RSMeans

Open Shop Building Construction Cost Data

16th Annual Edition

2000

Senior Editor
Phillip R. Waier, PE

Contributing Editors
Thomas J. Akins
Barbara Balboni
Howard M. Chandler
John H. Chiang, PE
Paul C. Crosscup
Jennifer L. Curran
Stephen E. Donnelly
J. Robert Lang
Robert C. McNichols
Robert W. Mewis
Melville J. Mossman, PE
John J. Moylan
Jeannene D. Murphy
Peter T. Nightingale
Stephen C. Plotner
Michael J. Regan
William R. Tennyson, II

Manager, Engineering Operations
John H. Ferguson, PE

President
Durwood S. Snead

Vice President and General Manager
Roger J. Grant

Vice President, Sales and Marketing
John M. Shea

Production Manager
Michael Kokernak

Production Coordinator
Marion E. Schofield

Technical Support
Wayne D. Anderson
Thomas J. Dion
Michael H. Donelan
Jonathan Forgit
Gary L. Hoitt
Paula Reale-Camelio
Kathryn S. Rodriguez
Sheryl A. Rose
James N. Wills

Art Director
Helen A. Marcella

Book & Cover Design
Norman R. Forgit

R.S. Means Company, Inc. ("R.S. Means"), its authors, editors and engineers, apply diligence and judgment in locating and using reliable sources for the information published. However, R.S. Means makes no express or implied warranty or guarantee in connection with the content of the information contained herein, including the accuracy, correctness, value, sufficiency, or completeness of the data, methods and other information contained herein. R.S. Means makes no express or implied warranty of merchantability or fitness for a particular purpose. R.S. Means shall have no liability to any customer or third party for any loss, expense, or damage including consequential, incidental, special or punitive damages, including lost profits or lost revenue, caused directly or indirectly by any error or omission, or arising out of, or in connection with, the information contained herein.

No part of this publication may be reproduced, stored in a retrieval system, or transmitted in any form or by any means without prior written permission of R.S. Means Company, Inc.

Editorial Advisory Board

James E. Armstrong
Energy Consultant
EnergyVision

William R. Barry
Chief Estimator
Mitchell Construction Company

Robert F. Cox, PhD
Assistant Professor
ME Rinker Sr. School of Bldg. Constr.
University of Florida

Roy F. Gilley, AIA
Principal
Gilley-Hinkel Architects

Kenneth K. Humphreys, PhD, PE, CCE
Secretary-Treasurer
International Cost Engineering Council

Patricia L. Jackson, PE
Vice President, Project Management
Aguirre Corporation

Martin F. Joyce
Vice President, Utility Division
Bond Brothers, Inc.

First Printing

Foreword

R.S. Means Co., Inc. is owned by CMD Group, a leading worldwide provider of proprietary construction information. CMD Group is comprised of three synergistic product groups crafted to be the complete resource for reliable, timely and actionable construction market data. In North America, CMD Group encompasses: Architects' First Source, an innovative product selection and specification solution in print and on the Internet; Construction Market Data (CMD), the source for construction activity information, as well as early planning reports for the design community; Associated Construction Publications, with 14 magazines, one of the largest editorial networks dedicated to U.S. highway and heavy construction coverage; Manufacturer's Survey Associates (MSA), a leading estimating and quantity survey firm in the U.S.; R.S. Means, the authority on construction cost data in North America; CMD Canada, the leading supplier of project information, industry news and forecasting data products for the Canadian construction industry; and BIMSA/Mexico, the dominant distributor of information on building projects and construction throughout Mexico. Worldwide, CMD Group includes Byggfakta Scandinavia, providing construction market data to Denmark, Estonia, Finland, Norway and Sweden; and Cordell Building Information Services, the market leader for construction and cost information in Australia.

Our Mission

Since 1942, R.S. Means Company, Inc. has been actively engaged in construction cost publishing and consulting throughout North America.

Today, over fifty years after the company began, our primary objective remains the same: to provide you, the construction and facilities professional, with the most current and comprehensive construction cost data possible.

Whether you are a contractor, an owner, an architect, an engineer, a facilities manager, or anyone else who needs a fast and reliable construction cost estimate, you'll find this publication to be a highly useful and necessary tool.

Today, with the constant flow of new construction methods and materials, it's difficult to find the time to look at and evaluate all the different construction cost possibilities. In addition, because labor and material costs keep changing, last year's cost information is not a reliable basis for today's estimate or budget.

That's why so many construction professionals turn to R.S. Means. We keep track of the costs for you, along with a wide range of other key information, from city cost indexes . . . to productivity rates . . . to crew composition . . . to contractor's overhead and profit rates.

R.S. Means performs these functions by collecting data from all facets of the industry, and organizing it in a format that is instantly accessible to you. From the preliminary budget to the detailed unit price estimate, you'll find the data in this book useful for all phases of construction cost determination.

The Staff, the Organization, and Our Services

When you purchase one of R.S. Means' publications, you are in effect hiring the services of a full-time staff of construction and engineering professionals.

Our thoroughly experienced and highly qualified staff works daily at collecting, analyzing, and disseminating comprehensive cost information for your needs. These staff members have years of practical construction experience and engineering training prior to joining the firm. As a result, you can count on them not only for the cost figures, but also for additional background reference information that will help you create a realistic estimate.

The Means organization is always prepared to help you solve construction problems through its five major divisions: Construction and Cost Data Publishing, Electronic Products and Services, Consulting Services, Insurance Division, and Educational Services.

Besides a full array of construction cost estimating books, Means also publishes a number of other reference works for the construction industry. Subjects include construction estimating and project and business management; special topics such as HVAC, roofing, plumbing, and hazardous waste remediation; and a library of facility management references.

In addition, you can access all of our construction cost data through your computer with Means CostWorks 2000 CD-ROM, an electronic tool that offers over 50,000 lines of Means construction cost data.

What's more, you can increase your knowledge and improve your construction estimating and management performance with a Means Construction Seminar or In-House Training Program. These two-day seminar programs offer unparalleled opportunities for everyone in your organization to get updated on a wide variety of construction-related issues.

Means also is a worldwide provider of construction cost management and analysis services for commercial and government owners and of claims and valuation services for insurers.

In short, R.S. Means can provide you with the tools and expertise for constructing accurate and dependable construction estimates and budgets in a variety of ways.

Robert Snow Means Established a Tradition of Quality That Continues Today

Robert Snow Means spent years building his company, making certain he always delivered a quality product.

Today, at R.S. Means, we do more than talk about the quality of our data and the usefulness of our books. We stand behind all of our data, from historical cost indexes... to construction materials and techniques... to current costs.

If you have any questions about our products or services, please call us toll-free at 1-800-334-3509. Our customer service representatives will be happy to assist you.

Table of Contents

Foreword	ii
How the Book Is Built: An Overview	iv
Quick Start	v
How To Use the Book: The Details	vi
Unit Price Section	1
Reference Section	479
Reference Numbers	480
Change Orders	573
Crew Listings	577
Historical Cost Indexes	600
City Cost Indexes	601
Location Factors	643
Abbreviations	649
Index	652
Other Means Publications and Reference Works	Yellow Pages
Installing Contractor's Overhead & Profit	Inside Back Cover

UNIT PRICES

GENERAL REQUIREMENTS	1
SITE CONSTRUCTION	2
CONCRETE	3
MASONRY	4
METALS	5
WOOD & PLASTICS	6
THERMAL & MOISTURE PROTECTION	7
DOORS & WINDOWS	8
FINISHES	9
SPECIALTIES	10
EQUIPMENT	11
FURNISHINGS	12
SPECIAL CONSTRUCTION	13
CONVEYING SYSTEMS	14
MECHANICAL	15
ELECTRICAL	16
SQUARE FOOT	17

REFERENCE INFORMATION

- REFERENCE NUMBERS
- CREWS
- COST INDEXES
- INDEX

How the Book Is Built: An Overview

A Powerful Construction Tool

You have in your hands one of the most powerful construction tools available today. A successful project is built on the foundation of an accurate and dependable estimate. This book will enable you to construct just such an estimate.

For the casual user the book is designed to be:

- quickly and easily understood so you can get right to your estimate
- filled with valuable information so you can understand the necessary factors that go into the cost estimate

For the regular user, the book is designed to be:

- a handy desk reference that can be quickly referred to for key costs
- a comprehensive, fully reliable source of current construction costs and productivity rates, so you'll be prepared to estimate any project
- a source book for preliminary project cost, product selections, and alternate materials and methods

To meet all of these requirements we have organized the book into the following clearly defined sections.

Quick Start

This one-page section (see following page) can quickly get you started on your estimate.

How To Use the Book: The Details

This section contains an in-depth explanation of how the book is arranged . . . and how you can use it to determine a reliable construction cost estimate. It includes information about how we develop our cost figures and how to completely prepare your estimate.

Unit Price Section

All cost data has been divided into the 16 divisions according to the MasterFormat system of classification and numbering as developed by the Construction Specifications Institute (CSI) and Construction Specifications Canada (CSC). For a listing of these divisions and an outline of their subdivisions, see the Unit Price Section Table of Contents.

Estimating tips are included at the beginning of each division.

Division 17: Quick Project Estimates: In addition to the 16 Unit Price Divisions there is a S.F. (Square Foot) and C.F. (Cubic Foot) Cost Division, Division 17. It contains costs for 58 different building types that allow you to quickly make a rough estimate for the overall cost of a project or its major components.

Reference Section

This section includes information on Reference Numbers, Change Orders, Crew Listings, Historical Cost Indexes, City Cost Indexes, Location Factors and a listing of Abbreviations. It is visually identified by a vertical gray bar on the edge of pages.

Reference Numbers: At the beginning of selected major classifications in the Unit Price Section are "reference numbers" shown in bold squares. These numbers refer you to related information in the Reference Section.

In this section, you'll find reference tables, explanations, and estimating information that support how we develop the unit price data. Also included are alternate pricing methods, technical data, and estimating procedures, along with information on design and economy in construction. You'll also find helpful tips on what to expect and what to avoid when estimating and constructing your project.

It is recommended that you refer to the Reference Section if a "reference number" appears within the section you are estimating.

Change Orders: This section includes information on the factors that influence the pricing of change orders.

Crew Listings: This section lists all the crews referenced in the book. For the purposes of this book, a crew is composed of more than one trade classification and/or the addition of power equipment to any trade classification. Power equipment is included in the cost of the crew. Costs are shown both with bare labor rates and with the installing contractor's overhead and profit added. For each, the total crew cost per eight-hour day and the composite cost per labor-hour are listed.

Historical Cost Indexes: These indexes provide you with data to adjust construction costs over time. If you know costs for a project completed in the past, you can use these indexes to calculate a rough estimate of what it would cost to construct the same project today.

City Cost Indexes: Obviously, costs vary depending on the regional economy. You can adjust the "national average" costs in this book to over 930 locations throughout the U.S. and Canada by using the data in this section. How to use information is included.

Location Factors, to quickly adjust the data to over 930 zip code areas, are included.

Abbreviations: A listing of the abbreviations used throughout this book, along with the terms they represent, is included.

Index

A comprehensive listing of all terms and subjects in this book to help you find what you need quickly when you are not sure where it falls in MasterFormat.

The Scope of This Book

This book is designed to be as comprehensive and as easy to use as possible. To that end we have made certain assumptions and limited its scope in three key ways:

1. We have established material prices based on a "national average."
2. We have computed labor costs based on a 30-city "national average" of open shop wages across the U.S.
3. We have targeted the data for projects of a certain size range.

For a more detailed explanation of how the cost data is developed, see "How To Use the Book: The Details."

Project Size

This book is aimed primarily at commercial and industrial projects costing $1,000,000 and up, or large multi-family housing projects. Costs are primarily for new construction or major renovation of buildings rather than repairs or minor alterations.

With reasonable exercise of judgment the figures can be used for any building work. *For civil engineering structures such as bridges, dams, highways, or the like, please refer to* **Means Heavy Construction Cost Data.**

Quick Start

If you feel you are ready to use this book and don't think you need the detailed instructions that begin on the following page, this Quick Start section is for you.

These steps will allow you to get started estimating in a matter of minutes.

1 Find each cost data section you need in the Unit Price Section Table of Contents.

The cost data has been divided into 16 divisions according to the CSI MasterFormat.

2 Turn to the indicated section and locate the line item you need for your estimate. Portions of a sample page layout appear here.

- If there is a reference number listed at the beginning of a section, for example, R03410-030, it refers to additional information you may find useful. See the Reference Section for detailed information.
- Note the crew code designation. You'll find full descriptions of crews in the Crews Section, including labor-hour and equipment costs.

3 Determine the total number of units your job will require. Note that the unit of measure for the material you're using is listed under "UNIT."

- Bare Costs: These figures show unit costs for materials and installation. Labor and equipment costs are calculated according to crew costs and average daily output. Bare costs do not contain allowances for overhead, profit, or taxes.
- "Labor-hours" allows you to calculate the total labor-hours to complete that task. Just multiply the quantity of work by this figure for an estimate of activity duration.

4 Then multiply the total units by the "Total Incl. O&P," which stands for the total cost including the installing contractor's overhead and profit. (See the next pages for a complete explanation.)

- If the work is to be subcontracted, add the general contractor's markup, approximately 10%.

5 The price you calculate will be an estimate for a completed item of work.

6 Compile a list of all items included in the total project. Summarize cost information, and add project overhead.

Localize costs using the City Cost Indexes or Location Factors found in the Reference Section.

For a more complete explanation of the way costs are derived, please see the following section.

Commonly Used Abbreviations

R.S. Means utilizes standard industry abbreviations. There is a complete glossary of abbreviations in the reference section. The following are a few of the most commonly used abbreviations you'll find in the book:

B.F.	Board Feet
C	Hundred; Centigrade
C.Y.	Cubic Yard (27 Cubic Feet)
Cwt	100 Pounds
Ea.	Each
Flr.	Floor
L.F.	Linear Foot
Lb.	Pound
MBF	Thousand Board Feet
Opng.	Opening
S.F.	Square Foot
SFCA	Square Foot Contact Area
S.Y.	Square Yard
Sq.	Square; 100 Square Feet
Sty.	Story
Surf.	Surface
V.L.F.	Vertical Linear Foot

Editors' Note: We urge you to spend time reading and understanding the supporting material. An accurate estimate requires experience, knowledge, and careful calculation. The more you know about how we at R.S. Means developed the data, the more accurate your estimate will be. In addition, it's important to take into consideration some of the reference material such as Crews Listing and the "reference numbers."

03400 | Precast Concrete

		03410	Plant Precast		CREW	DAILY OUTPUT	LABOR-HOURS	UNIT	MAT.	LABOR	EQUIP.	TOTAL	TOTAL INCL O&P	
620	0010		SLABS Prestressed roof/floor members, solid, grouted, 4" thick	R03410-030	C-11	3,600	.016	S.F.	4.65	.31	.43	5.39	6.10	620
	0050		6" thick			4,500	.012		4.48	.25	.35	5.08	5.75	
	0100		8" thick, hollow			5,600	.010		5	.20	.28	5.48	6.15	
	0150		10" thick			8,800	.006		5.70	.13	.18	6.01	6.65	
	0200		12" thick			8,000	.007		6.15	.14	.20	6.49	7.20	
650	0010		PRESTRESSED CONCRETE pretensioned, see division 03400	R03410-090										650
	0020		See also division 03230-600											
	0100		Post-tensioned in place, small job	R03410-030	C-17B	8.50	9.647	C.Y.	215	194	34	443	610	
	0200		Large job		"	10	8.200	"	161	165	29	355	490	
660	0010		PRESTRESSED Roof and floor members, see division 03400	R03410-030										660

How to Use the Book: The Details

What's Behind the Numbers? The Development of Cost Data

The staff at R.S. Means continuously monitors developments in the construction industry in order to ensure reliable, thorough and up-to-date cost information.

While *overall* construction costs may vary relative to general economic conditions, price fluctuations within the industry are dependent upon many factors. Individual price variations may, in fact, be opposite to overall economic trends. Therefore, costs are continually monitored and complete updates are published yearly. Also, new items are frequently added in response to changes in materials and methods.

Costs—$ (U.S.)

All costs represent U.S. national averages and are given in U.S. dollars. The Means City Cost Indexes can be used to adjust costs to a particular location. The City Cost Indexes for Canada can be used to adjust U.S. national averages to local costs in Canadian dollars.

Material Costs

The R.S. Means staff contacts manufacturers, dealers, distributors, and contractors all across the U.S. and Canada to determine national average material costs. If you have access to current material costs for your specific location, you may wish to make adjustments to reflect differences from the national average. Included within material costs are fasteners for a normal installation. R.S. Means engineers use manufacturers' recommendations, written specifications and/or standard construction practice for size and spacing of fasteners. Adjustments to material costs may be required for your specific application or location. Material costs do not include sales tax.

Labor Costs

Labor costs are based on the average of open shop wage rates across the U.S. for construction trades for the current year. Rates along with overhead and profit markups are listed on the inside back cover of this book.

- If wage rates in your area vary from those used in this book, or if rate increases are expected within a given year, labor costs should be adjusted accordingly.

Labor costs reflect productivity based on actual working conditions. These figures include time spent during a normal workday on tasks other than actual installation, such as material receiving and handling, mobilization at site, site movement, breaks, and cleanup.

Productivity data is developed over an extended period so as not to be influenced by abnormal variations and reflects a typical average.

Equipment Costs

Equipment costs include not only rental, but also operating costs for equipment under normal use. The operating costs include parts and labor for routine servicing such as repair and replacement of pumps, filters and worn lines. Normal operating expendables such as fuel, lubricants, tires and electricity (where applicable) are also included. Extraordinary operating expendables with highly variable wear patterns such as diamond bits and blades are excluded. These costs are included under materials. Equipment rental rates are obtained from industry sources throughout North America—contractors, suppliers, dealers, manufacturers, and distributors.

Crew Equipment Cost/Day—The power equipment required for each crew is included in the crew cost. The daily cost for crew equipment is based on dividing the weekly bare rental rate by 5 (number of working days per week), and then adding the hourly operating cost times 8 (hours per day). This "Crew Equipment Cost/Day" is listed in Subdivision 01590.

General Conditions

Cost data in this book is presented in two ways: Bare Costs and Total Cost including O&P (Overhead and Profit). General Conditions, when applicable, should also be added to the Total Cost including O&P. The costs for General Conditions are listed in Division 1 and the Reference Section of this book. General Conditions for the *Installing Contractor* may range from 0% to 10% of the Total Cost including O&P. For the *General* or *Prime Contractor*, costs for General Conditions may range from 5% to 15% of the Total Cost including O&P, with a figure of 10% as the most typical allowance.

Overhead and Profit

Total Cost including O&P for the *Installing Contractor* is shown in the last column on the Unit Price pages of this book. This figure is the sum of the bare material cost plus 10% for profit, the base labor cost plus total overhead and profit, and the bare equipment cost plus 10% for profit. Details for the calculation of Overhead and Profit on labor are shown on the inside back cover and in the Reference Section of this book. (See the "How To Use the Unit Price Pages" for an example of this calculation.)

Factors Affecting Costs

Costs can vary depending upon a number of variables. Here's how we have handled the main factors affecting costs.

Quality—The prices for materials and the workmanship upon which productivity is based represent sound construction work. They are also in line with U.S. government specifications.

Overtime — We have made no allowance for overtime. If you anticipate premium time or work beyond normal working hours, be sure to make an appropriate adjustment to your labor costs.

Productivity — The productivity, daily output, and labor-hour figures for each line item are based on working an eight-hour day in daylight hours in moderate temperatures. For work that extends beyond normal work hours or is performed under adverse conditions, productivity may decrease. (See the section in "How To Use the Unit Price Pages" for more on productivity.)

Size of Project — The size, scope of work, and type of construction project will have a significant impact on cost. Economies of scale can reduce costs for large projects. Unit costs can often run higher for small projects. Costs in this book are intended for the size and type of project as previously described in "How the Book Is Built: An Overview." Costs for projects of a significantly different size or type should be adjusted accordingly.

Location — Material prices in this book are for metropolitan areas. However, in dense urban areas, traffic and site storage limitations may increase costs. Beyond a 20-mile radius of large cities, extra trucking or transportation charges may also increase the material costs slightly. On the other hand, lower wage rates may be in effect. Be sure to consider both these factors when preparing an estimate, particularly if the job site is located in a central city or remote rural location.

In addition, highly specialized subcontract items may require travel and per diem expenses for mechanics.

Other factors —
- season of year
- contractor management
- weather conditions
- local union restrictions
- building code requirements
- availability of:
  - adequate energy
  - skilled labor
  - building materials
- owner's special requirements/restrictions
- safety requirements
- environmental considerations

Unpredictable Factors — General business conditions influence "in-place" costs of all items. Substitute materials and construction methods may have to be employed. These may affect the installed cost and/or life cycle costs. Such factors may be difficult to evaluate and cannot necessarily be predicted on the basis of the job's location in a particular section of the country. Thus, where these factors apply, you may find significant, but unavoidable cost variations for which you will have to apply a measure of judgment to your estimate.

Rounding of Costs

In general, all unit prices in excess of $5.00 have been rounded to make them easier to use and still maintain adequate precision of the results. The rounding rules we have chosen are in the following table.

Prices from . . .	Rounded to the nearest . . .
$.01 to $5.00	$.01
$5.01 to $20.00	$.05
$20.01 to $100.00	$.50
$100.01 to $300.00	$1.00
$300.01 to $1,000.00	$5.00
$1,000.01 to $10,000.00	$25.00
$10,000.01 to $50,000.00	$100.00
$50,000.01 and above	$500.00

Final Checklist

Estimating can be a straightforward process provided you remember the basics. Here's a checklist of some of the items you should remember to do before completing your estimate.

Did you remember to . . .

- factor in the City Cost Index for your locale
- take into consideration which items have been marked up and by how much
- mark up the entire estimate sufficiently for your purposes
- read the background information on techniques and technical matters that could impact your project time span and cost
- include all components of your project in the final estimate
- double check your figures to be sure of your accuracy
- call R.S. Means if you have any questions about your estimate or the data you've found in our publications

Remember, R.S. Means stands behind its publications. If you have any questions about your estimate . . . about the costs you've used from our books . . . or even about the technical aspects of the job that may affect your estimate, feel free to call the R.S. Means editors at 1-800-334-3509.

Unit Price Section

Table of Contents

Div No.		Page
	General Requirements	**5**
01100	Summary	6
01200	Price & Payment Procedures	7
01300	Administrative Requirements	8
01400	Quality Requirements	10
01500	Temporary Facilities & Controls	12
01590	Material & Equipment	19
01700	Execution Requirements	27
01800	Facility Operation	27
	Site Construction	**29**
02050	Basic Site Materials & Methods	30
02100	Site Remediation	30
02200	Site Preparation	32
02300	Earthwork	46
02400	Tunneling, Boring & Jacking	55
02450	Foundation & L. B. Elements	55
02500	Utility Services	60
02600	Drainage & Containment	66
02700	Bases, Ballasts, Pavements, & Appurtenances	69
02800	Site Improvements & Amenities	74
02900	Planting	84
02950	Site Restoration & Rehab.	87
	Concrete	**89**
03050	Basic Conc. Mat. & Methods	90
03100	Concrete Forms & Accessories	92
03200	Concrete Reinforcement	103
03300	Cast-In-Place Concrete	109
03400	Precast Concrete	114
03500	Cementitious Decks & Underlay	116
03600	Grouts	117
	Masonry	**119**
04050	Basic Masonry Mat. & Methods	120
04200	Masonry Units	122
04400	Stone	130
04500	Refractories	133
04700	Simulated Masonry	133
04800	Masonry Assemblies	134
04900	Masonry Restoration & Cleaning	136
	Metals	**137**
05050	Basic Materials & Methods	138
05100	Structural Metal Framing	145
05200	Metal Joists	154
05300	Metal Decking	156
05400	Cold Formed Metal Framing	157
05500	Metal Fabrications	163
05650	Railroad Track & Accessories	169
05700	Ornamental Metals	169
05800	Expansion Control	170
	Wood & Plastics	**171**
06050	Basic Wd. & Plastic Mat. & Meth.	172
06100	Rough Carpentry	176
06200	Finish Carpentry	186
06400	Architectural Woodwork	189
06500	Structural Plastics	195
06600	Plastic Fabrications	195
	Thermal & Moisture Protection	**199**
07100	Dampproofing & Waterproofing	200
07200	Thermal Protection	201
07300	Shingles, Roof Tiles & Roof Cov.	206

Div No.		Page
07400	Roofing & Siding Panels	209
07500	Membrane Roofing	213
07600	Flashing & Sheet Metal	217
07700	Roof Specialties & Accessories	219
07800	Fire & Smoke Protection	223
07900	Joint Sealers	225
	Doors & Windows	**227**
08100	Metal Doors & Frames	228
08200	Wood & Plastic Doors	230
08300	Specialty Doors	238
08400	Entrances & Storefronts	244
08500	Windows	246
08600	Skylights	252
08700	Hardware	253
08800	Glazing	259
08900	Glazed Curtain Wall	262
	Finishes	**265**
09100	Metal Support Assemblies	266
09200	Plaster & Gypsum Board	266
09300	Tile	273
09400	Terrazzo	275
09500	Ceilings	276
09600	Flooring	278
09700	Wall Finishes	283
09800	Acoustical Treatment	284
09900	Paints & Coatings	285
	Specialties	**297**
10100	Visual Display Boards	298
10150	Compartments & Cubicles	300
10200	Louvers & Vents	302
10260	Wall & Corner Guards	303
10270	Access Flooring	304
10300	Fireplaces & Stoves	304
10340	Manufactured Exterior Specialties	306
10350	Flagpoles	306
10400	Identification Devices	307
10450	Pedestrian Control Devices	308
10500	Lockers	309
10520	Fire Protection Specialties	309
10530	Protective Covers	311
10550	Postal Specialties	311
10600	Partitions	312
10670	Storage Shelving	314
10750	Telephone Specialties	315
10800	Toilet, Bath, & Laundry Access.	316
10880	Scales	317
10900	Wardrobe & Closet Specialties	318
	Equipment	**319**
11010	Maintenance Equipment	320
11020	Security & Vault Equipment	320
11030	Teller & Service Equipment	320
11040	Ecclesiastical Equipment	321
11050	Library Equipment	322
11060	Theater & Stage Equipment	323
11100	Mercantile Equipment	324
11110	Coml. Laun. & Dry Cln. Eq.	325
11130	Audio-Visual Equip.	325
11140	Vehicle Service Equip.	326
11150	Parking Control Equip.	327
11160	Loading Dock Equip.	327
11170	Solid Waste Handling Equip.	328
11190	Detention Equipment	329
11300	Fluid Waste Treat. & Disp. Equip.	330

Div No.		Page
11400	Food Service Equipment	330
11450	Residential Equipment	334
11470	Darkroom Equipment	335
11480	Athletic Rec. & Thera. Equip.	336
11500	Industrial & Process Equipment	338
11600	Laboratory Equipment	339
11700	Medical Equipment	340
	Furnishings	**343**
12300	Manufactured Casework	344
12400	Furnishings & Access.	345
12500	Furniture	347
12600	Multiple Seating	349
12800	Interior Plants & Planters	350
	Special Construction	**351**
13010	Air Supported Structures	352
13030	Special Purpose Rooms	353
13080	Sound, Vib. & Seismic Control	356
13090	Radiation Protection	357
13120	Pre-Engineered Structures	358
13150	Swimming Pools	364
13170	Tubs & Pools	365
13175	Ice Rinks	365
13200	Storage Tanks	365
13280	Hazardous Mat. Remediation	368
13600	Solar & Wind Energy Equipment	372
13700	Security Access & Surveillance	373
13800	Building Automation & Control	374
13900	Fire Suppression	376
	Conveying Systems	**379**
14100	Dumbwaiters	380
14200	Elevators	380
14300	Escalators & Moving Walks	384
14400	Lifts	384
14500	Material Handling	385
14600	Hoists & Cranes	386
	Mechanical	**389**
15050	Basic Materials & Methods	390
15100	Building Services Piping	392
15200	Process Piping	409
15400	Plumbing Fixtures & Equipment	409
15500	Heat Generation Equipment	419
15600	Refrigeration Equipment	423
15700	Heating, Ventilation, & A/C Equip.	425
15800	Air Distribution	430
15950	Testing/Adjusting/Balancing	436
	Electrical	**439**
16050	Basic Elect. Mat. & Methods	440
16100	Wiring Methods	442
16200	Electrical Power	456
16400	Low-Voltage Distribution	459
16500	Lighting	463
16700	Communications	467
16800	Sound & Video	467
	Square Foot	**469**
17100	S.F., C.F. & % of Total Costs	470

How to Use the Unit Price Pages

The following is a detailed explanation of a sample entry in the Unit Price Section. Next to each bold number below is the item being described with appropriate component of the sample entry following in parenthesis. Some prices are listed as bare costs, others as costs that include overhead and profit of the installing contractor. In most cases, if the work is to be subcontracted, the general contractor will need to add an additional markup (R.S. Means suggests using 10%) to the figures in the column "Total Incl. O&P."

1 Division Number/Title (03400/Precast Concrete)

Use the Unit Price Section Table of Contents to locate specific items. The sections are classified according to the CSI MasterFormat (1995 Edition).

2 Line Numbers (03410 620 0100)

Each unit price line item has been assigned a unique 12-digit code based on the CSI MasterFormat classification.

```
03400
03410-620-0100
```
- Level One - CSI-MasterFormat Division
- Level Two - CSI
- Level Three - CSI
- Level Four - Means
- Means 12-digit Line Number

3 Description (SLABS, etc.)

Each line item is described in detail. Sub-items and additional sizes are indented beneath the appropriate line items. The first line or two after the main item (in boldface) may contain descriptive information that pertains to all line items beneath this boldface listing.

4 Reference Number Information

R03410-030 You'll see reference numbers shown in bold rectangles at the beginning of some sections. These refer to related items in the Reference Section, visually identified by a vertical gray bar on the edge of pages.

The relation may be: (1) an estimating procedure that should be read before estimating, (2) an alternate pricing method, or (3) technical information.

The "R" designates the Reference Section. The numbers refer to the MasterFormat classification system.

It is strongly recommended that you review all reference numbers that appear within the section in which you are working.

Example: The rectangle number above is directing you to refer to the reference number R03410-030. This particular reference number shows delivered and erected costs for standard prestressed, precast concrete shapes and other detail information.

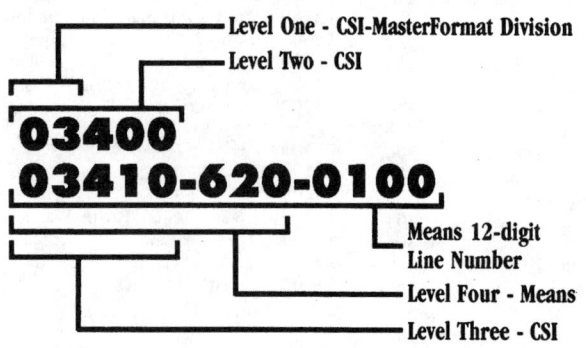

03400	Precast Concrete							2000 BARE COSTS				TOTAL	
03410	Plant Precast			CREW	DAILY OUTPUT	LABOR-HOURS	UNIT	MAT.	LABOR	EQUIP.	TOTAL	INCL O&P	
620	0010	**SLABS** Prestressed roof/floor members, solid, grouted		R03410-030	C-11	3,600	.016	S.F.	4.65	.31	.43	5.39	6.10	620
	0050	6" thick				4,500	.012		4.48	.25	.35	5.08	5.75	
	0100	8" thick, hollow				5,600	.010		5	.20	.28	5.48	6.15	
	0150	10" thick				8,800	.006		5.70	.13	.18	6.01	6.65	
	0200	12" thick			▼	8,	.007	▼	6.15	.14	.20	6.49	7.20	
650	0010	**PRESTRESSED CONCRETE** Tensioned, see division 03400												650
	0020	See also division 03230-600												
	0100	Post-tensioned in place, small job		R03410-030	C-17B	8.50	9.647	C.Y.	215	194	34	443	610	
	0200	Large job			"	10	8.200		161	165	29	355	490	
660	0010	**PRESTRESSED** Roof and floor members, see division 03400		R03410-030										660

Crew (C-11)

The "Crew" column designates the typical trade or crew used to install the item. If an installation can be accomplished by one trade and requires no power equipment, that trade and the number of workers are listed (for example, "2 Struc. Steel Workers"). If an installation requires a composite crew, a crew code designation is listed (for example, "C-11"). You'll find full details on all composite crews in the Crew Listings.
* For a complete list of all trades utilized in this book and their abbreviations, see the inside back cover.

Crews

Crew No.	Bare Costs		Incl. Subs O & P		Cost Per Labor-Hour	
Crew C-11	Hr.	Daily	Hr.	Daily	Bare Costs	Incl. O&P
1 Skilled Worker Foreman	$21.75	$174.00	$37.30	$298.40	$20.16	$34.44
5 Skilled Worker	19.75	790.00	33.90	1356.00		
1 Equip. Oper. (crane)	20.65	165.20	34.25	274.00		
1 Truck Crane, 150 Ton		1563.00		1719.30	27.91	30.70
56 L.H., Daily Totals		$2692.20		$3647.70	$48.07	$65.14

Productivity: Daily Output (5600)/Labor-Hours (.010)

The "Daily Output" represents the typical number of units the designated crew will install in a normal 8-hour day. To find out the number of days the given crew would require to complete the installation, divide your quantity by the daily output. For example:

Quantity	÷	Daily Output	=	Duration
15,000 S.F.	÷	5600 S.F./Crew Day	=	2.68 Crew Days

The "Labor-Hours" figure represents the number of labor-hours required to install one unit of work. To find out the number of labor-hours required for your particular task, multiply the quantity of the item times the number of labor-hours shown. For example:

Quantity	x	Productivity Rate	=	Duration
15,000 S.F.	x	.010 Labor-Hours/S.F.	=	150 Labor-Hours

Unit (S.F.)

The abbreviated designation indicates the unit of measure upon which the price, production, and crew are based (S.F. = Square Foot). For a complete listing of abbreviations refer to the Abbreviations Listing in the Reference Section of this book.

Bare Costs:

Mat. (Bare Material Cost) (5.00)

The unit material cost is the "bare" material cost with no overhead and profit included. *Costs shown reflect national average material prices for January of the current year and include delivery to the job site. No sales taxes are included.*

Labor (.20)

The unit labor cost is derived by multiplying bare labor-hour costs for Crew C-11 by labor-hour units. The bare labor-hour cost is found in the Crew Section under C-11. (If a trade is listed, the hourly labor cost—the wage rate—is found on the inside back cover.)

Labor-Hour Cost Crew C-11	x	Labor-Hour Units	=	Labor
$20.16	x	.010	=	$.20

Equip. (Equipment) (.28)

Equipment costs for each crew are listed in the description of each crew. Tools or equipment whose value justifies purchase or ownership by a contractor are considered overhead as shown on the inside back cover. The unit equipment cost is derived by multiplying the bare equipment hourly cost by the labor-hour units.

Equipment Cost Crew C-11	x	Labor-Hour Units	=	Equip.
$27.91	x	.010	=	$.28

Total (5.48)

The total of the bare costs is the arithmetic total of the three previous columns: mat., labor, and equip.

Material	+	Labor	+	Equip.	=	Total
$5.00	+	$.20	+	$.28	=	$5.48

Total Costs Including O&P

This figure is the sum of the bare material cost plus 10% for profit; the bare labor cost plus total overhead and profit (per the inside back cover or, if a crew is listed, from the crew listings); and the bare equipment cost plus 10% for profit.

Material is Bare Material Cost + 10% = 5.00 + .50	=	$5.50
Labor for Crew C-11 = Labor-Hour Cost (34.44) x Labor-Hour Units (.010)	=	$ .34
Equip. is Bare Equip. Cost + 10% = .28 + .03	=	$ .31
Total	=	$6.15

Division 1
General Requirements

Estimating Tips

The General Requirements of any contract are very important to both the bidder and the owner. These lay the ground rules under which the contract will be executed and have a significant influence on the cost of operations. Therefore, it is extremely important to thoroughly read and understand the General Requirements both before preparing an estimate and when the estimate is complete, to ascertain that nothing in the contract is overlooked. Caution should be exercised when applying items listed in Division 1 to an estimate. Many of the items are included in the unit prices listed in the other divisions such as mark-ups on labor and company overhead.

01300 Administrative Requirements
- Before determining a final cost estimate, it is a good practice to review all the items listed in subdivision 01300 to make final adjustments for items that may need customizing to specific job conditions.

01330 Submittal Procedures
- Requirements for initial and periodic submittals can represent a significant cost to the General Requirements of a job. Thoroughly check the submittal specifications when estimating a project to determine any costs that should be included.

01400 Quality Requirements
- All projects will require some degree of Quality Control. This cost is not included in the unit cost of construction listed in each division. Depending upon the terms of the contract, the various costs of inspection and testing can be the responsibility of either the owner or the contractor. Be sure to include the required costs in your estimate.

01500 Temporary Facilities & Controls
- Barricades, access roads, safety nets, scaffolding, security and many more requirements for the execution of a safe project are elements of direct cost. These costs can easily be overlooked when preparing an estimate. When looking through the major classifications of this subdivision, determine which items apply to each division in your estimate.

01559 Equipment Rental
- This subdivision contains transportation, handling, storage, protection and product options and substitutions. Listed in this cost manual are average equipment rental rates for all types of equipment. This is useful information when estimating the time and materials requirement of any particular operation in order to establish a unit or total cost.
- A good rule of thumb is that weekly rental is 3 times daily rental and that monthly rental is 3 times weekly rental.
- The figures in the column for Crew Equipment Cost represent the rental rate used in determining the daily cost of equipment in a crew. It is calculated by dividing the weekly rate by 5 days and adding the hourly operating cost times 8 hours.

01770 Closeout Procedures
- When preparing an estimate, read the specifications to determine the requirements for Contract Closeout thoroughly. Final cleaning, record documentation, operation and maintenance data, warranties and bonds, and spare parts and maintenance materials can all be elements of cost for the completion of a contract. Do not overlook these in your estimate.

01830 Operations & Maintenance
- If maintenance and repair are included in your contract, they require special attention. To estimate the cost to remove and replace any unit usually requires a site visit to determine the accessibility and the specific difficulty at that location. Obstructions, dust control, safety, and often overtime hours must be considered when preparing your estimate.

Reference Numbers

Reference numbers are shown in bold squares at the beginning of some major classifications. These numbers refer to related items in the Reference Section. The reference information may be an estimating procedure, an alternate pricing method or technical information.

Note: Not all subdivisions listed here necessarily appear in this publication.

01100 | Summary

01103 | Models & Renderings

			CREW	DAILY OUTPUT	LABOR-HOURS	UNIT	2000 BARE COSTS MAT.	LABOR	EQUIP.	TOTAL	TOTAL INCL O&P	
200	0010	**MODELS** Cardboard & paper, 1 building, minimum				Ea.	550			550	605	200
	0050	Maximum					1,250			1,250	1,375	
	0100	2 buildings, minimum					725			725	800	
	0150	Maximum					1,700			1,700	1,875	
	0200	Plexiglass and metal, basic layout				SF Flr.	.06			.06	.07	
	0210	Including equipment and personnel				"	.23			.23	.25	
	0300	Site plan layout, minimum				Ea.	1,050			1,050	1,150	
	0350	Maximum				"	1,800			1,800	1,975	
500	0010	**RENDERINGS** Color, matted, 20" x 30", eye level,										500
	0020	1 building, minimum				Ea.	1,525			1,525	1,675	
	0050	Average					2,550			2,550	2,800	
	0100	Maximum					3,600			3,600	3,950	
	1000	5 buildings, minimum					3,075			3,075	3,375	
	1100	Maximum					6,100			6,100	6,700	
	2000	Aerial perspective, color, 1 building, minimum					2,550			2,550	2,800	
	2100	Maximum					6,100			6,100	6,700	
	3000	5 buildings, minimum					3,075			3,075	3,375	
	3100	Maximum					10,200			10,200	11,200	

01107 | Professional Consultant

						UNIT	MAT.	LABOR	EQUIP.	TOTAL	TOTAL INCL O&P	
100	0011	**ARCHITECTURAL FEES**	R01107-010									100
	0020	For new construction										
	0060	Minimum				Project					4.90%	
	0090	Maximum									16%	
	0100	For alteration work, to $500,000, add to fee									50%	
	0150	Over $500,000, add to fee									25%	
200	0010	**CONSTRUCTION MANAGEMENT FEES** $1,000,000 job, minimum				Project					4.50%	200
	0050	Maximum									7.50%	
	0300	$5,000,000 job, minimum									2.50%	
	0350	Maximum									4%	
300	0010	**ENGINEERING FEES**	R01107-030									300
	0020	Educational planning consultant, minimum				Project					.50%	
	0100	Maximum				"					2.50%	
	0200	Electrical, minimum				Contrct					4.10%	
	0300	Maximum									10.10%	
	0400	Elevator & conveying systems, minimum									2.50%	
	0500	Maximum									5%	
	0600	Food service & kitchen equipment, minimum									8%	
	0700	Maximum									12%	
	0800	Landscaping & site development, minimum									2.50%	
	0900	Maximum									6%	
	1000	Mechanical (plumbing & HVAC), minimum									4.10%	
	1100	Maximum									10.10%	
	1200	Structural, minimum				Project					1%	
	1300	Maximum				"					2.50%	
700	0010	**SURVEYING** Conventional, topographical, minimum		A-7	3.30	7.273	Acre	16	148		164	271	700
	0100	Maximum		A-8	.60	53.333		48	1,075		1,123	1,875	
	0300	Lot location and lines, minimum, for large quantities		A-7	2	12		25	245		270	445	
	0320	Average		"	1.25	19.200		45	390		435	715	
	0400	Maximum, for small quantities		A-8	1	32		72	645		717	1,175	
	0600	Monuments, 3' long		A-7	10	2.400	Ea.	19	49		68	105	
	0800	Property lines, perimeter, cleared land		"	1,000	.024	L.F.	.03	.49		.52	.86	
	0900	Wooded land		A-8	875	.037	"	.05	.74		.79	1.31	
	1100	Crew for building layout, 2 person crew		A-6	1	16	Day		335		335	575	
	1200	3 person crew		A-7	1	24			490		490	835	

01100 | Summary

01107 | Professional Consultant

			CREW	DAILY OUTPUT	LABOR-HOURS	UNIT	2000 BARE COSTS MAT.	LABOR	EQUIP.	TOTAL	TOTAL INCL O&P	
700	1300	4 person crew	A-8	1	32	Day		645		645	1,100	700
	1500	Aerial surveying, including ground control, minimum fee, 10 acres				Total					5,500	
	1510	100 acres									9,100	
	1550	From existing photography, deduct				▼					1,340	
	1600	2' contours, 10 acres				Acre					440	
	1650	20 acres									300	
	1800	50 acres									90	
	1850	100 acres									80	
	2000	1000 acres									17.01	
	2050	10,000 acres				▼					11.01	
	2150	For 1' contours and										
	2160	dense urban areas, add to above				Acre					40%	
	3000	Inertial guidance system for										
	3010	locating coordinates, rent per day				Ea.					4,000	

01200 | Price & Payment Procedures

01250 | Contract Modification Procedures

			CREW	DAILY OUTPUT	LABOR-HOURS	UNIT	2000 BARE COSTS MAT.	LABOR	EQUIP.	TOTAL	TOTAL INCL O&P	
200	0010	**CONTINGENCIES** for estimate at conceptual stage				Project					15%	200
	0050	Schematic stage									10%	
	0100	Preliminary working drawing stage (Design Dev.)									7%	
	0150	Final working drawing stage				▼					2%	
300	0010	**CREWS** For building construction, see How To Use This Book										300
500	0010	**JOB CONDITIONS** Modifications to total										500
	0020	project cost summaries										
	0100	Economic conditions, favorable, deduct				Project					2%	
	0200	Unfavorable, add									5%	
	0300	Hoisting conditions, favorable, deduct									2%	
	0400	Unfavorable, add									5%	
	0500	General Contractor management, experienced, deduct									2%	
	0600	Inexperienced, add									10%	
	0700	Labor availability, surplus, deduct									1%	
	0800	Shortage, add									10%	
	0900	Material storage area, available, deduct									1%	
	1000	Not available, add									2%	
	1100	Subcontractor availability, surplus, deduct									5%	
	1200	Shortage, add									12%	
	1300	Work space, available, deduct									2%	
	1400	Not available, add				▼					5%	
600	0010	**OVERTIME** For early completion of projects or where	R01100 -110									600
	0020	labor shortages exist, add to usual labor, up to				Costs		100%				

01255 | Cost Indexes

			CREW	DAILY OUTPUT	LABOR-HOURS	UNIT	2000 BARE COSTS MAT.	LABOR	EQUIP.	TOTAL	TOTAL INCL O&P	
200	0010	**CONSTRUCTION COST INDEX** (Reference) over 930 zip code locations in										200
	0020	The U.S. and Canada, total bldg cost, min. (Fayetteville, AR)				%					67.20%	
	0050	Average									100%	
	0100	Maximum (New York, NY)				▼					133.80%	
400	0010	**HISTORICAL COST INDEXES** (Reference) Back to 1950										400

01200 | Price & Payment Procedures

01255 | Cost Indexes

			CREW	DAILY OUTPUT	LABOR-HOURS	UNIT	MAT.	LABOR	EQUIP.	TOTAL	TOTAL INCL O&P	
500	0010	**LABOR INDEX** (Reference) For over 930 zip code locations in										500
	0020	the U.S. and Canada, minimum (Statesboro, GA)				%		36.50%				
	0050	Average						100%				
	0100	Maximum (New York, NY)				↓		160.50%				
600	0010	**MATERIAL INDEX** (Reference) For over 930 zip code locations in										600
	0020	the U.S. and Canada, minimum (Elizabethtown, KY)				%	92.10%					
	0040	Average					100%					
	0060	Maximum (Ketchikan, AK)				↓	144%					

01290 | Payment Procedures

			CREW	DAILY OUTPUT	LABOR-HOURS	UNIT	MAT.	LABOR	EQUIP.	TOTAL	TOTAL INCL O&P	
800	0010	**TAXES** Sales tax, State, average	R01100-090			%	4.71%					800
	0050	Maximum					7.25%					
	0200	Social Security, on first $72,600 of wages	R01100-100					7.65%				
	0300	Unemployment, MA, combined Federal and State, minimum						2.10%				
	0350	Average						7%				
	0400	Maximum				↓		8%				

01300 | Administrative Requirements

01310 | Project Management/Coordination

			CREW	DAILY OUTPUT	LABOR-HOURS	UNIT	MAT.	LABOR	EQUIP.	TOTAL	TOTAL INCL O&P	
150	0010	**PERMITS** Rule of thumb, most cities, minimum				Job					.50%	150
	0100	Maximum				"					2%	
200	0010	**PERFORMANCE BOND** For buildings, minimum	R01100-080			Job					.60%	200
	0100	Maximum				"					2.50%	
300	0010	**CONSTRUCTION TIME** Requirements	R01100-020									300
350	0010	**INSURANCE** Builders risk, standard, minimum	R01100-040			Job					.22%	350
	0050	Maximum									.59%	
	0200	All-risk type, minimum	R01100-050								.25%	
	0250	Maximum				↓					.62%	
	0400	Contractor's equipment floater, minimum	R01100-060			Value					.50%	
	0450	Maximum				"					1.50%	
	0600	Public liability, average				Job					1.55%	
	0800	Workers' compensation & employer's liability, average										
	0850	by trade, carpentry, general				Payroll		19.85%				
	0900	Clerical						.53%				
	0950	Concrete						18.81%				
	1000	Electrical						7.03%				
	1050	Excavation						11.41%				
	1100	Glazing						14.43%				
	1150	Insulation						18.52%				
	1200	Lathing						12.26%				
	1250	Masonry						17.75%				
	1300	Painting & decorating						15.27%				
	1350	Pile driving						29.88%				
	1400	Plastering						15.72%				
	1450	Plumbing						8.82%				
	1500	Roofing						34.62%				
	1550	Sheet metal work (HVAC)						12.27%				
	1600	Steel erection, structural				↓		42.56%				

01300 | Administrative Requirements

01310 | Project Management/Coordination

			CREW	DAILY OUTPUT	LABOR-HOURS	UNIT	2000 BARE COSTS MAT.	LABOR	EQUIP.	TOTAL	TOTAL INCL O&P	
350	1650	Tile work, interior ceramic				Payroll		10.48%				350
	1700	Waterproofing, brush or hand caulking						8.32%				
	1800	Wrecking						43.48%				
	2000	Range of 35 trades in 50 states, excl. wrecking, minimum						2%				
	2100	Average						18.10%				
	2200	Maximum						132.92%				
400	0010	**MAIN OFFICE EXPENSE** Average for General Contractors										400
	0020	As a percentage of their annual volume										
	0125	Annual volume under 1 million dollars				% Vol.				13.60%		
	0145	Up to 2.5 million dollars								8%		
	0150	Up to 4.0 million dollars								6.80%		
	0200	Up to 7.0 million dollars								5.60%		
	0250	Up to 10 million dollars								5.10%		
	0300	Over 10 million dollars								3.90%		
500	0010	**MARK-UP** For General Contractors for change										500
	0100	of scope of job as bid										
	0200	Extra work, by subcontractors, add				%					10%	
	0250	By General Contractor, add									15%	
	0400	Omitted work, by subcontractors, deduct all but									5%	
	0450	By General Contractor, deduct all but									7.50%	
	0600	Overtime work, by subcontractors, add									15%	
	0650	By General Contractor, add									10%	
	1000	Installing contractors, on his own labor, minimum						48%				
	1100	Maximum						88%				
600	0010	**OVERHEAD** As percent of direct costs, minimum				%					5%	600
	0050	Average									12%	
	0100	Maximum									30%	
620	0010	**OVERHEAD & PROFIT** Allowance to add to items in this										620
	0020	book that do not include Subs O&P, average				%					25%	
	0100	Allowance to add to items in this book that										
	0110	do include Subs O&P, minimum				%					5%	
	0150	Average									10%	
	0200	Maximum									15%	
	0300	Typical, by size of project, under $100,000									30%	
	0350	$500,000 project									25%	
	0400	$2,000,000 project									20%	
	0450	Over $10,000,000 project									15%	
700	0010	**FIELD PERSONNEL** Clerk average				Week		270		270	425	700
	0100	Field engineer, minimum						650		650	1,020	
	0120	Average						845		845	1,330	
	0140	Maximum						970		970	1,530	
	0160	General purpose laborer, average						890		890	1,405	
	0180	Project manager, minimum						1,220		1,220	1,930	
	0200	Average						1,370		1,370	2,160	
	0220	Maximum						1,545		1,545	2,435	
	0240	Superintendent, minimum						1,165		1,165	1,840	
	0260	Average						1,290		1,290	2,035	
	0280	Maximum						1,455		1,455	2,295	
	0290	Timekeeper, average						750		750	1,185	

01320 | Construction Progress Documents

			CREW	DAILY OUTPUT	LABOR-HOURS	UNIT	MAT.	LABOR	EQUIP.	TOTAL	TOTAL INCL O&P	
200	0010	**SCHEDULING** Critical path, as % of architectural fee, minimum				%					2%	200
	0100	Maximum				"					4%	
	0300	Computer-update, micro, no plots, minimum				Ea.					220	
	0400	Including plots, maximum				"					2,300	

01300 | Administrative Requirements

01320 | Construction Progress Documents

			Crew	Daily Output	Labor-Hours	Unit	Mat.	Labor	Equip.	Total	Total Incl O&P	
200	0600	Rule of thumb, CPM scheduling, small job				Job					.10%	200
	0650	Large job									.05%	
	0700	Cost control, small job									.15%	
	0750	Large job									.04%	

01321 | Construction Photos

			Crew	Daily Output	Labor-Hours	Unit	Mat.	Labor	Equip.	Total	Total Incl O&P	
500	0010	PHOTOGRAPHS 8" x 10", 4 shots, 2 prints ea., std. mounting				Set	97			97	107	500
	0100	Hinged linen mounts					112			112	123	
	0200	8" x 10", 4 shots, 2 prints each, in color					194			194	213	
	0300	For I.D. slugs, add to all above					2.55			2.55	2.80	
	0500	Aerial photos, initial fly-over, 6 shots, 1 print ea., 8" x 10"					830			830	915	
	0550	11" x 14" prints					865			865	955	
	0600	16" x 20" prints					1,025			1,025	1,125	
	0700	For full color prints, add					40%				40%	
	0750	Add for traffic control area					265			265	292	
	0900	For over 30 miles from airport, add per				Mile	4.90			4.90	5.40	
	1000	Vertical photography, 4 to 6 shots with										
	1010	different scales, 1 print each				Set	1,000			1,000	1,100	
	1500	Time lapse equipment, camera and projector, buy					3,575			3,575	3,925	
	1550	Rent per month					530			530	585	
	1700	Cameraman and film, including processing, B.&W.				Day	590			590	650	
	1720	Color				"	665			665	730	

01400 | Quality Requirements

01450 | Quality Control

			Crew	Daily Output	Labor-Hours	Unit	Mat.	Labor	Equip.	Total	Total Incl O&P	
500	0010	FIELD TESTING	R02315-300									500
	0015	For concrete building costing $1,000,000, minimum				Project					5,000	
	0020	Maximum									40,000	
	0050	Steel building, minimum									5,000	
	0070	Maximum									15,000	
	0100	For building costing, $10,000,000, minimum									32,000	
	0150	Maximum									50,000	
	0200	Asphalt testing, compressive strength Marshall stability, set of 3				Ea.					165	
	0220	Density, set of 3									87	
	0250	Extraction, individual tests on sample									120	
	0300	Penetration									35	
	0350	Mix design, 5 specimens									200	
	0360	Additional specimen									40	
	0400	Specific gravity									45	
	0420	Swell test									70	
	0450	Water effect and cohesion, set of 6									200	
	0470	Water effect and plastic flow									70	
	0600	Concrete testing, aggregates, abrasion									120	
	0650	Absorption									46	
	0800	Petrographic analysis									850	
	0900	Specific gravity									45	
	1000	Sieve analysis, washed									65	
	1050	Unwashed									65	
	1200	Sulfate soundness									125	

Important: See the Reference Section for critical supporting data - Reference Nos., Crews, & City Cost Indexes

01400 | Quality Requirements

01450 | Quality Control

			CREW	DAILY OUTPUT	LABOR-HOURS	UNIT	2000 BARE COSTS				TOTAL INCL O&P
							MAT.	LABOR	EQUIP.	TOTAL	
500	1300	Weight per cubic foot	R02315 -300			Ea.					40
	1500	Cement, physical tests									300
	1600	Chemical tests									270
	1800	Compressive strength test, cylinder, delivered to lab/cyl									13
	1900	Picked up by lab, minimum									15
	1950	Average									20
	2000	Maximum									30
	2200	Compressive strength, cores (not incl. drilling)				↓					40
	2250	Core drilling, 4" diameter (plus technician)				Inch					25
	2260	Technician for core drilling				Hr.					50
	2300	Patching core holes				Ea.					24
	2400	Drying shrinkage at 28 days									260
	2500	Flexural test beams									65
	2600	Mix design, one batch mix									285
	2650	Added trial batches									132
	2800	Modulus of elasticity									150
	2900	Tensile test, cylinders									50
	3000	Water-Cement ratio curve, 3 batches									155
	3100	4 batches									205
	3300	Masonry testing, absorption, per 5 brick									50
	3350	Chemical resistance, per 2 brick									55
	3400	Compressive strength, per 5 brick									70
	3420	Efflorescence, per 5 brick									50
	3440	Imperviousness, per 5 brick									96
	3470	Modulus of rupture, per 5 brick									55
	3500	Moisture, block only									35
	3550	Mortar, compressive strength, set of 3									25
	4100	Reinforcing steel, bend test									61
	4200	Tensile test, up to #8 bar									35
	4220	#9 to #11 bar									40
	4240	#14 bar and larger									50
	4400	Soil testing, Atterberg limits, liquid and plastic limits									61
	4510	Hydrometer analysis									120
	4530	Specific gravity									48
	4600	Sieve analysis, washed									56.67
	4700	Unwashed									65
	4710	Consolidation test (ASTM D2435), minimum									320
	4715	Maximum									475
	4720	Density and classification of undisturbed sample									75
	4735	Soil density, nuclear method, ASTM D2922-71									38.67
	4740	Sand cone method ASTM D1556064									30.17
	4750	Moisture content									5.33
	4780	Permeability test, double ring infiltrometer									550
	4800	Permeability, variable or constant head, undisturbed									205
	4850	Recompacted									233.33
	4900	Proctor compaction, 4" standard mold									125
	4950	6" modified mold									75
	5100	Shear tests, triaxial, minimum									450
	5150	Maximum									600
	5300	Direct shear, minimum									250
	5350	Maximum									300
	5550	Technician for inspection, per day, earthwork									215
	5650	Bolting									270
	5750	Roofing									250
	5790	Welding				↓					260
	5820	Non-destructive testing, dye penetrant				Day					320

01400 | Quality Requirements

01450 | Quality Control

			CREW	DAILY OUTPUT	LABOR-HOURS	UNIT	2000 BARE COSTS MAT.	LABOR	EQUIP.	TOTAL	TOTAL INCL O&P	
500	5840	Magnetic particle				Day					320	500
	5860	Radiography	R02315-300								480	
	5880	Ultrasonic									330	
	6000	Welding certification, minimum				Ea.					100	
	6100	Maximum				"					275	
	7000	Underground storage tank										
	7500	Hydrostatic tank tightness test per tank, min.				Ea.					500	
	7510	Maximum				"					1,000	
	7600	Vadose zone (soil gas) sampling, 10-40 samples, min.				Day					1,500	
	7610	Maximum				"					2,500	
	7700	Ground water monitoring incl. drilling 3 wells, min.				Total					5,000	
	7710	Maximum				"					7,000	
	8000	X-ray concrete slabs				Ea.					200	

01500 | Temporary Facilities & Controls

01510 | Temporary Utilities

			CREW	DAILY OUTPUT	LABOR-HOURS	UNIT	2000 BARE COSTS MAT.	LABOR	EQUIP.	TOTAL	TOTAL INCL O&P	
800	0010	**TEMPORARY UTILITIES**										800
	0101	Heat, incl. fuel and operation, per week, 12 hrs. per day	1 Clab	100	.080	CSF Flr	5.10	1.16		6.26	7.65	
	0201	24 hrs. per day	"	4.50	1.778		7.70	25.50		33.20	52.50	
	0350	Lighting, incl. service lamps, wiring & outlets, minimum	1 Elec	34	.235		2.07	5.20		7.27	10.80	
	0360	Maximum	"	17	.471		4.51	10.40		14.91	22	
	0400	Power for temp lighting only, per month, min/month 6.6 KWH								.75	1.18	
	0450	Maximum/month 23.6 KWH								2.85	2.85	
	0600	Power for job duration incl. elevator, etc., minimum								47	51.70	
	0650	Maximum								110	121	
	1000	Toilet, portable, see division 01590-400										

01520 | Construction Facilities

			CREW	DAILY OUTPUT	LABOR-HOURS	UNIT	MAT.	LABOR	EQUIP.	TOTAL	INCL O&P	
500	0010	**OFFICE** Trailer, furnished, no hookups, 20' x 8', buy	2 Skwk	1	16	Ea.	5,075	315		5,390	6,125	500
	0250	Rent per month					134			134	148	
	0300	32' x 8', buy	2 Skwk	.70	22.857		7,825	450		8,275	9,400	
	0350	Rent per month					165			165	182	
	0400	50' x 10', buy	2 Skwk	.60	26.667		13,400	525		13,925	15,600	
	0450	Rent per month					293			293	325	
	0500	50' x 12', buy	2 Skwk	.50	32		15,800	630		16,430	18,500	
	0550	Rent per month					345			345	380	
	0700	For air conditioning, rent per month, add					35.50			35.50	39	
	0800	For delivery, add per mile				Mile	1.50			1.50	1.65	
	1000	Portable buildings, prefab, on skids, economy, 8' x 8'	2 Carp	265	.060	S.F.	80	1.19		81.19	90	
	1100	Deluxe, 8' x 12'	"	150	.107	"	87	2.10		89.10	99	
	1200	Storage boxes, 20' x 8', buy	2 Skwk	1.80	8.889	Ea.	3,200	176		3,376	3,825	
	1250	Rent per month					68.50			68.50	75	
	1300	40' x 8', buy	2 Skwk	1.40	11.429		3,325	226		3,551	4,050	
	1350	Rent per month					85			85	93.50	
	5000	Air supported structures, see division 13011-200										
550	0010	**FIELD OFFICE EXPENSE**										550
	0100	Field office expense, office equipment rental average				Month	135			135	149	

01500 | Temporary Facilities & Controls

01520 | Construction Facilities

		CREW	DAILY OUTPUT	LABOR-HOURS	UNIT	2000 BARE COSTS MAT.	LABOR	EQUIP.	TOTAL	TOTAL INCL O&P	
550	0120 Office supplies, average				Month	80			80	88	550
	0125 Office trailer rental, see division 01520-500										
	0140 Telephone bill; avg. bill/month incl. long dist.				Month	225			225	248	
	0160 Field office lights & HVAC				"	90			90	99	
900	0010 **WEATHER STATION** Remote recording, minimum				Ea.	3,000			3,000	3,300	900
	0100 Maximum				"	7,100			7,100	7,800	

01530 | Temporary Construction

		CREW	DAILY OUTPUT	LABOR-HOURS	UNIT	MAT.	LABOR	EQUIP.	TOTAL	INCL O&P	
700	0010 **PROTECTION** Stair tread, 2" x 12" planks, 1 use	1 Carp	75	.107	Tread	4.43	2.10		6.53	8.45	700
	0100 Exterior plywood, 1/2" thick, 1 use		65	.123		1.59	2.42		4.01	5.90	
	0200 3/4" thick, 1 use	↓	60	.133	↓	2.42	2.63		5.05	7.15	
900	0010 **WINTER PROTECTION** Reinforced plastic on wood										900
	0100 framing to close openings	2 Clab	750	.021	S.F.	.35	.31		.66	.92	
	0200 Tarpaulins hung over scaffolding, 8 uses, not incl. scaffolding		1,500	.011		.16	.15		.31	.44	
	0250 Tarpaulin polyester reinf. w/ integral fastening system 11 mils thick		1,600	.010		.73	.14		.87	1.05	
	0300 Prefab fiberglass panels, steel frame, 8 uses	↓	1,200	.013	↓	.70	.19		.89	1.10	

01540 | Construction Aids

		CREW	DAILY OUTPUT	LABOR-HOURS	UNIT	MAT.	LABOR	EQUIP.	TOTAL	INCL O&P	
500	0010 **PERSONNEL PROTECTIVE EQUIPMENT**										500
	0015 Hazardous waste protection										
	0020 Respirator mask only, full face, silicone				Ea.	199			199	219	
	0030 Half face, silicone					23			23	25.50	
	0040 Respirator cartridges, 2 reg'd/mask, dust or asbestos					34.50			34.50	38	
	0050 Chemical vapor					30			30	33	
	0060 Combination vapor and dust					33.50			33.50	37	
	0100 Emergency escape breathing apparatus, 5 min					405			405	445	
	0110 10 min					490			490	535	
	0150 Self contained breathing apparatus with full face piece, 30 min					1,825			1,825	2,000	
	0160 60 min					2,725			2,725	3,000	
	0200 Encapsulating suits, limited use, level A					690			690	760	
	0210 Level B				↓	148			148	163	
	0300 Over boots, latex				Pr.	5.35			5.35	5.90	
	0310 PVC					15.35			15.35	16.85	
	0320 Neoprene					37.50			37.50	41.50	
	0400 Gloves, nitrile/PVC					4.83			4.83	5.30	
	0410 Neoprene coated				↓	28			28	30.50	
550	0010 **PUMP STAGING**, Aluminum		R01540 -200								550
	0200 24' long pole section, buy				Ea.	335			335	365	
	0300 18' long pole section, buy					258			258	284	
	0400 12' long pole section, buy					174			174	192	
	0500 6' long pole section, buy					92			92	101	
	0600 6' long splice joint section, buy					68			68	75	
	0700 Pump jack					111			111	123	
	0900 Foldable brace					48.50			48.50	53	
	1000 Workbench/back safety rail support					59			59	64.50	
	1100 Scaffolding planks/workbench, 14" wide x 24' long					545			545	600	
	1200 Plank end safety rail					183			183	202	
	1250 Safety net, 22' long					267			267	294	
	1300 System in place, 50' working height, per use based on 50 uses	2 Carp	84.80	.189	C.S.F.	4.95	3.72		8.67	11.80	
	1400 100 uses		84.80	.189		2.48	3.72		6.20	9.05	
	1500 150 uses	↓	84.80	.189	↓	1.66	3.72		5.38	8.15	
700	0010 **SAFETY NETS** No supports, stock sizes, nylon, 4" mesh				S.F.	1.10			1.10	1.21	700
	0100 Polypropylene, 6" mesh				↓	1.55			1.55	1.70	

GENERAL REQUIREMENTS 1

01500 | Temporary Facilities & Controls

01540 | Construction Aids

		CREW	DAILY OUTPUT	LABOR-HOURS	UNIT	2000 BARE COSTS MAT.	LABOR	EQUIP.	TOTAL	TOTAL INCL O&P	
700	0200 Small mesh debris nets, 1/4" & 3/4" mesh, stock sizes				S.F.	.72			.72	.79	700
	0220 Combined 4" mesh and 1/4" mesh, stock sizes					1.80			1.80	1.98	
	0300 Monthly rental, 4" mesh, stock sizes, 1st month					.19			.19	.21	
	0320 2nd month rental					.15			.15	.17	
	0340 Maximum rental/year					.75			.75	.83	
750	0014 **SCAFFOLDING**				R01540 -100						750
	0015 Steel tubular, rent, 1 use/mo, no plank, erect & dismantle										
	0091 Building exterior, wall face, 1 to 5 stories	3 Clab	24	1	C.S.F.	24.50	14.45		38.95	51.50	
	0201 6 to 12 stories	4 Clab	21.20	1.509		24.50	22		46.50	64	
	0301 13 to 20 stories	5 Clab	20	2		24.50	29		53.50	76	
	0461 Building interior, walls face area, up to 16' high	3 Clab	25	.960		24.50	13.85		38.35	50.50	
	0561 16' to 40' high		23	1.043		24.50	15.10		39.60	52.50	
	0801 Building interior floor area, up to 30' high		312	.077	C.C.F.	1.87	1.11		2.98	3.96	
	0901 Over 30' high	4 Clab	275	.116	"	1.87	1.68		3.55	4.94	
	0910 Steel tubular, heavy duty shoring, buy										
	0920 Frames 5' high 2' wide				Ea.	75			75	82.50	
	0925 5' high 4' wide					85			85	93.50	
	0930 6' high 2' wide					86			86	94.50	
	0935 6' high 4' wide					101			101	111	
	0940 Accessories										
	0945 Cross braces				Ea.	16			16	17.60	
	0950 U-head, 8" x 8"					16			16	17.60	
	0955 J-head, 4" x 8"					12			12	13.20	
	0960 Base plate, 8" x 8"					13			13	14.30	
	0965 Leveling jack					30.50			30.50	33.50	
	1000 Steel tubular, regular, buy										
	1100 Frames 3' high 5' wide				Ea.	58			58	64	
	1150 5' high 5' wide					67			67	73.50	
	1200 6'-4" high 5' wide					84			84	92.50	
	1350 7'-6" high 6' wide					145			145	160	
	1500 Accessories cross braces					15			15	16.50	
	1550 Guardrail post					15			15	16.50	
	1600 Guardrail 7' section					7.25			7.25	8	
	1650 Screw jacks & plates					24			24	26.50	
	1700 Sidearm brackets					28			28	31	
	1750 8" casters					33			33	36.50	
	1800 Plank 2" x 10" x 16'-0"					42.50			42.50	47	
	1900 Stairway section					235			235	259	
	1910 Stairway starter bar					21			21	23.50	
	1920 Stairway inside handrail					53			53	58.50	
	1930 Stairway outside handrail					73			73	80.50	
	1940 Walk-thru frame guardrail					28			28	31	
	2000 Steel tubular, regular, rent/mo.										
	2100 Frames 3' high 5' wide				Ea.	3.75			3.75	4.13	
	2150 5' high 5' wide					3.75			3.75	4.13	
	2200 6'-4" high 5' wide					3.75			3.75	4.13	
	2250 7'-6" high 6' wide					7			7	7.70	
	2500 Accessories, cross braces					.60			.60	.66	
	2550 Guardrail post					1			1	1.10	
	2600 Guardrail 7' section					.75			.75	.83	
	2650 Screw jacks & plates					1.50			1.50	1.65	
	2700 Sidearm brackets					1.50			1.50	1.65	
	2750 8" casters					6			6	6.60	
	2800 Outrigger for rolling tower					3			3	3.30	
	2850 Plank 2" x 10" x 16'-0"					5			5	5.50	

01500 | Temporary Facilities & Controls

01540 | Construction Aids

			CREW	DAILY OUTPUT	LABOR-HOURS	UNIT	2000 BARE COSTS MAT.	LABOR	EQUIP.	TOTAL	TOTAL INCL O&P	
750	2900	Stairway section				Ea.	10			10	11	750
	2910	Stairway starter bar	R01540 -100				.10			.10	.11	
	2920	Stairway inside handrail					5			5	5.50	
	2930	Stairway outside handrail					5			5	5.50	
	2940	Walk-thru frame guardrail					2			2	2.20	
	3000	Steel tubular, heavy duty shoring, rent/mo.										
	3250	5' high 2' & 4' wide				Ea.	5			5	5.50	
	3300	6' high 2' & 4' wide					5			5	5.50	
	3500	Accessories, cross braces					1			1	1.10	
	3600	U - head, 8" x 8"					1			1	1.10	
	3650	J - head, 4" x 8"					1			1	1.10	
	3700	Base plate, 8" x 8"					1			1	1.10	
	3750	Leveling jack					2			2	2.20	
	5700	Planks, 2"x10"x16'-0" in place, up to 50' high	3 Carp	144	.167			3.28		3.28	5.65	
	5800	In place over 50' high	4 Carp	160	.200			3.94		3.94	6.75	
	6000	Heavy duty shoring for elevated slab forms to 8'-2" high, floor area										
	6010	Set up and take down										
	6101	1 use/month	4 Clab	36	.889	C.S.F.	29.50	12.85		42.35	54.50	
	6151	2 uses/month	"	36	.889	"	14.80	12.85		27.65	38.50	
	6500	To 14'-8" high										
	6601	1 use/month	4 Clab	18	1.778	C.S.F.	43	25.50		68.50	91.50	
	6651	2 uses/month	"	18	1.778	"	21.50	25.50		47	68	
755	0011	**SCAFFOLDING SPECIALTIES**										755
	1200	Sidewalk bridge, heavy duty steel posts & beams, including										
	1210	parapet protection & waterproofing										
	1221	8' to 10' wide, 2 posts	3 Clab	15	1.600	L.F.	42	23		65	85.50	
	1231	3 posts	"	10	2.400	"	63	34.50		97.50	129	
	1500	Sidewalk bridge using tubular steel										
	1511	scaffold frames, including planking	3 Clab	45	.533	L.F.	4.72	7.70		12.42	18.40	
	1600	For 2 uses per month, deduct from all above					50%					
	1700	For 1 use every 2 months, add to all above					100%					
	1900	Catwalks, 32" wide, no guardrails, 6' span, buy				Ea.	120			120	132	
	2000	10' span, buy				"	190			190	209	
	2800	Hand winch-operated masons										
	2810	scaffolding, no plank moving required										
	2900	98' long, 10'-6" high, buy				Ea.	19,700			19,700	21,700	
	3000	Rent per month					790			790	865	
	3100	28'-6" high, buy					25,700			25,700	28,200	
	3200	Rent per month					1,025			1,025	1,125	
	3400	196' long, 28'-6" high, buy					49,700			49,700	54,500	
	3500	Rent per month					2,000			2,000	2,175	
	3600	64'-6" high, buy					73,000			73,000	80,000	
	3700	Rent per month					2,900			2,900	3,200	
	3720	Putlog, standard, 8' span, with hangers, buy					61			61	67	
	3730	Rent per month					10			10	11	
	3750	12' span, buy					92			92	101	
	3755	Rent per month					15			15	16.50	
	3760	Trussed type, 16' span, buy					210			210	231	
	3770	Rent per month					20			20	22	
	3790	22' span, buy					252			252	277	
	3795	Rent per month					30			30	33	
	3800	Rolling ladders with handrails, 30" wide, buy, 2 step					142			142	156	
	4000	7 step					425			425	470	
	4050	10 step					640			640	705	
	4100	Rolling towers, buy, 5' wide, 7' long, 10' high					1,150			1,150	1,250	
	4200	For 5' high added sections, to buy, add					188			188	207	

01500 | Temporary Facilities & Controls

01540 | Construction Aids

			DAILY	LABOR-		2000 BARE COSTS				TOTAL
		CREW	OUTPUT	HOURS	UNIT	MAT.	LABOR	EQUIP.	TOTAL	INCL O&P
755	4300 Complete incl. wheels, railings, outriggers,									
	4350 21' high, to buy				Ea.	1,925			1,925	2,125
	4400 Rent/month				"	138			138	152
760	0010 **STAGING AIDS** and fall protection equipment									
	0100 Sidewall staging bracket, tubular, buy				Ea.	29			29	31.50
	0110 Cost each per day, based on 250 days use				Day	.11			.11	.13
	0200 Guard post, buy				Ea.	15			15	16.50
	0210 Cost each per day, based on 250 days use				Day	.06			.06	.07
	0300 End guard chains, buy per set				Ea.	25			25	27.50
	0310 Cost per set per day, based on 250 days use				Day	.11			.11	.13
	1000 Roof shingling bracket, steel, buy				Ea.	5.75			5.75	6.30
	1010 Cost each per day, based on 250 days use				Day	.02			.02	.03
	1100 Wood bracket, buy				Ea.	10.60			10.60	11.65
	1110 Cost each per day, based on 250 days use				Day	.04			.04	.05
	2000 Ladder jack, aluminum, buy per pair				Pair	76.50			76.50	84
	2010 Cost per pair per day, based on 250 days use				Day	.31			.31	.34
	2100 Steel siderail jack, buy per pair				Pair	69			69	76
	2110 Cost per pair per day, based on 250 days use				Day	.28			.28	.30
	3000 Laminated wood plank, 2x10x16', buy				Ea.	42.50			42.50	47
	3010 Cost each per day, based on 250 days use				Day	.17			.17	.19
	3100 Aluminum scaffolding plank, 20" wide x 24' long, buy				Ea.	660			660	730
	3110 Cost each per day, based on 250 days use				Day	2.65			2.65	2.91
	4000 Nylon full body harness, lanyard and rope grab				Ea.	225			225	248
	4010 Cost each per day, based on 250 days use				Day	.90			.90	.99
	4100 Rope for safety line, 5/8" x 100' nylon, buy				Ea.	35			35	38.50
	4110 Cost each per day, based on 250 days use				Day	.14			.14	.15
	4200 Permanent U-Bolt roof anchor, buy				Ea.	29.50			29.50	32.50
	4300 Temporary (one use) roof ridge anchor, buy				"	18.60			18.60	20.50
	5000 Installation (setup and removal) of staging aids									
	5010 Sidewall staging bracket	2 Carp	64	.250	Ea.		4.93		4.93	8.45
	5020 Guard post with 2 wood rails	"	64	.250			4.93		4.93	8.45
	5030 End guard chains, set	1 Carp	64	.125			2.46		2.46	4.22
	5100 Roof shingling bracket		96	.083			1.64		1.64	2.81
	5200 Ladder jack	↓	64	.125			2.46		2.46	4.22
	5300 Wood plank, 2x10x16'	2 Carp	80	.200			3.94		3.94	6.75
	5310 Aluminum scaffold plank, 20" x 24'	"	40	.400			7.90		7.90	13.50
	5410 Safety rope	1 Carp	40	.200			3.94		3.94	6.75
	5420 Permanent U-Bolt roof anchor (install only)	2 Carp	40	.400			7.90		7.90	13.50
	5430 Temporary roof ridge anchor (install only)	1 Carp	64	.125	↓		2.46		2.46	4.22
780	0010 **SWING STAGING**, 500 lb cap., 2' wide to 24' long, hand operated hoist									
	0020 steel cable type, with 60' cables, buy				Ea.	3,650			3,650	4,000
	0030 Rent per month				"	450			450	495
	0600 Lightweight (not for masons) 24' long for 150' height.									
	0610 manual type, buy				Ea.	3,900			3,900	4,300
	0620 Rent per month					520			520	575
	0700 Powered, electric or air, to 150' high, buy					13,500			13,500	14,800
	0710 Rent per month					840			840	925
	0780 To 300' high, buy					14,300			14,300	15,700
	0800 Rent per month					910			910	1,000
	1000 Bosun's chair or work basket 3' x 3.5', to 300' high, electric, buy					7,250			7,250	7,975
	1010 Rent per month				↓	450			450	495
	2200 Move swing staging (setup and remove)	E-4	2	16	Move		350	42	392	730
790	0010 **SURVEYOR STAKES** Hardwood, 1" x 1" x 48" long				C	41.50			41.50	46
	0100 2" x 2" x 18" long				↓	50			50	55

Important: See the Reference Section for critical supporting data - Reference Nos., Crews, & City Cost Indexes

01500 | Temporary Facilities & Controls

01540 | Construction Aids

			CREW	DAILY OUTPUT	LABOR-HOURS	UNIT	MAT.	LABOR	EQUIP.	TOTAL	TOTAL INCL O&P	
790	0150	2" x 2" x 24" long				C	62			62	68	790
	0200	2" x 2" x 30" long					71			71	78	
800	0010	TARPAULINS Cotton duck, 10 oz. to 13.13 oz. per S.Y., minimum				S.F.	.46			.46	.51	800
	0050	Maximum					.55			.55	.61	
	0100	Polyvinyl coated nylon, 14 oz. to 18 oz., minimum					.45			.45	.50	
	0150	Maximum					.65			.65	.71	
	0200	Reinforced polyethylene 3 mils thick, white					.10			.10	.11	
	0300	4 mils thick, white, clear or black					.12			.12	.13	
	0400	5.5 mils thick, clear					.09			.09	.10	
	0500	White, fire retardant					.16			.16	.18	
	0600	7.5 mils, oil resistant, fire retardant					.17			.17	.19	
	0700	8.5 mils, black					.22			.22	.24	
	0710	Woven polyethylene, 6 mils thick					.45			.45	.50	
	0730	Polyester reinforced w/ integral fastening system 11 mils thick					1			1	1.10	
	0740	Mylar polyester, non-reinforced, 7 mils thick					1.10			1.10	1.21	
820	0010	SMALL TOOLS As % of contractor's work, minimum R01100-050				Total					.50%	820
	0100	Maximum				"					2%	

01550 | Vehicular Access & Parking

			CREW	DAILY OUTPUT	LABOR-HOURS	UNIT	MAT.	LABOR	EQUIP.	TOTAL	TOTAL INCL O&P	
700	0010	ROADS AND SIDEWALKS Temporary										700
	0050	Roads, gravel fill, no surfacing, 4" gravel depth	B-14	715	.067	S.Y.	1.04	1.04	.28	2.36	3.23	
	0100	8" gravel depth	"	615	.078	"	2.07	1.21	.33	3.61	4.71	
	1001	Ramp, 3/4" plywood on 2" x 6" joists, 16" O.C.	2 Clab	300	.053	S.F.	1.58	.77		2.35	3.06	
	1101	On 2" x 10" joists, 16" O.C.	"	275	.058		1.91	.84		2.75	3.54	
	2201	Sidewalks, 2" x 12" planks, 2 uses	1 Clab	350	.023		.60	.33		.93	1.23	
	2301	Exterior plywood, 2 uses, 1/2" thick		750	.011		.33	.15		.48	.62	
	2401	5/8" thick		650	.012		.38	.18		.56	.72	
	2501	3/4" thick		600	.013		.46	.19		.65	.84	

01560 | Barriers & Enclosures

			CREW	DAILY OUTPUT	LABOR-HOURS	UNIT	MAT.	LABOR	EQUIP.	TOTAL	TOTAL INCL O&P	
100	0011	BARRICADES 5' high, 3 rail @ 2" x 8", fixed	2 Clab	30	.533	L.F.	10.80	7.70		18.50	25	100
	0200	Precast barrier walls, 10' sections	B-6	240	.100	"	26.50	1.60	.85	28.95	32.50	
	0300	Stock units, 6' high, 8' wide, plain, buy				Ea.	430			430	475	
	0350	With reflective tape, buy				"	525			525	580	
	0400	Break-a-way 3" PVC pipe barricade										
	0410	with 3 ea. 1' x 4' reflectorized panels, buy				Ea.	305			305	335	
	0500	Plywood with steel legs, 32" wide					70			70	77	
	0600	Telescoping Christmas tree, 9' high, 5 flags, buy					121			121	133	
	0800	Traffic cones, PVC, 18" high					5.80			5.80	6.40	
	0850	28" high					18.15			18.15	20	
	1001	Guardrail, wooden, 3' high, 1" x 6" on 2" x 4" posts	2 Clab	200	.080	L.F.	1.01	1.16		2.17	3.09	
	1101	2" x 6" on 4" x 4" posts	"	165	.097		2.04	1.40		3.44	4.65	
	1200	Portable metal with base pads, buy					15			15	16.50	
	1251	Typical installation, assume 10 reuses	2 Clab	600	.027		1.58	.39		1.97	2.40	
	5000	Barricades, see also division 01590-400										
250	0010	FENCING Chain link, 11 ga, 5' high	2 Clab	100	.160	L.F.	4.04	2.31		6.35	8.40	250
	0100	6' high		75	.213		3.37	3.08		6.45	9	
	0200	Rented chain link, 6' high, to 500' (up to 12 mo.)		100	.160		2.26	2.31		4.57	6.45	
	0250	Over 1000' (up to 12 mo.)		110	.145		1.64	2.10		3.74	5.40	
	0351	Plywood, painted, 2" x 4" frame, 4' high	3 Clab	135	.178		4.53	2.57		7.10	9.40	
	0401	4" x 4" frame, 8' high	"	110	.218		8.60	3.15		11.75	14.85	
	0501	Wire mesh on 4" x 4" posts, 4' high	2 Clab	100	.160		6.15	2.31		8.46	10.70	
	0551	8' high	"	80	.200		9.10	2.89		11.99	14.95	

GENERAL REQUIREMENTS 1

01500 | Temporary Facilities & Controls

01560 | Barriers & Enclosures

			CREW	DAILY OUTPUT	LABOR-HOURS	UNIT	2000 BARE COSTS				TOTAL INCL O&P	
							MAT.	LABOR	EQUIP.	TOTAL		
400	0010	**TEMPORARY CONSTRUCTION** See also division 01530										400
800	0010	**WATCHMAN** Service, monthly basis, uniformed person, minimum				Hr.					7.80	800
	0100	Maximum									14.15	
	0200	Person and command dog, minimum									10.30	
	0300	Maximum									15.30	
	0500	Sentry dog, leased, with job patrol (yard dog), 1 dog				Week					195	
	0600	2 dogs				"					275	
	0800	Purchase, trained sentry dog, minimum				Ea.					800	
	0900	Maximum				"					2,000	

01580 | Project Signs

700	0010	**SIGNS** Hi-intensity reflectorized, no posts, buy				S.F.	12.05			12.05	13.25	700

Important: See the Reference Section for critical supporting data - Reference Nos., Crews, & City Cost Indexes

01500 | Temporary Facilities & Controls

01590 | Equipment Rental

			Unit	Hourly Oper. Cost	Rent Per Day	Rent Per Week	Rent Per Month	Crew Equipment Cost/Day
100	0010	**CONCRETE EQUIPMENT RENTAL** R01590-100						
	0100	without operators						
	0200	Bucket, concrete lightweight, 1/2 C.Y. R03310-090	Ea.	.15	28.50	85	255	18.20
	0300	1 C.Y.		.20	42.50	128	385	27.20
	0400	1-1/2 C.Y.		.25	45	135	405	29
	0500	2 C.Y.		.26	58.50	175	525	37.10
	0600	Cart, concrete, self propelled, operator walking, 10 C.F.		1.25	70	210	630	52
	0700	Operator riding, 18 C.F.		2.40	107	320	960	83.20
	0800	Conveyer for concrete, portable, gas, 16" wide, 26' long		3.60	183	550	1,650	138.80
	0900	46' long		3.70	237	710	2,125	171.60
	1000	56' long		4	250	750	2,250	182
	1100	Core drill, electric, 2-1/2 H.P., 1" to 8" bit diameter		.50	73.50	220	660	48
	1150	11 H.P., 8" to 18" cores		.70	91.50	275	825	60.60
	1200	Finisher, concrete floor, gas, riding trowel, 48" diameter		2.70	117	350	1,050	91.60
	1300	Gas, manual, 3 blade, 36" trowel		2	53.50	160	480	48
	1400	4 blade, 48" trowel		2.25	58.50	175	525	53
	1500	Float, hand-operated (Bull float) 48" wide		.10	8.35	25	75	5.80
	1570	Curb builder, 14 H.P., gas, single screw		1.75	66.50	200	600	54
	1590	Double screw		2.35	66.50	200	600	58.80
	1600	Grinder, concrete and terrazzo, electric, floor		1.25	66.50	200	600	50
	1700	Wall grinder		.60	33.50	100	300	24.80
	1800	Mixer, powered, mortar and concrete, gas, 6 C.F., 18 H.P.		2.44	58.50	175	525	54.50
	1900	10 C.F., 25 H.P.		3.18	53.50	160	480	57.45
	2000	16 C.F.		3.65	75	225	675	74.20
	2100	Concrete, stationary, tilt drum, 2 C.Y.		7.75	267	800	2,400	222
	2120	Pump, concrete, truck mounted, 4" line, 80' boom		12	935	2,800	8,400	656
	2140	5" line, 110' boom		13.50	1,100	3,300	9,900	768
	2160	Mud jack, 50 C.F. per hr.		3.15	70	210	630	67.20
	2180	225 C.F. per hr.		6.25	335	1,000	3,000	250
	2600	Saw, concrete, manual, gas, 18 H.P.		2.75	80	240	720	70
	2650	Self-propelled, gas, 30 H.P.		5.50	110	330	990	110
	2700	Vibrators, concrete, electric, 60 cycle, 2 H.P.		.25	28.50	85	255	19
	2800	3 H.P.		.26	40	120	360	26.10
	2900	Gas engine, 5 H.P.		.65	40	120	360	29.20
	3000	8 H.P.		1	50	150	450	38
	3100	Concrete transit mixer, hydraulic drive						
	3120	6 x 4, 250 H.P., 8 C.Y., rear discharge		30	935	2,800	8,400	800
	3200	Front discharge		30.25	1,000	3,000	9,000	842
	3300	6 x 6, 285 H.P., 12 C.Y., rear discharge		32.50	1,050	3,150	9,450	890
	3400	Front discharge		32.75	1,100	3,300	9,900	922
200	0010	**EARTHWORK EQUIPMENT RENTAL** Without operators R01590-100						
	0040	Aggregate spreader, push type 8' to 12' wide	Ea.	1.10	108	325	975	73.80
	0050	Augers for truck or trailer mounting, vertical drilling R02315-300						
	0055	Fence post auger, truck mounted	Ea.	10	485	1,450	4,350	370
	0060	4" to 36" diam., 54 H.P., gas, 10' spindle travel R02315-400		16	590	1,775	5,325	483
	0070	14' spindle travel		16.40	665	2,000	6,000	531.20
	0075	Auger, truck mounted, vertical drilling, to 25' depth R02315-450		49.50	2,325	7,000	21,000	1,796
	0080	Auger, horizontal boring machine, 12" to 36" diameter, 45 H.P.		7.35	335	1,000	3,000	258.80
	0090	12" to 48" diameter, 65 H.P. R02455-900		12.25	665	2,000	6,000	498
	0100	Excavator, diesel hydraulic, crawler mounted, 1/2 C.Y. cap.		11.65	415	1,248	3,750	342.80
	0120	5/8 C.Y. capacity		15.50	480	1,438	4,325	411.60
	0140	3/4 C.Y. capacity		17.40	530	1,583	4,750	455.80
	0150	1 C.Y. capacity		22.25	600	1,800	5,400	538
	0200	1-1/2 C.Y. capacity		26.65	835	2,500	7,500	713.20
	0300	2 C.Y. capacity		42.15	1,125	3,400	10,200	1,017
	0320	2-1/2 C.Y. capacity		59.15	2,000	6,000	18,000	1,673
	0340	3-1/2 C.Y. capacity		78.25	2,500	7,500	22,500	2,126
	0341	Attachments						

GENERAL REQUIREMENTS 1

01500 | Temporary Facilities & Controls

01590 | Equipment Rental

			UNIT	HOURLY OPER. COST	RENT PER DAY	RENT PER WEEK	RENT PER MONTH	CREW EQUIPMENT COST/DAY
0342	Bucket thumbs	R01590-100	Ea.	.45	292	875	2,625	178.60
0345	Grapples			.35	300	900	2,700	182.80
0350	Gradall type, truck mounted, 3 ton @ 15' radius, 5/8 C.Y.	R02315-300		25.40	690	2,075	6,225	618.20
0370	1 C.Y. capacity			26.33	1,000	3,036	9,100	817.85
0400	Backhoe-loader, 40 to 45 H.P., 5/8 C.Y. capacity	R02315-400		7.15	217	650	1,950	187.20
0450	45 H.P. to 60 H.P., 3/4 C.Y. capacity			8.50	225	675	2,025	203
0460	80 H.P., 1-1/4 C.Y. capacity	R02315-450		12.22	292	877	2,625	273.15
0470	112 H.P., 1-1/2 C.Y. capacity			15.50	395	1,185	3,550	361
0480	Attachments	R02455-900						
0482	Compactor, 20,000 lb			1.35	192	575	1,725	125.80
0485	Hydraulic hammer, 750 ft-lbs			1.15	300	900	2,700	189.20
0486	Hydraulic hammer, 1200 ft-lbs			1.70	335	1,000	3,000	213.60
0500	Brush chipper, gas engine, 6" cutter head, 35 H.P.			3.85	142	425	1,275	115.80
0550	12" cutter head, 130 H.P.			11.88	193	580	1,750	211.05
0600	15" cutter head, 165 H.P.			15.48	267	800	2,400	283.85
0750	Bucket, clamshell, general purpose, 3/8 C.Y.			.75	61.50	185	555	43
0800	1/2 C.Y.			.85	70	210	630	48.80
0850	3/4 C.Y.			1.10	80	240	720	56.80
0900	1 C.Y.			1.25	100	300	900	70
0950	1-1/2 C.Y.			1.50	133	400	1,200	92
1000	2 C.Y.			1.75	150	450	1,350	104
1010	Bucket, dragline, medium duty, 1/2 C.Y.			.36	40	120	360	26.90
1020	3/4 C.Y.			.65	46.50	140	420	33.20
1030	1 C.Y.			.88	50	150	450	37.05
1040	1-1/2 C.Y.			1.12	66.50	200	600	48.95
1050	2 C.Y.			1.40	83.50	250	750	61.20
1070	3 C.Y.			1.45	117	350	1,050	81.60
1200	Compactor, roller, 2 drum, 2000 lb., operator walking			1.78	133	400	1,200	94.25
1250	Rammer compactor, gas, 1000 lb. blow			.75	50	150	450	36
1300	Vibratory plate, gas, 13" plate, 1000 lb. blow			.75	43.50	130	390	32
1350	24" plate, 5000 lb. blow			1.95	75	225	675	60.60
1370	Curb builder/extruder, 14 H.P., gas, single screw			1.50	50	150	450	42
1390	Double screw			2.25	91.50	275	825	73
1750	Extractor, piling, see lines 2500 to 2750							
1860	Grader, self-propelled, 25,000 lb.		Ea.	12.85	555	1,658	4,975	434.40
1910	30,000 lb.			14.50	665	2,000	6,000	516
1920	40,000 lb.			20.90	835	2,500	7,500	667.20
1930	55,000 lb.			29.90	1,400	4,183	12,500	1,076
1950	Hammer, pavement demo., hyd., gas, self-prop., 1000 to 1250 lb.			16.55	410	1,230	3,700	378.40
2000	Diesel 1300 to 1500 lb.			16.60	460	1,375	4,125	407.80
2050	Pile driving hammer, steam or air, 4150 ft.-lb. @ 225 BPM			1.40	293	880	2,650	187.20
2100	8750 ft.-lb. @ 145 BPM			1.85	420	1,260	3,775	266.80
2150	15,000 ft.-lb. @ 60 BPM			2.05	490	1,475	4,425	311.40
2200	24,450 ft.-lb. @ 111 BPM			2.45	600	1,800	5,400	379.60
2250	Leads, 15,000 ft.-lb. hammers		L.F.	.25	4	12	36	4.40
2300	24,450 ft.-lb. hammers and heavier		"	.28	5	15	45	5.25
2350	Diesel type hammer, 22,400 ft.-lb.		Ea.	6.35	490	1,468	4,400	344.40
2400	41,300 ft.-lb.			9.95	750	2,250	6,750	529.60
2450	141,000 ft.-lb.			19.96	1,475	4,455	13,400	1,051
2500	Vib. elec. hammer/extractor, 200 KW diesel generator, 34 H.P.			9.35	610	1,825	5,475	439.80
2550	80 H.P.			16.12	945	2,835	8,500	695.95
2600	150 H.P.			31.35	1,600	4,835	14,500	1,218
2700	Extractor, steam or air, 700 ft.-lb.			1.44	173	520	1,550	115.50
2750	1000 ft.-lb.			1.90	242	725	2,175	160.20
3000	Roller, tandem, gas, 3 to 5 ton			4.15	167	500	1,500	133.20
3050	Diesel, 8 to 12 ton			6	293	880	2,650	224
3100	Towed type, vibratory, gas 12.5 H.P., 2 ton			3.75	133	400	1,200	110
3150	Sheepsfoot, double 60" x 60"			4.80	147	440	1,325	126.40

Important: See the Reference Section for critical supporting data - Reference Nos., Crews, & City Cost Indexes

01500 | Temporary Facilities & Controls

01590 | Equipment Rental

			UNIT	HOURLY OPER. COST	RENT PER DAY	RENT PER WEEK	RENT PER MONTH	CREW EQUIPMENT COST/DAY
3200	Pneumatic tire diesel roller, 12 ton	R01590-100	Ea.	7.60	300	900	2,700	240.80
3250	21 to 25 ton			12.56	350	1,050	3,150	310.50
3300	Sheepsfoot roller, self-propelled, 4 wheel, 130 H.P.	R02315-300		20.95	665	2,000	6,000	567.60
3320	300 H.P.			26.70	785	2,350	7,050	683.60
3350	Vibratory steel drum & pneumatic tire, diesel, 18,000 lb.	R02315-400		11.50	465	1,400	4,200	372
3400	29,000 lb.			12.80	550	1,650	4,950	432.40
3450	Scrapers, towed type, 9 to 12 C.Y. capacity	R02315-450		3.25	91.50	275	825	81
3500	12 to 17 C.Y. capacity			6.10	260	780	2,350	204.80
3550	Scrapers, self-propelled, 4 x 4 drive, 2 engine, 14 C.Y. capacity	R02455-900		60.47	1,900	5,710	17,100	1,626
3600	2 engine, 24 C.Y. capacity			76.64	2,175	6,500	19,500	1,913
3650	Self-loading, 11 C.Y. capacity			27.32	800	2,400	7,200	698.55
3700	22 C.Y. capacity			33.10	1,200	3,600	10,800	984.80
3710	Screening plant 110 hp. w / 5' x 10' screen			16.28	450	1,350	4,050	400.25
3720	5' x 16' screen			17.65	520	1,560	4,675	453.20
3850	Shovels, see Cranes division 01590-600							
3860	Shovel/backhoe bucket, 1/2 C.Y.		Ea.	1.15	78.50	235	705	56.20
3870	3/4 C.Y.			3.15	127	380	1,150	101.20
3880	1 C.Y.			3.65	173	520	1,550	133.20
3890	1-1/2 C.Y.			4.25	212	635	1,900	161
3910	3 C.Y.			7.55	400	1,200	3,600	300.40
4110	Tractor, crawler, with bulldozer, torque converter, diesel 75 H.P.			9.95	335	1,000	3,000	279.60
4150	105 H.P.			10.75	510	1,535	4,600	393
4200	140 H.P.			17.55	650	1,950	5,850	530.40
4260	200 H.P.			25.85	1,000	3,000	9,000	806.80
4310	300 H.P.			34.70	1,375	4,150	12,500	1,108
4360	410 H.P.			47.10	1,700	5,100	15,300	1,397
4380	700 H.P.			94	3,325	10,000	30,000	2,752
4400	Loader, crawler, torque conv., diesel, 1-1/2 C.Y., 80 H.P.			9.87	420	1,255	3,775	329.95
4450	1-1/2 to 1-3/4 C.Y., 95 H.P.			10.23	475	1,425	4,275	366.85
4510	1-3/4 to 2-1/4 C.Y., 130 H.P.			12.50	585	1,750	5,250	450
4530	2-1/2 to 3-1/4 C.Y., 190 H.P.			19.25	1,050	3,150	9,450	784
4560	3-1/2 to 5 C.Y., 275 H.P.			26.90	1,375	4,100	12,300	1,035
4610	Tractor loader, wheel, torque conv., 4 x 4, 1 to 1-1/4 C.Y., 65 H.P.			8.45	277	830	2,500	233.60
4620	1-1/2 to 1-3/4 C.Y., 80 H.P.			9.75	375	1,125	3,375	303
4650	1-3/4 to 2 C.Y., 100 H.P.			11.35	405	1,215	3,650	333.80
4710	2-1/2 to 3-1/2 C.Y., 130 H.P.			12.60	500	1,500	4,500	400.80
4730	3 to 4-1/2 C.Y., 170 H.P.			15.30	700	2,100	6,300	542.40
4760	5-1/4 to 5-3/4 C.Y., 270 H.P.			30.15	1,025	3,100	9,300	861.20
4810	7 to 8 C.Y., 375 H.P.			45.65	1,425	4,300	12,900	1,225
4870	12-1/2 C.Y., 690 H.P.			91.25	2,500	7,500	22,500	2,230
4880	Wheeled, skid steer, 10 C.F., 30 H.P. gas			4.55	133	400	1,200	116.40
4890	1 C.Y., 78 H.P., diesel			6.35	292	875	2,625	225.80
4891	Attachments for all skid steer loaders							
4892	Auger		Ea.	.12	83.50	250	750	50.95
4893	Backhoe			.15	110	330	990	67.20
4894	Broom			.16	107	320	960	65.30
4895	Forks			.08	36.50	110	330	22.65
4896	Grapple			.12	83.50	250	750	50.95
4897	Concrete hammer			.25	180	540	1,625	110
4898	Tree spade			.36	153	460	1,375	94.90
4899	Trencher			.41	122	365	1,100	76.30
4900	Trencher, chain, boom type, gas, operator walking, 12 H.P.			2.16	142	426	1,275	102.50
4910	Operator riding, 40 H.P.			6.36	267	800	2,400	210.90
5000	Wheel type, diesel, 4' deep, 12" wide			12.42	515	1,550	4,650	409.35
5100	Diesel, 6' deep, 20" wide			15.65	735	2,200	6,600	565.20
5150	Ladder type, diesel, 5' deep, 8" wide			9.40	385	1,150	3,450	305.20
5200	Diesel, 8' deep, 16" wide			17.61	665	2,000	6,000	540.90
5250	Truck, dump, tandem, 12 ton payload			17.41	375	1,130	3,400	365.30

GENERAL REQUIREMENTS

01500 | Temporary Facilities & Controls

01590 | Equipment Rental

		UNIT	HOURLY OPER. COST	RENT PER DAY	RENT PER WEEK	RENT PER MONTH	CREW EQUIPMENT COST/DAY	
200 5300	Three axle dump, 16 ton payload	Ea.	20.28	475	1,425	4,275	447.25	200
5350	Dump trailer only, rear dump, 16-1/2 C.Y.		3.45	177	530	1,600	133.60	
5400	20 C.Y.		3.51	180	540	1,625	136.10	
5450	Flatbed, single axle, 1-1/2 ton rating		10.87	143	430	1,300	172.95	
5500	3 ton rating		11.43	148	445	1,325	180.45	
5550	Off highway rear dump, 25 ton capacity		20.50	895	2,680	8,050	700	
5600	35 ton capacity		32.27	1,325	4,000	12,000	1,058	
400 0010	**GENERAL EQUIPMENT RENTAL** Without operators							400
0150	Aerial lift, scissor type, to 15' high, 1000 lb. cap., electric	Ea.	1.18	83.50	250	750	59.45	
0160	To 25' high, 2000 lb. capacity		1.80	125	375	1,125	89.40	
0170	Telescoping boom to 40' high, 750 lb. capacity, gas		6.42	335	1,000	3,000	251.35	
0180	1000 lb. capacity		7.98	435	1,300	3,900	323.85	
0190	To 60' high, 750 lb. capacity		8.47	500	1,500	4,500	367.75	
0195	Air compressor, portable, 6.5 CFM, electric		.12	33.50	100	300	20.95	
0196	gasoline		.13	43.50	130	390	27.05	
0200	Air compressor, portable, gas engine, 60 C.F.M.		5.41	58.50	175	525	78.30	
0300	160 C.F.M.		6.80	73.50	220	660	98.40	
0400	Diesel engine, rotary screw, 250 C.F.M.		6.50	125	375	1,125	127	
0500	365 C.F.M.		9.30	167	500	1,500	174.40	
0600	600 C.F.M.		16	250	750	2,250	278	
0700	750 C.F.M.		17.60	267	800	2,400	300.80	
0800	For silenced models, small sizes, add		3%	5%	5%	5%		
0900	Large sizes, add		5%	7%	7%	7%		
0920	Air tools and accessories							
0930	Breaker, pavement, 60 lb.	Ea.	.19	30	90	270	19.50	
0940	80 lb.		.21	40	120	360	25.70	
0950	Drills, hand (jackhammer) 65 lb.		.23	28.50	85	255	18.85	
0960	Track or wagon, swing boom, 4" drifter		10.57	380	1,140	3,425	312.55	
0970	5" drifter		11.38	660	1,975	5,925	486.05	
0980	Dust control per drill		2.11	12.65	38	114	24.50	
0990	Hammer, chipping, 12 lb.		.12	26.50	80	240	16.95	
1000	Hose, air with couplings, 50' long, 3/4" diameter		.15	5	15	45	4.20	
1100	1" diameter		.15	6.65	20	60	5.20	
1200	1-1/2" diameter		.15	10	30	90	7.20	
1300	2" diameter		.21	16.65	50	150	11.70	
1400	2-1/2" diameter		.23	20	60	180	13.85	
1410	3" diameter		.24	25	75	225	16.90	
1450	Drill, steel, 7/8" x 2'			4	12	36	2.40	
1460	7/8" x 6'			5	15	45	3	
1520	Moil points		.82	3.33	10	30	8.55	
1525	Pneumatic nailer w/accessories		.12	30	90	270	18.95	
1530	Sheeting driver for 60 lb. breaker		.15	11.65	35	105	8.20	
1540	For 90 lb. breaker		.15	18.35	55	165	12.20	
1550	Spade, 25 lb.		.08	8.35	25	75	5.65	
1560	Tamper, single, 35 lb.		.10	25	75	225	15.80	
1570	Triple, 140 lb.		1.80	41.50	125	375	39.40	
1580	Wrenches, impact, air powered, up to 3/4" bolt		.25	22	66	198	15.20	
1590	Up to 1-1/4" bolt		.35	41.50	125	375	27.80	
1600	Barricades, barrels, reflectorized, 1 to 50 barrels			2.33	7	21	1.40	
1610	100 to 200 barrels			1.67	5	15	1	
1620	Barrels with flashers, 1 to 50 barrels			3.33	10	30	2	
1630	100 to 200 barrels			2.67	8	24	1.60	
1640	Barrels with steady burn type C lights			4	12	36	2.40	
1650	Illuminated board, trailer mounted, with generator		.78	91.50	275	825	61.25	
1670	Portable, stock, with flashers, 1 to 6 units			.83	2.50	7.50	.50	
1680	25 to 50 units			.77	2.30	6.90	.45	
1690	Butt fusion machine, electric		1.50	293	880	2,650	188	

Important: See the Reference Section for critical supporting data - Reference Nos., Crews, & City Cost Indexes

01500 | Temporary Facilities & Controls

01590 | Equipment Rental

			UNIT	HOURLY OPER. COST	RENT PER DAY	RENT PER WEEK	RENT PER MONTH	CREW EQUIPMENT COST/DAY
400	1695	Electro fusion machine	Ea.	1.25	127	380	1,150	86
	1700	Carts, brick, hand powered, 1000 lb. capacity		1.11	23.50	70	210	22.90
	1800	Gas engine, 1500 lb., 7-1/2' lift		1.65	91.50	275	825	68.20
	1830	Distributor, asphalt, trailer mtd, 2000 gal., 38 H.P. diesel		7.54	485	1,450	4,350	350.30
	1840	3000 gal., 38 H.P. diesel		8.12	515	1,550	4,650	374.95
	1850	Drill, rotary hammer, electric, 1-1/2" diameter		.15	25	75	225	16.20
	1860	Carbide bit for above			6	18	54	3.60
	1870	Emulsion sprayer, 65 gal., 5 H.P. gas engine		.63	56.50	170	510	39.05
	1880	200 gal., 5 H.P. engine		.67	60	180	540	41.35
	1900	Fencing, see division 015-304 & 028-300						
	1920	Floodlight, mercury, vapor or quartz, on tripod						
	1930	1000 watt	Ea.	.12	25	75	225	15.95
	1940	2000 watt		.13	41.50	125	375	26.05
	1960	Floodlights, trailer mounted with generator, 2-1000 watt lights		1.80	110	330	990	80.40
	2020	Forklift, wheeled, for brick, 18', 3000 lb., 2 wheel drive, gas		9.28	167	500	1,500	174.25
	2040	28', 4000 lb., 4 wheel drive, diesel		6.40	200	600	1,800	171.20
	2100	Generator, electric, gas engine, 1.5 KW to 3 KW		1.13	40	120	360	33.05
	2200	5 KW		1.49	58.50	175	525	46.90
	2300	10 KW		2.40	133	400	1,200	99.20
	2400	25 KW		6.80	153	460	1,375	146.40
	2500	Diesel engine, 20 KW		4.35	117	350	1,050	104.80
	2600	50 KW		6.90	133	400	1,200	135.20
	2700	100 KW		11.38	193	580	1,750	207.05
	2800	250 KW		29.05	335	1,000	3,000	432.40
	2850	Hammer, hydraulic, for mounting on boom, to 500 ft.-lb.		.99	133	400	1,200	87.90
	2860	500 to 1200 ft.-lb.		2.50	258	775	2,325	175
	2900	Heaters, space, oil or electric, 50 MBH		.11	26.50	80	240	16.90
	3000	100 MBH		.12	33.50	100	300	20.95
	3100	300 MBH		.13	50	150	450	31.05
	3150	500 MBH		.15	66.50	200	600	41.20
	3200	Hose, water, suction with coupling, 20' long, 2" diameter		.06	10	30	90	6.50
	3210	3" diameter		.06	15	45	135	9.50
	3220	4" diameter		.06	20	60	180	12.50
	3230	6" diameter		.06	36.50	110	330	22.50
	3240	8" diameter		.07	45	135	405	27.55
	3250	Discharge hose with coupling, 50' long, 2" diameter		.05	8.35	25	75	5.40
	3260	3" diameter		.05	10	30	90	6.40
	3270	4" diameter		.06	13.35	40	120	8.50
	3280	6" diameter		.07	33.50	100	300	20.55
	3290	8" diameter		.09	36.50	110	330	22.70
	3300	Ladders, extension type, 16' to 36' long			16.65	50	150	10
	3400	40' to 60' long			30	90	270	18
	3410	Level, laser type, for pipe laying, self leveling			93.50	281	845	56.20
	3430	Manual leveling			66.50	200	600	40
	3440	Rotary beacon with rod and sensor			91.50	275	825	55
	3460	Builders level with tripod and rod			26.50	80	240	16
	3500	Light towers, towable, with diesel generator, 2000 watt		1.60	125	375	1,125	87.80
	3600	4000 watt		1.95	142	425	1,275	100.60
	3700	Mixer, powered, plaster and mortar, 6 C.F., 7 H.P.		.85	56.50	170	510	40.80
	3800	10 C.F., 9 H.P.		1.25	78.50	235	705	57
	3850	Nailer, pneumatic		.12	28.50	85	255	17.95
	3900	Paint sprayers complete, 8 CFM		.08	45	135	405	27.65
	4000	17 CFM		.08	61.50	185	555	37.65
	4020	Pavers, bituminous, rubber tires, 8' wide, 52 H.P., gas		14.95	575	1,725	5,175	464.60
	4030	8' wide, 64 H.P., diesel		15.50	1,000	3,025	9,075	729
	4050	Crawler, 10' wide, 78 H.P., gas		22.25	1,400	4,200	12,600	1,018
	4060	10' wide, 87 H.P., diesel		22.75	1,075	3,200	9,600	822
	4070	Concrete paver, 12' to 24' wide, 250 H.P.		24.25	1,525	4,550	13,700	1,104

GENERAL REQUIREMENTS 1

01500 | Temporary Facilities & Controls
01590 | Equipment Rental

			UNIT	HOURLY OPER. COST	RENT PER DAY	RENT PER WEEK	RENT PER MONTH	CREW EQUIPMENT COST/DAY
4080	Placer-spreader-trimmer, 24' wide, 300 H.P.		Ea.	32.35	1,775	5,325	16,000	1,324
4100	Pump, centrifugal gas pump, 1-1/2", 4 MGPH	R02250-450		.45	35	105	315	24.60
4200	2", 8 MGPH			.55	36	108	325	26
4300	3", 15 MGPH	R02250-400		1.25	46.50	140	420	38
4400	6", 90 MGPH	R02315-300		9.65	133	400	1,200	157.20
4500	Submersible electric pump, 1-1/4", 55 GPM			.35	36.50	110	330	24.80
4600	1-1/2", 83 GPM			.41	40	120	360	27.30
4700	2", 120 GPM			.42	46.50	140	420	31.35
4800	3", 300 GPM			.75	58.50	175	525	41
4900	4", 560 GPM			1.21	70	210	630	51.70
5000	6", 1590 GPM			5.38	217	650	1,950	173.05
5100	Diaphragm pump, gas, single, 1-1/2" diameter			.50	26.50	80	240	20
5200	2" diameter			.55	41.50	125	375	29.40
5300	3" diameter			.75	47.50	143	430	34.60
5400	Double, 4" diameter			1.58	88.50	265	795	65.65
5500	Trash pump, self-priming, gas, 2" diameter			1.26	40	120	360	34.10
5600	Diesel, 4" diameter			1.99	91.50	275	825	70.90
5650	Diesel, 6" diameter			5.05	128	385	1,150	117.40
5660	Rollers, see division 016-408							
5700	Salamanders, L.P. gas fired, 100,000 B.T.U.		Ea.	.75	21.50	65	195	19
5705	50,000 BTU			.35	15	45	135	11.80
5720	Sandblaster, portable, open top, 3 C.F. capacity			.20	50	150	450	31.60
5730	6 C.F. capacity			.35	63.50	190	570	40.80
5740	Accessories for above			.07	16.65	50	150	10.55
5750	Sander, floor			.15	40.50	121	365	25.40
5760	Edger			.10	25	75	225	15.80
5800	Saw, chain, gas engine, 18" long			.55	40	120	360	28.40
5900	36" long			1.15	66.50	200	600	49.20
5950	60" long			1.20	83.50	250	750	59.60
6000	Masonry, table mounted, 14" diameter, 5 H.P.			1.80	53.50	160	480	46.40
6050	Saw, portable cut-off, 8 H.P.			.80	56.50	170	510	40.40
6100	Circular, hand held, electric, 7-1/4" diameter			.18	21.50	65	195	14.45
6200	12" diameter			.28	33.50	100	300	22.25
6275	Shot blaster, walk behind, 20" wide			1.50	217	650	1,950	142
6300	Steam cleaner, 100 gallons per hour			.45	56.50	170	510	37.60
6310	200 gallons per hour			.75	66.50	200	600	46
6340	Tar Kettle/Pot, 400 gallon			.50	75	225	675	49
6350	Torch, cutting, acetylene-oxygen, 150' hose			7.50	33.50	100	300	80
6360	Hourly operating cost includes tips and gas			7.65				61.20
6420	Recycle flush type				20	60	180	12
6430	Toilet, fresh water flush, garden hose,				22.50	68	204	13.60
6440	Hoisted, non-flush, for high rise				20	60	180	12
6450	Toilet, trailers, minimum				33.50	100	300	20
6460	Maximum				100	300	900	60
6470	Trailer, office, see division 015-904							
6500	Trailers, platform, flush deck, 2 axle, 25 ton capacity		Ea.	1.40	125	375	1,125	86.20
6600	40 ton capacity			1.75	230	690	2,075	152
6700	3 axle, 50 ton capacity			2.90	247	740	2,225	171.20
6800	75 ton capacity			3.75	325	980	2,950	226
6850	Trailer, storage, see division 015-904							
6900	Water tank, engine driven discharge, 5000 gallons		Ea.	8.24	233	700	2,100	205.90
7000	10,000 gallons			11	335	1,000	3,000	288
7020	Transit with tripod				33.50	100	300	20
7030	Trench box, 3000 lbs. 6'x8'			.80	100	300	900	66.40
7040	7200 lbs. 6'x20'			1.48	179	537	1,600	119.25
7050	8000 lbs., 8' x 16'			1.65	177	530	1,600	119.20
7060	9500 lbs., 8'x20'			1.80	190	570	1,700	128.40
7065	11,000 lbs., 8'x24'			1.85	241	724	2,175	159.60

Important: See the Reference Section for critical supporting data - Reference Nos., Crews, & City Cost Indexes

01500 | Temporary Facilities & Controls

01590 | Equipment Rental

			UNIT	HOURLY OPER. COST	RENT PER DAY	RENT PER WEEK	RENT PER MONTH	CREW EQUIPMENT COST/DAY	
400	7070	12,000 lbs., 10' x 20'	Ea.	2.10	267	800	2,400	176.80	400
	7100	Truck, pickup, 3/4 ton, 2 wheel drive		10.65	75	225	675	130.20	
	7200	4 wheel drive		12.08	80	240	720	144.65	
	7250	Crew carrier, 9 passenger		16.95	103	310	930	197.60	
	7290	Tool van, 24,000 G.V.W.		18.80	107	320	960	214.40	
	7300	Tractor, 4 x 2, 30 ton capacity, 195 H.P.		11.20	395	1,180	3,550	325.60	
	7410	250 H.P.		15.25	435	1,300	3,900	382	
	7500	6 x 2, 40 ton capacity, 240 H.P.		17.95	475	1,420	4,250	427.60	
	7600	6 x 4, 45 ton capacity, 240 H.P.		23.15	575	1,730	5,200	531.20	
	7620	Vacuum truck, hazardous material, 2500 gallon		13.95	305	910	2,725	293.60	
	7625	5,000 gallon		15	335	1,000	3,000	320	
	7650	Vacuum, H.E.P.A., 16 gal., wet/dry		.90	55	165	495	40.20	
	7655	55 gal, wet/dry		.85	50	150	450	36.80	
	7690	Large production vacuum loader, 3150 CFM		11.85	725	2,175	6,525	529.80	
	7700	Welder, electric, 200 amp		.81	39	117	350	29.90	
	7800	300 amp		1.10	66.50	200	600	48.80	
	7900	Gas engine, 200 amp		4.40	55	165	495	68.20	
	8000	300 amp		5.25	70	210	630	84	
	8100	Wheelbarrow, any size			8.35	25	75	5	
	8200	Wrecking ball, 4000 lb.		.41	66.50	200	600	43.30	
600	0010	**LIFTING AND HOISTING EQUIPMENT RENTAL**							600
	0100	without operators							
	0200	Crane, climbing, 106' jib, 6000 lb. capacity, 410 FPM	Ea.	26.58	1,225	3,675	11,000	947.65	
	0300	101' jib, 10,250 lb. capacity, 270 FPM	"	35.34	1,550	4,655	14,000	1,214	
	0400	Tower, static, 130' high, 106' jib,							
	0500	6200 lb. capacity at 400 FPM	Ea.	53.48	1,425	4,260	12,800	1,280	
	0600	Crawler, cable, 1/2 C.Y., 15 tons at 12' radius		19	540	1,625	4,875	477	
	0700	3/4 C.Y., 20 tons at 12' radius		20.20	550	1,655	4,975	492.60	
	0800	1 C.Y., 25 tons at 12' radius		20.85	585	1,750	5,250	516.80	
	0900	Crawler, cable, 1-1/2 C.Y., 40 tons at 12' radius		30.54	785	2,350	7,050	714.30	
	1000	2 C.Y., 50 tons at 12' radius		35.60	945	2,830	8,500	850.80	
	1100	3 C.Y., 75 tons at 12' radius		43.65	985	2,950	8,850	939.20	
	1200	100 ton capacity, standard boom		42.43	1,325	4,000	12,000	1,139	
	1300	165 ton capacity, standard boom		64.75	2,200	6,600	19,800	1,838	
	1400	200 ton capacity, 150' boom		120.75	2,375	7,100	21,300	2,386	
	1500	450' boom		135	3,025	9,100	27,300	2,900	
	1600	Truck mounted, cable operated, 6 x 4, 20 tons at 10' radius		14.07	700	2,100	6,300	532.55	
	1700	25 tons at 10' radius		20.80	1,075	3,250	9,750	816.40	
	1800	8 x 4, 30 tons at 10' radius		28.64	615	1,850	5,550	599.10	
	1900	40 tons at 12' radius		29.35	785	2,350	7,050	704.80	
	2000	8 x 4, 60 tons at 15' radius		44.87	915	2,750	8,250	908.95	
	2050	82 tons at 15' radius		45.57	1,675	5,000	15,000	1,365	
	2100	90 tons at 15' radius		49	1,050	3,150	9,450	1,022	
	2200	115 tons at 15' radius		51.20	1,875	5,660	17,000	1,542	
	2300	150 tons at 18' radius		76.61	1,575	4,750	14,300	1,563	
	2350	165 tons at 18' radius		77.65	2,225	6,700	20,100	1,961	
	2400	Truck mounted, hydraulic, 12 ton capacity		22.93	450	1,350	4,050	453.45	
	2500	25 ton capacity		23.65	615	1,850	5,550	559.20	
	2550	33 ton capacity		24.37	835	2,500	7,500	694.95	
	2600	55 ton capacity		34.15	890	2,675	8,025	808.20	
	2700	80 ton capacity		37.30	1,325	4,000	12,000	1,098	
	2800	Self-propelled, 4 x 4, with telescoping boom, 5 ton		10.18	325	975	2,925	276.45	
	2900	12-1/2 ton capacity		16.42	450	1,350	4,050	401.35	
	3000	15 ton capacity		18.25	490	1,465	4,400	439	
	3100	25 ton capacity		20.99	660	1,980	5,950	563.90	
	3200	Derricks, guy, 20 ton capacity, 60' boom, 75' mast		8.75	300	900	2,700	250	
	3300	100' boom, 115' mast		16.39	520	1,560	4,675	443.10	
	3400	Stiffleg, 20 ton capacity, 70' boom, 37' mast		11.56	385	1,160	3,475	324.50	

GENERAL REQUIREMENTS

01500 | Temporary Facilities & Controls

01590 | Equipment Rental

			UNIT	HOURLY OPER. COST	RENT PER DAY	RENT PER WEEK	RENT PER MONTH	CREW EQUIPMENT COST/DAY	
600	3500	100' boom, 47' mast	Ea.	18.06	630	1,885	5,650	521.50	600
	3550	Helicopter, small, lift to 1250 lbs. maximum, w/pilot		260	2,425	7,300	21,900	3,540	
	3600	Hoists, chain type, overhead, manual, 3/4 ton		.06	6.65	20	60	4.50	
	3900	10 ton		.25	28.50	85	255	19	
	4000	Hoist and tower, 5000 lb. cap., portable electric, 40' high		4	173	520	1,550	136	
	4100	For each added 10' section, add			13.35	40	120	8	
	4200	Hoist and single tubular tower, 5000 lb. electric, 100' high		5.45	242	725	2,175	188.60	
	4300	For each added 6'-6" section, add		.75	23.50	70	210	20	
	4400	Hoist and double tubular tower, 5000 lb., 100' high		5.75	267	800	2,400	206	
	4500	For each added 6'-6" section, add		.06	25	75	225	15.50	
	4550	Hoist and tower, mast type, 6000 lb., 100' high		5.35	277	830	2,500	208.80	
	4570	For each added 10' section, add		.15	16.65	50	150	11.20	
	4600	Hoist and tower, personnel, electric, 2000 lb., 100' @ 125 FPM		10	735	2,200	6,600	520	
	4700	3000 lb., 100' @ 200 FPM		10.79	835	2,500	7,500	586.30	
	4800	3000 lb., 150' @ 300 FPM		11.47	935	2,800	8,400	651.75	
	4900	4000 lb., 100' @ 300 FPM		12.10	950	2,850	8,550	666.80	
	5000	6000 lb., 100' @ 275 FPM		12.95	1,000	3,000	9,000	703.60	
	5100	For added heights up to 500', add	L.F.		1.67	5	15	1	
	5200	Jacks, hydraulic, 20 ton	Ea.	.15	14.65	44	132	10	
	5500	100 ton	"	.20	40.50	122	365	26	
	6000	Jacks, hydraulic, climbing with 50' jackrods							
	6010	and control consoles, minimum 3 mo. rental							
	6100	30 ton capacity	Ea.	.06	100	300	900	60.50	
	6150	For each added 10' jackrod section, add			3.33	10	30	2	
	6300	50 ton capacity			160	480	1,450	96	
	6350	For each added 10' jackrod section, add			4	12	36	2.40	
	6500	125 ton capacity			415	1,250	3,750	250	
	6550	For each added 10' jackrod section, add			28.50	85	255	17	
	6600	Cable jack, 10 ton capacity with 200' cable			83.50	250	750	50	
	6650	For each added 50' of cable, add			8.35	25	75	5	
700	0010	**WELLPOINT EQUIPMENT RENTAL** See also division 02240							700
	0020	Based on 2 months rental							
	0100	Combination jetting & wellpoint pump, 60 H.P. diesel	Ea.	3.50	247	740	2,225	176	
	0200	High pressure gas jet pump, 200 H.P., 300 psi	"	9.75	210	630	1,900	204	
	0300	Discharge pipe, 8" diameter	L.F.		.40	1.20	3.60	.25	
	0350	12" diameter			.58	1.75	5.25	.35	
	0400	Header pipe, flows up to 150 G.P.M., 4" diameter			.37	1.10	3.30	.20	
	0500	400 G.P.M., 6" diameter			.42	1.25	3.75	.25	
	0600	800 G.P.M., 8" diameter			.58	1.75	5.25	.35	
	0700	1500 G.P.M., 10" diameter			.62	1.85	5.55	.35	
	0800	2500 G.P.M., 12" diameter			1.17	3.50	10.50	.70	
	0900	4500 G.P.M., 16" diameter			1.50	4.50	13.50	.90	
	0950	For quick coupling aluminum and plastic pipe, add			1.55	4.65	13.95	.95	
	1100	Wellpoint, 25' long, with fittings & riser pipe, 1-1/2" or 2" diameter	Ea.		3.10	9.30	28	1.85	
	1200	Wellpoint pump, diesel powered, 4" diameter, 20 H.P.		3.49	142	425	1,275	112.90	
	1300	6" diameter, 30 H.P.		5.40	177	530	1,600	149.20	
	1400	8" suction, 40 H.P.		6.41	242	725	2,175	196.30	
	1500	10" suction, 75 H.P.		6.91	283	850	2,550	225.30	
	1600	12" suction, 100 H.P.		9.89	455	1,360	4,075	351.10	
	1700	12" suction, 175 H.P.		10.83	500	1,500	4,500	386.65	

Reference numbers appearing in column: R01559-150, R01590-100, R02315-450, R02240-900

Important: See the Reference Section for critical supporting data - Reference Nos., Crews, & City Cost Indexes

01700 | Execution Requirements

01740 | Cleaning

			CREW	DAILY OUTPUT	LABOR-HOURS	UNIT	MAT.	LABOR	EQUIP.	TOTAL	TOTAL INCL O&P	
500	0010	**CLEANING UP** After job completion, allow, minimum				Job					.30%	500
	0040	Maximum				"					1%	
	0050	Cleanup of floor area, continuous, per day	A-5	24	.750	M.S.F.	1.65	10.95	1.80	14.40	22.50	
	0100	Final	"	11.50	1.565	"	2.63	23	3.76	29.39	46	
	0200	Rubbish removal, see division 02225-730										

01800 | Facility Operation

01810 | Commissioning

			CREW	DAILY OUTPUT	LABOR-HOURS	UNIT	MAT.	LABOR	EQUIP.	TOTAL	TOTAL INCL O&P	
100	0010	**COMMISSIONING**										100
	0100	As-built drawings, punchlist, training, manuals, minimum				Project					.50%	
	0150	Maximum				"					2%	

For information about Means Estimating Seminars, see yellow pages 11 and 12 in back of book

Division Notes

		CREW	DAILY OUTPUT	LABOR-HOURS	UNIT	2000 BARE COSTS MAT.	LABOR	EQUIP.	TOTAL	TOTAL INCL O&P

Division 2
Site Construction

Estimating Tips
02200 Site Preparation
- If possible visit the site and take an inventory of the type, quantity and size of the trees. Certain trees may have a landscape resale value or firewood value. Stump disposal can be very expensive, particularly if they cannot be buried at the site. Consider using a bulldozer in lieu of hand cutting trees.
- Estimators should visit the site to determine the need for haul road, access, storage of materials, and security considerations. When estimating for access roads on unstable soil, consider using a geotextile stabilization fabric. It can greatly reduce the quantity of crushed stone or gravel. Sites of limited size and access can cause cost overruns due to lost productivity. Theft and damage is another consideration if the location is isolated. A temporary fence or security guards may be required. Investigate the site thoroughly.

02210 Subsurface Investigation
In preparing estimates on structures involving earthwork or foundations, all information concerning soil characteristics should be obtained. Look particularly for hazardous waste, evidence of prior dumping of debris, and previous stream beds.
- The costs shown for selective demolition do not include rubbish handling or disposal. These items should be estimated separately using Means data or other sources.

02300 Earthwork
- Estimating the actual cost of performing earthwork requires careful consideration of the variables involved. This includes items such as type of soil, whether or not water will be encountered, dewatering, whether or not banks need bracing, disposal of excavated earth, length of haul to fill or spoil sites, etc. If the project has large quantities of cut or fill, consider raising or lowering the site to reduce costs while paying close attention to the effect on site drainage and utilities if doing this.
- If the project has large quantities of fill, creating a borrow pit on the site can significantly lower the costs. It is very important to consider what time of year the project is scheduled for completion. Bad weather can create large cost overruns from dewatering, site repair and lost productivity from cold weather.

02700 Bases, Ballasts, Pavements/Appurtenances
- When estimating paving, keep in mind the project schedule. If an asphaltic paving project is in a colder climate and runs through to the spring, consider placing the base course in the autumn, then topping it in the spring just prior to completion. This could save considerable costs in spring repair. Keep in mind that prices for asphalt and concrete are generally higher in the cold seasons.

02500 Utility Services
02600 Drainage & Containment
- Never assume that the water, sewer and drainage lines will go in at the early stages of the project. Consider the site access needs before dividing the site in half with open trenches, loose pipe, and machinery obstructions. Always inspect the site to establish that the site drawings are complete. Check off all existing utilities on your drawing as you locate them. If you find any discrepancies, mark up the site plan for further research. Differing site conditions can be very costly if discovered later in the project.

02900 Planting
- The timing of planting and guarantee specifications often dictate the costs for establishing tree and shrub growth and a stand of grass or ground cover. Establish the work performance schedule to coincide with the local planting season. Maintenance and growth guarantees can add from 20% to 100% to the total landscaping cost. The cost to replace trees and shrubs can be as high as 5% of the total cost depending on the planting zone, soil conditions and time of year.

Reference Numbers
Reference numbers are shown in bold squares at the beginning of some major classifications. These numbers refer to related items in the Reference Section. The reference information may be an estimating procedure, an alternate pricing method or technical information.

Note: Not all subdivisions listed here necessarily appear in this publication.

02050 | Basic Site Materials & Methods

02060 | Aggregate

			CREW	DAILY OUTPUT	LABOR-HOURS	UNIT	MAT.	LABOR	EQUIP.	TOTAL	TOTAL INCL O&P	
150	0010	**BORROW**										150
	0020	and spread, with 200 H.P. dozer, no compaction	R02315 -400									
	0100	Bank run gravel	B-15	600	.047	C.Y.	5.20	.79	2.84	8.83	10.15	
	0200	Common borrow		600	.047		4.77	.79	2.84	8.40	9.70	
	0300	Crushed stone, (1.40 tons per CY), 1-1/2"		600	.047		17.05	.79	2.84	20.68	23.50	
	0320	3/4"		600	.047		18.80	.79	2.84	22.43	25	
	0340	1/2"		600	.047		18.25	.79	2.84	21.88	24.50	
	0360	3/8"		600	.047		15.75	.79	2.84	19.38	22	
	0400	Sand, washed, concrete		600	.047		10.65	.79	2.84	14.28	16.15	
	0500	Dead or bank sand		600	.047		3.50	.79	2.84	7.13	8.30	
	0600	Select structural fill		600	.047		7.50	.79	2.84	11.13	12.70	
	0700	Screened loam		600	.047		17.35	.79	2.84	20.98	23.50	
	0800	Topsoil, weed free		600	.047		13.10	.79	2.84	16.73	18.85	
	0900	For 5 mile haul, add	B-34B	200	.040			.65	2.24	2.89	3.54	

02065 | Cement & Concrete

						UNIT	MAT.	LABOR	EQUIP.	TOTAL	TOTAL INCL O&P	
300	0010	**ASPHALTIC CONCRETE** plant mix (145 lb. per C.F.)	R02065 -300			Ton	30			30	33	300
	0200	All weather patching mix, hot					31.50			31.50	34.50	
	0250	Cold patch					36			36	39.50	
	0300	Berm mix					31.50			31.50	34.50	

02080 | Utility Materials

			CREW	DAILY OUTPUT	LABOR-HOURS	UNIT	MAT.	LABOR	EQUIP.	TOTAL	TOTAL INCL O&P	
400	0010	**PIPING, WATER DISTRIBUTION** Mech. joints unless noted										400
	1000	Fire hydrants, two way; excavation and backfill not incl.										
	1100	4-1/2" valve size, depth 2'-0"	B-21	10	2.800	Ea.	730	44.50	13.85	788.35	890	
	1120	2'-6"		10	2.800		770	44.50	13.85	828.35	935	
	1140	3'-0"		10	2.800		815	44.50	13.85	873.35	985	
	1300	7'-0"		6	4.667		985	74.50	23	1,082.50	1,225	
	2400	Lower barrel extensions with stems, 1'-0"	B-20	14	1.714		242	26		268	310	
	2480	3'-0"	"	12	2		735	30		765	855	
790	0010	**UNDERGROUND MARKING TAPE**										790
	0400	Underground tape, detectable aluminum, 2"	1 Clab	150	.053	C.L.F.	3.70	.77		4.47	5.40	
	0500	6"	"	140	.057	"	9.25	.83		10.08	11.60	
800	0010	**UTILITY VAULTS** Precast concrete, 6" thick										800
	0050	5' x 10' x 6' high, I.D.	B-13	2	24	Ea.	1,400	380	280	2,060	2,475	
	0100	6' x 10' x 6' high, I.D.		2	24		1,425	380	280	2,085	2,525	
	0150	5' x 12' x 6' high, I.D.		2	24		1,525	380	280	2,185	2,625	
	0200	6' x 12' x 6' high, I.D.		1.80	26.667		1,700	420	310	2,430	2,925	
	0250	6' x 13' x 6' high, I.D.		1.50	32		2,250	505	375	3,130	3,725	
	0300	8' x 14' x 7' high, I.D.		1	48		2,425	760	560	3,745	4,575	
	0350	Hand hole, precast concrete, 1-1/2" thick										
	0400	1'-0" x 2'-0" x 1'-9", I.D., light duty	B-1	4	6	Ea.	355	90.50		445.50	545	
	0450	4'-6" x 3'-2" x 2'-0", O.D., heavy duty	B-6	3	8	"	880	128	67.50	1,075.50	1,250	

02100 | Site Remediation

02110 | Excavation, Removal & Handling

			CREW	DAILY OUTPUT	LABOR-HOURS	UNIT	MAT.	LABOR	EQUIP.	TOTAL	TOTAL INCL O&P	
300	0010	**HAZARDOUS WASTE CLEANUP/PICKUP/DISPOSAL**										300
	0100	For contractor equipment, i.e. dozer,										

02100 | Site Remediation

02110 | Excavation, Removal & Handling

			CREW	DAILY OUTPUT	LABOR-HOURS	UNIT	2000 BARE COSTS				TOTAL INCL O&P
							MAT.	LABOR	EQUIP.	TOTAL	
300	0110	Front end loader, dump truck, etc., see div. 01590-200										300
	1000	Solid pickup									
	1100	55 gal. drums				Ea.					200
	1120	Bulk material, minimum				Ton					150
	1130	Maximum				"					500
	1200	Transportation to disposal site									
	1220	Truckload = 80 drums or 25 C.Y. or 18 tons									
	1260	Minimum				Mile					2.30
	1270	Maximum				"					4
	3000	Liquid pickup, vacuum truck, stainless steel tank									
	3100	Minimum charge, 4 hours									
	3110	1 compartment, 2200 gallon				Hr.					100
	3120	2 compartment, 5000 gallon				"					100
	3400	Transportation in 6900 gallon bulk truck				Mile					4.30
	3410	In teflon lined truck				"					5
	5000	Heavy sludge or dry vacuumable material				Hr.					100
	6000	Dumpsite disposal charge, minimum				Ton					100
	6020	Maximum				"					400

02115 | Underground Tank Removal

			CREW	DAILY OUTPUT	LABOR-HOURS	UNIT	MAT.	LABOR	EQUIP.	TOTAL	TOTAL INCL O&P
200	0010	REMOVAL OF UNDERGROUND STORAGE TANKS	R02115-200									200
	0011	Petroleum storage tanks, non-leaking									
	0100	Excavate & load onto trailer									
	0110	3000 gal. to 5000 gal. tank	B-14	4	12	Ea.		187	51	238	375
	0120	6000 gal to 8000 gal tank	B-3A	3	13.333			207	238	445	610
	0130	9000 gal to 12000 gal tank	"	2	20	↓		310	355	665	920
	0190	Known leaking tank add				%				100%	100%
	0200	Remove sludge, water and remaining product from bottom									
	0201	of tank with vacuum truck									
	0300	3000 gal to 5000 gal tank	A-13	5	1.600	Ea.		30.50	106	136.50	168
	0310	6000 gal to 8000 gal tank	↓	4	2			38	132	170	210
	0320	9000 gal to 12000 gal tank		3	2.667	↓		51	177	228	279
	0390	Dispose of sludge off-site, average				Gal.					4
	0400	Insert solid carbon dioxide "dry ice" to produce inert gas									
	0401	For cleaning/transporting tanks (1.5 lbs./100 gal. cap)	1 Clab	500	.016	Lb.	1.20	.23		1.43	1.72
	1020	Haul tank to certified salvage dump, 100 miles round trip									
	1023	3000 gal. to 5000 gal. tank				Ea.				550	630
	1026	6000 gal. to 8000 gal. tank								650	750
	1029	9,000 gal. to 12,000 gal. tank				↓				875	1,000
	1100	Disposal of contaminated soil to landfill									
	1110	Minimum				C.Y.					100
	1111	Maximum				"					300
	1120	Disposal of contaminated soil to									
	1121	bituminous concrete batch plant									
	1130	Minimum				C.Y.					50
	1131	Maximum				"					100
	2010	Decontamination of soil on site incl poly tarp on top/bottom									
	2011	Soil containment berm, and chemical treatment									
	2020	Minimum	B-11C	100	.160	C.Y.	5.60	2.75	2.03	10.38	13
	2021	Maximum	"	100	.160		7.25	2.75	2.03	12.03	14.85
	2050	Disposal of decontaminated soil, minimum									60
	2055	Maximum				↓					125

02200 | Site Preparation

02210 | Subsurface Investigation

		CREW	DAILY OUTPUT	LABOR-HOURS	UNIT	2000 BARE COSTS				TOTAL INCL O&P
						MAT.	LABOR	EQUIP.	TOTAL	
310	0010	**BORINGS** Initial field stake out and determination of elevations	A-6	1	16	Day		335		335	575
	0100	Drawings showing boring details				Total		170		170	245
	0200	Report and recommendations from P.E.						375		375	540
	0300	Mobilization and demobilization, minimum	B-55	4	4			60.50	166	226.50	285
	0350	For over 100 miles, per added mile		450	.036	Mile		.54	1.47	2.01	2.53
	0600	Auger holes in earth, no samples, 2-1/2" diameter		78.60	.204	L.F.		3.08	8.45	11.53	14.50
	0650	4" diameter		67.50	.237			3.59	9.85	13.44	16.90
	0800	Cased borings in earth, with samples, 2-1/2" diameter		55.50	.288		12.35	4.37	11.95	28.67	34
	0850	4" diameter		32.60	.491		19.75	7.45	20.50	47.70	56.50
	1000	Drilling in rock, "BX" core, no sampling	B-56	34.90	.458			6.60	17.40	24	30.50
	1050	With casing & sampling		31.70	.505		12.35	7.30	19.15	38.80	47
	1200	"NX" core, no sampling		25.92	.617			8.90	23.50	32.40	41.50
	1250	With casing and sampling		25	.640		15.30	9.25	24.50	49.05	59
	1400	Drill rig and crew with truck mounted auger	B-55	1	16	Day		242	665	907	1,150
	1450	With crawler type drill	B-56	1	16	"		231	610	841	1,075
	1500	For inner city borings add, minimum									10%
	1510	Maximum									20%
320	0010	**DRILLING, CORE** Reinforced concrete slab, up to 6" thick slab									
	0020	Including bit, layout and set up									
	0100	1" diameter core	B-89A	28	.571	Ea.	2.30	9.75	2.17	14.22	21.50
	0150	Each added inch thick, add		300	.053		.41	.91	.20	1.52	2.23
	0300	3" diameter core		23	.696		5.10	11.90	2.64	19.64	29
	0350	Each added inch thick, add		186	.086		.92	1.47	.33	2.72	3.89
	0500	4" diameter core		19	.842		5.10	14.40	3.19	22.69	33.50
	0550	Each added inch thick, add		170	.094		1.17	1.61	.36	3.14	4.44
	0700	6" diameter core		14	1.143		8.45	19.55	4.33	32.33	47.50
	0750	Each added inch thick, add		140	.114		1.43	1.95	.43	3.81	5.40
	0900	8" diameter core		11	1.455		11.55	25	5.50	42.05	61.50
	0950	Each added inch thick, add		95	.168		1.94	2.88	.64	5.46	7.75
	1100	10" diameter core		10	1.600		15.40	27.50	6.05	48.95	70.50
	1150	Each added inch thick, add		80	.200		2.55	3.42	.76	6.73	9.50
	1300	12" diameter core		9	1.778		18.45	30.50	6.75	55.70	80
	1350	Each added inch thick, add		68	.235		3.06	4.02	.89	7.97	11.25
	1500	14" diameter core		7	2.286		22.50	39	8.65	70.15	101
	1550	Each added inch thick, add		55	.291		3.88	4.97	1.10	9.95	14.05
	1700	18" diameter core		4	4		29	68.50	15.15	112.65	166
	1750	Each added inch thick, add		28	.571		5.10	9.75	2.17	17.02	24.50
	1760	For horizontal holes, add to above								30%	30%
	1770	Prestressed hollow core plank, 6" thick									
	1780	1" diameter core	B-89A	52	.308	Ea.	1.53	5.25	1.17	7.95	11.95
	1790	Each added inch thick, add		350	.046		.26	.78	.17	1.21	1.82
	1800	3" diameter core		50	.320		3.37	5.45	1.21	10.03	14.45
	1810	Each added inch thick, add		240	.067		.56	1.14	.25	1.95	2.86
	1820	4" diameter core		48	.333		4.49	5.70	1.26	11.45	16.15
	1830	Each added inch thick, add		216	.074		.77	1.27	.28	2.32	3.33
	1840	6" diameter core		44	.364		5.55	6.20	1.38	13.13	18.25
	1850	Each added inch thick, add		175	.091		.92	1.56	.35	2.83	4.07
	1860	8" diameter core		32	.500		7.45	8.55	1.89	17.89	25
	1870	Each added inch thick, add		118	.136		1.28	2.32	.51	4.11	5.95
	1880	10" diameter core		28	.571		10.05	9.75	2.17	21.97	30
	1890	Each added inch thick, add		99	.162		1.38	2.76	.61	4.75	6.95
	1900	12" diameter core		22	.727		12.25	12.45	2.76	27.46	38
	1910	Each added inch thick, add		85	.188		2.04	3.22	.71	5.97	8.50
	1950	Minimum charge for above, 3" diameter core		7	2.286	Total		39	8.65	47.65	76.50
	2000	4" diameter core		6.80	2.353			40	8.90	48.90	79

02200 | Site Preparation

02210 | Subsurface Investigation

			CREW	DAILY OUTPUT	LABOR-HOURS	UNIT	MAT.	LABOR	EQUIP.	TOTAL	TOTAL INCL O&P	
320	2050	6" diameter core	B-89A	6	2.667	Total		45.50	10.10	55.60	89	320
	2100	8" diameter core		5.50	2.909			50	11.05	61.05	97.50	
	2150	10" diameter core		4.75	3.368			57.50	12.75	70.25	113	
	2200	12" diameter core		3.90	4.103			70	15.55	85.55	137	
	2250	14" diameter core		3.38	4.734			81	17.95	98.95	159	
	2300	18" diameter core	↓	3.15	5.079	↓		87	19.25	106.25	170	
	3010	Bits for core drill, diamond, premium, 1" diameter				Ea.	114			114	126	
	3020	3" diameter					280			280	310	
	3040	4" diameter					310			310	340	
	3050	6" diameter					500			500	550	
	3080	8" diameter					730			730	800	
	3120	12" diameter					1,125			1,125	1,225	
	3180	18" diameter					1,825			1,825	2,025	
	3240	24" diameter				↓	2,450			2,450	2,700	
900	0010	**TEST PITS** Hand digging, light soil	1 Clab	4.50	1.778	C.Y.		25.50		25.50	44	900
	0100	Heavy soil	"	2.50	3.200			46		46	79	
	0120	Loader-backhoe, light soil	B-11M	28	.571			9.80	9.75	19.55	27.50	
	0130	Heavy soil	"	20	.800	↓		13.75	13.65	27.40	38	
	1000	Subsurface exploration, mobilization				Mile				5	5.75	
	1010	Difficult access for rig, add				Hr.				100	115	
	1020	Auger borings, drill rig, incl. samples				L.F.				11.50	13.25	
	1030	Hand auger								17	19.55	
	1050	Drill and sample every 5', split spoon				↓				15	17.25	
	1060	Extra samples				Ea.				20	23	

02220 | Site Demolition

			CREW	DAILY OUTPUT	LABOR-HOURS	UNIT	MAT.	LABOR	EQUIP.	TOTAL	TOTAL INCL O&P	
100	0010	**BUILDING DEMOLITION** Large urban projects, incl. 20 Mi. haul										100
	0012	No foundation or dump fees, C.F. is volume of building standing, steel	B-8	21,500	.003	C.F.		.04	.10	.14	.18	
	0050	Concrete		15,300	.004			.06	.15	.21	.26	
	0080	Masonry		20,100	.003			.05	.11	.16	.20	
	0100	Mixture of types, average	↓	20,100	.003			.05	.11	.16	.20	
	0500	Small bldgs, or single bldgs, no salvage included, steel	B-3	14,800	.003			.05	.11	.16	.21	
	0600	Concrete		11,300	.004			.07	.15	.22	.28	
	0650	Masonry		14,800	.003			.05	.11	.16	.21	
	0700	Wood	↓	14,800	.003	↓		.05	.11	.16	.21	
	1000	Single family, one story house, wood, minimum				Ea.				2,300	2,700	
	1020	Maximum								4,000	4,800	
	1200	Two family, two story house, wood, minimum								3,000	3,600	
	1220	Maximum								5,800	7,000	
	1300	Three family, three story house, wood, minimum								4,000	4,800	
	1320	Maximum								7,000	8,400	
	5000	For buildings with no interior walls, deduct				↓				50%		
400	0010	**EXPLOSIVE/IMPLOSIVE DEMOLITION** Large projects, no disposal/fee										400
	0020	based on building volume, steel building	B-5B	16,900	.003	C.F.		.05	.11	.16	.21	
	0100	Concrete building		16,900	.003			.05	.11	.16	.21	
	0200	Masonry building	↓	16,900	.003	↓		.05	.11	.16	.21	
	0400	Disposal of material, minimum	B-3	445	.108	C.Y.		1.75	3.77	5.52	7.10	
	0500	Maximum	"	365	.132	"		2.14	4.60	6.74	8.65	
550	0010	**FOOTINGS AND FOUNDATIONS DEMOLITION**										550
	0200	Floors, concrete slab on grade,										
	0240	4" thick, plain concrete	B-9C	500	.080	S.F.		1.19	.36	1.55	2.44	
	0280	Reinforced, wire mesh		470	.085			1.26	.38	1.64	2.59	
	0300	Rods		400	.100			1.49	.45	1.94	3.04	
	0400	6" thick, plain concrete		375	.107			1.58	.48	2.06	3.24	
	0420	Reinforced, wire mesh		340	.118			1.75	.53	2.28	3.57	
	0440	Rods	↓	300	.133	↓		1.98	.60	2.58	4.05	

For expanded coverage of these items see *Means Heavy Construction Cost Data 2000*

02200 | Site Preparation

02220 | Site Demolition

		CREW	DAILY OUTPUT	LABOR-HOURS	UNIT	2000 BARE COSTS				TOTAL INCL O&P
						MAT.	LABOR	EQUIP.	TOTAL	
550	1000 Footings, concrete, 1' thick, 2' wide	B-5	300	.133	L.F.		2.13	3.21	5.34	7.15
	1080 1'-6" thick, 2' wide		250	.160			2.55	3.86	6.41	8.55
	1120 3' wide		200	.200			3.19	4.82	8.01	10.70
	1140 2' thick, 3' wide		175	.229			3.64	5.50	9.14	12.25
	1200 Average reinforcing, add								10%	10%
	1220 Heavy reinforcing, add								20%	20%
	2000 Walls, block, 4" thick	1 Clab	180	.044	S.F.		.64		.64	1.10
	2040 6" thick		170	.047			.68		.68	1.16
	2080 8" thick		150	.053			.77		.77	1.32
	2100 12" thick		150	.053			.77		.77	1.32
	2200 For horizontal reinforcing, add								10%	10%
	2220 For vertical reinforcing, add								20%	20%
	2400 Concrete, plain concrete, 6" thick	B-9	160	.250			3.71	1.13	4.84	7.60
	2420 8" thick		140	.286			4.24	1.29	5.53	8.65
	2440 10" thick		120	.333			4.95	1.50	6.45	10.15
	2500 12" thick		100	.400			5.95	1.80	7.75	12.20
	2600 For average reinforcing, add								10%	10%
	2620 For heavy reinforcing, add								20%	20%
	4000 For congested sites or small quantities, add up to								200%	200%
	4200 Add for disposal, on site	B-11A	232	.069	C.Y.		1.18	3.48	4.66	5.80
	4250 To five miles	B-30	220	.109	"		1.90	7.30	9.20	11.20
575	0010 HYDRODEMOLITION, concrete pavement, 4000 PSI, 2" depth	B-5	500	.080	S.F.		1.28	1.93	3.21	4.29
	0120 4" depth		450	.089			1.42	2.14	3.56	4.77
	0130 6" depth		400	.100			1.59	2.41	4	5.35
	0410 6000 PSI, 2" depth		410	.098			1.56	2.35	3.91	5.25
	0420 4" depth		350	.114			1.82	2.76	4.58	6.15
	0430 6" depth		300	.133			2.13	3.21	5.34	7.15
	0510 8000 PSI, 2" depth		330	.121			1.93	2.92	4.85	6.50
	0520 4" depth		280	.143			2.28	3.44	5.72	7.65
	0530 6" depth		240	.167			2.66	4.02	6.68	8.95
875	0010 SITE DEMOLITION No hauling, abandon catch basin or manhole	B-6	7	3.429	Ea.		55	29	84	125
	0020 Remove existing catch basin or manhole, masonry		4	6			96	51	147	218
	0030 Catch basin or manhole frames and covers, stored		13	1.846			29.50	15.60	45.10	67
	0040 Remove and reset		7	3.429			55	29	84	125
	0100 Roadside delineators, remove only	B-80	175	.137			2.07	3.15	5.22	7
	0110 Remove and reset	"	100	.240			3.63	5.50	9.13	12.25
	0600 Fencing, barbed wire, 3 strand	2 Clab	430	.037	L.F.		.54		.54	.92
	0650 5 strand	"	280	.057			.83		.83	1.41
	0700 Chain link, posts & fabric, remove only, 8' to 10' high	B-6	445	.054			.86	.46	1.32	1.96
	0750 Remove and reset	"	70	.343			5.50	2.90	8.40	12.50
	0800 Guiderail, corrugated steel, remove only	B-80A	100	.240			3.47	1.80	5.27	7.95
	0850 Remove and reset	"	40	.600			8.65	4.51	13.16	19.80
	0860 Guide posts, remove only	B-80B	120	.267	Ea.		4.16	2.36	6.52	9.65
	0870 Remove and reset	B-55	50	.320			4.85	13.25	18.10	23
	0901 Hydrants, fire, remove only	2 Skwk	4.70	3.404			67		67	115
	0951 Remove and reset	"	1.40	11.429			226		226	385
	1000 Masonry walls, block or tile, solid, remove	B-5	1,800	.022	C.F.		.35	.54	.89	1.19
	1100 Cavity wall		2,200	.018			.29	.44	.73	.97
	1200 Brick, solid		900	.044			.71	1.07	1.78	2.38
	1300 With block back-up		1,130	.035			.56	.85	1.41	1.90
	1400 Stone, with mortar		900	.044			.71	1.07	1.78	2.38
	1500 Dry set		1,500	.027			.43	.64	1.07	1.43
	1600 Median barrier, precast concrete, remove and store	B-3	430	.112	L.F.		1.82	3.90	5.72	7.35
	1610 Remove and reset	"	390	.123	"		2	4.30	6.30	8.10

02200 | Site Preparation

02220 | Site Demolition

			CREW	DAILY OUTPUT	LABOR-HOURS	UNIT	MAT.	LABOR	EQUIP.	TOTAL	TOTAL INCL O&P	
875	1710	Pavement removal, bituminous roads, 3" thick	B-38	690	.035	S.Y.		.56	.60	1.16	1.60	875
	1750	4" to 6" thick		420	.057			.91	.99	1.90	2.64	
	1800	Bituminous driveways		640	.038			.60	.65	1.25	1.74	
	1900	Concrete to 6" thick, hydraulic hammer, mesh reinforced		255	.094			1.51	1.63	3.14	4.35	
	2000	Rod reinforced		200	.120			1.92	2.08	4	5.55	
	2100	Concrete, 7" to 24" thick, plain		33	.727	C.Y.		11.65	12.65	24.30	33.50	
	2200	Reinforced		24	1	"		16	17.35	33.35	46	
	2300	With hand held air equipment, bituminous, to 6" thick	B-39	1,900	.025	S.F.		.37	.09	.46	.74	
	2320	Concrete to 6" thick, no reinforcing		1,200	.040			.59	.15	.74	1.18	
	2340	Mesh reinforced		1,400	.034			.51	.13	.64	1.01	
	2360	Rod reinforced		765	.063			.93	.24	1.17	1.85	
	2400	Curbs, concrete, plain	B-6	360	.067	L.F.		1.07	.56	1.63	2.42	
	2500	Reinforced		275	.087			1.40	.74	2.14	3.17	
	2600	Granite		360	.067			1.07	.56	1.63	2.42	
	2700	Bituminous		528	.045			.73	.38	1.11	1.65	
	2900	Pipe removal, sewer/water, no excavation, 12" diameter		175	.137			2.19	1.16	3.35	4.99	
	2930	15" diameter		150	.160			2.56	1.35	3.91	5.80	
	2960	24" diameter		120	.200			3.20	1.69	4.89	7.25	
	3000	36" diameter		90	.267			4.27	2.26	6.53	9.70	
	3200	Steel, welded connections, 4" diameter		160	.150			2.40	1.27	3.67	5.45	
	3300	10" diameter		80	.300			4.80	2.54	7.34	10.90	
	3500	Railroad track removal, ties and track	B-13	330	.145			2.30	1.69	3.99	5.75	
	3600	Ballast	B-14	500	.096	C.Y.		1.49	.41	1.90	2.99	
	3700	Remove and re-install, ties & track using new bolts & spikes		50	.960	L.F.		14.95	4.06	19.01	30	
	3800	Turnouts using new bolts and spikes		1	48	Ea.		745	203	948	1,500	
	4000	Sidewalk removal, bituminous, 2-1/2" thick	B-6	325	.074	S.Y.		1.18	.62	1.80	2.69	
	4050	Brick, set in mortar		185	.130			2.08	1.10	3.18	4.72	
	4100	Concrete, plain, 4"		160	.150			2.40	1.27	3.67	5.45	
	4200	Mesh reinforced		150	.160			2.56	1.35	3.91	5.80	
	5000	Slab on grade removal, plain	B-5	45	.889	C.Y.		14.15	21.50	35.65	47.50	
	5100	Mesh reinforced		33	1.212			19.30	29	48.30	65	
	5200	Rod reinforced		25	1.600			25.50	38.50	64	86	
	5500	For congested sites or small quantities, add up to								200%	200%	
	5550	For disposal on site, add	B-11A	232	.069			1.18	3.48	4.66	5.80	
	5600	To 5 miles, add	B-34D	76	.105			1.71	7.40	9.11	11	

02225 | Selective Demolition

			CREW	DAILY OUTPUT	LABOR-HOURS	UNIT	MAT.	LABOR	EQUIP.	TOTAL	TOTAL INCL O&P	
310	0010	**CEILING DEMOLITION**										310
	0200	Drywall, furred and nailed	2 Clab	800	.020	S.F.		.29		.29	.50	
	0220	On metal frame		760	.021			.30		.30	.52	
	0240	On suspension system, including system		720	.022			.32		.32	.55	
	1000	Plaster, lime and horse hair, on wood lath, incl. lath		700	.023			.33		.33	.57	
	1020	On metal lath		570	.028			.41		.41	.69	
	1100	Gypsum, on gypsum lath		720	.022			.32		.32	.55	
	1120	On metal lath		500	.032			.46		.46	.79	
	1200	Suspended ceiling, mineral fiber, 2'x2' or 2'x4'		1,500	.011			.15		.15	.26	
	1250	On suspension system, incl. system		1,200	.013			.19		.19	.33	
	1500	Tile, wood fiber, 12" x 12", glued		900	.018			.26		.26	.44	
	1540	Stapled		1,500	.011			.15		.15	.26	
	1580	On suspension system, incl. system		760	.021			.30		.30	.52	
	2000	Wood, tongue and groove, 1" x 4"		1,000	.016			.23		.23	.40	
	2040	1" x 8"		1,100	.015			.21		.21	.36	
	2400	Plywood or wood fiberboard, 4' x 8' sheets		1,200	.013			.19		.19	.33	
320	0010	**CUTOUT DEMOLITION** Conc., elev. slab, light reinf., under 6 C.F.	B-9C	65	.615	C.F.		9.15	2.78	11.93	18.70	320
	0050	Light reinforcing, over 6 C.F.	"	75	.533	"		7.90	2.41	10.31	16.20	

For expanded coverage of these items see Means Heavy Construction Cost Data 2000

02200 | Site Preparation

02225 | Selective Demolition

		CREW	DAILY OUTPUT	LABOR-HOURS	UNIT	2000 BARE COSTS MAT.	LABOR	EQUIP.	TOTAL	TOTAL INCL O&P
320 0200	Slab on grade to 6" thick, not reinforced, under 8 S.F.	B-9	85	.471	S.F.		7	2.12	9.12	14.30
0250	Not reinforced, over 8 S.F.		175	.229	"		3.39	1.03	4.42	6.95
0600	Walls, not reinforced, under 6 C.F.		60	.667	C.F.		9.90	3.01	12.91	20.50
0650	Not reinforced, over 6 C.F.		65	.615			9.15	2.78	11.93	18.70
1000	Concrete, elevated slab, bar reinforced, under 6 C.F.	B-9C	45	.889			13.20	4.01	17.21	27
1050	Bar reinforced, over 6 C.F.	"	50	.800			11.90	3.61	15.51	24.50
1200	Slab on grade to 6" thick, bar reinforced, under 8 S.F.	B-9	75	.533	S.F.		7.90	2.41	10.31	16.20
1250	Bar reinforced, over 8 S.F.	"	105	.381	"		5.65	1.72	7.37	11.60
1400	Walls, bar reinforced, under 6 C.F.	B-9C	50	.800	C.F.		11.90	3.61	15.51	24.50
1450	Bar reinforced, over 6 C.F.	"	55	.727	"		10.80	3.28	14.08	22
2000	Brick, to 4 S.F. opening, not including toothing									
2040	4" thick	B-9C	30	1.333	Ea.		19.80	6	25.80	40.50
2060	8" thick		18	2.222			33	10	43	67.50
2080	12" thick		10	4			59.50	18.05	77.55	122
2400	Concrete block, to 4 S.F. opening, 2" thick		35	1.143			16.95	5.15	22.10	34.50
2420	4" thick		30	1.333			19.80	6	25.80	40.50
2440	8" thick		27	1.481			22	6.70	28.70	45
2460	12" thick		24	1.667			25	7.50	32.50	51
2600	Gypsum block, to 4 S.F. opening, 2" thick	B-9	80	.500			7.45	2.26	9.71	15.20
2620	4" thick		70	.571			8.50	2.58	11.08	17.40
2640	8" thick		55	.727			10.80	3.28	14.08	22
2800	Terra cotta, to 4 S.F. opening, 4" thick		70	.571			8.50	2.58	11.08	17.40
2840	8" thick		65	.615			9.15	2.78	11.93	18.70
2880	12" thick		50	.800			11.90	3.61	15.51	24.50
3000	Toothing masonry cutouts, brick, soft old mortar	1 Brhe	40	.200	V.L.F.		3.14		3.14	5.30
3100	Hard mortar		30	.267			4.19		4.19	7.10
3200	Block, soft old mortar		70	.114			1.79		1.79	3.04
3400	Hard mortar		50	.160			2.51		2.51	4.26
6000	Walls, interior, not including re-framing,									
6010	openings to 5 S.F.									
6100	Drywall to 5/8" thick	A-1	24	.333	Ea.		4.82	2.92	7.74	11.45
6200	Paneling to 3/4" thick		20	.400			5.80	3.50	9.30	13.75
6300	Plaster, on gypsum lath		20	.400			5.80	3.50	9.30	13.75
6340	On wire lath		14	.571			8.25	5	13.25	19.65
7000	Wood frame, not including re-framing, openings to 5 S.F.									
7200	Floors, sheathing and flooring to 2" thick	A-1	5	1.600	Ea.		23	14	37	55
7310	Roofs, sheathing to 1" thick, not including roofing		6	1.333			19.25	11.65	30.90	46
7410	Walls, sheathing to 1" thick, not including siding		7	1.143			16.50	10	26.50	39.50
340 0010	**DOOR DEMOLITION**									
0200	Doors, exterior, 1-3/4" thick, single, 3' x 7' high	1 Clab	16	.500	Ea.		7.20		7.20	12.40
0220	Double, 6' x 7' high		12	.667			9.65		9.65	16.50
0500	Interior, 1-3/8" thick, single, 3' x 7' high		20	.400			5.80		5.80	9.90
0520	Double, 6' x 7' high		16	.500			7.20		7.20	12.40
0700	Bi-folding, 3' x 6'-8" high		20	.400			5.80		5.80	9.90
0720	6' x 6'-8" high		18	.444			6.40		6.40	11
0900	Bi-passing, 3' x 6'-8" high		16	.500			7.20		7.20	12.40
0940	6' x 6'-8" high		14	.571			8.25		8.25	14.15
1500	Remove and reset, minimum	1 Carp	8	1			19.70		19.70	34
1520	Maximum	"	6	1.333			26.50		26.50	45
2000	Frames, including trim, metal	A-1	8	1			14.45	8.75	23.20	34.50
2201	Alternate pricing method	"	200	.040	L.F.		.58	.35	.93	1.38
3001	Special doors, counter doors	2 Clab	6	2.667	Ea.		38.50		38.50	66
3101	Double acting		10	1.600			23		23	39.50
3201	Floor door (trap type)		8	2			29		29	49.50
3301	Glass, sliding, including frames		12	1.333			19.25		19.25	33
3401	Overhead, commercial, 12' x 12' high		4	4			58		58	99

02200 | Site Preparation

02225 | Selective Demolition

			CREW	DAILY OUTPUT	LABOR-HOURS	UNIT	MAT.	LABOR	EQUIP.	TOTAL	TOTAL INCL O&P	
340	3441	20' x 16' high	2 Clab	3	5.333	Ea.		77		77	132	340
	3501	Residential, 9' x 7' high		8	2			29		29	49.50	
	3541	16' x 7' high		7	2.286			33		33	56.50	
	3601	Remove and reset, minimum	1 Clab	2	4			58		58	99	
	3621	Maximum	"	1.25	6.400			92.50		92.50	158	
	3701	Roll-up grille	2 Clab	5	3.200			46		46	79	
	3801	Revolving door		2	8			116		116	198	
	3902	Swing door		3	5.333			77		77	132	
380	0010	**FLOORING DEMOLITION**										380
	0200	Brick with mortar	2 Clab	475	.034	S.F.		.49		.49	.83	
	0400	Carpet, bonded, including surface scraping		2,000	.008			.12		.12	.20	
	0440	Scrim applied		8,000	.002			.03		.03	.05	
	0480	Tackless		9,000	.002			.03		.03	.04	
	0601	Composition	1 Clab	200	.040			.58		.58	.99	
	0800	Resilient, sheet goods	2 Clab	1,400	.011			.17		.17	.28	
	0820	For gym floors		900	.018			.26		.26	.44	
	0900	Vinyl composition tile, 12" x 12"		1,000	.016			.23		.23	.40	
	2000	Tile, ceramic, thin set		675	.024			.34		.34	.59	
	2020	Mud set		625	.026			.37		.37	.63	
	2200	Marble, slate, thin set		675	.024			.34		.34	.59	
	2220	Mud set		625	.026			.37		.37	.63	
	2600	Terrazzo, thin set		450	.036			.51		.51	.88	
	2620	Mud set		425	.038			.54		.54	.93	
	2640	Cast in place		300	.053			.77		.77	1.32	
	3000	Wood, block, on end	1 Carp	400	.020			.39		.39	.68	
	3200	Parquet		450	.018			.35		.35	.60	
	3400	Strip flooring, interior, 2-1/4" x 25/32" thick		325	.025			.48		.48	.83	
	3500	Exterior, porch flooring, 1" x 4"		220	.036			.72		.72	1.23	
	3800	Subfloor, tongue and groove, 1" x 6"		325	.025			.48		.48	.83	
	3820	1" x 8"		430	.019			.37		.37	.63	
	3840	1" x 10"		520	.015			.30		.30	.52	
	4000	Plywood, nailed		600	.013			.26		.26	.45	
	4100	Glued and nailed		400	.020			.39		.39	.68	
	8000	Remove flooring, bead blast, minimum	A-1A	1,000	.008			.16	.14	.30	.43	
	8100	Maximum		400	.020			.39	.36	.75	1.07	
	8150	Mastic only		1,500	.005			.11	.09	.20	.28	
390	0010	**FRAMING DEMOLITION**										390
	1020	Concrete, average reinforcing, beams, 8" x 10"	B-9	120	.333	L.F.		4.95	1.50	6.45	10.15	
	1040	10" x 12"		110	.364			5.40	1.64	7.04	11.05	
	1060	12" x 14"		90	.444			6.60	2	8.60	13.50	
	1200	Columns, 8" x 8"		120	.333			4.95	1.50	6.45	10.15	
	1240	10" x 10"		120	.333			4.95	1.50	6.45	10.15	
	1280	12" x 12"		110	.364			5.40	1.64	7.04	11.05	
	1320	14" x 14"		100	.400			5.95	1.80	7.75	12.20	
	1400	Girders, 14" x 16"		55	.727			10.80	3.28	14.08	22	
	1440	16" x 18"		40	1			14.85	4.51	19.36	30.50	
	1600	Slabs, elevated, 6" thick		600	.067	S.F.		.99	.30	1.29	2.03	
	1640	8" thick		450	.089			1.32	.40	1.72	2.70	
	1680	10" thick		360	.111			1.65	.50	2.15	3.38	
	1900	Add for heavy reinforcement									25%	
	2000	Steel framing, beams, 4" x 6"	B-13	500	.096	L.F.		1.52	1.12	2.64	3.81	
	2020	4" x 8"		400	.120			1.90	1.40	3.30	4.77	
	2080	8" x 12"		250	.192			3.04	2.24	5.28	7.60	
	2200	Columns, 6" x 6"		400	.120			1.90	1.40	3.30	4.77	
	2240	8" x 8"		350	.137			2.17	1.60	3.77	5.45	
	2280	10" x 10"		320	.150			2.37	1.75	4.12	5.95	

For expanded coverage of these items see Means Heavy Construction Cost Data 2000

02200 | Site Preparation

02225 | Selective Demolition

		CREW	DAILY OUTPUT	LABOR-HOURS	UNIT	MAT.	LABOR	EQUIP.	TOTAL	TOTAL INCL O&P	
390	2400 Girders, 10" x 12"	B-13	225	.213	L.F.		3.37	2.49	5.86	8.50	390
	2440 10" x 14"		200	.240			3.80	2.80	6.60	9.55	
	2480 10" x 16"		165	.291			4.60	3.39	7.99	11.60	
	2520 10" x 24"		125	.384			6.05	4.47	10.52	15.25	
	3000 Wood framing, beams, 6" x 8"	B-2	275	.145			2.16		2.16	3.70	
	3040 6" x 10"		220	.182			2.70		2.70	4.63	
	3080 6" x 12"		185	.216			3.21		3.21	5.50	
	3120 8" x 12"		140	.286			4.24		4.24	7.25	
	3160 10" x 12"		110	.364			5.40		5.40	9.25	
	3400 Fascia boards, 1" x 6"	1 Clab	500	.016			.23		.23	.40	
	3440 1" x 8"		450	.018			.26		.26	.44	
	3480 1" x 10"		400	.020			.29		.29	.50	
	3800 Headers over openings, 2 @ 2" x 6"		110	.073			1.05		1.05	1.80	
	3840 2 @ 2" x 8"		100	.080			1.16		1.16	1.98	
	3880 2 @ 2" x 10"		90	.089			1.28		1.28	2.20	
	4230 Joists, 2" x 6"	2 Clab	970	.016			.24		.24	.41	
	4240 2" x 8"		940	.017			.25		.25	.42	
	4250 2" x 10"		910	.018			.25		.25	.44	
	4280 2" x 12"		880	.018			.26		.26	.45	
	5400 Posts, 4" x 4"		800	.020			.29		.29	.50	
	5440 6" x 6"		400	.040			.58		.58	.99	
	5480 8" x 8"		300	.053			.77		.77	1.32	
	5500 10" x 10"		240	.067			.96		.96	1.65	
	5800 Rafters, ordinary, 2" x 6"		850	.019			.27		.27	.47	
	5840 2" x 8"		837	.019			.28		.28	.47	
	6200 Stairs and stringers, minimum		40	.400	Riser		5.80		5.80	9.90	
	6240 Maximum		26	.615	"		8.90		8.90	15.25	
	6600 Studs, 2" x 4"		2,000	.008	L.F.		.12		.12	.20	
	6640 2" x 6"		1,600	.010	"		.14		.14	.25	
	9500 See Div. 02225-730 for rubbish handling										
400	0010 **GUTTING** Building interior, including disposal, dumpster fees not included										400
	0500 Residential building										
	0560 Minimum	B-16	400	.080	SF Flr.		1.23	1.12	2.35	3.33	
	0580 Maximum	"	360	.089	"		1.37	1.24	2.61	3.70	
	0900 Commercial building										
	1000 Minimum	B-16	350	.091	SF Flr.		1.41	1.28	2.69	3.80	
	1020 Maximum	"	250	.128	"		1.97	1.79	3.76	5.30	
610	0010 **MASONRY DEMOLITION**										610
	1000 Chimney, 16" x 16", soft old mortar	A-1	24	.333	V.L.F.		4.82	2.92	7.74	11.45	
	1020 Hard mortar		18	.444			6.40	3.89	10.29	15.30	
	1080 20" x 20", soft old mortar		12	.667			9.65	5.85	15.50	23	
	1100 Hard mortar		10	.800			11.55	7	18.55	27.50	
	1140 20" x 32", soft old mortar		10	.800			11.55	7	18.55	27.50	
	1160 Hard mortar		8	1			14.45	8.75	23.20	34.50	
	1200 48" x 48", soft old mortar		5	1.600			23	14	37	55	
	1220 Hard mortar		4	2			29	17.50	46.50	69	
	2000 Columns, 8" x 8", soft old mortar		48	.167			2.41	1.46	3.87	5.75	
	2020 Hard mortar		40	.200			2.89	1.75	4.64	6.90	
	2060 16" x 16", soft old mortar		16	.500			7.20	4.38	11.58	17.20	
	2100 Hard mortar		14	.571			8.25	5	13.25	19.65	
	2140 24" x 24", soft old mortar		8	1			14.45	8.75	23.20	34.50	
	2160 Hard mortar		6	1.333			19.25	11.65	30.90	46	
	2200 36" x 36", soft old mortar		4	2			29	17.50	46.50	69	
	2220 Hard mortar		3	2.667			38.50	23.50	62	91.50	
	3000 Copings, precast or masonry, to 8" wide										

Important: See the Reference Section for critical supporting data - Reference Nos., Crews, & City Cost Indexes

02200 | Site Preparation

02225 | Selective Demolition

			CREW	DAILY OUTPUT	LABOR-HOURS	UNIT	MAT.	LABOR	EQUIP.	TOTAL	TOTAL INCL O&P	
610	3020	Soft old mortar	A-1	180	.044	L.F.		.64	.39	1.03	1.53	610
	3040	Hard mortar	"	160	.050	"		.72	.44	1.16	1.72	
	3100	To 12" wide										
	3120	Soft old mortar	A-1	160	.050	L.F.		.72	.44	1.16	1.72	
	3140	Hard mortar	"	140	.057	"		.83	.50	1.33	1.96	
	4000	Fireplace, brick, 30" x 24" opening										
	4020	Soft old mortar	A-1	2	4	Ea.		58	35	93	138	
	4040	Hard mortar		1.25	6.400			92.50	56	148.50	220	
	4100	Stone, soft old mortar		1.50	5.333			77	46.50	123.50	184	
	4120	Hard mortar		1	8			116	70	186	275	
	5000	Veneers, brick, soft old mortar		140	.057	S.F.		.83	.50	1.33	1.96	
	5020	Hard mortar		125	.064			.92	.56	1.48	2.20	
	5100	Granite and marble, 2" thick		180	.044			.64	.39	1.03	1.53	
	5120	4" thick		170	.047			.68	.41	1.09	1.61	
	5140	Stone, 4" thick		180	.044			.64	.39	1.03	1.53	
	5160	8" thick		175	.046			.66	.40	1.06	1.57	
	5400	Alternate pricing method, stone, 4" thick		60	.133	C.F.		1.93	1.17	3.10	4.58	
	5420	8" thick		85	.094	"		1.36	.82	2.18	3.24	
620	0010	**MILLWORK AND TRIM DEMOLITION**										620
	1000	Cabinets, wood, base cabinets	2 Clab	80	.200	L.F.		2.89		2.89	4.95	
	1020	Wall cabinets	"	80	.200	"		2.89		2.89	4.95	
	1060	Remove and reset, base cabinets	2 Carp	18	.889	Ea.		17.50		17.50	30	
	1070	Wall cabinets	"	20	.800	"		15.75		15.75	27	
	1100	Steel, painted, base cabinets	2 Clab	60	.267	L.F.		3.85		3.85	6.60	
	1120	Wall cabinets		60	.267	"		3.85		3.85	6.60	
	1200	Casework, large area		320	.050	S.F.		.72		.72	1.24	
	1220	Selective		200	.080	"		1.16		1.16	1.98	
	1500	Counter top, minimum		200	.080	L.F.		1.16		1.16	1.98	
	1510	Maximum		120	.133			1.93		1.93	3.30	
	1550	Remove and reset, minimum	2 Carp	50	.320			6.30		6.30	10.80	
	1560	Maximum	"	40	.400			7.90		7.90	13.50	
	2000	Paneling, 4' x 8' sheets, 1/4" thick	2 Clab	2,000	.008	S.F.		.12		.12	.20	
	2100	Boards, 1" x 4"		700	.023			.33		.33	.57	
	2120	1" x 6"		750	.021			.31		.31	.53	
	2140	1" x 8"		800	.020			.29		.29	.50	
	3000	Trim, baseboard, to 6" wide		1,200	.013	L.F.		.19		.19	.33	
	3040	12" wide		1,000	.016			.23		.23	.40	
	3080	Remove and reset, minimum	2 Carp	400	.040			.79		.79	1.35	
	3090	Maximum	"	300	.053			1.05		1.05	1.80	
	3100	Ceiling trim	2 Clab	1,000	.016			.23		.23	.40	
	3120	Chair rail		1,200	.013			.19		.19	.33	
	3140	Railings with balusters		240	.067			.96		.96	1.65	
	3160	Wainscoting		700	.023	S.F.		.33		.33	.57	
690	0010	**ROOFING AND SIDING DEMOLITION**										690
	1000	Deck, roof, concrete plank	B-13	1,680	.029	S.F.		.45	.33	.78	1.14	
	1100	Gypsum plank		3,900	.012			.19	.14	.33	.49	
	1150	Metal decking		3,500	.014			.22	.16	.38	.55	
	1200	Wood, boards, tongue and groove, 2" x 6"	2 Clab	960	.017			.24		.24	.41	
	1220	2" x 10"		1,040	.015			.22		.22	.38	
	1280	Standard planks, 1" x 6"		1,080	.015			.21		.21	.37	
	1320	1" x 8"		1,160	.014			.20		.20	.34	
	1340	1" x 12"		1,200	.013			.19		.19	.33	
	1350	Plywood, to 1" thick		2,000	.008			.12		.12	.20	
	2000	Gutters, aluminum or wood, edge hung	1 Clab	240	.033	L.F.		.48		.48	.83	
	2100	Built-in		100	.080	"		1.16		1.16	1.98	

For expanded coverage of these items see *Means Heavy Construction Cost Data 2000*

02200 | Site Preparation

02225 | Selective Demolition

			CREW	DAILY OUTPUT	LABOR-HOURS	UNIT	MAT.	LABOR	EQUIP.	TOTAL	TOTAL INCL O&P	
690	2500	Roof accessories, plumbing vent flashing	1 Clab	14	.571	Ea.		8.25		8.25	14.15	690
	2600	Adjustable metal chimney flashing		9	.889	"		12.85		12.85	22	
	2650	Coping, sheet metal, up to 12" wide		240	.033	L.F.		.48		.48	.83	
	2660	Concrete, up to 12" wide	2 Clab	160	.100	"		1.44		1.44	2.48	
	3000	Roofing, built-up, 5 ply roof, no gravel	B-2	1,600	.025	S.F.		.37		.37	.64	
	3001	Including gravel		890	.045			.67		.67	1.14	
	3100	Gravel removal, minimum		5,000	.008			.12		.12	.20	
	3120	Maximum		2,000	.020			.30		.30	.51	
	3400	Roof insulation board, up to 2" thick		3,900	.010			.15		.15	.26	
	3450	Roll roofing, cold adhesive	1 Clab	12	.667	Sq.		9.65		9.65	16.50	
	4000	Shingles, asphalt strip, 1 layer	B-2	3,500	.011	S.F.		.17		.17	.29	
	4100	Slate		2,500	.016			.24		.24	.41	
	4300	Wood		2,200	.018			.27		.27	.46	
	4500	Skylight to 10 S.F.	1 Clab	8	1	Ea.		14.45		14.45	25	
	5000	Siding, metal, horizontal		444	.018	S.F.		.26		.26	.45	
	5020	Vertical		400	.020			.29		.29	.50	
	5200	Wood, boards, vertical		400	.020			.29		.29	.50	
	5220	Clapboards, horizontal		380	.021			.30		.30	.52	
	5240	Shingles		350	.023			.33		.33	.57	
	5260	Textured plywood		725	.011			.16		.16	.27	
720	0010	**DISPOSAL ONLY** Urban buildings with salvage value allowed										720
	0020	Including loading and 5 mile haul to dump										
	0200	Steel frame	B-3	430	.112	C.Y.		1.82	3.90	5.72	7.35	
	0300	Concrete frame		365	.132			2.14	4.60	6.74	8.65	
	0400	Masonry construction		445	.108			1.75	3.77	5.52	7.10	
	0500	Wood frame		247	.194			3.16	6.80	9.96	12.85	
730	0010	**RUBBISH HANDLING** The following are to be added to the										730
	0020	demolition prices										
	0400	Chute, circular, prefabricated steel, 18" diameter	B-1	40	.600	L.F.	19.35	9.05		28.40	37	
	0440	30" diameter	"	30	.800	"	25	12.10		37.10	48	
	0600	Dumpster, weekly rental, 1 dump/week, 6 C.Y. capacity (2 Tons)				Ea.					300	
	0700	10 C.Y. capacity (4 Tons)									375	
	0800	30 C.Y. capacity (10 Tons)									640	
	0840	40 C.Y. capacity (13 Tons)									775	
	1001	Dust partition, 6 mil polyethylene, 4' x 8' panels, 1" x 3" frame	2 Clab	2,000	.008	S.F.	.40	.12		.52	.64	
	1081	2" x 4" frame		2,000	.008	"	.50	.12		.62	.75	
	2000	Load, haul to chute & dumping into chute, 50' haul		24	.667	C.Y.		9.65		9.65	16.50	
	2040	100' haul		16.50	.970			14		14	24	
	2080	Over 100' haul, add per 100 L.F.		35.50	.451			6.50		6.50	11.15	
	2120	In elevators, per 10 floors, add		140	.114			1.65		1.65	2.83	
	3000	Loading & trucking, including 2 mile haul, chute loaded	B-16	45	.711			10.95	9.95	20.90	29.50	
	3040	Hand loading truck, 50' haul	"	48	.667			10.25	9.30	19.55	27.50	
	3080	Machine loading truck	B-17	120	.267			4.28	4.74	9.02	12.40	
	5000	Haul, per mile, up to 8 C.Y. truck	B-34B	1,165	.007			.11	.38	.49	.61	
	5100	Over 8 C.Y. truck	"	1,550	.005			.08	.29	.37	.46	
740	0010	**DUMP CHARGES** Typical urban city, tipping fees only										740
	0100	Building construction materials				Ton					55	
	0200	Trees, brush, lumber									45	
	0300	Rubbish only									50	
	0500	Reclamation station, usual charge									80	
760	0010	**SAW CUTTING**, Asphalt, up to 3" deep	B-89	1,050	.015	L.F.	.24	.26	.05	.55	.76	760
	0020	Each additional inch of depth		1,800	.009		.06	.15	.03	.24	.36	
	0400	Concrete slabs, mesh reinforcing, up to 3" deep		980	.016		.33	.28	.05	.66	.89	
	0420	Each additional inch of depth		1,600	.010		.44	.17	.03	.64	.80	

Important: See the Reference Section for critical supporting data - Reference Nos., Crews, & City Cost Indexes

02200 | Site Preparation

02225 | Selective Demolition

		CREW	DAILY OUTPUT	LABOR-HOURS	UNIT	MAT.	LABOR	EQUIP.	TOTAL	TOTAL INCL O&P	
760	0800	Concrete walls, hydraulic saw, plain, per inch of depth	B-89B	250	.064	L.F.	.30	1.12	2.14	3.56	4.54	760
	0820	Rod reinforcing, per inch of depth		150	.107		.42	1.86	3.56	5.84	7.45	
	1200	Masonry walls, hydraulic saw, brick, per inch of depth		300	.053		.30	.93	1.78	3.01	3.84	
	1220	Block walls, solid, per inch of depth		250	.064		.31	1.12	2.14	3.57	4.55	
	2000	Brick or masonry w/hand held saw, per inch of depth	A-1	125	.064		.25	.92	.56	1.73	2.47	
	5001	Wood sheathing, to 1" thick on walls	1 Clab	200	.040			.58		.58	.99	
	5021	On roof	"	250	.032			.46		.46	.79	
	9950	See also div. 02210-320 core drilling										
790	0010	**TORCH CUTTING** Steel, 1" thick plate	1 Clab	32	.250	L.F.		3.61		3.61	6.20	790
	0040	1" diameter bar	"	210	.038	Ea.		.55		.55	.94	
	1000	Oxygen lance cutting, reinforced concrete walls										
	1040	12" to 16" thick walls	1 Clab	10	.800	L.F.		11.55		11.55	19.80	
	1080	24" thick walls	"	6	1.333	"		19.25		19.25	33	
840	0010	**WALLS AND PARTITIONS DEMOLITION**										840
	0100	Brick, 4" to 12" thick	B-9C	220	.182	C.F.		2.70	.82	3.52	5.55	
	0200	Concrete block, 4" thick		1,000	.040	S.F.		.59	.18	.77	1.22	
	0280	8" thick		810	.049			.73	.22	.95	1.50	
	0300	Exterior stucco 1" thick over netting	B-9	3,200	.013			.19	.06	.25	.38	
	1000	Drywall, nailed	1 Clab	1,000	.008			.12		.12	.20	
	1020	Glued and nailed		900	.009			.13		.13	.22	
	1500	Fiberboard, nailed		900	.009			.13		.13	.22	
	1520	Glued and nailed		800	.010			.14		.14	.25	
	2000	Movable walls, metal, 5' high		300	.027			.39		.39	.66	
	2020	8' high		400	.020			.29		.29	.50	
	2200	Metal or wood studs, finish 2 sides, fiberboard	B-1	520	.046			.70		.70	1.20	
	2250	Lath and plaster		260	.092			1.40		1.40	2.39	
	2300	Plasterboard (drywall)		520	.046			.70		.70	1.20	
	2350	Plywood		450	.053			.81		.81	1.38	
	3000	Plaster, lime and horsehair, on wood lath	1 Clab	400	.020			.29		.29	.50	
	3020	On metal lath		335	.024			.35		.35	.59	
	3400	Gypsum or perlite, on gypsum lath		410	.020			.28		.28	.48	
	3420	On metal lath		300	.027			.39		.39	.66	
	3600	Plywood, one side	B-1	1,500	.016			.24		.24	.41	
	3750	Terra cotta block and plaster, to 6" thick	"	175	.137			2.07		2.07	3.55	
	3800	Toilet partitions, slate or marble	1 Clab	5	1.600	Ea.		23		23	39.50	
	3820	Hollow metal	"	8	1	"		14.45		14.45	25	
850	0010	**WINDOW DEMOLITION**										850
	0200	Aluminum, including trim, to 12 S.F.	1 Clab	16	.500	Ea.		7.20		7.20	12.40	
	0240	To 25 S.F.		11	.727			10.50		10.50	18	
	0280	To 50 S.F.		5	1.600			23		23	39.50	
	0320	Storm windows, to 12 S.F.		27	.296			4.28		4.28	7.35	
	0360	To 25 S.F.		21	.381			5.50		5.50	9.45	
	0400	To 50 S.F.		16	.500			7.20		7.20	12.40	
	0600	Glass, minimum		200	.040	S.F.		.58		.58	.99	
	0620	Maximum		150	.053	"		.77		.77	1.32	
	1000	Steel, including trim, to 12 S.F.		13	.615	Ea.		8.90		8.90	15.25	
	1020	To 25 S.F.		9	.889			12.85		12.85	22	
	1040	To 50 S.F.		4	2			29		29	49.50	
	2000	Wood, including trim, to 12 S.F.		22	.364			5.25		5.25	9	
	2020	To 25 S.F.		18	.444			6.40		6.40	11	
	2060	To 50 S.F.		13	.615			8.90		8.90	15.25	
	5020	Remove and reset window, minimum	1 Carp	6	1.333			26.50		26.50	45	
	5040	Average		4	2			39.50		39.50	67.50	
	5080	Maximum		2	4			79		79	135	

For expanded coverage of these items see *Means Heavy Construction Cost Data 2000*

02200 | Site Preparation

02230 | Site Clearing

		CREW	DAILY OUTPUT	LABOR-HOURS	UNIT	MAT.	2000 BARE COSTS LABOR	EQUIP.	TOTAL	TOTAL INCL O&P	
200	0010	**CLEAR AND GRUB** Cut & chip light, trees to 6" diam.	B-7	1	48	Acre		755	1,100	1,855	2,475	200
	0150	Grub stumps and remove	B-30	2	12			209	805	1,014	1,225	
	0200	Cut & chip medium, trees to 12" diam.	B-7	.70	68.571			1,075	1,550	2,625	3,550	
	0250	Grub stumps and remove	B-30	1	24			420	1,600	2,020	2,475	
	0300	Cut & chip heavy, trees to 24" diam.	B-7	.30	160			2,500	3,650	6,150	8,275	
	0350	Grub stumps and remove	B-30	.50	48			835	3,225	4,060	4,925	
	0400	If burning is allowed, reduce cut & chip									40%	
	3000	Chipping stumps, to 18" deep, 12" diam.	B-86	20	.400	Ea.		7.95	8.30	16.25	22.50	
	3040	18" diameter		16	.500			9.95	10.35	20.30	28	
	3080	24" diameter		14	.571			11.35	11.85	23.20	32	
	3100	30" diameter		12	.667			13.25	13.80	27.05	37	
	3120	36" diameter		10	.800			15.90	16.55	32.45	44.50	
	3160	48" diameter		8	1			19.90	20.50	40.40	56	
	5000	Tree thinning, feller buncher, conifer										
	5080	Up to 8" diameter	B-93	240	.033	Ea.		.66	1.81	2.47	3.09	
	5120	12" diameter		160	.050			.99	2.72	3.71	4.64	
	5240	Hardwood, up to 4" diameter		240	.033			.66	1.81	2.47	3.09	
	5280	8" diameter		180	.044			.88	2.41	3.29	4.12	
	5320	12" diameter		120	.067			1.33	3.62	4.95	6.20	
	7000	Tree removal, congested area, aerial lift truck										
	7040	8" diameter	B-85	7	5.714	Ea.		91	115	206	279	
	7080	12" diameter		6	6.667			106	134	240	325	
	7120	18" diameter		5	8			127	161	288	390	
	7160	24" diameter		4	10			159	201	360	490	
	7240	36" diameter		3	13.333			212	268	480	655	
	7280	48" diameter		2	20			320	400	720	975	
220	0010	**CLEARING** Brush with brush saw	A-1	.25	32	Acre		460	280	740	1,100	220
	0100	By hand	"	.12	66.667			965	585	1,550	2,300	
	0300	With dozer, ball and chain, light clearing	B-11A	2	8			137	405	542	675	
	0400	Medium clearing		1.50	10.667			183	540	723	900	
	0500	With dozer and brush rake, light		10	1.600			27.50	80.50	108	135	
	0550	Medium brush to 4" diameter		8	2			34.50	101	135.50	169	
	0600	Heavy brush to 4" diameter		6.40	2.500			43	126	169	211	
	1000	Brush mowing, tractor w/rotary mower, no removal										
	1020	Light density	B-84	2	4	Acre		79.50	105	184.50	247	
	1040	Medium density		1.50	5.333			106	140	246	330	
	1080	Heavy density		1	8			159	210	369	495	
250	0010	**FELLING TREES & PILING** With tractor, large tract, firm										250
	0020	level terrain, no boulders, less than 12" diam. trees										
	0300	300 HP dozer, up to 400 trees/acre, 0 to 25% hardwoods	B-10M	.75	10.667	Acre		212	1,475	1,687	1,975	
	0340	25% to 50% hardwoods		.60	13.333			265	1,850	2,115	2,475	
	0370	75% to 100% hardwoods		.45	17.778			355	2,450	2,805	3,275	
	0400	500 trees/acre, 0% to 25% hardwoods		.60	13.333			265	1,850	2,115	2,475	
	0440	25% to 50% hardwoods		.48	16.667			330	2,300	2,630	3,100	
	0470	75% to 100% hardwoods		.36	22.222			440	3,075	3,515	4,100	
	0500	More than 600 trees/acre, 0 to 25% hardwoods		.52	15.385			305	2,125	2,430	2,850	
	0540	25% to 50% hardwoods		.42	19.048			380	2,650	3,030	3,525	
	0570	75% to 100% hardwoods		.31	25.806			515	3,575	4,090	4,775	
	0900	Large tract clearing per tree										
	1500	300 HP dozer, to 12" diameter, softwood	B-10M	320	.025	Ea.		.50	3.46	3.96	4.63	
	1550	Hardwood		100	.080			1.59	11.10	12.69	14.85	
	1600	12" to 24" diameter, softwood		200	.040			.80	5.55	6.35	7.40	
	1650	Hardwood		80	.100			1.99	13.85	15.84	18.55	
	1700	24" to 36" diameter, softwood		100	.080			1.59	11.10	12.69	14.85	
	1750	Hardwood		50	.160			3.18	22	25.18	30	

02200 | Site Preparation

02230 | Site Clearing

			CREW	DAILY OUTPUT	LABOR-HOURS	UNIT	MAT.	LABOR	EQUIP.	TOTAL	TOTAL INCL O&P	
250	1800	36" to 48" diameter, softwood	B-10M	70	.114	Ea.		2.27	15.85	18.12	21	250
	1850	Hardwood	↓	35	.229	↓		4.55	31.50	36.05	42.50	
280	0010	**SELECTIVE CLEARING**										280
	1000	Stump removal on site by hydraulic backhoe, 1-1/2 C.Y.										
	1040	4" to 6" diameter	B-17	60	.533	Ea.		8.55	9.45	18	25	
	1050	8" to 12" diameter	B-30	33	.727			12.70	48.50	61.20	74.50	
	1100	14" to 24" diameter		25	.960			16.75	64.50	81.25	98.50	
	1150	26" to 36" diameter	↓	16	1.500			26	100	126	155	
	2000	Remove selective trees, on site using chain saws and chipper,										
	2050	not incl. stumps, up to 6" diameter	B-7	18	2.667	Ea.		42	61	103	138	
	2100	8" to 12" diameter		12	4			63	91	154	207	
	2150	14" to 24" diameter		10	4.800			75.50	109	184.50	248	
	2200	26" to 36" diameter	↓	8	6			94	137	231	310	
	2300	Machine load, 2 mile haul to dump, 12" diam. tree, add				↓			150		225	
880	0010	**STRIPPING** Topsoil, and stockpiling, sandy loam										880
	0020	200 H.P. dozer, ideal conditions	B-10B	2,300	.003	C.Y.		.07	.35	.42	.50	
	0100	Adverse conditions	"	1,150	.007			.14	.70	.84	1	
	0200	300 HP dozer, ideal conditions	B-10M	3,000	.003			.05	.37	.42	.50	
	0300	Adverse conditions	"	1,650	.005			.10	.67	.77	.90	
	0400	400 HP dozer, ideal conditions	B-10X	3,900	.002			.04	.36	.40	.46	
	0500	Adverse conditions	"	2,000	.004			.08	.70	.78	.90	
	0600	Clay, dry and soft, 200 HP dozer, ideal conditions	B-10B	1,600	.005			.10	.50	.60	.72	
	0601	Strip topsoil, clay, dry & soft, 200 HP dozer, ideal conditions		1,600	.005			.10	.50	.60	.72	
	0700	Adverse conditions	↓	800	.010			.20	1.01	1.21	1.44	
	1000	Medium hard, 300 HP dozer, ideal conditions	B-10M	2,000	.004			.08	.55	.63	.74	
	1100	Adverse conditions	"	1,100	.007			.14	1.01	1.15	1.35	
	1200	Very hard, 400 HP dozer, ideal conditions	B-10X	2,600	.003			.06	.54	.60	.69	
	1300	Adverse conditions	"	1,340	.006	↓		.12	1.04	1.16	1.35	

02240 | Dewatering

			CREW	DAILY OUTPUT	LABOR-HOURS	UNIT	MAT.	LABOR	EQUIP.	TOTAL	TOTAL INCL O&P	
500	0010	**DEWATERING** Excavate drainage trench, 2' wide, 2' deep	B-11C	90	.178	C.Y.		3.05	2.26	5.31	7.65	500
	0100	2' wide, 3' deep, with backhoe loader	"	135	.119			2.03	1.50	3.53	5.05	
	0200	Excavate sump pits by hand, light soil	1 Clab	7.10	1.127			16.30		16.30	28	
	0300	Heavy soil	"	3.50	2.286	↓		33		33	56.50	
	0500	Pumping 8 hr., attended 2 hrs. per day, including 20 L.F.										
	0550	of suction hose & 100 L.F. discharge hose										
	0601	2" diaphragm pump used for 8 hours	B-94A	4	2	Day		29	11.70	40.70	62.50	
	0652	4" diaphragm pump used for 8 hours	B-94B	4	2			29	24	53	75.50	
	0801	8 hrs. attended, 2" diaphragm pump	B-94A	1	8			116	46.50	162.50	250	
	0901	3" centrifugal pump	B-10J	1	8			159	60.50	219.50	330	
	1001	4" diaphragm pump	B-10I	1	8			159	95	254	370	
	1101	6" centrifugal pump	B-10K	1	8			159	221	380	505	
	1300	CMP, incl. excavation 3' deep, 12" diameter	B-6	115	.209	L.F.	9.05	3.34	1.77	14.16	17.55	
	1400	18" diameter		100	.240	"	8.60	3.84	2.03	14.47	18.20	
	1600	Sump hole construction, incl. excavation and gravel, pit		1,250	.019	C.F.	.56	.31	.16	1.03	1.32	
	1700	With 12" gravel collar, 12" pipe, corrugated, 16 ga.		70	.343	L.F.	11.90	5.50	2.90	20.30	25.50	
	1800	15" pipe, corrugated, 16 ga.		55	.436		14.65	7	3.69	25.34	32	
	1900	18" pipe, corrugated, 16 ga.		50	.480		11.50	7.70	4.06	23.26	30	
	2000	24" pipe, corrugated, 14 ga.		40	.600	↓	21.50	9.60	5.10	36.20	46	
	2200	Wood lining, up to 4' x 4', add	↓	300	.080	SFCA	12	1.28	.68	13.96	16.10	
	9950	See div. 02240-900 for wellpoints										
	9960	See div. 02240-700 for deep well systems										
700	0010	**WELLS** For dewatering 10' to 20' deep, 2' diameter										700
	0020	with steel casing, minimum	B-6	165	.145	V.L.F.	2	2.33	1.23	5.56	7.50	
	0050	Average		98	.245		4	3.92	2.07	9.99	13.35	
	0100	Maximum	↓	49	.490		10	7.85	4.14	21.99	29	

For expanded coverage of these items see *Means Heavy Construction Cost Data 2000*

02200 | Site Preparation

02240 | Dewatering

			CREW	DAILY OUTPUT	LABOR-HOURS	UNIT	2000 BARE COSTS MAT.	LABOR	EQUIP.	TOTAL	TOTAL INCL O&P	
700	0300	For pumps for dewatering, see division 01590-400-4100 to 4400										700
	0500	For domestic water wells, see division 02520-900										
900	0010	**WELLPOINTS** For wellpoint equipment rental, see div. 01590-700 R02240-900										900
	0100	Installation and removal of single stage system										
	0110	Labor only, .75 labor-hours per L.F., minimum	1 Clab	10.70	.748	LF Hdr		10.80		10.80	18.50	
	0200	2.0 labor-hours per L.F., maximum	"	4	2	"		29		29	49.50	
	0400	Pump operation, 4 @ 6 hr. shifts										
	0411	Per 24 hour day	4 Clab	1.27	25.197	Day		365		365	625	
	0501	Per 168 hour week, 160 hr. straight, 8 hr. double time	↓	.18	177	Week		2,575		2,575	4,400	
	0551	Per 4.3 week month		.04	800	Month		11,600		11,600	19,800	
	0600	Complete installation, operation, equipment rental, fuel &										
	0610	removal of system with 2" wellpoints 5' O.C.										
	0700	100' long header, 6" diameter, first month	4 Eqlt	3.23	9.907	LF Hdr	113	189		302	440	
	0800	Thereafter, per month		4.13	7.748		90	148		238	345	
	1000	200' long header, 8" diameter, first month		6	5.333		100	102		202	279	
	1100	Thereafter, per month		8.39	3.814		50.50	73		123.50	177	
	1300	500' long header, 8" diameter, first month		10.63	3.010		39.50	57.50		97	139	
	1400	Thereafter, per month		20.91	1.530		28	29		57	79.50	
	1600	1,000' long header, 10" diameter, first month		11.62	2.754		34	52.50		86.50	125	
	1700	Thereafter, per month	↓	41.81	.765	↓	16.90	14.60		31.50	43	
	1900	Note: above figures include pumping 168 hrs. per week										
	1910	and include the pump operator and one stand-by pump.										

02250 | Shoring & Underpinning

			CREW	DAILY OUTPUT	LABOR-HOURS	UNIT	MAT.	LABOR	EQUIP.	TOTAL	TOTAL INCL O&P	
050	0010	**GROUTING, PRESSURE** Cement and sand, 1:1 mix, minimum	B-61	124	.323	Bag	8.40	4.79	2.58	15.77	20.50	050
	0100	Maximum		51	.784	"	8.40	11.65	6.30	26.35	36	
	0200	Cement and sand, 1:1 mix, minimum		250	.160	C.F.	16.80	2.38	1.28	20.46	24	
	0300	Maximum		100	.400		25	5.95	3.20	34.15	41	
	0400	Epoxy cement grout, minimum		137	.292		100	4.34	2.34	106.68	120	
	0500	Maximum	↓	57	.702	↓	100	10.40	5.60	116	134	
	0600	Structural epoxy grout				Gal.	45			45	49.50	
	0700	Alternate pricing method: (Add for materials)										
	0710	5 person crew and equipment	B-61	1	40	Day		595	320	915	1,375	
100	0010	**UNDERPINNING FOUNDATIONS** Including excavation,										100
	0020	forming, reinforcing, concrete and equipment										
	0100	5' to 16' below grade, 100 to 500 C.Y.	B-52	2.30	24.348	C.Y.	179	400	170	749	1,075	
	0200	Over 500 C.Y.		2.50	22.400		161	365	157	683	980	
	0400	16' to 25' below grade, 100 to 500 C.Y.		2	28		197	460	196	853	1,225	
	0500	Over 500 C.Y.		2.10	26.667		186	435	187	808	1,150	
	0700	26' to 40' below grade, 100 to 500 C.Y.		1.60	35		215	570	245	1,030	1,500	
	0800	Over 500 C.Y.	↓	1.80	31.111		197	510	218	925	1,325	
	0900	For under 50 C.Y., add					10%	40%				
	1000	For 50 C.Y. to 100 C.Y., add				↓	5%	20%				
400	0010	**SHEET PILING** Steel, not incl. wales, 22 psf, 15' excav., left in place	B-40	10.81	5.180	Ton	780	99.50	179	1,058.50	1,225	400
	0100	Drive, extract & salvage		6	9.333		216	179	320	715	915	
	0300	20' deep excavation, 27 psf, left in place		12.95	4.324		780	83	149	1,012	1,175	
	0400	Drive, extract & salvage R02250-450		6.55	8.550		216	164	295	675	860	
	0600	25' deep excavation, 38 psf, left in place		19	2.947		780	56.50	102	938.50	1,075	
	0700	Drive, extract & salvage R02250-400		10.50	5.333		216	102	184	502	625	
	0900	40' deep excavation, 38 psf, left in place		21.20	2.642		780	50.50	91	921.50	1,050	
	1000	Drive, extract & salvage		12.25	4.571	↓	216	87.50	158	461.50	570	
	1200	15' deep excavation, 22 psf, left in place		983	.057	S.F.	9.05	1.09	1.97	12.11	14.10	
	1300	Drive, extract & salvage		656	.085		2.42	1.64	2.95	7.01	8.90	
	1500	20' deep excavation, 27 psf, left in place		960	.058		11.35	1.12	2.01	14.48	16.75	
	1600	Drive, extract & salvage	↓	640	.087	↓	3.15	1.68	3.02	7.85	9.85	

02200 | Site Preparation

02250 | Shoring & Underpinning

			CREW	DAILY OUTPUT	LABOR-HOURS	UNIT	MAT.	LABOR	EQUIP.	TOTAL	TOTAL INCL O&P	
400	1800	25' deep excavation, 38 psf, left in place	B-40	1,000	.056	S.F.	16.75	1.07	1.93	19.75	22.50	400
	1900	Drive, extract & salvage		670	.084		4.31	1.60	2.88	8.79	10.80	
	2100	Rent steel sheet piling and wales, first month				Ton	244			244	268	
	2200	Per added month					24.50			24.50	27	
	2300	Rental piling left in place, add to rental					475			475	525	
	2500	Wales, connections & struts, 2/3 salvage					183			183	201	
	2700	High strength piling, 50,000 psi, add					63.50			63.50	70	
	2800	55,000 psi, add					69			69	76	
	3000	Tie rod, not upset, 1-1/2" to 4" diameter with turnbuckle					1,275			1,275	1,400	
	3100	No turnbuckle					1,050			1,050	1,175	
	3300	Upset, 1-3/4" to 4" diameter with turnbuckle					1,600			1,600	1,750	
	3400	No turnbuckle					1,375			1,375	1,525	
	3600	Lightweight, 18" to 28" wide, 7 ga., 9.22 psf, and										
	3610	9 ga., 8.6 psf, minimum				Lb.	.56			.56	.62	
	3700	Average					.61			.61	.67	
	3750	Maximum					.69			.69	.76	
	3900	Wood, solid sheeting, incl. wales, braces and spacers,										
	3910	drive, extract & salvage, 8' deep excavation	B-31	330	.121	S.F.	1.73	1.80	.47	4	5.50	
	4000	10' deep, 50 S.F./hr. in & 150 S.F./hr. out		300	.133		1.78	1.98	.51	4.27	5.90	
	4100	12' deep, 45 S.F./hr. in & 135 S.F./hr. out		270	.148		1.83	2.20	.57	4.60	6.40	
	4200	14' deep, 42 S.F./hr. in & 126 S.F./hr. out		250	.160		1.88	2.38	.61	4.87	6.80	
	4300	16' deep, 40 S.F./hr. in & 120 S.F./hr. out		240	.167		1.94	2.48	.64	5.06	7.10	
	4400	18' deep, 38 S.F./hr. in & 114 S.F./hr. out		230	.174		2.01	2.58	.67	5.26	7.35	
	4500	20' deep, 35 S.F./hr. in & 105 S.F./hr. out		210	.190		2.07	2.83	.73	5.63	7.95	
	4520	Left in place, 8' deep, 55 S.F./hr.		440	.091		3.11	1.35	.35	4.81	6.10	
	4540	10' deep, 50 S.F./hr.		400	.100		3.27	1.49	.38	5.14	6.55	
	4560	12' deep, 45 S.F./hr.		360	.111		3.46	1.65	.43	5.54	7.10	
	4565	14' deep, 42 S.F./hr.		335	.119		3.66	1.77	.46	5.89	7.55	
	4570	16' deep, 40 S.F./hr.		320	.125		3.89	1.86	.48	6.23	8	
	4580	18' deep, 38 S.F./hr.		305	.131		4.15	1.95	.50	6.60	8.45	
	4590	20' deep, 35 S.F./hr.		280	.143		4.44	2.12	.55	7.11	9.10	
	4700	Alternate pricing, left in place, 8' deep		1.76	22.727	M.B.F.	700	340	87.50	1,127.50	1,450	
	4800	Drive, extract and salvage, 8' deep		1.32	30.303	"	620	450	116	1,186	1,575	
	5000	For treated lumber add cost of treatment to lumber										
	5010	See division 06073-400										
500	0010	**SHORING** Existing building, with timber, no salvage allowance	B-51	2.20	21.818	M.B.F.	740	330	78.50	1,148.50	1,450	500
	1000	With 35 ton screw jacks, per box and jack	"	3.60	13.333	Jack	43	200	48	291	440	
900	0010	**VIBROFLOTATION**										900
	0900	Vibroflotation compacted sand cylinder, minimum	B-60	750	.064	V.L.F.		1.06	1.49	2.55	3.44	
	0950	Maximum		325	.148			2.45	3.43	5.88	7.95	
	1100	Vibro replacement compacted stone cylinder, minimum		500	.096			1.59	2.23	3.82	5.15	
	1150	Maximum		250	.192			3.19	4.46	7.65	10.30	
	1300	Mobilization and demobilization, minimum		.47	102	Total		1,700	2,375	4,075	5,475	
	1400	Maximum		.14	342	"		5,700	7,975	13,675	18,400	

02260 | Excavation Support/Protection

			CREW	DAILY OUTPUT	LABOR-HOURS	UNIT	MAT.	LABOR	EQUIP.	TOTAL	TOTAL INCL O&P	
700	0010	**SLURRY TRENCH** Excavated slurry trench in wet soils										700
	0020	backfilled with 3000 PSI concrete, no reinforcing steel										
	0050	Minimum	C-7	333	.216	C.F.	4.65	3.47	2.74	10.86	13.95	
	0100	Maximum		200	.360	"	7.80	5.80	4.56	18.16	23.50	
	0200	Alternate pricing method, minimum		150	.480	S.F.	9.30	7.70	6.10	23.10	30	
	0300	Maximum		120	.600		13.95	9.65	7.60	31.20	40	
	0500	Reinforced slurry trench, minimum	B-48	177	.271		7	4.29	11.30	22.59	27.50	
	0600	Maximum	"	69	.696		23.50	11	29	63.50	76	

For expanded coverage of these items see *Means Heavy Construction Cost Data 2000*

02200 | Site Preparation

02260 | Excavation Support/Protection

			CREW	DAILY OUTPUT	LABOR-HOURS	UNIT	MAT.	LABOR	EQUIP.	TOTAL	TOTAL INCL O&P	
700	0800	Haul for disposal, 2 mile haul, excavated material, add	B-34B	99	.081	C.Y.		1.31	4.52	5.83	7.15	700
	0900	Haul bentonite castings for disposal, add	"	40	.200	"		3.24	11.20	14.44	17.70	
850	0010	**SOLDIER BEAMS & LAGGING** H piles with 3" wood sheeting										850
	0020	horizontal between piles, including removal of wales & braces										
	0100	No hydrostatic head, 15' deep, 1 line of braces, minimum	B-50	545	.191	S.F.	7.40	3.39	3.03	13.82	17.55	
	0200	Maximum		495	.210		8.20	3.73	3.33	15.26	19.35	
	0400	15' to 22' deep with 2 lines of braces, 10" H, minimum		360	.289		8.70	5.15	4.58	18.43	24	
	0500	Maximum		330	.315		9.85	5.60	5	20.45	26.50	
	0700	23' to 35' deep with 3 lines of braces, 12" H, minimum		325	.320		11.35	5.70	5.10	22.15	28.50	
	0800	Maximum		295	.353		12.30	6.25	5.60	24.15	31	
	1000	36' to 45' deep with 4 lines of braces, 14" H, minimum		290	.359		12.70	6.35	5.70	24.75	31.50	
	1100	Maximum		265	.392		13.40	6.95	6.25	26.60	34	
	1300	No hydrostatic head, left in place, 15' dp., 1 line of braces, min.		635	.164		9.85	2.91	2.60	15.36	18.90	
	1400	Maximum		575	.181		10.55	3.21	2.87	16.63	20.50	
	1600	15' to 22' deep with 2 lines of braces, minimum		455	.229		14.75	4.06	3.63	22.44	27.50	
	1700	Maximum		415	.251		16.40	4.45	3.98	24.83	30.50	
	1900	23' to 35' deep with 3 lines of braces, minimum		420	.248		17.55	4.40	3.93	25.88	31.50	
	2000	Maximum		380	.274		19.40	4.86	4.34	28.60	35	
	2200	36' to 45' deep with 4 lines of braces, minimum		385	.270		21	4.80	4.29	30.09	36.50	
	2300	Maximum		350	.297		24.50	5.30	4.72	34.52	41.50	
	2350	Lagging only, 3" thick wood between piles 8' O.C., minimum	B-46	400	.120		1.64	2.07	.12	3.83	5.70	
	2370	Maximum		250	.192		2.46	3.32	.20	5.98	8.95	
	2400	Open sheeting no bracing, for trenches to 10' deep, min.		1,736	.028		.74	.48	.03	1.25	1.70	
	2450	Maximum		1,510	.032		.82	.55	.03	1.40	1.93	
	2500	Tie-back method, add to open sheeting, add, minimum								20%	20%	
	2550	Maximum								60%	60%	
	2700	Tie-backs only, based on tie-backs total length, minimum	B-46	86.80	.553	L.F.	9.45	9.55	.57	19.57	28	
	2750	Maximum		38.50	1.247	"	16.60	21.50	1.28	39.38	58.50	
	3500	Tie-backs only, typical average, 25' long		2	24	Ea.	415	415	24.50	854.50	1,225	
	3600	35' long		1.58	30.380	"	555	525	31.50	1,111.50	1,600	

02300 | Earthwork

02305 | Equipment

			CREW	DAILY OUTPUT	LABOR-HOURS	UNIT	MAT.	LABOR	EQUIP.	TOTAL	TOTAL INCL O&P	
250	0010	**MOBILIZATION OR DEMOBILIZATION** Up to 50 miles	R01590 -100									250
	0020	Dozer, loader, backhoe or excavator, 70 H.P.- 250 H.P.	B-34K	6	1.333	Ea.		21.50	153	174.50	204	
	0100	Above 250 H.P		4	2			32.50	229	261.50	305	
	0300	Scraper, towed type (incl. tractor), 6 C.Y. capacity		3.75	2.133			34.50	244	278.50	325	
	0400	10 C.Y.		3.50	2.286			37	262	299	350	
	0600	Self-propelled scraper, 15 C.Y.		3.30	2.424			39.50	278	317.50	370	
	0700	24 C.Y.		3	2.667			43	305	348	405	
	0900	Shovel or dragline, 3/4 C.Y.		3.60	2.222			36	254	290	340	
	1000	1-1/2 C.Y.		3	2.667			43	305	348	405	
	1100	Delivery charge for small equipment on flatbed trailer, minimum									40	
	1150	Maximum									100	
	3000	For large pieces of equipment, allow for knockdown, assembly										
	3001	and lead and tail vehicles for highway transport										

02300 | Earthwork

02310 | Grading

		Description	CREW	DAILY OUTPUT	LABOR-HOURS	UNIT	MAT.	LABOR	EQUIP.	TOTAL	TOTAL INCL O&P	
440	0010	FINE GRADE Area to be paved with grader, small area	B-11L	400	.040	S.Y.		.69	1.29	1.98	2.58	440
	0100	Large area		2,000	.008			.14	.26	.40	.51	
	1100	Fine grade for slab on grade, machine		1,040	.015			.26	.50	.76	.99	
	1150	Hand grading	B-18	700	.034			.52	.09	.61	.99	
460	0010	LOAM OR TOPSOIL Remove and stockpile on site										460
	0020	6" deep, 200' haul	B-10B	865	.009	C.Y.		.18	.93	1.11	1.34	
	0100	300' haul		520	.015			.31	1.55	1.86	2.22	
	0150	500' haul		225	.036			.71	3.59	4.30	5.10	
	0200	Alternate method: 6" deep, 200' haul		5,090	.002	S.Y.		.03	.16	.19	.22	
	0250	500' haul		1,325	.006	"		.12	.61	.73	.87	
	0400	Spread from pile to rough finish grade, F.E. loader, 1.5 C.Y.	B-10S	200	.040	C.Y.		.80	1.52	2.32	2.99	
	0500	Up to 200' radius, by hand	1 Clab	14	.571			8.25		8.25	14.15	
	0600	Top dress by hand, 1 C.Y. for 600 S.F.	"	11.50	.696		17.35	10.05		27.40	36.50	
	0700	Furnish and place, truck dumped, screened, 4" deep	B-10S	1,300	.006	S.Y.	2.17	.12	.23	2.52	2.85	
	0800	6" deep	"	820	.010	"	2.78	.19	.37	3.34	3.78	

02315 | Excavation and Fill

		Description	CREW	DAILY OUTPUT	LABOR-HOURS	UNIT	MAT.	LABOR	EQUIP.	TOTAL	TOTAL INCL O&P	
100	0010	BACKFILL By hand, no compaction, light soil (R02315-300)	1 Clab	14	.571	C.Y.		8.25		8.25	14.15	100
	0100	Heavy soil		11	.727			10.50		10.50	18	
	0300	Compaction in 6" layers, hand tamp, add to above		20.60	.388			5.60		5.60	9.60	
	0400	Roller compaction operator walking, add	B-10A	100	.080			1.59	.94	2.53	3.68	
	0500	Air tamp, add	B-9C	190	.211			3.13	.95	4.08	6.40	
	0600	Vibrating plate, add	A-1	60	.133			1.93	1.17	3.10	4.58	
	0800	Compaction in 12" layers, hand tamp, add to above	1 Clab	34	.235			3.40		3.40	5.80	
	0900	Roller compaction operator walking, add	B-10A	150	.053			1.06	.63	1.69	2.45	
	1000	Air tamp, add	B-9	285	.140			2.08	.63	2.71	4.27	
	1100	Vibrating plate, add	A-1	90	.089			1.28	.78	2.06	3.06	
	1300	Dozer backfilling, bulk, up to 300' haul, no compaction	B-10B	1,200	.007			.13	.67	.80	.96	
	1400	Air tamped	B-11B	240	.067			1.14	4.21	5.35	6.55	
	1600	Compacting backfill, 6" to 12" lifts, vibrating roller	B-10C	800	.010			.20	1.15	1.35	1.59	
	1700	Sheepsfoot roller	B-10D	750	.011			.21	1.24	1.45	1.72	
	1900	Dozer backfilling, trench, up to 300' haul, no compaction	B-10B	900	.009			.18	.90	1.08	1.28	
	2000	Air tamped	B-11B	235	.068			1.17	4.30	5.47	6.70	
	2200	Compacting backfill, 6" to 12" lifts, vibrating roller	B-10C	700	.011			.23	1.31	1.54	1.82	
	2300	Sheepsfoot roller	B-10D	650	.012			.24	1.44	1.68	1.99	
130	0010	BEDDING For pipe and conduit, not incl. compaction										130
	0050	Crushed or screened bank run gravel	B-6	150	.160	C.Y.	6.50	2.56	1.35	10.41	12.90	
	0100	Crushed stone 3/4" to 1/2"		150	.160		17.05	2.56	1.35	20.96	24.50	
	0200	Sand, dead or bank		150	.160		3.50	2.56	1.35	7.41	9.65	
	0500	Compacting bedding in trench	A-1	90	.089			1.28	.78	2.06	3.06	
320	0010	COMPACTION, STRUCTURAL Steel wheel tandem roller, 5 tons (R02315-300)	B-10E	8	1	Hr.		19.90	16.65	36.55	51.50	320
	0100	10 tons	B-10F	8	1	"		19.90	28	47.90	64	
	0300	Sheepsfoot or wobbly wheel roller, 8" lifts, common fill	B-10G	1,300	.006	C.Y.		.12	.44	.56	.68	
	0400	Select fill	"	1,500	.005			.11	.38	.49	.60	
	0600	Vibratory plate, 8" lifts, common fill	A-1	200	.040			.58	.35	.93	1.38	
	0700	Select fill	"	216	.037			.54	.32	.86	1.28	
340	0010	DRILLING AND BLASTING Only, rock, open face, under 1500 C.Y.	B-47	225	.071	C.Y.	1.60	1.10	2.77	5.47	6.70	340
	0100	Over 1500 C.Y.		300	.053		1.60	.82	2.08	4.50	5.45	
	0200	Areas where blasting mats are required, under 1500 C.Y.		175	.091		1.60	1.41	3.57	6.58	8.10	
	0250	Over 1500 C.Y.		250	.064		1.60	.99	2.50	5.09	6.20	
	0300	Bulk drilling and blasting, can vary greatly, average									5	
	0500	Pits, average									20	
	1300	Deep hole method, up to 1500 C.Y.	B-47	50	.320		1.60	4.94	12.50	19.04	24	
	1400	Over 1500 C.Y.		66	.242		1.60	3.75	9.45	14.80	18.55	

For expanded coverage of these items see *Means Heavy Construction Cost Data 2000*

02300 | Earthwork

02315 | Excavation and Fill

			CREW	DAILY OUTPUT	LABOR-HOURS	UNIT	2000 BARE COSTS MAT.	LABOR	EQUIP.	TOTAL	TOTAL INCL O&P	
340	1900	Restricted areas, up to 1500 C.Y.	B-47	13	1.231	C.Y.	1.60	19	48	68.60	87.50	340
	2000	Over 1500 C.Y.		20	.800		1.60	12.35	31	44.95	57.50	
	2200	Trenches, up to 1500 C.Y.		22	.727		4.64	11.25	28.50	44.39	55.50	
	2300	Over 1500 C.Y.		26	.615		4.64	9.50	24	38.14	48	
	2500	Pier holes, up to 1500 C.Y.		22	.727		1.60	11.25	28.50	41.35	52	
	2600	Over 1500 C.Y.		31	.516		1.60	7.95	20	29.55	37.50	
	2800	Boulders under 1/2 C.Y., loaded on truck, no hauling	B-100	80	.100			1.99	5.65	7.64	9.50	
	2900	Boulders, drilled, blasted	B-47	100	.160		1.60	2.47	6.25	10.32	12.85	
	3100	Jackhammer operators with foreman compressor, air tools	B-9	1	40	Day		595	180	775	1,225	
	3300	Track drill, compressor, operator and foreman	B-47	1	16	"		247	625	872	1,100	
	3500	Blasting caps				Ea.	3			3	3.30	
	3700	Explosives					2.05			2.05	2.26	
	3900	Blasting mats, rent, for first day					90			90	99	
	4000	Per added day					30			30	33	
	4200	Preblast survey for 6 room house, individual lot, minimum	A-6	2.40	6.667			140		140	240	
	4300	Maximum	"	1.35	11.852			248		248	425	
	4500	City block within zone of influence, minimum	A-8	25,200	.001	S.F.		.03		.03	.04	
	4600	Maximum	"	15,100	.002	"		.04		.04	.07	
345	0010	**DRILLING ONLY** 2" hole for rock bolts, average	B-47	316	.051	L.F.		.78	1.98	2.76	3.51	345
	0800	2-1/2" hole for pre-splitting, average		600	.027			.41	1.04	1.45	1.85	
	1600	Quarry operations, 2-1/2" to 3-1/2" diameter		715	.022			.35	.87	1.22	1.55	
400	0010	**EXCAVATING, BULK BANK MEASURE** Common earth piled	R02315-400									400
	0020	For loading onto trucks, add								15%	15%	
	0050	For mobilization and demobilization, see division 02305-250	R02315-450									
	0100	For hauling, see division 02320-200										
	0200	Backhoe, hydraulic, crawler mtd., 1 C.Y. cap. = 75 C.Y./hr.	B-12A	600	.013	C.Y.		.28	.90	1.18	1.45	
	0250	1-1/2 C.Y. cap. = 100 C.Y./hr.	B-12B	800	.010			.21	.89	1.10	1.32	
	0260	2 C.Y. cap. = 130 C.Y./hr.	B-12C	1,040	.008			.16	.98	1.14	1.34	
	0300	3 C.Y. cap. = 160 C.Y./hr.	B-12D	1,280	.006			.13	1.66	1.79	2.04	
	0310	Wheel mounted, 1/2 C.Y. cap. = 30 C.Y./hr.	B-12E	240	.033			.69	1.43	2.12	2.71	
	0360	3/4 C.Y. cap. = 45 C.Y./hr.	B-12F	360	.022			.46	1.27	1.73	2.15	
	0500	Clamshell, 1/2 C.Y. cap. = 20 C.Y./hr.	B-12G	160	.050			1.03	3.29	4.32	5.35	
	0550	1 C.Y. cap. = 35 C.Y./hr.	B-12H	280	.029			.59	2.10	2.69	3.29	
	0950	Dragline, 1/2 C.Y. cap. = 30 C.Y./hr.	B-12I	240	.033			.69	2.19	2.88	3.55	
	1000	3/4 C.Y. cap. = 35 C.Y./hr.	"	280	.029			.59	1.88	2.47	3.05	
	1050	1-1/2 C.Y. cap. = 65 C.Y./hr.	B-12P	520	.015			.32	1.47	1.79	2.14	
	1200	Front end loader, track mtd., 1-1/2 C.Y. cap. = 70 C.Y./hr.	B-10N	560	.014			.28	.59	.87	1.12	
	1250	2-1/2 C.Y. cap. = 95 C.Y./hr.	B-10O	760	.011			.21	.59	.80	1	
	1300	3 C.Y. cap. = 130 C.Y./hr.	B-10P	1,040	.008			.15	.75	.90	1.08	
	1350	5 C.Y. cap. = 160 C.Y./hr.	B-10Q	1,280	.006			.12	.81	.93	1.10	
	1500	Wheel mounted, 3/4 C.Y. cap. = 45 C.Y./hr.	B-10R	360	.022			.44	.65	1.09	1.44	
	1550	1-1/2 C.Y. cap. = 80 C.Y./hr.	B-10S	640	.013			.25	.47	.72	.93	
	1600	2-1/4 C.Y. cap. = 100 C.Y./hr.	B-10T	800	.010			.20	.50	.70	.88	
	1650	5 C.Y. cap. = 185 C.Y./hr.	B-10U	1,480	.005			.11	.58	.69	.82	
	1800	Hydraulic excavator, truck mtd, 1/2 C.Y. = 30 C.Y./hr.	B-12J	240	.033			.69	2.58	3.27	3.97	
	1850	48 inch bucket, 1 C.Y. = 45 C.Y./hr.	B-12K	360	.022			.46	2.27	2.73	3.26	
	3700	Shovel, 1/2 C.Y. capacity = 55 C.Y./hr.	B-12L	440	.018			.38	1.21	1.59	1.95	
	3750	3/4 C.Y. capacity = 85 C.Y./hr.	B-12M	680	.012			.24	.87	1.11	1.36	
	3800	1 C.Y. capacity = 120 C.Y./hr.	B-12N	960	.008			.17	.68	.85	1.03	
	3850	1-1/2 C.Y. capacity = 160 C.Y./hr.	B-12O	1,280	.006			.13	.68	.81	.96	
	3900	3 C.Y. cap. = 250 C.Y./hr.	B-12T	2,000	.004			.08	.62	.70	.82	
	4000	For soft soil or sand, deduct								15%	15%	
	4100	For heavy soil or stiff clay, add								60%	60%	
	4200	For wet excavation with clamshell or dragline, add								100%	100%	
	4250	All other equipment, add								50%	50%	

02300 | Earthwork

02315 | Excavation and Fill

			CREW	DAILY OUTPUT	LABOR-HOURS	UNIT	MAT.	2000 BARE COSTS LABOR	EQUIP.	TOTAL	TOTAL INCL O&P	
400	4400	Clamshell in sheeting or cofferdam, minimum	B-12H	160	.050	C.Y.		1.03	3.67	4.70	5.75	400
	4450	Maximum	"	60	.133	↓		2.75	9.80	12.55	15.30	
	8000	For hauling excavated material, see div. 02320-200										
410	0010	**EXCAVATING, BULK, DOZER** Open site										410
	2000	75 H.P., 50' haul, sand & gravel	B-10L	460	.017	C.Y.		.35	.61	.96	1.24	
	2020	Common earth		400	.020			.40	.70	1.10	1.43	
	2040	Clay		250	.032			.64	1.12	1.76	2.29	
	2200	150' haul, sand & gravel		230	.035			.69	1.22	1.91	2.49	
	2220	Common earth		200	.040			.80	1.40	2.20	2.86	
	2240	Clay		125	.064			1.27	2.24	3.51	4.57	
	2400	300' haul, sand & gravel		120	.067			1.33	2.33	3.66	4.76	
	2440	Clay		65	.123			2.45	4.30	6.75	8.80	
	3000	105 H.P., 50' haul, sand & gravel	B-10W	700	.011			.23	.56	.79	1	
	3020	Common earth		610	.013			.26	.64	.90	1.14	
	3040	Clay		385	.021			.41	1.02	1.43	1.81	
	3200	150' haul, sand & gravel		310	.026			.51	1.27	1.78	2.24	
	3220	Common earth		270	.030			.59	1.46	2.05	2.58	
	3240	Clay		170	.047			.94	2.31	3.25	4.09	
	3300	300' haul, sand & gravel		140	.057			1.14	2.81	3.95	4.98	
	3320	Common earth		120	.067			1.33	3.28	4.61	5.80	
	3340	Clay		100	.080			1.59	3.93	5.52	6.95	
	4000	200 H.P., 50' haul, sand & gravel	B-10B	1,400	.006			.11	.58	.69	.82	
	4020	Common earth		1,230	.007			.13	.66	.79	.93	
	4040	Clay		770	.010			.21	1.05	1.26	1.49	
	4200	150' haul, sand & gravel		595	.013			.27	1.36	1.63	1.93	
	4220	Common earth		516	.016			.31	1.56	1.87	2.23	
	4240	Clay		325	.025			.49	2.48	2.97	3.54	
	4400	300' haul, sand & gravel		310	.026			.51	2.60	3.11	3.71	
	4420	Common earth		270	.030			.59	2.99	3.58	4.27	
	4440	Clay		170	.047			.94	4.75	5.69	6.75	
	5000	300 H.P., 50' haul, sand & gravel	B-10M	1,900	.004			.08	.58	.66	.78	
	5020	Common earth		1,650	.005			.10	.67	.77	.90	
	5040	Clay		1,025	.008			.16	1.08	1.24	1.45	
	5200	150' haul, sand & gravel		920	.009			.17	1.20	1.37	1.61	
	5220	Common earth		800	.010			.20	1.39	1.59	1.85	
	5240	Clay		500	.016			.32	2.22	2.54	2.97	
	5400	300' haul, sand & gravel		470	.017			.34	2.36	2.70	3.15	
	5420	Common earth		410	.020			.39	2.70	3.09	3.61	
	5440	Clay		250	.032			.64	4.43	5.07	5.95	
	5500	460 H.P., 50' haul, sand & gravel	B-10X	1,930	.004			.08	.72	.80	.94	
	5510	Common earth		1,680	.005			.09	.83	.92	1.07	
	5520	Clay		1,050	.008			.15	1.33	1.48	1.71	
	5530	150' haul, sand & gravel		1,290	.006			.12	1.08	1.20	1.39	
	5540	Common earth		1,120	.007			.14	1.25	1.39	1.61	
	5550	Clay		700	.011			.23	2	2.23	2.58	
	5560	300' haul, sand & gravel		660	.012			.24	2.12	2.36	2.73	
	5570	Common earth		575	.014			.28	2.43	2.71	3.13	
	5580	Clay		350	.023			.45	3.99	4.44	5.15	
	6000	700 H.P., 50' haul, sand & gravel	B-10V	3,500	.002			.05	.79	.84	.94	
	6010	Common earth		3,035	.003			.05	.91	.96	1.09	
	6020	Clay		1,925	.004			.08	1.43	1.51	1.71	
	6030	150' haul, sand & gravel		2,025	.004			.08	1.36	1.44	1.62	
	6040	Common earth		1,750	.005			.09	1.57	1.66	1.88	
	6050	Clay		1,100	.007			.14	2.50	2.64	2.99	
	6060	300' haul, sand & gravel		1,030	.008			.15	2.67	2.82	3.20	

For expanded coverage of these items see Means Heavy Construction Cost Data 2000

02300 | Earthwork

02315 | Excavation and Fill

			CREW	DAILY OUTPUT	LABOR-HOURS	UNIT	MAT.	LABOR	EQUIP.	TOTAL	TOTAL INCL O&P	
410	6070	Common earth	B-10V	900	.009	C.Y.	.18	3.06		3.24	3.65	410
	6080	Clay	↓	550	.015	↓	.29	5		5.29	6	
430	0010	**EXCAVATION, BULK, SCRAPERS**	R02315 -400									430
	0100	Elevating scraper 11 C.Y., sand & gravel 1500' haul	B-33F	690	.014	C.Y.		.29	1.41	1.70	2.04	
	0150	3000' haul		610	.016			.33	1.60	1.93	2.30	
	0200	5000' haul		505	.020			.39	1.93	2.32	2.77	
	0300	Common earth, 1500' haul		600	.017			.33	1.63	1.96	2.34	
	0350	3000' haul		530	.019			.38	1.84	2.22	2.64	
	0400	5000' haul		440	.023			.45	2.22	2.67	3.19	
	0500	Clay, 1500' haul		375	.027			.53	2.60	3.13	3.74	
	0550	3000' haul		330	.030			.60	2.96	3.56	4.25	
	0600	5000' haul	↓	275	.036	↓		.72	3.55	4.27	5.10	
	1000	Self propelled scraper, 14 C.Y. 1/4 push dozer, sand										
	1050	and gravel, 1500' haul	B-33D	920	.011	C.Y.		.22	2.07	2.29	2.64	
	1100	3000' haul		805	.012			.25	2.36	2.61	3.01	
	1200	5000' haul		645	.016			.31	2.95	3.26	3.76	
	1300	Common earth, 1500' haul		800	.013			.25	2.38	2.63	3.03	
	1350	3000' haul		700	.014			.28	2.72	3	3.46	
	1400	5000' haul		560	.018			.36	3.40	3.76	4.33	
	1500	Clay, 1500' haul		500	.020			.40	3.81	4.21	4.85	
	1550	3000' haul		440	.023			.45	4.33	4.78	5.50	
	1600	5000' haul	↓	350	.029			.57	5.45	6.02	6.95	
	2000	21 C.Y., 1/4 push dozer, sand & gravel, 1500' haul	B-33E	1,180	.008			.17	1.86	2.03	2.32	
	2100	3000' haul		910	.011			.22	2.41	2.63	3.01	
	2200	5000' haul		750	.013			.27	2.92	3.19	3.65	
	2300	Common earth, 1500' haul		1,030	.010			.19	2.13	2.32	2.66	
	2350	3000' haul		790	.013			.25	2.77	3.02	3.47	
	2400	5000' haul		650	.015			.31	3.37	3.68	4.22	
	2500	Clay, 1500' haul		645	.016			.31	3.40	3.71	4.24	
	2550	3000' haul		495	.020			.40	4.42	4.82	5.55	
	2600	5000' haul	↓	405	.025			.49	5.40	5.89	6.75	
	2700	Towed, 10 C.Y., 1/4 push dozer, sand & gravel, 1500' haul	B-33B	560	.018			.36	2.84	3.20	3.71	
	2720	3000' haul		450	.022			.44	3.53	3.97	4.62	
	2730	5000' haul		365	.027			.55	4.36	4.91	5.70	
	2750	Common earth, 1500' haul		420	.024			.47	3.79	4.26	4.95	
	2770	3000' haul		400	.025			.50	3.97	4.47	5.20	
	2780	5000' haul		310	.032			.64	5.15	5.79	6.70	
	2800	Clay, 1500' haul		315	.032			.63	5.05	5.68	6.60	
	2820	3000' haul		300	.033			.66	5.30	5.96	6.95	
	2840	5000' haul		225	.044			.88	7.05	7.93	9.20	
	2900	15 C.Y., 1/4 push dozer, sand & gravel, 1500' haul	B-33C	800	.013			.25	1.99	2.24	2.60	
	2920	3000' haul		640	.016			.31	2.48	2.79	3.25	
	2940	5000' haul		520	.019			.38	3.06	3.44	3.99	
	2960	Common earth, 1500' haul		600	.017			.33	2.65	2.98	3.46	
	2980	3000' haul		560	.018			.36	2.84	3.20	3.71	
	3000	5000' haul		440	.023			.45	3.61	4.06	4.72	
	3020	Clay, 1500' haul		450	.022			.44	3.53	3.97	4.62	
	3040	3000' haul		420	.024			.47	3.79	4.26	4.95	
	3060	5000' haul	↓	320	.031	↓		.62	4.97	5.59	6.50	
440	0010	**EXCAVATING, STRUCTURAL** Hand, pits to 6' deep, sandy soil	1 Clab	8	1	C.Y.		14.45		14.45	25	440
	0100	Heavy soil or clay		4	2			29		29	49.50	
	0300	Pits 6' to 12' deep, sandy soil		5	1.600			23		23	39.50	
	0500	Heavy soil or clay		3	2.667			38.50		38.50	66	
	0700	Pits 12' to 18' deep, sandy soil		4	2			29		29	49.50	
	0900	Heavy soil or clay	↓	2	4	↓		58		58	99	

02300 | Earthwork

02315 | Excavation and Fill

				DAILY	LABOR-			2000 BARE COSTS			TOTAL	
			CREW	OUTPUT	HOURS	UNIT	MAT.	LABOR	EQUIP.	TOTAL	INCL O&P	
440	1100	Hand loading trucks from stock pile, sandy soil	1 Clab	12	.667	C.Y.		9.65		9.65	16.50	440
	1300	Heavy soil or clay	↓	8	1	↓		14.45		14.45	25	
	1500	For wet or muck hand excavation, add to above				%				50%	50%	
	2000	Machine excavation, for spread and mat footings, elevator pits,										
	2001	and small building foundations										
	2035	Common earth, hydraulic backhoe, 3/4 C.Y. bucket	B-12F	90	.089	C.Y.		1.84	5.05	6.89	8.60	
	2040	1 C.Y. bucket	B-12A	108	.074			1.53	4.98	6.51	8.05	
	2050	1-1/2 C.Y. bucket	B-12B	144	.056			1.15	4.95	6.10	7.35	
	2060	2 C.Y. bucket	B-12C	200	.040			.83	5.10	5.93	6.95	
	2070	Sand and gravel, 3/4 C.Y. bucket	B-12F	100	.080			1.65	4.56	6.21	7.75	
	2080	1 C.Y. bucket	B-12A	120	.067			1.38	4.48	5.86	7.20	
	2090	1-1/2 C.Y. bucket	B-12B	160	.050			1.03	4.46	5.49	6.60	
	3000	2 C.Y. bucket	B-12C	220	.036			.75	4.62	5.37	6.35	
	3010	Clay, till, or blasted rock, 3/4 C.Y. bucket	B-12F	80	.100			2.07	5.70	7.77	9.65	
	3020	1 C.Y. bucket	B-12A	95	.084			1.74	5.65	7.39	9.15	
	3030	1-1/2 C.Y. bucket	B-12B	130	.062			1.27	5.50	6.77	8.15	
	3040	2 C.Y. bucket	B-12C	175	.046	↓		.94	5.80	6.74	7.95	
	9010	For mobilization or demobilization, see div. 02305-250										
	9020	For dewatering, see div. 02240-500										
	9022	For larger structures, see Bulk Excavation, div. 02315-400										
	9024	For loading onto trucks, add								15%		
	9026	For hauling, see div. 02320-200										
	9030	For sheeting or soldier beams/lagging, see div. 02250 & 02260										
	9040	For trench excavation of strip footings, see div. 02315-900										
500	0010	**FILL** Borrow, load, 1 mile haul, spread with dozer										500
	0020	for embankments	B-15	1,200	.023	C.Y.	4.77	.40	1.42	6.59	7.45	
	0100	Select fill for shoulders & embankments	"	1,200	.023	"	7.50	.40	1.42	9.32	10.45	
	0201	For hauling over 1 mile, add to above per C.Y., div. 02320-200				Mile				.62	.82	
505	0010	**FILL** Spread dumped material, by dozer, no compaction	B-10B	1,000	.008	C.Y.		.16	.81	.97	1.15	505
	0100	By hand	1 Clab	12	.667	"		9.65		9.65	16.50	
	0500	Gravel fill, compacted, under floor slabs, 4" deep	B-37	10,000	.005	S.F.	.15	.07	.01	.23	.31	
	0600	6" deep		8,600	.006		.23	.09	.02	.34	.42	
	0700	9" deep		7,200	.007		.38	.10	.02	.50	.61	
	0800	12" deep		6,000	.008	↓	.52	.12	.02	.66	.81	
	1000	Alternate pricing method, 4" deep		120	.400	C.Y.	11.25	6.20	1.11	18.56	24	
	1100	6" deep		160	.300		11.25	4.67	.83	16.75	21.50	
	1200	9" deep		200	.240		11.25	3.73	.67	15.65	19.50	
	1300	12" deep	↓	220	.218	↓	11.25	3.39	.61	15.25	18.85	
	1500	For fill under exterior paving, see division 02720-200										
900	0010	**EXCAVATING, TRENCH** or continuous footing, common earth										900
	0020	No sheeting or dewatering included										
	0050	1' to 4' deep, 3/8 C.Y. tractor loader/backhoe	B-11C	150	.107	C.Y.		1.83	1.35	3.18	4.57	
	0060	1/2 C.Y. tractor loader/backhoe	B-11M	200	.080			1.37	1.37	2.74	3.81	
	0090	4' to 6' deep, 1/2 C.Y. tractor loader/backhoe	"	200	.080			1.37	1.37	2.74	3.81	
	0101	5/8 C.Y. hydraulic backhoe	B-95A	250	.064			1.12	1.65	2.77	3.70	
	0300	1/2 C.Y. hydraulic excavator, truck mounted	B-12J	200	.040			.83	3.09	3.92	4.77	
	0500	6' to 10' deep, 3/4 C.Y. hydraulic backhoe, 6' to 10' deep	B-12F	225	.036			.73	2.03	2.76	3.45	
	0510	1 C.Y. hydraulic backhoe	B-12A	400	.020			.41	1.35	1.76	2.16	
	0600	1 C.Y. hydraulic excavator, truck mounted	B-12K	400	.020			.41	2.04	2.45	2.93	
	0610	1-1/2 C.Y. hydraulic backhoe	B-12B	600	.013			.28	1.19	1.47	1.77	
	0900	10' to 14' deep, 3/4 C.Y. hydraulic backhoe	B-12F	200	.040			.83	2.28	3.11	3.88	
	0910	1 C.Y. hydraulic backhoe	B-12A	360	.022			.46	1.49	1.95	2.40	
	1000	1-1/2 C.Y. hydraulic backhoe	B-12B	540	.015			.31	1.32	1.63	1.96	
	1300	14' to 20' deep, 1 C.Y. hydraulic backhoe	B-12A	320	.025			.52	1.68	2.20	2.71	
	1310	1-1/2 C.Y. hydraulic backhoe	B-12B	480	.017	↓		.34	1.49	1.83	2.20	

For expanded coverage of these items see *Means Heavy Construction Cost Data 2000*

02300 | Earthwork

02315 | Excavation and Fill

		CREW	DAILY OUTPUT	LABOR-HOURS	UNIT	MAT.	LABOR	EQUIP.	TOTAL	TOTAL INCL O&P	
900	1320	2-1/2 C.Y. hydraulic backhoe	B-12S	850	.009	C.Y.		.19	1.97	2.16	2.49	900
	1400	By hand with pick and shovel 2' to 6' deep, light soil	1 Clab	8	1			14.45		14.45	25	
	1500	Heavy soil	"	4	2			29		29	49.50	
	1700	For tamping backfilled trenches, air tamp, add	A-1	100	.080			1.16	.70	1.86	2.75	
	1900	Vibrating plate, add	B-18	230	.104	↓		1.58	.26	1.84	2.99	
	2100	Trim sides and bottom for concrete pours, common earth		1,500	.016	S.F.		.24	.04	.28	.45	
	2300	Hardpan	↓	600	.040	"		.60	.10	.70	1.15	
	2400	Pier and spread footing excavation, add to above				C.Y.				30%	30%	
	3000	Backfill trench, F.E. loader, wheel mtd., 1 C.Y. bucket										
	3020	Minimal haul	B-10R	400	.020	C.Y.		.40	.58	.98	1.30	
	3040	100' haul	"	200	.040			.80	1.17	1.97	2.60	
	3080	2-1/4 C.Y. bucket, minimum haul	B-10T	600	.013			.27	.67	.94	1.17	
	3090	100' haul	"	300	.027	↓		.53	1.34	1.87	2.35	
940	0010	**EXCAVATING, UTILITY TRENCH** Common earth										940
	0050	Trenching with chain trencher, 12 H.P., operator walking										
	0100	4" wide trench, 12" deep	B-53	800	.010	L.F.		.14	.13	.27	.39	
	0150	18" deep		750	.011			.15	.14	.29	.41	
	0200	24" deep		700	.011			.17	.15	.32	.44	
	0300	6" wide trench, 12" deep		650	.012			.18	.16	.34	.47	
	0350	18" deep		600	.013			.19	.17	.36	.52	
	0400	24" deep		550	.015			.21	.19	.40	.56	
	0450	36" deep		450	.018			.26	.23	.49	.69	
	0600	8" wide trench, 12" deep		475	.017			.24	.22	.46	.66	
	0650	18" deep		400	.020			.29	.26	.55	.78	
	0700	24" deep		350	.023			.33	.29	.62	.89	
	0750	36" deep	↓	300	.027	↓		.39	.34	.73	1.04	
	1000	Backfill by hand including compaction, add										
	1050	4" wide trench, 12" deep	A-1	800	.010	L.F.		.14	.09	.23	.35	
	1100	18" deep		530	.015			.22	.13	.35	.52	
	1150	24" deep		400	.020			.29	.17	.46	.69	
	1300	6" wide trench, 12" deep		540	.015			.21	.13	.34	.51	
	1350	18" deep		405	.020			.29	.17	.46	.68	
	1400	24" deep		270	.030			.43	.26	.69	1.02	
	1450	36" deep		180	.044			.64	.39	1.03	1.53	
	1600	8" wide trench, 12" deep		400	.020			.29	.17	.46	.69	
	1650	18" deep		265	.030			.44	.26	.70	1.04	
	1700	24" deep		200	.040			.58	.35	.93	1.38	
	1750	36" deep	↓	135	.059	↓		.86	.52	1.38	2.04	
	2000	Chain trencher, 40 H.P. operator riding										
	2050	6" wide trench and backfill, 12" deep	B-54	1,200	.007	L.F.		.13	.18	.31	.40	
	2100	18" deep		1,000	.008			.15	.21	.36	.48	
	2150	24" deep		975	.008			.16	.22	.38	.50	
	2200	36" deep		900	.009			.17	.23	.40	.54	
	2250	48" deep		750	.011			.20	.28	.48	.65	
	2300	60" deep		650	.012			.24	.32	.56	.75	
	2400	8" wide trench and backfill, 12" deep		1,000	.008			.15	.21	.36	.48	
	2450	18" deep		950	.008			.16	.22	.38	.51	
	2500	24" deep		900	.009			.17	.23	.40	.54	
	2550	36" deep		800	.010			.19	.26	.45	.61	
	2600	48" deep		650	.012			.24	.32	.56	.75	
	2700	12" wide trench and backfill, 12" deep		975	.008			.16	.22	.38	.50	
	2750	18" deep		860	.009			.18	.25	.43	.56	
	2800	24" deep		800	.010			.19	.26	.45	.61	
	2850	36" deep		725	.011			.21	.29	.50	.67	
	3000	16" wide trench and backfill, 12" deep		835	.010	↓		.18	.25	.43	.58	

02300 | Earthwork

02315 | Excavation and Fill

		CREW	DAILY OUTPUT	LABOR-HOURS	UNIT	MAT.	LABOR	EQUIP.	TOTAL	TOTAL INCL O&P	
940	3050	18" deep	B-54	750	.011	L.F.		.20	.28	.48	.65	940
	3100	24" deep	↓	700	.011	↓		.22	.30	.52	.69	
	3200	Compaction with vibratory plate, add								50%	50%	

02320 | Hauling

			CREW	DAILY OUTPUT	LABOR-HOURS	UNIT	MAT.	LABOR	EQUIP.	TOTAL	TOTAL INCL O&P	
200	0011	**HAULING** Excavated or borrow material, loose cubic yards	R02315 -400									200
	0015	no loading included, highway haulers										
	0020	6 C.Y. dump truck, 1/4 mile round trip, 5.0 loads/hr.	B-34A	195	.041	C.Y.		.66	1.87	2.53	3.17	
	0030	1/2 mile round trip, 4.1 loads/hr.		160	.050			.81	2.28	3.09	3.86	
	0040	1 mile round trip, 3.3 loads/hr.		130	.062			1	2.81	3.81	4.75	
	0100	2 mile round trip, 2.6 loads/hr.		100	.080			1.30	3.65	4.95	6.20	
	0150	3 mile round trip, 2.1 loads/hr.		80	.100			1.62	4.57	6.19	7.70	
	0200	4 mile round trip, 1.8 loads/hr.	↓	70	.114			1.85	5.20	7.05	8.85	
	0310	12 C.Y. dump truck, 1/4 mile round trip 3.7 loads/hr.	B-34B	288	.028			.45	1.55	2	2.46	
	0320	1/2 mile round trip, 3.2 loads/hr.		250	.032			.52	1.79	2.31	2.84	
	0330	1 mile round trip 2.7, loads/hr.		210	.038			.62	2.13	2.75	3.37	
	0400	2 mile round trip, 2.2 loads/hr.		180	.044			.72	2.48	3.20	3.93	
	0450	3 mile round trip, 1.9 loads/hr.		170	.047			.76	2.63	3.39	4.16	
	0500	4 mile round trip, 1.6 loads/hr.		125	.064			1.04	3.58	4.62	5.65	
	0540	5 mile round trip, 1 load/hr.		78	.103			1.66	5.75	7.41	9.05	
	0550	10 mile round trip, 0.60 load/hr.		58	.138			2.23	7.70	9.93	12.25	
	0560	20 mile round trip, 0.4 load/hr.	↓	39	.205			3.32	11.45	14.77	18.15	
	0600	16.5 C.Y. dump trailer, 1 mile round trip, 2.6 loads/hr.	B-34C	280	.029			.46	2	2.46	2.97	
	0700	2 mile round trip, 2.1 loads/hr.		225	.036			.58	2.49	3.07	3.70	
	1000	3 mile round trip, 1.8 loads/hr.		193	.041			.67	2.91	3.58	4.32	
	1100	4 mile round trip, 1.6 loads/hr.		172	.047			.75	3.26	4.01	4.85	
	1110	5 mile round trip, 1 load/hr.		108	.074			1.20	5.20	6.40	7.70	
	1120	10 mile round trip, .60 load/hr.		80	.100			1.62	7	8.62	10.40	
	1130	20 mile round trip, .4 load/hr.	↓	54	.148			2.40	10.40	12.80	15.45	
	1150	20 C.Y. dump trailer, 1 mile round trip, 2.5 loads/hr.	B-34D	325	.025			.40	1.73	2.13	2.58	
	1200	2 mile round trip, 2 loads/hr.		260	.031			.50	2.17	2.67	3.21	
	1220	3 mile round trip, 1.7 loads/hr.		221	.036			.59	2.55	3.14	3.79	
	1240	4 mile round trip, 1.5 loads/hr.		195	.041			.66	2.89	3.55	4.29	
	1245	5 mile round trip, 1.1 load/hr.		143	.056			.91	3.94	4.85	5.85	
	1250	10 mile round trip, .75 load/hr.		110	.073			1.18	5.10	6.28	7.60	
	1255	20 mile round trip, .5 load/hr.	↓	78	.103			1.66	7.25	8.91	10.70	
	1300	Hauling in medium traffic, add								20%	20%	
	1400	Heavy traffic, add								30%	30%	
	1600	Grading at dump, or embankment if required, by dozer	B-10B	1,000	.008	↓		.16	.81	.97	1.15	
	1800	Spotter at fill or cut, if required	1 Clab	8	1	Hr.		14.45		14.45	25	

02325 | Dredging

			CREW	DAILY OUTPUT	LABOR-HOURS	UNIT	MAT.	LABOR	EQUIP.	TOTAL	TOTAL INCL O&P	
250	0010	**DREDGING** Mobilization and demobilization., add to below, minimum	B-8	.53	105	Total		1,775	4,225	6,000	7,625	250
	0100	Maximum	"	.10	560	"		9,400	22,400	31,800	40,400	
	0300	Barge mounted clamshell excavation into scows,										
	0310	Dumped 20 miles at sea, minimum	B-57	310	.129	C.Y.		2.08	5.25	7.33	9.35	
	0400	Maximum	"	213	.188	"		3.02	7.65	10.67	13.55	
	0500	Barge mounted dragline or clamshell, hopper dumped,										
	0510	pumped 1000' to shore dump, minimum	B-57	340	.118	C.Y.		1.89	4.80	6.69	8.50	
	0525	All pumping uses 2000 gallons of water per cubic yard										
	0600	Maximum	B-57	243	.165	C.Y.		2.65	6.70	9.35	11.90	
	1000	Hydraulic method, pumped 1000' to shore dump, minimum		460	.087			1.40	3.55	4.95	6.30	
	1100	Maximum		310	.129			2.08	5.25	7.33	9.35	
	1400	Into scows dumped 20 miles, minimum	↓	425	.094	↓		1.51	3.84	5.35	6.80	

For expanded coverage of these items see *Means Heavy Construction Cost Data 2000*

02300 | Earthwork

02325 | Dredging

		CREW	DAILY OUTPUT	LABOR-HOURS	UNIT	2000 BARE COSTS				TOTAL INCL O&P	
						MAT.	LABOR	EQUIP.	TOTAL		
250	1500	Maximum	B-57	243	.165	C.Y.		2.65	6.70	9.35	11.90	250
	1600	For inland rivers and canals in South, deduct								30%	30%	

02340 | Soil Stabilization

| 160 | 0010 | CALCIUM CHLORIDE Delivered, 100 lb. bags, truckload lots | | | | Ton | 400 | | | 400 | 440 | 160 |
| 0200 | Solution, 4 lb. flake per gallon, tank truck delivery | | | | Gal. | .85 | | | .85 | .94 | | |

02360 | Soil Treatment

800	0010	TERMITE PRETREATMENT										800
	0020	Slab and walls, residential	1 Skwk	1,200	.007	SF Flr.	.22	.13		.35	.47	
	0100	Commercial, minimum		2,496	.003		.22	.06		.28	.35	
	0200	Maximum		1,645	.005		.15	.10		.25	.33	
	0400	Insecticides for termite control, minimum		14.20	.563	Gal.	10	11.15		21.15	30	
	0500	Maximum		11	.727	"	17.10	14.35		31.45	43.50	

02370 | Erosion & Sedimentation Control

300	0010	RIP-RAP Random, broken stone										300
	0100	Machine placed for slope protection	B-12G	62	.129	C.Y.	17.05	2.66	8.50	28.21	32.50	
	0110	3/8 to 1/4 C.Y. pieces, grouted	B-13	80	.600	S.Y.	32.50	9.50	7	49	59.50	
	0200	18" minimum thickness, not grouted	"	53	.906	"	13.15	14.35	10.55	38.05	50.50	
	0300	Dumped, 50 lb. average	B-11A	800	.020	Ton	10	.34	1.01	11.35	12.70	
	0350	100 lb. average		700	.023		14.30	.39	1.15	15.84	17.70	
	0370	300 lb. average		600	.027		16.65	.46	1.34	18.45	20.50	
	0400	Gabions, galvanized steel mesh mats or boxes, stone filled, 6" deep	B-13	200	.240	S.Y.	10.65	3.80	2.80	17.25	21.50	
	0500	9" deep		163	.294		11.90	4.66	3.43	19.99	25	
	0600	12" deep		153	.314		16.30	4.96	3.65	24.91	30.50	
	0700	18" deep		102	.471		22.50	7.45	5.50	35.45	43.50	
	0800	36" deep		60	.800		33	12.65	9.30	54.95	68.50	
550	0010	EROSION CONTROL Jute mesh, 100 S.Y. per roll, 4' wide, stapled	B-80A	2,400	.010	S.Y.	.62	.14	.08	.84	1.01	550
	0100	Plastic netting, stapled, 2" x 1" mesh, 20 mil	B-1	2,500	.010		.40	.15		.55	.69	
	0200	Polypropylene mesh, stapled, 6.5 oz./S.Y.		2,500	.010		1	.15		1.15	1.35	
	0300	Tobacco netting, or jute mesh #2, stapled		2,500	.010		.07	.15		.22	.33	
	1000	Silt fence, polypropylene, 3' high, ideal conditions	2 Clab	1,600	.010	L.F.	.30	.14		.44	.58	
	1100	Adverse conditions	"	950	.017	"	.30	.24		.54	.75	
	1200	Place and remove hay bales	A-2	3	8	Ton	50	119	57.50	226.50	320	
	1250	Hay bales, staked	"	2,500	.010	L.F.	2	.14	.07	2.21	2.52	

02390 | Shore Protect/Mooring Structures

220	0010	DOCKS Floating, recreational, prefabricated galvanized steel with										220
	0020	polyethylene encased polystyrene, no pilings included	F-3	330	.121	S.F.	26	2.17	1.37	29.54	33.50	
	0200	Pile supported, shore constructed, bare, 3" decking		130	.308		15.45	5.50	3.49	24.44	30	
	0250	4" decking		120	.333		15.45	6	3.78	25.23	31.50	
	0400	Floating, small boat, prefab, no shore facilities, minimum		250	.160		6.20	2.87	1.81	10.88	13.70	
	0500	Maximum		150	.267		26	4.78	3.02	33.80	40	
	0700	Per slip, minimum (180 S.F. each)		1.59	25.157	Ea.	1,350	450	285	2,085	2,550	
	0800	Maximum		1.40	28.571	"	5,150	510	325	5,985	6,900	

02400 | Tunneling, Boring & Jacking

02420 | Initial Tunnel Support Systems

			CREW	DAILY OUTPUT	LABOR-HOURS	UNIT	MAT.	LABOR	EQUIP.	TOTAL	TOTAL INCL O&P
700	0011	**ROCK BOLTS**									700
	2020	Hollow core, prestressable anchor, 1" diameter, 5' long	2 Skwk	32	.500	Ea.	70.50	9.90		80.40	95
	2025	10' long		24	.667		133	13.15		146.15	169
	2060	2" diameter, 5' long		32	.500		265	9.90		274.90	310
	2065	10' long		24	.667		510	13.15		523.15	585
	2100	Super high-tensile, 3/4" diameter, 5' long		32	.500		15.50	9.90		25.40	34
	2105	10' long		24	.667		28.50	13.15		41.65	54
	2160	2" diameter, 5' long		32	.500		137	9.90		146.90	168
	2165	10' long		24	.667		246	13.15		259.15	294
	4400	Drill hole for rock bolt, 1-3/4" diam., 5' long (for 3/4" bolt)	B-56	17	.941			13.60	35.50	49.10	63
	4405	10' long		9	1.778			25.50	67.50	93	118
	4420	2" diameter, 5' long (for 1" bolt)		13	1.231			17.80	46.50	64.30	82
	4425	10' long		7	2.286			33	87	120	152
	4460	3-1/2" diameter, 5' long (for 2" bolt)		10	1.600			23	61	84	107
	4465	10' long		5	3.200			46	122	168	213

02441 | Microtunneling

			CREW	DAILY OUTPUT	LABOR-HOURS	UNIT	MAT.	LABOR	EQUIP.	TOTAL	TOTAL INCL O&P
400	0010	**MICROTUNNELING** Not including excavation, backfill, shoring,									400
	0020	or dewatering, average 50'/day, slurry method									
	0100	24" to 48" outside diameter, minimum				L.F.					600
	0110	Adverse conditions, add				%					50%
	1000	Rent microtunneling machine, average monthly lease				Month					80,000
	1010	Operating technician				Day					600
	1100	Mobilization and demobilization, minimum				Job					40,000
	1110	Maximum				"					400,000

02445 | Boring or Jacking Conduits

			CREW	DAILY OUTPUT	LABOR-HOURS	UNIT	MAT.	LABOR	EQUIP.	TOTAL	TOTAL INCL O&P
300	0010	**HORIZONTAL BORING** Casing only, 100' minimum,									300
	0020	not incl. jacking pits or dewatering									
	0100	Roadwork, 1/2" thick wall, 24" diameter casing	B-42	20	2.800	L.F.	45	46.50	57	148.50	192
	0200	36" diameter		16	3.500		75	58.50	71.50	205	260
	0300	48" diameter		15	3.733		105	62.50	76	243.50	305
	0500	Railroad work, 24" diameter		15	3.733		45	62.50	76	183.50	238
	0600	36" diameter		14	4		75	67	81.50	223.50	285
	0700	48" diameter		12	4.667		105	78	95	278	355
	0900	For ledge, add								145	175

02450 | Foundation & Load Bearing Elements

02455 | Driven Piles

			CREW	DAILY OUTPUT	LABOR-HOURS	UNIT	MAT.	LABOR	EQUIP.	TOTAL	TOTAL INCL O&P
220	0010	**PILES, CONCRETE** 200 piles, 60' long									220
	0020	unless specified otherwise, not incl. pile caps or mobilization									
	0050	Cast in place augered piles, no casing or reinforcing									
	0060	8" diameter	B-43	540	.074	V.L.F.	2.03	1.10	3.33	6.46	7.75
	0065	10" diameter		480	.083		3.23	1.24	3.74	8.21	9.80
	0070	12" diameter		420	.095		4.55	1.41	4.28	10.24	12.10
	0075	14" diameter		360	.111		6.15	1.65	4.99	12.79	15.10
	0080	16" diameter		300	.133		8.25	1.98	6	16.23	19.10

For expanded coverage of these items see *Means Heavy Construction Cost Data 2000*

02450 | Foundation & Load Bearing Elements

02455 | Driven Piles

		CREW	DAILY OUTPUT	LABOR-HOURS	UNIT	MAT.	LABOR	EQUIP.	TOTAL	TOTAL INCL O&P
0085	18" diameter	B-43	240	.167	V.L.F.	10.25	2.48	7.50	20.23	23.50
0100	Cast in place, thin wall shell pile, straight sided,									
0110	not incl. reinforcing, 8" diam., 16 ga., 5.8 lb./L.F.	B-19	700	.080	V.L.F.	3.92	1.54	2.29	7.75	9.60
0200	10" diameter, 16 ga. corrugated, 7.3 lb./L.F.		650	.086		5.15	1.65	2.46	9.26	11.35
0300	12" diameter, 16 ga. corrugated, 8.7 lb./L.F.		600	.093		6.65	1.79	2.67	11.11	13.55
0400	14" diameter, 16 ga. corrugated, 10.0 lb./L.F.		550	.102		7.85	1.95	2.91	12.71	15.35
0500	16" diameter, 16 ga. corrugated, 11.6 lb./L.F.	▼	500	.112	▼	9.60	2.15	3.20	14.95	17.95
0800	Cast in place friction pile, 50' long, fluted,									
0810	tapered steel, 4000 psi concrete, no reinforcing									
0900	12" diameter, 7 ga.	B-19	600	.093	V.L.F.	12	1.79	2.67	16.46	19.40
1000	14" diameter, 7 ga.		560	.100		13.10	1.92	2.86	17.88	21
1100	16" diameter, 7 ga.		520	.108		15.40	2.07	3.08	20.55	24
1200	18" diameter, 7 ga.	▼	480	.117	▼	18.05	2.24	3.34	23.63	27.50
1300	End bearing, fluted, constant diameter,									
1320	4000 psi concrete, no reinforcing									
1340	12" diameter, 7 ga.	B-19	600	.093	V.L.F.	12.55	1.79	2.67	17.01	20
1360	14" diameter, 7 ga.		560	.100		15.70	1.92	2.86	20.48	24
1380	16" diameter, 7 ga.		520	.108		18.20	2.07	3.08	23.35	27
1400	18" diameter, 7 ga.	▼	480	.117	▼	20	2.24	3.34	25.58	29.50
1500	For reinforcing steel, add				Lb.	.55			.55	.60
1700	For ball or pedestal end, add	B-19	11	5.091	C.Y.	78.50	97.50	146	322	425
1900	For lengths above 60', concrete, add	"	11	5.091	"	81.50	97.50	146	325	425
2000	For steel thin shell, pipe only				Lb.	.52			.52	.57
2200	Precast, prestressed, 50' long, 12" diam., 2-3/8" wall	B-19	720	.078	V.L.F.	9.90	1.49	2.22	13.61	16.05
2300	14" diameter, 2-1/2" wall		680	.082		13	1.58	2.36	16.94	19.75
2500	16" diameter, 3" wall	▼	640	.087		18	1.68	2.50	22.18	25.50
2600	18" diameter, 3" wall	B-19A	600	.107		22.50	2.10	2.70	27.30	31.50
2800	20" diameter, 3-1/2" wall		560	.114		26	2.25	2.90	31.15	36
2900	24" diameter, 3-1/2" wall	▼	520	.123		32	2.42	3.12	37.54	43
3100	Precast, prestressed, 40' long, 10" thick, square	B-19	700	.080		6.60	1.54	2.29	10.43	12.55
3200	12" thick, square		680	.082		8.10	1.58	2.36	12.04	14.35
3400	14" thick, square		600	.093		9.90	1.79	2.67	14.36	17.10
3500	Octagonal		640	.087		14	1.68	2.50	18.18	21
3700	16" thick, square		560	.100		15.85	1.92	2.86	20.63	24
3800	Octagonal	▼	600	.093		16.80	1.79	2.67	21.26	24.50
4000	18" thick, square	B-19A	520	.123		18.60	2.42	3.12	24.14	28.50
4100	Octagonal	B-19	560	.100		19.90	1.92	2.86	24.68	28.50
4300	20" thick, square	B-19A	480	.133		25	2.63	3.38	31.01	35.50
4400	Octagonal	B-19	520	.108		22	2.07	3.08	27.15	31
4600	24" thick, square	B-19A	440	.145		35	2.86	3.69	41.55	47.50
4700	Octagonal	B-19	480	.117		31.50	2.24	3.34	37.08	42.50
4750	Mobilization for 10,000 L.F. pile job, add		3,300	.017			.33	.49	.82	1.12
4800	25,000 L.F. pile job, add	▼	8,500	.007	▼		.13	.19	.32	.44
0011	**PILING SPECIAL COSTS** pile caps, see Division 03310-240									
0500	Cutoffs, concrete piles, plain	1 Pile	5.50	1.455	Ea.		28.50		28.50	52.50
0600	With steel thin shell, add		38	.211			4.09		4.09	7.65
0700	Steel pile or "H" piles		19	.421			8.20		8.20	15.25
0800	Wood piles	▼	38	.211	▼		4.09		4.09	7.65
0900	Pre-augering up to 30' deep, average soil, 24" diameter	B-43	180	.222	L.F.		3.30	10	13.30	16.65
0920	36" diameter		115	.348			5.15	15.60	20.75	26
0960	48" diameter		70	.571			8.50	25.50	34	42.50
0980	60" diameter	▼	50	.800	▼		11.90	36	47.90	60
1000	Testing, any type piles, test load is twice the design load									
1050	50 ton design load, 100 ton test				Ea.				14,000	15,000
1100	100 ton design load, 200 ton test				▼				18,000	19,000

02450 | Foundation & Load Bearing Elements

02455 | Driven Piles

			CREW	DAILY OUTPUT	LABOR-HOURS	UNIT	2000 BARE COSTS MAT.	LABOR	EQUIP.	TOTAL	TOTAL INCL O&P	
350	1150	150 ton design load, 300 ton test				Ea.				22,500	24,000	350
	1200	200 ton design load, 400 ton test								24,500	27,000	
	1250	400 ton design load, 800 ton test				↓				28,350	31,500	
	1500	Wet conditions, soft damp ground										
	1600	Requiring mats for crane, add								40%	40%	
	1700	Barge mounted driving rig, add								30%	30%	
500	0010	**MOBILIZATION** Set up & remove, air compressor, 600 C.F.M. R02455-900	A-5	3.30	5.455	Ea.		79.50	13.10	92.60	150	500
	0100	1200 C.F.M.	"	2.20	8.182			120	19.65	139.65	226	
	0200	Crane, with pile leads and pile hammer, 75 ton	B-19	.60	93.333			1,800	2,675	4,475	6,175	
	0300	150 ton	"	.36	155			2,975	4,450	7,425	10,300	
	0500	Drill rig, for caissons, to 36", minimum	B-43	2	20			297	900	1,197	1,500	
	0600	Up to 84"	"	1	40			595	1,800	2,395	3,000	
	0800	Auxiliary boiler, for steam small	A-5	1.66	10.843			158	26	184	299	
	0900	Large	"	.83	21.687			315	52	367	600	
	1100	Rule of thumb: complete pile driving set up, small	B-19	.45	124			2,400	3,550	5,950	8,275	
	1200	Large	"	.27	207	↓		3,975	5,925	9,900	13,800	
850	0010	**PILES, STEEL** Not including mobilization or demobilization										850
	0100	Step tapered, round, concrete filled										
	0110	8" tip, 60 ton capacity, 30' depth	B-19	760	.074	V.L.F.	5	1.41	2.11	8.52	10.40	
	0120	60' depth		740	.076		5.65	1.45	2.16	9.26	11.20	
	0130	80' depth		700	.080		5.85	1.54	2.29	9.68	11.75	
	0150	10" tip, 90 ton capacity, 30' depth		700	.080		6.15	1.54	2.29	9.98	12.05	
	0160	60' depth		690	.081		6.35	1.56	2.32	10.23	12.35	
	0170	80' depth		670	.084		6.85	1.60	2.39	10.84	13.05	
	0190	12" tip, 120 ton capacity, 30' depth		660	.085		8.80	1.63	2.43	12.86	15.30	
	0200	60' depth, 12" diameter		630	.089		8	1.71	2.54	12.25	14.70	
	0210	80' depth		590	.095		7.75	1.82	2.71	12.28	14.80	
	0250	"H" Sections, 50' long, HP8 x 36		640	.087		9	1.68	2.50	13.18	15.70	
	0400	HP10 X 42		610	.092		10.50	1.76	2.63	14.89	17.65	
	0500	HP10 X 57		610	.092		14.25	1.76	2.63	18.64	22	
	0700	HP12 X 53		590	.095		13.25	1.82	2.71	17.78	21	
	0800	HP12 X 74	B-19A	590	.108		18.50	2.14	2.75	23.39	27.50	
	1000	HP14 X 73		540	.119		18.25	2.33	3	23.58	27.50	
	1100	HP14 X 89		540	.119		22.50	2.33	3	27.83	32	
	1300	HP14 X 102		510	.125		25.50	2.47	3.18	31.15	36	
	1400	HP14 X 117	↓	510	.125	↓	29.50	2.47	3.18	35.15	40.50	
	1601	Splice on standard points, not in leads, 8" or 10"	1 Pile	5	1.600	Ea.	55	31		86	119	
	1701	12" or 14"		4	2		80	39		119	161	
	1901	Heavy duty points, not in leads, 10" wide		4	2		85	39		124	166	
	2101	14" wide	↓	3.50	2.286	↓	110	44.50		154.50	204	
	2600	Pipe piles, 50' lg. 8" diam., 29 lb. per L.F., no concrete	B-19	500	.112	V.L.F.	8.75	2.15	3.20	14.10	17	
	2700	Concrete filled		460	.122		9.35	2.34	3.48	15.17	18.35	
	2900	10" diameter, 34 lb. per L.F., no concrete		500	.112		11.90	2.15	3.20	17.25	20.50	
	3000	Concrete filled		450	.124		13.20	2.39	3.56	19.15	23	
	3200	12" diameter, 44 lb. per L.F., no concrete		475	.118		13.20	2.26	3.37	18.83	22.50	
	3300	Concrete filled		415	.135		15.15	2.59	3.86	21.60	25.50	
	3500	14" diameter, 46 lb. per L.F., no concrete		430	.130		14.10	2.50	3.72	20.32	24	
	3600	Concrete filled		355	.158		16.80	3.03	4.51	24.34	29	
	3800	16" diameter, 52 lb. per L.F., no concrete		385	.145		15.65	2.79	4.16	22.60	27	
	3900	Concrete filled		335	.167		19.35	3.21	4.78	27.34	32.50	
	4100	18" diameter, 59 lb. per L.F., no concrete		355	.158		18.65	3.03	4.51	26.19	31	
	4200	Concrete filled	↓	310	.181	↓	23	3.47	5.15	31.62	37	
	4400	Splices for pipe piles, not in leads, 8" diameter	1 Sswl	4.67	1.713	Ea.	33	36.50		69.50	108	
	4500	14" diameter		3.79	2.111		43	45		88	136	
	4600	16" diameter		3.03	2.640		53.50	56		109.50	170	
	4800	Points, standard, 8" diameter	↓	4.61	1.735	↓	28	37		65	104	

For expanded coverage of these items see *Means Heavy Construction Cost Data 2000*

02450 | Foundation & Load Bearing Elements

02455 | Driven Piles

			CREW	DAILY OUTPUT	LABOR-HOURS	UNIT	MAT.	LABOR	EQUIP.	TOTAL	TOTAL INCL O&P	
850	4900	14" diameter	1 Sswl	4.05	1.975	Ea.	39	42		81	126	850
	5000	16" diameter		3.37	2.374		47.50	50.50		98	152	
	5200	Points, heavy duty, 10" diameter		2.89	2.768		33.50	59		92.50	153	
	5300	14" or 16" diameter		2.02	3.960		53	84		137	225	
	5500	For reinforcing steel, add		1,150	.007	Lb.	.40	.15		.55	.73	
	5700	For thick wall sections, add				"	.45			.45	.50	
900	0010	**PILES, WOOD** Friction or end bearing, not including										900
	0050	mobilization or demobilization										
	0100	Untreated piles, up to 30' long, 12" butts, 8" points	B-19	625	.090	V.L.F.	5.60	1.72	2.56	9.88	12.15	
	0200	30' to 39' long, 12" butts, 8" points		700	.080		5.60	1.54	2.29	9.43	11.50	
	0300	40' to 49' long, 12" butts, 7" points		720	.078		5.90	1.49	2.22	9.61	11.65	
	0400	50' to 59' long, 13" butts, 7" points		800	.070		5.35	1.34	2	8.69	10.55	
	0500	60' to 69' long, 13" butts, 7" points		840	.067		6	1.28	1.91	9.19	11	
	0600	70' to 80' long, 13" butts, 6" points		840	.067		6.70	1.28	1.91	9.89	11.75	
	0800	Treated piles, 12 lb. per C.F.,										
	0810	friction or end bearing, ASTM class B										
	1000	Up to 30' long, 12" butts, 8" points	B-19	625	.090	V.L.F.	9.30	1.72	2.56	13.58	16.20	
	1100	30' to 39' long, 12" butts, 8" points		700	.080		9.55	1.54	2.29	13.38	15.80	
	1200	40' to 49' long, 12" butts, 7" points		720	.078		9.55	1.49	2.22	13.26	15.65	
	1300	50' to 59' long, 13" butts, 7" points		800	.070		8.65	1.34	2	11.99	14.15	
	1400	60' to 69' long, 13" butts, 6" points	B-19A	840	.076		14.45	1.50	1.93	17.88	20.50	
	1500	70' to 80' long, 13" butts, 6" points	"	840	.076		17.25	1.50	1.93	20.68	24	
	1600	Treated piles, C.C.A., 2.5# per C.F.										
	1610	8" butts, 10' long	B-19	400	.140	V.L.F.	6.40	2.69	4	13.09	16.35	
	1620	11' to 16' long		500	.112		6.40	2.15	3.20	11.75	14.45	
	1630	17' to 20' long		575	.097		6.40	1.87	2.79	11.06	13.50	
	1640	10" butts, 10' to 16' long		500	.112		6.95	2.15	3.20	12.30	15.05	
	1650	17' to 20' long		575	.097		6.95	1.87	2.79	11.61	14.10	
	1660	21' to 40' long		700	.080		6.95	1.54	2.29	10.78	12.95	
	1670	12" butts, 10' to 20' long		575	.097		7.50	1.87	2.79	12.16	14.70	
	1680	21' to 35' long		650	.086		7.50	1.65	2.46	11.61	13.95	
	1690	36' to 40' long		700	.080		7.50	1.54	2.29	11.33	13.55	
	1695	14" butts. to 40' long		700	.080		10.70	1.54	2.29	14.53	17.05	
	1700	Boot for pile tip, minimum	1 Pile	27	.296	Ea.	15	5.75		20.75	27.50	
	1800	Maximum		21	.381		45	7.40		52.40	63.50	
	2000	Point for pile tip, minimum		20	.400		17.10	7.80		24.90	33.50	
	2100	Maximum		15	.533		61.50	10.35		71.85	87.50	
	2300	Splice for piles over 50' long, minimum	B-46	35	1.371		43	23.50	1.41	67.91	91	
	2400	Maximum		20	2.400		51.50	41.50	2.47	95.47	134	
	2600	Concrete encasement with wire mesh and tube		331	.145	V.L.F.	8.05	2.51	.15	10.71	13.50	
	2700	Mobilization for 10,000 L.F. pile job, add	B-19	3,300	.017			.33	.49	.82	1.12	
	2800	25,000 L.F. pile job, add	"	8,500	.007			.13	.19	.32	.44	

02465 | Bored Piles

			CREW	DAILY OUTPUT	LABOR-HOURS	UNIT	MAT.	LABOR	EQUIP.	TOTAL	TOTAL INCL O&P	
600	0010	**CAISSONS** Incl. excav., concrete, 50 lbs. reinf. per C.Y., not										600
	0020	incl. mobilization, boulder removal, disposal										
	0100	Open style, machine drilled, to 50' deep, in stable ground, no										
	0110	casings or ground water, 18" diam., 0.065 C.Y./L.F.	B-43	200	.200	V.L.F.	5.05	2.97	9	17.02	20.50	
	0200	24" diameter, 0.116 C.Y./L.F.		190	.211		9.05	3.13	9.45	21.63	25.50	
	0300	30" diameter, 0.182 C.Y./L.F.		150	.267		14.15	3.96	11.95	30.06	35.50	
	0400	36" diameter, 0.262 C.Y./L.F.		125	.320		20.50	4.75	14.35	39.60	46.50	
	0500	48" diameter, 0.465 C.Y./L.F.		100	.400		36	5.95	17.95	59.90	70	
	0600	60" diameter, 0.727 C.Y./L.F.		90	.444		56.50	6.60	19.95	83.05	95.50	
	0700	72" diameter, 1.05 C.Y./L.F.		80	.500		81.50	7.45	22.50	111.45	127	
	0800	84" diameter, 1.43 C.Y./L.F.		75	.533		111	7.90	24	142.90	162	
	1000	For bell excavation and concrete, add										

02450 | Foundation & Load Bearing Elements

02465 | Bored Piles

		CREW	DAILY OUTPUT	LABOR-HOURS	UNIT	MAT.	LABOR	EQUIP.	TOTAL	TOTAL INCL O&P
1020	4' bell diameter, 24" shaft, 0.444 C.Y.	B-43	20	2	Ea.	28	29.50	90	147.50	181
1040	6' bell diameter, 30" shaft, 1.57 C.Y.		5.70	7.018		98.50	104	315	517.50	630
1060	8' bell diameter, 36" shaft, 3.72 C.Y.		2.40	16.667		234	248	750	1,232	1,500
1080	9' bell diameter, 48" shaft, 4.48 C.Y.		2	20		281	297	900	1,478	1,800
1100	10' bell diameter, 60" shaft, 5.24 C.Y.		1.70	23.529		330	350	1,050	1,730	2,100
1120	12' bell diameter, 72" shaft, 8.74 C.Y.		1	40		550	595	1,800	2,945	3,600
1140	14' bell diameter, 84" shaft, 13.6 C.Y.		.70	57.143		855	850	2,575	4,280	5,225
1200	Open style, machine drilled, to 50' deep, in wet ground, pulled									
1300	casing and pumping, 18" diameter, 0.065 C.Y./L.F.	B-48	160	.300	V.L.F.	5.05	4.75	12.50	22.30	27.50
1400	24" diameter, 0.116 C.Y./L.F.		125	.384		9.05	6.05	15.95	31.05	38
1500	30" diameter, 0.182 C.Y./L.F.		85	.565		14.15	8.95	23.50	46.60	57
1600	36" diameter, 0.262 C.Y./L.F.		60	.800		20.50	12.65	33.50	66.65	80.50
1700	48" diameter, 0.465 C.Y./L.F.	B-49	55	1.309		36	21.50	46.50	104	129
1800	60" diameter, 0.727 C.Y./L.F.		35	2.057		56.50	34	73	163.50	202
1900	72" diameter, 1.05 C.Y./L.F.		30	2.400		81.50	39.50	85	206	253
2000	84" diameter, 1.43 C.Y./L.F.		25	2.880		111	47.50	102	260.50	315
2100	For bell excavation and concrete, add									
2120	4' bell diameter, 24" shaft, 0.444 C.Y.	B-48	19.80	2.424	Ea.	28	38.50	101	167.50	207
2140	6' bell diameter, 30" shaft, 1.57 C.Y.		5.70	8.421		98.50	133	350	581.50	720
2160	8' bell diameter, 36" shaft, 3.72 C.Y.		2.40	20		234	315	830	1,379	1,700
2180	9' bell diameter, 48" shaft, 4.48 C.Y.	B-49	3.30	21.818		281	360	775	1,416	1,775
2200	10' bell diameter, 60" shaft, 5.24 C.Y.		2.80	25.714		330	425	915	1,670	2,100
2220	12' bell diameter, 72" shaft, 8.74 C.Y.		1.60	45		550	740	1,600	2,890	3,650
2240	14' bell diameter, 84" shaft, 13.6 C.Y.		1	72		855	1,175	2,550	4,580	5,825
2300	Open style, machine drilled, to 50' deep, in soft rocks and									
2400	medium hard shales, 18" diameter, 0.065 C.Y./L.F.	B-49	50	1.440	V.L.F.	5.05	23.50	51	79.55	103
2500	24" diameter, 0.116 C.Y./L.F.		30	2.400		9.05	39.50	85	133.55	172
2600	30" diameter, 0.182 C.Y./L.F.		20	3.600		14.15	59.50	128	201.65	260
2700	36" diameter, 0.262 C.Y./L.F.		15	4.800		20.50	79	170	269.50	350
2800	48" diameter, 0.465 C.Y./L.F.		10	7.200		36	119	256	411	530
2900	60" diameter, 0.727 C.Y./L.F.		7	10.286		56.50	169	365	590.50	760
3000	72" diameter, 1.05 C.Y./L.F.		6	12		81.50	198	425	704.50	905
3100	84" diameter, 1.43 C.Y./L.F.		5	14.400		111	237	510	858	1,100
3200	For bell excavation and concrete, add									
3220	4' bell diameter, 24" shaft, 0.444 C.Y.	B-49	10.90	6.606	Ea.	28	109	234	371	480
3240	6' bell diameter, 30" shaft, 1.57 C.Y.		3.10	23.226		98.50	385	825	1,308.50	1,675
3260	8' bell diameter, 36" shaft, 3.72 C.Y.		1.30	55.385		234	910	1,975	3,119	4,000
3280	9' bell diameter, 48" shaft, 4.48 C.Y.		1.10	65.455		281	1,075	2,325	3,681	4,725
3300	10' bell diameter, 60" shaft, 5.24 C.Y.		.90	80		330	1,325	2,850	4,505	5,775
3320	12' bell diameter, 72" shaft, 8.74 C.Y.		.60	119		550	1,975	4,250	6,775	8,725
3340	14' bell diameter, 84" shaft, 13.6 C.Y.		.40	180		855	2,975	6,400	10,230	13,100
3600	For rock excavation, sockets, add, minimum		120	.600	C.F.		9.90	21.50	31.40	41
3650	Average		95	.758			12.50	27	39.50	51.50
3700	Maximum		48	1.500			24.50	53	77.50	102
3900	For 50' to 100' deep, add				V.L.F.				7%	7%
4000	For 100' to 150' deep, add								25%	25%
4100	For 150' to 200' deep, add								30%	30%
4200	For casings left in place, add				Lb.	.52			.52	.57
4300	For other than 50 lb. reinf. per C.Y., add or deduct				"	.52			.52	.57
4400	For steel "I" beam cores, add	B-49	8.30	8.675	Ton	960	143	310	1,413	1,650
4500	Load and haul excess excavation, 2 miles	B-34B	178	.045	C.Y.		.73	2.51	3.24	3.98
4600	For mobilization, 50 mile radius, rig to 36"	B-43	2	20	Ea.		297	900	1,197	1,500
4650	Rig to 84"	B-48	1.75	27.429	"		435	1,150	1,585	2,000
4700	For low headroom, add								50%	
5000	Bottom inspection	1 Skwk	1.20	6.667	Ea.		132		132	226

For expanded coverage of these items see *Means Heavy Construction Cost Data 2000*

02450 | Foundation & Load Bearing Elements

02465 | Bored Piles

			CREW	DAILY OUTPUT	LABOR-HOURS	UNIT	2000 BARE COSTS MAT.	LABOR	EQUIP.	TOTAL	TOTAL INCL O&P	
800	0010	**PRESSURE INJECTED FOOTINGS** or Displacement Caissons										800
	0100	incl. mobilization and demobilization, up to 50 miles	R02465-800									
	0200	Uncased shafts, 30 to 80 tons cap., 17" diam., 10' depth	B-44	88	.727	V.L.F.	12.55	13.55	10.35	36.45	49.50	
	0300	25' depth		165	.388		8.95	7.20	5.55	21.70	29	
	0400	80-150 ton capacity, 22" diameter, 10' depth		80	.800		15.70	14.90	11.40	42	57	
	0500	20' depth		130	.492		12.55	9.15	7	28.70	38	
	0700	Cased shafts, 10 to 30 ton capacity, 10-5/8" diam., 20' depth		175	.366		8.95	6.80	5.20	20.95	28	
	0800	30' depth		240	.267		8.35	4.96	3.80	17.11	22.50	
	0850	30 to 60 ton capacity, 12" diameter, 20' depth		160	.400		12.55	7.45	5.70	25.70	33.50	
	0900	40' depth		230	.278		9.65	5.20	3.97	18.82	24.50	
	1000	80 to 100 ton capacity, 16" diameter, 20' depth		160	.400		17.95	7.45	5.70	31.10	39.50	
	1100	40' depth		230	.278		16.75	5.20	3.97	25.92	32	
	1200	110 to 140 ton capacity, 17-5/8" diameter, 20' depth		160	.400		19.35	7.45	5.70	32.50	41	
	1300	40' depth		230	.278		17.95	5.20	3.97	27.12	33.50	
	1400	140 to 175 ton capacity, 19" diameter, 20' depth		130	.492		21	9.15	7	37.15	47.50	
	1500	40' depth		210	.305		19.35	5.65	4.34	29.34	36.50	
	1700	Over 30' long, L.F. cost tends to be lower										
	1900	Maximum depth is about 90'										

02500 | Utility Services

02510 | Water Distribution

			CREW	DAILY OUTPUT	LABOR-HOURS	UNIT	2000 BARE COSTS MAT.	LABOR	EQUIP.	TOTAL	TOTAL INCL O&P	
800	0010	**PIPING, WATER DISTRIBUTION SYSTEMS** Pipe laid in trench,										800
	0020	excavation and backfill not included										
	1400	Ductile Iron, cement lined, class 50 water pipe, 18' lengths										
	1410	Mechanical joint, 4" diameter	B-20	144	.167	L.F.	7.15	2.52		9.67	12.20	
	1420	6" diameter		126	.190		8.20	2.88		11.08	14	
	1430	8" diameter		108	.222		10.75	3.36		14.11	17.55	
	1440	10" diameter		90	.267		14.55	4.03		18.58	23	
	1450	12" diameter	B-21	72	.389		18	6.20	1.92	26.12	32.50	
	1460	14" diameter		54	.519		23	8.25	2.56	33.81	42	
	1470	16" diameter		46	.609		25	9.70	3.01	37.71	47.50	
	1480	18" diameter		42	.667		31.50	10.60	3.29	45.39	56.50	
	1490	24" diameter		35	.800		47.50	12.75	3.95	64.20	78	
	1550	Push on joint, 4" diameter	B-20	155	.155		6.35	2.34		8.69	10.95	
	1560	6" diameter		135	.178		7.25	2.69		9.94	12.60	
	1570	8" diameter		115	.209		9.95	3.16		13.11	16.35	
	1580	10" diameter		98	.245		15.60	3.70		19.30	23.50	
	1590	12" diameter		78	.308		16.45	4.65		21.10	26	
	1600	14" diameter	B-21	58	.483		18.05	7.70	2.38	28.13	35.50	
	1610	16" diameter		52	.538		25.50	8.55	2.66	36.71	45.50	
	1620	18" diameter		43	.651		28.50	10.35	3.22	42.07	52	
	1630	20" diameter		41	.683		31	10.85	3.37	45.22	56	
	1640	24" diameter		40	.700		40	11.15	3.46	54.61	67	
	1950	Butterfly valves with boxes, cast iron										
	1970	4" diameter	B-20	6	4	Ea.	465	60.50		525.50	615	
	1990	6" diameter	"	5	4.800		490	72.50		562.50	665	
	2010	8" diameter	B-21	4	7		675	111	34.50	820.50	975	
	2030	10" diameter		3.50	8		980	127	39.50	1,146.50	1,325	
	2050	12" diameter		3	9.333		1,275	148	46	1,469	1,700	

02500 | Utility Services

02510 | Water Distribution

		CREW	DAILY OUTPUT	LABOR-HOURS	UNIT	2000 BARE COSTS MAT.	LABOR	EQUIP.	TOTAL	TOTAL INCL O&P
2070	14" diameter	B-21	2	14	Ea.	2,325	223	69	2,617	3,000
2090	16" diameter		2	14		2,850	223	69	3,142	3,600
2650	Polyvinyl chloride pipe, class 160, S.D.R.-26, 1-1/2" diameter	B-20	300	.080	L.F.	.27	1.21		1.48	2.37
2700	2" diameter		250	.096		.38	1.45		1.83	2.91
2750	2-1/2" diameter		250	.096		.50	1.45		1.95	3.04
2800	3" diameter		200	.120		.75	1.81		2.56	3.94
2850	4" diameter		200	.120		1.18	1.81		2.99	4.41
2900	6" diameter		180	.133		2.50	2.02		4.52	6.20
2950	8" diameter	B-21	160	.175		4.25	2.78	.86	7.89	10.35
8000	Fittings, ductile iron, mechanical joint									
8010	90° bend 4" diameter	B-20	37	.649	Ea.	104	9.80		113.80	131
8020	6" diameter		25	.960		169	14.50		183.50	210
8040	8" diameter		21	1.143		245	17.30		262.30	300
8060	10" diameter	B-21	21	1.333		182	21	6.60	209.60	243
8080	12" diameter		18	1.556		430	25	7.70	462.70	525
8100	14" diameter		16	1.750		550	28	8.65	586.65	660
8120	16" diameter		14	2		635	32	9.90	676.90	765
8140	18" diameter		10	2.800		2,025	44.50	13.85	2,083.35	2,325
8160	20" diameter		8	3.500		2,200	55.50	17.30	2,272.80	2,550
8180	24" diameter		6	4.667		2,400	74.50	23	2,497.50	2,775
8200	Wye or tee, 4" diameter	B-20	25	.960		169	14.50		183.50	211
8220	6" diameter		17	1.412		178	21.50		199.50	233
8240	8" diameter		14	1.714		255	26		281	325
8260	10" diameter	B-21	14	2		550	32	9.90	591.90	670
8280	12" diameter		12	2.333		745	37	11.55	793.55	895
8300	14" diameter		10	2.800		875	44.50	13.85	933.35	1,050
8320	16" diameter		8	3.500		1,000	55.50	17.30	1,072.80	1,225
8340	18" diameter		6	4.667		2,075	74.50	23	2,172.50	2,425
8360	20" diameter		4	7		2,500	111	34.50	2,645.50	2,975
8380	24" diameter		3	9.333		2,950	148	46	3,144	3,550

02520 | Wells

		CREW	DAILY OUTPUT	LABOR-HOURS	UNIT	MAT.	LABOR	EQUIP.	TOTAL	TOTAL INCL O&P
0010	**WELLS** Domestic water									
0100	Drilled, 4" to 6" diameter	B-23	120	.333	L.F.		4.95	16.45	21.40	26.50
0200	8" diameter	"	95.20	.420	"		6.25	21	27.25	33.50
0400	Gravel pack well, 40' deep, incl. gravel & casing, complete									
0500	24" diameter casing x 18" diameter screen	B-23	.13	307	Total	20,000	4,575	15,200	39,775	46,500
0600	36" diameter casing x 18" diameter screen		.12	333	"	21,500	4,950	16,500	42,950	50,500
0800	Observation wells, 1-1/4" riser pipe		163	.245	V.L.F.	11	3.64	12.15	26.79	31.50
0900	For flush Buffalo roadway box, add	1 Skwk	16.60	.482	Ea.	30	9.50		39.50	49.50
1200	Test well, 2-1/2" diameter, up to 50' deep (15 to 50 GPM)	B-23	1.51	26.490	"	450	395	1,300	2,145	2,625
1300	Over 50' deep, add	"	121.80	.328	L.F.	12	4.88	16.25	33.13	39.50
1500	Pumps, installed in wells to 100' deep, 4" submersible									
1510	1/2 H.P.	Q-1	3.22	4.969	Ea.	425	98		523	630
1520	3/4 H.P.		2.66	6.015		475	119		594	720
1600	1 H.P.		2.29	6.987		525	138		663	810
1700	1-1/2 H.P.	Q-22	1.60	10		580	198	283	1,061	1,275
1800	2 H.P.		1.33	12.030		620	238	340	1,198	1,450
1900	3 H.P.		1.14	14.035		775	277	400	1,452	1,750
2000	5 H.P.		1.14	14.035		1,250	277	400	1,927	2,275
2050	Remove and install motor only, 4 H.P.		1.14	14.035		585	277	400	1,262	1,550
3000	Pump, 6" submersible, 25' to 150' deep, 25 H.P., 249 to 297 GPM		.89	17.978		3,775	355	510	4,640	5,300
3100	25' to 500' deep, 30 H.P., 100 to 300 GPM		.73	21.918		3,875	435	620	4,930	5,650
9950	See div. 02240-900 for wellpoints									
9960	See div. 02240-700 for drainage wells									

02500 | Utility Services

02520 | Wells

		CREW	DAILY OUTPUT	LABOR-HOURS	UNIT	2000 BARE COSTS MAT.	LABOR	EQUIP.	TOTAL	TOTAL INCL O&P	
910	0010	**PUMPS, WELL** Water system, with pressure control										910
	1000	Deep well, jet, 42 gal. galvanized tank										
	1040	3/4 HP	1 Plum	.80	10	Ea.	545	220		765	965	
	3000	Shallow well, jet, 30 gal. galvanized tank										
	3040	1/2 HP	1 Plum	2	4	Ea.	550	88		638	750	

02530 | Sanitary Sewerage

		CREW	DAILY OUTPUT	LABOR-HOURS	UNIT	MAT.	LABOR	EQUIP.	TOTAL	TOTAL INCL O&P	
100	0010	**SEWAGE TREATMENT** Plant, not incl. fencing or external piping										100
	0020	Steel packaged, blown air aeration plants										
	0100	1,000 GPD				Gal.				15	17.25	
	0200	5,000 GPD								10	11.50	
	0300	15,000 GPD								5.50	6.30	
	0400	30,000 GPD								5.20	6	
	0500	50,000 GPD								4	4.60	
	0600	100,000 GPD								3.50	4	
	0700	200,000 GPD								2.50	2.88	
	0800	500,000 GPD								2.45	2.80	
	1000	Concrete, extended aeration, primary and secondary treatment										
	1010	10,000 GPD				Gal.				11	12.65	
	1100	30,000 GPD								5.50	6.35	
	1200	50,000 GPD								4.50	5.18	
	1400	100,000 GPD								3.50	4.05	
	1500	500,000 GPD								2.50	2.90	
	1700	Municipal wastewater treatment facility										
	1720	1.0 MGD				Gal.				4.30	4.95	
	1740	1.5 MGD								4.25	4.90	
	1760	2.0 MGD								3.65	4.20	
	1780	3.0 MGD								2.85	3.30	
	1800	5.0 MGD								2.60	3	
	2000	Holding tank system, not incl. excavation or backfill										
	2010	Recirculating chemical water closet	2 Plum	4	4	Ea.	675	88		763	885	
	2100	For voltage converter, add	"	16	1		179	22		201	233	
	2200	For high level alarm, add	1 Plum	7.80	1.026		102	22.50		124.50	149	
730	0010	**PIPING, DRAINAGE & SEWAGE, CONCRETE**										730
	0020	Not including excavation or backfill	R02510 -810									
	1000	Non-reinforced pipe, extra strength, B&S or T&G joints										
	1010	6" diameter	B-14	265.04	.181	L.F.	3.47	2.82	.77	7.06	9.45	
	1020	8" diameter		224	.214		3.82	3.33	.91	8.06	10.85	
	1030	10" diameter		216	.222		4.23	3.46	.94	8.63	11.60	
	1040	12" diameter		200	.240		5.20	3.73	1.02	9.95	13.20	
	1050	15" diameter		180	.267		6.05	4.15	1.13	11.33	15	
	1060	18" diameter		144	.333		7.45	5.20	1.41	14.06	18.60	
	1070	21" diameter		112	.429		9.20	6.65	1.81	17.66	23.50	
	1080	24" diameter		100	.480		11.30	7.45	2.03	20.78	27.50	
	2000	Reinforced culvert, class 3, no gaskets										
	2010	12" diameter	B-14	210	.229	L.F.	8.35	3.56	.97	12.88	16.30	
	2020	15" diameter		175	.274		10.40	4.27	1.16	15.83	20	
	2030	18" diameter		130	.369		11.90	5.75	1.56	19.21	24.50	
	2035	21" diameter		120	.400		15.15	6.20	1.69	23.04	29	
	2040	24" diameter		100	.480		18.20	7.45	2.03	27.68	35	
	2045	27" diameter	B-13	92	.522		23	8.25	6.10	37.35	46.50	
	2050	30" diameter		88	.545		27	8.65	6.35	42	51	
	2060	36" diameter		72	.667		37	10.55	7.75	55.30	67	
	2070	42" diameter	B-13B	72	.778		47.50	12.45	11.20	71.15	86	
	2080	48" diameter		64	.875		59	14	12.65	85.65	102	

02500 | Utility Services

02530 | Sanitary Sewerage

		CREW	DAILY OUTPUT	LABOR-HOURS	UNIT	2000 BARE COSTS MAT.	LABOR	EQUIP.	TOTAL	TOTAL INCL O&P	
730	2090	60" diameter	B-13B	48	1.167	L.F.	70	18.65	16.85	105.50	127	730
	2100	72" diameter		40	1.400		108	22.50	20	150.50	179	
	2120	84" diameter		32	1.750		209	28	25.50	262.50	305	
	2140	96" diameter		24	2.333		251	37.50	33.50	322	375	
	2200	With gaskets, class 3, 12" diameter	B-21	168	.167		7.45	2.65	.82	10.92	13.65	
	2220	15" diameter		160	.175		8.95	2.78	.86	12.59	15.55	
	2230	18" diameter		152	.184		11.20	2.93	.91	15.04	18.35	
	2240	24" diameter		136	.206		16.80	3.28	1.02	21.10	25	
	2260	30" diameter	B-13	88	.545		22.50	8.65	6.35	37.50	46	
	2270	36" diameter	"	72	.667		33.50	10.55	7.75	51.80	63.50	
	2290	48" diameter	B-13B	64	.875		55	14	12.65	81.65	98	
	2310	72" diameter	"	40	1.400		142	22.50	20	184.50	216	
	2330	Flared ends, 6'-1" long, 12" diameter	B-21	190	.147		28	2.34	.73	31.07	36	
	2340	15" diameter		155	.181		31.50	2.87	.89	35.26	40.50	
	2400	6'-2" long, 18" diameter		122	.230		35	3.65	1.13	39.78	46	
	2420	24" diameter		88	.318		46.50	5.05	1.57	53.12	61.50	
	2440	36" diameter	B-13	60	.800		58.50	12.65	9.30	80.45	96.50	
	3040	Vitrified plate lined, add to above, 30" to 36" diameter				SFCA	3.09			3.09	3.40	
	3050	42" to 54" diameter, add					3.30			3.30	3.63	
	3060	60" to 72" diameter, add					3.86			3.86	4.25	
	3070	Over 72" diameter, add					4.12			4.12	4.53	
	3080	Radius pipe, add to pipe prices, 12" to 60" diameter				L.F.	50%					
	3090	Over 60" diameter, add				"	20%					
	3500	Reinforced elliptical, 8' lengths, C507 class 3										
	3520	14" x 23" inside, round equivalent 18" diameter	B-21	82	.341	L.F.	41	5.45	1.69	48.14	56.50	
	3530	24" x 38" inside, round equivalent 30" diameter	B-13	58	.828		51.50	13.10	9.65	74.25	89.50	
	3540	29" x 45" inside, round equivalent 36" diameter		52	.923		67	14.60	10.75	92.35	110	
	3550	38" x 60" inside, round equivalent 48" diameter		38	1.263		103	20	14.70	137.70	163	
	3560	48" x 76" inside, round equivalent 60" diameter		26	1.846		139	29	21.50	189.50	226	
	3570	58" x 91" inside, round equivalent 72" diameter		22	2.182		206	34.50	25.50	266	315	
	3780	Concrete slotted pipe, class 4 mortar joint										
	3800	12" diameter	B-21	168	.167	L.F.	10.80	2.65	.82	14.27	17.35	
	3840	18" diameter	"	152	.184	"	16.75	2.93	.91	20.59	24.50	
	3900	Class 4 O-ring										
	3940	12" diameter	B-21	168	.167	L.F.	12.35	2.65	.82	15.82	19.05	
	3960	18" diameter	"	152	.184	"	18.55	2.93	.91	22.39	26.50	
780	0010	**PIPING, DRAINAGE & SEWAGE, POLYVINYL CHLORIDE**										780
	0020	Not including excavation or backfill										
	2000	10' lengths, S.D.R. 35, B&S, 4" diameter	B-20	375	.064	L.F.	2.21	.97		3.18	4.09	
	2040	6" diameter		350	.069		2.87	1.04		3.91	4.94	
	2080	8" diameter		335	.072		4.40	1.08		5.48	6.70	
	2120	10" diameter	B-21	330	.085		4.84	1.35	.42	6.61	8.05	
	2160	12" diameter		320	.087		5.10	1.39	.43	6.92	8.50	
	2200	15" diameter		190	.147		11.15	2.34	.73	14.22	17.10	
790	0010	**PIPING, DRAINAGE & SEWAGE, VITRIFIED CLAY** C700										790
	0020	Not including excavation or backfill,										
	4030	Extra strength, compression joints, C425										
	5000	4" diameter x 4' long	B-20	265	.091	L.F.	1.79	1.37		3.16	4.32	
	5020	6" diameter x 5' long	"	200	.120		2.93	1.81		4.74	6.35	
	5040	8" diameter x 5' long	B-21	200	.140		4.14	2.23	.69	7.06	9.10	
	5060	10" diameter x 5' long		190	.147		6.80	2.34	.73	9.87	12.25	
	5080	12" diameter x 6' long		150	.187		8.95	2.97	.92	12.84	15.85	
	5100	15" diameter x 7' long		110	.255		16.25	4.05	1.26	21.56	26	
	5120	18" diameter x 7' long		88	.318		23.50	5.05	1.57	30.12	36.50	
	5140	24" diameter x 7' long		45	.622		48.50	9.90	3.07	61.47	73.50	
	5160	30" diameter x 7' long	B-22	31	.968		85	15.70	6.70	107.40	127	

For expanded coverage of these items see *Means Heavy Construction Cost Data 2000*

02500 | Utility Services

02530 | Sanitary Sewerage

		CREW	DAILY OUTPUT	LABOR-HOURS	UNIT	2000 BARE COSTS MAT.	LABOR	EQUIP.	TOTAL	TOTAL INCL O&P	
790	5180 36" diameter x 7' long	B-22	20	1.500	L.F.	125	24.50	10.35	159.85	191	790
	6000 For 3' lengths, add					30%	30%				
	6020 For 2' lengths, add					40%	60%				
	6060 For plain joints, deduct					25%					
	7060 2' lengths, add to above					40%					

02540 | Septic Tank Systems

		CREW	DAILY OUTPUT	LABOR-HOURS	UNIT	MAT.	LABOR	EQUIP.	TOTAL	TOTAL INCL O&P	
700	0010 SEPTIC TANKS Not incl. excav. or piping, precast, 1,000 gallon	B-21	8	3.500	Ea.	490	55.50	17.30	562.80	650	700
	0100 2,000 gallon	"	5	5.600		1,000	89	27.50	1,116.50	1,275	
	0200 5,000 gallon	B-13	3.50	13.714		4,775	217	160	5,152	5,800	
	0300 15,000 gallon, 4 piece	B-13B	1.70	32.941		11,000	525	475	12,000	13,500	
	0400 25,000 gallon, 4 piece		1.10	50.909		24,400	815	735	25,950	29,100	
	0500 40,000 gallon, 4 piece		.80	70		31,800	1,125	1,000	33,925	38,000	
	0520 50,000 gallon, 5 piece	B-13C	.60	93.333		36,600	1,500	1,900	40,000	44,800	
	0540 75,000 gallon, cast in place	C-14C	.25	448		44,500	8,275	152	52,927	63,500	
	0560 100,000 gallon	"	.15	746		55,000	13,800	254	69,054	84,500	
	0600 High density polyethylene, 1,000 gallon	B-21	6	4.667		800	74.50	23	897.50	1,025	
	0700 1,500 gallon	"	4	7		1,000	111	34.50	1,145.50	1,325	
	1000 Distribution boxes, concrete, 7 outlets	2 Clab	16	1		86.50	14.45		100.95	120	
	1100 9 outlets	"	8	2		225	29		254	298	
	1150 Leaching field chambers, 13' x 3'-7" x 1'-4", standard	B-13	16	3		665	47.50	35	747.50	850	
	1200 Heavy duty, 8' x 4' x 1'-6"		14	3.429		320	54	40	414	485	
	1300 13' x 3'-9" x 1'-6"		12	4		910	63.50	46.50	1,020	1,150	
	1350 20' x 4' x 1'-6"		5	9.600		750	152	112	1,014	1,200	
	1400 Leaching pit, precast concrete, 3' diameter, 3' deep	B-21	8	3.500		145	55.50	17.30	217.80	274	
	1500 6' diameter, 3' section		4.70	5.957		375	95	29.50	499.50	610	
	2000 Velocity reducing pit, precast conc., 6' diameter, 3' deep		4.70	5.957		225	95	29.50	349.50	440	
	2200 Excavation for septic tank, 3/4 C.Y. backhoe	B-12F	145	.055	C.Y.		1.14	3.14	4.28	5.35	
	2400 4' trench for disposal field, 3/4 C.Y. backhoe	"	335	.024	L.F.		.49	1.36	1.85	2.32	
	2600 Gravel fill, run of bank	B-6	150	.160	C.Y.	5.20	2.56	1.35	9.11	11.50	
	2800 Crushed stone, 3/4"	"	150	.160	"	21	2.56	1.35	24.91	29.50	

02550 | Piped Energy Distribution

		CREW	DAILY OUTPUT	LABOR-HOURS	UNIT	MAT.	LABOR	EQUIP.	TOTAL	TOTAL INCL O&P	
450	0010 **GAS STATION PRODUCT LINE**										450
	0020 Primary containment pipe, fiberglass-reinforced										
	0030 Plastic pipe 15' & 30' lengths										
	0040 2" diameter	Q-6	425	.056	L.F.	3.06	1.08		4.14	5.15	
	0050 3" diameter		400	.060		4.53	1.15		5.68	6.90	
	0060 4" diameter		375	.064		5.90	1.23		7.13	8.55	
	0100 Fittings										
	0110 Elbows, 90° & 45°, bell-ends, 2"	Q-6	24	1	Ea.	32	19.15		51.15	66.50	
	0120 3" diameter		22	1.091		36	21		57	74	
	0130 4" diameter		20	1.200		43.50	23		66.50	85.50	
	0200 Tees, bell ends, 2"		21	1.143		38	22		60	77.50	
	0210 3" diameter		18	1.333		41	25.50		66.50	87.50	
	0220 4" diameter		15	1.600		53.50	30.50		84	110	
	0230 Flanges bell ends, 2"		24	1		20.50	19.15		39.65	54	
	0240 3" diameter		22	1.091		22.50	21		43.50	59.50	
	0250 4" diameter		20	1.200		26	23		49	66.50	
	0260 Sleeve couplings, 2"		21	1.143		7.95	22		29.95	45	
	0270 3" diameter		18	1.333		11.05	25.50		36.55	54	
	0280 4" diameter		15	1.600		15.95	30.50		46.45	68	
	0290 Threaded adapters 2"		21	1.143		11.30	22		33.30	48.50	
	0300 3" diameter		18	1.333		17	25.50		42.50	60.50	
	0310 4" diameter		15	1.600		23.50	30.50		54	76.50	

02500 | Utility Services

02550 | Piped Energy Distribution

			CREW	DAILY OUTPUT	LABOR-HOURS	UNIT	2000 BARE COSTS				TOTAL INCL O&P
							MAT.	LABOR	EQUIP.	TOTAL	
450	0320	Reducers, 2"	Q-6	27	.889	Ea.	14.60	17.05		31.65	44
	0330	3" diameter		22	1.091		17	21		38	53
	0340	4" diameter	▼	20	1.200	▼	26	23		49	66.50
	1010	Gas station product line for secondary containment (double wall)									
	1100	Fiberglass reinforced plastic pipe 25' lengths									
	1120	Pipe, plain end, 3"	Q-6	375	.064	L.F.	4.10	1.23		5.33	6.55
	1130	4" diameter		350	.069		5.10	1.31		6.41	7.75
	1140	5" diameter		325	.074		5.65	1.42		7.07	8.60
	1150	6" diameter	▼	300	.080	▼	9.75	1.53		11.28	13.25
	1200	Fittings									
	1230	Elbows, 90° & 45°, 3"	Q-6	18	1.333	Ea.	38	25.50		63.50	84
	1240	4" diameter		16	1.500		72	29		101	127
	1250	5" diameter		14	1.714		149	33		182	219
	1260	6" diameter		12	2		155	38.50		193.50	234
	1270	Tees, 3"		15	1.600		55	30.50		85.50	111
	1280	4" diameter		12	2		87.50	38.50		126	160
	1290	5" diameter		9	2.667		163	51		214	264
	1300	6" diameter		6	4		170	76.50		246.50	315
	1310	Couplings, 3"		18	1.333		25	25.50		50.50	69.50
	1320	4" diameter		16	1.500		67	29		96	121
	1330	5" diameter		14	1.714		139	33		172	208
	1340	6" diameter		12	2		144	38.50		182.50	223
	1350	Cross-over nipples, 3"		18	1.333		5.90	25.50		31.40	48.50
	1360	4" diameter		16	1.500		6.95	29		35.95	55
	1370	5" diameter		14	1.714		10.30	33		43.30	66
	1380	6" diameter		12	2		10.80	38.50		49.30	75.50
	1400	Telescoping, reducers, concentric 4" x 3"		18	1.333		19.70	25.50		45.20	63.50
	1410	5" x 4"		17	1.412		51.50	27		78.50	101
	1420	6" x 5"	▼	16	1.500	▼	124	29		153	184
464	0010	**PIPING, GAS SERVICE & DISTRIBUTION, POLYETHYLENE**									464
	0020	not including excavation or backfill									
	1000	60 psi coils, comp cplg @ 100', 1/2" diameter, SDR 9.3	B-20A	608	.053	L.F.	.36	.93		1.29	1.96
	1040	1-1/4" diameter, SDR 11		544	.059		.59	1.04		1.63	2.38
	1100	2" diameter, SDR 11		488	.066		.74	1.15		1.89	2.75
	1160	3" diameter, SDR 11	▼	408	.078		1.55	1.38		2.93	4.02
	1500	60 PSI 40' joints with coupling, 3" diameter, SDR 11	B-21A	408	.098		1.55	1.79	.98	4.32	5.75
	1540	4" diameter, SDR 11		352	.114		3.44	2.07	1.14	6.65	8.50
	1600	6" diameter, SDR 11		328	.122		11.25	2.22	1.22	14.69	17.45
	1640	8" diameter, SDR 11	▼	272	.147	▼	15	2.68	1.48	19.16	22.50

02580 | Elec/Communication Structures

			CREW	DAILY OUTPUT	LABOR-HOURS	UNIT	MAT.	LABOR	EQUIP.	TOTAL	TOTAL INCL O&P
300	0010	**ELECTRIC & TELEPHONE SITE WORK** Not including excavation									300
	0200	backfill and cast in place concrete									
	0400	Hand holes, precast concrete, with concrete cover									
	0600	2' x 2' x 3' deep	R-3	2.40	8.333	Ea.	355	183	57.50	595.50	760
	0800	3' x 3' x 3' deep		1.90	10.526		465	232	72.50	769.50	970
	1000	4' x 4' x 4' deep	▼	1.40	14.286	▼	555	315	98.50	968.50	1,225
	1200	Manholes, precast with iron racks & pulling irons, C.I. frame									
	1400	and cover, 4' x 6' x 7' deep	B-13	2	24	Ea.	1,175	380	280	1,835	2,250
	1600	6' x 8' x 7' deep		1.90	25.263		1,450	400	294	2,144	2,600
	1800	6' x 10' x 7' deep	▼	1.80	26.667		1,625	420	310	2,355	2,850
	2000	Poles, wood, preservative treatment, see also div. 16520, 20' high	R-3	3.10	6.452		236	142	44.50	422.50	540
	2400	25' high		2.90	6.897		250	152	47.50	449.50	575
	2600	30' high		2.60	7.692		270	169	53	492	635
	2800	35' high		2.40	8.333		350	183	57.50	590.50	750

For expanded coverage of these items see *Means Heavy Construction Cost Data 2000*

02500 | Utility Services

02580 | Elec/Communication Structures

		CREW	DAILY OUTPUT	LABOR-HOURS	UNIT	2000 BARE COSTS MAT.	LABOR	EQUIP.	TOTAL	TOTAL INCL O&P
3000	40' high	R-3	2.30	8.696	Ea.	430	191	60	681	855
3200	45' high	↓	1.70	11.765	↓	525	259	81.50	865.50	1,100
3400	Cross arms with hardware & insulators									
3600	4' long	1 Elec	2.50	3.200	Ea.	110	70.50		180.50	237
3800	5' long		2.40	3.333		128	73.50		201.50	262
4000	6' long	↓	2.20	3.636	↓	147	80.50		227.50	293
4200	Underground duct, banks ready for concrete fill, min. of 7.5"									
4400	between conduits, ctr. to ctr.(for wire & cable see div. 16120)									
4580	PVC, type EB, 1 @ 2" diameter	2 Elec	480	.033	L.F.	.57	.74		1.31	1.84
4600	2 @ 2" diameter		240	.067		1.14	1.47		2.61	3.67
4800	4 @ 2" diameter		120	.133		2.29	2.95		5.24	7.35
5000	2 @ 3" diameter		200	.080		1.57	1.77		3.34	4.61
5200	4 @ 3" diameter		100	.160		3.13	3.54		6.67	9.25
5400	2 @ 4" diameter		160	.100		2.43	2.21		4.64	6.30
5600	4 @ 4" diameter		80	.200		4.86	4.42		9.28	12.60
5800	6 @ 4" diameter		54	.296		7.30	6.55		13.85	18.70
6200	Rigid galvanized steel, 2 @ 2" diameter		180	.089		9.30	1.96		11.26	13.40
6400	4 @ 2" diameter		90	.178		18.60	3.93		22.53	27
6800	2 @ 3" diameter		100	.160		20.50	3.54		24.04	28.50
7000	4 @ 3" diameter		50	.320		41	7.05		48.05	56.50
7200	2 @ 4" diameter		70	.229		29.50	5.05		34.55	41
7400	4 @ 4" diameter		34	.471		59	10.40		69.40	81.50
7600	6 @ 4" diameter	↓	22	.727	↓	88	16.05		104.05	124
0010	**RADIO TOWERS** Guyed, 50'h, 40 lb. sec., 70MPH basic wind spd.	2 Sswk	1	16	Ea.	1,575	340		1,915	2,400
0100	Wind load 90 MPH basic wind speed	"	1	16		1,575	340		1,915	2,400
0300	190' high, 40 lb. section, wind load 70 MPH basic wind speed	K-2	.33	72.727		4,225	1,475	545	6,245	8,025
0400	200' high, 70 lb. section, wind load 90 MPH basic wind speed		.33	72.727		8,525	1,475	545	10,545	12,800
0600	300' high, 70 lb. section, wind load 70 MPH basic wind speed		.20	120		12,100	2,425	900	15,425	18,900
0700	270' high, 90 lb. section, wind load 90 MPH basic wind speed		.20	120		14,000	2,425	900	17,325	21,000
0800	400' high, 100 lb. section, wind load 70 MPH basic wind speed		.14	171		20,400	3,450	1,300	25,150	30,400
0900	Self-supporting, 60' high, wind load 70 MPH basic wind speed		.80	30		3,275	605	226	4,106	5,000
0910	60' high, wind load 90MPH basic wind speed		.45	53.333		5,875	1,075	400	7,350	8,950
1000	120' high, wind load 70MPH basic wind speed		.40	60		8,025	1,200	450	9,675	11,600
1200	190' high, wind load 90 MPH basic wind speed	↓	.20	120	↓	19,500	2,425	900	22,825	27,000
2000	For states west of Rocky Mountains, add for shipping					10%				

02600 | Drainage & Containment

02620 | Subdrainage

		CREW	DAILY OUTPUT	LABOR-HOURS	UNIT	2000 BARE COSTS MAT.	LABOR	EQUIP.	TOTAL	TOTAL INCL O&P
0010	**PIPING, SUBDRAINAGE, CONCRETE**									
0021	Not including excavation and backfill	R02510-810								
3000	Porous wall concrete underdrain, std. strength, 4" diameter	B-20	335	.072	L.F.	1.78	1.08		2.86	3.82
3020	6" diameter	"	315	.076		2.31	1.15		3.46	4.52
3040	8" diameter	B-21	310	.090		2.86	1.44	.45	4.75	6.10
3060	12" diameter		285	.098		6.05	1.56	.49	8.10	9.85
3080	15" diameter		230	.122		6.95	1.94	.60	9.49	11.60
3100	18" diameter	↓	165	.170		9.15	2.70	.84	12.69	15.60
4000	Extra strength, 6" diameter	B-20	315	.076		2.29	1.15		3.44	4.49
4020	8" diameter	B-21	310	.090	↓	3.43	1.44	.45	5.32	6.70

Important: See the Reference Section for critical supporting data - Reference Nos., Crews, & City Cost Indexes

02600 | Drainage & Containment

02620 | Subdrainage

			CREW	DAILY OUTPUT	LABOR-HOURS	UNIT	2000 BARE COSTS MAT.	LABOR	EQUIP.	TOTAL	TOTAL INCL O&P	
210	4040	10" diameter	B-21	285	.098	L.F.	6.85	1.56	.49	8.90	10.75	210
	4060	12" diameter		230	.122		7.45	1.94	.60	9.99	12.15	
	4080	15" diameter		200	.140		8.25	2.23	.69	11.17	13.60	
	4100	18" diameter		165	.170		12	2.70	.84	15.54	18.70	
240	0010	**PIPING, SUBDRAINAGE, CORRUGATED METAL**										240
	0021	Not including excavation and backfill										
	2010	Aluminum, perforated										
	2020	6" diameter, 18 ga.	B-14	380	.126	L.F.	2.57	1.97	.53	5.07	6.75	
	2200	8" diameter, 16 ga.		370	.130		3.73	2.02	.55	6.30	8.15	
	2220	10" diameter, 16 ga.		360	.133		4.66	2.07	.56	7.29	9.30	
	2240	12" diameter, 16 ga.		285	.168		5.20	2.62	.71	8.53	11	
	2260	18" diameter, 16 ga.		205	.234		7.85	3.64	.99	12.48	15.90	
	3000	Uncoated galvanized, perforated										
	3020	6" diameter, 18 ga.	B-20	380	.063	L.F.	4	.95		4.95	6.05	
	3200	8" diameter, 16 ga.	"	370	.065		5.50	.98		6.48	7.75	
	3220	10" diameter, 16 ga.	B-21	360	.078		8.25	1.24	.38	9.87	11.60	
	3240	12" diameter, 16 ga.		285	.098		8.65	1.56	.49	10.70	12.70	
	3260	18" diameter, 16 ga.		205	.137		13.20	2.17	.67	16.04	18.95	
	4000	Steel, perforated, asphalt coated										
	4020	6" diameter 18 ga.	B-20	380	.063	L.F.	3.20	.95		4.15	5.15	
	4030	8" diameter 18 ga	"	370	.065		5	.98		5.98	7.20	
	4040	10" diameter 16 ga	B-21	360	.078		5.75	1.24	.38	7.37	8.90	
	4050	12" diameter 16 ga		285	.098		6.60	1.56	.49	8.65	10.45	
	4060	18" diameter 16 ga		205	.137		9	2.17	.67	11.84	14.35	
270	0010	**PIPING, SUBDRAINAGE, POLYVINYL CHLORIDE**										270
	0020	Perforated, price as solid pipe, division 02530-780										
280	0010	**PIPING, SUBDRAINAGE, VITRIFIED CLAY**										280
	0020	Not including excavation and backfill										
	3000	Perforated, 5' lengths, C700, 4" diameter	B-14	400	.120	L.F.	1.93	1.87	.51	4.31	5.85	
	3020	6" diameter		315	.152		3	2.37	.64	6.01	8.05	
	3040	8" diameter		290	.166		4.02	2.58	.70	7.30	9.55	
	3060	12" diameter		275	.175		8.15	2.72	.74	11.61	14.40	
	4000	Channel pipe, 4" diameter	B-20	430	.056		2	.84		2.84	3.65	
	4020	6" diameter		335	.072		3	1.08		4.08	5.15	
	4060	8" diameter		295	.081		4.50	1.23		5.73	7.05	
	4080	12" diameter	B-21	280	.100		9.50	1.59	.49	11.58	13.70	

02630 | Storm Drainage

			CREW	DAILY OUTPUT	LABOR-HOURS	UNIT	MAT.	LABOR	EQUIP.	TOTAL	TOTAL INCL O&P	
100	0010	**PIPING, STORM DRAINAGE, CORRUGATED METAL**										100
	0020	Not including excavation or backfill										
	2000	Corrugated metal pipe, galvanized and coated										
	2020	Bituminous coated with paved invert, 20' lengths										
	2040	8" diameter, 16 ga.	B-14	330	.145	L.F.	7.20	2.26	.62	10.08	12.50	
	2060	10" diameter, 16 ga.		260	.185		8.65	2.87	.78	12.30	15.30	
	2080	12" diameter, 16 ga.		210	.229		12.05	3.56	.97	16.58	20.50	
	2100	15" diameter, 16 ga.		200	.240		12.65	3.73	1.02	17.40	21.50	
	2120	18" diameter, 16 ga.		190	.253		14.35	3.93	1.07	19.35	23.50	
	2140	24" diameter, 14 ga.		160	.300		27	4.67	1.27	32.94	39	
	2160	30" diameter, 14 ga.	B-13	120	.400		34.50	6.35	4.66	45.51	54	
	2180	36" diameter, 12 ga.		120	.400		50	6.35	4.66	61.01	71	
	2200	48" diameter, 12 ga.		100	.480		64.50	7.60	5.60	77.70	90	
	2220	60" diameter, 10 ga.	B-13B	75	.747		83	11.95	10.75	105.70	123	
	2240	72" diameter, 8 ga.	"	45	1.244		117	19.90	17.95	154.85	182	
	2500	Galvanized, uncoated, 20' lengths										
	2520	8" diameter, 16 ga.	B-14	355	.135	L.F.	6.45	2.10	.57	9.12	11.25	
	2540	10" diameter, 16 ga.		280	.171		7.75	2.67	.73	11.15	13.90	

For expanded coverage of these items see *Means Heavy Construction Cost Data 2000*

02600 | Drainage & Containment

02630 | Storm Drainage

			CREW	DAILY OUTPUT	LABOR-HOURS	UNIT	2000 BARE COSTS				TOTAL INCL O&P
							MAT.	LABOR	EQUIP.	TOTAL	
100	2560	12" diameter, 16 ga.	B-14	220	.218	L.F.	8.30	3.39	.92	12.61	15.90
	2580	15" diameter, 16 ga.		220	.218		9.60	3.39	.92	13.91	17.35
	2600	18" diameter, 16 ga.		205	.234		12.30	3.64	.99	16.93	21
	2620	24" diameter, 14 ga.	↓	175	.274		18.35	4.27	1.16	23.78	28.50
	2640	30" diameter, 14 ga.	B-13	130	.369		23	5.85	4.30	33.15	39.50
	2660	36" diameter, 12 ga.		130	.369		38	5.85	4.30	48.15	56.50
	2680	48" diameter, 12 ga.	↓	110	.436		51	6.90	5.10	63	73.50
	2700	60" diameter, 10 ga.	B-13B	78	.718	↓	79.50	11.50	10.35	101.35	118
	2780	End sections, 8" diameter	B-14	24	2	Ea.	53.50	31	8.45	92.95	121
	2785	10" diameter		22	2.182		55	34	9.25	98.25	129
	2790	12" diameter		35	1.371		60.50	21.50	5.80	87.80	110
	2800	18" diameter	↓	30	1.600		74	25	6.75	105.75	131
	2810	24" diameter	B-13	25	1.920		107	30.50	22.50	160	193
	2820	30" diameter		25	1.920		212	30.50	22.50	265	310
	2825	36" diameter		20	2.400		305	38	28	371	430
	2830	48" diameter	↓	10	4.800		680	76	56	812	940
	2835	60" diameter	B-13B	5	11.200		1,200	179	162	1,541	1,800
	2840	72" diameter	"	4	14		1,400	224	202	1,826	2,150
	3000	Corrugated galvanized or alum. oval arch culverts, coated & paved									
	3020	17" x 13", 16 ga., 15" equivalent	B-14	200	.240	L.F.	11.95	3.73	1.02	16.70	20.50
	3040	21" x 15", 16 ga., 18" equivalent		150	.320		30.50	4.98	1.35	36.83	43.50
	3060	28" x 20", 14 ga., 24" equivalent		125	.384		24	6	1.62	31.62	38.50
	3080	35" x 24", 14 ga., 30" equivalent	↓	100	.480		50	7.45	2.03	59.48	70
	3100	42" x 29", 12 ga., 36" equivalent	B-13	100	.480		39.50	7.60	5.60	52.70	62.50
	3120	49" x 33", 12 ga., 42" equivalent		90	.533		45.50	8.45	6.20	60.15	71
	3140	57" x 38", 12 ga., 48" equivalent		75	.640		93	10.10	7.45	110.55	128
	3160	Steel, plain oval arch culverts, plain									
	3180	17" x 13", 16 ga., 15" equivalent	B-14	225	.213	L.F.	7.85	3.32	.90	12.07	15.30
	3200	21" x 15", 16 ga., 18" equivalent		175	.274		15.50	4.27	1.16	20.93	25.50
	3220	28" x 20", 14 ga., 24" equivalent	↓	150	.320		14.60	4.98	1.35	20.93	26
	3240	35" x 24", 14 ga., 30" equivalent	B-13	108	.444		30	7.05	5.20	42.25	50.50
	3260	42" x 29", 12 ga., 36" equivalent		108	.444		36	7.05	5.20	48.25	57
	3280	49" x 33", 12 ga., 42" equivalent		92	.522		42.50	8.25	6.10	56.85	67.50
	3300	57" x 38", 12 ga., 48" equivalent		75	.640		51	10.10	7.45	68.55	81.50
	3320	End sections, 17" x 13"		22	2.182	Ea.	34	34.50	25.50	94	124
	3340	42" x 29"	↓	17	2.824	"	172	44.50	33	249.50	300
	3360	Multi-plate arch, steel	B-20	1,690	.014	Lb.	.60	.21		.81	1.03
200	0010	**CATCH BASINS OR MANHOLES** not including footing, excavation,									
	0020	backfill, frame and cover									
	0050	Brick, 4' inside diameter, 4' deep	D-1	1	16	Ea.	264	286		550	775
	0100	6' deep		.70	22.857		370	410		780	1,100
	0150	8' deep		.50	32	↓	475	570		1,045	1,475
	0200	For depths over 8', add		4	4	V.L.F.	100	71.50		171.50	231
	0400	Concrete blocks (radial), 4' I.D., 4' deep		1.50	10.667	Ea.	300	190		490	650
	0500	6' deep		1	16		390	286		676	915
	0600	8' deep		.70	22.857	↓	510	410		920	1,250
	0700	For depths over 8', add		5.50	2.909	V.L.F.	76.50	52		128.50	172
	0800	Concrete, cast in place, 4' x 4', 8" thick, 4' deep	C-14H	2	24	Ea.	340	460	18.95	818.95	1,200
	0900	6' deep		1.50	32		495	615	25.50	1,135.50	1,650
	1000	8' deep		1	48	↓	650	925	38	1,613	2,350
	1100	For depths over 8', add		8	6	V.L.F.	84.50	116	4.74	205.24	298
	1110	Precast, 4' I.D., 4' deep	B-22	4.10	7.317	Ea.	315	119	50.50	484.50	605
	1120	6' deep		3	10		425	162	69	656	815
	1130	8' deep		2	15	↓	470	243	104	817	1,050
	1140	For depths over 8', add	↓	16	1.875	V.L.F.	78.50	30.50	12.95	121.95	152

02600 | Drainage & Containment

02630 | Storm Drainage

		CREW	DAILY OUTPUT	LABOR-HOURS	UNIT	2000 BARE COSTS MAT.	LABOR	EQUIP.	TOTAL	TOTAL INCL O&P
1150	5' I.D., 4' deep	B-6	3	8	Ea.	445	128	67.50	640.50	780
1160	6' deep		2	12		605	192	102	899	1,100
1170	8' deep		1.50	16		760	256	135	1,151	1,425
1180	For depths over 8', add		12	2	V.L.F.	99	32	16.90	147.90	182
1190	6' I.D., 4' deep		2	12	Ea.	730	192	102	1,024	1,250
1200	6' deep		1.50	16		950	256	135	1,341	1,625
1210	8' deep		1	24		1,175	385	203	1,763	2,150
1220	For depths over 8', add		8	3	V.L.F.	153	48	25.50	226.50	277
1250	Slab tops, precast, 8" thick									
1300	4' diameter manhole	B-6	8	3	Ea.	158	48	25.50	231.50	282
1400	5' diameter manhole		7.50	3.200		275	51	27	353	420
1500	6' diameter manhole		7	3.429		315	55	29	399	470
1600	Frames & covers, C.I., 24" square, 500 lb.		7.80	3.077		213	49	26	288	345
1700	26" D shape, 600 lb.		7	3.429		214	55	29	298	360
1800	Light traffic, 18" diameter, 100 lb.		10	2.400		76.50	38.50	20.50	135.50	172
1900	24" diameter, 300 lb.		8.70	2.759		138	44	23.50	205.50	251
2000	36" diameter, 900 lb.		5.80	4.138		385	66	35	486	570
2100	Heavy traffic, 24" diameter, 400 lb.		7.80	3.077		172	49	26	247	300
2200	36" diameter, 1150 lb.		3	8		510	128	67.50	705.50	850
2300	Mass. State standard, 26" diameter, 475 lb.		7	3.429		214	55	29	298	360
2400	30" diameter, 620 lb.		7	3.429		243	55	29	327	390
2500	Watertight, 24" diameter, 350 lb.		7.80	3.077		293	49	26	368	430
2600	26" diameter, 500 lb.		7	3.429		330	55	29	414	490
2700	32" diameter, 575 lb.		6	4		390	64	34	488	570
2800	3 piece cover & frame, 10" deep,									
2900	1200 lbs., for heavy equipment	B-6	3	8	Ea.	770	128	67.50	965.50	1,150
3000	Raised for paving 1-1/4" to 2" high,									
3100	4 piece expansion ring									
3200	20" to 26" diameter	1 Clab	3	2.667	Ea.	102	38.50		140.50	178
3300	30" to 36" diameter	"	3	2.667	"	143	38.50		181.50	223
3320	Frames and covers, existing, raised for paving 2", including									
3340	row of brick, concrete collar, up to 12" wide frame	B-6	18	1.333	Ea.	31	21.50	11.30	63.80	82.50
3360	20" to 26" wide frame		11	2.182		41	35	18.45	94.45	125
3380	30" to 36" wide frame		9	2.667		51	42.50	22.50	116	153
3400	Inverts, single channel brick	D-1	3	5.333		57	95		152	224
3500	Concrete		5	3.200		45	57		102	146
3600	Triple channel, brick		2	8		86.50	143		229.50	340
3700	Concrete		3	5.333		61	95		156	228
3800	Steps, heavyweight cast iron, 7" x 9"	1 Bric	40	.200		8.35	4		12.35	15.90
3900	8" x 9"		40	.200		12.50	4		16.50	20.50
3928	12" x 10-1/2"		40	.200		13	4		17	21
4000	Standard sizes, galvanized steel		40	.200		11.75	4		15.75	19.65
4100	Aluminum		40	.200		13	4		17	21

02700 | Bases, Ballasts, Pavements & Appurtenances

02720 | Unbound Base Courses & Ballasts

		CREW	DAILY OUTPUT	LABOR-HOURS	UNIT	2000 BARE COSTS MAT.	LABOR	EQUIP.	TOTAL	TOTAL INCL O&P
0010	BASE COURSE For roadways and large paved areas									
0050	Crushed 3/4" stone base, compacted, 3" deep	B-36B	5,200	.012	S.Y.	2.95	.22	.61	3.78	4.28

For expanded coverage of these items see *Means Heavy Construction Cost Data 2000*

02700 | Bases, Ballasts, Pavements & Appurtenances

02720 | Unbound Base Courses & Ballasts

			CREW	DAILY OUTPUT	LABOR-HOURS	UNIT	MAT.	LABOR	EQUIP.	TOTAL	TOTAL INCL O&P	
200	0100	6" deep	B-36B	5,000	.013	S.Y.	5.95	.23	.63	6.81	7.60	200
	0200	9" deep		4,600	.014		8.90	.25	.69	9.84	10.95	
	0300	12" deep		4,200	.015		12.10	.27	.75	13.12	14.60	
	0301	Crushed 1-1/2" stone base, compacted to 4" deep		6,000	.011		3.14	.19	.53	3.86	4.35	
	0302	6" deep		5,400	.012		4.81	.21	.59	5.61	6.30	
	0303	8" deep		4,500	.014		6.30	.25	.70	7.25	8.10	
	0304	12" deep		3,800	.017		9.60	.30	.83	10.73	12	
	0350	Bank run gravel, spread and compacted										
	0370	6" deep	B-32	6,000	.005	S.Y.	2.20	.10	.26	2.56	2.87	
	0390	9" deep		4,900	.007		3.24	.12	.32	3.68	4.11	
	0400	12" deep		4,200	.008		4.40	.14	.37	4.91	5.50	
	0700	Liquid application to gravel base, asphalt emulsion	B-45	6,000	.003	Gal.	2	.04	.06	2.10	2.34	
	0800	Prime and seal, cut back asphalt		6,000	.003	"	2.36	.04	.06	2.46	2.74	
	1000	Macadam penetration crushed stone, 2 gal. per S.Y., 4" thick		6,000	.003	S.Y.	4	.04	.06	4.10	4.54	
	1100	6" thick, 3 gal. per S.Y.		4,000	.004		6	.06	.09	6.15	6.80	
	1200	8" thick, 4 gal. per S.Y.		3,000	.005		8	.08	.12	8.20	9.10	
	6000	Stabilization fabric, polypropylene, 6 oz./S.Y.	B-6	10,000	.002		.84	.04	.02	.90	1	
	8900	For small and irregular areas, add						50%	50%			
215	0010	BASE Prepare and roll sub-base, small areas to 2500 S.Y.	B-32A	1,500	.016	S.Y.		.29	.63	.92	1.18	215
	0100	Large areas over 2500 S.Y.	B-32	3,700	.009	"		.16	.42	.58	.73	

02740 | Flexible Pavement

			CREW	DAILY OUTPUT	LABOR-HOURS	UNIT	MAT.	LABOR	EQUIP.	TOTAL	TOTAL INCL O&P	
300	0010	ASPHALTIC CONCRETE PAVEMENT for highways										300
	0020	and large paved areas			R02065-300							
	0080	Binder course, 1-1/2" thick	B-25	7,725	.011	S.Y.	1.87	.18	.23	2.28	2.62	
	0120	2" thick		6,345	.014		2.49	.22	.28	2.99	3.43	
	0160	3" thick		4,905	.018		3.70	.29	.36	4.35	4.96	
	0200	4" thick		4,140	.021		4.95	.34	.43	5.72	6.50	
	0300	Wearing course, 1" thick	B-25B	10,575	.009		1.47	.15	.19	1.81	2.07	
	0340	1-1/2" thick		7,725	.012		2.23	.20	.26	2.69	3.08	
	0380	2" thick		6,345	.015		3	.25	.31	3.56	4.07	
	0420	2-1/2" thick		5,480	.018		3.70	.29	.36	4.35	4.96	
	0460	3" thick		4,900	.020		4.41	.32	.41	5.14	5.85	
	0800	Alternate method of figuring paving costs										
	0810	Binder course, 1-1/2" thick	B-25	630	.140	Ton	24.50	2.25	2.82	29.57	34	
	0811	2" thick		690	.128		24.50	2.06	2.57	29.13	33.50	
	0812	3" thick		800	.110		24.50	1.77	2.22	28.49	32.50	
	0813	4" thick		850	.104		24.50	1.67	2.09	28.26	32	
	0850	Wearing course, 1" thick	B-25B	575	.167		27	2.74	3.47	33.21	38.50	
	0851	1-1/2" thick		630	.152		27	2.50	3.17	32.67	37.50	
	0852	2" thick		690	.139		27	2.29	2.90	32.19	37	
	0853	2-1/2" thick		745	.129		27	2.12	2.68	31.80	36.50	
	0854	3" thick		800	.120		27	1.97	2.50	31.47	36	
	1000	Pavement replacement over trench, 2" thick	B-37	90	.533	S.Y.	3.05	8.30	1.48	12.83	19.10	
	1050	4" thick		70	.686		6.05	10.65	1.91	18.61	27	
	1080	6" thick		55	.873		9.60	13.60	2.43	25.63	36	
315	0011	PAVING Asphaltic concrete, parking lots & driveways										315
	0020	6" stone base, 2" binder course, 1" topping	B-25C	9,000	.005	S.F.	1.07	.09	.17	1.33	1.52	
	0300	Binder course, 1-1/2" thick		35,000	.001		.27	.02	.04	.33	.39	
	0400	2" thick		25,000	.002		.35	.03	.06	.44	.51	
	0500	3" thick		15,000	.003		.54	.05	.10	.69	.80	
	0600	4" thick		10,800	.004		.71	.07	.14	.92	1.06	
	0800	Sand finish course, 3/4" thick		41,000	.001		.15	.02	.04	.21	.24	
	0900	1" thick		34,000	.001		.19	.02	.05	.26	.30	
	1000	Fill pot holes, hot mix, 2" thick	B-16	4,200	.008		.39	.12	.11	.62	.74	
	1100	4" thick		3,500	.009		.56	.14	.13	.83	1	

02700 | Bases, Ballasts, Pavements & Appurtenances

02740 | Flexible Pavement

			CREW	DAILY OUTPUT	LABOR-HOURS	UNIT	MAT.	LABOR	EQUIP.	TOTAL	TOTAL INCL O&P	
315	1120	6" thick	B-16	3,100	.010	S.F.	.76	.16	.14	1.06	1.26	315
	1140	Cold patch, 2" thick	B-51	3,000	.016		.45	.24	.06	.75	.97	
	1160	4" thick		2,700	.018		.86	.27	.06	1.19	1.47	
	1180	6" thick		1,900	.025		1.33	.38	.09	1.80	2.22	

02750 | Rigid Pavement

			CREW	DAILY OUTPUT	LABOR-HOURS	UNIT	MAT.	LABOR	EQUIP.	TOTAL	TOTAL INCL O&P	
100	0010	**CONCRETE PAVEMENT** Including joints, finishing, and curing										100
	0020	Fixed form, 12' pass, unreinforced, 6" thick	B-26	3,000	.029	S.Y.	15.85	.49	.61	16.95	18.95	
	0100	8" thick		2,750	.032		22	.53	.67	23.20	26	
	0200	9" thick		2,500	.035		25.50	.59	.74	26.83	30	
	0300	10" thick		2,100	.042		27.50	.70	.88	29.08	32.50	
	0400	12" thick		1,800	.049		29.50	.81	1.02	31.33	35	
	0500	15" thick		1,500	.059		33	.98	1.23	35.21	39.50	
	0510	For small irregular areas, add						100%				
	0700	Finishing, broom finish small areas	2 Cefi	120	.133	S.Y.		2.52		2.52	4.11	
	1000	Curing, with sprayed membrane by hand	2 Clab	1,500	.011	"	.32	.15		.47	.61	
	1650	For integral coloring, see div. 03310-220										

02766 | Pavement Markings

			CREW	DAILY OUTPUT	LABOR-HOURS	UNIT	MAT.	LABOR	EQUIP.	TOTAL	TOTAL INCL O&P	
550	0010	**LINES ON PAV'T** Acrylic waterborne, white or yellow, 4" wide	B-78	20,000	.002	L.F.	.12	.03	.03	.18	.21	550
	0200	6" wide		11,000	.004		.10	.05	.05	.20	.25	
	0500	8" wide		10,000	.004		.16	.06	.05	.27	.33	
	0600	12" wide		4,000	.010		.29	.15	.13	.57	.71	
	0620	Arrows or gore lines		2,300	.017	S.F.	.50	.26	.23	.99	1.24	
	0640	Temporary paint, white or yellow		15,000	.003	L.F.	.15	.04	.03	.22	.28	
	0660	Removal	1 Clab	300	.027			.39		.39	.66	
	0680	Temporary tape	2 Clab	1,500	.011		1.12	.15		1.27	1.49	
	0710	Thermoplastic, white or yellow, 4" wide	B-79	15,000	.002		.56	.03	.05	.64	.72	
	0730	6" wide		14,000	.002		.81	.03	.05	.89	1	
	0740	8" wide		12,000	.003		1.09	.04	.06	1.19	1.33	
	0750	12" wide		6,000	.005		1.62	.08	.11	1.81	2.06	
	0760	Arrows		660	.048	S.F.	1.50	.72	1.04	3.26	4.03	
	0770	Gore lines		2,500	.013		1	.19	.27	1.46	1.73	
	0780	Letters		660	.048		1.25	.72	1.04	3.01	3.76	
	0790	Layout of pavement marking	A-2	25,000	.001	L.F.		.01	.01	.02	.03	
	0800	Parking stall, paint, white	B-78	440	.091	Stall	3	1.35	1.18	5.53	6.90	
	1000	Street letters and numbers	"	1,600	.025	S.F.	.50	.37	.32	1.19	1.55	

02770 | Curbs and Gutters

			CREW	DAILY OUTPUT	LABOR-HOURS	UNIT	MAT.	LABOR	EQUIP.	TOTAL	TOTAL INCL O&P	
225	0010	**CURBS** Asphaltic, machine formed, 8" wide, 6" high, 40 L.F./ton	B-27	1,000	.032	L.F.	.54	.48	.07	1.09	1.50	225
	0100	8" wide, 8" high, 30 L.F. per ton		900	.036		.62	.53	.08	1.23	1.69	
	0150	Asphaltic berm, 12" W, 3"-6" H, 35 L.F./ton, before pavement		700	.046		.80	.68	.10	1.58	2.17	
	0200	12" W, 1-1/2" to 4" H, 60 L.F. per ton, laid with pavement	B-2	1,050	.038		.49	.57		1.06	1.51	
	0300	Concrete, wood forms, 6" x 18", straight	C-2A	500	.096		2.15	1.83		3.98	5.45	
	0400	6" x 18", radius	"	200	.240		2.26	4.56		6.82	10.25	
	0415	Machine formed, 6" x 18", straight	B-69A	2,000	.024		3.04	.39	.23	3.66	4.26	
	0416	6" x 18", radius		900	.053		3.17	.88	.50	4.55	5.50	
	0421	Curb and gutter, straight										
	0422	with 6" high curb and 6" thick gutter, wood forms										
	0430	24" wide, .055 C.Y. per L.F.	C-2A	375	.128	L.F.	10.65	2.43		13.08	15.85	
	0435	30" wide, .066 C.Y. per L.F.		340	.141		11.55	2.69		14.24	17.25	
	0440	Steel forms, 24" wide, straight		700	.069		4.95	1.30		6.25	7.65	
	0441	Radius		300	.160		4.95	3.04		7.99	10.60	
	0442	30" wide, straight		700	.069		5.95	1.30		7.25	8.75	
	0443	Radius		300	.160		5.95	3.04		8.99	11.70	

For expanded coverage of these items see *Means Heavy Construction Cost Data 2000*

02700 | Bases, Ballasts, Pavements & Appurtenances

02770 | Curbs and Gutters

			CREW	DAILY OUTPUT	LABOR-HOURS	UNIT	2000 BARE COSTS				TOTAL INCL O&P
							MAT.	LABOR	EQUIP.	TOTAL	
225	0445	Machine formed, 24" wide, straight	B-69A	2,000	.024	L.F.	4.95	.39	.23	5.57	6.35
	0446	Radius		900	.053		4.95	.88	.50	6.33	7.50
	0447	30" wide, straight		2,000	.024		5.95	.39	.23	6.57	7.45
	0448	Radius		900	.053		5.95	.88	.50	7.33	8.60
	0550	Precast, 6" x 18", straight	B-29	700	.069		6.25	1.08	.88	8.21	9.70
	0600	6" x 18", radius	"	325	.148		7.75	2.34	1.90	11.99	14.55
	1000	Granite, split face, straight, 5" x 16"	D-13	500	.096		14.10	1.80	.80	16.70	19.40
	1100	6" x 18"	"	450	.107		18.55	2	.89	21.44	25
	1300	Radius curbing, 6" x 18", over 10' radius	B-29	260	.185		22.50	2.92	2.38	27.80	32.50
	1400	Corners, 2' radius		80	.600	Ea.	76	9.50	7.75	93.25	109
	1600	Edging, 4-1/2" x 12", straight		300	.160	L.F.	7.05	2.53	2.06	11.64	14.35
	1800	Curb inlets, (guttermouth) straight		41	1.171	Ea.	169	18.50	15.10	202.60	234
	2000	Indian granite (belgian block)									
	2100	Jumbo, 10-1/2" x 7-1/2" x 4", grey	D-1	150	.107	L.F.	1.75	1.90		3.65	5.15
	2150	Pink		150	.107		2.15	1.90		4.05	5.60
	2200	Regular, 9" x 4-1/2" x 4-1/2", grey		160	.100		1.70	1.79		3.49	4.89
	2250	Pink		160	.100		2	1.79		3.79	5.20
	2300	Cubes, 4" x 4" x 4", grey		175	.091		1.65	1.63		3.28	4.58
	2350	Pink		175	.091		1.75	1.63		3.38	4.69
	2400	6" x 6" x 6", pink		155	.103		3.60	1.84		5.44	7.10
	2500	Alternate pricing method for indian granite									
	2550	Jumbo, 10-1/2" x 7-1/2" x 4" (30lb), grey				Ton	100			100	110
	2600	Pink					125			125	138
	2650	Regular, 9" x 4-1/2" x 4-1/2" (20lb), grey					120			120	132
	2700	Pink					140			140	154
	2750	Cubes, 4" x 4" x 4" (5lb), grey					200			200	220
	2800	Pink					225			225	248
	2850	6" x 6" x 6" (25lb), pink					140			140	154
	2900	For pallets, add					15			15	16.50

02775 | Sidewalks

			CREW	DAILY OUTPUT	LABOR-HOURS	UNIT	MAT.	LABOR	EQUIP.	TOTAL	TOTAL INCL O&P
275	0010	**SIDEWALKS, DRIVEWAYS, & PATIOS** No base (R02065-300)									
	0020	Asphaltic concrete, 2" thick	B-37	720	.067	S.Y.	3.16	1.04	.19	4.39	5.45
	0100	2-1/2" thick	"	660	.073	"	4	1.13	.20	5.33	6.55
	0300	Concrete, 3000 psi, CIP, 6 x 6 - W1.4 x W1.4 mesh,									
	0311	broomed finish, no base, 4" thick	3 Clab	600	.040	S.F.	1.05	.58		1.63	2.14
	0351	5" thick		545	.044		1.40	.64		2.04	2.63
	0401	6" thick		510	.047		1.63	.68		2.31	2.95
	0450	For bank run gravel base, 4" thick, add	B-18	2,500	.010		.13	.15	.02	.30	.43
	0520	8" thick, add	"	1,600	.015		.27	.23	.04	.54	.72
	0550	Exposed aggregate finish, add to above, minimum	B-24	1,875	.013		.09	.23		.32	.48
	0600	Maximum	"	455	.053		.29	.93		1.22	1.89
	1000	Crushed stone, 1" thick, white marble	2 Clab	1,700	.009		.23	.14		.37	.48
	1050	Bluestone	"	1,700	.009		.18	.14		.32	.43
	1700	Redwood, prefabricated, 4' x 4' sections	2 Carp	316	.051		7.15	1		8.15	9.55
	1750	Redwood planks, 1" thick, on sleepers	"	240	.067		4.99	1.31		6.30	7.75

02778 | Steps

			CREW	DAILY OUTPUT	LABOR-HOURS	UNIT	MAT.	LABOR	EQUIP.	TOTAL	TOTAL INCL O&P
280	0010	**STEPS** Incl. excav., borrow & concrete base, where applicable									
	0100	Brick steps	B-24	35	.686	LF Riser	7.60	12.10		19.70	29
	0300	Bluestone treads, 12" x 2" or 12" x 1-1/2"	"	30	.800	"	19.30	14.15		33.45	45
	0500	Concrete, cast in place, see division 03310-240									
	0600	Precast concrete, see division 03480-800									

02700 | Bases, Ballasts, Pavements & Appurtenances

02780 | Unit Pavers

			CREW	DAILY OUTPUT	LABOR-HOURS	UNIT	MAT.	LABOR	EQUIP.	TOTAL	TOTAL INCL O&P	
100	0010	**ASPHALT BLOCKS**, 6"x12"x1-1/4", w/bed & neopr. adhesive	D-1	135	.119	S.F.	2.84	2.12		4.96	6.70	100
	0100	3" thick		130	.123		3.98	2.20		6.18	8.10	
	0300	Hexagonal tile, 8" wide, 1-1/4" thick		135	.119		2.91	2.12		5.03	6.80	
	0400	2" thick		130	.123		4.07	2.20		6.27	8.20	
	0500	Square, 8" x 8", 1-1/4" thick		135	.119		2.89	2.12		5.01	6.75	
	0600	2" thick		130	.123		4.05	2.20		6.25	8.15	
	0900	For exposed aggregate (ground finish) add					.45			.45	.50	
	0910	For colors, add					.40			.40	.44	
200	0010	**BRICK PAVING** 4" x 8" x 1-1/2", without joints (4.5 brick/S.F.)	D-1	110	.145	S.F.	2.04	2.60		4.64	6.65	200
	0100	Grouted, 3/8" joint (3.9 brick/S.F.)		90	.178		2.39	3.17		5.56	8	
	0200	4" x 8" x 2-1/4", without joints (4.5 bricks/S.F.)		110	.145		2.64	2.60		5.24	7.30	
	0300	Grouted, 3/8" joint (3.9 brick/S.F.)		90	.178		2.44	3.17		5.61	8.05	
	0500	Bedding, asphalt, 3/4" thick	B-25	5,130	.017		.30	.28	.35	.93	1.18	
	0540	Course washed sand bed, 1" thick	B-18	5,000	.005		.15	.07	.01	.23	.29	
	0580	Mortar, 1" thick	D-1	300	.053		.36	.95		1.31	2	
	0620	2" thick		200	.080		.36	1.43		1.79	2.81	
	1500	Brick on 1" thick sand bed laid flat, 4.5 per S.F.		100	.160		3.02	2.86		5.88	8.15	
	2000	Brick pavers, laid on edge, 7.2 per S.F.		70	.229		1.99	4.08		6.07	9.10	
	2500	For 4" thick concrete bed and joints, add		595	.027		.75	.48		1.23	1.64	
	2800	For steam cleaning, add	A-1	950	.008		.05	.12	.07	.24	.35	
600	0010	**PRECAST CONCRETE PAVING SLABS**										600
	0750	Exposed local aggregate, natural	2 Bric	250	.064	S.F.	4.68	1.28		5.96	7.30	
	0800	Colors		250	.064		5	1.28		6.28	7.65	
	0850	Exposed granite or limestone aggregate		250	.064		5.60	1.28		6.88	8.30	
	0900	Exposed white tumblestone aggregate		250	.064		3.07	1.28		4.35	5.55	
650	0010	**PLANTER BLOCKS** Precast concrete, interlocking										650
	0020	"V" blocks for retaining soil	D-1	205	.078	S.F.	3.75	1.39		5.14	6.50	
800	0010	**STONE PAVERS**										800
	1100	Flagging, bluestone, irregular, 1" thick,	D-1	81	.198	S.F.	4	3.53		7.53	10.35	
	1150	Snapped random rectangular, 1" thick		92	.174		6.05	3.10		9.15	11.90	
	1200	1-1/2" thick		85	.188		7.30	3.36		10.66	13.70	
	1250	2" thick		83	.193		8.50	3.44		11.94	15.20	
	1300	Slate, natural cleft, irregular, 3/4" thick		92	.174		1.80	3.10		4.90	7.25	
	1350	Random rectangular, gauged, 1/2" thick		105	.152		3.90	2.72		6.62	8.90	
	1400	Random rectangular, butt joint, gauged, 1/4" thick		150	.107		4.19	1.90		6.09	7.85	
	1500	For interior setting, add								25%	25%	
	1550	Granite blocks, 3-1/2" x 3-1/2" x 3-1/2"	D-1	92	.174	S.F.	5.25	3.10		8.35	11.05	
	1600	4" to 12" long, 3" to 5" wide, 3" to 5" thick		98	.163		4.38	2.91		7.29	9.75	
	1650	6" to 15" long, 3" to 6" wide, 3" to 5" thick		105	.152		2.34	2.72		5.06	7.20	

02785 | Flexible Pavement Coating

			CREW	DAILY OUTPUT	LABOR-HOURS	UNIT	MAT.	LABOR	EQUIP.	TOTAL	TOTAL INCL O&P	
800	0010	**SEALCOATING** 2 coat coal tar pitch emulsion over 10,000 S.Y.	B-45	5,000	.003	S.Y.	.43	.05	.07	.55	.63	800
	0030	1000 to 10,000 S.Y.	"	3,000	.005		.43	.08	.12	.63	.75	
	0100	Under 1000 S.Y.	B-1	1,050	.023		.43	.35		.78	1.06	
	0300	Petroleum resistant, over 10,000 S.Y.	B-45	5,000	.003		.50	.05	.07	.62	.71	
	0320	1000 to 10,000 S.Y.	"	3,000	.005		.50	.08	.12	.70	.83	
	0400	Under 1000 S.Y.	B-1	1,050	.023		.50	.35		.85	1.14	
	0600	Non-skid pavement renewal, over 10,000 S.Y.	B-45	5,000	.003		.60	.05	.07	.72	.82	
	0620	1000 to 10,000 S.Y.	"	3,000	.005		.60	.08	.12	.80	.94	
	0700	Under 1000 S.Y.	B-1	1,050	.023		.60	.35		.95	1.25	
	0800	Prepare and clean surface for above	A-2	8,545	.003			.04	.02	.06	.09	
	1000	Hand seal asphalt curbing	B-1	4,420	.005	L.F.	.30	.08		.38	.47	
	1900	Asphalt surface treatment, single course, small area										
	1901	0.30 gal/S.Y. asphalt material, 20#/S.Y. aggregate	B-91	5,000	.013	S.Y.	.70	.23	.30	1.23	1.48	
	1910	Roadway or large area		10,000	.006		.64	.11	.15	.90	1.06	

For expanded coverage of these items see *Means Heavy Construction Cost Data 2000*

02700 | Bases, Ballasts, Pavements & Appurtenances

02785 | Flexible Pavement Coating

		CREW	DAILY OUTPUT	LABOR-HOURS	UNIT	MAT.	LABOR	EQUIP.	TOTAL	TOTAL INCL O&P
1950	Asphalt surface treatment, dbl. course for small area	B-91	3,000	.021	S.Y.	1.30	.38	.49	2.17	2.60
1960	Roadway or large area		6,000	.011		1.17	.19	.25	1.61	1.88
1980	Asphalt surface treatment, single course, for shoulders		7,500	.009		.75	.15	.20	1.10	1.30
2080	Sand sealing, sharp sand, asphalt emulsion, small area		10,000	.006		.50	.11	.15	.76	.90
2120	Roadway or large area		18,000	.004		.43	.06	.08	.57	.67

02790 | Athletic/Recreational Surfaces

		CREW	DAILY OUTPUT	LABOR-HOURS	UNIT	MAT.	LABOR	EQUIP.	TOTAL	TOTAL INCL O&P
0010	**TURF, ARTIFICIAL** Not including asphalt base or drainage, but									
0020	including cushion pad, over 50,000 S.F.									
0200	1/2" pile and 5/16" cushion pad, standard	C-17	3,200	.025	S.F.	5.60	.50		6.10	7
0300	Deluxe		2,560	.031		6.65	.63		7.28	8.40
0500	1/2" pile and 5/8" cushion pad, standard		2,844	.028		8.15	.57		8.72	9.95
0600	Deluxe		2,327	.034		8.95	.69		9.64	11
0800	For asphaltic concrete base, 2-1/2" thick,									
0900	with 6" crushed stone sub-base, add	B-25	12,000	.007	S.F.	.96	.12	.15	1.23	1.42
0010	**TENNIS COURT** Asphalt, incl. base, 2-1/2" thick, one court	B-37	450	.107	S.Y.	9.25	1.66	.30	11.21	13.35
0200	Two courts		675	.071		7.15	1.11	.20	8.46	9.95
0300	Clay courts		360	.133		26.50	2.07	.37	28.94	33
0400	Pulverized natural greenstone with 4" base, fast dry		250	.192		26	2.99	.53	29.52	34
0800	Rubber-acrylic base resilient pavement		600	.080		26.50	1.24	.22	27.96	31.50
1000	Colored sealer, acrylic emulsion, 3 coats	2 Clab	800	.020		3.80	.29		4.09	4.68
1100	3 coat, 2 colors	"	900	.018		4.60	.26		4.86	5.50
1200	For preparing old courts, add	1 Clab	825	.010			.14		.14	.24
1400	Posts for nets, 3-1/2" diameter with eye bolts	B-1	3.40	7.059	Pr.	140	107		247	335
1500	With pulley & reel		3.40	7.059	"	196	107		303	400
1700	Net, 42' long, nylon thread with binder		50	.480	Ea.	176	7.25		183.25	205
1800	All metal		6.50	3.692	"	360	56		416	490
2001	Paint markings on asphalt, 2 coat	1 Clab	2.50	3.200	Court	51	46		97	135
2200	Complete court with fence, etc., asphaltic conc., minimum	B-37	.20	240		11,800	3,725	665	16,190	20,000
2300	Maximum		.16	300		14,300	4,675	835	19,810	24,600
2800	Clay courts, minimum		.20	240		13,800	3,725	665	18,190	22,300
2900	Maximum		.16	300		16,900	4,675	835	22,410	27,500
0010	**RUNNING TRACK** Asphalt, incl base, 3" thick	B-37	300	.160	S.Y.	9.90	2.49	.44	12.83	15.65
0100	Surface, latex rubber system, 3/8" thick, black	B-20	125	.192		4.45	2.90		7.35	9.85
0150	Colors		125	.192		7.90	2.90		10.80	13.65
0300	Urethane rubber system, 3/8" thick, black		120	.200		11.90	3.02		14.92	18.25
0400	Color coating		115	.209		16.25	3.16		19.41	23.50

02800 | Site Improvements and Amenities

02810 | Irrigation System

		CREW	DAILY OUTPUT	LABOR-HOURS	UNIT	MAT.	LABOR	EQUIP.	TOTAL	TOTAL INCL O&P
0010	**SPRINKLER IRRIGATION SYSTEM** For lawns									
0100	Golf course with fully automatic system	C-17	.05	1,600	9 holes	75,000	32,200		107,200	138,000
0200	24' diam. head at 15' O.C incl. piping, auto oper., minimum	B-20	70	.343	Head	16.50	5.20		21.70	27
0300	Maximum		40	.600		38	9.05		47.05	57.50
0500	60' diam. head at 40' O.C incl. piping, auto oper., minimum		28	.857		50	12.95		62.95	77
0600	Maximum		23	1.043		140	15.80		155.80	181

02800 | Site Improvements and Amenities

02810 | Irrigation System

		CREW	DAILY OUTPUT	LABOR-HOURS	UNIT	2000 BARE COSTS MAT.	LABOR	EQUIP.	TOTAL	TOTAL INCL O&P
0800	Residential system, custom, 1" supply	B-20	2,000	.012	S.F.	.25	.18		.43	.59
0900	1-1/2" supply	↓	1,800	.013	"	.28	.20		.48	.66
1020	Pop up spray head w/risers, hi-pop, full circle pattern, 4"	2 Skwk	76	.211	Ea.	3.17	4.16		7.33	10.65
1030	1/2 circle pattern		76	.211		3.17	4.16		7.33	10.65
1040	6", full circle pattern		76	.211		7.45	4.16		11.61	15.30
1050	1/2 circle pattern		76	.211		7.45	4.16		11.61	15.30
1060	12", full circle pattern		76	.211		9.25	4.16		13.41	17.35
1070	1/2 circle pattern		76	.211		9.30	4.16		13.46	17.40
1080	Pop up bubbler head w/risers, hi-pop bubbler head, 4"		76	.211		3.27	4.16		7.43	10.75
1090	6"		76	.211		7.45	4.16		11.61	15.30
1100	12"		76	.211		9.25	4.16		13.41	17.35
1110	Impact full/part circle sprinklers, 28'-54' 25-60 PSI		37	.432		8.80	8.55		17.35	24.50
1120	Spaced 37'-49' @ 25-50 PSI		37	.432		20.50	8.55		29.05	37
1130	Spaced 43'-61' @ 30-60 PSI		37	.432		42.50	8.55		51.05	61
1140	Spaced 54'-78' @ 40-80 PSI	↓	37	.432	↓	92.50	8.55		101.05	117
1145	Impact rotor pop-up full/part commercial circle sprinklers									
1150	Spaced 42'-65' 35-80 PSI	2 Skwk	25	.640	Ea.	20.50	12.65		33.15	44
1160	Spaced 48'-76' 45-85 PSI	"	25	.640	"	6.20	12.65		18.85	28.50
1165	Impact rotor pop-up part. circle comm., 53'-75', 55-100 PSI, w/ acc									
1170	Plastic case, metal cover	2 Skwk	25	.640	Ea.	134	12.65		146.65	169
1180	Rubber cover		25	.640		102	12.65		114.65	134
1190	Iron case, metal cover		22	.727		127	14.35		141.35	164
1200	Rubber cover		22	.727		133	14.35		147.35	172
1250	Plastic case, 2 nozzle, metal cover		25	.640		101	12.65		113.65	133
1260	Rubber cover		25	.640		107	12.65		119.65	139
1270	Iron case, 2 nozzle, metal cover		22	.727		143	14.35		157.35	183
1280	Rubber cover	↓	22	.727	↓	152	14.35		166.35	193
1282	Impact rotor pop-up full circle comm., 39'-99', 30-100 PSI									
1284	Plastic case, metal cover	2 Skwk	25	.640	Ea.	101	12.65		113.65	133
1286	Rubber cover		25	.640		111	12.65		123.65	144
1288	Iron case, metal cover		22	.727		149	14.35		163.35	189
1290	Rubber cover		22	.727		154	14.35		168.35	195
1292	Plastic case, 2 nozzle, metal cover		22	.727		108	14.35		122.35	143
1294	Rubber cover		22	.727		113	14.35		127.35	149
1296	Iron case, 2 nozzle, metal cover		20	.800		144	15.80		159.80	185
1298	Rubber cover		20	.800		152	15.80		167.80	194
1305	Electric remote control valve, plastic, 3/4"		18	.889		15.55	17.55		33.10	47
1310	1"		18	.889		23	17.55		40.55	55.50
1320	1-1/2"		18	.889		49	17.55		66.55	84
1330	2"	↓	18	.889	↓	69.50	17.55		87.05	107
1335	Quick coupling valves, brass, locking cover									
1340	Inlet coupling valve, 3/4"	2 Skwk	18.75	.853	Ea.	38.50	16.85		55.35	71.50
1350	1"		18.75	.853		47.50	16.85		64.35	81
1360	Controller valve boxes, 6" round boxes		18.75	.853		3.27	16.85		20.12	32.50
1370	10" round boxes		14.25	1.123		7.20	22		29.20	46
1380	12" square box	↓	9.75	1.641	↓	13.90	32.50		46.40	71
1388	Electromech. control, 14 day 3-60 min, auto start to 23/day									
1390	4 station	2 Skwk	1.04	15.385	Ea.	160	305		465	695
1400	7 station		.64	25		159	495		654	1,025
1410	12 station		.40	40		243	790		1,033	1,625
1420	Dual programs, 18 station		.24	66.667		1,200	1,325		2,525	3,575
1430	23 station	↓	.16	100	↓	1,575	1,975		3,550	5,125
1435	Backflow preventer, bronze, 0-175 PSI, w/valves, test cocks									
1440	3/4"	2 Skwk	2	8	Ea.	97	158		255	380
1450	1"		2	8		99	158		257	380
1460	1-1/2"	↓	2	8	↓	200	158		358	490

For expanded coverage of these items see *Means Site Work and Landscape Cost Data 2000*

02800 | Site Improvements and Amenities

02810 | Irrigation System

			CREW	DAILY OUTPUT	LABOR-HOURS	UNIT	MAT.	LABOR	EQUIP.	TOTAL	TOTAL INCL O&P
800	1470	2"	2 Skwk	2	8	Ea.	214	158		372	505
	1475	Pressure vacuum breaker, brass, 15-150 PSI									
	1480	3/4"	2 Skwk	2	8	Ea.	54	158		212	330
	1490	1"		2	8		81.50	158		239.50	360
	1500	1-1/2"		2	8		161	158		319	450
	1510	2"	↓	2	8	↓	149	158		307	435

02815 | Fountains

			CREW	DAILY OUTPUT	LABOR-HOURS	UNIT	MAT.	LABOR	EQUIP.	TOTAL	TOTAL INCL O&P
225	0010	**FOUNTAINS/AERATORS**									
	0100	Pump w/controls									
	0200	Single phase, 100' chord, 1/2 H.P. pump	2 Skwk	4.40	3.636	Ea.	2,150	72		2,222	2,475
	0300	3/4 H.P. pump		4.30	3.721		2,450	73.50		2,523.50	2,825
	0400	1 H.P. pump		4.20	3.810		2,475	75		2,550	2,850
	0500	1-1/2 H.P. pump		4.10	3.902		2,575	77		2,652	2,950
	0600	2 H.P. pump		4	4		2,600	79		2,679	2,975
	0700	Three phase, 200' chord, 5 H.P. pump		3.90	4.103		6,500	81		6,581	7,300
	0800	7-1/2 H.P. pump		3.80	4.211		7,675	83		7,758	8,575
	0900	10 H.P. pump		3.70	4.324		8,450	85.50		8,535.50	9,450
	1000	15 H.P. pump		3.60	4.444		10,400	88		10,488	11,600
	1100	Nozzles, minimum		8	2		135	39.50		174.50	217
	1200	Maximum		8	2		246	39.50		285.50	340
	1300	Lights w/mounting kits, 200 watt		18	.889		297	17.55		314.55	355
	1400	300 watt		18	.889		330	17.55		347.55	395
	1500	500 watt		18	.889		360	17.55		377.55	430
	1600	Color blender	↓	12	1.333	↓	292	26.50		318.50	365

02820 | Fences & Gates

			CREW	DAILY OUTPUT	LABOR-HOURS	UNIT	MAT.	LABOR	EQUIP.	TOTAL	TOTAL INCL O&P
500	0010	**FENCE, MISC. METAL** Chicken wire, posts @ 4', 1" mesh, 4' high	B-80	410	.059	L.F.	1.10	.89	1.34	3.33	4.21
	0100	2" mesh, 6' high		350	.069		1	1.04	1.57	3.61	4.61
	0200	Galv. steel, 12 ga., 2" x 4" mesh, posts 5' O.C., 3' high		300	.080		1.50	1.21	1.84	4.55	5.75
	0300	5' high		300	.080		2	1.21	1.84	5.05	6.30
	0400	14 ga., 1" x 2" mesh, 3' high		300	.080		1.60	1.21	1.84	4.65	5.85
	0500	5' high	↓	300	.080	↓	2.20	1.21	1.84	5.25	6.50
	1000	Kennel fencing, 1-1/2" mesh, 6' long, 3'-6" wide, 6'-2" high	2 Clab	4	4	Ea.	250	58		308	375
	1050	12' long		4	4		300	58		358	430
	1200	Top covers, 1-1/2" mesh, 6' long		15	1.067		51	15.40		66.40	82.50
	1250	12' long	↓	12	1.333	↓	81	19.25		100.25	123
	1300	For kennel doors, see division 08344-350									
	4500	Security fence, prison grade, set in concrete, 12' high	B-80	25	.960	L.F.	20.50	14.50	22	57	71.50
	4600	16' high		20	1.200		24.50	18.15	27.50	70.15	88.50
	5300	Tubular picket, steel, 6' sections, 1-9/16" posts, 4' high		300	.080		15.65	1.21	1.84	18.70	21.50
	5400	2" posts, 5' high		240	.100		21.50	1.51	2.29	25.30	29
	5600	2" posts, 6' high		200	.120		24.50	1.81	2.75	29.06	33
	5700	Staggered picket 1-9/16" posts, 4' high		300	.080		14.15	1.21	1.84	17.20	19.65
	5800	2" posts, 5' high		240	.100		23	1.51	2.29	26.80	30.50
	5900	2" posts, 6' high	↓	200	.120		24	1.81	2.75	28.56	32.50
	6200	Gates, 4' high, 3' wide	B-1	10	2.400	Ea.	136	36.50		172.50	212
	6300	5' high, 3' wide		10	2.400		177	36.50		213.50	256
	6400	6' high, 3' wide		10	2.400		182	36.50		218.50	262
	6500	4' wide	↓	10	2.400	↓	212	36.50		248.50	295
528	0010	**FENCE, CHAIN LINK INDUSTRIAL**, schedule 40									
	0020	3 strands barb wire, 2" post @ 10' O.C., set in concrete, 6' H									
	0200	9 ga. wire, galv. steel	B-80	240	.100	L.F.	7.20	1.51	2.29	11	13.05
	0300	Aluminized steel	↓	240	.100		9.25	1.51	2.29	13.05	15.30

02800 | Site Improvements and Amenities

02820 | Fences & Gates

		CREW	DAILY OUTPUT	LABOR-HOURS	UNIT	MAT.	LABOR	EQUIP.	TOTAL	TOTAL INCL O&P
0500	6 ga. wire, galv. steel	B-80	240	.100	L.F.	11.70	1.51	2.29	15.50	17.95
0600	Aluminized steel		240	.100		13.40	1.51	2.29	17.20	19.85
0800	6 ga. wire, 6' high but omit barbed wire, galv. steel		250	.096		11.35	1.45	2.20	15	17.35
0900	Aluminized steel		250	.096		15.85	1.45	2.20	19.50	22.50
0920	8' H, 6 ga. wire, 2-1/2" line post, galv. steel		180	.133		18.45	2.02	3.06	23.53	27.50
0940	Aluminized steel		180	.133		23	2.02	3.06	28.08	32
1100	Add for corner posts, 3" diam., galv. steel		40	.600	Ea.	55	9.05	13.75	77.80	91
1200	Aluminized steel		40	.600		66	9.05	13.75	88.80	103
1300	Add for braces, galv. steel		80	.300		15	4.54	6.90	26.44	32
1350	Aluminized steel		80	.300		20	4.54	6.90	31.44	37.50
1400	Gate for 6' high fence, 1-5/8" frame, 3' wide, galv. steel		10	2.400		91	36.50	55	182.50	223
1500	Aluminized steel		10	2.400		111	36.50	55	202.50	245
2000	5'-0" high fence, 9 ga., no barbed wire, 2" line post,									
2010	10' O.C., 1-5/8" top rail									
2100	Galvanized steel	B-80	300	.080	L.F.	6	1.21	1.84	9.05	10.70
2200	Aluminized steel		300	.080	"	7.25	1.21	1.84	10.30	12.10
2400	Gate, 4' wide, 5' high, 2" frame, galv. steel		10	2.400	Ea.	100	36.50	55	191.50	233
2500	Aluminized steel		10	2.400	"	110	36.50	55	201.50	244
3100	Overhead slide gate, chain link, 6' high, to 18' wide		38	.632	L.F.	82.50	9.55	14.50	106.55	123
3110	Cantilever type		48	.500		38	7.55	11.45	57	67.50
3120	8' high		24	1		55	15.10	23	93.10	112
3130	10' high		18	1.333		65	20	30.50	115.50	140
5000	Double swing gates, incl. posts & hardware									
5010	5' high, 12' opening	B-80	3.40	7.059	Opng.	263	107	162	532	650
5020	20' opening		2.80	8.571		340	130	197	667	815
5060	6' high, 12' opening		3.20	7.500		475	113	172	760	910
5070	20' opening		2.60	9.231		655	140	212	1,007	1,200
5080	8' high, 12' opening		2.13	11.252		700	170	258	1,128	1,350
5090	20' opening		1.45	16.552		935	250	380	1,565	1,875
5100	10' high, 12' opening		1.31	18.321		800	277	420	1,497	1,825
5110	20' opening		1.03	23.301		1,200	350	535	2,085	2,525
5120	12' high, 12' opening		1.05	22.857		1,175	345	525	2,045	2,450
5130	20' opening		.85	28.235		1,500	425	650	2,575	3,100
5190	For aluminized steel add					20%				
7001	Snow fence on steel posts 10' O.C., 4' high	B-1	500	.048	L.F.	1.56	.73		2.29	2.95
0010	**FENCE, CHAIN LINK RESIDENTIAL**, sch. 20, 11 ga. wire, 1-5/8" post									
0020	10' O.C., 1-3/8" top rail, 2" corner post, galv. stl. 3' high	B-1	500	.048	L.F.	2.49	.73		3.22	3.98
0050	4' high		400	.060		2.83	.91		3.74	4.66
0100	6' high		200	.120		3.44	1.81		5.25	6.90
0150	Add for gate 3' wide, 1-3/8" frame, 3' high		12	2	Ea.	37.50	30		67.50	93
0170	4' high		10	2.400		43	36.50		79.50	110
0190	6' high		10	2.400		59	36.50		95.50	127
0200	Add for gate 4' wide, 1-3/8" frame, 3' high		9	2.667		42	40.50		82.50	115
0220	4' high		9	2.667		47.50	40.50		88	121
0240	6' high		8	3		55.50	45.50		101	139
0350	Aluminized steel, 11 ga. wire, 3' high		500	.048	L.F.	3.23	.73		3.96	4.79
0380	4' high		400	.060		4.15	.91		5.06	6.10
0400	6' high		200	.120		5.85	1.81		7.66	9.50
0450	Add for gate 3' wide, 1-3/8" frame, 3' high		12	2	Ea.	45	30		75	102
0470	4' high		10	2.400		74.50	36.50		111	144
0490	6' high		10	2.400		93.50	36.50		130	165
0500	Add for gate 4' wide, 1-3/8" frame, 3' high		10	2.400		51.50	36.50		88	119
0520	4' high		9	2.667		57	40.50		97.50	132
0540	6' high		8	3		65.50	45.50		111	150
0620	Vinyl covered, 9 ga. wire, 3' high		500	.048	L.F.	2.69	.73		3.42	4.20

For expanded coverage of these items see *Means Site Work and Landscape Cost Data 2000*

02800 | Site Improvements and Amenities

02820 | Fences & Gates

		CREW	DAILY OUTPUT	LABOR-HOURS	UNIT	2000 BARE COSTS MAT.	LABOR	EQUIP.	TOTAL	TOTAL INCL O&P	
530	0640	4' high	B-1	400	.060	L.F.	3.18	.91		4.09	5.05	530
	0660	6' high		200	.120		4.04	1.81		5.85	7.55	
	0720	Add for gate 3' wide, 1-3/8" frame, 3' high		12	2	Ea.	52.50	30		82.50	110	
	0740	4' high		10	2.400		59.50	36.50		96	128	
	0760	6' high		10	2.400		91.50	36.50		128	162	
	0780	Add for gate 4' wide, 1-3/8" frame, 3' high		10	2.400		59.50	36.50		96	128	
	0800	4' high		9	2.667		65	40.50		105.50	141	
	0820	6' high		8	3		73.50	45.50		119	159	
	0860	Tennis courts, 11 ga. wire, 2-1/2" post 10' O.C., 1-5/8" top rail										
	0900	10' high	B-1	155	.155	L.F.	7.05	2.34		9.39	11.75	
	0920	12' high		130	.185	"	8.55	2.79		11.34	14.20	
	1000	Add for gate 4' wide, 1-5/8" frame 7' high		10	2.400	Ea.	74	36.50		110.50	144	
	1040	Aluminized steel, 11 ga. wire 10' high		155	.155	L.F.	8.70	2.34		11.04	13.55	
	1100	12' high		130	.185	"	10.90	2.79		13.69	16.80	
	1140	Add for gate 4' wide, 1-5/8" frame, 7' high		10	2.400	Ea.	85	36.50		121.50	156	
	1250	Vinyl covered, 9 ga. wire, 10' high		155	.155	L.F.	8.05	2.34		10.39	12.85	
	1300	12' high		130	.185	"	13.95	2.79		16.74	20	
	1400	Add for gate 4' wide, 1-5/8" frame, 7' high		10	2.400	Ea.	87	36.50		123.50	158	
890	0005	**WIRE FENCING**										890
	0010	Barbed wire, galvanized, domestic steel, hi-tensile 15-1/2 ga.				M.L.F.	25			25	27.50	
	0020	standard, 12-3/4 ga.					32.50			32.50	36	
	0210	Barbless wire, 2-strand galvanized, 12-1/2 ga.					33			33	36.50	
	0500	Helical razor ribbon, stainless steel, 18" dia x 18" spacing				C.L.F.	102			102	113	
	0600	Hardware cloth galv., 1/4" mesh, 23 ga., 2' wide				C.S.F.	45			45	49.50	
	0700	3' wide					44			44	48.50	
	0900	1/2" mesh, 19 ga., 2' wide					38.50			38.50	42.50	
	1000	4' wide					38			38	41.50	
	1200	Chain link fabric, steel, 2" mesh, 6 ga, galvanized					119			119	131	
	1300	9 ga, galvanized					55			55	60.50	
	1350	Vinyl coated					45			45	49.50	
	1360	Aluminized					76			76	83.50	
	1400	2-1/4" mesh, 11.5 ga., galvanized					36			36	39.50	
	1600	1-3/4" mesh (tennis courts), 11.5 ga (core), vinyl coated					51.50			51.50	57	
	1700	9 ga, galvanized					67.50			67.50	74	
	2100	Welded wire fabric, galvanized, 1" x 2", 14 ga.					31			31	34	
	2200	2" x 4", 12-1/2 ga.					21.50			21.50	23.50	
925	0010	**FENCE, RAIL** Picket, No. 2 cedar, Gothic, 2 rail, 3' high	B-1	160	.150	L.F.	4.47	2.27		6.74	8.80	925
	0050	Gate, 3'-6" wide		9	2.667	Ea.	38.50	40.50		79	112	
	0400	3 rail, 4' high		150	.160	L.F.	5.15	2.42		7.57	9.80	
	0500	Gate, 3'-6" wide		9	2.667	Ea.	46	40.50		86.50	120	
	0600	Open rail, rustic, No. 1 cedar, 2 rail, 3' high		160	.150	L.F.	4.01	2.27		6.28	8.30	
	0650	Gate, 3' wide		9	2.667	Ea.	44.50	40.50		85	118	
	0700	3 rail, 4' high		150	.160	L.F.	4.58	2.42		7	9.20	
	0900	Gate, 3' wide		9	2.667	Ea.	57.50	40.50		98	132	
	1200	Stockade, No. 2 cedar, treated wood rails, 6' high		160	.150	L.F.	5.55	2.27		7.82	10.05	
	1250	Gate, 3' wide		9	2.667	Ea.	45.50	40.50		86	120	
	1300	No. 1 cedar, 3-1/4" cedar rails, 6' high		160	.150	L.F.	13.55	2.27		15.82	18.85	
	1500	Gate, 3' wide		9	2.667	Ea.	108	40.50		148.50	188	
	1520	Open rail, split, No. 1 cedar, 2 rail, 3' high		160	.150	L.F.	4.07	2.27		6.34	8.35	
	1540	3 rail, 4'-0" high		150	.160		5.30	2.42		7.72	9.95	
	3300	Board, shadow box, 1" x 6", treated pine, 6' high		160	.150		8.25	2.27		10.52	12.95	
	3400	No. 1 cedar, 6' high		150	.160		16.25	2.42		18.67	22	
	3900	Basket weave, No. 1 cedar, 6' high		160	.150		16.15	2.27		18.42	21.50	
	3950	Gate, 3'-6" wide		8	3	Ea.	103	45.50		148.50	191	
	4000	Treated pine, 6' high		150	.160	L.F.	8.80	2.42		11.22	13.85	
	4200	Gate, 3'-6" wide		9	2.667	Ea.	48	40.50		88.50	122	

02800 | Site Improvements and Amenities

02820 | Fences & Gates

							2000 BARE COSTS				TOTAL
			CREW	DAILY OUTPUT	LABOR-HOURS	UNIT	MAT.	LABOR	EQUIP.	TOTAL	INCL O&P
925	5000	Fence rail, redwood, 2" x 4", merch grade 8'	B-1	2,400	.010	L.F.	.87	.15		1.02	1.22
	5050	Select grade, 8'		2,400	.010	"	2.70	.15		2.85	3.23
	6000	Fence post, select redwood, earthpacked & treated, 4" x 4" x 6'		96	.250	Ea.	8.65	3.78		12.43	16
	6010	4" x 4" x 8'		96	.250		10.30	3.78		14.08	17.75
	6020	Set in concrete, 4" x 4" x 6'		50	.480		11.15	7.25		18.40	24.50
	6030	4" x 4" x 8'		50	.480		13.30	7.25		20.55	27
	6040	Wood post, 4' high, set in concrete, incl. concrete		50	.480		6.50	7.25		13.75	19.60
	6050	Earth packed		96	.250		4.44	3.78		8.22	11.35
	6060	6' high, set in concrete, incl. concrete		50	.480		8.35	7.25		15.60	21.50
	6070	Earth packed		96	.250		6.05	3.78		9.83	13.10

02830 | Retaining Walls

100	0010	**RETAINING WALLS** Aluminized steel bin, excavation										100
	0020	and backfill not included, 10' wide									
	0100	4' high, 5.5' deep	B-13	650	.074	S.F.	14.70	1.17	.86	16.73	19.10
	0200	8' high, 5.5' deep		615	.078		16.90	1.23	.91	19.04	21.50
	0300	10' high, 7.7' deep		580	.083		17.75	1.31	.96	20.02	23
	0400	12' high, 7.7' deep		530	.091		19.15	1.43	1.06	21.64	24.50
	0500	16' high, 7.7' deep		515	.093		20	1.47	1.09	22.56	26
	0600	16' high, 9.9' deep		500	.096		21.50	1.52	1.12	24.14	27.50
	0700	20' high, 9.9' deep		470	.102		24	1.62	1.19	26.81	30.50
	0800	20' high, 12.1' deep		460	.104		26	1.65	1.22	28.87	32.50
	0900	24' high, 12.1' deep		455	.105		27.50	1.67	1.23	30.40	34
	1000	24' high, 14.3' deep		450	.107		28.50	1.69	1.24	31.43	35.50
	1100	28' high, 14.3' deep		440	.109		29.50	1.73	1.27	32.50	37
	1300	For plain galvanized bin type walls, deduct					10%				
	1800	Concrete gravity wall with vertical face including excavation & backfill									
	1850	No reinforcing									
	1900	6' high, level embankment	C-17C	36	2.306	L.F.	45.50	46.50	11.45	103.45	143
	2000	33° slope embankment		32	2.594		42	52.50	12.85	107.35	150
	2200	8' high, no surcharge		27	3.074		57	62	15.25	134.25	186
	2300	33° slope embankment		24	3.458		69	70	17.15	156.15	215
	2500	10' high, level embankment		19	4.368		81.50	88	21.50	191	265
	2600	33° slope embankment		18	4.611		113	93	23	229	310
	2800	Reinforced concrete cantilever, incl. excavation, backfill & reinf.									
	2900	6' high, 33° slope embankment	C-17C	35	2.371	L.F.	42	48	11.75	101.75	141
	3000	8' high, 33° slope embankment		29	2.862		48.50	57.50	14.20	120.20	168
	3100	10' high, 33° slope embankment		20	4.150		63	83.50	20.50	167	235
	3200	20' high, 500 lb. per L.F. surcharge		7.50	11.067		188	223	55	466	655
	3500	Concrete cribbing, incl. excavation and backfill									
	3700	12' high, open face	B-13	210	.229	S.F.	21	3.62	2.66	27.28	32
	3900	Closed face	"	210	.229	"	19.65	3.62	2.66	25.93	30.50
	4100	Concrete filled slurry trench, see division 02260-700									
	4300	Stone filled gabions, not incl. excavation,									
	4310	Stone, delivered, 3' wide									
	4350	Galvanized, 6' high, 33° slope embankment	B-13	49	.980	L.F.	20	15.50	11.40	46.90	61
	4500	Highway surcharge		27	1.778		40	28	20.50	88.50	115
	4600	9' high, up to 33° slope embankment		24	2		45	31.50	23.50	100	129
	4700	Highway surcharge		16	3		70	47.50	35	152.50	196
	4900	12' high, up to 33° slope embankment		14	3.429		70	54	40	164	214
	5000	Highway surcharge		11	4.364		100	69	51	220	283
	5950	For PVC coating, add					20%				
	7100	Segmental wall system, incl backfill, compaction, to 8'									
	7120	interlocking pins, no scaffolding, base, or tiebacks									
	7140	8" x 18" x 21.5", 95 lbs	D-12	250	.128	S.F.	8.80	2.25		11.05	13.45
	7160	8" x 18" x 12.5", 85 lbs		330	.097		8.50	1.71		10.21	12.25

For expanded coverage of these items see *Means Site Work and Landscape Cost Data 2000*

02800 | Site Improvements and Amenities

02830 | Retaining Walls

		Description	CREW	DAILY OUTPUT	LABOR-HOURS	UNIT	MAT.	LABOR	EQUIP.	TOTAL	TOTAL INCL O&P	
100	7180	6" x 17.25" x 12", 72 lbs	D-12	375	.085	S.F.	8.10	1.50		9.60	11.45	100
	7200	Beveled, 6" x 6" x 12", 68 lbs		350	.091		7.75	1.61		9.36	11.20	
	7220	Step, 6" x 16" x 12", 85 lbs		300	.107		7	1.88		8.88	10.95	
	7240	Caps, to 40 lbs		600	.053	L.F.	8.80	.94		9.74	11.25	
	7260	For reinforcing, add				S.F.				2	2.50	
	8000	For higher walls, add components as necessary										
400	0010	**STONE WALL** Including excavation, concrete footing and										400
	0020	stone 3' below grade. Price is exposed face area.										
	0200	Decorative random stone, to 6' high, 1'-6" thick, dry set	D-1	35	.457	S.F.	7.60	8.15		15.75	22	
	0300	Mortar set		40	.400		9.25	7.15		16.40	22.50	
	0500	Cut stone, to 6' high, 1'-6" thick, dry set		35	.457		11.50	8.15		19.65	26.50	
	0600	Mortar set		40	.400		13.50	7.15		20.65	27	
	0800	Retaining wall, random stone, 6' to 10' high, 2' thick, dry set		45	.356		9.50	6.35		15.85	21	
	0900	Mortar set		50	.320		11.50	5.70		17.20	22.50	
	1100	Cut stone, 6' to 10' high, 2' thick, dry set		45	.356		14.75	6.35		21.10	27	
	1200	Mortar set		50	.320		15.75	5.70		21.45	27	

02840 | Walk/Road/Parking Appurtenances

		Description	CREW	DAILY OUTPUT	LABOR-HOURS	UNIT	MAT.	LABOR	EQUIP.	TOTAL	TOTAL INCL O&P	
155	0010	**BUMPER RAILS** For garages, 12 ga. rail, 6" wide, with steel										155
	0020	posts 12'-6" O.C., Minimum	E-4	190	.168	L.F.	8	3.66	.44	12.10	16.50	
	0030	Average		165	.194		10	4.22	.51	14.73	19.85	
	0100	Maximum		140	.229		12	4.97	.60	17.57	23.50	
	0300	12" channel rail, minimum		160	.200		10	4.35	.53	14.88	20	
	0400	Maximum		120	.267		15	5.80	.70	21.50	28.50	
200	0010	**TRAFFIC CONTROL DEVICES**										200
	0100	Traffic channelizing pavement markers, layout only	A-7	2,000	.012	Ea.		.24		.24	.42	
	0110	13" x 7-1/2" x 2-1/2" high, non-plowable install	2 Clab	96	.167		16.50	2.41		18.91	22.50	
	0200	8" x 8" x 3-1/4" high, non-plowable, install		96	.167		18	2.41		20.41	24	
	0230	4" x 4" x 3/4" high, non-plowable, install		120	.133		2	1.93		3.93	5.50	
	0240	9-1/4" x 5-7/8" x 1/4" high, plowable, concrete pav't	A-2A	70	.343		12	5.10	4.04	21.14	26.50	
	0250	9-1/4" x 5-7/8" x 1/4" high, plowable, asphalt pav't	"	120	.200		2.30	2.98	2.36	7.64	10.15	
	0300	Barrier and curb delineators, reflectorized, 2" x 4"	2 Clab	150	.107		1.30	1.54		2.84	4.07	
	0310	3" x 5"	"	150	.107		2.65	1.54		4.19	5.55	
	0500	Rumble strip, polycarbonate										
	0510	24" x 3-1/2" x 1/2" high	2 Clab	50	.320	Ea.	5	4.62		9.62	13.40	
500	0010	**GUIDE/GUARD RAIL** Corrugated steel, galv. steel posts, 6'-3" O.C.	B-80	850	.028	L.F.	10	.43	.65	11.08	12.45	500
	0200	End sections, galvanized, flared		50	.480	Ea.	40	7.25	11	58.25	68.50	
	0300	Wrap around end		50	.480	"	60	7.25	11	78.25	90.50	
	0400	Timber guide rail, 4" x 8" with 6" x 8" wood posts, treated		960	.025	L.F.	12	.38	.57	12.95	14.50	
	0600	Cable guide rail, 3 at 3/4" cables, steel posts, single face		900	.027		4.71	.40	.61	5.72	6.55	
	0700	Wood posts		950	.025		5.90	.38	.58	6.86	7.80	
	0900	Guide rail, steel box beam, 6" x 6"		120	.200		15.50	3.02	4.59	23.11	27.50	
	1100	Median barrier, steel box beam, 6" x 8"		215	.112		19.50	1.69	2.56	23.75	27	
	1200	Impact barrier, UTMCD, barrel type	B-16	30	1.067	Ea.	252	16.40	14.90	283.30	320	
	1400	Resilient guide fence and light shield, 6' high	B-2	130	.308	L.F.	17.15	4.57		21.72	27	
	1500	Concrete posts, individual, 6'-5", triangular	B-80	110	.218	Ea.	30.50	3.30	5	38.80	44.50	
	1550	Square	"	110	.218	"	33	3.30	5	41.30	47	
	2000	Median, precast concrete, 3'-6" high, 2' wide, single face	B-29	380	.126	L.F.	28.50	2	1.63	32.13	36	
	2200	Double face	"	340	.141	"	24	2.23	1.82	28.05	32	
	2400	Speed bumps, thermoplastic, 10-1/2" x 2-1/4" x 48" long	B-2	120	.333	Ea.	81	4.95		85.95	98	
700	0010	**PARKING BARRIERS** Timber with saddles, treated type										700
	0100	4" x 4" for cars	B-2	520	.077	L.F.	2.80	1.14		3.94	5.05	
	0200	6" x 6" for trucks		520	.077	"	6.05	1.14		7.19	8.60	
	0400	Folding with individual padlocks		50	.800	Ea.	345	11.90		356.90	400	

Important: See the Reference Section for critical supporting data - Reference Nos., Crews, & City Cost Indexes

02800 | Site Improvements and Amenities

02840 | Walk/Road/Parking Appurtenances

			CREW	DAILY OUTPUT	LABOR-HOURS	UNIT	MAT.	LABOR	EQUIP.	TOTAL	TOTAL INCL O&P	
700	0600	Flexible fixed stanchion, 2' high, 3" diameter	B-2	100	.400	Ea.	17.75	5.95		23.70	29.50	700
	1000	Wheel stops, precast concrete incl. dowels, 6" x 10" x 6'-0"		120	.333		26.50	4.95		31.45	37.50	
	1100	8" x 13" x 6'-0"		120	.333		30	4.95		34.95	41.50	
	1200	Thermoplastic, 6" x 10" x 6'-0"		120	.333		51	4.95		55.95	64.50	
	1300	Pipe bollards, conc filled/paint, 8' L x 4' D hole, 6" diam.	B-6	20	1.200		168	19.20	10.15	197.35	229	
	1400	8" diam.		15	1.600		255	25.50	13.55	294.05	340	
	1500	12" diam.		12	2		330	32	16.90	378.90	440	
	9000	Parking lot control, see division 11156-600										

02850 | Prefabricated Bridges

			CREW	DAILY OUTPUT	LABOR-HOURS	UNIT	MAT.	LABOR	EQUIP.	TOTAL	TOTAL INCL O&P	
210	0010	BRIDGES Pedestrian, spans over streams, roadways, etc.										210
	0020	including erection, not including foundations										
	0050	Precast concrete, complete in place, 8' wide, 60' span	E-2	215	.223	S.F.	25.50	4.80	4.75	35.05	42.50	
	0100	100' span		185	.259		27.50	5.55	5.50	38.55	47.50	
	0150	120' span		160	.300		30	6.45	6.40	42.85	52.50	
	0200	150' span		145	.331		31.50	7.10	7.05	45.65	56	
	0300	Steel, trussed or arch spans, compl. in place, 8' wide, 40' span		320	.150		35.50	3.22	3.19	41.91	48.50	
	0400	50' span		395	.122		31.50	2.61	2.59	36.70	43	
	0500	60' span		465	.103		31.50	2.22	2.20	35.92	41.50	
	0600	80' span		570	.084		38	1.81	1.79	41.60	47	
	0700	100' span		465	.103		53	2.22	2.20	57.42	65	
	0800	120' span		365	.132		67	2.82	2.80	72.62	82.50	
	0900	150' span		310	.155		71.50	3.33	3.30	78.13	88.50	
	1000	160' span		255	.188		71.50	4.04	4.01	79.55	90.50	
	1100	10' wide, 80' span		640	.075		36.50	1.61	1.60	39.71	45	
	1200	120' span		415	.116		47	2.48	2.46	51.94	59.50	
	1300	150' span		445	.108		53	2.32	2.30	57.62	65	
	1400	200' span		205	.234		56.50	5.05	4.98	66.53	77	
	1600	Wood, laminated type, complete in place, 80' span	C-12	203	.236		30.50	4.57	2.23	37.30	44	
	1700	130' span	"	153	.314		31.50	6.05	2.96	40.51	48	

02870 | Site Furnishings

			CREW	DAILY OUTPUT	LABOR-HOURS	UNIT	MAT.	LABOR	EQUIP.	TOTAL	TOTAL INCL O&P	
610	0010	BENCHES Park, precast concrete, w/backs, wood rails, 4' long	2 Clab	5	3.200	Ea.	370	46		416	485	610
	0100	8' long		4	4		765	58		823	940	
	0300	Fiberglass, without back, one piece, 4' long		10	1.600		405	23		428	490	
	0400	8' long		7	2.286		840	33		873	980	
	0500	Steel barstock pedestals w/backs, 2" x 3" wood rails, 4' long		10	1.600		755	23		778	870	
	0510	8' long		7	2.286		895	33		928	1,050	
	0520	3" x 8" wood plank, 4' long		10	1.600		760	23		783	880	
	0530	8' long		7	2.286		890	33		923	1,025	
	0540	Backless, 4" x 4" wood plank, 4' square		10	1.600		740	23		763	850	
	0550	8' long		7	2.286		700	33		733	825	
	0600	Aluminum pedestals, with backs, aluminum slats, 8' long		8	2		180	29		209	248	
	0610	15' long		5	3.200		300	46		346	410	
	0620	Portable, aluminum slats, 8' long		8	2		204	29		233	274	
	0630	15' long		5	3.200		330	46		376	445	
	0800	Cast iron pedestals, back & arms, wood slats, 4' long		8	2		460	29		489	555	
	0820	8' long		5	3.200		765	46		811	920	
	0840	Backless, wood slats, 4' long		8	2		400	29		429	490	
	0860	8' long		5	3.200		675	46		721	825	
	1700	Steel frame, fir seat, 10' long		10	1.600		150	23		173	205	
800	0010	TRASH RECEPTACLE Fiberglass, 2' square, 18" high	2 Clab	30	.533	Ea.	204	7.70		211.70	237	800
	0100	2' square, 2'-6" high		30	.533		286	7.70		293.70	330	
	0300	Circular, 2' diameter, 18" high		30	.533		184	7.70		191.70	215	
	0400	2' diameter, 2'-6" high		30	.533		245	7.70		252.70	282	

For expanded coverage of these items see *Means Site Work and Landscape Cost Data 2000*

02800 | Site Improvements and Amenities

02870 | Site Furnishings

			CREW	DAILY OUTPUT	LABOR-HOURS	UNIT	2000 BARE COSTS MAT.	LABOR	EQUIP.	TOTAL	TOTAL INCL O&P	
815	0010	**TRASH CLOSURE** Steel with pullover cover										815
	0020	2'-3" wide, 4'-7" high, 6'-2" long	2 Clab	5	3.200	Ea.	695	46		741	845	
	0100	10'-1" long		4	4		920	58		978	1,100	
	0300	Wood, 10' wide, 6' high, 10' long		1.20	13.333		725	193		918	1,125	

02880 | Playfield Equipment

			CREW	DAILY OUTPUT	LABOR-HOURS	UNIT	MAT.	LABOR	EQUIP.	TOTAL	TOTAL INCL O&P	
100	0010	**BLEACHERS** Outdoor, portable, 3 to 5 tiers, to 300' long, min.	2 Sswk	120	.133	Seat	24.50	2.83		27.33	32.50	100
	0100	Maximum, less than 15' long, prefabricated		80	.200		47	4.25		51.25	60	
	0200	6 to 20 tiers, minimum, up to 300' long		120	.133		30	2.83		32.83	38.50	
	0300	Max., under 15', (highly prefabricated, on wheels)		80	.200		68.50	4.25		72.75	84	
	0500	Permanent grandstands, wood seat, steel frame, 24" row										
	0600	3 to 15 tiers, minimum	2 Sswk	60	.267	Seat	59	5.65		64.65	76	
	0700	Maximum		48	.333		65.50	7.10		72.60	86	
	0900	16 to 30 tiers, minimum		60	.267		82.50	5.65		88.15	102	
	0950	Average		55	.291		109	6.20		115.20	132	
	1000	Maximum		48	.333		116	7.10		123.10	141	
	1200	Seat backs only, 30" row, fiberglass		160	.100		17.50	2.13		19.63	23.50	
	1300	Steel and wood		160	.100		21.50	2.13		23.63	28	
	1400	NOTE: average seating is 1.5' in width										
140	0010	**BACKSTOPS** Baseball, prefabricated, 30' wide, 12' high & 1 overhang	B-1	1	24	Ea.	1,875	365		2,240	2,700	140
	0100	40' wide, 12' high & 2 overhangs	"	.75	32		1,800	485		2,285	2,800	
	0300	Basketball, steel, single goal	B-13	3.04	15.789		670	250	184	1,104	1,350	
	0400	Double goal	"	1.92	25		535	395	291	1,221	1,575	
	0600	Tennis, wire mesh with pair of ends	B-1	2.48	9.677	Set	1,125	146		1,271	1,500	
	0700	Enclosed court	"	1.30	18.462	Ea.	3,200	279		3,479	4,000	
	0900	Handball or squash court, outdoor, wood	2 Carp	.50	32		2,750	630		3,380	4,100	
	1000	Masonry handball/squash court	D-1	.30	53.333		21,100	950		22,050	24,800	
225	0010	**GOAL POSTS** Steel, football, double post	B-1	1.50	16	Pr.	1,250	242		1,492	1,825	225
	0100	Deluxe, single post		1.50	16		1,750	242		1,992	2,350	
	0300	Football, convertible to soccer		1.50	16		1,775	242		2,017	2,375	
	0500	Soccer, regulation		2	12		1,425	181		1,606	1,875	
700	0010	**PLAYGROUND EQUIPMENT** See also individual items										700
	0200	Bike rack, 10' long, permanent	B-1	12	2	Ea.	315	30		345	400	
	0400	Horizontal monkey ladder, 14' long		4	6		1,225	90.50		1,315.50	1,475	
	0590	Parallel bars, 10' long		4	6		224	90.50		314.50	400	
	0600	Posts, tether ball set, 2-3/8" O.D.		12	2		133	30		163	198	
	0800	Poles, multiple purpose, 10'-6" long		12	2	Pr.	110	30		140	173	
	1000	Ground socket for movable posts, 2-3/8" post		10	2.400		74	36.50		110.50	144	
	1100	3-1/2" post		10	2.400		78.50	36.50		115	148	
	1300	See-saw, steel, 2 units		6	4	Ea.	230	60.50		290.50	355	
	1400	4 units		4	6		800	90.50		890.50	1,025	
	1500	6 units		3	8		855	121		976	1,150	
	1700	Shelter, fiberglass golf tee, 3 person		4.60	5.217		2,150	79		2,229	2,500	
	1900	Slides, stainless steel bed, 12' long, 6' high		3	8		1,600	121		1,721	1,975	
	2000	20' long, 10' high		2	12		1,900	181		2,081	2,375	
	2200	Swings, plain seats, 8' high, 4 seats		2	12		775	181		956	1,175	
	2300	8 seats		1.30	18.462		1,375	279		1,654	1,975	
	2500	12' high, 4 seats		2	12		1,075	181		1,256	1,475	
	2600	8 seats		1.30	18.462		1,675	279		1,954	2,325	
	2800	Whirlers, 8' diameter		3	8		1,575	121		1,696	1,925	
	2900	10' diameter		3	8		2,275	121		2,396	2,700	
710	0010	**MODULAR PLAYGROUND** Basic components										710
	0100	Deck, square, steel	B-1	1	24	Ea.	560	365		925	1,225	
	0110	Polyurethane		1	24		465	365		830	1,125	
	0120	Triangular, steel		1	24		400	365		765	1,050	

Important: See the Reference Section for critical supporting data - Reference Nos., Crews, & City Cost Indexes

02800 | Site Improvements and Amenities

02880 | Playfield Equipment

			CREW	DAILY OUTPUT	LABOR-HOURS	UNIT	MAT.	LABOR	EQUIP.	TOTAL	TOTAL INCL O&P	
710	0130	Post, steel, 5" square	B-1	18	1.333	Ea.	17.20	20		37.20	53.50	710
	0140	Aluminum, 2-3/8" square		20	1.200		15.50	18.15		33.65	48	
	0160	Roof, square poly, 54" side		18	1.333		835	20		855	950	
	0170	Transfer module		3	8		1,725	121		1,846	2,100	
	0180	Guardrail, pipe		60	.400	L.F.	260	6.05		266.05	296	
	0190	Steps, deck-to-deck		8	3	Ea.	1,375	45.50		1,420.50	1,575	
	0200	Activity panel, minimum		2	12		292	181		473	630	
	0210	Maximum		2	12		585	181		766	955	
	0360	With guardrails		3	8		1,650	121		1,771	2,000	
	0370	Crawl tunnel, straight, 56" long		4	6		845	90.50		935.50	1,075	
	0380	90°		4	6		1,175	90.50		1,265.50	1,450	
	1200	Slide, tunnel		8	3		1,100	45.50		1,145.50	1,275	
	1210	Straight, poly		8	3		174	45.50		219.50	269	
	1220	Stainless steel, 54" high		6	4		181	60.50		241.50	305	
	1230	Curved, poly, 40" high		6	4		735	60.50		795.50	910	
	1240	Spyroslide, 56" - 72" high		5	4.800		2,950	72.50		3,022.50	3,375	
	1300	Ladder, vertical		5	4.800		330	72.50		402.50	485	
	1310	Horizontal, 8' long		5	4.800		580	72.50		652.50	760	
	1320	Corkscrew climber		3	8		455	121		576	705	
	1330	Fire pole		6	4		217	60.50		277.50	345	
	1340	Bridge, ring		4	6		810	90.50		900.50	1,050	
	1350	Clatter		4	6	L.F.	285	90.50		375.50	470	
880	0010	PLATFORM/PADDLE TENNIS COURT Complete with lighting, etc.										880
	0100	Aluminum slat deck with aluminum frame	B-1	.08	300	Court	24,000	4,525		28,525	34,200	
	0500	Aluminum slat deck and wood frame	C-1	.12	266		24,400	4,575		28,975	34,700	
	0800	Aluminum deck heater, add	B-1	1.18	20.339		3,150	310		3,460	4,000	
	0900	Douglas fir planking and wood frame 2" x 6" x 30'	C-1	.12	266		22,500	4,575		27,075	32,700	
	1000	Plywood deck with steel frame		.12	266		23,500	4,575		28,075	33,800	
	1100	Steel slat deck with wood frame		.12	266		24,500	4,575		29,075	34,900	

02890 | Traffic Signs & Signals

			CREW	DAILY OUTPUT	LABOR-HOURS	UNIT	MAT.	LABOR	EQUIP.	TOTAL	TOTAL INCL O&P	
700	0010	SIGNS Stock, 24" x 24", no posts, .080" alum. reflectorized	B-80	70	.343	Ea.	25	5.20	7.85	38.05	45	700
	0100	High intensity		70	.343		45	5.20	7.85	58.05	67	
	0300	30" x 30", reflectorized		70	.343		48	5.20	7.85	61.05	70.50	
	0400	High intensity		70	.343		64	5.20	7.85	77.05	88	
	0600	Guide and directional signs, 12" x 18", reflectorized		70	.343		15.30	5.20	7.85	28.35	34.50	
	0700	High intensity		70	.343		15	5.20	7.85	28.05	34	
	0900	18" x 24", stock signs, reflectorized		70	.343		24	5.20	7.85	37.05	44	
	1000	High intensity		70	.343		38	5.20	7.85	51.05	59.50	
	1200	24" x 24", stock signs, reflectorized		70	.343		25	5.20	7.85	38.05	45	
	1300	High intensity		70	.343		45	5.20	7.85	58.05	67	
	1500	Add to above for steel posts, galvanized, 10'-0" upright, bolted		200	.120		28	1.81	2.75	32.56	37	
	1600	12'-0" upright, bolted		140	.171		30	2.59	3.93	36.52	42	
	1800	Highway road signs, aluminum, over 20 S. F., reflectorized		350	.069	S.F.	12	1.04	1.57	14.61	16.70	
	2000	High intensity		350	.069		15	1.04	1.57	17.61	20	
	2200	Highway, suspended over road, 80 S.F. min., reflectorized		165	.145		15	2.20	3.34	20.54	24	
	2300	High intensity		165	.145		20	2.20	3.34	25.54	29.50	
900	0010	TRAFFIC SIGNALS Mid block pedestrian crosswalk,										900
	0020	with pushbutton and mast arms	R-2	.30	186	Total	11,700	3,700	920	16,320	20,100	
	0100	Intersection, 8 signals w/three sect. (2 each direction), programmed	"	.15	373		24,500	7,425	1,850	33,775	41,300	
	0120	For each additional traffic phase controller, add	L-9	1.20	30		1,500	550		2,050	2,575	
	0200	Semi-actuated, detectors in side street only, add		.81	44.444		2,500	810		3,310	4,125	
	0300	Fully-actuated, detectors in all streets, add		.49	73.469		4,500	1,350		5,850	7,225	
	0400	For pedestrian pushbutton, add		.70	51.429		3,500	940		4,440	5,450	
	0500	Optically programmed signal only, add per head		1.64	21.951		2,500	400		2,900	3,425	

For expanded coverage of these items see *Means Site Work and Landscape Cost Data 2000*

02800 | Site Improvements and Amenities

02890 | Traffic Signs & Signals

			CREW	DAILY OUTPUT	LABOR-HOURS	UNIT	2000 BARE COSTS MAT.	LABOR	EQUIP.	TOTAL	TOTAL INCL O&P	
900	0600	School flashing system, programmed	L-9	.41	87.805	Signal	6,000	1,600		7,600	9,325	900

02900 | Planting

02905 | Transplanting

			CREW	DAILY OUTPUT	LABOR-HOURS	UNIT	2000 BARE COSTS MAT.	LABOR	EQUIP.	TOTAL	TOTAL INCL O&P	
725	0011	**PLANTING** Moving shrubs on site, 12" ball	3 Clab	28	.857	Ea.		12.40		12.40	21	725
	0100	24" ball	B-62	22	1.091			17.45	5.30	22.75	35.50	
	0300	Moving trees on site, 36" ball	B-6	3.75	6.400			102	54	156	233	
	0400	60" ball	"	1	24			385	203	588	875	

02910 | Plant Preparation

			CREW	DAILY OUTPUT	LABOR-HOURS	UNIT	MAT.	LABOR	EQUIP.	TOTAL	INCL O&P	
500	0010	**MULCH**										500
	0100	Aged barks, 3" deep, hand spread	1 Clab	100	.080	S.Y.	2	1.16		3.16	4.18	
	0150	Skid steer loader	B-63	13.50	2.963	M.S.F.	220	43	8.60	271.60	325	
	0200	Hay, 1" deep, hand spread	1 Clab	475	.017	S.Y.	.25	.24		.49	.70	
	0250	Power mulcher, small	B-64	180	.089	M.S.F.	18.50	1.35	1.60	21.45	24.50	
	0350	Large	B-65	530	.030	"	18.50	.46	.86	19.82	22	
	0400	Humus peat, 1" deep, hand spread	1 Clab	700	.011	S.Y.	1.11	.17		1.28	1.50	
	0450	Push spreader	A-1	2,500	.003	"	1.39	.05	.03	1.47	1.64	
	0550	Tractor spreader	B-66	700	.011	M.S.F.	104	.22	.27	104.49	115	
	0600	Oat straw, 1" deep, hand spread	1 Clab	475	.017	S.Y.	.28	.24		.52	.73	
	0650	Power mulcher, small	B-64	180	.089	M.S.F.	25	1.35	1.60	27.95	31.50	
	0700	Large	B-65	530	.030	"	25	.46	.86	26.32	29	
	0750	Add for asphaltic emulsion	B-45	1,770	.009	Gal.	1.60	.14	.21	1.95	2.22	
	0800	Peat moss, 1" deep, hand spread	1 Clab	900	.009	S.Y.	1.55	.13		1.68	1.92	
	0850	Push spreader	A-1	2,500	.003	"	1.60	.05	.03	1.68	1.87	
	0950	Tractor spreader	B-66	700	.011	M.S.F.	175	.22	.27	175.49	194	
	1000	Polyethylene film, 6 mil.	2 Clab	2,000	.008	S.Y.	.15	.12		.27	.37	
	1100	Redwood nuggets, 3" deep, hand spread	1 Clab	150	.053	"	6	.77		6.77	7.90	
	1150	Skid steer loader	B-63	13.50	2.963	M.S.F.	600	43	8.60	651.60	745	
	1200	Stone mulch, hand spread, ceramic chips, economy	1 Clab	125	.064	S.Y.	5.75	.92		6.67	7.95	
	1250	Deluxe	"	95	.084	"	8.60	1.22		9.82	11.55	
	1300	Granite chips	B-1	10	2.400	C.Y.	28	36.50		64.50	93	
	1400	Marble chips		10	2.400		105	36.50		141.50	178	
	1500	Onyx gemstone		10	2.400		310	36.50		346.50	400	
	1600	Pea gravel		28	.857		61.50	12.95		74.45	89.50	
	1700	Quartz		10	2.400		135	36.50		171.50	211	
	1800	Tar paper, 15 Lb. felt	1 Clab	800	.010	S.Y.	.40	.14		.54	.69	
	1900	Wood chips, 2" deep, hand spread	"	220	.036	"	1.65	.53		2.18	2.72	
	1950	Skid steer loader	B-63	20.30	1.970	M.S.F.	108	28.50	5.75	142.25	174	

02912 | General Planting

			CREW	DAILY OUTPUT	LABOR-HOURS	UNIT	MAT.	LABOR	EQUIP.	TOTAL	INCL O&P	
275	0010	**GROUND COVER** Plants, pachysandra, in prepared beds	B-1	15	1.600	C	15	24		39	58	275
	0200	Vinca minor, 1 yr, bare root		12	2	"	19	30		49	73	
	0600	Stone chips, in 50 lb. bags, Georgia marble		520	.046	Bag	3.36	.70		4.06	4.90	
	0700	Onyx gemstone		260	.092		13.10	1.40		14.50	16.85	
	0800	Quartz		260	.092		5.50	1.40		6.90	8.45	
	0900	Pea gravel, truckload lots		28	.857	Ton	20.50	12.95		33.45	44.50	

02900 | Planting

02920 | Lawns & Grasses

		CREW	DAILY OUTPUT	LABOR-HOURS	UNIT	2000 BARE COSTS MAT.	LABOR	EQUIP.	TOTAL	TOTAL INCL O&P	
500	0010	**SEEDING** Mechanical seeding, 215 lb./acre	B-66	1.50	5.333	Acre	485	102	125	712	840	500
	0100	44 lb./M.S.Y.	"	2,500	.003	S.Y.	.15	.06	.07	.28	.35	
	0300	Fine grading and seeding incl. lime, fertilizer & seed,										
	0310	with equipment	B-14	1,000	.048	S.Y.	.16	.75	.20	1.11	1.67	
	0600	Limestone hand push spreader, 50 lbs. per M.S.F.	1 Clab	180	.044	M.S.F.	3.25	.64		3.89	4.68	
	0800	Grass seed hand push spreader, 4.5 lbs. per M.S.F.	"	180	.044	"	10	.64		10.64	12.10	
	1000	Hydro or air seeding for large areas, incl. seed and fertilizer	B-81	8,900	.002	S.Y.	.15	.03	.07	.25	.30	
	1100	With wood fiber mulch added	"	8,900	.002	"	.25	.03	.07	.35	.41	
	1300	Seed only, over 100 lbs., field seed, minimum				Lb.	1			1	1.10	
	1400	Maximum					3.50			3.50	3.85	
	1500	Lawn seed, minimum					1.50			1.50	1.65	
	1600	Maximum					4.25			4.25	4.68	
	1800	Aerial operations, seeding only, field seed	B-58	50	.480	Acre	350	7.70	75	432.70	480	
	1900	Lawn seed		50	.480		510	7.70	75	592.70	655	
	2100	Seed and liquid fertilizer, field seed		50	.480		400	7.70	75	482.70	535	
	2200	Lawn seed		50	.480		550	7.70	75	632.70	700	
600	0010	**SODDING** 1" deep, bluegrass sod, on level ground, over 8 M.S.F.	B-63	22	1.818	M.S.F.	165	26.50	5.30	196.80	233	600
	0200	4 M.S.F.		17	2.353		230	34	6.85	270.85	320	
	0300	1000 S.F.		3.50	11.429		230	165	33.50	428.50	575	
	0500	Sloped ground, over 8 M.S.F.		6	6.667		165	96.50	19.40	280.90	370	
	0600	4 M.S.F.		5	8		230	116	23.50	369.50	475	
	0700	1000 S.F.		4	10		230	145	29	404	535	
	1000	Bent grass sod, on level ground, over 6 M.S.F.		20	2		315	29	5.80	349.80	400	
	1100	3 M.S.F.		18	2.222		230	32	6.45	268.45	315	
	1200	Sodding 1000 S.F. or less		14	2.857		345	41.50	8.30	394.80	460	
	1500	Sloped ground, over 6 M.S.F.		15	2.667		315	38.50	7.75	361.25	420	
	1600	3 M.S.F.		13.50	2.963		230	43	8.60	281.60	335	
	1700	1000 S.F.		12	3.333		188	48	9.70	245.70	300	

02930 | Exterior Plants

		CREW	DAILY OUTPUT	LABOR-HOURS	UNIT	MAT.	LABOR	EQUIP.	TOTAL	TOTAL INCL O&P	
050	0010	**SHRUBS AND TREES** Evergreen, in prepared beds, B & B										050
	0100	Arborvitae pyramidal, 4'-5'	B-17	30	1.067	Ea.	35	17.10	18.95	71.05	88.50	
	0150	Globe, 12"-15"	B-1	96	.250		10.05	3.78		13.83	17.50	
	0300	Cedar, blue, 8'-10'	B-17	18	1.778		146	28.50	31.50	206	244	
	0500	Hemlock, canadian, 2-1/2'-3'	B-1	36	.667		14.25	10.10		24.35	33	
	0550	Holly, Savannah, 8' - 10' H		9.68	2.479		500	37.50		537.50	615	
	0600	Juniper, andorra, 18"-24"		80	.300		14	4.54		18.54	23	
	0620	Wiltoni, 15"-18"		80	.300		14.50	4.54		19.04	23.50	
	0640	Skyrocket, 4-1/2'-5'	B-17	55	.582		38.50	9.35	10.35	58.20	69.50	
	0660	Blue pfitzer, 2'-2-1/2'	B-1	44	.545		16	8.25		24.25	32	
	0680	Ketleerie, 2-1/2'-3'		50	.480		28	7.25		35.25	43.50	
	0700	Pine, black, 2-1/2'-3'		50	.480		29.50	7.25		36.75	45	
	0720	Mugo, 18"-24"		60	.400		30	6.05		36.05	43.50	
	0740	White, 4'-5'	B-17	75	.427		45.50	6.85	7.60	59.95	70	
	0800	Spruce, blue, 18"-24"	B-1	60	.400		28	6.05		34.05	41.50	
	0840	Norway, 4'-5'	B-17	75	.427		56	6.85	7.60	70.45	81.50	
	0900	Yew, denisforma, 12"-15"	B-1	60	.400		22.50	6.05		28.55	35.50	
	1000	Capitata, 18"-24"		30	.800		21	12.10		33.10	43.50	
	1100	Hicksi, 2'-2-1/2'		30	.800		26	12.10		38.10	49	
410	0010	**SHRUBS** Broadleaf evergreen, planted in prepared beds										410
	0100	Andromeda, 15"-18", container	B-1	96	.250	Ea.	14	3.78		17.78	22	
	0200	Azalea, 15" - 18", container		96	.250		18.50	3.78		22.28	27	
	0300	Barberry, 9"-12", container		130	.185		11	2.79		13.79	16.90	
	0400	Boxwood, 15"-18", B & B		96	.250		17.50	3.78		21.28	25.50	
	0500	Euonymus, emerald gaiety, 12" to 15", container		115	.209		12.25	3.16		15.41	18.90	

For expanded coverage of these items see *Means Site Work and Landscape Cost Data 2000*

02900 | Planting

02930 | Exterior Plants

		CREW	DAILY OUTPUT	LABOR-HOURS	UNIT	2000 BARE COSTS				TOTAL INCL O&P
						MAT.	LABOR	EQUIP.	TOTAL	
410	0600 Holly, 15"-18", B & B	B-1	96	.250	Ea.	16.75	3.78		20.53	25	410
	0900 Mount laurel, 18"-24", B & B		80	.300		75	4.54		79.54	90.50
	1000 Paxistema, 9-12" high		130	.185		15	2.79		17.79	21.50
	1100 Rhododendron, 18"-24", container		48	.500		37.50	7.55		45.05	54.50
	1200 Rosemary, 1 gal container		600	.040		35	.60		35.60	39.50
	2000 Deciduous, amelanchier, 2'-3', B & B		57	.421		55	6.35		61.35	71.50
	2100 Azalea, 15"-18", B & B		96	.250		23	3.78		26.78	32
	2300 Bayberry, 2'-3', B & B		57	.421		22.50	6.35		28.85	35.50
	2600 Cotoneaster, 15"-18", B & B		80	.300		21	4.54		25.54	31
	2800 Dogwood, 3'-4', B & B	B-17	40	.800		30	12.85	14.20	57.05	70
	2900 Euonymus, alatus compacta, 15" to 18", container	B-1	80	.300		33	4.54		37.54	44.50
	3200 Forsythia, 2'-3', container	"	60	.400		24	6.05		30.05	37
	3300 Hibiscus, 3'-4', B & B	B-17	75	.427		36	6.85	7.60	50.45	59.50
	3400 Honeysuckle, 3'-4', B & B	B-1	60	.400		25.50	6.05		31.55	38.50
	3500 Hydrangea, 2'-3', B & B	"	57	.421		36	6.35		42.35	50.50
	3600 Lilac, 3'-4', B & B	B-17	40	.800		39	12.85	14.20	66.05	80
	3900 Privet, bare root, 18"-24"	B-1	80	.300		7.50	4.54		12.04	16
	4100 Quince, 2'-3', B & B	"	57	.421		19.50	6.35		25.85	32.50
	4200 Russian olive, 3'-4', B & B	B-17	75	.427		17	6.85	7.60	31.45	38.50
	4400 Spirea, 3'-4', B & B	B-1	70	.343		27	5.20		32.20	38.50
	4500 Viburnum, 3'-4', B & B	B-17	40	.800		24	12.85	14.20	51.05	63.50
680	0010 **PLANT BED PREPARATION**										680
	0100 Backfill planting pit, by hand, on site topsoil	2 Clab	18	.889	C.Y.		12.85		12.85	22
	0200 Prepared planting mix	"	24	.667			9.65		9.65	16.50
	0300 Skid steer loader, on site topsoil	B-62	340	.071			1.13	.34	1.47	2.29
	0400 Prepared planting mix	"	410	.059			.94	.28	1.22	1.89
	1000 Excavate planting pit, by hand, sandy soil	2 Clab	16	1			14.45		14.45	25
	1100 Heavy soil or clay	"	8	2			29		29	49.50
	1200 1/2 C.Y. backhoe, sandy soil	B-11C	150	.107			1.83	1.35	3.18	4.57
	1300 Heavy soil or clay	"	115	.139			2.39	1.77	4.16	5.95
	2000 Mix planting soil, incl. loam, manure, peat, by hand	2 Clab	60	.267		23	3.85		26.85	31.50
	2100 Skid steer loader	B-62	150	.160		23	2.56	.78	26.34	30
	3000 Pile sod, skid steer loader	"	2,800	.009	S.Y.		.14	.04	.18	.28
	3100 By hand	2 Clab	400	.040			.58		.58	.99
	4000 Remove sod, F.E. loader	B-10S	2,000	.004			.08	.15	.23	.30
	4100 Sod cutter	B-12K	3,200	.002			.05	.26	.31	.37
	4200 By hand	2 Clab	240	.067			.96		.96	1.65
900	0010 **TREES** Deciduous, in prep. beds, balled & burlapped (B&B) R02930-900										900
	0100 Ash, 2" caliper	B-17	8	4	Ea.	100	64	71	235	296
	0200 Beech, 5'-6'		50	.640		200	10.25	11.35	221.60	250
	0300 Birch, 6'-8', 3 stems		20	1.600		113	25.50	28.50	167	199
	0500 Crabapple, 6'-8'		20	1.600		150	25.50	28.50	204	240
	0600 Dogwood, 4'-5'		40	.800		58	12.85	14.20	85.05	101
	0700 Eastern redbud 4'-5'		40	.800		115	12.85	14.20	142.05	164
	0800 Elm, 8'-10'		20	1.600		85	25.50	28.50	139	169
	0900 Ginkgo, 6'-7'		24	1.333		100	21.50	23.50	145	172
	1000 Hawthorn, 8'-10', 1" caliper		20	1.600		120	25.50	28.50	174	207
	1100 Honeylocust, 10'-12', 1-1/2" caliper		10	3.200		120	51.50	57	228.50	281
	1300 Larch, 8'		32	1		75	16.05	17.75	108.80	129
	1400 Linden, 8'-10', 1" caliper		20	1.600		120	25.50	28.50	174	207
	1500 Magnolia, 4'-5'		20	1.600		55	25.50	28.50	109	135
	1600 Maple, red, 8'-10', 1-1/2" caliper		10	3.200		122	51.50	57	230.50	283
	1700 Mountain ash, 8'-10', 1" caliper		16	2		150	32	35.50	217.50	258
	1800 Oak, 2-1/2"-3" caliper		3	10.667		163	171	189	523	675
	2100 Planetree, 9'-11', 1-1/4" caliper		10	3.200		90	51.50	57	198.50	248

Important: See the Reference Section for critical supporting data - Reference Nos., Crews, & City Cost Indexes

02900 | Planting

02930 | Exterior Plants

			CREW	DAILY OUTPUT	LABOR-HOURS	UNIT	2000 BARE COSTS MAT.	LABOR	EQUIP.	TOTAL	TOTAL INCL O&P	
900	2200	Plum, 6'-8', 1" caliper	B-17	20	1.600	Ea.	75	25.50	28.50	129	158	900
	2300	Poplar, 9'-11', 1-1/4" caliper R02930-900		10	3.200		51.50	51.50	57	160	206	
	2500	Sumac, 2'-3'		75	.427		24	6.85	7.60	38.45	46.50	
	2700	Tulip, 5'-6'		40	.800		29.50	12.85	14.20	56.55	69.50	
	2800	Willow, 6'-8', 1" caliper		20	1.600		63	25.50	28.50	117	144	
910	0010	**TRAVEL** To all nursery items, for 10 to 20 miles, add				All					5%	910
	0100	30 to 50 miles, add				"					10%	

02945 | Planting Accessories

			CREW	DAILY OUTPUT	LABOR-HOURS	UNIT	MAT.	LABOR	EQUIP.	TOTAL	TOTAL INCL O&P	
500	0010	**PLANTERS** Concrete, sandblasted, precast, 48" diameter, 24" high	2 Clab	15	1.067	Ea.	425	15.40		440.40	495	500
	0100	Fluted, precast, 7' diameter, 36" high		10	1.600		875	23		898	1,000	
	0300	Fiberglass, circular, 36" diameter, 24" high		15	1.067		335	15.40		350.40	395	
	0400	60" diameter, 24" high		10	1.600		650	23		673	755	
	0600	Square, 24" side, 36" high		15	1.067		435	15.40		450.40	505	
	0700	48" side, 36" high		15	1.067		805	15.40		820.40	910	
	0900	Planter/bench, 72" square, 36" high		5	3.200		2,700	46		2,746	3,050	
	1000	96" square, 27" high		5	3.200		4,400	46		4,446	4,925	
	1200	Wood, square, 48" side, 24" high		15	1.067		755	15.40		770.40	855	
	1300	Circular, 48" diameter, 30" high		10	1.600		675	23		698	785	
	1500	72" diameter, 30" high		10	1.600		1,100	23		1,123	1,250	
	1600	Planter/bench, 72"		5	3.200		2,100	46		2,146	2,400	
775	0010	**TREE GUYING** Including stakes, guy wire and wrap										775
	0100	Less than 3" caliper, 2 stakes	2 Clab	35	.457	Ea.	15	6.60		21.60	28	
	0200	3" to 4" caliper, 3 stakes	"	21	.762	"	17.60	11		28.60	38	
	1000	Including arrowhead anchor, cable, turnbuckles and wrap										
	1100	Less than 3" caliper, 3 anchors	2 Clab	20	.800	Ea.	45	11.55		56.55	69.50	
	1200	3" to 6" caliper, 4 anchors		15	1.067		65	15.40		80.40	98	
	1300	6" caliper, 6" anchors		12	1.333		80	19.25		99.25	121	
	1400	8" caliper, 8" anchors		9	1.778		14.90	25.50		40.40	60.50	

02950 | Site Restoration & Rehabilitation

02955 | Restoration of Underground Piping

			CREW	DAILY OUTPUT	LABOR-HOURS	UNIT	MAT.	LABOR	EQUIP.	TOTAL	TOTAL INCL O&P	
700	0010	**LINING PIPE** with cement, incl. bypass and cleaning										700
	0020	Less than 10,000 L.F., urban, 6" to 10"	C-17E	130	.615	L.F.	5.80	12.40	.47	18.67	28.50	
	0200	24" to 36"		90	.889		9.25	17.90	.68	27.83	41.50	
	0300	48" to 72"		80	1		14.80	20	.76	35.56	51.50	
800	0010	**CORROSION RESISTANCE** Wrap & coat, add to pipe, 4" dia.				L.F.	1.31			1.31	1.44	800
	0040	6" diameter					1.38			1.38	1.52	
	0060	8" diameter					2.13			2.13	2.34	
	0100	12" diameter					3.17			3.17	3.49	
	0200	24" diameter					6			6	6.60	
	0500	Coating, bituminous, per diameter inch, 1 coat, add					.23			.23	.25	
	0540	3 coat					.37			.37	.41	
	0560	Coal tar epoxy, per diameter inch, 1 coat, add					.15			.15	.17	
	0600	3 coat					.29			.29	.32	

For expanded coverage of these items see Means Site Work and Landscape Cost Data 2000

02950 | Site Restoration & Rehabilitation

02990 | Structure Moving

		CREW	DAILY OUTPUT	LABOR-HOURS	UNIT	2000 BARE COSTS				TOTAL INCL O&P
						MAT.	LABOR	EQUIP.	TOTAL	
0010	**MOVING BUILDINGS** One day move, up to 24' wide									
0020	Reset on new foundation, patch & hook-up, average move				Total					8,700
0040	Wood or steel frame bldg., based on ground floor area	B-4	185	.259	S.F.		3.91	2.58	6.49	9.50
0060	Masonry bldg., based on ground floor area	"	137	.350			5.30	3.49	8.79	12.85
0200	For 24' to 42' wide, add									15%
0220	For each additional day on road, add	B-4	1	48	Day		725	480	1,205	1,750
0240	Construct new basement, move building, 1 day									
0300	move, patch & hook-up, based on ground floor area	B-3	155	.310	S.F.	5.85	5.05	10.85	21.75	27

For information about Means Estimating Seminars, see yellow pages 11 and 12 in back of book

Division 3 Concrete

Estimating Tips
General
- Carefully check all the plans and specifications. Concrete often appears on drawings other than structural drawings, including mechanical and electrical drawings for equipment pads. The cost of cutting and patching is often difficult to estimate. See Subdivision 02225 for demolition costs.
- Always obtain concrete prices from suppliers near the job site. A volume discount can often be negotiated depending upon competition in the area. Remember to add for waste, particularly for slabs and footings on grade.

03100 Concrete Forms & Accessories
- A primary cost for concrete construction is forming. Most jobs today are constructed with prefabricated forms. The selection of the forms best suited for the job and the total square feet of forms required for efficient concrete forming and placing are key elements in estimating concrete construction. Enough forms must be available for erection to make efficient use of the concrete placing equipment and crew.
- Concrete accessories for forming and placing depend upon the systems used. Study the plans and specifications to assure that all special accessory requirements have been included in the cost estimate such as anchor bolts, inserts and hangers.

03200 Concrete Reinforcement
- Ascertain that the reinforcing steel supplier has included all accessories, cutting, bending and an allowance for lapping, splicing and waste. A good rule of thumb is 10% for lapping, splicing and waste. Also, 10% waste should be allowed for welded wire fabric.

03300 Cast-in-Place Concrete
- When estimating structural concrete, pay particular attention to requirements for concrete additives, curing methods and surface treatments. Special consideration for climate, hot or cold, must be included in your estimate. Be sure to include requirements for concrete placing equipment and concrete finishing.

03400 Precast Concrete
03500 Cementitious Decks & Toppings
- The cost of hauling precast concrete structural members is often an important factor. For this reason, it is important to get a quote from the nearest supplier. It may become economically feasible to set up precasting beds on the site if the hauling costs are prohibitive.

Reference Numbers
Reference numbers are shown in bold squares at the beginning of some major classifications. These numbers refer to related items in the Reference Section. The reference information may be an estimating procedure, an alternate pricing method or technical information.

Note: Not all subdivisions listed here necessarily appear in this publication.

03050 | Basic Concrete Materials & Methods

03060 | Basic Concrete Materials

		CREW	DAILY OUTPUT	LABOR-HOURS	UNIT	2000 BARE COSTS MAT.	LABOR	EQUIP.	TOTAL	TOTAL INCL O&P
100	**0010 CONCRETE ADMIXTURES & SURFACE TREATMENTS**									
	0040 Abrasives, aluminum oxide, over 20 tons				Lb.	.90			.90	.99
	0070 Under 1 ton					.95			.95	1.05
	0100 Silicon carbide, black, over 20 tons					1.25			1.25	1.38
	0120 Under 1 ton					1.30			1.30	1.43
	0200 Air entraining agent, .7 to 1.5 oz. per bag, 55 gallon lots				Gal.	8.50			8.50	9.35
	0220 5 gallon lots					8.65			8.65	9.50
	0300 Bonding agent, acrylic latex (200-250 S.F. per gallon)					26.50			26.50	29
	0320 Epoxy resin (70-80 S.F. per gallon)					50.50			50.50	55.50
	0400 Calcium chloride, 100 lb. bags, FOB plant, truckload lots				Ton	335			335	370
	0420 Less than truckload lots				Bag	21.50			21.50	23.50
	0500 Carbon black, liquid, 2 to 8 lbs. per bag of cement				Lb.	2.90			2.90	3.19
	0600 Colors, integral, 2 to 10 lb. per bag of cement, minimum					1.38			1.38	1.52
	0610 Average					2.17			2.17	2.39
	0620 Maximum					3.97			3.97	4.37
	0700 Curing compound, (200 to 400 S.F. per gallon), 55 gal. lots				Gal.	10			10	11
	0720 5 gallon lots					8.50			8.50	9.35
	0800 Premium grade, (450 S.F. per gallon), 55 gallon lots					13			13	14.30
	0820 5 gallon lots					14.50			14.50	15.95
	0900 Dustproofing compound, (200-600 S.F./gal.), 55 gallon lots					7.85			7.85	8.65
	0920 5 gallon lots					9			9	9.90
	1000 Epoxy dustproof coating, colors, (300-400 S.F. per coat),									
	1010 or transparent, (400-600 S.F. per coat)				Gal.	56.50			56.50	62
	1100 Hardeners, metallic, 55 lb. bags, natural (grey)				Lb.	.75			.75	.83
	1200 Colors, average					.97			.97	1.07
	1300 Non-metallic, 55 lb. bags, natural (grey), minimum					.32			.32	.35
	1310 Maximum					.36			.36	.40
	1320 Non-metallic, colors, minimum					.44			.44	.48
	1340 Maximum					.70			.70	.77
	1400 Non-metallic, non-slip, 100 lb. bags, minimum					.44			.44	.48
	1420 Maximum					.88			.88	.97
	1500 Solution type, (300 to 400 S.F. per gallon)				Gal.	5.90			5.90	6.50
	1550 Release agent, for tilt slabs					6.50			6.50	7.15
	1570 For forms, average					4.45			4.45	4.90
	1600 Sealer, hardener and dustproofer, clear, 450 S.F., minimum					6.85			6.85	7.50
	1620 Maximum					14			14	15.40
	1700 Colors (300-400 S.F. per gallon)					36			36	39.50
	1800 Set accelerator for below freezing, 1 to 1-1/2 gal. per C.Y.					5.25			5.25	5.80
	1900 Set retarder, 2 to 4 fl. oz. per bag of cement					9.85			9.85	10.80
	2000 Waterproofing, integral 1 lb. per bag of cement				Lb.	.75			.75	.83
	2100 Powdered metallic, 40 lbs. per 100 S.F., minimum					.95			.95	1.05
	2120 Maximum					1.95			1.95	2.15
	2200 Water reducing admixture, average				Gal.	8.80			8.80	9.70
110	**0010 AGGREGATE** Expanded shale, C.L. lots, 52 lb. per C.F., minimum R03310-020				Ton	36			36	39.50
	0050 Maximum				"	46			46	50.50
	0100 Lightweight vermiculite or perlite, 4 C.F. bag, C.L. lots				Bag	5.45			5.45	6
	0150 L.C.L. lots				"	6			6	6.60
	0250 Sand & stone, loaded at pit, crushed bank gravel				Ton	11.65			11.65	12.80
	0350 Sand, washed, for concrete					9.45			9.45	10.40
	0400 For plaster or brick					11.75			11.75	12.90
	0450 Stone, 3/4" to 1-1/2"					12.50			12.50	13.75
	0500 3/8" roofing stone & 1/2" pea stone					13.25			13.25	14.60
	0550 For trucking 10 miles, add to the above					3.75			3.75	4.13
	0600 30 miles, add to the above					8			8	8.80
	0850 Sand & stone, loaded at pit, crushed bank gravel				C.Y.	16.10			16.10	17.70

03050 | Basic Concrete Materials & Methods

03060 | Basic Concrete Materials

			CREW	DAILY OUTPUT	LABOR-HOURS	UNIT	2000 BARE COSTS MAT.	LABOR	EQUIP.	TOTAL	TOTAL INCL O&P	
110	0950	Sand, washed, for concrete [R03310-020]				C.Y.	11.90			11.90	13.10	110
	1000	For plaster or brick					16.95			16.95	18.65	
	1050	Stone, 3/4" to 1-1/2"					21.50			21.50	23.50	
	1100	3/8" roofing stone & 1/2" pea stone					35			35	38.50	
	1150	For trucking 10 miles, add to the above					5.25			5.25	5.80	
	1200	30 miles, add to the above					11.20			11.20	12.30	
	1310	Quartz chips, 50 lb. bags				Cwt.	15.25			15.25	16.80	
	1330	Silica chips, 50 lb. bags					8.35			8.35	9.20	
	1410	White marble, 3/8" to 1/2", 50 lb. bags					12.60			12.60	13.85	
	1430	3/4" F.O.B. plant				Ton	26			26	28.50	
200	0010	**CEMENT** Material only [R03310-020]										200
	0050	Masonry, gray, T.L. or C.L. lots				Bag	5.90			5.90	6.50	
	0060	L.T.L. or L.C.L. lots [R03310-060]					5.70			5.70	6.25	
	0100	Masonry, white, T.L. or C.L. lots					16.05			16.05	17.65	
	0120	L.T.L. or L.C.L. lots					16.60			16.60	18.25	
	0240	Portland, type I, plain/air entrained, TL lots, 94 lb bags					7.30			7.30	8.05	
	0300	Trucked in bulk, per cwt				Cwt.	4.09			4.09	4.50	
	0400	Type III, high early strength, TL lots, 94 lb bags				Bag	8.40			8.40	9.25	
	0420	L.T.L. or L.C.L. lots					9.05			9.05	9.95	
	0500	White high early strength, T.L. or C.L. lots, bags					17.40			17.40	19.15	
	0520	L.T.L. or L.C.L. lots					21.50			21.50	23.50	
	0600	White, T.L. or C.L. lots, bags					16.75			16.75	18.45	
	0620	L.T.L. or L.C.L. lots					20.50			20.50	22.50	
210	0010	**CRIBBING** See under Retaining Walls, division 02370-700										210
220	0010	**CUTTING** Concrete see division 02225-760										220
250	0010	**DAMPPROOFING** See division 07110										250
300	0010	**EQUIPMENT** For placing conc. see div. 01590-100 [R01590-100]										300
400	0010	**LIFT SLAB** See division 03310-240 [R03310-120]										400
700	0010	**SAWING CONCRETE** See division 02225-760										700
850	0010	**WATERPROOFING AND DAMPPROOFING** See division 07110										850
	0050	Integral waterproofing, add to cost of regular concrete				C.Y.	4.25			4.25	4.68	
870	0010	**WINTER PROTECTION** For heated ready mix, add, minimum					3.90			3.90	4.29	870
	0050	Maximum					5			5	5.50	
	0100	Protecting concrete and temporary heat, add, minimum	2 Clab	6,000	.003	S.F.	.16	.04		.20	.25	
	0150	Maximum, see also division 01510-800	"	2,000	.008	"	.66	.12		.78	.93	
	0200	Temporary shelter for slab on grade, wood frame and polyethylene										
	0201	sheeting, minimum	2 Carp	10	1.600	M.S.F.	299	31.50		330.50	385	
	0210	Maximum	"	3	5.333	"	360	105		465	575	
	0300	See also Division 03390-200										

03100 | Concrete Forms & Accessories

03110 | Structural C.I.P. Forms

			CREW	DAILY OUTPUT	LABOR-HOURS	UNIT	2000 BARE COSTS MAT.	LABOR	EQUIP.	TOTAL	TOTAL INCL O&P	
300	0010	**EXPANSION JOINT** See division 03150-250										300
405	0010	**FORMS IN PLACE, BEAMS AND GIRDERS**										405
	0020	See also Elevated Slabs, division 03310-240										
	0500	Beams & girders, exterior spandrel, plywood, 12" wide, 1 use	C-2	225	.213	SFCA	2.40	3.74		6.14	9.05	
	0550	2 use		275	.175		1.25	3.06		4.31	6.60	
	0600	3 use		295	.163		.96	2.85		3.81	5.95	
	0650	4 use		310	.155		.78	2.71		3.49	5.50	
	1000	Exterior spandrel, 18" wide, 1 use		250	.192		2.17	3.36		5.53	8.15	
	1050	2 use		275	.175		1.19	3.06		4.25	6.55	
	1100	3 use		305	.157		.87	2.76		3.63	5.70	
	1150	4 use		315	.152		.71	2.67		3.38	5.35	
	1500	Exterior spandrel, 24" wide, 1 use		265	.181		1.97	3.17		5.14	7.60	
	1550	2 use		290	.166		1.11	2.90		4.01	6.20	
	1600	3 use		315	.152		.79	2.67		3.46	5.45	
	1650	4 use		325	.148		.64	2.59		3.23	5.15	
	2000	Interior beam, 12" wide, 1 use		300	.160		2.33	2.80		5.13	7.35	
	2050	2 use		340	.141		1.32	2.47		3.79	5.70	
	2100	3 use		364	.132		.93	2.31		3.24	4.98	
	2150	4 use		377	.127		.76	2.23		2.99	4.65	
	2500	Interior beam, 24" wide, 1 use		320	.150		2.02	2.63		4.65	6.70	
	2550	2 use		365	.132		1.14	2.30		3.44	5.20	
	2600	3 use		385	.125		.80	2.18		2.98	4.62	
	2650	4 use		395	.122		.65	2.13		2.78	4.37	
	3000	Beam and Girder, encasing steel frame, hung, plywood, 1 use		325	.148		2.29	2.59		4.88	6.95	
	3050	2 use		390	.123		1.26	2.16		3.42	5.10	
	3100	3 use		415	.116		.92	2.03		2.95	4.48	
	3150	4 use		430	.112		.74	1.96		2.70	4.17	
	3500	Beam bottoms only, to 30" wide, plywood, 1 use		230	.209		3.52	3.66		7.18	10.10	
	3550	2 use		265	.181		1.97	3.17		5.14	7.60	
	3600	3 use		280	.171		1.41	3		4.41	6.70	
	3650	4 use		290	.166		1.14	2.90		4.04	6.25	
	4000	Beam sides only, vertical, 36" high, plywood, 1 use		335	.143		3.52	2.51		6.03	8.15	
	4050	2 use		405	.119		1.94	2.08		4.02	5.70	
	4100	3 use		430	.112		1.41	1.96		3.37	4.90	
	4150	4 use		445	.108		1.14	1.89		3.03	4.50	
	4500	Sloped sides, 36" high, plywood, 1 use		305	.157		3.49	2.76		6.25	8.55	
	4550	2 use		370	.130		1.95	2.27		4.22	6.05	
	4600	3 use		405	.119		1.40	2.08		3.48	5.10	
	4650	4 use		425	.113		1.14	1.98		3.12	4.64	
	5000	Upstanding beams, 36" high, plywood, 1 use		225	.213		4.10	3.74		7.84	10.90	
	5050	2 use		255	.188		2.29	3.30		5.59	8.15	
	5100	3 use		275	.175		1.66	3.06		4.72	7.10	
	5150	4 use		280	.171		1.35	3		4.35	6.65	
410	0010	**FORMS IN PLACE, COLUMNS**										410
	0500	Round fiberglass, 4 use per mo., rent, 12" diameter	C-1	160	.200	L.F.	2.10	3.43		5.53	8.20	
	0550	16" diameter		150	.213		2.25	3.66		5.91	8.75	
	0600	18" diameter		140	.229		2.50	3.92		6.42	9.45	
	0650	24" diameter		135	.237		3.20	4.07		7.27	10.45	
	0700	28" diameter		130	.246		3.50	4.22		7.72	11.10	
	0800	30" diameter		125	.256		4.10	4.39		8.49	12.05	
	0850	36" diameter		120	.267		4.70	4.58		9.28	13	
	1500	Round fiber tube, 1 use, 8" diameter		155	.206		1.72	3.54		5.26	7.95	
	1550	10" diameter		155	.206		2.35	3.54		5.89	8.65	
	1600	12" diameter		150	.213		3.15	3.66		6.81	9.70	
	1650	14" diameter		145	.221		3.81	3.79		7.60	10.70	

03100 | Concrete Forms & Accessories

03110 | Structural C.I.P. Forms

			CREW	DAILY OUTPUT	LABOR-HOURS	UNIT	MAT.	LABOR	EQUIP.	TOTAL	TOTAL INCL O&P	
410	1700	16" diameter	C-1	140	.229	L.F.	4.80	3.92		8.72	12	410
	1750	20" diameter		135	.237		7.70	4.07		11.77	15.45	
	1800	24" diameter		130	.246		9.35	4.22		13.57	17.55	
	1850	30" diameter		125	.256		13.10	4.39		17.49	22	
	1900	36" diameter		115	.278		18.05	4.77		22.82	28	
	1950	42" diameter		100	.320		38.50	5.50		44	52	
	2000	48" diameter		85	.376		53	6.45		59.45	69.50	
	2200	For seamless type, add					15%					
	3000	Round, steel, 4 use per mo., rent, regular duty, 12" diameter	C-1	145	.221	L.F.	2.75	3.79		6.54	9.55	
	3050	16" diameter		125	.256		3	4.39		7.39	10.85	
	3100	Heavy duty, 20" diameter		105	.305		3.25	5.25		8.50	12.55	
	3150	24" diameter		85	.376		3.35	6.45		9.80	14.75	
	3200	30" diameter		70	.457		3.60	7.85		11.45	17.40	
	3250	36" diameter		60	.533		4.15	9.15		13.30	20.50	
	3300	48" diameter		50	.640		6.15	11		17.15	25.50	
	3350	60" diameter		45	.711		7.50	12.20		19.70	29.50	
	4000	Column capitals, steel, 4 uses/mo., 24" col, 4' cap diameter		12	2.667	Ea.	18.25	46		64.25	98.50	
	4050	5' cap diameter		11	2.909		19.80	50		69.80	108	
	4100	6' cap diameter		10	3.200		22.50	55		77.50	119	
	4150	7' cap diameter		9	3.556		25	61		86	133	
	4500	For second and succeeding months, deduct					50%					
	5000	Plywood, 8" x 8" columns, 1 use	C-1	165	.194	SFCA	1.65	3.33		4.98	7.50	
	5050	2 use		195	.164		.95	2.82		3.77	5.85	
	5100	3 use		210	.152		.66	2.61		3.27	5.20	
	5150	4 use		215	.149		.54	2.55		3.09	4.98	
	5500	12" x 12" columns, 1 use		180	.178		1.66	3.05		4.71	7.05	
	5550	2 use		210	.152		.91	2.61		3.52	5.50	
	5600	3 use		220	.145		.66	2.50		3.16	5	
	5650	4 use		225	.142		.54	2.44		2.98	4.77	
	6000	16" x 16" columns, 1 use		185	.173		1.71	2.97		4.68	7	
	6050	2 use		215	.149		.92	2.55		3.47	5.40	
	6100	3 use		230	.139		.69	2.39		3.08	4.84	
	6150	4 use		235	.136		.56	2.34		2.90	4.61	
	6500	24" x 24" columns, 1 use		190	.168		1.84	2.89		4.73	6.95	
	6550	2 use		216	.148		1.01	2.54		3.55	5.45	
	6600	3 use		230	.139		.70	2.39		3.09	4.85	
	6650	4 use		238	.134		.60	2.31		2.91	4.61	
	7000	36" x 36" columns, 1 use		200	.160		1.80	2.75		4.55	6.70	
	7050	2 use		230	.139		1.01	2.39		3.40	5.20	
	7100	3 use		245	.131		.72	2.24		2.96	4.63	
	7150	4 use		250	.128		.59	2.20		2.79	4.40	
	7500	Steel framed plywood, 4 use per mo., rent, 8" x 8"		340	.094		3.92	1.62		5.54	7.10	
	7550	10" x 10"		350	.091		2.36	1.57		3.93	5.30	
	7600	12" x 12"		370	.086		3.05	1.48		4.53	5.90	
	7650	16" x 16"		400	.080		2.49	1.37		3.86	5.10	
	7700	20" x 20"		420	.076		2.10	1.31		3.41	4.55	
	7750	24" x 24"		440	.073		1.74	1.25		2.99	4.05	
	7755	30" x 30"		440	.073		2.35	1.25		3.60	4.72	
415	0010	FORMS IN PLACE, CULVERT 5' to 8' square or rectangular, 1 use	C-1	170	.188	SFCA	2.70	3.23		5.93	8.55	415
	0050	2 use		180	.178		1.44	3.05		4.49	6.85	
	0100	3 use		190	.168		1.12	2.89		4.01	6.20	
	0150	4 use		200	.160		.97	2.75		3.72	5.75	
420	0010	FORMS IN PLACE, ELEVATED SLABS										420
	0050	See also corrugated form deck, division 05310-300										
	1000	Flat plate plywood to 15' high, 1 use	C-2	470	.102	S.F.	2.61	1.79		4.40	5.95	
	1050	2 use		520	.092		1.44	1.62		3.06	4.35	

03100 | Concrete Forms & Accessories

03110 | Structural C.I.P. Forms

			CREW	DAILY OUTPUT	LABOR-HOURS	UNIT	MAT.	LABOR	EQUIP.	TOTAL	TOTAL INCL O&P
1100		3 use	C-2	545	.088	S.F.	1.04	1.54		2.58	3.79
1150		4 use		560	.086		.85	1.50		2.35	3.50
1500		15' to 20' high ceilings, 4 use		495	.097		1.05	1.70		2.75	4.07
1600		21' to 35' high ceilings, 4 use		450	.107		1.25	1.87		3.12	4.58
2000		Flat slab with drop panels, to 15' high, 1 use		449	.107		2.73	1.87		4.60	6.20
2050		2 use		509	.094		1.69	1.65		3.34	4.69
2100		3 use		532	.090		1.39	1.58		2.97	4.24
2150		4 use		544	.088		1.22	1.55		2.77	3.99
2250		15' to 20' high ceilings, 4 use		480	.100		1.42	1.75		3.17	4.56
2350		20' to 35' high ceilings, 4 use		435	.110		1.56	1.93		3.49	5.05
3000		Floor slab hung from steel beams, 1 use		485	.099		1.65	1.73		3.38	4.78
3050		2 use		535	.090		1.20	1.57		2.77	4.01
3100		3 use		550	.087		1.05	1.53		2.58	3.77
3150		4 use		565	.085		.97	1.49		2.46	3.62
3500		Floor slab, with 20" metal pans, 1 use		415	.116		3.60	2.03		5.63	7.45
3550		2 use		445	.108		1.80	1.89		3.69	5.20
3600		3 use		475	.101		1.20	1.77		2.97	4.35
3650		4 use		500	.096		1.25	1.68		2.93	4.26
3700		Floor slab with 30" pans, 1 use		418	.115		7.50	2.01		9.51	11.70
3720		2 use		455	.105		3.91	1.85		5.76	7.45
3740		3 use		470	.102		3.25	1.79		5.04	6.65
3760		4 use		480	.100		2.77	1.75		4.52	6.05
4000		Floor slab with 19" metal domes, 1 use		405	.119		3.60	2.08		5.68	7.50
4050		2 use		435	.110		2.40	1.93		4.33	5.95
4100		3 use		465	.103		2	1.81		3.81	5.30
4150		4 use		495	.097		1.25	1.70		2.95	4.29
4500		With 30" fiberglass domes, 1 use		405	.119		4.57	2.08		6.65	8.60
4520		2 use		450	.107		3.19	1.87		5.06	6.70
4530		3 use		460	.104		2.62	1.83		4.45	6
4550		4 use		470	.102		2.22	1.79		4.01	5.50
5000		Box out for slab openings, over 16" deep, 1 use		190	.253	SFCA	3.01	4.43		7.44	10.90
5050		2 use		240	.200	"	1.66	3.50		5.16	7.80
5500		Shallow slab box outs, to 10 S.F.		42	1.143	Ea.	10.85	20		30.85	46.50
5550		Over 10 S.F. (use perimeter)		600	.080	L.F.	1.45	1.40		2.85	3.99
6000		Bulkhead forms for slab, with keyway, 1 use, 2 piece		500	.096		1.74	1.68		3.42	4.79
6100		3 piece (see also edge forms)		460	.104		2.31	1.83		4.14	5.65
6200		Bulkhead forms for slab, w/keyway expanded metal									
6210		In lieu of 2 piece form	C-1	1,100	.029	L.F.	.69	.50		1.19	1.62
6215		In lieu of 3 piece form		960	.033		.69	.57		1.26	1.74
6220		6" high, 4 uses		1,100	.029		.84	.50		1.34	1.78
6500		Curb forms, wood, 6" to 12" high, on elevated slabs, 1 use		180	.178	SFCA	1.58	3.05		4.63	7
6550		2 use		205	.156		1.27	2.68		3.95	6
6600		3 use		220	.145		.93	2.50		3.43	5.30
6650		4 use		225	.142		.75	2.44		3.19	5
7000		Edge forms to 6" high, on elevated slab, 4 use		500	.064	L.F.	.40	1.10		1.50	2.31
7500		Depressed area forms to 12" high, 4 use		300	.107		.67	1.83		2.50	3.88
7550		12" to 24" high, 4 use		175	.183		.91	3.14		4.05	6.40
8000		Perimeter deck and rail for elevated slabs, straight		90	.356		9.50	6.10		15.60	21
8050		Curved		65	.492		13.05	8.45		21.50	29
8500		Void forms, round fiber, 3" diameter		450	.071		1.08	1.22		2.30	3.28
8550		4" diameter		425	.075		1.41	1.29		2.70	3.76
8600		6" diameter		400	.080		2.13	1.37		3.50	4.69
8650		8" diameter		375	.085		3.46	1.46		4.92	6.30
8700		10" diameter		350	.091		4.12	1.57		5.69	7.20
8750		12" diameter		300	.107		4.95	1.83		6.78	8.60
8800		Metal end closures, loose, minimum				C	30			30	33

03100 | Concrete Forms & Accessories

03110 | Structural C.I.P. Forms

			CREW	DAILY OUTPUT	LABOR-HOURS	UNIT	2000 BARE COSTS MAT.	LABOR	EQUIP.	TOTAL	TOTAL INCL O&P	
420	8850	Maximum				C	154			154	169	420
425	0010	**FORMS IN PLACE, EQUIPMENT FOUNDATIONS** 1 use	C-2	160	.300	SFCA	2.41	5.25		7.66	11.65	425
	0050	2 use		190	.253		1.33	4.43		5.76	9.05	
	0100	3 use		200	.240		.96	4.20		5.16	8.25	
	0150	4 use		205	.234		.79	4.10		4.89	7.90	
430	0010	**FORMS IN PLACE, FOOTINGS** Continuous wall, plywood, 1 use	C-1	375	.085	SFCA	2.16	1.46		3.62	4.88	430
	0050	2 use		440	.073		1.19	1.25		2.44	3.44	
	0100	3 use		470	.068		.86	1.17		2.03	2.95	
	0150	4 use		485	.066		.70	1.13		1.83	2.72	
	0500	Dowel supports for footings or beams, 1 use		500	.064	L.F.	.89	1.10		1.99	2.86	
	1000	Integral starter wall, to 4" high, 1 use		400	.080		.89	1.37		2.26	3.32	
	1500	Keyway, 4 use, tapered wood, 2" x 4"	1 Carp	530	.015		.21	.30		.51	.74	
	1550	2" x 6"		500	.016		.29	.32		.61	.86	
	2000	Tapered plastic, 2" x 3"		530	.015		.50	.30		.80	1.06	
	2050	2" x 4"		500	.016		.65	.32		.97	1.25	
	2250	For keyway hung from supports, add		150	.053		.89	1.05		1.94	2.78	
	3000	Pile cap, square or rectangular, plywood, 1 use	C-1	290	.110	SFCA	1.94	1.89		3.83	5.35	
	3050	2 use		346	.092		1.07	1.59		2.66	3.89	
	3100	3 use		371	.086		.77	1.48		2.25	3.39	
	3150	4 use		383	.084		.63	1.43		2.06	3.15	
	4000	Triangular or hexagonal caps, plywood, 1 use		225	.142		2.29	2.44		4.73	6.70	
	4050	2 use		280	.114		1.26	1.96		3.22	4.74	
	4100	3 use		305	.105		.91	1.80		2.71	4.09	
	4150	4 use		315	.102		.74	1.74		2.48	3.81	
	5000	Spread footings, plywood, 1 use		305	.105		1.69	1.80		3.49	4.94	
	5050	2 use		371	.086		.93	1.48		2.41	3.57	
	5100	3 use		401	.080		.68	1.37		2.05	3.09	
	5150	4 use		414	.077		.55	1.33		1.88	2.88	
	6000	Supports for dowels, plinths or templates, 2' x 2'		25	1.280	Ea.	3.46	22		25.46	41.50	
	6050	4' x 4' footing		22	1.455		6.90	25		31.90	50.50	
	6100	8' x 8' footing		20	1.600		13.80	27.50		41.30	62	
	6150	12' x 12' footing		17	1.882		24	32.50		56.50	82	
	7000	Plinths, 1 use		250	.128	SFCA	2.50	2.20		4.70	6.50	
	7100	4 use		270	.119	"	.82	2.03		2.85	4.38	
435	0010	**FORMS IN PLACE, GRADE BEAM** Plywood, 1 use	C-2	530	.091	SFCA	1.53	1.59		3.12	4.41	435
	0050	2 use		580	.083		.84	1.45		2.29	3.42	
	0100	3 use		600	.080		.61	1.40		2.01	3.07	
	0150	4 use		605	.079		.50	1.39		1.89	2.93	
440	0010	**FORMS IN PLACE, MAT FOUNDATION** 1 use	C-2	290	.166	SFCA	1.69	2.90		4.59	6.85	440
	0050	2 use		310	.155		.93	2.71		3.64	5.65	
	0100	3 use		330	.145		.66	2.55		3.21	5.10	
	0120	4 use		350	.137		.55	2.40		2.95	4.72	
445	0010	**FORMS IN PLACE, SLAB ON GRADE**										445
	1000	Bulkhead forms with keyway, wood, 1 use, 2 piece	C-1	510	.063	L.F.	.70	1.08		1.78	2.61	
	1050	3 piece		400	.080		.85	1.37		2.22	3.29	
	1100	4 piece		350	.091		1.18	1.57		2.75	3.99	
	1400	Bulkhead forms w/keyway, 1 piece expanded metal, left in place										
	1410	In lieu of 2 piece form	C-1	1,375	.023	L.F.	1.15	.40		1.55	1.95	
	1420	In lieu of 3 piece form		1,200	.027		1.15	.46		1.61	2.05	
	1430	In lieu of 4 piece form		1,050	.030		1.15	.52		1.67	2.17	
	2000	Curb forms, wood, 6" to 12" high, on grade, 1 use		215	.149	SFCA	1.39	2.55		3.94	5.90	
	2050	2 use		250	.128		.77	2.20		2.97	4.60	
	2100	3 use		265	.121		.55	2.07		2.62	4.16	
	2150	4 use		275	.116		.45	2		2.45	3.92	

03100 | Concrete Forms & Accessories

03110 | Structural C.I.P. Forms

			CREW	DAILY OUTPUT	LABOR-HOURS	UNIT	2000 BARE COSTS				TOTAL INCL O&P
							MAT.	LABOR	EQUIP.	TOTAL	
445	3000	Edge forms, wood, 4 use, on grade, to 6" high	C-1	600	.053	L.F.	.32	.92		1.24	1.93
	3050	7" to 12" high		435	.074	SFCA	.75	1.26		2.01	2.99
	3500	For depressed slabs, 4 use, to 12" high		300	.107	L.F.	.68	1.83		2.51	3.89
	3550	To 24" high		175	.183		.89	3.14		4.03	6.40
	4000	For slab blockouts, to 12" high, 1 use		200	.160		.70	2.75		3.45	5.45
	4050	To 24" high, 1 use		120	.267		.89	4.58		5.47	8.80
	4100	Plastic (extruded), to 6" high, multiple use, on grade		800	.040		.23	.69		.92	1.43
	5000	Screed, 24 ga. metal key joint, see Div 03150-250									
	5020	Wood, incl. wood stakes, 1" x 3"	C-1	900	.036	L.F.	.44	.61		1.05	1.53
	5050	2" x 4"		900	.036	"	1.25	.61		1.86	2.43
	6000	Trench forms in floor, wood, 1 use		160	.200	SFCA	2.10	3.43		5.53	8.20
	6050	2 use		175	.183		1.26	3.14		4.40	6.80
	6100	3 use		180	.178		.92	3.05		3.97	6.25
	6150	4 use		185	.173		.74	2.97		3.71	5.90
450	0010	**FORMS IN PLACE, STAIRS** (Slant length x width), 1 use	C-2	165	.291	S.F.	3.68	5.10		8.78	12.80
	0050	2 use		170	.282		2.02	4.95		6.97	10.75
	0100	3 use		180	.267		1.47	4.67		6.14	9.60
	0150	4 use		190	.253		1.03	4.43		5.46	8.75
	1000	Alternate pricing method (0.7 L.F./S.F.), 1 use		100	.480	LF Rsr	4.80	8.40		13.20	19.70
	1050	2 use		105	.457		2.64	8		10.64	16.65
	1100	3 use		110	.436		1.92	7.65		9.57	15.20
	1150	4 use		115	.417		1.58	7.30		8.88	14.30
	2000	Stairs, cast on sloping ground (length x width), 1 use		220	.218	S.F.	2.45	3.82		6.27	9.25
	2100	4 use		240	.200	"	.81	3.50		4.31	6.90
455	0010	**FORMS IN PLACE, WALLS**									
	0100	Box out for wall openings, to 16" thick, to 10 S.F.	C-2	24	2	Ea.	24	35		59	86.50
	0150	Over 10 S.F. (use perimeter)	"	280	.171	L.F.	2.09	3		5.09	7.45
	0250	Brick shelf, 4" w, add to wall forms, use wall area abv shelf									
	0260	1 use	C-2	240	.200	SFCA	2.29	3.50		5.79	8.50
	0300	2 use		275	.175		1.26	3.06		4.32	6.65
	0350	4 use		300	.160		.92	2.80		3.72	5.80
	0500	Bulkhead forms, with keyway, 1 use, 2 piece		265	.181	L.F.	2.50	3.17		5.67	8.20
	0550	3 piece		175	.274	"	3.10	4.81		7.91	11.65
	0600	Bulkhead forms w/keyway, 1 piece expanded metal, left in place									
	0610	In lieu of 2 piece form	C-1	800	.040	L.F.	1.15	.69		1.84	2.45
	0620	In lieu of 3 piece form	"	525	.061	"	1.15	1.05		2.20	3.06
	0700	Buttress forms, to 8' high, 1 use	C-2	350	.137	SFCA	3.04	2.40		5.44	7.45
	0750	2 use		430	.112		1.67	1.96		3.63	5.20
	0800	3 use		460	.104		1.22	1.83		3.05	4.47
	0850	4 use		480	.100		1	1.75		2.75	4.10
	1000	Corbel (haunch) forms, to 12" wide, add to wall forms, 1 use		150	.320	L.F.	1.68	5.60		7.28	11.45
	1050	2 use		170	.282		.92	4.95		5.87	9.50
	1100	3 use		175	.274		.67	4.81		5.48	9
	1150	4 use		180	.267		.55	4.67		5.22	8.60
	2000	Job built plywood wall forms, to 8' high, 1 use, below grade		300	.160	SFCA	1.90	2.80		4.70	6.90
	2050	2 use, below grade		365	.132		1.06	2.30		3.36	5.10
	2100	3 use, below grade		425	.113		.76	1.98		2.74	4.23
	2150	4 use, below grade		435	.110		.63	1.93		2.56	4
	2400	Over 8' to 16' high, 1 use		280	.171		3.68	3		6.68	9.20
	2450	2 use		345	.139		1.17	2.44		3.61	5.45
	2500	3 use		375	.128		.84	2.24		3.08	4.76
	2550	4 use		395	.122		.69	2.13		2.82	4.40
	2700	Over 16' high, 1 use		235	.204		2.39	3.58		5.97	8.80
	2750	2 use		290	.166		1.31	2.90		4.21	6.40
	2800	3 use		315	.152		.95	2.67		3.62	5.65
	2850	4 use		330	.145		.78	2.55		3.33	5.20

Important: See the Reference Section for critical supporting data - Reference Nos., Crews, & City Cost Indexes

03100 | Concrete Forms & Accessories

03110 | Structural C.I.P. Forms

		CREW	DAILY OUTPUT	LABOR-HOURS	UNIT	2000 BARE COSTS MAT.	LABOR	EQUIP.	TOTAL	TOTAL INCL O&P
3000	For architectural finish, add	C-2	1,820	.026	SFCA	.60	.46		1.06	1.45
3500	Polystyrene (expanded) wall forms									
3510	To 8' high, 1 use, left in place	1 Carp	295	.027	SFCA	1.60	.53		2.13	2.68
4000	Radial wall forms, plywood, smooth curved, 1 use	C-2	245	.196		2.19	3.43		5.62	8.30
4050	2 use		300	.160		1.21	2.80		4.01	6.15
4100	3 use		325	.148		.88	2.59		3.47	5.40
4150	4 use		335	.143		.72	2.51		3.23	5.10
4200	Wall forms, smooth curved, below grade, job built plywood, 1 use		225	.213		2.85	3.74		6.59	9.55
4210	2 use		225	.213		1.57	3.74		5.31	8.15
4220	3 use		225	.213		1.29	3.74		5.03	7.80
4230	4 use		225	.213		.92	3.74		4.66	7.40
4300	Curved, plywood, with 2' chords, 1 use		290	.166		1.90	2.90		4.80	7.05
4350	2 use		355	.135		1.04	2.37		3.41	5.20
4400	3 use		385	.125		.76	2.18		2.94	4.58
4450	4 use		400	.120		.62	2.10		2.72	4.28
4500	Over 8' high, 1 use		290	.166		.81	2.90		3.71	5.85
4525	2 use		355	.135		.44	2.37		2.81	4.55
4550	3 use		385	.125		.32	2.18		2.50	4.09
4575	4 use		400	.120		.27	2.10		2.37	3.89
4600	Retaining wall forms, plywood, battered, to 8' high, 1 use		300	.160		1.78	2.80		4.58	6.75
4650	2 use		355	.135		.98	2.37		3.35	5.15
4700	3 use		375	.128		.71	2.24		2.95	4.62
4750	4 use		390	.123		.54	2.16		2.70	4.30
4900	Over 8' to 16' high, 1 use		240	.200		1.96	3.50		5.46	8.15
4950	2 use		295	.163		1.08	2.85		3.93	6.05
5000	3 use		305	.157		.78	2.76		3.54	5.60
5050	4 use		320	.150		.64	2.63		3.27	5.20
5500	For gang wall forming, 192 S.F. sections, deduct					10%	10%			
5550	384 S.F. sections, deduct					20%	20%			
5750	Liners for forms (add to wall forms), A.B.S. plastic									
5800	Aged wood, 4" wide, 1 use	1 Carp	250	.032	SFCA	4.85	.63		5.48	6.45
5820	2 use		400	.020		2.70	.39		3.09	3.65
5840	4 use		750	.011		1.65	.21		1.86	2.18
5900	Fractured rope rib, 1 use		250	.032		7.50	.63		8.13	9.35
6000	4 use		750	.011		2.45	.21		2.66	3.06
6100	Ribbed look, 1/2" & 3/4" deep, 1 use		300	.027		5.15	.53		5.68	6.55
6200	4 use		800	.010		1.65	.20		1.85	2.16
6300	Rustic brick pattern, 1 use		250	.032		4.80	.63		5.43	6.40
6400	4 use		750	.011		1.60	.21		1.81	2.12
6500	Striated, random, 3/8" x 3/8" deep, 1 use		300	.027		5.10	.53		5.63	6.50
6600	4 use		800	.010		1.65	.20		1.85	2.16
6800	Rustication strips, A.B.S. plastic, 2 piece snap-on									
6850	1" deep x 1-3/8" wide, 1 use	C-2	400	.120	L.F.	3.75	2.10		5.85	7.75
6900	2 use		600	.080		2.10	1.40		3.50	4.71
6950	4 use		800	.060		1.25	1.05		2.30	3.18
7050	Wood, beveled edge, 3/4" deep, 1 use		600	.080		.22	1.40		1.62	2.64
7100	1" deep, 1 use		450	.107		.35	1.87		2.22	3.59
7200	For solid board finish, uniform, 1 use, add to wall forms		300	.160	SFCA	.74	2.80		3.54	5.60
7300	Non-uniform finish		250	.192		.65	3.36		4.01	6.45
7500	Lintel or sill forms, 1 use	1 Carp	30	.267		2.43	5.25		7.68	11.65
7520	2 use		34	.235		1.34	4.64		5.98	9.40
7540	3 use		36	.222		.97	4.38		5.35	8.55
7560	4 use		37	.216		.79	4.26		5.05	8.15
7800	Modular prefabricated plywood, to 8' high, 1 use per month	C-2	1,180	.041		1.13	.71		1.84	2.46
7820	2 use per month		1,200	.040		.62	.70		1.32	1.88
7840	3 use per month		1,240	.039		.45	.68		1.13	1.66

03100 | Concrete Forms & Accessories

03110 | Structural C.I.P. Forms

			CREW	DAILY OUTPUT	LABOR-HOURS	UNIT	2000 BARE COSTS MAT.	LABOR	EQUIP.	TOTAL	TOTAL INCL O&P	
455	7860	4 use per month	C-2	1,260	.038	SFCA	.37	.67		1.04	1.55	455
	8000	To 16' high, 1 use per month		715	.067		1.38	1.18		2.56	3.54	
	8020	2 use per month		740	.065		.76	1.14		1.90	2.78	
	8040	3 use per month		770	.062		.55	1.09		1.64	2.48	
	8060	4 use per month		790	.061		.46	1.06		1.52	2.32	
	8100	Over 16' high, 1 use per month		715	.067		1.66	1.18		2.84	3.84	
	8120	2 use per month		740	.065		.91	1.14		2.05	2.95	
	8140	3 use per month		770	.062		.66	1.09		1.75	2.60	
	8160	4 use per month		790	.061		.55	1.06		1.61	2.43	
	8600	Pilasters, 1 use		270	.178		2.30	3.11		5.41	7.90	
	8620	2 use		330	.145		1.27	2.55		3.82	5.75	
	8640	3 use		370	.130		.92	2.27		3.19	4.91	
	8660	4 use		385	.125		.75	2.18		2.93	4.56	
	9000	Steel framed plywood, to 8' high, 1 use per month		600	.080		1.80	1.40		3.20	4.38	
	9020	2 use per month		640	.075		.99	1.31		2.30	3.34	
	9040	3 use per month		655	.073		.72	1.28		2	2.99	
	9060	4 use per month		665	.072		.59	1.26		1.85	2.82	
	9200	Over 8' to 16' high, 1 use per month		455	.105		2	1.85		3.85	5.35	
	9220	2 use per month		505	.095		1.10	1.67		2.77	4.06	
	9240	3 use per month		525	.091		.80	1.60		2.40	3.63	
	9260	4 use per month		530	.091		.66	1.59		2.25	3.45	
	9400	Over 16' to 20' high, 1 use per month		425	.113		1.68	1.98		3.66	5.25	
	9420	2 use per month		435	.110		.92	1.93		2.85	4.33	
	9440	3 use per month		455	.105		.67	1.85		2.52	3.91	
	9460	4 use per month		465	.103		.55	1.81		2.36	3.71	
	9475	For elevated walls, add						10%				
	9480	For battered walls, 1 side battered, add					10%	10%				
	9485	For battered walls, 2 sides battered, add					15%	15%				
500	0010	**GAS STATION FORMS** Curb fascia, with template,										500
	0050	12 ga. steel, left in place, 9" high	1 Carp	50	.160	L.F.	6.95	3.15		10.10	13.05	
	1000	Sign or light bases, 18" diameter, 9" high		9	.889	Ea.	39	17.50		56.50	73	
	1050	30" diameter, 13" high		8	1		66	19.70		85.70	107	
	2000	Island forms, 10' long, 9" high, 3'-6" wide	C-1	10	3.200		180	55		235	292	
	2050	4' wide		9	3.556		195	61		256	320	
	2500	20' long, 9" high, 4' wide		6	5.333		315	91.50		406.50	500	
	2550	5' wide		5	6.400		330	110		440	555	
750	0010	**REGLET** See division 07710-750										750
800	0010	**SCAFFOLDING** See division 01540-750										800
820	0010	**SLIPFORMS** Silos, minimum	C-17E	3,885	.021	SFCA	1.05	.41	.02	1.48	1.88	820
	0050	Maximum		1,095	.073		1.50	1.47	.06	3.03	4.24	
	1000	Buildings, minimum		3,660	.022		1.25	.44	.02	1.71	2.16	
	1050	Maximum		875	.091		2.20	1.84	.07	4.11	5.65	

03150 | Concrete Accessories

			CREW	DAILY OUTPUT	LABOR-HOURS	UNIT	MAT.	LABOR	EQUIP.	TOTAL	INCL O&P	
080	0010	**ACCESSORIES, ANCHOR BOLTS** J-type, incl. nut and washer										080
	0020	1/2" diameter, 6" long	1 Carp	90	.089	Ea.	1.11	1.75		2.86	4.22	
	0050	10" long		85	.094		1.22	1.85		3.07	4.52	
	0100	12" long		85	.094		1.27	1.85		3.12	4.58	
	0200	5/8" diameter, 12" long		80	.100		1.40	1.97		3.37	4.92	
	0250	18" long		70	.114		1.65	2.25		3.90	5.70	
	0300	24" long		60	.133		1.90	2.63		4.53	6.60	
	0350	3/4" diameter, 8" long		80	.100		1.71	1.97		3.68	5.25	
	0400	12" long		70	.114		2.11	2.25		4.36	6.20	
	0450	18" long		60	.133		2.58	2.63		5.21	7.35	

03100 | Concrete Forms & Accessories

03150 | Concrete Accessories

		Crew	Daily Output	Labor-Hours	Unit	Mat.	Labor	Equip.	Total	Total Incl O&P
080										
0500	24" long	1 Carp	50	.160	Ea.	3.43	3.15		6.58	9.15
0600	7/8" diameter, 12" long		60	.133		2.32	2.63		4.95	7.05
0650	18" long		50	.160		2.97	3.15		6.12	8.65
0700	24" long		40	.200		3.10	3.94		7.04	10.15
0800	1" diameter, 12" long		55	.145		3.20	2.87		6.07	8.45
0850	18" long		45	.178		3.61	3.50		7.11	9.95
0900	24" long		35	.229		4.25	4.50		8.75	12.40
0950	36" long		25	.320		5.60	6.30		11.90	16.95
1200	1-1/2" diameter, 18" long		22	.364		8.95	7.15		16.10	22
1250	24" long		18	.444		10.15	8.75		18.90	26
1300	36" long		12	.667		11.95	13.15		25.10	35.50
1350	Larger sizes		200	.040	Lb.	.88	.79		1.67	2.32
8000	Sleeves, see Division 03150-620									
160	**0010 ACCESSORIES, CHAMFER STRIPS**									
2000	Polyvinyl chloride, 1/2" wide with leg	1 Carp	535	.015	L.F.	6.55	.29		6.84	7.75
2200	3/4" wide with leg		525	.015		10.30	.30		10.60	11.80
2400	1" radius with leg		515	.016		.45	.31		.76	1.02
2800	1-1/2" radius with leg		500	.016		1.73	.32		2.05	2.44
5000	Wood, 1/2" wide		535	.015		.09	.29		.38	.60
5200	3/4" wide		525	.015		.24	.30		.54	.77
5400	1" wide		515	.016		.31	.31		.62	.86
170	**0010 ACCESSORIES, COLUMN FORM**									
1000	Column clamps, adjustable to 24" x 24", buy				Set	73			73	80.50
1100	Rent per month					7.75			7.75	8.55
1300	For sizes to 30" x 30", buy					84			84	92.50
1400	Rent per month					6.50			6.50	7.15
1600	For sizes to 36" x 36", buy					103			103	113
1700	Rent per month					10.25			10.25	11.30
2000	Bull winch (band iron) 36" x 36", buy					46.50			46.50	51
2100	Rent per month					4.25			4.25	4.68
2300	48" x 48", buy					48.50			48.50	53.50
2400	Rent per month					5.25			5.25	5.80
3000	Chain & wedge type 36" x 36", buy					67.50			67.50	74.50
3100	Rent per month					7.75			7.75	8.55
3300	60" x 60", buy					89			89	98
3400	Rent per month					10.25			10.25	11.30
4000	Friction collars 2'-6" dia., buy					675			675	745
4100	Rent per month					65			65	71.50
4300	4'-0" dia., buy					760			760	835
4400	Rent per month					83			83	91.50
200	**0010 ACCESSORIES, DOVETAIL ANCHOR SYSTEM**									
0500	Anchor slot, galv., filled, 24 ga.	1 Carp	425	.019	L.F.	.56	.37		.93	1.26
0600	20 ga.		400	.020		.73	.39		1.12	1.48
0800	16 oz. copper, foam filled		375	.021		1.63	.42		2.05	2.51
0900	26 ga. stainless steel, foam filled		375	.021		1.10	.42		1.52	1.93
1200	Brick anchor, corr., galv., 3-1/2" long, 16 ga.	1 Bric	10.50	.762	C	26.50	15.25		41.75	55
1300	12 ga.		10.50	.762		34.50	15.25		49.75	64
1500	Flat, galv., 3-1/2" long, 16 ga.		10.50	.762		23.50	15.25		38.75	52
1600	12 ga.		10.50	.762		45.50	15.25		60.75	76
2000	Cavity wall, corr., galv., 5" long, 16 ga.		10.50	.762		25	15.25		40.25	53.50
2100	12 ga.		10.50	.762		36	15.25		51.25	65.50
3000	Furring anchors, corr., galv., 1-1/2" long, 16 ga.		10.50	.762		9.50	15.25		24.75	36.50
3100	12 ga.		10.50	.762		16	15.25		31.25	43.50
6000	Stone anchors, 3-1/2" long, galv., 1/8" x 1" wide		10.50	.762		80.50	15.25		95.75	115

03100 | Concrete Forms & Accessories

03150 | Concrete Accessories

			CREW	DAILY OUTPUT	LABOR-HOURS	UNIT	2000 BARE COSTS MAT.	LABOR	EQUIP.	TOTAL	TOTAL INCL O&P	
200	6100	1/4" x 1" wide	1 Bric	10.50	.762	C	125	15.25		140.25	164	200
250	0010	**EXPANSION JOINT** Keyed, cold, 24 ga, incl. stakes, 3-1/2" high	1 Carp	200	.040	L.F.	.50	.79		1.29	1.90	250
	0050	4-1/2" high		200	.040		.62	.79		1.41	2.03	
	0100	5-1/2" high		195	.041		.69	.81		1.50	2.14	
	0150	7-1/2" high		190	.042		.84	.83		1.67	2.34	
	0300	Poured asphalt, plain, 1/2" x 1"	1 Clab	450	.018		.34	.26		.60	.81	
	0350	1" x 2"		400	.020		1.19	.29		1.48	1.81	
	0500	Neoprene, liquid, cold applied, 1/2" x 1"		450	.018		1.57	.26		1.83	2.17	
	0550	1" x 2"		400	.020		6.10	.29		6.39	7.20	
	0700	Polyurethane, poured, 2 part, 1/2" x 1"		400	.020		1.88	.29		2.17	2.57	
	0750	1" x 2"		350	.023		7.35	.33		7.68	8.65	
	0900	Rubberized asphalt, hot or cold applied, 1/2" x 1"		450	.018		.63	.26		.89	1.13	
	0950	1" x 2"		400	.020		1.21	.29		1.50	1.83	
	1100	Hot applied, fuel resistant, 1/2" x 1"		450	.018		1.05	.26		1.31	1.59	
	1150	1" x 2"		400	.020		1.46	.29		1.75	2.11	
	2000	Premolded, bituminous fiber, 1/2" x 6"	1 Carp	375	.021		.37	.42		.79	1.13	
	2050	1" x 12"		300	.027		1.58	.53		2.11	2.64	
	2250	Cork with resin binder, 1/2" x 6"		375	.021		1.59	.42		2.01	2.47	
	2300	1" x 12"		300	.027		4.87	.53		5.40	6.25	
	2500	Neoprene sponge, closed cell, 1/2" x 6"		375	.021		1.26	.42		1.68	2.11	
	2550	1" x 12"		300	.027		5.65	.53		6.18	7.15	
	2750	Polyethylene foam, 1/2" x 6"		375	.021		.37	.42		.79	1.13	
	2800	1" x 12"		300	.027		1.78	.53		2.31	2.86	
	3000	Polyethylene backer rod, 3/8" diameter		460	.017		.03	.34		.37	.62	
	3050	3/4" diameter		460	.017		.06	.34		.40	.66	
	3100	1" diameter		460	.017		.11	.34		.45	.71	
	3500	Polyurethane foam, with polybutylene, 1/2" x 1/2"		475	.017		.48	.33		.81	1.10	
	3550	1" x 1"		450	.018		1.47	.35		1.82	2.22	
	3750	Polyurethane foam, regular, closed cell, 1/2" x 6"		375	.021		1.11	.42		1.53	1.94	
	3800	1" x 12"		300	.027		1.78	.53		2.31	2.86	
	4000	Polyvinyl chloride foam, closed cell, 1/2" x 6"		375	.021		1.68	.42		2.10	2.57	
	4050	1" x 12"		300	.027		5.20	.53		5.73	6.65	
	4250	Rubber, gray sponge, 1/2" x 6"		375	.021		2.34	.42		2.76	3.29	
	4300	1" x 12"		300	.027		9.15	.53		9.68	11	
	4500	Lead wool for joints, 1 ton lots				Lb.	2.18			2.18	2.40	
	4550	Retail				"	2.33			2.33	2.56	
	5000	For installation in walls, add						75%				
	5250	For installation in boxouts, add						25%				
350	0010	**ACCESSORIES, HANGERS**										350
	0020	Slab and beam form										
	0500	Banding iron, 3/4" x 22 ga, 14 L.F. per lb or										
	0550	1/2" x 14 ga, 7 L.F. per lb.				Lb.	.92			.92	1.01	
	1000	Fascia ties, coil type add to frame ties below				C	56			56	61.50	
	1500	Frame ties to 8-1/8"					145			145	160	
	1550	8-1/8" to 10-1/8"					154			154	169	
	5000	Snap tie hanger, to 30" overall length, 4000 #					320			320	350	
	5050	30" to 36" overall length					345			345	380	
	5100	42" to 48" overall length					385			385	425	
	5500	Steel beam hanger										
	5600	Flange to 8-1/8"				C	280			280	310	
	5650	8-1/8" to 10-1/8"					259			259	285	
	6000	Tie hangers to 24" overall length, 6000 #					320			320	355	
	6100	30" to 36" overall length					375			375	415	
	6500	Tie back hanger, up to 12-1/8" flange					305			305	335	

03100 | Concrete Forms & Accessories

03150 | Concrete Accessories

			CREW	DAILY OUTPUT	LABOR-HOURS	UNIT	MAT.	LABOR	EQUIP.	TOTAL	TOTAL INCL O&P	
350	8500	Wire, black annealed, 9 ga				Cwt.	84			84	92.50	350
	8600	16 ga				"	89			89	98	
400	0010	**ACCESSORIES, INSERTS**										400
	1000	All size nut insert, 5/8" & 3/4", incl. nut	1 Carp	84	.095	Ea.	3	1.88		4.88	6.50	
	2000	Continuous slotted, 1-5/8" x 1-3/8"										
	2100	3" long, 12 ga.	1 Carp	65	.123	Ea.	2.75	2.42		5.17	7.20	
	2150	6" long, 12 ga.		65	.123		3.60	2.42		6.02	8.10	
	2200	12" long, 8 ga.		65	.123		8.80	2.42		11.22	13.85	
	2250	24" long, 8 ga.		65	.123		14	2.42		16.42	19.55	
	2300	36" long, 8 ga.		60	.133		19.50	2.63		22.13	26	
	2350	60" long, 8 ga.		55	.145		33	2.87		35.87	41.50	
	7000	Threaded cast										
	7100	1/4" diameter bolt	1 Carp	84	.095	Ea.	4.50	1.88		6.38	8.15	
	7350	7/8" diameter bolt	"	84	.095	"	7.50	1.88		9.38	11.45	
	9000	Wedge										
	9050	For 5/8" diameter bolt	1 Carp	60	.133	Ea.	3.25	2.63		5.88	8.10	
	9100	For 3/4" diameter bolt	"	60	.133	"	7.10	2.63		9.73	12.30	
	9800	Cut washers, "Black"										
	9850	5/8" bolt				Ea.	.20			.20	.22	
	9950	For galvanized inserts, add				C	30%					
600	0010	**SHORES** Erect and strip, by hand, horizontal members										600
	0501	Aluminum joists and stringers	L-2	60	.267	Ea.		4.60		4.60	7.90	
	0602	Steel, adjustable beams		45	.356			6.15		6.15	10.50	
	0701	Wood joists		50	.320			5.50		5.50	9.45	
	0801	Wood stringers		30	.533			9.20		9.20	15.75	
	1001	Vertical members to 10' high		55	.291			5		5	8.60	
	1051	To 13' high		50	.320			5.50		5.50	9.45	
	1101	To 16' high		45	.356			6.15		6.15	10.50	
	1501	Reshoring		1,400	.011	S.F.	.18	.20		.38	.54	
	1600	Flying truss system	C-17D	9,600	.009	SFCA		.18	.06	.24	.36	
	1760	Horizontal, aluminum joists, 6' to 30' spans, buy				L.F.	10.85			10.85	11.95	
	1770	Aluminum stringers, 12' & 16' spans				"	16.25			16.25	17.90	
	1810	Horizontal, steel beam, adjustable, 4' to 7' span				Ea.	112			112	123	
	1830	6' to 10' span					143			143	157	
	1920	9' to 15' span					265			265	292	
	1940	12' to 20' span					305			305	335	
	1970	Steel stringer, 6' to 15' span				L.F.	8			8	8.80	
	3000	Rent for job duration, aluminum, first month				SF Flr.	.25			.25	.28	
	3050	Steel				"	.20			.20	.22	
	3500	Vertical, adjustable steel, 5'-7" to 9'-6" high, 10,000# cap., buy				Ea.	59			59	65	
	3550	7'-3" to 12'-10" high, 7800# capacity					71			71	78	
	3600	8'-10" to 12'-4" high, 10,000# capacity					79			79	87	
	3650	8'-10" to 16'-1" high, 3800# capacity					88			88	97	
	4000	Frame shoring systems, aluminum, 10,000# per leg,										
	4050	6' wide, 5' & 6' high				Ea.	240			240	264	
	4100	5' to 7' post with base, jack screw & top plate				Set	53			53	58.50	
	5010	Steel, 10,000# per leg										
	5040	2' & 4' wide, 3', 4', 5' & 6' high				Ea.	97			97	107	
	5250	6' extension tube with adjusting collar					91			91	100	
	5550	Base plate					9.75			9.75	10.75	
	5600	12" adjustable leg					39			39	43	
	5650	Top plate					21			21	23	
	5750	Flying truss system				SFCA	9			9	9.90	
620	0010	**ACCESSORIES, SLEEVES AND CHASES**										620
	0100	Plastic, 1 use, 9" long, 2" diameter	1 Carp	100	.080	Ea.	.55	1.58		2.13	3.31	

03100 | Concrete Forms & Accessories

03150 | Concrete Accessories

			CREW	DAILY OUTPUT	LABOR-HOURS	UNIT	2000 BARE COSTS				TOTAL INCL O&P	
							MAT.	LABOR	EQUIP.	TOTAL		
620	0150	4" diameter	1 Carp	90	.089	Ea.	1.56	1.75		3.31	4.72	620
	0200	6" diameter		75	.107		2.75	2.10		4.85	6.65	
	0250	12" diameter	▼	60	.133	▼	9.70	2.63		12.33	15.20	
640	0010	**ACCESSORIES, SNAP TIES, FLAT WASHER**										640
	0100	3000 lb., to 8"				C	70			70	77	
	0200	11" & 12"					84			84	92.50	
	0250	16"					90			90	99	
	0300	18"					88.50			88.50	97.50	
	0500	With plastic cone, to 8"					62.50			62.50	68.50	
	0600	11" & 12"					85.50			85.50	94	
	0650	16"					89.50			89.50	98.50	
	0700	18"					93			93	102	
	1000	5000 lb., to 8"					99.50			99.50	110	
	1150	11" & 12"					113			113	124	
	1200	16"					125			125	138	
	1250	18"					122			122	135	
	1500	With plastic cone, to 8"					107			107	118	
	1600	11" & 12"					141			141	155	
	1650	16"					155			155	171	
	1700	18"				▼	161			161	177	
660	0010	**STAIR TREAD INSERTS** Cast iron, abrasive, 3" wide	1 Carp	90	.089	L.F.	5.75	1.75		7.50	9.35	660
	0020	4" wide		80	.100		7	1.97		8.97	11.10	
	0040	6" wide		75	.107		8.65	2.10		10.75	13.10	
	0050	9" wide		70	.114		12.75	2.25		15	17.90	
	0100	12" wide	▼	65	.123		19.15	2.42		21.57	25	
	0300	Cast aluminum, compared to cast iron, deduct					10%					
	0500	Extruded aluminum safety tread, 3" wide	1 Carp	75	.107		4.85	2.10		6.95	8.95	
	0550	4" wide		75	.107		6.55	2.10		8.65	10.80	
	0600	6" wide		75	.107		10.95	2.10		13.05	15.65	
	0650	9" wide to resurface stairs	▼	70	.114	▼	15.45	2.25		17.70	21	
	1700	Cement filled pan type, plain	1 Cefi	115	.070	S.F.	2.25	1.31		3.56	4.62	
	1750	Non-slip	"	100	.080	"	3.38	1.51		4.89	6.15	
850	0010	**ACCESSORIES, WALL AND FOUNDATION**										850
	2000	Footings, form braces, solid steel, adjustable				Ea.	12			12	13.20	
	2050	Spreaders for footer, adjustable				"	3.94			3.94	4.33	
	3000	Form oil, coverage varies greatly, minimum				Gal.	4.75			4.75	5.20	
	3050	Maximum				"	6.75			6.75	7.45	
	3500	Form patches, 1-3/4" diameter				C	8.80			8.80	9.70	
	3550	2-3/4" diameter				"	16			16	17.60	
	4000	Nail stakes, 3/4" diameter, 18" long				Ea.	1.17			1.17	1.29	
	4050	24" long					2.75			2.75	3.03	
	4200	30" long					2.85			2.85	3.14	
	4250	36" long				▼	3.10			3.10	3.41	
860	0010	**WATERSTOP** PVC, ribbed 3/16" thick, 4" wide	1 Carp	155	.052	L.F.	.56	1.02		1.58	2.36	860
	0050	6" wide		145	.055		1.21	1.09		2.30	3.19	
	0500	Ribbed, PVC, with center bulb, 9" wide, 3/16" thick		135	.059		1.70	1.17		2.87	3.87	
	0550	3/8" thick		130	.062		1.94	1.21		3.15	4.21	
	0800	Dumbbell type, PVC, 6" wide, 3/16" thick		150	.053		1	1.05		2.05	2.90	
	0850	3/8" thick		145	.055		1.40	1.09		2.49	3.40	
	1000	9" wide, 3/8" thick, PVC, plain		130	.062		2.73	1.21		3.94	5.10	
	1050	Center bulb		130	.062		5.70	1.21		6.91	8.40	
	1250	Split PVC, 3/8" thick, 6" wide		145	.055		3.85	1.09		4.94	6.10	
	1300	9" wide	▼	130	.062	▼	4.36	1.21		5.57	6.90	

03100 | Concrete Forms & Accessories

03150	Concrete Accessories	CREW	DAILY OUTPUT	LABOR-HOURS	UNIT	2000 BARE COSTS MAT.	LABOR	EQUIP.	TOTAL	TOTAL INCL O&P	
2000	Rubber, flat dumbbell, 3/8" thick, 6" wide	1 Carp	145	.055	L.F.	4.79	1.09		5.88	7.10	860
2050	9" wide		135	.059		7.90	1.17		9.07	10.70	
2500	Flat dumbbell split, 3/8" thick, 6" wide		145	.055		6.90	1.09		7.99	9.45	
2550	9" wide		135	.059		10.20	1.17		11.37	13.20	
3000	Center bulb, 1/4" thick, 6" wide		145	.055		3.64	1.09		4.73	5.85	
3050	9" wide		135	.059		7.10	1.17		8.27	9.80	
3500	Center bulb split, 3/8" thick, 6" wide		145	.055		4.33	1.09		5.42	6.60	
3550	9" wide		135	.059		7.50	1.17		8.67	10.25	
5000	Waterstop fittings, rubber, flat										
5010	Dumbbell or center bulb, 3/8" thick,										
5200	Field union, 6" wide	1 Carp	50	.160	Ea.	9.55	3.15		12.70	15.90	
5250	9" wide		50	.160		13.75	3.15		16.90	20.50	
5500	Flat cross, 6" wide		30	.267		35	5.25		40.25	47.50	
5550	9" wide		30	.267		58.50	5.25		63.75	73.50	
6000	Flat tee, 6" wide		30	.267		34.50	5.25		39.75	47	
6050	9" wide		30	.267		49	5.25		54.25	63	
6500	Flat ell, 6" wide		40	.200		30.50	3.94		34.44	40.50	
6550	9" wide		40	.200		46	3.94		49.94	57.50	
7000	Vertical tee, 6" wide		25	.320		28.50	6.30		34.80	42.50	
7050	9" wide		25	.320		43	6.30		49.30	58	
7500	Vertical ell, 6" wide		35	.229		26	4.50		30.50	36	
7550	9" wide		35	.229		43	4.50		47.50	54.50	

03200 | Concrete Reinforcement

03210	Reinforcing Steel	CREW	DAILY OUTPUT	LABOR-HOURS	UNIT	2000 BARE COSTS MAT.	LABOR	EQUIP.	TOTAL	TOTAL INCL O&P	
0010	**ACCESSORIES** Materials only										100
0020	See also Form Accessories, division 03150										
0100	Beam bolsters, (BB) standard, lower, up to 1-1/2" high, plain				C.L.F.	35			35	38.50	
0102	Galvanized					54			54	59.50	
0104	Stainless					102			102	112	
0106	Plastic					61			61	67	
0108	Epoxy					84			84	92.50	
0110	2-1/2" to 3" high, plain					46			46	50.50	
0120	Galvanized					61			61	67	
0140	Stainless					140			140	154	
0160	Plastic					64			64	70.50	
0162	Epoxy					89			89	98	
0200	Upper, standard (BBU) to 1-1/2" high, plain					101			101	111	
0210	2-1/2" to 3" high					182			182	200	
0300	Beam bolster with plate (BBP) to 1-1/2" high, plain					136			136	150	
0310	2-1/2" to 3" high					263			263	289	
0500	Slab bolsters, continuous, plain (SB) 3/4" to 1" high, plain					30			30	33	
0502	Galvanized					35			35	38.50	
0504	Stainless					57			57	62.50	
0506	Plastic					42			42	46	
0510	1" to 2" high, plain					36			36	39.50	
0515	Galvanized					42			42	46	
0520	Stainless					80			80	88	
0525	Plastic					47			47	51.50	

03200 | Concrete Reinforcement

03210 | Reinforcing Steel

			CREW	DAILY OUTPUT	LABOR-HOURS	UNIT	2000 BARE COSTS MAT.	LABOR	EQUIP.	TOTAL	TOTAL INCL O&P	
100	0530	For bolsters with wire runners (SBR), add				C.L.F.	41			41	45	100
	0540	For bolsters with plates (SBP), add				↓	105			105	116	
	0700	Clip or bar ties, 16 ga., plain, 3" long				C	8.75			8.75	9.65	
	0710	4" long					9.25			9.25	10.20	
	0720	6" long					10.75			10.75	11.80	
	0730	8" long				↓	11.25			11.25	12.40	
	0900	Flange clips, expandable flanges, 10 ga., 12" O.C., continuous,										
	0910	galvanized, over 500 L.F., 4" to 8"				C.L.F.	32			32	35	
	0920	9" to 12"					44			44	48.50	
	0930	17" to 24"				↓	45			45	49.50	
	1200	High chairs, individual, no plates (1 HC), to 3" high, plain				C	43.50			43.50	48	
	1202	Galvanized					60			60	66	
	1204	Stainless					124			124	136	
	1206	Plastic					58			58	64	
	1210	5" high, plain					66			66	72.50	
	1212	Galvanized					83			83	91.50	
	1214	Stainless					179			179	197	
	1216	Plastic					83			83	91.50	
	1220	8" high, plain					143			143	157	
	1222	Galvanized					183			183	201	
	1224	Stainless					300			300	330	
	1226	Plastic					180			180	198	
	1230	12" high, plain					298			298	330	
	1232	Galvanized					365			365	400	
	1234	Stainless					550			550	605	
	1236	Plastic					335			335	370	
	1240	15" high, plain					560			560	615	
	1242	Galvanized					620			620	680	
	1244	Stainless					915			915	1,000	
	1246	Plastic					605			605	665	
	1250	For each added 1" up to 24" high, plain, add					30			30	33	
	1252	Galvanized, add					36			36	39.50	
	1254	Stainless, add					43			43	47.50	
	1256	Plastic, add					35			35	38.50	
	1400	Individual high chairs, with plates, (HCP), to 5" high, add					159			159	175	
	1410	Over 5" high, add					208			208	229	
	1500	Bar chair (BC) for up to 1-3/4" high, plain					24			24	26.50	
	1520	Galvanized					27			27	29.50	
	1530	Stainless					72			72	79	
	1540	Plastic					47			47	51.50	
	1550	Joist chair (JC), joists up to 6", plain					30			30	33	
	1580	Galvanized					36			36	39.50	
	1600	Stainless					51			51	56	
	1620	Plastic					51			51	56	
	1630	Epoxy				↓	119			119	131	
	1700	Continuous high chairs, legs 8" O.C. (CHC) to 4" high, plain				C.L.F.	58			58	64	
	1705	Galvanized					67			67	73.50	
	1710	Stainless					143			143	157	
	1715	Plastic					71			71	78	
	1718	Epoxy					103			103	113	
	1720	6" high, plain					83			83	91.50	
	1725	Galvanized					113			113	124	
	1730	Stainless					167			167	184	
	1735	Plastic					113			113	124	
	1738	Epoxy					138			138	152	
	1740	8" high, plain				↓	121			121	133	

Important: See the Reference Section for critical supporting data - Reference Nos., Crews, & City Cost Indexes

03200 | Concrete Reinforcement

03210 | Reinforcing Steel

		CREW	DAILY OUTPUT	LABOR-HOURS	UNIT	2000 BARE COSTS				TOTAL INCL O&P
						MAT.	LABOR	EQUIP.	TOTAL	
1745	Galvanized				C.L.F.	159			159	175
1750	Stainless					198			198	218
1755	Plastic					158			158	174
1758	Epoxy					208			208	229
1760	12" high, plain					295			295	325
1765	Galvanized					340			340	375
1770	Stainless					475			475	520
1775	Plastic					335			335	370
1778	Epoxy					490			490	540
1780	15" high, plain					330			330	365
1785	Galvanized					380			380	420
1790	Stainless					380			380	415
1795	Plastic					380			380	415
1798	Epoxy					650			650	715
1800	For each added 1" up to 24" high, plain, add					27			27	29.50
1820	Galvanized, add					33			33	36.50
1840	Stainless, add					29			29	32
1860	Plastic, add					29			29	32
1900	For continuous bottom plate, (CHCP), add					129			129	142
1940	For upper continuous high chairs, (CHCU), add					129			129	142
1960	For galvanized wire runners, add					125			125	138
2100	Paper tubing, 4' lengths, for #2 & #3 bar					48			48	53
2120	For #6 bar					66			66	72.50
2200	Screed base, 1/2" diameter, 2-1/2" high, plain				C	135			135	149
2210	Galvanized					140			140	154
2220	5-1/2" high, plain					162			162	178
2250	Galvanized					170			170	187
2300	3/4" diameter, 2-1/2" high, plain					168			168	185
2310	Galvanized					177			177	195
2320	5-1/2" high, plain					208			208	229
2350	Galvanized					222			222	244
2400	Screed holder, 1/2" diam. for 1" I.D. pipe, plain, 6" long					140			140	154
2420	12" long					232			232	255
2500	3/4" diameter, for 1-1/2" I.D. pipe, 6" long					252			252	277
2520	12" long					415			415	455
2700	Screw anchor for bolts, plain, 1/2" diameter					92			92	101
2720	1" diameter					273			273	300
2740	1-1/2" diameter					455			455	500
2800	Screw eye bolts, 1/2" x 5" long					105			105	116
2820	1" x 9" long					415			415	455
2840	1-1/2" x 14" long					1,000			1,000	1,100
2900	Screw anchor bolts, 1/2" x up to 7" long					420			420	460
2920	1" x up to 12" long					1,350			1,350	1,475
3000	Slab lifting inserts, single, 3/4" dia., galv., 4" high					282			282	310
3010	6" high					345			345	380
3030	7" high					395			395	435
3100	1" diameter, 5" high					445			445	490
3120	7" high					470			470	515
3200	Double lifting inserts, 1" diameter, 5" high					880			880	970
3220	7" high					935			935	1,025
3330	1-1/4" diameter, 5" high					960			960	1,050
3500	Sleeper clips for wood sleepers, 20 ga., galv., 2" wide				M	335			335	370
3520	4" wide					415			415	455
3600	Spacers, plastic for 1" bar clearance, average					50			50	55
3620	For 2" bar clearance, average					60			60	66
3800	Subgrade chairs, 1/2" diameter, 3-1/2" high				C	275			275	305

03200 | Concrete Reinforcement

03210 | Reinforcing Steel

			CREW	DAILY OUTPUT	LABOR-HOURS	UNIT	2000 BARE COSTS MAT.	LABOR	EQUIP.	TOTAL	TOTAL INCL O&P	
100	3850	12" high				C	775			775	855	100
	3900	3/4" diameter, 3-1/2" high					355			355	390	
	3950	12" high					845			845	930	
	4200	Subgrade stakes, 3/4" diameter, 12" long					280			280	310	
	4250	24" long					380			380	420	
	4300	1" diameter, 12" long					425			425	470	
	4350	24" long					630			630	695	
	4500	Tie wire, 16 ga. annealed steel, under 500 lbs.				Cwt.	82			82	90	
	4520	2,000 to 4,000 lbs.				"	77			77	84.50	
	4550	Tie wire holder, plastic case				Ea.	32			32	35	
	4600	Aluminum case				"	38			38	42	
200	0010	**COATED REINFORCING** Add to material										200
	0100	Epoxy coated, A775				Cwt.	24			24	26.50	
	0150	Galvanized, #3					32			32	35	
	0200	#4					32			32	35	
	0250	#5					31.50			31.50	34.50	
	0300	#6 or over					31.50			31.50	34.50	
	1000	For over 20 tons, #6 or larger, minimum					29			29	32	
	1500	Maximum					35			35	38.50	
600	0010	**REINFORCING IN PLACE** A615 Grade 60										600
	0100	Beams & Girders, #3 to #7	4 Rodm	1.60	20	Ton	560	420		980	1,400	
	0150	#8 to #18		2.70	11.852		560	250		810	1,075	
	0200	Columns, #3 to #7		1.50	21.333		560	450		1,010	1,450	
	0250	#8 to #18		2.30	13.913		560	294		854	1,150	
	0300	Spirals, hot rolled, 8" to 15" diameter		2.20	14.545		955	305		1,260	1,625	
	0320	15" to 24" diameter		2.20	14.545		930	305		1,235	1,600	
	0330	24" to 36" diameter		2.30	13.913		910	294		1,204	1,550	
	0340	36" to 48" diameter		2.40	13.333		890	281		1,171	1,500	
	0360	48" to 64" diameter		2.50	12.800		975	270		1,245	1,575	
	0380	64" to 84" diameter		2.60	12.308		1,025	260		1,285	1,600	
	0390	84" to 96" diameter		2.70	11.852		1,025	250		1,275	1,600	
	0400	Elevated slabs, #4 to #7		2.90	11.034		590	233		823	1,075	
	0500	Footings, #4 to #7		2.10	15.238		530	320		850	1,175	
	0550	#8 to #18		3.60	8.889		500	188		688	900	
	0600	Slab on grade, #3 to #7		2.30	13.913		530	294		824	1,125	
	0700	Walls, #3 to #7		3	10.667		530	225		755	1,000	
	0750	#8 to #18		4	8		530	169		699	900	
	1000	Typical in place, 10 ton lots, average		1.70	18.824		560	395		955	1,350	
	1100	Over 50 ton lots, average		2.30	13.913		535	294		829	1,125	
	1200	High strength steel, Grade 75, #14 bars only, add					60			60	66	
	2000	Unloading & sorting, add to above	C-5	100	.480			9.20	5.60	14.80	22.50	
	2200	Crane cost for handling, add to above, minimum		135	.356			6.80	4.14	10.94	16.70	
	2210	Average		92	.522			10	6.10	16.10	24.50	
	2220	Maximum		35	1.371			26.50	16	42.50	64.50	
	2400	Dowels, 2 feet long, deformed, #3	2 Rodm	520	.031	Ea.	.23	.65		.88	1.46	
	2410	#4		480	.033		.41	.70		1.11	1.76	
	2420	#5		435	.037		.65	.78		1.43	2.15	
	2430	#6		360	.044		.92	.94		1.86	2.76	
	2450	Longer and heavier dowels		725	.022	Lb.	.28	.47		.75	1.18	
	2500	Smooth dowels, 12" long, 1/4" or 3/8" diameter		140	.114	Ea.	.43	2.41		2.84	4.96	
	2520	5/8" diameter		125	.128		.75	2.70		3.45	5.85	
	2530	3/4" diameter		110	.145		.93	3.07		4	6.70	
	2700	Dowel caps, 5" long, 1/2" to 3/4" diameter		800	.020		.21	.42		.63	1.02	
	2720	1-1/4" diameter		750	.021		.26	.45		.71	1.13	

Important: See the Reference Section for critical supporting data - Reference Nos., Crews, & City Cost Indexes

03200 | Concrete Reinforcement

03210 | Reinforcing Steel

			CREW	DAILY OUTPUT	LABOR-HOURS	UNIT	2000 BARE COSTS MAT.	LABOR	EQUIP.	TOTAL	TOTAL INCL O&P	
700	0010	**SPLICING REINFORCING BARS** Incl. holding bars in										700
	0020	place while splicing	R03210-070									
	0100	Butt weld columns #4 bars		C-5	190	.253	Ea.	1.05	4.84	2.94	8.83	13.05	
	0110	#6 bars			150	.320		1.50	6.10	3.73	11.33	16.70	
	0130	#10 bars			95	.505		2.05	9.65	5.90	17.60	26	
	0150	#14 bars			65	.738		2.55	14.15	8.60	25.30	38	
	0280	Column splice clamps, sleeve & wedge, or end bearing											
	0300	#7 or #8 bars		C-5	190	.253	Ea.	3.30	4.84	2.94	11.08	15.50	
	0310	#9 or #10 bars			170	.282		3.45	5.40	3.29	12.14	17.05	
	0320	#11 bars			160	.300		4.45	5.75	3.49	13.69	19	
	0330	#14 bars			150	.320		5.50	6.10	3.73	15.33	21	
	0340	#18 bars			140	.343		8.30	6.55	3.99	18.84	25.50	
	0500	Reducer inserts for above, #14 to #18 bar						3.20			3.20	3.52	
	0520	#14 to #11 bar						2.85			2.85	3.14	
	0550	#10 to #9 bar						.85			.85	.94	
	0560	#9 to #8 bar						.90			.90	.99	
	0580	#8 to #7 bar						.85			.85	.94	
	0600	For bolted speed sleeve type, deduct							15%				
	0800	Mechanical butt splice, sleeve type with filler metal, compression											
	0810	only, all grades, columns only #11 bars		C-5	68	.706	Ea.	13.50	13.50	8.20	35.20	48	
	0900	#14 bars			62	.774		14	14.80	9	37.80	52	
	0920	#18 bars			62	.774		17	14.80	9	40.80	55	
	1000	125% yield point, grade 60, columns only, #6 bars			68	.706		18.50	13.50	8.20	40.20	53.50	
	1020	#7 or #8 bars			68	.706		16	13.50	8.20	37.70	50.50	
	1030	#9 bars			68	.706		16	13.50	8.20	37.70	50.50	
	1040	#10 bars			68	.706		17	13.50	8.20	38.70	52	
	1050	#11 bars			68	.706		21.50	13.50	8.20	43.20	56.50	
	1060	#14 bars			62	.774		27	14.80	9	50.80	66	
	1070	#18 bars			62	.774		39.50	14.80	9	63.30	80	
	1200	Full tension, grade 60 steel, columns,											
	1220	slabs or beams, #6, #7, #8 bars		C-5	68	.706	Ea.	15.75	13.50	8.20	37.45	50.50	
	1230	#9 bars			68	.706		17.25	13.50	8.20	38.95	52	
	1240	#10 bars			68	.706		19	13.50	8.20	40.70	54	
	1250	#11 bars			68	.706		22.50	13.50	8.20	44.20	58	
	1260	#14 bars			62	.774		30.50	14.80	9	54.30	70	
	1270	#18 bars			62	.774		51	14.80	9	74.80	92.50	
	1400	If equipment handling not required, deduct							50%				
	1600	Mechanical threaded type, bar threading not included,											
	1700	Straight bars, #10 & #11		C-5	140	.343	Ea.	16.75	6.55	3.99	27.29	34.50	
	1750	#14 bars			130	.369		19	7.05	4.30	30.35	38.50	
	1800	#18 bars			75	.640		30	12.25	7.45	49.70	63	
	2100	#11 to #18 & #14 to #18 transition			75	.640		32	12.25	7.45	51.70	65	
	2400	Bent bars, #10 & #11			105	.457		28	8.75	5.35	42.10	52.50	
	2500	#14			90	.533		37	10.20	6.20	53.40	65.50	
	2600	#18			70	.686		54	13.10	8	75.10	92	
	2800	#11 to #14 transition			75	.640		39	12.25	7.45	58.70	73	
	2900	#11 to #18 & #14 to #18 transition			70	.686		54	13.10	8	75.10	92	

03220 | Welded Wire Fabric

			CREW	DAILY OUTPUT	LABOR-HOURS	UNIT	MAT.	LABOR	EQUIP.	TOTAL	TOTAL INCL O&P	
200	0010	**WELDED WIRE FABRIC** ASTM A185										200
	0050	Sheets										
	0100	6 x 6 - W1.4 x W1.4 (10 x 10) 21 lb. per C.S.F.	2 Rodm	35	.457	C.S.F.	6.90	9.65		16.55	25.50	
	0200	6 x 6 - W2.1 x W2.1 (8 x 8) 30 lb. per C.S.F.		31	.516		10.55	10.90		21.45	32	
	0300	6 x 6 - W2.9 x W2.9 (6 x 6) 42 lb. per C.S.F.		29	.552		15.45	11.65		27.10	38.50	
	0400	6 x 6 - W4 x W4 (4 x 4) 58 lb. per C.S.F.		27	.593		21.50	12.50		34	47	

03200 | Concrete Reinforcement

03220 | Welded Wire Fabric

		CREW	DAILY OUTPUT	LABOR-HOURS	UNIT	MAT.	LABOR	EQUIP.	TOTAL	TOTAL INCL O&P
0500	4 x 4 - W1.4 x W1.4 (10 x 10) 31 lb. per C.S.F.	2 Rodm	31	.516	C.S.F.	11.30	10.90		22.20	33
0600	4 x 4 - W2.1 x W2.1 (8 x 8) 44 lb. per C.S.F.		29	.552		13.65	11.65		25.30	36.50
0650	4 x 4 - W2.9 x W2.9 (6 x 6) 61 lb. per C.S.F.		27	.593		19.70	12.50		32.20	45
0700	4 x 4 - W4 x W4 (4 x 4) 85 lb. per C.S.F.		25	.640		27	13.50		40.50	55
0750	Rolls									
0800	2 x 2 - #14 galv. @ 21 lb., beam & column wrap	2 Rodm	6.50	2.462	C.S.F.	14.75	52		66.75	113
0900	2 x 2 - #12 galv. for gunite reinforcing	"	6.50	2.462	"	20.50	52		72.50	119
0950	Material prices for above include 10% lap									
1000	Specially fabricated heavier gauges in sheets	4 Rodm	50	.640	C.S.F.		13.50		13.50	25
1010	Material only, minimum				Ton	520			520	570
1020	Average					725			725	800
1030	Maximum					930			930	1,025

03230 | Stressing Tendons

		CREW	DAILY OUTPUT	LABOR-HOURS	UNIT	MAT.	LABOR	EQUIP.	TOTAL	TOTAL INCL O&P
0010	**PRESTRESSING STEEL** Post-tensioned in field R03410-090									
0100	Grouted strand, 50' span, 100 kip	C-3	1,200	.053	Lb.	2.29	.99	.17	3.45	4.49
0150	300 kip		2,700	.024		2.04	.44	.08	2.56	3.11
0300	100' span, 100 kip		1,700	.038		2.29	.70	.12	3.11	3.91
0350	300 kip		3,200	.020		2.23	.37	.06	2.66	3.19
0500	200' span, 100 kip		2,700	.024		2.29	.44	.08	2.81	3.38
0550	300 kip		3,500	.018		2.22	.34	.06	2.62	3.12
0800	Grouted bars, 50' span, 42 kip		2,600	.025		.37	.46	.08	.91	1.32
0850	143 kip		3,200	.020		.37	.37	.06	.80	1.14
1000	75' span, 42 kip		3,200	.020		.38	.37	.06	.81	1.15
1050	143 kip		4,200	.015		.35	.28	.05	.68	.94
1200	Ungrouted strand, 50' span, 100 kip	C-4	1,275	.025		1.92	.50	.06	2.48	3.10
1250	300 kip		1,475	.022		2.10	.43	.05	2.58	3.16
1400	100' span, 100 kip		1,500	.021		1.95	.43	.05	2.43	2.99
1450	300 kip		1,650	.019		2.10	.39	.05	2.54	3.07
1600	200' span, 100 kip		1,500	.021		1.95	.43	.05	2.43	2.99
1650	300 kip		1,700	.019		2.10	.38	.05	2.53	3.05
1800	Ungrouted bars, 50' span, 42 kip		1,400	.023		.28	.46	.06	.80	1.21
1850	143 kip		1,700	.019		.28	.38	.05	.71	1.05
2000	75' span, 42 kip		1,800	.018		.28	.35	.04	.67	1.01
2050	143 kip		2,200	.015		.28	.29	.04	.61	.88
2220	Ungrouted single strand, 100' slab, 25 kip		1,200	.027		2.09	.53	.07	2.69	3.34
2250	35 kip		1,475	.022		1.95	.43	.05	2.43	3

03240 | Fibrous Reinforcing

		CREW	DAILY OUTPUT	LABOR-HOURS	UNIT	MAT.	LABOR	EQUIP.	TOTAL	TOTAL INCL O&P
0010	**FIBROUS REINFORCING**									
0100	Synthetic fibers				Lb.	3.79			3.79	4.17
0110	1-1/2 lb. per C.Y., add to concrete				C.Y.	5.85			5.85	6.45
0150	Steel fibers				Lb.	.48			.48	.53
0155	25 lb. per C.Y., add to concrete				C.Y.	12			12	13.20
0160	50 lb. per C.Y., add to concrete					24			24	26.50
0170	75 lb. per C.Y., add to concrete					37			37	40.50
0180	100 lb. per C.Y., add to concrete					48			48	53

03300 | Cast-In-Place Concrete

03310 | Structural Concrete

			CREW	DAILY OUTPUT	LABOR-HOURS	UNIT	MAT.	LABOR	EQUIP.	TOTAL	TOTAL INCL O&P	
200	0010	**CONCRETE, FIELD MIX** FOB forms 2250 psi	R03310-080			C.Y.	64			64	70.50	200
	0020	3000 psi				"	67			67	73.50	
220	0010	**CONCRETE, READY MIX** Regular weight	R03310-050									220
	0020	2000 psi				C.Y.	60.50			60.50	66.50	
	0100	2500 psi	R03310-060				61			61	67	
	0150	3000 psi					63			63	69	
	0200	3500 psi	R03310-070				64.50			64.50	70.50	
	0300	4000 psi					67			67	73.50	
	0350	4500 psi					69			69	76	
	0400	5000 psi					71.50			71.50	78.50	
	0411	6000 psi					81.50			81.50	90	
	0412	8000 psi					133			133	146	
	0413	10,000 psi					189			189	208	
	0414	12,000 psi					228			228	251	
	1000	For high early strength cement, add					10%					
	1010	For structural lightweight with regular sand, add					25%					
	2000	For all lightweight aggregate, add					45%					
	3000	For integral colors, 2500 psi, 5 bag mix										
	3100	Red, yellow or brown, 1.8 lb. per bag, add				C.Y.	13.45			13.45	14.80	
	3200	9.4 lb. per bag, add					72.50			72.50	80	
	3400	Black, 1.8 lb. per bag, add					16			16	17.60	
	3500	7.5 lb. per bag, add					67			67	73.50	
	3700	Green, 1.8 lb. per bag, add					32			32	35	
	3800	7.5 lb. per bag, add					153			153	169	
240	0010	**CONCRETE IN PLACE** Including forms (4 uses), reinforcing	R03310-010									240
	0050	steel, including finishing unless otherwise indicated										
	0300	Beams, 5 kip per L.F., 10' span	R03310-100	C-14A	15.62	12.804	C.Y.	213	250	44.50	507.50	720
	0350	25' span		"	18.55	10.782		193	211	37.50	441.50	620
	0500	Chimney foundations, industrial, minimum	R03350-130	C-14C	32.22	3.476		130	64.50	1.18	195.68	254
	0510	Maximum		"	23.71	4.724		151	87.50	1.61	240.11	320
	0700	Columns, square, 12" x 12", minimum reinforcing	R04210-055	C-14A	11.96	16.722		229	325	58	612	880
	0720	Average reinforcing			10.13	19.743		320	385	68.50	773.50	1,100
	0740	Maximum reinforcing			9.03	22.148		400	435	77	912	1,275
	0800	16" x 16", minimum reinforcing			16.22	12.330		185	241	43	469	670
	0820	Average reinforcing			12.57	15.911		299	310	55	664	930
	0840	Maximum reinforcing			10.25	19.512		410	380	67.50	857.50	1,200
	0900	24" x 24", minimum reinforcing			23.66	8.453		155	165	29.50	349.50	490
	0920	Average reinforcing			17.71	11.293		246	221	39	506	700
	0940	Maximum reinforcing			14.15	14.134		335	276	49	660	905
	1000	36" x 36", minimum reinforcing			33.69	5.936		142	116	20.50	278.50	380
	1020	Average reinforcing			23.32	8.576		228	168	30	426	575
	1040	Maximum reinforcing			17.82	11.223		315	220	39	574	770
	1200	16" diameter, minimum reinforcing			31.49	6.351		203	124	22	349	465
	1220	Average reinforcing			19.12	10.460		330	205	36.50	571.50	760
	1240	Maximum reinforcing			13.77	14.524		460	284	50.50	794.50	1,050
	1300	20" diameter, minimum reinforcing			41.04	4.873		200	95.50	16.90	312.40	405
	1320	Average reinforcing			24.05	8.316		310	163	29	502	660
	1340	Maximum reinforcing			17.01	11.758		425	230	41	696	910
	1400	24" diameter, minimum reinforcing			51.85	3.857		188	75.50	13.40	276.90	355
	1420	Average reinforcing			27.06	7.391		305	145	25.50	475.50	615
	1440	Maximum reinforcing			18.29	10.935		415	214	38	667	870
	1500	36" diameter, minimum reinforcing			75.04	2.665		176	52	9.25	237.25	294
	1520	Average reinforcing			37.49	5.335		262	104	18.50	384.50	490
	1540	Maximum reinforcing			22.84	8.757		375	171	30.50	576.50	740
	1900	Elevated slabs, flat slab, 125 psf Sup. Load, 20' span		C-14B	38.45	5.410		143	106	18.05	267.05	360
	1950	30' span			50.99	4.079		131	79.50	13.60	224.10	297

03300 | Cast-In-Place Concrete

03310 | Structural Concrete

			CREW	DAILY OUTPUT	LABOR-HOURS	UNIT	2000 BARE COSTS MAT.	LABOR	EQUIP.	TOTAL	TOTAL INCL O&P
240	2100	Flat plate, 125 psf Sup. Load, 15' span	C-14B	30.24	6.878	C.Y.	152	134	23	309	425
	2150	25' span		49.60	4.194		125	82	14	221	294
	2300	Waffle const., 30" domes, 125 psf Sup. Load, 20' span		37.07	5.611		188	110	18.75	316.75	415
	2350	30' span		44.07	4.720		170	92	15.75	277.75	365
	2500	One way joists, 30" pans, 125 psf Sup. Load, 15' span		27.38	7.597		219	148	25.50	392.50	525
	2550	25' span		31.15	6.677		205	130	22.50	357.50	475
	2700	One way beam & slab, 125 psf Sup. Load, 15' span		20.59	10.102		173	197	33.50	403.50	570
	2750	25' span		28.36	7.334		158	143	24.50	325.50	450
	2900	Two way beam & slab, 125 psf Sup. Load, 15' span		24.04	8.652		163	169	29	361	505
	2950	25' span		35.87	5.799		137	113	19.35	269.35	370
	3100	Elevated slabs including finish, not									
	3110	including forms or reinforcing									
	3150	Regular concrete, 4" slab	C-8	2,613	.021	S.F.	.84	.36	.25	1.45	1.80
	3200	6" slab		2,585	.022		1.30	.36	.25	1.91	2.32
	3250	2-1/2" thick floor fill		2,685	.021		.57	.35	.24	1.16	1.49
	3300	Lightweight, 110# per C.F., 2-1/2" thick floor fill		2,585	.022		.67	.36	.25	1.28	1.63
	3400	Cellular concrete, 1-5/8" fill, under 5000 S.F.		2,000	.028		.46	.47	.33	1.26	1.66
	3450	Over 10,000 S.F.		2,200	.025		.36	.43	.30	1.09	1.45
	3500	Add per floor for 3 to 6 stories high		31,800	.002			.03	.02	.05	.07
	3520	For 7 to 20 stories high		21,200	.003			.04	.03	.07	.10
	3800	Footings, spread under 1 C.Y.	C-14C	38.07	2.942	C.Y.	92.50	54.50	1	148	197
	3850	Over 5 C.Y.		81.04	1.382		85	25.50	.47	110.97	138
	3900	Footings, strip, 18" x 9", plain		41.04	2.729		84	50.50	.93	135.43	181
	3950	36" x 12", reinforced		61.55	1.820		85.50	33.50	.62	119.62	153
	4000	Foundation mat, under 10 C.Y.		38.67	2.896		116	53.50	.98	170.48	222
	4050	Over 20 C.Y.		56.40	1.986		103	36.50	.68	140.18	177
	4200	Grade walls, 8" thick, 8' high	C-14D	45.83	4.364		105	85	15.15	205.15	278
	4250	14' high		27.26	7.337		131	143	25.50	299.50	415
	4260	12" thick, 8' high		64.32	3.109		95	60.50	10.80	166.30	220
	4270	14' high		40.01	4.999		105	97	17.35	219.35	300
	4300	15" thick, 8' high		80.02	2.499		90	48.50	8.65	147.15	192
	4350	12' high		51.26	3.902		93.50	76	13.55	183.05	249
	4500	18' high		48.85	4.094		104	79.50	14.20	197.70	267
	4520	Handicap access ramp, railing both sides, 3' wide	C-14H	14.58	3.292	L.F.	97	63.50	2.60	163.10	219
	4525	5' wide		12.22	3.928		111	75.50	3.10	189.60	256
	4530	With cheek walls and rails both sides, 3' wide		8.55	5.614		99.50	108	4.44	211.94	300
	4535	5' wide		7.31	6.566		100	126	5.20	231.20	335
	4650	Slab on grade, not including finish, 4" thick	C-14E	60.75	1.449	C.Y.	77	27.50	.62	105.12	134
	4700	6" thick	"	92	.957	"	74	18.05	.41	92.46	114
	4751	Slab on grade, incl. troweled finish, not incl. forms									
	4760	or reinforcing, over 10,000 S.F., 4" thick slab	C-14F	3,425	.021	S.F.	.84	.37	.01	1.22	1.54
	4820	6" thick slab		3,350	.021		1.22	.38	.01	1.61	1.98
	4840	8" thick slab		3,184	.023		1.67	.40	.01	2.08	2.51
	4900	12" thick slab		2,734	.026		2.51	.46	.01	2.98	3.55
	4950	15" thick slab		2,505	.029		3.15	.51	.02	3.68	4.32
	5000	Slab on grade, incl. textured finish, not incl. forms									
	5001	or reinforcing, 4" thick slab	C-14G	2,873	.019	S.F.	.82	.34	.01	1.17	1.47
	5010	6" thick		2,590	.022		1.28	.37	.01	1.66	2.04
	5020	8" thick		2,320	.024		1.66	.42	.02	2.10	2.54
	5200	Lift slab in place above the foundation, incl. forms,									
	5210	reinforcing, concrete and columns, minimum	C-14B	2,113	.098	S.F.	4.75	1.92	.33	7	8.90
	5250	Average		1,650	.126		5.40	2.46	.42	8.28	10.60
	5300	Maximum		1,500	.139		6.15	2.71	.46	9.32	11.95
	5500	Lightweight, ready mix, including screed finish only,									
	5510	not including forms or reinforcing									
	5550	1:4 for structural roof decks	C-14B	260	.800	C.Y.	94	15.60	2.67	112.27	133

Reference numbers in column: R03310-010, R03310-100, R03350-130, R04210-055

03300 | Cast-In-Place Concrete

03310 | Structural Concrete

			CREW	DAILY OUTPUT	LABOR-HOURS	UNIT	MAT.	2000 BARE COSTS LABOR	EQUIP.	TOTAL	TOTAL INCL O&P	
240	5600	1:6 for ground slab with radiant heat	C-14F	92	.783	C.Y.	89	13.80	.41	103.21	121	240
	5650	1:3:2 with sand aggregate, roof deck	C-14B	260	.800		94	15.60	2.67	112.27	133	
	5700	Ground slab	C-14F	107	.673		94	11.85	.36	106.21	123	
	5900	Pile caps, incl. forms and reinf., sq. or rect., under 5 C.Y.	C-14C	54.14	2.069		78	38.50	.70	117.20	153	
	5950	Over 10 C.Y.		75	1.493		80.50	27.50	.51	108.51	138	
	6000	Triangular or hexagonal, under 5 C.Y.		53	2.113		76	39	.72	115.72	152	
	6050	Over 10 C.Y.		85	1.318		83.50	24.50	.45	108.45	134	
	6200	Retaining walls, gravity, 4' high see division 02370-700	C-14D	66.20	3.021		84	59	10.50	153.50	205	
	6250	10' high		125	1.600		74	31	5.55	110.55	141	
	6300	Cantilever, level backfill loading, 8' high		70	2.857		93	55.50	9.90	158.40	208	
	6350	16' high		91	2.198		89	43	7.65	139.65	180	
	6800	Stairs, not including safety treads, free standing, 3'-6" wide	C-14H	83	.578	LF Nose	5.90	11.15	.46	17.51	26.50	
	6850	Cast on ground		125	.384	"	4.10	7.40	.30	11.80	17.60	
	7000	Stair landings, free standing		200	.240	S.F.	2.30	4.62	.19	7.11	10.75	
	7050	Cast on ground		475	.101	"	1.35	1.95	.08	3.38	4.94	
450	0010	**INSULATING CONCRETE** See division 03310 and 03500										450
700	0010	**PLACING CONCRETE** and vibrating, including labor & equipment										700
	0050	Beams, elevated, small beams, pumped	C-20	60	1.067	C.Y.		17	12.20	29.20	42.50	
	0100	With crane and bucket	C-7	45	1.600			25.50	20.50	46	66	
	0200	Large beams, pumped	C-20	90	.711			11.35	8.15	19.50	28	
	0250	With crane and bucket	C-7	65	1.108			17.80	14	31.80	45.50	
	0400	Columns, square or round, 12" thick, pumped	C-20	60	1.067			17	12.20	29.20	42.50	
	0450	With crane and bucket	C-7	40	1.800			29	23	52	74	
	0600	18" thick, pumped	C-20	90	.711			11.35	8.15	19.50	28	
	0650	With crane and bucket	C-7	55	1.309			21	16.55	37.55	53.50	
	0800	24" thick, pumped	C-20	92	.696			11.10	7.95	19.05	27.50	
	0850	With crane and bucket	C-7	70	1.029			16.50	13	29.50	42.50	
	1000	36" thick, pumped	C-20	140	.457			7.30	5.25	12.55	18.10	
	1050	With crane and bucket	C-7	100	.720			11.55	9.10	20.65	29.50	
	1400	Elevated slabs, less than 6" thick, pumped	C-20	140	.457			7.30	5.25	12.55	18.10	
	1450	With crane and bucket	C-7	95	.758			12.15	9.60	21.75	31	
	1500	6" to 10" thick, pumped	C-20	160	.400			6.40	4.58	10.98	15.85	
	1550	With crane and bucket	C-7	110	.655			10.50	8.30	18.80	27	
	1600	Slabs over 10" thick, pumped	C-20	180	.356			5.65	4.07	9.72	14.05	
	1650	With crane and bucket	C-7	130	.554			8.90	7	15.90	22.50	
	1900	Footings, continuous, shallow, direct chute	C-6	120	.400			6.20	.63	6.83	11.25	
	1950	Pumped	C-20	150	.427			6.80	4.88	11.68	16.85	
	2000	With crane and bucket	C-7	90	.800			12.85	10.15	23	32.50	
	2100	Footings, continuous, deep, direct chute	C-6	140	.343			5.30	.54	5.84	9.65	
	2150	Pumped	C-20	160	.400			6.40	4.58	10.98	15.85	
	2200	With crane and bucket	C-7	110	.655			10.50	8.30	18.80	27	
	2400	Footings, spread, under 1 C.Y., direct chute	C-6	55	.873			13.55	1.38	14.93	24.50	
	2450	Pumped	C-20	65	.985			15.70	11.25	26.95	39	
	2500	With crane and bucket	C-7	45	1.600			25.50	20.50	46	66	
	2600	Footings, spread, over 5 C.Y., direct chute	C-6	120	.400			6.20	.63	6.83	11.25	
	2650	Pumped	C-20	150	.427			6.80	4.88	11.68	16.85	
	2700	With crane and bucket	C-7	100	.720			11.55	9.10	20.65	29.50	
	2900	Foundation mats, over 20 C.Y., direct chute	C-6	350	.137			2.13	.22	2.35	3.85	
	2950	Pumped	C-20	400	.160			2.55	1.83	4.38	6.35	
	3000	With crane and bucket	C-7	300	.240			3.85	3.04	6.89	9.85	
	3200	Grade beams, direct chute	C-6	150	.320			4.97	.51	5.48	9	
	3250	Pumped	C-20	180	.356			5.65	4.07	9.72	14.05	
	3300	With crane and bucket	C-7	120	.600			9.65	7.60	17.25	24.50	
	3500	High rise, for more than 5 stories, pumped, add per story	C-20	2,100	.030			.49	.35	.84	1.20	

03300 | Cast-In-Place Concrete

03310 | Structural Concrete

			CREW	DAILY OUTPUT	LABOR-HOURS	UNIT	MAT.	LABOR	EQUIP.	TOTAL	TOTAL INCL O&P	
700	3510	With crane and bucket, add per story	C-7	2,100	.034	C.Y.		.55	.43	.98	1.41	700
	3700	Pile caps, under 5 C.Y., direct chute	C-6	90	.533			8.30	.84	9.14	15	
	3750	Pumped	C-20	110	.582			9.25	6.65	15.90	23	
	3800	With crane and bucket	C-7	80	.900			14.45	11.40	25.85	37	
	3850	Pile cap, 5 C.Y. to 10 C.Y., direct chute	C-6	175	.274			4.26	.43	4.69	7.70	
	3900	Pumped	C-20	200	.320			5.10	3.66	8.76	12.70	
	3950	With crane and bucket	C-7	150	.480			7.70	6.10	13.80	19.70	
	4000	Over 10 C.Y., direct chute	C-6	215	.223			3.47	.35	3.82	6.30	
	4050	Pumped	C-20	240	.267			4.25	3.05	7.30	10.55	
	4100	With crane and bucket	C-7	185	.389			6.25	4.93	11.18	15.95	
	4300	Slab on grade, 4" thick, direct chute	C-6	110	.436			6.80	.69	7.49	12.25	
	4350	Pumped	C-20	130	.492			7.85	5.65	13.50	19.50	
	4400	With crane and bucket	C-7	110	.655			10.50	8.30	18.80	27	
	4600	Over 6" thick, direct chute	C-6	165	.291			4.52	.46	4.98	8.15	
	4650	Pumped	C-20	185	.346			5.50	3.96	9.46	13.70	
	4700	With crane and bucket	C-7	145	.497			7.95	6.30	14.25	20.50	
	4900	Walls, 8" thick, direct chute	C-6	90	.533			8.30	.84	9.14	15	
	4950	Pumped	C-20	100	.640			10.20	7.30	17.50	25.50	
	5000	With crane and bucket	C-7	80	.900			14.45	11.40	25.85	37	
	5050	12" thick, direct chute	C-6	100	.480			7.45	.76	8.21	13.50	
	5100	Pumped	C-20	110	.582			9.25	6.65	15.90	23	
	5200	With crane and bucket	C-7	90	.800			12.85	10.15	23	32.50	
	5300	15" thick, direct chute	C-6	105	.457			7.10	.72	7.82	12.85	
	5350	Pumped	C-20	120	.533			8.50	6.10	14.60	21	
	5400	With crane and bucket	C-7	95	.758			12.15	9.60	21.75	31	
	5600	Wheeled concrete dumping, add to placing costs above										
	5610	Walking cart, 50' haul, add	C-18	32	.281	C.Y.		4.13	1.63	5.76	8.85	
	5620	150' haul, add		24	.375			5.50	2.17	7.67	11.80	
	5700	250' haul, add		18	.500			7.35	2.89	10.24	15.75	
	5800	Riding cart, 50' haul, add	C-19	80	.112			1.65	1.04	2.69	3.97	
	5810	150' haul, add		60	.150			2.20	1.39	3.59	5.30	
	5900	250' haul, add		45	.200			2.93	1.85	4.78	7.10	

03350 | Concrete Finishing

			CREW	DAILY OUTPUT	LABOR-HOURS	UNIT	MAT.	LABOR	EQUIP.	TOTAL	TOTAL INCL O&P	
300	0010	FINISHING FLOORS Monolithic, screed finish	1 Cefi	900	.009	S.F.		.17		.17	.27	300
	0050	Darby finish		750	.011			.20		.20	.33	
	0100	Screed and float finish		725	.011			.21		.21	.34	
	0150	Screed, float, and broom finish		630	.013			.24		.24	.39	
	0200	Screed, float, and hand trowel		600	.013			.25		.25	.41	
	0250	Machine trowel		550	.015			.27		.27	.45	
	0400	Integral topping and finish, using 1:1:2 mix, 3/16" thick	C-10	1,000	.024		.06	.42		.48	.75	
	0450	1/2" thick		950	.025		.15	.44		.59	.90	
	0500	3/4" thick		850	.028		.24	.49		.73	1.07	
	0600	1" thick		750	.032		.31	.56		.87	1.26	
	0800	Granolithic topping, laid after, 1:1:1-1/2 mix, 1/2" thick		590	.041		.16	.71		.87	1.35	
	0820	3/4" thick		580	.041		.25	.72		.97	1.46	
	0850	1" thick		575	.042		.33	.73		1.06	1.56	
	0950	2" thick		500	.048		.66	.84		1.50	2.10	
	1200	Heavy duty, 1:1:2, 3/4" thick, preshrunk, gray, 20 MSF		320	.075		.22	1.31		1.53	2.40	
	1300	100 MSF		380	.063		.22	1.10		1.32	2.06	
	1350	For colors, .50 psf, add, minimum		1,650	.015		.43	.25		.68	.89	
	1400	Maximum		1,500	.016		1.92	.28		2.20	2.57	
	1600	Exposed local aggregate finish, minimum	1 Cefi	625	.013		.37	.24		.61	.80	
	1650	Maximum		465	.017		1.18	.33		1.51	1.83	
	1800	Floor abrasives, .25 psf, add to above, aluminum oxide		850	.009		.15	.18		.33	.46	
	1850	Silicon carbide		850	.009		.18	.18		.36	.49	

03300 | Cast-In-Place Concrete

03350 | Concrete Finishing

				CREW	DAILY OUTPUT	LABOR-HOURS	UNIT	2000 BARE COSTS MAT.	LABOR	EQUIP.	TOTAL	TOTAL INCL O&P	
300	2000	Floor hardeners, metallic, light service, .50 psf, add	R03350-130	1 Cefi	850	.009	S.F.	.28	.18		.46	.60	300
	2050	Medium service, .75 psf, add			750	.011		.40	.20		.60	.77	
	2100	Heavy service, 1.0 psf, add	R03350-140		650	.012		.56	.23		.79	1	
	2150	Extra heavy, 1.5 psf, add			575	.014		.64	.26		.90	1.13	
	2300	Non-metallic, light service, .50 psf, add			850	.009		.17	.18		.35	.48	
	2350	Medium service, .75 psf, add			750	.011		.24	.20		.44	.59	
	2400	Heavy service, 1.00 psf, add			650	.012		.31	.23		.54	.72	
	2450	Extra heavy, 1.50 psf, add			575	.014		.48	.26		.74	.95	
	2600	Add for colored hardeners, metallic						50%					
	2650	Non-metallic						25%					
	2800	Trap rock wearing surface for monolithic floors											
	2810	2.0 psf, add to above		C-10	1,250	.019	S.F.	.78	.33		1.11	1.41	
	3000	Floor coloring, dusted on, 0.5 psf per S.F., add to above, min.		1 Cefi	1,300	.006		.58	.12		.70	.83	
	3050	Maximum		"	625	.013		2.03	.24		2.27	2.62	
	3100	Colors only, minimum					Lb.	.98			.98	1.08	
	3120	Maximum					"	3.43			3.43	3.77	
	3200	Integral colors, see division 03310-220											
	3600	1/2" topping using 5 lb. per bag, regular colors		C-10	590	.041	S.F.	.19	.71		.90	1.38	
	3650	Blue or green		"	590	.041		.19	.71		.90	1.38	
	3800	Dustproofing, silicate liquids, 1 coat		1 Cefi	1,900	.004		.07	.08		.15	.21	
	3850	2 coats			1,300	.006		.13	.12		.25	.33	
	4000	Epoxy coating, 1 coat, clear			1,500	.005		.38	.10		.48	.58	
	4050	Colors			1,500	.005		.38	.10		.48	.58	
	4400	Stair finish, float			275	.029			.55		.55	.90	
	4500	Steel trowel finish			200	.040			.76		.76	1.23	
	4600	Silicon carbide finish, .25 psf			150	.053		.52	1.01		1.53	2.21	
350	0010	**FINISHING WALLS** Break ties and patch voids		1 Cefi	540	.015	S.F.	.05	.28		.33	.52	350
	0050	Burlap rub with grout			450	.018		.05	.34		.39	.61	
	0100	Carborundum rub, dry			270	.030		.05	.56		.61	.97	
	0150	Wet rub			175	.046		.05	.86		.91	1.47	
	0300	Bush hammer, green concrete		B-39	1,000	.048		.05	.71	.18	.94	1.48	
	0350	Cured concrete		"	650	.074		.05	1.09	.28	1.42	2.23	
	0600	Float finish, 1/16" thick		1 Cefi	300	.027		.03	.50		.53	.85	
	0701	Sandblast, light penetration		D-4	1,835	.022		.20	.36	.06	.62	.89	
	0751	Heavy penetration		"	625	.064		.40	1.04	.17	1.61	2.40	
	0800	Board finish, see division 03110-455											
	0850	Rustication strips, see division 03110-455											
600	0010	**SLAB TEXTURE STAMPING,** buy											600
	0020	Approx. 3 S.F.- 5 S.F. each, minimum					Ea.	40			40	44	
	0030	Average					"	44			44	48.50	
	0120	Per S.F. of tool, average					S.F.	48			48	53	
	0200	Commonly used chemicals for texture systems											
	0210	Hardeners w/colors average					S.F.	.40			.40	.44	
	0220	Release agents w/colors, average						.15			.15	.17	
	0225	Clear, average						.10			.10	.11	
	0230	Sealers, clear, average						.10			.10	.11	
	0240	Colors, average						.12			.12	.13	

03370 | Specially Placed Concrete

			CREW	DAILY OUTPUT	LABOR-HOURS	UNIT	MAT.	LABOR	EQUIP.	TOTAL	TOTAL INCL O&P	
300	0010	**GUNITE,** dry mix										300
	0020	Applied in 1" layers, no mesh included	C-8	2,000	.028	S.F.	.94	.47	.33	1.74	2.18	
	0100	Mesh for gunite 2 x 2, #12, to 3" thick	2 Rodm	800	.020		.17	.42		.59	.98	
	0150	Over 3" thick	"	500	.032		.21	.68		.89	1.49	
	0300	Typical in place, including mesh, 2" thick, minimum	C-16	1,000	.072		1.75	1.28	.66	3.69	4.85	
	0350	Maximum		500	.144		2.71	2.55	1.31	6.57	8.85	

03300 | Cast-In-Place Concrete

03370 | Specially Placed Concrete

			CREW	DAILY OUTPUT	LABOR-HOURS	UNIT	2000 BARE COSTS MAT.	LABOR	EQUIP.	TOTAL	TOTAL INCL O&P	
300	0500	4" thick, minimum	C-16	750	.096	S.F.	2.59	1.70	.87	5.16	6.75	300
	0550	Maximum	↓	350	.206	↓	4.03	3.65	1.87	9.55	12.80	
	0901	Prepare old walls, no scaffolding, minimum	2 Clab	665	.024	S.Y.	.58	.35		.93	1.24	
	0951	Maximum	"	185	.086	"	1.85	1.25		3.10	4.18	
	1100	For high finish requirement or close tolerance, add, minimum				S.F.		50%				
	1150	Maximum				"		110%				

03390 | Concrete Curing

			CREW	DAILY OUTPUT	LABOR-HOURS	UNIT	2000 BARE COSTS MAT.	LABOR	EQUIP.	TOTAL	TOTAL INCL O&P	
200	0010	CURING Burlap, 4 uses assumed, 7.5 oz.	2 Clab	55	.291	C.S.F.	3.47	4.20		7.67	11	200
	0100	12 oz.		55	.291		5.65	4.20		9.85	13.45	
	0200	Waterproof curing paper, 2 ply, reinforced		70	.229		4.89	3.30		8.19	11.05	
	0300	Sprayed membrane curing compound	↓	95	.168	↓	3.52	2.43		5.95	8.05	
	0400	Curing blankets, 1" to 2" thick, buy, minimum				S.F.	1.10			1.10	1.21	
	0450	Maximum					2.75			2.75	3.03	
	0500	Electrically heated pads, 110 volts, 15 watts per S.F., buy					4.50			4.50	4.95	
	0600	20 watts per S.F., buy					6			6	6.60	
	0710	Electrically, heated pads, 15 watts/S.F., 20 uses, minimum					.16			.16	.18	
	0800	Maximum				↓	.27			.27	.29	

03400 | Precast Concrete

03410 | Plant Precast

			CREW	DAILY OUTPUT	LABOR-HOURS	UNIT	2000 BARE COSTS MAT.	LABOR	EQUIP.	TOTAL	TOTAL INCL O&P	
100	0011	BEAMS, "L" shaped, 20' span, 12" x 20"	C-11	32	1.750	Ea.	1,350	35.50	49	1,434.50	1,625	100
	1000	Inverted tee beams, add to above, small beams				L.F.	15%					
	1050	Large beams				"	5.55			5.55	6.10	
	1200	Rectangular, 20' span, 12" x 20"	C-11	32	1.750	Ea.	925	35.50	49	1,009.50	1,150	
	1250	18" x 36"		24	2.333		1,700	47	65	1,812	2,025	
	1300	24" x 44"		22	2.545		2,450	51.50	71	2,572.50	2,875	
	1400	30' span, 12" x 36"		24	2.333		2,175	47	65	2,287	2,525	
	1450	18" x 44"		20	2.800		3,050	56.50	78	3,184.50	3,525	
	1500	24" x 52"		16	3.500		4,325	70.50	97.50	4,493	4,975	
	1600	40' span, 12" x 52"		20	2.800		4,025	56.50	78	4,159.50	4,600	
	1650	18" x 52"		16	3.500		4,900	70.50	97.50	5,068	5,625	
	1700	24" x 52"		12	4.667		6,000	94	130	6,224	6,900	
	2000	"T" shaped, 20' span, 12" x 20"		32	1.750		1,600	35.50	49	1,684.50	1,900	
	2050	18" x 36"		24	2.333		2,550	47	65	2,662	2,975	
	2100	24" x 44"		22	2.545		3,600	51.50	71	3,722.50	4,125	
	2200	30' span, 12" x 36"		24	2.333		3,650	47	65	3,762	4,150	
	2250	18" x 44"		20	2.800		4,975	56.50	78	5,109.50	5,650	
	2300	24" x 52"		16	3.500		5,150	70.50	97.50	5,318	5,875	
	2500	40' span, 12" x 52"		20	2.800		6,850	56.50	78	6,984.50	7,725	
	2550	18" x 52"		16	3.500		7,500	70.50	97.50	7,668	8,475	
	2600	24" x 52"	↓	12	4.667	↓	9,150	94	130	9,374	10,400	
210	0010	COLUMNS Rectangular to 12' high, small columns	C-11	120	.467	L.F.	66.50	9.40	13	88.90	103	210
	0050	Large columns		96	.583		105	11.75	16.30	133.05	154	
	0300	24' high, small columns		192	.292		99	5.90	8.15	113.05	128	
	0350	Large columns	↓	144	.389		132	7.85	10.85	150.70	170	
400	0011	JOISTS 40 psf L.L., 6" deep for 12' spans	C-11	700	.080	L.F.	5.70	1.61	2.23	9.54	11.45	400
	0051	8" deep for 16' spans	↓	670	.084		6.70	1.69	2.33	10.72	12.80	

03400 | Precast Concrete

03410 | Plant Precast

			CREW	DAILY OUTPUT	LABOR-HOURS	UNIT	MAT.	LABOR	EQUIP.	TOTAL	TOTAL INCL O&P	
400	0101	10" deep for 20' spans	C-11	640	.087	L.F.	7.50	1.76	2.44	11.70	13.95	400
	0151	12" deep for 24' spans		610	.092		10.25	1.85	2.56	14.66	17.30	
620	0010	SLABS Prestressed roof/floor members, solid, grouted, 4" thick	C-11	3,600	.016	S.F.	4.65	.31	.43	5.39	6.10	620
	0050	6" thick		4,500	.012		4.48	.25	.35	5.08	5.75	
	0100	8" thick, hollow		5,600	.010		5	.20	.28	5.48	6.15	
	0150	10" thick		8,800	.006		5.70	.13	.18	6.01	6.65	
	0200	12" thick		8,000	.007		6.15	.14	.20	6.49	7.20	
650	0010	PRESTRESSED CONCRETE pretensioned, see division 03400										650
	0020	See also division 03230-600										
	0100	Post-tensioned in place, small job	C-17B	8.50	9.647	C.Y.	215	194	34	443	610	
	0200	Large job	"	10	8.200	"	161	165	29	355	490	
660	0010	PRESTRESSED Roof and floor members, see division 03400										660
750	0010	TEES Prestressed										750
	0020	Quad tee, short spans, roof	C-11	7,200	.008	S.F.	4.35	.16	.22	4.73	5.30	
	0050	Floor		7,200	.008		4.35	.16	.22	4.73	5.30	
	0200	Double tee, floor members, 60' span		8,400	.007		6.10	.13	.19	6.42	7.15	
	0250	80' span		8,000	.007		7.30	.14	.20	7.64	8.45	
	0300	Roof members, 30' span		4,800	.012		4.65	.24	.33	5.22	5.85	
	0350	50' span		6,400	.009		5.50	.18	.24	5.92	6.60	
	0400	Wall members, up to 55' high		3,600	.016		6.25	.31	.43	6.99	7.85	
	0500	Single tee roof members, 40' span		3,200	.017		5.75	.35	.49	6.59	7.45	
	0550	80' span		5,120	.011		7.15	.22	.31	7.68	8.55	
	0600	100' span		6,000	.009		11	.19	.26	11.45	12.70	
	0650	120' span		6,000	.009		11.50	.19	.26	11.95	13.25	
	1000	Double tees, floor members										
	1100	Lightweight, 20" x 8' wide, 45' span	C-11	20	2.800	Ea.	1,875	56.50	78	2,009.50	2,225	
	1150	24" x 8' wide, 50' span		18	3.111		2,025	62.50	87	2,174.50	2,425	
	1200	32" x 10' wide, 60' span		16	3.500		3,425	70.50	97.50	3,593	4,000	
	1250	Standard weight, 12" x 8' wide, 20' span		22	2.545		700	51.50	71	822.50	935	
	1300	16" x 8' wide, 25' span		20	2.800		930	56.50	78	1,064.50	1,200	
	1350	18" x 8' wide, 30' span		20	2.800		1,175	56.50	78	1,309.50	1,475	
	1400	20" x 8' wide, 45' span		18	3.111		1,325	62.50	87	1,474.50	1,675	
	1450	24" x 8' wide, 50' span		16	3.500		1,725	70.50	97.50	1,893	2,125	
	1500	32" x 10' wide, 60' span		14	4		3,150	80.50	112	3,342.50	3,725	
	2000	Roof members										
	2050	Lightweight, 20" x 8' wide, 40' span	C-11	20	2.800	Ea.	1,425	56.50	78	1,559.50	1,750	
	2100	24" x 8' wide, 50' span		18	3.111		1,900	62.50	87	2,049.50	2,300	
	2150	32" x 10' wide, 60' span		16	3.500		3,150	70.50	97.50	3,318	3,700	
	2200	Standard weight, 12" x 8' wide, 30' span		22	2.545		955	51.50	71	1,077.50	1,225	
	2250	16" x 8' wide, 30' span		20	2.800		1,000	56.50	78	1,134.50	1,275	
	2300	18" x 8' wide, 30' span		20	2.800		1,125	56.50	78	1,259.50	1,400	
	2350	20" x 8' wide, 40' span		18	3.111		1,150	62.50	87	1,299.50	1,450	
	2400	24" x 8' wide, 50' span		16	3.500		1,525	70.50	97.50	1,693	1,900	
	2450	32" x 10' wide, 60' span		14	4		2,675	80.50	112	2,867.50	3,200	

03450 | Architectural Precast

			CREW	DAILY OUTPUT	LABOR-HOURS	UNIT	MAT.	LABOR	EQUIP.	TOTAL	TOTAL INCL O&P	
850	0011	WALL PANELS Material only										850
	0050	Uninsulated 4" thick, smooth gray										
	0150	Low rise, 4' x 8'x 4" thick	C-11	320	.175	S.F.	8.70	3.53	4.88	17.11	21	
	0200	8'x 8' x 4" thick		576	.097		8.60	1.96	2.71	13.27	15.80	
	0250	8'x 16'x 4" thick		1,024	.055		8.55	1.10	1.53	11.18	12.95	
	0400	8'x 8', 4" thick, smooth gray		576	.097		8.60	1.96	2.71	13.27	15.80	
	0500	Exposed aggregate		576	.097		9.50	1.96	2.71	14.17	16.80	
	0600	High rise, 4' x 8' x 4" thick		288	.194		8.70	3.92	5.45	18.07	22	

03400 | Precast Concrete

03450 | Architectural Precast

			CREW	DAILY OUTPUT	LABOR-HOURS	UNIT	2000 BARE COSTS				TOTAL INCL O&P	
							MAT.	LABOR	EQUIP.	TOTAL		
850	0650	8' x 8' x 4" thick	C-11	512	.109	S.F.	8.60	2.21	3.05	13.86	16.60	850
	0700	8' x 16' x 4" thick	R03450-010	768	.073		8.55	1.47	2.04	12.06	14.15	
	0800	Insulated panel, 2" polystyrene, add					1.60			1.60	1.76	
	0850	2" urethane, add					2.10			2.10	2.31	
	1200	Finishes, white, add					1.70			1.70	1.87	
	1250	Exposed aggregate, add					.80			.80	.88	
	1300	Granite faced, domestic, add					27			27	29.50	
	2200	Fiberglass reinforced cement with urethane core										
	2210	R20, 8' x 8', minimum	E-2	750	.064	S.F.	8.30	1.37	1.36	11.03	13.30	
	2220	Maximum	"	600	.080	"	15.50	1.72	1.70	18.92	22	

03470 | Tilt-Up Precast

			CREW	DAILY OUTPUT	LABOR-HOURS	UNIT	MAT.	LABOR	EQUIP.	TOTAL	TOTAL INCL O&P	
600	0010	TILT-UP Wall panel construction, walls only, 5-1/2" thick	C-14	1,600	.085	S.F.	2.82	1.49	.69	5	6.40	600
	0100	7-1/2" thick	R03470-020	1,550	.088		3.02	1.54	.71	5.27	6.75	
	0500	Walls and columns, 5-1/2" thick walls, 12" x 12" columns		1,565	.087		3	1.53	.70	5.23	6.70	
	0550	7-1/2" thick wall, 12" x 12" columns		1,370	.099		3.90	1.74	.80	6.44	8.15	
	0800	Columns only, site precast, minimum		200	.680	L.F.	23	11.95	5.50	40.45	52	
	0850	Maximum		105	1.295	"	42	22.50	10.45	74.95	96.50	

03480 | Precast Specialties

			CREW	DAILY OUTPUT	LABOR-HOURS	UNIT	MAT.	LABOR	EQUIP.	TOTAL	TOTAL INCL O&P	
200	0010	CURBS Roadway type, see division 02770-225										200
400	0010	LINTELS										400
	0800	Precast concrete, 4" wide, 8" high, to 5' long	D-1	175	.091	L.F.	4.72	1.63		6.35	7.95	
	0850	5'-12' long	D-4	190	.211		4.93	3.43	.57	8.93	11.90	
	1000	6" wide, 8" high, to 5' long		185	.216		6.65	3.53	.59	10.77	14	
	1050	5'-12' long		190	.211		6.45	3.43	.57	10.45	13.60	
	1200	8" wide, 8" high, to 5' long		185	.216		9.80	3.53	.59	13.92	17.40	
	1250	5'-12' long		190	.211		9.05	3.43	.57	13.05	16.45	
	1400	10" wide, 8" high, to 14' long		180	.222		32.50	3.62	.61	36.73	42.50	
	1450	12" wide, 8" high, to 19' long		185	.216		34.50	3.53	.59	38.62	44.50	
800	0010	STAIRS, Precast concrete treads on steel stringers, 3' wide	C-12	75	.640	Riser	53.50	12.35	6.05	71.90	86.50	800
	0302	Front entrance, 5' wide with 48" platform, 2 risers	B-1	8	3	Flight	279	45.50		324.50	385	
	0351	5 risers		6	4		310	60.50		370.50	445	
	0501	6' wide, 2 risers		8	3		315	45.50		360.50	425	
	0551	5 risers		6	4		335	60.50		395.50	475	
	0702	7' wide, 2 risers		7	3.429		380	52		432	510	
	0751	5 risers		5	4.800		410	72.50		482.50	580	
	1201	Basement entrance stairs, steel bulkhead doors, minimum		11	2.182		460	33		493	560	
	1251	Maximum		5.50	4.364		800	66		866	995	

03500 | Cementitious Decks & Underlayments

03510 | Cementitious Roof Deck

			CREW	DAILY OUTPUT	LABOR-HOURS	UNIT	MAT.	LABOR	EQUIP.	TOTAL	TOTAL INCL O&P	
200	0010	WOOD FIBER Lightweight cement system										200
	0050	Plank, beveled, 1" thick	R05120-250 / 2 Carp	1,000	.016	S.F.	1.85	.32		2.17	2.58	
	0100	Plank, T & G, 1-1/2" thick		975	.016		2.15	.32		2.47	2.92	
	0150	2" thick		950	.017		2.90	.33		3.23	3.76	

03500 | Cementitious Decks & Underlayments

03510 | Cementitious Roof Deck

			CREW	DAILY OUTPUT	LABOR-HOURS	UNIT	2000 BARE COSTS				TOTAL INCL O&P	
							MAT.	LABOR	EQUIP.	TOTAL		
200	0200	2-1/2" thick	2 Carp	925	.017	S.F.	3	.34		3.34	3.88	200
	0250	3" thick		900	.018		3.50	.35		3.85	4.45	
	0300	3-1/2" thick		875	.018		5	.36		5.36	6.10	
	0350	4" thick		850	.019		5.50	.37		5.87	6.70	
	1000	Bulb tee, sub-purlin and grout, 6' span, add	E-1	5,000	.003		1.62	.07	.02	1.71	1.93	
	1100	8' span	"	4,200	.004		1.67	.08	.02	1.77	2.02	
250	0011	CONCRETE CHANNEL SLABS 2-3/4" or 3-1/2" thick, straight	C-11	1,840	.030	S.F.	3	.61	.85	4.46	5.30	250
	0051	Chopped up		915	.061		3	1.23	1.71	5.94	7.30	
	0201	6" thick, span to 20'		1,515	.037		3.50	.75	1.03	5.28	6.25	
	0301	8" thick, span to 24'		1,285	.044		4	.88	1.22	6.10	7.25	
270	0011	CONCRETE PLANK Lightweight, nailable, T&G, 2" thick	C-11	2,100	.027	S.F.	2.80	.54	.74	4.08	4.82	270
	0051	2-3/4" thick		1,840	.030		3.05	.61	.85	4.51	5.35	
	0101	3-3/4" thick		1,600	.035		3.80	.71	.98	5.49	6.45	
	0150	For premium ceiling finish, add					.33			.33	.36	
	0200	For sloping roofs, slope over 4 in 12, add						25%				
	0250	Slope over 6 in 12, add						150%				
350	0010	FORMBOARD Including sub-purlins										350
	0050	Non-asbestos fiber cement, 1/8" thick	C-13	2,950	.008	S.F.	2.10	.17	.03	2.30	2.66	
	0070	1/4" thick		2,950	.008		3.75	.17	.03	3.95	4.48	
	0100	Fiberglass, 1" thick, economy		2,700	.009		.95	.18	.03	1.16	1.43	
	1000	Poured gypsum, 2" thick, add to formboard above	C-8	6,000	.009		.75	.16	.11	1.02	1.21	
	1100	3" thick	"	4,800	.012		1.05	.20	.14	1.39	1.63	
900	0010	WOOD PLANK Roof decks, see division 06170-600 & 06170-550										900

03520 | Lightweight Concrete Roof Insul

			CREW	DAILY OUTPUT	LABOR-HOURS	UNIT	MAT.	LABOR	EQUIP.	TOTAL	TOTAL INCL O&P	
250	0010	INSULATING Lightweight cellular concrete roof fill										250
	0020	Portland cement and foaming agent	C-8	50	1.120	C.Y.	72	18.80	13.10	103.90	125	
	0100	Poured vermiculite or perlite, field mix,										
	0110	1:6 field mix	C-8	50	1.120	C.Y.	77.50	18.80	13.10	109.40	131	
	0200	Ready mix, 1:6 mix, roof fill, 2" thick		10,000	.006	S.F.	.43	.09	.07	.59	.70	
	0250	3" thick		7,700	.007		.65	.12	.09	.86	1	
	0401	Expanded volcanic glass rock, with binder, minimum	L-2	1,500	.011		.25	.18		.43	.60	
	0451	Maximum	"	1,200	.013		.75	.23		.98	1.22	

03600 | Grouts

03610 | Construction Grout

			CREW	DAILY OUTPUT	LABOR-HOURS	UNIT	MAT.	LABOR	EQUIP.	TOTAL	TOTAL INCL O&P	
400	0010	GROUT Column & machine bases, non-shrink, metallic, 1" deep	1 Cefi	35	.229	S.F.	2.52	4.32		6.84	9.80	400
	0050	2" deep		25	.320		5.05	6.05		11.10	15.40	
	0300	Non-shrink, non-metallic, 1" deep		35	.229		10.45	4.32		14.77	18.55	
	0350	2" deep		25	.320		20.50	6.05		26.55	32.50	
600	0010	PATCHING CONCRETE										600
	0100	Floors, 1/4" thick, small areas, regular grout	1 Cefi	170	.047	S.F.	.07	.89		.96	1.52	
	0150	Epoxy grout	"	100	.080	"	.12	1.51		1.63	2.59	
	0300	Slab on Grade, cut outs, up to 50 C.F.	2 Cefi	50	.320	C.F.	4.65	6.05		10.70	14.95	
	2000	Walls, including chipping, cleaning and epoxy grout										
	2100	Minimum	1 Cefi	65	.123	S.F.	.03	2.33		2.36	3.82	
	2150	Average		50	.160		.04	3.02		3.06	4.98	
	2200	Maximum		40	.200		.08	3.78		3.86	6.25	

03600 | Grouts

03610 | Construction Grout

		Crew	Daily Output	Labor-Hours	Unit	2000 Bare Costs				Total Incl O&P
						Mat.	Labor	Equip.	Total	
2510	Underlayment, P.C based self-leveling, 4100 psi, pumped, 1/4"	C-8	20,000	.003	S.F.	1.20	.05	.03	1.28	1.44
2520	1/2"		19,000	.003		2.16	.05	.03	2.24	2.50
2530	3/4"		18,000	.003		3.36	.05	.04	3.45	3.83
2540	1"		17,000	.003		4.56	.06	.04	4.66	5.15
2550	1-1/2"	↓	15,000	.004		6.95	.06	.04	7.05	7.80
2560	Hand mix, 1/2"	C-18	4,000	.002		2.16	.03	.01	2.20	2.45
2610	Topping, P.C. based self-level/dry 6100 psi, pumped, 1/4"	C-8	20,000	.003		1.80	.05	.03	1.88	2.10
2620	1/2"		19,000	.003		3.24	.05	.03	3.32	3.68
2630	3/4"		18,000	.003		5.05	.05	.04	5.14	5.70
2660	1"		17,000	.003		6.85	.06	.04	6.95	7.65
2670	1-1/2"	↓	15,000	.004		10.45	.06	.04	10.55	11.65
2680	Hand mix, 1/2"	C-18	4,000	.002	↓	3.24	.03	.01	3.28	3.63

For information about Means Estimating Seminars, see yellow pages 11 and 12 in back of book

Important: See the Reference Section for critical supporting data - Reference Nos., Crews, & City Cost Indexes

Division 4
Masonry

Estimating Tips
04050 Basic Masonry Materials & Methods

- The terms *mortar* and *grout* are often used interchangeably, and incorrectly. Mortar is used to bed masonry units, seal the entry of air and moisture, provide architectural appearance, and allow for size variations in the units. Grout is used primarily in reinforced masonry construction and is used to bond the masonry to the reinforcing steel. Common mortar types are M(2500 psi), S(1800 psi), N(750 psi), and O(350 psi), and conform to ASTM C270. Grout is either fine or coarse, conforms to ASTM C476, and in-place strengths generally exceed 2500 psi. Mortar and grout are different components of masonry construction and are placed by entirely different methods. An estimator should be aware of their unique uses and costs.

04200 Masonry Units

- The most common types of unit masonry are brick and concrete masonry. The major classifications of brick are building brick (ASTM C62), facing brick (ASTM C216) and glazed brick, fire brick and pavers. Many varieties of texture and appearance can exist within these classifications, and the estimator would be wise to check local custom and availability within the project area. On repair and remodeling jobs, matching the existing brick may be the most important criteria.
- Brick and concrete block are priced by the piece and then converted into a price per square foot of wall. Openings less than two square feet are generally ignored by the estimator because any savings in units used is offset by the cutting and trimming required.
- All masonry walls, whether interior or exterior, require bracing. The cost of bracing walls during construction should be included by the estimator and this bracing must remain in place until permanent bracing is complete. Permanent bracing of masonry walls is accomplished by masonry itself, in the form of pilasters or abutting wall corners, or by anchoring the walls to the structural frame. Accessories in the form of anchors, anchor slots and ties are used, but their supply and installation can be by different trades. For instance, anchor slots on spandrel beams and columns are supplied and welded in place by the steel fabricator, but the ties from the slots into the masonry are installed by the bricklayer. Regardless of the installation method the estimator must be certain that these accessories are accounted for in pricing.

Reference Numbers

Reference numbers are shown in bold squares at the beginning of some major classifications. These numbers refer to related items in the Reference Section. The reference information may be an estimating procedure, an alternate pricing method or technical information.

Note: Not all subdivisions listed here necessarily appear in this publication.

04050 | Basic Masonry Materials & Methods

04060 | Masonry Mortar

			CREW	DAILY OUTPUT	LABOR-HOURS	UNIT	2000 BARE COSTS MAT.	LABOR	EQUIP.	TOTAL	TOTAL INCL O&P	
200	0010	**CEMENT** Gypsum 80 lb. bag, T.L. lots				Bag	11.40			11.40	12.55	200
	0050	L.T.L. lots					11.90			11.90	13.10	
	0100	Masonry, 70 lb. bag, T.L. lots					5.90			5.90	6.50	
	0150	L.T.L. lots					5.70			5.70	6.25	
	0200	White, 70 lb. bag, T.L. lots					16.05			16.05	17.65	
	0250	L.T.L. lots					16.60			16.60	18.25	
400	0010	**LIME** Masons, hydrated, 50 lb. bag, T.L. lots				Bag	5.60			5.60	6.15	400
	0050	L.T.L. lots					5.80			5.80	6.40	
	0200	Finish, double hydrated, 50 lb. bag, T.L. lots					6.85			6.85	7.55	
	0250	L.T.L. lots					7.55			7.55	8.30	
500	0010	**MORTAR**	R04060-100									500
	0020	With masonry cement										
	0100	Type M, 1:1:6 mix	1 Brhe	143	.056	C.F.	3.08	.88		3.96	4.88	
	0200	Type N, 1:3 mix	"	143	.056	"	2.83	.88		3.71	4.61	
	2000	With portland cement and lime										
	2100	Type M, 1:1/4:3 mix	1 Brhe	143	.056	C.F.	3.77	.88		4.65	5.65	
	2200	Type N, 1:1:6 mix, 750 psi		143	.056		3.05	.88		3.93	4.84	
	2300	Type O, 1:2:9 mix (Pointing Mortar)		143	.056		2.97	.88		3.85	4.76	
	2700	Mortar for glass block		143	.056		7.35	.88		8.23	9.60	
	2800	Gypsum cement mortar					5.65			5.65	6.25	
	2900	Mortar for Fire Brick, 80 lb. bag, T.L. Lots				Bag	12.50			12.50	13.75	
520	0010	**POINTING MORTAR** See also Division 04060-500				Lb.	.85			.85	.94	520
	0050	White				"	1.03			1.03	1.13	
540	0010	**COLORS** 50 lb. bags (2 bags per M bricks),	R04060-100									540
	0020	range 2 to 10 lb. per bag of cement, minimum				Lb.	1.50			1.50	1.65	
	0050	Average					2.50			2.50	2.75	
	0100	Maximum					5			5	5.50	
750	0010	**SAND** For mortar, screened and washed, at the pit				Ton	11.40			11.40	12.55	750
	0050	With 10 mile haul					12.40			12.40	13.65	
	0100	With 30 mile haul					13.45			13.45	14.80	
	0200	Screened and washed, at the pit				C.Y.	15.85			15.85	17.45	
	0250	With 10 mile haul					17.25			17.25	19	
	0300	With 30 mile haul					18.70			18.70	20.50	
770	0010	**SURFACE BONDING** CMU walls with fiberglass mortar,										770
	0020	gray or white colors, not incl. block work	1 Bric	540	.015	S.F.	.52	.30		.82	1.07	
900	0010	**WATERPROOFING** Admixture, 1 qt. to 2 bags of masonry cement				Qt.	6			6	6.60	900

04070 | Masonry Grout

			CREW	DAILY OUTPUT	LABOR-HOURS	UNIT	MAT.	LABOR	EQUIP.	TOTAL	INCL O&P	
420	0010	**GROUTING** Bond bms. & lintels, 8" dp., pumped, not incl. block	R04060-100									420
	0020	8" thick, 0.2 C.F. per L.F.	D-4	1,400	.029	L.F.	.70	.47	.08	1.25	1.65	
	0050	10" thick, 0.25 C.F. per L.F.		1,200	.033		.88	.54	.09	1.51	1.99	
	0060	12" thick, 0.3 C.F. per L.F.		1,040	.038		1.05	.63	.10	1.78	2.34	
	0200	Concrete block cores, solid, 4" thk., by hand, 0.067 C.F./S.F.	D-8	1,100	.036	S.F.	.25	.66		.91	1.40	
	0210	6" thick, pumped, 0.175 C.F. per S.F.	D-4	720	.056		.60	.91	.15	1.66	2.37	
	0250	8" thick, pumped, 0.258 C.F. per S.F.		680	.059		.91	.96	.16	2.03	2.81	
	0300	10" thick, pumped, 0.340 C.F. per S.F.		660	.061		1.19	.99	.17	2.35	3.17	
	0350	12" thick, pumped, 0.422 C.F. per S.F.		640	.063		1.47	1.02	.17	2.66	3.54	
	0500	Cavity walls, 2" space, pumped, no shoring, 0.167 C.F./S.F.		1,700	.024		.60	.38	.06	1.04	1.38	
	0550	3" space, 0.250 C.F./S.F.		1,200	.033		.88	.54	.09	1.51	1.99	
	0600	4" space, 0.333 C.F. per S.F.		1,150	.035		1.16	.57	.09	1.82	2.33	
	0700	6" space, 0.500 C.F. per S.F.		800	.050		1.76	.82	.14	2.72	3.46	
	0800	Door frames, 3' x 7' opening, 2.5 C.F. per opening		60	.667	Opng.	8.80	10.85	1.82	21.47	30	
	0850	6' x 7' opening, 3.5 C.F. per opening		45	.889	"	12.30	14.50	2.43	29.23	40.50	
	2000	Grout, C476, for bond beams, lintels and CMU cores		350	.114	C.F.	3.51	1.86	.31	5.68	7.35	

04050 | Basic Masonry Materials & Methods

04080 | Masonry Anchor & Reinforcement

			CREW	DAILY OUTPUT	LABOR-HOURS	UNIT	2000 BARE COSTS MAT.	LABOR	EQUIP.	TOTAL	TOTAL INCL O&P	
070	0010	**ANCHOR BOLTS** Hooked type with nut and washer, 1/2" diam., 8" long	1 Bric	200	.040	Ea.	.47	.80		1.27	1.87	070
	0030	12" long		190	.042		1.22	.84		2.06	2.77	
	0040	5/8" diameter, 8" long		180	.044		1.25	.89		2.14	2.88	
	0050	12" long		170	.047		1.38	.94		2.32	3.10	
	0060	3/4" diameter, 8" long		160	.050		1.71	1		2.71	3.57	
	0070	12" long		150	.053		2.11	1.07		3.18	4.13	
200	0010	**REINFORCING** Steel bars A615, placed horiz., #3 & #4 bars	1 Bric	450	.018	Lb.	.28	.36		.64	.91	200
	0020	#5 & #6 bars		800	.010		.28	.20		.48	.65	
	0050	Placed vertical, #3 & #4 bars		350	.023		.28	.46		.74	1.08	
	0060	#5 & #6 bars		650	.012		.28	.25		.53	.73	
	0200	Joint reinforcing, regular truss, to 6" wide, mill std galvanized		30	.267	C.L.F.	10.20	5.35		15.55	20.50	
	0250	12" wide		20	.400		11.75	8		19.75	26.50	
	0400	Cavity truss with drip section, to 6" wide		30	.267		9.70	5.35		15.05	19.70	
	0450	12" wide		20	.400		11.05	8		19.05	25.50	
650	0010	**WALL TIES** To brick veneer, galv., corrugated, 7/8" x 7", 22 Ga.	1 Bric	10.50	.762	C	4.47	15.25		19.72	31	650
	0100	24 Ga.		10.50	.762		3.97	15.25		19.22	30.50	
	0150	16 Ga.		10.50	.762		13.45	15.25		28.70	41	
	0200	Buck anchors, galv., corrugated, 16 gauge, 2" bend. 8" x 2"		10.50	.762		114	15.25		129.25	151	
	0250	8" x 3"		10.50	.762		99	15.25		114.25	135	
	0600	Cavity wall, Z type, galvanized, 6" long, 1/4" diameter		10.50	.762		20	15.25		35.25	48	
	0650	3/16" diameter		10.50	.762		9.50	15.25		24.75	36.50	
	0800	8" long, 1/4" diameter		10.50	.762		24	15.25		39.25	52.50	
	0850	3/16" diameter		10.50	.762		10.45	15.25		25.70	37.50	
	1000	Rectangular type, galvanized, 1/4" diameter, 2" x 6"		10.50	.762		25.50	15.25		40.75	54.50	
	1050	2" x 8" or 4" x 6"		10.50	.762		29	15.25		44.25	57.50	
	1100	3/16" diameter, 2" x 6"		10.50	.762		16.95	15.25		32.20	44.50	
	1150	2" x 8" or 4" x 6"		10.50	.762		19.40	15.25		34.65	47.50	
	1500	Rigid partition anchors, plain, 8" long, 1" x 1/8"		10.50	.762		48	15.25		63.25	79	
	1550	1" x 1/4"		10.50	.762		94	15.25		109.25	129	
	1580	1-1/2" x 1/8"		10.50	.762		66.50	15.25		81.75	99	
	1600	1-1/2" x 1/4"		10.50	.762		158	15.25		173.25	199	
	1650	2" x 1/8"		10.50	.762		83	15.25		98.25	118	
	1700	2" x 1/4"		10.50	.762		225	15.25		240.25	274	

04090 | Masonry Accessories

			CREW	DAILY OUTPUT	LABOR-HOURS	UNIT	MAT.	LABOR	EQUIP.	TOTAL	TOTAL INCL O&P	
150	0010	**CAULKING** See division 07920-800										150
170	0010	**CONTROL JOINT** Rubber, 4" and wider wall	1 Bric	400	.020	L.F.	2.58	.40		2.98	3.52	170
	0050	PVC, 4" wall		400	.020		1.77	.40		2.17	2.63	
	0100	Rubber, 6" wall		320	.025		4.68	.50		5.18	6	
	0120	PVC, 6" wall		320	.025		1.81	.50		2.31	2.84	
	0140	Rubber, 8" and wider wall		280	.029		5.35	.57		5.92	6.85	
	0160	PVC, 8" wall		280	.029		2.02	.57		2.59	3.19	
	0180	12" wall		240	.033		2.93	.67		3.60	4.35	
420	0010	**INSULATION** See also division 07210-550										420
	0100	Inserts, styrofoam, plant installed, add to block prices										
	0200	8" x 16" units, 6" thick				S.F.	.75			.75	.83	
	0250	8" thick					.75			.75	.83	
	0300	10" thick					.90			.90	.99	
	0350	12" thick					.95			.95	1.05	
	0500	8" x 8" units, 8" thick					.62			.62	.68	
	0550	12" thick					.75			.75	.83	
700	0010	**SCAFFOLDING & SWING STAGING** See division 01540										700

04050 | Basic Masonry Materials & Methods

04090 | Masonry Accessories

			CREW	DAILY OUTPUT	LABOR-HOURS	UNIT	MAT.	LABOR	EQUIP.	TOTAL	TOTAL INCL O&P	
850	0010	**VENT BOX** See division 04090-860										850
860	0010	**VENT BOX** Extruded aluminum, 4" deep, 2-3/8" x 8-1/8"	1 Bric	30	.267	Ea.	30.50	5.35		35.85	42.50	860
	0050	5" x 8-1/8"		25	.320		42	6.40		48.40	57	
	0100	2-1/4" x 25"		25	.320		60	6.40		66.40	77	
	0150	5" x 16-1/2"		22	.364		44	7.25		51.25	61	
	0200	6" x 16-1/2"		22	.364		85	7.25		92.25	106	
	0250	7-3/4" x 16-1/2"		20	.400		80	8		88	102	
	0400	For baked enamel finish, add					35%					
	0500	For cast aluminum, painted, add					60%					
	1000	Stainless steel ventilators, 6" x 6"	1 Bric	25	.320		90	6.40		96.40	110	
	1050	8" x 8"		24	.333		95	6.65		101.65	116	
	1100	12" x 12"		23	.348		110	6.95		116.95	133	
	1150	12" x 6"		24	.333		95	6.65		101.65	116	
	1200	Foundation block vent, galv., 1-1/4" thk, 8" high, 16" long, no damper		30	.267		16	5.35		21.35	26.50	
	1250	For damper, add					5			5	5.50	
900	0010	**WALL PLUGS** For nailing to brickwork, 26 ga., galvanized, plain	1 Bric	10.50	.762	C	24	15.25		39.25	52.50	900
	0050	Wood filled	"	10.50	.762	"	82	15.25		97.25	116	

04200 | Masonry Units

04210 | Clay Masonry Units

			CREW	DAILY OUTPUT	LABOR-HOURS	UNIT	MAT.	LABOR	EQUIP.	TOTAL	TOTAL INCL O&P	
100	0010	**COMMON BUILDING BRICK** C62, TL lots, material only										100
	0020	Standard, minimum				M	250			250	275	
	0050	Average (select)				"	305			305	335	
120	0010	**BRICK VENEER** Scaffolding not included, truck load lots										120
	0015	Material costs incl. 3% brick and 25% mortar waste										
	0020	Standard, select common, 4" x 2-2/3" x 8" (6.75/S.F.)	D-8	1.50	26.667	M	350	485		835	1,200	
	0050	Red, 4" x 2-2/3" x 8", running bond		1.50	26.667		395	485		880	1,250	
	0100	Full header every 6th course (7.88/S.F.)		1.45	27.586		395	505		900	1,300	
	0150	English, full header every 2nd course (10.13/S.F.)		1.40	28.571		395	520		915	1,325	
	0200	Flemish, alternate header every course (9.00/S.F.)		1.40	28.571		395	520		915	1,325	
	0250	Flemish, alt. header every 6th course (7.13/S.F.)		1.45	27.586		395	505		900	1,300	
	0300	Full headers throughout (13.50/S.F.)		1.40	28.571		395	520		915	1,325	
	0350	Rowlock course (13.50/S.F.)		1.35	29.630		395	540		935	1,350	
	0400	Rowlock stretcher (4.50/S.F.)		1.40	28.571		400	520		920	1,325	
	0450	Soldier course (6.75/S.F.)		1.40	28.571		395	520		915	1,325	
	0500	Sailor course (4.50/S.F.)		1.30	30.769		400	560		960	1,400	
	0601	Buff or gray face, running bond, (6.75/S.F.)		1.50	26.667		395	485		880	1,250	
	0700	Glazed face, 4" x 2-2/3" x 8", running bond		1.40	28.571		1,525	520		2,045	2,550	
	0750	Full header every 6th course (7.88/S.F.)		1.35	29.630		1,475	540		2,015	2,550	
	1000	Jumbo, 6" x 4" x 12",(3.00/S.F.)		1.30	30.769		1,250	560		1,810	2,325	
	1051	Norman, 4" x 2-2/3" x 12" (4.50/S.F.)		1.45	27.586		735	505		1,240	1,675	
	1100	Norwegian, 4" x 3-1/5" x 12" (3.75/S.F.)		1.40	28.571		570	520		1,090	1,500	
	1150	Economy, 4" x 4" x 8" (4.50 per S.F.)		1.40	28.571		475	520		995	1,400	
	1201	Engineer, 4" x 3-1/5" x 8", (5.63/S.F.)		1.45	27.586		335	505		840	1,225	
	1251	Roman, 4" x 2" x 12", (6.00/S.F.)		1.50	26.667		760	485		1,245	1,675	
	1300	S.C.R. 6" x 2-2/3" x 12" (4.50/S.F.)		1.40	28.571		900	520		1,420	1,875	
	1350	Utility, 4" x 4" x 12" (3.00/S.F.)		1.35	29.630		1,000	540		1,540	2,025	

04200 | Masonry Units

04210 | Clay Masonry Units

			CREW	DAILY OUTPUT	LABOR-HOURS	UNIT	MAT.	LABOR	EQUIP.	TOTAL	TOTAL INCL O&P
120	1360	For less than truck load lots, add				M	10				
	1400	For battered walls, add						30%			
	1450	For corbels, add						75%			
	1500	For curved walls, add						30%			
	1550	For pits and trenches, deduct						20%			
	1999	Alternate method of figuring by square foot									
	2000	Standard, sel. common, 4" x 2-2/3" x 8", (6.75/S.F.)	D-8	230	.174	S.F.	2.68	3.18		5.86	8.35
	2020	Standard, red, 4" x 2-2/3" x 8", running bond (6.75/SF)		220	.182		2.68	3.32		6	8.60
	2050	Full header every 6th course (7.88/S.F.)		185	.216		3.13	3.95		7.08	10.15
	2100	English, full header every 2nd course (10.13/S.F.)		140	.286		4.01	5.20		9.21	13.25
	2150	Flemish, alternate header every course (9.00/S.F.)		150	.267		3.57	4.87		8.44	12.15
	2200	Flemish, alt. header every 6th course (7.13/S.F.)		205	.195		2.83	3.57		6.40	9.15
	2250	Full headers throughout (13.50/S.F.)		105	.381		5.35	6.95		12.30	17.65
	2300	Rowlock course (13.50/S.F.)		100	.400		5.35	7.30		12.65	18.25
	2350	Rowlock stretcher (4.50/S.F.)		310	.129		1.80	2.36		4.16	5.95
	2400	Soldier course (6.75/S.F.)		200	.200		2.68	3.66		6.34	9.15
	2450	Sailor course (4.50/S.F.)		290	.138		1.80	2.52		4.32	6.25
	2600	Buff or gray face, running bond, (6.75/S.F.)		220	.182		2.85	3.32		6.17	8.80
	2700	Glazed face brick, running bond		210	.190		10	3.48		13.48	16.90
	2750	Full header every 6th course (7.88/S.F.)		170	.235		11.65	4.30		15.95	20
	3000	Jumbo, 6" x 4" x 12" running bond (3.00/S.F.)		435	.092		3.53	1.68		5.21	6.75
	3050	Norman, 4" x 2-2/3" x 12" running bond, (4.5/S.F.)		320	.125		3.19	2.28		5.47	7.40
	3100	Norwegian, 4" x 3-1/5" x 12" (3.75/S.F.)		375	.107		2.08	1.95		4.03	5.60
	3150	Economy, 4" x 4" x 8" (4.50/S.F.)		310	.129		2.12	2.36		4.48	6.35
	3200	Engineer, 4" x 3-1/5" x 8" (5.63/S.F.)		260	.154		1.88	2.81		4.69	6.85
	3250	Roman, 4" x 2" x 12" (6.00/S.F.)		250	.160		4.52	2.92		7.44	9.90
	3300	SCR, 6" x 2-2/3" x 12" (4.50/S.F.)		310	.129		4.14	2.36		6.50	8.55
	3350	Utility, 4" x 4" x 12" (3.00/S.F.)		450	.089		2.96	1.62		4.58	6
	3400	For cavity wall construction, add						15%			
	3450	For stacked bond, add						10%			
	3500	For interior veneer construction, add						15%			
	3550	For curved walls, add						30%			
200	0011	**CAST CERAMIC FLOORING** See division 09600									
300	0010	**FACE BRICK** C216, TL lots, material only									
	0300	Standard modular, 4" x 2-2/3" x 8", minimum				M	350			350	385
	0350	Maximum					315			315	345
	0450	Economy, 4" x 4" x 8", minimum					420			420	460
	0500	Maximum					525			525	580
	0510	Economy, 4" x 4" x 12", minimum					420			420	460
	0520	Maximum					525			525	580
	0550	Jumbo, 6" x 4" x 12", minimum					1,100			1,100	1,200
	0600	Maximum					1,400			1,400	1,550
	0610	Jumbo, 8" x 4" x 12", minimum					1,100			1,100	1,200
	0620	Maximum					1,400			1,400	1,550
	0650	Norwegian, 4" x 3-1/5" x 12", minimum					500			500	550
	0700	Maximum					700			700	770
	0710	Norwegian, 6" x 3-1/5" x 12", minimum					500			500	550
	0720	Maximum					700			700	770
	0850	Standard glazed, plain colors, 4" x 2-2/3" x 8", minimum					1,400			1,400	1,550
	0900	Maximum					1,100			1,100	1,200
	1000	Deep trim shades, 4" x 2-2/3" x 8", minimum					1,125			1,125	1,250
	1050	Maximum					1,250			1,250	1,375
	1080	Jumbo utility, 4" x 4" x 12"					915			915	1,000
	1120	4" x 8" x 8"					1,050			1,050	1,150
	1140	4" x 8" x 16"					2,200			2,200	2,425

04200 | Masonry Units

04210 | Clay Masonry Units

			CREW	DAILY OUTPUT	LABOR-HOURS	UNIT	2000 BARE COSTS MAT.	LABOR	EQUIP.	TOTAL	TOTAL INCL O&P	
300	1260	Engineer, 4" x 3-1/5" x 8", minimum				M	288			288	315	300
	1270	Maximum	R04210-120				475			475	525	
	1350	King, 4" x 2-3/4" x 10", minimum					410			410	450	
	1360	Maximum					430			430	475	
	1400	Norman, 4" x 2-3/4" x 12"					410			410	450	
	1450	Roman, 4" x 2" x 12"					410			410	450	
	1500	SCR, 6" x 2-2/3" x 12"					410			410	450	
	1550	Double, 4" x 5-1/3" x 8"					410			410	450	
	1600	Triple, 4" x 5-1/3" x 12"					410			410	450	
	1770	Standard modular, double glazed, 4" x 2-2/3" x 8"					950			950	1,050	
	1850	Jumbo, colored glazed ceramic, 6" x 4" x 12"					1,400			1,400	1,550	
	2050	Jumbo utility, glazed, 4" x 4" x 12"					1,050			1,050	1,150	
	2100	4" x 8" x 8"					1,450			1,450	1,600	
	2150	4" x 16" x 8"					2,500			2,500	2,750	
	2160	Fire Brick, 2" x 2-2/3" x 9", minimum										
	2165	Maximum				M	900			900	990	
	2170	For less than truck load lots, add					10					
	2180	For buff or gray brick, add					15					
350	0010	**STRUCTURAL FACING TILE** Scaffolding not incl, standard colors										350
	0020	6T series, 5-1/3" x 12", 2.3 pieces per S.F., glazed 1 side, 2" thick	D-8	225	.178	S.F.	4.97	3.25		8.22	10.95	
	0100	4" thick		220	.182		5.55	3.32		8.87	11.75	
	0150	Glazed 2 sides		195	.205		9	3.75		12.75	16.25	
	0250	6" thick		210	.190		9.10	3.48		12.58	15.90	
	0300	Glazed 2 sides		185	.216		12.75	3.95		16.70	21	
	0400	8" thick		180	.222		11.80	4.06		15.86	19.90	
	0500	Special shapes, group 1		400	.100	Ea.	4.36	1.83		6.19	7.90	
	0550	Group 2		375	.107		6.85	1.95		8.80	10.80	
	0600	Group 3		350	.114		8.70	2.09		10.79	13.10	
	0650	Group 4		325	.123		12.45	2.25		14.70	17.50	
	0700	Group 5		300	.133		16.80	2.44		19.24	22.50	
	0750	Group 6		275	.145		28.50	2.66		31.16	35.50	
	1000	Fire rated, 4" thick, 1 hr. rating		210	.190	S.F.	11.05	3.48		14.53	18.05	
	1300	Acoustic, 4" thick		210	.190	"	12.05	3.48		15.53	19.20	
	1400	For designer colors, add					25%					
	2000	8W series, 8" x 16", 1.125 pieces per S.F.										
	2050	2" thick, glazed 1 side	D-8	360	.111	S.F.	5.25	2.03		7.28	9.20	
	2100	4" thick, glazed 1 side		345	.116		6.05	2.12		8.17	10.25	
	2150	Glazed 2 sides		325	.123		9.80	2.25		12.05	14.60	
	2200	6" thick, glazed 1 side		330	.121		8.60	2.22		10.82	13.25	
	2250	8" thick, glazed 1 side		310	.129		10.45	2.36		12.81	15.50	
	2500	Special shapes, group 1		300	.133	Ea.	7.05	2.44		9.49	11.95	
	2550	Group 2		280	.143		9.25	2.61		11.86	14.55	
	2600	Group 3		260	.154		9.75	2.81		12.56	15.50	
	2650	Group 4		250	.160		13.05	2.92		15.97	19.30	
	2700	Group 5		240	.167		42.50	3.05		45.55	52	
	2750	Group 6		230	.174		56.50	3.18		59.68	67.50	
	3000	4" thick, glazed 1 side		345	.116	S.F.	8.45	2.12		10.57	12.85	
	3100	Acoustic, 4" thick		345	.116	"	9.50	2.12		11.62	14.05	
	3120	4W series, 8" x 8", 2.25 pieces per S.F.										
	3125	2" thick, glazed 1 side	D-8	360	.111	S.F.	5.20	2.03		7.23	9.15	
	3130	4" thick, glazed 1 side		345	.116		6.30	2.12		8.42	10.55	
	3135	Glazed 2 sides		325	.123		8.60	2.25		10.85	13.25	
	3140	6" thick, glazed 1 side		330	.121		8.40	2.22		10.62	13	
	3150	8" thick, glazed 1 side		310	.129		12.60	2.36		14.96	17.90	
	3155	Special shapes, group I		300	.133	Ea.	4.03	2.44		6.47	8.55	
	3160	Group II		280	.143	"	4.91	2.61		7.52	9.80	

Important: See the Reference Section for critical supporting data - Reference Nos., Crews, & City Cost Indexes

04200 | Masonry Units

04210 | Clay Masonry Units

			Daily Output	Labor-Hours	Unit	Mat.	Labor	Equip.	Total	Total Incl O&P	
						2000 Bare Costs					
			Crew								
350	3200	For designer colors, add				25%					350
	3300	For epoxy mortar joints, add			S.F.	1			1	1.10	
810	0010	**TERRA COTTA** Coping, split type, not glazed, 9" wide	D-1	90	.178	L.F.	4.78	3.17		7.95	10.60	810
	0100	13" wide		80	.200		7.80	3.57		11.37	14.65	
	0200	Split type, glazed, 9" wide		90	.178		5.55	3.17		8.72	11.50	
	0250	13" wide		80	.200		9.05	3.57		12.62	16	
	0500	Partition or back-up blocks, scored, in C.L. lots										
	0700	Non-load bearing 12" x 12", 3" thick, Special Order Only	D-8	550	.073	S.F.	4.45	1.33		5.78	7.15	
	0750	4" thick, standard		500	.080		3.65	1.46		5.11	6.50	
	0800	6" thick		450	.089		3.70	1.62		5.32	6.80	
	0850	8" thick		400	.100		4.60	1.83		6.43	8.15	
	1000	Load bearing, 12" x 12", 4" thick, in walls		500	.080		4.03	1.46		5.49	6.90	
	1050	In floors		750	.053		4.03	.97		5	6.10	
	1200	6" thick, in walls		450	.089		3.85	1.62		5.47	7	
	1250	In floors		675	.059		3.85	1.08		4.93	6.05	
	1400	8" thick, in walls		400	.100		4.70	1.83		6.53	8.25	
	1450	In floors		575	.070		4.70	1.27		5.97	7.30	
	1600	10" thick, in walls, special order		350	.114		5.45	2.09		7.54	9.55	
	1650	In floors, special order		500	.080		5.45	1.46		6.91	8.50	
	1800	12" thick, in walls, special order		300	.133		6.50	2.44		8.94	11.30	
	1850	In floors, special order		450	.089		6.50	1.62		8.12	9.90	
	2000	For reinforcing with steel rods, add to above					15%	5%				
	2100	For smooth tile instead of scored, add					.10			.10	.11	
	2200	For L.C.L. quantities, add					10%	10%				
820	0010	**TERRA COTTA TILE** On walls, dry set, 1/2" thick										820
	0100	Square, hexagonal or lattice shapes, unglazed	1 Tilf	135	.059	S.F.	4.25	1.14		5.39	6.50	
	0300	Glazed, plain colors		130	.062		5.70	1.18		6.88	8.15	
	0400	Intense colors		125	.064		8.80	1.23		10.03	11.70	

04220 | Concrete Masonry Units

			Crew	Daily Output	Labor-Hours	Unit	Mat.	Labor	Equip.	Total	Total Incl O&P	
200	0010	**CHIMNEY BLOCK** Scaffolding not included										200
	0220	1 piece, with 8" x 8" flue, 16" x 16"	D-1	28	.571	V.L.F.	12.50	10.20		22.70	31	
	0230	2 piece, 16" x 16"		26	.615		14.10	11		25.10	34	
	0240	2 piece, with 8" x 12" flue, 16" x 20"		24	.667		17.50	11.90		29.40	39.50	
220	0010	**CONCRETE BLOCK, BACK-UP** Scaffolding not included	R04220-200									220
	0020	Sand aggregate, 8" x 16" units, tooled joint 1 side										
	1000	Reinforced, alternate courses, 4" thick	D-8	435	.092	S.F.	.88	1.68		2.56	3.81	
	1100	6" thick		415	.096		1.18	1.76		2.94	4.27	
	1150	8" thick		395	.101		1.38	1.85		3.23	4.65	
	1200	10" thick		385	.104		1.92	1.90		3.82	5.35	
	1250	12" thick	D-9	365	.132		2.05	2.35		4.40	6.25	
230	0010	**CONCRETE BLOCK BOND BEAM** Scaffolding not included										230
	0020	Not including grout or reinforcing										
	0100	Regular block, 8" high, 8" thick	D-8	565	.071	L.F.	1.62	1.29		2.91	3.98	
	0150	12" thick	D-9	510	.094		2.54	1.68		4.22	5.65	
	0500	Lightweight, 8" high, 8" thick	D-8	575	.070		1.58	1.27		2.85	3.88	
	0550	12" thick	D-9	520	.092		2.75	1.65		4.40	5.80	
	2000	Including grout and 2 #5 bars										
	2100	Regular block, 8" high, 8" thick	D-8	300	.133	L.F.	3.01	2.44		5.45	7.45	
	2150	12" thick	D-9	250	.192		4.36	3.43		7.79	10.60	
	2500	Lightweight, 8" high, 8" thick	D-8	305	.131		3.34	2.40		5.74	7.75	
	2550	12" thick	D-9	255	.188		4.57	3.36		7.93	10.75	
240	0010	**CONCRETE BLOCK, DECORATIVE** Scaffolding not included										240
	0020	Embossed, simulated brick face										

04200 | Masonry Units

04220 | Concrete Masonry Units

		CREW	DAILY OUTPUT	LABOR-HOURS	UNIT	2000 BARE COSTS MAT.	LABOR	EQUIP.	TOTAL	TOTAL INCL O&P
0100	8" x 16" units, 4" thick	D-8	400	.100	S.F.	2.15	1.83		3.98	5.45
0200	8" thick		340	.118		2.93	2.15		5.08	6.85
0250	12" thick	↓	300	.133	↓	3.78	2.44		6.22	8.30
0400	Embossed both sides									
0500	8" thick	D-8	300	.133	S.F.	3.27	2.44		5.71	7.70
0550	12" thick	"	275	.145	"	4.28	2.66		6.94	9.20
1000	Fluted high strength									
1100	Flutes 1 side, 8" x 16" x 4" thick	D-8	345	.116	S.F.	2.60	2.12		4.72	6.45
1150	Flutes 2 sides, 8" x 16" x 4" thick		335	.119		3.16	2.18		5.34	7.15
1200	8" thick	↓	300	.133		4.06	2.44		6.50	8.60
1250	For special colors, add				↓	.24			.24	.26
1400	Deep grooved, smooth face									
1450	8" x 16" x 4" thick	D-8	345	.116	S.F.	1.63	2.12		3.75	5.40
1500	8" thick	"	300	.133	"	2.80	2.44		5.24	7.20
2000	Formblock, incl. inserts & reinforcing									
2100	8" x 16" x 8" thick	D-8	345	.116	S.F.	2.93	2.12		5.05	6.80
2150	12" thick	"	310	.129	"	3.60	2.36		5.96	7.95
2500	Ground face									
2600	8" x 16" x 4" thick	D-8	345	.116	S.F.	3.68	2.12		5.80	7.65
2650	6" thick		310	.129		4.03	2.36		6.39	8.40
2700	8" thick	↓	290	.138		4.65	2.52		7.17	9.35
2750	12" thick	D-9	265	.181	↓	5.60	3.23		8.83	11.65
2900	For special colors, add, minimum					15%				
2950	For special colors, add, maximum					45%				
4000	Slump block									
4100	4" face height x 16" x 4" thick	D-1	165	.097	S.F.	3.76	1.73		5.49	7.05
4150	6" thick		160	.100		4.61	1.79		6.40	8.05
4200	8" thick		155	.103		6.25	1.84		8.09	9.95
4250	10" thick		140	.114		9.35	2.04		11.39	13.75
4300	12" thick		130	.123		9.95	2.20		12.15	14.65
4400	6" face height x 16" x 6" thick		155	.103		3.77	1.84		5.61	7.25
4450	8" thick		150	.107		4.94	1.90		6.84	8.65
4500	10" thick		130	.123		7.50	2.20		9.70	11.95
4550	12" thick	↓	120	.133	↓	7.75	2.38		10.13	12.60
5000	Split rib profile units, 1" deep ribs, 8 ribs									
5100	8" x 16" x 4" thick	D-8	345	.116	S.F.	1.89	2.12		4.01	5.65
5150	6" thick		325	.123		2.29	2.25		4.54	6.30
5200	8" thick	↓	305	.131		2.74	2.40		5.14	7.10
5250	12" thick	D-9	275	.175		3.35	3.12		6.47	9
5400	For special deeper colors, 4" thick, add					.17			.17	.19
5450	12" thick, add					.34			.34	.37
5600	For white, 4" thick, add					.70			.70	.77
5650	6" thick, add					.93			.93	1.02
5700	8" thick, add					1.13			1.13	1.24
5750	12" thick, add				↓	1.58			1.58	1.74
6000	Split face or scored split face									
6100	8" x 16" x 4" thick	D-8	350	.114	S.F.	1.95	2.09		4.04	5.70
6150	6" thick		315	.127		2.21	2.32		4.53	6.35
6200	8" thick	↓	295	.136		4.05	2.48		6.53	8.65
6250	12" thick	D-9	270	.178		5.30	3.17		8.47	11.20
6400	For special deeper colors, 4" thick, add					.23			.23	.25
6450	6" thick, add					.23			.23	.25
6500	8" thick, add					.24			.24	.26
6550	12" thick, add					.41			.41	.45
6650	For white, 4" thick, add					.73			.73	.81
6700	6" thick, add				↓	.93			.93	1.02

Important: See the Reference Section for critical supporting data - Reference Nos., Crews, & City Cost Indexes

04200 | Masonry Units

04220 | Concrete Masonry Units

			CREW	DAILY OUTPUT	LABOR-HOURS	UNIT	2000 BARE COSTS MAT.	LABOR	EQUIP.	TOTAL	TOTAL INCL O&P	
240	6750	8" thick, add				S.F.	1.19			1.19	1.31	240
	6800	12" thick, add					1.64			1.64	1.80	
	7000	Scored ground face, 2 to 5 scores										
	7100	8" x 16" x 4" thick	D-8	345	.116	S.F.	3.54	2.12		5.66	7.50	
	7150	6" thick		310	.129		4.25	2.36		6.61	8.65	
	7200	8" thick		290	.138		4.76	2.52		7.28	9.50	
	7250	12" thick	D-9	265	.181		6	3.23		9.23	12.10	
	8000	Hexagonal face profile units, 8" x 16" units										
	8100	4" thick, hollow	D-8	345	.116	S.F.	1.99	2.12		4.11	5.80	
	8200	Solid		345	.116		2.83	2.12		4.95	6.70	
	8300	6" thick, hollow		310	.129		2.33	2.36		4.69	6.55	
	8350	8" thick, hollow		290	.138		3.18	2.52		5.70	7.75	
	8500	For stacked bond, add						26%				
	8550	For high rise construction, add per story	D-8	67.80	.590	M.S.F.		10.80		10.80	18.25	
	8600	For scored block, add					10%					
	8650	For honed or ground face, per face, add				Ea.	.25			.25	.27	
	8700	For honed or ground end, per end, add				"	2.49			2.49	2.73	
	8750	For bullnose block, add					10%					
	8800	For special color, add					13%					
250	0010	**CONCRETE BLOCK, EXTERIOR** Not including scaffolding										250
	0020	Reinforced alt courses, tooled joints 2 sides, foam inserts										
	0100	Regular, 8" x 16" x 6" thick	D-8	390	.103	S.F.	1.17	1.87		3.04	4.45	
	0200	8" thick		365	.110		1.36	2		3.36	4.88	
	0250	10" thick		355	.113		1.60	2.06		3.66	5.25	
	0300	12" thick	D-9	330	.145		2.47	2.60		5.07	7.10	
	0500	Lightweight, 8" x 16" x 6" thick	D-8	410	.098		2.13	1.78		3.91	5.35	
	0600	8" thick		385	.104		3.09	1.90		4.99	6.60	
	0650	10" thick		370	.108		2.69	1.98		4.67	6.30	
	0700	12" thick	D-9	350	.137		3.38	2.45		5.83	7.85	
260	0010	**CONCRETE BLOCK FOUNDATION WALL** Scaffolding not included										260
	0050	Normal-weight, trowel cut joints, parged 1/2" thick, no reinforcing										
	0200	Hollow, 8" x 16" x 6" thick	D-8	450	.089	S.F.	1.30	1.62		2.92	4.18	
	0250	8" thick		430	.093		1.50	1.70		3.20	4.53	
	0300	10" thick		420	.095		2.04	1.74		3.78	5.20	
	0350	12" thick	D-9	395	.122		2.18	2.17		4.35	6.05	
	0500	Solid, 8" x 16" block, 6" thick	D-8	440	.091		1.65	1.66		3.31	4.62	
	0550	8" thick	"	415	.096		2.08	1.76		3.84	5.25	
	0600	12" thick	D-9	380	.126		3.24	2.25		5.49	7.40	
270	0010	**CONCRETE BLOCK, HIGH STRENGTH** Scaffolding not included										270
	0050	Hollow, reinforced alternate courses, 8" x 16" units										
	0200	3500 psi, 4" thick	D-8	430	.093	S.F.	.96	1.70		2.66	3.93	
	0250	6" thick		400	.100		1.13	1.83		2.96	4.35	
	0300	8" thick		375	.107		1.33	1.95		3.28	4.76	
	0350	12" thick	D-9	340	.141		2.43	2.52		4.95	6.95	
	0500	5000 psi, 4" thick	D-8	430	.093		1.55	1.70		3.25	4.59	
	0550	6" thick		400	.100		2.01	1.83		3.84	5.30	
	0600	8" thick		375	.107		2.97	1.95		4.92	6.55	
	0650	12" thick	D-9	340	.141		3.26	2.52		5.78	7.85	
	1000	For 75% solid block, add					30%					
	1050	For 100% solid block, add					50%					
280	0010	**CONCRETE BLOCK, LINTELS** Scaffolding not included										280
	0100	Including grout and horizontal reinforcing										
	0200	8" x 8" x 8", 1 #4 bar	D-4	300	.133	L.F.	3.38	2.17	.36	5.91	7.80	
	0250	2 #4 bars		295	.136		3.50	2.21	.37	6.08	8	

04200 | Masonry Units

04220	Concrete Masonry Units	CREW	DAILY OUTPUT	LABOR-HOURS	UNIT	2000 BARE COSTS MAT.	LABOR	EQUIP.	TOTAL	TOTAL INCL O&P	
280	0400	8" x 16" x 8", 1 #4 bar	D-4	275	.145	L.F.	5.95	2.37	.40	8.72	10.95	280
	0450	2 #4 bars		270	.148		6.05	2.42	.40	8.87	11.20	
	1000	12" x 8" x 8", 1 #4 bar		275	.145		4.79	2.37	.40	7.56	9.70	
	1100	2 #4 bars		270	.148		4.92	2.42	.40	7.74	9.95	
	1150	2 #5 bars		270	.148		5.05	2.42	.40	7.87	10.10	
	1200	2 #6 bars		265	.151		5.20	2.46	.41	8.07	10.40	
	1500	12" x 16" x 8", 1 #4 bar		250	.160		7.90	2.61	.44	10.95	13.55	
	1600	2 #3 bars		245	.163		7.90	2.66	.45	11.01	13.70	
	1650	2 #4 bars		245	.163		8	2.66	.45	11.11	13.80	
	1700	2 #5 bars		240	.167		8.15	2.72	.46	11.33	14.05	
300	0010	**CONCRETE BRICK** C55, grade N, type I, scaffolding not included										300
	0100	Regular, 4 x 2-1/4 x 8	D-8	220	.182	Ea.	.28	3.32		3.60	5.95	
	0125	Rusticated, 4 x 2-1/4 x 8		220	.182		.31	3.32		3.63	6	
	0150	Frog, 4 x 2-1/4 x 8		220	.182		.30	3.32		3.62	6	
	0200	Double, 4 x 4-7/8 x 8		180	.222		.49	4.06		4.55	7.45	
320	0010	**CONCRETE SCREEN BLOCK** Scaffolding not included										320
	0200	8" x 16", 4" thick	D-8	180	.222	S.F.	1.54	4.06		5.60	8.60	
	0300	8" thick		270	.148		2.26	2.71		4.97	7.10	
	0350	12" x 12", 4" thick		300	.133		1.88	2.44		4.32	6.20	
	0500	8" thick		330	.121		2.30	2.22		4.52	6.30	
340	0010	**COPING** Stock units										340
	0050	Precast concrete, 10" wide, 4" tapers to 3-1/2", 8" wall	D-1	75	.213	L.F.	10	3.81		13.81	17.45	
	0100	12" wide, 3-1/2" tapers to 3", 10" wall		70	.229		10.10	4.08		14.18	18	
	0110	14" wide, 4" tapers to 3-1/2", 12" wall		65	.246		9.50	4.39		13.89	17.90	
	0150	16" wide, 4" tapers to 3-1/2", 14" wall		60	.267		13.90	4.76		18.66	23.50	
	0250	Precast concrete corners		40	.400	Ea.	23.50	7.15		30.65	38	
	0300	Limestone for 12" wall, 4" thick		90	.178	L.F.	13	3.17		16.17	19.65	
	0350	6" thick		80	.200		15.25	3.57		18.82	23	
	0500	Marble, to 4" thick, no wash, 9" wide		90	.178		18.90	3.17		22.07	26.50	
	0550	12" wide		80	.200		28	3.57		31.57	37	
	0700	Terra cotta, 9" wide		90	.178		4.55	3.17		7.72	10.35	
	0750	12" wide		80	.200		7.60	3.57		11.17	14.40	
	0800	Aluminum, for 12" wall		80	.200		10.75	3.57		14.32	17.85	
500	0010	**CONCRETE BLOCK, INTERLOCKING** Scaffolding not incl., mortar incl.										500
	0100	Not including grout or reinforcing										
	0200	8" x 16" units, 2,000 psi, 8" thick	D-1	245	.065	S.F.	1.65	1.17		2.82	3.78	
	0300	12" thick		220	.073		2.48	1.30		3.78	4.93	
	0350	16" thick		185	.086		3.80	1.54		5.34	6.80	
	0400	Including grout & reinforcing, 8" thick	D-4	245	.163		5.25	2.66	.45	8.36	10.75	
	0450	12" thick		220	.182		6.20	2.97	.50	9.67	12.40	
	0500	16" thick		185	.216		7.60	3.53	.59	11.72	15.05	
700	0010	**GLAZED CONCRETE BLOCK** No scaffolding or reinforcing incl.										700
	0100	Single face, 8" x 16" units, 2" thick	D-8	360	.111	S.F.	5.90	2.03		7.93	9.95	
	0200	4" thick		345	.116		6.30	2.12		8.42	10.50	
	0250	6" thick		330	.121		6.60	2.22		8.82	11.05	
	0300	8" thick		310	.129		6.90	2.36		9.26	11.60	
	0350	10" thick		295	.136		8.30	2.48		10.78	13.30	
	0400	12" thick	D-9	280	.171		8.55	3.06		11.61	14.60	
	0700	Double face, 8" x 16" units, 4" thick	D-8	310	.129		10.25	2.36		12.61	15.25	
	0750	6" thick		290	.138		10.55	2.52		13.07	15.90	
	0800	8" thick		270	.148		10.95	2.71		13.66	16.65	
	1000	Jambs, bullnose or square, single face, 8" x 16", 2" thick		315	.127	Ea.	10	2.32		12.32	14.95	
	1050	4" thick		285	.140	"	9.85	2.57		12.42	15.20	

04200 | Masonry Units

04220 | Concrete Masonry Units

		Crew	Daily Output	Labor-Hours	Unit	Mat.	Labor	Equip.	Total	Total Incl O&P	
700	1200 Caps, bullnose or square, 8" x 16", 2" thick	D-8	420	.095	L.F.	6.85	1.74		8.59	10.50	700
	1250 4" thick		380	.105		12.60	1.92		14.52	17.10	
	1500 Cove base, 8" x 16", 2" thick		315	.127		6.70	2.32		9.02	11.30	
	1550 4" thick		285	.140		6.75	2.57		9.32	11.80	
	1600 6" thick		265	.151		6.90	2.76		9.66	12.25	
	1650 8" thick		245	.163		7.30	2.98		10.28	13.10	

04270 | Glass Masonry Units

		Crew	Daily Output	Labor-Hours	Unit	Mat.	Labor	Equip.	Total	Total Incl O&P	
200	0010 **GLASS BLOCK** Scaffolding not included (R04270-700)										200
	0100 Plain, 4" thick, under 1,000 S.F., 6" x 6" block	D-8	115	.348	S.F.	13	6.35		19.35	25	
	0150 8" x 8" block		160	.250		9.25	4.57		13.82	17.95	
	0200 12" x 12" block		175	.229		11.10	4.18		15.28	19.25	
	0300 1,000 to 5,000 S.F., 6" x 6" block		135	.296		12.30	5.40		17.70	22.50	
	0350 8" x 8" block		190	.211		8.45	3.85		12.30	15.80	
	0400 12" x 12" block		215	.186		9.45	3.40		12.85	16.15	
	0500 Over 5,000 S.F., 6" x 6" block		145	.276		12.35	5.05		17.40	22	
	0550 8" x 8" block		215	.186		8.30	3.40		11.70	14.85	
	0600 12" x 12" block		240	.167		9.25	3.05		12.30	15.35	
	0700 For solar reflective blocks, add					100%					
	0800 Under 1,000 S.F., 4" x 8" blocks	D-8	145	.276	S.F.	12.60	5.05		17.65	22.50	
	0850 Over 5,000 S.F.		170	.235		11.70	4.30		16	20	
	1000 Thinline, plain, 3-1/8" thick, under 1,000 S.F., 6" x 6" block		115	.348		9.35	6.35		15.70	21	
	1050 8" x 8" block		160	.250		5.80	4.57		10.37	14.10	
	1200 Over 5,000 S.F., 6" x 6" block		145	.276		8.35	5.05		13.40	17.75	
	1250 8" x 8" block		215	.186		5.15	3.40		8.55	11.45	
	1400 For cleaning block after installation (both sides), add		1,000	.040		.10	.73		.83	1.35	

04290 | Adobe Masonry Units

		Crew	Daily Output	Labor-Hours	Unit	Mat.	Labor	Equip.	Total	Total Incl O&P	
100	0010 **ADOBE BRICK** Unstabilized, with adobe mortar (Southwestern States)										100
	0060 Brick, 4" x 3" x 8" (6.0 per S.F.)	D-8	1.60	25	M	300	455		755	1,100	
	0080 4" x 4" x 8" (4.5 per S.F.)		1.50	26.667		405	485		890	1,275	
	0100 4" x 4" x 14" (2.5 per S.F.)		1.35	29.630		450	540		990	1,400	
	0120 8" x 3" x 16" (3.0 per S.F.)		1.40	28.571		505	520		1,025	1,450	
	0140 4" x 3" x 12" (4.0 per S.F.)		1.45	27.586		465	505		970	1,375	
	0160 6" x 3" x 12" (4.0 per S.F.)		1.45	27.586		415	505		920	1,325	
	0180 4" x 5" x 16" (1.80 per S.F.)		1.35	29.630		495	540		1,035	1,450	
	0200 8" x 4" x 16" (2.25 per S.F.)		1.30	30.769		650	560		1,210	1,675	
	0220 10" x 4" x 14" (2.57 per S.F.)		1.25	32		600	585		1,185	1,650	
	0260 Brick, 4" x 3" x 8" (6.0 per S.F.)		266	.150	S.F.	1.80	2.75		4.55	6.65	
	0280 4" x 4" x 8" (4.5 per S.F.)		333	.120		1.83	2.20		4.03	5.75	
	0300 4" x 4" x 14" (2.5 per S.F.)		540	.074		1.13	1.35		2.48	3.53	
	0320 8" x 3" x 16" (3.0 per S.F.)		466	.086		1.52	1.57		3.09	4.33	
	0340 4" x 3" x 12" (4.0 per S.F.)		362	.110		1.86	2.02		3.88	5.45	
	0360 6" x 3" x 12" (4.0 per S.F.)		362	.110		1.67	2.02		3.69	5.25	
	0380 4" x 5" x 16" (1.80 per S.F.)		600	.067		.89	1.22		2.11	3.04	
	0400 8" x 4" x 16" (2.25 per S.F.)		577	.069		1.46	1.27		2.73	3.76	
	0420 10" x 4" x 14" (2.57 per S.F.)		555	.072		1.54	1.32		2.86	3.93	
	0440 Adobe, partially stabilized, add					10%					
	0480 Fully stabilized, add					25%					

04400 | Stone

04412 | Bluestone

		CREW	DAILY OUTPUT	LABOR-HOURS	UNIT	2000 BARE COSTS MAT.	LABOR	EQUIP.	TOTAL	TOTAL INCL O&P
100	0010 **BLUESTONE** Cut to size									100
	0500 Sills, natural cleft, 10" wide to 6' long, 1-1/2" thick	D-11	70	.343	L.F.	9.30	6.25		15.55	21
	0550 2" thick	"	63	.381		11.05	6.95		18	24
	1000 Stair treads, natural cleft, 12" wide, 6' long, 1-1/2" thick	D-10	115	.348		14.05	6.45	3.49	23.99	30
	1050 2" thick		105	.381		15.70	7.10	3.82	26.62	33.50
	1100 Smooth finish, 1-1/2" thick		115	.348		18.40	6.45	3.49	28.34	35
	1300 Thermal finish, 1-1/2" thick		115	.348		20.50	6.45	3.49	30.44	37
	1350 2" thick		105	.381		23	7.10	3.82	33.92	41.50
800	0010 **WINDOW SILL** Bluestone, thermal top, 10" wide, 1-1/2" thick	D-1	85	.188	S.F.	12.75	3.36		16.11	19.75	800
	0050 2" thick		75	.213	"	14.75	3.81		18.56	22.50
	0100 Cut stone, 5" x 8" plain		48	.333	L.F.	10	5.95		15.95	21
	0200 Face brick on edge, brick, 8" wide		80	.200		1.90	3.57		5.47	8.15
	0400 Marble, 9" wide, 1" thick		85	.188		7.10	3.36		10.46	13.50
	0600 Precast concrete, 4" tapers to 3", 9" wide		70	.229		9.05	4.08		13.13	16.85
	0650 11" wide		60	.267		12.25	4.76		17.01	21.50
	0700 13" wide, 3 1/2" tapers to 2 1/2", 12" wall		50	.320		12.15	5.70		17.85	23
	0900 Slate, colored, unfading, honed, 12" wide, 1" thick		85	.188		15	3.36		18.36	22
	0950 2" thick		70	.229		21	4.08		25.08	30

04413 | Granite

		CREW	DAILY OUTPUT	LABOR-HOURS	UNIT	MAT.	LABOR	EQUIP.	TOTAL	TOTAL INCL O&P
300	0010 **GRANITE** Cut to size									300
	0050 Veneer, polished face, 3/4" to 1-1/2" thick									
	0150 Low price, gray, light gray, etc.	D-10	130	.308	S.F.	18.75	5.75	3.09	27.59	33.50
	0220 High price, red, black, etc.	"	130	.308	"	32	5.75	3.09	40.84	48
	0300 1-1/2" to 2-1/2" thick, veneer									
	0350 Low price, gray, light gray, etc.	D-10	130	.308	S.F.	21	5.75	3.09	29.84	36
	0550 High price, red, black, etc.	"	130	.308	"	36.50	5.75	3.09	45.34	53
	0700 2-1/2" to 4" thick, veneer									
	0750 Low price, gray, light gray, etc.	D-10	110	.364	S.F.	26	6.75	3.65	36.40	44
	0950 High price, red, black, etc.	"	110	.364		41.50	6.75	3.65	51.90	61.50
	1000 For bush hammered finish, deduct					5%				
	1050 Coarse rubbed finish, deduct					10%				
	1100 Honed finish, deduct					5%				
	1150 Thermal finish, deduct					18%				
	2450 For radius under 5', add				L.F.	100%				
	2500 Steps, copings, etc., finished on more than one surface									
	2550 Minimum	D-10	50	.800	C.F.	73	14.90	8	95.90	114
	2600 Maximum	"	50	.800	"	109	14.90	8	131.90	154
	2800 Pavers, 4" x 4" x 4" blocks, split face and joints									
	2850 Minimum	D-11	80	.300	S.F.	10.40	5.45		15.85	20.50
	2900 Maximum	"	80	.300	"	20.50	5.45		25.95	32
	3500 Curbing, city street type, See Division 02770-225									
	4000 Soffits, 2" thick, minimum	D-13	35	1.371	S.F.	31	25.50	11.45	67.95	90.50
	4100 Maximum		35	1.371		62.50	25.50	11.45	99.45	125
	4200 4" thick, minimum		35	1.371		42.50	25.50	11.45	79.45	103
	4300 Maximum		35	1.371		80	25.50	11.45	116.95	144

04414 | Limestone

		CREW	DAILY OUTPUT	LABOR-HOURS	UNIT	MAT.	LABOR	EQUIP.	TOTAL	TOTAL INCL O&P
400	0010 **LIMESTONE** See also Ashlar Veneer, division 04414-400									400
	0020 Veneer facing panels									
	0500 Texture finish, light stick, 4-1/2" thick, 5'x 12'	D-4	300	.133	S.F.	18.95	2.17	.36	21.48	25
	0750 5" thick, 5' x 14' panels	D-10	275	.145		22.50	2.71	1.46	26.67	31
	1000 Sugarcube finish, 2" Thick, 3' x 5' panels		275	.145		11.25	2.71	1.46	15.42	18.55
	1050 3" Thick, 4' x 9' panels		275	.145		11.70	2.71	1.46	15.87	19.05
	1200 4" Thick, 5' x 11' panels		275	.145		15.60	2.71	1.46	19.77	23.50
	1400 Sugarcube, textured finish, 4-1/2" thick, 5' x 12'		275	.145		20	2.71	1.46	24.17	28

04400 | Stone

04414 | Limestone

		Description	CREW	DAILY OUTPUT	LABOR-HOURS	UNIT	MAT.	LABOR	EQUIP.	TOTAL	TOTAL INCL O&P
400	1450	5" thick, 5' x 14' panels	D-10	275	.145	S.F.	23.50	2.71	1.46	27.67	31.50
	2000	Coping, sugarcube finish, top & 2 sides		30	1.333	C.F.	52	25	13.35	90.35	114
	2100	Sills, lintels, jambs, trim, stops, sugarcube finish, average		20	2		47	37	20	104	136
	2150	Detailed		20	2		61.50	37	20	118.50	153
	2300	Steps, extra hard, 14" wide, 6" rise		50	.800	L.F.	42.50	14.90	8	65.40	80.50
	3000	Quoins, plain finish, 6"x12"x12"	D-12	25	1.280	Ea.	102	22.50		124.50	150
	3050	6"x16"x24"	"	25	1.280	"	136	22.50		158.50	188

04415 | Marble

		Description	CREW	DAILY OUTPUT	LABOR-HOURS	UNIT	MAT.	LABOR	EQUIP.	TOTAL	TOTAL INCL O&P
500	0011	**MARBLE** Ashlar, split face, 4" + or - thick, random									
	0040	lengths 1' to 4' & heights 2" to 7-1/2", average	D-8	175	.229	S.F.	13.45	4.18		17.63	22
	0100	Base, polished, 3/4" or 7/8" thick, polished, 6" high	D-10	65	.615	L.F.	12.25	11.45	6.15	29.85	39.50
	0300	Carvings or bas relief, from templates, average		80	.500	S.F.	109	9.30	5	123.30	141
	0350	Maximum		80	.500	"	255	9.30	5	269.30	300
	0600	Columns, cornices, mouldings, etc.									
	0650	Hand or special machine cut, average	D-10	35	1.143	C.F.	109	21.50	11.45	141.95	169
	0700	Maximum	"	35	1.143	"	234	21.50	11.45	266.95	305
	1000	Facing, polished finish, cut to size, 3/4" to 7/8" thick									
	1050	Average	D-10	130	.308	S.F.	17.95	5.75	3.09	26.79	33
	1100	Maximum		130	.308		41.50	5.75	3.09	50.34	59
	1300	1-1/4" thick, average		125	.320		26	5.95	3.21	35.16	42
	1350	Maximum		125	.320		52	5.95	3.21	61.16	70.50
	1500	2" thick, average		120	.333		30	6.20	3.34	39.54	47
	1550	Maximum		120	.333		52	6.20	3.34	61.54	71.50
	2200	Window sills, 6" x 3/4" thick	D-1	85	.188	L.F.	6.75	3.36		10.11	13.15
	2500	Flooring, polished tiles, 12" x 12" x 3/8" thick									
	2510	Thin set, average	D-11	90	.267	S.F.	8.30	4.86		13.16	17.40
	2600	Maximum		90	.267		30	4.86		34.86	41.50
	2700	Mortar bed, average		65	.369		8.45	6.75		15.20	20.50
	2740	Maximum		65	.369		27.50	6.75		34.25	42
	2780	Travertine, 3/8" thick, average	D-10	130	.308		11.50	5.75	3.09	20.34	25.50
	2790	Maximum	"	130	.308		28	5.75	3.09	36.84	44
	2800	Patio tile, non-slip, 1/2" thick, flame finish	D-11	75	.320		11.70	5.85		17.55	23
	2900	Shower or toilet partitions, 7/8" thick partitions									
	3050	3/4" or 1-1/4" thick stiles, polished 2 sides, average	D-11	75	.320	S.F.	35	5.85		40.85	48.50
	3201	Soffits, add to above prices				"	20%	100%			
	3210	Stairs, risers, 7/8" thick x 6" high	D-10	115	.348	L.F.	11.75	6.45	3.49	21.69	27.50
	3360	Treads, 12" wide x 1-1/4" thick	"	115	.348	"	17.15	6.45	3.49	27.09	33.50
	3500	Thresholds, 3' long, 7/8" thick, 4" to 5" wide, plain	D-12	24	1.333	Ea.	12.95	23.50		36.45	54
	3550	Beveled		24	1.333	"	15.10	23.50		38.60	56
	3700	Window stools, polished, 7/8" thick, 5" wide		85	.376	L.F.	12.10	6.65		18.75	24.50

04417 | Sandstone

		Description	CREW	DAILY OUTPUT	LABOR-HOURS	UNIT	MAT.	LABOR	EQUIP.	TOTAL	TOTAL INCL O&P
700	0011	**SANDSTONE OR BROWNSTONE**									
	0100	Sawed face veneer, 2-1/2" thick, to 2' x 4' panels	D-10	130	.308	S.F.	15.35	5.75	3.09	24.19	30
	0150	4" thick, to 3'-6" x 8' panels		100	.400		15.35	7.45	4.01	26.81	34
	0300	Split face, random sizes		100	.400		9.05	7.45	4.01	20.51	27
	0350	Cut stone trim (limestone)									
	0360	Ribbon stone, 4" thick, 5' pieces	D-8	120	.333	Ea.	111	6.10		117.10	132
	0370	Cove stone, 4" thick, 5' pieces		105	.381		111	6.95		117.95	134
	0380	Cornice stone, 10" to 12" wide		90	.444		137	8.10		145.10	165
	0390	Band stone, 4" thick, 5' pieces		145	.276		71	5.05		76.05	86.50
	0410	Window and door trim, 3" to 4" wide		160	.250		60	4.57		64.57	74
	0420	Key stone, 18" long		60	.667		63.50	12.20		75.70	90.50

04400 | Stone

04418 | Slate

		CREW	DAILY OUTPUT	LABOR-HOURS	UNIT	MAT.	LABOR	EQUIP.	TOTAL	TOTAL INCL O&P
0010	**SLATE** Pennsylvania, blue gray to gray black; Vermont,									
0050	Unfading green, mottled green & purple, gray & purple									
0100	Virginia, blue black									
0200	Exterior paving, natural cleft, 1" thick									
0500	24" x 24", Pennsylvania	D-12	120	.267	S.F.	8.65	4.69		13.34	17.45
0550	Vermont		120	.267		11.30	4.69		15.99	20.50
0600	Virginia	▼	120	.267	▼	12.65	4.69		17.34	22
1000	Interior flooring, natural cleft, 1/2" thick									
1300	24" x 24" Pennsylvania	D-12	120	.267	S.F.	5.95	4.69		10.64	14.50
1350	Vermont		120	.267		8.90	4.69		13.59	17.75
1400	Virginia	▼	120	.267	▼	9.20	4.69		13.89	18.05
2000	Facing panels, 1-1/4" thick, to 4' x 4' panels									
2100	Natural cleft finish, Pennsylvania	D-10	180	.222	S.F.	17.60	4.14	2.23	23.97	29
2110	Vermont		180	.222		18.35	4.14	2.23	24.72	29.50
2120	Virginia	▼	180	.222		20.50	4.14	2.23	26.87	32
2150	Sand rubbed finish, surface, add					1.91			1.91	2.10
2200	Honed finish, add					3.77			3.77	4.15
2500	Ribbon, natural cleft finish, 1" thick, to 9 S.F.	D-10	80	.500		8.10	9.30	5	22.40	30
2700	1-1/2" thick		78	.513		10.55	9.55	5.15	25.25	33.50
2850	2" thick	▼	76	.526	▼	12.70	9.80	5.30	27.80	36.50
3000	Roofing, see division 07310-800									
3500	Stair treads, sand finish, 1" thick x 12" wide									
3600	3 L.F. to 6 L.F.	D-10	120	.333	L.F.	14.60	6.20	3.34	24.14	30
3700	Ribbon, sand finish, 1" thick x 12" wide									
3750	To 6 L.F.	D-10	120	.333	L.F.	9.75	6.20	3.34	19.29	25
4000	Stools or sills, sand finish, 1" thick, 6" wide	D-12	160	.200		7.05	3.52		10.57	13.70
4200	10" wide		90	.356		10.65	6.25		16.90	22.50
4400	2" thick, 6" wide		140	.229		11.35	4.02		15.37	19.30
4600	10" wide	▼	90	.356		16.95	6.25		23.20	29.50
4800	For lengths over 3', add				▼	25%				

04420 | Collected Stone

		CREW	DAILY OUTPUT	LABOR-HOURS	UNIT	MAT.	LABOR	EQUIP.	TOTAL	TOTAL INCL O&P
0011	**LIGHTWEIGHT NATURAL STONE** Lava type									
0100	Veneer, rubble face, sawed back, irregular shapes	D-10	130	.308	S.F.	5.10	5.75	3.09	13.94	18.65
0200	Sawed face and back, irregular shapes	"	130	.308	"	5.25	5.75	3.09	14.09	18.85
0011	**ROUGH STONE WALL**, Dry									
0100	Random fieldstone, under 18" thick	D-12	60	.533	C.F.	9.05	9.40		18.45	26
0150	Over 18" thick	"	63	.508	"	12.50	8.95		21.45	29

04430 | Quarried Stone

		CREW	DAILY OUTPUT	LABOR-HOURS	UNIT	MAT.	LABOR	EQUIP.	TOTAL	TOTAL INCL O&P
0011	**ASHLAR VENEER** 4" + or - thk, random or random rectangular (R04430-500)									
0150	Sawn face, split joints, low priced stone	D-8	140	.286	S.F.	4.86	5.20		10.06	14.20
0200	Medium priced stone		130	.308		7.35	5.60		12.95	17.60
0300	High priced stone		120	.333		9.85	6.10		15.95	21
0600	Seam face, split joints, medium price stone		125	.320		8.25	5.85		14.10	18.95
0700	High price stone		120	.333		10.50	6.10		16.60	22
1000	Split or rock face, split joints, medium price stone		125	.320		8.25	5.85		14.10	18.95
1100	High price stone	▼	120	.333	▼	10.75	6.10		16.85	22

04500 | Refractories

04550 | Flue Liners

			CREW	DAILY OUTPUT	LABOR-HOURS	UNIT	2000 BARE COSTS MAT.	LABOR	EQUIP.	TOTAL	TOTAL INCL O&P	
250	0010	**FLUE LINING** Including mortar joints, 8" x 8"	D-1	125	.128	V.L.F.	2.92	2.28		5.20	7.10	250
	0100	8" x 12"		103	.155		3.80	2.77		6.57	8.90	
	0200	12" x 12"		93	.172		5.70	3.07		8.77	11.45	
	0300	12" x 18"		84	.190		8.75	3.40		12.15	15.35	
	0400	18" x 18"		75	.213		12.05	3.81		15.86	19.70	
	0500	20" x 20"		66	.242		12.65	4.33		16.98	21.50	
	0600	24" x 24"		56	.286		14.85	5.10		19.95	25	
	1000	Round, 18" diameter		66	.242		14.25	4.33		18.58	23	
	1100	24" diameter	▼	47	.340	▼	46	6.10		52.10	61	

04580 | Refractory Brick

			CREW	DAILY OUTPUT	LABOR-HOURS	UNIT	MAT.	LABOR	EQUIP.	TOTAL	INCL O&P	
250	0010	**FIRE BRICK** 9" x 2-1/2" x 4-1/2", low duty, 2000° F	D-1	.60	26.667	M	760	475		1,235	1,650	250
	0050	High duty, 3000° F	"	.60	26.667	"	1,550	475		2,025	2,500	
260	0010	**FIRE CLAY** Gray, high duty, 100 lb. bag				Bag	40			40	44	260
	0050	100 lb. drum, premixed (400 brick per drum)				Drum	50			50	55	
270	0010	**FIREPLACE** For prefabricated fireplace, see div. 10305-100										270
	0100	Brick fireplace, not incl. foundations or chimneys										
	0110	30" x 29" opening, incl. chamber, plain brickwork	D-1	.40	40	Ea.	355	715		1,070	1,600	
	0200	Fireplace box only (110 brick)	"	2	8	"	115	143		258	370	
	0300	For elaborate brickwork and details, add					35%	35%				
	0400	For hearth, brick & stone, add	D-1	2	8	Ea.	132	143		275	385	
	0410	For steel angle, damper, cleanouts, add		4	4		95	71.50		166.50	226	
	0600	Plain brickwork, incl. metal circulator		.50	32	▼	700	570		1,270	1,725	
	0800	Face brick only, standard size, 8" x 2-2/3" x 4"	▼	.30	53.333	M	375	950		1,325	2,025	

04700 | Simulated Masonry

04710 | Simulated Brick

			CREW	DAILY OUTPUT	LABOR-HOURS	UNIT	MAT.	LABOR	EQUIP.	TOTAL	INCL O&P	
600	0010	**SIMULATED BRICK** Aluminum, baked on colors	1 Carp	200	.040	S.F.	2.15	.79		2.94	3.72	600
	0050	Fiberglass panels		200	.040		2.25	.79		3.04	3.83	
	0100	Urethane pieces cemented in mastic		150	.053		4.50	1.05		5.55	6.75	
	0150	Vinyl siding panels	▼	200	.040		1.80	.79		2.59	3.33	
	0160	Cement base, brick, incl. mastic	D-1	100	.160	▼	3	2.86		5.86	8.15	
	0170	Corner		50	.320	V.L.F.	7.25	5.70		12.95	17.65	
	0180	Stone face, incl. mastic		100	.160	S.F.	6.90	2.86		9.76	12.45	
	0190	Corner	▼	50	.320	V.L.F.	7.75	5.70		13.45	18.20	

04730 | Simulated Stone

			CREW	DAILY OUTPUT	LABOR-HOURS	UNIT	MAT.	LABOR	EQUIP.	TOTAL	INCL O&P	
600	0010	**SIMULATED STONE**										600
	0100	Insulated fiberglass panels, 5/8" ply backer	L-4	200	.080	S.F.	8.75	1.38		10.13	12	

MASONRY 4

04800 | Masonry Assemblies

04810 | Unit Masonry Assemblies

			CREW	DAILY OUTPUT	LABOR-HOURS	UNIT	MAT.	LABOR	EQUIP.	TOTAL	TOTAL INCL O&P
160	0010	**CHIMNEY** See Div. 03310 for foundation, add to prices below R04210-050										**160**
	0100	Brick, 16" x 16", 8" flue, scaff. not incl.	D-1	18.20	.879	V.L.F.	16.25	15.70		31.95	44.50
	0150	16" x 20" with one 8" x 12" flue		16	1		19.70	17.85		37.55	51.50
	0200	16" x 24" with two 8" x 8" flues		14	1.143		28	20.50		48.50	65.50
	0250	20" x 20" with one 12" x 12" flue		13.70	1.168		19.75	21		40.75	57
	0300	20" x 24" with two 8" x 12" flues		12	1.333		31.50	24		55.50	75.50
	0350	20" x 32" with two 12" x 12" flues		10	1.600		33	28.50		61.50	84.50
	1800	Metal, high temp. steel jacket, factory lining, 24" diam.	E-2	65	.738		174	15.85	15.70	205.55	239
	1900	60" diameter	"	30	1.600		630	34.50	34	698.50	800
	2100	Poured concrete, brick lining, 200' high x 10' diam.					5,350			5,350	5,875
	2200	200' high x 18' diameter									4,995
	2400	250' high x 9' diameter									3,920
	2500	300' x 14' diameter									4,690
	2700	400' high x 18' diameter									5,255
	2800	500' x 20' diameter					9,500			9,500	10,500
170	0010	**COLUMNS** Brick, scaffolding not included R04210-100										**170**
	0050	8" x 8", 9 brick	D-1	56	.286	V.L.F.	3.50	5.10		8.60	12.50
	0100	12" x 8", 13.5 brick		37	.432		5.25	7.70		12.95	18.80
	0200	12" x 12", 20 brick		25	.640		7.80	11.40		19.20	28
	0300	16" x 12", 27 brick		19	.842		10.50	15.05		25.55	37
	0400	16" x 16", 36 brick		14	1.143		14	20.50		34.50	50
	0500	20" x 16", 45 brick		11	1.455		17.50	26		43.50	63.50
	0600	20" x 20", 56 brick		9	1.778		22	31.50		53.50	77.50
	0700	24" x 20", 68 brick		7	2.286		26.50	41		67.50	98
	0800	24" x 24", 81 brick		6	2.667		31.50	47.50		79	115
	1000	36" x 36", 182 brick		3	5.333		71	95		166	239
180	0010	**CONCRETE BLOCK COLUMN** or pilaster. Scaffolding not included.										**180**
	0050	Including vertical reinforcing (4-#4 bars) and grout										
	0160	1 piece unit, 16" x 16"	D-1	26	.615	V.L.F.	8.90	11		19.90	28.50
	0170	2 piece units, 16" x 20"		24	.667		12.05	11.90		23.95	33.50
	0180	20" x 20"		22	.727		17.65	13		30.65	41.50
	0190	22" x 24"		18	.889		21.50	15.85		37.35	51
	0200	20" x 32"		14	1.143		22	20.50		42.50	59
210	0010	**CONCRETE BLOCK, PARTITIONS** Scaffolding not included R04220-200										**210**
	0100	Acoustical slotted block										
	0200	NRC .50, type A, 4" thick	D-8	315	.127	S.F.	2.34	2.32		4.66	6.50
	0210	NRC .65 type R, 4" thick		315	.127		2.85	2.32		5.17	7.05
	0250	NRC .50, type A, 6" thick		290	.138		2.63	2.52		5.15	7.15
	0260	NRC .65 type R, 6" thick		290	.138		3.54	2.52		6.06	8.15
	0400	NRC .45 type A-1, 8" thick		265	.151		4.59	2.76		7.35	9.70
	0410	NRC .55, type Q, 8" thick		265	.151		5.15	2.76		7.91	10.35
	0500	NRC .60 type R, 8" thick		265	.151		3.68	2.76		6.44	8.70
	0600	NRC .65 type RR, 8" thick		265	.151		10.50	2.76		13.26	16.20
	0700	NRC .65 type 4R-RF, 8" thick		265	.151		5.05	2.76		7.81	10.20
	0710	NRC .70 type R, 12" thick		245	.163		5.45	2.98		8.43	11.05
	1000	Lightweight block, tooled joints, 2 sides, hollow										
	1100	Not reinforced, 8" x 16" x 4" thick	D-8	440	.091	S.F.	.91	1.66		2.57	3.81
	1150	6" thick		410	.098		1.21	1.78		2.99	4.35
	1200	8" thick		385	.104		1.52	1.90		3.42	4.90
	1250	10" thick		370	.108		1.98	1.98		3.96	5.55
	1300	12" thick	D-9	350	.137		2.12	2.45		4.57	6.50
	2000	Not reinforced, 8" x 24" x 4" thick, hollow		460	.104		.65	1.86		2.51	3.87
	2100	6" thick		440	.109		.86	1.95		2.81	4.24

04800 | Masonry Assemblies

04810 | Unit Masonry Assemblies

			CREW	DAILY OUTPUT	LABOR-HOURS	UNIT	2000 BARE COSTS MAT.	LABOR	EQUIP.	TOTAL	TOTAL INCL O&P	
210	2150	8" thick	D-9	415	.116	S.F.	1.09	2.06		3.15	4.70	210
	2200	10" thick		385	.125		1.41	2.23		3.64	5.30	
	2250	12" thick		365	.132		1.52	2.35		3.87	5.65	
	2800	Solid, not reinforced, 8" x 16" x 2" thick	D-8	440	.091		.84	1.66		2.50	3.74	
	2900	4" thick		420	.095		1.06	1.74		2.80	4.12	
	2950	6" thick		390	.103		2.07	1.87		3.94	5.45	
	3000	8" thick		365	.110		3.03	2		5.03	6.70	
	3050	10" thick		350	.114		2.79	2.09		4.88	6.60	
	3100	12" thick	D-9	330	.145		3.31	2.60		5.91	8.05	
	4000	Regular block, tooled joints, 2 sides, hollow										
	4100	Not reinforced, 8" x 16" x 4" thick	D-8	430	.093	S.F.	.77	1.70		2.47	3.72	
	4150	6" thick		400	.100		1.06	1.83		2.89	4.27	
	4200	8" thick		375	.107		1.26	1.95		3.21	4.69	
	4250	10" thick		360	.111		1.80	2.03		3.83	5.40	
	4300	12" thick	D-9	340	.141		1.93	2.52		4.45	6.40	
	4500	Reinforced alternate courses, 8" x 16" x 4" thick	D-8	425	.094		.84	1.72		2.56	3.84	
	4550	6" thick		395	.101		1.14	1.85		2.99	4.39	
	4600	8" thick		370	.108		1.34	1.98		3.32	4.83	
	4650	10" thick		355	.113		1.96	2.06		4.02	5.65	
	4700	12" thick	D-9	335	.143		2.02	2.56		4.58	6.55	
	4900	Solid, not reinforced, 2" thick	D-8	435	.092		.69	1.68		2.37	3.61	
	5000	3" thick		430	.093		.86	1.70		2.56	3.83	
	5050	4" thick		415	.096		1.15	1.76		2.91	4.25	
	5100	6" thick		385	.104		1.41	1.90		3.31	4.78	
	5150	8" thick		360	.111		1.84	2.03		3.87	5.45	
	5200	12" thick	D-9	325	.148		2.99	2.64		5.63	7.75	
	5500	Solid, reinforced alternate courses, 4" thick	D-8	420	.095		1.20	1.74		2.94	4.27	
	5550	6" thick		380	.105		1.47	1.92		3.39	4.88	
	5600	8" thick		355	.113		1.90	2.06		3.96	5.60	
	5650	12" thick	D-9	320	.150		2.65	2.68		5.33	7.45	
300	0010	**FACING PANELS** Stone aggregate mounted on plywood,										300
	0020	see division 07440-200										
400	0010	**LINTELS** See division 05120-480										400
650	0010	**WALLS** Building brick, including mortar										650
	0060	Includes 3% brick waste and 25% mortar waste										
	0140	4" thick, facing, 4" x 2-2/3" x 8"	D-8	1.45	27.586	M	289	505		794	1,175	
	0150	4" thick, as back-up, 6.75 bricks per S.F.		1.60	25		289	455		744	1,100	
	0204	8" thick, 13.50 bricks per S.F.		1.80	22.222		296	405		701	1,025	
	0250	12" thick, 20.25 bricks per S.F.		1.90	21.053		298	385		683	980	
	0304	16" thick, 27.00 bricks per S.F.		2	20		300	365		665	950	
	0500	Reinforced, 4" wall, 4" x 2-2/3" x 8"		1.40	28.571		300	520		820	1,225	
	0550	8" thick, 13.50 bricks per S.F.		1.75	22.857		365	420		785	1,100	
	0600	12" thick, 20.25 bricks per S.F.		1.85	21.622		365	395		760	1,075	
	0650	16" thick, 27.00 bricks per S.F.		1.95	20.513		370	375		745	1,050	
	0790	Alternate method of figuring by square foot										
	0800	4" wall, face, 4" x 2-2/3" x 8"	D-8	215	.186	S.F.	2.65	3.40		6.05	8.65	
	0850	4" thick, as back up, 6.75 bricks per S.F.		240	.167		1.95	3.05		5	7.30	
	0900	8" thick wall, 13.50 brick per S.F.		135	.296		4	5.40		9.40	13.55	
	1000	12" thick wall, 20.25 bricks per S.F.		95	.421		6	7.70		13.70	19.65	
	1050	16" thick wall, 27.00 bricks per S.F.		75	.533		8.10	9.75		17.85	25.50	
	1200	Reinforced, 4" x 2-2/3" x 8", 4" wall		205	.195		1.97	3.57		5.54	8.20	
	1250	8" thick wall, 13.50 brick per S.F.		130	.308		4.01	5.60		9.61	13.90	
	1300	12" thick wall, 20.25 bricks per S.F.		90	.444		6.05	8.10		14.15	20.50	
	1350	16" thick wall, 27.00 bricks per S.F.		70	.571		8.15	10.45		18.60	26.50	

04800 | Masonry Assemblies

04810 | Unit Masonry Assemblies

			CREW	DAILY OUTPUT	LABOR-HOURS	UNIT	MAT.	LABOR	EQUIP.	TOTAL	TOTAL INCL O&P	
670	0010	**STEPS** With select common	D-1	.30	53.333	M	305	950		1,255	1,925	670

04840 | Prefabricated Masonry Panels

			CREW	DAILY OUTPUT	LABOR-HOURS	UNIT	MAT.	LABOR	EQUIP.	TOTAL	TOTAL INCL O&P	
900	0010	**WALL PANELS** Prefabricated, 4" thick, minimum	C-11	775	.072	S.F.	4.58	1.46	2.02	8.06	9.75	900
	0100	Maximum	"	500	.112		6.65	2.26	3.13	12.04	14.65	
	0200	4" brick & 2" concrete back-up, add					2.70					
	0300	4" brick & 1" urethane & 3" concrete back-up, add					3.50					

04900 | Masonry Restoration & Cleaning

04910 | Unit Masonry Restoration

			CREW	DAILY OUTPUT	LABOR-HOURS	UNIT	MAT.	LABOR	EQUIP.	TOTAL	TOTAL INCL O&P	
600	0010	**NEEDLE BEAM MASONRY** Incl. shoring 10' x 10' opening										600
	0400	Block, concrete, 8" thick	B-9	7.10	5.634	Ea.	32.50	83.50	25.50	141.50	207	
	0420	12" thick		6.70	5.970		46.50	88.50	27	162	233	
	0800	Brick, 4" thick with 8" backup block		5.70	7.018		46.50	104	31.50	182	265	
	1000	Brick, solid, 8" thick		6.20	6.452		32.50	96	29	157.50	232	
	1040	12" thick		4.90	8.163		46.50	121	37	204.50	300	
	1080	16" thick		4.50	8.889		75	132	40	247	350	
	2000	Add for additional floors of shoring	B-1	6	4		32.50	60.50		93	140	
750	0010	**SAWING** Brick or block by hand, per inch depth	D-5	300	.027	L.F.		.42		.42	.71	750

04930 | Unit Masonry Cleaning

			CREW	DAILY OUTPUT	LABOR-HOURS	UNIT	MAT.	LABOR	EQUIP.	TOTAL	TOTAL INCL O&P	
200	0010	**CLEAN AND POINT** Smooth brick	1 Bric	300	.027	S.F.	.20	.53		.73	1.12	200
	0100	Rough brick	"	265	.030	"	.21	.60		.81	1.25	
750	0010	**STEAM CLEAN** Building, not incl. scaffolding, minimum	B-9	3,000	.013	S.F.		.20	.06	.26	.41	750
	0100	Maximum		1,500	.027			.40	.12	.52	.81	
	0300	Common face brick		1,750	.023			.34	.10	.44	.69	
	0400	Wire cut face brick		1,250	.032			.48	.14	.62	.97	
900	0010	**WASHING BRICK** Acid wash, smooth brick R04930-100	1 Bric	560	.014	S.F.	.23	.29		.52	.73	900
	0050	Rough brick		400	.020		.23	.40		.63	.93	
	0060	Stone, acid wash		600	.013		.23	.27		.50	.70	
	1000	Muriatic acid, price per gallon in 5 gallon lots				Gal.	4			4	4.40	

For information about Means Estimating Seminars, see yellow pages 11 and 12 in back of book

Division 5 Metals

Estimating Tips

05050 Basic Metal Materials & Methods

- Nuts, bolts, washers, connection angles and plates can add a significant amount to both the tonnage of a structural steel job as well as the estimated cost. As a rule of thumb add 10% to the total weight to account for these accessories.
- Type 2 steel construction, commonly referred to as "simple construction," consists generally of field bolted connections with lateral bracing supplied by other elements of the building, such as masonry walls or x-bracing. The estimator should be aware, however, that shop connections may be accomplished by welding or bolting. The method may be particular to the fabrication shop and may have an impact on the estimated cost.

05200 Metal Joists

- In any given project the total weight of open web steel joists is determined by the loads to be supported and the design. However, economies can be realized in minimizing the amount of labor used to place the joists. This is done by maximizing the joist spacing and therefore minimizing the number of joists required to be installed on the job. Certain spacings and locations may be required by the design, but in other cases maximizing the spacing and keeping it as uniform as possible will keep the costs down.

05300 Metal Deck

- The takeoff and estimating of metal deck involves more than simply the area of the floor or roof and the type of deck specified or shown on the drawings. Many different sizes and types of openings may exist. Small openings for individual pipes or conduits may be drilled after the floor/roof is installed, but larger openings may require special deck lengths as well as reinforcing or structural support. The estimator should determine who will be supplying this reinforcing. Additionally, some deck terminations are part of the deck package, such as screed angles and pour stops, and others will be part of the steel contract, such as angles attached to structural members and cast-in-place angles and plates. The estimator must ensure that all pieces are accounted for in the complete estimate.

05500 Metal Fabrications

- The most economical steel stairs are those that use common materials, standard details and most importantly, a uniform and relatively simple method of field assembly. Commonly available A36 channels and plates are very good choices for the main stringers of the stairs, as are angles and tees for the carrier members. Risers and treads are usually made by specialty shops, and it is most economical to use a typical detail in as many places as possible. The stairs should be pre-assembled and shipped directly to the site. The field connections should be simple and straightforward to be accomplished efficiently and with a minimum of equipment and labor.

Reference Numbers

Reference numbers are shown in bold squares at the beginning of some major classifications. These numbers refer to related items in the Reference Section. The reference information may be an estimating procedure, an alternate pricing method or technical information.

Note: Not all subdivisions listed here necessarily appear in this publication.

05050 | Basic Materials & Methods

05090 | Metal Fastenings

			CREW	DAILY OUTPUT	LABOR-HOURS	UNIT	2000 BARE COSTS MAT.	LABOR	EQUIP.	TOTAL	TOTAL INCL O&P
080	0010	**ANCHOR BOLTS**									
	0020	See also divisions 03150 and 04080									
	0100	J-type, incl. nut, washer, 1/2" diameter x 6" long	2 Carp	70	.229	Ea.	1.11	4.50		5.61	8.90
	0110	12" long		65	.246		1.27	4.85		6.12	9.70
	0120	18" long		60	.267		1.73	5.25		6.98	10.90
	0130	3/4" diameter x 8" long		50	.320		1.71	6.30		8.01	12.70
	0140	12" long		45	.356		2.11	7		9.11	14.30
	0150	18" long		40	.400		2.58	7.90		10.48	16.35
	0160	1" diameter x 12" long		35	.457		3.20	9		12.20	18.95
	0170	18" long		30	.533		3.61	10.50		14.11	22
	0180	24" long		25	.640		4.25	12.60		16.85	26
	0190	36" long		20	.800		5.60	15.75		21.35	33
	0200	1-1/2" diameter x 18" long		22	.727		8.95	14.35		23.30	34.50
	0210	24" long		16	1		10.15	19.70		29.85	45
	0300	L-type, incl. hex nuts, 3/4" diameter x 12" long		45	.356		1.11	7		8.11	13.20
	0310	18" long		40	.400		1.40	7.90		9.30	15.05
	0320	24" long		35	.457		1.69	9		10.69	17.30
	0330	30" long		30	.533		2.18	10.50		12.68	20.50
	0340	36" long		25	.640		2.45	12.60		15.05	24
	0350	1" diameter x 12" long		35	.457		2.24	9		11.24	17.90
	0360	18" long		30	.533		2.74	10.50		13.24	21
	0370	24" long		25	.640		3.29	12.60		15.89	25
	0380	30" long		23	.696		3.82	13.70		17.52	27.50
	0390	36" long		20	.800		4.34	15.75		20.09	32
	0400	42" long		18	.889		5.50	17.50		23	36
	0410	48" long		15	1.067		6.30	21		27.30	43
	0420	1-1/4" diameter x 18" long		25	.640		5	12.60		17.60	27
	0430	24" long		20	.800		5.95	15.75		21.70	33.50
	0440	30" long		20	.800		6.75	15.75		22.50	34.50
	0450	36" long		18	.889		7.65	17.50		25.15	38.50
	0460	42" long		16	1		8.60	19.70		28.30	43.50
	0470	48" long		14	1.143		9.65	22.50		32.15	49
	0480	54" long		12	1.333		10.80	26.50		37.30	57
	0490	60" long		10	1.600		11.65	31.50		43.15	67
	0500	1-1/2" diameter x 18" long		22	.727		7.95	14.35		22.30	33.50
	0510	24" long		19	.842		9	16.60		25.60	38.50
	0520	30" long		17	.941		10.30	18.55		28.85	43.50
	0530	36" long		16	1		11.95	19.70		31.65	47
	0540	42" long		15	1.067		13.55	21		34.55	51
	0550	48" long		13	1.231		15.20	24.50		39.70	58
	0560	54" long		11	1.455		16.90	28.50		45.40	67.50
	0570	60" long		9	1.778		18.55	35		53.55	80.50
	0580	1-3/4" diameter x 18" long		20	.800		12.60	15.75		28.35	41
	0590	24" long		18	.889		14.45	17.50		31.95	46
	0600	30" long		17	.941		16.35	18.55		34.90	50
	0610	36" long		16	1		18.55	19.70		38.25	54.50
	0620	42" long		14	1.143		20.50	22.50		43	61
	0630	48" long		12	1.333		22.50	26.50		49	70
	0640	54" long		10	1.600		25	31.50		56.50	81.50
	0650	60" long		8	2		27	39.50		66.50	97
	0660	2" diameter x 24" long		17	.941		18.85	18.55		37.40	52.50
	0670	30" long		15	1.067		21	21		42	59
	0680	36" long		13	1.231		23	24.50		47.50	67
	0690	42" long		11	1.455		25.50	28.50		54	77
	0700	48" long		10	1.600		28.50	31.50		60	85
	0710	54" long		9	1.778		30	35		65	93

05050 | Basic Materials & Methods

05090 | Metal Fastenings

			CREW	DAILY OUTPUT	LABOR-HOURS	UNIT	2000 BARE COSTS MAT.	LABOR	EQUIP.	TOTAL	TOTAL INCL O&P	
080	0720	60" long	2 Carp	8	2	Ea.	32.50	39.50		72	103	080
	0730	66" long		7	2.286		34.50	45		79.50	115	
	0740	72" long	↓	6	2.667	↓	37.50	52.50		90	132	
150	0010	**BOLTS & HEX NUTS** Steel, A307										150
	0100	1/4" diameter, 1/2" long				Ea.	.05			.05	.06	
	0200	1" long					.06			.06	.07	
	0300	2" long					.07			.07	.08	
	0400	3" long					.10			.10	.11	
	0500	4" long					.15			.15	.17	
	0600	3/8" diameter, 1" long					.10			.10	.11	
	0700	2" long					.14			.14	.15	
	0800	3" long					.17			.17	.19	
	0900	4" long					.22			.22	.24	
	1000	5" long					.26			.26	.29	
	1100	1/2" diameter, 1-1/2" long					.22			.22	.24	
	1200	2" long					.25			.25	.28	
	1300	4" long					.37			.37	.41	
	1400	6" long					.49			.49	.54	
	1500	8" long					.66			.66	.73	
	1600	5/8" diameter, 1-1/2" long					.38			.38	.42	
	1700	2" long					.42			.42	.46	
	1800	4" long					.59			.59	.65	
	1900	6" long					.80			.80	.88	
	2000	8" long					1.06			1.06	1.17	
	2100	10" long					1.26			1.26	1.39	
	2200	3/4" diameter, 2" long					.66			.66	.73	
	2300	4" long					.97			.97	1.07	
	2400	6" long					1.23			1.23	1.35	
	2500	8" long					1.54			1.54	1.69	
	2600	10" long					2.07			2.07	2.28	
	2700	12" long					2.36			2.36	2.60	
	2800	1" diameter, 3" long					1.61			1.61	1.77	
	2900	6" long					2.36			2.36	2.60	
	3000	12" long					4.74			4.74	5.20	
	3100	For galvanized, add					75%					
	3200	For stainless, add				↓	350%					
340	0010	**DRILLING** For anchors, up to 4" deep, incl. bit and layout										340
	0050	in concrete or brick walls and floors, no anchor										
	0100	Holes, 1/4" diameter	1 Carp	75	.107	Ea.	.09	2.10		2.19	3.69	
	0150	For each additional inch of depth, add		430	.019		.02	.37		.39	.65	
	0200	3/8" diameter		63	.127		.08	2.50		2.58	4.37	
	0250	For each additional inch of depth, add		340	.024		.02	.46		.48	.81	
	0300	1/2" diameter		50	.160		.07	3.15		3.22	5.50	
	0350	For each additional inch of depth, add		250	.032		.02	.63		.65	1.10	
	0400	5/8" diameter		48	.167		.14	3.28		3.42	5.80	
	0450	For each additional inch of depth, add		240	.033		.03	.66		.69	1.17	
	0500	3/4" diameter		45	.178		.14	3.50		3.64	6.15	
	0550	For each additional inch of depth, add		220	.036		.04	.72		.76	1.27	
	0600	7/8" diameter		43	.186		.17	3.67		3.84	6.50	
	0650	For each additional inch of depth, add		210	.038		.04	.75		.79	1.34	
	0700	1" diameter		40	.200		.20	3.94		4.14	6.95	
	0750	For each additional inch of depth, add		190	.042		.05	.83		.88	1.48	
	0800	1-1/4" diameter		38	.211		.30	4.15		4.45	7.45	
	0850	For each additional inch of depth, add	↓	180	.044	↓	.07	.88		.95	1.58	

METALS

05050 | Basic Materials & Methods

05090 | Metal Fastenings

		CREW	DAILY OUTPUT	LABOR-HOURS	UNIT	2000 BARE COSTS MAT.	LABOR	EQUIP.	TOTAL	TOTAL INCL O&P
340 0900	1-1/2" diameter	1 Carp	35	.229	Ea.	.48	4.50		4.98	8.20	**340**
0950	For each additional inch of depth, add	↓	165	.048	↓	.12	.96		1.08	1.77
1000	For ceiling installations, add						40%			
1100	Drilling & layout for drywall or plaster walls, no anchor									
1200	Holes, 1/4" diameter	1 Carp	150	.053	Ea.	.01	1.05		1.06	1.81
1300	3/8" diameter		140	.057		.01	1.13		1.14	1.94
1400	1/2" diameter		130	.062		.01	1.21		1.22	2.09
1500	3/4" diameter		120	.067		.02	1.31		1.33	2.27
1600	1" diameter		110	.073		.03	1.43		1.46	2.48
1700	1-1/4" diameter		100	.080		.04	1.58		1.62	2.74
1800	1-1/2" diameter	↓	90	.089		.06	1.75		1.81	3.07
1900	For ceiling installations, add				↓		40%			
380 0010	**EXPANSION ANCHORS** & shields										**380**
0100	Bolt anchors for concrete, brick or stone, no layout and drilling									
0200	Expansion shields, zinc, 1/4" diameter, 1" long, single	1 Carp	90	.089	Ea.	.90	1.75		2.65	3.99
0300	1-3/8" long, double		85	.094		.99	1.85		2.84	4.27
0400	3/8" diameter, 2" long, single		85	.094		1.49	1.85		3.34	4.82
0500	2" long, double		80	.100		1.84	1.97		3.81	5.40
0600	1/2" diameter, 2-1/2" long, single		80	.100		2.46	1.97		4.43	6.10
0700	2-1/2" long, double		75	.107		2.38	2.10		4.48	6.20
0800	5/8" diameter, 2-5/8" long, single		75	.107		3.52	2.10		5.62	7.45
0900	3" long, double		70	.114		3.52	2.25		5.77	7.75
1000	3/4" diameter, 2-3/4" long, single		70	.114		5.25	2.25		7.50	9.60
1100	4" long, double		65	.123		7	2.42		9.42	11.85
1410	Concrete anchor, w/rod & epoxy cartridge, 1-3/4" diameter x 15" long	E-22	20	1.200		75.50	24.50		100	125
1415	18" long		17	1.412		91	29		120	150
1420	2" diameter x 18" long		16	1.500		116	30.50		146.50	180
1425	24" long		15	1.600		151	32.50		183.50	222
1430	Chemical anchor, w/rod & epoxy cartridge, 3/4" diam. x 9-1/2" long		27	.889		11.10	18.15		29.25	43
1435	1" diameter x 11-3/4" long		24	1		21	20.50		41.50	58
1440	1-1/4" diameter x 14" long	↓	21	1.143		40	23.50		63.50	84
1500	Self drilling anchor, snap-off, for 1/4" diameter bolt	1 Carp	26	.308		.94	6.05		6.99	11.45
1600	3/8" diameter bolt		23	.348		1.41	6.85		8.26	13.30
1700	1/2" diameter bolt		20	.400		2.13	7.90		10.03	15.85
1800	5/8" diameter bolt		18	.444		3.62	8.75		12.37	19
1900	3/4" diameter bolt	↓	16	.500	↓	6.55	9.85		16.40	24
2100	Hollow wall anchors for gypsum wall board, plaster or tile									
2300	1/8" diameter, short				Ea.	.25			.25	.28
2400	Long					.29			.29	.32
2500	3/16" diameter, short					.59			.59	.65
2600	Long					.65			.65	.71
2700	1/4" diameter, short					.75			.75	.83
2800	Long					.87			.87	.96
3000	Toggle bolts, bright steel, 1/8" diameter, 2" long	1 Carp	85	.094		.26	1.85		2.11	3.47
3100	4" long		80	.100		.33	1.97		2.30	3.74
3200	3/16" diameter, 3" long		80	.100		.37	1.97		2.34	3.79
3300	6" long		75	.107		.56	2.10		2.66	4.22
3400	1/4" diameter, 3" long		75	.107		.41	2.10		2.51	4.05
3500	6" long		70	.114		.61	2.25		2.86	4.53
3600	3/8" diameter, 3" long		70	.114		.90	2.25		3.15	4.85
3700	6" long		60	.133		1.32	2.63		3.95	5.95
3800	1/2" diameter, 4" long		60	.133		2.86	2.63		5.49	7.65
3900	6" long	↓	50	.160	↓	3.81	3.15		6.96	9.60
4000	Nailing anchors									
4100	Nylon nailing anchor, 1/4" diameter, 1" long				C	15.90			15.90	17.50
4200	1-1/2" long				↓	20.50			20.50	22.50

05050 | Basic Materials & Methods

05090 | Metal Fastenings

			CREW	DAILY OUTPUT	LABOR-HOURS	UNIT	2000 BARE COSTS MAT.	LABOR	EQUIP.	TOTAL	TOTAL INCL O&P	
380	4300	2" long				C	34			34	37.50	380
	4400	Metal nailing anchor, 1/4" diameter, 1" long					22			22	24	
	4500	1-1/2" long					30			30	33	
	4600	2" long					38.50			38.50	42	
	5000	Screw anchors for concrete, masonry,										
	5100	stone & tile, no layout or drilling included										
	5200	Jute fiber, #6, #8, & #10, 1" long				Ea.	.19			.19	.21	
	5300	#12, 1-1/2" long					.28			.28	.31	
	5400	#14, 2" long					.44			.44	.48	
	5500	#16, 2" long					.46			.46	.51	
	5600	#20, 2" long					.74			.74	.81	
	5700	Lag screw shields, 1/4" diameter, short					.43			.43	.47	
	5800	Long					.50			.50	.55	
	5900	3/8" diameter, short					.74			.74	.81	
	6000	Long					.84			.84	.92	
	6100	1/2" diameter, short					1.11			1.11	1.22	
	6200	Long					1.31			1.31	1.44	
	6300	3/4" diameter, short					2.42			2.42	2.66	
	6400	Long					3.09			3.09	3.40	
	6600	Lead, #6 & #8, 3/4" long					.18			.18	.20	
	6700	#10 - #14, 1-1/2" long					.27			.27	.30	
	6800	#16 & #18, 1-1/2" long					.37			.37	.41	
	6900	Plastic, #6 & #8, 3/4" long					.04			.04	.04	
	7000	#8 & #10, 7/8" long					.04			.04	.04	
	7100	#10 & #12, 1" long					.05			.05	.06	
	7200	#14 & #16, 1-1/2" long					.06			.06	.07	
	8000	Wedge anchors, not including layout or drilling										
	8050	Carbon steel, 1/4" diameter, 1-3/4" long	1 Carp	150	.053	Ea.	.33	1.05		1.38	2.16	
	8100	3 1/4" long		145	.055		.50	1.09		1.59	2.41	
	8150	3/8" diameter, 2-1/4" long		150	.053		.60	1.05		1.65	2.46	
	8200	5" long		145	.055		1.06	1.09		2.15	3.03	
	8250	1/2" diameter, 2-3/4" long		140	.057		.89	1.13		2.02	2.91	
	8300	7" long		130	.062		1.56	1.21		2.77	3.80	
	8350	5/8" diameter, 3-1/2" long		130	.062		1.68	1.21		2.89	3.93	
	8400	8-1/2" long		115	.070		3.29	1.37		4.66	5.95	
	8450	3/4" diameter, 4-1/4" long		115	.070		2.18	1.37		3.55	4.75	
	8500	10" long		100	.080		4.97	1.58		6.55	8.15	
	8550	1" diameter, 6" long		100	.080		7.45	1.58		9.03	10.90	
	8575	9" long		80	.100		9.70	1.97		11.67	14.10	
	8600	12" long		80	.100		10.50	1.97		12.47	14.95	
	8650	1-1/4" diameter, 9" long		70	.114		14.70	2.25		16.95	20	
	8700	12" long		60	.133		15.05	2.63		17.68	21	
	8750	For type 303 stainless steel, add					350%					
	8800	For type 316 stainless steel, add					450%					
420	0010	**HIGH STRENGTH BOLTS**	R05090-510									420
	0020	A325 Type 1, structural steel, bolt-nut-washer set										
	0100	1/2" diameter x 1-1/2" long	1 Sswk	120	.067	Ea.	.24	1.42		1.66	3.05	
	0120	2" long		120	.067		.27	1.42		1.69	3.09	
	0150	3" long		120	.067		.33	1.42		1.75	3.15	
	0170	5/8" diameter x 1-1/2" long		120	.067		.39	1.42		1.81	3.22	
	0180	2" long		120	.067		.43	1.42		1.85	3.26	
	0190	3" long		120	.067		.51	1.42		1.93	3.35	
	0200	3/4" diameter x 2" long		120	.067		.60	1.42		2.02	3.45	
	0220	3" long		115	.070		.70	1.48		2.18	3.68	
	0250	4" long		110	.073		.82	1.55		2.37	3.95	
	0300	6" long		105	.076		1.06	1.62		2.68	4.36	

05050 | Basic Materials & Methods

05090 | Metal Fastenings

		CREW	DAILY OUTPUT	LABOR-HOURS	UNIT	2000 BARE COSTS MAT.	LABOR	EQUIP.	TOTAL	TOTAL INCL O&P	
420	0350	8" long	1 Sswk	95	.084	Ea.	1.57	1.79		3.36	5.25	**420**
	0360	7/8" diameter x 2" long	R05090 -510	115	.070		.89	1.48		2.37	3.89	
	0365	3" long		110	.073		1.05	1.55		2.60	4.20	
	0370	4" long		105	.076		1.19	1.62		2.81	4.50	
	0380	6" long		100	.080		1.50	1.70		3.20	5	
	0390	8" long		90	.089		2.07	1.89		3.96	6	
	0400	1" diameter x 2" long		105	.076		1.26	1.62		2.88	4.58	
	0420	3" long		105	.076		1.45	1.62		3.07	4.79	
	0450	4" long		105	.076		1.60	1.62		3.22	4.95	
	0500	6" long		90	.089		1.95	1.89		3.84	5.85	
	0550	8" long		85	.094		2.82	2		4.82	7.05	
	0600	1-1/4" diameter x 3" long		85	.094		2.88	2		4.88	7.10	
	0650	4" long		80	.100		3.18	2.13		5.31	7.70	
	0700	6" long		75	.107		3.73	2.27		6	8.55	
	0750	8" long		70	.114		4.65	2.43		7.08	9.90	
	1020	A490, bolt-nut-washer set										
	1100	1/2" diameter x 1-1/2" long	1 Sswk	120	.067	Ea.	.30	1.42		1.72	3.12	
	1120	2" long		120	.067		.35	1.42		1.77	3.18	
	1150	3" long		120	.067		.41	1.42		1.83	3.24	
	1170	5/8" diameter x 1-1/2" long		120	.067		.42	1.42		1.84	3.25	
	1180	2" long		120	.067		.45	1.42		1.87	3.29	
	1190	3" long		120	.067		.56	1.42		1.98	3.41	
	1200	3/4" diameter x 2" long		120	.067		.69	1.42		2.11	3.55	
	1220	3" long		115	.070		1.06	1.48		2.54	4.08	
	1250	4" long		110	.073		.93	1.55		2.48	4.07	
	1300	6" long		105	.076		1.20	1.62		2.82	4.51	
	1350	8" long		95	.084		1.71	1.79		3.50	5.40	
	1360	7/8" diameter x 2" long		115	.070		1	1.48		2.48	4.01	
	1365	3" long		110	.073		1.14	1.55		2.69	4.30	
	1370	4" long		105	.076		1.37	1.62		2.99	4.70	
	1380	6" long		100	.080		1.83	1.70		3.53	5.35	
	1390	8" long		90	.089		2.42	1.89		4.31	6.40	
	1400	1" diameter x 2" long		105	.076		1.41	1.62		3.03	4.74	
	1420	3" long		105	.076		1.58	1.62		3.20	4.93	
	1450	4" long		105	.076		1.99	1.62		3.61	5.40	
	1500	6" long		90	.089		2.26	1.89		4.15	6.20	
	1550	8" long		85	.094		3.22	2		5.22	7.50	
	1600	1-1/4" diameter x 3" long		85	.094		3.15	2		5.15	7.40	
	1650	4" long		80	.100		3.49	2.13		5.62	8.05	
	1700	6" long		75	.107		4.38	2.27		6.65	9.30	
	1750	8" long		70	.114		5.35	2.43		7.78	10.65	
460	0005	**LAG SCREWS**										**460**
	0010	Steel, 1/4" diameter, 2" long	1 Carp	200	.040	Ea.	.07	.79		.86	1.43	
	0100	3/8" diameter, 3" long		150	.053		.18	1.05		1.23	2	
	0200	1/2" diameter, 3" long		130	.062		.30	1.21		1.51	2.41	
	0300	5/8" diameter, 3" long		120	.067		.59	1.31		1.90	2.90	
500	0005	**MACHINE SCREWS**										**500**
	0010	Steel, round head, #8 x 1" long				C	1.57			1.57	1.73	
	0110	#8 x 2" long					4			4	4.40	
	0200	#10 x 1" long					2.63			2.63	2.89	
	0300	#10 x 2" long					4.83			4.83	5.30	
540	0010	**MACHINERY ANCHORS** Standard, flush mounted,										**540**
	0020	incl. stud w/fiber plug, connecting nut, washer & bolt										

05050 | Basic Materials & Methods

05090 | Metal Fastenings

			CREW	DAILY OUTPUT	LABOR-HOURS	UNIT	2000 BARE COSTS MAT.	LABOR	EQUIP.	TOTAL	TOTAL INCL O&P	
540	0200	Material only, 1/2" diameter stud & bolt				Ea.	43.50			43.50	48	540
	0300	5/8" diameter					49.50			49.50	54.50	
	0500	3/4" diameter					54			54	59	
	0600	7/8" diameter					62			62	68	
	0800	1" diameter					68.50			68.50	75	
	0900	1-1/4" diameter					86.50			86.50	95	
580	0005	**POWDER ACTUATED** Tools & fasteners										580
	0010	Stud driver, .22 caliber, buy, minimum				Ea.	289			289	320	
	0100	Maximum				"	505			505	555	
	0300	Powder charges for above, low velocity				C	15.15			15.15	16.65	
	0400	Standard velocity					25.50			25.50	28	
	0600	Drive pins & studs, 1/4" & 3/8" diam., to 3" long, minimum					22			22	24	
	0700	Maximum					57.50			57.50	63	
	0800	Pneumatic stud driver for 1/8" diameter studs				Ea.	1,850			1,850	2,025	
	0900	Drive pins for above, 1/2" to 3/4" long				M	400			400	440	
600	0010	**RIVETS**										600
	0100	Aluminum rivet & mandrel, 1/2" grip length x 1/8" diameter				C	4.53			4.53	4.98	
	0200	3/16" diameter					7			7	7.70	
	0300	Aluminum rivet, steel mandrel, 1/8" diameter					4.23			4.23	4.65	
	0400	3/16" diameter					6.35			6.35	7	
	0500	Copper rivet, steel mandrel, 1/8" diameter					5.50			5.50	6.05	
	0600	Monel rivet, steel mandrel, 1/8" diameter					19.30			19.30	21	
	0700	3/16" diameter					40.50			40.50	44.50	
	0800	Stainless rivet & mandrel, 1/8" diameter					12			12	13.20	
	0900	3/16" diameter					18.20			18.20	20	
	1000	Stainless rivet, steel mandrel, 1/8" diameter					7.60			7.60	8.35	
	1100	3/16" diameter					13.75			13.75	15.10	
	1200	Steel rivet and mandrel, 1/8" diameter					4.74			4.74	5.20	
	1300	3/16" diameter					7.15			7.15	7.85	
	1400	Hand riveting tool, minimum				Ea.	76.50			76.50	84	
	1500	Maximum					460			460	505	
	1600	Power riveting tool, minimum					660			660	730	
	1700	Maximum					1,750			1,750	1,925	
820	0010	**VIBRATION PADS**										820
	0300	Laminated synthetic rubber impregnated cotton duck, 1/2" thick	2 Sswk	20	.800	S.F.	50	17		67	88.50	
	0400	1" thick		20	.800		101	17		118	145	
	0600	Neoprene bearing pads, 1/2" thick		24	.667		18.90	14.15		33.05	49	
	0700	1" thick		20	.800		39	17		56	76	
	0900	Fabric reinforced neoprene, 5000 psi, 1/2" thick		24	.667		8.50	14.15		22.65	37.50	
	1000	1" thick		20	.800		17	17		34	52	
	1200	Felt surfaced vinyl pads, cork and sisal, 5/8" thick		24	.667		22	14.15		36.15	52.50	
	1300	1" thick		20	.800		40	17		57	78	
	1500	Teflon bonded to 10 ga. carbon steel, 1/32" layer		24	.667		36	14.15		50.15	67.50	
	1600	3/32" layer		24	.667		54	14.15		68.15	87.50	
	1800	Bonded to 10 ga. stainless steel, 1/32" layer		24	.667		65	14.15		79.15	99.50	
	1900	3/32" layer		24	.667		83.50	14.15		97.65	120	
	2100	Circular, encased, rule of thumb				Kip	5.50			5.50	6.05	
840	0005	**WELD SHEAR CONNECTORS**										840
	0010	3/4" diameter, 3-3/16" long	E-10	1,030	.023	Ea.	.30	.51	.50	1.31	1.89	
	0020	3-3/8" long		1,030	.023		.51	.51	.50	1.52	2.12	
	0200	3-7/8" long		1,030	.023		.34	.51	.50	1.35	1.93	
	0300	4-3/16" long		1,030	.023		.35	.51	.50	1.36	1.95	
	0500	4-7/8" long		1,030	.023		.39	.51	.50	1.40	1.99	
	0600	5-3/16" long		1,030	.023		.41	.51	.50	1.42	2.01	
	0800	5-3/8" long		1,030	.023		.44	.51	.50	1.45	2.04	

05050 | Basic Materials & Methods

05090 | Metal Fastenings

			Crew	Daily Output	Labor-Hours	Unit	Mat.	Labor	Equip.	Total	Total Incl O&P	
840	0900	6-3/16" long	E-10	1,000	.024	Ea.	.45	.53	.52	1.50	2.11	840
	1000	7-3/16" long		1,000	.024		.57	.53	.52	1.62	2.24	
	1100	8-3/16" long		1,000	.024		.62	.53	.52	1.67	2.29	
	1500	7/8" diameter, 3-11/16" long		1,030	.023		.48	.51	.50	1.49	2.09	
	1600	4-3/16" long		1,030	.023		.52	.51	.50	1.53	2.13	
	1700	5-3/16" long		1,030	.023		.60	.51	.50	1.61	2.22	
	1800	6-3/16" long		1,000	.024		.67	.53	.52	1.72	2.35	
	1900	7-3/16" long		1,000	.024		.75	.53	.52	1.80	2.44	
	2000	8-3/16" long		1,000	.024		.82	.53	.52	1.87	2.51	
860	0005	**WELD STUDS**										860
	0010	1/4" diameter, 2-11/16" long	E-10	1,030	.023	Ea.	.19	.51	.50	1.20	1.77	
	0100	4-1/8" long		1,030	.023		.18	.51	.50	1.19	1.76	
	0200	3/8" diameter, 4-1/8" long		1,030	.023		.21	.51	.50	1.22	1.79	
	0300	6-1/8" long		1,030	.023		.27	.51	.50	1.28	1.86	
	0400	1/2" diameter, 2-1/8" long		1,030	.023		.20	.51	.50	1.21	1.78	
	0500	3-1/8" long		1,030	.023		.24	.51	.50	1.25	1.82	
	0600	4-1/8" long		1,030	.023		.28	.51	.50	1.29	1.87	
	0700	5-5/16" long		1,030	.023		.35	.51	.50	1.36	1.95	
	0800	6-1/8" long		1,000	.024		.38	.53	.52	1.43	2.03	
	0900	8-1/8" long		1,000	.024		.53	.53	.52	1.58	2.19	
	1000	5/8" diameter, 2-11/16" long		1,030	.023		.34	.51	.50	1.35	1.93	
	1010	4-3/16" long		1,030	.023		.43	.51	.50	1.44	2.03	
	1100	6-9/16" long		1,000	.024		.56	.53	.52	1.61	2.23	
	1200	8-3/16" long		1,000	.024		.75	.53	.52	1.80	2.44	
880	0005	**WELD ROD**										880
	0010	Steel, type E6011 (all purpose), 1/8" dia, less than 500#				Lb.	1.49			1.49	1.64	
	0100	500# to 2,000#					1.32			1.32	1.45	
	0200	2,000# to 5,000#					1.21			1.21	1.33	
	0300	5/32" diameter, less than 500#					1.35			1.35	1.49	
	0310	500# to 2,000#					1.19			1.19	1.31	
	0320	2,000# to 5,000#					1.10			1.10	1.21	
	0400	3/16" dia, less than 500#					1.32			1.32	1.45	
	0500	500# to 2,000#					1.16			1.16	1.28	
	0600	2,000# to 5,000#					1.07			1.07	1.18	
	0620	Steel, type E6010 (pipe), 1/8" dia, less than 500#					1.51			1.51	1.66	
	0630	500# to 2,000#					1.36			1.36	1.50	
	0640	2,000# to 5,000#					1.24			1.24	1.36	
	0650	Steel, type E7018 (low hydrogen), 1/8" dia, less than 500#					1.35			1.35	1.49	
	0660	500# to 2,000#					1.19			1.19	1.31	
	0670	2,000# to 5,000#					1.08			1.08	1.19	
	0700	Steel, type E7024 (jet weld), 1/8" dia, less than 500#					1.37			1.37	1.51	
	0710	500# to 2,000#					1.21			1.21	1.33	
	0720	2,000# to 5,000#					1.09			1.09	1.20	
	1550	Aluminum, type 4043, 1/8" dia, less than 10#					4.49			4.49	4.94	
	1560	10# to 60#					3.95			3.95	4.35	
	1570	Over 60#					3.46			3.46	3.81	
	1600	Aluminum, type 5356, 1/8" dia, less than 10#					5			5	5.50	
	1610	10# to 60#					4.38			4.38	4.82	
	1620	Over 60#					3.86			3.86	4.25	
	1900	Cast iron (cold welding), 1/8" dia, less than 500#					18.90			18.90	21	
	1910	500# to 1,000#					17.40			17.40	19.15	
	1920	Over 1,000#					16.20			16.20	17.85	
	2000	Stainless steel, type 308-16, 1/8" dia, less than 500#					7.40			7.40	8.10	
	2100	500# to 1000#					6.50			6.50	7.15	

05050 | Basic Materials & Methods

05090 | Metal Fastenings

			CREW	DAILY OUTPUT	LABOR-HOURS	UNIT	MAT.	LABOR	EQUIP.	TOTAL	TOTAL INCL O&P	
880	2220	Over 1000#				Lb.	5.85			5.85	6.40	880
900	0005	**WELDING STRUCTURAL**										900
	0010	Field welding, 1/8" E6011, cost per welder, no operating engr R05090-520	E-14	8	1	Hr.	2.98	21.50	10.50	34.98	57	
	0200	With 1/2 operating engineer	E-13	8	1.500		2.98	33	10.50	46.48	76.50	
	0300	With 1 operating engineer	E-12	8	2	↓	2.98	42.50	10.50	55.98	92.50	
	0500	With no operating engineer, 2# weld rod per ton	E-14	8	1	Ton	2.98	21.50	10.50	34.98	57	
	0600	8# E6011 per ton	"	2	4		11.90	85	42	138.90	227	
	0800	With one operating engineer per welder, 2# E6011 per ton	E-12	8	2		2.98	42.50	10.50	55.98	92.50	
	0900	8# E6011 per ton	"	2	8	↓	11.90	169	42	222.90	370	
	1200	Continuous fillet, stick welding, incl. equipment										
	1300	Single pass, 1/8" thick, 0.1#/L.F.	E-14	150	.053	L.F.	.15	1.13	.56	1.84	3.01	
	1400	3/16" thick, 0.2#/L.F.		75	.107		.30	2.27	1.12	3.69	6.05	
	1500	1/4" thick, 0.3#/L.F.		50	.160		.45	3.40	1.68	5.53	9.05	
	1610	5/16" thick, 0.4#/L.F.		38	.211		.60	4.47	2.21	7.28	11.90	
	1800	3 passes, 3/8" thick, 0.5#/L.F.		30	.267		.74	5.65	2.80	9.19	15.05	
	2010	4 passes, 1/2" thick, 0.7#/L.F.		22	.364		1.04	7.75	3.82	12.61	20.50	
	2200	5 to 6 passes, 3/4" thick, 1.3#/L.F.		12	.667		1.94	14.15	7	23.09	38	
	2400	8 to 11 passes, 1" thick, 2.4#/L.F.	↓	6	1.333		3.58	28.50	14	46.08	75.50	
	2600	For all position welding, add, minimum						20%				
	2700	Maximum						300%				
	2900	For semi-automatic welding, deduct, minimum						5%				
	3000	Maximum				↓		15%				
	4000	Cleaning and welding plates, bars, or rods										
	4010	to existing beams, columns, or trusses	E-14	12	.667	L.F.	.74	14.15	7	21.89	36.50	
920	0010	**STEEL CUTTING**										920
	0020	Hand burning, incl. preparation, torch cutting & grinding, no staging										
	0100	Steel to 1/2" thick	E-14	70	.114	L.F.		2.43	1.20	3.63	6.10	
	0150	3/4" thick		50	.160			3.40	1.68	5.08	8.55	
	0200	1" thick	↓	45	.178	↓		3.78	1.87	5.65	9.50	

05100 | Structural Metal Framing

05120 | Structural Steel

			CREW	DAILY OUTPUT	LABOR-HOURS	UNIT	MAT.	LABOR	EQUIP.	TOTAL	TOTAL INCL O&P	
140	0005	**SUBPURLINS**										140
	0006	Bulb tees, painted, 32-5/8" O.C., 40 psf L.L. R05120-250										
	0010	Up to 5'-6" span, 1.48 plf	E-1	5,900	.003	S.F.	.28	.06	.01	.35	.44	
	0050	6'-4" span, 1.68 plf		5,200	.003		.28	.07	.02	.37	.46	
	0060	7'-8" span, 1.87 plf		4,700	.003		.30	.07	.02	.39	.49	
	0100	8'-9" span, 2.15 plf		4,200	.004		.33	.08	.02	.43	.54	
	0200	10'-2" span, 3.19 plf		3,100	.005		.41	.11	.03	.55	.70	
	0300	12'-0" span, 3.87 plf	↓	2,700	.006	↓	.52	.13	.03	.68	.86	
	0600	For galvanizing, add				Lb.	.21			.21	.23	
	1000	Truss tees, painted, 32-5/8" O.C., 40 psf L.L.										
	1100	2" high, up to 5'-9" span, 1.01 plf	E-1	7,000	.002	S.F.	.43	.05	.01	.49	.59	
	1110	6'-0" span, 1.17 plf		6,800	.002		.45	.05	.01	.51	.61	
	1130	7'-6" span, 1.75 plf		4,800	.003		.53	.07	.02	.62	.74	
	1200	2-1/2" high, up to 6'-9" span, 1.06 plf	↓	7,000	.002	↓	.44	.05	.01	.50	.60	

05100 | Structural Metal Framing

05120 | Structural Steel

			CREW	DAILY OUTPUT	LABOR-HOURS	UNIT	2000 BARE COSTS MAT.	LABOR	EQUIP.	TOTAL	TOTAL INCL O&P	
140	1210	6'-9" span, 1.31 plf	E-1	6,500	.002	S.F.	.47	.05	.01	.53	.63	140
	1230	8'-9" span, 1.90 plf		4,700	.003		.55	.07	.02	.64	.76	
	1300	3" high, up to 7'-3" span, 1.06 plf		7,000	.002		.45	.05	.01	.51	.60	
	1310	9'-0" span, 1.90 plf		4,700	.003		.55	.07	.02	.64	.77	
	1400	3-1/2" high, up to 7'-9" span, 1.06 plf		6,900	.002		.46	.05	.01	.52	.61	
	1405	3-1/2" high, up to 8'-6" span, 1.31 plf		5,800	.003		.48	.06	.01	.55	.67	
	1410	10'-9" span, 1.90 plf		4,700	.003		.56	.07	.02	.65	.78	
	1420	For 24-5/8" spacing, add					33%	33%				
	1430	For 48-5/8" spacing, deduct					50%	50%				
180	0005	**CANOPY FRAMING**										180
	0010	6" and 8" members	E-4	3,000	.011	Lb.	.72	.23	.03	.98	1.28	
220	0010	**CEILING SUPPORTS**										220
	1000	Entrance door/folding partition supports	E-4	60	.533	L.F.	12	11.60	1.40	25	37.50	
	1100	Linear accelerator door supports		14	2.286		54.50	49.50	6	110	165	
	1200	Lintels or shelf angles, hung, exterior hot dipped galv.		267	.120		8.20	2.61	.32	11.13	14.50	
	1250	Two coats primer paint instead of galv.		267	.120		7.10	2.61	.32	10.03	13.30	
	1400	Monitor support, ceiling hung, expansion bolted		4	8	Ea.	190	174	21	385	575	
	1450	Hung from pre-set inserts		6	5.333		205	116	14.05	335.05	470	
	1600	Motor supports for overhead doors		4	8		96.50	174	21	291.50	475	
	1700	Partition support for heavy folding partitions, without pocket		24	1.333	L.F.	27.50	29	3.51	60.01	91	
	1750	Supports at pocket only		12	2.667		54.50	58	7	119.50	182	
	2000	Rolling grilles & fire door supports		34	.941		23.50	20.50	2.48	46.48	68.50	
	2100	Spider-leg light supports, expansion bolted to ceiling slab		8	4	Ea.	78	87	10.50	175.50	270	
	2150	Hung from pre-set inserts		12	2.667	"	84	58	7	149	214	
	2400	Toilet partition support		36	.889	L.F.	27.50	19.35	2.34	49.19	70.50	
	2500	X-ray travel gantry support		12	2.667	"	93.50	58	7	158.50	225	
260	0005	**COLUMNS**										260
	0010	Aluminum, extruded, stock units, 6" diameter	E-4	240	.133	L.F.	7.65	2.90	.35	10.90	14.55	
	0100	8" diameter		170	.188		10.10	4.09	.50	14.69	19.75	
	0200	10" diameter		150	.213		13.35	4.64	.56	18.55	24.50	
	0300	12" diameter		140	.229		21.50	4.97	.60	27.07	34	
	0400	15" diameter		120	.267		29	5.80	.70	35.50	44	
	0410	Caps and bases, plain, 6" diameter				Set	17.05			17.05	18.75	
	0420	8" diameter					21			21	23	
	0430	10" diameter					28			28	31	
	0440	12" diameter					59.50			59.50	65.50	
	0450	15" diameter					120			120	132	
	0460	Caps, ornamental, minimum					96.50			96.50	106	
	0470	Maximum					1,000			1,000	1,100	
	0500	For square columns, add to column prices above				L.F.	50%					
	0700	Residential, flat, 8' high, plain	E-4	20	1.600	Ea.	54	35	4.21	93.21	133	
	0720	Fancy		20	1.600		68.50	35	4.21	107.71	149	
	0740	Corner type, plain		20	1.600		94.50	35	4.21	133.71	177	
	0760	Fancy		20	1.600		123	35	4.21	162.21	209	
	0800	Steel (lally), concrete filled, extra strong pipe, 3-1/2" diameter	E-2	660	.073	L.F.	18.35	1.56	1.55	21.46	24.50	
	0930	6" diameter		1,200	.040		34.50	.86	.85	36.21	40.50	
	1000	Lightweight units, 3-1/2" diameter		780	.062		2.33	1.32	1.31	4.96	6.55	
	1050	4" diameter		900	.053		3.26	1.15	1.14	5.55	7.05	
	1100	For galvanizing, add				Lb.	.38			.38	.42	
	1300	For web ties, angles, etc., add per added lb.	1 Sswk	945	.008		.60	.18		.78	1.01	
	1500	Steel pipe, extra strong, no concrete, 3" to 5" diameter	E-2	16,000	.003		.69	.06	.06	.81	.95	
	1600	6" to 12" diameter		14,000	.003		.70	.07	.07	.84	.99	
	1700	Steel pipe, extra strong, no concrete, 3" diameter x 12'-0"		60	.800	Ea.	93	17.20	17.05	127.25	154	
	1750	4" diameter x 12'-0"		58	.828		138	17.80	17.60	173.40	205	

05100 | Structural Metal Framing

05120 | Structural Steel

		CREW	DAILY OUTPUT	LABOR-HOURS	UNIT	2000 BARE COSTS MAT.	LABOR	EQUIP.	TOTAL	TOTAL INCL O&P
260 1800	6" diameter x 12'-0"	E-2	54	.889	Ea.	270	19.10	18.90	308	355
1850	8" diameter x 14'-0"	R05120-210	50	.960		475	20.50	20.50	516	585
1900	10" diameter x 16'-0"		48	1		690	21.50	21.50	733	825
1950	12" diameter x 18'-0"		45	1.067		940	23	22.50	985.50	1,100
3300	Structural tubing, square, A500GrB, 4" to 6" square, light section		11,270	.004	Lb.	.63	.09	.09	.81	.97
3600	Heavy section		32,000	.002	"	.63	.03	.03	.69	.79
4000	Concrete filled, add				L.F.	2.90			2.90	3.19
4500	Structural tubing, sq, 4" x 4" x 1/4" x 12'-0"	E-2	58	.828	Ea.	104	17.80	17.60	139.40	167
4550	6" x 6" x 1/4" x 12'-0"		54	.889		170	19.10	18.90	208	245
4600	8" x 8" x 3/8" x 14'-0"		50	.960		370	20.50	20.50	411	465
4650	10" x 10" x 1/2" x 16'-0"		48	1		685	21.50	21.50	728	815
5100	Structural tubing, rect, 5" to 6" wide, light section		9,500	.005	Lb.	.63	.11	.11	.85	1.02
5200	Heavy section		31,200	.002		.63	.03	.03	.69	.79
5300	7" to 10" wide, light section		37,000	.001		.61	.03	.03	.67	.75
5400	Heavy section		68,000	.001		.61	.02	.02	.65	.72
5500	Structural tubing, rect, 5" x 3" x 1/4" x 12'-0"		58	.828	Ea.	101	17.80	17.60	136.40	164
5550	6" x 4" x 5/16" x 12'-0"		54	.889		158	19.10	18.90	196	231
5600	8" x 4" x 3/8" x 12'-0"		54	.889		223	19.10	18.90	261	305
5650	10" x 6" x 3/8" x 14'-0"		50	.960		355	20.50	20.50	396	455
5700	12" x 8" x 1/2" x 16'-0"		48	1		660	21.50	21.50	703	790
5800	Adjustable jack post, 8' maximum height, 2-3/4" diameter					21			21	23
5850	4" diameter					34			34	37.50
6000	Prefabricated, fireproof, with steel jackets and one coat									
6100	shop paint, 2 to 4 hour rated, minimum	E-2	27,000	.002	Lb.	.43	.04	.04	.51	.58
6200	Average		35,000	.001		.61	.03	.03	.67	.76
6250	Maximum		43,000	.001		1.01	.02	.02	1.05	1.19
6400	Mild steel, flat, 9" wide, stock units, painted, plain	E-4	160	.200	L.F.	4.61	4.35	.53	9.49	14.25
6450	Fancy		160	.200		9.10	4.35	.53	13.98	19.20
6500	Corner columns, painted, plain		160	.200		7.30	4.35	.53	12.18	17.25
6550	Fancy		160	.200		14.35	4.35	.53	19.23	25
6800	W Shape, A36 steel, 2 tier, W8 x 24	E-2	1,080	.044		20.50	.95	.95	22.40	25.50
6850	W8 x 31		1,080	.044		26.50	.95	.95	28.40	32.50
6900	W8 x 48		1,032	.047		41	1	.99	42.99	48.50
6950	W8 x 67		984	.049		57.50	1.05	1.04	59.59	66
7000	W10 x 45		1,032	.047		38.50	1	.99	40.49	45.50
7050	W10 x 68		984	.049		58.50	1.05	1.04	60.59	67
7100	W10 x 112		960	.050		96	1.07	1.06	98.13	109
7150	W12 x 50		1,032	.047		43	1	.99	44.99	50
7200	W12 x 87		984	.049		74.50	1.05	1.04	76.59	85
7250	W12 x 120		960	.050		103	1.07	1.06	105.13	116
7300	W12 x 190		912	.053		163	1.13	1.12	165.25	182
7350	W14 x 74		984	.049		63.50	1.05	1.04	65.59	73
7400	W14 x 120		960	.050		103	1.07	1.06	105.13	116
7450	W14 x 176		912	.053		151	1.13	1.12	153.25	169
300 0005	**CURB EDGING**									
0011	Steel angle w/anchors, on forms, 1" x 1", 0.8#/L.F.	2 Carp	175	.091	L.F.	1.24	1.80		3.04	4.45
0101	2" x 2" angles, 3.92#/L.F.		165	.097		3.48	1.91		5.39	7.10
0201	3" x 3" angles, 6.1#/L.F.		150	.107		6.70	2.10		8.80	11
0301	4" x 4" angles, 8.2#/L.F.		140	.114		7.10	2.25		9.35	11.65
1002	6" x 4" angles, 12.3#/L.F.		125	.128		10.05	2.52		12.57	15.35
1051	Steel channels with anchors, on forms, 3" channel, 5#/L.F.		145	.110		4.26	2.17		6.43	8.40
1101	4" channel, 5.4#/L.F.		135	.119		4.55	2.33		6.88	9
1201	6" channel, 8.2#/L.F.		130	.123		7.10	2.42		9.52	11.95
1301	8" channel, 11.5#/L.F.		115	.139		9.45	2.74		12.19	15.10
1401	10" channel, 15.3#/L.F.		90	.178		12.20	3.50		15.70	19.40
1501	12" channel, 20.7#/L.F.		70	.229		16.10	4.50		20.60	25.50

05100 | Structural Metal Framing

05120 | Structural Steel

		CREW	DAILY OUTPUT	LABOR-HOURS	UNIT	MAT.	LABOR	EQUIP.	TOTAL	TOTAL INCL O&P
300 2000	For curved edging, add				L.F.	35%	10%			
440 0010	**LIGHTWEIGHT FRAMING**									
0200	For load-bearing steel studs see division 05410-400									
0400	Angle framing, field fabricated, 4" and larger	E-3	440	.055	Lb.	.35	1.20	.19	1.74	2.95
0450	Less than 4" angles		265	.091	"	.36	1.99	.32	2.67	4.66
0460	1/2" x 1/2" x 1/8"		200	.120	L.F.	.07	2.63	.42	3.12	5.75
0462	3/4" x 3/4" x 1/8"		160	.150		.20	3.29	.53	4.02	7.30
0464	1" x 1" x 1/8"		135	.178		.29	3.90	.62	4.81	8.70
0466	1-1/4" x 1-1/4" x 3/16"		115	.209		.53	4.57	.73	5.83	10.40
0468	1-1/2" x 1-1/2" x 3/16"		100	.240		.65	5.25	.84	6.74	12
0470	2" x 2" x 1/4"		90	.267		1.15	5.85	.93	7.93	13.85
0472	2-1/2" x 2-1/2" x 1/4"		72	.333		1.48	7.30	1.17	9.95	17.30
0474	3" x 2" x 3/8"		65	.369		2.12	8.10	1.29	11.51	19.70
0476	3" x 3" x 3/8"		57	.421		2.59	9.25	1.47	13.31	22.50
0600	Channel framing, field fabricated, 8" and larger		500	.048	Lb.	.36	1.05	.17	1.58	2.65
0650	Less than 8" channels		335	.072	"	.36	1.57	.25	2.18	3.78
0660	C2 x 1.78		115	.209	L.F.	.64	4.57	.73	5.94	10.50
0662	C3 x 4.1		80	.300		1.48	6.60	1.05	9.13	15.70
0664	C4 x 5.4		66	.364		1.94	7.95	1.27	11.16	19.25
0666	C5 x 6.7		57	.421		2.41	9.25	1.47	13.13	22.50
0668	C6 x 8.2		55	.436		2.85	9.55	1.53	13.93	23.50
0670	C7 x 9.8		40	.600		3.53	13.15	2.10	18.78	32
0672	C8 x 11.5		36	.667		4.14	14.60	2.33	21.07	36
0710	Structural bar tee, field fabricated, 3/4" x 3/4" x 1/8"		160	.150		.20	3.29	.53	4.02	7.30
0712	1" x 1" x 1/8"		135	.178		.29	3.90	.62	4.81	8.70
0714	1-1/2" x 1-1/2" x 1/4"		114	.211		.84	4.61	.74	6.19	10.85
0716	2" x 2" x 1/4"		89	.270		1.15	5.90	.94	7.99	13.95
0718	2-1/2" x 2-1/2" x 3/8"		72	.333		2.12	7.30	1.17	10.59	18
0720	3" x 3" x 3/8"		57	.421		2.59	9.25	1.47	13.31	22.50
0730	Structural zee, field fabricated, 1-1/4" x 1-3/4" x 1-3/4"		114	.211		.27	4.61	.74	5.62	10.20
0732	2-11/16" x 3" x 2-11/16"		114	.211		.64	4.61	.74	5.99	10.60
0734	3-1/16" x 4" x 3-1/16"		133	.180		.97	3.96	.63	5.56	9.55
0736	3-1/4" x 5" x 3-1/4"		133	.180		1.32	3.96	.63	5.91	9.95
0738	3-1/2" x 6" x 3-1/2"		160	.150		1.99	3.29	.53	5.81	9.25
0740	Junior beam, field fabricated, 3"		80	.300		2.05	6.60	1.05	9.70	16.35
0742	4"		72	.333		2.77	7.30	1.17	11.24	18.75
0744	5"		67	.358		3.60	7.85	1.25	12.70	21
0746	6"		62	.387		4.50	8.50	1.35	14.35	23
0748	7"		57	.421		5.50	9.25	1.47	16.22	26
0750	8"		53	.453		6.60	9.95	1.58	18.13	28.50
1000	Continuous slotted channel framing system, minimum	2 Sswk	220	.073	Lb.	1.86	1.55		3.41	5.10
1200	Maximum	"	135	.119		2.10	2.52		4.62	7.30
1300	Cross bracing, rods, 3/4" diameter	E-3	700	.034		.72	.75	.12	1.59	2.40
1310	7/8" diameter		850	.028		.72	.62	.10	1.44	2.12
1320	1" diameter		1,000	.024		.72	.53	.08	1.33	1.92
1330	Angle, 5" x 5" x 3/8"		440	.055		.72	1.20	.19	2.11	3.36
1350	Hanging lintels, average		350	.069		.72	1.50	.24	2.46	4.01
1380	Roof frames, 3'-0" square, 5' span	E-2	4,200	.011		.72	.25	.24	1.21	1.53
1400	Tie rod, not upset, 1-1/2" to 4" diameter, with turnbuckle	2 Sswk	800	.020		.78	.43		1.21	1.70
1420	No turnbuckle		700	.023		.75	.49		1.24	1.79
1500	Upset, 1-3/4" to 4" diameter, with turnbuckle		800	.020		.78	.43		1.21	1.70
1520	No turnbuckle		700	.023		.75	.49		1.24	1.79
480 0005	**LINTELS**									
0010	Plain steel angles, under 500 lb.	1 Bric	550	.015	Lb.	.46	.29		.75	1

05100 | Structural Metal Framing

05120 | Structural Steel

			CREW	DAILY OUTPUT	LABOR-HOURS	UNIT	2000 BARE COSTS MAT.	LABOR	EQUIP.	TOTAL	TOTAL INCL O&P	
480	0100	500 to 1000 lb.	1 Bric	640	.013	Lb.	.45	.25		.70	.92	480
	0200	1,000 to 2,000 lb.		640	.013		.44	.25		.69	.90	
	0300	2,000 to 4,000 lb.		640	.013		.43	.25		.68	.89	
	0500	For built-up angles and plates, add to above					.15			.15	.16	
	0700	For engineering, add to above					.06			.06	.07	
	0900	For galvanizing, add to above, under 500 lb.					.36			.36	.40	
	0950	500 to 2,000 lb.					.35			.35	.38	
	1000	Over 2,000 lb.					.33			.33	.36	
	2000	Steel angles, 3-1/2" x 3", 1/4" thick, 2'-6" long	1 Bric	47	.170	Ea.	6.50	3.40		9.90	12.90	
	2100	4'-6" long		26	.308		11.65	6.15		17.80	23.50	
	2600	4" x 3-1/2", 1/4" thick, 5'-0" long		21	.381		14.90	7.60		22.50	29.50	
	2700	9'-0" long		12	.667		27	13.35		40.35	52	
	3500	For precast concrete lintels, see div. 03480-400										
520	0005	**PIPE SUPPORT FRAMING**										520
	0010	Under 10#/L.F.	E-4	3,900	.008	Lb.	.80	.18	.02	1	1.25	
	0200	10.1 to 15#/L.F.		4,300	.007		.79	.16	.02	.97	1.21	
	0400	15.1 to 20#/L.F.		4,800	.007		.78	.15	.02	.95	1.17	
	0600	Over 20#/L.F.		5,400	.006		.77	.13	.02	.92	1.11	
560	0010	**PLATES** Structural steel										560
	2010	48"-60" wide, 10'-60' long, 5-50 tons, mill prices										
	2100	1/4" thick				Cwt.	37.50			37.50	41.50	
	2150	5/16" thick					37.50			37.50	41.50	
	2200	3/8" thick					35.50			35.50	39	
	2250	1/2" thick					35.50			35.50	39	
	2300	3/4" thick					35.50			35.50	39	
	2350	1" thick					35.50			35.50	39	
	2400	2" thick					36.50			36.50	40.50	
	2450	4" thick					38.50			38.50	42.50	
	2500	6" thick					40.50			40.50	44.50	
	2550	8" thick					40.50			40.50	44.50	
600	0005	**STRESSED SKIN** Roof & ceiling system										600
	0010	Double panel flat roof, spans to 100'	E-2	1,150	.042	S.F.	4.80	.90	.89	6.59	8	
	0100	Double panel convex roof, spans to 200'		960	.050		7.80	1.07	1.06	9.93	11.85	
	0200	Double panel arched roof, spans to 300'		760	.063		12	1.36	1.34	14.70	17.30	
640	0010	**STRUCTURAL STEEL MEMBERS**										640
	0020	Shop fabricated for 1-2 story bldg., bolted conn's., 100 tons										
	0100	W 6 x 9	E-2	600	.080	L.F.	5.50	1.72	1.70	8.92	11.20	
	0120	x 16		600	.080		9.80	1.72	1.70	13.22	15.90	
	0140	x 20		600	.080		12.25	1.72	1.70	15.67	18.60	
	0300	W 8 x 10		600	.080		6.10	1.72	1.70	9.52	11.90	
	0320	x 15		600	.080		9.20	1.72	1.70	12.62	15.25	
	0350	x 21		600	.080		12.85	1.72	1.70	16.27	19.30	
	0360	x 24		550	.087		14.70	1.87	1.86	18.43	22	
	0370	x 28		550	.087		17.15	1.87	1.86	20.88	24.50	
	0500	x 31		550	.087		18.60	1.87	1.86	22.33	26	
	0520	x 35		550	.087		21	1.87	1.86	24.73	28.50	
	0540	x 48		550	.087		29	1.87	1.86	32.73	37	
	0600	W 10 x 12		600	.080		7.35	1.72	1.70	10.77	13.25	
	0620	x 15		600	.080		9.20	1.72	1.70	12.62	15.25	
	0700	x 22		600	.080		13.45	1.72	1.70	16.87	19.95	
	0720	x 26		600	.080		15.90	1.72	1.70	19.32	22.50	
	0740	x 33		550	.087		19.80	1.87	1.86	23.53	27.50	
	0900	x 49		550	.087		29.50	1.87	1.86	33.23	38	
	1100	W 12 x 14		880	.055		8.55	1.17	1.16	10.88	12.95	

05100 | Structural Metal Framing

05120 | Structural Steel

			CREW	DAILY OUTPUT	LABOR-HOURS	UNIT	2000 BARE COSTS				TOTAL INCL O&P
							MAT.	LABOR	EQUIP.	TOTAL	
1300	x 22		E-2	880	.055	L.F.	13.45	1.17	1.16	15.78	18.35
1500	x 26	R05120-210		880	.055		15.90	1.17	1.16	18.23	21
1520	x 35			810	.059		21	1.27	1.26	23.53	27
1560	x 50	R05120-240		750	.064		30	1.37	1.36	32.73	37
1580	x 58			750	.064		35	1.37	1.36	37.73	42.50
1700	x 72			640	.075		42.50	1.61	1.60	45.71	51.50
1740	x 87			640	.075		51	1.61	1.60	54.21	61.50
1900	W 14 x 26			990	.048		15.90	1.04	1.03	17.97	20.50
2100	x 30			900	.053		18.35	1.15	1.14	20.64	23.50
2300	x 34			810	.059		20.50	1.27	1.26	23.03	26.50
2320	x 43			810	.059		26	1.27	1.26	28.53	32.50
2340	x 53			800	.060		32	1.29	1.28	34.57	39
2360	x 74			760	.063		43.50	1.36	1.34	46.20	52
2380	x 90			740	.065		53	1.39	1.38	55.77	62
2500	x 120			720	.067		69	1.43	1.42	71.85	80.50
2700	W 16 x 26			1,000	.048		15.90	1.03	1.02	17.95	20.50
2900	x 31			900	.053		18.60	1.15	1.14	20.89	24
3100	x 40			800	.060		24	1.29	1.28	26.57	30.50
3120	x 50			800	.060		30	1.29	1.28	32.57	37
3140	x 67			760	.063		39.50	1.36	1.34	42.20	47.50
3300	W 18 x 35		E-5	960	.075		21	1.61	1.15	23.76	27.50
3500	x 40			960	.075		24	1.61	1.15	26.76	31
3520	x 46			960	.075		27.50	1.61	1.15	30.26	35
3700	x 50			912	.079		30	1.69	1.21	32.90	37.50
3900	x 55			912	.079		33	1.69	1.21	35.90	41
3920	x 65			900	.080		39	1.71	1.23	41.94	47.50
3940	x 76			900	.080		44.50	1.71	1.23	47.44	53.50
3960	x 86			900	.080		50.50	1.71	1.23	53.44	60
3980	x 106			900	.080		61	1.71	1.23	63.94	71.50
4100	W 21 x 44			1,064	.068		26.50	1.45	1.04	28.99	33
4300	x 50			1,064	.068		30	1.45	1.04	32.49	37
4500	x 62			1,036	.069		37	1.49	1.07	39.56	45
4700	x 68			1,036	.069		40	1.49	1.07	42.56	48
4720	x 83			1,000	.072		49	1.54	1.11	51.65	57.50
4740	x 93			1,000	.072		54.50	1.54	1.11	57.15	64
4760	x 101			1,000	.072		58	1.54	1.11	60.65	68
4780	x 122			1,000	.072		70.50	1.54	1.11	73.15	81.50
4900	W 24 x 55			1,110	.065		33	1.39	1	35.39	40.50
5100	x 62			1,110	.065		37	1.39	1	39.39	45
5300	x 68			1,110	.065		40	1.39	1	42.39	48
5500	x 76			1,110	.065		44.50	1.39	1	46.89	53
5700	x 84			1,080	.067		49.50	1.43	1.02	51.95	58.50
5720	x 94			1,080	.067		55.50	1.43	1.02	57.95	65
5740	x 104			1,050	.069		60	1.47	1.05	62.52	70
5760	x 117			1,050	.069		67.50	1.47	1.05	70.02	78
5780	x 146			1,050	.069		84	1.47	1.05	86.52	96.50
5800	W 27 x 84			1,190	.061		49.50	1.30	.93	51.73	58
5900	x 94			1,190	.061		55.50	1.30	.93	57.73	64.50
5920	x 114			1,150	.063		65.50	1.34	.96	67.80	75.50
5940	x 146			1,150	.063		84	1.34	.96	86.30	96
5960	x 161			1,150	.063		92.50	1.34	.96	94.80	106
6100	W 30 x 99			1,200	.060		58	1.28	.92	60.20	67.50
6300	x 108			1,200	.060		62	1.28	.92	64.20	72
6500	x 116			1,160	.062		67	1.33	.95	69.28	77
6520	x 132			1,160	.062		76	1.33	.95	78.28	87
6540	x 148			1,160	.062		85.50	1.33	.95	87.78	97.50

05100 | Structural Metal Framing

05120 | Structural Steel

			CREW	DAILY OUTPUT	LABOR-HOURS	UNIT	2000 BARE COSTS				TOTAL INCL O&P
							MAT.	LABOR	EQUIP.	TOTAL	
640	6560	x 173	E-5	1,120	.064	L.F.	99.50	1.38	.99	101.87	114
	6580	x 191	R05120-210	1,120	.064		110	1.38	.99	112.37	125
	6700	W 33 x 118		1,176	.061		68	1.31	.94	70.25	78.50
	6900	x 130	R05120-240	1,134	.063		75	1.36	.98	77.34	86
	7100	x 141		1,134	.063		81	1.36	.98	83.34	93
	7120	x 169		1,100	.065		97.50	1.40	1.01	99.91	111
	7140	x 201		1,100	.065		116	1.40	1.01	118.41	131
	7300	W 36 x 135		1,170	.062		78	1.32	.95	80.27	89
	7500	x 150		1,170	.062		86.50	1.32	.95	88.77	98.50
	7600	x 170		1,150	.063		98	1.34	.96	100.30	112
	7700	x 194		1,125	.064		112	1.37	.98	114.35	127
	7900	x 230		1,125	.064		132	1.37	.98	134.35	150
	7920	x 260		1,035	.070		150	1.49	1.07	152.56	169
	8100	x 300		1,035	.070		173	1.49	1.07	175.56	194
	8490	For jobs less than 100 tons, add				Ton	10%				
680	0010	**STRUCTURAL STEEL PROJECTS** Bolted, unless noted otherwise	R05080-310								
	0201	Apartments, nursing homes, etc., 1 to 2 stories		E-2	6.45	7.442	Ton	1,200	160	158	1,518	1,800
	0302	3 to 6 stories	R05090-510	"	6.30	7.619		1,225	164	162	1,551	1,850
	0402	7 to 15 stories		E-5	8.74	8.238		1,250	176	127	1,553	1,850
	0500	Over 15 stories	R05120-210	E-6	13.90	8.633		1,300	183	90.50	1,573.50	1,875
	0701	Offices, hospitals, etc., steel bearing, 1 to 2 stories		E-2	6.45	7.442		1,200	160	158	1,518	1,800
	0801	3 to 6 stories	R05120-220	"	6.30	7.619		1,225	164	162	1,551	1,850
	0901	7 to 15 stories		E-5	8.74	8.238		1,250	176	127	1,553	1,850
	1000	Over 15 stories	R05120-230	E-6	13.90	8.633		1,300	183	90.50	1,573.50	1,875
	1100	For multi-story masonry wall bearing construction, add							30%			
	1301	Industrial bldgs., 1 story, beams & girders, steel bearing	R05120-240	E-2	8.06	5.955		1,200	128	127	1,455	1,700
	1401	Masonry bearing		"	6.25	7.680		1,200	165	164	1,529	1,825
	1500	Industrial bldgs., 1 story, under 10 tons,										
	1510	steel from warehouse, trucked		E-2	7.50	6.400	Ton	1,450	137	136	1,723	2,000
	1601	1 story with roof trusses, steel bearing			6.63	7.240		1,425	156	154	1,735	2,025
	1701	Masonry bearing			6	8		1,425	172	170	1,767	2,075
	1901	Monumental structures, banks, stores, etc., minimum		E-5	8	9		1,200	193	138	1,531	1,850
	2000	Maximum		E-6	9	13.333		2,000	283	140	2,423	2,900
	2201	Churches, minimum		E-2	7.25	6.621		1,125	142	141	1,408	1,650
	2300	Maximum		E-5	5.20	13.846		1,500	296	213	2,009	2,425
	2800	Power stations, fossil fuels, minimum		E-6	11	10.909		1,200	231	115	1,546	1,900
	2900	Maximum			5.70	21.053		1,800	445	221	2,466	3,075
	2950	Nuclear fuels, non-safety steel, minimum			7	17.143		1,200	365	180	1,745	2,225
	3000	Maximum			5.50	21.818		1,800	465	229	2,494	3,125
	3040	Safety steel, minimum			2.50	48		1,750	1,025	505	3,280	4,450
	3070	Maximum			1.50	80		2,300	1,700	840	4,840	6,725
	3101	Roof trusses, minimum		E-2	8.13	5.904		1,675	127	126	1,928	2,225
	3200	Maximum		E-5	8.30	8.675		2,050	186	133	2,369	2,750
	3211	Schools, minimum		E-2	9	5.333		1,200	115	114	1,429	1,675
	3220	Maximum		E-5	8.30	8.675		1,750	186	133	2,069	2,425
	3400	Welded construction, simple commercial bldgs., 1 to 2 stories		E-7	7.60	9.474		1,225	203	157	1,585	1,925
	3501	7 to 15 stories		E-8	6.38	13.793		1,425	295	213	1,933	2,350
	3701	Welded rigid frame, 1 story, minimum		E-2	9.88	4.858		1,250	104	103	1,457	1,700
	3800	Maximum		E-7	5.50	13.091		1,625	280	216	2,121	2,550
	4000	High strength steels, add to A36 price, minimum						15			15	16.50
	4100	Maximum						154			154	169
	4300	Column base plates, light, up to 150 lb		2 Sswk	2,000	.008	Lb.	.66	.17		.83	1.07
	4400	Heavy, over 150 lb		E-2	7,500	.006	"	.69	.14	.14	.97	1.17
	4600	Castellated beams, light sections, to 50#/L.F., minimum			10.70	4.486	Ton	1,250	96.50	95.50	1,442	1,675
	4700	Maximum			7	6.857		1,375	147	146	1,668	1,975

05100 | Structural Metal Framing

05120 | Structural Steel

			CREW	DAILY OUTPUT	LABOR-HOURS	UNIT	2000 BARE COSTS MAT.	LABOR	EQUIP.	TOTAL	TOTAL INCL O&P	
680	4900	Heavy sections, over 50# per L.F., minimum	E-2	11.70	4.103	Ton	1,325	88	87.50	1,500.50	1,725	680
	5000	Maximum		7.80	6.154		1,450	132	131	1,713	1,975	
	5500	Steel domes - see R13128-310										
	5700	Steel estimating weights per S.F. - see R05120-220										
	5900	Galvanizing structural steel in shop, under 1 ton, add to above				Ton	775			775	855	
	5950	1 ton to 20 tons					730			730	805	
	6000	Over 20 tons					675			675	745	
	6101	Cold galvanizing, brush	1 Pord	1,100	.007	S.F.	.06	.13		.19	.28	
	6125	Steel surface treatments										
	6171	Wire brush, hand	1 Pord	695	.012	S.F.	.02	.21		.23	.36	
	6181	Power tool		533	.015		.05	.27		.32	.51	
	6215	Pressure washing, 2800-6000 S.F./day		3,500	.002			.04		.04	.07	
	6220	Steam cleaning, 2800-4000 S.F./day		2,400	.003			.06		.06	.10	
	6225	Water blasting		3,200	.002			.04		.04	.07	
	6230	Brush-off blast	2 Pord	3,800	.004		.07	.08		.15	.20	
	6235	Blast (SSPC-6), loose scale, fine pwdr rust, 2.0 #/S.F.		2,500	.006		.13	.11		.24	.34	
	6240	Tight mill scale, little/no rust, 3.0 #/S.F.		1,920	.008		.20	.15		.35	.47	
	6245	Exist coat blistered/pitted, 4.0 #/S.F.		1,280	.013		.27	.22		.49	.67	
	6250	Exist coat badly pitted/nodules, 6.7 #/S.F.		770	.021		.45	.37		.82	1.12	
	6255	Near white blast (SSPC-10), loose scale, fine rust, 5.6 #/S.F.		1,025	.016		.38	.28		.66	.88	
	6260	Tight mill scale, little/no rust, 6.9 #/S.F.		830	.019		.46	.35		.81	1.09	
	6265	Exist coat blistered/pitted, 9.0 #/S.F.		640	.025		.60	.45		1.05	1.42	
	6270	Exist coat badly pitted/nodules, 11.3 #/S.F.		510	.031		.76	.56		1.32	1.78	
	6510	Paints & protective coatings, sprayed										
	6520	Alkyds, primer	2 Psst	3,600	.004	S.F.	.05	.08		.13	.23	
	6540	Gloss topcoats		3,200	.005		.04	.09		.13	.24	
	6560	Silicone alkyd		3,200	.005		.09	.09		.18	.29	
	6610	Epoxy, primer		3,000	.005		.16	.10		.26	.37	
	6630	Intermediate or topcoat		2,800	.006		.13	.11		.24	.36	
	6650	Enamel coat		2,800	.006		.16	.11		.27	.39	
	6700	Epoxy ester, primer		2,800	.006		.33	.11		.44	.58	
	6720	Topcoats		2,800	.006		.10	.11		.21	.34	
	6810	Latex primer		3,600	.004		.05	.08		.13	.23	
	6830	Topcoats		3,200	.005		.05	.09		.14	.25	
	6910	Universal primers, one part, phenolic, modified alkyd		2,000	.008		.08	.15		.23	.39	
	6940	Two part, epoxy spray		2,000	.008		.19	.15		.34	.50	
	7000	Zinc rich primers, self cure, spray, inorganic		1,800	.009		.47	.17		.64	.86	
	7010	Epoxy, spray, organic		1,800	.009		.13	.17		.30	.48	
	7020	Above one story, simple structures, add						25%				
	7030	Intricate structures, add						50%				

05140 | Structural Aluminum

			CREW	DAILY OUTPUT	LABOR-HOURS	UNIT	MAT.	LABOR	EQUIP.	TOTAL	TOTAL INCL O&P	
080	0005	ALUMINUM										080
	0010	Structural shapes, 1" to 10" members, under 1 ton	E-2	1,050	.046	Lb.	1.82	.98	.97	3.77	4.96	
	0050	1 to 5 tons		1,330	.036		1.68	.78	.77	3.23	4.19	
	0100	Over 5 tons		1,330	.036		1.65	.78	.77	3.20	4.16	
	0300	Extrusions, over 5 tons, stock shapes		1,330	.036		1.61	.78	.77	3.16	4.11	
	0400	Custom shapes		1,330	.036		1.74	.78	.77	3.29	4.25	

05150 | Wire Rope Assemblies

			CREW	DAILY OUTPUT	LABOR-HOURS	UNIT	MAT.	LABOR	EQUIP.	TOTAL	TOTAL INCL O&P	
800	0005	STEEL WIRE ROPE										800
	0010	6 x 19, bright, fiber core, 5000' rolls, 1/2" diameter				L.F.	.60			.60	.66	
	0050	Steel core					.76			.76	.84	
	0100	Fiber core, 1" diameter					1.99			1.99	2.19	
	0150	Steel core					2.25			2.25	2.48	
	0300	6 x 19, galvanized, fiber core, 1/2" diameter					.84			.84	.92	

Important: See the Reference Section for critical supporting data - Reference Nos., Crews, & City Cost Indexes

05100 | Structural Metal Framing

05150 | Wire Rope Assemblies

		CREW	DAILY OUTPUT	LABOR-HOURS	UNIT	2000 BARE COSTS MAT.	LABOR	EQUIP.	TOTAL	TOTAL INCL O&P
0400	Fiber core, 1" diameter				L.F.	2.46			2.46	2.71
0450	Steel core					2.59			2.59	2.85
0500	6 x 7, bright, IPS, fiber core, <500 L.F. w/acc., 1/4" diameter	E-17	6,400	.002		.65	.06		.71	.82
0510	1/2" diameter		2,100	.008		1.60	.17		1.77	2.09
0520	3/4" diameter		960	.017		2.89	.37		3.26	3.91
0550	6 x 19, bright, IPS, IWRC, <500 L.F. w/acc., 1/4" diameter		5,760	.003		.60	.06		.66	.78
0560	1/2" diameter		1,730	.009		1.02	.21		1.23	1.53
0570	3/4" diameter		770	.021		1.76	.46		2.22	2.85
0580	1" diameter		420	.038		3.32	.85	.01	4.18	5.35
0590	1-1/4" diameter		290	.055		4.80	1.23	.01	6.04	7.75
0600	1-1/2" diameter		192	.083		6.95	1.85	.01	8.81	11.30
0610	1-3/4" diameter	E-18	240	.167		10.75	3.56	2.22	16.53	21
0620	2" diameter		160	.250		13.85	5.35	3.33	22.53	29
0630	2-1/4" diameter		160	.250		18.60	5.35	3.33	27.28	34.50
0650	6 x 37, bright, IPS, IWRC, <500 L.F. w/acc., 1/4" diameter	E-17	6,400	.002		.71	.06		.77	.89
0660	1/2" diameter		1,730	.009		1.20	.21		1.41	1.73
0670	3/4" diameter		770	.021		2.01	.46		2.47	3.12
0680	1" diameter		430	.037		3.20	.83	.01	4.04	5.15
0690	1-1/4" diameter		290	.055		4.94	1.23	.01	6.18	7.90
0700	1-1/2" diameter		190	.084		6.95	1.87	.01	8.83	11.35
0710	1-3/4" diameter	E-18	260	.154		10.45	3.29	2.05	15.79	20
0720	2" diameter		200	.200		13.50	4.28	2.66	20.44	26
0730	2-1/4" diameter		160	.250		17.50	5.35	3.33	26.18	33
0800	6 x 19 & 6 x 37, swaged, 1/2" diameter	E-17	1,220	.013		1.04	.29		1.33	1.72
0810	9/16" diameter		1,120	.014		1.18	.32		1.50	1.93
0820	5/8" diameter		930	.017		1.37	.38		1.75	2.26
0830	3/4" diameter		640	.025		1.92	.56		2.48	3.21
0840	7/8" diameter		480	.033		2.50	.74	.01	3.25	4.22
0850	1" diameter		350	.046		3.08	1.02	.01	4.11	5.40
0860	1-1/8" diameter		288	.056		3.72	1.24	.01	4.97	6.55
0870	1-1/4" diameter		230	.070		4.73	1.55	.01	6.29	8.25
0880	1-3/8" diameter		192	.083		5.45	1.85	.01	7.31	9.65
0890	1-1/2" diameter	E-18	300	.133		6.60	2.85	1.77	11.22	14.70

05160 | Metal Framing Systems

		CREW	DAILY OUTPUT	LABOR-HOURS	UNIT	MAT.	LABOR	EQUIP.	TOTAL	TOTAL INCL O&P
0005	**SPACE FRAME**									
0010	Steel collars and channel members, 4' modules, minimum	E-2	1,200	.040	S.F.	15	.86	.85	16.71	19.10
0200	Maximum		900	.053		30	1.15	1.14	32.29	36.50
0400	5' modules, minimum		1,300	.037		13	.79	.79	14.58	16.70
0500	Maximum		1,000	.048		19	1.03	1.02	21.05	24
0600	Steel collars and square tubular members, 5' modules, minimum		1,200	.040		15	.86	.85	16.71	19.10
0650	Maximum		1,000	.048		22	1.03	1.02	24.05	27
0700	4' modules, minimum		1,100	.044		16	.94	.93	17.87	20.50
0800	Maximum		900	.053		25	1.15	1.14	27.29	31
0900	Steel nodes & tubular members, 7' to 10' modules, minimum		1,650	.029		20	.62	.62	21.24	24
0950	Maximum		825	.058		50	1.25	1.24	52.49	59
1100	Less than 7' modules, minimum		1,200	.040		33	.86	.85	34.71	39
1200	Less than 7' modules, maximum		600	.080		65	1.72	1.70	68.42	76.50

05200 | Metal Joists

05210 | Steel Joists

		CREW	DAILY OUTPUT	LABOR-HOURS	UNIT	2000 BARE COSTS MAT.	LABOR	EQUIP.	TOTAL	TOTAL INCL O&P	
600	0010 **OPEN WEB JOISTS**, Truckload lots										600
	0020 K series, horizontal bridging, spans up to 30', minimum	E-7	15	4.800	Ton	810	103	79.50	992.50	1,175	
	0050 Average		12	6		955	128	99	1,182	1,400	
	0080 Maximum		9	8		1,275	171	132	1,578	1,875	
	0130 8K1		1,200	.060	L.F.	2.44	1.28	.99	4.71	6.25	
	0140 10K1		1,200	.060		2.39	1.28	.99	4.66	6.20	
	0160 12K3		1,500	.048		2.72	1.03	.79	4.54	5.85	
	0180 14K3		1,500	.048		2.86	1.03	.79	4.68	6	
	0200 16K3		1,800	.040		3.01	.86	.66	4.53	5.70	
	0220 16K6		1,800	.040		3.87	.86	.66	5.39	6.65	
	0240 18K5		2,000	.036		3.68	.77	.60	5.05	6.20	
	0260 18K9		2,000	.036		4.87	.77	.60	6.24	7.50	
	0410 Span 30' to 50', minimum		17	4.235	Ton	790	90.50	70	950.50	1,125	
	0460 Maximum		10	7.200	"	1,125	154	119	1,398	1,675	
	0500 20K5		2,000	.036	L.F.	3.69	.77	.60	5.06	6.20	
	0520 20K9		2,000	.036		4.86	.77	.60	6.23	7.50	
	0540 22K5		2,000	.036		3.96	.77	.60	5.33	6.50	
	0560 22K9		2,000	.036		5.10	.77	.60	6.47	7.75	
	0580 24K6		2,200	.033		4.36	.70	.54	5.60	6.75	
	0600 24K10		2,200	.033		5.90	.70	.54	7.14	8.45	
	0620 26K6		2,200	.033		4.77	.70	.54	6.01	7.20	
	0640 26K10		2,200	.033		6.20	.70	.54	7.44	8.80	
	0660 28K8		2,400	.030		5.70	.64	.50	6.84	8.10	
	0680 28K12		2,400	.030		7.70	.64	.50	8.84	10.25	
	0700 30K8		2,400	.030		5.95	.64	.50	7.09	8.35	
	0720 30K12		2,400	.030		7.90	.64	.50	9.04	10.50	
	1100 10CS2		1,200	.060		4.14	1.28	.99	6.41	8.15	
	1120 12CS2		1,500	.048		4.41	1.03	.79	6.23	7.70	
	1140 14CS2		1,500	.048		4.41	1.03	.79	6.23	7.70	
	1160 16CS2		1,800	.040		4.69	.86	.66	6.21	7.55	
	1180 16CS4		1,800	.040		8	.86	.66	9.52	11.20	
	1200 18CS2		2,000	.036		4.96	.77	.60	6.33	7.60	
	1220 18CS4		2,000	.036		8.25	.77	.60	9.62	11.25	
	1240 20CS2		2,000	.036		5.25	.77	.60	6.62	7.90	
	1260 20CS4		2,000	.036		9.10	.77	.60	10.47	12.15	
	1280 22CS2		2,000	.036		5.50	.77	.60	6.87	8.20	
	1300 22CS4		2,000	.036		9.10	.77	.60	10.47	12.15	
	1320 24CS2		2,200	.033		5.50	.70	.54	6.74	8	
	1340 24CS4		2,200	.033		9.10	.70	.54	10.34	11.95	
	1360 26CS2		2,200	.033		5.50	.70	.54	6.74	8	
	1380 26CS4		2,200	.033		9.10	.70	.54	10.34	11.95	
	1400 28CS2		2,400	.030		5.80	.64	.50	6.94	8.15	
	1420 28CS4		2,400	.030		9.10	.64	.50	10.24	11.80	
	1440 30CS2		2,400	.030		6.05	.64	.50	7.19	8.50	
	1460 30CS4		2,400	.030		9.10	.64	.50	10.24	11.80	
	2000 LH series, bolted cross bridging										
	2020 Spans to 96', minimum	E-7	16	4.500	Ton	960	96.50	74.50	1,131	1,325	
	2040 Average		13	5.538		1,125	119	91.50	1,335.50	1,550	
	2080 Maximum		11	6.545		1,475	140	108	1,723	2,025	
	2200 18LH04		1,400	.051	L.F.	6.70	1.10	.85	8.65	10.45	
	2220 18LH08		1,400	.051		10.60	1.10	.85	12.55	14.75	
	2240 20LH04		1,400	.051		6.70	1.10	.85	8.65	10.45	
	2260 20LH08		1,400	.051		10.60	1.10	.85	12.55	14.75	
	2280 24LH05		1,400	.051		7.25	1.10	.85	9.20	11.05	
	2300 24LH10		1,400	.051		12.85	1.10	.85	14.80	17.20	
	2320 28LH06		1,800	.040		8.95	.86	.66	10.47	12.25	

Important: See the Reference Section for critical supporting data - Reference Nos., Crews, & City Cost Indexes

05200 | Metal Joists

05210 | Steel Joists

		CREW	DAILY OUTPUT	LABOR-HOURS	UNIT	2000 BARE COSTS MAT.	LABOR	EQUIP.	TOTAL	TOTAL INCL O&P
2340	28LH11	E-7	1,800	.040	L.F.	13.95	.86	.66	15.47	17.75
2360	32LH08		1,800	.040		9.50	.86	.66	11.02	12.85
2380	32LH13		1,800	.040		16.75	.86	.66	18.27	21
2400	36LH09		1,800	.040		11.75	.86	.66	13.27	15.30
2420	36LH14		1,800	.040		20	.86	.66	21.52	24.50
2440	40LH10		2,200	.033		11.75	.70	.54	12.99	14.85
2460	40LH15		2,200	.033		20	.70	.54	21.24	24
2480	44LH11		2,200	.033		12.30	.70	.54	13.54	15.45
2500	44LH16		2,200	.033		23.50	.70	.54	24.74	28
2520	48LH11		2,200	.033		12.30	.70	.54	13.54	15.45
2540	48LH16	▼	2,200	.033	▼	23.50	.70	.54	24.74	28
3010	DLH series, bolted cross bridging									
3020	Spans to 144' (shipped in 2 pieces), minimum	E-7	16	4.500	Ton	1,050	96.50	74.50	1,221	1,425
3040	Average		13	5.538		1,325	119	91.50	1,535.50	1,775
3100	Maximum		11	6.545	▼	1,475	140	108	1,723	2,025
3200	52DLH11		2,000	.036	L.F.	16.55	.77	.60	17.92	20.50
3220	52DLH16		2,000	.036		30	.77	.60	31.37	35
3240	56DLH11		2,000	.036		17.20	.77	.60	18.57	21
3260	56DLH16		2,000	.036		30.50	.77	.60	31.87	35.50
3280	60DLH12		2,000	.036		19.20	.77	.60	20.57	23
3300	60DLH17		2,000	.036		34.50	.77	.60	35.87	40
3320	64DLH12		2,200	.033		20.50	.70	.54	21.74	24.50
3340	64DLH17		2,200	.033		34.50	.70	.54	35.74	40
3360	68DLH13		2,200	.033		24.50	.70	.54	25.74	29
3380	68DLH18		2,200	.033		40.50	.70	.54	41.74	46.50
3400	72DLH14		2,200	.033		27	.70	.54	28.24	32
3420	72DLH19	▼	2,200	.033	▼	46.50	.70	.54	47.74	53
4010	SLH series, bolted cross bridging									
4020	Spans to 200', minimum	E-7	16	4.500	Ton	1,225	96.50	74.50	1,396	1,625
4040	Average		13	5.538		1,500	119	91.50	1,710.50	1,975
4060	Maximum		11	6.545	▼	1,750	140	108	1,998	2,325
4200	80SLH15		1,500	.048	L.F.	30	1.03	.79	31.82	36
4220	80SLH20		1,500	.048		56.50	1.03	.79	58.32	65
4240	88SLH16		1,500	.048		34.50	1.03	.79	36.32	41
4260	88SLH21		1,500	.048		67	1.03	.79	68.82	76.50
4280	96SLH17		1,500	.048		39	1.03	.79	40.82	46
4300	96SLH22		1,500	.048		76.50	1.03	.79	78.32	87
4320	104SLH18		1,800	.040		44.50	.86	.66	46.02	51
4340	104SLH23		1,800	.040		82	.86	.66	83.52	92.50
4360	112SLH19		1,800	.040		50.50	.86	.66	52.02	58
4380	112SLH24		1,800	.040		98.50	.86	.66	100.02	110
4400	120SLH20		1,800	.040		58	.86	.66	59.52	66
4420	120SLH25	▼	1,800	.040	▼	114	.86	.66	115.52	127
6000	For welded cross bridging, add						30%			
6100	For L.T.L. lots, add					10%	15%			
6200	For shop prime paint other than mfrs. standard, add					20%				
6300	For bottom chord extensions, add per chord				Ea.	15.05			15.05	16.55
7000	Joist girders, minimum	E-5	15	4.800	Ton	840	103	73.50	1,016.50	1,200
8000	Trusses, factory fabricated WT chords, average	"	11	6.545	"	2,875	140	101	3,116	3,550

05300 | Metal Decking

05310 | Steel Deck

			CREW	DAILY OUTPUT	LABOR-HOURS	UNIT	2000 BARE COSTS				TOTAL INCL O&P
							MAT.	LABOR	EQUIP.	TOTAL	
300	0010	**METAL DECKING** Steel decking										300
	0200	Cellular units, galvanized, 2" deep, 20-20 gauge, over 15 squares	E-4	1,460	.022	S.F.	2.81	.48	.06	3.35	4.09
	0250	18-20 gauge		1,420	.023		3.27	.49	.06	3.82	4.64
	0300	18-18 gauge		1,390	.023		3.69	.50	.06	4.25	5.10
	0320	16-18 gauge		1,360	.024		4.16	.51	.06	4.73	5.65
	0340	16-16 gauge		1,330	.024		4.57	.52	.06	5.15	6.15
	0400	3" deep, galvanized, 20-20 gauge		1,375	.023		2.93	.51	.06	3.50	4.29
	0500	18-20 gauge		1,350	.024		3.40	.52	.06	3.98	4.83
	0600	18-18 gauge		1,290	.025		3.87	.54	.07	4.48	5.40
	0700	16-18 gauge		1,230	.026		4.27	.57	.07	4.91	5.90
	0800	16-16 gauge		1,150	.028		4.80	.61	.07	5.48	6.55
	1000	4-1/2" deep, galvanized, 20-18 gauge		1,100	.029		5.40	.63	.08	6.11	7.30
	1100	18-18 gauge		1,040	.031		5.95	.67	.08	6.70	7.95
	1200	16-18 gauge		980	.033		6.75	.71	.09	7.55	8.90
	1300	16-16 gauge		935	.034		6.95	.74	.09	7.78	9.20
	1500	For acoustical deck, add					15%				
	1700	For cells used for ventilation, add					15%				
	1900	For multi-story or congested site, add						50%			
	2100	Open type, galv., 1-1/2" deep wide rib, 22 gauge, under 50 squares	E-4	4,500	.007	S.F.	.80	.15	.02	.97	1.20
	2400	Over 500 squares		5,100	.006		.64	.14	.02	.80	.99
	2600	20 gauge, under 50 squares		3,865	.008		.94	.18	.02	1.14	1.41
	2700	Over 500 squares		4,300	.007		.74	.16	.02	.92	1.16
	2900	18 gauge, under 50 squares		3,800	.008		1.22	.18	.02	1.42	1.72
	3000	Over 500 squares		4,300	.007		.96	.16	.02	1.14	1.39
	3050	16 gauge, under 50 squares		3,700	.009		1.51	.19	.02	1.72	2.06
	3200	3" deep, 22 gauge, under 50 squares		3,600	.009		1.14	.19	.02	1.35	1.67
	3300	20 gauge, under 50 squares		3,400	.009		1.30	.20	.02	1.52	1.86
	3400	18 gauge, under 50 squares		3,200	.010		1.66	.22	.03	1.91	2.28
	3500	16 gauge, under 50 squares		3,000	.011		1.46	.23	.03	1.72	2.10
	3700	4-1/2" deep, long span roof, over 50 squares, 20 gauge		2,700	.012		2.18	.26	.03	2.47	2.94
	3800	18 gauge		2,460	.013		2.57	.28	.03	2.88	3.43
	3900	16 gauge		2,350	.014		3.25	.30	.04	3.59	4.20
	4100	6" deep, long span, 18 gauge		2,000	.016		3.60	.35	.04	3.99	4.70
	4200	16 gauge		1,930	.017		3.93	.36	.04	4.33	5.10
	4300	14 gauge		1,860	.017		4.68	.37	.05	5.10	5.95
	4500	7-1/2" deep, long span, 18 gauge		1,690	.019		3.93	.41	.05	4.39	5.20
	4600	16 gauge		1,590	.020		4.57	.44	.05	5.06	5.95
	4700	14 gauge		1,490	.021		5.45	.47	.06	5.98	7
	4800	For painted instead of galvanized, deduct					2%				
	5000	For acoustical perforated, with fiberglass, add				S.F.	1.12			1.12	1.23
	5200	Non-cellular composite deck, galv., 2" deep, 22 gauge	E-4	3,860	.008		.84	.18	.02	1.04	1.30
	5300	20 gauge		3,600	.009		.96	.19	.02	1.17	1.47
	5400	18 gauge		3,380	.009		1.26	.21	.02	1.49	1.83
	5500	16 gauge		3,200	.010		1.57	.22	.03	1.82	2.19
	5700	3" deep, galv., 22 gauge		3,200	.010		.92	.22	.03	1.17	1.47
	5800	20 gauge		3,000	.011		1.07	.23	.03	1.33	1.67
	5900	18 gauge		2,850	.011		1.31	.24	.03	1.58	1.95
	6000	16 gauge		2,700	.012		1.81	.26	.03	2.10	2.53
	6100	Slab form, steel, 28 gauge, 9/16" deep, uncoated		4,000	.008		.46	.17	.02	.65	.87
	6200	Galvanized		4,000	.008		.52	.17	.02	.71	.93
	6220	24 gauge, 1" deep, uncoated		3,900	.008		.45	.18	.02	.65	.87
	6240	Galvanized		3,900	.008		.56	.18	.02	.76	.99
	6300	24 gauge, 1-5/16" deep, uncoated		3,800	.008		.65	.18	.02	.85	1.09
	6400	Galvanized		3,800	.008		.71	.18	.02	.91	1.16
	6500	22 gauge, 1-5/16" deep, uncoated		3,700	.009		.57	.19	.02	.78	1.02
	6600	Galvanized		3,700	.009		.80	.19	.02	1.01	1.27

05300 | Metal Decking

05310 | Steel Deck

			CREW	DAILY OUTPUT	LABOR-HOURS	UNIT	2000 BARE COSTS				TOTAL INCL O&P	
							MAT.	LABOR	EQUIP.	TOTAL		
300	6700	22 gauge, 3" deep uncoated	E-4	3,600	.009	S.F.	.66	.19	.02	.87	1.14	300
	6800	Galvanized	↓	3,600	.009	↓	.77	.19	.02	.98	1.26	
	7000	Sheet metal edge closure form, 12" wide with 2 bends										
	7100	18 gauge	E-14	360	.022	L.F.	1.40	.47	.23	2.10	2.73	
	7200	16 gauge	"	360	.022	"	1.30	.47	.23	2	2.62	
	8000	Metal deck and trench, 2" thick, 20 gauge, combination										
	8010	60% cellular, 40% non-cellular, inserts and trench	R-4	1,100	.036	S.F.	5.60	.79	.08	6.47	7.80	

05400 | Cold Formed Metal Framing

05410 | Load-Bearing Metal Studs

			CREW	DAILY OUTPUT	LABOR-HOURS	UNIT	2000 BARE COSTS				TOTAL INCL O&P	
							MAT.	LABOR	EQUIP.	TOTAL		
100	0010	**BRACING**, shear wall X-bracing, per 10' x 10' bay, one face										100
	0120	Metal strap, 20 ga x 4" wide	2 Carp	18	.889	Ea.	13.95	17.50		31.45	45.50	
	0130	6" wide		18	.889		21.50	17.50		39	54	
	0160	18 ga x 4" wide		16	1		23.50	19.70		43.20	59.50	
	0170	6" wide	↓	16	1	↓	34	19.70		53.70	71.50	
	0410	Continuous strap bracing, per horizontal row on both faces										
	0420	Metal strap, 20 ga x 2" wide, studs 12" O.C.	1 Carp	7	1.143	C.L.F.	35	22.50		57.50	77	
	0430	16" O.C.		8	1		35	19.70		54.70	72.50	
	0440	24" O.C.		10	.800		35	15.75		50.75	65.50	
	0450	18 ga x 2" wide, studs 12" O.C.		6	1.333		57	26.50		83.50	108	
	0460	16" O.C.		7	1.143		57	22.50		79.50	102	
	0470	24" O.C.	↓	8	1	↓	57	19.70		76.70	97	
120	0010	**BRIDGING**, solid between studs w/ 1-1/4" leg track, per stud bay										120
	0200	Studs 12" O.C., 18 ga x 2-1/2" wide	1 Carp	125	.064	Ea.	.55	1.26		1.81	2.77	
	0210	3-5/8" wide		120	.067		.63	1.31		1.94	2.94	
	0220	4" wide		120	.067		.69	1.31		2	3.01	
	0230	6" wide		115	.070		.89	1.37		2.26	3.33	
	0240	8" wide		110	.073		1.18	1.43		2.61	3.75	
	0300	16 ga x 2-1/2" wide		115	.070		.68	1.37		2.05	3.10	
	0310	3-5/8" wide		110	.073		.80	1.43		2.23	3.33	
	0320	4" wide		110	.073		.86	1.43		2.29	3.40	
	0330	6" wide		105	.076		1.09	1.50		2.59	3.77	
	0340	8" wide		100	.080		1.48	1.58		3.06	4.33	
	1200	Studs 16" O.C., 18 ga x 2-1/2" wide		125	.064		.71	1.26		1.97	2.94	
	1210	3-5/8" wide		120	.067		.81	1.31		2.12	3.14	
	1220	4" wide		120	.067		.89	1.31		2.20	3.23	
	1230	6" wide		115	.070		1.14	1.37		2.51	3.60	
	1240	8" wide		110	.073		1.52	1.43		2.95	4.12	
	1300	16 ga x 2-1/2" wide		115	.070		.87	1.37		2.24	3.31	
	1310	3-5/8" wide		110	.073		1.02	1.43		2.45	3.58	
	1320	4" wide		110	.073		1.11	1.43		2.54	3.67	
	1330	6" wide		105	.076		1.40	1.50		2.90	4.11	
	1340	8" wide		100	.080		1.90	1.58		3.48	4.79	
	2200	Studs 24" O.C., 18 ga x 2-1/2" wide		125	.064		1.03	1.26		2.29	3.29	
	2210	3-5/8" wide		120	.067		1.17	1.31		2.48	3.54	
	2220	4" wide		120	.067		1.29	1.31		2.60	3.67	
	2230	6" wide		115	.070		1.65	1.37		3.02	4.16	
	2240	8" wide	↓	110	.073	↓	2.20	1.43		3.63	4.87	

05400 | Cold Formed Metal Framing

05410 | Load-Bearing Metal Studs

			CREW	DAILY OUTPUT	LABOR-HOURS	UNIT	2000 BARE COSTS MAT.	LABOR	EQUIP.	TOTAL	TOTAL INCL O&P	
120	2300	16 ga x 2-1/2" wide	1 Carp	115	.070	Ea.	1.27	1.37		2.64	3.74	120
	2310	3-5/8" wide		110	.073		1.48	1.43		2.91	4.08	
	2320	4" wide		110	.073		1.60	1.43		3.03	4.21	
	2330	6" wide		105	.076		2.03	1.50		3.53	4.80	
	2340	8" wide		100	.080		2.75	1.58		4.33	5.70	
	3000	Continuous bridging, per row										
	3100	16 ga x 1-1/2" channel thru studs 12" O.C.	1 Carp	6	1.333	C.L.F.	29	26.50		55.50	76.50	
	3110	16" O.C.		7	1.143		29	22.50		51.50	70	
	3120	24" O.C.		8.80	.909		29	17.90		46.90	62	
	4100	2" x 2" angle x 18 ga, studs 12" O.C.		7	1.143		59.50	22.50		82	104	
	4110	16" O.C.		9	.889		59.50	17.50		77	95.50	
	4120	24" O.C.		12	.667		59.50	13.15		72.65	88	
	4200	16 ga, studs 12" O.C.		5	1.600		75	31.50		106.50	137	
	4210	16" O.C.		7	1.143		75	22.50		97.50	121	
	4220	24" O.C.		10	.800		75	15.75		90.75	110	
300	0010	FRAMING, boxed headers/beams										300
	0200	Double, 18 ga x 6" deep	2 Carp	220	.073	L.F.	3.12	1.43		4.55	5.90	
	0210	8" deep		210	.076		3.58	1.50		5.08	6.50	
	0220	10" deep		200	.080		4.29	1.58		5.87	7.40	
	0230	12" deep		190	.084		4.93	1.66		6.59	8.25	
	0300	16 ga x 8" deep		180	.089		4.12	1.75		5.87	7.55	
	0310	10" deep		170	.094		4.93	1.85		6.78	8.60	
	0320	12" deep		160	.100		5.40	1.97		7.37	9.35	
	0400	14 ga x 10" deep		140	.114		5.85	2.25		8.10	10.30	
	0410	12" deep		130	.123		6.50	2.42		8.92	11.30	
	1210	Triple, 18 ga x 8" deep		170	.094		5.15	1.85		7	8.85	
	1220	10" deep		165	.097		6.15	1.91		8.06	10	
	1230	12" deep		160	.100		7.10	1.97		9.07	11.20	
	1300	16 ga x 8" deep		145	.110		5.95	2.17		8.12	10.25	
	1310	10" deep		140	.114		7.10	2.25		9.35	11.65	
	1320	12" deep		135	.119		7.80	2.33		10.13	12.55	
	1400	14 ga x 10" deep		115	.139		8.05	2.74		10.79	13.55	
	1410	12" deep		110	.145		9	2.87		11.87	14.80	
400	0010	FRAMING, STUD WALLS w/ top & bottom track, no openings,										400
	0020	headers, beams, bridging or bracing										
	4100	8' high walls, 18 ga x 2-1/2" wide, studs 12" O.C.	2 Carp	54	.296	L.F.	5.35	5.85		11.20	15.90	
	4110	16" O.C.		77	.208		4.29	4.09		8.38	11.70	
	4120	24" O.C.		107	.150		3.21	2.95		6.16	8.60	
	4130	3-5/8" wide, studs 12" O.C.		53	.302		6	5.95		11.95	16.80	
	4140	16" O.C.		76	.211		4.80	4.15		8.95	12.40	
	4150	24" O.C.		105	.152		3.60	3		6.60	9.10	
	4160	4" wide, studs 12" O.C.		52	.308		6.50	6.05		12.55	17.60	
	4170	16" O.C.		74	.216		5.20	4.26		9.46	13.05	
	4180	24" O.C.		103	.155		3.92	3.06		6.98	9.55	
	4190	6" wide, studs 12" O.C.		51	.314		8.10	6.20		14.30	19.50	
	4200	16" O.C.		73	.219		6.50	4.32		10.82	14.55	
	4210	24" O.C.		101	.158		4.89	3.12		8.01	10.75	
	4220	8" wide, studs 12" O.C.		50	.320		10.40	6.30		16.70	22.50	
	4230	16" O.C.		72	.222		8.35	4.38		12.73	16.70	
	4240	24" O.C.		100	.160		6.35	3.15		9.50	12.35	
	4300	16 ga x 2-1/2" wide, studs 12" O.C.		47	.340		6.15	6.70		12.85	18.30	
	4310	16" O.C.		68	.235		4.89	4.64		9.53	13.35	
	4320	24" O.C.		94	.170		3.61	3.35		6.96	9.70	
	4330	3-5/8" wide, studs 12" O.C.		46	.348		7.10	6.85		13.95	19.60	
	4340	16" O.C.		66	.242		5.65	4.78		10.43	14.40	

05400 | Cold Formed Metal Framing

05410 | Load-Bearing Metal Studs

		CREW	DAILY OUTPUT	LABOR-HOURS	UNIT	2000 BARE COSTS MAT.	LABOR	EQUIP.	TOTAL	TOTAL INCL O&P
4350	24" O.C.	2 Carp	92	.174	L.F.	4.16	3.43		7.59	10.45
4360	4" wide, studs 12" O.C.		45	.356		7.70	7		14.70	20.50
4370	16" O.C.		65	.246		6.10	4.85		10.95	15.05
4380	24" O.C.		90	.178		4.52	3.50		8.02	11
4390	6" wide, studs 12" O.C.		44	.364		9.55	7.15		16.70	23
4400	16" O.C.		64	.250		7.55	4.93		12.48	16.80
4410	24" O.C.		88	.182		5.60	3.58		9.18	12.30
4420	8" wide, studs 12" O.C.		43	.372		12.40	7.35		19.75	26
4430	16" O.C.		63	.254		9.85	5		14.85	19.40
4440	24" O.C.		86	.186		7.35	3.67		11.02	14.35
5100	10' high walls, 18 ga x 2-1/2" wide, studs 12" O.C.		54	.296		6.45	5.85		12.30	17.10
5110	16" O.C.		77	.208		5.10	4.09		9.19	12.60
5120	24" O.C.		107	.150		3.75	2.95		6.70	9.20
5130	3-5/8" wide, studs 12" O.C.		53	.302		7.20	5.95		13.15	18.10
5140	16" O.C.		76	.211		5.70	4.15		9.85	13.35
5150	24" O.C.		105	.152		4.20	3		7.20	9.75
5160	4" wide, studs 12" O.C.		52	.308		7.80	6.05		13.85	19
5170	16" O.C.		74	.216		6.20	4.26		10.46	14.10
5180	24" O.C.		103	.155		4.57	3.06		7.63	10.30
5190	6" wide, studs 12" O.C.		51	.314		9.70	6.20		15.90	21.50
5200	16" O.C.		73	.219		7.70	4.32		12.02	15.85
5210	24" O.C.		101	.158		5.70	3.12		8.82	11.60
5220	8" wide, studs 12" O.C.		50	.320		12.45	6.30		18.75	24.50
5230	16" O.C.		72	.222		9.90	4.38		14.28	18.40
5240	24" O.C.		100	.160		7.35	3.15		10.50	13.50
5300	16 ga x 2-1/2" wide, studs 12" O.C.		47	.340		7.45	6.70		14.15	19.70
5310	16" O.C.		68	.235		5.85	4.64		10.49	14.40
5320	24" O.C.		94	.170		4.25	3.35		7.60	10.45
5330	3-5/8" wide, studs 12" O.C.		46	.348		8.60	6.85		15.45	21
5340	16" O.C.		66	.242		6.75	4.78		11.53	15.65
5350	24" O.C.		92	.174		4.90	3.43		8.33	11.25
5360	4" wide, studs 12" O.C.		45	.356		9.30	7		16.30	22.50
5370	16" O.C.		65	.246		7.30	4.85		12.15	16.35
5380	24" O.C.		90	.178		5.30	3.50		8.80	11.85
5390	6" wide, studs 12" O.C.		44	.364		11.50	7.15		18.65	25
5400	16" O.C.		64	.250		9.05	4.93		13.98	18.40
5410	24" O.C.		88	.182		6.60	3.58		10.18	13.40
5420	8" wide, studs 12" O.C.		43	.372		14.95	7.35		22.30	29
5430	16" O.C.		63	.254		11.80	5		16.80	21.50
5440	24" O.C.		86	.186		8.60	3.67		12.27	15.75
6190	12' high walls, 18 ga x 6" wide, studs 12" O.C.		41	.390		11.30	7.70		19	25.50
6200	16" O.C.		58	.276		8.90	5.45		14.35	19.10
6210	24" O.C.		81	.198		6.50	3.89		10.39	13.80
6220	8" wide, studs 12" O.C.		40	.400		14.50	7.90		22.40	29.50
6230	16" O.C.		57	.281		11.45	5.55		17	22
6240	24" O.C.		80	.200		8.35	3.94		12.29	15.95
6390	16 ga x 6" wide, studs 12" O.C.		35	.457		13.45	9		22.45	30.50
6400	16" O.C.		51	.314		10.50	6.20		16.70	22
6410	24" O.C.		70	.229		7.55	4.50		12.05	16.05
6420	8" wide, studs 12" O.C.		34	.471		17.50	9.25		26.75	35
6430	16" O.C.		50	.320		13.70	6.30		20	26
6440	24" O.C.		69	.232		9.85	4.57		14.42	18.70
6530	14 ga x 3-5/8" wide, studs 12" O.C.		34	.471		13.30	9.25		22.55	30.50
6540	16" O.C.		48	.333		10.35	6.55		16.90	22.50
6550	24" O.C.		65	.246		7.40	4.85		12.25	16.45
6560	4" wide, studs 12" O.C.		33	.485		14.10	9.55		23.65	32

05400 | Cold Formed Metal Framing

05410 | Load-Bearing Metal Studs

		CREW	DAILY OUTPUT	LABOR-HOURS	UNIT	2000 BARE COSTS				TOTAL INCL O&P
						MAT.	LABOR	EQUIP.	TOTAL	
400 6570	16" O.C.	2 Carp	47	.340	L.F.	11	6.70		17.70	23.50	400
6580	24" O.C.		64	.250		7.90	4.93		12.83	17.10
6730	12 ga x 3-5/8" wide, studs 12" O.C.		31	.516		19.75	10.15		29.90	39
6740	16" O.C.		43	.372		15.20	7.35		22.55	29.50
6750	24" O.C.		59	.271		10.65	5.35		16	21
6760	4" wide, studs 12" O.C.		30	.533		21.50	10.50		32	41.50
6770	16" O.C.		42	.381		16.60	7.50		24.10	31
6780	24" O.C.		58	.276		11.60	5.45		17.05	22
7390	16' high walls, 16 ga x 6" wide, studs 12" O.C.		33	.485		17.35	9.55		26.90	35.50
7400	16" O.C.		48	.333		13.45	6.55		20	26
7410	24" O.C.		67	.239		9.55	4.70		14.25	18.55
7420	8" wide, studs 12" O.C.		32	.500		22.50	9.85		32.35	42
7430	16" O.C.		47	.340		17.50	6.70		24.20	31
7440	24" O.C.		66	.242		12.40	4.78		17.18	22
7560	14 ga x 4" wide, studs 12" O.C.		31	.516		18.30	10.15		28.45	37.50
7570	16" O.C.		45	.356		14.10	7		21.10	27.50
7580	24" O.C.		61	.262		9.95	5.15		15.10	19.80
7590	6" wide, studs 12" O.C.		30	.533		23	10.50		33.50	43.50
7600	16" O.C.		44	.364		17.90	7.15		25.05	32
7610	24" O.C.		60	.267		12.65	5.25		17.90	23
7760	12 ga x 4" wide, studs 12" O.C.		29	.552		28	10.85		38.85	49.50
7770	16" O.C.		40	.400		21.50	7.90		29.40	37
7780	24" O.C.		55	.291		14.90	5.75		20.65	26
7790	6" wide, studs 12" O.C.		28	.571		35	11.25		46.25	58
7800	16" O.C.		39	.410		27	8.10		35.10	43.50
7810	24" O.C.		54	.296		18.55	5.85		24.40	30.50
8590	20' high walls, 14 ga x 6" wide, studs 12" O.C.		29	.552		28.50	10.85		39.35	50
8600	16" O.C.		42	.381		22	7.50		29.50	37
8610	24" O.C.		57	.281		15.30	5.55		20.85	26.50
8620	8" wide, studs 12" O.C.		28	.571		35	11.25		46.25	58
8630	16" O.C.		41	.390		27	7.70		34.70	43
8640	24" O.C.		56	.286		19	5.65		24.65	30.50
8790	12 ga x 6" wide, studs 12" O.C.		27	.593		43.50	11.65		55.15	67.50
8800	16" O.C.		37	.432		33	8.50		41.50	51
8810	24" O.C.		51	.314		22.50	6.20		28.70	35.50
8820	8" wide, studs 12" O.C.		26	.615		53	12.10		65.10	79
8830	16" O.C.		36	.444		40.50	8.75		49.25	59.50
8840	24" O.C.		50	.320		28	6.30		34.30	41.50

05420 | Cold-Formed Metal Joists

		CREW	DAILY OUTPUT	LABOR-HOURS	UNIT	MAT.	LABOR	EQUIP.	TOTAL	TOTAL INCL O&P
100 0010	BRACING, continuous, per row, top & bottom										100
0120	Flat strap, 20 ga x 2" wide, joists at 12" O.C.	1 Carp	4.67	1.713	C.L.F.	37	34		71	98.50
0130	16" O.C.		5.33	1.501		35.50	29.50		65	89.50
0140	24" O.C.		6.66	1.201		34	23.50		57.50	78
0150	18 ga x 2" wide, joists at 12" O.C.		4	2		56.50	39.50		96	130
0160	16" O.C.		4.67	1.713		55.50	34		89.50	119
0170	24" O.C.		5.33	1.501		54.50	29.50		84	111
120 0010	BRIDGING, solid between joists w/ 1-1/4" leg track, per joist bay										120
0230	Joists 12" O.C., 18 ga track x 6" wide	1 Carp	80	.100	Ea.	.89	1.97		2.86	4.36
0240	8" wide		75	.107		1.18	2.10		3.28	4.90
0250	10" wide		70	.114		1.52	2.25		3.77	5.55
0260	12" wide		65	.123		1.75	2.42		4.17	6.10
0330	16 ga track x 6" wide		70	.114		1.09	2.25		3.34	5.05
0340	8" wide		65	.123		1.48	2.42		3.90	5.80
0350	10" wide		60	.133		1.81	2.63		4.44	6.50

05400 | Cold Formed Metal Framing

05420 | Cold-Formed Metal Joists

			CREW	DAILY OUTPUT	LABOR-HOURS	UNIT	2000 BARE COSTS MAT.	LABOR	EQUIP.	TOTAL	TOTAL INCL O&P	
120	0360	12" wide	1 Carp	55	.145	Ea.	2.15	2.87		5.02	7.25	120
	0440	14 ga track x 8" wide		60	.133		1.90	2.63		4.53	6.60	
	0450	10" wide		55	.145		2.34	2.87		5.21	7.50	
	0460	12" wide		50	.160		2.78	3.15		5.93	8.45	
	0550	12 ga track x 10" wide		45	.178		3.55	3.50		7.05	9.90	
	0560	12" wide		40	.200		4.18	3.94		8.12	11.35	
	1230	16" O.C., 18 ga track x 6" wide		80	.100		1.14	1.97		3.11	4.63	
	1240	8" wide		75	.107		1.52	2.10		3.62	5.25	
	1250	10" wide		70	.114		1.95	2.25		4.20	6	
	1260	12" wide		65	.123		2.24	2.42		4.66	6.60	
	1330	16 ga track x 6" wide		70	.114		1.40	2.25		3.65	5.40	
	1340	8" wide		65	.123		1.90	2.42		4.32	6.25	
	1350	10" wide		60	.133		2.33	2.63		4.96	7.05	
	1360	12" wide		55	.145		2.76	2.87		5.63	7.95	
	1440	14 ga track x 8" wide		60	.133		2.44	2.63		5.07	7.20	
	1450	10" wide		55	.145		3	2.87		5.87	8.20	
	1460	12" wide		50	.160		3.56	3.15		6.71	9.30	
	1550	12 ga track x 10" wide		45	.178		4.55	3.50		8.05	11	
	1560	12" wide		40	.200		5.35	3.94		9.29	12.65	
	2230	24" O.C., 18 ga track x 6" wide		80	.100		1.65	1.97		3.62	5.20	
	2240	8" wide		75	.107		2.20	2.10		4.30	6	
	2250	10" wide		70	.114		2.82	2.25		5.07	6.95	
	2260	12" wide		65	.123		3.25	2.42		5.67	7.70	
	2330	16 ga track x 6" wide		70	.114		2.03	2.25		4.28	6.10	
	2340	8" wide		65	.123		2.75	2.42		5.17	7.15	
	2350	10" wide		60	.133		3.37	2.63		6	8.20	
	2360	12" wide		55	.145		3.99	2.87		6.86	9.30	
	2440	14 ga track x 8" wide		60	.133		3.53	2.63		6.16	8.40	
	2450	10" wide		55	.145		4.34	2.87		7.21	9.70	
	2460	12" wide		50	.160		5.15	3.15		8.30	11.05	
	2550	12 ga track x 10" wide		45	.178		6.60	3.50		10.10	13.25	
	2560	12" wide		40	.200		7.75	3.94		11.69	15.30	
200	0010	**FRAMING, BAND JOIST** (track) fastened to bearing wall										200
	0220	18 ga track x 6" deep	2 Carp	1,000	.016	L.F.	.72	.32		1.04	1.34	
	0230	8" deep		920	.017		.97	.34		1.31	1.65	
	0240	10" deep		860	.019		1.24	.37		1.61	1.99	
	0320	16 ga track x 6" deep		900	.018		.89	.35		1.24	1.58	
	0330	8" deep		840	.019		1.21	.38		1.59	1.97	
	0340	10" deep		780	.021		1.48	.40		1.88	2.32	
	0350	12" deep		740	.022		1.75	.43		2.18	2.66	
	0430	14 ga track x 8" deep		750	.021		1.55	.42		1.97	2.43	
	0440	10" deep		720	.022		1.91	.44		2.35	2.85	
	0450	12" deep		700	.023		2.27	.45		2.72	3.26	
	0540	12 ga track x 10" deep		670	.024		2.90	.47		3.37	4	
	0550	12" deep		650	.025		3.41	.48		3.89	4.58	
300	0010	**FRAMING, BOXED HEADERS/BEAMS**										300
	0200	Double, 18 ga x 6" deep	2 Carp	220	.073	L.F.	3.12	1.43		4.55	5.90	
	0210	8" deep		210	.076		3.58	1.50		5.08	6.50	
	0220	10" deep		200	.080		4.29	1.58		5.87	7.40	
	0230	12" deep		190	.084		4.93	1.66		6.59	8.25	
	0300	16 ga x 8" deep		180	.089		4.12	1.75		5.87	7.55	
	0310	10" deep		170	.094		4.93	1.85		6.78	8.60	
	0320	12" deep		160	.100		5.40	1.97		7.37	9.35	
	0400	14 ga x 10" deep		140	.114		5.85	2.25		8.10	10.30	
	0410	12" deep		130	.123		6.50	2.42		8.92	11.30	

05400 | Cold Formed Metal Framing

05420 | Cold-Formed Metal Joists

			CREW	DAILY OUTPUT	LABOR-HOURS	UNIT	2000 BARE COSTS MAT.	LABOR	EQUIP.	TOTAL	TOTAL INCL O&P	
300	0500	12 ga x 10" deep	2 Carp	110	.145	L.F.	7.85	2.87		10.72	13.55	300
	0510	12" deep		100	.160		8.75	3.15		11.90	15	
	1210	Triple, 18 ga x 8" deep		170	.094		5.15	1.85		7	8.85	
	1220	10" deep		165	.097		6.15	1.91		8.06	10	
	1230	12" deep		160	.100		7.10	1.97		9.07	11.20	
	1300	16 ga x 8" deep		145	.110		5.95	2.17		8.12	10.25	
	1310	10" deep		140	.114		7.10	2.25		9.35	11.65	
	1320	12" deep		135	.119		7.80	2.33		10.13	12.55	
	1400	14 ga x 10" deep		115	.139		8.50	2.74		11.24	14.05	
	1410	12" deep		110	.145		9.45	2.87		12.32	15.30	
	1500	12 ga x 10" deep		90	.178		11.50	3.50		15	18.65	
	1510	12" deep		85	.188		12.85	3.71		16.56	20.50	
410	0010	**FRAMING, JOISTS**, no band joists (track), web stiffeners, headers,										410
	0020	beams, bridging or bracing										
	0030	Joists (2" flange) and fasteners, materials only										
	0220	18 ga x 6" deep				L.F.	.96			.96	1.05	
	0230	8" deep					1.20			1.20	1.32	
	0240	10" deep					1.41			1.41	1.55	
	0320	16 ga x 6" deep					1.23			1.23	1.35	
	0330	8" deep					1.48			1.48	1.63	
	0340	10" deep					1.74			1.74	1.92	
	0350	12" deep					1.98			1.98	2.18	
	0430	14 ga x 8" deep					1.88			1.88	2.07	
	0440	10" deep					2.24			2.24	2.46	
	0450	12" deep					2.56			2.56	2.82	
	0540	12 ga x 10" deep					3.29			3.29	3.62	
	0550	12" deep					3.75			3.75	4.12	
	1010	Installation of joists to band joists, beams & headers, labor only										
	1220	18 ga x 6" deep	2 Carp	110	.145	Ea.		2.87		2.87	4.91	
	1230	8" deep		90	.178			3.50		3.50	6	
	1240	10" deep		80	.200			3.94		3.94	6.75	
	1320	16 ga x 6" deep		95	.168			3.32		3.32	5.70	
	1330	8" deep		70	.229			4.50		4.50	7.70	
	1340	10" deep		60	.267			5.25		5.25	9	
	1350	12" deep		55	.291			5.75		5.75	9.80	
	1430	14 ga x 8" deep		65	.246			4.85		4.85	8.30	
	1440	10" deep		45	.356			7		7	12	
	1450	12" deep		35	.457			9		9	15.45	
	1540	12 ga x 10" deep		40	.400			7.90		7.90	13.50	
	1550	12" deep		30	.533			10.50		10.50	18	
500	0010	**FRAMING, WEB STIFFENERS** at joist bearing, fabricated from										500
	0020	stud piece (1-5/8" flange) to stiffen joist (2" flange)										
	2120	For 6" deep joist, with 18 ga x 2-1/2" stud	1 Carp	120	.067	Ea.	1.19	1.31		2.50	3.56	
	2130	3-5/8" stud		110	.073		1.23	1.43		2.66	3.80	
	2140	4" stud		105	.076		1.23	1.50		2.73	3.93	
	2150	6" stud		100	.080		1.32	1.58		2.90	4.15	
	2160	8" stud		95	.084		1.43	1.66		3.09	4.41	
	2220	8" deep joist, with 2-1/2" stud		120	.067		1.30	1.31		2.61	3.68	
	2230	3-5/8" stud		110	.073		1.33	1.43		2.76	3.91	
	2240	4" stud		105	.076		1.35	1.50		2.85	4.06	
	2250	6" stud		100	.080		1.45	1.58		3.03	4.29	
	2260	8" stud		95	.084		1.64	1.66		3.30	4.64	
	2320	10" deep joist, with 2-1/2" stud		110	.073		1.84	1.43		3.27	4.47	
	2330	3-5/8" stud		100	.080		1.89	1.58		3.47	4.78	
	2340	4" stud		95	.084		1.94	1.66		3.60	4.98	
	2350	6" stud		90	.089		2.06	1.75		3.81	5.25	

05400 | Cold Formed Metal Framing

05420 | Cold-Formed Metal Joists

		CREW	DAILY OUTPUT	LABOR-HOURS	UNIT	2000 BARE COSTS MAT.	LABOR	EQUIP.	TOTAL	TOTAL INCL O&P
2360	8" stud	1 Carp	85	.094	Ea.	2.20	1.85		4.05	5.60
2420	12" deep joist, with 2-1/2" stud		110	.073		1.94	1.43		3.37	4.59
2430	3-5/8" stud		100	.080		1.98	1.58		3.56	4.88
2440	4" stud		95	.084		2.02	1.66		3.68	5.05
2450	6" stud		90	.089		2.16	1.75		3.91	5.40
2460	8" stud		85	.094		2.45	1.85		4.30	5.85
3130	For 6" deep joist, with 16 ga x 3-5/8" stud		100	.080		1.30	1.58		2.88	4.12
3140	4" stud		95	.084		1.32	1.66		2.98	4.29
3150	6" stud		90	.089		1.42	1.75		3.17	4.56
3160	8" stud		85	.094		1.59	1.85		3.44	4.93
3230	8" deep joist, with 3-5/8" stud		100	.080		1.44	1.58		3.02	4.28
3240	4" stud		95	.084		1.45	1.66		3.11	4.43
3250	6" stud		90	.089		1.58	1.75		3.33	4.73
3260	8" stud		85	.094		1.79	1.85		3.64	5.15
3330	10" deep joist, with 3-5/8" stud		85	.094		1.97	1.85		3.82	5.35
3340	4" stud		80	.100		2.06	1.97		4.03	5.65
3350	6" stud		75	.107		2.20	2.10		4.30	6
3360	8" stud		70	.114		2.42	2.25		4.67	6.55
3430	12" deep joist, with 3-5/8" stud		85	.094		2.15	1.85		4	5.55
3440	4" stud		80	.100		2.16	1.97		4.13	5.75
3450	6" stud		75	.107		2.35	2.10		4.45	6.20
3460	8" stud		70	.114		2.67	2.25		4.92	6.80
4230	For 8" deep joist, with 14 ga x 3-5/8" stud		90	.089		1.97	1.75		3.72	5.15
4240	4" stud		85	.094		2.02	1.85		3.87	5.40
4250	6" stud		80	.100		2.21	1.97		4.18	5.80
4260	8" stud		75	.107		2.39	2.10		4.49	6.25
4330	10" deep joist, with 3-5/8" stud		75	.107		2.77	2.10		4.87	6.65
4340	4" stud		70	.114		2.76	2.25		5.01	6.90
4350	6" stud		65	.123		3.07	2.42		5.49	7.50
4360	8" stud		60	.133		3.23	2.63		5.86	8.05
4430	12" deep joist, with 3-5/8" stud		75	.107		2.94	2.10		5.04	6.85
4440	4" stud		70	.114		3.02	2.25		5.27	7.20
4450	6" stud		65	.123		3.30	2.42		5.72	7.80
4460	8" stud		60	.133		3.56	2.63		6.19	8.40
5330	For 10" deep joist, with 12 ga x 3-5/8" stud		65	.123		3.15	2.42		5.57	7.60
5340	4" stud		60	.133		3.31	2.63		5.94	8.15
5350	6" stud		55	.145		3.59	2.87		6.46	8.85
5360	8" stud		50	.160		3.94	3.15		7.09	9.75
5430	12" deep joist, with 3-5/8" stud		65	.123		3.50	2.42		5.92	8
5440	4" stud		60	.133		3.49	2.63		6.12	8.35
5450	6" stud		55	.145		3.91	2.87		6.78	9.20
5460	8" stud		50	.160		4.50	3.15		7.65	10.35

05500 | Metal Fabrications

05514 | Ladders

		CREW	DAILY OUTPUT	LABOR-HOURS	UNIT	2000 BARE COSTS MAT.	LABOR	EQUIP.	TOTAL	TOTAL INCL O&P
0005	**LADDER**									
0011	Steel, 20" wide, bolted to concrete, with cage	2 Carp	25	.640	V.L.F.	53	12.60		65.60	80
0101	Without cage		45	.356		24.50	7		31.50	39
0301	Aluminum, bolted to concrete, with cage		25	.640		100	12.60		112.60	132

05500 | Metal Fabrications

05514 | Ladders

		CREW	DAILY OUTPUT	LABOR-HOURS	UNIT	2000 BARE COSTS MAT.	LABOR	EQUIP.	TOTAL	TOTAL INCL O&P	
500	0401 Without cage	2 Carp	45	.356	V.L.F.	55.50	7		62.50	73	500
	1350 Alternating tread stair, 56/68°, steel, standard paint color	2 Sswk	50	.320		132	6.80		138.80	158	
	1360 Non-standard paint color		50	.320		149	6.80		155.80	177	
	1370 Galvanized steel		50	.320		149	6.80		155.80	177	
	1380 Stainless steel		50	.320		221	6.80		227.80	256	
	1390 68°, aluminum		50	.320		165	6.80		171.80	194	

05517 | Metal Stairs

		CREW	DAILY OUTPUT	LABOR-HOURS	UNIT	MAT.	LABOR	EQUIP.	TOTAL	TOTAL INCL O&P	
300	0010 **FIRE ESCAPE**										300
	0200 2' wide balcony, 1" x 1/4" bars 1-1/2" O.C.	1 Sswk	5	1.600	L.F.	34	34		68	105	
	0400 1st story cantilevered stair, standard		.09	88.889	Ea.	1,425	1,900		3,325	5,300	
	0700 Platform & fixed stair, 36" x 40"		.17	47.059	Flight	630	1,000		1,630	2,675	
	0900 For 3'-6" wide escapes, add to above					100%	150%				
350	0005 **FIRE ESCAPE STAIRS**										350
	0010 One story, disappearing, stainless steel	2 Sswk	20	.800	V.L.F.	142	17		159	190	
	0100 Portable ladder				Ea.	50			50	55	
700	0010 **STAIR** Steel, safety nosing, steel stringers										700
	0020 Grating tread and pipe railing, 3'-6" wide	E-4	35	.914	Riser	100	19.90	2.40	122.30	152	
	0100 4'-0" wide		30	1.067		130	23	2.81	155.81	192	
	0200 Cement fill metal pan, picket rail, 3'-6" wide		35	.914		150	19.90	2.40	172.30	207	
	0300 4'-0" wide		30	1.067		170	23	2.81	195.81	236	
	0350 Wall rail, both sides, 3'-6" wide		53	.604		115	13.15	1.59	129.74	155	
	0400 Cast iron tread and pipe rail, 3'-6" wide		35	.914		160	19.90	2.40	182.30	218	
	0500 Checkered plate tread, industrial, 3'-6" wide		28	1.143		100	25	3.01	128.01	162	
	0550 Circular, for tanks, 3'-0" wide		33	.970		110	21	2.55	133.55	165	
	0600 For isolated stairs, add						100%				
	0800 Custom steel stairs, 3'-6" wide, minimum	E-4	35	.914		150	19.90	2.40	172.30	207	
	0810 Average		30	1.067		200	23	2.81	225.81	269	
	0900 Maximum		20	1.600		250	35	4.21	289.21	350	
	1100 For 4' wide stairs, add					10%	5%				
	1300 For 5' wide stairs, add					20%	10%				
	1500 Landing, steel pan, conventional	E-4	160	.200	S.F.	20	4.35	.53	24.88	31	
	1600 Pre-erected	"	255	.125	"	35	2.73	.33	38.06	44.50	
	1700 Pre-erected, steel pan tread, 3'-6" wide, 2 line pipe rail	E-2	87	.552	Riser	165	11.85	11.75	188.60	218	
	1810 Spiral aluminum, 5'-0" diameter, stock units	E-4	45	.711		185	15.45	1.87	202.32	237	
	1820 Custom units		45	.711		350	15.45	1.87	367.32	420	
	1900 Spiral, cast iron, 4'-0" diameter, ornamental, minimum		45	.711		165	15.45	1.87	182.32	215	
	1920 Maximum		25	1.280		225	28	3.37	256.37	305	
	2000 Spiral, steel, industrial checkered plate, 4' diameter		45	.711		165	15.45	1.87	182.32	215	
	2200 Stock units, 6'-0" diameter		40	.800		200	17.40	2.10	219.50	257	
	3110 Spiral steel, stock units, primed, flat metal tread, 3'-6" dia	2 Carp	1.60	10	Flight	840	197		1,037	1,275	
	3120 4'-0" dia		1.45	11.034		965	217		1,182	1,425	
	3130 4'-6" dia		1.35	11.852		1,050	233		1,283	1,575	
	3140 5'-0" dia		1.25	12.800		1,150	252		1,402	1,700	
	3210 Galvanized, 3'-6" dia		1.60	10		1,425	197		1,622	1,925	
	3220 4'-0" dia		1.45	11.034		1,600	217		1,817	2,150	
	3230 4'-6" dia		1.35	11.852		1,725	233		1,958	2,300	
	3240 5'-0" dia		1.25	12.800		1,900	252		2,152	2,500	
	3310 Checkered plate tread, 3'-6" dia		1.45	11.034		1,025	217		1,242	1,500	
	3320 4'-0" dia		1.35	11.852		1,175	233		1,408	1,700	
	3330 4'-6" dia		1.25	12.800		1,275	252		1,527	1,825	
	3340 5'-0" dia		1.15	13.913		1,375	274		1,649	2,000	
	3410 Galvanized, 3'-6" dia		1.45	11.034		1,625	217		1,842	2,175	
	3420 4'-0" dia		1.35	11.852		1,825	233		2,058	2,425	

05500 | Metal Fabrications

05517 | Metal Stairs

			CREW	DAILY OUTPUT	LABOR-HOURS	UNIT	2000 BARE COSTS MAT.	LABOR	EQUIP.	TOTAL	TOTAL INCL O&P	
700	3430	4'-6" dia	2 Carp	1.25	12.800	Flight	2,000	252		2,252	2,625	700
	3440	5'-0" dia		1.15	13.913		2,125	274		2,399	2,800	
	3510	Red oak tread on flat metal, 3'-6" dia		1.35	11.852		1,400	233		1,633	1,950	
	3520	4'-0" dia		1.25	12.800		1,550	252		1,802	2,125	
	3530	4'-6" dia		1.15	13.913		1,700	274		1,974	2,325	
	3540	5'-0" dia		1.05	15.238		1,825	300		2,125	2,525	
	3900	Industrial ships ladder, 3' W, grating treads, 2 line pipe rail	E-4	30	1.067	Riser	65	23	2.81	90.81	121	
	4000	Aluminum	"	30	1.067	"	100	23	2.81	125.81	159	

05520 | Handrails & Railings

			CREW	DAILY OUTPUT	LABOR-HOURS	UNIT	MAT.	LABOR	EQUIP.	TOTAL	TOTAL INCL O&P	
700	0005	**RAILING, PIPE**										700
	0010	Aluminum, 2 rail, satin finish, 1-1/4" diameter	E-4	160	.200	L.F.	13	4.35	.53	17.88	23.50	
	0030	Clear anodized		160	.200		16.10	4.35	.53	20.98	27	
	0040	Dark anodized		160	.200		18.15	4.35	.53	23.03	29	
	0080	1-1/2" diameter, satin finish		160	.200		15.55	4.35	.53	20.43	26.50	
	0090	Clear anodized		160	.200		17.35	4.35	.53	22.23	28.50	
	0100	Dark anodized		160	.200		19.25	4.35	.53	24.13	30	
	0140	Aluminum, 3 rail, 1-1/4" diam., satin finish		137	.234		19.90	5.10	.61	25.61	32.50	
	0150	Clear anodized		137	.234		25	5.10	.61	30.71	38	
	0160	Dark anodized		137	.234		27.50	5.10	.61	33.21	41	
	0200	1-1/2" diameter, satin finish		137	.234		24	5.10	.61	29.71	37	
	0210	Clear anodized		137	.234		27	5.10	.61	32.71	40.50	
	0220	Dark anodized		137	.234		29.50	5.10	.61	35.21	43	
	0700	Stainless steel, 2 rail, 1-1/4" diam. #4 finish		137	.234		32	5.10	.61	37.71	45.50	
	0720	High polish		137	.234		51.50	5.10	.61	57.21	67	
	0740	Mirror polish		137	.234		64.50	5.10	.61	70.21	81.50	
	0760	Stainless steel, 3 rail, 1-1/2" diam., #4 finish		120	.267		48	5.80	.70	54.50	65	
	0770	High polish		120	.267		79.50	5.80	.70	86	99.50	
	0780	Mirror finish		120	.267		97	5.80	.70	103.50	119	
	0900	Wall rail, alum. pipe, 1-1/4" diam., satin finish		213	.150		7.45	3.27	.40	11.12	15.05	
	0905	Clear anodized		213	.150		9.05	3.27	.40	12.72	16.90	
	0910	Dark anodized		213	.150		11	3.27	.40	14.67	19	
	0915	1-1/2" diameter, satin finish		213	.150		8.25	3.27	.40	11.92	15.95	
	0920	Clear anodized		213	.150		10.35	3.27	.40	14.02	18.30	
	0925	Dark anodized		213	.150		12.80	3.27	.40	16.47	21	
	0930	Steel pipe, 1-1/4" diameter, primed		213	.150		5.70	3.27	.40	9.37	13.15	
	0935	Galvanized		213	.150		8.25	3.27	.40	11.92	15.95	
	0940	1-1/2" diameter		176	.182		5.85	3.95	.48	10.28	14.75	
	0945	Galvanized		213	.150		8.30	3.27	.40	11.97	16	
	0955	Stainless steel pipe, 1-1/2" diam., #4 finish		107	.299		25.50	6.50	.79	32.79	41.50	
	0960	High polish		107	.299		52	6.50	.79	59.29	70.50	
	0965	Mirror polish		107	.299		61	6.50	.79	68.29	81	
780	0005	**RAILINGS, INDUSTRIAL** Welded										780
	0010	2 rail, 3'-6" high, 1-1/2" pipe	E-4	255	.125	L.F.	13.70	2.73	.33	16.76	21	
	0100	2" angle rail	"	255	.125		12.50	2.73	.33	15.56	19.50	
	0200	For 4" high kick plate, 10 gauge, add					2.86			2.86	3.15	
	0300	1/4" thick, add					3.67			3.67	4.04	
	0500	For curved rails, add					30%	30%				

05530 | Gratings

			CREW	DAILY OUTPUT	LABOR-HOURS	UNIT	MAT.	LABOR	EQUIP.	TOTAL	TOTAL INCL O&P	
300	0010	**FLOOR GRATING, ALUMINUM**										300
	0200	For straight cuts, add				L.F.	2.31			2.31	2.54	
	0300	For curved cuts, add					3.04			3.04	3.34	
	0400	For straight banding, add					2.89			2.89	3.18	
	0500	For curved banding, add					3.62			3.62	3.98	
	0600	For aluminum checkered plate nosings, add					3.38			3.38	3.72	

05500 | Metal Fabrications

05530 | Gratings

			Crew	Daily Output	Labor-Hours	Unit	2000 Bare Costs Mat.	Labor	Equip.	Total	Total Incl O&P	
300	0700	For straight toe plate, add				L.F.	6.25			6.25	6.90	300
	0800	For curved toe plate, add					7.70			7.70	8.45	
	1000	For cast aluminum abrasive nosings, add					4.99			4.99	5.50	
	1200	Expanded aluminum, .65# per S.F.	E-4	1,050	.030	S.F.	3.58	.66	.08	4.32	5.35	
	1400	Extruded I bars are 10% less than 3/16" bars										
	1600	Heavy duty, all extruded plank, 3/4" deep, 1.8 # per S.F.	E-4	1,100	.029	S.F.	8.70	.63	.08	9.41	10.95	
	1700	1-1/4" deep, 2.9# per S.F.		1,000	.032		12.80	.70	.08	13.58	15.55	
	1800	1-3/4" deep, 4.2# per S.F.		925	.035		17.35	.75	.09	18.19	20.50	
	1900	2-1/4" deep, 5.0# per S.F.		875	.037		19.65	.80	.10	20.55	23	
	2100	For safety serrated surface, add					15%					
320	0010	**FLOOR GRATING PLANKS**										320
	0020	Aluminum, 9-1/2" wide, 14 ga., 2" rib	E-4	950	.034	L.F.	11.20	.73	.09	12.02	13.85	
	0200	Galvanized steel, 9-1/2" wide, 14 ga., 2-1/2" rib		950	.034		7	.73	.09	7.82	9.25	
	0300	4" rib		950	.034		8.25	.73	.09	9.07	10.60	
	0500	12 gauge, 2-1/2" rib		950	.034		8.60	.73	.09	9.42	11.05	
	0600	3" rib		950	.034		9.65	.73	.09	10.47	12.20	
	0800	Stainless steel, type 304, 16 ga., 2" rib		950	.034		18.25	.73	.09	19.07	21.50	
	0900	Type 316		950	.034		22	.73	.09	22.82	26	
340	0010	**FLOOR GRATING, STEEL**										340
	0050	Labor for installing, from ground/floor	E-4	845	.038	S.F.		.82	.10	.92	1.73	
	0100	Elevated		460	.070	"		1.51	.18	1.69	3.18	
	0300	Platforms, to 12' high, rectangular		3,150	.010	Lb.	1.07	.22	.03	1.32	1.65	
	0400	Circular		2,300	.014	"	1.18	.30	.04	1.52	1.94	
	0800	For straight cuts, add				L.F.	3.74			3.74	4.11	
	0900	For curved cuts, add					4.70			4.70	5.15	
	1000	For straight banding, add					3.59			3.59	3.95	
	1100	For curved banding, add					4.70			4.70	5.15	
	1200	For checkered plate nosings, add					4.04			4.04	4.44	
	1300	For straight toe or kick plate, add					7.25			7.25	8	
	1400	For curved toe or kick plate, add					8.10			8.10	8.90	
	1500	For abrasive nosings, add					5.90			5.90	6.50	
	1510	For stair treads, see division 05550-700										
	1600	For safety serrated surface, minimum, add					15%					
	1700	Maximum, add					25%					
	2000	Stainless steel gratings, close spaced, 1" x 1/8" bars, up to 300 S.F.	E-4	450	.071	S.F.	82	1.55	.19	83.74	93.50	
	2100	Standard spacing, 3/4" x 1/8" bars		500	.064		67	1.39	.17	68.56	76.50	
	2200	1-1/4" x 3/16" bars		400	.080		97	1.74	.21	98.95	111	
	2400	Expanded steel grating, at ground, 3.0# per S.F.		900	.036		2.49	.77	.09	3.35	4.36	
	2500	3.14# per S.F.		900	.036		2.55	.77	.09	3.41	4.42	
	2600	4.0# per S.F.		850	.038		3.18	.82	.10	4.10	5.20	
	2650	4.27# per S.F.		850	.038		3.44	.82	.10	4.36	5.50	
	2700	5.0# per S.F.		800	.040		4.14	.87	.11	5.12	6.40	
	2800	6.25# per S.F.		750	.043		4.99	.93	.11	6.03	7.45	
	2900	7.0# per S.F.		700	.046		5.85	.99	.12	6.96	8.50	
	3100	For flattened expanded steel grating, add					8%					
	3300	For elevated installation, add						15%				
360	0005	**GRATING FRAME**										360
	0011	Aluminum, for gratings 1" to 1-1/2" deep	1 Carp	70	.114	L.F.	6.20	2.25		8.45	10.70	
	0100	For each corner, add				Ea.	4.90			4.90	5.40	

05540 | Floor Plates

			Crew	Daily Output	Labor-Hours	Unit	Mat.	Labor	Equip.	Total	Total Incl O&P	
200	0005	**CHECKERED PLATE**										200
	0010	1/4" & 3/8", 2000 to 5000 S.F., bolted	E-4	2,900	.011	Lb.	.59	.24	.03	.86	1.15	
	0100	Welded		4,400	.007	"	.55	.16	.02	.73	.94	
	0300	Pit or trench cover and frame, 1/4" plate, 2' to 3' wide		100	.320	S.F.	15.50	6.95	.84	23.29	31.50	

05500 | Metal Fabrications

05540 | Floor Plates

			Crew	Daily Output	Labor-Hours	Unit	Mat.	Labor	Equip.	Total	Total Incl O&P	
200	0400	For galvanizing, add				Lb.	.34			.34	.37	200
	0500	Platforms, 1/4" plate, no handrails included, rectangular	E-4	4,200	.008		.93	.17	.02	1.12	1.37	
	0600	Circular	"	2,500	.013		1.30	.28	.03	1.61	2.02	
700	0010	**TRENCH COVER**										700
	0021	Cast iron grating with bar stops and angle frame, to 18" wide	1 Carp	20	.400	L.F.	41	7.90		48.90	58.50	
	0101	Frame only (both sides of trench), 1" grating		45	.178		9	3.50		12.50	15.90	
	0151	2" grating		35	.229		14	4.50		18.50	23	
	0200	Aluminum, stock units, including frames and										
	0211	3/8" plain cover plate, 4" opening	2 Carp	100	.160	L.F.	29.50	3.15		32.65	38	
	0301	6" opening		90	.178		36.50	3.50		40	46	
	0401	10" opening		85	.188		50	3.71		53.71	61.50	
	0501	16" opening		80	.200		69	3.94		72.94	82.50	
	0700	Add per inch for additional widths to 24"					2.73			2.73	3	
	0900	For custom fabrication, add					50%					
	1100	For 1/4" plain cover plate, deduct					12%					
	1500	For cover recessed for tile, 1/4" thick, deduct					12%					
	1600	3/8" thick, add					5%					
	1800	For checkered plate cover, 1/4" thick, deduct					12%					
	1900	3/8" thick, add					2%					
	2100	For slotted or round holes in cover, 1/4" thick, add					3%					
	2200	3/8" thick, add					4%					
	2300	For abrasive cover, add					12%					

05550 | Stair Treads & Nosings

			Crew	Daily Output	Labor-Hours	Unit	Mat.	Labor	Equip.	Total	Total Incl O&P	
700	0005	**STAIR TREADS**										700
	0010	Aluminum grating, 3' long, 1-1/2" x 3/16" rect. bars, 6" wide	1 Sswk	24	.333	Ea.	25	7.10		32.10	41.50	
	0100	12" wide		22	.364		43.50	7.75		51.25	63.50	
	0200	1-1/2" x 3/16" I-bars, 6" wide		24	.333		34.50	7.10		41.60	52	
	0300	12" wide		22	.364		58.50	7.75		66.25	79.50	
	0400	For abrasive nosings, add					13.50			13.50	14.85	
	0500	For narrow mesh, add					60%					
	0600	Treads, 12" x 3'-6", not incl. stringers. See also div. 03150-660										
	0701	Cast aluminum, abrasive, 5/16" thick	1 Carp	15	.533	Ea.	99.50	10.50		110	128	
	0801	3/8" thick		15	.533		104	10.50		114.50	132	
	0901	1/2" thick		15	.533		118	10.50		128.50	148	
	1001	Cast bronze, abrasive, 3/8" thick		8	1		400	19.70		419.70	475	
	1101	1/2" thick		8	1		525	19.70		544.70	610	
	1201	Cast iron, abrasive, 3/8" thick		15	.533		74	10.50		84.50	99.50	
	1301	1/2" thick		15	.533		85.50	10.50		96	112	
	1400	Fiberglass reinforced plastic with safety nosing,										
	1501	1-1/2" thick, 12" wide, 24" long	1 Carp	22	.364	Ea.	87.50	7.15		94.65	108	
	1601	30" long		22	.364		91.50	7.15		98.65	113	
	1701	36" long		22	.364		100	7.15		107.15	122	
	2000	Steel grating, painted, 3' long, 1-1/4" x 3/16" bars, 6" wide	1 Sswk	20	.400		26	8.50		34.50	45.50	
	2010	12" wide	"	18	.444		26	9.45		35.45	47	
	2100	Add for abrasive nosing, 3' long, painted					12.25			12.25	13.50	
	2200	Galvanized					20.50			20.50	22.50	
	2300	Painting 3' treads, in shop, nonstandard paint					.93			.93	1.02	
	2400	Added coats, standard paint					.48			.48	.53	
	2501	Expanded steel, 2-1/2" deep, 9" x 3' long, 18 gauge	1 Carp	20	.400		17.75	7.90		25.65	33	
	2601	14 gauge	"	20	.400		22	7.90		29.90	37.50	

05500 | Metal Fabrications

05560 | Metal Castings

			CREW	DAILY OUTPUT	LABOR-HOURS	UNIT	2000 BARE COSTS MAT.	LABOR	EQUIP.	TOTAL	TOTAL INCL O&P	
200	0010	**CONSTRUCTION CASTINGS**										200
	0020	Manhole covers and frames see Division 02630-200										
	0101	Column bases, cast iron, 16" x 16", approx. 65 lbs.	2 Carp	23	.696	Ea.	79	13.70		92.70	111	
	0201	32" x 32", approx. 256 lbs.	"	11.50	1.391		295	27.50		322.50	370	
	0400	Cast aluminum for wood columns, 8" x 8"	1 Carp	32	.250		28	4.93		32.93	39	
	0500	12" x 12"	"	32	.250	↓	60	4.93		64.93	74.50	
	0601	Miscellaneous C.I. castings, light sections	2 Carp	1,600	.010	Lb.	1.21	.20		1.41	1.67	
	1101	Heavy sections		2,100	.008		.60	.15		.75	.92	
	1301	Special low volume items	↓	1,600	.010	↓	2.12	.20		2.32	2.67	
	1500	For ductile iron, add					100%					

05580 | Formed Metal Fabrications

			CREW	DAILY OUTPUT	LABOR-HOURS	UNIT	MAT.	LABOR	EQUIP.	TOTAL	TOTAL INCL O&P	
150	0005	**ALLOY STEEL CHAIN**										150
	0010	Self-colored, cut lengths, w/accessories, 1/4"	E-17	4	4	C.L.F.	320	89	.60	409.60	530	
	0020	3/8"		2	8		420	178	1.20	599.20	815	
	0030	1/2"		1.20	13.333		620	297	2	919	1,275	
	0040	5/8"		.72	22.222		1,200	495	3.33	1,698.33	2,300	
	0050	3/4"		.48	33.333		1,675	740	5	2,420	3,325	
	0060	7/8"		.40	40		2,525	890	6	3,421	4,550	
	0070	1"		.35	45.714		3,825	1,025	6.85	4,856.85	6,200	
	0080	1-1/4"	↓	.24	66.667	↓	6,875	1,475	10	8,360	10,500	
	0110	Clevis slip hook, 1/4"				Ea.	7.55			7.55	8.30	
	0120	3/8"					10.95			10.95	12.05	
	0130	1/2"					20			20	22	
	0140	5/8"					35.50			35.50	39	
	0150	3/4"					75			75	82.50	
	0160	Eye/sling hook w/ hammerlock coupling, 7/8"					185			185	203	
	0170	1"					305			305	335	
	0180	1-1/4"				↓	495			495	545	
600	0005	**LAMP POSTS**										600
	0010	Aluminum, 7' high, stock units, post only	1 Carp	16	.500	Ea.	65	9.85		74.85	88.50	
	0100	Mild steel, plain	"	16	.500	"	39	9.85		48.85	60	
900	0005	**WINDOW GUARDS**										900
	0010	Expanded metal, steel angle frame, permanent	E-4	350	.091	S.F.	14.35	1.99	.24	16.58	19.95	
	0020	Steel bars, 1/2" x 1/2", spaced 5" O.C.	"	290	.110	"	9.90	2.40	.29	12.59	15.95	
	0030	Hinge mounted, add				Opng.	28.50			28.50	31.50	
	0040	Removable type, add				"	18.25			18.25	20	
	0050	For galvanized guards, add				S.F.	35%					
	0070	For pivoted or projected type, add					105%	40%				
	0100	Mild steel, stock units, economy	E-4	405	.079		3.89	1.72	.21	5.82	7.90	
	0200	Deluxe		405	.079	↓	8	1.72	.21	9.93	12.40	
	0400	Woven wire, stock units, 3/8" channel frame, 3' x 5' opening		40	.800	Opng.	105	17.40	2.10	124.50	153	
	0500	4' x 6' opening	↓	38	.842		168	18.30	2.21	188.51	223	
	0800	Basket guards for above, add					144			144	158	
	1000	Swinging guards for above, add				↓	49.50			49.50	54.50	

Important: See the Reference Section for critical supporting data - Reference Nos., Crews, & City Cost Indexes

05650 | Railroad Track & Accessories

05655 | Railroad Trackwork

			CREW	DAILY OUTPUT	LABOR-HOURS	UNIT	MAT.	LABOR	EQUIP.	TOTAL	TOTAL INCL O&P	
700	0010	**RAILROAD** Car bumpers, standard	B-14	2	24	Ea.	2,050	375	102	2,527	3,025	700
	0100	Heavy duty		2	24		3,900	375	102	4,377	5,050	
	0200	Derails hand throw (sliding)		10	4.800		720	74.50	20.50	815	945	
	0300	Hand throw with standard timbers, open stand & target		8	6		780	93.50	25.50	899	1,050	
	0400	Resurface and realign existing track		200	.240	L.F.		3.73	1.02	4.75	7.45	
	0600	For crushed stone ballast, add		500	.096	"	9.05	1.49	.41	10.95	12.95	
	0800	Siding, yard spur, level grade										
	0810	100 lb. rail, new material on wood ties	B-14	57	.842	L.F.	57	13.10	3.56	73.66	89	
	1000	Steel ties in concrete w/100# rail, fasteners & plates		22	2.182	"	101	34	9.25	144.25	179	
	1200	Switch timber, for a #8 switch, pressure treated		3.70	12.973	M.B.F.	620	202	55	877	1,075	
	1300	Complete set of timbers, 3.7 M.B.F. for #8 switch		1	48	Total	2,675	745	203	3,623	4,450	
	1400	Ties, concrete, 8'-6" long, 30" O.C.		80	.600	Ea.	70	9.35	2.54	81.89	95.50	
	1600	Wood, pressure treated, 6" x 8" x 8'-6", C.L. lots		90	.533		28.50	8.30	2.26	39.06	47.50	
	1700	L.C.L. lots		90	.533		29.50	8.30	2.26	40.06	49	
	1900	Heavy duty, 7" x 9" x 8'-6", C.L. lots		70	.686		31	10.65	2.90	44.55	55.50	
	2000	L.C.L. lots		70	.686		29	10.65	2.90	42.55	53.50	
	2200	Turnouts, #8, incl. 100 lb. rails, plates, bars, frog, switch pt.										
	2300	Timbers and ballast 6" below bottom of tie	B-14	.50	96	Ea.	18,800	1,500	405	20,705	23,600	
	2400	Wheel stops, fixed		18	2.667	Pr.	460	41.50	11.30	512.80	590	
	2450	Hinged		14	3.429	"	505	53.50	14.50	573	660	
750	0005	**RAILROAD TRACK**										750
	0010	Track bolts				Ea.	1.73			1.73	1.90	
	0100	Joint bars				Pr.	26			26	28.50	
	0200	Spikes				Ea.	.41			.41	.45	
	0300	Tie plates				"	1.50			1.50	1.65	

05700 | Ornamental Metal

05720 | Ornamental Railings

			CREW	DAILY OUTPUT	LABOR-HOURS	UNIT	MAT.	LABOR	EQUIP.	TOTAL	TOTAL INCL O&P	
700	0005	**RAILINGS, ORNAMENTAL**										700
	0010	Aluminum, bronze or stainless, minimum	1 Sswk	24	.333	L.F.	44	7.10		51.10	62.50	
	0100	Maximum		9	.889		545	18.90		563.90	635	
	0200	Aluminum ornamental rail, minimum		15	.533		44	11.35		55.35	71	
	0300	Maximum		8	1		59.50	21.50		81	108	
	0400	Hand-forged wrought iron, minimum		12	.667		95	14.15		109.15	132	
	0500	Maximum		8	1		405	21.50		426.50	485	
	0600	Composite metal and wood or glass, minimum		6	1.333		50	28.50		78.50	111	
	0700	Maximum		5	1.600		150	34		184	232	

05800 | Expansion Control

05810 | Exp. Joint Cover Assemblies

		CREW	DAILY OUTPUT	LABOR-HOURS	UNIT	2000 BARE COSTS				TOTAL INCL O&P
						MAT.	LABOR	EQUIP.	TOTAL	
0010	**EXPANSION JOINT ASSEMBLIES** Custom units									
0201	Floor cover assemblies, 1" space, aluminum	1 Carp	38	.211	L.F.	13.35	4.15		17.50	22
0301	Bronze or stainless		38	.211		26	4.15		30.15	35.50
0501	2" space, aluminum		38	.211		16.10	4.15		20.25	25
0601	Bronze or stainless		38	.211		28	4.15		32.15	37.50
0801	Wall and ceiling assemblies, 1" space, aluminum		38	.211		7.95	4.15		12.10	15.85
0901	Bronze or stainless		38	.211		27.50	4.15		31.65	37
1101	2" space, aluminum		38	.211		13.50	4.15		17.65	22
1201	Bronze or stainless		38	.211		25	4.15		29.15	34.50
1401	Floor to wall assemblies, 1" space, aluminum		38	.211		12.15	4.15		16.30	20.50
1501	Bronze or stainless		38	.211		29.50	4.15		33.65	39
1701	Gym floor angle covers, aluminum, 3" x 3" angle		46	.174		10.50	3.43		13.93	17.40
1801	3" x 4" angle		46	.174		12.40	3.43		15.83	19.50
2001	Roof closures, aluminum, 1" space, flat roof, low profile		57	.140		24.50	2.76		27.26	31
2101	High profile		57	.140		30	2.76		32.76	37.50
2302	Roof to wall, 1" space, low profile		57	.140		13.40	2.76		16.16	19.50
2401	High profile	▼	57	.140	▼	17.30	2.76		20.06	23.50

For information about Means Estimating Seminars, see yellow pages 11 and 12 in back of book

Division 6
Wood & Plastics

Estimating Tips

06050 Basic Wood & Plastic Materials & Methods

- Common to any wood framed structure are the accessory connector items such as screws, nails, adhesives, hangers, connector plates, straps, angles and holdowns. For typical wood framed buildings, such as residential projects, the aggregate total for these items can be significant, especially in areas where seismic loading is a concern. For floor and wall framing, nail quantities can be figured on a "pounds per thousand board feet basis", with 10 to 25 lbs. per MBF the range. Holdowns, hangers and other connectors should be taken off by the piece.

06100 Rough Carpentry

- Lumber is a traded commodity and therefore sensitive to supply and demand in the marketplace. Even in "budgetary" estimating of wood framed projects, it is advisable to call local suppliers for the latest market pricing.
- Common quantity units for wood framed projects are "thousand board feet" (MBF). A board foot is a volume of wood, 1" x 1' x 1', or 144 cubic inches. Board foot quantities are generally calculated using nominal material dimensions—dressed sizes are ignored. Board foot per lineal foot of any stick of lumber can be calculated by dividing the nominal cross sectional area by 12. As an example, 2,000 lineal feet of 2 x 12 equates to 4 MBF by dividing the nominal area, 2 x 12, by 12, which equals 2, and multiplying by 2,000 to give 4,000 board feet. This simple rule applies to all nominal dimensioned lumber.
- Waste is an issue of concern at the quantity takeoff for any area of construction. Framing lumber is sold in even foot lengths, i.e., 10', 12', 14', 16', and depending on spans, wall heights and the grade of lumber, waste is inevitable. A rule of thumb for lumber waste is 5% to 10% depending on material quality and the complexity of the framing.
- Wood in various forms and shapes is used in many projects, even where the main structural framing is steel, concrete or masonry. Plywood as a back-up partition material and 2x boards used as blocking and cant strips around roof edges are two common examples. The estimator should ensure that the costs of all wood materials are included in the final estimate.

06200 Finish Carpentry

- It is necessary to consider the grade of workmanship when estimating labor costs for erecting millwork and interior finish. In practice, there are three grades: premium, custom and economy. The Means daily output for base and case moldings is in the range of 200 to 250 L.F. per carpenter per day. This is appropriate for most average custom grade projects. For premium projects an adjustment to productivity of 25% to 50% should be made depending on the complexity of the job.

Reference Numbers

Reference numbers are shown in bold squares at the beginning of some major classifications. These numbers refer to related items in the Reference Section. The reference information may be an estimating procedure, an alternate pricing method or technical information.

Note: Not all subdivisions listed here necessarily appear in this publication.

06050 | Basic Wood / Plastic Materials / Methods

06055 | Wood & Plastic Laminate

			DAILY	LABOR-		2000 BARE COSTS				TOTAL
		CREW	OUTPUT	HOURS	UNIT	MAT.	LABOR	EQUIP.	TOTAL	INCL O&P
720	0010	**CONVECTOR COVERS** Laminated plastic on 3/4"									720
	0020	thick particle board, 12" wide, minimum	1 Carp	16	.500	L.F.	20	9.85		29.85	39
	0050	Average		13	.615		25.50	12.10		37.60	49
	0100	Maximum		13	.615	↓	35	12.10		47.10	59.50
	0300	Add to above for grille, minimum		150	.053	S.F.	1.60	1.05		2.65	3.56
	0400	Maximum	↓	75	.107	"	4.50	2.10		6.60	8.55
740	0010	**COUNTER TOP** Stock, plastic lam., 24" wide w/backsplash, min.	1 Carp	30	.267	L.F.	4.89	5.25		10.14	14.40	740
	0100	Maximum		25	.320		14.25	6.30		20.55	26.50
	0300	Custom plastic, 7/8" thick, aluminum molding, no splash		30	.267		15.70	5.25		20.95	26.50
	0400	Cove splash		30	.267		20.50	5.25		25.75	31.50
	0600	1-1/4" thick, no splash		28	.286		18.50	5.65		24.15	30
	0700	Square splash		28	.286		23	5.65		28.65	35
	0900	Square edge, plastic face, 7/8" thick, no splash		30	.267		19.80	5.25		25.05	31
	1000	With splash		30	.267		26	5.25		31.25	37.50
	1200	For stainless channel edge, 7/8" thick, add					2.21			2.21	2.43
	1300	1-1/4" thick, add					2.58			2.58	2.84
	1500	For solid color suede finish, add				↓	1.96			1.96	2.16
	1700	For end splash, add				Ea.	12.35			12.35	13.60
	1900	For cut outs, standard, add, minimum	1 Carp	32	.250		2.63	4.93		7.56	11.35
	2000	Maximum		8	1	↓	3.14	19.70		22.84	37.50
	2100	Postformed, including backsplash and front edge		30	.267	L.F.	8.55	5.25		13.80	18.40
	2110	Mitred, add		12	.667	Ea.		13.15		13.15	22.50
	2200	Built-in place, 25" wide, plastic laminate		25	.320	L.F.	10.80	6.30		17.10	22.50
	2300	Ceramic tile mosaic	↓	25	.320		24.50	6.30		30.80	38
	2500	Marble, stock, with splash, 1/2" thick, minimum	1 Bric	17	.471		30	9.40		39.40	49
	2700	3/4" thick, maximum	"	13	.615		76	12.30		88.30	105
	2900	Maple, solid, laminated, 1-1/2" thick, no splash	1 Carp	28	.286		31	5.65		36.65	43.50
	3000	With square splash		28	.286	↓	35	5.65		40.65	48
	3200	Stainless steel		24	.333	S.F.	73	6.55		79.55	92
	3400	Recessed cutting block with trim, 16" x 20" x 1"		8	1	Ea.	41	19.70		60.70	79.50
	3600	Table tops, plastic laminate, square edge, 7/8" thick		45	.178	S.F.	7	3.50		10.50	13.70
	3700	1-1/8" thick	↓	40	.200	"	7.20	3.94		11.14	14.70

06073 | Fire Retardant Treatment

400	0011	**LUMBER TREATMENT**									400
	0400	Fire retardant, wet				M.B.F.	282			282	310
	0500	KDAT					255			255	281
	0700	Salt treated, water borne, .40 lb. retention					122			122	134
	0800	Oil borne, 8 lb. retention					143			143	157
	1000	Kiln dried lumber, 1" & 2" thick, softwoods					81.50			81.50	90
	1100	Hardwoods					87			87	95.50
	1500	For small size 1" stock, add				↓	11			11	12.10
	1700	For full size rough lumber, add					20%				
600	0010	**PLYWOOD TREATMENT** Fire retardant, 1/4" thick				M.S.F.	204			204	224	600
	0030	3/8" thick					224			224	246
	0050	1/2" thick					240			240	264
	0070	5/8" thick					255			255	281
	0100	3/4" thick					280			280	310
	0200	For KDAT, add					61			61	67
	0500	Salt treated water borne, .25 lb., wet, 1/4" thick					112			112	123
	0530	3/8" thick					117			117	129
	0550	1/2" thick					122			122	134
	0570	5/8" thick					133			133	146
	0600	3/4" thick					138			138	152
	0800	For KDAT add				↓	61			61	67

06050 | Basic Wood / Plastic Materials / Methods

	06073	Fire Retardant Treatment	CREW	DAILY OUTPUT	LABOR-HOURS	UNIT	2000 BARE COSTS				TOTAL INCL O&P	
							MAT.	LABOR	EQUIP.	TOTAL		
600	0900	For .40 lb., per C.F. retention, add				M.S.F.	51			51	56	600
	1000	For certification stamp, add				↓	30			30	33	
	06090	**Wood & Plastic Fastenings**										
600	0010	**NAILS** Prices of material only, based on 50# box purchase, copper, plain				Lb.	4.10			4.10	4.51	600
	0400	Stainless steel, plain					5.40			5.40	5.95	
	0500	Box, 3d to 20d, bright					1.13			1.13	1.24	
	0520	Galvanized					1.31			1.31	1.44	
	0600	Common, 3d to 60d, plain					.77			.77	.85	
	0700	Galvanized					.99			.99	1.09	
	0800	Aluminum					3.31			3.31	3.64	
	1000	Annular or spiral thread, 4d to 60d, plain					.66			.66	.73	
	1200	Galvanized					.83			.83	.91	
	1400	Drywall nails, plain					.77			.77	.85	
	1600	Galvanized					1.10			1.10	1.21	
	1800	Finish nails, 4d to 10d, plain					.90			.90	.99	
	2000	Galvanized					1.04			1.04	1.14	
	2100	Aluminum					4.85			4.85	5.35	
	2300	Flooring nails, hardened steel, 2d to 10d, plain					1.22			1.22	1.34	
	2400	Galvanized					1.34			1.34	1.47	
	2500	Gypsum lath nails, 1-1/8", 13 ga. flathead, blued					1.33			1.33	1.46	
	2600	Masonry nails, hardened steel, 3/4" to 3" long, plain					1.61			1.61	1.77	
	2700	Galvanized					1.44			1.44	1.58	
	2900	Roofing nails, threaded, galvanized					1.30			1.30	1.43	
	3100	Aluminum					4.70			4.70	5.15	
	3300	Compressed lead head, threaded, galvanized					1.44			1.44	1.58	
	3600	Siding nails, plain shank, galvanized					1.33			1.33	1.46	
	3800	Aluminum					4			4	4.40	
	5000	Add to prices above for cement coating					.07			.07	.08	
	5200	Zinc or tin plating					.12			.12	.13	
	5500	Vinyl coated sinkers, 8d to 16d				↓	.55			.55	.61	
650	0010	**NAILS** mat. only, for pneumatic tools, framing, per carton of 5000, 2"				Ea.	36.50			36.50	40	650
	0100	2-3/8"					41.50			41.50	45.50	
	0200	Per carton of 4000, 3"					37			37	40.50	
	0300	3-1/4"					39			39	43	
	0400	Per carton of 5000, 2-3/8", galv.					56.50			56.50	62	
	0500	Per carton of 4000, 3", galv.					63.50			63.50	70	
	0600	3-1/4", galv.					79			79	87	
	0700	Roofing, per carton of 7200, 1"					34.50			34.50	38	
	0800	1-1/4"					32			32	35.50	
	0900	1-1/2"					37			37	40.50	
	1000	1-3/4"				↓	44.50			44.50	49	
700	0010	**SHEET METAL SCREWS** Steel, standard, #8 x 3/4", plain				C	2.69			2.69	2.96	700
	0100	Galvanized					3.37			3.37	3.71	
	0300	#10 x 1", plain					3.69			3.69	4.06	
	0400	Galvanized					4.26			4.26	4.69	
	0600	With washers, #14 x 1", plain					10.85			10.85	11.95	
	0700	Galvanized					12.05			12.05	13.30	
	0900	#14 x 2", plain					15.80			15.80	17.40	
	1000	Galvanized					17.80			17.80	19.55	
	1500	Self-drilling, with washers, (pinch point) #8 x 3/4", plain					4.68			4.68	5.15	
	1600	Galvanized					7			7	7.70	
	1800	#10 x 3/4", plain					6.80			6.80	7.50	
	1900	Galvanized				↓	7.75			7.75	8.55	

WOOD & PLASTICS 6

06050 | Basic Wood / Plastic Materials / Methods

	06090	Wood & Plastic Fastenings	CREW	DAILY OUTPUT	LABOR-HOURS	UNIT	2000 BARE COSTS MAT.	LABOR	EQUIP.	TOTAL	TOTAL INCL O&P	
700	3000	Stainless steel w/aluminum or neoprene washers, #14 x 1", plain				C	18			18	19.80	700
	3100	#14 x 2", plain					24.50			24.50	27	
750	0010	**WOOD SCREWS** #8, 1" long, steel				C	3.29			3.29	3.62	750
	0100	Brass					11			11	12.10	
	0200	#8, 2" long, steel					3.69			3.69	4.06	
	0300	Brass					11.50			11.50	12.65	
	0400	#10, 1" long, steel					4.22			4.22	4.64	
	0500	Brass					22.50			22.50	24.50	
	0600	#10, 2" long, steel					7.50			7.50	8.25	
	0700	Brass					39.50			39.50	43.50	
	0800	#10, 3" long, steel					13.75			13.75	15.10	
	1000	#12, 2" long, steel					4.79			4.79	5.25	
	1100	Brass					16			16	17.60	
	1500	#12, 3" long, steel					15.90			15.90	17.50	
	2000	#12, 4" long, steel					28.50			28.50	31	
800	0010	**TIMBER CONNECTORS** Add up cost of each part for total										800
	0020	cost of connection										
	0100	Connector plates, steel, with bolts, straight	2 Carp	75	.213	Ea.	16.35	4.20		20.55	25	
	0110	Tee	"	50	.320		24	6.30		30.30	37.50	
	0200	Bolts, machine, sq. hd. with nut & washer, 1/2" diameter, 4" long	1 Carp	140	.057		.74	1.13		1.87	2.74	
	0300	7-1/2" long		130	.062		.94	1.21		2.15	3.11	
	0500	3/4" diameter, 7-1/2" long		130	.062		1.69	1.21		2.90	3.94	
	0600	15" long		95	.084		2.16	1.66		3.82	5.20	
	0800	Drilling bolt holes in timber, 1/2" diameter		450	.018	Inch		.35		.35	.60	
	0900	1" diameter		350	.023	"		.45		.45	.77	
	1100	Framing anchors, 2 or 3 dimensional, 10 gauge, no nails incl.		175	.046	Ea.	.39	.90		1.29	1.97	
	1250	Holdowns, 3 gauge base, 10 gauge body		8	1		13.60	19.70		33.30	49	
	1300	Joist and beam hangers, 18 ga. galv., for 2" x 4" joist		175	.046		.49	.90		1.39	2.08	
	1400	2" x 6" to 2" x 10" joist		165	.048		.42	.96		1.38	2.10	
	1600	16 ga. galv., 3" x 6" to 3" x 10" joist		160	.050		2.29	.99		3.28	4.21	
	1700	3" x 10" to 3" x 14" joist		160	.050		2.65	.99		3.64	4.61	
	1800	4" x 6" to 4" x 10" joist		155	.052		2	1.02		3.02	3.94	
	1900	4" x 10" to 4" x 14" joist		155	.052		2.72	1.02		3.74	4.73	
	2000	Two-2" x 6" to two-2" x 10" joists		150	.053		2.19	1.05		3.24	4.21	
	2100	Two-2" x 10" to two-2" x 14" joists		150	.053		2.19	1.05		3.24	4.21	
	2300	3/16" thick, 6" x 8" joist		145	.055		4.80	1.09		5.89	7.15	
	2400	6" x 10" joist		140	.057		5.65	1.13		6.78	8.20	
	2500	6" x 12" joist		135	.059		6.80	1.17		7.97	9.50	
	2700	1/4" thick, 6" x 14" joist		130	.062		8.45	1.21		9.66	11.40	
	2800	Joist anchors, 1/4" x 1-1/4" x 18"		140	.057		3.29	1.13		4.42	5.55	
	2900	Plywood clips, extruded aluminum H clip, for 3/4" panels					.12			.12	.13	
	3000	Galvanized 18 ga. back-up clip					.11			.11	.12	
	3200	Post framing, 16 ga. galv. for 4" x 4" base, 2 piece	1 Carp	130	.062		4.74	1.21		5.95	7.30	
	3300	Cap		130	.062		2.33	1.21		3.54	4.64	
	3500	Rafter anchors, 18 ga. galv., 1-1/2" wide, 5-1/4" long		145	.055		.38	1.09		1.47	2.28	
	3600	10-3/4" long		145	.055		.78	1.09		1.87	2.72	
	3800	Shear plates, 2-5/8" diameter		120	.067		1.41	1.31		2.72	3.80	
	3900	4" diameter		115	.070		3.26	1.37		4.63	5.95	
	4000	Sill anchors, embedded in concrete or block, 18-5/8" long		115	.070		.95	1.37		2.32	3.40	
	4100	Spike grids, 4" x 4", flat or curved		120	.067		.34	1.31		1.65	2.62	
	4400	Split rings, 2-1/2" diameter		120	.067		1.15	1.31		2.46	3.52	
	4500	4" diameter		110	.073		1.77	1.43		3.20	4.40	
	4700	Strap ties, 16 ga., 1-3/8" wide, 12" long		180	.044		.92	.88		1.80	2.51	
	4800	24" long		160	.050		1.42	.99		2.41	3.25	
	5000	Toothed rings, 2-5/8" or 4" diameter		90	.089		.97	1.75		2.72	4.07	

06050 | Basic Wood / Plastic Materials / Methods
06090 | Wood & Plastic Fastenings

		CREW	DAILY OUTPUT	LABOR-HOURS	UNIT	MAT.	LABOR	EQUIP.	TOTAL	TOTAL INCL O&P
5200	Truss plates, nailed, 20 gauge, up to 32' span	1 Carp	17	.471	Truss	7	9.25		16.25	23.50
5400	Washers, 2" x 2" x 1/8"				Ea.	.22			.22	.24
5500	3" x 3" x 3/16"				"	.57			.57	.63
6101	Beam hangers, polymer painted									
6102	Bolted, 3 ga., (W x H x L)									
6104	3-1/4" x 9" x 12" top flange	1 Carp	1	8	C	5,575	158		5,733	6,400
6106	5-1/4" x 9" x 12" top flange		1	8		5,800	158		5,958	6,650
6108	5-1/4" x 11" x 11-3/4" top flange		1	8		13,000	158		13,158	14,600
6110	6-7/8" x 9" x 12" top flange		1	8		5,975	158		6,133	6,850
6112	6-7/8" x 11" x 13-1/2" top flange		1	8		13,700	158		13,858	15,400
6114	8-7/8" x 11" x 15-1/2" top flange	▼	1	8	▼	14,600	158		14,758	16,400
6116	Nailed, 3 ga., (W x H x L)									
6118	3-1/4" x 10-1/2" x 10" top flange	1 Carp	1.80	4.444	C	3,850	87.50		3,937.50	4,375
6120	3-1/4" x 10-1/2" x 12" top flange		1.80	4.444		4,450	87.50		4,537.50	5,050
6122	5-1/4" x 9-1/2" x 10" top flange		1.80	4.444		4,150	87.50		4,237.50	4,700
6124	5-1/4" x 9-1/2" x 12" top flange		1.80	4.444		4,700	87.50		4,787.50	5,325
6126	5-1/2" x 9-1/2" x 12" top flange		1.80	4.444		4,150	87.50		4,237.50	4,700
6128	6-7/8" x 8-1/2" x 12" top flange		1.80	4.444		4,275	87.50		4,362.50	4,875
6130	7-1/2" x 8-1/2" x 12" top flange		1.80	4.444		4,350	87.50		4,437.50	4,925
6132	8-7/8" x 7-1/2" x 14" top flange	▼	1.80	4.444	▼	4,800	87.50		4,887.50	5,425
6201	Beam and purlin hangers, galvanized, 12 ga.									
6202	Purlin or joist size, 3" x 8"	1 Carp	1.70	4.706	C	705	92.50		797.50	935
6204	3" x 10"		1.70	4.706		790	92.50		882.50	1,025
6206	3" x 12"		1.65	4.848		935	95.50		1,030.50	1,200
6208	3" x 14"		1.65	4.848		1,100	95.50		1,195.50	1,375
6210	3" x 16"		1.65	4.848		1,250	95.50		1,345.50	1,550
6212	4" x 8"		1.65	4.848		705	95.50		800.50	940
6214	4" x 10"		1.65	4.848		820	95.50		915.50	1,075
6216	4" x 12"		1.60	5		880	98.50		978.50	1,150
6218	4" x 14"		1.60	5		955	98.50		1,053.50	1,225
6220	4" x 16"		1.60	5		1,100	98.50		1,198.50	1,375
6224	6" x 10"		1.55	5.161		1,200	102		1,302	1,500
6226	6" x 12"		1.55	5.161		1,275	102		1,377	1,600
6228	6" x 14"		1.50	5.333		1,400	105		1,505	1,700
6230	6" x 16"	▼	1.50	5.333	▼	1,575	105		1,680	1,925
6300	Column bases									
6302	4 x 4, 16 ga.	1 Carp	1.80	4.444	C	980	87.50		1,067.50	1,225
6306	7 ga.		1.80	4.444		1,825	87.50		1,912.50	2,150
6314	6 x 6, 16 ga.		1.75	4.571		1,250	90		1,340	1,525
6318	7 ga.		1.75	4.571		2,550	90		2,640	2,950
6326	8 x 8, 7 ga.		1.65	4.848		4,250	95.50		4,345.50	4,850
6330	8 x 10, 7 ga.	▼	1.65	4.848	▼	4,550	95.50		4,645.50	5,175
6590	Joist hangers, heavy duty 12 ga., galvanized									
6592	2" x 4"	1 Carp	1.75	4.571	C	800	90		890	1,025
6594	2" x 6"		1.65	4.848		855	95.50		950.50	1,100
6596	2" x 8"		1.65	4.848		910	95.50		1,005.50	1,175
6598	2" x 10"		1.65	4.848		975	95.50		1,070.50	1,250
6600	2" x 12"		1.65	4.848		1,100	95.50		1,195.50	1,375
6622	(2) 2" x 6"		1.60	5		1,075	98.50		1,173.50	1,350
6624	(2) 2" x 8"		1.60	5		1,150	98.50		1,248.50	1,450
6626	(2) 2" x 10"		1.55	5.161		1,300	102		1,402	1,600
6628	(2) 2" x 12"	▼	1.55	5.161	▼	1,575	102		1,677	1,900
6890	Purlin hangers, painted									
6892	12 ga., 2" x 6"	1 Carp	1.80	4.444	C	975	87.50		1,062.50	1,225
6894	2" x 8"		1.80	4.444		1,025	87.50		1,112.50	1,275
6896	2" x 10"	▼	1.80	4.444	▼	1,050	87.50		1,137.50	1,300

06050 | Basic Wood / Plastic Materials / Methods

06090 | Wood & Plastic Fastenings

			CREW	DAILY OUTPUT	LABOR-HOURS	UNIT	2000 BARE COSTS				TOTAL INCL O&P	
							MAT.	LABOR	EQUIP.	TOTAL		
800	6898	2" x 12"	1 Carp	1.75	4.571	C	1,150	90		1,240	1,400	800
	6934	(2) 2" x 6"		1.70	4.706		985	92.50		1,077.50	1,225	
	6936	(2) 2" x 8"		1.70	4.706		1,075	92.50		1,167.50	1,350	
	6938	(2) 2" x 10"		1.70	4.706		1,200	92.50		1,292.50	1,475	
	6940	(2) 2" x 12"	↓	1.65	4.848	↓	1,275	95.50		1,370.50	1,575	
825	0010	**ROUGH HARDWARE** Average % of carpentry material, minimum					.50%					825
	0200	Maximum					1.50%					
850	0010	**BRACING**										850
	0301	Let-in, "T" shaped, 22 ga. galv. steel, studs at 16" O.C.	2 Carp	11.60	1.379	C.L.F.	37.50	27		64.50	88	
	0401	Studs at 24" O.C.		12	1.333		37.50	26.50		64	86.50	
	0501	16 ga. galv. steel straps, studs at 16" O.C.		12	1.333		49	26.50		75.50	99	
	0601	Studs at 24" O.C.	↓	12.40	1.290	↓	49	25.50		74.50	97.50	

06100 | Rough Carpentry

06110 | Wood Framing

			CREW	DAILY OUTPUT	LABOR-HOURS	UNIT	2000 BARE COSTS				TOTAL INCL O&P	
							MAT.	LABOR	EQUIP.	TOTAL		
100	0010	**BLOCKING**										100
	2600	Miscellaneous, to wood construction										
	2621	2" x 4"	2 Carp	.34	47.059	M.B.F.	590	925		1,515	2,250	
	2626	Pneumatic nailed		.42	38.095		590	750		1,340	1,925	
	2661	2" x 8"		.54	29.630		620	585		1,205	1,675	
	2666	Pneumatic nailed	↓	.66	24.242	↓	620	480		1,100	1,500	
	2720	To steel construction										
	2741	2" x 4"	2 Carp	.28	57.143	M.B.F.	590	1,125		1,715	2,575	
	2781	2" x 8"	"	.42	38.095	"	620	750		1,370	1,950	
150	0010	**BRACING** Let-in, with 1" x 6" boards, studs @ 16" O.C.	1 Carp	1.50	5.333	C.L.F.	49	105		154	234	150
	0200	Studs @ 24" O.C.	"	2.30	3.478	"	49	68.50		117.50	171	
200	0010	**BRIDGING** Wood, for joists 16" O.C., 1" x 3"	1 Carp	1.30	6.154	C.Pr.	26.50	121		147.50	238	200
	0015	Pneumatic nailed		1.70	4.706		26.50	92.50		119	189	
	0100	2" x 3" bridging		1.30	6.154		46	121		167	259	
	0105	Pneumatic nailed		1.70	4.706		46	92.50		138.50	210	
	0300	Steel, galvanized, 18 ga., for 2" x 10" joists at 12" O.C.		1.30	6.154		58.50	121		179.50	273	
	0400	24" O.C.		1.40	5.714		220	113		333	435	
	0600	For 2" x 14" joists at 16" O.C.		1.30	6.154		54.50	121		175.50	268	
	0700	24" O.C.		1.40	5.714		54.50	113		167.50	253	
	0900	Compression type, 16" O.C., 2" x 8" joists		2	4		126	79		205	274	
	1000	2" x 12" joists	↓	2	4	↓	141	79		220	290	
505	0010	**FRAMING, BEAMS & GIRDERS**	R06100-010									505
	3500	Single, 2" x 6"	2 Carp	.70	22.857	M.B.F.	590	450		1,040	1,425	
	3505	Pneumatic nailed	R06110-030	.81	19.704		590	390		980	1,325	
	3520	2" x 8"		.86	18.605		620	365		985	1,325	
	3525	Pneumatic nailed		1	16.048		620	315		935	1,225	
	3540	2" x 10"		1	16		605	315		920	1,200	
	3545	Pneumatic nailed		1.16	13.793		605	272		877	1,125	
	3560	2" x 12"		1.10	14.545		740	287		1,027	1,300	
	3565	Pneumatic nailed		1.28	12.539		740	247		987	1,225	
	3580	2" x 14"	↓	1.17	13.675	↓	885	269		1,154	1,425	

06100 | Rough Carpentry

06110 | Wood Framing

			Crew	Daily Output	Labor-Hours	Unit	Mat.	Labor	Equip.	Total	Total Incl O&P	
505	3585	Pneumatic nailed	2 Carp	1.36	11.791	M.B.F.	885	232		1,117	1,375	505
	3600	3" x 8"		1.10	14.545		1,175	287		1,462	1,775	
	3620	3" x 10"		1.25	12.800		1,175	252		1,427	1,700	
	3640	3" x 12"		1.35	11.852		1,175	233		1,408	1,675	
	3660	3" x 14"		1.40	11.429		1,175	225		1,400	1,650	
	3680	4" x 8"	F-3	2.66	15.038		1,325	270	171	1,766	2,100	
	3700	4" x 10"		3.16	12.658		1,325	227	144	1,696	2,000	
	3720	4" x 12"		3.60	11.111		1,325	199	126	1,650	1,925	
	3740	4" x 14"		3.96	10.101		1,325	181	115	1,621	1,875	
	4000	Double, 2" x 6"	2 Carp	1.25	12.800		590	252		842	1,075	
	4005	Pneumatic nailed		1.45	11.034		590	217		807	1,025	
	4020	2" x 8"		1.60	10		620	197		817	1,025	
	4025	Pneumatic nailed		1.86	8.621		620	170		790	975	
	4040	2" x 10"		1.92	8.333		605	164		769	950	
	4045	Pneumatic nailed		2.23	7.185		605	142		747	910	
	4060	2" x 12"		2.20	7.273		740	143		883	1,050	
	4065	Pneumatic nailed		2.55	6.275		740	124		864	1,025	
	4080	2" x 14"		2.45	6.531		885	129		1,014	1,200	
	4085	Pneumatic nailed		2.84	5.634		885	111		996	1,150	
	5000	Triple, 2" x 6"		1.65	9.697		590	191		781	975	
	5005	Pneumatic nailed		1.91	8.377		590	165		755	935	
	5020	2" x 8"		2.10	7.619		620	150		770	940	
	5025	Pneumatic nailed		2.44	6.568		620	129		749	905	
	5040	2" x 10"		2.50	6.400		605	126		731	885	
	5045	Pneumatic nailed		2.90	5.517		605	109		714	855	
	5060	2" x 12"		2.85	5.614		740	111		851	1,000	
	5065	Pneumatic nailed		3.31	4.840		740	95.50		835.50	975	
	5080	2" x 14"		3.15	5.079		885	100		985	1,150	
	5085	Pneumatic nailed		3.35	4.770		885	94		979	1,125	
510	0010	**FRAMING, CEILINGS**										510
	6400	Suspended, 2" x 3"	2 Carp	.50	32	M.B.F.	610	630		1,240	1,750	
	6450	2" x 4"		.59	27.119		590	535		1,125	1,575	
	6500	2" x 6"		.80	20		590	395		985	1,325	
	6550	2" x 8"		.86	18.605		620	365		985	1,325	
515	0010	**FRAMING, COLUMNS**										515
	0400	4" x 4"	2 Carp	.52	30.769	M.B.F.	830	605		1,435	1,975	
	0420	4" x 6"		.55	29.091		1,325	575		1,900	2,425	
	0440	4" x 8"		.59	27.119		1,325	535		1,860	2,375	
	0460	6" x 6"		.65	24.615		1,825	485		2,310	2,825	
	0480	6" x 8"		.70	22.857		1,425	450		1,875	2,350	
	0500	6" x 10"		.75	21.333		1,425	420		1,845	2,300	
520	0010	**FRAMING, HEAVY** Mill timber, beams, single 6" x 10"	2 Carp	1.10	14.545	M.B.F.	1,425	287		1,712	2,075	520
	0100	Single 8" x 16"		1.20	13.333		1,625	263		1,888	2,250	
	0200	Built from 2" lumber, multiple 2" x 14"		.90	17.778		885	350		1,235	1,575	
	0210	Built from 3" lumber, multiple 3" x 6"		.70	22.857		1,175	450		1,625	2,050	
	0220	Multiple 3" x 8"		.80	20		1,175	395		1,570	1,950	
	0230	Multiple 3" x 10"		.90	17.778		1,175	350		1,525	1,875	
	0240	Multiple 3" x 12"		1	16		1,175	315		1,490	1,825	
	0250	Built from 4" lumber, multiple 4" x 6"		.80	20		1,325	395		1,720	2,125	
	0260	Multiple 4" x 8"		.90	17.778		1,325	350		1,675	2,050	
	0270	Multiple 4" x 10"		1	16		1,325	315		1,640	2,000	
	0280	Multiple 4" x 12"		1.10	14.545		1,325	287		1,612	1,950	
	0281											

WOOD & PLASTICS 6

06100 | Rough Carpentry

06110 | Wood Framing

			CREW	DAILY OUTPUT	LABOR-HOURS	UNIT	MAT.	LABOR	EQUIP.	TOTAL	TOTAL INCL O&P	
520	0290	Columns, structural grade, 1500f, 4" x 4"	2 Carp	.60	26.667	M.B.F.	1,300	525		1,825	2,325	520
	0300	6" x 6"		.65	24.615		1,800	485		2,285	2,800	
	0400	8" x 8"		.70	22.857		1,950	450		2,400	2,925	
	0500	10" x 10"		.75	21.333		2,025	420		2,445	2,950	
	0600	12" x 12"		.80	20		2,025	395		2,420	2,900	
	0800	Floor planks, 2" thick, T & G, 2" x 6"		1.05	15.238		1,600	300		1,900	2,275	
	0900	2" x 10"		1.10	14.545		760	287		1,047	1,325	
	1100	3" thick, 3" x 6"		1.05	15.238		1,150	300		1,450	1,800	
	1200	3" x 10"		1.10	14.545		1,150	287		1,437	1,775	
	1400	Girders, structural grade, 12" x 12"		.80	20		1,650	395		2,045	2,500	
	1500	10" x 16"		1	16		1,600	315		1,915	2,325	
	2050	Roof planks, see division 06150-600										
	2300	Roof purlins, 4" thick, structural grade	2 Carp	1.05	15.238	M.B.F.	1,125	300		1,425	1,775	
	2500	Roof trusses, add timber connectors, division 06090-800	"	.45	35.556	"	1,100	700		1,800	2,400	
530	0010	**FRAMING, JOISTS**										530
	2650	Joists, 2" x 4" R06100-010	2 Carp	.83	19.277	M.B.F.	590	380		970	1,300	
	2655	Pneumatic nailed		.96	16.667		590	330		920	1,225	
	2680	2" x 6" R06110-030		1.25	12.800		590	252		842	1,075	
	2685	Pneumatic nailed		1.44	11.111		590	219		809	1,025	
	2700	2" x 8"		1.46	10.959		620	216		836	1,050	
	2705	Pneumatic nailed		1.68	9.524		620	188		808	1,000	
	2720	2" x 10"		1.49	10.738		605	212		817	1,025	
	2725	Pneumatic nailed		1.71	9.357		605	184		789	985	
	2740	2" x 12"		1.75	9.143		740	180		920	1,125	
	2745	Pneumatic nailed		2.01	7.960		740	157		897	1,075	
	2760	2" x 14"		1.79	8.939		885	176		1,061	1,275	
	2765	Pneumatic nailed		2.06	7.767		885	153		1,038	1,225	
	2780	3" x 6"		1.39	11.511		1,175	227		1,402	1,675	
	2790	3" x 8"		1.90	8.421		1,175	166		1,341	1,550	
	2800	3" x 10"		1.95	8.205		1,175	162		1,337	1,550	
	2820	3" x 12"		1.80	8.889		1,175	175		1,350	1,575	
	2840	4" x 6"		1.60	10		1,325	197		1,522	1,800	
	2851	4" x 8"	F-3	4.15	9.639		1,325	173	109	1,607	1,875	
	2861	4" x 10"		5	8		1,325	143	90.50	1,558.50	1,800	
	2881	4" x 12"		4.50	8.889		1,325	159	101	1,585	1,825	
	3000	Composite wood joist 9-1/2" deep	2 Carp	.90	17.778	M.L.F.	1,400	350		1,750	2,150	
	3010	11-1/2" deep		.88	18.182		1,525	360		1,885	2,300	
	3020	14" deep		.82	19.512		1,700	385		2,085	2,525	
	3030	16" deep		.78	20.513		2,225	405		2,630	3,150	
	4000	Open web joist 12" deep		.88	18.182		1,575	360		1,935	2,350	
	4010	14" deep		.82	19.512		1,825	385		2,210	2,650	
	4020	16" deep		.78	20.513		1,900	405		2,305	2,800	
	4030	18" deep		.74	21.622		1,975	425		2,400	2,900	
545	0010	**FRAMING, MISCELLANEOUS**										545
	8500	Firestops, 2" x 4"	2 Carp	.51	31.373	M.B.F.	590	620		1,210	1,700	
	8505	Pneumatic nailed		.62	25.806		590	510		1,100	1,525	
	8520	2" x 6"		.60	26.667		590	525		1,115	1,550	
	8525	Pneumatic nailed		.73	21.858		590	430		1,020	1,400	
	8540	2" x 8"		.60	26.667		620	525		1,145	1,575	
	8560	2" x 12"		.70	22.857		740	450		1,190	1,575	
	8600	Nailers, treated, wood construction, 2" x 4"		.53	30.189		985	595		1,580	2,100	
	8605	Pneumatic nailed		.64	25.157		985	495		1,480	1,925	
	8620	2" x 6"		.75	21.333		1,050	420		1,470	1,875	
	8625	Pneumatic nailed		.90	17.778		1,050	350		1,400	1,750	
	8640	2" x 8"		.93	17.204		1,100	340		1,440	1,800	

06100 | Rough Carpentry

06110 | Wood Framing

			CREW	DAILY OUTPUT	LABOR-HOURS	UNIT	2000 BARE COSTS MAT.	LABOR	EQUIP.	TOTAL	TOTAL INCL O&P	
545	8645	Pneumatic nailed	2 Carp	1.12	14.337	M.B.F.	1,100	282		1,382	1,700	545
	8660	Steel construction, 2" x 4"		.50	32		985	630		1,615	2,150	
	8680	2" x 6"		.70	22.857		1,050	450		1,500	1,925	
	8700	2" x 8"		.87	18.391		1,100	360		1,460	1,850	
	8760	Rough bucks, treated, for doors or windows, 2" x 6"		.40	40		1,050	790		1,840	2,500	
	8765	Pneumatic nailed		.48	33.333		1,050	655		1,705	2,275	
	8780	2" x 8"		.51	31.373		1,100	620		1,720	2,275	
	8785	Pneumatic nailed		.61	26.144		1,100	515		1,615	2,100	
	8800	Stair stringers, 2" x 10"		.22	72.727		605	1,425		2,030	3,125	
	8820	2" x 12"		.26	61.538		740	1,200		1,940	2,875	
	8840	3" x 10"		.31	51.613		1,175	1,025		2,200	3,025	
	8860	3" x 12"		.38	42.105		1,175	830		2,005	2,700	
550	0010	**PARTITIONS** Wood stud with single bottom plate and										550
	0020	double top plate, no waste, std. & better lumber										
	0180	2" x 4" studs, 8' high, studs 12" O.C.	2 Carp	80	.200	L.F.	4.34	3.94		8.28	11.55	
	0185	12" O.C., pneumatic nailed		96	.167		4.34	3.28		7.62	10.45	
	0200	16" O.C.		100	.160		3.55	3.15		6.70	9.30	
	0205	16" O.C., pneumatic nailed		120	.133		3.55	2.63		6.18	8.40	
	0300	24" O.C.		125	.128		2.76	2.52		5.28	7.35	
	0305	24" O.C., pneumatic nailed		150	.107		2.76	2.10		4.86	6.65	
	0380	10' high, studs 12" O.C.		80	.200		5.15	3.94		9.09	12.40	
	0385	12" O.C., pneumatic nailed		96	.167		5.15	3.28		8.43	11.30	
	0400	16" O.C.		100	.160		4.14	3.15		7.29	9.95	
	0405	16" O.C., pneumatic nailed		120	.133		4.14	2.63		6.77	9.05	
	0500	24" O.C.		125	.128		3.16	2.52		5.68	7.80	
	0505	24" O.C., pneumatic nailed		150	.107		3.16	2.10		5.26	7.05	
	0580	12' high, studs 12" O.C.		65	.246		5.90	4.85		10.75	14.80	
	0585	12" O.C., pneumatic nailed		78	.205		5.90	4.04		9.94	13.40	
	0600	16" O.C.		80	.200		4.74	3.94		8.68	11.95	
	0605	16" O.C., pneumatic nailed		96	.167		4.74	3.28		8.02	10.85	
	0700	24" O.C.		100	.160		3.55	3.15		6.70	9.30	
	0705	24" O.C., pneumatic nailed		120	.133		3.55	2.63		6.18	8.40	
	0780	2" x 6" studs, 8' high, studs 12" O.C.		70	.229		6.50	4.50		11	14.85	
	0785	12" O.C., pneumatic nailed		84	.190		6.50	3.75		10.25	13.60	
	0800	16" O.C.		90	.178		5.30	3.50		8.80	11.85	
	0805	16" O.C., pneumatic nailed		108	.148		5.30	2.92		8.22	10.85	
	0900	24" O.C.		115	.139		4.13	2.74		6.87	9.25	
	0905	24" O.C., pneumatic nailed		138	.116		4.13	2.28		6.41	8.45	
	0980	10' high, studs 12" O.C.		70	.229		7.65	4.50		12.15	16.15	
	0985	12" O.C., pneumatic nailed		84	.190		7.65	3.75		11.40	14.90	
	1000	16" O.C.		90	.178		6.20	3.50		9.70	12.80	
	1005	16" O.C., pneumatic nailed		108	.148		6.20	2.92		9.12	11.80	
	1100	24" O.C.		115	.139		4.72	2.74		7.46	9.90	
	1105	24" O.C., pneumatic nailed		138	.116		4.72	2.28		7	9.10	
	1180	12' high, studs 12" O.C.		55	.291		8.85	5.75		14.60	19.55	
	1185	12" O.C., pneumatic nailed		66	.242		8.85	4.78		13.63	17.95	
	1200	16" O.C.		70	.229		7.10	4.50		11.60	15.50	
	1205	16" O.C., pneumatic nailed		84	.190		7.10	3.75		10.85	14.25	
	1300	24" O.C.		90	.178		5.30	3.50		8.80	11.85	
	1305	24" O.C., pneumatic nailed		108	.148		5.30	2.92		8.22	10.85	
	1400	For horizontal blocking, 2" x 4", add		600	.027		.39	.53		.92	1.33	
	1500	2" x 6", add		600	.027		.59	.53		1.12	1.55	
	1600	For openings, add		250	.064			1.26		1.26	2.16	
	1700	Headers for above openings, material only, add				M.B.F.	620			620	685	
555	0010	**FRAMING, ROOFS**										555
	6070	Fascia boards, 2" x 8"	2 Carp	.30	53.333	M.B.F.	620	1,050		1,670	2,475	

06100 | Rough Carpentry

06110 | Wood Framing

			CREW	DAILY OUTPUT	LABOR-HOURS	UNIT	2000 BARE COSTS MAT.	LABOR	EQUIP.	TOTAL	TOTAL INCL O&P	
555	6080	2" x 10"	2 Carp	.30	53.333	M.B.F.	605	1,050		1,655	2,475	555
	7000	Rafters, to 4 in 12 pitch, 2" x 6"		1	16		590	315		905	1,200	
	7060	2" x 8"		1.26	12.698		620	250		870	1,125	
	7300	Hip and valley rafters, 2" x 6"		.76	21.053		590	415		1,005	1,350	
	7360	2" x 8"		.96	16.667		620	330		950	1,250	
	7540	Hip and valley jacks, 2" x 6"		.60	26.667		590	525		1,115	1,550	
	7600	2" x 8"		.65	24.615		620	485		1,105	1,525	
	7780	For slopes steeper than 4 in 12, add						30%				
	7790	For dormers or complex roofs, add						50%				
	7800	Rafter tie, 1" x 4", #3	2 Carp	.27	59.259	M.B.F.	1,025	1,175		2,200	3,125	
	7820	Ridge board, #2 or better, 1" x 6"		.30	53.333		1,800	1,050		2,850	3,775	
	7840	1" x 8"		.37	43.243		1,800	850		2,650	3,425	
	7860	1" x 10"		.42	38.095		1,800	750		2,550	3,250	
	7880	2" x 6"		.50	32		590	630		1,220	1,725	
	7900	2" x 8"		.60	26.667		620	525		1,145	1,575	
	7920	2" x 10"		.66	24.242		605	480		1,085	1,500	
	7940	Roof cants, split, 4" x 4"		.86	18.605		830	365		1,195	1,550	
	7960	6" x 6"		1.80	8.889		1,825	175		2,000	2,300	
	7980	Roof curbs, untreated, 2" x 6"		.52	30.769		590	605		1,195	1,700	
	8000	2" x 12"		.80	20		740	395		1,135	1,475	
560	0010	**FRAMING, SILLS**										560
	4482	Ledgers, nailed, 2" x 4"	2 Carp	.50	32	M.B.F.	590	630		1,220	1,725	
	4484	2" x 6"		.60	26.667		590	525		1,115	1,550	
	4486	Bolted, not including bolts, 3" x 8"		.65	24.615		1,175	485		1,660	2,100	
	4488	3" x 12"		.70	22.857		1,175	450		1,625	2,050	
	4490	Mud sills, redwood, construction grade, 2" x 4"		.59	27.119		4,000	535		4,535	5,325	
	4492	2" x 6"		.78	20.513		4,000	405		4,405	5,100	
	4500	Sills, 2" x 4"		.40	40		590	790		1,380	2,000	
	4520	2" x 6"		.55	29.091		590	575		1,165	1,625	
	4540	2" x 8"		.67	23.881		620	470		1,090	1,500	
	4600	Treated, 2" x 4"		.36	44.444		985	875		1,860	2,575	
	4620	2" x 6"		.50	32		1,050	630		1,680	2,225	
	4640	2" x 8"		.60	26.667		1,100	525		1,625	2,125	
	4700	4" x 4"		.60	26.667		1,325	525		1,850	2,375	
	4720	4" x 6"		.70	22.857		1,275	450		1,725	2,175	
	4740	4" x 8"		.80	20		1,275	395		1,670	2,075	
	4760	4" x 10"		.87	18.391		1,375	360		1,735	2,150	
565	0010	**FRAMING, SLEEPERS**										565
	0300	On concrete, treated, 1" x 2"	2 Carp	.39	41.026	M.B.F.	1,200	810		2,010	2,700	
	0320	1" x 3"		.50	32		805	630		1,435	1,950	
	0340	2" x 4"		.99	16.162		985	320		1,305	1,625	
	0360	2" x 6"		1.30	12.308		1,050	242		1,292	1,575	
570	0010	**FRAMING, SOFFITS & CANOPIES**										570
	1300	Canopy or soffit framing, 1" x 4"	2 Carp	.30	53.333	M.B.F.	1,800	1,050		2,850	3,775	
	1340	1" x 8"		.50	32		1,800	630		2,430	3,050	
	1360	2" x 4"		.41	39.024		590	770		1,360	1,975	
	1400	2" x 8"		.67	23.881		620	470		1,090	1,500	
	1420	3" x 4"		.50	32		1,075	630		1,705	2,250	
	1460	3" x 8"		.60	26.667		1,175	525		1,700	2,175	
575	0010	**FRAMING, TREATED LUMBER**										575
	0020	Water-borne salt, C.C.A., A.C.A., wet, .40 P.C.F. retention										
	0100	2" x 4"				M.B.F.	985			985	1,075	
	0110	2" x 6"					1,050			1,050	1,150	

06100 | Rough Carpentry

06110 | Wood Framing

		Crew	Daily Output	Labor-Hours	Unit	Mat.	Labor	Equip.	Total	Total Incl O&P	
575											575
0120	2" x 8"				M.B.F.	1,100			1,100	1,225	
0130	2" x 10"					1,075			1,075	1,175	
0140	2" x 12"					1,175			1,175	1,275	
0200	4" x 4"					1,325			1,325	1,475	
0210	4" x 6"					1,275			1,275	1,400	
0220	4" x 8"					1,275			1,275	1,400	
0250	Add for .60 P.C.F. retention					40%					
0260	Add for 2.5 P.C.F. retention					200%					
0270	Add for K.D.A.T.					20%					
590											590
0010	**FRAMING, WALLS**										
5860	Headers over openings, 2" x 6"	2 Carp	.36	44.444	M.B.F.	590	875		1,465	2,150	
5865	2" x 6", pneumatic nailed		.43	37.209		590	735		1,325	1,900	
5880	2" x 8"		.45	35.556		620	700		1,320	1,875	
5885	2" x 8", pneumatic nailed		.54	29.630		620	585		1,205	1,675	
5900	2" x 10"		.53	30.189		605	595		1,200	1,700	
5905	2" x 10", pneumatic nailed		.67	23.881		605	470		1,075	1,475	
5920	2" x 12"		.60	26.667		740	525		1,265	1,700	
5925	2" x 12", pneumatic nailed		.72	22.222		740	440		1,180	1,550	
5940	4" x 12"		.76	21.053		1,325	415		1,740	2,150	
5945	4" x 12", pneumatic nailed		.92	17.391		1,325	345		1,670	2,025	
5960	6" x 12"		.84	19.048		1,875	375		2,250	2,700	
5965	6" x 12", pneumatic nailed		1.01	15.873		1,875	315		2,190	2,575	
6000	Plates, untreated, 2" x 3"		.43	37.209		610	735		1,345	1,925	
6005	2" x 3", pneumatic nailed		.52	30.769		610	605		1,215	1,725	
6020	2" x 4"		.53	30.189		590	595		1,185	1,675	
6025	2" x 4", pneumatic nailed		.67	23.881		590	470		1,060	1,450	
6040	2" x 6"		.75	21.333		590	420		1,010	1,375	
6045	2" x 6", pneumatic nailed		.90	17.778		590	350		940	1,250	
6120	Studs, 8' high wall, 2" x 3"		.60	26.667		610	525		1,135	1,575	
6125	2" x 3", pneumatic nailed		.72	22.222		610	440		1,050	1,425	
6140	2" x 4"		.92	17.391		590	345		935	1,225	
6145	2" x 4", pneumatic nailed		1.10	14.493		590	286		876	1,150	
6160	2" x 6"		1	16		590	315		905	1,200	
6165	2" x 6", pneumatic nailed		1.20	13.333		590	263		853	1,100	
6180	3" x 4"		.80	20		1,075	395		1,470	1,850	
6185	3" x 4", pneumatic nailed		.96	16.667		1,075	330		1,405	1,750	
8200	For 12' high walls, deduct						5%				
8220	For stub wall, 6' high, add						20%				
8240	3' high, add						40%				
8250	For second story & above, add						5%				
8300	For dormer & gable, add						15%				
600											600
0010	**FURRING** Wood strips, 1" x 2", on walls, on wood	1 Carp	550	.015	L.F.	.18	.29		.47	.69	
0015	On wood, pneumatic nailed		710	.011		.18	.22		.40	.58	
0300	On masonry		495	.016		.18	.32		.50	.75	
0400	On concrete		260	.031		.18	.61		.79	1.24	
0600	1" x 3", on walls, on wood		550	.015		.18	.29		.47	.69	
0605	On wood, pneumatic nailed		710	.011		.18	.22		.40	.58	
0700	On masonry		495	.016		.18	.32		.50	.75	
0800	On concrete		260	.031		.18	.61		.79	1.24	
0850	On ceilings, on wood		350	.023		.18	.45		.63	.97	
0855	On wood, pneumatic nailed		450	.018		.18	.35		.53	.80	
0900	On masonry		320	.025		.18	.49		.67	1.04	
0950	On concrete		210	.038		.18	.75		.93	1.49	
700											700
0010	**GROUNDS** For casework, 1" x 2" wood strips, on wood	1 Carp	330	.024	L.F.	.18	.48		.66	1.02	
0100	On masonry		285	.028		.18	.55		.73	1.15	

06100 | Rough Carpentry

06110 | Wood Framing

		CREW	DAILY OUTPUT	LABOR-HOURS	UNIT	2000 BARE COSTS MAT.	LABOR	EQUIP.	TOTAL	TOTAL INCL O&P	
700	0200	On concrete	1 Carp	250	.032	L.F.	.18	.63		.81	1.28	700
	0400	For plaster, 3/4" deep, on wood		450	.018		.18	.35		.53	.80	
	0500	On masonry		225	.036		.18	.70		.88	1.40	
	0600	On concrete		175	.046		.18	.90		1.08	1.74	
	0700	On metal lath		200	.040		.18	.79		.97	1.55	

06120 | Structural Panels

			CREW	DAILY OUTPUT	LABOR-HOURS	UNIT	MAT.	LABOR	EQUIP.	TOTAL	TOTAL INCL O&P	
200	0010	**MINERAL FIBER CEMENT PANELS** Including panels, fasteners,										200
	0100	accessories, trim & sealant										
	0130	Architectural, textured finish, 1/8" thick, minimum	G-3	500	.064	S.F.	5.75	1.15		6.90	8.30	
	0140	Maximum		500	.064		7.25	1.15		8.40	9.95	
	0150	1/4" thick, minimum		500	.064		6.75	1.15		7.90	9.40	
	0200	Maximum		500	.064		8.85	1.15		10	11.70	
	0250	3/8" thick, minimum		500	.064		8.90	1.15		10.05	11.75	
	0300	Maximum		500	.064		11.75	1.15		12.90	14.90	
	0350	5/8" thick, minimum		300	.107		11.75	1.92		13.67	16.20	
	0400	Maximum		300	.107		17.10	1.92		19.02	22	
	2000	Flat sheets, 1/8" thick		1,200	.027		4.05	.48		4.53	5.25	
	2100	1/4" thick		1,020	.031		5.80	.56		6.36	7.35	
	2200	3/8" thick		885	.036		7.30	.65		7.95	9.15	
	2300	5/8" thick		795	.040		9.95	.72		10.67	12.20	
	3000	Glasweld, mineral enamel coating, 1/8" thick		600	.053		3.50	.96		4.46	5.50	
	3100	1/4" thick		322	.099		4.45	1.79		6.24	7.95	
	4000	Sandwich panel, Glasweld face and back										
	4100	1" thick, perlite core	G-3	322	.099	S.F.	7.25	1.79		9.04	11.05	
	4200	Polyurethane core		322	.099		7.65	1.79		9.44	11.45	
	4500	2" thick, perlite core		322	.099		7.95	1.79		9.74	11.80	
	4600	Polyurethane core		322	.099		8.35	1.79		10.14	12.25	
800	0010	**STRESSED SKIN PLYWOOD ROOF PANELS** 3/8" group 1 top										800
	0020	skin, 3/8" exterior AD bottom skin										
	0030	1150f stringers, 4' x 8' panels										
	0100	4-1/4" deep	F-3	2,075	.019	SF Roof	3.04	.35	.22	3.61	4.17	
	0200	6-1/8" deep		1,725	.023		3.29	.42	.26	3.97	4.62	
	0300	8-1/8" deep		1,475	.027		3.81	.49	.31	4.61	5.35	
	0500	3/8" top skin, no bottom skin, 5-3/4" deep		1,725	.023		2.88	.42	.26	3.56	4.17	
	0600	7-3/4" deep		1,475	.027		3.29	.49	.31	4.09	4.79	
	0800	For 3-1/2" factory fiberglass insulation, add					.35			.35	.39	
	1000	For 1/2" thick top skin, add					.35			.35	.39	
	1500	Floor panels, substitute 5/8" underlayment as										
	1510	top skin, add to roof panels above				SF Flr.	.41			.41	.45	
	2000	Curved roof panels, 3/8" structural 1 top skin,										
	2010	3/8" exterior AC bottom skin, laminated ribs										
	2200	8' radius, 2-1/4" deep, tie rods not req'd.	F-3	1,150	.035	SF Flr.	5.70	.62	.39	6.71	7.80	
	2400	10' radius, 1-1/2" deep, tie rods are included		950	.042		4.48	.75	.48	5.71	6.75	
	2600	10' radius, 3-3/8" deep, tie rods not req'd.		1,150	.035		7.80	.62	.39	8.81	10.05	
	2800	12' radius, 2" deep, tie rods are included		950	.042		4.63	.75	.48	5.86	6.90	
	3000	12' radius, 4-1/2" deep, tie rods not req'd.		1,150	.035		8.10	.62	.39	9.11	10.40	
	4000	Folded plate roofs, structural 1 top skin with intermediate										
	4010	rafters and end chord. Cost of tie rods included										
	4200	Slope 7 in 12, 4' fold, 2" thick, 32' span	F-3	850	.047	SF Flr.	3.71	.84	.53	5.08	6.10	
	4400	Slope 8-1/2 in 12, 5' fold, 4" thick, 56' span		950	.042		7.40	.75	.48	8.63	9.95	
	4600	Slope 10 in 12, 8' fold, 4" thick, 52' span		950	.042		4.43	.75	.48	5.66	6.70	
	4800	Slope 10 in 12, 8' fold, 4" thick, 72' span		1,025	.039		7.80	.70	.44	8.94	10.25	
	6000	Box beams, structural 1 web										

Important: See the Reference Section for critical supporting data - Reference Nos., Crews, & City Cost Indexes

06100 | Rough Carpentry

06120 | Structural Panels

		CREW	DAILY OUTPUT	LABOR-HOURS	UNIT	MAT.	LABOR	EQUIP.	TOTAL	TOTAL INCL O&P	
800	6200	24" deep, 2-2" x 4" flanges, 2 webs @ 3/8"	F-3	295	.136	L.F.	8.90	2.43	1.54	12.87	15.60	800
	6400	24" deep, 3-2" x 4" flanges, 2 webs @ 1/2"		260	.154		11.05	2.76	1.74	15.55	18.80	
	6600	48" deep, 3-2" x 6" flanges, 2 webs @ 3/4"		140	.286		19.15	5.10	3.24	27.49	33.50	
	6800	48" deep, 6-2" x 6" flanges, 4 webs @ 3/8",										
	6810	including 2 interior webs	F-3	115	.348	L.F.	39	6.25	3.94	49.19	58	
	7000	For exterior AC outer webs, add					1.03			1.03	1.13	
	7200	For medium density overlaid outer webs, add					1.54			1.54	1.69	

06150 | Wood Decking

		CREW	DAILY OUTPUT	LABOR-HOURS	UNIT	MAT.	LABOR	EQUIP.	TOTAL	TOTAL INCL O&P	
600	0010	**ROOF DECKS**										600
	0020	For laminated decks, see division 06170-550										
	0200	For cementitious decks, see division 03450-000										
	0400	Cedar planks, 3" thick	2 Carp	320	.050	S.F.	5.65	.99		6.64	7.95	
	0500	4" thick		250	.064		7.65	1.26		8.91	10.55	
	0700	Douglas fir, 3" thick		320	.050		2.12	.99		3.11	4.02	
	0800	4" thick		250	.064		2.84	1.26		4.10	5.30	
	1000	Hemlock, 3" thick		320	.050		2.12	.99		3.11	4.02	
	1100	4" thick		250	.064		2.83	1.26		4.09	5.25	
	1300	Western white spruce, 3" thick		320	.050		2.05	.99		3.04	3.95	
	1400	4" thick		250	.064		2.73	1.26		3.99	5.15	

06160 | Sheathing

		CREW	DAILY OUTPUT	LABOR-HOURS	UNIT	MAT.	LABOR	EQUIP.	TOTAL	TOTAL INCL O&P	
800	0010	**SHEATHING** Plywood on roof, CDX R06160-020										800
	0030	5/16" thick	2 Carp	1,600	.010	S.F.	.36	.20		.56	.74	
	0035	Pneumatic nailed R06110-030		1,952	.008		.36	.16		.52	.68	
	0050	3/8" thick		1,525	.010		.40	.21		.61	.80	
	0055	Pneumatic nailed		1,860	.009		.40	.17		.57	.74	
	0100	1/2" thick		1,400	.011		.53	.23		.76	.97	
	0105	Pneumatic nailed		1,708	.009		.53	.18		.71	.90	
	0200	5/8" thick		1,300	.012		.67	.24		.91	1.15	
	0205	Pneumatic nailed		1,586	.010		.67	.20		.87	1.07	
	0300	3/4" thick		1,200	.013		.81	.26		1.07	1.34	
	0305	Pneumatic nailed		1,464	.011		.81	.22		1.03	1.26	
	0500	Plywood on walls with exterior CDX, 3/8" thick		1,200	.013		.40	.26		.66	.90	
	0505	Pneumatic nailed		1,488	.011		.40	.21		.61	.81	
	0600	1/2" thick		1,125	.014		.53	.28		.81	1.06	
	0605	Pneumatic nailed		1,395	.011		.53	.23		.76	.97	
	0700	5/8" thick		1,050	.015		.67	.30		.97	1.24	
	0705	Pneumatic nailed		1,302	.012		.67	.24		.91	1.14	
	0800	3/4" thick		975	.016		.81	.32		1.13	1.44	
	0805	Pneumatic nailed		1,209	.013		.81	.26		1.07	1.34	
	1000	For shear wall construction, add						20%				
	1200	For structural 1 exterior plywood, add				S.F.	10%					
	1400	With boards, on roof 1" x 6" boards, laid horizontal	2 Carp	725	.022		1.30	.43		1.73	2.17	
	1500	Laid diagonal		650	.025		1.30	.48		1.78	2.26	
	1700	1" x 8" boards, laid horizontal		875	.018		1.30	.36		1.66	2.05	
	1800	Laid diagonal		725	.022		1.30	.43		1.73	2.17	
	2000	For steep roofs, add						40%				
	2200	For dormers, hips and valleys, add					5%	50%				
	2400	Boards on walls, 1" x 6" boards, laid regular	2 Carp	650	.025		1.30	.48		1.78	2.26	
	2500	Laid diagonal		585	.027		1.30	.54		1.84	2.35	
	2700	1" x 8" boards, laid regular		765	.021		1.30	.41		1.71	2.14	
	2800	Laid diagonal		650	.025		1.30	.48		1.78	2.26	
	2850	Gypsum, weatherproof, 1/2" thick		1,125	.014		.28	.28		.56	.79	

06100 | Rough Carpentry

06160 | Sheathing

			CREW	DAILY OUTPUT	LABOR-HOURS	UNIT	2000 BARE COSTS				TOTAL INCL O&P	
							MAT.	LABOR	EQUIP.	TOTAL		
800	2900	Sealed, 4/10" thick	2 Carp	1,100	.015	S.F.	.44	.29		.73	.97	800
	3000	Wood fiber, regular, no vapor barrier, 1/2" thick		1,200	.013		.53	.26		.79	1.03	
	3100	5/8" thick		1,200	.013		.71	.26		.97	1.23	
	3300	No vapor barrier, in colors, 1/2" thick		1,200	.013		.77	.26		1.03	1.30	
	3400	5/8" thick		1,200	.013		.95	.26		1.21	1.50	
	3600	With vapor barrier one side, white, 1/2" thick		1,200	.013		.54	.26		.80	1.04	
	3700	Vapor barrier 2 sides, 1/2" thick		1,200	.013		.82	.26		1.08	1.35	
	3800	Asphalt impregnated, 25/32" thick		1,200	.013		.35	.26		.61	.84	
	3850	Intermediate, 1/2" thick		1,200	.013		.29	.26		.55	.77	
850	0010	**SUBFLOOR** Plywood, CDX, 1/2" thick	2 Carp	1,500	.011	SF Flr.	.53	.21		.74	.94	850
	0015	Pneumatic nailed		1,860	.009		.53	.17		.70	.87	
	0100	5/8" thick		1,350	.012		.67	.23		.90	1.13	
	0105	Pneumatic nailed		1,674	.010		.67	.19		.86	1.05	
	0200	3/4" thick		1,250	.013		.81	.25		1.06	1.32	
	0205	Pneumatic nailed		1,550	.010		.81	.20		1.01	1.24	
	0300	1-1/8" thick, 2-4-1 including underlayment		1,050	.015		2	.30		2.30	2.71	
	0500	With boards, 1" x 10" S4S, laid regular		1,100	.015		1.01	.29		1.30	1.60	
	0600	Laid diagonal		900	.018		1.02	.35		1.37	1.72	
	0800	1" x 8" S4S, laid regular		1,000	.016		.91	.32		1.23	1.54	
	0900	Laid diagonal		850	.019		.91	.37		1.28	1.64	
	1100	Wood fiber, T&G, 2' x 8' planks, 1" thick		1,000	.016		1.26	.32		1.58	1.93	
	1200	1-3/8" thick		900	.018		1.55	.35		1.90	2.30	
900	0010	**UNDERLAYMENT** Plywood, underlayment grade, 3/8" thick	2 Carp	1,500	.011	SF Flr.	.62	.21		.83	1.04	900
	0015	Pneumatic nailed		1,860	.009		.62	.17		.79	.97	
	0100	1/2" thick		1,450	.011		.80	.22		1.02	1.25	
	0105	Pneumatic nailed		1,798	.009		.80	.18		.98	1.18	
	0200	5/8" thick		1,400	.011		.76	.23		.99	1.23	
	0205	Pneumatic nailed		1,736	.009		.76	.18		.94	1.15	
	0300	3/4" thick		1,300	.012		.97	.24		1.21	1.49	
	0305	Pneumatic nailed		1,612	.010		.97	.20		1.17	1.40	
	0500	Particle board, 3/8" thick		1,500	.011		.38	.21		.59	.78	
	0505	Pneumatic nailed		1,860	.009		.38	.17		.55	.71	
	0600	1/2" thick		1,450	.011		.40	.22		.62	.81	
	0605	Pneumatic nailed		1,798	.009		.40	.18		.58	.74	
	0800	5/8" thick		1,400	.011		.46	.23		.69	.90	
	0805	Pneumatic nailed		1,736	.009		.46	.18		.64	.82	
	0900	3/4" thick		1,300	.012		.57	.24		.81	1.05	
	0905	Pneumatic nailed		1,612	.010		.57	.20		.77	.96	
	1100	Hardboard, underlayment grade, 4' x 4', .215" thick		1,500	.011		.41	.21		.62	.81	

06170 | Prefabricated Structural Wood

			CREW	DAILY OUTPUT	LABOR-HOURS	UNIT	MAT.	LABOR	EQUIP.	TOTAL	TOTAL INCL O&P	
550	0010	**LAMINATED ROOF DECK** Pine or hemlock, 3" thick	2 Carp	425	.038	S.F.	2.84	.74		3.58	4.39	550
	0100	4" thick		325	.049		3.78	.97		4.75	5.80	
	0300	Cedar, 3" thick		425	.038		3.47	.74		4.21	5.10	
	0400	4" thick		325	.049		4.42	.97		5.39	6.50	
	0600	Fir, 3" thick		425	.038		2.89	.74		3.63	4.45	
	0700	4" thick		325	.049		3.63	.97		4.60	5.65	
600	0010	**STRUCTURAL JOISTS** Fabricated "I" joists with wood flanges,										600
	0100	Plywood webs, incl. bridging & blocking, panels 24" O.C.										
	1200	15' to 24' span, 50 psf live load	F-5	2,400	.013	SF Flr.	1.44	.23		1.67	1.97	
	1300	55 psf live load		2,250	.014		1.55	.25		1.80	2.12	
	1400	24' to 30' span, 45 psf live load		2,600	.012		1.80	.21		2.01	2.34	
	1500	55 psf live load		2,400	.013		1.80	.23		2.03	2.37	

Important: See the Reference Section for critical supporting data - Reference Nos., Crews, & City Cost Indexes

06100 | Rough Carpentry

06170 | Prefabricated Structural Wood

			CREW	DAILY OUTPUT	LABOR-HOURS	UNIT	MAT.	LABOR	EQUIP.	TOTAL	TOTAL INCL O&P	
600	1600	Tubular steel open webs, 45 psf, 24" O.C., 40' span	F-3	6,250	.006	SF Flr.	1.80	.11	.07	1.98	2.26	600
	1700	55' span		7,750	.005		1.75	.09	.06	1.90	2.15	
	1800	70' span		9,250	.004		2.27	.08	.05	2.40	2.68	
	1900	85 psf live load, 26' span		2,300	.017		2.11	.31	.20	2.62	3.07	
980	0010	**ROOF TRUSSES**										980
	0020	For timber connectors, see div. 06090-800										
	0100	Fink (W) or King post type, 2'-0" O.C.										
	0200	Metal plate connected, 4 in 12 slope										
	0210	24' to 29' span	F-3	3,000	.013	SF Flr.	1.46	.24	.15	1.85	2.19	
	0300	30' to 43' span		3,000	.013		1.62	.24	.15	2.01	2.36	
	0400	44' to 60' span		3,000	.013		1.79	.24	.15	2.18	2.55	
	0600	For change in roof pitch, subtract					.06			.06	.07	
	0700	Glued and nailed, add					50%					

06180 | Glued-Laminated Construction

			CREW	DAILY OUTPUT	LABOR-HOURS	UNIT	MAT.	LABOR	EQUIP.	TOTAL	TOTAL INCL O&P	
400	0010	**LAMINATED FRAMING** Not including decking										400
	0020	30 lb., short term live load, 15 lb. dead load										
	0200	Straight roof beams, 20' clear span, beams 8' O.C.	F-3	2,560	.016	SF Flr.	1.45	.28	.18	1.91	2.27	
	0300	Beams 16' O.C.		3,200	.013		1.06	.22	.14	1.42	1.71	
	0500	40' clear span, beams 8' O.C.		3,200	.013		2.79	.22	.14	3.15	3.61	
	0600	Beams 16' O.C.		3,840	.010		2.28	.19	.12	2.59	2.96	
	0800	60' clear span, beams 8' O.C.	F-4	2,880	.014		4.79	.25	.28	5.32	6	
	0900	Beams 16' O.C.	"	3,840	.010		3.57	.19	.21	3.97	4.48	
	1100	Tudor arches, 30' to 40' clear span, frames 8' O.C.	F-3	1,680	.024		6.25	.43	.27	6.95	7.90	
	1200	Frames 16' O.C.	"	2,240	.018		4.89	.32	.20	5.41	6.15	
	1400	50' to 60' clear span, frames 8' O.C.	F-4	2,200	.018		6.75	.33	.37	7.45	8.35	
	1500	Frames 16' O.C.		2,640	.015		5.75	.27	.31	6.33	7.10	
	1700	Radial arches, 60' clear span, frames 8' O.C.		1,920	.021		6.30	.37	.42	7.09	8.05	
	1800	Frames 16' O.C.		2,880	.014		4.84	.25	.28	5.37	6.05	
	2000	100' clear span, frames 8' O.C.		1,600	.025		6.50	.45	.51	7.46	8.45	
	2100	Frames 16' O.C.		2,400	.017		5.75	.30	.34	6.39	7.20	
	2300	120' clear span, frames 8' O.C.		1,440	.028		8.70	.50	.56	9.76	11	
	2400	Frames 16' O.C.		1,920	.021		7.90	.37	.42	8.69	9.80	
	2600	Bowstring trusses, 20' O.C., 40' clear span	F-3	2,400	.017		3.91	.30	.19	4.40	5	
	2700	60' clear span	F-4	3,600	.011		3.51	.20	.22	3.93	4.45	
	2800	100' clear span		4,000	.010		4.97	.18	.20	5.35	5.95	
	2900	120' clear span		3,600	.011		5.35	.20	.22	5.77	6.45	
	3000	For less than 1000 B.F., add					20%					
	3050	For over 5000 B.F., deduct					10%					
	3100	For premium appearance, add to S.F. prices					5%					
	3300	For industrial type, deduct					15%					
	3500	For stain and varnish, add					5%					
	3900	For 3/4" laminations, add to straight					25%					
	4100	Add to curved					15%					
	4300	Alternate pricing method: (use nominal footage of										
	4310	components). Straight beams, camber less than 6"	F-3	3.50	11.429	M.B.F.	2,175	205	130	2,510	2,875	
	4400	Columns, including hardware		2	20		2,325	360	227	2,912	3,400	
	4600	Curved members, radius over 32'		2.50	16		2,375	287	181	2,843	3,325	
	4700	Radius 10' to 32'		3	13.333		2,350	239	151	2,740	3,175	
	4900	For complicated shapes, add maximum					100%					
	5100	For pressure treating, add to straight					35%					
	5200	Add to curved					45%					
	6000	Laminated veneer members, southern pine or western species										
	6050	1-3/4" wide x 5-1/2" deep	2 Carp	480	.033	L.F.	3.69	.66		4.35	5.20	
	6100	9-1/2" deep		480	.033		3.86	.66		4.52	5.40	

06100 | Rough Carpentry

06180 | Glued-Laminated Construction

		CREW	DAILY OUTPUT	LABOR-HOURS	UNIT	MAT.	LABOR	EQUIP.	TOTAL	TOTAL INCL O&P
400							2000 BARE COSTS				400
6150	14" deep	2 Carp	450	.036	L.F.	5.50	.70		6.20	7.25
6200	18" deep	↓	450	.036	↓	7.15	.70		7.85	9.10
6300	Parallel strand members, southern pine or western species									
6350	1-3/4" wide x 9-1/4" deep	2 Carp	480	.033	L.F.	4.42	.66		5.08	6
6400	11-1/4" deep		450	.036		5.45	.70		6.15	7.15
6450	14" deep		400	.040		6.45	.79		7.24	8.45
6500	3-1/2" wide x 9-1/4" deep		480	.033		8.60	.66		9.26	10.60
6550	11-1/4" deep		450	.036		10.60	.70		11.30	12.85
6600	14" deep		400	.040		12.65	.79		13.44	15.25
6650	7" wide x 9-1/4" deep		450	.036		16.95	.70		17.65	19.85
6700	11-1/4" deep		420	.038		21	.75		21.75	24.50
6750	14" deep	↓	400	.040	↓	25	.79		25.79	29

06200 | Finish Carpentry

06220 | Millwork

		CREW	DAILY OUTPUT	LABOR-HOURS	UNIT	MAT.	LABOR	EQUIP.	TOTAL	TOTAL INCL O&P
200	**0010 MOLDINGS, BASE**										**200**
0500	Base, stock pine, 9/16" x 3-1/2"	1 Carp	240	.033	L.F.	1	.66		1.66	2.23
0550	9/16" x 4-1/2"		200	.040		1.28	.79		2.07	2.76
0561	Base shoe, oak, 3/4" x 1"	↓	240	.033	↓	.76	.66		1.42	1.97
400	**0010 MOLDINGS, CASINGS**										**400**
0090	Apron, stock pine, 5/8" x 2"	1 Carp	250	.032	L.F.	.96	.63		1.59	2.14
0110	5/8" x 3-1/2"		220	.036		1.40	.72		2.12	2.77
0300	Band, stock pine, 11/16" x 1-1/8"		270	.030		.38	.58		.96	1.42
0350	11/16" x 1-3/4"		250	.032		.60	.63		1.23	1.74
0700	Casing, stock pine, 11/16" x 2-1/2"		240	.033		.77	.66		1.43	1.98
0750	11/16" x 3-1/2"	↓	215	.037	↓	1.48	.73		2.21	2.89
450	**0010 MOLDINGS, CEILINGS**										**450**
0600	Bed, stock pine, 9/16" x 1-3/4"	1 Carp	270	.030	L.F.	.60	.58		1.18	1.66
0650	9/16" x 2"		240	.033		.58	.66		1.24	1.77
1200	Cornice molding, stock pine, 9/16" x 1-3/4"		330	.024		.62	.48		1.10	1.50
1300	9/16" x 2-1/4"		300	.027		.85	.53		1.38	1.84
2400	Cove scotia, stock pine, 9/16" x 1-3/4"		270	.030		.54	.58		1.12	1.59
2500	11/16" x 2-3/4"		255	.031		1.15	.62		1.77	2.33
2600	Crown, stock pine, 9/16" x 3-5/8"		250	.032		1.51	.63		2.14	2.74
2700	11/16" x 4-5/8"	↓	220	.036	↓	1.91	.72		2.63	3.33
500	**0010 MOLDINGS, EXTERIOR**										**500**
1500	Cornice, boards, pine, 1" x 2"	1 Carp	330	.024	L.F.	.24	.48		.72	1.08
1700	1" x 6"		250	.032		.86	.63		1.49	2.03
2000	1" x 12"		180	.044		1.14	.88		2.02	2.75
2200	Three piece, built-up, pine, minimum		80	.100		1.47	1.97		3.44	5
2300	Maximum		65	.123		5.05	2.42		7.47	9.70
3000	Corner board, sterling pine, 1" x 4"		200	.040		.66	.79		1.45	2.08
3100	1" x 6"		200	.040		.96	.79		1.75	2.41
3350	Fascia, sterling pine, 1" x 6"		250	.032		.96	.63		1.59	2.14
3370	1" x 8"		225	.036		1.50	.70		2.20	2.85
3400	Trim, exterior, sterling pine, back band		250	.032		.57	.63		1.20	1.71
3500	Casing	↓	250	.032	↓	.60	.63		1.23	1.74

Important: See the Reference Section for critical supporting data - Reference Nos., Crews, & City Cost Indexes

06200 | Finish Carpentry

06220 | Millwork

			CREW	DAILY OUTPUT	LABOR-HOURS	UNIT	2000 BARE COSTS MAT.	LABOR	EQUIP.	TOTAL	TOTAL INCL O&P	
500	3600	Crown	1 Carp	250	.032	L.F.	1.20	.63		1.83	2.40	500
	3700	Porch rail with balusters		22	.364		9	7.15		16.15	22	
	3800	Screen		395	.020		.32	.40		.72	1.03	
	4100	Verge board, sterling pine, 1" x 4"		200	.040		.47	.79		1.26	1.87	
	4200	1" x 6"		200	.040		.72	.79		1.51	2.14	
	4300	2" x 6"		165	.048		1.40	.96		2.36	3.18	
	4400	2" x 8"	▼	165	.048	▼	1.86	.96		2.82	3.69	
	4700	For redwood trim, add					200%					
700	0010	**MOLDINGS, TRIM**										700
	0200	Astragal, stock pine, 11/16" x 1-3/4"	1 Carp	255	.031	L.F.	.78	.62		1.40	1.92	
	0250	1-5/16" x 2-3/16"		240	.033		2.73	.66		3.39	4.13	
	0800	Chair rail, stock pine, 5/8" x 2-1/2"		270	.030		.84	.58		1.42	1.92	
	0900	5/8" x 3-1/2"		240	.033		1.26	.66		1.92	2.52	
	1000	Closet pole, stock pine, 1-1/8" diameter		200	.040		.80	.79		1.59	2.23	
	1100	Fir, 1-5/8" diameter		200	.040		1.19	.79		1.98	2.66	
	3300	Half round, stock pine, 1/4" x 1/2"		270	.030		.19	.58		.77	1.21	
	3350	1/2" x 1"	▼	255	.031	▼	.34	.62		.96	1.43	
	3400	Handrail, fir, single piece, stock, hardware not included										
	3450	1-1/2" x 1-3/4"	1 Carp	80	.100	L.F.	1.21	1.97		3.18	4.71	
	3470	Pine, 1-1/2" x 1-3/4"		80	.100		1.06	1.97		3.03	4.55	
	3500	1-1/2" x 2-1/2"		76	.105		1.43	2.07		3.50	5.10	
	3600	Lattice, stock pine, 1/4" x 1-1/8"		270	.030		.32	.58		.90	1.35	
	3700	1/4" x 1-3/4"		250	.032		.35	.63		.98	1.47	
	3800	Miscellaneous, custom, pine, 1" x 1"		270	.030		.98	.58		1.56	2.08	
	3900	1" x 3"		240	.033		.78	.66		1.44	1.99	
	4100	Birch or oak, nominal 1" x 1"		240	.033		.46	.66		1.12	1.64	
	4200	Nominal 1" x 3"		215	.037		1.59	.73		2.32	3.01	
	4400	Walnut, nominal 1" x 1"		215	.037		.76	.73		1.49	2.10	
	4500	Nominal 1" x 3"		200	.040		2.29	.79		3.08	3.87	
	4700	Teak, nominal 1" x 1"		215	.037		1.05	.73		1.78	2.41	
	4800	Nominal 1" x 3"		200	.040		3	.79		3.79	4.65	
	4900	Quarter round, stock pine, 1/4" x 1/4"		275	.029		.20	.57		.77	1.20	
	4950	3/4" x 3/4"		255	.031	▼	.41	.62		1.03	1.51	
	5600	Wainscot moldings, 1-1/8" x 9/16", 2' high, minimum		76	.105	S.F.	5.90	2.07		7.97	10.05	
	5700	Maximum	▼	65	.123	"	13.45	2.42		15.87	18.95	
800	0010	**MOLDINGS, WINDOW AND DOOR**										800
	2800	Door moldings, stock, decorative, 1-1/8" wide, plain	1 Carp	17	.471	Set	28.50	9.25		37.75	47	
	2900	Detailed		17	.471	"	68	9.25		77.25	91	
	3150	Door trim set, 1 head and 2 sides, pine, 2-1/2 wide		5.90	1.356	Opng.	12.70	26.50		39.20	60	
	3170	4-1/2" wide		5.30	1.509	"	24	29.50		53.50	77	
	3200	Glass beads, stock pine, 1/4" x 11/16"		285	.028	L.F.	.30	.55		.85	1.28	
	3250	3/8" x 1/2"		275	.029		.36	.57		.93	1.38	
	3270	3/8" x 7/8"		270	.030		.40	.58		.98	1.44	
	4850	Parting bead, stock pine, 3/8" x 3/4"		275	.029		.29	.57		.86	1.30	
	4870	1/2" x 3/4"		255	.031		.36	.62		.98	1.46	
	5000	Stool caps, stock pine, 11/16" x 3-1/2"		200	.040		1.31	.79		2.10	2.79	
	5100	1-1/16" x 3-1/4"		150	.053	▼	1.97	1.05		3.02	3.97	
	5300	Threshold, oak, 3' long, inside, 5/8" x 3-5/8"		32	.250	Ea.	6.20	4.93		11.13	15.30	
	5400	Outside, 1-1/2" x 7-5/8"	▼	16	.500	"	25	9.85		34.85	44.50	
	5900	Window trim sets, including casings, header, stops,										
	5910	stool and apron, 2-1/2" wide, minimum	1 Carp	13	.615	Opng.	15.30	12.10		27.40	38	
	5950	Average		10	.800		26.50	15.75		42.25	56	
	6000	Maximum	▼	6	1.333	▼	38	26.50		64.50	86.50	
900	0010	**SOFFITS** Wood fiber, no vapor barrier, 15/32" thick	2 Carp	525	.030	S.F.	.75	.60		1.35	1.86	900
	0100	5/8" thick	▼	525	.030		.81	.60		1.41	1.92	

06200 | Finish Carpentry

06220 | Millwork

		CREW	DAILY OUTPUT	LABOR-HOURS	UNIT	2000 BARE COSTS				TOTAL INCL O&P	
						MAT.	LABOR	EQUIP.	TOTAL		
900	0300 As above, 5/8" thick, with factory finish	2 Carp	525	.030	S.F.	.83	.60		1.43	1.94	900
	0500 Hardboard, 3/8" thick, slotted		525	.030		1	.60		1.60	2.13	
	1000 Exterior AC plywood, 1/4" thick		420	.038		.70	.75		1.45	2.06	
	1100 1/2" thick		420	.038		.97	.75		1.72	2.36	
	1150 For aluminum soffit, see division 07460-750										

06250 | Prefinished Paneling

		CREW	DAILY OUTPUT	LABOR-HOURS	UNIT	MAT.	LABOR	EQUIP.	TOTAL	INCL O&P	
200	0010 **PANELING, HARDBOARD**										200
	0050 Not incl. furring or trim, hardboard, tempered, 1/8" thick	2 Carp	500	.032	S.F.	.30	.63		.93	1.41	
	0100 1/4" thick		500	.032		.38	.63		1.01	1.50	
	0300 Tempered pegboard, 1/8" thick		500	.032		.38	.63		1.01	1.50	
	0400 1/4" thick		500	.032		.41	.63		1.04	1.53	
	0600 Untempered hardboard, natural finish, 1/8" thick		500	.032		.32	.63		.95	1.43	
	0700 1/4" thick		500	.032		.31	.63		.94	1.42	
	0900 Untempered pegboard, 1/8" thick		500	.032		.32	.63		.95	1.43	
	1000 1/4" thick		500	.032		.36	.63		.99	1.48	
	1200 Plastic faced hardboard, 1/8" thick		500	.032		.53	.63		1.16	1.66	
	1300 1/4" thick		500	.032		.71	.63		1.34	1.86	
	1500 Plastic faced pegboard, 1/8" thick		500	.032		.50	.63		1.13	1.63	
	1600 1/4" thick		500	.032		.62	.63		1.25	1.76	
	1800 Wood grained, plain or grooved, 1/4" thick, minimum		500	.032		.47	.63		1.10	1.60	
	1900 Maximum		425	.038		.89	.74		1.63	2.25	
	2100 Moldings for hardboard, wood or aluminum, minimum		500	.032	L.F.	.32	.63		.95	1.43	
	2200 Maximum		425	.038	"	.90	.74		1.64	2.26	
500	0010 **PANELING, PLYWOOD** R06160-020										500
	2400 Plywood, prefinished, 1/4" thick, 4' x 8' sheets										
	2410 with vertical grooves. Birch faced, minimum	2 Carp	500	.032	S.F.	.72	.63		1.35	1.87	
	2420 Average		420	.038		1.10	.75		1.85	2.50	
	2430 Maximum		350	.046		1.62	.90		2.52	3.32	
	2600 Mahogany, African		400	.040		2.08	.79		2.87	3.64	
	2700 Philippine (Lauan)		500	.032		.88	.63		1.51	2.05	
	2900 Oak or Cherry, minimum		500	.032		1.74	.63		2.37	2.99	
	3000 Maximum		400	.040		2.68	.79		3.47	4.30	
	3200 Rosewood		320	.050		3.81	.99		4.80	5.90	
	3400 Teak		400	.040		2.68	.79		3.47	4.30	
	3600 Chestnut		375	.043		3.97	.84		4.81	5.80	
	3800 Pecan		400	.040		1.70	.79		2.49	3.22	
	3900 Walnut, minimum		500	.032		2.27	.63		2.90	3.58	
	3950 Maximum		400	.040		4.33	.79		5.12	6.10	
	4000 Plywood, prefinished, 3/4" thick, stock grades, minimum		320	.050		1.03	.99		2.02	2.82	
	4100 Maximum		224	.071		4.48	1.41		5.89	7.35	
	4300 Architectural grade, minimum		224	.071		3.30	1.41		4.71	6.05	
	4400 Maximum		160	.100		5.05	1.97		7.02	8.95	
	4600 Plywood, "A" face, birch, V.C., 1/2" thick, natural		450	.036		1.55	.70		2.25	2.90	
	4700 Select		450	.036		1.70	.70		2.40	3.07	
	4900 Veneer core, 3/4" thick, natural		320	.050		1.65	.99		2.64	3.51	
	5000 Select		320	.050		1.85	.99		2.84	3.73	
	5200 Lumber core, 3/4" thick, natural		320	.050		2.47	.99		3.46	4.41	
	5500 Plywood, knotty pine, 1/4" thick, A2 grade		450	.036		1.34	.70		2.04	2.67	
	5600 A3 grade		450	.036		1.70	.70		2.40	3.07	
	5800 3/4" thick, veneer core, A2 grade		320	.050		1.75	.99		2.74	3.62	
	5900 A3 grade		320	.050		1.96	.99		2.95	3.85	
	6100 Aromatic cedar, 1/4" thick, plywood		400	.040		1.72	.79		2.51	3.24	
	6200 1/4" thick, particle board		400	.040		.83	.79		1.62	2.26	

06200 | Finish Carpentry

06260 | Board Paneling

		Crew	Daily Output	Labor-Hours	Unit	Mat.	Labor	Equip.	Total	Total Incl O&P
0010	**PANELING, BOARDS**									
6400	Wood board paneling, 3/4" thick, knotty pine	2 Carp	300	.053	S.F.	1.25	1.05		2.30	3.18
6500	Rough sawn cedar		300	.053		1.60	1.05		2.65	3.56
6700	Redwood, clear, 1" x 4" boards		300	.053		3.75	1.05		4.80	5.95
6900	Aromatic cedar, closet lining, boards		275	.058		2.89	1.15		4.04	5.15

06270 | Closet/Utility Wood Shelving

		Crew	Daily Output	Labor-Hours	Unit	Mat.	Labor	Equip.	Total	Total Incl O&P
0010	**SHELVING** Pine, clear grade, no edge band, 1" x 8"	1 Carp	115	.070	L.F.	1.71	1.37		3.08	4.23
0100	1" x 10"		110	.073		2.31	1.43		3.74	4.99
0200	1" x 12"		105	.076		4.18	1.50		5.68	7.15
0400	For lumber edge band, by hand, add					1.46			1.46	1.61
0420	By machine, add					.94			.94	1.03
0600	Plywood, 3/4" thick with lumber edge, 12" wide	1 Carp	75	.107		1.24	2.10		3.34	4.96
0700	24" wide		70	.114		2.35	2.25		4.60	6.45
0900	Bookcase, clear grade pine, shelves 12" O.C., 8" deep		70	.114	S.F.	3.35	2.25		5.60	7.55
1000	12" deep shelves		65	.123	"	4.19	2.42		6.61	8.75
1200	Adjustable closet rod and shelf, 12" wide, 3' long		20	.400	Ea.	40.50	7.90		48.40	58
1300	8' long		15	.533	"	57	10.50		67.50	80.50
1500	Prefinished shelves with supports, stock, 8" wide		75	.107	L.F.	3.65	2.10		5.75	7.60
1600	10" wide		70	.114	"	4.07	2.25		6.32	8.35
1800	Custom, high quality dadoed pine shelving units, minimum				S.F.					28
1900	Maximum				"					40

06400 | Architectural Woodwork

06410 | Custom Cabinets

		Crew	Daily Output	Labor-Hours	Unit	Mat.	Labor	Equip.	Total	Total Incl O&P
0010	**CABINETS** Corner china cabinets, stock pine,									
0020	80" high, unfinished, minimum	2 Carp	6.60	2.424	Ea.	430	48		478	550
0100	Maximum	"	4.40	3.636	"	950	71.50		1,021.50	1,175
0300	Built-in drawer units, pine, 18" deep, 32" high, unfinished									
0400	Minimum	2 Carp	53	.302	L.F.	138	5.95		143.95	162
0500	Maximum	"	40	.400	"	109	7.90		116.90	134
0700	Kitchen base cabinets, hardwood, not incl. counter tops,									
0710	24" deep, 35" high, prefinished									
0800	One top drawer, one door below, 12" wide	2 Carp	24.80	.645	Ea.	125	12.70		137.70	160
0840	18" wide		23.30	.687		188	13.55		201.55	230
0880	24" wide		22.30	.717		223	14.15		237.15	269
1000	Four drawers, 12" wide		24.80	.645		305	12.70		317.70	355
1040	18" wide		23.30	.687		330	13.55		343.55	385
1060	24" wide		22.30	.717		355	14.15		369.15	415
1200	Two top drawers, two doors below, 27" wide		22	.727		266	14.35		280.35	315
1260	36" wide		20.30	.788		305	15.55		320.55	360
1300	48" wide		18.90	.847		340	16.70		356.70	405
1500	Range or sink base, two doors below, 30" wide		21.40	.748		222	14.75		236.75	270
1540	36" wide		20.30	.788		250	15.55		265.55	300
1580	48" wide		18.90	.847		281	16.70		297.70	340
1800	For sink front units, deduct					53			53	58.50
2000	Corner base cabinets, 36" wide, standard	2 Carp	18	.889		195	17.50		212.50	245

06400 | Architectural Woodwork

06410 | Custom Cabinets

			CREW	DAILY OUTPUT	LABOR-HOURS	UNIT	MAT.	LABOR	EQUIP.	TOTAL	TOTAL INCL O&P
100	2100	Lazy Susan with revolving door	2 Carp	16.50	.970	Ea.	320	19.10		339.10	385
	4000	Kitchen wall cabinets, hardwood, 12" deep with two doors									
	4050	12" high, 30" wide	2 Carp	24.80	.645	Ea.	122	12.70		134.70	157
	4100	36" wide		24	.667		144	13.15		157.15	182
	4400	15" high, 30" wide		24	.667		144	13.15		157.15	181
	4440	36" wide		22.70	.705		157	13.90		170.90	197
	4700	24" high, 30" wide		23.30	.687		166	13.55		179.55	206
	4720	36" wide		22.70	.705		183	13.90		196.90	225
	5000	30" high, one door, 12" wide		22	.727		107	14.35		121.35	143
	5040	18" wide		20.90	.766		139	15.10		154.10	179
	5060	24" wide		20.30	.788		152	15.55		167.55	194
	5300	Two doors, 27" wide		19.80	.808		209	15.90		224.90	258
	5340	36" wide		18.80	.851		211	16.75		227.75	262
	5380	48" wide		18.40	.870		260	17.15		277.15	315
	6000	Corner wall, 30" high, 24" wide		18	.889		132	17.50		149.50	175
	6050	30" wide		17.20	.930		165	18.35		183.35	214
	6100	36" wide		16.50	.970		176	19.10		195.10	227
	6500	Revolving Lazy Susan		15.20	1.053		249	20.50		269.50	310
	7000	Broom cabinet, 84" high, 24" deep, 18" wide		10	1.600		350	31.50		381.50	440
	7500	Oven cabinets, 84" high, 24" deep, 27" wide		8	2		530	39.50		569.50	655
	7750	Valance board trim		396	.040	L.F.	7.60	.80		8.40	9.70
	9000	For deluxe models of all cabinets, add					40%				
	9500	For custom built in place, add					25%	10%			
	9550	Rule of thumb, kitchen cabinets not including									
	9560	appliances & counter top, minimum	2 Carp	30	.533	L.F.	80.50	10.50		91	107
	9600	Maximum	"	25	.640	"	225	12.60		237.60	270
	9610	For metal cabinets, see division 12310-750									
210	0010	**CASEWORK, FRAMES**									
	0050	Base cabinets, counter storage, 36" high, one bay									
	0100	18" wide	1 Carp	2.70	2.963	Ea.	91	58.50		149.50	200
	0400	Two bay, 36" wide		2.20	3.636		139	71.50		210.50	276
	1100	Three bay, 54" wide		1.50	5.333		165	105		270	360
	2800	Book cases, one bay, 7' high, 18" wide		2.40	3.333		107	65.50		172.50	231
	3500	Two bay, 36" wide		1.60	5		155	98.50		253.50	340
	4100	Three bay, 54" wide		1.20	6.667		257	131		388	510
	5100	Coat racks, one bay, 7' high, 24" wide		4.50	1.778		107	35		142	178
	5300	Two bay, 48" wide		2.75	2.909		149	57.50		206.50	262
	5800	Three bay, 72" wide		2.10	3.810		219	75		294	370
	6100	Wall mounted cabinet, one bay, 24" high, 18" wide		3.60	2.222		59	44		103	140
	6800	Two bay, 36" wide		2.20	3.636		86	71.50		157.50	218
	7400	Three bay, 54" wide		1.70	4.706		107	92.50		199.50	277
	8400	30" high, one bay, 18" wide		3.60	2.222		64	44		108	146
	9000	Two bay, 36" wide		2.15	3.721		85	73.50		158.50	220
	9400	Three bay, 54" wide		1.60	5		106	98.50		204.50	286
	9800	Wardrobe, 7' high, single, 24" wide		2.70	2.963		118	58.50		176.50	230
	9880	Partition & adjustable shelves, 48" wide		1.70	4.706		150	92.50		242.50	325
	9950	Partition, adjustable shelves & drawers, 48" wide		1.40	5.714		225	113		338	440
220	0010	**CABINET DOORS**									
	2000	Glass panel, hardwood frame									
	2200	12" wide, 18" high	1 Carp	34	.235	Ea.	14.85	4.64		19.49	24.50
	2600	30" high		32	.250		24.50	4.93		29.43	35.50
	4450	18" wide, 18" high		32	.250		15.30	4.93		20.23	25.50
	4550	30" high		29	.276		24.50	5.45		29.95	36.50
	5000	Hardwood, raised panel									
	5100	12" wide, 18" high	1 Carp	16	.500	Ea.	20.50	9.85		30.35	39.50

06400 | Architectural Woodwork

06410 | Custom Cabinets

			CREW	DAILY OUTPUT	LABOR-HOURS	UNIT	2000 BARE COSTS MAT.	LABOR	EQUIP.	TOTAL	TOTAL INCL O&P	
220	5200	30" high	1 Carp	15	.533	Ea.	32.50	10.50		43	54	220
	5500	18" wide, 18" high		15	.533		29.50	10.50		40	50.50	
	5600	30" high	▼	14	.571	▼	50	11.25		61.25	74.50	
	6000	Plastic laminate on particle board										
	6100	12" wide, 18" high	1 Carp	25	.320	Ea.	13	6.30		19.30	25	
	6140	30" high		23	.348		21	6.85		27.85	35	
	6500	18" wide, 18" high		24	.333		19	6.55		25.55	32.50	
	6600	30" high	▼	22	.364	▼	32	7.15		39.15	47.50	
	7000	Plywood, with edge band										
	7010	12" wide, 18" high	1 Carp	27	.296	Ea.	17.50	5.85		23.35	29.50	
	7120	30" high		25	.320		30	6.30		36.30	44	
	7650	18" wide, 18" high		26	.308		25.50	6.05		31.55	38.50	
	7750	30" high	▼	24	.333	▼	44	6.55		50.55	60	
230	0010	**CABINET HARDWARE**										230
	1000	Catches, minimum	1 Carp	235	.034	Ea.	.72	.67		1.39	1.94	
	1040	Maximum	"	80	.100	"	4.19	1.97		6.16	8	
	2000	Door/drawer pulls, handles										
	2200	Handles and pulls, projecting, metal, minimum	1 Carp	160	.050	Ea.	1.54	.99		2.53	3.38	
	2240	Maximum		68	.118		7.35	2.32		9.67	12.05	
	2300	Wood, minimum		160	.050		1.54	.99		2.53	3.38	
	2340	Maximum		68	.118		4.20	2.32		6.52	8.60	
	2600	Flush, metal, minimum		160	.050		1.47	.99		2.46	3.31	
	2640	Maximum		68	.118	▼	10.50	2.32		12.82	15.50	
	3000	Drawer tracks/glides, minimum		48	.167	Pr.	5.75	3.28		9.03	12	
	3040	Maximum		24	.333		16.80	6.55		23.35	30	
	4000	Cabinet hinges, minimum		160	.050		1.43	.99		2.42	3.26	
	4040	Maximum	▼	68	.118	▼	6.30	2.32		8.62	10.90	
240	0010	**DRAWERS**										240
	0100	Solid hardwood front										
	1000	4" high, 12" wide	1 Carp	17	.471	Ea.	19.10	9.25		28.35	37	
	1200	18" wide	"	16	.500	"	25	9.85		34.85	44.50	
	2800	Plastic laminate on particle board front										
	3000	4" high, 12" wide	1 Carp	17	.471	Ea.	19.95	9.25		29.20	38	
	3200	18" wide	"	16	.500	"	23	9.85		32.85	42.50	
	5400	Plywood, flush panel front										
	6000	4" high, 12" wide	1 Carp	17	.471	Ea.	20.50	9.25		29.75	38.50	
	6200	18" wide	"	16	.500	"	25	9.85		34.85	44.50	
400	0010	**VANITIES**										400
	8000	Vanity bases, 2 doors, 30" high, 21" deep, 24" wide	2 Carp	20	.800	Ea.	124	15.75		139.75	163	
	8050	30" wide		16	1		130	19.70		149.70	177	
	8100	36" wide		13.33	1.200		146	23.50		169.50	201	
	8150	48" wide	▼	11.43	1.400		184	27.50		211.50	249	
	9000	For deluxe models of all vanities, add to above					40%					
	9500	For custom built in place, add to above				▼	25%	10%				

06430 | Stairs & Railings

			CREW	DAILY OUTPUT	LABOR-HOURS	UNIT	MAT.	LABOR	EQUIP.	TOTAL	TOTAL INCL O&P	
500	0010	**RAILING** Custom design, architectural grade, hardwood, minimum	1 Carp	38	.211	L.F.	12	4.15		16.15	20.50	500
	0100	Maximum		30	.267		46	5.25		51.25	60	
	0300	Stock interior railing with spindles 6" O.C., 4' long		40	.200		29	3.94		32.94	38.50	
	0400	8' long	▼	48	.167	▼	27	3.28		30.28	35	
505	0010	**DECK, WOOD, PRESSURE TREATED LUMBER**										505
	0100	Railings and trim, 1" x 4"	1 Carp	300	.027	L.F.	.62	.53		1.15	1.58	
	0200	2" x 4"		300	.027		.64	.53		1.17	1.60	
	0300	2" x 6"	▼	300	.027	▼	.85	.53		1.38	1.84	

06400 | Architectural Woodwork

06430 | Stairs & Railings

			CREW	DAILY OUTPUT	LABOR-HOURS	UNIT	2000 BARE COSTS MAT.	LABOR	EQUIP.	TOTAL	TOTAL INCL O&P	
505	0400	Decking, 1" x 4"	1 Carp	275	.029	S.F.	1.56	.57		2.13	2.70	505
	0500	2" x 4"		300	.027		1.62	.53		2.15	2.68	
	0600	2" x 6"		320	.025		1.61	.49		2.10	2.61	
	0650	5/4" x 6"		320	.025		1.68	.49		2.17	2.69	
	0700	Redwood decking, 1" x 4"		275	.029		4	.57		4.57	5.40	
	0800	2" x 6"		340	.024		11.40	.46		11.86	13.35	
	0900	5/4" x 6"		320	.025		7.15	.49		7.64	8.70	
620	0011	**STAIRS, PREFABRICATED**										620
	0100	Box stairs, prefabricated, 3'-0" wide										
	0110	Oak treads, no handrails, 2' high	2 Carp	5	3.200	Flight	216	63		279	345	
	0200	4' high		4	4		430	79		509	610	
	0300	6' high		3.50	4.571		620	90		710	835	
	0400	8' high		3	5.333		775	105		880	1,025	
	0600	With pine treads for carpet, 2' high		5	3.200		88.50	63		151.50	205	
	0700	4' high		4	4		164	79		243	315	
	0800	6' high		3.50	4.571		240	90		330	420	
	0900	8' high		3	5.333		274	105		379	480	
	1100	For 4' wide stairs, add					25%					
	1500	Prefabricated stair rail with balusters, 5 risers	2 Carp	15	1.067	Ea.	229	21		250	288	
	1700	Basement stairs, prefabricated, soft wood,										
	1710	open risers, 3' wide, 8' high	2 Carp	4	4	Flight	575	79		654	765	
	1900	Open stairs, prefabricated prefinished poplar, metal stringers,										
	1910	treads 3'-6" wide, no railings										
	2000	3' high	2 Carp	5	3.200	Flight	229	63		292	360	
	2100	4' high		4	4		485	79		564	670	
	2200	6' high		3.50	4.571		555	90		645	770	
	2300	8' high		3	5.333		730	105		835	980	
	2500	For prefab. 3 piece wood railings & balusters, add for										
	2600	3' high stairs	2 Carp	15	1.067	Ea.	31.50	21		52.50	70.50	
	2700	4' high stairs		14	1.143		51	22.50		73.50	95	
	2800	6' high stairs		13	1.231		63	24.50		87.50	111	
	2900	8' high stairs		12	1.333		96.50	26.50		123	151	
	3100	For 3'-6" x 3'-6" platform, add		4	4		72.50	79		151.50	215	
	3300	Curved stairways, 3'-3" wide, prefabricated, oak, unfinished,										
	3310	incl. curved balustrade system, open one side										
	3400	9' high	2 Carp	.70	22.857	Flight	6,375	450		6,825	7,800	
	3500	10' high		.70	22.857		7,200	450		7,650	8,700	
	3700	Open two sides, 9' high		.50	32		10,000	630		10,630	12,100	
	3800	10' high		.50	32		10,800	630		11,430	13,000	
	4000	Residential, wood, oak treads, prefabricated		1.50	10.667		930	210		1,140	1,375	
	4200	Built in place		.44	36.364		1,325	715		2,040	2,675	
	4400	Spiral, oak, 4'-6" diameter, unfinished, prefabricated,										
	4500	incl. railing, 9' high	2 Carp	1.50	10.667	Flight	4,000	210		4,210	4,750	
630	0010	**STAIR PARTS** Balusters, turned, 30" high, pine, minimum	1 Carp	28	.286	Ea.	4.10	5.65		9.75	14.15	630
	0100	Maximum	R06430-100	26	.308		9	6.05		15.05	20.50	
	0300	30" high birch balusters, minimum		28	.286		6.30	5.65		11.95	16.60	
	0400	Maximum		26	.308		10.20	6.05		16.25	21.50	
	0600	42" high, pine balusters, minimum		27	.296		6	5.85		11.85	16.60	
	0700	Maximum		25	.320		13	6.30		19.30	25	
	0900	42" high birch balusters, minimum		27	.296		7.65	5.85		13.50	18.40	
	1000	Maximum		25	.320		27	6.30		33.30	40.50	
	1050	Baluster, stock pine, 1-1/16" x 1-1/16"		240	.033	L.F.	1.97	.66		2.63	3.30	
	1100	1-5/8" x 1-5/8"		220	.036	"	2.18	.72		2.90	3.63	
	1200	Newels, 3-1/4" wide, starting, minimum		7	1.143	Ea.	33.50	22.50		56	75.50	
	1300	Maximum		6	1.333		126	26.50		152.50	184	

06400 | Architectural Woodwork

06430 | Stairs & Railings

			CREW	DAILY OUTPUT	LABOR-HOURS	UNIT	2000 BARE COSTS MAT.	LABOR	EQUIP.	TOTAL	TOTAL INCL O&P	
630	1500	Landing, minimum	1 Carp	5	1.600	Ea.	73.50	31.50		105	135	630
	1600	Maximum		4	2	↓	179	39.50		218.50	264	
	1800	Railings, oak, built-up, minimum		60	.133	L.F.	5.55	2.63		8.18	10.60	
	1900	Maximum		55	.145		15.75	2.87		18.62	22.50	
	2100	Add for sub rail		110	.073		4.20	1.43		5.63	7.05	
	2300	Risers, beech, 3/4" x 7-1/2" high		64	.125		5.75	2.46		8.21	10.55	
	2400	Fir, 3/4" x 7-1/2" high		64	.125		1.58	2.46		4.04	5.95	
	2600	Oak, 3/4" x 7-1/2" high		64	.125		4.99	2.46		7.45	9.70	
	2800	Pine, 3/4" x 7-1/2" high		66	.121		1.57	2.39		3.96	5.80	
	2850	Skirt board, pine, 1" x 10"		55	.145		1.73	2.87		4.60	6.80	
	2900	1" x 12"		52	.154	↓	2.10	3.03		5.13	7.50	
	3000	Treads, 1-1/16" x 9-1/2" wide, 3' long, oak		18	.444	Ea.	23	8.75		31.75	40.50	
	3100	4' long, oak		17	.471		29	9.25		38.25	48	
	3300	1-1/16" x 11-1/2" wide, 3' long, oak		18	.444		23.50	8.75		32.25	40.50	
	3400	6' long, oak	↓	14	.571	↓	56	11.25		67.25	81	
	3600	Beech treads, add					40%					
	3800	For mitered return nosings, add	↓			L.F.	8.40			8.40	9.25	

06440 | Wood Ornaments

			CREW	DAILY OUTPUT	LABOR-HOURS	UNIT	MAT.	LABOR	EQUIP.	TOTAL	TOTAL INCL O&P	
150	0010	**BEAMS, DECORATIVE** Rough sawn cedar, non-load bearing, 4" x 4"	2 Carp	180	.089	L.F.	1.30	1.75		3.05	4.43	150
	0100	4" x 6"		170	.094		2.50	1.85		4.35	5.95	
	0200	4" x 8"		160	.100		3.21	1.97		5.18	6.90	
	0300	4" x 10"		150	.107		4.46	2.10		6.56	8.50	
	0400	4" x 12"		140	.114		5.40	2.25		7.65	9.80	
	0500	8" x 8"	↓	130	.123	↓	7.55	2.42		9.97	12.45	
	1100	Beam connector plates see div. 06090-800										
350	0010	**GRILLES** and panels, hardwood, sanded										350
	0020	2' x 4' to 4' x 8', custom designs, unfinished, minimum	1 Carp	38	.211	S.F.	12	4.15		16.15	20.50	
	0050	Average		30	.267		26	5.25		31.25	37.50	
	0100	Maximum		19	.421		40	8.30		48.30	58	
	0300	As above, but prefinished, minimum		38	.211		12	4.15		16.15	20.50	
	0400	Maximum	↓	19	.421	↓	45	8.30		53.30	63.50	
400	0010	**LOUVERS** Redwood, 2'-0" diameter, full circle	1 Carp	16	.500	Ea.	105	9.85		114.85	133	400
	0100	Half circle		16	.500		104	9.85		113.85	131	
	0200	Octagonal		16	.500		86	9.85		95.85	111	
	0300	Triangular, 5/12 pitch, 5'-0" at base	↓	16	.500	↓	180	9.85		189.85	215	
500	0010	**FIREPLACE MANTELS** 6" molding, 6' x 3'-6" opening, minimum	1 Carp	5	1.600	Opng.	130	31.50		161.50	196	500
	0100	Maximum		5	1.600		157	31.50		188.50	227	
	0300	Prefabricated pine, colonial type, stock, deluxe		2	4		760	79		839	970	
	0400	Economy	↓	3	2.667	↓	256	52.50		308.50	370	
550	0010	**FIREPLACE MANTEL BEAMS** Rough texture wood, 4" x 8"	1 Carp	36	.222	L.F.	4.24	4.38		8.62	12.15	550
	0100	4" x 10"		35	.229	"	5.30	4.50		9.80	13.55	
	0300	Laminated hardwood, 2-1/4" x 10-1/2" wide, 6' long		5	1.600	Ea.	95.50	31.50		127	159	
	0400	8' long		5	1.600	"	133	31.50		164.50	200	
	0600	Brackets for above, rough sawn		12	.667	Pr.	8.75	13.15		21.90	32	
	0700	Laminated	↓	12	.667	"	13.25	13.15		26.40	37	
700	0010	**COLUMNS** For base plates, see division 05560-200										700
	0050	Aluminum, round colonial, 6" diameter	2 Carp	80	.200	V.L.F.	16	3.94		19.94	24.50	
	0100	8" diameter		62.25	.257		19	5.05		24.05	29.50	
	0200	10" diameter		55	.291		23.50	5.75		29.25	36	
	0250	Fir, stock units, hollow round, 6" diameter		80	.200		13	3.94		16.94	21	
	0300	8" diameter		80	.200		15	3.94		18.94	23.50	
	0350	10" diameter		70	.229		19	4.50		23.50	28.50	
	0400	Solid turned, to 8' high, 3-1/2" diameter	↓	80	.200	↓	7	3.94		10.94	14.45	

06400 | Architectural Woodwork

06440 | Wood Ornaments

		CREW	DAILY OUTPUT	LABOR-HOURS	UNIT	2000 BARE COSTS MAT.	LABOR	EQUIP.	TOTAL	TOTAL INCL O&P	
700	0500	4-1/2" diameter	2 Carp	75	.213	V.L.F.	10	4.20		14.20	18.20	700
	0600	5-1/2" diameter		70	.229		14	4.50		18.50	23	
	0800	Square columns, built-up, 5" x 5"		65	.246		13	4.85		17.85	22.50	
	0900	Solid, 3-1/2" x 3-1/2"		130	.123		6	2.42		8.42	10.75	
	1600	Hemlock, tapered, T & G, 12" diam, 10' high		100	.160		30	3.15		33.15	38.50	
	1700	16' high		65	.246		53	4.85		57.85	67	
	1900	10' high, 14" diameter		100	.160		77	3.15		80.15	90	
	2000	18' high		65	.246		73	4.85		77.85	89	
	2200	18" diameter, 12' high		65	.246		103	4.85		107.85	121	
	2300	20' high		50	.320		99	6.30		105.30	120	
	2500	20" diameter, 14' high		40	.400		122	7.90		129.90	148	
	2600	20' high		35	.457		126	9		135	154	
	2800	For flat pilasters, deduct					33%					
	3000	For splitting into halves, add				Ea.	60			60	66	
	4000	Rough sawn cedar posts, 4" x 4"	2 Carp	250	.064	V.L.F.	2.43	1.26		3.69	4.83	
	4100	4" x 6"		235	.068		3.62	1.34		4.96	6.30	
	4200	6" x 6"		220	.073		5.45	1.43		6.88	8.45	
	4300	8" x 8"		200	.080		5.60	1.58		7.18	8.85	

06445 | Simulated Wood Ornaments

100	0010	MILLWORK, HIGH DENSITY POLYMER										100
	0100	Base, 9/16" x 3-3/16"	1 Carp	230	.035	L.F.	1.22	.69		1.91	2.51	
	0200	Casing, fluted, 5/8" x 3-1/4"		215	.037		1.22	.73		1.95	2.60	
	0300	Chair rail, 9/16" x 2-1/4"		260	.031		.62	.61		1.23	1.72	
	0400	5/8" x 3-1/8"		230	.035		1.17	.69		1.86	2.46	
	0500	Corner, inside, 1/2" x 1-1/8"		220	.036		.61	.72		1.33	1.90	
	0600	Cove, 13/16" x 3-3/4"		260	.031		1.22	.61		1.83	2.38	
	0700	Crown, 3/4" x 3-13/16"		260	.031		1.22	.61		1.83	2.38	
	0800	Half round, 15/16" x 2"		240	.033		.67	.66		1.33	1.87	

06470 | Screen, Blinds & Shutters

100	0010	SHUTTERS, EXTERIOR Aluminum, louvered, 1'-4" wide, 3'-0" long	1 Carp	10	.800	Pr.	29.50	15.75		45.25	59.50	100
	0400	6'-8" long		9	.889		48	17.50		65.50	82.50	
	1000	Pine, louvered, primed, each 1'-2" wide, 3'-3" long		10	.800		41	15.75		56.75	72	
	1001	Pine, louvered, primed, each 1'-2" wide, 3'-3" long		20	.400	Ea.	24	7.90		31.90	40	
	1100	4'-7" long		10	.800	Pr.	49	15.75		64.75	81	
	1101	4'-7" long		20	.400	Ea.	34.50	7.90		42.40	51.50	
	1500	Each 1'-6" wide, 3'-3" long		10	.800	Pr.	45	15.75		60.75	76.50	
	1620	Hemlock, louvered, 1'-2" wide, 5'-7" long		10	.800		60.50	15.75		76.25	93.50	
	1630	Each 1'-4" wide, 2'-2" long		10	.800		38	15.75		53.75	68.50	
	1670	4'-3" long		10	.800		50	15.75		65.75	81.50	
	1690	5'-11" long		10	.800		61.50	15.75		77.25	95	
	1700	Door blinds, 6'-9" long, each 1'-3" wide		9	.889		71.50	17.50		89	109	
	1710	1'-6" wide		9	.889		90	17.50		107.50	129	
	1720	Hemlock, solid raised panel, each 1'-4" wide, 3'-3" long		10	.800		61.50	15.75		77.25	95	
	1740	4'-3" long		10	.800		78	15.75		93.75	113	
	1770	5'-11" long		10	.800		111	15.75		126.75	149	
	1800	Door blinds, 6'-9" long, each 1'-3" wide		9	.889		119	17.50		136.50	161	
	1900	1'-6" wide		9	.889		124	17.50		141.50	167	
	2500	Polystyrene, solid raised panel, each 1'-4" wide, 3'-3" long		10	.800		52	15.75		67.75	84	
	2700	4'-7" long		10	.800		62.50	15.75		78.25	96	
	4500	Polystyrene, louvered, each 1'-2" wide, 3'-3" long		10	.800		40	15.75		55.75	71	
	4750	5'-3" long		10	.800		53	15.75		68.75	85.50	
	6000	Vinyl, louvered, each 1'-2" x 4'-7" long		10	.800		52	15.75		67.75	84	
	6200	Each 1'-4" x 6'-8" long		9	.889		79	17.50		96.50	117	

06500 | Structural Plastics

06510 | Struct Plastic Shapes & Plates

			CREW	DAILY OUTPUT	LABOR-HOURS	UNIT	2000 BARE COSTS MAT.	LABOR	EQUIP.	TOTAL	TOTAL INCL O&P
400	0010	**CASTINGS, FIBERGLASS**									
	0100	Angle, 1" x 1" x 1/8" thick	2 Sswk	240	.067	L.F.	1.02	1.42		2.44	3.91
	0120	3" x 3" x 1/4" thick		200	.080		4.26	1.70		5.96	8.05
	0140	4" x 4" x 1/4" thick		200	.080		5.45	1.70		7.15	9.35
	0160	4" x 4" x 3/8" thick		200	.080		8.50	1.70		10.20	12.70
	0180	6" x 6" x 1/2" thick		160	.100		16.65	2.13		18.78	22.50
	1000	Flat sheet, 1/8" thick		140	.114	S.F.	4.34	2.43		6.77	9.55
	1020	1/4" thick		120	.133		8.35	2.83		11.18	14.80
	1040	3/8" thick		100	.160		14.40	3.40		17.80	22.50
	1060	1/2" thick		80	.200		18.45	4.25		22.70	29
	2000	Handrail, 42" high, 2" diam. rails pickets 5' O.C.		32	.500	L.F.	42.50	10.65		53.15	67.50
	3000	Round bar, 1/4" diam.		240	.067		.47	1.42		1.89	3.31
	3020	1/2" diam.		200	.080		1.12	1.70		2.82	4.58
	3040	3/4" diam.		200	.080		1.62	1.70		3.32	5.15
	3060	1" diam.		160	.100		2.89	2.13		5.02	7.35
	3080	1-1/4" diam.		160	.100		3.41	2.13		5.54	7.95
	3100	1-1/2" diam.		140	.114		4.19	2.43		6.62	9.40
	3500	Round tube, 1" diam. x 1/8" thick		240	.067		1.86	1.42		3.28	4.84
	3520	2" diam. x 1/4" thick		200	.080		4.54	1.70		6.24	8.35
	3540	3" diam. x 1/4" thick		160	.100		8.80	2.13		10.93	13.85
	4000	Square bar, 1/2" square		240	.067		3.68	1.42		5.10	6.85
	4020	1" square		200	.080		3.79	1.70		5.49	7.50
	4040	1-1/2" square		160	.100		7	2.13		9.13	11.90
	4500	Square tube, 1" x 1" x 1/8" thick		240	.067		1.64	1.42		3.06	4.59
	4520	2" x 2" x 1/8" thick		200	.080		3.09	1.70		4.79	6.75
	4540	3" x 3" x 1/4" thick		160	.100		8.45	2.13		10.58	13.50
	5000	Threaded rod, 3/8" diam.		320	.050		2.86	1.06		3.92	5.25
	5020	1/2" diam.		320	.050		3.37	1.06		4.43	5.80
	5040	5/8" diam.		280	.057		3.73	1.21		4.94	6.50
	5060	3/4" diam.		280	.057		4.24	1.21		5.45	7.05
	6000	Wide flange beam, 4" x 4" x 1/4" thick		120	.133		7.65	2.83		10.48	14
	6020	6" x 6" x 1/4" thick		100	.160		13.85	3.40		17.25	22
	6040	8" x 8" x 3/8" thick		80	.200		26.50	4.25		30.75	37.50

06520 | Plastic Struct Assemblies

			CREW	DAILY OUTPUT	LABOR-HOURS	UNIT	MAT.	LABOR	EQUIP.	TOTAL	TOTAL INCL O&P
100	0010	**STAIR TREAD, FIBERGLASS** Isophthalic Resin, 10-1/2" deep									
	0100	24" wide	2 Sswk	52	.308	Ea.	38.50	6.55		45.05	55.50
	0140	30" wide		52	.308		48	6.55		54.55	66
	0180	36" wide		52	.308		58	6.55		64.55	76.50
	0220	42" wide		52	.308		67.50	6.55		74.05	87

06600 | Plastic Fabrications

06610 | Fiberglass

			CREW	DAILY OUTPUT	LABOR-HOURS	UNIT	2000 BARE COSTS MAT.	LABOR	EQUIP.	TOTAL	TOTAL INCL O&P
300	0010	**GRATING, FIBERGLASS**									
	0100	Molded, green (for mod. corrosive environment)									
	0140	1" x 4" mesh, 1" thick	2 Sswk	400	.040	S.F.	10.25	.85		11.10	13
	0180	1-1/2" square mesh, 1" thick		400	.040		15.30	.85		16.15	18.55

06600 | Plastic Fabrications

06610 | Fiberglass

		Crew	Daily Output	Labor-Hours	Unit	Mat.	Labor	Equip.	Total	Total Incl O&P
300	0220 1-1/4" thick	2 Sswk	400	.040	S.F.	12.10	.85		12.95	15
	0260 1-1/2" thick		400	.040		21.50	.85		22.35	25.50
	0300 2" square mesh, 2" thick	▼	320	.050	▼	20.50	1.06		21.56	24.50
	1000 Orange (for highly corrosive environment)									
	1040 1" x 4" mesh, 1" thick	2 Sswk	400	.040	S.F.	12.60	.85		13.45	15.60
	1080 1-1/2" square mesh, 1" thick		400	.040		17.15	.85		18	20.50
	1120 1-1/4" thick		400	.040		17.55	.85		18.40	21
	1160 1-1/2" thick		400	.040		19	.85		19.85	22.50
	1200 2" square mesh, 2" thick	▼	320	.050	▼	22.50	1.06		23.56	27
	3000 Pultruded, green (for mod. corrosive environment)									
	3040 1" O.C. bar spacing, 1" thick	2 Sswk	400	.040	S.F.	16	.85		16.85	19.30
	3080 1-1/2" thick		320	.050		14.95	1.06		16.01	18.50
	3120 1-1/2" O.C. bar spacing, 1" thick		400	.040		11.85	.85		12.70	14.75
	3160 1-1/2" thick	▼	400	.040	▼	12.85	.85		13.70	15.85
	4000 Grating support legs, fixed height, no base				Ea.	40			40	44
	4040 With base					35.50			35.50	39
	4080 Adjustable to 60"				▼	52.50			52.50	57.50

06620 | Non-Structural Plastics

		Crew	Daily Output	Labor-Hours	Unit	Mat.	Labor	Equip.	Total	Total Incl O&P
200	0010 **FLOOR GRATING, FIBERGLASS**									200
	0101 1" x 4" grid, 1" thick	2 Carp	255	.063	S.F.	13.75	1.24		14.99	17.25
	0201 1-1/2" x 6" mesh, 1-1/2" thick		250	.064		16.85	1.26		18.11	20.50
	0301 With grit surface, 1-1/2" x 6" grid, 1-1/2" thick	▼	250	.064		17.75	1.26		19.01	21.50
600	0010 **NETTING, FLEXIBLE PLASTIC** 1/8" square mesh	4 Clab	4,000	.008	S.F.	.61	.12		.73	.87	600
	0100 1/4" square mesh		4,000	.008		.10	.12		.22	.31
	0120 1/2" square mesh		4,000	.008		.07	.12		.19	.28
	0140 5/8" x 3/4" mesh		4,000	.008		.03	.12		.15	.23
	0160 1-1/4" x 1-1/2" mesh		4,000	.008		.26	.12		.38	.49
	0200 4" square mesh	▼	4,000	.008	▼	.04	.12		.16	.24
	1000 Poly clips				Ea.	.29			.29	.32
810	0010 **SOLID SURFACE COUNTERTOPS**, Acrylic polymer										810
	0020 Pricing for orders of 100 L.F. or greater									
	0100 25" wide, solid colors	2 Carp	28	.571	L.F.	42	11.25		53.25	65.50
	0200 Patterned colors		28	.571		53	11.25		64.25	78
	0300 Premium patterned colors		28	.571		66.50	11.25		77.75	92.50
	0400 With silicone attached 4" backsplash, solid colors		27	.593		46	11.65		57.65	70.50
	0500 Patterned colors		27	.593		58	11.65		69.65	84
	0600 Premium patterned colors		27	.593		72.50	11.65		84.15	99.50
	0700 With hard seam attached 4" backsplash, solid colors		23	.696		46	13.70		59.70	74
	0800 Patterned colors		23	.696		58	13.70		71.70	87.50
	0900 Premium patterned colors	▼	23	.696	▼	72.50	13.70		86.20	103
	1000 Pricing for order of 51 - 99 L.F.									
	1100 25" wide, solid colors	2 Carp	24	.667	L.F.	48	13.15		61.15	75.50
	1200 Patterned colors		24	.667		61	13.15		74.15	89.50
	1300 Premium patterned colors		24	.667		76.50	13.15		89.65	107
	1400 With silicone attached 4" backsplash, solid colors		23	.696		53	13.70		66.70	81.50
	1500 Patterned colors		23	.696		67	13.70		80.70	97
	1600 Premium patterned colors		23	.696		83.50	13.70		97.20	115
	1700 With hard seam attached 4" backsplash, solid colors		20	.800		53	15.75		68.75	85
	1800 Patterned colors		20	.800		67	15.75		82.75	101
	1900 Premium patterned colors	▼	20	.800	▼	83.50	15.75		99.25	119
	2000 Pricing for order of 1 - 50 L.F.									
	2100 25" wide, solid colors	2 Carp	20	.800	L.F.	56.50	15.75		72.25	89
	2200 Patterned colors	▼	20	.800	▼	71.50	15.75		87.25	106

06600 | Plastic Fabrications

06620 | Non-Structural Plastics

			CREW	DAILY OUTPUT	LABOR-HOURS	UNIT	2000 BARE COSTS MAT.	LABOR	EQUIP.	TOTAL	TOTAL INCL O&P	
810	2300	Premium patterned colors	2 Carp	20	.800	L.F.	89.50	15.75		105.25	126	810
	2400	With silicone attached 4" backsplash, solid colors		19	.842		62	16.60		78.60	96.50	
	2500	Patterned colors		19	.842		78.50	16.60		95.10	115	
	2600	Premium patterned colors		19	.842		98	16.60		114.60	137	
	2700	With hard seam attached 4" backsplash, solid colors		4	4		62	79		141	203	
	2800	Patterned colors		15	1.067		78.50	21		99.50	123	
	2900	Premium patterned colors	▼	15	1.067	▼	98	21		119	144	
	3000	Sinks, pricing for order of 100 or greater units										
	3100	Single bowl, hard seamed, solid colors, 13" x 17"	1 Carp	3	2.667	Ea.	283	52.50		335.50	400	
	3200	10" x 15"		7	1.143		131	22.50		153.50	183	
	3300	Cutouts for sinks	▼	8	1	▼		19.70		19.70	34	
	3400	Sinks, pricing for order of 51 - 99 units										
	3500	Single bowl, hard seamed, solid colors, 13" x 17"	1 Carp	2.55	3.137	Ea.	325	62		387	460	
	3600	10" x 15"		6	1.333		150	26.50		176.50	210	
	3700	Cutouts for sinks	▼	7	1.143	▼		22.50		22.50	38.50	
	3800	Sinks, pricing for order of 1 - 50 units										
	3900	Single bowl, hard seamed, solid colors, 13" x 17"	1 Carp	2	4	Ea.	380	79		459	555	
	4000	10" x 15"		4.55	1.758		176	34.50		210.50	254	
	4100	Cutouts for sinks		5.25	1.524			30		30	51.50	
	4200	Cooktop cutouts, pricing for 100 or greater units		4	2		20.50	39.50		60	90	
	4300	51 - 99 units		3.40	2.353		23.50	46.50		70	106	
	4400	1 - 50 units	▼	3	2.667	▼	27.50	52.50		80	121	
850	0010	**VANITY TOPS**										850
	0015	Solid surface, center bowl, 17" x 19"	1 Carp	12	.667	Ea.	168	13.15		181.15	208	
	0020	19" x 25"		12	.667		203	13.15		216.15	246	
	0030	19" x 31"		12	.667		247	13.15		260.15	295	
	0040	19" x 37"		12	.667		287	13.15		300.15	340	
	0050	22" x 25"		10	.800		229	15.75		244.75	278	
	0060	22" x 31"		10	.800		267	15.75		282.75	320	
	0070	22" x 37"		10	.800		310	15.75		325.75	365	
	0080	22" x 43"		10	.800		355	15.75		370.75	415	
	0090	22" x 49"		10	.800		390	15.75		405.75	455	
	0110	22" x 55"		8	1		445	19.70		464.70	525	
	0120	22" x 61"		8	1		510	19.70		529.70	595	
	0130	22" x 68"		8	1		650	19.70		669.70	750	
	0140	22" x 73"		8	1		735	19.70		754.70	845	
	0150	22" x 85"		8	1		850	19.70		869.70	970	
	0160	Offset bowl, left or right, 22" x 49"		10	.800		470	15.75		485.75	540	
	0170	22" x 55"		8	1		535	19.70		554.70	620	
	0180	22" x 61"		8	1		605	19.70		624.70	700	
	0190	22" x 68"		8	1		710	19.70		729.70	815	
	0200	22" x 73"		8	1		875	19.70		894.70	995	
	0210	22" x 85"		8	1		1,025	19.70		1,044.70	1,150	
	0220	Double bowl, 22" x 61"		8	1		555	19.70		574.70	650	
	0230	Corner top/bowl, 22" x 49"	▼	8	1	▼	435	19.70		454.70	510	
	0240	For aggregate colors, add					35%					
	0250	For faucets and fittings see 15410-300										

For information about Means Estimating Seminars, see yellow pages 11 and 12 in back of book

WOOD & PLASTICS 6

Division Notes

		CREW	DAILY OUTPUT	LABOR-HOURS	UNIT	2000 BARE COSTS				TOTAL INCL O&P
						MAT.	LABOR	EQUIP.	TOTAL	

Division 7
Thermal & Moisture Protection

Estimating Tips

07100 Dampproofing & Waterproofing
- Be sure of the job specifications before pricing this subdivision. The difference in cost between waterproofing and dampproofing can be great. Waterproofing will hold back standing water. Dampproofing prevents the transmission of water vapor. Also included in this section are vapor retarding membranes.

07200 Thermal Protection
- Insulation and fireproofing products are measured by area, thickness, volume or R value. Specifications may only give what the specific R value should be in a certain situation. The estimator may need to choose the type of insulation to meet that R value.

07300 Shingles, Roof Tiles & Roof Coverings
07400 Roofing & Siding Panels
- Many roofing and siding products are bought and sold by the square. One square is equal to an area that measures 100 square feet. This simple change in unit of measure could create a large error if the estimator is not observant. Accessories and fasteners necessary for a complete installation must be figured into any calculations for both material and labor.

07500 Membrane Roofing
07600 Flashing & Sheet Metal
07700 Roof Specialties & Accessories
- The items in these subdivisions compose a roofing system. No one component completes the installation and all must be estimated. Built-up or single ply membrane roofing systems are made up of many products and installation trades. Wood blocking at roof perimeters or penetrations, parapet coverings, reglets, roof drains, gutters, downspouts, sheet metal flashing, skylights, smoke vents or roof hatches all need to be considered along with the roofing material. Several different installation trades will need to work together on the roofing system. Inherent difficulties in the scheduling and coordination of various trades must be accounted for when estimating labor costs.

07900 Joint Sealers
- To complete the weather-tight shell the sealants and caulkings must be estimated. Where different materials meet—at expansion joints, at flashing penetrations, and at hundreds of other locations throughout a construction project—they provide another line of defense against water penetration. Often, an entire system is based on the proper location and placement of caulking or sealants. The detail drawings that are included as part of a set of architectural plans, show typical locations for these materials. When caulking or sealants are shown at typical locations, this means the estimator must include them for all the locations where this detail is applicable. Be careful to keep different types of sealants separate, and remember to consider backer rods and primers if necessary.

Reference Numbers
Reference numbers are shown in bold squares at the beginning of some major classifications. These numbers refer to related items in the Reference Section. The reference information may be an estimating procedure, an alternate pricing method or technical information.

Note: Not all subdivisions listed here necessarily appear in this publication.

07100 | Dampproofing & Waterproofing

07110 | Dampproofing

		Description	CREW	DAILY OUTPUT	LABOR-HOURS	UNIT	MAT.	LABOR	EQUIP.	TOTAL	TOTAL INCL O&P	
100	0010	**BITUMINOUS ASPHALT COATING** For foundation										100
	0030	Brushed on, below grade, 1 coat	1 Rofc	665	.012	S.F.	.06	.21		.27	.45	
	0100	2 coat		500	.016		.10	.27		.37	.62	
	0300	Sprayed on, below grade, 1 coat, 25.6 S.F./gal.		830	.010		.07	.16		.23	.39	
	0400	2 coat, 20.5 S.F./gal.		500	.016		.14	.27		.41	.67	
	0500	Asphalt coating, with fibers				Gal.	3.68			3.68	4.05	
	0600	Troweled on, asphalt with fibers, 1/16" thick	1 Rofc	500	.016	S.F.	.15	.27		.42	.68	
	0700	1/8" thick		400	.020		.28	.34		.62	.94	
	1000	1/2" thick		350	.023		.92	.39		1.31	1.74	
200	0010	**CEMENT PARGING** 2 coats, 1/2" thick, regular P.C.	R07110-010 D-1	250	.064	S.F.	.15	1.14		1.29	2.10	200
	0100	Waterproofed Portland cement	"	250	.064	"	.17	1.14		1.31	2.12	

07130 | Sheet Waterproofing

		Description	CREW	DAILY OUTPUT	LABOR-HOURS	UNIT	MAT.	LABOR	EQUIP.	TOTAL	TOTAL INCL O&P	
200	0010	**ELASTOMERIC WATERPROOFING**										200
	0050	Acrylic rubber, fluid applied, 20 mils thick	3 Rofc	1,000	.024	S.F.	1.52	.41		1.93	2.43	
	0060	50 mil, reinforced, stucco texture	"	600	.040		2.46	.68		3.14	3.98	
	0090	EPDM, plain, 45 mils thick	2 Rofc	580	.028		.74	.47		1.21	1.69	
	0100	60 mils thick		570	.028		1.08	.48		1.56	2.08	
	0300	Nylon reinforced sheets, 45 mils thick		580	.028		1.04	.47		1.51	2.02	
	0400	60 mils thick		570	.028		1.44	.48		1.92	2.47	
	0600	Vulcanizing splicing tape for above, 2" wide				C.L.F.	33.50			33.50	36.50	
	0700	4" wide				"	67			67	73.50	
	0900	Adhesive, bonding, 60 SF per gal				Gal.	13.60			13.60	14.95	
	1000	Splicing, 75 SF per gal				"	21			21	23	
	1200	Neoprene sheets, plain, 45 mils thick	2 Rofc	580	.028	S.F.	1.09	.47		1.56	2.08	
	1300	60 mils thick		570	.028		1.82	.48		2.30	2.89	
	1500	Nylon reinforced, 45 mils thick		580	.028		1.26	.47		1.73	2.27	
	1600	60 mils thick		570	.028		1.28	.48		1.76	2.30	
	1800	120 mils thick		500	.032		2.54	.55		3.09	3.81	
	1900	Adhesive, splicing, 150 S.F. per gal. per coat				Gal.	18.15			18.15	19.95	
	2100	Fiberglass reinforced, fluid applied, 1/8" thick	2 Rofc	500	.032	S.F.	1.50	.55		2.05	2.67	
	2200	Polyethylene and rubberized asphalt sheets, 1/8" thick		550	.029		.54	.50		1.04	1.51	
	2210	Asphaltic hardboard protection board, 1/8" thick		500	.032		.30	.55		.85	1.35	
	2220	Asphaltic hardboard protection board, 1/4" thick		450	.036		.52	.61		1.13	1.70	
	2400	Polyvinyl chloride sheets, plain, 10 mils thick		580	.028		.14	.47		.61	1.03	
	2500	20 mils thick		570	.028		.22	.48		.70	1.13	
	2700	30 mils thick		560	.029		.30	.49		.79	1.24	
	3000	Adhesives, trowel grade, 40-100 SF per gal				Gal.	20			20	22	
	3100	Brush grade, 100-250 SF per gal.				"	20			20	22	
	3300	Bitumen modified polyurethane, fluid applied, 55 mils thick	2 Rofc	665	.024	S.F.	.65	.41		1.06	1.47	
	3600	Vinyl plastic, sprayed on, 25 to 40 mils thick	"	475	.034	"	.98	.57		1.55	2.15	
500	0010	**MEMBRANE WATERPROOFING** On slabs, 1 ply, felt	G-1	3,000	.019	S.F.	.13	.30	.14	.57	.85	500
	0100	Glass fiber fabric		2,100	.027		.13	.43	.20	.76	1.16	
	0300	2 ply, felt		2,500	.022		.26	.36	.17	.79	1.13	
	0400	Glass fiber fabric		1,650	.034		.30	.55	.25	1.10	1.64	
	0600	3 ply, felt		2,100	.027		.38	.43	.20	1.01	1.44	
	0700	Glass fiber fabric		1,550	.036		.40	.58	.27	1.25	1.81	
	0900	For installation on walls, add						15%				
	1000	For adhered 1/4" EPS protection board, add	2 Rofc	3,500	.005		.14	.08		.22	.30	
	1050	3/8" thick, add		3,500	.005		.16	.08		.24	.33	
	1060	1/2" thick, add		3,500	.005		.18	.08		.26	.35	

07100 | Dampproofing & Waterproofing

07130 | Sheet Waterproofing

			CREW	DAILY OUTPUT	LABOR-HOURS	UNIT	2000 BARE COSTS MAT.	LABOR	EQUIP.	TOTAL	TOTAL INCL O&P	
500	1070	Fiberglass fabric, black, 20/10 mesh	2 Rofc	116	.138	Sq.	9.95	2.35		12.30	15.35	500
	1080	White, 20/10 mesh		116	.138	"	10.25	2.35		12.60	15.70	
	1100	1/16" urethane, troweled		200	.080	S.F.	.69	1.36		2.05	3.30	
	1200	Roller applied		120	.133	"	.60	2.27		2.87	4.89	

07160 | Cementitious Waterproofing

			CREW	DAILY OUTPUT	LABOR-HOURS	UNIT	MAT.	LABOR	EQUIP.	TOTAL	TOTAL INCL O&P	
150	0010	**CEMENTITIOUS WATERPROOFING** One coat cement base										150
	0020	1/8" application, sprayed on	G-2	1,000	.024	S.F.	1.50	.39	.25	2.14	2.59	
	0030	2 coat, cementitious/metallic slurry, troweled, 1/4" thick	1 Cefi	2.48	3.226	C.S.F.	35	61		96	138	
	0040	3 coat, 3/8" thick		1.84	4.348		52.50	82		134.50	192	
	0050	4 coat, 1/2" thick		1.20	6.667		70	126		196	282	

07170 | Bentonite Waterproofing

			CREW	DAILY OUTPUT	LABOR-HOURS	UNIT	MAT.	LABOR	EQUIP.	TOTAL	TOTAL INCL O&P	
700	0010	**BENTONITE**, Panels, 4' x 4', 3/16" thick	1 Rofc	625	.013	S.F.	.70	.22		.92	1.18	700
	0100	Rolls, 3/8" thick, with geotextile fabric both sides	"	550	.015	"	.84	.25		1.09	1.38	
	0300	Granular bentonite, 50 lb. bags (.625 C.F.)				Bag	11.65			11.65	12.85	
	0400	3/8" thick, troweled on	1 Rofc	475	.017	S.F.	.58	.29		.87	1.17	
	0500	Drain board, expanded polystyrene, binder encapsulated, 1-1/2" thick	1 Rohe	1,600	.005		.62	.06		.68	.80	
	0510	2" thick		1,600	.005		.83	.06		.89	1.03	
	0520	3" thick		1,600	.005		1.26	.06		1.32	1.51	
	0530	4" thick		1,600	.005		1.68	.06		1.74	1.97	
	0600	With filter fabric, 1-1/2" thick		1,600	.005		.78	.06		.84	.98	
	0625	2" thick		1,600	.005		1.03	.06		1.09	1.25	
	0650	3" thick		1,600	.005		1.46	.06		1.52	1.73	
	0675	4" thick		1,600	.005		1.88	.06		1.94	2.19	
	0700	Vapor retarder, see polyethelene, 07260-100										

07190 | Water Repellents

			CREW	DAILY OUTPUT	LABOR-HOURS	UNIT	MAT.	LABOR	EQUIP.	TOTAL	TOTAL INCL O&P	
700	0010	**RUBBER COATING** Water base liquid, roller applied	2 Rofc	7,000	.002	S.F.	.55	.04		.59	.68	700
	0200	Silicone or stearate, sprayed on CMU, 1 coat	1 Rofc	4,000	.002		.27	.03		.30	.36	
	0300	2 coats	"	3,000	.003		.54	.05		.59	.68	

07200 | Thermal Protection

07210 | Building Insulation

			CREW	DAILY OUTPUT	LABOR-HOURS	UNIT	MAT.	LABOR	EQUIP.	TOTAL	TOTAL INCL O&P	
150	0010	**BLOWN-IN INSULATION** Ceilings, with open access										150
	0020	Cellulose, 3-1/2" thick, R13	G-4	5,000	.005	S.F.	.13	.07	.05	.25	.32	
	0030	5-3/16" thick, R19		3,800	.006		.19	.10	.07	.36	.45	
	0050	6-1/2" thick, R22		3,000	.008		.24	.12	.09	.45	.57	
	0100	8-11/16" thick, R30		2,600	.009		.32	.14	.10	.56	.70	
	0120	10-7/8" thick, R38		1,800	.013		.41	.20	.15	.76	.97	
	1000	Fiberglass, 5" thick, R11		3,800	.006		.16	.10	.07	.33	.42	
	1050	6" thick, R13		3,000	.008		.17	.12	.09	.38	.50	
	1100	8-1/2" thick, R19		2,200	.011		.23	.16	.12	.51	.67	
	1200	10" thick, R22		1,800	.013		.27	.20	.15	.62	.82	
	1300	12" thick, R26		1,500	.016		.33	.24	.18	.75	.97	
	2000	Mineral wool, 4" thick, R12		3,500	.007		.16	.10	.08	.34	.45	

07200 | Thermal Protection

07210 | Building Insulation

			CREW	DAILY OUTPUT	LABOR-HOURS	UNIT	2000 BARE COSTS MAT.	LABOR	EQUIP.	TOTAL	TOTAL INCL O&P	
150	2050	6" thick, R17	G-4	2,500	.010	S.F.	.18	.15	.11	.44	.57	150
	2100	9" thick, R23	↓	1,750	.014	↓	.27	.21	.16	.64	.83	
	2500	Wall installation, incl. drilling & patching from outside, two 1"										
	2510	diam. holes @ 16" O.C., top & mid-point of wall, add to above										
	2700	For masonry	G-4	415	.058	S.F.	.06	.87	.65	1.58	2.29	
	2800	For wood siding	↓	840	.029		.06	.43	.32	.81	1.17	
	2900	For stucco/plaster	↓	665	.036	↓	.06	.55	.41	1.02	1.45	
400	0010	**INSULATION FASTENERS**										400
	0050	Clips, galv, 1" or 2" insulation				C	17.50			17.50	19.25	
	0100	6" insulation					30			30	33	
	0300	Screws with plates, galv, for 1" or 2" insulation					15.75			15.75	17.35	
	0500	For 6" insulation				↓	28.50			28.50	31.50	
500	0010	**POURED INSULATION** Cellulose fiber, R3.8 per inch	1 Carp	200	.040	C.F.	.45	.79		1.24	1.85	500
	0040	Ceramic type (perlite), R3.2 per inch		200	.040		1.45	.79		2.24	2.95	
	0080	Fiberglass wool, R4 per inch		200	.040		.33	.79		1.12	1.71	
	0100	Mineral wool, R3 per inch		200	.040		.30	.79		1.09	1.68	
	0300	Polystyrene, R4 per inch		200	.040		1.93	.79		2.72	3.47	
	0400	Vermiculite or perlite, R2.7 per inch		200	.040		1.45	.79		2.24	2.95	
	0700	Wood fiber, R3.85 per inch	↓	200	.040	↓	.57	.79		1.36	1.98	
550	0010	**MASONRY INSULATION** Vermiculite or perlite, poured										550
	0100	In cores of concrete block, 4" thick wall, .115 CF/SF	D-1	4,800	.003	S.F.	.17	.06		.23	.28	
	0200	6" thick wall, .175 CF/SF		3,000	.005		.25	.10		.35	.44	
	0300	8" thick wall, .258 CF/SF		2,400	.007		.37	.12		.49	.61	
	0400	10" thick wall, .340 CF/SF		1,850	.009		.49	.15		.64	.80	
	0500	12" thick wall, .422 CF/SF	↓	1,200	.013	↓	.61	.24		.85	1.07	
	0550	For sand fill, deduct from above					70%					
	0600	Poured cavity wall, vermiculite or perlite, water repellant	D-1	250	.064	C.F.	1.45	1.14		2.59	3.53	
	0700	Foamed in place, urethane in 2-5/8" cavity	G-2	1,035	.023	S.F.	.38	.38	.24	1	1.32	
	0800	For each 1" added thickness, add	"	2,372	.010	"	.12	.16	.11	.39	.53	
600	0600	**PERIMETER INSULATION**, polystyrene, expanded, 1" thick, R4	1 Carp	680	.012	S.F.	.17	.23		.40	.59	600
	0700	2" thick, R8		675	.012	"	.32	.23		.55	.75	
700	0010	**REFLECTIVE INSULATION**, aluminum foil on reinforced scrim		19	.421	C.S.F.	13.90	8.30		22.20	29.50	700
	0100	Reinforced with woven polyolefin		19	.421		16.90	8.30		25.20	33	
	0500	With single bubble air space, R8.8		15	.533		26	10.50		36.50	46.50	
	0600	With double bubble air space, R9.8	↓	15	.533	↓	28	10.50		38.50	49	
800	0010	**SPRAYED** Fibrous/cementitious, finished wall, 1" thick, R3.7	G-2	2,050	.012	S.F.	.22	.19	.12	.53	.69	800
	0100	Attic, 5.2" thick, R19	"	1,550	.015	"	.36	.25	.16	.77	1	
	0300	Foam type, incl. preparation										
	0600	3 #/CF, 1" thick, R3.8	G-2	770	.031	S.F.	.42	.51	.32	1.25	1.67	
	0700	2" thick, R7.5	"	475	.051	"	.84	.82	.53	2.19	2.88	
900	0010	**WALL INSULATION, RIGID**										900
	0040	Fiberglass, 1.5#/CF, unfaced, 1" thick, R4.1	1 Carp	1,000	.008	S.F.	.25	.16		.41	.55	
	0060	1-1/2" thick, R6.2		1,000	.008		.34	.16		.50	.64	
	0080	2" thick, R8.3		1,000	.008		.41	.16		.57	.72	
	0120	3" thick, R12.4		800	.010		.49	.20		.69	.88	
	0370	3#/CF, unfaced, 1" thick, R4.3		1,000	.008		.32	.16		.48	.62	
	0390	1-1/2" thick, R6.5		1,000	.008		.63	.16		.79	.96	
	0400	2" thick, R8.7		890	.009		.77	.18		.95	1.15	
	0420	2-1/2" thick, R10.9		800	.010		.96	.20		1.16	1.40	
	0440	3" thick, R13		800	.010		1.14	.20		1.34	1.59	
	0520	Foil faced, 1" thick, R4.3		1,000	.008		.76	.16		.92	1.11	
	0540	1-1/2" thick, R6.5	↓	1,000	.008	↓	1.02	.16		1.18	1.39	

07200 | Thermal Protection

07210 | Building Insulation

		Crew	Daily Output	Labor-Hours	Unit	Mat.	Labor	Equip.	Total	Total Incl O&P	
900	0560	2" thick, R8.7	1 Carp	890	.009	S.F.	1.27	.18		1.45	1.70	900
	0580	2-1/2" thick, R10.9		800	.010		1.50	.20		1.70	1.99	
	0600	3" thick, R13		800	.010		1.64	.20		1.84	2.14	
	0670	6#/CF, unfaced, 1" thick, R4.3		1,000	.008		.73	.16		.89	1.07	
	0690	1-1/2" thick, R6.5		890	.009		1.13	.18		1.31	1.54	
	0700	2" thick, R8.7		800	.010		1.58	.20		1.78	2.08	
	0721	2-1/2" thick, R10.9		800	.010		1.74	.20		1.94	2.25	
	0741	3" thick, R13		730	.011		2.08	.22		2.30	2.66	
	0821	Foil faced, 1" thick, R4.3		1,000	.008		1.03	.16		1.19	1.40	
	0840	1-1/2" thick, R6.5		890	.009		1.48	.18		1.66	1.93	
	0850	2" thick, R8.7		800	.010		1.93	.20		2.13	2.46	
	0880	2-1/2" thick, R10.9		800	.010		2.32	.20		2.52	2.89	
	0900	3" thick, R13		730	.011		2.77	.22		2.99	3.42	
	1500	Foamglass, 1-1/2" thick, R4.5		800	.010		1.45	.20		1.65	1.94	
	1550	3" thick, R9		730	.011		2.71	.22		2.93	3.35	
	1600	Isocyanurate, 4' x 8' sheet, foil faced, both sides										
	1610	1/2" thick, R3.9	1 Carp	800	.010	S.F.	.27	.20		.47	.64	
	1620	5/8" thick, R4.5		800	.010		.28	.20		.48	.65	
	1630	3/4" thick, R5.4		800	.010		.29	.20		.49	.66	
	1640	1" thick, R7.2		800	.010		.32	.20		.52	.69	
	1650	1-1/2" thick, R10.8		730	.011		.35	.22		.57	.76	
	1660	2" thick, R14.4		730	.011		.44	.22		.66	.85	
	1670	3" thick, R21.6		730	.011		1.05	.22		1.27	1.52	
	1680	4" thick, R28.8		730	.011		1.29	.22		1.51	1.79	
	1700	Perlite, 1" thick, R2.77		800	.010		.24	.20		.44	.60	
	1750	2" thick, R5.55		730	.011		.48	.22		.70	.90	
	1900	Extruded polystyrene, 25 PSI compressive strength, 1" thick, R5		800	.010		.30	.20		.50	.67	
	1940	2" thick R10		730	.011		.59	.22		.81	1.02	
	1960	3" thick, R15		730	.011		.86	.22		1.08	1.32	
	2100	Expanded polystyrene, 1" thick, R3.85		800	.010		.12	.20		.32	.47	
	2120	2" thick, R7.69		730	.011		.32	.22		.54	.72	
	2140	3" thick, R11.49		730	.011		.49	.22		.71	.91	
950	0010	**WALL OR CEILING INSUL., NON-RIGID**										950
	0040	Fiberglass, kraft faced, batts or blankets										
	0060	3-1/2" thick, R11, 11" wide	1 Carp	1,150	.007	S.F.	.22	.14		.36	.47	
	0080	15" wide		1,600	.005		.22	.10		.32	.41	
	0100	23" wide		1,600	.005		.22	.10		.32	.41	
	0140	6" thick, R19, 11" wide		1,000	.008		.31	.16		.47	.61	
	0160	15" wide		1,350	.006		.31	.12		.43	.54	
	0180	23" wide		1,600	.005		.31	.10		.41	.51	
	0200	9" thick, R30, 15" wide		1,150	.007		.56	.14		.70	.85	
	0220	23" wide		1,350	.006		.56	.12		.68	.82	
	0240	12" thick, R38, 15" wide		1,000	.008		.72	.16		.88	1.06	
	0260	23" wide		1,350	.006		.72	.12		.84	.99	
	0400	Fiberglass, foil faced, batts or blankets										
	0420	3-1/2" thick, R11, 15" wide	1 Carp	1,600	.005	S.F.	.32	.10		.42	.52	
	0440	23" wide		1,600	.005		.32	.10		.42	.52	
	0460	6" thick, R19, 15" wide		1,350	.006		.39	.12		.51	.63	
	0480	23" wide		1,600	.005		.39	.10		.49	.60	
	0500	9" thick, R30, 15" wide		1,150	.007		.67	.14		.81	.97	
	0550	23" wide		1,350	.006		.67	.12		.79	.94	
	0800	Fiberglass, unfaced, batts or blankets										
	0820	3-1/2" thick, R11, 15" wide	1 Carp	1,350	.006	S.F.	.20	.12		.32	.42	
	0830	23" wide		1,600	.005		.19	.10		.29	.38	
	0860	6" thick, R19, 15" wide		1,150	.007		.33	.14		.47	.59	
	0880	23" wide		1,350	.006		.32	.12		.44	.55	

07200 | Thermal Protection

07210 | Building Insulation

		CREW	DAILY OUTPUT	LABOR-HOURS	UNIT	2000 BARE COSTS MAT.	LABOR	EQUIP.	TOTAL	TOTAL INCL O&P	
950	0900	9" thick, R30, 15" wide	1 Carp	1,000	.008	S.F.	.56	.16		.72	.89	950
	0920	23" wide		1,150	.007		.56	.14		.70	.85	
	0940	12" thick, R38, 15" wide		1,000	.008		.72	.16		.88	1.06	
	0960	23" wide	▼	1,150	.007	▼	.72	.14		.86	1.02	
	1300	Mineral fiber batts, kraft faced										
	1320	3-1/2" thick, R12	1 Carp	1,600	.005	S.F.	.24	.10		.34	.43	
	1340	6" thick, R19		1,600	.005		.37	.10		.47	.58	
	1380	10" thick, R30		1,350	.006	▼	.58	.12		.70	.84	
	1850	Friction fit wire insulation supports, 16" O.C.	▼	960	.008	Ea.	.05	.16		.21	.34	
	1900	For foil backing, add				S.F.	.04			.04	.04	

07220 | Roof & Deck Insulation

		CREW	DAILY OUTPUT	LABOR-HOURS	UNIT	2000 BARE COSTS MAT.	LABOR	EQUIP.	TOTAL	TOTAL INCL O&P	
700	0010	ROOF DECK INSULATION										700
	0020	Fiberboard low density, 1/2" thick R1.39	1 Rofc	1,000	.008	S.F.	.17	.14		.31	.44	
	0030	1" thick R2.78		800	.010		.32	.17		.49	.67	
	0080	1 1/2" thick R4.17		800	.010		.48	.17		.65	.85	
	0100	2" thick R5.56		800	.010		.64	.17		.81	1.02	
	0110	Fiberboard high density, 1/2" thick R1.3		1,000	.008		.18	.14		.32	.45	
	0120	1" thick R2.5		800	.010		.34	.17		.51	.69	
	0130	1-1/2" thick R3.8		800	.010		.55	.17		.72	.93	
	0200	Fiberglass, 3/4" thick R2.78		1,000	.008		.44	.14		.58	.73	
	0400	15/16" thick R3.70		1,000	.008		.57	.14		.71	.88	
	0460	1-1/16" thick R4.17		1,000	.008		.72	.14		.86	1.04	
	0600	1-5/16" thick R5.26		1,000	.008		.99	.14		1.13	1.34	
	0650	2-1/16" thick R8.33		800	.010		1.06	.17		1.23	1.49	
	0700	2-7/16" thick R10		800	.010		1.20	.17		1.37	1.64	
	1500	Foamglass, 1-1/2" thick R4.5		800	.010		1.41	.17		1.58	1.87	
	1530	3" thick R9		700	.011	▼	2.81	.19		3	3.45	
	1600	Tapered for drainage		600	.013	B.F.	.91	.23		1.14	1.42	
	1650	Perlite, 1/2" thick R1.32		1,050	.008	S.F.	.25	.13		.38	.52	
	1655	3/4" thick R2.08		800	.010		.30	.17		.47	.65	
	1660	1" thick R2.78		800	.010		.24	.17		.41	.58	
	1670	1-1/2" thick R4.17		800	.010		.36	.17		.53	.72	
	1680	2" thick R5.56		700	.011		.48	.19		.67	.89	
	1685	2-1/2" thick R6.67		700	.011	▼	.71	.19		.90	1.14	
	1690	Tapered for drainage		800	.010	B.F.	.54	.17		.71	.91	
	1700	Polyisocyanurate, 2#/CF density, 3/4" thick, R5.1		1,500	.005	S.F.	.29	.09		.38	.49	
	1705	1" thick R7.14		1,400	.006		.30	.10		.40	.51	
	1715	1-1/2" thick R10.87		1,250	.006		.34	.11		.45	.57	
	1725	2" thick R14.29		1,100	.007		.42	.12		.54	.69	
	1735	2-1/2" thick R16.67		1,050	.008		.48	.13		.61	.77	
	1745	3" thick R21.74		1,000	.008		.59	.14		.73	.90	
	1755	3-1/2" thick R25		1,000	.008	▼	.70	.14		.84	1.02	
	1765	Tapered for drainage		1,400	.006	B.F.	.36	.10		.46	.58	
	1900	Extruded Polystyrene										
	1910	15 PSI compressive strength, 1" thick, R5	1 Rofc	1,500	.005	S.F.	.21	.09		.30	.40	
	1920	2" thick, R10		1,250	.006		.33	.11		.44	.56	
	1930	3" thick R15		1,000	.008		.73	.14		.87	1.05	
	1932	4" thick R20		1,000	.008	▼	1.01	.14		1.15	1.36	
	1934	Tapered for drainage		1,500	.005	B.F.	.33	.09		.42	.53	
	1940	25 PSI compressive strength, 1" thick R5		1,500	.005	S.F.	.31	.09		.40	.51	
	1942	2" thick R10		1,250	.006		.63	.11		.74	.89	
	1944	3" thick R15		1,000	.008		.94	.14		1.08	1.28	
	1946	4" thick R20		1,000	.008	▼	1.05	.14		1.19	1.40	
	1948	Tapered for drainage		1,500	.005	B.F.	.38	.09		.47	.59	
	1950	40 psi compressive strength, 1" thick R5	▼	1,500	.005	S.F.	.33	.09		.42	.53	

07200 | Thermal Protection

07220 | Roof & Deck Insulation

		CREW	DAILY OUTPUT	LABOR-HOURS	UNIT	MAT.	LABOR	EQUIP.	TOTAL	TOTAL INCL O&P	
1952	2" thick R10	1 Rofc	1,250	.006	S.F.	.65	.11		.76	.91	700
1954	3" thick R15		1,000	.008		.95	.14		1.09	1.30	
1956	4" thick R20		1,000	.008	↓	1.27	.14		1.41	1.65	
1958	Tapered for drainage		1,400	.006	B.F.	.48	.10		.58	.71	
1960	60 PSI compressive strength, 1" thick R5		1,450	.006	S.F.	.40	.09		.49	.62	
1962	2" thick R10		1,200	.007		.71	.11		.82	.99	
1964	3" thick R15		975	.008		1.06	.14		1.20	1.43	
1966	4" thick R20		950	.008	↓	1.47	.14		1.61	1.89	
1968	Tapered for drainage		1,400	.006	B.F.	.57	.10		.67	.81	
2010	Expanded polystyrene, 1#/CF density, 3/4" thick R2.89		1,500	.005	S.F.	.17	.09		.26	.36	
2020	1" thick R3.85		1,500	.005		.17	.09		.26	.36	
2100	2" thick R7.69		1,250	.006		.32	.11		.43	.55	
2110	3" thick R11.49		1,250	.006		.48	.11		.59	.73	
2120	4" thick R15.38		1,200	.007		.53	.11		.64	.79	
2130	5" thick R19.23		1,150	.007		.67	.12		.79	.96	
2140	6" thick R23.26		1,150	.007	↓	.79	.12		.91	1.09	
2150	Tapered for drainage	↓	1,500	.005	B.F.	.32	.09		.41	.52	
2400	Composites with 2" EPS										
2410	1" fiberboard	1 Rofc	950	.008	S.F.	.74	.14		.88	1.08	
2420	7/16" oriented strand board		800	.010		.87	.17		1.04	1.28	
2430	1/2" plywood		800	.010		.94	.17		1.11	1.35	
2440	1" perlite	↓	800	.010	↓	.78	.17		.95	1.18	
2450	Composites with 1 1/2" polyisocyanurate										
2460	1" fiberboard	1 Rofc	800	.010	S.F.	.80	.17		.97	1.20	
2470	1" perlite		850	.009		.83	.16		.99	1.21	
2480	7/16" oriented strand board	↓	800	.010	↓	.95	.17		1.12	1.37	

07240 | Ext. Insulation/Finish Systems

		CREW	DAILY OUTPUT	LABOR-HOURS	UNIT	MAT.	LABOR	EQUIP.	TOTAL	TOTAL INCL O&P	
0010	**EXTERIOR INSULATION FINISH SYSTEM**										100
0095	Field applied, 1" EPS insulation	J-1	295	.136	S.F.	1.70	2.37	.18	4.25	6.05	
0100	With 1/2" cement board sheathing		220	.182		2.74	3.17	.25	6.16	8.60	
0105	2" EPS insulation		295	.136		1.90	2.37	.18	4.45	6.25	
0110	With 1/2" cement board sheathing		220	.182		2.94	3.17	.25	6.36	8.80	
0115	3" EPS insulation		295	.136		2.07	2.37	.18	4.62	6.45	
0120	With 1/2" cement board sheathing		220	.182		3.11	3.17	.25	6.53	9	
0125	4" EPS insulation		295	.136		2.22	2.37	.18	4.77	6.60	
0130	With 1/2" cement board sheathing		220	.182		4.30	3.17	.25	7.72	10.30	
0140	Premium finish add		1,265	.032		.27	.55	.04	.86	1.27	
0150	Heavy duty reinforcement add	↓	914	.044	↓	1.65	.76	.06	2.47	3.17	
0160	2.5#/S.Y. metal lath substrate add	1 Lath	75	.107	S.Y.	2.12	2.05		4.17	5.70	
0170	3.4#/S.Y. metal lath substrate add	"	75	.107	"	2.21	2.05		4.26	5.80	
0180	Color or texture change,	J-1	1,265	.032	S.F.	.72	.55	.04	1.31	1.76	
0190	With substrate leveling base coat	1 Plas	530	.015		.72	.28		1	1.26	
0210	With substrate sealing base coat	1 Pord	1,224	.007		.07	.12		.19	.28	
0220	Prefab. panels, with hat channels 2 1/2" x 3/4", 2" EPS insul.	L-5	1,800	.031		7.70	.67	.31	8.68	10.10	
0240	3" EPS insulation		1,800	.031		8.55	.67	.31	9.53	11.05	
0250	4" EPS insulation		1,800	.031		9.55	.67	.31	10.53	12.15	
0260	1-1/2" profile lightweight metal backing, 1" EPS insulation		1,800	.031		11.75	.67	.31	12.73	14.60	
0270	2" EPS insulation		1,800	.031		13.10	.67	.31	14.08	16.10	
0280	3" EPS insulation		1,800	.031		14.65	.67	.31	15.63	17.75	
0290	4" EPS insulation		1,800	.031		16.35	.67	.31	17.33	19.65	
0310	6"-16 Ga. steel stud back-up, 1" EPS insulation		1,800	.031		13.55	.67	.31	14.53	16.55	
0320	2" EPS insulation		1,800	.031		15.15	.67	.31	16.13	18.30	
0330	3" EPS insulation		1,800	.031		16.85	.67	.31	17.83	20	
0340	4" EPS insulation	↓	1,800	.031		18.85	.67	.31	19.83	22	
0350	Premium finish add	J-1	1,265	.032	↓	.27	.55	.04	.86	1.27	

07200 | Thermal Protection

07240 | Ext. Insulation/Finish Systems

			CREW	DAILY OUTPUT	LABOR-HOURS	UNIT	MAT.	LABOR	EQUIP.	TOTAL	TOTAL INCL O&P	
100	0360	Heavy duty reinforcement add	J-1	914	.044	S.F.	1.65	.76	.06	2.47	3.17	100
	0370	V groove shape in panel face				L.F.	.52			.52	.57	
	0380	U groove shape in panel face				"	.69			.69	.76	
	0390	Architectural features,										
	0410	Crown moulding 8" high x 4" wide	1 Plas	150	.053	L.F.	1.15	.99		2.14	2.93	
	0420	Crown moulding 12" high x 6" wide		150	.053		2.15	.99		3.14	4.03	
	0430	Crown moulding 16" high x 12" wide	↓	150	.053	↓	5.65	.99		6.64	7.90	
	0440	For higher than one story, add						25%				

07260 | Vapor Retarders

			CREW	DAILY OUTPUT	LABOR-HOURS	UNIT	MAT.	LABOR	EQUIP.	TOTAL	TOTAL INCL O&P	
100	0010	**BUILDING PAPER** Aluminum and kraft laminated, foil 1 side	1 Carp	37	.216	Sq.	3.55	4.26		7.81	11.20	100
	0100	Foil 2 sides		37	.216		5.70	4.26		9.96	13.55	
	0300	Asphalt, two ply, 30#, for subfloors		19	.421		10.90	8.30		19.20	26	
	0400	Asphalt felt sheathing paper, 15#	↓	37	.216	↓	2.59	4.26		6.85	10.15	
	0450	Housewrap, exterior, spun bonded polypropylene										
	0470	Small roll	1 Carp	3,800	.002	S.F.	.10	.04		.14	.18	
	0480	Large roll	"	4,000	.002	"	.09	.04		.13	.17	
	0500	Material only, 3' x 111.1' roll				Ea.	33			33	36.50	
	0520	9' x 111.1' roll				"	94			94	103	
	0600	Polyethylene vapor barrier, standard, .002" thick	1 Carp	37	.216	Sq.	1.09	4.26		5.35	8.50	
	0700	.004" thick		37	.216		2.36	4.26		6.62	9.90	
	0900	.006" thick		37	.216		2.86	4.26		7.12	10.45	
	1200	.010" thick		37	.216		6.15	4.26		10.41	14.05	
	1300	Clear reinforced, fire retardant, .008" thick		37	.216		8.40	4.26		12.66	16.55	
	1350	Cross laminated type, .003" thick		37	.216		6.50	4.26		10.76	14.45	
	1400	.004" thick		37	.216		7.25	4.26		11.51	15.30	
	1500	Red rosin paper, 5 sq rolls, 4 lb per square		37	.216		1.55	4.26		5.81	9	
	1600	5 lbs. per square		37	.216		2	4.26		6.26	9.50	
	1800	Reinf. waterproof, .002" polyethylene backing, 1 side		37	.216		4.92	4.26		9.18	12.70	
	1900	2 sides	↓	37	.216		6.50	4.26		10.76	14.45	
	2100	Roof deck vapor barrier, class 1 metal decks	1 Rofc	37	.216		8.30	3.69		11.99	16	
	2200	For all other decks	"	37	.216		5.70	3.69		9.39	13.15	
	2400	Waterproofed kraft with sisal or fiberglass fibers, minimum	1 Carp	37	.216		5.30	4.26		9.56	13.15	
	2500	Maximum	"	37	.216		13.25	4.26		17.51	22	

07300 | Shingles, Roof Tiles & Roof Coverings

07310 | Shingles

			CREW	DAILY OUTPUT	LABOR-HOURS	UNIT	MAT.	LABOR	EQUIP.	TOTAL	TOTAL INCL O&P	
050	0010	**ALUMINUM** Shingles, mill finish, .019" thick	1 Carp	5	1.600	Sq.	150	31.50		181.50	219	050
	0100	.020" thick	"	5	1.600		145	31.50		176.50	213	
	0300	For colors, add				↓	15.15			15.15	16.65	
	0600	Ridge cap, .024" thick	1 Carp	170	.047	L.F.	1.85	.93		2.78	3.63	
	0700	End wall flashing, .024" thick		170	.047		1.26	.93		2.19	2.98	
	0900	Valley section, .024" thick		170	.047		2.30	.93		3.23	4.12	
	1000	Starter strip, .024" thick		400	.020		1.20	.39		1.59	2	
	1200	Side wall flashing, .024" thick		170	.047		1.25	.93		2.18	2.97	
	1500	Gable flashing, .024" thick	↓	400	.020	↓	1.35	.39		1.74	2.17	
100	0010	**ASPHALT SHINGLES**										100
	0100	Standard strip shingles										

Important: See the Reference Section for critical supporting data - Reference Nos., Crews, & City Cost Indexes

07300 | Shingles, Roof Tiles & Roof Coverings

07310 | Shingles

			CREW	DAILY OUTPUT	LABOR-HOURS	UNIT	MAT.	LABOR	EQUIP.	TOTAL	TOTAL INCL O&P	
100	0150	Inorganic, class A, 210-235 lb/sq	1 Rofc	5.50	1.455	Sq.	26	25		51	74.50	100
	0155	Pneumatic nailed		7	1.143		26	19.50		45.50	65	
	0200	Organic, class C, 235-240 lb/sq		5	1.600		35.50	27.50		63	90	
	0205	Pneumatic nailed		6.25	1.280		35.50	22		57.50	79.50	
	0250	Standard, laminated multi-layered shingles										
	0300	Class A, 240-260 lb/sq	1 Rofc	4.50	1.778	Sq.	34	30.50		64.50	94	
	0305	Pneumatic nailed		5.63	1.422		34	24.50		58.50	82.50	
	0350	Class C, 260-300 lb/square, 4 bundles/square		4	2		49	34		83	118	
	0355	Pneumatic nailed		5	1.600		49	27.50		76.50	105	
	0400	Premium, laminated multi-layered shingles										
	0450	Class A, 260-300 lb, 4 bundles/sq	1 Rofc	3.50	2.286	Sq.	43	39		82	120	
	0455	Pneumatic nailed		4.37	1.831		43	31		74	106	
	0500	Class C, 300-385 lb/square, 5 bundles/square		3	2.667		65.50	45.50		111	157	
	0505	Pneumatic nailed		3.75	2.133		65.50	36.50		102	140	
	0800	#15 felt underlayment		64	.125		2.59	2.13		4.72	6.80	
	0825	#30 felt underlayment		58	.138		5.45	2.35		7.80	10.40	
	0850	Self adhering polyethylene and rubberized asphalt underlayment		22	.364		37	6.20		43.20	52.50	
	0900	Ridge shingles		330	.024	L.F.	.72	.41		1.13	1.56	
	0905	Pneumatic nailed		412.50	.019	"	.72	.33		1.05	1.41	
	1000	For steep roofs (7 to 12 pitch or greater), add						50%				
500	0010	FIBER CEMENT shingles, 16" x 9.35", 500 lb per square	1 Carp	4	2	Sq.	244	39.50		283.50	335	500
	0110	Starters, 16" x 9.35"		3	2.667	C.L.F.	83	52.50		135.50	181	
	0120	Hip & ridge, 4.75" x 14"		1	8	"	600	158		758	930	
	0200	Shakes, 16" x 9.35", 550 lb per square		2.20	3.636	Sq.	221	71.50		292.50	365	
	0300	Hip & ridge, 4.75" x 14"		1	8	C.L.F.	600	158		758	930	
	0400	Hexagonal, 16" x 16"		3	2.667	Sq.	165	52.50		217.50	272	
	0500	Square, 16" x 16"		3	2.667	"	148	52.50		200.50	253	
800	0010	SLATE, Buckingham, Virginia, black										800
	0102	3/16" - 1/4" thick	1 Rots	1.75	4.571	Sq.	540	78		618	740	
	0202	1/4" thick		1.75	4.571		720	78		798	935	
	0901	Pennsylvania black, Bangor, #1 clear		1.75	4.571		435	78		513	625	
	1201	Vermont, unfading, green, mottled green		1.75	4.571		395	78		473	580	
	1302	Semi-weathering green & gray		1.75	4.571		296	78		374	470	
	1401	Purple		1.75	4.571		390	78		468	575	
	1501	Black or gray		1.75	4.571		355	78		433	535	
	1601	Red		1.75	4.571		1,125	78		1,203	1,375	
	1700	Variegated purple		1.75	4.571		375	78		453	555	
900	0011	STEEL Shingles, galvanized, 26 gauge	1 Rofc	2.20	3.636	Sq.	149	62		211	279	900
	0201	24 gauge	"	2.20	3.636		157	62		219	287	
	0300	For colored galvanized shingles, add					41			41	45	
	0500	For 1" factory applied polystyrene insulation, add					30			30	33	
980	0010	WOOD 16" No. 1 red cedar shingles, 5" exposure, on roof	1 Carp	2.50	3.200	Sq.	151	63		214	274	980
	0015	Pneumatic nailed		3.25	2.462		151	48.50		199.50	249	
	0200	7-1/2" exposure, on walls		2.05	3.902		101	77		178	243	
	0205	Pneumatic nailed		2.67	2.996		101	59		160	212	
	0300	18" No. 1 red cedar perfections, 5-1/2" exposure, on roof		2.75	2.909		164	57.50		221.50	279	
	0305	Pneumatic nailed		3.57	2.241		164	44		208	257	
	0500	7-1/2" exposure, on walls		2.25	3.556		121	70		191	253	
	0505	Pneumatic nailed		2.92	2.740		121	54		175	226	
	0600	Resquared, and rebutted, 5-1/2" exposure, on roof		3	2.667		199	52.50		251.50	310	
	0605	Pneumatic nailed		3.90	2.051		199	40.50		239.50	288	
	0900	7-1/2" exposure, on walls		2.45	3.265		146	64.50		210.50	271	
	0905	Pneumatic nailed		3.18	2.516		146	49.50		195.50	246	
	1000	Add to above for fire retardant shingles, 16" long					30			30	33	
	1050	18" long					28.50			28.50	31.50	

THERMAL & MOISTURE PROTECTION 7

07300 | Shingles, Roof Tiles & Roof Coverings

07310 | Shingles

			CREW	DAILY OUTPUT	LABOR-HOURS	UNIT	2000 BARE COSTS MAT.	LABOR	EQUIP.	TOTAL	TOTAL INCL O&P	
980	1060	Preformed ridge shingles	1 Carp	400	.020	L.F.	1.65	.39		2.04	2.50	980
	1100	Hand-split red cedar shakes, 1/2" thick x 24" long, 10" exp. on roof		2.50	3.200	Sq.	138	63		201	260	
	1105	Pneumatic nailed		3.25	2.462		138	48.50		186.50	235	
	1110	3/4" thick x 24" long, 10" exp. on roof		2.25	3.556		138	70		208	272	
	1115	Pneumatic nailed		2.92	2.740		138	54		192	245	
	1200	1/2" thick, 18" long, 8-1/2" exp. on roof		2	4		97.50	79		176.50	242	
	1205	Pneumatic nailed		2.60	3.077		97.50	60.50		158	211	
	1210	3/4" thick x 18" long, 8 1/2" exp. on roof		1.80	4.444		97.50	87.50		185	257	
	1215	Pneumatic nailed		2.34	3.419		97.50	67.50		165	222	
	1255	10" exp. on walls		2	4		110	79		189	256	
	1260	10" exposure on walls, pneumatic nailed		2.60	3.077		110	60.50		170.50	225	
	1700	Add to above for fire retardant shakes, 24" long					30			30	33	
	1800	18" long					30			30	33	
	1810	Ridge shakes	1 Carp	350	.023	L.F.	2.35	.45		2.80	3.35	
	2000	White cedar shingles, 16" long, extras, 5" exposure, on roof		2.40	3.333	Sq.	112	65.50		177.50	237	
	2005	Pneumatic nailed		3.12	2.564		112	50.50		162.50	211	
	2050	5" exposure on walls		2	4		112	79		191	259	
	2055	Pneumatic nailed		2.60	3.077		112	60.50		172.50	228	
	2100	7-1/2" exposure, on walls		2	4		80	79		159	224	
	2105	Pneumatic nailed		2.60	3.077		80	60.50		140.50	193	
	2150	"B" grade, 5" exposure on walls		2	4		117	79		196	264	
	2155	Pneumatic nailed		2.60	3.077		117	60.50		177.50	233	
	2300	For 15# organic felt underlayment on roof, 1 layer, add		64	.125		2.59	2.46		5.05	7.05	
	2400	2 layers, add		32	.250		5.20	4.93		10.13	14.15	
	2600	For steep roofs (7/12 pitch or greater), add to above						50%				
	2700	Panelized systems, No.1 cedar shingles on 5/16" CDX plywood										
	2800	On walls, 8' strips, 7" or 14" exposure	2 Carp	700	.023	S.F.	3.20	.45		3.65	4.29	
	3500	On roofs, 8' strips, 7" or 14" exposure	1 Carp	3	2.667	Sq.	320	52.50		372.50	440	
	3505	Pneumatic nailed	"	4	2	"	320	39.50		359.50	420	

07320 | Roof Tiles

			CREW	DAILY OUTPUT	LABOR-HOURS	UNIT	MAT.	LABOR	EQUIP.	TOTAL	TOTAL INCL O&P	
100	0010	**ALUMINUM** Tiles with accessories, .032" thick, mission tile	1 Carp	2.50	3.200	Sq.	355	63		418	500	100
	0200	Spanish tiles	"	3	2.667	"	355	52.50		407.50	480	
200	0010	**CLAY TILE** ASTM C1167, GR 1, severe weathering, acces. incl.										200
	0202	Lanai tile or Classic tile, 158 pc per sq	1 Rofc	1.65	4.848	Sq.	490	82.50		572.50	695	
	0302	Americana, 158 pc per sq, most colors		1.65	4.848		490	82.50		572.50	695	
	0351	Green, gray or brown		1.65	4.848		490	82.50		572.50	695	
	0401	Blue		1.65	4.848		490	82.50		572.50	695	
	0602	Spanish tile, 171 pc per sq, red		1.80	4.444		365	76		441	540	
	0802	Buff, green, gray, brown		1.80	4.444		420	76		496	600	
	0902	Glazed white		1.80	4.444		500	76		576	690	
	1101	Mission tile, 166 pc per sq, scored finish, red		1.15	6.957		515	119		634	785	
	1701	French tile, 133 pc per sq, smooth finish, red		1.35	5.926		495	101		596	735	
	1751	Blue or green		1.35	5.926		685	101		786	945	
	1801	Norman tile, 317 pc per sq		1	8		805	136		941	1,150	
	2201	Williamsburg tile, 158 pc per sq, aged cedar		1.35	5.926		465	101		566	705	
	2251	Gray or green		1.35	5.926		490	101		591	730	
	2510	One piece mission tile, natural red, 75 pc per square	1 Rots	1.65	4.848		133	83		216	300	
	2530	Mission Tile, 134 pc per square	"	1.15	6.957		178	119		297	415	
300	0010	**CONCRETE TILE** Including installation of accessories										300
	0020	Corrugated, 13" x 16-1/2", 90 per sq, 950 lb per sq										
	0051	Earthtone colors, nailed to wood deck	1 Rofc	1.35	5.926	Sq.	103	101		204	300	
	0152	Blues		1.35	5.926		114	101		215	315	
	0202	Greens		1.35	5.926		114	101		215	315	
	0500	Shakes, 13" x 16-1/2", 90 per sq, 950 lb per sq										

07300 | Shingles, Roof Tiles & Roof Coverings

07320 | Roof Tiles

			CREW	DAILY OUTPUT	LABOR-HOURS	UNIT	2000 BARE COSTS				TOTAL INCL O&P	
							MAT.	LABOR	EQUIP.	TOTAL		
300	0601	All colors, nailed to wood deck	1 Rofc	1.50	5.333	Sq.	185	91		276	375	300
	1500	Accessory pieces, ridge & hip, 10" x 16-1/2", 8 lbs. each				Ea.	2.25			2.25	2.48	
	1700	Rake, 6-1/2" x 16-3/4", 9 lbs. each					2.25			2.25	2.48	
	1800	Mansard hip, 10" x 16-1/2", 9.2 lbs. each					2.25			2.25	2.48	
	1900	Hip starter, 10" x 16-1/2", 10.5 lbs. each					9.50			9.50	10.45	
	2000	3 or 4 way apex, 10" each side, 11.5 lbs. each					10.25			10.25	11.30	

07400 | Roofing & Siding Panels

07410 | Metal Roof & Wall Panels

			CREW	DAILY OUTPUT	LABOR-HOURS	UNIT	2000 BARE COSTS				TOTAL INCL O&P	
							MAT.	LABOR	EQUIP.	TOTAL		
100	0010	ALUMINUM ROOFING Corrugated or ribbed, .0155" thick, natural	G-3	1,200	.027	S.F.	.61	.48		1.09	1.48	100
	0300	Painted		1,200	.027		.87	.48		1.35	1.77	
	0400	Corrugated, .018" thick, on steel frame, natural finish		1,200	.027		.78	.48		1.26	1.67	
	0600	Painted		1,200	.027		.97	.48		1.45	1.88	
	0700	Corrugated, on steel frame, natural, .024" thick		1,200	.027		1.14	.48		1.62	2.06	
	0800	Painted, .024" thick		1,200	.027		1.37	.48		1.85	2.32	
	0900	.032" thick, natural		1,200	.027		1.21	.48		1.69	2.14	
	1200	painted		1,200	.027		1.60	.48		2.08	2.57	
	1300	V-Beam, on steel frame construction, .032" thick, natural		1,200	.027		1.37	.48		1.85	2.32	
	1500	Painted		1,200	.027		1.75	.48		2.23	2.74	
	1600	.040" thick, natural		1,200	.027		1.70	.48		2.18	2.68	
	1800	Painted		1,200	.027		2.10	.48		2.58	3.12	
	1900	.050" thick, natural		1,200	.027		2.07	.48		2.55	3.09	
	2100	Painted		1,200	.027		2.52	.48		3	3.58	
	2200	For roofing on wood frame, deduct		4,600	.007		.05	.13		.18	.27	
	2400	Ridge cap, .032" thick, natural		800	.040	L.F.	1.53	.72		2.25	2.90	
500	0010	MANSARD Colored aluminum, with battens, .032" thick										500
	0600	Stock units, straight surfaces	1 Shee	115	.070	S.F.	2.13	1.50		3.63	4.87	
	0700	Concave or convex surfaces		75	.107	"	2.35	2.29		4.64	6.45	
	0800	For framing, to 5' high, add		115	.070	L.F.	2.35	1.50		3.85	5.10	
	0900	Soffits, to 1' wide		125	.064	S.F.	1.16	1.38		2.54	3.60	
690	0010	METAL FACING PANELS										690
	0400	Textured aluminum, 4' x 8' x 5/16" plywood backing, single face	2 Shee	375	.043	S.F.	2.18	.92		3.10	3.95	
	0600	Double face		375	.043		3.28	.92		4.20	5.15	
	0700	4' x 10' x 5/16" plywood backing, single face		375	.043		2.08	.92		3	3.84	
	0900	Double face		375	.043		3.15	.92		4.07	5	
	1000	4' x 12' x 5/15" plywood backing, single face		375	.043		2.99	.92		3.91	4.84	
	1300	Smooth al, 1/4" panel, fluoropolymer finish, double face		375	.043		5.35	.92		6.27	7.45	
	1350	Clear anodized finish, double face		375	.043		5.95	.92		6.87	8.10	
	1400	Double face textured aluminum, structural panel, 1" EPS insulation		375	.043		3.60	.92		4.52	5.50	
	1500	Accessories, outside corner	1 Shee	175	.046	L.F.	1.54	.98		2.52	3.35	
	1600	Inside corner		175	.046		1.04	.98		2.02	2.80	
	1800	Batten mounting clip		200	.040		.35	.86		1.21	1.84	
	1900	Low profile batten		480	.017		.72	.36		1.08	1.39	
	2100	High profile batten		480	.017		1.21	.36		1.57	1.93	
	2200	Water table		200	.040		1.28	.86		2.14	2.86	
	2400	Horizontal joint connector		200	.040		1.05	.86		1.91	2.60	

07400 | Roofing & Siding Panels

07410 | Metal Roof & Wall Panels

			CREW	DAILY OUTPUT	LABOR-HOURS	UNIT	MAT.	LABOR	EQUIP.	TOTAL	TOTAL INCL O&P	
690	2500	Corner cap	1 Shee	200	.040	L.F.	1.28	.86		2.14	2.86	690
	2700	H - moulding	↓	480	.017	↓	.68	.36		1.04	1.35	
700	0010	STEEL ROOFING on steel frame, corrugated or ribbed, 30 ga galv	G-3	1,100	.029	S.F.	.75	.52		1.27	1.72	700
	0100	28 ga		1,050	.030		.79	.55		1.34	1.80	
	0300	26 ga		1,000	.032		.86	.58		1.44	1.93	
	0400	24 ga		950	.034		1.02	.61		1.63	2.15	
	0600	Colored, 28 ga		1,050	.030		1.05	.55		1.60	2.08	
	0700	26 ga		1,000	.032		1.12	.58		1.70	2.21	
	0710	Flat profile, 1-3/4" standing seams, 10" wide, standard finish, 26 ga		1,000	.032		2.30	.58		2.88	3.51	
	0715	24 ga		950	.034		2.67	.61		3.28	3.97	
	0720	22 ga		900	.036		3.29	.64		3.93	4.71	
	0725	Zinc aluminum alloy finish, 26 ga		1,000	.032		1.80	.58		2.38	2.96	
	0730	24 ga		950	.034		2.15	.61		2.76	3.40	
	0735	22 ga		900	.036		2.47	.64		3.11	3.81	
	0740	12" wide, standard finish, 26 ga		1,000	.032		2.29	.58		2.87	3.50	
	0745	24 ga		950	.034		2.66	.61		3.27	3.96	
	0750	Zinc aluminum alloy finish, 26 ga		1,000	.032		1.70	.58		2.28	2.85	
	0755	24 ga		950	.034		2.03	.61		2.64	3.26	
	0840	Flat profile, 1" x 3/8" batten, 12" wide, standard finish, 26 ga		1,000	.032		1.79	.58		2.37	2.95	
	0845	24 ga		950	.034		2.10	.61		2.71	3.34	
	0850	22 ga		900	.036		2.52	.64		3.16	3.86	
	0855	Zinc aluminum alloy finish, 26 ga		1,000	.032		1.51	.58		2.09	2.64	
	0860	24 ga		950	.034		1.68	.61		2.29	2.88	
	0865	22 ga		900	.036		1.95	.64		2.59	3.24	
	0870	16-1/2" wide, standard finish, 24 ga		950	.034		2.13	.61		2.74	3.37	
	0875	22 ga		900	.036		2.39	.64		3.03	3.72	
	0880	Zinc aluminum alloy finish, 24 ga		950	.034		1.60	.61		2.21	2.79	
	0885	22 ga		900	.036		1.78	.64		2.42	3.05	
	0890	Flat profile, 2" x 2" batten, 12" wide, standard finish, 26 ga		1,000	.032		2.05	.58		2.63	3.24	
	0895	24 ga		950	.034		2.45	.61		3.06	3.73	
	0900	22 ga		900	.036		3	.64		3.64	4.39	
	0905	Zinc aluminum alloy finish, 26 ga		1,000	.032		1.69	.58		2.27	2.84	
	0910	24 ga		950	.034		1.83	.61		2.44	3.04	
	0915	22 ga		900	.036		2.25	.64		2.89	3.57	
	0920	16-1/2" wide, standard finish, 24 ga		950	.034		2.27	.61		2.88	3.53	
	0925	22 ga		900	.036		2.64	.64		3.28	3.99	
	0930	Zinc aluminum alloy finish, 24 ga		950	.034		1.80	.61		2.41	3.01	
	0935	22 ga		900	.036		2.05	.64		2.69	3.35	
	1200	Ridge, galvanized, 10" wide		800	.040	L.F.	1.68	.72		2.40	3.07	
	1210	20" wide	↓	750	.043	"	2.95	.77		3.72	4.55	

07420 | Plastic Roof & Wall Panels

			CREW	DAILY OUTPUT	LABOR-HOURS	UNIT	MAT.	LABOR	EQUIP.	TOTAL	TOTAL INCL O&P	
770	0010	FIBERGLASS Corrugated panels, roofing, 8 oz per SF	G-3	1,000	.032	S.F.	2.17	.58		2.75	3.37	770
	0100	12 oz per SF		1,000	.032		3.76	.58		4.34	5.10	
	0300	Corrugated siding, 6 oz per SF		880	.036		1.88	.65		2.53	3.18	
	0400	8 oz per SF		880	.036		2.17	.65		2.82	3.50	
	0500	Fire retardant		880	.036		3	.65		3.65	4.41	
	0600	12 oz. siding, textured		880	.036		3.68	.65		4.33	5.15	
	0700	Fire retardant		880	.036		4.02	.65		4.67	5.55	
	0900	Flat panels, 6 oz per SF, clear or colors		880	.036		1.68	.65		2.33	2.96	
	1100	Fire retardant, class A		880	.036		2.91	.65		3.56	4.31	
	1300	8 oz per SF, clear or colors		880	.036		2.17	.65		2.82	3.50	
	1700	Sandwich panels, fiberglass, 1-9/16" thick, panels to 20 SF		180	.178		21	3.20		24.20	28.50	
	1900	As above, but 2-3/4" thick, panels to 100 SF	↓	265	.121	↓	15.60	2.17		17.77	21	

07400 | Roofing & Siding Panels

07440 | Faced Panels

		CREW	DAILY OUTPUT	LABOR-HOURS	UNIT	MAT.	LABOR	EQUIP.	TOTAL	TOTAL INCL O&P
0010	**EXPOSED AGGREGATE PANELS**									
1400	Fiberglass polymer back-up,									
1500	Small size aggregate	F-3	445	.090	S.F.	4.62	1.61	1.02	7.25	8.95
1600	Medium size aggregate		445	.090		5.20	1.61	1.02	7.83	9.55
1700	Large size aggregate	↓	445	.090	↓	5.35	1.61	1.02	7.98	9.75

07460 | Siding

		CREW	DAILY OUTPUT	LABOR-HOURS	UNIT	MAT.	LABOR	EQUIP.	TOTAL	TOTAL INCL O&P
0010	**ALUMINUM SIDING** .019" thick, on steel construction, natural	G-3	775	.041	S.F.	.65	.74		1.39	1.97
0100	Painted		775	.041		.84	.74		1.58	2.18
0400	Farm type, .021" thick on steel frame, natural		775	.041		.75	.74		1.49	2.09
0600	Painted		775	.041		.92	.74		1.66	2.27
0700	Industrial type, corrugated, on steel, .024" thick, mill		775	.041		.95	.74		1.69	2.31
0900	Painted		775	.041		1.15	.74		1.89	2.53
1000	.032" thick, mill		775	.041		1.25	.74		1.99	2.64
1200	Painted		775	.041		1.60	.74		2.34	3.02
1300	V-Beam, on steel frame, .032" thick, mill		775	.041		1.47	.74		2.21	2.88
1500	Painted		775	.041		1.75	.74		2.49	3.19
1600	.040" thick, mill		775	.041		1.78	.74		2.52	3.22
1800	Painted		775	.041		2.10	.74		2.84	3.57
1900	.050" thick, mill		775	.041		2.11	.74		2.85	3.58
2100	Painted		775	.041		2.52	.74		3.26	4.03
2200	Ribbed, 3" profile, on steel frame, .032" thick, natural		775	.041		1.35	.74		2.09	2.75
2400	Painted		775	.041		1.61	.74		2.35	3.03
2500	.040" thick, natural		775	.041		1.60	.74		2.34	3.02
2700	Painted		775	.041		1.87	.74		2.61	3.32
2750	.050" thick, natural		775	.041		1.84	.74		2.58	3.28
2760	Painted		775	.041		2.10	.74		2.84	3.57
3300	For siding on wood frame, deduct from above	↓	2,800	.011	↓	.06	.21		.27	.42
3400	Screw fasteners, aluminum, self tapping, neoprene washer, 1"				M	123			123	135
3600	Stitch screws, self tapping, with neoprene washer, 5/8"				"	92.50			92.50	101
3630	Flashing, sidewall, .032" thick	G-3	800	.040	L.F.	1.75	.72		2.47	3.15
3650	End wall, .040" thick		800	.040		2.05	.72		2.77	3.48
3670	Closure strips, corrugated, .032" thick		800	.040		.52	.72		1.24	1.79
3680	Ribbed, 4" or 8", .032" thick		800	.040		.52	.72		1.24	1.79
3690	V-beam, .040" thick	↓	800	.040	↓	.71	.72		1.43	2
3800	Horizontal, colored clapboard, 8" wide, plain	2 Carp	515	.031	S.F.	1.23	.61		1.84	2.40
3900	Insulated		515	.031		1.49	.61		2.10	2.69
4000	Vertical board & batten, colored, non-insulated	↓	515	.031		1.06	.61		1.67	2.22
4200	For simulated wood design, add				↓	.08			.08	.09
4300	Corners for above, outside	2 Carp	515	.031	V.L.F.	1.67	.61		2.28	2.89
4500	Inside corners	"	515	.031	"	.94	.61		1.55	2.08
4600	Sandwich panels, 1" insulation, single story	G-3	395	.081	S.F.	4.52	1.46		5.98	7.45
4900	Multi-story	"	345	.093		6.35	1.67		8.02	9.80
5100	For baked enamel finish 1 side, add				↓	.24			.24	.26
5200	See also Metal Facing Panels, division 07410-690									
0010	**FASCIA** Aluminum, reverse board and batten,									
0100	.032" thick, colored, no furring included	1 Shee	145	.055	S.F.	2.10	1.19		3.29	4.31
0300	Steel, galv and enameled, stock, no furring, long panels		145	.055		2.19	1.19		3.38	4.41
0600	Short panels	↓	115	.070	↓	3.31	1.50		4.81	6.15
0010	**FIBER CEMENT SIDING**									
0020	Lap siding, 5/16" thick, 6" wide, smooth texture	2 Carp	415	.039	S.F.	.88	.76		1.64	2.27
0025	Woodgrain texture		415	.039		.88	.76		1.64	2.27
0030	7-1/2" wide, smooth texture		425	.038		.88	.74		1.62	2.24
0035	Woodgrain texture		425	.038		.88	.74		1.62	2.24
0040	8" wide, smooth texture	↓	425	.038	↓	.87	.74		1.61	2.23

THERMAL & MOISTURE PROTECTION 7

07400 | Roofing & Siding Panels

07460 | Siding

			CREW	DAILY OUTPUT	LABOR-HOURS	UNIT	2000 BARE COSTS MAT.	LABOR	EQUIP.	TOTAL	TOTAL INCL O&P	
500	0045	Roughsawn texture	2 Carp	425	.038	S.F.	.87	.74		1.61	2.23	500
	0050	9-1/2" wide, smooth texture		440	.036		.84	.72		1.56	2.16	
	0055	Woodgrain texture		440	.036		.84	.72		1.56	2.16	
	0060	12" wide, smooth texture		455	.035		.81	.69		1.50	2.09	
	0065	Woodgrain texture		455	.035		.81	.69		1.50	2.09	
	0070	Panel siding, 5/16" thick, smooth texture		750	.021		.73	.42		1.15	1.52	
	0075	Stucco texture		750	.021		.73	.42		1.15	1.52	
	0080	Grooved woodgrain texture		750	.021		.73	.42		1.15	1.52	
	0085	V - grooved woodgrain texture		750	.021		.73	.42		1.15	1.52	
	0090	Wood starter strip		400	.040	L.F.	.20	.79		.99	1.57	
600	0010	**VINYL SIDING** Solid PVC panels, 8" to 10" wide, plain	1 Carp	255	.031	S.F.	.62	.62		1.24	1.74	600
	0100	with 3/8" insulation		255	.031		.76	.62		1.38	1.90	
	0200	Soffit and fascia		205	.039		1.40	.77		2.17	2.86	
	0300	Window and door trim moldings		185	.043	L.F.	.31	.85		1.16	1.80	
	0500	Corner posts, outside corner		205	.039		1.10	.77		1.87	2.53	
	0600	Inside corner		205	.039		.58	.77		1.35	1.96	
750	0010	**SOFFIT** Aluminum, residential, stock units, .020" thick	1 Carp	210	.038	S.F.	.99	.75		1.74	2.38	750
	0100	Baked enamel on steel, 16 or 18 gauge		105	.076		3.74	1.50		5.24	6.70	
	0300	Polyvinyl chloride, white, solid		230	.035		.65	.69		1.34	1.88	
	0400	Perforated		230	.035		.65	.69		1.34	1.88	
	0500	For colors, add					.06			.06	.07	
800	0010	**STEEL SIDING**, Beveled, vinyl coated, 8" wide, including fasteners	1 Carp	265	.030	S.F.	1.12	.59		1.71	2.25	800
	0050	10" wide	"	275	.029		1.19	.57		1.76	2.29	
	0080	Galv, corrugated or ribbed, on steel frame, 30 gauge	G-3	800	.040		.75	.72		1.47	2.05	
	0100	28 gauge		795	.040		.79	.72		1.51	2.10	
	0300	26 gauge		790	.041		.86	.73		1.59	2.19	
	0400	24 gauge		785	.041		1.03	.73		1.76	2.37	
	0600	22 gauge		770	.042		1.17	.75		1.92	2.56	
	0700	Colored, corrugated/ribbed, on steel frame, 10 yr fnsh, 28 ga.		800	.040		1.08	.72		1.80	2.41	
	0900	26 gauge		795	.040		.94	.72		1.66	2.26	
	1000	24 gauge		790	.041		1.11	.73		1.84	2.46	
	1020	20 gauge		785	.041		1.37	.73		2.10	2.75	
	1200	Factory sandwich panel, 26 ga., 1" insulation, galvanized		380	.084		2.41	1.51		3.92	5.20	
	1300	Colored 1 side		380	.084		3.50	1.51		5.01	6.40	
	1500	Galvanized 2 sides		380	.084		4.35	1.51		5.86	7.35	
	1600	Colored 2 sides		380	.084		4.50	1.51		6.01	7.50	
	1800	Acrylic paint face, regular paint liner		380	.084		3.40	1.51		4.91	6.30	
	1900	For 2" thick polystyrene, add					.52			.52	.57	
	2000	22 ga, galv, 2" insulation, baked enamel exterior	G-3	360	.089		6.65	1.60		8.25	10	
	2100	P.V.F. exterior finish	"	360	.089		7	1.60		8.60	10.40	
900	0010	**WOOD SIDING, BOARDS**										900
	3201	Wood, cedar bevel, short lengths, A grade, 1/2" x 6"	2 Carp	500	.032	S.F.	1.90	.63		2.53	3.17	
	3301	1/2" x 8"		550	.029		1.56	.57		2.13	2.70	
	3501	3/4" x 10", clear grade, 3' to 16'		600	.027		2.68	.53		3.21	3.85	
	3601	"B" grade		600	.027		2.89	.53		3.42	4.08	
	3801	Cedar, rough sawn, 1" x 4", A grade/ natural		480	.033		2.75	.66		3.41	4.16	
	3901	Stained		480	.033		3.10	.66		3.76	4.54	
	4101	1" x 12", board & batten, #3 & Btr., natural		520	.031		2.08	.61		2.69	3.33	
	4201	Stained		520	.031		2.42	.61		3.03	3.70	
	4401	1" x 8" channel siding, #3 & Btr., natural		500	.032		1.73	.63		2.36	2.98	
	4501	Stained		500	.032		2.30	.63		2.93	3.61	
	4701	Redwood, clear, beveled, vertical grain, 1/2" x 4"		400	.040		3.22	.79		4.01	4.89	
	4801	1/2" x 8"		500	.032		2.19	.63		2.82	3.49	
	5001	3/4" x 10"		600	.027		3.57	.53		4.10	4.83	

Important: See the Reference Section for critical supporting data - Reference Nos., Crews, & City Cost Indexes

07400 | Roofing & Siding Panels

07460 | Siding

			CREW	DAILY OUTPUT	LABOR-HOURS	UNIT	2000 BARE COSTS MAT.	LABOR	EQUIP.	TOTAL	TOTAL INCL O&P	
900	5201	Channel siding, 1" x 10", B grade	2 Carp	570	.028	S.F.	2.30	.55		2.85	3.48	900
	5250	Redwood, T&G boards, B grade, 1" x 4"		300	.053		2.74	1.05		3.79	4.81	
	5270	1" x 8"		375	.043		2.36	.84		3.20	4.04	
	5402	White pine, rough sawn, 1" x 8", natural		550	.029		.69	.57		1.26	1.74	
	5501	Stained		550	.029		1.02	.57		1.59	2.10	
950	0010	**WOOD PRODUCT SIDING**										950
	0030	Lap siding, hardboard, 7/16" x 8", primed										
	0050	Wood grain texture finish	2 Carp	650	.025	S.F.	1.04	.48		1.52	1.97	
	0100	Panels, 7/16" thick, smooth, textured or grooved, primed		700	.023		.74	.45		1.19	1.58	
	0200	Stained		700	.023		.88	.45		1.33	1.74	
	0700	Particle board, overlaid, 3/8" thick		750	.021		.63	.42		1.05	1.41	
	0900	Plywood, medium density overlaid, 3/8" thick		750	.021		1.02	.42		1.44	1.84	
	1000	1/2" thick		700	.023		1.19	.45		1.64	2.08	
	1100	3/4" thick		650	.025		1.56	.48		2.04	2.55	
	1600	Texture 1-11, cedar, 5/8" thick, natural		675	.024		1.09	.47		1.56	2	
	1700	Factory stained		675	.024		1.74	.47		2.21	2.71	
	1900	Texture 1-11, fir, 5/8" thick, natural		675	.024		1.01	.47		1.48	1.91	
	2000	Factory stained		675	.024		1.14	.47		1.61	2.05	
	2050	Texture 1-11, S.Y.P., 5/8" thick, natural		675	.024		.85	.47		1.32	1.74	
	2100	Factory stained		675	.024		.93	.47		1.40	1.82	
	2200	Rough sawn cedar, 3/8" thick, natural		675	.024		1.14	.47		1.61	2.05	
	2300	Factory stained		675	.024		1.26	.47		1.73	2.19	
	2500	Rough sawn fir, 3/8" thick, natural		675	.024		.61	.47		1.08	1.47	
	2600	Factory stained		675	.024		.68	.47		1.15	1.55	
	2800	Redwood, textured siding, 5/8" thick		675	.024		1.89	.47		2.36	2.88	
	3000	Polyvinyl chloride coated, 3/8" thick		750	.021		.93	.42		1.35	1.74	

07500 | Membrane Roofing

07510 | Built-Up Bituminous Roofing

			CREW	DAILY OUTPUT	LABOR-HOURS	UNIT	2000 BARE COSTS MAT.	LABOR	EQUIP.	TOTAL	TOTAL INCL O&P	
050	0010	**ASPHALT** Coated felt, #30, 2 sq per roll, not mopped	1 Rofc	58	.138	Sq.	5.45	2.35		7.80	10.40	050
	0200	#15, 4 sq per roll, plain or perforated, not mopped		58	.138		2.59	2.35		4.94	7.25	
	0300	Roll roofing, smooth, #65		15	.533		12	9.10		21.10	30	
	0500	#90		15	.533		13.65	9.10		22.75	32	
	0520	Mineralized		15	.533		14.75	9.10		23.85	33	
	0540	D.C. (Double coverage), 19" selvage edge		10	.800		27.50	13.65		41.15	55.50	
	0580	Adhesive (lap cement)				Gal.	3.68			3.68	4.05	
	0600	Steep, flat or dead level asphalt, 10 ton lots, bulk				Ton	260			260	286	
	0800	Packaged				"	260			260	286	
300	0010	**BUILT-UP ROOFING**	R07510 -020									300
	0120	Asphalt flood coat with gravel/slag surfacing, not including										
	0140	Insulation, flashing or wood nailers										
	0200	Asphalt base sheet, 3 plies #15 asphalt felt, mopped	G-1	22	2.545	Sq.	35.50	41	19	95.50	137	
	0350	On nailable decks		21	2.667		38.50	43	19.90	101.40	145	
	0500	4 plies #15 asphalt felt, mopped		20	2.800		49.50	45	21	115.50	161	
	0550	On nailable decks		19	2.947		45	47.50	22	114.50	162	
	0700	Coated glass base sheet, 2 plies glass (type IV), mopped		22	2.545		37.50	41	19	97.50	139	

07500 | Membrane Roofing

07510 | Built-Up Bituminous Roofing

			CREW	DAILY OUTPUT	LABOR-HOURS	UNIT	2000 BARE COSTS				TOTAL INCL O&P
							MAT.	LABOR	EQUIP.	TOTAL	
300	0850	3 plies glass, mopped	G-1	20	2.800	Sq.	44	45	21	110	156
	0950	On nailable decks		19	2.947		42	47.50	22	111.50	159
	1100	4 plies glass fiber felt (type IV), mopped		20	2.800		53	45	21	119	166
	1150	On nailable decks		19	2.947		48.50	47.50	22	118	166
	1200	Coated & saturated base sheet, 3 plies #15 asph. felt, mopped		20	2.800		42	45	21	108	153
	1250	On nailable decks		19	2.947		39.50	47.50	22	109	156
	1300	4 plies #15 asphalt felt, mopped		22	2.545		48	41	19	108	151
	2000	Asphalt flood coat, smooth surface									
	2200	Asphalt base sheet & 3 plies #15 asphalt felt, mopped	G-1	24	2.333	Sq.	35.50	37.50	17.45	90.45	128
	2400	On nailable decks		23	2.435		33.50	39	18.20	90.70	130
	2600	4 plies #15 asphalt felt, mopped		24	2.333		41.50	37.50	17.45	96.45	135
	2700	On nailable decks		23	2.435		39.50	39	18.20	96.70	137
	2900	Coated glass fiber base sheet, mopped, and 2 plies of									
	2910	glass fiber felt (type IV)	G-1	25	2.240	Sq.	32	36	16.75	84.75	121
	3100	On nailable decks		24	2.333		30.50	37.50	17.45	85.45	123
	3200	3 plies, mopped		23	2.435		39	39	18.20	96.20	136
	3300	On nailable decks		22	2.545		36.50	41	19	96.50	138
	3800	4 plies glass fiber felt (type IV), mopped		23	2.435		45.50	39	18.20	102.70	143
	3900	On nailable decks		22	2.545		43.50	41	19	103.50	145
	4000	Coated & saturated base sheet, 3 plies #15 asph. felt, mopped		24	2.333		36.50	37.50	17.45	91.45	129
	4200	On nailable decks		23	2.435		34.50	39	18.20	91.70	131
	4300	4 plies #15 organic felt, mopped		22	2.545		42.50	41	19	102.50	145
	4500	Coal tar pitch with gravel/slag surfacing									
	4600	4 plies #15 tarred felt, mopped	G-1	21	2.667	Sq.	101	43	19.90	163.90	213
	4800	3 plies glass fiber felt (type IV), mopped	"	19	2.947	"	83	47.50	22	152.50	204
	5000	Coated glass fiber base sheet, and 2 plies of									
	5010	glass fiber felt, (type IV), mopped	G-1	19	2.947	Sq.	82.50	47.50	22	152	204
	5300	On nailable decks		18	3.111		73	50	23	146	200
	5600	4 plies glass fiber felt (type IV), mopped		21	2.667		113	43	19.90	175.90	226
	5800	On nailable decks		20	2.800		103	45	21	169	221
400	0010	**CANTS** 4" x 4", treated timber, cut diagonally	1 Rofc	325	.025	L.F.	.80	.42		1.22	1.66
	0100	Foamglass		325	.025		1.92	.42		2.34	2.89
	0300	Mineral or fiber, trapezoidal, 1" x 4" x 48"		325	.025		.17	.42		.59	.97
	0400	1-1/2" x 5-5/8" x 48"		325	.025		.29	.42		.71	1.10
700	0010	**FELT** Glass fibered, #15, no mopping	1 Rofc	58	.138	Sq.	3	2.35		5.35	7.70
	0300	Base sheet, #45, channel vented		58	.138		16.50	2.35		18.85	22.50
	0400	#50, coated		58	.138		7.65	2.35		10	12.80
	0500	Cap, mineral surfaced		58	.138		16.50	2.35		18.85	22.50
	0600	Flashing membrane, #65		16	.500		21	8.50		29.50	39.50
	0800	Coal tar fibered, #15, no mopping		58	.138		7.90	2.35		10.25	13.10
	0900	Asphalt felt, #15, 4 sq per roll, no mopping		58	.138		2.59	2.35		4.94	7.25
	1100	#30, 2 sq per roll		58	.138		5.45	2.35		7.80	10.40
	1200	Double coated, #33		58	.138		5.90	2.35		8.25	10.90
	1400	#40, base sheet		58	.138		5.50	2.35		7.85	10.45
	1450	Coated and saturated		58	.138		6.45	2.35		8.80	11.50
	1500	Tarred felt, organic, #15, 4 sq rolls		58	.138		7.80	2.35		10.15	13
	1550	#30, 2 sq roll		58	.138		15.60	2.35		17.95	21.50
	1700	Add for mopping above felts, per ply, asphalt, 24 lb per sq	G-1	192	.292		3.12	4.70	2.18	10	14.55
	1800	Coal tar mopping, 30 lb per sq		186	.301		8.45	4.85	2.25	15.55	21
	1900	Flood coat, with asphalt, 60 lb per sq		60	.933		7.80	15.05	6.95	29.80	44.50
	2000	With coal tar, 75 lb per sq		56	1		21	16.10	7.45	44.55	61.50

07520 | Cold Applied Bituminous Roofing

200	0010	**COLD APPLIED** 3-ply system (components listed below)	G-5	50	.800	Sq.		12.60	3.43	16.03	27.50
	0100	Spunbond poly. fabric, 1.35 oz/SY, 36"W, 10.8 Sq/roll				Ea.	123			123	135

07500 | Membrane Roofing

07520 | Cold Applied Bituminous Roofing

			CREW	DAILY OUTPUT	LABOR-HOURS	UNIT	MAT.	LABOR	EQUIP.	TOTAL	TOTAL INCL O&P	
200	0200	49" wide, 14.6 Sq./roll				Ea.	169			169	186	200
	0300	2.10 oz./S.Y., 36" wide, 10.8 Sq./roll					185			185	203	
	0400	49" wide, 14.6 Sq./roll				▼	250			250	275	
	0500	Base & finish coat, 3 gal./Sq., 5 gal./can				Gal.	3.25			3.25	3.58	
	0600	Coating, ceramic granules, 1/2 Sq./bag				Ea.	11.50			11.50	12.65	
	0700	Aluminum, 2 gal./Sq.				Gal.	9.25			9.25	10.20	
	0800	Emulsion, fibered or non-fibered, 4 gal./Sq.				"	4.25			4.25	4.68	

07530 | Elastomeric Membrane Roofing

			CREW	DAILY OUTPUT	LABOR-HOURS	UNIT	MAT.	LABOR	EQUIP.	TOTAL	TOTAL INCL O&P	
350	0010	**ELASTOMERIC ROOFING**										350
	0100	For Elastomeric waterproofing, see division 07130-200										
	0110	Acrylic rubber, fluid applied, 20 mils thick	G-5	2,000	.020	S.F.	1.80	.31	.09	2.20	2.66	
	0120	50 mils, reinforced		1,200	.033		2.80	.52	.14	3.46	4.22	
	0130	For walking surface, add	▼	900	.044		.85	.70	.19	1.74	2.45	
	0300	Hypalon neoprene, fluid applied, 20 mil thick, not-reinforced	G-1	1,135	.049		2.05	.79	.37	3.21	4.15	
	0600	Non-woven polyester, reinforced		960	.058		2.07	.94	.44	3.45	4.51	
	0700	5 coat neoprene deck, 60 mil thick, under 10,000 SF		325	.172		4.56	2.78	1.29	8.63	11.55	
	0900	Over 10,000 SF		625	.090		4.25	1.44	.67	6.36	8.10	
	1300	Vinyl plastic traffic deck, sprayed, 2 to 4 mils thick		625	.090		1.33	1.44	.67	3.44	4.89	
	1500	Vinyl and neoprene membrane traffic deck	▼	1,550	.036	▼	1.41	.58	.27	2.26	2.93	
	1600	Polyurethane spray-on with 20 mil silicone rubber coating applied										
	1700	1" thick, R7, minimum	G-2	875	.027	S.F.	1.34	.45	.29	2.08	2.53	
	1800	Maximim		805	.030		1.80	.48	.31	2.59	3.14	
	1900	2" thick, R14, minimum		685	.035		1.90	.57	.37	2.84	3.45	
	2000	Maximum		575	.042		2.41	.68	.43	3.52	4.27	
	2100	3" thick, R21, minimum		500	.048		2.50	.78	.50	3.78	4.61	
	2200	Maximum	▼	440	.055	▼	3.20	.89	.57	4.66	5.65	
800	0010	**SINGLE-PLY MEMBRANE**										800
	0800	Chlorosulfonated polyethylene-hypalon (CSPE), 45 mils,										
	0900	0.29 P.S.F., fully adhered	G-5	26	1.538	Sq.	122	24	6.60	152.60	186	
	1100	Loose-laid & ballasted with stone (10 P.S.F.)		51	.784		129	12.35	3.36	144.71	169	
	1200	Partially adhered with fastening strips		35	1.143		124	18	4.90	146.90	175	
	1300	Plates with adhesive attachment	▼	35	1.143	▼	122	18	4.90	144.90	173	
	3500	Ethylene propylene diene monomer (EPDM), 45 mils, 0.28 P.S.F.										
	3600	Loose-laid & ballasted with stone (10 P.S.F.)	G-5	51	.784	Sq.	62.50	12.35	3.36	78.21	95	
	3700	Partially adhered		35	1.143		53.50	18	4.90	76.40	98	
	3800	Fully adhered with adhesive		26	1.538		76	24	6.60	106.60	136	
	4500	60 mils, 0.40 P.S.F.										
	4600	Loose-laid & ballasted with stone (10 P.S.F.)	G-5	51	.784	Sq.	74.50	12.35	3.36	90.21	109	
	4700	Partially adhered		35	1.143		65	18	4.90	87.90	110	
	4800	Fully adhered with adhesive		26	1.538		87	24	6.60	117.60	148	
	4810	45 mil, .28 PSF, membrane only					34			34	37	
	4820	60 mil, .40 PSF, membrane only				▼	44			44	48.50	
	4850	Seam tape for membrane, 4" x 100' roll				Ea.	67			67	73.50	
	4900	Batten strips, 10' sections					2.45			2.45	2.70	
	4910	Cover tape for batten strips, 6" x 100' roll				▼	126			126	139	
	4930	Plate anchors				M	123			123	135	
	4970	Adhesive for fully adhered systems, 60 S.F./gal.				Gal.	13.60			13.60	14.95	
	7500	Polyisobutylene (PIB), 100 mils, 0.57 P.S.F.										
	7600	Loose-laid & ballasted with stone/gravel (10 P.S.F.)	G-5	51	.784	Sq.	123	12.35	3.36	138.71	162	
	7700	Partially adhered with adhesive		35	1.143		154	18	4.90	176.90	208	
	7800	Hot asphalt attachment		35	1.143		147	18	4.90	169.90	201	
	7900	Fully adhered with contact cement	▼	26	1.538	▼	159	24	6.60	189.60	227	
	8200	Polyvinyl chloride (PVC), heat welded seams										
	8700	Reinforced, 48 mils, 0.33 P.S.F.										

07500 | Membrane Roofing

07530 | Elastomeric Membrane Roofing

			CREW	DAILY OUTPUT	LABOR-HOURS	UNIT	MAT.	LABOR	EQUIP.	TOTAL	TOTAL INCL O&P	
800	8750	Loose-laid & ballasted with stone/gravel (12 P.S.F.)	G-5	51	.784	Sq.	93.50	12.35	3.36	109.21	130	800
	8800	Partially adhered with mechanical fasteners		35	1.143		83.50	18	4.90	106.40	131	
	8850	Fully adhered with adhesive		26	1.538		116	24	6.60	146.60	179	
	8860	Reinforced, 60 mils, .40 P.S.F.										
	8870	Loose-laid & ballasted with stone/gravel (12 P.S.F.)	G-5	51	.784	Sq.	89.50	12.35	3.36	105.21	125	
	8880	Partially adhered with mechanical fasteners		35	1.143		80	18	4.90	102.90	127	
	8890	Fully adhered with adhesive		26	1.538		112	24	6.60	142.60	175	

07550 | Modified Bit. Membrane Roofing

			CREW	DAILY OUTPUT	LABOR-HOURS	UNIT	MAT.	LABOR	EQUIP.	TOTAL	TOTAL INCL O&P	
500	0010	**MODIFIED BITUMEN ROOFING**	R07550-030									500
	0020	Base sheet, #15 glass fiber felt, nailed to deck	1 Rofc	58	.138	Sq.	3.73	2.35		6.08	8.50	
	0030	Spot mopped to deck	G-1	295	.190		4.56	3.06	1.42	9.04	12.25	
	0040	Fully mopped to deck	"	192	.292		6.10	4.70	2.18	12.98	17.90	
	0050	#15 organic felt, nailed to deck	1 Rofc	58	.138		3.32	2.35		5.67	8.05	
	0060	Spot mopped to deck	G-1	295	.190		4.15	3.06	1.42	8.63	11.85	
	0070	Fully mopped to deck	"	192	.292		5.70	4.70	2.18	12.58	17.45	
	0080	SBS modified, granule surf cap sheet, polyester rein., mopped										
	0600	150 mils	G-1	2,000	.028	S.F.	.38	.45	.21	1.04	1.49	
	1100	160 mils		2,000	.028		.57	.45	.21	1.23	1.70	
	1500	Glass fiber reinforced, mopped, 160 mils		2,000	.028		.36	.45	.21	1.02	1.47	
	1600	Smooth surface cap sheet, mopped, 145 mils		2,100	.027		.36	.43	.20	.99	1.42	
	1700	Smooth surface flashing, 145 mils		1,260	.044		.36	.72	.33	1.41	2.09	
	1800	150 mils		1,260	.044		.35	.72	.33	1.40	2.08	
	1900	Granular surface flashing, 150 mils		1,260	.044		.38	.72	.33	1.43	2.11	
	2000	160 mils		1,260	.044		.57	.72	.33	1.62	2.32	
	2100	APP mod., smooth surf. cap sheet, poly. reinf., torched, 160 mils	G-5	2,100	.019		.35	.30	.08	.73	1.04	
	2150	170 mils		2,100	.019		.39	.30	.08	.77	1.08	
	2200	Granule surface cap sheet, poly. reinf., torched, 180 mils		2,000	.020		.45	.31	.09	.85	1.18	
	2250	Smooth surface flashing, torched, 160 mils		1,260	.032		.35	.50	.14	.99	1.47	
	2300	170 mils		1,260	.032		.39	.50	.14	1.03	1.51	
	2350	Granule surface flashing, torched, 180 mils		1,260	.032		.45	.50	.14	1.09	1.58	
	2400	Fibrated aluminum coating	1 Rofc	3,800	.002		.09	.04		.13	.17	

07580 | Roll Roofing

			CREW	DAILY OUTPUT	LABOR-HOURS	UNIT	MAT.	LABOR	EQUIP.	TOTAL	TOTAL INCL O&P	
200	0010	**ROLL ROOFING**										200
	0100	Asphalt, mineral surface										
	0200	1 ply #15 organic felt, 1 ply mineral surfaced										
	0300	Selvage roofing, lap 19", nailed & mopped	G-1	27	2.074	Sq.	33.50	33.50	15.50	82.50	116	
	0400	3 plies glass fiber felt (type IV), 1 ply mineral surfaced										
	0500	Selvage roofing, lapped 19", mopped	G-1	25	2.240	Sq.	48.50	36	16.75	101.25	139	
	0600	Coated glass fiber base sheet, 2 plies of glass fiber										
	0700	Felt (type IV), 1 ply mineral surfaced selvage										
	0800	Roofing, lapped 19", mopped	G-1	25	2.240	Sq.	53.50	36	16.75	106.25	144	
	0900	On nailable decks	"	24	2.333	"	50.50	37.50	17.45	105.45	145	
	1000	3 plies glass fiber felt (type III), 1 ply mineral surfaced										
	1100	Selvage roofing, lapped 19", mopped	G-1	25	2.240	Sq.	48.50	36	16.75	101.25	139	

07590 | Roof Maintenance & Repairs

			CREW	DAILY OUTPUT	LABOR-HOURS	UNIT	MAT.	LABOR	EQUIP.	TOTAL	TOTAL INCL O&P	
300	0010	**ROOF COATINGS** Asphalt				Gal.	2.95			2.95	3.25	300
	0200	Asphalt base, fibered aluminum coating					8.90			8.90	9.75	
	0300	Asphalt primer, 5 gallon					3.27			3.27	3.60	
	0600	Coal tar pitch, 200 lb. barrels				Ton	565			565	620	
	0700	Tar roof cement, 5 gal. lots				Gal.	5.75			5.75	6.35	
	0800	Glass fibered roof & patching cement, 5 gallon				"	3.50			3.50	3.85	

07500 | Membrane Roofing

07590 | Roof Maintenance & Repairs

		CREW	DAILY OUTPUT	LABOR-HOURS	UNIT	MAT.	LABOR	EQUIP.	TOTAL	TOTAL INCL O&P	
300	0900 Reinforcing glass membrane, 450 S.F./roll				Ea.	45.50			45.50	50	300
	1000 Neoprene roof coating, 5 gal, 2 gal/sq				Gal.	19.95			19.95	22	
	1100 Roof patch & flashing cement, 5 gallon					17.45			17.45	19.20	
	1200 Roof resaturant, glass fibered, 3 gal/sq					7.30			7.30	8.05	
	1300 Mineral rubber, 3 gal/sq				↓	4.71			4.71	5.20	

07600 | Flashing & Sheet Metal

07610 | Sheet Metal Roofing

		CREW	DAILY OUTPUT	LABOR-HOURS	UNIT	MAT.	LABOR	EQUIP.	TOTAL	TOTAL INCL O&P	
300	0010 **COPPER ROOFING** Batten seam, over 10 sq, 16 oz, 130 lb/sq	1 Shee	1.10	7.273	Sq.	395	156		551	700	300
	0200 18 oz, 145 lb per sq		1	8		440	172		612	775	
	0300 20 oz, 160 lb per sq		1	8		490	172		662	825	
	0400 Standing seam, over 10 squares, 16 oz, 125 lb per sq		1.30	6.154		380	132		512	645	
	0600 18 oz, 140 lb per sq		1.20	6.667		425	143		568	710	
	0700 20 oz, 150 lb per sq		1.10	7.273		460	156		616	770	
	0900 Flat seam, over 10 squares, 16 oz, 115 lb per sq		1.20	6.667		350	143		493	625	
	1000 20 oz, 145 lb per sq	↓	1.10	7.273		440	156		596	750	
	1200 For abnormal conditions or small areas, add					25%	100%				
	1300 For lead-coated copper, add				↓	25%					
500	0010 **LEAD ROOFING** 5 lb. per SF, batten seam	1 Shee	1.20	6.667	Sq.	375	143		518	655	500
	0100 Flat seam	"	1.30	6.154	"	375	132		507	640	
700	0010 **STAINLESS STEEL ROOFING** Type 304, batten seam, 28 gauge	1 Shee	1.20	6.667	Sq.	300	143		443	570	700
	0100 26 gauge	"	1.15	6.957		375	150		525	665	
	0200 For standing seam construction, deduct					2%					
	0500 For flat seam construction, deduct					3%					
	0800 For lead or terne coated stainless, 28 gauge, add					68.50			68.50	75.50	
	0900 For 26 gauge, add				↓	91			91	100	
900	0010 **ZINC** Copper alloy roofing, batten seam, .020" thick	1 Shee	1.20	6.667	Sq.	510	143		653	800	900
	0100 .027" thick		1.15	6.957		615	150		765	935	
	0300 .032" thick		1.10	7.273		695	156		851	1,025	
	0400 .040" thick	↓	1.05	7.619		820	164		984	1,175	
	0600 For standing seam construction, deduct					2%					
	0700 For flat seam construction, deduct				↓	3%					

07650 | Flexible Flashing

		CREW	DAILY OUTPUT	LABOR-HOURS	UNIT	MAT.	LABOR	EQUIP.	TOTAL	TOTAL INCL O&P	
600	0010 **FLASHING** Aluminum, mill finish, .013" thick	1 Shee	145	.055	S.F.	.34	1.19		1.53	2.37	600
	0030 .016" thick		145	.055		.50	1.19		1.69	2.55	
	0060 .019" thick		145	.055		.75	1.19		1.94	2.83	
	0100 .032" thick		145	.055		.99	1.19		2.18	3.09	
	0200 .040" thick		145	.055		1.36	1.19		2.55	3.50	
	0300 .050" thick		145	.055	↓	1.80	1.19		2.99	3.98	
	0325 Mill finish 5" x 7" step flashing, .016" thick		1,920	.004	Ea.	.10	.09		.19	.26	
	0350 Mill finish 12" x 12" step flashing, .016" thick	↓	1,600	.005	"	.40	.11		.51	.62	
	0400 Painted finish, add				S.F.	.24			.24	.26	
	0500 Fabric-backed 2 sides, .004" thick	1 Shee	330	.024		.93	.52		1.45	1.90	
	0700 .005" thick		330	.024		1.09	.52		1.61	2.08	
	0750 Mastic-backed, self adhesive		460	.017		2.22	.37		2.59	3.07	
	0800 Mastic-coated 2 sides, .004" thick		330	.024		.93	.52		1.45	1.90	
	1000 .005" thick	↓	330	.024	↓	1.09	.52		1.61	2.08	

07600 | Flashing & Sheet Metal

07650 | Flexible Flashing

			CREW	DAILY OUTPUT	LABOR-HOURS	UNIT	2000 BARE COSTS				TOTAL INCL O&P
							MAT.	LABOR	EQUIP.	TOTAL	
600	1100	.016" thick	1 Shee	330	.024	S.F.	1.20	.52		1.72	2.20
	1300	Asphalt flashing cement, 5 gallon				Gal.	3.34			3.34	3.67
	1600	Copper, 16 oz, sheets, under 1000 lbs.	1 Shee	115	.070	S.F.	3.05	1.50		4.55	5.90
	1700	Over 4000 lbs.		155	.052		2.85	1.11		3.96	5
	1900	20 oz sheets, under 1000 lbs.		110	.073		3.82	1.56		5.38	6.85
	2000	Over 4000 lbs.		145	.055		3.55	1.19		4.74	5.90
	2200	24 oz sheets, under 1000 lbs.		105	.076		4.60	1.64		6.24	7.80
	2300	Over 4000 lbs.		135	.059		4.25	1.27		5.52	6.85
	2500	32 oz sheets, under 1000 lbs.		100	.080		6.10	1.72		7.82	9.60
	2600	Over 4000 lbs.		130	.062		5.70	1.32		7.02	8.50
	2700	W shape for valleys, 16 oz, 24" wide		100	.080	L.F.	5.90	1.72		7.62	9.40
	2800	Copper, paperbacked 1 side, 2 oz		330	.024	S.F.	.86	.52		1.38	1.83
	2900	3 oz		330	.024		1.12	.52		1.64	2.11
	3100	Paperbacked 2 sides, 2 oz		330	.024		.88	.52		1.40	1.85
	3150	3 oz		330	.024		1.11	.52		1.63	2.10
	3200	5 oz		330	.024		1.70	.52		2.22	2.75
	3250	7 oz		330	.024		2.72	.52		3.24	3.87
	3400	Mastic-backed 2 sides, copper, 2 oz		330	.024		1.02	.52		1.54	2
	3500	3 oz		330	.024		1.25	.52		1.77	2.26
	3700	5 oz		330	.024		1.85	.52		2.37	2.92
	3800	Fabric-backed 2 sides, copper, 2 oz		330	.024		1.09	.52		1.61	2.08
	4000	3 oz		330	.024		1.35	.52		1.87	2.37
	4100	5 oz		330	.024		1.90	.52		2.42	2.97
	4300	Copper-clad stainless steel, .015" thick, under 500 lbs.		115	.070		3.10	1.50		4.60	5.95
	4400	Over 2000 lbs.		155	.052		2.99	1.11		4.10	5.15
	4600	.018" thick, under 500 lbs.		100	.080		4.10	1.72		5.82	7.40
	4700	Over 2000 lbs.		145	.055		3	1.19		4.19	5.30
	4900	Fabric, asphalt-saturated cotton, specification grade	1 Rofc	35	.229	S.Y.	1.93	3.90		5.83	9.35
	5000	Utility grade		35	.229		1.22	3.90		5.12	8.60
	5200	Open-mesh fabric, saturated, 40 oz per S.Y.		35	.229		1.35	3.90		5.25	8.75
	5300	Close-mesh fabric, saturated, 17 oz per S.Y.		35	.229		1.42	3.90		5.32	8.80
	5500	Fiberglass, resin-coated		35	.229		1.16	3.90		5.06	8.55
	5600	Asphalt-coated, 40 oz per S.Y.		35	.229		7.90	3.90		11.80	15.95
	5800	Lead, 2.5 lb. per SF, up to 12" wide		135	.059	S.F.	3.25	1.01		4.26	5.45
	5900	Over 12" wide		135	.059		3.25	1.01		4.26	5.45
	6100	Lead-coated copper, fabric-backed, 2 oz	1 Shee	330	.024		1.41	.52		1.93	2.43
	6200	5 oz		330	.024		1.66	.52		2.18	2.71
	6400	Mastic-backed 2 sides, 2 oz		330	.024		1.10	.52		1.62	2.09
	6500	5 oz		330	.024		1.39	.52		1.91	2.41
	6700	Paperbacked 1 side, 2 oz		330	.024		.95	.52		1.47	1.93
	6800	3 oz		330	.024		1.12	.52		1.64	2.11
	7000	Paperbacked 2 sides, 2 oz		330	.024		.98	.52		1.50	1.96
	7100	5 oz		330	.024		1.58	.52		2.10	2.62
	7300	Polyvinyl chloride, black, .010" thick	1 Rofc	285	.028		.14	.48		.62	1.04
	7400	.020" thick		285	.028		.19	.48		.67	1.10
	7600	.030" thick		285	.028		.29	.48		.77	1.21
	7700	.056" thick		285	.028		.70	.48		1.18	1.66
	7900	Black or white for exposed roofs, .060" thick		285	.028		1.55	.48		2.03	2.59
	8060	PVC tape, 5" x 45 mils, for joint covers, 100 L.F./roll				Ea.	79.50			79.50	87
	8100	Rubber, butyl, 1/32" thick	1 Rofc	285	.028	S.F.	.70	.48		1.18	1.66
	8200	1/16" thick		285	.028		1.05	.48		1.53	2.04
	8300	Neoprene, cured, 1/16" thick		285	.028		1.48	.48		1.96	2.52
	8400	1/8" thick		285	.028		2.99	.48		3.47	4.18
	8500	Shower pan, bituminous membrane, 7 oz	1 Shee	155	.052		1.08	1.11		2.19	3.06
	8550	3 ply copper and fabric, 3 oz		155	.052		1.60	1.11		2.71	3.63
	8600	7 oz		155	.052		3.30	1.11		4.41	5.50

07600 | Flashing & Sheet Metal

07650 | Flexible Flashing

			CREW	DAILY OUTPUT	LABOR-HOURS	UNIT	2000 BARE COSTS MAT.	LABOR	EQUIP.	TOTAL	TOTAL INCL O&P	
600	8650	Copper, 16 oz	1 Shee	100	.080	S.F.	3.05	1.72		4.77	6.25	600
	8700	Lead on copper and fabric, 5 oz		155	.052		1.66	1.11		2.77	3.70	
	8800	7 oz		155	.052		2.87	1.11		3.98	5.05	
	8850	Polyvinyl chloride, .030" thick		160	.050		.30	1.08		1.38	2.14	
	8900	Stainless steel sheets, 32 ga, .010" thick		155	.052		2.16	1.11		3.27	4.25	
	9000	28 ga, .015" thick		155	.052		2.55	1.11		3.66	4.67	
	9100	26 ga, .018" thick		155	.052		3.16	1.11		4.27	5.35	
	9200	24 ga, .025" thick		155	.052		4.10	1.11		5.21	6.40	
	9290	For mechanically keyed flashing, add					40%					
	9300	Stainless steel, paperbacked 2 sides, .005" thick	1 Shee	330	.024	S.F.	1.97	.52		2.49	3.05	
	9400	Terne coated stainless steel, .015" thick, 28 ga		155	.052		3.97	1.11		5.08	6.25	
	9500	.018" thick, 26 ga		155	.052		4.48	1.11		5.59	6.80	
	9600	Zinc and copper alloy (brass), .020" thick		155	.052		3.20	1.11		4.31	5.40	
	9700	.027" thick		155	.052		4.28	1.11		5.39	6.60	
	9800	.032" thick		155	.052		5	1.11		6.11	7.35	
	9900	.040" thick		155	.052		6.10	1.11		7.21	8.55	

07700 | Roof Specialties & Accessories

07710 | Manufactured Roof Specialties

			CREW	DAILY OUTPUT	LABOR-HOURS	UNIT	2000 BARE COSTS MAT.	LABOR	EQUIP.	TOTAL	TOTAL INCL O&P	
400	0010	DOWNSPOUTS Aluminum 2" x 3", .020" thick, embossed	1 Shee	190	.042	L.F.	.71	.91		1.62	2.31	400
	0100	Enameled		190	.042		.69	.91		1.60	2.29	
	0300	Enameled, .024" thick, 2" x 3"		180	.044		1.09	.96		2.05	2.81	
	0400	3" x 4"		140	.057		1.56	1.23		2.79	3.79	
	0600	Round, corrugated aluminum, 3" diameter, .020" thick		190	.042		.85	.91		1.76	2.47	
	0700	4" diameter, .025" thick		140	.057		1.41	1.23		2.64	3.62	
	0900	Wire strainer, round, 2" diameter		155	.052	Ea.	1.75	1.11		2.86	3.80	
	1000	4" diameter		155	.052		1.82	1.11		2.93	3.87	
	1200	Rectangular, perforated, 2" x 3"		145	.055		2.15	1.19		3.34	4.37	
	1300	3" x 4"		145	.055		3.10	1.19		4.29	5.40	
	1500	Copper, round, 16 oz., stock, 2" diameter		190	.042	L.F.	4.94	.91		5.85	7	
	1600	3" diameter		190	.042		3.79	.91		4.70	5.70	
	1800	4" diameter		145	.055		4.60	1.19		5.79	7.05	
	1900	5" diameter		130	.062		6.55	1.32		7.87	9.45	
	2100	Rectangular, corrugated copper, stock, 2" x 3"		190	.042		3.50	.91		4.41	5.40	
	2200	3" x 4"		145	.055		4.36	1.19		5.55	6.80	
	2400	Rectangular, plain copper, stock, 2" x 3"		190	.042		4.71	.91		5.62	6.75	
	2500	3" x 4"		145	.055		6.20	1.19		7.39	8.80	
	2700	Wire strainers, rectangular, 2" x 3"		145	.055	Ea.	2.76	1.19		3.95	5.05	
	2800	3" x 4"		145	.055		4.37	1.19		5.56	6.80	
	3000	Round, 2" diameter		145	.055		2.59	1.19		3.78	4.85	
	3100	3" diameter		145	.055		3.62	1.19		4.81	6	
	3300	4" diameter		145	.055		5.60	1.19		6.79	8.15	
	3400	5" diameter		115	.070		8.10	1.50		9.60	11.45	
	3600	Lead-coated copper, round, stock, 2" diameter		190	.042	L.F.	4.95	.91		5.86	7	
	3700	3" diameter		190	.042		4.51	.91		5.42	6.50	
	3900	4" diameter		145	.055		5.65	1.19		6.84	8.25	
	4000	5" diameter, corrugated		130	.062		5.70	1.32		7.02	8.50	
	4200	6" diameter, corrugated		105	.076		9.70	1.64		11.34	13.40	
	4300	Rectangular, corrugated, stock, 2" x 3"		190	.042		4.09	.91		5	6.05	

07700 | Roof Specialties & Accessories

07710 | Manufactured Roof Specialties

			CREW	DAILY OUTPUT	LABOR-HOURS	UNIT	MAT.	LABOR	EQUIP.	TOTAL	TOTAL INCL O&P	
400	4500	Plain, stock, 2" x 3"	1 Shee	190	.042	L.F.	6.95	.91		7.86	9.20	**400**
	4600	3" x 4"		145	.055		7.55	1.19		8.74	10.30	
	4800	Steel, galvanized, round, corrugated, 2" or 3" diam, 28 ga		190	.042		.68	.91		1.59	2.28	
	4900	4" diameter, 28 gauge		145	.055		.90	1.19		2.09	2.99	
	5100	5" diameter, 28 gauge		130	.062		1.37	1.32		2.69	3.74	
	5200	26 gauge		130	.062		1.34	1.32		2.66	3.70	
	5400	6" diameter, 28 gauge		105	.076		2.15	1.64		3.79	5.15	
	5500	26 gauge		105	.076		1.40	1.64		3.04	4.31	
	5700	Rectangular, corrugated, 28 gauge, 2" x 3"		190	.042		.53	.91		1.44	2.11	
	5800	3" x 4"		145	.055		1.48	1.19		2.67	3.63	
	6000	Rectangular, plain, 28 gauge, galvanized, 2" x 3"		190	.042		.80	.91		1.71	2.41	
	6100	3" x 4"		145	.055		1.23	1.19		2.42	3.35	
	6300	Epoxy painted, 24 gauge, corrugated, 2" x 3"		190	.042		1.01	.91		1.92	2.64	
	6400	3" x 4"		145	.055		1.95	1.19		3.14	4.15	
	6600	Wire strainers, rectangular, 2" x 3"		145	.055	Ea.	1.62	1.19		2.81	3.78	
	6700	3" x 4"		145	.055		2.61	1.19		3.80	4.87	
	6900	Round strainers, 2" or 3" diameter		145	.055		1.20	1.19		2.39	3.32	
	7000	4" diameter		145	.055		1.42	1.19		2.61	3.56	
	7200	5" diameter		145	.055		2.20	1.19		3.39	4.42	
	7300	6" diameter		115	.070		2.63	1.50		4.13	5.40	
	7500	Steel pipe, black, extra heavy, 4" diameter		20	.400	L.F.	5.30	8.60		13.90	20.50	
	7600	6" diameter		18	.444		14.30	9.55		23.85	32	
	7800	Stainless steel tubing, schedule 5, 2" x 3" or 3" diameter		190	.042		12.80	.91		13.71	15.65	
	7900	3" x 4" or 4" diameter		145	.055		16.30	1.19		17.49	19.95	
	8100	4" x 5" or 5" diameter		135	.059		33.50	1.27		34.77	39	
	8200	Vinyl, rectangular, 2" x 3"		210	.038		.72	.82		1.54	2.17	
	8300	Round, 2-1/2"		220	.036		.72	.78		1.50	2.11	
450	0010	**DRIP EDGE**, aluminum, .016" thick, 5" wide, mill finish	1 Carp	400	.020	L.F.	.20	.39		.59	.90	**450**
	0100	White finish		400	.020		.22	.39		.61	.92	
	0200	8" wide, mill finish		400	.020		.30	.39		.69	1.01	
	0300	Ice belt, 28" wide, mill finish		100	.080		3.42	1.58		5	6.45	
	0310	Vented, mill finish		400	.020		1.38	.39		1.77	2.20	
	0320	Painted finish		400	.020		1.50	.39		1.89	2.33	
	0400	Galvanized, 5" wide		400	.020		.22	.39		.61	.92	
	0500	8" wide, mill finish		400	.020		.33	.39		.72	1.04	
	0510	Rake edge, aluminum, 1-1/2" x 1-1/2"		400	.020		.13	.39		.52	.82	
	0520	3-1/2" x 1-1/2"		400	.020		.19	.39		.58	.89	
500	0010	**ELBOWS** Aluminum, 2" x 3", embossed	1 Shee	100	.080	Ea.	.90	1.72		2.62	3.89	**500**
	0100	Enameled		100	.080		1.63	1.72		3.35	4.69	
	0200	3" x 4", .025" thick, embossed		100	.080		3.15	1.72		4.87	6.35	
	0300	Enameled		100	.080		3.15	1.72		4.87	6.35	
	0400	Round corrugated, 3", embossed, .020" thick		100	.080		1.95	1.72		3.67	5.05	
	0500	4", .025" thick		100	.080		2.95	1.72		4.67	6.15	
	0600	Copper, 16 oz. round, 2" diameter		100	.080		11	1.72		12.72	15	
	0700	3" diameter		100	.080		5.05	1.72		6.77	8.45	
	0800	4" diameter		100	.080		9.50	1.72		11.22	13.35	
	1000	2" x 3" corrugated		100	.080		5.05	1.72		6.77	8.50	
	1100	3" x 4" corrugated		100	.080		9.75	1.72		11.47	13.65	
	1300	Vinyl, 2-1/2" diameter, 45° or 75°		100	.080		2	1.72		3.72	5.10	
	1400	Tee Y junction		75	.107		8.50	2.29		10.79	13.20	
550	0010	**GRAVEL STOP** Aluminum, .050" thick, 4" face height, mill finish	1 Shee	145	.055	L.F.	2.72	1.19		3.91	4.99	**550**
	0080	Duranodic finish		145	.055		3.69	1.19		4.88	6.05	
	0100	Painted		145	.055		4.26	1.19		5.45	6.70	
	0300	6" face height		135	.059		3.22	1.27		4.49	5.70	

07700 | Roof Specialties & Accessories

07710 | Manufactured Roof Specialties

			CREW	DAILY OUTPUT	LABOR-HOURS	UNIT	2000 BARE COSTS MAT.	LABOR	EQUIP.	TOTAL	TOTAL INCL O&P	
550	0350	Duranodic finish	1 Shee	135	.059	L.F.	4.35	1.27		5.62	6.95	**550**
	0400	Painted		135	.059		5.05	1.27		6.32	7.70	
	0600	8" face height		125	.064		3.98	1.38		5.36	6.70	
	0650	Duranodic finish		125	.064		4.98	1.38		6.36	7.80	
	0700	Painted		125	.064		5.05	1.38		6.43	7.90	
	0900	12" face height, .080 thick, 2 piece		100	.080		6.45	1.72		8.17	10	
	0950	Duranodic finish		100	.080		6.30	1.72		8.02	9.80	
	1000	Painted		100	.080		7.40	1.72		9.12	11.05	
	1350	Galv steel, 24 ga., 4" leg, plain, with continuous cleat, 4" face		145	.055		1.50	1.19		2.69	3.65	
	1360	6" face height		145	.055		2.27	1.19		3.46	4.50	
	1500	Polyvinyl chloride, 6" face height		135	.059		3.28	1.27		4.55	5.75	
	1600	9" face height		125	.064		3.87	1.38		5.25	6.60	
	1800	Stainless steel, 24 ga., 6" face height		135	.059		7.15	1.27		8.42	10	
	1900	12" face height		100	.080		15	1.72		16.72	19.40	
	2100	20 ga., 6" face height		135	.059		8.10	1.27		9.37	11.05	
	2200	12" face height		100	.080		17.10	1.72		18.82	21.50	
650	0010	**GUTTERS** Aluminum, stock units, 5" box, .027" thick, plain	1 Shee	120	.067	L.F.	.97	1.43		2.40	3.49	**650**
	0100	Enameled		120	.067		1.07	1.43		2.50	3.60	
	0300	5" box type, .032" thick, plain		120	.067		1.08	1.43		2.51	3.61	
	0400	Enameled		120	.067		1.20	1.43		2.63	3.74	
	0600	5" x 6" combination fascia & gutter, .032" thick, enameled		60	.133		3.45	2.87		6.32	8.65	
	0700	Copper, half round, 16 oz, stock units, 4" wide		120	.067		3.18	1.43		4.61	5.90	
	0900	5" wide		120	.067		3.87	1.43		5.30	6.70	
	1000	6" wide		115	.070		4.91	1.50		6.41	7.95	
	1200	K type, 16 oz, stock, 4" wide		120	.067		3.59	1.43		5.02	6.35	
	1300	5" wide		120	.067		3.61	1.43		5.04	6.40	
	1500	Lead coated copper, half round, stock, 4" wide		120	.067		4.58	1.43		6.01	7.45	
	1600	6" wide		115	.070		5.60	1.50		7.10	8.70	
	1800	K type, stock, 4" wide		120	.067		6.95	1.43		8.38	10.05	
	1900	5" wide		120	.067		6.20	1.43		7.63	9.25	
	2100	Stainless steel, half round or box, stock, 4" wide		120	.067		4.65	1.43		6.08	7.50	
	2200	5" wide		120	.067		5	1.43		6.43	7.90	
	2400	Steel, galv, half round or box, 28 ga, 5" wide, plain		120	.067		.90	1.43		2.33	3.41	
	2500	Enameled		120	.067		.95	1.43		2.38	3.47	
	2700	26 ga, stock, 5" wide		120	.067		.95	1.43		2.38	3.47	
	2800	6" wide		120	.067		1.61	1.43		3.04	4.19	
	3000	Vinyl, O.G., 4" wide	1 Carp	110	.073		.85	1.43		2.28	3.39	
	3100	5" wide		110	.073		1	1.43		2.43	3.55	
	3200	4" half round, stock units		110	.073		.68	1.43		2.11	3.20	
	3250	Joint connectors				Ea.	1.36			1.36	1.50	
	3300	Wood, clear treated cedar, fir or hemlock, 3" x 4"	1 Carp	100	.080	L.F.	6.30	1.58		7.88	9.60	
	3400	4" x 5"	"	100	.080	"	7.30	1.58		8.88	10.70	
700	0010	**GUTTER GUARD** 6" wide strip, aluminum mesh	1 Carp	500	.016	L.F.	.37	.32		.69	.95	**700**
	0100	Vinyl mesh		500	.016	"	.22	.32		.54	.78	
750	0010	**REGLET** Aluminum, .025" thick, in concrete parapet		225	.036	L.F.	.92	.70		1.62	2.21	**750**
	0100	Copper, 10 oz.		225	.036		1.59	.70		2.29	2.95	
	0300	16 oz.		225	.036		2.12	.70		2.82	3.53	
	0400	Galvanized steel, 24 gauge		225	.036		.78	.70		1.48	2.06	
	0600	Stainless steel, .020" thick		225	.036		1.57	.70		2.27	2.93	
	0700	Zinc and copper alloy, 20 oz.		225	.036		1.75	.70		2.45	3.13	
	0900	Counter flashing for above, 12" wide, .032" aluminum	1 Shee	150	.053		1.19	1.15		2.34	3.25	
	1000	Copper, 10 oz.		150	.053		3.32	1.15		4.47	5.60	
	1200	16 oz.		150	.053		3.70	1.15		4.85	6	
	1300	Galvanized steel, .020" thick		150	.053		.62	1.15		1.77	2.62	
	1500	Stainless steel, .020" thick		150	.053		2.76	1.15		3.91	4.98	
	1600	Zinc and copper alloy, 20 oz.		150	.053		3.11	1.15		4.26	5.35	

07700 | Roof Specialties & Accessories

07710 | Manufactured Roof Specialties

			CREW	DAILY OUTPUT	LABOR-HOURS	UNIT	MAT.	LABOR	EQUIP.	TOTAL	TOTAL INCL O&P	
800	0010	**EXPANSION JOINT**										800
	0300	Butyl or neoprene center with foam insulation, metal flanges										
	0400	Aluminum, .032" thick for openings to 2-1/2"	1 Rofc	165	.048	L.F.	7.40	.83		8.23	9.70	
	0600	For joint openings to 3-1/2"		165	.048		8.65	.83		9.48	11.05	
	0610	For joint openings to 5"		165	.048		10.40	.83		11.23	13	
	0620	For joint openings to 8"		165	.048		16.40	.83		17.23	19.60	
	0700	Copper, 16 oz. for openings to 2-1/2"		165	.048		10.40	.83		11.23	13	
	0900	For joint openings to 3-1/2"		165	.048		11.95	.83		12.78	14.70	
	0910	For joint openings to 5"		165	.048		14	.83		14.83	16.95	
	0920	For joint openings to 8"		165	.048		21	.83		21.83	24.50	
	1000	Galvanized steel, 26 ga. for openings to 2-1/2"		165	.048		6.25	.83		7.08	8.45	
	1200	For joint openings to 3-1/2"		165	.048		7.40	.83		8.23	9.70	
	1210	For joint openings to 5"		165	.048		9.25	.83		10.08	11.75	
	1220	For joint openings to 8"		165	.048		15.75	.83		16.58	18.90	
	1300	Lead-coated copper, 16 oz. for openings to 2-1/2"		165	.048		18.50	.83		19.33	22	
	1500	For joint openings to 3-1/2"		165	.048		22	.83		22.83	25.50	
	1600	Stainless steel, .018", for openings to 2-1/2"		165	.048		9.25	.83		10.08	11.75	
	1800	For joint openings to 3-1/2"		165	.048		10.55	.83		11.38	13.15	
	1810	For joint openings to 5"		165	.048		13.05	.83		13.88	15.90	
	1820	For joint openings to 8"		165	.048		19.85	.83		20.68	23.50	
	1900	Neoprene, double-seal type with thick center, 4-1/2" wide		125	.064		8.90	1.09		9.99	11.85	
	1950	Polyethylene bellows, with galv steel flat flanges		100	.080		3.45	1.36		4.81	6.35	
	1960	With galvanized angle flanges		100	.080		3.80	1.36		5.16	6.70	
	2000	Roof joint with extruded aluminum cover, 2"	1 Shee	115	.070		24	1.50		25.50	28.50	
	2100	Roof joint, plastic curbs, foam center, standard	1 Rofc	100	.080		8.70	1.36		10.06	12.10	
	2200	Large		100	.080		11.60	1.36		12.96	15.30	
	2300	Transitions, regular, minimum		10	.800	Ea.	75	13.65		88.65	108	
	2350	Maximum		4	2		95.50	34		129.50	169	
	2400	Large, minimum		9	.889		110	15.15		125.15	149	
	2450	Maximum		3	2.667		115	45.50		160.50	212	
	2500	Roof to wall joint with extruded aluminum cover	1 Shee	115	.070	L.F.	20.50	1.50		22	25	
	2600	See also divisions 03150 & 05810										
	2700	Wall joint, closed cell foam on PVC cover, 9" wide	1 Rofc	125	.064	L.F.	3.10	1.09		4.19	5.45	
	2800	12" wide	"	115	.070	"	3.50	1.19		4.69	6.05	

07720 | Roof Accessories

			CREW	DAILY OUTPUT	LABOR-HOURS	UNIT	MAT.	LABOR	EQUIP.	TOTAL	TOTAL INCL O&P	
480	0010	**PITCH POCKETS**										480
	0100	Adjustable, 4" to 7", welded corners, 4" deep	1 Rofc	48	.167	Ea.	10.50	2.84		13.34	16.85	
	0200	Side extenders, 6"	"	240	.033	"	1.70	.57		2.27	2.93	
500	0010	**ROOF VENTS** Mushroom for built-up roofs, aluminum	1 Rofc	30	.267	Ea.	24	4.55		28.55	35	500
	0100	PVC, 6" high	"	30	.267	"	27.50	4.55		32.05	39	
550	0010	**RIDGE VENT**										550
	0100	Aluminum strips, mill finish	1 Rofc	160	.050	L.F.	1.19	.85		2.04	2.90	
	0150	Painted finish		160	.050	"	2.12	.85		2.97	3.93	
	0200	Connectors		48	.167	Ea.	1.81	2.84		4.65	7.30	
	0300	End caps		48	.167	"	.76	2.84		3.60	6.15	
	0400	Galvanized strips, with damper and bird screen		160	.050	L.F.	20.50	.85		21.35	24.50	
	0430	Molded polyethylene, shingles not included		160	.050	"	2.50	.85		3.35	4.34	
	0440	End plugs		48	.167	Ea.	.76	2.84		3.60	6.15	
	0450	Flexible roll, shingles not included		160	.050	L.F.	1.95	.85		2.80	3.73	
700	0010	**ROOF HATCHES** With curb, 1" fiberglass insulation, 2'-6" x 3'-0"										700
	0500	Aluminum curb and cover	G-3	10	3.200	Ea.	405	57.50		462.50	545	
	0520	Galvanized steel curb and aluminum cover		10	3.200		340	57.50		397.50	475	
	0540	Galvanized steel curb and cover		10	3.200		405	57.50		462.50	545	

07700 | Roof Specialties & Accessories

07720 | Roof Accessories

					DAILY	LABOR-			2000 BARE COSTS			TOTAL	
				CREW	OUTPUT	HOURS	UNIT	MAT.	LABOR	EQUIP.	TOTAL	INCL O&P	
700	0600	2'-6" x 4'-6", aluminum curb and cover		G-3	9	3.556	Ea.	560	64		624	725	700
	0800	Galvanized steel curb and aluminum cover			9	3.556		475	64		539	635	
	0900	Galvanized steel curb and cover			9	3.556		615	64		679	785	
	1200	2'-6" x 8'-0", aluminum curb and cover			6.60	4.848		995	87		1,082	1,250	
	1400	Galvanized steel curb and aluminum cover			6.60	4.848		920	87		1,007	1,150	
	1500	Galvanized steel curb and cover			6.60	4.848		905	87		992	1,150	
	1800	For plexiglass panels, 2'-6" x 3'-0", add to above						345			345	380	
800	0010	**WALKWAY** For built-up roofs, asphalt impregnated, 3' x 6' x 1/2" thk		1 Rofc	400	.020	S.F.	.87	.34		1.21	1.59	800
	0100	3' x 3' x 3/4" thick		"	400	.020		2.07	.34		2.41	2.91	
	0300	Concrete patio blocks, 2" thick, natural		1 Clab	115	.070		1.39	1.01		2.40	3.25	
	0400	Colors		"	115	.070		1.75	1.01		2.76	3.65	
850	0010	**SMOKE HATCHES** Unlabeled, not including hand winch operator											850
	0200	For 3'-0" long, add to roof hatches from division 07720-700					Ea.	25%	5%				
	0300	For 8'-0" long, add to roof hatches from division 07720-700					"	10%	5%				
860	0010	**SMOKE VENT**, insulated, 4' x 4'											860
	0100	Aluminum cover and frame		G-3	13	2.462	Ea.	1,050	44.50		1,094.50	1,225	
	0200	Galvanized steel cover and frame			13	2.462		950	44.50		994.50	1,125	
	0300	4' x 8' aluminum cover and frame			8	4		1,425	72		1,497	1,700	
	0400	Galvanized steel cover and frame			8	4		1,250	72		1,322	1,500	
870	0010	**VENTS, ONE-WAY** For insul. decks, 1 per M.S.F., plastic, min.		1 Rofc	40	.200	Ea.	12.50	3.41		15.91	20	870
	0100	Maximum			20	.400		28.50	6.80		35.30	44	
	0300	Aluminum			30	.267		12.50	4.55		17.05	22	
	0800	Polystyrene baffles, 12" wide for 16" O.C. rafter spacing		1 Carp	90	.089		.45	1.75		2.20	3.50	
	0900	For 24" O.C. rafter spacing		"	110	.073		1.05	1.43		2.48	3.60	

07800 | Fire & Smoke Protection

07812 | Cementitious Fireproofing

			CREW	DAILY OUTPUT	LABOR-HOURS	UNIT	MAT.	LABOR	EQUIP.	TOTAL	TOTAL INCL O&P	
600	0010	**SPRAYED** Mineral fiber or cementitious for fireproofing,										600
	0050	not incl tamping or canvas protection										
	0100	1" thick, on flat plate steel	G-2	3,000	.008	S.F.	.42	.13	.08	.63	.77	
	0200	Flat decking		2,400	.010		.42	.16	.10	.68	.84	
	0400	Beams		1,500	.016		.42	.26	.17	.85	1.08	
	0500	Corrugated or fluted decks		1,250	.019		.63	.31	.20	1.14	1.44	
	0700	Columns, 1-1/8" thick		1,100	.022		.47	.35	.23	1.05	1.37	
	0800	2-3/16" thick		700	.034		.88	.56	.36	1.80	2.30	
	0850	For tamping, add							10%				
	0900	For canvas protection, add	G-2	5,000	.005	S.F.	.06	.08	.05	.19	.26	
	1000	Acoustical sprayed, 1" thick, finished, straight work, minimum		520	.046		.44	.75	.48	1.67	2.27	
	1100	Maximum		200	.120		.47	1.95	1.25	3.67	5.20	
	1300	Difficult access, minimum		225	.107		.47	1.73	1.11	3.31	4.66	
	1400	Maximum		130	.185		.51	3	1.92	5.43	7.75	
	1500	Intumescent epoxy fireproofing on wire mesh, 3/16" thick										
	1550	1 hour rating, exterior use	G-2	136	.176	S.F.	5.40	2.87	1.84	10.11	12.80	
	1600	Magnesium oxychloride, 35# to 40# density, 1/4" thick		3,000	.008		1.14	.13	.08	1.35	1.56	
	1650	1/2" thick		2,000	.012		2.28	.19	.13	2.60	2.98	

07800 | Fire & Smoke Protection

07812 | Cementitious Fireproofing

		Crew	Daily Output	Labor-Hours	Unit	Mat.	Labor	Equip.	Total	Total Incl O&P
1700	60# to 70# density, 1/4" thick	G-2	3,000	.008	S.F.	1.50	.13	.08	1.71	1.96
1750	1/2" thick		2,000	.012		3.03	.19	.13	3.35	3.80
2000	Vermiculite cement, troweled or sprayed, 1/4" thick		3,000	.008		1.03	.13	.08	1.24	1.44
2050	1/2" thick		2,000	.012		2.04	.19	.13	2.36	2.71

07840 | Firestopping

		Crew	Daily Output	Labor-Hours	Unit	Mat.	Labor	Equip.	Total	Total Incl O&P
0010	**FIRESTOPPING**									
0100	Metallic piping, non insulated									
0110	Through walls, 2" diameter	1 Carp	16	.500	Ea.	9.60	9.85		19.45	27.50
0120	4" diameter		14	.571		14.65	11.25		25.90	35.50
0130	6" diameter		12	.667		19.70	13.15		32.85	44
0140	12" diameter		10	.800		35	15.75		50.75	65.50
0150	Through floors, 2" diameter		32	.250		5.80	4.93		10.73	14.85
0160	4" diameter		28	.286		8.35	5.65		14	18.85
0170	6" diameter		24	.333		11	6.55		17.55	23.50
0180	12" diameter		20	.400		18.50	7.90		26.40	34
0190	Metallic piping, insulated									
0200	Through walls, 2" diameter	1 Carp	16	.500	Ea.	13.60	9.85		23.45	32
0210	4" diameter		14	.571		18.65	11.25		29.90	40
0220	6" diameter		12	.667		23.50	13.15		36.65	48.50
0230	12" diameter		10	.800		39	15.75		54.75	69.50
0240	Through floors, 2" diameter		32	.250		9.80	4.93		14.73	19.25
0250	4" diameter		28	.286		12.35	5.65		18	23.50
0260	6" diameter		24	.333		15	6.55		21.55	28
0270	12" diameter		20	.400		18.50	7.90		26.40	34
0280	Non metallic piping, non insulated									
0290	Through walls, 2" diameter	1 Carp	12	.667	Ea.	39.50	13.15		52.65	66
0300	4" diameter		10	.800		49.50	15.75		65.25	81.50
0310	6" diameter		8	1		69	19.70		88.70	110
0330	Through floors, 2" diameter		16	.500		31	9.85		40.85	51
0340	4" diameter		6	1.333		38.50	26.50		65	87
0350	6" diameter		6	1.333		46	26.50		72.50	95.50
0370	Ductwork, insulated & non insulated, round									
0380	Through walls, 6" diameter	1 Carp	12	.667	Ea.	20	13.15		33.15	44.50
0390	12" diameter		10	.800		40	15.75		55.75	71
0400	18" diameter		8	1		65	19.70		84.70	106
0410	Through floors, 6" diameter		16	.500		11	9.85		20.85	29
0420	12" diameter		14	.571		20	11.25		31.25	41.50
0430	18" diameter		12	.667		35	13.15		48.15	61
0440	Ductwork, insulated & non insulated, rectangular									
0450	With stiffener/closure angle, through walls, 6" x 12"	1 Carp	8	1	Ea.	16.65	19.70		36.35	52.50
0460	12" x 24"		6	1.333		22	26.50		48.50	69.50
0470	24" x 48"		4	2		63	39.50		102.50	137
0480	With stiffener/closure angle, through floors, 6" x 12"		10	.800		9	15.75		24.75	37
0490	12" x 24"		8	1		16.20	19.70		35.90	52
0500	24" x 48"		6	1.333		32	26.50		58.50	80
0510	Multi trade openings									
0520	Through walls, 6" x 12"	1 Carp	2	4	Ea.	35	79		114	174
0530	12" x 24"	"	1	8		141	158		299	425
0540	24" x 48"	2 Carp	1	16		565	315		880	1,150
0550	48" x 96"	"	.75	21.333		2,275	420		2,695	3,225
0560	Through floors, 6" x 12"	1 Carp	2	4		35	79		114	174
0570	12" x 24"	"	1	8		141	158		299	425
0580	24" x 48"	2 Carp	.75	21.333		565	420		985	1,350
0590	48" x 96"	"	.50	32		2,275	630		2,905	3,575
0600	Structural penetrations, through walls									

07800 | Fire & Smoke Protection

07840 | Firestopping

			CREW	DAILY OUTPUT	LABOR-HOURS	UNIT	MAT.	LABOR	EQUIP.	TOTAL	TOTAL INCL O&P	
100	0610	Steel beams, W8 x 10	1 Carp	8	1	Ea.	22	19.70		41.70	58	100
	0620	W12 x 14		6	1.333		35	26.50		61.50	83.50	
	0630	W21 x 44		5	1.600		70	31.50		101.50	131	
	0640	W36 x 135		3	2.667		170	52.50		222.50	277	
	0650	Bar joists, 18" deep		6	1.333		32	26.50		58.50	80	
	0660	24" deep		6	1.333		40	26.50		66.50	89	
	0670	36" deep		5	1.600		60	31.50		91.50	120	
	0680	48" deep		4	2		70	39.50		109.50	145	
	0690	Construction joints, floor slab at exterior wall										
	0700	Precast, brick, block or drywall exterior										
	0710	2" wide joint	1 Carp	125	.064	L.F.	5	1.26		6.26	7.65	
	0720	4" wide joint	"	75	.107	"	10	2.10		12.10	14.60	
	0730	Metal panel, glass or curtain wall exterior										
	0740	2" wide joint	1 Carp	40	.200	L.F.	11.85	3.94		15.79	19.75	
	0750	4" wide joint	"	25	.320	"	16.15	6.30		22.45	28.50	
	0760	Floor slab to drywall partition										
	0770	Flat joint	1 Carp	100	.080	L.F.	4.90	1.58		6.48	8.10	
	0780	Fluted joint		50	.160		10	3.15		13.15	16.40	
	0790	Etched fluted joint		75	.107		6.50	2.10		8.60	10.75	
	0800	Floor slab to concrete/masonry partition										
	0810	Flat joint	1 Carp	75	.107	L.F.	11	2.10		13.10	15.70	
	0820	Fluted joint	"	50	.160	"	13	3.15		16.15	19.70	
	0830	Concrete/CMU wall joints										
	0840	1" wide	1 Carp	100	.080	L.F.	6	1.58		7.58	9.30	
	0850	2" wide		75	.107		11	2.10		13.10	15.70	
	0860	4" wide		50	.160		21	3.15		24.15	28.50	
	0870	Concrete/CMU floor joints										
	0880	1" wide	1 Carp	200	.040	L.F.	3	.79		3.79	4.65	
	0890	2" wide		150	.053		5.50	1.05		6.55	7.85	
	0900	4" wide		100	.080		10.50	1.58		12.08	14.25	

07900 | Joint Sealers

07920 | Joint Sealants

			CREW	DAILY OUTPUT	LABOR-HOURS	UNIT	MAT.	LABOR	EQUIP.	TOTAL	TOTAL INCL O&P	
800	0010	CAULKING AND SEALANTS										800
	0020	Acoustical sealant, elastomeric, cartridges				Ea.	2.05			2.05	2.26	
	0030	Backer rod, polyethylene, 1/4" diameter	1 Bric	4.60	1.739	C.L.F.	1.28	35		36.28	60.50	
	0050	1/2" diameter		4.60	1.739		2.54	35		37.54	62	
	0070	3/4" diameter		4.60	1.739		4.90	35		39.90	64.50	
	0090	1" diameter		4.60	1.739		9	35		44	69	
	0100	Acrylic latex caulk, white										
	0200	11 fl. oz cartridge				Ea.	1.99			1.99	2.19	
	0500	1/4" x 1/2"	1 Bric	248	.032	L.F.	.16	.65		.81	1.27	
	0600	1/2" x 1/2"		250	.032		.32	.64		.96	1.44	
	0800	3/4" x 3/4"		230	.035		.73	.70		1.43	1.98	
	0900	3/4" x 1"		200	.040		.98	.80		1.78	2.42	
	1000	1" x 1"		180	.044		1.22	.89		2.11	2.84	
	1400	Butyl based, bulk				Gal.	22			22	24	
	1500	Cartridges				"	26.50			26.50	29.50	
	1700	Bulk, in place 1/4" x 1/2", 154 L.F./gal.	1 Bric	230	.035	L.F.	.14	.70		.84	1.34	

07900 | Joint Sealers

07920 | Joint Sealants

		CREW	DAILY OUTPUT	LABOR-HOURS	UNIT	2000 BARE COSTS MAT.	LABOR	EQUIP.	TOTAL	TOTAL INCL O&P
1800	1/2" x 1/2", 77 L.F./gal.	1 Bric	180	.044	L.F.	.29	.89		1.18	1.81
2000	Latex acrylic based, bulk				Gal.	23			23	25.50
2100	Cartridges					26			26	28.50
2300	Polysulfide compounds, 1 component, bulk					43.50			43.50	47.50
2600	1 or 2 component, in place, 1/4" x 1/4", 308 L.F./gal.	1 Bric	145	.055	L.F.	.14	1.10		1.24	2.02
2700	1/2" x 1/4", 154 L.F./gal.		135	.059		.28	1.19		1.47	2.32
2900	3/4" x 3/8", 68 L.F./gal.		130	.062		.64	1.23		1.87	2.78
3000	1" x 1/2", 38 L.F./gal.		130	.062		1.14	1.23		2.37	3.33
3200	Polyurethane, 1 or 2 component				Gal.	49			49	54
3500	Bulk, in place, 1/4" x 1/4"	1 Bric	150	.053	L.F.	.16	1.07		1.23	1.98
3600	1/2" x 1/4"		145	.055		.32	1.10		1.42	2.22
3800	3/4" x 3/8", 68 L.F./gal.		130	.062		.72	1.23		1.95	2.87
3900	1" x 1/2"		110	.073		1.27	1.45		2.72	3.86
4100	Silicone rubber, bulk				Gal.	34			34	37.50
4200	Cartridges				"	40			40	44
4400	Neoprene gaskets, closed cell, adhesive, 1/8" x 3/8"	1 Bric	240	.033	L.F.	.20	.67		.87	1.35
4500	1/4" x 3/4"		215	.037		.48	.74		1.22	1.79
4700	1/2" x 1"		200	.040		1.40	.80		2.20	2.89
4800	3/4" x 1-1/2"		165	.048		2.91	.97		3.88	4.84
5500	Resin epoxy coating, 2 component, heavy duty				Gal.	26			26	28.50
5800	Tapes, sealant, P.V.C. foam adhesive, 1/16" x 1/4"				C.L.F.	4.60			4.60	5.05
5900	1/16" x 1/2"					6.80			6.80	7.50
5950	1/16" x 1"					11.30			11.30	12.45
6000	1/8" x 1/2"					7.65			7.65	8.40
6200	Urethane foam, 2 component, handy pack, 1 C.F.				Ea.	27.50			27.50	30.50
6300	50.0 C.F. pack				C.F.	14.05			14.05	15.45

For information about Means Estimating Seminars, see yellow pages 11 and 12 in back of book

Important: See the Reference Section for critical supporting data - Reference Nos., Crews, & City Cost Indexes

Division 8
Doors & Windows

Estimating Tips

08100 Metal Doors & Frames
- Most metal doors and frames look alike, but there may be significant differences among them. When estimating these items be sure to choose the line item that most closely compares to the specification or door schedule requirements regarding:
  - type of metal
  - metal gauge
  - door core material
  - fire rating
  - finish

08200 Wood & Plastic Doors
- Wood and plastic doors vary considerably in price. The primary determinant is the veneer material. Lauan, birch and oak are the most common veneers. Other variables include the following:
  - hollow or solid core
  - fire rating
  - flush or raised panel
  - finish
- If the specifications require compliance with AWI (Architectural Woodwork Institute) standards or acoustical standards, the cost of the door may increase substantially. All wood doors are priced pre-mortised for hinges and predrilled for cylindrical locksets.

08300 Specialty Doors
- There are many varieties of special doors, and they are usually priced per each. Add frames, hardware or operators required for a complete installation.

08510 Steel Windows
- Most metal windows are delivered preglazed. However, some metal windows are priced without glass. Refer to 08800 Glazing for glass pricing. The grade C indicates commercial grade windows, usually ASTM C-35.

08550 Wood Windows
- All wood windows are priced preglazed. The two glazing options priced are single pane float glass and insulating glass 1/2" thick. Add the cost of screens and grills if required.

08700 Hardware
- Hardware costs add considerably to the cost of a door. The most efficient method to determine the hardware requirements for a project is to review the door schedule. This schedule, in conjunction with the specifications, is all you should need to take off the door hardware.
- Door hinges are priced by the pair, with most doors requiring 1-1/2 pairs per door. The hinge prices do not include installation labor because it is included in door installation. Hinges are classified according to the frequency of use.

08800 Glazing
- Different openings require different types of glass. The three most common types are:
  - float
  - tempered
  - insulating
- Most exterior windows are glazed with insulating glass. Entrance doors and window walls, where the glass is less than 18" from the floor, are generally glazed with tempered glass. Interior windows and some residential windows are glazed with float glass.

08900 Glazed Curtain Wall
- Glazed curtain walls consist of the metal tube framing and the glazing material. The cost data in this subdivision is presented for the metal tube framing alone or the composite wall. If your estimate requires a detailed takeoff of the framing, be sure to add the glazing cost.

Reference Numbers
Reference numbers are shown in bold squares at the beginning of some major classifications. These numbers refer to related items in the Reference Section. The reference information may be an estimating procedure, an alternate pricing method or technical information.

Note: Not all subdivisions listed here necessarily appear in this publication.

08100 | Metal Doors & Frames

08110 | Steel Doors & Frames

			CREW	DAILY OUTPUT	LABOR-HOURS	UNIT	MAT.	LABOR	EQUIP.	TOTAL	TOTAL INCL O&P
200	0010	**COMMERCIAL STEEL DOORS**									200
	0015	Flush, full panel, hollow core	R08110-120								
	0020	1-3/8" thick, 20 ga., 2'-0" x 6'-8"	2 Carp	20	.800	Ea.	150	15.75		165.75	192
	0040	2'-8" x 6'-8"		18	.889		155	17.50		172.50	201
	0060	3'-0" x 6'-8"		17	.941		158	18.55		176.55	205
	0100	3'-0" x 7'-0"		17	.941		163	18.55		181.55	211
	0120	For vision lite, add					51.50			51.50	57
	0140	For narrow lite, add					62.50			62.50	69
	0160	For bottom louver, add					64			64	70.50
	0230	For baked enamel finish, add					30%	15%			
	0260	For galvanizing, add					15%				
	0320	Half glass, 20 ga., 2'-0" x 6'-8"	2 Carp	20	.800	Ea.	202	15.75		217.75	249
	0340	2'-8" x 6'-8"		18	.889		215	17.50		232.50	267
	0360	3'-0" x 6'-8"		17	.941		221	18.55		239.55	275
	0400	3'-0" x 7'-0"		17	.941		219	18.55		237.55	273
	0500	Hollow core, 1-3/4" thick, full panel, 20 ga., 2'-8" x 6'-8"		18	.889		162	17.50		179.50	209
	0520	3'-0" x 6'-8"		17	.941		165	18.55		183.55	214
	0640	3'-0" x 7'-0"		17	.941		170	18.55		188.55	219
	0680	4'-0" x 7'-0"		15	1.067		340	21		361	410
	0700	4'-0" x 8'-0"		13	1.231		360	24.50		384.50	435
	1000	18 ga., 2'-8" x 6'-8"		17	.941		187	18.55		205.55	238
	1020	3'-0" x 6'-8"		16	1		185	19.70		204.70	237
	1120	3'-0" x 7'-0"		17	.941		189	18.55		207.55	240
	1180	4'-0" x 7'-0"		14	1.143		246	22.50		268.50	310
	1200	4'-0" x 8'-0"		17	.941		291	18.55		309.55	350
	1230	Half glass, 20 ga., 2'-8" x 6'-8"		20	.800		235	15.75		250.75	286
	1240	3'-0" x 6'-8"		18	.889		235	17.50		252.50	289
	1260	3'-0" x 7'-0"		18	.889		246	17.50		263.50	300
	1280	4'-0" x 7'-0"		16	1		535	19.70		554.70	625
	1300	4'-0" x 8'-0"		13	1.231		560	24.50		584.50	655
	1320	18 ga., 2'-8" x 6'-8"		18	.889		158	17.50		175.50	204
	1340	3'-0" x 6'-8"		17	.941		253	18.55		271.55	310
	1360	3'-0" x 7'-0"		17	.941		263	18.55		281.55	320
	1380	4'-0" x 7'-0"		15	1.067		370	21		391	445
	1400	4'-0" x 8'-0"		14	1.143		430	22.50		452.50	510
	1720	Insulated, 1-3/4" thick, full panel, 18 ga., 3'-0" x 6'-8"		15	1.067		245	21		266	305
	1740	2'-8" x 7'-0"		16	1		256	19.70		275.70	315
	1760	3'-0" x 7'-0"		15	1.067		252	21		273	315
	1800	4'-0" x 8'-0"		13	1.231		305	24.50		329.50	375
	1820	Half glass, 18 ga., 3'-0" x 6'-8"		16	1		315	19.70		334.70	380
	1840	2'-8" x 7'-0"		17	.941		330	18.55		348.55	390
	1860	3'-0" x 7'-0"		16	1		260	19.70		279.70	320
	1900	4'-0" x 8'-0"		14	1.143		490	22.50		512.50	575
250	0010	**DOOR FRAMES**									250
	0020	Steel channels with anchors and bar stops									
	0100	6" channel @ 8.2#/L.F., 3' x 7' door, weighs 150#	E-4	13	2.462	Ea.	108	53.50	6.45	167.95	232
	0200	8" channel @ 11.5#/L.F., 6' x 8' door, weighs 275#		9	3.556		198	77.50	9.35	284.85	380
	0300	8' x 12' door, weighs 400#		6.50	4.923		288	107	12.95	407.95	540
	0400	10" channel @ 15.3#/L.F., 10' x 10' door, weighs 500#		6	5.333		360	116	14.05	490.05	640
	0500	12' x 12' door, weighs 600#		5.50	5.818		430	127	15.30	572.30	740
	0600	12" channel @ 20.7#/L.F., 12' x 12' door, weighs 825#		4.50	7.111		595	155	18.70	768.70	980
	0700	12' x 16' door, weighs 1000#		4	8		720	174	21	915	1,150
	0800	For frames without bar stops, light sections, deduct					15%				
	0900	Heavy sections, deduct					10%				

08100 | Metal Doors & Frames

08110 | Steel Doors & Frames

			CREW	DAILY OUTPUT	LABOR-HOURS	UNIT	2000 BARE COSTS MAT.	LABOR	EQUIP.	TOTAL	TOTAL INCL O&P
300	0010	**FIRE DOOR**									
	0015	Steel, flush, "B" label, 90 minute									
	0020	Full panel, 20 ga., 2'-0" x 6'-8"	2 Carp	20	.800	Ea.	173	15.75		188.75	217
	0040	2'-8" x 6'-8"		18	.889		180	17.50		197.50	228
	0060	3'-0" x 6'-8"		17	.941		180	18.55		198.55	230
	0080	3'-0" x 7'-0"		17	.941		187	18.55		205.55	238
	0140	18 ga., 3'-0" x 6'-8"		16	1		200	19.70		219.70	253
	0160	2'-8" x 7'-0"		17	.941		213	18.55		231.55	266
	0180	3'-0" x 7'-0"		16	1		207	19.70		226.70	261
	0200	4'-0" x 7'-0"		15	1.067		261	21		282	325
	0220	For "A" label, 3 hour, 18 ga., use same price as "B" label									
	0240	For vision lite, add				Ea.	35			35	38.50
	0520	Flush, "B" label 90 min., composite, 20 ga., 2'-0" x 6'-8"	2 Carp	18	.889		233	17.50		250.50	286
	0540	2'-8" x 6'-8"		17	.941		237	18.55		255.55	293
	0560	3'-0" x 6'-8"		16	1		240	19.70		259.70	298
	0580	3'-0" x 7'-0"		16	1		248	19.70		267.70	305
	0640	Flush, "A" label 3 hour, composite, 18 ga., 3'-0" x 6'-8"		15	1.067		260	21		281	320
	0660	2'-8" x 7'-0"		16	1		270	19.70		289.70	330
	0680	3'-0" x 7'-0"		15	1.067		267	21		288	330
	0700	4'-0" x 7'-0"		14	1.143		320	22.50		342.50	395
600	0010	**RESIDENTIAL STEEL DOOR**									
	0020	Prehung, insulated, exterior									
	0030	Embossed, full panel, 2'-8" x 6'-8"	2 Carp	17	.941	Ea.	178	18.55		196.55	228
	0040	3'-0" x 6'-8"		15	1.067		179	21		200	233
	0060	3'-0" x 7'-0"		15	1.067		233	21		254	292
	0070	5'-4" x 6'-8", double		8	2		315	39.50		354.50	420
	0220	Half glass, 2'-8" x 6'-8"		17	.941		226	18.55		244.55	280
	0240	3'-0" x 6'-8"		16	1		229	19.70		248.70	286
	0260	3'-0" x 7'-0"		16	1		277	19.70		296.70	340
	0270	5'-4" x 6'-8", double		8	2		485	39.50		524.50	600
	0720	Raised plastic face, full panel, 2'-8" x 6'-8"		16	1		221	19.70		240.70	277
	0740	3'-0" x 6'-8"		15	1.067		225	21		246	283
	0760	3'-0" x 7'-0"		15	1.067		239	21		260	298
	0780	5'-4" x 6'-8", double		8	2		440	39.50		479.50	555
	0820	Half glass, 2'-8" x 6'-8"		17	.941		263	18.55		281.55	320
	0840	3'-0" x 6'-8"		16	1		266	19.70		285.70	325
	0860	3'-0" x 7'-0"		16	1		291	19.70		310.70	355
	0880	5'-4" x 6'-8", double		8	2		550	39.50		589.50	675
	1320	Flush face, full panel, 2'-6" x 6'-8"		16	1		200	19.70		219.70	254
	1340	3'-0" x 6'-8"		15	1.067		202	21		223	259
	1360	3'-0" x 7'-0"		15	1.067		256	21		277	320
	1380	5'-4" x 6'-8", double		8	2		400	39.50		439.50	510
	1420	Half glass, 2'-8" x 6'-8"		17	.941		247	18.55		265.55	305
	1440	3'-0" x 6'-8"		16	1		250	19.70		269.70	310
	1460	3'-0" x 7'-0"		16	1		305	19.70		324.70	370
	1480	5'-4" x 6'-8", double		8	2		495	39.50		534.50	615
	2300	Interior, residential, closet, bi-fold, 6'-8" x 2'-0" wide		16	1		128	19.70		147.70	175
	2330	3'-0" wide		16	1		144	19.70		163.70	192
	2360	4'-0" wide		15	1.067		218	21		239	276
	2400	5'-0" wide		14	1.143		253	22.50		275.50	315
	2420	6'-0" wide		13	1.231		283	24.50		307.50	350
820	0010	**STEEL FRAMES, KNOCK DOWN** R08110-100									
	0020	18 ga., up to 5-3/4" deep									
	0025	6'-8" high, 3'-0" wide, single	2 Carp	16	1	Ea.	61.50	19.70		81.20	102
	0040	6'-0" wide, double		14	1.143		74.50	22.50		97	121

08100 | Metal Doors & Frames

08110 | Steel Doors & Frames

		CREW	DAILY OUTPUT	LABOR-HOURS	UNIT	MAT.	LABOR	EQUIP.	TOTAL	TOTAL INCL O&P
0100	7'-0" high, 3'-0" wide, single	2 Carp	16	1	Ea.	63.50	19.70		83.20	104
0140	6'-0" wide, double	R08110-100	14	1.143		77.50	22.50		100	124
1000	18 ga., up to 4-7/8" deep, 7'-0" H, 3'-0" W, single		16	1		65.50	19.70		85.20	106
1140	6'-0" wide, double		14	1.143		79	22.50		101.50	126
2800	16 ga., up to 3-7/8" deep, 7'-0" high, 3'-0" wide, single		16	1		70	19.70		89.70	111
2840	6'-0" wide, double		14	1.143		86	22.50		108.50	133
3600	5-3/4" deep, 7'-0" high, 4'-0" wide, single		15	1.067		73	21		94	117
3640	8'-0" wide, double		12	1.333		93.50	26.50		120	148
3700	8'-0" high, 4'-0" wide, single		15	1.067		91	21		112	136
3740	8'-0" wide, double		12	1.333		92.50	26.50		119	147
4000	6-3/4" deep, 7'-0" high, 4'-0" wide, single		15	1.067		82	21		103	127
4040	8'-0"		12	1.333		101	26.50		127.50	156
4100	8'-0" high, 4'-0" wide, single		15	1.067		95	21		116	141
4140	8'-0" wide, double		12	1.333		113	26.50		139.50	169
4400	8-3/4" deep, 7'-0" high, 4'-0" wide, single		15	1.067		93	21		114	138
4440	8'-0" wide, double		12	1.333		120	26.50		146.50	177
4500	8'-0" high, 4'-0" wide, single		15	1.067		104	21		125	150
4540	8'-0" wide, double		12	1.333		113	26.50		139.50	170
4900	For welded frames, add					32			32	35
5400	16 ga., "B" label, up to 5-3/4" deep, 7'-0" high, 4'-0" wide, single	2 Carp	15	1.067		80.50	21		101.50	125
5440	8'-0" wide, double		12	1.333		122	26.50		148.50	180
5800	6-3/4" deep, 7'-0" high, 4'-0" wide, single		15	1.067		86.50	21		107.50	132
5840	8'-0" wide, double		12	1.333		116	26.50		142.50	172
6200	8-3/4" deep, 7'-0" high, 4'-0" wide, single		15	1.067		96.50	21		117.50	142
6240	8'-0" wide, double		12	1.333		138	26.50		164.50	197
6300	For "A" label use same price as "B" label									
6400	For baked enamel finish, add					30%	15%			
6500	For galvanizing, add					15%				
7900	Transom lite frames, fixed, add	2 Carp	155	.103	S.F.	25	2.03		27.03	31
8000	Movable, add	"	130	.123	"	30	2.42		32.42	37

08160 | Sliding Metal Doors & Grilles

		CREW	DAILY OUTPUT	LABOR-HOURS	UNIT	MAT.	LABOR	EQUIP.	TOTAL	TOTAL INCL O&P
0010	**STEEL, SLIDING**									
0021	Up to 50' x 18', electric, standard duty, minimum	E-2	360	.133	S.F.	18	2.86	2.84	23.70	28.50
0101	Maximum		340	.141		30	3.03	3.01	36.04	42
0501	Heavy duty, minimum		297	.162		24.50	3.47	3.44	31.41	37.50
0601	Maximum		277	.173		64.50	3.72	3.69	71.91	82

08200 | Wood & Plastic Doors

08210 | Wood Doors

		CREW	DAILY OUTPUT	LABOR-HOURS	UNIT	MAT.	LABOR	EQUIP.	TOTAL	TOTAL INCL O&P
0010	**KALAMEIN**									
0021	Interior, flush type, 3' x 7'	2 Carp	4.30	3.721	Opng.	173	73.50		246.50	315
0010	**PRE-HUNG DOORS**									
0300	Exterior, wood, comb. storm & screen, 6'-9" x 2'-6" wide	2 Carp	15	1.067	Ea.	208	21		229	265
0320	2'-8" wide		15	1.067		208	21		229	265
0340	3'-0" wide		15	1.067		218	21		239	276
0360	For 7'-0" high door, add					28			28	30.50
0370	For aluminum storm doors, see division 08280-800									

08200 | Wood & Plastic Doors

08210 | Wood Doors

			CREW	DAILY OUTPUT	LABOR-HOURS	UNIT	2000 BARE COSTS MAT.	LABOR	EQUIP.	TOTAL	TOTAL INCL O&P
720	1600	Entrance door, flush, birch, solid core										720
	1620	4-5/8" solid jamb, 1-3/4" x 6'-8" x 2'-8" wide	2 Carp	16	1	Ea.	191	19.70		210.70	244
	1640	3'-0" wide	"	16	1		199	19.70		218.70	252
	1680	For 7'-0" high door, add					13.25			13.25	14.60
	2000	Entrance door, colonial, 6 panel pine									
	2020	4-5/8" solid jamb, 1-3/4" x 6'-8" x 2'-8" wide	2 Carp	16	1	Ea.	440	19.70		459.70	520
	2040	3'-0" wide	"	16	1		465	19.70		484.70	545
	2060	For 7'-0" high door, add					13.25			13.25	14.60
	2200	For 5-5/8" solid jamb, add					13.25			13.25	14.60
	4000	Interior, passage door, 4-5/8" solid jamb									
	4400	Lauan, flush, solid core, 1-3/8" x 6'-8" x 2'-6" wide	2 Carp	20	.800	Ea.	135	15.75		150.75	175
	4420	2'-8" wide		20	.800		138	15.75		153.75	179
	4440	3'-0" wide		19	.842		161	16.60		177.60	206
	4600	Hollow core, 1-3/8" x 6'-8" x 2'-6" wide		20	.800		102	15.75		117.75	139
	4620	2'-8" wide		20	.800		102	15.75		117.75	139
	4640	3'-0" wide		19	.842		104	16.60		120.60	143
	4700	For 7'-0" high door, add					8.95			8.95	9.80
	5000	Birch, flush, solid core, 1-3/8" x 6'-8" x 2'-6" wide	2 Carp	20	.800		153	15.75		168.75	195
	5020	2'-8" wide		20	.800		156	15.75		171.75	198
	5040	3'-0" wide		19	.842		162	16.60		178.60	208
	5200	Hollow core, 1-3/8" x 6'-8" x 2'-6" wide		20	.800		122	15.75		137.75	161
	5220	2'-8" wide		20	.800		146	15.75		161.75	188
	5240	3'-0" wide		19	.842		127	16.60		143.60	169
	5280	For 7'-0" high door, add					9.40			9.40	10.35
	5500	Hardboard paneled, 1-3/8" x 6'-8" x 2'-6" wide	2 Carp	20	.800		123	15.75		138.75	162
	5520	2'-8" wide		20	.800		125	15.75		140.75	164
	5540	3'-0" wide		19	.842		127	16.60		143.60	169
	6000	Pine paneled, 1-3/8" x 6'-8" x 2'-6" wide		20	.800		210	15.75		225.75	258
	6020	2'-8" wide		20	.800		221	15.75		236.75	270
	6040	3'-0" wide		19	.842		229	16.60		245.60	281
	6500	For 5-5/8" solid jamb, add					21.50			21.50	24
	6520	For split jamb, deduct					12.05			12.05	13.25
850	0010	**TIN CLAD** R08110-100										850
	0021	3 ply, 6' x 7', double sliding, manual with hardware	2 Carp	1	16	Opng.	1,325	315		1,640	2,000
	1000	For electric operator, add	1 Elec	2	4	"	2,350	88.50		2,438.50	2,725
900	0010	**WOOD DOOR, ARCHITECTURAL** R08210-100										900
	0015	Flush, int., 1-3/4", 7 ply, hollow core,									
	0020	Lauan face, 2'-0" x 6'-8"	2 Carp	17	.941	Ea.	31.50	18.55		50.05	66.50
	0040	2'-6" x 6'-8"		17	.941		49.50	18.55		68.05	86.50
	0080	3'-0" x 6'-8"		17	.941		45	18.55		63.55	81.50
	0100	4'-0" x 6'-8"		16	1		72	19.70		91.70	113
	0120	Birch face, 2'-0" x 6'-8"		17	.941		46	18.55		64.55	82.50
	0140	2'-6" x 6'-8"		17	.941		60	18.55		78.55	98
	0180	3'-0" x 6'-8"		17	.941		63	18.55		81.55	102
	0200	4'-0" x 6'-8"		16	1		80.50	19.70		100.20	123
	0220	Oak face, 2'-0" x 6'-8"		17	.941		61.50	18.55		80.05	99.50
	0240	2'-6" x 6'-8"		17	.941		65.50	18.55		84.05	104
	0280	3'-0" x 6'-8"		17	.941		70	18.55		88.55	109
	0300	4'-0" x 6'-8"		16	1		89	19.70		108.70	132
	0320	Walnut face, 2'-0" x 6'-8"		17	.941		124	18.55		142.55	169
	0340	2'-6" x 6'-8"		17	.941		127	18.55		145.55	171
	0380	3'-0" x 6'-8"		17	.941		131	18.55		149.55	177
	0400	4'-0" x 6'-8"		16	1		149	19.70		168.70	198
	0430	For 7'-0" high, add					13.65			13.65	15
	0440	For 8'-0" high, add					19.10			19.10	21

08200 | Wood & Plastic Doors

08210 | Wood Doors

		Crew	Daily Output	Labor Hours	Unit	2000 Bare Costs Mat.	2000 Bare Costs Labor	2000 Bare Costs Equip.	2000 Bare Costs Total	Total Incl O&P
0460	For 8'-0" high walnut, add				Ea.	11.45			11.45	12.60
0480	For prefinishing, clear, add					30			30	33
0500	For prefinishing, stain, add					41			41	45
1320	M.D. overlay on hardboard, 2'-0" x 6'-8"	2 Carp	17	.941		84.50	18.55		103.05	125
1340	2'-6" x 6'-8"		17	.941		84.50	18.55		103.05	125
1380	3'-0" x 6'-8"		17	.941		100	18.55		118.55	142
1400	4'-0" x 6'-8"		16	1		130	19.70		149.70	177
1420	For 7'-0" high, add					7.50			7.50	8.25
1440	For 8'-0" high, add					20			20	22
1720	H.P. plastic laminate, 2'-0" x 6'-8"	2 Carp	16	1		189	19.70		208.70	242
1740	2'-6" x 6'-8"		16	1		189	19.70		208.70	242
1780	3'-0" x 6'-8"		15	1.067		215	21		236	272
1800	4'-0" x 6'-8"		14	1.143		300	22.50		322.50	370
1820	For 7'-0" high, add					7.50			7.50	8.25
1840	For 8'-0" high, add					19.10			19.10	21
2020	5 ply particle core, lauan face, 2'-6" x 6'-8"	2 Carp	15	1.067		67	21		88	110
2040	3'-0" x 6'-8"		14	1.143		73	22.50		95.50	119
2080	3'-0" x 7'-0"		13	1.231		76.50	24.50		101	126
2100	4'-0" x 7'-0"		12	1.333		94.50	26.50		121	149
2120	Birch face, 2'-6" x 6'-8"		15	1.067		80	21		101	124
2140	3'-0" x 6'-8"		14	1.143		87.50	22.50		110	135
2180	3'-0" x 7'-0"		13	1.231		89.50	24.50		114	140
2200	4'-0" x 7'-0"		12	1.333		109	26.50		135.50	165
2220	Oak face, 2'-6" x 6'-8"		15	1.067		88	21		109	133
2240	3'-0" x 6'-8"		14	1.143		97	22.50		119.50	146
2280	3'-0" x 7'-0"		13	1.231		99.50	24.50		124	152
2300	4'-0" x 7'-0"		12	1.333		122	26.50		148.50	179
2320	Walnut face, 2'-0" x 6'-8"		15	1.067		98	21		119	144
2340	2'-6" x 6'-8"		14	1.143		112	22.50		134.50	162
2380	3'-0" x 7'-0"		13	1.231		126	24.50		150.50	180
2400	4'-0" x 6'-8"		12	1.333		157	26.50		183.50	217
2440	For 8'-0" high, add					23			23	25.50
2460	For 8'-0" high walnut, add					13.40			13.40	14.75
2480	For solid wood core, add					30			30	33
2720	For prefinishing, clear, add					19.55			19.55	21.50
2740	For prefinishing, stain, add					40.50			40.50	45
3320	M.D. overlay on hardboard, 2'-6" x 6'-8"	2 Carp	14	1.143		73.50	22.50		96	120
3340	3'-0" x 6'-8"		13	1.231		81	24.50		105.50	131
3380	3'-0" x 7'-0"		12	1.333		82.50	26.50		109	136
3400	4'-0" x 7'-0"		10	1.600		101	31.50		132.50	165
3440	For 8'-0" height, add					23			23	25.50
3460	For solid wood core, add					30			30	33
3720	H.P. plastic laminate, 2'-6" x 6'-8"	2 Carp	13	1.231		113	24.50		137.50	167
3740	3'-0" x 6'-8"		12	1.333		128	26.50		154.50	186
3780	3'-0" x 7'-0"		11	1.455		133	28.50		161.50	195
3800	4'-0" x 7'-0"		8	2		162	39.50		201.50	246
3840	For 8'-0" height, add					23			23	25.50
3860	For solid wood core, add					28			28	31
4000	Exterior, flush, solid wood stave core, birch, 1-3/4" x 7'-0" x 2'-6"	2 Carp	15	1.067		142	21		163	192
4020	2'-8" wide		15	1.067		148	21		169	199
4040	3'-0" wide		14	1.143		157	22.50		179.50	212
4100	Oak faced 1-3/4" x 7'-0" x 2'-6" wide		15	1.067		156	21		177	207
4120	2'-8" wide		15	1.067		167	21		188	220
4140	3'-0" wide		14	1.143		178	22.50		200.50	235
4200	Walnut faced, 1-3/4" x 7'-0" x 2'-6" wide		15	1.067		229	21		250	288
4220	2'-8" wide		15	1.067		239	21		260	299

08200 | Wood & Plastic Doors

08210 | Wood Doors

			CREW	DAILY OUTPUT	LABOR-HOURS	UNIT	2000 BARE COSTS MAT.	LABOR	EQUIP.	TOTAL	TOTAL INCL O&P	
900	4240	3'-0" wide	2 Carp	14	1.143	Ea.	250	22.50		272.50	315	900
	4300	For 6'-8" high door, deduct from 7'-0" door	R08210 -100				12.75			12.75	14.05	
910	0010	**WOOD DOORS, DECORATOR**										910
	3000	Solid wood, 1-3/4" thick stile and rail										
	3020	Mahogany, 3'-0" x 7'-0", minimum	2 Carp	14	1.143	Ea.	440	22.50		462.50	525	
	3030	Maximum		10	1.600		640	31.50		671.50	760	
	3040	3'-6" x 8'-0", minimum		10	1.600		645	31.50		676.50	765	
	3050	Maximum		8	2		750	39.50		789.50	895	
	3100	Pine, 3'-0" x 7'-0", minimum		14	1.143		289	22.50		311.50	355	
	3110	Maximum		10	1.600		520	31.50		551.50	625	
	3120	3'-6" x 8'-0", minimum		10	1.600		570	31.50		601.50	685	
	3130	Maximum		8	2		1,550	39.50		1,589.50	1,775	
	3200	Red oak, 3'-0" x 7'-0", minimum		14	1.143		650	22.50		672.50	755	
	3210	Maximum		10	1.600		1,150	31.50		1,181.50	1,325	
	3220	3'-6" x 8'-0", minimum		10	1.600		825	31.50		856.50	960	
	3230	Maximum		8	2		1,350	39.50		1,389.50	1,575	
	4000	Hand carved door, mahogany										
	4020	3'-0" x 7'-0", minimum	2 Carp	14	1.143	Ea.	645	22.50		667.50	750	
	4030	Maximum		11	1.455		1,550	28.50		1,578.50	1,750	
	4040	3'-6" x 8'-0", minimum		10	1.600		1,050	31.50		1,081.50	1,200	
	4050	Maximum		8	2		2,200	39.50		2,239.50	2,500	
	4200	Red oak, 3'-0" x 7'-0", minimum		14	1.143		1,275	22.50		1,297.50	1,450	
	4210	Maximum		11	1.455		2,975	28.50		3,003.50	3,325	
	4220	3'-6" x 8'-0", minimum		10	1.600		2,550	31.50		2,581.50	2,850	
	4280	For 6'-8" high door, deduct from 7'-0" door					23			23	25.50	
	4400	For custom finish, add					101			101	112	
	4600	Side light, mahogany, 7'-0" x 1'-6" wide, minimum	2 Carp	18	.889		255	17.50		272.50	310	
	4610	Maximum		14	1.143		695	22.50		717.50	800	
	4620	8'-0" x 1'-6" wide, minimum		14	1.143		315	22.50		337.50	385	
	4630	Maximum		10	1.600		805	31.50		836.50	940	
	4640	Side light, oak, 7'-0" x 1'-6" wide, minimum		18	.889		345	17.50		362.50	410	
	4650	Maximum		14	1.143		800	22.50		822.50	920	
	4660	8'-0" x 1-6" wide, minimum		14	1.143		415	22.50		437.50	500	
	4670	Maximum		10	1.600		935	31.50		966.50	1,075	
	6520	Interior cafe doors, 2'-6" opening, stock, panel pine		16	1		146	19.70		165.70	194	
	6540	3'-0" opening		16	1		151	19.70		170.70	200	
	6550	Louvered pine										
	6560	2'-6" opening	2 Carp	16	1	Ea.	127	19.70		146.70	173	
	8000	3'-0" opening		16	1		135	19.70		154.70	183	
	8010	2'-6" opening, hardwood		16	1		171	19.70		190.70	222	
	8020	3'-0" opening		16	1		201	19.70		220.70	255	
	8800	Pre-hung doors, see division 08210-720										
920	0010	**WOOD DOORS, PANELED**										920
	0020	Interior, six panel, hollow core, 1-3/8" thick										
	0040	Molded hardboard, 2'-0" x 6'-8"	2 Carp	17	.941	Ea.	41	18.55		59.55	77	
	0060	2'-6" x 6'-8"		17	.941		44.50	18.55		63.05	80.50	
	0080	3'-0" x 6'-8"		17	.941		49	18.55		67.55	85.50	
	0140	Embossed print, molded hardboard, 2'-0" x 6'-8"		17	.941		44.50	18.55		63.05	80.50	
	0160	2'-6" x 6'-8"		17	.941		44.50	18.55		63.05	80.50	
	0180	3'-0" x 6'-8"		17	.941		49	18.55		67.55	85.50	
	0540	Six panel, solid, 1-3/8" thick, pine, 2'-0" x 6'-8"		15	1.067		107	21		128	154	
	0560	2'-6" x 6'-8"		14	1.143		120	22.50		142.50	171	
	0580	3'-0" x 6'-8"		13	1.231		138	24.50		162.50	194	
	1020	Two panel, bored rail, solid, 1-3/8" thick, pine, 1'-6" x 6'-8"		16	1		195	19.70		214.70	249	
	1040	2'-0" x 6'-8"		15	1.067		257	21		278	320	
	1060	2'-6" x 6'-8"		14	1.143		293	22.50		315.50	365	

08200 | Wood & Plastic Doors

08210 | Wood Doors

			CREW	DAILY OUTPUT	LABOR-HOURS	UNIT	2000 BARE COSTS MAT.	LABOR	EQUIP.	TOTAL	TOTAL INCL O&P	
920	1340	Two panel, solid, 1-3/8" thick, fir, 2'-0" x 6'-8"	2 Carp	15	1.067	Ea.	107	21		128	154	920
	1360	2'-6" x 6'-8"		14	1.143		120	22.50		142.50	171	
	1380	3'-0" x 6'-8"		13	1.231		293	24.50		317.50	365	
	1740	Five panel, solid, 1-3/8" thick, fir, 2'-0" x 6'-8"		15	1.067		192	21		213	247	
	1760	2'-6" x 6'-8"		14	1.143		293	22.50		315.50	365	
	1780	3'-0" x 6'-8"		13	1.231		294	24.50		318.50	365	
930	0010	**WOOD DOORS, RESIDENTIAL**										930
	0200	Exterior, combination storm & screen, pine										
	0260	2'-8" wide	2 Carp	10	1.600	Ea.	199	31.50		230.50	273	
	0280	3'-0" wide		9	1.778		206	35		241	286	
	0300	7'-1" x 3'-0" wide		9	1.778		209	35		244	290	
	0400	Full lite, 6'-9" x 2'-6" wide		11	1.455		204	28.50		232.50	274	
	0420	2'-8" wide		10	1.600		204	31.50		235.50	279	
	0440	3'-0" wide		9	1.778		211	35		246	292	
	0500	7'-1" x 3'-0" wide		9	1.778		230	35		265	315	
	0700	Dutch door, pine, 1-3/4" x 6'-8" x 2'-8" wide, minimum		12	1.333		550	26.50		576.50	650	
	0720	Maximum		10	1.600		585	31.50		616.50	695	
	0800	3'-0" wide, minimum		12	1.333		575	26.50		601.50	680	
	0820	Maximum		10	1.600		625	31.50		656.50	740	
	1000	Entrance door, colonial, 1-3/4" x 6'-8" x 2'-8" wide		16	1		298	19.70		317.70	365	
	1020	6 panel pine, 3'-0" wide		15	1.067		325	21		346	390	
	1100	8 panel pine, 2'-8" wide		16	1		445	19.70		464.70	525	
	1120	3'-0" wide		15	1.067		480	21		501	565	
	1200	For tempered safety glass lites, add					23.50			23.50	26	
	1300	Flush, birch, solid core, 1-3/4" x 6'-8" x 2'-8" wide	2 Carp	16	1		80	19.70		99.70	123	
	1320	3'-0" wide		15	1.067		76	21		97	120	
	1350	7'-0" x 2'-8" wide		16	1		77.50	19.70		97.20	119	
	1360	3'-0" wide		15	1.067		91.50	21		112.50	137	
	1380	For tempered safety glass lites, add					49.50			49.50	54.50	
	2700	Interior, closet, bi-fold, w/hardware, no frame or trim incl.										
	2720	Flush, birch, 6'-6" or 6'-8" x 2'-6" wide	2 Carp	13	1.231	Ea.	43	24.50		67.50	89	
	2740	3'-0" wide		13	1.231		46.50	24.50		71	93	
	2760	4'-0" wide		12	1.333		78.50	26.50		105	132	
	2780	5'-0" wide		11	1.455		85.50	28.50		114	143	
	2800	6'-0" wide		10	1.600		93	31.50		124.50	156	
	3000	Raised panel pine, 6'-6" or 6'-8" x 2'-6" wide		13	1.231		119	24.50		143.50	173	
	3020	3'-0" wide		13	1.231		134	24.50		158.50	189	
	3040	4'-0" wide		12	1.333		203	26.50		229.50	268	
	3060	5'-0" wide		11	1.455		234	28.50		262.50	305	
	3080	6'-0" wide		10	1.600		263	31.50		294.50	345	
	3200	Louvered, pine 6'-6" or 6'-8" x 2'-6" wide		13	1.231		98	24.50		122.50	150	
	3220	3'-0" wide		13	1.231		108	24.50		132.50	160	
	3240	4'-0" wide		12	1.333		169	26.50		195.50	231	
	3260	5'-0" wide		11	1.455		153	28.50		181.50	217	
	3280	6'-0" wide		10	1.600		208	31.50		239.50	283	
	4400	Bi-passing closet, incl. hardware and frame, no trim incl.										
	4420	Flush, lauan, 6'-8" x 4'-0" wide	2 Carp	12	1.333	Opng.	150	26.50		176.50	210	
	4440	5'-0" wide		11	1.455		157	28.50		185.50	222	
	4460	6'-0" wide		10	1.600		163	31.50		194.50	233	
	4600	Flush, birch, 6'-8" x 4'-0" wide		12	1.333		165	26.50		191.50	227	
	4620	5'-0" wide		11	1.455		183	28.50		211.50	250	
	4640	6'-0" wide		10	1.600		198	31.50		229.50	272	
	4800	Louvered, pine, 6'-8" x 4'-0" wide		12	1.333		330	26.50		356.50	410	
	4820	5'-0" wide		11	1.455		360	28.50		388.50	450	
	4840	6'-0" wide		10	1.600		405	31.50		436.50	500	
	5000	Paneled, pine, 6'-8" x 4'-0" wide		12	1.333		315	26.50		341.50	395	

08200 | Wood & Plastic Doors

08210 | Wood Doors

		CREW	DAILY OUTPUT	LABOR-HOURS	UNIT	2000 BARE COSTS MAT.	LABOR	EQUIP.	TOTAL	TOTAL INCL O&P	
930	5020 5'-0" wide	2 Carp	11	1.455	Opng.	350	28.50		378.50	435	930
	5040 6'-0" wide	↓	10	1.600	↓	390	31.50		421.50	485	
	6100 Folding accordion, closet, including track and frame										
	6120 Vinyl, 2 layer, stock (see also division 10651-100)	2 Carp	400	.040	S.F.	2.76	.79		3.55	4.39	
	6140 Woven mahogany and vinyl, stock		400	.040		1.41	.79		2.20	2.90	
	6160 Wood slats with vinyl overlay, stock		400	.040		9.65	.79		10.44	11.95	
	6180 Economy vinyl, stock		400	.040		1.48	.79		2.27	2.98	
	6200 Rigid PVC	↓	400	.040	↓	4.17	.79		4.96	5.95	
	6220 For custom partition, add					25%	10%				
	7310 Passage doors, flush, no frame included										
	7320 Hardboard, hollow core, 1-3/8" x 6'-8" x 1'-6" wide	2 Carp	18	.889	Ea.	36	17.50		53.50	69.50	
	7330 2'-0" wide		18	.889		37	17.50		54.50	70.50	
	7340 2'-6" wide		18	.889		41	17.50		58.50	75	
	7350 2'-8" wide		18	.889		42	17.50		59.50	76	
	7360 3'-0" wide		17	.941		34.50	18.55		53.05	70	
	7420 Lauan, hollow core, 1-3/8" x 6'-8" x 1'-6" wide		18	.889		21.50	17.50		39	54	
	7440 2'-0" wide		18	.889		23	17.50		40.50	55.50	
	7450 2'-4" wide		18	.889		26.50	17.50		44	59	
	7460 2'-6" wide		18	.889		26.50	17.50		44	59	
	7480 2'-8" wide		18	.889		27.50	17.50		45	60	
	7500 3'-0" wide		17	.941		29.50	18.55		48.05	64.50	
	7700 Birch, hollow core, 1-3/8" x 6'-8" x 1'-6" wide		18	.889		31	17.50		48.50	64	
	7720 2'-0" wide		18	.889		33.50	17.50		51	67	
	7740 2'-6" wide		18	.889		39	17.50		56.50	73	
	7760 2'-8" wide		18	.889		40.50	17.50		58	75	
	7780 3'-0" wide		17	.941		42.50	18.55		61.05	79	
	8000 Pine louvered, 1-3/8" x 6'-8" x 1'-6" wide		19	.842		99.50	16.60		116.10	138	
	8020 2'-0" wide		18	.889		115	17.50		132.50	157	
	8040 2'-6" wide		18	.889		128	17.50		145.50	171	
	8060 2'-8" wide		18	.889		139	17.50		156.50	183	
	8080 3'-0" wide		17	.941		147	18.55		165.55	194	
	8300 Pine paneled, 1-3/8" x 6'-8" x 1'-6" wide		19	.842		99.50	16.60		116.10	138	
	8320 2'-0" wide		18	.889		115	17.50		132.50	157	
	8330 2'-4" wide		18	.889		122	17.50		139.50	164	
	8340 2'-6" wide		18	.889		128	17.50		145.50	171	
	8360 2'-8" wide		18	.889		139	17.50		156.50	183	
	8380 3'-0" wide	↓	17	.941	↓	147	18.55		165.55	194	
	8550 For over 20 doors, deduct					15%					
950	0010 **WOOD FIRE DOORS**	R08210 -100									950
	0020 Particle core, 7 face plys, "B" label,										
	0040 1 hour, birch face, 1-3/4" x 2'-6" x 6'-8"	2 Carp	14	1.143	Ea.	167	22.50		189.50	223	
	0080 3'-0" x 6'-8"		13	1.231		173	24.50		197.50	232	
	0090 3'-0" x 7'-0"		12	1.333		193	26.50		219.50	257	
	0100 4'-0" x 7'-0"		12	1.333		270	26.50		296.50	340	
	0140 Oak face, 2'-6" x 6'-8"		14	1.143		190	22.50		212.50	248	
	0180 3'-0" x 6'-8"		13	1.231		198	24.50		222.50	259	
	0190 3'-0" x 7'-0"		12	1.333		208	26.50		234.50	274	
	0200 4'-0" x 7'-0"		12	1.333		282	26.50		308.50	355	
	0240 Walnut face, 2'-6" x 6'-8"		14	1.143		238	22.50		260.50	300	
	0280 3'-0" x 6'-8"		13	1.231		245	24.50		269.50	310	
	0290 3'-0" x 7'-0"		12	1.333		257	26.50		283.50	330	
	0300 4'-0" x 7'-0"		12	1.333		360	26.50		386.50	440	
	0440 M.D. overlay on hardboard, 2'-6" x 6'-8"		15	1.067		174	21		195	228	
	0480 3'-0" x 6'-8"		14	1.143		180	22.50		202.50	237	
	0490 3'-0" x 7'-0"		13	1.231		188	24.50		212.50	249	
	0500 4'-0" x 7'-0"	↓	12	1.333		263	26.50		289.50	335	

DOORS & WINDOWS 8

08200 | Wood & Plastic Doors

08210 | Wood Doors

			CREW	DAILY OUTPUT	LABOR-HOURS	UNIT	2000 BARE COSTS MAT.	LABOR	EQUIP.	TOTAL	TOTAL INCL O&P	
950	0540	H.P. plastic laminate, 2'-6" x 6'-8"	2 Carp	13	1.231	Ea.	228	24.50		252.50	293	950
	0580	3'-0" x 6'-8"	R08210 -100	12	1.333		242	26.50		268.50	310	
	0590	3'-0" x 7'-0"		11	1.455		245	28.50		273.50	320	
	0600	4'-0" x 7'-0"		10	1.600		320	31.50		351.50	410	
	0740	90 minutes, birch face, 1-3/4" x 2'-6" x 6'-8"		14	1.143		247	22.50		269.50	310	
	0780	3'-0" x 6'-8"		13	1.231		200	24.50		224.50	262	
	0790	3'-0" x 7'-0"		12	1.333		206	26.50		232.50	271	
	0800	4'-0" x 7'-0"		12	1.333		300	26.50		326.50	380	
	0840	Oak face, 2'-6" x 6'-8"		14	1.143		201	22.50		223.50	260	
	0880	3'-0" x 6'-8"		13	1.231		209	24.50		233.50	272	
	0890	3'-0" x 7'-0"		12	1.333		219	26.50		245.50	286	
	0900	4'-0" x 7'-0"		12	1.333		305	26.50		331.50	380	
	0940	Walnut face, 2'-6" x 6'-8"		14	1.143		276	22.50		298.50	345	
	0980	3'-0" x 6'-8"		13	1.231		283	24.50		307.50	350	
	0990	3'-0" x 7'-0"		12	1.333		297	26.50		323.50	370	
	1000	4'-0" x 7'-0"		12	1.333		425	26.50		451.50	515	
	1140	M.D. overlay on hardboard, 2'-6" x 6'-8"		15	1.067		185	21		206	240	
	1180	3'-0" x 6'-8"		14	1.143		191	22.50		213.50	249	
	1190	3'-0" x 7'-0"		13	1.231		200	24.50		224.50	261	
	1200	4'-0" x 7'-0"	▼	12	1.333		284	26.50		310.50	355	
	1240	For 8'-0" height, add					40			40	44	
	1260	For 8'-0" height walnut, add					50			50	55	
	1340	H.P. plastic laminate, 2'-6" x 6'-8"	2 Carp	13	1.231		239	24.50		263.50	305	
	1380	3'-0" x 6'-8"		12	1.333		252	26.50		278.50	325	
	1390	3'-0" x 7'-0"		11	1.455		256	28.50		284.50	330	
	1400	4'-0" x 7'-0"	▼	10	1.600	▼	340	31.50		371.50	430	
	2200	Custom architectural "B" label, flush, 1-3/4" thick, birch,										
	2210	Solid core										
	2220	2'-6" x 7'-0"	2 Carp	15	1.067	Ea.	325	21		346	395	
	2260	3'-0" x 7'-0"		14	1.143		330	22.50		352.50	405	
	2300	4'-0" x 7'-0"		13	1.231		455	24.50		479.50	540	
	2420	4'-0" x 8'-0"	▼	11	1.455		530	28.50		558.50	630	
	2460	For 6'-8" high door, deduct from 7'-0" door					13.15			13.15	14.45	
	2480	For oak veneer, add					50%					
	2500	For walnut veneer, add	▼			▼	75%					
960	0010	**WOOD FRAMES**										960
	0400	Exterior frame, incl. ext. trim, pine, 5/4 x 4-9/16" deep	2 Carp	375	.043	L.F.	3.89	.84		4.73	5.70	
	0420	5-3/16" deep		375	.043		6.60	.84		7.44	8.75	
	0440	6-9/16" deep		375	.043		5.70	.84		6.54	7.70	
	0600	Oak, 5/4 x 4-9/16" deep		350	.046		8.35	.90		9.25	10.70	
	0620	5-3/16" deep		350	.046		9.40	.90		10.30	11.90	
	0640	6-9/16" deep		350	.046		10.45	.90		11.35	13.05	
	0800	Walnut, 5/4 x 4-9/16" deep		350	.046		9.80	.90		10.70	12.35	
	0820	5-3/16" deep		350	.046		14.25	.90		15.15	17.20	
	0840	6-9/16" deep		350	.046		16.80	.90		17.70	20	
	1000	Sills, 8/4 x 8" deep, oak, no horns		100	.160		11.85	3.15		15	18.40	
	1020	2" horns		100	.160		13.20	3.15		16.35	19.90	
	1040	3" horns		100	.160		15.20	3.15		18.35	22	
	1100	8/4 x 10" deep, oak, no horns		90	.178		15.95	3.50		19.45	23.50	
	1120	2" horns		90	.178		17.75	3.50		21.25	25.50	
	1140	3" horns		90	.178	▼	19.35	3.50		22.85	27.50	
	2000	Exterior, colonial, frame & trim, 3' opng., in-swing, minimum		22	.727	Ea.	276	14.35		290.35	330	
	2010	Average		21	.762		235	15		250	284	
	2020	Maximum		20	.800		930	15.75		945.75	1,050	
	2100	5'-4" opening, in-swing, minimum	▼	17	.941	▼	310	18.55		328.55	375	

08200 | Wood & Plastic Doors

08210 | Wood Doors

		CREW	DAILY OUTPUT	LABOR-HOURS	UNIT	MAT.	LABOR	EQUIP.	TOTAL	TOTAL INCL O&P	
960	2120	Maximum	2 Carp	15	1.067	Ea.	930	21		951	1,050	960
	2140	Out-swing, minimum		17	.941		320	18.55		338.55	380	
	2160	Maximum		15	1.067		965	21		986	1,075	
	2400	6'-0" opening, in-swing, minimum		16	1		299	19.70		318.70	365	
	2420	Maximum		10	1.600		965	31.50		996.50	1,100	
	2460	Out-swing, minimum		16	1		320	19.70		339.70	385	
	2480	Maximum		10	1.600		1,200	31.50		1,231.50	1,375	
	2600	For two sidelights, add, minimum		30	.533	Opng.	310	10.50		320.50	360	
	2620	Maximum		20	.800	"	985	15.75		1,000.75	1,100	
	2700	Custom birch frame, 3'-0" opening		16	1	Ea.	177	19.70		196.70	229	
	2750	6'-0" opening		16	1		252	19.70		271.70	310	
	2900	Exterior, modern, plain trim, 3' opng., in-swing, minimum		26	.615		26.50	12.10		38.60	50.50	
	2920	Average		24	.667		34	13.15		47.15	59.50	
	2940	Maximum		22	.727		38.50	14.35		52.85	67	
	3000	Interior frame, pine, 11/16" x 3-5/8" deep		375	.043	L.F.	3.81	.84		4.65	5.65	
	3020	4-9/16" deep		375	.043		5.10	.84		5.94	7.10	
	3200	Oak, 11/16" x 3-5/8" deep		350	.046		4.73	.90		5.63	6.75	
	3220	4-9/16" deep		350	.046		5.10	.90		6	7.15	
	3240	5-3/16" deep		350	.046		5.25	.90		6.15	7.35	
	3400	Walnut, 11/16" x 3-5/8" deep		350	.046		8.40	.90		9.30	10.80	
	3420	4-9/16" deep		350	.046		8.40	.90		9.30	10.80	
	3440	5-3/16" deep		350	.046		8.75	.90		9.65	11.15	
	3600	Pocket door frame		16	1	Ea.	71	19.70		90.70	112	
	3800	Threshold, oak, 5/8" x 3-5/8" deep		200	.080	L.F.	3.07	1.58		4.65	6.10	
	3820	4-5/8" deep		190	.084		3.91	1.66		5.57	7.15	
	3840	5-5/8" deep		180	.089		4.69	1.75		6.44	8.15	
	4000	For casing see division 06220-400										

08260 | Sliding Wood & Plastic Doors

		CREW	DAILY OUTPUT	LABOR-HOURS	UNIT	MAT.	LABOR	EQUIP.	TOTAL	TOTAL INCL O&P	
700	0010	**GLASS, SLIDING**										700
	0013	1" insulated glass, 6'-0" x 6'-10" high	2 Carp	4	4	Opng.	750	79		829	955	
	0101	8'-0" x 6'-10" high		4	4		1,675	79		1,754	1,975	
	0501	3 leaf, 9'-0" x 6'-10" high		3	5.333		1,525	105		1,630	1,850	
	0601	12'-0" x 6'-10" high		3	5.333		1,875	105		1,980	2,250	

08280 | Wood/Plastic Storm/Screen Doors

		CREW	DAILY OUTPUT	LABOR-HOURS	UNIT	MAT.	LABOR	EQUIP.	TOTAL	TOTAL INCL O&P	
800	0010	**STORM DOORS & FRAMES** Aluminum, residential,										800
	0020	combination storm and screen										
	0400	Clear anodic coating, 6'-8" x 2'-6" wide	2 Carp	15	1.067	Ea.	138	21		159	187	
	0420	2'-8" wide		14	1.143		165	22.50		187.50	220	
	0440	3'-0" wide		14	1.143		165	22.50		187.50	220	
	0500	For 7' door height, add					5%					
	1000	Mill finish, 6'-8" x 2'-6" wide	2 Carp	15	1.067	Ea.	194	21		215	249	
	1020	2'-8" wide		14	1.143		194	22.50		216.50	252	
	1040	3'-0" wide		14	1.143		210	22.50		232.50	270	
	1100	For 7'-0" door, add					5%					
	1500	White painted, 6'-8" x 2'-6" wide	2 Carp	15	1.067	Ea.	189	21		210	244	
	1520	2'-8" wide		14	1.143		177	22.50		199.50	233	
	1540	3'-0" wide		14	1.143		196	22.50		218.50	255	
	1600	For 7'-0" door, add					5%					
	2000	Wood door & screen, see division 08210-930										
	2020											

08300 | Specialty Doors

08310 | Access Doors & Panels

		Crew	Daily Output	Labor-Hours	Unit	2000 Bare Costs Mat.	Labor	Equip.	Total	Total Incl O&P
100	**0010 ACCESS DOORS**									**100**
	1000 Fire rated door with lock									
	1100 Metal, 12" x 12"	1 Carp	10	.800	Ea.	107	15.75		122.75	145
	1150 18" x 18"		9	.889		135	17.50		152.50	179
	1200 24" x 24"		9	.889		162	17.50		179.50	209
	1250 24" x 36"		8	1		214	19.70		233.70	269
	1300 24" x 48"		8	1		261	19.70		280.70	320
	1350 36" x 36"		7.50	1.067		320	21		341	385
	1400 48" x 48"		7.50	1.067		380	21		401	450
	1600 Stainless steel, 12" x 12"		10	.800		172	15.75		187.75	217
	1650 18" x 18"		9	.889		250	17.50		267.50	305
	1700 24" x 24"		9	.889		300	17.50		317.50	360
	1750 24" x 36"		8	1		365	19.70		384.70	435
	2000 Flush door for finishing									
	2100 Metal 8" x 8"	1 Carp	10	.800	Ea.	32.50	15.75		48.25	62.50
	2150 12" x 12"	"	10	.800	"	35	15.75		50.75	65.50
	3000 Recessed door for acoustic tile									
	3100 Metal, 12" x 12"	1 Carp	4.50	1.778	Ea.	40.50	35		75.50	105
	3150 12" x 24"		4.50	1.778		56.50	35		91.50	122
	3200 24" x 24"		4	2		72.50	39.50		112	147
	3250 24" x 36"		4	2		92.50	39.50		132	169
	4000 Recessed door for drywall									
	4100 Metal 12" x 12"	1 Carp	6	1.333	Ea.	47	26.50		73.50	96.50
	4150 12" x 24"		5.50	1.455		66	28.50		94.50	122
	4200 24" x 36"		5	1.600		101	31.50		132.50	166
	6000 Standard door									
	6100 Metal, 8" x 8"	1 Carp	10	.800	Ea.	28.50	15.75		44.25	58
	6150 12" x 12"		10	.800		31	15.75		46.75	61
	6200 18" x 18"		9	.889		39	17.50		56.50	73
	6250 24" x 24"		9	.889		51	17.50		68.50	86
	6300 24" x 36"		8	1		71	19.70		90.70	113
	6350 36" x 36"		8	1		87	19.70		106.70	130
	6500 Stainless steel, 8" x 8"		10	.800		54	15.75		69.75	86.50
	6550 12" x 12"		10	.800		69	15.75		84.75	103
	6600 18" x 18"		9	.889		120	17.50		137.50	162
	6650 24" x 24"		9	.889		156	17.50		173.50	201
	7010 Aluminum cover	G-3	11	2.909		455	52.50		507.50	590
150	**0010 BULKHEAD CELLAR DOORS**									**150**
	0020 Steel, not incl. sides, 44" x 62"	1 Carp	5.50	1.455	Ea.	187	28.50		215.50	255
	0100 52" x 73"		5.10	1.569		208	31		239	282
	0500 With sides and foundation plates, 57" x 45" x 24"		4.70	1.702		244	33.50		277.50	325
	0600 42" x 49" x 51"		4.30	1.860		294	36.50		330.50	390
300	**0010 FLOOR, COMMERCIAL**									**300**
	0021 Aluminum tile, steel frame, one leaf, 2'x2' opng.	L-4	3.50	4.571	Opng.	350	79		429	520
	0051 3'-6" x 3'-6" opening		3.50	4.571		635	79		714	830
	0501 Double leaf, 4' x 4' opening		3	5.333		930	92		1,022	1,175
	0551 5' x 5' opening		3	5.333		1,375	92		1,467	1,650
350	**0010 FLOOR, INDUSTRIAL**									**350**
	0021 Steel 300 psf L.L., single leaf, 2' x 2', 175#	L-4	6	2.667	Opng.	465	46		511	590
	0051 3' x 3' opening, 300#		5.50	2.909		640	50		690	790
	0301 Double leaf, 4' x 4' opening, 455#		5	3.200		960	55.50		1,015.50	1,150
	0351 5' x 5' opening, 645#		4.50	3.556		1,250	61.50		1,311.50	1,475
	1001 Aluminum, 300 psf L.L., single leaf, 2' x 2', 60#		6	2.667		455	46		501	580

Important: See the Reference Section for critical supporting data - Reference Nos., Crews, & City Cost Indexes

08300 | Specialty Doors

08310 | Access Doors & Panels

			CREW	DAILY OUTPUT	LABOR-HOURS	UNIT	MAT.	LABOR	EQUIP.	TOTAL	TOTAL INCL O&P	
350	1051	3' x 3' opening, 100#	L-4	5.50	2.909	Opng.	710	50		760	865	350
	1501	Double leaf, 4' x 4' opening, 160#		5	3.200		1,125	55.50		1,180.50	1,325	
	1551	5' x 5' opening, 235#		4.50	3.556		1,475	61.50		1,536.50	1,725	
	2001	Aluminum, 150 psf L.L., single leaf, 2' x 2', 60#		6	2.667		415	46		461	540	
	2051	3' x 3' opening, 95#		5.50	2.909		620	50		670	765	
	2501	Double leaf, 4' x 4' opening, 150#		5	3.200		990	55.50		1,045.50	1,200	
	2551	5' x 5' opening, 230#		4.50	3.556		1,325	61.50		1,386.50	1,575	

08320 | Detention Doors & Frames

			CREW	DAILY OUTPUT	LABOR-HOURS	UNIT	MAT.	LABOR	EQUIP.	TOTAL	TOTAL INCL O&P	
900	0010	**SECURITY GATES** See division 10610-100										900
950	0010	**VAULT FRONT**										950
	0020	Door and frame, 32" x 78", clear opening										
	0101	1 hour test, weighs 750 lbs.	L-4	1.50	10.667	Opng.	2,650	184		2,834	3,225	
	0201	2 hour test, 32" door, weighs 950 lbs.		1.30	12.308		3,150	213		3,363	3,825	
	0251	40" door, weighs 1130 lbs.		1	16		3,300	276		3,576	4,100	
	0301	4 hour test, 32" door, weighs 1025 lbs.		1.20	13.333		3,175	230		3,405	3,900	
	0351	40" door, weighs 1140 lbs.		.90	17.778		3,700	305		4,005	4,575	
	0500	For stainless steel front, including frame, add to above					1,625			1,625	1,800	
	0550	Back, add					1,625			1,625	1,800	
	0600	For time lock, two movement, add	1 Elec	2	4	Ea.	1,300	88.50		1,388.50	1,575	
	0650	Three movement, add	"	2	4		1,700	88.50		1,788.50	2,025	
	0801	Day gate, painted, wire mesh, 32" wide	L-4	1.50	10.667		1,475	184		1,659	1,950	
	0851	40" wide		1.40	11.429		1,650	197		1,847	2,150	
	0901	Aluminum, 32" wide		1.50	10.667		2,250	184		2,434	2,800	
	0951	40" wide		1.40	11.429		2,500	197		2,697	3,100	
	2051	Security vault door, class I	E-24	.19	166	Opng.	13,500	3,475	2,925	19,900	24,600	
	2101	Class II		.19	166		16,100	3,475	2,925	22,500	27,500	
	2151	Class III		.13	250		21,900	5,225	4,375	31,500	38,800	

08330 | Coiling Doors & Grilles

			CREW	DAILY OUTPUT	LABOR-HOURS	UNIT	MAT.	LABOR	EQUIP.	TOTAL	TOTAL INCL O&P	
130	0010	**COUNTER DOORS**										130
	0020	Manual, incl. frm and hdwe, galv. stl., 4' roll-up, 6' long	2 Carp	2	8	Opng.	630	158		788	965	
	0301	Galvanized steel, UL label		1.80	8.889		765	175		940	1,150	
	0601	Stainless steel, 4' high roll-up, 6' long		2	8		1,050	158		1,208	1,425	
	0701	10' long		1.80	8.889		1,550	175		1,725	2,000	
	2000	Aluminum, 4' high, 4' long		2.20	7.273		700	143		843	1,025	
	2020	6' long		2	8		775	158		933	1,125	
	2040	8' long		1.90	8.421		895	166		1,061	1,275	
	2060	10' long		1.80	8.889		1,075	175		1,250	1,475	
	2080	14' long		1.40	11.429		1,550	225		1,775	2,075	
	2100	6' high, 4' long		2	8		775	158		933	1,125	
	2120	6' long		1.60	10		910	197		1,107	1,350	
	2140	10' long		1.40	11.429		1,250	225		1,475	1,750	
640	0010	**COILING GRILLE**										640
	2021	Aluminum, manual operated, mill finish	L-4	82	.195	S.F.	16.90	3.37		20.27	24.50	
	2041	Bronze anodized		82	.195	"	27	3.37		30.37	35.50	
	2061	Steel, manual operated, 10' x 10' high		1	16	Opng.	1,500	276		1,776	2,125	
	2081	15' x 8' high		.80	20	"	1,775	345		2,120	2,550	
	3000	For safety edge bottom bar, electric, add				L.F.	31			31	34.50	
	3101	For motor operation, add	L-4	5	3.200	Opng.	760	55.50		815.50	935	
	8000	For motor operation, add	2 Sswk	5	3.200	"	760	68		828	975	
700	0010	**ROLLING GRILLE SUPPORTS** Overhead framed	E-4	36	.889	L.F.	16.65	19.35	2.34	38.34	59	700

08300 | Specialty Doors

08330 | Coiling Doors & Grilles

		CREW	DAILY OUTPUT	LABOR-HOURS	UNIT	MAT.	LABOR	EQUIP.	TOTAL	TOTAL INCL O&P
0010	**ROLLING SERVICE DOORS** Steel, manual, 20 ga., incl. hardware									
0051	8' x 8' high	L-4	1.60	10	Ea.	690	173		863	1,050
0101	10' x 10' high		1.40	11.429		910	197		1,107	1,350
0201	20' x 10' high		1	16		1,875	276		2,151	2,525
0301	12' x 12' high		1.20	13.333		1,175	230		1,405	1,700
0401	20' x 12' high		.90	17.778		1,925	305		2,230	2,650
0501	14' x 14' high		.80	20		1,575	345		1,920	2,325
0601	20' x 16' high		.60	26.667		2,225	460		2,685	3,250
0701	10' x 20' high		.50	32		1,350	555		1,905	2,425
1001	12' x 12', crank operated, crank on door side		.80	20		1,050	345		1,395	1,750
1101	Crank thru wall		.70	22.857		1,175	395		1,570	1,975
1300	For vision panel, add					192			192	211
1400	For 22 ga., deduct				S.F.	.66			.66	.73
1600	3' x 7' pass door within rolling steel door, new construction				Ea.	1,050			1,050	1,150
1701	Existing construction	L-4	2	8		1,050	138		1,188	1,375
2001	Class A fire doors, manual, 20 ga., 8' x 8' high		1.40	11.429		905	197		1,102	1,325
2101	10' x 10' high		1.10	14.545		1,200	251		1,451	1,750
2201	20' x 10' high		.80	20		2,400	345		2,745	3,250
2301	12' x 12' high		1	16		1,600	276		1,876	2,225
2401	20' x 12' high		.80	20		2,775	345		3,120	3,650
2501	14' x 14' high		.60	26.667		2,075	460		2,535	3,075
2601	20' x 16' high		.50	32		3,450	555		4,005	4,750
2701	10' x 20' high		.40	40		2,200	690		2,890	3,575
3000	For 18 ga. doors, add				S.F.	.69			.69	.76
3300	For enamel finish, add				"	.81			.81	.89
3600	For safety edge bottom bar, pneumatic, add				L.F.	10.65			10.65	11.70
3700	Electric, add					20			20	22
4000	For weatherstripping, extruded rubber, jambs, add					6.90			6.90	7.60
4100	Hood, add					4.38			4.38	4.82
4200	Sill, add					2.50			2.50	2.75
4501	Motor operators, to 14' x 14' opening	L-4	5	3.200	Ea.	665	55.50		720.50	825
4601	Over 14' x 14', jack shaft type	"	5	3.200		595	55.50		650.50	750
4700	For fire door, additional fusible link, add					12.50			12.50	13.75

08341 | Cold Storage Doors

		CREW	DAILY OUTPUT	LABOR-HOURS	UNIT	MAT.	LABOR	EQUIP.	TOTAL	TOTAL INCL O&P
0010	**COLD STORAGE**									
0020	Single, 20 ga. galvanized steel									
0300	Horizontal sliding, 5' x 7', manual operation, 2" thick	2 Carp	2	8	Ea.	2,050	158		2,208	2,525
0400	4" thick		2	8		2,700	158		2,858	3,250
0500	6" thick		2	8		2,625	158		2,783	3,175
0800	5' x 7', power operation, 2" thick		1.90	8.421		4,375	166		4,541	5,075
0900	4" thick		1.90	8.421		4,450	166		4,616	5,150
1000	6" thick		1.90	8.421		5,050	166		5,216	5,850
1300	9' x 10', manual operation, 2" insulation		1.70	9.412		3,525	185		3,710	4,200
1400	4" insulation		1.70	9.412		3,650	185		3,835	4,325
1500	6" insulation		1.70	9.412		4,400	185		4,585	5,150
1800	Power operation, 2" insulation		1.60	10		6,075	197		6,272	7,025
1900	4" insulation		1.60	10		6,225	197		6,422	7,200
2000	6" insulation		1.70	9.412		7,050	185		7,235	8,075
2300	For stainless steel face, add					20%				
3001	Hinged, lightweight, 3' x 7'-0", galvanized, 2" thick	2 Carp	2	8	Ea.	1,000	158		1,158	1,375
3051	4" thick		1.90	8.421		1,150	166		1,316	1,550
3301	Aluminum doors, 3' x 7'-0", 4" thick		1.90	8.421		1,050	166		1,216	1,425
3351	6" thick		1.40	11.429		1,850	225		2,075	2,425
3601	Stainless steel, 3' x 7'-0", 4" thick		1.90	8.421		1,325	166		1,491	1,750

08300 | Specialty Doors

08341 | Cold Storage Doors

		Crew	Daily Output	Labor-Hours	Unit	Mat.	Labor	Equip.	Total	Total Incl O&P	
200	3651	6" thick	2 Carp	1.40	11.429	Ea.	2,250	225		2,475	2,850	200
	3901	Painted, 3' x 7'-0", 4" thick		1.90	8.421		955	166		1,121	1,325	
	3951	6" thick		1.40	11.429		1,825	225		2,050	2,375	
	5000	Bi-parting, electric operated										
	5011	6' x 8' opening, galv. faces, 4" thick for cooler	2 Carp	.80	20	Opng.	5,700	395		6,095	6,950	
	5051	For freezer		.80	20		6,300	395		6,695	7,600	
	5301	For door buck framing and door protection, add		2.50	6.400		465	126		591	730	
	6001	Galvanized batten door, galvanized hinges, 4' x 7'		2	8		1,400	158		1,558	1,825	
	6051	6' x 8'		1.80	8.889		1,925	175		2,100	2,400	
	6501	Fire door, 3 hr., 6' x 8', single slide		.80	20		6,600	395		6,995	7,925	
	6551	Double, bi-parting		.70	22.857		8,575	450		9,025	10,200	

08343 | Hangar Doors

		Crew	Daily Output	Labor-Hours	Unit	Mat.	Labor	Equip.	Total	Total Incl O&P	
400	0010	**HANGAR DOOR**										400
	0020	Bi-fold, ovhd., 20 PSF wind load, incl. elec. oper.										
	0101	Without sheeting, 12' high x 40'	L-4	240	.067	S.F.	9.40	1.15		10.55	12.35	
	0201	16' high x 60'		230	.070		17.05	1.20		18.25	21	
	0301	20' high x 80'		220	.073		19.35	1.26		20.61	23.50	

08344 | Industrial Doors

		Crew	Daily Output	Labor-Hours	Unit	Mat.	Labor	Equip.	Total	Total Incl O&P	
120	0010	**AIR CURTAINS** Incl. motor starters, transformers,										120
	0050	door switches & temperature controls										
	0100	Shipping and receiving doors, unheated, minimal wind stoppage										
	0150	8' high, multiples of 3' wide	2 Shee	3.60	4.444	L.F.	173	95.50		268.50	350	
	0160	5' wide		5	3.200		184	69		253	320	
	0210	10' high, multiples of 4' wide		3.60	4.444		172	95.50		267.50	350	
	0250	12' high, 3'-6" wide		3.30	4.848		171	104		275	365	
	0260	12' wide		3.30	4.848		201	104		305	395	
	0350	16' high, 3'-6" wide		3.30	4.848		335	104		439	540	
	0360	12' wide		3.30	4.848		166	104		270	360	
	0500	Maximum wind stoppage										
	0550	10' high, multiples of 4' wide	2 Shee	4.40	3.636	L.F.	127	78		205	272	
	0650	14' high, multiples of 4' wide		3.25	4.923		166	106		272	360	
	0750	20' high, multiples of 8' wide		3	5.333		360	115		475	590	
	1100	Heated, maximum wind stoppage, steam heat										
	1150	10' high, multiples of 4' wide	2 Shee	3	5.333	L.F.	465	115		580	705	
	1250	14' high, multiples of 4' wide		2.80	5.714		565	123		688	830	
	1350	20' high, multiples of 8' wide		2.60	6.154		565	132		697	850	
	1500	Customer entrance doors, unheated, minimal wind stoppage										
	1550	10' high, multiples of 3' wide	2 Shee	3.30	4.848	L.F.	173	104		277	365	
	1560	5' wide		5	3.200		187	69		256	320	
	1650	Maximum wind stoppage, 12' high, multiples of 4' wide		3.30	4.848		185	104		289	380	
	1700	Heated, minimal wind stoppage, electric heat										
	1750	8' high, multiples of 3' wide	2 Shee	3.10	5.161	L.F.	510	111		621	745	
	1760	Multiples of 5' wide		3.10	5.161		360	111		471	580	
	1850	10' high, multiples of 3' wide		3.10	5.161		455	111		566	685	
	1860	Multiples of 5' wide		3.10	5.161		360	111		471	580	
	1950	Maximum wind stoppage, steam heat										
	1960	12' high, multiples of 4' wide	2 Shee	2.80	5.714	L.F.	288	123		411	520	
	2000	Walk-in coolers and freezers, ambient air, minimal wind stoppage										
	2050	8' high, multiples of 3' wide	2 Shee	3.60	4.444	L.F.	155	95.50		250.50	330	
	2060	Multiples of 5' wide		3.60	4.444		155	95.50		250.50	330	
	2250	Maximum wind stoppage, 12' high, multiples of 3' wide		3.30	4.848		165	104		269	360	
	2450	Conveyor openings or service windows, unheated, 5' high		4	4		200	86		286	365	

08300 | Specialty Doors

08344 | Industrial Doors

				DAILY	LABOR-		2000 BARE COSTS				TOTAL	
			CREW	OUTPUT	HOURS	UNIT	MAT.	LABOR	EQUIP.	TOTAL	INCL O&P	
120	2460	Heated, electric, 5' high, 2'-6" wide	2 Shee	3.30	4.848	L.F.	205	104		309	400	120
	2700	Entrance with recirculating system, unheated, add		↓		↓	173			173	191	
	2800	Installation of each doorway	2 Shee	.10	160	Total		3,450		3,450	5,800	
200	0010	**DOUBLE ACTING, SWING**										200
	1001	.063" aluminum, 7'-2" high, 3'-4" wide	2 Carp	4.20	3.810	Pr.	355	75		430	520	
	1051	6'-8" wide	"	4	4	"	490	79		569	670	
	2000	Solid core wood, 3/4" thick, metal frame, stainless steel										
	2011	base plate, 7' high opening, 4' wide	2 Carp	4	4	Pr.	720	79		799	925	
	2051	7' wide	"	3.80	4.211	"	1,025	83		1,108	1,275	
300	0010	**GLASS DOOR, SWING**										300
	0020	Including hardware, 1/2" thick, tempered, 3' x 7' opening	2 Glaz	2	8	Opng.	1,600	155		1,755	2,000	
	0100	6' x 7' opening	"	1.40	11.429	"	3,175	221		3,396	3,875	
350	0010	**KENNEL DOORS**										350
	0021	2 way, swinging type, 13" x 19" opening	2 Carp	11	1.455	Opng.	197	28.50		225.50	266	
	0101	17" x 29" opening	"	11	1.455	"	197	28.50		225.50	266	
600	0010	**SHOCK ABSORBING DOORS**										600
	0011	Rigid, no frame, insulated, 1-13/16" thick, 5' x 7'	L-4	1.90	8.421	Opng.	1,125	145		1,270	1,500	
	0021	Rigid, no frame, insulated, 1-13/16" thick, 5' x 7'		1.90	8.421		1,125	145		1,270	1,500	
	0101	8' x 8'		1.80	8.889		1,600	154		1,754	2,025	
	0501	Flexible, frame not incl., 5' x 7' opening, economy		2	8		1,375	138		1,513	1,750	
	0601	Deluxe		1.90	8.421		2,100	145		2,245	2,550	
	1001	8' x 8' opening, economy		2	8		2,150	138		2,288	2,600	
	1101	Deluxe	↓	1.90	8.421	↓	2,775	145		2,920	3,300	
650	0010	**TUBULAR STEEL SWING DOORS**										650
	0021	Tubular steel, 7' high, single, 3'-4" opening	L-4	2.50	6.400	Ea.	465	111		576	700	
	0101	Double, 6'-0" opening	"	2	8	Pr.	590	138		728	885	

08348 | Sound Control Doors

100	0010	**ACOUSTICAL DOORS**										100
	0020	Including framed seals, 3' x 7', wood, 27 STC rating	2 Carp	1.50	10.667	Ea.	330	210		540	725	
	0100	Steel, 40 STC rating		1.50	10.667		1,275	210		1,485	1,750	
	0200	45 STC rating		1.50	10.667		1,675	210		1,885	2,200	
	0300	48 STC rating		1.50	10.667		2,125	210		2,335	2,700	
	0400	52 STC rating	↓	1.50	10.667	↓	2,550	210		2,760	3,150	

08360 | Overhead Doors

550	0010	**OVERHEAD, COMMERCIAL** Frames not included										550
	1001	Stock, sectional, heavy duty, wood, 1-3/4" thick, 8' x 8' high	2 Carp	2	8	Ea.	460	158		618	775	
	1101	10' x 10' high		1.80	8.889		690	175		865	1,050	
	1201	12' x 12' high		1.50	10.667		960	210		1,170	1,400	
	1301	Chain hoist, 14' x 14' high		1.30	12.308		1,525	242		1,767	2,100	
	1401	12' x 16' high		1	16		1,350	315		1,665	2,025	
	1501	20' x 8' high		.80	20		1,250	395		1,645	2,050	
	1601	20' x 16' high		.60	26.667		3,150	525		3,675	4,375	
	1801	Center mullion openings, 8' high		4	4		490	79		569	675	
	1901	20' high	↓	2	8	↓	915	158		1,073	1,275	
	2100	For medium duty custom door, deduct					5%	5%				
	2150	For medium duty stock doors, deduct					10%	5%				
	2301	Fiberglass and aluminum, heavy duty, sectional, 12' x 12' high	2 Carp	1.50	10.667	Ea.	1,275	210		1,485	1,750	
	2451	Chain hoist, 20' x 20' high		.50	32		4,400	630		5,030	5,900	
	2601	Steel, 24 ga. sectional, manual, 8' x 8' high		2	8		375	158		533	680	
	2651	10' x 10' high	↓	1.80	8.889	↓	515	175		690	865	

08300 | Specialty Doors

08360 | Overhead Doors

		Crew	Daily Output	Labor-Hours	Unit	2000 Bare Costs Mat.	Labor	Equip.	Total	Total Incl O&P	
550	2701	12' x 12' high	2 Carp	1.50	10.667	Ea.	690	210		900	1,125	550
	2801	Chain hoist, 20' x 14' high	↓	.70	22.857	↓	1,825	450		2,275	2,775	
	2850	For 1-1/4" rigid insulation and 26 ga. galv.										
	2860	back panel, add				S.F.	1.78			1.78	1.96	
	2900	For electric trolley operator, 1/3 H.P., to 12' x 12', add	1 Carp	2	4	Ea.	2,950	79		3,029	3,375	
	2950	Over 12' x 12', 1/2 H.P., add	"	1	8	"	745	158		903	1,100	
600	0010	**RESIDENTIAL GARAGE DOORS** Including hardware, no frame										600
	0051	Hinged, wood, custom, double door, 9' x 7'	2 Carp	3	5.333	Ea.	310	105		415	520	
	0071	16' x 7'		2	8		525	158		683	850	
	0201	Overhead, sectional, incl. hardware, fiberglass, 9' x 7', standard		8	2		510	39.50		549.50	630	
	0221	Deluxe		8	2		595	39.50		634.50	725	
	0301	16' x 7', standard		6	2.667		790	52.50		842.50	955	
	0321	Deluxe		6	2.667		985	52.50		1,037.50	1,175	
	0501	Hardboard, 9' x 7', standard		8	2		325	39.50		364.50	425	
	0521	Deluxe		8	2		390	39.50		429.50	500	
	0601	16' x 7', standard		6	2.667		570	52.50		622.50	720	
	0621	Deluxe		6	2.667		665	52.50		717.50	825	
	0701	Metal, 9' x 7', standard		8	2		257	39.50		296.50	350	
	0721	Deluxe		8	2		510	39.50		549.50	630	
	0801	16' x 7', standard		6	2.667		505	52.50		557.50	645	
	0821	Deluxe		6	2.667		700	52.50		752.50	860	
	0901	Wood, 9' x 7', standard		8	2		360	39.50		399.50	470	
	0921	Deluxe		8	2		1,025	39.50		1,064.50	1,200	
	1001	16' x 7', standard		6	2.667		725	52.50		777.50	890	
	1021	Deluxe	↓	6	2.667		1,525	52.50		1,577.50	1,775	
	1800	Door hardware, sectional	1 Carp	4	2		159	39.50		198.50	243	
	1810	Door tracks only		4	2		74.50	39.50		114	150	
	1820	One side only	↓	7	1.143		51.50	22.50		74	95	
	3001	Swing-up, including hardware, fiberglass, 9' x 7', standard	2 Carp	8	2		500	39.50		539.50	620	
	3021	Deluxe		8	2		590	39.50		629.50	720	
	3101	16' x 7' standard		6	2.667		630	52.50		682.50	780	
	3121	Deluxe		6	2.667		735	52.50		787.50	895	
	3201	Hardboard, 9' x 7', standard		8	2		258	39.50		297.50	350	
	3221	Deluxe		8	2		340	39.50		379.50	445	
	3301	16' x 7', standard		6	2.667		360	52.50		412.50	485	
	3321	Deluxe		6	2.667		535	52.50		587.50	680	
	3401	Metal, 9' x 7', standard		8	2		283	39.50		322.50	380	
	3421	Deluxe		8	2		445	39.50		484.50	560	
	3501	16' x 7', standard		6	2.667		445	52.50		497.50	580	
	3522	Deluxe		6	2.667		715	52.50		767.50	875	
	3601	Wood, 9' x 7', standard		8	2		310	39.50		349.50	410	
	3621	Deluxe		8	2		495	39.50		534.50	615	
	3701	16' x 7', standard		6	2.667		540	52.50		592.50	680	
	3721	Deluxe	↓	6	2.667		760	52.50		812.50	925	
	3900	Door hardware only, swing up	1 Carp	4	2		78.50	39.50		118	154	
	3920	One side only		7	1.143		50.50	22.50		73	94	
	4000	For electric operator, economy, add		8	1		208	19.70		227.70	262	
	4100	Deluxe, including remote control	↓	8	1	↓	298	19.70		317.70	365	
	4500	For transmitter/receiver control, add to operator				Total	69.50			69.50	76.50	
	4600	Transmitters, additional				"	22.50			22.50	24.50	
800	0010	**TELESCOPING STEEL DOORS** Baked enamel finish										800
	1000	Overhead, .03" thick, electric operated, 10' x 10'	E-3	.80	30	Ea.	8,825	660	105	9,590	11,100	
	2000	20' x 10'		.60	40		11,700	875	140	12,715	14,800	
	3000	20' x 16'	↓	.40	60	↓	13,500	1,325	210	15,035	17,700	

243

08300 | Specialty Doors

08370 | Vertical Lift Doors

		CREW	DAILY OUTPUT	LABOR-HOURS	UNIT	MAT.	LABOR	EQUIP.	TOTAL	TOTAL INCL O&P	
950	0010 **VERTICAL LIFT DOORS**										950
	0020 Motorized, 14 ga. stl, incl., frm and ctrl pnl										
	0050 16' x 16' high	L-10	.50	48	Ea.	15,200	1,050	905	17,155	19,600	
	0100 10' x 20' high		1.30	18.462		17,000	400	350	17,750	19,800	
	0120 15' x 20' high		1.30	18.462		21,100	400	350	21,850	24,300	
	0140 20' x 20' high		1	24		25,500	520	455	26,475	29,500	
	0160 25' x 20' high		1	24		28,700	520	455	29,675	33,100	
	0170 32' x 24' high		.75	32		29,800	695	605	31,100	34,700	
	0180 20' x 25' high		1	24		30,200	520	455	31,175	34,700	
	0200 25' x 25' high		.70	34.286		34,200	745	650	35,595	39,800	
	0220 25' x 30' high		.70	34.286		36,800	745	650	38,195	42,600	
	0240 30' x 30' high		.70	34.286		42,200	745	650	43,595	48,500	
	0260 35' x 30' high	↓	.70	34.286	↓	46,600	745	650	47,995	53,500	

08380 | Traffic Doors

		CREW	DAILY OUTPUT	LABOR-HOURS	UNIT	MAT.	LABOR	EQUIP.	TOTAL	TOTAL INCL O&P	
480	0010 **FLEXIBLE TRANSPARENT STRIP ENTRANCE**										480
	0100 12" strip width, 2/3 overlap	3 Shee	135	.178	SF Surf	3.73	3.82		7.55	10.55	
	0200 Full overlap		115	.209		4.61	4.49		9.10	12.65	
	0220 8" strip width, 1/2 overlap		140	.171		2.73	3.69		6.42	9.20	
	0240 Full overlap	↓	120	.200	↓	3.62	4.30		7.92	11.25	
	0300 Add for suspension system, header mount				L.F.	7.20			7.20	7.95	
	0400 Wall mount				"	7.30			7.30	8	

08400 | Entrances & Storefronts

08411 | Aluminum Framed Storefront

		CREW	DAILY OUTPUT	LABOR-HOURS	UNIT	MAT.	LABOR	EQUIP.	TOTAL	TOTAL INCL O&P	
100	0010 **ALUMINUM FRAMES**										100
	0021 Entrance, 3' x 7' opening, clear finish	2 Sswk	7	2.286	Opng.	192	48.50		240.50	305	
	0060 Black finish		7	2.286		310	48.50		358.50	435	
	0101 Bronze finish		7	2.286		298	48.50		346.50	425	
	0501 6' x 7' opening, clear finish		6	2.667		273	56.50		329.50	410	
	0521 Bronze finish		6	2.667		310	56.50		366.50	450	
	0540 Black finish		6	2.667		365	56.50		421.50	510	
	1001 With 3' high transoms, 3' x 10' opening, clear finish		6.50	2.462		335	52.50		387.50	470	
	1051 Bronze finish		6.50	2.462		370	52.50		422.50	510	
	1101 Black finish		6.50	2.462		435	52.50		487.50	585	
	1501 With 3' high transoms, 6' x 10' opening, clear finish		5.50	2.909		405	62		467	565	
	1551 Bronze finish		5.50	2.909		435	62		497	600	
	1601 Black finish	↓	5.50	2.909	↓	515	62		577	685	
120	0010 **ALUMINUM DOORS** Commercial entrance										120
	2000 Medium stile, pair of 2'-6" x 7'-0"	2 Carp	1.70	9.412	Pr.	695	185		880	1,075	
	2100 3'-0" x 7'-0", single		3	5.333	Ea.	550	105		655	785	
	2200 Pair of 3'-0" x 7'-0"		1.70	9.412	Pr.	1,050	185		1,235	1,475	
	2300 3'-6" x 7'-0", single	↓	5.33	3.002	Ea.	645	59		704	810	
	5000 Flush panel doors, pair of 2'-6" x 7'-0"	2 Sswk	2	8	Pr.	845	170		1,015	1,275	
	5050 3'-0" x 7'-0", single		2.50	6.400	Ea.	425	136		561	735	
	5100 Pair of 3'-0" x 7'-0"	↓	2	8	Pr.	870	170		1,040	1,300	

08400 | Entrances & Storefronts

08411 | Aluminum Framed Storefront

				CREW	DAILY OUTPUT	LABOR-HOURS	UNIT	2000 BARE COSTS MAT.	LABOR	EQUIP.	TOTAL	TOTAL INCL O&P	
120	5150		3'-6" x 7'-0", single	2 Sswk	2.50	6.400	Ea.	520	136		656	840	120
140	0010	**ALUMINUM DOORS & FRAMES** Entrance, narrow stile, including											140
	0021		3' x 7' opening	2 Sswk	2	8	Ea.	415	170		585	795	
	0101		3'-0" x 10'-0" opening, 3' high transom		1.80	8.889		680	189		869	1,125	
	0201		3'-6" x 10'-0" opening, 3' high transom		1.80	8.889		670	189		859	1,100	
	0301		6'-0" x 7'-0" opening		1.30	12.308	Pr.	690	262		952	1,275	
	0401		6' x 10' opening, 3' high transom		1.10	14.545	"	625	310		935	1,300	
	0500		Wide stile, 2'-6" x 7'-0" opening		2	8	Ea.	700	170		870	1,100	
	0520		3'-0" x 7'-0" opening		2	8		625	170		795	1,025	
	0540		3'-6" x 7'-0" opening		2	8		655	170		825	1,050	
	0560		5'-0" x 7'-0" opening		2	8		1,000	170		1,170	1,425	
	0580		6'-0" x 7'-0" opening		1.30	12.308	Pr.	960	262		1,222	1,575	
	0600		7'-0" x 7'-0" opening		1	16	"	1,100	340		1,440	1,875	
	1100		For full vision doors, with 1/2" glass, add				Leaf	55%					
	1200		For non-standard size, add					67%					
	1300		Light bronze finish, add					36%					
	1400		Dark bronze finish, add					18%					
	1500		For black finish, add					36%					
	1600		Concealed panic device, add					895			895	985	
	1700		Electric striker release, add				Opng.	230			230	253	
	1800		Floor check, add				Leaf	685			685	750	
	1900		Concealed closer, add				"	455			455	500	
600	0010	**STAINLESS STEEL AND GLASS** Entrance unit, narrow stiles											600
	0021		3' x 7' opening, including hardware, minimum	L-4	1.60	10	Opng.	4,425	173		4,598	5,175	
	0051		Average		1.40	11.429		4,800	197		4,997	5,625	
	0101		Maximum		1.20	13.333		5,150	230		5,380	6,050	
	1000		For solid bronze entrance units, statuary finish, add					60%					
	1100		Without statuary finish, add					45%					
	2001		Balanced doors, 3' x 7', economy	L-4	.90	17.778	Ea.	6,000	305		6,305	7,125	
	2101		Premium	"	.70	22.857	"	10,300	395		10,695	12,100	
650	0010	**STOREFRONT SYSTEMS** Aluminum frame, clear 3/8" plate glass,											650
	0020		incl. 3' x 7' door with hardware (400 sq. ft. max. wall)										
	0500		Wall height to 12' high, commercial grade	2 Glaz	150	.107	S.F.	11.45	2.06		13.51	16	
	0600		Institutional grade		130	.123		15.25	2.38		17.63	20.50	
	0700		Monumental grade		115	.139		22	2.69		24.69	28.50	
	1000		6' x 7' door with hardware, commercial grade		135	.119		11.70	2.29		13.99	16.65	
	1100		Institutional grade		115	.139		16	2.69		18.69	22	
	1200		Monumental grade		100	.160		29.50	3.10		32.60	37.50	
	1500		For bronze anodized finish, add					15%					
	1600		For black anodized finish, add					30%					
	1700		For stainless steel framing, add to monumental					75%					

08460 | Automatic Entrance Doors

				CREW	DAILY OUTPUT	LABOR-HOURS	UNIT	MAT.	LABOR	EQUIP.	TOTAL	INCL O&P	
600	0010	**SLIDING ENTRANCE** 12' x 7'-6" opng., 5' x 7' door, 2 way traf.,											600
	0020		mat activated, panic pushout, incl. operator & hardware,										
	0030		not including glass or glazing	2 Glaz	.70	22.857	Opng.	5,625	440		6,065	6,900	
650	0010	**SLIDING PANELS**											650
	0020		Mall fronts, aluminum & glass, 15' x 9' high	2 Glaz	1.30	12.308	Opng.	2,200	238		2,438	2,825	
	0100		24' x 9' high		.70	22.857		3,200	440		3,640	4,250	
	0200		48' x 9' high, with fixed panels		.90	17.778		5,950	345		6,295	7,125	
	0500		For bronze finish, add					17%					

08400 | Entrances & Storefronts

08470 | Revolving Entrance Doors

			CREW	DAILY OUTPUT	LABOR-HOURS	UNIT	2000 BARE COSTS MAT.	LABOR	EQUIP.	TOTAL	TOTAL INCL O&P	
600	0010	**REVOLVING DOORS** Aluminum, 6'-6" to 7'-0" diameter,										600
	0021	6'-10" to 7' high, stock units, minimum	L-4	.38	42.105	Opng.	14,500	725		15,225	17,300	
	0051	Average		.30	53.333		17,900	920		18,820	21,300	
	0101	Maximum		.23	69.565		23,600	1,200		24,800	28,100	
	1001	Stainless steel		.15	106		26,400	1,850		28,250	32,300	
	1101	Solid bronze		.07	228		30,800	3,950		34,750	40,700	
	1500	For automatic controls, add	2 Elec	2	8		10,600	177		10,777	12,000	

08480 | Balanced Entrance Doors

			CREW	DAILY OUTPUT	LABOR-HOURS	UNIT	MAT.	LABOR	EQUIP.	TOTAL	INCL O&P	
150	0010	**BALANCED DOORS**										150
	0021	Hardware & frame, alum. & glass, 3' x 7', econ.	L-4	.90	17.778	Ea.	4,550	305		4,855	5,525	
	0151	Premium	"	.70	22.857	"	5,650	395		6,045	6,900	

08500 | Windows

08510 | Steel Windows

			CREW	DAILY OUTPUT	LABOR-HOURS	UNIT	MAT.	LABOR	EQUIP.	TOTAL	INCL O&P	
700	0010	**SCREENS**										700
	0021	For metal sash, aluminum or bronze mesh, flat screen	L-4	1,200	.013	S.F.	2.94	.23		3.17	3.63	
	0501	Wicket screen, inside window		1,000	.016		4.29	.28		4.57	5.20	
	0801	Security screen, aluminum frame with stainless steel cloth		1,200	.013		15.25	.23		15.48	17.15	
	0901	Steel grate, painted, on steel frame		1,600	.010		8.20	.17		8.37	9.30	
	1001	For solar louvers, add		160	.100		15.80	1.73		17.53	20.50	
	4000	See also division 05580-900										
750	0010	**STEEL SASH** Custom units, glazing and trim not included	R08510-100									750
	0103	Casement, 100% vented	L-4	200	.080	S.F.	34	1.38		35.38	40	
	0201	50% vented		200	.080		29	1.38		30.38	34.50	
	0301	Fixed		200	.080		19.35	1.38		20.73	24	
	1001	Projected, commercial, 40% vented		200	.080		35	1.38		36.38	41	
	1101	Intermediate, 50% vented		200	.080		36.50	1.38		37.88	42.50	
	1501	Industrial, horizontally pivoted		200	.080		37.50	1.38		38.88	44	
	1601	Fixed		200	.080		21	1.38		22.38	25.50	
	2001	Industrial security sash, 50% vented		200	.080		39	1.38		40.38	45.50	
	2101	Fixed		200	.080		32.50	1.38		33.88	38.50	
	2501	Picture window		200	.080		20.50	1.38		21.88	25	
	3001	Double hung		200	.080		38	1.38		39.38	44.50	
	5001	Mullions for above, open interior face		240	.067	L.F.	6.90	1.15		8.05	9.60	
	5101	With interior cover		240	.067	"	11.65	1.15		12.80	14.80	
770	0010	**STEEL WINDOWS** Stock, including frame, trim and insul. glass										770
	0020	See also division 13128										
	1001	Custom units, double hung, 2'-8" x 4'-6" opening	L-4 R08510-100	12	1.333	Ea.	485	23		508	575	
	1101	2'-4" x 3'-9" opening		12	1.333		400	23		423	480	
	1501	Commercial projected, 3'-9" x 5'-5" opening		10	1.600		850	27.50		877.50	985	
	1601	6'-9" x 4'-1" opening		7	2.286		1,125	39.50		1,164.50	1,300	
	2001	Intermediate projected, 2'-9" 4'-1" opening		12	1.333		475	23		498	565	
	2101	4'-1" x 5'-5" opening		10	1.600		960	27.50		987.50	1,100	

08500 | Windows

08520 | Aluminum Windows

			CREW	DAILY OUTPUT	LABOR-HOURS	UNIT	2000 BARE COSTS MAT.	LABOR	EQUIP.	TOTAL	TOTAL INCL O&P	
100	0010	**ALUMINUM SASH**										100
	0021	Stock, grade C, glaze & trim not incl., csmt.	L-4	200	.080	S.F.	26.50	1.38		27.88	31.50	
	0051	Double hung		200	.080		25.50	1.38		26.88	30.50	
	0101	Fixed casement		200	.080		9.65	1.38		11.03	13	
	0151	Picture window		200	.080		11	1.38		12.38	14.45	
	0201	Projected window		200	.080		22.50	1.38		23.88	27	
	0251	Single hung		200	.080		11.50	1.38		12.88	15	
	0301	Sliding		200	.080		14.85	1.38		16.23	18.70	
	1001	Mullions for above, tubular		240	.067	L.F.	3.94	1.15		5.09	6.30	
	2001	Custom aluminum sash, grade 3, glazing not included, minimum		200	.080	S.F.	27.50	1.38		28.88	33	
	2101	Maximum		85	.188	"	36	3.25		39.25	45	
120	0010	**ALUMINUM WINDOWS** Incl. frame and glazing, grade C										120
	0020	See also division 05120-440										
	1001	Stock units, casement, 3'-1" x 3'-2" opening	L-4	10	1.600	Ea.	258	27.50		285.50	330	
	1050	Add for storms					52			52	57	
	1601	Projected, with screen, 3'-1" x 3'-2" opening	L-4	10	1.600		184	27.50		211.50	251	
	1700	Add for storms					49			49	54	
	2001	4'-5" x 5'-3" opening	L-4	8	2		260	34.50		294.50	345	
	2100	Add for storms					68			68	74.50	
	2501	Enamel finish windows, 3'-1" x 3'-2"	L-4	10	1.600		165	27.50		192.50	230	
	2601	4'-5" x 5'-3"		8	2		248	34.50		282.50	330	
	3001	Single hung, 2' x 3' opening, enameled, standard glazed		10	1.600		124	27.50		151.50	184	
	3101	Insulating glass		10	1.600		150	27.50		177.50	213	
	3301	2'-8" x 6'-8" opening, standard glazed		8	2		263	34.50		297.50	350	
	3401	Insulating glass		8	2		340	34.50		374.50	430	
	3701	3'-4" x 5'-0" opening, standard glazed		9	1.778		170	30.50		200.50	240	
	3801	Insulating glass		9	1.778		238	30.50		268.50	315	
	3821	Folding type, 3' x 4' opening, standard glass		10	1.600		186	27.50		213.50	253	
	3841	Insulating glass		10	1.600		233	27.50		260.50	305	
	3861	4' x 5' opening, standard glass		10	1.600		260	27.50		287.50	335	
	3881	Insulating glass		10	1.600		340	27.50		367.50	425	
	3891	Awning type, 3' x 3' opening standard glass		14	1.143		305	19.75		324.75	370	
	3901	Insulating glass		14	1.143		325	19.75		344.75	390	
	3911	3' x 4' opening, standard glass		10	1.600		355	27.50		382.50	440	
	3921	Insulating glass		10	1.600		410	27.50		437.50	500	
	3931	3' x 5'-4" opening, standard glass		10	1.600		430	27.50		457.50	520	
	3941	Insulating glass		10	1.600		505	27.50		532.50	605	
	3951	4' x 5'-4" opening, standard glass		9	1.778		470	30.50		500.50	575	
	3961	Insulating glass		9	1.778		565	30.50		595.50	675	
	4001	Sliding aluminum, 3' x 2' opening, standard glazed		10	1.600		139	27.50		166.50	201	
	4101	Insulating glass		10	1.600		154	27.50		181.50	218	
	4301	5' x 3' opening, standard glazed		9	1.778		177	30.50		207.50	248	
	4401	Insulating glass		9	1.778		247	30.50		277.50	325	
	4601	8' x 4' opening, standard glazed		6	2.667		255	46		301	360	
	4701	Insulating glass		6	2.667		410	46		456	530	
	5001	9' x 5' opening, standard glazed		4	4		385	69		454	545	
	5101	Insulating glass		4	4		615	69		684	800	
	5501	Sliding, with thermal barrier and screen, 6' x 4', 2 track		8	2		525	34.50		559.50	635	
	5701	4 track		8	2		640	34.50		674.50	765	
	6000	For above units with bronze finish, add					12%					
	6200	For installation in concrete openings, add					5%					
500	0010	**JALOUSIES**										500
	0021	Aluminum incl. glazing & screens, stock, 1'-7" x 3'-2"	L-4	10	1.600	Ea.	113	27.50		140.50	172	
	0101	2'-3" x 4'-0"		10	1.600		162	27.50		189.50	226	
	0201	3'-1" x 2'-0"		10	1.600		111	27.50		138.50	170	

08500 | Windows

08520 | Aluminum Windows

			CREW	DAILY OUTPUT	LABOR-HOURS	UNIT	2000 BARE COSTS				TOTAL INCL O&P	
							MAT.	LABOR	EQUIP.	TOTAL		
500	0301	3'-1" x 5'-3"	L-4	10	1.600	Ea.	231	27.50		258.50	300	500
	1001	Mullions for above, 2'-0" long		80	.200		8.70	3.45		12.15	15.50	
	1101	5'-3" long		80	.200		14.90	3.45		18.35	22.50	
600	0010	**LOUVERS** See division 04090, 06440 & 10210										600

08550 | Wood Windows

			CREW	DAILY OUTPUT	LABOR-HOURS	UNIT	MAT.	LABOR	EQUIP.	TOTAL	INCL O&P	
100	0010	**AWNING WINDOW** Including frame, screen, and exterior trim										100
	0100	Average quality, builders model, 34" x 22", double insulated glass	1 Carp	10	.800	Ea.	220	15.75		235.75	269	
	0200	Low E glass		10	.800		220	15.75		235.75	269	
	0300	40" x 28", double insulated glass		9	.889		370	17.50		387.50	440	
	0400	Low E Glass		9	.889		370	17.50		387.50	440	
	0500	48" x 36", double insulated glass		8	1		685	19.70		704.70	790	
	0600	Low E glass		8	1		685	19.70		704.70	790	
	1000	34" x 22"		10	.800		193	15.75		208.75	239	
	1100	40" x 22"		10	.800		213	15.75		228.75	261	
	1200	36" x 28"		9	.889		227	17.50		244.50	280	
	1300	36" x 36"		9	.889		286	17.50		303.50	345	
	1400	48" x 28"		8	1		270	19.70		289.70	330	
	1500	60" x 36"		8	1		370	19.70		389.70	445	
	2000	Metal clad, deluxe, double insulated glass, 34" x 22"		10	.800		184	15.75		199.75	230	
	2100	40" x 22"		10	.800		201	15.75		216.75	249	
	2200	36" x 25"		9	.889		205	17.50		222.50	256	
	2300	40" x 30"		9	.889		267	17.50		284.50	325	
	2400	48" x 28"		8	1		241	19.70		260.70	299	
	2500	60" x 36"		8	1		237	19.70		256.70	295	
150	0010	**BOW-BAY WINDOW** Including frame, screens and grills,										150
	0020	end panels operable										
	1001	Bow type, casement, wood, bldrs mdl 8'x 5' dbl insltd glass, 4 panel	2 Carp	10	1.600	Ea.	850	31.50		881.50	990	
	1052	Low E glass		10	1.600		1,125	31.50		1,156.50	1,300	
	1101	10'-0" x 5'-0", double insulated glass, 6 panels		6	2.667		1,125	52.50		1,177.50	1,350	
	1201	Low E glass, 6 panels		6	2.667		1,200	52.50		1,252.50	1,425	
	1300	Vinyl clad, bldrs model, double insulated glass, 6'-0" x 4'-0", 3 panel		10	1.600		925	31.50		956.50	1,075	
	1340	9'-0" x 4'-0", 4 panel		8	2		1,225	39.50		1,264.50	1,425	
	1380	10'-0" x 6'-0", 5 panels		7	2.286		2,075	45		2,120	2,350	
	1420	12'-0" x 6'-0", 6 panels		6	2.667		2,650	52.50		2,702.50	3,000	
	1601	Metal clad, bldrs model, 6'-0" x 4'-0" dbl insltd glass, 3 panels		10	1.600		770	31.50		801.50	900	
	1641	9'-0" x 4'-0", 4 panels		8	2		1,075	39.50		1,114.50	1,275	
	1681	10'-0" x 5'-0", 5 panels		7	2.286		1,475	45		1,520	1,700	
	1721	12'-0" x 6'-0", 6 panels		6	2.667		2,075	52.50		2,127.50	2,375	
	2001	Bay window, casement, bldrs mdl, 8' x 5', dbl insul gls, 4 panels		10	1.600		1,300	31.50		1,331.50	1,475	
	2051	Low E glass		10	1.600		1,575	31.50		1,606.50	1,775	
	2101	12'-0" x 6'-0", double insulated glass, 6 panels		6	2.667		1,625	52.50		1,677.50	1,875	
	2201	Low E glass		6	2.667		1,675	52.50		1,727.50	1,950	
	2280	6'-0" x 4'-0"		11	1.455		955	28.50		983.50	1,100	
	2300	Vinyl clad, premium, insulating glass, 8'-0" x 5'-0"		10	1.600		1,125	31.50		1,156.50	1,275	
	2340	10'-0" x 5'-0"		8	2		1,600	39.50		1,639.50	1,825	
	2380	10'-0" x 6'-0"		7	2.286		1,650	45		1,695	1,900	
	2420	12'-0" x 6'-0"		6	2.667		1,975	52.50		2,027.50	2,275	
	2601	Metal clad, deluxed, insul. glass, 8'-0" x 5'-0" high, 4 panels		10	1.600		1,150	31.50		1,181.50	1,300	
	2641	10'-0" x 5'-0" high, 5 panels		8	2		1,225	39.50		1,264.50	1,425	
	2681	10'-0" x 6'-0" high, 5 panels		7	2.286		1,450	45		1,495	1,675	
	2721	12'-0" x 6'-0" high, 6 panels		6	2.667		2,025	52.50		2,077.50	2,325	
	3001	Double hung, bldrs. model, 8' x 4' high, std. glazed		10	1.600		910	31.50		941.50	1,050	

Important: See the Reference Section for critical supporting data - Reference Nos., Crews, & City Cost Indexes

08500 | Windows

08550 | Wood Windows

			Crew	Daily Output	Labor-Hours	Unit	Mat.	Labor	Equip.	Total	Total Incl O&P	
150	3051	Insulating glass	2 Carp	10	1.600	Ea.	975	31.50		1,006.50	1,125	150
	3101	9'-0" x 5'-0" high, 4 panels, standard glazed		6	2.667		980	52.50		1,032.50	1,175	
	3201	Insulating glass		6	2.667		1,025	52.50		1,077.50	1,225	
	3300	Vinyl clad, premium, insulating glass, 7'-0" x 4'-6"		10	1.600		945	31.50		976.50	1,075	
	3340	8'-0" x 4'-6"		8	2		960	39.50		999.50	1,125	
	3380	8'-0" x 5'-0"		7	2.286		1,000	45		1,045	1,175	
	3420	9'-0" x 5'-0"		6	2.667		1,025	52.50		1,077.50	1,225	
	3601	Metal clad, deluxe, insul. glass, 7'-0" x 4'-0" high, 3 panels		10	1.600		870	31.50		901.50	1,025	
	3641	8'-0" x 4'-0" high, 4 panels		8	2		900	39.50		939.50	1,075	
	3681	8'-0" x 5'-0" high, 4 panels		7	2.286		935	45		980	1,100	
	3721	9'-0" x 5'-0" high, 4 panels		6	2.667		990	52.50		1,042.50	1,200	
200	0010	**CASEMENT WINDOW** Including frame, screen, and exterior trim										200
	0100	Avg. quality, bldrs. model, 2'-0" x 3'-0" H, standard glazed	1 Carp	10	.800	Ea.	159	15.75		174.75	202	
	0150	Insulating glass		10	.800		196	15.75		211.75	242	
	0200	2'-0" x 4'-6" high, standard glazed		9	.889		207	17.50		224.50	257	
	0250	Insulating glass		9	.889		266	17.50		283.50	325	
	0300	2'-3" x 6'-0" high, standard glazed		8	1		226	19.70		245.70	282	
	0350	Insulating glass		8	1		299	19.70		318.70	365	
	0522	Vinyl clad, premium, insulating glass, 2'-0" x 3'-0"		10	.800		194	15.75		209.75	241	
	0524	2'-0" x 4'-0"		9	.889		231	17.50		248.50	284	
	0525	2'-0" x 5'-0"		8	1		266	19.70		285.70	325	
	0528	2'-0" x 6'-0"		8	1		315	19.70		334.70	385	
	8000	Solid vinyl, premium, insulating glass, 2'-0" x 3'-0" high		10	.800		162	15.75		177.75	206	
	8020	2'-0" x 4'-0" high		9	.889		199	17.50		216.50	249	
	8040	2'-0" x 5'-0" high		8	1		229	19.70		248.70	286	
	8100	Metal clad, deluxe, insulating glass, 2'-0" x 3'-0" high		10	.800		160	15.75		175.75	203	
	8120	2'-0" x 4'-0" high		9	.889		193	17.50		210.50	242	
	8140	2'-0" x 5'-0" high		8	1		219	19.70		238.70	275	
	8160	2'-0" x 6'-0" high		8	1		252	19.70		271.70	310	
	8200	For multiple leaf units, deduct for stationary sash										
	8220	2' high				Ea.	16.65			16.65	18.30	
	8240	4'-6" high					19.20			19.20	21	
	8260	6' high					25.50			25.50	28	
	8300	For installation, add per leaf						15%				
250	0010	**DOUBLE HUNG** Including frame, screen, and exterior trim R08550-010										250
	0100	Avg. quality, bldrs. model, 2'-0" x 3'-0" high, standard glazed	1 Carp	10	.800	Ea.	107	15.75		122.75	144	
	0150	Insulating glass		10	.800		151	15.75		166.75	194	
	0200	3'-0" x 4'-0" high, standard glazed		9	.889		140	17.50		157.50	184	
	0250	Insulating glass		9	.889		185	17.50		202.50	233	
	0300	4'-0" x 4'-6" high, standard glazed		8	1		172	19.70		191.70	223	
	0350	Insulating glass		8	1		227	19.70		246.70	284	
	1000	Vinyl clad, premium, insulating glass, 2'-6" x 3'-0"		10	.800		240	15.75		255.75	291	
	1100	3'-0" x 3'-6"		10	.800		291	15.75		306.75	345	
	1200	3'-0" x 4'-0"		9	.889		315	17.50		332.50	375	
	1300	3'-0" x 4'-6"		9	.889		330	17.50		347.50	390	
	1400	3'-0" x 5'-0"		8	1		340	19.70		359.70	410	
	1500	3'-6" x 6'-0"		8	1		570	19.70		589.70	660	
	1540	Solid vinyl, average quality, insulated glass, 2'-0" x 3'-0"		10	.800		121	15.75		136.75	160	
	1542	3'-0" x 4'-0"		9	.889		146	17.50		163.50	191	
	1544	4'-0" x 4'-6"		8	1		175	19.70		194.70	227	
	1560	Premium, insulating glass, 2'-6" x 3'-0"		10	.800		137	15.75		152.75	178	
	1562	3'-0" x 3'-6"		9	.889		159	17.50		176.50	205	
	1564	3'-0" x 4'-0"		9	.889		169	17.50		186.50	216	
	1566	3'-0" x 4'-6"		9	.889		173	17.50		190.50	221	

08500 | Windows

08550 | Wood Windows

			CREW	DAILY OUTPUT	LABOR-HOURS	UNIT	2000 BARE COSTS				TOTAL INCL O&P
							MAT.	LABOR	EQUIP.	TOTAL	
250	1568	3'-0" x 5'-0"	1 Carp	8	1	Ea.	178	19.70		197.70	230
	1570	3'-6" x 6'-0"		8	1		195	19.70		214.70	249
	2000	Metal clad, deluxe, insulating glass, 2'-6" x 3'-0" high		10	.800		170	15.75		185.75	214
	2100	3'-0" x 3'-6" high		10	.800		201	15.75		216.75	248
	2200	3'-0" x 4'-0" high		9	.889		214	17.50		231.50	265
	2300	3'-0" x 4'-6" high		9	.889		231	17.50		248.50	284
	2400	3'-0" x 5'-0" high		8	1		247	19.70		266.70	305
	2500	3'-6" x 6'-0" high		8	1		300	19.70		319.70	365
650	0010	**PALLADIAN WINDOWS**									
	0020	Aluminum clad, including frame and exterior trim									
	0040	3'-2" x 2'-0" high	2 Carp	11	1.455	Ea.	905	28.50		933.50	1,050
	0060	3'-2" x 4'-10"		11	1.455		1,025	28.50		1,053.50	1,175
	0080	3'-2" x 6'-4"		10	1.600		1,225	31.50		1,256.50	1,400
	0100	4'-0" x 4'-0"		10	1.600		1,000	31.50		1,031.50	1,150
	0120	4'-0" x 5'-4"	3 Carp	10	2.400		1,150	47.50		1,197.50	1,350
	0140	4'-0" x 6'-0"		9	2.667		1,200	52.50		1,252.50	1,425
	0160	4'-0" x 7'-4"		9	2.667		1,325	52.50		1,377.50	1,550
	0180	5'-5" x 4'-10"		9	2.667		1,300	52.50		1,352.50	1,525
	0200	5'-5" x 6'-10"		9	2.667		1,550	52.50		1,602.50	1,800
	0220	5'-5" x 7'-9"		9	2.667		1,675	52.50		1,727.50	1,950
	0240	6'-0" x 7'-11"		8	3		2,225	59		2,284	2,550
	0260	8'-0" x 6'-0"		8	3		1,900	59		1,959	2,200
670	0010	**PICTURE WINDOW** Including frame and exterior trim									
	0101	Average quality, bldrs. model, 3'-6" x 4' high, standard glazed	2 Carp	12	1.333	Ea.	238	26.50		264.50	305
	0151	Insulating glass		12	1.333		200	26.50		226.50	265
	0201	4'-0" x 4'-6" high, standard glazed		11	1.455		198	28.50		226.50	267
	0251	Insulating glass		11	1.455		202	28.50		230.50	271
	0301	5'-0" x 4'-0" high, standard glazed		11	1.455		200	28.50		228.50	269
	0351	Insulating glass		11	1.455		250	28.50		278.50	325
	0401	6'-0" x 4'-6" high, standard glazed		10	1.600		280	31.50		311.50	365
	0451	Insulating glass		10	1.600		345	31.50		376.50	435
	1000	Vinyl clad, premium, insulating glass, 4'-0" x 4'-0"		12	1.333		420	26.50		446.50	510
	1100	4'-0" x 6'-0"		11	1.455		575	28.50		603.50	685
	1200	5'-0" x 6'-0"		10	1.600		750	31.50		781.50	880
	1300	6'-0" x 6'-0"		10	1.600		765	31.50		796.50	895
	2001	Metal clad, deluxe, insulating glass, 4'-0" x 4'-0" high		12	1.333		252	26.50		278.50	320
	2101	4'-0" x 6'-0" high		11	1.455		370	28.50		398.50	460
	2201	5'-0" x 6'-0" high		10	1.600		410	31.50		441.50	505
	2301	6'-0" x 6'-0" high		10	1.600		470	31.50		501.50	570
750	0010	**SLIDING WINDOW** Including frame, screen, and exterior trim									
	0100	Average quality, bldrs. model, 3'-0" x 3'-0" high, standard glazed	1 Carp	10	.800	Ea.	128	15.75		143.75	168
	0120	Insulating glass		10	.800		161	15.75		176.75	205
	0200	4'-0" x 3'-6" high, standard glazed		9	.889		152	17.50		169.50	197
	0220	Insulating glass		9	.889		190	17.50		207.50	239
	0300	6'-0" x 5'-0" high, standard glazed		8	1		279	19.70		298.70	340
	0320	Insulating glass		8	1		335	19.70		354.70	400
	1000	Vinyl clad, premium, insulating glass, 3'-0" x 3'-0"		10	.800		465	15.75		480.75	540
	1050	4'-0" x 3'-6"		9	.889		560	17.50		577.50	645
	1100	5'-0" x 4'-0"		9	.889		645	17.50		662.50	740
	1150	6'-0" x 5'-0"		8	1		805	19.70		824.70	920
	2000	Metal clad, deluxe, insulating glass, 3'-0" x 3'-0" high		10	.800		260	15.75		275.75	315
	2050	4'-0" x 3'-6" high		9	.889		320	17.50		337.50	380
	2100	5'-0" x 4'-0" high		9	.889		385	17.50		402.50	450
	2150	6'-0" x 5'-0" high		8	1		470	19.70		489.70	550

Important: See the Reference Section for critical supporting data - Reference Nos., Crews, & City Cost Indexes

08500 | Windows

08550 | Wood Windows

		CREW	DAILY OUTPUT	LABOR-HOURS	UNIT	2000 BARE COSTS MAT.	LABOR	EQUIP.	TOTAL	TOTAL INCL O&P
770	0010 **WEATHERSTRIPPING** See division 08720									
800	0010 **WINDOW GRILLE OR MUNTIN** Snap-in type									
	0020 Colonial or diamond pattern									
	2000 Wood, awning window, glass size 28" x 16" high	1 Carp	30	.267	Ea.	22	5.25		27.25	33
	2060 44" x 24" high		32	.250		28	4.93		32.93	39.50
	2100 Casement, glass size, 20" x 36" high		30	.267		27.50	5.25		32.75	39
	2180 20" x 56" high		32	.250		36.50	4.93		41.43	49
	2200 Double hung, glass size, 16" x 24" high		24	.333	Set	57.50	6.55		64.05	75
	2280 32" x 32" high		34	.235	"	104	4.64		108.64	122
	2500 Picture, glass size, 48" x 48" high		30	.267	Ea.	110	5.25		115.25	130
	2580 60" x 68" high		28	.286	"	122	5.65		127.65	144
	2600 Sliding, glass size, 14" x 36" high		24	.333	Set	31.50	6.55		38.05	46
	2680 36" x 36" high		22	.364	"	49.50	7.15		56.65	67
820	0010 **WOOD SASH** Including glazing but not including trim									
	0051 Custom, 5'-0" x 4'-0", 1" dbl. glazed, 3/16" thick lites	2 Carp	3.20	5	Ea.	187	98.50		285.50	375
	0101 1/4" thick lites		5	3.200		192	63		255	320
	0201 1" thick, triple glazed		5	3.200		440	63		503	595
	0301 7'-0" x 4'-6" high, 1" double glazed, 3/16" thick lites		4.30	3.721		445	73.50		518.50	615
	0401 1/4" thick lites		4.30	3.721		505	73.50		578.50	680
	0501 1" thick, triple glazed		4.30	3.721		575	73.50		648.50	760
	0601 8'-6" x 5'-0" high, 1" double glazed, 3/16" thick lites		3.50	4.571		605	90		695	820
	0701 1/4" thick lites		3.50	4.571		660	90		750	880
	0801 1" thick, triple glazed		3.50	4.571		665	90		755	885
	0900 Window frames only, based on perimeter length				L.F.	3.69			3.69	4.06
	1200 Window sill, stock, per lineal foot					8.45			8.45	9.25
	1250 Casing, stock					2.93			2.93	3.22
	3000 Replacement sash, double hung, double glazing, to 12 S.F.	1 Carp	64	.125	S.F.	18	2.46		20.46	24
	3100 12 S.F. to 20 S.F.		94	.085		15.05	1.68		16.73	19.40
	3200 20 S.F. and over		106	.075		13.10	1.49		14.59	17
	3800 Triple glazing for above, add					2.71			2.71	2.98
	7000 Sash, single lite, 2'-0" x 2'-0" high	1 Carp	20	.400	Ea.	35.50	7.90		43.40	52.50
	7050 2'-6" x 2'-0" high		19	.421		38	8.30		46.30	56
	7100 2'-6" x 2'-6" high		18	.444		40.50	8.75		49.25	59.50
	7150 3'-0" x 2'-0" high		17	.471		51	9.25		60.25	72
840	0010 **WOOD SCREENS**									
	0021 Over 3 S.F., 3/4" frames	2 Carp	375	.043	S.F.	3.03	.84		3.87	4.77
	0101 1-1/8" frames	"	375	.043	"	2.98	.84		3.82	4.72

08580 | Special Function Windows

		CREW	DAILY OUTPUT	LABOR-HOURS	UNIT	MAT.	LABOR	EQUIP.	TOTAL	TOTAL INCL O&P
900	0010 **STORM WINDOWS** Aluminum, residential									
	0300 Basement, mill finish, incl. fiberglass screen									
	0320 1'-10" x 1'-0" high	2 Carp	30	.533	Ea.	25	10.50		35.50	45
	0340 2'-9" x 1'-6" high	"	30	.533	"	27	10.50		37.50	47.50
	1600 Double-hung, combination, storm & screen									
	2000 Average quality, clear anodic coating, 2'-0" x 3'-5" high	2 Carp	30	.533	Ea.	65	10.50		75.50	89.50
	2020 2'-6" x 5'-0" high		28	.571		82	11.25		93.25	110
	2040 4'-0" x 6'-0" high		25	.640		96.50	12.60		109.10	128
	2400 White painted, 2'-0" x 3'-5" high		30	.533		64	10.50		74.50	88.50
	2420 2'-6" x 5'-0" high		28	.571		71	11.25		82.25	97.50
	2440 4'-0" x 6'-0" high		25	.640		77.50	12.60		90.10	107
	2600 Mill finish, 2'-0" x 3'-5" high		30	.533		58.50	10.50		69	82.50
	2620 2'-6" x 5'-0" high		28	.571		65	11.25		76.25	91.50
	2640 4'-0" x 6-8" high		25	.640		73	12.60		85.60	102

08600 | Skylights

08620 | Unit Skylights

			CREW	DAILY OUTPUT	LABOR-HOURS	UNIT	2000 BARE COSTS MAT.	LABOR	EQUIP.	TOTAL	TOTAL INCL O&P	
400	0010	**PREFABRICATED** Glass block with metal frame, minimum	G-3	265	.121	S.F.	39.50	2.17		41.67	47	400
	0100	Maximum	"	160	.200	"	79.50	3.60		83.10	93	
800	0010	**SKYLIGHT** Plastic domes, flush or curb mounted, ten or										800
	0100	more units, curb not included										
	0300	Nominal size under 10 S.F., double	G-3	130	.246	S.F.	12	4.43		16.43	20.50	
	0400	Single		160	.200		8.05	3.60		11.65	14.95	
	0600	10 S.F. to 20 S.F., double		315	.102		9.80	1.83		11.63	13.85	
	0700	Single		395	.081		7.40	1.46		8.86	10.60	
	0900	20 S.F. to 30 S.F., double		395	.081		9.80	1.46		11.26	13.25	
	1000	Single		465	.069		8.75	1.24		9.99	11.70	
	1200	30 S.F. to 65 S.F., double		465	.069		9.20	1.24		10.44	12.20	
	1300	Single		610	.052		13.15	.94		14.09	16.05	
	1500	For insulated 4" curbs, double, add					25%					
	1600	Single, add					30%					
	1800	For integral insulated 9" curbs, double, add					30%					
	1900	Single, add					40%					
	2120	Ventilating insulated plexiglass dome with										
	2130	curb mounting, 36" x 36"	G-3	12	2.667	Ea.	340	48		388	455	
	2150	52" x 52"		12	2.667		510	48		558	645	
	2160	28" x 52"		10	3.200		400	57.50		457.50	540	
	2170	36" x 52"		10	3.200		430	57.50		487.50	575	
	2180	For electric opening system, add					256			256	281	
	2200	Field fabricated, factory type, aluminum and wire glass	G-3	120	.267	S.F.	13.20	4.79		17.99	22.50	
	2300	Insulated safety glass with aluminum frame		160	.200		78.50	3.60		82.10	92.50	
	2400	Sandwich panels, fiberglass, for walls, 1-9/16" thick, to 250 SF		200	.160		13.95	2.88		16.83	20	
	2500	250 SF and up		265	.121		12.50	2.17		14.67	17.45	
	2700	As above, but for roofs, 2-3/4" thick, to 250 SF		295	.108		20	1.95		21.95	25.50	
	2800	250 SF and up		330	.097		16.50	1.74		18.24	21	

08700 | Hardware

08710 | Door Hardware

			CREW	DAILY OUTPUT	LABOR-HOURS	UNIT	2000 BARE COSTS MAT.	LABOR	EQUIP.	TOTAL	TOTAL INCL O&P	
100	0010	**AUTOMATIC OPENERS COMMERCIAL** Swing doors, single	2 Skwk	.80	20	Ea.	2,225	395		2,620	3,125	100
	0100	Single operating pair		.50	32	Pr.	4,100	630		4,730	5,600	
	0400	For double simultaneous doors, one way, add		1.20	13.333		345	263		608	830	
	0500	Two way, add		.90	17.778		430	350		780	1,075	
	1000	Sliding doors, 3' wide, including track & hanger, single		.60	26.667	Opng.	3,775	525		4,300	5,075	
	1300	Bi-parting		.50	32		4,575	630		5,205	6,100	
	1450	Activating carpet, single door, one way add		2.20	7.273		700	144		844	1,025	
	1550	Two way, add		1.30	12.308		1,000	243		1,243	1,525	
	1751	Handicap opener, button operating	2 Carp	1.50	10.667	Ea.	1,300	210		1,510	1,775	
120	0010	**AUTOMATIC OPENERS INDUSTRIAL**, sliding doors, to 6' wide	2 Skwk	.60	26.667	Opng.	4,550	525		5,075	5,900	120
	0200	To 12' wide	"	.40	40	"	5,550	790		6,340	7,450	
	0400	Over 12' wide, add per L.F. of excess				L.F.	605			605	670	
	1000	Swing doors, to 5' wide	2 Skwk	.80	20	Ea.	2,650	395		3,045	3,600	
	1860	Add for controls, wall pushbutton, 3 button		4	4		162	79		241	315	
	1870	Ceiling pull cord		4.30	3.721		139	73.50		212.50	279	
150	0010	**AVERAGE** Percentage for hardware, total job cost, minimum									.75%	150
	0050	Maximum									3.50%	

08700 | Hardware

08710 | Door Hardware

		CREW	DAILY OUTPUT	LABOR-HOURS	UNIT	2000 BARE COSTS MAT.	LABOR	EQUIP.	TOTAL	TOTAL INCL O&P	
150	0500 Total hardware for building, average distribution					85%	15%				150
	1000 Door hardware, apartment, interior				Door	107			107	118	
	1500 Hospital bedroom, minimum					243			243	268	
	2000 Maximum				↓	535			535	585	
	2100 Pocket door				Ea.	109			109	120	
	2250 School, single exterior, incl. lever, not incl. panic device				Door	360			360	395	
	2500 Single interior, regular use, no lever included					240			240	264	
	2550 Including handicap lever					325			325	360	
	2600 Heavy use, incl. lever and closer					420			420	460	
	2850 Stairway, single interior				↓	600			600	660	
	3100 Double exterior, with panic device				Pr.	845			845	930	
	3600 Toilet, public, single interior				Door	132			132	145	
200	0010 **BOLTS, FLUSH**										200
	0020 Standard, concealed	1 Carp	7	1.143	Ea.	17.55	22.50		40.05	58	
	0800 Automatic fire exit	"	5	1.600		273	31.50		304.50	355	
	1600 For electric release, add	1 Elec	3	2.667		94	59		153	200	
	3000 Barrel, brass, 2" long	1 Carp	40	.200		2.16	3.94		6.10	9.15	
	3020 4" long		40	.200		2.35	3.94		6.29	9.35	
	3060 6" long	↓	40	.200	↓	4.78	3.94		8.72	12	
220	0010 **BUMPER PLATES**										220
	0021 1-1/2" x 3/4" U channel	2 Carp	80	.200	L.F.	3.15	3.94		7.09	10.20	
	1001 Tear drop, spring-steel 4" high		15	1.067	Ea.	40.50	21		61.50	80.50	
	1101 8" high		15	1.067		81	21		102	125	
	1201 10" high	↓	15	1.067	↓	101	21		122	148	
300	0010 **DOOR CLOSER** Rack and pinion	1 Carp	6.50	1.231	Ea.	108	24.50		132.50	161	300
	0020 Adjustable backcheck, 3 way mount, all sizes, regular arm		6	1.333		114	26.50		140.50	171	
	0040 Hold open arm		6	1.333		125	26.50		151.50	182	
	0100 Fusible link		6.50	1.231		97	24.50		121.50	149	
	0200 Non sized, regular arm		6	1.333		108	26.50		134.50	164	
	0240 Hold open arm		6	1.333		135	26.50		161.50	193	
	0400 4 way mount, non sized, regular arm		6	1.333		148	26.50		174.50	208	
	0440 Hold open arm	↓	6	1.333	↓	160	26.50		186.50	221	
	2000 Backcheck and adjustable power, hinge face mount										
	2010 All sizes, regular arm	1 Carp	6.50	1.231	Ea.	137	24.50		161.50	192	
	2040 Hold open arm		6.50	1.231		147	24.50		171.50	203	
	2400 Top jamb mount, all sizes, regular arm		6	1.333		137	26.50		163.50	195	
	2440 Hold open arm		6	1.333		147	26.50		173.50	207	
	2800 Top face mount, all sizes, regular arm		6.50	1.231		136	24.50		160.50	192	
	2840 Hold open arm		6.50	1.231		147	24.50		171.50	203	
	4000 Backcheck, overhead concealed, all sizes, regular arm		5.50	1.455		144	28.50		172.50	208	
	4040 Concealed arm		5	1.600		154	31.50		185.50	223	
	4400 Compact overhead, concealed, all sizes, regular arm		5.50	1.455		263	28.50		291.50	340	
	4440 Concealed arm		5	1.600		274	31.50		305.50	355	
	4800 Concealed in door, all sizes, regular arm		5.50	1.455		97.50	28.50		126	156	
	4840 Concealed arm		5	1.600		105	31.50		136.50	169	
	4900 Floor concealed, all sizes, single acting		2.20	3.636		124	71.50		195.50	259	
	4940 Double acting	↓	2.20	3.636		159	71.50		230.50	298	
	5000 For cast aluminum cylinder, deduct					13.10			13.10	14.40	
	5040 For delayed action, add					23			23	25	
	5080 For fusible link arm, add					9.50			9.50	10.45	
	5120 For shock absorbing arm, add					28.50			28.50	31	
	5160 For spring power adjustment, add					22			22	24	
	6000 Closer-holder, hinge face mount, all sizes, exposed arm	1 Carp	6.50	1.231		100	24.50		124.50	152	
	7000 Electronic closer-holder, hinge facemount, concealed arm	↓	5	1.600	↓	153	31.50		184.50	222	

08700 | Hardware

08710 | Door Hardware

			CREW	DAILY OUTPUT	LABOR-HOURS	UNIT	2000 BARE COSTS MAT.	LABOR	EQUIP.	TOTAL	TOTAL INCL O&P	
300	7400	With built-in detector	1 Carp	5	1.600	Ea.	460	31.50		491.50	560	300
320	0010	**DEADLOCKS** Mortise, heavy duty, outside key	1 Carp	9	.889	Ea.	112	17.50		129.50	153	320
	0020	Double cylinder		9	.889		124	17.50		141.50	166	
	0100	Medium duty, outside key		10	.800		87	15.75		102.75	123	
	0110	Double cylinder		10	.800		109	15.75		124.75	147	
	1000	Tubular, standard duty, outside key		10	.800		47	15.75		62.75	78.50	
	1010	Double cylinder		10	.800		60	15.75		75.75	93	
	1200	Night latch, outside key		10	.800		59	15.75		74.75	92	
340	0010	**DOORSTOPS** Holder and bumper, floor or wall	1 Carp	32	.250	Ea.	27	4.93		31.93	38.50	340
	1300	Wall bumper, 4" diameter, with rubber pad, aluminum		32	.250		6.60	4.93		11.53	15.70	
	1600	Door bumper, floor type, aluminum		32	.250		3.77	4.93		8.70	12.60	
	1900	Plunger type, door mounted		32	.250		23	4.93		27.93	34	
400	0010	**ENTRANCE LOCKS** Cylinder, grip handle, deadlocking latch	1 Carp	9	.889	Ea.	103	17.50		120.50	143	400
	0020	Deadbolt		8	1		125	19.70		144.70	171	
	0100	Push and pull plate, dead bolt		8	1		119	19.70		138.70	165	
	0900	For handicapped lever, add					130			130	143	
450	0010	**FLOOR CHECKS** For over 3' wide doors, single acting	1 Carp	2.50	3.200	Ea.	400	63		463	550	450
	0500	Double acting	"	2.50	3.200	"	505	63		568	670	
500	0010	**HASP** Steel, 3" assembly	1 Carp	26	.308	Ea.	2.39	6.05		8.44	13.05	500
	0020	4-1/2"		13	.615		3.04	12.10		15.14	24.50	
	0040	6"		12.50	.640		4.81	12.60		17.41	27	
520	0010	**HINGES** Full mortise, avg. freq., steel base, 4-1/2" x 4-1/2", USP R08700-100				Pr.	17.90			17.90	19.70	520
	0100	5" x 5", USP					29			29	32	
	0200	6" x 6", USP					60.50			60.50	67	
	0400	Brass base, 4-1/2" x 4-1/2", US10					37			37	40.50	
	0500	5" x 5", US10					52.50			52.50	57.50	
	0600	6" x 6", US10					87			87	96	
	0800	Stainless steel base, 4-1/2" x 4-1/2", US32					61			61	67.50	
	0900	For non removable pin, add				Ea.	2.14			2.14	2.35	
	0910	For floating pin, driven tips, add					3.90			3.90	4.29	
	0930	For hospital type tip on pin, add					11.45			11.45	12.60	
	0940	For steeple type tip on pin, add					7.65			7.65	8.40	
	0950	Full mortise, high frequency, steel base, 3-1/2" x 3-1/2", US26D				Pr.	19.40			19.40	21.50	
	1000	4-1/2" x 4-1/2", USP					37.50			37.50	41.50	
	1100	5" x 5", USP					40			40	44	
	1200	6" x 6", USP					97			97	107	
	1400	Brass base, 3-1/2" x 3-1/2", US4					35			35	38.50	
	1430	4-1/2" x 4-1/2", US10					41.50			41.50	46	
	1500	5" x 5", US10					74.50			74.50	82	
	1600	6" x 6", US10					124			124	136	
	1800	Stainless steel base, 4-1/2" x 4-1/2", US32					94.50			94.50	104	
	1930	For hospital type tip on pin, add				Ea.	15.05			15.05	16.55	
	1950	Full mortise, low frequency, steel base, 3-1/2" x 3-1/2", US26D				Pr.	12.25			12.25	13.45	
	2000	4-1/2" x 4-1/2", USP					13.45			13.45	14.80	
	2100	5" x 5", USP					21.50			21.50	23.50	
	2200	6" x 6", USP					42			42	46.50	
	2300	4-1/2" x 4-1/2", US3					9.85			9.85	10.80	
	2310	5" x 5", US3					30.50			30.50	33.50	
	2400	Brass bass, 4-1/2" x 4-1/2", US10					30.50			30.50	34	
	2500	5" x 5", US10					46			46	50.50	
	2800	Stainless steel base, 4-1/2" x 4-1/2", US32					51			51	56	
550	0010	**KICK PLATE** 6" high, for 3' door, stainless steel	1 Carp	15	.533	Ea.	14.25	10.50		24.75	33.50	550
	0500	Bronze		15	.533		20	10.50		30.50	40	

08700 | Hardware

08710 | Door Hardware

		CREW	DAILY OUTPUT	LABOR-HOURS	UNIT	2000 BARE COSTS MAT.	LABOR	EQUIP.	TOTAL	TOTAL INCL O&P
550										**550**
2000	Aluminum, .050, with 3 beveled edges, 10" x 28"	1 Carp	15	.533	Ea.	14.70	10.50		25.20	34
2010	10" x 30"		15	.533		15.75	10.50		26.25	35.50
2020	10" x 34"		15	.533		17.80	10.50		28.30	37.50
2040	10" x 38"	▼	15	.533	▼	19.65	10.50		30.15	39.50
650 0010	**LOCKSET** Standard duty, cylindrical, with sectional trim									**650**
0020	Non-keyed, passage	1 Carp	12	.667	Ea.	37	13.15		50.15	63.50
0100	Privacy		12	.667		45	13.15		58.15	72.50
0400	Keyed, single cylinder function		10	.800		66	15.75		81.75	100
0420	Hotel		8	1		90	19.70		109.70	133
0500	Lever handled, keyed, single cylinder function		10	.800		91	15.75		106.75	127
1000	Heavy duty with sectional trim, non-keyed, passages		12	.667		105	13.15		118.15	138
1100	Privacy		12	.667		132	13.15		145.15	169
1400	Keyed, single cylinder function		10	.800		156	15.75		171.75	199
1420	Hotel		8	1		233	19.70		252.70	290
1600	Communicating	▼	10	.800		185	15.75		200.75	230
1690	For re-core cylinder, add					26.50			26.50	29
1700	Residential, interior door, minimum	1 Carp	16	.500		12.60	9.85		22.45	31
1720	Maximum		8	1		33.50	19.70		53.20	71
1800	Exterior, minimum		14	.571		28	11.25		39.25	50.50
1810	Average		8	1		59.50	19.70		79.20	99.50
1820	Maximum	▼	8	1	▼	118	19.70		137.70	164
700 0010	**MORTISE LOCKSET** Comm., wrought knobs & full escutcheon trim									**700**
0020	Non-keyed, passage, minimum	1 Carp	9	.889	Ea.	139	17.50		156.50	183
0030	Maximum		8	1		225	19.70		244.70	281
0040	Privacy, minimum		9	.889		148	17.50		165.50	193
0050	Maximum		8	1		243	19.70		262.70	300
0100	Keyed, office/entrance/apartment, minimum		8	1		170	19.70		189.70	221
0110	Maximum		7	1.143		292	22.50		314.50	360
0120	Single cylinder, typical, minimum		8	1		146	19.70		165.70	195
0130	Maximum		7	1.143		270	22.50		292.50	335
0200	Hotel, minimum		7	1.143		176	22.50		198.50	232
0210	Maximum		6	1.333		286	26.50		312.50	360
0300	Communication, double cylinder, minimum		8	1		176	19.70		195.70	227
0310	Maximum		7	1.143		227	22.50		249.50	289
1000	Wrought knobs and sectional trim, non-keyed, passage, minimum		10	.800		92.50	15.75		108.25	129
1010	Maximum		9	.889		182	17.50		199.50	231
1040	Privacy, minimum		10	.800		108	15.75		123.75	146
1050	Maximum		9	.889		195	17.50		212.50	244
1100	Keyed, entrance, office/apartment, minimum		9	.889		160	17.50		177.50	206
1110	Maximum		8	1		232	19.70		251.70	289
1120	Single cylinder, typical, minimum		9	.889		154	17.50		171.50	200
1130	Maximum	▼	8	1	▼	225	19.70		244.70	281
2000	Cast knobs and full escutcheon trim									
2010	Non-keyed, passage, minimum	1 Carp	9	.889	Ea.	196	17.50		213.50	246
2020	Maximum		8	1		320	19.70		339.70	385
2040	Privacy, minimum		9	.889		236	17.50		253.50	290
2050	Maximum		8	1		345	19.70		364.70	410
2120	Keyed, single cylinder, typical, minimum		8	1		236	19.70		255.70	294
2130	Maximum		7	1.143		375	22.50		397.50	450
2200	Hotel, minimum		7	1.143		271	22.50		293.50	340
2210	Maximum		6	1.333		505	26.50		531.50	600
3000	Cast knob and sectional trim, non-keyed, passage, minimum		10	.800		153	15.75		168.75	195
3010	Maximum		10	.800		305	15.75		320.75	360
3040	Privacy, minimum		10	.800		174	15.75		189.75	218
3050	Maximum	▼	10	.800	▼	305	15.75		320.75	360

08700 | Hardware

08710 | Door Hardware

			CREW	DAILY OUTPUT	LABOR-HOURS	UNIT	2000 BARE COSTS MAT.	LABOR	EQUIP.	TOTAL	TOTAL INCL O&P	
700	3100	Keyed, office/entrance/apartment, minimum	1 Carp	9	.889	Ea.	197	17.50		214.50	247	700
	3110	Maximum		9	.889		310	17.50		327.50	370	
	3120	Single cylinder, typical, minimum		9	.889		197	17.50		214.50	247	
	3130	Maximum		9	.889		380	17.50		397.50	450	
	3190	For re-core cylinder, add					26.50			26.50	29.50	
	3900	Keyless, pushbutton type										
	4000	Residential/light commercial, deadbolt, standard	1 Carp	9	.889	Ea.	89	17.50		106.50	128	
	4010	Heavy duty		9	.889		106	17.50		123.50	146	
	4020	Industrial, heavy duty, with deadbolt		9	.889		221	17.50		238.50	273	
	4030	Key override		9	.889		245	17.50		262.50	299	
	4040	Lever activated handle		9	.889		268	17.50		285.50	325	
	4050	Key override		9	.889		297	17.50		314.50	355	
	4060	Double sided pushbutton type		8	1		490	19.70		509.70	575	
	4070	Key override		8	1		530	19.70		549.70	615	
750	0010	**PANIC DEVICE** For rim locks, single door, exit only	1 Carp	6	1.333	Ea.	330	26.50		356.50	410	750
	0020	Outside key and pull		5	1.600		380	31.50		411.50	475	
	0200	Bar and vertical rod, exit only		5	1.600		485	31.50		516.50	590	
	0210	Outside key and pull		4	2		580	39.50		619.50	710	
	0400	Bar and concealed rod		4	2		490	39.50		529.50	610	
	0600	Touch bar, exit only		6	1.333		390	26.50		416.50	475	
	0610	Outside key and pull		5	1.600		470	31.50		501.50	570	
	0700	Touch bar and vertical rod, exit only		5	1.600		530	31.50		561.50	640	
	0710	Outside key and pull		4	2		620	39.50		659.50	750	
	1000	Mortise, bar, exit only		4	2		435	39.50		474.50	550	
	1600	Touch bar, exit only		4	2		500	39.50		539.50	620	
	2000	Narrow stile, rim mounted, bar, exit only		6	1.333		525	26.50		551.50	625	
	2010	Outside key and pull		5	1.600		570	31.50		601.50	680	
	2200	Bar and vertical rod, exit only		5	1.600		540	31.50		571.50	650	
	2210	Outside key and pull		4	2		540	39.50		579.50	665	
	2400	Bar and concealed rod, exit only		3	2.667		630	52.50		682.50	780	
	3000	Mortise, bar, exit only		4	2		440	39.50		479.50	555	
	3600	Touch bar, exit only		4	2		645	39.50		684.50	775	
780	0010	**PUSH-PULL PLATE**										780
	0100	Push plate, .050 thick, 4" x 16", aluminum	1 Carp	12	.667	Ea.	5.05	13.15		18.20	28	
	0500	Bronze		12	.667		11.30	13.15		24.45	35	
	1500	Pull handle and push bar, aluminum		11	.727		108	14.35		122.35	144	
	2000	Bronze		10	.800		140	15.75		155.75	181	
	3000	Push plate both sides, aluminum		14	.571		13.25	11.25		24.50	34	
	3500	Bronze		13	.615		33.50	12.10		45.60	57.50	
	4000	Door pull, designer style, cast aluminum, minimum		12	.667		61	13.15		74.15	89.50	
	5000	Maximum		8	1		282	19.70		301.70	345	
	6000	Cast bronze, minimum		12	.667		65.50	13.15		78.65	94.50	
	7000	Maximum		8	1		305	19.70		324.70	370	
	8000	Walnut, minimum		12	.667		50	13.15		63.15	77.50	
	9000	Maximum		8	1		281	19.70		300.70	345	
800	0010	**SPECIAL HINGES**										800
	0015	Paumelle, high frequency										
	0020	Steel base, 6" x 4-1/2", US10				Pr.	107			107	118	
	0100	Bronze base, 5" x 4-1/2", US10					144			144	158	
	0200	Paumelle, average frequency, steel base, 4-1/2" x 3-1/2", US10					72.50			72.50	79.50	
	0400	Olive knuckle, low frequency, brass base, 6" x 4-1/2", US10					131			131	145	
	1000	Electric hinge with concealed conductor, average frequency										
	1010	Steel base, 4-1/2" x 4-1/2", US26D				Pr.	240			240	264	
	1100	Bronze base, 4-1/2" x 4-1/2", US26D				"	249			249	274	
	1200	Electric hinge with concealed conductor, high frequency										

08700 | Hardware

08710 | Door Hardware

		CREW	DAILY OUTPUT	LABOR-HOURS	UNIT	MAT.	LABOR	EQUIP.	TOTAL	TOTAL INCL O&P
800	1210	Steel base, 4-1/2" x 4-1/2", US26D				Pr.	179			179	197
	1600	Double weight, 800 lb., steel base, removable pin, 5" x 6", USP					112			112	124
	1700	Steel base-welded pin, 5" x 6", USP					124			124	137
	1800	Triple weight, 2000 lb., steel base, welded pin, 5" x 6", USP					130			130	142
	2000	Pivot reinf., high frequency, steel base, 7-3/4" door plate, USP					144			144	159
	2200	Bronze base, 7-3/4" door plate, US10				↓	175			175	192
	3000	Swing clear, full mortise, full or half surface, high frequency,									
	3010	Steel base, 5" high, USP				Pr.	121			121	133
	3200	Swing clear, full mortise, average frequency									
	3210	Steel base, 4-1/2" high, USP				Pr.	96			96	106
	4000	Wide throw, average frequency, steel base, 4-1/2" x 6", USP					73.50			73.50	81
	4200	High frequency, steel base, 4-1/2" x 6", USP				↓	112			112	124
	4600	Spring hinge, single acting, 6" flange, steel				Ea.	41.50			41.50	45.50
	4700	Brass					73			73	80.50
	4900	Double acting, 6" flange, steel					75.50			75.50	83
	4950	Brass				↓	123			123	136
	9001	Continuous hinge, steel, full mortise, heavy duty 501-85	2 Carp	64	.250	L.F.	10.40	4.93		15.33	19.85

08720 | Weatherstripping & Seals

		CREW	DAILY OUTPUT	LABOR-HOURS	UNIT	MAT.	LABOR	EQUIP.	TOTAL	TOTAL INCL O&P
100	0010	**ASTRAGALS** One piece overlapping									
	0400	Cadmium plated steel, flat, 3/16" x 2"	1 Carp	90	.089	L.F.	2.59	1.75		4.34	5.85
	0600	Prime coated steel, flat, 1/8" x 3"		90	.089		3.60	1.75		5.35	6.95
	0800	Stainless steel, flat, 3/32" x 1-5/8"		90	.089		13.50	1.75		15.25	17.85
	1000	Aluminum, flat, 1/8" x 2"		90	.089		2.60	1.75		4.35	5.85
	1200	Nail on, "T" extrusion		120	.067		.57	1.31		1.88	2.88
	1300	Vinyl bulb insert		105	.076		.93	1.50		2.43	3.59
	1600	Screw on, "T" extrusion		90	.089		3.79	1.75		5.54	7.15
	1700	Vinyl insert		75	.107		2.55	2.10		4.65	6.40
	2000	"L" extrusion, neoprene bulbs		75	.107		1.48	2.10		3.58	5.25
	2100	Neoprene sponge insert		75	.107		4.49	2.10		6.59	8.55
	2200	Magnetic		75	.107		7.40	2.10		9.50	11.75
	2400	Spring hinged security seal, with cam		75	.107		4.76	2.10		6.86	8.85
	2600	Spring loaded locking bolt, vinyl insert		45	.178		6.50	3.50		10	13.15
	2800	Neoprene sponge strip, "Z" shaped, aluminum		60	.133		3.08	2.63		5.71	7.90
	2900	Solid neoprene strip, nail on aluminum strip	↓	90	.089	↓	2.58	1.75		4.33	5.85
	3000	One piece stile protection									
	3020	Neoprene fabric loop, nail on aluminum strips	1 Carp	60	.133	L.F.	.49	2.63		3.12	5.05
	3110	Flush mounted aluminum extrusion, 1/2" x 1-1/4"		60	.133		2.46	2.63		5.09	7.20
	3140	3/4" x 1-3/8"		60	.133		3.11	2.63		5.74	7.90
	3160	1-1/8" x 1-3/4"		60	.133		4.97	2.63		7.60	9.95
	3300	Mortise, 9/16" x 3/4"		60	.133		2.64	2.63		5.27	7.40
	3320	13/16" x 1-3/8"		60	.133		2.86	2.63		5.49	7.65
	3600	Spring bronze strip, nail on type		105	.076		2.25	1.50		3.75	5.05
	3620	Screw on, with retainer		75	.107		1.84	2.10		3.94	5.60
	3800	Flexible stainless steel housing, pile insert, 1/2" door		105	.076		5.10	1.50		6.60	8.15
	3820	3/4" door		105	.076		5.75	1.50		7.25	8.85
	4000	Extruded aluminum retainer, flush mount, pile insert		105	.076		1.78	1.50		3.28	4.53
	4080	Mortise, felt insert		90	.089		3.17	1.75		4.92	6.50
	4160	Mortise with spring, pile insert		90	.089		2.48	1.75		4.23	5.75
	4400	Rigid vinyl retainer, mortise, pile insert		105	.076		1.77	1.50		3.27	4.52
	4600	Wool pile filler strip, aluminum backing	↓	105	.076	↓	1.78	1.50		3.28	4.53
	5000	Two piece overlapping astragal, extruded aluminum retainer									
	5010	Pile insert	1 Carp	60	.133	L.F.	2.31	2.63		4.94	7.05
	5020	Vinyl bulb insert		60	.133		1.49	2.63		4.12	6.15
	5040	Vinyl flap insert	↓	60	.133	↓	4.63	2.63		7.26	9.60

08700 | Hardware

08720 | Weatherstripping & Seals

			CREW	DAILY OUTPUT	LABOR-HOURS	UNIT	2000 BARE COSTS MAT.	LABOR	EQUIP.	TOTAL	TOTAL INCL O&P	
100	5060	Solid neoprene flap insert	1 Carp	60	.133	L.F.	4.58	2.63		7.21	9.55	100
	5080	Hypalon rubber flap insert		60	.133		4.68	2.63		7.31	9.65	
	5090	Snap on cover, pile insert		60	.133		5.35	2.63		7.98	10.40	
	5400	Magnetic aluminum, surface mounted		60	.133		18.10	2.63		20.73	24.50	
	5500	Interlocking aluminum, 5/8" x 1" neoprene bulb insert		45	.178		2.87	3.50		6.37	9.15	
	5600	Adjustable aluminum, 9/16" x 21/32", pile insert		45	.178		13.65	3.50		17.15	21	
	5790	For vinyl bulb, deduct					.37			.37	.41	
	5800	Magnetic, adjustable, 9/16" x 21/32"	1 Carp	45	.178		17.50	3.50		21	25.50	
	6000	Two piece stile protection										
	6010	Cloth backed rubber loop, 1" gap, nail on aluminum strips	1 Carp	45	.178	L.F.	2.94	3.50		6.44	9.25	
	6040	Screw on aluminum strips		45	.178		4.58	3.50		8.08	11.05	
	6100	1-1/2" gap, screw on aluminum extrusion		45	.178		4.12	3.50		7.62	10.55	
	6240	Vinyl fabric loop, slotted aluminum extrusion, 1" gap		45	.178		1.46	3.50		4.96	7.60	
	6300	1-1/4" gap		45	.178		4.34	3.50		7.84	10.75	
300	0010	**WEATHERSTRIPPING** Window, double hung, 3' x 5', zinc	1 Carp	7.20	1.111	Opng.	10.70	22		32.70	49.50	300
	0100	Bronze		7.20	1.111		19.45	22		41.45	59	
	0500	As above but heavy duty, zinc		4.60	1.739		13	34.50		47.50	73	
	0600	Bronze		4.60	1.739		22.50	34.50		57	83.50	
	1000	Doors, wood frame, interlocking, for 3' x 7' door, zinc		3	2.667		11.65	52.50		64.15	103	
	1100	Bronze		3	2.667		18.35	52.50		70.85	110	
	1300	6' x 7' opening, zinc		2	4		12.75	79		91.75	149	
	1400	Bronze		2	4		24	79		103	162	
	1700	Wood frame, spring type, bronze										
	1800	3' x 7' door	1 Carp	7.60	1.053	Opng.	15.45	20.50		35.95	52.50	
	1900	6' x 7' door	"	7	1.143	"	16.45	22.50		38.95	56.50	
	2200	Metal frame, spring type, bronze										
	2300	3' x 7' door	1 Carp	3	2.667	Opng.	25.50	52.50		78	118	
	2400	6' x 7' door	"	2.50	3.200	"	35	63		98	146	
	2500	For stainless steel, spring type, add					133%					
	2700	Metal frame, extruded sections, 3' x 7' door, aluminum	1 Carp	2	4	Opng.	34	79		113	172	
	2800	Bronze		2	4		85.50	79		164.50	229	
	3100	6' x 7' door, aluminum		1.20	6.667		43	131		174	273	
	3200	Bronze		1.20	6.667		101	131		232	335	
	3500	Threshold weatherstripping										
	3650	Door sweep, flush mounted, aluminum	1 Carp	25	.320	Ea.	10	6.30		16.30	22	
	3700	Vinyl		25	.320		11.85	6.30		18.15	24	
	5000	Garage door bottom weatherstrip, 12' aluminum, clear		14	.571		16.05	11.25		27.30	37	
	5010	Bronze		14	.571		61	11.25		72.25	87	
	5050	Bottom protection, 12' aluminum, clear		14	.571		19.45	11.25		30.70	41	
	5100	Bronze		14	.571		76	11.25		87.25	103	
800	0010	**THRESHOLD** 3' long door saddles, aluminum	1 Carp	48	.167	L.F.	3.38	3.28		6.66	9.35	800
	0100	Aluminum, 8" wide, 1/2" thick		12	.667	Ea.	29	13.15		42.15	54.50	
	0500	Bronze		60	.133	L.F.	27.50	2.63		30.13	34.50	
	0600	Bronze, panic threshold, 5" wide, 1/2" thick		12	.667	Ea.	57	13.15		70.15	85	
	0700	Rubber, 1/2" thick, 5-1/2" wide		20	.400		28.50	7.90		36.40	45	
	0800	2-3/4" wide		20	.400		13.25	7.90		21.15	28	

08770 | Door/Window Accessories

			CREW	DAILY OUTPUT	LABOR-HOURS	UNIT	MAT.	LABOR	EQUIP.	TOTAL	TOTAL INCL O&P	
100	0010	**AREA WALL** Galvanized steel, 20 ga., 3'-2" wide, 1' deep	1 Sswk	29	.276	Ea.	12.20	5.85		18.05	25	100
	0100	2' deep		23	.348		21.50	7.40		28.90	38.50	
	0300	16 ga., galv., 3'-2" wide, 1' deep		29	.276		17.20	5.85		23.05	30.50	
	0400	3' deep		23	.348		35	7.40		42.40	53	
	0600	Welded grating for above, 15 lbs., painted		45	.178		33	3.78		36.78	44	
	0700	Galvanized		45	.178		63	3.78		66.78	77	
	0900	Translucent plastic cap for above		60	.133		12.65	2.83		15.48	19.50	

08700 | Hardware

08770 | Door/Window Accessories

			CREW	DAILY OUTPUT	LABOR-HOURS	UNIT	2000 BARE COSTS				TOTAL INCL O&P	
							MAT.	LABOR	EQUIP.	TOTAL		
200	0010	DOOR PROTECTION Acrylic and neoprene cover for door										200
	0020	frame, high impact type	1 Carp	100	.080	L.F.	7.05	1.58		8.63	10.50	
550	0010	DETECTION SYSTEMS See division 13851-065										550

08800 | Glazing

08810 | Glass

			CREW	DAILY OUTPUT	LABOR-HOURS	UNIT	2000 BARE COSTS				TOTAL INCL O&P	
							MAT.	LABOR	EQUIP.	TOTAL		
100	0010	ACOUSTICAL GLASS UNITS 1 lite at 3/8", 1 lite at 3/16", for 1" thick	2 Glaz	100	.160	S.F.	18.45	3.10		21.55	25.50	100
	0100	For 4" thick	"	80	.200	"	31.50	3.87		35.37	41	
160	0010	BEVELED GLASS With design patterns, 1/4" thick, 1/2" bevel, minimum	2 Glaz	150	.107	S.F.	31.50	2.06		33.56	38	160
	0050	Average		125	.128		74	2.48		76.48	85.50	
	0100	Maximum		100	.160		131	3.10		134.10	149	
205	0010	CURTAIN WALL See division 08900										205
250	0010	FACETED Color tinted glass, 3/4" thick, minimum	2 Glaz	95	.168	S.F.	26.50	3.26		29.76	35	250
	0100	Maximum	"	75	.213	"	46.50	4.13		50.63	58.50	
260	0010	FLOAT GLASS 3/16" thick, clear, plain R08810-010	2 Glaz	130	.123	S.F.	3.55	2.38		5.93	7.85	260
	0200	Tempered, clear		130	.123		4.24	2.38		6.62	8.60	
	0300	Tinted		130	.123		5.30	2.38		7.68	9.80	
	0600	1/4" thick, clear, plain		120	.133		4.47	2.58		7.05	9.20	
	0700	Tinted		120	.133		4.24	2.58		6.82	8.95	
	0800	Tempered, clear		120	.133		5.30	2.58		7.88	10.15	
	0900	Tinted		120	.133		7.35	2.58		9.93	12.40	
	1600	3/8" thick, clear, plain		75	.213		7.05	4.13		11.18	14.60	
	1700	Tinted		75	.213		8.50	4.13		12.63	16.20	
	1800	Tempered, clear		75	.213		10.60	4.13		14.73	18.50	
	1900	Tinted		75	.213		13.20	4.13		17.33	21.50	
	2200	1/2" thick, clear, plain		55	.291		13.80	5.65		19.45	24.50	
	2300	Tinted		55	.291		14.85	5.65		20.50	25.50	
	2400	Tempered, clear		55	.291		15.90	5.65		21.55	27	
	2500	Tinted		55	.291		19.85	5.65		25.50	31.50	
	2800	5/8" thick, clear, plain		45	.356		14.85	6.90		21.75	27.50	
	2900	Tempered, clear		45	.356		16.95	6.90		23.85	30	
	3200	3/4" thick, clear, plain		35	.457		19.10	8.85		27.95	35.50	
	3300	Tempered, clear		35	.457		22.50	8.85		31.35	39	
	3600	1" thick, clear, plain		30	.533		32	10.30		42.30	52	
	8900	For low emissivity coating for 3/16" and 1/4" only, add to above					3					
270	0010	FULL VISION Window system with 3/4" glass mullions, 10' high	H-2	130	.185	S.F.	41.50	3.27		44.77	51.50	270
	0100	10' to 20' high, minimum		110	.218		44	3.87		47.87	55	
	0150	Average		100	.240		47.50	4.25		51.75	59.50	
	0200	Maximum		80	.300		53.50	5.30		58.80	68	
280	0010	GLASS BLOCK See division 04270-200 R04270-700										280
300	0010	GLAZING VARIABLES R08810-010										300
	0500	For high rise glazing, exterior, add per S.F. per story				S.F.					.08	
	0600	For glass replacement, add				"		100%				
	0700	For gasket settings, add				L.F.	3.18			3.18	3.50	

08800 | Glazing

08810 | Glass

			CREW	DAILY OUTPUT	LABOR-HOURS	UNIT	2000 BARE COSTS MAT.	LABOR	EQUIP.	TOTAL	TOTAL INCL O&P	
300	0800	For concrete reglet settings, add				S.F.	20%	25%				300
	0900	For sloped glazing, add				"		25%				
	2000	Fabrication, polished edges, 1/4" thick				Inch	.27			.27	.30	
	2100	1/2" thick					.70			.70	.77	
	2500	Mitered edges, 1/4" thick					.70			.70	.77	
	2600	1/2" thick					1.12			1.12	1.23	
460	0010	**INSULATING GLASS** 2 lites 1/8" float, 1/2" thk, under 15 S.F.										460
	0020	Clear	2 Glaz	95	.168	S.F.	6.15	3.26		9.41	12.20	
	0100	Tinted		95	.168		9.15	3.26		12.41	15.45	
	0200	2 lites 3/16" float, for 5/8" thk unit, 15 to 30 S.F., clear		90	.178		7.40	3.44		10.84	13.85	
	0300	Tinted		90	.178		7.65	3.44		11.09	14.10	
	0400	1" thk, dbl. glazed, 1/4" float, 30-70 S.F., clear		75	.213		10.50	4.13		14.63	18.40	
	0500	Tinted		75	.213		12.70	4.13		16.83	21	
	0600	1" thick double glazed, 1/4" float, 1/4" wire		75	.213		15.90	4.13		20.03	24.50	
	0700	1/4" float, 1/4" tempered		75	.213		15.40	4.13		19.53	24	
	0800	1/4" wire, 1/4" tempered		75	.213		21.50	4.13		25.63	30.50	
	0900	Both lites, 1/4" wire		75	.213		20	4.13		24.13	29	
	2000	Both lites, light & heat reflective		85	.188		16.95	3.64		20.59	24.50	
	2500	Heat reflective, film inside, 1" thick unit, clear		85	.188		14.85	3.64		18.49	22.50	
	2600	Tinted		85	.188		16	3.64		19.64	23.50	
	3000	Film on weatherside, clear, 1/2" thick unit		95	.168		10.60	3.26		13.86	17.05	
	3100	5/8" thick unit		90	.178		13.45	3.44		16.89	20.50	
	3200	1" thick unit		85	.188		14.60	3.64		18.24	22	
500	0010	**LAMINATED GLASS** Clear float, .03" vinyl, 1/4" thick	2 Glaz	90	.178	S.F.	7.15	3.44		10.59	13.55	500
	0100	3/8" thick		78	.205		11.60	3.97		15.57	19.35	
	0200	.06" vinyl, 1/2" thick		65	.246		13.55	4.76		18.31	23	
	1000	5/8" thick		90	.178		15.90	3.44		19.34	23	
	2000	Bullet-resisting, 1-3/16" thick, to 15 S.F.		16	1		40	19.35		59.35	76	
	2100	Over 15 S.F.		16	1		36	19.35		55.35	71.50	
	2500	2-1/4" thick, to 15 S.F.		12	1.333		51	26		77	99.50	
	2600	Over 15 S.F.		12	1.333		45.50	26		71.50	93	
550	0010	**LEAD GLASS** For X-ray, see division 13091-600										550
600	0010	**OBSCURE GLASS** 1/8" thick, minimum	2 Glaz	140	.114	S.F.	5.90	2.21		8.11	10.10	600
	0100	Maximum		125	.128		7.05	2.48		9.53	11.85	
	0300	7/32" thick, minimum		120	.133		6.50	2.58		9.08	11.45	
	0400	Maximum		105	.152		8.30	2.95		11.25	14.05	
650	0010	**PATTERNED GLASS** Colored, 1/8" thick, minimum	2 Glaz	140	.114	S.F.	6.95	2.21		9.16	11.30	650
	0100	Maximum		125	.128		8.20	2.48		10.68	13.15	
	0300	7/32" thick, minimum		120	.133		8.10	2.58		10.68	13.20	
	0400	Maximum		105	.152		8.80	2.95		11.75	14.60	
675	0010	**REFLECTIVE GLASS** 1/4" float with fused metallic oxide, tinted	2 Glaz	115	.139	S.F.	8.80	2.69		11.49	14.15	675
	0500	1/4" float glass with reflective applied coating		115	.139		7.35	2.69		10.04	12.55	
	2000	Solar film on glass, not including glass, minimum		180	.089		3.55	1.72		5.27	6.75	
	2050	Maximum		225	.071		8.20	1.38		9.58	11.35	
740	0010	**SANDBLASTED GLASS** Float glass, 1/8" thick	2 Glaz	160	.100	S.F.	5.90	1.94		7.84	9.65	740
	0100	3/16" thick		130	.123		6.50	2.38		8.88	11.10	
	0500	Plate glass, 1/4" thick		120	.133		6.75	2.58		9.33	11.75	
	0600	3/8" thick		75	.213		7.35	4.13		11.48	14.95	
760	0010	**SHEET GLASS** Gray, 1/8" thick	2 Glaz	160	.100	S.F.	2.92	1.94		4.86	6.40	760
	0200	1/4" thick		130	.123	"	3.83	2.38		6.21	8.15	
780	0010	**SPANDREL GLASS** 1/4" thick, standard colors, up to 1,000 S.F.		110	.145	S.F.	10	2.81		12.81	15.65	780
	0200	1,000 to 2,000 S.F.		120	.133	"	9.40	2.58		11.98	14.60	
	0300	For custom colors, add				Total	10%					
	0500	For 3/8" thick, add				S.F.	6.10			6.10	6.70	

08800 | Glazing

08810 | Glass

			CREW	DAILY OUTPUT	LABOR-HOURS	UNIT	2000 BARE COSTS MAT.	LABOR	EQUIP.	TOTAL	TOTAL INCL O&P	
780	1000	For double coated, 1/4" thick, add				S.F.	2.28			2.28	2.51	780
	1200	For insulation on panels, add					3.76			3.76	4.14	
	2000	Panels, insulated, with aluminum backed fiberglass, 1" thick	2 Glaz	120	.133		9.05	2.58		11.63	14.25	
	2100	2" thick	"	120	.133		10.55	2.58		13.13	15.90	
	2500	With galvanized steel backing, add					3.13			3.13	3.44	
850	0010	**WINDOW GLASS** Clear float, stops, putty bed, 1/8" thick	2 Glaz	480	.033	S.F.	2.92	.65		3.57	4.28	850
	0500	3/16" thick, clear		480	.033		3.60	.65		4.25	5.05	
	0600	Tinted		480	.033		4.08	.65		4.73	5.55	
	0700	Tempered		480	.033		4.94	.65		5.59	6.50	
860	0010	**WINDOW WALL** See division 08911-900										860
900	0010	**WIRE GLASS** 1/4" thick, rough obscure (chicken wire)	2 Glaz	135	.119	S.F.	9.95	2.29		12.24	14.75	900
	1000	Polished wire, 1/4" thick, diamond, clear		135	.119		12.85	2.29		15.14	17.95	
	1500	Pinstripe, obscure		135	.119		12.05	2.29		14.34	17.05	

08830 | Mirrors

			CREW	DAILY OUTPUT	LABOR-HOURS	UNIT	MAT.	LABOR	EQUIP.	TOTAL	TOTAL INCL O&P	
100	0010	**MIRRORS** No frames, wall type, 1/4" plate glass, polished edge										100
	0100	Up to 5 S.F.	2 Glaz	125	.128	S.F.	5.85	2.48		8.33	10.50	
	0200	Over 5 S.F.		160	.100		5.65	1.94		7.59	9.40	
	0500	Door type, 1/4" plate glass, up to 12 S.F.		160	.100		6.05	1.94		7.99	9.85	
	1000	Float glass, up to 10 S.F., 1/8" thick		160	.100		3.57	1.94		5.51	7.15	
	1100	3/16" thick		150	.107		4.15	2.06		6.21	8	
	1500	12" x 12" wall tiles, square edge, clear		195	.082		1.52	1.59		3.11	4.30	
	1600	Veined		195	.082		3.87	1.59		5.46	6.90	
	2000	1/4" thick, stock sizes, one way transparent		125	.128		13.20	2.48		15.68	18.60	
	2010	Bathroom, unframed, laminated		160	.100		10.10	1.94		12.04	14.30	

08840 | Plastic Glazing

			CREW	DAILY OUTPUT	LABOR-HOURS	UNIT	MAT.	LABOR	EQUIP.	TOTAL	TOTAL INCL O&P	
600	0010	**PLEXIGLASS ACRYLIC** Clear, masked, 1/8" thick, cut sheets	2 Glaz	170	.094	S.F.	2.93	1.82		4.75	6.25	600
	0200	Full sheets		195	.082		1.53	1.59		3.12	4.31	
	0500	1/4" thick, cut sheets		165	.097		5.15	1.88		7.03	8.80	
	0600	Full sheets		185	.086		2.82	1.67		4.49	5.90	
	0900	3/8" thick, cut sheets		155	.103		9.50	2		11.50	13.75	
	1000	Full sheets		180	.089		5.10	1.72		6.82	8.45	
	1300	1/2" thick, cut sheets		135	.119		10.90	2.29		13.19	15.80	
	1400	Full sheets		150	.107		10.60	2.06		12.66	15.05	
	1700	3/4" thick, cut sheets		115	.139		39	2.69		41.69	47	
	1800	Full sheets		130	.123		22.50	2.38		24.88	28.50	
	2100	1" thick, cut sheets		105	.152		43.50	2.95		46.45	53	
	2200	Full sheets		125	.128		27	2.48		29.48	33.50	
	3000	Colored, 1/8" thick, cut sheets		170	.094		9	1.82		10.82	12.90	
	3200	Full sheets		195	.082		5.85	1.59		7.44	9.05	
	3500	1/4" thick, cut sheets		165	.097		10.05	1.88		11.93	14.20	
	3600	Full sheets		185	.086		6.90	1.67		8.57	10.40	
	4000	Mirrors, untinted, cut sheets, 1/8" thick		185	.086		4.24	1.67		5.91	7.45	
	4200	1/4" thick		180	.089		7.70	1.72		9.42	11.30	
650	0010	**POLYCARBONATE** Clear, masked, cut sheets, 1/8" thick	2 Glaz	170	.094	S.F.	5.05	1.82		6.87	8.55	650
	0500	3/16" thick		165	.097		6.10	1.88		7.98	9.80	
	1000	1/4" thick		155	.103		6.75	2		8.75	10.70	
	1500	3/8" thick		150	.107		12.40	2.06		14.46	17.05	
900	0010	**VINYL GLASS** Steel mesh reinforced, stock sizes, .090" thick	2 Glaz	170	.094	S.F.	7.05	1.82		8.87	10.75	900
	0500	.120" thick		170	.094		8.90	1.82		10.72	12.80	
	1000	.250" thick		155	.103		12.40	2		14.40	16.95	
	1500	For non-standard sizes, add					15%					

08900 | Glazed Curtain Wall

08911 | Glazed Aluminum Curtain Wall

			CREW	DAILY OUTPUT	LABOR-HOURS	UNIT	2000 BARE COSTS MAT.	LABOR	EQUIP.	TOTAL	TOTAL INCL O&P	
200	0010	**CURTAIN WALLS** Aluminum, stock, including glazing, minimum	H-1	205	.156	S.F.	19.30	3.17		22.47	27	200
	0050	Average, single glazed		195	.164		26	3.33		29.33	35	
	0150	Average, double glazed		180	.178		36.50	3.61		40.11	47	
	0200	Maximum		160	.200		98.50	4.06		102.56	115	
700	0010	**TUBE FRAMING** For window walls and store fronts, aluminum, stock										700
	0050	Plain tube frame, mill finish, 1-3/4" x 1-3/4"	2 Glaz	103	.155	L.F.	6.05	3.01		9.06	11.65	
	0150	1-3/4" x 4"		98	.163		7.60	3.16		10.76	13.60	
	0200	1-3/4" x 4-1/2"		95	.168		8.40	3.26		11.66	14.65	
	0250	2" x 6"		89	.180		12.75	3.48		16.23	19.80	
	0350	4" x 4"		87	.184		12.45	3.56		16.01	19.60	
	0400	4-1/2" x 4-1/2"		85	.188		13.95	3.64		17.59	21.50	
	0450	Glass bead		240	.067		1.61	1.29		2.90	3.91	
	1000	Flush tube frame, mill finish, 1/4" glass, 1-3/4" x 4", open header		80	.200		7.45	3.87		11.32	14.60	
	1050	Open sill		82	.195		6.50	3.78		10.28	13.40	
	1100	Closed back header		83	.193		10.45	3.73		14.18	17.65	
	1150	Closed back sill		85	.188		9.90	3.64		13.54	16.95	
	1200	Vertical mullion, one piece		75	.213		11.05	4.13		15.18	19	
	1250	Two piece		73	.219		11.85	4.24		16.09	20	
	1300	90° or 180° vertical corner post		75	.213		18.60	4.13		22.73	27.50	
	1400	1-3/4" x 4-1/2", open header		80	.200		9.10	3.87		12.97	16.40	
	1450	Open sill		82	.195		7.50	3.78		11.28	14.50	
	1500	Closed back header		83	.193		11.50	3.73		15.23	18.85	
	1550	Closed back sill		85	.188		10.70	3.64		14.34	17.80	
	1600	Vertical mullion, one piece		75	.213		11.95	4.13		16.08	20	
	1650	Two piece		73	.219		12.65	4.24		16.89	21	
	1700	90° or 180° vertical corner post		75	.213		12.95	4.13		17.08	21	
	2000	Flush tube frame, mill fin. for ins. glass, 2" x 4-1/2", open header		75	.213		10.60	4.13		14.73	18.50	
	2050	Open sill		77	.208		9.30	4.02		13.32	16.85	
	2100	Closed back header		78	.205		11.55	3.97		15.52	19.30	
	2150	Closed back sill		80	.200		11.40	3.87		15.27	18.95	
	2200	Vertical mullion, one piece		70	.229		12.80	4.42		17.22	21.50	
	2250	Two piece		68	.235		13.70	4.55		18.25	22.50	
	2300	90° or 180° vertical corner post		70	.229		12.95	4.42		17.37	21.50	
	5000	Flush tube frame, mill fin., thermal brk., 2-1/4"x 4-1/2", open header		74	.216		11.70	4.18		15.88	19.80	
	5050	Open sill		75	.213		10.20	4.13		14.33	18.05	
	5100	Vertical mullion, one piece		69	.232		14.10	4.49		18.59	23	
	5150	Two piece		67	.239		15.05	4.62		19.67	24	
	5200	90° or 180° vertical corner post		69	.232		13.55	4.49		18.04	22.50	
	6980	Door stop (snap in)		380	.042		2.18	.81		2.99	3.75	
	7000	For joints, 90°, clip type, add				Ea.	18.65			18.65	20.50	
	7050	Screw spline joint, add					14.05			14.05	15.45	
	7100	For joint other than 90°, add					29.50			29.50	32.50	
	8000	For bronze anodized aluminum, add					15%					
	8020	For black finish, add					27%					
	8050	For stainless steel materials, add					350%					
	8100	For monumental grade, add					50%					
	8150	For steel stiffener, add	2 Glaz	200	.080	L.F.	7.15	1.55		8.70	10.45	
	8200	For 2 to 5 stories, add per story				Story		5%				
900	0010	**WINDOW WALLS** Aluminum, stock, including glazing, minimum	H-2	160	.150	S.F.	25	2.66		27.66	32	900
	0050	Average		140	.171	"	30.50	3.04		33.54	38.50	
	0100	Maximum		110	.218	S.F.	95.50	3.87		99.37	111	
	0500	For translucent sandwich wall systems, see div. 07420-770										
	0850	Cost of the above walls depends on material,										
	0860	finish, repetition, and size of units.										
	0870	The larger the opening, the lower the S.F. cost										
	1200	Double glazed acoustical window wall for airports,										

08900 | Glazed Curtain Wall

08911 | Glazed Aluminum Curtain Wall

			CREW	DAILY OUTPUT	LABOR-HOURS	UNIT	2000 BARE COSTS MAT.	LABOR	EQUIP.	TOTAL	TOTAL INCL O&P	
900	1220	including 1" thick glass with 2" x 4-1/2" tube frame	H-2	40	.600	S.F.	62.50	10.65		73.15	86.50	900

08950 | Translucent Wall/Roof Assemblies

			CREW	DAILY OUTPUT	LABOR-HOURS	UNIT	MAT.	LABOR	EQUIP.	TOTAL	TOTAL INCL O&P	
100	0010	**SKYROOFS** Translucent panels, 2-3/4" thick, under 5000 SF	G-3	395	.081	SF Hor.	18.95	1.46		20.41	23.50	100
	0100	Over 5000 SF		465	.069		16.90	1.24		18.14	20.50	
	0300	Continuous vaulted, semi-circular, to 8' wide, double glazed		145	.221		42.50	3.97		46.47	53.50	
	0400	Single glazed		160	.200		28	3.60		31.60	37	
	0600	To 20' wide, single glazed		175	.183		31	3.29		34.29	39.50	
	0700	Over 20' wide, single glazed		200	.160		36	2.88		38.88	44.50	
	0900	Motorized opening type, single glazed, 1/3 opening		145	.221		38	3.97		41.97	48.50	
	1000	Full opening		130	.246		43.50	4.43		47.93	55	
	1200	Pyramid type units, self-supporting, to 30' clear opening,										
	1300	square or circular, single glazed, minimum	G-3	200	.160	SF Hor.	20.50	2.88		23.38	27.50	
	1310	Average		165	.194		29	3.49		32.49	38	
	1400	Maximum		130	.246		41.50	4.43		45.93	53.50	
	1500	Grid type, 4' to 10' modules, single glass glazed, minimum		200	.160		26.50	2.88		29.38	34	
	1550	Maximum		128	.250		43	4.49		47.49	55	
	1600	Preformed acrylic, minimum		300	.107		31	1.92		32.92	38	
	1650	Maximum		175	.183		44	3.29		47.29	54	
	1800	Skyroofs, dome type, self-supporting, to 100' clear opening, circular										
	1900	Rise to span ratio = 0.20										
	1920	Minimum	G-3	197	.162	SF Hor.	13	2.92		15.92	19.25	
	1950	Maximum		113	.283		43.50	5.10		48.60	56	
	2100	Rise to span ratio = 0.33, minimum		169	.189		26	3.40		29.40	34.50	
	2150	Maximum		101	.317		51	5.70		56.70	66	
	2200	Rise to span ratio = 0.50, minimum		148	.216		39	3.89		42.89	49.50	
	2250	Maximum		87	.368		57	6.60		63.60	74	
	2400	Ridge units, continuous, to 8' wide, double		130	.246		99.50	4.43		103.93	117	
	2500	Single		200	.160		68	2.88		70.88	80	
	2700	Ridge and furrow units, over 4' O.C., double, minimum		200	.160		21.50	2.88		24.38	28.50	
	2750	Maximum		120	.267		39.50	4.79		44.29	51.50	
	2800	Single, minimum		214	.150		18.20	2.69		20.89	24.50	
	2850	Maximum		153	.209		38.50	3.76		42.26	49	
	3000	Rolling roof, translucent panels, flat roof, residential, minimum		253	.126	S.F.	16.50	2.27		18.77	22	
	3030	Maximum		160	.200		33	3.60		36.60	42.50	
	3100	Lean-to skyroof, long span, double, minimum		197	.162		22	2.92		24.92	29.50	
	3150	Maximum		101	.317		44	5.70		49.70	58	
	3300	Single, minimum		321	.100		16.55	1.79		18.34	21	
	3350	Maximum		160	.200		27.50	3.60		31.10	36.50	

For information about Means Estimating Seminars, see yellow pages 11 and 12 in back of book

Division Notes

	CREW	DAILY OUTPUT	LABOR-HOURS	UNIT	2000 BARE COSTS				TOTAL INCL O&P
					MAT.	LABOR	EQUIP.	TOTAL	

Division 9 Finishes

Estimating Tips

General
- Room Finish Schedule: A complete set of plans should contain a room finish schedule. If one is not available, it would be well worth the time and effort to put one together. A room finish schedule should contain the room number, room name (for clarity), floor materials, base materials, wainscot materials, wainscot height, wall materials (for each wall), ceiling materials and special instructions.
- Surplus Finishes: Review the specifications to determine if there is any requirement to provide certain amounts of extra materials for the owner's maintenance department. In some cases the owner may require a substantial amount of materials, especially when it is a special order item or long lead time item.

09200 Plaster & Gypsum Board
- Lath is estimated by the square yard for both gypsum and metal lath, plus usually 5% allowance for waste. Furring, channels and accessories are measured by the linear foot. An extra foot should be allowed for each accessory miter or stop.
- Plaster is also estimated by the square yard. Deductions for openings vary by preference, from zero deduction to 50% of all openings over 2 feet in width. Some estimators deduct a percentage of the total yardage for openings. The estimator should allow one extra square foot for each linear foot of horizontal interior or exterior angle located below the ceiling level. Also, double the areas of small radius work.
- Each room should be measured, perimeter times maximum wall height. Ceiling areas are equal to length times width.
- Drywall accessories, studs, track, and acoustical caulking are all measured by the linear foot. Drywall taping is figured by the square foot. Gypsum wallboard is estimated by the square foot. No material deductions should be made for door or window openings under 32 S.F. Coreboard can be obtained in a 1" thickness for solid wall and shaft work. Additions should be made to price out the inside or outside corners.
- Different types of partition construction should be listed separately on the quantity sheets. There may be walls with studs of various widths, double studded, and similar or dissimilar surface materials. Shaft work is usually different construction from surrounding partitions requiring separate quantities and pricing of the work.

09300 Tile
09400 Terrazzo
- Tile and terrazzo areas are taken off on a square foot basis. Trim and base materials are measured by the linear foot. Accent tiles are listed per each. Two basic methods of installation are used. Mud set is approximately 30% more expensive than the thin set. In terrazzo work, be sure to include the linear footage of embedded decorative strips, grounds, machine rubbing and power cleanup.

09600 Flooring
- Wood flooring is available in strip, parquet, or block configuration. The latter two types are set in adhesives with quantities estimated by the square foot. The laying pattern will influence labor costs and material waste. In addition to the material and labor for laying wood floors, the estimator must make allowances for sanding and finishing these areas unless the flooring is prefinished.
- Most of the various types of flooring are all measured on a square foot basis. Base is measured by the linear foot. If adhesive materials are to be quantified, they are estimated at a specified coverage rate by the gallon depending upon the specified type and the manufacturer's recommendations.
- Sheet flooring is measured by the square yard. Roll widths vary, so consideration should be given to use the most economical width, as waste must be figured into the total quantity. Consider also the installation methods available, direct glue down or stretched.

09700 Wall Finishes
- Wall coverings are estimated by the square foot. The area to be covered is measured, length by height of wall above baseboards, to calculate the square footage of each wall. This figure is divided by the number of square feet in the single roll which is being used. Deduct, in full, the areas of openings such as doors and windows. Where a pattern match is required allow 25%-30% waste. One gallon of paste should be sufficient to hang 12 single rolls of light to medium weight paper.

09800 Acoustical Treatment
- Acoustical systems fall into several categories. The takeoff of these materials is by the square foot of area with a 5% allowance for waste. Do not forget about scaffolding, if applicable, when estimating these systems.

09900 Paints & Coatings
- Painting is one area where bids vary to a greater extent than almost any other section of a project. This arises from the many methods of measuring surfaces to be painted. The estimator should check the plans and specifications carefully to be sure of the required number of coats.
- Protection of adjacent surfaces is not included in painting costs. When considering the method of paint application, an important factor is the amount of protection and masking required. These must be estimated separately and may be the determining factor in choosing the method of application.

Reference Numbers
Reference numbers are shown in bold squares at the beginning of some major classifications. These numbers refer to related items in the Reference Section. The reference information may be an estimating procedure, an alternate pricing method or technical information.

Note: Not all subdivisions listed here necessarily appear in this publication.

09100 | Metal Support Assemblies

09110 | Non-Load Bearing Wall Framing

			CREW	DAILY OUTPUT	LABOR-HOURS	UNIT	MAT.	LABOR	EQUIP.	TOTAL	TOTAL INCL O&P	
100	0010	**METAL STUDS, PARTITIONS,** 10' high, with runners										100
	0050	See also Studding, division 05120-440	R09110-610									
	2000	Non-load bearing, galvanized, 25 ga. 1-5/8" wide, 16" O.C.	1 Carp	450	.018	S.F.	.15	.35		.50	.76	
	2100	24" O.C.		520	.015		.11	.30		.41	.64	
	2200	2-1/2" wide, 16" O.C.		440	.018		.16	.36		.52	.79	
	2250	24" O.C.		510	.016		.12	.31		.43	.66	
	2300	3-5/8" wide, 16" O.C.		430	.019		.19	.37		.56	.84	
	2350	24" O.C.		500	.016		.14	.32		.46	.69	
	2400	4" wide, 16" O.C.		420	.019		.23	.38		.61	.89	
	2450	24" O.C.		490	.016		.17	.32		.49	.73	
	2500	6" wide, 16" O.C.		410	.020		.30	.38		.68	.99	
	2550	24" O.C.		480	.017		.22	.33		.55	.80	
	2600	20 ga. studs, 1-5/8" wide, 16" O.C.		450	.018		.25	.35		.60	.88	
	2650	24" O.C.		520	.015		.19	.30		.49	.73	
	2700	2-1/2" wide, 16" O.C.		440	.018		.27	.36		.63	.91	
	2750	24" O.C.		510	.016		.20	.31		.51	.75	
	2800	3-5/8" wide, 16" O.C.		430	.019		.29	.37		.66	.95	
	2850	24" O.C.		500	.016		.22	.32		.54	.78	
	2900	4" wide, 16" O.C.		420	.019		.37	.38		.75	1.04	
	2950	24" O.C.		490	.016		.27	.32		.59	.85	
	3000	6" wide, 16" O.C.		410	.020		.43	.38		.81	1.13	
	3050	24" O.C.		480	.017		.32	.33		.65	.91	
	5000	Load bearing studs, see division 05410-400										

09130 | Acoustical Suspension

			CREW	DAILY OUTPUT	LABOR-HOURS	UNIT	MAT.	LABOR	EQUIP.	TOTAL	TOTAL INCL O&P	
100	0010	**CEILING SUSPENSION SYSTEMS** For boards and tile										100
	0050	Class A suspension system, 15/16" T bar, 2' x 4' grid	1 Carp	800	.010	S.F.	.32	.20		.52	.69	
	0300	2' x 2' grid	"	650	.012		.40	.24		.64	.86	
	0350	For 9/16" grid, add					.13			.13	.14	
	0360	For fire rated grid, add					.07			.07	.08	
	0370	For colored grid, add					.15			.15	.17	
	0400	Concealed Z bar suspension system, 12" module	1 Carp	520	.015		.35	.30		.65	.91	
	0600	1-1/2" carrier channels, 4' O.C., add	"	470	.017		.21	.34		.55	.80	
	0700	Carrier channels for ceilings with										
	0900	recessed lighting fixtures, add	1 Carp	460	.017	S.F.	.34	.34		.68	.96	
	1040	Hanging wire, 12 ga., 4' long		65	.123	C.S.F.	1.56	2.42		3.98	5.85	
	1080	8' long		65	.123	"	3.05	2.42		5.47	7.50	

09200 | Plaster & Gypsum Board

09205 | Furring & Lathing

			CREW	DAILY OUTPUT	LABOR-HOURS	UNIT	MAT.	LABOR	EQUIP.	TOTAL	TOTAL INCL O&P	
530	0010	**FURRING** Beams & columns, 7/8" galvanized channels,										530
	0030	12" O.C.	1 Lath	155	.052	S.F.	.20	.99		1.19	1.84	
	0050	16" O.C.		170	.047		.16	.90		1.06	1.66	
	0070	24" O.C.		185	.043		.11	.83		.94	1.48	
	0100	Ceilings, on steel, 7/8" channels, galvanized, 12" O.C.		210	.038		.18	.73		.91	1.40	
	0300	16" O.C.		290	.028		.16	.53		.69	1.05	
	0400	24" O.C.		420	.019		.11	.37		.48	.72	
	0600	1-5/8" channels, galvanized, 12" O.C.		190	.042		.28	.81		1.09	1.62	

Important: See the Reference Section for critical supporting data - Reference Nos., Crews, & City Cost Indexes

09200 | Plaster & Gypsum Board

09205 | Furring & Lathing

			CREW	DAILY OUTPUT	LABOR-HOURS	UNIT	2000 BARE COSTS MAT.	LABOR	EQUIP.	TOTAL	TOTAL INCL O&P	
530	0700	16" O.C.	1 Lath	260	.031	S.F.	.25	.59		.84	1.24	530
	0900	24" O.C.		390	.021		.17	.39		.56	.83	
	1000	Walls, 7/8" channels, galvanized, 12" O.C.		235	.034		.18	.65		.83	1.27	
	1200	16" O.C.		265	.030		.16	.58		.74	1.13	
	1300	24" O.C.		350	.023		.11	.44		.55	.84	
	1500	1-5/8" channels, galvanized, 12" O.C.		210	.038		.28	.73		1.01	1.50	
	1600	16" O.C.		240	.033		.25	.64		.89	1.32	
	1800	24" O.C.		305	.026		.17	.50		.67	1	
	8000	Suspended ceilings, including carriers										
	8200	1-1/2" carriers, 24" O.C. with:										
	8300	7/8" channels, 16" O.C.	1 Lath	165	.048	S.F.	.58	.93		1.51	2.16	
	8320	24" O.C.		200	.040		.53	.77		1.30	1.84	
	8400	1-5/8" channels, 16" O.C.		155	.052		.67	.99		1.66	2.35	
	8420	24" O.C.		190	.042		.59	.81		1.40	1.96	
	8600	2" carriers, 24" O.C. with:										
	8700	7/8" channels, 16" O.C.	1 Lath	155	.052	S.F.	.16	.99		1.15	1.80	
	8720	24" O.C.		190	.042		.11	.81		.92	1.44	
	8800	1-5/8" channels, 16" O.C.		145	.055		.63	1.06		1.69	2.43	
	8820	24" O.C.		180	.044		.54	.85		1.39	2	
540	0010	GYPSUM LATH Plain or perforated, nailed, 3/8" thick R09205-050	1 Lath	85	.094	S.Y.	3.42	1.81		5.23	6.70	540
	0100	1/2" thick, nailed		80	.100		3.51	1.92		5.43	7	
	0300	Clipped to steel studs, 3/8" thick		75	.107		3.42	2.05		5.47	7.10	
	0400	1/2" thick		70	.114		3.60	2.19		5.79	7.55	
	0600	Firestop gypsum base, to steel studs, 3/8" thick		70	.114		3.51	2.19		5.70	7.45	
	0700	1/2" thick		65	.123		3.96	2.36		6.32	8.25	
	0900	Foil back, to steel studs, 3/8" thick		75	.107		3.60	2.05		5.65	7.30	
	1000	1/2" thick		70	.114		3.96	2.19		6.15	7.95	
	1500	For ceiling installations, add		216	.037			.71		.71	1.16	
	1600	For columns and beams, add		170	.047			.90		.90	1.48	
560	0010	METAL LATH Diamond, expanded, 2.5 lb. per S.Y., painted R09205-060				S.Y.	1.47			1.47	1.62	560
	0100	Galvanized, 2.5 lb. per S.Y.					1.47			1.47	1.62	
	0300	3.4 lb. per S.Y., painted					2.18			2.18	2.40	
	0400	Galvanized					2.14			2.14	2.35	
	0600	For 15# asphalt sheathing paper, add					.23			.23	.26	
	0900	Flat rib, 1/8" high, 2.75 lb., painted					2.28			2.28	2.51	
	1000	Foil backed					2.65			2.65	2.92	
	1200	3.4 lb. per S.Y., painted					2.26			2.26	2.49	
	1300	Galvanized					3			3	3.30	
	1500	For 15# asphalt sheating paper, add					.23			.23	.26	
	1800	High rib, 3/8" high, 3.4 lb. per S.Y., painted					3.20			3.20	3.52	
	1900	Galvanized					2.67			2.67	2.94	
	2400	High rib, 3/4" high, painted, .60 lb. per S.F.				S.F.	.30			.30	.33	
	2500	.75 lb. per S.F.				"	.66			.66	.73	
	2700	Stucco mesh, painted, 1.8 lb.				S.Y.	3.03			3.03	3.33	
	2800	3.6 lb.					1.62			1.62	1.78	
	3000	K-lath, perforated, absorbent paper, regular					2.33			2.33	2.56	
	3100	Heavy duty					2.75			2.75	3.03	
	3300	Waterproof, heavy duty, grade B backing					2.70			2.70	2.97	
	3400	Fire resistant backing					2.98			2.98	3.28	
	3600	2.5 lb. diamond painted, on wood framing, on walls	1 Lath	85	.094		1.47	1.81		3.28	4.58	
	3700	On ceilings		75	.107		1.47	2.05		3.52	4.97	
	3900	3.4 lb. diamond painted, on wood framing, on walls		80	.100		2.26	1.92		4.18	5.65	
	4000	On ceilings		70	.114		2.26	2.19		4.45	6.10	
	4200	3.4 lb. diamond painted, wired to steel framing		75	.107		2.26	2.05		4.31	5.85	
	4300	On ceilings		60	.133		2.26	2.56		4.82	6.70	

09200 | Plaster & Gypsum Board

09205 | Furring & Lathing

			CREW	DAILY OUTPUT	LABOR-HOURS	UNIT	2000 BARE COSTS MAT.	LABOR	EQUIP.	TOTAL	TOTAL INCL O&P	
560	4500	Columns and beams, wired to steel	1 Lath	40	.200	S.Y.	2.26	3.84		6.10	8.80	560
	4600	Cornices, wired to steel		35	.229		2.26	4.39		6.65	9.70	
	4800	Screwed to steel studs, 2.5 lb.		80	.100		1.47	1.92		3.39	4.77	
	4900	3.4 lb.		75	.107		2.18	2.05		4.23	5.75	
	5100	Rib lath, painted, wired to steel, on walls, 2.5 lb.		75	.107		2.28	2.05		4.33	5.85	
	5200	3.4 lb.		70	.114		3.20	2.19		5.39	7.10	
	5400	4.0 lb.		65	.123		3.28	2.36		5.64	7.50	
	5500	For self-furring lath, add					.06			.06	.07	
	5700	Suspended ceiling system, incl. 3.4 lb. diamond lath, painted	1 Lath	15	.533		8.65	10.25		18.90	26.50	
	5800	Galvanized	"	15	.533		8.90	10.25		19.15	26.50	
	6000	Hollow metal stud partitions, 3.4 lb. painted lath both sides										
	6010	Non-load bearing, 25 ga., w/rib lath 2-1/2" studs, 12" O.C.	1 Lath	20.30	.394	S.Y.	8.80	7.55		16.35	22	
	6300	16" O.C.		21.10	.379		8.20	7.30		15.50	21	
	6350	24" O.C.		22.70	.352		7.75	6.75		14.50	19.65	
	6400	3-5/8" studs, 16" O.C.		19.50	.410		8.40	7.90		16.30	22	
	6600	24" O.C.		20.40	.392		7.90	7.55		15.45	21	
	6700	4" studs, 16" O.C.		20.40	.392		9.10	7.55		16.65	22.50	
	6900	24" O.C.		21.60	.370		8	7.10		15.10	20.50	
	7000	6" studs, 16" O.C.		19.50	.410		9.10	7.90		17	23	
	7100	24" O.C.		21.10	.379		8.45	7.30		15.75	21	
	7200	L.B. partitions, 16 ga., w/rib lath, 2-1/2" studs, 16" O.C.		20	.400		8.60	7.70		16.30	22	
	7300	3-5/8" studs, 16 ga.		19.70	.406		10.15	7.80		17.95	24	
	7500	4" studs, 16 ga.		19.50	.410		10.70	7.90		18.60	24.50	
	7600	6" studs, 16 ga.		18.70	.428		12.40	8.20		20.60	27	
700	0010	**ACCESSORIES, PLASTER** Casing bead, expanded flange, galvanized	1 Lath	2.70	2.963	C.L.F.	27	57		84	123	700
	0900	Channels, cold rolled, 16 ga., 3/4" deep, galvanized					18			18	19.80	
	1200	1-1/2" deep, 16 ga., galvanized					27.50			27.50	30.50	
	1620	Corner bead, expanded bullnose, 3/4" radius, #10, galvanized	1 Lath	2.60	3.077		20	59		79	119	
	1650	#1, galvanized		2.55	3.137		35	60		95	137	
	1670	Expanded wing, 2-3/4" wide, galv. #1		2.65	3.019		20	58		78	117	
	1700	Inside corner, (corner rite) 3" x 3", painted		2.60	3.077		17.65	59		76.65	116	
	1750	Strip-ex, 4" wide, painted		2.55	3.137		14.75	60		74.75	115	
	1800	Expansion joint, 3/4" grounds, limited expansion, galv., 1 piece		2.70	2.963		65	57		122	165	
	2100	Extreme expansion, galvanized, 2 piece		2.60	3.077		120	59		179	229	

09210 | Gypsum Plaster

			CREW	DAILY OUTPUT	LABOR-HOURS	UNIT	MAT.	LABOR	EQUIP.	TOTAL	TOTAL INCL O&P	
100	0010	**GYPSUM PLASTER** 80# bag, less than 1 ton				Bag	14.20			14.20	15.65	100
	0100	Over 1 ton				"	12.90			12.90	14.20	
	0300	2 coats, no lath included, on walls	J-1	105	.381	S.Y.	3.42	6.65	.52	10.59	15.45	
	0400	On ceilings	"	92	.435		3.42	7.60	.59	11.61	17.10	
	0600	2 coats on and incl. 3/8" gypsum lath on steel, on walls	J-2	97	.495		6.85	8.80	.56	16.21	23	
	0700	On ceilings	"	83	.578		6.85	10.25	.66	17.76	25.50	
	0900	3 coats, no lath included, on walls	J-1	87	.460		4.77	8	.63	13.40	19.35	
	1000	On ceilings	"	78	.513		4.77	8.95	.70	14.42	21	
	1200	3 coats on and including painted metal lath, on wood studs	J-2	86	.558		7.05	9.90	.64	17.59	25	
	1300	On ceilings	"	76.50	.627		7.05	11.15	.72	18.92	27	
	1600	For irregular or curved surfaces, add						30%				
	1800	For columns & beams, add						50%				
200	0010	**GAUGING PLASTER** 100 lb. bags, less than 1 ton				Bag	21			21	23	200
	0100	Over 1 ton				"	20			20	22	
300	0010	**KEENES CEMENT** In 100 lb. bags, less than 1 ton				Bag	23			23	25.50	300
	0100	Over 1 ton				"	22.50			22.50	24.50	
	0300	Finish only, add to plaster prices, standard	J-1	215	.186	S.Y.	2.16	3.25	.25	5.66	8.10	
	0400	High quality	"	144	.278	"	2.19	4.85	.38	7.42	10.90	
500	0010	**PERLITE OR VERMICULITE PLASTER** 100 lb. bags Under 200 bags				Bag	13.05			13.05	14.35	500
	0100	Over 200 bags				"	12			12	13.20	

09200 | Plaster & Gypsum Board

09210 | Gypsum Plaster

			CREW	DAILY OUTPUT	LABOR-HOURS	UNIT	MAT.	LABOR	EQUIP.	TOTAL	TOTAL INCL O&P	
500	0300	2 coats, no lath included, on walls	J-1	92	.435	S.Y.	3.24	7.60	.59	11.43	16.90	500
	0400	On ceilings	"	79	.506		3.24	8.85	.69	12.78	19.05	
	0600	2 coats, on and incl. 3/8" gypsum lath, on metal studs	J-2	84	.571		5.90	10.15	.65	16.70	24	
	0700	On ceilings	"	70	.686		5.90	12.15	.78	18.83	28	
	0900	3 coats, no lath included, on walls	J-1	74	.541		5.30	9.45	.74	15.49	22.50	
	1000	On ceilings	"	63	.635		5.30	11.10	.86	17.26	25.50	
	1200	3 coats, on and incl. painted metal lath, on metal studs	J-2	72	.667		8.50	11.85	.76	21.11	30	
	1300	On ceilings		61	.787		8.50	13.95	.90	23.35	34	
	1500	3 coats, on and incl. suspended metal lath ceiling		37	1.297		13.95	23	1.48	38.43	55.50	
	1700	For irregular or curved surfaces, add to above						30%				
	1800	For columns and beams, add to above						50%				
	1900	For soffits, add to ceiling prices						40%				
650	0010	**PLASTER PARTITION WALL**										650
	0400	Stud walls, 3.4 lb. metal lath, 3 coat gypsum plaster, 2 sides										
	0600	2" x 4" wood studs, 16" O.C.	J-2	315	.152	S.F.	2.98	2.70	.17	5.85	7.95	
	0700	2-1/2" metal studs, 25 ga., 12" O.C.		325	.148		2.76	2.62	.17	5.55	7.60	
	0800	3-5/8" metal studs, 25 ga., 16" O.C.		320	.150		2.78	2.66	.17	5.61	7.70	
	0900	Gypsum lath, 2 coat vermiculite plaster, 2 sides										
	1000	2" x 4" wood studs, 16" O.C.	J-2	355	.135	S.F.	3.32	2.40	.15	5.87	7.80	
	1200	2-1/2" metal studs, 25 ga., 12" O.C.		365	.132		2.94	2.33	.15	5.42	7.30	
	1300	3-5/8" metal studs, 25 ga., 16" O.C.		360	.133		3.03	2.37	.15	5.55	7.45	
900	0010	**THIN COAT** Plaster, 1 coat veneer, not incl. lath	J-1	3,600	.011	S.F.	.07	.19	.02	.28	.42	900
	1000	In 50 lb. bags				Bag	8.60			8.60	9.45	

09220 | Portland Cement Plaster

			CREW	DAILY OUTPUT	LABOR-HOURS	UNIT	MAT.	LABOR	EQUIP.	TOTAL	TOTAL INCL O&P	
200	0010	**STUCCO**, 3 coats 1" thick, float finish, with mesh, on wood frame	J-2	135	.356	S.Y.	3.65	6.30	.41	10.36	14.95	200
	0100	On masonry construction, no mesh incl.	J-1	200	.200		2.03	3.49	.27	5.79	8.40	
	0300	For trowel finish, add	1 Plas	170	.047			.88		.88	1.47	
	0400	For 3/4" thick, on masonry, deduct	J-1	880	.045		.56	.79	.06	1.41	2.02	
	0600	For coloring and special finish, add, minimum		685	.058		.36	1.02	.08	1.46	2.19	
	0700	Maximum		200	.200		1.26	3.49	.27	5.02	7.55	
	0900	For soffits, add	J-2	155	.310		1.94	5.50	.35	7.79	11.65	
	1000	Exterior stucco, with bonding agent, 3 coats, on walls, no mesh incl.	J-1	200	.200		3.25	3.49	.27	7.01	9.75	
	1200	Ceilings		180	.222		3.25	3.88	.30	7.43	10.40	
	1300	Beams		80	.500		3.25	8.75	.68	12.68	18.95	
	1500	Columns		100	.400		3.25	7	.54	10.79	15.85	
	1600	Mesh, painted, nailed to wood, 1.8 lb.	1 Lath	60	.133		3.03	2.56		5.59	7.50	
	1800	3.6 lb.		55	.145		1.62	2.79		4.41	6.35	
	1900	Wired to steel, painted, 1.8 lb.		53	.151		3.03	2.90		5.93	8.10	
	2100	3.6 lb.		50	.160		1.62	3.07		4.69	6.85	

09250 | Gypsum Board

			CREW	DAILY OUTPUT	LABOR-HOURS	UNIT	MAT.	LABOR	EQUIP.	TOTAL	TOTAL INCL O&P	
200	0010	**CEMENTITIOUS BACKERBOARD**										200
	0070	Cementitious backerboard, on floor, 3' x 4' x 1/2" sheets	2 Carp	525	.030	S.F.	1.04	.60		1.64	2.18	
	0080	3' x 5' x 1/2" sheets		525	.030		1.04	.60		1.64	2.18	
	0090	3' x 6' x 1/2" sheets		525	.030		1.04	.60		1.64	2.17	
	0100	3' x 4' x 5/8" sheets		525	.030		1.08	.60		1.68	2.22	
	0110	3' x 5' x 5/8" sheets		525	.030		1.06	.60		1.66	2.20	
	0120	3' x 6' x 5/8" sheets		525	.030		1.07	.60		1.67	2.20	
	0150	On wall, 3' x 4' x 1/2" sheets		350	.046		1.04	.90		1.94	2.69	
	0160	3' x 5' x 1/2" sheets		350	.046		1.04	.90		1.94	2.69	
	0170	3' x 6' x 1/2" sheets		350	.046		1.04	.90		1.94	2.68	
	0180	3' x 4' x 5/8" sheets		350	.046		1.08	.90		1.98	2.73	
	0190	3' x 5' x 5/8" sheets		350	.046		1.06	.90		1.96	2.71	

FINISHES

09200 | Plaster & Gypsum Board

09250 | Gypsum Board

			CREW	DAILY OUTPUT	LABOR-HOURS	UNIT	2000 BARE COSTS				TOTAL INCL O&P	
							MAT.	LABOR	EQUIP.	TOTAL		
200	0200	3' x 6' x 5/8" sheets	2 Carp	350	.046	S.F.	1.07	.90		1.97	2.71	200
	0250	On counter, 3' x 4'x 1/2" sheets		180	.089		1.04	1.75		2.79	4.15	
	0260	3' x 5' x 1/2" sheets		180	.089		1.04	1.75		2.79	4.15	
	0270	3' x 6' x 1/2" sheets		180	.089		1.04	1.75		2.79	4.14	
	0300	3' x 4'x 5/8" sheets		180	.089		1.08	1.75		2.83	4.19	
	0310	3' x 5' x 5/8" sheets		180	.089		1.06	1.75		2.81	4.17	
	0320	3' x 6' x 5/8" sheets		180	.089		1.07	1.75		2.82	4.17	
300	0010	**BLUEBOARD** For use with thin coat										300
	0100	plaster application (see division 09210-900)										
	1000	3/8" thick, on walls or ceilings, standard, no finish included	2 Carp	1,900	.008	S.F.	.27	.17		.44	.58	
	1100	With thin coat plaster finish		875	.018		.34	.36		.70	.99	
	1400	On beams, columns, or soffits, standard, no finish included		675	.024		.31	.47		.78	1.14	
	1450	With thin coat plaster finish		475	.034		.39	.66		1.05	1.57	
	3000	1/2" thick, on walls or ceilings, standard, no finish included		1,900	.008		.27	.17		.44	.58	
	3100	With thin coat plaster finish		875	.018		.34	.36		.70	.99	
	3300	Fire resistant, no finish included		1,900	.008		.27	.17		.44	.58	
	3400	With thin coat plaster finish		875	.018		.34	.36		.70	.99	
	3450	On beams, columns, or soffits, standard, no finish included		675	.024		.31	.47		.78	1.14	
	3500	With thin coat plaster finish		475	.034		.39	.66		1.05	1.57	
	3700	Fire resistant, no finish included		675	.024		.31	.47		.78	1.14	
	3800	With thin coat plaster finish		475	.034		.39	.66		1.05	1.57	
	5000	5/8" thick, on walls or ceilings, fire resistant, no finish included		1,900	.008		.30	.17		.47	.61	
	5100	With thin coat plaster finish		875	.018		.37	.36		.73	1.03	
	5500	On beams, columns, or soffits, no finish included		675	.024		.35	.47		.82	1.18	
	5600	With thin coat plaster finish		475	.034		.43	.66		1.09	1.61	
	6000	For high ceilings, over 8' high, add		3,060	.005			.10		.10	.18	
	6500	For over 3 stories high, add per story		6,100	.003			.05		.05	.09	
500	0010	**CEILINGS** Gypsum drywall, fire rated, finished										500
	0100	Screwed to grid, channel or joists, 1/2" thick	2 Carp	765	.021	S.F.	.25	.41		.66	.99	
	0200	5/8" thick		765	.021		.29	.41		.70	1.03	
	0300	Over 8' high, 1/2" thick		615	.026		.25	.51		.76	1.16	
	0400	5/8" thick		615	.026		.29	.51		.80	1.20	
	0600	Grid suspension system, direct hung										
	0700	1-1/2" C.R.C., with 7/8" hi hat furring channel, 16" O.C.	2 Carp	600	.027	S.F.	.84	.53		1.37	1.82	
	0800	24" O.C.		900	.018		.77	.35		1.12	1.45	
	0900	3-5/8" C.R.C., with 7/8" hi hat furring channel, 16" O.C.		600	.027		.89	.53		1.42	1.88	
	1000	24" O.C.		900	.018		.78	.35		1.13	1.46	
700	0010	**DRYWALL** Gypsum plasterboard, nailed or screwed to studs										700
	0100	unless otherwise noted										
	0150	3/8" thick, on walls, standard, no finish included	2 Carp	2,000	.008	S.F.	.26	.16		.42	.56	
	0200	On ceilings, standard, no finish included		1,800	.009		.26	.18		.44	.59	
	0250	On beams, columns, or soffits, no finish included		675	.024		.30	.47		.77	1.13	
	0300	1/2" thick, on walls, standard, no finish included		2,000	.008		.26	.16		.42	.56	
	0350	Taped and finished (level 4 finish)		965	.017		.30	.33		.63	.89	
	0390	With compound skim coat (level 5 finish)		775	.021		.32	.41		.73	1.05	
	0400	Fire resistant, no finish included		2,000	.008		.25	.16		.41	.55	
	0450	Taped and finished (level 4 finish)		965	.017		.29	.33		.62	.88	
	0490	With compound skim coat (level 5 finish)		775	.021		.31	.41		.72	1.04	
	0500	Water resistant, no finish included		2,000	.008		.32	.16		.48	.62	
	0550	Taped and finished (level 4 finish)		965	.017		.36	.33		.69	.96	
	0590	With compound skim coat (level 5 finish)		775	.021		.38	.41		.79	1.12	
	0600	Prefinished, vinyl, clipped to studs		900	.018		.56	.35		.91	1.22	
	1000	On ceilings, standard, no finish included		1,800	.009		.26	.18		.44	.59	
	1050	Taped and finished (level 4 finish)		765	.021		.30	.41		.71	1.04	
	1090	With compound skim coat (level 5 finish)		610	.026		.32	.52		.84	1.24	

09200 | Plaster & Gypsum Board

09250 | Gypsum Board

		Crew	Daily Output	Labor-Hours	Unit	Mat.	Labor	Equip.	Total	Total Incl O&P
1100	Fire resistant, no finish included	2 Carp	1,800	.009	S.F.	.25	.18		.43	.58
1150	Taped and finished (level 4 finish)		765	.021		.29	.41		.70	1.03
1195	With compound skim coat (level 5 finish)		610	.026		.31	.52		.83	1.23
1200	Water resistant, no finish included		1,800	.009		.32	.18		.50	.65
1250	Taped and finished (level 4 finish)		765	.021		.36	.41		.77	1.11
1290	With compound skim coat (level 5 finish)		610	.026		.38	.52		.90	1.31
1500	On beams, columns, or soffits, standard, no finish included		675	.024		.30	.47		.77	1.13
1550	Taped and finished (level 4 finish)		475	.034		.35	.66		1.01	1.52
1590	With compound skim coat (level 5 finish)		540	.030		.37	.58		.95	1.40
1600	Fire resistant, no finish included		675	.024		.29	.47		.76	1.12
1650	Taped and finished (level 4 finish)		475	.034		.33	.66		.99	1.51
1690	With compound skim coat (level 5 finish)		540	.030		.36	.58		.94	1.39
1700	Water resistant, no finish included		675	.024		.37	.47		.84	1.20
1750	Taped and finished (level 4 finish)		475	.034		.41	.66		1.07	1.60
1790	With compound skim coat (level 5 finish)		540	.030		.44	.58		1.02	1.48
2000	5/8" thick, on walls, standard, no finish included		2,000	.008		.29	.16		.45	.59
2050	Taped and finished (level 4 finish)		965	.017		.33	.33		.66	.92
2090	With compound skim coat (level 5 finish)		775	.021		.35	.41		.76	1.09
2100	Fire resistant, no finish included		2,000	.008		.29	.16		.45	.59
2150	Taped and finished (level 4 finish)		965	.017		.33	.33		.66	.92
2195	With compound skim coat (level 5 finish)		775	.021		.35	.41		.76	1.09
2200	Water resistant, no finish included		2,000	.008		.35	.16		.51	.66
2250	Taped and finished (level 4 finish)		965	.017		.39	.33		.72	.99
2290	With compound skim coat (level 5 finish)		775	.021		.41	.41		.82	1.15
2300	Prefinished, vinyl, clipped to studs		900	.018		.65	.35		1	1.31
3000	On ceilings, standard, no finish included		1,800	.009		.29	.18		.47	.62
3050	Taped and finished (level 4 finish)		765	.021		.33	.41		.74	1.07
3090	With compound skim coat (level 5 finish)		615	.026		.35	.51		.86	1.27
3100	Fire resistant, no finish included		1,800	.009		.29	.18		.47	.62
3150	Taped and finished (level 4 finish)		765	.021		.33	.41		.74	1.07
3190	With compound skim coat (level 5 finish)		615	.026		.35	.51		.86	1.27
3200	Water resistant, no finish included		1,800	.009		.35	.18		.53	.69
3250	Taped and finished (level 4 finish)		765	.021		.39	.41		.80	1.14
3290	With compound skim coat (level 5 finish)		615	.026		.41	.51		.92	1.33
3500	On beams, columns, or soffits, no finish included		675	.024		.33	.47		.80	1.17
3550	Taped and finished (level 4 finish)		475	.034		.38	.66		1.04	1.56
3590	With compound skim coat (level 5 finish)		380	.042		.40	.83		1.23	1.86
3600	Fire resistant, no finish included		675	.024		.33	.47		.80	1.17
3650	Taped and finished (level 4 finish)		475	.034		.38	.66		1.04	1.56
3690	With compound skim coat (level 5 finish)		380	.042		.40	.83		1.23	1.86
3700	Water resistant, no finish included		675	.024		.40	.47		.87	1.24
3750	Taped and finished (level 4 finish)		475	.034		.45	.66		1.11	1.63
3790	With compound skim coat (level 5 finish)		380	.042		.47	.83		1.30	1.94
4000	Fireproofing, beams or columns, 2 layers, 1/2" thick, incl finish		330	.048		.54	.96		1.50	2.23
4050	5/8" thick		300	.053		.66	1.05		1.71	2.53
4100	3 layers, 1/2" thick		225	.071		.87	1.40		2.27	3.36
4150	5/8" thick		210	.076		.99	1.50		2.49	3.66
5050	For 1" thick coreboard on columns		480	.033		.69	.66		1.35	1.89
5100	For foil-backed board, add					.10			.10	.11
5200	For high ceilings, over 8' high, add	2 Carp	3,060	.005			.10		.10	.18
5270	For textured spray, add	2 Lath	1,600	.010		.05	.19		.24	.37
5300	For over 3 stories high, add per story	2 Carp	6,100	.003			.05		.05	.09
5350	For finishing corners, inside or outside, add	"	1,100	.015	L.F.	.06	.29		.35	.56
5500	For acoustical sealant, add per bead	1 Carp	500	.016	"	.03	.32		.35	.57
5550	Sealant, 1 quart tube				Ea.	4.89			4.89	5.40
5600	Sound deadening board, 1/4" gypsum	2 Carp	1,800	.009	S.F.	.25	.18		.43	.58

09200 | Plaster & Gypsum Board

09250 | Gypsum Board

			CREW	DAILY OUTPUT	LABOR-HOURS	UNIT	2000 BARE COSTS MAT.	LABOR	EQUIP.	TOTAL	TOTAL INCL O&P	
700	5650	1/2" wood fiber	2 Carp	1,800	.009	S.F.	.34	.18		.52	.67	700

09260 | Gypsum Board Systems

			CREW	DAILY OUTPUT	LABOR-HOURS	UNIT	MAT.	LABOR	EQUIP.	TOTAL	TOTAL INCL O&P	
100	0010	**PARTITION WALL** Stud wall, 8' to 12' high										100
	0050	1/2", interior, gypsum drywall, standard, taped both sides										
	0500	Installed on and incl., 2" x 4" wood studs, 16" O.C.	2 Carp	310	.052	S.F.	1.04	1.02		2.06	2.89	
	1000	Metal studs, NLB, 25 ga., 16" O.C., 3-5/8" wide		350	.046		.82	.90		1.72	2.44	
	1200	6" wide		330	.048		.90	.96		1.86	2.63	
	1400	Water resistant, on 2" x 4" wood studs, 16" O.C.		310	.052		1.16	1.02		2.18	3.02	
	1600	Metal studs, NLB, 25 ga., 16" O.C., 3-5/8" wide		350	.046		.94	.90		1.84	2.57	
	1800	6" wide		330	.048		1.02	.96		1.98	2.76	
	2000	Fire res., 2 layers, 1-1/2 hr., on 2" x 4" wood studs, 16" O.C.		210	.076		1.52	1.50		3.02	4.25	
	2200	Metal studs, NLB, 25 ga., 16" O.C., 3-5/8" wide		250	.064		1.30	1.26		2.56	3.59	
	2400	6" wide		230	.070		1.38	1.37		2.75	3.87	
	2600	Fire & water res., 2 layers, 1-1/2 hr., 2"x4" studs, 16" O.C.		210	.076		1.52	1.50		3.02	4.25	
	2800	Metal studs, NLB, 25 ga., 16" O.C., 3-5/8" wide		250	.064		1.30	1.26		2.56	3.59	
	3000	6" wide		230	.070		1.38	1.37		2.75	3.87	
	3200	5/8", interior, gypsum drywall, standard, taped both sides										
	3400	Installed on and including 2" x 4" wood studs, 16" O.C.	2 Carp	300	.053	S.F.	1.10	1.05		2.15	3.01	
	3600	24" O.C.		330	.048		1.01	.96		1.97	2.75	
	3800	Metal studs, NLB, 25 ga., 16" O.C., 3-5/8" wide		340	.047		.88	.93		1.81	2.56	
	4000	6" wide		320	.050		.96	.99		1.95	2.75	
	4200	24" O.C., 3-5/8" wide		360	.044		.88	.88		1.76	2.47	
	4400	6" wide		340	.047		.96	.93		1.89	2.65	
	4800	Water resistant, on 2" x 4" wood studs, 16" O.C.		300	.053		1.22	1.05		2.27	3.15	
	5000	24" O.C.		330	.048		1.13	.96		2.09	2.88	
	5200	Metal studs, NLB, 25 ga. 16" O.C., 3-5/8" wide		340	.047		1	.93		1.93	2.69	
	5400	6" wide		320	.050		1.08	.99		2.07	2.88	
	5600	24" O.C., 3-5/8" wide		360	.044		1	.88		1.88	2.60	
	5800	6" wide		340	.047		1.08	.93		2.01	2.78	
	6000	Fire res., 2 layers, 2 hr., on 2" x 4" wood studs, 16" O.C.		205	.078		1.59	1.54		3.13	4.37	
	6200	24" O.C.		235	.068		1.59	1.34		2.93	4.04	
	6400	Metal studs, NLB, 25 ga., 16" O.C., 3-5/8" wide		245	.065		1.46	1.29		2.75	3.81	
	6600	6" wide		225	.071		1.54	1.40		2.94	4.09	
	6800	24" O.C., 3-5/8" wide		265	.060		1.46	1.19		2.65	3.65	
	7000	6" wide		245	.065		1.54	1.29		2.83	3.89	
	7200	Fire & water res., 2 layers, 2 hr., 2" x 4" studs, 16" O.C.		205	.078		1.68	1.54		3.22	4.48	
	7400	24" O.C.		235	.068		1.59	1.34		2.93	4.04	
	7600	Metal studs, NLB, 25 ga., 16" O.C., 3-5/8" wide		245	.065		1.46	1.29		2.75	3.81	
	7800	6" wide		225	.071		1.54	1.40		2.94	4.09	
	8000	24" O.C., 3-5/8" wide		265	.060		1.46	1.19		2.65	3.65	
	8200	6" wide		245	.065		1.54	1.29		2.83	3.89	
	8600	1/2" blueboard, mesh tape both sides										
	8620	Installed on and including 2" x 4" wood studs, 16" O.C.	2 Carp	300	.053	S.F.	1.08	1.05		2.13	2.99	
	8640	Metal studs, NLB, 25 ga., 16" O.C., 3-5/8" wide		340	.047		.86	.93		1.79	2.54	
	8660	6" wide		320	.050		.94	.99		1.93	2.72	
	9000	Exterior, 1/2" gypsum sheathing, 1/2" gypsum finished, interior,										
	9100	including foil faced insulation, metal studs, 20 ga.										
	9200	16" O.C., 3-5/8" wide	2 Carp	290	.055	S.F.	1.44	1.09		2.53	3.44	
	9400	6" wide	"	270	.059	"	1.44	1.17		2.61	3.58	
800	0010	**SHAFT WALL** Cavity type on 25 ga. J track & C-H studs, 24" O.C.										800
	0030	1" thick coreboard wall liner on shaft side										
	0040	2-hour assembly with double layer										
	0060	5/8" fire rated gypsum board on room side	2 Carp	220	.073	S.F.	1.27	1.43		2.70	3.85	

09200 | Plaster & Gypsum Board

09260 | Gypsum Board Systems

			DAILY	LABOR-		\multicolumn{4}{c}{2000 BARE COSTS}	TOTAL
		CREW	OUTPUT	HOURS	UNIT	MAT.	LABOR	EQUIP.	TOTAL	INCL O&P
800	0100	3-hour assembly with triple layer									800
	0300	5/8" fire rated gypsum board on room side	2 Carp	180	.089	S.F.	1.56	1.75		3.31	4.72
	0400	4-hour assembly, 1" coreboard, 5/8" fire rated gypsum board									
	0600	and 3/4" galv. metal furring channels, 24" O.C., with									
	0700	Double layer 5/8" fire rated gypsum board on room side	2 Carp	110	.145	S.F.	1.38	2.87		4.25	6.45
	0900	For taping & finishing, add per side	1 Carp	1,050	.008	"	.04	.15		.19	.30
	1000	For insulation, see div. 07210-000									

09270 | Drywall Accessories

			DAILY	LABOR-		MAT.	LABOR	EQUIP.	TOTAL	INCL O&P
100	0010	ACCESSORIES, DRYWALL Casing bead, galvanized steel R09110-610	1 Carp	2.90	2.759	C.L.F.	13.80	54.50		68.30	108	100
	0100	Vinyl		3	2.667		16	52.50		68.50	108
	0400	Corner bead, galvanized steel, 1-1/4" x 1-1/4"		3.50	2.286		9	45		54	87
	0600	Vinyl corner bead		4	2		21	39.50		60.50	90.50
	0900	Furring channel, galv. steel, 7/8" deep, standard		2.60	3.077		17.35	60.50		77.85	123
	1000	Resilient		2.55	3.137		18.65	62		80.65	127
	1100	J trim, galvanized steel, 1/2" wide		3	2.667		11.85	52.50		64.35	103
	1120	5/8" wide		2.95	2.712		12.50	53.50		66	105
	1140	L trim, galvanized		3	2.667		15	52.50		67.50	107
	1150	U trim, galvanized		2.95	2.712		15.65	53.50		69.15	109
	1160	Screws #6 x 1" A				M	6.65			6.65	7.30
	1170	#6 x 1-5/8" A				"	8.65			8.65	9.55
	1200	Studs and runners for partitions, see also div. 05120-440									
	1500	Z stud, galvanized steel, 1-1/2" wide	1 Carp	2.60	3.077	C.L.F.	24	60.50		84.50	130
	1600	2" wide	"	2.55	3.137	"	28.50	62		90.50	138

09300 | Tile

09310 | Ceramic Tile

			DAILY	LABOR-		\multicolumn{4}{c}{2000 BARE COSTS}	TOTAL
		CREW	OUTPUT	HOURS	UNIT	MAT.	LABOR	EQUIP.	TOTAL	INCL O&P
100	0010	CERAMIC TILE									100
	0050	Base, using 1' x 4" high pc. with 1" x 1" tiles, mud set	D-7	82	.195	L.F.	3.99	3.38		7.37	9.90
	0100	Thin set	"	128	.125		3.79	2.17		5.96	7.70
	0300	For 6" high base, 1" x 1" tile face, add					.63			.63	.69
	0400	For 2" x 2" tile face, add to above					.33			.33	.36
	0600	Cove base, 4-1/4" x 4-1/4" high, mud set	D-7	91	.176		3.02	3.05		6.07	8.25
	0700	Thin set		128	.125		3.02	2.17		5.19	6.85
	0900	6" x 4-1/4" high, mud set		100	.160		2.54	2.77		5.31	7.30
	1000	Thin set		137	.117		2.54	2.02		4.56	6.05
	1200	Sanitary cove base, 6" x 4-1/4" high, mud set		93	.172		3.52	2.98		6.50	8.70
	1300	Thin set		124	.129		3.52	2.24		5.76	7.50
	1500	6" x 6" high, mud set		84	.190		3	3.30		6.30	8.65
	1600	Thin set		117	.137		3	2.37		5.37	7.15
	1800	Bathroom accessories, average		82	.195	Ea.	9.50	3.38		12.88	15.95
	1900	Bathtub, 5', rec. 4-1/4" x 4-1/4" tile wainscot, adhesive set 6' high		2.90	5.517		140	95.50		235.50	310
	2100	7' high wainscot		2.50	6.400		160	111		271	355
	2200	8' high wainscot		2.20	7.273		170	126		296	390
	2400	Bullnose trim, 4-1/4" x 4-1/4", mud set		82	.195	L.F.	2.74	3.38		6.12	8.50
	2500	Thin set		128	.125		2.74	2.17		4.91	6.50
	2700	6" x 4-1/4" bullnose trim, mud set		84	.190		2.30	3.30		5.60	7.90

09300 | Tile

09310 | Ceramic Tile

			CREW	DAILY OUTPUT	LABOR-HOURS	UNIT	MAT.	LABOR	EQUIP.	TOTAL	TOTAL INCL O&P	
100	2800	Thin set	D-7	124	.129	L.F.	2.30	2.24		4.54	6.15	100
	3000	Floors, natural clay, random or uniform, thin set, color group 1		183	.087	S.F.	3.55	1.52		5.07	6.35	
	3100	Color group 2		183	.087		3.82	1.52		5.34	6.65	
	3300	Porcelain type, 1 color, color group 2, 1" x 1"		183	.087		4.11	1.52		5.63	7	
	3400	2" x 2" or 2" x 1", thin set		190	.084		3.97	1.46		5.43	6.75	
	3600	For random blend, 2 colors, add					.75			.75	.83	
	3700	4 colors, add					1.08			1.08	1.19	
	3900	For color group 3, add					.44			.44	.48	
	4000	For abrasive non-slip tile, add					.43			.43	.47	
	4300	Specialty tile, 4-1/4" x 4-1/4" x 1/2", decorator finish	D-7	183	.087		8.95	1.52		10.47	12.30	
	4500	Add for epoxy grout, 1/16" joint, 1" x 1" tile		800	.020		.53	.35		.88	1.14	
	4600	2" x 2" tile		820	.020		.50	.34		.84	1.10	
	4800	Pregrouted sheets, walls, 4-1/4" x 4-1/4", 6" x 4-1/4"										
	4810	and 8-1/2" x 4-1/4", 4 S.F. sheets, silicone grout	D-7	240	.067	S.F.	4.07	1.16		5.23	6.35	
	5100	Floors, unglazed, 2 S.F. sheets,										
	5110	urethane adhesive	D-7	180	.089	S.F.	4.05	1.54		5.59	6.95	
	5400	Walls, interior, thin set, 4-1/4" x 4-1/4" tile		190	.084		2	1.46		3.46	4.56	
	5500	6" x 4-1/4" tile		190	.084		2.25	1.46		3.71	4.84	
	5700	8-1/2" x 4-1/4" tile		190	.084		3.19	1.46		4.65	5.85	
	5800	6" x 6" tile		200	.080		2.63	1.39		4.02	5.15	
	5810	8" x 8" tile		225	.071		3.08	1.23		4.31	5.40	
	5820	12" x 12" tile		300	.053		2.92	.92		3.84	4.71	
	5830	16" x 16" tile		500	.032		3.17	.55		3.72	4.39	
	6000	Decorated wall tile, 4-1/4" x 4-1/4", minimum		270	.059		4	1.03		5.03	6.05	
	6100	Maximum		180	.089		42	1.54		43.54	48.50	
	6300	Exterior walls, frostproof, mud set, 4-1/4" x 4-1/4"		102	.157		4.02	2.72		6.74	8.80	
	6400	1-3/8" x 1-3/8"		93	.172		3.71	2.98		6.69	8.90	
	6600	Crystalline glazed, 4-1/4" x 4-1/4", mud set, plain		100	.160		3.15	2.77		5.92	7.95	
	6700	4-1/4" x 4-1/4", scored tile		100	.160		3.97	2.77		6.74	8.85	
	6900	6" x 6" plain		93	.172		4.02	2.98		7	9.25	
	7000	For epoxy grout, 1/16" joints, 4-1/4" tile, add		800	.020		.33	.35		.68	.92	
	7200	For tile set in dry mortar, add		1,735	.009			.16		.16	.26	
	7300	For tile set in portland cement mortar, add		290	.055			.96		.96	1.55	
	9000	Regrout tile 4-1/2 x 4-1/2, or larger, wall	1 Tilf	100	.080		.13	1.54		1.67	2.63	
	9220	Floor	"	125	.064		.14	1.23		1.37	2.14	
300	0010	**CERAMIC TILE PANELS** Insulated, over 1000 S.F., 1-1/2" thick	D-7	220	.073	S.F.	8.65	1.26		9.91	11.60	300
	0100	2-1/2" thick	"	220	.073	"	9.30	1.26		10.56	12.25	

09330 | Quarry Tile

			CREW	DAILY OUTPUT	LABOR-HOURS	UNIT	MAT.	LABOR	EQUIP.	TOTAL	TOTAL INCL O&P	
100	0010	**QUARRY TILE** Base, cove or sanitary, 2" or 5" high, mud set										100
	0100	1/2" thick	D-7	110	.145	L.F.	3.70	2.52		6.22	8.15	
	0300	Bullnose trim, red, mud set, 6" x 6" x 1/2" thick		120	.133		3.81	2.31		6.12	7.95	
	0400	4" x 4" x 1/2" thick		110	.145		3.83	2.52		6.35	8.30	
	0600	4" x 8" x 1/2" thick, using 8" as edge		130	.123		3.63	2.13		5.76	7.45	
	0700	Floors, mud set, 1,000 S.F. lots, red, 4" x 4" x 1/2" thick		120	.133	S.F.	3.74	2.31		6.05	7.85	
	0900	6" x 6" x 1/2" thick		140	.114		2.70	1.98		4.68	6.20	
	1000	4" x 8" x 1/2" thick		130	.123		3.74	2.13		5.87	7.55	
	1300	For waxed coating, add					.60			.60	.66	
	1500	For colors other than green, add					.35			.35	.39	
	1600	For abrasive surface, add					.42			.42	.46	
	1800	Brown tile, imported, 6" x 6" x 3/4"	D-7	120	.133		4.44	2.31		6.75	8.60	
	1900	8" x 8" x 1"		110	.145		5.05	2.52		7.57	9.65	
	2100	For thin set mortar application, deduct		700	.023			.40		.40	.64	
	2200	For epoxy grout & mortar, 6" x 6" x 1/2", add		350	.046		1.58	.79		2.37	3.02	
	2700	Stair tread, 6" x 6" x 3/4", plain		50	.320		4.04	5.55		9.59	13.45	

09300 | Tile

09330 | Quarry Tile

			CREW	DAILY OUTPUT	LABOR-HOURS	UNIT	2000 BARE COSTS MAT.	LABOR	EQUIP.	TOTAL	TOTAL INCL O&P	
100	2800	Abrasive	D-7	47	.340	S.F.	4.59	5.90		10.49	14.60	100
	3000	Wainscot, 6" x 6" x 1/2", thin set, red		105	.152		3.42	2.64		6.06	8.05	
	3100	Colors other than green		105	.152		3.81	2.64		6.45	8.45	
	3300	Window sill, 6" wide, 3/4" thick		90	.178	L.F.	4.41	3.08		7.49	9.85	
	3400	Corners		80	.200	Ea.	4.81	3.47		8.28	10.90	

09350 | Glass Mosaics

			CREW	DAILY OUTPUT	LABOR-HOURS	UNIT	MAT.	LABOR	EQUIP.	TOTAL	INCL O&P	
100	0010	**GLASS MOSAICS** 3/4" tile on 12" sheets, standard grout										100
	0300	Color group 1 & 2	D-7	73	.219	S.F.	14.95	3.80		18.75	22.50	
	0350	Color group 3		73	.219		15.75	3.80		19.55	23.50	
	0400	Color group 4		73	.219		21.50	3.80		25.30	29.50	
	0450	Color group 5		73	.219		24	3.80		27.80	32.50	
	0500	Color group 6		73	.219		29	3.80		32.80	38	
	0600	Color group 7		73	.219		34	3.80		37.80	43	
	0700	Color group 8, golds, silvers & specialties		64	.250		49	4.33		53.33	61	

09370 | Metal Tile

			CREW	DAILY OUTPUT	LABOR-HOURS	UNIT	MAT.	LABOR	EQUIP.	TOTAL	INCL O&P	
100	0010	**METAL TILE** 4' x 4' sheet, 24 ga., tile pattern, nailed										100
	0200	Stainless steel	2 Carp	512	.031	S.F.	21.50	.62		22.12	25	
	0400	Aluminized steel	"	512	.031	"	11.70	.62		12.32	13.90	

09400 | Terrazzo

09420 | Precast Terrazzo

			CREW	DAILY OUTPUT	LABOR-HOURS	UNIT	MAT.	LABOR	EQUIP.	TOTAL	INCL O&P	
900	0011	**TERRAZZO, PRECAST** Base, 6" high, straight	1 Mstz	70	.114	L.F.	8	2.18		10.18	12.35	900
	0101	Cove		60	.133		10.75	2.55		13.30	15.95	
	0301	8" high, straight		60	.133		9.10	2.55		11.65	14.15	
	0401	Cove		50	.160		12.75	3.06		15.81	19	
	0600	For white cement, add					.33			.33	.36	
	0700	For 16 ga. zinc toe strip, add					.99			.99	1.09	
	0901	Curbs, 4" x 4" high	J-3	38	.421		23	7.30	3.27	33.57	40.50	
	1001	8" x 8" high	1 Mstz	15	.533		30	10.20		40.20	49.50	
	1200	Floor tiles, non-slip, 1" thick, 12" x 12"	D-1	29	.552	S.F.	14.60	9.85		24.45	33	
	1300	1-1/4" thick, 12" x 12"		29	.552		16.75	9.85		26.60	35	
	1500	16" x 16"		23	.696		18.20	12.40		30.60	41	
	1600	1-1/2" thick, 16" x 16"		21	.762		16.65	13.60		30.25	41.50	
	1800	For Venetian terrazzo, add					4.99			4.99	5.50	
	1900	For white cement, add					.47			.47	.52	
	2401	Stair treads, 1-1/2" thick, non-slip, three line pattern	2 Mstz	70	.229	L.F.	33.50	4.37		37.87	44	
	2501	Nosing and two lines	J-3	70	.229		33.50	3.96	1.77	39.23	45.50	
	2701	2" thick treads, straight	2 Mstz	60	.267		32.50	5.10		37.60	44	
	2801	Curved		50	.320		41	6.10		47.10	55	
	3001	Stair risers, straight sections, 1" thick, to 6" high		60	.267		8	5.10		13.10	17.05	
	3101	Cove		50	.320		12.15	6.10		18.25	23.50	
	3301	Curved, 1" thick, to 6" high, vertical		48	.333		20	6.35		26.35	32.50	
	3402	Cove		38	.421		33.50	8.05		41.55	49.50	
	3601	Stair tread and riser, single piece, straight, minimum		60	.267		43	5.10		48.10	55.50	
	3701	Maximum		40	.400		55.50	7.65		63.15	73.50	

09400 | Terrazzo

09420 | Precast Terrazzo

			CREW	DAILY OUTPUT	LABOR-HOURS	UNIT	2000 BARE COSTS MAT.	LABOR	EQUIP.	TOTAL	TOTAL INCL O&P	
900	3901	Curved tread and riser, minimum	J-3	40	.400	L.F.	60	6.95	3.10	70.05	80.50	900
	4001	Maximum	"	32	.500		75.50	8.65	3.88	88.03	101	
	4201	Stair stringers, notched, 1" thick	2 Mstz	25	.640		33.50	12.20		45.70	56.50	
	4301	2" thick		22	.727		40.50	13.90		54.40	67	
	4501	Stair landings, structural, non-slip, 1-1/2" thick		85	.188	S.F.	21	3.60		24.60	29.50	
	4601	3" thick		14	1.143		33.50	22		55.50	72.50	
	4801	Wainscot, 1" thick, 12" x 12" tiles	1 Mstz	24	.333		8.65	6.35		15	19.85	
	4901	16" x 16" x 1-1/2" tiles	J-3	16	1		12.30	17.35	7.75	37.40	50	

09450 | Cast-in-Place Terrazzo

			CREW	DAILY OUTPUT	LABOR-HOURS	UNIT	MAT.	LABOR	EQUIP.	TOTAL	TOTAL INCL O&P	
100	0011	**TERRAZZO, CAST IN PLACE** Cove base, 6" high R09400-100	J-3	30	.533	L.F.	2.64	9.25	4.14	16.03	22.50	100
	0101	Curb, 6" high and 6" wide	"	12	1.333		4.12	23	10.35	37.47	53.50	
	0300	Divider strip for floors, 14 ga., 1-1/4" deep, zinc	1 Mstz	375	.021		.91	.41		1.32	1.66	
	0400	Brass		375	.021		1.70	.41		2.11	2.53	
	0600	Heavy top strip 1/4" thick, 1-1/4" deep, zinc		300	.027		1.54	.51		2.05	2.52	
	0900	Galv. bottoms, brass		300	.027		2.61	.51		3.12	3.70	
	1200	For thin set floors, 16 ga., 1/2" x 1/2", zinc		350	.023		.80	.44		1.24	1.59	
	1300	Brass		350	.023		1.51	.44		1.95	2.37	
	1500	Floor, bonded to concrete, 1-3/4" thick, gray cement	J-3	130	.123	S.F.	2.37	2.13	.96	5.46	7.10	
	1600	White cement, mud set		130	.123		2.70	2.13	.96	5.79	7.50	
	1800	Not bonded, 3" total thickness, gray cement		115	.139		2.96	2.41	1.08	6.45	8.35	
	1900	White cement, mud set		115	.139		3.23	2.41	1.08	6.72	8.65	
	2100	For Venetian terrazzo, 1" topping, add					50%	50%				
	2200	For heavy duty abrasive terrazzo, add					50%	50%				
	2400	Bonded conductive floor for hospitals	J-3	90	.178	S.F.	3.23	3.08	1.38	7.69	10.05	
	2500	Epoxy terrazzo, 1/4" thick, minimum		100	.160		3.66	2.77	1.24	7.67	9.90	
	2550	Average		75	.213		4.03	3.70	1.66	9.39	12.25	
	2600	Maximum		60	.267		4.42	4.62	2.07	11.11	14.65	
	2700	Monolithic terrazzo, 1/2" thick										
	2710	10' panels	J-3	125	.128	S.F.	2.15	2.22	.99	5.36	7.05	
	3000	Stairs, cast in place, pan filled treads		30	.533	L.F.	2.10	9.25	4.14	15.49	22	
	3100	Treads and risers		14	1.143	"	4.41	19.80	8.85	33.06	46.50	
	3300	For stair landings, add to floor prices						50%				
	3400	Stair stringers and fascia	J-3	30	.533	S.F.	2.98	9.25	4.14	16.37	23	
	3600	For abrasive metal nosings on stairs, add		150	.107	L.F.	6.05	1.85	.83	8.73	10.60	
	3700	For abrasive surface finish, add		600	.027	S.F.	.74	.46	.21	1.41	1.79	
	3900	For raised abrasive strips, add		150	.107	L.F.	.69	1.85	.83	3.37	4.67	
	4000	Wainscot, bonded, 1-1/2" thick		30	.533	S.F.	2.70	9.25	4.14	16.09	22.50	
	4200	Epoxy terrazzo, 1/4" thick		40	.400	"	4.03	6.95	3.10	14.08	19.10	
200	0010	**TILE OR TERRAZZO BASE** Scratch coat only	1 Mstz	150	.053	S.F.	.31	1.02		1.33	1.99	200
	0500	Scratch and brown coat only	"	75	.107	"	.56	2.04		2.60	3.92	

09500 | Ceilings

09510 | Acoustical Ceilings

			CREW	DAILY OUTPUT	LABOR-HOURS	UNIT	MAT.	LABOR	EQUIP.	TOTAL	TOTAL INCL O&P	
700	0010	**SUSPENDED ACOUSTIC CEILING TILES**, Not including										700
	0100	suspension system										
	0300	Fiberglass boards, film faced, 2' x 2' or 2' x 4', 5/8" thick	1 Carp	625	.013	S.F.	.51	.25		.76	.99	
	0400	3/4" thick		600	.013		1.06	.26		1.32	1.62	

09500 | Ceilings

09510 | Acoustical Ceilings

		CREW	DAILY OUTPUT	LABOR-HOURS	UNIT	2000 BARE COSTS MAT.	LABOR	EQUIP.	TOTAL	TOTAL INCL O&P	
700	0500	3" thick, thermal, R11	1 Carp	450	.018	S.F.	1.25	.35		1.60	1.98	700
	0600	Glass cloth faced fiberglass, 3/4" thick		500	.016		1.65	.32		1.97	2.36	
	0700	1" thick		485	.016		1.86	.32		2.18	2.61	
	0820	1-1/2" thick, nubby face		475	.017		2.21	.33		2.54	3	
	1110	Mineral fiber tile, lay-in, 2' x 2' or 2' x 4', 5/8" thick, fine texture		625	.013		.41	.25		.66	.88	
	1115	Rough textured		625	.013		1	.25		1.25	1.53	
	1125	3/4" thick, fine textured		600	.013		1.12	.26		1.38	1.68	
	1130	Rough textured		600	.013		1.40	.26		1.66	1.99	
	1135	Fissured		600	.013		1.65	.26		1.91	2.27	
	1150	Tegular, 5/8" thick, fine textured		470	.017		.99	.34		1.33	1.66	
	1155	Rough textured		470	.017		1.29	.34		1.63	1.99	
	1165	3/4" thick, fine textured		450	.018		1.40	.35		1.75	2.14	
	1170	Rough textured		450	.018		1.58	.35		1.93	2.34	
	1175	Fissured		450	.018		2.46	.35		2.81	3.31	
	1180	For aluminum face, add					4.49			4.49	4.94	
	1185	For plastic film face, add					.69			.69	.76	
	1190	For fire rating, add					.34			.34	.37	
	1300	Mirror faced panels, 15/16" thick, 2' x 2'	1 Carp	500	.016		9.10	.32		9.42	10.55	
	1900	Eggcrate, acrylic, 1/2" x 1/2" x 1/2" cubes		500	.016		1.36	.32		1.68	2.04	
	2100	Polystyrene eggcrate, 3/8" x 3/8" x 1/2" cubes		510	.016		1.13	.31		1.44	1.77	
	2200	1/2" x 1/2" x 1/2" cubes		500	.016		1.52	.32		1.84	2.21	
	2400	Luminous panels, prismatic, acrylic		400	.020		1.64	.39		2.03	2.48	
	2500	Polystyrene		400	.020		.84	.39		1.23	1.60	
	2700	Flat white acrylic		400	.020		2.87	.39		3.26	3.84	
	2800	Polystyrene		400	.020		1.96	.39		2.35	2.84	
	3000	Drop pan, white, acrylic		400	.020		4.20	.39		4.59	5.30	
	3100	Polystyrene		400	.020		3.51	.39		3.90	4.54	
	3600	Perforated aluminum sheets, .024" thick, corrugated, painted		490	.016		1.67	.32		1.99	2.39	
	3700	Plain		500	.016		1.50	.32		1.82	2.19	
	3720	Mineral fiber, 24" x 24" or 48", reveal edge, painted, 5/8" thick		600	.013		.93	.26		1.19	1.47	
	3740	3/4" thick		575	.014		1.52	.27		1.79	2.14	
760	0010	**SUSPENDED CEILINGS, COMPLETE** Including standard										760
	0100	suspension system but not incl. 1-1/2" carrier channels										
	0600	Fiberglass ceiling board, 2' x 4' x 5/8", plain faced,	1 Carp	500	.016	S.F.	1.02	.32		1.34	1.66	
	0700	Offices, 2' x 4' x 3/4"		380	.021		1.13	.41		1.54	1.95	
	0800	Mineral fiber, on 15/16" T bar susp. 2' x 2' x 3/4" lay-in board		345	.023		1.61	.46		2.07	2.55	
	0810	2' x 4' x 5/8" tile		380	.021		1.41	.41		1.82	2.26	
	0820	Tegular, 2' x 2' x 5/8" tile on 9/16" grid		250	.032		1.90	.63		2.53	3.17	
	0830	2' x 4' x 3/4" tile		225	.036		1.98	.70		2.68	3.38	
	0900	Luminous panels, prismatic, acrylic		255	.031		2.54	.62		3.16	3.85	
	1200	Metal pan with acoustic pad, steel		75	.107		3.29	2.10		5.39	7.20	
	1300	Painted aluminum		75	.107		5.85	2.10		7.95	10	
	1500	Aluminum, degreased finish		75	.107		3.79	2.10		5.89	7.75	
	1600	Stainless steel		75	.107		6.95	2.10		9.05	11.25	
	1800	Tile, Z bar suspension, 5/8" mineral fiber tile		150	.053		1.47	1.05		2.52	3.42	
	1900	3/4" mineral fiber tile		150	.053		1.57	1.05		2.62	3.53	
	2400	For strip lighting, see division 16510-440										
	2500	For rooms under 500 S.F., add				S.F.		25%				
900	0010	**CEILING TILE**, Stapled or cemented										900
	0100	12" x 12" or 12" x 24", not including furring										
	0600	Mineral fiber, vinyl coated, 5/8" thick	1 Carp	1,000	.008	S.F.	.72	.16		.88	1.06	
	0700	3/4" thick		1,000	.008		1.17	.16		1.33	1.56	
	0900	Fire rated, 3/4" thick, plain faced		1,000	.008		1	.16		1.16	1.37	
	1000	Plastic coated face		1,000	.008		1.04	.16		1.20	1.41	

09500 | Ceilings

09510 | Acoustical Ceilings

			CREW	DAILY OUTPUT	LABOR-HOURS	UNIT	MAT.	LABOR	EQUIP.	TOTAL	TOTAL INCL O&P	
900	1200	Aluminum faced, 5/8" thick, plain	1 Carp	1,000	.008	S.F.	1.09	.16		1.25	1.47	900
	3700	Wall application of above, add	↓	3,100	.003			.05		.05	.09	
	3900	For ceiling primer, add					.12			.12	.13	
	4000	For ceiling cement, add				↓	.32			.32	.35	

09600 | Flooring

09631 | Brick Flooring

			CREW	DAILY OUTPUT	LABOR-HOURS	UNIT	MAT.	LABOR	EQUIP.	TOTAL	TOTAL INCL O&P	
100	0010	**FLOORING**										100
	0020	Acid proof shales, red, 8" x 3-3/4" x 1-1/4" thick	D-7	.43	37.209	M	695	645		1,340	1,825	
	0050	2-1/4" thick	D-1	.40	40		755	715		1,470	2,025	
	0200	Acid proof clay brick, 8" x 3-3/4" x 2-1/4" thick	"	.40	40	↓	755	715		1,470	2,025	
	0260	Cast ceramic, pressed, 4" x 8" x 1/2", unglazed	D-7	100	.160	S.F.	4.94	2.77		7.71	9.95	
	0270	Glazed		100	.160		6.60	2.77		9.37	11.75	
	0280	Hand molded flooring, 4" x 8" x 3/4", unglazed		95	.168		6.55	2.92		9.47	11.95	
	0290	Glazed		95	.168		8.20	2.92		11.12	13.75	
	0300	8" hexagonal, 3/4" thick, unglazed		85	.188		7.15	3.26		10.41	13.20	
	0310	Glazed	↓	85	.188		12.95	3.26		16.21	19.50	
	0400	Heavy duty industrial, cement mortar bed, 2" thick, not incl. brick	D-1	80	.200		.71	3.57		4.28	6.85	
	0450	Acid proof joints, 1/4" wide	"	65	.246		1.13	4.39		5.52	8.70	
	0500	Pavers, 8" x 4", 1" to 1-1/4" thick, red	D-7	95	.168		2.88	2.92		5.80	7.90	
	0510	Ironspot	"	95	.168		4.07	2.92		6.99	9.20	
	0540	1-3/8" to 1-3/4" thick, red	D-1	95	.168		2.78	3.01		5.79	8.15	
	0560	Ironspot		95	.168		4.02	3.01		7.03	9.50	
	0580	2-1/4" thick, red		90	.178		2.83	3.17		6	8.45	
	0590	Ironspot	↓	90	.178	↓	4.38	3.17		7.55	10.15	
	0800	For sidewalks and patios with pavers, see division 02780-200										
	0870	For epoxy joints, add	D-1	600	.027	S.F.	2.15	.48		2.63	3.18	
	0880	For Furan underlayment, add	"	600	.027		1.78	.48		2.26	2.77	
	0890	For waxed surface, steam cleaned, add	D-5	1,000	.008	↓	.15	.13		.28	.38	

09635 | Marble Flooring

			CREW	DAILY OUTPUT	LABOR-HOURS	UNIT	MAT.	LABOR	EQUIP.	TOTAL	TOTAL INCL O&P	
100	0010	**MARBLE** Thin gauge tile, 12" x 6", 3/8", White Carara	D-7	60	.267	S.F.	8.75	4.62		13.37	17.15	100
	0100	Travertine		60	.267		9.65	4.62		14.27	18.10	
	0200	12" x 12" x 3/8", thin set, floors		60	.267		6.70	4.62		11.32	14.85	
	0300	On walls	↓	52	.308	↓	8.75	5.35		14.10	18.30	

09637 | Stone Flooring

			CREW	DAILY OUTPUT	LABOR-HOURS	UNIT	MAT.	LABOR	EQUIP.	TOTAL	TOTAL INCL O&P	
100	0010	**SLATE TILE** Vermont, 6" x 6" x 1/4" thick, thin set	D-7	180	.089	S.F.	4.14	1.54		5.68	7.05	100
	0200	See also division 02775-275										
200	0010	**SLATE & STONE FLOORS** See division 02780-800										200
300	0010	**CONCRETE FLOORS** And toppings, see division 03350-300										300

09643 | Wood Block Flooring

			CREW	DAILY OUTPUT	LABOR-HOURS	UNIT	MAT.	LABOR	EQUIP.	TOTAL	TOTAL INCL O&P	
100	0010	**WOOD BLOCK FLOORING** End grain flooring, coated, 2" thick	1 Carp	295	.027	S.F.	2.70	.53		3.23	3.89	100
	0400	Natural finish, 1" thick, fir		125	.064		2.80	1.26		4.06	5.25	
	0600	1-1/2" thick, pine		125	.064		2.75	1.26		4.01	5.20	
	0700	2" thick, pine	↓	125	.064	↓	2.72	1.26		3.98	5.15	

09600 | Flooring

09644 | Wood Comp. Flooring

			CREW	DAILY OUTPUT	LABOR-HOURS	UNIT	2000 BARE COSTS MAT.	LABOR	EQUIP.	TOTAL	TOTAL INCL O&P	
100	0010	**WOOD COMPOSITION** Gym floors										100
	0100	2-1/4" x 6-7/8" x 3/8", on 2" grout setting bed	D-7	150	.107	S.F.	4.60	1.85		6.45	8.05	
	0200	Thin set, on concrete	"	250	.064		4.20	1.11		5.31	6.40	
	0300	Sanding and finishing, add	1 Carp	200	.040		.62	.79		1.41	2.03	

09648 | Wood Strip Flooring

			CREW	DAILY OUTPUT	LABOR-HOURS	UNIT	MAT.	LABOR	EQUIP.	TOTAL	TOTAL INCL O&P	
100	0010	**WOOD** Fir, vertical grain, 1" x 4", not incl. finish, B & better	1 Carp	255	.031	S.F.	2.30	.62		2.92	3.59	100
	0100	C grade & better	"	255	.031	"	2.16	.62		2.78	3.44	
	0600	Gym floor, in mastic, over 2 ply felt, #2 & better										
	0700	25/32" thick maple	1 Carp	100	.080	S.F.	3.25	1.58		4.83	6.30	
	0900	33/32" thick maple		98	.082		3.75	1.61		5.36	6.90	
	1000	For 1/2" corkboard underlayment, add		750	.011		.73	.21		.94	1.16	
	1300	For #1 grade maple, add					.40			.40	.44	
	1600	Maple flooring, over sleepers, #2 & better										
	1700	25/32" thick	1 Carp	85	.094	S.F.	3.35	1.85		5.20	6.85	
	1900	33/32" thick	"	83	.096		3.90	1.90		5.80	7.55	
	2000	For #1 grade, add					.40			.40	.44	
	2200	For 3/4" subfloor, add	1 Carp	350	.023		.87	.45		1.32	1.73	
	2300	With two 1/2" subfloors, 25/32" thick	"	69	.116		4.20	2.28		6.48	8.55	
	2500	Maple, incl. finish, #2 & btr., 25/32" thick, on rubber										
	2600	Sleepers, with two 1/2" subfloors	1 Carp	76	.105	S.F.	4.50	2.07		6.57	8.50	
	2800	With steel spline, double connection to channels	"	73	.110		4.80	2.16		6.96	9	
	2900	For 33/32" maple, add					.52			.52	.57	
	3100	For #1 grade maple, add					.40			.40	.44	
	3500	For termite proofing all of the above, add					.20			.20	.22	
	3700	Portable hardwood, prefinished panels	1 Carp	83	.096		6	1.90		7.90	9.85	
	3720	Insulated with polystyrene, 1" thick, add		165	.048		.52	.96		1.48	2.21	
	3750	Running tracks, Sitka spruce surface, 25/32" x 2-1/4"		62	.129		11	2.54		13.54	16.45	
	3770	3/4" plywood surface, finished		100	.080		2.60	1.58		4.18	5.55	
	3800	See also resilient gym floors, division 09658-100										
	4000	Maple, strip, 25/32" x 2-1/4", not incl. finish, select	1 Carp	170	.047	S.F.	3.10	.93		4.03	5	
	4100	#2 & better		170	.047		2.65	.93		3.58	4.51	
	4300	33/32" x 3-1/4", not incl. finish, #1 grade		170	.047		3.25	.93		4.18	5.15	
	4400	#2 & better		170	.047		2.90	.93		3.83	4.78	
	4600	Oak, white or red, 25/32" x 2-1/4", not incl. finish										
	4700	#1 common	1 Carp	170	.047	S.F.	2.70	.93		3.63	4.56	
	4900	Select quartered, 2-1/4" wide		170	.047		3.02	.93		3.95	4.91	
	5000	Clear		170	.047		3.25	.93		4.18	5.15	
	5200	Parquetry, standard, 5/16" thick, not incl. finish, oak, minimum		160	.050		3.04	.99		4.03	5.05	
	5300	Maximum		100	.080		5.10	1.58		6.68	8.35	
	5500	Teak, minimum		160	.050		4.15	.99		5.14	6.25	
	5600	Maximum		100	.080		7.25	1.58		8.83	10.70	
	5650	13/16" thick, select grade oak, minimum		160	.050		8	.99		8.99	10.50	
	5700	Maximum		100	.080		12.15	1.58		13.73	16.05	
	5800	Custom parquetry, including finish, minimum		100	.080		13.40	1.58		14.98	17.45	
	5900	Maximum		50	.160		17.75	3.15		20.90	25	
	6100	Prefinished, white oak, prime grade, 2-1/4" wide		170	.047		5.85	.93		6.78	8.05	
	6200	3-1/4" wide		185	.043		7.50	.85		8.35	9.70	
	6400	Ranch plank		145	.055		7.25	1.09		8.34	9.85	
	6500	Hardwood blocks, 9" x 9", 25/32" thick		160	.050		4.85	.99		5.84	7.05	
	6700	Parquetry, 5/16" thick, oak, minimum		160	.050		3.30	.99		4.29	5.30	
	6800	Maximum		100	.080		8.10	1.58		9.68	11.60	
	7000	Walnut or teak, parquetry, minimum		160	.050		4.50	.99		5.49	6.65	
	7100	Maximum		100	.080		7.85	1.58		9.43	11.35	
	7200	Acrylic wood parquet blocks, 12" x 12" x 5/16",										
	7210	irradiated, set in epoxy	1 Carp	160	.050	S.F.	6.55	.99		7.54	8.90	

09600 | Flooring

09648 | Wood Strip Flooring

			CREW	DAILY OUTPUT	LABOR-HOURS	UNIT	MAT.	LABOR	EQUIP.	TOTAL	TOTAL INCL O&P	
100	7400	Yellow pine, 3/4" x 3-1/8", T & G, C & better, not incl. finish	1 Carp	200	.040	S.F.	2.15	.79		2.94	3.72	100
	7500	Refinish wood floor, sand, 2 cts poly, wax, soft wood, min.	1 Clab	400	.020		.65	.29		.94	1.21	
	7600	Hard wood, max		130	.062		.98	.89		1.87	2.60	
	7800	Sanding and finishing, 2 coats polyurethane		295	.027		.65	.39		1.04	1.38	
	7900	Subfloor and underlayment, see division 06160-850 & 06160-791										
	8015	Transition molding, 2 1/4" wide, 5' long	1 Carp	19.20	.417	Ea.	10.35	8.20		18.55	25.50	

09651 | Resilient Base & Access.

			CREW	DAILY OUTPUT	LABOR-HOURS	UNIT	MAT.	LABOR	EQUIP.	TOTAL	TOTAL INCL O&P	
100	0010	**STAIR TREADS AND RISERS** See index for materials other										100
	0100	than rubber and vinyl										
	0300	Rubber, molded tread, 12" wide, 5/16" thick, black	1 Tilf	115	.070	L.F.	6.35	1.34		7.69	9.15	
	0400	Colors		115	.070		6.75	1.34		8.09	9.60	
	0600	1/4" thick, black		115	.070		6.25	1.34		7.59	9.05	
	0700	Colors		115	.070		6.80	1.34		8.14	9.60	
	0900	Grip strip safety tread, colors, 5/16" thick		115	.070		10.30	1.34		11.64	13.50	
	1000	3/16" thick		120	.067		7.80	1.28		9.08	10.60	
	1200	Landings, smooth sheet rubber, 1/8" thick		120	.067	S.F.	2.47	1.28		3.75	4.79	
	1300	3/16" thick		120	.067	"	3.59	1.28		4.87	6	
	1500	Nosings, 3" wide, 3/16" thick, black		140	.057	L.F.	2.59	1.10		3.69	4.63	
	1600	Colors		140	.057		2.53	1.10		3.63	4.56	
	1800	Risers, 7" high, 1/8" thick, flat		250	.032		2.90	.61		3.51	4.19	
	1900	Coved		250	.032		2.09	.61		2.70	3.30	
	2100	Vinyl, molded tread, 12" wide, colors, 1/8" thick		115	.070		2.63	1.34		3.97	5.05	
	2200	1/4" thick		115	.070		4.33	1.34		5.67	6.90	
	2300	Landing material, 1/8" thick		200	.040	S.F.	2.73	.77		3.50	4.24	
	2400	Riser, 7" high, 1/8" thick, coved		175	.046	L.F.	1.61	.88		2.49	3.19	
	2500	Tread and riser combined, 1/8" thick		80	.100	"	4.06	1.92		5.98	7.60	

09658 | Resilient Tile Flooring

			CREW	DAILY OUTPUT	LABOR-HOURS	UNIT	MAT.	LABOR	EQUIP.	TOTAL	TOTAL INCL O&P	
100	0010	**RESILIENT FLOORING**	R09658-600									100
	0800	Base, cove, rubber or vinyl, .080" thick										
	1100	Standard colors, 2-1/2" high	1 Tilf	315	.025	L.F.	.40	.49		.89	1.23	
	1150	4" high		315	.025		.44	.49		.93	1.27	
	1200	6" high		315	.025		.73	.49		1.22	1.59	
	1450	1/8" thick, standard colors, 2-1/2" high		315	.025		.45	.49		.94	1.29	
	1500	4" high		315	.025		.65	.49		1.14	1.50	
	1550	6" high		315	.025		.81	.49		1.30	1.68	
	1600	Corners, 2-1/2" high		315	.025	Ea.	1.02	.49		1.51	1.91	
	1630	4" high		315	.025		1.07	.49		1.56	1.97	
	1660	6" high		315	.025		1.38	.49		1.87	2.31	
	1700	Conductive flooring, rubber tile, 1/8" thick		315	.025	S.F.	2.50	.49		2.99	3.54	
	1800	Homogeneous vinyl tile, 1/8" thick		315	.025		3.42	.49		3.91	4.55	
	2200	Cork tile, standard finish, 1/8" thick		315	.025		3.41	.49		3.90	4.54	
	2250	3/16" thick		315	.025		3.64	.49		4.13	4.79	
	2300	5/16" thick		315	.025		4.56	.49		5.05	5.80	
	2350	1/2" thick		315	.025		5.45	.49		5.94	6.80	
	2500	Urethane finish, 1/8" thick		315	.025		4.33	.49		4.82	5.55	
	2550	3/16" thick		315	.025		5.15	.49		5.64	6.50	
	2600	5/16" thick		315	.025		6.05	.49		6.54	7.45	
	2650	1/2" thick		315	.025		8.10	.49		8.59	9.70	
	3700	Polyethylene, in rolls, no base incl., landscape surfaces		275	.029		2.24	.56		2.80	3.36	
	3800	Nylon action surface, 1/8" thick		275	.029		2.41	.56		2.97	3.55	
	3900	1/4" thick		275	.029		3.47	.56		4.03	4.72	
	4000	3/8" thick		275	.029		4.36	.56		4.92	5.70	
	4100	Golf tee surface with foam back		235	.034		4.31	.65		4.96	5.80	

09600 | Flooring

09658 | Resilient Tile Flooring

			CREW	DAILY OUTPUT	LABOR-HOURS	UNIT	MAT.	LABOR	EQUIP.	TOTAL	TOTAL INCL O&P	
100	4200	Practice putting, knitted nylon surface	1 Tilf	235	.034	S.F.	3.66	.65		4.31	5.10	100
	4400	Polyurethane, thermoset, prefabricated in place, indoor	R09658-600									
	4500	3/8" thick for basketball, gyms, etc.	1 Tilf	100	.080	S.F.	3.45	1.54		4.99	6.30	
	4600	1/2" thick for professional sports		95	.084		4.03	1.62		5.65	7.05	
	4700	Outdoor, 1/4" thick, smooth, for tennis		100	.080		3.09	1.54		4.63	5.90	
	4800	Rough, for track, 3/8" thick		95	.084		3.58	1.62		5.20	6.55	
	5000	Poured in place, indoor, with finish, 1/4" thick		80	.100		2.54	1.92		4.46	5.90	
	5050	3/8" thick		65	.123		3.09	2.36		5.45	7.25	
	5100	1/2" thick		50	.160		4.10	3.07		7.17	9.50	
	5500	Polyvinyl chloride, sheet goods for gyms, 1/4" thick		80	.100		3.52	1.92		5.44	7	
	5600	3/8" thick		60	.133		3.97	2.56		6.53	8.50	
	5900	Rubber, sheet goods, 36" wide, 1/8" thick		120	.067		3.09	1.28		4.37	5.45	
	5950	3/16" thick		100	.080		4.39	1.54		5.93	7.30	
	6000	1/4" thick		90	.089		5.05	1.71		6.76	8.35	
	6050	Tile, marbleized colors, 12" x 12", 1/8" thick		400	.020		3.50	.38		3.88	4.47	
	6100	3/16" thick		400	.020		4.75	.38		5.13	5.80	
	6300	Special tile, plain colors, 1/8" thick		400	.020		3.83	.38		4.21	4.83	
	6350	3/16" thick		400	.020		5.15	.38		5.53	6.25	
	6410	Raised, radial or square, minimum		400	.020		6	.38		6.38	7.20	
	6430	Maximum		400	.020		6	.38		6.38	7.20	
	6450	For golf course, skating rink, etc., 1/4" thick		275	.029		5.90	.56		6.46	7.35	
	6700	Synthetic turf, 3/8" thick		90	.089		3.16	1.71		4.87	6.25	
	6750	Interlocking 2' x 2' squares, 1/2" thick, not										
	6810	cemented, for playgrounds, minimum	1 Tilf	210	.038	S.F.	2.61	.73		3.34	4.05	
	6850	Maximum		190	.042		6.70	.81		7.51	8.65	
	7000	Vinyl composition tile, 12" x 12", 1/16" thick		500	.016		.69	.31		1	1.26	
	7050	Embossed		500	.016		.86	.31		1.17	1.45	
	7100	Marbleized		500	.016		.86	.31		1.17	1.45	
	7150	Solid		500	.016		.97	.31		1.28	1.57	
	7200	3/32" thick, embossed		500	.016		.85	.31		1.16	1.44	
	7250	Marbleized		500	.016		.98	.31		1.29	1.58	
	7300	Solid		500	.016		1.43	.31		1.74	2.07	
	7350	1/8" thick, marbleized		500	.016		.94	.31		1.25	1.53	
	7400	Solid		500	.016		2.01	.31		2.32	2.71	
	7450	Conductive		500	.016		3.33	.31		3.64	4.16	
	7500	Vinyl tile, 12" x 12", .050" thick, minimum		500	.016		1.54	.31		1.85	2.19	
	7550	Maximum		500	.016		3.01	.31		3.32	3.81	
	7600	1/8" thick, minimum		500	.016		1.94	.31		2.25	2.63	
	7650	Solid colors		500	.016		4.35	.31		4.66	5.30	
	7700	Marbleized or Travertine pattern		500	.016		3.11	.31		3.42	3.92	
	7750	Florentine pattern		500	.016		3.57	.31		3.88	4.43	
	7800	Maximum		500	.016		7.35	.31		7.66	8.55	
	8000	Vinyl sheet goods, backed, .065" thick, minimum		250	.032		1.25	.61		1.86	2.38	
	8050	Maximum		200	.040		2.32	.77		3.09	3.79	
	8100	.080" thick, minimum		230	.035		1.50	.67		2.17	2.73	
	8150	Maximum		200	.040		2.65	.77		3.42	4.16	
	8200	.125" thick, minimum		230	.035		1.63	.67		2.30	2.87	
	8250	Maximum		200	.040		3.57	.77		4.34	5.15	
	8700	Adhesive cement, 1 gallon does 200 to 300 S.F.				Gal.	13.85			13.85	15.20	
	8800	Asphalt primer, 1 gallon per 300 S.F.					8.40			8.40	9.25	
	8900	Emulsion, 1 gallon per 140 S.F.					9			9	9.90	
	8950	Latex underlayment, liquid, fortified					27.50			27.50	30	

09673 | Composition Flooring

			CREW	DAILY OUTPUT	LABOR-HOURS	UNIT	MAT.	LABOR	EQUIP.	TOTAL	TOTAL INCL O&P	
100	0010	**COMPOSITION FLOORING** Acrylic, 1/4" thick	C-6	520	.092	S.F.	1.21	1.43	.15	2.79	3.92	100
	0100	3/8" thick		450	.107		1.57	1.66	.17	3.40	4.73	

09600 | Flooring

09673 | Composition Flooring

		CREW	DAILY OUTPUT	LABOR-HOURS	UNIT	2000 BARE COSTS				TOTAL INCL O&P
						MAT.	LABOR	EQUIP.	TOTAL	
0300	Cupric oxychloride, on bond coat, minimum	C-6	480	.100	S.F.	2.66	1.55	.16	4.37	5.75
0400	Maximum		420	.114		4.45	1.77	.18	6.40	8.10
0600	Epoxy, with colored quartz chips, broadcast, minimum		675	.071		2.11	1.10	.11	3.32	4.31
0700	Maximum		490	.098		2.56	1.52	.15	4.23	5.55
0900	Trowelled, minimum		560	.086		2.72	1.33	.14	4.19	5.40
1000	Maximum		480	.100		3.96	1.55	.16	5.67	7.15
1200	Heavy duty epoxy topping, 1/4" thick,									
1300	500 to 1,000 S.F.	C-6	420	.114	S.F.	4.09	1.77	.18	6.04	7.70
1500	1,000 to 2,000 S.F.		450	.107		2.99	1.66	.17	4.82	6.30
1600	Over 10,000 S.F.		480	.100		2.67	1.55	.16	4.38	5.75
1800	Epoxy terrazzo, 1/4" thick, chemical resistant, minimum	J-3	200	.080		4.35	1.39	.62	6.36	7.70
1900	Maximum		150	.107		7.10	1.85	.83	9.78	11.75
2100	Conductive, minimum		210	.076		5.50	1.32	.59	7.41	8.85
2200	Maximum		160	.100		7.70	1.73	.78	10.21	12.10
2400	Mastic, hot laid, 2 coat, 1-1/2" thick, standard, minimum	C-6	690	.070		3.09	1.08	.11	4.28	5.35
2500	Maximum		520	.092		3.96	1.43	.15	5.54	6.95
2700	Acidproof, minimum		605	.079		3.96	1.23	.13	5.32	6.60
2800	Maximum		350	.137		5.50	2.13	.22	7.85	9.90
3000	Neoprene, trowelled on, 1/4" thick, minimum		545	.088		3.04	1.37	.14	4.55	5.80
3100	Maximum		430	.112		4.18	1.73	.18	6.09	7.75
3150	Polyacrylate terrazzo, 1/4" thick, minimum		735	.065		2.42	1.01	.10	3.53	4.49
3170	Maximum		480	.100		3.21	1.55	.16	4.92	6.35
3200	3/8" thick, minimum		620	.077		2.94	1.20	.12	4.26	5.40
3220	Maximum		480	.100		4.69	1.55	.16	6.40	7.95
3300	Conductive terrazzo, 1/4" thick, minimum		450	.107		5.80	1.66	.17	7.63	9.40
3330	Maximum		305	.157		7.45	2.44	.25	10.14	12.60
3350	3/8" thick, minimum		365	.132		7.85	2.04	.21	10.10	12.35
3370	Maximum		255	.188		9.95	2.92	.30	13.17	16.25
3450	Granite, conductive, 1/4" thick, minimum		695	.069		5.95	1.07	.11	7.13	8.45
3470	Maximum		420	.114		7.90	1.77	.18	9.85	11.90
3500	3/8" thick, minimum		695	.069		8.60	1.07	.11	9.78	11.40
3520	Maximum		380	.126		10.75	1.96	.20	12.91	15.35
3600	Polyester, with colored quartz chips, 1/16" thick, minimum		1,065	.045		2.14	.70	.07	2.91	3.62
3700	Maximum		560	.086		3.23	1.33	.14	4.70	5.95
3900	1/8" thick, minimum		810	.059		2.58	.92	.09	3.59	4.50
4000	Maximum		675	.071		3.73	1.10	.11	4.94	6.10
4200	Polyester, heavy duty, compared to epoxy, add		2,590	.019		.72	.29	.03	1.04	1.31
4300	Polyurethane, with suspended vinyl chips, minimum		1,065	.045		5.90	.70	.07	6.67	7.75
4500	Maximum		860	.056		8.55	.87	.09	9.51	10.95

09680 | Carpet

		CREW	DAILY OUTPUT	LABOR-HOURS	UNIT	MAT.	LABOR	EQUIP.	TOTAL	TOTAL INCL O&P
0010	**CARPET** Commercial grades, direct cement									
0700	Nylon, level loop, 26 oz., light to medium traffic	1 Tilf	75	.107	S.Y.	12.90	2.05		14.95	17.50
0720	28 oz., light to medium traffic		75	.107		14.25	2.05		16.30	19
0900	32 oz., medium traffic		75	.107		18.30	2.05		20.35	23.50
1100	40 oz., medium to heavy traffic		75	.107		27	2.05		29.05	33.50
2920	Nylon plush, 30 oz., medium traffic		57	.140		13.50	2.69		16.19	19.20
3000	36 oz., medium traffic		75	.107		17	2.05		19.05	22
3100	42 oz., medium to heavy traffic		70	.114		16.70	2.19		18.89	22
3200	46 oz., medium to heavy traffic		70	.114		23.50	2.19		25.69	29.50
3300	54 oz., heavy traffic		70	.114		26.50	2.19		28.69	32.50
3340	60 oz., heavy traffic		70	.114		37.50	2.19		39.69	44.50
3665	Olefin, 24 oz., light to medium traffic		75	.107		7.90	2.05		9.95	12
3670	26 oz., medium traffic		75	.107		8.20	2.05		10.25	12.30
3680	28 oz., medium to heavy traffic		75	.107		9.55	2.05		11.60	13.80

09600 | Flooring

09680 | Carpet

			CREW	DAILY OUTPUT	LABOR-HOURS	UNIT	2000 BARE COSTS MAT.	LABOR	EQUIP.	TOTAL	TOTAL INCL O&P	
800	3700	32 oz., medium to heavy traffic	1 Tilf	75	.107	S.Y.	11.75	2.05		13.80	16.20	800
	3730	42 oz., heavy traffic		70	.114		17.30	2.19		19.49	22.50	
	4110	Wool, level loop, 40 oz., medium traffic		75	.107		58.50	2.05		60.55	68	
	4500	50 oz., medium to heavy traffic		75	.107		60	2.05		62.05	69.50	
	4700	Patterned, 32 oz., medium to heavy traffic		70	.114		59	2.19		61.19	68.50	
	4900	48 oz., heavy traffic		70	.114		60.50	2.19		62.69	70	
	5000	For less than full roll, add					25%					
	5100	For small rooms, less than 12' wide, add						25%				
	5200	For large open areas (no cuts), deduct						25%				
	5600	For bound carpet baseboard, add	1 Tilf	300	.027	L.F.	1.02	.51		1.53	1.95	
	5610	For stairs, not incl. price of carpet, add	"	30	.267	Riser		5.10		5.10	8.30	
	5620	For borders and patterns, add to labor						18%				
	8950	For tackless, stretched installation, add padding to above										
	9000	Sponge rubber pad, minimum	1 Tilf	150	.053	S.Y.	2.61	1.02		3.63	4.53	
	9100	Maximum		150	.053		6.95	1.02		7.97	9.30	
	9200	Felt pad, minimum		150	.053		2.78	1.02		3.80	4.72	
	9300	Maximum		150	.053		5.20	1.02		6.22	7.35	
	9400	Bonded urethane pad, minimum		150	.053		3.15	1.02		4.17	5.15	
	9500	Maximum		150	.053		5.60	1.02		6.62	7.80	
	9600	Prime urethane pad, minimum		150	.053		1.83	1.02		2.85	3.67	
	9700	Maximum		150	.053		3.35	1.02		4.37	5.35	
	9850	For "branded" fiber, add					25%					
900	0010	**CARPET TILE**										900
	0100	Tufted nylon, 18" x 18", hard back, 20 oz.	1 Tilf	150	.053	S.Y.	16.85	1.02		17.87	20	
	0110	26 oz.		150	.053		29	1.02		30.02	33	
	0200	Cushion back, 20 oz.		150	.053		21	1.02		22.02	25	
	0210	26 oz.		150	.053		33	1.02		34.02	38	
910	0010	**CARPET MAINTENANCE** Steam clean, per cleaning, minimum	1 Clab	3,000	.003	S.F.	.07	.04		.11	.15	910
	0500	Maximum	"	2,000	.004	"	.10	.06		.16	.21	

09700 | Wall Finishes

09710 | Acoustical Wall Treatment

			CREW	DAILY OUTPUT	LABOR-HOURS	UNIT	2000 BARE COSTS MAT.	LABOR	EQUIP.	TOTAL	TOTAL INCL O&P	
100	0010	**WALL COVERING** Including sizing, add 10%-30% waste at takeoff R09700-700										100
	0051	Aluminum foil	1 Pord	275	.029	S.F.	.78	.52		1.30	1.73	
	0101	Copper sheets, .025" thick, vinyl backing		240	.033		4.15	.60		4.75	5.55	
	0301	Phenolic backing		240	.033		5.35	.60		5.95	6.90	
	0601	Cork tiles, light or dark, 12" x 12" x 3/16"		240	.033		2.67	.60		3.27	3.94	
	0701	5/16" thick		235	.034		2.77	.61		3.38	4.07	
	0901	1/4" basketweave		240	.033		4.25	.60		4.85	5.70	
	1001	1/2" natural, non-directional pattern		240	.033		4.18	.60		4.78	5.60	
	1201	Granular surface, 12" x 36", 1/2" thick		385	.021		.92	.37		1.29	1.63	
	1301	1" thick		370	.022		1.19	.39		1.58	1.96	
	1501	Polyurethane coated, 12" x 12" x 3/16" thick		240	.033		2.87	.60		3.47	4.16	
	1601	5/16" thick		235	.034		4.08	.61		4.69	5.50	
	1802	Cork wallpaper, paperbacked, natural		480	.017		1.63	.30		1.93	2.29	
	1901	Colors		480	.017		2.02	.30		2.32	2.72	
	2101	Flexible wood veneer, 1/32" thick, plain woods		100	.080		1.64	1.44		3.08	4.20	
	2201	Exotic woods		95	.084		2.49	1.51		4	5.25	

09700 | Wall Finishes

09710 | Acoustical Wall Treatment

			Crew	Daily Output	Labor-Hours	Unit	Mat.	Labor	Equip.	Total	Total Incl O&P
100	2400	Gypsum-based, fabric-backed, fire										100
	2501	resistant for masonry walls, minimum	1 Pord	800	.010	S.F.	.60	.18		.78	.96
	2701	Maximum, (small quantities)	"	300	.027		1	.48		1.48	1.90
	2750	Acrylic, modified, semi-rigid PVC, .028" thick	2 Carp	330	.048		.82	.96		1.78	2.54
	2800	.040" thick	"	320	.050		1.08	.99		2.07	2.88
	3001	Vinyl wall covering, fabric-backed, lightweight	1 Pord	640	.013		.53	.22		.75	.95
	3301	Medium weight		480	.017		.84	.30		1.14	1.42
	3401	Heavy weight	↓	435	.018		1.16	.33		1.49	1.83
	3600	Adhesive, 5 gal. lots, (18SY/Gal.)				Gal.	8.30			8.30	9.15
	3702	Wallpaper, average workmanship, solid pattern, low cost paper	1 Pord	640	.013	S.F.	.27	.22		.49	.67
	3902	Basic patterns (matching req.), avg. cost paper		535	.015		.48	.27		.75	.98
	4002	Paper at $50 per double roll, quality workmanship	↓	435	.018	↓	.92	.33		1.25	1.56
	4100	Linen wall covering, paper backed									
	4150	Flame treatment, minimum				S.F.	.65			.65	.71
	4180	Maximum					1.17			1.17	1.29
	4201	Grass cloths with lining paper, minimum	1 Pord	400	.020		.59	.36		.95	1.25
	4301	Maximum	"	350	.023	↓	1.88	.41		2.29	2.75

09770 | Special Wall Surfaces

			Crew	Daily Output	Labor-Hours	Unit	Mat.	Labor	Equip.	Total	Total Incl O&P
400	0010	**FIBERGLASS REINFORCED PLASTIC** panels, .090" thick, on walls										400
	0020	Adhesive mounted, embossed surface	2 Carp	640	.025	S.F.	1.21	.49		1.70	2.17
	0030	Smooth surface		640	.025		1.38	.49		1.87	2.36
	0040	Fire rated, embossed surface		640	.025		1.59	.49		2.08	2.59
	0050	Nylon rivet mounted, on drywall, embossed surface		480	.033		1.11	.66		1.77	2.35
	0060	Smooth surface		480	.033		1.28	.66		1.94	2.54
	0070	Fire rated, embossed surface		480	.033		1.49	.66		2.15	2.77
	0080	On masonry, embossed surface		320	.050		1.11	.99		2.10	2.91
	0090	Smooth surface		320	.050		1.28	.99		2.27	3.10
	0100	Fire rated, embossed surface		320	.050		1.49	.99		2.48	3.33
	0110	Nylon rivet and adhesive mounted, on drywall, embossed surface		240	.067		1.30	1.31		2.61	3.68
	0120	Smooth surface		240	.067		1.47	1.31		2.78	3.87
	0130	Fire rated, embossed surface		240	.067		1.68	1.31		2.99	4.10
	0140	On masonry, embossed surface		190	.084		1.30	1.66		2.96	4.27
	0150	Smooth surface		190	.084		1.47	1.66		3.13	4.46
	0160	Fire rated, embossed surface	↓	190	.084	↓	1.68	1.66		3.34	4.69
	0170	For moldings add	1 Carp	250	.032	L.F.	.22	.63		.85	1.32
	0180	On ceilings, for lay in grid system, embossed surface		400	.020	S.F.	1.21	.39		1.60	2.01
	0190	Smooth surface		400	.020		1.38	.39		1.77	2.20
	0200	Fire rated, embossed surface	↓	400	.020	↓	1.59	.39		1.98	2.43

09800 | Acoustical Treatment

09820 | Acoustical Insul/Sealants

			Crew	Daily Output	Labor-Hours	Unit	Mat.	Labor	Equip.	Total	Total Incl O&P
500	0010	**SOUND ATTENUATION** Blanket, 1" thick	1 Carp	925	.009	S.F.	.21	.17		.38	.52	500
	0500	1-1/2" thick		920	.009		.24	.17		.41	.55
	1000	2" thick		915	.009		.31	.17		.48	.64
	1500	3" thick		910	.009		.40	.17		.57	.74
	3000	Thermal or acoustical batt above ceiling, 2" thick		900	.009		.42	.18		.60	.76
	3100	3" thick	↓	900	.009	↓	.63	.18		.81	.99

09800 | Acoustical Treatment

09820 | Acoustical Insul/Sealants

		CREW	DAILY OUTPUT	LABOR-HOURS	UNIT	MAT.	LABOR	EQUIP.	TOTAL	TOTAL INCL O&P	
500	3200	4" thick	1 Carp	900	.009	S.F.	.77	.18		.95	1.15	500
	3400	Urethane plastic foam, open cell, on wall, 2" thick	2 Carp	2,050	.008		2.56	.15		2.71	3.08	
	3500	3" thick		1,550	.010		3.40	.20		3.60	4.09	
	3600	4" thick		1,050	.015		4.77	.30		5.07	5.75	
	3700	On ceiling, 2" thick		1,700	.009		2.55	.19		2.74	3.12	
	3800	3" thick		1,300	.012		3.40	.24		3.64	4.16	
	3900	4" thick	▼	900	.018	▼	4.77	.35		5.12	5.85	
	4000	Nylon matting 0.4" thick, with carbon black spinerette										
	4010	plus polyester fabric, on floor	J-4	4,000	.004	S.F.	1.90	.07		1.97	2.20	
	4200	Fiberglass reinf. backer board underlayment, 7/16" thick, on floor	"	800	.020	"	1.59	.35		1.94	2.31	

09830 | Acoustical Barriers

		CREW	DAILY OUTPUT	LABOR-HOURS	UNIT	MAT.	LABOR	EQUIP.	TOTAL	TOTAL INCL O&P	
100	0011	BARRIERS Plenum										100
	0600	Aluminum foil, fiberglass reinf., parallel with joists	1 Carp	275	.029	S.F.	.74	.57		1.31	1.79	
	0700	Perpendicular to joists		180	.044		.74	.88		1.62	2.31	
	0900	Aluminum mesh, kraft paperbacked		275	.029		.69	.57		1.26	1.74	
	0970	Fiberglass batts, kraft faced, 3-1/2" thick		1,400	.006		.25	.11		.36	.47	
	0980	6" thick		1,300	.006		.42	.12		.54	.67	
	1000	Sheet lead, 1 lb., 1/64" thick, perpendicular to joists		150	.053		1.55	1.05		2.60	3.50	
	1100	Vinyl foam reinforced, 1/8" thick, 1.0 lb. per S.F.	▼	150	.053	▼	2.70	1.05		3.75	4.77	

09840 | Acoustical Wall Treatment

		CREW	DAILY OUTPUT	LABOR-HOURS	UNIT	MAT.	LABOR	EQUIP.	TOTAL	TOTAL INCL O&P	
100	0010	SOUND ABSORBING PANELS Perforated steel facing, painted with										100
	0100	fiberglass or mineral filler, no backs, 2-1/4" thick, modular										
	0200	space units, ceiling or wall hung, white or colored	1 Carp	100	.080	S.F.	7.45	1.58		9.03	10.85	
	0300	Fiberboard sound deadening panels, 1/2" thick	"	600	.013	"	.28	.26		.54	.76	
	0500	Fiberglass panels, 4' x 8' x 1" thick, with										
	0600	glass cloth face for walls, cemented	1 Carp	155	.052	S.F.	4.69	1.02		5.71	6.90	
	0700	1-1/2" thick, dacron covered, inner aluminum frame,										
	0710	wall mounted	1 Carp	300	.027	S.F.	7.15	.53		7.68	8.80	
	0900	Mineral fiberboard panels, fabric covered, 30"x 108",										
	1000	3/4" thick, concealed spline, wall mounted	1 Carp	150	.053	S.F.	5.15	1.05		6.20	7.45	

09900 | Paints & Coatings

09910 | Paints & Coatings

		CREW	DAILY OUTPUT	LABOR-HOURS	UNIT	MAT.	LABOR	EQUIP.	TOTAL	TOTAL INCL O&P	
100	0010	CABINETS AND CASEWORK										100
	1000	Primer coat, oil base, brushwork	1 Pord	650	.012	S.F.	.04	.22		.26	.42	
	2000	Paint, oil base, brushwork, 1 coat		650	.012		.05	.22		.27	.43	
	3000	Stain, brushwork, wipe off		650	.012		.04	.22		.26	.42	
	4000	Shellac, 1 coat, brushwork		650	.012		.05	.22		.27	.42	
	4500	Varnish, 3 coats, brushwork, sand after 1st coat	▼	325	.025		.16	.44		.60	.92	
	5000	For latex paint, deduct				▼	10%					
300	0010	DOORS AND WINDOWS, EXTERIOR	R09910-100									300
	0100	Door frames & trim, only										
	0110	Brushwork, primer	R09910-220	1 Pord	512	.016	L.F.	.05	.28		.33	.53	
	0120	Finish coat, exterior latex		512	.016		.05	.28		.33	.53	
	0130	Primer & 1 coat, exterior latex		300	.027	▼	.10	.48		.58	.91	
	0135	2 coats, exterior latex, both sides	▼	15	.533	Ea.	4.70	9.55		14.25	21	

09900 | Paints & Coatings

09910 | Paints & Coatings

			CREW	DAILY OUTPUT	LABOR-HOURS	UNIT	MAT.	LABOR	EQUIP.	TOTAL	TOTAL INCL O&P
300	0140	Primer & 2 coats, exterior latex	1 Pord	265	.030	L.F.	.16	.54		.70	1.07
	0150	Doors, flush, both sides, incl. frame & trim	R09910-100								
	0160	Roll & brush, primer	1 Pord	10	.800	Ea.	3.83	14.35		18.18	28
	0170	Finish coat, exterior latex	R09910-220	10	.800		4	14.35		18.35	28.50
	0180	Primer & 1 coat, exterior latex		7	1.143		7.85	20.50		28.35	42.50
	0190	Primer & 2 coats, exterior latex		5	1.600		11.85	28.50		40.35	61
	0200	Brushwork, stain, sealer & 2 coats polyurethane		4	2		14.25	36		50.25	75.50
	0210	Doors, French, both sides, 10-15 lite, incl. frame & trim									
	0220	Brushwork, primer	1 Pord	6	1.333	Ea.	1.91	24		25.91	42
	0230	Finish coat, exterior latex		6	1.333		2	24		26	42
	0240	Primer & 1 coat, exterior latex		3	2.667		3.92	48		51.92	84.50
	0250	Primer & 2 coats, exterior latex		2	4		5.80	72		77.80	126
	0260	Brushwork, stain, sealer & 2 coats polyurethane		2.50	3.200		5.10	57.50		62.60	102
	0270	Doors, louvered, both sides, incl. frame & trim									
	0280	Brushwork, primer	1 Pord	7	1.143	Ea.	3.83	20.50		24.33	38
	0290	Finish coat, exterior latex		7	1.143		4	20.50		24.50	38.50
	0300	Primer & 1 coat, exterior latex		4	2		7.85	36		43.85	68.50
	0310	Primer & 2 coats, exterior latex		3	2.667		11.60	48		59.60	93
	0320	Brushwork, stain, sealer & 2 coats polyurethane		4.50	1.778		14.25	32		46.25	68.50
	0330	Doors, panel, both sides, incl. frame & trim									
	0340	Roll & brush, primer	1 Pord	6	1.333	Ea.	3.83	24		27.83	44
	0350	Finish coat, exterior latex		6	1.333		4	24		28	44.50
	0360	Primer & 1 coat, exterior latex		3	2.667		7.85	48		55.85	88.50
	0370	Primer & 2 coats, exterior latex		2.50	3.200		11.60	57.50		69.10	109
	0380	Brushwork, stain, sealer & 2 coats polyurethane		3	2.667		14.25	48		62.25	95.50
	0400	Windows, per ext. side, based on 15 SF									
	0410	1 to 6 lite									
	0420	Brushwork, primer	1 Pord	13	.615	Ea.	.76	11.05		11.81	19.30
	0430	Finish coat, exterior latex		13	.615		.79	11.05		11.84	19.30
	0440	Primer & 1 coat, exterior latex		8	1		1.55	17.95		19.50	31.50
	0450	Primer & 2 coats, exterior latex		6	1.333		2.29	24		26.29	42.50
	0460	Stain, sealer & 1 coat varnish		7	1.143		2.01	20.50		22.51	36
	0470	7 to 10 lite									
	0480	Brushwork, primer	1 Pord	11	.727	Ea.	.76	13.05		13.81	23
	0490	Finish coat, exterior latex		11	.727		.79	13.05		13.84	23
	0500	Primer & 1 coat, exterior latex		7	1.143		1.55	20.50		22.05	35.50
	0510	Primer & 2 coats, exterior latex		5	1.600		2.29	28.50		30.79	50.50
	0520	Stain, sealer & 1 coat varnish		6	1.333		2.01	24		26.01	42
	0530	12 lite									
	0540	Brushwork, primer	1 Pord	10	.800	Ea.	.76	14.35		15.11	25
	0550	Finish coat, exterior latex		10	.800		.79	14.35		15.14	25
	0560	Primer & 1 coat, exterior latex		6	1.333		1.55	24		25.55	41.50
	0570	Primer & 2 coats, exterior latex		5	1.600		2.29	28.50		30.79	50.50
	0580	Stain, sealer & 1 coat varnish		6	1.333		2	24		26	42
	0590	For oil base paint, add					10%				
310	0010	**DOORS & WINDOWS, INTERIOR LATEX**	R09910-100								
	0100	Doors flush, both sides, incl. frame & trim									
	0110	Roll & brush, primer	1 Pord	10	.800	Ea.	3.32	14.35		17.67	27.50
	0120	Finish coat, latex	R09910-220	10	.800		3.73	14.35		18.08	28
	0130	Primer & 1 coat latex		7	1.143		7.05	20.50		27.55	42
	0140	Primer & 2 coats latex		5	1.600		10.60	28.50		39.10	59.50
	0160	Spray, both sides, primer		20	.400		3.50	7.20		10.70	15.85
	0170	Finish coat, latex		20	.400		3.92	7.20		11.12	16.30
	0180	Primer & 1 coat latex		11	.727		7.45	13.05		20.50	30
	0190	Primer & 2 coats latex		8	1		11.20	17.95		29.15	42.50

09900 | Paints & Coatings

09910 | Paints & Coatings

		CREW	DAILY OUTPUT	LABOR-HOURS	UNIT	2000 BARE COSTS MAT.	LABOR	EQUIP.	TOTAL	TOTAL INCL O&P
310										
0200	Doors, French, both sides, 10-15 lite, incl. frame & trim (R09910-100)									
0210	Roll & brush, primer	1 Pord	6	1.333	Ea.	1.66	24		25.66	42
0220	Finish coat, latex (R09910-220)		6	1.333		1.87	24		25.87	42
0230	Primer & 1 coat latex		3	2.667		3.53	48		51.53	84
0240	Primer & 2 coats latex		2	4		5.30	72		77.30	126
0260	Doors, louvered, both sides, incl. frame & trim									
0270	Roll & brush, primer	1 Pord	7	1.143	Ea.	3.32	20.50		23.82	37.50
0280	Finish coat, latex		7	1.143		3.73	20.50		24.23	38
0290	Primer & 1 coat, latex		4	2		6.85	36		42.85	67.50
0300	Primer & 2 coats, latex		3	2.667		10.80	48		58.80	92
0320	Spray, both sides, primer		20	.400		3.50	7.20		10.70	15.85
0330	Finish coat, latex		20	.400		3.92	7.20		11.12	16.30
0340	Primer & 1 coat, latex		11	.727		7.45	13.05		20.50	30
0350	Primer & 2 coats, latex		8	1		11.45	17.95		29.40	42.50
0360	Doors, panel, both sides, incl. frame & trim									
0370	Roll & brush, primer	1 Pord	6	1.333	Ea.	3.50	24		27.50	44
0380	Finish coat, latex		6	1.333		3.73	24		27.73	44
0390	Primer & 1 coat, latex		3	2.667		7.05	48		55.05	88
0400	Primer & 2 coats, latex		2.50	3.200		10.80	57.50		68.30	108
0420	Spray, both sides, primer		10	.800		3.50	14.35		17.85	28
0430	Finish coat, latex		10	.800		3.92	14.35		18.27	28.50
0440	Primer & 1 coat, latex		5	1.600		7.45	28.50		35.95	56
0450	Primer & 2 coats, latex		4	2		11.45	36		47.45	72.50
0460	Windows, per interior side, base on 15 SF									
0470	1 to 6 lite									
0480	Brushwork, primer	1 Pord	13	.615	Ea.	.66	11.05		11.71	19.15
0490	Finish coat, enamel		13	.615		.74	11.05		11.79	19.25
0500	Primer & 1 coat enamel		8	1		1.39	17.95		19.34	31.50
0510	Primer & 2 coats enamel		6	1.333		2.13	24		26.13	42.50
0530	7 to 10 lite									
0540	Brushwork, primer	1 Pord	11	.727	Ea.	.66	13.05		13.71	22.50
0550	Finish coat, enamel		11	.727		.74	13.05		13.79	23
0560	Primer & 1 coat enamel		7	1.143		1.39	20.50		21.89	35.50
0570	Primer & 2 coats enamel		5	1.600		2.13	28.50		30.63	50.50
0590	12 lite									
0600	Brushwork, primer	1 Pord	10	.800	Ea.	.66	14.35		15.01	24.50
0610	Finish coat, enamel		10	.800		.74	14.35		15.09	25
0620	Primer & 1 coat enamel		6	1.333		1.39	24		25.39	41.50
0630	Primer & 2 coats enamel		5	1.600		2.13	28.50		30.63	50.50
0650	For oil base paint, add					10%				
320										
0010	**DOORS AND WINDOWS, INTERIOR ALKYD (OIL BASE)**									
0500	Flush door & frame, 3' x 7', oil, primer, brushwork	1 Pord	10	.800	Ea.	1.94	14.35		16.29	26
1000	Paint, 1 coat		10	.800		1.86	14.35		16.21	26
1400	Stain, brushwork, wipe off		18	.444		.90	8		8.90	14.30
1600	Shellac, 1 coat, brushwork		25	.320		.96	5.75		6.71	10.65
1800	Varnish, 3 coats, brushwork, sand after 1st coat		9	.889		3.36	15.95		19.31	30
2000	Panel door & frame, 3' x 7', oil, primer, brushwork		6	1.333		1.63	24		25.63	42
2200	Paint, 1 coat		6	1.333		1.86	24		25.86	42
2600	Stain, brushwork, panel door, 3' x 7', not incl. frame		16	.500		.90	9		9.90	16
2800	Shellac, 1 coat, brushwork		22	.364		.96	6.55		7.51	11.95
3000	Varnish, 3 coats, brushwork, sand after 1st coat		7.50	1.067		3.36	19.15		22.51	35.50
4400	Windows, including frame and trim, per side									
4600	Colonial type, 6/6 lites, 2' x 3', oil, primer, brushwork	1 Pord	14	.571	Ea.	.26	10.25		10.51	17.40
5800	Paint, 1 coat		14	.571		.29	10.25		10.54	17.40
6200	3' x 5' opening, 6/6 lites, primer coat, brushwork		12	.667		.64	11.95		12.59	20.50
6400	Paint, 1 coat		12	.667		.73	11.95		12.68	21

09900 | Paints & Coatings

09910 | Paints & Coatings

			CREW	DAILY OUTPUT	LABOR-HOURS	UNIT	2000 BARE COSTS MAT.	LABOR	EQUIP.	TOTAL	TOTAL INCL O&P	
320	6800	4' x 8' opening, 6/6 lites, primer coat, brushwork	1 Pord	8	1	Ea.	1.37	17.95		19.32	31.50	**320**
	7000	Paint, 1 coat		8	1		1.57	17.95		19.52	31.50	
	8000	Single lite type, 2' x 3', oil base, primer coat, brushwork		33	.242		.26	4.35		4.61	7.55	
	8200	Paint, 1 coat		33	.242		.29	4.35		4.64	7.55	
	8600	3' x 5' opening, primer coat, brushwork		20	.400		.64	7.20		7.84	12.70	
	8800	Paint, 1 coat		20	.400		.73	7.20		7.93	12.80	
	9200	4' x 8' opening, primer coat, brushwork		14	.571		1.37	10.25		11.62	18.60	
	9400	Paint, 1 coat		14	.571		1.57	10.25		11.82	18.80	
400	0010	**FENCES** R09910-100										**400**
	0100	Chain link or wire metal, one side, water base										
	0110	Roll & brush, first coat R09910-220	1 Pord	960	.008	S.F.	.05	.15		.20	.31	
	0120	Second coat		1,280	.006		.05	.11		.16	.24	
	0130	Spray, first coat		2,275	.004		.05	.06		.11	.17	
	0140	Second coat		2,600	.003		.05	.06		.11	.15	
	0150	Picket, water base										
	0160	Roll & brush, first coat	1 Pord	865	.009	S.F.	.06	.17		.23	.34	
	0170	Second coat		1,050	.008		.06	.14		.20	.29	
	0180	Spray, first coat		2,275	.004		.06	.06		.12	.17	
	0190	Second coat		2,600	.003		.06	.06		.12	.15	
	0200	Stockade, water base										
	0210	Roll & brush, first coat	1 Pord	1,040	.008	S.F.	.06	.14		.20	.29	
	0220	Second coat		1,200	.007		.06	.12		.18	.26	
	0230	Spray, first coat		2,275	.004		.06	.06		.12	.17	
	0240	Second coat		2,600	.003		.06	.06		.12	.15	
500	0010	**FLOORS**, Interior										**500**
	0100	Concrete										
	0120	1st coat	1 Pord	975	.008	S.F.	.11	.15		.26	.37	
	0130	2nd coat		1,150	.007		.07	.12		.19	.29	
	0140	3rd coat		1,300	.006		.06	.11		.17	.24	
	0150	Roll, latex, block filler										
	0160	1st coat	1 Pord	2,600	.003	S.F.	.14	.06		.20	.25	
	0170	2nd coat		3,250	.002		.08	.04		.12	.16	
	0180	3rd coat		3,900	.002		.06	.04		.10	.13	
	0190	Spray, latex, block filler										
	0200	1st coat	1 Pord	2,600	.003	S.F.	.12	.06		.18	.22	
	0210	2nd coat		3,250	.002		.07	.04		.11	.14	
	0220	3rd coat		3,900	.002		.05	.04		.09	.12	
620	0010	**MISCELLANEOUS, EXTERIOR**										**620**
	0015	For painting metals, see Div. 05120-680										
	0100	Railing, ext., decorative wood, incl. cap & baluster R09910-220										
	0110	newels & spindles @ 12" O.C.										
	0120	Brushwork, stain, sand, seal & varnish										
	0130	First coat	1 Pord	90	.089	L.F.	.40	1.60		2	3.10	
	0140	Second coat	"	120	.067	"	.40	1.20		1.60	2.44	
	0150	Rough sawn wood, 42" high, 2"x2" verticals, 6" O.C.										
	0160	Brushwork, stain, each coat	1 Pord	90	.089	L.F.	.13	1.60		1.73	2.80	
	0170	Wrought iron, 1" rail, 1/2" sq. verticals										
	0180	Brushwork, zinc chromate, 60" high, bars 6" O.C.										
	0190	Primer	1 Pord	130	.062	L.F.	.47	1.10		1.57	2.36	
	0200	Finish coat		130	.062		.15	1.10		1.25	2	
	0210	Additional coat		190	.042		.17	.76		.93	1.45	
	0220	Shutters or blinds, single panel, 2'x4', paint all sides										
	0230	Brushwork, primer	1 Pord	20	.400	Ea.	.55	7.20		7.75	12.60	
	0240	Finish coat, exterior latex		20	.400		.42	7.20		7.62	12.45	
	0250	Primer & 1 coat, exterior latex		13	.615		.85	11.05		11.90	19.40	

09900 | Paints & Coatings

09910 | Paints & Coatings

			Crew	Daily Output	Labor-Hours	Unit	Mat.	Labor	Equip.	Total	Total Incl O&P
620	0260	Spray, primer	1 Pord	35	.229	Ea.	.81	4.10		4.91	7.75
	0270	Finish coat, exterior latex		35	.229		.90	4.10		5	7.85
	0280	Primer & 1 coat, exterior latex		20	.400		.86	7.20		8.06	12.95
	0290	For louvered shutters, add				S.F.	10%				
	0300	Stair stringers, exterior, metal									
	0310	Roll & brush, zinc chromate, to 14", each coat	1 Pord	320	.025	L.F.	.05	.45		.50	.80
	0320	Rough sawn wood, 4" x 12"									
	0330	Roll & brush, exterior latex, each coat	1 Pord	215	.037	L.F.	.06	.67		.73	1.18
	0340	Trellis/lattice, 2"x2" @ 3" O.C. with 2"x8" supports									
	0350	Spray, latex, per side, each coat	1 Pord	475	.017	S.F.	.06	.30		.36	.57
	0450	Decking, Ext., sealer, alkyd, brushwork, sealer coat		1,140	.007		.05	.13		.18	.26
	0460	1st coat		1,140	.007		.05	.13		.18	.26
	0470	2nd coat		1,300	.006		.04	.11		.15	.22
	0500	Paint, alkyd, brushwork, primer coat		1,140	.007		.07	.13		.20	.28
	0510	1st coat		1,140	.007		.06	.13		.19	.28
	0520	2nd coat		1,300	.006		.05	.11		.16	.23
	0600	Sand paint, alkyd, brushwork, 1 coat		150	.053		.08	.96		1.04	1.69
630	0010	**MISCELLANEOUS, INTERIOR**									
	2400	Floors, conc./wood, oil base, primer/sealer coat, brushwork	2 Pord	1,950	.008	S.F.	.05	.15		.20	.31
	2450	Roller		5,200	.003		.06	.06		.12	.15
	2600	Spray		6,000	.003		.06	.05		.11	.14
	2650	Paint 1 coat, brushwork		1,950	.008		.05	.15		.20	.30
	2800	Roller		5,200	.003		.05	.06		.11	.15
	2850	Spray		6,000	.003		.05	.05		.10	.14
	3000	Stain, wood floor, brushwork, 1 coat		4,550	.004		.04	.06		.10	.16
	3200	Roller		5,200	.003		.05	.06		.11	.14
	3250	Spray		6,000	.003		.04	.05		.09	.13
	3400	Varnish, wood floor, brushwork		4,550	.004		.05	.06		.11	.17
	3450	Roller		5,200	.003		.06	.06		.12	.15
	3600	Spray		6,000	.003		.06	.05		.11	.15
	3650	For dust proofing or anti skid, see division 03350-300									
	3800	Grilles, per side, oil base, primer coat, brushwork	1 Pord	520	.015	S.F.	.09	.28		.37	.55
	3850	Spray		1,140	.007		.09	.13		.22	.31
	3920	Paint 2 coats, brushwork		325	.025		.19	.44		.63	.95
	3940	Spray		650	.012		.22	.22		.44	.61
	5000	Pipe, to 4" diameter, primer or sealer coat, oil base, brushwork	2 Pord	1,250	.013	L.F.	.06	.23		.29	.44
	5100	Spray		2,165	.007		.05	.13		.18	.28
	5350	Paint 2 coats, brushwork		775	.021		.10	.37		.47	.73
	5400	Spray		1,240	.013		.11	.23		.34	.51
	6300	To 16" diameter, primer or sealer coat, brushwork		310	.052		.22	.93		1.15	1.80
	6350	Spray		540	.030		.25	.53		.78	1.16
	6500	Paint 2 coats, brushwork		195	.082		.41	1.47		1.88	2.91
	6550	Spray		310	.052		.45	.93		1.38	2.05
	7000	Trim, wood, incl. puttying, under 6" wide									
	7200	Primer coat, oil base, brushwork	1 Pord	650	.012	L.F.	.02	.22		.24	.39
	7250	Paint, 1 coat, brushwork		650	.012		.02	.22		.24	.40
	7450	3 coats		325	.025		.07	.44		.51	.82
	7500	Over 6" wide, primer coat, brushwork		650	.012		.04	.22		.26	.42
	7550	Paint, 1 coat, brushwork		650	.012		.05	.22		.27	.42
	7650	3 coats		325	.025		.14	.44		.58	.90
	8000	Cornice, simple design, primer coat, oil base, brushwork		650	.012	S.F.	.04	.22		.26	.42
	8250	Paint, 1 coat		650	.012		.05	.22		.27	.42
	8350	Ornate design, primer coat		350	.023		.04	.41		.45	.73
	8400	Paint, 1 coat		350	.023		.05	.41		.46	.73
	8600	Balustrades, primer coat, oil base, brushwork		520	.015		.04	.28		.32	.51

09900 | Paints & Coatings

09910 | Paints & Coatings

		CREW	DAILY OUTPUT	LABOR-HOURS	UNIT	2000 BARE COSTS				TOTAL INCL O&P
						MAT.	LABOR	EQUIP.	TOTAL	
630 8650	Paint, 1 coat	1 Pord R09910-100	520	.015	S.F.	.05	.28		.33	.51	**630**
8900	Trusses and wood frames, primer coat, oil base, brushwork		800	.010		.04	.18		.22	.35
8950	Spray		1,200	.007		.04	.12		.16	.25
9220	Paint 2 coats, brushwork		500	.016		.10	.29		.39	.58
9240	Spray		600	.013		.11	.24		.35	.52
9260	Stain, brushwork, wipe off		600	.013		.04	.24		.28	.45
9280	Varnish, 3 coats, brushwork		275	.029		.16	.52		.68	1.05
9350	For latex paint, deduct					10%				
700 0010	**SIDING EXTERIOR**, Alkyd (oil base)	R09910-100									**700**
0500	Spray	2 Pord	4,550	.004	S.F.	.08	.06		.14	.19
0800	Paint 2 coats, brushwork		1,300	.012		.10	.22		.32	.48
1000	Spray		4,550	.004		.14	.06		.20	.26
1200	Stucco, rough, oil base, paint 2 coats, brushwork		1,300	.012		.10	.22		.32	.48
1400	Roller		1,625	.010		.11	.18		.29	.41
1600	Spray		2,925	.005		.11	.10		.21	.28
1800	Texture 1-11 or clapboard, oil base, primer coat, brushwork		1,300	.012		.09	.22		.31	.46
2000	Spray		4,550	.004		.09	.06		.15	.20
2400	Paint 2 coats, brushwork		810	.020		.15	.35		.50	.75
2600	Spray		2,600	.006		.16	.11		.27	.36
3400	Stain 2 coats, brushwork		950	.017		.09	.30		.39	.59
4000	Spray		3,050	.005		.10	.09		.19	.26
4200	Wood shingles, oil base primer coat, brushwork		1,300	.012		.08	.22		.30	.46
4400	Spray		3,900	.004		.07	.07		.14	.20
5000	Paint 2 coats, brushwork		810	.020		.12	.35		.47	.72
5200	Spray		2,275	.007		.12	.13		.25	.34
6500	Stain 2 coats, brushwork		950	.017		.09	.30		.39	.59
7000	Spray		2,660	.006		.12	.11		.23	.31
8000	For latex paint, deduct					10%				
710 0010	**SIDING, MISC.**	R09910-100									**710**
0100	Aluminum siding									
0110	Brushwork, primer	2 Pord R09910-220	2,275	.007	S.F.	.05	.13		.18	.27
0120	Finish coat, exterior latex		2,275	.007		.04	.13		.17	.25
0130	Primer & 1 coat exterior latex		1,300	.012		.10	.22		.32	.48
0140	Primer & 2 coats exterior latex		975	.016		.14	.29		.43	.64
0150	Mineral Fiber shingles									
0160	Brushwork, primer	2 Pord	1,495	.011	S.F.	.09	.19		.28	.41
0170	Finish coat, industrial enamel		1,495	.011		.09	.19		.28	.42
0180	Primer & 1 coat enamel		810	.020		.18	.35		.53	.79
0190	Primer & 2 coats enamel		540	.030		.27	.53		.80	1.19
0200	Roll, primer		1,625	.010		.09	.18		.27	.39
0210	Finish coat, industrial enamel		1,625	.010		.10	.18		.28	.40
0220	Primer & 1 coat enamel		975	.016		.20	.29		.49	.70
0230	Primer & 2 coats enamel		650	.025		.30	.44		.74	1.06
0240	Spray, primer		3,900	.004		.07	.07		.14	.20
0250	Finish coat, industrial enamel		3,900	.004		.08	.07		.15	.21
0260	Primer & 1 coat enamel		2,275	.007		.16	.13		.29	.38
0270	Primer & 2 coats enamel		1,625	.010		.24	.18		.42	.55
0280	Waterproof sealer, first coat		4,485	.004		.06	.06		.12	.18
0290	Second coat		5,235	.003		.06	.05		.11	.16
0300	Rough wood incl. shingles, shakes or rough sawn siding									
0310	Brushwork, primer	2 Pord	1,280	.013	S.F.	.11	.22		.33	.49
0320	Finish coat, exterior latex		1,280	.013		.07	.22		.29	.45
0330	Primer & 1 coat exterior latex		960	.017		.18	.30		.48	.70
0340	Primer & 2 coats exterior latex		700	.023		.25	.41		.66	.95
0350	Roll, primer		2,925	.005		.15	.10		.25	.32
0360	Finish coat, exterior latex		2,925	.005		.08	.10		.18	.25

Important: See the Reference Section for critical supporting data - Reference Nos., Crews, & City Cost Indexes

09900 | Paints & Coatings

09910 | Paints & Coatings

		CREW	DAILY OUTPUT	LABOR-HOURS	UNIT	2000 BARE COSTS MAT.	LABOR	EQUIP.	TOTAL	TOTAL INCL O&P
0370	Primer & 1 coat exterior latex	2 Pord	1,790	.009	S.F.	.23	.16		.39	.53
0380	Primer & 2 coats exterior latex		1,300	.012		.31	.22		.53	.72
0390	Spray, primer		3,900	.004		.13	.07		.20	.26
0400	Finish coat, exterior latex		3,900	.004		.06	.07		.13	.19
0410	Primer & 1 coat exterior latex		2,600	.006		.19	.11		.30	.39
0420	Primer & 2 coats exterior latex		2,080	.008		.25	.14		.39	.51
0430	Waterproof sealer, first coat		4,485	.004		.11	.06		.17	.23
0440	Second coat		4,485	.004		.06	.06		.12	.18
0450	Smooth wood incl. butt, T&G, beveled, drop or B&B siding									
0460	Brushwork, primer	2 Pord	2,325	.007	S.F.	.08	.12		.20	.30
0470	Finish coat, exterior latex		1,280	.013		.07	.22		.29	.45
0480	Primer & 1 coat exterior latex		800	.020		.15	.36		.51	.76
0490	Primer & 2 coats exterior latex		630	.025		.22	.46		.68	1
0500	Roll, primer		2,275	.007		.09	.13		.22	.31
0510	Finish coat, exterior latex		2,275	.007		.07	.13		.20	.29
0520	Primer & 1 coat exterior latex		1,300	.012		.16	.22		.38	.55
0530	Primer & 2 coats exterior latex		975	.016		.24	.29		.53	.75
0540	Spray, primer		4,550	.004		.07	.06		.13	.19
0550	Finish coat, exterior latex		4,550	.004		.06	.06		.12	.18
0560	Primer & 1 coat exterior latex		2,600	.006		.13	.11		.24	.33
0570	Primer & 2 coats exterior latex		1,950	.008		.20	.15		.35	.47
0580	Waterproof sealer, first coat		5,230	.003		.06	.05		.11	.16
0590	Second coat		5,980	.003		.06	.05		.11	.15
0600	For oil base paint, add					10%				
0010	**TRIM, EXTERIOR**									
0100	Door frames & trim (see Doors, interior or exterior)									
0110	Fascia, latex paint, one coat coverage									
0120	1" x 4", brushwork	1 Pord	640	.013	L.F.	.02	.22		.24	.39
0130	Roll		1,280	.006		.02	.11		.13	.21
0140	Spray		2,080	.004		.01	.07		.08	.13
0150	1" x 6" to 1" x 10", brushwork		640	.013		.06	.22		.28	.43
0160	Roll		1,230	.007		.06	.12		.18	.25
0170	Spray		2,100	.004		.05	.07		.12	.16
0180	1" x 12", brushwork		640	.013		.06	.22		.28	.43
0190	Roll		1,050	.008		.06	.14		.20	.29
0200	Spray		2,200	.004		.05	.07		.12	.16
0210	Gutters & downspouts, metal, zinc chromate paint									
0220	Brushwork, gutters, 5", first coat	1 Pord	640	.013	L.F.	.05	.22		.27	.43
0230	Second coat		960	.008		.05	.15		.20	.30
0240	Third coat		1,280	.006		.04	.11		.15	.23
0250	Downspouts, 4", first coat		640	.013		.05	.22		.27	.43
0260	Second coat		960	.008		.05	.15		.20	.30
0270	Third coat		1,280	.006		.04	.11		.15	.23
0280	Gutters & downspouts, wood									
0290	Brushwork, gutters, 5", primer	1 Pord	640	.013	L.F.	.05	.22		.27	.43
0300	Finish coat, exterior latex		640	.013		.05	.22		.27	.42
0310	Primer & 1 coat exterior latex		400	.020		.10	.36		.46	.71
0320	Primer & 2 coats exterior latex		325	.025		.16	.44		.60	.91
0330	Downspouts, 4", primer		640	.013		.05	.22		.27	.43
0340	Finish coat, exterior latex		640	.013		.05	.22		.27	.42
0350	Primer & 1 coat exterior latex		400	.020		.10	.36		.46	.71
0360	Primer & 2 coats exterior latex		325	.025		.08	.44		.52	.83
0370	Molding, exterior, up to 14" wide									
0380	Brushwork, primer	1 Pord	640	.013	L.F.	.06	.22		.28	.44
0390	Finish coat, exterior latex		640	.013		.06	.22		.28	.43
0400	Primer & 1 coat exterior latex		400	.020		.12	.36		.48	.74

FINISHES

09900 | Paints & Coatings

09910 | Paints & Coatings

			CREW	DAILY OUTPUT	LABOR-HOURS	UNIT	2000 BARE COSTS MAT.	LABOR	EQUIP.	TOTAL	TOTAL INCL O&P	
800	0410	Primer & 2 coats exterior latex	1 Pord	315	.025	L.F.	.12	.46		.58	.90	**800**
	0420	Stain & fill		1,050	.008		.05	.14		.19	.29	
	0430	Shellac		1,850	.004		.05	.08		.13	.19	
	0440	Varnish		1,275	.006		.06	.11		.17	.26	
910	0350	**WALLS, MASONRY (CMU), EXTERIOR**										**910**
	0360	Concrete masonry units (CMU), smooth surface										
	0370	Brushwork, latex, first coat	1 Pord	640	.013	S.F.	.03	.22		.25	.41	
	0380	Second coat		960	.008		.03	.15		.18	.28	
	0390	Waterproof sealer, first coat		736	.011		.06	.20		.26	.40	
	0400	Second coat		1,104	.007		.04	.13		.17	.26	
	0410	Roll, latex, paint, first coat		1,465	.005		.04	.10		.14	.20	
	0420	Second coat		1,790	.004		.03	.08		.11	.16	
	0430	Waterproof sealer, first coat		1,680	.005		.06	.09		.15	.21	
	0440	Second coat		2,060	.004		.04	.07		.11	.16	
	0450	Spray, latex, paint, first coat		1,950	.004		.03	.07		.10	.15	
	0460	Second coat		2,600	.003		.02	.06		.08	.12	
	0470	Waterproof sealer, first coat		2,245	.004		.08	.06		.14	.19	
	0480	Second coat		2,990	.003		.03	.05		.08	.11	
	0490	Concrete masonry unit (CMU), porous										
	0500	Brushwork, latex, first coat	1 Pord	640	.013	S.F.	.07	.22		.29	.44	
	0510	Second coat		960	.008		.03	.15		.18	.29	
	0520	Waterproof sealer, first coat		736	.011		.08	.20		.28	.41	
	0530	Second coat		1,104	.007		.04	.13		.17	.26	
	0540	Roll latex, first coat		1,465	.005		.05	.10		.15	.21	
	0550	Second coat		1,790	.004		.03	.08		.11	.17	
	0560	Waterproof sealer, first coat		1,680	.005		.08	.09		.17	.22	
	0570	Second coat		2,060	.004		.04	.07		.11	.17	
	0580	Spray latex, first coat		1,950	.004		.04	.07		.11	.16	
	0590	Second coat		2,600	.003		.02	.06		.08	.12	
	0600	Waterproof sealer, first coat		2,245	.004		.06	.06		.12	.18	
	0610	Second coat		2,990	.003		.03	.05		.08	.12	
920	0010	**WALLS AND CEILINGS, Interior**										**920**
	0100	Concrete, dry wall or plaster, oil base, primer or sealer coat										
	0200	Smooth finish, brushwork	1 Pord	1,150	.007	S.F.	.04	.12		.16	.26	
	0240	Roller		2,040	.004		.04	.07		.11	.16	
	0300	Sand finish, brushwork		975	.008		.04	.15		.19	.29	
	0340	Roller		1,150	.007		.05	.12		.17	.26	
	0380	Spray		2,275	.004		.03	.06		.09	.15	
	0800	Paint 2 coats, smooth finish, brushwork		680	.012		.08	.21		.29	.44	
	0840	Roller		800	.010		.08	.18		.26	.39	
	0880	Spray		1,625	.005		.07	.09		.16	.23	
	0900	Sand finish, brushwork		605	.013		.08	.24		.32	.49	
	0940	Roller		1,020	.008		.08	.14		.22	.32	
	0980	Spray		1,700	.005		.07	.08		.15	.22	
	1200	Paint 3 coats, smooth finish, brushwork		510	.016		.12	.28		.40	.60	
	1240	Roller		650	.012		.13	.22		.35	.51	
	1280	Spray		1,625	.005		.11	.09		.20	.27	
	1600	Glaze coating, 5 coats, spray, clear		900	.009		.60	.16		.76	.93	
	1640	Multicolor		900	.009		.81	.16		.97	1.16	
	1700	For latex paint, deduct					10%					
	1800	For ceiling installations, add						25%				
	2000	Masonry or concrete block, oil base, primer or sealer coat										
	2100	Smooth finish, brushwork	1 Pord	1,224	.007	S.F.	.04	.12		.16	.25	
	2180	Spray		2,400	.003		.06	.06		.12	.17	
	2200	Sand finish, brushwork		1,089	.007		.07	.13		.20	.29	

09900 | Paints & Coatings

09910 | Paints & Coatings

		Crew	Daily Output	Labor-Hours	Unit	Mat.	Labor	Equip.	Total	Total Incl O&P	
920	2280 Spray	1 Pord	2,400	.003	S.F.	.06	.06		.12	.17	920
	2800 Paint 2 coats, smooth finish, brushwork		756	.011		.13	.19		.32	.47	
	2880 Spray		1,360	.006		.12	.11		.23	.32	
	2900 Sand finish, brushwork		672	.012		.13	.21		.34	.51	
	2980 Spray		1,360	.006		.12	.11		.23	.32	
	3600 Glaze coating, 5 coats, spray, clear		900	.009		.60	.16		.76	.93	
	3620 Multicolor		900	.009		.81	.16		.97	1.16	
	4000 Block filler, 1 coat, brushwork		425	.019		.12	.34		.46	.69	
	4100 Silicone, water repellent, 2 coats, spray		2,000	.004		.24	.07		.31	.38	
	4120 For latex paint, deduct					10%					

09930 | Stains/Transp. Finishes

		Crew	Daily Output	Labor-Hours	Unit	Mat.	Labor	Equip.	Total	Total Incl O&P	
100	0010 VARNISH 1 coat + sealer, on wood trim, no sanding included	1 Pord	400	.020	S.F.	.06	.36		.42	.67	100
	0100 Hardwood floors, 2 coats, no sanding included, roller	"	1,890	.004	"	.12	.08		.20	.26	

09963 | Glazed Coatings

		Crew	Daily Output	Labor-Hours	Unit	Mat.	Labor	Equip.	Total	Total Incl O&P	
200	0010 WALL COATINGS Acrylic glazed coatings, minimum	1 Pord	525	.015	S.F.	.24	.27		.51	.72	200
	0100 Maximum		305	.026		.50	.47		.97	1.34	
	0300 Epoxy coatings, minimum		525	.015		.31	.27		.58	.80	
	0400 Maximum		170	.047		.96	.84		1.80	2.47	
	0600 Exposed aggregate, troweled on, 1/16" to 1/4", minimum		235	.034		.48	.61		1.09	1.55	
	0700 Maximum (epoxy or polyacrylate)		130	.062		1.03	1.10		2.13	2.97	
	0900 1/2" to 5/8" aggregate, minimum		130	.062		.95	1.10		2.05	2.89	
	1000 Maximum		80	.100		1.62	1.80		3.42	4.78	
	1500 Exposed aggregate, sprayed on, 1/8" aggregate, minimum		295	.027		.45	.49		.94	1.31	
	1600 Maximum		145	.055		.82	.99		1.81	2.55	
	1800 High build epoxy, 50 mil, minimum		390	.021		.52	.37		.89	1.18	
	1900 Maximum		95	.084		.90	1.51		2.41	3.51	
	2100 Laminated epoxy with fiberglass, minimum		295	.027		.57	.49		1.06	1.44	
	2200 Maximum		145	.055		1.02	.99		2.01	2.77	
	2400 Sprayed perlite or vermiculite, 1/16" thick, minimum		2,935	.003		.20	.05		.25	.30	
	2500 Maximum		640	.013		.58	.22		.80	1.01	
	2700 Vinyl plastic wall coating, minimum		735	.011		.26	.20		.46	.62	
	2800 Maximum		240	.033		.64	.60		1.24	1.70	
	3000 Urethane on smooth surface, 2 coats, minimum		1,135	.007		.19	.13		.32	.42	
	3100 Maximum		665	.012		.44	.22		.66	.84	
	3600 Ceramic-like glazed coating, cementitious, minimum		440	.018		.38	.33		.71	.96	
	3700 Maximum		345	.023		.63	.42		1.05	1.38	
	3900 Resin base, minimum		640	.013		.26	.22		.48	.66	
	4000 Maximum		330	.024		.43	.44		.87	1.20	
800	0010 SANDING and puttying interior trim, compared to										800
	0100 Painting 1 coat, on quality work				L.F.		100%				
	0300 Medium work						50%				
	0400 Industrial grade						25%				
	0500 Surface protection, placement and removal										
	0510 Basic drop cloths	1 Pord	6,400	.001	S.F.		.02		.02	.04	
	0520 Masking with paper		800	.010		.02	.18		.20	.32	
	0530 Volume cover up (using plastic sheathing, or building paper)		16,000	.001			.01		.01	.01	
900	0010 SURFACE PREPARATION, EXTERIOR										900
	0015 Doors, per side, not incl. frames or trim										
	0020 Scrape & sand										
	0030 Wood, flush	1 Pord	616	.013	S.F.		.23		.23	.39	
	0040 Wood, detail		496	.016			.29		.29	.48	
	0050 Wood, louvered		280	.029			.51		.51	.86	
	0060 Wood, overhead		616	.013			.23		.23	.39	
	0070 Wire brush										

09900 | Paints & Coatings

09963 | Glazed Coatings

		Crew	Daily Output	Labor-Hours	Unit	Mat.	Labor	Equip.	Total	Total Incl O&P
900	0080	Metal, flush	1 Pord	640	.013	S.F.		.22		.22	.37
	0090	Metal, detail		520	.015			.28		.28	.46
	0100	Metal, louvered		360	.022			.40		.40	.67
	0110	Metal or fibr., overhead		640	.013			.22		.22	.37
	0120	Metal, roll up		560	.014			.26		.26	.43
	0130	Metal, bulkhead		640	.013			.22		.22	.37
	0140	Power wash, based on 2500 lb. operating pressure									
	0150	Metal, flush	B-9	2,240	.018	S.F.		.27	.08	.35	.54
	0160	Metal, detail		2,120	.019			.28	.09	.37	.57
	0170	Metal, louvered		2,000	.020			.30	.09	.39	.61
	0180	Metal or fibr., overhead		2,400	.017			.25	.08	.33	.50
	0190	Metal, roll up		2,400	.017			.25	.08	.33	.50
	0200	Metal, bulkhead		2,200	.018			.27	.08	.35	.55
	0400	Windows, per side, not incl. trim									
	0410	Scrape & sand									
	0420	Wood, 1-2 lite	1 Pord	320	.025	S.F.		.45		.45	.75
	0430	Wood, 3-6 lite		280	.029			.51		.51	.86
	0440	Wood, 7-10 lite		240	.033			.60		.60	1
	0450	Wood, 12 lite		200	.040			.72		.72	1.20
	0460	Wood, Bay / Bow		320	.025			.45		.45	.75
	0470	Wire brush									
	0480	Metal, 1-2 lite	1 Pord	480	.017	S.F.		.30		.30	.50
	0490	Metal, 3-6 lite		400	.020			.36		.36	.60
	0500	Metal, Bay / Bow		480	.017			.30		.30	.50
	0510	Power wash, based on 2500 lb. operating pressure									
	0520	1-2 lite	B-9	4,400	.009	S.F.		.14	.04	.18	.28
	0530	3-6 lite		4,320	.009			.14	.04	.18	.29
	0540	7-10 lite		4,240	.009			.14	.04	.18	.29
	0550	12 lite		4,160	.010			.14	.04	.18	.29
	0560	Bay / Bow		4,400	.009			.14	.04	.18	.28
	0600	Siding, scrape and sand, light=10-30%, med.=30-70%									
	0610	Heavy=70-100%, % of surface to sand									
	0620	For Chemical Washing, see Division 04930-900									
	0630	For Steam Cleaning, see Division 04930-750									
	0640	For Sand Blasting, see Division 04930-700									
	0650	Texture 1-11, light	1 Pord	480	.017	S.F.		.30		.30	.50
	0660	Med.		440	.018			.33		.33	.54
	0670	Heavy		360	.022			.40		.40	.67
	0680	Wood shingles, shakes, light		440	.018			.33		.33	.54
	0690	Med.		360	.022			.40		.40	.67
	0700	Heavy		280	.029			.51		.51	.86
	0710	Clapboard, light		520	.015			.28		.28	.46
	0720	Med.		480	.017			.30		.30	.50
	0730	Heavy		400	.020			.36		.36	.60
	0740	Wire brush									
	0750	Aluminum, light	1 Pord	600	.013	S.F.		.24		.24	.40
	0760	Med.		520	.015			.28		.28	.46
	0770	Heavy		440	.018			.33		.33	.54
	0780	Pressure wash, based on 2500 lb.. operating pressure									
	0790	Stucco	B-9	3,080	.013	S.F.		.19	.06	.25	.39
	0800	Aluminum or vinyl		3,200	.013			.19	.06	.25	.38
	0810	Siding, masonry, brick & block		2,400	.017			.25	.08	.33	.50
	1300	Miscellaneous, wire brush									
	1310	Metal, pedestrian gate	1 Pord	100	.080	S.F.		1.44		1.44	2.40
910	0010	**SURFACE PREPARATION, INTERIOR**									
	0020	Doors									

09900 | Paints & Coatings

09963 | Glazed Coatings

		CREW	DAILY OUTPUT	LABOR-HOURS	UNIT	2000 BARE COSTS				TOTAL INCL O&P
						MAT.	LABOR	EQUIP.	TOTAL	
0030	Scrape & sand									
0040	Wood, flush	1 Pord	616	.013	S.F.		.23		.23	.39
0050	Wood, detail		496	.016			.29		.29	.48
0060	Wood, louvered	↓	280	.029	↓		.51		.51	.86
0070	Wire brush									
0080	Metal, flush	1 Pord	640	.013	S.F.		.22		.22	.37
0090	Metal, detail		520	.015			.28		.28	.46
0100	Metal, louvered	↓	360	.022			.40		.40	.67
0110	Hand wash									
0120	Wood, flush	1 Pord	2,160	.004	S.F.		.07		.07	.11
0130	Wood, detailed		2,000	.004			.07		.07	.12
0140	Wood, louvered		1,360	.006			.11		.11	.18
0150	Metal, flush		2,160	.004			.07		.07	.11
0160	Metal, detail		2,000	.004			.07		.07	.12
0170	Metal, louvered	↓	1,360	.006	↓		.11		.11	.18
0400	Windows, per side, not incl. trim									
0410	Scrape & sand									
0420	Wood, 1-2 lite	1 Pord	360	.022	S.F.		.40		.40	.67
0430	Wood, 3-6 lite		320	.025			.45		.45	.75
0440	Wood, 7-10 lite		280	.029			.51		.51	.86
0450	Wood, 12 lite		240	.033			.60		.60	1
0460	Wood, Bay / Bow	↓	360	.022	↓		.40		.40	.67
0470	Wire brush									
0480	Metal, 1-2 lite	1 Pord	520	.015	S.F.		.28		.28	.46
0490	Metal, 3-6 lite		440	.018			.33		.33	.54
0500	Metal, Bay / Bow	↓	520	.015	↓		.28		.28	.46
0600	Walls, sanding, light=10-30%									
0610	Med.=30-70%, heavy=70-100%, % of surface to sand									
0620	For Chemical Washing, see Division 04930-900									
0630	For Steam Cleaning, see Division 04930-750									
0640	For Sand Blasting, see Division 04930-700									
0650	Walls, sand									
0660	Drywall, gypsum, plaster, light	1 Pord	3,077	.003	S.F.		.05		.05	.08
0670	Drywall, gypsum, plaster, med.		2,160	.004			.07		.07	.11
0680	Drywall, gypsum, plaster, heavy		923	.009			.16		.16	.26
0690	Wood, T&G, light		2,400	.003			.06		.06	.10
0700	Wood, T&G, med.		1,600	.005			.09		.09	.15
0710	Wood, T&G, heavy	↓	800	.010	↓		.18		.18	.30
0720	Walls, wash									
0730	Drywall, gypsum, plaster	1 Pord	3,200	.002	S.F.		.04		.04	.07
0740	Wood, T&G		3,200	.002			.04		.04	.07
0750	Masonry, brick & block, smooth		2,800	.003			.05		.05	.09
0760	Masonry, brick & block, coarse	↓	2,000	.004	↓		.07		.07	.12

For information about Means Estimating Seminars, see yellow pages 11 and 12 in back of book

Division Notes

Division 10 Specialties

Estimating Tips
General
- The items in this division are usually priced per square foot or each.
- Many items in Division 10 require some type of support system that is not usually furnished with the item. Examples of these systems include blocking for the attachment of grab bars and support angles for ceiling hung toilet partitions. The required blocking or supports must be added to the estimate in the appropriate division.
- Some items in Division 10, such as lockers, may require assembly before installation. Verify the amount of assembly required. Assembly can often exceed installation time.

10100 Visual Display Boards
- Toilet partitions are priced by the stall. A stall consists of a side wall, pilaster and door with hardware. Toilet tissue holders and grab bars are extra.

10600 Partitions
- The required acoustical rating of a folding partition can have a significant impact on costs. Verify the sound transmission coefficient rating of the panel priced to the specification requirements.

Reference Numbers
Reference numbers are shown in bold squares at the beginning of some major classifications. These numbers refer to related items in the Reference Section. The reference information may be an estimating procedure, an alternate pricing method or technical information.

Note: Not all subdivisions listed here necessarily appear in this publication.

10100 | Visual Display Boards

10110 | Chalkboards

		CREW	DAILY OUTPUT	LABOR-HOURS	UNIT	2000 BARE COSTS MAT.	LABOR	EQUIP.	TOTAL	TOTAL INCL O&P
0010	**CHALKBOARDS** Porcelain enamel steel									
0100	Freestanding, reversible									
0120	Economy, wood frame, 4' x 6'									
0140	Chalkboard both sides				Ea.	460			460	510
0160	Chalkboard one side, cork other side				"	400			400	440
0200	Standard, lightweight satin finished aluminum, 4' x 6'									
0220	Chalkboard both sides				Ea.	475			475	525
0240	Chalkboard one side, cork other side				"	420			420	460
0300	Deluxe, heavy duty extruded aluminum, 4' x 6'									
0320	Chalkboard both sides				Ea.	1,025			1,025	1,150
0340	Chalkboard one side, cork other side				"	1,200			1,200	1,325
0400	Sliding chalkboards									
0450	Vertical, one sliding board with back panel, wall mounted									
0500	8' x 4'	2 Carp	8	2	Ea.	1,275	39.50		1,314.50	1,475
0520	8' x 8'		7.50	2.133		1,775	42		1,817	2,025
0540	8' x 12'		7	2.286		2,225	45		2,270	2,500
0600	Two sliding boards, with back panel									
0620	8' x 4'	2 Carp	8	2	Ea.	1,950	39.50		1,989.50	2,225
0640	8' x 8'		7.50	2.133		2,725	42		2,767	3,075
0660	8' x 12'		7	2.286		3,375	45		3,420	3,775
0700	Horizontal, two track									
0800	4' x 8', 2 sliding panels	2 Carp	8	2	Ea.	1,125	39.50		1,164.50	1,300
0820	4' x 12', 2 sliding panels		7.50	2.133		1,375	42		1,417	1,575
0840	4' x 16', 4 sliding panels		7	2.286		1,850	45		1,895	2,125
0900	Four track, four sliding panels									
0920	4' x 8'	2 Carp	8	2	Ea.	1,800	39.50		1,839.50	2,075
0940	4' x 12'		7.50	2.133		2,225	42		2,267	2,525
0960	4' x 16'		7	2.286		2,775	45		2,820	3,125
1200	Vertical, motor operated									
1400	One sliding panel with back panel									
1450	10' x 4'	2 Carp	4	4	Ea.	4,975	79		5,054	5,600
1500	10' x 10'		3.75	4.267		5,750	84		5,834	6,475
1550	10' x 16'		3.50	4.571		6,525	90		6,615	7,350
1700	Two sliding panels with back panel									
1750	10' x 4'	2 Carp	4	4	Ea.	7,975	79		8,054	8,900
1800	10' x 10'		3.75	4.267		9,075	84		9,159	10,100
1850	10' x 16'		3.50	4.571		10,200	90		10,290	11,400
2000	Three sliding panels with back panel									
2100	10' x 4'	2 Carp	4	4	Ea.	10,600	79		10,679	11,700
2150	10' x 10'		3.75	4.267		12,200	84		12,284	13,500
2200	10' x 16'		3.50	4.571		13,800	90		13,890	15,400
2400	For projection screen, glass beaded, add				S.F.	4.50			4.50	4.95
2500	For remote control, 1 panel control, add				Ea.	395			395	435
2600	2 panel control, add				"	695			695	765
2800	For units without back panels, deduct				S.F.	4.91			4.91	5.40
2850	For liquid chalk porcelain panels, add				"	3.64			3.64	4
3000	Swing leaf, any comb. of chalkboard & cork, aluminum frame									
3100	Floor style, 5 panels									
3150	30" x 40" panels				Ea.	1,150			1,150	1,250
3200	48" x 40" panels				"	1,300			1,300	1,450
3300	Wall mounted, 5 panels									
3400	30" x 40" panels	2 Carp	16	1	Ea.	1,225	19.70		1,244.70	1,375
3450	48" x 40" panels	"	16	1	"	1,400	19.70		1,419.70	1,575
3600	Extra panels for swing leaf units									
3700	30" x 40" panels				Ea.	201			201	221
3750	48" x 40" panels				"	237			237	260

10100 | Visual Display Boards

10110 | Chalkboards

		Description	CREW	DAILY OUTPUT	LABOR-HOURS	UNIT	MAT.	LABOR	EQUIP.	TOTAL	TOTAL INCL O&P
240	3900	Wall hung									240
	4000	Aluminum frame and chalktrough									
	4200	3' x 4'	2 Carp	16	1	Ea.	128	19.70		147.70	175
	4300	3' x 5'		15	1.067		186	21		207	241
	4500	4' x 8'		14	1.143		285	22.50		307.50	355
	4600	4' x 12'	↓	13	1.231	↓	410	24.50		434.50	495
	4700	Wood frame and chalktrough									
	4800	3' x 4'	2 Carp	16	1	Ea.	130	19.70		149.70	177
	5000	3' x 5'		15	1.067		184	21		205	238
	5100	4' x 5'		14	1.143		186	22.50		208.50	243
	5300	4' x 8'	↓	13	1.231	↓	264	24.50		288.50	330
	5400	Liquid chalk, white porcelain enamel, wall hung									
	5420	Deluxe units, aluminum trim and chalktrough									
	5450	4' x 4'	2 Carp	16	1	Ea.	181	19.70		200.70	234
	5500	4' x 8'		14	1.143		320	22.50		342.50	390
	5550	4' x 12'	↓	12	1.333	↓	435	26.50		461.50	525
	5700	Wood trim and chalktrough									
	5900	4' x 4'	2 Carp	16	1	Ea.	375	19.70		394.70	450
	6000	4' x 6'		15	1.067		465	21		486	545
	6200	4' x 8'	↓	14	1.143		540	22.50		562.50	635
	6300	Liquid chalk, felt tip markers					1.46			1.46	1.61
	6500	Erasers					1.35			1.35	1.49
	6600	Board cleaner, 8 oz. bottle				↓	3.95			3.95	4.35

10120 | Tack & Visual Aid Boards

		Description	CREW	DAILY OUTPUT	LABOR-HOURS	UNIT	MAT.	LABOR	EQUIP.	TOTAL	TOTAL INCL O&P
350	0010	CONTROL BOARDS Magnetic, porcelain finish, 18" x 24", framed	2 Carp	8	2	Ea.	145	39.50		184.50	228	350
	0100	24" x 36"		7.50	2.133		205	42		247	298
	0200	36" x 48"		7	2.286		325	45		370	435
	0300	48" x 72"		6	2.667		310	52.50		362.50	430
	0400	48" x 96"	↓	5	3.200	↓	650	63		713	825
940	0010	BULLETIN BOARD Cork sheets, unbacked, no frame, 1/4" thick	2 Carp	290	.055	S.F.	5.65	1.09		6.74	8.05	940
	0100	1/2" thick		290	.055		2.52	1.09		3.61	4.63
	0300	Fabric-face, no frame, on 7/32" cork underlay		290	.055		4.11	1.09		5.20	6.40
	0400	On 1/4" cork on 1/4" hardboard		290	.055		5.50	1.09		6.59	7.90
	0600	With edges wrapped		290	.055		5.85	1.09		6.94	8.30
	0700	On 7/16" fire retardant core		290	.055		4.10	1.09		5.19	6.35
	0900	With edges wrapped	↓	290	.055		4.40	1.09		5.49	6.70
	1000	Designer fabric only, cut to size					1.82			1.82	2
	1200	1/4" vinyl cork, on 1/4" hardboard, no frame	2 Carp	290	.055		4.48	1.09		5.57	6.80
	1300	On 1/4" coreboard		290	.055	↓	6	1.09		7.09	8.45
	2000	For map and display rail, economy, add		385	.042	L.F.	1.44	.82		2.26	2.98
	2100	Deluxe, add		350	.046	"	2.74	.90		3.64	4.55
	2120	Prefabricated, 1/4" cork, 3' x 5' with aluminum frame		16	1	Ea.	97.50	19.70		117.20	141
	2140	Wood frame		16	1		126	19.70		145.70	172
	2160	4' x 4' with aluminum frame		16	1		81	19.70		100.70	123
	2180	Wood frame		16	1		170	19.70		189.70	221
	2200	4' x 8' with aluminum frame		14	1.143		154	22.50		176.50	208
	2210	With wood frame		14	1.143		170	22.50		192.50	226
	2220	4' x 12' with aluminum frame	↓	12	1.333	↓	224	26.50		250.50	291
	2230	Bulletin board case, single glass door, with lock									
	2240	36" x 24", economy	2 Carp	12	1.333	Ea.	245	26.50		271.50	315
	2250	Deluxe		12	1.333		375	26.50		401.50	460
	2260	42" x 30", economy		12	1.333		275	26.50		301.50	350
	2270	Deluxe	↓	12	1.333	↓	410	26.50		436.50	500

SPECIALTIES 10

10100 | Visual Display Boards

10120 | Tack & Visual Aid Boards

		CREW	DAILY OUTPUT	LABOR-HOURS	UNIT	2000 BARE COSTS MAT.	LABOR	EQUIP.	TOTAL	TOTAL INCL O&P
2300	Glass enclosed cabinets, alum., cork panel, hinged doors									
2400	3' x 3', 1 door	2 Carp	12	1.333	Ea.	420	26.50		446.50	510
2500	4' x 4', 2 door		11	1.455		700	28.50		728.50	820
2600	4' x 7', 3 door		10	1.600		1,150	31.50		1,181.50	1,325
2800	4' x 10', 4 door	↓	8	2		1,425	39.50		1,464.50	1,650
2900	For lights, add per door opening	1 Elec	13	.615		138	13.60		151.60	174
3100	Horizontal sliding units, 4 doors, 4' x 8' 8' x 4'	2 Carp	9	1.778		1,225	35		1,260	1,400
3200	4' x 12'		7	2.286		1,550	45		1,595	1,775
3400	8 doors, 4' x 16'		5	3.200		3,250	63		3,313	3,675
3500	4' x 24'	↓	4	4	↓	4,025	79		4,104	4,550

10150 | Compartments & Cubicles

10155 | Toilet Compartments

		CREW	DAILY OUTPUT	LABOR-HOURS	UNIT	2000 BARE COSTS MAT.	LABOR	EQUIP.	TOTAL	TOTAL INCL O&P
0010	**PARTITIONS, TOILET**									
0101	Cubicles, ceiling hung, marble	2 Ston	2	8	Ea.	1,300	156		1,456	1,725
0200	Painted metal	2 Carp	4	4		385	79		464	555
0250	Phenolic		4	4		730	79		809	935
0300	Plastic laminate on particle board		4	4		495	79		574	680
0500	Stainless steel	↓	4	4		965	79		1,044	1,200
0600	For handicap units, incl. 52" grab bars, add					277			277	305
0801	Floor & ceiling anchored, marble	2 Ston	2.50	6.400		1,450	125		1,575	1,775
1000	Painted metal	2 Carp	5	3.200		355	63		418	500
1050	Phenolic		5	3.200		785	63		848	975
1100	Plastic laminate on particle board		5	3.200		470	63		533	625
1300	Stainless steel	↓	5	3.200		1,125	63		1,188	1,325
1400	For handicap units, incl. 52" grab bars, add					253			253	278
1601	Floor mounted, marble	2 Ston	3	5.333		850	104		954	1,100
1700	Painted metal	2 Carp	7	2.286		325	45		370	430
1750	Phenolic		7	2.286		730	45		775	875
1800	Plastic laminate on particle board		7	2.286		495	45		540	620
2000	Stainless steel	↓	7	2.286		1,050	45		1,095	1,225
2100	For handicap units, incl. 52" grab bars, add					253			253	278
2200	For juvenile units, deduct					35			35	38.50
2401	Floor mounted, headrail braced, marble	2 Ston	3	5.333		990	104		1,094	1,275
2500	Painted metal	2 Carp	6	2.667		355	52.50		407.50	480
2550	Phenolic		6	2.667		760	52.50		812.50	925
2600	Plastic laminate on particle board		6	2.667		525	52.50		577.50	665
2800	Stainless steel	↓	6	2.667		1,100	52.50		1,152.50	1,300
2900	For handicap units, incl. 52" grab bars, add					253			253	278
3000	Wall hung partitions, painted metal	2 Carp	7	2.286		450	45		495	570
3300	Stainless steel	"	7	2.286		1,075	45		1,120	1,275
3400	For handicap units, incl. 52" grab bars, add					253			253	278
4000	Screens, entrance, floor mounted, 58" high, 48" wide				↓					
4100	Marble	2 Marb	9	1.778	Ea.	535	35.50		570.50	645
4200	Painted metal	2 Carp	15	1.067		166	21		187	218
4300	Plastic laminate on particle board		15	1.067		294	21		315	360
4500	Stainless steel	↓	15	1.067		585	21		606	680
4600	Urinal screen, 18" wide, ceiling braced, marble	D-1	6	2.667		535	47.50		582.50	665
4700	Painted metal	2 Carp	8	2	↓	126	39.50		165.50	207

10150 | Compartments & Cubicles

10155 | Toilet Compartments

		Crew	Daily Output	Labor-Hours	Unit	Mat.	Labor	Equip.	Total	Total Incl O&P
4800	Plastic laminate on particle board	2 Carp	8	2	Ea.	235	39.50		274.50	325
5000	Stainless steel	↓	8	2	↓	420	39.50		459.50	530
5100	Floor mounted, head rail braced									
5200	Marble	D-1	6	2.667	Ea.	495	47.50		542.50	625
5300	Painted metal	2 Carp	8	2		166	39.50		205.50	250
5400	Plastic laminate on particle board		8	2		212	39.50		251.50	300
5600	Stainless steel	↓	8	2		450	39.50		489.50	565
5700	Pilaster, flush, marble	D-1	9	1.778		605	31.50		636.50	720
5800	Painted metal	2 Carp	10	1.600		216	31.50		247.50	292
5900	Plastic laminate on particle board		10	1.600		270	31.50		301.50	350
6100	Stainless steel	↓	10	1.600		370	31.50		401.50	460
6200	Post braced, marble	D-1	9	1.778		595	31.50		626.50	710
6300	Painted metal	2 Carp	10	1.600		222	31.50		253.50	299
6400	Plastic laminate on particle board		10	1.600		276	31.50		307.50	360
6600	Stainless steel	↓	10	1.600	↓	365	31.50		396.50	460
6700	Wall hung, bracket supported									
6800	Painted metal	2 Carp	10	1.600	Ea.	225	31.50		256.50	300
6900	Plastic laminate on particle board		10	1.600		108	31.50		139.50	173
7100	Stainless steel		10	1.600		335	31.50		366.50	425
7400	Flange supported, painted metal		10	1.600		165	31.50		196.50	236
7500	Plastic laminate on particle board		10	1.600		232	31.50		263.50	310
7700	Stainless steel		10	1.600		390	31.50		421.50	485
7800	Wedge type, painted metal		10	1.600		195	31.50		226.50	268
8100	Stainless steel	↓	10	1.600	↓	400	31.50		431.50	495

10185 | Shower/Dressing Compartments

		Crew	Daily Output	Labor-Hours	Unit	Mat.	Labor	Equip.	Total	Total Incl O&P
0010	**PARTITIONS, SHOWER** Floor mounted, no plumbing									
0100	Cabinet, incl. base, no door, painted steel, 1" thick walls	2 Shee	5	3.200	Ea.	600	69		669	775
0300	With door, fiberglass		4.50	3.556		490	76.50		566.50	670
0600	Galvanized and painted steel, 1" thick walls		5	3.200		635	69		704	815
0800	Stall, 1" thick wall, no base, enameled steel		5	3.200		610	69		679	790
1200	Stainless steel	↓	5	3.200		1,075	69		1,144	1,300
1400	For double entry type, no doors, deduct					10%				
1500	Circular fiberglass, cabinet 36" diameter,	2 Shee	4	4		505	86		591	700
1700	One piece, 36" diameter, less door		4	4		430	86		516	620
1800	With door		3.50	4.571		695	98.50		793.50	930
2000	Curved shell shower, no door needed	↓	3	5.333		640	115		755	895
2300	For fiberglass seat, add to both above					90			90	99
2400	Glass stalls, with doors, no receptors, chrome on brass	2 Shee	3	5.333		1,050	115		1,165	1,350
2700	Anodized aluminum	"	4	4		720	86		806	940
2901	Marble shower stall, stock design, with shower door	2 Ston	1.20	13.333		1,325	260		1,585	1,925
3001	With curtain		1.30	12.308		1,175	240		1,415	1,700
3201	Receptors/precast terrazzo, 32" x 32"		14	1.143		204	22.50		226.50	262
3301	48" x 34"		9.50	1.684		270	33		303	355
3501	Plastic, simulated terrazzo receptor, 32" x 32"		14	1.143		85.50	22.50		108	132
3601	32" x 48"		12	1.333		101	26		127	155
3801	Precast concrete, colors, 32" x 32"		14	1.143		134	22.50		156.50	185
3901	48" x 48"	↓	12	1.333		155	26		181	215
4100	Shower doors, economy plastic, 24" wide	1 Shee	9	.889		90	19.10		109.10	132
4200	Tempered glass door, economy		8	1		174	21.50		195.50	228
4400	Folding, tempered glass, aluminum frame		6	1.333		213	28.50		241.50	283
4500	Sliding, tempered glass, 48" opening		6	1.333		180	28.50		208.50	248
4700	Deluxe, tempered glass, chrome on brass frame, minimum		8	1		220	21.50		241.50	279
4800	Maximum		1	8		640	172		812	990
4850	On anodized aluminum frame, minimum		2	4		103	86		189	258
4900	Maximum	↓	1	8	↓	350	172		522	675

10150 | Compartments & Cubicles

10185 | Shower/Dressing Compartments

		CREW	DAILY OUTPUT	LABOR-HOURS	UNIT	2000 BARE COSTS MAT.	LABOR	EQUIP.	TOTAL	TOTAL INCL O&P
5100	Shower enclosure, tempered glass, anodized alum. frame									
5120	2 panel & door, corner unit, 32" x 32"	1 Shee	2	4	Ea.	350	86		436	530
5140	Neo-angle corner unit, 16" x 24" x 16"	"	2	4		610	86		696	820
5200	Shower surround, 3 wall, polypropylene, 32" x 32"	1 Carp	4	2		258	39.50		297.50	350
5220	PVC, 32" x 32"		4	2		226	39.50		265.50	315
5240	Fiberglass		4	2		271	39.50		310.50	365
5250	2 wall, polypropylene, 32" x 32"		4	2		187	39.50		226.50	274
5270	PVC		4	2		226	39.50		265.50	315
5290	Fiberglass		4	2		251	39.50		290.50	345
5300	Tub doors, tempered glass & frame, minimum	1 Shee	8	1		148	21.50		169.50	200
5400	Maximum		6	1.333		340	28.50		368.50	425
5600	Chrome plated, brass frame, minimum		8	1		193	21.50		214.50	250
5700	Maximum		6	1.333		385	28.50		413.50	475
5900	Tub/shower enclosure, temp. glass, alum. frame, minimum		2	4		264	86		350	435
6200	Maximum		1.50	5.333		515	115		630	760
6500	On chrome-plated brass frame, minimum		2	4		365	86		451	550
6600	Maximum		1.50	5.333		755	115		870	1,025
6800	Tub surround, 3 wall, polypropylene	1 Carp	4	2		148	39.50		187.50	231
6900	PVC		4	2		226	39.50		265.50	315
7000	Fiberglass, minimum		4	2		251	39.50		290.50	345
7100	Maximum		3	2.667		430	52.50		482.50	565

10190 | Cubicles

		CREW	DAILY OUTPUT	LABOR-HOURS	UNIT	MAT.	LABOR	EQUIP.	TOTAL	INCL O&P
0010	**PARTITIONS, HOSPITAL** Curtain track, box channel, ceiling mounted	1 Carp	135	.059	L.F.	3.76	1.17		4.93	6.15
0100	Suspended	"	100	.080	"	5.35	1.58		6.93	8.60
0300	Curtains, nylon mesh tops, fire resistant, 11 oz. per lineal yard									
0310	Polyester oxford cloth, 9' ceiling height	1 Carp	425	.019	L.F.	7.75	.37		8.12	9.20
0500	8' ceiling height		425	.019		7	.37		7.37	8.35
0700	Designer oxford cloth		425	.019		17	.37		17.37	19.35
0800	I.V. track systems									
0820	I.V. track, oval	1 Carp	135	.059	L.F.	3	1.17		4.17	5.30
0830	I.V. trolley		32	.250	Ea.	25.50	4.93		30.43	36.50
0840	I.V. pendent, (tree, 5 hook)		32	.250	"	94.50	4.93		99.43	112

10200 | Louvers & Vents

10210 | Wall Louvers

		CREW	DAILY OUTPUT	LABOR-HOURS	UNIT	MAT.	LABOR	EQUIP.	TOTAL	INCL O&P
0010	**LOUVERS** Aluminum with screen, residential, 8" x 8"	1 Carp	38	.211	Ea.	6.95	4.15		11.10	14.70
0100	12" x 12"		38	.211		8.75	4.15		12.90	16.70
0200	12" x 18"		35	.229		12.70	4.50		17.20	21.50
0250	14" x 24"		30	.267		16.35	5.25		21.60	27
0300	18" x 24"		27	.296		18.85	5.85		24.70	31
0500	24" x 30"		24	.333		27.50	6.55		34.05	42
0700	Triangle, adjustable, small		20	.400		23	7.90		30.90	38.50
0800	Large		15	.533		61.50	10.50		72	85.50
1200	Extruded aluminum, see division 15850-600									
2100	Midget, aluminum, 3/4" deep, 1" diameter	1 Carp	85	.094	Ea.	.56	1.85		2.41	3.80
2150	3" diameter		60	.133		1.32	2.63		3.95	5.95
2200	4" diameter		50	.160		1.66	3.15		4.81	7.25

10200 | Louvers & Vents

10210 | Wall Louvers

			CREW	DAILY OUTPUT	LABOR-HOURS	UNIT	2000 BARE COSTS MAT.	LABOR	EQUIP.	TOTAL	TOTAL INCL O&P	
800	2250	6" diameter	1 Carp	30	.267	Ea.	2.25	5.25		7.50	11.50	800
	2300	Ridge vent strip, mill finish	1 Shee	155	.052	L.F.	1.95	1.11		3.06	4.02	
	2330	Soffit vent, continuous, 3" wide, aluminum, mill finish	1 Carp	200	.040		.44	.79		1.23	1.83	
	2340	Baked enamel finish		200	.040		.52	.79		1.31	1.92	
	2400	Under eaves vent, aluminum, mill finish, 16" x 4"		48	.167	Ea.	1.49	3.28		4.77	7.30	
	2500	16" x 8"		48	.167		1.79	3.28		5.07	7.60	
	7000	Vinyl gable vent, 8" x 8"		38	.211		9.50	4.15		13.65	17.55	
	7020	12" x 12"		38	.211		17	4.15		21.15	26	
	7080	12" x 18"		35	.229		22	4.50		26.50	31.50	
	7200	18" x 24"		30	.267		26	5.25		31.25	37.50	

10260 | Wall & Corner Guards

10265 | Wall & Corner Guards

			CREW	DAILY OUTPUT	LABOR-HOURS	UNIT	2000 BARE COSTS MAT.	LABOR	EQUIP.	TOTAL	TOTAL INCL O&P	
200	0011	CORNER GUARDS Steel angle w/anchors, 1" x 1" x 1/4", 1.5#/L.F.	2 Carp	160	.100	L.F.	2.72	1.97		4.69	6.35	200
	0101	2" x 2" x 1/4" angles, 3.2#/L.F.		150	.107		6.45	2.10		8.55	10.70	
	0201	3" x 3" x 5/16" angles, 6.1#/L.F.		140	.114		8.35	2.25		10.60	13	
	0301	4" x 4" x 5/16" angles, 8.2#/L.F.		120	.133		8.40	2.63		11.03	13.75	
	0350	For angles drilled and anchored to masonry, add					15%	120%				
	0370	Drilled and anchored to concrete, add					20%	170%				
	0400	For galvanized angles, add					35%					
	0450	For stainless steel angles, add				L.F.	100%					
	0501	Steel door track/wheel guard, 4' - 0" high	2 Carp	12	1.333	Ea.	129	26.50		155.50	187	
	0801	Pipe bumper for truck doors, conc. filled, 8' long, 6" diam.		10	1.600		183	31.50		214.50	255	
	0901	8" diameter		10	1.600		281	31.50		312.50	365	
250	0010	CORNER PROTECTION										250
	0101	Stainless steel, 16 ga., adhesive mount, 3-1/2" leg	1 Carp	80	.100	L.F.	16.20	1.97		18.17	21	
	0201	12 ga. stainless, adhesive mount	"	80	.100		24.50	1.97		26.47	30.50	
	0300	For screw mount, add						10%				
	0500	Vinyl acrylic, adhesive mount, 3" leg	1 Carp	128	.063		4.80	1.23		6.03	7.40	
	0550	1-1/2" leg		160	.050		3.70	.99		4.69	5.75	
	0600	Screw mounted, 3" leg		80	.100		6.15	1.97		8.12	10.15	
	0650	1-1/2" leg		100	.080		3.60	1.58		5.18	6.65	
	0700	Clear plastic, screw mounted, 2-1/2"		60	.133		2.75	2.63		5.38	7.55	
	1000	Vinyl cover, alum. retainer, surface mount, 3" x 3"		48	.167		6.10	3.28		9.38	12.35	
	1050	2" x 2"		48	.167		4.43	3.28		7.71	10.50	
	1100	Flush mounted, 3" x 3"		32	.250		12.35	4.93		17.28	22	
	1150	2" x 2"		32	.250		9.10	4.93		14.03	18.45	
500	0010	WALLGUARD										500
	0400	Rub rail, vinyl, adhesive mounted	1 Carp	185	.043	L.F.	3.67	.85		4.52	5.50	
	0500	Neoprene, aluminum backing, 1-1/2" x 2"		110	.073		6.15	1.43		7.58	9.20	
	1000	Trolley rail, PVC, clipped to wall, 5" high		185	.043		5.10	.85		5.95	7.05	
	1050	8" high		180	.044		5.95	.88		6.83	8.05	
	1200	Bed bumper, vinyl acrylic, alum. retainer, 21" long		10	.800	Ea.	33.50	15.75		49.25	64	
	1300	53" long with aligner		9	.889	"	100	17.50		117.50	140	
	1400	Bumper, vinyl cover, alum. retain., cush. mnt., 1-1/2" x 2-3/4"		80	.100	L.F.	10.10	1.97		12.07	14.50	
	1500	2" x 4-1/4"		80	.100		13.50	1.97		15.47	18.25	
	1600	Surface mounted, 1-3/4" x 3-5/8"		80	.100		9.10	1.97		11.07	13.40	

10260 | Wall & Corner Guards

10265 | Wall & Corner Guards

		CREW	DAILY OUTPUT	LABOR-HOURS	UNIT	MAT.	LABOR	EQUIP.	TOTAL	TOTAL INCL O&P
2000	Crash rail, vinyl cover, alum. retainer, 1" x 4"	1 Carp	110	.073	L.F.	10.30	1.43		11.73	13.80
2100	1" x 8"		90	.089		12.85	1.75		14.60	17.15
2150	Vinyl inserts, aluminum plate, 1" x 2-1/2"		110	.073		10.05	1.43		11.48	13.55
2200	1" x 5"	↓	90	.089	↓	14.50	1.75		16.25	18.95
3000	Handrail/bumper, vinyl cover, alum. retainer									
3010	Bracket mounted, flat rail, 5-1/2"	1 Carp	80	.100	L.F.	15.50	1.97		17.47	20.50
3100	6-1/2"		80	.100		19.05	1.97		21.02	24.50
3200	Bronze bracket, 1-3/4" diam. rail	↓	80	.100	↓	15.50	1.97		17.47	20.50

10270 | Access Flooring

10275 | Access Flooring

		CREW	DAILY OUTPUT	LABOR-HOURS	UNIT	MAT.	LABOR	EQUIP.	TOTAL	TOTAL INCL O&P
0010	PEDESTAL ACCESS FLOORS Computer room application, metal									
0020	Particle board or steel panels, no covering, under 6,000 S.F.	2 Carp	400	.040	S.F.	5.80	.79		6.59	7.75
0300	Metal covered, over 6,000 S.F.		450	.036		5.90	.70		6.60	7.70
0400	Aluminum, 24" panels	↓	500	.032		21	.63		21.63	24
0600	For carpet covering, add					4.45			4.45	4.90
0700	For vinyl floor covering, add					4.30			4.30	4.73
0900	For high pressure laminate covering, add					3.01			3.01	3.31
0910	For snap on stringer system, add	2 Carp	1,000	.016	↓	1.05	.32		1.37	1.69
0950	Office applications, to 8" high, steel panels,									
0960	no covering, over 6,000 S.F.	2 Carp	500	.032	S.F.	6.85	.63		7.48	8.65
1000	Machine cutouts after initial installation	1 Carp	10	.800	Ea.	33.50	15.75		49.25	64
1050	Pedestals, 6" to 12"	2 Carp	85	.188		5.10	3.71		8.81	11.95
1100	Air conditioning grilles, 4" x 12"	1 Carp	17	.471		51.50	9.25		60.75	73
1150	4" x 18"	"	14	.571	↓	68.50	11.25		79.75	95
1200	Approach ramps, minimum	2 Carp	85	.188	S.F.	15.80	3.71		19.51	23.50
1300	Maximum	"	60	.267	"	26.50	5.25		31.75	38
1500	Handrail, 2 rail, aluminum	1 Carp	15	.533	L.F.	75	10.50		85.50	101

10300 | Fireplaces & Stoves

10305 | Manufactured Fireplaces

		CREW	DAILY OUTPUT	LABOR-HOURS	UNIT	MAT.	LABOR	EQUIP.	TOTAL	TOTAL INCL O&P
0010	FIREPLACE, PREFABRICATED Free standing or wall hung									
0100	with hood & screen, minimum	1 Carp	1.30	6.154	Ea.	930	121		1,051	1,225
0150	Average		1	8		1,125	158		1,283	1,500
0200	Maximum		.90	8.889	↓	2,675	175		2,850	3,250
0500	Chimney dbl. wall, all stainless, over 8'-6", 7" diam., add		33	.242	V.L.F.	31.50	4.78		36.28	42.50
0600	10" diameter, add		32	.250		45	4.93		49.93	58
0700	12" diameter, add		31	.258		58.50	5.10		63.60	73
0800	14" diameter, add		30	.267	↓	74	5.25		79.25	90.50
1000	Simulated brick chimney top, 4' high, 16" x 16"		10	.800	Ea.	160	15.75		175.75	203
1100	24" x 24"	↓	7	1.143	"	298	22.50		320.50	370

10300 | Fireplaces & Stoves

10305 | Manufactured Fireplaces

			DAILY	LABOR-		2000 BARE COSTS				TOTAL
		CREW	OUTPUT	HOURS	UNIT	MAT.	LABOR	EQUIP.	TOTAL	INCL O&P
100	1500	Simulated logs, gas fired, 40,000 BTU, 2' long, minimum	1 Carp	7	1.143	Set	415	22.50		437.50	500
	1600	Maximum		6	1.333		605	26.50		631.50	715
	1700	Electric, 1,500 BTU, 1'-6" long, minimum		7	1.143		110	22.50		132.50	161
	1800	11,500 BTU, maximum		6	1.333		239	26.50		265.50	310
	2000	Fireplace, built-in, 36" hearth, radiant		1.30	6.154	Ea.	480	121		601	735
	2100	Recirculating, small fan		1	8		685	158		843	1,025
	2150	Large fan		.90	8.889		1,275	175		1,450	1,700
	2200	42" hearth, radiant		1.20	6.667		610	131		741	895
	2300	Recirculating, small fan		.90	8.889		800	175		975	1,175
	2350	Large fan		.80	10		1,525	197		1,722	2,025
	2400	48" hearth, radiant		1.10	7.273		1,125	143		1,268	1,500
	2500	Recirculating, small fan		.80	10		1,400	197		1,597	1,900
	2550	Large fan		.70	11.429		2,175	225		2,400	2,775
	3000	See through, including doors		.80	10		1,800	197		1,997	2,325
	3200	Corner (2 wall)		1	8		890	158		1,048	1,250

10310 | Fireplace Specialties & Accessories

			DAILY	LABOR-		2000 BARE COSTS				TOTAL
		CREW	OUTPUT	HOURS	UNIT	MAT.	LABOR	EQUIP.	TOTAL	INCL O&P
100	0010	FIREPLACE ACCESSORIES Chimney screens, galv., 13" x 13" flue	1 Bric	8	1	Ea.	47.50	20		67.50	86.50
	0050	Galv., 24" x 24" flue		5	1.600		125	32		157	191
	0200	Stainless steel, 13" x 13" flue		8	1		264	20		284	325
	0250	20" x 20" flue		5	1.600		360	32		392	450
	0400	Cleanout doors and frames, cast iron, 8" x 8"		12	.667		17.50	13.35		30.85	42
	0450	12" x 12"		10	.800		33.50	16		49.50	63.50
	0500	18" x 24"		8	1		105	20		125	149
	0550	Cast iron frame, steel door, 24" x 30"		5	1.600		227	32		259	305
	0800	Damper, rotary control, steel, 30" opening		6	1.333		63.50	26.50		90	115
	0850	Cast iron, 30" opening		6	1.333		71	26.50		97.50	123
	0880	36" opening		6	1.333		76.50	26.50		103	130
	0900	48" opening		6	1.333		105	26.50		131.50	161
	0920	60" opening		6	1.333		250	26.50		276.50	320
	0950	72" opening		5	1.600		305	32		337	390
	1000	84" opening, special order		5	1.600		635	32		667	755
	1050	96" opening, special order		4	2		645	40		685	780
	1200	Steel plate, poker control, 60" opening		8	1		225	20		245	281
	1250	84" opening, special opening		5	1.600		410	32		442	505
	1400	"Universal" type, chain operated, 32" x 20" opening		8	1		153	20		173	203
	1450	48" x 24" opening		5	1.600		271	32		303	350
	1600	Dutch Oven door and frame, cast iron, 12" x 15" opening		13	.615		92	12.30		104.30	122
	1650	Copper plated, 12" x 15" opening		13	.615		177	12.30		189.30	215
	1800	Fireplace forms, no accessories, 32" opening		3	2.667		465	53.50		518.50	605
	1900	36" opening		2.50	3.200		565	64		629	730
	2000	40" opening		2	4		680	80		760	880
	2100	78" opening		1.50	5.333		970	107		1,077	1,250
	2400	Squirrel and bird screens, galvanized, 8" x 8" flue		16	.500		44.50	10		54.50	66
	2450	13" x 13" flue		12	.667		52	13.35		65.35	80

10320 | Stoves

			DAILY	LABOR-		2000 BARE COSTS				TOTAL
		CREW	OUTPUT	HOURS	UNIT	MAT.	LABOR	EQUIP.	TOTAL	INCL O&P
100	0010	WOODBURNING STOVES Cast iron, minimum	2 Carp	1.30	12.308	Ea.	765	242		1,007	1,250
	0020	Average		1	16		1,225	315		1,540	1,900
	0030	Maximum		.80	20		2,075	395		2,470	2,950
	0050	For gas log lighter, add					36			36	39.50

10340 | Manufactured Exterior Specialties

10342 | Cupolas

			CREW	DAILY OUTPUT	LABOR-HOURS	UNIT	2000 BARE COSTS MAT.	LABOR	EQUIP.	TOTAL	TOTAL INCL O&P	
100	0010	CUPOLA Stock units, pine, painted, 18" sq., 28" high, alum. roof	1 Carp	4.10	1.951	Ea.	99	38.50		137.50	175	100
	0100	Copper roof		3.80	2.105		141	41.50		182.50	226	
	0300	23" square, 33" high, aluminum roof		3.70	2.162		232	42.50		274.50	330	
	0400	Copper roof		3.30	2.424		233	48		281	340	
	0600	30" square, 37" high, aluminum roof		3.70	2.162		355	42.50		397.50	465	
	0700	Copper roof		3.30	2.424		365	48		413	480	
	0900	Hexagonal, 31" wide, 46" high, copper roof		4	2		470	39.50		509.50	590	
	1000	36" wide, 50" high, copper roof		3.50	2.286		560	45		605	695	
	1200	For deluxe stock units, add to above					25%					
	1400	For custom built units, add to above					50%	50%				
	1600	Fiberglass, 5'-0" base, 63' high minimum	F-3	6	6.667		2,325	120	75.50	2,520.50	2,825	
	1650	Maximum		4	10		2,725	179	113	3,017	3,400	
	1700	6'-0" base, 63' high, minimum		5	8		3,625	143	90.50	3,858.50	4,325	
	1750	Maximum		3	13.333		3,725	239	151	4,115	4,675	

10344 | Weathervanes

			CREW	DAILY OUTPUT	LABOR-HOURS	UNIT	MAT.	LABOR	EQUIP.	TOTAL	TOTAL INCL O&P	
800	0005	WEATHERVANES										800
	0010	Residential types, minimum	1 Carp	8	1	Ea.	40	19.70		59.70	78	
	0100	Maximum	"	2	4	"	950	79		1,029	1,175	

10350 | Flagpoles

10355 | Flagpoles

			CREW	DAILY OUTPUT	LABOR-HOURS	UNIT	2000 BARE COSTS MAT.	LABOR	EQUIP.	TOTAL	TOTAL INCL O&P	
400	0010	FLAGPOLE, Ground set										400
	0050	Not including base or foundation										
	0100	Aluminum, tapered, ground set 20' high	K-1	2	8	Ea.	600	142	90	832	1,000	
	0200	25' high		1.70	9.412		725	167	106	998	1,200	
	0300	30' high		1.50	10.667		995	190	120	1,305	1,550	
	0400	35' high		1.40	11.429		1,375	203	129	1,707	2,000	
	0500	40' high		1.20	13.333		1,775	237	150	2,162	2,550	
	0600	50' high		1	16		2,675	284	180	3,139	3,600	
	0700	60' high		.90	17.778		4,675	315	201	5,191	5,900	
	0800	70' high		.80	20		5,900	355	226	6,481	7,350	
	1100	Counterbalanced, internal halyard, 20' high		1.80	8.889		1,900	158	100	2,158	2,450	
	1200	30' high		1.50	10.667		2,350	190	120	2,660	3,025	
	1300	40' high		1.30	12.308		4,075	219	139	4,433	5,000	
	1400	50' high		1	16		7,950	284	180	8,414	9,425	
	2820	Aluminum, electronically operated, 30' high		1.40	11.429		3,900	203	129	4,232	4,775	
	2840	35' high		1.30	12.308		4,100	219	139	4,458	5,050	
	2860	39' high		1.10	14.545		4,625	259	164	5,048	5,725	
	2880	45' high		1	16		5,800	284	180	6,264	7,050	
	2900	50' high		.90	17.778		6,375	315	201	6,891	7,750	
	3000	Fiberglass, tapered, ground set, 23' high		2	8		860	142	90	1,092	1,275	
	3100	29"-7" high		1.50	10.667		1,225	190	120	1,535	1,800	
	3200	36'-1" high		1.40	11.429		1,575	203	129	1,907	2,225	
	3300	39'-5" high		1.20	13.333		1,825	237	150	2,212	2,575	
	3400	49'-2" high		1	16		4,300	284	180	4,764	5,400	
	3500	59' high		.90	17.778		5,475	315	201	5,991	6,775	
	4300	Concrete, direct imbedded installation										

10350 | Flagpoles

10355 | Flagpoles

		CREW	DAILY OUTPUT	LABOR-HOURS	UNIT	2000 BARE COSTS MAT.	LABOR	EQUIP.	TOTAL	TOTAL INCL O&P
400 4400	Internal halyard, 20' high	K-1	2.50	6.400	Ea.	1,250	114	72	1,436	1,650
4500	25' high		2.50	6.400		1,400	114	72	1,586	1,825
4600	30' high		2.30	6.957		1,625	124	78.50	1,827.50	2,075
4700	40' high		2.10	7.619		1,950	135	86	2,171	2,475
4800	50' high		1.90	8.421		3,500	150	95	3,745	4,200
5000	60' high		1.80	8.889		5,475	158	100	5,733	6,400
5100	70' high		1.60	10		6,575	178	113	6,866	7,650
5200	80' high		1.40	11.429		8,525	203	129	8,857	9,850
5300	90' high		1.20	13.333		11,200	237	150	11,587	12,900
5500	100' high		1	16		11,600	284	180	12,064	13,500
6400	Wood poles, tapered, clear vertical grain fir with tilting									
6410	base, not incl. foundation, 4" butt, 25' high	K-1	1.90	8.421	Ea.	480	150	95	725	890
6800	6" butt, 30' high	"	1.30	12.308	"	595	219	139	953	1,175
7300	Foundations for flagpoles, including									
7400	excavation and concrete, to 35' high poles	C-1	10	3.200	Ea.	400	55		455	535
7600	40' to 50' high		3.50	9.143		730	157		887	1,075
7700	Over 60' high		2	16		815	275		1,090	1,375
900 0010	FLAGPOLE, Structure mounted									
0100	Fiberglass, vertical wall set, 19'-8" long	K-1	1.50	10.667	Ea.	1,025	190	120	1,335	1,575
0200	23' long		1.40	11.429		1,025	203	129	1,357	1,600
0300	26'-3" long		1.30	12.308		1,475	219	139	1,833	2,125
0800	19'-8" long outrigger		1.30	12.308		1,225	219	139	1,583	1,875
1300	Aluminum, vertical wall set, tapered, with base, 20' high		1.20	13.333		1,025	237	150	1,412	1,700
1400	29'-6" high		1	16		1,600	284	180	2,064	2,450
2400	Outrigger poles with base, 12' long		1.30	12.308		1,125	219	139	1,483	1,775
2500	14' long		1	16		1,150	284	180	1,614	1,950

10400 | Identification Devices

10410 | Directories

		CREW	DAILY OUTPUT	LABOR-HOURS	UNIT	2000 BARE COSTS MAT.	LABOR	EQUIP.	TOTAL	TOTAL INCL O&P
100 0010	DIRECTORY BOARDS									
0050	Plastic, glass covered, 30" x 20"	2 Carp	3	5.333	Ea.	289	105		394	500
0100	36" x 48"		2	8		565	158		723	895
0300	Grooved cork, 30" x 20"		3	5.333		297	105		402	505
0400	36" x 48"		2	8		223	158		381	515
0600	Black felt, 30" x 20"		3	5.333		305	105		410	515
0700	36" x 48"		2	8		615	158		773	945
0900	Outdoor, weatherproof, black plastic, 36" x 24"		2	8		315	158		473	620
1000	36" x 36"		1.50	10.667		665	210		875	1,100
1800	Indoor, economy, open face, 18" x 24"		7	2.286		92.50	45		137.50	179
1900	24" x 36"		7	2.286		126	45		171	216
2000	36" x 24"		6	2.667		145	52.50		197.50	250
2100	36" x 48"		6	2.667		223	52.50		275.50	335
2400	Building directory, alum., black felt panels, 1 door, 24" x 18"		4	4		278	79		357	440
2500	36" x 24"		3.50	4.571		340	90		430	530
2600	48" x 32"		3	5.333		405	105		510	630
2700	36" x 48", 2 door		2.50	6.400		665	126		791	945
2800	36" x 60"		2	8		720	158		878	1,050

10400 | Identification Devices

10410 | Directories

			CREW	DAILY OUTPUT	LABOR-HOURS	UNIT	MAT.	LABOR	EQUIP.	TOTAL	TOTAL INCL O&P	
100	2900	48" x 60"	2 Carp	1	16	Ea.	770	315		1,085	1,375	100
	3100	For bronze enamel finish, add					15%					
	3200	For bronze anodized finish, add					25%					
	3400	For illuminated directory, single door unit, add					144			144	159	
	3500	For 6" header panel, 6 letters per foot, add				L.F.	23.50			23.50	26	

10430 | Exterior Signage

			CREW	DAILY OUTPUT	LABOR-HOURS	UNIT	MAT.	LABOR	EQUIP.	TOTAL	TOTAL INCL O&P	
200	0010	SIGNS Letters, 2" high, 3/8" deep, cast bronze	1 Carp	24	.333	Ea.	15.65	6.55		22.20	28.50	200
	0140	1/2" deep, cast aluminum		18	.444		15.65	8.75		24.40	32	
	0160	Cast bronze		32	.250		22	4.93		26.93	33	
	0300	6" high, 5/8" deep, cast aluminum		24	.333		17.50	6.55		24.05	30.50	
	0400	Cast bronze		24	.333		43	6.55		49.55	59	
	0600	8" high, 3/4" deep, cast aluminum		14	.571		28	11.25		39.25	50.50	
	0700	Cast bronze		20	.400		65.50	7.90		73.40	85.50	
	0900	10" high, 1" deep, cast aluminum		18	.444		38	8.75		46.75	57	
	1000	Bronze		18	.444		72	8.75		80.75	94	
	1200	12" high, 1-1/4" deep, cast aluminum		12	.667		46	13.15		59.15	73	
	1500	Cast bronze		18	.444		105	8.75		113.75	131	
	1600	14" high, 2-5/16" deep, cast aluminum		12	.667		58.50	13.15		71.65	86.50	
	1800	Fabricated stainless steel, 6" high, 2" deep		20	.400		91.50	7.90		99.40	114	
	1900	12" high, 3" deep		18	.444		108	8.75		116.75	134	
	2100	18" high, 3" deep		12	.667		163	13.15		176.15	203	
	2200	24" high, 4" deep		10	.800		228	15.75		243.75	278	
	2700	Acrylic, on high density foam,		20	.400		18.30	7.90		26.20	33.50	
	2800	12" high, 2" deep		18	.444		44	8.75		52.75	63.50	
	3900	Plaques, custom, 20" x 30", for up to 450 letters, cast aluminum	2 Carp	4	4		755	79		834	965	
	4000	Cast bronze		4	4		1,000	79		1,079	1,225	
	4200	30" x 36", up to 900 letters cast aluminum		3	5.333		1,525	105		1,630	1,850	
	4300	Cast bronze		3	5.333		1,950	105		2,055	2,325	
	4500	36" x 48", for up to 1300 letters, cast bronze		2	8		2,950	158		3,108	3,525	
	4800	Signs, reflective alum. street signs, dbl. face, 2-way, w/bracket		30	.533		40	10.50		50.50	62	
	4900	4-way		30	.533		71	10.50		81.50	96	
	5100	Exit signs, 24 ga. alum., 14" x 12" surface mounted	1 Carp	30	.267		12.95	5.25		18.20	23.50	
	5200	10" x 7"		20	.400		7.90	7.90		15.80	22	
	5400	Bracket mounted, double face, 12" x 10"		30	.267		22.50	5.25		27.75	34	
	5500	Sticky back, stock decals, 14" x 10"	1 Clab	50	.160		4.74	2.31		7.05	9.15	
	6000	Interior elec., wall mount, fiberglass panels, 2 lamps, 6"	1 Elec	8	1		48	22		70	88.50	
	6100	8"	"	8	1		40	22		62	80	
	6400	Replacement sign faces, 6" or 8"	1 Clab	50	.160		21	2.31		23.31	27.50	

10450 | Pedestrian Control Devices

10455 | Turnstiles

			CREW	DAILY OUTPUT	LABOR-HOURS	UNIT	MAT.	LABOR	EQUIP.	TOTAL	TOTAL INCL O&P	
900	0010	TURNSTILES One way, 4 arm, 46" diameter, economy, manual	2 Carp	5	3.200	Ea.	254	63		317	390	900
	0100	Electric		1.20	13.333		850	263		1,113	1,375	
	0300	High security, galv., 5'-5" diameter, 7' high, manual		1	16		2,300	315		2,615	3,075	
	0350	Electric		.60	26.667		2,850	525		3,375	4,025	
	0420	Three arm, 24" opening, light duty, manual		2	8		595	158		753	925	
	0450	Heavy duty		1.50	10.667		610	210		820	1,025	

10450 | Pedestrian Control Devices

10455 | Turnstiles

		Crew	Daily Output	Labor-Hours	Unit	Mat.	Labor	Equip.	Total	Total Incl O&P
0460	Manual, with registering & controls, light duty	2 Carp	2	8	Ea.	695	158		853	1,025
0470	Heavy duty		1.50	10.667		1,225	210		1,435	1,700
0480	Electric, heavy duty	↓	1.10	14.545		1,025	287		1,312	1,625
0500	For coin or token operating, add				↓	244			244	268
1200	One way gate with horizontal bars, 5'-5" diameter									
1300	7' high, recreation or transit type	2 Carp	.80	20	Ea.	2,600	395		2,995	3,525
1500	For electronic counter, add				"	173			173	191

10500 | Lockers

10505 | Metal Lockers

		Crew	Daily Output	Labor-Hours	Unit	Mat.	Labor	Equip.	Total	Total Incl O&P
0011	**LOCKERS** Steel, baked enamel	R10505-050								
0110	Single tier box locker, 12" x 15" x 72"	1 Shee	8	1	Ea.	132	21.50		153.50	182
0120	18" x 15" x 72"		8	1		154	21.50		175.50	207
0130	12" x 18" x 72"		8	1		136	21.50		157.50	187
0140	18" x 18" x 72"		8	1		136	21.50		157.50	187
0410	Double tier, 12" x 15" x 36"		21	.381		114	8.20		122.20	139
0420	18" x 15" x 36"		21	.381		116	8.20		124.20	141
0430	12" x 18" x 36"		21	.381		99	8.20		107.20	123
0440	18" x 18" x 36"		21	.381		96.50	8.20		104.70	120
0500	Two person, 18" x 15" x 72"		8	1		182	21.50		203.50	237
0510	18" x 18" x 72"		8	1		201	21.50		222.50	258
0520	Duplex, 15" x 15" x 72"		8	1		216	21.50		237.50	275
0530	15" x 21" x 72"		8	1	↓	259	21.50		280.50	320
0600	5 tier box lockers, minimum		30	.267	Opng.	23.50	5.75		29.25	35
0700	Maximum		24	.333		41	7.15		48.15	57
0900	6 tier box lockers, minimum		36	.222		24.50	4.78		29.28	35
1000	Maximum		30	.267	↓	32	5.75		37.75	44.50
1100	Wire meshed wardrobe, floor. mtd., open front varsity type		7.50	1.067	Ea.	147	23		170	201
2100	Locker bench, laminated maple, top only		100	.080	L.F.	12.20	1.72		13.92	16.35
2200	Pedestals, steel pipe	↓	25	.320	Ea.	31	6.90		37.90	45.50
2400	16-person locker unit with clothing rack									
2500	72 wide x 15" deep x 72" high	1 Shee	8	1	Ea.	355	21.50		376.50	425
2550	18" deep	"	8	1	"	350	21.50		371.50	420
3000	Wall mounted lockers, 4 person, with coat bar									
3100	48" wide x 18" deep x 12" high	1 Shee	8	1	Ea.	155	21.50		176.50	207
3250	Rack w/ 24 wire mesh baskets		1.50	5.333	Set	270	115		385	490
3260	30 baskets		1.25	6.400		232	138		370	485
3270	36 baskets		.95	8.421		325	181		506	665
3280	42 baskets	↓	.80	10	↓	365	215		580	770
3300	For built-in lock with 2 keys, add				Ea.	5.15			5.15	5.70
3600	For hanger rods, add				"	1.65			1.65	1.82

10520 | Fire Protection Specialties

10525 | Fire Prot. Specialties

			DAILY	LABOR-		2000 BARE COSTS				TOTAL
		CREW	OUTPUT	HOURS	UNIT	MAT.	LABOR	EQUIP.	TOTAL	INCL O&P
200	0010	**FIRE EQUIPMENT CABINETS** Not equipped, 20 ga. steel box,									
	0040	recessed, D.S. glass in door, box size given									
	1000	Portable extinguisher, single, 8" x 12" x 27", alum. door & frame	Q-12	8	2	Ea.	75.50	39.50		115	149
	1100	Steel door and frame	"	8	2	"	46	39.50		85.50	117
	3000	Hose rack assy., 1-1/2" valve & 100' hose, 24" x 40" x 5-1/2"									
	3100	Aluminum door and frame	Q-12	6	2.667	Ea.	159	53		212	262
	3200	Steel door and frame		6	2.667		101	53		154	199
	3300	Stainless steel door and frame		6	2.667		244	53		297	355
	4000	Hose rack assy., 2-1/2" x 1-1/2" valve, 100' hose, 24" x 40" x 8"									
	4100	Aluminum door and frame	Q-12	6	2.667	Ea.	160	53		213	264
	4200	Steel door and frame		6	2.667		108	53		161	207
	4300	Stainless steel door and frame		6	2.667		221	53		274	330
	5000	Hose rack assy., 2-1/2" x 1-1/2" valve, 100' hose									
	5100	Aluminum door and frame	Q-12	5	3.200	Ea.	204	63.50		267.50	330
	5200	Steel door and frame		5	3.200		129	63.50		192.50	247
	5300	Stainless steel door and frame		5	3.200		285	63.50		348.50	420
	8000	Valve cabinet for 2-1/2" FD angle valve, 18" x 18" x 8"									
	8100	Aluminum door and frame	Q-12	12	1.333	Ea.	70.50	26.50		97	122
	8200	Steel door and frame		12	1.333		54	26.50		80.50	104
	8300	Stainless steel door and frame		12	1.333		98.50	26.50		125	153
300	0010	**FIRE EXTINGUISHERS**									
	0120	CO2, portable with swivel horn, 5 lb.				Ea.	101			101	111
	0140	With hose and "H" horn, 10 lb.					150			150	165
	0160	15 lb.					172			172	189
	0180	20 lb.					207			207	228
	0360	Wheeled type, cart mounted, 50 lb.					1,200			1,200	1,325
	1000	Dry chemical, pressurized									
	1040	Standard type, portable, painted, 2-1/2 lb.				Ea.	27.50			27.50	30.50
	1060	5 lb.					42			42	46
	1080	10 lb.					67			67	73.50
	1100	20 lb.					90			90	99
	1120	30 lb.					145			145	160
	1300	Standard type, wheeled, 150 lb.					1,400			1,400	1,550
	2000	ABC all purpose type, portable, 2-1/2 lb.					27.50			27.50	30.50
	2060	5 lb.					40			40	44
	2080	9-1/2 lb.					65			65	71.50
	2100	20 lb.					88.50			88.50	97.50
	2300	Wheeled, 45 lb.					650			650	715
	2360	150 lb.					1,450			1,450	1,600
	3000	Dry chemical, outside cartridge to -65°F, painted, 9 lb.					200			200	220
	3060	26 lb.					250			250	275
	5000	Pressurized water, 2-1/2 gallon, stainless steel					75			75	82.50
	5060	With anti-freeze					110			110	121
	6000	Soda & acid, 2-1/2 gallon, stainless steel					75			75	82.50
	9400	Installation of extinguishers, 12 or more, on wood	1 Carp	30	.267			5.25		5.25	9
	9420	On masonry or concrete	"	15	.533			10.50		10.50	18

10530 | Protective Covers

10535 | Awning & Canopies

			CREW	DAILY OUTPUT	LABOR-HOURS	UNIT	2000 BARE COSTS MAT.	LABOR	EQUIP.	TOTAL	TOTAL INCL O&P	
200	0010	**CANOPIES** Wall hung, .032", aluminum, prefinished, 8' x 10'	K-2	1.30	18.462	Ea.	1,325	370	139	1,834	2,300	200
	0300	8' x 20'		1.10	21.818		2,625	440	164	3,229	3,900	
	0500	10' x 10'		1.30	18.462		1,475	370	139	1,984	2,475	
	0700	10' x 20'		1.10	21.818		2,725	440	164	3,329	4,000	
	1000	12' x 20'		1	24		3,350	485	180	4,015	4,825	
	1360	12' x 30'		.80	30		4,925	605	226	5,756	6,825	
	1700	12' x 40'		.60	40		5,250	805	300	6,355	7,625	
	1900	For free standing units, add					20%	10%				
	2300	Aluminum entrance canopies, flat soffit, .032"										
	2500	3'-6" x 4'-0", clear anodized	2 Carp	4	4	Ea.	550	79		629	740	
	2700	Bronze anodized		4	4		970	79		1,049	1,200	
	3000	Polyurethane painted		4	4		790	79		869	1,000	
	3300	4'-6" x 10'-0", clear anodized		2	8		1,525	158		1,683	1,950	
	3500	Bronze anodized		2	8		1,950	158		2,108	2,400	
	3700	Polyurethane painted		2	8		1,625	158		1,783	2,075	
	4000	Wall downspout, 10 L.F., clear anodized	1 Carp	7	1.143		94.50	22.50		117	143	
	4300	Bronze anodized		7	1.143		165	22.50		187.50	220	
	4500	Polyurethane painted		7	1.143		142	22.50		164.50	195	
	4700	Canvas awnings, including canvas, frame & lettering										
	5000	Minimum	2 Carp	100	.160	S.F.	38	3.15		41.15	47	
	5300	Average		90	.178		45.50	3.50		49	56.50	
	5500	Maximum		80	.200		92.50	3.94		96.44	109	
	7000	Carport, baked vinyl finish, .032", 20' x 10', no foundations, min.	K-2	4	6	Car	2,375	121	45	2,541	2,875	
	7250	Maximum		2	12	"	5,625	241	90	5,956	6,725	
	7500	Walkway cover, to 12' wide, stl., vinyl finish, .032",no fndtns., min.		250	.096	S.F.	13.85	1.93	.72	16.50	19.70	
	7750	Maximum		200	.120	"	15.05	2.41	.90	18.36	22	

10550 | Postal Specialties

10555 | Mail Delivery Systems

			CREW	DAILY OUTPUT	LABOR-HOURS	UNIT	2000 BARE COSTS MAT.	LABOR	EQUIP.	TOTAL	TOTAL INCL O&P	
600	0010	**MAIL BOXES** Horiz., key lock, 5"H x 6"W x 15"D, alum., rear load	1 Carp	34	.235	Ea.	34.50	4.64		39.14	46	600
	0100	Front loading		34	.235		38.50	4.64		43.14	50.50	
	0200	Double, 5"H x 12"W x 15"D, rear loading		26	.308		61	6.05		67.05	77.50	
	0300	Front loading		26	.308		66	6.05		72.05	83.50	
	0500	Quadruple, 10"H x 12"W x 15"D, rear loading		20	.400		110	7.90		117.90	135	
	0600	Front loading		20	.400		121	7.90		128.90	148	
	0800	Vertical, front load, 15"H x 5"W x 6"D, alum., per compartment		34	.235		29	4.64		33.64	40	
	0900	Bronze, duranodic finish		34	.235		46	4.64		50.64	59	
	1000	Steel, enameled		34	.235		26	4.64		30.64	36.50	
	1700	Alphabetical directories, 120 names		10	.800		140	15.75		155.75	181	
	1800	Letter collection box		6	1.333		575	26.50		601.50	680	
	1900	Letter slot, residential		20	.400		50.50	7.90		58.40	69	
	2000	Post office type		8	1		191	19.70		210.70	244	
	2200	Post office counter window, with grille		2	4		515	79		594	700	
	2250	Key keeper, single key, aluminum		26	.308		39.50	6.05		45.55	54	
	2300	Steel, enameled		26	.308		70	6.05		76.05	87.50	
700	0010	**MAIL CHUTES** Aluminum & glass, 14-1/4" wide, 4-5/8" deep	2 Shee	4	4	Floor	635	86		721	845	700
	0100	8-5/8" deep		3.80	4.211		705	90.50		795.50	930	
	0300	8-3/4" x 3-1/2", aluminum		5	3.200		540	69		609	705	
	0400	Bronze or stainless		4.50	3.556		835	76.50		911.50	1,050	

10550 | Postal Specialties

10555 | Mail Delivery Systems

		CREW	DAILY OUTPUT	LABOR-HOURS	UNIT	2000 BARE COSTS MAT.	LABOR	EQUIP.	TOTAL	TOTAL INCL O&P	
700	0600 Lobby collection boxes, aluminum	2 Shee	5	3.200	Ea.	1,750	69		1,819	2,050	700
	0700 Bronze or stainless	↓	4.50	3.556	"	2,150	76.50		2,226.50	2,500	

10600 | Partitions

10605 | Wire Mesh Partitions

		CREW	DAILY OUTPUT	LABOR-HOURS	UNIT	2000 BARE COSTS MAT.	LABOR	EQUIP.	TOTAL	TOTAL INCL O&P	
100	0010 **PARTITIONS, WOVEN WIRE** For tool or stockroom enclosures										100
	0100 Channel frame, 1-1/2" diamond mesh, 10 ga. wire, painted										
	0300 Wall panels, 4'-0" wide, 7' high	2 Carp	25	.640	Ea.	83.50	12.60		96.10	114	
	0400 8' high		23	.696		95.50	13.70		109.20	129	
	0600 10' high	↓	18	.889		113	17.50		130.50	154	
	0700 For 5' wide panels, add					5%					
	0900 Ceiling panels, 10' long, 2' wide	2 Carp	25	.640		80.50	12.60		93.10	110	
	1000 4' wide		15	1.067		95.50	21		116.50	141	
	1200 Panel with service window & shelf, 5' wide, 7' high		20	.800		265	15.75		280.75	320	
	1300 8' high		15	1.067		269	21		290	330	
	1500 Sliding doors, full height, 3' wide, 7' high		6	2.667		255	52.50		307.50	370	
	1600 10' high		5	3.200		292	63		355	430	
	1800 6' wide sliding door, 7' full height		5	3.200		330	63		393	470	
	1900 10' high		4	4		380	79		459	555	
	2100 Swinging doors, 3' wide, 7' high, no transom		6	2.667		178	52.50		230.50	286	
	2200 7' high, 3' transom	↓	5	3.200	↓	227	63		290	360	

10610 | Folding Gates

		CREW	DAILY OUTPUT	LABOR-HOURS	UNIT	2000 BARE COSTS MAT.	LABOR	EQUIP.	TOTAL	TOTAL INCL O&P	
100	0010 **SECURITY GATES** For roll up type, see division 08330-130										100
	0300 Scissors type folding gate, ptd. steel, single, 6' high, 5-1/2' wide	2 Sswk	4	4	Opng.	108	85		193	287	
	0350 6-1/2' wide		4	4		126	85		211	305	
	0400 7-1/2' wide		4	4		155	85		240	340	
	0600 Double gate, 7-1/2' high, 8' wide		2.50	6.400		198	136		334	485	
	0650 10' wide		2.50	6.400		244	136		380	535	
	0700 12' wide		2	8		330	170		500	695	
	0750 14' wide		2	8		440	170		610	820	
	0900 Door gate, folding steel, 4' wide, 61" high		4	4		54.50	85		139.50	228	
	1000 71" high		4	4		56.50	85		141.50	230	
	1200 81" high		4	4		61	85		146	235	
	1300 Window gates, 2' to 4' wide, 31" high		4	4		29.50	85		114.50	200	
	1500 55" high		3.75	4.267		41	90.50		131.50	224	
	1600 79" high	↓	3.50	4.571	↓	46.50	97		143.50	244	

10615 | Demountable Partitions

		CREW	DAILY OUTPUT	LABOR-HOURS	UNIT	2000 BARE COSTS MAT.	LABOR	EQUIP.	TOTAL	TOTAL INCL O&P	
100	0010 **PARTITIONS, MOVABLE OFFICE** Demountable, add for doors										100
	0100 Do not deduct door openings from total L.F.										
	0900 Demountable gypsum system on 2" to 2-1/2"										
	1000 steel studs, 9' high, 3" to 3-3/4" thick										
	1200 Vinyl clad gypsum	2 Carp	48	.333	L.F.	23.50	6.55		30.05	37.50	
	1300 Fabric clad gypsum		44	.364		73.50	7.15		80.65	93	
	1500 Steel clad gypsum	↓	40	.400	↓	78	7.90		85.90	99.50	
	1600 1.75 system, aluminum framing, vinyl clad hardboard,										
	1800 paper honeycomb core panel, 1-3/4" to 2-1/2" thick										
	1900 9' high	2 Carp	48	.333	L.F.	49.50	6.55		56.05	66	

10600 | Partitions

10615 | Demountable Partitions

		CREW	DAILY OUTPUT	LABOR-HOURS	UNIT	2000 BARE COSTS MAT.	LABOR	EQUIP.	TOTAL	TOTAL INCL O&P
2100	7' high	2 Carp	60	.267	L.F.	45.50	5.25		50.75	59
2200	5' high	↓	80	.200	↓	39	3.94		42.94	50
2250	Unitized gypsum system									
2300	Unitized panel, 9' high, 2" to 2-1/2" thick									
2350	Vinyl clad gypsum	2 Carp	48	.333	L.F.	67	6.55		73.55	85
2400	Fabric clad gypsum	"	44	.364	"	120	7.15		127.15	144
2500	Unitized mineral fiber system									
2510	Unitized panel, 9' high, 2-1/4" thick, aluminum frame									
2550	Vinyl clad mineral fiber	2 Carp	48	.333	L.F.	85	6.55		91.55	105
2600	Fabric clad mineral fiber	"	44	.364	"	109	7.15		116.15	131
2800	Movable steel walls, modular system									
2900	Unitized panels, 9' high, 48" wide									
3100	Baked enamel, pre-finished	2 Carp	60	.267	L.F.	78	5.25		83.25	95
3200	Fabric clad steel		56	.286	"	119	5.65		124.65	141
5310	Trackless wall, cork finish, semi-acoustic, 1-5/8" thick, minimum		325	.049	S.F.	19	.97		19.97	22.50
5320	Maximum		190	.084		23	1.66		24.66	28
5330	Acoustic, 2" thick, minimum		305	.052		19.60	1.03		20.63	23.50
5340	Maximum	↓	225	.071		29	1.40		30.40	34.50
5500	For acoustical partitions, add, minimum					1.38			1.38	1.52
5550	Maximum				↓	5.05			5.05	5.55
5700	For doors, see div. 08100 & 08200									
5800	For door hardware, see div. 08700									
6100	In-plant modular office system, w/prehung steel door									
6200	3" thick honeycomb core panels									
6250	12' x 12', 2 wall	2 Clab	3.80	4.211	Ea.	3,225	61		3,286	3,650
6300	4 wall		1.90	8.421		4,975	122		5,097	5,675
6350	16' x 16', 2 wall		3.60	4.444		4,825	64		4,889	5,400
6400	4 wall	↓	1.80	8.889	↓	6,925	128		7,053	7,850

10630 | Port. Partitions/Screens/Panels

		CREW	DAILY OUTPUT	LABOR-HOURS	UNIT	MAT.	LABOR	EQUIP.	TOTAL	TOTAL INCL O&P
0010	**PARTITIONS, PORTABLE** Divider panels, free standing, fiber core									
0020	Fabric face straight									
0100	3'-0" long, 4'-0" high	2 Carp	100	.160	L.F.	92.50	3.15		95.65	107
0200	5'-0" high		90	.178		93.50	3.50		97	109
0500	6'-0" high		75	.213		106	4.20		110.20	124
0900	5'-0" long, 4'-0" high		175	.091		70	1.80		71.80	80
1000	5'-0" high		150	.107		86	2.10		88.10	98
1500	6"-0" high		125	.128		82.50	2.52		85.02	95.50
1600	6'-0" long, 5'-0" high		162	.099		69	1.95		70.95	79
3100	Curved, 3'-0" long, 5'-0" high		90	.178		85	3.50		88.50	99.50
3150	6'-0" high		75	.213		98.50	4.20		102.70	115
3200	Economical panels, fabric face, 4'-0" long, 5'-0" high		132	.121		135	2.39		137.39	153
3250	6'-0" high		112	.143		23	2.81		25.81	30
3300	5'-0" long, 5'-0" high		150	.107		27.50	2.10		29.60	34
3350	6'-0" high		125	.128		33.50	2.52		36.02	41.50
3380	3'-0" curved, 5'-0" high		90	.178		85	3.50		88.50	99.50
3390	6'-0" high		75	.213		98.50	4.20		102.70	115
3450	Acoustical panels, 60 to 90 NRC, 3'-0" long, 5'-0" high		90	.178		79	3.50		82.50	93
3550	6'-0" high		75	.213		77	4.20		81.20	91.50
3600	5'-0" long, 5'-0" high		150	.107		55	2.10		57.10	64
3650	6'-0" high		125	.128		61	2.52		63.52	72
3700	6'-0" long, 5'-0" high		162	.099		53.50	1.95		55.45	62
3750	6'-0" high		138	.116		56.50	2.28		58.78	66.50
3800	Economy acoustical panels, 40 NRC, 4'-0" long, 5'-0" high		132	.121		34	2.39		36.39	41
3850	6'-0" high		112	.143		38	2.81		40.81	47
3900	5'-0" long, 6'-0" high	↓	125	.128	↓	33.50	2.52		36.02	41.50

10600 | Partitions

10630 | Port. Partitions/Screens/Panels

			CREW	DAILY OUTPUT	LABOR-HOURS	UNIT	2000 BARE COSTS MAT.	LABOR	EQUIP.	TOTAL	TOTAL INCL O&P	
100	3950	6'-0" long, 5'-0" high	2 Carp	162	.099	L.F.	33	1.95		34.95	40	100
	4000	Metal chalkboard, 6'-6" high, chalkboard, 1 side		125	.128		78	2.52		80.52	90.50	
	4100	Metal chalkboard, 2 sides		120	.133		89.50	2.63		92.13	103	
	4300	Tackboard, both sides		123	.130		70.50	2.56		73.06	82	

10651 | Accordion Folding Partitions

			CREW	DAILY OUTPUT	LABOR-HOURS	UNIT	MAT.	LABOR	EQUIP.	TOTAL	TOTAL INCL O&P	
100	0010	**PARTITIONS, FOLDING ACCORDION**										100
	0100	Vinyl covered, over 150 S.F., frame not included										
	0300	Residential, 1.25 lb. per S.F., 8' maximum height	2 Carp	300	.053	S.F.	12	1.05		13.05	15	
	0400	Commercial, 1.75 lb. per S.F., 8' maximum height		225	.071		12	1.40		13.40	15.60	
	0600	2 lb. per S.F., 17' maximum height		150	.107		14.60	2.10		16.70	19.65	
	0700	Industrial, 4 lb. per S.F., 20' maximum height		75	.213		24.50	4.20		28.70	34	
	0900	Acoustical, 3 lb. per S.F., 17' maximum height		100	.160		18.10	3.15		21.25	25.50	
	1200	5 lb. per S.F., 20' maximum height		95	.168		26.50	3.32		29.82	35	
	1300	5.5 lb. per S.F., 17' maximum height		90	.178		30	3.50		33.50	39	
	1400	Fire rated, 4.5 psf, 20' maximum height		160	.100		30	1.97		31.97	36.50	
	1500	Vinyl clad wood or steel, electric operation, 5.0 psf		160	.100		30	1.97		31.97	36.50	
	1900	Wood, non-acoustic, birch or mahogany, to 10' high		300	.053		16.10	1.05		17.15	19.50	

10653 | Folding Panel Partitions

			CREW	DAILY OUTPUT	LABOR-HOURS	UNIT	MAT.	LABOR	EQUIP.	TOTAL	TOTAL INCL O&P	
200	0010	**PARTITIONS, FOLDING LEAF** Acoustic, wood										200
	0100	Vinyl faced, to 18' high, 6 psf, minimum	2 Carp	60	.267	S.F.	35.50	5.25		40.75	48	
	0150	Average		45	.356		42.50	7		49.50	58.50	
	0200	Maximum		30	.533		55	10.50		65.50	78.50	
	0400	Formica or hardwood finish, minimum		60	.267		36.50	5.25		41.75	49.50	
	0500	Maximum		30	.533		39	10.50		49.50	61	
	0600	Wood, low acoustical type, 4.5 psf, to 14' high		50	.320		26.50	6.30		32.80	40.50	
	1100	Steel, acoustical, 9 to 12 lb. per S.F., vinyl faced, minimum		60	.267		38	5.25		43.25	50.50	
	1200	Maximum		30	.533		46	10.50		56.50	69	
	1700	Aluminum framed, acoustical, to 12' high, 5.5 psf, minimum		60	.267		25.50	5.25		30.75	37	
	1800	Maximum		30	.533		31	10.50		41.50	52	
	2000	6.5 lb. per S.F., minimum		60	.267		27	5.25		32.25	38.50	
	2100	Maximum		30	.533		33	10.50		43.50	54.50	

10658 | Acoustic Air Wall

			CREW	DAILY OUTPUT	LABOR-HOURS	UNIT	MAT.	LABOR	EQUIP.	TOTAL	TOTAL INCL O&P	
100	0010	**PARTITIONS, OPERABLE** Acoustic air wall, 1-5/8" thick, minimum	2 Carp	375	.043	S.F.	21.50	.84		22.34	25.50	100
	0100	Maximum		365	.044		37	.86		37.86	42.50	
	0300	2-1/4" thick, minimum		360	.044		24.50	.88		25.38	28.50	
	0400	Maximum		330	.048		42.50	.96		43.46	48.50	
	0600	For track type, add to above				L.F.	79.50			79.50	87.50	
	0700	Overhead track type, acoustical, 3" thick, 11 psf, minimum	2 Carp	350	.046	S.F.	54.50	.90		55.40	61.50	
	0800	Maximum	"	300	.053	"	66	1.05		67.05	74.50	

10670 | Storage Shelving

10674 | Storage Shelving

			CREW	DAILY OUTPUT	LABOR-HOURS	UNIT	MAT.	LABOR	EQUIP.	TOTAL	TOTAL INCL O&P	
500	0010	**SHELVING** Metal, industrial, cross-braced, 3' wide, 12" deep	1 Sswk	175	.046	SF Shlf	8.80	.97		9.77	11.60	500
	0100	24" deep		330	.024		6.85	.52		7.37	8.55	

10670 | Storage Shelving

10674 | Storage Shelving

			CREW	DAILY OUTPUT	LABOR-HOURS	UNIT	2000 BARE COSTS				TOTAL INCL O&P
							MAT.	LABOR	EQUIP.	TOTAL	
500	0300	4' wide, 12" deep	1 Sswk	185	.043	SF Shlf	6.60	.92		7.52	9.05
	0400	24" deep		380	.021		6.80	.45		7.25	8.40
	1200	Enclosed sides, cross-braced back, 3' wide, 12" deep		175	.046		12.85	.97		13.82	16.05
	1300	24" deep		290	.028		8.15	.59		8.74	10.10
	1500	Fully enclosed, sides and back, 3' wide, 12" deep		150	.053		15.05	1.13		16.18	18.80
	1600	24" deep		255	.031		9.45	.67		10.12	11.70
	1800	4' wide, 12" deep		150	.053		14.25	1.13		15.38	17.95
	1900	24" deep		290	.028		9.15	.59		9.74	11.25
	2200	Wide span, 1600 lb. capacity per shelf, 6' wide, 24" deep		380	.021		6.55	.45		7	8.10
	2400	36" deep		440	.018		5.80	.39		6.19	7.10
	2600	8' wide, 24" deep		440	.018		5.95	.39		6.34	7.30
	2800	36" deep		520	.015		4.99	.33		5.32	6.15
	4000	Pallet racks, steel frame 5,000 lb. capacity, 8' long, 36" deep	2 Sswk	450	.036		4.23	.76		4.99	6.15
	4200	42" deep		500	.032		3.71	.68		4.39	5.40
	4400	48" deep		520	.031		3.30	.65		3.95	4.92
600	0010	**PARTS BINS** Metal, gray baked enamel finish									
	0100	6'-3" high, 3' wide									
	0300	12 bins, 18" wide x 12" high, 12" deep	2 Clab	10	1.600	Ea.	239	23		262	300
	0400	24" deep		10	1.600		283	23		306	350
	0600	72 bins, 6" wide x 6" high, 12" deep		8	2		460	29		489	555
	0700	24" deep		8	2		775	29		804	900
	1000	7'-3" high, 3' wide									
	1200	14 bins, 18" wide x 12" high, 12" deep	2 Clab	10	1.600	Ea.	247	23		270	310
	1300	24" deep		10	1.600		340	23		363	410
	1500	84 bins, 6" wide x 6" high, 12" deep		8	2		535	29		564	635
	1600	24" deep		8	2		900	29		929	1,050

10750 | Telephone Specialties

10755 | Telephone Enclosures

			CREW	DAILY OUTPUT	LABOR-HOURS	UNIT	2000 BARE COSTS				TOTAL INCL O&P
							MAT.	LABOR	EQUIP.	TOTAL	
400	0010	**TELEPHONE ENCLOSURE**									
	0300	Shelf type, wall hung, minimum	2 Carp	5	3.200	Ea.	705	63		768	890
	0400	Maximum		5	3.200		2,425	63		2,488	2,775
	0600	Booth type, painted steel, indoor or outdoor, minimum		1.50	10.667		2,975	210		3,185	3,625
	0700	Maximum (stainless steel)		1.50	10.667		9,850	210		10,060	11,200
	1300	Outdoor, acoustical, on post		3	5.333		1,250	105		1,355	1,550
	1400	Phone carousel, pedestal mounted with dividers		.60	26.667		4,800	525		5,325	6,175
	1900	Outdoor, drive-up type, wall mounted		4	4		775	79		854	990
	2000	Post mounted, stainless steel posts		3	5.333		1,200	105		1,305	1,500
	2200	Directory shelf, wall mounted, stainless steel									
	2300	3 binders	2 Carp	8	2	Ea.	910	39.50		949.50	1,075
	2500	4 binders		7	2.286		1,050	45		1,095	1,225
	2600	5 binders		6	2.667		1,275	52.50		1,327.50	1,500
	2800	Table type, stainless steel, 4 binders		8	2		1,125	39.50		1,164.50	1,300
	2900	7 binders		7	2.286		1,375	45		1,420	1,575

10800 | Toilet/Bath/Laundry Accessories

10820 | Bath Accessories

		Description	CREW	DAILY OUTPUT	LABOR-HOURS	UNIT	MAT.	LABOR	EQUIP.	TOTAL	TOTAL INCL O&P
100	0010	**BATH ACCESSORIES**										100
	0200	Curtain rod, stainless steel, 5' long, 1" diameter	1 Carp	13	.615	Ea.	20.50	12.10		32.60	43.50
	0300	1-1/4" diameter	"	13	.615	"	24.50	12.10		36.60	48
	0500	Dispenser units, combined soap & towel dispensers,									
	0510	mirror and shelf, flush mounted	1 Carp	10	.800	Ea.	271	15.75		286.75	325
	0600	Towel dispenser and waste receptacle,									
	0610	18 gallon capacity	1 Carp	10	.800	Ea.	271	15.75		286.75	325
	0800	Grab bar, straight, 1-1/4" diameter, stainless steel, 18" long		24	.333		35.50	6.55		42.05	50.50
	0900	24" long		23	.348		32.50	6.85		39.35	48
	1000	30" long		22	.364		42	7.15		49.15	58.50
	1100	36" long		20	.400		45	7.90		52.90	63
	1200	1-1/2" diameter, 24" long		23	.348		41.50	6.85		48.35	58
	1300	36" long		20	.400		48.50	7.90		56.40	67
	1500	Tub bar, 1-1/4" diameter, 24" x 36"		14	.571		86.50	11.25		97.75	115
	1600	Plus vertical arm		12	.667		75.50	13.15		88.65	106
	1900	End tub bar, 1" diameter, 90° angle, 16" x 32"		12	.667		105	13.15		118.15	139
	2300	Hand dryer, surface mounted, electric, 115 volt, 20 amp		4	2		485	39.50		524.50	605
	2400	230 volt, 10 amp		4	2		485	39.50		524.50	605
	2600	Hat and coat strip, stainless steel, 4 hook, 36" long		24	.333		50	6.55		56.55	66.50
	2700	6 hook, 60" long		20	.400		78	7.90		85.90	99.50
	3000	Mirror, with stainless steel 3/4" square frame, 18" x 24"		20	.400		61	7.90		68.90	80.50
	3100	36" x 24"		15	.533		108	10.50		118.50	137
	3200	48" x 24"		10	.800		144	15.75		159.75	185
	3300	72" x 24"		6	1.333		196	26.50		222.50	261
	3500	With 5" stainless steel shelf, 18" x 24"		20	.400		90.50	7.90		98.40	114
	3600	36" x 24"		15	.533		132	10.50		142.50	163
	3700	48" x 24"		10	.800		165	15.75		180.75	209
	3800	72" x 24"		6	1.333		295	26.50		321.50	370
	4100	Mop holder strip, stainless steel, 5 holders, 48" long		20	.400		63	7.90		70.90	83
	4200	Napkin/tampon dispenser, recessed		15	.533		365	10.50		375.50	420
	4300	Robe hook, single, regular		36	.222		9.60	4.38		13.98	18.05
	4400	Heavy duty, concealed mounting		36	.222		14.20	4.38		18.58	23
	4600	Soap dispenser, chrome, surface mounted, liquid		20	.400		41	7.90		48.90	59
	4700	Powder		20	.400		59	7.90		66.90	78.50
	5000	Recessed stainless steel, liquid		10	.800		66.50	15.75		82.25	101
	5100	Powder		10	.800		255	15.75		270.75	310
	5300	Soap tank, stainless steel, 1 gallon		10	.800		159	15.75		174.75	202
	5400	5 gallon		5	1.600		219	31.50		250.50	295
	5600	Shelf, stainless steel, 5" wide, 18 ga., 24" long		24	.333		35.50	6.55		42.05	50.50
	5700	48" long		16	.500		63	9.85		72.85	86.50
	5800	8" wide shelf, 18 ga., 24" long		22	.364		44	7.15		51.15	61
	5900	48" long		14	.571		88	11.25		99.25	116
	6000	Toilet seat cover dispenser, stainless steel, recessed		20	.400		94	7.90		101.90	117
	6050	Surface mounted		15	.533		26.50	10.50		37	47
	6100	Toilet tissue dispenser, surface mounted, SS, single roll		30	.267		10.60	5.25		15.85	20.50
	6200	Double roll		24	.333		15.65	6.55		22.20	28.50
	6400	Towel bar, stainless steel, 18" long		23	.348		30	6.85		36.85	45
	6500	30" long		21	.381		34.50	7.50		42	51
	6700	Towel dispenser, stainless steel, surface mounted		16	.500		36	9.85		45.85	56.50
	6800	Flush mounted, recessed		10	.800		138	15.75		153.75	179
	7000	Towel holder, hotel type, 2 guest size		20	.400		12.40	7.90		20.30	27
	7200	Towel shelf, stainless steel, 24" long, 8" wide		20	.400		45	7.90		52.90	63
	7400	Tumbler holder, tumbler only		30	.267		20.50	5.25		25.75	31.50
	7500	Soap, tumbler & toothbrush		30	.267		20.50	5.25		25.75	31.50
	7700	Wall urn ash receiver, surface mount, 11" long		12	.667		104	13.15		117.15	137
	7800	7-1/2", long		18	.444		72	8.75		80.75	94

10800 | Toilet/Bath/Laundry Accessories

10820 | Bath Accessories

			CREW	DAILY OUTPUT	LABOR-HOURS	UNIT	MAT.	LABOR	EQUIP.	TOTAL	TOTAL INCL O&P	
100	8000	Waste receptacles, stainless steel, with top, 13 gallon	1 Carp	10	.800	Ea.	188	15.75		203.75	234	100
	8100	36 gallon		8	1		300	19.70		319.70	365	
400	0010	**MEDICINE CABINETS** With mirror, st. st. frame, 16" x 22", unlighted	1 Carp	14	.571	Ea.	65	11.25		76.25	91	400
	0100	Wood frame		14	.571		95	11.25		106.25	124	
	0300	Sliding mirror doors, 20" x 16" x 4-3/4", unlighted		7	1.143		85	22.50		107.50	132	
	0400	24" x 19" x 8-1/2", lighted		5	1.600		143	31.50		174.50	211	
	0600	Triple door, 30" x 32", unlighted, plywood body		7	1.143		211	22.50		233.50	271	
	0700	Steel body		7	1.143		278	22.50		300.50	345	
	0900	Oak door, wood body, beveled mirror, single door		7	1.143		124	22.50		146.50	175	
	1000	Double door		6	1.333		315	26.50		341.50	390	
	1200	Hotel cabinets, stainless, with lower shelf, unlighted		10	.800		170	15.75		185.75	214	
	1300	Lighted		5	1.600		252	31.50		283.50	330	

10880 | Scales

10885 | Scales

			CREW	DAILY OUTPUT	LABOR-HOURS	UNIT	MAT.	LABOR	EQUIP.	TOTAL	TOTAL INCL O&P	
100	0010	**SCALES** Built-in floor scale, not incl. foundations										100
	0100	Dial type, 5 ton capacity, 8' x 6' platform	3 Carp	.50	48	Ea.	4,400	945		5,345	6,450	
	0300	9' x 7' platform		.40	60		7,325	1,175		8,500	10,100	
	0400	10 ton capacity, steel platform, 8' x 6' platform		.40	60		7,125	1,175		8,300	9,875	
	0600	9' x 7' platform		.35	68.571		8,325	1,350		9,675	11,500	
	0700	Truck scales, incl. steel weigh bridge,										
	0800	not including foundation, pits										
	1550	Digital, electronic, 100 ton capacity, steel deck 12' x 10' platform	3 Carp	.20	120	Ea.	8,725	2,375		11,100	13,700	
	1600	40' x 10' platform		.14	171		13,800	3,375		17,175	21,000	
	1640	60' x 10' platform		.13	184		23,400	3,625		27,025	32,000	
	1680	70' x 10' platform		.12	200		27,600	3,950		31,550	37,200	
	2000	For standard automatic printing device, add					1,925			1,925	2,100	
	2100	For remote reading electronic system, add					2,025			2,025	2,225	
	2300	Concrete foundation pits for above, 8' x 6', 5 C.Y. required	C-1	.50	64		715	1,100		1,815	2,675	
	2400	14' x 6' platform, 10 C.Y. required		.35	91.429		1,025	1,575		2,600	3,825	
	2600	50' x 10' platform, 30 C.Y. required		.25	128		1,425	2,200		3,625	5,325	
	2700	70' x 10' platform, 40 C.Y. required		.15	213		2,375	3,650		6,025	8,875	
	2750	Crane scales, dial, 1 ton capacity					815			815	895	
	2780	5 ton capacity					970			970	1,075	
	2800	Digital, 1 ton capacity					1,800			1,800	2,000	
	2850	10 ton capacity					3,450			3,450	3,800	
	2900	Low profile electronic warehouse scale,										
	3000	not incl. printer, 4' x 4' platform, 10,000 lb. capacity	2 Carp	.30	53.333	Ea.	2,250	1,050		3,300	4,275	
	3300	5' x 7' platform, 10,000 lb. capacity		.25	64		3,425	1,250		4,675	5,925	
	3400	20,000 lb. capacity		.20	80		4,000	1,575		5,575	7,100	
	3500	For printers, incl. time, date & numbering, add					915			915	1,000	
	3800	Portable, beam type, capacity 1000#, platform 18" x 24"					495			495	545	
	3900	Dial type, capacity 2000#, platform 24" x 24"					855			855	945	
	4000	Digital type, capacity 1000#, platform 24" x 30"					1,475			1,475	1,600	
	4100	Portable contractor truck scales, 50 ton cap., 40' x 10' platform					25,600			25,600	28,100	
	4200	60' x 10' platform					28,600			28,600	31,400	

10900 | Wardrobe & Closet Specialties

10905 | Coat Racks/Wardrobes

			CREW	DAILY OUTPUT	LABOR-HOURS	UNIT	2000 BARE COSTS MAT.	LABOR	EQUIP.	TOTAL	TOTAL INCL O&P	
500	0010	**COAT RACKS & WARDROBES**										500
	0020	Floor model hat & coat racks, 6 hangers										
	0050	Standing, beech wood, 21" x 21" x 72", chrome				Ea.	98.50			98.50	108	
	0100	18 gauge tubular steel, 21" x 21" x 69", wood walnut				"	224			224	246	
	0500	16 gauge steel frame, 22 gauge steel shelves										
	0650	Single pedestal, 30" x 18" x 63"				Ea.	181			181	199	
	0800	Single face rack, 29" x 18-1/2" x 62"					263			263	289	
	0900	51" x 18-1/2" x 70"					224			224	246	
	0910	Double face rack, 39" x 26" x 70"					395			395	435	
	0920	63" x 26" x 70"				↓	440			440	480	
	0940	For 2" ball casters, add				Set	47.50			47.50	52.50	
	1400	Utility hook strips, 3/8" x 2-1/2" x 18", 6 hooks	1 Carp	48	.167	Ea.	29.50	3.28		32.78	38	
	1500	34" long, 12 hooks	"	48	.167	"	47	3.28		50.28	57	
	1650	Wall mounted racks, 16 gauge steel frame, 22 gauge steel shelves										
	1850	12" x 15" x 26", 6 hangers	1 Carp	32	.250	Ea.	113	4.93		117.93	132	
	2000	12" x 15" x 50", 12 hangers	"	32	.250	"	171	4.93		175.93	196	
	2150	Wardrobe cabinet, steel, baked enamel finish										
	2300	36" x 21" x 78", incl. top shelf & hanger rod				Ea.	375			375	410	

For information about Means Estimating Seminars, see yellow pages 11 and 12 in back of book

Division 11 Equipment

Estimating Tips
General
- The items in this division are usually priced per square foot or each. Many of these items are purchased by the owner for installation by the contractor. Check the specifications for responsibilities, and include time for receiving, storage, installation and mechanical and electrical hook-ups in the appropriate divisions.
- Many items in Division 11 require some type of support system that is not usually furnished with the item. Examples of these systems include blocking for the attachment of casework and support angles for ceiling hung projection screens. The required blocking or supports must be added to the estimate in the appropriate division.
- Some items in Division 11 may require assembly or electrical hook-ups. Verify the amount of assembly required or the need for a hard electrical connection and add the appropriate costs.

Reference Numbers
Reference numbers are shown in bold squares at the beginning of some major classifications. These numbers refer to related items in the Reference Section. The reference information may be an estimating procedure, an alternate pricing method or technical information.

Note: Not all subdivisions listed here necessarily appear in this publication.

11010 | Maintenance Equipment

11012 | Contractor Equipment

		CREW	DAILY OUTPUT	LABOR-HOURS	UNIT	2000 BARE COSTS MAT.	LABOR	EQUIP.	TOTAL	TOTAL INCL O&P	
100	0010	CONTRACTOR EQUIPMENT See division 01590 R01590-100										100

11013 | Floor/Wall Cleaning Equipment

			CREW	DAILY OUTPUT	LABOR-HOURS	UNIT	MAT.	LABOR	EQUIP.	TOTAL	TOTAL INCL O&P	
800	0010	VACUUM CLEANING										800
	0020	Central, 3 inlet, residential	1 Skwk	.90	8.889	Total	685	176		861	1,050	
	0200	Commercial		.70	11.429		790	226		1,016	1,250	
	0400	5 inlet system, residential		.50	16		840	315		1,155	1,450	
	0600	7 inlet system, commercial		.40	20		1,050	395		1,445	1,825	
	0800	9 inlet system, residential		.30	26.667		1,200	525		1,725	2,225	
	4010	Rule of thumb: First 1200 S.F., installed									1,044	
	4020	For each additional S.F., add				S.F.					.17	

11020 | Security & Vault Equipment

11021 | Safes

			CREW	DAILY OUTPUT	LABOR-HOURS	UNIT	MAT.	LABOR	EQUIP.	TOTAL	TOTAL INCL O&P	
600	0010	SAFE										600
	0015	Office, 4 hr. rating, 30" x 18" x 18" inside				Ea.	3,575			3,575	3,925	
	0100	60" x 36" x 18"					7,750			7,750	8,525	
	0200	1 hr. rating, 30" x 18" x 18"					2,000			2,000	2,200	
	0250	40" x 18" x 18"					2,250			2,250	2,475	
	0300	60" x 36" x 18", double door					5,750			5,750	6,325	
	0400	Data, 4 hr. rating, 23-1/2" x 19-1/2" x 17" inside					3,025			3,025	3,325	
	0450	52" x 19" x 17"					7,375			7,375	8,100	
	0500	63" x 34" x 16", double door					11,500			11,500	12,600	
	0550	52" x 34" x 16", inside					10,800			10,800	11,900	
	0600	1 hr. rating, 27" x 19" x 16"					6,025			6,025	6,625	
	0700	63" x 34" x 16"					10,200			10,200	11,200	
	0750	Diskette, 1 hr., 14" x 12" x 11", inside					2,825			2,825	3,125	
	0800	Money, "B" label, 9" x 14" x 14"					445			445	490	
	0900	Tool resistive, 24" x 24" x 20"					3,025			3,025	3,325	
	1050	Tool and torch resistive, 24" x 24" x 20"					7,675			7,675	8,450	
	1150	Jewelers, 23" x 20" x 18"					9,825			9,825	10,800	
	1200	63" x 25" x 18"					19,700			19,700	21,600	
	1300	For handling into building, add, minimum	A-2	8.50	2.824			42	20.50	62.50	94	
	1400	Maximum	"	.78	30.769			460	222	682	1,025	

11030 | Teller & Service Equipment

11038 | Bank Equipment

			CREW	DAILY OUTPUT	LABOR-HOURS	UNIT	MAT.	LABOR	EQUIP.	TOTAL	TOTAL INCL O&P	
150	0010	BANK EQUIPMENT										150
	0020	Alarm system, police	2 Elec	1.60	10	Ea.	3,600	221		3,821	4,300	
	0100	With vault alarm	"	.40	40		14,300	885		15,185	17,200	
	0400	Bullet resistant teller window, 44" x 60"	1 Glaz	.60	13.333		2,725	258		2,983	3,425	

11030 | Teller & Service Equipment

11038 | Bank Equipment

			CREW	DAILY OUTPUT	LABOR-HOURS	UNIT	MAT.	LABOR	EQUIP.	TOTAL	TOTAL INCL O&P
150	0500	48" x 60"	1 Glaz	.60	13.333	Ea.	3,975	258		4,233	4,800
	3000	Counters for banks, frontal only	2 Carp	1	16	Station	1,350	315		1,665	2,025
	3100	Complete with steel undercounter	"	.50	32	"	2,625	630		3,255	3,950
	4600	Door and frame, bullet-resistant, with vision panel, minimum	2 Sswk	1.10	14.545	Ea.	2,200	310		2,510	3,025
	4700	Maximum		1.10	14.545		4,200	310		4,510	5,225
	4800	Drive-up window, drawer & mike, not incl. glass, minimum		1	16		4,200	340		4,540	5,300
	4900	Maximum		.50	32		6,925	680		7,605	8,975
	5000	Night depository, with chest, minimum		1	16		6,175	340		6,515	7,475
	5100	Maximum		.50	32		9,775	680		10,455	12,200
	5200	Package receiver, painted		3.20	5		1,000	106		1,106	1,300
	5300	Stainless steel		3.20	5		1,700	106		1,806	2,075
	5400	Partitions, bullet-resistant, 1-3/16" glass, 8' high	2 Carp	10	1.600	L.F.	145	31.50		176.50	214
	5450	Acrylic	"	10	1.600	"	275	31.50		306.50	360
	5500	Pneumatic tube systems, 2 lane drive-up, complete	L-3	.25	40	Total	18,900	805		19,705	22,200
	5550	With T.V. viewer	"	.20	50	"	36,600	1,000		37,600	42,000
	5570	Safety deposit boxes, minimum	1 Sswk	44	.182	Opng.	42	3.86		45.86	53.50
	5580	Maximum, 10" x 15" opening		19	.421		89	8.95		97.95	116
	5590	Teller locker, average		15	.533		1,150	11.35		1,161.35	1,300
	5600	Pass thru, bullet-resist. window, painted steel, 24" x 36"	2 Sswk	1.60	10	Ea.	1,400	213		1,613	1,975
	5700	48" x 48"		1.20	13.333		1,950	283		2,233	2,700
	5800	72" x 40"		.80	20		2,350	425		2,775	3,425
	5900	For stainless steel frames, add					20%				
	6000	Surveillance system, 16 mm film camera, complete	2 Elec	1	16	Ea.	3,750	355		4,105	4,700
	6010	For each additional camera, add					1,575			1,575	1,725
	6100	Surveillance system, video camera, complete	2 Elec	1	16	Ea.	11,800	355		12,155	13,600
	6110	For each additional camera, add					725			725	800
	6200	Twenty-four hour teller, single unit,									
	6300	automated deposit, cash and memo	L-3	.25	40	Ea.	35,600	805		36,405	40,600
	7000	Vault front, see division 08320-950									

11040 | Ecclesiastical Equipment

11041 | Ecclesiastical Equipment

			CREW	DAILY OUTPUT	LABOR-HOURS	UNIT	MAT.	LABOR	EQUIP.	TOTAL	TOTAL INCL O&P
250	0010	**CHURCH EQUIPMENT**									
	0020	Altar, wood, custom design, plain	1 Carp	1.40	5.714	Ea.	1,575	113		1,688	1,925
	0050	Deluxe	"	.20	40		7,600	790		8,390	9,700
	0071	Granite or marble, average	2 Ston	.50	32		6,100	625		6,725	7,750
	0091	Deluxe	"	.20	80		20,000	1,550		21,550	24,700
	0100	Arks, prefabricated, plain	2 Carp	.80	20		6,000	395		6,395	7,275
	0130	Deluxe, maximum	"	.20	80		75,000	1,575		76,575	85,000
	0150	Baptistry, fiberglass, 3'-6" deep, x 13'-7" long,									
	0160	steps at both ends, incl. plumbing, minimum	L-8	1	20	Ea.	2,300	365		2,665	3,150
	0200	Maximum	"	.70	28.571		4,500	520		5,020	5,825
	0250	Add for filter, heater and lights					985			985	1,075
	0300	Carillon, 4 octave (48 bells), with keyboard				System	500,000			500,000	550,000
	0320	2 octave (24 bells)					175,000			175,000	192,500
	0340	3 to 4 bell peal, minimum					45,000			45,000	49,500
	0360	Maximum					250,000			250,000	275,000
	0380	Cast bronze bell, average				Ea.	35,000			35,000	38,500

11040 | Ecclesiastical Equipment

11041 | Ecclesiastical Equipment

		CREW	DAILY OUTPUT	LABOR-HOURS	UNIT	MAT.	LABOR	EQUIP.	TOTAL	TOTAL INCL O&P
0400	Electronic, digital, minimum				Ea.	11,000			11,000	12,100
0410	With keyboard, maximum					25,000			25,000	27,500
0500	Reconciliation room, wood, prefabricated, single, plain	1 Carp	.60	13.333		2,000	263		2,263	2,650
0550	Deluxe		.40	20		5,500	395		5,895	6,725
0650	Double, plain		.40	20		4,000	395		4,395	5,075
0700	Deluxe		.20	40		12,500	790		13,290	15,200
1000	Lecterns, wood, plain		5	1.600		450	31.50		481.50	550
1100	Deluxe		2	4		2,250	79		2,329	2,600
1500	Pews, bench type, hardwood, minimum		20	.400	L.F.	55	7.90		62.90	74
1550	Maximum		15	.533		110	10.50		120.50	139
1570	For kneeler, add					12.50			12.50	13.75
2000	Pulpits, hardwood, prefabricated, plain	1 Carp	2	4	Ea.	1,000	79		1,079	1,225
2100	Deluxe		1.60	5	"	6,600	98.50		6,698.50	7,425
2500	Railing, hardwood, average		25	.320	L.F.	125	6.30		131.30	149
2700	Safes, see division 11021-600									
3000	Seating, individual, oak, contour, laminated	1 Carp	21	.381	Person	110	7.50		117.50	134
3100	Cushion seat		21	.381		100	7.50		107.50	123
3200	Fully upholstered		21	.381		90	7.50		97.50	112
3300	Combination, self-rising		21	.381		275	7.50		282.50	320
3500	For cherry, add					30%				
4000	Steeples, translucent fiberglass, 30" square, 15' high	F-3	2	20	Ea.	3,000	360	227	3,587	4,150
4150	25' high		1.80	22.222		3,500	400	252	4,152	4,800
4350	Opaque fiberglass, 24" square, 14' high		2	20		2,500	360	227	3,087	3,600
4500	28' high		1.80	22.222		2,975	400	252	3,627	4,225
4600	Aluminum, baked finish, 14' high, 16" square					1,550			1,550	1,700
4620	20' high, 3'-6" base					4,125			4,125	4,525
4640	35' high, 8' base					15,600			15,600	17,100
4660	60' high, 14' base					35,100			35,100	38,600
4680	152' high, custom					345,500			345,500	380,000
4700	Porcelain enamel steeples, custom, 40' high	F-3	.50	80		10,400	1,425	905	12,730	14,900
4800	60' high	"	.30	133		18,000	2,400	1,500	21,900	25,600
5000	Wall cross, aluminum, extruded, 2" x 2" section	1 Carp	34	.235	L.F.	32	4.64		36.64	43
5150	4" x 4" section		29	.276		46	5.45		51.45	60
5300	Bronze, extruded, 1" x 2" section		31	.258		63	5.10		68.10	78
5350	2-1/2" x 2-1/2" section		34	.235		95.50	4.64		100.14	113
5450	Solid bar stock, 1/2" x 3" section		29	.276		125	5.45		130.45	147
5600	Fiberglass, stock		34	.235		26	4.64		30.64	36.50
5700	Stainless steel, 4" deep, channel section		29	.276		101	5.45		106.45	120
5800	4" deep box section		29	.276		125	5.45		130.45	147

11050 | Library Equipment

11051 | Library Equipment

		CREW	DAILY OUTPUT	LABOR-HOURS	UNIT	MAT.	LABOR	EQUIP.	TOTAL	TOTAL INCL O&P
0010	**LIBRARY EQUIPMENT**									
0020	Bookshelf, mtl, 90" high, 10" shelf, dbl face	1 Carp	11.50	.696	L.F.	89.50	13.70		103.20	122
0300	Single face	"	12	.667	"	65	13.15		78.15	94
0600	For 8" shelving, subtract from above					10%				
0700	For 12" shelving, add to above					10%				
0800	For 42" high with countertop, subtract from above					20%				

11050 | Library Equipment

11051 | Library Equipment

			CREW	DAILY OUTPUT	LABOR-HOURS	UNIT	MAT.	LABOR	EQUIP.	TOTAL	TOTAL INCL O&P
400	1100	Magazine shelving, 90" high, 12" deep, single face	1 Carp	11.50	.696	L.F.	83	13.70		96.70	115
	1400	Double face		11.50	.696	"	161	13.70		174.70	201
	2500	Carrels, hardwood, 36" x 24", minimum		5	1.600	Ea.	560	31.50		591.50	675
	2650	Maximum		4	2		735	39.50		774.50	880
	2700	Card catalog file, 60 trays, complete					3,775			3,775	4,150
	2720	Alternate method: each tray					90			90	99
	3100	Chairs, wood					115			115	127
	3500	Charging desk, built-in, with counter, plastic laminated top	1 Carp	7	1.143	L.F.	395	22.50		417.50	475
	3700	Reading table, laminated top, 60" x 36"				Ea.	600			600	660
	3800	Mobile compacted shelving, hand crank, 9'-0" high									
	3820	Double face, including track, 3' section				Ea.	820			820	900
	3840	For electrical operation, add					25%				

11060 | Theater & Stage Equipment

11063 | Stage Equipment

			CREW	DAILY OUTPUT	LABOR-HOURS	UNIT	MAT.	LABOR	EQUIP.	TOTAL	TOTAL INCL O&P
600	0010	**STAGE EQUIPMENT**									
	0050	Control boards with dimmers and breakers, minimum	1 Elec	1	8	Ea.	3,200	177		3,377	3,825
	0100	Average		.50	16		8,525	355		8,880	9,950
	0150	Maximum		.20	40		39,000	885		39,885	44,400
	0500	Curtain track, straight, light duty	2 Carp	20	.800	L.F.	18.35	15.75		34.10	47
	0600	Heavy duty		18	.889		37.50	17.50		55	71.50
	0700	Curved sections		12	1.333		110	26.50		136.50	166
	1000	Curtains, velour, medium weight		600	.027	S.F.	6	.53		6.53	7.50
	1150	Silica based yarn, fireproof		50	.320		12	6.30		18.30	24
	1500	Flooring, portable oak parquet, 3' x 3' sections					10.05			10.05	11.05
	1600	Cart to carry 225 S.F. of flooring				Ea.	305			305	335
	2000	Lights, border, quartz, reflector, vented,									
	2100	colored or white	1 Elec	20	.400	L.F.	138	8.85		146.85	166
	2500	Spotlight, follow spot, with transformer, 2,100 watt	"	4	2	Ea.	1,325	44		1,369	1,525
	2600	For no transformer, deduct					475			475	525
	3000	Stationary spot, fresnel quartz, 6" lens	1 Elec	4	2		115	44		159	200
	3100	8" lens		4	2		181	44		225	272
	3500	Ellipsoidal quartz, 1,000W, 6" lens		4	2		239	44		283	335
	3600	12" lens		4	2		420	44		464	535
	4000	Strobe light, 1 to 15 flashes per second, quartz		3	2.667		540	59		599	690
	4500	Color wheel, portable, five hole, motorized		4	2		115	44		159	200
	5000	Stages, portable with steps, folding legs, stock, 8" high				SF Stg.	7.50			7.50	8.25
	5100	16" high					37			37	40.50
	5200	32" high					29			29	32
	5300	40" high					30			30	33
	6000	Telescoping platforms, extruded alum., straight, minimum	4 Carp	157	.204		28	4.02		32.02	37.50
	6100	Maximum		77	.416		43	8.20		51.20	61.50
	6500	Pie-shaped, minimum		150	.213		31	4.20		35.20	41
	6600	Maximum		70	.457		68.50	9		77.50	91
	6800	For 3/4" plywood covered deck, deduct					2.73			2.73	3
	7000	Band risers, steel frame, plywood deck, minimum	4 Carp	275	.116		22	2.29		24.29	28
	7100	Maximum	"	138	.232		41.50	4.57		46.07	53.50
	7500	Chairs for above, self-storing, minimum	2 Carp	43	.372	Ea.	80	7.35		87.35	101
	7600	Maximum	"	40	.400	"	113	7.90		120.90	138

11060 | Theater & Stage Equipment

11063 | Stage Equipment

		CREW	DAILY OUTPUT	LABOR-HOURS	UNIT	2000 BARE COSTS				TOTAL INCL O&P	
						MAT.	LABOR	EQUIP.	TOTAL		
600	8000 Rule of thumb: total stage equipment, minimum	4 Carp	100	.320	SF Stg.	72.50	6.30		78.80	90.50	600
	8100 Maximum	"	25	1.280	"	400	25		425	485	

11100 | Mercantile Equipment

11102 | Barber Shop Equipment

		CREW	DAILY OUTPUT	LABOR-HOURS	UNIT	MAT.	LABOR	EQUIP.	TOTAL	INCL O&P	
150	0010 **BARBER EQUIPMENT**										150
	0020 Chair, hydraulic, movable, minimum	1 Carp	24	.333	Ea.	410	6.55		416.55	460	
	0050 Maximum	"	16	.500		2,700	9.85		2,709.85	3,000	
	0200 Wall hung styling station with mirrors, minimum	L-2	8	2		240	34.50		274.50	320	
	0300 Maximum	"	4	4		1,500	69		1,569	1,775	
	0500 Sink, hair washing basin, rough plumbing not incl.	1 Plum	8	1		273	22		295	335	
	1000 Sterilizer, liquid solution for tools					125			125	138	
	1100 Total equipment, rule of thumb, per chair, minimum	L-8	1	20		1,475	365		1,840	2,250	
	1150 Maximum	"	1	20		4,000	365		4,365	5,025	

11103 | Cash Register/Checking

		CREW	DAILY OUTPUT	LABOR-HOURS	UNIT	MAT.	LABOR	EQUIP.	TOTAL	INCL O&P	
200	0010 **CHECKOUT COUNTER**										200
	0020 Supermarket conveyor, single belt	2 Clab	10	1.600	Ea.	1,925	23		1,948	2,175	
	0100 Double belt, power take-away		9	1.778		3,300	25.50		3,325.50	3,675	
	0400 Double belt, power take-away, incl. side scanning		7	2.286		4,100	33		4,133	4,550	
	0800 Warehouse or bulk type		6	2.667		4,800	38.50		4,838.50	5,350	
	1000 Scanning system, 2 lanes, w/registers, scan gun & memory				System	12,000			12,000	13,200	
	1100 10 lanes, single processor, full scan, with scales				"	114,000			114,000	125,500	
	2000 Register, restaurant, minimum				Ea.	500			500	550	
	2100 Maximum					2,200			2,200	2,425	
	2150 Store, minimum					500			500	550	
	2200 Maximum					2,200			2,200	2,425	

11104 | Display Cases & Systems

		CREW	DAILY OUTPUT	LABOR-HOURS	UNIT	MAT.	LABOR	EQUIP.	TOTAL	INCL O&P	
700	0010 **REFRIGERATED FOOD CASES**										700
	0030 Dairy, multi-deck, 12' long	Q-5	3	5.333	Ea.	6,800	106		6,906	7,650	
	0100 For rear sliding doors, add					980			980	1,075	
	0200 Delicatessen case, service deli, 12' long, single deck	Q-5	3.90	4.103		4,575	81.50		4,656.50	5,150	
	0300 Multi-deck, 18 S.F. shelf display		3	5.333		5,600	106		5,706	6,325	
	0400 Freezer, self-contained, chest-type, 30 C.F.		3.90	4.103		3,325	81.50		3,406.50	3,800	
	0500 Glass door, upright, 78 C.F.		3.30	4.848		6,350	96.50		6,446.50	7,150	
	0600 Frozen food, chest type, 12' long		3.30	4.848		4,600	96.50		4,696.50	5,200	
	0700 Glass door, reach-in, 5 door		3	5.333		8,800	106		8,906	9,850	
	0800 Island case, 12' long, single deck		3.30	4.848		5,200	96.50		5,296.50	5,875	
	0900 Multi-deck		3	5.333		11,000	106		11,106	12,300	
	1000 Meat case, 12' long, single deck		3.30	4.848		3,775	96.50		3,871.50	4,300	
	1050 Multi-deck		3.10	5.161		6,500	103		6,603	7,325	
	1100 Produce, 12' long, single deck		3.30	4.848		5,000	96.50		5,096.50	5,650	
	1200 Multi-deck		3.10	5.161		5,575	103		5,678	6,300	

11110 | Commercial Laundry & Dry Cleaning Equipment

11119 | Laundry Cleaning

			CREW	DAILY OUTPUT	LABOR-HOURS	UNIT	MAT.	LABOR	EQUIP.	TOTAL	TOTAL INCL O&P	
450	0010	**LAUNDRY EQUIPMENT** Not incl. rough-in										450
	0500	Dryers, gas fired residential, 16 lb. capacity, average	1 Plum	3	2.667	Ea.	505	58.50		563.50	655	
	1000	Commercial, 30 lb. capacity, coin operated, single		3	2.667		2,325	58.50		2,383.50	2,650	
	1100	Double stacked		2	4		4,975	88		5,063	5,625	
	1500	Industrial, 30 lb. capacity		2	4		1,975	88		2,063	2,325	
	1600	50 lb. capacity		1.70	4.706		2,450	103		2,553	2,875	
	2000	Dry cleaners, electric, 20 lb. capacity	L-1	.20	50		27,900	1,100		29,000	32,500	
	2050	25 lb. capacity		.17	58.824		36,300	1,300		37,600	42,100	
	2100	30 lb. capacity		.15	66.667		38,300	1,475		39,775	44,500	
	2150	60 lb. capacity		.09	111		60,000	2,450		62,450	70,000	
	3500	Folders, blankets & sheets, minimum	1 Elec	.17	47.059		22,800	1,050		23,850	26,700	
	3700	King size with automatic stacker		.10	80		43,500	1,775		45,275	50,500	
	3800	For conveyor delivery, add		.45	17.778		5,175	395		5,570	6,350	
	4500	Ironers, institutional, 110", single roll		.20	40		23,800	885		24,685	27,700	
	4700	Lint collector, ductwork not included, 8,000 to 10,000 C.F.M.	Q-10	.30	80		6,825	1,600		8,425	10,200	
	5000	Washers, residential, 4 cycle, average	1 Plum	3	2.667		590	58.50		648.50	740	
	5300	Commercial, coin operated, average	"	3	2.667		945	58.50		1,003.50	1,150	
	6000	Combination washer/extractor, 20 lb. capacity	L-6	1.50	8		3,375	176		3,551	4,000	
	6100	30 lb. capacity		.80	15		7,000	330		7,330	8,250	
	6200	50 lb. capacity		.68	17.647		7,975	390		8,365	9,425	
	6300	75 lb. capacity		.30	40		15,700	880		16,580	18,700	
	6350	125 lb. capacity		.16	75		20,000	1,650		21,650	24,700	

11130 | Audio-Visual Equipment

11136 | Projection Screens

			CREW	DAILY OUTPUT	LABOR-HOURS	UNIT	MAT.	LABOR	EQUIP.	TOTAL	TOTAL INCL O&P	
500	0010	**PROJECTION SCREENS** Wall or ceiling hung, matte white										500
	0100	Manually operated, economy	2 Carp	500	.032	S.F.	4.41	.63		5.04	5.95	
	0300	Intermediate		450	.036		5.15	.70		5.85	6.85	
	0400	Deluxe		400	.040		7.15	.79		7.94	9.20	
	0600	Electric operated, matte white, 25 S.F., economy		5	3.200	Ea.	645	63		708	820	
	0700	Deluxe		4	4		1,450	79		1,529	1,700	
	0900	50 S.F., economy		3	5.333		325	105		430	535	
	1000	Deluxe		2	8		1,500	158		1,658	1,925	
	1200	Heavy duty, electric operated, 200 S.F.		1.50	10.667		2,950	210		3,160	3,600	
	1300	400 S.F.		1	16		3,650	315		3,965	4,550	
	1500	Rigid acrylic in wall, for rear projection, 1/4" thick	2 Glaz	30	.533	S.F.	35.50	10.30		45.80	56.50	
	1600	1/2" thick (maximum size 10' x 20')	"	25	.640	"	63	12.40		75.40	90	
600	0010	**MOVIE EQUIPMENT**										600
	0020	Changeover, minimum				Ea.	350			350	385	
	0100	Maximum					685			685	750	
	0400	Film transport, incl. platters and autowind, minimum					5,000			5,000	5,500	
	0500	Maximum					10,800			10,800	11,900	
	0800	Lamphouses, incl. rectifiers, xenon, 1,000 watt	1 Elec	2	4		5,875	88.50		5,963.50	6,625	
	0900	1,600 watt		2	4		6,625	88.50		6,713.50	7,425	
	1000	2,000 watt		1.50	5.333		6,825	118		6,943	7,725	
	1100	4,000 watt		1.50	5.333		9,675	118		9,793	10,800	
	1400	Lenses, anamorphic, minimum					895			895	980	
	1500	Maximum					2,100			2,100	2,300	
	1800	Flat 35 mm, minimum					800			800	880	

11130 | Audio-Visual Equipment

11136 | Projection Screens

			CREW	DAILY OUTPUT	LABOR-HOURS	UNIT	2000 BARE COSTS MAT.	LABOR	EQUIP.	TOTAL	TOTAL INCL O&P	
600	1900	Maximum				Ea.	1,125			1,125	1,250	600
	2200	Pedestals, for projectors					560			560	615	
	2300	Console type					10,000			10,000	11,000	
	2600	Projector mechanisms, incl. soundhead, 35 mm, minimum					9,925			9,925	10,900	
	2700	Maximum					13,200			13,200	14,500	
	3000	Projection screens, rigid, in wall, acrylic, 1/4" thick	2 Glaz	195	.082	S.F.	31.50	1.59		33.09	37.50	
	3100	1/2" thick	"	130	.123	"	36.50	2.38		38.88	44.50	
	3300	Electric operated, heavy duty, 400 S.F.	2 Carp	1	16	Ea.	2,225	315		2,540	3,000	
	3320	Theatre projection screens, matte white, including frames	"	200	.080	S.F.	5	1.58		6.58	8.20	
	3400	Also see division 11136-500										
	3700	Sound systems, incl. amplifier, mono, minimum	1 Elec	.90	8.889	Ea.	2,500	196		2,696	3,075	
	3800	Dolby/Super Sound, maximum		.40	20		13,700	440		14,140	15,700	
	4100	Dual system, 2 channel, front surround, minimum		.70	11.429		3,500	253		3,753	4,275	
	4200	Dolby/Super Sound, 4 channel, maximum		.40	20		12,500	440		12,940	14,500	
	4500	Sound heads, 35 mm					4,000			4,000	4,400	
	4900	Splicer, wet type, minimum					560			560	615	
	5000	Tape type, maximum					1,000			1,000	1,100	
	5300	Speakers, recessed behind screen, minimum	1 Elec	2	4		1,500	88.50		1,588.50	1,800	
	5400	Maximum	"	1	8		2,325	177		2,502	2,875	
	5700	Seating, painted steel, upholstered, minimum	2 Carp	35	.457		100	9		109	125	
	5800	Maximum	"	28	.571		237	11.25		248.25	279	
	6100	Rewind tables, minimum					2,000			2,000	2,200	
	6200	Maximum					3,550			3,550	3,900	
	7000	For automation, varying sophistication, minimum	1 Elec	1	8	System	1,800	177		1,977	2,275	
	7100	Maximum	2 Elec	.30	53.333	"	4,200	1,175		5,375	6,550	

11140 | Vehicle Service Equipment

11141 | Service Station Equipment

			CREW	DAILY OUTPUT	LABOR-HOURS	UNIT	2000 BARE COSTS MAT.	LABOR	EQUIP.	TOTAL	TOTAL INCL O&P	
150	0010	**AUTOMOTIVE**										150
	0030	Compressors, electric, 1-1/2 H.P., standard controls	L-4	1.50	10.667	Ea.	320	184		504	670	
	0550	Dual controls		1.50	10.667		420	184		604	780	
	0600	5 H.P., 115/230 volt, standard controls		1	16		1,525	276		1,801	2,175	
	0650	Dual controls		1	16		1,625	276		1,901	2,275	
	1100	Product dispenser with vapor recovery for 6 nozzles, installed, not										
	1110	including piping to storage tanks				Ea.	15,000			15,000	16,500	
	2200	Hoists, single post, 8,000# capacity, swivel arms	L-4	.40	40		3,500	690		4,190	5,025	
	2400	Two posts, adjustable frames, 11,000# capacity		.25	64		4,500	1,100		5,600	6,850	
	2500	24,000# capacity		.15	106		6,000	1,850		7,850	9,750	
	2700	7,500# capacity, frame supports		.50	32		5,000	555		5,555	6,450	
	2800	Four post, roll on ramp		.50	32		4,500	555		5,055	5,900	
	2810	Hydraulic lifts, above ground, 2 post, clear floor, 6000 lb cap		2.67	5.993		4,700	103		4,803	5,350	
	2815	9000 lb capacity		2.29	6.987		11,200	121		11,321	12,500	
	2820	15,000 lb capacity		2	8		25,900	138		26,038	28,700	
	2825	30,000 lb capacity		1.60	10		27,900	173		28,073	30,900	
	2830	4 post, ramp style, 25,000 lb capacity		2	8		10,900	138		11,038	12,200	
	2835	35,000 lb capacity		1	16		51,000	276		51,276	56,500	
	2840	50,000 lb capacity		1	16		57,000	276		57,276	63,000	
	2845	75,000 lb capacity		1	16		66,000	276		66,276	73,000	

11140 | Vehicle Service Equipment

11141 | Service Station Equipment

		CREW	DAILY OUTPUT	LABOR-HOURS	UNIT	2000 BARE COSTS MAT.	LABOR	EQUIP.	TOTAL	TOTAL INCL O&P	
150	2850 For drive thru tracks, add, minimum				Ea.	700			700	770	150
	2855 Maximum					1,200			1,200	1,325	
	2860 Ramp extensions, 3' (set of 2)					575			575	635	
	2865 Rolling jack platform					2,000			2,000	2,200	
	2870 Elec/hyd jacking beam					5,350			5,350	5,875	
	2880 Scissor lift, portable, 6000 lb capacity					5,250			5,250	5,775	
	3000 Lube equipment, 3 reel type, with pumps, not including piping	L-4	.50	32	Set	6,000	555		6,555	7,550	
	4000 Spray painting booth, 26' long, complete	"	.40	40	Ea.	12,000	690		12,690	14,400	

11150 | Parking Control Equipment

11156 | Parking Equipment

		CREW	DAILY OUTPUT	LABOR-HOURS	UNIT	2000 BARE COSTS MAT.	LABOR	EQUIP.	TOTAL	TOTAL INCL O&P	
600	0010 **PARKING EQUIPMENT**										600
	5000 Barrier gate with programmable controller	2 Elec	3	5.333	Ea.	2,550	118		2,668	3,000	
	5020 Industrial	"	3	5.333		3,575	118		3,693	4,125	
	5100 Card reader	1 Elec	2	4		1,775	88.50		1,863.50	2,125	
	5120 Proximity with customer display	2 Elec	1	16		5,100	355		5,455	6,175	
	5200 Cashier booth, average	B-22	1	30		9,175	485	207	9,867	11,200	
	5300 Collector station, pay on foot	2 Elec	.20	80		112,000	1,775		113,775	126,500	
	5320 Credit card only		.50	32		45,900	705		46,605	51,500	
	5500 Exit verifier		1	16		22,500	355		22,855	25,400	
	5600 Fee computer	1 Elec	1.50	5.333		13,500	118		13,618	15,100	
	5700 Full sign, 4" letters	"	2	4		1,225	88.50		1,313.50	1,500	
	5800 Inductive loop	2 Elec	4	4		510	88.50		598.50	705	
	5900 Ticket spitter with time/date stamp, standard		2	8		6,000	177		6,177	6,900	
	5920 Mag stripe encoding		2	8		16,300	177		16,477	18,300	
	5950 Vehicle detector, microprocessor based	1 Elec	3	2.667		410	59		469	545	
	6000 Parking control software, minimum		.50	16		18,900	355		19,255	21,400	
	6020 Maximum		.20	40		79,000	885		79,885	88,000	

11160 | Loading Dock Equipment

11161 | Loading Dock Equipment

		CREW	DAILY OUTPUT	LABOR-HOURS	UNIT	2000 BARE COSTS MAT.	LABOR	EQUIP.	TOTAL	TOTAL INCL O&P	
200	0011 **DOCK BUMPERS** Bolts not included, 2" x 6" to 4" x 8", average	2 Carp	.60	26.667	M.B.F.	480	525		1,005	1,425	200
400	0010 **LOADING DOCK**										400
	0020 Bumpers, rubber blocks 4-1/2" thk, 10" H, 14" long	1 Carp	26	.308	Ea.	38.50	6.05		44.55	52.50	
	0200 24" long		22	.364		60.50	7.15		67.65	79.50	
	0300 36" long		17	.471		70.50	9.25		79.75	93.50	
	0500 12" high, 14" long		25	.320		64	6.30		70.30	81.50	
	0550 24" long		20	.400		71	7.90		78.90	91.50	
	0600 36" long		15	.533		80	10.50		90.50	106	
	0800 Rubber blocks 6" thick, 10" high, 14" long		22	.364		59.50	7.15		66.65	78	

11160 | Loading Dock Equipment

11161 | Loading Dock Equipment

		CREW	DAILY OUTPUT	LABOR-HOURS	UNIT	2000 BARE COSTS MAT.	LABOR	EQUIP.	TOTAL	TOTAL INCL O&P
400	0850	24" long	1 Carp	18	.444	Ea.	77.50	8.75		86.25	101
	0900	36" long		13	.615		97	12.10		109.10	128
	0910	20" high, 11" long		13	.615		99	12.10		111.10	130
	0920	Extruded rubber bumpers, T section, 22" x 22" x 3" thick		41	.195		41	3.84		44.84	52
	0940	Molded rubber bumpers, 24" x 12" x 3" thick	▼	20	.400		39	7.90		46.90	56.50
	1000	Welded installation of above bumpers	E-14	8	1		2.25	21.50	10.50	34.25	56
	1100	For drilled anchors, add per anchor	1 Carp	36	.222	▼	4.05	4.38		8.43	11.95
	1300	Steel bumpers, see division 10265 & 08770									
	1350	Wood bumpers, see division 11161-200									
	2200	Dock boards, heavy duty, 60" x 60", aluminum, 5,000 lb. cap.				Ea.	1,025			1,025	1,125
	2700	9,000 lb. capacity					1,125			1,125	1,250
	3200	15,000 lb. capacity				▼	1,250			1,250	1,375
	3600	Door seal for door perimeter, 12" x 12", vinyl covered	1 Carp	26	.308	L.F.	16.95	6.05		23	29
	3900	Folding gates, see division 10610-100									
	4200	Platform lifter, 6' x 6', portable, 3,000 lb. capacity				Ea.	5,850			5,850	6,425
	4250	4,000 lb. capacity					8,200			8,200	9,025
	4400	Fixed, 6' x 8', 5,000 lb. capacity	E-16	.70	22.857		7,075	510	120	7,705	8,925
	4500	Levelers, hinged for trucks, 10 ton capacity, 6' x 8'		1.08	14.815		3,025	330	78	3,433	4,075
	4650	7' x 8'		1.08	14.815		3,150	330	78	3,558	4,200
	4670	Air bag power operated, 10 ton cap., 6'x8'		1.08	14.815		3,475	330	78	3,883	4,550
	4680	7' x 8'		1.08	14.815		3,700	330	78	4,108	4,800
	4700	Hydraulic, 10 ton capacity, 6' x 8'		1.08	14.815		5,550	330	78	5,958	6,825
	4800	7' x 8'	▼	1.08	14.815		5,950	330	78	6,358	7,275
	5000	Lights for loading docks, single arm, 24" long	1 Elec	3.80	2.105		110	46.50		156.50	197
	5700	Double arm, 60" long	"	3.80	2.105		135	46.50		181.50	224
	5800	Loading dock safety restraints, manual style	E-16	1.08	14.815		2,025	330	78	2,433	2,950
	5900	Automatic style	"	1.08	14.815		3,600	330	78	4,008	4,675
	6200	Shelters, fabric, for truck or train, scissor arms, minimum	1 Carp	1	8		980	158		1,138	1,350
	6300	Maximum	"	.50	16	▼	1,475	315		1,790	2,175

11170 | Solid Waste Handling Equipment

11179 | Waste Handling Equipment

		CREW	DAILY OUTPUT	LABOR-HOURS	UNIT	2000 BARE COSTS MAT.	LABOR	EQUIP.	TOTAL	TOTAL INCL O&P
150	0010	**WASTE HANDLING**									
	0020	Compactors, 115 volt, 250#/hr., chute fed	L-4	1	16	Ea.	8,225	276		8,501	9,525
	0100	Hand fed		2.40	6.667		5,675	115		5,790	6,450
	0300	Multi-bag, 230 volt, 600#/hr, chute fed		1	16		7,025	276		7,301	8,200
	0400	Hand fed		1	16		6,375	276		6,651	7,500
	0500	Containerized, hand fed, 2 to 6 C.Y. containers, 250#/hr.		1	16		7,475	276		7,751	8,700
	0550	For chute fed, add per floor		1	16		860	276		1,136	1,425
	1000	Heavy duty industrial compactor, 0.5 C.Y. capacity		1	16		5,050	276		5,326	6,025
	1050	1.0 C.Y. capacity		1	16		7,525	276		7,801	8,750
	1100	3 C.Y. capacity		.50	32		10,500	555		11,055	12,600
	1150	5.0 C.Y. capacity		.50	32		14,000	555		14,555	16,400
	1200	Combination shredder/compactor (5,000 lbs./hr.)	▼	.50	32		25,200	555		25,755	28,700
	1400	For handling hazardous waste materials, 55 gallon drum packer, std.					11,700			11,700	12,900
	1410	55 gallon drum packer w/HEPA filter					14,900			14,900	16,400
	1420	55 gallon drum packer w/charcoal & HEPA filter					19,000			19,000	20,900
	1430	All of the above made explosion proof, add				▼	9,075			9,075	9,975

BONUS OFFER INCLUDED on CD

FREE: The Means CostWorks Dictionary. Contains all of the approximately 17,000 construction terms found in the Means Illustrated Construction Dictionary. A $99.95 value.

Computer Specifications for Means CostWorks 2000:

Minimum Requirements: Windows 95, Windows 98 or Windows NT; 8 megabytes RAM; 640 x 480 Monitor, 16 Colors, 486 - 33 Mhz Computer; 2X CD-ROM

Also INCLUDED on CD

Architects' First Source®, featuring thousands of products listed by manufacturer!

The Construction Specifications Institute's SPEC-DATA®, for technical product information!

CSI's MANU-SPEC®, for manufacturers' complete product specifications!

Web-Enabled, launch to the Web to get updated construction data instantly!

Ordering Information

Try Means CostWorks FREE at Absolutely No Risk, With Our 30-Day Guarantee! Order By Check or Credit Card! Call 1-800-334-3509.

The Means CostWorks CD-ROM contains cost information from 18 digital construction cost data books. When you place your order, we'll send you one Means CostWorks CD-ROM and a special access code for the specific digital cost book(s) that you selected. If you want to order additional cost titles later, we'll give you the appropriate access codes designed to work with the CD-ROM you already have.

Your annual Means CostWorks license includes four Free quarterly data updates, unlimited access to Means CostWorks Homeport website, full use of the R.S. Means cost data that you purchase, and access to Means CostWorks navigation utilities.

Digital Cost Book	Regular Price	Cost
Assemblies Cost Data	$149.95	
Building Construction Cost Data	$ 99.95	
Concrete & Masonry Cost Data	$ 99.95	
Electrical Cost Data	$ 99.95	
Heavy Construction Cost Data	$ 99.95	
Interior Cost Data	$ 99.95	
Light Commercial Cost Data	$129.95	
Mechanical Cost Data	$ 99.95	
Open Shop Building Construction Cost Data	$ 99.95	
Plumbing Cost Data	$ 99.95	
Site Work & Landscape Cost Data	$ 99.95	
Wage Rate Utility *(Available for above titles only)*	$ 99.95	
Facilities Construction Cost Data	$249.95	
Repair & Remodeling Cost Data	$ 99.95	
Building Construction Cost Data/Metric	$ 99.95	
Heavy Construction Cost Data/Metric	$ 99.95	
Commercial Square Foot Model Costs	$149.95	
Residential Cost Data	$129.95	
Facilities Maintenance & Repair Cost Data	$249.95	
Total Cost	$	
Shipping and handling:	$	5.00
MA residents please add 5% sales tax:	$	
Grand Total Cost:	$	

[ ] Check enclosed
[ ] Please charge my credit card
    [ ] VISA [ ] MasterCard [ ] American Express [ ] Discover
    Card # _____
 Expiration date _____
 Cardholder signature _____

Name _____
Company _____
Street _____
City _____ State _____ Zip _____
Phone(___)_____ Fax(___)_____
E-mail _____

Complete this form and mail today or fax toll free 1-800-632-6732 CWCW-1000

Visit the R.S. Means CostWorks Website at www.rsmeans.com

RSMeans, Construction Plaza, 63 Smiths Lane, P.O. Box 800, Kingston, MA 02364-0800

Means CostWorks 2000

- **Search and Organize Cost Data**
- **Create and Save Project-Specific Documents**
- **Localize Costs to Your Area**
- **Calculate Estimates Automatically**
- **Export Data to Spreadsheets**
- **And Much, Much More!**

Search by Division. Locate cost data by CSI MasterFormat Division.

Open Titles. View Means CostWorks titles, then point and click to make your selections!

Search by Keyword. Enter a keyword and Means CostWorks takes you to that line item.

Select Location. Use this feature to modify costs to over 930 geographic areas in the U.S. and Canada.

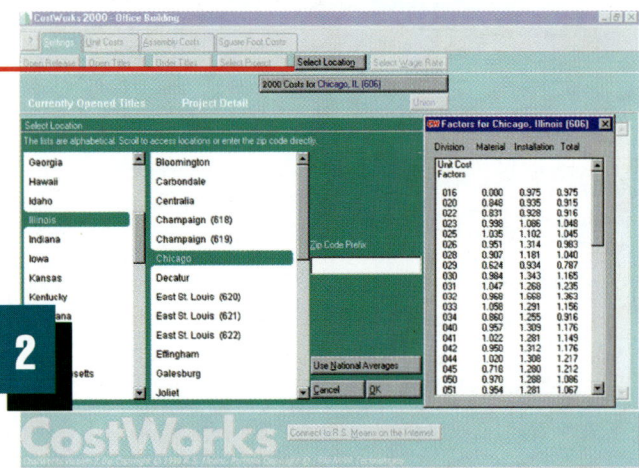

Architects' First Source®, MANU-SPEC® and SPEC-DATA®. Find product listings and specifications by manufacturer.

Wage Rate Utility. Allows you to select Open Shop or Union Wage Rates. (Available Option)

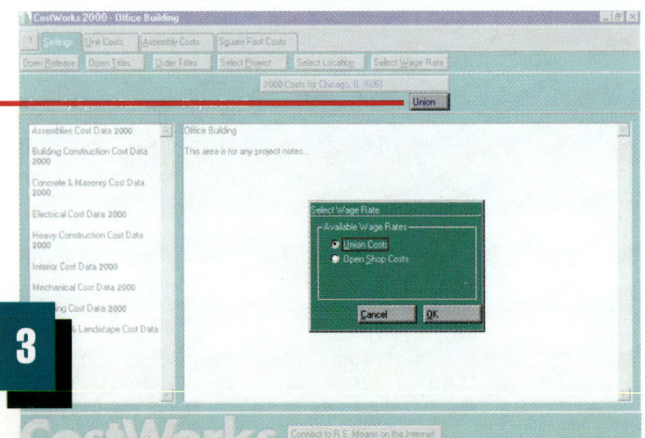

CostCalc. Figure costs automatically for entered quantity. Add selected line to your project list.

EASY TO USE - Just "Point & Click"

1 **Open Titles.** Make your selection, then open the cost book title to bring up more information. Open one or more titles within a group, and use the "Hot Keys" to toggle between screens.

2 **Select Location.** Costs vary by region, so you'll need to localize your data. With this smart program, all the numbers you see will be specific to the geographic area you've selected!

3 **Wage Rate Utility.** Click the Select Wage Rate button on the Settings screen, then look in the dialog box to utilize Union or Open Shop wage rates. *(Option available with selected titles - See order form)*

4 **Search by Division.** You can locate cost data by division..."Glazing"— "Plumbing Fixtures"— "Railroad and Marine Work" — whatever category applies, using convenient pulldown menus.

5 **Search by Keyword.** Enter a Keyword in the dialog box. When the keyword appears, press enter or point and click. Means CostWorks will jump to the corresponding information.

6 **Architects' First Source®, MANU-SPEC® and SPEC-DATA®.** Instantly access manufacturer product listings and specifications included on the Means CostWorks CD-ROM. Launch to the web for additional, updated building product information on-line.

7 **CostCalc.** Find out "how much for how many" by using the CostCalc Toolbar. Enter the required quantity of the highlighted line item. CostCalc will calculate values and totals for one or more items.

8 **Crew Lookup & Print.** This function will let you display AND PRINT a list of the labor and equipment you need for a specific task, wages and more.

9 **Graphics.** One picture is worth a thousand words! Highlighted line items are matched to realistic graphics.

10 **Reference Information.** Quick link to estimating procedures, alternate pricing methods, technical information and more.

11 **CostLists.** Build your CostList with Model, Assembly or Unit Price Data. Means CostWorks will save it automatically. When your CostList is complete, you can easily save, print, or export it to a Spreadsheet file or estimating form.

12 **Assembly Costs Screen.** Shows Division, Unit of Measure, Materials, Installation Costs and Total Costs.

13 **Assembly Components.** To display each factor that went into determining the assembly cost, click on the "Components" tab.

14 **Square Foot Model Costs.** Select your project's building type, exterior wall construction and building framing options, total square foot of floor area and base perimeter, and common additives. And Means CostWorks calculates the square foot cost localized to your geographic area.

Crew Lookup & Print. View and print crew lists, wages and equipment.

Assembly Costs Screen. Access cost data for Assemblies, like walls and foundations, lighting systems and more.

Graphics. View detailed graphic illustrations for Assemblies and Unit Price Lines.

Assembly Components. View each factor that went into the total cost.

Reference Information. Quick link to access additional technical information and estimating procedures about selected items.

Square Foot Models. Instantly generate sample estimates and alternative pricing methods for nearly 100 standard structures, and more than 6,000 building variations.

CostList. View custom project lists that you create using the line items you've selected. Then file, print or export your CostList to fit your project needs.

In minutes, Means CostWorks will give you the power of technology & the reliability of Means construction cost data. Means CostWorks is easy to use, saves time and money, and features NEW enhancements for 2000, to make your job even easier!

Call 1-800-334-3509.

For system requirements and HOW TO ORDER, SEE BACK PAGE

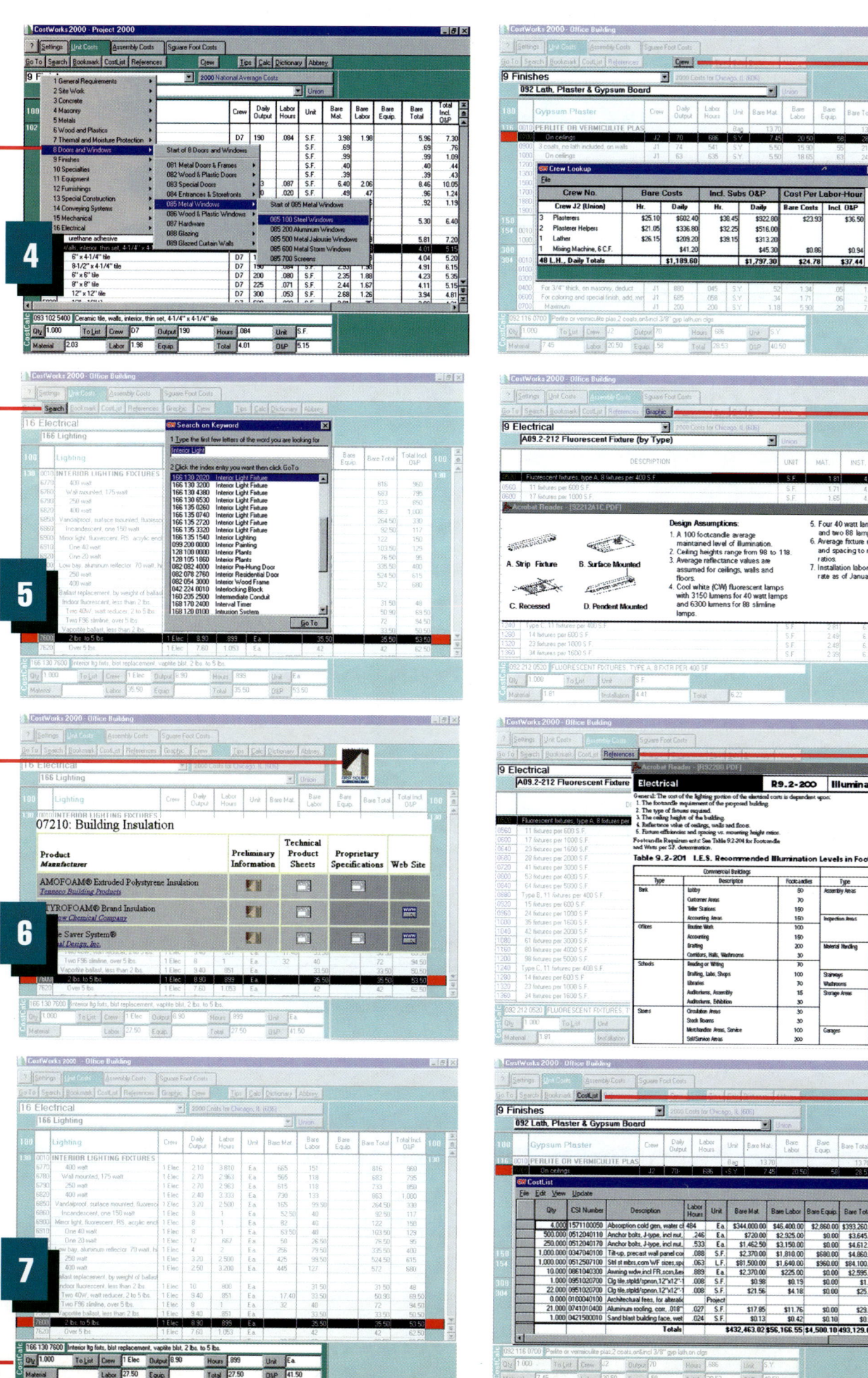

2000 Means CostWorks CD-ROM

Features 2000

- Now Available: Light Commercial Cost Data! Residential Cost Data! Facilities Maintenance & Repair Cost Data!
- Building Product Directory and CSI Formatted Specifications: Architects' First Source®, MANU-SPEC® and SPEC-DATA®
- More Hot Keys For Quick Moves Between Screens!
- Merge Multiple CostLists Into One File!
- Enhanced Search Screens Reveal More Key Words at a Glance!
- Export Data to Lotus, Excel and Windows Applications!
- Print Detailed Lists of Crews and System Components!
- Enhanced Bookmarks, Auto-Save Feature and Much More!

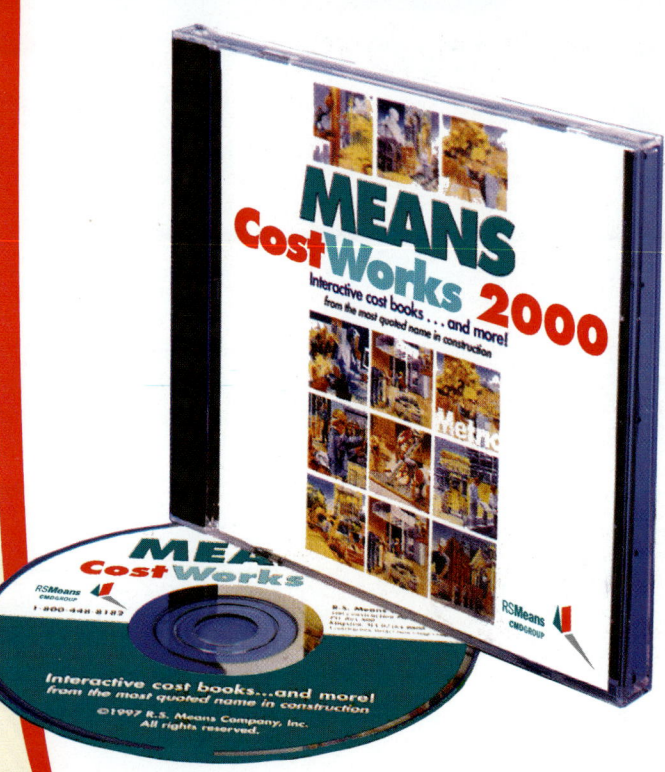

Try Means CostWorks FREE at Absolutely No Risk, With Our 30-Day Guarantee! Call 1-800-334-3509.

11170 | Solid Waste Handling Equipment

11179 | Waste Handling Equipment

			CREW	DAILY OUTPUT	LABOR-HOURS	UNIT	2000 BARE COSTS MAT.	LABOR	EQUIP.	TOTAL	TOTAL INCL O&P	
150	1500	Crematory, not including building, 1 place	Q-3	.20	160	Ea.	42,400	3,150		45,550	52,000	150
	1750	2 place		.10	320		60,500	6,325		66,825	77,000	
	4400	Incinerator, gas, not incl. chimney, elec. or pipe, 50#/hr., minimum		.80	40		16,200	790		16,990	19,100	
	4420	Maximum		.70	45.714		21,100	905		22,005	24,700	
	4440	200 lb. per hr., minimum (batch type)		.60	53.333		21,100	1,050		22,150	25,000	
	4460	Maximum (with feeder)		.50	64		41,100	1,275		42,375	47,300	
	4480	400 lb. per hr., minimum (batch type)		.30	106		25,000	2,100		27,100	31,000	
	4500	Maximum (with feeder)		.25	128		47,200	2,525		49,725	56,000	
	4520	800 lb. per hr., with feeder, minimum		.20	160		61,500	3,150		64,650	72,500	
	4540	Maximum		.17	188		84,000	3,725		87,725	98,500	
	4560	1,200 lb. per hr., with feeder, minimum		.15	213		89,000	4,225		93,225	105,000	
	4580	Maximum		.11	290		107,000	5,750		112,750	127,000	
	4600	2,000 lb. per hr., with feeder, minimum		.10	320		155,500	6,325		161,825	181,500	
	4620	Maximum		.05	640		261,500	12,600		274,100	308,500	
	4700	For heat recovery system, add, minimum		.25	128		50,500	2,525		53,025	59,500	
	4710	Add, maximum		.11	290		161,500	5,750		167,250	187,000	
	4720	For automatic ash conveyer, add		.50	64		20,200	1,275		21,475	24,300	
	4750	Large municipal incinerators, incl. stack, minimum		.25	128	Ton/day	12,900	2,525		15,425	18,400	
	4850	Maximum		.10	320	"	34,300	6,325		40,625	48,100	
	5500	Shredder, municipal use, 35 ton per hour				Ea.	192,000			192,000	211,000	
	5600	60 ton per hour					409,000			409,000	450,000	
	5750	Shredder & baler, 50 ton per day					383,500			383,500	422,000	
	5800	Shredder, industrial, minimum					15,100			15,100	16,600	
	5850	Maximum					81,000			81,000	89,000	
	5900	Baler, industrial, minimum					6,050			6,050	6,650	
	5950	Maximum					353,500			353,500	389,000	
	6000	Transfer station compactor, with power unit										
	6050	and pedestal, not including pit, 50 ton per hour				Ea.	121,000			121,000	133,000	

11190 | Detention Equipment

11191 | Detention Equipment

			CREW	DAILY OUTPUT	LABOR-HOURS	UNIT	2000 BARE COSTS MAT.	LABOR	EQUIP.	TOTAL	TOTAL INCL O&P	
150	0010	**DETENTION EQUIPMENT**										150
	0500	Bar front, rolling, 7/8" bars, 4" O.C., 7' high, 5' wide, with hardware	E-4	2	16	Ea.	4,175	350	42	4,567	5,325	
	1000	Doors & frames, 3' x 7', complete, with hardware, single plate		4	8		3,100	174	21	3,295	3,800	
	1650	Double plate		4	8		3,850	174	21	4,045	4,600	
	2000	Cells, prefab., 5' to 6' wide, 7' to 8' high, 7' to 8' deep,										
	2010	bar front, cot, not incl. plumbing	E-4	1.50	21.333	Ea.	7,400	465	56	7,921	9,125	
	2500	Cot, bolted, single, painted steel		20	1.600		279	35	4.21	318.21	380	
	2700	Stainless steel		20	1.600		725	35	4.21	764.21	875	
	3000	Toilet apparatus including wash basin, average	L-8	1.50	13.333		2,075	243		2,318	2,675	
	4000	Visitor cubicle, vision panel, no intercom	E-4	2	16		2,350	350	42	2,742	3,325	

11300 | Fluid Waste Treatment & Disposal Equipment

11310 | Sewage & Sludge Pumps

		CREW	DAILY OUTPUT	LABOR-HOURS	UNIT	2000 BARE COSTS				TOTAL INCL O&P
						MAT.	LABOR	EQUIP.	TOTAL	
700	0010 **SEWAGE PUMPING STATIONS** Prefabricated steel, concrete										700
	0020 or fiberglass, 200 GPM	C-17D	.17	494	Total	27,500	9,975	3,225	40,700	51,000
	0200 1,000 GPM		.07	1,200		40,800	24,200	7,850	72,850	95,000
	0500 Add for generator unit, 200 GPM, steel		.34	247		20,900	4,975	1,625	27,500	33,300
	0600 Concrete		.51	164		13,300	3,325	1,075	17,700	21,500
	1000 Add for generator unit, 1,000 GPM, steel		.30	280		22,200	5,650	1,825	29,675	36,200
	1200 Concrete	↓	.38	221		18,800	4,450	1,450	24,700	29,900
	1500 For drilled water well, if required, add	B-23	.50	80	↓	5,400	1,200	3,950	10,550	12,300

11390 | Pkg Sewage Treat Plants

		CREW	DAILY OUTPUT	LABOR-HOURS	UNIT	MAT.	LABOR	EQUIP.	TOTAL	TOTAL INCL O&P
900	0010 **WASTEWATER TREATMENT SYSTEM** Fiberglass, 1,000 gallon	B-21	1.29	21.705	Ea.	2,650	345	107	3,102	3,625	900
	0100 1,500 gallon	"	1.03	27.184	"	5,800	435	134	6,369	7,250

11400 | Food Service Equipment

11405 | Food Storage Equipment

		CREW	DAILY OUTPUT	LABOR-HOURS	UNIT	2000 BARE COSTS				TOTAL INCL O&P
						MAT.	LABOR	EQUIP.	TOTAL	
800	0010 **WINE CELLAR**, refrigerated, Redwood interior, carpeted, walk-in type										800
	0020 6'-8" high, including racks										
	0200 80"W x 48"D for 900 bottles	2 Carp	1.50	10.667	Ea.	2,800	210		3,010	3,425
	0250 80" W x 72" D for 1300 bottles		1.33	12.030		3,700	237		3,937	4,475
	0300 80" W x 94" D for 1900 bottles		1.17	13.675		4,800	269		5,069	5,725
	0400 80" W x 124" D for 2500 bottles	↓	1	16	↓	5,800	315		6,115	6,925
	0600 Portable cabinets, red oak, reach-in temp.& humidity controlled										
	0650 26-5/8"W x 26-1/2"D x 68"H for 235 bottles				Ea.	2,150			2,150	2,375
	0660 32"W x 21-1/2"D x 73-1/2"H for 144 bottles					2,925			2,925	3,225
	0670 32"W x 29-1/2"D x 73-1/2"H for 288 bottles					2,950			2,950	3,250
	0680 39-1/2"W x 29-1/2"D x 86-1/2"H for 440 bottles					3,350			3,350	3,675
	0690 52-1/2"W x 29-1/2"D x 73-1/2"H for 468 bottles					3,850			3,850	4,225
	0700 52-1/2"W x 29-1/2"D x 86-1/2"H for 572 bottles				↓	4,050			4,050	4,450
	0730 Portable, red oak, can be built-in with glass door										
	0750 23-7/8"W x 24"D x 34-1/2"H for 50 bottles				Ea.	895			895	985

11410 | Food Prep Equipment

		CREW	DAILY OUTPUT	LABOR-HOURS	UNIT	MAT.	LABOR	EQUIP.	TOTAL	TOTAL INCL O&P
150	0010 **COMMERCIAL KITCHEN EQUIPMENT**										150
	0020 Bake oven, gas, one section	Q-1	8	2	Ea.	3,650	39.50		3,689.50	4,100
	0300 Two sections		7	2.286		7,300	45		7,345	8,100
	0600 Three sections	↓	6	2.667		10,400	52.50		10,452.50	11,600
	0900 Electric convection, single deck	L-7	4	6.500		4,950	110		5,060	5,625
	1050 Butter pat dispenser	1 Clab	13	.615		410	8.90		418.90	470
	1100 Bread dispenser, counter top	"	13	.615		480	8.90		488.90	545
	1300 Broiler, without oven, standard	Q-1	8	2		3,525	39.50		3,564.50	3,950
	1550 Infra-red	L-7	4	6.500		4,500	110		4,610	5,125
	1650 Cabinet, heated, 1 compartment, reach-in	R-18	5.60	4.643		2,175	82		2,257	2,500
	1655 Pass-thru roll-in		5.60	4.643		3,025	82		3,107	3,450
	1660 2 compartment, reach-in	↓	4.80	5.417		4,575	95.50		4,670.50	5,175
	1670 Mobile					2,175			2,175	2,375
	1700 Choppers, 5 pounds	R-18	7	3.714		1,550	65.50		1,615.50	1,800
	1720 16 pounds		5	5.200		1,925	92		2,017	2,275
	1740 35 to 40 pounds	↓	4	6.500	↓	3,800	115		3,915	4,375

11400 | Food Service Equipment

11410 | Food Prep Equipment

			CREW	DAILY OUTPUT	LABOR-HOURS	UNIT	2000 BARE COSTS MAT.	LABOR	EQUIP.	TOTAL	TOTAL INCL O&P	
150	1840	Coffee brewer, 5 burners	1 Plum	3	2.667	Ea.	950	58.50		1,008.50	1,150	150
	1850	Coffee urn, twin 6 gallon urns		2	4		6,000	88		6,088	6,750	
	1860	Single, 3 gallon	▼	3	2.667		4,375	58.50		4,433.50	4,925	
	1900	Cup and glass dispenser, drop in	1 Clab	4	2		850	29		879	985	
	1920	Disposable cup, drop in	"	16	.500		240	7.20		247.20	275	
	2350	Cooler, reach-in, beverage, 6' long	Q-1	6	2.667		2,725	52.50		2,777.50	3,075	
	2650	Dish dispenser, drop in, 12"	1 Clab	11	.727		1,000	10.50		1,010.50	1,125	
	2660	Mobile	"	10	.800	▼	1,675	11.55		1,686.55	1,875	
	2700	Dishwasher, commercial, rack type										
	2720	10 to 12 racks per hour	Q-1	3.20	5	Ea.	2,500	99		2,599	2,925	
	2750	Semi-automatic 38 to 50 racks per hour	"	1.30	12.308		5,250	243		5,493	6,175	
	2800	Automatic, 190 to 230 racks per hour	Q-2	.70	34.286		10,700	650		11,350	12,900	
	2820	235 to 275 racks per hour		.50	48		21,600	915		22,515	25,300	
	2840	8,750 to 12,500 dishes per hour	▼	.20	120	▼	40,700	2,275		42,975	48,600	
	2950	Dishwasher hood, canopy type	L-3A	10	1.200	L.F.	3,600	26		3,626	4,025	
	2960	Pant leg type	"	2.50	4.800	Ea.	4,800	104		4,904	5,450	
	2970	Exhaust hood, sst, gutter on all sides, 4' x 4' x 2'	1 Carp	1.80	4.444		2,650	87.50		2,737.50	3,050	
	2980	4' x 4' x 7'	"	1.60	5		4,150	98.50		4,248.50	4,725	
	3000	Fast food equipment, total package, minimum	6 Skwk	.08	600		110,500	11,900		122,400	142,000	
	3100	Maximum	"	.07	685		150,000	13,500		163,500	188,500	
	3300	Food warmer, counter, 1.2 KW					635			635	700	
	3550	1.6 KW					810			810	890	
	3600	Well, hot food, built-in, rectangular, 12" x 20"	R-30	10	2.600		256	45.50		301.50	360	
	3610	Circular, 7 qt		10	2.600		214	45.50		259.50	315	
	3620	Refrigerated, 2 compartments		10	2.600		1,550	45.50		1,595.50	1,775	
	3630	3 compartments		9	2.889		1,550	50.50		1,600.50	1,775	
	3640	4 compartments	▼	8	3.250		2,125	57		2,182	2,425	
	3800	Food mixers, 20 quarts	L-7	7	3.714		3,150	62.50		3,212.50	3,550	
	3850	40 quarts		5.40	4.815		7,225	81		7,306	8,100	
	3900	60 quarts		5	5.200		9,850	87.50		9,937.50	11,000	
	4040	80 quarts		3.90	6.667		11,300	112		11,412	12,600	
	4080	130 quarts		2.20	11.818		16,900	199		17,099	18,900	
	4100	Floor type, 20 quarts		15	1.733		2,975	29		3,004	3,325	
	4120	60 quarts		14	1.857		8,650	31.50		8,681.50	9,575	
	4140	80 quarts		12	2.167		11,200	36.50		11,236.50	12,400	
	4160	140 quarts	▼	8.60	3.023		19,900	51		19,951	22,000	
	4300	Freezers, reach-in, 44 C.F.	Q-1	4	4		7,200	79		7,279	8,050	
	4500	68 C.F.	"	3	5.333		8,725	105		8,830	9,750	
	4600	Freezer, pre-fab, 8' x 8' w/refrigeration	2 Carp	.45	35.556		6,750	700		7,450	8,625	
	4620	8' x 12'		.35	45.714		8,900	900		9,800	11,400	
	4640	8' x 16'		.25	64		11,400	1,250		12,650	14,800	
	4660	8' x 20'	▼	.17	94.118		13,000	1,850		14,850	17,500	
	4680	Reach-in, 1 compartment	Q-1	4	4		1,950	79		2,029	2,275	
	4700	2 compartment	"	3	5.333		3,175	105		3,280	3,675	
	4720	Frost cold plate	R-30	9	2.889		10,600	50.50		10,650.50	11,700	
	4750	Fryer, with twin baskets, modular model	Q-1	7	2.286		2,850	45		2,895	3,200	
	5000	Floor model on 6" legs	"	5	3.200		3,875	63		3,938	4,350	
	5100	Extra single basket, large					87.50			87.50	96	
	5200	Garbage disposal 1.5 HP, 100 GPH	L-1	4.80	2.083		1,575	46		1,621	1,800	
	5210	3 HP, 120 GPH		4.60	2.174		2,500	48		2,548	2,825	
	5220	5 HP, 250 GPH	▼	4.50	2.222		3,075	49		3,124	3,475	
	5300	3' Long SST griddle, 24" plate dp w/4" legs, elec, 208V 3Ph.	Q-1	7	2.286		1,625	45		1,670	1,875	
	5550	4' long	"	6	2.667		2,800	52.50		2,852.50	3,150	
	5700	Hot chocolate dispenser	1 Plum	4	2		715	44		759	860	
	5800	Ice cube maker, 50 pounds per day	Q-1	6	2.667		1,500	52.50		1,552.50	1,725	
	5900	250 pounds per day		1.20	13.333	▼	1,925	263		2,188	2,550	

11400 | Food Service Equipment

11410 | Food Prep Equipment

		CREW	DAILY OUTPUT	LABOR-HOURS	UNIT	MAT.	LABOR	EQUIP.	TOTAL	TOTAL INCL O&P	
150	6050	500 pounds per day	Q-1	4	4	Ea.	2,725	79		2,804	3,125	150
	6060	With bin		1.20	13.333		3,525	263		3,788	4,325	
	6090	1000 pounds per day, with bin		1	16		7,650	315		7,965	8,925	
	6100	Ice flakers, 300 pounds per day		1.60	10		3,750	198		3,948	4,450	
	6120	600 pounds per day		.95	16.842		3,625	335		3,960	4,525	
	6130	1000 pounds per day		.75	21.333		6,300	420		6,720	7,625	
	6140	2000 pounds per day	▼	.65	24.615		11,800	485		12,285	13,800	
	6160	Ice storage bin, 500 pound capacity	Q-5	1	16		1,500	320		1,820	2,175	
	6180	1000 pound	"	.56	28.571		2,175	570		2,745	3,350	
	6200	Iced tea brewer	1 Plum	3.44	2.326		640	51		691	790	
	6250	Jet spray dispenser	R-18	4.50	5.778		2,150	102		2,252	2,550	
	6300	Juice dispenser, concentrate	"	4.50	5.778		815	102		917	1,075	
	6350	20 Gal kettle w/steam jacket, tilting w/positive lock, S/S	L-7	7	3.714		5,350	62.50		5,412.50	5,975	
	6600	60 gallons	"	6	4.333		5,875	73		5,948	6,575	
	6690	Milk dispenser, bulk, 2 flavor	R-30	8	3.250		925	57		982	1,125	
	6695	3 flavor	"	8	3.250		1,375	57		1,432	1,625	
	6700	Peelers, small	R-18	8	3.250		1,725	57.50		1,782.50	1,975	
	6720	Large	"	6	4.333		3,000	76.50		3,076.50	3,425	
	6750	Pot sink, 3 compartment	1 Plum	7.25	1.103	L.F.	435	24		459	515	
	6760	Pot washer, small		1.60	5	Ea.	12,600	110		12,710	14,100	
	6770	Large		1.20	6.667		31,500	146		31,646	34,900	
	6800	Pulper/extractor, close coupled, 5 HP	▼	1.90	4.211		2,775	92.50		2,867.50	3,200	
	6850	Mobile rack w/pan slide					850			850	935	
	6900	Range, restaurant type, 6 burners and 1 standard oven, 36" wide	Q-1	7	2.286		2,050	45		2,095	2,350	
	6950	Convection		7	2.286		2,775	45		2,820	3,125	
	7150	2 standard ovens, 24" griddle, 60" wide		6	2.667		4,350	52.50		4,402.50	4,875	
	7200	1 standard, 1 convection oven		6	2.667		5,575	52.50		5,627.50	6,225	
	7450	Heavy duty, single 34" standard oven, open top		5	3.200		3,775	63		3,838	4,250	
	7500	Convection oven		5	3.200		5,050	63		5,113	5,650	
	7700	Griddle top		6	2.667		4,475	52.50		4,527.50	5,000	
	7750	Convection oven		6	2.667		5,500	52.50		5,552.50	6,125	
	7950	Hood fire protection system, minimum		3	5.333		2,825	105		2,930	3,275	
	8050	Maximum		1	16		20,100	315		20,415	22,600	
	8300	Refrigerators, reach-in type, 44 C.F.		5	3.200		4,825	63		4,888	5,425	
	8310	With glass doors, 68 C.F.	▼	4	4		6,300	79		6,379	7,050	
	8320	Refrigerator, reach-in, 1 compartment	R-18	7.80	3.333		1,750	59		1,809	2,000	
	8330	2 compartment		6.20	4.194		2,300	74		2,374	2,650	
	8340	3 compartment	▼	5.60	4.643		3,375	82		3,457	3,825	
	8350	Pre-fab, with refrigeration, 8' x 8'	2 Carp	.45	35.556		5,475	700		6,175	7,225	
	8360	8' x 12'		.35	45.714		6,725	900		7,625	8,950	
	8370	8' x 16'		.25	64		7,700	1,250		8,950	10,600	
	8380	8' x 20'	▼	.17	94.118		10,100	1,850		11,950	14,300	
	8390	Pass-thru/roll-in, 1 compartment	R-18	7.80	3.333		2,925	59		2,984	3,325	
	8400	2 compartment		6.24	4.167		4,050	73.50		4,123.50	4,600	
	8410	3 compartment	▼	5.60	4.643		5,550	82		5,632	6,250	
	8420	Walk-in, alum, door & floor only, no refrig, 6' x 6' x 7'-6"	2 Carp	1.40	11.429		7,050	225		7,275	8,125	
	8430	10' x 6' x 7'-6"		.55	29.091		10,100	575		10,675	12,100	
	8440	12' x 14' x 7'-6"		.25	64		13,400	1,250		14,650	17,000	
	8450	12' x 20' x 7'-6"	▼	.17	94.118		16,200	1,850		18,050	21,000	
	8460	Refrigerated cabinets, mobile					2,150			2,150	2,350	
	8470	Refrigerator/freezer, reach-in, 1 compartment	R-18	5.60	4.643		4,475	82		4,557	5,050	
	8480	2 compartment	▼	4.80	5.417		6,150	95.50		6,245.50	6,900	
	8580	Slicer with table		9	2.889		2,800	51		2,851	3,175	
	8600	Stainless steel shelving, louvered 4-tier, 20" x 3'	1 Clab	6	1.333		1,025	19.25		1,044.25	1,175	
	8605	20" x 4'		6	1.333		1,225	19.25		1,244.25	1,375	
	8610	20" x 6'		6	1.333		1,675	19.25		1,694.25	1,875	

11400 | Food Service Equipment

11410 | Food Prep Equipment

		CREW	DAILY OUTPUT	LABOR-HOURS	UNIT	2000 BARE COSTS MAT.	LABOR	EQUIP.	TOTAL	TOTAL INCL O&P	
150	8615	24" x 3'	1 Clab	6	1.333	Ea.	1,075	19.25		1,094.25	1,225	150
	8620	24" x 4'		6	1.333		1,300	19.25		1,319.25	1,450	
	8625	24" x 6'		6	1.333		1,800	19.25		1,819.25	2,000	
	8630	Flat 4-tier, 20" x 3'		6	1.333		955	19.25		974.25	1,075	
	8635	20" x 4'		6	1.333		1,150	19.25		1,169.25	1,300	
	8640	20" x 5'		6	1.333		1,325	19.25		1,344.25	1,475	
	8645	24" x 3'		6	1.333		1,025	19.25		1,044.25	1,150	
	8650	24" x 4'		6	1.333		1,275	19.25		1,294.25	1,425	
	8655	24" x 6'		6	1.333		2,425	19.25		2,444.25	2,700	
	8700	Galvanized shelving, louvered 4-tier, 20" x 3'		6	1.333		365	19.25		384.25	440	
	8705	20" x 4'		6	1.333		415	19.25		434.25	490	
	8710	20" x 6'		6	1.333		615	19.25		634.25	710	
	8715	24" x 3'		6	1.333		400	19.25		419.25	475	
	8720	24" x 4'		6	1.333		455	19.25		474.25	535	
	8725	24" x 6'		6	1.333		625	19.25		644.25	725	
	8730	Flat 4-tier, 20" x 3'		6	1.333		375	19.25		394.25	445	
	8735	20" x 4'		6	1.333		425	19.25		444.25	505	
	8740	20" x 6'		6	1.333		575	19.25		594.25	670	
	8745	24" x 3'		6	1.333		395	19.25		414.25	470	
	8750	24" x 4'		6	1.333		455	19.25		474.25	535	
	8755	24" x 6'		6	1.333		625	19.25		644.25	720	
	8760	Stainless steel dunnage rack, 24" x 3'		8	1		470	14.45		484.45	540	
	8765	24" x 4'		8	1		545	14.45		559.45	625	
	8770	Galvanized dunnage rack, 24" x 3'		8	1		128	14.45		142.45	166	
	8775	24" x 4'		8	1		145	14.45		159.45	185	
	8800	Serving counter, straight	1 Carp	40	.200	L.F.	425	3.94		428.94	475	
	8820	Curved section	"	30	.267	"	605	5.25		610.25	680	
	8825	Solid surface, see section 06610-810										
	8830	Soft serve ice cream machine, medium	R-18	11	2.364	Ea.	6,300	41.50		6,341.50	7,025	
	8840	Large	"	9	2.889		9,500	51		9,551	10,500	
	8850	Steamer, electric 27 KW	L-7	7	3.714		7,425	62.50		7,487.50	8,275	
	9100	Electric, 10 KW or gas 100,000 BTU	"	5	5.200		3,425	87.50		3,512.50	3,900	
	9150	Toaster, conveyor type, 16-22 slices per minute					910			910	1,000	
	9160	Pop-up, 2 slot					455			455	500	
	9170	Trash compactor, small, up to 125 lb. compacted weight	L-4	4	4		14,300	69		14,369	15,900	
	9175	Large, up to 175 lb. compacted weight	"	3	5.333		17,500	92		17,592	19,500	
	9180	Tray and silver dispenser, mobile	1 Clab	16	.500		630	7.20		637.20	705	
	9200	For deluxe models of above equipment, add					75%					
	9400	Rule of thumb: Equipment cost based										
	9410	on kitchen work area										
	9420	Office buildings, minimum	L-7	77	.338	S.F.	48	5.70		53.70	62.50	
	9450	Maximum		58	.448		81.50	7.55		89.05	102	
	9550	Public eating facilities, minimum		77	.338		63	5.70		68.70	79	
	9600	Maximum		46	.565		103	9.55		112.55	129	
	9750	Hospitals, minimum		58	.448		65	7.55		72.55	84.50	
	9800	Maximum		39	.667		109	11.25		120.25	138	

11450 | Residential Equipment

11454 | Residential Appliances

		CREW	DAILY OUTPUT	LABOR-HOURS	UNIT	2000 BARE COSTS MAT.	LABOR	EQUIP.	TOTAL	TOTAL INCL O&P
0010	**RESIDENTIAL APPLIANCES**									
0020	Cooking range, 30" free standing, 1 oven, minimum	2 Clab	10	1.600	Ea.	259	23		282	325
0050	Maximum		4	4		1,175	58		1,233	1,400
0150	2 oven, minimum		10	1.600		1,125	23		1,148	1,275
0200	Maximum	↓	10	1.600		1,500	23		1,523	1,700
0350	Built-in, 30" wide, 1 oven, minimum	1 Elec	6	1.333		259	29.50		288.50	335
0400	Maximum	2 Carp	2	8		795	158		953	1,150
0500	2 oven, conventional, minimum		4	4		880	79		959	1,100
0550	1 conventional, 1 microwave, maximum	↓	2	8		1,550	158		1,708	1,975
0700	Free-standing, 1 oven, 21" wide range, minimum	2 Clab	10	1.600		281	23		304	350
0750	21" wide, maximum	"	4	4		465	58		523	615
0900	Counter top cook tops, 4 burner, standard, minimum	1 Elec	6	1.333		172	29.50		201.50	237
0950	Maximum		3	2.667		430	59		489	570
1050	As above, but with grille and griddle attachment, minimum		6	1.333		400	29.50		429.50	490
1100	Maximum		3	2.667		515	59		574	660
1200	Induction cooktop, 30" wide		3	2.667		500	59		559	645
1250	Microwave oven, minimum		4	2		87.50	44		131.50	169
1300	Maximum	↓	2	4		365	88.50		453.50	550
1500	Combination range, refrigerator and sink, 30" wide, minimum	L-1	2	5		635	110		745	880
1550	Maximum		1	10		1,275	220		1,495	1,775
1570	60" wide, average		1.40	7.143		2,100	157		2,257	2,575
1590	72" wide, average		1.20	8.333		2,375	183		2,558	2,925
1600	Office model, 48" wide		2	5		1,800	110		1,910	2,150
1620	Refrigerator and sink only	↓	2.40	4.167	↓	1,800	91.50		1,891.50	2,125
1640	Combination range, refrigerator, sink, microwave									
1660	oven and ice maker	L-1	.80	12.500	Ea.	3,525	275		3,800	4,325
1750	Compactor, residential size, 4 to 1 compaction, minimum	1 Carp	5	1.600		335	31.50		366.50	425
1800	Maximum	"	3	2.667		370	52.50		422.50	495
2000	Deep freeze, 15 to 23 C.F., minimum	2 Clab	10	1.600		370	23		393	445
2050	Maximum		5	3.200		510	46		556	640
2200	30 C.F., minimum		8	2		730	29		759	855
2250	Maximum	↓	3	5.333		850	77		927	1,075
2450	Dehumidifier, portable, automatic, 15 pint					185			185	204
2550	40 pint					220			220	242
2750	Dishwasher, built-in, 2 cycles, minimum	L-1	4	2.500		247	55		302	360
2800	Maximum		2	5		258	110		368	465
2950	4 or more cycles, minimum		4	2.500		239	55		294	355
2960	Average		4	2.500		320	55		375	440
3000	Maximum	↓	2	5		495	110		605	725
3200	Dryer, automatic, minimum	L-2	3	5.333		251	92		343	435
3250	Maximum	"	2	8		545	138		683	835
3300	Garbage disposer, sink type, minimum	L-1	10	1		37.50	22		59.50	77.50
3350	Maximum	"	10	1		174	22		196	228
3550	Heater, electric, built-in, 1250 watt, ceiling type, minimum	1 Elec	4	2		68	44		112	148
3600	Maximum		3	2.667		111	59		170	219
3700	Wall type, minimum		4	2		96.50	44		140.50	179
3750	Maximum		3	2.667		128	59		187	238
3900	1500 watt wall type, with blower		4	2		119	44		163	204
3950	3000 watt	↓	3	2.667		244	59		303	365
4150	Hood for range, 2 speed, vented, 30" wide, minimum	L-3	5	2		35	40.50		75.50	107
4200	Maximum		3	3.333		430	67.50		497.50	590
4300	42" wide, minimum		5	2		164	40.50		204.50	250
4330	Custom		5	2		575	40.50		615.50	700
4350	Maximum	↓	3	3.333		700	67.50		767.50	885
4500	For ventless hood, 2 speed, add					13.40			13.40	14.75
4650	For vented 1 speed, deduct from maximum				↓	26			26	28.50

11450 | Residential Equipment

11454 | Residential Appliances

			DAILY	LABOR-		2000 BARE COSTS				TOTAL
		CREW	OUTPUT	HOURS	UNIT	MAT.	LABOR	EQUIP.	TOTAL	INCL O&P
500	4850	Humidifier, portable, 8 gallons per day				Ea.	92.50			92.50	102
	5000	15 gallons per day					149			149	164
	5200	Icemaker, automatic, 20 lb. per day	1 Plum	7	1.143		505	25		530	595
	5350	51 lb. per day	"	2	4		785	88		873	1,000
	5500	Refrigerator, no frost, 10 C.F. to 12 C.F. minimum	2 Clab	10	1.600		420	23		443	505
	5600	Maximum		6	2.667		640	38.50		678.50	770
	5750	14 C.F. to 16 C.F., minimum		9	1.778		440	25.50		465.50	530
	5800	Maximum		5	3.200		520	46		566	650
	5950	18 C.F. to 20 C.F., minimum		8	2		485	29		514	585
	6000	Maximum		4	4		805	58		863	985
	6150	21 C.F. to 29 C.F., minimum		7	2.286		790	33		823	925
	6200	Maximum		3	5.333		2,600	77		2,677	2,975
	6400	Sump pump cellar drainer, pedestal, 1/3 H.P., molded PVC base	1 Plum	3	2.667		86.50	58.50		145	192
	6450	Solid brass	"	2	4		179	88		267	340
	6460	Sump pump, see also division 15440-940									
	6650	Washing machine, automatic, minimum	1 Plum	3	2.667	Ea.	320	58.50		378.50	445
	6700	Maximum	"	1	8		620	176		796	970
	6900	Water heater, electric, glass lined, 30 gallon, minimum	L-1	5	2		215	44		259	310
	6950	Maximum		3	3.333		299	73.50		372.50	450
	7100	80 gallon, minimum		2	5		415	110		525	635
	7150	Maximum		1	10		575	220		795	995
	7180	Water heater, gas, glass lined, 30 gallon, minimum	2 Plum	5	3.200		284	70		354	425
	7220	Maximum		3	5.333		395	117		512	630
	7260	50 gallon, minimum		2.50	6.400		515	140		655	800
	7300	Maximum		1.50	10.667		720	234		954	1,175
	7310	Water heater, see also division 15480-200									
	7350	Water softener, automatic, to 30 grains per gallon	2 Plum	5	3.200	Ea.	420	70		490	575
	7400	To 100 grains per gallon	"	4	4		510	88		598	710
	7450	Vent kits for dryers	1 Carp	10	.800		9.55	15.75		25.30	37.50
550	0010	**DISAPPEARING STAIRWAY** No trim included									
	0100	Custom grade, pine, 8'-6" ceiling, minimum	1 Carp	4	2	Ea.	100	39.50		139.50	178
	0150	Average		3.50	2.286		105	45		150	193
	0200	Maximum		3	2.667		185	52.50		237.50	294
	0500	Heavy duty, pivoted, from 7'-7" to 12'-10" floor to floor		3	2.667		315	52.50		367.50	440
	0600	16'-0" ceiling		2	4		1,025	79		1,104	1,275
	0800	Economy folding, pine, 8'-6" ceiling		4	2		83	39.50		122.50	159
	0900	9'-6" ceiling		4	2		93.50	39.50		133	171
	1000	Fire escape, galvanized steel, 8'-0" to 10'-4" ceiling	2 Carp	1	16		1,125	315		1,440	1,775
	1010	10'-6" to 13'-6" ceiling		1	16		1,400	315		1,715	2,100
	1100	Automatic electric, aluminum, floor to floor height, 8' to 9'		1	16		5,500	315		5,815	6,625
	1400	11' to 12'		.90	17.778		5,925	350		6,275	7,125
	1700	14' to 15'		.70	22.857		6,325	450		6,775	7,725

11470 | Darkroom Equipment

11471 | Darkroom Processing

			DAILY	LABOR-		2000 BARE COSTS				TOTAL
		CREW	OUTPUT	HOURS	UNIT	MAT.	LABOR	EQUIP.	TOTAL	INCL O&P
700	0010	**DARKROOM EQUIPMENT**									
	0020	Developing sink, 5" deep, 24" x 48"	Q-1	2	8	Ea.	3,425	158		3,583	4,025

11470 | Darkroom Equipment

11471 | Darkroom Processing

		CREW	DAILY OUTPUT	LABOR-HOURS	UNIT	2000 BARE COSTS MAT.	LABOR	EQUIP.	TOTAL	TOTAL INCL O&P
0050	48" x 52"	Q-1	1.70	9.412	Ea.	3,725	186		3,911	4,375
0200	10" deep, 24" x 48"		1.70	9.412		4,825	186		5,011	5,625
0250	24" x 108"	▼	1.50	10.667		1,750	211		1,961	2,275
0500	Dryers, dehumidified filtered air, 36" x 25" x 68" high	L-7	6	4.333		3,550	73		3,623	4,050
0550	48" x 25" x 68" high		5	5.200		6,750	87.50		6,837.50	7,575
2000	Processors, automatic, color print, minimum		4	6.500		9,775	110		9,885	11,000
2050	Maximum		.60	43.333		9,950	730		10,680	12,200
2300	Black and white print, minimum		2	13		7,425	219		7,644	8,525
2350	Maximum		.80	32.500		48,400	550		48,950	54,500
2600	Manual processor, 16" x 20" maximum print size		2	13		7,475	219		7,694	8,600
2650	20" x 24" maximum print size		1	26		7,025	440		7,465	8,475
3000	Viewing lites, 20" x 24"		6	4.333		310	73		383	465
3100	20" x 24" with color correction	▼	6	4.333		435	73		508	605
3500	Washers, round, minimum sheet 11" x 14"	Q-1	2	8		2,250	158		2,408	2,725
3550	Maximum sheet 20" x 24"		1	16		2,450	315		2,765	3,225
3800	Square, minimum sheet 20" x 24"		1	16		2,200	315		2,515	2,950
3900	Maximum sheet 50" x 56"	▼	.80	20	▼	3,425	395		3,820	4,425
4500	Combination tank sink, tray sink, washers, with									
4510	dry side tables, average	Q-1	.45	35.556	Ea.	7,450	700		8,150	9,350

11472 | Revolving Darkroom Doors

		CREW	DAILY OUTPUT	LABOR-HOURS	UNIT	MAT.	LABOR	EQUIP.	TOTAL	TOTAL INCL O&P
0011	DARKROOM DOORS Revolving, standard, 2 way, 36" diameter	2 Carp	3.10	5.161	Opng.	1,575	102		1,677	1,900
0021	41" diameter		3.10	5.161		1,725	102		1,827	2,075
0051	3 way, 51" diameter		1.40	11.429		2,275	225		2,500	2,875
1001	4 way, 49" diameter		1.40	11.429		2,675	225		2,900	3,325
2001	Hinged safety, 2 way, 41" diameter		2.30	6.957		2,100	137		2,237	2,550
2501	3 way, 51" diameter		1.40	11.429		2,725	225		2,950	3,375
3001	Pop out safety, 2 way, 41" diameter		3.10	5.161		2,550	102		2,652	2,975
4001	3 way, 51" diameter		1.40	11.429		2,575	225		2,800	3,200
5001	Wheelchair-type, pop out, 51" diameter	▼	1.40	11.429	▼	2,850	225		3,075	3,500
9300	For complete dark rooms, see division 13035-300									

11480 | Athletic Recreational & Therapeutic Equipment

11483 | Bowling Alleys

		CREW	DAILY OUTPUT	LABOR-HOURS	UNIT	MAT.	LABOR	EQUIP.	TOTAL	TOTAL INCL O&P
0010	BOWLING ALLEYS Including alley, pinsetter, scorer,									
0020	counters and misc. supplies, minimum	4 Carp	.20	160	Lane	32,200	3,150		35,350	40,900
0150	Average		.19	168		34,500	3,325		37,825	43,700
0300	Maximum	▼	.18	177		36,700	3,500		40,200	46,400
0400	Combo table ball rack, add					550			550	605
0600	For automatic scorer, add, minimum					6,000			6,000	6,600
0700	Maximum				▼	6,800			6,800	7,475

11484 | Exercise Equipment

		CREW	DAILY OUTPUT	LABOR-HOURS	UNIT	MAT.	LABOR	EQUIP.	TOTAL	TOTAL INCL O&P
0010	HEALTH CLUB EQUIPMENT									
0020	Abdominal rack, 2 board capacity				Ea.	380			380	415
0050	Abdominal board, upholstered					425			425	465
0200	Bicycle trainer, minimum				▼	665			665	730

11480 | Athletic Recreational & Therapeutic Equipment

11484 | Exercise Equipment

		CREW	DAILY OUTPUT	LABOR-HOURS	UNIT	MAT.	LABOR	EQUIP.	TOTAL	TOTAL INCL O&P	
400	0300	Deluxe, electric				Ea.	3,375			3,375	3,725	400
	0400	Bar bell set, chrome plated steel, 25 lbs.					199			199	219	
	0420	100 lbs.					300			300	330	
	0450	200 lbs.					580			580	640	
	0500	Weight plates, cast iron, per lb.				Lb.	4.08			4.08	4.49	
	0520	Storage rack, 10 station				Ea.	645			645	705	
	0600	Circuit training apparatus, 12 machines minimum	2 Clab	1.25	12.800	Set	22,400	185		22,585	25,000	
	0700	Average		1	16		27,500	231		27,731	30,700	
	0800	Maximum		.75	21.333		32,600	310		32,910	36,400	
	0820	Dumbbell set, cast iron, with rack and 5 pair					625			625	690	
	0900	Squat racks	2 Clab	5	3.200	Ea.	620	46		666	765	
	1200	Multi-station gym machine, 5 station					6,475			6,475	7,125	
	1250	9 station					9,075			9,075	9,975	
	1280	Rowing machine, hydraulic					1,200			1,200	1,325	
	1300	Treadmill, manual					1,175			1,175	1,300	
	1320	Motorized					3,625			3,625	4,000	
	1340	Electronic					4,425			4,425	4,875	
	1360	Cardio-testing					6,325			6,325	6,950	
	1400	Treatment/massage tables, minimum					420			420	465	
	1420	Deluxe, with accessories					625			625	685	
	1600	For saunas, see division 13035-800										
	1640	For steam baths, see division 13035-940										
	1680	For whirlpool baths, see division 15418-100										
	1720	For additional exercise equipment, see div. 11486-700										

11486 | Gymnasium Equipment

		CREW	DAILY OUTPUT	LABOR-HOURS	UNIT	MAT.	LABOR	EQUIP.	TOTAL	TOTAL INCL O&P	
700	0010	**SCHOOL EQUIPMENT**										700
	0200	For exterior equipment, see division 02800										
	0201	For chairs & desks, see division 12560-200										
	0300	For chalkboards & bulletin boards, see div. 10110-940 & 10110-240										
	0400	For lockers, see division 10505-500										
	1000	Basketball backstops, wall mtd., 6' extended, fixed, minimum	L-2	1	16	Ea.	535	276		811	1,075	
	1100	Maximum		1	16		700	276		976	1,250	
	1200	Swing up, minimum		1	16		1,075	276		1,351	1,650	
	1250	Maximum		1	16		1,925	276		2,201	2,575	
	1300	Portable, manual, heavy duty, spring operated		1.90	8.421		9,000	145		9,145	10,100	
	1400	Ceiling suspended, stationary, minimum		.78	20.513		1,325	355		1,680	2,050	
	1450	Fold up, with accessories, maximum		.40	40		4,825	690		5,515	6,475	
	1600	For electrically operated, add	1 Elec	1	8		1,575	177		1,752	2,050	
	2000	Benches, folding, in wall, 14' table, 2 benches	L-4	2	8	Set	510	138		648	795	
	3000	Bleachers, telescoping, manual to 15 tier, minimum	F-5	65	.492	Seat	51	8.50		59.50	70.50	
	3100	Maximum		60	.533		76.50	9.20		85.70	100	
	3300	16 to 20 tier, minimum		60	.533		122	9.20		131.20	151	
	3400	Maximum		55	.582		153	10.05		163.05	185	
	3600	21 to 30 tier, minimum		50	.640		128	11.05		139.05	159	
	3700	Maximum		40	.800		153	13.80		166.80	192	
	3900	For integral power operation, add, minimum	2 Elec	300	.053		25.50	1.18		26.68	30	
	4000	Maximum	"	250	.064		41	1.41		42.41	47.50	
	4100	Boxing ring, elevated, 22' x 22'	L-4	.10	160	Ea.	8,150	2,775		10,925	13,700	
	4110	For cellular plastic foam padding, add		.10	160		3,900	2,775		6,675	9,050	
	4120	Floor level, including posts and ropes only, 20' x 20'		.80	20		1,200	345		1,545	1,925	
	4130	Canvas, 30' x 30'		5	3.200		1,100	55.50		1,155.50	1,300	
	4150	Exercise equipment, bicycle trainer					350			350	385	
	4180	Chinning bar, adjustable, wall mounted	1 Carp	5	1.600		250	31.50		281.50	330	
	4200	Exercise ladder, 16' x 1'-7", suspended	L-2	3	5.333		1,075	92		1,167	1,325	
	4210	High bar, floor plate attached	1 Carp	4	2		925	39.50		964.50	1,100	

11480 | Athletic Recreational & Therapeutic Equipment

11486 | Gymnasium Equipment

		CREW	DAILY OUTPUT	LABOR-HOURS	UNIT	MAT.	LABOR	EQUIP.	TOTAL	TOTAL INCL O&P	
700	4240	Parallel bars, adjustable	1 Carp	4	2	Ea.	2,050	39.50		2,089.50	2,325	700
	4270	Uneven parallel bars, adjustable	↓	4	2	↓	2,425	39.50		2,464.50	2,750	
	4280	Wall mounted, adjustable	L-2	1.50	10.667	Set	700	184		884	1,075	
	4300	Rope, ceiling mounted, 18' long	1 Carp	3.66	2.186	Ea.	149	43		192	238	
	4330	Side horse, vaulting		5	1.600		1,475	31.50		1,506.50	1,675	
	4360	Treadmill, motorized, deluxe, training type	↓	5	1.600		4,750	31.50		4,781.50	5,275	
	4390	Weight lifting multi-station, minimum	2 Clab	1	16		2,825	231		3,056	3,500	
	4450	Maximum	"	.50	32	↓	12,200	460		12,660	14,300	
	4480	For additional exercise equipment, see div. 11484-400										
	4500	Gym divider curtain, mesh top, vinyl bottom, manual	L-4	500	.032	S.F.	6.10	.55		6.65	7.70	
	4700	Electric roll up	L-7	400	.065		9.20	1.10		10.30	11.95	
	5500	Gym mats, 2" thick, naugahyde covered					3.10			3.10	3.41	
	5600	Vinyl/nylon covered					5.15			5.15	5.65	
	5800	Wall pads, 1-1/2" thick	2 Carp	640	.025		5.80	.49		6.29	7.25	
	6000	Wrestling mats, 1" thick, heavy duty				↓	5.30			5.30	5.85	
	7000	Scoreboards, baseball, minimum	R-3	1.30	15.385	Ea.	2,350	340	106	2,796	3,275	
	7200	Maximum		.05	400		17,100	8,800	2,775	28,675	36,300	
	7300	Football, minimum		.86	23.256		5,775	510	161	6,446	7,375	
	7400	Maximum		.20	100		53,500	2,200	690	56,390	63,500	
	7500	Basketball (one side), minimum		2.07	9.662		2,475	213	67	2,755	3,125	
	7600	Maximum		.30	66.667		17,100	1,475	460	19,035	21,700	
	7700	Hockey-basketball (four sides), minimum		.25	80		8,775	1,750	555	11,080	13,100	
	7800	Maximum	↓	.15	133	↓	26,800	2,925	920	30,645	35,300	

11488 | Shooting Ranges

		CREW	DAILY OUTPUT	LABOR-HOURS	UNIT	MAT.	LABOR	EQUIP.	TOTAL	TOTAL INCL O&P	
700	0010	**SHOOTING RANGE** Incl. bullet traps, target provisions, controls,										700
	0100	separators, ceiling system, etc. Not incl. structural shell										
	0200	Commercial	L-9	.64	56.250	Point	7,275	1,025		8,300	9,750	
	0300	Law enforcement		.28	128		15,500	2,350		17,850	21,000	
	0400	National Guard armories		.71	50.704		4,975	925		5,900	7,050	
	0500	Reserve training centers		.71	50.704		5,075	925		6,000	7,150	
	0600	Schools and colleges		.32	112		10,900	2,050		12,950	15,500	
	0700	Major acadamies	↓	.19	189		20,300	3,450		23,750	28,200	
	0800	For acoustical treatment, add					10%	10%				
	0900	For lighting, add					28%	25%				
	1000	For plumbing, add					5%	5%				
	1100	For ventilating system, add, minimum					40%	40%				
	1200	Add, average					25%	25%				
	1300	Add, maximum				↓	35%	35%				

11500 | Industrial & Process Equipment

11520 | Industrial Equipment

		CREW	DAILY OUTPUT	LABOR-HOURS	UNIT	MAT.	LABOR	EQUIP.	TOTAL	TOTAL INCL O&P	
300	0010	**EQUIPMENT INSTALLATION**										300
	0020	Industrial equipment, minimum	E-2	12	4	Ton		86	85	171	259	
	0200	Maximum	"	2	24	"		515	510	1,025	1,550	
850	0010	**VOCATIONAL SHOP EQUIPMENT**										850
	0020	Benches, work, wood, average	2 Carp	5	3.200	Ea.	400	63		463	550	

11500 | Industrial & Process Equipment

11520 | Industrial Equipment

			CREW	DAILY OUTPUT	LABOR-HOURS	UNIT	MAT.	LABOR	EQUIP.	TOTAL	TOTAL INCL O&P
850	0100	Metal, average	2 Carp	5	3.200	Ea.	225	63		288	355
	0400	Combination belt & disc sander, 6"		4	4		965	79		1,044	1,175
	0700	Drill press, floor mounted, 12", 1/2 H.P.	↓	4	4		400	79		479	575
	0800	Dust collector, not incl. ductwork, 6" diameter	1 Shee	1.10	7.273		2,375	156		2,531	2,900
	1000	Grinders, double wheel, 1/2 H.P.	2 Carp	5	3.200		230	63		293	360
	1300	Jointer, 4", 3/4 H.P.		4	4		1,050	79		1,129	1,275
	1600	Kilns, 16 C.F., to 2000°		4	4		2,100	79		2,179	2,425
	1900	Lathe, woodworking, 10", 1/2 H.P.		4	4		550	79		629	740
	2200	Planer, 13" x 6"		4	4		1,025	79		1,104	1,275
	2500	Potter's wheel, motorized		4	4		525	79		604	715
	2800	Saws, band, 14", 3/4 H.P.		4	4		480	79		559	665
	3100	Metal cutting band saw, 14"		4	4		765	79		844	980
	3400	Radial arm saw, 10", 2 H.P.		4	4		800	79		879	1,025
	3700	Scroll saw, 24"		4	4		1,350	79		1,429	1,600
	4000	Table saw, 10", 3 H.P.		4	4		1,675	79		1,754	1,950
	4300	Welder AC arc, 30 amp capacity	↓	4	4	↓	2,025	79		2,104	2,350

11600 | Laboratory Equipment

11620 | Laboratory Equipment

			CREW	DAILY OUTPUT	LABOR-HOURS	UNIT	MAT.	LABOR	EQUIP.	TOTAL	TOTAL INCL O&P
350	0010	**LABORATORY EQUIPMENT**									
	0020	Cabinets, base, door units, metal	2 Carp	18	.889	L.F.	112	17.50		129.50	153
	0300	Drawer units		18	.889		245	17.50		262.50	299
	0700	Tall storage cabinets, open, 7' high		20	.800		246	15.75		261.75	297
	0900	With glazed doors		20	.800		277	15.75		292.75	330
	1300	Wall cabinets, metal, 12-1/2" deep, open		20	.800		70	15.75		85.75	104
	1500	With doors		20	.800	↓	145	15.75		160.75	187
	1550	Counter tops, not incl. base cabinets, acidproof, minimum		82	.195	S.F.	15.05	3.84		18.89	23
	1600	Maximum		70	.229		24	4.50		28.50	34
	1650	Stainless steel	↓	82	.195	↓	60	3.84		63.84	72.50
	2000	Fume hood, with countertop & base, not including HVAC									
	2020	Simple, minimum	2 Carp	5.40	2.963	L.F.	505	58.50		563.50	655
	2050	Complex, including fixtures		2.40	6.667		1,200	131		1,331	1,525
	2100	Special, maximum	↓	1.70	9.412	↓	1,250	185		1,435	1,700
	2200	Ductwork, minimum	2 Shee	1	16	Hood	835	345		1,180	1,500
	2250	Maximum	"	.50	32	"	4,225	690		4,915	5,800
	2300	Service fixtures, average				Ea.	43			43	47
	2500	For sink assembly with hot and cold water, add	1 Plum	1.40	5.714		500	125		625	755
	2550	Glassware washer, undercounter, minimum	L-1	1.80	5.556		4,800	122		4,922	5,475
	2600	Maximum	"	1	10		6,750	220		6,970	7,800
	2650	Glove box, fiberglass, bacteriological					11,000			11,000	12,100
	2700	Controlled atmosphere					8,500			8,500	9,350
	2750	Radioisotope					11,000			11,000	12,100
	2770	Carcinogenic					11,000			11,000	12,100
	2800	Sink, one piece plastic, flask wash, hose, free standing	1 Plum	1.60	5		1,275	110		1,385	1,575
	2850	Epoxy resin sink, 25" x 16" x 10"	"	2	4	↓	140	88		228	299
	3000	Utility table, acid resistant top with drawers	2 Carp	30	.533	L.F.	104	10.50		114.50	132
	3800	Titration unit, four 2000 ml reservoirs				Ea.	8,350			8,350	9,200
	4200	Alternate pricing method: as percent of lab furniture									
	4400	Installation, not incl. plumbing & duct work				% Furn.					22%

11600 | Laboratory Equipment

11620 | Laboratory Equipment

		CREW	DAILY OUTPUT	LABOR-HOURS	UNIT	2000 BARE COSTS MAT.	LABOR	EQUIP.	TOTAL	TOTAL INCL O&P
4800	Plumbing, final connections, simple system				% Furn.					10%
5000	Moderately complex system									15%
5200	Complex system									20%
5400	Electrical, simple system									10%
5600	Moderately complex system									20%
5800	Complex system									35%
6000	Safety equipment, eye wash, hand held				Ea.	210			210	231
6200	Deluge shower				"	500			500	550
6300	Rule of thumb: lab furniture including installation & connection									
6320	High school				S.F.					27.75
6340	College									40.75
6360	Clinical, health care									35
6380	Industrial									56.50

11700 | Medical Equipment

11710 | Medical Equipment

		CREW	DAILY OUTPUT	LABOR-HOURS	UNIT	2000 BARE COSTS MAT.	LABOR	EQUIP.	TOTAL	TOTAL INCL O&P
0010	**MEDICAL EQUIPMENT**									
0015	Autopsy table, standard	1 Plum	1	8	Ea.	6,300	176		6,476	7,225
0020	Deluxe	"	.60	13.333		7,875	293		8,168	9,150
0300	Blood pressure unit, mercurial, wall					158			158	174
0400	Diagnostic set, wall					460			460	510
0700	Distiller, water, steam heated, 50 gal. capacity	1 Plum	1.40	5.714		13,100	125		13,225	14,600
0750	Exam room furnishings, average per room					4,600			4,600	5,050
1800	Heat therapy unit, humidified, 26" x 78" x 28"					2,425			2,425	2,650
2100	Hubbard tank with accessories, stainless steel,									
2110	125 GPM at 45 psi water pressure				Ea.	18,700			18,700	20,600
2150	For electric overhead hoist, add					2,050			2,050	2,250
2500	Incubators, minimum					2,600			2,600	2,875
2600	Maximum					14,300			14,300	15,700
2900	K-Module for heat therapy, 20 oz. capacity, 75° to 110°F					289			289	320
3200	Mortuary refrigerator, end operated, 2 capacity					10,200			10,200	11,200
3300	6 capacity					18,900			18,900	20,800
3600	Paraffin bath, 126°F, auto controlled					1,100			1,100	1,225
3900	Parallel bars for walking training, 12'-0"					910			910	1,000
4200	Refrigerator, blood bank, 28.6 C.F. emergency signal					5,700			5,700	6,250
4210	Reach-in, 16.9 C.F.					6,075			6,075	6,675
4400	Scale, physician's, with height rod					250			250	275
4600	Station, dietary, medium, with ice					11,600			11,600	12,700
4700	Medicine					5,225			5,225	5,750
5000	Scrub, surgical, stainless steel, single station, minimum	1 Plum	3	2.667		970	58.50		1,028.50	1,175
5100	Maximum					5,650			5,650	6,200
5600	Sterilizers, floor loading, 26" x 62" x 42", single door, steam					114,500			114,500	126,000
5650	Double door, steam					147,000			147,000	161,500
5800	General purpose, 20" x 20" x 38", single door					11,000			11,000	12,100
6000	Portable, counter top, steam, minimum					2,750			2,750	3,025
6020	Maximum					4,300			4,300	4,725
6050	Portable, counter top, gas, 17"x15"x32-1/2"					28,400			28,400	31,200
6150	Manual washer/sterilizer, 16"x16"x26"	1 Plum	2	4		38,900	88		38,988	42,800

11700 | Medical Equipment

11710 | Medical Equipment

		CREW	DAILY OUTPUT	LABOR-HOURS	UNIT	2000 BARE COSTS MAT.	LABOR	EQUIP.	TOTAL	TOTAL INCL O&P
6200	Steam generators, electric 10 KW to 180 KW, freestanding									
6250	Minimum	1 Elec	3	2.667	Ea.	5,775	59		5,834	6,450
6300	Maximum	"	.70	11.429		20,800	253		21,053	23,300
6500	Surgery table, minor minimum	1 Sswk	.70	11.429		10,000	243		10,243	11,500
6520	Maximum		.50	16		18,800	340		19,140	21,400
6550	Major surgery table, minimum		.50	16		22,700	340		23,040	25,600
6570	Maximum		.50	16		75,000	340		75,340	83,000
6700	Surgical lights, doctor's office, single arm	2 Elec	2	8		1,050	177		1,227	1,450
6750	Dual arm		1	16		4,000	355		4,355	4,975
6800	Major operating room, dual head, minimum		1	16		12,600	355		12,955	14,500
6850	Maximum		1	16		21,000	355		21,355	23,700
7000	Tables, physical therapy, walk off, electric	2 Carp	3	5.333		2,300	105		2,405	2,725
7150	Standard, vinyl top with base cabinets, minimum		3	5.333		940	105		1,045	1,200
7200	Maximum		2	8		2,775	158		2,933	3,325
7500	Thermometer, electric, portable					288			288	315
8100	Utensil washer-sanitizer	1 Plum	2	4		7,875	88		7,963	8,825
8200	Bed pan washer-sanitizer	"	2	4		5,150	88		5,238	5,800
8400	Whirlpool bath, mobile, sst, 18" x 24" x 60"					3,425			3,425	3,750
8450	Fixed, incl. mixing valves	1 Plum	2	4		6,650	88		6,738	7,450
8700	X-ray, mobile, minimum					9,650			9,650	10,600
8750	Maximum					54,500			54,500	60,000
8900	Stationary, minimum					30,200			30,200	33,200
8950	Maximum					157,000			157,000	172,500
9150	Developing processors, minimum					8,400			8,400	9,250
9200	Maximum					22,100			22,100	24,300
9800	For hospital partitions, see division 10190-200									
9900	For casework & furniture, see division 12310-750 & 12560-400									

11740 | Dental Equipment

		CREW	DAILY OUTPUT	LABOR-HOURS	UNIT	MAT.	LABOR	EQUIP.	TOTAL	TOTAL INCL O&P
0010	**DENTAL EQUIPMENT**									
0020	Central suction system, minimum	1 Plum	1.20	6.667	Ea.	2,700	146		2,846	3,225
0100	Maximum	"	.90	8.889		20,900	195		21,095	23,300
0300	Air compressor, minimum	1 Skwk	.80	10		2,975	198		3,173	3,625
0400	Maximum		.50	16		20,600	315		20,915	23,200
0600	Chair, electric or hydraulic, minimum		.50	16		3,400	315		3,715	4,300
0700	Maximum		.25	32		7,200	630		7,830	9,000
0800	Doctor's/assistant's stool, minimum					385			385	425
0850	Maximum					755			755	835
1000	Drill console with accessories, minimum	1 Skwk	1.60	5		3,075	99		3,174	3,550
1100	Maximum		1.60	5		8,750	99		8,849	9,800
2000	Light, ceiling mounted, minimum		8	1		1,775	19.75		1,794.75	1,975
2100	Maximum		8	1		3,575	19.75		3,594.75	3,950
2200	Unit light, minimum	2 Skwk	5.33	3.002		1,325	59.50		1,384.50	1,550
2210	Maximum		5.33	3.002		4,325	59.50		4,384.50	4,850
2220	Track light, minimum		3.20	5		2,350	99		2,449	2,750
2230	Maximum		3.20	5		5,150	99		5,249	5,850
2300	Sterilizers, steam portable, minimum					2,975			2,975	3,275
2350	Maximum					10,300			10,300	11,300
2600	Steam, institutional					12,800			12,800	14,100
2650	Dry heat, electric, portable, 3 trays					1,000			1,000	1,100
2700	Ultra-sonic cleaner, portable, minimum					810			810	890
2750	Maximum (institutional)					1,800			1,800	2,000
3000	X-ray unit, wall, minimum	1 Skwk	4	2		4,100	39.50		4,139.50	4,575
3010	Maximum		4	2		7,725	39.50		7,764.50	8,575
3100	Panoramic unit		.60	13.333		18,500	263		18,763	20,900

11700 | Medical Equipment

11740 | Dental Equipment

		CREW	DAILY OUTPUT	LABOR-HOURS	UNIT	2000 BARE COSTS				TOTAL INCL O&P	
						MAT.	LABOR	EQUIP.	TOTAL		
3105	Deluxe, minimum	2 Skwk	1.60	10	Ea.	26,800	198		26,998	29,800	200
3110	Maximum	"	1.60	10		62,000	198		62,198	68,500	
3500	Developers, X-ray, average	1 Plum	5.33	1.501		3,600	33		3,633	4,025	
3600	Maximum	"	5.33	1.501		6,700	33		6,733	7,425	

For information about Means Estimating Seminars, see yellow pages 11 and 12 in back of book

Division 12 Furnishings

Estimating Tips
General
- The items in this division are usually priced per square foot or each. Most of these items are purchased by the owner and placed by the supplier. Do not assume the items in Division 12 will be purchased and installed by the supplier. Check the specifications for responsibilities and include receiving, storage, installation and mechanical and electrical hook-ups in the appropriate divisions.

- Some items in this division require some type of support system that is not usually furnished with the item. Examples of these systems include blocking for the attachment of casework and heavy drapery rods. The required blocking must be added to the estimate in the appropriate division.

Reference Numbers
Reference numbers are shown in bold squares at the beginning of some major classifications. These numbers refer to related items in the Reference Section. The reference information may be an estimating procedure, an alternate pricing method or technical information.

Note: Not all subdivisions listed here necessarily appear in this publication.

12300 | Manufactured Casework

12310 | Metal Casework

		CREW	DAILY OUTPUT	LABOR-HOURS	UNIT	2000 BARE COSTS MAT.	LABOR	EQUIP.	TOTAL	TOTAL INCL O&P	
100	0010	**KEY CABINETS**									100
	0020	Wall mounted, 60 key capacity	1 Carp	20	.400	Ea.	166	7.90		173.90	197	
	0200	Drawer type, 600 key capacity	1 Clab	15	.533		2,125	7.70		2,132.70	2,350	
	0300	2,400 key capacity		20	.400		4,775	5.80		4,780.80	5,250	
	0400	Tray type, 20 key capacity		50	.160		96	2.31		98.31	110	
	0500	50 key capacity		40	.200		127	2.89		129.89	145	
200	0010	**DISPLAY CASES** Free standing, all glass										200
	0020	Aluminum frame, 42" high x 36" x 12" deep	2 Carp	8	2	Ea.	460	39.50		499.50	580	
	0100	70" high x 48" x 18" deep	"	6	2.667		1,125	52.50		1,177.50	1,350	
	0500	For wood bases, add					9%					
	0600	For hardwood frames, deduct					8%					
	0700	For bronze, baked enamel finish, add					10%					
	2000	Wall mounted, glass front, aluminum frame										
	2010	Non-illuminated, one section 3' x 4' x 1'-4"	2 Carp	5	3.200	Ea.	965	63		1,028	1,150	
	2100	5' x 4' x 1'-4"		5	3.200		1,150	63		1,213	1,375	
	2200	6' x 4' x 1'-4"		4	4		1,250	79		1,329	1,500	
	2500	Two sections, 8' x 4' x 1'-4"		2	8		1,700	158		1,858	2,125	
	2600	10' x 4' x 1'-4"		2	8		1,825	158		1,983	2,275	
	3000	Three sections, 16' x 4' x 1'-4"		1.50	10.667		2,700	210		2,910	3,325	
	3500	For fluorescent lights, add				Section	101			101	111	
	4000	Table exhibit cases, 2' wide, 3' high, 4' long, flat top	2 Carp	5	3.200	Ea.	695	63		758	875	
	4100	3' wide, 3' high, 4' long, sloping top	"	3	5.333	"	1,050	105		1,155	1,325	
560	0010	**IRONING CENTER**										560
	0020	Including cabinet, board & light, minimum	1 Carp	2	4	Ea.	240	79		319	400	
	0100	Maximum, see also division 11119-450	"	1.50	5.333	"	465	105		570	695	
750	0010	**CASEWORK**										750
	0500	Hospital, base cabinets, laminated plastic	2 Carp	10	1.600	L.F.	180	31.50		211.50	252	
	1000	Stainless steel	"	10	1.600		225	31.50		256.50	300	
	1200	For all drawers, add					9.20			9.20	10.10	
	1300	Cabinet base trim, 4" high, enameled steel	2 Carp	200	.080		27.50	1.58		29.08	32.50	
	1400	Stainless steel		200	.080		41	1.58		42.58	48	
	1450	Counter top, laminated plastic, no backsplash		40	.400		31	7.90		38.90	47.50	
	1650	With backsplash		40	.400		38.50	7.90		46.40	56	
	1800	For sink cutout, add		12.20	1.311	Ea.		26		26	44.50	
	1900	Stainless steel counter top		40	.400	L.F.	86	7.90		93.90	109	
	2000	For drop-in stainless 43" x 21" sink, add				Ea.	440			440	485	
	2050	For solid surface countertops see 06610-810										
	2100	Nurses station, door type, laminated plastic	2 Carp	10	1.600	L.F.	209	31.50		240.50	284	
	2200	Enameled steel		10	1.600		181	31.50		212.50	253	
	2300	Stainless steel		10	1.600		241	31.50		272.50	320	
	2400	For drawer type, add					84.50			84.50	93	
	2500	Wall cabinets, laminated plastic	2 Carp	15	1.067		135	21		156	184	
	2600	Enameled steel		15	1.067		151	21		172	202	
	2700	Stainless steel		15	1.067		234	21		255	293	
	2800	For glass doors, add					21.50			21.50	23.50	
	3500	Kitchen, base cabinets, metal, minimum	2 Carp	30	.533		44	10.50		54.50	66.50	
	3600	Maximum		25	.640		112	12.60		124.60	145	
	3700	Wall cabinets, metal, minimum		30	.533		44	10.50		54.50	66.50	
	3800	Maximum		25	.640		101	12.60		113.60	134	
	5000	School, 24" deep, metal, 84" high units		15	1.067		221	21		242	279	
	5150	Counter height units		20	.800		172	15.75		187.75	216	
	5450	Wood, custom fabricated, 32" high counter		20	.800		143	15.75		158.75	184	
	5600	Add for counter top		56	.286		16.70	5.65		22.35	28	
	5800	84" high wall units		15	1.067		149	21		170	200	
	6000	Laminated plastic finish is same price as wood										

12400 | Furnishings & Accessories

12411 | Furniture Accessories

			CREW	DAILY OUTPUT	LABOR-HOURS	UNIT	MAT.	LABOR	EQUIP.	TOTAL	TOTAL INCL O&P
900	0010	**ASH/TRASH RECEIVERS**									900
	1000	Ash urn, cylindrical metal									
	1020	8" diameter, 20" high	1 Clab	60	.133	Ea.	110	1.93		111.93	124
	1060	10" diameter, 26" high	"	60	.133	"	155	1.93		156.93	173
	2000	Combination ash/trash urn, metal									
	2020	8" diameter, 20" high	1 Clab	60	.133	Ea.	101	1.93		102.93	114
	2050	10" diameter, 26" high	"	60	.133	"	174	1.93		175.93	194
	4000	Trash receptacle, metal									
	4020	8" diameter, 15" high	1 Clab	60	.133	Ea.	58	1.93		59.93	67
	4040	10" diameter, 18" high		60	.133		99	1.93		100.93	112
	5040	16" x 8" x 14" high		60	.133		29	1.93		30.93	35.50
	5500	Trash receptacle, plastic, with lid									
	5520	35 gallon	1 Clab	60	.133	Ea.	121	1.93		122.93	137
	5540	45 gallon	"	60	.133	"	149	1.93		150.93	167

12483 | Floor Mats & Frames

			CREW	DAILY OUTPUT	LABOR-HOURS	UNIT	MAT.	LABOR	EQUIP.	TOTAL	TOTAL INCL O&P
200	0010	**FLOOR MATS**									200
	0020	Recessed, in-laid black rubber, 3/8" thick, solid	1 Clab	155	.052	S.F.	19.20	.75		19.95	22.50
	0050	Perforated		155	.052		25	.75		25.75	29
	0100	1/2" thick, solid		155	.052		20.50	.75		21.25	24
	0150	Perforated		155	.052		23	.75		23.75	26.50
	0200	In colors, 3/8" thick, solid		155	.052		22	.75		22.75	25.50
	0250	Perforated		155	.052		24.50	.75		25.25	28.50
	0300	1/2" thick, solid		155	.052		30.50	.75		31.25	35
	0350	Perforated		155	.052		31.50	.75		32.25	36.50
	0500	Link mats, including nosings, aluminum, 3/8" thick		155	.052		16.85	.75		17.60	19.85
	0550	Black rubber with galvanized tie rods		155	.052		13.25	.75		14	15.90
	0600	Steel, galvanized, 3/8" thick		155	.052		6.40	.75		7.15	8.35
	0650	Vinyl, in colors		155	.052		17.15	.75		17.90	20
	0750	Add for nosings, rubber				L.F.	6.15			6.15	6.75
	0850	Recess frames for above mats, aluminum	1 Carp	100	.080		3.81	1.58		5.39	6.90
	0870	Bronze	"	100	.080		8.05	1.58		9.63	11.55
	0900	Skate lock tile, 24" x 24" x 1/2" thick, rubber, black	1 Clab	125	.064	S.F.	12.95	.92		13.87	15.80
	0950	Color		125	.064	"	14.90	.92		15.82	18
	1000	12" x 24" border, black		75	.107	L.F.	21.50	1.54		23.04	26
	1100	Color		75	.107	"	25.50	1.54		27.04	31
	1150	12" x 12" outside corner, black		100	.080	S.F.	20.50	1.16		21.66	25
	1200	Color		100	.080		24	1.16		25.16	28.50
	1500	Duckboard, aluminum slats		155	.052		17.80	.75		18.55	21
	1700	Hardwood strips on rubber base, to 54" wide		155	.052		12.50	.75		13.25	15.05
	1800	Assembled with brass rods and vinyl spacers, to 48" wide		155	.052		16.85	.75		17.60	19.85
	1850	Tire fabric, 3/4" thick		155	.052		10.25	.75		11	12.60
	1900	Vinyl, 36" wide, in colors, hollow top & bottoms		155	.052		4.24	.75		4.99	5.95
	1950	Solid top & bottom members		155	.052		7.75	.75		8.50	9.85

12492 | Blinds and Shades

			CREW	DAILY OUTPUT	LABOR-HOURS	UNIT	MAT.	LABOR	EQUIP.	TOTAL	TOTAL INCL O&P
100	0010	**BLINDS, INTERIOR**									100
	0020	Horizontal, 1" aluminum slats, solid color, stock	1 Carp	590	.014	S.F.	2.60	.27		2.87	3.32
	0090	Custom, minimum		590	.014		2.06	.27		2.33	2.73
	0100	Maximum		440	.018		6.50	.36		6.86	7.75
	0250	2" aluminum slats, custom, minimum		590	.014		2.71	.27		2.98	3.44
	0350	Maximum		440	.018		6.40	.36		6.76	7.60
	0450	Stock, minimum		590	.014		2.66	.27		2.93	3.39
	0500	Maximum		440	.018		4.94	.36		5.30	6.05
	0600	2" steel slats, stock, minimum		590	.014		1.50	.27		1.77	2.11
	0630	Maximum		440	.018		4.18	.36		4.54	5.20

12400 | Furnishings & Accessories

12492 | Blinds and Shades

			CREW	DAILY OUTPUT	LABOR-HOURS	UNIT	MAT.	LABOR	EQUIP.	TOTAL	TOTAL INCL O&P	
100	0750	Custom, minimum	1 Carp	590	.014	S.F.	1.35	.27		1.62	1.95	100
	0850	Maximum		400	.020		6.70	.39		7.09	8.05	
	1500	Vertical, 3" to 5" PVC or cloth strips, minimum		460	.017		6.55	.34		6.89	7.85	
	1600	Maximum		400	.020		8.75	.39		9.14	10.30	
	1800	4" aluminum slats, minimum		460	.017		2.93	.34		3.27	3.81	
	1900	Maximum		400	.020		6.95	.39		7.34	8.30	
	1950	Mylar mirror-finish strips, to 8" wide, minimum		460	.017		11.60	.34		11.94	13.35	
	1970	Maximum		400	.020		17.40	.39		17.79	19.85	
	3000	Wood folding panels with movable louvers, 7" x 20" each		17	.471	Pr.	36	9.25		45.25	55.50	
	3300	8" x 28" each		17	.471		52	9.25		61.25	73	
	3450	9" x 36" each		17	.471		62	9.25		71.25	84	
	3600	10" x 40" each		17	.471		70	9.25		79.25	93	
	4000	Fixed louver type, stock units, 8" x 20" each		17	.471		54	9.25		63.25	75.50	
	4150	10" x 28" each		17	.471		74	9.25		83.25	97.50	
	4300	12" x 36" each		17	.471		94	9.25		103.25	119	
	4450	18" x 40" each		17	.471		112	9.25		121.25	139	
	5000	Insert panel type, stock, 7" x 20" each		17	.471		13.90	9.25		23.15	31	
	5150	8" x 28" each		17	.471		25.50	9.25		34.75	44	
	5300	9" x 36" each		17	.471		32.50	9.25		41.75	51.50	
	5450	10" x 40" each		17	.471		35	9.25		44.25	54	
	5600	Raised panel type, stock, 10" x 24" each		17	.471		93	9.25		102.25	118	
	5650	12" x 26" each		17	.471		108	9.25		117.25	135	
	5700	14" x 30" each		17	.471		122	9.25		131.25	150	
	5750	16" x 36" each		17	.471		137	9.25		146.25	167	
	6000	For custom built pine, add					22%					
	6500	For custom built hardwood blinds, add					42%					
600	0010	**SHADES**										600
	0020	Basswood, roll-up, stain finish, 3/8" slats	1 Carp	300	.027	S.F.	9	.53		9.53	10.80	
	0200	7/8" slats		300	.027		8.50	.53		9.03	10.25	
	0300	Vertical side slide, stain finish, 3/8" slats		300	.027		12.50	.53		13.03	14.65	
	0400	7/8" slats		300	.027		12.50	.53		13.03	14.65	
	0500	For fire retardant finishes, add					16%					
	0600	For "B" rated finishes, add					20%					
	0900	Mylar, single layer, non-heat reflective	1 Carp	685	.012		5	.23		5.23	5.90	
	1000	Double layered, heat reflective		685	.012		6.90	.23		7.13	8	
	1100	Triple layered, heat reflective		685	.012		6.90	.23		7.13	8	
	1200	For metal roller instead of wood, add per				Shade	2.63			2.63	2.89	
	1300	Vinyl coated cotton, standard	1 Carp	685	.012	S.F.	1	.23		1.23	1.49	
	1400	Lightproof decorator shades		685	.012		1.03	.23		1.26	1.52	
	1500	Vinyl, lightweight, 4 gauge		685	.012		.37	.23		.60	.80	
	1600	Heavyweight, 6 gauge		685	.012		1.15	.23		1.38	1.66	
	1700	Vinyl laminated fiberglass, 6 ga., translucent		685	.012		1.08	.23		1.31	1.58	
	1800	Lightproof		685	.012		2.63	.23		2.86	3.28	
	3000	Woven aluminum, 3/8" thick, lightproof and fireproof		350	.023		3.79	.45		4.24	4.94	

12493 | Curtains and Drapes

			CREW	DAILY OUTPUT	LABOR-HOURS	UNIT	MAT.	LABOR	EQUIP.	TOTAL	TOTAL INCL O&P	
200	0010	**DRAPERY HARDWARE**										200
	0030	Standard traverse, per foot, minimum	1 Carp	59	.136	L.F.	1.40	2.67		4.07	6.10	
	0100	Maximum		51	.157	"	7.50	3.09		10.59	13.55	
	4000	Traverse rods, adjustable, 28" to 48"		22	.364	Ea.	15.20	7.15		22.35	29	
	4020	48" to 84"		20	.400		24.50	7.90		32.40	40.50	
	4040	66" to 120"		18	.444		26	8.75		34.75	43.50	
	4060	84" to 156"		16	.500		30	9.85		39.85	49.50	
	4080	100" to 180"		14	.571		29.50	11.25		40.75	51.50	
	4100	228" to 312"		13	.615		45.50	12.10		57.60	71	
	4600	Valance, pinch pleated fabric, 12" deep, up to 54" long, minimum					28			28	31	

12400 | Furnishings & Accessories

12493 | Curtains and Drapes

		CREW	DAILY OUTPUT	LABOR-HOURS	UNIT	2000 BARE COSTS				TOTAL INCL O&P	
						MAT.	LABOR	EQUIP.	TOTAL		
200	4610	Maximum				Ea.	70			70	77	200
	4620	Up to 77" long, minimum					43			43	47.50	
	4630	Maximum					113			113	124	
	5000	Stationary rods, first 2 feet				▼	6.95			6.95	7.65	
	5020	Each additional foot, add				L.F.	2.79			2.79	3.07	

12500 | Furniture

12510 | Office Furniture

		CREW	DAILY OUTPUT	LABOR-HOURS	UNIT	2000 BARE COSTS				TOTAL INCL O&P	
						MAT.	LABOR	EQUIP.	TOTAL		
580	0010	**MULTI-MEDIA BOARDS** See division 10110-240										580
600	0010	**OFFICE CASE GOODS**										600
	0020	Desks, 29" high, double pedestal, 30" x 60", metal, minimum				Ea.	340			340	375	
	0030	Maximum					965			965	1,050	
	0160	Wood, minimum					425			425	465	
	0180	Maximum					1,750			1,750	1,925	
	0600	Desks, single pedestal, 30" x 60", metal, minimum					283			283	310	
	0620	Maximum					735			735	805	
	0640	Wood, minimum					365			365	405	
	0650	Maximum					550			550	605	
	0670	Executive return, 24" x 42", with box, file, wood, minimum					252			252	277	
	0680	Maximum					550			550	605	
	0720	Desks, secretarial, 30" x 60", metal, minimum					194			194	213	
	0730	Maximum					550			550	605	
	0740	Return, 20" x 42", minimum					183			183	202	
	0750	Maximum					365			365	405	
	0800	Wood, minimum					510			510	560	
	0810	Maximum					1,775			1,775	1,950	
	0820	Return, 20" x 42", minimum					325			325	355	
	0830	Maximum					945			945	1,025	
	0900	Desktop organizer, 72" x 14" x 36" high, wood, minimum					105			105	115	
	0920	Maximum					236			236	259	
	0940	59" x 12" x 23" high, steel, minimum					136			136	150	
	0960	Maximum					210			210	231	
	2000	Chairs, office type, executive, minimum					252			252	277	
	2150	Maximum					1,600			1,600	1,775	
	2200	Management, minimum					209			209	230	
	2250	Maximum					1,200			1,200	1,325	
	2280	Task, minimum					136			136	150	
	2290	Maximum					465			465	515	
	2300	Arm kit, minimum					43.50			43.50	47.50	
	2320	Maximum				▼	89			89	98	
	6000	Table, conference										
	6050	Boat, 96" x 42", minimum				Ea.	565			565	620	
	6150	Maximum					3,150			3,150	3,450	
	6720	Rectangle, 96" x 42", minimum					625			625	690	
	6740	Maximum				▼	2,600			2,600	2,850	
675	0010	**POSTS**										675
	0020	Portable for pedestrian traffic control, standard, minimum				Ea.	89			89	97.50	

12500 | Furniture

12510 | Office Furniture

			CREW	DAILY OUTPUT	LABOR-HOURS	UNIT	2000 BARE COSTS MAT.	LABOR	EQUIP.	TOTAL	TOTAL INCL O&P	
675	0100	Maximum				Ea.	136			136	150	675
	0300	Deluxe posts, minimum					150			150	165	
	0400	Maximum				↓	285			285	315	
	0600	Ropes for above posts, plastic covered, 1-1/2" diameter				L.F.	8.10			8.10	8.95	
	0700	Chain core				"	8.25			8.25	9.05	

12520 | Seating

			CREW	DAILY OUTPUT	LABOR-HOURS	UNIT	MAT.	LABOR	EQUIP.	TOTAL	INCL O&P	
550	0010	**SEATING**										550
	0100	Bleachers, See div. 02880 & 11486										
	0101	Benches, See 02880, 10505 & 11486										
	0500	Classroom, movable chair & desk type, minimum				Set					65	
	0600	Maximum				"					120	
	1000	Lecture hall, pedestal type, minimum	2 Carp	22	.727	Ea.	102	14.35		116.35	137	
	1200	Maximum		14.50	1.103		365	21.50		386.50	435	
	2000	Auditorium chair, all veneer construction		22	.727		115	14.35		129.35	151	
	2200	Veneer back, padded seat		22	.727		145	14.35		159.35	184	
	2350	Fully upholstered, spring seat	↓	22	.727		146	14.35		160.35	186	
	2450	For tablet arms, add					38			38	42	
	2500	For fire retardancy, CATB-133, add				↓	15.25			15.25	16.75	

12540 | Hospitality Furniture

			CREW	DAILY OUTPUT	LABOR-HOURS	UNIT	MAT.	LABOR	EQUIP.	TOTAL	INCL O&P	
100	0010	**TABLES, FOLDING** Laminated plastic tops										100
	1000	Tubular steel legs with glides										
	1020	18" x 60", minimum				Ea.	164			164	181	
	1040	Maximum					570			570	625	
	1840	36" x 96", minimum					267			267	294	
	1860	Maximum					1,150			1,150	1,275	
	2000	Round, wood stained, plywood top, 60" diameter, minimum					247			247	271	
	2020	Maximum				↓	1,325			1,325	1,475	
500	0010	**FURNITURE, HOTEL**										500
	0020	Standard quality set, minimum				Room	1,475			1,475	1,625	
	0200	Maximum				"	7,175			7,175	7,875	
700	0010	**FURNITURE, RESTAURANT**										700
	0020	Bars, built-in, front bar	1 Carp	5	1.600	L.F.	189	31.50		220.50	262	
	0200	Back bar	"	5	1.600	"	137	31.50		168.50	205	
	0300	Booth seating see division 12640-150										
	2000	Chair, bentwood side chair, metal, minimum				Ea.	66.50			66.50	73.50	
	2020	Maximum					85			85	93.50	
	2600	Upholstered seat & back, arms, minimum					190			190	208	
	2620	Maximum					490			490	540	

12560 | Institutional Furniture

			CREW	DAILY OUTPUT	LABOR-HOURS	UNIT	MAT.	LABOR	EQUIP.	TOTAL	INCL O&P	
100	0010	**BANK FURNITURE** See division 12510-600										100
150	0010	**CHURCH FURNITURE** See division 11041-250										150
200	0010	**FURNITURE, SCHOOL**										200
	1000	Chair, molded plastic,										
	1100	Integral tablet arm, minimum				Ea.	52.50			52.50	58	
	1150	Maximum					212			212	233	
	2000	Desk, single pedestal, top book compartment, minimum					48			48	53	
	2020	Maximum					52.50			52.50	57.50	
	2200	Flip top, minimum					62			62	68	
	2220	Maximum				↓	67			67	73.50	

12500 | Furniture

12560 | Institutional Furniture

			CREW	DAILY OUTPUT	LABOR-HOURS	UNIT	2000 BARE COSTS				TOTAL INCL O&P
							MAT.	LABOR	EQUIP.	TOTAL	
300	0010	**FURNITURE, DORMITORY**										300
	1000	Chest, four drawer, minimum				Ea.	350			350	385
	1020	Maximum				"	495			495	545
	1050	Built-in, minimum	2 Carp	13	1.231	L.F.	105	24.50		129.50	158
	1150	Maximum		10	1.600		208	31.50		239.50	283
	1200	Desk top, built-in, laminated plastic, 24" deep, minimum		50	.320		23	6.30		29.30	36.50
	1300	Maximum		40	.400		69.50	7.90		77.40	89.50
	1450	30" deep, minimum		50	.320		29	6.30		35.30	43
	1550	Maximum		40	.400		128	7.90		135.90	155
	1750	Dressing unit, built-in, minimum		12	1.333		172	26.50		198.50	234
	1850	Maximum	▼	8	2		515	39.50		554.50	635
	8000	Rule of thumb: total cost for furniture, minimum				Student					1,900
	8050	Maximum				"					3,600
400	0010	**FURNITURE, HOSPITAL**										400
	0020	Beds, manual, minimum				Ea.	810			810	890
	0100	Maximum					1,425			1,425	1,550
	0300	Manual and electric beds, minimum					1,025			1,025	1,125
	0400	Maximum					2,175			2,175	2,400
	0600	All electric hospital beds, minimum					1,350			1,350	1,475
	0700	Maximum					3,350			3,350	3,700
	0900	Manual, nursing home beds, minimum					640			640	705
	1000	Maximum					1,325			1,325	1,450
	1020	Overbed table, laminated top, minimum					272			272	300
	1040	Maximum				▼	680			680	745
	1100	Patient wall systems, not incl. plumbing, minimum				Room	815			815	895
	1200	Maximum				"	1,500			1,500	1,650
	2000	Geriatric chairs, minimum				Ea.	263			263	289
	2020	Maximum				"	615			615	675
700	0010	**FURNITURE, LIBRARY**										700
	0100	Attendant desk, 36" x 62" x 29" high	1 Carp	16	.500	Ea.	2,200	9.85		2,209.85	2,425
	0200	Book display, "A" frame display, both sides, 42" x 42" x 60" high		16	.500		2,350	9.85		2,359.85	2,600
	0220	Table with bulletin board, 42" x 24" x 49" high		16	.500		1,200	9.85		1,209.85	1,350
	0800	Card catalogue, 30 tray unit		16	.500		2,375	9.85		2,384.85	2,650
	0840	60 tray unit	▼	16	.500		4,075	9.85		4,084.85	4,500
	0880	72 tray unit	2 Carp	16	1		4,425	19.70		4,444.70	4,875
	1000	Carrels, single face, initial unit	1 Carp	16	.500		725	9.85		734.85	815
	1500	Double face, initial unit	2 Carp	16	1		1,100	19.70		1,119.70	1,225
	4000	Dictionary stand, stationary	1 Carp	16	.500		670	9.85		679.85	755
	4020	Revolving		16	.500		233	9.85		242.85	273
	4200	Exhibit case, table style, 60" x 28" x 36"		11	.727		3,325	14.35		3,339.35	3,675
	7000	Tables, card catalog reference, 24" x 60" x 42"	▼	16	.500	▼	750	9.85		759.85	840

12600 | Multiple Seating

12640 | Booths & Tables

			CREW	DAILY OUTPUT	LABOR-HOURS	UNIT	2000 BARE COSTS				TOTAL INCL O&P
							MAT.	LABOR	EQUIP.	TOTAL	
150	0010	**BOOTHS**										150
	1000	Banquet, upholstered seat and back, custom									

12600 | Multiple Seating

12640 | Booths & Tables

		CREW	DAILY OUTPUT	LABOR-HOURS	UNIT	2000 BARE COSTS MAT.	LABOR	EQUIP.	TOTAL	TOTAL INCL O&P	
150	1500 Straight, minimum	2 Carp	40	.400	L.F.	109	7.90		116.90	134	150
	1520 Maximum		36	.444		212	8.75		220.75	248	
	1600 "L" or "U" shape, minimum		35	.457		111	9		120	137	
	1620 Maximum		30	.533		197	10.50		207.50	235	
	1800 Upholstered outside finished backs for										
	1810 single booths and custom banquets										
	1820 Minimum	2 Carp	44	.364	L.F.	12.55	7.15		19.70	26	
	1840 Maximum	"	40	.400	"	37.50	7.90		45.40	55	
	3000 Fixed seating, one piece plastic chair and										
	3010 plastic laminate table top										
	3100 Two seat, 24" x 24" table, minimum	F-7	30	1.067	Ea.	440	18.20		458.20	510	
	3120 Maximum		26	1.231		435	21		456	515	
	3200 Four seat, 24" x 48" table, minimum		28	1.143		435	19.50		454.50	515	
	3220 Maximum		24	1.333		740	23		763	855	
	5000 Mount in floor, wood fiber core with										
	5010 plastic laminate face, single booth										
	5050 24" wide	F-7	30	1.067	Ea.	168	18.20		186.20	216	
	5100 48" wide	"	28	1.143	"	213	19.50		232.50	269	

12800 | Interior Plants & Planters

12830 | Interior Planters

		CREW	DAILY OUTPUT	LABOR-HOURS	UNIT	2000 BARE COSTS MAT.	LABOR	EQUIP.	TOTAL	TOTAL INCL O&P	
600	0010 PLANTERS										600
	1000 Fiberglass, hanging, 12" diameter, 7" high				Ea.	38			38	41.50	
	1500 Rectangular, 48" long, 16" high x 15" wide					310			310	340	
	1650 60" long, 30" high, 28" wide					660			660	725	
	2000 Round, 12" diameter, 13" high					66			66	72.50	
	2050 25" high					78.50			78.50	86.50	
	5000 Square, 10" side, 20" high					91			91	100	
	5100 14" side, 15" high					92.50			92.50	102	
	6000 Metal bowl, 32" diameter, 8" high, minimum					310			310	340	
	6050 Maximum					465			465	510	
	8750 Wood, fiberglass liner, square										
	8780 14" square, 15" high, minimum				Ea.	171			171	188	
	8800 Maximum					210			210	232	
	9400 Plastic cylinder, molded, 10" diameter, 10" high					6.95			6.95	7.65	
	9500 11" diameter, 11" high					22			22	24	

For information about Means Estimating Seminars, see yellow pages 11 and 12 in back of book

Division 13
Special Construction

Estimating Tips
General
- The items and systems in this division are usually estimated, purchased, supplied and installed as a unit by one or more subcontractors. The estimator must ensure that all parties are operating from the same set of specifications and assumptions and that all necessary items are estimated and will be provided. Many times the complex items and systems are covered but the more common ones such as excavation or a crane are overlooked for the very reason that everyone assumes nobody could miss them. The estimator should be the central focus and be able to ensure that all systems are complete.
- Another area where problems can develop in this division is at the interface between systems. The estimator must ensure, for instance, that anchor bolts, nuts and washers are estimated and included for the air-supported structures and pre-engineered buildings to be bolted to their foundations.

Utility supply is a common area where essential items or pieces of equipment can be missed or overlooked due to the fact that each subcontractor may feel it is the others' responsibility. The estimator should also be aware of certain items which may be supplied as part of a package but installed by others, and ensure that the installing contractor's estimate includes the cost of installation. Conversely, the estimator must also ensure that items are not costed by two different subcontractors, resulting in an inflated overall estimate.

13120 Pre-Engineered Structures
- The foundations and floor slab, as well as rough mechanical and electrical, should be estimated, as this work is required for the assembly and erection of the structure. Generally, as noted in the book, the pre-engineered building comes as a shell and additional features must be included by the estimator. Here again, the estimator must have a clear understanding of the scope of each portion of the work and all the necessary interfaces.

13200 Storage Tanks
- The prices in this subdivision for above and below ground storage tanks do not include foundations or hold-down slabs. The estimator should refer to Divisions 2 and 3 for foundation system pricing. In addition to the foundations, required tank accessories such as tank gauges, leak detection devices, and additional manholes and piping must be added to the tank prices.

Reference Numbers
Reference numbers are shown in bold squares at the beginning of some major classifications. These numbers refer to related items in the Reference Section. The reference information may be an estimating procedure, an alternate pricing method or technical information.

Note: Not all subdivisions listed here necessarily appear in this publication.

13010 | Air Supported Structures

13011 | Air Supported Structures

		CREW	DAILY OUTPUT	LABOR-HOURS	UNIT	2000 BARE COSTS				TOTAL INCL O&P
						MAT.	LABOR	EQUIP.	TOTAL	
100	**0010 AIR SUPPORTED STORAGE TANK COVERS** Vinyl polyester									**100**
	0100 scrim, double layer, with hardware, blower, standby & controls									
	0200 Round, 75' diameter	B-2	4,500	.009	S.F.	5.60	.13		5.73	6.40
	0300 100' diameter		5,000	.008		5	.12		5.12	5.70
	0400 150' diameter		5,000	.008		3.55	.12		3.67	4.11
	0500 Rectangular, 20' x 20'		4,500	.009		12	.13		12.13	13.45
	0600 30' x 40'		4,500	.009		12	.13		12.13	13.45
	0700 50' x 60'		4,500	.009		12	.13		12.13	13.45
	0800 For single wall construction, deduct, minimum					.27			.27	.30
	0900 Maximum					1.06			1.06	1.17
	1000 For maximum resistance to atmosphere or cold, add					.33			.33	.36
	1100 For average shipping charges, add				Total	1,050			1,050	1,150
200	**0010 AIR SUPPORTED STRUCTURES**	R13011-110								**200**
	0020 Site preparation, incl. anchor placement and utilities	B-11B	2,000	.008	SF Flr.	.66	.14	.51	1.31	1.52
	0030 For concrete curb, see division 03310-240									
	0050 Warehouse, polyester/vinyl fabric, 28 oz., over 10 yr. life, welded									
	0060 Seams, tension cables, primary & auxiliary inflation system,									
	0070 airlock, personnel doors and liner									
	0100 5000 S.F.	4 Clab	5,000	.006	SF Flr.	15.40	.09		15.49	17.10
	0250 12,000 S.F.	"	6,000	.005		10.20	.08		10.28	11.35
	0400 24,000 S.F.	8 Clab	12,000	.005		7.80	.08		7.88	8.75
	0500 50,000 S.F.	"	12,500	.005		6.20	.07		6.27	6.95
	0700 12 oz. reinforced vinyl fabric, 5 yr. life, sewn seams,									
	0710 accordian door, including liner									
	0750 3000 S.F.	4 Clab	3,000	.011	SF Flr.	6.40	.15		6.55	7.30
	0800 12,000 S.F.	"	6,000	.005		5.10	.08		5.18	5.75
	0850 24,000 S.F.	8 Clab	12,000	.005		4.15	.08		4.23	4.70
	0950 Deduct for single layer					.58			.58	.64
	1000 Add for welded seams					.90			.90	.99
	1050 Add for double layer, welded seams included					1.85			1.85	2.04
	1250 Tedlar/vinyl fabric, 28 oz., with liner, over 10 yr. life,									
	1260 incl. overhead and personnel doors									
	1300 3000 S.F.	4 Clab	3,000	.011	SF Flr.	13.50	.15		13.65	15.10
	1450 12,000 S.F.	"	6,000	.005		9.45	.08		9.53	10.55
	1550 24,000 S.F.	8 Clab	12,000	.005		7.40	.08		7.48	8.30
	1700 Deduct for single layer					1.29			1.29	1.42
	2250 Greenhouse/shelter, woven polyethylene with liner, 2 yr. life,									
	2260 sewn seams, including doors									
	2300 3000 S.F.	4 Clab	3,000	.011	SF Flr.	5.90	.15		6.05	6.75
	2350 12,000 S.F.	"	6,000	.005		3.09	.08		3.17	3.53
	2450 24,000 S.F.	8 Clab	12,000	.005		2.69	.08		2.77	3.09
	2550 Deduct for single layer					.56			.56	.62
	2600 Tennis/gymnasium, polyester/vinyl fabric, 28 oz., over 10 yr. life,									
	2610 including thermal liner, heat and lights									
	2650 7200 S.F.	4 Clab	6,000	.005	SF Flr.	13	.08		13.08	14.45
	2750 13,000 S.F.	"	6,500	.005		9.50	.07		9.57	10.55
	2850 Over 24,000 S.F.	8 Clab	12,000	.005		8.50	.08		8.58	9.50
	2860 For low temperature conditions, add					.53			.53	.58
	2870 For average shipping charges, add				Total	2,000			2,000	2,200
	2900 Thermal liner, translucent reinforced vinyl				SF Flr.	.75			.75	.83
	2950 Metalized mylar fabric and mesh, double liner				"	1.30			1.30	1.43
	3050 Stadium/convention center, teflon coated fiberglass, heavy weight,									
	3060 over 20 yr. life, incl. thermal liner and heating system									
	3100 Minimum	9 Clab	26,000	.003	SF Flr.	35	.04		35.04	38.50
	3110 Maximum	"	19,000	.004	"	40	.05		40.05	44
	3400 Doors, air lock, 15' long, 10' x 10'	2 Carp	.80	20	Ea.	12,500	395		12,895	14,500

13010 | Air Supported Structures

13011 | Air Supported Structures

		CREW	DAILY OUTPUT	LABOR-HOURS	UNIT	MAT.	LABOR	EQUIP.	TOTAL	TOTAL INCL O&P	
200	3600	15' x 15'	2 Carp	.50	32	Ea.	18,500	630		19,130	21,500	200
	3700	For each added 5' length, add	R13011-110				2,000			2,000	2,200	
	3900	Revolving personnel door, 6' diameter, 6'-6" high	2 Carp	.80	20		8,800	395		9,195	10,400	
	4200	Double wall, self supporting, shell only, minimum				SF Flr.					21	
	4300	Maximum				"					39	

13030 | Special Purpose Rooms

13035 | Special Purpose Rooms

			CREW	DAILY OUTPUT	LABOR-HOURS	UNIT	MAT.	LABOR	EQUIP.	TOTAL	TOTAL INCL O&P	
100	0010	**INTEGRATED CEILINGS** Lighting, ventilating & acoustical										100
	0100	Luminaire, incl. connections, 5' x 5' modules, 50% lighted	L-3	90	.111	S.F.	3.53	2.24		5.77	7.70	
	0200	100% lighted	"	50	.200		5.25	4.04		9.29	12.65	
	0400	For ventilating capacity with perforations, add									.27	
	0600	For supply air diffuser, add				Ea.					64	
	0700	Dimensionaire, 2' x 4' board system, see also div. 09510-760	L-3	50	.200	L.F.	8.20	4.04		12.24	15.85	
	0900	Tile system	1 Carp	250	.032	S.F.	2.14	.63		2.77	3.43	
	1000	For air bar suspension, add, minimum									.61	
	1100	Maximum									1.05	
	1200	For vaulted coffer, including fixture, stock, add				Ea.	80			80	88	
	1300	Custom, add, minimum									133	
	1400	Average									345	
	1500	Maximum									575	
	1600	Grid sys. & ceil. tile only, flat profile, 5'x5' mod., 50% lghtd.										
	1620	Mineral fiber panels	1 Carp	700	.011	S.F.	3.23	.23		3.46	3.94	
	1630	Glass fiber panels	"	700	.011		3.47	.23		3.70	4.21	
	1640	100% lighted, add					1.68			1.68	1.85	
	1650	For supply air linear diffuser, add					.53			.53	.58	
	1660	For vaulted coffer, 5' x 5' modules, 50% lighted	1 Carp	493	.016		4.12	.32		4.44	5.10	
	1800	Radiant hot water system with finished acoustic ceiling,										
	1810	not including supply piping. Heating only (gross S.F.)										
	2100	Elementary schools, minimum				S.F.					5.30	
	2200	Maximum									6.65	
	2400	High schools and colleges, minimum									4.75	
	2500	Maximum									6.65	
	2700	Libraries, minimum									4.80	
	2800	Maximum									5.75	
	3000	Hospitals, minimum									6.50	
	3100	Maximum									8.15	
	3300	Office buildings, minimum									4.60	
	3400	Maximum									6.15	
	3600	For combined heating and cooling, add, minimum									30%	
	3700	Maximum									40%	
	4000	Radiant electric ceiling board, strapped between joists	1 Elec	250	.032		1.74	.71		2.45	3.07	
	4100	2' x 4' heating panel for grid system, manila finish		25	.320	Ea.	80	7.05		87.05	99.50	
	4200	Textured epoxy finish		22	.364		82	8.05		90.05	103	
	4300	Vinyl finish		19	.421		84	9.30		93.30	108	
	4400	Hair cell, ABS plastic finish		13	.615		94	13.60		107.60	126	
	4500	2' x 4' alternate blank panel, for use with above										
	4600	Manila finish	1 Elec	50	.160	Ea.	25	3.54		28.54	33.50	

13030 | Special Purpose Rooms

13035 | Special Purpose Rooms

			CREW	DAILY OUTPUT	LABOR-HOURS	UNIT	MAT.	LABOR	EQUIP.	TOTAL	TOTAL INCL O&P	
100	4700	Textured epoxy finish	1 Elec	45	.178	Ea.	30	3.93		33.93	39.50	100
	4800	Vinyl finish		40	.200		32	4.42		36.42	42.50	
	4900	Hair cell, ABS plastic finish	↓	25	.320	↓	64	7.05		71.05	82	
150	0010	**ANECHOIC CHAMBERS** Standard units, 7' ceiling heights										150
	0100	Area for pricing is net inside dimensions										
	0300	200 cycles per second cutoff, 25 S.F. floor area				SF Flr.					1,375	
	0400	50 S.F.									1,100	
	0600	75 S.F.									1,050	
	0700	100 S.F.									1,025	
	0900	For 150 cycles per second cutoff, add					30%	30%				
	1000	For 100 cycles per second cutoff, add				↓	45%	45%				
180	0010	**AUDIOMETRIC ROOMS** Under 500 S.F. surface	4 Carp	98	.327	SF Surf	40	6.45		46.45	55	180
	0100	Over 500 S.F. surface	"	120	.267	"	38	5.25		43.25	51	
200	0010	**CLEAN ROOMS**										200
	1100	Clean room, soft wall, 12' x 12', Class 100	1 Carp	.18	44.444	Ea.	12,200	875		13,075	14,900	
	1110	Class 1,000		.18	44.444		9,450	875		10,325	11,900	
	1120	Class 10,000		.21	38.095		8,050	750		8,800	10,100	
	1130	Class 100,000	↓	.21	38.095	↓	8,050	750		8,800	10,100	
	2800	Ceiling grid support, slotted channel struts 4'-0" O.C., ea. way				S.F.					6.50	
	3000	Ceiling panel, vinyl coated foil on mineral substrate										
	3020	Sealed, non-perforated				S.F.					1.40	
	4000	Ceiling panel seal, silicone sealant, 150 L.F./gal.	1 Carp	150	.053	L.F.	.22	1.05		1.27	2.04	
	4100	Two sided adhesive tape	"	240	.033	"	.10	.66		.76	1.24	
	4200	Clips, one per panel				Ea.	.85			.85	.94	
	6000	HEPA filter, 2'x4', 99.97% eff., 3" dp beveled frame (silicone seal)					260			260	286	
	6040	6" deep skirted frame (channel seal)					300			300	330	
	6100	99.99% efficient, 3" deep beveled frame (silicone seal)					280			280	310	
	6140	6" deep skirted frame (channel seal)					320			320	350	
	6200	99.999% efficient, 3" deep beveled frame (silicone seal)					320			320	350	
	6240	6" deep skirted frame (channel seal)				↓	360			360	395	
	7000	Wall panel systems, including channel strut framing										
	7020	Polyester coated aluminum, particle board				S.F.					20	
	7100	Porcelain coated aluminum, particle board									35	
	7400	Wall panel support, slotted channel struts, to 12' high				↓					18	
300	0010	**DARKROOMS** Shell, complete except for door, 64 S.F., 8' high	2 Carp	128	.125	SF Flr.	65.50	2.46		67.96	76	300
	0100	12' high		64	.250		78.50	4.93		83.43	95	
	0500	120 S.F. floor, 8' high		120	.133		48	2.63		50.63	57.50	
	0600	12' high		60	.267		59	5.25		64.25	74	
	0800	240 S.F. floor, 8' high		120	.133		34	2.63		36.63	42	
	0900	12' high		60	.267	↓	47	5.25		52.25	60.50	
	1200	Mini-cylindrical, revolving, unlined, 4' diameter		3.50	4.571	Ea.	3,250	90		3,340	3,725	
	1400	5'-6" diameter	↓	2.50	6.400		5,125	126		5,251	5,875	
	1600	Add for lead lining, inner cylinder, 1/32" thick					1,300			1,300	1,450	
	1700	1/16" thick					1,525			1,525	1,675	
	1800	Add for lead lining, inner and outer cylinder, 1/32" thick					2,425			2,425	2,675	
	1900	1/16" thick				↓	2,525			2,525	2,800	
	2000	For darkroom door, see division 11472-370										
500	0010	**MUSIC** Practice room, modular, perforated steel, under 500 S.F.	2 Carp	70	.229	SF Surf	21.50	4.50		26	31	500
	0100	Over 500 S.F.	"	80	.200	"	16.30	3.94		20.24	24.50	
700	0010	**REFRIGERATION** Curbs, 12" high, 4" thick, concrete	2 Carp	58	.276	L.F.	3.02	5.45		8.47	12.60	700
	1000	Doors, see division 08341-200										
	2400	Finishes, 2 coat portland cement plaster, 1/2" thick	1 Plas	48	.167	S.F.	.83	3.11		3.94	6.10	
	2500	For galvanized reinforcing mesh, add	1 Lath	335	.024	↓	.56	.46		1.02	1.37	

Important: See the Reference Section for critical supporting data - Reference Nos., Crews, & City Cost Indexes

13030 | Special Purpose Rooms

13035 | Special Purpose Rooms

		CREW	DAILY OUTPUT	LABOR-HOURS	UNIT	MAT.	LABOR	EQUIP.	TOTAL	TOTAL INCL O&P
2700	3/16" thick latex cement	1 Plas	88	.091	S.F.	1.42	1.70		3.12	4.40
2900	For glass cloth reinforced ceilings, add	"	450	.018		.34	.33		.67	.92
3100	Fiberglass panels, 1/8" thick	1 Carp	149.45	.054		1.88	1.05		2.93	3.88
3200	Polystyrene, plastic finish ceiling, 1" thick		274	.029		1.74	.58		2.32	2.90
3400	2" thick		274	.029		1.97	.58		2.55	3.16
3500	4" thick		219	.037		2.20	.72		2.92	3.65
3800	Floors, concrete, 4" thick	1 Cefi	93	.086		.82	1.63		2.45	3.55
3900	6" thick	"	85	.094		1.24	1.78		3.02	4.26
4000	Insulation, 1" to 6" thick, cork				B.F.	.80			.80	.88
4100	Urethane					.79			.79	.87
4300	Polystyrene, regular					.54			.54	.59
4400	Bead board					.41			.41	.45
4600	Installation of above, add per layer	2 Carp	657.60	.024	S.F.	.25	.48		.73	1.10
4700	Wall and ceiling juncture		298.90	.054	L.F.	1.28	1.05		2.33	3.22
4900	Partitions, galvanized sandwich panels, 4" thick, stock		219.20	.073	S.F.	5.40	1.44		6.84	8.40
5000	Aluminum or fiberglass		219.20	.073	"	5.95	1.44		7.39	8.95
5200	Prefab walk-in, 7'-6" high, aluminum, incl. door & floors,									
5210	not incl. partitions or refrigeration, 6' x 6' O.D. nominal	2 Carp	54.80	.292	SF Flr.	97	5.75		102.75	117
5500	10' x 10' O.D. nominal		82.20	.195		78	3.83		81.83	92.50
5700	12' x 14' O.D. nominal		109.60	.146		70	2.88		72.88	82
5800	12' x 20' O.D. nominal		109.60	.146		61	2.88		63.88	72
6100	For 8'-6" high, add					5%				
6300	Rule of thumb for complete units, w/o doors & refrigeration, cooler	2 Carp	146	.110		88	2.16		90.16	101
6400	Freezer		109.60	.146		104	2.88		106.88	119
6600	Shelving, plated or galvanized, steel wire type		360	.044	SF Hor.	7.70	.88		8.58	9.95
6700	Slat shelf type		375	.043		9.50	.84		10.34	11.90
6900	For stainless steel shelving, add					300%				
7000	Vapor barrier, on wood walls	2 Carp	1,644	.010	S.F.	.11	.19		.30	.45
7200	On masonry walls	"	1,315	.012	"	.29	.24		.53	.73
7500	For air curtain doors, see division 08344-120									
0010	**SAUNA** Prefabricated, incl. heater & controls, 7' high, 6' x 4', C/C	L-7	2.20	11.818	Ea.	3,200	199		3,399	3,875
0050	6' x 4', C/P		2	13		2,525	219		2,744	3,150
0400	6' x 5', C/C		2	13		3,200	219		3,419	3,900
0450	6' x 5', C/P		2	13		2,875	219		3,094	3,525
0600	6' x 6', C/C		1.80	14.444		3,750	244		3,994	4,550
0650	6' x 6', C/P		1.80	14.444		3,000	244		3,244	3,725
0800	6' x 9', C/C		1.60	16.250		4,675	274		4,949	5,625
0850	6' x 9', C/P		1.60	16.250		4,675	274		4,949	5,625
1000	8' x 12', C/C		1.10	23.636		6,300	400		6,700	7,600
1050	8' x 12', C/P		1.10	23.636		5,275	400		5,675	6,475
1200	8' x 8', C/C		1.40	18.571		4,925	315		5,240	5,950
1250	8' x 8', C/P		1.40	18.571		4,350	315		4,665	5,300
1400	8' x 10', C/C		1.20	21.667		5,400	365		5,765	6,575
1450	8' x 10', C/P		1.20	21.667		4,725	365		5,090	5,825
1600	10' x 12', C/C		1	26		7,200	440		7,640	8,675
1650	10' x 12', C/P		1	26		6,375	440		6,815	7,775
1700	Door only, cedar, 2'x6', with tempered insulated glass window	2 Carp	3.40	4.706		275	92.50		367.50	465
1800	Prehung, incl. jambs, pulls & hardware	"	12	1.333		325	26.50		351.50	405
2500	Heaters only (incl. above), wall mounted, to 200 C.F.					440			440	485
2750	To 300 C.F.					465			465	515
3000	Floor standing, to 720 C.F., 10,000 watts, w/controls	1 Elec	3	2.667		890	59		949	1,075
3250	To 1,000 C.F., 16,000 watts	"	3	2.667		1,575	59		1,634	1,825
0010	**SPORT COURT** Floors, No. 2 & better maple, 25/32" thick				SF Flr.					6.65
0100	Walls, laminated plastic bonded to galv. steel studs				SF Wall					8.45
0150	Laminated fiberglass surfacing, minimum									2.10
0180	Maximum									2.37

13030 | Special Purpose Rooms

13035 | Special Purpose Rooms

			CREW	DAILY OUTPUT	LABOR-HOURS	UNIT	2000 BARE COSTS MAT.	LABOR	EQUIP.	TOTAL	TOTAL INCL O&P	
900	0300	Squash, regulation court in existing building, minimum				Court					14,500	900
	0400	Maximum				"					27,100	
	0450	Rule of thumb for components:										
	0470	Walls	3 Carp	.15	160	Court	10,100	3,150		13,250	16,500	
	0500	Floor	"	.25	96		4,525	1,900		6,425	8,225	
	0550	Lighting	2 Elec	.60	26.667		1,425	590		2,015	2,550	
	0600	Handball, racquetball court in existing building, minimum	C-1	.20	160		1,750	2,750		4,500	6,625	
	0800	Maximum	"	.10	320		23,400	5,500		28,900	35,100	
	0900	Rule of thumb for components: walls	3 Carp	.12	200		9,225	3,950		13,175	16,900	
	1000	Floor		.25	96		4,800	1,900		6,700	8,525	
	1100	Ceiling		.33	72.727		2,000	1,425		3,425	4,650	
	1200	Lighting	2 Elec	.60	26.667		2,475	590		3,065	3,700	
940	0010	STEAM BATH Heater, timer & head, single, to 140 C.F.	1 Plum	1.20	6.667	Ea.	740	146		886	1,050	940
	0500	To 300 C.F.	"	1.10	7.273		800	160		960	1,150	
	1000	Commercial size, with blow-down assembly, to 800 C.F.	1 Plum	.90	8.889		2,700	195		2,895	3,300	
	1500	To 2500 C.F.	"	.80	10		4,250	220		4,470	5,050	
	2000	Multiple, motels, apts., 2 baths, w/ blow-down assm., 500 C.F.	Q-1	1.30	12.308		2,875	243		3,118	3,575	
	2500	4 baths	"	.70	22.857		3,825	450		4,275	4,950	
	2700	Conversion unit for residential tub, including door					2,225			2,225	2,450	

13039 | Vaults

900	0010	VAULT FRONT See division 08320-950										900

13080 | Sound, Vibration & Seismic Control

13081 | Sound Control

			CREW	DAILY OUTPUT	LABOR-HOURS	UNIT	2000 BARE COSTS MAT.	LABOR	EQUIP.	TOTAL	TOTAL INCL O&P	
100	0010	ACOUSTICAL Enclosure, 4" thick wall and ceiling panels										100
	0020	8# per S.F., up to 12' span	3 Carp	72	.333	SF Surf	25	6.55		31.55	38.50	
	0300	Better quality panels, 10.5# per S.F.		64	.375		28	7.40		35.40	43.50	
	0400	Reverb-chamber, 4" thick, parallel walls		60	.400		30	7.90		37.90	46.50	
	0600	Skewed wall, parallel roof, 4" thick panels		55	.436		30	8.60		38.60	48	
	0700	Skewed walls, skewed roof, 4" layers, 4" air space		48	.500		35.50	9.85		45.35	56	
	0900	Sound-absorbing panels, pntd mtl, 2'-6" x 8', under 1,000 S.F.		215	.112		9.05	2.20		11.25	13.70	
	1100	Over 1000 S.F.		240	.100		8.70	1.97		10.67	12.95	
	1200	Fabric faced		240	.100		7.25	1.97		9.22	11.40	
	1500	Flexible transparent curtain, clear	3 Shee	215	.112		5.60	2.40		8	10.20	
	1600	50% foam		215	.112		7.85	2.40		10.25	12.65	
	1700	75% foam		215	.112		7.85	2.40		10.25	12.65	
	1800	100% foam		215	.112		7.85	2.40		10.25	12.65	
	3100	Audio masking system, including speakers, amplification										
	3110	and signal generator										
	3200	Ceiling mounted, 5,000 S.F.	2 Elec	2,400	.007	S.F.	.96	.15		1.11	1.30	
	3300	10,000 S.F.		2,800	.006		.78	.13		.91	1.07	
	3400	Plenum mounted, 5,000 S.F.		3,800	.004		.81	.09		.90	1.04	
	3500	10,000 S.F.		4,400	.004		.54	.08		.62	.72	

13090 | Radiation Protection

13091 | X-Ray/Radio Freq Protection

		CREW	DAILY OUTPUT	LABOR-HOURS	UNIT	MAT.	LABOR	EQUIP.	TOTAL	TOTAL INCL O&P
0010	**SHIELDING LEAD**									
0100	Laminated lead in wood doors, 1/16" thick, no hardware				S.F.	26			26	28.50
0200	Lead lined door frame, not incl. hardware,									
0210	1/16" thick lead, butt prepared for hardware	1 Lath	2.40	3.333	Ea.	299	64		363	435
0300	Lead sheets, 1/16" thick	2 Lath	135	.119	S.F.	4.57	2.28		6.85	8.80
0400	1/8" thick		120	.133		9.55	2.56		12.11	14.70
0500	Lead shielding, 1/4" thick		135	.119		15.25	2.28		17.53	20.50
0550	1/2" thick		120	.133		30.50	2.56		33.06	37.50
0600	Lead glass, 1/4" thick, 2.0 mm LE, 12" x 16"	2 Glaz	13	1.231	Ea.	247	24		271	310
0700	24" x 36"		8	2		970	38.50		1,008.50	1,150
0800	36" x 60"		2	8		1,850	155		2,005	2,275
0850	Window frame with 1/16" lead and voice passage, 36" x 60"		2	8		380	155		535	670
0870	24" x 36" frame		8	2		288	38.50		326.50	380
0900	Lead gypsum board, 5/8" thick with 1/16" lead		160	.100	S.F.	4.78	1.94		6.72	8.45
0910	1/8" lead		140	.114		9.75	2.21		11.96	14.40
0930	1/32" lead	2 Lath	200	.080		2.94	1.54		4.48	5.75
0950	Lead headed nails (average 1 lb. per sheet)				Lb.	5.25			5.25	5.80
1000	Butt joints in 1/8" lead or thicker, 2" batten strip x 7' long	2 Lath	240	.067	Ea.	13.15	1.28		14.43	16.55
1200	X-ray protection, average radiography or fluoroscopy									
1210	room, up to 300 S.F. floor, 1/16" lead, minimum	2 Lath	.25	64	Total	4,600	1,225		5,825	7,075
1500	Maximum, 7'-0" walls	"	.15	106	"	6,000	2,050		8,050	9,950
1600	Deep therapy X-ray room, 250 KV capacity,									
1800	up to 300 S.F. floor, 1/4" lead, minimum	2 Lath	.08	200	Total	15,500	3,850		19,350	23,400
1900	Maximum, 7'-0" walls	"	.06	266	"	21,000	5,125		26,125	31,500
2000	X-ray viewing panels, clear lead plastic									
2010	7 mm thick, 0.3 mm LE, 2.3 lbs/S.F.	H-3	139	.115	S.F.	122	1.97		123.97	137
2020	12 mm thick, 0.5 mm LE, 3.9 lbs/S.F.		82	.195		165	3.33		168.33	188
2030	18 mm thick, 0.8mm LE, 5.9 lbs/S.F.		54	.296		179	5.05		184.05	206
2040	22 mm thick, 1.0 mm LE, 7.2 lbs/S.F.		44	.364		183	6.20		189.20	211
2050	35 mm thick, 1.5 mm LE, 11.5 lbs/S.F.		28	.571		205	9.75		214.75	242
2060	46 mm thick, 2.0 mm LE, 15.0 lbs/S.F.		21	.762		270	13		283	320
2090	For panels 12 S.F. to 48 S.F., add crating charge				Ea.					50
4000	X-ray barriers, modular, panels mounted within framework for									
4002	attaching to floor, wall or ceiling, upper portion is clear lead									
4005	plastic window panels 48"H, lower portion is opaque leaded									
4008	steel panels 36"H, structural supports not incl.									
4010	1-section barrier, 36"W x 84"H overall									
4020	0.5 mm LE panels	H-3	6.40	2.500	Ea.	2,750	42.50		2,792.50	3,100
4030	0.8 mm LE panels		6.40	2.500		2,925	42.50		2,967.50	3,300
4040	1.0 mm LE panels		5.33	3.002		3,025	51.50		3,076.50	3,400
4050	1.5 mm LE panels		5.33	3.002		3,225	51.50		3,276.50	3,625
4060	2-section barrier, 72"W x 84"H overall									
4070	0.5 mm LE panels	H-3	4	4	Ea.	5,700	68.50		5,768.50	6,400
4080	0.8 mm LE panels		4	4		6,025	68.50		6,093.50	6,750
4090	1.0 mm LE panels		3.56	4.494		6,150	77		6,227	6,900
5000	1.5 mm LE panels		3.20	5		6,650	85.50		6,735.50	7,475
5010	3-section barrier, 108"W x 84"H overall									
5020	0.5 mm LE panels	H-3	3.20	5	Ea.	8,150	85.50		8,235.50	9,125
5030	0.8 mm LE panels		3.20	5		8,550	85.50		8,635.50	9,550
5040	1.0 mm LE panels		2.67	5.993		8,750	102		8,852	9,800
5050	1.5 mm LE panels		2.46	6.504		9,500	111		9,611	10,700
7000	X-ray barriers, mobile, mounted within framework w/casters on									
7005	bottom, clear lead plastic window panels on upper portion,									
7010	opaque on lower, 30"W x 75"H overall, incl. framework									
7020	24"H upper w/0.5 mm LE, 48"H lower w/0.8 mm LE	1 Carp	16	.500	Ea.	1,875	9.85		1,884.85	2,075
7030	48"W x 75"H overall, incl. framework									

13090 | Radiation Protection

13091 | X-Ray/Radio Freq Protection

			CREW	DAILY OUTPUT	LABOR-HOURS	UNIT	2000 BARE COSTS MAT.	LABOR	EQUIP.	TOTAL	TOTAL INCL O&P	
600	7040	36"H upper w/0.5 mm LE, 36"H lower w/0.8 mm LE	1 Carp	16	.500	Ea.	3,300	9.85		3,309.85	3,650	600
	7050	36"H upper w/1.0 mm LE, 36"H lower w/1.5 mm LE	"	16	.500	"	4,000	9.85		4,009.85	4,425	
	7060	72"W x 75"H overall, incl. framework										
	7070	36"H upper w/0.5 mm LE, 36"H lower w/0.8 mm LE	1 Carp	16	.500	Ea.	3,925	9.85		3,934.85	4,325	
	7080	36"H upper w/1.0 mm LE, 36"H lower w/1.5 mm LE	"	16	.500	"	4,950	9.85		4,959.85	5,475	
700	0010	**SHIELDING, RADIO FREQUENCY**										700
	0020	Prefabricated or screen-type copper or steel, minimum	2 Carp	180	.089	SF Surf	24	1.75		25.75	29.50	
	0100	Average		155	.103		26	2.03		28.03	32	
	0150	Maximum		145	.110		31	2.17		33.17	37.50	

13120 | Pre-Engineered Structures

13128 | Pre-Eng. Structures

			CREW	DAILY OUTPUT	LABOR-HOURS	UNIT	2000 BARE COSTS MAT.	LABOR	EQUIP.	TOTAL	TOTAL INCL O&P	
060	0010	**PORTABLE BOOTHS** Prefab aluminum with doors, windows, ext. roof										060
	0100	lights wiring & insulation, 15 S.F. building, O.D., painted, minimum				S.F.	255			255	281	
	0300	30 S.F. building, minimum					175			175	193	
	0400	50 S.F. building, minimum					130			130	143	
	0600	80 S.F. building, minimum					105			105	116	
	0700	100 S.F. building, minimum					97			97	107	
	0900	Acoustical booth, 27 Db @ 1,000 Hz, 15 S.F. floor				Ea.	2,875			2,875	3,175	
	1000	7' x 7'-6", including light & ventilation					5,900			5,900	6,500	
	1200	Ticket booth, galv. steel, not incl. foundations., 4' x 4'					4,425			4,425	4,875	
	1300	4' x 6'					5,175			5,175	5,700	
070	0010	**CONTROL TOWERS**, 12' x 10', incl. instruments, min.				Ea.					450,000	070
	0100	Maximum									1,000,000	
	0500	With standard 40' tower, average									600,000	
	1000	Temporary portable control towers, 8' x 12',										
	1010	complete with one position communications, minimum				Ea.					275,000	
	2000	For fixed facilities, depending on height, minimum									50,000	
	2010	Maximum									100,000	
160	0010	**TENSION STRUCTURES** Rigid steel/alum. frame, vyl. coated polyester										160
	0100	fabric shell, 60' clear span, not incl. foundations or floors										
	0200	6,000 S.F.	B-41	1,000	.044	SF Flr.	10.40	.67	.18	11.25	12.80	
	0300	12,000 S.F.		1,100	.040		8.75	.61	.16	9.52	10.85	
	0400	80' clear span, 20,800 S.F.		1,220	.036		9.40	.55	.15	10.10	11.45	
	0410	100' clear span, 10,000 S.F.	L-5	2,175	.026		10.25	.55	.26	11.06	12.60	
	0430	26,000 S.F.		2,300	.024		8.80	.52	.24	9.56	11	
	0450	36,000 S.F.		2,500	.022		8.65	.48	.22	9.35	10.75	
	0460	120' clear span, 24,000 S.F.		3,000	.019		11.55	.40	.19	12.14	13.65	
	0470	150' clear span, 30,000 S.F.		6,000	.009		11.10	.20	.09	11.39	12.70	
	0480	200' clear span, 40,000 S.F.	E-6	8,000	.015		14.60	.32	.16	15.08	16.85	
	0500	For roll-up door, 12' x 14', add	L-2	1	16	Ea.	3,625	276		3,901	4,475	
	0600	For personnel doors, add, minimum				SF Flr.	5%					
	0700	Add, maximum					15%					
	0800	For site work, simple foundation, etc., add, minimum								1.25	1.95	
	0900	Add, maximum								2.75	3.05	
200	0010	**COMFORT STATIONS** Prefab., stock, w/doors, windows & fixt.										200
	0100	Not incl. interior finish or electrical										

13120 | Pre-Engineered Structures

13128 | Pre-Eng. Structures

			CREW	DAILY OUTPUT	LABOR-HOURS	UNIT	MAT.	LABOR	EQUIP.	TOTAL	TOTAL INCL O&P
200	0300	Mobile, on steel frame, minimum				S.F.	37.50			37.50	41
	0350	Maximum					63.50			63.50	70
	0400	Permanent, including concrete slab, minimum	B-12J	50	.160		138	3.30	12.35	153.65	171
	0500	Maximum	"	43	.186		200	3.84	14.40	218.24	242
	0600	Alternate pricing method, mobile, minimum				Fixture	1,725			1,725	1,900
	0650	Maximum					2,575			2,575	2,825
	0700	Permanent, minimum	B-12J	.70	11.429		9,900	236	885	11,021	12,300
	0750	Maximum	"	.50	16		16,500	330	1,225	18,055	20,100
300	0010	**DOMES**									
	0020	Revolv alum, elec drv for astronomy observation, shell only, stk units									
	0600	10'-0" diameter, 800#, dome	2 Carp	.25	64	Ea.	9,400	1,250		10,650	12,500
	0700	Base		.67	23.881		3,250	470		3,720	4,375
	0900	18'-0" diameter, 2,500#, dome		.17	94.118		27,500	1,850		29,350	33,500
	1000	Base		.33	48.485		9,100	955		10,055	11,600
	1200	24'-0" diameter, 4,500#, dome		.08	200		51,000	3,950		54,950	63,000
	1300	Base		.25	64		16,400	1,250		17,650	20,200
	1500	Bulk storage, shell only, dual radius hemispher. arch, steel									
	1600	framing, corrugated steel covering, 150' diameter	E-2	550	.087	SF Flr.	27	1.87	1.86	30.73	35
	1700	400' diameter	"	720	.067		22	1.43	1.42	24.85	28.50
	1800	Wood framing, wood decking, to 400' diameter	F-4	400	.100		20	1.79	2.02	23.81	27.50
	1900	Radial framed wood (2" x 6"), 1/2" thick									
	2000	plywood, asphalt shingles, 50' diameter	F-3	2,000	.020	SF Flr.	25	.36	.23	25.59	28.50
	2100	60' diameter		1,900	.021		19	.38	.24	19.62	22
	2200	72' diameter		1,800	.022		17	.40	.25	17.65	19.65
	2300	116' diameter		1,730	.023		15	.41	.26	15.67	17.50
	2400	150' diameter		1,500	.027		13	.48	.30	13.78	15.45
340	0010	**GEODESIC DOME** Shell only, interlocking plywood panels R13128-310									
	0400	30' diameter	F-5	1.60	20	Ea.	9,950	345		10,295	11,500
	0500	34' diameter		1.14	28.070		12,200	485		12,685	14,200
	0600	39' diameter		1	32		14,200	550		14,750	16,500
	0700	45' diameter	F-3	1.13	35.556		17,100	640	405	18,145	20,300
	0750	55' diameter		1	40		25,500	715	455	26,670	29,700
	0800	60' diameter		1	40		33,200	715	455	34,370	38,200
	0850	65' diameter		.80	50		39,400	895	565	40,860	45,500
	1100	Aluminum panel, with 6" insulation									
	1200	100' diameter				SF Flr.					30
	1300	500' diameter				"					25
	1600	Aluminum framed, plexiglass closure panels									
	1700	40' diameter				SF Flr.					80
	1801	200' diameter				"					70
	2100	Aluminum framed, aluminum closure panels									
	2200	40' diameter				SF Flr.					50
	2300	100' diameter									30
	2400	200' diameter									25
	2500	For VRP faced bonded fiberglass insulation, add									10
	2700	Aluminum framed, fiberglass sandwich panel closure									
	2800	6' diameter	2 Carp	150	.107	SF Flr.	33	2.10		35.10	40
	2900	28' diameter	"	350	.046	"	25	.90		25.90	29
360	0010	**GARAGE COSTS** R13128-910									
	0020	Public parking, average				Car					10,300
	0100	See also division 17100-410									
	0300	Residential, prefab shell, stock, wood, single car, minimum	2 Carp	1	16	Total	2,550	315		2,865	3,350
	0350	Maximum		.67	23.881		5,775	470		6,245	7,150
	0400	Two car, minimum		.67	23.881		3,825	470		4,295	5,000
	0450	Maximum		.50	32		7,625	630		8,255	9,475

SPECIAL CONSTRUCTION 13

13120 | Pre-Engineered Structures

13128	Pre-Eng. Structures	CREW	DAILY OUTPUT	LABOR-HOURS	UNIT	2000 BARE COSTS MAT.	LABOR	EQUIP.	TOTAL	TOTAL INCL O&P
380 0010	**GARDEN HOUSE** Prefab wood, no floors or foundations									
0100	32 to 200 S.F., minimum	2 Carp	200	.080	SF Flr.	13.45	1.58		15.03	17.50
0300	Maximum	"	48	.333	"	24.50	6.55		31.05	38.50
500 0010	**GRANDSTANDS** Permanent, municipal, including foundation									
0050	Steel understructure w/aluminum closed deck, minimum				Seat					130
0100	Maximum									225
0300	Steel, minimum									25
0400	Maximum									80
0600	Aluminum, extruded, stock design, minimum									70
0700	Maximum									115
0900	Composite, steel, wood and plastic, stock design, minimum									30
1000	Maximum									90
540 0010	**GREENHOUSE** Shell only, stock units, not incl. 2' stub walls,									
0020	foundation, floors, heat or compartments									
0300	Residential type, free standing, 8'-6" long x 7'-6" wide	2 Carp	59	.271	SF Flr.	35	5.35		40.35	47.50
0400	10'-6" wide		85	.188		27	3.71		30.71	36
0600	13'-6" wide		108	.148		24	2.92		26.92	31.50
0700	17'-0" wide		160	.100		27	1.97		28.97	33
0900	Lean-to type, 3'-10" wide		34	.471		31	9.25		40.25	50
1000	6'-10" wide		58	.276		24	5.45		29.45	36
1500	Commercial, custom, truss frame, incl. equip., plumbing, elec.,									
1510	benches and controls, under 2,000 S.F., minimum				SF Flr.					30
1550	Maximum									38
1700	Over 5,000 S.F., minimum									23
1750	Maximum									30
2000	Institutional, custom, rigid frame, including compartments and									
2010	multi-controls, under 500 S.F., minimum				SF Flr.					77
2050	Maximum									99
2150	Over 2,000 S.F., minimum									39
2200	Maximum									65
2400	Concealed rigid frame, under 500 S.F., minimum									90
2450	Maximum									110
2550	Over 2,000 S.F., minimum									68
2600	Maximum									79
2800	Lean-to type, under 500 S.F., minimum									80
2850	Maximum									120
3000	Over 2,000 S.F., minimum									44
3050	Maximum									73
3600	For 1/4" clear plate glass, add				SF Surf	1.40			1.40	1.54
3700	For 1/4" tempered glass, add				"	3.15			3.15	3.47
3900	For cooling, add, minimum				SF Flr.	2.10			2.10	2.31
4000	Maximum					5.20			5.20	5.70
4200	For heaters, 13.6 MBH, add					4			4	4.40
4300	60 MBH, add					1.50			1.50	1.65
4500	For benches, 2' x 3'-6", add				SF Hor.	17.90			17.90	19.70
4600	3' x 10', add				S.F.	9.70			9.70	10.65
4800	For controls, add, minimum				Total	1,800			1,800	1,975
4900	Maximum				"	10,700			10,700	11,800
5100	For humidification equipment, add				M.C.F.	4.60			4.60	5.05
5200	For vinyl shading, add				S.F.	.97			.97	1.07
6000	Geodesic hemisphere, 1/8" plexiglass glazing									
6050	8' diameter	2 Carp	2	8	Ea.	2,025	158		2,183	2,500
6150	24' diameter		.35	45.714		10,400	900		11,300	13,000
6250	48' diameter		.20	80		28,200	1,575		29,775	33,700

13120 | Pre-Engineered Structures

13128 | Pre-Eng. Structures

		CREW	DAILY OUTPUT	LABOR-HOURS	UNIT	MAT.	LABOR	EQUIP.	TOTAL	TOTAL INCL O&P
						\multicolumn{4}{c	}{2000 BARE COSTS}	

		CREW	DAILY OUTPUT	LABOR-HOURS	UNIT	MAT.	LABOR	EQUIP.	TOTAL	TOTAL INCL O&P
580	**0010 HANGARS** Prefabricated steel T hangars, Galv. steel roof &									
0100	walls, incl. electric bi-folding doors, 4 or more units,									
0110	not including floors or foundations, minimum	E-2	1,275	.038	SF Flr.	8.85	.81	.80	10.46	12.20
0130	Maximum		1,063	.045		9.50	.97	.96	11.43	13.35
0900	With bottom rolling doors, minimum		1,386	.035		8.05	.74	.74	9.53	11.10
1000	Maximum		966	.050		9.50	1.07	1.06	11.63	13.65
1200	Alternate pricing method:									
1300	Galv. roof and walls, electric bi-folding doors, minimum	E-2	1.06	45.283	Plane	10,600	975	965	12,540	14,500
1500	Maximum		.91	52.747		11,700	1,125	1,125	13,950	16,300
1600	With bottom rolling doors, minimum		1.25	38.400		8,400	825	820	10,045	11,700
1800	Maximum		.97	49.485		10,000	1,075	1,050	12,125	14,200
2000	Circular type, prefab., steel frame, plastic skin, electric									
2010	door, including foundations, 80' diameter,									
2020	for up to 5 light planes, minimum	E-2	.50	96	Total	81,000	2,050	2,050	85,100	95,000
2200	Maximum	"	.25	192	"	91,000	4,125	4,100	99,225	113,000
600	**0010 KIOSKS** Round, 5' diameter, 8' high, 1/4" fiberglass wall				Ea.	5,500			5,500	6,050
0100	1" insulated double wall, fiberglass					6,250			6,250	6,875
0500	Rectangular, 5' x 9', 7'-6" high, 1/4" fiberglass wall					8,000			8,000	8,800
0600	1" insulated double wall, fiberglass					9,500			9,500	10,500
700	**0010 PRE-ENGINEERED STEEL BUILDINGS** R13128-210									
0100	Clear span rigid frame, 26 ga. colored roofing and siding									
0200	30' to 40' wide, 10' eave height	E-2	1,800	.027	SF Flr.	3.92	.57	.57	5.06	6.05
0300	14' eave height		1,800	.027		4.13	.57	.57	5.27	6.25
0400	16' eave height		1,400	.034		4.34	.74	.73	5.81	7
0500	20' eave height		1,000	.048		4.76	1.03	1.02	6.81	8.35
0600	24' eave height		1,000	.048		5.35	1.03	1.02	7.40	9
0700	50' to 100' wide, 10' eave height		1,800	.027		3.43	.57	.57	4.57	5.50
0800	14' eave height		1,800	.027		3.64	.57	.57	4.78	5.70
0900	16' eave height		1,400	.034		3.85	.74	.73	5.32	6.45
1000	20' eave height		1,000	.048		4.10	1.03	1.02	6.15	7.60
1100	24' eave height		1,000	.048		4.49	1.03	1.02	6.54	8.05
1200	Clear span tapered beam frame, 26 ga. colored roofing and siding									
1300	30' wide, 10' eave height	E-2	1,800	.027	SF Flr.	4.37	.57	.57	5.51	6.55
1400	14' eave height		1,800	.027		4.84	.57	.57	5.98	7
1500	16' eave height		1,400	.034		5.20	.74	.73	6.67	7.95
1600	20' eave height		1,000	.048		5.75	1.03	1.02	7.80	9.45
1700	40' wide, 10' eave height		1,800	.027		4.01	.57	.57	5.15	6.15
1800	14' eave height		1,800	.027		4.34	.57	.57	5.48	6.50
1900	16' eave height		1,400	.034		4.57	.74	.73	6.04	7.25
2000	20' eave height		1,000	.048		4.99	1.03	1.02	7.04	8.60
2100	50' to 80' wide, 10' eave height		1,800	.027		3.83	.57	.57	4.97	5.95
2200	14' eave height		1,800	.027		4.14	.57	.57	5.28	6.25
2300	16' eave height		1,400	.034		4.32	.74	.73	5.79	6.95
2400	20' eave height		1,000	.048		4.80	1.03	1.02	6.85	8.40
2500	Single post 2-span frame, 26 ga. colored roofing and siding									
2600	80' wide, 14' eave height	E-2	1,800	.027	SF Flr.	3.23	.57	.57	4.37	5.25
2700	16' eave height		1,400	.034		3.44	.74	.73	4.91	6
2800	20' eave height		1,000	.048		3.75	1.03	1.02	5.80	7.25
2900	24' eave height		1,000	.048		4.04	1.03	1.02	6.09	7.55
3000	100' wide, 14' eave height		1,800	.027		3.14	.57	.57	4.28	5.15
3100	16' eave height		1,400	.034		3.30	.74	.73	4.77	5.85
3200	20' eave height		1,000	.048		3.61	1.03	1.02	5.66	7.05
3300	24' eave height		1,000	.048		3.89	1.03	1.02	5.94	7.40
3400	120' wide, 14' eave height		1,800	.027		3.11	.57	.57	4.25	5.15
3500	16' eave height		1,400	.034		3.22	.74	.73	4.69	5.75

13120 | Pre-Engineered Structures

13128 | Pre-Eng. Structures

		CREW	DAILY OUTPUT	LABOR-HOURS	UNIT	2000 BARE COSTS MAT.	LABOR	EQUIP.	TOTAL	TOTAL INCL O&P
3600	20' eave height	E-2	1,000	.048	SF Flr.	3.51	1.03	1.02	5.56	6.95
3700	24' eave height		1,000	.048		3.81	1.03	1.02	5.86	7.30
3800	Double post 3-span frame, 26 ga. colored roofing and siding									
3900	150' wide, 14' eave height	E-2	1,800	.027	SF Flr.	2.87	.57	.57	4.01	4.88
4000	16' eave height		1,400	.034		2.97	.74	.73	4.44	5.50
4100	20' eave height		1,000	.048		3.18	1.03	1.02	5.23	6.60
4200	24' eave height		1,000	.048		3.73	1.03	1.02	5.78	7.20
4300	Triple post 4-span frame, 26 ga. colored roofing and siding									
4400	160' wide, 14' eave height	E-2	1,800	.027	SF Flr.	2.78	.57	.57	3.92	4.78
4500	16' eave height		1,400	.034		2.91	.74	.73	4.38	5.40
4600	20' eave height		1,000	.048		3.10	1.03	1.02	5.15	6.50
4700	24' eave height		1,000	.048		3.61	1.03	1.02	5.66	7.05
4800	200' wide, 14' eave height		1,800	.027		2.80	.57	.57	3.94	4.80
4900	16' eave height		1,400	.034		2.87	.74	.73	4.34	5.40
5000	20' eave height		1,000	.048		3.10	1.03	1.02	5.15	6.50
5100	24' eave height		1,000	.048		3.62	1.03	1.02	5.67	7.10
5200	Accessory items: add to the basic building cost above									
5250	Eave overhang, 2' wide, 26 ga., with soffit	E-2	360	.133	L.F.	14.65	2.86	2.84	20.35	24.50
5300	4' wide, without soffit		300	.160		13.75	3.44	3.41	20.60	25.50
5350	With soffit		250	.192		19	4.12	4.09	27.21	33.50
5400	6' wide, without soffit		250	.192		18.10	4.12	4.09	26.31	32.50
5450	With soffit		200	.240		24.50	5.15	5.10	34.75	42.50
5500	Entrance canopy, incl. frame, 4' x 4'		25	1.920	Ea.	127	41	41	209	265
5550	4' x 8'		19	2.526	"	237	54.50	54	345.50	425
5600	End wall roof overhang, 4' wide, without soffit		850	.056	L.F.	9.20	1.21	1.20	11.61	13.80
5650	With soffit		500	.096	"	13.30	2.06	2.04	17.40	21
5700	Doors, H.M. self-framing, incl. butts, lockset and trim									
5750	Single leaf, 3070 (3' x 7'), economy	2 Sswk	5	3.200	Opng.	234	68		302	390
5800	Deluxe		4	4		315	85		400	515
5825	Glazed		4	4		310	85		395	510
5850	3670 (3'-6" x 7')		4	4		410	85		495	625
5900	4070 (4' x 7')		3	5.333		340	113		453	600
5950	Double leaf, 6070 (6' x 7')		2	8		425	170		595	805
6000	Glazed		2	8		575	170		745	970
6050	Framing only, for openings, 3' x 7'	E-2	25	1.920		83.50	41	41	165.50	217
6100	10' x 10'		21	2.286		275	49	48.50	372.50	455
6150	For windows below, 2020 (2' x 2')		25	1.920		88.50	41	41	170.50	222
6200	4030 (4' x 3')		22	2.182		108	47	46.50	201.50	260
6250	Flashings, 26 ga., corner or eave, painted	2 Sswk	240	.067	L.F.	2.41	1.42		3.83	5.45
6300	Galvanized		240	.067		1.60	1.42		3.02	4.55
6350	Rake flashing, painted		240	.067		2.41	1.42		3.83	5.45
6400	Galvanized		240	.067		1.80	1.42		3.22	4.77
6450	Ridge flashing, 18" wide, painted		240	.067		2.70	1.42		4.12	5.75
6500	Galvanized		240	.067		1.70	1.42		3.12	4.66
6550	Gutter, eave type, 26 ga., painted		320	.050		3.13	1.06		4.19	5.55
6600	Galvanized		320	.050		1.70	1.06		2.76	3.97
6650	Valley type, between buildings, painted		120	.133		3.35	2.83		6.18	9.30
6700	Galvanized		120	.133		5.10	2.83		7.93	11.20
6750	Insulation, rated .6 lb density, vinyl faced									
6801	1-1/2" thick, R5	2 Clab	2,300	.007	S.F.	.19	.10		.29	.38
6851	3" thick, R10		2,300	.007		.26	.10		.36	.46
6901	4" thick, R11		2,300	.007		.32	.10		.42	.52
6951	Foil faced, 1-1/2" thick, R5		2,300	.007		.21	.10		.31	.40
7001	2" thick, R6		2,300	.007		.25	.10		.35	.45
7051	3" thick, R10		2,300	.007		.30	.10		.40	.50
7101	4" thick, R13		2,300	.007		.36	.10		.46	.57

13120 | Pre-Engineered Structures

13128 | Pre-Eng. Structures

			CREW	DAILY OUTPUT	LABOR-HOURS	UNIT	MAT.	LABOR	EQUIP.	TOTAL	TOTAL INCL O&P	
700	7151	Metalized polyester facing, 1-1/2" thick, R5	2 Clab	2,300	.007	S.F.	.32	.10		.42	.52	700
	7201	2" thick, R6		2,300	.007		.37	.10		.47	.58	
	7251	3" thick, R11		2,300	.007		.38	.10		.48	.59	
	7301	4" thick, R13		2,300	.007		.45	.10		.55	.67	
	7351	Vinyl, scrim foil (SD), 1-1/2" thick, R5		2,300	.007		.35	.10		.45	.56	
	7401	2" thick, R6		2,300	.007		.35	.10		.45	.56	
	7451	3" thick, R10		2,300	.007		.43	.10		.53	.64	
	7501	4" thick, R13		2,300	.007		.48	.10		.58	.70	
	7650	Sash, single slide, glazed, with screens, 2020 (2'x 2')	E-1	22	.727	Opng.	71	15.45	3.82	90.27	113	
	7700	3030 (3' x 3')		14	1.143		160	24.50	6	190.50	231	
	7750	4030 (4' x 3')		13	1.231		213	26	6.45	245.45	293	
	7800	6040 (6' x 4')		12	1.333		425	28.50	7	460.50	535	
	7850	Double slide sash, 3030 (3' x 3')		14	1.143		136	24.50	6	166.50	205	
	7900	6040 (6' x 4')		12	1.333		365	28.50	7	400.50	465	
	7950	Fixed glass, no screens, 3030 (3' x 3')		14	1.143		126	24.50	6	156.50	194	
	8000	6040 (6' x 4')		12	1.333		335	28.50	7	370.50	435	
	8050	Prefinished storm sash, 3030 (3' x 3')		70	.229		54.50	4.86	1.20	60.56	71	
	8100	Siding and roofing, see division 07300 & 07400										
	8200	Skylight, fiberglass panels, to 30 S.F.	E-1	10	1.600	Ea.	113	34	8.40	155.40	200	
	8250	Larger sizes, add for excess over 30 S.F.	"	300	.053	S.F.	3.76	1.13	.28	5.17	6.70	
	8300	Roof vents, circular with damper, birdscreen										
	8350	and operator hardware, painted										
	8400	26 ga., 12" diameter	1 Sswk	4	2	Ea.	70.50	42.50		113	162	
	8450	20" diameter		3	2.667		145	56.50		201.50	271	
	8500	24 ga., 24" diameter		2	4		287	85		372	485	
	8550	Galvanized		2	4		241	85		326	435	
	8600	Continuous, 26 ga., 10' long, 9" wide	2 Sswk	4	4		207	85		292	395	
	8650	12" wide		4	4		248	85		333	440	
800	0010	**SHELTERS** Aluminum frame, acrylic glazing, 3' x 9' x 8' high		1.14	14.035	Ea.	2,550	298		2,848	3,400	800
	0100	9' x 12' x 8' high		.73	21.918	"	4,250	465		4,715	5,600	
840	0010	**SILOS** Concrete stave industrial, not incl. foundations, conical or										840
	0100	sloping bottoms, 12' diameter, 35' high	D-8	.11	363	Ea.	15,700	6,650		22,350	28,600	
	0200	16' diameter, 45' high		.08	500		21,000	9,150		30,150	38,600	
	0400	25' diameter, 75' high		.05	800		51,500	14,600		66,100	81,500	
	0501	Steel, factory fab., 30,000 gallon cap., painted, minimum	E-2	1	48		10,200	1,025	1,025	12,250	14,300	
	0701	Maximum		.50	96		15,900	2,050	2,050	20,000	23,700	
	0801	Epoxy lined, minimum		1	48		16,400	1,025	1,025	18,450	21,100	
	1001	Maximum		.50	96		20,500	2,050	2,050	24,600	28,800	
880	0010	**SWIMMING POOL ENCLOSURE** Translucent, free standing,										880
	0020	not including foundations, heat or light										
	0200	Economy, minimum	2 Carp	200	.080	SF Hor.	10	1.58		11.58	13.70	
	0300	Maximum		100	.160		21	3.15		24.15	28.50	
	0400	Deluxe, minimum		100	.160		23	3.15		26.15	31	
	0600	Maximum		70	.229		250	4.50		254.50	283	
	0700	For motorized roof, 40% opening, solid roof, add					3			3	3.30	
	0800	Skylight type roof, add					4.50			4.50	4.95	
	0900	Air-inflated, including blowers and heaters, minimum									3.50	
	1000	Maximum									6.70	

Note: Row 7151 references R13128-210.

13150 | Swimming Pools

13151 | Swimming Pools

							2000 BARE COSTS				TOTAL
			CREW	DAILY OUTPUT	LABOR-HOURS	UNIT	MAT.	LABOR	EQUIP.	TOTAL	INCL O&P
200	0010	**SWIMMING POOLS** Residential in-ground, vinyl lined, concrete sides									200
	0020	Sides including equipment, sand bottom	B-52	300	.187	SF Surf	10.40	3.05	1.31	14.76	18.15
	0100	Metal or polystyrene sides	B-14	410	.117		8.70	1.82	.50	11.02	13.20
	0200	Add for vermiculite bottom	R13128-520				.67			.67	.74
	0500	Gunite bottom and sides, white plaster finish									
	0600	12' x 30' pool	B-52	145	.386	SF Surf	17.25	6.30	2.70	26.25	33
	0720	16' x 32' pool		155	.361		15.50	5.90	2.53	23.93	30
	0750	20' x 40' pool		250	.224		13.85	3.66	1.57	19.08	23.50
	0810	Concrete bottom and sides, tile finish									
	0820	12' x 30' pool	B-52	80	.700	SF Surf	17.40	11.45	4.90	33.75	44
	0830	16' x 32' pool		95	.589		14.40	9.65	4.13	28.18	37
	0840	20' x 40' pool		130	.431		11.45	7.05	3.02	21.52	28
	1100	Motel, gunite with plaster finish, incl. medium									
	1150	capacity filtration & chlorination	B-52	115	.487	SF Surf	21	7.95	3.41	32.36	41
	1200	Municipal, gunite with plaster finish, incl. high									
	1250	capacity filtration & chlorination	B-52	100	.560	SF Surf	27.50	9.15	3.92	40.57	50
	1350	Add for formed gutters				L.F.	47.50			47.50	52.50
	1360	Add for stainless steel gutters				"	141			141	155
	1600	For water heating system, see division 15510-880									
	1700	Filtration and deck equipment only, as % of total				Total				20%	20%
	1800	Deck equipment, rule of thumb, 20' x 40' pool				SF Pool					1.30
	1900	5000 S.F. pool				"					1.90
	3000	Painting pools, preparation + 3 coats, 20' x 40' pool, epoxy	2 Pord	.33	48.485	Total	595	870		1,465	2,100
	3100	Rubber base paint, 18 gallons	"	.33	48.485		450	870		1,320	1,950
	3500	42' x 82' pool, 75 gallons, epoxy paint	3 Pord	.14	171		2,500	3,075		5,575	7,875
	3600	Rubber base paint	"	.14	171		1,925	3,075		5,000	7,250
700	0010	**SWIMMING POOL EQUIPMENT** Diving stand, stainless steel, 3 meter	2 Carp	.40	40	Ea.	4,725	790		5,515	6,525	700
	0300	1 meter		2.70	5.926		2,650	117		2,767	3,125
	0600	Diving boards, 16' long, aluminum		2.70	5.926		2,100	117		2,217	2,525
	0700	Fiberglass		2.70	5.926		1,575	117		1,692	1,925
	0900	Filter system, sand or diatomite type, incl. pump, 6,000 gal./hr.	2 Plum	1.80	8.889	Total	900	195		1,095	1,325
	1020	Add for chlorination system, 800 S.F. pool		3	5.333	Ea.	196	117		313	410
	1040	5,000 S.F. pool		3	5.333	"	1,150	117		1,267	1,450
	1100	Gutter system, stainless steel, with grating, stock,									
	1110	contains supply and drainage system	E-1	20	.800	L.F.	160	17	4.20	181.20	214
	1120	Integral gutter and 5' high wall system, stainless steel	"	10	1.600	"	240	34	8.40	282.40	340
	1200	Ladders, heavy duty, stainless steel, 2 tread	2 Carp	7	2.286	Ea.	790	45		835	940
	1500	4 tread		6	2.667		1,350	52.50		1,402.50	1,600
	1800	Lifeguard chair, stainless steel, fixed		2.70	5.926		930	117		1,047	1,225
	1900	Portable					1,700			1,700	1,875
	2100	Lights, underwater, 12 volt, with transformer, 300 watt	1 Elec	.40	20		150	440		590	890
	2200	110 volt, 500 watt, standard		.40	20		148	440		588	890
	2400	Low water cutoff type		.40	20		128	440		568	865
	2800	Heaters, see division 15510-880									
	3000	Pool covers, reinforced vinyl	3 Clab	1,800	.013	S.F.	.35	.19		.54	.72
	3050	Automatic, electric									9.65
	3100	Vinyl water tube	3 Clab	1,800	.013		.25	.19		.44	.61
	3200	Maximum	"	1,600	.015		.45	.22		.67	.87
	3250	Sealed air bubble polyethylene solar blanket					.17			.17	.19
	3300	Slides, tubular, fiberglass, aluminum handrails & ladder, 5'-0", straight	2 Carp	1.60	10	Ea.	1,025	197		1,222	1,475
	3320	8'-0", curved		3	5.333		11,000	105		11,105	12,300
	3400	10'-0", curved		1	16		13,400	315		13,715	15,300
	3420	12'-0", straight with platform		1.20	13.333		950	263		1,213	1,500
	4500	Hydraulic lift, movable pool bottom, single ram									
	4520	Under 1,000 S.F. area	L-9	.03	1,200	Ea.	80,000	21,900		101,900	125,500
	4600	Four ram lift, over 1,000 S.F.	"	.02	1,800		96,500	32,900		129,400	162,000

13150 | Swimming Pools

13151 | Swimming Pools

			CREW	DAILY OUTPUT	LABOR-HOURS	UNIT	MAT.	LABOR	EQUIP.	TOTAL	TOTAL INCL O&P	
700	5000	Removable access ramp, stainless steel	2 Clab	2	8	Ea.	5,875	116		5,991	6,675	700
	5500	Removable stairs, stainless steel, collapsible	"	2	8	↓	2,475	116		2,591	2,925	

13170 | Tubs & Pools

13171 | Therapeutic Pools

			CREW	DAILY OUTPUT	LABOR-HOURS	UNIT	MAT.	LABOR	EQUIP.	TOTAL	TOTAL INCL O&P	
800	0010	THERAPEUTIC POOLS See division 15418										800

13175 | Ice Rinks

13176 | Ice Rinks

			CREW	DAILY OUTPUT	LABOR-HOURS	UNIT	MAT.	LABOR	EQUIP.	TOTAL	TOTAL INCL O&P	
500	0010	ICE SKATING Equipment incl. refrigeration and plumbing, not										500
	0020	including building or slab, 85' x 200' rink										
	0300	55° system, 5 mos., 100 ton				Total					192,500	
	0700	90° system, 12 mos., 135 ton				"					253,000	
	1000	Dasher boards, 1/2" H.D. polyethylene faced steel frame, 3' acrylic										
	1020	screen at sides, 5' acrylic ends, 85' x 200'	F-5	.06	533	Ea.	115,000	9,200		124,200	142,500	
	1100	Fiberglass & aluminum construction, same sides and ends	"	.06	533		150,000	9,200		159,200	181,000	
	1200	Subsoil heating system (recycled from compressor), 85' x 200'	Q-7	.27	118		25,000	2,350		27,350	31,400	
	1300	Subsoil insulation, 2 lb. polystyrene with vapor barrier, 85' x 200'	2 Carp	.14	114	↓	30,000	2,250		32,250	36,900	

13200 | Storage Tanks

13201 | Storage Tanks

			CREW	DAILY OUTPUT	LABOR-HOURS	UNIT	MAT.	LABOR	EQUIP.	TOTAL	TOTAL INCL O&P	
200	0010	ELEVATED STORAGE TANKS Not incl. pipe or pumps										200
	3000	Elevated water tanks, 100' to bottom capacity line, incl painting										
	3010	50,000 gallons				Ea.					166,000	
	3300	100,000 gallons									229,000	
	3400	250,000 gallons									316,000	
	3600	500,000 gallons									510,000	
	3700	750,000 gallons									703,000	
	3900	1,000,000 gallons				↓					812,000	
300	0010	GROUND TANKS Not incl. pipe or pumps, prestress conc., 250,000 gal.				Ea.					280,000	300
	0100	500,000 gallons									380,000	
	0300	1,000,000 gallons									540,000	
	0400	2,000,000 gallons				↓					816,000	

13200 | Storage Tanks

13201	Storage Tanks	CREW	DAILY OUTPUT	LABOR-HOURS	UNIT	2000 BARE COSTS MAT.	LABOR	EQUIP.	TOTAL	TOTAL INCL O&P
0600	4,000,000 gallons				Ea.					1,290,000
0700	6,000,000 gallons									1,760,000
0750	8,000,000 gallons									2,240,000
0800	10,000,000 gallons									2,700,000
0900	Steel, ground level, not incl. foundations, 100,000 gallons									98,000
1000	250,000 gallons									137,000
1200	500,000 gallons									209,000
1250	750,000 gallons									249,000
1300	1,000,000 gallons									336,000
1500	2,000,000 gallons									615,000
1600	4,000,000 gallons									955,000
1800	6,000,000 gallons									1,325,000
1850	8,000,000 gallons									1,735,000
1900	10,000,000 gallons									1,945,000
2100	Steel standpipes, hgt/dial, 100' to overflow, no foundations									
2200	500,000 gallons				Ea.					275,000
2400	750,000 gallons									335,000
2500	1,000,000 gallons									408,000
2700	1,500,000 gallons									570,000
2800	2,000,000 gallons									700,000
3000	Steel, storage, above ground, including cradles, coating,									
3020	fittings, not including fdn, pumps or piping									
3040	Single wall, interior, 275 gallon	Q-5	5	3.200	Ea.	231	63.50		294.50	360
3060	550 gallon	"	2.70	5.926		815	118		933	1,100
3080	1,000 gallon	Q-7	5	6.400		1,025	127		1,152	1,325
3100	1,500 gallon		4.75	6.737		1,525	134		1,659	1,900
3120	2,000 gallon		4.60	6.957		1,975	138		2,113	2,375
3140	5,000 gallon		3.20	10		4,200	199		4,399	4,950
3150	10,000 gallon		2	16		6,900	320		7,220	8,125
3160	15,000 gallon		1.70	18.824		9,650	375		10,025	11,200
3170	20,000 gallon		1.45	22.069		11,300	440		11,740	13,100
3180	25,000 gallon capacity		1.30	24.615		14,100	490		14,590	16,300
3190	30,000 gallon		1.10	29.091		17,000	580		17,580	19,700
3320	Double wall, 500 gallon capacity	Q-5	2.40	6.667		2,200	133		2,333	2,650
3330	2000 gallon capacity	Q-7	4.15	7.711		4,575	153		4,728	5,275
3340	4000 gallon capacity		3.60	8.889		8,125	177		8,302	9,225
3350	6000 gallon capacity		2.40	13.333		9,675	265		9,940	11,000
3360	8000 gallon capacity		2	16		11,100	320		11,420	12,700
3370	10000 gallon capacity		1.80	17.778		12,200	355		12,555	14,000
3380	15000 gallon capacity		1.50	21.333		18,200	425		18,625	20,700
3390	20000 gallon capacity		1.30	24.615		20,700	490		21,190	23,600
3400	25000 gallon capacity		1.15	27.826		25,100	555		25,655	28,600
3410	30000 gallon capacity		1	32		27,500	635		28,135	31,400
4000	Fixed roof oil storage tanks, steel, (1 BBL=42 GAL w/ fdn 3'D x 1'W)									
4200	5,000 barrels				Ea.					144,000
4300	25,000 barrels									202,000
4500	55,000 barrels									342,000
4600	100,000 barrels									550,000
4800	150,000 barrels									775,000
4900	225,000 barrels									1,168,000
5100	Floating roof gasoline tanks, steel, 5,000 barrels									122,000
5200	25,000 barrels									273,000
5400	55,000 barrels									413,000
5500	100,000 barrels									625,000
5700	150,000 barrels									846,000
5800	225,000 barrels									1,173,000

13200 | Storage Tanks

13201 | Storage Tanks

			CREW	DAILY OUTPUT	LABOR-HOURS	UNIT	MAT.	LABOR	EQUIP.	TOTAL	TOTAL INCL O&P	
300	6000	Wood tanks, ground level, 2" cypress, 3,000 gallons	C-1	.19	168	Ea.	5,250	2,900		8,150	10,700	300
	6100	2-1/2" cypress, 10,000 gallons		.12	266		14,200	4,575		18,775	23,500	
	6300	3" redwood or 3" fir, 20,000 gallons		.10	320		21,700	5,500		27,200	33,200	
	6400	30,000 gallons		.08	400		26,800	6,875		33,675	41,300	
	6600	45,000 gallons	▼	.07	457	▼	40,600	7,850		48,450	58,000	
	6700	Larger sizes, minimum				Gal.					.66	
	6900	Maximum				"					.81	
	7000	Vinyl coated fabric pillow tanks, freestanding, 5,000 gallons	4 Clab	4	8	Ea.	1,900	116		2,016	2,300	
	7100	Supporting embankment not included, 25,000 gallons	6 Clab	2	24		6,700	345		7,045	7,975	
	7200	50,000 gallons	8 Clab	1.50	42.667		9,750	615		10,365	11,800	
	7300	100,000 gallons	9 Clab	.90	80		15,100	1,150		16,250	18,600	
	7400	150,000 gallons		.50	144		19,100	2,075		21,175	24,600	
	7500	200,000 gallons		.40	180		22,300	2,600		24,900	29,000	
	7600	250,000 gallons	▼	.30	239		26,800	3,475		30,275	35,500	
800	0010	**UNDERGROUND STORAGE TANKS**										800
	0210	Fiberglass, underground, single wall, U.L. listed, not including										
	0220	manway or hold-down strap										
	0240	2,000 gallon capacity	Q-7	4.57	7.002	Ea.	2,375	139		2,514	2,825	
	0250	4,000 gallon capacity		3.55	9.014		3,425	179		3,604	4,075	
	0260	6,000 gallon capacity		2.67	11.985		3,825	239		4,064	4,600	
	0280	10,000 gallon capacity		2	16		5,200	320		5,520	6,250	
	0284	15,000 gallon capacity		1.68	19.048		7,575	380		7,955	8,975	
	0290	20,000 gallon capacity	▼	1.45	22.069	▼	10,500	440		10,940	12,200	
	0500	For manway, fittings and hold-downs, add					20%	15%				
	1020	Fiberglass, underground, double wall, U.L. listed										
	1030	includes manways, not incl. hold-down straps										
	1040	600 gallon capacity	Q-5	2.42	6.612	Ea.	3,150	132		3,282	3,700	
	1050	1,000 gallon capacity	"	2.25	7.111		4,175	142		4,317	4,825	
	1060	2,500 gallon capacity	Q-7	4.16	7.692		5,950	153		6,103	6,800	
	1070	3,000 gallon capacity		3.90	8.205		6,400	163		6,563	7,325	
	1080	4,000 gallon capacity		3.64	8.791		7,000	175		7,175	8,000	
	1090	6,000 gallon capacity		2.42	13.223		8,100	263		8,363	9,350	
	1100	8,000 gallon capacity		2.08	15.385		10,200	305		10,505	11,700	
	1110	10,000 gallon capacity		1.82	17.582		10,700	350		11,050	12,300	
	1120	12,000 gallon capacity	▼	1.70	18.824	▼	13,000	375		13,375	14,900	
	2210	Fiberglass, underground, single wall, U.L. listed, including										
	2220	hold-down straps, no manways										
	2240	2,000 gallon capacity	Q-7	3.55	9.014	Ea.	2,525	179		2,704	3,075	
	2250	4,000 gallon capacity		2.90	11.034		3,575	220		3,795	4,325	
	2260	6,000 gallon capacity		2	16		4,150	320		4,470	5,075	
	2280	10,000 gallon capacity		1.60	20		5,525	400		5,925	6,725	
	2284	15,000 gallon capacity		1.39	23.022		7,900	460		8,360	9,450	
	2290	20,000 gallon capacity	▼	1.14	28.070	▼	11,000	560		11,560	13,000	
	3020	Fiberglass, underground, double wall, U.L. listed										
	3030	includes manways and hold-down straps										
	3040	600 gallon capacity	Q-5	1.86	8.602	Ea.	3,325	171		3,496	3,925	
	3050	1,000 gallon capacity	"	1.70	9.412		4,350	187		4,537	5,075	
	3060	2,500 gallon capacity	Q-7	3.29	9.726		6,125	194		6,319	7,050	
	3070	3,000 gallon capacity		3.13	10.224		6,575	203		6,778	7,550	
	3080	4,000 gallon capacity		2.93	10.922		7,150	217		7,367	8,225	
	3090	6,000 gallon capacity		1.86	17.204		8,425	340		8,765	9,850	
	3100	8,000 gallon capacity		1.65	19.394		10,500	385		10,885	12,200	
	3110	10,000 gallon capacity		1.48	21.622		11,000	430		11,430	12,800	
	3120	12,000 gallon capacity	▼	1.40	22.857	▼	13,400	455		13,855	15,500	
	5000	Steel underground, sti-P3, set in place, not incl. hold-down bars.										
	5500	Excavation, pad, pumps and piping not included										

SPECIAL CONSTRUCTION 13

13200 | Storage Tanks

13201 | Storage Tanks

		CREW	DAILY OUTPUT	LABOR-HOURS	UNIT	2000 BARE COSTS MAT.	LABOR	EQUIP.	TOTAL	TOTAL INCL O&P
5510	Single wall, 500 gallon capacity, 7 gauge shell	Q-5	2.70	5.926	Ea.	820	118		938	1,100
5520	1,000 gallon capacity, 7 gauge shell	"	2.50	6.400		1,350	127		1,477	1,700
5530	2,000 gallon capacity, 1/4" thick shell	Q-7	4.60	6.957		2,175	138		2,313	2,625
5535	2,500 gallon capacity, 7 gauge shell	Q-5	3	5.333		2,375	106		2,481	2,800
5540	5,000 gallon capacity, 1/4" thick shell	Q-7	3.20	10		4,450	199		4,649	5,225
5580	15,000 gallon capacity, 5/16" thick shell		1.70	18.824		10,200	375		10,575	11,900
5600	20,000 gallon capacity, 5/16" thick shell		1.50	21.333		14,000	425		14,425	16,100
5610	25,000 gallon capacity, 3/8" thick shell		1.30	24.615		17,600	490		18,090	20,200
5620	30,000 gallon capacity, 3/8" thick shell		1.10	29.091		20,400	580		20,980	23,500
5630	40,000 gallon capacity, 3/8" thick shell		.90	35.556		30,700	710		31,410	34,900
5640	50,000 gallon capacity, 3/8" thick shell		.80	40		38,300	795		39,095	43,500

13280 | Hazardous Material Remediation

13281 | Hazardous Material Remediation

		CREW	DAILY OUTPUT	LABOR-HOURS	UNIT	2000 BARE COSTS MAT.	LABOR	EQUIP.	TOTAL	TOTAL INCL O&P
0010	**BULK ASBESTOS REMOVAL**									
0020	Includes disposable tools and 1 suit and respirator/day/worker									
0100	Beams, W 10 x 19	A-9	235	.272	L.F.		5.60		5.60	9.80
0110	W 12 x 22		210	.305			6.25		6.25	10.95
0120	W 14 x 26		180	.356			7.30		7.30	12.80
0130	W 16 x 31		160	.400			8.20		8.20	14.40
0140	W 18 x 40		140	.457			9.40		9.40	16.45
0150	W 24 x 55		110	.582			11.95		11.95	21
0160	W 30 x 108		85	.753			15.50		15.50	27
0170	W 36 x 150		72	.889			18.30		18.30	32
0200	Boiler insulation		480	.133	S.F.	.06	2.74		2.80	4.87
0210	With metal lath add				%				50%	
0300	Boiler breeching or flue insulation	A-9	520	.123	S.F.		2.53		2.53	4.43
0310	For active boiler, add				%				100%	
0400	Duct or AHU insulation	A-10B	440	.073	S.F.		1.50		1.50	2.63
0500	Duct vibration isolation joints, up to 24 Sq. In. duct	A-9	56	1.143	Ea.		23.50		23.50	41
0520	25 Sq. In. to 48 Sq. In. duct		48	1.333			27.50		27.50	48
0530	49 Sq. In. to 76 Sq. In. duct		40	1.600			33		33	57.50
0600	Pipe insulation, air cell type, up to 4" diameter pipe		900	.071	L.F.		1.46		1.46	2.56
0610	4" to 8" diameter pipe		800	.080			1.64		1.64	2.88
0620	10" to 12" diameter pipe		700	.091			1.88		1.88	3.29
0630	14" to 16" diameter pipe		550	.116			2.39		2.39	4.19
0650	Over 16" diameter pipe		650	.098	S.F.		2.02		2.02	3.55
0700	With glove bag up to 3" diameter pipe		100	.640	L.F.	3.15	13.15		16.30	26.50
1000	Pipe fitting insulation up to 4" diameter pipe		320	.200	Ea.		4.11		4.11	7.20
1100	6" to 8" diameter pipe		304	.211			4.33		4.33	7.60
1110	10" to 12" diameter pipe		192	.333			6.85		6.85	12
1120	14" to 16" diameter pipe		128	.500			10.30		10.30	18
1130	Over 16" diameter pipe		176	.364	S.F.		7.50		7.50	13.10
1200	With glove bag, up to 8" diameter pipe		40	1.600	Ea.	6.55	33		39.55	64.50
2000	Scrape foam fireproofing from flat surface		2,400	.027	S.F.		.55		.55	.96
2100	Irregular surfaces		1,200	.053			1.10		1.10	1.92
3000	Remove cementitious material from flat surface		1,800	.036			.73		.73	1.28
3100	Irregular surface		1,400	.046			.94		.94	1.65

13280 | Hazardous Material Remediation

13281 | Hazardous Material Remediation

			CREW	DAILY OUTPUT	LABOR-HOURS	UNIT	2000 BARE COSTS				TOTAL INCL O&P
							MAT.	LABOR	EQUIP.	TOTAL	
120	4000	Scrape acoustical coating/fireproofing, from ceiling	A-9	3,200	.020	S.F.		.41		.41	.72
	5000	Remove VAT from floor by hand	↓	2,400	.027			.55		.55	.96
	5100	By machine	A-11	4,800	.013	↓		.27	.01	.28	.49
	5150	For 2 layers, add				%				50%	
	6000	Remove contaminated soil from crawl space by hand	A-9	400	.160	C.F.		3.29		3.29	5.75
	6100	With large production vacuum loader	A-12	700	.091	"		1.88	.76	2.64	4.12
	7000	Radiator backing, not including radiator removal	A-9	1,200	.053	S.F.		1.10		1.10	1.92
	8000	Cement-asbestos transite board	2 Asbe	2,000	.008		.08	.16		.24	.38
	8100	Transite shingle siding	A-10D	750	.043		.08	.84	.93	1.85	2.54
	8200	Shingle roofing	A-10B	2,000	.016		.07	.33		.40	.66
	8250	Built-up, no gravel, non-friable	B-2	1,400	.029		.07	.42		.49	.81
	8300	Asbestos millboard	2 Asbe	1,000	.016	↓	.08	.33		.41	.66
	9000	For type C (supplied air) respirator equipment, add				%					10%
125	0010	**ASBESTOS ABATEMENT WORK AREA** Containment and preparation.									
	0100	Pre-cleaning, HEPA vacuum and wet wipe, flat surfaces	A-10	12,000	.005	S.F.		.11		.11	.19
	0200	Protect carpeted area, 2 layers 6 mil poly on 3/4" plywood	"	1,000	.064		1.50	1.32		2.82	3.95
	0300	Separation barrier, 2" x 4" @ 16", 1/2" plywood ea. side, 8' high	2 Carp	400	.040		1.25	.79		2.04	2.73
	0310	12' high		320	.050		1.40	.99		2.39	3.23
	0320	16' high		200	.080		1.50	1.58		3.08	4.35
	0400	Personnel decontam. chamber, 2" x 4" @ 16", 3/4" ply ea. side		280	.057		2.50	1.13		3.63	4.68
	0450	Waste decontam. chamber, 2" x 4" studs @ 16", 3/4" ply ea. side	↓	360	.044		3	.88		3.88	4.80
	0500	Cover surfaces with polyethelene sheeting									
	0501	Including glue and tape									
	0550	Floors, each layer, 6 mil	A-10	8,000	.008	S.F.	.10	.16		.26	.40
	0551	4 mil		9,000	.007		.06	.15		.21	.33
	0560	Walls, each layer, 6 mil		6,000	.011		.09	.22		.31	.48
	0561	4 mil	↓	7,000	.009		.07	.19		.26	.41
	0570	For heights above 12', add						20%			
	0575	For heights above 20', add						30%			
	0580	For fire retardant poly, add					100%				
	0590	For large open areas, deduct					10%	20%			
	0600	Seal floor penetrations with foam firestop to 36 Sq. In.	2 Carp	200	.080	Ea.	6.25	1.58		7.83	9.60
	0610	36 Sq. In. to 72 Sq. In.		125	.128		12.50	2.52		15.02	18.05
	0615	72 Sq. In. to 144 Sq. In.		80	.200		25	3.94		28.94	34.50
	0620	Wall penetrations, to 36 square inches		180	.089		6.25	1.75		8	9.90
	0630	36 Sq. In. to 72 Sq. In.		100	.160		12.50	3.15		15.65	19.15
	0640	72 Sq. In. to 144 Sq. In.	↓	60	.267	↓	25	5.25		30.25	36.50
	0800	Caulk seams with latex	1 Carp	230	.035	L.F.	.15	.69		.84	1.34
130	0010	**DEMOLITION IN ASBESTOS CONTAMINATED AREA**									
	0200	Ceiling, including suspension system, plaster and lath	A-9	2,100	.030	S.F.		.63		.63	1.10
	0210	Finished plaster, leaving wire lath		585	.109			2.25		2.25	3.94
	0220	Suspended acoustical tile		3,500	.018			.38		.38	.66
	0230	Concealed tile grid system		3,000	.021			.44		.44	.77
	0240	Metal pan grid system		1,500	.043			.88		.88	1.54
	0250	Gypsum board		2,500	.026	↓		.53		.53	.92
	0260	Lighting fixtures up to 2' x 4'		72	.889	Ea.		18.30		18.30	32
	0400	Partitions, non load bearing									
	0410	Plaster, lath, and studs	A-9	690	.093	S.F.	.55	1.91		2.46	3.95
	0450	Gypsum board and studs	"	1,390	.046	"		.95		.95	1.66
	9000	For type C (supplied air) respirator equipment, add				%					10%
135	0010	**ASBESTOS ABATEMENT EQUIPMENT** and supplies, buy	R13281-120								
	0200	Air filtration device, 2000 C.F.M.				Ea.	2,500			2,500	2,750
	0250	Large volume air sampling pump, minimum					450			450	495
	0260	Maximum	↓			↓	1,000			1,000	1,100

13280 | Hazardous Material Remediation

13281 | Hazardous Material Remediation

			CREW	DAILY OUTPUT	LABOR-HOURS	UNIT	MAT.	LABOR	EQUIP.	TOTAL	TOTAL INCL O&P	
135	0300	Airless sprayer unit, 2 gun				Ea.	2,000			2,000	2,200	135
	0350	Light stand, 500 watt	R13281-120			↓	250			250	275	
	0400	Personal respirators										
	0410	Negative pressure, 1/2 face, dual operation, min.				Ea.	22.50			22.50	25	
	0420	Maximum					24			24	26.50	
	0450	P.A.P.R., full face, minimum					400			400	440	
	0460	Maximum					700			700	770	
	0470	Supplied air, full face, incl. air line, minimum					450			450	495	
	0480	Maximum					600			600	660	
	0500	Personnel sampling pump, minimum					450			450	495	
	0510	Maximum					750			750	825	
	1500	Power panel, 20 unit, incl. G.F.I.					1,800			1,800	1,975	
	1600	Shower unit, including pump and filters					1,125			1,125	1,250	
	1700	Supplied air system (type C)					10,000			10,000	11,000	
	1750	Vacuum cleaner, HEPA, 16 gal., stainless steel, wet/dry					1,000			1,000	1,100	
	1760	55 gallon					2,200			2,200	2,425	
	1800	Vacuum loader, 9-18 ton/hr					90,000			90,000	99,000	
	1900	Water atomizer unit, including 55 gal. drum					230			230	253	
	2000	Worker protection, whole body, foot, head cover & gloves, plastic					30			30	33	
	2500	Respirator, single use					10.50			10.50	11.55	
	2550	Cartridge for respirator					37.50			37.50	41.50	
	2570	Glove bag, 7 mil, 50" x 64"					8.50			8.50	9.35	
	2580	10 mil, 44" x 60"					8.40			8.40	9.25	
	3000	HEPA vacuum for work area, minimun					1,000			1,000	1,100	
	3050	Maximum					2,775			2,775	3,050	
	6000	Disposable polyethelene bags, 6 mil, 3 C.F.					1.15			1.15	1.27	
	6300	Disposable fiber drums, 3 C.F.					6.50			6.50	7.15	
	6400	Pressure sensitive caution lables, 3" x 5"					8.90			8.90	9.80	
	6450	11" x 17"					6.50			6.50	7.15	
	6500	Negative air machine, 1800 C.F.M.		↓		↓	775			775	855	
140	0010	**DECONTAMINATION CONTAINMENT AREA DEMOLITION** and clean-up										140
	0100	Spray exposed substrate with surfactant (bridging)										
	0200	Flat surfaces	A-9	6,000	.011	S.F.	.35	.22		.57	.77	
	0250	Irregular surfaces		4,000	.016	"	.30	.33		.63	.91	
	0300	Pipes, beams, and columns		2,000	.032	L.F.	.55	.66		1.21	1.76	
	1000	Spray encapsulate polyethelene sheeting		8,000	.008	S.F.	.30	.16		.46	.62	
	1100	Roll down polyethelene sheeting		8,000	.008	"		.16		.16	.29	
	1500	Bag polyethelene sheeting		400	.160	Ea.	.75	3.29		4.04	6.60	
	2000	Fine clean exposed substrate, with nylon brush		2,400	.027	S.F.		.55		.55	.96	
	2500	Wet wipe substrate		4,800	.013			.27		.27	.48	
	2600	Vacuum surfaces, fine brush	↓	6,400	.010	↓		.21		.21	.36	
	3000	Structural demolition										
	3100	Wood stud walls	A-9	2,800	.023	S.F.		.47		.47	.82	
	3500	Window manifolds, not incl. window replacement		4,200	.015			.31		.31	.55	
	3600	Plywood carpet protection	↓	2,000	.032	↓		.66		.66	1.15	
	4000	Remove custom decontamination facility	A-10A	8	3	Ea.	15	62		77	126	
	4100	Remove portable decontamination facility	3 Asbe	12	2	"	12.50	41		53.50	86	
	5000	HEPA vacuum, shampoo carpeting	A-9	4,800	.013	S.F.	.05	.27		.32	.54	
	9000	Final cleaning of protected surfaces	A-10A	8,000	.003	"		.06		.06	.11	
145	0010	**OSHA TESTING**										145
	0100	Certified technician, minimum				Day					300	
	0110	Maximum				"					500	
	0200	Personal sampling, PCM analysis, minimum	1 Asbe	8	1	Ea.	2.75	20.50		23.25	39	
	0210	Maximum	"	4	2	"	3	41		44	75.50	
	0300	Industrial hygienist, minimum				Day					400	

13280 | Hazardous Material Remediation

13281 | Hazardous Material Remediation

			CREW	DAILY OUTPUT	LABOR-HOURS	UNIT	2000 BARE COSTS MAT.	LABOR	EQUIP.	TOTAL	TOTAL INCL O&P	
145	0310	Maximum				Day					550	145
	1000	Cleaned area samples	1 Asbe	8	1	Ea.	2.63	20.50		23.13	39	
	1100	PCM air sample analysis, minimum		8	1		30	20.50		50.50	69	
	1110	Maximum		4	2		3.09	41		44.09	75.50	
	1200	TEM air sample analysis, minimum								100	125	
	1210	Maximum								400	500	
150	0010	**ENCAPSULATION WITH SEALANTS**										150
	0100	Ceilings and walls, minimum	A-9	21,000	.003	S.F.	.25	.06		.31	.39	
	0110	Maximum		10,600	.006		.40	.12		.52	.66	
	0200	Columns and beams, minimum		13,300	.005		.25	.10		.35	.45	
	0210	Maximum		5,325	.012		.45	.25		.70	.93	
	0300	Pipes to 12" diameter including minor repairs, minimum		800	.080	L.F.	.35	1.64		1.99	3.27	
	0310	Maximum		400	.160	"	1	3.29		4.29	6.85	
155	0010	**WASTE PACKAGING, HANDLING, & DISPOSAL**										155
	0100	Collect and bag bulk material, 3 C.F. bags, by hand	A-9	400	.160	Ea.	1.15	3.29		4.44	7	
	0200	Large production vacuum loader	A-12	880	.073		.80	1.50	.60	2.90	4.16	
	1000	Double bag and decontaminate	A-9	960	.067		2.30	1.37		3.67	4.93	
	2000	Containerize bagged material in drums, per 3 C.F. drum	"	800	.080		6.50	1.64		8.14	10.05	
	3000	Cart bags 50' to dumpster	2 Asbe	400	.040			.82		.82	1.44	
	5000	Disposal charges, not including haul, minimum				C.Y.					50	
	5020	Maximum				"					175	
	5100	Remove refrigerant from system	1 Plum	40	.200	Lb.		4.39		4.39	7.25	
	9000	For type C (supplied air) respirator equipment, add				%					10%	
440	0010	**REMOVAL** Existing lead paint, by chemicals, per application	R13281-460									440
	0020	See also, Div. 13280, Haz. Mat'l. Abatement										
	0050	Baseboard, to 6" wide	1 Pord	64	.125	L.F.	1.42	2.24		3.66	5.30	
	0070	To 12" wide		32	.250	"	2.78	4.49		7.27	10.55	
	0200	Balustrades, one side		28	.286	S.F.	3.16	5.15		8.31	12.05	
	1400	Cabinets, simple design		32	.250		2.76	4.49		7.25	10.55	
	1420	Ornate design		25	.320		3.55	5.75		9.30	13.50	
	1600	Cornice, simple design		60	.133		1.48	2.39		3.87	5.60	
	1620	Ornate design		20	.400		4.37	7.20		11.57	16.80	
	2800	Doors, one side, flush		84	.095		1.07	1.71		2.78	4.03	
	2820	Two panel		80	.100		1.11	1.80		2.91	4.22	
	2840	Four panel		45	.178		1.96	3.19		5.15	7.45	
	2880	For trim, one side, add		64	.125	L.F.	1.42	2.24		3.66	5.30	
	3000	Fence, picket, one side		30	.267	S.F.	2.97	4.79		7.76	11.25	
	3200	Grilles, one side, simple design		30	.267		2.97	4.79		7.76	11.25	
	3220	Ornate design		25	.320		3.55	5.75		9.30	13.50	
	4400	Pipes, to 4" diameter		90	.089	L.F.	1.01	1.60		2.61	3.77	
	4420	To 8" diameter		50	.160		1.76	2.87		4.63	6.75	
	4440	To 12" diameter		36	.222		2.46	3.99		6.45	9.35	
	4460	To 16" diameter		20	.400		4.40	7.20		11.60	16.85	
	4500	For hangers, add		40	.200	Ea.	2.21	3.59		5.80	8.45	
	4800	Siding		90	.089	S.F.	1.01	1.60		2.61	3.77	
	5000	Trusses, open		55	.145	SF Face	1.61	2.61		4.22	6.15	
	6200	Windows, one side only, double hung, 1/1 light, 24" x 48" high		4	2	Ea.	22	36		58	84.50	
	6220	30" x 60" high		3	2.667		29.50	48		77.50	113	
	6240	36" x 72" high		2.50	3.200		35.50	57.50		93	135	
	6280	40" x 80" high		2	4		44.50	72		116.50	169	
	6400	Colonial window, 6/6 light, 24" x 48" high		2	4		44.50	72		116.50	169	
	6420	30" x 60" high		1.50	5.333		59.50	95.50		155	225	
	6440	36" x 72" high		1	8		89	144		233	340	
	6480	40" x 80" high		1	8		89	144		233	340	
	6600	8/8 light, 24" x 48" high		2	4		44.50	72		116.50	169	

13280 | Hazardous Material Remediation

13281 | Hazardous Material Remediation

			CREW	DAILY OUTPUT	LABOR-HOURS	UNIT	MAT.	LABOR	EQUIP.	TOTAL	TOTAL INCL O&P
440	6620	40" x 80" high	1 Pord	1	8	Ea.	89	144		233	340
	6800	12/12 light, 24" x 48" high		1	8		89	144		233	340
	6820	40" x 80" high		.75	10.667		119	191		310	450
	6840	Window frame & trim items, included in pricing above									
460	0010	**LEAD PAINT ENCAPSULATION**, water based polymer coating, 14 mil DFT									
	0020	Interior, brushwork, trim, under 6"	1 Pord	240	.033	L.F.	2.09	.60		2.69	3.30
	0030	6" to 12" wide		180	.044		2.77	.80		3.57	4.38
	0040	Balustrades		300	.027		1.67	.48		2.15	2.64
	0050	Pipe to 4" diameter		500	.016		1	.29		1.29	1.58
	0060	To 8" diameter		375	.021		1.33	.38		1.71	2.10
	0070	To 12" diameter		250	.032		2	.57		2.57	3.16
	0080	To 16" diameter		170	.047		2.94	.84		3.78	4.64
	0090	Cabinets, ornate design		200	.040	S.F.	2.50	.72		3.22	3.95
	0100	Simple design		250	.032	"	2	.57		2.57	3.16
	0110	Doors, 3'x 7', both sides, incl. frame & trim									
	0120	Flush	1 Pord	6	1.333	Ea.	25.50	24		49.50	68
	0130	French, 10-15 lite		3	2.667		5.10	48		53.10	85.50
	0140	Panel		4	2		30.50	36		66.50	94
	0150	Louvered		2.75	2.909		28	52		80	118
	0160	Windows, per interior side, per 15 S.F.									
	0170	1 to 6 lite	1 Pord	14	.571	Ea.	17.70	10.25		27.95	36.50
	0180	7 to 10 lite		7.50	1.067		19.50	19.15		38.65	53.50
	0190	12 lite		5.75	1.391		26	25		51	70.50
	0200	Radiators		8	1		62.50	17.95		80.45	98.50
	0210	Grilles, vents		275	.029	S.F.	1.82	.52		2.34	2.87
	0220	Walls, roller, drywall or plaster		1,000	.008		.49	.14		.63	.78
	0230	With spunbonded reinforcing fabric		720	.011		.56	.20		.76	.95
	0240	Wood		800	.010		.62	.18		.80	.98
	0250	Ceilings, roller, drywall or plaster		900	.009		.56	.16		.72	.89
	0260	Wood		700	.011		.71	.21		.92	1.12
	0270	Exterior, brushwork, gutters and downspouts		300	.027	L.F.	1.67	.48		2.15	2.64
	0280	Columns		400	.020	S.F.	1.25	.36		1.61	1.98
	0290	Spray, siding		600	.013	"	.83	.24		1.07	1.31
	0300	Miscellaneous									
	0310	Electrical conduit, brushwork, to 2" diameter	1 Pord	500	.016	L.F.	1	.29		1.29	1.58
	0320	Brick, block or concrete, spray		500	.016	S.F.	1	.29		1.29	1.58
	0330	Steel, flat surfaces and tanks to 12"		500	.016		1	.29		1.29	1.58
	0340	Beams, brushwork		400	.020		1.25	.36		1.61	1.98
	0350	Trusses		400	.020		1.25	.36		1.61	1.98

13600 | Solar and Wind Energy Equipment

13630 | Solar Collector Components

			CREW	DAILY OUTPUT	LABOR-HOURS	UNIT	MAT.	LABOR	EQUIP.	TOTAL	TOTAL INCL O&P
200	0010	**SOLAR ENERGY**									
	0020	System/Package prices, not including connecting									
	0030	pipe, insulation, or special heating/plumbing fixtures									
	0500	Hot water, standard package, low temperature									
	0540	1 collector, circulator, fittings, 65 gal. tank	Q-1	.50	32	Ea.	1,350	630		1,980	2,550
	0580	2 collectors, circulator, fittings, 120 gal. tank		.40	40		1,925	790		2,715	3,425

13600 | Solar and Wind Energy Equipment

13630 | Solar Collector Components

		Crew	Daily Output	Labor-Hours	Unit	Mat.	Labor	Equip.	Total	Total Incl O&P
0620	3 collectors, circulator, fittings, 120 gal. tank	Q-1	.40	40	Ea.	2,350	790		3,140	3,875
0700	Medium temperature package									
0720	1 collector, circulator, fittings, 80 gal. tank	Q-1	.50	32	Ea.	1,475	630		2,105	2,675
0740	2 collectors, circulator, fittings, 120 gal. tank		.40	40		2,075	790		2,865	3,600
0780	3 collectors, circulator, fittings, 120 gal. tank		.30	53.333		3,775	1,050		4,825	5,900
0980	For each additional 120 gal. tank, add					720			720	795
2300	Circulators, air									
2310	Blowers									
2400	Reversible fan, 20" diameter, 2 speed	Q-9	18	.889	Ea.	102	17.20		119.20	141
2520	Space & DHW system, less duct work	"	.50	32		1,400	620		2,020	2,600
2870	1/12 HP, 30 GPM	Q-1	10	1.600		315	31.50		346.50	395
3000	Collector panels, air with aluminum absorber plate									
3010	Wall or roof mount									
3040	Flat black, plastic glazing									
3080	4' x 8'	Q-9	6	2.667	Ea.	600	51.50		651.50	745
3200	Flush roof mount, 10' to 16' x 22" wide	"	96	.167	L.F.	146	3.23		149.23	166
3300	Collector panels, liquid with copper absorber plate									
3330	Alum. frame, 4' x 8', 5/32" single glazing	Q-1	9.50	1.684	Ea.	545	33.50		578.50	655
3390	Alum. frame, 4' x 10', 5/32" single glazing	"	6	2.667	"	650	52.50		702.50	795
3440	Flat black									
3450	Alum. frame, 3' x 8'	Q-1	9	1.778	Ea.	430	35		465	535
3500	Alum. frame, 4' x 8.5'		5.50	2.909		490	57.50		547.50	635
3520	Alum. frame, 4' x 10.5'		10	1.600		595	31.50		626.50	705
3540	Alum. frame, 4' x 12.5'		5	3.200		735	63		798	915
3600	Liquid, full wetted, plastic, alum. frame, 3' x 10'		5	3.200		195	63		258	320
3650	Collector panel mounting, flat roof or ground rack		7	2.286		64.50	45		109.50	146
3670	Roof clamps		70	.229	Set	1.61	4.51		6.12	9.20
3700	Roof strap, teflon	1 Plum	205	.039	L.F.	9.50	.86		10.36	11.85
3900	Differential controller with two sensors									
3930	Thermostat, hard wired	1 Plum	8	1	Ea.	92	22		114	138
4100	Five station with digital read-out	"	3	2.667	"	202	58.50		260.50	320
4300	Heat exchanger									
4580	Fluid to fluid package includes two circulating pumps									
4590	expansion tank, check valve, relief valve									
4600	controller, high temperature cutoff and sensors	Q-1	2.50	6.400	Ea.	695	126		821	975
4650	Heat transfer fluid									
4700	Propylene glycol, inhibited anti-freeze	1 Plum	28	.286	Gal.	8.80	6.25		15.05	20
8250	Water storage tank with heat exchanger and electric element									
8280	80 gal. with 2" x 1/2 lb. density insulation	1 Plum	1.60	5	Ea.	360	110		470	575
8300	80 gal. with 2" x 2 lb. density insulation		1.60	5		745	110		855	1,000
8350	120 gal. with 2" x 1/2 lb. density insulation		1.40	5.714		390	125		515	630
8380	120 gal. with 2" x 2 lb. density insulation		1.40	5.714		840	125		965	1,125
8400	120 gal. with 2" x 2 lb. density insul., 40 S.F. heat coil		1.40	5.714		965	125		1,090	1,275

13700 | Security Access and Surveillance

13710 | Security Access

		Crew	Daily Output	Labor-Hours	Unit	Mat.	Labor	Equip.	Total	Total Incl O&P
0010	**ACCESS CONTROL**									
0020	Card type, 1 time zone, minimum				Ea.	300			300	330

13700 | Security Access and Surveillance

13710 | Security Access

			CREW	DAILY OUTPUT	LABOR-HOURS	UNIT	2000 BARE COSTS MAT.	LABOR	EQUIP.	TOTAL	TOTAL INCL O&P	
300	0040	Maximum				Ea.	1,025			1,025	1,125	300
	0060	3 time zones, minimum					740			740	815	
	0080	Maximum				↓	1,700			1,700	1,875	
	0100	System with printer, and control console, 3 zones				Total	8,300			8,300	9,125	
	0120	6 zones				"	10,900			10,900	12,000	
	0140	For each door, minimum, add				Ea.	1,225			1,225	1,350	
	0160	Maximum, add				"	1,800			1,800	2,000	

13800 | Building Automation & Control

13834 | Electric/Electronic Control

			CREW	DAILY OUTPUT	LABOR-HOURS	UNIT	2000 BARE COSTS MAT.	LABOR	EQUIP.	TOTAL	TOTAL INCL O&P	
200	0010	CONTROL SYSTEMS, ELECTRONIC										200
	0020	For electronic costs, add to division 13836-200				Ea.					15%	

13836 | Pneumatic Controls

			CREW	DAILY OUTPUT	LABOR-HOURS	UNIT	MAT.	LABOR	EQUIP.	TOTAL	TOTAL INCL O&P	
200	0010	CONTROL SYSTEMS, PNEUMATIC (Sub's quote incl. mat. & labor)										200
	0011	and nominal 50 Ft. of tubing. Add control panelboard if req'd.										
	0100	Heating and Ventilating, split system										
	0200	Mixed air control, economizer cycle, panel readout, tubing										
	0220	Up to 10 tons	Q-19	.68	35.294	Ea.	2,650	730		3,380	4,125	
	0240	For 10 to 20 tons		.63	37.915		2,875	780		3,655	4,450	
	0260	For over 20 tons	↓	.58	41.096	↓	3,100	850		3,950	4,825	
	0300	Heating coil, hot water, 3 way valve,										
	0320	Freezestat, limit control on discharge, readout	Q-5	.69	23.088	Ea.	2,000	460		2,460	2,950	
	0500	Cooling coil, chilled water, room										
	0520	Thermostat, 3 way valve	Q-5	2	8	Ea.	885	159		1,044	1,250	
	0600	Cooling tower, fan cycle, damper control,										
	0620	Control system including water readout in/out at panel	Q-19	.67	35.821	Ea.	3,525	740		4,265	5,100	
	1000	Unit ventilator, day/night operation,										
	1100	freezestat, ASHRAE, cycle 2	Q-19	.91	26.374	Ea.	1,950	545		2,495	3,050	
	2000	Compensated hot water from boiler, valve control,										
	2100	readout and reset at panel, up to 60 GPM	Q-19	.55	43.956	Ea.	3,650	905		4,555	5,525	
	2120	For 120 GPM		.51	47.059		3,900	970		4,870	5,900	
	2140	For 240 GPM		.49	49.180		4,100	1,025		5,125	6,175	
	3000	Boiler room combustion air, damper to 5 SF, controls		1.36	17.582		1,750	365		2,115	2,550	
	3500	Fan coil, heating and cooling valves, 4 pipe control system		3	8		795	165		960	1,150	
	3600	Heat exchanger system controls	↓	.86	27.907		1,700	575		2,275	2,825	
	4000	Pneumatic thermostat, including controlling room radiator valve	Q-5	2.43	6.593		530	131		661	800	
	4060	Pump control system	Q-19	3	8	↓	815	165		980	1,175	
	4500	Air supply for pneumatic control system										
	4600	Tank mounted duplex compressor, starter, alternator,										
	4620	piping, dryer, PRV station and filter										
	4630	1/2 HP	Q-19	.68	35.139	Ea.	6,550	725		7,275	8,400	
	4660	1-1/2 HP		.57	41.739		8,000	860		8,860	10,200	
	4690	5 HP	↓	.42	57.143	↓	19,000	1,175		20,175	22,900	
050	0010	CLOCKS										050
	0080	12" diameter, single face	1 Elec	8	1	Ea.	75.50	22		97.50	119	
	0100	Double face	"	6.20	1.290	"	144	28.50		172.50	205	

13800 | Building Automation & Control

13836 | Pneumatic Controls

		CREW	DAILY OUTPUT	LABOR-HOURS	UNIT	MAT.	LABOR	EQUIP.	TOTAL	TOTAL INCL O&P
055 0010	**CLOCK SYSTEMS**, not including wires & conduits									
0100	Time system components, master controller	1 Elec	.33	24.242	Ea.	1,625	535		2,160	2,675
0200	Program bell		8	1		53.50	22		75.50	95
0400	Combination clock & speaker		3.20	2.500		189	55.50		244.50	299
0600	Frequency generator		2	4		6,350	88.50		6,438.50	7,125
0800	Job time automatic stamp recorder, minimum		4	2		445	44		489	565
1000	Maximum		4	2		680	44		724	825
1200	Time stamp for correspondence, hand operated					340			340	375
1400	Fully automatic					495			495	545
1600	Master time clock system, clocks & bells, 20 room	4 Elec	.20	160		4,075	3,525		7,600	10,300
1800	50 room	"	.08	400		9,400	8,850		18,250	24,800
2000	Time clock, 100 cards in & out, 1 color	1 Elec	3.20	2.500		950	55.50		1,005.50	1,150
2200	2 colors		3.20	2.500		1,025	55.50		1,080.50	1,225
2400	With 3 circuit program device, minimum		2	4		375	88.50		463.50	560
2600	Maximum		2	4		545	88.50		633.50	745
2800	Metal rack for 25 cards		7	1.143		60	25.50		85.50	108
3000	Watchman's tour station		8	1		63	22		85	106
3200	Annunciator with zone indication		1	8		230	177		407	540
3400	Time clock with tape		1	8		610	177		787	960
065 0010	**DETECTION SYSTEMS**, not including wires & conduits									
0100	Burglar alarm, battery operated, mechanical trigger	1 Elec	4	2	Ea.	249	44		293	345
0200	Electrical trigger		4	2		297	44		341	400
0400	For outside key control, add		8	1		70.50	22		92.50	114
0600	For remote signaling circuitry, add		8	1		112	22		134	159
0800	Card reader, flush type, standard		2.70	2.963		835	65.50		900.50	1,025
1000	Multi-code		2.70	2.963		1,075	65.50		1,140.50	1,275
1200	Door switches, hinge switch		5.30	1.509		52.50	33.50		86	113
1400	Magnetic switch		5.30	1.509		62	33.50		95.50	123
1600	Exit control locks, horn alarm		4	2		310	44		354	415
1800	Flashing light alarm		4	2		350	44		394	460
2000	Indicating panels, 1 channel		2.70	2.963		330	65.50		395.50	470
2200	10 channel	2 Elec	3.20	5		1,125	111		1,236	1,425
2400	20 channel		2	8		2,200	177		2,377	2,725
2600	40 channel		1.14	14.035		4,000	310		4,310	4,900
2800	Ultrasonic motion detector, 12 volt	1 Elec	2.30	3.478		206	77		283	355
3000	Infrared photoelectric detector	"	2.30	3.478		170	77		247	315
3594	Fire, alarm control panel									
3600	4 zone	2 Elec	2	8	Ea.	920	177		1,097	1,300
3800	8 zone		1	16		1,400	355		1,755	2,125
4000	12 zone		.67	23.988		1,825	530		2,355	2,875
4020	Alarm device	1 Elec	8	1		122	22		144	170
4050	Actuating device		8	1		292	22		314	355
4200	Battery and rack		4	2		690	44		734	835
4400	Automatic charger		8	1		445	22		467	525
4600	Signal bell		8	1		49.50	22		71.50	90.50
4800	Trouble buzzer or manual station		8	1		37	22		59	76.50
5000	Detector, rate of rise		8	1		33.50	22		55.50	73
5200	Smoke detector, ceiling type		6.20	1.290		75	28.50		103.50	129
5400	Duct type		3.20	2.500		250	55.50		305.50	365
5600	Strobe and horn		5.30	1.509		95	33.50		128.50	160
5800	Fire alarm horn		6.70	1.194		36.50	26.50		63	83
6000	Door holder, electro-magnetic		4	2		77.50	44		121.50	158
6200	Combination holder and closer		3.20	2.500		430	55.50		485.50	565
6400	Code transmitter		4	2		690	44		734	835
6600	Drill switch		8	1		86.50	22		108.50	131

13800 | Building Automation & Control

13836 | Pneumatic Controls

		CREW	DAILY OUTPUT	LABOR-HOURS	UNIT	2000 BARE COSTS MAT.	LABOR	EQUIP.	TOTAL	TOTAL INCL O&P	
065	6800	Master box	1 Elec	2.70	2.963	Ea.	3,100	65.50		3,165.50	3,500	065
	7000	Break glass station		8	1		50	22		72	91	
	7800	Remote annunciator, 8 zone lamp	↓	1.80	4.444		175	98		273	355	
	8000	12 zone lamp	2 Elec	2.60	6.154		300	136		436	550	
	8200	16 zone lamp	"	2.20	7.273	↓	300	161		461	595	

13900 | Fire Suppression

13910 | Basic Fire Protection Matl/Methd

		CREW	DAILY OUTPUT	LABOR-HOURS	UNIT	2000 BARE COSTS MAT.	LABOR	EQUIP.	TOTAL	TOTAL INCL O&P	
400	0010	**FIRE HOSE AND EQUIPMENT**										400
	0200	Adapters, rough brass, straight hose threads										
	0220	One piece, female to male, rocker lugs										
	0240	1" x 1"				Ea.	17.85			17.85	19.65	
	2200	Hose, less couplings										
	2260	Synthetic jacket, lined, 300 lb. test, 1-1/2" diameter	Q-12	2,600	.006	L.F.	1.33	.12		1.45	1.66	
	2280	2-1/2" diameter		2,200	.007		2.17	.14		2.31	2.63	
	2360	High strength, 500 lb. test, 1-1/2" diameter		2,600	.006		1.59	.12		1.71	1.95	
	2380	2-1/2" diameter	↓	2,200	.007	↓	2.75	.14		2.89	3.27	
	2600	Hose rack, swinging, for 1-1/2" diameter hose,										
	2620	Enameled steel, 50' & 75' lengths of hose	Q-12	20	.800	Ea.	32.50	15.90		48.40	62	
	2640	100' and 125' lengths of hose	"	20	.800	"	32.50	15.90		48.40	62	
	3750	Hydrants, wall, w/caps, single, flush, polished brass										
	3800	2-1/2" x 2-1/2"	Q-12	5	3.200	Ea.	91.50	63.50		155	205	
	3840	2-1/2" x 3"	"	5	3.200	"	122	63.50		185.50	239	
	3950	Double, flush, polished brass										
	4000	2-1/2" x 2-1/2" x 4"	Q-12	5	3.200	Ea.	245	63.50		308.50	375	
	4040	2-1/2" x 2-1/2" x 6"	"	4.60	3.478	↓	355	69		424	505	
	4200	For polished chrome, add					10%					
	4350	Double, projecting, polished brass										
	4400	2-1/2" x 2-1/2" x 4"	Q-12	5	3.200	Ea.	113	63.50		176.50	229	
	4450	2-1/2" x 2-1/2" x 6"	"	4.60	3.478	"	218	69		287	355	
	4460	Valve control, dbl. flush/projecting hydrant, cap &										
	4470	chain, ext. rod & cplg., escutcheon, polished brass	Q-12	8	2	Ea.	133	39.50		172.50	213	
	5600	Nozzles, brass										
	5620	Adjustable fog, 3/4" booster line				Ea.	69			69	76	
	5630	1" booster line					80.50			80.50	88.50	
	5640	1-1/2" leader line					84.50			84.50	93	
	5660	2-1/2" direct connection				↓	165			165	182	
	5780	For chrome plated, add					8%					
	5850	Electrical fire, adjustable fog, no shock										
	5900	1-1/2"				Ea.	263			263	289	
	5920	2-1/2"					360			360	395	
	5980	For polished chrome, add				↓	6%					
	6200	Heavy duty, comb. adj. fog and str. stream, with handle										
	6210	1" booster line				Ea.	251			251	277	
	7140	Standpipe connections, wall, w/plugs & chains										
	7160	Single, flush, brass, 2-1/2" x 2-1/2"	Q-12	5	3.200	Ea.	65	63.50		128.50	177	
	7180	2-1/2" x 3"	"	5	3.200	"	71	63.50		134.50	183	
	7240	For polished chrome, add					15%					

13900 | Fire Suppression

13910 | Basic Fire Protection Matl/Methd

			CREW	DAILY OUTPUT	LABOR-HOURS	UNIT	MAT.	LABOR	EQUIP.	TOTAL	TOTAL INCL O&P	
400	7280	Double, flush, polished brass										400
	7300	2-1/2" x 2-1/2" x 4"	Q-12	5	3.200	Ea.	219	63.50		282.50	345	
	7330	2-1/2" x 2-1/2" x 6"	"	4.60	3.478	"	315	69		384	460	
	7400	For polished chrome, add					15%					
	7440	For sill cock combination, add				Ea.	38.50			38.50	42.50	
	7900	Three way, flush, polished brass										
	7920	2-1/2" (3) x 4"	Q-12	4.80	3.333	Ea.	715	66		781	900	
	7930	2-1/2" (3) x 6"	"	4.80	3.333		725	66		791	905	
	8000	For polished chrome, add				▼	9%					
	8020	Three way, projecting, polished brass										
	8040	2-1/2"(3) x 4"	Q-12	4.80	3.333	Ea.	475	66		541	630	
800	0010	**FIRE VALVES**										800
	0080	Wheel handle, 300 lb., 1-1/2"	1 Spri	12	.667	Ea.	22.50	14.70		37.20	49	
	0090	2-1/2"	"	7	1.143	"	55	25		80	102	
	0100	For polished brass, add					35%					
	0110	For polished chrome, add					50%					

13920 | Fire Pumps

			CREW	DAILY OUTPUT	LABOR-HOURS	UNIT	MAT.	LABOR	EQUIP.	TOTAL	TOTAL INCL O&P	
400	0010	**FIRE PUMPS** Including controller, fittings and relief valve										400
	0030	Diesel										
	0050	500 GPM, 50 psi, 27 HP, 4" pump	Q-13	.64	50	Ea.	46,600	995		47,595	53,000	
	0200	750 GPM, 50 psi, 44 HP, 5" pump		.60	53.333		49,700	1,050		50,750	56,500	
	0400	1000 GPM, 100 psi, 89 HP, 4" pump		.56	57.143		53,500	1,125		54,625	60,500	
	0700	2,000 GPM, 100 psi, 167 HP, 6" pump		.34	94.118		56,500	1,875		58,375	65,000	
	0950	3500 GPM, 100 psi, 300 HP, 10" pump	▼	.24	133	▼	99,500	2,650		102,150	114,000	
	3000	Electric										
	3100	250 GPM, 55 psi, 15 HP, 3,550 RPM, 2" pump	Q-13	.70	45.714	Ea.	10,900	905		11,805	13,500	
	3200	500 GPM, 50 psi, 27 HP, 1770 RPM, 4" pump		.68	47.059		13,500	935		14,435	16,500	
	3350	750 GPM, 50 psi, 44 HP, 1,770 RPM, 5" pump		.64	50		19,800	995		20,795	23,500	
	3400	750 GPM, 100 psi, 66 HP, 3550 RPM, 4" pump	▼	.58	55.172		17,500	1,100		18,600	21,100	
	5000	For jockey pump 1", 3 HP, with control, add	Q-12	2	8	▼	2,425	159		2,584	2,950	

13930 | Wet-Pipe Fire Supp. Sprinklers

			CREW	DAILY OUTPUT	LABOR-HOURS	UNIT	MAT.	LABOR	EQUIP.	TOTAL	TOTAL INCL O&P	
400	0010	**SPRINKLER SYSTEM COMPONENTS**										400
	0600	Accelerator	1 Spri	8	1	Ea.	335	22		357	405	
	2600	Sprinkler heads, not including supply piping										
	2640	Dry, pendent, 1/2" orifice, 3/4" or 1" NPT										
	2700	11" to 12-3/4" length	1 Spri	14	.571	Ea.	33.50	12.60		46.10	58	
	2710	13" to 14-3/4" length		13	.615		37.50	13.55		51.05	64	
	2720	15" to 16-3/4" length		13	.615		37.50	13.55		51.05	64	
	2730	17" to 18-3/4" length		13	.615		40.50	13.55		54.05	67	
	3600	Foam-water, pendent or upright, 1/2" NPT	▼	12	.667	▼	32	14.70		46.70	59.50	
	3700	Standard spray, pendent or upright, brass, 135° to 286°F										
	3730	1/2" NPT, 7/16" orifice	1 Spri	16	.500	Ea.	5.30	11		16.30	24	
	3740	1/2" NPT, 1/2" orifice	"	16	.500		3.64	11		14.64	22.50	
	3860	For wax and lead coating, add					4.25			4.25	4.68	
	3880	For wax coating, add					1.50			1.50	1.65	
	3900	For lead coating, add				▼	3			3	3.30	
	3920	For 360°F, same cost										
	3930	For 400°F, add				Ea.	5.50			5.50	6.05	
	3940	For 500°F, add				"	5.50			5.50	6.05	
	4500	Sidewall, horizontal, brass, 135° to 286°F										
	4520	1/2" NPT, 1/2" orifice	1 Spri	16	.500	Ea.	5.25	11		16.25	24	

13900 | Fire Suppression

13930 | Wet-Pipe Fire Supp. Sprinklers

			CREW	DAILY OUTPUT	LABOR-HOURS	UNIT	2000 BARE COSTS MAT.	LABOR	EQUIP.	TOTAL	TOTAL INCL O&P	
400	4540	For 360°F, same cost										400
	4800	Recessed pendent, brass, 135° to 286°F										
	4820	1/2" NPT, 3/8" orifice	1 Spri	10	.800	Ea.	5.30	17.65		22.95	35	
	4830	1/2" NPT, 7/16" orifice		10	.800		5.30	17.65		22.95	35	
	4840	1/2" NPT, 1/2" orifice	↓	10	.800	↓	3.64	17.65		21.29	33	
	6200	Valves										
	6500	Check, swing, C.I. body, brass fittings, auto. ball drip										
	6520	4" size	Q-12	3	5.333	Ea.	95.50	106		201.50	280	
	6800	Check, wafer, butterfly type, C.I. body, bronze fittings										
	6820	4" size	Q-12	4	4	Ea.	152	79.50		231.50	298	
	7000	Deluge, assembly, incl. trim, pressure										
	7020	operated relief, emergency release, gauges										
	7040	2" size	Q-12	2	8	Ea.	930	159		1,089	1,300	
	7060	3" size	"	1.50	10.667	"	1,100	212		1,312	1,550	

13960 | CO2 Fire Extinguishing

			CREW	DAILY OUTPUT	LABOR-HOURS	UNIT	MAT.	LABOR	EQUIP.	TOTAL	TOTAL INCL O&P	
200	0010	**AUTOMATIC FIRE SUPPRESSION SYSTEMS**										200
	0040	For detectors and control stations, see division 13850										
	1000	Dispersion nozzle, CO2, 3" x 5"	1 Plum	18	.444	Ea.	50	9.75		59.75	71	
	1100	FM200, 1-1/2"	"	14	.571		80	12.55		92.55	109	
	2000	Extinguisher, CO2 system, high pressure, 75 lb. cylinder	Q-1	6	2.667	↓	975	52.50		1,027.50	1,150	
	2400	FM200 system, filled, with mounting bracket										
	2540	196 lb. container	Q-1	4	4	Ea.	5,800	79		5,879	6,500	
	3000	Electro/mechanical release	L-1	4	2.500	↓	115	55		170	218	
	3400	Manual pull station	1 Plum	6	1.333	↓	45	29.50		74.50	98	
	6000	Average FM200 system, minimum				C.F.	1.25			1.25	1.38	
	6020	Maximum				"	2.50			2.50	2.75	

For information about Means Estimating Seminars, see yellow pages 11 and 12 in back of book

Division 14
Conveying Systems

Estimating Tips
General
- Many products in Division 14 will require some type of support or blocking for installation not included with the item itself. Examples are supports for conveyors or tube systems, attachment points for lifts, and footings for hoists or cranes. Add these supports in the appropriate division.

14100 Dumbwaiters
14200 Elevators
- Dumbwaiters and elevators are estimated and purchased in a method similar to buying a car. The manufacturer has a base unit with standard features. Added to this base unit price will be whatever options the owner or specifications require. Increased load capacity, additional vertical travel, additional stops, higher speed, and cab finish options are items to be considered. When developing an estimate for dumbwaiters and elevators, remember that some items needed by the installers may have to be included as part of the general contract.

Examples are:
- shaftway
- rail support brackets
- machine room
- electrical supply
- sill angles
- electrical connections
- pits
- roof penthouses
- pit ladders

Check the job specifications and drawings before pricing.

14300 Escalators & Moving Walks
- Escalators and moving walks are specialty items installed by specialty contractors. There are numerous options associated with these items. For specific options contact a manufacturer or contractor. In a method similar to estimating dumbwaiters and elevators, you should verify the extent of general contract work and add items as necessary.

14400 Lifts
14500 Material Handling
14600 Hoists & Cranes
- Products such as correspondence lifts, conveyors, chutes, pneumatic tube systems, material handling cranes and hoists as well as other items specified in this subdivision may require trained installers. The general contractor might not have any choice as to who will perform the installation or when it will be performed. Long lead times are often required for these products, making early decisions in scheduling necessary.

Reference Numbers
Reference numbers are shown in bold squares at the beginning of some major classifications. These numbers refer to related items in the Reference Section. The reference information may be an estimating procedure, an alternate pricing method or technical information.

Note: Not all subdivisions listed here necessarily appear in this publication.

14100 | Dumbwaiters

14110	Manual Dumbwaiters	CREW	DAILY OUTPUT	LABOR-HOURS	UNIT	2000 BARE COSTS				TOTAL INCL O&P
						MAT.	LABOR	EQUIP.	TOTAL	
0010	**DUMBWAITERS** 2 stop, hand, minimum	2 Elev	.75	21.333	Ea.	2,225	490		2,715	3,250
0100	Maximum		.50	32	"	5,050	735		5,785	6,750
0300	For each additional stop, add	↓	.75	21.333	Stop	800	490		1,290	1,700

14120	Electric Dumbwaiters									
0010	**DUMBWAITERS** 2 stop, electric, minimum	2 Elev	.13	123	Ea.	5,600	2,825		8,425	10,800
0100	Maximum		.11	145	"	16,800	3,325		20,125	24,000
0600	For each additional stop, add	↓	.54	29.630	Stop	2,475	680		3,155	3,825

14200 | Elevators

14210	Electric Traction Elevators	CREW	DAILY OUTPUT	LABOR-HOURS	UNIT	2000 BARE COSTS				TOTAL INCL O&P
						MAT.	LABOR	EQUIP.	TOTAL	
0012	**ELEVATORS OR LIFTS**									
7000	Residential, cab type, 1 floor, 2 stop, minimum	2 Elev	.20	80	Ea.	7,850	1,825		9,675	11,700
7100	Maximum		.10	160		13,300	3,675		16,975	20,700
7200	2 floor, 3 stop, minimum		.12	133		11,700	3,050		14,750	17,900
7300	Maximum		.06	266		19,000	6,100		25,100	31,000
7700	Stair climber (chair lift), single seat, minimum		1	16		3,800	365		4,165	4,775
7800	Maximum		.20	80		5,225	1,825		7,050	8,775
8000	Wheelchair lift, minimum		1	16		5,200	365		5,565	6,325
8500	Maximum		.50	32		12,300	735		13,035	14,800
8700	Stair lift, minimum		1	16		10,300	365		10,665	11,900
8900	Maximum	↓	.20	80	↓	16,300	1,825		18,125	20,900
0010	**ELEVATORS**									
0020	For multi-story buildings, housing project, minimum	R14200-100			% total					2.50%
0100	Maximum									4.50%
0300	Office building, minimum	R14200-200								2.50%
0400	Maximum				↓					10%
0410	Holeless hydraulic passenger, 2000 lb capacity, 2 story	R14200-300	2 Elev	.10	160	Ea.	26,800	3,675		30,475	35,500
0420	2500 lb capacity			.10	160		29,400	3,675		33,075	38,500
0425	Electric freight, base unit, 4000 lb, 200 fpm, 4 stop, std. fin.	R14200-400	↓	.05	320		61,500	7,325		68,825	80,000
0450	For 5000 lb capacity, add						4,425			4,425	4,875
0500	For 6000 lb capacity, add						7,800			7,800	8,575
0525	For 7000 lb capacity, add						10,500			10,500	11,500
0550	For 8000 lb capacity, add						14,600			14,600	16,000
0575	For 10000 lb capacity, add						17,400			17,400	19,100
0600	For 12000 lb capacity, add						21,200			21,200	23,300
0625	For 16000 lb capacity, add						25,500			25,500	28,000
0650	For 20000 lb capacity, add						28,000			28,000	30,700
0675	For increased speed, 250 fpm, add						8,800			8,800	9,700
0700	300 fpm, geared electric, add						11,000			11,000	12,100
0725	350 fpm, geared electric, add						13,000			13,000	14,300
0750	400 fpm, geared electric, add						14,400			14,400	15,800
0775	500 fpm, gearless electric, add						18,000			18,000	19,800
0800	600 fpm, gearless electric, add						19,900			19,900	21,900
0825	700 fpm, gearless electric, add						23,200			23,200	25,500
0850	800 fpm, gearless electric, add	↓			↓		25,800			25,800	28,300

Important: See the Reference Section for critical supporting data - Reference Nos., Crews, & City Cost Indexes

14200 | Elevators

14210 | Electric Traction Elevators

		CREW	DAILY OUTPUT	LABOR-HOURS	UNIT	MAT.	LABOR	EQUIP.	TOTAL	TOTAL INCL O&P
0875	For class "B" loading, add				Ea.	1,550			1,550	1,700
0900	For class "C-1" loading, add					3,825			3,825	4,200
0925	For class "C-2" loading, add					4,550			4,550	5,000
0950	For class "C-3" loading, add					6,250			6,250	6,875
0975	For travel over 40 V.L.F., add	2 Elev	7.25	2.207	V.L.F.	94	50.50		144.50	187
1000	For number of stops over 4, add		.27	59.259	Stop	1,650	1,350		3,000	4,075
1025	Hydraulic freight, base unit, 2000 lb, 50 fpm, 2 stop, std. fin.		.10	160	Ea.	30,400	3,675		34,075	39,600
1050	For 2500 lb capacity, add					2,050			2,050	2,250
1075	For 3000 lb capacity, add					3,525			3,525	3,875
1100	For 3500 lb capacity, add					5,325			5,325	5,850
1125	For 4000 lb capacity, add					5,700			5,700	6,275
1150	For 4500 lb capacity, add					6,750			6,750	7,425
1175	For 5000 lb capacity, add					9,225			9,225	10,100
1200	For 6000 lb capacity, add					9,500			9,500	10,400
1225	For 7000 lb capacity, add					15,000			15,000	16,500
1250	For 8000 lb capacity, add					16,300			16,300	18,000
1275	For 10000 lb capacity, add					17,400			17,400	19,100
1300	For 12000 lb capacity, add					20,700			20,700	22,700
1325	For 16000 lb capacity, add					26,900			26,900	29,600
1350	For 20000 lb capacity, add					29,900			29,900	32,900
1375	For increased speed, 100 fpm, add					660			660	730
1400	125 fpm, add					1,225			1,225	1,350
1425	150 fpm, add					2,275			2,275	2,525
1450	175 fpm, add					3,575			3,575	3,950
1475	For class "B" loading, add					1,525			1,525	1,675
1500	For class "C-1" loading, add					3,800			3,800	4,175
1525	For class "C-2" loading, add					4,550			4,550	5,000
1550	For class "C-3" loading, add					6,225			6,225	6,825
1575	For travel over 20 V.L.F., add	2 Elev	7.25	2.207	V.L.F.	206	50.50		256.50	310
1600	For number of stops over 2, add		.27	59.259	Stop	470	1,350		1,820	2,775
1625	Electric pass., base unit, 2000 lb, 200 fpm, 4 stop, std. fin.		.05	320	Ea.	62,000	7,325		69,325	80,500
1650	For 2500 lb capacity, add					2,525			2,525	2,775
1675	For 3000 lb capacity, add					3,750			3,750	4,125
1700	For 3500 lb capacity, add					5,325			5,325	5,875
1725	For 4000 lb capacity, add					5,425			5,425	5,975
1750	For 4500 lb capacity, add					7,150			7,150	7,850
1775	For 5000 lb capacity, add					9,050			9,050	9,950
1800	For increased speed, 250 fpm, geared electric, add					2,075			2,075	2,275
1825	300 fpm, geared electric, add					4,225			4,225	4,625
1850	350 fpm, geared electric, add					4,950			4,950	5,450
1875	400 fpm, geared electric, add					7,075			7,075	7,775
1900	500 fpm, gearless electric, add					32,900			32,900	36,200
1925	600 fpm, gearless electric, add					34,700			34,700	38,200
1950	700 fpm, gearless electric, add					38,000			38,000	41,800
1975	800 fpm, gearless electric, add					41,900			41,900	46,100
2000	For travel over 40 V.L.F., add	2 Elev	7.25	2.207	V.L.F.	94	50.50		144.50	187
2025	For number of stops over 4, add		.27	59.259	Stop	2,225	1,350		3,575	4,700
2050	Hyd. pass., base unit, 1500 lb, 100 fpm, 2 stop, std. fin.		.10	160	Ea.	27,100	3,675		30,775	35,900
2075	For 2000 lb capacity, add					475			475	525
2100	For 2500 lb capacity, add					1,050			1,050	1,150
2125	For 3000 lb capacity, add					2,600			2,600	2,875
2150	For 3500 lb capacity, add					4,450			4,450	4,900
2175	For 4000 lb capacity, add					5,125			5,125	5,650
2200	For 4500 lb capacity, add					6,125			6,125	6,750
2225	For 5000 lb capacity, add					8,525			8,525	9,375
2250	For increased speed, 125 fpm, add					725			725	800

14200 | Elevators

14210 | Electric Traction Elevators

		CREW	DAILY OUTPUT	LABOR-HOURS	UNIT	2000 BARE COSTS MAT.	LABOR	EQUIP.	TOTAL	TOTAL INCL O&P	
200	2275	150 fpm, add				Ea.	1,575			1,575	1,725	200
	2300	175 fpm, add					2,725			2,725	2,975	
	2325	200 fpm, add					4,400			4,400	4,850	
	2350	For travel over 12 V.L.F., add	2 Elev	7.25	2.207	V.L.F.	206	50.50		256.50	310	
	2375	For number of stops over 2, add		.27	59.259	Stop	415	1,350		1,765	2,700	
	2400	Electric hospital, base unit, 4000 lb, 200 fpm, 4 stop, std fin.		.05	320	Ea.	66,500	7,325		73,825	85,000	
	2425	For 4500 lb capacity, add					4,400			4,400	4,850	
	2450	For 5000 lb capacity, add					5,750			5,750	6,325	
	2475	For increased speed, 250 fpm, geared electric, add					2,225			2,225	2,450	
	2500	300 fpm, geared electric, add					4,150			4,150	4,550	
	2525	350 fpm, geared electric, add					5,050			5,050	5,575	
	2550	400 fpm, geared electric, add					7,050			7,050	7,750	
	2575	500 fpm, gearless electric, add					31,400			31,400	34,600	
	2600	600 fpm, gearless electric, add					34,700			34,700	38,200	
	2625	700 fpm, gearless electric, add					37,900			37,900	41,700	
	2650	800 fpm, gearless electric, add					41,900			41,900	46,100	
	2675	For travel over 40 V.L.F., add	2 Elev	7.25	2.207	V.L.F.	94	50.50		144.50	187	
	2700	For number of stops over 4, add		.27	59.259	Stop	2,625	1,350		3,975	5,150	
	2725	Hydraulic hospital, base unit, 4000 lb, 100 fpm, 2 stop, std. fin.		.10	160	Ea.	37,200	3,675		40,875	47,100	
	2750	For 4000 lb capacity, add					4,500			4,500	4,950	
	2775	For 4500 lb capacity, add					5,250			5,250	5,775	
	2800	For 5000 lb capacity, add					7,650			7,650	8,400	
	2825	For increased speed, 125 fpm, add					1,250			1,250	1,375	
	2850	150 fpm, add					2,100			2,100	2,325	
	2875	175 fpm, add					3,325			3,325	3,675	
	2900	200 fpm, add					4,875			4,875	5,350	
	2925	For travel over 12 V.L.F., add	2 Elev	7.25	2.207	V.L.F.	206	50.50		256.50	310	
	2950	For number of stops over 2, add	"	.27	59.259	Stop	2,725	1,350		4,075	5,250	
	2975	Passenger elevator options										
	3000	2 car group automatic controls	2 Elev	.66	24.242	Ea.	2,150	555		2,705	3,300	
	3025	3 car group automatic controls		.44	36.364		3,200	835		4,035	4,900	
	3050	4 car group automatic controls		.33	48.485		5,300	1,100		6,400	7,650	
	3075	5 car group automatic controls		.26	61.538		7,125	1,400		8,525	10,200	
	3100	6 car group automatic controls		.22	72.727		10,800	1,675		12,475	14,600	
	3125	Intercom service		3	5.333		296	122		418	525	
	3150	Duplex car selective collective		.66	24.242		2,425	555		2,980	3,575	
	3175	Center opening 1 speed doors		2	8		1,225	183		1,408	1,650	
	3200	Center opening 2 speed doors		2	8		1,600	183		1,783	2,050	
	3225	Rear opening doors (opposite front)		2	8		3,375	183		3,558	4,025	
	3250	Side opening 2 speed doors		2	8		5,225	183		5,408	6,050	
	3275	Automatic emergency power switching		.66	24.242		820	555		1,375	1,825	
	3300	Manual emergency power switching		8	2		270	46		316	375	
	3325	Cab finishes (based on 3500 lb cab size)										
	3350	Acrylic panel ceiling				Ea.	365			365	400	
	3375	Aluminum eggcrate ceiling					430			430	475	
	3400	Stainless steel doors					745			745	820	
	3425	Carpet flooring					310			310	340	
	3450	Epoxy flooring					255			255	281	
	3475	Quarry tile flooring					390			390	430	
	3500	Slate flooring					545			545	600	
	3525	Textured rubber flooring					103			103	113	
	3550	Stainless steel walls					2,275			2,275	2,500	
	3575	Stainless steel returns at door					445			445	490	
	3625	Hall finishes, stainless steel doors					745			745	820	
	3650	Stainless steel frames					745			745	820	
	3675	12 month maintenance contract									2,640	

14200 | Elevators

14210 | Electric Traction Elevators

			CREW	DAILY OUTPUT	LABOR-HOURS	UNIT	2000 BARE COSTS MAT.	LABOR	EQUIP.	TOTAL	TOTAL INCL O&P
3700	Signal devices, hall lanterns	R14200-100	2 Elev	8	2	Ea.	300	46		346	405
3725	Position indicators, up to 3			9.40	1.702		210	39		249	296
3750	Position indicators, per each over 3	R14200-200		32	.500		58	11.45		69.45	83
3775	High speed heavy duty door opener						1,425			1,425	1,550
3800	Variable voltage, O.H. gearless machine, min.	R14200-300	2 Elev	.16	100		21,900	2,300		24,200	27,900
3815	Maximum			.07	228		48,300	5,225		53,525	61,500
3825	Basement installed geared machine	R14200-400		.33	48.485		10,800	1,100		11,900	13,700
3850	Freight elevator options										
3875	Doors, bi-parting		2 Elev	.66	24.242	Ea.	3,600	555		4,155	4,875
3900	Power operated door and gate		"	.66	24.242		14,500	555		15,055	16,800
3925	Finishes, steel plate floor						625			625	685
3950	14 ga. 1/4" x 4' steel plate walls						1,525			1,525	1,675
3975	12 month maintenance contract										1,992
4000	Signal devices, hall lanterns		2 Elev	8	2		300	46		346	405
4025	Position indicators, up to 3			9.40	1.702		210	39		249	296
4050	Position indicators, per each over 3			32	.500		58	11.45		69.45	83
4075	Variable voltage basement installed geared machine			.66	24.242		12,000	555		12,555	14,100
4100	Hospital elevator options										
4125	2 car group automatic controls		2 Elev	.66	24.242	Ea.	2,150	555		2,705	3,300
4150	3 car group automatic controls			.44	36.364		3,200	835		4,035	4,900
4175	4 car group automatic controls			.33	48.485		5,300	1,100		6,400	7,650
4200	5 car group automatic controls			.26	61.538		7,150	1,400		8,550	10,200
4225	6 car group automatic controls			.22	72.727		10,900	1,675		12,575	14,800
4250	Intercom service			3	5.333		296	122		418	525
4275	Duplex car selective collective			.66	24.242		2,425	555		2,980	3,575
4300	Center opening 1 speed doors			2	8		1,225	183		1,408	1,650
4325	Center opening 2 speed doors			2	8		1,600	183		1,783	2,050
4350	Rear opening doors (opposite front)			2	8		3,375	183		3,558	4,025
4375	Side opening 2 speed doors			2	8		5,225	183		5,408	6,050
4400	Automatic emergency power switching			.66	24.242		820	555		1,375	1,825
4425	Manual emergency power switching			8	2		270	46		316	375
4450	Cab finishes (based on 3500# cab size)										
4475	Aluminum eggcrate ceiling					Ea.	430			430	475
4500	Stainless steel doors						745			745	820
4525	Epoxy flooring						255			255	281
4550	Quarry tile flooring						395			395	435
4575	Textured rubber flooring						103			103	113
4600	Stainless steel walls						2,275			2,275	2,500
4625	Stainless steel returns at door						475			475	525
4675	Hall finishes, stainless steel doors						745			745	820
4700	Stainless steel frames						745			745	820
4725	12 month maintenance contract										3,985
4750	Signal devices, hall lanterns		2 Elev	8	2		300	46		346	405
4775	Position indicators, up to 3			9.40	1.702		210	39		249	296
4800	Position indicators, per each over 3			32	.500		58	11.45		69.45	83
4825	High speed heavy duty door opener						1,425			1,425	1,550
4850	Variable voltage, O.H. gearless machine, min.		2 Elev	.16	100		21,900	2,300		24,200	27,900
4865	Maximum			.07	228		48,300	5,225		53,525	61,500
4875	Basement installed geared machine			.33	48.485		10,800	1,100		11,900	13,700
5000	Drilling for piston, casing included, 18" diameter		B-48	80	.600	V.L.F.	24.50	9.50	25	59	70.50

14300 | Escalators & Moving Walks

14320 | Escalators

			CREW	DAILY OUTPUT	LABOR-HOURS	UNIT	2000 BARE COSTS MAT.	LABOR	EQUIP.	TOTAL	TOTAL INCL O&P
300	0010	**ESCALATORS**										300
	1000	Glass, 32" wide x 10' floor to floor height R14320-100	M-1	.07	457	Ea.	57,500	9,950	1,350	68,800	81,500
	1010	48" wide x 10' floor to floor height		.07	457		61,500	9,950	1,350	72,800	86,000
	1020	32" wide x 15' floor to floor height		.06	533		59,000	11,600	1,575	72,175	86,000
	1030	48" wide x 15' floor to floor height		.06	533		62,500	11,600	1,575	75,675	89,500
	1040	32" wide x 20' floor to floor height		.05	653		59,500	14,200	1,950	75,650	91,000
	1050	48" wide x 20' floor to floor height		.05	653		63,500	14,200	1,950	79,650	95,500
	1060	32" wide x 25' floor to floor height		.04	800		63,000	17,400	2,375	82,775	101,000
	1070	48" wide x 25' floor to floor height		.04	800		68,000	17,400	2,375	87,775	106,500
	1080	Enameled steel, 32" wide x 10' floor to floor height		.07	457		57,500	9,950	1,350	68,800	81,500
	1090	48" wide x 10' floor to floor height		.07	457		61,500	9,950	1,350	72,800	86,000
	1110	32" wide x 15' floor to floor height		.06	533		59,000	11,600	1,575	72,175	86,000
	1120	48" wide x 15' floor to floor height		.06	533		62,500	11,600	1,575	75,675	89,500
	1130	32" wide x 20' floor to floor height		.05	653		59,500	14,200	1,950	75,650	91,000
	1140	48" wide x 20' floor to floor height		.05	653		63,500	14,200	1,950	79,650	95,500
	1150	32" wide x 25' floor to floor height		.04	800		63,000	17,400	2,375	82,775	101,000
	1160	48" wide x 25' floor to floor height		.04	800		68,000	17,400	2,375	87,775	106,500
	1170	Stainless steel, 32" wide x 10' floor to floor height		.07	457		63,500	9,950	1,350	74,800	88,000
	1180	48" wide x 10' floor to floor height		.07	457		66,500	9,950	1,350	77,800	91,000
	1500	32" wide x 15' floor to floor height		.06	533		64,500	11,600	1,575	77,675	92,000
	1700	48" wide x 15' floor to floor height		.06	533		67,500	11,600	1,575	80,675	95,000
	2300	32" wide x 25' floor to floor height		.04	800		72,500	17,400	2,375	92,275	111,000
	2500	48" wide x 25' floor to floor height		.04	800		76,500	17,400	2,375	96,275	115,500

14360 | Moving Walks

			CREW	DAILY OUTPUT	LABOR-HOURS	UNIT	MAT.	LABOR	EQUIP.	TOTAL	TOTAL INCL O&P
500	0010	**MOVING RAMPS AND WALKS** Walk, 27" tread width, minimum R14360-200	M-1	6.50	4.923	L.F.	690	107	14.60	811.60	955	500
	0100	300' to 500', maximum		4.43	7.223		955	157	21.50	1,133.50	1,325
	0300	48" tread width walk, minimum		4.43	7.223		1,550	157	21.50	1,728.50	1,975
	0400	100' to 350', maximum		3.82	8.377		1,825	182	25	2,032	2,325
	0600	Ramp, 12° incline, 36" tread width, minimum		5.27	6.072		1,275	132	18.05	1,425.05	1,650
	0700	70' to 90' maximum		3.82	8.377		1,825	182	25	2,032	2,325
	0900	48" tread width, minimum		3.57	8.964		1,850	195	26.50	2,071.50	2,400
	1000	40' to 70', maximum		2.91	10.997		2,350	239	32.50	2,621.50	3,000

14400 | Lifts

14460 | Correspondence & Parcel Lifts

			CREW	DAILY OUTPUT	LABOR-HOURS	UNIT	MAT.	LABOR	EQUIP.	TOTAL	TOTAL INCL O&P
200	0010	**CORRESPONDENCE LIFT** 1 floor 2 stop, 25 lb capacity, electric	2 Elev	.20	80	Ea.	4,825	1,825		6,650	8,325	200
	0100	Hand, 5 lb capacity	"	.20	80	"	1,825	1,825		3,650	5,025
600	0010	**PARCEL LIFT** 20" x 20", 100 lb capacity, electric, per floor	2 Mill	.25	64	Ea.	7,475	1,325		8,800	10,400	600

14500 | Material Handling

14510 | Material Handling

			CREW	DAILY OUTPUT	LABOR-HOURS	UNIT	MAT.	LABOR	EQUIP.	TOTAL	TOTAL INCL O&P
900	0010	**MOTORIZED CAR** Distribution systems, single track,										900
	0100	20 lb. per car capacity, material handling									
	0200	Minimum, 50' to 100'	4 Mill	.19	168	Station	23,000	3,475		26,475	31,000
	0300	Maximum, 100' to 200'	"	.15	213	"	31,000	4,400		35,400	41,300
	0400	Larger system, incl. hospital transport, track type,									
	0500	fully automated material handling system									
	0600	Minimum, complete system	4 Mill	.05	640	Ea.	102,000	13,200		115,200	133,500
	0700	Maximum, with several stations	E-6	.05	2,400	"	594,500	51,000	25,200	670,700	780,000

14550 | Conveyors

			CREW	DAILY OUTPUT	LABOR-HOURS	UNIT	MAT.	LABOR	EQUIP.	TOTAL	TOTAL INCL O&P
350	0010	**MATERIAL HANDLING** Conveyors, gravity type 2" rollers, 3" O.C.										350
	0050	10' sections with 2 supports, 600 lb. capacity, 18" wide				Ea.	355			355	390
	0100	24" wide					410			410	450
	0150	1400 lb. capacity, 18" wide					325			325	360
	0200	24" wide				↓	365			365	400
	0350	Horizontal belt, center drive and takeup, 60 fpm									
	0400	16" belt, 26.5' length	2 Mill	.50	32	Ea.	2,500	660		3,160	3,825
	0450	24" belt, 41.5' length		.40	40		3,700	825		4,525	5,425
	0500	61.5' length	↓	.30	53.333	↓	4,800	1,100		5,900	7,075
	0600	Inclined belt, 10' rise with horizontal loader and									
	0620	End idler assembly, 27.5' length, 18" belt	2 Mill	.30	53.333	Ea.	7,300	1,100		8,400	9,825
	0700	24" belt	"	.15	106	"	8,200	2,200		10,400	12,600
	3600	Monorail, overhead, manual, channel type									
	3700	125 lb. per L.F.	1 Mill	26	.308	L.F.	8.10	6.35		14.45	19.25
	3900	500 lb. per L.F.	"	21	.381	"	7.35	7.85		15.20	21
	4000	Trolleys for above, 2 wheel, 125 lb. capacity				Ea.	45			45	49.50
	4200	4 wheel, 250 lb. capacity					165			165	181
	4300	8 wheel, 1,000 lb. capacity				↓	360			360	395
900	0010	**VERTICAL CONVEYOR** Automatic selective										900
	0100	to 10 floors, base price	2 Mill	.13	120	Floor	22,900	2,475		25,375	29,300
	0200	Add for electrical hook-up and testing	2 Elec	.50	32	"		705		705	1,150

14560 | Chutes

			CREW	DAILY OUTPUT	LABOR-HOURS	UNIT	MAT.	LABOR	EQUIP.	TOTAL	TOTAL INCL O&P
250	0010	**CHUTES** Linen or refuse, incl. sprinklers, 12' floor height										250
	0050	Aluminized steel, 16 ga., 18" diameter	2 Shee	3.50	4.571	Floor	735	98.50		833.50	975
	0100	24" diameter		3.20	5		760	108		868	1,025
	0200	30" diameter		3	5.333		840	115		955	1,125
	0300	36" diameter		2.80	5.714		985	123		1,108	1,275
	0400	Galvanized steel, 16 ga., 18" diameter		3.50	4.571		700	98.50		798.50	935
	0500	24" diameter		3.20	5		790	108		898	1,050
	0600	30" diameter		3	5.333		885	115		1,000	1,175
	0700	36" diameter		2.80	5.714		1,050	123		1,173	1,350
	0800	Stainless steel, 18" diameter		3.50	4.571		920	98.50		1,018.50	1,175
	0900	24" diameter		3.20	5		970	108		1,078	1,250
	1000	30" diameter		3	5.333		1,150	115		1,265	1,475
	1005	36" diameter		2.80	5.714	↓	1,725	123		1,848	2,100
	1200	Linen chute bottom collector, aluminized steel		4	4	Ea.	980	86		1,066	1,225
	1300	Stainless steel		4	4		1,250	86		1,336	1,525
	1500	Refuse, bottom hopper, aluminized steel, 18" diameter		3	5.333		600	115		715	855
	1600	24" diameter		3	5.333		665	115		780	930
	1800	36" diameter		3	5.333	↓	895	115		1,010	1,175
	2900	Package chutes, spiral type, minimum		4.50	3.556	Floor	1,700	76.50		1,776.50	2,000
	3000	Maximum	↓	1.50	10.667	"	4,425	229		4,654	5,250

14500 | Material Handling

14580 | Pneumatic Tube Systems

			CREW	DAILY OUTPUT	LABOR-HOURS	UNIT	MAT.	LABOR	EQUIP.	TOTAL	TOTAL INCL O&P	
800	0010	**PNEUMATIC TUBE SYSTEM** Single tube, 2 stations, blower										800
	0020	100' long, stock										
	0100	3" diameter	2 Stpi	.12	133	Total	4,750	2,950		7,700	10,100	
	0300	4" diameter	"	.09	177	"	5,350	3,925		9,275	12,400	
	0400	Twin tube, two stations or more, conventional system										
	0600	2-1/2" round	2 Stpi	62.50	.256	L.F.	9.55	5.65		15.20	19.90	
	0700	3" round		46	.348		11	7.70		18.70	25	
	0900	4" round		49.60	.323		12.15	7.15		19.30	25	
	1000	4" x 7" oval		37.60	.426		17.75	9.40		27.15	35	
	1050	Add for blower		2	8	System	3,750	177		3,927	4,425	
	1110	Plus for each round station, add		7.50	2.133	Ea.	420	47		467	545	
	1150	Plus for each oval station, add		7.50	2.133	"	420	47		467	545	
	1200	Alternate pricing method: base cost, minimum		.75	21.333	Total	4,600	470		5,070	5,825	
	1300	Maximum		.25	64	"	9,125	1,425		10,550	12,400	
	1500	Plus total system length, add, minimum		93.40	.171	L.F.	5.90	3.79		9.69	12.75	
	1600	Maximum		37.60	.426		17.45	9.40		26.85	35	
	1800	Completely automatic system, 4" round, 15 to 50 stations		.29	55.172	Station	14,500	1,225		15,725	17,900	
	2200	51 to 144 stations		.32	50		11,200	1,100		12,300	14,200	
	2400	6" round or 4" x 7" oval, 15 to 50 stations		.24	66.667		18,100	1,475		19,575	22,300	
	2800	51 to 144 stations		.23	69.565		15,200	1,525		16,725	19,300	

14600 | Hoists & Cranes

14610 | Fixed Hoists

			CREW	DAILY OUTPUT	LABOR-HOURS	UNIT	MAT.	LABOR	EQUIP.	TOTAL	TOTAL INCL O&P	
500	0010	**MATERIAL HANDLING**										500
	1500	Cranes, portable hydraulic, floor type, 2,000 lb capacity				Ea.	1,750			1,750	1,925	
	1600	4,000 lb capacity					3,025			3,025	3,325	
	1800	Movable gantry type, 12' to 15' range, 2,000 lb capacity					2,700			2,700	2,975	
	1900	6,000 lb capacity					3,900			3,900	4,300	
	2100	Hoists, electric overhead, chain, hook hung, 15' lift, 1 ton cap.					1,750			1,750	1,925	
	2200	3 ton capacity					2,175			2,175	2,400	
	2500	5 ton capacity					4,850			4,850	5,325	
	2600	For hand-pushed trolley, add					15%					
	2700	For geared trolley, add					30%					
	2800	For motor trolley, add					75%					
	3000	For lifts over 15', 1 ton, add				L.F.	16.60			16.60	18.25	
	3100	5 ton, add				"	38.50			38.50	42.50	
	3300	Lifts, scissor type, portable, electric, 36" high, 2,000 lb				Ea.	3,150			3,150	3,475	
	3400	48" high, 4,000 lb				"	3,600			3,600	3,950	

14630 | Bridge Cranes

			CREW	DAILY OUTPUT	LABOR-HOURS	UNIT	MAT.	LABOR	EQUIP.	TOTAL	TOTAL INCL O&P	
300	0010	**CRANE RAIL** Box beam bridge, no equipment included	E-4	3,400	.009	Lb.	.87	.20	.02	1.09	1.39	300
	0200	Running track only, 104 lb per yard	"	5,600	.006	"	.46	.12	.02	.60	.78	
700	0010	**OVERHEAD BRIDGE CRANES**										700
	0100	1 girder, 20' span, 3 ton	M-3	1	34	Ea.	14,700	680	119	15,499	17,500	
	0125	5 ton		1	34		16,100	680	119	16,899	19,000	
	0150	7.5 ton		1	34		19,200	680	119	19,999	22,500	
	0175	10 ton		.80	42.500		25,200	845	148	26,193	29,300	
	0200	15 ton		.80	42.500		32,300	845	148	33,293	37,100	

14600 | Hoists & Cranes

14630 | Bridge Cranes

		CREW	DAILY OUTPUT	LABOR-HOURS	UNIT	2000 BARE COSTS				TOTAL INCL O&P
						MAT.	LABOR	EQUIP.	TOTAL	
0225	30' span, 3 ton	M-3	1	34	Ea.	15,300	680	119	16,099	18,200
0250	5 ton		1	34		16,800	680	119	17,599	19,800
0275	7.5 ton		1	34		20,200	680	119	20,999	23,500
0300	10 ton		.80	42.500		26,100	845	148	27,093	30,300
0325	15 ton		.80	42.500		33,700	845	148	34,693	38,700
0350	2 girder, 40' span, 3 ton	M-4	.50	72		25,500	1,425	390	27,315	30,900
0375	5 ton		.50	72		26,700	1,425	390	28,515	32,100
0400	7.5 ton		.50	72		29,200	1,425	390	31,015	34,900
0425	10 ton		.40	90		33,900	1,775	490	36,165	40,800
0450	15 ton		.40	90		46,100	1,775	490	48,365	54,000
0475	25 ton		.30	119		55,000	2,375	655	58,030	65,000
0500	50' span, 3 ton		.50	72		28,900	1,425	390	30,715	34,600
0525	5 ton		.50	72		30,000	1,425	390	31,815	35,800
0550	7.5 ton		.50	72		32,200	1,425	390	34,015	38,200
0575	10 ton		.40	90		36,900	1,775	490	39,165	44,100
0600	15 ton		.40	90		48,300	1,775	490	50,565	56,500
0625	25 ton		.30	119		57,500	2,375	655	60,530	67,500

For information about Means Estimating Seminars, see yellow pages 11 and 12 in back of book

Division Notes

	CREW	DAILY OUTPUT	LABOR-HOURS	UNIT	2000 BARE COSTS				TOTAL INCL O&P
					MAT.	LABOR	EQUIP.	TOTAL	

Division 15 Mechanical

Estimating Tips

15100 Building Services Piping
This subdivision is primarily basic pipe and related materials. The pipe may be used by any of the mechanical disciplines, i.e., plumbing, fire protection, heating, and air conditioning.

- The piping section lists the add to labor for elevated pipe installation. These adds apply to all elevated pipe, fittings, valves, insulation, etc., that are placed above 10' high. CAUTION: the correct percentage may vary for the same pipe. For example, the percentage add for the basic pipe installation should be based on the maximum height that the craftsman must install for that particular section. If the pipe is to be located 14' above the floor but it is suspended on threaded rod from beams, the bottom flange of which is 18' high (4' rods), then the height is actually 18' and the add is 20%. The pipe coverer, however, does not have to go above the 14' and so his add should be 10%.

- Most pipe is priced first as straight pipe with a joint (coupling, weld, etc.) every 10' and a hanger usually every 10'. There are exceptions with hanger spacing such as: for cast iron pipe (5') and plastic pipe (3 per 10'). Following each type of pipe there are several lines listing sizes and the amount to be subtracted to delete couplings and hangers. This is for pipe that is to be buried or supported together on trapeze hangers. The reason that the couplings are deleted is that these runs are usually long and frequently longer lengths of pipe are used. By deleting the couplings the estimator is expected to look up and add back the correct reduced number of couplings.

- When preparing an estimate it may be necessary to approximate the fittings. Fittings usually run between 25% and 50% of the cost of the pipe. The lower percentage is for simpler runs, and the higher number is for complex areas like mechanical rooms.

15400 Plumbing Fixtures & Equipment
- Plumbing fixture costs usually require two lines, the fixture itself and its "rough-in, supply and waste".
- Remember that gas and oil fired units need venting.

15500 Heat Generation Equipment
- When estimating the cost of an HVAC system, check to see who is responsible for providing and installing the temperature control system. It is possible to overlook controls, assuming that they would be included in the electrical estimate.
- When looking up a boiler be careful on specified capacity. Some manufacturers rate their products on output while others use input.
- Include HVAC insulation for pipe, boiler and duct (wrap and liner).
- Be careful when looking up mechanical items to get the correct pressure rating and connection type (thread, weld, flange).

15700 Heating/Ventilation/ Air Conditioning Equipment
- Combination heating and cooling units are sized by the air conditioning requirements. (See Reference No. R15710-020 for preliminary sizing guide.)
- A ton of air conditioning is nominally 400 CFM.

- Rectangular duct is taken off by the linear foot for each size, but its cost is usually estimated by the pound. Remember that SMACNA standards now base duct on internal pressure.
- Prefabricated duct is estimated and purchased like pipe: straight sections and fittings.
- Note that cranes or other lifting equipment are not included on any lines in Division 15. For example, if a crane is required to lift a heavy piece of pipe into place high above a gym floor, or to put a rooftop unit on the roof of a four-story building, etc., it must be added. Due to the potential for extreme variation—from nothing additional required, to a major crane or helicopter—we feel that including a nominal amount for "lifting contingency" would be useless and detract from the accuracy of the estimate. When using equipment rental from Means do not forget to include the cost of the operator(s).

Reference Numbers
Reference numbers are shown in bold squares at the beginning of some major classifications. These numbers refer to related items in the Reference Section. The reference information may be an estimating procedure, an alternate pricing method or technical information.

Note: Not all subdivisions listed here necessarily appear in this publication.

15050 | Basic Materials & Methods

15051 | Mechanical General

			CREW	DAILY OUTPUT	LABOR-HOURS	UNIT	2000 BARE COSTS MAT.	LABOR	EQUIP.	TOTAL	TOTAL INCL O&P	
100	0010	**BOILERS, GENERAL** Prices do not include flue piping, elec. wiring,										100
	0020	gas or oil piping, boiler base, pad, or tankless unless noted										
	0100	Boiler H.P.: 10 KW = 34 lbs/steam/hr = 33,475 BTU/hr.										
	0120											
	0150	To convert SFR to BTU rating: Hot water, 150 x SFR;	R15500 -050									
	0160	Forced hot water, 180 x SFR; steam, 240 x SFR										
700	0010	**PIPING** See also divisions 02500 & 02600 for site work										700
	1000	Add to labor for elevated installation										
	1080	10' to 15' high						10%				
	1100	15' to 20' high						20%				
	1120	20' to 25' high						25%				
	1140	25' to 30' high						35%				
	1160	30' to 35' high						40%				
	1180	35' to 40' high						50%				
	1200	Over 40' high						55%				

15055 | Selective Mech Demolition

			CREW	DAILY OUTPUT	LABOR-HOURS	UNIT	MAT.	LABOR	EQUIP.	TOTAL	INCL O&P	
300	0010	**HVAC DEMOLITION**										300
	0100	Air conditioner, split unit, 3 ton	Q-5	2	8	Ea.		159		159	263	
	0150	Package unit, 3 ton	Q-6	3	8			153		153	253	
	0200	Air conditioner, rooftop, gas heat, remove & reset, 10 ton		.30	80			1,525		1,525	2,525	
	0240	Remove & reset, 20 ton		.15	160			3,075		3,075	5,075	
	0300	Boiler, electric	Q-19	2	12			248		248	410	
	0340	Gas or oil, steel, under 150 MBH	Q-6	3	8			153		153	253	
	0380	Over 150 MBH	"	2	12			230		230	380	
	1000	Ductwork, 4" high, 8" wide	1 Clab	200	.040	L.F.		.58		.58	.99	
	1100	6" high, 8" wide		165	.048			.70		.70	1.20	
	1200	10" high, 12" wide		125	.064			.92		.92	1.58	
	1300	12"-14" high, 16"-18" wide		85	.094			1.36		1.36	2.33	
	1400	18" high, 24" wide		67	.119			1.73		1.73	2.96	
	1500	30" high, 36" wide		56	.143			2.06		2.06	3.54	
	1540	72" wide		50	.160			2.31		2.31	3.96	
	3000	Mechanical equipment, light items. Unit is weight, not cooling.	Q-5	.90	17.778	Ton		355		355	585	
	3600	Heavy items	"	1.10	14.545	"		289		289	480	
	3700	Deduct for salvage (when applicable), minimum				Job					52	
	3750	Maximum				"					420	
600	0010	**PLUMBING DEMOLITION**										600
	1020	Fixtures, including 10' piping										
	1101	Bath tubs, cast iron	1 Clab	4	2	Ea.		29		29	49.50	
	1121	Fiberglass		6	1.333			19.25		19.25	33	
	1141	Steel		5	1.600			23		23	39.50	
	1201	Lavatory, wall hung		10	.800			11.55		11.55	19.80	
	1221	Counter top		16	.500			7.20		7.20	12.40	
	1301	Sink, steel or cast iron, single		16	.500			7.20		7.20	12.40	
	1321	Double		10	.800			11.55		11.55	19.80	
	1401	Water closet, floor mounted		16	.500			7.20		7.20	12.40	
	1421	Wall mounted		7	1.143			16.50		16.50	28.50	
	1501	Urinal, floor mounted		4	2			29		29	49.50	
	1521	Wall mounted		7	1.143			16.50		16.50	28.50	
	1601	Water fountains, free standing		8	1			14.45		14.45	25	
	1621	Recessed		6	1.333			19.25		19.25	33	
	2001	Piping, metal, to 2" diameter		200	.040	L.F.		.58		.58	.99	
	2051	To 4" diameter		150	.053			.77		.77	1.32	
	2101	To 8" diameter		100	.080			1.16		1.16	1.98	

Important: See the Reference Section for critical supporting data - Reference Nos., Crews, & City Cost Indexes

15050 | Basic Materials & Methods

15055 | Selective Mech Demolition

		CREW	DAILY OUTPUT	LABOR-HOURS	UNIT	MAT.	LABOR	EQUIP.	TOTAL	TOTAL INCL O&P
2151	To 16" diameter	1 Clab	60	.133	L.F.		1.93		1.93	3.30
2160	Deduct for salvage, aluminum scrap				Ton					525
2170	Brass scrap									420
2180	Copper scrap									1,025
2200	Lead scrap									235
2220	Steel scrap				↓					65
2251	Water heater, 40 gal.	1 Clab	6	1.333	Ea.		19.25		19.25	33

15082 | Duct Insulation

		CREW	DAILY OUTPUT	LABOR-HOURS	UNIT	MAT.	LABOR	EQUIP.	TOTAL	TOTAL INCL O&P
0010	**INSULATION**									
0100	Rule of thumb, as a percentage of total mechanical costs				Job					10%
0110	Insulation req'd is based on the surface size/area to be covered									
1000	Boiler, 1-1/2" calcium silicate, 1/2" cement finish	Q-14	50	.320	S.F.	2.53	5.90		8.43	13.15
1020	2" fiberglass	"	80	.200	"	1.49	3.69		5.18	8.10
2000	Breeching, 2" calcium silicate with 1/2" cement finish, no lath									
2020	Rectangular	Q-14	42	.381	S.F.	2.86	7.05		9.91	15.45
2040	Round	"	38.70	.413	"	3.21	7.65		10.86	16.90
3000	Ductwork									
3020	Blanket type, fiberglass, flexible									
3030	Fire resistant liner, black coating one side									
3050	1/2" thick, 2 lb. density	Q-14	380	.042	S.F.	.29	.78		1.07	1.68
3060	1" thick, 1-1/2 lb. density		350	.046		.39	.84		1.23	1.91
3070	1-1/2" thick, 1-1/2 lb. density		320	.050		.68	.92		1.60	2.36
3080	2" thick, 1-1/2 lb. density	↓	300	.053	↓	.80	.98		1.78	2.60
3140	FRK vapor barrier wrap, .75 lb. density									
3160	1" thick	Q-14	350	.046	S.F.	.27	.84		1.11	1.78
3170	1-1/2" thick		320	.050		.24	.92		1.16	1.87
3180	2" thick		300	.053		.29	.98		1.27	2.04
3190	3" thick		260	.062		.43	1.14		1.57	2.46
3200	4" thick	↓	242	.066	↓	.58	1.22		1.80	2.78
3210	Vinyl jacket, same as FRK									
3280	Unfaced, 1 lb. density									
3310	1" thick	Q-14	360	.044	S.F.	.24	.82		1.06	1.70
3320	1-1/2" thick		330	.048		.33	.89		1.22	1.93
3330	2" thick	↓	310	.052	↓	.39	.95		1.34	2.10
3490	Board type, fiberglass liner, 3 lb. density									
3500	Fire resistant, black pigmented, 1 side									
3520	1" thick	Q-14	150	.107	S.F.	1.16	1.97		3.13	4.73
3540	1-1/2" thick		130	.123		1.43	2.27		3.70	5.55
3560	2" thick	↓	120	.133	↓	1.72	2.46		4.18	6.20
3600	FRK vapor barrier									
3620	1" thick	Q-14	150	.107	S.F.	1.27	1.97		3.24	4.85
3630	1-1/2" thick		130	.123		1.64	2.27		3.91	5.80
3640	2" thick	↓	120	.133	↓	1.99	2.46		4.45	6.50
3680	No finish									
3700	1" thick	Q-14	170	.094	S.F.	.68	1.74		2.42	3.79
3710	1-1/2" thick		140	.114		1.02	2.11		3.13	4.81
3720	2" thick		130	.123		1.30	2.27		3.57	5.40
3800	1/2" cement over 1" wire mesh, incl. corner bead		100	.160		1.23	2.95		4.18	6.50
3820	Glass cloth, pasted on		170	.094		2	1.74		3.74	5.25
3900	Weatherproof, non-metallic, 2 lb. per S.F.	↓	100	.160	↓	2.86	2.95		5.81	8.30
4000	Pipe covering (price copper tube one size less than IPS)									
6600	Fiberglass, with all service jacket									
6840	1" wall, 1/2" iron pipe size	Q-14	240	.067	L.F.	1.04	1.23		2.27	3.29
6870	1" iron pipe size	↓	220	.073	↓	1.21	1.34		2.55	3.68

15050 | Basic Materials & Methods

15082 | Duct Insulation

		CREW	DAILY OUTPUT	LABOR-HOURS	UNIT	2000 BARE COSTS MAT.	LABOR	EQUIP.	TOTAL	TOTAL INCL O&P
6900	2" iron pipe size	Q-14	200	.080	L.F.	1.62	1.48		3.10	4.36
6920	3" iron pipe size		180	.089		2.01	1.64		3.65	5.10
6940	4" iron pipe size		150	.107		2.58	1.97		4.55	6.30
7320	2" wall, 1/2" iron pipe size		220	.073		3.43	1.34		4.77	6.10
7440	6" iron pipe size		100	.160		6.55	2.95		9.50	12.35
7460	8" iron pipe size		80	.200		8.10	3.69		11.79	15.35
7480	10" iron pipe size		70	.229		9.70	4.22		13.92	18.05
7490	12" iron pipe size		65	.246		10.65	4.54		15.19	19.70
7600	For fiberglass with standard canvas jacket, deduct					5%				
7660	For fittings, add 3 L.F. for each fitting									
7680	plus 4 L.F. for each flange of the fitting									
7700	For equipment or congested areas, add						20%			
7720	Finishes, for .010" aluminum jacket, add	Q-14	200	.080	S.F.	.30	1.48		1.78	2.91
7740	For .016" aluminum jacket, add		200	.080		.39	1.48		1.87	3.01
7760	For .010" stainless steel, add		160	.100		1.18	1.85		3.03	4.53
7780	For single layer of felt, add					10%	10%			
7879	Rubber tubing, flexible closed cell foam									
7880	3/8" wall, 1/4" iron pipe size	1 Asbe	120	.067	L.F.	.35	1.37		1.72	2.78
7910	1/2" iron pipe size		115	.070		.45	1.43		1.88	3
7920	3/4" iron pipe size		115	.070		.52	1.43		1.95	3.07
7930	1" iron pipe size		110	.073		.62	1.49		2.11	3.29
7950	1-1/2" iron pipe size		110	.073		.86	1.49		2.35	3.56
8100	1/2" wall, 1/4" iron pipe size		90	.089		.57	1.82		2.39	3.82
8130	1/2" iron pipe size		89	.090		.70	1.84		2.54	4
8140	3/4" iron pipe size		89	.090		.79	1.84		2.63	4.10
8150	1" iron pipe size		88	.091		.87	1.86		2.73	4.22
8170	1-1/2" iron pipe size		87	.092		1.27	1.89		3.16	4.70
8180	2" iron pipe size		86	.093		1.56	1.91		3.47	5.05
8200	3" iron pipe size		85	.094		2.09	1.93		4.02	5.70
8300	3/4" wall, 1/4" iron pipe size		90	.089		.86	1.82		2.68	4.14
8330	1/2" iron pipe size		89	.090		1.12	1.84		2.96	4.46
8340	3/4" iron pipe size		89	.090		1.38	1.84		3.22	4.75
8350	1" iron pipe size		88	.091		1.56	1.86		3.42	4.98
8370	1-1/2" iron pipe size		87	.092		2.35	1.89		4.24	5.90
8380	2" iron pipe size		86	.093		2.77	1.91		4.68	6.40
8400	3" iron pipe size		85	.094		4.23	1.93		6.16	8.05
8444	1" wall, 1/2" iron pipe size		86	.093		2.18	1.91		4.09	5.75
8445	3/4" iron pipe size		84	.095		2.64	1.95		4.59	6.30
8446	1" iron pipe size		84	.095		3.07	1.95		5.02	6.80
8447	1-1/4" iron pipe size		82	.098		3.47	2		5.47	7.30
8448	1-1/2" iron pipe size		82	.098		4.03	2		6.03	7.95
8449	2" iron pipe size		80	.100		5.40	2.05		7.45	9.55
8450	2-1/2" iron pipe size		80	.100		7	2.05		9.05	11.30
8456	Rubber insulation tape, 1/8" x 2" x 30'				Ea.	12.95			12.95	14.25

15100 | Building Services Piping

15106 | Glass Pipe & Fittings

		CREW	DAILY OUTPUT	LABOR-HOURS	UNIT	2000 BARE COSTS MAT.	LABOR	EQUIP.	TOTAL	TOTAL INCL O&P
0010	**PIPE, GLASS** Borosilicate, couplings & hangers 10' O.C.									
0020	Drainage									

15100 | Building Services Piping

15106 | Glass Pipe & Fittings

			CREW	DAILY OUTPUT	LABOR-HOURS	UNIT	2000 BARE COSTS MAT.	LABOR	EQUIP.	TOTAL	TOTAL INCL O&P	
120	1100	1-1/2" diameter	Q-1	52	.308	L.F.	8.45	6.10		14.55	19.35	120
	1120	2" diameter		44	.364		11.05	7.20		18.25	24	
	1140	3" diameter		39	.410		14.55	8.10		22.65	29.50	
	1160	4" diameter		30	.533		26	10.55		36.55	46	
	1180	6" diameter	▼	26	.615	▼	46.50	12.15		58.65	71.50	

15107 | Metal Pipe & Fittings

			CREW	DAILY OUTPUT	LABOR-HOURS	UNIT	MAT.	LABOR	EQUIP.	TOTAL	TOTAL INCL O&P	
220	0010	**PIPE, BRASS** Plain end,										220
	0900	Field threaded, coupling & clevis hanger 10' O.C.										
	0920	Regular weight										
	0980	1/8" diameter	1 Plum	62	.129	L.F.	2.35	2.83		5.18	7.25	
	1120	1/2" diameter		48	.167		4.13	3.66		7.79	10.60	
	1140	3/4" diameter		46	.174		5.55	3.82		9.37	12.40	
	1160	1" diameter	▼	43	.186		7.95	4.08		12.03	15.50	
	1180	1-1/4" diameter	Q-1	72	.222		11.75	4.39		16.14	20	
	1200	1-1/2" diameter		65	.246		13.85	4.86		18.71	23.50	
	1220	2" diameter	▼	53	.302		18.70	5.95		24.65	30.50	
320	0010	**PIPE, CAST IRON** Soil, on hangers 5' O.C.										320
	0020	Single hub, service wt., lead & oakum joints 10' O.C.										
	2120	2" diameter	Q-1	63	.254	L.F.	4.92	5		9.92	13.70	
	2140	3" diameter		60	.267		6.35	5.25		11.60	15.65	
	2160	4" diameter	▼	55	.291		6.25	5.75		12	16.35	
	2180	5" diameter	Q-2	76	.316		10.95	6		16.95	22	
	2200	6" diameter	"	73	.329		12.90	6.25		19.15	24.50	
	2220	8" diameter	Q-3	59	.542		20.50	10.70		31.20	40	
	2240	10" diameter		54	.593		34	11.70		45.70	57	
	2260	12" diameter	▼	48	.667		47.50	13.15		60.65	74.50	
	2320	For service weight, double hub, add					10%					
	2340	For extra heavy, single hub, add					48%	4%				
	2360	For extra heavy, double hub, add				▼	71%	4%				
	2400	Lead for caulking, (1#/dia in)				Lb.	.98			.98	1.08	
	2420	Oakum for caulking, (1/8#/dia in)				"	4.66			4.66	5.15	
	4000	No hub, couplings 10' O.C.										
	4100	1-1/2" diameter	Q-1	71	.225	L.F.	5.60	4.45		10.05	13.50	
	4120	2" diameter		67	.239		5.70	4.72		10.42	14.10	
	4160	4" diameter	▼	58	.276	▼	9.30	5.45		14.75	19.20	
420	0010	**PIPE, COPPER** Solder joints										420
	1000	Type K tubing, couplings & clevis hangers 10' O.C.										
	1100	1/4" diameter	1 Plum	84	.095	L.F.	1.03	2.09		3.12	4.60	
	1200	1" diameter		66	.121		2.37	2.66		5.03	7	
	1260	2" diameter	▼	40	.200	▼	4.90	4.39		9.29	12.65	
	2000	Type L tubing, couplings & hangers 10' O.C.										
	2100	1/4" diameter	1 Plum	88	.091	L.F.	.88	2		2.88	4.26	
	2120	3/8" diameter		84	.095		1.02	2.09		3.11	4.58	
	2140	1/2" diameter		81	.099		1.14	2.17		3.31	4.84	
	2160	5/8" diameter		79	.101		1.39	2.22		3.61	5.20	
	2180	3/4" diameter		76	.105		1.50	2.31		3.81	5.45	
	2200	1" diameter		68	.118		2.03	2.58		4.61	6.50	
	2220	1-1/4" diameter		58	.138		2.42	3.03		5.45	7.65	
	2240	1-1/2" diameter		52	.154		2.93	3.38		6.31	8.80	
	2260	2" diameter	▼	42	.190		4.20	4.18		8.38	11.50	
	2280	2-1/2" diameter	Q-1	62	.258		5.90	5.10		11	14.95	
	2300	3" diameter		56	.286		7.85	5.65		13.50	18	
	2320	3-1/2" diameter	▼	43	.372	▼	9.95	7.35		17.30	23	

15100 | Building Services Piping

15107 | Metal Pipe & Fittings

		CREW	DAILY OUTPUT	LABOR-HOURS	UNIT	2000 BARE COSTS MAT.	LABOR	EQUIP.	TOTAL	TOTAL INCL O&P	
420	2340	4" diameter	Q-1	39	.410	L.F.	12.40	8.10		20.50	27	**420**
	2360	5" diameter		34	.471		34.50	9.30		43.80	53	
	2380	6" diameter	Q-2	40	.600		42.50	11.40		53.90	66	
	2400	8" diameter	"	36	.667		74	12.70		86.70	103	
	2410	For other than full hard temper, add					21%					
	2590	For silver solder, add						15%				
	4000	Type DWV tubing, couplings & hangers 10' O.C.										
	4100	1-1/4" diameter	1 Plum	60	.133	L.F.	2.06	2.93		4.99	7.10	
	4120	1-1/2" diameter		54	.148		2.45	3.25		5.70	8.10	
	4140	2" diameter		44	.182		3.04	3.99		7.03	9.95	
	4160	3" diameter	Q-1	58	.276		4.74	5.45		10.19	14.20	
	4180	4" diameter		40	.400		8.25	7.90		16.15	22	
	4200	5" diameter		36	.444		23.50	8.80		32.30	40.50	
	4220	6" diameter	Q-2	42	.571		32.50	10.85		43.35	54	
500	0010	**PIPE, CORROSION RESISTANT** No couplings or hangers										**500**
	0020	Iron alloy, drain, mechanical joint										
	1000	1-1/2" diameter	Q-1	70	.229	L.F.	22.50	4.51		27.01	32.50	
	1100	2" diameter		66	.242		26	4.79		30.79	36.50	
	1120	3" diameter		60	.267		36	5.25		41.25	48.50	
	1140	4" diameter		52	.308		46.50	6.10		52.60	61	
	2980	Plastic, epoxy, fiberglass filament wound										
	3000	2" diameter	Q-1	62	.258	L.F.	6.15	5.10		11.25	15.20	
	3100	3" diameter		51	.314		9.90	6.20		16.10	21	
	3120	4" diameter		45	.356		11.80	7		18.80	24.50	
	3140	6" diameter		32	.500		17.50	9.90		27.40	35.50	
	3980	Polyester, fiberglass filament wound										
	4000	2" diameter	Q-1	62	.258	L.F.	9.90	5.10		15	19.30	
	4100	3" diameter		51	.314		13.20	6.20		19.40	25	
	4120	4" diameter		45	.356		14.90	7		21.90	28	
	4140	6" diameter		32	.500		23.50	9.90		33.40	42.50	
	4980	Polypropylene, acid resistant, fire retardant, schedule 40										
	5000	1-1/2" diameter	Q-1	68	.235	L.F.	.88	4.65		5.53	8.65	
	5100	2" diameter		62	.258		1.18	5.10		6.28	9.75	
	5120	3" diameter		51	.314		2.48	6.20		8.68	13	
	5140	4" diameter		45	.356		3.53	7		10.53	15.50	
	5980	Proxylene, fire retardant, Schedule 40										
	6000	1-1/2" diameter	Q-1	68	.235	L.F.	2.81	4.65		7.46	10.80	
	6100	2" diameter		62	.258		3.81	5.10		8.91	12.65	
	6120	3" diameter		51	.314		6.90	6.20		13.10	17.80	
	6140	4" diameter		45	.356		9.75	7		16.75	22.50	
620	0010	**PIPE, STEEL**	R15100-050									**620**
	0020	All pipe sizes are to Spec. A-53 unless noted otherwise										
	0050	Schedule 40, threaded, with couplings, and clevis type										
	0060	hangers sized for covering, 10' O.C.										
	0540	Black, 1/4" diameter	1 Plum	66	.121	L.F.	1.45	2.66		4.11	6	
	0550	3/8" diameter		65	.123		1.61	2.70		4.31	6.25	
	0560	1/2" diameter		63	.127		1.40	2.79		4.19	6.15	
	0570	3/4" diameter		61	.131		1.54	2.88		4.42	6.45	
	0580	1" diameter		53	.151		1.83	3.31		5.14	7.50	
	0590	1-1/4" diameter	Q-1	89	.180		2.17	3.55		5.72	8.25	
	0600	1-1/2" diameter		80	.200		2.42	3.95		6.37	9.20	
	0610	2" diameter		64	.250		3.28	4.94		8.22	11.75	
	0620	2-1/2" diameter		50	.320		5.05	6.30		11.35	16	
	0630	3" diameter		43	.372		6.30	7.35		13.65	19.05	
	0640	3-1/2" diameter		40	.400		8.10	7.90		16	22	
	0650	4" diameter		36	.444		8.95	8.80		17.75	24.50	

15100 | Building Services Piping

15107 | Metal Pipe & Fittings

			CREW	DAILY OUTPUT	LABOR-HOURS	UNIT	2000 BARE COSTS MAT.	LABOR	EQUIP.	TOTAL	TOTAL INCL O&P	
620	0660	5" diameter	Q-1	26	.615	L.F.	11.95	12.15		24.10	33	620
	0670	6" diameter	Q-2	31	.774		19	14.75		33.75	45.50	
	1290	Galvanized, 1/4" diameter	1 Plum	66	.121		1.71	2.66		4.37	6.30	
	1300	3/8" diameter		65	.123		1.88	2.70		4.58	6.55	
	1310	1/2" diameter		63	.127		1.59	2.79		4.38	6.35	
	1320	3/4" diameter		61	.131		1.78	2.88		4.66	6.70	
	1330	1" diameter		53	.151		1.98	3.31		5.29	7.65	
	1340	1-1/4" diameter	Q-1	89	.180		2.39	3.55		5.94	8.50	
	1350	1-1/2" diameter		80	.200		2.66	3.95		6.61	9.45	
	1360	2" diameter		64	.250		3.57	4.94		8.51	12.10	
	1370	2-1/2" diameter		50	.320		5.55	6.30		11.85	16.60	
	1380	3" diameter		43	.372		6.90	7.35		14.25	19.75	
	1390	3-1/2" diameter		40	.400		8.95	7.90		16.85	23	
	1400	4" diameter		36	.444		9.90	8.80		18.70	25.50	
	1410	5" diameter		26	.615		14.50	12.15		26.65	36	
	1420	6" diameter	Q-2	31	.774		19.15	14.75		33.90	45.50	
	1430	8" diameter		27	.889		31.50	16.90		48.40	62.50	
	1440	10" diameter		23	1.043		54	19.85		73.85	92	
	1450	12" diameter		18	1.333		64	25.50		89.50	112	
	2000	Welded, sch. 40, on yoke & roll hangers, sized for covering,										
	2010	10' O.C. (no hangers incl. for 14" diam. and up)										
	2040	Black, 1" diameter	Q-15	93	.172	L.F.	2.30	3.40	.52	6.22	8.70	
	2070	2" diameter		61	.262		3.20	5.20	.80	9.20	12.95	
	2090	3" diameter		43	.372		5.05	7.35	1.13	13.53	18.95	
	2110	4" diameter		37	.432		6.80	8.55	1.32	16.67	23	
	2120	5" diameter		32	.500		9.80	9.90	1.52	21.22	29	
	2130	6" diameter	Q-16	36	.667		16.60	13.65	1.35	31.60	42	
	2140	8" diameter		29	.828		22	16.95	1.68	40.63	54.50	
	2150	10" diameter		24	1		42	20.50	2.03	64.53	82	
	2160	12" diameter		19	1.263		51.50	26	2.56	80.06	102	
690	0010	**PIPE, GROOVED-JOINT STEEL FITTINGS & VALVES**										690
	0020	Pipe includes coupling & clevis type hanger 10' O.C.										
	1000	Schedule 40, black										
	1040	3/4" diameter	1 Plum	71	.113	L.F.	2.04	2.47		4.51	6.35	
	1050	1" diameter		63	.127		2.17	2.79		4.96	7	
	1060	1-1/4" diameter		58	.138		2.64	3.03		5.67	7.90	
	1070	1-1/2" diameter		51	.157		2.97	3.44		6.41	8.95	
	1080	2" diameter		40	.200		3.46	4.39		7.85	11.05	
	1090	2-1/2" diameter	Q-1	57	.281		4.68	5.55		10.23	14.30	
	1100	3" diameter		50	.320		5.70	6.30		12	16.70	
	1110	4" diameter		45	.356		8.05	7		15.05	20.50	
	1120	5" diameter		37	.432		12.40	8.55		20.95	27.50	
	1130	6" diameter	Q-2	42	.571		15.90	10.85		26.75	35.50	
	1800	Galvanized										
	1840	3/4" diameter	1 Plum	71	.113	L.F.	2.27	2.47		4.74	6.60	
	1850	1" diameter		63	.127		2.34	2.79		5.13	7.20	
	1860	1-1/4" diameter		58	.138		2.85	3.03		5.88	8.15	
	1870	1-1/2" diameter		51	.157		3.23	3.44		6.67	9.25	
	1880	2" diameter		40	.200		3.81	4.39		8.20	11.45	
	1890	2-1/2" diameter	Q-1	57	.281		5.25	5.55		10.80	14.90	
	1900	3" diameter		50	.320		6.40	6.30		12.70	17.50	
	1910	4" diameter		45	.356		9.10	7		16.10	21.50	
	1920	5" diameter		37	.432		15.25	8.55		23.80	31	
	1930	6" diameter	Q-2	42	.571		19.80	10.85		30.65	40	
	3990	Fittings: cplg. & labor required at joints not incl. in fitting										
	3994	price. Add 1 per joint for installed price.										

15100 | Building Services Piping

15107 | Metal Pipe & Fittings

		CREW	DAILY OUTPUT	LABOR-HOURS	UNIT	2000 BARE COSTS MAT.	LABOR	EQUIP.	TOTAL	TOTAL INCL O&P
4000	Elbow, 90° or 45°, painted									
4030	3/4" diameter	1 Plum	50	.160	Ea.	12.10	3.51		15.61	19.10
4040	1" diameter		50	.160		12.10	3.51		15.61	19.10
4050	1-1/4" diameter		40	.200		12.10	4.39		16.49	20.50
4060	1-1/2" diameter		33	.242		12.10	5.30		17.40	22
4070	2" diameter		25	.320		12.10	7		19.10	25
4080	2-1/2" diameter	Q-1	40	.400		12.10	7.90		20	26.50
4090	3" diameter		33	.485		21.50	9.60		31.10	39.50
4100	4" diameter		25	.640		23.50	12.65		36.15	47
4110	5" diameter		20	.800		56.50	15.80		72.30	88.50
4120	6" diameter	Q-2	25	.960		66.50	18.25		84.75	103
4250	For galvanized elbows, add					26%				
4690	Tee, painted									
4700	3/4" diameter	1 Plum	38	.211	Ea.	18.65	4.62		23.27	28
4740	1" diameter		33	.242		18.65	5.30		23.95	29.50
4750	1-1/4" diameter		27	.296		18.65	6.50		25.15	31.50
4760	1-1/2" diameter		22	.364		18.65	8		26.65	33.50
4770	2" diameter		17	.471		18.65	10.35		29	37.50
4780	2-1/2" diameter	Q-1	27	.593		18.65	11.70		30.35	40
4790	3" diameter		22	.727		26	14.35		40.35	53
4800	4" diameter		17	.941		40	18.60		58.60	74.50
4810	5" diameter		13	1.231		93.50	24.50		118	143
4820	6" diameter	Q-2	17	1.412		108	27		135	164
4900	For galvanized tees, add					24%				
4906	Couplings, rigid style, painted									
4908	1" diameter	1 Plum	100	.080	Ea.	8.80	1.76		10.56	12.55
4909	1-1/4" diameter		100	.080		8.80	1.76		10.56	12.55
4910	1-1/2" diameter		67	.119		8.80	2.62		11.42	14
4912	2" diameter		50	.160		9	3.51		12.51	15.70
4914	2-1/2" diameter	Q-1	80	.200		10.40	3.95		14.35	17.95
4916	3" diameter		67	.239		12.15	4.72		16.87	21
4918	4" diameter		50	.320		17.20	6.30		23.50	29.50
4920	5" diameter		40	.400		22.50	7.90		30.40	37.50
4922	6" diameter	Q-2	50	.480		30	9.15		39.15	48
4924	8" diameter		42	.571		47	10.85		57.85	69.50
4926	10" diameter		35	.686		84	13.05		97.05	114
4928	12" diameter		32	.750		94.50	14.25		108.75	128
4930	14" diameter		24	1		123	19		142	167
4931	16" diameter		20	1.200		161	23		184	215
4932	18" diameter		18	1.333		186	25.50		211.50	246
4933	20" diameter		16	1.500		221	28.50		249.50	290
4934	24" diameter	Q-9	13	1.231		325	24		349	400
4940	Flexible, standard, painted									
4950	3/4" diameter	1 Plum	100	.080	Ea.	6.30	1.76		8.06	9.85
4960	1" diameter		100	.080		6.30	1.76		8.06	9.85
4970	1-1/4" diameter		80	.100		8.40	2.20		10.60	12.90
4980	1-1/2" diameter		67	.119		9.20	2.62		11.82	14.45
4990	2" diameter		50	.160		9.70	3.51		13.21	16.45
5000	2-1/2" diameter	Q-1	80	.200		11.60	3.95		15.55	19.30
5010	3" diameter		67	.239		12.80	4.72		17.52	22
5020	3-1/2" diameter		57	.281		18.70	5.55		24.25	29.50
5030	4" diameter		50	.320		18.70	6.30		25	31
5040	5" diameter		40	.400		28.50	7.90		36.40	44.50
5050	6" diameter	Q-2	50	.480		34	9.15		43.15	52.50
0010	**PIPE, STAINLESS STEEL**									
3500	Threaded, couplings and hangers 10' O.C.									

15100 | Building Services Piping

15107 | Metal Pipe & Fittings

			CREW	DAILY OUTPUT	LABOR-HOURS	UNIT	MAT.	LABOR	EQUIP.	TOTAL	TOTAL INCL O&P	
920	3520	Schedule 40, type 304										920
	3540	1/4" diameter	1 Plum	54	.148	L.F.	3.72	3.25		6.97	9.50	
	3550	3/8" diameter		53	.151		4.22	3.31		7.53	10.15	
	3560	1/2" diameter		52	.154		5.45	3.38		8.83	11.60	
	3580	1" diameter		45	.178		8.30	3.90		12.20	15.55	
	3610	2" diameter	Q-1	57	.281		15.70	5.55		21.25	26.50	
	3640	4" diameter	Q-2	51	.471		44.50	8.95		53.45	64	
	3740	For small quantities, add					10%					
	4250	Schedule 40, type 316										
	4290	1/4" diameter	1 Plum	54	.148	L.F.	4.64	3.25		7.89	10.50	
	4300	3/8" diameter		53	.151		6.10	3.31		9.41	12.20	
	4310	1/2" diameter		52	.154		6.80	3.38		10.18	13.05	
	4320	3/4" diameter		51	.157		7.90	3.44		11.34	14.40	
	4330	1" diameter		45	.178		10.50	3.90		14.40	17.95	
	4360	2" diameter	Q-1	57	.281		20.50	5.55		26.05	31.50	
	4390	4" diameter	Q-2	51	.471		59	8.95		67.95	80	
	4490	For small quantities, add					10%					

15108 | Plastic Pipe & Fittings

			CREW	DAILY OUTPUT	LABOR-HOURS	UNIT	MAT.	LABOR	EQUIP.	TOTAL	TOTAL INCL O&P	
520	0010	**PIPE, PLASTIC**										520
	0020	Fiberglass reinforced, couplings 10' O.C., hangers 3 per 10'										
	0200	High strength										
	0240	2" diameter	Q-1	58	.276	L.F.	10.75	5.45		16.20	21	
	0260	3" diameter		51	.314		14.65	6.20		20.85	26.50	
	0280	4" diameter		47	.340		17.65	6.70		24.35	30.50	
	0300	6" diameter		38	.421		26	8.30		34.30	42.50	
	0320	8" diameter	Q-2	48	.500		41	9.50		50.50	60.50	
	1800	PVC, couplings 10' O.C., hangers 3 per 10'										
	1820	Schedule 40										
	1860	1/2" diameter	1 Plum	54	.148	L.F.	1.90	3.25		5.15	7.50	
	1870	3/4" diameter		51	.157		2.06	3.44		5.50	7.95	
	1880	1" diameter		46	.174		2.57	3.82		6.39	9.15	
	1890	1-1/4" diameter		42	.190		2.61	4.18		6.79	9.75	
	1900	1-1/2" diameter		36	.222		2.81	4.88		7.69	11.15	
	1910	2" diameter	Q-1	59	.271		2.83	5.35		8.18	11.95	
	1920	2-1/2" diameter		56	.286		3.52	5.65		9.17	13.20	
	1930	3" diameter		53	.302		4.05	5.95		10	14.30	
	1940	4" diameter		48	.333		5.50	6.60		12.10	16.95	
	1950	5" diameter		43	.372		7.50	7.35		14.85	20.50	
	1960	6" diameter		39	.410		8.30	8.10		16.40	22.50	
	4100	DWW type, schedule 40, couplings 10' O.C., hangers 3 per 10'										
	4120	ABS										
	4140	1-1/4" diameter	1 Plum	42	.190	L.F.	2.58	4.18		6.76	9.75	
	4150	1-1/2" diameter	"	36	.222		2.56	4.88		7.44	10.85	
	4160	2" diameter	Q-1	59	.271		2.66	5.35		8.01	11.75	
	4400	PVC										
	4410	1-1/4" diameter	1 Plum	42	.190	L.F.	2.66	4.18		6.84	9.80	
	4420	1-1/2" diameter	"	36	.222		2.73	4.88		7.61	11.05	
	4460	2" diameter	Q-1	59	.271		2.91	5.35		8.26	12.05	
	4470	3" diameter		53	.302		3.79	5.95		9.74	14	
	4480	4" diameter		48	.333		5.10	6.60		11.70	16.55	
	4490	6" diameter		39	.410		7.40	8.10		15.50	21.50	
	5360	CPVC, couplings 10' O.C., hangers 3 per 10'										
	5380	Schedule 40										
	5460	1/2" diameter	1 Plum	54	.148	L.F.	2.63	3.25		5.88	8.30	

15100 | Building Services Piping

15108 | Plastic Pipe & Fittings

			CREW	DAILY OUTPUT	LABOR-HOURS	UNIT	2000 BARE COSTS MAT.	LABOR	EQUIP.	TOTAL	TOTAL INCL O&P	
520	5470	3/4" diameter	1 Plum	51	.157	L.F.	3.04	3.44		6.48	9.05	520
	5480	1" diameter		46	.174		3.98	3.82		7.80	10.70	
	5490	1-1/4" diameter		42	.190		4.57	4.18		8.75	11.95	
	5500	1-1/2" diameter	▼	36	.222		5.20	4.88		10.08	13.75	
	5510	2" diameter	Q-1	59	.271		6.05	5.35		11.40	15.55	
	5520	2-1/2" diameter		56	.286		8.85	5.65		14.50	19.10	
	5530	3" diameter	▼	53	.302	▼	10.55	5.95		16.50	21.50	

15110 | Valves

			CREW	DAILY OUTPUT	LABOR-HOURS	UNIT	MAT.	LABOR	EQUIP.	TOTAL	TOTAL INCL O&P	
100	0010	**VALVES, BRASS**										100
	0500	Gas cocks, threaded										
	0530	1/2" size	1 Plum	24	.333	Ea.	7.65	7.30		14.95	20.50	
	0540	3/4" size		22	.364		10.30	8		18.30	24.50	
	0550	1" size		19	.421		12.60	9.25		21.85	29	
	0560	1-1/4" size	▼	15	.533	▼	36	11.70		47.70	59	
160	0010	**VALVES, BRONZE**										160
	1020	Angle, 150 lb., rising stem, threaded										
	1030	1/8" size	1 Plum	24	.333	Ea.	39	7.30		46.30	55	
	1040	1/4" size		24	.333		39	7.30		46.30	55	
	1050	3/8" size		24	.333		39	7.30		46.30	55	
	1060	1/2" size		22	.364		39	8		47	56	
	1070	3/4" size		20	.400		52.50	8.80		61.30	72.50	
	1080	1" size		19	.421		76	9.25		85.25	99	
	1100	1-1/2" size		13	.615		127	13.50		140.50	162	
	1110	2" size	▼	11	.727		206	15.95		221.95	254	
	1380	Ball, 150 psi, threaded										
	1400	1/4" size	1 Plum	24	.333	Ea.	6.40	7.30		13.70	19.15	
	1430	3/8" size		24	.333		6.40	7.30		13.70	19.15	
	1450	1/2" size		22	.364		6.40	8		14.40	20.50	
	1460	3/4" size		20	.400		10.60	8.80		19.40	26	
	1470	1" size		19	.421		13.30	9.25		22.55	30	
	1480	1-1/4" size		15	.533		22.50	11.70		34.20	44.50	
	1490	1-1/2" size		13	.615		28.50	13.50		42	54	
	1500	2" size	▼	11	.727	▼	35.50	15.95		51.45	66	
	1750	Check, swing, class 150, regrinding disc, threaded										
	1800	1/8" size	1 Plum	24	.333	Ea.	21.50	7.30		28.80	35.50	
	1830	1/4" size		24	.333		21.50	7.30		28.80	35.50	
	1840	3/8" size		24	.333		21.50	7.30		28.80	35.50	
	1850	1/2" size		24	.333		21.50	7.30		28.80	35.50	
	1860	3/4" size		20	.400		28.50	8.80		37.30	46	
	1870	1" size		19	.421		43	9.25		52.25	63	
	1880	1-1/4" size		15	.533		59.50	11.70		71.20	85	
	1890	1-1/2" size		13	.615		70	13.50		83.50	99.50	
	1900	2" size	▼	11	.727		103	15.95		118.95	140	
	1910	2-1/2" size	Q-1	15	1.067	▼	180	21		201	233	
	2000	For 200 lb, add					5%	10%				
	2040	For 300 lb, add					15%	15%				
	2850	Gate, N.R.S., soldered, 125 psi										
	2900	3/8" size	1 Plum	24	.333	Ea.	16.20	7.30		23.50	30	
	2920	1/2" size		24	.333		14.05	7.30		21.35	27.50	
	2940	3/4" size		20	.400		16.20	8.80		25	32.50	
	2950	1" size		19	.421		23	9.25		32.25	41	
	2960	1-1/4" size		15	.533		40.50	11.70		52.20	64	
	2970	1-1/2" size		13	.615		45.50	13.50		59	72.50	
	2980	2" size	▼	11	.727	▼	54.50	15.95		70.45	86.50	

Important: See the Reference Section for critical supporting data - Reference Nos., Crews, & City Cost Indexes

15100 | Building Services Piping

15110 | Valves

		Crew	Daily Output	Labor-Hours	Unit	Mat.	Labor	Equip.	Total	Total Incl O&P
2990	2-1/2" size	Q-1	15	1.067	Ea.	138	21		159	186
3000	3" size	"	13	1.231	↓	180	24.50		204.50	238
3850	Rising stem, soldered, 300 psi									
3950	1" size	1 Plum	19	.421	Ea.	43.50	9.25		52.75	63.50
3980	2" size	"	11	.727		104	15.95		119.95	141
4000	3" size	Q-1	13	1.231	↓	325	24.50		349.50	395
4250	Threaded, class 150									
4310	1/4" size	1 Plum	24	.333	Ea.	21	7.30		28.30	35
4320	3/8" size		24	.333		21	7.30		28.30	35
4330	1/2" size		24	.333		19.45	7.30		26.75	33.50
4340	3/4" size		20	.400		23	8.80		31.80	40
4350	1" size		19	.421		30.50	9.25		39.75	49
4360	1-1/4" size		15	.533		41	11.70		52.70	64.50
4370	1-1/2" size		13	.615		51	13.50		64.50	78.50
4380	2" size	↓	11	.727		70	15.95		85.95	104
4390	2-1/2" size	Q-1	15	1.067		162	21		183	213
4400	3" size	"	13	1.231		230	24.50		254.50	293
4500	For 300 psi, threaded, add					100%	15%			
4540	For chain operated type, add				↓	15%				
4850	Globe, class 150, rising stem, threaded									
4920	1/4" size	1 Plum	24	.333	Ea.	30	7.30		37.30	45
4940	3/8" size		24	.333		30	7.30		37.30	45
4950	1/2" size		24	.333		29.50	7.30		36.80	44
4960	3/4" size		20	.400		40.50	8.80		49.30	59
4970	1" size		19	.421		63	9.25		72.25	85
4980	1-1/4" size		15	.533		100	11.70		111.70	129
4990	1-1/2" size		13	.615		121	13.50		134.50	156
5000	2" size	↓	11	.727		182	15.95		197.95	227
5010	2-1/2" size	Q-1	15	1.067		370	21		391	440
5020	3" size	"	13	1.231	↓	520	24.50		544.50	615
5120	For 300 lb threaded, add					50%	15%			
5600	Relief, pressure & temperature, self-closing, ASME, threaded									
5640	3/4" size	1 Plum	28	.286	Ea.	63	6.25		69.25	80
5650	1" size		24	.333		92	7.30		99.30	113
5660	1-1/4" size		20	.400		184	8.80		192.80	217
5670	1-1/2" size		18	.444		350	9.75		359.75	400
5680	2" size	↓	16	.500	↓	385	11		396	440
5950	Pressure, poppet type, threaded									
6000	1/2" size	1 Plum	30	.267	Ea.	13.15	5.85		19	24
6040	3/4" size	"	28	.286	"	39	6.25		45.25	53.50
6400	Pressure, water, ASME, threaded									
6440	3/4" size	1 Plum	28	.286	Ea.	45	6.25		51.25	60
6450	1" size		24	.333		91	7.30		98.30	112
6460	1-1/4"size		20	.400		131	8.80		139.80	159
6470	1-1/2" size		18	.444		206	9.75		215.75	242
6480	2" size		16	.500		298	11		309	345
6490	2-1/2" size	↓	15	.533	↓	830	11.70		841.70	935
6900	Reducing, water pressure									
6920	300 psi to 25-75 psi, threaded or sweat									
6940	1/2" size	1 Plum	24	.333	Ea.	103	7.30		110.30	126
6950	3/4" size		20	.400		103	8.80		111.80	129
6960	1" size		19	.421		160	9.25		169.25	191
6970	1-1/4" size		15	.533		288	11.70		299.70	335
6980	1-1/2" size	↓	13	.615	↓	430	13.50		443.50	500
8350	Tempering, water, sweat connections									
8400	1/2" size	1 Plum	24	.333	Ea.	39	7.30		46.30	55

15100 | Building Services Piping

15110 | Valves

			CREW	DAILY OUTPUT	LABOR-HOURS	UNIT	2000 BARE COSTS MAT.	LABOR	EQUIP.	TOTAL	TOTAL INCL O&P	
160	8440	3/4" size	1 Plum	20	.400	Ea.	48	8.80		56.80	67.50	160
	8650	Threaded connections										
	8700	1/2" size	1 Plum	24	.333	Ea.	48	7.30		55.30	65	
	8740	3/4" size		20	.400		182	8.80		190.80	215	
	8750	1" size		19	.421		201	9.25		210.25	236	
	8760	1-1/4" size		15	.533		320	11.70		331.70	370	
	8770	1-1/2" size		13	.615		350	13.50		363.50	410	
	8780	2" size		11	.727		525	15.95		540.95	600	
200	0010	**VALVES, IRON BODY**										200
	1020	Butterfly, wafer type, gear actuator, 200 lb.										
	1030	2" size	1 Plum	14	.571	Ea.	109	12.55		121.55	140	
	1040	2-1/2" size	Q-1	9	1.778		112	35		147	181	
	1050	3" size		8	2		116	39.50		155.50	194	
	1060	4" size		5	3.200		145	63		208	264	
	1070	5" size	Q-2	5	4.800		175	91.50		266.50	345	
	1080	6" size	"	5	4.800		198	91.50		289.50	370	
	1650	Gate, 125 lb., N.R.S.										
	2150	Flanged										
	2200	2" size	1 Plum	5	1.600	Ea.	263	35		298	345	
	2240	2-1/2" size	Q-1	5	3.200		270	63		333	400	
	2260	3" size		4.50	3.556		300	70		370	450	
	2280	4" size		3	5.333		435	105		540	655	
	2300	6" size	Q-2	3	8		740	152		892	1,050	
	3550	OS&Y, 125 lb., flanged										
	3600	2" size	1 Plum	5	1.600	Ea.	198	35		233	276	
	3660	3" size	Q-1	4.50	3.556		231	70		301	370	
	3680	4" size	"	3	5.333		248	105		353	445	
	3700	6" size	Q-2	3	8		545	152		697	850	
	3900	For 250 lb, flanged, add					200%	10%				
	5450	Swing check, 125 lb., threaded										
	5500	2" size	1 Plum	11	.727	Ea.	320	15.95		335.95	375	
	5540	2-1/2" size	Q-1	15	1.067		415	21		436	490	
	5550	3" size		13	1.231		445	24.50		469.50	530	
	5560	4" size		10	1.600		715	31.50		746.50	835	
	5950	Flanged										
	6000	2" size	1 Plum	5	1.600	Ea.	140	35		175	212	
	6040	2-1/2" size	Q-1	5	3.200		168	63		231	289	
	6050	3" size		4.50	3.556		181	70		251	315	
	6060	4" size		3	5.333		286	105		391	490	
	6070	6" size	Q-2	3	8		505	152		657	805	
500	0010	**VALVES, PLASTIC**										500
	1100	Angle, PVC, threaded										
	1110	1/4" size	1 Plum	26	.308	Ea.	47.50	6.75		54.25	63	
	1120	1/2" size		26	.308		47.50	6.75		54.25	63	
	1130	3/4" size		25	.320		55.50	7		62.50	72.50	
	1140	1" size		23	.348		66.50	7.65		74.15	85.50	
	1150	Ball, PVC, socket or threaded, single union										
	1200	1/4" size	1 Plum	26	.308	Ea.	22	6.75		28.75	35	
	1220	3/8" size		26	.308		22	6.75		28.75	35	
	1230	1/2" size		26	.308		22	6.75		28.75	35	
	1240	3/4" size		25	.320		26.50	7		33.50	40.50	
	1250	1" size		23	.348		31	7.65		38.65	46.50	
	1260	1-1/4" size		21	.381		41.50	8.35		49.85	60	
	1270	1-1/2" size		20	.400		52	8.80		60.80	71.50	
	1280	2" size		17	.471		74.50	10.35		84.85	98.50	
	1360	For PVC, flanged, add					100%	15%				

15100 | Building Services Piping

15110 | Valves

		Crew	Daily Output	Labor-Hours	Unit	2000 Bare Costs Mat.	Labor	Equip.	Total	Total Incl O&P
1650	CPVC, socket or threaded, single union									
1700	1/2" size	1 Plum	26	.308	Ea.	36.50	6.75		43.25	51
1720	3/4" size		25	.320		43.50	7		50.50	59
1730	1" size		23	.348		51	7.65		58.65	68.50
1750	1-1/4" size		21	.381		88	8.35		96.35	111
1760	1-1/2" size		20	.400		88	8.80		96.80	112
1840	For CPVC, flanged, add					65%	15%			
1880	For true union, socket or threaded, add					50%	5%			
2050	Polypropylene, threaded									
2100	1/4" size	1 Plum	26	.308	Ea.	30	6.75		36.75	44
2120	3/8" size		26	.308		30	6.75		36.75	44
2130	1/2" size		26	.308		30	6.75		36.75	44
2140	3/4" size		25	.320		37.50	7		44.50	52.50
2150	1" size		23	.348		44.50	7.65		52.15	61.50
2160	1-1/4" size		21	.381		64.50	8.35		72.85	85
2170	1-1/2" size		20	.400		74	8.80		82.80	96
2180	2" size		17	.471		101	10.35		111.35	128
4850	Foot valve, PVC, socket or threaded									
4900	1/2" size	1 Plum	34	.235	Ea.	36.50	5.15		41.65	48.50
4930	3/4" size		32	.250		41.50	5.50		47	54.50
4940	1" size		28	.286		54	6.25		60.25	70
4950	1-1/4" size		27	.296		104	6.50		110.50	125
4960	1-1/2" size		26	.308		104	6.75		110.75	125
6350	Y sediment strainer, PVC, socket or threaded									
6400	1/2" size	1 Plum	26	.308	Ea.	29.50	6.75		36.25	43.50
6440	3/4" size		24	.333		32	7.30		39.30	47.50
6450	1" size		23	.348		39	7.65		46.65	55.50
6460	1-1/4" size		21	.381		64.50	8.35		72.85	85
6470	1-1/2" size		20	.400		67.50	8.80		76.30	89
0010	**VALVES, STEEL**									
0800	Cast									
1350	Check valve, swing type, 150 lb., flanged									
1400	2" size	1 Plum	8	1	Ea.	465	22		487	550
1440	2-1/2" size	Q-1	5	3.200		595	63		658	760
1450	3" size		4.50	3.556		595	70		665	770
1460	4" size		3	5.333		895	105		1,000	1,150
1540	For 300 lb., flanged, add					50%	15%			
1548	For 600 lb., flanged, add					110%	20%			
1950	Gate valve, 150 lb., flanged									
2000	2" size	1 Plum	8	1	Ea.	515	22		537	600
2040	2-1/2" size	Q-1	5	3.200		725	63		788	905
2050	3" size		4.50	3.556		725	70		795	915
2060	4" size		3	5.333		895	105		1,000	1,150
2070	6" size	Q-2	3	8		1,425	152		1,577	1,825
3650	Globe valve, 150 lb., flanged									
3700	2" size	1 Plum	8	1	Ea.	675	22		697	780
3740	2-1/2" size	Q-1	5	3.200		850	63		913	1,050
3750	3" size		4.50	3.556		850	70		920	1,050
3760	4" size		3	5.333		1,250	105		1,355	1,550
3770	6" size	Q-2	3	8		1,950	152		2,102	2,400
5150	Forged									
5650	Check valve, class 800, horizontal, socket									
5698	Threaded									
5700	1/4" size	1 Plum	24	.333	Ea.	41	7.30		48.30	57
5720	3/8" size		24	.333		41	7.30		48.30	57

15100 | Building Services Piping

15110 | Valves

			CREW	DAILY OUTPUT	LABOR-HOURS	UNIT	2000 BARE COSTS MAT.	LABOR	EQUIP.	TOTAL	TOTAL INCL O&P	
700	5730	1/2" size	1 Plum	24	.333	Ea.	41	7.30		48.30	57	700
	5740	3/4" size		20	.400		41	8.80		49.80	59.50	
	5750	1" size		19	.421		52	9.25		61.25	72.50	
	5760	1-1/4" size		15	.533		99.50	11.70		111.20	128	

15120 | Piping Specialties

			CREW	DAILY OUTPUT	LABOR-HOURS	UNIT	MAT.	LABOR	EQUIP.	TOTAL	TOTAL INCL O&P	
120	0010	**AIR CONTROL**										120
	0030	Air separator, with strainer										
	0040	2" diameter	Q-5	6	2.667	Ea.	505	53		558	645	
	0080	2-1/2" diameter		5	3.200		575	63.50		638.50	735	
	0100	3" diameter		4	4		875	79.50		954.50	1,100	
	0120	4" diameter		3	5.333		1,275	106		1,381	1,575	
	0130	5" diameter	Q-6	3.60	6.667		1,625	128		1,753	1,975	
	0140	6" diameter	"	3.40	7.059		1,925	135		2,060	2,350	
160	0010	**AUTOMATIC AIR VENT**										160
	0020	Cast iron body, stainless steel internals, float type										
	0060	1/2" NPT inlet, 300 psi	1 Stpi	12	.667	Ea.	77	14.75		91.75	109	
	0220	3/4" NPT inlet, 250 psi	"	10	.800		246	17.70		263.70	299	
	0340	1-1/2" NPT inlet, 250 psi	Q-5	12	1.333		765	26.50		791.50	885	
300	0010	**EXPANSION JOINTS**										300
	0100	Bellows type, neoprene cover, flanged spool										
	0120	6" face to face, 1" diameter	1 Stpi	13	.615	Ea.	142	13.60		155.60	179	
	0140	1-1/4" diameter		11	.727		145	16.05		161.05	187	
	0160	1-1/2" diameter		10.60	.755		145	16.70		161.70	188	
	0180	2" diameter	Q-5	13.30	1.203		147	24		171	202	
	0190	2-1/2" diameter		12.40	1.290		153	25.50		178.50	211	
	0200	3" diameter		11.40	1.404		174	28		202	237	
	0480	10" face to face, 2" diameter		13	1.231		216	24.50		240.50	279	
	0500	2-1/2" diameter		12	1.333		228	26.50		254.50	295	
	0520	3" diameter		11	1.455		231	29		260	300	
	0540	4" diameter		8	2		258	40		298	350	
	0560	5" diameter		7	2.286		315	45.50		360.50	420	
	0580	6" diameter		6	2.667		320	53		373	445	
320	0010	**EXPANSION TANKS**										320
	1505	Fiberglass and steel single / double wall storage, see Div 13201										
	1510	Tank leak detection systems, see Div 13851-350										
	2000	Steel, liquid expansion, ASME, painted, 15 gallon capacity	Q-5	17	.941	Ea.	330	18.75		348.75	395	
	2020	24 gallon capacity		14	1.143		335	22.50		357.50	405	
	2040	30 gallon capacity		12	1.333		370	26.50		396.50	450	
	2060	40 gallon capacity		10	1.600		420	32		452	515	
	2080	60 gallon capacity		8	2		505	40		545	620	
	2100	80 gallon capacity		7	2.286		520	45.50		565.50	645	
	2120	100 gallon capacity		6	2.667		670	53		723	825	
	3000	Steel ASME expansion, rubber diaphragm, 19 gal. cap. accept.		12	1.333		1,250	26.50		1,276.50	1,425	
	3020	31 gallon capacity		8	2		1,400	40		1,440	1,600	
	3040	61 gallon capacity		6	2.667		1,950	53		2,003	2,250	
	3080	119 gallon capacity		4	4		2,200	79.50		2,279.50	2,550	
	3100	158 gallon capacity		3.80	4.211		3,075	84		3,159	3,550	
	3140	317 gallon capacity		2.80	5.714		4,650	114		4,764	5,325	
	3180	528 gallon capacity		2.40	6.667		7,550	133		7,683	8,525	
	5950											
350	0010	**FLEXIBLE CONNECTORS**, Corrugated, 7/8" O.D., 1/2" I.D.										350
	0050	Gas, seamless brass, steel fittings										
	0200	12" long	1 Plum	36	.222	Ea.	9.45	4.88		14.33	18.45	
	0220	18" long		36	.222		11.75	4.88		16.63	21	

15100 | Building Services Piping

15120 | Piping Specialties

			CREW	DAILY OUTPUT	LABOR-HOURS	UNIT	2000 BARE COSTS MAT.	LABOR	EQUIP.	TOTAL	TOTAL INCL O&P	
350	0240	24" long	1 Plum	34	.235	Ea.	13.90	5.15		19.05	24	**350**
	0280	36" long		32	.250		16.55	5.50		22.05	27.50	
	0340	60" long	↓	30	.267	↓	25	5.85		30.85	37	
	2000	Water, copper tubing, dielectric separators										
	2100	12" long	1 Plum	36	.222	Ea.	8.10	4.88		12.98	16.95	
	2260	24" long	"	34	.235	"	12	5.15		17.15	22	
370	0010	**FLEXIBLE METAL HOSE** Connectors, standard lengths										**370**
	0100	Bronze braided, bronze ends										
	0120	3/8" diameter x 12"	1 Stpi	26	.308	Ea.	14.40	6.80		21.20	27	
	0160	3/4" diameter x 12"		20	.400		20	8.85		28.85	36.50	
	0180	1" diameter x 18"		19	.421		26	9.30		35.30	44	
	0200	1-1/2" diameter x 18"		13	.615		41	13.60		54.60	67.50	
	0220	2" diameter x 18"	↓	11	.727	↓	49.50	16.05		65.55	81	
520	0010	**HYDRONIC HEATING CONTROL VALVES**										**520**
	0100	Radiator supply, 1/2" diameter	1 Stpi	24	.333	Ea.	35.50	7.35		42.85	51	
	0120	3/4" diameter		20	.400		36	8.85		44.85	54	
	0140	1" diameter		19	.421		44.50	9.30		53.80	64.50	
	0160	1-1/4" diameter	↓	15	.533	↓	63	11.80		74.80	88.50	
	0500	For low pressure steam, add					25%					
670	0010	**PRESSURE REGULATOR**										**670**
	3000	Steam, high capacity, bronze body, stainless steel trim										
	3020	Threaded, 1/2" diameter	1 Stpi	24	.333	Ea.	660	7.35		667.35	735	
	3030	3/4" diameter		24	.333		660	7.35		667.35	735	
	3040	1" diameter		19	.421		735	9.30		744.30	825	
	3060	1-1/4" diameter		15	.533		815	11.80		826.80	915	
	3080	1-1/2" diameter		13	.615		930	13.60		943.60	1,050	
	3100	2" diameter	↓	11	.727		1,150	16.05		1,166.05	1,275	
	3120	2-1/2" diameter	Q-5	12	1.333		1,500	26.50		1,526.50	1,700	
	3140	3" diameter	"	11	1.455	↓	1,625	29		1,654	1,850	
	3500	Flanged connection, iron body, 125 lb. W.S.P.										
	3520	3" diameter	Q-5	11	1.455	Ea.	1,775	29		1,804	2,025	
	3540	4" diameter	"	5	3.200	"	2,250	63.50		2,313.50	2,575	
760	0010	**STEAM TRAP**										**760**
	0030	Cast iron body, threaded										
	0040	Inverted bucket										
	0050	1/2" pipe size	1 Stpi	12	.667	Ea.	98.50	14.75		113.25	133	
	0070	3/4" pipe size		10	.800		169	17.70		186.70	215	
	0100	1" pipe size		9	.889		260	19.65		279.65	320	
	0120	1-1/4" pipe size	↓	8	1	↓	390	22		412	465	
	1000	Float & thermostatic, 15 psi										
	1010	3/4" pipe size	1 Stpi	16	.500	Ea.	96	11.05		107.05	124	
	1020	1" pipe size		15	.533		113	11.80		124.80	144	
	1040	1-1/2" pipe size		9	.889		254	19.65		273.65	310	
	1060	2" pipe size	↓	6	1.333	↓	460	29.50		489.50	560	
820	0010	**STRAINERS, Y TYPE** Bronze body										**820**
	0050	Screwed, 150 lb., 1/4" pipe size	1 Stpi	24	.333	Ea.	11.40	7.35		18.75	25	
	0070	3/8" pipe size		24	.333		15.90	7.35		23.25	29.50	
	0100	1/2" pipe size		20	.400		15.90	8.85		24.75	32	
	0140	1" pipe size		17	.471		18.70	10.40		29.10	37.50	
	0160	1-1/2" pipe size		14	.571		40.50	12.65		53.15	65.50	
	0180	2" pipe size		13	.615		53.50	13.60		67.10	81.50	
	0182	3" pipe size	↓	12	.667		286	14.75		300.75	340	
	0220	3" pipe size	Q-5	16	1		565	19.90		584.90	660	
	0240	4" pipe size	"	15	1.067	↓	1,300	21		1,321	1,450	

15100 | Building Services Piping

15120 | Piping Specialties

			CREW	DAILY OUTPUT	LABOR-HOURS	UNIT	2000 BARE COSTS MAT.	LABOR	EQUIP.	TOTAL	TOTAL INCL O&P	
820	0500	For 300 lb rating 1/4" thru 2", add					15%					820
	1000	Flanged, 150 lb., 1-1/2" pipe size	1 Stpi	11	.727	Ea.	305	16.05		321.05	360	
	1020	2" pipe size	"	8	1		390	22		412	465	
	1030	2-1/2" pipe size	Q-5	5	3.200		505	63.50		568.50	660	
	1040	3" pipe size		4.50	3.556		620	71		691	800	
	1060	4" pipe size		3	5.333		945	106		1,051	1,200	
	1100	6" pipe size	Q-6	3	8		2,500	153		2,653	3,000	
	1106	8" pipe size	"	2.60	9.231		3,900	177		4,077	4,600	
	1500	For 300 lb rating, add					40%					
840	0010	**STRAINERS, Y TYPE** Iron body										840
	0050	Screwed, 250 lb., 1/4" pipe size	1 Stpi	20	.400	Ea.	6.65	8.85		15.50	22	
	0070	3/8" pipe size		20	.400		6.30	8.85		15.15	21.50	
	0100	1/2" pipe size		20	.400		6.65	8.85		15.50	22	
	0140	1" pipe size		16	.500		10.75	11.05		21.80	30	
	0160	1-1/2" pipe size		12	.667		17.75	14.75		32.50	44	
	0180	2" pipe size		8	1		27.50	22		49.50	66.50	
	0220	3" pipe size	Q-5	11	1.455		151	29		180	214	
	0240	4" pipe size	"	5	3.200		255	63.50		318.50	385	
	0500	For galvanized body, add					50%					
	1000	Flanged, 125 lb., 1-1/2" pipe size	1 Stpi	11	.727	Ea.	86	16.05		102.05	121	
	1020	2" pipe size	"	8	1		64	22		86	107	
	1040	3" pipe size	Q-5	4.50	3.556		84.50	71		155.50	210	
	1060	4" pipe size	"	3	5.333		155	106		261	345	
	1080	5" pipe size	Q-6	3.40	7.059		242	135		377	490	
	1100	6" pipe size	"	3	8		295	153		448	580	
	1500	For 250 lb rating, add					20%					
	2000	For galvanized body, add					50%					
	2500	For steel body, add					40%					
920	0010	**VENTURI FLOW** Measuring device										920
	0050	1/2" diameter	1 Stpi	24	.333	Ea.	105	7.35		112.35	128	
	0120	1" diameter		19	.421		94	9.30		103.30	118	
	0140	1-1/4" diameter		15	.533		134	11.80		145.80	167	
	0160	1-1/2" diameter		13	.615		140	13.60		153.60	177	
	0180	2" diameter		11	.727		166	16.05		182.05	210	
	0220	3" diameter	Q-5	14	1.143		335	22.50		357.50	410	
	0240	4" diameter	"	11	1.455		375	29		404	465	
	0280	6" diameter	Q-6	3.50	6.857		660	131		791	940	
	0500	For meter, add					1,175			1,175	1,300	
940	0010	**WATER SUPPLY METERS**										940
	2000	Domestic/commercial, bronze										
	2020	Threaded										
	2060	5/8" diameter, to 20 GPM	1 Plum	16	.500	Ea.	66.50	11		77.50	91.50	
	2080	3/4" diameter, to 30 GPM		14	.571		115	12.55		127.55	148	
	2100	1" diameter, to 50 GPM		12	.667		157	14.65		171.65	196	
	2300	Threaded/flanged										
	2340	1-1/2" diameter, to 100 GPM	1 Plum	8	1	Ea.	485	22		507	570	
	2360	2" diameter, to 160 GPM	"	6	1.333	"	645	29.50		674.50	760	
	2600	Flanged, compound										
	2640	3" diameter, 320 GPM	Q-1	3	5.333	Ea.	2,900	105		3,005	3,375	
	2660	4" diameter, to 500 GPM		1.50	10.667		4,525	211		4,736	5,325	
	2680	6" diameter, to 1,000 GPM		1	16		6,500	315		6,815	7,700	
	2700	8" diameter, to 1,800 GPM		.80	20		12,800	395		13,195	14,800	

15100 | Building Services Piping

15140 | Domestic Water Piping

		CREW	DAILY OUTPUT	LABOR-HOURS	UNIT	2000 BARE COSTS MAT.	LABOR	EQUIP.	TOTAL	TOTAL INCL O&P
100 0010	**BACKFLOW PREVENTER** Includes valves									
0020	and four test cocks, corrosion resistant, automatic operation									
4000	Reduced pressure principle									
4100	Threaded, valves are ball									
4120	3/4" pipe size	1 Plum	16	.500	Ea.	685	11		696	775
4140	1" pipe size		14	.571		730	12.55		742.55	820
4150	1-1/4" pipe size		12	.667		1,025	14.65		1,039.65	1,150
4160	1-1/2" pipe size		10	.800		990	17.55		1,007.55	1,100
4180	2" pipe size		7	1.143		1,050	25		1,075	1,200
5000	Flanged, bronze, valves are OS&Y									
5060	2-1/2" pipe size	Q-1	5	3.200	Ea.	2,725	63		2,788	3,100
5080	3" pipe size		4.50	3.556		3,375	70		3,445	3,850
5100	4" pipe size		3	5.333		4,800	105		4,905	5,450
5120	6" pipe size	Q-2	3	8		10,200	152		10,352	11,500
5600	Flanged, iron, valves are OS&Y									
5660	2-1/2" pipe size	Q-1	5	3.200	Ea.	1,825	63		1,888	2,100
5680	3" pipe size		4.50	3.556		1,950	70		2,020	2,250
5700	4" pipe size		3	5.333		2,575	105		2,680	3,000
5720	6" pipe size	Q-2	3	8		4,050	152		4,202	4,700
5740	8" pipe size		2	12		8,175	228		8,403	9,375
5760	10" pipe size		1	24		11,200	455		11,655	13,100
600 0010	**VACUUM BREAKERS** Hot or cold water									
1030	Anti-siphon, brass									
1040	1/4" size	1 Plum	24	.333	Ea.	18.10	7.30		25.40	32
1050	3/8" size		24	.333		18.10	7.30		25.40	32
1060	1/2" size		24	.333		20.50	7.30		27.80	34.50
1080	3/4" size		20	.400		24.50	8.80		33.30	41.50
1100	1" size		19	.421		38	9.25		47.25	57.50
1120	1-1/4" size		15	.533		66.50	11.70		78.20	92.50
1140	1-1/2" size		13	.615		78	13.50		91.50	109
1160	2" size		11	.727		122	15.95		137.95	161
1300	For polished chrome, (1/4" thru 1"), add					50%				
1900	Vacuum relief, water service, bronze									
2000	1/2" size	1 Plum	30	.267	Ea.	17.20	5.85		23.05	28.50
800 0010	**WATER HAMMER ARRESTORS / SHOCK ABSORBERS**									
0490	Copper									
0500	3/4" male I.P.S. For 1 to 11 fixtures	1 Plum	12	.667	Ea.	16.05	14.65		30.70	41.50
0600	1" male I.P.S., For 12 to 32 fixtures		8	1		41	22		63	81.50
0700	1-1/4" male I.P.S. For 33 to 60 fixtures		8	1		47.50	22		69.50	89
0800	1-1/2" male I.P.S. For 61 to 113 fixtures		8	1		65	22		87	108
0900	2" male I.P.S. For 114 to 154 fixtures		8	1		104	22		126	151
1000	2-1/2" male I.P.S. For 155 to 330 fixtures		4	2		231	44		275	325

15155 | Drainage Specialties

		CREW	DAILY OUTPUT	LABOR-HOURS	UNIT	MAT.	LABOR	EQUIP.	TOTAL	TOTAL INCL O&P
160 0010	**CLEANOUTS**									
0060	Floor type									
0080	Round or square, scoriated nickel bronze top									
0100	2" pipe size	1 Plum	10	.800	Ea.	84	17.55		101.55	122
0120	3" pipe size		8	1		113	22		135	162
0140	4" pipe size		6	1.333		113	29.50		142.50	174
0980	Round top, recessed for terrazzo									
1000	2" pipe size	1 Plum	9	.889	Ea.	84	19.50		103.50	125
1080	3" pipe size		6	1.333		113	29.50		142.50	174
1100	4" pipe size		4	2		113	44		157	198

15100 | Building Services Piping

15155 | Drainage Specialties

		CREW	DAILY OUTPUT	LABOR-HOURS	UNIT	MAT.	LABOR	EQUIP.	TOTAL	TOTAL INCL O&P	
160	1120	5" pipe size	Q-1	6	2.667	Ea.	137	52.50		189.50	238	160
170	0010	**CLEANOUT TEE**										170
	0100	Cast iron, B&S, with countersunk plug										
	0200	2" pipe size	1 Plum	4	2	Ea.	79.50	44		123.50	160
	0220	3" pipe size		3.60	2.222		87	49		136	176
	0240	4" pipe size		3.30	2.424		108	53		161	207
	0280	6" pipe size	Q-1	5	3.200		291	63		354	425
	0500	For round smooth access cover, same price									
	4000	Plastic, tees and adapters. Add plugs									
	4010	ABS, DWV									
	4020	Cleanout tee, 1-1/2" pipe size	1 Plum	15	.533	Ea.	5.45	11.70		17.15	25.50
	4030	2" pipe size	Q-1	27	.593		7.20	11.70		18.90	27.50
	4040	3" pipe size		21	.762		14.45	15.05		29.50	41
	4050	4" pipe size		16	1		27	19.75		46.75	62
	4100	Cleanout plug, 1-1/2" pipe size	1 Plum	32	.250		1	5.50		6.50	10.15
	4110	2" pipe size	Q-1	56	.286		1.32	5.65		6.97	10.80
	4120	3" pipe size		36	.444		2.21	8.80		11.01	16.95
	4130	4" pipe size		30	.533		4.62	10.55		15.17	22.50
	4180	Cleanout adapter fitting, 1-1/2" pipe size	1 Plum	32	.250		1.77	5.50		7.27	11
	4190	2" pipe size	Q-1	56	.286		2.48	5.65		8.13	12.10
	4200	3" pipe size		36	.444		6.95	8.80		15.75	22
	4210	4" pipe size		30	.533		11.50	10.55		22.05	30
	5000	PVC, DWV									
	5010	Cleanout tee, 1-1/2" pipe size	1 Plum	15	.533	Ea.	1.67	11.70		13.37	21
	5020	2" pipe size	Q-1	27	.593		2.30	11.70		14	22
	5030	3" pipe size		21	.762		6.30	15.05		21.35	32
	5040	4" pipe size		16	1		12.05	19.75		31.80	46
	5090	Cleanout plug, 1-1/2" pipe size	1 Plum	32	.250		.98	5.50		6.48	10.15
	5100	2" pipe size	Q-1	56	.286		1.22	5.65		6.87	10.70
	5110	3" pipe size		36	.444		2.57	8.80		11.37	17.35
	5120	4" pipe size		30	.533		3.79	10.55		14.34	21.50
	5130	6" pipe size		24	.667		11.55	13.15		24.70	35
	5170	Cleanout adapter fitting, 1-1/2" pipe size	1 Plum	32	.250		.65	5.50		6.15	9.75
	5180	2" pipe size	Q-1	56	.286		.92	5.65		6.57	10.35
	5190	3" pipe size		36	.444		2.59	8.80		11.39	17.35
	5200	4" pipe size		30	.533		4.25	10.55		14.80	22
	5210	6" pipe size		24	.667		9.55	13.15		22.70	32.50
300	0010	**DRAINS**										300
	0140	Cornice, C.I., 45° or 90° outlet									
	0180	1-1/2" & 2" pipe size	Q-1	14	1.143	Ea.	72	22.50		94.50	117
	0200	3" and 4" pipe size	"	12	1.333		98	26.50		124.50	152
	0260	For galvanized body, add					19.60			19.60	21.50
	0280	For polished bronze dome, add					16.65			16.65	18.30
	0400	Deck, auto park, C.I., 13" top									
	0440	3", 4", 5", and 6" pipe size	Q-1	8	2	Ea.	425	39.50		464.50	535
	0480	For galvanized body, add				"	202			202	223
	2000	Floor, medium duty, C.I., deep flange, 7" dia top									
	2040	2" and 3" pipe size	Q-1	12	1.333	Ea.	60	26.50		86.50	110
	2080	For galvanized body, add					25			25	27.50
	2120	For polished bronze top, add					33			33	36.50
	2400	Heavy duty, with sediment bucket, C.I., 12" dia. loose grate									
	2420	2", 3", 4", 5", and 6" pipe size	Q-1	9	1.778	Ea.	205	35		240	284
	2460	For polished bronze top, add				"	85			85	93.50
	2500	Heavy duty, cleanout & trap w/bucket, C.I., 15" top									
	2540	2", 3", and 4" pipe size	Q-1	6	2.667	Ea.	1,475	52.50		1,527.50	1,700

15100 | Building Services Piping

15155 | Drainage Specialties

			CREW	DAILY OUTPUT	LABOR-HOURS	UNIT	MAT.	LABOR	EQUIP.	TOTAL	TOTAL INCL O&P	
300	2560	For galvanized body, add				Ea.	440			440	485	300
	2580	For polished bronze top, add					465			465	510	
	2780	Shower, with strainer, uniform diam. trap, bronze top										
	2800	1-1/2", 2" and 3" pipe size	Q-1	8	2	Ea.	126	39.50		165.50	205	
	2820	4" pipe size	"	7	2.286		139	45		184	228	
	2840	For galvanized body, add					49.50			49.50	54.50	
	3860	Roof, flat metal deck, C.I. body, 12" C.I. dome										
	3890	3" pipe size	Q-1	14	1.143	Ea.	117	22.50		139.50	167	
	3920	6" pipe size	"	10	1.600	"	85.50	31.50		117	146	
	4620	Main, all aluminum, 12" low profile dome										
	4640	2", 3" and 4" pipe size	Q-1	14	1.143	Ea.	130	22.50		152.50	181	
	4980	Scupper floor, oblique strainer, C.I.										
	5000	6" x 7" top, 2", 3" and 4" pipe size	Q-1	16	1	Ea.	84	19.75		103.75	125	
	5100	8" x 12" top, 5" and 6" pipe size	"	14	1.143		165	22.50		187.50	220	
	5160	For galvanized body, add					40%					
	5200	For polished bronze strainer, add					85%					
	5980	Trench, floor, hvy duty, modular, C.I., 12" x 12" top										
	6000	2", 3", 4", 5", & 6" pipe size	Q-1	8	2	Ea.	266	39.50		305.50	360	
	6100	For polished bronze top, add				"	130			130	143	
	6960	Backwater valve, soil pipe, C.I. body										
	6980	Bronze gate and automatic flapper valves										
	7000	3" and 4" pipe size	Q-1	13	1.231	Ea.	530	24.50		554.50	620	
	7100	5" and 6" pipe size	"	13	1.231	"	850	24.50		874.50	975	
	7240	Bronze flapper valve, bolted cover										
	7260	2" pipe size	Q-1	16	1	Ea.	177	19.75		196.75	228	
	7300	4" pipe size	"	13	1.231		340	24.50		364.50	415	
	7340	6" pipe size	Q-2	17	1.412		490	27		517	585	
400	0010	**INTERCEPTORS**										400
	0150	Grease, cast iron, 4 GPM, 8 lb. fat capacity	1 Plum	4	2	Ea.	355	44		399	465	
	0200	7 GPM, 14 lb. fat capacity		4	2		490	44		534	615	
	1000	10 GPM, 20 lb. fat capacity		4	2		575	44		619	710	
	1040	15 GPM, 30 lb. fat capacity		4	2		855	44		899	1,025	
	1060	20 GPM, 40 lb. fat capacity		3	2.667		1,050	58.50		1,108.50	1,250	
	1160	100 GPM, 200 lb. fat capacity	Q-1	2	8		4,125	158		4,283	4,775	
	1560	For chemical add-port, add					77.50			77.50	85	
	1580	For seepage pan, add					7%					
	3000	Hair, cast iron, 1-1/4" and 1-1/2" pipe connection	1 Plum	8	1	Ea.	132	22		154	183	
	3100	For chrome-plated cast iron, add					81.50			81.50	89.50	
	4000	Oil, fabricated steel, 10 GPM, 2" pipe size	1 Plum	4	2		745	44		789	895	
	4100	15 GPM, 2" or 3" pipe size		4	2		1,025	44		1,069	1,200	
	4120	20 GPM, 2" or 3" pipe size		3	2.667		1,225	58.50		1,283.50	1,450	
	4220	100 GPM, 3" pipe size	Q-1	2	8		4,125	158		4,283	4,775	
	6000	Solids, precious metals recovery, C.I., 1-1/4" to 2" pipe	1 Plum	4	2		200	44		244	292	
	6100	Dental Lab., large, C.I., 1-1/2" to 2" pipe	"	3	2.667		700	58.50		758.50	865	
780	0010	**TRAPS**										780
	0030	Cast iron, service weight										
	0050	Running P trap, without vent										
	1100	2"	Q-1	16	1	Ea.	15.95	19.75		35.70	50	
	1140	3"		14	1.143		27	22.50		49.50	67.50	
	1150	4"		13	1.231		50	24.50		74.50	95.50	
	1160	6"	Q-2	17	1.412		232	27		259	300	
	1180	Running trap, single hub, with vent										
	2080	3" pipe size, 3" vent	Q-1	14	1.143	Ea.	44	22.50		66.50	86	
	2120	4" pipe size, 4" vent	"	13	1.231		60	24.50		84.50	106	

15100 | Building Services Piping

15155 | Drainage Specialties

			Crew	Daily Output	Labor-Hours	Unit	Mat.	Labor	Equip.	Total	Total Incl O&P	
780	2300	For double hub, vent, add				Ea.	10%	20%				780
	3000	P trap, B&S, 2" pipe size	Q-1	16	1		13.20	19.75		32.95	47	
	3040	3" pipe size	"	14	1.143		19.75	22.50		42.25	59.50	
	3350	Deep seal trap, B&S										
	3400	1-1/4" pipe size	Q-1	14	1.143	Ea.	17.10	22.50		39.60	56.50	
	3410	1-1/2" pipe size		14	1.143		17.10	22.50		39.60	56.50	
	3420	2" pipe size		14	1.143		17.05	22.50		39.55	56.50	
	3440	3" pipe size		12	1.333		27.50	26.50		54	74	
	3800	Drum trap, 4" x 5", 1-1/2" tapping	Q-2	17	1.412		13.50	27		40.50	59.50	
	3820	2" tapping	"	17	1.412		13.50	27		40.50	59.50	
	3840	For galvanized, add					60%					
	4700	Copper, drainage, drum trap										
	4800	3" x 5" solid, 1-1/2" pipe size	1 Plum	16	.500	Ea.	15.05	11		26.05	34.50	
	4840	3" x 6" swivel, 1-1/2" pipe size	"	16	.500	"	22	11		33	42.50	
	5100	P trap, standard pattern										
	5200	1-1/4" pipe size	1 Plum	18	.444	Ea.	10.75	9.75		20.50	28	
	5240	1-1/2" pipe size		17	.471		10.75	10.35		21.10	29	
	5260	2" pipe size		15	.533		16.60	11.70		28.30	37.50	
	5280	3" pipe size		11	.727		40	15.95		55.95	70.50	
	5340	With cleanout and slip joint										
	5360	1-1/4" pipe size	1 Plum	18	.444	Ea.	15.90	9.75		25.65	33.50	
	5400	1-1/2" pipe size	"	17	.471	"	15.55	10.35		25.90	34	
940	0010	**VENT FLASHING**										940
	1000	Aluminum with lead ring										
	1020	1-1/4" pipe	1 Plum	20	.400	Ea.	6.40	8.80		15.20	21.50	
	1030	1-1/2" pipe		20	.400		6.40	8.80		15.20	21.50	
	1040	2" pipe		18	.444		6.85	9.75		16.60	23.50	
	1050	3" pipe		17	.471		7.60	10.35		17.95	25.50	
	1060	4" pipe		16	.500		9.15	11		20.15	28.50	
	1350	Copper with neoprene ring										
	1400	1-1/4" pipe	1 Plum	20	.400	Ea.	14.05	8.80		22.85	30	
	1430	1-1/2" pipe		20	.400		14.05	8.80		22.85	30	
	1440	2" pipe		18	.444		14.85	9.75		24.60	32.50	
	1450	3" pipe		17	.471		17.40	10.35		27.75	36.50	
	1460	4" pipe		16	.500		19.20	11		30.20	39	

15180 | Heating and Cooling Piping

			Crew	Daily Output	Labor-Hours	Unit	Mat.	Labor	Equip.	Total	Total Incl O&P	
200	0010	**PUMPS, CIRCULATING** Heated or chilled water application										200
	0600	Bronze, sweat connections, 1/40 HP, in line										
	0640	3/4" size	Q-1	16	1	Ea.	110	19.75		129.75	154	
	1000	Flange connection, 3/4" to 1-1/2" size										
	1040	1/12 HP	Q-1	6	2.667	Ea.	300	52.50		352.50	415	
	1060	1/8 HP		6	2.667		520	52.50		572.50	655	
	1100	1/3 HP		6	2.667		570	52.50		622.50	710	
	1140	2" size, 1/6 HP		5	3.200		740	63		803	920	
	1180	2-1/2" size, 1/4 HP		5	3.200		965	63		1,028	1,175	
	2000	Cast iron, flange connection										
	2040	3/4" to 1-1/2" size, in line, 1/12 HP	Q-1	6	2.667	Ea.	197	52.50		249.50	305	
	2100	1/3 HP		6	2.667		365	52.50		417.50	490	
	2140	2" size, 1/6 HP		5	3.200		400	63		463	545	
	2180	2-1/2" size, 1/4 HP		5	3.200		525	63		588	685	
	2220	3" size, 1/4 HP		4	4		535	79		614	715	
	2600	For non-ferrous impeller, add					3%					
300	0010	**PUMPS, CONDENSATE RETURN SYSTEM**										300
	0200	Simplex, 3/4 H.P. mtr, float switch, controls, 10 Gal. C.I. rcvr, 6-15GPM	Q-1	1	16	Ea.	1,250	315		1,565	1,900	

Important: See the Reference Section for critical supporting data - Reference Nos., Crews, & City Cost Indexes

15100 | Building Services Piping

15180 | Heating and Cooling Piping

			CREW	DAILY OUTPUT	LABOR-HOURS	UNIT	MAT.	LABOR	EQUIP.	TOTAL	TOTAL INCL O&P	
300	1000	Duplex, 2 pumps, 3/4 H.P. motors, float switch,										309
	1060	alternator asssembly, 15 Gal. C.I. receiver	Q-1	.50	32	Ea.	1,750	630		2,380	2,975	
800	0010	**STEAM CONDENSATE METER**										800
	0100	500 lb. per hour	1 Stpi	14	.571	Ea.	2,300	12.65		2,312.65	2,550	
	0140	1500 lb. per hour	"	7	1.143	"	2,475	25.50		2,500.50	2,775	

15200 | Process Piping

15230 | Industrial Process Pipe

			CREW	DAILY OUTPUT	LABOR-HOURS	UNIT	MAT.	LABOR	EQUIP.	TOTAL	TOTAL INCL O&P	
500	0010	**PUMPS, GENERAL UTILITY** With motor										500
	0200	Multi-stage, horizontal split, for boiler feed applications										
	0300	Two stage, 3" discharge x 4" suction, 75 HP	Q-7	.30	106	Ea.	14,900	2,125		17,025	19,900	
	0340	Four stage, 3" discharge x 4" suction, 150 HP	"	.18	177	"	24,500	3,550		28,050	32,900	
	2000	Single stage										
	2060	End suction, 1"D. x 2"S., 3 HP	Q-1	.50	32	Ea.	3,650	630		4,280	5,050	
	2100	1-1/2"D. x 3"S., 10 HP	"	.40	40		3,925	790		4,715	5,625	
	2140	2"D. x 3"S., 15 HP	Q-2	.60	40		4,275	760		5,035	5,950	
	3000	Double suction, 2"D. x 2-1/2"S., 10 HP	Q-1	.30	53.333		5,050	1,050		6,100	7,300	
	3060	3"D. x 4"S., 15 HP	Q-2	.46	52.174		5,200	990		6,190	7,375	
	3180	6"D. x 8"S., 60 HP	Q-3	.30	106		8,475	2,100		10,575	12,800	
	3190	75 HP, to 2500 GPM		.28	114		10,700	2,250		12,950	15,500	
	3220	100 HP, to 3000 GPM		.26	123		14,400	2,425		16,825	19,900	
	3240	150 HP, to 4000 GPM		.24	133		17,700	2,625		20,325	23,900	

15400 | Plumbing Fixtures & Equipment

15410 | Plumbing Fixtures

			CREW	DAILY OUTPUT	LABOR-HOURS	UNIT	MAT.	LABOR	EQUIP.	TOTAL	TOTAL INCL O&P	
040	0010	**FIXTURES** Includes trim fittings unless otherwise noted	R15100-420									040
	0080	For rough-in, supply, waste, and vent, see add for each type										
	0120	For electric water coolers, see division 15413										
	0160	For color, unless otherwise noted, add				Ea.	20%					
200	0010	**CARRIERS/SUPPORTS** For plumbing fixtures										200
	0500	Drinking fountain, wall mounted										
	0600	Plate type with studs, top back plate	1 Plum	7	1.143	Ea.	26.50	25		51.50	70.50	
	0700	Top front and back plate		7	1.143		33	25		58	77.50	
	0800	Top & bottom, front & back plates, w/bearing jacks		7	1.143		47.50	25		72.50	94	
	3000	Lavatory, concealed arm										
	3050	Floor mounted, single										
	3100	High back fixture	1 Plum	6	1.333	Ea.	128	29.50		157.50	190	
	3200	Flat slab fixture		6	1.333		149	29.50		178.50	213	
	3220	Paraplegic		6	1.333		136	29.50		165.50	199	
	3250	Floor mounted, back to back										
	3300	High back fixtures	1 Plum	5	1.600	Ea.	182	35		217	258	

15400 | Plumbing Fixtures & Equipment

15410 | Plumbing Fixtures

		CREW	DAILY OUTPUT	LABOR-HOURS	UNIT	2000 BARE COSTS				TOTAL INCL O&P
						MAT.	LABOR	EQUIP.	TOTAL	
3400	Flat slab fixtures	1 Plum	5	1.600	Ea.	185	35		220	261
3430	Paraplegic	↓	5	1.600	↓	191	35		226	268
3500	Wall mounted, in stud or masonry									
3600	High back fixture	1 Plum	6	1.333	Ea.	75.50	29.50		105	132
3700	Flat slab fixture	"	6	1.333	"	96.50	29.50		126	155
4600	Sink, floor mounted									
4650	Exposed arm system									
4700	Single heavy fixture	1 Plum	5	1.600	Ea.	298	35		333	385
4750	Single heavy sink with slab		5	1.600		201	35		236	279
4800	Back to back, standard fixtures		5	1.600		230	35		265	310
4850	Back to back, heavy fixtures		5	1.600		268	35		303	355
4900	Back to back, heavy sink with slab	↓	5	1.600	↓	268	35		303	355
4950	Exposed offset arm system									
5000	Single heavy deep fixture	1 Plum	5	1.600	Ea.	203	35		238	282
5100	Plate type system									
5200	With bearing jacks, single fixture	1 Plum	5	1.600	Ea.	271	35		306	355
5300	With exposed arms, single heavy fixture		5	1.600		350	35		385	440
5400	Wall mounted, exposed arms, single heavy fixture		5	1.600		126	35		161	197
6000	Urinal, floor mounted, 2" or 3" coupling, blowout type		6	1.333		136	29.50		165.50	198
6100	With fixture or hanger bolts, blowout or washout		6	1.333		96.50	29.50		126	155
6200	With bearing plate		6	1.333		108	29.50		137.50	168
6300	Wall mounted, plate type system	↓	6	1.333	↓	97	29.50		126.50	155
6980	Water closet, siphon jet									
7000	Horizontal, adjustable, caulk									
7040	Single, 4" pipe size	1 Plum	6	1.333	Ea.	205	29.50		234.50	275
7050	4" pipe size, paraplegic		6	1.333		205	29.50		234.50	275
7060	5" pipe size		6	1.333		271	29.50		300.50	345
7100	Double, 4" pipe size		5	1.600		385	35		420	480
7110	4" pipe size, paraplegic		5	1.600		205	35		240	284
7120	5" pipe size	↓	5	1.600	↓	271	35		306	355
7160	Horizontal, adjustable, extended, caulk									
7180	Single, 4" pipe size	1 Plum	6	1.333	Ea.	205	29.50		234.50	275
7200	5" pipe size		6	1.333		365	29.50		394.50	455
7240	Double, 4" pipe size		5	1.600		277	35		312	365
7260	5" pipe size	↓	5	1.600	↓	365	35		400	465
7400	Vertical, adjustable, caulk or thread									
7440	Single, 4" pipe size	1 Plum	6	1.333	Ea.	242	29.50		271.50	315
7460	5" pipe size		6	1.333		305	29.50		334.50	385
7480	6" pipe size		5	1.600		360	35		395	455
7520	Double, 4" pipe size		5	1.600		420	35		455	525
7540	5" pipe size		5	1.600		485	35		520	595
7560	6" pipe size	↓	4	2	↓	535	44		579	665
7600	Vertical, adjustable, extended, caulk									
7620	Single, 4" pipe size	1 Plum	6	1.333	Ea.	242	29.50		271.50	315
7640	5" pipe size		6	1.333		305	29.50		334.50	385
7680	6" pipe size		5	1.600		360	35		395	455
7720	Double, 4" pipe size		5	1.600		420	35		455	525
7740	5" pipe size		5	1.600		485	35		520	595
7760	6" pipe size	↓	4	2	↓	535	44		579	665
7780	Water closet, blow out									
7800	Vertical offset, caulk or thread									
7820	Single, 4" pipe size	1 Plum	6	1.333	Ea.	184	29.50		213.50	251
7840	Double, 4" pipe size	"	5	1.600	"	315	35		350	405
7880	Vertical offset, extended, caulk									
7900	Single, 4" pipe size	1 Plum	6	1.333	Ea.	231	29.50		260.50	305
7920	Double, 4" pipe size	"	5	1.600	"	360	35		395	460

15400 | Plumbing Fixtures & Equipment

15410 | Plumbing Fixtures

			CREW	DAILY OUTPUT	LABOR-HOURS	UNIT	MAT.	LABOR	EQUIP.	TOTAL	TOTAL INCL O&P	
200	7960	Vertical, for floor mounted back-outlet										200
	7980	Single, 4" thread, 2" vent	1 Plum	6	1.333	Ea.	187	29.50		216.50	255	
	8000	Double, 4" thread, 2" vent	"	6	1.333	"	335	29.50		364.50	420	
	8040	Vertical, for floor mounted back-outlet, extended										
	8060	Single, 4" caulk, 2" vent	1 Plum	6	1.333	Ea.	179	29.50		208.50	246	
	8080	Double, 4" caulk, 2" vent	"	6	1.333	"	278	29.50		307.50	355	
	8200	Water closet, residential										
	8220	Vertical centerline, floor mount										
	8240	Single, 3" caulk, 2" or 3" vent	1 Plum	6	1.333	Ea.	147	29.50		176.50	211	
	8260	4" caulk, 2" or 4" vent		6	1.333		185	29.50		214.50	253	
	8280	3" copper sweat, 3" vent		6	1.333		147	29.50		176.50	211	
	8300	4" copper sweat, 4" vent		6	1.333		185	29.50		214.50	252	
	8400	Vertical offset, floor mount										
	8420	Single, 3" or 4" caulk, vent	1 Plum	4	2	Ea.	184	44		228	275	
	8440	3" or 4" copper sweat, vent		5	1.600		184	35		219	260	
	8460	Double, 3" or 4" caulk, vent		4	2		315	44		359	420	
	8480	3" or 4" copper sweat, vent		5	1.600		315	35		350	405	
	9000	Water cooler (electric), floor mounted										
	9100	Plate type with bearing plate, single	1 Plum	6	1.333	Ea.	108	29.50		137.50	168	
300	0010	**FAUCETS/FITTINGS**										300
	0150	Bath, faucets, diverter spout combination, sweat	1 Plum	8	1	Ea.	65	22		87	108	
	0200	For integral stops, IPS unions, add				"	68.50			68.50	75.50	
	0810	Bidet										
	0812	Fitting, over the rim, swivel spray/pop-up drain	1 Plum	8	1	Ea.	119	22		141	168	
	0840	Flush valves, with vacuum breaker										
	0850	Water closet										
	0860	Exposed, rear spud	1 Plum	8	1	Ea.	105	22		127	152	
	0870	Top spud		8	1		96	22		118	142	
	0880	Concealed, rear spud		8	1		138	22		160	188	
	0890	Top spud		8	1		138	22		160	189	
	0900	Wall hung		8	1		121	22		143	170	
	0920	Urinal										
	0930	Exposed, stall	1 Plum	8	1	Ea.	96	22		118	142	
	0940	Wall, (washout)		8	1		96	22		118	143	
	0950	Pedestal, top spud		8	1		97.50	22		119.50	144	
	0960	Concealed, stall		8	1		110	22		132	158	
	0970	Wall (washout)		8	1		115	22		137	164	
	0971	Automatic flush sensor and operator for										
	0972	urinals or water closets	1 Plum	16	.500	Ea.	315	11		326	365	
	1000	Kitchen sink faucets, top mount, cast spout		10	.800		48	17.55		65.55	82	
	1100	For spray, add		24	.333		12	7.30		19.30	25.50	
	2000	Laundry faucets, shelf type, IPS or copper unions		12	.667		39.50	14.65		54.15	67.50	
	2100	Lavatory faucet, centerset, without drain		10	.800		51	17.55		68.55	85	
	2810	Automatic sensor and operator, with faucet head		6.15	1.301		279	28.50		307.50	350	
	3000	Service sink faucet, cast spout, pail hook, hose end		14	.571		67.50	12.55		80.05	94.50	
	4000	Shower by-pass valve with union		18	.444		47.50	9.75		57.25	68	
	4200	Shower thermostatic mixing valve, concealed		8	1		226	22		248	286	
	5000	Sillcock, compact, brass, IPS or copper to hose		24	.333		4.94	7.30		12.24	17.55	

15411 | Commercial/Indust Fixtures

			CREW	DAILY OUTPUT	LABOR-HOURS	UNIT	MAT.	LABOR	EQUIP.	TOTAL	TOTAL INCL O&P	
400	0010	**HOT WATER DISPENSERS**										400
	0160	Commercial, 100 cup, 11.3 amp	1 Plum	14	.571	Ea.	320	12.55		332.55	375	
	3180	Household, 60 cup	"	14	.571	"	172	12.55		184.55	210	

15400 | Plumbing Fixtures & Equipment
15411 | Commercial/Indust Fixtures

		CREW	DAILY OUTPUT	LABOR-HOURS	UNIT	2000 BARE COSTS MAT.	LABOR	EQUIP.	TOTAL	TOTAL INCL O&P
500 0010	**HYDRANTS**									**500**
0050	Wall type, moderate climate, bronze, encased									
0200	3/4" IPS connection	1 Plum	16	.500	Ea.	226	11		237	267
0300	1" IPS connection	"	14	.571		226	12.55		238.55	270
0500	Anti-siphon type					30.50			30.50	33.50
1000	Non-freeze, bronze, exposed									
1100	3/4" IPS connection, 4" to 9" thick wall	1 Plum	14	.571	Ea.	153	12.55		165.55	190
1120	10" to 14" thick wall		12	.667		166	14.65		180.65	207
1140	15" to 19" thick wall		12	.667		185	14.65		199.65	228
1160	20" to 24" thick wall		10	.800		201	17.55		218.55	250
1200	For 1" IPS connection, add					15%	10%			
1240	For 3/4" adapter type vacuum breaker, add				Ea.	19.25			19.25	21
1280	For anti-siphon type				"	40.50			40.50	44.50
2000	Non-freeze bronze, encased									
2100	3/4" IPS connection, 5" to 9" thick wall	1 Plum	14	.571	Ea.	264	12.55		276.55	310
2140	15" to 19" thick wall	"	12	.667		297	14.65		311.65	350
2280	Anti-siphon type					124			124	136
3000	Ground box type, bronze frame, 3/4" IPS connection									
3080	Non-freeze, all bronze, polished face, set flush									
3100	2 feet depth of bury	1 Plum	8	1	Ea.	282	22		304	345
3180	6 feet depth of bury		7	1.143		370	25		395	445
3220	8 feet depth of bury		5	1.600		415	35		450	515
3400	For 1" IPS connection, add					15%	10%			
3550	For 2" connection, add					445%	24%			
3600	For tapped drain port in box, add					25.50			25.50	28.50
5000	Moderate climate, all bronze, polished face									
5020	and scoriated cover, set flush									
5100	3/4" IPS connection	1 Plum	16	.500	Ea.	201	11		212	239
5120	1" IPS connection	"	14	.571		200	12.55		212.55	241
5200	For tapped drain port in box, add					25.50			25.50	28.50
700 0010	**URINALS**									**700**
3000	Wall hung, vitreous china, with hanger & self-closing valve									
3100	Siphon jet type	Q-1	3	5.333	Ea.	278	105		383	480
3120	Blowout type		3	5.333		315	105		420	525
3300	Rough-in, supply, waste & vent		2.83	5.654		83	112		195	277
5000	Stall type, vitreous china, includes valve		2.50	6.400		450	126		576	705
5100	3" seam cover, add		12	1.333		118	26.50		144.50	174
5200	6" seam cover, add		12	1.333		164	26.50		190.50	224
6980	Rough-in, supply, waste and vent		1.99	8.040		99.50	159		258.50	375
840 0010	**WASH FOUNTAINS** Rigging not included									**840**
1900	Group, foot control									
2000	Precast terrazzo, circular, 36" diam., 5 or 6 persons	Q-2	3	8	Ea.	2,225	152		2,377	2,700
2100	54" diameter for 8 or 10 persons		2.50	9.600		2,725	183		2,908	3,300
2400	Semi-circular, 36" diam. for 3 persons		3	8		2,050	152		2,202	2,525
2500	54" diam. for 4 or 5 persons		2.50	9.600		2,450	183		2,633	3,000
2700	Quarter circle (corner), 54" for 3 persons		3.50	6.857		2,400	130		2,530	2,850
3000	Stainless steel, circular, 36" diameter		3.50	6.857		2,575	130		2,705	3,050
3100	54" diameter		2.80	8.571		2,475	163		2,638	2,975
3400	Semi-circular, 36" diameter		3.50	6.857		1,775	130		1,905	2,175
3500	54" diameter		2.80	8.571		2,225	163		2,388	2,700
5610	Group, infrared control, barrier free									
5614	Precast terrazzo									
5620	Semi-circular 36" diam. for 3 persons	Q-2	3	8	Ea.	2,850	152		3,002	3,400
5630	46" diam. for 4 persons		2.80	8.571		3,075	163		3,238	3,675
5640	Circular, 54" diam. for 8 persons, button control		2.50	9.600		4,750	183		4,933	5,525

15400 | Plumbing Fixtures & Equipment

15411 | Commercial/Indust Fixtures

		CREW	DAILY OUTPUT	LABOR-HOURS	UNIT	MAT.	LABOR	EQUIP.	TOTAL	TOTAL INCL O&P	
840	5700 Rough-in, supply, waste and vent for above wash fountains	Q-1	1.82	8.791	Ea.	75.50	174		249.50	370	840
	6200 Duo for small washrooms, stainless steel		2	8		1,575	158		1,733	1,975	
	6500 Rough-in, supply, waste & vent for duo fountains		2.02	7.921		42	156		198	305	

15412 | Drinking Fountains

		CREW	DAILY OUTPUT	LABOR-HOURS	UNIT	MAT.	LABOR	EQUIP.	TOTAL	TOTAL INCL O&P	
200	0010 **DRINKING FOUNTAIN** For connection to cold water supply R15100-410										200
	1000 Wall mounted, non-recessed										
	1400 Bronze, with no back	1 Plum	4	2	Ea.	965	44		1,009	1,125	
	1800 Cast aluminum, enameled, for correctional institutions		4	2		570	44		614	705	
	2000 Fiberglass, 12" back, single bubbler unit		4	2		510	44		554	635	
	2040 Dual bubbler		3.20	2.500		740	55		795	905	
	2400 Precast stone, no back		4	2		465	44		509	585	
	2700 Stainless steel, single bubbler, no back		4	2		795	44		839	950	
	2740 With back		4	2		865	44		909	1,025	
	2780 Dual handle & wheelchair projection type		4	2		480	44		524	605	
	2820 Dual level for handicapped type		3.20	2.500		965	55		1,020	1,150	
	3300 Vitreous china										
	3340 7" back	1 Plum	4	2	Ea.	395	44		439	505	
	3940 For vandal-resistant bottom plate, add					49			49	54	
	3960 For freeze-proof valve system, add	1 Plum	2	4		385	88		473	570	
	3980 For rough-in, supply and waste, add	"	2.21	3.620		32	79.50		111.50	167	
	4000 Wall mounted, semi-recessed										
	4200 Poly-marble, single bubbler	1 Plum	4	2	Ea.	620	44		664	755	
	4600 Stainless steel, satin finish, single bubbler		4	2		720	44		764	865	
	4900 Vitreous china, single bubbler		4	2		435	44		479	550	
	5980 For rough-in, supply and waste, add		1.83	4.372		32	96		128	195	
	6000 Wall mounted, fully recessed										
	6400 Poly-marble, single bubbler	1 Plum	4	2	Ea.	705	44		749	850	
	6800 Stainless steel, single bubbler		4	2		570	44		614	705	
	7560 For freeze-proof valve system, add		2	4		470	88		558	660	
	7580 For rough-in, supply and waste, add		1.83	4.372		32	96		128	195	
	7600 Floor mounted, pedestal type										
	7700 Aluminum, architectural style, C.I. base	1 Plum	2	4	Ea.	425	88		513	615	
	7780 Wheelchair handicap unit		2	4		1,000	88		1,088	1,275	
	8400 Stainless steel, architectural style		2	4		1,100	88		1,188	1,350	
	8600 Enameled iron, heavy duty service, 2 bubblers		2	4		1,025	88		1,113	1,275	
	8660 4 bubblers		2	4		1,075	88		1,163	1,350	
	8880 For freeze-proof valve system, add		2	4		310	88		398	490	
	8900 For rough-in, supply and waste, add		1.83	4.372		32	96		128	195	
	9100 Deck mounted										
	9500 Stainless steel, circular receptor	1 Plum	4	2	Ea.	283	44		327	385	
	9760 White enameled steel, 14" x 9" receptor		4	2		234	44		278	330	
	9860 White enameled cast iron, 24" x 16" receptor		3	2.667		223	58.50		281.50	340	
	9980 For rough-in, supply and waste, add		1.83	4.372		32	96		128	195	

15413 | Electric Water Coolers

		CREW	DAILY OUTPUT	LABOR-HOURS	UNIT	MAT.	LABOR	EQUIP.	TOTAL	TOTAL INCL O&P	
900	0010 **WATER COOLER** R15100-430										900
	0030 See line 15413-900-9800 for rough-in, waste & vent										
	0040 for all water coolers										
	0100 Wall mounted, non-recessed										
	0140 4 GPH	Q-1	4	4	Ea.	440	79		519	615	
	0160 8 GPH, barrier free, sensor operated		4	4		800	79		879	1,000	
	0180 8.2 GPH		4	4		520	79		599	700	
	0600 For hot and cold water, add					133			133	146	

15400 | Plumbing Fixtures & Equipment

15413 | Electric Water Coolers

			CREW	DAILY OUTPUT	LABOR-HOURS	UNIT	MAT.	LABOR	EQUIP.	TOTAL	TOTAL INCL O&P	
900	0640	For stainless steel cabinet, add				Ea.	68.50			68.50	75.50	900
	1000	Dual height, 8.2 GPH	Q-1	3.80	4.211		715	83		798	920	
	1040	14.3 GPH	"	3.80	4.211		765	83		848	975	
	1240	For stainless steel cabinet, add					110			110	121	
	2600	Wheelchair type, 8 GPH	Q-1	4	4		1,225	79		1,304	1,475	
	3300	Semi-recessed, 8.1 GPH		4	4		715	79		794	915	
	3320	12 GPH		4	4		760	79		839	965	
	4600	Floor mounted, flush-to-wall										
	4640	4 GPH	1 Plum	3	2.667	Ea.	500	58.50		558.50	645	
	4680	8.2 GPH		3	2.667		530	58.50		588.50	675	
	4720	14.3 GPH		3	2.667		550	58.50		608.50	700	
	4960	For hot and cold water, add					160			160	176	
	4980	For stainless steel cabinet, add					93			93	103	
	5000	Dual height, 8.2 GPH	1 Plum	2	4		775	88		863	1,000	
	5040	14.3 GPH	"	2	4		790	88		878	1,025	
	5120	For stainless steel cabinet, add					136			136	150	
	9800	For supply, waste & vent, all coolers	1 Plum	2.21	3.620		32	79.50		111.50	167	

15414 | Emergency Fixtures

			CREW	DAILY OUTPUT	LABOR-HOURS	UNIT	MAT.	LABOR	EQUIP.	TOTAL	TOTAL INCL O&P	
200	0010	**INDUSTRIAL SAFETY FIXTURES** Rough-in not included										200
	1000	Eye wash fountain										
	1400	Plastic bowl, pedestal mounted	Q-1	4	4	Ea.	178	79		257	325	
	1600	Unmounted		4	4		127	79		206	271	
	1800	Wall mounted		4	4		131	79		210	275	
	2000	Stainless steel, pedestal mounted		4	4		266	79		345	425	
	2200	Unmounted		4	4		177	79		256	325	
	2400	Wall mounted		4	4		170	79		249	320	
	3000	Eye wash, portable, self-contained					365			365	400	
	4000	Eye and face wash, combination fountain										
	4200	Stainless steel, pedestal mounted	Q-1	4	4	Ea.	380	79		459	550	
	4400	Unmounted		4	4		287	79		366	445	
	4600	Wall mounted		4	4		294	79		373	455	
	5000	Shower, single head, drench, ball valve, pull, freestanding		4	4		246	79		325	400	
	5200	Horizontal or vertical supply		4	4		221	79		300	375	
	6000	Multi-nozzle, eye/face wash combination		4	4		480	79		559	660	
	6400	Multi-nozzle, 12 spray, shower only		4	4		1,075	79		1,154	1,300	
	6600	For freeze-proof, add		6	2.667		261	52.50		313.50	375	

15418 | Resi/Comm/Industrial Fixtures

			CREW	DAILY OUTPUT	LABOR-HOURS	UNIT	MAT.	LABOR	EQUIP.	TOTAL	TOTAL INCL O&P	
100	0010	**BATHS**										100
	0100	Tubs, recessed porcelain enamel on cast iron, with trim										
	0140	42" x 37"	Q-1	5	3.200	Ea.	670	63		733	845	
	0180	48" x 42"		4	4		1,025	79		1,104	1,250	
	0220	72" x 36"		3	5.333		1,125	105		1,230	1,400	
	2000	Enameled formed steel, 4'-6" long		5.80	2.759		240	54.50		294.50	355	
	2200	5' long		5.50	2.909		224	57.50		281.50	340	
	4000	Soaking, acrylic with pop-up drain, 60" x 32" x 21" deep		5.50	2.909		680	57.50		737.50	840	
	4100	60" x 48" x 18-1/2" deep		5	3.200		550	63		613	710	
	6000	Whirlpool, bath with vented overflow, molded fiberglass										
	6100	66" x 48" x 24"	Q-1	1	16	Ea.	1,975	315		2,290	2,700	
	9600	Rough-in, supply, waste and vent, for all above tubs, add	"	2.07	7.729	"	102	153		255	365	
400	0010	**LAUNDRY SINKS** With trim										400
	0020	Porcelain enamel on cast iron, black iron frame										
	0050	24" x 20", single compartment	Q-1	6	2.667	Ea.	340	52.50		392.50	455	
	0100	24" x 23", single compartment	"	6	2.667	"	365	52.50		417.50	490	

15400 | Plumbing Fixtures & Equipment

15418 | Resi/Comm/Industrial Fixtures

			CREW	DAILY OUTPUT	LABOR-HOURS	UNIT	MAT.	LABOR	EQUIP.	TOTAL	TOTAL INCL O&P	
400	2000	Molded stone, on wall hanger or legs										400
	2020	22" x 23", single compartment	Q-1	6	2.667	Ea.	117	52.50		169.50	216	
	2100	45" x 21", double compartment	"	5	3.200	"	242	63		305	370	
	3000	Plastic, on wall hanger or legs										
	3020	18" x 23", single compartment	Q-1	6.50	2.462	Ea.	82	48.50		130.50	171	
	3300	40" x 24", double compartment		5.50	2.909		187	57.50		244.50	300	
	5000	Stainless steel, counter top, 22" x 17" single compartment		6	2.667		253	52.50		305.50	365	
	5100	22" x 22", single compartment		6	2.667		320	52.50		372.50	435	
	5200	33" x 22", double compartment		5	3.200		295	63		358	430	
	9600	Rough-in, supply, waste and vent, for all laundry sinks		2.14	7.477		60	148		208	310	
450	0010	**LAVATORIES** With trim, white unless noted otherwise	R15100-420									450
	0500	Vanity top, porcelain enamel on cast iron										
	0600	20" x 18"	Q-1	6.40	2.500	Ea.	199	49.50		248.50	300	
	0640	33" x 19" oval		6.40	2.500		345	49.50		394.50	455	
	0720	18" round		6.40	2.500		180	49.50		229.50	280	
	0860	For color, add					25%					
	1000	Cultured marble, 19" x 17", single bowl	Q-1	6.40	2.500	Ea.	117	49.50		166.50	211	
	1040	25" x 19", single bowl	"	6.40	2.500	"	144	49.50		193.50	241	
	1580	For color, same price										
	1900	Stainless steel, self-rimming, 25" x 22", single bowl, ledge	Q-1	6.40	2.500	Ea.	155	49.50		204.50	253	
	1960	17" x 22", single bowl		6.40	2.500		128	49.50		177.50	223	
	2600	Steel, enameled, 20" x 17", single bowl		5.80	2.759		101	54.50		155.50	201	
	2660	19" round		5.80	2.759		99	54.50		153.50	199	
	2900	Vitreous china, 20" x 16", single bowl		5.40	2.963		184	58.50		242.50	299	
	2960	20" x 17", single bowl		5.40	2.963		164	58.50		222.50	277	
	3020	19" round, single bowl		5.40	2.963		156	58.50		214.50	268	
	3200	22" x 13", single bowl		5.40	2.963		193	58.50		251.50	310	
	3560	For color, add					50%					
	3580	Rough-in, supply, waste and vent for all above lavatories	Q-1	2.30	6.957	Ea.	52.50	137		189.50	285	
	4000	Wall hung										
	4040	Porcelain enamel on cast iron, 16" x 14", single bowl	Q-1	8	2	Ea.	345	39.50		384.50	445	
	4180	20" x 18", single bowl		8	2		200	39.50		239.50	286	
	4240	22" x 19", single bowl		8	2		350	39.50		389.50	455	
	4580	For color, add					30%					
	6000	Vitreous china, 18" x 15", single bowl with backsplash	Q-1	7	2.286	Ea.	197	45		242	291	
	6180	19" x 19", corner style	"	8	2	"	380	39.50		419.50	480	
	6500	For color, add					30%					
	6960	Rough-in, supply, waste and vent for above lavatories	Q-1	1.66	9.639	Ea.	160	190		350	490	
500	0010	**SHOWERS**										500
	1500	Stall, with drain only. Add for valve and door/curtain										
	3000	Fiberglass, one piece, with 3 walls, 32" x 32" square	Q-1	2.40	6.667	Ea.	290	132		422	540	
	3100	36" x 36" square		2.40	6.667		330	132		462	585	
	3250	64" x 65-3/4" x 81-1/2" fold. seat, whlchr.		1.80	8.889		2,675	176		2,851	3,250	
	4000	Polypropylene, with molded-stone floor, 30" x 30"		2	8		370	158		528	670	
	4200	Rough-in, supply, waste and vent for above showers		2.05	7.805		53.50	154		207.50	315	
	5000	Built-in, head, arm, 4 GPM valve	1 Plum	4	2		71	44		115	151	
	5200	Head, arm, by-pass, integral stops, handles	"	3.60	2.222		179	49		228	278	
	6000	Group, w/pressure balancing valve, rough-in and rigging not included										
	6800	Column, 6 heads, no receptors, less partitions	Q-1	3	5.333	Ea.	2,200	105		2,305	2,600	
	6900	With stainless steel partitions		1	16		5,325	315		5,640	6,375	
	7600	5 heads, no receptors, less partitions		3	5.333		1,850	105		1,955	2,200	
	7700	With stainless steel partitions		1	16		1,675	315		1,990	2,375	
	8000	Wall, 2 heads, no receptors, less partitions		4	4		920	79		999	1,150	
	8100	With stainless steel partitions		2	8		2,275	158		2,433	2,750	
600	0010	**SINKS** With faucets and drain	R15100-420									600
	2000	Kitchen, counter top style, P.E. on C.I., 24" x 21" single bowl	Q-1	5.60	2.857	Ea.	185	56.50		241.50	298	

15400 | Plumbing Fixtures & Equipment

15418 | Resi/Comm/Industrial Fixtures

			CREW	DAILY OUTPUT	LABOR-HOURS	UNIT	MAT.	LABOR	EQUIP.	TOTAL	TOTAL INCL O&P	
600	2100	31" x 22" single bowl	Q-1	5.60	2.857	Ea.	315	56.50		371.50	440	600
	2200	32" x 21" double bowl		4.80	3.333		276	66		342	415	
	2300	42" x 21" double bowl		4.80	3.333		430	66		496	580	
	3000	Stainless steel, self rimming, 19" x 18" single bowl		5.60	2.857		263	56.50		319.50	385	
	3100	25" x 22" single bowl		5.60	2.857		290	56.50		346.50	415	
	4000	Steel, enameled, with ledge, 24" x 21" single bowl		5.60	2.857		99	56.50		155.50	203	
	4100	32" x 21" double bowl		4.80	3.333		116	66		182	237	
	4960	For color sinks except stainless steel, add					10%					
	4980	For rough-in, supply, waste and vent, counter top sinks	Q-1	2.14	7.477		60	148		208	310	
	5000	Kitchen, raised deck, P.E. on C.I.										
	5100	32" x 21", dual level, double bowl	Q-1	2.60	6.154	Ea.	350	122		472	585	
	5700	For color, add					20%					
	5790	For rough-in, supply, waste & vent, sinks	Q-1	1.85	8.649		60	171		231	350	
	6650	Service, floor, corner, P.E. on C.I., 28" x 28"		4.40	3.636		475	72		547	640	
	6790	For rough-in, supply, waste & vent, floor service sinks		1.64	9.756		114	193		307	445	
	7000	Service, wall, P.E. on C.I., roll rim, 22" x 18"		4	4		405	79		484	575	
	7100	24" x 20"		4	4		445	79		524	620	
	8600	Vitreous china, 22" x 20"		4	4		380	79		459	545	
	8960	For stainless steel rim guard, front or side, add					30			30	33	
	8980	For rough-in, supply, waste & vent, wall service sinks	Q-1	1.30	12.308		233	243		476	655	
900	0010	**WATER CLOSETS**										900
	0030	For automatic flush, see 15410-300-0972										
	0150	Tank type, vitreous china, incl. seat, supply pipe w/stop										
	0200	Wall hung, one piece	Q-1	5.30	3.019	Ea.	445	59.50		504.50	590	
	0400	Two piece, close coupled		5.30	3.019		385	59.50		444.50	525	
	0960	For rough-in, supply, waste, vent and carrier		2.73	5.861		255	116		371	470	
	1000	Floor mounted, one piece		5.30	3.019		435	59.50		494.50	580	
	1100	Two piece, close coupled, water saver		5.30	3.019		126	59.50		185.50	238	
	1960	For color, add					30%					
	1980	For rough-in, supply, waste and vent	Q-1	3.05	5.246	Ea.	125	104		229	310	
	3000	Bowl only, with flush valve, seat										
	3100	Wall hung	Q-1	5.80	2.759	Ea.	335	54.50		389.50	455	
	3200	For rough-in, supply, waste and vent, single WC		2.56	6.250		261	123		384	490	
	3300	Floor mounted		5.80	2.759		299	54.50		353.50	420	
	3350	With wall outlet		5.80	2.759		445	54.50		499.50	580	
	3400	For rough-in, supply, waste and vent, single WC		2.84	5.634		131	111		242	330	

15440 | Plumbing Pumps

			CREW	DAILY OUTPUT	LABOR-HOURS	UNIT	MAT.	LABOR	EQUIP.	TOTAL	TOTAL INCL O&P	
240	0010	**PUMPS, PRESSURE BOOSTER SYSTEM**										240
	0200	Pump system, with diaphragm tank, control, press. switch										
	0300	1 HP pump	Q-1	1.30	12.308	Ea.	2,950	243		3,193	3,650	
	0400	1-1/2 HP pump		1.25	12.800		2,975	253		3,228	3,700	
	0420	2 HP pump		1.20	13.333		3,050	263		3,313	3,775	
	0440	3 HP pump		1.10	14.545		3,100	287		3,387	3,875	
	0460	5 HP pump	Q-2	1.50	16		3,425	305		3,730	4,275	
	0480	7-1/2 HP pump		1.42	16.901		3,825	320		4,145	4,725	
	0500	10 HP pump		1.34	17.910		4,000	340		4,340	4,975	
	1000	Pump/ energy storage system, diaphragm tank, 3 HP pump										
	1100	motor, PRV, switch, gauge, control center, flow switch										
	1200	125 lb. working pressure	Q-2	.70	34.286	Ea.	8,575	650		9,225	10,500	
	1300	250 lb. working pressure	"	.64	37.500	"	9,375	715		10,090	11,500	
400	0010	**PUMPS, GRINDER SYSTEM** Complete, incl. check valve, tank, std.										400
	0020	controls incl. alarm/disconnect panel w/wire. Excavation not included										
	0260	Simplex, 9 GPM at 60 PSIG, 60 gal. tank				Ea.	2,225			2,225	2,450	
	0300	For manway, 26" I.D., 18" high, add					330			330	360	

15400 | Plumbing Fixtures & Equipment

15440 | Plumbing Pumps

			CREW	DAILY OUTPUT	LABOR-HOURS	UNIT	MAT.	LABOR	EQUIP.	TOTAL	TOTAL INCL O&P	
400	0340	26" I.D., 36" high, add				Ea.	370			370	405	400
	0380	43" I.D., 4' high, add					420			420	460	
800	0010	**PUMPS, SEWAGE EJECTOR** With operating and level controls										800
	0100	Simplex system incl. tank, cover, pump 15' head										
	0500	37 gal PE tank, 12 GPM, 1/2 HP, 2" discharge	Q-1	3.20	5	Ea.	365	99		464	565	
	0510	3" discharge		3.10	5.161		390	102		492	600	
	0530	87 GPM, .7 HP, 2" discharge		3.20	5		560	99		659	780	
	0540	3" discharge		3.10	5.161		610	102		712	840	
	0600	45 gal. coated stl tank, 12 GPM, 1/2 HP, 2" discharge		3	5.333		645	105		750	885	
	0610	3" discharge		2.90	5.517		675	109		784	920	
	0630	87 GPM, .7 HP, 2" discharge		3	5.333		830	105		935	1,075	
	0640	3" discharge		2.90	5.517		875	109		984	1,150	
	0660	134 GPM, 1 HP, 2" discharge		2.80	5.714		895	113		1,008	1,175	
	0680	3" discharge		2.70	5.926		945	117		1,062	1,250	
	0700	70 gal. PE tank, 12 GPM, 1/2 HP, 2" discharge		2.60	6.154		730	122		852	1,000	
	0710	3" discharge		2.40	6.667		780	132		912	1,075	
	0730	87 GPM, 0.7 HP, 2" discharge		2.50	6.400		980	126		1,106	1,275	
	0740	3" discharge		2.30	6.957		1,000	137		1,137	1,325	
	0760	134 GPM, 1 HP, 2" discharge		2.20	7.273		1,025	144		1,169	1,350	
	0770	3" discharge		2	8		1,100	158		1,258	1,450	
900	0010	**PUMPS, PEDESTAL SUMP** With float control										900
	0400	Molded PVC base, 21 GPM at 15' head, 1/3 HP	1 Plum	5	1.600	Ea.	86.50	35		121.50	153	
	0800	Iron base, 21 GPM at 15' head, 1/3 HP		5	1.600		110	35		145	179	
	1200	Solid brass, 21 GPM at 15' head, 1/3 HP		5	1.600		179	35		214	255	
940	0010	**PUMPS, SUBMERSIBLE** Sump										940
	7000	Sump pump, automatic										
	7100	Plastic, 1-1/4" discharge, 1/4 HP	1 Plum	6	1.333	Ea.	105	29.50		134.50	165	
	7140	1/3 HP		5	1.600		127	35		162	198	
	7160	1/2 HP		5	1.600		153	35		188	226	
	7180	1-1/2" discharge, 1/2 HP		4	2		171	44		215	261	
	7500	Cast iron, 1-1/4" discharge, 1/4 HP		6	1.333		125	29.50		154.50	186	
	7540	1/3 HP		6	1.333		140	29.50		169.50	203	
	7560	1/2 HP		5	1.600		176	35		211	251	

15460 | Domestic Water Cond Equipment

			CREW	DAILY OUTPUT	LABOR-HOURS	UNIT	MAT.	LABOR	EQUIP.	TOTAL	TOTAL INCL O&P	
900	0010	**WATER SOFTENER**										900
	5800	Softener systems, automatic, intermediate sizes										
	5820	available, may be used in multiples.										
	6000	Hardness capacity between regenerations and flow										
	6100	150,000 grains, 37 GPM cont., 51 GPM peak	Q-1	1.20	13.333	Ea.	3,425	263		3,688	4,200	
	6200	300,000 grains, 81 GPM cont., 113 GPM peak		1	16		5,075	315		5,390	6,125	
	6300	750,000 grains, 160 GPM cont., 230 GPM peak		.80	20		6,625	395		7,020	7,950	
	6400	900,000 grains, 185 GPM cont., 270 GPM peak		.70	22.857		10,700	450		11,150	12,400	

15480 | Domestic Water Heaters

			CREW	DAILY OUTPUT	LABOR-HOURS	UNIT	MAT.	LABOR	EQUIP.	TOTAL	TOTAL INCL O&P	
200	0010	**WATER HEATERS**										200
	0020	For solar, see division 13600										
	1000	Residential, electric, glass lined tank, 5 yr, 10 gal., single element	1 Plum	2.30	3.478	Ea.	188	76.50		264.50	335	
	1040	20 gallon, single element		2.20	3.636		222	80		302	375	
	1060	30 gallon, double element		2.20	3.636		239	80		319	395	
	1080	40 gallon, double element		2	4		258	88		346	430	
	1100	52 gallon, double element		2	4		296	88		384	470	
	1180	120 gallon, double element		1.40	5.714		675	125		800	945	
	2000	Gas fired, glass lined tank, 5 yr, vent not incl., 20 gallon		2.10	3.810		265	83.50		348.50	430	
	2040	30 gallon		2	4		315	88		403	490	

15400 | Plumbing Fixtures & Equipment

15480 | Domestic Water Heaters

			CREW	DAILY OUTPUT	LABOR-HOURS	UNIT	2000 BARE COSTS				TOTAL INCL O&P	
							MAT.	LABOR	EQUIP.	TOTAL		
200	2100	75 gallon	1 Plum	1.50	5.333	Ea.	805	117		922	1,075	200
	2120	100 gallon		1.30	6.154		940	135		1,075	1,250	
	3000	Oil fired, glass lined tank, 5 yr, vent not included, 30 gallon		2	4		745	88		833	965	
	3040	50 gallon		1.80	4.444		1,000	97.50		1,097.50	1,250	
	3060	70 gallon		1.50	5.333		1,450	117		1,567	1,800	
	3080	85 gallon		1.40	5.714		3,325	125		3,450	3,850	
	4000	Commercial, 100° rise. NOTE: for each size tank, a range of										
	4010	heaters between the ones shown are available										
	4020	Electric										
	4100	5 gal., 3 kW, 12 GPH, 208V	1 Plum	2	4	Ea.	1,250	88		1,338	1,525	
	4120	10 gal., 6 kW, 25 GPH, 208V		2	4		1,375	88		1,463	1,675	
	4140	50 gal., 9 kW, 37 GPH, 208V		1.80	4.444		1,900	97.50		1,997.50	2,250	
	4160	50 gal., 36 kW, 148 GPH, 208V		1.80	4.444		2,900	97.50		2,997.50	3,350	
	4300	200 gal., 15 kW, 61 GPH, 480V	Q-1	1.70	9.412		8,975	186		9,161	10,200	
	4320	200 gal., 120 kW, 490 GPH, 480V		1.70	9.412		12,300	186		12,486	13,800	
	4460	400 gal., 30 kW, 123 GPH, 480V		1	16		12,200	315		12,515	13,900	
	5400	Modulating step control, 2-5 steps	1 Elec	5.30	1.509		1,175	33.50		1,208.50	1,350	
	5440	6-10 steps		3.20	2.500		1,500	55.50		1,555.50	1,750	
	5460	11-15 steps		2.70	2.963		1,875	65.50		1,940.50	2,150	
	5480	16-20 steps		1.60	5		2,200	111		2,311	2,600	
	6000	Gas fired, flush jacket, std. controls, vent not incl.										
	6040	75 MBH input, 73 GPH	1 Plum	1.40	5.714	Ea.	950	125		1,075	1,250	
	6060	98 MBH input, 95 GPH		1.40	5.714		1,625	125		1,750	1,975	
	6080	120 MBH input, 110 GPH		1.20	6.667		1,775	146		1,921	2,200	
	6180	200 MBH input, 192 GPH		.60	13.333		3,175	293		3,468	3,975	
	6200	250 MBH input, 245 GPH		.50	16		3,375	350		3,725	4,275	
	6900	For low water cutoff, add		8	1		189	22		211	245	
	6960	For bronze body hot water circulator, add		4	2		720	44		764	870	
	8000	Oil fired, flush jacket, std. controls, vent not incl.										
	8060	103 MBH gross output, 116 GPH	1 Plum	1.10	7.273	Ea.	895	160		1,055	1,250	
	8080	122 MBH gross output, 141 GPH		1	8		1,875	176		2,051	2,375	
	8100	137 MBH gross output, 161 GPH		.80	10		1,950	220		2,170	2,500	
	8160	225 MBH gross output, 256 GPH	Q-1	.80	20		2,750	395		3,145	3,675	
	8180	262 MBH gross output, 315 GPH		.70	22.857		2,650	450		3,100	3,675	
	8280	735 MBH gross output, 880 GPH		.40	40		6,525	790		7,315	8,475	
	8300	840 MBH gross output, 1000 GPH		.40	40		7,525	790		8,315	9,575	
	8900	For low water cutoff, add	1 Plum	8	1		189	22		211	245	
	8960	For bronze body hot water circulator, add	"	4	2		720	44		764	870	
900	0010	**HEAT TRANSFER PACKAGES** Complete, controls,										900
	0020	expansion tank, converter, air separator										
	1000	Hot water, 180°F enter, 200°F leaving, 15# steam										
	1010	One pump system, 28 GPM	Q-6	.75	32	Ea.	10,500	615		11,115	12,500	
	1020	35 GPM		.70	34.286		11,600	655		12,255	13,800	
	1040	55 GPM		.65	36.923		13,200	710		13,910	15,700	
	1060	130 GPM		.55	43.636		16,300	835		17,135	19,400	
	1080	255 GPM		.40	60		22,400	1,150		23,550	26,600	
	1100	550 GPM		.30	80		28,500	1,525		30,025	33,900	

15500 | Heat Generation Equipment

15510 | Heating Boilers and Accessories

			CREW	DAILY OUTPUT	LABOR-HOURS	UNIT	MAT.	LABOR	EQUIP.	TOTAL	TOTAL INCL O&P	
300	0010	**BOILERS, ELECTRIC, ASME** Standard controls and trim										300
	1000	Steam, 6 KW, 20.5 MBH	Q-19	1.20	20	Ea.	7,900	415		8,315	9,375	
	1160	60 KW, 205 MBH		1	24		9,775	495		10,270	11,500	
	1220	120 KW, 409 MBH		.75	32		11,200	660		11,860	13,400	
	1280	210 KW, 716 MBH		.55	43.636		15,300	900		16,200	18,300	
	1380	510 KW, 1740 MBH	Q-21	.36	88.889		20,600	1,875		22,475	25,700	
	1480	1,080 KW, 3685 MBH		.25	128		31,500	2,700		34,200	39,100	
	1600	2,340 KW, 7984 MBH		.16	200		60,000	4,200		64,200	72,500	
	2000	Hot water, 12 KW, 41 MBH	Q-19	1.30	18.462		3,400	380		3,780	4,350	
	2100	60 KW, 205 MBH		1.10	21.818		4,725	450		5,175	5,950	
	2220	240 KW, 820 MBH		.55	43.636		10,600	900		11,500	13,200	
	2320	480 KW, 1636 MBH	Q-21	.46	69.565		18,400	1,450		19,850	22,600	
	2500	1,200 KW, 4095 MBH		.34	94.118		33,600	1,975		35,575	40,300	
	2680	2,400 KW, 8191 MBH		.25	128		57,500	2,700		60,200	67,500	
	2820	3,600 KW, 12,283 MBH		.16	200		72,000	4,200		76,200	86,500	
400	0010	**BOILERS, GAS FIRED** Natural or propane, standard controls										400
	1000	Cast iron, with insulated jacket										
	2000	Steam, gross output, 81 MBH	Q-7	1.40	22.857	Ea.	1,125	455		1,580	2,000	
	2080	203 MBH		.90	35.556		1,900	710		2,610	3,275	
	2180	400 MBH		.56	56.838		3,150	1,125		4,275	5,325	
	2240	765 MBH		.40	79.602		5,025	1,575		6,600	8,150	
	2320	1,875 MBH		.22	142		10,700	2,825		13,525	16,500	
	2440	4,720 MBH		.13	250		20,500	4,975		25,475	30,800	
	2480	6,100 MBH		.11	283		49,000	5,625		54,625	63,500	
	2540	6,970 MBH		.08	410		55,000	8,175		63,175	74,000	
	3000	Hot water, gross output, 80 MBH		1.46	21.918		1,000	435		1,435	1,850	
	3140	320 MBH		.80	40		2,450	795		3,245	4,000	
	3260	1,088 MBH		.40	80		6,325	1,600		7,925	9,575	
	3380	3,264 MBH		.18	179		16,100	3,575		19,675	23,600	
	3480	6,100 MBH		.13	250		49,500	4,975		54,475	62,500	
	3540	6,970 MBH		.09	359		55,500	7,150		62,650	73,000	
	4000	Steel, insulating jacket										
	4500	Steam, not including burner, gross output										
	5500	1,440 MBH	Q-6	.35	68.571	Ea.	15,000	1,325		16,325	18,700	
	5580	3,065 MBH	"	.18	133		21,400	2,550		23,950	27,800	
	5640	7,200 MBH	Q-7	.18	177		39,300	3,550		42,850	49,100	
	5720	17,990 MBH	"	.10	320		73,500	6,375		79,875	91,500	
	6000	Hot water, including burner & one zone valve, gross output										
	6020	72 MBH	Q-6	2	12	Ea.	1,750	230		1,980	2,300	
	6120	227 MBH		1.30	18.462		4,050	355		4,405	5,025	
	6180	480 MBH		.70	34.286		6,600	655		7,255	8,350	
	6240	960 MBH		.45	53.333		11,400	1,025		12,425	14,200	
	6340	3,000 MBH		.15	160		31,600	3,075		34,675	39,800	
	7000	For tankless water heater on smaller gas units, add					10%					
460	0010	**BOILERS, GAS/OIL** Combination with burners and controls										460
	1000	Cast iron with insulated jacket										
	2000	Steam, gross output, 720 MBH	Q-7	.43	74.074	Ea.	6,375	1,475		7,850	9,450	
	2080	1,600 MBH		.30	107		9,800	2,125		11,925	14,300	
	2140	2,700 MBH		.19	165		14,000	3,300		17,300	20,900	
	2280	5,520 MBH		.14	235		51,000	4,675		55,675	64,000	
	2340	6,390 MBH		.11	296		54,500	5,900		60,400	70,000	
	2380	6,970 MBH		.09	372		58,000	7,400		65,400	75,500	
	2900	Hot water, gross output										
	2910	200 MBH	Q-6	.61	39.024	Ea.	4,150	750		4,900	5,800	

15500 | Heat Generation Equipment

15510 | Heating Boilers and Accessories

			CREW	DAILY OUTPUT	LABOR-HOURS	UNIT	2000 BARE COSTS MAT.	LABOR	EQUIP.	TOTAL	TOTAL INCL O&P	
460	2920	300 MBH	Q-6	.49	49.080	Ea.	4,150	940		5,090	6,125	460
	2930	400 MBH		.41	57.971		4,850	1,100		5,950	7,175	
	2940	500 MBH		.36	67.039		5,225	1,275		6,500	7,875	
	3000	584 MBH	Q-7	.44	72.072		8,100	1,425		9,525	11,300	
	3060	1,460 MBH		.28	113		20,100	2,250		22,350	25,800	
	3160	4,088 MBH		.16	195		56,000	3,875		59,875	68,000	
	3300	13,500 MBH, 403.3 BHP		.04	727		117,000	14,500		131,500	153,000	
	4000	Steel, insulated jacket, skid base, tubeless										
	4500	Steam, 150 psi gross output, 335 MBH, 10 BHP	Q-6	.54	44.037	Ea.	9,250	845		10,095	11,600	
	4560	670 MBH, 20 BHP		.36	65.934		11,500	1,275		12,775	14,700	
	4640	1,339 MBH, 40 BHP		.29	81.911		18,100	1,575		19,675	22,600	
	4720	2,511 MBH, 75 BHP		.17	141		24,700	2,700		27,400	31,700	
	5000	Hot water, gross output, 525 MBH		.45	52.980		9,500	1,025		10,525	12,200	
	5080	1,050 MBH		.35	67.989		15,300	1,300		16,600	19,100	
	5140	2,310 MBH		.20	120		28,500	2,300		30,800	35,200	
	5180	3,150 MBH		.16	146		37,900	2,800		40,700	46,300	
500	0010	**BOILERS, OIL FIRED** Standard controls, flame retention burner										500
	1000	Cast iron, with insulated flush jacket										
	2000	Steam, gross output, 109 MBH	Q-7	1.20	26.667	Ea.	1,325	530		1,855	2,350	
	2060	207 MBH		.90	35.556		1,825	710		2,535	3,175	
	2180	1,084 MBH		.38	85.106		6,500	1,700		8,200	9,950	
	2280	3,000 MBH		.19	170		13,900	3,375		17,275	20,900	
	2380	5,520 MBH		.14	235		42,200	4,675		46,875	54,000	
	2460	6,970 MBH		.09	363		54,500	7,225		61,725	72,000	
	3000	Hot water, same price as steam										
	4000	For tankless coil in smaller sizes, add					15%					
	5000	Steel, insulated jacket, burner										
	6000	Steam, full water leg construction, gross output										
	6020	144 MBH	Q-6	1.33	18.045	Ea.	3,200	345		3,545	4,100	
	6120	468 MBH		.70	34.483		5,900	660		6,560	7,600	
	6180	1,008 MBH		.39	61.381		9,150	1,175		10,325	12,100	
	6280	2,400 MBH		.19	125		17,500	2,400		19,900	23,200	
	6400	Larger sizes are same as steel, gas fired										
	7000	Hot water, gross output, 103 MBH	Q-6	1.60	15	Ea.	1,300	288		1,588	1,900	
	7120	420 MBH		.70	34.483		5,175	660		5,835	6,775	
	7200	840 MBH		.43	55.172		10,200	1,050		11,250	13,000	
	7260	1,680 MBH		.29	83.624		18,500	1,600		20,100	23,000	
	7320	3,150 MBH		.13	184		24,400	3,550		27,950	32,700	
	7340	For tankless coil in steam or hot water, add					7%					
880	0010	**SWIMMING POOL HEATERS** Not including wiring, external										880
	0020	piping, base or pad,										
	0060	Gas fired, input, 115 MBH	Q-6	3	8	Ea.	1,500	153		1,653	1,900	
	0100	135 MBH		2	12		1,675	230		1,905	2,225	
	0160	155 MBH		1.50	16		1,775	305		2,080	2,450	
	0200	190 MBH		1	24		2,300	460		2,760	3,300	
	0280	500 MBH		.40	60		5,600	1,150		6,750	8,050	
	0400	1,800 MBH		.14	171		15,600	3,275		18,875	22,500	
	0440	3,750 MBH		.09	266		31,400	5,100		36,500	43,100	
	2000	Electric, 12 KW, 4,800 gallon pool	Q-19	3	8		1,475	165		1,640	1,900	
	2020	15 KW, 7,200 gallon pool		2.80	8.571		1,475	177		1,652	1,925	
	2040	24 KW, 9,600 gallon pool		2.40	10		1,975	206		2,181	2,525	
	2100	55 KW, 24,000 gallon pool		1.20	20		2,825	415		3,240	3,775	

15500 | Heat Generation Equipment

15530 | Furnaces

			CREW	DAILY OUTPUT	LABOR-HOURS	UNIT	2000 BARE COSTS MAT.	LABOR	EQUIP.	TOTAL	TOTAL INCL O&P	
400	0010	**FURNACES** Hot air heating, blowers, standard controls										400
	0020	not including gas, oil or flue piping										
	1000	Electric, UL listed										
	1020	10.2 MBH	Q-20	5	4	Ea.	315	79.50		394.50	480	
	1040	17.1 MBH		4.80	4.167		330	83		413	500	
	1060	27.3 MBH		4.60	4.348		350	86.50		436.50	530	
	1100	34.1 MBH		4.40	4.545		405	90.50		495.50	595	
	3000	Gas, AGA certified, direct drive models										
	3020	45 MBH input	Q-9	4	4	Ea.	435	77.50		512.50	610	
	3040	60 MBH input		3.80	4.211		580	81.50		661.50	780	
	3060	75 MBH input		3.60	4.444		615	86		701	825	
	3100	100 MBH input		3.20	5		655	97		752	885	
	6000	Oil, UL listed, atomizing gun type burner										
	6020	56 MBH output	Q-9	3.60	4.444	Ea.	745	86		831	965	
	6030	84 MBH output		3.50	4.571		775	88.50		863.50	1,000	
	6040	95 MBH output		3.40	4.706		790	91		881	1,025	
	6060	134 MBH output		3.20	5		1,100	97		1,197	1,375	
	6080	151 MBH output		3	5.333		1,225	103		1,328	1,500	

15540 | Fuel-Fired Heaters

			CREW	DAILY OUTPUT	LABOR-HOURS	UNIT	MAT.	LABOR	EQUIP.	TOTAL	TOTAL INCL O&P	
300	0010	**DUCT FURNACES** Includes burner, controls, stainless steel										300
	0020	heat exchanger. Gas fired, electric ignition										
	0030	Indoor installation										
	0100	120 MBH output	Q-5	4	4	Ea.	1,150	79.50		1,229.50	1,375	
	0130	200 MBH output		2.70	5.926		1,575	118		1,693	1,925	
	0140	240 MBH output		2.30	6.957		1,650	138		1,788	2,025	
	0180	320 MBH output		1.60	10		2,000	199		2,199	2,525	
	0300	For powered venter and adapter, add					250			250	275	
	0500	For required flue pipe, see 15550										
	1000	Outdoor installation, with vent cap										
	1020	75 MBH output	Q-5	4	4	Ea.	1,500	79.50		1,579.50	1,775	
	1060	120 MBH output		4	4		1,925	79.50		2,004.50	2,250	
	1100	187 MBH output		3	5.333		2,750	106		2,856	3,200	
	1140	300 MBH output		1.80	8.889		3,600	177		3,777	4,250	
	1180	450 MBH output		1.40	11.429		4,250	227		4,477	5,050	
	1400	For powered venter, add					10%					
900	0010	**SPACE HEATERS** Cabinet, grilles, fan, controls, burner,										900
	0020	thermostat, no piping. For flue see 15550										
	1000	Gas fired, floor mounted										
	1100	60 MBH output	Q-5	10	1.600	Ea.	530	32		562	640	
	1140	100 MBH output		8	2		580	40		620	700	
	1180	180 MBH output		6	2.667		790	53		843	960	
	2000	Suspension mounted, propeller fan, 20 MBH output		8.50	1.882		400	37.50		437.50	500	
	2040	60 MBH output		7	2.286		495	45.50		540.50	615	
	2060	80 MBH output		6	2.667		550	53		603	695	
	2100	130 MBH output		5	3.200		730	63.50		793.50	910	
	2240	320 MBH output		2	8		1,500	159		1,659	1,925	
	2500	For powered venter and adapter, add					243			243	267	
	5000	Wall furnace, 17.5 MBH output	Q-5	6	2.667		470	53		523	605	
	5020	24 MBH output		5	3.200		485	63.50		548.50	635	
	5040	35 MBH output		4	4		650	79.50		729.50	845	
	6000	Oil fired, suspension mounted, 94 MBH output		4	4		1,850	79.50		1,929.50	2,150	
	6040	140 MBH output		3	5.333		1,975	106		2,081	2,350	
	6060	184 MBH output		3	5.333		2,125	106		2,231	2,525	

15500 | Heat Generation Equipment
15550 | Breechings, Chimneys & Stacks

		CREW	DAILY OUTPUT	LABOR-HOURS	UNIT	MAT.	LABOR	EQUIP.	TOTAL	TOTAL INCL O&P
0010	**VENT CHIMNEY** Prefab metal, U.L. listed									
0020	Gas, double wall, galvanized steel									
0080	3" diameter	Q-9	72	.222	V.L.F.	3.38	4.30		7.68	10.95
0100	4" diameter		68	.235		4.14	4.55		8.69	12.25
0120	5" diameter		64	.250		4.91	4.84		9.75	13.55
0140	6" diameter		60	.267		5.70	5.15		10.85	15
0160	7" diameter		56	.286		8.40	5.55		13.95	18.60
0180	8" diameter		52	.308		9.35	5.95		15.30	20.50
0200	10" diameter		48	.333		19.75	6.45		26.20	32.50
0220	12" diameter		44	.364		26.50	7.05		33.55	41
0260	16" diameter		40	.400		73.50	7.75		81.25	93.50
0300	20" diameter	Q-10	36	.667		92	13.40		105.40	124
0480	38" diameter	"	24	1		292	20		312	355
7800	All fuel, double wall, stainless steel, 6" diameter	Q-9	60	.267		28	5.15		33.15	39
7802	7" diameter		56	.286		36	5.55		41.55	49
7804	8" diameter		52	.308		42	5.95		47.95	56
7806	10" diameter		48	.333		61	6.45		67.45	78.50
7808	12" diameter		44	.364		82	7.05		89.05	102
7810	14" diameter		42	.381		108	7.35		115.35	130
9000	High temp. (2000° F), steel jacket, acid resist. refractory lining									
9010	11 ga. galvanized jacket, U.L. listed									
9020	Straight section, 48" long, 10" diameter	Q-10	13.30	1.805	Ea.	335	36		371	425
9040	18" diameter		7.40	3.243		510	65		575	670
9070	36" diameter		2.70	8.889		1,075	178		1,253	1,500
9090	48" diameter	Q-11	2.70	11.852		1,575	229		1,804	2,125
9120	Tee section, 10" diameter	Q-10	4.40	5.455		705	109		814	960
9140	18" diameter		2.40	10		1,000	201		1,201	1,450
9170	36" diameter		.80	30		3,300	600		3,900	4,650
9190	48" diameter	Q-11	.90	35.556		5,600	690		6,290	7,325
9220	Cleanout pier section, 10" diameter	Q-10	3.50	6.857		545	138		683	830
9240	18" diameter		1.90	12.632		730	254		984	1,225
9270	36" diameter		.75	32		1,350	640		1,990	2,550
9290	48" diameter	Q-11	.75	42.667		2,050	825		2,875	3,650
9320	For drain, add					53%				
9330	Elbow, 30° and 45°, 10" diameter	Q-10	6.60	3.636		460	73		533	630
9350	18" diameter		3.70	6.486		745	130		875	1,025
9380	36" diameter		1.30	18.462		1,850	370		2,220	2,650
9400	48" diameter	Q-11	1.35	23.704		3,225	460		3,685	4,325
9430	For 60° and 90° elbow, add					112%				
9440	End cap, 10" diameter	Q-10	26	.923		790	18.55		808.55	895
9460	18" diameter		15	1.600		1,125	32		1,157	1,275
9490	36" diameter		5	4.800		1,950	96.50		2,046.50	2,325
9510	48" diameter	Q-11	5.30	6.038		2,875	117		2,992	3,350
9540	Increaser (1 diameter), 10" diameter	Q-10	6.60	3.636		390	73		463	555
9560	18" diameter		3.70	6.486		610	130		740	890
9590	36" diameter		1.30	18.462		1,200	370		1,570	1,950
9592	42" diameter	Q-11	1.60	20		1,325	385		1,710	2,100
9594	48" diameter		1.35	23.704		1,575	460		2,035	2,525
9596	54" diameter		1.10	29.091		1,925	565		2,490	3,050
9600	For expansion joints, add to straight section					7%				
9610	For 1/4" hot rolled steel jacket, add					157%				
9620	For 2950°F very high temperature, add					89%				
9630	26 ga. aluminized jacket, straight section, 48" long									
9640	10" diameter	Q-10	15.30	1.569	V.L.F.	197	31.50		228.50	269
9660	18" diameter		8.50	2.824		300	56.50		356.50	425
9690	36" diameter		3.10	7.742		695	155		850	1,025

15500 | Heat Generation Equipment

15550 | Breechings, Chimneys & Stacks

		CREW	DAILY OUTPUT	LABOR-HOURS	UNIT	MAT.	LABOR	EQUIP.	TOTAL	TOTAL INCL O&P	
440	9810	Draw band, galv. stl., 11 gauge, 10" diameter	Q-10	32	.750	Ea.	53	15.05		68.05	84	440
	9820	12" diameter		30	.800		57	16.05		73.05	89.50	
	9830	18" diameter		26	.923		71.50	18.55		90.05	110	
	9860	36" diameter		18	1.333		121	27		148	178	
	9880	48" diameter	Q-11	20	1.600		207	31		238	281	
600	0010	**INDUCED DRAFT FANS**										600
	1000	Breeching installation										
	1800	Hot gas, 600°F, variable pitch pulley and motor										
	1860	8" diam. inlet, 1/4 H.P., 1 phase, 1120 CFM	Q-9	4	4	Ea.	1,575	77.50		1,652.50	1,850	
	1900	12" diam. inlet, 3/4 H.P., 3 phase, 2960 CFM		3	5.333		2,150	103		2,253	2,525	
	1940	18" diam. inlet, 3 H.P., 3 phase, 9120 CFM		2	8		3,325	155		3,480	3,900	
	1980	24" diam. inlet, 7-1/2 H.P., 3 phase, 17,760 CFM		.80	20		5,200	385		5,585	6,375	
	2300	For multi-blade damper at fan inlet, add					20%					

15600 | Refrigeration Equipment

15620 | Packaged Water Chillers

		CREW	DAILY OUTPUT	LABOR-HOURS	UNIT	MAT.	LABOR	EQUIP.	TOTAL	TOTAL INCL O&P	
100	0010	**ABSORPTION WATER CHILLERS**										100
	0020	Steam or hot water, water cooled										
	0400	420 ton	Q-7	.10	323	Ea.	187,500	6,425		193,925	216,500	
600	0010	**CENTRIFUGAL/RECIP. WATER CHILLERS,** With standard controls										600
	0020	Centrifugal liquid chiller, water cooled										
	0030	not including water tower										
	0100	Open drive, 2000 ton	Q-7	.07	477	Ea.	483,500	9,500		493,000	547,000	
	0490	Reciprocating, packaged w/integral air cooled condenser, 15 ton cool		.37	86.486		12,300	1,725		14,025	16,400	
	0500	20 ton cooling		.34	94.955		15,500	1,900		17,400	20,200	
	0520	40 ton cooling		.30	108		24,900	2,150		27,050	31,000	
	0600	100 ton cooling		.25	129		53,000	2,575		55,575	62,500	
	0680	Water cooled, single compressor, semi-hermetic, tower not incl.										
	0700	2 to 5 ton cooling	Q-5	.57	28.021	Ea.	5,725	560		6,285	7,225	
	0740	8 ton cooling	"	.31	52.117		9,825	1,025		10,850	12,500	
	0760	10 ton cooling	Q-6	.36	67.039		10,200	1,275		11,475	13,300	
	0800	20 ton cooling	Q-7	.38	83.990		12,800	1,675		14,475	16,800	
	0820	30 ton cooling	"	.33	96.096		15,400	1,900		17,300	20,100	
	0980	Water cooled, multiple compress., semi-hermetic, tower not incl.										
	1000	15 ton cooling	Q-6	.36	65.934	Ea.	16,000	1,275		17,275	19,600	
	1020	20 ton cooling	Q-7	.41	78.049		17,400	1,550		18,950	21,800	
	1060	30 ton cooling		.31	101		19,900	2,025		21,925	25,300	
	1100	50 ton cooling		.28	113		25,100	2,275		27,375	31,400	
	1160	100 ton cooling		.18	179		48,700	3,575		52,275	59,500	
	1180	120 ton cooling		.16	196		52,000	3,900		55,900	63,500	
	1200	140 ton cooling		.16	202		56,500	4,025		60,525	68,500	
	1450	Water cooled, dual compressors, direct drive, tower not incl.										
	1500	80 ton cooling	Q-7	.14	222	Ea.	46,400	4,425		50,825	58,500	
	1520	100 ton cooling		.14	228		54,000	4,550		58,550	67,000	
	1540	120 ton cooling		.14	231		59,500	4,625		64,125	73,000	
	1580	150 ton cooling		.13	240		72,500	4,800		77,300	87,500	
	1620	200 ton cooling		.13	250		88,500	4,975		93,475	105,000	

15600 | Refrigeration Equipment

15620 | Packaged Water Chillers

			CREW	DAILY OUTPUT	LABOR-HOURS	UNIT	2000 BARE COSTS MAT.	LABOR	EQUIP.	TOTAL	TOTAL INCL O&P	
600	1660	250 ton cooling	Q-7	.12	260	Ea.	102,000	5,175		107,175	120,500	600

15640 | Packaged Cooling Towers

			CREW	DAILY OUTPUT	LABOR-HOURS	UNIT	MAT.	LABOR	EQUIP.	TOTAL	INCL O&P	
400	0010	**COOLING TOWERS** Packaged units										400
	0070	Galvanized steel										
	0080	Draw thru, single flow										
	0100	Belt drive, 60 tons	Q-6	90	.267	TonAC	76	5.10		81.10	92	
	0150	95 tons		100	.240		67	4.60		71.60	81	
	0200	110 tons		109	.220		65.50	4.22		69.72	79	
	0250	125 tons		120	.200		63.50	3.83		67.33	76.50	
	1500	Induced air, double flow										
	1900	Gear drive, 150 ton	Q-6	126	.190	TonAC	88.50	3.65		92.15	103	
	2000	300 ton		129	.186		60	3.57		63.57	72	
	2100	600 ton		132	.182		56.50	3.49		59.99	68	
	2150	840 ton		142	.169		45.50	3.24		48.74	55.50	
	2200	Up to 1,000 tons		150	.160		42.50	3.07		45.57	51.50	
	3000	For higher capacities, use multiples										
	3500	For pumps and piping, add	Q-6	38	.632	TonAC	38	12.10		50.10	62	
	4000	For absorption systems, add				"	75%	75%				
	4500	For rigging, see division 01559										
	5000	Fiberglass										
	5010	Draw thru										
	5100	60 tons	Q-6	1.50	16	Ea.	2,775	305		3,080	3,550	
	5120	125 tons		.99	24.242		5,675	465		6,140	7,025	
	5140	300 tons		.43	55.814		13,300	1,075		14,375	16,400	
	5160	600 tons		.22	109		24,200	2,100		26,300	30,100	
	5180	1000 tons		.15	160		41,500	3,075		44,575	50,500	
	6000	Stainless steel										
	6010	Draw thru										
	6100	60 tons	Q-6	1.50	16	Ea.	7,200	305		7,505	8,425	
	6120	110 tons		.99	24.242		11,500	465		11,965	13,500	
	6140	300 tons		.43	55.814		31,900	1,075		32,975	36,900	
	6160	600 tons		.22	109		49,400	2,100		51,500	58,000	
	6180	1000 tons		.15	160		80,500	3,075		83,575	93,500	

15660 | Liquid Coolers/Evap Condensers

			CREW	DAILY OUTPUT	LABOR-HOURS	UNIT	MAT.	LABOR	EQUIP.	TOTAL	INCL O&P	
100	0010	**CONDENSERS** Ratings are for 30°F TD, R-22										100
	0080	Air cooled, belt drive, propeller fan										
	0240	50 ton	Q-6	.69	34.985	Ea.	10,200	670		10,870	12,300	
	0280	59 ton		.58	41.308		11,300	790		12,090	13,700	
	0320	73 ton		.47	51.173		13,100	980		14,080	16,100	
	0360	86 ton		.40	60.302		16,400	1,150		17,550	19,900	
	0380	88 ton		.39	61.697		17,500	1,175		18,675	21,200	
	1550	Air cooled, direct drive, propeller fan										
	1590	1 ton	Q-5	3.80	4.211	Ea.	535	84		619	730	
	1600	1-1/2 ton		3.60	4.444		635	88.50		723.50	845	
	1620	2 ton		3.20	5		700	99.50		799.50	940	
	1640	5 ton		2	8		1,425	159		1,584	1,850	
	1660	10 ton		1.40	11.429		2,200	227		2,427	2,800	
	1690	16 ton		1.10	14.545		3,100	289		3,389	3,900	
	1720	26 ton		.84	19.002		3,975	380		4,355	4,975	
	1760	41 ton	Q-6	.77	31.008		7,250	595		7,845	8,975	
	1800	63 ton	"	.54	44.037		11,500	845		12,345	14,000	

15700 | Heating/Ventilating/Air Conditioning Equipment

15710 | Heat Exchangers

			CREW	DAILY OUTPUT	LABOR-HOURS	UNIT	2000 BARE COSTS MAT.	LABOR	EQUIP.	TOTAL	TOTAL INCL O&P	
900	0010	**HEAT EXCHANGERS**										900
	0016	Shell & tube type, 4 pass, 3/4" O.D. copper tubes,										
	0020	C.I. heads, C.I. tube sheet, steel shell										
	0100	Hot water 40°F to 180°F, by steam at 10 PSI										
	0120	8 GPM	Q-5	6	2.667	Ea.	875	53		928	1,050	
	0140	10 GPM		5	3.200		1,325	63.50		1,388.50	1,550	
	0160	40 GPM		4	4		2,050	79.50		2,129.50	2,375	
	0180	64 GPM		2	8		3,125	159		3,284	3,725	
	0200	96 GPM	↓	1	16		4,150	320		4,470	5,075	
	0220	120 GPM	Q-6	1.50	16		5,450	305		5,755	6,475	
	1000	Hot water 40°F to 140°F, by water at 200°F										
	1020	7 GPM	Q-5	6	2.667	Ea.	1,075	53		1,128	1,275	
	1040	16 GPM		5	3.200		1,525	63.50		1,588.50	1,775	
	1060	34 GPM		4	4		2,325	79.50		2,404.50	2,675	
	1100	74 GPM	↓	1.50	10.667	↓	4,125	212		4,337	4,900	
	3000	Plate type,										
	3100	400 GPM	Q-6	.80	30	Ea.	18,400	575		18,975	21,200	
	3120	800 GPM	"	.50	48		32,300	920		33,220	37,100	
	3140	1200 GPM	Q-7	.34	94.118		66,500	1,875		68,375	76,000	
	3160	1800 GPM	"	.24	133	↓	99,500	2,650		102,150	114,000	

15720 | Air Handling Units

			CREW	DAILY OUTPUT	LABOR-HOURS	UNIT	MAT.	LABOR	EQUIP.	TOTAL	TOTAL INCL O&P	
200	0010	**CENTRAL STATION AIR-HANDLING UNIT** Chilled water										200
	1000	Modular, capacity at 700 FPM face velocity										
	1100	1300 CFM	Q-5	1.20	13.333	Ea.	2,325	265		2,590	3,025	
	1300	3200 CFM	"	.80	20		3,075	400		3,475	4,050	
	1500	8000 CFM	Q-6	.60	40		5,425	765		6,190	7,225	
	1700	20,000 CFM	↓	.31	77.419		11,100	1,475		12,575	14,700	
	1900	33,500 CFM	↓	.19	126		19,200	2,425		21,625	25,200	
	2100	63,000 CFM	Q-7	.13	246		37,400	4,900		42,300	49,200	
	2200	For hot water heating coil, add				↓	21%	5%				
500	0010	**MAKE-UP AIR UNIT**										500
	0020	Indoor suspension, natural/LP gas, direct fired,										
	0030	standard control. For flue see division 15550-440										
	0040	70°F temperature rise, MBH is input										
	0100	2000 CFM, 168 MBH	Q-6	3	8	Ea.	4,800	153		4,953	5,550	
	0160	6000 CFM, 502 MBH	↓	1.50	16		6,050	305		6,355	7,175	
	0220	12,000 CFM, 1005 MBH	↓	1	24		7,550	460		8,010	9,050	
	0300	24,000 CFM, 2007 MBH	Q-7	1	32		9,825	635		10,460	11,900	
	0400	50,000 CFM, 4180 MBH	"	.80	40		13,500	795		14,295	16,100	
	0600	For discharge louver assembly, add					10%					
	0700	For filters, add					20%					
	0800	For air shut-off damper section, add					10%					
	0900	For vertical unit, add				↓	10%					

15730 | Unitary Air Conditioning Equip

			CREW	DAILY OUTPUT	LABOR-HOURS	UNIT	MAT.	LABOR	EQUIP.	TOTAL	TOTAL INCL O&P	
200	0010	**COMPUTER ROOM UNITS**										200
	1000	Air cooled, includes remote condenser but not										
	1020	interconnecting tubing or refrigerant										
	1080	3 ton	Q-5	.50	32	Ea.	7,025	635		7,660	8,775	
	1120	5 ton		.45	35.556		9,400	710		10,110	11,500	
	1160	6 ton		.30	53.333		17,300	1,050		18,350	20,900	
	1200	8 ton		.27	59.259		19,600	1,175		20,775	23,600	
	1240	10 ton	↓	.25	64	↓	20,400	1,275		21,675	24,500	

15700 | Heating/Ventilating/Air Conditioning Equipment

15730 | Unitary Air Conditioning Equip

			CREW	DAILY OUTPUT	LABOR-HOURS	UNIT	MAT.	LABOR	EQUIP.	TOTAL	TOTAL INCL O&P	
200	1280	15 ton	Q-5	.22	72.727	Ea.	22,800	1,450		24,250	27,400	200
	1290	18 ton	↓	.20	80		25,900	1,600		27,500	31,100	
	1320	20 ton	Q-6	.29	82.759		27,200	1,575		28,775	32,500	
	1360	23 ton	"	.28	85.714		28,600	1,650		30,250	34,100	
500	0010	**PACKAGED TERMINAL AIR CONDITIONER** Cabinet, wall sleeve,										500
	0100	louver, electric heat, thermostat, manual changeover, 208 V										
	0200	6,000 BTUH cooling, 8800 BTU heat	Q-5	6	2.667	Ea.	915	53		968	1,100	
	0220	9,000 BTUH cooling, 13,900 BTU heat		5	3.200		935	63.50		998.50	1,125	
	0240	12,000 BTUH cooling, 13,900 BTU heat	↓	4	4		970	79.50		1,049.50	1,200	
	0260	15,000 BTUH cooling, 13,900 BTU heat		3	5.333		1,050	106		1,156	1,325	
	0500	For hot water coil, increase heat by 10%, add					5%	10%				
	1000	For steam, increase heat output by 30%, add					8%	10%				
600	0010	**ROOF TOP AIR CONDITIONERS** Standard controls, curb, economizer										600
	1000	Single zone, electric cool, gas heat										
	1090	2 ton cooling, 55 MBH heating	Q-5	.93	17.204	Ea.	4,050	340		4,390	5,025	
	1100	3 ton cooling, 60 MBH heating		.70	22.857		4,225	455		4,680	5,375	
	1120	4 ton cooling, 95 MBH heating		.61	26.403		4,500	525		5,025	5,825	
	1140	5 ton cooling, 112 MBH heating		.56	28.520		4,675	570		5,245	6,100	
	1145	6 ton cooling, 140 MBH heating		.52	30.769		5,350	610		5,960	6,900	
	1150	7.5 ton cooling, 170 MBH heating	↓	.50	32.258		7,325	640		7,965	9,100	
	1160	10 ton cooling, 200 MBH heating	Q-6	.67	35.982		9,500	690		10,190	11,600	
	1170	12.5 ton cooling, 230 MBH heating		.63	37.975		10,800	730		11,530	13,100	
	1190	18 ton cooling, 330 MBH heating	↓	.52	45.889		15,100	880		15,980	18,100	
	1200	20 ton cooling, 360 MBH heating	Q-7	.67	47.976		20,600	955		21,555	24,200	
	1210	25 ton cooling, 450 MBH heating		.56	57.554		24,600	1,150		25,750	29,000	
	1220	30 ton cooling, 540 MBH heating		.47	68.376		29,100	1,350		30,450	34,300	
	1260	50 ton cooling, 810 MBH heating		.28	113		46,000	2,275		48,275	54,500	
	1265	60 ton cooling, 900 MBH heating		.23	136		52,000	2,725		54,725	62,000	
	1270	80 ton cooling, 1000 MBH heating		.17	182		86,500	3,650		90,150	101,500	
	1275	90 ton cooling, 1200 MBH heating		.16	205		102,500	4,075		106,575	119,500	
	1280	100 ton cooling, 1350 MBH heating	↓	.14	228	↓	110,000	4,550		114,550	128,500	
	2000	Multizone, electric cool, gas heat, economizer										
	2100	15 cooling, 360 MBH heating	Q-7	.61	52.545	Ea.	39,900	1,050		40,950	45,600	
	2120	20 cooling, 360 MBH heating		.53	60.038		45,500	1,200		46,700	52,000	
	2200	40 ton cooling, 540 MBH heating		.28	113		66,500	2,275		68,775	77,500	
	2210	50 ton cooling, 540 MBH heating		.22	142		80,500	2,825		83,325	93,000	
	2220	70 ton cooling, 1500 MBH heating		.16	198		112,500	3,950		116,450	130,000	
	2240	80 ton cooling, 1500 MBH heating		.14	228		128,500	4,550		133,050	149,000	
	2260	90 ton cooling, 1500 MBH heating		.13	256		144,500	5,100		149,600	167,500	
	2280	105 ton cooling, 1500 MBH heating	↓	.11	290		168,500	5,800		174,300	195,000	
	2400	For hot water heat coil, deduct					5%					
	2500	For steam heat coil, deduct					2%					
	2600	For electric heat, deduct	↓			↓	3%	5%				
840	0010	**SELF-CONTAINED SINGLE PACKAGE**										840
	0100	Air cooled, for free blow or duct, including remote condenser										
	0200	3 ton cooling	Q-5	1	16	Ea.	4,400	320		4,720	5,350	
	0220	5 ton cooling	Q-6	1.20	20		5,525	385		5,910	6,700	
	0240	10 ton cooling	Q-7	1	32		9,400	635		10,035	11,400	
	0260	20 ton cooling		.90	35.556		15,300	710		16,010	18,000	
	0280	30 ton cooling	↓	.80	40		19,300	795		20,095	22,500	
	0340	60 ton cooling	Q-8	.40	80	↓	38,400	1,675	122	40,197	45,100	
	0490	For duct mounting no price change										
	0500	For hot water or steam heating coils, add				Ea.	10%	10%				
	1000	Water cooled for free blow or duct, not including tower										
	1010	Constant volume										

15700 | Heating/Ventilating/Air Conditioning Equipment

15730 | Unitary Air Conditioning Equip

		CREW	DAILY OUTPUT	LABOR-HOURS	UNIT	MAT.	LABOR	EQUIP.	TOTAL	TOTAL INCL O&P	
1100	3 ton cooling	Q-6	1	24	Ea.	3,075	460		3,535	4,125	
1120	5 ton cooling	"	1	24		3,650	460		4,110	4,775	
1140	10 ton cooling	Q-7	.90	35.556		7,425	710		8,135	9,350	
1160	20 ton cooling		.80	40		13,300	795		14,095	15,900	
1180	30 ton cooling		.70	45.714		17,600	910		18,510	20,900	

15740 | Heat Pumps

		CREW	DAILY OUTPUT	LABOR-HOURS	UNIT	MAT.	LABOR	EQUIP.	TOTAL	TOTAL INCL O&P
0010	**HEAT PUMPS** (Not including interconnecting tubing)									
1000	Air to air, split system, not including curbs, pads, or ductwork									
1020	2 ton cooling, 8.5 MBH heat @ 0°F	Q-5	1.20	13.333	Ea.	2,100	265		2,365	2,750
1060	5 ton cooling, 27 MBH heat @ 0°F		.50	32		4,275	635		4,910	5,750
1080	7.5 ton cooling, 33 MBH heat @ 0°F		.30	53.333		6,575	1,050		7,625	8,975
1100	10 ton cooling, 50 MBH heat @ 0°F	Q-6	.38	63.158		8,500	1,200		9,700	11,400
1120	15 ton cooling, 64 MBH heat @ 0°F	"	.26	92.308		11,700	1,775		13,475	15,800
1180	40 ton cooling, 193 MBH heat @ 0°F	Q-7	.12	266		33,900	5,300		39,200	46,000
1300	Supplementary electric heat coil, included									
1500	Single package, not including curbs, pads, or plenums									
1520	2 ton cooling, 6.5 MBH heat @ 0°F	Q-5	1.50	10.667	Ea.	2,525	212		2,737	3,125
1580	4 ton cooling, 13 MBH heat @ 0°F		.96	16.667		3,800	330		4,130	4,725
1640	7.5 ton cooling, 35 MBH heat @ 0°F		.40	40		8,225	795		9,020	10,400
1660	15 ton cooling, 56 MBH heat @ 0°F	Q-6	.30	80		16,100	1,525		17,625	20,200
2000	Water source to air, single package									
2100	1 ton cooling, 13 MBH heat @ 75°F	Q-5	2	8	Ea.	965	159		1,124	1,350
2140	2 ton cooling, 19 MBH heat @ 75°F		1.70	9.412		1,225	187		1,412	1,625
2220	5 ton cooling, 29 MBH heat @ 75°F		.90	17.778		2,125	355		2,480	2,925
3960	For supplementary heat coil, add					10%				
4000	For increase in capacity thru use									
4020	of solar collector, size boiler at 60%									

15750 | Humidity Control Equipment

		CREW	DAILY OUTPUT	LABOR-HOURS	UNIT	MAT.	LABOR	EQUIP.	TOTAL	TOTAL INCL O&P
0010	**HUMIDIFIERS**									
0520	Steam, room or duct, filter, regulators, auto. controls, 220 V									
0540	11 lb. per hour	Q-5	6	2.667	Ea.	2,000	53		2,053	2,300
0560	22 lb. per hour		5	3.200		2,200	63.50		2,263.50	2,525
0580	33 lb. per hour		4	4		2,250	79.50		2,329.50	2,600
0600	50 lb. per hour		4	4		2,750	79.50		2,829.50	3,150
0620	100 lb. per hour		3	5.333		3,300	106		3,406	3,825

15761 | Air Coils

		CREW	DAILY OUTPUT	LABOR-HOURS	UNIT	MAT.	LABOR	EQUIP.	TOTAL	TOTAL INCL O&P
0010	**DUCT HEATERS** Electric, 480 V, 3 Ph									
0020	Finned tubular insert, 500°F									
0100	8" wide x 6" high, 4.0 kW	Q-20	16	1.250	Ea.	510	25		535	605
0120	12" high, 8.0 kW		15	1.333		815	26.50		841.50	945
0140	18" high, 12.0 kW		14	1.429		1,150	28.50		1,178.50	1,300
0160	24" high, 16.0 kW		13	1.538		1,475	30.50		1,505.50	1,675
0180	30" high, 20.0 kW		12	1.667		1,800	33		1,833	2,025
0300	12" wide x 6" high, 6.7 kW		15	1.333		525	26.50		551.50	620
0320	12" high, 13.3 kW		14	1.429		845	28.50		873.50	980
0340	18" high, 20.0 kW		13	1.538		1,175	30.50		1,205.50	1,350
0360	24" high, 26.7 kW		12	1.667		1,525	33		1,558	1,725
0700	24" wide x 6" high, 17.8 kW		13	1.538		630	30.50		660.50	740
0760	24" high, 71.1 kW		10	2		1,900	40		1,940	2,175
8000	To obtain BTU multiply kW by 3413									

15700 | Heating/Ventilating/Air Conditioning Equipment

15761 | Air Coils

		CREW	DAILY OUTPUT	LABOR-HOURS	UNIT	MAT.	LABOR	EQUIP.	TOTAL	TOTAL INCL O&P
700 0010	**COILS, FLANGED**									
0500	Chilled water cooling, 6 rows, 24" x 48"	Q-5	3.20	5	Ea.	2,575	99.50		2,674.50	3,000
1000	Direct expansion cooling, 6 rows, 24" x 48"		2.80	5.714		2,775	114		2,889	3,250
1500	Hot water heating, 1 row, 24" x 48"		4	4		1,025	79.50		1,104.50	1,250
2000	Steam heating, 1 row, 24" x 48"	↓	3.06	5.229	↓	1,450	104		1,554	1,775

15765 | Fan Coil Unit/Unit Ventilators

		CREW	DAILY OUTPUT	LABOR-HOURS	UNIT	MAT.	LABOR	EQUIP.	TOTAL	TOTAL INCL O&P
200 0010	**FAN COIL AIR CONDITIONING** Cabinet mounted, filters, controls									
0100	Chilled water, 1/2 ton cooling	Q-5	8	2	Ea.	730	40		770	870
0120	1 ton cooling		6	2.667		830	53		883	1,000
0180	3 ton cooling	↓	4	4		1,675	79.50		1,754.50	1,950
0200	10 ton cooling	Q-6	2.80	8.571		2,400	164		2,564	2,925
0220	15 ton cooling		1.50	16		3,350	305		3,655	4,175
0240	20 ton cooling		.80	30		4,300	575		4,875	5,675
0260	30 ton cooling	↓	.60	40		6,350	765		7,115	8,275
0940	Direct expansion, for use w/air cooled condensing, 1.5 ton cooling	Q-5	5	3.200		440	63.50		503.50	590
1000	5 ton cooling	"	3	5.333		1,000	106		1,106	1,275
1040	10 ton cooling	Q-6	2.60	9.231		2,250	177		2,427	2,775
1060	20 ton cooling		.70	34.286		4,175	655		4,830	5,675
1100	40 ton cooling	↓	.45	53.333		8,775	1,025		9,800	11,400
1510	For condensing unit add see division 15670.									
600 0010	**HEATING & VENTILATING UNITS** Classroom									
0020	Includes filter, heating/cooling coils, standard controls									
0080	750 CFM, 2 tons cooling	Q-6	2	12	Ea.	2,950	230		3,180	3,625
0120	1250 CFM, 3 tons cooling		1.40	17.143		3,600	330		3,930	4,500
0140	1500 CFM, 4 tons cooling	↓	.80	30		3,850	575		4,425	5,175
0500	For electric heat, add					35%				
1000	For no cooling, deduct				↓	25%	10%			

15766 | Fin Tube Radiation

		CREW	DAILY OUTPUT	LABOR-HOURS	UNIT	MAT.	LABOR	EQUIP.	TOTAL	TOTAL INCL O&P
190 0010	**HYDRONIC HEATING** Terminal units, not incl. main supply pipe									
1000	Radiation									
1100	Panel, baseboard, C.I., including supports, no covers	Q-5	46	.348	L.F.	21	6.90		27.90	34.50
1150	Fin tube, wall hung, 14" slope top cover, with damper									
1200	1-1/4" copper tube, 4-1/4" alum. fin	Q-5	38	.421	L.F.	34.50	8.40		42.90	52
1250	1-1/4" steel tube, 4-1/4" steel fin	"	36	.444	"	31	8.85		39.85	48.50
1500	Note: fin tube may also require corners, caps, etc.									
1990	Convector unit, floor recessed, flush, with trim									
2000	for under large glass wall areas, no damper	Q-5	20	.800	L.F.	24.50	15.90		40.40	53.50
2100	For unit with damper, add				"	9.95			9.95	10.95
3000	Radiators, cast iron									
3100	Free standing or wall hung, 6 tube, 25" high	Q-5	96	.167	Section	21.50	3.32		24.82	29
3200	4 tube, 19" high	"	96	.167	"	14.80	3.32		18.12	22
3250	Adj. brackets, 2 per wall radiator up to 30 sections	1 Stpi	32	.250	Ea.	32	5.55		37.55	44
3950	Unit heaters, propeller, 115 V 2 psi steam, 60°F entering air									
4000	Horizontal, 12 MBH	Q-5	12	1.333	Ea.	330	26.50		356.50	410
4060	43.9 MBH		8	2		440	40		480	550
4140	96.8 MBH		6	2.667		540	53		593	685
4180	157.6 MBH		4	4		700	79.50		779.50	900
4240	286.9 MBH		2	8		1,175	159		1,334	1,575
4250	326.0 MBH		1.90	8.421		1,375	168		1,543	1,800
4260	364 MBH		1.80	8.889		1,525	177		1,702	1,975
4270	404 MBH	↓	1.60	10		1,925	199		2,124	2,450
4300	For vertical diffuser, add					136			136	149

15700 | Heating/Ventilating/Air Conditioning Equipment

15766 | Fin Tube Radiation

		CREW	DAILY OUTPUT	LABOR-HOURS	UNIT	2000 BARE COSTS MAT.	LABOR	EQUIP.	TOTAL	TOTAL INCL O&P	
190	4310	Vertical flow, 40 MBH	Q-5	11	1.455	Ea.	365	29		394	450	190
	4314	58.5 MBH		8	2		415	40		455	520	
	4326	131.0 MBH		4	4		685	79.50		764.50	880	
	4346	297.0 MBH		1.80	8.889		1,850	177		2,027	2,350	
	4354	420 MBH, (460 V)	Q-6	1.80	13.333		2,425	256		2,681	3,075	
	4358	500 MBH, (460 V)		1.71	14.035		2,750	269		3,019	3,475	
	4362	570 MBH, (460 V)		1.40	17.143		2,850	330		3,180	3,700	
	4366	620 MBH, (460 V)		1.30	18.462		3,275	355		3,630	4,175	
	4370	960 MBH, (460 V)		1.10	21.818		5,750	420		6,170	7,025	
	9500	To convert SFR to BTU rating: Hot water, 150 x SFR										
	9510	Forced hot water, 180 x SFR; steam, 240 x SFR										

15768 | Infrared Heaters

			CREW	DAILY OUTPUT	LABOR-HOURS	UNIT	MAT.	LABOR	EQUIP.	TOTAL	INCL O&P	
600	0010	**INFRA-RED UNIT**										600
	0020	Gas fired, unvented, electric ignition, 100% shutoff. Piping and										
	0030	wiring not included										
	0060	Input, 15 MBH	Q-5	7	2.286	Ea.	345	45.50		390.50	455	
	0120	45 MBH		5	3.200		390	63.50		453.50	530	
	0160	60 MBH		4	4		455	79.50		534.50	630	
	0240	120 MBH		2	8		805	159		964	1,150	

15770 | Floor-Heating & Snow-Melting Eq.

			CREW	DAILY OUTPUT	LABOR-HOURS	UNIT	MAT.	LABOR	EQUIP.	TOTAL	INCL O&P	
200	0010	**ELECTRIC HEATING**, equipments are not including wires & conduits										200
	0200	Snow melting for paved surface embedded mat heaters & controls	1 Elec	130	.062	S.F.	6.10	1.36		7.46	8.90	
	0400	Cable heating, radiant heat plaster, no controls, in South		130	.062		7.25	1.36		8.61	10.20	
	0600	In North		90	.089		7.25	1.96		9.21	11.20	
	0800	Cable on 1/2" board not incl. controls, tract housing		90	.089		6.10	1.96		8.06	9.90	
	1000	Custom housing		80	.100		7.10	2.21		9.31	11.40	
	1100	Rule of thumb: Baseboard units, including control		4.40	1.818	kW	75	40		115	148	
	1200	Duct heaters, including controls		5.30	1.509	"	61.50	33.50		95	123	
	1300	Baseboard heaters, 2' long, 375 watt		8	1	Ea.	34	22		56	73.50	
	1400	3' long, 500 watt		8	1		41	22		63	81	
	1600	4' long, 750 watt		6.70	1.194		48	26.50		74.50	96	
	1800	5' long, 935 watt		5.70	1.404		57	31		88	113	
	2000	6' long, 1125 watt		5	1.600		65	35.50		100.50	130	
	2400	8' long, 1500 watt		4	2		81	44		125	162	
	2950	Wall heaters with fan, 120 to 277 volt										
	2970	surface mounted, residential, 750 watt	1 Elec	7	1.143	Ea.	105	25.50		130.50	158	
	3000	1500 watt		5	1.600		120	35.50		155.50	190	
	3040	2250 watt		4	2		255	44		299	355	
	3070	4000 watt		3.50	2.286		260	50.50		310.50	370	
	3600	Thermostats, integral		16	.500		25	11.05		36.05	45.50	
	3800	Line voltage, 1 pole		8	1		25	22		47	63.50	

15780 | Energy Recovery Equipment

			CREW	DAILY OUTPUT	LABOR-HOURS	UNIT	MAT.	LABOR	EQUIP.	TOTAL	INCL O&P	
100	0010	**HEAT RECOVERY PACKAGES**										100
	0100	Air to air										
	2000	Kitchen exhaust, commercial, heat pipe exchanger										
	2040	Combined supply/exhaust air volume										
	2080	2.5 to 6.0 MCFM	Q-10	2.80	8.571	MCFM	4,725	172		4,897	5,500	
	2120	6 to 16 MCFM		5	4.800		2,900	96.50		2,996.50	3,350	
	2160	16 to 22 MCFM		6	4		2,250	80.50		2,330.50	2,625	
	7950											

15800 | Air Distribution

15810 | Ducts

		CREW	DAILY OUTPUT	LABOR-HOURS	UNIT	MAT.	LABOR	EQUIP.	TOTAL	TOTAL INCL O&P
0010	**DUCTWORK**									
0020	Fabricated rectangular, includes fittings, joints, supports,			R15810-050						
0030	allowance for flexible connections, no insulation									
0031	NOTE: Fabrication and installation are combined									
0040	as LABOR cost. Approx. 25% fittings assumed.									
0100	Aluminum, alloy 3003-H14, under 100 lb.	Q-10	75	.320	Lb.	2.80	6.40		9.20	13.95
0110	100 to 500 lb.		80	.300		2.60	6		8.60	13
0120	500 to 1,000 lb.		95	.253		1.50	5.05		6.55	10.20
0140	1,000 to 2,000 lb.		120	.200		1.20	4.01		5.21	8.10
0150	2,000 to 5,000 lb.		130	.185		1.15	3.71		4.86	7.50
0160	Over 5,000 lb.		145	.166		1.10	3.32		4.42	6.80
0500	Galvanized steel, under 200 lb.		235	.102		1.05	2.05		3.10	4.61
0520	200 to 500 lb.		245	.098		.85	1.97		2.82	4.26
0540	500 to 1,000 lb.		255	.094		.41	1.89		2.30	3.64
0560	1,000 to 2,000 lb.		265	.091		.41	1.82		2.23	3.52
0580	Over 5,000 lb.		285	.084		.31	1.69		2	3.19
1000	Stainless steel, type 304, under 100 lb.		165	.145		2.15	2.92		5.07	7.30
1020	100 to 500 lb.		175	.137		2.05	2.75		4.80	6.90
1030	500 to 1,000 lb.		190	.126		1.35	2.54		3.89	5.75
1040	1,000 to 2,000 lb.		200	.120		1.15	2.41		3.56	5.35
1050	2,000 to 5,000 lb.		225	.107		1.05	2.14		3.19	4.76
1060	Over 5,000 lb.		235	.102		.98	2.05		3.03	4.54
1100	For medium pressure ductwork, add						15%			
1200	For high pressure ductwork, add						40%			
1300	Flexible, coated fiberglass fabric on corr. resist. metal helix									
1400	pressure to 12" (WG) UL-181									
1500	Non-insulated, 3" diameter	Q-9	400	.040	L.F.	.95	.77		1.72	2.36
1540	5" diameter		320	.050		1.11	.97		2.08	2.85
1560	6" diameter		280	.057		1.25	1.11		2.36	3.25
1580	7" diameter		240	.067		1.64	1.29		2.93	3.98
1600	8" diameter		200	.080		1.69	1.55		3.24	4.47
1640	10" diameter		160	.100		2.17	1.94		4.11	5.65
1660	12" diameter		120	.133		2.57	2.58		5.15	7.20
1900	Insulated, 1" thick with 3/4 lb., PE jacket, 3" diameter		380	.042		1.64	.81		2.45	3.18
1910	4" diameter		340	.047		1.64	.91		2.55	3.34
1920	5" diameter		300	.053		1.92	1.03		2.95	3.85
1940	6" diameter		260	.062		2.15	1.19		3.34	4.38
1960	7" diameter		220	.073		2.57	1.41		3.98	5.20
1980	8" diameter		180	.089		2.63	1.72		4.35	5.80
2020	10" diameter		140	.114		3.42	2.21		5.63	7.50
2040	12" diameter		100	.160		3.98	3.10		7.08	9.65
3490	Rigid fiberglass duct board, foil reinf. kraft facing									
3500	Rectangular, 1" thick, alum. faced, (FRK), std. weight	Q-10	350	.069	SF Surf	.81	1.38		2.19	3.21

15820 | Duct Accessories

		CREW	DAILY OUTPUT	LABOR-HOURS	UNIT	MAT.	LABOR	EQUIP.	TOTAL	TOTAL INCL O&P
0010	**DUCT ACCESSORIES**									
0050	Air extractors, 12" x 4"	1 Shee	24	.333	Ea.	14.45	7.15		21.60	28
0100	8" x 6"		22	.364		14.45	7.80		22.25	29
0200	20" x 8"		16	.500		32.50	10.75		43.25	54
0280	24" x 12"		10	.800		44.50	17.20		61.70	78
1000	Duct access door, insulated, 6" x 6"		14	.571		13.65	12.30		25.95	35.50
1020	10" x 10"		11	.727		16.25	15.65		31.90	44.50
1040	12" x 12"		10	.800		17.55	17.20		34.75	48.50
1050	12" x 18"		9	.889		27.50	19.10		46.60	62.50
1070	18" x 18"		8	1		31	21.50		52.50	71

15800 | Air Distribution

15820 | Duct Accessories

			CREW	DAILY OUTPUT	LABOR-HOURS	UNIT	MAT.	LABOR	EQUIP.	TOTAL	TOTAL INCL O&P
300	1074	24" x 18"	1 Shee	8	1	Ea.	37	21.50		58.50	77.50
	2000	Fabrics for flexible connections, with metal edge		100	.080	L.F.	1.55	1.72		3.27	4.60
	2100	Without metal edge		160	.050	"	1.17	1.08		2.25	3.10
	3000	Fire damper, curtain type, 1-1/2 hr rated, vertical, 6" x 6"		24	.333	Ea.	19.70	7.15		26.85	33.50
	3020	8" x 6"		22	.364		19.70	7.80		27.50	34.50
	3240	16" x 14"		18	.444		31.50	9.55		41.05	50.50
	3400	24" x 20"		8	1		39.50	21.50		61	80
	5990	Multi-blade dampers, opposed blade, 8" x 6"		24	.333		18.40	7.15		25.55	32
	5994	8" x 8"		22	.364		19.20	7.80		27	34
	5996	10" x 10"		21	.381		21.50	8.20		29.70	38
	6000	12" x 12"		21	.381		25	8.20		33.20	41.50
	6020	12" x 18"		18	.444		33.50	9.55		43.05	53
	6030	14" x 10"		20	.400		21.50	8.60		30.10	38.50
	6031	14" x 14"		17	.471		29.50	10.10		39.60	49.50
	6033	16" x 12"		17	.471		29.50	10.10		39.60	49.50
	6035	16" x 16"		16	.500		37	10.75		47.75	58.50
	6037	18" x 14"		16	.500		37	10.75		47.75	58.50
	6038	18" x 18"		15	.533		44	11.45		55.45	68
	6070	20" x 16"		14	.571		44	12.30		56.30	69
	6072	20" x 20"		13	.615		53	13.25		66.25	80.50
	6074	22" x 18"		14	.571		53	12.30		65.30	78.50
	6076	24" x 16"		11	.727		52	15.65		67.65	83.50
	6078	24" x 20"		8	1		61.50	21.50		83	105
	6080	24" x 24"		8	1		72	21.50		93.50	116
	6110	26" x 26"		6	1.333		84.50	28.50		113	142
	6133	30" x 30"	Q-9	6.60	2.424		112	47		159	203
	6135	32" x 32"		6.40	2.500		128	48.50		176.50	223
	6180	48" x 36"		5.60	2.857		215	55.50		270.50	330
	7000	Splitter damper assembly, self-locking, 1' rod	1 Shee	24	.333		15.80	7.15		22.95	29.50
	7020	3' rod		22	.364		20.50	7.80		28.30	35.50
	7040	4' rod		20	.400		22.50	8.60		31.10	39.50
	7060	6' rod		18	.444		27.50	9.55		37.05	46.50
	8000	Multi-blade dampers, parallel blade									
	8100	8" x 8"	1 Shee	24	.333	Ea.	52.50	7.15		59.65	70
	8140	16" x 10"		20	.400		55.50	8.60		64.10	75.50
	8200	24" x 16"		11	.727		74	15.65		89.65	108
	8260	30" x 18"		7	1.143		86	24.50		110.50	136
	9000	Silencers, noise control for air flow, duct				MCFM	41.50			41.50	45.50
	9100	Louvers, galvanized				"	60			60	66
	9200	Plenums, measured by panel surface				S.F.	8.25			8.25	9.10
	9300	Vent, air transfer				Ea.	139			139	153

15830 | Fans

			CREW	DAILY OUTPUT	LABOR-HOURS	UNIT	MAT.	LABOR	EQUIP.	TOTAL	TOTAL INCL O&P
100	0010	**FANS** R15700-040									100
	0020	Air conditioning and process air handling									
	0030	Axial flow, compact, low sound, 2.5" S.P.									
	0050	3,800 CFM, 5 HP	Q-20	3.40	5.882	Ea.	3,450	117		3,567	4,000
	0080	6,400 CFM, 5 HP		2.80	7.143		3,850	142		3,992	4,475
	0100	10,500 CFM, 7-1/2 HP		2.40	8.333		4,800	166		4,966	5,550
	0120	15,600 CFM, 10 HP		1.60	12.500		6,050	249		6,299	7,075
	0200	In-line centrifugal, supply/exhaust booster									
	0220	aluminum wheel/hub, disconnect switch, 1/4" S.P.									
	0240	500 CFM, 10" diameter connection	Q-20	3	6.667	Ea.	700	133		833	990
	0260	1,380 CFM, 12" diameter connection		2	10		800	199		999	1,225
	0280	1,520 CFM, 16" diameter connection		2	10		915	199		1,114	1,325

15800 | Air Distribution

15830 | Fans

			CREW	DAILY OUTPUT	LABOR-HOURS	UNIT	MAT.	LABOR	EQUIP.	TOTAL	TOTAL INCL O&P	
100	0300	2,560 CFM, 18" diameter connection R15700-040	Q-20	1	20	Ea.	1,000	400		1,400	1,775	100
	0320	3,480 CFM, 20" diameter connection		.80	25		1,200	500		1,700	2,150	
	0326	5,080 CFM, 20" diameter connection		.75	26.667		1,325	530		1,855	2,375	
	1500	Vaneaxial, low pressure, 2000 CFM, 1/2 HP		3.60	5.556		1,575	111		1,686	1,925	
	1520	4,000 CFM, 1 HP		3.20	6.250		1,625	124		1,749	1,975	
	1540	8,000 CFM, 2 HP		2.80	7.143		1,900	142		2,042	2,350	
	2500	Ceiling fan, right angle, extra quiet, 0.10" S.P.										
	2520	95 CFM	Q-20	20	1	Ea.	93	19.90		112.90	136	
	2540	210 CFM		19	1.053		120	21		141	167	
	2560	385 CFM		18	1.111		137	22		159	187	
	2580	885 CFM		16	1.250		305	25		330	375	
	2600	1,650 CFM		13	1.538		390	30.50		420.50	480	
	2620	2,960 CFM		11	1.818		635	36		671	760	
	2640	For wall or roof cap, add	1 Shee	16	.500		65	10.75		75.75	89.50	
	2660	For straight thru fan, add					10%					
	2680	For speed control switch, add	1 Elec	16	.500		31.50	11.05		42.55	52.50	
	3000	Paddle blade air circulator, 3 speed switch										
	3020	42", 5,000 CFM high, 3000 CFM low	1 Elec	2.40	3.333	Ea.	73	73.50		146.50	201	
	3040	52", 6,500 CFM high, 4000 CFM low	"	2.20	3.636	"	90	80.50		170.50	230	
	3100	For antique white motor, same cost										
	3200	For brass plated motor, same cost										
	3300	For light adaptor kit, add				Ea.	26			26	29	
	3500	Centrifugal, airfoil, motor and drive, complete										
	3520	1000 CFM, 1/2 HP	Q-20	2.50	8	Ea.	870	159		1,029	1,225	
	3540	2,000 CFM, 1 HP		2	10		935	199		1,134	1,350	
	3560	4,000 CFM, 3 HP		1.80	11.111		1,200	221		1,421	1,700	
	3580	8,000 CFM, 7-1/2 HP		1.40	14.286		1,775	284		2,059	2,425	
	3600	12,000 CFM, 10 HP		1	20		2,600	400		3,000	3,525	
	4500	Corrosive fume resistant, plastic										
	4600	roof ventilators, centrifugal, V belt drive, motor										
	4620	1/4" S.P., 250 CFM, 1/4 HP	Q-20	6	3.333	Ea.	2,225	66.50		2,291.50	2,550	
	4640	895 CFM, 1/3 HP		5	4		2,425	79.50		2,504.50	2,800	
	4660	1630 CFM, 1/2 HP		4	5		2,875	99.50		2,974.50	3,325	
	4680	2240 CFM, 1 HP		3	6.667		3,000	133		3,133	3,525	
	5000	Utility set, centrifugal, V belt drive, motor										
	5020	1/4" S.P., 1200 CFM, 1/4 HP	Q-20	6	3.333	Ea.	2,925	66.50		2,991.50	3,325	
	5040	1520 CFM, 1/3 HP		5	4		2,925	79.50		3,004.50	3,350	
	5060	1850 CFM, 1/2 HP		4	5		2,950	99.50		3,049.50	3,400	
	5080	2180 CFM, 3/4 HP		3	6.667		2,950	133		3,083	3,450	
	5100	1/2" S.P., 3600 CFM, 1 HP		2	10		3,950	199		4,149	4,675	
	5120	4250 CFM, 1-1/2 HP		1.60	12.500		3,975	249		4,224	4,800	
	5140	4800 CFM, 2 HP		1.40	14.286		4,000	284		4,284	4,875	
	6000	Propeller exhaust, wall shutter, 1/4" S.P.										
	6020	Direct drive, two speed										
	6100	375 CFM, 1/10 HP	Q-20	10	2	Ea.	198	40		238	284	
	6120	730 CFM, 1/7 HP		9	2.222		220	44		264	315	
	6140	1000 CFM, 1/8 HP		8	2.500		305	50		355	420	
	6200	4720 CFM, 1 HP		5	4		490	79.50		569.50	675	
	6300	V-belt drive, 3 phase										
	6320	6175 CFM, 3/4 HP	Q-20	5	4	Ea.	410	79.50		489.50	585	
	6340	7500 CFM, 3/4 HP		5	4		435	79.50		514.50	610	
	6360	10,100 CFM, 1 HP		4.50	4.444		530	88.50		618.50	730	
	6380	14,300 CFM, 1-1/2 HP		4	5		620	99.50		719.50	850	
	6650	Residential, bath exhaust, grille, back draft damper										
	6660	50 CFM	Q-20	24	.833	Ea.	13.25	16.60		29.85	42.50	
	6670	110 CFM		22	.909		47	18.10		65.10	82.50	

Important: See the Reference Section for critical supporting data - Reference Nos., Crews, & City Cost Indexes

15800 | Air Distribution

15830 | Fans

			CREW	DAILY OUTPUT	LABOR-HOURS	UNIT	2000 BARE COSTS MAT.	LABOR	EQUIP.	TOTAL	TOTAL INCL O&P
6680	Light combination, squirrel cage, 100 watt, 70 CFM	R15700-040	Q-20	24	.833	Ea.	58	16.60		74.60	92
6700	Light/heater combination, ceiling mounted										
6710	70 CFM, 1450 watt		Q-20	24	.833	Ea.	70.50	16.60		87.10	106
6800	Heater combination, recessed, 70 CFM			24	.833		33.50	16.60		50.10	64.50
6820	With 2 infrared bulbs			23	.870		50.50	17.30		67.80	84.50
6900	Kitchen exhaust, grille, complete, 160 CFM			22	.909		58.50	18.10		76.60	95
6910	180 CFM			20	1		49.50	19.90		69.40	88
6920	270 CFM			18	1.111		91	22		113	137
6930	350 CFM			16	1.250		70	25		95	119
6940	Residential roof jacks and wall caps										
6944	Wall cap with back draft damper										
6946	3" & 4" dia. round duct		1 Shee	11	.727	Ea.	12.05	15.65		27.70	40
6948	6" dia. round duct		"	11	.727	"	29	15.65		44.65	58.50
6958	Roof jack with bird screen and back draft damper										
6960	3" & 4" dia. round duct		1 Shee	11	.727	Ea.	11.60	15.65		27.25	39.50
6962	3-1/4" x 10" rectangular duct		"	10	.800	"	21.50	17.20		38.70	52.50
6980	Transition										
6982	3-1/4" x 10" to 6" dia. round		1 Shee	20	.400	Ea.	13.05	8.60		21.65	29
7000	Roof exhauster, centrifugal, aluminum housing, 12" galvanized										
7020	curb, bird screen, back draft damper, 1/4" S.P.										
7100	Direct drive, 320 CFM, 11" sq. damper		Q-20	7	2.857	Ea.	380	57		437	515
7120	600 CFM, 11" sq. damper			6	3.333		390	66.50		456.50	540
7140	815 CFM, 13" sq. damper			5	4		390	79.50		469.50	565
7160	1450 CFM, 13" sq. damper			4.20	4.762		510	95		605	720
7180	2050 CFM, 16" sq. damper			4	5		510	99.50		609.50	725
7200	V-belt drive, 1650 CFM, 12" sq. damper			6	3.333		650	66.50		716.50	825
7220	2750 CFM, 21" sq. damper			5	4		750	79.50		829.50	960
7230	3500 CFM, 21" sq. damper			4.50	4.444		830	88.50		918.50	1,050
7240	4910 CFM, 23" sq. damper			4	5		1,025	99.50		1,124.50	1,300
7260	8525 CFM, 28" sq. damper			3	6.667		1,275	133		1,408	1,625
7280	13,760 CFM, 35" sq. damper			2	10		1,725	199		1,924	2,225
7300	20,558 CFM, 43" sq. damper			1	20		3,750	400		4,150	4,825
7320	For 2 speed winding, add						15%				
7340	For explosionproof motor, add					Ea.	273			273	300
7360	For belt driven, top discharge, add						15%				
7500	Utility set, steel construction, pedestal, 1/4" S.P.										
7520	Direct drive, 150 CFM, 1/8 HP		Q-20	6.40	3.125	Ea.	580	62		642	740
7540	485 CFM, 1/6 HP			5.80	3.448		730	68.50		798.50	915
7560	1950 CFM, 1/2 HP			4.80	4.167		855	83		938	1,075
7580	2410 CFM, 3/4 HP			4.40	4.545		1,575	90.50		1,665.50	1,875
7600	3328 CFM, 1-1/2 HP			3	6.667		1,750	133		1,883	2,150
7680	V-belt drive, drive cover, 3 phase										
7700	800 CFM, 1/4 HP		Q-20	6	3.333	Ea.	695	66.50		761.50	875
7720	1,300 CFM, 1/3 HP			5	4		745	79.50		824.50	950
7740	2,000 CFM, 1 HP			4.60	4.348		765	86.50		851.50	985
7760	2,900 CFM, 3/4 HP			4.20	4.762		1,125	95		1,220	1,400
8500	Wall exhausters, centrifugal, auto damper, 1/8" S.P.										
8520	Direct drive, 610 CFM, 1/20 HP		Q-20	14	1.429	Ea.	160	28.50		188.50	224
8540	796 CFM, 1/12 HP			13	1.538		165	30.50		195.50	234
8560	822 CFM, 1/6 HP			12	1.667		261	33		294	345
8580	1,320 CFM, 1/4 HP			12	1.667		263	33		296	345
9500	V-belt drive, 3 phase										
9520	2,800 CFM, 1/4 HP		Q-20	9	2.222	Ea.	725	44		769	875
9540	3,740 CFM, 1/2 HP		"	8	2.500	"	755	50		805	915

15800 | Air Distribution

15850 | Air Outlets & Inlets

			CREW	DAILY OUTPUT	LABOR-HOURS	UNIT	2000 BARE COSTS MAT.	LABOR	EQUIP.	TOTAL	TOTAL INCL O&P	
300	0010	**DIFFUSERS** Aluminum, opposed blade damper unless noted										300
	0100	Ceiling, linear, also for sidewall										
	0500	Perforated, 24" x 24" lay-in panel size, 6" x 6"	1 Shee	16	.500	Ea.	79.50	10.75		90.25	105	
	0520	8" x 8"		15	.533		82	11.45		93.45	109	
	0530	9" x 9"		14	.571		84	12.30		96.30	113	
	0540	10" x 10"		14	.571		86	12.30		98.30	115	
	0560	12" x 12"		12	.667		86	14.35		100.35	119	
	0590	16" x 16"		11	.727		107	15.65		122.65	145	
	0600	18" x 18"		10	.800		114	17.20		131.20	155	
	0610	20" x 20"		10	.800		130	17.20		147.20	172	
	0620	24" x 24"		9	.889		163	19.10		182.10	212	
	1000	Rectangular, 1 to 4 way blow, 6" x 6"		16	.500		46	10.75		56.75	69	
	1010	8" x 8"		15	.533		54	11.45		65.45	79	
	1014	9" x 9"		15	.533		56	11.45		67.45	81	
	1016	10" x 10"		15	.533		67	11.45		78.45	93	
	1020	12" x 6"		15	.533		57	11.45		68.45	82	
	1040	12" x 9"		14	.571		69.50	12.30		81.80	97	
	1060	12" x 12"		12	.667		82	14.35		96.35	114	
	1070	14" x 6"		13	.615		61	13.25		74.25	90	
	1074	14" x 14"		12	.667		101	14.35		115.35	135	
	1150	18" x 18"		9	.889		149	19.10		168.10	197	
	1160	21" x 21"		8	1		189	21.50		210.50	244	
	1170	24" x 12"		10	.800		134	17.20		151.20	177	
	1500	Round, butterfly damper, 6" diameter		18	.444		15.45	9.55		25	33	
	1520	8" diameter		16	.500		16.35	10.75		27.10	36	
	1540	10" diameter		14	.571		20.50	12.30		32.80	43	
	1560	12" diameter		12	.667		27	14.35		41.35	53.50	
	1580	14" diameter		10	.800		33.50	17.20		50.70	66	
	2000	T bar mounting, 24" x 24" lay-in frame, 6" x 6"		16	.500		75.50	10.75		86.25	101	
	2020	9" x 9"		14	.571		84	12.30		96.30	113	
	2040	12" x 12"		12	.667		111	14.35		125.35	146	
	2060	15" x 15"		11	.727		143	15.65		158.65	184	
	2080	18" x 18"		10	.800		149	17.20		166.20	193	
	6000	For steel diffusers instead of aluminum, deduct					10%					
500	0010	**GRILLES**										500
	0020	Aluminum										
	1000	Air return, 6" x 6"	1 Shee	26	.308	Ea.	12.35	6.60		18.95	25	
	1020	10" x 6"		24	.333		14.95	7.15		22.10	28.50	
	1080	16" x 8"		22	.364		21.50	7.80		29.30	36.50	
	1100	12" x 12"		22	.364		21.50	7.80		29.30	36.50	
	1120	24" x 12"		18	.444		38.50	9.55		48.05	58	
	1220	24" x 18"		16	.500		45.50	10.75		56.25	68	
	1280	36" x 24"		14	.571		90.50	12.30		102.80	120	
	3000	Filter grille with filter, 12" x 12"		24	.333		35	7.15		42.15	50.50	
	3020	18" x 12"		20	.400		51.50	8.60		60.10	71	
	3040	24" x 18"		18	.444		74	9.55		83.55	97.50	
	3060	24" x 24"		16	.500		97	10.75		107.75	125	
	6000	For steel grilles instead of aluminum in above, deduct					10%					
600	0010	**LOUVERS**										600
	0100	Aluminum, extruded, with screen, mill finish										
	1000	Brick vent, (see also division 04090)										
	1100	Standard, 4" deep, 8" wide, 5" high	1 Shee	24	.333	Ea.	22.50	7.15		29.65	37	
	1200	Modular, 4" deep, 7-3/4" wide, 5" high		24	.333		23.50	7.15		30.65	38	
	1300	Speed brick, 4" deep, 11-5/8" wide, 3-7/8" high		24	.333		23.50	7.15		30.65	38	
	1400	Fuel oil brick, 4" deep, 8" wide, 5" high		24	.333		41	7.15		48.15	57	
	2000	Cooling tower and mechanical equip., screens, light weight		40	.200	S.F.	9.75	4.30		14.05	18	

15800 | Air Distribution

15850 | Air Outlets & Inlets

		CREW	DAILY OUTPUT	LABOR-HOURS	UNIT	2000 BARE COSTS MAT.	LABOR	EQUIP.	TOTAL	TOTAL INCL O&P
2020	Standard weight	1 Shee	35	.229	S.F.	27	4.91		31.91	38
2500	Dual combination, automatic, intake or exhaust		20	.400		37.50	8.60		46.10	55.50
2520	Manual operation		20	.400		27.50	8.60		36.10	45
2540	Electric or pneumatic operation		20	.400		27.50	8.60		36.10	45
2560	Motor, for electric or pneumatic		14	.571	Ea.	325	12.30		337.30	380
3000	Fixed blade, continuous line									
3100	Mullion type, stormproof	1 Shee	28	.286	S.F.	27.50	6.15		33.65	41
3200	Stormproof		28	.286		27.50	6.15		33.65	41
3300	Vertical line		28	.286		36.50	6.15		42.65	50.50
3500	For damper to use with above, add					50%	30%			
3520	Motor, for damper, electric or pneumatic	1 Shee	14	.571	Ea.	325	12.30		337.30	380
4000	Operating, 45°, manual, electric or pneumatic		24	.333	S.F.	27.50	7.15		34.65	42.50
4100	Motor, for electric or pneumatic		14	.571	Ea.	325	12.30		337.30	380
4200	Penthouse, roof		56	.143	S.F.	16	3.07		19.07	23
4300	Walls		40	.200		38.50	4.30		42.80	50
5000	Thinline, under 4" thick, fixed blade		40	.200		15.50	4.30		19.80	24.50
5010	Finishes, applied by mfr. at additional cost, available in colors									
5020	Prime coat only, add				S.F.	2.25			2.25	2.48
5040	Baked enamel finish coating, add					4.15			4.15	4.57
5060	Anodized finish, add					4.50			4.50	4.95
5080	Duranodic finish, add					8.15			8.15	8.95
5100	Fluoropolymer finish coating, add					12.75			12.75	14.05
9980	For small orders (under 10 pieces), add					25%				
0010	**REGISTERS**									
0980	Air supply									
1000	Ceiling/wall, O.B. damper, anodized aluminum									
1010	One or two way deflection, adj. curved face bars									
1020	8" x 4"	1 Shee	26	.308	Ea.	18.05	6.60		24.65	31
1120	12" x 12"		18	.444		33	9.55		42.55	52.50
1240	20" x 6"		18	.444		29.50	9.55		39.05	48.50
1340	24" x 8"		13	.615		40	13.25		53.25	66.50
1350	24" x 18"		12	.667		74	14.35		88.35	106
2700	Above registers in steel instead of aluminum, deduct					10%				
4000	Floor, toe operated damper, enameled steel									
4020	4" x 8"	1 Shee	32	.250	Ea.	15.90	5.40		21.30	26.50
4100	8" x 10"		22	.364		18.85	7.80		26.65	34
4140	10" x 10"		20	.400		22.50	8.60		31.10	39.50
4220	14" x 14"		16	.500		75.50	10.75		86.25	101
4240	14" x 20"		15	.533		100	11.45		111.45	129
4980	Air return									
5000	Ceiling or wall, fixed 45° face blades									
5010	Adjustable O.B. damper, anodized aluminum									
5020	4" x 8"	1 Shee	26	.308	Ea.	21.50	6.60		28.10	34.50
5060	6" x 10"		19	.421		24	9.05		33.05	42
5280	24" x 24"		11	.727		103	15.65		118.65	140
5300	24" x 36"		8	1		159	21.50		180.50	212
6000	For steel construction instead of aluminum, deduct					10%				

15854 | Ventilators

		CREW	DAILY OUTPUT	LABOR-HOURS	UNIT	MAT.	LABOR	EQUIP.	TOTAL	TOTAL INCL O&P
0010	**VENTILATORS** Base, damper & bird screen, CFM in 5 MPH wind									
0500	Rotary syphon, galvanized, 6" neck diameter, 185 CFM	Q-9	16	1	Ea.	50	19.35		69.35	87.50
0560	12" neck diameter, 310 CFM		10	1.600		70.50	31		101.50	130
0660	24" neck diameter, 1,530 CFM		8	2		287	38.50		325.50	380
0700	36" neck diameter, 3,800 CFM		6	2.667		780	51.50		831.50	945
0720	42" neck diameter, 4,500 CFM		4	4		1,275	77.50		1,352.50	1,525

15800 | Air Distribution

15854 | Ventilators

			Crew	Daily Output	Labor-Hours	Unit	2000 Bare Costs Mat.	Labor	Equip.	Total	Total Incl O&P
300	1280	Spinner ventilators, wind driven, galvanized R15700-040									
	1300	4" neck diameter, 180 CFM	Q-9	20	.800	Ea.	51	15.50		66.50	82
	1340	6" neck diameter, 250 CFM		16	1		57	19.35		76.35	95
	1400	12" neck diameter, 770 CFM		10	1.600		82.50	31		113.50	144
	1500	24" neck diameter, 3,100 CFM		8	2		241	38.50		279.50	330
	1540	36" neck diameter, 5,500 CFM	↓	6	2.667	↓	705	51.50		756.50	860
	2000	Stationary, gravity, syphon, galvanized									
	2160	6" neck diameter, 66 CFM	Q-9	16	1	Ea.	44	19.35		63.35	81
	2240	12" neck diameter, 160 CFM		10	1.600		92.50	31		123.50	155
	2340	24" neck diameter, 900 CFM		8	2		252	38.50		290.50	345
	2380	36" neck diameter, 2,000 CFM		6	2.667		790	51.50		841.50	955
	4200	Stationary mushroom, aluminum, 16" orifice diameter		10	1.600		246	31		277	325
	4220	26" orifice diameter		9	1.778		365	34.50		399.50	460
	4230	30" orifice diameter		8.60	1.860		530	36		566	645
	4240	38" orifice diameter		8	2		760	38.50		798.50	900
	4250	42" orifice diameter		7.60	2.105		1,000	40.50		1,040.50	1,175
	4260	50" orifice diameter	↓	7	2.286	↓	1,200	44		1,244	1,400
	5000	Relief vent									
	5500	Rectangular, aluminum, galvanized curb									
	5510	intake/exhaust, 0.05" SP									
	5600	500 CFM, 12" x 16"	Q-9	8	2	Ea.	355	38.50		393.50	455
	5640	1000 CFM, 12" x 24"		6.60	2.424		420	47		467	545
	5680	3000 CFM, 20" x 42"	↓	4	4	↓	730	77.50		807.50	930

15860 | Air Cleaning Devices

			Crew	Daily Output	Labor-Hours	Unit	Mat.	Labor	Equip.	Total	Total Incl O&P
100	0010	**AIR FILTERS**										100
	0050	Activated charcoal type, full flow				MCFM	600			600	660
	2000	Electronic air cleaner, duct mounted									
	2150	400 - 1000 CFM	1 Shee	2.30	3.478	Ea.	545	75		620	720
	2200	1000 - 1400 CFM		2.20	3.636		700	78		778	895
	2250	1400 - 2000 CFM	↓	2.10	3.810	↓	760	82		842	980
	2950	Mechanical media filtration units									
	3000	High efficiency type, with frame, non-supported				MCFM	45			45	49.50
	3100	Supported type					55			55	60.50
	4000	Medium efficiency, extended surface					5			5	5.50
	4500	Permanent washable					20			20	22
	5000	Renewable disposable roll				↓	120			120	132
	5500	Throwaway glass or paper media type				Ea.	4.60			4.60	5.05

15950 | Testing/Adjusting/Balancing

15955 | HVAC Test/Adjust/Balance

			Crew	Daily Output	Labor-Hours	Unit	Mat.	Labor	Equip.	Total	Total Incl O&P
100	0010	**BALANCING, AIR** (Subcontractor's quote incl. material & labor)										100
	0900	Heating and ventilating equipment									
	1000	Centrifugal fans, utility sets				Ea.					241.92
	1100	Heating and ventilating unit									362.88
	1200	In-line fan									362.88
	1300	Propeller and wall fan				↓					68.54

15950 | Testing/Adjusting/Balancing

	15955	HVAC Test/Adjust/Balance	CREW	DAILY OUTPUT	LABOR-HOURS	UNIT	MAT.	LABOR	EQUIP.	TOTAL	TOTAL INCL O&P	
							\multicolumn{4}{c}{2000 BARE COSTS}		
100	1400	Roof exhaust fan				Ea.					161.28	100
	2000	Air conditioning equipment, central station									524.16	
	2100	Built-up low pressure unit									483.84	
	2200	Built-up high pressure unit									564.48	
	2500	Multi-zone A.C. and heating unit									362.88	
	2600	For each zone over one, add									80.64	
	2700	Package A.C. unit									201.60	
	2800	Rooftop heating and cooling unit									282.64	
	3000	Supply, return, exhaust, registers & diffusers, avg. height ceiling									48.38	
	3100	High ceiling									72.58	
	3200	Floor height				▼					40.32	
900	0010	BALANCING, WATER (Subcontractor's quote incl. material & labor)										900
	0050	Air cooled condenser				Ea.					141.96	
	0100	Cabinet unit heater									48.67	
	0200	Chiller									344.76	
	0300	Convector									40.56	
	0500	Cooling tower									263.64	
	0600	Fan coil unit, unit ventilator									73.01	
	0700	Fin tube and radiant panels									81.12	
	0800	Main and duct re-heat coils									75.04	
	0810	Heat exchanger									75.04	
	1000	Pumps									178.46	
	1100	Unit heater				▼					56.78	

For information about Means Estimating Seminars, see yellow pages 11 and 12 in back of book

Division Notes

	CREW	DAILY OUTPUT	LABOR-HOURS	UNIT	2000 BARE COSTS				TOTAL INCL O&P
					MAT.	LABOR	EQUIP.	TOTAL	

Division 16 Electrical

Estimating Tips

16060 Grounding & Bonding
- When taking off grounding system, identify separately the type and size of wire and list each unique type of ground connection.

16100 Wiring Methods
- Conduit should be taken off in three main categories: power distribution, branch power, and branch lighting, so the estimator can concentrate on systems and components, therefore making it easier to ensure all items have been accounted for.
- For cost modifications for elevated conduit installation, add the percentages to labor according to the height of installation and only the quantities exceeding the different height levels, not to the total conduit quantities.
- Remember that aluminum wiring of equal ampacity is larger in diameter than copper and may require larger conduit.
- If more than three wires at a time are being pulled, deduct percentages from the labor hours of that grouping of wires.
- The estimator should take the weights of materials into consideration when completing a takeoff. Topics to consider include: How will the materials be supported? What methods of support are available? How high will the support structure have to reach? Will the final support structure be able to withstand the total burden? Is the support material included or separate from the fixture, equipment and material specified?

16200 Electrical Power
- Do not overlook the costs for equipment used in the installation. If scaffolding or highlifts are available in the field, contractors may use them in lieu of the proposed ladders and rolling staging.

16400 Low-Voltage Distribution
- Supports and concrete pads may be shown on drawings for the larger equipment, or the support system may be just a piece of plywood for the back of a panelboard. In either case, it must be included in the costs.

16500 Lighting
- Fixtures should be taken off room by room, using the fixture schedule, specifications, and the ceiling plan. For large concentrations of lighting fixtures in the same area deduct the percentages from labor hours.

16700 Communications
16800 Sound & Video
- When estimating material costs for special systems, it is always prudent to obtain manufacturers' quotations for equipment prices and special installation requirements which will affect the total costs.

Reference Numbers
Reference numbers are shown in bold squares at the beginning of some major classifications. These numbers refer to related items in the Reference Section. The reference information may be an estimating procedure, an alternate pricing method or technical information.

Note: Not all subdivisions listed here necessarily appear in this publication.

16050 | Basic Electrical Materials & Methods

16055 | Selective Demolition

		CREW	DAILY OUTPUT	LABOR-HOURS	UNIT	2000 BARE COSTS				TOTAL INCL O&P
						MAT.	LABOR	EQUIP.	TOTAL	
0010	**ELECTRICAL DEMOLITION**									
0020	Conduit to 15' high, including fittings & hangers									
0100	Rigid galvanized steel, 1/2" to 1" diameter	1 Elec	242	.033	L.F.		.73		.73	1.20
0120	1-1/4" to 2"		200	.040			.88		.88	1.45
0140	2" to 4"		151	.053			1.17		1.17	1.92
0160	4" to 6"		80	.100			2.21		2.21	3.62
0200	Electric metallic tubing (EMT) 1/2" to 1"		394	.020			.45		.45	.73
0220	1-1/4" to 1-1/2"		326	.025			.54		.54	.89
0260	3-1/2" to 4"		155	.052			1.14		1.14	1.87
0270	Armored cable, (BX) avg. 50' runs									
0280	#14, 2 wire	1 Elec	690	.012	L.F.		.26		.26	.42
0290	#14, 3 wire		571	.014			.31		.31	.51
0300	#12, 2 wire		605	.013			.29		.29	.48
0310	#12, 3 wire		514	.016			.34		.34	.56
0320	#10, 2 wire		514	.016			.34		.34	.56
0330	#10, 3 wire		425	.019			.42		.42	.68
0340	#8, 3 wire		342	.023			.52		.52	.85
0350	Non metallic sheathed cable (Romex)									
0360	#14, 2 wire	1 Elec	720	.011	L.F.		.25		.25	.40
0370	#14, 3 wire		657	.012			.27		.27	.44
0380	#12, 2 wire		629	.013			.28		.28	.46
0390	#10, 3 wire		450	.018			.39		.39	.64
0400	Wiremold raceway, including fittings & hangers									
0420	No. 3000	1 Elec	250	.032	L.F.		.71		.71	1.16
0440	No. 4000		217	.037			.81		.81	1.33
0460	No. 6000		166	.048			1.07		1.07	1.74
0500	Channels, steel, including fittings & hangers									
0520	3/4" x 1-1/2"	1 Elec	308	.026	L.F.		.57		.57	.94
0540	1-1/2" x 1-1/2"		269	.030			.66		.66	1.08
0560	1-1/2" x 1-7/8"		229	.035			.77		.77	1.26
0600	Copper bus duct, indoor, 3 phase									
0610	Including hangers & supports									
0620	225 amp	2 Elec	135	.119	L.F.		2.62		2.62	4.28
0640	400 amp		106	.151			3.34		3.34	5.45
0660	600 amp		86	.186			4.11		4.11	6.75
0680	1000 amp		60	.267			5.90		5.90	9.65
0700	1600 amp		40	.400			8.85		8.85	14.45
0720	3000 amp		10	1.600			35.50		35.50	58
1300	Transformer, dry type, 1 ph, incl. removal of									
1320	supports, wire & pipe terminations									
1340	1 kVA	1 Elec	7.70	1.039	Ea.		23		23	37.50
1420	75 kVA	"	1.25	6.400	"		141		141	231
1440	3 Phase to 600V, primary									
1460	3 kVA	1 Elec	3.85	2.078	Ea.		46		46	75
1520	75 kVA	"	1.35	5.926			131		131	214
1550	300 kVA	R-3	1.80	11.111			245	77	322	485
1570	750 kVA	"	1.10	18.182			400	126	526	795
1800	Wire, THW-THWN-THHN, removed from									
1810	in place conduit, to 15' high									
1830	#14	1 Elec	65	.123	C.L.F.		2.72		2.72	4.45
1840	#12		55	.145			3.21		3.21	5.25
1850	#10		45.50	.176			3.89		3.89	6.35
1860	#8		40.40	.198			4.38		4.38	7.15
1870	#6		32.60	.245			5.40		5.40	8.85
1880	#4	2 Elec	53	.302			6.65		6.65	10.90
1890	#3		50	.320			7.05		7.05	11.55

16050 | Basic Electrical Materials & Methods

16055 | Selective Demolition

		CREW	DAILY OUTPUT	LABOR-HOURS	UNIT	2000 BARE COSTS MAT.	2000 BARE COSTS LABOR	2000 BARE COSTS EQUIP.	2000 BARE COSTS TOTAL	TOTAL INCL O&P
1900	#2	2 Elec	44.60	.359	C.L.F.		7.95		7.95	12.95
1910	1/0		33.20	.482			10.65		10.65	17.40
1920	2/0		29.20	.548			12.10		12.10	19.80
1930	3/0		25	.640			14.15		14.15	23
1940	4/0		22	.727			16.05		16.05	26.50
1950	250 kcmil		20	.800			17.70		17.70	29
1960	300 kcmil		19	.842			18.60		18.60	30.50
1970	350 kcmil		18	.889			19.65		19.65	32
1980	400 kcmil		17	.941			21		21	34
1990	500 kcmil		16.20	.988			22		22	35.50
2000	Interior fluorescent fixtures, incl. supports									
2010	& whips, to 15' high									
2100	Recessed drop-in 2' x 2', 2 lamp	2 Elec	35	.457	Ea.		10.10		10.10	16.55
2120	2' x 4', 2 lamp		33	.485			10.70		10.70	17.55
2140	2' x 4', 4 lamp		30	.533			11.80		11.80	19.30
2160	4' x 4', 4 lamp		20	.800			17.70		17.70	29
2180	Surface mount, acrylic lens & hinged frame									
2200	1' x 4', 2 lamp	2 Elec	44	.364	Ea.		8.05		8.05	13.15
2220	2' x 2', 2 lamp		44	.364			8.05		8.05	13.15
2260	2' x 4', 4 lamp		33	.485			10.70		10.70	17.55
2280	4' x 4', 4 lamp		40	.400			8.85		8.85	14.45
2300	Strip fixtures, surface mount									
2320	4' long, 1 lamp	2 Elec	53	.302	Ea.		6.65		6.65	10.90
2340	4' long, 2 lamp		50	.320			7.05		7.05	11.55
2360	8' long, 1 lamp		42	.381			8.40		8.40	13.75
2380	8' long, 2 lamp		40	.400			8.85		8.85	14.45
2400	Pendant mount, industrial, incl. removal									
2410	of chain or rod hangers, to 15' high									
2420	4' long, 2 lamp	2 Elec	35	.457	Ea.		10.10		10.10	16.55
2440	8' long, 2 lamp	"	27	.593	"		13.10		13.10	21.50

16060 | Grounding & Bonding

		CREW	DAILY OUTPUT	LABOR-HOURS	UNIT	MAT.	LABOR	EQUIP.	TOTAL	INCL O&P
0010	**GROUNDING**									
0030	Rod, copper clad, 8' long, 1/2" diameter	1 Elec	5.50	1.455	Ea.	15.85	32		47.85	70
0050	3/4" diameter		5.30	1.509		30	33.50		63.50	87.50
0080	10' long, 1/2" diameter		4.80	1.667		17.50	37		54.50	80
0100	3/4" diameter		4.40	1.818		29.50	40		69.50	98
0130	15' long, 3/4" diameter		4	2		79	44		123	160
0390	Bare copper wire, #8 stranded		11	.727	C.L.F.	14	16.05		30.05	42
0400	#6		10	.800		21	17.70		38.70	52.50
0600	#2	2 Elec	10	1.600		51	35.50		86.50	115
0800	3/0		6.60	2.424		135	53.50		188.50	236
1000	4/0		5.70	2.807		168	62		230	286
1200	250 kcmil	3 Elec	7.20	3.333		199	73.50		272.50	340
1800	Water pipe ground clamps, heavy duty									
2000	Bronze, 1/2" to 1" diameter	1 Elec	8	1	Ea.	11.70	22		33.70	49
2100	1-1/4" to 2" diameter		8	1		15.30	22		37.30	53
2200	2-1/2" to 3" diameter		6	1.333		41	29.50		70.50	93
2800	Brazed connections, #6 wire		12	.667		10.75	14.75		25.50	36
3000	#2 wire		10	.800		14.40	17.70		32.10	45
3100	3/0 wire		8	1		21.50	22		43.50	60
3200	4/0 wire		7	1.143		24.50	25.50		50	68.50
3400	250 kcmil wire		5	1.600		29	35.50		64.50	89.50
3600	500 kcmil wire		4	2		35.50	44		79.50	112

16100 | Wiring Methods
16120 | Conductors & Cables

			CREW	DAILY OUTPUT	LABOR-HOURS	UNIT	2000 BARE COSTS MAT.	LABOR	EQUIP.	TOTAL	TOTAL INCL O&P	
120	0010	**ARMORED CABLE**										120
	0050	600 volt, copper (BX), #14, 2 conductor, solid	1 Elec	2.40	3.333	C.L.F.	46.50	73.50		120	173	
	0100	3 conductor, solid		2.20	3.636		61	80.50		141.50	198	
	0150	#12, 2 conductor, solid		2.30	3.478		48.50	77		125.50	179	
	0200	3 conductor, solid		2	4		72.50	88.50		161	225	
	0250	#10, 2 conductor, solid		2	4		85.50	88.50		174	239	
	0300	3 conductor, solid		1.60	5		112	111		223	305	
	0350	#8, 3 conductor, solid		1.30	6.154		183	136		319	425	
	0400	3 conductor with PVC jacket, in cable tray, #6		3.10	2.581		249	57		306	370	
	0450	#4	2 Elec	5.40	2.963		320	65.50		385.50	460	
	0500	#2		4.60	3.478		425	77		502	595	
	0550	#1		4	4		575	88.50		663.50	780	
	0600	1/0		3.60	4.444		680	98		778	910	
	0650	2/0		3.40	4.706		825	104		929	1,075	
	0700	3/0		3.20	5		965	111		1,076	1,225	
	0750	4/0		3	5.333		1,100	118		1,218	1,425	
	0800	250 kcmil	3 Elec	3.60	6.667		1,250	147		1,397	1,625	
	0850	350 kcmil		3.30	7.273		1,650	161		1,811	2,100	
	0900	500 kcmil		3	8		2,150	177		2,327	2,675	
	1050	5 kV, copper, 3 conductor with PVC jacket,										
	1060	non-shielded, in cable tray, #4	2 Elec	380	.042	L.F.	4.90	.93		5.83	6.90	
	1100	#2		360	.044		6.40	.98		7.38	8.65	
	1200	#1		300	.053		8.10	1.18		9.28	10.85	
	1400	1/0		290	.055		9.35	1.22		10.57	12.30	
	1600	2/0		260	.062		10.85	1.36		12.21	14.15	
	2000	4/0		240	.067		14.45	1.47		15.92	18.30	
	2100	250 kcmil	3 Elec	330	.073		19.75	1.61		21.36	24	
	2150	350 kcmil		315	.076		24.50	1.68		26.18	30	
	2200	500 kcmil		270	.089		30	1.96		31.96	35.50	
	2400	15 kV, copper, 3 conductor with PVC jacket galv steel armored										
	2500	grounded neutral, in cable tray, #2	2 Elec	300	.053	L.F.	10.30	1.18		11.48	13.30	
	2600	#1		280	.057		11	1.26		12.26	14.15	
	2800	1/0		260	.062		12.65	1.36		14.01	16.10	
	2900	2/0		220	.073		16.65	1.61		18.26	21	
	3000	4/0		190	.084		18.95	1.86		20.81	24	
	3100	250 kcmil	3 Elec	270	.089		21	1.96		22.96	26	
	3150	350 kcmil		240	.100		25	2.21		27.21	31	
	3200	500 kcmil		210	.114		33.50	2.53		36.03	40.50	
	3400	15 kV, copper, 3 conductor with PVC jacket,										
	3450	ungrounded neutral, in cable tray, #2	2 Elec	260	.062	L.F.	11.10	1.36		12.46	14.40	
	3500	#1		230	.070		12.30	1.54		13.84	16.05	
	3600	1/0		200	.080		14.05	1.77		15.82	18.35	
	3700	2/0		190	.084		17.30	1.86		19.16	22	
	3800	4/0		160	.100		21	2.21		23.21	26.50	
	4000	250 kcmil	3 Elec	210	.114		24.50	2.53		27.03	31	
	4050	350 kcmil		195	.123		32	2.72		34.72	39.50	
	4100	500 kcmil		180	.133		39	2.95		41.95	48	
	9010	600 volt, copper (MC) steel clad, #14, 2 wire	1 Elec	2.40	3.333	C.L.F.	46.50	73.50		120	172	
	9020	3 wire		2.20	3.636		60	80.50		140.50	197	
	9040	#12, 2 wire		2.30	3.478		49.50	77		126.50	181	
	9050	3 wire		2	4		73	88.50		161.50	226	
	9070	#10, 2 wire		2	4		85	88.50		173.50	239	
	9080	3 wire		1.60	5		128	111		239	320	
	9100	#8, 2 wire, stranded		1.80	4.444		162	98		260	340	
	9110	3 wire, stranded		1.30	6.154		230	136		366	475	
	9200	600 volt, copper (MC) alum. clad, #14, 2 wire		2.65	3.019		46.50	66.50		113	160	

16100 | Wiring Methods

16120 | Conductors & Cables

			CREW	DAILY OUTPUT	LABOR-HOURS	UNIT	2000 BARE COSTS MAT.	LABOR	EQUIP.	TOTAL	TOTAL INCL O&P	
120	9210	3 wire	1 Elec	2.45	3.265	C.L.F.	60	72		132	184	120
	9220	4 wire		2.20	3.636		83.50	80.50		164	223	
	9230	#12, 2 wire		2.55	3.137		49.50	69.50		119	168	
	9240	3 wire		2.20	3.636		73	80.50		153.50	211	
	9250	4 wire		2	4		95	88.50		183.50	250	
	9260	#10, 2 wire		2.20	3.636		85	80.50		165.50	225	
	9270	3 wire		1.80	4.444		128	98		226	300	
	9280	4 wire		1.55	5.161		197	114		311	405	
230	0010	**CABLE TERMINATIONS**										230
	0015	Wire connectors, screw type, #22 to #14	1 Elec	260	.031	Ea.	.05	.68		.73	1.17	
	0020	#18 to #12		240	.033		.06	.74		.80	1.28	
	0025	#18 to #10		240	.033		.09	.74		.83	1.31	
	0030	Screw-on connectors, insulated, #18 to #12		240	.033		.08	.74		.82	1.30	
	0035	#16 to #10		230	.035		.10	.77		.87	1.37	
	0040	#14 to #8		210	.038		.22	.84		1.06	1.62	
	0045	#12 to #6		180	.044		.35	.98		1.33	2	
	0050	Terminal lugs, solderless, #16 to #10		50	.160		.44	3.54		3.98	6.30	
	0100	#8 to #4		30	.267		.55	5.90		6.45	10.25	
	0150	#2 to #1		22	.364		.74	8.05		8.79	13.95	
	0200	1/0 to 2/0		16	.500		2.05	11.05		13.10	20.50	
	0250	3/0		12	.667		2.50	14.75		17.25	27	
	0300	4/0		11	.727		2.50	16.05		18.55	29.50	
	0350	250 kcmil		9	.889		2.50	19.65		22.15	35	
	0400	350 kcmil		7	1.143		3.25	25.50		28.75	45	
	0450	500 kcmil		6	1.333		6.30	29.50		35.80	55	
	1600	Crimp 1 hole lugs, copper or aluminum, 600 volt										
	1620	#14	1 Elec	60	.133	Ea.	.27	2.95		3.22	5.10	
	1630	#12		50	.160		.38	3.54		3.92	6.20	
	1640	#10		45	.178		.38	3.93		4.31	6.85	
	1780	#8		36	.222		1.42	4.91		6.33	9.60	
	1800	#6		30	.267		2.48	5.90		8.38	12.40	
	2000	#4		27	.296		2.79	6.55		9.34	13.75	
	2200	#2		24	.333		3.54	7.35		10.89	15.95	
	2400	#1		20	.400		4.06	8.85		12.91	18.90	
	2600	2/0		15	.533		4.65	11.80		16.45	24.50	
	2800	3/0		12	.667		5.45	14.75		20.20	30	
	3000	4/0		11	.727		5.95	16.05		22	33	
	3200	250 kcmil		9	.889		7.20	19.65		26.85	40	
	3400	300 kcmil		8	1		8.05	22		30.05	45	
	3500	350 kcmil		7	1.143		8.20	25.50		33.70	50.50	
	3600	400 kcmil		6.50	1.231		9.70	27		36.70	55	
	3800	500 kcmil		6	1.333		11.75	29.50		41.25	61	
280	0010	**CONTROL CABLE**										280
	0020	600 volt, copper, #14 THWN wire with PVC jacket, 2 wires	1 Elec	9	.889	C.L.F.	13	19.65		32.65	46.50	
	0100	4 wires		7	1.143		20.50	25.50		46	64	
	0200	6 wires		6	1.333		33	29.50		62.50	84.50	
	0300	8 wires		5.30	1.509		39	33.50		72.50	97.50	
	0400	10 wires		4.80	1.667		47	37		84	112	
	0500	12 wires		4.30	1.860		55	41		96	128	
	0600	14 wires		3.80	2.105		65	46.50		111.50	148	
	0700	16 wires		3.50	2.286		76	50.50		126.50	166	
	0800	18 wires		3.30	2.424		79	53.50		132.50	175	
	0900	20 wires		3	2.667		97	59		156	204	
	1000	22 wires		2.80	2.857		99	63		162	212	
400	0010	**FIBER OPTICS**										400
	0020	Fiber optics cable only. Added costs depend on the type of fiber										

16100 | Wiring Methods
16120 | Conductors & Cables

			CREW	DAILY OUTPUT	LABOR-HOURS	UNIT	2000 BARE COSTS MAT.	LABOR	EQUIP.	TOTAL	TOTAL INCL O&P	
400	0030	special connectors, optical modems, and networking parts.										400
	0040	Specialized tools & techniques cause installation costs to vary.										
	0070	Cable, minimum, bulk simplex	1 Elec	8	1	C.L.F.	24	22		46	62.50	
	0080	Cable, maximum, bulk plenum quad	"	2.29	3.493	"	117	77		194	255	
	0150	Fiber optic jumper				Ea.	55.50			55.50	61	
	0200	Fiber optic pigtail					30			30	33	
	0300	Fiber optic connector	1 Elec	24	.333		16.90	7.35		24.25	30.50	
	0350	Fiber optic finger splice		32	.250		32	5.55		37.55	44	
	0400	Transceiver (low cost bi-directional)		8	1		375	22		397	450	
	0450	Multi-channel rack enclosure (10 modules)		2	4		550	88.50		638.50	750	
	0500	Fiber optic patch panel (12 ports)	↓	6	1.333	↓	178	29.50		207.50	244	
500	0010	**MINERAL INSULATED CABLE** 600 volt										500
	0100	1 conductor, #12	1 Elec	1.60	5	C.L.F.	253	111		364	460	
	0200	#10		1.60	5		264	111		375	470	
	0400	#8		1.50	5.333		299	118		417	525	
	0500	#6		1.40	5.714		335	126		461	575	
	0600	#4	2 Elec	2.40	6.667		465	147		612	750	
	0800	#2		2.20	7.273		620	161		781	945	
	0900	#1		2.10	7.619		710	168		878	1,050	
	1000	1/0		2	8		815	177		992	1,175	
	1100	2/0		1.90	8.421		975	186		1,161	1,375	
	1200	3/0		1.80	8.889		1,150	196		1,346	1,600	
	1400	4/0	↓	1.60	10		1,325	221		1,546	1,800	
	1410	250 kcmil	3 Elec	2.40	10		1,500	221		1,721	2,000	
	1420	350 kcmil		1.95	12.308		1,675	272		1,947	2,300	
	1430	500 kcmil	↓	1.95	12.308	↓	2,100	272		2,372	2,750	
550	0010	**NON-METALLIC SHEATHED CABLE** 600 volt										550
	0100	Copper with ground wire, (Romex)										
	0150	#14, 2 conductor	1 Elec	2.70	2.963	C.L.F.	13.40	65.50		78.90	122	
	0200	3 conductor		2.40	3.333		22.50	73.50		96	146	
	0250	#12, 2 conductor		2.50	3.200		18.90	70.50		89.40	137	
	0300	3 conductor		2.20	3.636		32	80.50		112.50	166	
	0350	#10, 2 conductor		2.20	3.636		32	80.50		112.50	166	
	0400	3 conductor		1.80	4.444		48.50	98		146.50	214	
	0450	#8, 3 conductor		1.50	5.333		97.50	118		215.50	300	
	0500	#6, 3 conductor	↓	1.40	5.714	↓	150	126		276	370	
	0550	SE type SER aluminum cable, 3 RHW and										
	0600	1 bare neutral, 3 #8 & 1 #8	1 Elec	1.60	5	C.L.F.	96	111		207	287	
	0650	3 #6 & 1 #6	"	1.40	5.714		109	126		235	325	
	0700	3 #4 & 1 #6	2 Elec	2.40	6.667		122	147		269	375	
	0750	3 #2 & 1 #4		2.20	7.273		179	161		340	460	
	0800	3 #1/0 & 1 #2		2	8		271	177		448	585	
	0850	3 #2/0 & 1 #1		1.80	8.889		320	196		516	670	
	0900	3 #4/0 & 1 #2/0	↓	1.60	10	↓	455	221		676	860	
700	0010	**SHIELDED CABLE** Splicing & terminations not included										700
	0040	Copper, XLP shielding, 5 kV, #6	2 Elec	4.40	3.636	C.L.F.	85	80.50		165.50	225	
	0050	#4		4.40	3.636		110	80.50		190.50	252	
	0100	#2		4	4		130	88.50		218.50	288	
	0200	#1		4	4		150	88.50		238.50	310	
	0400	1/0		3.80	4.211		165	93		258	335	
	0600	2/0		3.60	4.444		205	98		303	385	
	0800	4/0	↓	3.20	5		270	111		381	480	
	1000	250 kcmil	3 Elec	4.50	5.333		325	118		443	555	
	1200	350 kcmil	↓	3.90	6.154	↓	420	136		556	680	

16100 | Wiring Methods

16120 | Conductors & Cables

			CREW	DAILY OUTPUT	LABOR-HOURS	UNIT	2000 BARE COSTS				TOTAL INCL O&P
							MAT.	LABOR	EQUIP.	TOTAL	
700	1400	500 kcmil	3 Elec	3.60	6.667	C.L.F.	520	147		667	810
	1600	15 KV, ungrounded neutral, #1	2 Elec	4	4		184	88.50		272.50	345
	1800	1/0		3.80	4.211		220	93		313	395
	2000	2/0		3.60	4.444		250	98		348	435
	2200	4/0		3.20	5		330	111		441	545
	2400	250 kcmil	3 Elec	4.50	5.333		365	118		483	595
	2600	350 kcmil		3.90	6.154		460	136		596	725
	2800	500 kcmil		3.60	6.667		570	147		717	865
900	0010	**WIRE**									
	0020	600 volt type THW, copper, solid, #14	1 Elec	13	.615	C.L.F.	4.08	13.60		17.68	27
	0030	#12		11	.727		5.25	16.05		21.30	32.50
	0040	#10		10	.800		8.40	17.70		26.10	38.50
	0050	Stranded, #14		13	.615		4.40	13.60		18	27.50
	0100	#12		11	.727		6.05	16.05		22.10	33
	0120	#10		10	.800		9.30	17.70		27	39.50
	0140	#8		8	1		15.60	22		37.60	53
	0160	#6		6.50	1.231		25.50	27		52.50	72.50
	0180	#4	2 Elec	10.60	1.509		41.50	33.50		75	100
	0200	#3		10	1.600		47.50	35.50		83	111
	0220	#2		9	1.778		59.50	39.50		99	130
	0240	#1		8	2		75	44		119	155
	0260	1/0		6.60	2.424		92.50	53.50		146	190
	0280	2/0		5.80	2.759		115	61		176	226
	0300	3/0		5	3.200		145	70.50		215.50	275
	0350	4/0		4.40	3.636		180	80.50		260.50	330
	0400	250 kcmil	3 Elec	6	4		216	88.50		304.50	385
	0420	300 kcmil		5.70	4.211		257	93		350	435
	0450	350 kcmil		5.40	4.444		297	98		395	485
	0480	400 kcmil		5.10	4.706		345	104		449	550
	0490	500 kcmil		4.80	5		420	111		531	640
	0530	Aluminum, stranded, #8	1 Elec	9	.889		12.35	19.65		32	45.50
	0540	#6	"	8	1		15.75	22		37.75	53.50
	0560	#4	2 Elec	13	1.231		19.70	27		46.70	66
	0580	#2		10.60	1.509		26.50	33.50		60	84
	0600	#1		9	1.778		39	39.50		78.50	108
	0620	1/0		8	2		46.50	44		90.50	124
	0640	2/0		7.20	2.222		55	49		104	141
	0680	3/0		6.60	2.424		68.50	53.50		122	163
	0700	4/0		6.20	2.581		76	57		133	177
	0720	250 kcmil	3 Elec	8.70	2.759		92.50	61		153.50	202
	0740	300 kcmil		8.10	2.963		128	65.50		193.50	248
	0760	350 kcmil		7.50	3.200		130	70.50		200.50	259
	0780	400 kcmil		6.90	3.478		152	77		229	293
	0800	500 kcmil		6	4		168	88.50		256.50	330
	0850	600 kcmil		5.70	4.211		213	93		306	385
	0880	700 kcmil		5.10	4.706		245	104		349	440
	0900	750 kcmil		4.80	5		248	111		359	455
	0920	Type THWN-THHN, copper, solid, #14	1 Elec	13	.615		4.08	13.60		17.68	27
	0940	#12		11	.727		5.25	16.05		21.30	32.50
	0960	#10		10	.800		8.40	17.70		26.10	38.50
	1000	Stranded, #14		13	.615		4.40	13.60		18	27.50
	1200	#12		11	.727		6.05	16.05		22.10	33
	1250	#10		10	.800		9.30	17.70		27	39.50
	1300	#8		8	1		15.60	22		37.60	53
	1350	#6		6.50	1.231		25.50	27		52.50	72.50
	1400	#4	2 Elec	10.60	1.509		41.50	33.50		75	100

R16132-210

16100 | Wiring Methods

16131 | Cable Trays

			CREW	DAILY OUTPUT	LABOR-HOURS	UNIT	2000 BARE COSTS				TOTAL INCL O&P
							MAT.	LABOR	EQUIP.	TOTAL	
105	0010	**CABLE TRAY LADDER TYPE** w/ ftngs & supports, 4" dp., to 15' elev.									105
	0160	Galvanized steel tray									
	0170	4" rung spacing, 6" wide	2 Elec	98	.163	L.F.	8.15	3.61		11.76	14.85
	0200	12" wide		86	.186		9.80	4.11		13.91	17.55
	0400	18" wide		82	.195		11.40	4.31		15.71	19.60
	0600	24" wide		78	.205		13.10	4.53		17.63	22
	3200	Aluminum tray, 4" deep, 6" rung spacing, 6" wide		134	.119		10.25	2.64		12.89	15.55
	3220	12" wide		124	.129		11.45	2.85		14.30	17.25
	3230	18" wide		114	.140		12.75	3.10		15.85	19.10
	3240	24" wide	▼	106	.151	▼	14.90	3.34		18.24	22
	9980	Allow. for tray ftngs., 5% min.-20% max.									

16132 | Conduit & Tubing

			CREW	DAILY OUTPUT	LABOR-HOURS	UNIT	MAT.	LABOR	EQUIP.	TOTAL	TOTAL INCL O&P
205	0010	**CONDUIT** To 15' high, includes 2 terminations, 2 elbows and									205
	0020	11 beam clamps per 100 L.F. R16132-210									
	0300	Aluminum, 1/2" diameter	1 Elec	100	.080	L.F.	1.18	1.77		2.95	4.19
	0500	3/4" diameter		90	.089		1.58	1.96		3.54	4.95
	0700	1" diameter		80	.100		2.14	2.21		4.35	6
	1000	1-1/4" diameter		70	.114		2.77	2.53		5.30	7.15
	1030	1-1/2" diameter		65	.123		3.43	2.72		6.15	8.20
	1050	2" diameter		60	.133		4.54	2.95		7.49	9.80
	1070	2-1/2" diameter	▼	50	.160		7.05	3.54		10.59	13.55
	1100	3" diameter	2 Elec	90	.178		9.45	3.93		13.38	16.85
	1130	3-1/2" diameter		80	.200		11.60	4.42		16.02	20
	1140	4" diameter	▼	70	.229		14.10	5.05		19.15	24
	1750	Rigid galvanized steel, 1/2" diameter	1 Elec	90	.089		1.52	1.96		3.48	4.88
	1770	3/4" diameter		80	.100		1.81	2.21		4.02	5.60
	1800	1" diameter		65	.123		2.53	2.72		5.25	7.25
	1830	1-1/4" diameter		60	.133		3.38	2.95		6.33	8.55
	1850	1-1/2" diameter		55	.145		4.02	3.21		7.23	9.65
	1870	2" diameter		45	.178		5.35	3.93		9.28	12.35
	1900	2-1/2" diameter	▼	35	.229		8.90	5.05		13.95	18
	1930	3" diameter	2 Elec	50	.320		11.30	7.05		18.35	24
	1950	3-1/2" diameter		44	.364		13.75	8.05		21.80	28.50
	1970	4" diameter	▼	40	.400		16.50	8.85		25.35	32.50
	2500	Steel, intermediate conduit (IMC), 1/2" diameter	1 Elec	100	.080		1.21	1.77		2.98	4.22
	2530	3/4" diameter		90	.089		1.44	1.96		3.40	4.79
	2550	1" diameter		70	.114		1.98	2.53		4.51	6.30
	2570	1-1/4" diameter		65	.123		2.61	2.72		5.33	7.35
	2600	1-1/2" diameter		60	.133		2.98	2.95		5.93	8.10
	2630	2" diameter		50	.160		3.87	3.54		7.41	10.05
	2650	2-1/2" diameter	▼	40	.200		7.65	4.42		12.07	15.70
	2670	3" diameter	2 Elec	60	.267		9.90	5.90		15.80	20.50
	2700	3-1/2" diameter		54	.296		12.05	6.55		18.60	24
	2730	4" diameter	▼	50	.320		14	7.05		21.05	27
	5000	Electric metallic tubing (EMT), 1/2" diameter	1 Elec	170	.047		.36	1.04		1.40	2.10
	5020	3/4" diameter		130	.062		.55	1.36		1.91	2.82
	5040	1" diameter		115	.070		.93	1.54		2.47	3.53
	5060	1-1/4" diameter		100	.080		1.39	1.77		3.16	4.42
	5080	1-1/2" diameter		90	.089		1.76	1.96		3.72	5.15
	5100	2" diameter		80	.100		2.23	2.21		4.44	6.10
	5120	2-1/2" diameter	▼	60	.133		5.95	2.95		8.90	11.35
	5140	3" diameter	2 Elec	100	.160		6.45	3.54		9.99	12.90
	5160	3-1/2" diameter		90	.178		8.35	3.93		12.28	15.65
	5180	4" diameter	▼	80	.200	▼	9.45	4.42		13.87	17.65

16100 | Wiring Methods

16132 | Conduit & Tubing

			CREW	DAILY OUTPUT	LABOR-HOURS	UNIT	2000 BARE COSTS				TOTAL INCL O&P	
							MAT.	LABOR	EQUIP.	TOTAL		
205	9900	Add to labor for higher elevated installation										205
	9910	15' to 20' high, add	R16132-210					10%				
	9920	20' to 25' high, add						20%				
	9930	25' to 30' high, add						25%				
	9940	30' to 35' high, add						30%				
	9950	35' to 40' high, add						35%				
	9960	Over 40' high, add						40%				
	9980	Allow. for cond. ftngs., 5% min.-20% max.										
230	0010	**CONDUIT IN CONCRETE SLAB** Including terminations,										230
	0020	fittings and supports										
	3230	PVC, schedule 40, 1/2" diameter	1 Elec	270	.030	L.F.	.35	.65		1	1.45	
	3250	3/4" diameter		230	.035		.42	.77		1.19	1.73	
	3270	1" diameter		200	.040		.56	.88		1.44	2.07	
	3300	1-1/4" diameter		170	.047		.78	1.04		1.82	2.56	
	3330	1-1/2" diameter		140	.057		.96	1.26		2.22	3.13	
	3350	2" diameter		120	.067		1.23	1.47		2.70	3.76	
	4350	Rigid galvanized steel, 1/2" diameter		200	.040		1.27	.88		2.15	2.84	
	4400	3/4" diameter		170	.047		1.56	1.04		2.60	3.41	
	4450	1" diameter		130	.062		2.28	1.36		3.64	4.73	
	4500	1-1/4" diameter		110	.073		2.96	1.61		4.57	5.90	
	4600	1-1/2" diameter		100	.080		3.59	1.77		5.36	6.85	
	4800	2" diameter		90	.089		4.79	1.96		6.75	8.45	
240	0010	**CONDUIT IN TRENCH** Includes terminations and fittings										240
	0020	Does not include excavation or backfill, see div. 02315										
	0200	Rigid galvanized steel, 2" diameter	1 Elec	150	.053	L.F.	4.63	1.18		5.81	7.05	
	0400	2-1/2" diameter	"	100	.080		8.05	1.77		9.82	11.75	
	0600	3" diameter	2 Elec	160	.100		10.40	2.21		12.61	15.05	
	0800	3-1/2" diameter		140	.114		13.20	2.53		15.73	18.65	
	1000	4" diameter		100	.160		15.15	3.54		18.69	22.50	
	1200	5" diameter		80	.200		31.50	4.42		35.92	42	
	1400	6" diameter		60	.267		45	5.90		50.90	58.50	
320	0010	**FLEXIBLE METALLIC CONDUIT**										320
	0050	Steel, 3/8" diameter	1 Elec	200	.040	L.F.	.23	.88		1.11	1.70	
	0100	1/2" diameter		200	.040		.32	.88		1.20	1.80	
	0200	3/4" diameter		160	.050		.41	1.11		1.52	2.26	
	0250	1" diameter		100	.080		.85	1.77		2.62	3.83	
	0300	1-1/4" diameter		70	.114		1.09	2.53		3.62	5.35	
	0350	1-1/2" diameter		50	.160		1.40	3.54		4.94	7.35	
	0370	2" diameter		40	.200		2.05	4.42		6.47	9.50	
	0420	Connectors, plain, 3/8" diameter		100	.080	Ea.	.95	1.77		2.72	3.94	
	0430	1/2" diameter		80	.100		1.44	2.21		3.65	5.20	
	0440	3/4" diameter		70	.114		1.60	2.53		4.13	5.90	
	0450	1" diameter		50	.160		3.93	3.54		7.47	10.10	
	0452	1-1/4" diameter		45	.178		6.65	3.93		10.58	13.75	
	0454	1-1/2" diameter		40	.200		9.35	4.42		13.77	17.55	
	0456	2" diameter		28	.286		13.30	6.30		19.60	25	
	0490	Insulated, 1" diameter		40	.200		4.65	4.42		9.07	12.35	
	0500	1-1/4" diameter		40	.200		8.60	4.42		13.02	16.70	
	0550	1-1/2" diameter		32	.250		12.70	5.55		18.25	23	
	0600	2" diameter		23	.348		18.15	7.70		25.85	32.50	

16133 | Multi-outlet Assemblies

540	0010	**TRENCH DUCT** Steel with cover										540
	0020	Standard adjustable, depths to 4"										

16100 | Wiring Methods

16133 | Multi-outlet Assemblies

			CREW	DAILY OUTPUT	LABOR-HOURS	UNIT	MAT.	LABOR	EQUIP.	TOTAL	TOTAL INCL O&P	
540	0100	Straight, single compartment, 9" wide	2 Elec	40	.400	L.F.	66	8.85		74.85	87	540
	0200	12" wide		32	.500		75	11.05		86.05	101	
	0400	18" wide		26	.615		100	13.60		113.60	133	
	0600	24" wide		22	.727		128	16.05		144.05	168	
	0800	30" wide		20	.800		155	17.70		172.70	200	
	1000	36" wide		16	1		186	22		208	241	
	1200	Horizontal elbow, 9" wide		5.40	2.963	Ea.	245	65.50		310.50	375	
	1400	12" wide		4.60	3.478		282	77		359	435	
	1600	18" wide		4	4		365	88.50		453.50	545	
	1800	24" wide		3.20	5		510	111		621	740	
	2000	30" wide		2.60	6.154		680	136		816	970	
	2200	36" wide		2.40	6.667		890	147		1,037	1,225	
	2400	Vertical elbow, 9" wide		5.40	2.963		85	65.50		150.50	201	
	2600	12" wide		4.60	3.478		93	77		170	228	
	2800	18" wide		4	4		107	88.50		195.50	263	
	3000	24" wide		3.20	5		132	111		243	325	
	3200	30" wide		2.60	6.154		147	136		283	385	
	3400	36" wide		2.40	6.667		161	147		308	420	
	3600	Cross, 9" wide		4	4		405	88.50		493.50	590	
	3800	12" wide		3.20	5		425	111		536	650	
	4000	18" wide		2.60	6.154		510	136		646	780	
	4200	24" wide		2.20	7.273		645	161		806	975	
	4400	30" wide		2	8		840	177		1,017	1,225	
	4600	36" wide		1.80	8.889		1,050	196		1,246	1,475	
	4800	End closure, 9" wide		14.40	1.111		25	24.50		49.50	67.50	
	5000	12" wide		12	1.333		29	29.50		58.50	80	
	5200	18" wide		10	1.600		44	35.50		79.50	107	
	5400	24" wide		8	2		58	44		102	137	
	5600	30" wide		6.60	2.424		72	53.50		125.50	167	
	5800	36" wide		5.80	2.759		86	61		147	194	
	6000	Tees, 9" wide		4	4		244	88.50		332.50	415	
	6200	12" wide		3.60	4.444		284	98		382	470	
	6400	18" wide		3.20	5		365	111		476	580	
	6600	24" wide		3	5.333		520	118		638	770	
	6800	30" wide		2.60	6.154		680	136		816	970	
	7000	36" wide		2	8		890	177		1,067	1,275	
	7200	Riser, and cabinet connector, 9" wide		5.40	2.963		107	65.50		172.50	225	
	7400	12" wide		4.60	3.478		124	77		201	262	
	7600	18" wide		4	4		153	88.50		241.50	315	
	7800	24" wide		3.20	5		184	111		295	385	
	8000	30" wide		2.60	6.154		213	136		349	455	
	8200	36" wide		2	8		248	177		425	560	
	8400	Insert assembly, cell to conduit adapter, 1-1/4"	1 Elec	16	.500		42	11.05		53.05	64	
560	0010	**UNDERFLOOR DUCT**										560
	0100	Duct, 1-3/8" x 3-1/8" blank, standard	2 Elec	160	.100	L.F.	7.25	2.21		9.46	11.55	
	0200	1-3/8" x 7-1/4" blank, super duct		120	.133		14.45	2.95		17.40	20.50	
	0400	7/8" or 1-3/8" insert type, 24" O.C., 1-3/8" x 3-1/8", std.		140	.114		9.70	2.53		12.23	14.80	
	0600	1-3/8" x 7-1/4", super duct		100	.160		16.90	3.54		20.44	24.50	
	0800	Junction box, single duct, 1 level, 3-1/8"	1 Elec	4	2	Ea.	218	44		262	315	
	1000	Junction box, single duct, 1 level, 7-1/4"		2.70	2.963		238	65.50		303.50	370	
	1200	1 level, 2 duct, 3-1/8"		3.20	2.500		290	55.50		345.50	410	
	1400	Junction box, 1 level, 2 duct, 7-1/4"		2.30	3.478		785	77		862	990	
	1600	Triple duct, 3-1/8"		2.30	3.478		495	77		572	670	
	1800	Insert to conduit adapter, 3/4" & 1"		32	.250		9.95	5.55		15.50	20	
	2000	Support, single cell		27	.296		15.80	6.55		22.35	28	

16100 | Wiring Methods

16133 | Multi-outlet Assemblies

			CREW	DAILY OUTPUT	LABOR-HOURS	UNIT	2000 BARE COSTS				TOTAL INCL O&P	
							MAT.	LABOR	EQUIP.	TOTAL		
560	2200	Super duct	1 Elec	16	.500	Ea.	15.80	11.05		26.85	35.50	560
	2400	Double cell		16	.500		15.80	11.05		26.85	35.50	
	2600	Triple cell		11	.727		14.90	16.05		30.95	43	
	2800	Vertical elbow, standard duct		10	.800		47.50	17.70		65.20	81.50	
	3000	Super duct		8	1		47.50	22		69.50	88.50	
	3200	Cabinet connector, standard duct		32	.250		36	5.55		41.55	49	
	3400	Super duct		27	.296		36	6.55		42.55	50.50	
	3600	Conduit adapter, 1" to 1-1/4"		32	.250		36	5.55		41.55	49	
	3800	2" to 1-1/4"		27	.296		36	6.55		42.55	50.50	
	4000	Outlet, low tension (tele, computer, etc.)		8	1		38	22		60	77.50	
	4200	High tension, receptacle (120 volt)		8	1		41.50	22		63.50	81.50	
800	0010	**SURFACE RACEWAY**										800
	0090	Metal, straight section										
	0100	No. 500	1 Elec	100	.080	L.F.	.63	1.77		2.40	3.58	
	0110	No. 700		100	.080		.71	1.77		2.48	3.67	
	0400	No. 1500, small pancake		90	.089		1.33	1.96		3.29	4.67	
	0600	No. 2000, base & cover, blank		90	.089		1.29	1.96		3.25	4.63	
	0800	No. 3000, base & cover, blank		75	.107		2.53	2.36		4.89	6.65	
	1000	No. 4000, base & cover, blank		65	.123		4.30	2.72		7.02	9.20	
	1200	No. 6000, base & cover, blank		50	.160		7.05	3.54		10.59	13.55	
	2400	Fittings, elbows, No. 500		40	.200	Ea.	1.14	4.42		5.56	8.50	
	2800	Elbow cover, No. 2000		40	.200		2.30	4.42		6.72	9.80	
	2880	Tee, No. 500		42	.190		2.21	4.21		6.42	9.35	
	2900	No. 2000		27	.296		7.05	6.55		13.60	18.45	
	3000	Switch box, No. 500		16	.500		8.10	11.05		19.15	27	
	3400	Telephone outlet, No. 1500		16	.500		8.60	11.05		19.65	27.50	
	3600	Junction box, No. 1500		16	.500		6	11.05		17.05	24.50	
	3800	Plugmold wired sections, No. 2000										
	4000	1 circuit, 6 outlets, 3 ft. long	1 Elec	8	1	Ea.	21.50	22		43.50	59.50	
	4100	2 circuits, 8 outlets, 6 ft. long	"	5.30	1.509	"	36	33.50		69.50	94	

16134 | Wireway & Aux Gutters

			CREW	DAILY OUTPUT	LABOR-HOURS	UNIT	MAT.	LABOR	EQUIP.	TOTAL	TOTAL INCL O&P	
150	0010	**WIREWAY** to 15' high										150
	0100	Screw cover, with fittings and supports, 2-1/2" x 2-1/2"	1 Elec	45	.178	L.F.	8.65	3.93		12.58	15.95	
	0200	4" x 4"	"	40	.200		9.55	4.42		13.97	17.75	
	0400	6" x 6"	2 Elec	60	.267		16.10	5.90		22	27.50	
	0600	8" x 8"	"	40	.400		29	8.85		37.85	46.50	
	9980	Allow. for wireway ftngs., 5% min.-20% max.										

16136 | Boxes

			CREW	DAILY OUTPUT	LABOR-HOURS	UNIT	MAT.	LABOR	EQUIP.	TOTAL	TOTAL INCL O&P	
600	0010	**OUTLET BOXES**										600
	0020	Pressed steel, octagon, 4"	1 Elec	20	.400	Ea.	1.46	8.85		10.31	16.05	
	0060	Covers, blank		64	.125		.60	2.76		3.36	5.20	
	0100	Extension		40	.200		2.33	4.42		6.75	9.80	
	0150	Square, 4"		20	.400		1.98	8.85		10.83	16.65	
	0200	Extension		40	.200		2.44	4.42		6.86	9.95	
	0250	Covers, blank		64	.125		.69	2.76		3.45	5.30	
	0300	Plaster rings		64	.125		1.11	2.76		3.87	5.75	
	0650	Switchbox		27	.296		2.26	6.55		8.81	13.20	
	1100	Concrete, floor, 1 gang		5.30	1.509		57.50	33.50		91	118	
	2000	Poke-thru fitting, fire rated, for 3-3/4" floor		6.80	1.176		90	26		116	142	
	2040	For 7" floor		6.80	1.176		90	26		116	142	
	2100	Pedestal, 15 amp, duplex receptacle & blank plate		5.25	1.524		93	33.50		126.50	157	
	2120	Duplex receptacle and telephone plate		5.25	1.524		93.50	33.50		127	158	

16100 | Wiring Methods

16136 | Boxes

			CREW	DAILY OUTPUT	LABOR-HOURS	UNIT	MAT.	LABOR	EQUIP.	TOTAL	TOTAL INCL O&P	
600	2140	Pedestal, 20 amp, duplex recept. & phone plate	1 Elec	5	1.600	Ea.	94	35.50		129.50	161	600
	2200	Abandonment plate	↓	32	.250	↓	27.50	5.55		33.05	39	
700	0010	**PULL BOXES & CABINETS**										700
	0100	Sheet metal, pull box, NEMA 1, type SC, 6" W x 6" H x 4" D	1 Elec	8	1	Ea.	8.35	22		30.35	45	
	0200	8" W x 8" H x 4" D		8	1		11.45	22		33.45	48.50	
	0300	10" W x 12" H x 6" D		5.30	1.509		20	33.50		53.50	76.50	
	0400	16" W x 20" H x 8" D		4	2		76	44		120	156	
	0500	20" W x 24" H x 8" D		3.20	2.500		89	55.50		144.50	189	
	0600	24" W x 36" H x 8" D		2.70	2.963		125	65.50		190.50	245	
	0650	Hinged cabinets, NEMA 1, 6" W x 6" H x 4" D		8	1		8.50	22		30.50	45.50	
	0800	12" W x 16" H x 6" D		4.70	1.702		28	37.50		65.50	92	
	1000	20" W x 20" H x 6" D		3.60	2.222		56	49		105	142	
	1200	20" W x 20" H x 8" D		3.20	2.500		113	55.50		168.50	215	
	1400	24" W x 36" H x 8" D		2.70	2.963		180	65.50		245.50	305	
	1600	24" W x 42" H x 8" D	↓	2	4	↓	273	88.50		361.50	445	
	2100	Pull box, NEMA 3R, type SC, raintight & weatherproof										
	2150	6" L x 6" W x 6" D	1 Elec	10	.800	Ea.	14.30	17.70		32	45	
	2200	8" L x 6" W x 6" D		8	1		17.85	22		39.85	55.50	
	2250	10" L x 6" W x 6" D		7	1.143		23.50	25.50		49	67.50	
	2300	12" L x 12" W x 6" D		5	1.600		33.50	35.50		69	95	
	2350	16" L x 16" W x 6" D		4.50	1.778		67	39.50		106.50	138	
	2400	20" L x 20" W x 6" D		4	2		92	44		136	174	
	2450	24" L x 18" W x 8" D		3	2.667		100	59		159	207	
	2500	24" L x 24" W x 10" D		2.50	3.200		239	70.50		309.50	380	
	2550	30" L x 24" W x 12" D		2	4		246	88.50		334.50	415	
	2600	36" L x 36" W x 12" D	↓	1.50	5.333	↓	325	118		443	550	
	2800	Cast iron, pull boxes for surface mounting										
	3000	NEMA 4, watertight & dust tight										
	3050	6" L x 6" W x 6" D	1 Elec	4	2	Ea.	139	44		183	226	
	3100	8" L x 6" W x 6" D		3.20	2.500		189	55.50		244.50	299	
	3150	10" L x 6" W x 6" D		2.50	3.200		234	70.50		304.50	375	
	3200	12" L x 12" W x 6" D		2.30	3.478		395	77		472	560	
	3250	16" L x 16" W x 6" D		1.30	6.154		810	136		946	1,100	
	3300	20" L x 20" W x 6" D		.80	10		1,500	221		1,721	2,000	
	3350	24" L x 18" W x 8" D		.70	11.429		1,575	253		1,828	2,175	
	3400	24" L x 24" W x 10" D		.50	16		2,425	355		2,780	3,250	
	3450	30" L x 24" W x 12" D		.40	20		4,075	440		4,515	5,200	
	3500	36" L x 36" W x 12" D	↓	.20	40	↓	5,025	885		5,910	6,975	
	4000	NEMA 7, explosionproof										
	4050	6" L x 6" W x 6" D	1 Elec	2	4	Ea.	465	88.50		553.50	655	
	4100	8" L x 6" W x 6" D		1.80	4.444		530	98		628	745	
	4150	10" L x 6" W x 6" D		1.60	5		650	111		761	895	
	4200	12" L x 12" W x 6" D		1	8		1,075	177		1,252	1,475	
	4250	16" L x 14" W x 6" D		.60	13.333		1,475	295		1,770	2,100	
	4300	18" L x 18" W x 8" D		.50	16		2,600	355		2,955	3,450	
	4350	24" L x 18" W x 8" D		.40	20		3,475	440		3,915	4,550	
	4400	24" L x 24" W x 10" D		.30	26.667		4,525	590		5,115	5,950	
	4450	30" L x 24" W x 12" D	↓	.20	40	↓	6,900	885		7,785	9,050	
	6000	J.I.C. wiring boxes, NEMA 12, dust tight & drip tight										
	6050	6" L x 8" W x 4" D	1 Elec	10	.800	Ea.	33	17.70		50.70	65	
	6100	8" L x 10" W x 4" D		8	1		41	22		63	81	
	6150	12" L x 14" W x 6" D		5.30	1.509		67	33.50		100.50	128	
	6200	14" L x 16" W x 6" D		4.70	1.702		79.50	37.50		117	149	
	6250	16" L x 20" W x 6" D		4.40	1.818		170	40		210	253	
	6300	24" L x 30" W x 6" D		3.20	2.500		250	55.50		305.50	365	
	6350	24" L x 30" W x 8" D		2.90	2.759	↓	267	61		328	395	

16100 | Wiring Methods

16136 | Boxes

		CREW	DAILY OUTPUT	LABOR-HOURS	UNIT	MAT.	LABOR	EQUIP.	TOTAL	TOTAL INCL O&P
6400	24" L x 36" W x 8" D	1 Elec	2.70	2.963	Ea.	295	65.50		360.50	430
6450	24" L x 42" W x 8" D		2.30	3.478		320	77		397	480
6500	24" L x 48" W x 8" D	▼	2	4	▼	350	88.50		438.50	530
7000	Cabinets, current transformer									
7050	Single door, 24" H x 24" W x 10" D	1 Elec	1.60	5	Ea.	107	111		218	299
7100	30" H x 24" W x 10" D		1.30	6.154		120	136		256	355
7150	36" H x 24" W x 10" D		1.10	7.273		134	161		295	410
7200	30" H x 30" W x 10" D		1	8		140	177		317	445
7250	36" H x 30" W x 10" D		.90	8.889		189	196		385	530
7300	36" H x 36" W x 10" D		.80	10		206	221		427	585
7500	Double door, 48" H x 36" W x 10" D		.60	13.333		360	295		655	875
7550	24" H x 24" W x 12" D	▼	1	8	▼	193	177		370	500

16139 | Residential Wiring

		CREW	DAILY OUTPUT	LABOR-HOURS	UNIT	MAT.	LABOR	EQUIP.	TOTAL	TOTAL INCL O&P
0010	**RESIDENTIAL WIRING**									
0020	20' avg. runs and #14/2 wiring incl. unless otherwise noted									
1000	Service & panel, includes 24' SE-AL cable, service eye, meter,									
1010	Socket, panel board, main bkr., ground rod, 15 or 20 amp									
1020	1-pole circuit breakers, and misc. hardware									
1100	100 amp, with 10 branch breakers	1 Elec	1.19	6.723	Ea.	400	149		549	680
1110	With PVC conduit and wire		.92	8.696		425	192		617	785
1120	With RGS conduit and wire		.73	10.959		545	242		787	995
1150	150 amp, with 14 branch breakers		1.03	7.767		615	172		787	960
1170	With PVC conduit and wire		.82	9.756		675	216		891	1,100
1180	With RGS conduit and wire	▼	.67	11.940		885	264		1,149	1,400
1200	200 amp, with 18 branch breakers	2 Elec	1.80	8.889		810	196		1,006	1,200
1220	With PVC conduit and wire		1.46	10.959		870	242		1,112	1,350
1230	With RGS conduit and wire	▼	1.24	12.903		1,150	285		1,435	1,750
1800	Lightning surge suppressor for above services, add	1 Elec	32	.250	▼	39.50	5.55		45.05	52.50
2000	Switch devices									
2100	Single pole, 15 amp, Ivory, with a 1-gang box, cover plate,									
2110	Type NM (Romex) cable	1 Elec	17.10	.468	Ea.	6.65	10.35		17	24.50
2120	Type MC (BX) cable		14.30	.559		16.80	12.35		29.15	38.50
2130	EMT & wire		5.71	1.401		17.45	31		48.45	69.50
2150	3-way, #14/3, type NM cable		14.55	.550		9.85	12.15		22	30.50
2170	Type MC cable		12.31	.650		21	14.35		35.35	46.50
2180	EMT & wire		5	1.600		19.65	35.50		55.15	79.50
2200	4-way, #14/3, type NM cable		14.55	.550		22.50	12.15		34.65	45
2220	Type MC cable		12.31	.650		34	14.35		48.35	60.50
2230	EMT & wire		5	1.600		32.50	35.50		68	93.50
2250	S.P., 20 amp, #12/2, type NM cable		13.33	.600		11.65	13.25		24.90	34.50
2270	Type MC cable		11.43	.700		21	15.45		36.45	48.50
2280	EMT & wire		4.85	1.649		23	36.50		59.50	84.50
2290	S.P. rotary dimmer, 600W, no wiring		17	.471		15.55	10.40		25.95	34
2300	S.P. rotary dimmer, 600W, type NM cable		14.55	.550		18.25	12.15		30.40	40
2320	Type MC cable		12.31	.650		28.50	14.35		42.85	54.50
2330	EMT & wire		5	1.600		30	35.50		65.50	91
2350	3-way rotary dimmer, type NM cable		13.33	.600		15.65	13.25		28.90	38.50
2370	Type MC cable		11.43	.700		26	15.45		41.45	54
2380	EMT & wire	▼	4.85	1.649	▼	27.50	36.50		64	89.50
2400	Interval timer wall switch, 20 amp, 1-30 min., #12/2									
2410	Type NM cable	1 Elec	14.55	.550	Ea.	28.50	12.15		40.65	51
2420	Type MC cable		12.31	.650		35.50	14.35		49.85	62.50
2430	EMT & wire	▼	5	1.600	▼	39.50	35.50		75	102
2500	Decorator style									
2510	S.P., 15 amp, type NM cable	1 Elec	17.10	.468	Ea.	10.05	10.35		20.40	28

16100 | Wiring Methods

16139 | Residential Wiring

		CREW	DAILY OUTPUT	LABOR-HOURS	UNIT	2000 BARE COSTS MAT.	LABOR	EQUIP.	TOTAL	TOTAL INCL O&P
2520	Type MC cable	1 Elec	14.30	.559	Ea.	20	12.35		32.35	42
2530	EMT & wire		5.71	1.401		21	31		52	73.50
2550	3-way, #14/3, type NM cable		14.55	.550		13.25	12.15		25.40	34.50
2570	Type MC cable		12.31	.650		24.50	14.35		38.85	50.50
2580	EMT & wire		5	1.600		23	35.50		58.50	83.50
2600	4-way, #14/3, type NM cable		14.55	.550		26	12.15		38.15	48.50
2620	Type MC cable		12.31	.650		37	14.35		51.35	64.50
2630	EMT & wire		5	1.600		36	35.50		71.50	97.50
2650	S.P., 20 amp, #12/2, type NM cable		13.33	.600		15.05	13.25		28.30	38
2670	Type MC cable		11.43	.700		24.50	15.45		39.95	52.50
2680	EMT & wire		4.85	1.649		26	36.50		62.50	88.50
2700	S.P., slide dimmer, type NM cable		17.10	.468		25.50	10.35		35.85	45
2720	Type MC cable		14.30	.559		35.50	12.35		47.85	59
2730	EMT & wire		5.71	1.401		37	31		68	91.50
2750	S.P., touch dimmer, type NM cable		17.10	.468		21.50	10.35		31.85	40.50
2770	Type MC cable		14.30	.559		31.50	12.35		43.85	54.50
2780	EMT & wire		5.71	1.401		33	31		64	87
2800	3-way touch dimmer, type NM cable		13.33	.600		39	13.25		52.25	64
2820	Type MC cable		11.43	.700		49	15.45		64.45	79.50
2830	EMT & wire		4.85	1.649		50.50	36.50		87	115
3000	Combination devices									
3100	S.P. switch/15 amp recpt., Ivory, 1-gang box, plate									
3110	Type NM cable	1 Elec	11.43	.700	Ea.	14.45	15.45		29.90	41.50
3120	Type MC cable		10	.800		24.50	17.70		42.20	56
3130	EMT & wire		4.40	1.818		26	40		66	94
3150	S.P. switch/pilot light, type NM cable		11.43	.700		15.05	15.45		30.50	42
3170	Type MC cable		10	.800		25	17.70		42.70	56.50
3180	EMT & wire		4.43	1.806		26.50	40		66.50	95
3190	2-S.P. switches, 2-#14/2, no wiring		14	.571		5.45	12.65		18.10	26.50
3200	2-S.P. switches, 2-#14/2, type NM cables		10	.800		16.45	17.70		34.15	47
3220	Type MC cable		8.89	.900		33	19.90		52.90	69
3230	EMT & wire		4.10	1.951		28	43		71	101
3250	3-way switch/15 amp recpt., #14/3, type NM cable		10	.800		20.50	17.70		38.20	51.50
3270	Type MC cable		8.89	.900		31.50	19.90		51.40	67.50
3280	EMT & wire		4.10	1.951		30.50	43		73.50	104
3300	2-3 way switches, 2-#14/3, type NM cables		8.89	.900		27	19.90		46.90	62
3320	Type MC cable		8	1		46	22		68	86.50
3330	EMT & wire		4	2		35	44		79	111
3350	S.P. switch/20 amp recpt., #12/2, type NM cable		10	.800		24.50	17.70		42.20	55.50
3370	Type MC cable		8.89	.900		31	19.90		50.90	67
3380	EMT & wire		4.10	1.951		35.50	43		78.50	110
3400	Decorator style									
3410	S.P. switch/15 amp recpt., type NM cable	1 Elec	11.43	.700	Ea.	17.85	15.45		33.30	45
3420	Type MC cable		10	.800		28	17.70		45.70	60
3430	EMT & wire		4.40	1.818		29.50	40		69.50	98
3450	S.P. switch/pilot light, type NM cable		11.43	.700		18.45	15.45		33.90	46
3470	Type MC cable		10	.800		28.50	17.70		46.20	60.50
3480	EMT & wire		4.40	1.818		30	40		70	98.50
3500	2-S.P. switches, 2-#14/2, type NM cables		10	.800		19.85	17.70		37.55	51
3520	Type MC cable		8.89	.900		36.50	19.90		56.40	73
3530	EMT & wire		4.10	1.951		31	43		74	105
3550	3-way/15 amp recpt., #14/3, type NM cable		10	.800		24	17.70		41.70	55.50
3570	Type MC cable		8.89	.900		35	19.90		54.90	71
3580	EMT & wire		4.10	1.951		34	43		77	108
3650	2-3 way switches, 2-#14/3, type NM cables		8.89	.900		30.50	19.90		50.40	66
3670	Type MC cable		8	1		49.50	22		71.50	90.50

16100 | Wiring Methods

16139 | Residential Wiring

		CREW	DAILY OUTPUT	LABOR-HOURS	UNIT	2000 BARE COSTS				TOTAL INCL O&P
						MAT.	LABOR	EQUIP.	TOTAL	
3680	EMT & wire	1 Elec	4	2	Ea.	38	44		82	115
3700	S.P. switch/20 amp recpt., #12/2, type NM cable		10	.800		27.50	17.70		45.20	59.50
3720	Type MC cable		8.89	.900		34.50	19.90		54.40	70.50
3730	EMT & wire	▼	4.10	1.951	▼	39	43		82	114
4000	Receptacle devices									
4010	Duplex outlet, 15 amp recpt., Ivory, 1-gang box, plate									
4015	Type NM cable	1 Elec	14.55	.550	Ea.	5.20	12.15		17.35	25.50
4020	Type MC cable		12.31	.650		15.30	14.35		29.65	40.50
4030	EMT & wire		5.33	1.501		16	33		49	72
4050	With #12/2, type NM cable		12.31	.650		6.30	14.35		20.65	30.50
4070	Type MC cable		10.67	.750		15.65	16.55		32.20	44
4080	EMT & wire		4.71	1.699		17.50	37.50		55	81
4100	20 amp recpt., #12/2, type NM cable		12.31	.650		9.50	14.35		23.85	34
4120	Type MC cable		10.67	.750		18.85	16.55		35.40	48
4130	EMT & wire	▼	4.71	1.699	▼	20.50	37.50		58	84.50
4140	For GFI see line 4300 below									
4150	Decorator style, 15 amp recpt., type NM cable	1 Elec	14.55	.550	Ea.	8.60	12.15		20.75	29.50
4170	Type MC cable		12.31	.650		18.70	14.35		33.05	44
4180	EMT & wire		5.33	1.501		19.40	33		52.40	76
4200	With #12/2, type NM cable		12.31	.650		9.70	14.35		24.05	34
4220	Type MC cable		10.67	.750		19.05	16.55		35.60	48
4230	EMT & wire		4.71	1.699		21	37.50		58.50	84.50
4250	20 amp recpt. #12/2, type NM cable		12.31	.650		12.90	14.35		27.25	37.50
4270	Type MC cable		10.67	.750		22.50	16.55		39.05	51.50
4280	EMT & wire		4.71	1.699		24	37.50		61.50	88
4300	GFI, 15 amp recpt., type NM cable		12.31	.650		33	14.35		47.35	59.50
4320	Type MC cable		10.67	.750		43	16.55		59.55	74.50
4330	EMT & wire		4.71	1.699		43.50	37.50		81	110
4350	GFI with #12/2, type NM cable		10.67	.750		34	16.55		50.55	64.50
4370	Type MC cable		9.20	.870		43.50	19.20		62.70	79
4380	EMT & wire		4.21	1.900		45	42		87	118
4400	20 amp recpt., #12/2 type NM cable		10.67	.750		35.50	16.55		52.05	66
4420	Type MC cable		9.20	.870		45	19.20		64.20	81
4430	EMT & wire		4.21	1.900		47	42		89	120
4500	Weather-proof cover for above receptacles, add	▼	32	.250	▼	4.40	5.55		9.95	13.90
4550	Air conditioner outlet, 20 amp-240 volt recpt.									
4560	30' of #12/2, 2 pole circuit breaker									
4570	Type NM cable	1 Elec	10	.800	Ea.	39.50	17.70		57.20	72
4580	Type MC cable		9	.889		51.50	19.65		71.15	88.50
4590	EMT & wire		4	2		50	44		94	128
4600	Decorator style, type NM cable		10	.800		43	17.70		60.70	76.50
4620	Type MC cable		9	.889		55.50	19.65		75.15	93
4630	EMT & wire	▼	4	2	▼	54	44		98	132
4650	Dryer outlet, 30 amp-240 volt recpt., 20' of #10/3									
4660	2 pole circuit breaker									
4670	Type NM cable	1 Elec	6.41	1.248	Ea.	47.50	27.50		75	97.50
4680	Type MC cable		5.71	1.401		57	31		88	114
4690	EMT & wire	▼	3.48	2.299	▼	53	51		104	142
4700	Range outlet, 50 amp-240 volt recpt., 30' of #8/3									
4710	Type NM cable	1 Elec	4.21	1.900	Ea.	69	42		111	145
4720	Type MC cable		4	2		98.50	44		142.50	181
4730	EMT & wire		2.96	2.703		67.50	59.50		127	172
4750	Central vacuum outlet, Type NM cable		6.40	1.250		41.50	27.50		69	90.50
4770	Type MC cable		5.71	1.401		58.50	31		89.50	115
4780	EMT & wire	▼	3.48	2.299	▼	52.50	51		103.50	141
4800	30 amp-110 volt locking recpt., #10/2 circ. bkr.									

16100 | Wiring Methods

16139 | Residential Wiring

		CREW	DAILY OUTPUT	LABOR-HOURS	UNIT	2000 BARE COSTS				TOTAL INCL O&P
						MAT.	LABOR	EQUIP.	TOTAL	
4810	Type NM cable	1 Elec	6.20	1.290	Ea.	48	28.50		76.50	99
4820	Type MC cable		5.40	1.481		70.50	32.50		103	131
4830	EMT & wire	↓	3.20	2.500	↓	61	55.50		116.50	158
4900	Low voltage outlets									
4910	Telephone recpt., 20' of 4/C phone wire	1 Elec	26	.308	Ea.	6.75	6.80		13.55	18.50
4920	TV recpt., 20' of RG59U coax wire, F type connector	"	16	.500	"	11.45	11.05		22.50	30.50
4950	Door bell chime, transformer, 2 buttons, 60' of bellwire									
4970	Economy model	1 Elec	11.50	.696	Ea.	48.50	15.35		63.85	78.50
4980	Custom model		11.50	.696		83.50	15.35		98.85	117
4990	Luxury model, 3 buttons	↓	9.50	.842	↓	228	18.60		246.60	282
6000	Lighting outlets									
6050	Wire only (for fixture), type NM cable	1 Elec	32	.250	Ea.	3.51	5.55		9.06	12.90
6070	Type MC cable		24	.333		11.40	7.35		18.75	24.50
6080	EMT & wire		10	.800		11.15	17.70		28.85	41.50
6100	Box (4"), and wire (for fixture), type NM cable		25	.320		7.50	7.05		14.55	19.80
6120	Type MC cable		20	.400		15.40	8.85		24.25	31.50
6130	EMT & wire	↓	11	.727	↓	15.10	16.05		31.15	43
6200	Fixtures (use with lines 6050 or 6100 above)									
6210	Canopy style, economy grade	1 Elec	40	.200	Ea.	25.50	4.42		29.92	35.50
6220	Custom grade		40	.200		46	4.42		50.42	58
6250	Dining room chandelier, economy grade		19	.421		76	9.30		85.30	98.50
6260	Custom grade		19	.421		225	9.30		234.30	263
6270	Luxury grade		15	.533		495	11.80		506.80	565
6310	Kitchen fixture (fluorescent), economy grade		30	.267		51.50	5.90		57.40	66
6320	Custom grade		25	.320		161	7.05		168.05	189
6350	Outdoor, wall mounted, economy grade		30	.267		27	5.90		32.90	39
6360	Custom grade		30	.267		101	5.90		106.90	121
6370	Luxury grade		25	.320		227	7.05		234.05	262
6410	Outdoor PAR floodlights, 1 lamp, 150 watt		20	.400		20	8.85		28.85	36.50
6420	2 lamp, 150 watt each		20	.400		33.50	8.85		42.35	51.50
6430	For infrared security sensor, add		32	.250		87	5.55		92.55	105
6450	Outdoor, quartz-halogen, 300 watt flood		20	.400		38	8.85		46.85	56.50
6600	Recessed downlight, round, pre-wired, 50 or 75 watt trim		30	.267		35	5.90		40.90	48
6610	With shower light trim		30	.267		42	5.90		47.90	55.50
6620	With wall washer trim		28	.286		52.50	6.30		58.80	68.50
6630	With eye-ball trim	↓	28	.286		49	6.30		55.30	64.50
6640	For direct contact with insulation, add					1.60			1.60	1.76
6700	Porcelain lamp holder	1 Elec	40	.200		3.50	4.42		7.92	11.10
6710	With pull switch		40	.200		3.75	4.42		8.17	11.40
6750	Fluorescent strip, 1-20 watt tube, wrap around diffuser, 24"		24	.333		51.50	7.35		58.85	68.50
6760	1-40 watt tube, 48"		24	.333		65	7.35		72.35	83.50
6770	2-40 watt tubes, 48"		20	.400		78	8.85		86.85	100
6780	With residential ballast		20	.400		89.50	8.85		98.35	113
6800	Bathroom heat lamp, 1-250 watt		28	.286		30.50	6.30		36.80	44
6810	2-250 watt lamps	↓	28	.286	↓	51	6.30		57.30	66.50
6820	For timer switch, see line 2400									
6900	Outdoor post lamp, incl. post, fixture, 35' of #14/2									
6910	Type NMC cable	1 Elec	3.50	2.286	Ea.	165	50.50		215.50	265
6920	Photo-eye, add		27	.296		29	6.55		35.55	42.50
6950	Clock dial time switch, 24 hr., w/enclosure, type NM cable		11.43	.700		50.50	15.45		65.95	81
6970	Type MC cable		11	.727		60.50	16.05		76.55	93.50
6980	EMT & wire	↓	4.85	1.649	↓	61.50	36.50		98	127
7000	Alarm systems									
7050	Smoke detectors, box, #14/3, type NM cable	1 Elec	14.55	.550	Ea.	28	12.15		40.15	50.50
7070	Type MC cable		12.31	.650		36.50	14.35		50.85	64
7080	EMT & wire	↓	5	1.600	↓	35.50	35.50		71	97

16100 | Wiring Methods

16139 | Residential Wiring

		CREW	DAILY OUTPUT	LABOR-HOURS	UNIT	MAT.	LABOR	EQUIP.	TOTAL	TOTAL INCL O&P	
700	7090	For relay output to security system, add				Ea.	11.75			11.75	12.95	700
	8000	Residential equipment										
	8050	Disposal hook-up, incl. switch, outlet box, 3' of flex										
	8060	20 amp-1 pole circ. bkr., and 25' of #12/2										
	8070	Type NM cable	1 Elec	10	.800	Ea.	19.10	17.70		36.80	50	
	8080	Type MC cable		8	1		30	22		52	69	
	8090	EMT & wire	↓	5	1.600	↓	31.50	35.50		67	93	
	8100	Trash compactor or dishwasher hook-up, incl. outlet box,										
	8110	3' of flex, 15 amp-1 pole circ. bkr., and 25' of #14/2										
	8120	Type NM cable	1 Elec	10	.800	Ea.	13.40	17.70		31.10	44	
	8130	Type MC cable		8	1		25.50	22		47.50	64	
	8140	EMT & wire	↓	5	1.600	↓	26	35.50		61.50	86.50	
	8150	Hot water sink dispensor hook-up, use line 8100										
	8200	Vent/exhaust fan hook-up, type NM cable	1 Elec	32	.250	Ea.	3.51	5.55		9.06	12.90	
	8220	Type MC cable		24	.333		11.40	7.35		18.75	24.50	
	8230	EMT & wire		10	.800		11.15	17.70		28.85	41.50	
	8250	Bathroom vent fan, 50 CFM (use with above hook-up)										
	8260	Economy model	1 Elec	15	.533	Ea.	21.50	11.80		33.30	43	
	8270	Low noise model		15	.533		30	11.80		41.80	52.50	
	8280	Custom model	↓	12	.667	↓	111	14.75		125.75	146	
	8300	Bathroom or kitchen vent fan, 110 CFM										
	8310	Economy model	1 Elec	15	.533	Ea.	53	11.80		64.80	78	
	8320	Low noise model	"	15	.533	"	72	11.80		83.80	98.50	
	8350	Paddle fan, variable speed (w/o lights)										
	8360	Economy model (AC motor)	1 Elec	10	.800	Ea.	94.50	17.70		112.20	133	
	8370	Custom model (AC motor)		10	.800		165	17.70		182.70	211	
	8380	Luxury model (DC motor)		8	1		325	22		347	395	
	8390	Remote speed switch for above, add	↓	12	.667	↓	21	14.75		35.75	47	
	8500	Whole house exhaust fan, ceiling mount, 36", variable speed										
	8510	Remote switch, incl. shutters, 20 amp-1 pole circ. bkr.										
	8520	30' of #12/2, type NM cable	1 Elec	4	2	Ea.	675	44		719	820	
	8530	Type MC cable		3.50	2.286		690	50.50		740.50	845	
	8540	EMT & wire	↓	3	2.667	↓	690	59		749	855	
	8600	Whirlpool tub hook-up, incl. timer switch, outlet box										
	8610	3' of flex, 20 amp-1 pole GFI circ. bkr.										
	8620	30' of #12/2, type NM cable	1 Elec	5	1.600	Ea.	74.50	35.50		110	140	
	8630	Type MC cable		4.20	1.905		82	42		124	160	
	8640	EMT & wire	↓	3.40	2.353	↓	84	52		136	178	
	8650	Hot water heater hook-up, incl. 1-2 pole circ. bkr., box;										
	8660	3' of flex, 20' of #10/2, type NM cable	1 Elec	5	1.600	Ea.	19.20	35.50		54.70	79	
	8670	Type MC cable		4.20	1.905		33.50	42		75.50	106	
	8680	EMT & wire	↓	3.40	2.353	↓	28	52		80	116	
	9000	Heating/air conditioning										
	9050	Furnace/boiler hook-up, incl. firestat, local on-off switch										
	9060	Emergency switch, and 40' of type NM cable	1 Elec	4	2	Ea.	42	44		86	119	
	9070	Type MC cable		3.50	2.286		59	50.50		109.50	148	
	9080	EMT & wire	↓	1.50	5.333	↓	60	118		178	259	
	9100	Air conditioner hook-up, incl. local 60 amp disc. switch										
	9110	3' sealtite, 40 amp, 2 pole circuit breaker										
	9130	40' of #8/2, type NM cable	1 Elec	3.50	2.286	Ea.	136	50.50		186.50	232	
	9140	Type MC cable		3	2.667		182	59		241	298	
	9150	EMT & wire	↓	1.30	6.154	↓	146	136		282	380	
	9200	Heat pump hook-up, 1-40 & 1-100 amp 2 pole circ. bkr.										
	9210	Local disconnect switch, 3' sealtite										
	9220	40' of #8/2 & 30' of #3/2										
	9230	Type NM cable	1 Elec	1.30	6.154	Ea.	325	136		461	575	

16100 | Wiring Methods

16139 | Residential Wiring

			CREW	DAILY OUTPUT	LABOR-HOURS	UNIT	2000 BARE COSTS MAT.	LABOR	EQUIP.	TOTAL	TOTAL INCL O&P	
700	9240	Type MC cable	1 Elec	1.08	7.407	Ea.	445	164		609	760	700
	9250	EMT & wire	↓	.94	8.511	↓	335	188		523	680	
	9500	Thermostat hook-up, using low voltage wire										
	9520	Heating only	1 Elec	24	.333	Ea.	5.75	7.35		13.10	18.35	
	9530	Heating/cooling	"	20	.400	"	7.10	8.85		15.95	22.50	

16140 | Wiring Devices

			CREW	DAILY OUTPUT	LABOR-HOURS	UNIT	MAT.	LABOR	EQUIP.	TOTAL	INCL O&P	
500	0010	**LOW VOLTAGE SWITCHING**										500
	3600	Relays, 120 V or 277 V standard	1 Elec	12	.667	Ea.	26	14.75		40.75	52.50	
	3800	Flush switch, standard		40	.200		9.05	4.42		13.47	17.20	
	4000	Interchangeable		40	.200		11.80	4.42		16.22	20.50	
	4100	Surface switch, standard		40	.200		6.60	4.42		11.02	14.55	
	4200	Transformer 115 V to 25 V		12	.667		93	14.75		107.75	126	
	4400	Master control, 12 circuit, manual		4	2		94	44		138	176	
	4500	25 circuit, motorized		4	2		102	44		146	185	
	4600	Rectifier, silicon		12	.667		30.50	14.75		45.25	57.50	
	4800	Switchplates, 1 gang, 1, 2 or 3 switch, plastic		80	.100		3	2.21		5.21	6.90	
	5000	Stainless steel		80	.100		8.10	2.21		10.31	12.50	
	5400	2 gang, 3 switch, stainless steel		53	.151		15.65	3.34		18.99	22.50	
	5500	4 switch, plastic		53	.151		6.70	3.34		10.04	12.80	
	5800	3 gang, 9 switch, stainless steel	↓	32	.250	↓	50	5.55		55.55	64	
910	0010	**WIRING DEVICES**										910
	0200	Toggle switch, quiet type, single pole, 15 amp	1 Elec	40	.200	Ea.	4.63	4.42		9.05	12.35	
	0600	3 way, 15 amp		23	.348		6.70	7.70		14.40	19.90	
	0900	4 way, 15 amp		15	.533		19.40	11.80		31.20	41	
	1650	Dimmer switch, 120 volt, incandescent, 600 watt, 1 pole		16	.500		10.80	11.05		21.85	30	
	2460	Receptacle, duplex, 120 volt, grounded, 15 amp		40	.200		1.14	4.42		5.56	8.50	
	2470	20 amp		27	.296		9.50	6.55		16.05	21	
	2490	Dryer, 30 amp		15	.533		11.70	11.80		23.50	32	
	2500	Range, 50 amp		11	.727		10.45	16.05		26.50	38	
	2600	Wall plates, stainless steel, 1 gang		80	.100		2.05	2.21		4.26	5.90	
	2800	2 gang		53	.151		4.10	3.34		7.44	9.95	
	3200	Lampholder, keyless		26	.308		8.80	6.80		15.60	21	
	3400	Pullchain with receptacle	↓	22	.364	↓	8.90	8.05		16.95	23	

16150 | Wiring Connections

			CREW	DAILY OUTPUT	LABOR-HOURS	UNIT	MAT.	LABOR	EQUIP.	TOTAL	INCL O&P	
275	0010	**MOTOR CONNECTIONS**										275
	0020	Flexible conduit and fittings, 115 volt, 1 phase, up to 1 HP motor	1 Elec	8	1	Ea.	4.30	22		26.30	40.50	
	0200	25 HP motor		2.70	2.963		17.50	65.50		83	126	
	0400	50 HP motor		2.20	3.636		45.50	80.50		126	181	
	0600	100 HP motor	↓	1.50	5.333	↓	114	118		232	320	

16200 | Electrical Power

16230 | Generator Assemblies

			CREW	DAILY OUTPUT	LABOR-HOURS	UNIT	MAT.	LABOR	EQUIP.	TOTAL	INCL O&P	
450	0010	**GENERATOR SET**										450
	0020	Gas or gasoline operated, includes battery,										

16200 | Electrical Power

16230 | Generator Assemblies

			CREW	DAILY OUTPUT	LABOR-HOURS	UNIT	MAT.	LABOR	EQUIP.	TOTAL	TOTAL INCL O&P
450	0050	charger, muffler & transfer switch									450
	0200	3 phase 4 wire, 277/480 volt, 7.5 kW	R-3	.83	24.096	Ea.	6,000	530	167	6,697	7,650
	0300	11.5 kW		.71	28.169		8,500	620	195	9,315	10,600
	0400	20 kW		.63	31.746		10,000	700	219	10,919	12,400
	0500	35 kW		.55	36.364		12,000	800	251	13,051	14,800
	0600	80 kW		.40	50		20,500	1,100	345	21,945	24,800
	0700	100 kW		.33	60.606		22,500	1,325	420	24,245	27,400
	0800	125 kW		.28	71.429		46,000	1,575	495	48,070	53,500
	0900	185 kW		.25	80		61,000	1,750	555	63,305	70,500
	2000	Diesel engine, including battery, charger,									
	2010	muffler, automatic transfer switch & day tank, 30 kW	R-3	.55	36.364	Ea.	16,000	800	251	17,051	19,200
	2100	50 kW		.42	47.619		19,700	1,050	330	21,080	23,700
	2200	75 kW		.35	57.143		25,700	1,250	395	27,345	30,800
	2300	100 kW		.31	64.516		28,500	1,425	445	30,370	34,200
	2400	125 kW		.29	68.966		30,400	1,525	475	32,400	36,400
	2500	150 kW		.26	76.923		35,000	1,700	530	37,230	41,900
	2600	175 kW		.25	80		36,500	1,750	555	38,805	43,700
	2700	200 kW		.24	83.333		38,000	1,825	575	40,400	45,400
	2800	250 kW		.23	86.957		41,500	1,925	600	44,025	49,500
	2900	300 kW		.22	90.909		49,800	2,000	630	52,430	59,000
	3000	350 kW		.20	100		53,500	2,200	690	56,390	63,500
	3100	400 kW		.19	105		65,500	2,325	725	68,550	76,500
	3200	500 kW		.18	111		76,000	2,450	770	79,220	88,500

16270 | Transformers

			CREW	DAILY OUTPUT	LABOR-HOURS	UNIT	MAT.	LABOR	EQUIP.	TOTAL	TOTAL INCL O&P
200	0010	DRY TYPE TRANSFORMER									200
	0050	Single phase, 240/480 volt primary, 120/240 volt secondary									
	0100	1 kVA	1 Elec	2	4	Ea.	147	88.50		235.50	305
	0300	2 kVA		1.60	5		221	111		332	425
	0500	3 kVA		1.40	5.714		274	126		400	505
	0700	5 kVA		1.20	6.667		375	147		522	655
	0900	7.5 kVA	2 Elec	2.20	7.273		525	161		686	840
	1100	10 kVA		1.60	10		670	221		891	1,100
	1300	15 kVA		1.20	13.333		915	295		1,210	1,475
	1500	25 kVA		1	16		1,175	355		1,530	1,850
	1700	37.5 kVA		.80	20		1,525	440		1,965	2,400
	1900	50 kVA		.70	22.857		1,800	505		2,305	2,800
	2100	75 kVA		.65	24.615		2,375	545		2,920	3,525
	2190	480V primary 120/240V secondary, nonvent., 15 kVA		1.20	13.333		855	295		1,150	1,425
	2200	25 kVA		.90	17.778		1,350	395		1,745	2,150
	2210	37 kVA		.75	21.333		1,625	470		2,095	2,550
	2220	50 kVA		.65	24.615		1,925	545		2,470	3,025
	2300	3 phase, 480 volt primary 120/208 volt secondary									
	2310	Ventilated, 3 kVA	1 Elec	1	8	Ea.	425	177		602	760
	2700	6 kVA		.80	10		630	221		851	1,050
	2900	9 kVA		.70	11.429		715	253		968	1,200
	3100	15 kVA	2 Elec	1.10	14.545		930	320		1,250	1,550
	3300	30 kVA		.90	17.778		1,100	395		1,495	1,875
	3500	45 kVA		.80	20		1,300	440		1,740	2,150
	3700	75 kVA		.70	22.857		2,025	505		2,530	3,050
	3900	112.5 kVA	R-3	.90	22.222		2,675	490	154	3,319	3,925
	4100	150 kVA		.85	23.529		3,500	520	163	4,183	4,875
	4300	225 kVA		.65	30.769		4,725	675	213	5,613	6,525
	4500	300 kVA		.55	36.364		5,975	800	251	7,026	8,150
	4700	500 kVA		.45	44.444		9,900	980	305	11,185	12,800

16200 | Electrical Power

16270 | Transformers

		CREW	DAILY OUTPUT	LABOR-HOURS	UNIT	2000 BARE COSTS				TOTAL INCL O&P
						MAT.	LABOR	EQUIP.	TOTAL	
200 4800	750 kVA	R-3	.35	57.143	Ea.	17,400	1,250	395	19,045	21,600
600 0010	**OIL FILLED TRANSFORMER** Pad mounted, primary delta or Y,									
0050	5 kV or 15 kV, with taps, 277/480 V secondary, 3 phase									
0100	150 kVA	R-3	.65	30.769	Ea.	5,975	675	213	6,863	7,900
0200	300 kVA		.45	44.444		7,875	980	305	9,160	10,600
0300	500 kVA		.40	50		10,500	1,100	345	11,945	13,800
0400	750 kVA		.38	52.632		14,200	1,150	365	15,715	17,900
0500	1000 kVA		.26	76.923		16,800	1,700	530	19,030	21,900
0600	1500 kVA		.23	86.957		20,000	1,925	600	22,525	25,700
0700	2000 kVA		.20	100		25,200	2,200	690	28,090	32,100
0800	3750 kVA		.16	125		47,300	2,750	865	50,915	57,500
300 0010	**CAPACITORS** Indoor									
0020	240 volts, single & 3 phase, 0.5 kVAR	1 Elec	2.70	2.963	Ea.	262	65.50		327.50	395
0100	1.0 kVAR		2.70	2.963		315	65.50		380.50	455
0150	2.5 kVAR		2	4		375	88.50		463.50	560
0200	5.0 kVAR		1.80	4.444		445	98		543	650
0250	7.5 kVAR		1.60	5		515	111		626	745
0300	10 kVAR		1.50	5.333		615	118		733	875
0350	15 kVAR		1.30	6.154		850	136		986	1,150
0400	20 kVAR		1.10	7.273		1,050	161		1,211	1,450
0450	25 kVAR		1	8		1,250	177		1,427	1,675
1000	480 volts, single & 3 phase, 1 kVAR		2.70	2.963		246	65.50		311.50	380
1050	2 kVAR		2.70	2.963		283	65.50		348.50	415
1100	5 kVAR		2	4		360	88.50		448.50	540
1150	7.5 kVAR		2	4		390	88.50		478.50	575
1200	10 kVAR		2	4		430	88.50		518.50	620
1250	15 kVAR		2	4		535	88.50		623.50	730
1300	20 kVAR		1.60	5		595	111		706	835
1350	30 kVAR		1.50	5.333		745	118		863	1,000
1400	40 kVAR		1.20	6.667		945	147		1,092	1,300
1450	50 kVAR		1.10	7.273		1,075	161		1,236	1,450

16290 | Power Measure & Control

		CREW	DAILY OUTPUT	LABOR-HOURS	UNIT	MAT.	LABOR	EQUIP.	TOTAL	TOTAL INCL O&P
800 0010	**SWITCHBOARD INSTRUMENTS** 3 phase, 4 wire									
0100	AC indicating, ammeter & switch	1 Elec	8	1	Ea.	1,200	22		1,222	1,325
0200	Voltmeter & switch		8	1		1,200	22		1,222	1,325
0300	Wattmeter		8	1		2,350	22		2,372	2,625
0400	AC recording, ammeter		4	2		4,200	44		4,244	4,700
0500	Voltmeter		4	2		4,200	44		4,244	4,700
0600	Ground fault protection, zero sequence		2.70	2.963		3,700	65.50		3,765.50	4,175
0700	Ground return path		2.70	2.963		3,700	65.50		3,765.50	4,175
0800	3 current transformers, 5 to 800 amp		2	4		1,550	88.50		1,638.50	1,850
0900	1000 to 1500 amp		1.30	6.154		2,225	136		2,361	2,675
1200	2000 to 4000 amp		1	8		2,625	177		2,802	3,200
1300	Fused potential transformer, maximum 600 volt		8	1		705	22		727	810

Important: See the Reference Section for critical supporting data - Reference Nos., Crews, & City Cost Indexes

16400 | Low-Voltage Distribution

16410 | Encl Switches & Circuit Breakers

			CREW	DAILY OUTPUT	LABOR-HOURS	UNIT	MAT.	LABOR	EQUIP.	TOTAL	TOTAL INCL O&P	
200	0010	**CIRCUIT BREAKERS** (in enclosure)										200
	0100	Enclosed (NEMA 1), 600 volt, 3 pole, 30 amp	1 Elec	3.20	2.500	Ea.	365	55.50		420.50	490	
	0200	60 amp		2.80	2.857		365	63		428	505	
	0400	100 amp		2.30	3.478		415	77		492	585	
	0600	225 amp		1.50	5.333		965	118		1,083	1,275	
	0700	400 amp	2 Elec	1.60	10		1,650	221		1,871	2,175	
	0800	600 amp		1.20	13.333		2,400	295		2,695	3,100	
	1000	800 amp		.94	17.021		3,100	375		3,475	4,050	
800	0010	**SAFETY SWITCHES**										800
	0100	General duty 240 volt, 3 pole NEMA 1, fusible, 30 amp	1 Elec	3.20	2.500	Ea.	65.50	55.50		121	163	
	0200	60 amp		2.30	3.478		111	77		188	248	
	0300	100 amp		1.90	4.211		191	93		284	360	
	0400	200 amp		1.30	6.154		410	136		546	670	
	0500	400 amp	2 Elec	1.80	8.889		1,025	196		1,221	1,450	
	0600	600 amp	"	1.20	13.333		1,925	295		2,220	2,600	
	2900	Heavy duty, 240 volt, 3 pole NEMA 1 fusible										
	2910	30 amp	1 Elec	3.20	2.500	Ea.	106	55.50		161.50	208	
	3000	60 amp		2.30	3.478		179	77		256	325	
	3300	100 amp		1.90	4.211		282	93		375	460	
	3500	200 amp		1.30	6.154		485	136		621	755	
	3700	400 amp	2 Elec	1.80	8.889		1,125	196		1,321	1,550	
	3900	600 amp	"	1.20	13.333		2,025	295		2,320	2,700	

16420 | Enclosed Controllers

			CREW	DAILY OUTPUT	LABOR-HOURS	UNIT	MAT.	LABOR	EQUIP.	TOTAL	TOTAL INCL O&P	
220	0010	**CONTROL STATIONS**										220
	0050	NEMA 1, heavy duty, stop/start	1 Elec	8	1	Ea.	113	22		135	160	
	0100	Stop/start, pilot light		6.20	1.290		153	28.50		181.50	215	
	0200	Hand/off/automatic		6.20	1.290		83.50	28.50		112	139	
	0400	Stop/start/reverse		5.30	1.509		153	33.50		186.50	223	

16440 | Swbds, Panels & Cont Centers

			CREW	DAILY OUTPUT	LABOR-HOURS	UNIT	MAT.	LABOR	EQUIP.	TOTAL	TOTAL INCL O&P	
660	0010	**MOTOR STARTERS & CONTROLS**										660
	0050	Magnetic, FVNR, with enclosure and heaters, 480 volt	R16220-610									
	0100	5 HP, size 0	1 Elec	2.30	3.478	Ea.	199	77		276	345	
	0200	10 HP, size 1	"	1.60	5		224	111		335	425	
	0300	25 HP, size 2	2 Elec	2.20	7.273		420	161		581	730	
	0400	50 HP, size 3		1.80	8.889		685	196		881	1,075	
	0500	100 HP, size 4		1.20	13.333		1,525	295		1,820	2,150	
	0600	200 HP, size 5		.90	17.778		3,550	395		3,945	4,575	
	0700	Combination, with motor circuit protectors, 5 HP, size 0	1 Elec	1.80	4.444		645	98		743	870	
	0800	10 HP, size 1	"	1.30	6.154		670	136		806	960	
	0900	25 HP, size 2	2 Elec	2	8		940	177		1,117	1,325	
	1000	50 HP, size 3		1.32	12.121		1,350	268		1,618	1,950	
	1200	100 HP, size 4		.80	20		2,950	440		3,390	3,975	
	1400	Combination, with fused switch, 5 HP, size 0	1 Elec	1.80	4.444		495	98		593	705	
	1600	10 HP, size 1	"	1.30	6.154		530	136		666	800	
	1800	25 HP, size 2	2 Elec	2	8		860	177		1,037	1,225	
	2000	50 HP, size 3		1.32	12.121		1,450	268		1,718	2,050	
	2200	100 HP, size 4		.80	20		2,525	440		2,965	3,500	
700	0010	**PANELBOARD & LOAD CENTER CIRCUIT BREAKERS**										700
	0050	Bolt-on, 10,000 amp IC, 120 volt, 1 pole										
	0100	15 to 50 amp	1 Elec	10	.800	Ea.	10	17.70		27.70	40	
	0200	60 amp		8	1		10	22		32	47	
	0300	70 amp		8	1		19	22		41	57	
	0350	240 volt, 2 pole										

16400 | Low-Voltage Distribution

16440 | Swbds, Panels & Cont Centers

			CREW	DAILY OUTPUT	LABOR-HOURS	UNIT	MAT.	LABOR	EQUIP.	TOTAL	TOTAL INCL O&P	
700	0400	15 to 50 amp	1 Elec	8	1	Ea.	22	22		44	60	700
	0500	60 amp		7.50	1.067		22	23.50		45.50	62.50	
	0600	80 to 100 amp		5	1.600		56.50	35.50		92	120	
	0700	3 pole, 15 to 60 amp		6.20	1.290		69.50	28.50		98	123	
	0800	70 amp		5	1.600		87.50	35.50		123	155	
	0900	80 to 100 amp		3.60	2.222		99.50	49		148.50	190	
	1000	22,000 amp I.C., 240 volt, 2 pole, 70 - 225 amp		2.70	2.963		430	65.50		495.50	575	
	1100	3 pole, 70 - 225 amp		2.30	3.478		475	77		552	650	
	1200	14,000 amp I.C., 277 volts, 1 pole, 15 - 30 amp		8	1		26.50	22		48.50	65	
	1300	22,000 amp I.C., 480 volts, 2 pole, 70 - 225 amp		2.70	2.963		430	65.50		495.50	575	
	1400	3 pole, 70 - 225 amp	↓	2.30	3.478	↓	475	77		552	650	
720	0010	**PANELBOARDS** (Commercial use)										720
	0050	NQOD, w/20 amp 1 pole bolt-on circuit breakers										
	0100	3 wire, 120/240 volts, 100 amp main lugs										
	0150	10 circuits	1 Elec	1	8	Ea.	390	177		567	715	
	0200	14 circuits		.88	9.091		455	201		656	830	
	0250	18 circuits		.75	10.667		500	236		736	930	
	0300	20 circuits	↓	.65	12.308		560	272		832	1,050	
	0350	225 amp main lugs, 24 circuits	2 Elec	1.20	13.333		635	295		930	1,175	
	0400	30 circuits		.90	17.778		740	395		1,135	1,450	
	0450	36 circuits		.80	20		845	440		1,285	1,650	
	0500	38 circuits		.72	22.222		910	490		1,400	1,800	
	0550	42 circuits	↓	.66	24.242		950	535		1,485	1,925	
	0600	4 wire, 120/208 volts, 100 amp main lugs, 12 circuits	1 Elec	1	8		440	177		617	775	
	0650	16 circuits		.75	10.667		505	236		741	940	
	0700	20 circuits		.65	12.308		590	272		862	1,100	
	0750	24 circuits		.60	13.333		640	295		935	1,175	
	0800	30 circuits	↓	.53	15.094		740	335		1,075	1,350	
	0850	225 amp main lugs, 32 circuits	2 Elec	.90	17.778		830	395		1,225	1,550	
	0900	34 circuits		.84	19.048		850	420		1,270	1,625	
	0950	36 circuits		.80	20		870	440		1,310	1,675	
	1000	42 circuits		.68	23.529	↓	975	520		1,495	1,925	
	1200	NEHB,w/20 amp, 1 pole bolt-on circuit breakers										
	1250	4 wire, 277/480 volts, 100 amp main lugs, 12 circuits	1 Elec	.88	9.091	Ea.	840	201		1,041	1,250	
	1300	20 circuits	"	.60	13.333		1,250	295		1,545	1,850	
	1350	225 amp main lugs, 24 circuits	2 Elec	.90	17.778		1,425	395		1,820	2,225	
	1400	30 circuits		.80	20		1,725	440		2,165	2,625	
	1450	36 circuits	↓	.72	22.222	↓	2,000	490		2,490	3,000	
	1600	NQOD panel, w/20 amp, 1 pole, circuit breakers										
	1650	3 wire, 120/240 volt with main circuit breaker										
	1700	100 amp main, 12 circuits	1 Elec	.80	10	Ea.	540	221		761	950	
	1750	20 circuits	"	.60	13.333		690	295		985	1,250	
	1800	225 amp main, 30 circuits	2 Elec	.68	23.529		1,325	520		1,845	2,300	
	1850	42 circuits		.52	30.769		1,525	680		2,205	2,775	
	1900	400 amp main, 30 circuits		.54	29.630		1,825	655		2,480	3,100	
	1950	42 circuits	↓	.50	32	↓	2,050	705		2,755	3,400	
	2000	4 wire, 120/208 volts with main circuit breaker										
	2050	100 amp main, 24 circuits	1 Elec	.47	17.021	Ea.	805	375		1,180	1,500	
	2100	30 circuits	"	.40	20		910	440		1,350	1,725	
	2200	225 amp main, 32 circuits	2 Elec	.72	22.222		1,550	490		2,040	2,500	
	2250	42 circuits		.56	28.571		1,675	630		2,305	2,875	
	2300	400 amp main, 42 circuits		.48	33.333		2,275	735		3,010	3,700	
	2350	600 amp main, 42 circuits	↓	.40	40	↓	3,375	885		4,260	5,175	
	2400	NEHB, with 20 amp, 1 pole circuit breaker										
	2450	4 wire, 277/480 volts with main circuit breaker										

16400 | Low-Voltage Distribution

16440 | Swbds, Panels & Cont Centers

			CREW	DAILY OUTPUT	LABOR-HOURS	UNIT	MAT.	LABOR	EQUIP.	TOTAL	TOTAL INCL O&P	
720	2500	100 amp main, 24 circuits	1 Elec	.42	19.048	Ea.	1,650	420		2,070	2,525	720
	2550	30 circuits	"	.38	21.053		1,950	465		2,415	2,875	
	2600	225 amp main, 30 circuits	2 Elec	.72	22.222		2,450	490		2,940	3,500	
	2650	42 circuits	"	.56	28.571		3,025	630		3,655	4,350	
800	0010	**SWITCHBOARDS** distribution section										800
	0100	Aluminum bus bars, not including breakers										
	0200	120/208 or 277/480 volt, 4 wire, 600 amp	2 Elec	1	16	Ea.	1,150	355		1,505	1,825	
	0300	800 amp		.88	18.182		1,450	400		1,850	2,250	
	0400	1000 amp		.80	20		1,725	440		2,165	2,625	
	0500	1200 amp		.72	22.222		2,375	490		2,865	3,400	
	0600	1600 amp		.66	24.242		2,775	535		3,310	3,925	
	0700	2000 amp		.62	25.806		3,250	570		3,820	4,500	
	0800	2500 amp		.60	26.667		3,675	590		4,265	5,000	
	0900	3000 amp		.56	28.571		4,450	630		5,080	5,900	
	0950	4000 amp		.52	30.769		6,475	680		7,155	8,225	
820	0010	**SWITCHBOARDS** feeder section group mounted devices										820
	0030	Circuit breakers										
	0160	FA frame, 15 to 60 amp, 240 volt, 1 pole	1 Elec	8	1	Ea.	63	22		85	106	
	0280	FA frame, 70 to 100 amp, 240 volt, 1 pole		7	1.143		83	25.50		108.50	133	
	0420	KA frame, 70 to 225 amp		3.20	2.500		695	55.50		750.50	855	
	0430	LA frame, 125 to 400 amp		2.30	3.478		1,575	77		1,652	1,850	
	0460	MA frame, 450 to 600 amp		1.60	5		2,600	111		2,711	3,025	
	0470	700 to 800 amp		1.30	6.154		3,375	136		3,511	3,925	
	0480	MAL frame, 1000 amp		1	8		3,500	177		3,677	4,150	
	0490	PA frame, 1200 amp		.80	10		7,100	221		7,321	8,175	
	0500	Branch circuit, fusible switch, 600 volt, double 30/30 amp		4	2		700	44		744	845	
	0550	60/60 amp		3.20	2.500		705	55.50		760.50	865	
	0600	100/100 amp		2.70	2.963		965	65.50		1,030.50	1,150	
	0650	Single, 30 amp		5.30	1.509		445	33.50		478.50	545	
	0700	60 amp		4.70	1.702		445	37.50		482.50	550	
	0750	100 amp		4	2		655	44		699	795	
	0800	200 amp		2.70	2.963		945	65.50		1,010.50	1,150	
	0850	400 amp		2.30	3.478		1,825	77		1,902	2,125	
	0900	600 amp		1.80	4.444		2,150	98		2,248	2,525	
	0950	800 amp		1.30	6.154		3,650	136		3,786	4,250	
	1000	1200 amp		.80	10		4,200	221		4,421	4,975	
840	0010	**SWITCHBOARDS** Incoming main service section										840
	0100	Aluminum bus bars, not including CT's or PT's										
	0200	No main disconnect, includes CT compartment										
	0300	120/208 volt, 4 wire, 600 amp	2 Elec	1	16	Ea.	2,650	355		3,005	3,500	
	0400	800 amp		.88	18.182		2,650	400		3,050	3,575	
	0500	1000 amp		.80	20		3,175	440		3,615	4,225	
	0600	1200 amp		.72	22.222		3,175	490		3,665	4,300	
	0700	1600 amp		.66	24.242		3,175	535		3,710	4,375	
	0800	2000 amp		.62	25.806		3,425	570		3,995	4,700	
	1000	3000 amp		.56	28.571		4,525	630		5,155	6,000	
	2000	Fused switch & CT compartment										
	2100	120/208 volt, 4 wire, 400 amp	2 Elec	1.12	14.286	Ea.	3,550	315		3,865	4,425	
	2200	600 amp		.94	17.021		4,200	375		4,575	5,250	
	2300	800 amp		.84	19.048		7,175	420		7,595	8,575	
	2400	1200 amp		.68	23.529		9,325	520		9,845	11,200	
	2900	Pressure switch & CT compartment										
	3000	120/208 volt, 4 wire, 800 amp	2 Elec	.80	20	Ea.	6,450	440		6,890	7,800	
	3100	1200 amp		.66	24.242		12,500	535		13,035	14,600	

16400 | Low-Voltage Distribution

16440 | Swbds, Panels & Cont Centers

			CREW	DAILY OUTPUT	LABOR-HOURS	UNIT	2000 BARE COSTS MAT.	LABOR	EQUIP.	TOTAL	TOTAL INCL O&P	
840	3200		1600 amp	2 Elec	.62	25.806	Ea.	13,300	570		13,870	15,500	840
	3300		2000 amp	↓	.56	28.571	↓	14,100	630		14,730	16,500	
	4400		Circuit breaker, molded case & CT compartment										
	4600		3 pole, 4 wire, 600 amp	2 Elec	.94	17.021	Ea.	5,550	375		5,925	6,725	
	4800		800 amp		.84	19.048		6,650	420		7,070	8,000	
	5000		1200 amp	↓	.68	23.529	↓	9,050	520		9,570	10,800	
	5100		Copper bus bars, not incl. CT's or PT's, add, minimum					15%					

16450 | Enclosed Bus Assemblies

			CREW	DAILY OUTPUT	LABOR-HOURS	UNIT	MAT.	LABOR	EQUIP.	TOTAL	TOTAL INCL O&P	
320	0010	**COPPER BUS DUCT** 10 ft long											320
	0050	3 pole 4 wire, plug-in/indoor, straight section, 225 amp		2 Elec	40	.400	L.F.	136	8.85		144.85	164	
	1000	400 amp			32	.500		136	11.05		147.05	168	
	1500	600 amp			26	.615		136	13.60		149.60	173	
	2400	800 amp			20	.800		162	17.70		179.70	207	
	2450	1000 amp			18	.889		179	19.65		198.65	228	
	2500	1350 amp			16	1		255	22		277	315	
	2510	1600 amp			12	1.333		289	29.50		318.50	370	
	2520	2000 amp			10	1.600		365	35.50		400.50	460	
	2550	Feeder, 600 amp			28	.571		119	12.65		131.65	152	
	2600	800 amp			22	.727		145	16.05		161.05	186	
	2700	1000 amp			20	.800		162	17.70		179.70	207	
	2800	1350 amp			18	.889		238	19.65		257.65	294	
	2900	1600 amp			14	1.143		272	25.50		297.50	340	
	3000	2000 amp			12	1.333	↓	350	29.50		379.50	435	
	3100	Elbows, 225 amp			4	4	Ea.	805	88.50		893.50	1,025	
	3200	400 amp			3.60	4.444		805	98		903	1,050	
	3300	600 amp			3.20	5		805	111		916	1,075	
	3400	800 amp			2.80	5.714		875	126		1,001	1,175	
	3500	1000 amp			2.60	6.154		975	136		1,111	1,300	
	3600	1350 amp			2.40	6.667		1,150	147		1,297	1,500	
	3700	1600 amp			2.20	7.273		1,250	161		1,411	1,650	
	3800	2000 amp			1.80	8.889		1,550	196		1,746	2,025	
	4000	End box, 225 amp			34	.471		109	10.40		119.40	137	
	4100	400 amp			32	.500		109	11.05		120.05	138	
	4200	600 amp			28	.571		109	12.65		121.65	141	
	4300	800 amp			26	.615		109	13.60		122.60	143	
	4400	1000 amp			24	.667		109	14.75		123.75	144	
	4500	1350 amp			22	.727		109	16.05		125.05	147	
	4600	1600 amp			20	.800		109	17.70		126.70	149	
	4700	2000 amp			18	.889		133	19.65		152.65	179	
	4800	Cable tap box end, 225 amp			3.20	5		780	111		891	1,025	
	5000	400 amp			2.60	6.154		780	136		916	1,075	
	5100	600 amp			2.20	7.273		780	161		941	1,125	
	5200	800 amp			2	8		875	177		1,052	1,250	
	5300	1000 amp			1.60	10		960	221		1,181	1,400	
	5400	1350 amp			1.40	11.429		1,150	253		1,403	1,700	
	5500	1600 amp			1.20	13.333		1,300	295		1,595	1,900	
	5600	2000 amp			1	16		1,475	355		1,830	2,200	
	5700	Switchboard stub, 225 amp			5.40	2.963		815	65.50		880.50	1,000	
	5800	400 amp			4.60	3.478		815	77		892	1,025	
	5900	600 amp			4	4		815	88.50		903.50	1,050	
	6000	800 amp			3.20	5		985	111		1,096	1,250	
	6100	1000 amp			3	5.333		1,150	118		1,268	1,450	
	6200	1350 amp			2.60	6.154		1,475	136		1,611	1,850	
	6300	1600 amp		↓	2.40	6.667	↓	1,650	147		1,797	2,075	

16400 | Low-Voltage Distribution

16450 | Enclosed Bus Assemblies

		CREW	DAILY OUTPUT	LABOR-HOURS	UNIT	MAT.	LABOR	EQUIP.	TOTAL	TOTAL INCL O&P
6400	2000 amp	2 Elec	2	8	Ea.	2,025	177		2,202	2,525
6490	Tee fittings, 225 amp		2.40	6.667		1,100	147		1,247	1,475
6500	400 amp		2	8		1,100	177		1,277	1,525
6600	600 amp		1.80	8.889		1,100	196		1,296	1,550
6700	800 amp		1.60	10		1,275	221		1,496	1,750
6800	1350 amp		1.20	13.333		1,850	295		2,145	2,500
7000	1600 amp		1	16		2,100	355		2,455	2,875
7100	2000 amp		.80	20		2,475	440		2,915	3,450
7200	Plug-in switches, 600 volt, 3 pole, 30 amp	1 Elec	4	2		360	44		404	470
7300	60 amp		3.60	2.222		390	49		439	510
7400	100 amp		2.70	2.963		780	65.50		845.50	965
7500	200 amp	2 Elec	3.20	5		985	111		1,096	1,250
7600	400 amp		1.40	11.429		2,575	253		2,828	3,250
7700	600 amp		.90	17.778		3,350	395		3,745	4,350
7800	800 amp		.66	24.242		5,800	535		6,335	7,250
7900	1200 amp		.50	32		10,900	705		11,605	13,100
8000	Plug-in circuit breakers, molded case, 15 to 50 amp	1 Elec	4.40	1.818		460	40		500	570
8100	70 to 100 amp	"	3.10	2.581		510	57		567	655
8200	150 to 225 amp	2 Elec	3.40	4.706		1,375	104		1,479	1,700
8300	250 to 400 amp		1.40	11.429		2,425	253		2,678	3,100
8400	500 to 600 amp		1	16		3,275	355		3,630	4,175
8500	700 to 800 amp		.64	25		4,025	555		4,580	5,350
8600	900 to 1000 amp		.56	28.571		5,775	630		6,405	7,375
8700	1200 amp		.44	36.364		6,950	805		7,755	8,950

16500 | Lighting

16510 | Interior Luminaires

		CREW	DAILY OUTPUT	LABOR-HOURS	UNIT	MAT.	LABOR	EQUIP.	TOTAL	TOTAL INCL O&P
0010	**INTERIOR LIGHTING FIXTURES** Including lamps, mounting									
0030	hardware and connections									
0100	Fluorescent, C.W. lamps, troffer, recess mounted in grid, RS									
0200	Acrylic lens, 1'W x 4'L, two 40 watt	1 Elec	5.70	1.404	Ea.	46	31		77	101
0210	1'W x 4'L, three 40 watt		5.40	1.481		52	32.50		84.50	111
0300	2'W x 2'L, two U40 watt		5.70	1.404		50	31		81	106
0400	2'W x 4'L, two 40 watt		5.30	1.509		49	33.50		82.50	109
0500	2'W x 4'L, three 40 watt		5	1.600		53	35.50		88.50	117
0600	2'W x 4'L, four 40 watt		4.70	1.702		56	37.50		93.50	123
0700	4'W x 4'L, four 40 watt	2 Elec	6.40	2.500		285	55.50		340.50	405
0800	4'W x 4'L, six 40 watt		6.20	2.581		295	57		352	420
0900	4'W x 4'L, eight 40 watt		5.80	2.759		305	61		366	435
0910	Acrylic lens, 1'W x 4'L, two 32 watt	1 Elec	5.70	1.404		55	31		86	111
0930	2'W x 2'L, two U32 watt		5.70	1.404		75	31		106	133
0940	2'W x 4'L, two 32 watt		5.30	1.509		67	33.50		100.50	128
0950	2'W x 4'L, three 32 watt		5	1.600		72	35.50		107.50	137
0960	2'W x 4'L, four 32 watt		4.70	1.702		75	37.50		112.50	144
1000	Surface mounted, RS									
1030	Acrylic lens with hinged & latched door frame									
1100	1'W x 4'L, two 40 watt	1 Elec	7	1.143	Ea.	70	25.50		95.50	119
1110	1'W x 4'L, three 40 watt		6.70	1.194		80	26.50		106.50	131
1200	2'W x 2'L, two U40 watt		7	1.143		85	25.50		110.50	135

16500 | Lighting

16510 | Interior Luminaires

		CREW	DAILY OUTPUT	LABOR-HOURS	UNIT	2000 BARE COSTS MAT.	LABOR	EQUIP.	TOTAL	TOTAL INCL O&P
1300	2'W x 4'L, two 40 watt	1 Elec	6.20	1.290	Ea.	78	28.50		106.50	133
1400	2'W x 4'L, three 40 watt		5.70	1.404		88	31		119	148
1500	2'W x 4'L, four 40 watt		5.30	1.509		90	33.50		123.50	154
1600	4'W x 4'L, four 40 watt	2 Elec	7.20	2.222		385	49		434	505
1700	4'W x 4'L, six 40 watt		6.60	2.424		420	53.50		473.50	550
1800	4'W x 4'L, eight 40 watt		6.20	2.581		430	57		487	570
1900	2'W x 8'L, four 40 watt		6.40	2.500		140	55.50		195.50	245
2000	2'W x 8'L, eight 40 watt		6.20	2.581		170	57		227	281
2100	Strip fixture									
2130	Surface mounted									
2200	4' long, one 40 watt RS	1 Elec	8.50	.941	Ea.	26	21		47	62.50
2300	4' long, two 40 watt RS		8	1		28	22		50	67
2400	4' long, one 40 watt, SL		8	1		38	22		60	78
2500	4' long, two 40 watt, SL		7	1.143		52	25.50		77.50	98.50
2600	8' long, one 75 watt, SL	2 Elec	13.40	1.194		39	26.50		65.50	86
2700	8' long, two 75 watt, SL	"	12.40	1.290		47	28.50		75.50	98
2800	4' long, two 60 watt, HO	1 Elec	6.70	1.194		76	26.50		102.50	127
2900	8' long, two 110 watt, HO	2 Elec	10.60	1.509		80	33.50		113.50	143
3000	Pendent mounted, industrial, white porcelain enamel									
3100	4' long, two 40 watt, RS	1 Elec	5.70	1.404	Ea.	43	31		74	98
3200	4' long, two 60 watt, HO	"	5	1.600		68	35.50		103.50	133
3300	8' long, two 75 watt, SL	2 Elec	8.80	1.818		81	40		121	155
3400	8' long, two 110 watt, HO	"	8	2		104	44		148	187
3470	Troffer, air handling, 2'W x 4'L with four 40 watt, RS	1 Elec	4	2		76	44		120	156
3480	2'W x 2'L with two U40 watt RS		5.50	1.455		66	32		98	125
3490	Air connector insulated, 5" diameter		20	.400		52	8.85		60.85	71.50
3500	6" diameter		20	.400		53	8.85		61.85	73
3580	Mercury vapor, integral ballast, ceiling, recess mounted,									
3590	prismatic glass lens, floating door									
3600	2'W x 2'L, 250 watt DX lamp	1 Elec	3.20	2.500	Ea.	250	55.50		305.50	365
3700	2'W x 2'L, 400 watt DX lamp		2.90	2.759		255	61		316	380
3800	Surface mtd., prismatic lens, 2'W x 2'L, 250 watt DX lamp		2.70	2.963		250	65.50		315.50	380
3900	2'W x 2'L, 400 watt DX lamp		2.40	3.333		283	73.50		356.50	430
4000	High bay, aluminum reflector									
4030	Single unit, 400 watt DX lamp	2 Elec	4.60	3.478	Ea.	248	77		325	400
4100	Single unit, 1000 watt DX lamp		4	4		390	88.50		478.50	575
4200	Twin unit, two 400 watt DX lamps		3.20	5		420	111		531	640
4450	Incandescent, high hat can, round alzak reflector, prewired									
4470	100 watt	1 Elec	8	1	Ea.	53	22		75	94.50
4480	150 watt		8	1		77	22		99	121
4500	300 watt		6.70	1.194		175	26.50		201.50	236
4600	Square glass lens with metal trim, prewired									
4630	100 watt	1 Elec	6.70	1.194	Ea.	38	26.50		64.50	85
4700	200 watt		6.70	1.194		58	26.50		84.50	107
4800	300 watt		5.70	1.404		90	31		121	150
4900	Ceiling/wall, surface mounted, metal cylinder, 75 watt		10	.800		33	17.70		50.70	65.50
4920	150 watt		10	.800		51	17.70		68.70	85
5200	Ceiling, surface mounted, opal glass drum									
5300	8", one 60 watt lamp	1 Elec	10	.800	Ea.	32	17.70		49.70	64
5400	10", two 60 watt lamps		8	1		36	22		58	75.50
5500	12", four 60 watt lamps		6.70	1.194		50	26.50		76.50	98
6010	Vapor tight, incandescent, ceiling mounted, 200 watt		6.20	1.290		46	28.50		74.50	97
6100	Fluorescent, surface mounted, 2 lamps, 4'L, RS, 40 watt		3.20	2.500		86	55.50		141.50	185
6200	Mercury vapor with ballast, 175 watt		3.20	2.500		252	55.50		307.50	370
6300	Explosionproof									
6510	Incandescent, ceiling mounted, 200 watt	1 Elec	4	2	Ea.	630	44		674	770

16500 | Lighting

16510 | Interior Luminaires

		CREW	DAILY OUTPUT	LABOR-HOURS	UNIT	2000 BARE COSTS MAT.	LABOR	EQUIP.	TOTAL	TOTAL INCL O&P
6600	Fluorescent, RS, 4' long, ceiling mounted, two 40 watt	1 Elec	2.70	2.963	Ea.	1,575	65.50		1,640.50	1,850
6700	Mercury vapor with ballast, surface mounted, 175 watt		2.70	2.963		525	65.50		590.50	685
6850	Vandalproof, surface mounted, fluorescent, two 40 watt		3.20	2.500		171	55.50		226.50	279
6860	Incandescent, one 150 watt		8	1		48	22		70	89
7500	Ballast replacement, by weight of ballast, to 15' high									
7520	Indoor fluorescent, less than 2 lbs.	1 Elec	10	.800	Ea.		17.70		17.70	29
7540	Two 40W, watt reducer, 2 to 5 lbs.		9.40	.851		18	18.80		36.80	51
7560	Two F96 slimline, over 5 lbs.		8	1		31	22		53	70
7580	Vaportite ballast, less than 2 lbs.		9.40	.851			18.80		18.80	31
7600	2 lbs. to 5 lbs.		8.90	.899			19.85		19.85	32.50
7620	Over 5 lbs.		7.60	1.053			23.50		23.50	38
7630	Electronic ballast for two tubes		8	1		30	22		52	69
7640	Dimmable ballast one lamp		8	1		43	22		65	83
7650	Dimmable ballast two-lamp		7.60	1.053		67	23.50		90.50	112

16520 | Exterior Luminaires

		CREW	DAILY OUTPUT	LABOR-HOURS	UNIT	MAT.	LABOR	EQUIP.	TOTAL	TOTAL INCL O&P
0010	**EXTERIOR FIXTURES** With lamps									
0200	Wall mounted, incandescent, 100 watt	1 Elec	8	1	Ea.	28	22		50	67
0400	Quartz, 500 watt		5.30	1.509		51	33.50		84.50	111
0600	Mercury vapor, 100 watt		5.30	1.509		218	33.50		251.50	295
0800	Wall pack, mercury vapor, 175 watt		4	2		210	44		254	305
1000	250 watt		4	2		213	44		257	305
1100	Low pressure sodium, 35 watt		4	2		208	44		252	300
1150	55 watt		4	2		280	44		324	385
1200	Floodlights with ballast and lamp,									
1400	pole mounted, pole not included									
1500	Mercury vapor, 250 watt	1 Elec	2.40	3.333	Ea.	200	73.50		273.50	340
1600	400 watt	2 Elec	4.40	3.636		265	80.50		345.50	425
1800	1000 watt	"	4	4		365	88.50		453.50	545
1950	Metal halide, 175 watt	1 Elec	2.70	2.963		285	65.50		350.50	420
2000	400 watt	2 Elec	4.40	3.636		350	80.50		430.50	515
2200	1000 watt	"	4	4		450	88.50		538.50	640
2340	High pressure sodium, 70 watt	1 Elec	2.70	2.963		175	65.50		240.50	300
2400	400 watt	2 Elec	4.40	3.636		335	80.50		415.50	500
2600	1000 watt	"	4	4		500	88.50		588.50	695
2650	Roadway area luminaire, low pressure sodium, 135 watt	1 Elec	2	4		535	88.50		623.50	735
2700	180 watt	"	2	4		565	88.50		653.50	765
2720	Mercury vapor, 400 watt	2 Elec	4.40	3.636		340	80.50		420.50	505
2730	1000 watt		4	4		425	88.50		513.50	615
2750	Metal halide, 400 watt		4.40	3.636		410	80.50		490.50	580
2760	1000 watt		4	4		490	88.50		578.50	685
2780	High pressure sodium, 400 watt		4.40	3.636		465	80.50		545.50	640
2790	1000 watt		4	4		515	88.50		603.50	710
2800	Light poles, anchor base									
2820	not including concrete bases									
2840	Aluminum pole, 8' high	1 Elec	4	2	Ea.	445	44		489	565
3000	20' high	R-3	2.90	6.897		605	152	47.50	804.50	965
3200	30' high		2.60	7.692		1,125	169	53	1,347	1,575
3400	35' high		2.30	8.696		1,225	191	60	1,476	1,725
3600	40' high		2	10		1,400	220	69	1,689	1,950
3800	Bracket arms, 1 arm	1 Elec	8	1		76	22		98	120
4000	2 arms		8	1		153	22		175	204
4200	3 arms		5.30	1.509		229	33.50		262.50	305
4400	4 arms		5.30	1.509		305	33.50		338.50	390
4500	Steel pole, galvanized, 8' high		3.80	2.105		420	46.50		466.50	535
4600	20' high	R-3	2.60	7.692		745	169	53	967	1,150

16500 | Lighting

16520 | Exterior Luminaires

			CREW	DAILY OUTPUT	LABOR-HOURS	UNIT	2000 BARE COSTS				TOTAL INCL O&P
							MAT.	LABOR	EQUIP.	TOTAL	
300	4800	30' high	R-3	2.30	8.696	Ea.	875	191	60	1,126	1,350
	5000	35' high		2.20	9.091		960	200	63	1,223	1,450
	5200	40' high		1.70	11.765		1,175	259	81.50	1,515.50	1,825
	5400	Bracket arms, 1 arm	1 Elec	8	1		122	22		144	170
	5600	2 arms		8	1		189	22		211	244
	5800	3 arms		5.30	1.509		205	33.50		238.50	281
	6000	4 arms	▼	5.30	1.509	▼	285	33.50		318.50	370

16530 | Emergency Lighting

			CREW	DAILY OUTPUT	LABOR-HOURS	UNIT	MAT.	LABOR	EQUIP.	TOTAL	TOTAL INCL O&P
320	0010	**EXIT AND EMERGENCY LIGHTING**									
	0080	Exit light ceiling or wall mount, incandescent, single face	1 Elec	8	1	Ea.	35.50	22		57.50	75.50
	0100	Double face		6.70	1.194		44	26.50		70.50	91.50
	0200	L.E.D. standard, single face		8	1		62	22		84	104
	0220	double face		6.70	1.194		65	26.50		91.50	115
	0240	L.E.D. w/ battery unit, single face		4.40	1.818		105	40		145	182
	0260	double face	▼	4	2	▼	107	44		151	191
	0300	Emergency light units, battery operated									
	0350	Twin sealed beam light, 25 watt, 6 volt each									
	0500	Lead battery operated	1 Elec	4	2	Ea.	160	44		204	249
	0700	Nickel cadmium battery operated		4	2		480	44		524	605
	0900	Self-contained fluorescent lamp pack	▼	10	.800	▼	111	17.70		128.70	151

16585 | Lamps

			CREW	DAILY OUTPUT	LABOR-HOURS	UNIT	MAT.	LABOR	EQUIP.	TOTAL	TOTAL INCL O&P
600	0010	**LAMPS**									
	0080	Fluorescent, rapid start, cool white, 2' long, 20 watt	1 Elec	1	8	C	274	177		451	590
	0100	4' long, 40 watt		.90	8.889		253	196		449	600
	0200	Slimline, 4' long, 40 watt		.90	8.889		830	196		1,026	1,225
	0400	High output, 4' long, 60 watt		.90	8.889		985	196		1,181	1,400
	0500	8' long, 110 watt		.80	10		905	221		1,126	1,350
	0560	Twin tube compact lamp		.90	8.889		405	196		601	765
	0570	Double twin tube compact lamp		.80	10		880	221		1,101	1,325
	0600	Mercury vapor, mogul base, deluxe white, 100 watt		.30	26.667		3,325	590		3,915	4,625
	0700	250 watt		.30	26.667		4,375	590		4,965	5,800
	0800	400 watt		.30	26.667		3,525	590		4,115	4,850
	0900	1000 watt		.20	40		8,175	885		9,060	10,500
	1000	Metal halide, mogul base, 175 watt		.30	26.667		4,225	590		4,815	5,625
	1200	400 watt		.30	26.667		4,550	590		5,140	5,975
	1300	1000 watt		.20	40		12,300	885		13,185	15,100
	1350	High pressure sodium, 70 watt		.30	26.667		4,225	590		4,815	5,600
	1380	250 watt		.30	26.667		4,800	590		5,390	6,250
	1400	400 watt		.30	26.667		4,925	590		5,515	6,400
	1450	1000 watt		.20	40		17,600	885		18,485	20,900
	3000	Guards, fluorescent lamp, 4' long		1	8		580	177		757	930
	3200	8' long	▼	.90	8.889	▼	830	196		1,026	1,225

16700 | Communications

16720 | Tel and Intercomm Equipment

			CREW	DAILY OUTPUT	LABOR-HOURS	UNIT	2000 BARE COSTS MAT.	LABOR	EQUIP.	TOTAL	TOTAL INCL O&P	
300	0010	**DOCTORS IN-OUT REGISTER**										300
	0050	Register, 200 names	4 Elec	.64	50	Ea.	9,475	1,100		10,575	12,200	
	0100	Combination control and recall, 200 names	"	.64	50		12,400	1,100		13,500	15,400	
	0200	Recording register	1 Elec	.50	16		4,825	355		5,180	5,875	
	0300	Transformers	"	4	2		162	44		206	251	
	0400	Pocket pages					835			835	920	
600	0010	**NURSE CALL SYSTEMS**										600
	0100	Single bedside call station	1 Elec	8	1	Ea.	170	22		192	222	
	0200	Ceiling speaker station		8	1		49	22		71	90	
	0400	Emergency call station		8	1		83	22		105	128	
	0600	Pillow speaker		8	1		153	22		175	204	
	0800	Double bedside call station		4	2		258	44		302	355	
	1000	Duty station		4	2		127	44		171	212	
	1200	Standard call button		8	1		56.50	22		78.50	98	
	1400	Lights, corridor, dome or zone indicator		8	1		52.50	22		74.50	94	
	1600	Master control station for 20 stations	2 Elec	.65	24.615	Total	3,000	545		3,545	4,200	

16800 | Sound & Video

16820 | Sound Reinforcement

			CREW	DAILY OUTPUT	LABOR-HOURS	UNIT	2000 BARE COSTS MAT.	LABOR	EQUIP.	TOTAL	TOTAL INCL O&P	
300	0010	**DOORBELL SYSTEM** Incl. transformer, button & signal										300
	0100	6" bell	1 Elec	4	2	Ea.	55.50	44		99.50	134	
	0200	Buzzer	"	4	2	"	54.50	44		98.50	133	
800	0010	**PUBLIC ADDRESS SYSTEM**										800
	0100	Conventional, office	1 Elec	5.33	1.501	Speaker	95.50	33		128.50	160	
	0200	Industrial	"	2.70	2.963	"	185	65.50		250.50	310	
	0400	Explosionproof system is 3 times cost of central control										
	0600	Installation costs run about 120% of material cost										
840	0010	**SOUND SYSTEM** not including rough-in wires, cables & conduits										840
	0100	Components, outlet, projector	1 Elec	8	1	Ea.	42	22		64	82	
	0200	Microphone		4	2		46.50	44		90.50	124	
	0400	Speakers, ceiling or wall		8	1		81	22		103	125	
	0600	Trumpets		4	2		147	44		191	235	
	0800	Privacy switch		8	1		58	22		80	99.50	
	1000	Monitor panel		4	2		263	44		307	360	
	1200	Antenna, AM/FM		4	2		147	44		191	235	
	1400	Volume control		8	1		58.50	22		80.50	101	
	1600	Amplifier, 250 watts		1	8		1,100	177		1,277	1,500	
	1800	Cabinets		1	8		570	177		747	920	
	2000	Intercom, 25 station capacity, master station	2 Elec	2	8		1,375	177		1,552	1,825	
	2200	Remote station	1 Elec	8	1		110	22		132	157	
	2400	Intercom outlets		8	1		64.50	22		86.50	107	
	2600	Handset		4	2		200	44		244	293	
	2800	Emergency call system, 12 zones, annunciator		1.30	6.154		645	136		781	930	
	3000	Bell		5.30	1.509		67	33.50		100.50	128	
	3200	Light or relay		8	1		33.50	22		55.50	73	
	3400	Transformer		4	2		147	44		191	235	
	3600	House telephone, talking station		1.60	5		315	111		426	525	

16800 | Sound & Video

16820 | Sound Reinforcement

			CREW	DAILY OUTPUT	LABOR-HOURS	UNIT	MAT.	LABOR	EQUIP.	TOTAL	TOTAL INCL O&P	
840	3800	Press to talk, release to listen	1 Elec	5.30	1.509	Ea.	73.50	33.50		107	136	840
	4000	System-on button					44			44	48.50	
	4200	Door release	1 Elec	4	2		77.50	44		121.50	158	
	4400	Combination speaker and microphone		8	1		134	22		156	183	
	4600	Termination box		3.20	2.500		42	55.50		97.50	137	
	4800	Amplifier or power supply		5.30	1.509		485	33.50		518.50	585	
	5000	Vestibule door unit		16	.500	Name	89	11.05		100.05	116	
	5200	Strip cabinet		27	.296	Ea.	168	6.55		174.55	196	
	5400	Directory		16	.500	"	79	11.05		90.05	105	

16850 | Television Equipment

			CREW	DAILY OUTPUT	LABOR-HOURS	UNIT	MAT.	LABOR	EQUIP.	TOTAL	TOTAL INCL O&P	
600	0010	**T.V. SYSTEMS** not including rough-in wires, cables & conduits										600
	0100	Master TV antenna system										
	0200	VHF reception & distribution, 12 outlets	1 Elec	6	1.333	Outlet	152	29.50		181.50	215	
	0400	30 outlets		10	.800		100	17.70		117.70	139	
	0600	100 outlets		13	.615		102	13.60		115.60	135	
	0800	VHF & UHF reception & distribution, 12 outlets		6	1.333		151	29.50		180.50	214	
	1000	30 outlets		10	.800		100	17.70		117.70	139	
	1200	100 outlets		13	.615		102	13.60		115.60	135	
	1400	School and deluxe systems, 12 outlets		2.40	3.333		199	73.50		272.50	340	
	1600	30 outlets		4	2		175	44		219	266	
	1800	80 outlets		5.30	1.509		168	33.50		201.50	240	
	2000	Closed circuit, surveillance, one station (camera & monitor)	2 Elec	2.60	6.154	Total	965	136		1,101	1,275	
	2200	For additional camera stations, add	1 Elec	2.70	2.963	Ea.	540	65.50		605.50	700	
	2400	Industrial quality, one station (camera & monitor)	2 Elec	2.60	6.154	Total	2,000	136		2,136	2,425	
	2600	For additional camera stations, add	1 Elec	2.70	2.963	Ea.	1,225	65.50		1,290.50	1,450	
	2610	For low light, add		2.70	2.963		985	65.50		1,050.50	1,175	
	2620	For very low light, add		2.70	2.963		7,250	65.50		7,315.50	8,075	
	2800	For weatherproof camera station, add		1.30	6.154		755	136		891	1,050	
	3000	For pan and tilt, add		1.30	6.154		1,950	136		2,086	2,375	
	3200	For zoom lens - remote control, add, minimum		2	4		1,800	88.50		1,888.50	2,125	
	3400	Maximum		2	4		6,575	88.50		6,663.50	7,375	
	3410	For automatic iris for low light, add		2	4		1,575	88.50		1,663.50	1,875	
	3600	Educational T.V. studio, basic 3 camera system, black & white,										
	3800	electrical & electronic equip. only, minimum	4 Elec	.80	40	Total	9,350	885		10,235	11,800	
	4000	Maximum (full console)		.28	114		39,800	2,525		42,325	47,900	
	4100	As above, but color system, minimum		.28	114		52,500	2,525		55,025	62,000	
	4120	Maximum		.12	266		229,000	5,900		234,900	261,500	
	4200	For film chain, black & white, add	1 Elec	1	8	Ea.	10,700	177		10,877	12,100	
	4250	Color, add		.25	32		13,000	705		13,705	15,500	
	4400	For video tape recorders, add, minimum		1	8		2,250	177		2,427	2,775	
	4600	Maximum	4 Elec	.40	80		18,700	1,775		20,475	23,500	

For information about Means Estimating Seminars, see yellow pages 11 and 12 in back of book

Division 17
Square Foot & Cubic Foot Costs

Estimating Tips

- The cost figures in Division 17 were derived from approximately 10,200 projects contained in the Means database of completed construction projects, and include the contractor's overhead and profit, but do not generally include architectural fees or land costs. The figures have been adjusted to January of the current year. New projects are added to our files each year, and outdated projects are discarded. For this reason, certain costs may not show a uniform annual progression. In no case are all subdivisions of a project listed.

- These projects were located throughout the U.S. and reflect a tremendous variation in square foot (S.F.) and cubic foot (C.F.) costs. This is due to differences, not only in labor and material costs, but also in individual owners' requirements. For instance, a bank in a large city would have different features than one in a rural area. This is true of all the different types of buildings analyzed. Therefore, caution should be exercised when using Division 17 costs. For example, for court houses, costs in the database are local court house costs and will not apply to the larger, more elaborate federal court houses. As a general rule, the projects in the 1/4 column do not include any site work or equipment, while the projects in the 3/4 column may include both equipment and site work. The median figures do not generally include site work.

- None of the figures "go with" any others. All individual cost items were computed and tabulated separately. Thus the sum of the median figures for Plumbing, HVAC and Electrical will not normally total up to the total Mechanical and Electrical costs arrived at by separate analysis and tabulation of the projects.

- Each building was analyzed as to total and component costs and percentages. The figures were arranged in ascending order with the results tabulated as shown. The 1/4 column shows that 25% of the projects had lower costs, 75% higher. The 3/4 column shows that 75% of the projects had lower costs, 25% had higher. The median column shows that 50% of the projects had lower costs, 50% had higher.

- There are two times when square foot costs are useful. The first is in the conceptual stage when no details are available. Then square foot costs make a useful starting point. The second is after the bids are in and the costs can be worked back into their appropriate units for information purposes. As soon as details become available in the project design, the square foot approach should be discontinued and the project priced as to its particular components. When more precision is required or for estimating the replacement cost of specific buildings, the current edition of *Means Square Foot Costs* should be used.

- In using the figures in Division 17, it is recommended that the median column be used for preliminary figures if no additional information is available. The median figures, when multiplied by the total city construction cost index figures (see City Cost Indexes) and then multiplied by the project size modifier in Reference Number R17100-100, should present a fairly accurate base figure, which would then have to be adjusted in view of the estimator's experience, local economic conditions, code requirements and the owner's particular requirements. There is no need to factor the percentage figures, as these should remain constant from city to city. All tabulations mentioning air conditioning had at least partial air conditioning.

- The editors of this book would greatly appreciate receiving cost figures on one or more of your recent projects which would then be included in the averages for next year. All cost figures received will be kept confidential except that they will be averaged with other similar projects to arrive at S.F. and C.F. cost figures for next year's book. See the last page of the book for details and the discount available for submitting one or more of your projects.

17100 | S.F., C.F. and % of Total Costs

17100 | S.F. & C.F. Costs

				UNIT COSTS			% OF TOTAL		
			UNIT	1/4	MEDIAN	3/4	1/4	MEDIAN	3/4
010	0010	**APARTMENTS** Low Rise (1 to 3 story)	S.F.	43.15	54.05	72.30			
	0020	Total project cost	C.F.	3.87	5.15	6.35			
	0100	Site work	S.F.	3.69	5.40	8.55	6.30%	11%	14.10%
	0500	Masonry		.80	2.09	3.43	1.50%	4%	6.50%
	1500	Finishes		4.54	5.90	7.65	9%	10.70%	12.90%
	1800	Equipment		1.41	2.13	3.09	2.70%	4.10%	6.20%
	2720	Plumbing		3.32	4.32	5.45	6.70%	9%	10.10%
	2770	Heating, ventilating, air conditioning		2.14	2.64	3.88	4.20%	5.80%	7.70%
	2900	Electrical		2.49	3.25	4.44	5.20%	6.70%	8.40%
	3100	Total: Mechanical & Electrical		8.60	10.95	13.75	16%	18.20%	23%
	9000	Per apartment unit, total cost	Apt.	39,800	60,600	90,500			
	9500	Total: Mechanical & Electrical	"	7,600	12,000	15,600			
020	0010	**APARTMENTS** Mid Rise (4 to 7 story)	S.F.	56.15	68.25	83.50			
	0020	Total project costs	C.F.	4.47	6.15	8.40			
	0100	Site work	S.F.	2.29	4.54	8.15	5.20%	6.70%	9.10%
	0500	Masonry		3.80	5.20	7.45	5.90%	7.60%	10.60%
	1500	Finishes		7.20	9.20	11.70	10.50%	13.10%	17.70%
	1800	Equipment		1.88	2.68	3.44	2.80%	3.50%	4.70%
	2500	Conveying equipment		1.30	1.60	1.95	2%	2.30%	2.70%
	2720	Plumbing		3.36	5.40	5.95	6.30%	7.40%	10%
	2900	Electrical		3.94	5.40	6.30	6.70%	7.50%	8.90%
	3100	Total: Mechanical & Electrical		11.40	15.40	18.95	18.80%	21.60%	25.10%
	9000	Per apartment unit, total cost	Apt.	64,700	76,400	126,400			
	9500	Total: Mechanical & Electrical	"	12,200	14,100	17,900			
030	0010	**APARTMENTS** High Rise (8 to 24 story)	S.F.	64.85	78.30	95.70			
	0020	Total project costs	C.F.	5.50	7.70	9.35			
	0100	Site work	S.F.	1.99	3.80	5.30	2.60%	4.80%	6.20%
	0500	Masonry		3.75	6.85	8.70	4.70%	9.60%	11.10%
	1500	Finishes		7.20	9	10.65	9.70%	11.80%	13.70%
	1800	Equipment		2.09	2.60	3.40	2.80%	3.50%	4.30%
	2500	Conveying equipment		1.48	2.24	3.19	2.20%	2.80%	3.40%
	2720	Plumbing		4.80	5.65	6.90	7%	9.10%	10.60%
	2900	Electrical		4.46	5.65	7.60	6.40%	7.70%	8.90%
	3100	Total: Mechanical & Electrical		13.35	17.05	20.55	18%	22.30%	24.40%
	9000	Per apartment unit, total cost	Apt.	67,100	72,900	103,000			
	9500	Total: Mechanical & Electrical	"	14,600	16,700	17,600			
040	0010	**AUDITORIUMS**	S.F.	66.70	90.35	122			
	0020	Total project costs	C.F.	4.22	5.85	8.85			
	2720	Plumbing	S.F.	4.27	5.90	7.45	6.30%	7.20%	9.05%
	2900	Electrical		5.45	7.50	9.40	6.80%	9%	10.60%
	3100	Total: Mechanical & Electrical		10.40	14	24.35	14.40%	18.50%	23.60%
050	0010	**AUTOMOTIVE SALES**	S.F.	48.60	55.90	82.30			
	0020	Total project costs	C.F.	3.24	3.94	5.10			
	2720	Plumbing	S.F.	2.43	3.95	4.30	4.70%	6.50%	7%
	2770	Heating, ventilating, air conditioning		3.51	5.45	5.80	6.40%	10.30%	10.40%
	2900	Electrical		3.87	6	6.80	7.30%	10%	12.40%
	3100	Total: Mechanical & Electrical		8.75	12.60	15.75	17%	20.40%	26%
060	0010	**BANKS**	S.F.	96.85	121	153			
	0020	Total project costs	C.F.	7	9.50	12.45			
	0100	Site work	S.F.	10.50	18	25.65	7.50%	13.90%	17.60%
	0500	Masonry		4.94	8.40	18.35	2.90%	6.20%	11.40%
	1500	Finishes		8.35	11.55	14.90	5.50%	7.70%	10.30%
	1800	Equipment		3.90	8.10	18.10	3.30%	7.70%	12.50%
	2720	Plumbing		3.10	4.41	6.40	2.80%	3.90%	4.90%
	2770	Heating, ventilating, air conditioning		5.95	7.80	10.35	5%	7.20%	8.50%
	2900	Electrical		9.40	12.40	16.10	8.30%	10.30%	12.20%
	3100	Total: Mechanical & Electrical		22.25	29.90	36.35	16.60%	20.40%	24.40%
	3500	See also division 11020 & 11030							

See R17100-100 and City Cost Indexes in the Reference Section

17100 | S.F., C.F. and % of Total Costs

17100 | S.F. & C.F. Costs

			UNIT	UNIT COSTS 1/4	UNIT COSTS MEDIAN	UNIT COSTS 3/4	% OF TOTAL 1/4	% OF TOTAL MEDIAN	% OF TOTAL 3/4	
130	0010	**CHURCHES**	S.F.	65.05	81.60	106				130
	0020	Total project costs	C.F.	4.08	5.15	6.80				
	1800	Equipment	S.F.	.79	1.87	4.14	1%	2.20%	4.70%	
	2720	Plumbing		2.56	3.59	5.30	3.60%	5%	6.30%	
	2770	Heating, ventilating, air conditioning		6	7.80	11.50	7.60%	10%	12.20%	
	2900	Electrical		5.40	7.45	9.65	7.30%	8.80%	10.90%	
	3100	Total: Mechanical & Electrical		14.05	21.05	30.60	19.60%	22.90%	25.40%	
	3500	See also division 11040								
150	0010	**CLUBS, COUNTRY**	S.F.	68.60	81.40	107				150
	0020	Total project costs	C.F.	5.70	6.95	9.60				
	2720	Plumbing	S.F.	4.53	6.35	14.40	6.40%	9.60%	10%	
	2900	Electrical		5.55	7.95	10.45	7%	9.30%	11.40%	
	3100	Total: Mechanical & Electrical		12	15.45	28.65	16.10%	22.30%	30.90%	
170	0010	**CLUBS, SOCIAL** Fraternal	S.F.	55.45	79.15	104				170
	0020	Total project costs	C.F.	3.33	5.35	6.20				
	2720	Plumbing	S.F.	3.19	4.40	5.35	5.20%	6.80%	8.30%	
	2770	Heating, ventilating, air conditioning		5.10	6.20	7.95	8.70%	11%	14.40%	
	2900	Electrical		4.50	7	8.45	7.30%	9.50%	11.40%	
	3100	Total: Mechanical & Electrical		11.95	16.20	23.20	19.30%	23%	33.10%	
180	0010	**CLUBS, Y.M.C.A.**	S.F.	64.55	81.60	109				180
	0020	Total project costs	C.F.	3.26	5.45	8.15				
	2720	Plumbing	S.F.	4.70	8.90	10	6%	9.70%	16%	
	2900	Electrical		5.35	7.05	10.40	6.40%	9.20%	10.80%	
	3100	Total: Mechanical & Electrical		11.75	16.80	25.05	16.20%	21.90%	28.50%	
190	0010	**COLLEGES** Classrooms & Administration	S.F.	80.75	107	143				190
	0020	Total project costs	C.F.	5.85	8.20	13.30				
	0500	Masonry	S.F.	5.15	10.35	11.50	5.10%	11.90%	12.10%	
	2720	Plumbing		3.96	7.75	14.40	4.80%	8.90%	11.80%	
	2900	Electrical		6.65	10.20	12.60	8.40%	10%	12.10%	
	3100	Total: Mechanical & Electrical		14.25	25.90	40.20	14.20%	24.70%	32.50%	
210	0010	**COLLEGES** Science, Engineering, Laboratories	S.F.	128	156	189				210
	0020	Total project costs	C.F.	7.65	11.15	12.65				
	1800	Equipment	S.F.	7.40	16.75	18.30	5.30%	9.70%	15%	
	2900	Electrical		11	15.85	23.90	7.10%	9.60%	15.40%	
	3100	Total: Mechanical & Electrical		40.55	48.30	74.80	28%	31.60%	39.80%	
	3500	See also division 11600								
230	0010	**COLLEGES** Student Unions	S.F.	84.80	119	139				230
	0020	Total project costs	C.F.	4.73	6.20	8				
	3100	Total: Mechanical & Electrical	S.F.	22.30	31.85	34.50	17.40%	23.90%	28.90%	
250	0010	**COMMUNITY CENTERS**	"	63.35	85.05	113				250
	0020	Total project costs	C.F.	4.55	6.55	8.45				
	1800	Equipment	S.F.	1.76	3	4.85	2%	3.30%	6.40%	
	2720	Plumbing		3.40	5.80	8.45	4.90%	7.10%	9.50%	
	2770	Heating, ventilating, air conditioning		5.60	8.05	11.10	7.70%	10.80%	13%	
	2900	Electrical		5.60	7.60	11.50	7.40%	9.40%	11.50%	
	3100	Total: Mechanical & Electrical		20.80	24.80	35.75	25.10%	27.90%	34.40%	
280	0010	**COURT HOUSES**	S.F.	100	116	133				280
	0020	Total project costs	C.F.	7.65	9.25	12.85				
	2720	Plumbing	S.F.	4.81	6.70	9.65	5.90%	7.40%	9.60%	
	2900	Electrical		8.80	10.75	14.10	8.40%	9.80%	11.60%	
	3100	Total: Mechanical & Electrical		19.80	27.20	34.95	20.10%	25.70%	28.70%	
300	0010	**DEPARTMENT STORES**	S.F.	37.35	50.50	63.80				300
	0020	Total project costs	C.F.	1.97	2.59	3.52				
	2720	Plumbing	S.F.	1.22	1.48	2.23	3.10%	4.20%	5.90%	
	2770	Heating, ventilating, air conditioning		3.40	5.24	7.90	8.30%	12.40%	14.80%	

For expanded coverage of these items see *Means Square Foot Costs 2000*

17100 | S.F., C.F. and % of Total Costs

17100 | S.F. & C.F. Costs

			UNIT	UNIT COSTS 1/4	MEDIAN	3/4	% OF TOTAL 1/4	MEDIAN	3/4	
300	2900	Electrical	S.F.	4.29	5.90	6.95	9.10%	12.20%	14.90%	300
	3100	Total: Mechanical & Electrical (R17100-100)		7.55	9.75	15.85	20%	19%	26.50%	
310	0010	**DORMITORIES** Low Rise (1 to 3 story)	S.F.	61.10	87.20	105				310
	0020	Total project costs	C.F.	4.23	6.45	9.70				
	2720	Plumbing	S.F.	4.26	5.70	7.20	8%	9%	9.60%	
	2770	Heating, ventilating, air conditioning		4.50	5.40	7.20	4.60%	8%	10%	
	2900	Electrical		4.50	6.85	8.60	6.60%	8.90%	9.60%	
	3100	Total: Mechanical & Electrical		23.65	24.80	25.85	22.50%	24.10%	27.10%	
	9000	Per bed, total cost	Bed	30,000	33,300	71,500				
320	0010	**DORMITORIES** Mid Rise (4 to 8 story)	S.F.	86.95	113	140				320
	0020	Total project costs	C.F.	9.60	10.55	12.60				
	2900	Electrical	S.F.	6.45	10.10	12.90	7.40%	8.90%	10.40%	
	3100	Total: Mechanical & Electrical	"	19.30	25.85	33.75	18.10%	22.20%	30.60%	
	9000	Per bed, total cost	Bed	12,400	28,300	58,200				
340	0010	**FACTORIES**	S.F.	33.20	48.95	75.75				340
	0020	Total project costs	C.F.	2.11	3.13	5.20				
	0100	Site work	S.F.	3.75	6.85	11.20	5%	10.70%	18.20%	
	2720	Plumbing		1.96	3.30	5.60	4.60%	6.20%	8.70%	
	2770	Heating, ventilating, air conditioning		3.46	4.95	6.70	5.20%	8.50%	11.40%	
	2900	Electrical		4.04	6.45	9.90	8.10%	10.60%	14.20%	
	3100	Total: Mechanical & Electrical		9.75	15.60	23.70	20.40%	28.10%	34.50%	
360	0010	**FIRE STATIONS**	S.F.	63.65	85.45	110				360
	0020	Total project costs	C.F.	3.82	5.20	6.90				
	0500	Masonry	S.F.	9.60	17.20	22.80	8.70%	14.10%	19.30%	
	1140	Roofing		2.13	5.75	6.55	1.90%	4.90%	5%	
	1580	Painting		1.65	2.47	2.53	1.40%	2.10%	2.20%	
	1800	Equipment		1.41	1.96	4.70	1.10%	2.50%	4.80%	
	2720	Plumbing		4.13	6.10	8.95	6.20%	7.70%	9.70%	
	2770	Heating, ventilating, air conditioning		3.56	5.85	8.95	4.90%	7.50%	9.30%	
	2900	Electrical		4.52	7.80	10.65	7%	9.20%	11.30%	
	3100	Total: Mechanical & Electrical		20.10	24.80	29.35	18.80%	23.80%	27.30%	
370	0010	**FRATERNITY HOUSES** and Sorority Houses	S.F.	65.35	84.10	119				370
	0020	Total project costs	C.F.	6.25	6.80	8.70				
	2720	Plumbing	S.F.	4.90	5.65	10.35	5.20%	8%	10.90%	
	2900	Electrical		4.31	9.30	11.40	6.60%	9.90%	10.70%	
	3100	Total: Mechanical & Electrical		11.95	16.90	20.35	14.60%	19.90%	20.70%	
380	0010	**FUNERAL HOMES**	S.F.	68.60	94	171				380
	0020	Total project costs	C.F.	7.05	7.85	15.05				
	2900	Electrical	S.F.	3.04	5.60	6.60	3.60%	4.40%	11%	
	3100	Total: Mechanical & Electrical	"	10.95	16.35	21.25	12.90%	17.80%	18.80%	
390	0010	**GARAGES, COMMERCIAL** (Service)	S.F.	37.65	59.70	82.05				390
	0020	Total project costs	C.F.	2.39	3.58	5.15				
	1800	Equipment	S.F.	2.26	4.95	7.60	3.70%	6.80%	8.30%	
	2720	Plumbing		2.54	3.81	7.45	4.70%	7.40%	11%	
	2730	Heating & ventilating		3.55	4.85	6.70	5.30%	6.90%	9.60%	
	2900	Electrical		3.59	5.50	7.60	7.20%	9%	11.20%	
	3100	Total: Mechanical & Electrical		8.40	15.70	23.10	14.10%	22.50%	26.90%	
400	0010	**GARAGES, MUNICIPAL** (Repair)	S.F.	57.35	76.40	109				400
	0020	Total project costs	C.F.	3.58	4.54	7.40				
	0500	Masonry	S.F.	5.40	10.50	16.30	6%	10%	15.50%	
	2720	Plumbing		2.57	4.93	9.30	3.60%	6.70%	8%	
	2730	Heating & ventilating		4.40	6.35	12.30	6.20%	7.40%	13.50%	
	2900	Electrical		4.23	6.25	9.60	6.30%	8%	11.70%	
	3100	Total: Mechanical & Electrical		10.80	19.35	31.90	15.20%	25.50%	32.70%	

See R17100-100 and City Cost Indexes in the Reference Section

17100 | S.F., C.F. and % of Total Costs

17100 | S.F. & C.F. Costs

			UNIT	UNIT COSTS 1/4	MEDIAN	3/4	% OF TOTAL 1/4	MEDIAN	3/4	
410	0010	**GARAGES, PARKING**	S.F.	21.60	31.30	55.90				410
	0020	Total project costs	C.F.	2.10	2.85	4.14				
	2720	Plumbing	S.F.	.58	.98	1.51	2.20%	3.40%	3.90%	
	2900	Electrical		.91	1.38	2.03	4.30%	5.20%	6.30%	
	3100	Total: Mechanical & Electrical		1.22	3.57	4.58	6.50%	9.40%	12.80%	
	3200									
	9000	Per car, total cost	Car	9,400	11,800	15,300				
430	0010	**GYMNASIUMS**	S.F.	61.85	78.95	101				430
	0020	Total project costs	C.F.	3.09	3.95	5.15				
	1800	Equipment	S.F.	1.31	2.66	5.30	2.10%	3.40%	6.70%	
	2720	Plumbing		3.92	4.84	5.80	5.40%	7.30%	7.90%	
	2770	Heating, ventilating, air conditioning		4.22	6.45	12.90	9%	11.10%	22.60%	
	2900	Electrical		4.70	5.90	7.90	6.60%	8.30%	10.70%	
	3100	Total: Mechanical & Electrical		14.55	20	26.85	20.60%	26.20%	29.40%	
	3500	See also division 11480								
460	0010	**HOSPITALS**	S.F.	128	149	221				460
	0020	Total project costs	C.F.	9	11.15	16.10				
	1800	Equipment	S.F.	3.10	5.80	10	2.50%	3.80%	6%	
	2720	Plumbing		10.65	14.35	18.45	7.80%	9.40%	11.80%	
	2770	Heating, ventilating, air conditioning		15.05	19.95	26.90	8.40%	14.60%	17%	
	2900	Electrical		12.80	16.65	26.20	9.90%	12%	14.50%	
	3100	Total: Mechanical & Electrical		36.45	48.35	79.20	26.60%	33.10%	39%	
	9000	Per bed or person, total cost	Bed	45,900	97,500	153,800				
	9900	See also division 11700 & 11780								
480	0010	**HOUSING** For the Elderly	S.F.	58.55	73.90	90.80				480
	0020	Total project costs	C.F.	4.16	5.80	7.40				
	0100	Site work	S.F.	4.08	6.60	9.30	5.10%	8.20%	12.10%	
	0500	Masonry		1.80	6.65	9.55	2.10%	7.10%	12.20%	
	1800	Equipment		1.40	1.91	3.10	1.90%	3.20%	4.40%	
	2510	Conveying systems		1.42	1.91	2.60	1.80%	2.30%	2.90%	
	2720	Plumbing		4.35	5.70	7.65	8.10%	9.70%	10.90%	
	2730	Heating, ventilating, air conditioning		2.23	3.16	4.72	3.30%	5.60%	7.20%	
	2900	Electrical		4.36	5.90	7.55	7.30%	8.50%	10.50%	
	3100	Total: Mechanical & Electrical		15	17.65	23.30	18.10%	22%	29.10%	
	9000	Per rental unit, total cost	Unit	54,500	63,400	70,800				
	9500	Total: Mechanical & Electrical	"	12,000	13,000	16,300				
500	0010	**HOUSING** Public (Low Rise)	S.F.	50.95	68.45	89				500
	0020	Total project costs	C.F.	3.91	5.45	6.80				
	0100	Site work	S.F.	6.25	9	14.60	8.40%	11.70%	16.50%	
	1800	Equipment		1.34	2.18	3.58	2.30%	3%	5.10%	
	2720	Plumbing		3.34	4.69	5.95	6.80%	9%	11.60%	
	2730	Heating, ventilating, air conditioning		1.78	3.46	3.79	4.20%	6%	6.40%	
	2900	Electrical		2.98	4.44	6.15	5.10%	6.60%	8.30%	
	3100	Total: Mechanical & Electrical		14.15	18.20	20.40	14.50%	17.60%	26.50%	
	9000	Per apartment, total cost	Apt.	54,100	61,500	77,200				
	9500	Total: Mechanical & Electrical	"	9,800	12,800	14,700				
510	0010	**ICE SKATING RINKS**	S.F.	43.80	76	108				510
	0020	Total project costs	C.F.	3.09	3.16	3.65				
	2720	Plumbing	S.F.	1.57	2.95	3.02	3.10%	5.60%	6.70%	
	2900	Electrical		4.51	6.90	7.30	6.70%	15%	15.80%	
	3100	Total: Mechanical & Electrical		7.55	10.90	13.60	9.90%	25.90%	29.80%	
520	0010	**JAILS**	S.F.	129	165	213				520
	0020	Total project costs	C.F.	12.10	14.80	18.90				
	1800	Equipment	S.F.	5.35	14.80	25.10	4%	9.40%	15.20%	
	2720	Plumbing		13.05	16.55	21.85	7%	8.90%	13.80%	
	2770	Heating, ventilating, air conditioning		11.55	15.40	29.85	7.50%	9.40%	17.70%	
	2900	Electrical		14.10	18.10	22.40	8.10%	11.40%	12.40%	

For expanded coverage of these items see *Means Square Foot Costs 2000*

17100 | S.F., C.F. and % of Total Costs

17100 | S.F. & C.F. Costs

			UNIT	UNIT COSTS 1/4	MEDIAN	3/4	% OF TOTAL 1/4	MEDIAN	3/4	
520	3100	Total: Mechanical & Electrical [R17100-100]	S.F.	36	63.85	75.55	29.20%	31.10%	34.10%	520
530	0010	**LIBRARIES**	S.F.	76.40	97.30	124				530
	0020	Total project costs	C.F.	5.35	6.70	8.65				
	0500	Masonry	S.F.	4.37	9.75	16.80	5.80%	9.50%	11.90%	
	1800	Equipment		1.11	2.98	4.65	1.20%	2.80%	4.50%	
	2720	Plumbing		3.09	4.35	5.90	3.60%	4.90%	5.70%	
	2770	Heating, ventilating, air conditioning		6.60	11.20	14.60	8%	11%	14.60%	
	2900	Electrical		7.85	10.15	12.60	8.30%	11%	12.10%	
	3100	Total: Mechanical & Electrical		22.65	30.95	38.75	18.90%	25.30%	27.60%	
550	0010	**MEDICAL CLINICS**	S.F.	75.20	93.15	117				550
	0020	Total project costs	C.F.	5.60	7.25	9.70				
	1800	Equipment	S.F.	2.06	4.33	6.75	1.80%	5.20%	7.40%	
	2720	Plumbing		5.05	7.15	9.55	6.10%	8.40%	10%	
	2770	Heating, ventilating, air conditioning		6.15	7.90	11.65	6.70%	9%	11.30%	
	2900	Electrical		6.40	9.10	12.05	8.10%	10%	12.20%	
	3100	Total: Mechanical & Electrical		20.10	28.25	39.50	22%	27.60%	34.30%	
	3500	See also division 11700								
570	0010	**MEDICAL OFFICES**	S.F.	70.60	87.50	108				570
	0020	Total project costs	C.F.	5.25	7.20	9.85				
	1800	Equipment	S.F.	2.45	4.70	6.70	3%	5.80%	7.20%	
	2720	Plumbing		3.96	6.10	8.30	5.70%	6.80%	8.60%	
	2770	Heating, ventilating, air conditioning		4.79	7.05	9.10	6.20%	8%	9.70%	
	2900	Electrical		5.60	8.15	11.40	7.60%	9.80%	11.40%	
	3100	Total: Mechanical & Electrical		13.90	20	29.70	18.50%	22%	24.90%	
590	0010	**MOTELS**	S.F.	45.20	67	86.40				590
	0020	Total project costs	C.F.	3.95	5.55	9.10				
	2720	Plumbing	S.F.	4.59	5.85	6.95	9.40%	10.60%	12.50%	
	2770	Heating, ventilating, air conditioning		2.79	4.17	7.45	5.60%	5.60%	10%	
	2900	Electrical		4.27	5.45	7.10	7.10%	8.20%	10.40%	
	3100	Total: Mechanical & Electrical		14.50	18.20	31.20	18.50%	21%	24.40%	
	5000									
	9000	Per rental unit, total cost	Unit	23,000	43,800	47,300				
	9500	Total: Mechanical & Electrical	"	4,500	6,800	7,900				
600	0010	**NURSING HOMES**	S.F.	68	89.95	110				600
	0020	Total project costs	C.F.	5.45	7	9.50				
	1800	Equipment	S.F.	2.28	3.04	4.90	2.40%	3.70%	6%	
	2720	Plumbing		6.40	8.15	11.30	9.40%	10.70%	14.20%	
	2770	Heating, ventilating, air conditioning		6.35	8.85	11.30	9.30%	11.40%	11.80%	
	2900	Electrical		7.05	8.80	11.80	9.70%	11%	13%	
	3100	Total: Mechanical & Electrical		16.75	23.45	34.35	26%	29.90%	30.50%	
	3200									
	9000	Per bed or person, total cost	Bed	29,400	36,200	48,200				
610	0010	**OFFICES** Low Rise (1 to 4 story)	S.F.	57.30	73	97.15				610
	0020	Total project costs	C.F.	4.15	5.80	7.85				
	0100	Site work	S.F.	4.32	7.35	11.40	5.30%	9.70%	14%	
	0500	Masonry		1.99	4.66	8.80	2.90%	5.80%	8.70%	
	1800	Equipment		.71	1.30	3.57	1.20%	1.50%	4%	
	2720	Plumbing		2.18	3.30	4.67	3.70%	4.50%	6.10%	
	2770	Heating, ventilating, air conditioning		4.71	6.50	9.65	7.20%	10.50%	11.90%	
	2900	Electrical		4.86	6.70	9.40	7.50%	9.60%	11.10%	
	3100	Total: Mechanical & Electrical		11.40	15.85	23.15	18%	21.80%	26.50%	
620	0010	**OFFICES** Mid Rise (5 to 10 story)	S.F.	63.20	76.65	104				620
	0020	Total project costs	C.F.	4.42	5.60	8.10				
	2720	Plumbing	S.F.	1.91	2.96	4.26	2.80%	3.70%	4.50%	
	2770	Heating, ventilating, air conditioning		4.80	6.85	10.95	7.60%	9.40%	11%	

17100 | S.F., C.F. and % of Total Costs

17100 | S.F. & C.F. Costs

				UNIT COSTS			% OF TOTAL			
			UNIT	1/4	MEDIAN	3/4	1/4	MEDIAN	3/4	
620	2900	Electrical	S.F.	4.69	6	9.10	6.50%	8.20%	10%	620
	3100	Total: Mechanical & Electrical		12	15.30	30.35	17.90%	22.30%	29.90%	
630	0010	**OFFICES** High Rise (11 to 20 story)	S.F.	76.45	98.05	121				630
	0020	Total project costs	C.F.	4.89	6.75	9.75				
	2900	Electrical	S.F.	3.91	5.65	8.55	5.80%	7%	10.50%	
	3100	Total: Mechanical & Electrical	"	14.25	18.25	33.45	15.40%	17%	34.10%	
640	0010	**POLICE STATIONS**	S.F.	91.65	122	154				640
	0020	Total project costs	C.F.	7.40	8.95	12.45				
	0500	Masonry	S.F.	10.90	16.10	19.40	7.80%	10.60%	11.30%	
	1800	Equipment		1.51	6.50	10.45	2.10%	5.20%	9.80%	
	2720	Plumbing		5.20	8.40	11.75	5.60%	6.90%	11.30%	
	2770	Heating, ventilating, air conditioning		8.15	10.80	14.70	7%	10.70%	12%	
	2900	Electrical		10.15	15.20	19.30	9.70%	11.90%	14.90%	
	3100	Total: Mechanical & Electrical		32.35	38.60	54.35	25.10%	31.30%	36.40%	
650	0010	**POST OFFICES**	S.F.	71.40	90.40	113				650
	0020	Total project costs	C.F.	4.36	5.60	6.65				
	2720	Plumbing	S.F.	3.31	4.25	5.20	4.40%	5.60%	5.60%	
	2770	Heating, ventilating, air conditioning		5.20	6.40	7.10	7%	8.40%	10.20%	
	2900	Electrical		6.05	8.40	9.90	7.20%	9.60%	11%	
	3100	Total: Mechanical & Electrical		10.65	20.60	26	16.40%	20.40%	22.30%	
660	0010	**POWER PLANTS**	S.F.	532	678	1,241				660
	0020	Total project costs	C.F.	14.05	28.45	65.50				
	2900	Electrical	S.F.	36.05	76.45	114	9.50%	12.80%	18.40%	
	8100	Total: Mechanical & Electrical	"	89.85	292	655	28.80%	32.50%	52.60%	
670	0010	**RELIGIOUS EDUCATION**	S.F.	57.70	74.85	87.30				670
	0020	Total project costs	C.F.	3.29	4.71	6.10				
	2720	Plumbing	S.F.	2.47	3.50	4.87	4.30%	4.90%	8.40%	
	2770	Heating, ventilating, air conditioning		5.80	6.55	7.20	9.80%	10%	11.80%	
	2900	Electrical		4.70	6.20	8.20	7.90%	9.10%	10.30%	
	3100	Total: Mechanical & Electrical		12.15	21.15	25.10	17.60%	20.50%	22%	
690	0010	**RESEARCH** Laboratories and facilities	S.F.	88.55	127	185				690
	0020	Total project costs	C.F.	6.60	11.10	14.95				
	1800	Equipment	S.F.	3.94	7.75	18.90	1.60%	5.20%	9.30%	
	2720	Plumbing		8.90	11.65	18.30	7.30%	9.20%	13.30%	
	2770	Heating, ventilating, air conditioning		8	26.85	31.70	7.20%	16.50%	17.50%	
	2900	Electrical		10.20	17.10	29.15	9.60%	12%	15.80%	
	3100	Total: Mechanical & Electrical		23.55	44.70	85.20	23.90%	35.60%	41.70%	
700	0010	**RESTAURANTS**	S.F.	85.35	110	143				700
	0020	Total project costs	C.F.	7.15	9.40	12.40				
	1800	Equipment	S.F.	5.10	13.65	20.65	6.10%	13.40%	15.60%	
	2720	Plumbing		6.75	8.30	11.75	6.10%	8.30%	9.10%	
	2770	Heating, ventilating, air conditioning		8.95	11.95	15.60	9.40%	12.20%	13%	
	2900	Electrical		9.10	11.20	14.60	8.50%	10.60%	11.60%	
	3100	Total: Mechanical & Electrical		28.45	29.70	37.55	21.10%	24%	29.50%	
	5000									
	9000	Per seat unit, total cost	Seat	3,200	4,200	5,300				
	9500	Total: Mechanical & Electrical	"	663	943	1,240				
720	0010	**RETAIL STORES**	S.F.	39.80	53.70	70.55				720
	0020	Total project costs	C.F.	2.71	3.87	5.35				
	2720	Plumbing	S.F.	1.46	2.43	4.15	3.20%	4.60%	6.80%	
	2770	Heating, ventilating, air conditioning		3.15	4.31	6.45	6.70%	8.70%	10.10%	
	2900	Electrical		3.59	4.87	7.10	7.30%	9.90%	11.60%	
	3100	Total: Mechanical & Electrical		9.60	12.30	16	17.10%	21.40%	23.80%	
740	0010	**SCHOOLS** Elementary	S.F.	63.35	78.20	94.25				740
	0020	Total project costs	C.F.	4.29	5.50	7.10				
	0500	Masonry	S.F.	5.85	9.25	12.95	7.10%	10.50%	15.20%	
	1800	Equipment		2	3.38	5.95	2.50%	4.20%	7.60%	
	2720	Plumbing		3.67	5.30	7.10	5.70%	7.10%	9.40%	
	2730	Heating, ventilating, air conditioning		5.65	9	12.55	8.10%	10.80%	15.20%	

For expanded coverage of these items see *Means Square Foot Costs 2000*

17100 | S.F., C.F. and % of Total Costs

17100 | S.F. & C.F. Costs

				UNIT COSTS			% OF TOTAL			
			UNIT	1/4	MEDIAN	3/4	1/4	MEDIAN	3/4	
740	2900	Electrical	S.F.	5.90	7.55	9.65	8.40%	10%	11.60%	**740**
	3100	Total: Mechanical & Electrical	↓ [R17100-100]	20.45	22.65	31.95	22.20%	28%	30.40%	
	9000	Per pupil, total cost	Ea.	6,300	10,500	34,100				
	9500	Total: Mechanical & Electrical	"	2,125	2,675	10,800				
760	0010	**SCHOOLS** Junior High & Middle	S.F.	64.40	79.65	93.25				**760**
	0020	Total project costs	C.F.	4.23	5.55	6.25				
	0500	Masonry	S.F.	7.25	9.70	11.95	8.80%	11.10%	14%	
	1800	Equipment		2.16	3.48	5.35	2.60%	4.30%	5.90%	
	2720	Plumbing		4.29	4.75	6.30	5.60%	6.90%	8.10%	
	2770	Heating, ventilating, air conditioning		4.94	9.50	13.30	8.70%	12.70%	17.40%	
	2900	Electrical		6.25	7.80	9.40	7.80%	9.40%	10.60%	
	3100	Total: Mechanical & Electrical	↓	18	22.60	30	23.30%	25.70%	27.30%	
	9000	Per pupil, total cost	Ea.	8,100	9,300	11,600				
780	0010	**SCHOOLS** Senior High	S.F.	69.55	79.65	112				**780**
	0020	Total project costs	C.F.	4.45	5.80	8.15				
	1800	Equipment	S.F.	1.85	4.35	6.45	2.30%	3.70%	5.40%	
	2720	Plumbing		3.60	6.55	10.80	5%	6.90%	12%	
	2770	Heating, ventilating, air conditioning		8.05	9.15	17.60	8.90%	11%	15%	
	2900	Electrical		6.85	9	15	8.30%	10.10%	12.60%	
	3100	Total: Mechanical & Electrical	↓	18.70	24.60	46	19.80%	23.10%	27.80%	
	9000	Per pupil, total cost	Ea.	6,625	11,300	16,900				
800	0010	**SCHOOLS** Vocational	S.F.	56.75	79.35	101				**800**
	0020	Total project costs	C.F.	3.48	5.05	7				
	0500	Masonry	S.F.	3.35	8.25	12.60	4%	10.90%	19.90%	
	1800	Equipment	"	1.78	2.41	6.10	2.90%	3.40%	4.70%	
	2720	Plumbing	S.F.	3.64	5.55	8	5.40%	7%	8.50%	
	2770	Heating, ventilating, air conditioning		5.10	9.50	15.85	8.80%	11.90%	14.60%	
	2900	Electrical		5.85	8.10	11.45	8.40%	11.20%	13.80%	
	3100	Total: Mechanical & Electrical	↓	15.40	20.55	27.10	21.70%	27.30%	34.70%	
	9000	Per pupil, total cost	Ea.	7,900	21,200	31,700				
830	0010	**SPORTS ARENAS**	S.F.	48.40	66.45	97.55				**830**
	0020	Total project costs	C.F.	2.70	4.79	6.25				
	2720	Plumbing	S.F.	2.57	4.28	8.05	4.50%	6.30%	9.40%	
	2770	Heating, ventilating, air conditioning		5.30	7.35	9.55	5.80%	10.20%	13.50%	
	2900	Electrical		4.27	6.75	8.45	7.70%	9.80%	12.30%	
	3100	Total: Mechanical & Electrical	↓	12.55	22.40	28.85	13.40%	22.50%	30.80%	
850	0010	**SUPERMARKETS**	S.F.	45.85	53.85	62.50				**850**
	0020	Total project costs	C.F.	2.56	3.09	4.69				
	2720	Plumbing	S.F.	2.57	3.23	3.76	5.40%	6%	7.50%	
	2770	Heating, ventilating, air conditioning		3.78	4.64	5.50	8.60%	8.60%	9.60%	
	2900	Electrical		5.35	6.60	7.85	10.40%	12.40%	13.60%	
	3100	Total: Mechanical & Electrical	↓	15	16.20	22.40	17.40%	20.40%	28.40%	
860	0010	**SWIMMING POOLS**	S.F.	81.10	125	174				**860**
	0020	Total project costs	C.F.	5.95	7.45	8.10				
	2720	Plumbing	S.F.	6.90	7.85	10.95	9.70%	12.30%	12.30%	
	2900	Electrical		5.60	8.70	11.30	7.50%	8%	8%	
	3100	Total: Mechanical & Electrical	↓	11.90	25.15	46.55	17.70%	24.90%	31.60%	
870	0010	**TELEPHONE EXCHANGES**	S.F.	98.80	145	183				**870**
	0020	Total project costs	C.F.	6.10	9.40	13.30				
	2720	Plumbing	S.F.	4.17	6.45	9.45	4.50%	5.80%	6.90%	
	2770	Heating, ventilating, air conditioning		8.50	19.10	24.20	11.80%	11.80%	18.40%	
	2900	Electrical		10.05	15.60	28.20	10.90%	14%	17.80%	
	3100	Total: Mechanical & Electrical	↓	21.60	29.70	73.65	27.30%	33.40%	44.40%	
910	0010	**THEATERS**	S.F.	62.15	79.70	117				**910**
	0020	Total project costs	C.F.	2.87	4.24	6.25				
	2720	Plumbing	S.F.	1.94	2.25	6.75	2.90%	4.70%	6.80%	
	2770	Heating, ventilating, air conditioning	↓	6.05	7.30	9.05	8%	12.20%	13.40%	

See R17100-100 and City Cost Indexes in the Reference Section

17100 | S.F., C.F. and % of Total Costs

17100 | S.F. & C.F. Costs

				UNIT	UNIT COSTS 1/4	UNIT COSTS MEDIAN	UNIT COSTS 3/4	% OF TOTAL 1/4	% OF TOTAL MEDIAN	% OF TOTAL 3/4	
910	2900	Electrical	R17100 -100	S.F.	5.45	7.35	14.95	8%	10%	12.40%	910
	3100	Total: Mechanical & Electrical			14	20.90	42.95	22.90%	26.60%	27.50%	
940	0010	**TOWN HALLS** City Halls & Municipal Buildings		S.F.	69.35	87.65	114				940
	0020	Total project costs		C.F.	5.20	7.25	9.70				
	2720	Plumbing		S.F.	2.47	5.10	8.45	4.20%	6.10%	7.90%	
	2770	Heating, ventilating, air conditioning			5.24	10.40	15.20	7%	9%	13.50%	
	2900	Electrical			6.25	8.40	11.90	8.20%	9.50%	11.70%	
	3100	Total: Mechanical & Electrical			22	22.85	29	22%	25.10%	30.40%	
970	0010	**WAREHOUSES** And Storage Buildings		S.F.	25.85	36.15	55.65				970
	0020	Total project costs		C.F.	1.30	2.12	3.52				
	0100	Site work		S.F.	2.59	5.30	8	6%	12.80%	19.80%	
	0500	Masonry			1.61	3.68	7.95	5%	7.70%	13.20%	
	1800	Equipment			.42	.90	5.05	1%	2.40%	6.50%	
	2720	Plumbing			.86	1.55	2.89	2.90%	4.80%	6.60%	
	2730	Heating, ventilating, air conditioning			.99	2.78	3.73	2.40%	5%	8.90%	
	2900	Electrical			1.53	2.88	4.78	5%	7.30%	10%	
	3100	Total: Mechanical & Electrical			4.27	6.30	14.35	12.80%	18.40%	26.20%	
990	0010	**WAREHOUSE & OFFICES** Combination		S.F.	31.15	41.75	55.40				990
	0020	Total project costs		C.F.	1.62	2.36	3.50				
	1800	Equipment		S.F.	.55	1.07	1.65	1.20%	2.40%	2.70%	
	2720	Plumbing			1.20	2.07	3.27	3.60%	4.70%	6.30%	
	2770	Heating, ventilating, air conditioning			1.91	3.02	4.25	5%	5.60%	9.60%	
	2900	Electrical			2.11	3.11	4.88	5.70%	7.70%	10%	
	3100	Total: Mechanical & Electrical			4.98	8.25	10.30	13.80%	18.60%	23%	

For information about Means Estimating Seminars, see yellow pages 11 and 12 in back of book

For expanded coverage of these items see *Means Square Foot Costs 2000*

Division Notes

	CREW	DAILY OUTPUT	LABOR-HOURS	UNIT	2000 BARE COSTS				TOTAL INCL O&P
					MAT.	LABOR	EQUIP.	TOTAL	

Reference Section

All the reference information is in one section making it easy to find what you need to know... and easy to use the book on a daily basis. This section is visually identified by a vertical gray bar on the edge of pages.

In the reference number information that follows, you'll see the background that relates to the "reference numbers" that appeared in the Unit Price Sections. You'll find reference tables, explanations and estimating information that support how we arrived at the unit price data. Also included are alternate pricing methods, technical data and estimating procedures along with information on design and economy in construction.

Also in this Reference Section, we've included Change Orders, information on pricing changes to contract documents; Crew Listings, a full listing of all the crews, equipment and their costs; Historical Cost Indexes for cost comparisons over time; City Cost Indexes and Location Factors for adjusting costs to the region you are in; and an explanation of all Abbreviations used in the book.

Table of Contents

Reference Numbers

R011	Overhead & Misc. Data	480
R015	Construction Aids	489
R020	Paving & Surfacing	492
R021	Subsurface Investigation & Demolition	493
R022	Site Preparation & Excavation Support	493
R023	Earthwork	496
R024	Foundation & L.B. Elements	499
R025	Sewerage & Drainage	500
R029	Landscaping	500
R031	Concrete Formwork	501
R032	Concrete Reinforcement	506
R033	Cast-in-Place Concrete	510
R034	Precast Concrete	521
R035	Cementitious Decks & Toppings	525
R040	Mortar & Masonry Accessories	526
R042	Unit Masonry	527
R044	Stone	532
R049	Restoration & Cleaning	532
R050	Materials, Coatings & Fastenings	533
R051	Structural Metal Framing	534
R056	Railroad & Marine Work	543
R061	Rough Carpentry	544
R064	Architectural Woodwork	547
R071	Waterproofing & Dampproofing	547
R073	Shingles & Roofing Tiles	548

Reference Numbers (cont.)

R075	Membrane Roofing	548
R078	Fire & Smoke Protection	549
R081	Metal Doors & Frames	549
R082	Wood & Plastic Doors	550
R085	Metal Windows	550
R087	Hardware	551
R088	Glazing	552
R091	Metal Support Assemblies	553
R092	Plaster & Gypsum Board	553
R094	Terrazzo	555
R096	Flooring	555
R097	Wall Finishes	555
R099	Paints & Coatings	556
R105	Lockers	557
R130	Air Supported Structures	557
R131	Pre-Engineered Structures & Aquatic Facilities	557
R132	Hazardous Material Remediation	559
R136	Solar & Wind Energy Equip.	560
R142	Elevators	560
R143	Escalators & Moving Walks	563
R150	Basic Materials & Methods	564
R151	Plumbing	564
R155	Heating	565
R157	Air Conditioning & Ventilation	567
R158	Air Distribution	568
R161	Wiring Methods	569
R162	Electrical Power	571
R171	Information	572

Change Orders	573
Crew Listings	577
Historical Cost Indexes	600
City Cost Indexes	601
Location Factors	643
Abbreviations	649

General Requirements | R011 | Overhead & Miscellaneous Data

R01100-005 Tips for Accurate Estimating

1. Use pre-printed or columnar forms for orderly sequence of dimensions and locations and for recording telephone quotations.
2. Use only the front side of each paper or form except for certain pre-printed summary forms.
3. Be consistent in listing dimensions: For example, length x width x height. This helps in rechecking to ensure that, the total length of partitions is appropriate for the building area.
4. Use printed (rather than measured) dimensions where given.
5. Add up multiple printed dimensions for a single entry where possible.
6. Measure all other dimensions carefully.
7. Use each set of dimensions to calculate multiple related quantities.
8. Convert foot and inch measurements to decimal feet when listing. Memorize decimal equivalents to .01 parts of a foot (1/8" equals approximately .01').
9. Do not "round off" quantities until the final summary.
10. Mark drawings with different colors as items are taken off.
11. Keep similar items together, different items separate.
12. Identify location and drawing numbers to aid in future checking for completeness.
13. Measure or list everything on the drawings or mentioned in the specifications.
14. It may be necessary to list items not called for to make the job complete.
15. Be alert for: Notes on plans such as N.T.S. (not to scale); changes in scale throughout the drawings; reduced size drawings; discrepancies between the specifications and the drawings.
16. Develop a consistent pattern of performing an estimate. For example:
    a. Start the quantity takeoff at the lower floor and move to the next higher floor.
    b. Proceed from the main section of the building to the wings.
    c. Proceed from south to north or vice versa, clockwise or counterclockwise.
    d. Take off floor plan quantities first, elevations next, then detail drawings.
17. List all gross dimensions that can be either used again for different quantities, or used as a rough check of other quantities for verification (exterior perimeter, gross floor area, individual floor areas, etc.).
18. Utilize design symmetry or repetition (repetitive floors, repetitive wings, symmetrical design around a center line, similar room layouts, etc.). Note: Extreme caution is needed here so as not to omit or duplicate an area.
19. Do not convert units until the final total is obtained. For instance, when estimating concrete work, keep all units to the nearest cubic foot, then summarize and convert to cubic yards.
20. When figuring alternatives, it is best to total all items involved in the basic system, then total all items involved in the alternates. Therefore you work with positive numbers in all cases. When adds and deducts are used, it is often confusing whether to add or subtract a portion of an item; especially on a complicated or involved alternate.

R01100-020 Construction Time Requirements

Table at left is average construction time in months for different types of building projects. Table at right is the construction time in months for different size projects. Design time runs 25% to 40% of construction time.

Type Building	Construction Time	Project Value	Construction Time
Industrial Buildings	12 Months	Under $1,400,000	10 Months
Commercial Buildings	15 Months	Up to $3,800,000	15 Months
Research & Development	18 Months	Up to $19,000,000	21 Months
Institutional Buildings	20 Months	over $19,000,000	28 Months

R01100-040 Builder's Risk Insurance

Builder's Risk Insurance is insurance on a building during construction. Premiums are paid by the owner or the contractor. Blasting, collapse and underground insurance would raise total insurance costs above those listed. Floater policy for materials delivered to the job runs $.75 to $1.25 per $100 value. Contractor equipment insurance runs $.50 to $1.50 per $100 value. Insurance for miscellaneous tools to $1,500 value runs from $3.00 to $7.50 per $100 value.

Tabulated below are New England Builder's Risk insurance rates in dollars per $100 value for $1,000 deductible. For $25,000 deductible, rates can be reduced 13% to 34%. On contracts over $1,000,000, rates may be lower than those tabulated. Policies are written annually for the total completed value in place. For "all risk" insurance (excluding flood, earthquake and certain other perils) add $.025 to total rates below.

Coverage	Frame Construction (Class 1)		Brick Construction (Class 4)		Fire Resistive (Class 6)	
	Range	Average	Range	Average	Range	Average
Fire Insurance	$.300 to $.420	$.394	$.132 to $.189	$.174	$.052 to $.080	$.070
Extended Coverage	.115 to .150	.144	.080 to .105	.101	.081 to .105	.100
Vandalism	.012 to .016	.015	.008 to .011	.011	.008 to .011	.010
Total Annual Rate	$.427 to $.586	$.553	$.220 to $.305	$.286	$.141 to $.196	$.180

General Requirements — R011 Overhead & Miscellaneous Data

R01100-050 General Contractor's Overhead

The table below shows a contractor's overhead as a percentage of direct cost in two ways. The figures on the right are for the overhead, markup based on both material and labor. The figures on the left are based on the entire overhead applied only to the labor. This figure would be used if the owner supplied the materials or if a contract is for labor only. Note: Some of these markups are included in the labor rates shown on Reference Table R01100-070.

Items of General Contractor's Indirect Costs	% of Direct Costs As a Markup of Labor Only	% of Direct Costs As a Markup of Both Material and Labor
Field Supervision	6.0%	2.9%
Main Office Expense (see details below)	16.2	7.7
Tools and Minor Equipment	1.0	0.5
Workers' Compensation & Employers' Liability. See R01100-060	18.1	8.6
Field Office, Sheds, Photos, Etc.	1.5	0.7
Performance and Payment Bond, 0.7% to 1.5%. See R01100-080	2.3	1.1
Unemployment Tax See R01100-100 (Combined Federal and State)	7.0	3.3
Social Security and Medicare, See R01100-100	7.7	3.7
Sales Tax — add if applicable 42/80 x % as markup of total direct costs including both material and labor. See R01100-090		
Sub Total	59.8%	28.5%
*Builder's Risk Insurance ranges from .141% to .586%. See R01100-040	0.6	0.3
*Public Liability Insurance	3.2	1.5
Grand Total	63.6%	30.3%

*Paid by Owner or Contractor

Main Office Expense

A General Contractor's main office expense consists of many items not detailed in the front portion of the book. The percentage of main office expense declines with increased annual volume of the contractor. Typical main office expense ranges from 2% to 20% with the median about 7.2% of total volume. This equals about 7.7% of direct costs. The following are approximate percentages of total overhead for different items usually included in a General Contractor's main office overhead. With different accounting procedures, these percentages may vary.

Item	Typical Range	Average
Managers', clerical and estimators' salaries	40% to 55%	48%
Profit sharing, pension and bonus plans	2 to 20	12
Insurance	5 to 8	6
Estimating and project management (not including salaries)	5 to 9	7
Legal, accounting and data processing	0.5 to 5	3
Automobile and light truck expense	2 to 8	5
Depreciation of overhead capital expenditures	2 to 6	4
Maintenance of office equipment	0.1 to 1.5	1
Office rental	3 to 5	4
Utilities including phone and light	1 to 3	2
Miscellaneous	5 to 15	8
Total		100%

General Requirements — R011 Overhead & Miscellaneous Data

R01100-060 Workers' Compensation Insurance Rates by Trade

The table below tabulates the national averages for Workers' Compensation insurance rates by trade and type of building. The average "Insurance Rate" is multiplied by the "% of Building Cost" for each trade. This produces the "Workers' Compensation Cost" by % of total labor cost, to be added for each trade by building type to determine the weighted average Workers' Compensation rate for the building types analyzed.

Trade	Insurance Rate (% Labor Cost) Range	Insurance Rate (% Labor Cost) Average	% of Building Cost Office Bldgs.	% of Building Cost Schools & Apts.	% of Building Cost Mfg.	Workers' Compensation Office Bldgs.	Workers' Compensation Schools & Apts.	Workers' Compensation Mfg.
Excavation, Grading, etc.	4.0% to 26.6%	11.4%	4.8%	4.9%	4.5%	.55%	.56%	.51%
Piles & Foundations	8.1 to 80.1	29.9	7.1	5.2	8.7	2.12	1.55	2.60
Concrete	7.5 to 39.4	18.8	5.0	14.8	3.7	.94	2.78	.70
Masonry	5.8 to 48.6	17.8	6.9	7.5	1.9	1.23	1.34	.34
Structural Steel	8.1 to 132.9	42.6	10.7	3.9	17.6	4.56	1.66	7.50
Miscellaneous & Ornamental Metals	5.4 to 34.0	13.8	2.8	4.0	3.6	.39	.55	.50
Carpentry & Millwork	7.0 to 49.9	19.9	3.7	4.0	0.5	.74	.80	.10
Metal or Composition Siding	6.9 to 36.7	18.1	2.3	0.3	4.3	.42	.05	.78
Roofing	8.1 to 88.8	34.6	2.3	2.6	3.1	.80	.90	1.07
Doors & Hardware	3.8 to 28.8	11.7	0.9	1.4	0.4	.11	.16	.05
Sash & Glazing	5.7 to 30.0	14.4	3.5	4.0	1.0	.50	.58	.14
Lath & Plaster	5.3 to 37.4	15.7	3.3	6.9	0.8	.52	1.08	.13
Tile, Marble & Floors	3.5 to 29.5	10.5	2.6	3.0	0.5	.27	.32	.05
Acoustical Ceilings	4.0 to 26.7	12.3	2.4	0.2	0.3	.30	.02	.04
Painting	6.1 to 41.1	15.3	1.5	1.6	1.6	.23	.24	.24
Interior Partitions	7.0 to 49.9	19.9	3.9	4.3	4.4	.78	.86	.88
Miscellaneous Items	2.6 to 112.5	19.0	5.2	3.7	9.7	.99	.70	1.84
Elevators	2.0 to 19.1	8.5	2.1	1.1	2.2	.18	.09	.19
Sprinklers	2.8 to 20.6	9.0	0.5	—	2.0	.05	—	.18
Plumbing	3.0 to 15.9	8.8	4.9	7.2	5.2	.43	.63	.46
Heat., Vent., Air Conditioning	3.9 to 25.8	12.3	13.5	11.0	12.9	1.66	1.35	1.59
Electrical	2.9 to 11.6	7.0	10.1	8.4	11.1	.71	.59	.78
Total	2.0% to 132.9%	—	100.0%	100.0%	100.0%	18.48%	16.81%	20.67%
Overall Weighted Average							18.65%	

Workers' Compensation Insurance Rates by States

The table below lists the weighted average Workers' Compensation base rate for each state with a factor comparing this with the national average of 18.1%.

State	Weighted Average	Factor	State	Weighted Average	Factor	State	Weighted Average	Factor
Alabama	30.4%	168	Kentucky	19.8%	109	North Dakota	16.7%	92
Alaska	11.3	62	Louisiana	27.3	151	Ohio	16.2	90
Arizona	14.5	80	Maine	21.8	120	Oklahoma	21.9	121
Arkansas	14.2	78	Maryland	11.9	66	Oregon	19.3	107
California	18.6	103	Massachusetts	26.6	147	Pennsylvania	22.9	127
Colorado	26.5	146	Michigan	20.6	114	Rhode Island	22.4	124
Connecticut	21.1	117	Minnesota	37.7	208	South Carolina	13.8	76
Delaware	12.3	68	Mississippi	23.8	131	South Dakota	17.9	99
District of Columbia	25.2	139	Missouri	15.8	87	Tennessee	13.0	72
Florida	28.0	155	Montana	37.4	207	Texas	24.7	136
Georgia	24.9	138	Nebraska	15.7	87	Utah	13.0	72
Hawaii	16.7	92	Nevada	19.4	107	Vermont	14.9	82
Idaho	11.9	66	New Hampshire	23.2	128	Virginia	11.0	61
Illinois	23.3	129	New Jersey	11.0	61	Washington	12.0	66
Indiana	7.5	41	New Mexico	23.1	128	West Virginia	14.1	78
Iowa	12.1	67	New York	17.0	94	Wisconsin	13.9	77
Kansas	10.1	56	North Carolina	14.1	78	Wyoming	8.9	49

Weighted Average for U.S. is 18.7% of payroll = 100%

Rates in the following table are the base or manual costs per $100 of payroll for Workers' Compensation in each state. Rates are usually applied to straight time wages only and not to premium time wages and bonuses.

The weighted average skilled worker rate for 35 trades is 18.1%. For bidding purposes, apply the full value of Workers' Compensation directly to total labor costs, or if labor is 38%, materials 42% and overhead and profit 20% of total cost, carry 38/80 x 18.1% = 8.6% of cost (before overhead and profit) into overhead. Rates vary not only from state to state but also with the experience rating of the contractor.

Rates are the most current available at the time of publication.

General Requirements — R011 Overhead & Miscellaneous Data

R01100-060 Workers' Compensation Insurance Rates by Trade and State (cont.)

State	Carpentry — 3 stories or less 5651	Carpentry — interior cab. work 5437	Carpentry — general 5403	Concrete Work — NOC 5213	Concrete Work — flat (flr, sdwk.) 5221	Electrical Wiring — inside 5190	Excavation — earth NOC 6217	Excavation — rock 6217	Glaziers 5462	Insulation Work 5479	Lathing 5443	Masonry 5022	Painting & Decorating 5474	Pile Driving 6003	Plastering 5480	Plumbing 5183	Roofing 5551	Sheet Metal Work (HVAC) 5538	Steel Erection — door & sash 5102	Steel Erection — inter., ornam. 5102	Steel Erection — structure 5040	Steel Erection — NOC 5057	Tile Work — (interior ceramic) 5348	Waterproofing 9014	Wrecking 5701
AL	36.41	14.51	34.52	32.46	16.19	9.57	20.98	20.98	24.55	20.09	15.28	29.78	21.80	64.38	33.79	11.96	57.62	25.78	16.41	16.41	54.68	66.67	16.19	6.92	54.68
AK	9.30	6.63	9.58	9.26	5.69	5.31	8.16	8.16	9.44	8.72	6.37	9.33	8.41	36.08	11.46	4.46	16.70	6.03	10.43	10.43	18.69	18.69	6.28	5.07	25.07
AZ	14.89	9.11	26.40	13.69	8.09	6.28	8.66	8.66	12.90	15.63	7.83	13.69	10.23	19.88	12.27	7.77	19.67	10.13	10.02	10.02	36.62	23.31	7.24	4.94	50.31
AR	16.12	10.60	15.38	16.45	5.97	7.93	9.21	9.21	12.58	15.29	12.66	9.32	11.10	19.14	13.01	6.58	30.74	10.28	8.19	8.19	27.11	25.39	7.80	4.58	27.11
CA	28.28	9.07	28.28	13.18	13.18	9.93	7.88	7.88	16.37	23.34	10.90	14.55	19.45	21.18	18.12	11.59	39.79	15.33	13.55	13.55	23.93	21.91	7.74	19.45	21.91
CO	32.22	13.54	20.95	20.96	15.13	9.20	15.99	15.99	15.98	21.31	14.68	31.06	21.85	41.53	37.40	14.50	59.10	12.54	11.83	11.83	73.46	43.49	15.04	12.26	73.46
CT	19.25	13.14	24.07	20.48	14.17	6.68	8.42	8.42	21.25	34.89	13.64	25.45	18.84	24.54	17.60	12.25	33.98	13.93	13.67	13.67	50.86	35.83	13.94	5.02	50.86
DE	13.83	13.83	11.56	9.28	6.29	5.82	8.54	8.54	11.18	11.56	10.51	9.85	13.41	11.64	10.51	5.41	22.89	10.29	10.21	10.21	26.69	10.21	7.54	9.85	25.69
DC	17.51	10.80	18.88	29.27	16.85	8.76	17.96	17.96	28.00	22.14	12.07	29.67	16.85	42.12	17.06	12.83	22.39	15.78	33.66	33.66	66.43	42.52	16.42	4.66	66.43
FL	36.73	23.28	29.09	29.92	16.15	10.51	15.27	15.27	17.81	25.18	24.27	28.72	28.92	49.35	28.79	13.37	53.20	17.12	18.67	18.67	43.71	50.61	12.56	7.89	43.71
GA	31.04	16.02	32.71	20.67	15.12	9.82	19.55	19.55	21.97	19.15	26.66	23.03	21.20	40.88	19.50	11.14	40.72	18.65	13.74	13.74	36.03	53.09	13.08	10.16	36.03
HI	17.38	11.62	28.20	13.74	12.27	7.10	7.73	7.73	20.55	21.19	10.94	17.87	11.33	20.15	15.61	6.06	33.24	8.02	11.11	11.11	31.71	21.70	9.78	11.54	31.71
ID	12.68	6.50	16.29	9.46	9.04	5.37	6.87	6.87	8.58	11.01	8.97	9.38	8.83	18.27	8.94	5.04	21.34	7.76	7.51	7.51	26.68	24.89	5.35	7.44	26.68
IL	20.76	12.31	21.64	35.48	12.83	8.86	8.69	8.69	23.09	24.90	14.24	20.34	13.15	37.63	13.78	11.68	35.88	14.58	21.53	21.53	58.63	52.47	13.87	6.26	58.63
IN	6.86	3.76	6.99	7.50	3.50	2.89	4.01	4.01	5.67	7.82	3.97	5.79	6.06	12.19	5.30	2.96	12.65	4.33	5.55	5.55	27.19	12.61	3.49	3.67	27.19
IA	10.22	5.19	12.01	9.18	6.42	3.84	5.56	5.56	11.62	16.49	5.95	10.93	7.58	16.30	9.26	5.01	17.38	6.09	9.80	9.80	45.02	27.20	5.60	4.16	45.02
KS	11.44	7.93	9.21	9.68	6.17	4.12	4.14	4.14	6.93	16.00	7.06	11.86	8.63	13.22	9.21	5.11	22.98	5.57	5.38	5.38	16.49	22.99	3.90	3.77	16.49
KY	16.47	15.26	24.34	22.12	9.53	8.95	13.67	13.67	15.96	14.68	10.62	22.00	19.83	30.44	13.94	10.38	32.83	15.92	13.21	13.21	45.21	33.73	10.60	8.85	45.21
LA	29.63	22.95	26.01	20.86	14.80	10.48	24.73	24.73	23.88	20.82	14.00	21.27	27.63	54.39	21.42	13.78	49.11	21.35	15.51	15.51	66.41	36.94	12.82	10.26	66.41
ME	13.43	10.42	43.59	24.29	11.12	10.02	14.75	14.75	13.35	15.39	14.03	17.26	16.07	31.44	17.95	11.12	32.07	17.70	15.08	15.08	37.84	61.81	11.66	9.37	37.84
MD	10.55	5.95	10.55	11.35	5.05	5.15	9.25	9.25	13.20	13.05	6.45	11.35	6.75	27.45	6.65	5.55	22.80	7.00	9.15	9.15	26.80	18.50	6.35	3.50	26.80
MA	15.25	11.38	22.82	39.38	16.79	6.14	10.72	10.72	16.78	27.76	14.02	29.03	14.95	29.58	13.75	8.49	67.48	13.20	16.07	16.07	80.76	79.51	16.46	6.62	62.87
MI	16.08	9.61	17.85	24.50	12.37	5.71	10.99	10.99	9.62	20.95	11.61	19.18	17.86	64.20	18.33	8.19	43.08	10.91	14.52	14.52	36.41	31.20	13.08	10.42	41.58
MN	27.66	28.76	49.94	30.20	22.59	8.28	19.73	19.73	30.01	37.20	23.93	29.89	19.91	80.08	23.93	15.88	88.83	13.91	27.21	27.21	132.92	31.83	29.54	8.71	132.92
MS	21.77	16.71	28.09	17.88	11.02	10.84	13.03	13.03	18.18	21.21	14.41	22.67	17.90	77.98	20.06	10.84	36.25	20.16	14.52	14.52	37.42	42.77	12.76	8.68	37.42
MO	18.72	7.52	11.84	13.28	10.66	5.79	11.05	11.05	8.97	19.99	9.67	13.57	11.77	16.88	14.22	7.10	25.74	9.76	13.82	13.82	47.98	31.96	5.83	5.71	47.98
MT	30.63	15.19	47.53	32.78	24.74	11.01	26.63	26.63	25.04	38.54	19.54	48.55	41.06	53.68	27.00	15.03	85.06	16.93	33.95	33.95	86.98	55.25	13.30	12.10	86.98
NE	16.19	9.23	15.94	17.31	12.73	5.43	10.19	10.19	12.36	16.79	9.34	18.07	11.58	22.01	14.62	9.06	33.91	13.86	10.75	10.75	25.03	26.96	7.88	5.47	25.03
NV	19.86	10.88	14.94	14.89	11.46	9.64	9.26	9.26	13.64	22.78	14.60	14.48	16.01	27.25	17.93	10.76	33.53	18.03	16.53	16.53	45.36	32.56	18.03	7.43	45.36
NH	16.38	9.82	19.68	28.99	9.29	7.10	13.87	13.87	12.87	43.98	10.90	25.97	17.94	28.09	17.33	9.74	58.76	12.48	13.27	13.27	55.90	49.67	12.82	7.37	55.90
NJ	10.40	8.85	10.40	10.83	8.37	4.23	7.95	7.95	7.36	12.78	7.88	10.77	11.49	11.32	7.88	5.48	29.41	6.20	8.03	8.03	22.52	12.96	5.74	5.37	29.92
NM	25.35	13.09	20.45	19.66	13.39	6.34	10.95	10.95	14.58	31.49	16.25	26.97	16.87	33.27	15.64	12.10	43.41	16.46	21.26	21.26	67.82	29.91	9.67	10.65	67.82
NY	15.39	6.82	14.50	24.92	15.16	6.89	9.69	9.69	16.92	15.82	13.00	18.61	12.55	27.49	13.03	9.03	32.78	14.13	14.12	14.12	27.11	24.09	10.56	7.48	32.53
NC	14.43	9.91	15.37	16.82	6.37	7.64	7.73	7.73	9.52	12.92	8.00	12.94	10.40	17.95	14.47	6.93	25.39	10.26	13.32	13.32	39.03	16.68	7.68	3.86	39.03
ND	13.89	13.89	13.89	12.17	12.17	6.16	9.91	9.91	9.85	9.38	13.16	12.20	11.42	29.82	13.16	8.05	31.37	8.05	13.89	13.89	29.82	29.82	9.63	31.37	29.82
OH	13.27	13.27	13.27	13.45	13.45	6.02	13.45	13.45	18.28	16.31	16.31	15.62	18.28	13.45	16.31	7.96	28.62	24.12	13.45	13.45	13.45	13.45	13.08	13.45	13.45
OK	26.16	11.79	19.33	15.51	11.50	7.41	18.45	18.45	13.55	18.00	12.05	16.80	17.10	36.26	18.29	8.72	42.26	12.45	11.17	11.17	69.45	46.76	11.42	8.20	69.45
OR	29.51	9.28	20.03	16.50	10.61	6.63	12.94	12.94	14.95	19.68	12.27	12.43	19.83	22.05	13.85	8.38	38.11	11.85	11.01	11.01	61.69	25.54	9.92	12.10	61.69
PA	15.80	15.80	19.85	28.18	11.84	7.80	12.74	12.74	15.69	19.85	18.94	18.96	22.05	32.12	18.94	11.57	43.30	12.58	24.71	24.71	59.05	24.71	11.83	18.96	81.34
RI	21.32	11.02	18.77	22.11	15.55	5.86	12.16	12.16	17.02	20.95	11.89	20.66	19.86	41.76	14.19	7.25	29.48	10.52	13.18	13.18	78.79	49.67	14.34	9.90	78.79
SC	19.10	13.03	19.24	12.88	6.19	7.02	8.32	8.32	12.22	10.69	7.27	9.55	11.19	16.64	18.89	6.88	28.98	11.42	9.49	9.49	21.80	23.66	6.23	4.65	21.80
SD	14.61	9.94	20.92	21.64	9.04	5.72	11.73	11.73	14.77	17.73	10.89	13.30	13.25	32.14	16.96	9.50	29.39	11.35	11.51	11.51	46.57	33.20	9.85	5.59	46.57
TN	13.22	9.52	11.99	12.39	7.53	5.00	8.68	8.68	8.96	15.04	8.51	12.73	10.88	15.62	12.19	5.91	24.81	8.76	7.36	7.36	45.06	13.95	5.89	4.98	45.06
TX	28.29	18.95	28.29	25.59	19.49	11.60	18.16	18.16	14.17	25.84	13.47	23.12	18.29	43.91	20.99	12.88	47.24	22.92	14.29	14.29	50.47	31.01	9.94	11.38	54.81
UT	11.17	11.17	11.17	17.68	7.99	7.05	6.38	6.38	9.55	10.95	12.32	13.75	15.29	17.24	10.60	6.41	26.27	6.30	8.99	8.99	23.51	23.51	6.06	5.83	26.53
VT	11.93	7.61	14.88	34.92	9.35	4.51	7.91	7.91	11.95	11.32	8.84	13.76	7.82	19.81	10.86	6.71	24.32	10.29	10.29	10.29	32.12	35.79	6.45	8.56	32.12
VA	10.00	6.46	8.65	9.29	5.47	3.45	6.13	6.13	10.17	8.77	7.59	8.07	8.11	24.38	8.20	5.60	17.80	6.74	10.42	10.42	19.28	30.25	6.78	2.57	19.28
WA	8.78	8.78	8.78	8.53	8.53	3.68	8.82	8.82	8.71	9.93	8.78	11.54	10.23	20.99	11.78	5.41	23.39	3.94	16.21	16.21	16.21	16.21	9.81	10.23	16.21
WV	15.43	15.43	15.43	20.87	20.87	5.01	9.14	9.14	7.33	16.88	16.02	16.88	10.56	16.88	7.40	12.04	7.33	14.98	14.98	14.98	13.27	14.98	16.02	4.87	13.27
WI	10.47	10.13	19.98	8.91	8.50	5.73	7.00	7.00	9.67	13.88	13.48	15.43	12.10	17.04	11.81	5.95	29.35	8.64	11.52	11.52	36.51	18.72	8.45	3.92	36.51
WY	8.13	8.13	8.13	8.13	8.13	8.13	8.13	8.13	8.13	8.13	8.13	8.13	8.13	8.13	8.13	8.13	8.13	8.13	8.13	8.13	8.13	8.13	8.13	8.13	8.13
AVG.	18.12	11.65	19.85	18.81	11.46	7.03	11.41	11.41	14.43	18.52	12.26	17.75	15.27	29.88	15.72	8.82	34.62	12.27	13.77	13.77	42.56	31.55	10.48	8.32	43.48

General Requirements — R011 Overhead & Miscellaneous Data

R01100-060 Workers' Compensation (cont.) (Canada in Canadian dollars)

Province		Alberta	British Columbia	Manitoba	Ontario	New Brunswick	Newfndld. & Labrador	Northwest Territories	Nova Scotia	Prince Edward Island	Quebec	Saskatchewan	Yukon
Carpentry—3 stories or less	Rate	2.65	6.89	5.35	8.19	3.76	5.36	2.75	6.34	5.68	13.85	5.38	2.35
	Code	25401	70600	40102	723	422	403	4-41	4013	401	80110	B12-02	4-042
Carpentry—interior cab. work	Rate	2.48	2.68	5.35	8.19	3.85	5.36	2.75	6.34	3.74	13.85	3.86	2.35
	Code	42133	60412	40102	723	427	403	4-41	4013	402	80110	B11-25	4-042
CARPENTRY—general	Rate	2.65	6.89	5.35	8.19	3.76	5.36	2.75	6.34	5.68	13.85	5.38	2.35
	Code	25401	70600	40102	723	422	403	4-41	4013	401	80110	B12-02	4-042
CONCRETE WORK—NOC	Rate	3.26	6.89	6.72	14.29	3.76	5.36	2.75	7.87	5.68	16.29	7.99	3.25
	Code	42104	70604	40110	745	422	403	4-41	4222	401	80100	B14-04	2-032
CONCRETE WORK—flat (flr. sidewalk)	Rate	3.26	6.89	6.72	14.29	3.76	5.36	2.75	7.87	5.68	16.29	7.99	3.25
	Code	42104	70604	40110	745	422	403	4-41	4222	401	80100	B14-04	2-032
ELECTRICAL Wiring—inside	Rate	1.61	4.61	3.50	3.61	1.49	4.34	2.00	3.19	3.74	7.41	3.86	2.35
	Code	42124	71100	40203	704	426	400	4-46	4261	402	80170	B11-05	4-041
EXCAVATION—earth NOC	Rate	2.03	4.13	5.35	4.95	2.72	5.36	2.50	3.92	3.95	7.87	4.69	3.25
	Code	40604	72607	40706	711	421	403	4-43	4214	404	80030	R11-06	2-016
EXCAVATION—rock	Rate	2.03	4.13	5.35	4.95	2.72	5.36	2.50	3.92	3.95	7.87	4.69	3.25
	Code	40604	72607	40706	711	421	403	4-43	4214	404	80030	R11-06	2-016
GLAZIERS	Rate	1.93	2.03	3.50	10.27	3.85	4.09	2.75	7.75	3.74	17.29	7.99	2.35
	Code	42121	60236	40109	751	423	402	4-41	4233	402	80150	B13-04	4-042
INSULATION WORK	Rate	2.28	6.86	5.35	10.27	3.85	4.09	2.75	7.75	5.68	15.05	5.38	3.25
	Code	42184	70504	40102	751	423	402	4-41	4234	401	80120	B12-07	2-035
LATHING	Rate	5.05	6.86	5.35	8.19	3.85	4.09	2.75	6.14	3.74	15.05	7.99	3.25
	Code	42135	70500	40102	723	427	402	4-41	4271	402	80120	B13-02	2-036
MASONRY	Rate	3.26	6.89	5.35	16.52	3.85	5.36	2.75	7.75	5.68	16.29	7.99	3.25
	Code	42102	70602	40102	741	423	403	4-41	4231	401	80100	B15-01	2-032
PAINTING & DECORATING	Rate	2.63	6.86	6.72	9.76	3.85	4.09	2.75	6.14	3.74	15.05	5.38	3.25
	Code	42111	70501	40105	719	427	402	4-41	4275	402	80120	B12-01	2-036
PILE DRIVING	Rate	3.26	17.50	5.35	5.76	3.76	9.60	2.50	7.87	5.68	7.87	5.38	3.25
	Code	42159	72502	40706	732	422	404	4-43	4221	401	80030	B12-10	2-030
PLASTERING	Rate	5.05	6.86	6.72	9.76	3.85	4.09	2.75	6.14	3.74	15.05	7.99	3.25
	Code	42135	70502	40108	719	427	402	4-41	4271	402	80120	B13-02	2-036
PLUMBING	Rate	1.61	4.07	3.50	4.42	1.88	3.57	2.00	3.70	3.74	8.03	3.86	2.35
	Code	42122	70712	40204	707	424	401	4-46	4241	402	80160	B11-01	4-039
ROOFING	Rate	6.56	6.89	6.72	12.05	7.72	5.36	2.75	9.30	5.68	22.52	7.99	3.25
	Code	42118	70600	40403	728	430	403	4-41	4235	401	80130	B15-02	2-031
SHEET METAL WORK (HVAC)	Rate	1.61	4.07	5.35	4.42	1.88	3.57	2.00	7.75	3.74	8.03	3.86	2.35
	Code	42117	70714	40402	707	424	401	4-46	4236	402	80160	B11-07	4-040
STEEL ERECTION—door & sash	Rate	2.28	17.50	5.35	20.96	3.76	9.60	2.75	7.87	5.68	32.75	7.99	3.25
	Code	42106	72509	40502	748	422	404	4-41	4223	401	80080	B15-03	2-012
STEEL ERECTION—inter., ornam.	Rate	2.28	17.50	5.35	20.96	3.76	5.36	2.75	7.87	5.68	32.75	7.99	3.25
	Code	42106	72509	40502	748	422	403	4-41	4223	401	80080	B15-03	2-012
STEEL ERECTION—structure	Rate	2.28	17.50	5.35	20.96	3.76	9.60	2.75	10.79	5.68	32.75	7.99	3.25
	Code	42106	72509	40502	748	422	404	4-41	4227	401	80080	B15-04	2-012
STEEL ERECTION—NOC	Rate	2.28	17.50	5.35	20.96	3.76	9.60	2.75	10.79	5.68	32.75	7.99	3.25
	Code	42106	72509	40502	748	422	404	4-41	4227	401	80080	B15-03	2-012
TILE WORK—inter. (ceramic)	Rate	2.53	6.86	2.31	9.76	3.85	4.09	2.75	6.14	3.74	15.05	7.99	3.25
	Code	42113	70506	40103	719	427	402	4-41	4276	402	80120	B13-01	2-034
WATERPROOFING	Rate	2.63	6.89	5.35	8.19	3.85	5.36	2.75	7.75	3.74	22.52	3.86	3.25
	Code	42139	70620	40102	723	423	403	4-41	4239	402	80130	B11-17	2-030
WRECKING	Rate	12.86	6.89	6.72	20.96	2.72	5.36	2.50	3.92	5.68	26.98	7.99	3.25
	Code	42108	70600	40106	748	421	403	4-43	4211	401	80220	B14-07	2-030

General Requirements — R011 Overhead & Miscellaneous Data

R01100-070 Contractor's Overhead & Profit

Below are the **average** installing contractor's percentage mark-ups applied to base labor rates to arrive at typical billing rates.

Column A: Labor rates are based on average open shop wages for 7 major U.S. regions. Base rates including fringe benefits are listed hourly and daily. These figures are the sum of the wage rate and employer-paid fringe benefits such as vacation pay, and employer-paid health costs.

Column B: Workers' Compensation rates are the national average of state rates established for each trade.

Column C: Column C lists average fixed overhead figures for all trades. Included are Federal and State Unemployment costs set at 7.0%; Social Security Taxes (FICA) set at 7.65%; Builder's Risk Insurance costs set at 0.34%; and Public Liability costs set at 1.55%. All the percentages except those for Social Security Taxes vary from state to state as well as from company to company.

Columns D and E: Percentages in Columns D and E are based on the presumption that the installing contractor has annual billing of $1,000,000 and up. Overhead percentages may increase with smaller annual billing. The overhead percentages for any given contractor may vary greatly and depend on a number of factors, such as the contractor's annual volume, engineering and logistical support costs, and staff requirements. The figures for overhead and profit will also vary depending on the type of job, the job location, and the prevailing economic conditions. All factors should be examined very carefully for each job.

Column F: Column F lists the total of Columns B, C, D, and E.

Column G: Column G is Column A (hourly base labor rate) multiplied by the percentage in Column F (O&P percentage).

Column H: Column H is the total of Column A (hourly base labor rate) plus Column G (Total O&P).

Column I: Column I is Column H multiplied by eight hours.

		A		B	C	D	E	F	G	H	I
		Base Rate Incl. Fringes		Workers' Comp. Ins.	Average Fixed Overhead	Overhead	Profit	Total Overhead & Profit		Rate with O & P	
Abbr.	Trade	Hourly	Daily					%	Amount	Hourly	Daily
Skwk	Skilled Workers Average (35 trades)	$19.75	$158.00	18.1%	16.5%	27.0%	10.0%	71.6%	$14.15	$33.90	$271.20
	Helpers Average (5 trades)	14.80	118.40	19.7		25.0		71.2	10.55	25.35	202.80
	Foreman Average, Inside ($.50 over trade)	20.25	162.00	18.1		27.0		71.6	14.50	34.75	278.00
	Foreman Average, Outside ($2.00 over trade)	21.75	174.00	18.1		27.0		71.6	15.55	37.30	298.40
Clab	Common Building Laborers	14.45	115.60	19.9		25.0		71.4	10.30	24.75	198.00
Asbe	Asbestos Workers	20.50	164.00	18.5		30.0		75.0	15.40	35.90	287.20
Boil	Boilermakers	22.10	176.80	16.2		30.0		72.7	16.05	38.15	305.20
Bric	Bricklayers	20.00	160.00	17.8		25.0		69.3	13.85	33.85	270.80
Brhe	Bricklayer Helpers	15.70	125.60	17.8		25.0		69.3	10.90	26.60	212.80
Carp	Carpenters	19.70	157.60	19.9		25.0		71.4	14.05	33.75	270.00
Cefi	Cement Finishers	18.90	151.20	11.5		25.0		63.0	11.90	30.80	246.40
Elec	Electricians	22.10	176.80	7.0		30.0		63.5	14.05	36.15	289.20
Elev	Elevator Constructors	22.90	183.20	8.5		30.0		65.0	14.90	37.80	302.40
Eqhv	Equipment Operators, Crane or Shovel	20.65	165.20	11.4		28.0		65.9	13.60	34.25	274.00
Eqmd	Equipment Operators, Medium Equipment	19.90	159.20	11.4		28.0		65.9	13.10	33.00	264.00
Eqlt	Equipment Operators, Light Equipment	19.10	152.80	11.4		28.0		65.9	12.60	31.70	253.60
Eqol	Equipment Operators, Oilers	17.00	136.00	11.4		28.0		65.9	11.20	28.20	225.60
Eqmm	Equipment Operators, Master Mechanics	21.05	168.40	11.4		28.0		65.9	13.85	34.90	279.20
Glaz	Glaziers	19.35	154.80	14.4		25.0		65.9	12.75	32.10	256.80
Lath	Lathers	19.20	153.60	12.3		25.0		63.8	12.25	31.45	251.60
Marb	Marble Setters	19.85	158.80	17.8		25.0		69.3	13.75	33.60	268.80
Mill	Millwrights	20.60	164.80	11.7		25.0		63.2	13.00	33.60	268.80
Mstz	Mosaic and Terrazzo Workers	19.10	152.80	10.5		25.0		62.0	11.85	30.95	247.60
Pord	Painters, Ordinary	17.95	143.60	15.3		25.0		66.8	12.00	29.95	239.60
Psst	Painters, Structural Steel	18.65	149.20	51.1		25.0		102.6	19.15	37.80	302.40
Pape	Paper Hangers	18.05	144.40	15.3		25.0		66.8	12.05	30.10	240.80
Pile	Pile Drivers	19.45	155.60	29.9		30.0		86.4	16.80	36.25	290.00
Plas	Plasterers	18.65	149.20	15.7		25.0		67.2	12.55	31.20	249.60
Plah	Plasterer Helpers	15.65	125.20	15.7		25.0		67.2	10.50	26.15	209.20
Plum	Plumbers	21.95	175.60	8.8		30.0		65.3	14.35	36.30	290.40
Rodm	Rodmen (Reinforcing)	21.10	168.80	31.6		28.0		86.1	18.15	39.25	314.00
Rofc	Roofers, Composition	17.05	136.40	34.6		25.0		86.1	14.70	31.75	254.00
Rots	Roofers, Tile and Slate	17.10	136.80	34.6		25.0		86.1	14.70	31.80	254.40
Rohe	Roofer Helpers (Composition)	12.75	102.00	34.6		25.0		86.1	11.00	23.75	190.00
Shee	Sheet Metal Workers	21.50	172.00	12.3		30.0		68.8	14.80	36.30	290.40
Spri	Sprinkler Installers	22.05	176.40	9.0		30.0		65.5	14.45	36.50	292.00
Stpi	Steamfitters or Pipefitters	22.10	176.80	8.8		30.0		65.3	14.45	36.55	292.40
Ston	Stone Masons	19.50	156.00	17.8		25.0		69.3	13.50	33.00	264.00
Sswk	Structural Steel Workers	21.25	170.00	42.6		28.0		97.1	20.65	41.90	335.20
Tilf	Tile Layers (Floor)	19.20	153.60	10.5		25.0		62.0	11.90	31.10	248.80
Tilh	Tile Layer Helpers	15.45	123.60	10.5		25.0		62.0	9.60	25.05	200.40
Trlt	Truck Drivers, Light	15.85	126.80	15.6		25.0		67.1	10.65	26.50	212.00
Trhv	Truck Drivers, Heavy	16.20	129.60	15.6		25.0		67.1	10.85	27.05	216.40
Sswl	Welders, Structural Steel	21.25	170.00	42.6		28.0		97.1	20.65	41.90	335.20
Wrck	*Wrecking	14.90	119.20	43.5		25.0		95.0	14.15	29.05	232.40

*Not included in Averages.

General Requirements — R011 Overhead & Miscellaneous Data

R01100-080 Performance Bond

This table shows the cost of a Performance Bond for a construction job scheduled to be completed in 12 months. Add 1% of the premium cost per month for jobs requiring more than 12 months to complete. The rates are "standard" rates offered to contractors that the bonding company considers financially sound and capable of doing the work. Preferred rates are offered by some bonding companies based upon financial strength of the contractor. Actual rates vary from contractor to contractor and from bonding company to bonding company. Contractors should prequalify through a bonding agency before submitting a bid on a contract that requires a bond.

Contract Amount	Building Construction Class B Projects	Highways & Bridges Class A New Construction	Highways & Bridges Class A-1 Highway Resurfacing
First $ 100,000 bid	$25.00 per M	$15.00 per M	$9.40 per M
Next 400,000 bid	$ 2,500 plus $15.00 per M	$ 1,500 plus $10.00 per M	$ 940 plus $7.20 per M
Next 2,000,000 bid	8,500 plus 10.00 per M	5,500 plus 7.00 per M	3,820 plus 5.00 per M
Next 2,500,000 bid	28,500 plus 7.50 per M	19,500 plus 5.50 per M	15,820 plus 4.50 per M
Next 2,500,000 bid	47,250 plus 7.00 per M	33,250 plus 5.00 per M	28,320 plus 4.50 per M
Over 7,500,000 bid	64,750 plus 6.00 per M	45,750 plus 4.50 per M	39,570 plus 4.00 per M

R01100-090 Sales Tax by State (United States)

State sales tax on materials is tabulated below (5 states have no sales tax). Many states allow local jurisdictions, such as a county or city, to levy additional sales tax.

Some projects may be sales tax exempt, particularly those constructed with public funds.

State	Tax (%)	State	Tax (%)	State	Tax (%)	State	Tax (%)
Alabama	4	Illinois	6.25	Montana	0	Rhode Island	7
Alaska	0	Indiana	5	Nebraska	4.5	South Carolina	5
Arizona	5	Iowa	5	Nevada	6.875	South Dakota	4
Arkansas	4.625	Kansas	4.9	New Hampshire	0	Tennessee	6
California	7.25	Kentucky	6	New Jersey	6	Texas	6.25
Colorado	3	Louisiana	4	New Mexico	5	Utah	4.75
Connecticut	6	Maine	6	New York	4	Vermont	5
Delaware	0	Maryland	5	North Carolina	4	Virginia	4.5
District of Columbia	5.75	Massachusetts	5	North Dakota	5	Washington	6.5
Florida	6	Michigan	6	Ohio	5	West Virginia	6
Georgia	4	Minnesota	6.5	Oklahoma	4.5	Wisconsin	5
Hawaii	4	Mississippi	7	Oregon	0	Wyoming	4
Idaho	5	Missouri	4.225	Pennsylvania	6	Average	4.71 %

Sales Tax by Province (Canada)

GST - a value-added tax, which the federal government imposes on most goods and services provided in or imported into Canada.
PST - a retail sales tax, which five of the provinces impose on the price of most goods and some services.

QST - a value-added tax, similar to the federal GST, which Quebec imposes.
HST - Three provinces have combined their retail sales tax with the federal GST into the Harmonized Sales Tax.

Province	PST (%)	QST (%)	GST (%)	HST (%)
Alberta	0	0	7	0
British Columbia	7	0	7	0
Manitoba	7	0	7	0
New Brunswick	0	0	0	15
Newfoundland	0	0	0	15
Northwest Territories	0	0	7	0
Nova Scotia	0	0	0	15
Ontario	8	0	7	0
Prince Edward Island	10	0	7	0
Quebec	0	6.5	7	0
Saskatchewan	7	0	7	0
Yukon	0	0	7	0

R01100-100 Unemployment Taxes and Social Security Taxes

Mass. State Unemployment tax ranges from 1.325% to 7.225% plus an experience rating assessment the following year, on the first $10,800 of wages. Federal Unemployment tax is 6.2% of the first $7,000 of wages. This is reduced by a credit for payment to the state. The minimum Federal Unemployment tax is .8% after all credits.

Combined rates in Mass. thus vary from 2.125% to 8.025% of the first $10,800 of wages. Combined average U.S. rate is about 7.0% of the first $7,000. Contractors with permanent workers will pay less since the average annual wages for skilled workers is $28.75 x 2,000 hours or about $57,500 per year. The average combined rate for U.S. would thus be 7.0% x $7,000 ÷ $57,500 = 0.9% of total wages for permanent employees.

Rates vary not only from state to state but also with the experience rating of the contractor.

Social Security (FICA) for 2000 is estimated at time of publication to be 7.65% of wages up to $72,600.

General Requirements — R011 Overhead & Miscellaneous Data

R01100-110 Overtime

One way to improve the completion date of a project or eliminate negative float from a schedule is to compress activity duration times. This can be achieved by increasing the crew size or working overtime with the proposed crew.

To determine the costs of working overtime to compress activity duration times, consider the following examples. Below is an overtime efficiency and cost chart based on a five, six, or seven day week with an eight through twelve hour day. Payroll percentage increases for time and one half and double time are shown for the various working days.

Days per Week	Hours per Day	Production Efficiency					Payroll Cost Factors	
		1 Week	2 Weeks	3 Weeks	4 Weeks	Average 4 Weeks	@ 1-1/2 Times	@ 2 Times
5	8	100%	100%	100%	100%	100 %	100 %	100 %
	9	100	100	95	90	96.25	105.6	111.1
	10	100	95	90	85	91.25	110.0	120.0
	11	95	90	75	65	81.25	113.6	127.3
	12	90	85	70	60	76.25	116.7	133.3
6	8	100	100	95	90	96.25	108.3	116.7
	9	100	95	90	85	92.50	113.0	125.9
	10	95	90	85	80	87.50	116.7	133.3
	11	95	85	70	65	78.75	119.7	139.4
	12	90	80	65	60	73.75	122.2	144.4
7	8	100	95	85	75	88.75	114.3	128.6
	9	95	90	80	70	83.75	118.3	136.5
	10	90	85	75	65	78.75	121.4	142.9
	11	85	80	65	60	72.50	124.0	148.1
	12	85	75	60	55	68.75	126.2	152.4

R01107-010 Architectural Fees

Tabulated below are typical percentage fees by project size, for good professional architectural service. Fees may vary from those listed depending upon degree of design difficulty and economic conditions in any particular area.

Rates can be interpolated horizontally and vertically. Various portions of the same project requiring different rates should be adjusted proportionately. For alterations, add 50% to the fee for the first $500,000 of project cost and add 25% to the fee for project cost over $500,000.

Architectural fees tabulated below include Structural, Mechanical and Electrical Engineering Fees. They do not include the fees for special consultants such as kitchen planning, security, acoustical, interior design, etc.

Building Types	Total Project Size in Thousands of Dollars						
	100	250	500	1,000	5,000	10,000	50,000
Factories, garages, warehouses, repetitive housing	9.0%	8.0%	7.0%	6.2%	5.3%	4.9%	4.5%
Apartments, banks, schools, libraries, offices, municipal buildings	12.2	12.3	9.2	8.0	7.0	6.6	6.2
Churches, hospitals, homes, laboratories, museums, research	15.0	13.6	12.7	11.9	9.5	8.8	8.0
Memorials, monumental work, decorative furnishings	—	16.0	14.5	13.1	10.0	9.0	8.3

General Requirements — R011 Overhead & Miscellaneous Data

R01107-030 Engineering Fees

Typical **Structural Engineering Fees** based on type of construction and total project size. These fees are included in Architectural Fees.

Type of Construction	Total Project Size (in thousands of dollars)			
	$500	$500-$1,000	$1,000-$5,000	Over $5000
Industrial buildings, factories & warehouses	Technical payroll times 2.0 to 2.5	1.60%	1.25%	1.00%
Hotels, apartments, offices, dormitories, hospitals, public buildings, food stores		2.00%	1.70%	1.20%
Museums, banks, churches and cathedrals		2.00%	1.75%	1.25%
Thin shells, prestressed concrete, earthquake resistive		2.00%	1.75%	1.50%
Parking ramps, auditoriums, stadiums, convention halls, hangars & boiler houses		2.50%	2.00%	1.75%
Special buildings, major alterations, underpinning & future expansion		Add to above 0.5%	Add to above 0.5%	Add to above 0.5%

For complex reinforced concrete or unusually complicated structures, add 20% to 50%.

Typical **Mechanical and Electrical Engineering Fees** are based on the size of the subcontract. The fee structure for Mechanical Engineering is shown below; Electrical Engineering Fees range from 2 to 5%. These fees are included in Architectural Fees.

Type of Construction	Subcontract Size							
	$25,000	$50,000	$100,000	$225,000	$350,000	$500,000	$750,000	$1,000,000
Simple structures	6.4%	5.7%	4.8%	4.5%	4.4%	4.3%	4.2%	4.1%
Intermediate structures	8.0	7.3	6.5	5.6	5.1	5.0	4.9	4.8
Complex structures	12.0	9.0	9.0	8.0	7.5	7.5	7.0	7.0

For renovations, add 15% to 25% to applicable fee.

General Requirements | R015 | Construction Aids

R01540-100 Steel Tubular Scaffolding

On new construction, tubular scaffolding is efficient up to 60' high or five stories. Above this it is usually better to use a hung scaffolding if construction permits. Swing scaffolding operations may interfere with tenants. In this case, the tubular is more practical at all heights.

In repairing or cleaning the front of an existing building the cost of tubular scaffolding per S.F. of building front increases as the height increases above the first tier. The first tier cost is relatively high due to leveling and alignment.

The minimum efficient crew for erection is three workers. For heights over 50', a crew of four is more efficient. Use two or more on top and two at the bottom for handing up or hoisting. Four workers can erect and dismantle about nine frames per hour up to five stories. From five to eight stories they will average six frames per hour. With 7' horizontal spacing this will run about 400 S.F. and 265 S.F. of wall surface, respectively. Time for placing planks must be added to the above. On heights above 50', five planks can be placed per labor-hour.

The cost per 1,000 S.F. of building front in the table below was developed by pricing the materials required for a typical tubular scaffolding system eleven frames long and two frames high. Planks were figured five wide for standing plus two wide for materials.

Frames are 5' wide and usually spaced 7' O.C. horizontally. Sidewalk frames are 6' wide. Rental rates will be lower for jobs over three months duration.

For jobs under twenty-five frames, add 50% to rental cost. These figures do not include accessories which are listed separately below. Large quantities for long periods can reduce rental rates by 20%.

Item	Unit	Monthly Rent	Per 1,000 S. F. of Building Front	
			No. of Pieces	Rental per Month
5' Wide Standard Frame, 6'-4" High	Ea.	$ 3.75	24	$ 90.00
Leveling Jack & Plate		1.50	24	36.00
Cross Brace		.60	44	26.40
Side Arm Bracket, 21"		1.50	12	18.00
Guardrail Post		1.00	12	12.00
Guardrail, 7' section		.75	22	16.50
Stairway Section		10.00	2	20.00
Stairway Starter Bar		.10	1	.10
Stairway Inside Handrail		5.00	2	10.00
Stairway Outside Handrail		5.00	2	10.00
Walk-Thru Frame Guardrail		2.00	2	4.00
			Total	$243.00
			Per C.S.F., 1 Use/Mo.	$ 24.30

Scaffolding is often used as falsework over 15' high during construction of cast-in-place concrete beams and slabs. Two foot wide scaffolding is generally used for heavy beam construction. The span between frames depends upon the load to be carried with a maximum span of 5'.

Heavy duty shoring frames with a capacity of 10,000#/leg can be spaced up to 10' O. C. depending upon form support design and loading.

Scaffolding used as horizontal shoring requires less than half the material required with conventional shoring.

On new construction, erection is done by carpenters.

Rolling towers supporting horizontal shores can reduce labor and speed the job. For maintenance work, catwalks with spans up to 70' can be supported by the rolling towers.

R01540-200 Pump Staging

Pump staging is generally not available for rent. Purchase prices for individual items are shown in the chart below.

Item	Unit	Purchase, Each	Per 2,400 S.F. of Building Front	
			No. of Pieces	Cost to Buy
Aluminum pole section, 24' long	Ea.	$335.00	6	$2,010.00
Aluminum splice joint, 6' long		68.00	3	204.00
Aluminum foldable brace		48.50	3	145.50
Aluminum pump jack		111.00	3	333.00
Aluminum support for workbench/back safety rail		59.00	3	177.00
Aluminum scaffold plank/workbench, 14" wide x 24' long		545.00	4	2,180.00
Safety net, 22' long		267.00	2	534.00
Aluminum plank end safety rail		183.00	2	366.00
			Total System	$5,949.50
			Per C.S.F., 1 Use	$247.90

This system in place will cover up to 2,400 square feet of wall area. The cost in place will depend on how many uses are realized during the life of the equipment. Several options are given in Division 01540-550. The above prices are bare costs.

General Requirements — R015 Construction Aids

R01559-100   Contractor Equipment

Rental Rates shown in the front of the book pertain to late model high quality machines in excellent working condition, rented from equipment dealers. Rental rates from contractors may be substantially lower than the rental rates from equipment dealers depending upon economic conditions. For older, less productive machines, reduce rates by a maximum of 15%. Any overtime must be added to the base rates. For shift work, rates are lower. Usual rule of thumb is 150% of one shift rate for two shifts; 200% for three shifts.

For periods of less than one week, operated equipment is usually more economical to rent than renting bare equipment and hiring an operator.

Equipment moving and mobilization costs must be added to rental rates where applicable. A large crane, for instance, may take two days to erect and two days to dismantle.

Rental rates vary throughout the country with larger cities generally having lower rates. Lease plans for new equipment are available for periods in excess of six months with a percentage of payments applying toward purchase.

Monthly rental rates vary from 2% to 5% of the cost of the equipment depending on the anticipated life of the equipment and its wearing parts. Weekly rates are about 1/3 the monthly rates and daily rental rates about 1/3 the weekly rate.

The hourly operating costs for each piece of equipment include costs to the user such as fuel, oil, lubrication, normal expendables for the equipment, and a percentage of mechanic's wages chargeable to maintenance. The hourly operating costs listed do not include the operator's wages.

The daily cost for equipment used in the standard crews is figured by dividing the weekly rate by five, then adding eight times the hourly operating cost to give the total daily equipment cost, not including the operator. This figure is in the right hand column of Division 01559 under Crew Equip. Cost./Day

Pile Driving rates shown for pile hammer and extractor do not include leads, crane, boiler or compressor. Vibratory pile driving requires an added field specialist during set-up and pile driving operation for the electric model. The hydraulic model requires a field specialist for set-up only. Up to 125 reuses of sheet piling are possible using vibratory drivers. For normal conditions, crane capacity for hammer type and size are as follows.

Crane Capacity	Hammer Type and Size		
	Air or Steam	Diesel	Vibratory
25 ton	to 8,750 ft.-lb.		70 H.P.
40 ton	15,000 ft.-lb.	to 32,000 ft.-lb.	170 H.P.
60 ton	25,000 ft.-lb.		300 H.P.
100 ton		112,000 ft.-lb.	

Cranes should be specified for the job by size, building and site characteristics, availability, performance characteristics, and duration of time required.

Backhoes & Shovels rent for about the same as equivalent size cranes but maintenance and operating expense is higher. Crane operators rate must be adjusted for high boom heights. Average adjustments: for 150' boom add $.80 per hour; over 185', add $1.30 per hour; over 210', add $1.80 per hour; over 250', add $2.50 per hour and over 295', add $4.00 per hour.

Tower Cranes

Capacity In Kip-Feet	Typical Jib Length in Feet	Speed at Maximum Reach and Load	Purchase Price (New)		Monthly Rental, to 6 mo.	
			Crane & 80' Mast	Mast Sections	Crane & 80' Mast	Mast Sections
725	100	350 FPM	$250,900	$660 /L.F.	$ 8,270	$17.60 /L.F.
900	100	500	297,500	770	9,510	19.00
*1100	130	1000	409,600	890	10,340	23.80
1450	150	1000	568,300	1,270	14,680	28.20
2150	200	1000	901,100	1,450	18,410	32.60
3000	200	1000	998,400	1,480	24,200	38.90

*Most widely used.

Tower Cranes of the climbing or static type have jibs from 50' to 200' and capacities at maximum reach range from 4,000 to 14,000 pounds. Lifting capacities increase up to maximum load as the hook radius decreases.

Typical rental rates, based on purchase price are about 2% to 3% per month.

Erection and dismantling runs between $12,500 and $80,000. Climbing operation takes ten labor hours per 20' climb. Crane dead time is about five hours per 40' climb. If crane is bolted to side of the building add cost of ties and extra mast sections. Mast sections cost $500 to $1,500 per vertical foot or can be rented at 2% to 3% of purchase price per month. Contractors using climbers claim savings of $1.50 per C.Y. of concrete placed, plus $.20 per S.F. of formwork. Climbing cranes have from 80' to 180' of mast while static cranes have 80' to 800' of mast.

Truck Cranes can be converted to tower cranes by using tower attachments. Mast heights over 400' have been used. See Division 01559-600 for rental rates of high boom cranes.

A single 100' high material **Hoist and Tower** can be erected and dismantled for about $15,000; a double 100' high hoist and tower for about $20,000. Erection costs for additional heights are $100 and $125 per vertical foot respectively up to 150' and $100 to $150 per vertical foot over 150' high. A 40' high portable Buck hoist costs about $5,000 to erect and dismantle. Additional heights run $80 per vertical foot to 80' and $100 per vertical foot for the next 100'. Most material hoists do not meet local code requirements for carrying personnel.

A 150' high **Personnel Hoist** requires about 500 to 800 labor hours to erect and dismantle with costs ranging from $12,000 to $28,000. Budget erection cost is $150 per vertical foot for all trades. Local code requirements or labor scarcity requiring overtime can add up to 50% to any of the above erection costs.

Earthmoving Equipment: The selection of earthmoving equipment depends upon the type and quantity of material, moisture content, haul distance, haul road, time available, and equipment available. Short haul cut and fill operations may require dozers only, while another operation may require excavators, a fleet of trucks, and spreading and compaction equipment. Stockpiled material and granular material are easily excavated with front end loaders. Scrapers are most economically used with hauls between 300' and 1-1/2 miles if adequate haul roads can be maintained. Shovels are often used for blasted rock and any material where a vertical face of 8' or more can be excavated. Special conditions may dictate the use of draglines, clamshells, or backhoes. Spreading and compaction equipment must be matched to the soil characteristics, the compaction required and the rate the fill is being supplied.

General Requirements | R015 Construction Aids

R01559-150 Heavy Lifting

Hydraulic Climbing Jacks

The use of hydraulic heavy lift systems is an alternative to conventional type crane equipment. The lifting, lowering, pushing, or pulling mechanism is a hydraulic climbing jack moving on a square steel jackrod from 1-5/8" to 4" square, or a steel cable. The jackrod or cable can be vertical or horizontal, stationary or movable, depending on the individual application. When the jackrod is stationary, the climbing jack will climb the rod and push or pull the load along with itself. When the climbing jack is stationary, the jackrod is movable with the load attached to the end and the climbing jack will lift or lower the jackrod with the attached load. The heavy lift system is normally operated by a single control lever located at the hydraulic pump.

The system is flexible in that one or more climbing jacks can be applied wherever a load support point is required, and the rate of lift synchronized.

Economic benefits have been demonstrated on projects such as: erection of ground assembled roofs and floors, complete bridge spans, girders and trusses, towers, chimney liners and steel vessels, storage tanks, and heavy machinery. Other uses are raising and lowering offshore work platforms, caissons, tunnel sections and pipelines.

Site Construction — R020 Paving & Surfacing

R02065-300 Bituminous Paving

City	Bituminous Asphalt per Ton*	Pavement (3") 6.13 S.Y./ton				Sidewalks (2") 9.2 S.Y./ton			
		Cost per S.Y.			Per Ton	Cost per S.Y.			Per Ton
		Material*	Installation	Total	Total	Material*	Installation	Total	Total
Atlanta	$25.00	$4.08	$ .68	$4.76	$29.15	$2.72	$1.11	$3.83	$35.24
Baltimore	30.50	4.98	.72	5.70	34.94	3.32	1.25	4.57	42.04
Boston	36.25	5.91	.89	6.80	41.69	3.94	1.83	5.77	53.08
Buffalo	31.80	5.19	.86	6.05	37.09	3.46	1.73	5.19	47.75
Chicago	32.50	5.30	.91	6.21	38.06	3.53	1.89	5.42	49.86
Cincinnati	33.25	5.42	.76	6.18	37.88	3.61	1.39	5.00	46.00
Cleveland	30.00	4.89	.82	5.71	34.98	3.26	1.58	4.84	44.53
Columbus	26.50	4.32	.76	5.08	31.14	2.88	1.39	4.27	39.28
Dallas	25.00	4.08	.69	4.77	29.23	2.72	1.15	3.87	35.60
Denver	25.75	4.20	.70	4.90	30.03	2.80	1.18	3.98	36.62
Detroit	29.00	4.73	.82	5.55	34.02	3.15	1.59	4.74	43.61
Houston	30.50	4.98	.68	5.66	34.72	3.32	1.13	4.45	40.94
Indianapolis	27.80	4.54	.77	5.31	32.52	3.02	1.41	4.43	40.76
Kansas City	24.50	4.00	.78	4.78	29.29	2.66	1.45	4.11	37.81
Los Angeles	30.00	4.89	.88	5.77	35.38	3.26	1.80	5.06	46.55
Memphis	29.30	4.78	.65	5.43	33.29	3.18	1.02	4.20	38.64
Milwaukee	26.50	4.32	.86	5.18	31.77	2.88	1.74	4.62	42.50
Minneapolis	27.25	4.45	.82	5.27	32.28	2.96	1.58	4.54	41.77
Nashville	28.00	4.57	.68	5.25	32.15	3.04	1.11	4.15	38.18
New Orleans	37.00	6.04	.64	6.68	40.96	4.02	.99	5.01	46.09
New York City	44.00	7.18	1.04	8.22	50.38	4.78	2.33	7.11	65.41
Philadelphia	27.75	4.53	.82	5.35	32.82	3.02	1.60	4.62	42.50
Phoenix	23.75	3.87	.69	4.56	27.95	2.58	1.15	3.73	34.32
Pittsburgh	32.25	5.26	.80	6.06	37.12	3.51	1.51	5.02	46.18
St. Louis	25.50	4.16	.84	5.00	30.64	2.77	1.65	4.42	40.66
San Antonio	27.25	4.45	.63	5.08	31.14	2.96	.95	3.91	35.97
San Diego	28.50	4.65	.88	5.53	33.91	3.10	1.80	4.90	45.08
San Francisco	31.50	5.14	.91	6.05	37.08	3.42	1.89	5.31	48.85
Seattle	32.75	5.34	.84	6.18	37.87	3.56	1.65	5.21	47.93
Washington, D.C.	35.00	5.71	.71	6.42	39.37	3.80	1.23	5.03	46.28
Average	$29.80	$4.87	$ .78	$5.65	$34.63	$3.24	$1.47	$4.71	$43.34

Assumed density is 145 lb. per C.F.
*Includes delivery within 20 miles

Table below shows quantities and bare costs for 1000 S.Y. of Bituminous Paving.

Item	Roads and Parking Areas, 3" Thick (02740-300-0460)		Sidewalks, 2" Thick (02775-275-0010)	
	Quantities	Cost	Quantities	Cost
Bituminous asphalt	163 tons @ $29.80 per ton	$4,857.40	109 tons @ $29.80 per ton	$3,248.20
Installation using	Crew B-25B @ $3,575.40 /4900SY/ day x 1000	729.67	Crew B-37 @ $880.00 /720 SY/day x 1000	1,222.22
Total per 1000 S.Y.		$5,587.07		$4,470.42
Total per S.Y.		$ 5.65		$ 4.47
Total per Ton		$ 34.28		$ 41.01

Site Construction — R021 Subsurface Invest. & Demol.

R02115-200 Underground Storage Tank Removal

Underground Storage Tank Removal can be divided into two categories: Non-Leaking and Leaking. Prior to removing an underground storage tank, tests should be made, with the proper authorities present, to determine whether a tank has been leaking or the surrounding soil has been contaminated.

To safely remove Liquid Underground Storage Tanks:
1. Excavate to the top of the tank.
2. Disconnect all piping.
3. Open all tank vents and access ports.
4. Remove all liquids and/or sludge.
5. Purge the tank with an inert gas.
6. Provide access to the inside of the tank and clean out the interior using proper personal protective equipment (PPE).
7. Excavate soil surrounding the tank using proper PPE for on-site personnel.
8. Pull and properly dispose of the tank.
9. Clean up the site of all contaminated material.
10. Install new tanks or close the excavation.

Site Construction — R022 Site Prep. & Excav. Support

R02240-900 Wellpoints

A single stage wellpoint system is usually limited to dewatering an average 15' depth below normal ground water level. Multi-stage systems are employed for greater depth with the pumping equipment installed only at the lowest header level. Ejectors, with unlimited lift capacity, can be economical when two or more stages of wellpoints can be replaced or when horizontal clearance is restricted, such as in deep trenches or tunneling projects, and where low water flows are expected. Wellpoints are usually spaced on 2-1/2' to 10' centers along a header pipe. Wellpoint spacing, header size, and pump size are all determined by the expected flow, as dictated by soil conditions.

In almost all soils encountered in wellpoint dewatering, the wellpoints may be jetted into place. Cemented soils and stiff clays may require sand wicks about 12" in diameter around each wellpoint to increase efficiency and eliminate weeping into the excavation. These sand wicks require 1/2 to 3 C.Y. of washed filter sand and are installed by using a 12" diameter steel casing and hole puncher jetted into the ground 2' deeper than the wellpoint. Rock may require predrilled holes.

Labor required for the complete installation and removal of a single stage wellpoint system is in the range of 3/4 to 2 labor-hours per linear foot of header, depending upon jetting conditions, wellpoint spacing, etc.

Continuous pumping is necessary except in some free draining soil where temporary flooding is permissible (as in trenches which are backfilled after each day's work). Good practice requires provision of a stand-by pump during the continuous pumping operation.

Systems for continuous trenching below the water table should be installed three to four times the length of expected daily progress to insure uninterrupted digging, and header pipe size should not be changed during the job.

For pervious free draining soils, deep wells in place of wellpoints may be economical because of lower installation and maintenance costs. Daily production ranges between two to three wells per day, for 25' to 40' depths, to one well per day for depths over 50'.

Detailed analysis and estimating for any dewatering problem is available at no cost from wellpoint manufacturers. Major firms will quote "sufficient equipment" quotes or their affiliates offer lump sum proposals to cover complete dewatering responsibility.

	Description for 200' System with 8" Header	Quantities	1st Month Unit	1st Month Total	Thereafter Unit	Thereafter Total
Equipment & Material	Wellpoints 25' long, 2" diameter @ 5' O.C.	40 Each	$ 28.00	$ 1,120.00	$ 21.00	$ 840.00
	Header pipe, 8" diameter	200 L.F.	5.10	1,020.00	2.55	510.00
	Discharge pipe, 8" diameter	100 L.F.	3.45	345.00	2.24	224.00
	8" valves	3 Each	140.00	420.00	91.00	273.00
	Combination Jetting & Wellpoint pump (standby)	1 Each	2,175.00	2,175.00	1,522.50	1,522.50
	Wellpoint pump, 8" diameter	1 Each	2,150.00	2,150.00	1,505.00	1,505.00
	Transportation to and from site	1 Day	708.40	708.40	—	—
	Fuel 30 days x 60 gal./day	1800 Gallons	1.40	2,520.00	1.40	2,520.00
	Lubricants for 30 days x 16 lbs./day	480 Lbs.	1.00	480.00	1.00	480.00
	Sand for points	40 C.Y.	15.60	624.00	—	—
	Equipment & Materials Subtotal			$11,562.40		$ 7,874.50
Labor	Technician to supervise installation	1 Week	$ 21.05	$ 842.00	—	—
	Labor for installation and removal of system	300 Labor-hours	14.45	4,335.00	—	—
	4 Operators straight time 40 hrs./wk. for 4.33 wks.	693 Hrs.	19.10	13,236.30	$ 19.10	$13,236.30
	4 Operators overtime 2 hrs./wk. for 4.33 wks.	35 Hrs.	38.20	1,337.00	38.20	1,337.00
	Labor Subtotal			$19,750.30		$14,573.30
	Total Cost			$31,312.70		$22,447.80
	Monthly Cost/L.F. Header			$ 156.56		$ 112.24

Site Construction — R022 Site Prep. & Excav. Support

R02250-400 Wood Sheet Piling

Wood sheet piling may be used for depths to 20' where there is no ground water. If moderate ground water is encountered Tongue & Groove sheeting will help to keep it out. When considerable ground water is present, steel sheeting must be used.

For estimating purposes on trench excavation, sizes are as follows:

Depth	Sheeting	Wales	Braces	B.F. per S.F.
To 8'	3 x 12's	6 x 8's, 2 line	6 x 8's, @ 10'	4.0 @ 8'
8' x 12'	3 x 12's	10 x 10's, 2 line	10 x 10's, @ 9'	5.0 average
12' to 20'	3 x 12's	12 x 12's, 3 line	12 x 12's, @ 8'	7.0 average

Sheeting to be toed in at least 2' depending upon soil conditions. A five person crew with an air compressor and sheeting driver can drive and brace 440 SF/day at 8' deep, 360 SF/day at 12' deep, and 320 SF/day at 16' deep. For normal soils, piling can be pulled in 1/3 the time to install. Pulling difficulty increases with the time in the ground. Production can be increased by high pressure jetting. Figures below assume 50% of lumber is salvaged and includes pulling costs. Some jurisdictions require an equipment operator in addition to Crew B-31.

Sheeting Pulled

Daily Cost Crew B-31	L.H./Day	Hourly Cost	Daily Cost	8' Depth, 440 S.F./Day — To Drive (1 Day)	8' Depth — To Pull (1/3 Day)	16' Depth, 320 S.F./Day — To Drive (1 Day)	16' Depth — To Pull (1/3 Day)
1 Foreman	8	$16.45	$131.60	$131.60	$43.82	$131.60	$43.82
4 Laborers	32	14.45	462.40	462.40	153.98	462.40	153.98
1 Air Compressor			127.00	127.00	42.29	127.00	42.29
1 Sheeting Driver			12.20	12.20	4.06	12.20	4.06
2 -50 Ft. Air Hoses, 1-1/2" Diam.			14.40	14.40	4.80	14.40	4.80
Lumber (50% salvage)				1.76 MBF 614.96		2.24 MBF 782.68	
Total			$747.60	$1,362.56	$248.95	$1,530.28	$248.95
Total/S.F.				$ 3.10	$ .57	$ 4.78	$ .78
Total (Drive and Pull)/S.F.				$ 3.67		$ 5.56	

Sheeting Left in Place

Daily Cost	8' Depth 440 S.F./Day	10' Depth 400 S.F./Day	12' Depth 360 S.F./Day	16' Depth 320 S.F./Day	18' Depth 305 S.F./Day	20' Depth 280 S.F./Day
Crew B-31	$ 747.60	$ 747.60	$ 747.60	$ 747.60	$ 747.60	$ 747.60
Lumber	1.76 M 1,229.92	1.8 M 1,257.88	1.8 M 1,257.88	2.24 M 1,565.36	2.1 M 1,467.52	1.9 M 1,327.76
Total in Place	$1,977.52	$2,005.48	$2,005.48	$2,312.96	$2,215.12	$2,075.36
Total/S.F.	$ 4.49	$ 5.01	$ 5.57	$ 7.23	$ 7.26	$ 7.41
Total/M.B.F.	$1,123.59	$1,114.15	$1,114.15	$1,032.57	$1,054.82	$1,092.29

Site Construction — R022 — Site Prep. & Excav. Support

R02250-450 Steel Sheet Piling

Limiting weights are 22 to 38#/S.F. of wall surface with 27#/S.F. average for usual types and sizes. (Weights of piles themselves are from 30.7#/L.F. to 57#/L.F. but they are 15″ to 21″ wide.) Lightweight sections 12″ to 28″ wide from 3 ga. to 12 ga. thick are also available for shallow excavations. Piles may be driven two at a time with an impact or vibratory hammer (use vibratory to pull) hung from a crane without leads. A reasonable estimate of the life of steel sheet piling is 10 uses with up to 125 uses possible if a vibratory hammer is used. Used piling costs from 50% to 80% of new piling depending on location and market conditions. Sheet piling and H piles can be rented for about 30% of the delivered mill price for the first month and 5% per month thereafter. Allow 1 labor-hour per pile for cleaning and trimming after driving. These costs increase with depth and hydrostatic head. Vibratory drivers are faster in wet granular soils and are excellent for pile extraction. Pulling difficulty increases with the time in the ground and may cost more than driving. It is often economical to abandon the sheet piling, especially if it can be used as the outer wall form. Allow about 1/3 additional length or more, for toeing into ground. Add bracing, waler and strut costs. Waler costs can equal the cost per ton of sheeting.

Cost of Sheet Piling & Production Rate by Ton & S.F.						
Depth of Excavation	15′ Depth		20′ Depth		25′ Depth	
Description of Pile	22 psf = 90.9 S.F./Ton		27 psf = 74 S.F./Ton		38 psf = 52.6 S.F./Ton	
Type of operation: Left in Place or Removed	Drive & Left	Drive & Extract*	Drive & Left	Drive & Extract*	Drive & Left	Drive & Extract*
Labor & Equip. to Drive 1 ton	$ 278.18	$416.50	$ 232.03	$348.04	$ 158.27	$236.22
Piling (75% Salvage)	779.10	194.78	779.10	194.78	779.10	194.78
Cost/Ton in Place	$1,057.28	$611.27	$1,011.13	$542.82	$937.37	$431.00
Production Rate, Tons/Day	10.81	7.22	12.96	8.64	19.00	12.73
Cost, S.F. in Place (incl. 33% toe in)	$ 11.63	$ 6.72	$ 13.66	$ 7.34	$ 17.82	$ 8.19
Production Rate, S.F./Day (incl. 33% toe in)	983	656	960	640	1000	670

Crew B-40	Bare Costs		Inc. Subs O&P		Installation Cost per Ton
	Hr.	Daily	Hr.	Daily	
1 Pile Driver Foreman	$21.45	$ 171.60	$40.00	$ 320.00	for 15′ Deep Excavation plus 33% toe-in
4 Pile Drivers	19.45	622.40	36.25	1,160.00	$3,043.80 per day/10.81 tons = $281.57 per ton
1 Building Laborer	14.45	115.60	24.75	198.00	Installation Cost per S.F.
2 Equip. Oper. (crane)	20.65	330.40	34.25	548.00	$\frac{2000\#}{22\#/S.F.}$ = 90.9 S.F. x 10.81 tons = 983 S.F. per day
1 Crane, 40 Ton		714.30		785.75	
Vibratory Hammer & Gen.		1,218.00		1,339.80	$3,043.80 / 983 S.F. per day = $3.10 per S.F.
56 L.H., Daily Totals		$3,007.10		$4,077.55	

*For driving & extracting, two mobilizations & demobilizations will be necessary

R02250-900 Vibroflotation and Vibro Replacement Soil Compaction

Vibroflotation is a proprietary system of compacting sandy soils in place to increase relative density to about 70%. Typical bearing capacities attained will be 6000 psf for saturated sand and 12,000 psf for dry sand. Usual range is 4000 to 8000 psf capacity. Costs in the front of the book are for a vertical foot of compacted cylinder 6′ to 10′ in diameter.

Vibro replacement is a proprietary system of improving cohesive soils in place to increase bearing capacity. Most silts and clays above or below the water table can be strengthened by installation of stone columns.

The process consists of radial displacement of the soil by vibration. The created hole is then backfilled in stages with coarse granular fill which is thoroughly compacted and displaced into the surrounding soil in the form of a column.

The total project cost would depend on the number and depth of the compacted cylinders. The installing company guarantees relative soil density of the sand cylinders after compaction and the bearing capacity of the soil after the replacement process. Detailed estimating information is available from the installer at no cost.

Site Construction | R023 Earthwork

R02315-300 Compacting Backfill

Compaction of fill in embankments, around structures, in trenches, and under slabs is important to control settlement. Factors affecting compaction are:

1. Soil gradation
2. Moisture content
3. Equipment used
4. Depth of fill per lift
5. Density required

The costs for testing and soil analyses are listed in Division 01450-500. Also, see Division 02315 for further backfill, borrow, and compaction costs.

Example:

Compact granular fill around a building foundation using a 21" wide x 24" vibratory plate in 8" lifts. Operator moves at 50 FPM working a 50 minute hour to develop 95% Modified Proctor Density with 4 passes.

Production Rate:

$$\frac{1.75\text{" plate wide} \times 50 \text{ F.P.M.} \times 50 \text{ min./hr} \times 67\text{" life}}{27 \text{ C.F. per C.Y.}} = 108.5 \text{ C.Y./hr.}$$

Production Rate for 4 Passes:

$$\frac{108.5 \text{ C.Y.}}{4 \text{ passes}} = 27.125 \text{ C.Y./hr.} \times 8 \text{ hrs.} = 217 \text{ C.Y./day}$$

	Compacting 217 C.Y. with 21" Wide Vibratory Plate	L.H./Day	Hourly Cost	Daily Cost	CY Cost
1	Laborer	8	$14.45	$115.60	$.54
1	Vibratory Plate Compactor			60.60	.28
Total for 217 C.Y./day				$176.20	$.82

Site Construction | R023 Earthwork

R02315-400 Excavating

The selection of equipment used for structural excavation and bulk excavation or for grading is determined by the following factors.
1. Quantity of material.
2. Type of material.
3. Depth or height of cut.
4. Length of haul.
5. Condition of haul road.
6. Accessibility of site.
7. Moisture content and dewatering requirements.
8. Availability of excavating and hauling equipment.

Some additional costs must be allowed for hand trimming the sides and bottom of concrete pours and other excavation below the general excavation.

When planning excavation and fill, the following should also be considered.
1. Swell factor.
2. Compaction factor.
3. Moisture content.
4. Density requirements.

A typical example for scheduling and estimating the cost of excavation of a 15′ deep basement on a dry site when the material must be hauled off the site, is outlined below.

Assumptions:
1. Swell factor, 18%.
2. No mobilization or demobilization.
3. Allowance included for idle time and moving on job.
4. No dewatering, sheeting, or bracing.
5. No truck spotter or hand trimming.

Number of B.C.Y. per truck = 1.5 C.Y. bucket × 8 passes = 12 loose C.Y.

$$= 12 \times \frac{100}{118} = 10.2 \text{ B.C.Y. per truck}$$

Truck Haul Cycle:
Load truck 8 passes	=	4 minutes
Haul distance 1 mile	=	9 minutes
Dump time	=	2 minutes
Return 1 mile	=	7 minutes
Spot under machine	=	1 minute
		23 minute cycle

Fleet Haul Production per day in B.C.Y.

$$4 \text{ trucks} \times \frac{50 \text{ min. hour}}{23 \text{ min. haul cycle}} \times 8 \text{ hrs.} \times 10.2 \text{ B.C.Y.}$$

$$= 4 \times 2.2 \times 8 \times 10.2 = 718 \text{ B.C.Y./day}$$

	Excavating Cost with a 1-1/2 C.Y. Hydraulic Excavator 15′ Deep, 2 Mile Round Trip Haul	L.H./Day	Hourly Cost	Daily Cost	Subtotal	Unit Price
1	Equipment Operator	8	$20.65	$ 165.20		
1	Oiler	8	17.00	136.00		
4	Truck Drivers	32	16.20	518.40	$ 819.60	$1.14
1	Hydraulic Excavator			713.20		
4	Dump Trucks			1,789.00	2,502.20	3.48
	Total for 720 BCY			$3,321.80	$3,321.80	$4.61

Description		1-1/2 C.Y. Hyd. Backhoe 15′ Deep		1-1/2 C.Y. Power Shovel 7′ Bank		1-1/2 C.Y. Dragline 7′ Deep		2-1/2 C.Y. Trackloader Stockpile
Operator (and Oiler, if required)		$ 301.20		$ 301.20		$ 301.20		$ 301.20
Truck Drivers	3 Ea.	388.80	4 Ea.	518.40	3 Ea.	388.80	4 Ea.	518.40
Equipment Rental		713.20		875.30		899.75		865.10
20 C.Y. Trailer Dump Trucks	3 Ea.	1,341.75	4 Ea.	1,789.00	3 Ea.	1,341.75	4 Ea.	1,789.00
Total Cost per Day		$2,744.95		$3,483.90		$2,931.50		$3,473.70
Daily Production, C.Y. Bank Measure		720.00		960.00		640.00		1000.00
Cost per C.Y.		$ 3.81		$ 3.63		$ 4.58		$ 3.47

Add the mobilization and demobilization costs to the total excavation costs. When equipment is rented for more than three days, there is often no mobilization charge by the equipment dealer. On larger jobs outside of urban areas, scrapers can move earth economically provided a dump site or fill area and adequate haul roads are available. Excavation within sheeting bracing or cofferdam bracing is usually done with a clamshell and production is low, since the clamshell may have to be guided by hand between the bracing. When excavating or filling an area enclosed with a wellpoint system, add 10% to 15% to the cost to allow for restricted access. When estimating earth excavation quantities for structures, allow work space outside the building footprint for construction of the foundation, and a slope of 1:1 unless sheeting is used.

Site Construction — R023 Earthwork

R02315-450 Excavating Equipment

The table below lists THEORETICAL hourly production in C.Y./hr. bank measure for some typical excavation equipment. Figures assume 50 minute hours, 83% job efficiency, 100% operator efficiency, 90° swing and properly sized hauling units, which must be modified for adverse digging and loading conditions. Actual production costs in the front of the book average about 50% of the theoretical values listed here.

Equipment	Soil Type	B.C.Y. Weight	% Swell	1 C.Y.	1-1/2 C.Y	2 C.Y.	2-1/2 C.Y.	3 C.Y.	3-1/2 C.Y.	4 C.Y.
Hydraulic Excavator "Backhoe" 15' Deep Cut	Moist loam, sandy clay	3400 lb.	40%	85	125	175	220	275	330	380
	Sand and gravel	3100	18	80	120	160	205	260	310	365
	Common earth	2800	30	70	105	150	190	240	280	330
	Clay, hard, dense	3000	33	65	100	130	170	210	255	300
Power Shovel Optimum Cut (Ft.)	Moist loam, sandy clay	3400	40	170 (6.0)	245 (7.0)	295 (7.8)	335 (8.4)	385 (8.8)	435 (9.1)	475 (9.4)
	Sand and gravel	3100	18	165 (6.0)	225 (7.0)	275 (7.8)	325 (8.4)	375 (8.8)	420 (9.1)	460 (9.4)
	Common earth	2800	30	145 (7.8)	200 (9.2)	250 (10.2)	295 (11.2)	335 (12.1)	375 (13.0)	425 (13.8)
	Clay, hard, dense	3000	33	120 (9.0)	175 (10.7)	220 (12.2)	255 (13.3)	300 (14.2)	335 (15.1)	375 (16.0)
Drag Line Optimum Cut (Ft.)	Moist loam, sandy clay	3400	40	130 (6.6)	180 (7.4)	220 (8.0)	250 (8.5)	290 (9.0)	325 (9.5)	385 (10.0)
	Sand and gravel	3100	18	130 (6.6)	175 (7.4)	210 (8.0)	245 (8.5)	280 (9.0)	315 (9.5)	375 (10.0)
	Common earth	2800	30	110 (8.0)	160 (9.0)	190 (9.9)	220 (10.5)	250 (11.0)	280 (11.5)	310 (12.0)
	Clay, hard, dense	3000	33	90 (9.3)	130 (10.7)	160 (11.8)	190 (12.3)	225 (12.8)	250 (13.3)	280 (12.0)

Equipment	Soil Type	B.C.Y. Weight	% Swell	Wheel Loaders				Track Loaders		
				3 C.Y.	4 C.Y.	6 C.Y.	8 C.Y.	2-1/4 C.Y.	3 C.Y.	4 C.Y.
Loading Tractors	Moist loam, sandy clay	3400	40	260	340	510	690	135	180	250
	Sand and gravel	3100	18	245	320	480	650	130	170	235
	Common earth	2800	30	230	300	460	620	120	155	220
	Clay, hard, dense	3000	33	200	270	415	560	110	145	200
	Rock, well-blasted	4000	50	180	245	380	520	100	130	180

Site Construction — R024 Foundation & Load Bearing Elements

R02455-900 Wood Bearing Piles

Untreated Southern Yellow Pine is most generally used for pile foundations cut off below the low water line. These are driven with the bark on. All piles cut off above the low water line should be treated with a preservative or encased in concrete for the section above water.

Item	Unit Cost	Units	Quantity	Total Cost	Cost/Pile	Cost/L.F.
50' Treated Piles, 13" Butt, 7" Tip	$ 8.63	L.F.	200	$ 86,300.00	$431.50	$ 8.63
Installation, Crew B-19	2,676.30	Day	13	34,791.90	173.96	3.48
Mobilization & Demobilization, Crew B-19	2,676.30	Day	3	8,028.90	40.14	.80
Transportation of Eq. One Way	1,045.55	Day	1	1,045.55	5.23	.10
Totals				$130,166.35	$650.83	$13.01

The above figures are based on driving 800 L.F. daily which can be considered average. Time is included for moving rig, cutoff & ordinary delays. A general observation is that the cost of a pile in place complete is about two times the cost of pile only. See also equipment rental division 01559-000 and 01559-100 for equipment capacities.

R02465-600 Caissons

The three principal types of caissons are:

(1) **Belled Caissons**, which except for shallow depths and poor soil conditions, are generally recommended. They provide more bearing than shaft area. Because of its conical shape, no horizontal reinforcement of the bell is required.

(2) **Straight Shaft Caissons** are used where relatively light loads are to be supported by caissons that rest on high value bearing strata. While the shaft is larger in diameter than for belled types this is more than offset by the saving in time and labor.

(3) **Keyed Caissons** are used when extremely heavy loads are to be carried. A keyed or socketed caisson transfers its load into rock by a combination of end-bearing and shear reinforcing of the shaft. The most economical shaft often consists of a steel casing, a steel wide flange core and concrete. Allowable compressive stresses of .225 f'c for concrete, 16,000 psi for the wide flange core, and 9,000 psi for the steel casing are commonly used. The usual range of shaft diameter is 18" to 84". The number of sizes specified for any one project should be limited due to the problems of casing and auger storage. When hand work is to be performed, shaft diameters should not be less than 32". When inspection of borings is required a minimum shaft diameter of 30" is recommended. Concrete caissons are intended to be poured against earth excavation so permanent forms which add to cost should not be used if the excavation is clean and the earth sufficiently impervious to prevent excessive loss of concrete.

Soil Conditions for Belling		
Good	Requires Handwork	Not Recommended
Clay	Hard Shale	Silt
Sandy Clay	Limestone	Sand
Silty Clay	Sandstone	Gravel
Clayey Silt	Weathered Mica	Igneous Rock
Hard-pan		
Soft Shale		
Decomposed Rock		

R02465-800 Pressure Injected Footings

Pressure Injected Footings are end bearing foundation units consisting of expanded bulbous bases formed by high energy blows, and concrete shafts either cased or uncased, to transfer the load.

The bulb must be formed in granular, rock or hardpan soils. Bearing is achieved at a predetermined depth so that the length of the shaft is usually less than for friction piles. High load capacity reduces the number of units and caps required.

Mobilization and demobilization costs below assume maximum distance of 50 miles.

100 Uncased Pressure Injected Footings 25' Long	Quantity Unit	Unit Cost	Total	Cost/Pile	Cost/L.F.
Crew B-44, 8 piles/day	13 Days	$2,102.70	$27,335.10	$273.35	$10.93
Mobilization & Demobilization	2 Days	2,102.70	4,205.40	42.05	1.68
Transportation one way	1.5 Days	1,045.55	1,568.33	15.68	.63
Total Installation			$33,108.83	$331.09	$13.24
Shafts: 100 ea. 17" diam., 25' deep	155 C.Y.	$ 83.60	$12,958.00	$129.58	$ 5.18
Pressure Bulbs: 100 ea. @ .75 C.Y.	75 C.Y.	83.60	6,270.00	62.70	2.51
Waste: 15% of above	35 C.Y.	83.60	2,926.00	29.26	1.17
Total Material Cost			$22,154.00	$221.54	$ 8.86
Total Cost in Place			$55,262.83	$552.63	$22.10

Site Construction — R025 — Sewerage & Drainage

R02510-810 Concrete Pipe

Prices given are for inside 20 mile delivery zone. Add $1.50 per ton of pipe for each additional 10 miles. Minimum truckload is 10 tons. The non-reinforced pipe listed in the front of the book is designation ASTM C14-59 extra strength. The reinforced pipe listed is ASTM C76-65T class 3, no gaskets. The installation cost given includes shaping bottom of the trench, placing the pipe, and backfilling and tamping to the top of the pipe only.

Site Construction — R029 — Landscaping

R02920-500 Seeding

The type of grass is determined by light, shade and moisture content of soil plus intended use. Fertilizer should be disked 4" before seeding. For steep slopes disk five tons of mulch and lay two tons of hay or straw on surface per acre after seeding. Surface mulch can be staked, lightly disked or tar emulsion sprayed. Material for mulch can be wood chips, peat moss, partially rotted hay or straw, wood fibers and sprayed emulsions. Hemp seed blankets with fertilizer are also available. For spring seeding, watering is necessary. Late fall seeding may have to be reseeded in the spring. Hydraulic seeding, power mulching, and aerial seeding can be used on large areas.

R02930-900 Cost of Trees: Based on Pin Oak (Quercus palustris)

Tree Diameter	Normal Height	Catalog List Price of Tree	Guying Material	Equipment Charge	Installation Labor	Total
2 to 3 inch	14 feet	$ 140	$15.00	$ 56.83	$ 51.36	$ 263.59
3 to 4 inch	16 feet	231	18.00	94.72	85.60	429.44
4 to 5 inch	18 feet	371	55.00	113.66	102.72	642.36
6 to 7 inch	22 feet	822	75.00	142.08	128.40	1,167.36
8 to 9 inch	26 feet	1053	95.00	189.43	171.20	1,508.63

Installation Time & Cost for Planting Trees, Bare Costs

Ball Size Diam. X Depth Inches	Soil in Ball C.F.	Weight of Ball Lbs.	Hole Diam. Req'd Feet	Hole Excavation C.F.	Amount of Soil Displ. C.F.	Topsoil Handled C.F.	Dig & Lace	Handle Ball	Dig Hole	Plant & Prune	Water & Guy	Total L.H.	Crew	Total per Tree
12 x 12	.70	56.00	2.00	4.00	3.00	11.00	.25	.17	.33	.25	.07	1.10	1 Clab	$ 15.90
18 x 16	2.00	160.00	2.50	8.00	6.00	21.00	.50	.33	.47	.35	.08	1.70	2 Clab	24.57
24 x 18	4.00	320.00	3.00	13.00	9.00	38.00	1.00	.67	1.08	.82	.20	3.80	3 Clab	54.91
30 x 21	7.50	600.00	4.00	27.00	19.50	76.00	.82	.71	.79	1.22	.26	3.80		92.95
36 x 24	12.50	980.00	4.50	38.00	25.50	114.00	1.08	.95	1.11	1.32	.30	4.76		116.43
42 x 27	19.00	1,520.00	5.50	64.00	45.00	185.00	1.90	1.27	1.87	1.43	.34	6.80	B-6	166.33
48 x 30	28.00	2,040.00	6.00	85.00	57.00	254.00	2.41	1.60	2.06	1.55	.39	8.00	@	195.68
54 x 33	38.50	3,060.00	7.00	127.00	88.50	370.00	2.86	1.90	2.39	1.76	.45	9.40	$24.46	229.92
60 x 36	52.00	4,160.00	7.50	159.00	107.00	474.00	3.26	2.17	2.73	2.00	.51	10.70	per	261.72
66 x 39	68.00	5,440.00	8.00	196.00	128.00	596.00	3.61	2.41	3.07	2.26	.58	11.90	Labor-	291.07
72 x 42	87.00	7,160.00	9.00	267.00	180.00	785.00	3.90	2.60	3.71	2.78	.70	13.70	hour	335.10

Concrete — R031 Concrete Formwork

R03110-010 Wall Form Materials

Aluminum Forms

Approximate weight is 3 lbs. per S.F.C.A. Standard widths are available from 4″ to 36″ with 36″ most common. Standard lengths of 2′, 4′, 6′ to 8′ are available. Forms are lightweight and fewer ties are needed with the wider widths. The form face is either smooth or textured.

Cost of bare forms per S.F.C.A. with different facing surface is listed below. Forms may also be rented.

	Purchase Cost Per S.F.							
Finish	3′ x 8′	2′ x 8′	12″ x 8′	6″ x 8′	3′ x 4′	2′ x 4′	12″ x 4′	6″ x 4′
Smooth Aluminum (.096″ Face Sheet)	$12.78	$15.78	$21.38	$35.41	$13.69	$18.84	$24.72	$38.64
Textured Brick Aluminum	$14.88	$18.95	$25.57	$40.67	$17.08	$22.57	$28.96	$46.10

Metal Framed Plywood Forms

Manufacturers claim over 75 reuses of plywood and over 300 reuses of steel frames. Sale price for steel framed forms is from $8.50 per S.F. for 2′ x 8′ to $12.50 per S.F. for 2′ x 3′. Narrower forms range between $12.00 per S.F. to $22.00 per S.F. for 1′ x 3′. Many specials such as corners, fillers, pilasters, etc. are available. Monthly rental is generally about 15% of purchase price for first month and 9% per month thereafter with 90% of rental applied to purchase for the first month and decreasing percentages thereafter. Aluminum framed forms cost 25% to 30% more than steel framed.

Rule of thumb purchase price including corners, specials, etc.; steel framed $12.50 per S.F.; aluminum framed $17.50 per S.F.

Reconditioned steel framed forms are rented for an average of $1.35 per S.F per month. Aluminum framed forms are rented for $2.25 per S.F. for the first month and $1.50 per S.F. per month thereafter.

After the first month, extra days may be prorated from the monthly charge. Rental rates do not include ties, accessories, cleaning, loss of hardware or freight in and out. Approximate weight is 5 lbs. per S.F. for steel; 3 lbs. per S.F. for aluminum.

Forms can be rented with option to buy.

Plywood Forms, Job Fabricated

There are two types of plywood used for concrete forms.
1. Exterior plyform which is completely waterproof. This is face oiled to facilitate stripping. Ten reuses can be expected with this type with 25 reuses possible.
2. An overlaid type consists of a resin fiber fused to exterior plyform. No oiling is required except to facilitate cleaning. This is available in both high density (HDO) and medium density overlaid (MDO). Using HDO, 50 reuses can be expected with 200 possible.

Plyform is available in 5/8″ and 3/4″ thickness. High density overlaid is available in 3/8″, 1/2″, 5/8″ and 3/4″ thickness.

5/8″ thick is sufficient for most building forms, while 3/4″ is best on heavy construction.

For prices on plywood and framing lumber see R06100-010 and R06160-020.

Plywood Forms, Modular, Prefabricated

There are many plywood forming systems without frames. Most of these are manufactured from 1-1/8″ (HDO) plywood and have some hardware attached. These are used principally for foundation walls 8′ or less high. With care and maintenance, 100 reuses can be attained with decreasing quality of surface finish. Sale price of 2′ x 8′ panels is $6.50 per S.F. with 4″ x 8′ fillers costing $25.00 per S.F. Typical forms and accessories for 1000 S.F. of form area cost $7000 not including form ties.

Steel Forms

Approximate weight is 6-1/2 lbs. per S.F.C.A. including accessories. Standard widths are available from 2″ to 24″, with 24″ most common. Standard lengths are from 2′ to 8′, with 4′ the most common. Forms are easily ganged into modular units.

Sale price for typical hand set job is $12.50 per S.F.C.A. including all usual accessories, specials, corners, etc., but not including ties or freight. Cost of bare forms runs from $10.00 for wide forms to over $22.00 per S.F.C.A. for narrow widths and/or short lengths. Forms are usually leased for 15% of the purchase price per month prorated daily over 30 days. Standard 6000 lb. wall ties for 12″ walls cost $84 per hundred and 24″ long ties cost $121 per hundred.

Rental may be applied to sale price and usually rental forms are bought. With careful handling and cleaning 200 to 400 reuses are possible.

Straight wall gang forms up to 12′ x 20′ or 8′ x 30′ can be fabricated. These crane handled forms usually cost from $21.00 to $28.00 per S.F.C.A. or can be leased for approx. 9% per month. Straight wall gang forms utilizing 40,000 lb. and 60,000 lb. ties with sizes from 24′ x 10′ to 24′ x 24′ cost from $22.00 to $30.00 per S.F. including all accessories and taper ties. Rental rates on these gang forms run from $.30 to $.75 per week per S.F.C.A.

Individual job analysis is available from the manufacturer at no charge.

Concrete | R031 | Concrete Formwork

R03110-020 Floor Pans and Domes

For 8' to 15' Ceiling Heights Using Crew C-2 at $17.52 per Labor-Hour	20" Pans (031-150)				19" Domes (031-150)			
	(Line 3500)	1 Use	(Line 3650)	4 Use	(Line 4000)	1 Use	(Line 4150)	4 Use
Rent 20" pans or 19" domes		$ 84.00		$ 26.00		$ 84.00		$ 26.00
Adjustable shores and accessories		46.00		25.25		46.00		25.25
110 S.F. 3/4" plyform at $1036 per M.S.F.		113.96		37.04		113.96		37.04
210 B.F. supporting joists, girts and braces at $600 per MBF		126.00		40.95		126.00		40.95
Labor handle, place, strip, oil and clean pans and domes	1.5 LH	26.28	1.0 LH	17.52	1.8 LH	31.54	1.2 LH	21.02
Make up, erect, remove wood decking	10 LH	175.20	8.5 LH	148.92	10.0 LH	175.20	8.5 LH	148.92
Total per 100 S.F. of floor area		$571.44		$295.68		$576.70		$299.18

The figures above are for closed deck forming. For open deck forming deduct $25 per 100 S.F. for four uses. For pan rental, figure $.84 per S.F. of total area for one use; $.42 for two uses; $.30 for three uses; and $.25 for four uses, which is about the most that can be expected for any one project. The purchase price of fiberglass long pans is $6.00 per S.F.C.A. or an average $12.00 per S.F. of floor area.*

For two-way grid system, 19" steel domes are leased for $.25 to $.75 and 30" fiberglass domes are leased for $.50 to $2.25 per S.F. of floor area. The purchase price of custom size fiberglass forms runs from $6.00 to $18.00 per S.F.C.A.*

For slab height from 15' to 20' add $55 per 100 S.F. and for 20' to 35' add $75 per 100 S.F., both for four uses.

*It is necessary to divide the purchase cost by the number of expected uses to determine the cost per use.

R03110-030 Slipforms

The slipform method of forming may be used for forming circular silo and multi-celled storage bin type structures over 30' high, and building core shear walls over eight stories high. The shear walls, usually enclose elevator shafts, stairwells, mechanical spaces, and toilet rooms. Reuse of the form on duplicate structures will reduce the height necessary and spread the cost of building the form. Slipform systems can be used to cast chimneys, towers, piers, dams, underground shafts or other structures capable of being extruded.

Slipforms are usually 4' high and are raised semi-continuously by jacks climbing on rods which are embedded in the concrete. The jacks are powered by a hydraulic, pneumatic, or electric source and are available in 3, 6, and 22 ton capacities. Interior work decks and exterior scaffolds must be provided for placing inserts, embedded items, reinforcing steel, and concrete. Scaffolds below the form for finishers may be required. The interior work decks are often used as roof slab forms on silos and bin work. Form raising rates will range from 6" to 20" per hour for silos; 6" to 30" per hour for buildings; and 6" to 48" per hour for shaft work.

Reinforcing bars and stressing strands are usually hoisted by crane or gin pole and the concrete material can be hoisted by crane, winch-powered skip, or pumps. The slipform system is operated on a continuous 24 hour day when a monolithic structure is desired. For least cost, the system is operated only during normal working hours.

The cost of the form, including equipment setting, will range from $8.00 to $20 per square foot of actual form built. The cost of work decks and scaffolds will range from $2.50 to $10 per square foot of total deck area. Slipform equipment rental will range from $.20 to $.60 per S.F. of surface formed. Operating costs of slipform systems will range from $.07 to $.40 per S.F. of surface formed.

Placing concrete will range from 0.5 to 1.5 labor-hours per C.Y. Finishing (if required) and curing costs will range from $.06 to $.30 per S.F. of surface. Bucks, blockouts, keyways, weldplates, etc. are extra. Recent in-place total costs have ranged from $275 to $900 per C.Y.

Concrete — R031 Concrete Formwork

R03110-040 Forms for Reinforced Concrete

Design Economy
Avoid many sizes in proportioning beams and columns.

From story to story avoid changing column dimensions. Gain strength by adding steel or using a richer mix. If a change in size of column is necessary, vary one dimension only to minimize form alterations. Keep beams and columns the same width.

From floor to floor in a multi-story building vary beam depth, not width, as that will leave slab panel form unchanged. It is cheaper to vary the strength of a beam from floor to floor by means of steel area than by 2" changes in either width or depth.

Cost Factors
Material includes the cost of lumber, cost of rent for metal pans or forms if used, nails, form ties, form oil, bolts and accessories.

Labor includes the cost of carpenters to make up, erect, remove and repair, plus common labor to clean and move. Having carpenters remove forms minimizes repairs.

Improper alignment and condition of forms will increase finishing cost. When forms are heavily oiled, concrete surfaces must be neutralized before finishing. Special curing compounds will cause spillages to spall off in first frost. Gang forming methods will reduce costs on large projects.

Materials Used
Boards are seldom used unless their architectural finish is required. Generally, steel, fiberglass and plywood are used for contact surfaces. Labor on plywood is 10% less than with boards. The plywood is backed up with 2 x 4's at 12" to 32" O.C. Walers are generally 2 - 2 x 4's. Column forms are held together with steel yokes or bands. Shoring is with adjustable shoring or scaffolding for high ceilings.

Reuse
Floor and column forms can be reused four or possibly five times without excessive repair. Remember to allow for 10% waste on each reuse.

When modular sized wall forms are made, up to twenty uses can be expected with exterior plyform.

When forms are reused, the cost to erect, strip, clean and move will not be affected. 10% replacement of lumber should be included and about one hour of carpenter time for repairs on each reuse per 100 S.F.

The reuse cost for certain accessory items normally rented on a monthly basis will be lower than the cost for the first use.

After fifth use, new material required plus time needed for repair prevent form cost from dropping further and it may go up. Much depends on care in stripping, the number of special bays, changes in beam or column sizes and other factors.

Costs for multiple use of formwork may be developed as follows:

2 Uses
$$\frac{(\text{1st Use + Reuse})}{2} = \text{avg. cost/2 uses}$$

3 Uses
$$\frac{(\text{1st Use + 2 Reuse})}{3} = \text{avg. cost/3 uses}$$

4 Uses
$$\frac{(\text{1st use + 3 Reuse})}{4} = \text{avg. cost/4 uses}$$

R03110-050 Forms In Place

This section assumes that all cuts are made with power saws, that adjustable shores are employed and that maximum use is made of commercial form ties and accessories. Bare costs are used in the table below.

BEAM AND GIRDER, INTERIOR, 12" Wide (Line 031-138-2000)			First Use			Reuse		
Item	Cost	Unit	Quantity	Material	Installation	Quantity	Material	Installation
5/8" exterior plyform	$966.00	M.S.F.	115 S.F.	$111.10		11.5 S.F.	$11.10	
Lumber	600.00	M.B.F.	200 B.F.	120.00		20.0 B.F.	12.00	
Accessories, incl. adjustable shores			Allow	23.15		Allow	23.15	
Make up, crew C-2	17.52	L.H.	6.4 L.H.		$112.15	1.0 L.H.		$ 17.50
Erect and strip			8.3 L.H.		145.40	8.3 L.H.		145.40
Clean and move			1.3 L.H.		22.80	1.3 L.H.		22.80
Total per 100 S.F.C.A.			16.0 L.H.	$254.25	$280.35	10.6 L.H.	$46.25	$185.70

For structural steel frame with beams encased, subtract 1.2 labor-hours, and 50 B.F. lumber or about $48.50 per 100 S.F.C.A. for the first use and $31 for each reuse.

BOX CULVERT, 5' to 8' Square or Rectangular (Line 031-146-0010)			First Use			Reuse		
Item	Cost	Unit	Quantity	Material	Installation	Quantity	Material	Installation
3/4" exterior plyform	$1,036.00	M.S.F.	110 S.F.	$113.95		11.0 S.F.	$11.40	
Lumber	600.00	M.B.F.	170 B.F.	102.00		17.0 B.F.	10.20	
Accessories			Allow	18.25		Allow	18.25	
Build in place, crew C-1	17.16	L.H.	14.5 L.H.		$248.80	14.5 L.H.		$248.80
Strip and salvage			4.3 L.H.		73.80	4.3 L.H.		73.80
Total per 100 S.F.C.A.			18.8 L.H.	$234.20	$322.60	18.8 L.H.	$39.85	$322.60

Concrete | R031 | Concrete Formwork

R03110-050 Forms In Place (cont.)

COLUMNS, 24" x 24" (Line 031-142-6500)

Item	Cost	Unit	First Use Quantity	Material	Installation	Reuse Quantity	Material	Installation
5/8" exterior plyform	$966.00	M.S.F.	120 S.F.	$115.90		12.0 S.F.	$11.60	
Lumber	600.00	M.B.F.	125 B.F.	75.00		12.5 B.F.	7.50	
Clamps, chamfer strips and accessories			Allow	22.60		Allow	22.60	
Make up, crew C-1	17.16	L.H.	5.8 L.H.		$ 99.55	1.0 L.H.		$ 17.15
Erect and strip			9.8 L.H.		168.15	9.8 L.H.		168.15
Clean and move			1.2 L.H.		20.60	1.2 L.H.		20.60
Total per 100 S.F.C.A.			16.8 L.H.	$213.50	$288.30	12.0 L.H.	$41.70	$205.90

FLAT SLAB WITH DROP PANELS (Line 031-150-2000)

Item	Cost	Unit	First Use Quantity	Material	Installation	Reuse Quantity	Material	Installation
5/8" exterior plyform	$966.00	M.S.F.	115 S.F.	$111.10		11.5 S.F.	$11.10	
Lumber	600.00	M.B.F.	210 B.F.	126.00		21.0 B.F.	12.60	
Accessories, incl. adjustable shores			Allow	23.30		Allow	23.30	
Make up, crew C-2	17.52	L.H.	3.5 L.H.		$ 61.30	1.0 L.H.		$ 17.50
Erect and strip			6.0 L.H.		105.10	6.0 L.H.		105.10
Clean and move			1.2 L.H.		21.00	1.2 L.H.		21.00
Total per 100 S.F.C.A.			10.7 L.H.	$260.40	$187.40	8.2 L.H.	$47.00	$143.60

Drop panels included but column caps figure with columns.

FOOTINGS, SPREAD (Line 031-158-5000)

Item	Cost	Unit	First Use Quantity	Materials	Installation	Reuse Quantity	Material	Installation
Lumber	$600.00	M.B.F.	260 B.F.	$156.00		26 B.F.	$15.60	
Accessories			Allow	7.35		Allow	7.35	
Make up, crew C-1	17.16	L.H.	4.7 L.H.		$ 80.65	1.0 L.H.		$ 17.15
Erect and strip			4.2 L.H.		72.05	4.2 L.H.		72.05
Clean and move			1.6 L.H.		27.45	1.6 L.H.		27.45
Total per 100 S.F.C.A.			10.5 L.H.	$163.35	$180.15	6.8 L.H.	$22.95	$116.65

FOUNDATION WALL, 8' High (Line 031-182-2000)

Item	Cost	Unit	First Use Quantity	Material	Installation	Reuse Quantity	Material	Installation
5/8" exterior plyform	$966.00	M.S.F.	110 S.F.	$106.25		11.0 S.F.	$10.65	
Lumber	600.00	M.B.F.	140 B.F.	84.00		14.0 B.F.	8.40	
Accessories			Allow	23.50		Allow	23.50	
Make up, crew C-2	17.52	L.H.	5.0 L.H.		$ 87.60	1.0 L.H.		$ 17.50
Erect and strip			6.5 L.H.		113.90	6.5 L.H.		113.90
Clean and move			1.5 L.H.		26.30	1.5 L.H.		26.30
Total per 100 S.F.C.A.			13.0 L.H.	$213.75	$227.80	9.0 L.H.	$42.55	$157.70

PILE CAPS, Square or Rectangular (Line 031-158-3000)

Item	Cost	Unit	First Use Quantity	Material	Installation	Reuse Quantity	Material	Installation
5/8" exterior plyform	$966.00	M.S.F.	110 S.F.	$106.25		11.0 S.F.	$10.65	
Lumber	600.00	M.B.F.	160 B.F.	96.00		16.0 B.F.	9.60	
Accessories			Allow	7.00		Allow	7.00	
Make up, crew C-1	17.16	L.H.	4.5 L.H.		$ 77.20	1.0 L.H.		$ 17.15
Erect and strip			5.0 L.H.		85.80	5.0 L.H.		85.80
Clean and move			1.5 L.H.		25.75	1.5 L.H.		25.75
Total per 100 S.F.C.A.			11.0 L.H.	$209.25	$188.75	7.5 L.H.	$27.25	$128.70

STAIRS, Average Run (Inclined Length x Width) (Line 031-174-0010)

Item	Cost	Unit	First Use Quantity	Materials	Installation	Reuse Quantity	Material	Installation
5/8" exterior plyform	$966.00	M.S.F.	110 S.F.	$106.25		11.0 S.F.	$10.65	
Lumber	600.00	M.B.F.	425 B.F.	255.00		42.5 B.F.	25.50	
Accessories			Allow	18.25		Allow	18.25	
Build in place, crew C-2	17.52	L.H.	25.0 L.H.		$438.00	25.0 L.H.		$438.00
Strip and salvage			4.0 L.H.		70.10	4.0 L.H.		70.10
Total per 100 S.F.			29.0 L.H.	$379.50	$508.10	29.0 L.H.	$54.40	$508.10

Concrete — R031 Concrete Formwork

R03110-060 Formwork Labor Hours

Item	Unit	Hours Required — Fabricate	Hours Required — Erect & Strip	Hours Required — Clean & Move	Total Hours 1 Use	Multiple Use 2 Use	Multiple Use 3 Use	Multiple Use 4 Use
Beam and Girder, interior beams, 12" wide	100 S.F.	6.4	8.3	1.3	16.0	13.3	12.4	12.0
Hung from steel beams		5.8	7.7	1.3	14.8	12.4	11.6	11.2
Beam sides only, 36" high		5.8	7.2	1.3	14.3	11.9	11.1	10.7
Beam bottoms only, 24" wide		6.6	13.0	1.3	20.9	18.1	17.2	16.7
Box out for openings		9.9	10.0	1.1	21.0	16.6	15.1	14.3
Buttress forms, to 8' high		6.0	6.5	1.2	13.7	11.2	10.4	10.0
Centering, steel, 3/4" rib lath			1.0		1.0			
3/8" rib lath or slab form			0.9		0.9			
Chamfer strip or keyway	100 L.F.		1.5		1.5	1.5	1.5	1.5
Columns, fiber tube 8" diameter			20.6		20.6			
12"			21.3		21.3			
16"			22.9		22.9			
20"			23.7		23.7			
24"			24.6		24.6			
30"			25.6		25.6			
Round Steel, 12" diameter			22.0		22.0	22.0	22.0	22.0
16"			25.6		25.6	25.6	25.6	25.6
20"			30.5		30.5	30.5	30.5	30.5
24"			37.7		37.7	37.7	37.7	37.7
Plywood 8" x 8"	100 S.F.	7.0	11.0	1.2	19.2	16.2	15.2	14.7
12" x 12"		6.0	10.5	1.2	17.7	15.2	14.4	14.0
16" x 16"		5.9	10.0	1.2	17.1	14.7	13.8	13.4
24" x 24"		5.8	9.8	1.2	16.8	14.4	13.6	13.2
Steel framed plywood 8" x 8"			10.0	1.0	11.0	11.0	11.0	11.0
12" x 12"			9.3	1.0	10.3	10.3	10.3	10.3
16" x 16"			8.5	1.0	9.5	9.5	9.5	9.5
24" x 24"			7.8	1.0	8.8	8.8	8.8	8.8
Drop head forms, plywood		9.0	12.5	1.5	23.0	19.0	17.7	17.0
Coping forms		8.5	15.0	1.5	25.0	21.3	20.0	19.4
Culvert, box			14.5	4.3	18.8	18.8	18.8	18.8
Curb forms, 6" to 12" high, on grade		5.0	8.5	1.2	14.7	12.7	12.1	11.7
On elevated slabs		6.0	10.8	1.2	18.0	15.5	14.7	14.3
Edge forms to 6" high, on grade	100 L.F.	2.0	3.5	0.6	6.1	5.6	5.4	5.3
7" to 12" high	100 S.F.	2.5	5.0	1.0	8.5	7.8	7.5	7.4
Equipment foundations		10.0	18.0	2.0	30.0	25.5	24.0	23.3
Flat slabs, including drops		3.5	6.0	1.2	10.7	9.5	9.0	8.8
Hung from steel		3.0	5.5	1.2	9.7	8.7	8.4	8.2
Closed deck for domes		3.0	5.8	1.2	10.0	9.0	8.7	8.5
Open deck for pans		2.2	5.3	1.0	8.5	7.9	7.7	7.6
Footings, continuous, 12" high		3.5	3.5	1.5	8.5	7.3	6.8	6.6
Spread, 12" high		4.7	4.2	1.6	10.5	8.7	8.0	7.7
Pile caps, square or rectangular		4.5	5.0	1.5	11.0	9.3	8.7	8.4
Grade beams, 24" deep		2.5	5.3	1.2	9.0	8.3	8.0	7.9
Lintel or Sill forms		8.0	17.0	2.0	27.0	23.5	22.3	21.8
Spandrel beams, 12" wide		9.0	11.2	1.3	21.5	17.5	16.2	15.5
Stairs			25.0	4.0	29.0	29.0	29.0	29.0
Trench forms in floor		4.5	14.0	1.5	20.0	18.3	17.7	17.4
Walls, Plywood, at grade, to 8' high		5.0	6.5	1.5	13.0	11.0	9.7	9.5
8' to 16'		7.5	8.0	1.5	17.0	13.8	12.7	12.1
16' to 20'		9.0	10.0	1.5	20.5	16.5	15.2	14.5
Foundation walls, to 8' high		4.5	6.5	1.0	12.0	10.3	9.7	9.4
8' to 16' high		5.5	7.5	1.0	14.0	11.8	11.0	10.6
Retaining wall to 12' high, battered		6.0	8.5	1.5	16.0	13.5	12.7	12.3
Radial walls to 12' high, smooth		8.0	9.5	2.0	19.5	16.0	14.8	14.3
2' chords		7.0	8.0	1.5	16.5	13.5	12.5	12.0
Prefabricated modular, to 8' high		—	4.3	1.0	5.3	5.3	5.3	5.3
Steel, to 8' high		—	6.8	1.2	8.0	8.0	8.0	8.0
8' to 16' high		—	9.1	1.5	10.6	10.3	10.2	10.2
Steel framed plywood to 8' high		—	6.8	1.2	8.0	7.5	7.3	7.2
8' to 16' high		—	9.3	1.2	10.5	9.5	9.2	9.0

Concrete — R032 Concrete Reinforcement

R03210-010 Reinforcing Steel Weights and Measures

Bar Designation No.**	Nominal Weight Lb./Ft.	U.S. Customary Units Nominal Dimensions*			SI Units Nominal Dimensions*			
		Diameter in.	Cross Sectional Area, in.²	Perimeter in.	Nominal Weight kg/m	Diameter mm	Cross Sectional Area, cm²	Perimeter mm
3	.376	.375	.11	1.178	.560	9.52	.71	29.9
4	.668	.500	.20	1.571	.994	12.70	1.29	39.9
5	1.043	.625	.31	1.963	1.552	15.88	2.00	49.9
6	1.502	.750	.44	2.356	2.235	19.05	2.84	59.8
7	2.044	.875	.60	2.749	3.042	22.22	3.87	69.8
8	2.670	1.000	.79	3.142	3.973	25.40	5.10	79.8
9	3.400	1.128	1.00	3.544	5.059	28.65	6.45	90.0
10	4.303	1.270	1.27	3.990	6.403	32.26	8.19	101.4
11	5.313	1.410	1.56	4.430	7.906	35.81	10.06	112.5
14	7.650	1.693	2.25	5.320	11.384	43.00	14.52	135.1
18	13.600	2.257	4.00	7.090	20.238	57.33	25.81	180.1

* The nominal dimensions of a deformed bar are equivalent to those of a plain round bar having the same weight per foot as the deformed bar.
** Bar numbers are based on the number of eighths of an inch included in the nominal diameter of the bars.

R03210-020 Metric Rebar Specification - ASTM A615-81

Grade 300 (300 MPa* = 43,560 psi; +8.7% vs. Grade 40)
Grade 400 (400 MPa* = 58,000 psi; −3.4% vs. Grade 60)

Bar No.	Diameter mm	Area mm²	Equivalent in.²	Comparison with U.S. Customary Bars
10	11.3	100	.16	Between #3 & #4
15	16.0	200	.31	#5 (.31 in.²)
20	19.5	300	.47	#6 (.44 in.²)
25	25.2	500	.78	#8 (.79 in.²)
30	29.9	700	1.09	#9 (1.00 in.²)
35	35.7	1000	1.55	#11 (1.56 in.²)
45	43.7	1500	2.33	#14 (2.25 in.²)
55	56.4	2500	3.88	#18 (4.00 in.²)

* MPa = megapascals

R03210-040 Weight of Steel Reinforcing Per Square Foot of Wall (PSF)

Reinforced Weights: The table below suggests the weights per square foot for reinforcing steel in walls. Weights are approximate and will be the same for all grades of steel bars. For bars in two directions, add weights for each size and spacing.

C/C Spacing in Inches	#3 Wt. (PSF)	#4 Wt. (PSF)	#5 Wt. (PSF)	#6 Wt. (PSF)	#7 Wt. (PSF)	#8 Wt. (PSF)	#9 Wt. (PSF)	#10 Wt. (PSF)	#11 Wt. (PSF)
2"	2.26	4.01	6.26	9.01	12.27				
3"	1.50	2.67	4.17	6.01	8.18	10.68	13.60	17.21	21.25
4"	1.13	2.01	3.13	4.51	6.13	8.10	10.20	12.91	15.94
5"	.90	1.60	2.50	3.60	4.91	6.41	8.16	10.33	12.75
6"	.752	1.34	2.09	3.00	4.09	5.34	6.80	8.61	10.63
8"	.564	1.00	1.57	2.25	3.07	4.01	5.10	6.46	7.97
10"	.451	.802	1.25	1.80	2.45	3.20	4.08	5.16	6.38
12"	.376	.668	1.04	1.50	2.04	2.67	3.40	4.30	5.31
18"	.251	.445	.695	1.00	1.32	1.78	2.27	2.86	3.54
24"	.188	.334	.522	.751	1.02	1.34	1.70	2.15	2.66
30"	.150	.267	.417	.600	.817	1.07	1.36	1.72	2.13
36"	.125	.223	.348	.501	.681	.890	1.13	1.43	1.77
42"	.107	.191	.298	.429	.584	.753	.97	1.17	1.52
48"	.094	.167	.261	.376	.511	.668	.85	1.08	1.33

Concrete | R032 | Concrete Reinforcement

R03210-050 Minimum Wall Reinforcement Weight (PSF)

This table lists the approximate minimum wall reinforcement weights per S.F. according to the specification of .12% of gross area for vertical bars and .20% of gross area for horizontal bars.

Location	Wall Thickness	Bar Size	Horizontal Steel Spacing C/C	Sq. In. Req'd per S.F.	Total Wt. per S.F.	Bar Size	Vertical Steel Spacing C/C	Sq. In. Req'd per S.F.	Total Wt. per S.F.	Horizontal & Vertical Steel Total Weight per S.F.
Both Faces	10"	#4	18"	.24	.89#	#3	18"	.14	.50#	1.39#
	12"	#4	16"	.29	1.00	#3	16"	.17	.60	1.60
	14"	#4	14"	.34	1.14	#3	13"	.20	.69	1.84
	16"	#4	12"	.38	1.34	#3	11"	.23	.82	2.16
	18"	#5	17"	.43	1.47	#4	18"	.26	.89	2.36
One Face	6"	#3	9"	.15	.50	#3	18"	.09	.25	.75
	8"	#4	12"	.19	.67	#3	11"	.12	.41	1.08
	10"	#5	15"	.24	.83	#4	16"	.14	.50	1.34

R03210-060 Reinforcing Extras

There are two "base prices" used, the price for bars shipped directly from the mill and the base price for the bars shipped from the steel fabricator's shops. The extras for both methods of pricing are listed below and are listed in dollars per ton for A615 steel.

Extras	Item	From Mill	From Fabricator	Extras	Item	From Mill	From Fabricator
Base Price		$267	$460				$500 minimum
Size Extras	#3	$100	$102	Quantity Extras	Under 20 tons	$12	$30
	#4	40	50		20-49 tons	—	20
	#5	30	35		50-99 tons	—	15
	#6	25	25		100-300 tons	—	10
	#7 to 11	20	20		Over 300 tons	—	—
	#14	40	40	Bending	Light #2 thru #11	N/A	85
	#18	40	40		Heavy #4 up		44
Grade Extras	Grade 40	—	—	Detailing	Under 50 tons		46
	Grade 60	$ 13	$ 13		50-100 tons		35
	Grade 75	N/A	N/A		150-500 tons		31
					Over 500 tons		21

"Listing only" runs $4 to $5 per ton.

R03210-070 Bend, Place and Tie Reinforcing

See R03210-060 for shop bending costs. Field bending costs are usually higher.

Placing and tying by rodmen for footings and slabs runs from nine hrs. per ton for heavy bars to fifteen hrs. per ton for light bars. For beams, columns, and walls, production runs from eight hrs. per ton for heavy bars to twenty hrs. per ton for light bars. Overall average for typical reinforced concrete buildings is about fourteen hrs. per ton. These production figures include placing of accessories and usual inserts. Equipment handling is necessary for the larger size bars so that installation costs for the very heavy bars will not decrease proportionately.

For tie wire, figure 5 lbs. 16 ga. black annealed wire at $.80 per lb. for each ton of bars. Use of commercial accessories improves placing accuracy and costs of $60 per ton of bars is offset by reduced labor time.

Installation costs for splicing reinforcing bars includes allowance for equipment to hold the bars in place while splicing as well as necessary scaffolding for iron workers.

Concrete | R032 Concrete Reinforcement

R03210-080 Reinforcing Key Prices

Costs given below are for over 50 tons delivered and bent from fabricators in metropolitan area. Over 500 tons, subtract $10 per ton. Figures are for domestic A615, grade 60. Economic conditions can alter the prices substantially.

City	Bare Costs Per Ton					Costs Incl. Subs O & P		Total per Ton
	Material		Installation			Material	Installation	
	50 Tons Fabricated	Allow for Accessories	Unload, Sort and Pile	Place & Tie	in Place	Incl. 10% Mark-Up	Include Mark-up	
Atlanta	$475	$60	$13	$336	$884	$589	$416	$1,005
Baltimore	480	$60	$15	$377	932	594	466	1,060
Boston	465	$60	$19	$469	1013	578	581	1,159
Buffalo	490	$60	$18	$435	1002	605	538	1,143
Chicago	465	$60	$20	$505	1050	578	625	1,203
Cincinnati	420	$60	$15	$385	880	528	476	1,004
Cleveland	480	$60	$18	$435	993	594	539	1,133
Columbus	480	$60	$15	$391	946	594	483	1,077
Dallas	440	$60	$12	$300	812	550	372	922
Denver	450	$60	$13	$323	846	561	400	961
Detroit	410	$60	$18	$435	922	517	538	1,055
Houston	430	$60	$12	$302	804	539	373	912
Indianapolis	470	$60	$15	$389	934	583	481	1,064
Kansas City	440	$60	$15	$384	900	550	476	1,026
Los Angeles	465	$60	$19	$474	1018	578	586	1,164
Memphis	470	$60	$12	$304	846	583	376	959
Milwaukee	485	$60	$18	$431	994	600	534	1,134
Minneapolis	460	$60	$18	$434	972	572	537	1,109
Nashville	450	$60	$13	$316	838	561	391	952
New Orleans	440	$60	$11	$287	798	550	355	905
New York	460	$60	$27	$657	1203	572	813	1,385
Philadelphia	485	$60	$19	$472	1036	600	584	1,184
Phoenix	480	$60	$14	$349	903	594	431	1,025
Pittsburgh	500	$60	$18	$437	1014	616	541	1,157
St. Louis	480	$60	$17	$428	985	594	529	1,123
San Antonio	405	$60	$11	$277	753	512	342	854
San Diego	490	$60	$19	$474	1043	605	586	1,191
San Francisco	500	$60	$19	$477	1056	616	590	1,206
Seattle	450	$60	$17	$427	954	561	528	1,089
Washington, D.C.	455	$60	$14	$358	887	567	442	1,009
Average per ton	$460	$60	$16	$405	941	$575	$501	$1,072

Note: For field fabrication, add $75 to $100 per ton to shop fabrication costs.
Base price of A615, Grade 40 is $284 per ton.

Place and tie costs do not include crane charge.

Bare Material Costs		Total Installed Costs		
Item	Cost per Ton	Description	Bare Costs	Cost Incl. Subs O&P
Base price at mill	$267	Unload, sort & pile, C-5 crew	$ 16	$ 25
Grade 60 extra	13	Place & tie, 14 hours at $30.79 per hour	$405	476
Size and length extras	30			
Freight from mill	26	Total Installation Cost	$421	$ 501
Warehouse handling & storage	22			
Shearing and shop bending	44	Bare Material Cost	$520	
Drafting, detailing and listing	34	Material Cost + 10%		$ 572
Trucking to job site	24	Total Installed Cost	$941	$1,073
Reinforcing Delivered	$460			
Accessories Delivered	60			
Total Material Delivered	$520			

Engineering or design fees are not included above.

Material and erection costs can be considerably higher than those above for small jobs consisting primarily of smaller bars.

Concrete | R032 | Concrete Reinforcement

R03220-030 Common Stock Styles of Welded Wire Fabric

This table provides some of the basic specifications, sizes, and weights of welded wire fabric used for reinforcing concrete.

	New Designation Spacing — Cross Sectional Area (in.) — (Sq. in. 100)	Old Designation Spacing — Wire Gauge (in.) — (AS & W)		Steel Area per Foot				Approximate Weight per 100 S.F.	
				Longitudinal		Transverse			
				in.	cm	in.	cm	lbs	kg
Rolls	6 x 6 — W1.4 x W1.4	6 x 6 — 10 x 10		.028	.071	.028	.071	21	9.53
	6 x 6 — W2.0 x W2.0	6 x 6 — 8 x 8	1	.040	.102	.040	.102	29	13.15
	6 x 6 — W2.9 x W2.9	6 x 6 — 6 x 6		.058	.147	.058	.147	42	19.05
	6 x 6 — W4.0 x W4.0	6 x 6 — 4 x 4		.080	.203	.080	.203	58	26.91
	4 x 4 — W1.4 x W1.4	4 x 4 — 10 x 10		.042	.107	.042	.107	31	14.06
	4 x 4 — W2.0 x W2.0	4 x 4 — 8 x 8	1	.060	.152	.060	.152	43	19.50
	4 x 4 — W2.9 x W2.9	4 x 4 — 6 x 6		.087	.227	.087	.227	62	28.12
	4 x 4 — W4.0 x W4.0	4 x 4 — 4 x 4		.120	.305	.120	.305	85	38.56
Sheets	6 x 6 — W2.9 x W2.9	6 x 6 — 6 x 6		.058	.147	.058	.147	42	19.05
	6 x 6 — W4.0 x W4.0	6 x 6 — 4 x 4		.080	.203	.080	.203	58	26.31
	6 x 6 — W5.5 x W5.5	6 x 6 — 2 x 2	2	.110	.279	.110	.279	80	36.29
	4 x 4 — W1.4 x W1.4	4 x 4 — 4 x 4		.120	.305	.120	.305	85	38.56

NOTES: 1. Exact W—number size for 8 gauge is W2.1
2. Exact W—number size for 2 gauge is W5.4

Concrete | R033 | Cast-In-Place Concrete

R03310-010 Proportionate Quantities

The tables below show both quantities per S.F. of floor areas as well as form and reinforcing quantities per C.Y. Unusual structural requirements would increase the ratios below. High strength reinforcing would reduce the steel weights. Figures are for 3000 psi concrete and 60,000 psi reinforcing unless specified otherwise.

Type of Construction	Live Load	Span	Per S.F. of Floor Area				Per C.Y. of Concrete		
			Concrete	Forms	Reinf.	Pans	Forms	Reinf.	Pans
Flat Plate	50 psf	15 Ft.	.46 C.F.	1.06 S.F.	1.71 lb.		62 S.F.	101 lb.	
		20	.63	1.02	2.40		44	104	
		25	.79	1.02	3.03		35	104	
	100	15	.46	1.04	2.14		61	126	
		20	.71	1.02	2.72		39	104	
		25	.83	1.01	3.47		33	113	
Flat Plate (waffle construction) 20" domes	50	20	.43	1.00	2.10	.84 S.F.	63	135	53 S.F.
		25	.52	1.00	2.90	.89	52	150	46
		30	.64	1.00	3.70	.87	42	155	37
	100	20	.51	1.00	2.30	.84	53	125	45
		25	.64	1.00	3.20	.83	42	135	35
		30	.76	1.00	4.40	.81	36	160	29
Waffle Construction 30" domes	50	25	.69	1.06	1.83	.68	42	72	40
		30	.74	1.06	2.39	.69	39	87	39
		35	.86	1.05	2.71	.69	33	85	39
		40	.78	1.00	4.80	.68	35	165	40
Flat Slab (two way with drop panels)	50	20	.62	1.03	2.34		45	102	
		25	.77	1.03	2.99		36	105	
		30	.95	1.03	4.09		29	116	
	100	20	.64	1.03	2.83		43	119	
		25	.79	1.03	3.88		35	133	
		30	.96	1.03	4.66		29	131	
	200	20	.73	1.03	3.03		38	112	
		25	.86	1.03	4.23		32	133	
		30	1.06	1.03	5.30		26	135	
One Way Joists 20" Pans	50	15	.36	1.04	1.40	.93	78	105	70
		20	.42	1.05	1.80	.94	67	120	60
		25	.47	1.05	2.60	.94	60	150	54
	100	15	.38	1.07	1.90	.93	77	140	66
		20	.44	1.08	2.40	.94	67	150	58
		25	.52	1.07	3.50	.94	55	185	49
One Way Joists 8" x 16" filler blocks	50	15	.34	1.06	1.80	.81 Ea.	84	145	64 Ea.
		20	.40	1.08	2.20	.82	73	145	55
		25	.46	1.07	3.20	.83	63	190	49
	100	15	.39	1.07	1.90	.81	74	130	56
		20	.46	1.09	2.80	.82	64	160	48
		25	.53	1.10	3.60	.83	56	190	42
One Way Beam & Slab	50	15	.42	1.30	1.73		84	111	
		20	.51	1.28	2.61		68	138	
		25	.64	1.25	2.78		53	117	
	100	15	.42	1.30	1.90		84	122	
		20	.54	1.35	2.69		68	154	
		25	.69	1.37	3.93		54	145	
	200	15	.44	1.31	2.24		80	137	
		20	.58	1.40	3.30		65	163	
		25	.69	1.42	4.89		53	183	
Two Way Beam & Slab	100	15	.47	1.20	2.26		69	130	
		20	.63	1.29	3.06		55	131	
		25	.83	1.33	3.79		43	123	
	200	15	.49	1.25	2.70		41	149	
		20	.66	1.32	4.04		54	165	
		25	.88	1.32	6.08		41	187	

Concrete

R033 Cast-In-Place Concrete

R03310-010 Proportionate Quantities (cont.)

4000 psi Concrete and 60,000 psi Reinforcing—Form and Reinforcing Quantities per C.Y.					
Item	Size	Forms	Reinforcing	Minimum	Maximum
Columns (square tied)	10" x 10"	130 S.F.C.A.	#5 to #11	220 lbs.	875 lbs.
	12" x 12"	108	#6 to #14	200	955
	14" x 14"	92	#7 to #14	190	900
	16" x 16"	81	#6 to #14	187	1082
	18" x 18"	72	#6 to #14	170	906
	20" x 20"	65	#7 to #18	150	1080
	22" x 22"	59	#8 to #18	153	902
	24" x 24"	54	#8 to #18	164	884
	26" x 26"	50	#9 to #18	169	994
	28" x 28"	46	#9 to #18	147	864
	30" x 30"	43	#10 to #18	146	983
	32" x 32"	40	#10 to #18	175	866
	34" x 34"	38	#10 to #18	157	772
	36" x 36"	36	#10 to #18	175	852
	38" x 38"	34	#10 to #18	158	765
	40" x 40"	32	#10 to #18	143	692

Item	Size	Form	Spiral	Reinforcing	Minimum	Maximum
Columns (spirally reinforced)	12" diameter	34.5 L.F.	190 lbs.	#4 to #11	165 lbs.	1505 lb.
		34.5	190	#14 & #18	—	1100
	14"	25	170	#4 to #11	150	970
		25	170	#14 & #18	800	1000
	16"	19	160	#4 to #11	160	950
		19	160	#14 & #18	605	1080
	18"	15	150	#4 to #11	160	915
		15	150	#14 & #18	480	1075
	20"	12	130	#4 to #11	155	865
		12	130	#14 & #18	385	1020
	22"	10	125	#4 to #11	165	775
		10	125	#14 & #18	320	995
	24"	9	120	#4 to #11	195	800
		9	120	#14 & #18	290	1150
	26"	7.3	100	#4 to #11	200	729
		7.3	100	#14 & #18	235	1035
	28"	6.3	95	#4 to #11	175	700
		6.3	95	#14 & #18	200	1075
	30"	5.5	90	#4 to #11	180	670
		5.5	90	#14 & #18	175	1015
	32"	4.8	85	#4 to #11	185	615
		4.8	85	#14 & #18	155	955
	34"	4.3	80	#4 to #11	180	600
		4.3	80	#14 & #18	170	855
	36"	3.8	75	#4 to #11	165	570
		3.8	75	#14 & #18	155	865
	40"	3.0	70	#4 to #11	165	500
		3.0	70	#14 & #18	145	765

Concrete R033 Cast-In-Place Concrete

R03310-010 Proportionate Quantities (cont.)

3000 psi Concrete and 60,000 psi Reinforcing—Form and Reinforcing Quantities per C.Y.

Item	Type	Loading	Height	C.Y./L.F.	Forms/C.Y.	Reinf./C.Y.
Retaining Walls	Cantilever	Level Backfill	4 Ft.	0.2 C.Y.	49 S.F.	35 lbs.
			8	0.5	42	45
			12	0.8	35	70
			16	1.1	32	85
			20	1.6	28	105
		Highway Surcharge	4	0.3	41	35
			8	0.5	36	55
			12	0.8	33	90
			16	1.2	30	120
			20	1.7	27	155
		Railroad Surcharge	4	0.4	28	45
			8	0.8	25	65
			12	1.3	22	90
			16	1.9	20	100
			20	2.6	18	120
	Gravity, with Vertical Face	Level Backfill	4	0.4	37	None
			7	0.6	27	
			10	1.2	20	
		Sloping Backfill	4	0.3	31	
			7	0.8	21	
			10	1.6	15	↓

		Live Load in Kips per Linear Foot						
	Span	Under 1 Kip		2 to 3 Kips		4 to 5 Kips		6 to 7 Kips	
		Forms	Reinf.	Forms	Reinf.	Forms	Reinf.	Forms	Reinf.
Beams	10 Ft.	—	—	90 S.F.	170 #	85 S.F.	175 #	75 S.F.	185 #
	16	130 S.F.	165 #	85	180	75	180	65	225
	20	110	170	75	185	62	200	51	200
	26	90	170	65	215	62	215	—	—
	30	85	175	60	200	—	—	—	—

Item	Size	Type	Forms per C.Y.	Reinforcing per C.Y.
Spread Footings	Under 1 C.Y.	1,000 psf soil	24 S.F.	44 lbs.
		5,000	24	42
		10,000	24	52
	1 C.Y. to 5 C.Y.	1,000	14	49
		5,000	14	50
		10,000	14	50
	Over 5 C.Y.	1,000	9	54
		5,000	9	52
		10,000	9	56
Pile Caps (30 Ton Concrete Piles)	Under 5 C.Y.	shallow caps	20	65
		medium	20	50
		deep	20	40
	5 C.Y. to 10 C.Y.	shallow	14	55
		medium	15	45
		deep	15	40
	10 C.Y. to 20 C.Y.	shallow	11	60
		medium	11	45
		deep	12	35
	Over 20 C.Y.	shallow	9	60
		medium	9	45
		deep	10	40

Concrete — R033 Cast-In-Place Concrete

R03310-010 Proportionate Quantities (cont.)

	3000 psi Concrete and 60,000 psi Reinforcing — Form and Reinforcing Quantities per C.Y.					
Item	Pile Spacing		50 T Pile	100 T Pile	50 T Pile	100 T Pile
Pile Caps (Steel H Piles)	Under 5 C.Y.	24" O.C.	24 S.F.	24 S.F.	75 lbs.	90 lbs.
		30"	25	25	80	100
		36"	24	24	80	110
	5 C.Y. to 10 C.Y.	24"	15	15	80	110
		30"	15	15	85	110
		36"	15	15	75	90
	Over 10 C.Y.	24"	13	13	85	90
		30"	11	11	85	95
		36"	10	10	85	90

		8" Thick		10" Thick		12" Thick		15" Thick	
	Height	Forms	Reinf.	Forms	Reinf.	Forms	Reinf.	Forms	Reinf.
Basement Walls	7 Ft.	81 S.F.	44 lbs.	65 S.F.	45 lbs.	54 S.F.	44 lbs.	41 S.F.	43 lbs.
	8	↓	44	↓	45	↓	44	↓	43
	9		46		45		44		43
	10		57		45		44		43
	12		83		50		52		43
	14		116		65		64		51
	16				86		90		65
	18						106	↓	70

R03310-020 Materials for One C.Y. of Concrete

This is an approximate method of figuring quantities of cement, sand and coarse aggregate for a field mix with waste allowance included.

With crushed gravel as coarse aggregate, to determine barrels of cement required, divide 10 by total mix; that is, for 1:2:4 mix, 10 divided by 7 = 1-3/7 barrels.

If the coarse aggregate is crushed stone, use 10-1/2 instead of 10 as given for gravel.

To determine tons of sand required, multiply barrels of cement by parts of sand and then by 0.2; that is, for the 1:2:4 mix, as above, 1-3/7 x 2 x .2 = .57 tons.

Tons of crushed gravel are in the same ratio to tons of sand as parts in the mix, or 4/2 x .57 = 1.14 tons.

1 bag cement = 94#	1 C.Y. sand or crushed gravel = 2700#	1 C.Y. crushed stone = 2575#
4 bags = 1 barrel	1 ton sand or crushed gravel = 20 C.F.	1 ton crushed stone = 21 C.F.

Average carload of cement is 692 bags; of sand or gravel is 56 tons.
Do not stack stored cement over 10 bags high.

R03310-030 Metric Equivalents of Cement Content for Concrete Mixes

94 Pound Bags per Cubic Yard	Kilograms per Cubic Meter	94 Pound Bags per Cubic Yard	Kilograms per Cubic Meter
1.0	55.77	7.0	390.4
1.5	83.65	7.5	418.3
2.0	111.5	8.0	446.2
2.5	139.4	8.5	474.0
3.0	167.3	9.0	501.9
3.5	195.2	9.5	529.8
4.0	223.1	10.0	557.7
4.5	251.0	10.5	585.6
5.0	278.8	11.0	613.5
5.5	306.7	11.5	641.3
6.0	334.6	12.0	669.2
6.5	362.5	12.5	697.1

a. If you know the cement content in pounds per cubic yard, multiply by .5933 to obtain kilograms per cubic meter.

b. If you know the cement content in 94 pound bags per cubic yard, multiply by 55.77 to obtain kilograms per cubic meter.

Concrete — R033 Cast-In-Place Concrete

R03310-040 Metric Equivalents of Common Concrete Strengths
(to convert other psi values to megapascals, multiply by 0.006895)

U.S. Values psi	SI Value Megapascals	Non-SI Metric Value kgf/cm²*
2000	14	140
2500	17	175
3000	21	210
3500	24	245
4000	28	280
4500	31	315
5000	34	350
6000	41	420
7000	48	490
8000	55	560
9000	62	630
10,000	69	705

* kilograms force per square centimeter

R03310-050 Quantities of Cement, Sand and Stone for One C.Y. of Concrete per Various Mixes

This table can be used to determine the quantities of the ingredients for smaller quantities of site mixed concrete.

Concrete (C.Y.)	Mix = 1:1:1-3/4			Mix = 1:2:2.25			Mix = 1:2.25:3			Mix = 1:3:4		
	Cement (sacks)	Sand (C.Y.)	Stone (C.Y.)	Cement (sacks)	Sand (C.Y.)	Stone (C.Y.)	Cement (sacks)	Sand (C.Y.)	Stone (C.Y.)	Cement (sacks)	Sand (C.Y.)	Stone (C.Y.)
1	10	.37	.63	7.75	.56	.65	6.25	.52	.70	5	.56	.74
2	20	.74	1.26	15.50	1.12	1.30	12.50	1.04	1.40	10	1.12	1.48
3	30	1.11	1.89	23.25	1.68	1.95	18.75	1.56	2.10	15	1.68	2.22
4	40	1.48	2.52	31.00	2.24	2.60	25.00	2.08	2.80	20	2.24	2.96
5	50	1.85	3.15	38.75	2.80	3.25	31.25	2.60	3.50	25	2.80	3.70
6	60	2.22	3.78	46.50	3.36	3.90	37.50	3.12	4.20	30	3.36	4.44
7	70	2.59	4.41	54.25	3.92	4.55	43.75	3.64	4.90	35	3.92	5.18
8	80	2.96	5.04	62.00	4.48	5.20	50.00	4.16	5.60	40	4.48	5.92
9	90	3.33	5.67	69.75	5.04	5.85	56.25	4.68	6.30	45	5.04	6.66
10	100	3.70	6.30	77.50	5.60	6.50	62.50	5.20	7.00	50	5.60	7.40
11	110	4.07	6.93	85.25	6.16	7.15	68.75	5.72	7.70	55	6.16	8.14
12	120	4.44	7.56	93.00	6.72	7.80	75.00	6.24	8.40	60	6.72	8.88
13	130	4.82	8.20	100.76	7.28	8.46	81.26	6.76	9.10	65	7.28	9.62
14	140	5.18	8.82	108.50	7.84	9.10	87.50	7.28	9.80	70	7.84	10.36
15	150	5.56	9.46	116.26	8.40	9.76	93.76	7.80	10.50	75	8.40	11.10
16	160	5.92	10.08	124.00	8.96	10.40	100.00	8.32	11.20	80	8.96	11.84
17	170	6.30	10.72	131.76	9.52	11.06	106.26	8.84	11.90	85	9.52	12.58
18	180	6.66	11.34	139.50	10.08	11.70	112.50	9.36	12.60	90	10.08	13.32
19	190	7.04	11.98	147.26	10.64	12.36	118.76	9.84	13.30	95	10.64	14.06
20	200	7.40	12.60	155.00	11.20	13.00	125.00	10.40	14.00	100	11.20	14.80
21	210	7.77	13.23	162.75	11.76	13.65	131.25	10.92	14.70	105	11.76	15.54
22	220	8.14	13.86	170.05	12.32	14.30	137.50	11.44	15.40	110	12.32	16.28
23	230	8.51	14.49	178.25	12.88	14.95	143.75	11.96	16.10	115	12.88	17.02
24	240	8.88	15.12	186.00	13.44	15.60	150.00	12.48	16.80	120	13.44	17.76
25	250	9.25	15.75	193.75	14.00	16.25	156.25	13.00	17.50	125	14.00	18.50
26	260	9.64	16.40	201.52	14.56	16.92	162.52	13.52	18.20	130	14.56	19.24
27	270	10.00	17.00	209.26	15.12	17.56	168.76	14.04	18.90	135	15.02	20.00
28	280	10.36	17.64	217.00	15.68	18.20	175.00	14.56	19.60	140	15.68	20.72
29	290	10.74	18.28	224.76	16.24	18.86	181.26	15.08	20.30	145	16.24	21.46

Concrete | R033 | Cast-In-Place Concrete

R03310-060 Concrete Material Net Prices

Costs below are C.Y. of concrete delivered; per ton of bulk cement; per bag cement delivered T.L.L.; per ton for stone and sand aggregates loaded at plant (no trucking included) and per 4 C.F. bag for perlite or vermiculite aggregate delivered T.L.L.

City	Ready Mix Concrete Regular Weight		Cement T.L. Lots		Aggregates per Ton			Vermiculite or Perlite 4 C.F. Bag
	3000 psi	5000 psi	Bulk per Ton	Bags per Bag	Crushed Stone 1-1/2"	3/4"	Sand	
Atlanta	$64.50	$69.50	$73.25	$6.45	$13.95	$14.25	$12.84	$10.35
Baltimore	66.50	71.50	78.10	6.85	9.45	10.00	9.07	8.85
Boston	64.50	70.50	85.20	7.50	11.60	12.80	10.02	9.75
Buffalo	83.50	89.50	90.45	7.75	11.85	12.30	11.74	10.10
Chicago	61.00	67.00	84.50	7.20	9.40	9.70	7.64	8.70
Cincinnati	56.00	61.00	79.45	6.75	8.45	10.45	6.68	10.15
Cleveland	60.00	66.00	85.75	7.30	13.65	14.20	9.45	9.95
Columbus	55.00	59.00	83.25	7.40	12.25	12.50	6.44	10.10
Dallas	66.00	71.00	78.20	6.60	16.35	16.60	6.73	9.80
Denver	63.00	71.00	83.25	8.10	10.30	10.50	7.64	10.20
Detroit	64.00	69.00	82.00	7.20	8.95	9.15	8.30	10.00
Houston	62.00	68.00	86.05	7.35	18.50	18.70	10.02	8.95
Indianapolis	61.00	69.00	80.70	6.90	8.00	8.55	9.45	9.95
Kansas City	64.00	69.00	75.65	6.45	7.95	8.30	6.68	10.20
Los Angeles	53.00	61.00	88.30	7.90	13.10	13.70	11.93	10.40
Memphis	62.00	66.00	78.20	6.75	11.95	12.30	8.92	9.60
Milwaukee	64.00	69.00	88.45	7.55	8.85	9.25	8.16	9.40
Minneapolis	68.00	73.00	85.50	7.30	14.85	15.10	9.40	9.55
Nashville	56.00	61.00	75.65	6.50	9.40	9.60	11.07	9.10
New Orleans	60.00	68.00	79.10	6.70	15.40	16.00	12.46	9.80
New York City	83.00	91.00	89.55	7.70	24.70	26.30	13.84	10.35
Philadelphia	55.50	63.50	82.00	7.15	11.10	11.70	7.16	9.15
Phoenix	61.00	66.00	80.00	7.25	11.35	11.85	10.98	9.70
Pittsburgh	58.00	62.00	76.95	6.65	15.05	15.20	10.02	9.55
St. Louis	53.00	59.00	78.20	6.65	9.85	10.75	9.12	9.80
San Antonio	52.00	56.00	75.65	6.45	8.00	8.20	5.63	8.90
San Diego	64.00	70.00	85.75	7.30	9.30	9.50	7.73	8.95
San Francisco	78.00	83.00	89.25	7.65	18.80	20.85	11.93	9.55
Seattle	64.00	70.00	83.25	7.15	15.05	15.50	11.36	9.95
Washington, D.C.	68.00	78.00	82.00	7.05	14.00	15.00	10.74	10.20
Average	$63.00	$68.90	$82.10	$7.10	$12.40	$12.95	$ 9.45	$ 9.70

R03310-070 Ready Mix Material Prices

Table below lists national average prices per C.Y. of concrete. Prices in the key cities for different strengths can be closely estimated by factoring against the 3000 psi or 5000 psi price from R03310-060 above or from the City Cost Indexes, cast-in-place concrete material factor.

Strength in psi	Design Mix Nominal Mix	Bags per C.Y.	Heavy Weight Regular	High Early	Light Weight 110# per C.F.	All Light Weight	Admixtures and Special Items Add to Each C.Y. of Concrete for the Following Items:		
2,000	1:3:5	4.5	$59.85	$65.70	$74.80	$ 89.80	Calcium chloride, 1%	$1.75	per C.Y.
2,500	1:2-½:4-½	5	61.10	67.60	76.40	91.65	" 2%	3.00	per C.Y.
3,000	1:2:4	5.5	63.00	70.15	78.75	94.50	Water reducing agent	.25	per bag
3,500	1:2:3-½	6	64.60	72.40	80.75	96.90	Set retarder	.62	per bag
3,750	1:2:3-¼	6.3	65.50	73.70	81.90	98.25	High early cement	1.30	per bag
4,000	1:2:3	6.5	66.15	74.60	82.70	99.25	White cement	9.60	per bag
4,500	1:2:2-½	7	68.05	77.15	85.05	102.10	Pump aid	2.15	per C.Y.
5,000	1:1-½:2-½	7.5	69.00	78.75	86.25	103.50	Winter concrete	4.60	per C.Y.
	1:1:2	8	73.85	81.25	—	Perlite	Synthetic Fiber Reinforcing	8.50	per C.Y.
	1:4 topping	7.5	70.15	77.15	—	1:6=			
	1:3 topping	8.5	75.35	82.90	—	$ 92.05			
	1:2 topping	11	81.95	90.15	—				

Concrete — R033 Cast-In-Place Concrete

R03310-080 Field Mix Concrete

Presently most building jobs are built with ready mix concrete except for isolated locations and some larger jobs requiring over 10,000 C.Y. where land is readily available for setting up a temporary batch plant.

The most economical mix is a controlled mix using local aggregate proportioned by trial to give the required strength with the least cost of material.

Costs tabulated below are based on a job that requires 10,000 C.Y. of concrete to be placed within an eight month period.

28 Day Strength	3000 psi		2250 psi	
Mix Proportions	1:2-½:3		1:2-¾:4	
Cement, trucked in bulk @ $4.10 per Cwt.	6-1/4 bags @ 94 lb./bag	$24.09	5 bags @ 94 lb./bag	$19.27
Sand @ $9.45 per ton	.68 tons	6.43	.7 tons	6.62
Stone @ $14.25 per ton, maximum size aggregate 3/4"	.9 tons	12.83	1.02 tons	14.54
Set up plant, foundations, piping, etc.		2.00		2.00
Plant, trucks and equipment rental		9.41		9.41
Labor to mix and transport to forms		12.82		12.82
Supervision		2.78		2.78
Cost per C.Y. FOB Forms		$70.35		$67.42

See R03310-060 for material prices in the major cities.

R03310-090 Placing Ready Mixed Concrete

For ground pours allow for 5% waste when figuring quantities.

Prices in the front of the book assume normal deliveries. If deliveries are made before 8 A.M. or after 5 P.M. or on Saturday afternoons add $25 per C.Y. Large volume discounts are not included in prices in front of book.

For the lower floors without truck access, concrete may be wheeled in rubber tired buggies, conveyer handled, crane handled or pumped. Pumping is economical if there is top steel. Conveyers are more efficient for thick slabs. Concrete pump with an operator can be rented from $620 per day for up to 25 C.Y. to $1,500 per day for a 400 C.Y. pour. Figures include travel time if done at straight time. Pumping lightweight concrete costs an extra $2.40 per C.Y.

At higher floors the rubber tired buggies may be hoisted by a hoisting tower then wheeled to location. Placement by a conveyer is limited to three floors and is best for high volume pours. Pumped concrete is best when building has no crane access. Concrete may be pumped directly as high as thirty-six stories using special pumping techniques. Normal maximum height is about fifteen stories.

Best pumping aggregate is screened and graded bank gravel rather than crushed stone.

Pumping downward is more difficult than pumping upwards. Horizontal distance from pump to pour may increase preparation time prior to pour. Placing by cranes, either mobile, climbing or tower types continues as the most efficient method for high rise concrete buildings.

	Cost per C.Y. for Wheeled Concrete, Dumped Only (Add to appropriate placing cost)							
	10 C.F. Walking Cart				18 C.F. Riding Cart			
Item	Hourly Cost	Wheeled up to 50 ft.	Wheeled up to 150 ft.	Wheeled up to 250 ft.	Hourly Cost	Wheeled up to 50 ft.	Wheeled up to 150 ft.	Wheeled up to 250 ft.
Laborer	$14.45	$3.61	$4.83	$ 6.43	$14.45	$1.45	$1.88	$2.46
.125 Labor foreman	2.06	.51	.69	.92	2.06	.21	.27	.35
Concrete cart	6.50	1.63	2.17	2.89	10.40	1.04	1.35	1.77
Total Cost/C.Y.		$5.75	$7.69	$10.24		$2.70	$3.50	$4.58
Hourly production		4 C.Y.				10 C.Y.		

Concrete | R033 | Cast-In-Place Concrete

R03310-100 Average C.Y. of Concrete

Rubbing and floor finish not included — 4 uses of forms assumed, 5 story building.

Item	Description	Strength in psi	Ready Mix	Place	Forms	Reinforcing	Total
Beams	10' span	4000	$66.00	$43.80	$295.32	$ 82.25	$487.37
5 kip/L.F.	25'		66.00	29.20	269.64	101.05	465.89
Beam & Slab, 1 way	15'	4000	66.00	13.48	277.20	46.36	403.04
125 psf Sup. L.	25'		66.00	11.68	178.20	58.52	314.40
Beam & Slab, 2 way	15'	4000	66.00	13.48	229.08	49.40	357.96
125 psf Sup. L.	25'		66.00	11.68	142.76	46.74	267.18
Columns,	16" x 16"	4000	66.00	29.20	259.20	259.94	614.34
square tied	24" x 24"		66.00	19.04	172.26	214.84	472.14
Columns, Tied	16" diameter	4000	66.00	29.20	178.98	248.46	522.64
reinforced	24" diameter		66.00	19.04	127.98	246.00	459.02
Flat Plate,	15' span	4000	66.00	15.93	141.52	47.88	271.33
125 psf Sup. L.	25'		66.00	13.48	76.56	42.94	198.98
Flat Slab with drops,	20'	4000	66.00	15.93	110.08	45.22	237.23
125 psf Sup. L.	30'		66.00	13.48	74.24	49.78	203.50
Grade Wall,	8" thick	3000	63.00	9.12	211.41	16.28	299.81
8' high	15" thick		63.00	7.82	107.01	15.91	193.74
Metal Pan Joists,	15' span	4000	66.00	13.48	203.91	17.39	300.78
125 psf Sup. L.	25' span		66.00	11.68	168.35	29.97	276.00
Pile Caps	under 5 C.Y.	3000	66.15	9.12	43.40	18.20	136.87
"	over 10 C.Y.		66.15	3.82	21.70	16.80	108.47
Slab	4" thick	3500	68.25	7.47	11.31	12.15	99.18
on Grade	6" thick		68.25	4.98	7.38	8.10	88.71
Spread	under 1 C.Y.	3000	63.00	14.93	45.84	19.32	143.09
Footings	over 5 C.Y.		63.00	7.47	17.19	22.68	110.34
Strip	9" x 18" plain	3000	63.00	7.47	56.52	—	126.99
Footings	12" x 36" reinforced		63.00	5.30	28.26	16.80	113.36
Waffle 30" Domes	20' span	4000	66.00	13.48	170.82	52.43	302.73
125 psf Sup. L.	30' span		66.00	11.68	135.78	49.98	263.44

*Placement by direct chute assumed. All others, placement by pump assumed.

When form and reinforcing quantity ratios are available compute the C.Y. costs directly from the unit prices in the front of the book. See also R03310-010 for relative quantities per C.Y. for additional items. Higher strength concrete would change the ratios below. The tables below show typical examples of how the total costs per C.Y. have been derived. Ready mix concrete prices are U.S. average and equipment handling is assumed.

Beams 5 kip/L.F.	10' Span				25' Span			
	Material		Installation		Material		Installation	
4000 psi concrete	1 C.Y.	$ 66.00	1 C.Y.	$ 43.80	1 C.Y.	$ 66.00	1 C.Y.	$ 29.20
Formwork, 4 uses	69 S.F. @ .99	68.31	69 S.F. @ 3.29	227.01	63 S.F. @ .99	62.37	63 S.F.@ 3.29	207.27
Reinforcing steel	175 # @ .26	45.50	175 # @ .21	36.75	215 # @ .26	55.90	215 # @ .21	45.15
Total per C.Y.		$179.81		$307.56		$184.27		$281.62

Beam & Slab, One Way	125 psf, 15' Span				125 psf, 25' Span			
	Material		Installation		Material		Installation	
4000 psi concrete	1 C.Y.	$ 66.00	1 C.Y.	$ 13.48	1 C.Y.	$ 66.00	1 C.Y.	$ 11.68
Formwork, 4 uses	84 S.F. @ 1.04	87.36	84 S.F. @ 2.26	189.84	54 S.F. @ 1.04	56.16	54 S.F. @ 2.26	122.04
Reinforcing steel	122 # @ .26	31.72	122 # @ .12	14.64	154 # @ .26	40.04	154 # @ .12	18.48
Total per C.Y.		$185.08		$217.96		$162.20		$152.20

Beam & Slab, Two Way	125 psf, 15' Span				125 psf, 25' Span			
	Material		Installation		Material		Installation	
4000 psi concrete	1 C.Y.	$ 66.00	1 C.Y.	$ 13.48	1 C.Y.	$ 66.00	1 C.Y.	$ 11.68
Formwork, 4 uses	69 S.F. @ .95	65.55	69 S.F. @ 2.37	163.53	43 S.F. @ .95	40.85	43 S.F. @ 2.37	101.91
Reinforcing steel	130 # @ .26	33.80	130 # @ .12	15.60	123 # @ .26	31.98	123 # @ .12	14.76
Total per C.Y.		$165.35		$192.61		$138.83		$128.35

Concrete | R033 | Cast-In-Place Concrete

R03310-100 Average C.Y. of Concrete (cont.)

Columns, Square Tied	16" Square				24" Square			
	Material		Installation		Material		Installation	
4000 psi concrete	1 C.Y.		$ 66.00	1 C.Y.		$ 29.20	1 C.Y.		$ 66.00	1 C.Y.		$ 19.04
Formwork, 4 uses	81 S.F. @	.86	69.66	81 S.F. @	2.34	189.54	54 S.F. @	.86	46.44	54 S.F. @	2.33	125.82
Reinforcing steel, avg.	634 # @	.26	164.84	634 # @	.15	95.10	524 # @	.26	136.24	524 # @	.15	78.60
Total per C.Y.			$300.50			$313.84			$248.68			$223.46

Note: Reinforcing of 16" and 24" square columns can vary from 144 lb. to 972 lb. per C.Y.

Columns, Round Tied Reinforced	16" Diameter						24" Diameter					
	Material			Installation			Material			Installation		
4000 psi concrete	1 C.Y.		$ 66.00	1 C.Y.		$ 29.20	1 C.Y.		$ 66.00	1 C.Y.		$ 19.04
Formwork, fiber forms	19 L.F. @	5.50	104.50	19 L.F. @	3.92	74.48	9 L.F. @	10.00	90.00	9 L.F. @	4.22	37.98
Reinforcing steel, avg.	606 # @	.26	157.56	606 # @	.15	90.90	600 # @	.26	156.00	600 # @	.15	90.00
Ties	70 # @	.52	36.40	70 # @	.23	16.10	18 # @	.52	9.36	18 # @	.23	4.14
Total per C.Y.			$364.46			$210.68			$321.36			$151.16

Note: Reinforcing of 16" and 24" diameter columns vary from 160 lb. to 1150 lb. per C.Y.

Flat Plate	125 psf, 15' Span						125 psf, 25' Span					
	Material			Installation			Material			Installation		
4000 psi concrete	1 C.Y.		$ 66.00	1 C.Y.		$ 15.93	1 C.Y.		$ 66.00	1 C.Y.		$13.48
Formwork, 4 uses	61 S.F. @	.82	50.02	61 S.F. @	1.50	91.50	33 S.F. @	.82	27.06	33 S.F. @	1.50	49.50
Reinforcing steel	126 # @	.26	32.76	126 # @	.12	15.12	113 # @	.26	29.38	113 # @	.12	13.56
Total per C.Y.			$148.78			$122.55			$122.44			$76.54

Flat Slab with drops	125 psf, 20' Span						125 psf, 30' Span					
	Material			Installation			Material			Installation		
4000 psi concrete	1 C.Y.		$ 66.00	1 C.Y.		$15.93	1 C.Y.		$ 66.00	1 C.Y.		$13.48
Formwork, 4 uses	43 S.F. @	1.01	43.43	43 S.F. @	1.55	66.65	29 S.F. @	1.01	29.29	29 S.F. @	1.55	44.95
Reinforcing steel	119 # @	.26	30.94	119 # @	.12	14.28	131 # @	.26	34.06	131 # @	.12	15.72
Total per C.Y.			$140.37			$96.86			$129.35			$74.15

Grade Wall, 8' High	8" Thick						15" Thick					
	Material			Installation			Material			Installation		
3000 psi concrete	1 C.Y.		$ 63.00	1 C.Y.		$ 9.12	1 C.Y.		$ 63.00	1 C.Y.		$ 7.82
Formwork, 4 uses	81 S.F. @	.86	69.66	81 S.F. @	1.75	141.75	41 S.F. @	.86	35.26	41 S.F. @	1.75	71.75
Reinforcing steel	44 # @	.26	11.44	44 # @	.11	4.84	43 # @	.26	11.18	43 # @	.11	4.73
Total per C.Y.			$144.10			$155.71			$109.44			$84.30

Metal Pan Joists, 30" One Way	125 psf, 15' Span						125 psf, 25' Span					
	Material			Installation			Material			Installation		
4000 psi concrete	1 C.Y.		$ 66.00	1 C.Y.		$ 13.48	1 C.Y.		$ 66.00	1 C.Y.		$11.68
Formwork, 4 uses	49 S.F. @	.84	41.16	49 S.F. @	1.75	85.75	40 S.F. @	.84	33.60	40 S.F. @	1.75	70.00
Reinforcing steel	47 # @	.26	12.22	47 # @	.11	5.17	81 # @	.26	21.06	81 # @	.11	8.91
30" metal pans	44 S.F. @	1.50	66.00	44 S.F. @	.25	11.00	37 S.F. @	1.50	55.50	37 # @	.25	9.25
Total per C.Y.			$185.38			$115.40			$176.16			$99.84

Pile Caps	Under 5 C.Y.						Over 10 C.Y.					
	Material			Installation			Material			Installation		
3000 psi concrete	1.05 C.Y.		$66.15	1 C.Y.		$ 9.12	1.05 C.Y.		$66.15	1 C.Y.		$ 3.82
Formwork, 4 uses	20 S.F. @	.73	14.60	20 S.F. @	1.44	28.80	10 S.F. @	.73	7.30	10 S.F. @	1.44	14.40
Reinforcing steel	52 # @	.26	13.52	52 # @	.09	4.68	48 # @	.26	12.48	48 # @	.09	4.32
Total per C.Y.			$94.27			$42.60			$85.93			$22.54

Slab on Grade	4" Thick						6" Thick					
	Material			Installation			Material			Installation		
3500 psi concrete	1.05 C.Y.		$68.25	1 C.Y.		$ 7.47	1.05 C.Y.		$68.25	1 C.Y.		$ 4.98
Formwork, 4 uses	9.2 L.F. @	.31	2.85	9.2 L.F. @	.92	8.46	6 L.F. @	.31	1.86	6 L.F. @	.92	5.52
W.W.Fabric	81 # @	.07	5.67	81 # @	.08	6.48	54 # @	.07	3.78	54 # @	.08	4.32
Total per C.Y.			$76.77			$22.41			$73.89			$14.82

Concrete | R033 | Cast-In-Place Concrete

R03310-100 Average C.Y. of Concrete (cont.)

Spread Footings	Under 1 C.Y. Material			Under 1 C.Y. Installation			Over 5 C.Y. Material			Over 5 C.Y. Installation		
3000 psi concrete	1.05 C.Y.		$63.00	1.05 C.Y.		$14.93	1.05 C.Y.		$63.00	1 C.Y.		$ 7.47
Formwork, 4 uses	24 S.F. @	.58	13.92	24 S.F. @	1.33	31.92	9 S.F. @	.58	5.22	9 S.F. @	1.33	11.97
Reinforcing steel	46 # @	.26	11.96	46 # @	.16	7.36	54 # @	.26	14.04	54 # @	.16	8.64
Total per C.Y.			$88.88			$54.21			$82.26			$28.08

Strip Footings	9' x 18" Plain Material			9' x 18" Plain Installation			12" x 36" Reinforced Material			12" x 36" Reinforced Installation		
3000 psi concrete	1.05 C.Y.		$63.00	1.05 C.Y.		$ 7.47	1.05 C.Y.		$63.00	1 C.Y.		$ 5.30
Formwork, 4 uses	36 S.F. @	.44	15.84	36 S.F. @	1.13	40.68	18 S.F. @	.44	7.92	18 S.F. @	1.13	20.34
Reinforcing steel			—			—	40 # @	.26	10.40	40 # @	.16	6.40
Total per C.Y.			$78.84			$48.15			$81.32			$32.04

Waffle, 30" Domes	125 psf, 20' Span Material			125 psf, 20' Span Installation			125 psf, 30' Span Material			125 psf, 30' Span Installation		
4000 psi concrete	1 C.Y.		$ 66.00	1 C.Y.		$ 13.48	1 C.Y.		$ 66.00	1 C.Y.		$11.68
Formwork, 4 uses	39 S.F. @	.84	32.76	39 S.F. @	1.79	69.81	31 S.F. @	.84	26.04	31 S.F. @	1.79	55.49
Reinforcing steel & mesh	107 # @	.33	35.31	107 # @	.16	17.12	102 # @	.33	33.66	102 # @	.16	16.32
Metal Domes	39 S.F. @	1.50	58.50	39 S.F. @	.25	9.75	31 S.F. @	1.50	46.50	31 S.F. @	.25	7.75
Total per C.Y.			$192.57			$110.16			$172.20			$91.24

Concrete | R033 | Cast-In-Place Concrete

R03310-120 Lift Slabs

The cost advantage of the lift slab method is due to placing all concrete, reinforcing steel, inserts and electrical conduit at ground level and in reduction of formwork. Minimum economical project size is about 30,000 S.F. Slabs may be tilted for parking garage ramps.

It is now used in all types of buildings and has gone up to 22 stories high in apartment buildings. Current trend is to use post-tensioned flat plate slabs with spans from 22' to 35'. Cylindrical void forms are used when deep slabs are required. One pound of prestressing steel is about equal to seven pounds of conventional reinforcing.

To be considered cured for stressing and lifting, a slab must have attained 75% of design strength. Seven days are usually sufficient with four to five days possible if high early strength cement is used. Slabs can be stacked using two coats of a non-bonding agent to insure that slabs do not stick to each other. Lifting is done by companies specializing in this work. Lift rate is 5' to 15' per hour with an average of 10' per hour. Total areas up to 33,000 S.F. have been lifted at one time. 24 to 36 jacking columns are common. Most economical bay sizes are 24' to 28' with four to fourteen stories most efficient. Continuous design reduces reinforcing steel cost. Use of post-tensioned slabs allows larger bay sizes. Supplementary reinforcing, post-tensioned tendons, accessories and field labor cost about $2.40 per pound of tendon.

Table below shows the usual range and average S.F. cost of a typical lift slab project that has been designed to take advantage of lift slab techniques. Figures include Subs Overhead and Profit.

Typical Cost per S.F. for Lift Slabs Project Incl. Subs O & P

Item	Description	Typical Range			Average	Sub Total
Concrete Slabs	Edge and bulkhead forms	.05	to	.38	$ .22	
	Reinforcing and post-tensioning steel	1.47	to	2.63	2.05	
	Reinforcing and accessories	.06	to	.15	.11	$4.86
	Concrete cast at ground level	1.41	to	2.38	1.90	
	Float or trowel finish	.47	to	.68	.58	
Lifting Slabs	Separating and curing compound	.03	to	.07	.05	
	Lifting collars	.24	to	0.5	.37	2.54
	Lifting and welding	1.24	to	2.99	2.12	
Columns	Fabricated columns & accessories	.68	to	2	1.34	
	Set initial stage	.05	to	.21	.13	1.72
	Column splicing	.13	to	.37	.25	
Miscellaneous	Grout plates, patching, bolts, etc.	.05	to	.22	.14	.14
Total in place per S.F.		$5.88	to	$12.58	$9.26	$9.26

R03350-130 Granolithic Finish and Base

	Granolithic Topping, Mix 1:1:1-1/2				Granolithic Base	
Description	(Line 033-454-0850)	1" Topping	(Line 033-454-0950)	2" Topping	(Line 033-130-0200)	1" x 5" Straight
Cement @ $7.10 per bag	3.6 bags	$ 25.56	7.2 bags	$ 51.12	1.7 bags	$ 12.07
Sand @ $12.75 per C.Y.	3.6 C.F.	1.70	7.2 C.F.	3.40	1.7 C.F.	.80
Peastone @ $20.50 per C.Y.	5.4 C.F.	4.10	10.8 C.F.	8.20	1.85 C.F.	1.40
Clean, mix, place forms and finish. Crew C-10 @ $17.42/L.H.	4.07 L.H.	70.90	4.80 L.H.	83.62	13.70 L.H.	238.65
Total	100 S.F.	$102.26	100 S.F.	$146.34	100 L.F.	$252.92

R03350-140 Integral Floor Finish

Mix 1:1:2		(Line 033-454-0450)	1/2" Thick	(Line 033-454-0600)	1" Thick
Cement @	$ 7.10 per bag	1.7 bag	$12.07	3.4 bag	$24.14
Sand @	$12.75 per C.Y.	0.8 C.F.	.38	1.6 C.F.	.76
Gravel @	$18.85 per C.Y.	3.2 C.F.	2.23	6.4 C.F.	4.47
Mix, place and finish using Crew C-10 @	$17.42 L.H.	2.52 L.H.	43.90	3.21 L.H.	55.92
Total for 100 S.F.			$58.58		$85.29

Concrete — R034 Precast Concrete

R03410-030 Prestressed Precast Concrete Structural Units

See also R03410-090 for post-tensioned prestressed concrete.

Type	Location	Depth	Span in Ft.		Live Load Lb. per S.F.	Cost per S.F. Incl. Subs O & P	
						Delivered	Erected
Double Tee	Floor	28" to 34"	60 to 80		50 to 80	$ 5.85 to 8.15	$ 7.00 to 9.80
	Roof	12" to 24"	30 to 50		40	$ 3.90 to 6.20	$ 4.70 to 7.45
	Wall	Width 8'	Up to 55' high		Wind	$4.10 to $ 8.25	$4.90 to $ 9.90
Multiple Tee	Roof	8" to 12"	15 to 40		40	$ 4.10	$ 4.90
	Floor	8" to 12"	15 to 30		100	$ 4.30	$ 4.95
Plank	Roof or Floor		Roof	Floor			
		4"	13	12	40 for Roof	$ 3.75	$ 4.70
		6"	22	18		4.00	5.20
		8"	26	25		4.25	5.30
		10"	33	29	100 for Floor	4.75	5.95
		12"	42	32		5.00	6.25
Single Tee	Roof	28"	40			$ 6.75 *	$ 7.75 *
		32"	80		40	8.00 *	9.20 *
		36"	100			10.70 *	12.30 *
		48"	120			11.25 *	12.95 *
AASHO Girder	Bridges	Type 4	100			$107/L.F.	$161/L.F.
		5	110		Highway	142	178
		6	125			178	214
Box Beam	Bridges	15"	40			$115/L.F.	$173/L.F.
		27"	to		Highway	153	191
		33"	100			176	211

*Costs are for 10' wide members; for 8' wide members add $.55 per S.F.

The costs above are based on a project of 10,000 to 20,000 S.F. with a haul distance of 25 to 50 miles.

The majority of precast projects today utilize double tees rather than single tees because of speed and ease of installation. As a result casting beds at manufacturing plants are normally formed for double tees. Single tee projects will therefore require an initial set up charge of approximately $7,000 to be spread over the individual single tee costs. The prices above for single tees includes this cost based on a project size of 15,000 square feet.

For floors, a 2" to 3" topping is field cast over the shapes. For roofs, insulating concrete or rigid insulation is placed over the shapes. Topping is not included in the above costs.

Hauling costs (incl. above) run from $.50 per S.F. for short haul to $.90 for 50 mile haul. Member lengths up to 40' are standard haul, 40' to 60' require special permits and lengths over 60' must be escorted. Over width and/or over length can add up to 100% on hauling costs.

Multiple tee erection runs between $.60 to $1.75 per S.F. and single tee erection runs between $.75 to $1.75 per S.F. Large heavy members may require two cranes for lifting which would increase these erection costs by about 45%. An eight man erection crew and crane will run about $3,200 to $5,800 per day depending on location and size of crane. The crew can install 12 to 20 double tees, or 45 to 70 quad tees or planks per day.

The cost of supporting beams must be added to the above costs. Simple support beams run from $42 to $110 per L.F. delivered. Inverted tee beams run from $72 to $260 per L.F. delivered. Standard sized columns run $40 to $160 per L.F. delivered.

Grouting of connections must be added to the above. Typical costs are about $25 per connection but can go as high as $50 per connection.

Grouting planks run between $.45 to $.65 per S.F.

Single story buildings, including double tee roof members, supporting columns and girders, but no foundations, cost from $9.00 to $30.00 per S.F. of floor area. Parking garages run from $11.00 to $22.00 per S.F. above the foundations. Optimum parking garage design runs .02 C.Y. per S.F with overall costs for concrete in place running between $540 to $750 per C.Y.

Several system buildings utilizing precast members are available. Heights can go to 22 stories for apartment buildings with costs ranging from $11.00 to $40.00 per S.F. depending on the system, its components, its location and the degree of interior finish supplied. Optimum design ratio is 3 S.F. of surface to 1 S.F. of floor area.

Concrete | R034 | Precast Concrete

R03410-090 Prestressed Concrete, Post-tensioned

In post-tensioned concrete the steel tendons are tensioned after the concrete has reached about 3/4 of its ultimate strength. The cableways are grouted after tensioning to provide bond between the steel and concrete. If bond is to be prevented, the tendons are coated with a corrosion-preventative grease and wrapped with waterproof paper or plastic. Bonded tendons are usually used when ultimate strength (beams & girders) is a controlling factor.

High strength concrete is used to fully utilize the steel, thereby reducing the size and weight of the member. A plasticizing agent may be added to reduce water content. Maximum size aggregate ranges from 1/2" to 1-1/2" depending on the spacing of the tendons.

The types of steel commonly used are bars and strands. Job conditions determine which is best suited. Bars are best for vertical prestresses since they are easy to support. The trend is for steel manufacturers to supply a finished package, cut to length, which reduces field preparation to a minimum.

Bars vary from 3/4" to 1-3/8" diameter. Table below gives time in labor-hours per tendon and the labor cost per pound for placing, tensioning and grouting (if required) a 75' beam. Tendons used in buildings are not usually grouted; tendons for bridges usually are grouted. For strands the table indicates the labor-hours and labor cost per pound for typical prestressed units 100' long. Simple span beams usually require one end stressing regardless of lengths. Continuous beams are usually stressed from two ends. Long slabs are poured from the center outward and stressed in 75' increments after the initial 150' center pour. Prices below do not include subcontractor's overhead and profit.

Labor Hours per Tendon and Labor Costs per Pound of Prestress Steel						
Length	100' Beam		75' Beam		100' Slab	
Type Steel	Strand		Bars		Strand	
Diameter	0.5"		3/4"	1-3/8"	0.5"	0.6"
Number	12	24	1	1	1	1
Force in Kips	298	595	42	143	25	35
Preparation & Placing Cables	6.6	10.0	0.7	2.3	0.8	0.8
Stressing Cables	2.4	3.0	0.6	1.3	0.4	0.4
Grouting, if required	3.0	3.5	0.5	1.0		
Total Labor Hours	12.0	16.5	1.8	4.6	1.2	1.2
Prestressing Steel Weights	640#	1280#	115#	380#	53#	74#
Labor cost per lb. Bonded	$.70 to $.90	$.55 to $.65	$.55 to $.65	$.40 to $.45		
Non-bonded			$.48 to $.56	$.42 to $.50	$.88 to $1.10	$.88 to $1.10

Flat Slab construction — 4000 psi concrete with span to depth ratio between 36 and 44. Two way post-tensioned steel averages 1.0 lb. per S.F. for 24' to 28' bays (usually strand) and additional reinforcing steel averages .5 lb. per S.F.

Pan and Joist construction — 4000 psi concrete with span to depth ratio 28 to 30. Post-tensioned steel averages .8 lb. per S.F. and reinforcing steel about 1.0 lb. per S.F. Placing and stressing averages 40 hours per ton of total material.

Beam construction — 4000 to 5000 psi concrete. Steel weights vary greatly. Tendon material cost runs from $1.00 to $1.90 per lb. for grouted and $.85 to $1.18 per lb. for non-grouted strand. Bar costs are from $.85 to $1.35 per lb.

From the chart above placing and stressing cost for grouted tendons runs from $.40 to $.90 per lb; non-grouted placing costs run from $.41 to $1.10 per lb. Labor cost per pound goes down as the size and length of the tendon increases. The primary economic consideration is the cost per kip for the member.

Post-tensioning becomes feasible for beams and girders over 30' long; for continuous two-way slabs over 20' clear; also in transferring upper building loads over longer spans at lower levels. Post-tension suppliers will provide engineering services at no cost to the user. Substantial economies are possible by using post-tensioned Lift Slabs. See R03310-120 for Lift Slabs.

Concrete — R034 Precast Concrete

R03450-010 Precast Concrete Wall Panels

Panels are either solid or insulated with plain, colored or textured finishes. Transportation is an important cost factor. Prices below are based on delivery within 50 miles of a plant including fabricators' overhead and profit. Engineering data is available from fabricators to assist with construction details. Usual minimum job size for economical use of panels is about 5000 S.F. Small jobs can double the prices below. For large, highly repetitive jobs, deduct up to 15% from the prices below.

Panel Cost and Maximum Size Base price for panels based on 50 S.F. or larger panels is as follows:

Thickness	Cost per S.F.	Maximum Size	Thickness	Cost per S.F.	Maximum Size
3"	$ 7.90	50 S.F.	6"	$12.00	300 S.F.
4"	8.70	150 S.F.	7"	12.95	300 S.F.
5"	10.45	200 S.F.	8"	13.45	300 S.F.

The above prices are for gray smooth form or board finish one side. Add $1.60 per S.F. for white facing; $8.50 per C.F. for solid white for full thickness. For fluted surface (not broken) add $.50 to $1.05 per S.F. For broken rib finish add $.80 to $1.60 to fluted surface. Add $.35 to $1.35 per S.F. for exposed local aggregate and $1.00 to $4.00 per S.F. for special facing aggregates.

Sandblasting runs $.55 to $1.00 per S.F., and bush hammering runs $1.75 to $3.45 per S.F. Loose hardware (not included above) runs $1.00 to $2.00 per S.F.

2" thick panels cost about the same as 3" thick panels and maximum panel size is less. For building panels faced with granite, marble or stone, add the material prices from Division 04400 to the plain panel price above. There is a growing trend toward aggregate facings and broken rib finish rather than plain gray concrete panels.

Composite Panels Add to above panel prices for core insulation, mesh and shear ties per S.F.

Type	1" Thick	1-1/2" Thick	2" Thick
Fiberglass	$1.65	$2.30	$2.95
Polystyrene (E.P.S. Board)	1.10	1.55	2.00

Erection Table shows cost ranges for erection including Subcontractor's O & P using a six man crew, crane, operator & oiler.

Panel Size L x H	Area per Panel	Low Rise Hyd. 55 Ton Crane @ $3626 Daily Cost				High Rise							
		Daily Production Range		Erection Cost		Daily Production Range		55 Ton Crane $3,626 Daily Cost		90 Ton Crane $3385 Daily Cost		150 Ton Crane $3983 Daily Cost	
		Pieces	Area	Per Piece	Per S.F.	Pieces	Area	Piece Cost	S.F. Cost	Piece Cost	S.F. Cost	Piece Cost	S.F. Cost
4' x 4'	16 S.F.	10	160 S.F.	$339	$21.15	9	144 S.F.	$376	$23.51	$403	$25.18	$443	$27.66
		20	320	169	10.60	18	288	188	11.75	201	12.59	221	13.83
4' x 8'	32	10	320	339	10.60	9	288	376	11.75	403	12.59	443	13.83
		19	608	178	5.55	17	544	199	6.22	213	6.67	234	7.32
8' x 8'	64	9	576	376	5.90	8	512	423	6.61	453	7.08	498	7.78
		18	1152	188	2.95	16	1024	212	3.31	227	3.54	249	3.89
10' x 10'	100	8	800	423	4.25	7	700	484	4.84	518	5.18	569	5.69
		15	1500	226	2.25	14	1400	242	2.42	259	2.59	285	2.85
15' x 10'	150	7	1050	484	3.20	6	900	—	—	604	4.03	664	4.43
		12	1800	282	1.90	11	1650	—	—	330	2.20	362	2.41
20 x 10'	200	6	1200	564	2.80	5	1000	—	—	725	3.63	797	3.98
		8	1600	423	2.10	7	1400	—	—	518	2.59	569	2.85
30' x 10'	300	5	1500	677	2.25	4	1200	—	—	907	3.02	996	3.32
		7	2100	484	1.60	7	2100	—	—	518	1.73	569	1.90

Total Cost in Place for Low Rise Construction Including Subcontractor's O & P

Description	Gray		White Face		Exposed Aggregate	
	4' x 8' x 4"	20' x 10' x 6"	4' x 8' x 4"	20' x 10' x 6"	4' x 8' x 4"	20' x 10' x 6"
Panel, steel form, broomed finish	$ 8.70	$12.00	$10.25	$13.55	$10.70	$13.80
Caulking, grout, etc. (runs about $2.50 per L.F.)	.95	.36	.90	.36	.90	.36
Erect, plumb, align (from above)	10.60	2.80	10.60	2.80	10.60	2.80
Total in place per S.F.	$20.25	$15.16	$21.75	$16.71	$22.20	$16.96

No allowance has been made for supporting steel framework. On one story buildings, panels may rest on grade beams and require only wind bracing and fasteners. On multi-story buildings panels can span from column to column and floor to floor. Plastic designed steel framed structures may have large deflections which slow down erection and raise costs.

Large panels are more economical than small panels on a S.F. basis. When figuring areas include all protrusions, returns, etc. Overhangs can triple erection costs. Panels over 45' have been produced. Larger flat units should be prestressed. Vacuum lifting of smooth finish panels eliminates inserts and can speed erection.

Concrete — R034 Precast Concrete

R03470-020 Tilt Up Concrete Panels

The advantage of tilt up construction is in the low cost of forms and the placing of concrete and reinforcing. Panels up to 75' high and 5-1/2" thick have been tilted using strongbacks. Tilt up has been used for one to five story buildings and is well suited for warehouses, stores, offices, schools and residences.

The panels are cast in forms on the floor slab. Most jobs use 5-1/2" thick solid reinforced concrete panels. Sandwich panels with a layer of insulating materials are also used. Where dampness is a factor, lightweight aggregate is used at an added cost of $.50 per square foot. Optimum panel size is 300 to 500 S.F.

Slabs are usually poured with 3000 psi concrete which permits tilting seven days after pouring. Slabs may be stacked on top of each other and are separated from each other by either two coats of bond breaker or a film of polyethylene. Use of high early strength cement allows tilting two days after a pour. Tilting up is done with a roller outrigger crane with a capacity of at least 1-1/2 times the weight of the panel at the required reach. Exterior precast columns can be set at the same time as the panels; interior precast columns can be set first and the panels clipped directly to them. The use of cast-in-place concrete columns is diminishing due to shrinkage problems. Structural steel columns are sometimes used if crane rails are planned. Panels can be clipped to the columns or lowered between the flanges. Steel channels with anchors may be used as edge forms for the slab. When the panels are lifted the channels form an integral steel column to take structural loads. Roof loads can be carried directly by the panels for wall heights to 14'. For soft ground requiring mats for cranes, add 100% to costs below.

Below are typical costs per S.F., for panels of 300 to 500 S.F. 20' high, not including contractor's overhead and profit.

Item	5-1/2" Thick Material	5-1/2" Thick Installation	5-1/2" Thick Total	7-1/2" Thick Material	7-1/2" Thick Installation	7-1/2" Thick Total
Prepare pouring surface	$ .02	$ .05	$ .07	$ .02	$ .05	$ .07
Erect, strip side forms	.08	.24	.32	.14	.33	.47
Place concrete, 3000 psi	1.06	.09	1.15	1.44	.33	1.77
Steel trowel finish & curing	.02	.30	.32	.02	.30	.32
Reinforcing, inserts & misc. items	1.02	.42	1.44	1.41	.56	1.97
Panel erection and aligning	—	.68	.68	—	.68	.68
Total per S.F. of Wall	$ 2.20	$ 1.78	$ 3.98	$ 3.03	$ 2.25	$ 5.28
Site precast concrete columns, add	1.10	.98	2.08	1.06	.98	2.04
Total per S.F. of Wall	$ 3.30	$ 2.76	$ 6.06	$ 4.09	$ 3.23	$ 7.32
Panels only, per C.Y.	$118.80	$96.12	$214.92	$130.90	$97.20	$228.10

Requirements of local building codes may be a limiting factor and should be checked. Building floor slabs should be poured first and should be a minimum of 5" thick with 100% compaction of soil or 6" thick with less than 100% compaction.

Setting times as fast as nine minutes per panel have been observed, but a safer expectation would be four panels per hour with a crane and a four person setting crew. If crane erects from inside building, some provision must be made to get crane out after walls are erected. Good yarding procedure is important to minimize delays. Equalizing three point lifting beams and self-releasing pick-up hooks speed erection. If panels must be carried to their final location, setting time per panel will be increased and erection costs may fall in the erection cost range of architectural precast wall panels. Placing panels into slots formed in continuous footers will speed erection.

Reinforcing should be with #5 bars with vertical bars on the bottom. If surface is to be sandblasted, stainless steel chairs should be used to prevent rust staining.

Use of a broom finish is popular since the unavoidable surface blemishes are concealed. Many West Coast jobs have used exposed aggregate panels which cost an additional $.50 to $3.60 per S.F. depending on the aggregate and construction method employed. Eastern prices for exposed aggregate tend to be considerably higher. Panel connections run $1.60 to $6.00 per L.F. Precast columns run from three to five times the C.Y. price of the panels only.

Concrete | R035 Cementitious Decks & Toppings

R03520-010 Lightweight Concrete

Vermiculite or Perlite come in bags of 4 C.F. under various trade names. Weight is about 8 lbs. per C.F. For insulating roof fill use 1:6 mix. For structural deck use 1:4 mix over gypsum boards, steeltex, steel centering, etc. supported by closely spaced joists or bulb trees. For structural slabs use 1:3:2 vermiculite sand concrete over steeltex, metal lath, steel centering, etc. on joists spaced 2'-0" O.C. for maximum L.L. of 80#/S.F. Use same mix for slab base fill over steel flooring or regular reinforced concrete slab when tile, terrazzo or other finish is to be laid over.

For slabs on grade use 1:3:2 mix when tile, etc. finish is to be laid over. If radiant heating units are installed use a 1:6 mix for a base. After coils are in place, cover with a regular granolithic finish (mix 1:3:2) to a minimum depth of 1-1/2" over top of units.

Reinforce all slabs with 6 x 6 or 10 x 10 welded wire mesh.

Vermiculite concrete can be purchased ready mixed but the following breakdown is included for field mix. Prices given below are for a one story building and assume 50 C.Y. or more. For less than 50 C.Y. add 10%. For over one story add $3.50 per C.Y. Screed finish cost is included below. Ready mix 1:6 costs $92.05 per C.Y. delivered in truckload lots.

See R3310-070 for prices of ready mix lightweight concrete.

Quantities per C.Y., Field Mix			1:6 Mix Insulating Roof Fill		1:3:2 Mix Lightweight Structural Concrete	
Portland cement @	$ 7.10	per bag	5.0 bags	$ 35.50	6.2 bags	$ 44.02
Vermiculite or Perlite @	9.85	per bag	7.5 bags	73.88		
treated type @	11.50	per bag			4.7 bags	54.05
Sand @	12.75	per C.Y.			12.5 C.F.	5.90
Plant and Water				6.46		6.46
Labor, machine mix, hoist and place, Crew C-8 @	$28.50	per L.H.	1.12 L.H.	31.92	.91 L.H.	25.94
Total in place, per C.Y.				$147.76		$136.37

Masonry | R040 | Mortar & Masonry Accessories

R04060-100 Cement Mortar (material only)

Type N - 1:1:6 mix by volume. Use everywhere above grade except as noted below.
- 1:3 mix using conventional masonry cement which saves handling two separate bagged materials.

Type M - 1:1/4:3 mix by volume, or 1 part cement, 1/4 (10% by wt.) lime, 3 parts sand. Use for heavy loads and where earthquakes or hurricanes may occur. Also for reinforced brick, sewers, manholes and everywhere below grade.

Cost and Mix Proportions of Various Types of Mortar

Components	Type Mortar and Mix Proportions by Volume										
	M		S		N		O		K	PM	PL
	1:1:6	1:1/4:3	1/2:1:4	1:1/2:4	1:3	1:1:6	1:3	1:2:9	1:3:12	1:1:6	1:1/2:4
Portland cement @ $7.10 per bag	$ 7.10	$ 7.10	$ 3.55	$ 7.10	—	$ 7.10	—	$ 7.10	$ 7.10	$ 7.10	$ 7.10
Masonry cement @ $5.90 per bag	5.90	—	5.90	—	5.90	—	$5.90	—	—	$ 5.90	—
Lime @ $5.60 per 50 lb. bag	—	1.40	—	2.80	—	5.60	—	11.20	16.80	—	2.80
Masonry sand @ $17.50 per C.Y.*	3.89	1.94	2.59	2.59	1.94	3.89	1.94	5.83	7.78	3.89	2.59
Mixing machine incl. fuel**	1.72	.86	1.15	1.15	.86	1.72	.86	2.58	3.45	1.72	1.15
Total for Materials	$18.61	$11.30	$13.19	$13.64	$8.70	$18.31	$8.70	$26.71	$35.13	$18.61	$13.64
Total C.F.	6	3	4	4	3	6	3	9	12	6	4
Approximate Cost per C.F.	$ 3.10	$ 3.77	$ 3.30	$ 3.41	$2.90	$ 3.05	$2.90	$ 2.97	$ 2.93	$ 3.10	$ 3.41

*Includes 10 mile haul
**Based on a daily rental, 10 C.F., 25 H.P. mixer, mix 200 C.F./Day

Mix Proportions by Volume, Compressive Strength and Cost of Mortar

Where Used	Mortar Type	Allowable Proportions by Volume				Compressive Strength @ 28 days	Cost per Cubic Foot
		Portland Cement	Masonry Cement	Hydrated Lime	Masonry Sand		
Plain Masonry	M	1	1	—	6		$3.10
		1	—	1/4	3	2500 psi	3.77
	S	1/2	1	—	4		3.30
		1	—	1/4 to 1/2	4	1800 psi	3.41
	N	—	1	—	3		2.90
		1	—	1/2 to 1-1/4	6	750 psi	3.05
	O	—	1	—	3		2.90
		1	—	1-1/4 to 2-1/2	9	350 psi	2.97
	K	1	—	2-1/2 to 4	12	75 psi	2.93
Reinforced Masonry	PM	1	1	—	6	2500 psi	3.10
	PL	1	—	1/4 to 1/2	4	2500 psi	3.41

Note: The total aggregate should be between 2.25 to 3 times the sum of the cement and lime used.

The labor cost to mix the mortar is included in the labor cost on brickwork.

Machine mixing is usually specified on jobs of any size. There is a large price saving over hand mixing and mortar is more uniform.

There are two types of mortar color used. Prices in Section 04060 are for the inert additive type with about 100 lbs. per M brick as the typical quantity required. These colors are also available in smaller batch size bags (1 lb. to 15 lb.) which can be placed directly into the mixer without measuring. The other type is premixed and replaces the masonry cement with ranges in price from $5 to $15 per 70 lb. bag. Dark green color has the highest cost.

R04060-200 Miscellaneous Mortar (material only)

Quantities	Glass Block Mortar		Gypsum Cement Mortar	
White Portland cement at $17.75 per bag	7 bags	$124.25		
Gypsum cement at $11.40 per 80 lb. bag			11.25 bags	$128.25
Lime at $5.60 per 50 lb. bag	280 lbs.	31.36		
Sand at $17.50 per C.Y.*	1 C.Y.	17.50	1 C.Y.	17.50
Mixing machine and fuel		7.75		7.75
Total per C.Y.		$180.86		$153.50
Approximate Total per C.F.		$ 6.70		$ 5.69

* Includes 10 mile haul

Masonry | R040 | Mortar & Masonry Accessories

R04080-500 Masonry Reinforcing

Horizontal joint reinforcing helps prevent wall cracks where wall movement may occur and in many locations is required by code. Horizontal joint reinforcing is generally not considered to be structural reinforcing and an unreinforced wall may still contain joint reinforcing.

Reinforcing strips come in 10' and 12' lengths and in truss and ladder shapes, with and without drips. Field labor runs between 2.7 to 5.3 hours per 1000 L.F. for wall thicknesses up to 12".

The wire meets ASTM A82 for cold drawn steel wire and the typical size is 9 ga. sides and ties with 3/16" diameter also available. Typical finish is mill galvanized with zinc coating at .10 oz. per S.F. Class I (.40 oz. per S.F.) and Class III (.80 oz. per S.F.) are also available, as is hot dipped galvanizing at 1.50 oz. per S.F.

Masonry | R042 | Unit Masonry

R04210-050 Brick Chimneys

Quantities	16" x 16"		20" x 20"		20" x 24"		20" x 32"	
Brick at $300 per M	28 brick	$ 8.40	37 brick	$11.10	42 brick	$12.60	51 brick	$15.30
Type M mortar at $3.70 per C.F.	.5 C.F.	1.89	.6 C.F.	2.26	1.0 C.F.	3.77	1.3 C.F.	4.90
Flue tile (square)(per foot)	8" x 8"	3.00	12" x 12"	5.70	2 @ 8" x 12"	7.54	2 @ 12" x 12"	11.40
Install tile & brick, crew D-1	.055 day	15.71	.073 day	20.85	.083 day	23.70	.10 day	28.56
Total per L.F. high		$29.00		$39.91		$47.61		$60.16

Material costs include 3% waste for brick and 25% waste for mortar.

Labor costs are bare costs and do not include contractor's O&P.

Labor for chimney brick using D-1 crew is 31 hours per thousand brick or about $530 per thousand brick. An 8" x 12" flue takes 33 brick and two 8" x 8" flues take 37 brick.

R04210-055 Industrial Chimneys

Foundation requirements in C.Y. of concrete for various sized chimneys.

Size Chimney	2 Ton Soil	3 Ton Soil	Size Chimney	2 Ton Soil	3 Ton Soil	Size Chimney	2 Ton Soil	3 Ton Soil
75' x 3'-0"	13 C.Y.	11 C.Y.	160' x 6'-6"	86 C.Y.	76 C.Y.	300' x 10'-0"	325 C.Y.	245 C.Y.
85' x 5'-6"	19	16	175' x 7'-0"	108	95	350' x 12'-0"	422	320
100' x 5'-0"	24	20	200' x 6'-0"	125	105	400' x 14'-0"	520	400
125' x 5'-6"	43	36	250' x 8'-0"	230	175	500' x 18'-0"	725	575

R04210-100 Economy in Bricklaying

Have adequate supervision. Be sure bricklayers are always supplied with materials so there is no waiting. Place best bricklayers at corners and openings.

Use only screened sand for mortar. Otherwise, labor time will be wasted picking out pebbles. Use seamless metal tubs for mortar as they do not leak or catch the trowel. Locate stack and mortar for easy wheeling.

Have brick delivered for stacking. This makes for faster handling, reduces chipping and breakage, and requires less storage space. Many dealers will deliver select common in 2' x 3' x 4' pallets or face brick packaged. This affords quick handling with a crane or forklift and easy tonging in units of ten, which reduces waste.

Use wider bricks for one wythe wall construction. Keep scaffolding away from wall to allow mortar to fall clear and not stain wall.

On large jobs develop specialized crews for each type of masonry unit.

Consider designing for prefabricated panel construction on high rise projects.

Avoid excessive corners or openings. Each opening adds about 50% to labor cost for area of opening.

Bolting stone panels and using window frames as stops reduces labor costs and speeds up erection.

Masonry — R042 Unit Masonry

R04210-120 Common and Face Brick Prices

Prices are based on truckload lot purchases for Common Brick and carload lots for Face Brick. Prices are per M, (thousand), brick.

City	Material — Brick per M Delivered — Face	Material — 3/8" Joint	Mortar Bare Costs	Installation — Common in 8" Wall — Incl. O & P	Installation — Common in 8" Wall — Bare Costs	Installation — Face Brick, 4" Veneer — Incl. O & P	Installation — Face Brick, 4" Veneer — Bare Costs	Total — Common in 8" Wall — Incl. O & P	Total — Common in 8" Wall — Bare Costs	Total — Face Brick, 4" Veneer — Incl. O & P	Total — Face Brick, 4" Veneer — Bare Costs
Atlanta	$190	$250	$38.10 for 8" Wall and $31.40 for 4" Wall	$278	$ 471	$334	$ 565	$512	$ 728	$ 623	$ 883
Baltimore	235	280		320	542	384	651	600	850	704	1,002
Boston	305	480		573	970	687	1,164	925	1,357	1,213	1,742
Buffalo	260	350		491	832	590	998	797	1,168	981	1,429
Chicago	260	380		520	881	624	1,057	826	1,218	1,047	1,522
Cincinnati	230	330		384	650	461	781	659	953	832	1,189
Cleveland	235	325		464	785	556	942	744	1,093	923	1,345
Columbus	235	350		379	641	454	769	659	949	846	1,200
Dallas	200	275		273	463	328	555	517	731	643	901
Denver	210	310		318	538	381	645	572	818	732	1,031
Detroit	235	275		487	825	585	990	768	1,133	900	1,336
Houston	250	275		284	481	341	578	580	806	656	924
Indianapolis	255	265		388	657	466	788	689	988	770	1,123
Kansas City	260	300		416	704	499	845	722	1,041	840	1,220
Los Angeles	250	325		483	818	580	982	779	1,144	946	1,385
Memphis	190	235		273	463	328	555	507	720	601	856
Milwaukee	280	365		458	776	550	931	785	1,135	957	1,379
Minneapolis	255	400		473	801	568	961	774	1,132	1,011	1,449
Nashville	185	245		267	452	321	543	496	704	604	855
New Orleans	210	380		236	400	283	480	491	680	706	945
New York City	300	355		649	1,098	778	1,318	996	1,480	1,176	1,755
Philadelphia	245	345		516	873	619	1,048	806	1,193	1,006	1,473
Phoenix	340	350		296	501	355	601	684	928	747	1,032
Pittsburgh	210	275		420	711	504	854	675	991	819	1,200
St. Louis	235	265		249	422	299	506	529	730	603	841
San Antonio	260	325		429	726	514	871	735	1,062	880	1,274
San Diego	285	360		562	951	674	1,141	893	1,316	1,076	1,584
San Francisco	350	435		470	795	563	954	868	1,233	1,043	1,481
Seattle	340	390		473	801	568	961	861	1,228	1,001	1,438
Washington, D.C.	210	240		330	558	396	670	584	838	674	976
Average	$235	$305	▼	$410	$ 690	$490	$ 823	$700	$1,010	$ 850	$1,226

Common building brick manufactured according to ASTM C62 and facing brick manufactured according to ASTM C216 are the two standard bricks available for general building use.

Building brick is made in three grades; SW, where high resistance to damage caused by cyclic freezing is required; MW, where moderate resistance to cyclic freezing is needed; and NW, where little resistance to cyclic freezing is needed. Facing brick is made in only the two grades SW and MW. Additionally, facing brick is available in three types; FBS, for general use; FBX, for general use where a higher degree of precision and lower permissible variation in size than FBS is needed; and FBA, for general use to produce characteristic architectural effects resulting from non-uniformity in size and texture of the units.

In figuring above installation costs, a D-8 Crew (with a daily output of 1.5 M) was used for the 4" veneer. A D-8 Crew (with a daily output of 1.8 M) was used for the 8" solid wall.

In figuring the total cost including overhead and profit, an allowance of 10% was added to the sum of the cost of the brick and mortar. Also, 3% breakage was included for both the bare costs and the costs with overhead and profit. If bricks are delivered palletized with 280 to 300 per pallet, or packaged, allow only 1-1/2% for breakage. Then add $10 per M to the cost of brick and deduct two hours helper time. The net result is a savings of $30 to $40 per M in place. Packaged or palletized delivery is practical when a job is big enough to have a crane or other equipment available to handle a package of brick. This is so on all industrial work but not always true on small commercial buildings.

There are many types of red face brick. The prices above are for the most usual type used in commercial, apartment house or industrial construction. If it is possible to obtain the price of the actual brick to be used, it should be done and substituted in the table. The use of buff and gray face is increasing, and there is a continuing trend to the Norman, Roman, Jumbo and SCR brick.

See R04210-500 for brick quantities per S.F. and mortar quantities per M brick. (Average prices for the various sizes are listed in Division 4)

Common red clay brick for backup is not used that often. Concrete block is the most usual backup material with occasional use of sand lime or cement brick. Sand lime cost about $15 per M less than red clay and cement brick are about $5 per M less than red clay. These figures may be substituted in the common brick breakdown for the cost of these items in place, as labor is about the same. Occasionally common brick is being used in solid walls for strength and as a fire stop.

Brick panels built on the ground floor and then crane erected to the upper floors have proven to be economical. This allows the work to be done under cover and without scaffolding.

Masonry | R042 | Unit Masonry

R04210-180 Brick in Place

Table below is for common bond with 3/8" concave joints and includes 3% waste for brick and 25% waste for mortar. Crew costs are bare costs.

Item	8" Common Brick Wall 8" x 2-2/3" x 4"		Select Common Face 8" x 2-2/3" x 4"		Red Face Brick 8" x 2-2/3" x 4"	
1030 brick delivered	$250 per M	$257.50	$300 per M	$309.00	$325 per M	$334.75
Type N mortar @ $2.99 per C.F.	12.5 C.F.	38.13	10.3 C.F.	31.42	10.3 C.F.	31.42
Installation using indicated crew	Crew D-8 @ .556 Days	406.55	Crew D-8 @ .667 Days	487.71	Crew D-8 @ .667 Days	487.71
Total per M in place		$702.18		$828.13		$853.88
Total per S.F. of wall	13.5 bricks/S.F.	$ 9.48	6.75 bricks/S.F.	$ 5.59	6.75 bricks/S.F.	$ 5.76

R04210-185 Reinforced Brick in Walls

Table below is for common bond with 3/8" concave joints and includes 3% waste. Standard 8" x 2-2/3" x 4" bricks.

Item	8" to 9" Thick Wall		16" to 17" Thick Wall	
1030 brick (select common)	$305 per M	$309.00	$305 per M	$309.00
Type PL mortar @ $3.41 per C.F.	12.5 C.F.	42.63	13.9 C.F.	47.40
Reinforcing bars delivered	50 lb. @ .27 per lb.	13.00	50 lb. @ .27 per lb.	13.00
Installation incl. reinforcing using indicated crew	Crew D-8 @ .517 Days	378.03	Crew D-8 @ .513 Days	375.11
Total per M in place		$742.66		$744.50
Total per S.F. of wall	8" wall, 13.5 brick/S.F.	$ 10.03	16" wall, 27.0 brick/S.F.	$ 20.10

Masonry — R042 Unit Masonry

R04210-500 Brick, Block & Mortar Quantities

Type Brick	Nominal Size (incl. mortar) L × H × W	Modular Coursing	Number of Brick per S.F.	C.F. of Mortar per M Bricks, Waste Included 3/8" Joint	C.F. of Mortar per M Bricks, Waste Included 1/2" Joint
Standard	8 × 2-2/3 × 4	3C=8"	6.75	10.3	12.9
Economy	8 × 4 × 4	1C=4"	4.50	11.4	14.6
Engineer	8 × 3-1/5 × 4	5C=16"	5.63	10.6	13.6
Fire	9 × 2-1/2 × 4-1/2	2C=5"	6.40	550 # Fireclay	—
Jumbo	12 × 4 × 6 or 8	1C=4"	3.00	23.8	30.8
Norman	12 × 2-2/3 × 4	3C=8"	4.50	14.0	17.9
Norwegian	12 × 3-1/5 × 4	5C=16"	3.75	14.6	18.6
Roman	12 × 2 × 4	2C=4"	6.00	13.4	17.0
SCR	12 × 2-2/3 × 6	3C=8"	4.50	21.8	28.0
Utility	12 × 4 × 4	1C=4"	3.00	15.4	19.6

For Other Bonds Standard Size — Add to S.F. Quantities in Table to Left

Bond Type	Description	Factor
Common	full header every fifth course	+20%
	full header every sixth course	+16.7%
English	full header every second course	+50%
Flemish	alternate headers every course	+33.3%
	every sixth course	+5.6%
Header = W × H exposed		+100%
Rowlock = H × W exposed		+100%
Rowlock stretcher = L × W exposed		+33.3%
Soldier = H × L exposed		—
Sailor = W × L exposed		-33.3%

Concrete Blocks Nominal Size	Approximate Weight per S.F. Standard	Approximate Weight per S.F. Lightweight	Blocks per 100 S.F.	Mortar per M block Partitions	Mortar per M block Back up
2" × 8" × 16"	20 PSF	15 PSF	113	16 C.F.	36 C.F.
4"	30	20		31	51
6"	42	30		46	66
8"	55	38		62	82
10"	70	47		77	97
12"	85	55		92	112

R04210-550 Brick Veneer in Place

Table below is for running bond with 3/8" concave joints and includes 3% waste for brick and 25% waste for mortar.

Item	Buff Face Brick 8" × 2-2/3" × 4"		Red Norman Face Brick 12" × 2-2/3" × 4"		Buff Roman Face Brick 12" × 2" × 4"	
1030 brick delivered	$390 per M	$401.70	$675 per M	$ 695.25	$700 per M	$ 721.00
Type N mortar @ $3.05 per C.F.	10.3 C.F.	31.42	14.0 C.F.	42.70	13.4 C.F.	40.87
Installation using indicated crew	Crew D-8 @ .667 days	487.71	Crew D-8 @ .690 days	504.53	Crew D-8 @ .667 days	487.71
Total per M in place		$920.83		$1,242.48		$1,249.58
Total per S.F. of wall	6.75 brick/S.F.	$ 6.22	4.50 brick/S.F.	$ 5.59	6.00 brick/S.F.	$ 7.50

Masonry | R042 | Unit Masonry

R04220-200 Concrete Block

8" x 16" block, sand aggregate blocks with 3/8" joints for partitions.

	Material					Material			
	Per Block, Delivered		113 Block, Delivered			Per Block, Delivered		113 Block, Delivered	
City	8" Thick	4" Thick	8" Thick	City	4" Thick	8" Thick	4" Thick	8" Thick	
Atlanta	$.74	$1.00	$83.62	$113.00	Memphis	.68	.94	$76.84	$106.22
Baltimore	.65	.85	73.45	96.05	Milwaukee	.76	1.10	85.88	124.30
Boston	.60	.85	67.80	96.05	Minneapolis	.75	1.01	84.75	114.13
Buffalo	.70	1.15	79.10	129.95	Nashville	.77	1.02	87.01	115.26
Chicago	.60	.92	67.80	103.96	New Orleans	.82	1.15	92.66	129.95
Cincinnati	.60	.85	67.80	96.05	New York City	.77	1.03	87.01	116.39
Cleveland	.66	.93	74.58	105.09	Philadelphia	.70	.80	79.10	90.40
Columbus	.69	.83	77.97	93.79	Phoenix	.62	.86	70.06	97.18
Dallas	.61	1.02	68.93	115.26	Pittsburgh	.72	.92	81.36	103.96
Denver	.73	1.05	82.49	118.65	St. Louis	.76	1.10	85.88	124.30
Detroit	.77	1.05	87.01	118.65	San Antonio	.75	.96	84.75	108.48
Houston	.75	.97	84.75	109.61	San Diego	.63	.86	71.19	97.18
Indianapolis	.70	1.00	79.10	113.00	San Francisco	.88	1.45	99.44	163.85
Kansas City	.72	1.20	81.36	135.60	Seattle	.88	1.27	99.44	143.51
Los Angeles	.81	.95	91.53	107.35	Washington D.C.	.66	1.07	74.58	120.91
					Average	$.72	$1.01	$80.91	$113.60

The cost of sand aggregate block is based on the delivered price of 113 blocks, which is the equivalent of a 100 S.F. wall area. Mortar for a 100 S.F. wall area will average $10.50 for 4" thick block and $21.10 for 8" thick block.

Cost for 100 S.F. of 8" x 16" Concrete Block Partitions to Four Stories High, Tooled Joint One Side

8" x 16" Sand Aggregate	4" Thick Block		8" Thick Block		12" Thick Block	
113 block delivered	$.72 Ea.	$ 79.10	$1.01 Ea.	$113.00	$1.60 Ea.	$175.15
Type N mortar at $3.05 per C.F.	3.5 C.F.	10.36	7.0 C.F.	20.72	10.4 C.F.	30.78
Installation Crew (Bare Costs)	D-2 @ 9.32 L.H.	164.22	D-2 @ 10.68 L.H.	188.18	D-6 @ 14.70 L.H.	303.41
Total per 100 S.F.		$253.68		$321.90		$509.34
Add for filling cores solid	6.7 C.F.	$ 85.58	25.8 C.F.	$183.19	42.2 C.F.	$251.22

Cost for 100 S.F. of 8" x 16" Concrete Block Backup with Trowel Cut Joints

8" x 16" Sand Aggregate	4" Thick Block		8" Thick Block		12" Thick Block	
113 block delivered	$.72 Ea.	$ 79.10	$1.01 Ea.	$113.00	$1.60 Ea.	$175.15
Type N mortar at $3.05 per C.F.	5.8 C.F.	17.17	9.3 C.F.	27.53	12.7 C.F.	37.59
Installation Crew (Bare Costs)	D-2 @ 9.20 L.H.	162.10	D-2 @ 10.10 L.H.	177.96	D-6 @ 13.20 L.H.	272.45
Total per 100 S.F.		$258.37		$318.49		$485.19

Special block: corner, jamb and head block are same price as ordinary block of same size. Tabulated on the next page are national average prices per block. Labor on specials is about the same as equal sized regular block. Bond beam and 16" high lintel blocks cost 30% more than regular units of equal size. Lintel blocks are 8" long and 8" or 16" high. Costs in individual cities may be factored from the table above.

Use of motorized mortar spreader box will speed construction of continuous walls.

R04220-220 Interlocking Grout Block Walls

Cost per 100 S.F. of 8" x 16" Interlocking Grout Block, "Open End" Type

8" x 16" Sand Aggregate	8" Thick Block		12" Thick Block		16" Thick Block	
113 block delivered	$1.55 ea.	$175.15	$2.35 ea.	$265.55	$3.65 ea.	$ 412.45
Type M mortar @ $3.77 per C.F.	7.0 C.F.	26.39	10.4 C.F.	39.21	13.8 C.F.	52.03
Minimum reinf. req.@ $ .26 per lb.						
8" block #4 @ 24" horiz., 1 face	33.4 lb.	8.68				
32" vert., 1 face	40.0 lb.	10.40				
12" to 16" block @ 8" horiz., 1 face			99.7 lb.	25.92	99.7 lb.	25.92
16" vert., 1 face			50.2 lb.	13.05	50.2 lb.	13.05
Type PM mortar@ $3.10 C.F.	25.8 C.F.	79.98	42.2 C.F.	130.82	56.1 C.F.	173.91
Installation Crew D-4 @ $761.75 per day	.408 days	310.79	.455 days	346.60	.541 days	412.11
Total per 100 S.F.		$611.39		$821.15		$1,089.47
1 S.F.		6.11		8.21		10.89

| Masonry | **R042** | **Unit Masonry** |

R04270-700 Glass Block

Nominal Size (Including Mortar)	Cost of Blocks Each, Zone 1 Contractor Prices							
	Truckload or Carload				Less Than Truckload or Carload			
	Regular	Thinline	Essex	Solar Reflective	Regular	Thinline	Essex	Solar Reflective
6" x 6"	$ 3.25	$ 2.10	—	$ 7.00	$ 3.50	$ 2.35	—	$ 8.00
8" x 8"	4.10	2.35	4.70	$ 9.00	4.25	2.75	4.85	10.00
Solid 8" x 8"	22.00	—	—	—	24.00	—	—	—
4" x 8"	2.75	2.50	—	—	3.00	3.00	—	—
12" x 12"	9.25	—	—	—	11.00	—	—	—

Size	Per 100 S.F.		Per 1000 Block				
	No. of Block	Mortar " Joint	Asphalt Emulsion	Caulk	Expansion Joint	Panel Anchors	Wall Mesh
6" x 6"	410 ea.	5.0 C.F.	.17 gal.	1.5 gal.	80 L.F.	20 ea.	500 L.F.
8" x 8"	230	3.6	.33	2.8	140	36	670
12" x 12"	102	2.3	.67	6.0	312	80	1000
Approximate quantity per 100 S.F.			.07 gal.	0.6 gal.	32 L.F.	9 ea.	51, 68, 102 L.F.

Accessories required for installation — Wall ties: run galvanized double steel mesh full length of joint at $.32 per L.F.— fiberglass expansion joints at sides and top at $.40 per L.F.— silicone caulking one gallon at $50.00 does 95 L.F. asphalt emulsion one gallon does 600 L.F. at $4.50 per gallon in one gallon lots. If blocks are not set in wall chase use 2'-0" long wall anchors at $1.50 each at 2'-0" O.C.

Quantities	Cost per 100 S.F. (Truckload or Carload Price)					
	6" x 6" Block		8" x 8" Block		12" x 12" Block	
Block, delivered	410 @ $3.25	$1,332.50	230 @ $4.10	$ 943.00	102 @ $9.25	$ 943.50
Mortar @ $6.30 per C.F.	5.0 C.F.	31.50	3.6 C.F.	22.68	2.3 C.F.	14.49
Ties, expansion joints, etc.		92.37		92.82		102.00
Crew D-8 @	.690 days	504.53	.465 days	340.01	.417 days	304.91
Total per 100 S.F.		$1,960.90		$1,398.51		$1,364.90

| Masonry | **R044** | **Stone** |

R04430-500 Ashlar Veneer

Quantities			Low Price Stone		High Price Stone	
4" Stone, random ashlar,	2.5	tons average	$275 per ton	$ 687.50	$375 per ton	$ 937.50
Type N mortar @	$3.05	C.F.	20 C.F.	61.00	20 C.F.	61.00
50 - 1" x 1/4" x 6" galv. stone anchors @	$1.10	each		55.00		55.00
Using crew D-8 @	$731.20	per day	.714 days	522.00	.833 days	609.09
Total per 100 S.F.				$1,325.50		$1,662.59

Stone coverage varies from 30 S.F. to 50 S.F. per ton. Mortar quantities include 1" backing bed.

| Masonry | **R049** | **Restoration & Cleaning** |

R04930-100 Cleaning Face Brick

On smooth brick a person can clean 70 S.F. an hour; on rough brick 50 S.F. per hour. Use one gallon muriatic acid to 20 gallons of water for 1000 S.F. Do not use acid solution until wall is at least seven days old, but a mild soap solution may be used after two days. Commercial cleaners cost from $9 to $12 per gallon.

Time has been allowed for clean-up in brick prices.

Metals | R050 | Materials, Coatings, & Fastenings

R05080-310 Coating Structural Steel

On field-welded jobs, the shop-applied primer coat is necessarily omitted. All painting must be done in the field and usually consists of red oxide rust inhibitive paint or an aluminum paint (see Division 09910-650 for paint material costs). The table below shows paint coverage and daily production for field painting.

See Division 05120-680 for hot-dipped galvanizing and for field-applied cold galvanizing and other paints and protective coatings.

See division 05120-680 for steel surface preparation treatments such as wire brushing, pressure washing and sand blasting.

Type Construction	Surface Area per Ton	Coat	One Gallon Covers		In 8 Hrs. Person Covers		Average per Ton Spray	
			Brush	Spray	Brush	Spray	Gallons	Labor-hours
Light Structural	300 S.F. to 500 S.F.	1st	500 S.F.	455 S.F.	640 S.F.	2000 S.F.	0.9 gals.	1.6 L.H.
		2nd	450	410	800	2400	1.0	1.3
		3rd	450	410	960	3200	1.0	1.0
Medium	150 S.F. to 300 S.F.	All	400	365	1600	3200	0.6	0.6
Heavy Structural	50 S.F. to 150 S.F.	1st	400	365	1920	4000	0.2	0.2
		2nd	400	365	2000	4000	0.2	0.2
		3rd	400	365	2000	4000	0.2	0.2
Weighted Average	225 S.F.	All	400	365	1350	3000	0.6	0.6

R05090-510 High Strength Bolts

Common bolts (A307) are usually used in secondary connections (see Division 05090-150).

High strength bolts (A325 and A490) are usually specified for primary connections such as column splices, beam and girder connections to columns, column bracing, connections for supports of operating equipment or of other live loads which produce impact or reversal of stress, and in structures carrying cranes of over 5-ton capacity.

Allow 20 field bolts per ton of steel for a 6 story office building, apartment house or light industrial building. For 6 to 12 stories allow 25 bolts per ton, and above 12 stories, 30 bolts per ton. On power stations, 20 to 25 bolts per ton are needed.

R05090-520 Welded Structural Steel

Usual weight reductions with welded design run 10% to 20% compared with bolted or riveted connections. This amounts to about the same total cost compared with bolted structures since field welding is more expensive than bolts. For normal spans of 18' to 24' figure 6 to 7 connections per ton.

Trusses — For welded trusses add 4% to weight of main members for connections. Up to 15% less steel can be expected in a welded truss compared to one that is shop bolted. Cost of erection is the same whether shop bolted or welded.

General — Typical electrodes for structural steel welding are E6010, E6011, E60T and E70T. Typical buildings vary between 2# to 8# of weld rod per ton of steel. Buildings utilizing continuous design require about three times as much welding as conventional welded structures. In estimating field erection by welding, it is best to use the average linear feet of weld per ton to arrive at the welding cost per ton. The type, size and position of the weld will have a direct bearing on the cost per linear foot. A typical field welder will deposit 1.8# to 2# of weld rod per hour manually. Using semiautomatic methods can increase production by as much as 50% to 75%. Below is the cost per hour for manual welding.

Welded Structural Steel in Field				
Item		No Operating Engr.	1/2 Operating Engr.	1 Operating Engr.
2 lb. weld rod	$1.32 per lb.	$ 2.64	$ 2.64	$ 2.64
Equipment (for welding only)	$210 /40 hrs.	5.25	5.25	5.25
Operating cost @	$5.25 per hour	5.25	5.25	5.25
Welder 1 hr. @	$21.25 per hour	21.25	21.25	21.25
Operating engineer @	$19.10 per hour	—	9.55	19.10
Total per Welder Hour (Bare Costs)		$34.39	$43.94	$53.49

The welding costs per ton of structural steel will vary from $35 to $215 per ton depending on Union requirements, crew size, design and inspection required.

Metals — R051 Structural Metal Framing

R05120-210 Structural Steel

1 to 2 Story Building	Bare Costs for 100 Ton Job with Common Bolts					Including Subs O&P		
City	Fabricated and Delivered	Unloading and Sorting	Erection Equipment	Field Erection Labor	Total per Ton in Place	Material	Installation	Total
Atlanta	$1,115	$20	48	$127	1,310	$1,227	$ 438	$1,665
Baltimore	1,065	24	59	152	1,300	1,172	528	1,700
Boston	1,225	31	75	198	1,529	1,348	684	2,032
Buffalo	1,300	28	57	174	1,559	1,430	591	2,021
Chicago	1,150	34	77	215	1,476	1,265	737	2,002
Cincinnati	1,090	24	51	155	1,320	1,199	524	1,723
Cleveland	1,100	28	50	180	1,358	1,210	596	1,806
Columbus	1,095	25	48	156	1,324	1,205	524	1,729
Dallas	1,200	16	53	102	1,371	1,320	373	1,693
Denver	1,260	18	50	114	1,442	1,386	404	1,790
Detroit	1,240	30	61	193	1,524	1,364	649	2,013
Houston	1,160	16	56	103	1,335	1,276	379	1,655
Indianapolis	1,300	25	47	156	1,528	1,430	523	1,953
Kansas City	1,300	24	63	149	1,536	1,430	527	1,957
Los Angeles	1,400	31	58	195	1,684	1,540	652	2,192
Memphis	1,140	17	56	111	1,324	1,254	401	1,655
Milwaukee	1,200	27	53	173	1,453	1,320	579	1,899
Minneapolis	1,115	28	73	180	1,396	1,227	628	1,855
Nashville	1,265	17	54	110	1,446	1,392	396	1,788
New Orleans	1,250	16	46	98	1,410	1,375	352	1,727
New York City	1,370	48	85	306	1,809	1,507	1,014	2,521
Philadelphia	1,195	33	71	212	1,511	1,315	719	2,034
Phoenix	1,110	21	46	136	1,313	1,221	461	1,682
Pittsburgh	1,200	28	75	180	1,483	1,320	631	1,951
St. Louis	1,275	27	74	172	1,548	1,403	608	2,011
San Antonio	1,120	15	43	95	1,273	1,232	338	1,570
San Diego	1,350	31	58	195	1,634	1,485	652	2,137
San Francisco	1,315	31	63	195	1,604	1,447	660	2,107
Seattle	1,165	27	54	169	1,415	1,282	570	1,852
Washington, DC	1,110	23	66	144	1,343	1,221	515	1,736
U.S. Average	$1,200	$36	$59	$223	$1,518	$1,320	$ 555	$1,875

Adjustments to the Base Price Above	Bare Costs	Incl. Subs O&P
For 3 to 6 story building with high strength bolting, add per ton	$30	$ 50
For 7 to 15 stories with high strength bolting, add per ton	35	50
For over 15 stories with high strength bolting, add per ton	55	75
For field welded connections, add to these prices per ton	65	125
For multi-story masonry wall bearing construction, add to erection costs	30%	30%

General Average per Ton in Place for A36 Steel (High Strength Steel Base Price May Be Substituted)			Bare Costs		Incl. Subs O&P
Item	Description		Itemized	Summary	
Material	Base price		$ 330		
	Extras and delivery to shop (with common bolts)		140		
	Drafting		90	$1,200	$1,320
	Shop fabrication & warehouse rehandling		490		
	Shop coat paint		60		
	Trucking to job site		90		
Installation	Unload and shake out, 1.7 hours @ $20.50		36		
	Erect and plumb, 7.7 hours @ $20.50		164	$ 318	$ 555
	Field bolt, 2.8 hours @ $20.50		59		
	Crane & minor erection equipment (includes operator)	20 tons/day	59		
	Total per ton in place		$1,518	$1,518	$1,875

Metals | R051 | Structural Metal Framing

R05120-220 Steel Estimating Quantities

One estimate on erection is that a crane can handle 35 to 60 pieces per day. Say the average is 45. With usual sizes of beams, girders, and columns, this would amount to about 20 tons per day. Allowing a bare cost of $1,187 per day for the crane including crew equipment cost per day plus operator and oiler, this would amount amount to $59 per ton for the crane charge. The type of connection greatly affects the speed of erection. Moment connections for continuous design slow down production and increase erection costs.

Short open web bar joists can be set at the rate of 75 to 80 per day, with 50 per day being the average for setting long span joists.

After main members are calculated, add the following for usual allowances: base plates 2% to 3%; column splices 4% to 5%; and miscellaneous details 4% to 5%, for a total of 10% to 13% in addition to main members.

The ratio of column to beam tonnage varies depending on type of steels used, typical spans, story heights and live loads.

It is more economical to keep the column size constant and to vary the strength of the column by using high strength steels. This also saves floor space. Buildings have recently gone as high as ten stories with 8″ high strength columns. For light columns under W8X31 lb. sections, concrete filled steel columns are economical.

High strength steels may be used in columns and beams to save floor space and to meet head room requirements. High strength steels in some sizes sometimes require long lead times.

Round, square and rectangular columns, both plain and concrete filled, are readily available and save floor area, but are higher in cost per pound than rolled columns. For high unbraced columns, tube columns may be less expensive.

Below are average minimum figures for the weights of the structural steel frame for different types of buildings using A36 steel, rolled shapes and simple joints. For economy in domes, rise to span ratio = .13. Open web joist framing systems will reduce weights by 10% to 40%. Composite design can reduce steel weight by up to 25% but additional concrete floor slab thickness may be required. Continuous design can reduce the weights up to 20%. There are many building codes with different live load requirements and different structural requirements, such as hurricane and earthquake loadings which can alter the figures.

*See R13128-310 for Domes and Thin Shell Structures.

Structural Steel Weights per S.F. of Floor Area									
Type of Building	No. of Stories	Avg. Spans	L.L. #/S.F.	Lbs. Per S.F.	Type of Building	No. of Stories	Avg. Spans	L.L. #/S.F.	Lbs. Per S.F.
Steel Frame Mfg.	1	20'x20'	40	8	Apartments	2-8	20'x20'	40	8
		30'x30'		13		9-25			14
		40'x40'		18	Office	to 10	Various	80	10
Parking garage	4	Various	80	8.5		20			18
Domes (Schwedler)*	1	200'	30	10		30			26
		300'		15		over 50			35

Metals | R051 | Structural Metal Framing

R05120-225 Common Structural Steel Specifications

ASTM A36 is the all-purpose carbon grade steel widely used in building and bridge construction.

The other high-strength steels listed below may each have certain advantages over ASTM A36 stuctural carbon steel, depending on the application. They have proven to be economical choices where, due to lighter members, the reduction of dead load and the associated savings in shipping cost can be significant.

ASTM A588 atmospheric weathering, high-strength low-alloy steels can be used in the bare (uncoated) condition, where exposure to normal atmosphere causes a tightly adherant oxide to form on the surface protecting the steel from further oxidation. ASTM A242 corrosion-resistant, high-strength low-alloy steels have enhanced atmospheric corrosion resistance of at least two times that of carbon structural steels with copper, or four times that of carbon structural steels without copper. The reduction or elimination of maintenance resulting from the use of these steels often offsets their higher initial cost.

Steel Type	ASTM Designation	Minimum Yield Stress in KSI	Shapes Available
Carbon	A36	36	All structural shape groups, and plates & bars up thru 8" thick
	A529	42	Structural shape group 1, and plates & bars up thru 1/2" thick
High-Strength Low-Alloy Manganese-Vanadium	A441	40	Plates & bars over 4" thru 8" thick
		42	Structural shape groups 4 & 5, and plates & bars over 1-1/2" up thru 4" thick
		46	Structural shape group 3, and plates & bars over 3/4" up thru 1-1/2" thick
		50	Structural shape groups 1 & 2, and plates & bars up thru 3/4" thick
High-Strength Low-Alloy Columbium-Vanadium	A572	42	All structural shape groups, and plates & bars up thru 6" thick
		50	All structural shape groups, and plates & bars up thru 4" thick
		60	Structural shape groups 1 & 2, and plates & bars up thru 1-1/4" thick
		65	Structural shape group 1, and plates & bars up thru 1-1/4" thick
High-Strength Low-Alloy Columbium-Vanadium	A992	50	All structural shape groups
Corrosion-Resistant High-Strength Low-Alloy	A242	42	Structural shape groups 4 & 5, and plates & bars over 1-1/2" up thru 4" thick
		46	Structural shape group 3, and plates & bars over 3/4" up thru 1-1/2" thick
		50	Structural shape groups 1 & 2, and plates & bars up thru 3/4" thick
Weathering High-Strength Low-Alloy	A588	42	Plates & bars over 5" up thru 8" thick
		46	Plates & bars over 4" up thru 5" thick
		50	All structural shape groups, and plates & bars up thru 4" thick
Quenched and Tempered Low-Alloy	A852	70	Plates & bars up thru 4" thick
Quenched and Tempered Alloy	A514	90	Plates & bars over 2-1/2" up thru 6" thick
		100	Plates & bars up thru 2-1/2" thick

Metals — R051 Structural Metal Framing

R05120-230 High Strength Steels

The mill price of these steels may be higher than A36 carbon steel but their proper use can achieve overall savings thru total reduced weights. For columns with L/r over 100, A36 steel is best; under 100, high strength steels are economical. For heavy columns high strength steels are economical when cover plates are eliminated. There is no economy using high strength steels for clip angles or supports or for beams where deflection governs. Thinner members are more economical than thick.

Below is a table of the high strength steels available and their base price with specification extras. A36 is included for comparison purposes. See also R05120-210 for typical in place costs of A36 steel. The per ton erection and fabricating costs of the high strength steels will be higher than for A36 since the same number of pieces, but less weight, will be installed. See R05120-240 for extras for Jumbo 14″ A36 columns.

Mill Base Prices and Range of Size and Specification Extras per Ton for Structural Shapes						Special Column Sections W 14 x 455 Thru 805		
High Strength Low Alloy Steels									
Type of Steel	Minimum Yield Stress in KSI	Base	Size Extra	Spec. Extra	Base + Size & Spec. Extra		Size Extra	Spec. Extra	Base + Size Spec Extra
A36 Carbon Structural	36	$330	$0 — $140	—	$330 — $470		$120	—	$450
A529 Carbon Structural	42			—	330 — 470			—	450
A441 Manganese-Vanadium	40 to 50			—	330 — 470			—	450
A992 Columbium-Vanadium	50			—	330 — 470			—	450
A572 Columbiun-Vanadium	42			—	330 — 470			—	450
	50			$40	370 — 510		$45	495
	60			35	365 — 505		40	490
	65			40	370 — 510		50	500
A242 Corrosion-Resistant	42 to 50			—	330 — 470			—	450
A588 Weathering	42 to 50	▼	▼	60	390 — 530		▼	60	510

Metals — R051 Structural Metal Framing

R05120-235 Common Steel Sections

The upper portion of this table shows the name, shape, common designation and basic characteristics of commonly used steel sections. The lower portion explains how to read the designations used for the above illustrated common sections.

Shape & Designation	Name & Characteristics	Shape & Designation	Name & Characteristics
W	W Shape — Parallel flange surfaces	MC	Miscellaneous Channel — Infrequently rolled by some producers
S	American Standard Beam (I Beam) — Sloped inner flange	L	Angle — Equal or unequal legs, constant thickness
M	Miscellaneous Beams — Cannot be classified as W, HP or S; infrequently rolled by some producers	T	Structural Tee — Cut from W, M or S on center of web
C	American Standard Channel — Sloped inner flange	HP	Bearing Pile — Parallel flanges and equal flange and web thickness

Common drawing designations follow:

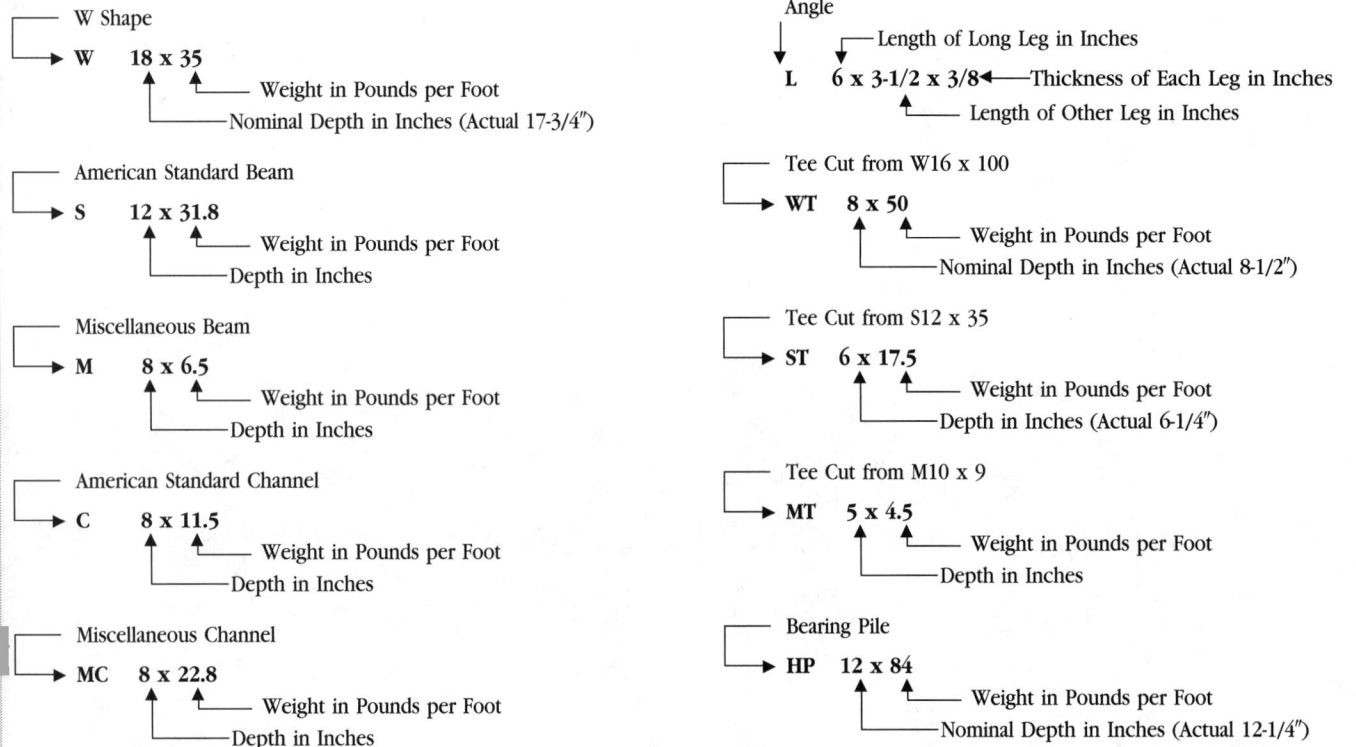

W Shape
W 18 x 35
— Weight in Pounds per Foot
— Nominal Depth in Inches (Actual 17-3/4")

American Standard Beam
S 12 x 31.8
— Weight in Pounds per Foot
— Depth in Inches

Miscellaneous Beam
M 8 x 6.5
— Weight in Pounds per Foot
— Depth in Inches

American Standard Channel
C 8 x 11.5
— Weight in Pounds per Foot
— Depth in Inches

Miscellaneous Channel
MC 8 x 22.8
— Weight in Pounds per Foot
— Depth in Inches

Angle
L 6 x 3-1/2 x 3/8
— Length of Long Leg in Inches
— Length of Other Leg in Inches
— Thickness of Each Leg in Inches

Tee Cut from W16 x 100
WT 8 x 50
— Weight in Pounds per Foot
— Nominal Depth in Inches (Actual 8-1/2")

Tee Cut from S12 x 35
ST 6 x 17.5
— Weight in Pounds per Foot
— Depth in Inches (Actual 6-1/4")

Tee Cut from M10 x 9
MT 5 x 4.5
— Weight in Pounds per Foot
— Depth in Inches

Bearing Pile
HP 12 x 84
— Weight in Pounds per Foot
— Nominal Depth in Inches (Actual 12-1/4")

Metals | R051 Structural Metal Framing

R05120-240 Structural Steel Extras

Principal Extras in Dollars Per Ton

Item quantity — using 5 tons per size as base price. Under 5 tons to 3 tons inclusive add $6 per ton. Under 3 tons to 2 tons inclusive add $11 per ton. Under 2 tons to 1 ton inclusive add $16 per ton. Under 1 ton to 1/2 ton add $55 per ton. Under 1/2 ton add $100 per ton.

Size Extras:

W Shapes: W36 x 393-135, W33 x 354-118, W30 x 326-90, W27 x 217-84, W24 x 176; W21 x 166, W18 x 311-130, W14 x 426-145, W12 x 336-136, add $20; W24 x 162-55, W21 x 147-44, W18 x 119-35, W16 x 100-26, W14 x 132-22, W12 x 120-14, W10 x 112-12, W8 x 67-10, W6 x 25-9, W5 x 19-16, and W4 x 13, no add; W14 x 808-455 add $124. For specification extras see R05120-230.

Miscellaneous Shapes: M14 x 18 to M5 x 18.9 no add. Standard Beams: S24 x 121-80, S20 x 96-66 add $140; S18 x 70 & 54.7, S15 x 50 & 42.9, S12 x 50-31.8, S10 x 35 & 25.4, S8 x 23 & 18.4, S7 x 20 & 15.3 and S6 x 17.25 & 12.5 add $80.

Standard Channels: C15 x 50-33.9, C12 x 30-20.7, C10 x 30-15.3, C9 x 20-13.4, C8 x 18.75-11.5, C7 x 14.75-9.8 add $60.

Miscellaneous Channels: MC18 x 58 to MC13 x 31.8 add $80; MC12 x 50 to MC6 x 18 add $80. Car Building Sections: add $120. Bulb Angles: add $140. Angles Equal & Unequal Leg: add $80.

Cambering: Channels, standard beams, tees & wide flange shapes to 300 lbs./ft. add $60.

Galvanizing under 1 ton, $600; over 20 tons, $500 per ton. For color coating of galvanizing add 40% to prices.

Government Specifications: MIL-S-20166 B, class U, type 1, Grade M-medium add $20; Grade HT-high tensile add $15. American Association of State Highway & Transportation Officials (AASHTO): Without Charpy Impacts, Grade M 188 no add; Grade M 222 add $65; M 223, Gr. 50 add $15.

Cut Lengths: 10' to 20' add $20; 20' to 25' add $20; 25' to 65' no add; over 65' to 90' no add; over 90' add $5.

Milling One or Two Ends: 10 lbs./ft. to 100 lbs./ft., over 10' add $70; over 100 lbs./ft., over 10' add $60.

Handling & Loading: Under 5 tons add $10. Banding or Wiring: add $10. Special loading: add $10. Special marking: add $10.

Special Testing: Bend Tests, $10, Charpy Impact Testing: one set of tests — impact strength values only, $5; one set of tests — impact strength values, lateral expansion, percent shear fracture, $7.

Special Straightening: For tolerances not more restrictive than 50% of standard camber or sweep, $7.

Splitting Beams to Produce Tees: over 15 lbs./ft.-30 lbs./ft., $60, over 30 lbs./ft.-150 lbs./ft. $40, over 150 lbs./ft., $35.

Chemistry: Ladle analysis limits for carbon, manganese, silicon & copper tests run from $10 to $30 per ton.

Metals R051 Structural Metal Framing

R05120-245 Installation Time for Structural Steel Building Components

The following tables show the expected average installation times for various structural steel shapes. Table A presents installation times for columns, Table B for beams, Table C for light framing and bolts, and Table D for structural steel for various project types.

Table A

Description	Labor-Hours	Unit
Columns		
Steel, Concrete Filled		
3-1/2" Diameter	.933	Ea.
6-5/8" Diameter	1.120	Ea.
Steel Pipe		
3" Diameter	.933	Ea.
8" Diameter	1.120	Ea.
12" Diameter	1.244	Ea.
Structural Tubing		
4" x 4"	.966	Ea.
8" x 8"	1.120	Ea.
12" x 8"	1.167	Ea.
W Shape 2 Tier		
W8 x 31	.052	L.F.
W8 x 67	.057	L.F.
W10 x 45	.054	L.F.
W10 x 112	.058	L.F.
W12 x 50	.054	L.F.
W12 x 190	.061	L.F.
W14 x 74	.057	L.F.
W14 x 176	.061	L.F.

Table B

Description	Labor-Hours	Unit	Labor-Hours	Unit
Beams, W Shape				
W6 x 9	.949	Ea.	.093	L.F.
W10 x 22	1.037	Ea.	.085	L.F.
W12 x 26	1.037	Ea.	.064	L.F.
W14 x 34	1.333	Ea.	.069	L.F.
W16 x 31	1.333	Ea.	.062	L.F.
W18 x 50	2.162	Ea.	.088	L.F.
W21 x 62	2.222	Ea.	.077	L.F.
W24 x 76	2.353	Ea.	.072	L.F.
W27 x 94	2.581	Ea.	.067	L.F.
W30 x 108	2.857	Ea.	.067	L.F.
W33 x 130	3.200	Ea.	.071	L.F.
W36 x 300	3.810	Ea.	.077	L.F.

Table C

Description	Labor-Hours	Unit
Light Framing		
Angles 4" and Larger	.055	lbs.
Less than 4"	.091	lbs.
Channels 8" and Larger	.048	lbs.
Less than 8"	.072	lbs.
Cross Bracing Angles	.055	lbs.
Rods	.034	lbs.
Hanging Lintels	.069	lbs.
High Strength Bolts in Place		
3/4" Bolts	.070	Ea.
7/8" Bolts	.076	Ea.

Table D

Description	Labor-Hours	Unit	Labor-Hours	Unit
Apartments, Nursing Homes, etc.				
1-2 Stories	4.211	Piece	7.767	Ton
3-6 Stories	4.444	Piece	7.921	Ton
7-15 Stories	4.923	Piece	9.014	Ton
Over 15 Stories	5.333	Piece	9.209	Ton
Offices, Hospitals, etc.				
1-2 Stories	4.211	Piece	7.767	Ton
3-6 Stories	4.741	Piece	8.889	Ton
7-15 Stories	4.923	Piece	9.014	Ton
Over 15 Stories	5.120	Piece	9.209	Ton
Industrial Buildings				
1 Story	3.478	Piece	6.202	Ton

Metals — R051 Structural Metal Framing

R05120-250 Subpurlins

Bulb tee and truss tee subpurlins are structural members designed to support and reinforce a variety of roof deck systems such as precast cement fiber roof deck tiles, monolithic roof deck systems, and gypsum or lightweight concrete over formboard. Other uses include interstitial service ceiling systems, wall panel systems, and joist anchoring in bond beams.

The table below shows bare costs for material only for subpurlins spaced at 32-5/8" O.C., purchased in lots of 6,000 to 10,000 L.F. (see Division 05120-140 for pricing on a square foot basis). Maximum span is based on a 3-span condition with a total allowable vertical load of 55 psf.

	Bulb Tees, Painted			Truss Tees, Painted							
Type	Wt. per L.F.	Cost per L.F.	Max. Span	Size	Wt. per L.F.	Cost per L.F.	Max. Span	Size	Wt. per L.F.	Cost per L.F.	Max. Span
112	1.48 #	$ .85	5'-6"	2"	1.01 #	$1.32	5'-9"	3"	1.06 #	$1.37	7'-3"
158	1.68	.86	6'-4"		1.17	1.38	6'-0"				
168	1.87	.91	7'-8"		1.75	1.61	7'-6"		1.90	1.69	9'-0"
178	2.15	1.00	8'-9"	2-1/2"	1.06	1.35	6'-9"	3-1/2"	1.06	1.39	7'-9"
218	3.19	1.26	10'-2"		1.31	1.44	6'-9"		1.31	1.46	8'-6"
228	3.87	1.60	12'-0"		1.90	1.67	8'-9"		1.90	1.71	10'-9"

Metals — R051 Structural Metal Framing

R05120-810 Dimensions and Weights of Sheet Steel

Gauge No.	Approximate Thickness				Weight		
	Inches (in fractions)	Inches (in decimal parts)		Millimeters	per S.F. in Ounces	per S.F. in Lbs.	per Square Meter in Kg.
	Wrought Iron	Wrought Iron	Steel	Steel			
0000000	1/2"	.5	.4782	12.146	320	20.000	97.650
000000	15/32"	.46875	.4484	11.389	300	18.750	91.550
00000	7/16"	.4375	.4185	10.630	280	17.500	85.440
0000	13/32"	.40625	.3886	9.870	260	16.250	79.330
000	3/8"	.375	.3587	9.111	240	15.000	73.240
00	11/32"	.34375	.3288	8.352	220	13.750	67.130
0	5/16"	.3125	.2989	7.592	200	12.500	61.030
1	9/32"	.28125	.2690	6.833	180	11.250	54.930
2	17/64"	.265625	.2541	6.454	170	10.625	51.880
3	1/4"	.25	.2391	6.073	160	10.000	48.820
4	15/64"	.234375	.2242	5.695	150	9.375	45.770
5	7/32"	.21875	.2092	5.314	140	8.750	42.720
6	13/64"	.203125	.1943	4.935	130	8.125	39.670
7	3/16"	.1875	.1793	4.554	120	7.500	36.320
8	11/64"	.171875	.1644	4.176	110	6.875	33.570
9	5/32"	.15625	.1495	3.797	100	6.250	30.520
10	9/64"	.140625	.1345	3.416	90	5.625	27.460
11	1/8"	.125	.1196	3.038	80	5.000	24.410
12	7/64"	.109375	.1046	2.657	70	4.375	21.360
13	3/32"	.09375	.0897	2.278	60	3.750	18.310
14	5/64"	.078125	.0747	1.897	50	3.125	15.260
15	9/128"	.0713125	.0673	1.709	45	2.813	13.730
16	1/16"	.0625	.0598	1.519	40	2.500	12.210
17	9/160"	.05625	.0538	1.367	36	2.250	10.990
18	1/20"	.05	.0478	1.214	32	2.000	9.765
19	7/160"	.04375	.0418	1.062	28	1.750	8.544
20	3/80"	.0375	.0359	.912	24	1.500	7.324
21	11/320"	.034375	.0329	.836	22	1.375	6.713
22	1/32"	.03125	.0299	.759	20	1.250	6.103
23	9/320"	.028125	.0269	.683	18	1.125	5.490
24	1/40"	.025	.0239	.607	16	1.000	4.882
25	7/320"	.021875	.0209	.531	14	.875	4.272
26	3/160"	.01875	.0179	.455	12	.750	3.662
27	11/640"	.0171875	.0164	.417	11	.688	3.357
28	1/64"	.015625	.0149	.378	10	.625	3.052

Metals — R056 Railroad & Marine Work

R05650-100 Single Track R.R. Siding

Description and Bare Costs	Quantity Unit	Unit Cost	Bare Cost/L.F.	Cost with O&P/L.F.
Rail @ 100 lb/yard	.033 Ton	$800.00	$26.40	$29.04
Spikes, Plates, and Bolts	.200 CWT	55.00	11.00	12.10
6" x 8" x 8'-6" Treated Timber Ties, 22" O.C.	.550 Ea.	28.25	15.54	17.09
Ballast, 1-1/2" Stone @ .67 Ton/L.F.	.667 Ton	12.50	8.34	9.17
Crew B-14 @ 57 L.F./Day	.018 Day	949.80	17.10	26.90
Total Cost in Place			$78.38	$94.31

R05650-200 Single Track, Steel Ties, Concrete Bed

Description and Bare Costs	Quantity	Unit Cost	Bare Cost/L.F.	Cost with O&P/L.F.
Rail, Prime Grade @ 100 lb/yard	.033 Ton	$800.00	$ 26.40	$ 29.04
Fasteners and Plates	.200 CWT	55.00	11.00	12.10
6" WF Beams @ 30" OC 6'-6" Long	2.600 L.F.	6.12	15.91	17.50
Concrete, 9' wide, 10" thick	.280 C.Y.	73.92	20.70	22.77
Crew B-14 @ 22 L.F./Day	.045 Day	949.80	42.74	67.25
Total Cost in Place			$116.75	$148.67

For information about Means Estimating Seminars, see yellow pages 11 and 12 in back of book

Wood & Plastics — R061 Rough Carpentry

R06100-010 Thirty City Lumber Prices

Prices for boards are for #2 or better or sterling, whichever is in best supply. Dimension lumber is "Standard or Better" either Southern Yellow Pine (S.Y.P.), Spruce-Pine-Fir (S.P.F.), Hem-Fir (H.F.) or Douglas Fir (D.F.). The species of lumber used in a geographic area is listed by city. Plyform is 3/4" BB oil sealed fir or S.Y.P. whichever prevails locally, 3/4" CDX is S.Y.P. or Fir.

These are prices at the time of publication and should be checked against the current market price. Relative differences between cities will stay approximately constant.

City	Species	Contractor Purchases per M.B.F. S4S Dimensions						Boards		Contractor Purchases per M.S.F.	
		2"x4"	2"x6"	2"x8"	2"x10"	2"x12"	4"x4"	1"x6"	1"x12"	3/4" Ext. Plyform	3/4" Thick CDX T&G
Atlanta	S.P.F.	$603	$577	$644	$819	$915	$824	$757	$1,289	$900	$804
Baltimore	S.P.F.	596	528	642	703	681	943	1000	1500	746	789
Boston	S.P.F.	600	568	630	790	830	1042	1920	1550	1032	831
Buffalo	H.F.	513	553	579	642	650	755	1280	1610	635	515
Chicago	S.P.F.	608	586	614	748	685	690	936	1543	779	774
Cincinnati	S.P.F.	599	683	691	795	882	863	1200	1446	824	904
Cleveland	S.P.F.	673	631	650	892	893	787	984	1910	1030	837
Columbus	S.P.F.	566	722	696	894	908	744	1419	1947	1186	775
Dallas	S.P.F.	771	656	609	690	731	668	728	1029	718	718
Denver	H.F.	631	631	646	766	658	958	771	1131	1196	900
Detroit	S.P.F.	550	589	614	851	886	594	1147	1859	804	889
Houston	S.Y.P.	528	536	536	592	641	802	707	855	985	703
Indianapolis	S.Y.F.	633	615	705	795	835	888	917	1177	825	840
Kansas City	D.F.	590	570	570	645	670	864	1050	1300	785	830
Los Angeles	D.F.	525	536	625	642	654	696	671	1385	690	790
Memphis	S.P.F.	698	654	693	834	770	1068	909	1314	954	1045
Milwaukee	S.P.F.	785	750	862	900	995	842	850	1237	828	859
Minneapolis	S.P.F.	559	545	536	675	754	674	1197	1409	1138	753
Nashville	S.P.F.	609	625	642	804	753	864	831	1543	910	754
New Orleans	S.Y.P.	577	577	561	612	570	732	591	1547	870	717
New York City	S.P.F.	539	522	605	666	577	1044	800	1101	1064	796
Philadelphia	H.F.	576	561	599	704	676	804	1223	1468	749	796
Phoenix	D.F.	614	658	715	838	768	1134	1178	1455	1308	924
Pittsburgh	H.F.	584	624	679	749	780	799	1300	1490	780	830
St. Louis	S.P.F.	488	470	412	438	506	858	910	654	742	691
San Antonio	S.Y.P.	527	531	618	662	737	616	743	712	753	802
San Diego	D.F.	522	522	550	618	588	768	1059	1178	1284	904
San Francisco	D.F.	490	490	487	630	625	840	760	1355	906	937
Seattle	S.P.F.	638	627	673	794	801	817	666	910	800	792
Washington, DC	S.P.F.	567	548	586	752	716	908	819	1210	1126	820
Average		$592	$590	$622	$731	$738	$830	$977	$1,337	$912	$811

To convert square feet of surface to board feet, 4% waste included.

S4S Size	Multiply S.F. by	T & G Size	Multiply S.F. by	Flooring Size	Multiply S.F. by
				25/32" x 2-1/4"	1.37
1 x 4	1.18	1 x 4	1.27	25/32" x 3-1/4"	1.29
1 x 6	1.13	1 x 6	1.18	15/32" x 1-1/2"	1.54
1 x 8	1.11	1 x 8	1.14	1" x 3"	1.28
1 x 10	1.09	2 x 6	2.36	1" x 4"	1.24

Wood & Plastics — R061 Rough Carpentry

R06110-030 Lumber Product Material Prices

The price of forest products fluctuates widely from location to location and from season to season depending upon economic conditions. The table below indicates National Average material prices in effect Jan. 1, 2000. The table shows relative differences between various sizes, grades and species. These percentage differentials remain fairly constant even though lumber prices in general may change significantly during the year.

Availability of certain items depends upon geographic location and must be checked prior to firm price bidding.

	National Average Contractor Price, Quantity Purchase					Heavy Timbers, Fir	
	Dimension Lumber, S4S, #2 & Better, KD						
	Species	2"x4"	2"x6"	2"x8"	2"x10"	2"x12"	
Framing Lumber per MBF	Douglas Fir	$548	$555	$589	$681	$661	3" x 4" thru 3" x 12" — $1,164
	Spruce	614	604	635	767	780	4" x 4" thru 4" x 12" — 1,326
	Southern Yellow Pine	544	548	571	622	649	6" x 6" thru 6" x 12" — 2,000
	Hem-Fir	576	592	625	715	691	8" x 8" thru 8" x 12" — 2,055
							10" x 10" and 10" x 12" — 2,086

	S4S "D" Quality or Clear, KD					S4S #2 & Better or Sterling, KD					
	Species	1"x4"	1"x6"	1"x8"	1"x10"	1"x12"	Species	1"x4"	1"x6"	1"x8"	1"x10"	1"x12"
Boards per MBF *See also Cedar Siding	Sugar Pine	$1,491	$1,645	$1,757	$2,079	$2,506	Sugar Pine	$602	$805	$861	$987	$1,288
	Idaho Pine	1,288	1,645	1,624	1,834	2,394	Idaho Pine	1,029	1,036	1,022	1,043	1,253
	Engleman Spruce	1,260	1,722	1,750	1,841	2,275	Engleman Spruce	630	756	847	966	1,295
	So. Yellow Pine	1,275	1,498	1,589	1,505	1,695	So. Yellow Pine	504	728	790	616	833
	Ponderosa Pine	1,300	1,624	1,645	1,869	2,394	Ponderosa Pine	553	756	812	945	1,218
	Redwood, CVG	3,940	3,800	3,750	3,770	5,940						

Flooring per MSF	1" x 4" Vertical grain, Fir "B" & better	$2,400	2-1/4" x 25/32" Maple, select — $2,970
	2-1/4" x 25/32", Oak, clear	3,200	#2 & better — 2,700
	Select	3,050	2-1/4" x 33/32" Maple, #2 & better — 3,132
	#1 common	2,700	3-1/4" x 33/32" Maple, #2 & better — 2,916
	Oak, prefinished, standard & better	3,750	Parquet, unfinished, 5/16", minimum — 2,700
	Standard	3,200	Maximum — 5,000

Siding per MBF	Clapboard, Cedar, beveled		*Rough sawn, Cedar, T&G, "A" grade 1" x 4" — $2,434
	1/2" x 6" thru 1/2" x 8", clear	$1,344	1" x 6" — 3,201
	"A" grade	1,246	"STK" grade, 1" x 6" — 1,614
	"B" grade	933	Board, "STK" grade, 1" x 8" — 1,561
	3/4" x 10" "clear"	2,592	Board & Batten 1" x 12" — 1,561
	"A" grade	2,334	Cedar channel siding 1" x 8", #3 & better — $1,092
	Redwood, beveled		Factory stained — 1,292
	1/2" x 6" thru 1/2" x 8", vertical grain, clear	1,907	White Pine siding, T&G, rough sawn — $637
	3/4" x 10" vertical grain, clear	2,889	Factory stained — 837

Shingles per CSF	Red Cedar		White Cedar shingles
	5X—16" long #1 regular	$151	16" long, extra grade (East Coast) — $125
	#2	170	Clear, 1st grade (East Coast) — 115
	18" long perfections #1	184	
	#2	128	Fire retardant Red Cedar
	Resquared & Rebutted #1	116	5X — 16" long — $251
			18" long perfections — 284
	Handsplit shakes, resawn		Handsplit & resawn
	24" long, 1/2" to 3/4"	145	24" long, 1/2" to 3/4" — 245
	18" long, 1/2" to 3/4"	125	18" long, 1/2" to 3/4" — 225

545

Wood & Plastics — R061 Rough Carpentry

R06160-020 Plywood

There are two types of plywood used in construction: interior, which is moisture resistant but not waterproofed, and exterior, which is waterproofed.

The grade of the exterior surface of the plywood sheets is designated by the first letter: A, for smooth surface with patches allowed; B, for solid surface with patches and plugs allowed; C, which may be surface plugged or may have knot holes up to 1" wide; and D, which is used only for interior type plywood and may have knot holes up to 2-1/2" wide. "Structural Grade" is specifically designed for engineered applications such as box beams. All CC & DD grades have roof and floor spans marked on them.

Underlayment grade plywood runs from 1/4" to 1-1/4" thick. Thicknesses 5/8" and over have optional tongue and groove joints which eliminates the need for blocking the edges. Underlayment 19/32" and over may be referred to as Sturd-i-Floor.

The price of plywood can fluctuate widely due to geographic and economic conditions. When one or two local prices are known, the relative prices for other types and sizes may be found by direct factoring of the prices in the table below.

Typical uses for various plywood grades are as follows:

AA-AD Interior — cupboards, shelving, paneling, furniture
BB Plyform — concrete form plywood
CDX — wall and roof sheathing
Structural — box beams, girders, stressed skin panels
AA-AC Exterior — fences, signs, siding, soffits, etc.
Underlayment — base for resilient floor coverings
Overlaid HDO — high density for concrete forms & highway signs
Overlaid MDO — medium density for painting, siding, soffits & signs
303 Siding — exterior siding, textured, striated, embossed, etc.

Grade	Type	4'x8'	Type	4'x8'	4'x10'
			National Average Price in Lots of 10 MSF, per MSF-January 2000		
Sanded Grade	1/4" Interior AD	$ 609	1/4" Exterior AC	$ 616	$ 696
	3/8"	693	3/8"	707	727
	1/2"	805	1/2"	819	871
	5/8"	959	5/8"	937	1,045
	3/4"	1,050	3/4"	1,064	1,239
	1"	1,221	1"	1,380	1,427
	1-1/4"	1,464	Exterior AA, add	155	160
	Interior AA, add	155	Exterior AB, add	130	135
			CD Structural 1	**Underlayment**	
Unsanded Grade 4' x 8' Sheets	5/16" CDX	$ 400	5/16", 4'x8' sheets $365	3/8", 4'x8' sheets	$ 572
	3/8"	405	3/8" 459	1/2"	733
	1/2"	531	1/2" 647	5/8"T&G	756
	5/8"	666	5/8" 777	3/4"T&G	1,211
	3/4"	806	3/4" 889	1-1/8" 2-4-1T&G	1,897
	3/4" T&G	831			
Form Plywood	5/8" Exterior, oiled BB, plyform	$ 966	5/8" HDO (overlay 2 sides)		$2,148
	3/4" Exterior, oiled BB, plyform	1,036	3/4" HDO (overlay 2 sides)		2,214
Overlaid 4'x8' Sheets	Overlay 2 Sides MDO		Overlay 1 Side MDO		
	3/8" thick	$1,212	3/8" thick		$1,029
	1/2"	1,418	1/2"		1,186
	5/8"	1,600	5/8"		1,332
	3/4"	1,830	3/4"		1,562
303 Siding	Fir, rough sawn, natural finish, 3/8" thick	$ 707	Texture 1-11	5/8" thick, Fir	$1,093
	Redwood	1,890		Redwood	1,890
	Cedar	1,890		Cedar	1,537
	Southern Yellow Pine	539		Southern Yellow	850
Waferboard/O.S.B.	1/4" sheathing	$ 298	19/32" T&G		$ 407
	7/16" sheathing	435	23/32" T&G		594

For 2 MSF to 10 MSF, add 10%. For less than 2 MSF, add 15%.

Wood & Plastics — R061 Rough Carpentry

R06170-100 Wood Roof Trusses

Loading figures represent live load. An additional load of 10 psf on the top chord and 10 psf on the bottom chord is included in the truss design. Spacing is 24" O.C.

Span in Feet	Cost per Truss for Different Live Loads and Roof Pitches					
	Flat	4 in 12 Pitch		5 in 12 Pitch		8 in 12 Pitch
	40 psf	30 psf	40 psf	30 psf	40 psf	30 psf
20	$ 66	$ 48	$ 50	$ 48	$ 52	$ 58
22	73	53	55	53	57	63
24	79	58	60	58	62	69
26	86	62	64	63	67	75
28	92	67	69	68	72	81
30	99	72	74	73	77	86
32	123	86	88	87	91	101
34	131	92	94	93	97	107
36	139	97	99	99	103	113
38	146	103	105	104	108	120
40	154	108	110	110	114	126

Wood & Plastics — R064 Architectural Woodwork

R06430-100 Wood Stair, Residential

Item					Quantity	Unit Cost	Bare Costs	Costs Incl. Subs O & P
One Flight with 8'-6" Story Height, 3'-6" Wide Oak Treads Open One Side, Built in Place								
Treads 10-1/2" x 1-1/16" thick					11 Ea.	$ 27.50	$ 302.50	$ 332.80
Landing tread nosing, rabbeted					1 Ea.	12.50	12.50	13.80
Risers 3/4" thick					12 Ea.	12.40	148.80	163.60
Single end starting step	(range	$175	to	$210)	1 Ea.	193.00	193.00	212.40
Balusters	(range	$4.50	to	$25.00)	22 Ea.	7.75	170.50	187.60
Newels, starting & landing	(range	$40	to	$120)	2 Ea.	73.00	146.00	160.60
Rail starter	(range	$39	to	$190)	1 Ea.	67.00	67.00	73.80
Handrail	(range	$4.50	to	$10.30)	26 L.F.	7.00	182.00	200.20
Cove trim					50 L.F.	.78	39.00	43.00
Rough stringers three - 2 x 12's, 14' long					84 B.F.	.74	62.16	68.40
Carpenter's installation: Bare Cost					36 Hrs.	$ 19.70	$ 709.20	
Incl. Subs O & P						$ 33.75		$1,215.00
					Total per Flight		$2,032.66	$2,671.20

Add for rail return on second floor and for varnishing or other finish. Adjoining walls or landings must be figured separately.

Thermal & Moist. Protection — R071 Waterproofing & Dampproofing

R07110-010 1/2" Pargeting (rough dampproofing plaster)

1:2-1/2 Mix, 4.5 C.F. Covers 100 S.F., 2 Coats, Waste Included			Regular Portland Cement		Waterproofed Portland Cement	
1.7	lbs. integral waterproofing admixture				$.65 per lb.	$ 1.11
1.7	bags Portland cement		$7.10 per bag	$ 12.07	$7.10 per bag	12.07
4.25	C.F. sand @	$18.70 per C.Y.		2.94		2.94
Labor mix, apply, crew D-1 @		$285.60 per day	.40 days	114.24	.40 days	114.24
Total Bare Cost per 100 S.F.				$129.25		$130.36

For information about Means Estimating Seminars, see yellow pages 11 and 12 in back of book

Thermal & Moist. Protection — R073 Shingles & Roofing Tiles

R07310-020 Roof Slate

16", 18" and 20" are standard lengths and slate usually comes in random widths. For standard 3/16" thickness use 1-1/2" copper nails. Allow for 3% breakage.

Quantities per Square	Unfading Vermont Colored	Weathering Sea Green	Buckingham, Virginia Black
Slate delivered (incl. punching)	$385.00	$286.00	$470.00
# 30 Felt, 2.5 lbs. copper nails	14.49	14.49	14.49
Slate roofer 4.6 hrs. @ $17.10 per hr.	78.66	78.66	78.66
Total Bare Cost per Square	$478.15	$379.15	$563.15

Thermal & Moist. Protection — R075 Membrane Roofing

R07510-020 Built-Up Roofing

Asphalt is available in kegs of 100 lbs. each; coal tar pitch in 560 lb. kegs. Prepared roofing felts are available in a wide range of sizes, weights & characteristics. However, the most commonly used are #15 (432 S.F. per roll, 13 lbs. per square) and #30 (216 S.F. per roll, 27 lbs. per square).

Inter-ply bitumen varies from 24 lbs. per sq. (asphalt) to 30 lbs. per sq. (coal tar) per ply, ± 25%. Flood coat bitumen also varies from 60 lbs. per sq. (asphalt) to 75 lbs. per sq. (coal tar), ± 25%. Expendable equipment (mops, brooms, screeds, etc.) runs about 16% of the bitumen cost. For new, inexperienced crews this factor may be much higher.

Rigid insulation board is typically applied in two layers. The first is mechanically attached to nailable decks or spot or solid mopped to non-nailable decks; the second layer is then spot or solid mopped to the first layer. Membrane application follows the insulation, except in protected membrane roofs, where the membrane goes down first and the insulation on top, followed with ballast (stone or concrete pavers). Insulation and related labor costs are NOT included in Div. 07510-300; they can be found in Div. 07510-700.

Prices shown are bare costs only.

4-Ply Felt on Insulated Deck	Asphalt/Glass Fiber Felt		Coal Tar/Organic Felt	
4 ply #15 felt (1/4 roll per ply)	$3.70 per ply	$ 14.80	$7.50 per ply	$ 30.00
156 lbs. hot asphalt/195 lbs. hot tar	$260 per ton	20.24	$550 per ton	53.63
500 lbs. roofing stone (1/4" to 1/2")	$21.50 per ton	5.38	$21.50 per ton	5.38
Installation, crew G-1, @ $1,320.20 per day	.050 days	66.01	.048 days	63.37
Total Bare Cost per Square in-place		$106.43		$152.38

R07550-030 Modified Bitumen Roofing

The cost of modified bitumen roofing is highly dependent on the type of installation that is planned. Installation is based on the type of modifier used in the bitumen. The two most popular modifiers are atactic polypropylene (APP) and styrene butadiene styrene (SBS). The modifiers are added to heated bitumen during the manufacturing process to change its characteristics. A polyethylene, polyester or fiberglass reinforcing sheet is then sandwiched between layers of this bitumen. When completed, the result is a pre-assembled, built-up roof that has increased elasticity and weatherablility. Some manufacturers include a surfacing material such as ceramic or mineral granules, metal particles or sand.

The preferred method of adhering SBS-modified bitumen roofing to the substrate is with hot-mopped asphalt (much the same as built-up roofing). This installation method requires a tar kettle/pot to heat the asphalt, as well as the labor, tools and equipment necessary to distribute and spread the hot asphalt.

The alternative method for applying APP and SBS modified bitumen is as follows. A skilled installer uses a torch to melt a small pool of bitumen off the membrane. This pool must form across the entire roll for proper adhesion. The installer must unroll the roofing at a pace slow enough to melt the bitumen, but fast enough to prevent damage to the rest of the membrane.

Modified bitumen roofing provides the advantages of both built-up and single-ply roofing. Labor costs are reduced over those of built-up roofing because only a single ply is necessary. The elasticity of single-ply roofing is attained with the reinforcing sheet and polymer modifiers. Modifieds have some self-healing characteristics and because of their multi-layer construction, they offer the reliability and safety of built-up roofing.

Thermal & Moist. Protection — R078 Fire and Smoke Protection

R07800-030 Firestopping

Firestopping is the sealing of structural, mechanical, electrical and other penetrations through fire-rated assemblies. The basic components of firestop systems are safing insulation and firestop sealant on both sides of wall penetrations and the top side of floor penetrations.

Pipe penetrations are assumed to be through concrete, grout, or joint compound and can be sleeved or unsleeved. Costs for the penetrations and sleeves are not included. An annular space of 1″ is assumed. Escutcheons are not included.

Metallic pipe is assumed to be copper, aluminum, cast iron or similar metallic material. Insulated metallic pipe is assumed to be covered with a thermal insulating jacket of varying thickness and materials.

Non-metallic pipe is assumed to be PVC, CPVC, FR Polypropylene or similar plastic piping material. Intumescent firestop sealant or wrap strips are included. Collars on both sides of wall penetrations and a sheet metal plate on the underside of floor penetrations are included.

Ductwork is assumed to be sheet metal, stainless steel or similar metallic material. Duct penetrations are assumed to be through concrete, grout or joint compound. Costs for penetrations and sleeves are not included. An annular space of 1/2″ is assumed.

Multi-trade openings include costs for sheet metal forms, firestop mortar, wrap strips, collars and sealants as necessary.

Structural penetrations joints are assumed to be 1/2″ or less. CMU walls are assumed to be within 1-1/2″ of metal deck. Drywall walls are assumed to be tight to the underside of metal decking.

Metal panel, glass or curtain wall systems include a spandrel area of 5′ filled with mineral wool foil-faced insulation. Fasteners and stiffeners are included.

Doors and Windows — R081 Metal Doors and Frames

R08110-100 Hollow Metal Frames

Table below lists material prices only.

Frame for Opening Size	16 Ga. Frames		16 Ga. UL Frames		16 Ga. Drywall Frames		16 Ga. UL Drywall Frames	
	6-3/4″ Deep	8-3/4″ Deep	6-3/4″ Deep	8-3/4″ Deep	7-1/8″ Deep	8-1/4″ Deep	7-1/8″ Deep	8-1/4″ Deep
2′-0″ x 6′-8″	$69	$84	$78	$93	$83	$89	$92	$98
2′-6″ x 6′-8″	69	84	78	93	83	89	92	98
3′-0″ x 7′-0″	75	85	84	94	84	90	93	99
3′-6″ x 7′-0″	80	87	89	96	88	94	97	103
4′-0″ x 7′-0″	85	97	94	104	89	94	98	103
6′-0″ x 7′-0″	90	101	108	126	99	106	117	124
8′-0″ x 8′-0″	107	124	116	134	119	126	137	144

For welded frames, add $26.00.
For galvanized frames, add $22.00.

R08110-120 Hollow Metal Doors

Table below lists material prices only, not including hardware or labor.

Door Thickness and Size		Full Flush Doors, 18 Ga.				Flush Fire Doors		
		Hollow Metal		Composite Core		Hollow Metal "B"		Hollow Metal Class "A"
	Plain	Glazed	Plain	Glazed	20 Ga.	18 Ga.	16 Ga.	18 Ga.
1-3/4″	3′-0″ x 6′-8″	$189	$246	$214	$271	$179	$205	$255	$217
	3′-0″ x 7′-0″	199	255	224	280	192	211	265	222
	3′-6″ x 7′-0″	227	300	252	325	215	244	285	258
	3′-0″ x 8′-0″	247	330	272	355	249	269	310	283
	4′-0″ x 8′-0″	285	380	310	405	320	316	349	321
1-3/8″	2′-0″ x 6′-8″	146	203	—	—	160	—	—	—
	2′-6″ x 6′-8″	155	212	—	—	170	—	—	—
	3′-0″ x 7′-0″	165	222	—	—	180	—	—	—

*Indicates 20 gauge doors.

Doors & Windows — R081 Metal Doors & Frames

R08110-130 Tin Clad Fire Door

6' x 7' Opening, A Label	Double Sliding Door		Double Swing Door	
Doors, 3 Ply		$ 510.00		$ 510.00
Frame, 8" channel (11.5 lb. per L.F.), installed	275 lb.	250.25	275 lb.	250.25
Track, hangers, hardware	13.75 L.F.	693.00	Straps and hinges	326.00
2 Carpenters @ $19.70 per hour each	16 hrs. total	315.20	16 hrs. total	315.20
Complete in place		$1,768.45		$1,401.45

Doors & Windows — R082 Wood & Plastic Doors

R08210-100 Wood Doors

Table below, except where indicated, lists price per door only, not including frame, hardware or labor. For pre-hung exterior door units up to 3' x 7', add $125 per door for wood frame and hardware for types not listed under pre-hung. Pricing is for ten or more doors. Doors are factory trimmed for butts and locksets.

Door Thickness and Size		Flush Type Doors (Interior)					Architectural (1-3/4" Ext., 1-3/8" Int.)				Pre-hung	
		Hollow Core		Solid Particle Core			Pine		Fir			
		Lauan	Birch	Lauan	Birch	Oak	Panel	Glazed	Panel	Glazed	Pine Panel	Flush Birch S. C.
1-3/4"	2'-6" x 6'-8"	$36	$56	$ 73	$ 91 *	$ 97	$289	$304	$242	$257	$394	$250
	3'-0" x 6'-8"	41	64	79	96 *	106	311	326	247	262	417	260
	3'-0" x 7'-0"	52	71	97	102 *	113	354	369	260	270	464	274
	3'-6" x 7'-0"	63	82	104	125 *	137	—	—	—	—	—	—
	4'-0" x 7'-0"	69	97	112	136 *	147	—	—	—	—	—	—
1-3/8"	2'-0" x 6'-8"	33	47	69	76 *	86	110	—	140	—	177	140
	2'-6" x 6'-8"	35	49	73	83 *	91	124	139	159	—	190	147
	3'-0" x 6'-8"	40	51	77	94 *	98	148	163	182	—	220	157
	3'-0" x 7'-0"	46	56	82	99 *	106	166	—	206	—	—	—

*Add to the above for the following birch face door types:

Solid wood stave core, add $25
3/4 hour label door, add $61
1 hour label door, add $72
1-1/2 hour label door, add $82
8' high door, add 31%

Doors & Windows — R085 Metal Windows

R08510-100 Steel Sash

Ironworker crew will erect 25 S.F. or 1.3 sash unit per hour, whichever is less.

Mechanic will point 30 L.F. per hour.

Painter will paint 90 S.F. per coat per hour.

Glazier production depends on light size.

Allow 1 lb. special steel sash putty per 16" x 20" light.

Doors & Windows — R085 | Metal Windows

R08550-010 D. H. Wood Window (ready hung)

Ponderosa pine and vinyl clad sash, exterior primed with insulating glass.

Description	2'-0" x 3'-0"				3'-0" x 4'-0"			
	Wood		Vinyl Clad		Wood		Vinyl Clad	
Window and frame, 2 lights		$122.50		$177.00		$171.00		$236.00
Removable grilles		13.50		13.50		15.60		15.60
Aluminum storm/screen		57.00		60.00		72.00		72.00
Interior trim set		15.00		15.00		19.00		19.00
Carpenter @ $19.70 per hr.	1.7 hr.	33.49	1.7 hr.	33.49	2 hr.	39.40	2 hr.	39.40
Complete in place		$241.49		$298.99		$317.00		$382.00

Mullions for above windows are about $28 each. Aluminum clad double-hung windows run about $28.60 per S.F. of glass.

Doors & Windows — R087 | Hardware

R08700-100 Hinges

All closer equipped doors should have ball bearing hinges. Lead lined or extremely heavy doors require special strength hinges. Usually 1-1/2 pair of hinges are used per door up to 7'-6" high openings. Table below shows typical hinge requirements.

Use Frequency	Type Hinge Required	Type of Opening	Type of Structure
High	Heavy weight ball bearing	Entrances	Banks, Office buildings, Schools, Stores & Theaters
		Toilet Rooms	Office buildings and Schools
Average	Standard weight ball bearing	Entrances	Dwellings
		Corridors	Office buildings and Schools
		Toilet Rooms	Stores
Low	Plain bearing	Interior	Dwellings

Door Thickness	Weight of Doors in Pounds per Square Foot				
	White Pine	Oak	Hollow Core	Solid Core	Hollow Metal
1-3/8"	3 psf	6 psf	1-1/2 psf	3-1/2 — 4 psf	6-1/2 psf
1-3/4"	3-1/2	7	2-1/2	4-1/2 — 5-1/4	6-1/2
2-1/4"	4-1/2	9	—	5-1/2 — 6-3/4	6-1/2

Doors & Windows — R088 Glazing

R08810-010 Glazing Labor

Glass sizes are estimated by the "united inch" (height + width). Table below shows the number of lights glazed in an eight hour period by the crew size indicated, for glass up to 1/4" thick. Square or nearly square lights are more economical on a S.F. basis. Long slender lights will have a high S.F. installation cost. For insulated glass reduce production by 33%. For 1/2" float glass reduce production by 50%. Production time for glazing with two glaziers per day averages: 1/4" float glass 120 S.F.; 1/2" float glass 55 S.F.; 1/2" insulated glass 95 S.F.; 3/4" insulated glass 75 S.F.

Glazing Method	United Inches per Light							
	40"	60"	80"	100"	135"	165"	200"	240"
Number of Men in Crew	1	1	1	1	2	3	3	4
Industrial sash, putty	60	45	24	15	18	—	—	—
With stops, putty bed	50	36	21	12	16	8	4	3
Wood stops, rubber	40	27	15	9	11	6	3	2
Metal stops, rubber	30	24	14	9	9	6	3	2
Structural glass	10	7	4	3	—	—	—	—
Corrugated glass	12	9	7	4	4	4	3	—
Storefronts	16	15	13	11	7	6	4	4
Skylights, putty glass	60	36	21	12	16	—	—	—
Thiokol set	15	15	11	9	9	6	3	2
Vinyl set, snap on	18	18	13	12	12	7	5	4
Maximum area per light	2.8 S.F.	6.3 S.F.	11.1 S.F.	17.4 S.F.	31.6 S.F.	47 S.F.	69 S.F.	100 S.F.
Daily Bare Crew Cost	$154.80	$154.80	$154.80	$154.80	$309.60	$464.40	$464.40	$619.20

R08810-100 Commercial Window Glass

Description		1/2" Insulated Glass		1/4" Plate Glass	
Glass per S.F.		$10.00	R Value	$3.75	R Value
Allow for breakage 10%		1.00	3.00	.38	.93
Glazing compound 1/2 lb. per S.F. @	$1.10 per lb.	.55	Average	.55	Average
Labor, 2 glaziers @	$19.35 per hour each (75 S.F./day)	4.13		4.13	
Total in place, per S.F.		$15.68		$8.81	

Finishes | R091 | Metal Support Assemblies

R09110-610　Studs, Joists and Track

Bare material prices per 1000 L.F., for galvanized studs, joists and tracks. Panhead framing screws, 7/16" long are $13.00 per thousand; #6 x 1-5/8" long drywall screws are $10.00 per thousand.

Non-load bearing, 20 ga. stud and track are primarily used for curtain wall. (C.W.)

	Non-Load Bearing				Load Bearing 1-5/8" Flange—Light Gauge Structural					
	25 Ga.		20 Ga. (C.W.)		18 Ga.		16 Ga.		14 Ga.	
Size	Stud	Track	Stud	Track	Stud	Track	Stud	Track	Stud	Track
1-5/8"	$155	$150	$267	$266						
2-1/2"	170	163	285	289	$ 527	$441	$ 646	$ 549	—	—
3-5/8"	197	195	308	318	586	501	729	632	$ 980	—
4"	238	237	384	390	647	554	796	692	1,050	—
6"	315	310	447	458	785	705	969	870	1,331	—
8"					1,019	925	1,281	1,175	1,635	$1,516

	Non-Load Bearing				Load Bearing—Extra Wide Flange — 2" Flange					
	25 Ga.		22 Ga.		18 Ga.		16 Ga.		14 Ga.	12 Ga.
Size	C-H Stud	J-Track	C-H Stud	J-Track	Joist	Track	Joist	Track	Joist	Track	Joist	Track
2-1/2"	$647	$497	$788	$750	—	—	—	—	—	—	—	—
4"	799	745	866	709	—	—	—	—	—	—	—	—
6"					$ 912	$ 705	$1,201	$ 870	$1,537	$1,162	—	—
8"					1,152	925	1,434	1,175	1,826	1,516	—	—
10"					1,359	1,222	1,707	1,467	2,195	1,873	$3,231	$2,867
12"					1,660	1,430	1,944	1,729	2,512	2,247	3,669	3,382

Finishes | R092 | Plaster & Gypsum Board

R09205-050　Gypsum Lath

		Nailed to Wood Studs		Clipped to Steel Studs	
Item		Quantities and Bare Cost	Incl. O & P	Quantities and Bare Cost	Incl. O & P
105 S.Y. 3/8" gypsum lath		$3.73 per S.Y.	$391.65	$430.82	$3.73 per S.Y.	$391.65	$430.82
Fasteners, nails & clips		6 lb. @ $1.20/lb.	7.20	7.92	600 @ $.07 ea.	42.00	46.20
Lather @ $19.20 and $31.45 per hr.		9.4 hours	180.48	295.63	10.7 hours	205.44	336.52
Total per 100 S.Y. in place			$579.33	$734.37		$639.09	$813.54

Regular lath comes in 16" x 32" and 16" x 48" sheets. Firestop gypsum base comes in 4' x 8' sheets. For nailing use 1-1/8" No. 13 ga. flathead blued nails.

R09205-060　Metal Lath

	Nailed to Wood Studs		Screwed to Steel Studs	
Painted Diamond Expanded	Quantity and Bare Cost	Incl. O & P	Quantity and Bare Cost	Incl. O & P
105 S.Y. 3.4 lb. lath	$2.31 per S.Y.	$242.55	$266.81	$2.31 per S.Y.	$242.55	$266.81
Fasteners, corner beads, etc.		4.38	4.82		9.88	10.87
Lather @ $19.20 and $31.45 per hr.	10.0 hrs.	192.00	314.50	10.7 hrs.	205.44	336.52
Total per 100 S.Y. in place		$438.93	$586.13		$457.87	$614.20

Material prices of other types of lath in Div. 092 may be substituted in the table above to arrive at their costs in place. For nailing on studs use a 1" roofing nail with 7/16" head. For nailing on ceiling use 1-1/2" No. 11 ga. barbed roofers nail with 7/16" head for holding power.

Finishes | R092 Plaster & Gypsum Board

R09210-105 Gypsum Plaster

Quantities for 100 S.Y.	2 Coat, 5/8" Thick		3 Coat, 3/4" Thick		
	Base 1:3 Mix	Finish 2:1 Mix	Scratch 1:2 Mix	Brown 1:3 Mix	Finish 2:1 Mix
Gypsum plaster	lb.	1350	lb. 650	lb.	
Sand	1.75 C.Y.		1.16 C.Y.	.84 C.Y.	
Finish hydrated lime		340 lb.			340 lb.
Gauging plaster		170 lb.			170 lb.

Total, in Place for 100 S.Y. on Walls				2 Coat, 5/8" Thick			3 Coat, 3/4" Thick		
				Quantities	Bare Cost	Incl. O & P	Quantities	Bare Cost	Incl. O & P
Gypsum plaster @	$13.02	per 80	lb. bag	1300 lb.	$211.58	$232.74	2000 lb.	$325.50	$358.05
Finish hydrated lime @	$5.60	per 50	lb. bag	340 lb.	38.08	41.89	340 lb.	38.08	41.89
Gauging plaster @	$19.02	per 100	lb. bag	170 lb.	32.33	35.56	170 lb.	32.33	35.56
Sand @	$17.25	per C.Y.		1.7 C.Y.	29.33	32.26	2.0 C.Y.	34.50	37.95
J-1 crew @	$17.45	& $29.18	per L.H.	34.8 L.H.	607.26	1,015.46	42.0 L.H.	732.90	1,225.56
Cleaning, staging, handling, patching				3.6 L.H.	62.82	105.05	4.0 L.H.	69.80	116.72
Total per 100 S.Y. in place					$981.40	$1,462.96		$1,233.11	$1,815.73

R09210-115 Vermiculite or Perlite Plaster

Proportions: Over lath, scratch coat and brown coat,
100# gypsum plaster to 2 C.F. aggregate; over masonry,
100# gypsum plaster to 3 C.F. aggregate.

Quantities for 100 S.Y.				2 Coat, 5/8" Thick			3 Coat, 3/4" Thick		
				Quantities	Bare Cost	Incl. O & P	Quantities	Bare Cost	Incl. O & P
Gypsum plaster @	$13.02	per 80	lb. bag	1250 lb.	$203.44	$223.78	2250 lb.	$366.19	$402.81
Vermiculite or perlite @	$13.03	per	bag	7.8 bags	101.63	111.79	11.3 bags	147.24	161.96
Finish hydrated lime @	$5.60	per 50	lb. bag	340 lb.	38.08	41.89	340 lb.	38.08	41.89
Gauging plaster @	$19.02	per 100	lb. bag	170 lb.	32.33	35.56	170 lb.	32.33	35.56
J-1 crew @	$18.81	& $30.68	per L.H.	40.0 L.H.	752.50	1,227.15	50.0 L.H.	940.63	1,533.94
Cleaning, staging, handling, patching				3.6 L.H.	67.73	110.44	4.0 L.H.	75.25	122.72
Total per 100 S.Y. in place					$1,195.71	$1,750.61		$1,599.72	$2,298.88

R09220-300 Stucco

Quantities for 100 S.Y. 3 Coats, 1" Thick			On Wood Frame			On Masonry		
			Quantities	Bare Cost	Incl. O & P	Quantities	Bare Cost	Incl. O & P
Portland cement @	$7.10 per bag		29 bags	$205.90	$226.49	21 bags	$149.10	$164.01
Sand @	$15.85 per C.Y.		2.6 C.Y.	41.21	45.33	2 C.Y.	31.70	34.87
Hydrated lime @	$5.60 per 50 lb. bag		180 lb.	20.16	22.18	120 lb.	13.44	14.78
Painted stucco mesh @	$2.32 per S.Y., 3.6#		105 S.Y.	243.60	267.96	—	—	—
Furring nails and jute fiber				13.84	15.22	—	—	—
Mix and install with crew indicated			J-2 @ .74 days	670.51	1,094.28	J-1 @ 0.5 days	376.25	613.58
Cleaning, staging, handling, patching			.10 days	85.16	143.63	.10 days	85.16	143.63
Total per 100 S.Y. in place				$1,280.38	$1,815.09		$655.65	$970.87

Finishes | R094 | Terrazzo

R09400-100 Terrazzo Floor

5000 S.F. 5/8" Terrazzo Topping Quantities for 100 S.F.	Bonded to Concrete 1-1/8" Bed, 1:4 Mix			Not Bonded 2-1/8" Bed and 1/4" Sand		
	Quantities	Bare Cost	Incl. O & P	Quantities	Bare Cost	Incl. O & P
Portland cement @ $7.10 per bag	6 bags	$ 42.60	$ 46.86	8 bags	$ 56.80	$ 62.48
Sand @ $15.85 per C.Y.	10 C.F.	5.87	6.46	20 C.F.	11.74	12.91
Divider strips for 4' square panels, zinc, 14 ga.	55 L.F.	50.60	55.66	55 L.F.	50.60	55.66
Terrazzo fill, domestic aggregates	600 lb.	180.00	198.00	600 lb.	180.00	198.00
15 lb. tarred felt	—	—	—		2.00	2.20
Mesh 2 x 2 #14 galvanized	—	—	—		14.75	16.23
Crew J-3 @ $25.09 & $36.62 per day	12.30 L.H.	308.61	450.43	13.90 L.H.	348.75	509.02
Total per 100 S.F. in place	Bonded	$587.68	$757.41	Not Bonded	$664.64	$856.50

2' x 2' panels require 1 L.F. divider strip per S.F.; 6' x 6' panels need .33 L.F. per S.F.

Finishes | R096 | Flooring

R09658-600 Resilient Flooring and Base

Description	12" x 12" x 3/32" V.C. Tile			4" x 1/8" Vinyl Base		
	Quantities	Bare Cost	Incl. O & P	Quantities	Bare Cost	Incl. O & P
Vinyl composition, tile, standard line	100 S.F.	$112.00	$123.20	100 L.F.	$ 63.00	$ 69.30
Vinyl cement @ $10.95 per gallon	0.8 gallon	8.76	9.64	0.5 gallon	5.48	6.03
Tile layer @ $19.20 and $31.10 per L.H.	1.5 L.H.	28.80	46.65	2.7 L.H.	51.84	83.97
Total per 100 S.F. in place		$149.56	$179.49	100 L.F.	$120.32	$159.30

Finishes | R097 | Wall Finishes

R09700-700 Wall Covering

Quantities for 100 S.F.	Medium Price Paper			Expensive Paper		
	Quantities	Bare Cost	Incl. O & P	Quantities	Bare Cost	Incl. O & P
Paper @ $30.00 and $57.00 per double roll	1.6 dbl. rolls	$48.00	$ 52.80	1.6 dbl. rolls	$ 91.20	$100.32
Wall sizing @ $19.60 per gallon	.25 gallon	4.90	5.39	.25 gallon	4.90	5.39
Vinyl wall paste @ $ 9.00 per gallon	0.4 gallon	3.60	3.96	0.4 gallon	3.60	3.96
Apply sizing @ $17.95 and $29.95 per hour	0.3 hour	5.39	8.99	0.3 hour	5.39	8.99
Apply paper @ $17.95 and $29.95 per hour	1.2 hours	21.54	35.94	1.5 hours	26.93	44.93
Total including waste allowance per 100 S.F.		$83.43	$107.08		$132.02	$163.59

This is equivalent to about $66.95 and $102.25 per double roll complete in place. Most wallpapers now come in double rolls only. To remove old paper allow 1.3 hours per 100 S.F.

Finishes | R099 | Paints & Coatings

R09910-100 Paint

Material prices per gallon in 5 gallon lots, up to 25 gallons. For 100 gallons, deduct 10%.

Exterior, Alkyd (oil base)		Masonry, Exterior		Rust inhibitor ferrous metal	$23.50
Flat	$23.00	Alkali resistant primer	$19.25	Zinc chromate	20.00
Gloss	22.25	Block filler, epoxy	20.25	**Heavy Duty Coatings**	
Primer	22.00	Block filler, latex	10.00	Acrylic urethane	51.50
		Latex, flat or semi-gloss	17.00	Chlorinated rubber	29.75
Exterior, Latex (water base)		**Masonry, Interior**		Coal tar epoxy	29.25
Acrylic stain	19.00	Alkali resistant primer	15.75	Metal pretreatment (polyvinyl butyral)	23.50
Gloss enamel	23.00	Block filler, epoxy	23.25	Polyamide epoxy finish	31.50
Flat	19.50	Block filler, latex	10.25	Polyamide epoxy primer	31.50
Primer	21.75	Floor, alkyd	21.00	Silicone alkyd	32.00
Semi-gloss	22.50	Floor, latex	21.25	2 component solvent based acrylic epoxy	59.50
		Latex, flat acrylic	11.25	2 component solvent based polyester epoxy	88.50
Interior, Alkyd (oil base)		Latex, flat emulsion	11.00	Vinyl	27.00
Enamel undercoater	20.75	Latex, sealer	12.25	Zinc rich primer	94.50
Flat	18.50	Latex, semi-gloss	18.75		
Gloss	23.00			**Special Coatings/Miscellaneous**	
Primer sealer	17.25	**Varnish and Stain**		Aluminum	22.75
Semi-gloss	22.25	Alkyd clear	21.25	Dry fall out, flat	12.25
		Polyurethane, clear	21.25	Fire retardant, intumescent	36.50
		Primer sealer	15.00	Linseed oil	10.25
		Semi-transparent stain	15.75	Shellac	18.25
Interior, Latex (water base)		Solid color stain	15.75	Swimming pool, epoxy or urethane base	36.00
Enamel undercoater	15.50			Swimming pool, rubber base	27.75
Flat	15.75	**Metal Coatings**		Texture paint	11.25
Floor and deck	21.00	Galvanized	24.50	Turpentine	10.75
Gloss	22.50	High heat	48.50	Water repellent 5% silicone	17.75
Primer sealer	18.00	Machinery enamel, alkyd	25.00		
Semi-gloss	21.50	Normal heat	25.50		

R09910-220 Painting

Item	Coat	One Gallon Covers			In 8 Hours Man Covers			Man-Hours per 100 S.F.		
		Brush	Roller	Spray	Brush	Roller	Spray	Brush	Roller	Spray
Paint wood siding	prime	250 S.F.	225 S.F.	290 S.F.	1150 S.F.	1300 S.F.	2275 S.F.	.695	.615	.351
	others	270	250	290	1300	1625	2600	.615	.492	.307
Paint exterior trim	prime	400	—	—	650	—	—	1.230	—	—
	1st	475	—	—	800	—	—	1.000	—	—
	2nd	520	—	—	975	—	—	.820	—	—
Paint shingle siding	prime	270	255	300	650	975	1950	1.230	.820	.410
	others	360	340	380	800	1150	2275	1.000	.695	.351
Stain shingle siding	1st	180	170	200	750	1125	2250	1.068	.711	.355
	2nd	270	250	290	900	1325	2600	.888	.603	.307
Paint brick masonry	prime	180	135	160	750	800	1800	1.066	1.000	.444
	1st	270	225	290	815	975	2275	.981	.820	.351
	2nd	340	305	360	815	1150	2925	.981	.695	.273
Paint interior plaster or drywall	prime	400	380	495	1150	2000	3250	.695	.400	.246
	others	450	425	495	1300	2300	4000	.615	.347	.200
Paint interior doors and windows	prime	400	—	—	650	—	—	1.230	—	—
	1st	425	—	—	800	—	—	1.000	—	—
	2nd	450	—	—	975	—	—	.820	—	—

Specialties — R105 Lockers

R10505-050 Steel Lockers

Price per Opening, Material Only, Based on 100 Openings, Not Including Locks					
Single Tier	1-Wide	3-Wide	Double Tier	1-Wide	3-Wide
12" x 12" x 60"	$128.83	$112.35	12" x 12" x 30"	$75.26	$42.56
12" x 15" x 60"	135.37	117.13	12" x 12" x 36"	79.84	45.05
12" x 18" x 60"	141.54	121.03	12" x 15" x 36"	82.36	47.28
12" x 12" x 72"	139.65	123.27	12" x 18" x 36"	89.26	49.32
12" x 15" x 72"	142.08	122.62	15" x 15" x 36"	91.78	51.00
12" x 18" x 72"	145.94	121.76	Multiple Tier	1-Wide	3-Wide
15" x 18" x 72"	174.02	134.84	5-High 12" x 12" x 12"	$33.61	$29.54
18" x 18" x 72"	182.36	162.74	12" x 15" x 12"	36.25	32.02
18" x 21" x 72"	194.00	170.85	15" x 15" x 12"	39.71	36.05
			6-High 12" x 12" x 12"	31.92	28.90
			12" x 15" x 12"	36.15	31.04

Special Construction — R130 Air Supported Structures

R13011-110 Air Supported Structures

Air supported structures are made from fabrics that can be classified into two groups: temporary and permanent. Temporary fabrics include nylon, woven polyethylene, vinyl film, and vinyl coated dacron. These have lifespans that range from five to fifteen plus years, and costs that vary accordingly from $7 to $15 per S.F. floor area. This is a package cost for a fabric shell, tension cables, primary and back-up inflation systems and doors. The lower cost structures are used for construction shelters, bulk storage and pond covers. The more expensive are used for recreational structures and warehouses.

Permanent fabrics are teflon coated fiberglass. The package cost runs $35 to $40 per S.F. floor area. The life of this structure is twenty plus years. The high cost limits its application to architectural designed structures which call for a clear span covered area, such as stadiums and convention centers. Both temporary and permanent structures are available in translucent fabrics which eliminates the need for daytime lighting.

Areas to be covered vary from 10,000 S.F. to any area up to 1000 foot wide by any lenght. Height restrictions range from a maximum of 1/2 of width to a minimum of 1/6 of the width. Erection of even the largest of the temporary structures requires no more than a week ($.07 to $.25 per S.F. floor area).

Centrifugal fans provide the inflation necessary to support the structure during application of live loads. Airlocks are usually used at large entrances to prevent loss of static pressure. Some manufacturers employ propeller fans which generate sufficient airflow (30,000 CFM) to eliminate the need for airlocks. These fans may also be automatically controlled to resist high wind conditions, regulate humidity (air changes), and provide cooling and heat.

Insulation can be provided with the addition of a second or even third interior liner, creating a dead air space with an "R" value of four to nine. Some structures allow for the liner to be collapsed into the outer shell to enable the internal heat to melt accumulated snow. For cooling or air conditioning, the exterior face of the liner can be aluminized to reflect the sun's heat. Costs for the liner run about $.80 to $1.65 per S.F. floor area.

Special Construction — R131 Pre-Engin. Struct. & Aquatic Facil.

R13128-210 Pre-engineered Steel Buildings

These buildings are manufactured by many companies and normally erected by franchised dealers throughout the U.S. The four basic types are: Rigid Frames, Truss type, Post and Beam and the Sloped Beam type. Most popular roof slope is low pitch of 1" in 12". The minimum economical area of these buildings is about 3000 S.F. of floor area. Bay sizes are usually 20' to 24' but can go as high as 30' with heavier girts and purlins. Eave heights are usually 12' to 24' with 18' to 20' most typical.

Material prices shown in Division 13128-700 are bare costs for the building shell only and do not include floors, foundations, interior finishes or utilities. Costs assume at least three bays of 24' each, a 1" in 12" roof slope, and they are based on 30 psf roof load and 20 psf wind load and no unusual requirements. Costs include the structural frame, 26 ga. colored steel roofing, 26 ga. colored steel siding, fasteners, closures, trim and flashing but no allowance for doors, windows, skylights, gutters or downspouts. Very large projects would generally cost less for materials than the prices shown.

Conditions at the site, weather, shape and size of the building, and labor availability will affect the erection cost of the building.

Special Construction | R131 | Pre-Engin. Struct. & Aquatic Facil.

R13128-310   Dome Structures

Steel — The four types are Lamella, Schwedler, Arch and Geodesic. For maximum economy, rise should be about 15 to 20% of diameter. Most common diameters are in the 200' to 300' range. Lamella domes weigh about 5 P.S.F. of floor area less than Schwedler domes. Schwedler dome weight in lbs. per S.F. approaches .046 times the diameter. Domes below 125' diameter weigh .07 times diameter and the cost per ton of steel is higher. Recent costs per ton erected range from $1,800 to $2,800 per ton of steel. See R05120-220 for estimating weights.

Wood — Small domes are of sawn lumber, larger ones are laminated. In larger sizes, triaxial and triangular cost about the same; radial domes cost more. Radial domes are economical in the 60' to 70' diameter range. Most economical range of all types is 80' to 200' diameters. Diameters can run over 400'. Costs above foundations run $31 per S.F. for under 50' diameter. At 77' diameter, radial cost $20 and triaxial cost $23 per S.F. Cost is $13.80 for 120'; $19.50 for 200'; $25.00 for 300' and $32.00 for 400' diameter. Prices include 2" decking and tension tie ring in place.

Plywood — See division 06120-200 for stressed skin and folded plates. Stock prefab geodesic domes are available with diameters from 24' to 60'. Shells cost between $8.30 to $20.00 per S.F. of floor with erection running between $.58 and $2.60 per S.F. Canopies without sides run about $1.20 per S.F. less.

Fiberglass — Aluminum framed translucent sandwich panels with spans from 5' to 45' are commercially available.

Aluminum — Stressed skin aluminum panels form geodesic domes with spans ranging from 82' to 232'. Total in-place cost for the entire range is $35.00 per S.F. of floor. For vinyl faced fiberglass insulation with an R rating of 20, add $10.00 per S.F. of floor. An aluminum space truss, triangulated or nontriangulated, with aluminum or clear acrylic closure panels can be used for clear spans of 40' to 415'. The aluminum panel costs range from $22 to $24 per S.F. floor, plus $5 per S.F. for R 19 bonded thermal and acoustical insulation. The total in-place cost for the acrylic panel is from $70 to $75 per S.F. of floor.

R13128-520   Swimming Pools

Pool prices given per square foot of surface area include pool structure, filter and chlorination equipment where required, pumps, related piping, diving boards, ladders, maintenance kit, skimmer and vacuum system. Decks and electrical service to equipment are not included.

Residential in-ground pool construction can be divided into two categories: vinyl lined and gunite. Vinyl lined pool walls are constructed of different materials including wood, concrete, plastic or metal. The bottom is often graded with sand over which the vinyl liner is installed. Costs are generally in the $16 to $23 per S.F. range. Vermiculite or soil cement bottoms may be substituted for an added cost.

Gunite pool construction is used both in residential and municipal installations. These structures are steel reinforced for strength and finished with a white cement limestone plaster. Residential costs run from $26 to $38 per S.F. surface. Municipal costs vary from $50 to $75 because plumbing codes require more expensive materials, chlorination equipment and higher filtration rates.

Municipal pools greater than 1,800 S.F. require gutter systems to control waves. This gutter may be formed into the concrete wall. Often a vinyl, stainless steel gutter or gutter and wall system is specified, which will raise the pool cost an additional $51 to $150 per L.F. of gutter installed and up to $290 per L.F. if a gutter and wall system is installed.

Competition pools usually require tile bottoms and sides with contrasting lane striping. Add $12 per S.F. of wall or bottom to be tiled.

R13128-910   Public Garage Costs

Cars	Per Car	Levels	Per Car	Levels	Per Car	Levels	Per S.F.
200 to 400	$ 9,775	3	$10,225	6	$10,025	Surface	$27.00
400 to 600	$10,050	4	$10,225	7	$10,375	2	$27.60
600 to 800	$10,475	5	$10,025	8 to 12	$10,875	3 or 4	$33.75
800 to 1200	$10,875					5 or 6	$33.30
Average	$10,300	Average per Car (310 to 320 S.F. per car)			$10,300	7 or 8	$35.00

The above costs refer to open wall construction. For enclosed structures, costs must be added for sprinkler and ventilation systems.

Special Construction — R132 | Hazardous Material Remediation

R13281-120 Asbestos Removal Process

Asbestos removal is accomplished by a specialty contractor who understands the federal and state regulations regarding the handling and disposal of the material. The process of asbestos removal is divided into many individual steps. An accurate estimate can be calculated only after all the steps have been priced.

The steps are generally as follows:
1. Obtain an asbestos abatement plan from an industrial hygienist.
2. Monitor the air quality in and around the removal area and along the path of travel between the removal area and transport area. This establishes the background contamination.
3. Construct a two part decontamination chamber at entrance to removal area.
4. Install a HEPA filter to create a negative pressure in the removal area.
5. Install wall, floor and ceiling protection as required by the plan, usually 2 layers of fireproof 6 mil polyethylene.
6. Industrial hygienist visually inspects work area to verify compliance with plan.
7. Provide temporary supports for conduit and piping affected by the removal process.
8. Proceed with asbestos removal and bagging process. Monitor air quality as described in Step #2. Discontinue operations when contaminate levels exceed applicable standards.
9. Document the legal disposal of materials in accordance with EPA standards.
10. Thoroughly clean removal area including all ledges, crevices and surfaces.
11. Post abatement inspection by industrial hygienist to verify plan compliance.
12. Provide a certificate from a licensed industrial hygienist attesting that contaminate levels are within acceptable standards before returning area to regular use.

R13281-460 Lead Paint Remediation Methods

Lead paint remediation can be accomplished by the following methods.
1. Abrasive blast
2. Chemical stripping
3. Power tool cleaning with vacuum collection system
4. Encapsulation
5. Remove and replace
6. Enclosure

Each of these methods has strengths and weakness depending on the specific circumstances of the project. The following is an overview of each method.

1. **Abrasive blasting** is usually accomplished with sand or recyclable metallic blast. Before work can begin, the area must be contained to ensure the blast material with lead does not escape to the atmosphere. The use of vacuum blast greatly reduces the containment requirements. Lead abatement equipment that may be associated with this work includes a negative air machine. In addition, it is necessary to have an industrial hygienist monitor the project on a continual basis. When the work is complete, the spent blast sand with lead must be disposed of as a hazardous material. If metallic shot was used, the lead is separated from the shot and disposed of as hazardous material. Worker protection includes disposable clothing and respiratory protection.

2. **Chemical stripping** requires strong chemicals be applied to the surface to remove the lead paint. Before the work can begin, the area under/adjacent to the work area must be covered to catch the chemical and removed lead. After the chemical is applied to the painted surface it is usually covered with paper. The chemical is left in place for the specified period, then the paper with lead paint is pulled or scraped off. The process may require several chemical applications. The paper with chemicals and lead paint adhered to it, plus the containment and loose scrapings collected by a HEPA (High Efficiency Particulate Air Filter) vac, must be disposed of as a hazardous material. The chemical stripping process usually requires a neutralizing agent and several wash downs after the paint is removed. Worker protection includes a neoprene or other compatible protective clothing and respiratory protection with face shield. An industrial hygienist is required intermittently during the process.

3. **Power tool cleaning** is accomplished using shrouded needle blasting guns. The shrouding with different end configurations is held up against the surface to be cleaned. The area is blasted with hardened needles and the shroud captures the lead with a HEPA vac and deposits it in a holding tank. An industrial hygienist monitors the project, protective clothing and a respirator is required until air samples prove otherwise. When the work is complete the lead must be disposed of as a hazardous material.

4. **Encapsulation** is a method that leaves the well bonded lead paint in place after the peeling paint has been removed. Before the work can begin, the area under/adjacent to the work must be covered to catch the scrapings. The scraped surface is then washed with a detergent and rinsed. The prepared surface is covered with approximately 10 mils of paint. A reinforcing fabric can also be embedded in the paint covering. The scraped paint and containment must be disposed of as a hazardous material. Workers must wear protective clothing and respirators.

5. **Remove and replace** is an effective way to remove lead paint from windows, gypsum walls and concrete masonry surfaces. The painted materials are removed and new materials are installed. Workers should wear a respirator and tyvek suit. The demolished materials must be disposed of as hazardous waste if it fails the TCLP (Toxicity Characteristic Leachate Process) test.

6. **Enclosure** is the process that permanently seals lead painted materials in place. This process has many applications such as covering lead painted drywall with new drywall, covering exterior construction with tyvek paper then residing, or covering lead painted structural members with aluminum or plastic. The seams on all enclosing materials must be securely sealed. An industrial hygienist monitors the project, and protective clothing and a respirator is required until air samples prove otherwise.

All the processes require clearance monitoring and wipe testing as required by the hygienist.

Special Construction | R136 | Solar & Wind Energy Equip.

R13600-610 Solar Heating (Space and Hot Water)

Collectors should face as close to due South as possible, however, variations of up to 20 degrees on either side of true South are acceptable. Local climate and collector type may influence the choice between east or west deviations. Obviously they should be located so they are not shaded from the sun's rays. Incline collectors at a slope of latitude minus 5 degrees for domestic hot water and latitude plus 15 degrees for space heating.

Flat plate collectors consist of a number of components as follows: Insulation to reduce heat loss through the bottom and sides of the collector. The enclosure which contains all the components in this assembly is usually weatherproof and prevents dust, wind and water from coming in contact with the absorber plate. The cover plate usually consists of one or more layers of a variety of glass or plastic and reduces the reradiation by creating an air space which traps the heat between the cover and the absorber plates.

The absorber plate must have a good thermal bond with the fluid passages. The absorber plate is usually metallic and treated with a surface coating which improves absorptivity. Black or dark paints or selective coatings are used for this purpose, and the design of this passage and plate combination helps determine a solar system's effectiveness.

Heat transfer fluid passage tubes are attached above and below or integral with an absorber plate for the purpose of transferring thermal energy from the absorber plate to a heat transfer medium. The heat exchanger is a device for transferring thermal energy from one fluid to another.

Piping and storage tanks should be well insulated to minimize heat losses.

Size domestic water heating storage tanks to hold 20 gallons of water per user, minimum, plus 10 gallons per dishwasher or washing machine. For domestic water heating an optimum collector size is approximately 3/4 square foot of area per gallon of water storage. For space heating of residences and small commercial applications the collector is commonly sized between 30% and 50% of the internal floor area. For space heating of large commercial applications, collector areas less than 30% of the internal floor area can still provide significant heat reductions.

A supplementary heat source is recommended for Northern states for December through February.

The solar energy transmission per square foot of collector surface varies greatly with the material used. Initial cost, heat transmittance and useful life are obviously interrelated.

Conveying Systems | R142 | Elevators

R14200-100 Freight Elevators

Capacities run from 2,000 lbs. to over 100,000 lbs. with 3,000 lbs. to 10,000 lbs. most common. Travel speeds are generally lower and control less intricate than on passenger elevators. Unit prices in division 14210 are for hydraulic and geared elevators.

Conveying Systems — R142 Elevators

R14200-200 Elevator Selective Costs
See R14200-400 for cost development.

	Passenger		Freight		Hospital	
A. Base Unit	**Hydraulic**	**Electric**	**Hydraulic**	**Electric**	**Hydraulic**	**Electric**
Capacity	1,500 Lb.	2,000 Lb.	2,000 Lb.	4,000 Lb.	4,000 lb.	4,000 lb.
Speed	100 F.P.M.	200 F.P.M.	50 F.P.M.	200 F.P.M.	100 F.P.M.	200 F.P.M.
#Stops/Travel Ft.	2/12	4/40	2/20	4/40	2/20	4/40
Push Button Oper.	Yes	Yes	Yes	Yes	Yes	Yes
Telephone Box & Wire	"	"	"	"	"	"
Emergency Lighting	"	"	No	No	"	"
Cab	Plastic Lam. Walls	Plastic Lam. Walls	Painted Steel	Painted Steel	Plastic Lam. Walls	Plastic Lam. Walls
Cove Lighting	Yes	Yes	No	No	Yes	Yes
Floor	V.C.T.	V.C.T.	Wood w/Safety Treads	Wood w/Safety Treads	V.C.T.	V.C.T.
Doors, & Speedside Slide	Yes	Yes	Yes	Yes	Yes	Yes
Gates, Manual	No	No	No	No	No	No
Signals, Lighted Buttons	Car and Hall	Car and Hall	Car and Hall	Car and Hall	Car and Hall	Car and Hall
O.H. Geared Machine	N.A.	Yes	N.A.	Yes	N.A.	Yes
Variable Voltage Contr.	"	"	N.A.	"	"	"
Emergency Alarm	Yes	"	Yes	"	Yes	"
Class "A" Loading	N.A.	N.A.	"	"	N.A.	N.A.
Base Cost	$38,070	$84,943	$43,160	$84,313	$49,205	$89,441
B. Capacity Adjustment						
2,000 Lb.	$ 525					
2,500	1,150	$ 2,766	$ 2,242			
3,000	2,928	4,120	3,870			
3,500	5,012	5,871	5,853			
4,000	5,781	5,977	6,263			
4,500	6,919	7,858	7,435		5,769	$ 4,844
5,000	9,661	9,947	10,137	$ 4,869	8,403	6,326
6,000			10,440	8,586		
7,000			16,496	11,495		
8,000			17,976	16,023		
10,000			19,088	19,097		
12,000			22,721	23,320		
16,000			29,627	28,007		
20,000			32,877	30,748		
C. Travel Over Base	$ 340 V.L.F.	$ 217 V.L.F.	$ 340 V.L.F.	$ 217 V.L.F.	$ 340 V.L.F.	$ 217 V.L.F.
D. Additional Stops	$ 3,512 Ea.	$ 5,500 Ea.	$ 3,573 Ea.	$ 4,881 Ea.	$ 6,052 Ea.	$ 5,948 Ea.
E. Speed Adjustment						
100 F.P.M.			$ 728			
125	$ 800		1,348		$ 1,375	
150	1,725		2,515		2,315	
175	2,986		3,944		3,670	
200	4,841				5,355	
Geared 250		$ 2,275		$ 9,691		$ 2,446
4 Flrs. 300		4,637		12,090		4,560
Min. 350		5,449		14,269		5,563
400		7,774		15,818		7,748
Gearless 500		36,171		19,812		34,585
4 Flrs. 600		38,198		21,892		38,199
Min. 700		41,810		25,503		41,702
800		46,072		28,348		46,077
1,000		Spec. Applic.		Spec. Applic.		Spec. Applic.
1,200		"		"		"
F. Other Than Class "A" Loading						
"B"			$ 1,664	$ 1,698		
"C-1"			4,174	4,199		
"C-2"			4,997	5,012		
"C-3"			6,836	6,881		

Conveying Systems — R142 Elevators

R14200-200 Elevator Selective Costs (cont.)

	Passenger	Freight	Hospital
G. Options			
1. Controls			
Automatic, 2 car group	$3,628		$3,628
3 car group	5,399		5,395
4 car group	8,319		8,330
5 car group	11,004		11,033
6 car group	15,580		15,740
Emergency, fireman service	Yes		Yes
Intercom service	600		600
Selective collective, single car	—		—
Duplex car	3,908		3,908
2. Doors			
Center opening, 1 speed	$1,763		$1,763
2 speed	2,174		2,174
Rear opening-opposite front	4,132		4,132
Side opening, 2 speed	6,159		6,159
Freight, bi-parting	—	$ 5,200	—
Power operated door and gate	—	17,160	—
3. Emergency power switching, automatic	$2,150		$2,150
Manual	400		400
4. Finishes based on 3500# cab			
Ceilings, acrylic panel	$400		
Aluminum egg crate	475		$475
Doors, stainless steel	820 Ea.		820 Ea.
Floors, carpet, class "A" — (Finish flooring by others)	342		—
Epoxy	281		281
Quarry tile	431		433
Slate	602		—
Steel plate	—	$ 685	
Textured rubber	113		113
Walls, plastic laminate	Std.		Std.
Stainless steel	2,500		2,500
Return at door	525		525
Steel plate, 1/4" x 4' high, 14 ga. above	—	$ 1,665	—
Entrance, doors, baked enamel	Std.		Std.
Stainless steel	820 Ea.		820 Ea.
Frames, baked enamel	Std.		Std.
Stainless steel	820 Ea.		820 Ea.
5. Maintenance contract - 12 months	$2,925	$ 2,186	$4,249
6. Signal devices			
Hall lantern, each	$435	$ 435	$435
Position indicator, car or lobby	319	319	319
Add for over three each, per floor	90	90	90
7. Specialties			
High speed, heavy duty door opener	$1,560		$1,560
Variable voltage, O.H. gearless machine	29,224-64,896		29,8224-64,896
Basement installed geared machine	$14,400	$14,450	$14,400

Conveying Systems — R142 Elevators

R14200-300 Passenger Elevators

Electric elevators are used generally but hydraulic elevators can be used for lifts up to 70' and where large capacities are required. Hydraulic speeds are limited to 200 F.P.M. but cars are self leveling at the stops. On low rises, hydraulic installation runs about 15% less than standard electric types but on higher rises this installation cost advantage is reduced. Maintenance of hydraulic elevators is about the same as electric type but underground portion is not included in the maintenance contract.

In electric elevators there are several control systems available, the choice of which will be based upon elevator use, size, speed and cost criteria. The two types of drives are geared for low speeds and gearless for 450 F.P.M. and over.

The tables on the preceding pages illustrate typical installed costs of the various types of elevators available.

R14200-400 Elevator Cost Development

Requirement: One passenger elevator, five story hydraulic, 2,500 lb. capacity, 12' floor to floor, speed 150 F.P.M., emergency power switching and maintenance contract.

Description	Total Cost
A. Base Elevator, (Hydraulic Passenger)	$38,070
B. Capacity Adjustment (2,500 lb.)	1,150
C. Excess Travel Over Base (4 x 12') = 48' — (12' for Base Unit) = 36' x $340 per V.L.F.	12,240
D. Stops Over Base 5 — (2 for base unit) = 3 x $3,512 Ea.	10,536
E. Speed Adjustment (150 F.P.M.)	1,725
F. Options:	
1. Intercom Service	600
2. Emergency power switching, automatic	2,150
3. Stainless steel entrance doors 5 x $820 Ea.	4,100
4. Maintenance Contract (12 months)	2,925
5. Position indicators 2 x $319	638
Total Cost	$74,134

Conveying Systems — R143 Escalators & Moving Walks

R14320-100 Escalators

Moving stairs can be used for buildings where 600 or more people are to be carried to the second floor or beyond. Freight cannot be carried on escalators and at least one elevator must be available for this function. Carrying capacity is 5,000 to 8,000 people per hour. Power requirement is 2 to 3 KW per hour and incline angle is 30°.

Table below indicates approximate installed costs per unit, with O & P. Installation outside major cities would increase the costs below, due to mechanics additional travel expense.

| Vertical Rise | Stainless Steel Balustrade | | Enameled Steel or Glass Balustrade | |
	32" Wide	48" Wide	32" Wide	48" Wide
10'	$ 93,373	$ 96,483	$ 88,112	$ 92,729
15'	98,393	101,707	93,647	97,524
20'	107,738	111,798	100,637	105,153
25'	120,719	125,443	113,117	118,155

R14360-200 Moving Ramps and Walks

These are a specialized form of conveyor 3' to 6' wide with capacities of 3,600 to 18,000 persons per hour. Maximum speed is 140 F.P.M. and normal incline is 0° to 15°.

Local codes will determine the maximum angle. A 48" wide, 150' long indoor moving ramp costs between $290,000 to $330,000 per unit installed, including O & P. Outdoor units would require additional weather protection. Approximate cost add is $250 per foot.

Mechanical — R150 — Basic Materials & Methods

R15050-720 Demolition (Selective vs. Removal for Replacement)

Demolition can be divided into two basic categories.

One type of demolition involves the removal of material with no concern for its replacement. The labor-hours to estimate this work are found in Div. 15055 under "Selective Demolition". It is selective in that individual items or all the material installed as a system or trade grouping such as plumbing or heating systems are removed. This may be accomplished by the easiest way possible, such as sawing, torch cutting, or sledge hammer as well as simple unbolting.

The second type of demolition is the removal of some item for repair or replacement. This removal may involve careful draining, opening of unions, disconnecting and tagging of electrical connections, capping of pipes/ducts to prevent entry of debris or leakage of the material contained as well as transport of the item away from its in-place location to a truck/dumpster. An approximation of the time required to accomplish this type of demolition is to use half of the time indicated as necessary to install a new unit. For example; installation of a new pump might be listed as requiring 6 labor-hours so if we had to estimate the removal of the old pump we would allow an additional 3 hours for a total of 9 hours. That is, the complete replacement of a defective pump with a new pump would be estimated to take 9 labor-hours.

Mechanical — R151 — Plumbing

R15100-050 Pipe Material Considerations

1. Malleable fittings should be used for gas service.
2. Malleable fittings are used where there are stresses/strains due to expansion and vibration.
3. Cast fittings may be broken as an aid to disassembling of heating lines frozen by long use, temperature and minerals.
4. Cast iron pipe is extensively used for underground and submerged service.
5. Type M (light wall) copper tubing is available in hard temper only and is used for nonpressure and less severe applications than K and L.
6. Type L (medium wall) copper tubing, available hard or soft for interior service.
7. Type K (heavy wall) copper tubing, available in hard or soft temper for use where conditions are severe. For underground and interior service.
8. Hard drawn tubing requires fewer hangers or supports but should not be bent. Silver brazed fittings are recommended, however soft solder is normally used.
9. Type DMV (very light wall) copper tubing designed for drainage, waste and vent plus other non-critical pressure services.

Domestic/Imported Pipe and Fittings Cost

The prices shown in this publication for steel/cast iron pipe and steel, cast iron, malleable iron fittings are based on domestic production sold at the normal trade discounts. The above listed items of foreign manufacture may be available at prices of 1/3 to 1/2 those shown. Some imported items after minor machining or finishing operations are being sold as domestic to further complicate the system.

Caution: Most pipe prices in this book also include a coupling and pipe hangers which for the larger sizes can add significantly to the per foot cost and should be taken into account when comparing "book cost" with quoted supplier's cost.

Mechanical | R151 Plumbing

R15100-420 Plumbing Fixture Installation Time

Item	Rough-In	Set	Total Hours	Item	Rough-In	Set	Total Hours
Bathtub	5	5	10	Shower head only	2	1	3
Bathtub and shower, cast iron	6	6	12	Shower drain	3	1	4
Fire hose reel and cabinet	4	2	6	Shower stall, slate		15	15
Floor drain to 4 inch diameter	3	1	4	Slop sink	5	3	8
Grease trap, single, cast iron	5	3	8	Test 6 fixtures			14
Kitchen gas range		4	4	Urinal, wall	6	2	8
Kitchen sink, single	4	4	8	Urinal, pedestal or floor	6	4	10
Kitchen sink, double	6	6	12	Water closet and tank	4	3	7
Laundry tubs	4	2	6	Water closet and tank, wall hung	5	3	8
Lavatory wall hung	5	3	8	Water heater, 45 gals. gas, automatic	5	2	7
Lavatory pedestal	5	3	8	Water heaters, 65 gals. gas, automatic	5	2	7
Shower and stall	6	4	10	Water heaters, electric, plumbing only	4	2	6

Fixture prices in front of book are based on the cost per fixture set in place. The rough-in cost, which must be added for each fixture, includes carrier, if required, some supply, waste and vent pipe connecting fittings and stops. The lengths of rough-in pipe are nominal runs which would connect to the larger runs and stacks. The supply runs and DWV runs and stacks must be accounted for in separate entries. In the eastern half of the United States it is common for the plumber to carry these to a point 5′ outside the building.

R15100-430 Water Cooler Application

Type of Service	Requirement
Office, School or Hospital	12 persons per gallon per hour
Office, Lobby or Department Store	4 or 5 gallons per hour per fountain
Light manufacturing	7 persons per gallon per hour
Heavy manufacturing	5 persons per gallon per hour
Hot heavy manufacturing	4 persons per gallon per hour
Hotel	.08 gallons per hour per room
Theatre	1 gallon per hour per 100 seats

Mechanical | R155 Heating

R15500-035 Heating (42° degrees latitude)

$$\text{Approximate S.F. radiation} = \frac{\text{S.F. sash}}{2} + \frac{\text{S.F. wall + roof} - \text{sash}}{20} + \frac{\text{C.F. building}}{200}$$

Mechanical | R155 | Heating

R15500-050 Factor for Determining Heat Loss for Various Types of Buildings

General: While the most accurate estimates of heating requirements would naturally be based on detailed information about the building being considered, it is possible to arrive at a reasonable approximation using the following procedure:

1. Calculate the cubic volume of the room or building.
2. Select the appropriate factor from Table 1 below. Note that the factors apply only to inside temperatures listed in the first column and to 0°F outside temperature.
3. If the building has bad north and west exposures, multiply the heat loss factor by 1.1.
4. If the outside design temperature is other than 0°F, multiply the factor from Table 1 by the factor from Table 2.
5. Multiply the cubic volume by the factor selected from Table 1. This will give the estimated BTUH heat loss which must be made up to maintain inside temperature.

Table 1 — Building Type	Conditions	Qualifications	Loss Factor*
Factories & Industrial Plants General Office Areas at 70°F	One Story	Skylight in Roof	6.2
		No Skylight in Roof	5.7
	Multiple Story	Two Story	4.6
		Three Story	4.3
		Four Story	4.1
		Five Story	3.9
		Six Story	3.6
	All Walls Exposed	Flat Roof	6.9
		Heated Space Above	5.2
	One Long Warm Common Wall	Flat Roof	6.3
		Heated Space Above	4.7
	Warm Common Walls on Both Long Sides	Flat Roof	5.8
		Heated Space Above	4.1
Warehouses at 60°F	All Walls Exposed	Skylights in Roof	5.5
		No Skylight in Roof	5.1
		Heated Space Above	4.0
	One Long Warm Common Wall	Skylight in Roof	5.0
		No Skylight in Roof	4.9
		Heated Space Above	3.4
	Warm Common Walls on Both Long Sides	Skylight in Roof	4.7
		No Skylight in Roof	4.4
		Heated Space Above	3.0

*Note: This table tends to be conservative particularly for new buildings designed for minimum energy consumption.

Table 2 — Outside Design Temperature Correction Factor (for Degrees Fahrenheit)									
Outside Design Temperature	50	40	30	20	10	0	-10	-20	-30
Correction Factor	.29	.43	.57	.72	.86	1.00	1.14	1.28	1.43

Mechanical — R157 Air Conditioning & Ventilation

R15700-020 Air Conditioning Requirements

BTU's per hour per S.F. of floor area and S.F. per ton of air conditioning.

Type of Building	BTU per S.F.	S.F. per Ton	Type of Building	BTU per S.F.	S.F. per Ton	Type of Building	BTU per S.F.	S.F. per Ton
Apartments, Individual	26	450	Dormitory, Rooms	40	300	Libraries	50	240
Corridors	22	550	Corridors	30	400	Low Rise Office, Exterior	38	320
Auditoriums & Theaters	40	300/18*	Dress Shops	43	280	Interior	33	360
Banks	50	240	Drug Stores	80	150	Medical Centers	28	425
Barber Shops	48	250	Factories	40	300	Motels	28	425
Bars & Taverns	133	90	High Rise Office—Ext. Rms.	46	263	Office (small suite)	43	280
Beauty Parlors	66	180	Interior Rooms	37	325	Post Office, Individual Office	42	285
Bowling Alleys	68	175	Hospitals, Core	43	280	Central Area	46	260
Churches	36	330/20*	Perimeter	46	260	Residences	20	600
Cocktail Lounges	68	175	Hotel, Guest Rooms	44	275	Restaurants	60	200
Computer Rooms	141	85	Corridors	30	400	Schools & Colleges	46	260
Dental Offices	52	230	Public Spaces	55	220	Shoe Stores	55	220
Dept. Stores, Basement	34	350	Industrial Plants, Offices	38	320	Shop'g. Ctrs., Supermarkets	34	350
Main Floor	40	300	General Offices	34	350	Retail Stores	48	250
Upper Floor	30	400	Plant Areas	40	300	Specialty	60	200

*Persons per ton
12,000 BTU = 1 ton of air conditioning

R15700-040 Recommended Ventilation Air Changes

Table below lists range of time in minutes per change for various types of facilities.

Facility	Min/Change	Facility	Min/Change	Facility	Min/Change
Assembly Halls	2-10	Dance Halls	2-10	Laundries	1-3
Auditoriums	2-10	Dining Rooms	3-10	Markets	2-10
Bakeries	2-3	Dry Cleaners	1-5	Offices	2-10
Banks	3-10	Factories	2-5	Pool Rooms	2-5
Bars	2-5	Garages	2-10	Recreation Rooms	2-10
Beauty Parlors	2-5	Generator Rooms	2-5	Sales Rooms	2-10
Boiler Rooms	1-5	Gymnasiums	2-10	Theaters	2-8
Bowling Alleys	2-10	Kitchens-Hospitals	2-5	Toilets	2-5
Churches	5-10	Kitchens-Restaurant	1-3	Transformer Rooms	1-5

CFM air required for changes = Volume of room in cubic feet ÷ Minutes per change.

Mechanical

R158 | Air Distribution

R15810-050 Ductwork

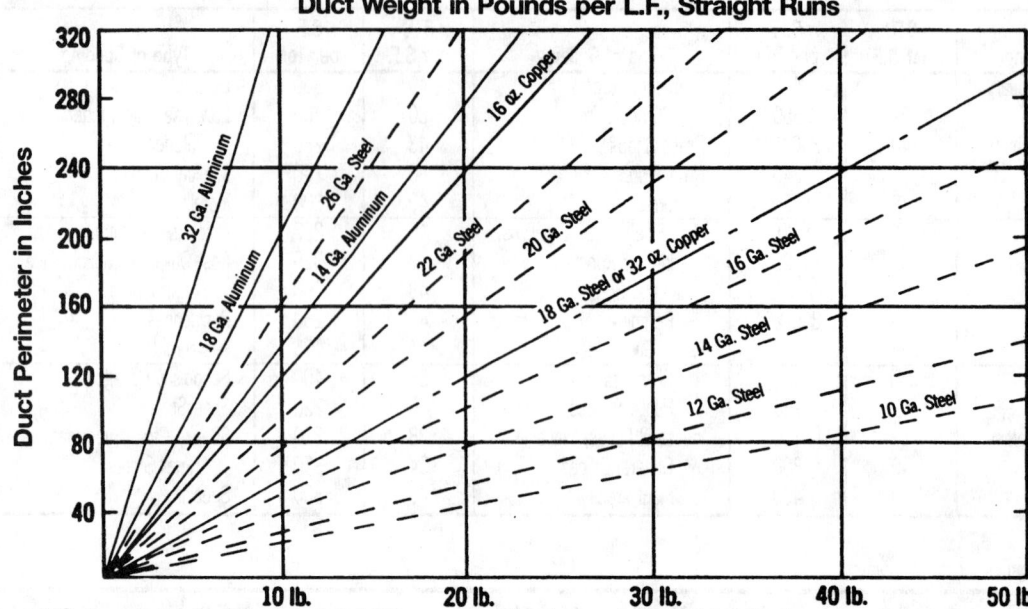

Duct Weight in Pounds per L.F., Straight Runs

Add to the above for fittings; 90° elbow is 3 L.F.; 45° elbow is 2.5 L.F.; offset is 4 L.F.; transition offset is 6 L.F.; square-to-round transition is 4 L.F.; 90° reducing elbow is 5 L.F. For bracing and waste, add 20% to aluminum and copper, 15% to steel.

Electrical — R161 Wiring Methods

R16120-910 Minimum Copper and Aluminum Wire Size Allowed for Various Types of Insulation

Minimum Wire Sizes

Amperes	Copper THW THWN or XHHW	Copper THHN XHHW *	Aluminum THW XHHW	Aluminum THHN XHHW *	Amperes	Copper THW THWN or XHHW	Copper THHN XHHW *	Aluminum THW XHHW	Aluminum THHN XHHW *
15A	#14	#14	#12	#12	195	3/0	2/0	250kcmil	4/0
20	#12	#12	#10	#10	200	3/0	3/0	250kcmil	4/0
25	#10	#10	#10	#10	205	4/0	3/0	250kcmil	4/0
30	#10	#10	#8	#8	225	4/0	3/0	300kcmil	250kcmil
40	#8	#8	#8	#8	230	4/0	4/0	300kcmil	250kcmil
45	#8	#8	#6	#8	250	250kcmil	4/0	350kcmil	300kcmil
50	#8	#8	#6	#6	255	250kcmil	4/0	400kcmil	300kcmil
55	#6	#8	#4	#6	260	300kcmil	4/0	400kcmil	350kcmil
60	#6	#6	#4	#6	270	300kcmil	250kcmil	400kcmil	350kcmil
65	#6	#6	#4	#4	280	300kcmil	250kcmil	500kcmil	350kcmil
75	#4	#6	#3	#4	285	300kcmil	250kcmil	500kcmil	400kcmil
85	#4	#4	#2	#3	290	350kcmil	250kcmil	500kcmil	400kcmil
90	#3	#4	#2	#2	305	350kcmil	300kcmil	500kcmil	400kcmil
95	#3	#4	#1	#2	310	350kcmil	300kcmil	500kcmil	500kcmil
100	#3	#3	#1	#2	320	400kcmil	300kcmil	600kcmil	500kcmil
110	#2	#3	1/0	#1	335	400kcmil	350kcmil	600kcmil	500kcmil
115	#2	#2	1/0	#1	340	500kcmil	350kcmil	600kcmil	500kcmil
120	#1	#2	1/0	1/0	350	500kcmil	350kcmil	700kcmil	500kcmil
130	#1	#2	2/0	1/0	375	500kcmil	400kcmil	700kcmil	600kcmil
135	1/0	#1	2/0	1/0	380	500kcmil	400kcmil	750kcmil	600kcmil
150	1/0	#1	3/0	2/0	385	600kcmil	500kcmil	750kcmil	600kcmil
155	2/0	1/0	3/0	3/0	420	600kcmil	500kcmil		700kcmil
170	2/0	1/0	4/0	3/0	430		500kcmil		750kcmil
175	2/0	2/0	4/0	3/0	435		600kcmil		750kcmil
180	3/0	2/0	4/0	4/0	475		600kcmil		

*Dry Locations Only

Notes:
1. Size #14 to 4/0 is in AWG units (American Wire Gauge).
2. Size 250 to 750 is in kcmil units (Thousand Circular Mils).
3. Use next higher ampere value if exact value is not listed in table.
4. For loads that operate continuously increase ampere value by 25% to obtain proper wire size.
5. Refer to Table R16120-905 for the maximum circuit length for the various size wires.
6. Table R16120-910 has been written for estimating purpose only, based on ambient temperature of 30° C (86° F); for ambient temperature other than 30° C (86° F), ampacity correction factors will be applied.

Electrical — R161 Wiring Methods

R16132-210 Conductors in Conduit

Table below lists maximum number of conductors for various sized conduit using THW, TW or THWN insulations.

Copper Wire Size	1/2" TW	1/2" THW	1/2" THWN	3/4" TW	3/4" THW	3/4" THWN	1" TW	1" THW	1" THWN	1-1/4" TW	1-1/4" THW	1-1/4" THWN	1-1/2" TW	1-1/2" THW	1-1/2" THWN	2" TW	2" THW	2" THWN	2-1/2" TW	2-1/2" THW	2-1/2" THWN	3" THW	3" THWN	3-1/2" THW	3-1/2" THWN	4" THW	4" THWN
#14	9	6	13	15	10	24	25	16	39	44	29	69	60	40	94	99	65	154	142	93		143		192			
#12	7	4	10	12	8	18	19	13	29	35	24	51	47	32	70	78	53	114	111	76	164	117		157			
#10	5	4	6	9	6	11	15	11	18	26	19	32	36	26	44	60	43	73	85	61	104	95	160	127		163	
#8	2	1	3	4	3	5	7	5	9	12	10	16	17	13	22	28	22	36	40	32	51	49	79	66	106	85	136
#6		1	1		2	4		4	6		7	11		10	15		16	26		23	37	36	57	48	76	62	98
#4		1	1		1	2		3	4		5	7		7	9		12	16		17	22	27	35	36	47	47	60
#3		1	1		1	1		2	3		4	6		6	8		10	13		15	19	23	29	31	39	40	51
#2		1	1		1	1		2	3		4	5		5	7		9	11		13	16	20	25	27	33	34	43
#1					1	1		1	1		3	3		4	5		6	8		9	12	14	18	19	25	25	32
1/0					1	1		1	1		2	3		3	4		5	7		8	10	12	15	16	21	21	27
2/0					1	1		1	1		1	2		3	3		5	6		7	8	10	13	14	17	18	22
3/0					1	1		1	1		1	1		2	3		4	5		6	7	9	11	12	14	15	18
4/0						1		1	1		1	1		1	2		3	4		5	6	7	9	10	12	13	15
250 kcmil								1	1		1	1		1	1		2	3		4	4	6	7	8	10	10	12
300								1	1		1	1		1	1		2	3		3	4	5	6	7	8	9	11
350									1		1	1		1	1		1	2		3	3	4	5	6	7	8	9
400											1	1		1	1		1	1		2	3	4	5	5	6	7	8
500											1	1		1	1		1	1		1	2	3	4	4	5	6	7
600												1		1	1		1	1		1	1	3	3	4	4	5	5
700														1	1		1	1		1	1	2	3	3	4	4	5
750														1	1		1	1		1	1	2	2	3	3	4	4

Electrical | R162 | Electrical Power

R16220-610 Motors

230/460 Volt A.C., three phase 60 cycle ball bearing squirrel cage induction motors, NEMA design B Standard Line, including installation and Subs O & P. (No conduit, wire, or terminations)

Horsepower	Synchronous Speed in RPM	Drip Proof, Class B Insulation 1.15 Service Factor			Totally Enclosed, Class B Insulation, 1.0 or 1.15 Service Factor		
		Motor Only	With Manual Starter	With Magnetic Starter	Motor Only	With Manual Starter	With Magnetic Starter
1	1,200	$ 287	$512	$ 552	$ 354	$ 582	$ 617
	1,800	257	482	522	320	547	582
2	1,200	379	607	725	409	637	755
	1,800	291	517	635	369	597	715
	3,600	276	502	620	307	532	650
3	1,200	489	717	835	524	752	870
	1,800	292	517	635	374	602	720
	3,600	314	542	660	314	542	660
5	1,200	574	889	1,000	734	1049	1,160
	1,800	358	674	785	414	729	840
	3,600	358	674	785	409	724	835
7.5	1,800	499	813	1,228	574	888	1,303
10		592	907	1,322	682	997	1,412
15		785	—	1,505	775	—	1,505
20		946	—	2,032	946	—	2,032
25		1,101	—	2,190	1,086	—	2,165
30		1,271	—	2,343	1,246	—	2,318
40		1,819	—	3,976	1,544	—	3,701
50		2,081	—	4,238	1,806	—	3,963
60		2,707	—	7,267	2,207	—	6,767
75		3,067	—	7,629	2,692	—	7,254
100		4,022	—	8,596	3,422	—	7,996
125		4,714	—	16,435	4,139	—	15,835
150		6,208	—	17,909	5,858	—	17,609
200		8,129	—	19,900	7,379	—	19,100

S.F. & C.F. Costs — R171 Information

R17100-100 Square Foot Project Size Modifier

One factor that affects the S.F. cost of a particular building is the size. In general, for buildings built to the same specifications in the same locality, the larger building will have the lower S.F. cost. This is due mainly to the decreasing contribution of the exterior walls plus the economy of scale usually achievable in larger buildings. The Area Conversion Scale shown below will give a factor to convert costs for the typical size building to an adjusted cost for the particular project.

The Square Foot Base Size lists the median costs, most typical project size in our accumulated data and the range in size of the projects.

The Size Factor for your project is determined by dividing your project area in S.F. by the typical project size for the particular Building Type. With this factor, enter the Area Conversion Scale at the appropriate Size Factor and determine the appropriate cost multiplier for your building size.

Example: Determine the cost per S.F. for a 100,000 S.F. Mid-rise apartment building.

$$\frac{\text{Proposed building area} = 100{,}000 \text{ S.F.}}{\text{Typical size from below} = 50{,}000 \text{ S.F.}} = 2.00$$

Enter Area Conversion scale at 2.0, intersect curve, read horizontally the appropriate cost multiplier of .94. Size adjusted cost becomes .94 x $68.25 = $64.15 based on national average costs.

Note: For Size Factors less than .50, the Cost Multiplier is 1.1
For Size Factors greater than 3.5, the Cost Multiplier is .90

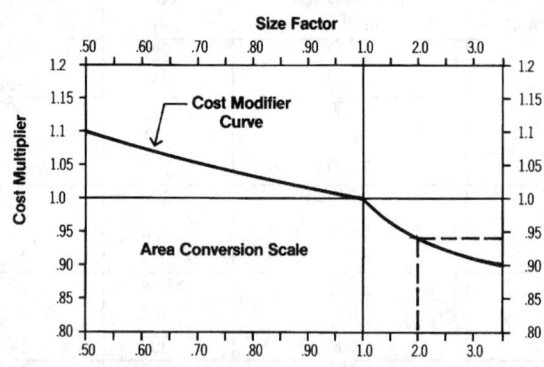

Square Foot Base Size

Building Type	Median Cost per S.F.	Typical Size Gross S.F.	Typical Range Gross S.F.	Building Type	Median Cost per S.F.	Typical Size Gross S.F.	Typical Range Gross S.F.
Apartments, Low Rise	$ 54.05	21,000	9,700 - 37,200	Jails	$165.00	13,700	7,500 - 28,000
Apartments, Mid Rise	68.25	50,000	32,000 - 100,000	Libraries	97.30	12,000	7,000 - 31,000
Apartments, High Rise	78.30	310,000	100,000 - 650,000	Medical Clinics	93.15	7,200	4,200 - 15,700
Auditoriums	90.35	25,000	7,600 - 39,000	Medical Offices	87.50	6,000	4,000 - 15,000
Auto Sales	55.90	20,000	10,800 - 28,600	Motels	67.00	27,000	15,800 - 51,000
Banks	121.00	4,200	2,500 - 7,500	Nursing Homes	89.95	23,000	15,000 - 37,000
Churches	81.60	9,000	5,300 - 13,200	Offices, Low Rise	73.00	8,600	4,700 - 19,000
Clubs, Country	81.40	6,500	4,500 - 15,000	Offices, Mid Rise	76.65	52,000	31,300 - 83,100
Clubs, Social	79.15	10,000	6,000 - 13,500	Offices, High Rise	98.05	260,000	151,000 - 468,000
Clubs, YMCA	81.60	28,300	12,800 - 39,400	Police Stations	122.00	10,500	4,000 - 19,000
Colleges (Class)	107.00	50,000	23,500 - 98,500	Post Offices	90.40	12,400	6,800 - 30,000
Colleges (Science Lab)	156.00	45,600	16,600 - 80,000	Power Plants	678.00	7,500	1,000 - 20,000
College (Student Union)	119.00	33,400	16,000 - 85,000	Religious Education	74.85	9,000	6,000 - 12,000
Community Center	85.05	9,400	5,300 - 16,700	Research	127.00	19,000	6,300 - 45,000
Court Houses	116.00	32,400	17,800 - 106,000	Restaurants	110.00	4,400	2,800 - 6,000
Dept. Stores	50.50	90,000	44,000 - 122,000	Retail Stores	53.70	7,200	4,000 - 17,600
Dormitories, Low Rise	87.20	24,500	13,400 - 40,000	Schools, Elementary	78.20	41,000	24,500 - 55,000
Dormitories, Mid Rise	113.00	55,600	36,100 - 90,000	Schools, Jr. High	79.65	92,000	52,000 - 119,000
Factories	48.95	26,400	12,900 - 50,000	Schools, Sr. High	79.65	101,000	50,500 - 175,000
Fire Stations	85.45	5,800	4,000 - 8,700	Schools, Vocational	79.35	37,000	20,500 - 82,000
Fraternity Houses	84.10	12,500	8,200 - 14,800	Sports Arenas	66.45	15,000	5,000 - 40,000
Funeral Homes	94.00	7,800	4,500 - 11,000	Supermarkets	53.85	20,000	12,000 - 30,000
Garages, Commercial	59.70	9,300	5,000 - 13,600	Swimming Pools	125.00	13,000	7,800 - 22,000
Garages, Municipal	76.40	8,300	4,500 - 12,600	Telephone Exchange	145.00	4,500	1,200 - 10,600
Garages, Parking	31.30	163,000	76,400 - 225,300	Theaters	79.70	10,500	8,800 - 17,500
Gymnasiums	78.95	19,200	11,600 - 41,000	Town Halls	87.65	10,800	4,800 - 23,400
Hospitals	149.00	55,000	27,200 - 125,000	Warehouses	36.15	25,000	8,000 - 72,000
House (Elderly)	73.90	37,000	21,000 - 66,000	Warehouse & Office	41.75	25,000	8,000 - 72,000
Housing (Public)	68.45	36,000	14,400 - 74,400				
Ice Rinks	76.00	29,000	27,200 - 33,600				

Change Orders

Change Order Considerations

A Change Order is a written document, usually prepared by the design professional, and signed by the owner, the architect/engineer and the contractor. A change order states the agreement of the parties to: an addition, deletion, or revision in the work; an adjustment in the contract sum, if any; or an adjustment in the contract time, if any. Change orders, or "extras" in the construction process occur after execution of the construction contract and impact architects/engineers, contractors and owners.

Change orders that are properly recognized and managed can ensure orderly, professional and profitable progress for all who are involved in the project. There are many causes for change orders and change order requests. In all cases, change orders or change order requests should be addressed promptly and in a precise and prescribed manner. The following paragraphs include information regarding change order pricing and procedures.

The Causes of Change Orders

Reasons for issuing change orders include:
- Unforeseen field conditions that require a change in the work
- Correction of design discrepancies, errors or omissions in the contract documents
- Owner-requested changes, either by design criteria, scope of work, or project objectives
- Completion date changes for reasons unrelated to the construction process
- Changes in building code interpretations, or other public authority requirements that require a change in the work
- Changes in availability of existing or new materials and products

Procedures

Properly written contract documents must include the correct change order procedures for all parties—owners, design professionals and contractors—to follow in order to avoid costly delays and litigation.

Being "in the right" is not always a sufficient or acceptable defense. The contract provisions requiring notification and documentation must be adhered to within a defined or reasonable time frame.

The appropriate method of handling change orders is by a written proposal and acceptance by all parties involved. Prior to starting work on a project, all parties should identify their authorized agents who may sign and accept change orders, as well as any limits placed on their authority.

Time may be a critical factor when the need for a change arises. For such cases, the contractor might be directed to proceed on a "time and materials" basis, rather than wait for all paperwork to be processed—a delay that could impede progress. In this situation, the contractor must still follow the prescribed change order procedures, including but not limited to, notification and documentation.

All forms used for change orders should be dated and signed by the proper authority. Lack of documentation can be very costly, especially if legal judgments are to be made and if certain field personnel are no longer available. For time and material change orders, the contractor should keep accurate daily records of all labor and material allocated to the change. Forms that can be used to document change order work are available in *Means Forms for Building Construction Professionals*.

Owners or awarding authorities who do considerable and continual building construction (such as the federal government) realize the inevitability of change orders for numerous reasons, both predictable and unpredictable. As a result, the federal government, the American Institute of Architects (AIA), the Engineers Joint Contract Documents Committee (EJCDC) and other contractor, legal and technical organizations have developed standards and procedures to be followed by all parties to achieve contract continuance and timely completion, while being financially fair to all concerned.

In addition to the change order standards put forth by industry associations, there are also many books available on the subject.

Pricing Change Orders

When pricing change orders, regardless of their cause, the most significant factor is *when* the change occurs. The need for a change may be perceived in the field or requested by the architect/engineer *before* any of the actual installation has begun, or may evolve or appear *during* construction when the item of work in question is partially installed. In the latter cases, the original sequence of construction is disrupted, along with all contiguous and supporting systems. Change orders cause the greatest impact when they occur *after* the installation has been completed and must be uncovered, or even replaced. Post-completion changes may be caused by necessary design changes, product failure, or changes in the owner's requirements that are not discovered until the building or the systems begin to function.

Specified procedures of notification and record keeping must be adhered to and enforced regardless of the stage of construction: *before*, *during*, or *after* installation. Some bidding documents anticipate change orders by requiring that unit prices including overhead and profit percentages—for additional as well as deductible changes—be listed. Generally these unit prices do not fully take into account the ripple effect, or impact on other trades, and should be used for general guidance only.

When pricing change orders, it is important to classify the time frame in which the change occurs. There are two basic time frames for change orders: *pre-installation change orders*, which occur before the start of construction, and *post-installation change orders*, which involve reworking after the original installation. Change orders that occur between these stages may be priced according to the extent of work completed using a combination of techniques developed for pricing *pre-* and *post-installation* changes.

The following factors are the basis for a check list to use when preparing a change order estimate.

Factors To Consider When Pricing Change Orders

As an estimator begins to prepare a change order, the following questions should be reviewed to determine their impact on the final price.

General

- Is the change order work *pre-installation* or *post-installation*?

    Change order work costs vary according to how much of the installation has been completed. Once workers have the project scoped in their mind, even though they have not started, it can be difficult to refocus. Consequently they may spend more than the normal amount of time understanding the change. Also, modifications to work in place such as trimming or refitting usually take more time than was initially estimated. The greater the amount of work in place, the more reluctant workers are to change it. Psychologically they may resent the change and as a result the rework takes longer than normal. Post-installation change order estimates must include demolition of existing work as required to accomplish the change. If the work is performed at a later time, additional obstacles such as building finishes may be present which must be protected. Regardless of whether the change occurs pre-installation or post-installation, attempt to isolate the identifiable factors and price them separately. For example, add shipping costs that may be required pre-installation or any demolition required post-installation. Then analyze the potential impact on productivity of psychological and/or learning curve factors and adjust the output rates accordingly. One approach is to break down the typical workday into segments and quantify the impact on each segment. The following chart may be useful as a guide:

	Activities (Productivity) Expressed as Percentages of a Workday		
Task	Means Mechanical Cost Data (for New Construction)	Pre-Installation Change Orders	Post-Installation Change Orders
1. Study plans	3%	6%	6%
2. Material procurement	3%	3%	3%
3. Receiving and storing	3%	3%	3%
4. Mobilization	5%	5%	5%
5. Site movement	5%	5%	8%
6. Layout and marking	8%	10%	12%
7. Actual installation	64%	59%	54%
8. Clean-up	3%	3%	3%
9. Breaks—non-productive	6%	6%	6%
Total	100%	100%	100%

Change Order Installation Efficiency

The labor-hours expressed (for new construction) are based on average installation time, using an efficiency level of approximately 60-65%. For change order situations, adjustments to this efficiency level should reflect the daily labor-hour allocation for that particular occurrence.

If any of the specific percentages expressed in the above chart do not apply to a particular project situation, then those percentage points should be reallocated to the appropriate task(s). Example: Using data for new construction, assume there is no new material being utilized. The percentages for Tasks 2 and 3 would therefore be reallocated to other tasks. If the time required for Tasks 2 and 3 can now be applied to installation, we can add the time allocated for *Material Procurement* and *Receiving and Storing* to the *Actual Installation* time for new construction, thereby increasing the Actual Installation percentage.

This chart shows that, due to reduced productivity, labor costs will be higher than those for new construction by 5% to 15% for pre-installation change orders and by 15% to 25% for post-installation change orders. Each job and change order is unique and must be examined individually. Many factors, covered elsewhere in this section, can each have a significant impact on productivity and change order costs. All such factors should be considered in every case.

- Will the change substantially delay the original completion date?

  A significant change in the project may cause the original completion date to be extended. The extended schedule may subject the contractor to new wage rates dictated by relevant labor contracts. Project supervision and other project overhead must also be extended beyond the original completion date. The schedule extension may also put installation into a new weather season. For example, underground piping scheduled for October installation was delayed until January. As a result, frost penetrated the trench area, thereby changing the degree of difficulty of the task. Changes and delays may have a ripple effect throughout the project. This effect must be analyzed and negotiated with the owner.

- What is the net effect of a deduct change order?

  In most cases, change orders resulting in a deduction or credit reflect only bare costs. The contractor may retain the overhead and profit based on the original bid.

Materials

- Will you have to pay more or less for the new material, required by the change order, than you paid for the original purchase?

  The same material prices or discounts will usually apply to materials purchased for change orders as new construction. In some instances, however, the contractor may forfeit the advantages of competitive pricing for change orders. Consider the following example:

  A contractor purchased over $20,000 worth of fan coil units for an installation, and obtained the maximum discount. Some time later it was determined the project required an additional matching unit. The contractor has to purchase this unit from the original supplier to ensure a match. The supplier at this time may not discount the unit because of the small quantity, and the fact that he is no longer in a competitive situation. The impact of quantity on purchase can add between 0% and 25% to material prices and/or subcontractor quotes.

- If materials have been ordered or delivered to the job site, will they be subject to a cancellation charge or restocking fee?

  Check with the supplier to determine if ordered materials are subject to a cancellation charge. Delivered materials not used as result of a change order may be subject to a restocking fee if returned to the supplier. Common restocking charges run between 20% and 40%. Also, delivery charges to return the goods to the supplier must be added.

Labor

- How efficient is the existing crew at the actual installation?

  Is the same crew that performed the initial work going to do the change order? Possibly the change consists of the installation of a unit identical to one already installed; therefore the change should take less time. Be sure to consider this potential productivity increase and modify the productivity rates accordingly.

- If the crew size is increased, what impact will that have on supervision requirements?

  Under most bargaining agreements or management practices, there is a point at which a working foreman is replaced by a nonworking foreman. This replacement increases project overhead by adding a nonproductive worker. If additional workers are added to accelerate the project or to perform changes while maintaining the schedule, be sure to add additional supervision time if warranted. Calculate the hours involved and the additional cost directly if possible.

- What are the other impacts of increased crew size?

  The larger the crew, the greater the potential for productivity to decrease. Some of the factors that cause this productivity loss are: overcrowding (producing restrictive conditions in the working space), and possibly a shortage of any special tools and equipment required. Such factors affect not only the crew working on the elements directly involved in the change order, but other crews whose movement may also be hampered.

  As the crew increases, check its basic composition for changes by the addition or deletion of apprentices or nonworking foreman and quantify the potential effects of equipment shortages or other logistical factors.

- As new crews, unfamiliar with the project, are brought onto the site, how long will it take them to become oriented to the project requirements?

  The orientation time for a new crew to become 100% effective varies with the site and type of project. Orientation is easiest at a new construction site, and most difficult at existing, very restrictive renovation sites. The type of work also affects orientation time. When all elements of the work are exposed, such as concrete or masonry work, orientation is decreased. When the work is concealed or less visible, such as existing electrical systems, orientation takes longer. Usually orientation can be accomplished in one day or less. Costs for added orientation should be itemized and added to the total estimated cost.

- How much actual production can be gained by working overtime?

  Short term overtime can be used effectively to accomplish more work in a day. However, as overtime is scheduled to run beyond several weeks, studies have shown marked decreases in output. The following chart shows the effect of long term overtime on worker efficiency. If the anticipated change requires extended overtime to keep the job on schedule, these factors can be used as a guide to predict the impact on time and cost. Add project overhead, particularly supervision, that may also be incurred.

| | | Production Efficiency | | | | | Payroll Cost Factors | |
Days per Week	Hours per Day	1 Week	2 Weeks	3 Weeks	4 Weeks	Average 4 Weeks	@ 1-1/2 Times	@ 2 Times
5	8	100%	100%	100%	100%	100%	100%	100%
	9	100	100	95	90	96.25	105.6	111.1
	10	100	95	90	85	91.25	110.0	120.0
	11	95	90	75	65	81.25	113.6	127.3
	12	90	85	70	60	76.25	116.7	133.3
6	8	100	100	95	90	96.25	108.3	116.7
	9	100	95	90	85	92.50	113.0	125.9
	10	95	90	85	80	87.50	116.7	133.3
	11	95	85	70	65	78.75	119.7	139.4
	12	90	80	65	60	73.75	122.2	144.4
7	8	100	95	85	75	88.75	114.3	128.6
	9	95	90	80	70	83.75	118.3	136.5
	10	90	85	75	65	78.75	121.4	142.9
	11	85	80	65	60	72.50	124.0	148.1
	12	85	75	60	55	68.75	126.2	152.4

Effects of Overtime

Caution: Under many labor agreements, Sundays and holidays are paid at a higher premium than the normal overtime rate.

The use of long-term overtime is counterproductive on almost any construction job; that is, the longer the period of overtime, the lower the actual production rate. Numerous studies have been conducted, and while they have resulted in slightly different numbers, all reach the same conclusion. The figure above tabulates the effects of overtime work on efficiency.

As illustrated, there can be a difference between the *actual* payroll cost per hour and the *effective* cost per hour for overtime work. This is due to the reduced production efficiency with the increase in weekly hours beyond 40. This difference between actual and effective cost results from overtime work over a prolonged period. Short-term overtime work does not result in as great a reduction in efficiency, and in such cases, effective cost may not vary significantly from the actual payroll cost. As the total hours per week are increased on a regular basis, more time is lost because of fatigue, lowered morale, and an increased accident rate.

As an example, assume a project where workers are working 6 days a week, 10 hours per day. From the figure above (based on productivity studies), the average effective productive hours over a four-week period are:

$$0.875 \times 60 = 52.5$$

Depending upon the locale and day of week, overtime hours may be paid at time and a half or double time. For time and a half, the overall (average) *actual* payroll cost (including regular and overtime hours) is determined as follows:

$$\frac{40 \text{ reg. hrs.} + (20 \text{ overtime hrs.} \times 1.5)}{60 \text{ hrs.}} = 1.167$$

Based on 60 hours, the payroll cost per hour will be 116.7% of the normal rate at 40 hours per week. However, because the effective production (efficiency) for 60 hours is reduced to the equivalent of 52.5 hours, the effective cost of overtime is calculated as follows:

For time and a half:

$$\frac{40 \text{ reg. hrs.} + (20 \text{ overtime hrs.} \times 1.5)}{52.5 \text{ hrs.}} = 1.33$$

Installed cost will be 133% of the normal rate (for labor).

Thus, when figuring overtime, the actual cost per unit of work will be higher than the apparent overtime payroll dollar increase, due to the reduced productivity of the longer workweek. These efficiency calculations are true only for those cost factors determined by hours worked. Costs that are applied weekly or monthly, such as equipment rentals, will not be similarly affected.

Equipment

- What equipment is required to complete the change order?

  Change orders may require extending the rental period of equipment already on the job site, or the addition of special equipment brought in to accomplish the change work. In either case, the additional rental charges and operator labor charges must be added.

Summary

The preceding considerations and others you deem appropriate should be analyzed and applied to a change order estimate. The impact of each should be quantified and listed on the estimate to form an audit trail.

Change orders that are properly identified, documented, and managed help to ensure the orderly, professional and profitable progress of the work. They also minimize potential claims or disputes at the end of the project.

Crews

Crew No.	Bare Costs		Incl. Subs O & P		Cost Per Labor-Hour	
Crew A-1	Hr.	Daily	Hr.	Daily	Bare Costs	Incl. O&P
1 Building Laborer	$14.45	$115.60	$24.75	$198.00	$14.45	$24.75
1 Gas Eng. Power Tool		70.00		77.00	8.75	9.63
8 L.H., Daily Totals		$185.60		$275.00	$23.20	$34.38
Crew A-1A	Hr.	Daily	Hr.	Daily	Bare Costs	Incl. O&P
1 Skilled Worker	$19.75	$158.00	$33.90	$271.20	$19.75	$33.90
1 Shot Blaster, 20"		142.00		156.20	17.75	19.53
8 L.H., Daily Totals		$300.00		$427.40	$37.50	$53.43
Crew A-2	Hr.	Daily	Hr.	Daily	Bare Costs	Incl. O&P
2 Laborers	$14.45	$231.20	$24.75	$396.00	$14.92	$25.33
1 Truck Driver (light)	15.85	126.80	26.50	212.00		
1 Light Truck, 1.5 Ton		172.95		190.25	7.21	7.93
24 L.H., Daily Totals		$530.95		$798.25	$22.13	$33.26
Crew A-2A	Hr.	Daily	Hr.	Daily	Bare Costs	Incl. O&P
2 Laborers	$14.45	$231.20	$24.75	$396.00	$14.92	$25.33
1 Truck Driver (light)	15.85	126.80	26.50	212.00		
1 Light Truck, 1.5 ton		172.95		190.25		
1 Concrete Saw		110.00		121.00	11.79	12.97
24 L.H., Daily Totals		$640.95		$919.25	$26.71	$38.30
Crew A-3	Hr.	Daily	Hr.	Daily	Bare Costs	Incl. O&P
1 Truck Driver (heavy)	$16.20	$129.60	$27.05	$216.40	$16.20	$27.05
1 Dump Truck, 12 Ton		365.30		401.85	45.66	50.23
8 L.H., Daily Totals		$494.90		$618.25	$61.86	$77.28
Crew A-3A	Hr.	Daily	Hr.	Daily	Bare Costs	Incl. O&P
1 Truck Driver (light)	$15.85	$126.80	$26.50	$212.00	$15.85	$26.50
1 Pickup truck (4x4)		144.65		159.10	18.08	19.89
8 L.H., Daily Totals		$271.45		$371.10	$33.93	$46.39
Crew A-3B	Hr.	Daily	Hr.	Daily	Bare Costs	Incl. O&P
1 Equip. Oper. (medium)	$19.90	$159.20	$33.00	$264.00	$18.05	$30.02
1 Truck Driver (heavy)	16.20	129.60	27.05	216.40		
1 Dump Truck, 16 Ton		447.25		492.00		
1 F.E. Loader, 3 C.Y.		400.80		440.90	53.00	58.30
16 L.H., Daily Totals		$1136.85		$1413.30	$71.05	$88.32
Crew A-3C	Hr.	Daily	Hr.	Daily	Bare Costs	Incl. O&P
1 Equip. Oper. (light)	$19.10	$152.80	$31.70	$253.60	$19.10	$31.70
1 Wheeled Skid Steer Loader		225.80		248.40	28.23	31.05
8 L.H., Daily Totals		$378.60		$502.00	$47.33	$62.75
Crew A-4	Hr.	Daily	Hr.	Daily	Bare Costs	Incl. O&P
2 Carpenters	$19.70	$315.20	$33.75	$540.00	$19.12	$32.48
1 Painter, Ordinary	17.95	143.60	29.95	239.60		
24 L.H., Daily Totals		$458.80		$779.60	$19.12	$32.48
Crew A-5	Hr.	Daily	Hr.	Daily	Bare Costs	Incl. O&P
2 Laborers	$14.45	$231.20	$24.75	$396.00	$14.61	$24.94
.25 Truck Driver (light)	15.85	31.70	26.50	53.00		
.25 Light Truck, 1.5 Ton		43.24		47.55	2.40	2.64
18 L.H., Daily Totals		$306.14		$496.55	$17.01	$27.58
Crew A-6	Hr.	Daily	Hr.	Daily	Bare Costs	Incl. O&P
1 Chief Of Party	$22.10	$176.80	$38.15	$305.20	$20.93	$36.03
1 Instrument Man	19.75	158.00	33.90	271.20		
16 L.H., Daily Totals		$334.80		$576.40	$20.93	$36.03
Crew A-7	Hr.	Daily	Hr.	Daily	Bare Costs	Incl. O&P
1 Chief Of Party	$22.10	$176.80	$38.15	$305.20	$20.40	$34.72
1 Instrument Man	19.75	158.00	33.90	271.20		
1 Rodman/Chainman	19.35	154.80	32.10	256.80		
24 L.H., Daily Totals		$489.60		$833.20	$20.40	$34.72
Crew A-8	Hr.	Daily	Hr.	Daily	Bare Costs	Incl. O&P
1 Chief Of Party	$22.10	$176.80	$38.15	$305.20	$20.14	$34.06
1 Instrument Man	19.75	158.00	33.90	271.20		
2 Rodmen/Chainmen	19.35	309.60	32.10	513.60		
32 L.H., Daily Totals		$644.40		$1090.00	$20.14	$34.06
Crew A-9	Hr.	Daily	Hr.	Daily	Bare Costs	Incl. O&P
1 Asbestos Foreman	$21.00	$168.00	$36.75	$294.00	$20.56	$36.01
7 Asbestos Workers	20.50	1148.00	35.90	2010.40		
64 L.H., Daily Totals		$1316.00		$2304.40	$20.56	$36.01
Crew A-10	Hr.	Daily	Hr.	Daily	Bare Costs	Incl. O&P
1 Asbestos Foreman	$21.00	$168.00	$36.75	$294.00	$20.56	$36.01
7 Asbestos Workers	20.50	1148.00	35.90	2010.40		
64 L.H., Daily Totals		$1316.00		$2304.40	$20.56	$36.01
Crew A-10A	Hr.	Daily	Hr.	Daily	Bare Costs	Incl. O&P
1 Asbestos Foreman	$21.00	$168.00	$36.75	$294.00	$20.67	$36.18
2 Asbestos Workers	20.50	328.00	35.90	574.40		
24 L.H., Daily Totals		$496.00		$868.40	$20.67	$36.18
Crew A-10B	Hr.	Daily	Hr.	Daily	Bare Costs	Incl. O&P
1 Asbestos Foreman	$21.00	$168.00	$36.75	$294.00	$20.63	$36.11
3 Asbestos Workers	20.50	492.00	35.90	861.60		
32 L.H., Daily Totals		$660.00		$1155.60	$20.63	$36.11
Crew A-10C	Hr.	Daily	Hr.	Daily	Bare Costs	Incl. O&P
3 Asbestos Workers	$20.50	$492.00	$35.90	$861.60	$20.50	$35.90
1 Flatbed truck		172.95		190.25	7.21	7.93
24 L.H., Daily Totals		$664.95		$1051.85	$27.71	$43.83
Crew A-10D	Hr.	Daily	Hr.	Daily	Bare Costs	Incl. O&P
2 Asbestos Workers	$20.50	$328.00	$35.90	$574.40	$19.66	$33.56
1 Equip. Oper. (crane)	20.65	165.20	34.25	274.00		
1 Equip. Oper. Oiler	17.00	136.00	28.20	225.60		
1 Hydraulic crane, 33 ton		694.95		764.45	21.72	23.89
32 L.H., Daily Totals		$1324.15		$1838.45	$41.38	$57.45
Crew A-11	Hr.	Daily	Hr.	Daily	Bare Costs	Incl. O&P
1 Asbestos Foreman	$21.00	$168.00	$36.75	$294.00	$20.56	$36.01
7 Asbestos Workers	20.50	1148.00	35.90	2010.40		
2 Chipping Hammers		33.90		37.30	.53	.58
64 L.H., Daily Totals		$1349.90		$2341.70	$21.09	$36.59
Crew A-12	Hr.	Daily	Hr.	Daily	Bare Costs	Incl. O&P
1 Asbestos Foreman	$21.00	$168.00	$36.75	$294.00	$20.56	$36.01
7 Asbestos Workers	20.50	1148.00	35.90	2010.40		
1 Large Prod. Vac. Loader		529.80		582.80	8.28	9.11
64 L.H., Daily Totals		$1845.80		$2887.20	$28.84	$45.12
Crew A-13	Hr.	Daily	Hr.	Daily	Bare Costs	Incl. O&P
1 Equip. Oper. (light)	$19.10	$152.80	$31.70	$253.60	$19.10	$31.70
1 Large Prod. Vac. Loader		529.80		582.80	66.23	72.85
8 L.H., Daily Totals		$682.60		$836.40	$85.33	$104.55

Crews

Crew No.		Bare Costs		Incl. Subs O & P	Cost Per Labor-Hour	
Crew B-1	Hr.	Daily	Hr.	Daily	Bare Costs	Incl. O&P
1 Labor Foreman (outside)	$16.45	$131.60	$28.20	$225.60	$15.12	$25.90
2 Laborers	14.45	231.20	24.75	396.00		
24 L.H., Daily Totals		$362.80		$621.60	$15.12	$25.90
Crew B-2	Hr.	Daily	Hr.	Daily	Bare Costs	Incl. O&P
1 Labor Foreman (outside)	$16.45	$131.60	$28.20	$225.60	$14.85	$25.44
4 Laborers	14.45	462.40	24.75	792.00		
40 L.H., Daily Totals		$594.00		$1017.60	$14.85	$25.44
Crew B-3	Hr.	Daily	Hr.	Daily	Bare Costs	Incl. O&P
1 Labor Foreman (outside)	$16.45	$131.60	$28.20	$225.60	$16.27	$27.47
2 Laborers	14.45	231.20	24.75	396.00		
1 Equip. Oper. (med.)	19.90	159.20	33.00	264.00		
2 Truck Drivers (heavy)	16.20	259.20	27.05	432.80		
1 F.E. Loader, T.M., 2.5 C.Y.		784.00		862.40		
2 Dump Trucks, 16 Ton		894.50		983.95	34.97	38.47
48 L.H., Daily Totals		$2459.70		$3164.75	$51.24	$65.94
Crew B-3A	Hr.	Daily	Hr.	Daily	Bare Costs	Incl. O&P
4 Laborers	$14.45	$462.40	$24.75	$792.00	$15.54	$26.40
1 Equip. Oper. (med.)	19.90	159.20	33.00	264.00		
1 Hyd. Excavator, 1.5 C.Y.		713.20		784.50	17.83	19.61
40 L.H., Daily Totals		$1334.80		$1840.50	$33.37	$46.01
Crew B-3B	Hr.	Daily	Hr.	Daily	Bare Costs	Incl. O&P
2 Laborers	$14.45	$231.20	$24.75	$396.00	$16.25	$27.39
1 Equip. Oper. (med.)	19.90	159.20	33.00	264.00		
1 Truck Drivers (heavy)	16.20	129.60	27.05	216.40		
1 Backhoe Loader, 80 H.P.		273.15		300.45		
1 Dump Trucks, 16 Ton		447.25		492.00	22.51	24.76
32 L.H., Daily Totals		$1240.40		$1668.85	$38.76	$52.15
Crew B-3C	Hr.	Daily	Hr.	Daily	Bare Costs	Incl. O&P
3 Laborers	$14.45	$346.80	$24.75	$594.00	$15.81	$26.81
1 Equip. Oper. (med.)	19.90	159.20	33.00	264.00		
1 F.E. Loader, 3.75 C.Y.		542.40		596.65	16.95	18.65
32 L.H., Daily Totals		$1048.40		$1454.65	$32.76	$45.46
Crew B-4	Hr.	Daily	Hr.	Daily	Bare Costs	Incl. O&P
1 Labor Foreman (outside)	$16.45	$131.60	$28.20	$225.60	$15.08	$25.71
4 Laborers	14.45	462.40	24.75	792.00		
1 Truck Driver (heavy)	16.20	129.60	27.05	216.40		
1 Tractor, 4 x 2, 195 H.P.		325.60		358.15		
1 Platform Trailer		152.00		167.20	9.95	10.95
48 L.H., Daily Totals		$1201.20		$1759.35	$25.03	$36.66
Crew B-5	Hr.	Daily	Hr.	Daily	Bare Costs	Incl. O&P
1 Labor Foreman (outside)	$16.45	$131.60	$28.20	$225.60	$15.94	$27.09
3 Laborers	14.45	346.80	24.75	594.00		
1 Equip. Oper. (med.)	19.90	159.20	33.00	264.00		
1 Air Compr. 250 C.F.M.		127.00		139.70		
2 Air Tools & Accessories		39.00		42.90		
2-50 Ft. Air Hoses, 1.5" Dia.		14.40		15.85		
1 F.E. Loader, T.M., 2.5 C.Y.		784.00		862.40	24.11	26.52
40 L.H., Daily Totals		$1602.00		$2144.45	$40.05	$53.61

Crew No.		Bare Costs		Incl. Subs O & P	Cost Per Labor-Hour	
Crew B-5A	Hr.	Daily	Hr.	Daily	Bare Costs	Incl. O&P
1 Foreman	$16.45	$131.60	$28.20	$225.60	$16.20	$27.38
6 Laborers	14.45	693.60	24.75	1188.00		
2 Equip. Oper. (med.)	19.90	318.40	33.00	528.00		
1 Equip. Oper. (light)	19.10	152.80	31.70	253.60		
2 Truck Drivers (heavy)	16.20	259.20	27.05	432.80		
1 Air Compr. 365 C.F.M.		174.40		191.85		
2 Pavement Breakers		39.00		42.90		
8 Air Hoses w/Coup.,1"		41.60		45.75		
2 Dump Trucks, 12 Ton		730.60		803.65	10.27	11.29
96 L.H., Daily Totals		$2541.20		$3712.15	$26.47	$38.67
Crew B-5B	Hr.	Daily	Hr.	Daily	Bare Costs	Incl. O&P
1 Powderman	$19.75	$158.00	$33.90	$271.20	$18.02	$30.18
2 Equip. Oper. (med.)	19.90	318.40	33.00	528.00		
3 Truck Drivers (heavy)	16.20	388.80	27.05	649.20		
1 F.E. Ldr. 2-1/2 CY		400.80		440.90		
3 Dump Trucks, 16 Ton		1341.75		1475.95		
1 Air Compr. 365 C.F.M.		174.40		191.85	39.94	43.93
48 L.H., Daily Totals		$2782.15		$3557.10	$57.96	$74.11
Crew B-5C	Hr.	Daily	Hr.	Daily	Bare Costs	Incl. O&P
3 Laborers	$14.45	$346.80	$24.75	$594.00	$16.66	$27.98
1 Equip. Oper. (medium)	19.90	159.20	33.00	264.00		
2 Truck Drivers (heavy)	16.20	259.20	27.05	432.80		
1 Equip. Oper. (crane)	20.65	165.20	34.25	274.00		
1 Equip. Oper. Oiler	17.00	136.00	28.20	225.60		
2 Dump Truck, 16 Ton		894.50		983.95		
1 F.E. Lder, 3.75 C.Y.		1035.00		1138.50		
1 Hyd. Crane, 25 Ton		563.90		620.30	38.96	42.86
64 L.H., Daily Totals		$3559.80		$4533.15	$55.62	$70.84
Crew B-6	Hr.	Daily	Hr.	Daily	Bare Costs	Incl. O&P
2 Laborers	$14.45	$231.20	$24.75	$396.00	$16.00	$27.07
1 Equip. Oper. (light)	19.10	152.80	31.70	253.60		
1 Backhoe Loader, 48 H.P.		203.00		223.30	8.46	9.30
24 L.H., Daily Totals		$587.00		$872.90	$24.46	$36.37
Crew B-7	Hr.	Daily	Hr.	Daily	Bare Costs	Incl. O&P
1 Labor Foreman (outside)	$16.45	$131.60	$28.20	$225.60	$15.69	$26.70
4 Laborers	14.45	462.40	24.75	792.00		
1 Equip. Oper. (med.)	19.90	159.20	33.00	264.00		
1 Chipping Machine		211.05		232.15		
1 F.E. Loader, T.M., 2.5 C.Y.		784.00		862.40		
2 Chain Saws, 36"		98.40		108.25	22.78	25.06
48 L.H., Daily Totals		$1846.65		$2484.40	$38.47	$51.76
Crew B-7A	Hr.	Daily	Hr.	Daily	Bare Costs	Incl. O&P
2 Laborers	$14.45	$231.20	$24.75	$396.00	$16.00	$27.07
1 Equip. Oper. (light)	19.10	152.80	31.70	253.60		
1 Rake w/Tractor		219.05		240.95		
2 Chain Saws, 18"		56.80		62.50	11.49	12.64
24 L.H., Daily Totals		$659.85		$953.05	$27.49	$39.71

Crews

Crew No.	Bare Costs		Incl. Subs O & P		Cost Per Labor-Hour	
Crew B-8	Hr.	Daily	Hr.	Daily	Bare Costs	Incl. O&P
1 Labor Foreman (outside)	$16.45	$131.60	$28.20	$225.60	$16.79	$28.26
2 Laborers	14.45	231.20	24.75	396.00		
2 Equip. Oper. (med.)	19.90	318.40	33.00	528.00		
2 Truck Drivers (heavy)	16.20	259.20	27.05	432.80		
1 Hyd. Crane, 25 Ton		559.20		615.10		
1 F.E. Loader, T.M., 2.5 C.Y.		784.00		862.40		
2 Dump Trucks, 16 Ton		894.50		983.95	39.96	43.95
56 L.H., Daily Totals		$3178.10		$4043.85	$56.75	$72.21
Crew B-9	Hr.	Daily	Hr.	Daily	Bare Costs	Incl. O&P
1 Labor Foreman (outside)	$16.45	$131.60	$28.20	$225.60	$14.85	$25.44
4 Laborers	14.45	462.40	24.75	792.00		
1 Air Compr., 250 C.F.M.		127.00		139.70		
2 Air Tools & Accessories		39.00		42.90		
2-50 Ft. Air Hoses, 1.5" Dia.		14.40		15.85	4.51	4.96
40 L.H., Daily Totals		$774.40		$1216.05	$19.36	$30.40
Crew B-9A	Hr.	Daily	Hr.	Daily	Bare Costs	Incl. O&P
2 Laborers	$14.45	$231.20	$24.75	$396.00	$15.03	$25.52
1 Truck Driver (heavy)	16.20	129.60	27.05	216.40		
1 Water Tanker		205.90		226.50		
1 Tractor		325.60		358.15		
2-50 Ft. Disch. Hoses		12.80		14.10	22.68	24.95
24 L.H., Daily Totals		$905.10		$1211.15	$37.71	$50.47
Crew B-9B	Hr.	Daily	Hr.	Daily	Bare Costs	Incl. O&P
2 Laborers	$14.45	$231.20	$24.75	$396.00	$15.03	$25.52
1 Truck Driver (heavy)	16.20	129.60	27.05	216.40		
2-50 Ft. Disch. Hoses		12.80		14.10		
1 Water Tanker		205.90		226.50		
1 Tractor		325.60		358.15		
1 Pressure Washer		60.40		66.45	25.20	27.72
24 L.H., Daily Totals		$965.50		$1277.60	$40.23	$53.24
Crew B-9C	Hr.	Daily	Hr.	Daily	Bare Costs	Incl. O&P
1 Labor Foreman (outside)	$16.45	$131.60	$28.20	$225.60	$14.85	$25.44
4 Laborers	14.45	462.40	24.75	792.00		
1 Air Compr., 250 C.F.M.		127.00		139.70		
2-50 Ft. Air Hoses, 1.5" Dia.		14.40		15.85		
2 Breaker, Pavement, 60 lb.		39.00		42.90	4.51	4.96
40 L.H., Daily Totals		$774.40		$1216.05	$19.36	$30.40
Crew B-10	Hr.	Daily	Hr.	Daily	Bare Costs	Incl. O&P
1 Equip. Oper. (med.)	$19.90	$159.20	$33.00	$264.00	$19.90	$33.00
8 L.H., Daily Totals		$159.20		$264.00	$19.90	$33.00
Crew B-10A	Hr.	Daily	Hr.	Daily	Bare Costs	Incl. O&P
1 Equip. Oper. (med.)	$19.90	$159.20	$33.00	$264.00	$19.90	$33.00
1 Roll. Compact., 2K Lbs.		94.25		103.70	11.78	12.96
8 L.H., Daily Totals		$253.45		$367.70	$31.68	$45.96
Crew B-10B	Hr.	Daily	Hr.	Daily	Bare Costs	Incl. O&P
1 Equip. Oper. (med.)	$19.90	$159.20	$33.00	$264.00	$19.90	$33.00
1 Dozer, 200 H.P.		806.80		887.50	100.85	110.94
8 L.H., Daily Totals		$966.00		$1151.50	$120.75	$143.94

Crew No.	Bare Costs		Incl. Subs O & P		Cost Per Labor-Hour	
Crew B-10C	Hr.	Daily	Hr.	Daily	Bare Costs	Incl. O&P
1 Equip. Oper. (med.)	$19.90	$159.20	$33.00	$264.00	$19.90	$33.00
1 Dozer, 200 H.P.		806.80		887.50		
1 Vibratory Roller, Towed		110.00		121.00	114.60	126.06
8 L.H., Daily Totals		$1076.00		$1272.50	$134.50	$159.06
Crew B-10D	Hr.	Daily	Hr.	Daily	Bare Costs	Incl. O&P
1 Equip. Oper. (med.)	$19.90	$159.20	$33.00	$264.00	$19.90	$33.00
1 Dozer, 200 H.P.		806.80		887.50		
1 Sheepsft. Roller, Towed		126.40		139.05	116.65	128.32
8 L.H., Daily Totals		$1092.40		$1290.55	$136.55	$161.32
Crew B-10E	Hr.	Daily	Hr.	Daily	Bare Costs	Incl. O&P
1 Equip. Oper. (med.)	$19.90	$159.20	$33.00	$264.00	$19.90	$33.00
1 Tandem Roller, 5 Ton		133.20		146.50	16.65	18.32
8 L.H., Daily Totals		$292.40		$410.50	$36.55	$51.32
Crew B-10F	Hr.	Daily	Hr.	Daily	Bare Costs	Incl. O&P
1 Equip. Oper. (med.)	$19.90	$159.20	$33.00	$264.00	$19.90	$33.00
1 Tandem Roller, 10 Ton		224.00		246.40	28.00	30.80
8 L.H., Daily Totals		$383.20		$510.40	$47.90	$63.80
Crew B-10G	Hr.	Daily	Hr.	Daily	Bare Costs	Incl. O&P
1 Equip. Oper. (med.)	$19.90	$159.20	$33.00	$264.00	$19.90	$33.00
1 Sheepsft. Roll., 130 H.P.		567.60		624.35	70.95	78.05
8 L.H., Daily Totals		$726.80		$888.35	$90.85	$111.05
Crew B-10H	Hr.	Daily	Hr.	Daily	Bare Costs	Incl. O&P
1 Equip. Oper. (med.)	$19.90	$159.20	$33.00	$264.00	$19.90	$33.00
1 Diaphr. Water Pump, 2"		29.40		32.35		
1-20 Ft. Suction Hose, 2"		6.50		7.15		
2-50 Ft. Disch. Hoses, 2"		10.80		11.90	5.84	6.42
8 L.H., Daily Totals		$205.90		$315.40	$25.74	$39.42
Crew B-10I	Hr.	Daily	Hr.	Daily	Bare Costs	Incl. O&P
1 Equip. Oper. (med.)	$19.90	$159.20	$33.00	$264.00	$19.90	$33.00
1 Diaphr. Water Pump, 4"		65.65		72.20		
1-20 Ft. Suction Hose, 4"		12.50		13.75		
2-50 Ft. Disch. Hoses, 4"		17.00		18.70	11.89	13.08
8 L.H., Daily Totals		$254.35		$368.65	$31.79	$46.08
Crew B-10J	Hr.	Daily	Hr.	Daily	Bare Costs	Incl. O&P
1 Equip. Oper. (med.)	$19.90	$159.20	$33.00	$264.00	$19.90	$33.00
1 Centr. Water Pump, 3"		38.00		41.80		
1-20 Ft. Suction Hose, 3"		9.50		10.45		
2-50 Ft. Disch. Hoses, 3"		12.80		14.10	7.54	8.29
8 L.H., Daily Totals		$219.50		$330.35	$27.44	$41.29
Crew B-10K	Hr.	Daily	Hr.	Daily	Bare Costs	Incl. O&P
1 Equip. Oper. (med.)	$19.90	$159.20	$33.00	$264.00	$19.90	$33.00
1 Centr. Water Pump, 6"		157.20		172.90		
1-20 Ft. Suction Hose, 6"		22.50		24.75		
2-50 Ft. Disch. Hoses, 6"		41.10		45.20	27.60	30.36
8 L.H., Daily Totals		$380.00		$506.85	$47.50	$63.36
Crew B-10L	Hr.	Daily	Hr.	Daily	Bare Costs	Incl. O&P
1 Equip. Oper. (med.)	$19.90	$159.20	$33.00	$264.00	$19.90	$33.00
1 Dozer, 75 H.P.		279.60		307.55	34.95	38.45
8 L.H., Daily Totals		$438.80		$571.55	$54.85	$71.45

Crews

Crew No.	Bare Costs		Incl. Subs O & P		Cost Per Labor-Hour	
Crew B-10M	Hr.	Daily	Hr.	Daily	Bare Costs	Incl. O&P
1 Equip. Oper. (med.)	$19.90	$159.20	$33.00	$264.00	$19.90	$33.00
1 Dozer, 300 H.P.		1108.00		1218.80	138.50	152.35
8 L.H., Daily Totals		$1267.20		$1482.80	$158.40	$185.35
Crew B-10N	Hr.	Daily	Hr.	Daily	Bare Costs	Incl. O&P
1 Equip. Oper. (med.)	$19.90	$159.20	$33.00	$264.00	$19.90	$33.00
1 F.E. Loader, T.M., 1.5 C.Y.		329.95		362.95	41.24	45.37
8 L.H., Daily Totals		$489.15		$626.95	$61.14	$78.37
Crew B-10O	Hr.	Daily	Hr.	Daily	Bare Costs	Incl. O&P
1 Equip. Oper. (med.)	$19.90	$159.20	$33.00	$264.00	$19.90	$33.00
1 F.E. Loader, T.M., 2.25 C.Y.		450.00		495.00	56.25	61.88
8 L.H., Daily Totals		$609.20		$759.00	$76.15	$94.88
Crew B-10P	Hr.	Daily	Hr.	Daily	Bare Costs	Incl. O&P
1 Equip. Oper. (med.)	$19.90	$159.20	$33.00	$264.00	$19.90	$33.00
1 F.E. Loader, T.M., 2.5 C.Y.		784.00		862.40	98.00	107.80
8 L.H., Daily Totals		$943.20		$1126.40	$117.90	$140.80
Crew B-10Q	Hr.	Daily	Hr.	Daily	Bare Costs	Incl. O&P
1 Equip. Oper. (med.)	$19.90	$159.20	$33.00	$264.00	$19.90	$33.00
1 F.E. Loader, T.M., 5 C.Y.		1035.00		1138.50	129.38	142.31
8 L.H., Daily Totals		$1194.20		$1402.50	$149.28	$175.31
Crew B-10R	Hr.	Daily	Hr.	Daily	Bare Costs	Incl. O&P
1 Equip. Oper. (med.)	$19.90	$159.20	$33.00	$264.00	$19.90	$33.00
1 F.E. Loader, W.M., 1 C.Y.		233.60		256.95	29.20	32.12
8 L.H., Daily Totals		$392.80		$520.95	$49.10	$65.12
Crew B-10S	Hr.	Daily	Hr.	Daily	Bare Costs	Incl. O&P
1 Equip. Oper. (med.)	$19.90	$159.20	$33.00	$264.00	$19.90	$33.00
1 F.E. Loader, W.M., 1.5 C.Y.		303.00		333.30	37.88	41.66
8 L.H., Daily Totals		$462.20		$597.30	$57.78	$74.66
Crew B-10T	Hr.	Daily	Hr.	Daily	Bare Costs	Incl. O&P
1 Equip. Oper. (med.)	$19.90	$159.20	$33.00	$264.00	$19.90	$33.00
1 F.E. Ldr, W.M.,2.5CY		400.80		440.90	50.10	55.11
8 L.H., Daily Totals		$560.00		$704.90	$70.00	$88.11
Crew B-10U	Hr.	Daily	Hr.	Daily	Bare Costs	Incl. O&P
1 Equip. Oper. (med.)	$19.90	$159.20	$33.00	$264.00	$19.90	$33.00
1 F.E. Loader, W.M., 5.5 C.Y.		861.20		947.30	107.65	118.42
8 L.H., Daily Totals		$1020.40		$1211.30	$127.55	$151.42
Crew B-10V	Hr.	Daily	Hr.	Daily	Bare Costs	Incl. O&P
1 Equip. Oper. (med.)	$19.90	$159.20	$33.00	$264.00	$19.90	$33.00
1 Dozer, 700 H.P.		2752.00		3027.20	344.00	378.40
8 L.H., Daily Totals		$2911.20		$3291.20	$363.90	$411.40
Crew B-10W	Hr.	Daily	Hr.	Daily	Bare Costs	Incl. O&P
1 Equip. Oper. (med.)	$19.90	$159.20	$33.00	$264.00	$19.90	$33.00
1 Dozer, 105 H.P.		393.00		432.30	49.13	54.04
8 L.H., Daily Totals		$552.20		$696.30	$69.03	$87.04
Crew B-10X	Hr.	Daily	Hr.	Daily	Bare Costs	Incl. O&P
1 Equip. Oper. (med.)	$19.90	$159.20	$33.00	$264.00	$19.90	$33.00
1 Dozer, 410 H.P.		1397.00		1536.70	174.63	192.09
8 L.H., Daily Totals		$1556.20		$1800.70	$194.53	$225.09
Crew B-10Y	Hr.	Daily	Hr.	Daily	Bare Costs	Incl. O&P
1 Equip. Oper. (med.)	$19.90	$159.20	$33.00	$264.00	$19.90	$33.00
1 Vibratory Drum Roller		372.00		409.20	46.50	51.15
8 L.H., Daily Totals		$531.20		$673.20	$66.40	$84.15
Crew B-11	Hr.	Daily	Hr.	Daily	Bare Costs	Incl. O&P
1 Equipment Oper. (med.)	$19.90	$159.20	$33.00	$264.00	$17.17	$28.88
1 Laborer	14.45	115.60	24.75	198.00		
16 L.H., Daily Totals		$274.80		$462.00	$17.17	$28.88
Crew B-11A	Hr.	Daily	Hr.	Daily	Bare Costs	Incl. O&P
1 Equipment Oper. (med.)	$19.90	$159.20	$33.00	$264.00	$17.17	$28.88
1 Laborer	14.45	115.60	24.75	198.00		
1 Dozer, 200 H.P.		806.80		887.50	50.43	55.47
16 L.H., Daily Totals		$1081.60		$1349.50	$67.60	$84.35
Crew B-11B	Hr.	Daily	Hr.	Daily	Bare Costs	Incl. O&P
1 Equipment Oper. (med.)	$19.90	$159.20	$33.00	$264.00	$17.17	$28.88
1 Laborer	14.45	115.60	24.75	198.00		
1 Dozer, 200 H.P.		806.80		887.50		
1 Air Powered Tamper		15.80		17.40		
1 Air Compr. 365 C.F.M.		174.40		191.85		
2-50 Ft. Air Hoses, 1.5" Dia.		14.40		15.85	63.21	69.53
16 L.H., Daily Totals		$1286.20		$1574.60	$80.38	$98.41
Crew B-11C	Hr.	Daily	Hr.	Daily	Bare Costs	Incl. O&P
1 Equipment Oper. (med.)	$19.90	$159.20	$33.00	$264.00	$17.17	$28.88
1 Laborer	14.45	115.60	24.75	198.00		
1 Backhoe Loader, 48 H.P.		203.00		223.30	12.69	13.96
16 L.H., Daily Totals		$477.80		$685.30	$29.86	$42.84
Crew B-11K	Hr.	Daily	Hr.	Daily	Bare Costs	Incl. O&P
1 Equipment Oper. (med.)	$19.90	$159.20	$33.00	$264.00	$17.17	$28.88
1 Laborer	14.45	115.60	24.75	198.00		
1 Trencher, 8' D., 16" W.		540.90		595.00	33.81	37.19
16 L.H., Daily Totals		$815.70		$1057.00	$50.98	$66.07
Crew B-11L	Hr.	Daily	Hr.	Daily	Bare Costs	Incl. O&P
1 Equipment Oper. (med.)	$19.90	$159.20	$33.00	$264.00	$17.17	$28.88
1 Laborer	14.45	115.60	24.75	198.00		
1 Grader, 30,000 Lbs.		516.00		567.60	32.25	35.48
16 L.H., Daily Totals		$790.80		$1029.60	$49.42	$64.36
Crew B-11M	Hr.	Daily	Hr.	Daily	Bare Costs	Incl. O&P
1 Equipment Oper. (med.)	$19.90	$159.20	$33.00	$264.00	$17.17	$28.88
1 Laborer	14.45	115.60	24.75	198.00		
1 Backhoe Loader, 80 H.P.		273.15		300.45	17.07	18.78
16 L.H., Daily Totals		$547.95		$762.45	$34.24	$47.66
Crew B-12	Hr.	Daily	Hr.	Daily	Bare Costs	Incl. O&P
1 Equip. Oper. (crane)	$20.65	$165.20	$34.25	$274.00	$20.65	$34.25
8 L.H., Daily Totals		$165.20		$274.00	$20.65	$34.25
Crew B-12A	Hr.	Daily	Hr.	Daily	Bare Costs	Incl. O&P
1 Equip. Oper. (crane)	$20.65	$165.20	$34.25	$274.00	$20.65	$34.25
1 Hyd. Excavator, 1 C.Y.		538.00		591.80	67.25	73.98
8 L.H., Daily Totals		$703.20		$865.80	$87.90	$108.23

Crews

Crew No.	Bare Costs		Incl. Subs O & P		Cost Per Labor-Hour	
Crew B-12B	Hr.	Daily	Hr.	Daily	Bare Costs	Incl. O&P
1 Equip. Oper. (crane)	$20.65	$165.20	$34.25	$274.00	$20.65	$34.25
1 Hyd. Excavator, 1.5 C.Y.		713.20		784.50	89.15	98.07
8 L.H., Daily Totals		$878.40		$1058.50	$109.80	$132.32
Crew B-12C	Hr.	Daily	Hr.	Daily	Bare Costs	Incl. O&P
1 Equip. Oper. (crane)	$20.65	$165.20	$34.25	$274.00	$20.65	$34.25
1 Hyd. Excavator, 2 C.Y.		1017.00		1118.70	127.13	139.84
8 L.H., Daily Totals		$1182.20		$1392.70	$147.78	$174.09
Crew B-12D	Hr.	Daily	Hr.	Daily	Bare Costs	Incl. O&P
1 Equip. Oper. (crane)	$20.65	$165.20	$34.25	$274.00	$20.65	$34.25
1 Hyd. Excavator, 3.5 C.Y.		2126.00		2338.60	265.75	292.33
8 L.H., Daily Totals		$2291.20		$2612.60	$286.40	$326.58
Crew B-12E	Hr.	Daily	Hr.	Daily	Bare Costs	Incl. O&P
1 Equip. Oper. (crane)	$20.65	$165.20	$34.25	$274.00	$20.65	$34.25
1 Hyd. Excavator, .5 C.Y.		342.80		377.10	42.85	47.14
8 L.H., Daily Totals		$508.00		$651.10	$63.50	$81.39
Crew B-12F	Hr.	Daily	Hr.	Daily	Bare Costs	Incl. O&P
1 Equip. Oper. (crane)	$20.65	$165.20	$34.25	$274.00	$20.65	$34.25
1 Hyd. Excavator, .75 C.Y.		455.80		501.40	56.98	62.67
8 L.H., Daily Totals		$621.00		$775.40	$77.63	$96.92
Crew B-12G	Hr.	Daily	Hr.	Daily	Bare Costs	Incl. O&P
1 Equip. Oper. (crane)	$20.65	$165.20	$34.25	$274.00	$20.65	$34.25
1 Power Shovel, .5 C.Y.		477.00		524.70		
1 Clamshell Bucket, .5 C.Y		48.80		53.70	65.73	72.30
8 L.H., Daily Totals		$691.00		$852.40	$86.38	$106.55
Crew B-12H	Hr.	Daily	Hr.	Daily	Bare Costs	Incl. O&P
1 Equip. Oper. (crane)	$20.65	$165.20	$34.25	$274.00	$20.65	$34.25
1 Power Shovel, 1 C.Y.		516.80		568.50		
1 Clamshell Bucket, 1 C.Y.		70.00		77.00	73.35	80.69
8 L.H., Daily Totals		$752.00		$919.50	$94.00	$114.94
Crew B-12I	Hr.	Daily	Hr.	Daily	Bare Costs	Incl. O&P
1 Equip. Oper. (crane)	$20.65	$165.20	$34.25	$274.00	$20.65	$34.25
1 Power Shovel, .75 C.Y.		492.60		541.85		
1 Dragline Bucket, .75 C.Y.		33.20		36.50	65.73	72.30
8 L.H., Daily Totals		$691.00		$852.35	$86.38	$106.55
Crew B-12J	Hr.	Daily	Hr.	Daily	Bare Costs	Incl. O&P
1 Equip. Oper. (crane)	$20.65	$165.20	$34.25	$274.00	$20.65	$34.25
1 Gradall, 3 Ton, .5 C.Y.		618.20		680.00	77.28	85.00
8 L.H., Daily Totals		$783.40		$954.00	$97.93	$119.25
Crew B-12K	Hr.	Daily	Hr.	Daily	Bare Costs	Incl. O&P
1 Equip. Oper. (crane)	$20.65	$165.20	$34.25	$274.00	$20.65	$34.25
1 Gradall, 3 Ton, 1 C.Y.		817.85		899.65	102.23	112.45
8 L.H., Daily Totals		$983.05		$1173.65	$122.88	$146.70
Crew B-12L	Hr.	Daily	Hr.	Daily	Bare Costs	Incl. O&P
1 Equip. Oper. (crane)	$20.65	$165.20	$34.25	$274.00	$20.65	$34.25
1 Power Shovel, .5 C.Y.		477.00		524.70		
1 F.E. Attachment, .5 C.Y.		56.20		61.80	66.65	73.32
8 L.H., Daily Totals		$698.40		$860.50	$87.30	$107.57

Crew No.	Bare Costs		Incl. Subs O & P		Cost Per Labor-Hour	
Crew B-12M	Hr.	Daily	Hr.	Daily	Bare Costs	Incl. O&P
1 Equip. Oper. (crane)	$20.65	$165.20	$34.25	$274.00	$20.65	$34.25
1 Power Shovel, .75 C.Y		492.60		541.85		
1 F.E. Attachment, .75 C.Y.		101.20		111.30	74.23	81.65
8 L.H., Daily Totals		$759.00		$927.15	$94.88	$115.90
Crew B-12N	Hr.	Daily	Hr.	Daily	Bare Costs	Incl. O&P
1 Equip. Oper. (crane)	$20.65	$165.20	$34.25	$274.00	$20.65	$34.25
1 Power Shovel, 1 C.Y.		516.80		568.50		
1 F.E. Attachment, 1 C.Y.		133.20		146.50	81.25	89.38
8 L.H., Daily Totals		$815.20		$989.00	$101.90	$123.63
Crew B-12O	Hr.	Daily	Hr.	Daily	Bare Costs	Incl. O&P
1 Equip. Oper. (crane)	$20.65	$165.20	$34.25	$274.00	$20.65	$34.25
1 Power Shovel, 1.5 C.Y.		714.30		785.75		
1 F.E. Attachment, 1.5 C.Y.		161.00		177.10	109.41	120.35
8 L.H., Daily Totals		$1040.50		$1236.85	$130.06	$154.60
Crew B-12P	Hr.	Daily	Hr.	Daily	Bare Costs	Incl. O&P
1 Equip. Oper. (crane)	$20.65	$165.20	$34.25	$274.00	$20.65	$34.25
1 Crawler Crane, 40 Ton		714.30		785.75		
1 Dragline Bucket, 1.5 C.Y.		48.95		53.85	95.41	104.95
8 L.H., Daily Totals		$928.45		$1113.60	$116.06	$139.20
Crew B-12Q	Hr.	Daily	Hr.	Daily	Bare Costs	Incl. O&P
1 Equip. Oper. (crane)	$20.65	$165.20	$34.25	$274.00	$20.65	$34.25
1 Hyd. Excavator, 5/8 C.Y.		411.60		452.75	51.45	56.60
8 L.H., Daily Totals		$576.80		$726.75	$72.10	$90.85
Crew B-12R	Hr.	Daily	Hr.	Daily	Bare Costs	Incl. O&P
1 Equip. Oper. (crane)	$20.65	$165.20	$34.25	$274.00	$20.65	$34.25
1 Hyd. Excavator, 1.5 C.Y.		713.20		784.50	89.15	98.07
8 L.H., Daily Totals		$878.40		$1058.50	$109.80	$132.32
Crew B-12S	Hr.	Daily	Hr.	Daily	Bare Costs	Incl. O&P
1 Equip. Oper. (crane)	$20.65	$165.20	$34.25	$274.00	$20.65	$34.25
1 Hyd. Excavator, 2.5 C.Y.		1673.00		1840.30	209.13	230.04
8 L.H., Daily Totals		$1838.20		$2114.30	$229.78	$264.29
Crew B-12T	Hr.	Daily	Hr.	Daily	Bare Costs	Incl. O&P
1 Equip. Oper. (crane)	$20.65	$165.20	$34.25	$274.00	$20.65	$34.25
1 Crawler Crane, 75 Ton		939.20		1033.10		
1 F.E. Attachment, 3 C.Y.		300.40		330.45	154.95	170.45
8 L.H., Daily Totals		$1404.80		$1637.55	$175.60	$204.70
Crew B-12V	Hr.	Daily	Hr.	Daily	Bare Costs	Incl. O&P
1 Equip. Oper. (crane)	$20.65	$165.20	$34.25	$274.00	$20.65	$34.25
1 Crawler Crane, 75 Ton		939.20		1033.10		
1 Dragline Bucket, 3 C.Y.		81.60		89.75	127.60	140.36
8 L.H., Daily Totals		$1186.00		$1396.85	$148.25	$174.61
Crew B-13	Hr.	Daily	Hr.	Daily	Bare Costs	Incl. O&P
1 Labor Foreman (outside)	$16.45	$131.60	$28.20	$225.60	$15.82	$26.91
4 Laborers	14.45	462.40	24.75	792.00		
1 Equip. Oper. (crane)	20.65	165.20	34.25	274.00		
1 Hyd. Crane, 25 Ton		559.20		615.10	11.65	12.82
48 L.H., Daily Totals		$1318.40		$1906.70	$27.47	$39.73

Crews

Crew No.	Bare Costs		Incl. Subs O & P		Cost Per Labor-Hour	
Crew B-13A	Hr.	Daily	Hr.	Daily	Bare Costs	Incl. O&P
1 Foreman	$16.45	$131.60	$28.20	$225.60	$16.79	$28.26
2 Laborers	14.45	231.20	24.75	396.00		
2 Equipment Operator	19.90	318.40	33.00	528.00		
2 Truck Drivers (heavy)	16.20	259.20	27.05	432.80		
1 Crane, 75 Ton		939.20		1033.10		
1 F.E. Lder, 3.75 C.Y.		1035.00		1138.50		
2 Dump Trucks, 12 Ton		730.60		803.65	48.30	53.13
56 L.H., Daily Totals		$3645.20		$4557.65	$65.09	$81.39
Crew B-13B	Hr.	Daily	Hr.	Daily	Bare Costs	Incl. O&P
1 Labor Foreman (outside)	$16.45	$131.60	$28.20	$225.60	$15.99	$27.09
4 Laborers	14.45	462.40	24.75	792.00		
1 Equip. Oper. (crane)	20.65	165.20	34.25	274.00		
1 Equip. Oper. Oiler	17.00	136.00	28.20	225.60		
1 Hyd. Crane, 55 Ton		808.20		889.00	14.43	15.88
56 L.H., Daily Totals		$1703.40		$2406.20	$30.42	$42.97
Crew B-13C	Hr.	Daily	Hr.	Daily	Bare Costs	Incl. O&P
1 Labor Foreman (outside)	$16.45	$131.60	$28.20	$225.60	$15.99	$27.09
4 Laborers	14.45	462.40	24.75	792.00		
1 Equip. Oper. (crane)	20.65	165.20	34.25	274.00		
1 Equip. Oper. Oiler	17.00	136.00	28.20	225.60		
1 Crawler Crane, 100 Ton		1139.00		1252.90	20.34	22.37
56 L.H., Daily Totals		$2034.20		$2770.10	$36.33	$49.46
Crew B-14	Hr.	Daily	Hr.	Daily	Bare Costs	Incl. O&P
1 Labor Foreman (outside)	$16.45	$131.60	$28.20	$225.60	$15.56	$26.48
4 Laborers	14.45	462.40	24.75	792.00		
1 Equip. Oper. (light)	19.10	152.80	31.70	253.60		
1 Backhoe Loader, 48 H.P.		203.00		223.30	4.23	4.65
48 L.H., Daily Totals		$949.80		$1494.50	$19.79	$31.13
Crew B-15	Hr.	Daily	Hr.	Daily	Bare Costs	Incl. O&P
1 Equipment Oper. (med)	$19.90	$159.20	$33.00	$264.00	$17.01	$28.42
.5 Laborer	14.45	57.80	24.75	99.00		
2 Truck Drivers (heavy)	16.20	259.20	27.05	432.80		
2 Dump Trucks, 16 Ton		894.50		983.95		
1 Dozer, 200 H.P.		806.80		887.50	60.76	66.84
28 L.H., Daily Totals		$2177.50		$2667.25	$77.77	$95.26
Crew B-16	Hr.	Daily	Hr.	Daily	Bare Costs	Incl. O&P
1 Labor Foreman (outside)	$16.45	$131.60	$28.20	$225.60	$15.39	$26.19
2 Laborers	14.45	231.20	24.75	396.00		
1 Truck Driver (heavy)	16.20	129.60	27.05	216.40		
1 Dump Truck, 16 Ton		447.25		492.00	13.98	15.37
32 L.H., Daily Totals		$939.65		$1330.00	$29.37	$41.56
Crew B-17	Hr.	Daily	Hr.	Daily	Bare Costs	Incl. O&P
2 Laborers	$14.45	$231.20	$24.75	$396.00	$16.05	$27.06
1 Equip. Oper. (light)	19.10	152.80	31.70	253.60		
1 Truck Driver (heavy)	16.20	129.60	27.05	216.40		
1 Backhoe Loader, 48 H.P.		203.00		223.30		
1 Dump Truck, 12 Ton		365.30		401.85	17.76	19.54
32 L.H., Daily Totals		$1081.90		$1491.15	$33.81	$46.60
Crew B-18	Hr.	Daily	Hr.	Daily	Bare Costs	Incl. O&P
1 Labor Foreman (outside)	$16.45	$131.60	$28.20	$225.60	$15.12	$25.90
2 Laborers	14.45	231.20	24.75	396.00		
1 Vibrating Compactor		60.60		66.65	2.53	2.78
24 L.H., Daily Totals		$423.40		$688.25	$17.65	$28.68

Crew No.	Bare Costs		Incl. Subs O & P		Cost Per Labor-Hour	
Crew B-19	Hr.	Daily	Hr.	Daily	Bare Costs	Incl. O&P
1 Pile Driver Foreman	$21.45	$171.60	$40.00	$320.00	$19.19	$34.86
4 Pile Drivers	19.45	622.40	36.25	1160.00		
1 Equip. Oper. (crane)	20.65	165.20	34.25	274.00		
1 Building Laborer	14.45	115.60	24.75	198.00		
1 Crane, 40 Ton & Access.		714.30		785.75		
60 L.F. Leads, 15K Ft. Lbs.		264.00		290.40		
1 Hammer, 15K Ft. Lbs.		311.40		342.55		
1 Air Compr., 600 C.F.M.		278.00		305.80		
2-50 Ft. Air Hoses, 3" Dia.		33.80		37.20	28.60	31.46
56 L.H., Daily Totals		$2676.30		$3713.70	$47.79	$66.32
Crew B-19A	Hr.	Daily	Hr.	Daily	Bare Costs	Incl. O&P
1 Pile Driver Foreman	$21.45	$171.60	$40.00	$320.00	$19.69	$35.21
4 Pile Drivers	19.45	622.40	36.25	1160.00		
2 Equip. Oper. (crane)	20.65	330.40	34.25	548.00		
1 Equip. Oper. Oiler	17.00	136.00	28.20	225.60		
1 Crawler Crane, 75 Ton		939.20		1033.10		
60 L.F. Leads, 25K Ft. Lbs.		315.00		346.50		
1 Air Compressor, 750 CFM		300.80		330.90		
4-50 Ft. Air Hose, 3" Dia.		67.60		74.35	25.35	27.89
64 L.H., Daily Totals		$2883.00		$4038.45	$45.04	$63.10
Crew B-20	Hr.	Daily	Hr.	Daily	Bare Costs	Incl. O&P
1 Labor Foreman (out)	$16.45	$131.60	$28.20	$225.60	$15.12	$25.90
2 Laborer	14.45	231.20	24.75	396.00		
24 L.H., Daily Totals		$362.80		$621.60	$15.12	$25.90
Crew B-20A	Hr.	Daily	Hr.	Daily	Bare Costs	Incl. O&P
1 Labor Foreman	$16.45	$131.60	$28.20	$225.60	$17.60	$29.56
1 Laborer	14.45	115.60	24.75	198.00		
1 Plumber	21.95	175.60	36.30	290.40		
1 Plumber Apprentice	17.55	140.40	29.00	232.00		
32 L.H., Daily Totals		$563.20		$946.00	$17.60	$29.56
Crew B-21	Hr.	Daily	Hr.	Daily	Bare Costs	Incl. O&P
1 Labor Foreman (out)	$16.45	$131.60	$28.20	$225.60	$15.91	$27.09
2 Laborer	14.45	231.20	24.75	396.00		
.5 Equip. Oper. (crane)	20.65	82.60	34.25	137.00		
.5 S.P. Crane, 5 Ton		138.23		152.05	4.94	5.43
28 L.H., Daily Totals		$583.63		$910.65	$20.85	$32.52
Crew B-21A	Hr.	Daily	Hr.	Daily	Bare Costs	Incl. O&P
1 Labor Foreman	$16.45	$131.60	$28.20	$225.60	$18.21	$30.50
1 Laborer	14.45	115.60	24.75	198.00		
1 Plumber	21.95	175.60	36.30	290.40		
1 Plumber Apprentice	17.55	140.40	29.00	232.00		
1 Equip. Oper. (crane)	20.65	165.20	34.25	274.00		
1 S.P. Crane, 12 Ton		401.35		441.50	10.03	11.04
40 L.H., Daily Totals		$1129.75		$1661.50	$28.24	$41.54
Crew B-22	Hr.	Daily	Hr.	Daily	Bare Costs	Incl. O&P
1 Labor Foreman (out)	$16.45	$131.60	$28.20	$225.60	$16.22	$27.57
2 Laborer	14.45	231.20	24.75	396.00		
.75 Equip. Oper. (crane)	20.65	123.90	34.25	205.50		
.75 S.P. Crane, 5 Ton		207.34		228.05	6.91	7.60
30 L.H., Daily Totals		$694.04		$1055.15	$23.13	$35.17

Crews

Crew No.	Bare Costs		Incl. Subs O & P		Cost Per Labor-Hour	
Crew B-22A	**Hr.**	**Daily**	**Hr.**	**Daily**	**Bare Costs**	**Incl. O&P**
1 Labor Foreman (out)	$16.45	$131.60	$28.20	$225.60	$16.97	$28.90
1 Skilled Worker	19.75	158.00	33.90	271.20		
2 Laborers	14.45	231.20	24.75	396.00		
.75 Equipment Oper. (crane)	20.65	123.90	34.25	205.50		
.75 Crane, 5 Ton		207.34		228.05		
1 Generator, 5 KW		46.90		51.60		
1 Butt Fusion Machine		188.00		206.80	11.64	12.80
38 L.H., Daily Totals		$1086.94		$1584.75	$28.61	$41.70
Crew B-22B	**Hr.**	**Daily**	**Hr.**	**Daily**	**Bare Costs**	**Incl. O&P**
1 Skilled Worker	$19.75	$158.00	$33.90	$271.20	$17.10	$29.33
1 Laborer	14.45	115.60	24.75	198.00		
1 Electro Fusion Machine		86.00		94.60	5.38	5.91
16 L.H., Daily Totals		$359.60		$563.80	$22.48	$35.24
Crew B-23	**Hr.**	**Daily**	**Hr.**	**Daily**	**Bare Costs**	**Incl. O&P**
1 Labor Foreman (outside)	$16.45	$131.60	$28.20	$225.60	$14.85	$25.44
4 Laborers	14.45	462.40	24.75	792.00		
1 Drill Rig, Wells		1796.00		1975.60		
1 Light Truck, 3 Ton		180.45		198.50	49.41	54.35
40 L.H., Daily Totals		$2570.45		$3191.70	$64.26	$79.79
Crew B-23A	**Hr.**	**Daily**	**Hr.**	**Daily**	**Bare Costs**	**Incl. O&P**
1 Labor Foreman (outside)	$16.45	$131.60	$28.20	$225.60	$16.93	$28.65
1 Laborers	14.45	115.60	24.75	198.00		
1 Equip. Operator, medium	19.90	159.20	33.00	264.00		
1 Drill Rig, Wells		1796.00		1975.60		
1 Pickup Truck, 3/4 Ton		130.20		143.20	80.26	88.28
24 L.H., Daily Totals		$2332.60		$2806.40	$97.19	$116.93
Crew B-23B	**Hr.**	**Daily**	**Hr.**	**Daily**	**Bare Costs**	**Incl. O&P**
1 Labor Foreman (outside)	$16.45	$131.60	$28.20	$225.60	$16.93	$28.65
1 Laborer	14.45	115.60	24.75	198.00		
1 Equip. Operator, medium	19.90	159.20	33.00	264.00		
1 Drill Rig, Wells		1796.00		1975.60		
1 Pickup Truck, 3/4 Ton		130.20		143.20		
1 Pump, Cntfgl, 6"		157.20		172.90	86.81	95.49
24 L.H., Daily Totals		$2489.80		$2979.30	$103.74	$124.14
Crew B-24	**Hr.**	**Daily**	**Hr.**	**Daily**	**Bare Costs**	**Incl. O&P**
1 Cement Finisher	$18.90	$151.20	$30.80	$246.40	$17.68	$29.77
1 Laborer	14.45	115.60	24.75	198.00		
1 Carpenter	19.70	157.60	33.75	270.00		
24 L.H., Daily Totals		$424.40		$714.40	$17.68	$29.77
Crew B-25	**Hr.**	**Daily**	**Hr.**	**Daily**	**Bare Costs**	**Incl. O&P**
1 Labor Foreman	$16.45	$131.60	$28.20	$225.60	$16.12	$27.31
7 Laborers	14.45	809.20	24.75	1386.00		
3 Equip. Oper. (med.)	19.90	477.60	33.00	792.00		
1 Asphalt Paver, 130 H.P.		1309.00		1439.90		
1 Tandem Roller, 10 Ton		224.00		246.40		
1 Roller, Pneumatic Wheel		240.80		264.90	20.16	22.17
88 L.H., Daily Totals		$3192.20		$4354.80	$36.28	$49.48

Crew No.	Bare Costs		Incl. Subs O & P		Cost Per Labor-Hour	
Crew B-25B	**Hr.**	**Daily**	**Hr.**	**Daily**	**Bare Costs**	**Incl. O&P**
1 Labor Foreman	$16.45	$131.60	$28.20	$225.60	$16.43	$27.79
7 Laborers	14.45	809.20	24.75	1386.00		
4 Equip. Oper. (medium)	19.90	636.80	33.00	1056.00		
1 Asphalt Paver, 130 H.P.		1309.00		1439.90		
2 Rollers, Steel Wheel		448.00		492.80		
1 Roller, Pneumatic Wheel		240.80		264.90	20.81	22.89
96 L.H., Daily Totals		$3575.40		$4865.20	$37.24	$50.68
Crew B-26	**Hr.**	**Daily**	**Hr.**	**Daily**	**Bare Costs**	**Incl. O&P**
1 Labor Foreman (outside)	$16.45	$131.60	$28.20	$225.60	$16.63	$28.43
6 Laborers	14.45	693.60	24.75	1188.00		
2 Equip. Oper. (med.)	19.90	318.40	33.00	528.00		
1 Rodman (reinf.)	21.10	168.80	39.25	314.00		
1 Cement Finisher	18.90	151.20	30.80	246.40		
1 Grader, 30,000 Lbs.		516.00		567.60		
1 Paving Mach. & Equip.		1324.00		1456.40	20.91	23.00
88 L.H., Daily Totals		$3303.60		$4526.00	$37.54	$51.43
Crew B-27	**Hr.**	**Daily**	**Hr.**	**Daily**	**Bare Costs**	**Incl. O&P**
1 Labor Foreman (outside)	$16.45	$131.60	$28.20	$225.60	$14.95	$25.61
3 Laborers	14.45	346.80	24.75	594.00		
1 Berm Machine		73.00		80.30	2.28	2.51
32 L.H., Daily Totals		$551.40		$899.90	$17.23	$28.12
Crew B-28	**Hr.**	**Daily**	**Hr.**	**Daily**	**Bare Costs**	**Incl. O&P**
2 Carpenters	$19.70	$315.20	$33.75	$540.00	$17.95	$30.75
1 Laborer	14.45	115.60	24.75	198.00		
24 L.H., Daily Totals		$430.80		$738.00	$17.95	$30.75
Crew B-29	**Hr.**	**Daily**	**Hr.**	**Daily**	**Bare Costs**	**Incl. O&P**
1 Labor Foreman (outside)	$16.45	$131.60	$28.20	$225.60	$15.82	$26.91
4 Laborers	14.45	462.40	24.75	792.00		
1 Equip. Oper. (crane)	20.65	165.20	34.25	274.00		
1 Gradall, 3 Ton, 1/2 C.Y.		618.20		680.00	12.88	14.17
48 L.H., Daily Totals		$1377.40		$1971.60	$28.70	$41.08
Crew B-30	**Hr.**	**Daily**	**Hr.**	**Daily**	**Bare Costs**	**Incl. O&P**
1 Equip. Oper. (med.)	$19.90	$159.20	$33.00	$264.00	$17.43	$29.03
2 Truck Drivers (heavy)	16.20	259.20	27.05	432.80		
1 Hyd. Excavator, 1.5 C.Y.		713.20		784.50		
2 Dump Trucks, 16 Ton		894.50		983.95	66.99	73.69
24 L.H., Daily Totals		$2026.10		$2465.25	$84.42	$102.72
Crew B-31	**Hr.**	**Daily**	**Hr.**	**Daily**	**Bare Costs**	**Incl. O&P**
1 Labor Foreman (outside)	$16.45	$131.60	$28.20	$225.60	$14.85	$25.44
4 Laborers	14.45	462.40	24.75	792.00		
1 Air Compr., 250 C.F.M.		127.00		139.70		
1 Sheeting Driver		12.20		13.40		
2-50 Ft. Air Hoses, 1.5" Dia.		14.40		15.85	3.84	4.22
40 L.H., Daily Totals		$747.60		$1186.55	$18.69	$29.66
Crew B-32	**Hr.**	**Daily**	**Hr.**	**Daily**	**Bare Costs**	**Incl. O&P**
1 Laborer	$14.45	$115.60	$24.75	$198.00	$18.54	$30.94
3 Equip. Oper. (med.)	19.90	477.60	33.00	792.00		
1 Grader, 30,000 Lbs.		516.00		567.60		
1 Tandem Roller, 10 Ton		224.00		246.40		
1 Dozer, 200 H.P.		806.80		887.50	48.34	53.17
32 L.H., Daily Totals		$2140.00		$2691.50	$66.88	$84.11

Crews

Crew No.	Bare Costs		Incl. Subs O & P		Cost Per Labor-Hour	
	Hr.	Daily	Hr.	Daily	Bare Costs	Incl. O&P
Crew B-32A						
1 Laborer	$14.45	$115.60	$24.75	$198.00	$18.08	$30.25
2 Equip. Oper. (medium)	19.90	318.40	33.00	528.00		
1 Grader, 30,000 Lbs.		516.00		567.60		
1 Roller, Vibratory, 29,000 Lbs.		432.40		475.65	39.52	43.47
24 L.H., Daily Totals		$1382.40		$1769.25	$57.60	$73.72
Crew B-32B	Hr.	Daily	Hr.	Daily	Bare Costs	Incl. O&P
1 Laborer	$14.45	$115.60	$24.75	$198.00	$18.08	$30.25
2 Equip. Oper. (medium)	19.90	318.40	33.00	528.00		
1 Dozer, 200 H.P.		806.80		887.50		
1 Roller, Vibratory, 29,000 Lbs.		432.40		475.65	51.63	56.80
24 L.H., Daily Totals		$1673.20		$2089.15	$69.71	$87.05
Crew B-32C	Hr.	Daily	Hr.	Daily	Bare Costs	Incl. O&P
1 Labor Foreman	$16.45	$131.60	$28.20	$225.60	$17.51	$29.45
2 Laborers	14.45	231.20	24.75	396.00		
3 Equip. Oper. (medium)	19.90	477.60	33.00	792.00		
1 Grader, 30,000 Lbs.		516.00		567.60		
1 Roller, Steel Wheel		224.00		246.40		
1 Dozer, 200 H.P.		806.80		887.50	32.23	35.45
48 L.H., Daily Totals		$2387.20		$3115.10	$49.74	$64.90
Crew B-33	Hr.	Daily	Hr.	Daily	Bare Costs	Incl. O&P
1 Equip. Oper. (med.)	$19.90	$159.20	$33.00	$264.00	$19.90	$33.00
.25 Equip. Oper. (med.)	19.90	39.80	33.00	66.00		
10 L.H., Daily Totals		$199.00		$330.00	$19.90	$33.00
Crew B-33A	Hr.	Daily	Hr.	Daily	Bare Costs	Incl. O&P
1 Equip. Oper. (med.)	$19.90	$159.20	$33.00	$264.00	$19.90	$33.00
.25 Equip. Oper. (med.)	19.90	39.80	33.00	66.00		
1 Scraper, Towed, 7 C.Y.		81.00		89.10		
1.25 Dozer, 300 H.P.		1385.00		1523.50	146.60	161.26
10 L.H., Daily Totals		$1665.00		$1942.60	$166.50	$194.26
Crew B-33B	Hr.	Daily	Hr.	Daily	Bare Costs	Incl. O&P
1 Equip. Oper. (med.)	$19.90	$159.20	$33.00	$264.00	$19.90	$33.00
.25 Equip. Oper. (med.)	19.90	39.80	33.00	66.00		
1 Scraper, Towed, 10 C.Y.		204.80		225.30		
1.25 Dozer, 300 H.P.		1385.00		1523.50	158.98	174.88
10 L.H., Daily Totals		$1788.80		$2078.80	$178.88	$207.88
Crew B-33C	Hr.	Daily	Hr.	Daily	Bare Costs	Incl. O&P
1 Equip. Oper. (med.)	$19.90	$159.20	$33.00	$264.00	$19.90	$33.00
.25 Equip. Oper. (med.)	19.90	39.80	33.00	66.00		
1 Scraper, Towed, 12 C.Y.		204.80		225.30		
1.25 Dozer, 300 H.P.		1385.00		1523.50	158.98	174.88
10 L.H., Daily Totals		$1788.80		$2078.80	$178.88	$207.88
Crew B-33D	Hr.	Daily	Hr.	Daily	Bare Costs	Incl. O&P
1 Equip. Oper. (med.)	$19.90	$159.20	$33.00	$264.00	$19.90	$33.00
.25 Equip. Oper. (med.)	19.90	39.80	33.00	66.00		
1 S.P. Scraper, 14 C.Y.		1626.00		1788.60		
.25 Dozer, 300 H.P.		277.00		304.70	190.30	209.33
10 L.H., Daily Totals		$2102.00		$2423.30	$210.20	$242.33

Crew No.	Bare Costs		Incl. Subs O & P		Cost Per Labor-Hour	
Crew B-33E	Hr.	Daily	Hr.	Daily	Bare Costs	Incl. O&P
1 Equip. Oper. (med.)	$19.90	$159.20	$33.00	$264.00	$19.90	$33.00
.25 Equip. Oper. (med.)	19.90	39.80	33.00	66.00		
1 S.P. Scraper, 24 C.Y.		1913.00		2104.30		
.25 Dozer, 300 H.P.		277.00		304.70	219.00	240.90
10 L.H., Daily Totals		$2389.00		$2739.00	$238.90	$273.90
Crew B-33F	Hr.	Daily	Hr.	Daily	Bare Costs	Incl. O&P
1 Equip. Oper. (med.)	$19.90	$159.20	$33.00	$264.00	$19.90	$33.00
.25 Equip. Oper. (med.)	19.90	39.80	33.00	66.00		
1 Elev. Scraper, 11 C.Y.		698.55		768.40		
.25 Dozer, 300 H.P.		277.00		304.70	97.56	107.31
10 L.H., Daily Totals		$1174.55		$1403.10	$117.46	$140.31
Crew B-33G	Hr.	Daily	Hr.	Daily	Bare Costs	Incl. O&P
1 Equip. Oper. (med.)	$19.90	$159.20	$33.00	$264.00	$19.90	$33.00
.25 Equip. Oper. (med.)	19.90	39.80	33.00	66.00		
1 Elev. Scraper, 20 C.Y.		984.80		1083.30		
.25 Dozer, 300 H.P.		277.00		304.70	126.18	138.80
10 L.H., Daily Totals		$1460.80		$1718.00	$146.08	$171.80
Crew B-34A	Hr.	Daily	Hr.	Daily	Bare Costs	Incl. O&P
1 Truck Driver (heavy)	$16.20	$129.60	$27.05	$216.40	$16.20	$27.05
1 Dump Truck, 12 Ton		365.30		401.85	45.66	50.23
8 L.H., Daily Totals		$494.90		$618.25	$61.86	$77.28
Crew B-34B	Hr.	Daily	Hr.	Daily	Bare Costs	Incl. O&P
1 Truck Driver (heavy)	$16.20	$129.60	$27.05	$216.40	$16.20	$27.05
1 Dump Truck, 16 Ton		447.25		492.00	55.91	61.50
8 L.H., Daily Totals		$576.85		$708.40	$72.11	$88.55
Crew B-34C	Hr.	Daily	Hr.	Daily	Bare Costs	Incl. O&P
1 Truck Driver (heavy)	$16.20	$129.60	$27.05	$216.40	$16.20	$27.05
1 Truck Tractor, 40 Ton		427.60		470.35		
1 Dump Trailer, 16.5 C.Y.		133.60		146.95	70.15	77.17
8 L.H., Daily Totals		$690.80		$833.70	$86.35	$104.22
Crew B-34D	Hr.	Daily	Hr.	Daily	Bare Costs	Incl. O&P
1 Truck Driver (heavy)	$16.20	$129.60	$27.05	$216.40	$16.20	$27.05
1 Truck Tractor, 40 Ton		427.60		470.35		
1 Dump Trailer, 20 C.Y.		136.10		149.70	70.46	77.51
8 L.H., Daily Totals		$693.30		$836.45	$86.66	$104.56
Crew B-34E	Hr.	Daily	Hr.	Daily	Bare Costs	Incl. O&P
1 Truck Driver (heavy)	$16.20	$129.60	$27.05	$216.40	$16.20	$27.05
1 Truck, Off Hwy., 25 Ton		700.00		770.00	87.50	96.25
8 L.H., Daily Totals		$829.60		$986.40	$103.70	$123.30
Crew B-34F	Hr.	Daily	Hr.	Daily	Bare Costs	Incl. O&P
1 Truck Driver (heavy)	$16.20	$129.60	$27.05	$216.40	$16.20	$27.05
1 Truck, Off Hwy., 22 C.Y.		1058.00		1163.80	132.25	145.48
8 L.H., Daily Totals		$1187.60		$1380.20	$148.45	$172.53
Crew B-34G	Hr.	Daily	Hr.	Daily	Bare Costs	Incl. O&P
1 Truck Driver (heavy)	$16.20	$129.60	$27.05	$216.40	$16.20	$27.05
1 Truck, Off Hwy., 34 C.Y.		1405.00		1545.50	175.63	193.19
8 L.H., Daily Totals		$1534.60		$1761.90	$191.83	$220.24

Crews

Crew No.	Bare Costs		Incl. Subs O & P		Cost Per Labor-Hour	
Crew B-34H	**Hr.**	**Daily**	**Hr.**	**Daily**	**Bare Costs**	**Incl. O&P**
1 Truck Driver (heavy)	$16.20	$129.60	$27.05	$216.40	$16.20	$27.05
1 Truck, Off Hwy., 42 C.Y.		1599.00		1758.90	199.88	219.86
8 L.H., Daily Totals		$1728.60		$1975.30	$216.08	$246.91
Crew B-34J	**Hr.**	**Daily**	**Hr.**	**Daily**	**Bare Costs**	**Incl. O&P**
1 Truck Driver (heavy)	$16.20	$129.60	$27.05	$216.40	$16.20	$27.05
1 Truck, Off Hwy., 60 C.Y.		2240.00		2464.00	280.00	308.00
8 L.H., Daily Totals		$2369.60		$2680.40	$296.20	$335.05
Crew B-34K	**Hr.**	**Daily**	**Hr.**	**Daily**	**Bare Costs**	**Incl. O&P**
1 Truck Driver (heavy)	$16.20	$129.60	$27.05	$216.40	$16.20	$27.05
1 Truck Tractor, 240 H.P.		531.20		584.30		
1 Low Bed Trailer		384.75		423.25	114.49	125.94
8 L.H., Daily Totals		$1045.55		$1223.95	$130.69	$152.99
Crew B-35	**Hr.**	**Daily**	**Hr.**	**Daily**	**Bare Costs**	**Incl. O&P**
1 Laborer Foreman (out)	$16.45	$131.60	$28.20	$225.60	$18.65	$31.48
1 Skilled Worker	19.75	158.00	33.90	271.20		
1 Welder (plumber)	21.95	175.60	36.30	290.40		
1 Laborer	14.45	115.60	24.75	198.00		
1 Equip. Oper. (crane)	20.65	165.20	34.25	274.00		
1 Electric Welding Mach.		48.80		53.70		
1 Hyd. Excavator, .75 C.Y.		455.80		501.40	12.62	13.88
40 L.H., Daily Totals		$1250.60		$1814.30	$31.27	$45.36
Crew B-35A	**Hr.**	**Daily**	**Hr.**	**Daily**	**Bare Costs**	**Incl. O&P**
1 Laborer Foreman (out)	$16.45	$131.60	$28.20	$225.60	$17.81	$30.05
2 Laborers	14.45	231.20	24.75	396.00		
1 Skilled Worker	19.75	158.00	33.90	271.20		
1 Welder (plumber)	21.95	175.60	36.30	290.40		
1 Equip. Oper. (crane)	20.65	165.20	34.25	274.00		
1 Equip. Oper. Oiler	17.00	136.00	28.20	225.60		
1 Welder, 300 amp		84.00		92.40		
1 Crane, 75 Ton		939.20		1033.10	18.27	20.10
56 L.H., Daily Totals		$2020.80		$2808.30	$36.08	$50.15
Crew B-36	**Hr.**	**Daily**	**Hr.**	**Daily**	**Bare Costs**	**Incl. O&P**
1 Labor Foreman (outside)	$16.45	$131.60	$28.20	$225.60	$17.03	$28.74
2 Laborers	14.45	231.20	24.75	396.00		
2 Equip. Oper. (med.)	19.90	318.40	33.00	528.00		
1 Dozer, 200 H.P.		806.90		887.50		
1 Aggregate Spreader		73.80		81.20		
1 Tandem Roller, 10 Ton		224.00		246.40	27.62	30.38
40 L.H., Daily Totals		$1785.80		$2364.70	$44.65	$59.12
Crew B-36A	**Hr.**	**Daily**	**Hr.**	**Daily**	**Bare Costs**	**Incl. O&P**
1 Labor Foreman (outside)	$16.45	$131.60	$28.20	$225.60	$17.85	$29.96
2 Laborers	14.45	231.20	24.75	396.00		
4 Equip. Oper. (med.)	19.90	636.80	33.00	1056.00		
1 Dozer, 200 H.P.		806.90		887.50		
1 Aggregate Spreader		73.80		81.20		
1 Roller, Steel Wheel		224.00		246.40		
1 Roller, Pneumatic Wheel		240.80		264.90	24.03	26.43
56 L.H., Daily Totals		$2345.00		$3157.60	$41.88	$56.39

Crew No.	Bare Costs		Incl. Subs O & P		Cost Per Labor-Hour	
Crew B-36B	**Hr.**	**Daily**	**Hr.**	**Daily**	**Bare Costs**	**Incl. O&P**
1 Labor Foreman (outside)	$16.45	$131.60	$28.20	$225.60	$17.64	$29.59
2 Laborers	14.45	231.20	24.75	396.00		
4 Equip. Oper. (medium)	19.90	636.80	33.00	1056.00		
1 Truck Driver, Heavy	16.20	129.60	27.05	216.40		
1 Grader, 30,000 Lbs.		516.00		567.60		
1 F.E. Loader, crl, 1.5 C.Y.		366.85		403.55		
1 Dozer, 300 H.P.		1108.00		1218.80		
1 Roller, Vibratory		432.40		475.65		
1 Truck, Tractor, 240 H.P.		531.20		584.30		
1 Water Tanker, 5000 Gal.		205.90		226.50	49.38	54.32
64 L.H., Daily Totals		$4289.55		$5370.40	$67.02	$83.91
Crew B-37	**Hr.**	**Daily**	**Hr.**	**Daily**	**Bare Costs**	**Incl. O&P**
1 Labor Foreman (outside)	$16.45	$131.60	$28.20	$225.60	$15.56	$26.48
4 Laborers	14.45	462.40	24.75	792.00		
1 Equip. Oper. (light)	19.10	152.80	31.70	253.60		
1 Tandem Roller, 5 Ton		133.20		146.50	2.78	3.05
48 L.H., Daily Totals		$880.00		$1417.70	$18.34	$29.53
Crew B-38	**Hr.**	**Daily**	**Hr.**	**Daily**	**Bare Costs**	**Incl. O&P**
2 Laborers	$14.45	$231.20	$24.75	$396.00	$16.00	$27.07
1 Equip. Oper. (light)	19.10	152.80	31.70	253.60		
1 Backhoe Loader, 48 H.P.		203.00		223.30		
1 Hyd. Hammer,(1200 lb)		213.60		234.95	17.36	19.09
24 L.H., Daily Totals		$800.60		$1107.85	$33.36	$46.16
Crew B-39	**Hr.**	**Daily**	**Hr.**	**Daily**	**Bare Costs**	**Incl. O&P**
1 Labor Foreman (outside)	$16.45	$131.60	$28.20	$225.60	$14.78	$25.32
5 Laborers	14.45	578.00	24.75	990.00		
1 Air Compr., 250 C.F.M.		127.00		139.70		
2 Air Tools & Accessories		39.00		42.90		
2-50 Ft. Air Hoses, 1.5" Dia.		14.40		15.85	3.76	4.13
48 L.H., Daily Totals		$890.00		$1414.05	$18.54	$29.45
Crew B-40	**Hr.**	**Daily**	**Hr.**	**Daily**	**Bare Costs**	**Incl. O&P**
1 Pile Driver Foreman (out)	$21.45	$171.60	$40.00	$320.00	$19.19	$34.86
4 Pile Drivers	19.45	622.40	36.25	1160.00		
1 Building Laborer	14.45	115.60	24.75	198.00		
1 Equip. Oper. (crane)	20.65	165.20	34.25	274.00		
1 Crane, 40 Ton		714.30		785.75		
1 Vibratory Hammer & Gen.		1218.00		1339.80	34.51	37.96
56 L.H., Daily Totals		$3007.10		$4077.55	$53.70	$72.82
Crew B-41	**Hr.**	**Daily**	**Hr.**	**Daily**	**Bare Costs**	**Incl. O&P**
1 Labor Foreman (outside)	$16.45	$131.60	$28.20	$225.60	$15.21	$25.97
4 Laborers	14.45	462.40	24.75	792.00		
.25 Equip. Oper. (crane)	20.65	41.30	34.25	68.50		
.25 Equip. Oper. Oiler	17.00	34.00	28.20	56.40		
.25 Crawler Crane, 40 Ton		178.57		196.45	4.06	4.46
44 L.H., Daily Totals		$847.87		$1338.95	$19.27	$30.43
Crew B-42	**Hr.**	**Daily**	**Hr.**	**Daily**	**Bare Costs**	**Incl. O&P**
1 Labor Foreman (outside)	$16.45	$131.60	$28.20	$225.60	$16.69	$28.25
4 Laborers	14.45	462.40	24.75	792.00		
1 Equip. Oper. (crane)	20.65	165.20	34.25	274.00		
1 Welder	21.95	175.60	36.30	290.40		
1 Hyd. Crane, 25 Ton		559.20		615.10		
1 Gas Welding Machine		84.00		92.40		
1 Horz. Boring Csg. Mch.		498.00		547.80	20.38	22.42
56 L.H., Daily Totals		$2076.00		$2837.30	$37.07	$50.67

Crews

Crew No.	Bare Costs		Incl. Subs O & P		Cost Per Labor-Hour	
Crew B-43	Hr.	Daily	Hr.	Daily	Bare Costs	Incl. O&P
1 Labor Foreman (outside)	$16.45	$131.60	$28.20	$225.60	$14.85	$25.44
4 Laborers	14.45	462.40	24.75	792.00		
1 Drill Rig & Augers		1796.00		1975.60	44.90	49.39
40 L.H., Daily Totals		$2390.00		$2993.20	$59.75	$74.83
Crew B-44	Hr.	Daily	Hr.	Daily	Bare Costs	Incl. O&P
1 Pile Driver Foreman	$21.45	$171.60	$40.00	$320.00	$18.60	$33.59
4 Pile Drivers	19.45	622.40	36.25	1160.00		
1 Equip. Oper. (crane)	20.65	165.20	34.25	274.00		
2 Laborer	14.45	231.20	24.75	396.00		
1 Crane, 40 Ton, & Access.		714.30		785.75		
45 L.F. Leads, 15K Ft. Lbs.		198.00		217.80	14.25	15.68
64 L.H., Daily Totals		$2102.70		$3153.55	$32.85	$49.27
Crew B-45	Hr.	Daily	Hr.	Daily	Bare Costs	Incl. O&P
1 Building Laborer	$14.45	$115.60	$24.75	$198.00	$15.33	$25.90
1 Truck Driver (heavy)	16.20	129.60	27.05	216.40		
1 Dist. Tank Truck, 3K Gal.		374.95		412.45	23.43	25.78
16 L.H., Daily Totals		$620.15		$826.85	$38.76	$51.68
Crew B-46	Hr.	Daily	Hr.	Daily	Bare Costs	Incl. O&P
1 Pile Driver Foreman	$21.45	$171.60	$40.00	$320.00	$17.28	$31.13
2 Pile Drivers	19.45	311.20	36.25	580.00		
3 Laborers	14.45	346.80	24.75	594.00		
1 Chain Saw, 36" Long		49.20		54.10	1.03	1.13
48 L.H., Daily Totals		$878.80		$1548.10	$18.31	$32.26
Crew B-47	Hr.	Daily	Hr	Daily	Bare Costs	Incl. O&P
1 Blast Foreman	$16.45	$131.60	$28.20	$225.60	$15.45	$26.48
1 Driller	14.45	115.60	24.75	198.00		
1 Crawler Type Drill, 4"		312.55		343.80		
1 Air Compr., 600 C.F.M.		278.00		305.80		
2-50 Ft. Air Hoses, 3" Dia.		33.80		37.20	39.02	42.92
16 L.H., Daily Totals		$871.55		$1110.40	$54.47	$69.40
Crew B-47A	Hr.	Daily	Hr.	Daily	Bare Costs	Incl. O&P
1 Drilling Foreman	$16.45	$131.60	$28.20	$225.60	$18.03	$30.22
1 Equip. Oper. (heavy)	20.65	165.20	34.25	274.00		
1 Oiler	17.00	136.00	28.20	225.60		
1 Quarry Drill		486.05		534.65	20.25	22.28
24 L.H., Daily Totals		$918.85		$1259.85	$38.28	$52.50
Crew B-47C	Hr.	Daily	Hr.	Daily	Bare Costs	Incl. O&P
1 Laborer	$14.45	$115.60	$24.75	$198.00	$16.77	$28.23
1 Equip. Oper. (light)	19.10	152.80	31.70	253.60		
1 Air Compressor, 750 CFM		300.80		330.90		
2-50' Air Hose, 3"		33.80		37.20		
1 Air Track Drill, 4"		312.55		343.80	40.45	44.49
16 L.H., Daily Totals		$915.55		$1163.50	$57.22	$72.72
Crew B-47E	Hr.	Daily	Hr.	Daily	Bare Costs	Incl. O&P
1 Laborer Forman	$16.45	$131.60	$28.20	$225.60	$14.95	$25.61
3 Laborers	14.45	346.80	24.75	594.00		
1 Truck, Flatbed, 3 ton		180.45		198.50	5.64	6.20
32 L.H., Daily Totals		$658.85		$1018.10	$20.59	$31.81

Crew No.	Bare Costs		Incl. Subs O & P		Cost Per Labor-Hour	
Crew B-48	Hr.	Daily	Hr.	Daily	Bare Costs	Incl. O&P
1 Labor Foreman (outside)	$16.45	$131.60	$28.20	$225.60	$15.82	$26.91
4 Laborers	14.45	462.40	24.75	792.00		
1 Equip. Oper. (crane)	20.65	165.20	34.25	274.00		
1 Centr. Water Pump, 6"		157.20		172.90		
1-20 Ft. Suction Hose, 6"		22.50		24.75		
1-50 Ft. Disch. Hose, 6"		20.55		22.60		
1 Drill Rig & Augers		1796.00		1975.60	41.59	45.75
48 L.H., Daily Totals		$2755.45		$3487.45	$57.41	$72.66
Crew B-49	Hr.	Daily	Hr.	Daily	Bare Costs	Incl. O&P
1 Labor Foreman (outside)	$16.45	$131.60	$28.20	$225.60	$16.47	$28.74
5 Laborers	14.45	578.00	24.75	990.00		
1 Equip. Oper. (crane)	20.65	165.20	34.25	274.00		
2 Pile Drivers	19.45	311.20	36.25	580.00		
1 Hyd. Crane, 25 Ton		559.20		615.10		
1 Centr. Water Pump, 6"		157.20		172.90		
1-20 Ft. Suction Hose, 6"		22.50		24.75		
1-50 Ft. Disch. Hose, 6"		20.55		22.60		
1 Drill Rig & Augers		1796.00		1975.60	35.49	39.04
72 L.H., Daily Totals		$3741.45		$4880.55	$51.96	$67.78
Crew B-50	Hr.	Daily	Hr.	Daily	Bare Costs	Incl. O&P
1 Pile Driver Foremen	$21.45	$171.60	$40.00	$320.00	$17.77	$31.96
6 Pile Drivers	19.45	933.60	36.25	1740.00		
1 Equip. Oper. (crane)	20.65	165.20	34.25	274.00		
5 Laborers	14.45	578.00	24.75	990.00		
1 Crane, 40 Ton		714.30		785.75		
60 L.F. Leads, 15K Ft. Lbs.		264.00		290.40		
1 Hammer, 15K Ft. Lbs.		311.40		342.55		
1 Air Compr., 600 C.F.M.		278.00		305.80		
2-50 Ft. Air Hoses, 3" Dia.		33.80		37.20		
1 Chain Saw, 36" Long		49.20		54.10	15.87	17.46
104 L.H., Daily Totals		$3499.10		$5139.80	$33.64	$49.42
Crew B-51	Hr.	Daily	Hr.	Daily	Bare Costs	Incl. O&P
1 Labor Foreman (outside)	$16.45	$131.60	$28.20	$225.60	$15.02	$25.62
4 Laborers	14.45	462.40	24.75	792.00		
1 Truck Driver (light)	15.85	126.80	26.50	212.00		
1 Light Truck, 1.5 Ton		172.95		190.25	3.60	3.96
48 L.H., Daily Totals		$893.75		$1419.85	$18.62	$29.58
Crew B-52	Hr.	Daily	Hr.	Daily	Bare Costs	Incl. O&P
1 Labor Foreman	$16.45	$131.60	$28.20	$225.60	$16.35	$28.15
1 Carpenter	19.70	157.60	33.75	270.00		
4 Laborers	14.45	462.40	24.75	792.00		
.5 Rodman (reinf.)	21.10	84.40	39.25	157.00		
.5 Equip. Oper. (med.)	19.90	79.60	33.00	132.00		
.5 F.E. Ldr., T.M., 2.5 C.Y.		392.00		431.20	7.00	7.70
56 L.H., Daily Totals		$1307.60		$2007.80	$23.35	$35.85
Crew B-53	Hr.	Daily	Hr.	Daily	Bare Costs	Incl. O&P
1 Building Laborer	$14.45	$115.60	$24.75	$198.00	$14.45	$24.75
1 Trencher, Chain, 12 H.P.		102.50		112.75	12.81	14.09
8 L.H., Daily Totals		$218.10		$310.75	$27.26	$38.84
Crew B-54	Hr.	Daily	Hr.	Daily	Bare Costs	Incl. O&P
1 Equip. Oper. (light)	$19.10	$152.80	$31.70	$253.60	$19.10	$31.70
1 Trencher, Chain, 40 H.P.		210.90		232.00	26.36	29.00
8 L.H., Daily Totals		$363.70		$485.60	$45.46	$60.70

Crews

Crew No.		Bare Costs		Incl. Subs O & P		Cost Per Labor-Hour	
Crew B-54A	Hr.	Daily	Hr.	Daily	Bare Costs	Incl. O&P
.17 Labor Foreman (outside)	$16.45	$22.37	$28.20	$38.35	$19.40	$32.30
1 Equipment Operator (med.)	19.90	159.20	33.00	264.00		
1 Wheel Trencher, 67HP		409.35		450.30	43.73	48.11
9.36 L.H., Daily Totals		$590.92		$752.65	$63.13	$80.41
Crew B-54B	Hr.	Daily	Hr.	Daily	Bare Costs	Incl. O&P
.25 Labor Foreman (outside)	$16.45	$32.90	$28.20	$56.40	$19.21	$32.04
1 Equipment Operator (med.)	19.90	159.20	33.00	264.00		
1 Wheel Trencher, 150HP		565.20		621.70	56.52	62.17
10 L.H., Daily Totals		$757.30		$942.10	$75.73	$94.21
Crew B-55	Hr.	Daily	Hr.	Daily	Bare Costs	Incl. O&P
1 Laborers	$14.45	$115.60	$24.75	$198.00	$15.15	$25.63
1 Truck Driver (light)	15.85	126.80	26.50	212.00		
1 Auger, 4" to 36" Dia		483.00		531.30		
1 Flatbed 3 Ton Truck		180.45		198.50	41.47	45.61
16 L.H., Daily Totals		$905.85		$1139.80	$56.62	$71.24
Crew B-56	Hr.	Daily	Hr.	Daily	Bare Costs	Incl. O&P
2 Laborer	$14.45	$231.20	$24.75	$396.00	$14.45	$24.75
1 Crawler Type Drill, 4"		312.55		343.80		
1 Air Compr., 600 C.F.M.		278.00		305.80		
1-50 Ft. Air Hose, 3" Dia.		16.90		18.60	37.97	41.76
16 L.H., Daily Totals		$838.65		$1064.20	$52.42	$66.51
Crew B-57	Hr.	Daily	Hr.	Daily	Bare Costs	Incl. O&P
1 Labor Foreman (outside)	$16.45	$131.60	$28.20	$225.60	$16.09	$27.34
3 Laborers	14.45	346.80	24.75	594.00		
1 Equip. Oper. (crane)	20.65	165.20	34.25	274.00		
1 Barge, 400 Ton		453.65		499.00		
1 Power Shovel, 1 C.Y.		516.80		568.50		
1 Clamshell Bucket, 1 C.Y.		70.00		77.00		
1 Centr. Water Pump, 6"		157.20		172.90		
1-20 Ft. Suction Hose, 6"		22.50		24.75		
20-50 Ft. Disch. Hoses, 6"		411.00		452.10	40.78	44.86
40 L.H., Daily Totals		$2274.75		$2887.85	$56.87	$72.20
Crew B-58	Hr.	Daily	Hr.	Daily	Bare Costs	Incl. O&P
2 Laborers	$14.45	$231.20	$24.75	$396.00	$16.00	$27.07
1 Equip. Oper. (light)	19.10	152.80	31.70	253.60		
1 Backhoe Loader, 48 H.P.		203.00		223.30		
1 Small Helicopter, w/pilot		3540.00		3894.00	155.96	171.55
24 L.H., Daily Totals		$4127.00		$4766.90	$171.96	$198.62
Crew B-59	Hr.	Daily	Hr.	Daily	Bare Costs	Incl. O&P
1 Truck Driver (heavy)	$16.20	$129.60	$27.05	$216.40	$16.20	$27.05
1 Truck, 30 Ton		325.60		358.15		
1 Water tank, 5000 Gal.		205.90		226.50	66.44	73.08
8 L.H., Daily Totals		$661.10		$801.05	$82.64	$100.13
Crew B-60	Hr.	Daily	Hr.	Daily	Bare Costs	Incl. O&P
1 Labor Foreman (outside)	$16.45	$131.60	$28.20	$225.60	$16.59	$28.07
3 Laborers	14.45	346.80	24.75	594.00		
1 Equip. Oper. (crane)	20.65	165.20	34.25	274.00		
1 Equip. Oper. (light)	19.10	152.80	31.70	253.60		
1 Crawler Crane, 40 Ton		714.30		785.75		
45 L.F. Leads, 15K Ft. Lbs.		198.00		217.80		
1 Backhoe Loader, 48 H.P.		203.00		223.30	23.24	25.56
48 L.H., Daily Totals		$1911.70		$2574.05	$39.83	$53.63

Crew No.		Bare Costs		Incl. Subs O & P		Cost Per Labor-Hour	
Crew B-61	Hr.	Daily	Hr.	Daily	Bare Costs	Incl. O&P
1 Labor Foreman (outside)	$16.45	$131.60	$28.20	$225.60	$14.85	$25.44
4 Laborers	14.45	462.40	24.75	792.00		
1 Cement Mixer, 2 C.Y.		222.00		244.20		
1 Air Compr., 160 C.F.M.		98.40		108.25	8.01	8.81
40 L.H., Daily Totals		$914.40		$1370.05	$22.86	$34.25
Crew B-62	Hr.	Daily	Hr.	Daily	Bare Costs	Incl. O&P
2 Laborers	$14.45	$231.20	$24.75	$396.00	$16.00	$27.07
1 Equip. Oper. (light)	19.10	152.80	31.70	253.60		
1 Loader, Skid Steer		116.40		128.05	4.85	5.34
24 L.H., Daily Totals		$500.40		$777.65	$20.85	$32.41
Crew B-63	Hr.	Daily	Hr.	Daily	Bare Costs	Incl. O&P
5 Laborers	$14.45	$578.00	$24.75	$990.00	$14.45	$24.75
1 Loader, Skid Steer		116.40		128.05	2.91	3.20
40 L.H., Daily Totals		$694.40		$1118.05	$17.36	$27.95
Crew B-64	Hr.	Daily	Hr.	Daily	Bare Costs	Incl. O&P
1 Laborer	$14.45	$115.60	$24.75	$198.00	$15.15	$25.63
1 Truck Driver (light)	15.85	126.80	26.50	212.00		
1 Power Mulcher (small)		115.80		127.40		
1 Light Truck, 1.5 Ton		172.95		190.25	18.05	19.85
16 L.H., Daily Totals		$531.15		$727.65	$33.20	$45.48
Crew B-65	Hr.	Daily	Hr.	Daily	Bare Costs	Incl. O&P
1 Laborer	$14.45	$115.60	$24.75	$198.00	$15.15	$25.63
1 Truck Driver (light)	15.85	126.80	26.50	212.00		
1 Power Mulcher (large)		283.85		312.25		
1 Light Truck, 1.5 Ton		172.95		190.25	28.55	31.41
16 L.H., Daily Totals		$699.20		$912.50	$43.70	$57.04
Crew B-66	Hr.	Daily	Hr.	Daily	Bare Costs	Incl. O&P
1 Equip. Oper. (light)	$19.10	$152.80	$31.70	$253.60	$19.10	$31.70
1 Backhoe Ldr. w/Attchmt.		187.20		205.90	23.40	25.74
8 L.H., Daily Totals		$340.00		$459.50	$42.50	$57.44
Crew B-67	Hr.	Daily	Hr.	Daily	Bare Costs	Incl. O&P
1 Millwright	$20.60	$164.80	$33.60	$268.80	$19.85	$32.65
1 Equip. Oper. (light)	19.10	152.80	31.70	253.60		
1 Forklift		171.20		188.30	10.70	11.77
16 L.H., Daily Totals		$488.80		$710.70	$30.55	$44.42
Crew B-68	Hr.	Daily	Hr.	Daily	Bare Costs	Incl. O&P
2 Millwrights	$20.60	$329.60	$33.60	$537.60	$20.10	$32.97
1 Equip. Oper. (light)	19.10	152.80	31.70	253.60		
1 Forklift		171.20		188.30	7.13	7.85
24 L.H., Daily Totals		$653.60		$979.50	$27.23	$40.82
Crew B-69	Hr.	Daily	Hr.	Daily	Bare Costs	Incl. O&P
1 Labor Foreman (outside)	$16.45	$131.60	$28.20	$225.60	$16.24	$27.48
3 Laborers	14.45	346.80	24.75	594.00		
1 Equip Oper. (crane)	20.65	165.20	34.25	274.00		
1 Equip Oper. Oiler	17.00	136.00	28.20	225.60		
1 Truck Crane, 80 Ton		1098.00		1207.80	22.88	25.16
48 L.H., Daily Totals		$1877.60		$2527.00	$39.12	$52.64

Crews

Crew No.	Bare Costs		Incl. Subs O & P		Cost Per Labor-Hour	
Crew B-69A	Hr.	Daily	Hr.	Daily	Bare Costs	Incl. O&P
1 Labor Foreman	$16.45	$131.60	$28.20	$225.60	$16.43	$27.71
3 Laborers	14.45	346.80	24.75	594.00		
1 Equip. Oper. (medium)	19.90	159.20	33.00	264.00		
1 Concrete Finisher	18.90	151.20	30.80	246.40		
1 Curb Paver		454.00		499.40	9.46	10.40
48 L.H., Daily Totals		$1242.80		$1829.40	$25.89	$38.11
Crew B-69B	Hr.	Daily	Hr.	Daily	Bare Costs	Incl. O&P
1 Labor Foreman	$16.45	$131.60	$28.20	$225.60	$16.43	$27.71
3 Laborers	14.45	346.80	24.75	594.00		
1 Equip. Oper. (medium)	19.90	159.20	33.00	264.00		
1 Cement Finisher	18.90	151.20	30.80	246.40		
1 Curb/Gutter Paver		911.85		1003.05	19.00	20.90
48 L.H., Daily Totals		$1700.65		$2333.05	$35.43	$48.61
Crew B-70	Hr.	Daily	Hr.	Daily	Bare Costs	Incl. O&P
1 Labor Foreman (outside)	$16.45	$131.60	$28.20	$225.60	$17.07	$28.78
3 Laborers	14.45	346.80	24.75	594.00		
3 Equip. Oper. (med.)	19.90	477.60	33.00	792.00		
1 Motor Grader, 30,000 Lb.		516.00		567.60		
1 Grader Attach., Ripper		64.00		70.40		
1 Road Sweeper, S.P.		218.40		240.25		
1 F.E. Loader, 1-3/4 C.Y.		303.00		333.30	19.67	21.63
56 L.H., Daily Totals		$2057.40		$2823.15	$36.74	$50.41
Crew B-71	Hr.	Daily	Hr.	Daily	Bare Costs	Incl. O&P
1 Labor Foreman (outside)	$16.45	$131.60	$28.20	$225.60	$17.07	$28.78
3 Laborers	14.45	346.80	24.75	594.00		
3 Equip. Oper. (med.)	19.90	477.60	33.00	792.00		
1 Pvmt. Profiler, 450 H.P.		3442.00		3786.20		
1 Road Sweeper, S.P.		218.40		240.25		
1 F.E. Loader, 1-3/4 C.Y.		303.00		333.30	70.78	77.85
56 L.H., Daily Totals		$4919.40		$5971.35	$87.85	$106.63
Crew B-72	Hr.	Daily	Hr.	Daily	Bare Costs	Incl. O&P
1 Labor Foreman (outside)	$16.45	$131.60	$28.20	$225.60	$17.42	$29.31
3 Laborers	14.45	346.80	24.75	594.00		
4 Equip. Oper. (med.)	19.90	636.80	33.00	1056.00		
1 Pvmt. Profiler, 450 H.P.		3442.00		3786.20		
1 Hammermill, 250 H.P.		1177.00		1294.70		
1 Windrow Loader		929.85		1022.85		
1 Mix Paver 165 H.P.		1440.00		1584.00		
1 Roller, Pneu. Tire, 12 T.		240.80		264.90	112.96	124.26
64 L.H., Daily Totals		$8344.85		$9828.25	$130.38	$153.57
Crew B-73	Hr.	Daily	Hr.	Daily	Bare Costs	Incl. O&P
1 Labor Foreman (outside)	$16.45	$131.60	$28.20	$225.60	$18.11	$30.34
2 Laborers	14.45	231.20	24.75	396.00		
5 Equip. Oper. (med.)	19.90	796.00	33.00	1320.00		
1 Road Mixer, 310 H.P.		1008.00		1108.80		
1 Roller, Tandem, 12 Ton		224.00		246.40		
1 Hammermill, 250 H.P.		1177.00		1294.70		
1 Motor Grader, 30,000 Lb.		516.00		567.60		
.5 F.E. Loader, 1-3/4 C.Y.		151.50		166.65		
.5 Truck, 30 Ton		162.80		179.10		
.5 Water Tank 5000 Gal.		102.95		113.25	52.22	57.44
64 L.H., Daily Totals		$4501.05		$5618.10	$70.33	$87.78

Crew No.	Bare Costs		Incl. Subs O & P		Cost Per Labor-Hour	
Crew B-74	Hr.	Daily	Hr.	Daily	Bare Costs	Incl. O&P
1 Labor Foreman (outside)	$16.45	$131.60	$28.20	$225.60	$17.86	$29.88
1 Laborer	14.45	115.60	24.75	198.00		
4 Equip. Oper. (med.)	19.90	636.80	33.00	1056.00		
2 Truck Drivers (heavy)	16.20	259.20	27.05	432.80		
1 Motor Grader, 30,000 Lb.		516.00		567.60		
1 Grader Attach., Ripper		64.00		70.40		
2 Stabilizers, 310 H.P.		1394.10		1533.50		
1 Flatbed Truck, 3 Ton		180.45		198.50		
1 Chem. Spreader, Towed		101.20		111.30		
1 Vibr. Roller, 29,000 Lb.		432.40		475.65		
1 Water Tank 5000 Gal.		205.90		226.50		
1 Truck, 30 Ton		325.60		358.15	50.31	55.34
64 L.H., Daily Totals		$4362.85		$5454.00	$68.17	$85.22
Crew B-75	Hr.	Daily	Hr.	Daily	Bare Costs	Incl. O&P
1 Labor Foreman (outside)	$16.45	$131.60	$28.20	$225.60	$18.10	$30.29
1 Laborer	14.45	115.60	24.75	198.00		
4 Equip. Oper. (med.)	19.90	636.80	33.00	1056.00		
1 Truck Driver (heavy)	16.20	129.60	27.05	216.40		
1 Motor Grader, 30,000 Lb.		516.00		567.60		
1 Grader Attach., Ripper		64.00		70.40		
2 Stabilizers, 310 H.P.		1394.10		1533.50		
1 Dist. Truck, 3000 Gal.		374.95		412.45		
1 Vibr. Roller, 29,000 Lb.		432.40		475.65	49.67	54.64
56 L.H., Daily Totals		$3795.05		$4755.60	$67.77	$84.93
Crew B-76	Hr.	Daily	Hr.	Daily	Bare Costs	Incl. O&P
1 Dock Builder Foreman	$21.45	$171.60	$40.00	$320.00	$19.67	$35.33
5 Dock Builders	19.45	778.00	36.25	1450.00		
2 Equip. Oper. (crane)	20.65	330.40	34.25	548.00		
1 Equip. Oper. Oiler	17.00	136.00	28.20	225.60		
1 Crawler Crane, 50 Ton		850.80		935.90		
1 Barge, 400 Ton		453.65		499.00		
1 Hammer, 15K Ft. Lbs.		311.40		342.55		
60 L.F. Leads, 15K Ft. Lbs.		264.00		290.40		
1 Air Compr., 600 C.F.M.		278.00		305.80		
2-50 Ft. Air Hoses, 3" Dia.		33.80		37.20	30.44	33.48
72 L.H., Daily Totals		$3607.65		$4954.45	$50.11	$68.81
Crew B-77	Hr.	Daily	Hr.	Daily	Bare Costs	Incl. O&P
1 Labor Foreman	$16.45	$131.60	$28.20	$225.60	$15.13	$25.79
3 Laborers	14.45	346.80	24.75	594.00		
1 Truck Driver (light)	15.85	126.80	26.50	212.00		
1 Crack Cleaner, 25 H.P.		76.95		84.65		
1 Crack Filler, Trailer Mtd.		145.20		159.70		
1 Flatbed Truck, 3 Ton		180.45		198.50	10.07	11.07
40 L.H., Daily Totals		$1007.80		$1474.45	$25.20	$36.86
Crew B-78	Hr.	Daily	Hr.	Daily	Bare Costs	Incl. O&P
1 Labor Foreman	$16.45	$131.60	$28.20	$225.60	$14.85	$25.44
4 Laborers	14.45	462.40	24.75	792.00		
1 Paint Striper, S.P.		209.00		229.90		
1 Flatbed Truck, 3 Ton		180.45		198.50		
1 Pickup Truck, 3/4 Ton		130.20		143.20	12.99	14.29
40 L.H., Daily Totals		$1113.65		$1589.20	$27.84	$39.73

Crews

Crew No.	Bare Costs Hr.	Bare Costs Daily	Incl. Subs O & P Hr.	Incl. Subs O & P Daily	Cost Per Labor-Hour Bare Costs	Cost Per Labor-Hour Incl. O&P
Crew B-79						
1 Labor Foreman	$16.45	$131.60	$28.20	$225.60	$14.95	$25.61
3 Laborers	14.45	346.80	24.75	594.00		
1 Thermo. Striper, T.M.		245.05		269.55		
1 Flatbed Truck, 3 Ton		180.45		198.50		
2 Pickup Truck, 3/4 Ton		260.40		286.45	21.43	23.58
32 L.H., Daily Totals		$1164.30		$1574.10	$36.38	$49.19
Crew B-80						
1 Labor Foreman	$16.45	$131.60	$28.20	$225.60	$15.12	$25.90
2 Laborer	14.45	231.20	24.75	396.00		
1 Flatbed Truck, 3 Ton		180.45		198.50		
1 Fence Post Auger, TM		370.00		407.00	22.94	25.23
24 L.H., Daily Totals		$913.25		$1227.10	$38.06	$51.13
Crew B-80A						
3 Laborers	$14.45	$346.80	$24.75	$594.00	$14.45	$24.75
1 Flatbed Truck, 3 Ton		180.45		198.50	7.52	8.27
24 L.H., Daily Totals		$527.25		$792.50	$21.97	$33.02
Crew B-80B						
3 Laborers	$14.45	$346.80	$24.75	$594.00	$15.61	$26.49
1 Equip. Oper. (light)	19.10	152.80	31.70	253.60		
1 Crane, flatbed mnt		283.20		311.50	8.85	9.74
32 L.H., Daily Totals		$782.80		$1159.10	$24.46	$36.23
Crew B-81						
1 Laborer	$14.45	$115.60	$24.75	$198.00	$15.33	$25.90
1 Truck Driver (heavy)	16.20	129.60	27.05	216.40		
1 Hydromulcher, T.M.		301.75		331.95		
1 Tractor Truck, 4x2		325.60		358.15	39.21	43.13
16 L.H., Daily Totals		$872.55		$1104.50	$54.54	$69.03
Crew B-82						
1 Laborer	$14.45	$115.60	$24.75	$198.00	$16.77	$28.23
1 Equip. Oper. (light)	19.10	152.80	31.70	253.60		
1 Horiz. Borer, 6 H.P.		49.80		54.80	3.11	3.42
16 L.H., Daily Totals		$318.20		$506.40	$19.88	$31.65
Crew B-83						
1 Tugboat Captain	$19.90	$159.20	$33.00	$264.00	$17.17	$28.88
1 Tugboat Hand	14.45	115.60	24.75	198.00		
1 Tugboat, 250 H.P.		444.80		489.30	27.80	30.58
16 L.H., Daily Totals		$719.60		$951.30	$44.97	$59.46
Crew B-84						
1 Equip. Oper. (med.)	$19.90	$159.20	$33.00	$264.00	$19.90	$33.00
1 Rotary Mower/Tractor		209.50		230.45	26.19	28.81
8 L.H., Daily Totals		$368.70		$494.45	$46.09	$61.81
Crew B-85						
3 Laborers	$14.45	$346.80	$24.75	$594.00	$15.89	$26.86
1 Equip. Oper. (med.)	19.90	159.20	33.00	264.00		
1 Truck Driver (heavy)	16.20	129.60	27.05	216.40		
1 Aerial Lift Truck, 80'		569.60		626.55		
1 Brush Chipper, 130 H.P.		211.05		232.15		
1 Pruning Saw, Rotary		22.25		24.50	20.07	22.08
40 L.H., Daily Totals		$1438.50		$1957.60	$35.96	$48.94

Crew No.	Bare Costs Hr.	Bare Costs Daily	Incl. Subs O & P Hr.	Incl. Subs O & P Daily	Cost Per Labor-Hour Bare Costs	Cost Per Labor-Hour Incl. O&P
Crew B-86						
1 Equip. Oper. (med.)	$19.90	$159.20	$33.00	$264.00	$19.90	$33.00
1 Stump Chipper, S.P.		165.60		182.15	20.70	22.77
8 L.H., Daily Totals		$324.80		$446.15	$40.60	$55.77
Crew B-86A						
1 Equip. Oper. (medium)	$19.90	$159.20	$33.00	$264.00	$19.90	$33.00
1 Grader, 30,000 Lbs.		516.00		567.60	64.50	70.95
8 L.H., Daily Totals		$675.20		$831.60	$84.40	$103.95
Crew B-86B						
1 Equip. Oper. (medium)	$19.90	$159.20	$33.00	$264.00	$19.90	$33.00
1 Dozer, 200 H.P.		806.80		887.50	100.85	110.94
8 L.H., Daily Totals		$966.00		$1151.50	$120.75	$143.94
Crew B-87						
1 Laborer	$14.45	$115.60	$24.75	$198.00	$18.81	$31.35
4 Equip. Oper. (med.)	19.90	636.80	33.00	1056.00		
2 Feller Bunchers, 50 H.P.		868.80		955.70		
1 Log Chipper, 22" Tree		2017.00		2218.70		
1 Dozer, 105 H.P.		279.60		307.55		
1 Chainsaw, Gas, 36" Long		49.20		54.10	80.37	88.40
40 L.H., Daily Totals		$3967.00		$4790.05	$99.18	$119.75
Crew B-88						
1 Laborer	$14.45	$115.60	$24.75	$198.00	$19.12	$31.82
6 Equip. Oper. (med.)	19.90	955.20	33.00	1584.00		
2 Feller Bunchers, 50 H.P.		868.80		955.70		
1 Log Chipper, 22" Tree		2017.00		2218.70		
2 Log Skidders, 50 H.P.		784.00		862.40		
1 Dozer, 105 H.P.		279.60		307.55		
1 Chainsaw, Gas, 36" Long		49.20		54.10	71.40	78.54
56 L.H., Daily Totals		$5069.40		$6180.45	$90.52	$110.36
Crew B-89						
1 Skilled Worker	$19.75	$158.00	$33.90	$271.20	$17.10	$29.33
1 Building Laborer	14.45	115.60	24.75	198.00		
1 Cutting Machine		48.00		52.80	3.00	3.30
16 L.H., Daily Totals		$321.60		$522.00	$20.10	$32.63
Crew B-89A						
1 Skilled Worker	$19.75	$158.00	$33.90	$271.20	$17.10	$29.33
1 Laborer	14.45	115.60	24.75	198.00		
1 Core Drill (large)		60.60		66.65	3.79	4.17
16 L.H., Daily Totals		$334.20		$535.85	$20.89	$33.50
Crew B-89B						
1 Equip. Oper. (light)	$19.10	$152.80	$31.70	$253.60	$17.48	$29.10
1 Truck Driver, Light	15.85	126.80	26.50	212.00		
1 Wall Saw, Hydraulic, 10 H.P.		133.20		146.50		
1 Generator, Diesel, 100 KW		207.05		227.75		
1 Water Tank, 65 Gal.		13.00		14.30		
1 Flatbed Truck, 3 Ton		180.45		198.50	33.36	36.69
16 L.H., Daily Totals		$813.30		$1052.65	$50.84	$65.79

Crews

Crew No.	Bare Costs		Incl. Subs O & P		Cost Per Labor-Hour	
Crew B-90	Hr.	Daily	Hr.	Daily	Bare Costs	Incl. O&P
1 Labor Foreman (outside)	$16.45	$131.60	$28.20	$225.60	$16.30	$27.49
3 Laborers	14.45	346.80	24.75	594.00		
2 Equip. Oper. (light)	19.10	305.60	31.70	507.20		
2 Truck Drivers (heavy)	16.20	259.20	27.05	432.80		
1 Road Mixer, 310 H.P.		1008.00		1108.80		
1 Dist. Truck, 2000 Gal.		350.30		385.35	21.22	23.35
64 L.H., Daily Totals		$2401.50		$3253.75	$37.52	$50.84
Crew B-90A	Hr.	Daily	Hr.	Daily	Bare Costs	Incl. O&P
1 Labor Foreman	$16.45	$131.60	$28.20	$225.60	$17.85	$29.96
2 Laborers	14.45	231.20	24.75	396.00		
4 Equip. Oper. (medium)	19.90	636.80	33.00	1056.00		
2 Graders, 30,000 Lbs.		1032.00		1135.20		
1 Roller, Steel Wheel		224.00		246.40		
1 Roller, Pneumatic Wheel		240.80		264.90	26.73	29.40
56 L.H., Daily Totals		$2496.40		$3324.10	$44.58	$59.36
Crew B-90B	Hr.	Daily	Hr.	Daily	Bare Costs	Incl. O&P
1 Labor Foreman	$16.45	$131.60	$28.20	$225.60	$17.51	$29.45
2 Laborers	14.45	231.20	24.75	396.00		
3 Equip. Oper. (medium)	19.90	477.60	33.00	792.00		
1 Roller, Steel Wheel		224.00		246.40		
1 Roller, Pneumatic Wheel		240.80		264.90		
1 Road Mixer, 310 H.P.		1008.00		1108.80	30.68	33.75
48 L.H., Daily Totals		$2313.20		$3033.70	$48.19	$63.20
Crew B-91	Hr.	Daily	Hr.	Daily	Bare Costs	Incl. O&P
1 Labor Foreman (outside)	$16.45	$131.60	$28.20	$225.60	$17.64	$29.59
2 Laborers	14.45	231.20	24.75	396.00		
4 Equip. Oper. (med.)	19.90	636.80	33.00	1056.00		
1 Truck Driver (heavy)	16.20	129.60	27.05	216.40		
1 Dist. Truck, 3000 Gal.		374.95		412.45		
1 Aggreg. Spreader, S.P.		640.40		704.45		
1 Roller, Pneu. Tire, 12 Ton		240.80		264.90		
1 Roller, Steel, 10 Ton		224.00		246.40	23.13	25.44
64 L.H., Daily Totals		$2609.35		$3522.20	$40.77	$55.03
Crew B-92	Hr.	Daily	Hr.	Daily	Bare Costs	Incl. O&P
1 Labor Foreman (outside)	$16.45	$131.60	$28.20	$225.60	$14.95	$25.61
3 Laborers	14.45	346.80	24.75	594.00		
1 Crack Cleaner, 25 H.P.		76.95		84.65		
1 Air Compressor		78.30		86.15		
1 Tar Kettle, T.M.		23.20		25.50		
1 Flatbed Truck, 3 Ton		180.45		198.50	11.22	12.34
32 L.H., Daily Totals		$837.30		$1214.40	$26.17	$37.95
Crew B-93	Hr.	Daily	Hr.	Daily	Bare Costs	Incl. O&P
1 Equip. Oper. (med.)	$19.90	$159.20	$33.00	$264.00	$19.90	$33.00
1 Feller Buncher, 50 H.P.		434.40		477.85	54.30	59.73
8 L.H., Daily Totals		$593.60		$741.85	$74.20	$92.73
Crew B-94A	Hr.	Daily	Hr.	Daily	Bare Costs	Incl. O&P
1 Laborer	$14.45	$115.60	$24.75	$198.00	$14.45	$24.75
1 Diaph. Water Pump, 2"		29.40		32.35		
1-20 Ft. Suction Hose, 2"		6.50		7.15		
2-50 Ft. Disch. Hoses, 2"		10.80		11.90	5.84	6.42
8 L.H., Daily Totals		$162.30		$249.40	$20.29	$31.17

Crew No.	Bare Costs		Incl. Subs O & P		Cost Per Labor-Hour	
Crew B-94B	Hr.	Daily	Hr.	Daily	Bare Costs	Incl. O&P
1 Laborer	$14.45	$115.60	$24.75	$198.00	$14.45	$24.75
1 Diaph. Water Pump, 4"		65.65		72.20		
1-20 Ft. Suction Hose, 4"		12.50		13.75		
2-50 Ft. Disch. Hoses, 4"		17.00		18.70	11.89	13.08
8 L.H., Daily Totals		$210.75		$302.65	$26.34	$37.83
Crew B-94C	Hr.	Daily	Hr.	Daily	Bare Costs	Incl. O&P
1 Laborer	$14.45	$115.60	$24.75	$198.00	$14.45	$24.75
1 Centr. Water Pump, 3"		38.00		41.80		
1-20 Ft. Suction Hose, 3"		9.50		10.45		
2-50 Ft. Disch. Hoses, 3"		12.80		14.10	7.54	8.29
8 L.H., Daily Totals		$175.90		$264.35	$21.99	$33.04
Crew B-94D	Hr.	Daily	Hr.	Daily	Bare Costs	Incl. O&P
1 Laborer	$14.45	$115.60	$24.75	$198.00	$14.45	$24.75
1 Centr. Water Pump, 6"		157.20		172.90		
1-20 Ft. Suction Hose, 6"		22.50		24.75		
2-50 Ft. Disch. Hoses, 6"		41.10		45.20	27.60	30.36
8 L.H., Daily Totals		$336.40		$440.85	$42.05	$55.11
Crew B-95	Hr.	Daily	Hr.	Daily	Bare Costs	Incl. O&P
1 Equip. Oper. (crane)	$20.65	$165.20	$34.25	$274.00	$17.55	$29.50
1 Laborer	14.45	115.60	24.75	198.00		
16 L.H., Daily Totals		$280.80		$472.00	$17.55	$29.50
Crew B-95A	Hr.	Daily	Hr.	Daily	Bare Costs	Incl. O&P
1 Equip. Oper. (crane)	$20.65	$165.20	$34.25	$274.00	$17.55	$29.50
1 Laborer	14.45	115.60	24.75	198.00		
1 Hyd. Excavator, 5/8 C.Y.		411.60		452.75	25.73	28.30
16 L.H., Daily Totals		$692.40		$924.75	$43.28	$57.80
Crew B-95B	Hr.	Daily	Hr.	Daily	Bare Costs	Incl. O&P
1 Equip. Oper. (crane)	$20.65	$165.20	$34.25	$274.00	$17.55	$29.50
1 Laborer	14.45	115.60	24.75	198.00		
1 Hyd. Excavator, 1.5 C.Y.		713.20		784.50	44.58	49.03
16 L.H., Daily Totals		$994.00		$1256.50	$62.13	$78.53
Crew B-95C	Hr.	Daily	Hr.	Daily	Bare Costs	Incl. O&P
1 Equip. Oper. (crane)	$20.65	$165.20	$34.25	$274.00	$17.55	$29.50
1 Laborer	14.45	115.60	24.75	198.00		
1 Hyd. Excavator, 2.5 C.Y.		1673.00		1840.30	104.56	115.02
16 L.H., Daily Totals		$1953.80		$2312.30	$122.11	$144.52
Crew C-1	Hr.	Daily	Hr.	Daily	Bare Costs	Incl. O&P
2 Carpenters	$19.70	$315.20	$33.75	$540.00	$17.16	$29.40
1 Carpenter Helper	14.80	118.40	25.35	202.80		
1 Laborer	14.45	115.60	24.75	198.00		
32 L.H., Daily Totals		$549.20		$940.80	$17.16	$29.40
Crew C-2	Hr.	Daily	Hr.	Daily	Bare Costs	Incl. O&P
1 Carpenter Foreman (out)	$21.70	$173.60	$37.20	$297.60	$17.52	$30.03
2 Carpenters	19.70	315.20	33.75	540.00		
2 Carpenter Helpers	14.80	236.80	25.35	405.60		
1 Laborer	14.45	115.60	24.75	198.00		
48 L.H., Daily Totals		$841.20		$1441.20	$17.52	$30.03

Crews

Crew No.	Bare Costs		Incl. Subs O & P		Cost Per Labor-Hour	
Crew C-2A	**Hr.**	**Daily**	**Hr.**	**Daily**	**Bare Costs**	**Incl. O&P**
1 Carpenter Foreman (out)	$21.70	$173.60	$37.20	$297.60	$19.02	$32.33
3 Carpenters	19.70	472.80	33.75	810.00		
1 Cement Finisher	18.90	151.20	30.80	246.40		
1 Laborer	14.45	115.60	24.75	198.00		
48 L.H., Daily Totals		$913.20		$1552.00	$19.02	$32.33
Crew C-3	**Hr.**	**Daily**	**Hr.**	**Daily**	**Bare Costs**	**Incl. O&P**
1 Rodman Foreman	$23.10	$184.80	$43.00	$344.00	$18.61	$33.34
3 Rodmen (reinf.)	21.10	506.40	39.25	942.00		
1 Equip. Oper. (light)	19.10	152.80	31.70	253.60		
3 Laborers	14.45	346.80	24.75	594.00		
3 Stressing Equipment		78.00		85.80		
.5 Grouting Equipment		125.00		137.50	3.17	3.49
64 L.H., Daily Totals		$1393.80		$2356.90	$21.78	$36.83
Crew C-4	**Hr.**	**Daily**	**Hr.**	**Daily**	**Bare Costs**	**Incl. O&P**
1 Rodman Foreman	$23.10	$184.80	$43.00	$344.00	$19.94	$36.56
2 Rodmen (reinf.)	21.10	337.60	39.25	628.00		
1 Building Laborer	14.45	115.60	24.75	198.00		
3 Stressing Equipment		78.00		85.80	2.44	2.68
32 L.H., Daily Totals		$716.00		$1255.80	$22.38	$39.24
Crew C-5	**Hr.**	**Daily**	**Hr.**	**Daily**	**Bare Costs**	**Incl. O&P**
1 Rodman Foreman	$23.10	$184.80	$43.00	$344.00	$19.14	$34.21
2 Rodmen (reinf.)	21.10	337.60	39.25	628.00		
1 Equip. Oper. (crane)	20.65	165.20	34.25	274.00		
2 Building Laborers	14.45	231.20	24.75	396.00		
1 Hyd. Crane, 25 Ton		559.20		615.10	11.65	12.82
48 L.H., Daily Totals		$1478.00		$2257.10	$30.79	$47.03
Crew C-6	**Hr.**	**Daily**	**Hr.**	**Daily**	**Bare Costs**	**Incl. O&P**
1 Labor Foreman (outside)	$16.45	$131.60	$28.20	$225.60	$15.53	$26.33
4 Laborers	14.45	462.40	24.75	792.00		
1 Cement Finisher	18.90	151.20	30.80	246.40		
2 Gas Engine Vibrators		76.00		83.60	1.58	1.74
48 L.H., Daily Totals		$821.20		$1347.60	$17.11	$28.07
Crew C-7	**Hr.**	**Daily**	**Hr.**	**Daily**	**Bare Costs**	**Incl. O&P**
1 Labor Foreman (outside)	$16.45	$131.60	$28.20	$225.60	$16.06	$27.11
5 Laborers	14.45	578.00	24.75	990.00		
1 Cement Finisher	18.90	151.20	30.80	246.40		
1 Equip. Oper. (med.)	19.90	159.20	33.00	264.00		
1 Equip. Oper. (oiler)	17.00	136.00	28.20	225.60		
2 Gas Engine Vibrators		76.00		83.60		
1 Concrete Bucket, 1 C.Y.		27.20		29.90		
1 Hyd. Crane, 55 Ton		808.20		889.00	12.66	13.92
72 L.H., Daily Totals		$2067.40		$2954.10	$28.72	$41.03
Crew C-8	**Hr.**	**Daily**	**Hr.**	**Daily**	**Bare Costs**	**Incl. O&P**
1 Labor Foreman (outside)	$16.45	$131.60	$28.20	$225.60	$16.79	$28.15
3 Laborers	14.45	346.80	24.75	594.00		
2 Cement Finishers	18.90	302.40	30.80	492.80		
1 Equip. Oper. (med.)	19.90	159.20	33.00	264.00		
1 Concrete Pump (small)		656.00		721.60	11.71	12.89
56 L.H., Daily Totals		$1596.00		$2298.00	$28.50	$41.04
Crew C-8A	**Hr.**	**Daily**	**Hr.**	**Daily**	**Bare Costs**	**Incl. O&P**
1 Labor Foreman (outside)	$16.45	$131.60	$28.20	$225.60	$16.27	$27.34
3 Laborers	14.45	346.80	24.75	594.00		
2 Cement Finishers	18.90	302.40	30.80	492.80		
48 L.H., Daily Totals		$780.80		$1312.40	$16.27	$27.34

Crew No.	Bare Costs		Incl. Subs O & P		Cost Per Labor-Hour	
Crew C-8B	**Hr.**	**Daily**	**Hr.**	**Daily**	**Bare Costs**	**Incl. O&P**
1 Labor Foreman (outside)	$16.45	$131.60	$28.20	$225.60	$15.94	$27.09
3 Laborers	14.45	346.80	24.75	594.00		
1 Equipment Operator	19.90	159.20	33.00	264.00		
1 Vibrating Screed		49.20		54.10		
1 Vibratory Roller		432.40		475.65		
1 Dozer, 200 HP		806.80		887.50	32.21	35.43
40 L.H., Daily Totals		$1926.00		$2500.85	$48.15	$62.52
Crew C-8C	**Hr.**	**Daily**	**Hr.**	**Daily**	**Bare Costs**	**Incl. O&P**
1 Labor Foreman (outside)	$16.45	$131.60	$28.20	$225.60	$16.43	$27.71
3 Laborers	14.45	346.80	24.75	594.00		
1 Cement Finisher	18.90	151.20	30.80	246.40		
1 Equipment Operator (med.)	19.90	159.20	33.00	264.00		
1 Shotcrete Rig, 12 CY/hr		346.00		380.60	7.21	7.93
48 L.H., Daily Totals		$1134.80		$1710.60	$23.64	$35.64
Crew C-8D	**Hr.**	**Daily**	**Hr.**	**Daily**	**Bare Costs**	**Incl. O&P**
1 Labor Foreman (outside)	$16.45	$131.60	$28.20	$225.60	$17.23	$28.86
1 Laborers	14.45	115.60	24.75	198.00		
1 Cement Finisher	18.90	151.20	30.80	246.40		
1 Equipment Operator (light)	19.10	152.80	31.70	253.60		
1 Compressor, 250 CFM		127.00		139.70		
2 Hoses, 1", 50'		10.40		11.45	4.29	4.72
32 L.H., Daily Totals		$688.60		$1074.75	$21.52	$33.58
Crew C-8E	**Hr.**	**Daily**	**Hr.**	**Daily**	**Bare Costs**	**Incl. O&P**
1 Labor Foreman (outside)	$16.45	$131.60	$28.20	$225.60	$17.23	$28.86
1 Laborers	14.45	115.60	24.75	198.00		
1 Cement Finisher	18.90	151.20	30.80	246.40		
1 Equipment Operator (light)	19.10	152.80	31.70	253.60		
1 Compressor, 250 CFM		127.00		139.70		
2 Hoses, 1", 50'		10.40		11.45		
1 Concrete Pump (small)		656.00		721.60	24.79	27.27
32 L.H., Daily Totals		$1344.60		$1796.35	$42.02	$56.13
Crew C-10	**Hr.**	**Daily**	**Hr.**	**Daily**	**Bare Costs**	**Incl. O&P**
1 Laborer	$14.45	$115.60	$24.75	$198.00	$17.42	$28.78
2 Cement Finishers	18.90	302.40	30.80	492.80		
24 L.H., Daily Totals		$418.00		$690.80	$17.42	$28.78
Crew C-11	**Hr.**	**Daily**	**Hr.**	**Daily**	**Bare Costs**	**Incl. O&P**
1 Skilled Worker Foreman	$21.75	$174.00	$37.30	$298.40	$20.16	$34.44
5 Skilled Worker	19.75	790.00	33.90	1356.00		
1 Equip. Oper. (crane)	20.65	165.20	34.25	274.00		
1 Truck Crane, 150 Ton		1563.00		1719.30	27.91	30.70
56 L.H., Daily Totals		$2692.20		$3647.70	$48.07	$65.14
Crew C-12	**Hr.**	**Daily**	**Hr.**	**Daily**	**Bare Costs**	**Incl. O&P**
1 Carpenter Foreman (out)	$21.70	$173.60	$37.20	$297.60	$19.32	$32.91
3 Carpenters	19.70	472.80	33.75	810.00		
1 Laborer	14.45	115.60	24.75	198.00		
1 Equip. Oper. (crane)	20.65	165.20	34.25	274.00		
1 Hyd. Crane, 12 Ton		453.45		498.80	9.45	10.39
48 L.H., Daily Totals		$1380.65		$2078.40	$28.77	$43.30
Crew C-13	**Hr.**	**Daily**	**Hr.**	**Daily**	**Bare Costs**	**Incl. O&P**
2 Struc. Steel Worker	$21.25	$340.00	$41.90	$670.40	$20.73	$39.18
1 Carpenter	19.70	157.60	33.75	270.00		
1 Gas Welding Machine		84.00		92.40	3.50	3.85
24 L.H., Daily Totals		$581.60		$1032.80	$24.23	$43.03

Crews

Crew C-14	Hr.	Daily	Hr.	Daily	Bare Costs	Incl. O&P
1 Carpenter Foreman (out)	$21.70	$173.60	$37.20	$297.60	$17.56	$30.19
3 Carpenters	19.70	472.80	33.75	810.00		
2 Carpenter Helpers	14.80	236.80	25.35	405.60		
4 Laborers	14.45	462.40	24.75	792.00		
2 Rodmen (reinf.)	21.10	337.60	39.25	628.00		
2 Rodman Helpers	14.80	236.80	25.35	405.60		
2 Cement Finishers	18.90	302.40	30.80	492.80		
1 Equip. Oper. (crane)	20.65	165.20	34.25	274.00		
1 Crane, 80 Ton, & Tools		1098.00		1207.80	8.07	8.88
136 L.H., Daily Totals		$3485.60		$5313.40	$25.63	$39.07

Crew C-14A	Hr.	Daily	Hr.	Daily	Bare Costs	Incl. O&P
1 Carpenter Foreman (out)	$21.70	$173.60	$37.20	$297.60	$19.56	$33.90
16 Carpenters	19.70	2521.60	33.75	4320.00		
4 Rodmen (reinf.)	21.10	675.20	39.25	1256.00		
2 Laborers	14.45	231.20	24.75	396.00		
1 Cement Finisher	18.90	151.20	30.80	246.40		
1 Equip. Oper. (med.)	19.90	159.20	33.00	264.00		
1 Gas Engine Vibrator		38.00		41.80		
1 Concrete Pump (small)		656.00		721.60	3.47	3.82
200 L.H., Daily Totals		$4606.00		$7543.40	$23.03	$37.72

Crew C-14B	Hr.	Daily	Hr.	Daily	Bare Costs	Incl. O&P
1 Carpenter Foreman (out)	$21.70	$173.60	$37.20	$297.60	$19.53	$33.78
16 Carpenters	19.70	2521.60	33.75	4320.00		
4 Rodmen (reinf.)	21.10	675.20	39.25	1256.00		
2 Laborers	14.45	231.20	24.75	396.00		
2 Cement Finishers	18.90	302.40	30.80	492.80		
1 Equip. Oper. (med.)	19.90	159.20	33.00	264.00		
1 Gas Engine Vibrator		38.00		41.80		
1 Concrete Pump (small)		656.00		721.60	3.34	3.67
208 L.H., Daily Totals		$4757.20		$7789.80	$22.87	$37.45

Crew C-14C	Hr.	Daily	Hr.	Daily	Bare Costs	Incl. O&P
1 Carpenter Foreman (out)	$21.70	$173.60	$37.20	$297.60	$18.49	$32.00
6 Carpenters	19.70	945.60	33.75	1620.00		
2 Rodmen (reinf)	21.10	337.60	39.25	628.00		
4 Laborers	14.45	462.40	24.75	792.00		
1 Cement Finisher	18.90	151.20	30.80	246.40		
1 Gas Engine Vibrator		38.00		41.80	.34	.37
112 L.H., Daily Totals		$2108.40		$3625.80	$18.83	$32.37

Crew C-14D	Hr.	Daily	Hr.	Daily	Bare Costs	Incl. O&P
1 Carpenter Foreman (out)	$21.70	$173.60	$37.20	$297.60	$19.45	$33.46
18 Carpenters	19.70	2836.80	33.75	4860.00		
2 Rodmen (reinf.)	21.10	337.60	39.25	628.00		
2 Laborers	14.45	231.20	24.75	396.00		
1 Cement Finisher	18.90	151.20	30.80	246.40		
1 Equip. Oper. (med.)	19.90	159.20	33.00	264.00		
1 Gas Engine Vibrator		38.00		41.80		
1 Concrete Pump (small)		656.00		721.60	3.47	3.82
200 L.H., Daily Totals		$4583.60		$7455.40	$22.92	$37.28

Crew C-14E	Hr.	Daily	Hr.	Daily	Bare Costs	Incl. O&P
1 Carpenter Foreman (out)	$21.70	$173.60	$37.20	$297.60	$18.89	$33.34
2 Carpenters	19.70	315.20	33.75	540.00		
4 Rodmen (reinf.)	21.10	675.20	39.25	1256.00		
3 Laborers	14.45	346.80	24.75	594.00		
1 Cement Finisher	18.90	151.20	30.80	246.40		
1 Gas Engine Vibrator		38.00		41.80	.43	.48
88 L.H., Daily Totals		$1700.00		$2975.80	$19.32	$33.82

Crew C-14F	Hr.	Daily	Hr.	Daily	Bare Costs	Incl. O&P
1 Laborer Foreman (out)	$16.45	$131.60	$28.20	$225.60	$17.64	$29.17
2 Laborers	14.45	231.20	24.75	396.00		
6 Cement Finishers	18.90	907.20	30.80	1478.40		
1 Gas Engine Vibrator		38.00		41.80	.53	.58
72 L.H., Daily Totals		$1308.00		$2141.80	$18.17	$29.75

Crew C-14G	Hr.	Daily	Hr.	Daily	Bare Costs	Incl. O&P
1 Laborer Foreman (out)	$16.45	$131.60	$28.20	$225.60	$17.28	$28.70
2 Laborers	14.45	231.20	24.75	396.00		
4 Cement Finishers	18.90	604.80	30.80	985.60		
1 Gas Engine Vibrator		38.00		41.80	.68	.75
56 L.H., Daily Totals		$1005.60		$1649.00	$17.96	$29.45

Crew C-14H	Hr.	Daily	Hr.	Daily	Bare Costs	Incl. O&P
1 Carpenter Foreman (out)	$21.70	$173.60	$37.20	$297.60	$19.26	$33.25
2 Carpenters	19.70	315.20	33.75	540.00		
1 Rodman (reinf.)	21.10	168.80	39.25	314.00		
1 Laborer	14.45	115.60	24.75	198.00		
1 Cement Finisher	18.90	151.20	30.80	246.40		
1 Gas Engine Vibrator		38.00		41.80	.79	.87
48 L.H., Daily Totals		$962.40		$1637.80	$20.05	$34.12

Crew C-15	Hr.	Daily	Hr.	Daily	Bare Costs	Incl. O&P
1 Carpenter Foreman (out)	$21.70	$173.60	$37.20	$297.60	$18.15	$31.09
2 Carpenters	19.70	315.20	33.75	540.00		
3 Laborers	14.45	346.80	24.75	594.00		
2 Cement Finishers	18.90	302.40	30.80	492.80		
1 Rodman (reinf.)	21.10	168.80	39.25	314.00		
72 L.H., Daily Totals		$1306.80		$2238.40	$18.15	$31.09

Crew C-16	Hr.	Daily	Hr.	Daily	Bare Costs	Incl. O&P
1 Labor Foreman (outside)	$16.45	$131.60	$28.20	$225.60	$17.74	$30.62
3 Laborers	14.45	346.80	24.75	594.00		
2 Cement Finishers	18.90	302.40	30.80	492.80		
1 Equip. Oper. (med.)	19.90	159.20	33.00	264.00		
2 Rodmen (reinf.)	21.10	337.60	39.25	628.00		
1 Concrete Pump (small)		656.00		721.60	9.11	10.02
72 L.H., Daily Totals		$1933.60		$2926.00	$26.85	$40.64

Crew C-17	Hr.	Daily	Hr.	Daily	Bare Costs	Incl. O&P
2 Skilled Worker Foremen	$21.75	$348.00	$37.30	$596.80	$20.15	$34.58
8 Skilled Workers	19.75	1264.00	33.90	2169.60		
80 L.H., Daily Totals		$1612.00		$2766.40	$20.15	$34.58

Crew C-17A	Hr.	Daily	Hr.	Daily	Bare Costs	Incl. O&P
2 Skilled Worker Foremen	$21.75	$348.00	$37.30	$596.80	$20.16	$34.58
8 Skilled Workers	19.75	1264.00	33.90	2169.60		
.125 Equip. Oper. (crane)	20.65	20.65	34.25	34.25		
.125 Crane, 80 Ton, & Tools		137.25		151.00	1.69	1.86
81 L.H., Daily Totals		$1769.90		$2951.65	$21.85	$36.44

Crew C-17B	Hr.	Daily	Hr.	Daily	Bare Costs	Incl. O&P
2 Skilled Worker Foremen	$21.75	$348.00	$37.30	$596.80	$20.16	$34.57
8 Skilled Workers	19.75	1264.00	33.90	2169.60		
.25 Equip. Oper. (crane)	20.65	41.30	34.25	68.50		
.25 Crane, 80 Ton, & Tools		274.50		301.95		
.25 Walk Behind Power Tools		13.25		14.60	3.51	3.86
82 L.H., Daily Totals		$1941.05		$3151.45	$23.67	$38.43

Crews

Crew No.	Bare Costs		Incl. Subs O & P		Cost Per Labor-Hour	
Crew C-17C	Hr.	Daily	Hr.	Daily	Bare Costs	Incl. O&P
2 Skilled Worker Foremen	$21.75	$348.00	$37.30	$596.80	$20.17	$34.57
8 Skilled Workers	19.75	1264.00	33.90	2169.60		
.375 Equip. Oper. (crane)	20.65	61.95	34.25	102.75		
.375 Crane, 80 Ton & Tools		411.75		452.95	4.96	5.46
83 L.H., Daily Totals		$2085.70		$3322.10	$25.13	$40.03
Crew C-17D	Hr.	Daily	Hr.	Daily	Bare Costs	Incl. O&P
2 Skilled Worker Foremen	$21.75	$348.00	$37.30	$596.80	$20.17	$34.56
8 Skilled Workers	19.75	1264.00	33.90	2169.60		
.5 Equip. Oper. (crane)	20.65	82.60	34.25	137.00		
.5 Crane, 80 Ton & Tools		549.00		603.90	6.54	7.19
84 L.H., Daily Totals		$2243.60		$3507.30	$26.71	$41.75
Crew C-17E	Hr.	Daily	Hr.	Daily	Bare Costs	Incl. O&P
2 Skilled Worker Foremen	$21.75	$348.00	$37.30	$596.80	$20.15	$34.58
8 Skilled Workers	19.75	1264.00	33.90	2169.60		
1 Hyd. Jack with Rods		60.50		66.55	.76	.83
80 L.H., Daily Totals		$1672.50		$2832.95	$20.91	$35.41
Crew C-18	Hr.	Daily	Hr.	Daily	Bare Costs	Incl. O&P
.125 Labor Foreman (out)	$16.45	$16.45	$28.20	$28.20	$14.67	$25.13
1 Laborer	14.45	115.60	24.75	198.00		
1 Concrete Cart, 10 C.F.		52.00		57.20	5.78	6.36
9 L.H., Daily Totals		$184.05		$283.40	$20.45	$31.49
Crew C-19	Hr.	Daily	Hr.	Daily	Bare Costs	Incl. O&P
.125 Labor Foreman (out)	$16.45	$16.45	$28.20	$28.20	$14.67	$25.13
1 Laborer	14.45	115.60	24.75	198.00		
1 Concrete Cart, 18 C.F.		83.20		91.50	9.24	10.17
9 L.H., Daily Totals		$215.25		$317.70	$23.91	$35.30
Crew C-20	Hr.	Daily	Hr.	Daily	Bare Costs	Incl. O&P
1 Labor Foreman (outside)	$16.45	$131.60	$28.20	$225.60	$15.94	$26.97
5 Laborers	14.45	578.00	24.75	990.00		
1 Cement Finisher	18.90	151.20	30.80	246.40		
1 Equip. Oper. (med.)	19.90	159.20	33.00	264.00		
2 Gas Engine Vibrators		76.00		83.60		
1 Concrete Pump (small)		656.00		721.60	11.44	12.58
64 L.H., Daily Totals		$1752.00		$2531.20	$27.38	$39.55
Crew C-21	Hr.	Daily	Hr.	Daily	Bare Costs	Incl. O&P
1 Labor Foreman (outside)	$16.45	$131.60	$28.20	$225.60	$15.94	$26.97
5 Laborers	14.45	578.00	24.75	990.00		
1 Cement Finisher	18.90	151.20	30.80	246.40		
1 Equip. Oper. (med.)	19.90	159.20	33.00	264.00		
2 Gas Engine Vibrators		76.00		83.60		
1 Concrete Conveyer		182.00		200.20	4.03	4.43
64 L.H., Daily Totals		$1278.00		$2009.80	$19.97	$31.40
Crew C-22	Hr.	Daily	Hr.	Daily	Bare Costs	Incl. O&P
1 Rodman Foreman	$23.10	$184.80	$43.00	$344.00	$21.37	$39.58
4 Rodmen (reinf.)	21.10	675.20	39.25	1256.00		
.125 Equip. Oper. (crane)	20.65	20.65	34.25	34.25		
.125 Equip. Oper. Oiler	17.00	17.00	28.20	28.20		
.125 Hyd. Crane, 25 Ton		69.90		76.90	1.66	1.83
42 L.H., Daily Totals		$967.55		$1739.35	$23.03	$41.41
Crew C-23	Hr.	Daily	Hr.	Daily	Bare Costs	Incl. O&P
2 Skilled Worker Foremen	$21.75	$348.00	$37.30	$596.80	$19.97	$34.05
6 Skilled Workers	19.75	948.00	33.90	1627.20		
1 Equip. Oper. (crane)	20.65	165.20	34.25	274.00		
1 Equip. Oper. Oiler	17.00	136.00	28.20	225.60		
1 Crane, 90 Ton		1022.00		1124.20	12.78	14.05
80 L.H., Daily Totals		$2619.20		$3847.80	$32.75	$48.10
Crew C-24	Hr.	Daily	Hr.	Daily	Bare Costs	Incl. O&P
2 Skilled Worker Foremen	$21.75	$348.00	$37.30	$596.80	$19.97	$34.05
6 Skilled Workers	19.75	948.00	33.90	1627.20		
1 Equip. Oper. (crane)	20.65	165.20	34.25	274.00		
1 Equip. Oper. Oiler	17.00	136.00	28.20	225.60		
1 Truck Crane, 150 Ton		1563.00		1719.30	19.54	21.49
80 L.H., Daily Totals		$3160.20		$4442.90	$39.51	$55.54
Crew C-25	Hr.	Daily	Hr.	Daily	Bare Costs	Incl. O&P
2 Rodmen (reinf.)	$21.10	$337.60	$39.25	$628.00	$16.93	$31.50
2 Rodmen Helpers	12.75	204.00	23.75	380.00		
32 L.H., Daily Totals		$541.60		$1008.00	$16.93	$31.50
Crew D-1	Hr.	Daily	Hr.	Daily	Bare Costs	Incl. O&P
1 Bricklayer	$20.00	$160.00	$33.85	$270.80	$17.85	$30.23
1 Bricklayer Helper	15.70	125.60	26.60	212.80		
16 L.H., Daily Totals		$285.60		$483.60	$17.85	$30.23
Crew D-2	Hr.	Daily	Hr.	Daily	Bare Costs	Incl. O&P
3 Bricklayers	$20.00	$480.00	$33.85	$812.40	$18.28	$30.95
2 Bricklayer Helpers	15.70	251.20	26.60	425.60		
40 L.H., Daily Totals		$731.20		$1238.00	$18.28	$30.95
Crew D-3	Hr.	Daily	Hr.	Daily	Bare Costs	Incl. O&P
3 Bricklayers	$20.00	$480.00	$33.85	$812.40	$18.35	$31.08
2 Bricklayer Helpers	15.70	251.20	26.60	425.60		
.25 Carpenter	19.70	39.40	33.75	67.50		
42 L.H., Daily Totals		$770.60		$1305.50	$18.35	$31.08
Crew D-4	Hr.	Daily	Hr.	Daily	Bare Costs	Incl. O&P
1 Bricklayer	$20.00	$160.00	$33.85	$270.80	$16.31	$27.68
3 Bricklayer Helpers	15.70	376.80	26.60	638.40		
1 Building Laborer	14.45	115.60	24.75	198.00		
1 Grout Pump, 50 C.F./hr		67.20		73.90		
1 Hoses & Hopper		31.60		34.75		
1 Accessories		10.55		11.60	2.73	3.01
40 L.H., Daily Totals		$761.75		$1227.45	$19.04	$30.69
Crew D-5	Hr.	Daily	Hr.	Daily	Bare Costs	Incl. O&P
1 Block Mason Helper	$15.70	$125.60	$26.60	$212.80	$15.70	$26.60
8 L.H., Daily Totals		$125.60		$212.80	$15.70	$26.60
Crew D-6	Hr.	Daily	Hr.	Daily	Bare Costs	Incl. O&P
3 Bricklayers	$20.00	$480.00	$33.85	$812.40	$17.85	$30.23
3 Bricklayer Helpers	15.70	376.80	26.60	638.40		
48 L.H., Daily Totals		$856.80		$1450.80	$17.85	$30.23
Crew D-7	Hr.	Daily	Hr.	Daily	Bare Costs	Incl. O&P
1 Tile Layer	$19.20	$153.60	$31.10	$248.80	$17.33	$28.08
1 Tile Layer Helper	15.45	123.60	25.05	200.40		
16 L.H., Daily Totals		$277.20		$449.20	$17.33	$28.08

Crews

Crew No.	Bare Costs		Incl. Subs O & P		Cost Per Labor-Hour	
	Hr.	Daily	Hr.	Daily	Bare Costs	Incl. O&P
Crew D-8						
3 Bricklayers	$20.00	$480.00	$33.85	$812.40	$18.28	$30.95
2 Bricklayer Helpers	15.70	251.20	26.60	425.60		
40 L.H., Daily Totals		$731.20		$1238.00	$18.28	$30.95
Crew D-9	Hr.	Daily	Hr.	Daily	Bare Costs	Incl. O&P
3 Block Masons	$20.00	$480.00	$33.85	$812.40	$17.85	$30.23
3 Block Mason Helpers	15.70	376.80	26.60	638.40		
48 L.H., Daily Totals		$856.80		$1450.80	$17.85	$30.23
Crew D-10	Hr.	Daily	Hr.	Daily	Bare Costs	Incl. O&P
1 Stone Mason Foreman	$21.50	$172.00	$36.40	$291.20	$18.61	$31.37
1 Stone Mason	19.50	156.00	33.00	264.00		
2 Stone Mason Helpers	15.70	251.20	26.60	425.60		
1 Equip. Oper. (crane)	20.65	165.20	34.25	274.00		
1 Truck Crane, 12.5 Ton		401.35		441.50	10.03	11.04
40 L.H., Daily Totals		$1145.75		$1696.30	$28.64	$42.41
Crew D-11	Hr.	Daily	Hr.	Daily	Bare Costs	Incl. O&P
2 Stone Masons	$19.50	$312.00	$33.00	$528.00	$18.23	$30.87
1 Stone Mason Helper	15.70	125.60	26.60	212.80		
24 L.H., Daily Totals		$437.60		$740.80	$18.23	$30.87
Crew D-12	Hr.	Daily	Hr.	Daily	Bare Costs	Incl. O&P
2 Stone Masons	$19.50	$312.00	$33.00	$528.00	$17.60	$29.80
2 Stone Mason Helpers	15.70	251.20	26.60	425.60		
32 L.H., Daily Totals		$563.20		$953.60	$17.60	$29.80
Crew D-13	Hr.	Daily	Hr.	Daily	Bare Costs	Incl. O&P
1 Stone Mason Foreman	$21.50	$172.00	$36.40	$291.20	$18.76	$31.64
2 Stone Masons	19.50	312.00	33.00	528.00		
2 Stone Mason Helpers	15.70	251.20	26.60	425.60		
1 Equip. Oper. (crane)	20.65	165.20	34.25	274.00		
1 Truck Crane, 12.5 Ton		401.35		441.50	8.36	9.20
48 L.H., Daily Totals		$1301.75		$1960.30	$27.12	$40.84
Crew E-1	Hr.	Daily	Hr.	Daily	Bare Costs	Incl. O&P
2 Struc. Steel Workers	$21.25	$340.00	$41.90	$670.40	$21.25	$41.90
1 Gas Welding Machine		84.00		92.40	5.25	5.78
16 L.H., Daily Totals		$424.00		$762.80	$26.50	$47.68
Crew E-2	Hr.	Daily	Hr.	Daily	Bare Costs	Incl. O&P
1 Struc. Steel Foreman	$23.25	$186.00	$45.85	$366.80	$21.48	$41.28
4 Struc. Steel Workers	21.25	680.00	41.90	1340.80		
1 Equip. Oper. (crane)	20.65	165.20	34.25	274.00		
1 Crane, 90 Ton		1022.00		1124.20	21.29	23.42
48 L.H., Daily Totals		$2053.20		$3105.80	$42.77	$64.70
Crew E-3	Hr.	Daily	Hr.	Daily	Bare Costs	Incl. O&P
1 Struc. Steel Foreman	$23.25	$186.00	$45.85	$366.80	$21.92	$43.22
2 Struc. Steel Worker	21.25	340.00	41.90	670.40		
1 Gas Welding Machine		84.00		92.40	3.50	3.85
24 L.H., Daily Totals		$610.00		$1129.60	$25.42	$47.07
Crew E-4	Hr.	Daily	Hr.	Daily	Bare Costs	Incl. O&P
1 Struc. Steel Foreman	$23.25	$186.00	$45.85	$366.80	$21.75	$42.89
3 Struc. Steel Workers	21.25	510.00	41.90	1005.60		
1 Gas Welding Machine		84.00		92.40	2.63	2.89
32 L.H., Daily Totals		$780.00		$1464.80	$24.38	$45.78
Crew E-5	Hr.	Daily	Hr.	Daily	Bare Costs	Incl. O&P
1 Struc. Steel Foremen	$23.25	$186.00	$45.85	$366.80	$21.41	$41.49
7 Struc. Steel Workers	21.25	1190.00	41.90	2346.40		
1 Equip. Oper. (crane)	20.65	165.20	34.25	274.00		
1 Crane, 90 Ton		1022.00		1124.20		
1 Gas Welding Machine		84.00		92.40	15.36	16.90
72 L.H., Daily Totals		$2647.20		$4203.80	$36.77	$58.39
Crew E-6	Hr.	Daily	Hr.	Daily	Bare Costs	Incl. O&P
1 Struc. Steel Foreman	$23.25	$186.00	$45.85	$366.80	$21.20	$40.97
12 Struc. Steel Workers	21.25	2040.00	41.90	4022.40		
1 Equip. Oper. (crane)	20.65	165.20	34.25	274.00		
1 Equip. Oper. (light)	19.10	152.80	31.70	253.60		
1 Crane, 90 Ton		1022.00		1124.20		
1 Gas Welding Machine		84.00		92.40		
1 Air Compr., 160 C.F.M.		98.40		108.25		
2 Impact Wrenches		55.60		61.15	10.50	11.55
120 L.H., Daily Totals		$3804.00		$6302.80	$31.70	$52.52
Crew E-7	Hr.	Daily	Hr.	Daily	Bare Costs	Incl. O&P
1 Struc. Steel Foreman	$23.25	$186.00	$45.85	$366.80	$21.41	$41.49
7 Struc. Steel Workers	21.25	1190.00	41.90	2346.40		
1 Equip. Oper. (crane)	20.65	165.20	34.25	274.00		
1 Crane, 90 Ton		1022.00		1124.20		
2 Gas Welding Machines		168.00		184.80	16.53	18.18
72 L.H., Daily Totals		$2731.20		$4296.20	$37.94	$59.67
Crew E-8	Hr.	Daily	Hr.	Daily	Bare Costs	Incl. O&P
1 Struc. Steel Foreman	$23.25	$186.00	$45.85	$366.80	$21.38	$41.56
9 Struc. Steel Workers	21.25	1530.00	41.90	3016.80		
1 Equip. Oper. (crane)	20.65	165.20	34.25	274.00		
1 Crane, 90 Ton		1022.00		1124.20		
4 Gas Welding Machines		336.00		369.60	15.43	16.98
88 L.H., Daily Totals		$3239.20		$5151.40	$36.81	$58.54
Crew E-9	Hr.	Daily	Hr.	Daily	Bare Costs	Incl. O&P
2 Struc. Steel Foremen	$23.25	$372.00	$45.85	$733.60	$21.19	$40.67
5 Struc. Steel Workers	21.25	850.00	41.90	1676.00		
1 Welder Foreman	23.25	186.00	45.85	366.80		
5 Welders	21.25	850.00	41.90	1676.00		
1 Equip. Oper. (crane)	20.65	165.20	34.25	274.00		
1 Equip. Oper. Oiler	17.00	136.00	28.20	225.60		
1 Equip. Oper. (light)	19.10	152.80	31.70	253.60		
1 Crane, 90 Ton		1022.00		1124.20		
5 Gas Welding Machines		420.00		462.00	11.27	12.39
128 L.H., Daily Totals		$4154.00		$6791.80	$32.46	$53.06
Crew E-10	Hr.	Daily	Hr.	Daily	Bare Costs	Incl. O&P
1 Struc. Steel Foreman	$23.25	$186.00	$45.85	$366.80	$21.92	$43.22
2 Struc. Steel Workers	21.25	340.00	41.90	670.40		
4 Gas Welding Machines		336.00		369.60		
1 Truck, 3 Ton		180.45		198.50	21.52	23.67
24 L.H., Daily Totals		$1042.45		$1605.30	$43.44	$66.89
Crew E-11	Hr.	Daily	Hr.	Daily	Bare Costs	Incl. O&P
2 Painters, Struc. Steel	$18.65	$298.40	$37.80	$604.80	$17.25	$33.45
1 Building Laborer	14.45	115.60	24.75	198.00		
1 Air Compressor 250 C.F.M.		127.00		139.70		
1 Sand Blaster		31.60		34.75		
1 Sand Blasting Accessories		10.55		11.60	7.05	7.75
24 L.H., Daily Totals		$583.15		$988.85	$24.30	$41.20

Crews

Crew No.		Bare Costs		Incl. Subs O & P		Cost Per Labor-Hour	
Crew E-12	Hr.	Daily	Hr.	Daily	Bare Costs	Incl. O&P
1 Welder Foreman	$23.25	$186.00	$45.85	$366.80	$21.18	$38.78
1 Equip. Oper. (light)	19.10	152.80	31.70	253.60		
1 Gas Welding Machine		84.00		92.40	5.25	5.78
16 L.H., Daily Totals		$422.80		$712.80	$26.43	$44.56
Crew E-13	Hr.	Daily	Hr.	Daily	Bare Costs	Incl. O&P
1 Welder Foreman	$23.25	$186.00	$45.85	$366.80	$21.87	$41.13
.5 Equip. Oper. (light)	19.10	76.40	31.70	126.80		
1 Gas Welding Machine		84.00		92.40	7.00	7.70
12 L.H., Daily Totals		$346.40		$586.00	$28.87	$48.83
Crew E-14	Hr.	Daily	Hr.	Daily	Bare Costs	Incl. O&P
1 Struc. Steel Worker	$21.25	$170.00	$41.90	$335.20	$21.25	$41.90
1 Gas Welding Machine		84.00		92.40	10.50	11.55
8 L.H., Daily Totals		$254.00		$427.60	$31.75	$53.45
Crew E-16	Hr.	Daily	Hr.	Daily	Bare Costs	Incl. O&P
1 Welder Foreman	$23.25	$186.00	$45.85	$366.80	$22.25	$43.88
1 Welder	21.25	170.00	41.90	335.20		
1 Gas Welding Machine		84.00		92.40	5.25	5.78
16 L.H., Daily Totals		$440.00		$794.40	$27.50	$49.66
Crew E-17	Hr.	Daily	Hr.	Daily	Bare Costs	Incl. O&P
1 Structural Steel Foreman	$23.25	$186.00	$45.85	$366.80	$22.25	$43.88
1 Structural Steel Worker	21.25	170.00	41.90	335.20		
1 Power Tool		2.40		2.65	.15	.17
16 L.H., Daily Totals		$358.40		$704.65	$22.40	$44.05
Crew E-18	Hr.	Daily	Hr.	Daily	Bare Costs	Incl. O&P
1 Structural Steel Foreman	$23.25	$186.00	$45.85	$366.80	$21.38	$40.91
3 Structural Steel Workers	21.25	510.00	41.90	1005.60		
1 Equipment Operator (med.)	19.90	159.20	33.00	264.00		
1 Crane, 20 Ton		532.55		585.80	13.31	14.65
40 L.H., Daily Totals		$1387.75		$2222.20	$34.69	$55.56
Crew E-19	Hr.	Daily	Hr.	Daily	Bare Costs	Incl. O&P
1 Structural Steel Worker	$21.25	$170.00	$41.90	$335.20	$21.20	$39.82
1 Structural Steel Foreman	23.25	186.00	45.85	366.80		
1 Equip. Oper. (light)	19.10	152.80	31.70	253.60		
1 Power Tool		2.40		2.65		
1 Crane, 20 ton		532.55		585.80	22.29	24.52
24 L.H., Daily Totals		$1043.75		$1544.05	$43.49	$64.34
Crew E-20	Hr.	Daily	Hr.	Daily	Bare Costs	Incl. O&P
1 Structural Steel Foreman	$23.25	$186.00	$45.85	$366.80	$20.89	$39.73
5 Structural Steel Workers	21.25	850.00	41.90	1676.00		
1 Equip. Oper. (crane)	20.65	165.20	34.25	274.00		
1 Oiler	17.00	136.00	28.20	225.60		
1 Power Tool		2.40		2.65		
1 Crane, 40 ton		704.80		775.30	11.05	12.16
64 L.H., Daily Totals		$2044.40		$3320.35	$31.94	$51.89
Crew E-22	Hr.	Daily	Hr.	Daily	Bare Costs	Incl. O&P
1 Skilled Worker Foreman	$21.75	$174.00	$37.30	$298.40	$20.42	$35.03
2 Skilled Worker	19.75	316.00	33.90	542.40		
24 L.H., Daily Totals		$490.00		$840.80	$20.42	$35.03
Crew E-24	Hr.	Daily	Hr.	Daily	Bare Costs	Incl. O&P
3 Structural Steel Worker	$21.25	$510.00	$41.90	$1005.60	$20.91	$39.68
1 Equipment Operator (medium)	19.90	159.20	33.00	264.00		
1-25 ton crane		559.20		615.10	17.48	19.22
32 L.H., Daily Totals		$1228.40		$1884.70	$38.39	$58.90
Crew F-3	Hr.	Daily	Hr.	Daily	Bare Costs	Incl. O&P
2 Carpenters	$19.70	$315.20	$33.75	$540.00	$17.93	$30.49
2 Carpenter Helpers	14.80	236.80	25.35	405.60		
1 Equip. Oper. (crane)	20.65	165.20	34.25	274.00		
1 Hyd. Crane, 12 Ton		453.45		498.80	11.34	12.47
40 L.H., Daily Totals		$1170.65		$1718.40	$29.27	$42.96
Crew F-4	Hr.	Daily	Hr.	Daily	Bare Costs	Incl. O&P
2 Carpenters	$19.70	$315.20	$33.75	$540.00	$17.93	$30.49
2 Carpenter Helpers	14.80	236.80	25.35	405.60		
1 Equip. Oper. (crane)	20.65	165.20	34.25	274.00		
1 Hyd. Crane, 55 Ton		808.20		889.00	20.21	22.23
40 L.H., Daily Totals		$1525.40		$2108.60	$38.14	$52.72
Crew F-5	Hr.	Daily	Hr.	Daily	Bare Costs	Incl. O&P
2 Carpenters	$19.70	$315.20	$33.75	$540.00	$17.25	$29.55
2 Carpenter Helpers	14.80	236.80	25.35	405.60		
32 L.H., Daily Totals		$552.00		$945.60	$17.25	$29.55
Crew F-6	Hr.	Daily	Hr.	Daily	Bare Costs	Incl. O&P
2 Carpenters	$19.70	$315.20	$33.75	$540.00	$17.79	$30.25
2 Building Laborers	14.45	231.20	24.75	396.00		
1 Equip. Oper. (crane)	20.65	165.20	34.25	274.00		
1 Hyd. Crane, 12 Ton		453.45		498.80	11.34	12.47
40 L.H., Daily Totals		$1165.05		$1708.80	$29.13	$42.72
Crew F-7	Hr.	Daily	Hr.	Daily	Bare Costs	Incl. O&P
2 Carpenters	$19.70	$315.20	$33.75	$540.00	$17.08	$29.25
2 Building Laborers	14.45	231.20	24.75	396.00		
32 L.H., Daily Totals		$546.40		$936.00	$17.08	$29.25
Crew G-1	Hr.	Daily	Hr.	Daily	Bare Costs	Incl. O&P
1 Roofer Foreman	$19.05	$152.40	$35.45	$283.60	$16.11	$29.99
4 Roofers, Composition	17.05	545.60	31.75	1016.00		
2 Roofer Helpers	12.75	204.00	23.75	380.00		
1 Application Equipment		171.60		188.75		
1 Tar Kettle/Pot		49.00		53.90		
1 Crew Truck		197.60		217.35	7.47	8.21
56 L.H., Daily Totals		$1320.20		$2139.60	$23.58	$38.20
Crew G-2	Hr.	Daily	Hr.	Daily	Bare Costs	Incl. O&P
1 Plasterer	$18.65	$149.20	$31.20	$249.60	$16.25	$27.37
1 Plasterer Helper	15.65	125.20	26.15	209.20		
1 Building Laborer	14.45	115.60	24.75	198.00		
1 Grouting Equipment		250.00		275.00	10.42	11.46
24 L.H., Daily Totals		$640.00		$931.80	$26.67	$38.83
Crew G-3	Hr.	Daily	Hr.	Daily	Bare Costs	Incl. O&P
2 Sheet Metal Workers	$21.50	$344.00	$36.30	$580.80	$17.98	$30.52
2 Building Laborers	14.45	231.20	24.75	396.00		
32 L.H., Daily Totals		$575.20		$976.80	$17.98	$30.52

Crews

Crew No.	Bare Costs		Incl. Subs O & P		Cost Per Labor-Hour	
Crew G-4	Hr.	Daily	Hr.	Daily	Bare Costs	Incl. O&P
1 Labor Foreman (outside)	$16.45	$131.60	$28.20	$225.60	$15.12	$25.90
2 Building Laborers	14.45	231.20	24.75	396.00		
1 Light Truck, 1.5 Ton		172.95		190.25		
1 Air Compr., 160 C.F.M.		98.40		108.25	11.31	12.44
24 L.H., Daily Totals		$634.15		$920.10	$26.43	$38.34
Crew G-5	Hr.	Daily	Hr.	Daily	Bare Costs	Incl. O&P
1 Roofer Foreman	$19.05	$152.40	$35.45	$283.60	$15.73	$29.29
2 Roofers, Composition	17.05	272.80	31.75	508.00		
2 Roofer Helpers	12.75	204.00	23.75	380.00		
1 Application Equipment		171.60		188.75	4.29	4.72
40 L.H., Daily Totals		$800.80		$1360.35	$20.02	$34.01
Crew G-6A	Hr.	Daily	Hr.	Daily	Bare Costs	Incl. O&P
2 Roofers Composition	$17.05	$272.80	$31.75	$508.00	$17.05	$31.75
1 Small Compressor		20.95		23.05		
2 Pneumatic Nailers		37.90		41.70	3.68	4.05
16 L.H., Daily Totals		$331.65		$572.75	$20.73	$35.80
Crew G-7	Hr.	Daily	Hr.	Daily	Bare Costs	Incl. O&P
1 Carpenter	$19.70	$157.60	$33.75	$270.00	$19.70	$33.75
1 Small Compressor		20.95		23.05		
1 Pneumatic Nailer		18.95		20.85	4.99	5.49
8 L.H., Daily Totals		$197.50		$313.90	$24.69	$39.24
Crew H-1	Hr.	Daily	Hr.	Daily	Bare Costs	Incl. O&P
2 Glaziers	$19.35	$309.60	$32.10	$513.60	$20.30	$37.00
2 Struc. Steel Workers	21.25	340.00	41.90	670.40		
32 L.H., Daily Totals		$649.60		$1184.00	$20.30	$37.00
Crew H-2	Hr.	Daily	Hr.	Daily	Bare Costs	Incl. O&P
2 Glaziers	$19.35	$309.60	$32.10	$513.60	$17.72	$29.65
1 Building Laborer	14.45	115.60	24.75	198.00		
24 L.H., Daily Totals		$425.20		$711.60	$17.72	$29.65
Crew H-3	Hr.	Daily	Hr.	Daily	Bare Costs	Incl. O&P
1 Glazier	$19.35	$154.80	$32.10	$256.80	$17.08	$28.73
1 Helper	14.80	118.40	25.35	202.80		
16 L.H., Daily Totals		$273.20		$459.60	$17.08	$28.73
Crew J-1	Hr.	Daily	Hr.	Daily	Bare Costs	Incl. O&P
3 Plasterers	$18.65	$447.60	$31.20	$748.80	$17.45	$29.18
2 Plasterer Helpers	15.65	250.40	26.15	418.40		
1 Mixing Machine, 6 C.F.		54.50		59.95	1.36	1.50
40 L.H., Daily Totals		$752.50		$1227.15	$18.81	$30.68
Crew J-2	Hr.	Daily	Hr.	Daily	Bare Costs	Incl. O&P
3 Plasterers	$18.65	$447.60	$31.20	$748.80	$17.74	$29.56
2 Plasterer Helpers	15.65	250.40	26.15	418.40		
1 Lather	19.20	153.60	31.45	251.60		
1 Mixing Machine, 6 C.F.		54.50		59.95	1.14	1.25
48 L.H., Daily Totals		$906.10		$1478.75	$18.88	$30.81
Crew J-3	Hr.	Daily	Hr.	Daily	Bare Costs	Incl. O&P
1 Terrazzo Worker	$19.10	$152.80	$30.95	$247.60	$17.33	$28.08
1 Terrazzo Helper	15.55	124.40	25.20	201.60		
1 Terrazzo Grinder, Electric		50.00		55.00		
1 Terrazzo Mixer		74.20		81.60	7.76	8.54
16 L.H., Daily Totals		$401.40		$585.80	$25.09	$36.62

Crew No.	Bare Costs		Incl. Subs O & P		Cost Per Labor-Hour	
Crew J-4	Hr.	Daily	Hr.	Daily	Bare Costs	Incl. O&P
1 Tile Layer	$19.20	$153.60	$31.10	$248.80	$17.33	$28.08
1 Tile Layer Helper	15.45	123.60	25.05	200.40		
16 L.H., Daily Totals		$277.20		$449.20	$17.33	$28.08
Crew K-1	Hr.	Daily	Hr.	Daily	Bare Costs	Incl. O&P
1 Carpenter	$19.70	$157.60	$33.75	$270.00	$17.78	$30.13
1 Truck Driver (light)	15.85	126.80	26.50	212.00		
1 Truck w/Power Equip.		180.45		198.50	11.28	12.41
16 L.H., Daily Totals		$464.85		$680.50	$29.06	$42.54
Crew K-2	Hr.	Daily	Hr.	Daily	Bare Costs	Incl. O&P
1 Struc. Steel Foreman	$23.25	$186.00	$45.85	$366.80	$20.12	$38.08
1 Struc. Steel Worker	21.25	170.00	41.90	335.20		
1 Truck Driver (light)	15.85	126.80	26.50	212.00		
1 Truck w/Power Equip.		180.45		198.50	7.52	8.27
24 L.H., Daily Totals		$663.25		$1112.50	$27.64	$46.35
Crew L-1	Hr.	Daily	Hr.	Daily	Bare Costs	Incl. O&P
.25 Electrician	$22.10	$44.20	$36.15	$72.30	$21.98	$36.27
1 Plumber	21.95	175.60	36.30	290.40		
10 L.H., Daily Totals		$219.80		$362.70	$21.98	$36.27
Crew L-2	Hr.	Daily	Hr.	Daily	Bare Costs	Incl. O&P
1 Carpenter	$19.70	$157.60	$33.75	$270.00	$17.25	$29.55
1 Carpenter Helper	14.80	118.40	25.35	202.80		
16 L.H., Daily Totals		$276.00		$472.80	$17.25	$29.55
Crew L-3	Hr.	Daily	Hr.	Daily	Bare Costs	Incl. O&P
1 Carpenter	$19.70	$157.60	$33.75	$270.00	$20.18	$34.23
.25 Electrician	22.10	44.20	36.15	72.30		
10 L.H., Daily Totals		$201.80		$342.30	$20.18	$34.23
Crew L-3A	Hr.	Daily	Hr.	Daily	Bare Costs	Incl. O&P
1 Carpenter Foreman (outside)	$21.70	$173.60	$37.20	$297.60	$21.63	$36.90
.5 Sheet Metal Worker	21.50	86.00	36.30	145.20		
12 L.H., Daily Totals		$259.60		$442.80	$21.63	$36.90
Crew L-4	Hr.	Daily	Hr.	Daily	Bare Costs	Incl. O&P
1 Skilled Workers	$19.75	$158.00	$33.90	$271.20	$17.27	$29.63
1 Helper	14.80	118.40	25.35	202.80		
16 L.H., Daily Totals		$276.40		$474.00	$17.27	$29.63
Crew L-5	Hr.	Daily	Hr.	Daily	Bare Costs	Incl. O&P
1 Struc. Steel Foreman	$23.25	$186.00	$45.85	$366.80	$21.45	$41.37
5 Struc. Steel Workers	21.25	850.00	41.90	1676.00		
1 Equip. Oper. (crane)	20.65	165.20	34.25	274.00		
1 Hyd. Crane, 25 Ton		559.20		615.10	9.99	10.98
56 L.H., Daily Totals		$1760.40		$2931.90	$31.44	$52.35
Crew L-5A	Hr.	Daily	Hr.	Daily	Bare Costs	Incl. O&P
1 Structural Steel Foreman	$23.25	$186.00	$45.85	$366.80	$21.60	$40.98
2 Structural Steel Workers	21.25	340.00	41.90	670.40		
1 Equip. Oper. (crane)	20.65	165.20	34.25	274.00		
1 Crane, SP, 25 Ton		563.90		620.30	17.62	19.38
32 L.H., Daily Totals		$1255.10		$1931.50	$39.22	$60.36

Crews

Crew No.		Bare Costs		Incl. Subs O & P		Cost Per Labor-Hour	
Crew L-6	Hr.	Daily	Hr.	Daily	Bare Costs	Incl. O&P
1 Plumber	$21.95	$175.60	$36.30	$290.40	$22.00	$36.25
.5 Electrician	22.10	88.40	36.15	144.60		
12 L.H., Daily Totals		$264.00		$435.00	$22.00	$36.25

Crew L-7	Hr.	Daily	Hr.	Daily	Bare Costs	Incl. O&P
1 Carpenters	$19.70	$157.60	$33.75	$270.00	$16.87	$28.77
2 Carpenter Helpers	14.80	236.80	25.35	405.60		
.25 Electrician	22.10	44.20	36.15	72.30		
26 L.H., Daily Totals		$438.60		$747.90	$16.87	$28.77

Crew L-8	Hr.	Daily	Hr.	Daily	Bare Costs	Incl. O&P
1 Carpenters	$19.70	$157.60	$33.75	$270.00	$18.19	$30.90
1 Carpenter Helper	14.80	118.40	25.35	202.80		
.5 Plumber	21.95	87.80	36.30	145.20		
20 L.H., Daily Totals		$363.80		$618.00	$18.19	$30.90

Crew L-9	Hr.	Daily	Hr.	Daily	Bare Costs	Incl. O&P
1 Skilled Worker Foreman	$21.75	$174.00	$37.30	$298.40	$18.26	$31.11
1 Skilled Worker	19.75	158.00	33.90	271.20		
2 Helpers	14.80	236.80	25.35	405.60		
.5 Electrician	22.10	88.40	36.15	144.60		
36 L.H., Daily Totals		$657.20		$1119.80	$18.26	$31.11

Crew L-10	Hr.	Daily	Hr.	Daily	Bare Costs	Incl. O&P
1 Structural Steel Foreman	$23.25	$186.00	$45.85	$366.80	$21.72	$40.67
1 Structural Steel Worker	21.25	170.00	41.90	335.20		
1 Equip. Oper. (crane)	20.65	165.20	34.25	274.00		
1 Hyd. Crane, 12 Ton		453.45		498.80	18.89	20.78
24 L.H., Daily Totals		$974.65		$1474.80	$40.61	$61.45

Crew M-1	Hr.	Daily	Hr.	Daily	Bare Costs	Incl. O&P
3 Elevator Constructors	$22.90	$549.60	$37.80	$907.20	$21.75	$35.90
1 Elevator Apprentice	18.30	146.40	30.20	241.60		
5 Hand Tools		95.00		104.50	2.97	3.27
32 L.H., Daily Totals		$791.00		$1253.30	$24.72	$39.17

Crew M-3	Hr.	Daily	Hr.	Daily	Bare Costs	Incl. O&P
1 Electrician Foreman (out)	$24.10	$192.80	$39.40	$315.20	$19.94	$33.04
1 Common Laborer	14.45	115.60	24.75	198.00		
.25 Equipment Operator, Medium	19.90	39.80	33.00	66.00		
1 Elevator Constructor	22.90	183.20	37.80	302.40		
1 Elevator Apprentice	18.30	146.40	30.20	241.60		
.25 Crane, SP, 4 x 4, 20 ton		118.83		130.70	3.49	3.84
34 L.H., Daily Totals		$796.63		$1253.90	$23.43	$36.88

Crew M-4	Hr.	Daily	Hr.	Daily	Bare Costs	Incl. O&P
1 Electrician Foreman (out)	$24.10	$192.80	$39.40	$315.20	$19.81	$32.84
1 Common Laborer	14.45	115.60	24.75	198.00		
.25 Equipment Operator, Crane	20.65	41.30	34.25	68.50		
.25 Equipment Operator, Oiler	17.00	34.00	28.20	56.40		
1 Elevator Constructor	22.90	183.20	37.80	302.40		
1 Elevator Apprentice	18.30	146.40	30.20	241.60		
.25 Crane, hyd, SP, 4WD, 40 ton		195.95		215.55	5.44	5.99
36 L.H., Daily Totals		$909.25		$1397.65	$25.25	$38.83

Crew Q-1	Hr.	Daily	Hr.	Daily	Bare Costs	Incl. O&P
1 Plumber	$21.95	$175.60	$36.30	$290.40	$19.75	$32.65
1 Plumber Apprentice	17.55	140.40	29.00	232.00		
16 L.H., Daily Totals		$316.00		$522.40	$19.75	$32.65

Crew Q-1C	Hr.	Daily	Hr.	Daily	Bare Costs	Incl. O&P
1 Plumber	$21.95	$175.60	$36.30	$290.40	$19.80	$32.77
1 Plumber Apprentice	17.55	140.40	29.00	232.00		
1 Equip. Oper. (medium)	19.90	159.20	33.00	264.00		
1 Trencher, Chain		540.90		595.00	22.54	24.79
24 L.H., Daily Totals		$1016.10		$1381.40	$42.34	$57.56

Crew Q-2	Hr.	Daily	Hr.	Daily	Bare Costs	Incl. O&P
1 Plumber	$21.95	$175.60	$36.30	$290.40	$19.02	$31.43
2 Plumber Apprentices	17.55	280.80	29.00	464.00		
24 L.H., Daily Totals		$456.40		$754.40	$19.02	$31.43

Crew Q-3	Hr.	Daily	Hr.	Daily	Bare Costs	Incl. O&P
2 Plumbers	$21.95	$351.20	$36.30	$580.80	$19.75	$32.65
2 Plumber Apprentices	17.55	280.80	29.00	464.00		
32 L.H., Daily Totals		$632.00		$1044.80	$19.75	$32.65

Crew Q-4	Hr.	Daily	Hr.	Daily	Bare Costs	Incl. O&P
2 Plumbers	$21.95	$351.20	$36.30	$580.80	$20.85	$34.47
1 Welder (plumber)	21.95	175.60	36.30	290.40		
1 Plumber Apprentice	17.55	140.40	29.00	232.00		
1 Electric Welding Mach.		48.80		53.70	1.53	1.68
32 L.H., Daily Totals		$716.00		$1156.90	$22.38	$36.15

Crew Q-5	Hr.	Daily	Hr.	Daily	Bare Costs	Incl. O&P
1 Steamfitter	$22.10	$176.80	$36.55	$292.40	$19.90	$32.90
1 Steamfitter Apprentice	17.70	141.60	29.25	234.00		
16 L.H., Daily Totals		$318.40		$526.40	$19.90	$32.90

Crew Q-6	Hr.	Daily	Hr.	Daily	Bare Costs	Incl. O&P
1 Steamfitters	$22.10	$176.80	$36.55	$292.40	$19.17	$31.68
2 Steamfitter Apprentices	17.70	283.20	29.25	468.00		
24 L.H., Daily Totals		$460.00		$760.40	$19.17	$31.68

Crew Q-7	Hr.	Daily	Hr.	Daily	Bare Costs	Incl. O&P
2 Steamfitters	$22.10	$353.60	$36.55	$584.80	$19.90	$32.90
2 Steamfitter Apprentices	17.70	283.20	29.25	468.00		
32 L.H., Daily Totals		$636.80		$1052.80	$19.90	$32.90

Crew Q-8	Hr.	Daily	Hr.	Daily	Bare Costs	Incl. O&P
2 Steamfitters	$22.10	$353.60	$36.55	$584.80	$21.00	$34.72
1 Welder (steamfitter)	22.10	176.80	36.55	292.40		
1 Steamfitter Apprentice	17.70	141.60	29.25	234.00		
1 Electric Welding Mach.		48.80		53.70	1.53	1.68
32 L.H., Daily Totals		$720.80		$1164.90	$22.53	$36.40

Crew Q-9	Hr.	Daily	Hr.	Daily	Bare Costs	Incl. O&P
1 Sheet Metal Worker	$21.50	$172.00	$36.30	$290.40	$19.35	$32.67
1 Sheet Metal Apprentice	17.20	137.60	29.05	232.40		
16 L.H., Daily Totals		$309.60		$522.80	$19.35	$32.67

Crew Q-10	Hr.	Daily	Hr.	Daily	Bare Costs	Incl. O&P
2 Sheet Metal Workers	$21.50	$344.00	$36.30	$580.80	$20.07	$33.88
1 Sheet Metal Apprentice	17.20	137.60	29.05	232.40		
24 L.H., Daily Totals		$481.60		$813.20	$20.07	$33.88

Crew Q-11	Hr.	Daily	Hr.	Daily	Bare Costs	Incl. O&P
2 Sheet Metal Workers	$21.50	$344.00	$36.30	$580.80	$19.35	$32.67
2 Sheet Metal Apprentices	17.20	275.20	29.05	464.80		
32 L.H., Daily Totals		$619.20		$1045.60	$19.35	$32.67

Crews

Crew No.	Bare Costs		Incl. Subs O & P		Cost Per Labor-Hour	
Crew Q-12	Hr.	Daily	Hr.	Daily	Bare Costs	Incl. O&P
1 Sprinkler Installer	$22.05	$176.40	$36.50	$292.00	$19.85	$32.85
1 Sprinkler Apprentice	17.65	141.20	29.20	233.60		
16 L.H., Daily Totals		$317.60		$525.60	$19.85	$32.85
Crew Q-13	Hr.	Daily	Hr.	Daily	Bare Costs	Incl. O&P
2 Sprinkler Installers	$22.05	$352.80	$36.50	$584.00	$19.85	$32.85
2 Sprinkler Apprentices	17.65	282.40	29.20	467.20		
32 L.H., Daily Totals		$635.20		$1051.20	$19.85	$32.85
Crew Q-14	Hr.	Daily	Hr.	Daily	Bare Costs	Incl. O&P
1 Asbestos Worker	$20.50	$164.00	$35.90	$287.20	$18.45	$32.30
1 Asbestos Apprentice	16.40	131.20	28.70	229.60		
16 L.H., Daily Totals		$295.20		$516.80	$18.45	$32.30
Crew Q-15	Hr.	Daily	Hr.	Daily	Bare Costs	Incl. O&P
1 Plumber	$21.95	$175.60	$36.30	$290.40	$19.75	$32.65
1 Plumber Apprentice	17.55	140.40	29.00	232.00		
1 Electric Welding Mach.		48.80		53.70	3.05	3.36
16 L.H., Daily Totals		$364.80		$576.10	$22.80	$36.01
Crew Q-16	Hr.	Daily	Hr.	Daily	Bare Costs	Incl. O&P
2 Plumbers	$21.95	$351.20	$36.30	$580.80	$20.48	$33.87
1 Plumber Apprentice	17.55	140.40	29.00	232.00		
1 Electric Welding Mach.		48.80		53.70	2.03	2.24
24 L.H., Daily Totals		$540.40		$866.50	$22.51	$36.11
Crew Q-17	Hr.	Daily	Hr.	Daily	Bare Costs	Incl. O&P
1 Steamfitter	$22.10	$176.80	$36.55	$292.40	$19.90	$32.90
1 Steamfitter Apprentice	17.70	141.60	29.25	234.00		
1 Electric Welding Mach.		48.80		53.70	3.05	3.36
16 L.H., Daily Totals		$367.20		$580.10	$22.95	$36.26
Crew Q-17A	Hr.	Daily	Hr.	Daily	Bare Costs	Incl. O&P
1 Steamfitter	$22.10	$176.80	$36.55	$292.40	$20.15	$33.35
1 Steamfitter Apprentice	17.70	141.60	29.25	234.00		
1 Equip. Oper. (crane)	20.65	165.20	34.25	274.00		
1 Truck Crane, 12 Ton		453.45		498.80		
1 Electric Welding Mach.		48.80		53.70	20.93	23.02
24 L.H., Daily Totals		$985.85		$1352.90	$41.08	$56.37
Crew Q-18	Hr.	Daily	Hr.	Daily	Bare Costs	Incl. O&P
2 Steamfitters	$22.10	$353.60	$36.55	$584.80	$20.63	$34.12
1 Steamfitter Apprentice	17.70	141.60	29.25	234.00		
1 Electric Welding Mach.		48.80		53.70	2.03	2.24
24 L.H., Daily Totals		$544.00		$872.50	$22.66	$36.36
Crew Q-19	Hr.	Daily	Hr.	Daily	Bare Costs	Incl. O&P
1 Steamfitter	$22.10	$176.80	$36.55	$292.40	$20.63	$33.98
1 Steamfitter Apprentice	17.70	141.60	29.25	234.00		
1 Electrician	22.10	176.80	36.15	289.20		
24 L.H., Daily Totals		$495.20		$815.60	$20.63	$33.98
Crew Q-20	Hr.	Daily	Hr.	Daily	Bare Costs	Incl. O&P
1 Sheet Metal Worker	$21.50	$172.00	$36.30	$290.40	$19.90	$33.37
1 Sheet Metal Apprentice	17.20	137.60	29.05	232.40		
.5 Electrician	22.10	88.40	36.15	144.60		
20 L.H., Daily Totals		$398.00		$667.40	$19.90	$33.37

Crew No.	Bare Costs		Incl. Subs O & P		Cost Per Labor-Hour	
Crew Q-21	Hr.	Daily	Hr.	Daily	Bare Costs	Incl. O&P
2 Steamfitters	$22.10	$353.60	$36.55	$584.80	$21.00	$34.63
1 Steamfitter Apprentice	17.70	141.60	29.25	234.00		
1 Electrician	22.10	176.80	36.15	289.20		
32 L.H., Daily Totals		$672.00		$1108.00	$21.00	$34.63
Crew Q-22	Hr.	Daily	Hr.	Daily	Bare Costs	Incl. O&P
1 Plumber	$21.95	$175.60	$36.30	$290.40	$19.75	$32.65
1 Plumber Apprentice	17.55	140.40	29.00	232.00		
1 Truck Crane, 12 Ton		453.45		498.80	28.34	31.17
16 L.H., Daily Totals		$769.45		$1021.20	$48.09	$63.82
Crew Q-22A	Hr.	Daily	Hr.	Daily	Bare Costs	Incl. O&P
1 Plumber	$21.95	$175.60	$36.30	$290.40	$18.65	$31.08
1 Plumber Apprentice	17.55	140.40	29.00	232.00		
1 Laborer	14.45	115.60	24.75	198.00		
1 Equip. Oper. (crane)	20.65	165.20	34.25	274.00		
1 Truck Crane, 12 Ton		453.45		498.80	14.17	15.59
32 L.H., Daily Totals		$1050.25		$1493.20	$32.82	$46.67
Crew Q-23	Hr.	Daily	Hr.	Daily	Bare Costs	Incl. O&P
1 Plumber Foreman	$23.95	$191.60	$39.60	$316.80	$21.93	$36.30
1 Plumber	21.95	175.60	36.30	290.40		
1 Equip. Oper. (medium)	19.90	159.20	33.00	264.00		
1 Power Tools		2.40		2.65		
1 Crane, 20 Ton		532.55		585.80	22.29	24.52
24 L.H., Daily Totals		$1061.35		$1459.65	$44.22	$60.82
Crew R-1	Hr.	Daily	Hr.	Daily	Bare Costs	Incl. O&P
1 Electrician Foreman	$22.60	$180.80	$36.95	$295.60	$19.75	$32.68
3 Electricians	22.10	530.40	36.15	867.60		
2 Helpers	14.80	236.80	25.35	405.60		
48 L.H., Daily Totals		$948.00		$1568.80	$19.75	$32.68
Crew R-1A	Hr.	Daily	Hr.	Daily	Bare Costs	Incl. O&P
1 Electrician	$22.10	$176.80	$36.15	$289.20	$18.45	$30.75
1 Helper	14.80	118.40	25.35	202.80		
16 L.H., Daily Totals		$295.20		$492.00	$18.45	$30.75
Crew R-2	Hr.	Daily	Hr.	Daily	Bare Costs	Incl. O&P
1 Electrician Foreman	$22.60	$180.80	$36.95	$295.60	$19.88	$32.91
3 Electricians	22.10	530.40	36.15	867.60		
2 Helpers	14.80	236.80	25.35	405.60		
1 Equip. Oper. (crane)	20.65	165.20	34.25	274.00		
1 S.P. Crane, 5 Ton		276.45		304.10	4.94	5.43
56 L.H., Daily Totals		$1389.65		$2146.90	$24.82	$38.34
Crew R-3	Hr.	Daily	Hr.	Daily	Bare Costs	Incl. O&P
1 Electrician Foreman	$22.60	$180.80	$36.95	$295.60	$22.01	$36.09
1 Electrician	22.10	176.80	36.15	289.20		
.5 Equip. Oper. (crane)	20.65	82.60	34.25	137.00		
.5 S.P. Crane, 5 Ton		138.23		152.05	6.91	7.60
20 L.H., Daily Totals		$578.43		$873.85	$28.92	$43.69
Crew R-4	Hr.	Daily	Hr.	Daily	Bare Costs	Incl. O&P
1 Struc. Steel Foreman	$23.25	$186.00	$45.85	$366.80	$21.82	$41.54
3 Struc. Steel Workers	21.25	510.00	41.90	1005.60		
1 Electrician	22.10	176.80	36.15	289.20		
1 Gas Welding Machine		84.00		92.40	2.10	2.31
40 L.H., Daily Totals		$956.80		$1754.00	$23.92	$43.85

Crews

Crew No.	Bare Costs		Incl. Subs O & P		Cost Per Labor-Hour	
Crew R-5	**Hr.**	**Daily**	**Hr.**	**Daily**	**Bare Costs**	**Incl. O&P**
1 Electrician Foreman	$22.60	$180.80	$36.95	$295.60	$19.49	$32.30
4 Electrician Linemen	22.10	707.20	36.15	1156.80		
2 Electrician Operators	22.10	353.60	36.15	578.40		
4 Electrician Groundmen	14.80	473.60	25.35	811.20		
1 Crew Truck		197.60		217.35		
1 Tool Van		214.40		235.85		
1 Pickup Truck, 3/4 Ton		130.20		143.20		
.2 Crane, 55 Ton		161.64		177.80		
.2 Crane, 12 Ton		90.69		99.75		
.2 Auger, Truck Mtd.		359.20		395.10		
1 Tractor w/Winch		298.80		328.70	16.51	18.16
88 L.H., Daily Totals		$3167.73		$4439.75	$36.00	$50.46
Crew R-6	**Hr.**	**Daily**	**Hr.**	**Daily**	**Bare Costs**	**Incl. O&P**
1 Electrician Foreman	$22.60	$180.80	$36.95	$295.60	$19.49	$32.30
4 Electrician Linemen	22.10	707.20	36.15	1156.80		
2 Electrician Operators	22.10	353.60	36.15	578.40		
4 Electrician Groundmen	14.80	473.60	25.35	811.20		
1 Crew Truck		197.60		217.35		
1 Tool Van		214.40		235.85		
1 Pickup Truck, 3/4 Ton		130.20		143.20		
.2 Crane, 55 Ton		161.64		177.80		
.2 Crane, 12 Ton		90.69		99.75		
.2 Auger, Truck Mtd.		359.20		395.10		
1 Tractor w/Winch		298.80		328.70		
3 Cable Trailers		422.40		464.65		
.5 Tensioning Rig		141.20		155.30		
.5 Cable Pulling Rig		819.50		901.45	32.22	35.45
88 L.H., Daily Totals		$4550.83		$5961.15	$51.71	$67.75
Crew R-7	**Hr.**	**Daily**	**Hr.**	**Daily**	**Bare Costs**	**Incl. O&P**
1 Electrician Foreman	$22.60	$180.80	$36.95	$295.60	$16.10	$27.28
5 Electrician Groundmen	14.80	592.00	25.35	1014.00		
1 Crew Truck		197.60		217.35	4.12	4.53
48 L.H., Daily Totals		$970.40		$1526.95	$20.22	$31.81
Crew R-8	**Hr.**	**Daily**	**Hr.**	**Daily**	**Bare Costs**	**Incl. O&P**
1 Electrician Foreman	$22.60	$180.80	$36.95	$295.60	$19.75	$32.68
3 Electrician Linemen	22.10	530.40	36.15	867.60		
2 Electrician Groundmen	14.80	236.80	25.35	405.60		
1 Pickup Truck, 3/4 Ton		130.20		143.20		
1 Crew Truck		197.60		217.35	6.83	7.51
48 L.H., Daily Totals		$1275.80		$1929.35	$26.58	$40.19
Crew R-9	**Hr.**	**Daily**	**Hr.**	**Daily**	**Bare Costs**	**Incl. O&P**
1 Electrician Foreman	$22.60	$180.80	$36.95	$295.60	$18.51	$30.85
1 Electrician Lineman	22.10	176.80	36.15	289.20		
2 Electrician Operators	22.10	353.60	36.15	578.40		
4 Electrician Groundmen	14.80	473.60	25.35	811.20		
1 Pickup Truck, 3/4 Ton		130.20		143.20		
1 Crew Truck		197.60		217.35	5.12	5.63
64 L.H., Daily Totals		$1512.60		$2334.95	$23.63	$36.48
Crew R-10	**Hr.**	**Daily**	**Hr.**	**Daily**	**Bare Costs**	**Incl. O&P**
1 Electrician Foreman	$22.60	$180.80	$36.95	$295.60	$20.97	$34.48
4 Electrician Linemen	22.10	707.20	36.15	1156.80		
1 Electrician Groundman	14.80	118.40	25.35	202.80		
1 Crew Truck		197.60		217.35		
3 Tram Cars		541.20		595.30	15.39	16.93
48 L.H., Daily Totals		$1745.20		$2467.85	$36.36	$51.41

Crew No.	Bare Costs		Incl. Subs O & P		Cost Per Labor-Hour	
Crew R-11	**Hr.**	**Daily**	**Hr.**	**Daily**	**Bare Costs**	**Incl. O&P**
1 Electrician Foreman	$22.60	$180.80	$36.95	$295.60	$20.04	$33.09
4 Electricians	22.10	707.20	36.15	1156.80		
1 Helper	14.80	118.40	25.35	202.80		
1 Common Laborer	14.45	115.60	24.75	198.00		
1 Crew Truck		197.60		217.35		
1 Crane, 12 Ton		453.45		498.80	11.63	12.79
56 L.H., Daily Totals		$1773.05		$2569.35	$31.67	$45.88
Crew R-12	**Hr.**	**Daily**	**Hr.**	**Daily**	**Bare Costs**	**Incl. O&P**
1 Carpenter Foreman	$20.20	$161.60	$34.60	$276.80	$18.00	$31.23
4 Carpenters	19.70	630.40	33.75	1080.00		
4 Common Laborers	14.45	462.40	24.75	792.00		
1 Equip. Oper. (med.)	19.90	159.20	33.00	264.00		
1 Steel Worker	21.25	170.00	41.90	335.20		
1 Dozer, 200 H.P.		806.80		887.50		
1 Pickup Truck, 3/4 Ton		130.20		143.20	10.65	11.71
88 L.H., Daily Totals		$2520.60		$3778.70	$28.65	$42.94
Crew R-15	**Hr.**	**Daily**	**Hr.**	**Daily**	**Bare Costs**	**Incl. O&P**
1 Electrician Foreman	$22.60	$180.80	$36.95	$295.60	$21.68	$35.54
4 Electricians	22.10	707.20	36.15	1156.80		
1 Equipment Operator	19.10	152.80	31.70	253.60		
1 Aerial Lift Truck		251.35		276.50	5.24	5.76
48 L.H., Daily Totals		$1292.15		$1982.50	$26.92	$41.30
Crew R-18	**Hr.**	**Daily**	**Hr.**	**Daily**	**Bare Costs**	**Incl. O&P**
.25 Electrician Foreman	$22.60	$45.20	$36.95	$73.90	$17.65	$29.57
1 Electrician	22.10	176.80	36.15	289.20		
2 Helpers	14.80	236.80	25.35	405.60		
26 L.H., Daily Totals		$458.80		$768.70	$17.65	$29.57
Crew R-19	**Hr.**	**Daily**	**Hr.**	**Daily**	**Bare Costs**	**Incl. O&P**
.5 Electrician Foreman	$22.60	$90.40	$36.95	$147.80	$22.20	$36.31
2 Electricians	22.10	353.60	36.15	578.40		
20 L.H., Daily Totals		$444.00		$726.20	$22.20	$36.31
Crew R-21	**Hr.**	**Daily**	**Hr.**	**Daily**	**Bare Costs**	**Incl. O&P**
1 Electrician Foreman	$22.60	$180.80	$36.95	$295.60	$22.17	$36.27
3 Electricians	22.10	530.40	36.15	867.60		
.1 Equip. Oper. (med.)	19.90	15.92	33.00	26.40		
.1 Hyd. Crane 25 Ton		56.39		62.05	1.72	1.89
32. L.H., Daily Totals		$783.51		$1251.65	$23.89	$38.16
Crew R-22	**Hr.**	**Daily**	**Hr.**	**Daily**	**Bare Costs**	**Incl. O&P**
.66 Electrician Foreman	$22.60	$119.33	$36.95	$195.10	$19.04	$31.63
2 Helpers	14.80	236.80	25.35	405.60		
2 Electricians	22.10	353.60	36.15	578.40		
37.28 L.H., Daily Totals		$709.73		$1179.10	$19.04	$31.63
Crew R-30	**Hr.**	**Daily**	**Hr.**	**Daily**	**Bare Costs**	**Incl. O&P**
.25 Electricians	$24.10	$48.20	$39.40	$78.80	$17.55	$29.38
1 Electricians	22.10	176.80	36.15	289.20		
2 Laborers, (Semi-Skilled)	14.45	231.20	24.75	396.00		
26 L.H., Daily Totals		$456.20		$764.00	$17.55	$29.38

Historical Cost Indexes

The table below lists both the Means Historical Cost Index based on Jan. 1, 1993 = 100 as well as the computed value of an index based on Jan. 1, 2000 costs. Since the Jan. 1, 2000 figure is estimated, space is left to write in the actual index figures as they become available through either the quarterly "Means Construction Cost Indexes" or as printed in the "Engineering News-Record." To compute the actual index based on Jan. 1, 2000 = 100, divide the Historical Cost Index for a particular year by the actual Jan. 1, 2000 Construction Cost Index. Space has been left to advance the index figures as the year progresses.

Year	Historical Cost Index Jan. 1, 1993 = 100		Current Index Based on Jan. 1, 2000 = 100		Year	Historical Cost Index Jan. 1, 1993 = 100	Current Index Based on Jan. 1, 2000 = 100		Year	Historical Cost Index Jan. 1, 1993 = 100	Current Index Based on Jan. 1, 2000 = 100	
	Est.	Actual	Est.	Actual		Actual	Est.	Actual		Actual	Est.	Actual
Oct 2000					July 1985	82.6	69.1		July 1967	23.5	19.6	
July 2000					1984	82.0	68.5		1966	22.7	19.0	
April 2000					1983	80.2	67.0		1965	21.7	18.1	
Jan 2000	119.6		100.0	100.0	1982	76.1	63.7		1964	21.2	17.7	
July 1999		117.6	98.3		1981	70.0	58.5		1963	20.7	17.3	
1998		115.1	96.2		1980	62.9	52.6		1962	20.2	16.9	
1997		112.8	94.3		1979	57.8	48.3		1961	19.8	16.6	
1996		110.2	92.1		1978	53.5	44.7		1960	19.7	16.5	
1995		107.6	90.0		1977	49.5	41.4		1959	19.3	16.1	
1994		104.4	87.3		1976	46.9	39.2		1958	18.8	15.7	
1993		101.7	85.0		1975	44.8	37.5		1957	18.4	15.4	
1992		99.4	83.1		1974	41.4	34.6		1956	17.6	14.7	
1991		96.8	80.9		1973	37.7	31.5		1955	16.6	13.9	
1990		94.3	78.9		1972	34.8	29.1		1954	16.0	13.4	
1989		92.1	77.0		1971	32.1	26.8		1953	15.8	13.2	
1988		89.9	75.1		1970	28.7	24.0		1952	15.4	12.9	
1987		87.7	73.3		1969	26.9	22.5		1951	15.0	12.5	
1986		84.2	70.4		1968	24.9	20.8		1950	13.7	11.5	

Adjustments to Costs

The Historical Cost Index can be used to convert National Average building costs at a particular time to the approximate building costs for some other time.

Example:

Estimate and compare construction costs for different years in the same city.

To estimate the National Average construction cost of a building in 1970, knowing that it cost $900,000 in 2000:

INDEX in 1970 = 28.7
INDEX in 2000 = 119.6

Note: The City Cost Indexes for Canada can be used to convert U.S. National averages to local costs in Canadian dollars.

Time Adjustment using the Historical Cost Indexes:

$$\frac{\text{Index for Year A}}{\text{Index for Year B}} \times \text{Cost in Year B} = \text{Cost in Year A}$$

$$\frac{\text{INDEX 1970}}{\text{INDEX 2000}} \times \text{Cost 2000} = \text{Cost 1970}$$

$$\frac{28.7}{119.6} \times \$900{,}000 = .240 \times \$900{,}000 = \$216{,}000$$

The construction cost of the building in 1970 is $216,000.

How to Use the City Cost Indexes

What you should know before you begin

Means City Cost Indexes (CCI) are an extremely useful tool to use when you want to compare costs from city to city and region to region.

This publication contains average construction cost indexes for 718 U.S. and Canadian cities covering over 930 three-digit zip code locations, as listed directly under each city.

Keep in mind that a City Cost Index number is a *percentage ratio* of a specific city's cost to the national average cost of the same item at a stated time period.

In other words, these index figures represent relative construction *factors* (or, if you prefer, multipliers) for Material and Installation costs, as well as the weighted average for Total In Place costs for each CSI MasterFormat division. Installation costs include both labor and equipment rental costs.

The 30 City Average Index is the average of 30 major U.S. cities and serves as a National Average.

Index figures for both material and installation are based on the 30 major city average of 100 and represent the cost relationship as of July 1, 1999. The index for each division is computed from representative material and labor quantities for that division. The weighted average for each city is a weighted total of the components listed above it, but does not include relative productivity between trades or cities.

As changes occur in local material prices, labor rates and equipment rental rates, the impact of these changes should be accurately measured by the change in the City Cost Index for each particular city (as compared to the 30 City Average).

> Therefore, if you know (or have estimated) building costs in one city today, you can easily convert those costs to expected building costs in another city.
>
> In addition, by using the Historical Cost Index, you can easily convert National Average building costs at a particular time to the approximate building costs for some other time. The City Cost Indexes can then be applied to calculate the costs for a particular city.

Quick Calculations

Location Adjustment Using the City Cost Indexes:

$$\frac{\text{Index for City A}}{\text{Index for City B}} \times \text{Cost in City B} = \text{Cost in City A}$$

Time Adjustment for the National Average Using the Historical Cost Index:

$$\frac{\text{Index for Year A}}{\text{Index for Year B}} \times \text{Cost in Year B} = \text{Cost in Year A}$$

Adjustment from the National Average:

$$\frac{\text{Index for City A}}{100} \times \text{National Average Cost} = \text{Cost in City A}$$

Since each of the other R.S. Means publications contains many different items, any *one* item multiplied by the particular city index may give incorrect results. However, the larger the number of items compiled, the closer the results should be to actual costs for that particular city.

The City Cost Indexes for Canadian cities are calculated using Canadian material and equipment prices and labor rates, in Canadian dollars. Therefore, indexes for Canadian cities can be used to convert U.S. National Average prices to local costs in Canadian dollars.

How to use this section

1. Compare costs from city to city.

In using the Means Indexes, remember that an index number is not a fixed number but a *ratio:* It's a percentage ratio of a building component's cost at any stated time to the National Average cost of that same component at the same time period. Put in the form of an equation:

$$\frac{\text{Specific City Cost}}{\text{National Average Cost}} \times 100 = \text{City Index Number}$$

Therefore, when making cost comparisons between cities, do not subtract one city's index number from the index number of another city and read the result as a percentage difference. Instead, divide one city's index number by that of the other city. The resulting number may then be used as a multiplier to calculate cost differences from city to city.

The formula used to find cost differences between cities for the purpose of comparison is as follows:

$$\frac{\text{City A Index}}{\text{City B Index}} \times \text{City B Cost (Known)} = \text{City A Cost (Unknown)}$$

In addition, you can use *Means CCI* to calculate and compare costs division by division between cities using the same basic formula. (Just be sure that you're comparing similar divisions.)

2. Compare a specific city's construction costs with the National Average.

When you're studying construction location feasibility, it's advisable to compare a prospective project's cost index with an index of the National Average cost.

For example, divide the weighted average index of construction costs of a specific city by that of the 30 City Average, which = 100.

$$\frac{\text{City Index}}{100} = \% \text{ of National Average}$$

As a result, you get a ratio that indicates the relative cost of construction in that city in comparison with the National Average.

3. Covert U.S. National Average to actual costs in Canadian City.

$$\frac{\text{Index for Canadian City}}{100} \times \text{National Average Cost} = \text{Cost in Canadian City in \$ CAN}$$

4. Adjust construction cost data based on a National Average.

When you use a source of construction cost data which is based on a National Average (such as *Means cost data publications*), it is necessary to adjust those costs to a specific location.

$$\frac{\text{City Index}}{100} \times \text{"Book" Cost Based on National Average Costs} = \text{City Cost (Unknown)}$$

5. When applying the City Cost Indexes to demolition projects, use the appropriate division installation index. For example, for removal of existing doors and windows, use the Division 8 index.

What you might like to know about how we developed the Indexes

To create a reliable index, R.S. Means researched the building type most often constructed in the United States and Canada. Because it was concluded that no one type of building completely represented the building construction industry, nine different types of buildings were combined to create a composite model.

The exact material, labor and equipment quantities are based on detailed analysis of these nine building types, then each quantity is weighted in proportion to expected usage. These various material items, labor hours, and equipment rental rates are thus combined to form a composite building representing as closely as possible the actual usage of materials, labor and equipment used in the North American Building Construction Industry.

The following structures were chosen to make up that composite model:

1. Factory, 1 story
2. Office, 2–4 story
3. Store, Retail
4. Town Hall, 2–3 story
5. High School, 2–3 story
6. Hospital, 4–8 story
7. Garage, Parking
8. Apartment, 1–3 story
9. Hotel/Motel, 2–3 story

For the purposes of ensuring the timeliness of the data, the components of the index for the composite model have been streamlined. They currently consist of:

- specific quantities of 66 commonly used construction materials;
- specific labor-hours for 21 building construction trades; and
- specific days of equipment rental for 6 types of construction equipment (normally used to install the 66 material items by the 21 trades.)

A sophisticated computer program handles the updating of all costs for each city on a quarterly basis. Material and equipment price quotations are gathered quarterly from 718 cities in the United States and Canada. These prices and the latest negotiated labor wage rates for 21 different building trades are used to compile the quarterly update of the City Cost Index.

The 30 major U.S. cities used to calculate the National Average are:

Atlanta, GA
Baltimore, MD
Boston, MA
Buffalo, NY
Chicago, IL
Cincinnati, OH
Cleveland, OH
Columbus, OH
Dallas, TX
Denver, CO
Detroit, MI
Houston, TX
Indianapolis, IN
Kansas City, MO
Los Angeles, CA
Memphis, TN
Milwaukee, WI
Minneapolis, MN
Nashville, TN
New Orleans, LA
New York, NY
Philadelphia, PA
Phoenix, AZ
Pittsburgh, PA
St. Louis, MO
San Antonio, TX
San Diego, CA
San Francisco, CA
Seattle, WA
Washington, DC

F.Y.I.: The CSI MasterFormat Divisions

1. General Requirements
2. Site Work
3. Concrete
4. Masonry
5. Metals
6. Wood & Plastics
7. Thermal & Moisture Protection
8. Doors & Windows
9. Finishes
10. Specialties
11. Equipment
12. Furnishings
13. Special Construction
14. Conveying Systems
15. Mechanical
16. Electrical

The information presented in the CCI is organized according to the Construction Specifications Institute (CSI) MasterFormat.

What the CCI does not indicate

The weighted average for each city is a total of the components listed above weighted to reflect typical usage, but it does *not* include the productivity variations between trades or cities.

In addition, the CCI does not take into consideration factors such as the following:

- managerial efficiency
- competitive conditions
- automation
- restrictive union practices
- unique local requirements
- regional variations due to specific building codes

City Cost Indexes

	DIVISION	UNITED STATES 30 CITY AVERAGE			ALABAMA ANNISTON 362			BIRMINGHAM 350 - 352			BUTLER 369			DECATUR 356			DOTHAN 363		
		MAT.	INST.	TOTAL	MAT.	INST.	TOTAL	MAT.	INST.	TOTAL	MAT.	INST.	TOTAL	MAT.	INST.	TOTAL	MAT.	INST.	TOTAL
02	SITE CONSTRUCTION	100.0	100.0	100.0	95.7	92.5	93.3	89.0	94.2	92.9	110.5	86.8	92.3	88.4	93.0	91.9	107.8	86.8	91.7
03100	CONCRETE FORMS & ACCESSORIES	100.0	100.0	100.0	96.5	42.8	50.4	95.2	78.9	81.2	85.6	56.6	60.9	96.7	60.5	65.6	98.4	56.8	62.7
03200	CONCRETE REINFORCEMENT	100.0	100.0	100.0	92.7	74.1	82.2	92.7	91.5	92.0	103.5	74.0	86.8	92.7	70.8	80.3	103.5	74.0	86.8
03300	CAST-IN-PLACE CONCRETE	100.0	100.0	100.0	93.6	47.5	74.3	95.8	68.6	84.5	91.3	57.2	77.1	93.1	66.2	81.8	91.2	57.2	77.1
03	CONCRETE	100.0	100.0	100.0	93.7	52.7	73.0	90.4	78.7	84.5	94.8	61.8	78.2	89.2	66.0	77.5	94.4	61.8	78.0
04	MASONRY	100.0	100.0	100.0	83.0	41.8	57.4	84.1	72.9	77.2	86.4	43.7	59.8	84.2	62.0	70.3	87.4	43.7	60.2
05	METALS	100.0	100.0	100.0	97.0	93.1	95.6	99.4	96.2	98.2	95.9	86.7	92.6	99.0	88.1	95.1	96.0	86.8	92.7
06	WOOD & PLASTICS	100.0	100.0	100.0	88.4	43.0	64.9	95.9	79.8	87.6	82.6	59.4	70.6	94.9	60.8	77.0	96.2	59.4	77.2
07	THERMAL & MOISTURE PROTECTION	100.0	100.0	100.0	97.0	45.0	72.6	97.2	69.4	84.2	97.1	54.1	77.0	96.9	60.8	80.0	97.2	54.1	77.0
08	DOORS & WINDOWS	100.0	100.0	100.0	93.6	51.0	83.2	97.3	78.2	92.7	93.6	60.1	85.5	97.3	58.5	87.9	93.7	60.1	85.5
09200	PLASTER & GYPSUM BOARD	100.0	100.0	100.0	105.1	41.7	65.9	111.3	79.7	91.7	104.2	58.6	76.0	108.3	59.5	78.0	108.8	58.6	77.7
09500	CEILINGS	100.0	100.0	100.0	92.1	41.7	59.0	96.3	79.7	85.4	92.1	58.6	70.1	96.3	59.5	72.1	92.1	58.6	70.1
09600	FLOORING	100.0	100.0	100.0	97.4	42.9	84.1	100.4	60.5	90.7	103.6	33.9	86.6	100.4	63.1	91.3	111.3	41.0	94.2
09900	PAINTS & COATINGS	100.0	100.0	100.0	93.7	38.0	61.2	93.7	57.8	72.7	93.7	64.5	76.6	93.7	60.3	74.2	93.7	64.5	76.6
09	FINISHES	100.0	100.0	100.0	96.9	42.1	68.9	99.1	73.4	86.0	100.3	53.3	76.3	98.6	60.8	79.3	103.3	54.5	78.4
10 - 14	TOTAL DIV. 10000 - 14000	100.0	100.0	100.0	100.0	62.9	92.2	100.0	85.9	97.0	100.0	66.8	93.0	100.0	69.5	93.5	100.0	66.8	93.0
15	MECHANICAL	100.0	100.0	100.0	98.9	46.8	75.3	99.9	72.5	87.5	98.2	50.4	76.5	99.9	61.2	82.4	98.2	50.4	76.5
16	ELECTRICAL	100.0	100.0	100.0	93.2	33.0	52.2	95.4	68.8	77.3	96.4	46.5	62.4	95.8	61.2	72.2	94.5	46.5	61.8
01 - 16	WEIGHTED AVERAGE	100.0	100.0	100.0	95.7	53.5	75.2	96.7	77.7	87.5	96.6	58.9	78.3	96.4	67.5	82.4	96.9	59.0	78.6

	DIVISION	ALABAMA EVERGREEN 364			GADSDEN 359			HUNTSVILLE 357 - 358			JASPER 355			MOBILE 365 - 366			MONTGOMERY 360 - 361		
		MAT.	INST.	TOTAL	MAT.	INST.	TOTAL	MAT.	INST.	TOTAL	MAT.	INST.	TOTAL	MAT.	INST.	TOTAL	MAT.	INST.	TOTAL
02	SITE CONSTRUCTION	111.1	87.1	92.7	94.4	93.3	93.6	88.1	93.4	92.2	94.0	92.5	92.8	99.5	87.3	90.1	99.9	87.6	90.5
03100	CONCRETE FORMS & ACCESSORIES	80.8	58.9	62.0	91.7	54.3	59.9	96.7	60.3	65.5	99.0	44.5	52.2	96.7	63.9	68.5	94.9	57.4	62.7
03200	CONCRETE REINFORCEMENT	104.1	74.0	87.0	101.5	90.9	95.5	92.7	89.0	90.6	92.7	74.2	82.2	95.8	69.1	80.7	95.8	90.5	92.8
03300	CAST-IN-PLACE CONCRETE	91.3	60.3	78.3	93.1	60.7	79.6	90.6	66.3	80.5	103.1	53.9	82.6	95.7	64.3	82.6	97.4	57.7	80.8
03	CONCRETE	94.9	63.7	79.2	93.7	65.3	79.4	88.0	69.3	78.6	97.1	55.3	76.1	90.9	66.5	78.6	91.6	65.3	78.4
04	MASONRY	86.4	49.2	63.2	82.3	51.9	63.3	84.0	62.0	70.3	79.9	45.2	58.3	84.7	62.4	70.8	85.1	43.4	59.1
05	METALS	96.0	86.7	92.6	97.1	94.9	96.3	99.0	94.4	97.4	96.9	87.1	93.4	97.6	87.2	93.9	97.1	94.0	96.4
06	WOOD & PLASTICS	77.6	59.4	68.2	83.2	54.2	68.2	94.9	59.4	76.5	90.9	44.0	66.7	94.9	64.8	79.4	92.6	59.4	75.4
07	THERMAL & MOISTURE PROTECTION	97.0	56.5	78.0	96.9	59.2	79.3	96.9	63.1	81.1	97.0	50.1	75.0	96.9	64.3	81.7	96.7	56.9	78.0
08	DOORS & WINDOWS	93.7	60.1	85.5	93.6	58.3	85.0	97.3	62.1	88.8	93.6	52.6	83.6	97.3	62.5	88.8	97.3	63.7	89.2
09200	PLASTER & GYPSUM BOARD	102.4	58.6	75.3	103.3	53.2	72.3	108.3	58.6	77.5	105.3	42.8	66.6	108.3	64.2	81.0	108.3	58.6	77.5
09500	CEILINGS	92.1	58.6	70.1	92.1	53.2	66.6	96.3	58.6	71.6	90.8	42.8	59.3	96.3	64.2	75.2	96.3	58.6	71.6
09600	FLOORING	100.9	33.9	84.6	95.4	54.1	85.3	100.4	52.3	88.7	98.5	41.3	84.5	108.8	65.6	98.2	108.8	33.9	90.5
09900	PAINTS & COATINGS	93.7	64.5	76.6	93.7	65.5	77.2	93.7	56.7	72.1	93.7	46.6	66.2	93.7	65.5	77.3	93.7	64.5	76.6
09	FINISHES	99.2	54.7	76.4	95.8	54.9	74.9	98.6	57.7	77.7	96.8	43.9	69.8	102.4	64.3	83.0	102.5	53.2	77.3
10 - 14	TOTAL DIV. 10000 - 14000	100.0	69.1	93.5	100.0	77.4	95.2	100.0	80.0	95.8	100.0	63.9	92.4	100.0	77.9	95.3	100.0	77.2	95.2
15	MECHANICAL	98.2	53.4	77.9	106.3	53.6	82.4	99.9	56.2	80.1	106.3	40.0	76.3	99.9	66.9	85.0	99.9	50.6	77.6
16	ELECTRICAL	92.2	46.5	61.0	94.9	68.8	77.1	95.8	75.3	81.8	95.2	33.4	53.1	95.8	63.6	73.9	95.4	46.5	62.1
01 - 16	WEIGHTED AVERAGE	96.2	60.6	78.9	97.2	66.3	82.2	96.3	70.1	83.6	97.6	52.8	75.9	97.2	69.4	83.7	97.2	60.6	79.5

	DIVISION	ALABAMA PHENIX CITY 368			SELMA 367			TUSCALOOSA 354			ALASKA ANCHORAGE 995 - 996			FAIRBANKS 997			JUNEAU 998		
		MAT.	INST.	TOTAL	MAT.	INST.	TOTAL	MAT.	INST.	TOTAL	MAT.	INST.	TOTAL	MAT.	INST.	TOTAL	MAT.	INST.	TOTAL
02	SITE CONSTRUCTION	115.3	87.1	94.1	107.6	86.8	91.6	88.6	93.1	92.0	128.6	135.7	134.1	114.3	135.7	130.7	125.6	135.7	133.4
03100	CONCRETE FORMS & ACCESSORIES	90.3	50.6	56.3	87.1	56.8	61.1	96.7	47.2	54.2	130.6	116.6	118.5	132.1	122.5	123.8	131.8	116.6	118.7
03200	CONCRETE REINFORCEMENT	102.9	97.3	99.7	103.5	74.0	86.8	92.7	90.9	91.7	142.4	108.3	123.1	119.7	108.4	113.3	105.9	108.3	107.3
03300	CAST-IN-PLACE CONCRETE	91.3	58.4	77.6	91.2	57.2	77.1	94.4	55.0	78.0	180.2	117.4	154.0	152.6	118.0	138.1	181.0	117.4	154.5
03	CONCRETE	97.5	63.9	80.6	93.5	61.8	77.5	89.8	60.0	74.8	151.2	114.5	132.7	129.3	117.3	123.3	146.9	114.5	130.6
04	MASONRY	86.4	49.5	63.4	91.4	43.7	61.7	84.3	45.8	60.3	195.4	122.0	149.7	183.0	122.0	145.0	184.8	122.0	145.7
05	METALS	95.8	96.5	96.1	95.8	86.7	92.6	98.1	94.8	96.9	129.9	99.9	119.1	130.0	100.2	119.3	130.2	99.9	119.3
06	WOOD & PLASTICS	87.8	49.5	68.0	84.3	59.4	71.4	94.9	47.1	70.2	122.9	114.4	118.5	123.2	121.1	122.6	122.9	114.4	118.5
07	THERMAL & MOISTURE PROTECTION	97.5	57.7	78.9	96.9	54.1	76.9	96.9	55.4	77.4	200.6	115.3	160.7	196.8	118.0	159.9	197.4	115.3	159.0
08	DOORS & WINDOWS	93.6	59.8	85.4	93.6	60.1	85.5	97.3	60.5	88.4	132.1	108.6	126.4	129.1	113.1	125.3	129.2	108.6	124.1
09200	PLASTER & GYPSUM BOARD	106.1	48.4	70.4	104.7	58.6	76.2	108.3	46.0	69.7	134.2	114.8	122.2	134.2	122.8	127.1	134.2	114.8	122.2
09500	CEILINGS	92.1	48.4	63.4	92.1	58.6	70.1	96.3	46.0	63.3	129.2	114.8	119.8	129.2	122.8	125.0	129.2	114.8	119.8
09600	FLOORING	106.5	33.9	88.8	104.5	33.9	87.3	100.4	49.7	88.0	128.5	126.2	127.9	128.6	126.2	128.0	128.5	126.2	127.9
09900	PAINTS & COATINGS	93.7	64.5	76.6	93.7	64.5	76.6	93.7	55.5	71.4	120.4	114.3	116.9	120.4	119.9	120.1	120.4	114.3	116.9
09	FINISHES	101.8	47.7	74.2	100.4	53.3	76.3	98.6	47.4	72.5	143.3	118.3	130.5	141.3	123.4	132.2	142.0	118.3	129.9
10 - 14	TOTAL DIV. 10000 - 14000	100.0	76.5	95.0	100.0	66.8	93.0	100.0	74.1	94.5	100.0	116.1	103.4	100.0	117.2	103.6	100.0	116.1	103.4
15	MECHANICAL	98.2	51.3	76.9	98.2	50.4	76.5	99.9	44.4	74.8	106.6	109.6	108.0	106.6	118.5	112.0	106.6	110.6	108.4
16	ELECTRICAL	95.5	46.5	62.1	94.0	46.5	61.6	95.8	68.8	77.4	161.0	114.7	129.5	164.0	114.7	130.4	164.0	114.7	130.4
01 - 16	WEIGHTED AVERAGE	97.3	60.4	79.4	96.5	58.9	78.3	96.4	62.2	79.8	134.4	115.4	125.2	130.2	118.5	124.5	132.9	115.6	124.5

City Cost Indexes

DIVISION		ALASKA KETCHIKAN 999			ARIZONA CHAMBERS 865			ARIZONA FLAGSTAFF 860			ARIZONA GLOBE 855			ARIZONA KINGMAN 864			ARIZONA MESA/TEMPE 852		
		MAT.	INST.	TOTAL	MAT.	INST.	TOTAL	MAT.	INST.	TOTAL	MAT.	INST.	TOTAL	MAT.	INST.	TOTAL	MAT.	INST.	TOTAL
02	SITE CONSTRUCTION	178.2	135.6	145.6	69.3	99.7	92.6	89.6	99.9	97.5	90.6	99.9	97.7	69.2	99.7	92.6	81.1	99.5	95.2
03100	CONCRETE FORMS & ACCESSORIES	122.6	116.4	117.3	100.0	71.6	75.6	107.3	77.3	81.5	97.6	71.5	75.2	97.1	71.6	75.2	101.7	72.8	76.9
03200	CONCRETE REINFORCEMENT	110.4	108.3	109.2	104.5	69.0	84.4	103.7	79.5	90.0	105.6	74.6	88.0	105.3	74.9	88.1	106.4	75.0	88.6
03300	CAST-IN-PLACE CONCRETE	291.8	117.3	219.0	96.4	75.9	87.8	96.5	78.2	88.9	98.3	75.7	88.9	96.1	75.9	87.7	99.1	70.0	86.9
03	CONCRETE	222.8	114.4	168.3	101.6	72.3	86.9	122.9	77.7	100.1	110.4	73.4	91.8	101.2	73.5	87.3	102.2	72.1	87.1
04	MASONRY	193.8	122.0	149.1	102.1	66.9	80.2	101.6	68.0	80.6	108.5	66.8	82.5	102.1	73.2	84.1	108.6	58.5	77.3
05	METALS	130.0	99.6	119.1	97.9	67.4	87.0	98.4	72.4	89.1	99.8	71.7	89.8	98.5	71.5	88.8	100.1	73.0	90.4
06	WOOD & PLASTICS	113.2	114.4	113.8	101.9	70.9	85.9	109.5	78.2	93.3	96.7	71.0	83.4	98.4	70.9	84.2	100.9	79.8	90.0
07	THERMAL & MOISTURE PROTECTION	201.2	115.3	161.0	109.0	75.1	93.1	111.2	76.4	94.9	107.8	72.2	91.1	108.9	74.6	92.8	106.9	71.0	90.1
08	DOORS & WINDOWS	129.2	108.6	124.1	102.1	68.8	94.0	102.2	75.6	95.7	100.8	71.4	93.6	102.3	71.3	94.7	100.9	76.0	94.8
09200	PLASTER & GYPSUM BOARD	131.4	114.8	121.2	92.2	70.1	78.5	94.8	77.5	84.1	94.2	70.1	79.3	90.7	70.1	77.9	95.6	79.1	85.4
09500	CEILINGS	129.2	114.8	119.8	97.7	70.1	79.5	99.0	77.5	84.9	103.4	70.1	81.5	99.0	70.1	80.0	103.4	79.1	87.4
09600	FLOORING	125.0	126.2	125.3	97.8	63.6	89.5	100.4	63.9	91.5	101.5	63.6	92.3	96.8	72.5	90.9	103.0	72.5	95.5
09900	PAINTS & COATINGS	120.4	108.7	113.6	101.6	61.2	78.0	101.6	61.2	78.0	112.6	61.2	82.6	101.6	61.2	78.0	112.6	65.7	85.2
09	FINISHES	143.2	117.7	130.2	95.5	68.9	81.9	98.5	73.2	85.6	99.9	69.0	84.1	95.2	70.4	82.5	99.8	72.3	85.8
10 - 14	TOTAL DIV. 10000 - 14000	100.0	116.1	103.4	100.0	79.9	95.7	100.0	80.9	96.0	100.0	80.2	95.8	100.0	79.9	95.7	100.0	75.3	94.8
15	MECHANICAL	105.4	110.6	107.7	97.7	81.6	90.4	100.0	82.4	92.0	96.6	72.8	85.8	97.7	77.5	88.6	99.8	69.5	86.1
16	ELECTRICAL	164.0	114.7	130.4	103.1	58.0	72.4	101.5	60.3	73.4	104.8	58.0	72.9	103.1	58.0	72.4	99.5	40.7	59.4
01 - 16	WEIGHTED AVERAGE	144.0	115.4	130.1	98.9	73.2	86.9	103.0	75.9	89.8	101.3	72.1	87.1	98.9	73.8	86.7	100.4	68.2	84.8

DIVISION		ARIZONA PHOENIX 850,853			ARIZONA PRESCOTT 863			ARIZONA SHOW LOW 859			ARIZONA TUCSON 856 - 857			ARKANSAS BATESVILLE 725			ARKANSAS CAMDEN 717		
		MAT.	INST.	TOTAL	MAT.	INST.	TOTAL	MAT.	INST.	TOTAL	MAT.	INST.	TOTAL	MAT.	INST.	TOTAL	MAT.	INST.	TOTAL
02	SITE CONSTRUCTION	81.5	100.3	96.0	75.6	99.7	94.1	92.7	99.9	98.2	77.3	100.1	94.8	73.8	83.0	80.9	74.7	82.6	80.7
03100	CONCRETE FORMS & ACCESSORIES	103.0	78.7	82.2	101.8	71.4	75.7	107.3	71.7	76.7	102.5	78.4	81.8	79.2	53.4	57.0	78.8	37.5	43.4
03200	CONCRETE REINFORCEMENT	104.5	79.9	90.5	103.7	68.9	84.0	106.4	69.0	85.2	103.4	79.5	89.9	94.6	54.8	72.0	95.9	30.2	58.6
03300	CAST-IN-PLACE CONCRETE	99.2	78.2	90.4	96.4	75.8	87.8	98.3	75.8	88.9	91.8	78.0	86.1	82.6	52.4	70.0	84.6	38.6	65.4
03	CONCRETE	101.8	78.4	90.1	107.2	72.4	89.7	112.8	72.4	92.4	98.1	78.1	88.0	80.0	54.0	67.0	81.8	37.7	59.7
04	MASONRY	95.5	73.2	81.6	102.1	66.9	80.1	108.5	66.8	82.5	95.6	67.9	78.4	97.5	51.2	68.7	108.5	44.2	68.5
05	METALS	101.4	74.5	91.8	98.4	70.9	88.5	99.6	68.2	88.4	100.7	73.1	90.8	94.9	65.0	84.2	95.0	55.1	80.7
06	WOOD & PLASTICS	102.3	79.8	90.7	103.8	70.9	86.8	106.9	71.0	88.4	102.3	79.8	90.7	78.2	55.0	66.2	77.9	38.8	57.7
07	THERMAL & MOISTURE PROTECTION	106.8	77.1	92.9	109.6	73.0	92.4	108.1	74.3	92.3	107.5	73.1	91.4	95.0	53.7	75.7	94.7	40.5	69.3
08	DOORS & WINDOWS	100.0	76.5	94.3	102.2	71.3	94.7	99.8	68.8	92.3	98.7	76.5	93.3	97.4	50.8	86.1	93.6	33.6	79.0
09200	PLASTER & GYPSUM BOARD	96.1	79.1	85.6	93.0	70.1	78.8	97.2	70.1	80.4	96.3	79.1	85.7	88.9	54.6	67.6	88.9	37.9	57.3
09500	CEILINGS	103.4	79.1	87.4	99.0	70.1	80.0	102.0	70.1	81.0	104.7	79.1	87.9	98.3	54.6	69.6	98.3	37.9	58.7
09600	FLOORING	103.3	72.5	95.8	98.7	63.6	90.1	104.8	63.6	94.8	102.4	63.9	93.0	109.7	60.5	97.7	109.6	20.0	87.8
09900	PAINTS & COATINGS	112.6	62.3	83.2	101.6	61.2	78.0	112.6	61.2	82.6	109.9	61.2	81.5	95.9	53.6	71.2	95.9	41.8	64.3
09	FINISHES	100.0	76.0	87.7	96.6	68.9	82.4	101.3	69.0	84.8	99.7	74.1	86.6	97.1	55.0	75.6	97.1	35.4	65.6
10 - 14	TOTAL DIV. 10000 - 14000	100.0	81.3	96.1	100.0	79.9	95.7	100.0	80.2	95.8	100.0	81.3	96.1	100.0	58.8	91.3	100.0	53.8	90.2
15	MECHANICAL	99.8	82.4	91.9	100.0	75.0	88.7	96.6	81.6	89.8	99.8	75.5	88.8	96.8	51.4	76.2	96.8	39.6	70.9
16	ELECTRICAL	108.6	64.4	78.5	101.0	58.0	71.7	100.1	58.0	71.4	103.0	63.9	76.4	100.0	59.5	72.4	94.2	49.5	63.8
01 - 16	WEIGHTED AVERAGE	100.5	77.8	89.5	100.3	72.3	86.7	101.5	73.2	87.8	99.2	75.4	87.7	94.2	58.2	76.7	94.2	46.8	71.2

DIVISION		ARKANSAS FAYETTEVILLE 727			ARKANSAS FORT SMITH 729			ARKANSAS HARRISON 726			ARKANSAS HOT SPRINGS 719			ARKANSAS JONESBORO 724			ARKANSAS LITTLE ROCK 720 - 722		
		MAT.	INST.	TOTAL	MAT.	INST.	TOTAL	MAT.	INST.	TOTAL	MAT.	INST.	TOTAL	MAT.	INST.	TOTAL	MAT.	INST.	TOTAL
02	SITE CONSTRUCTION	73.4	82.2	80.2	76.7	83.0	81.5	78.7	83.0	82.0	78.3	82.3	81.4	101.7	98.0	98.8	76.6	83.0	81.5
03100	CONCRETE FORMS & ACCESSORIES	73.7	31.0	37.1	98.8	49.3	56.3	84.9	53.4	57.8	75.6	35.1	40.9	83.1	57.1	60.7	93.0	54.1	59.6
03200	CONCRETE REINFORCEMENT	94.7	26.2	55.9	95.7	71.8	82.2	94.1	54.8	71.8	94.0	28.7	57.0	91.6	55.6	71.2	95.9	65.2	78.5
03300	CAST-IN-PLACE CONCRETE	82.6	40.4	65.0	85.5	52.7	71.8	91.7	52.4	75.3	86.7	37.5	66.2	90.1	64.8	79.5	86.5	52.7	72.4
03	CONCRETE	79.6	34.8	57.1	84.2	55.5	69.8	87.3	54.0	70.6	84.9	36.0	60.3	85.0	61.1	73.0	84.3	56.4	70.3
04	MASONRY	88.0	34.0	54.4	96.8	58.9	73.2	97.7	51.2	68.7	80.9	34.3	51.9	89.9	56.6	69.1	94.8	58.9	72.4
05	METALS	94.9	52.1	79.6	96.8	72.1	87.9	95.9	65.0	84.9	94.9	53.6	80.1	91.1	82.1	87.9	96.4	69.7	86.9
06	WOOD & PLASTICS	72.8	30.4	50.9	99.8	48.6	73.3	84.6	55.0	69.3	74.4	35.2	54.1	81.6	58.5	69.7	96.1	55.0	74.8
07	THERMAL & MOISTURE PROTECTION	96.0	42.2	70.8	96.6	55.3	77.3	95.4	53.7	75.9	95.0	41.6	70.0	106.2	60.6	84.9	95.2	57.6	76.8
08	DOORS & WINDOWS	97.4	28.2	80.6	98.2	50.9	86.7	98.2	50.8	86.6	93.6	33.6	79.0	99.5	56.5	89.1	98.2	53.8	87.4
09200	PLASTER & GYPSUM BOARD	86.6	29.2	51.1	94.8	48.0	65.8	92.0	54.6	68.8	88.0	34.1	54.6	99.1	57.7	73.5	94.8	54.6	69.9
09500	CEILINGS	98.3	29.2	52.9	102.5	48.0	66.7	102.5	54.6	71.0	98.3	34.1	56.2	103.1	57.7	73.3	102.5	54.6	71.0
09600	FLOORING	106.0	60.5	94.9	118.9	74.2	108.0	112.9	60.5	100.1	107.9	60.5	96.3	79.1	54.7	73.1	120.2	74.2	109.0
09900	PAINTS & COATINGS	95.9	30.4	57.6	95.9	70.0	80.8	95.9	53.6	71.2	95.9	44.7	66.0	84.2	63.0	71.8	95.9	53.6	71.2
09	FINISHES	95.6	36.3	65.3	101.9	55.9	78.4	99.6	55.0	76.8	96.6	40.8	68.1	92.2	57.3	74.4	102.5	57.8	79.5
10 - 14	TOTAL DIV. 10000 - 14000	100.0	52.7	90.0	100.0	68.6	93.4	100.0	58.8	91.3	100.0	53.4	90.2	100.0	67.7	93.2	100.0	69.5	93.5
15	MECHANICAL	96.8	34.6	68.6	100.0	52.3	78.4	96.8	51.4	76.2	96.8	40.9	71.5	100.2	52.4	78.5	100.0	58.0	81.0
16	ELECTRICAL	89.5	26.7	46.8	95.5	70.2	78.3	97.6	59.5	71.6	97.8	48.5	64.2	100.5	59.5	72.6	96.6	75.7	82.3
01 - 16	WEIGHTED AVERAGE	92.7	40.1	67.2	96.3	62.0	79.7	95.7	58.2	77.5	93.3	46.1	70.5	95.5	63.8	80.1	96.2	64.4	80.8

City Cost Indexes

		ARKANSAS												CALIFORNIA						
		PINE BLUFF			RUSSELLVILLE			TEXARKANA			WEST MEMPHIS			ALHAMBRA			ANAHEIM		
	DIVISION	716			728			718			723			917 - 918			928		
		MAT.	INST.	TOTAL	MAT.	INST.	TOTAL	MAT.	INST.	TOTAL	MAT.	INST.	TOTAL	MAT.	INST.	TOTAL	MAT.	INST.	TOTAL
02	SITE CONSTRUCTION	79.8	83.0	82.3	75.3	83.0	81.2	96.1	83.8	86.6	109.9	98.0	100.7	88.3	109.5	104.6	90.0	110.7	105.9
03100	CONCRETE FORMS & ACCESSORIES	75.1	54.0	57.0	80.2	48.4	53.0	84.1	47.3	52.5	90.2	57.1	61.7	113.9	118.7	118.0	103.8	119.4	117.2
03200	CONCRETE REINFORCEMENT	95.9	65.2	78.5	95.2	57.0	73.6	95.3	53.9	71.8	91.6	55.6	71.2	112.5	116.0	114.5	107.5	116.0	112.3
03300	CAST-IN-PLACE CONCRETE	86.7	52.7	72.5	86.7	52.4	72.4	86.6	50.5	71.6	93.2	64.8	81.4	90.6	114.6	100.6	98.8	118.8	107.1
03	CONCRETE	85.6	56.4	70.9	83.3	52.3	67.7	81.2	50.5	65.8	90.8	61.1	75.9	111.8	115.6	113.7	113.1	117.6	115.3
04	MASONRY	112.4	58.9	79.1	96.9	51.2	68.4	96.9	39.6	61.2	79.5	56.6	65.2	130.7	118.5	123.1	88.0	120.0	108.0
05	METALS	95.5	69.7	86.3	94.9	65.6	84.4	87.9	64.2	79.4	90.4	82.1	87.4	92.0	97.9	94.1	110.5	101.3	107.2
06	WOOD & PLASTICS	73.8	55.0	64.1	79.9	48.6	63.7	84.2	50.7	66.9	89.2	58.5	73.3	101.1	117.0	109.3	99.0	117.2	108.4
07	THERMAL & MOISTURE PROTECTION	95.1	55.9	76.8	96.3	53.1	76.1	96.1	49.5	74.3	96.8	60.6	85.2	105.7	114.4	109.8	121.7	118.9	120.4
08	DOORS & WINDOWS	93.6	53.8	83.9	97.4	47.9	85.4	98.6	48.4	86.4	99.5	56.5	89.1	98.3	114.6	102.3	105.8	114.7	107.9
09200	PLASTER & GYPSUM BOARD	87.5	54.6	67.1	88.9	48.0	63.6	90.7	50.1	65.6	101.4	57.7	74.4	90.9	117.8	107.6	94.1	117.8	108.8
09500	CEILINGS	98.3	54.6	69.6	98.3	48.0	65.3	105.2	50.1	69.1	103.1	57.7	73.3	110.0	117.8	115.2	116.8	117.8	117.5
09600	FLOORING	107.7	74.2	99.5	109.3	60.5	97.4	110.7	52.2	96.4	82.1	54.7	75.4	119.4	112.5	117.7	125.4	113.6	122.5
09900	PAINTS & COATINGS	95.9	53.6	71.2	95.9	53.6	71.2	95.9	44.1	65.6	84.2	63.0	71.8	117.0	118.5	117.9	114.3	118.5	116.8
09	FINISHES	96.6	57.8	76.8	97.1	51.3	73.7	99.7	48.0	73.3	94.0	57.3	75.3	111.7	117.4	114.6	114.7	118.1	116.4
10 - 14	TOTAL DIV. 10000 - 14000	100.0	69.5	93.5	100.0	58.0	91.1	100.0	48.2	89.1	100.0	67.7	93.2	100.0	116.2	103.4	100.0	117.4	103.7
15	MECHANICAL	100.0	55.2	79.7	96.8	44.9	73.3	100.0	46.3	75.7	97.0	52.4	76.8	96.8	115.2	105.1	100.2	116.6	107.6
16	ELECTRICAL	94.4	75.7	81.6	95.5	59.5	70.9	97.6	49.5	64.8	103.3	59.5	73.4	116.3	118.2	117.6	94.8	105.9	102.3
01 - 16	WEIGHTED AVERAGE	95.7	63.8	80.3	94.4	56.1	75.8	94.9	52.6	74.4	95.5	63.8	80.2	103.3	114.2	108.6	104.8	113.5	109.0

		CALIFORNIA																	
		BAKERSFIELD			BERKELEY			EUREKA			FRESNO			INGLEWOOD			LONG BEACH		
	DIVISION	932 - 933			947			955			936 - 938			903 - 905			906 - 908		
		MAT.	INST.	TOTAL	MAT.	INST.	TOTAL	MAT.	INST.	TOTAL	MAT.	INST.	TOTAL	MAT.	INST.	TOTAL	MAT.	INST.	TOTAL
02	SITE CONSTRUCTION	95.5	107.6	104.8	121.8	104.6	108.6	106.2	106.3	106.3	97.0	107.7	105.2	80.9	106.5	100.5	87.8	106.5	102.1
03100	CONCRETE FORMS & ACCESSORIES	94.9	118.8	115.5	123.5	137.2	135.3	115.4	119.8	119.2	100.5	120.9	118.0	110.5	118.4	117.3	103.1	118.4	116.3
03200	CONCRETE REINFORCEMENT	107.1	115.6	111.9	96.2	116.6	107.8	103.7	116.6	111.0	107.6	115.9	112.3	110.7	116.0	113.7	109.8	116.0	113.3
03300	CAST-IN-PLACE CONCRETE	94.3	117.5	104.0	129.8	120.6	125.9	102.3	115.6	107.9	103.3	116.9	108.9	80.6	114.8	94.8	91.8	114.8	101.3
03	CONCRETE	110.2	116.8	113.5	125.1	126.3	125.7	125.4	116.7	121.0	115.1	117.5	116.3	102.8	115.7	109.2	113.4	115.7	114.5
04	MASONRY	107.1	112.6	110.5	133.6	125.2	128.3	113.5	128.7	123.0	109.7	114.3	112.6	84.4	118.6	105.7	93.7	118.6	109.2
05	METALS	105.0	99.9	103.2	107.0	105.8	106.6	110.0	101.1	106.8	106.9	100.9	104.7	103.9	99.0	102.1	103.8	99.0	102.1
06	WOOD & PLASTICS	87.9	117.3	103.1	122.9	139.2	131.3	113.8	120.0	117.0	100.7	120.0	110.6	102.8	116.7	110.0	95.0	116.7	106.3
07	THERMAL & MOISTURE PROTECTION	105.3	109.6	107.3	113.4	131.2	121.7	112.4	112.7	112.6	101.8	111.6	106.4	114.6	115.8	115.2	114.8	115.8	115.3
08	DOORS & WINDOWS	104.6	113.8	106.8	106.9	130.0	112.5	105.6	111.9	107.1	104.6	115.2	107.2	95.0	114.4	99.7	95.0	114.4	99.7
09200	PLASTER & GYPSUM BOARD	94.1	117.8	108.8	95.5	140.1	123.1	99.0	120.6	112.4	93.2	120.6	110.2	90.6	117.8	107.5	88.3	117.8	106.6
09500	CEILINGS	116.8	117.8	117.5	110.0	140.1	129.8	125.1	120.6	122.1	116.8	120.6	119.3	108.7	117.8	114.7	108.7	117.8	114.7
09600	FLOORING	121.0	75.4	109.9	117.2	113.5	116.3	125.7	128.8	126.5	121.7	113.5	119.7	116.5	112.5	115.5	113.5	112.5	113.3
09900	PAINTS & COATINGS	117.4	109.6	112.9	119.1	126.5	123.4	116.9	73.6	91.6	116.9	102.6	108.5	114.2	118.5	116.7	114.2	118.5	116.7
09	FINISHES	116.2	110.5	113.3	116.0	132.6	124.5	120.7	117.2	118.9	116.3	117.5	116.9	109.5	117.2	113.5	108.6	117.2	113.0
10 - 14	TOTAL DIV. 10000 - 14000	100.0	135.4	107.5	100.0	139.1	108.3	100.0	134.4	107.3	100.0	135.0	107.4	100.0	115.7	103.3	100.0	115.7	103.3
15	MECHANICAL	100.2	103.3	101.6	96.9	137.6	115.3	97.0	98.9	97.8	100.2	116.6	107.7	96.9	115.1	105.1	96.9	115.1	105.1
16	ELECTRICAL	95.2	93.7	94.2	113.9	135.3	128.5	105.0	102.5	103.3	93.9	99.0	97.4	114.6	118.2	117.1	114.0	118.2	116.9
01 - 16	WEIGHTED AVERAGE	104.3	106.9	105.5	110.1	127.3	118.5	108.6	109.8	109.2	105.3	111.7	108.4	100.8	114.0	107.2	102.6	114.0	108.1

		CALIFORNIA																	
		LOS ANGELES			MARYSVILLE			MODESTO			MOJAVE			OAKLAND			OXNARD		
	DIVISION	900 - 902			959			953			935			946			930		
		MAT.	INST.	TOTAL	MAT.	INST.	TOTAL	MAT.	INST.	TOTAL	MAT.	INST.	TOTAL	MAT.	INST.	TOTAL	MAT.	INST.	TOTAL
02	SITE CONSTRUCTION	87.0	109.5	104.2	101.9	106.6	105.5	95.2	106.9	104.2	90.3	107.5	103.5	129.8	104.6	110.5	96.5	105.3	103.2
03100	CONCRETE FORMS & ACCESSORIES	106.7	119.0	117.2	101.6	120.9	118.1	96.1	121.0	117.4	110.6	118.4	117.3	108.0	137.3	133.1	101.7	119.4	116.9
03200	CONCRETE REINFORCEMENT	110.7	116.0	113.7	103.7	115.8	110.6	107.6	116.0	112.4	109.0	115.6	112.8	98.4	116.6	108.7	107.1	115.7	111.9
03300	CAST-IN-PLACE CONCRETE	87.8	116.0	99.5	114.4	116.6	115.3	102.3	116.9	108.4	86.0	118.7	99.7	126.3	120.6	123.9	100.1	117.8	107.5
03	CONCRETE	109.7	116.3	113.0	125.3	117.4	121.3	114.3	117.6	115.9	103.9	117.0	110.5	127.1	126.3	126.7	113.5	117.2	115.4
04	MASONRY	99.9	120.0	112.4	114.4	113.9	114.1	111.9	117.5	115.4	107.4	116.8	113.3	141.6	125.2	131.4	112.6	112.7	112.7
05	METALS	110.0	99.3	106.2	109.3	100.4	106.1	106.5	101.4	104.6	106.3	99.2	103.7	102.3	106.3	103.7	104.4	100.5	103.0
06	WOOD & PLASTICS	98.5	116.7	107.9	98.8	120.0	109.8	92.3	120.0	106.6	103.6	115.2	109.6	106.7	139.2	123.5	97.2	117.3	107.6
07	THERMAL & MOISTURE PROTECTION	115.0	117.0	116.0	111.6	114.4	112.9	111.0	111.9	111.4	105.9	111.0	108.3	113.5	131.3	121.9	110.9	113.9	112.3
08	DOORS & WINDOWS	99.2	114.4	102.9	104.8	115.3	107.4	103.8	115.3	106.6	100.1	112.6	103.1	107.0	130.0	112.6	103.3	114.8	106.1
09200	PLASTER & GYPSUM BOARD	90.8	117.8	107.6	94.1	120.6	110.5	95.0	120.6	110.9	101.5	115.7	110.3	90.7	140.1	121.3	94.1	117.8	108.8
09500	CEILINGS	116.9	117.8	117.5	123.7	120.6	121.7	116.8	120.6	119.3	123.7	115.7	118.4	111.4	140.1	130.3	116.8	117.8	117.5
09600	FLOORING	114.9	113.6	114.6	120.3	113.5	118.7	121.0	113.5	119.2	128.0	74.6	115.0	111.4	113.5	111.9	121.0	113.6	119.2
09900	PAINTS & COATINGS	114.2	118.7	116.9	116.9	103.3	108.9	116.9	97.7	105.7	116.9	101.7	108.0	119.1	124.9	122.5	116.9	113.0	114.6
09	FINISHES	110.8	117.7	114.3	117.5	117.4	117.5	116.2	116.4	116.3	120.5	109.0	114.6	114.2	132.4	123.5	116.0	117.5	116.8
10 - 14	TOTAL DIV. 10000 - 14000	100.0	116.2	103.4	100.0	135.0	107.4	100.0	135.1	107.4	100.0	136.4	107.7	100.0	139.1	108.3	100.0	117.8	103.8
15	MECHANICAL	100.1	116.5	107.5	97.0	116.8	105.9	100.2	116.8	107.7	97.0	111.9	103.8	100.1	137.6	117.1	100.2	116.7	107.6
16	ELECTRICAL	109.1	118.3	115.3	99.3	103.6	102.2	103.2	117.4	112.8	93.4	93.7	93.6	112.7	135.3	128.1	99.0	115.2	110.1
01 - 16	WEIGHTED AVERAGE	104.5	114.9	109.6	107.4	112.3	109.8	106.0	114.9	110.3	102.9	108.7	105.7	110.5	127.4	118.7	105.3	113.5	109.3

City Cost Indexes

CALIFORNIA

DIVISION		PALM SPRINGS 922			PALO ALTO 943			PASADENA 910 - 912			REDDING 960			RICHMOND 948			RIVERSIDE 925		
		MAT.	INST.	TOTAL	MAT.	INST.	TOTAL	MAT.	INST.	TOTAL	MAT.	INST.	TOTAL	MAT.	INST.	TOTAL	MAT.	INST.	TOTAL
02	SITE CONSTRUCTION	82.7	107.9	102.0	116.8	104.6	107.4	85.4	109.2	103.7	101.6	106.6	105.4	129.2	104.6	110.3	88.9	108.3	103.8
03100	CONCRETE FORMS & ACCESSORIES	99.8	118.6	115.9	105.2	137.3	132.8	99.0	118.5	115.7	101.3	120.8	118.0	127.6	137.1	135.8	104.9	119.2	117.2
03200	CONCRETE REINFORCEMENT	109.2	115.7	112.9	96.2	117.0	108.0	113.5	116.0	114.9	103.7	115.7	110.5	96.2	116.6	107.7	106.1	115.7	111.5
03300	CAST-IN-PLACE CONCRETE	90.0	117.6	101.5	109.7	120.6	114.3	86.0	114.4	97.8	113.4	116.6	114.7	126.3	120.5	123.9	97.8	118.7	106.5
03	CONCRETE	105.2	116.7	111.0	112.5	126.4	119.5	106.4	115.4	110.9	123.9	117.3	120.6	128.2	126.2	127.2	112.5	117.4	115.0
04	MASONRY	84.5	114.1	102.9	111.2	133.0	124.8	111.5	118.4	115.8	114.3	112.4	113.1	133.3	125.2	128.2	85.4	114.0	103.2
05	METALS	110.2	100.0	106.6	100.3	106.5	102.5	92.0	97.2	93.9	108.9	100.2	105.8	100.3	105.7	102.2	110.5	100.8	107.0
06	WOOD & PLASTICS	93.6	117.2	105.8	103.6	139.2	122.0	85.2	117.0	101.6	98.4	120.0	109.6	127.8	139.2	133.7	99.0	117.2	108.4
07	THERMAL & MOISTURE PROTECTION	120.3	115.5	118.1	112.9	132.6	122.1	105.2	114.4	109.5	111.6	114.0	112.8	113.9	131.2	122.0	120.8	115.9	118.5
08	DOORS & WINDOWS	102.5	114.7	105.5	107.0	130.0	112.6	98.3	113.1	101.9	106.1	112.8	107.7	107.0	130.0	112.6	105.8	114.7	107.9
09200	PLASTER & GYPSUM BOARD	92.0	117.8	108.0	89.8	140.1	120.9	86.3	117.8	105.8	94.4	120.6	110.6	97.1	140.1	123.7	93.9	117.8	108.7
09500	CEILINGS	115.4	117.8	117.0	111.4	140.1	130.3	110.0	117.8	115.2	125.1	120.6	122.1	111.4	140.1	130.3	115.4	117.8	117.0
09600	FLOORING	123.1	107.9	119.4	110.4	113.5	111.2	113.6	112.5	113.3	120.2	113.5	118.6	119.2	113.5	117.8	125.4	113.6	122.5
09900	PAINTS & COATINGS	114.3	118.5	116.8	119.1	125.4	122.8	117.0	118.5	117.9	116.9	103.3	108.9	119.1	126.5	123.4	114.3	118.5	116.8
09	FINISHES	113.0	116.8	114.9	113.1	132.5	123.0	109.0	117.4	113.3	117.7	117.4	117.6	117.6	132.6	125.3	114.3	118.1	116.2
10 - 14	TOTAL DIV. 10000 - 14000	100.0	116.8	103.6	100.0	139.1	108.3	100.0	116.2	103.4	100.0	135.0	107.4	100.0	139.1	108.3	100.0	117.4	103.7
15	MECHANICAL	97.0	115.9	105.6	96.9	147.5	119.8	96.8	115.2	105.1	100.2	116.7	107.7	96.9	136.2	114.7	100.1	116.6	107.6
16	ELECTRICAL	100.8	98.9	99.5	112.4	129.8	124.3	111.7	118.2	116.2	104.2	103.6	103.8	113.9	124.4	121.0	94.9	98.9	97.6
01 - 16	WEIGHTED AVERAGE	102.5	110.8	106.5	105.5	129.2	117.0	100.7	114.0	107.1	108.4	112.0	110.2	109.9	125.2	117.3	104.5	111.3	107.8

CALIFORNIA

DIVISION		SACRAMENTO 942, 956 - 958			SALINAS 939			SAN BERNARDINO 923 - 924			SAN DIEGO 919 - 921			SAN FRANCISCO 940 - 941			SAN JOSE 951		
		MAT.	INST.	TOTAL	MAT.	INST.	TOTAL	MAT.	INST.	TOTAL	MAT.	INST.	TOTAL	MAT.	INST.	TOTAL	MAT.	INST.	TOTAL
02	SITE CONSTRUCTION	95.4	111.1	107.4	112.6	107.7	108.8	68.3	108.0	98.7	90.6	102.6	99.8	131.9	111.2	116.0	131.7	101.0	108.1
03100	CONCRETE FORMS & ACCESSORIES	106.5	121.1	119.1	106.4	125.3	122.6	109.8	118.3	117.1	104.6	116.6	114.9	108.3	138.3	134.1	104.1	137.4	132.7
03200	CONCRETE REINFORCEMENT	99.5	115.9	108.8	107.6	116.4	112.6	106.1	115.7	111.5	109.9	115.6	113.1	112.1	117.2	115.0	104.0	117.0	111.4
03300	CAST-IN-PLACE CONCRETE	109.6	116.8	112.6	98.9	117.2	106.5	67.5	117.1	88.2	94.2	107.2	99.6	126.2	122.3	124.6	119.0	121.0	119.8
03	CONCRETE	118.9	117.6	118.2	127.4	119.7	123.5	82.5	116.4	99.5	112.6	112.2	112.4	128.8	127.5	128.2	122.5	126.7	124.6
04	MASONRY	114.1	113.9	114.0	107.0	121.8	116.2	92.8	111.8	104.6	103.4	113.2	109.5	141.9	134.6	137.4	149.2	133.1	139.2
05	METALS	98.3	100.2	99.0	109.5	102.2	106.9	110.4	99.9	106.6	109.4	100.0	106.0	107.5	108.3	107.8	111.0	108.6	110.1
06	WOOD & PLASTICS	101.7	120.1	111.2	103.1	125.0	114.4	104.0	117.2	110.8	98.6	114.3	106.7	106.7	139.4	123.6	104.9	139.1	122.6
07	THERMAL & MOISTURE PROTECTION	123.2	114.0	118.9	107.5	121.0	113.8	119.2	114.7	117.1	114.7	108.5	111.8	113.5	135.0	123.6	107.5	133.1	119.5
08	DOORS & WINDOWS	118.7	115.3	117.9	104.7	122.4	109.0	102.6	114.7	105.5	104.2	112.7	106.3	109.2	130.2	114.3	95.5	130.0	103.9
09200	PLASTER & GYPSUM BOARD	89.3	120.6	108.7	95.5	125.7	114.2	95.5	117.8	109.3	90.2	114.7	105.3	92.6	140.1	122.0	94.6	140.1	122.8
09500	CEILINGS	119.7	120.6	120.3	123.7	125.7	125.0	116.8	117.8	117.5	118.3	114.7	115.9	121.1	140.1	133.6	108.5	140.1	129.3
09600	FLOORING	113.8	113.5	113.7	121.8	113.5	119.8	127.4	108.1	122.7	123.1	112.9	120.6	111.4	114.9	112.3	118.0	114.9	117.2
09900	PAINTS & COATINGS	117.2	105.3	110.3	116.9	126.0	122.2	114.3	118.5	116.8	114.3	115.2	114.8	119.1	138.7	130.5	118.8	126.0	123.0
09	FINISHES	114.6	117.7	116.2	119.2	123.8	121.5	114.1	116.6	115.4	114.8	115.9	115.4	116.3	134.5	125.6	115.2	132.7	124.2
10 - 14	TOTAL DIV. 10000 - 14000	100.0	135.2	107.4	100.0	135.8	107.6	100.0	116.5	103.5	100.0	117.3	103.7	100.0	139.8	108.4	100.0	138.9	108.2
15	MECHANICAL	100.0	119.7	108.9	97.0	116.9	106.0	97.0	114.7	105.0	100.1	114.8	106.7	100.1	164.6	129.3	100.2	147.6	121.7
16	ELECTRICAL	103.2	104.5	104.1	93.9	121.2	112.5	100.8	108.9	106.3	94.0	94.3	94.2	108.5	146.1	134.1	109.6	129.8	123.4
01 - 16	WEIGHTED AVERAGE	107.4	113.5	110.4	107.3	117.9	112.4	100.0	112.0	105.8	105.0	108.2	106.6	111.8	136.6	123.8	110.1	129.1	119.3

CALIFORNIA

DIVISION		SAN LUIS OBISPO 934			SAN MATEO 944			SAN RAFAEL 949			SANTA ANA 926 - 927			SANTA BARBARA 931			SANTA CRUZ 950		
		MAT.	INST.	TOTAL	MAT.	INST.	TOTAL	MAT.	INST.	TOTAL	MAT.	INST.	TOTAL	MAT.	INST.	TOTAL	MAT.	INST.	TOTAL
02	SITE CONSTRUCTION	103.7	107.3	106.5	126.0	104.6	109.6	108.8	111.1	110.5	80.7	108.0	101.7	96.9	107.6	105.1	132.4	100.8	108.2
03100	CONCRETE FORMS & ACCESSORIES	113.1	118.8	118.0	113.7	137.3	133.9	121.9	136.9	134.8	110.7	118.6	117.5	102.3	119.2	116.8	104.1	125.6	122.5
03200	CONCRETE REINFORCEMENT	109.0	115.6	112.8	96.2	117.0	108.0	97.0	116.9	108.3	109.8	115.7	113.2	107.1	115.7	111.9	121.2	116.4	118.5
03300	CAST-IN-PLACE CONCRETE	106.3	116.5	110.6	122.2	120.6	121.5	142.3	119.1	132.6	86.3	117.6	99.3	99.7	117.6	107.2	118.1	119.7	118.8
03	CONCRETE	123.5	116.4	119.9	123.7	126.4	125.0	151.4	125.4	138.3	102.6	116.7	109.7	113.4	117.0	115.2	124.4	120.9	122.6
04	MASONRY	109.0	111.3	110.4	133.0	132.2	132.5	108.9	132.2	123.4	81.3	113.2	101.2	107.5	113.2	111.0	149.5	122.0	132.4
05	METALS	106.7	99.7	104.2	100.1	106.4	102.4	101.6	102.0	101.7	110.5	100.0	106.7	104.3	100.3	102.9	112.2	107.0	110.4
06	WOOD & PLASTICS	106.2	117.3	112.0	112.8	139.2	126.4	114.6	138.9	127.1	106.2	117.2	111.9	97.2	117.3	107.6	104.9	125.2	115.3
07	THERMAL & MOISTURE PROTECTION	107.2	112.5	109.7	113.5	131.9	122.1	130.2	132.0	131.0	120.9	116.4	118.8	106.4	112.5	109.3	117.5	124.8	115.3
08	DOORS & WINDOWS	102.4	113.8	105.2	107.0	130.0	112.6	117.5	129.9	120.5	101.7	114.7	104.9	104.6	114.8	107.1	97.5	122.4	103.6
09200	PLASTER & GYPSUM BOARD	102.4	117.8	112.0	92.5	140.1	122.0	93.9	140.1	122.5	96.4	117.8	109.7	94.1	117.8	108.8	96.8	125.7	114.7
09500	CEILINGS	123.7	117.8	119.9	111.4	140.1	130.3	119.7	140.1	133.1	116.8	117.8	117.5	116.8	117.8	117.5	119.5	125.7	123.6
09600	FLOORING	129.4	101.9	122.7	113.6	113.5	113.6	123.5	135.5	126.4	128.2	111.1	124.2	121.0	101.9	116.3	118.0	113.5	116.9
09900	PAINTS & COATINGS	116.9	113.0	114.6	119.1	125.4	122.8	114.6	122.6	119.3	114.3	118.5	116.8	116.9	113.0	114.6	118.8	126.0	123.0
09	FINISHES	121.9	115.2	118.5	115.0	132.5	123.9	116.9	135.7	126.5	115.5	117.4	116.5	116.2	115.5	115.9	117.6	123.9	120.8
10 - 14	TOTAL DIV. 10000 - 14000	100.0	134.9	107.4	100.0	139.1	108.3	100.0	138.1	108.1	100.0	116.9	103.6	100.0	117.7	103.8	100.0	136.3	107.7
15	MECHANICAL	97.0	115.3	105.3	96.9	137.0	115.1	96.8	145.2	118.7	97.0	115.2	105.2	100.2	116.7	107.6	100.2	117.0	107.8
16	ELECTRICAL	93.4	108.5	103.7	112.4	140.6	131.6	106.0	105.8	105.9	100.8	103.3	102.5	91.7	110.4	104.5	107.4	121.2	116.8
01 - 16	WEIGHTED AVERAGE	106.3	112.0	109.1	108.6	128.8	118.4	111.9	125.1	118.3	102.3	111.5	106.8	104.5	112.6	108.4	110.9	118.0	114.4

City Cost Indexes

		CALIFORNIA															COLORADO		
		SANTA ROSA			STOCKTON			SUSANVILLE			VALLEJO			VAN NUYS			ALAMOSA		
DIVISION		954			952			961			945			913 - 916			811		
		MAT.	INST.	TOTAL	MAT.	INST.	TOTAL	MAT.	INST.	TOTAL	MAT.	INST.	TOTAL	MAT.	INST.	TOTAL	MAT.	INST.	TOTAL
02	SITE CONSTRUCTION	95.6	107.1	104.4	94.8	106.9	104.1	109.3	106.7	107.3	91.6	111.3	106.7	101.3	109.3	107.4	145.5	94.0	106.0
03100	CONCRETE FORMS & ACCESSORIES	101.3	136.5	131.6	96.1	120.9	117.4	103.2	120.8	118.3	108.4	136.3	132.3	107.7	118.6	117.0	102.6	58.2	64.5
03200	CONCRETE REINFORCEMENT	104.8	117.1	111.7	107.6	116.0	112.4	103.7	115.8	110.6	98.2	116.9	108.8	113.5	116.0	114.9	109.9	64.9	84.4
03300	CAST-IN-PLACE CONCRETE	112.3	118.7	115.0	99.7	116.8	106.8	103.2	116.6	108.8	113.3	118.4	115.4	90.7	114.4	100.6	107.5	59.6	87.5
03	CONCRETE	125.3	125.3	125.3	113.0	117.6	115.3	127.8	117.3	122.6	120.7	124.9	122.8	124.5	115.5	120.0	121.2	60.7	90.8
04	MASONRY	112.9	132.8	125.3	112.2	117.5	115.5	113.4	113.9	113.7	81.4	130.5	112.0	130.7	118.4	123.1	137.3	65.3	92.5
05	METALS	110.9	104.3	108.5	106.5	101.4	104.6	109.4	100.3	106.2	101.6	102.7	102.0	91.2	97.4	93.4	100.7	76.9	92.2
06	WOOD & PLASTICS	95.1	138.8	117.7	92.3	120.0	106.6	100.5	120.0	110.6	100.6	138.9	120.4	94.3	117.0	106.0	99.9	55.8	77.1
07	THERMAL & MOISTURE PROTECTION	121.5	131.3	126.1	111.0	111.2	111.1	112.5	114.4	113.4	128.0	131.4	129.6	106.6	114.4	110.3	106.0	75.7	91.8
08	DOORS & WINDOWS	103.4	129.8	109.8	103.8	115.3	106.6	106.0	115.3	108.2	120.0	129.6	122.4	98.1	114.6	102.1	93.6	55.7	84.4
09200	PLASTER & GYPSUM BOARD	92.8	140.1	122.1	95.0	120.6	110.9	94.6	120.6	110.7	89.7	140.1	120.9	88.3	117.8	106.6	88.9	54.0	67.3
09500	CEILINGS	116.8	140.1	132.1	116.8	120.6	119.3	123.7	120.6	121.7	119.7	140.1	133.1	108.7	117.8	114.7	94.9	54.0	68.1
09600	FLOORING	123.7	114.9	121.5	121.0	113.5	119.2	121.0	113.5	119.2	118.2	114.9	117.4	117.1	112.5	115.9	118.1	85.3	110.1
09900	PAINTS & COATINGS	114.3	125.7	121.0	116.9	102.2	108.3	116.9	95.5	104.4	115.3	124.9	120.9	117.0	118.5	117.9	124.9	37.5	73.9
09	FINISHES	114.3	132.1	123.4	116.2	117.0	116.6	118.5	116.4	117.4	113.2	132.1	122.8	111.1	117.4	114.3	106.4	60.4	82.9
10 - 14	TOTAL DIV. 10000 - 14000	100.0	137.2	107.9	100.0	135.1	107.4	100.0	135.0	107.4	100.0	137.5	107.9	100.0	116.2	103.4	100.0	71.3	93.9
15	MECHANICAL	97.0	163.4	127.1	100.2	116.8	107.7	96.9	115.9	105.5	100.0	134.4	115.6	96.8	114.0	104.6	96.6	67.0	83.2
16	ELECTRICAL	101.3	105.9	104.4	103.2	112.5	109.5	104.8	103.6	104.0	99.2	122.8	115.3	111.7	118.2	116.2	96.4	75.6	82.2
01 - 16	WEIGHTED AVERAGE	107.3	127.9	117.3	105.8	114.1	109.8	108.6	112.0	110.2	106.1	125.2	115.3	104.6	113.9	109.1	105.1	70.0	88.1

		COLORADO																	
		BOULDER			COLORADO SPRINGS			DENVER			DURANGO			FORT COLLINS			FORT MORGAN		
DIVISION		803			808 - 809			800 - 802			813			805			807		
		MAT.	INST.	TOTAL	MAT.	INST.	TOTAL	MAT.	INST.	TOTAL	MAT.	INST.	TOTAL	MAT.	INST.	TOTAL	MAT.	INST.	TOTAL
02	SITE CONSTRUCTION	115.2	100.7	104.1	116.4	98.6	102.8	114.7	108.0	109.6	137.7	94.0	104.1	128.0	101.9	108.0	116.7	101.8	105.3
03100	CONCRETE FORMS & ACCESSORIES	111.9	57.4	65.1	92.2	75.5	77.8	102.4	78.1	81.6	115.9	57.1	65.4	101.8	74.3	78.2	113.5	74.2	79.7
03200	CONCRETE REINFORCEMENT	100.4	59.3	77.1	96.7	78.9	86.6	96.2	79.5	86.7	109.9	68.3	86.2	100.8	78.6	88.2	101.4	69.3	83.2
03300	CAST-IN-PLACE CONCRETE	103.5	70.0	89.5	107.6	83.1	97.4	100.6	82.7	93.1	123.6	58.9	96.7	116.9	81.3	102.0	101.6	70.0	88.4
03	CONCRETE	109.6	62.6	86.0	101.6	79.1	90.3	106.3	80.3	93.2	124.4	60.8	92.4	120.5	77.9	99.1	108.2	72.2	90.1
04	MASONRY	102.1	49.1	69.1	107.8	71.5	85.2	106.9	82.0	91.4	124.0	65.3	87.4	119.2	70.4	88.8	117.8	70.4	88.3
05	METALS	98.8	67.5	87.6	100.4	85.0	94.9	103.9	85.9	97.4	100.7	80.7	93.5	99.7	83.2	93.8	98.6	79.0	91.5
06	WOOD & PLASTICS	111.4	58.3	84.0	90.6	75.5	82.8	101.1	78.4	89.4	114.5	54.9	83.7	100.8	75.2	87.6	113.1	75.2	93.5
07	THERMAL & MOISTURE PROTECTION	105.7	66.3	87.2	106.1	81.6	94.6	105.5	84.3	95.6	106.0	68.9	88.6	106.1	76.2	92.1	105.7	76.2	91.9
08	DOORS & WINDOWS	94.2	51.1	83.7	96.6	78.8	92.3	96.0	81.1	92.4	110.9	58.1	90.5	94.2	73.9	89.2	94.2	71.2	88.6
09200	PLASTER & GYPSUM BOARD	101.9	57.2	74.2	91.5	74.7	81.1	99.7	78.0	86.2	94.5	53.1	68.9	98.6	74.7	83.8	102.3	74.7	85.2
09500	CEILINGS	88.9	57.2	68.1	95.8	74.7	81.9	94.4	78.0	83.6	94.9	53.1	67.5	88.9	74.7	79.5	88.9	74.7	79.5
09600	FLOORING	118.9	85.3	110.7	113.7	69.2	102.9	114.7	89.1	108.5	123.1	85.3	113.9	114.6	85.3	107.5	119.5	85.3	111.2
09900	PAINTS & COATINGS	113.9	39.8	70.6	114.0	64.5	85.1	113.9	77.2	92.5	124.9	37.5	73.9	113.9	64.0	84.7	113.9	76.4	92.0
09	FINISHES	104.4	60.4	81.9	102.5	72.5	87.2	103.6	79.7	91.4	108.1	59.9	83.5	103.1	75.3	88.9	104.6	76.7	90.4
10 - 14	TOTAL DIV. 10000 - 14000	100.0	67.5	93.1	100.0	85.5	96.9	100.0	85.4	96.9	100.0	71.2	93.9	100.0	73.5	94.4	100.0	73.6	94.4
15	MECHANICAL	96.6	77.0	87.7	99.9	80.6	91.2	99.8	85.0	93.1	96.6	47.2	74.2	99.8	82.5	92.0	96.6	82.3	90.1
16	ELECTRICAL	90.8	73.3	78.9	95.9	85.3	88.7	98.0	86.6	90.3	95.5	72.5	79.8	90.8	73.3	78.9	91.0	85.8	87.4
01 - 16	WEIGHTED AVERAGE	100.2	69.5	85.3	100.8	81.6	91.5	102.1	85.9	94.2	105.7	65.9	86.4	103.4	79.5	91.8	100.9	80.5	91.0

		COLORADO																	
		GLENWOOD SPRINGS			GOLDEN			GRAND JUNCTION			GREELEY			MONTROSE			PUEBLO		
DIVISION		816			804			815			806			814			810		
		MAT.	INST.	TOTAL	MAT.	INST.	TOTAL	MAT.	INST.	TOTAL	MAT.	INST.	TOTAL	MAT.	INST.	TOTAL	MAT.	INST.	TOTAL
02	SITE CONSTRUCTION	156.0	101.4	114.2	133.2	101.1	108.6	135.9	100.5	108.8	112.5	100.7	103.4	148.4	96.8	108.9	126.8	95.2	102.6
03100	CONCRETE FORMS & ACCESSORIES	98.4	67.5	71.8	95.7	74.3	77.3	111.2	52.8	61.0	98.8	57.3	63.2	97.8	53.3	59.6	105.3	75.9	80.0
03200	CONCRETE REINFORCEMENT	107.6	78.2	90.9	101.4	61.2	78.6	109.0	75.0	89.8	100.4	59.2	77.1	107.5	75.1	89.1	101.7	78.8	88.7
03300	CAST-IN-PLACE CONCRETE	107.5	59.4	87.4	101.6	81.2	93.1	119.0	52.1	91.1	97.7	70.0	86.1	107.5	53.0	84.8	106.7	83.9	97.2
03	CONCRETE	126.6	67.3	96.8	119.8	75.0	97.2	120.1	57.6	88.7	103.6	62.6	83.0	116.9	58.2	87.4	108.8	79.6	94.2
04	MASONRY	110.8	61.3	79.9	121.6	60.7	83.7	146.3	58.2	91.4	113.0	49.1	73.2	115.8	58.3	80.0	107.0	69.3	83.5
05	METALS	100.4	83.4	94.3	98.7	82.6	92.9	101.7	79.7	93.8	99.7	67.5	88.1	99.5	80.3	92.6	103.2	86.1	97.1
06	WOOD & PLASTICS	94.1	67.7	80.5	94.1	75.2	84.4	108.1	54.4	80.4	97.6	58.3	77.3	93.5	54.7	73.4	102.6	75.8	88.7
07	THERMAL & MOISTURE PROTECTION	105.7	72.9	90.3	106.6	73.4	91.1	104.6	62.7	84.9	105.2	66.4	87.1	105.9	65.6	87.0	103.9	80.3	92.8
08	DOORS & WINDOWS	100.1	74.6	93.9	94.2	78.7	90.4	100.8	57.9	90.4	94.2	51.1	83.7	101.0	58.0	90.6	95.2	79.0	91.2
09200	PLASTER & GYPSUM BOARD	99.4	66.8	79.2	96.4	74.7	82.9	104.4	53.1	72.7	97.3	57.2	72.6	84.8	53.1	65.2	91.5	74.7	81.1
09500	CEILINGS	93.5	66.8	76.0	88.9	74.7	79.5	93.5	53.1	67.0	88.9	57.2	68.1	94.9	53.1	67.5	103.2	74.7	84.5
09600	FLOORING	117.3	78.6	107.9	112.0	85.3	105.5	122.6	85.3	113.5	113.2	85.3	106.4	120.5	70.6	108.4	119.0	89.1	111.7
09900	PAINTS & COATINGS	124.9	76.4	96.5	113.9	76.4	92.0	124.9	37.5	73.8	113.9	39.8	70.6	124.9	37.5	73.9	124.9	59.7	86.8
09	FINISHES	107.8	70.4	88.7	102.6	76.7	89.4	108.7	57.3	82.4	101.7	60.4	80.6	106.6	54.4	79.9	107.0	76.2	91.3
10 - 14	TOTAL DIV. 10000 - 14000	100.0	71.6	94.0	100.0	73.6	94.4	100.0	66.2	92.9	100.0	67.5	93.1	100.0	66.8	93.0	100.0	86.4	97.1
15	MECHANICAL	96.6	76.9	87.7	96.6	82.6	90.3	99.9	41.4	73.4	99.8	76.9	89.4	96.6	42.1	71.9	99.9	78.7	90.3
16	ELECTRICAL	91.5	73.3	79.1	91.0	85.8	87.4	95.0	72.5	79.7	90.8	73.3	78.9	95.0	72.5	79.7	96.4	75.6	82.3
01 - 16	WEIGHTED AVERAGE	105.1	75.4	90.7	102.7	80.5	91.9	107.1	63.6	86.0	100.4	69.5	85.4	104.0	63.3	84.3	102.8	79.6	91.6

City Cost Indexes

DIVISION		COLORADO – SALIDA 812			CONNECTICUT – BRIDGEPORT 066			CONNECTICUT – BRISTOL 060			CONNECTICUT – HARTFORD 061			CONNECTICUT – MERIDEN 064			CONNECTICUT – NEW BRITAIN 060		
		MAT.	INST.	TOTAL	MAT.	INST.	TOTAL	MAT.	INST.	TOTAL	MAT.	INST.	TOTAL	MAT.	INST.	TOTAL	MAT.	INST.	TOTAL
02	SITE CONSTRUCTION	137.3	97.7	106.9	110.2	99.0	101.6	109.3	99.0	101.4	109.6	99.0	101.5	108.7	99.9	101.9	109.5	99.0	101.5
03100	CONCRETE FORMS & ACCESSORIES	110.2	57.9	65.3	98.2	94.2	94.7	98.2	102.4	101.8	97.3	102.4	101.7	98.0	102.5	101.9	98.6	102.4	101.9
03200	CONCRETE REINFORCEMENT	106.4	65.0	82.9	115.5	125.6	121.2	115.5	125.6	121.2	115.5	125.6	121.2	115.5	125.6	121.2	115.5	125.6	121.2
03300	CAST-IN-PLACE CONCRETE	123.1	58.7	96.2	110.3	113.8	111.8	103.3	113.7	107.6	103.0	113.7	107.5	99.4	113.8	105.4	105.0	113.7	108.6
03	CONCRETE	118.7	60.3	89.3	114.4	106.5	110.4	111.0	110.0	110.5	110.8	110.0	110.4	109.0	110.1	109.6	111.8	110.0	110.9
04	MASONRY	147.8	67.5	97.8	105.4	113.2	110.3	95.3	113.2	106.4	95.3	113.2	106.4	95.1	113.2	106.3	95.4	113.2	106.5
05	METALS	99.2	76.6	91.1	101.0	104.7	102.3	101.0	104.4	102.2	101.6	104.4	102.6	100.9	104.5	102.2	97.7	104.4	100.1
06	WOOD & PLASTICS	106.7	55.5	80.2	100.1	87.8	93.8	100.1	99.4	99.7	100.1	99.4	99.7	100.1	99.4	99.7	100.1	99.4	99.7
07	THERMAL & MOISTURE PROTECTION	104.7	76.2	91.4	100.5	111.7	105.8	100.7	106.6	103.4	99.5	106.6	102.8	100.7	106.6	103.5	100.7	106.6	103.4
08	DOORS & WINDOWS	93.7	55.5	84.4	108.3	103.2	107.0	108.3	104.5	107.4	108.3	104.5	107.4	110.8	104.9	110.5	108.3	104.5	107.4
09200	PLASTER & GYPSUM BOARD	88.5	54.0	67.1	98.0	86.5	90.9	98.0	98.4	98.3	98.0	98.4	98.3	100.5	98.4	99.2	98.0	98.4	98.3
09500	CEILINGS	94.9	54.0	68.1	99.4	86.5	90.9	99.4	98.4	98.8	99.4	98.4	98.8	111.8	98.4	103.0	99.4	98.4	98.8
09600	FLOORING	126.1	70.6	112.6	96.4	115.3	101.0	96.4	116.4	101.3	96.4	116.4	101.3	96.4	115.3	101.0	96.4	116.4	101.3
09900	PAINTS & COATINGS	124.9	37.5	73.9	93.5	117.4	107.4	93.5	115.3	106.3	93.5	115.3	106.3	93.5	115.3	106.3	93.5	115.3	106.3
09	FINISHES	108.1	57.2	82.1	97.9	99.1	98.5	98.0	105.9	102.0	97.9	105.9	102.0	100.6	105.6	103.2	98.0	105.9	102.0
10 - 14	TOTAL DIV. 10000 - 14000	100.0	70.7	93.8	100.0	113.3	102.8	100.0	114.8	103.1	100.0	114.8	103.1	100.0	114.8	103.1	100.0	114.8	103.1
15	MECHANICAL	96.6	66.9	83.2	99.9	105.7	102.5	99.9	105.7	102.5	99.9	105.7	102.5	96.7	105.7	100.8	99.9	105.7	102.5
16	ELECTRICAL	95.4	75.6	81.9	104.4	101.0	102.1	104.4	99.0	100.7	103.6	102.4	102.8	104.4	99.0	100.7	104.4	99.0	100.7
01 - 16	WEIGHTED AVERAGE	105.1	70.1	88.1	103.6	104.2	103.9	102.6	105.2	103.9	105.8	104.1		102.2	105.5	103.8	102.2	105.2	103.7

DIVISION		CONNECTICUT – NEW HAVEN 065			CONNECTICUT – NEW LONDON 063			CONNECTICUT – NORWALK 068			CONNECTICUT – STAMFORD 069			CONNECTICUT – WATERBURY 067			CONNECTICUT – WILLIMANTIC 062		
		MAT.	INST.	TOTAL	MAT.	INST.	TOTAL	MAT.	INST.	TOTAL	MAT.	INST.	TOTAL	MAT.	INST.	TOTAL	MAT.	INST.	TOTAL
02	SITE CONSTRUCTION	109.3	99.9	102.1	101.8	99.8	100.3	110.0	99.0	101.6	110.7	99.0	101.8	109.7	99.0	101.5	110.1	98.9	101.5
03100	CONCRETE FORMS & ACCESSORIES	98.0	102.5	101.9	98.0	102.4	101.8	98.2	94.2	94.7	98.2	94.5	95.0	98.2	102.6	102.0	98.2	102.3	101.7
03200	CONCRETE REINFORCEMENT	115.5	125.6	121.2	90.6	125.5	110.4	115.5	125.7	121.3	115.5	125.8	121.3	115.5	125.6	121.2	115.5	125.5	121.2
03300	CAST-IN-PLACE CONCRETE	106.8	113.8	109.7	91.0	113.7	100.5	108.5	115.8	111.5	110.3	116.0	112.7	110.3	113.8	111.7	103.0	113.7	107.4
03	CONCRETE	113.6	110.1	111.8	98.4	110.0	104.3	113.5	107.2	110.3	114.4	107.4	110.9	114.4	110.1	112.3	110.8	109.9	110.4
04	MASONRY	95.6	113.2	106.5	94.0	113.2	105.9	95.7	115.5	108.0	95.8	115.5	108.1	95.8	113.2	106.6	95.3	113.2	106.4
05	METALS	97.9	104.5	100.3	97.7	104.4	100.1	101.0	104.9	102.4	101.0	105.2	102.5	101.0	104.6	102.3	100.8	104.3	102.0
06	WOOD & PLASTICS	100.1	99.4	99.7	100.1	99.4	99.7	100.1	87.8	93.8	100.1	87.8	93.8	100.1	99.4	99.7	100.1	99.4	99.7
07	THERMAL & MOISTURE PROTECTION	100.8	109.2	104.7	100.7	106.6	103.4	100.7	112.6	106.3	100.7	112.6	106.3	100.7	109.2	104.7	100.9	106.6	103.6
08	DOORS & WINDOWS	108.3	109.4	108.6	111.7	103.8	109.8	108.3	103.2	107.0	108.3	103.2	107.0	108.3	109.4	108.6	111.7	104.7	110.0
09200	PLASTER & GYPSUM BOARD	98.0	98.4	98.3	98.0	98.4	98.3	98.0	86.5	90.9	98.0	86.5	90.9	98.0	98.4	98.3	98.0	98.4	98.3
09500	CEILINGS	99.4	98.4	98.8	99.4	98.4	98.8	99.4	86.5	90.9	99.4	86.5	90.9	99.4	98.4	98.8	99.4	98.4	98.8
09600	FLOORING	96.4	115.3	101.0	96.4	116.4	101.3	96.4	116.3	101.3	96.4	116.3	101.3	96.4	115.3	101.0	96.4	110.4	99.8
09900	PAINTS & COATINGS	93.5	105.6	100.6	93.5	112.1	104.4	93.5	117.4	107.4	93.5	117.4	107.4	93.5	115.3	106.3	93.5	115.3	106.3
09	FINISHES	98.0	104.4	101.2	97.5	105.5	101.6	98.0	99.3	98.6	98.0	99.3	98.7	97.8	105.6	101.8	98.0	104.6	101.4
10 - 14	TOTAL DIV. 10000 - 14000	100.0	114.8	103.1	100.0	114.8	103.1	100.0	113.3	102.8	100.0	113.5	102.9	100.0	114.8	103.1	100.0	114.9	103.1
15	MECHANICAL	99.9	105.7	102.5	96.7	105.7	100.8	99.9	105.7	102.5	99.9	105.7	102.6	99.9	105.7	102.5	99.9	101.2	100.5
16	ELECTRICAL	104.4	99.0	100.7	99.0	99.0	99.0	104.4	90.5	94.9	104.4	140.4	128.9	103.6	101.0	101.8	104.4	102.4	103.1
01 - 16	WEIGHTED AVERAGE	102.4	105.4	103.9	99.5	105.2	102.3	102.9	102.8	102.9	103.1	111.3	107.0	103.0	105.8	104.3	103.0	104.8	103.8

DIVISION		D.C. – WASHINGTON 200-205			DELAWARE – DOVER 199			DELAWARE – NEWARK 197			DELAWARE – WILMINGTON 198			FLORIDA – DAYTONA BEACH 321			FLORIDA – FORT LAUDERDALE 333		
		MAT.	INST.	TOTAL	MAT.	INST.	TOTAL	MAT.	INST.	TOTAL	MAT.	INST.	TOTAL	MAT.	INST.	TOTAL	MAT.	INST.	TOTAL
02	SITE CONSTRUCTION	103.5	91.0	93.9	107.6	113.4	112.0	107.6	113.4	112.0	91.4	113.7	108.5	122.3	87.5	95.6	107.6	74.0	81.8
03100	CONCRETE FORMS & ACCESSORIES	96.9	80.0	82.4	102.2	93.2	94.5	102.2	93.2	94.5	102.2	93.2	94.4	97.3	67.9	72.0	94.4	64.5	68.8
03200	CONCRETE REINFORCEMENT	101.0	98.3	99.5	96.5	94.6	95.4	97.4	94.6	95.8	97.4	94.6	95.8	95.8	81.4	87.6	95.8	82.2	88.1
03300	CAST-IN-PLACE CONCRETE	116.4	85.9	103.7	84.2	93.2	87.9	84.2	93.2	87.9	79.1	93.2	84.9	92.9	72.0	84.2	97.4	71.1	86.5
03	CONCRETE	108.0	87.0	97.5	100.3	94.8	97.5	100.4	94.8	97.6	97.9	94.8	96.3	89.6	73.2	81.3	91.6	71.6	81.6
04	MASONRY	90.9	83.5	86.3	99.5	88.1	92.4	99.5	88.1	92.4	103.4	88.1	93.9	85.1	68.4	74.7	85.3	65.4	72.9
05	METALS	97.3	113.3	103.0	97.6	114.6	103.7	97.6	114.6	103.7	97.3	114.6	103.5	99.9	93.0	97.4	99.3	94.3	97.5
06	WOOD & PLASTICS	95.8	79.9	87.6	106.4	92.3	99.1	106.4	92.3	99.1	106.4	92.3	99.1	95.6	68.1	81.4	90.3	65.7	77.6
07	THERMAL & MOISTURE PROTECTION	94.1	84.5	89.6	99.3	102.2	100.7	99.4	102.2	100.7	99.0	102.2	100.5	96.9	70.5	84.5	96.9	69.0	83.9
08	DOORS & WINDOWS	100.7	89.2	97.9	98.7	97.9	98.5	98.7	97.9	98.5	98.4	97.9	98.3	99.7	66.3	91.5	97.3	65.3	89.5
09200	PLASTER & GYPSUM BOARD	100.4	79.1	87.2	102.7	91.9	96.0	102.7	91.9	96.0	102.4	91.9	95.9	108.3	67.6	83.1	107.8	65.2	81.4
09500	CEILINGS	102.5	79.1	87.1	104.0	91.9	96.1	104.0	91.9	96.1	102.6	91.9	95.6	96.3	67.6	77.4	96.3	65.2	75.9
09600	FLOORING	93.4	88.7	92.3	80.4	90.6	82.9	80.4	90.6	82.9	80.2	90.6	82.7	114.5	70.0	103.6	114.5	72.8	104.3
09900	PAINTS & COATINGS	98.9	95.9	97.2	86.3	96.1	92.0	86.3	96.1	92.0	86.3	96.1	92.0	110.0	90.5	98.6	106.3	62.6	80.8
09	FINISHES	95.1	82.7	88.8	95.3	92.5	93.9	95.4	92.5	93.9	95.0	92.5	93.7	107.1	70.5	88.4	105.2	65.4	84.8
10 - 14	TOTAL DIV. 10000 - 14000	100.0	94.5	98.8	100.0	109.6	102.0	100.0	109.6	102.0	100.0	109.6	102.0	100.0	82.2	96.2	100.0	80.3	95.8
15	MECHANICAL	100.0	91.7	96.3	100.3	100.1	100.2	100.3	100.1	100.2	100.3	100.1	100.2	99.9	72.7	87.6	99.9	68.3	85.6
16	ELECTRICAL	98.5	98.4	98.4	100.5	100.1	100.3	100.5	100.1	100.3	100.3	100.1	100.2	95.8	68.3	77.1	95.8	87.0	89.8
01 - 16	WEIGHTED AVERAGE	99.4	91.9	95.7	99.3	100.1	99.7	99.4	100.1	99.7	98.7	100.2	99.4	98.8	74.6	87.1	98.0	74.4	86.6

City Cost Indexes

DIVISION		FLORIDA																	
		FORT MYERS 339,341			GAINESVILLE 326,344			JACKSONVILLE 320,322			LAKELAND 338			MELBOURNE 329			MIAMI 330 - 332,340		
		MAT.	INST.	TOTAL	MAT.	INST.	TOTAL	MAT.	INST.	TOTAL	MAT.	INST.	TOTAL	MAT.	INST.	TOTAL	MAT.	INST.	TOTAL
02	SITE CONSTRUCTION	119.8	87.2	94.8	133.7	86.8	97.8	122.5	87.9	96.0	122.0	86.9	95.1	130.5	87.5	97.5	106.8	73.9	81.6
03100	CONCRETE FORMS & ACCESSORIES	89.0	57.1	61.6	90.3	60.2	64.5	97.0	60.8	65.9	84.0	57.2	61.0	90.9	67.5	70.8	94.2	64.5	68.7
03200	CONCRETE REINFORCEMENT	96.7	71.4	82.4	101.6	63.1	79.8	95.8	63.7	77.6	99.2	71.9	83.7	96.7	81.4	88.1	95.8	82.2	88.1
03300	CAST-IN-PLACE CONCRETE	101.6	63.7	85.8	106.6	54.8	85.0	93.8	59.8	79.6	103.9	64.9	87.6	107.8	71.5	92.7	94.9	71.5	85.1
03	CONCRETE	92.6	63.8	78.1	100.4	60.5	80.3	90.0	62.6	76.2	94.2	64.3	79.2	99.0	72.8	85.9	90.3	71.7	81.0
04	MASONRY	79.3	52.0	62.3	96.9	48.8	66.9	84.3	58.4	68.2	90.9	60.7	72.1	80.5	67.5	72.4	81.7	65.9	71.8
05	METALS	101.5	88.5	96.9	97.8	84.7	93.1	99.5	85.9	94.7	101.3	88.3	96.6	108.0	92.9	102.6	99.8	94.0	97.7
06	WOOD & PLASTICS	86.9	57.2	71.5	88.3	61.5	74.4	95.6	61.5	78.0	81.5	57.2	68.9	88.8	68.1	78.1	90.3	65.7	77.6
07	THERMAL & MOISTURE PROTECTION	96.6	57.3	78.2	97.2	62.0	80.7	97.2	65.5	82.4	96.6	59.4	79.2	97.1	70.1	84.5	100.0	70.6	86.2
08	DOORS & WINDOWS	98.4	57.8	88.5	97.8	57.2	87.9	99.7	57.9	89.5	98.4	55.1	87.8	98.9	66.3	91.0	97.3	66.0	89.7
09200	PLASTER & GYPSUM BOARD	104.7	56.3	74.7	104.9	60.8	77.6	108.3	60.8	78.9	102.6	56.3	73.9	105.1	67.6	81.9	107.8	65.2	81.4
09500	CEILINGS	92.1	56.3	68.6	90.8	60.8	71.1	96.3	60.8	73.0	90.8	56.3	68.2	92.1	67.6	76.0	96.3	65.2	75.9
09600	FLOORING	110.3	50.0	95.6	110.9	42.8	94.3	114.5	59.8	101.1	107.5	63.4	96.7	111.2	70.0	101.1	122.8	73.7	110.8
09900	PAINTS & COATINGS	110.0	58.4	79.8	110.0	56.2	78.6	110.0	60.4	81.0	110.0	58.4	79.8	110.0	89.4	97.9	106.3	62.6	80.8
09	FINISHES	104.3	55.4	79.4	105.3	56.3	80.3	107.2	60.3	83.2	103.0	58.2	80.1	105.4	70.2	87.4	107.8	65.5	86.2
10 - 14	TOTAL DIV. 10000 - 14000	100.0	74.4	94.6	100.0	79.8	95.7	100.0	74.2	94.6	100.0	74.4	94.6	100.0	81.6	96.2	100.0	80.3	95.8
15	MECHANICAL	98.2	59.1	80.5	99.6	64.7	83.8	99.9	61.3	82.4	98.2	62.2	81.9	99.9	72.2	87.4	99.9	68.3	85.6
16	ELECTRICAL	99.5	50.2	65.9	96.4	52.0	66.1	95.4	68.9	77.3	96.3	58.2	70.3	95.6	65.2	74.8	96.4	79.0	84.6
01 - 16	WEIGHTED AVERAGE	98.4	62.8	81.2	100.3	63.5	82.4	98.7	67.5	83.6	98.8	65.9	82.8	100.8	73.7	87.7	98.2	73.1	86.0

DIVISION		FLORIDA																	
		ORLANDO 327 - 328,347			PANAMA CITY 324			PENSACOLA 325			SARASOTA 342			ST. PETERSBURG 337			TALLAHASSEE 323		
		MAT.	INST.	TOTAL	MAT.	INST.	TOTAL	MAT.	INST.	TOTAL	MAT.	INST.	TOTAL	MAT.	INST.	TOTAL	MAT.	INST.	TOTAL
02	SITE CONSTRUCTION	123.0	87.2	95.5	137.8	85.0	97.3	135.3	87.2	98.4	124.0	87.0	95.6	123.3	87.0	95.4	123.8	86.6	95.3
03100	CONCRETE FORMS & ACCESSORIES	97.0	63.2	68.0	95.9	33.5	42.3	85.2	61.4	64.8	97.3	57.1	62.8	94.3	57.1	62.4	97.0	46.9	54.0
03200	CONCRETE REINFORCEMENT	95.8	81.4	87.6	99.9	62.4	78.7	102.4	62.8	79.9	95.8	71.8	82.2	99.2	71.8	83.7	95.8	63.1	77.2
03300	CAST-IN-PLACE CONCRETE	100.9	71.9	88.8	98.5	40.3	74.2	98.5	63.8	84.0	106.3	64.8	89.0	105.0	64.8	88.2	97.2	56.2	80.1
03	CONCRETE	93.5	71.1	82.2	98.1	43.1	70.5	96.9	64.1	80.4	96.2	64.2	80.1	95.8	64.2	79.9	91.7	55.1	73.3
04	MASONRY	82.8	68.4	73.8	88.9	34.0	54.7	86.4	61.3	70.8	86.1	60.7	70.3	126.0	60.7	85.3	87.1	47.7	62.5
05	METALS	109.2	92.5	103.3	98.2	71.8	88.7	98.1	85.6	93.6	102.9	88.2	97.7	102.0	88.2	97.1	99.9	84.3	94.3
06	WOOD & PLASTICS	95.6	62.3	78.4	94.2	33.7	62.9	82.8	62.3	72.2	95.6	57.2	75.7	92.5	57.2	74.2	95.6	45.3	69.6
07	THERMAL & MOISTURE PROTECTION	97.2	69.9	84.4	97.5	36.8	69.1	97.2	62.0	80.7	96.9	59.4	79.4	96.8	58.3	78.8	97.3	54.6	77.3
08	DOORS & WINDOWS	99.7	62.3	90.6	97.4	31.6	81.4	97.4	59.4	88.1	99.7	54.4	88.6	98.4	54.9	87.8	99.7	48.4	87.2
09200	PLASTER & GYPSUM BOARD	108.3	61.6	79.4	106.7	32.1	60.5	103.0	61.6	77.4	108.3	56.3	76.1	106.3	56.3	75.3	108.3	44.1	68.5
09500	CEILINGS	96.3	61.6	73.5	90.8	32.1	52.3	90.8	61.6	71.6	96.3	56.3	70.1	90.8	56.3	68.2	96.3	44.1	62.0
09600	FLOORING	114.5	70.0	103.6	113.8	23.0	91.7	108.2	63.5	97.3	114.5	63.1	101.9	112.9	63.4	100.8	114.5	46.5	97.9
09900	PAINTS & COATINGS	110.0	69.3	86.2	110.0	30.9	63.7	110.0	70.1	86.7	110.0	58.4	79.8	110.0	58.4	79.8	110.0	49.7	74.7
09	FINISHES	107.2	64.7	85.5	106.8	30.7	67.9	104.3	62.8	83.1	107.1	58.1	82.1	105.3	58.2	81.3	107.3	46.3	76.1
10 - 14	TOTAL DIV. 10000 - 14000	100.0	81.4	96.1	100.0	57.8	91.1	100.0	64.6	92.5	100.0	74.3	94.6	100.0	67.7	93.2	100.0	71.9	94.1
15	MECHANICAL	99.9	64.0	83.7	99.9	31.3	68.9	99.9	62.3	82.9	99.8	59.7	81.6	99.9	62.1	82.8	99.9	49.6	77.1
16	ELECTRICAL	96.4	55.3	68.4	94.1	41.4	58.2	99.5	61.9	73.9	95.8	45.4	61.5	96.3	59.7	71.4	96.4	51.2	65.6
01 - 16	WEIGHTED AVERAGE	100.7	69.4	85.5	99.8	44.6	73.1	99.4	67.0	83.7	100.2	63.2	82.3	101.8	65.9	84.4	99.3	57.4	79.0

DIVISION		FLORIDA						GEORGIA											
		TAMPA 335 - 336,346			WEST PALM BEACH 334,349			ALBANY 317			ATHENS 306			ATLANTA 300 - 303,399			AUGUSTA 308 - 309		
		MAT.	INST.	TOTAL	MAT.	INST.	TOTAL	MAT.	INST.	TOTAL	MAT.	INST.	TOTAL	MAT.	INST.	TOTAL	MAT.	INST.	TOTAL
02	SITE CONSTRUCTION	123.6	87.0	95.5	104.3	73.9	81.0	107.8	76.5	84.0	118.8	92.8	98.9	113.4	95.5	99.8	109.3	93.2	97.0
03100	CONCRETE FORMS & ACCESSORIES	98.6	57.2	63.1	98.9	63.8	68.8	96.6	51.1	57.5	93.1	45.6	52.4	98.0	78.2	81.0	94.6	61.8	66.4
03200	CONCRETE REINFORCEMENT	95.8	71.9	82.2	98.3	63.7	78.7	95.8	89.3	92.1	106.9	82.8	93.2	103.0	90.7	96.0	108.8	82.5	93.9
03300	CAST-IN-PLACE CONCRETE	102.7	64.9	86.9	92.6	66.2	81.6	99.2	48.3	77.9	103.7	50.9	81.7	103.7	74.0	91.3	98.0	56.9	80.8
03	CONCRETE	94.5	64.3	79.3	88.9	66.2	77.5	92.6	59.0	75.7	100.3	55.3	77.7	97.7	79.0	88.3	93.9	64.3	79.0
04	MASONRY	86.0	60.7	70.2	85.0	57.8	68.1	87.2	38.7	57.0	74.2	55.1	62.3	89.6	71.0	78.0	89.7	48.9	64.3
05	METALS	102.9	88.4	97.7	98.1	87.1	94.2	97.2	92.0	95.4	92.6	76.7	86.9	94.0	80.5	89.2	92.7	74.5	86.2
06	WOOD & PLASTICS	97.3	57.2	76.6	94.2	65.7	79.8	94.9	51.4	72.4	93.7	44.3	68.2	98.7	80.5	89.3	95.3	64.5	79.4
07	THERMAL & MOISTURE PROTECTION	97.2	59.4	79.5	96.6	65.6	82.1	97.0	56.9	78.2	93.9	53.5	75.0	94.0	75.9	85.5	93.5	58.6	77.1
08	DOORS & WINDOWS	99.7	55.2	88.8	96.6	61.2	87.9	97.3	56.1	87.3	93.0	52.3	83.1	96.6	77.2	91.9	93.0	62.0	85.5
09200	PLASTER & GYPSUM BOARD	108.3	56.3	76.1	108.4	65.2	81.6	108.3	50.4	72.5	115.8	42.9	70.6	117.1	80.3	94.3	116.2	63.8	83.7
09500	CEILINGS	96.3	56.3	70.1	92.1	65.2	74.4	96.3	50.4	66.2	103.8	42.9	63.8	103.8	80.3	88.4	103.8	63.8	77.5
09600	FLOORING	114.5	63.4	102.0	117.1	60.8	103.4	114.5	38.8	96.0	84.8	69.9	81.2	86.5	74.3	83.5	85.3	49.4	76.6
09900	PAINTS & COATINGS	110.0	58.4	79.8	106.3	57.9	78.0	106.3	49.8	73.3	92.9	41.5	62.8	92.9	77.6	84.0	92.9	47.3	66.2
09	FINISHES	107.2	58.2	82.2	105.2	62.4	83.3	105.2	47.6	75.8	93.5	48.6	70.6	93.8	77.4	85.4	93.1	58.1	75.2
10 - 14	TOTAL DIV. 10000 - 14000	100.0	74.4	94.6	100.0	80.2	95.8	100.0	75.0	94.7	100.0	61.8	91.9	100.0	83.2	96.5	100.0	72.4	94.2
15	MECHANICAL	99.9	62.2	82.8	98.2	67.7	84.4	99.9	55.0	79.6	96.9	70.7	85.0	100.1	79.7	90.8	100.1	52.3	78.4
16	ELECTRICAL	95.3	59.8	71.1	95.9	64.5	74.5	89.9	66.4	73.9	93.9	78.6	83.5	92.7	84.7	87.3	94.9	56.3	68.6
01 - 16	WEIGHTED AVERAGE	100.0	66.2	83.6	97.0	67.8	82.8	97.5	61.1	79.9	95.1	66.9	81.4	96.8	80.9	89.1	95.7	62.2	79.5

City Cost Indexes

		GEORGIA																	
		COLUMBUS			DALTON			GAINESVILLE			MACON			SAVANNAH			STATESBORO		
DIVISION		318 - 319			307			305			310 - 312			313 - 314			304		
		MAT.	INST.	TOTAL	MAT.	INST.	TOTAL	MAT.	INST.	TOTAL	MAT.	INST.	TOTAL	MAT.	INST.	TOTAL	MAT.	INST.	TOTAL
02	SITE CONSTRUCTION	107.7	76.9	84.1	119.7	95.9	101.5	117.6	91.1	97.3	108.5	94.9	98.0	108.3	77.9	85.0	119.8	74.3	84.9
03100	CONCRETE FORMS & ACCESSORIES	98.2	65.4	70.1	82.9	24.9	33.1	97.6	43.1	50.8	95.6	66.1	70.3	96.5	61.3	66.3	75.7	26.8	33.8
03200	CONCRETE REINFORCEMENT	95.8	89.5	92.2	105.0	65.3	82.5	105.8	80.9	91.7	98.1	89.5	93.2	101.4	82.8	90.9	106.8	23.6	59.6
03300	CAST-IN-PLACE CONCRETE	98.8	48.9	78.0	100.7	35.6	73.6	109.0	35.5	78.4	97.5	52.7	78.8	95.7	57.0	79.6	103.5	33.4	74.3
03	CONCRETE	92.5	65.6	79.0	99.0	38.1	68.4	102.0	48.4	75.1	92.0	67.2	79.5	91.6	65.3	78.4	99.3	31.3	65.1
04	MASONRY	87.2	42.6	59.4	77.0	30.9	48.3	85.5	27.4	49.3	102.3	46.6	67.6	90.9	59.7	71.5	76.2	26.9	45.5
05	METALS	97.6	92.7	95.8	95.3	68.6	85.7	92.6	69.4	84.3	92.5	93.1	92.7	97.9	90.5	95.3	98.0	69.7	87.8
06	WOOD & PLASTICS	96.8	70.9	83.4	76.4	23.6	49.1	98.4	44.3	70.4	100.6	69.6	84.6	95.0	60.7	77.3	68.6	26.3	46.7
07	THERMAL & MOISTURE PROTECTION	96.7	59.9	79.4	99.1	31.6	67.5	93.9	42.0	69.6	95.5	64.2	80.9	97.1	60.0	79.7	94.7	28.7	63.8
08	DOORS & WINDOWS	97.3	66.9	89.9	93.9	21.6	76.3	93.0	45.0	81.3	95.8	67.2	88.8	97.3	59.3	88.1	94.8	23.5	77.4
09200	PLASTER & GYPSUM BOARD	108.3	70.5	84.9	100.0	21.6	51.4	117.1	42.9	71.2	113.6	69.1	86.1	108.3	60.0	78.4	102.3	24.4	54.0
09500	CEILINGS	96.3	70.5	79.4	116.9	21.6	54.3	103.8	42.9	63.8	88.6	69.1	75.8	96.3	60.0	72.5	113.3	24.4	54.9
09600	FLOORING	114.5	45.2	97.6	85.8	16.7	68.9	86.4	16.7	69.4	89.4	45.6	78.7	114.5	58.2	100.7	99.9	58.2	89.8
09900	PAINTS & COATINGS	106.3	47.7	72.1	83.3	23.9	48.6	92.9	41.5	62.8	108.1	58.3	79.0	106.3	59.1	78.8	91.4	21.4	50.5
09	FINISHES	105.1	60.0	82.1	100.5	22.7	60.8	94.0	37.8	65.3	91.5	61.4	76.1	105.4	60.4	82.4	103.7	32.1	67.1
10 - 14	TOTAL DIV. 10000 - 14000	100.0	77.5	95.2	100.0	23.7	83.9	100.0	61.7	91.9	100.0	79.0	95.6	100.0	73.7	94.4	100.0	56.8	90.9
15	MECHANICAL	99.9	45.0	75.0	97.0	26.3	65.0	96.9	22.5	63.2	99.9	50.7	77.6	99.9	55.3	79.7	97.2	23.9	64.0
16	ELECTRICAL	91.2	48.6	62.2	101.3	19.9	45.8	93.9	78.6	83.5	89.4	68.1	74.9	92.6	65.3	74.0	94.1	25.0	47.0
01 - 16	WEIGHTED AVERAGE	97.7	59.8	79.3	97.0	37.5	68.1	96.0	51.3	74.4	95.9	66.9	81.8	98.0	65.6	82.3	97.1	36.5	67.7

		GEORGIA						HAWAII						IDAHO		
		VALDOSTA			WAYCROSS			HILO			HONOLULU			STATES & POSS., GUAM			BOISE		
DIVISION		316			315			967			968			969			836 - 837		
		MAT.	INST.	TOTAL	MAT.	INST.	TOTAL	MAT.	INST.	TOTAL	MAT.	INST.	TOTAL	MAT.	INST.	TOTAL	MAT.	INST.	TOTAL
02	SITE CONSTRUCTION	118.8	77.2	86.9	115.5	75.4	84.8	119.5	108.5	111.1	120.9	108.5	111.4	119.5	101.3	105.5	84.5	102.0	97.9
03100	CONCRETE FORMS & ACCESSORIES	81.8	52.8	56.9	84.4	34.2	41.3	105.1	150.6	144.2	105.5	150.6	144.2	105.3	38.5	48.0	99.9	88.9	90.5
03200	CONCRETE REINFORCEMENT	101.6	52.0	73.5	101.6	78.5	88.5	108.0	124.9	117.6	107.1	124.9	117.2	107.1	27.6	62.0	100.4	83.8	91.0
03300	CAST-IN-PLACE CONCRETE	97.1	57.7	80.6	109.6	38.3	79.9	190.8	129.3	165.1	183.9	129.3	161.1	192.5	47.9	132.2	100.8	95.2	98.5
03	CONCRETE	96.5	56.2	76.2	100.8	46.2	73.3	158.3	136.6	147.4	154.8	136.6	145.7	159.0	41.0	99.7	104.5	89.9	97.1
04	MASONRY	93.8	53.8	68.9	94.7	34.4	57.2	132.4	134.9	134.0	132.4	134.9	134.0	131.5	35.3	71.6	136.9	85.9	105.1
05	METALS	97.4	80.0	91.2	96.7	86.6	93.1	117.0	109.1	114.2	116.2	109.1	113.7	115.9	56.5	94.6	113.4	81.5	101.9
06	WOOD & PLASTICS	79.0	50.1	64.1	81.7	32.9	56.5	101.6	154.5	128.9	101.5	154.5	128.9	101.5	38.6	69.0	96.1	88.3	92.1
07	THERMAL & MOISTURE PROTECTION	96.7	62.6	80.7	96.5	39.8	70.0	111.3	133.6	121.7	111.3	133.6	122.8	111.6	43.0	79.5	97.1	85.0	91.4
08	DOORS & WINDOWS	93.0	46.5	81.7	93.0	42.0	80.6	103.5	144.4	113.5	108.9	144.4	117.5	103.3	35.2	86.7	94.8	82.2	91.7
09200	PLASTER & GYPSUM BOARD	102.1	49.1	69.3	103.0	31.3	58.6	94.7	156.2	132.8	112.3	156.2	139.5	94.1	36.7	58.6	90.1	87.7	88.6
09500	CEILINGS	90.8	49.1	63.4	90.8	31.3	51.7	119.5	156.2	143.6	116.8	156.2	142.6	116.8	36.7	64.2	103.2	87.7	93.0
09600	FLOORING	106.5	46.5	91.9	107.9	38.8	91.1	123.4	131.9	125.5	121.9	131.9	124.3	121.9	36.5	101.1	105.2	67.5	96.0
09900	PAINTS & COATINGS	106.3	43.2	69.4	106.3	41.5	68.4	117.5	152.3	137.8	116.9	152.3	137.6	116.9	43.3	73.9	111.0	63.2	83.0
09	FINISHES	101.4	50.0	75.1	101.7	34.7	67.5	118.9	148.7	134.1	120.4	148.7	134.9	117.4	38.6	77.1	100.1	82.5	91.1
10 - 14	TOTAL DIV. 10000 - 14000	100.0	72.2	94.1	100.0	58.4	91.2	100.0	131.2	106.6	100.0	131.2	106.6	100.0	83.1	96.4	100.0	85.0	96.8
15	MECHANICAL	99.9	48.4	76.6	97.9	31.1	67.6	100.2	125.6	111.7	100.2	125.6	111.7	100.2	33.9	70.2	99.9	86.5	93.8
16	ELECTRICAL	88.2	37.5	53.6	95.0	68.3	76.8	110.8	124.3	120.0	114.0	124.3	121.0	111.2	35.4	59.6	82.3	79.7	80.5
01 - 16	WEIGHTED AVERAGE	97.5	55.0	76.9	98.0	50.9	75.2	115.6	128.8	122.0	116.2	128.8	122.3	115.4	46.1	81.8	102.1	86.1	94.3

		IDAHO												ILLINOIS		
		COEUR D'ALENE			IDAHO FALLS			LEWISTON			POCATELLO			TWIN FALLS			BLOOMINGTON		
DIVISION		838			834			835			832			833			617		
		MAT.	INST.	TOTAL	MAT.	INST.	TOTAL	MAT.	INST.	TOTAL	MAT.	INST.	TOTAL	MAT.	INST.	TOTAL	MAT.	INST.	TOTAL
02	SITE CONSTRUCTION	84.6	94.2	91.9	82.3	100.6	96.3	89.8	95.5	94.2	86.7	102.0	98.4	94.3	100.0	98.7	88.6	95.2	93.7
03100	CONCRETE FORMS & ACCESSORIES	103.6	47.4	55.4	93.0	44.0	51.0	110.8	79.8	84.2	99.9	88.6	90.2	103.3	42.9	51.5	84.4	106.9	103.7
03200	CONCRETE REINFORCEMENT	113.3	94.1	102.4	108.1	67.1	84.9	113.3	101.2	106.5	100.7	83.3	90.8	109.4	60.0	81.4	99.8	112.7	107.1
03300	CAST-IN-PLACE CONCRETE	105.2	82.9	95.9	92.6	64.5	80.9	109.3	93.5	102.7	100.1	95.1	98.0	102.7	51.6	81.4	93.8	99.9	96.4
03	CONCRETE	112.3	69.5	90.8	97.1	56.5	76.7	116.3	88.6	102.4	104.1	89.6	96.8	113.5	50.2	81.7	90.5	105.8	98.2
04	MASONRY	136.9	66.3	92.9	131.0	39.4	73.9	137.0	95.5	111.1	131.4	77.3	97.7	135.0	48.0	80.8	107.1	106.0	106.4
05	METALS	98.8	87.2	94.7	113.0	72.9	98.6	97.0	89.6	94.3	113.7	80.6	101.8	113.6	69.2	97.7	94.1	111.4	100.3
06	WOOD & PLASTICS	93.4	41.9	66.8	86.8	41.6	63.5	101.1	74.3	87.2	96.1	88.3	92.1	99.6	43.0	70.4	83.0	105.0	94.3
07	THERMAL & MOISTURE PROTECTION	167.1	72.5	122.8	96.7	57.7	78.4	167.0	87.2	129.6	97.3	79.5	89.0	98.3	55.1	78.1	95.4	101.4	98.2
08	DOORS & WINDOWS	115.4	55.8	100.9	98.1	48.1	85.9	115.4	77.5	106.2	94.8	76.9	90.4	98.2	43.6	84.9	92.5	104.8	95.5
09200	PLASTER & GYPSUM BOARD	152.8	40.1	83.0	86.0	39.6	57.3	155.0	73.4	104.5	90.1	87.7	88.6	89.9	41.0	59.6	102.6	104.7	103.9
09500	CEILINGS	148.8	40.1	77.4	96.3	39.6	59.1	148.8	73.4	99.3	103.2	87.7	93.0	97.7	41.0	60.5	90.9	104.7	100.0
09600	FLOORING	143.2	71.8	125.8	102.1	67.5	93.6	146.5	99.2	135.0	105.5	67.5	96.3	106.9	67.5	97.3	94.2	103.4	96.5
09900	PAINTS & COATINGS	138.8	79.0	103.9	111.0	52.3	76.7	138.8	79.0	103.9	111.0	66.2	84.8	111.0	39.0	68.9	87.6	96.9	93.0
09	FINISHES	169.7	53.4	110.3	96.8	48.2	72.0	171.2	82.4	125.9	100.2	82.8	91.3	100.2	46.7	72.9	91.8	105.1	98.6
10 - 14	TOTAL DIV. 10000 - 14000	100.0	74.1	94.5	100.0	61.5	91.9	100.0	97.3	99.4	100.0	85.0	96.8	100.0	60.2	91.6	100.0	102.9	100.6
15	MECHANICAL	109.4	58.3	86.2	102.2	61.6	83.8	100.9	94.8	98.1	99.9	86.4	93.8	99.9	44.9	75.0	97.1	109.0	102.4
16	ELECTRICAL	87.7	46.8	59.8	85.9	82.0	83.3	84.6	91.5	89.3	85.4	82.1	83.1	88.0	39.4	54.9	95.7	95.5	95.5
01 - 16	WEIGHTED AVERAGE	115.3	64.7	90.8	101.4	64.7	83.6	113.8	90.5	102.5	102.1	85.2	93.9	104.3	53.5	79.7	95.0	103.8	99.3

City Cost Indexes

ILLINOIS

DIVISION		CARBONDALE 629			CENTRALIA 628			CHAMPAIGN 618 - 619			CHICAGO 606			DECATUR 625			EAST ST. LOUIS 620 - 622		
		MAT.	INST.	TOTAL	MAT.	INST.	TOTAL	MAT.	INST.	TOTAL	MAT.	INST.	TOTAL	MAT.	INST.	TOTAL	MAT.	INST.	TOTAL
02	SITE CONSTRUCTION	98.1	96.2	96.6	98.5	96.7	97.1	97.9	96.3	96.7	83.3	93.7	91.3	81.9	96.2	92.9	100.2	96.9	97.7
03100	CONCRETE FORMS & ACCESSORIES	96.0	96.5	96.4	99.0	105.9	104.9	93.2	104.2	102.6	104.7	129.0	125.5	100.4	103.5	103.0	92.6	106.0	104.1
03200	CONCRETE REINFORCEMENT	100.1	94.5	96.9	100.1	112.5	107.1	99.8	109.8	105.5	94.8	158.2	130.7	93.9	105.0	100.2	99.4	112.5	106.9
03300	CAST-IN-PLACE CONCRETE	87.8	97.0	91.6	88.3	96.0	91.5	108.9	107.4	108.3	100.2	129.2	112.3	95.9	94.3	95.2	89.8	110.6	98.5
03	CONCRETE	82.6	97.6	90.2	83.2	104.7	94.0	102.4	106.6	104.5	97.1	133.6	115.4	91.9	100.9	96.4	83.7	109.9	96.9
04	MASONRY	73.9	99.8	90.0	73.9	113.7	98.7	129.3	102.3	112.5	94.2	131.0	117.2	70.2	102.8	90.5	74.0	113.8	98.8
05	METALS	93.2	115.2	101.1	93.3	123.1	104.0	94.1	109.1	99.5	96.9	125.0	107.0	96.7	108.1	100.7	94.1	124.0	104.8
06	WOOD & PLASTICS	99.7	94.4	96.9	103.0	103.8	103.4	92.1	102.6	97.5	101.4	127.4	114.8	102.5	102.6	102.5	96.1	103.8	100.1
07	THERMAL & MOISTURE PROTECTION	91.6	93.6	92.5	91.7	102.8	96.9	96.2	104.1	99.9	101.1	127.6	113.5	96.6	94.9	95.8	91.6	104.5	97.6
08	DOORS & WINDOWS	88.6	103.5	92.3	88.6	110.8	94.0	93.3	102.0	95.4	104.6	135.6	112.2	99.5	101.7	100.0	88.7	107.5	93.3
09200	PLASTER & GYPSUM BOARD	107.2	93.8	98.9	108.1	103.5	105.3	104.9	102.2	103.2	98.2	127.7	116.5	109.2	102.2	104.9	106.2	103.5	104.6
09500	CEILINGS	90.9	93.8	92.8	90.9	103.5	99.2	90.9	102.2	98.3	92.0	127.7	115.4	96.4	102.2	100.2	90.9	103.5	99.2
09600	FLOORING	112.3	89.6	106.7	113.6	78.3	105.0	97.6	98.3	97.7	85.0	126.5	95.1	101.5	99.4	101.0	110.9	63.3	99.3
09900	PAINTS & COATINGS	95.3	89.6	92.0	95.3	93.9	94.5	87.6	103.0	96.6	76.5	133.7	109.9	87.6	98.2	93.8	95.3	100.4	98.3
09	FINISHES	96.4	93.6	95.0	97.0	99.7	98.4	93.7	103.0	98.4	86.2	128.8	108.0	96.1	102.7	99.5	95.9	96.9	96.4
10 - 14	TOTAL DIV. 10000 - 14000	100.0	105.2	101.1	100.0	108.5	101.8	100.0	101.3	100.3	100.0	126.6	105.6	100.0	101.4	100.3	100.0	106.1	101.3
15	MECHANICAL	97.0	96.3	96.7	97.0	93.7	95.5	97.1	102.6	99.6	100.2	123.5	110.8	100.3	98.4	99.4	100.2	100.7	100.4
16	ELECTRICAL	95.1	98.9	97.6	97.4	101.6	100.3	101.1	87.5	91.9	100.6	127.3	118.8	100.3	87.0	91.3	96.7	106.3	103.3
01 - 16	WEIGHTED AVERAGE	92.4	99.0	95.6	92.7	103.5	98.0	98.7	100.6	99.6	97.6	124.7	110.7	95.9	98.6	97.2	93.4	105.9	99.5

ILLINOIS

DIVISION		EFFINGHAM 624			GALESBURG 614			JOLIET 604			KANKAKEE 609			LA SALLE 613			NORTH SUBURBAN 600 - 603		
		MAT.	INST.	TOTAL	MAT.	INST.	TOTAL	MAT.	INST.	TOTAL	MAT.	INST.	TOTAL	MAT.	INST.	TOTAL	MAT.	INST.	TOTAL
02	SITE CONSTRUCTION	89.8	96.2	94.7	90.9	95.2	94.2	83.3	92.7	90.5	78.0	92.3	89.0	90.4	95.4	94.2	82.6	92.7	90.3
03100	CONCRETE FORMS & ACCESSORIES	106.2	102.8	103.2	93.0	106.4	104.5	105.7	121.1	118.9	96.2	110.8	108.7	112.1	95.1	97.5	104.6	123.5	120.6
03200	CONCRETE REINFORCEMENT	99.8	105.8	103.2	97.5	113.1	106.3	94.8	141.4	121.2	99.0	123.6	112.9	98.2	121.6	111.5	94.8	151.2	126.7
03300	CAST-IN-PLACE CONCRETE	95.6	84.3	90.9	96.6	97.1	96.8	100.1	115.5	106.5	93.4	113.1	101.6	96.6	105.8	100.4	100.2	124.0	110.1
03	CONCRETE	93.3	97.3	95.3	92.9	104.7	98.8	97.1	122.1	109.7	91.6	113.5	102.6	94.3	104.3	99.3	97.1	127.8	112.5
04	MASONRY	77.7	95.9	89.0	107.3	107.8	107.6	95.8	123.3	112.9	93.7	112.9	105.7	107.3	106.5	106.8	94.2	125.4	113.6
05	METALS	94.0	108.0	99.0	94.1	111.2	100.2	95.0	115.4	102.3	95.0	109.5	100.2	94.1	115.1	101.7	96.1	118.6	104.1
06	WOOD & PLASTICS	105.3	102.6	103.9	91.9	105.3	98.8	103.3	120.1	112.0	93.7	109.6	101.9	111.4	90.2	100.4	101.4	122.0	112.1
07	THERMAL & MOISTURE PROTECTION	96.0	96.4	96.2	95.7	102.1	98.7	100.9	122.4	110.9	99.7	116.1	107.4	96.0	100.1	97.9	101.1	121.0	110.4
08	DOORS & WINDOWS	93.2	105.0	96.0	92.5	102.4	94.9	104.5	127.6	110.1	95.4	116.7	100.6	92.5	94.3	93.0	104.6	128.8	110.5
09200	PLASTER & GYPSUM BOARD	108.5	102.2	104.6	104.9	105.3	105.0	96.4	120.3	111.2	94.1	109.5	103.6	110.4	89.5	97.4	98.2	122.3	113.1
09500	CEILINGS	90.9	102.2	98.3	90.9	105.0	100.2	92.0	120.3	110.6	92.0	109.5	103.5	90.9	89.5	90.0	92.0	122.3	111.9
09600	FLOORING	102.6	98.3	101.6	97.5	101.2	98.4	84.5	121.3	93.5	81.8	113.3	89.5	104.8	86.2	100.3	85.0	121.3	93.9
09900	PAINTS & COATINGS	87.6	89.1	88.5	87.6	103.1	96.7	74.1	107.5	93.6	74.1	95.8	86.8	87.6	87.0	87.2	76.5	118.0	100.7
09	FINISHES	95.4	100.8	98.2	93.2	105.4	99.5	85.7	119.6	103.0	84.2	108.7	96.7	96.4	91.4	93.8	86.2	122.2	104.6
10 - 14	TOTAL DIV. 10000 - 14000	100.0	100.3	100.1	100.0	102.6	100.6	100.0	123.6	105.0	100.0	118.3	103.9	100.0	100.2	100.0	100.0	122.8	104.8
15	MECHANICAL	97.1	101.1	98.9	97.1	104.4	100.4	100.4	117.7	108.2	97.1	106.0	101.1	97.1	108.2	102.1	100.2	112.2	105.6
16	ELECTRICAL	96.8	98.8	98.2	97.2	78.1	84.2	100.2	95.7	97.2	92.1	90.6	91.1	92.4	90.6	91.2	99.4	119.3	112.9
01 - 16	WEIGHTED AVERAGE	94.5	99.9	97.1	95.7	100.0	97.8	97.3	113.4	105.1	93.7	105.4	99.4	96.1	100.7	98.3	97.3	117.8	107.3

ILLINOIS

DIVISION		PEORIA 615 - 616			QUINCY 623			ROCKFORD 610 - 611			ROCK ISLAND 612			SOUTH SUBURBAN 605			SPRINGFIELD 626 - 627		
		MAT.	INST.	TOTAL	MAT.	INST.	TOTAL	MAT.	INST.	TOTAL	MAT.	INST.	TOTAL	MAT.	INST.	TOTAL	MAT.	INST.	TOTAL
02	SITE CONSTRUCTION	89.9	95.0	93.8	88.7	95.8	94.2	90.1	95.6	94.3	89.1	94.6	93.3	82.6	92.5	90.2	88.5	96.2	94.4
03100	CONCRETE FORMS & ACCESSORIES	98.4	107.2	106.0	103.1	100.7	101.0	103.4	108.2	107.5	95.5	99.8	99.2	104.6	122.5	119.9	102.7	104.2	104.0
03200	CONCRETE REINFORCEMENT	93.9	112.8	104.6	98.2	98.9	98.6	93.9	137.7	118.7	97.5	112.9	106.2	94.8	151.2	126.7	93.9	105.1	100.2
03300	CAST-IN-PLACE CONCRETE	93.7	107.3	99.4	95.8	96.9	96.2	95.9	105.3	99.8	94.6	99.2	96.5	100.2	124.0	110.1	89.1	102.9	94.9
03	CONCRETE	90.7	108.5	99.7	92.6	99.4	96.0	92.1	112.9	102.6	91.5	102.4	97.0	97.1	127.4	112.4	88.7	104.2	96.5
04	MASONRY	113.0	109.4	110.8	96.5	92.0	93.7	84.6	113.3	102.5	107.1	100.1	102.8	94.2	123.3	112.3	68.9	103.5	90.4
05	METALS	96.7	111.5	102.0	94.0	104.1	97.6	96.7	121.6	105.6	94.1	109.5	99.6	96.1	118.6	104.1	96.7	107.5	100.6
06	WOOD & PLASTICS	102.5	105.1	103.9	102.3	102.6	102.4	102.5	104.9	103.7	94.5	99.0	96.8	101.4	122.0	112.1	102.3	102.6	102.4
07	THERMAL & MOISTURE PROTECTION	96.4	102.9	99.4	96.0	90.4	93.4	96.2	111.2	103.2	95.6	98.5	97.0	101.1	120.1	110.0	96.1	101.3	98.6
08	DOORS & WINDOWS	99.5	104.9	100.8	94.1	99.3	95.4	99.5	111.7	102.4	92.5	98.8	94.1	104.6	127.9	110.3	99.5	100.8	99.8
09200	PLASTER & GYPSUM BOARD	109.2	104.9	106.5	108.1	102.2	104.5	109.2	104.6	106.3	105.8	98.5	101.3	98.2	122.3	113.1	109.2	102.2	104.9
09500	CEILINGS	96.4	104.9	102.0	90.9	102.2	98.3	96.4	104.6	101.8	90.9	98.5	95.9	92.0	122.3	111.9	96.4	102.2	100.2
09600	FLOORING	101.5	101.2	101.4	101.5	93.4	99.5	101.5	98.3	100.7	98.6	92.9	97.2	85.0	121.3	93.9	101.7	99.5	101.2
09900	PAINTS & COATINGS	87.6	97.2	93.2	87.6	63.7	73.6	87.6	106.3	98.5	87.6	95.4	92.1	76.5	117.2	100.3	87.6	101.1	95.5
09	FINISHES	96.1	105.0	100.7	95.0	95.8	95.4	96.1	105.5	100.9	93.7	98.0	95.9	86.2	121.6	104.3	96.2	103.2	99.8
10 - 14	TOTAL DIV. 10000 - 14000	100.0	103.3	100.7	100.0	98.3	99.6	100.0	112.2	102.6	100.0	96.1	99.2	100.0	121.9	104.6	100.0	101.6	100.3
15	MECHANICAL	100.3	101.9	101.0	97.1	99.8	98.3	100.3	109.2	104.3	97.1	102.0	99.3	100.2	111.1	105.1	100.3	99.5	99.9
16	ELECTRICAL	99.2	94.0	95.7	92.6	81.3	84.9	99.2	112.1	108.0	84.6	93.4	90.6	99.4	119.5	112.9	100.3	95.5	97.0
01 - 16	WEIGHTED AVERAGE	98.2	103.0	100.5	95.2	95.2	95.2	96.8	110.1	103.3	94.6	99.5	97.0	97.3	117.2	106.9	95.6	100.9	98.2

City Cost Indexes

DIVISION		INDIANA																	
		ANDERSON			BLOOMINGTON			COLUMBUS			EVANSVILLE			FORT WAYNE			GARY		
		460			474			472			476 - 477			467 - 468			463 - 464		
		MAT.	INST.	TOTAL	MAT.	INST.	TOTAL	MAT.	INST.	TOTAL	MAT.	INST.	TOTAL	MAT.	INST.	TOTAL	MAT.	INST.	TOTAL
02	SITE CONSTRUCTION	84.3	95.1	92.6	77.5	95.7	91.5	75.0	95.6	90.8	83.0	130.8	119.6	84.7	95.3	92.9	84.9	97.1	94.3
03100	CONCRETE FORMS & ACCESSORIES	90.3	82.5	83.6	103.3	78.1	81.7	92.2	79.5	81.3	91.0	85.1	86.0	89.9	77.8	79.5	90.8	101.1	99.7
03200	CONCRETE REINFORCEMENT	97.3	85.6	90.6	89.7	83.4	86.1	90.0	83.4	86.3	98.2	82.0	89.0	97.3	82.3	88.8	97.3	100.6	99.2
03300	CAST-IN-PLACE CONCRETE	103.1	84.0	95.1	104.1	81.4	94.6	103.7	81.9	94.6	99.4	90.0	95.5	109.5	85.6	99.5	107.6	102.0	105.3
03	CONCRETE	96.9	84.1	90.5	103.2	79.9	91.5	102.0	80.7	91.3	105.8	86.3	96.0	100.0	82.0	91.0	99.2	101.5	100.3
04	MASONRY	92.6	84.5	87.6	92.1	83.5	86.7	92.0	83.4	86.7	88.1	86.6	87.2	90.2	85.9	87.5	94.0	101.2	98.5
05	METALS	99.3	92.3	96.8	96.2	76.8	89.2	96.2	76.7	89.2	89.8	88.1	89.2	99.3	91.0	96.3	99.3	103.0	100.6
06	WOOD & PLASTICS	99.6	81.9	90.4	116.5	76.0	95.6	104.7	77.7	90.8	92.1	84.7	88.3	99.2	75.6	87.0	98.6	100.6	99.6
07	THERMAL & MOISTURE PROTECTION	98.5	82.5	91.0	92.8	85.0	89.2	92.2	85.3	89.0	96.2	89.4	93.0	97.9	84.0	91.4	97.7	102.0	99.7
08	DOORS & WINDOWS	102.2	84.4	97.8	105.8	80.1	99.5	101.1	82.1	96.5	98.4	83.8	94.8	102.2	78.4	96.4	102.2	100.1	101.7
09200	PLASTER & GYPSUM BOARD	110.3	81.8	92.7	107.9	76.0	88.2	103.8	77.8	87.7	103.3	83.1	90.8	105.7	75.4	86.9	106.6	101.2	103.2
09500	CEILINGS	92.0	81.8	85.3	83.5	76.0	78.6	83.5	77.8	79.7	95.1	83.1	87.2	92.0	75.4	81.1	92.0	101.2	98.0
09600	FLOORING	89.4	79.5	87.0	102.9	84.5	98.4	97.8	84.5	94.5	96.7	83.2	93.4	89.4	84.7	88.3	89.4	108.2	94.0
09900	PAINTS & COATINGS	89.1	80.2	83.9	90.8	90.1	90.4	90.8	90.1	90.4	97.4	92.2	94.4	89.1	80.2	83.9	89.1	94.2	92.1
09	FINISHES	92.8	81.8	87.2	97.2	80.1	88.5	94.9	81.1	87.9	96.5	85.4	90.9	92.0	78.9	85.3	92.2	101.6	97.0
10 - 14	TOTAL DIV. 10000 - 14000	100.0	89.5	97.8	100.0	88.2	97.5	100.0	88.5	97.6	100.0	94.1	98.8	100.0	89.0	97.7	100.0	107.5	101.6
15	MECHANICAL	99.8	87.8	94.4	99.8	86.5	93.8	96.5	87.9	92.6	100.0	84.9	93.2	99.8	86.7	93.8	99.8	100.7	100.2
16	ELECTRICAL	82.1	93.4	89.8	101.0	88.2	92.3	99.7	93.4	95.4	95.9	89.1	91.3	82.5	87.7	86.1	89.1	105.5	100.3
01 - 16	WEIGHTED AVERAGE	96.7	88.1	92.5	99.1	84.5	92.0	97.1	86.0	91.7	96.9	91.2	94.2	96.9	86.0	91.6	97.5	101.8	99.6

DIVISION		INDIANA																	
		INDIANAPOLIS			KOKOMO			LAFAYETTE			LAWRENCEBURG			MUNCIE			NEW ALBANY		
		461 - 462			469			479			470			473			471		
		MAT.	INST.	TOTAL	MAT.	INST.	TOTAL	MAT.	INST.	TOTAL	MAT.	INST.	TOTAL	MAT.	INST.	TOTAL	MAT.	INST.	TOTAL
02	SITE CONSTRUCTION	84.9	99.0	95.8	81.6	94.8	91.7	75.7	95.4	90.8	73.3	116.0	106.0	77.1	96.0	91.6	71.3	100.5	93.7
03100	CONCRETE FORMS & ACCESSORIES	90.8	87.1	87.6	100.0	75.3	78.8	89.2	80.5	81.7	87.8	74.4	76.3	88.5	82.2	83.1	87.2	75.9	77.5
03200	CONCRETE REINFORCEMENT	96.6	85.6	90.4	88.0	77.0	81.8	89.7	81.7	85.2	89.0	75.0	81.1	99.1	85.6	91.4	90.3	75.6	81.9
03300	CAST-IN-PLACE CONCRETE	99.2	87.4	94.3	102.0	88.8	96.5	104.2	82.8	95.3	97.4	75.7	88.4	109.4	79.8	97.1	100.6	75.4	90.1
03	CONCRETE	99.3	86.4	92.8	94.6	81.2	87.8	102.2	81.1	91.6	95.3	75.9	85.5	101.5	82.6	92.0	100.7	76.0	88.3
04	MASONRY	99.9	85.8	91.1	92.4	83.3	86.7	99.0	83.1	89.1	77.9	78.9	78.5	94.1	84.6	88.2	84.5	77.4	80.1
05	METALS	101.3	79.8	93.6	96.2	90.7	94.2	94.7	75.3	87.7	90.8	89.3	90.3	97.8	92.2	95.8	93.1	82.3	89.2
06	WOOD & PLASTICS	97.1	87.3	92.1	110.0	72.8	90.7	101.4	79.6	90.1	90.8	72.6	81.4	101.9	81.7	91.5	93.2	76.0	84.3
07	THERMAL & MOISTURE PROTECTION	97.2	87.9	92.8	98.2	81.5	90.3	92.2	84.7	88.7	92.2	78.7	85.9	94.5	82.6	88.9	87.3	76.4	82.2
08	DOORS & WINDOWS	108.3	87.3	103.2	96.2	78.4	91.9	99.6	82.1	95.4	100.5	75.3	94.3	100.5	84.3	96.6	98.0	76.9	92.9
09200	PLASTER & GYPSUM BOARD	106.5	87.2	94.6	112.4	72.4	87.7	101.6	79.7	88.0	95.0	72.2	80.9	102.2	81.8	89.6	101.6	75.5	85.4
09500	CEILINGS	90.6	87.2	88.4	86.4	72.4	77.3	79.4	79.7	79.6	95.1	72.2	80.1	84.9	81.8	82.9	90.9	75.5	80.8
09600	FLOORING	89.2	94.1	90.4	93.5	80.8	90.4	96.5	89.6	94.8	71.7	83.7	74.6	95.9	79.5	91.9	94.8	76.8	90.4
09900	PAINTS & COATINGS	89.1	90.1	89.7	89.1	77.1	82.1	90.8	83.2	86.4	92.8	82.1	86.5	90.8	80.2	84.6	97.4	71.1	82.0
09	FINISHES	92.3	88.9	90.6	93.4	75.9	84.4	93.4	82.3	87.8	87.3	76.8	81.9	94.1	81.6	87.7	95.1	75.5	85.1
10 - 14	TOTAL DIV. 10000 - 14000	100.0	91.0	98.1	100.0	88.1	97.5	100.0	88.4	97.6	100.0	67.5	93.1	100.0	88.9	97.7	100.0	67.9	93.2
15	MECHANICAL	99.8	88.4	94.6	96.5	85.2	91.4	96.5	85.6	91.6	97.1	81.7	90.1	99.7	87.8	94.3	96.8	75.5	87.2
16	ELECTRICAL	99.8	93.4	95.4	89.4	91.0	90.5	98.8	81.7	87.2	92.8	78.7	83.2	89.6	86.1	87.2	93.3	83.3	86.5
01 - 16	WEIGHTED AVERAGE	99.6	89.1	94.5	95.2	85.4	90.4	96.9	83.6	90.4	93.3	82.9	88.2	97.3	86.6	92.1	95.0	80.1	87.7

DIVISION		INDIANA						IOWA					
		SOUTH BEND			TERRE HAUTE			WASHINGTON			BURLINGTON			CARROLL			CEDAR RAPIDS		
		465 - 466			478			475			526			514			522 - 524		
		MAT.	INST.	TOTAL	MAT.	INST.	TOTAL	MAT.	INST.	TOTAL	MAT.	INST.	TOTAL	MAT.	INST.	TOTAL	MAT.	INST.	TOTAL
02	SITE CONSTRUCTION	84.5	95.4	92.9	84.2	130.7	119.9	85.0	130.3	119.8	81.6	96.3	92.9	73.3	95.5	90.3	82.6	94.8	91.9
03100	CONCRETE FORMS & ACCESSORIES	95.4	76.1	78.8	92.6	79.4	81.2	93.5	82.9	84.4	94.1	70.6	73.9	77.7	60.2	62.7	101.5	79.4	82.5
03200	CONCRETE REINFORCEMENT	97.3	77.6	86.1	98.2	81.9	88.9	90.8	76.7	82.8	97.0	71.1	82.3	97.8	61.8	77.4	97.7	81.0	88.2
03300	CAST-IN-PLACE CONCRETE	100.3	88.5	95.4	96.2	86.6	92.2	104.8	90.2	98.7	103.8	54.4	83.2	103.8	68.9	89.3	104.1	85.8	96.5
03	CONCRETE	91.6	82.5	87.0	105.4	82.6	93.9	111.3	84.3	97.7	95.9	65.9	80.8	94.3	64.4	79.2	96.1	82.3	89.2
04	MASONRY	87.5	82.9	84.6	95.1	84.4	88.4	88.4	82.7	84.8	100.0	58.0	73.8	103.9	58.4	75.6	107.8	80.0	90.5
05	METALS	99.3	104.9	101.3	90.5	88.0	89.6	83.2	83.8	83.4	86.2	84.1	85.5	86.3	75.2	82.3	88.5	86.4	87.7
06	WOOD & PLASTICS	99.8	73.9	86.4	94.7	77.5	85.8	95.0	81.6	88.1	98.0	74.1	85.6	80.2	63.0	71.3	105.8	78.7	91.8
07	THERMAL & MOISTURE PROTECTION	97.5	86.1	92.2	96.3	84.9	91.0	96.5	85.6	91.4	96.0	69.3	83.5	96.3	64.8	81.6	97.3	79.4	88.9
08	DOORS & WINDOWS	95.2	77.3	90.9	99.0	81.0	94.6	95.9	84.5	93.1	96.8	70.4	90.4	102.6	56.9	91.5	103.6	78.4	97.5
09200	PLASTER & GYPSUM BOARD	109.8	73.6	87.4	103.3	75.7	86.2	101.8	79.9	88.3	94.1	73.5	81.3	88.5	62.0	72.1	97.2	78.6	85.6
09500	CEILINGS	92.0	73.6	79.9	95.1	75.7	82.3	82.6	79.9	80.9	107.3	73.5	85.1	107.3	62.0	77.6	111.4	78.6	89.8
09600	FLOORING	89.4	80.8	87.3	96.7	87.6	94.4	97.9	76.5	92.6	110.2	58.5	97.6	103.4	51.5	90.7	127.7	69.0	113.4
09900	PAINTS & COATINGS	89.1	82.6	85.3	97.4	89.4	92.7	97.4	93.0	94.8	109.3	103.9	106.2	109.3	48.7	73.9	114.0	79.7	93.9
09	FINISHES	92.8	77.1	84.8	96.5	81.6	88.9	94.6	82.5	88.4	104.6	72.3	88.1	100.7	57.4	78.6	111.8	76.7	93.9
10 - 14	TOTAL DIV. 10000 - 14000	100.0	84.8	96.8	100.0	92.8	98.5	100.0	94.2	98.8	100.0	82.3	96.2	100.0	59.5	91.4	100.0	85.7	97.0
15	MECHANICAL	99.8	79.1	90.4	100.0	85.8	93.6	96.8	87.1	92.4	97.1	77.5	88.2	97.1	64.7	82.4	100.3	84.9	93.3
16	ELECTRICAL	95.0	83.0	86.8	93.0	87.6	89.3	93.9	90.3	91.5	97.7	78.3	84.5	98.8	51.9	66.8	94.0	86.9	89.2
01 - 16	WEIGHTED AVERAGE	95.9	84.5	90.4	97.3	89.6	93.6	95.4	90.3	92.9	96.0	75.7	86.2	96.0	64.6	80.8	99.0	84.1	91.8

City Cost Indexes

DIVISION		IOWA																	
		COUNCIL BLUFFS			CRESTON			DAVENPORT			DECORAH			DES MOINES			DUBUQUE		
		515			508			527 - 528			521			500 - 503,509			520		
		MAT.	INST.	TOTAL	MAT.	INST.	TOTAL	MAT.	INST.	TOTAL	MAT.	INST.	TOTAL	MAT.	INST.	TOTAL	MAT.	INST.	TOTAL
02	SITE CONSTRUCTION	86.1	92.4	90.9	73.2	96.6	91.1	81.4	99.2	95.0	80.4	96.0	92.3	75.2	100.0	94.2	80.8	92.3	89.6
03100	CONCRETE FORMS & ACCESSORIES	76.8	65.9	67.5	78.6	70.0	71.2	101.0	89.8	91.4	90.6	57.0	61.7	103.2	79.7	83.0	78.7	65.8	67.7
03200	CONCRETE REINFORCEMENT	99.7	74.8	85.6	97.1	65.7	79.3	97.7	82.7	89.2	97.0	62.1	77.2	97.7	77.1	86.0	96.3	80.7	87.4
03300	CAST-IN-PLACE CONCRETE	108.3	74.2	94.1	103.5	71.9	90.4	100.3	89.4	95.8	101.1	65.1	86.0	104.4	86.2	96.8	101.9	67.7	87.7
03	CONCRETE	97.7	71.3	84.4	94.1	70.6	82.3	94.3	88.7	91.5	93.8	61.7	77.6	96.4	81.9	89.1	92.4	70.1	81.2
04	MASONRY	108.5	59.5	78.0	103.9	63.3	78.6	107.8	81.7	91.5	122.8	54.3	80.1	102.1	83.1	90.3	108.8	71.5	85.5
05	METALS	93.6	82.1	89.5	86.2	81.8	84.6	88.5	93.2	90.2	86.3	75.1	82.3	88.4	86.3	87.6	87.0	85.3	86.4
06	WOOD & PLASTICS	78.6	64.6	71.4	80.8	72.6	76.6	105.8	90.1	97.7	94.2	58.5	75.7	107.5	78.3	92.4	81.0	64.5	72.5
07	THERMAL & MOISTURE PROTECTION	96.4	63.6	81.1	97.5	67.9	83.6	96.7	85.8	91.6	96.3	54.4	76.7	97.4	79.4	89.0	96.7	69.9	84.1
08	DOORS & WINDOWS	102.6	67.0	93.9	116.7	71.6	105.7	103.6	90.2	100.3	100.7	59.4	90.7	103.6	77.2	97.2	102.6	72.7	95.3
09200	PLASTER & GYPSUM BOARD	88.5	64.0	73.4	88.5	72.0	78.3	97.2	90.1	92.8	93.1	57.4	71.0	93.9	77.9	84.0	89.0	63.9	73.4
09500	CEILINGS	107.3	64.0	78.9	107.3	72.0	84.1	111.4	90.1	97.4	107.3	57.4	74.5	110.0	77.9	88.9	107.3	63.9	78.8
09600	FLOORING	102.1	56.3	90.9	103.7	51.5	90.9	113.0	86.1	106.5	109.7	73.1	100.8	112.8	51.5	97.8	117.0	48.0	100.2
09900	PAINTS & COATINGS	104.7	60.1	78.6	109.3	60.1	80.6	109.3	103.9	106.2	109.3	51.6	75.6	109.3	80.4	92.4	112.7	63.2	83.7
09	FINISHES	100.8	62.8	81.4	100.8	65.5	82.8	106.7	90.8	98.6	104.3	59.6	81.5	105.2	74.0	89.3	106.2	61.2	83.2
10 - 14	TOTAL DIV. 10000 - 14000	100.0	82.3	96.3	100.0	83.0	96.4	100.0	89.1	97.7	100.0	66.8	93.0	100.0	87.4	97.3	100.0	81.4	96.1
15	MECHANICAL	100.3	80.1	91.1	97.1	67.5	83.7	100.3	87.7	94.6	97.1	56.4	78.6	100.3	84.3	93.0	100.3	72.7	87.8
16	ELECTRICAL	102.2	84.8	90.3	90.7	67.7	75.0	91.5	89.1	89.9	94.0	51.6	65.1	94.7	89.6	91.2	99.9	77.8	84.8
01 - 16	WEIGHTED AVERAGE	99.1	76.0	87.9	97.0	72.1	85.0	98.0	89.6	93.9	97.1	62.5	80.3	97.8	84.9	91.5	97.7	74.8	86.6

DIVISION		IOWA																	
		FORT DODGE			MASON CITY			OTTUMWA			SHENANDOAH			SIBLEY			SIOUX CITY		
		505			504			525			516			512			510 - 511		
		MAT.	INST.	TOTAL	MAT.	INST.	TOTAL	MAT.	INST.	TOTAL	MAT.	INST.	TOTAL	MAT.	INST.	TOTAL	MAT.	INST.	TOTAL
02	SITE CONSTRUCTION	78.0	95.1	91.1	78.0	96.0	91.8	81.6	91.3	89.0	85.0	91.2	89.8	88.9	95.4	93.9	89.9	96.6	95.1
03100	CONCRETE FORMS & ACCESSORIES	79.4	32.2	38.9	85.7	57.3	61.3	88.1	63.4	66.9	79.1	55.2	58.5	79.8	51.2	55.3	101.5	54.5	61.2
03200	CONCRETE REINFORCEMENT	97.1	61.3	76.8	97.0	76.1	85.2	97.0	71.4	82.5	99.7	50.8	72.0	99.7	49.7	71.4	97.7	70.6	82.3
03300	CAST-IN-PLACE CONCRETE	97.2	43.1	74.6	97.2	65.2	83.8	104.5	70.4	90.3	104.7	59.4	85.8	102.6	59.3	84.6	103.2	57.8	84.3
03	CONCRETE	89.8	43.1	66.3	90.3	64.5	77.3	95.3	68.2	81.7	95.2	57.3	76.1	94.3	54.8	74.5	95.7	59.8	77.6
04	MASONRY	101.5	34.3	59.6	116.3	54.3	77.7	105.6	63.2	79.2	108.3	44.2	68.4	126.4	48.3	77.7	101.9	58.7	75.0
05	METALS	86.3	73.3	81.6	86.3	80.9	84.4	86.2	83.7	85.3	92.8	78.5	87.7	86.4	69.1	80.2	88.5	80.8	85.7
06	WOOD & PLASTICS	81.6	29.7	54.8	88.6	58.5	73.0	91.4	64.3	77.4	81.2	59.0	69.7	82.3	51.9	66.6	105.8	54.5	79.3
07	THERMAL & MOISTURE PROTECTION	96.6	39.1	69.7	95.8	57.9	78.1	97.0	67.1	83.0	95.4	52.3	75.2	95.9	56.9	77.6	96.7	59.3	79.2
08	DOORS & WINDOWS	108.5	36.5	91.0	97.9	63.1	89.4	102.6	69.8	94.6	91.1	59.6	83.4	98.3	50.5	86.6	103.6	60.5	93.1
09200	PLASTER & GYPSUM BOARD	89.0	27.7	51.1	91.3	57.4	70.3	92.2	63.7	74.6	89.5	58.2	70.1	89.5	50.6	65.4	97.2	53.3	70.0
09500	CEILINGS	107.3	27.7	55.1	107.3	57.4	74.5	107.3	63.7	78.7	107.3	58.2	75.1	107.3	50.6	70.1	111.4	53.3	73.2
09600	FLOORING	105.3	73.1	97.4	107.8	73.1	99.4	120.6	74.8	109.5	102.9	53.0	90.8	104.4	52.6	91.8	113.6	70.8	103.2
09900	PAINTS & COATINGS	109.3	22.1	58.4	109.3	47.1	73.0	112.7	71.2	88.4	104.7	33.3	63.0	109.3	48.7	73.9	110.7	63.8	83.3
09	FINISHES	102.1	37.7	69.2	103.3	59.1	80.7	107.9	65.8	86.4	101.1	52.7	76.4	102.9	51.2	76.5	107.9	58.1	82.5
10 - 14	TOTAL DIV. 10000 - 14000	100.0	60.2	91.6	100.0	66.8	93.0	100.0	67.7	93.2	100.0	76.8	95.1	100.0	50.8	89.6	100.0	77.5	95.2
15	MECHANICAL	97.1	65.4	82.7	97.1	56.7	78.8	97.1	77.0	88.0	97.1	42.7	72.4	97.1	50.9	76.2	100.3	51.4	78.1
16	ELECTRICAL	101.5	68.9	79.3	99.9	51.6	67.0	97.4	71.6	79.8	94.0	41.1	57.9	94.0	51.9	65.3	94.0	81.2	85.3
01 - 16	WEIGHTED AVERAGE	96.5	58.0	77.8	96.2	63.6	80.4	97.2	73.4	85.7	96.0	55.8	76.5	97.1	57.6	77.9	98.4	67.4	83.4

DIVISION		IOWA						KANSAS											
		SPENCER			WATERLOO			BELLEVILLE			COLBY			DODGE CITY			EMPORIA		
		513			506 - 507			669			677			678			668		
		MAT.	INST.	TOTAL	MAT.	INST.	TOTAL	MAT.	INST.	TOTAL	MAT.	INST.	TOTAL	MAT.	INST.	TOTAL	MAT.	INST.	TOTAL
02	SITE CONSTRUCTION	89.0	95.4	93.9	81.9	96.4	93.0	106.1	94.4	97.2	103.4	94.4	96.5	105.4	94.4	97.0	96.4	91.9	92.9
03100	CONCRETE FORMS & ACCESSORIES	87.9	51.1	56.3	102.1	57.5	63.8	96.4	52.1	58.4	105.2	52.1	59.6	95.8	52.1	58.3	84.7	47.4	52.7
03200	CONCRETE REINFORCEMENT	99.7	49.7	71.4	97.7	80.5	88.0	101.6	74.6	86.3	104.1	74.6	87.4	101.6	74.6	86.3	100.2	65.5	80.6
03300	CAST-IN-PLACE CONCRETE	102.6	59.3	84.5	104.1	54.7	83.5	113.7	67.1	94.3	109.4	67.1	91.8	111.4	67.1	92.9	109.9	50.7	85.2
03	CONCRETE	94.9	54.8	74.7	96.2	61.9	78.9	109.3	63.1	86.1	106.9	63.1	84.9	107.6	63.1	85.2	102.0	53.6	77.7
04	MASONRY	126.4	48.3	77.7	102.0	62.8	77.6	95.1	52.5	68.6	96.9	52.5	69.3	103.1	52.5	71.6	101.2	52.0	70.5
05	METALS	86.4	69.1	80.2	88.5	84.9	87.2	92.5	87.2	90.6	92.8	87.2	90.8	93.9	87.2	91.5	92.2	82.9	88.9
06	WOOD & PLASTICS	91.2	51.9	70.9	106.6	58.5	81.7	94.4	52.2	72.6	103.2	52.2	76.8	93.7	52.2	72.2	83.0	47.8	64.8
07	THERMAL & MOISTURE PROTECTION	97.3	53.7	76.9	96.3	57.4	78.1	97.5	58.5	79.2	97.6	58.5	79.3	97.5	58.5	79.2	95.7	49.8	74.2
08	DOORS & WINDOWS	113.0	50.5	97.8	98.9	66.9	91.1	99.4	55.9	88.8	99.4	55.9	88.8	99.4	55.9	88.8	96.8	50.6	85.6
09200	PLASTER & GYPSUM BOARD	92.2	50.6	66.5	97.2	57.4	72.5	107.4	50.3	72.0	110.1	50.3	73.1	106.9	50.3	71.8	103.2	45.7	67.6
09500	CEILINGS	107.3	50.6	70.1	111.4	57.4	75.9	89.5	50.3	63.8	89.5	50.3	63.8	89.5	50.3	63.8	89.5	45.7	60.8
09600	FLOORING	107.6	52.6	94.2	114.8	73.1	104.6	101.0	60.8	91.2	104.7	60.8	94.0	100.8	60.8	91.0	95.3	56.5	85.8
09900	PAINTS & COATINGS	109.3	48.7	73.9	110.7	43.6	71.5	87.6	61.4	72.3	87.6	61.4	72.3	87.6	61.4	72.3	87.6	61.4	72.3
09	FINISHES	104.4	51.2	77.2	107.4	58.6	82.5	95.4	54.2	74.3	96.9	54.2	75.1	95.2	54.2	74.2	92.2	50.0	70.6
10 - 14	TOTAL DIV. 10000 - 14000	100.0	50.8	89.6	100.0	78.0	95.4	100.0	62.5	92.1	100.0	62.5	92.1	100.0	62.5	92.1	100.0	57.0	90.9
15	MECHANICAL	97.1	50.9	76.1	100.3	56.6	80.5	97.1	58.8	79.7	97.1	58.8	79.7	100.3	58.8	81.5	97.1	64.2	82.2
16	ELECTRICAL	96.6	51.9	66.1	94.0	62.0	72.2	100.3	62.6	74.6	100.3	62.6	74.6	95.0	62.6	72.9	95.6	69.5	77.8
01 - 16	WEIGHTED AVERAGE	99.3	57.5	79.1	97.6	66.6	82.6	98.6	64.9	82.2	98.6	64.9	82.3	99.3	64.9	82.6	96.5	64.0	80.8

City Cost Indexes

		KANSAS																	
		FORT SCOTT			HAYS			HUTCHINSON			INDEPENDENCE			KANSAS CITY			LIBERAL		
	DIVISION	667			676			675			673			660 - 662			679		
		MAT.	INST.	TOTAL	MAT.	INST.	TOTAL	MAT.	INST.	TOTAL	MAT.	INST.	TOTAL	MAT.	INST.	TOTAL	MAT.	INST.	TOTAL
02	SITE CONSTRUCTION	93.4	93.5	93.5	109.0	94.4	97.8	86.0	94.8	92.7	106.8	94.8	97.6	85.9	91.1	89.9	109.4	94.8	98.2
03100	CONCRETE FORMS & ACCESSORIES	107.3	68.7	74.2	101.8	52.1	59.1	88.3	36.0	43.4	117.5	36.4	47.9	102.6	88.1	90.2	96.4	35.6	44.2
03200	CONCRETE REINFORCEMENT	99.6	71.2	83.5	101.6	74.6	86.3	101.6	62.1	79.2	101.0	62.1	79.0	96.6	80.9	87.7	103.0	62.0	79.8
03300	CAST-IN-PLACE CONCRETE	102.0	71.9	89.5	86.2	67.1	78.2	79.8	49.3	67.1	111.8	50.0	86.0	87.3	87.2	87.3	86.2	49.1	70.7
03	CONCRETE	98.1	71.4	84.6	98.2	63.1	80.5	82.1	47.6	64.7	109.5	48.0	78.6	90.1	87.1	88.6	99.8	47.3	73.4
04	MASONRY	101.5	60.2	75.8	103.7	52.5	71.8	97.1	50.6	68.1	93.6	52.0	67.7	104.2	83.5	91.3	101.2	50.6	69.7
05	METALS	92.2	86.5	90.1	92.4	87.2	90.6	92.3	82.8	88.9	92.2	82.7	88.8	98.8	95.5	97.6	92.7	82.4	89.0
06	WOOD & PLASTICS	106.5	71.0	88.2	99.7	52.2	75.1	86.3	33.9	59.2	115.4	33.9	73.3	102.3	88.6	95.3	94.4	33.9	63.1
07	THERMAL & MOISTURE PROTECTION	96.8	69.2	83.8	97.9	58.5	79.4	96.2	49.6	74.4	97.7	48.7	74.7	95.1	89.6	92.5	98.0	45.2	73.3
08	DOORS & WINDOWS	96.8	70.7	90.4	99.3	55.9	88.8	99.3	43.1	85.6	96.8	43.1	83.7	98.2	85.3	95.0	99.4	43.1	85.7
09200	PLASTER & GYPSUM BOARD	108.7	69.7	84.5	108.7	50.3	72.5	104.6	31.4	59.3	113.8	31.4	62.8	103.1	87.9	93.7	107.4	31.4	60.3
09500	CEILINGS	89.5	69.7	76.5	89.5	50.3	63.8	89.5	31.4	51.4	89.5	31.4	51.4	96.4	87.9	90.8	89.5	31.4	51.4
09600	FLOORING	111.8	61.2	99.4	103.3	60.8	92.9	97.5	60.8	88.5	109.7	60.8	97.8	90.4	64.9	84.2	101.0	60.8	91.2
09900	PAINTS & COATINGS	90.2	61.4	73.4	87.6	61.4	72.3	87.6	61.4	72.3	87.6	61.4	72.3	95.3	89.4	91.9	87.6	61.4	72.3
09	FINISHES	98.6	66.8	82.4	96.6	54.2	74.9	92.7	42.1	66.9	99.1	42.5	70.2	94.2	84.0	89.0	95.8	42.1	68.4
10 - 14	TOTAL DIV. 10000 - 14000	100.0	74.3	94.6	100.0	62.5	92.1	100.0	57.9	91.1	100.0	58.5	91.2	100.0	93.8	98.7	100.0	57.8	91.1
15	MECHANICAL	97.1	56.5	78.7	97.1	58.8	79.7	97.1	48.0	74.8	97.1	46.2	74.0	100.1	92.2	96.5	97.1	40.6	71.5
16	ELECTRICAL	94.3	69.5	77.4	98.5	62.6	74.0	90.0	67.9	74.9	93.7	75.3	81.1	104.2	101.2	102.1	95.0	49.5	64.0
01 - 16	WEIGHTED AVERAGE	96.9	70.2	83.9	97.8	64.9	81.8	93.6	58.6	76.6	98.3	59.7	79.6	97.8	91.1	94.5	97.5	53.9	76.4

		KANSAS									KENTUCKY								
		SALINA			TOPEKA			WICHITA			ASHLAND			BOWLING GREEN			CAMPTON		
	DIVISION	674			664 - 666			670 - 672			411 - 412			421 - 422			413 - 414		
		MAT.	INST.	TOTAL	MAT.	INST.	TOTAL	MAT.	INST.	TOTAL	MAT.	INST.	TOTAL	MAT.	INST.	TOTAL	MAT.	INST.	TOTAL
02	SITE CONSTRUCTION	95.3	94.4	94.6	88.1	91.7	90.9	89.5	94.7	93.5	100.8	85.5	89.1	70.6	100.6	93.6	78.5	101.1	95.8
03100	CONCRETE FORMS & ACCESSORIES	90.9	54.5	59.7	101.4	52.6	59.5	98.0	55.9	61.9	82.4	106.6	103.2	82.2	81.3	81.4	84.6	40.5	46.7
03200	CONCRETE REINFORCEMENT	101.0	87.5	93.4	93.9	88.8	91.0	93.9	88.4	90.8	91.5	104.5	98.9	89.0	69.7	78.0	89.8	75.6	81.8
03300	CAST-IN-PLACE CONCRETE	96.6	56.6	79.9	91.0	61.5	78.7	85.5	64.2	76.6	91.5	102.3	96.0	90.8	76.9	85.0	101.3	47.0	78.7
03	CONCRETE	95.3	62.9	79.0	89.6	64.0	76.7	86.6	66.3	76.4	95.9	105.4	100.7	94.4	77.7	86.0	98.4	50.5	74.3
04	MASONRY	117.7	50.4	75.8	97.6	63.1	76.1	92.5	67.9	77.2	93.5	103.9	100.0	94.9	77.8	84.3	93.7	37.3	58.6
05	METALS	93.7	91.1	92.8	96.7	93.4	95.5	96.7	93.2	95.4	92.1	113.6	99.8	93.6	78.3	88.1	93.1	76.6	87.2
06	WOOD & PLASTICS	88.8	56.2	71.9	98.5	49.7	73.3	96.0	53.5	74.0	75.1	107.1	91.6	87.9	81.8	84.7	86.0	39.8	62.2
07	THERMAL & MOISTURE PROTECTION	96.8	63.8	81.4	96.8	72.7	85.5	96.4	71.3	84.7	88.1	98.3	92.9	87.1	76.5	82.1	96.7	49.2	74.4
08	DOORS & WINDOWS	99.3	58.9	89.5	99.5	61.8	90.3	99.5	61.4	90.2	96.3	101.7	97.5	98.1	72.9	92.0	99.7	47.2	86.9
09200	PLASTER & GYPSUM BOARD	105.5	54.4	73.9	109.2	47.7	71.1	109.2	51.6	73.5	70.0	107.3	93.1	99.7	81.4	88.4	99.7	36.8	60.8
09500	CEILINGS	89.5	54.4	66.5	96.4	47.7	64.4	96.4	51.6	67.0	83.4	107.3	99.1	90.9	81.4	84.7	90.9	36.8	55.4
09600	FLOORING	98.7	44.4	85.5	102.4	56.5	91.2	101.5	78.1	95.8	76.7	99.6	82.3	92.6	75.2	88.4	94.0	52.2	83.8
09900	PAINTS & COATINGS	87.6	48.3	64.6	87.6	70.2	77.4	87.6	61.6	72.4	99.9	99.3	99.6	97.4	81.5	88.1	97.4	40.5	64.1
09	FINISHES	93.8	51.4	72.2	96.4	53.6	74.5	96.2	59.8	77.6	81.6	105.2	93.6	94.0	80.5	87.1	94.6	41.9	67.7
10 - 14	TOTAL DIV. 10000 - 14000	100.0	71.2	93.9	100.0	82.7	96.3	100.0	73.2	94.3	100.0	101.9	100.4	100.0	73.3	94.4	100.0	69.8	93.6
15	MECHANICAL	100.3	44.3	74.9	100.3	72.3	87.6	100.3	72.4	87.6	96.7	95.5	96.1	100.2	82.7	92.3	96.8	43.1	72.5
16	ELECTRICAL	94.5	71.0	78.5	101.2	75.3	83.5	99.0	71.0	79.9	89.2	91.3	90.6	93.3	86.0	88.3	88.9	34.3	51.7
01 - 16	WEIGHTED AVERAGE	98.0	63.8	81.4	97.3	72.0	85.1	96.5	72.9	85.1	93.4	99.3	96.3	95.4	82.4	89.1	95.4	51.7	74.2

		KENTUCKY																	
		CORBIN			COVINGTON			ELIZABETHTOWN			FRANKFORT			HAZARD			HENDERSON		
	DIVISION	407 - 409			410			427			406			417 - 418			424		
		MAT.	INST.	TOTAL	MAT.	INST.	TOTAL	MAT.	INST.	TOTAL	MAT.	INST.	TOTAL	MAT.	INST.	TOTAL	MAT.	INST.	TOTAL
02	SITE CONSTRUCTION	75.8	101.1	95.2	74.9	116.6	106.9	65.7	100.5	92.4	76.5	103.1	96.9	76.4	101.1	95.4	73.0	130.7	117.2
03100	CONCRETE FORMS & ACCESSORIES	81.1	40.5	46.2	80.0	79.8	79.8	76.3	78.3	78.0	93.0	71.1	74.2	81.1	40.5	46.2	88.7	82.4	83.3
03200	CONCRETE REINFORCEMENT	89.4	75.6	81.6	88.6	98.6	94.3	89.4	94.0	92.0	90.7	88.8	89.4	90.3	75.6	81.9	89.1	94.2	92.0
03300	CAST-IN-PLACE CONCRETE	95.9	47.0	75.5	97.0	85.6	92.3	82.1	75.4	79.3	95.9	79.7	89.1	97.4	47.0	76.4	80.3	94.1	86.0
03	CONCRETE	93.5	50.5	71.9	96.6	85.8	91.2	86.4	80.5	83.4	94.6	77.7	86.1	95.0	50.5	72.6	91.3	88.7	90.0
04	MASONRY	93.3	37.3	58.4	109.2	87.7	95.9	79.5	76.0	77.3	93.3	67.8	77.4	91.5	37.3	57.8	99.5	94.0	96.1
05	METALS	93.0	76.6	87.1	90.8	95.5	92.5	93.0	88.3	91.3	93.1	87.0	90.9	93.1	76.6	87.2	83.1	91.1	86.0
06	WOOD & PLASTICS	82.3	39.8	60.4	82.7	77.3	79.9	81.7	78.4	80.0	94.7	69.7	81.8	82.3	39.8	60.4	89.9	78.9	84.2
07	THERMAL & MOISTURE PROTECTION	96.3	49.2	74.3	92.4	89.7	91.1	86.6	79.6	83.3	96.5	75.3	86.6	96.5	49.2	74.3	95.8	90.2	93.2
08	DOORS & WINDOWS	99.0	47.2	86.3	101.6	81.7	96.8	98.1	84.2	94.7	98.2	75.8	92.7	100.1	47.2	87.2	96.4	84.3	93.4
09200	PLASTER & GYPSUM BOARD	98.8	36.8	60.4	92.7	77.1	83.0	98.4	77.9	85.7	102.5	67.7	80.9	98.8	36.8	60.4	100.0	77.1	85.8
09500	CEILINGS	90.9	36.8	55.4	95.1	77.1	83.3	90.9	77.9	82.4	90.9	67.7	75.7	90.9	36.8	55.4	82.6	77.1	79.0
09600	FLOORING	92.3	52.2	82.5	69.1	91.3	74.5	89.9	73.9	86.0	97.5	63.9	89.3	92.3	52.2	82.5	95.8	89.0	94.1
09900	PAINTS & COATINGS	97.4	40.5	64.1	92.8	87.4	89.6	97.4	77.4	85.7	97.4	61.1	76.2	97.4	40.5	64.1	97.4	98.0	97.7
09	FINISHES	93.7	41.9	67.3	86.3	82.0	84.1	92.7	77.0	84.6	96.0	68.0	81.7	93.8	41.9	67.3	93.1	84.8	88.9
10 - 14	TOTAL DIV. 10000 - 14000	100.0	69.8	93.6	100.0	91.4	98.2	100.0	89.7	97.8	100.0	85.1	96.8	100.0	69.8	93.6	100.0	94.3	98.8
15	MECHANICAL	96.8	43.1	72.5	97.2	90.3	94.1	97.0	81.4	89.9	100.0	76.2	89.3	96.8	43.1	72.5	97.0	84.0	91.1
16	ELECTRICAL	89.4	34.3	51.8	95.4	83.4	87.2	88.6	86.0	86.8	93.5	75.7	81.3	88.9	34.3	51.7	92.4	83.3	86.2
01 - 16	WEIGHTED AVERAGE	94.5	51.7	73.8	95.3	89.9	92.7	92.1	83.7	88.0	96.0	78.3	87.4	94.7	51.7	73.9	92.9	91.2	92.1

City Cost Indexes

		KENTUCKY																
		LEXINGTON 403 - 405			LOUISVILLE 400 - 402			OWENSBORO 423			PADUCAH 420			PIKEVILLE 415 - 416			SOMERSET 425 - 426		
DIVISION		MAT.	INST.	TOTAL	MAT.	INST.	TOTAL	MAT.	INST.	TOTAL	MAT.	INST.	TOTAL	MAT.	INST.	TOTAL	MAT.	INST.	TOTAL
02	SITE CONSTRUCTION	77.2	103.0	97.0	67.7	100.6	92.9	82.2	130.5	119.2	76.4	130.5	117.9	111.9	82.9	89.7	69.0	101.1	93.6
03100	CONCRETE FORMS & ACCESSORIES	93.1	73.1	76.0	91.8	81.3	82.8	86.7	77.1	78.5	83.9	80.0	80.6	93.5	67.0	70.7	81.8	40.5	46.3
03200	CONCRETE REINFORCEMENT	98.2	96.7	97.4	98.2	97.5	97.8	89.1	95.4	92.6	89.6	83.0	85.9	92.0	66.2	77.3	89.4	75.6	81.6
03300	CAST-IN-PLACE CONCRETE	98.1	77.0	89.3	94.9	76.1	87.1	93.7	89.2	91.8	85.5	88.5	86.7	100.6	67.7	86.9	80.3	47.0	66.4
03	CONCRETE	96.6	79.3	87.9	95.0	82.7	88.8	103.4	85.0	94.2	95.7	83.7	89.7	109.6	68.8	89.1	81.3	50.5	65.8
04	MASONRY	92.1	58.2	71.0	92.5	79.9	84.6	92.3	63.1	74.1	95.0	78.0	84.4	93.4	60.0	72.6	85.9	37.3	55.7
05	METALS	95.1	89.8	93.2	95.1	89.8	93.2	85.3	90.1	87.0	83.0	85.6	83.9	92.0	93.6	92.6	93.0	76.6	87.1
06	WOOD & PLASTICS	94.7	75.2	84.6	98.0	81.8	89.6	87.8	77.0	82.2	85.0	78.9	81.8	86.6	67.4	76.7	83.2	39.8	60.8
07	THERMAL & MOISTURE PROTECTION	96.7	69.1	83.7	87.2	81.2	84.4	95.3	77.9	87.7	95.8	80.0	88.4	89.0	61.9	76.3	95.8	49.2	74.0
08	DOORS & WINDOWS	99.0	83.3	95.2	99.0	86.9	96.0	96.4	84.2	93.4	95.5	81.6	92.1	97.1	65.8	89.5	99.0	47.2	86.3
09200	PLASTER & GYPSUM BOARD	103.3	73.3	84.7	106.3	81.4	90.9	99.5	75.1	84.4	98.6	77.1	85.3	73.2	66.3	68.9	99.3	36.8	60.6
09500	CEILINGS	95.1	73.3	80.8	95.1	81.4	86.1	82.6	75.1	77.7	82.6	77.1	79.0	83.4	66.3	72.2	90.9	36.8	55.4
09600	FLOORING	97.5	52.2	86.5	96.0	73.9	90.6	94.9	75.2	90.1	93.7	75.2	89.2	81.3	62.1	76.6	92.5	52.2	82.7
09900	PAINTS & COATINGS	97.4	71.9	82.5	97.4	77.4	85.7	97.4	104.1	101.3	97.4	75.0	84.3	99.9	44.7	67.6	97.4	40.5	64.1
09	FINISHES	96.9	68.8	82.5	96.5	79.2	87.7	93.1	79.4	86.1	92.4	78.3	85.2	84.2	63.6	73.7	93.4	41.9	67.1
10 - 14	TOTAL DIV. 10000 - 14000	100.0	90.3	97.9	100.0	90.6	98.0	100.0	97.3	99.4	100.0	85.9	97.0	100.0	71.6	94.0	100.0	69.8	93.6
15	MECHANICAL	100.0	59.1	81.5	100.0	82.8	92.2	100.2	65.0	84.3	97.0	83.7	90.9	96.7	64.9	82.3	97.0	43.1	72.6
16	ELECTRICAL	94.1	63.0	72.9	94.1	88.2	90.1	92.5	81.0	84.6	96.5	85.2	88.8	94.2	61.2	71.7	89.6	34.3	51.9
01 - 16	WEIGHTED AVERAGE	96.7	72.8	85.2	96.0	85.7	91.0	95.3	82.7	89.2	93.4	87.4	90.5	96.2	68.7	82.9	92.4	51.7	72.7

		LOUISIANA																
		ALEXANDRIA 713 - 714			BATON ROUGE 707 - 708			HAMMOND 704			LAFAYETTE 705			LAKE CHARLES 706			MONROE 712		
DIVISION		MAT.	INST.	TOTAL	MAT.	INST.	TOTAL	MAT.	INST.	TOTAL	MAT.	INST.	TOTAL	MAT.	INST.	TOTAL	MAT.	INST.	TOTAL
02	SITE CONSTRUCTION	104.3	84.5	89.1	113.5	87.4	93.5	114.3	89.0	94.9	114.6	88.8	94.8	115.2	87.1	93.7	104.3	84.3	89.0
03100	CONCRETE FORMS & ACCESSORIES	78.5	51.9	55.7	97.7	57.6	63.3	75.9	63.7	65.5	97.2	48.1	55.1	97.8	58.6	64.1	77.7	52.8	56.3
03200	CONCRETE REINFORCEMENT	97.2	64.4	78.6	99.3	60.3	77.2	97.9	65.5	79.5	99.3	59.7	76.9	99.3	60.2	77.1	96.2	64.3	78.1
03300	CAST-IN-PLACE CONCRETE	98.8	49.5	78.2	91.4	60.0	78.3	92.4	58.7	78.4	91.9	59.2	78.2	96.8	61.7	82.2	98.8	54.5	80.3
03	CONCRETE	90.0	54.5	72.1	92.9	59.8	76.2	91.7	63.1	77.3	93.1	55.3	74.1	95.5	60.8	78.0	89.8	56.5	73.1
04	MASONRY	113.7	52.3	75.5	101.1	62.5	77.1	102.2	59.2	75.4	102.2	50.4	69.9	100.1	62.9	76.9	108.8	50.3	72.4
05	METALS	86.8	74.5	82.4	98.1	73.0	89.1	98.6	76.0	90.5	97.6	72.5	88.6	97.6	73.1	88.8	86.8	73.7	82.1
06	WOOD & PLASTICS	78.2	52.7	65.0	103.8	58.2	79.9	79.5	65.1	72.0	103.8	46.3	73.8	101.8	59.1	79.3	77.3	53.3	64.9
07	THERMAL & MOISTURE PROTECTION	96.6	56.3	77.7	97.8	63.9	81.9	99.9	64.7	83.4	100.7	58.6	80.9	100.5	63.8	83.3	96.6	57.8	78.4
08	DOORS & WINDOWS	100.3	57.0	89.8	103.1	58.4	92.2	98.7	68.9	91.4	103.1	51.9	90.6	103.1	60.7	92.8	100.3	62.3	91.0
09200	PLASTER & GYPSUM BOARD	88.5	52.2	66.1	101.1	57.7	74.2	94.5	64.8	76.1	101.1	45.4	66.6	101.1	58.6	74.8	88.1	52.8	66.2
09500	CEILINGS	101.1	52.2	69.0	100.8	57.7	72.5	102.1	64.8	77.6	100.8	45.4	64.4	100.8	58.6	73.1	101.1	52.8	69.4
09600	FLOORING	107.9	59.9	96.2	108.2	72.2	99.4	97.0	72.2	90.9	108.4	54.0	95.1	108.4	63.0	97.3	107.5	52.0	94.0
09900	PAINTS & COATINGS	95.9	50.7	69.5	94.8	58.1	73.4	94.8	53.0	70.4	94.8	58.1	73.4	94.8	53.5	70.7	95.9	44.4	65.8
09	FINISHES	98.1	53.0	75.1	102.3	60.4	80.9	97.9	64.2	80.7	102.3	49.2	75.2	102.3	58.8	80.1	97.9	51.2	74.1
10 - 14	TOTAL DIV. 10000 - 14000	100.0	67.0	93.0	100.0	69.3	93.5	100.0	71.2	93.9	100.0	67.3	93.1	100.0	69.5	93.6	100.0	67.4	93.1
15	MECHANICAL	100.0	46.7	75.9	100.1	53.2	78.9	96.9	59.4	79.9	100.1	58.4	81.2	100.1	64.1	83.8	100.0	53.5	78.9
16	ELECTRICAL	95.4	55.1	68.0	97.5	62.5	73.7	96.1	61.5	72.5	98.1	74.6	82.1	97.5	62.5	73.7	98.5	59.3	71.8
01 - 16	WEIGHTED AVERAGE	96.8	58.0	78.0	99.7	63.6	82.2	97.7	66.1	82.4	99.7	63.0	82.0	100.0	65.7	83.4	96.7	60.0	78.9

		LOUISIANA						MAINE								
		NEW ORLEANS 700 - 701			SHREVEPORT 710 - 711			THIBODAUX 703			AUGUSTA 043			BANGOR 044			BATH 045		
DIVISION		MAT.	INST.	TOTAL	MAT.	INST.	TOTAL	MAT.	INST.	TOTAL	MAT.	INST.	TOTAL	MAT.	INST.	TOTAL	MAT.	INST.	TOTAL
02	SITE CONSTRUCTION	117.2	89.5	96.0	103.0	84.2	88.6	116.9	88.9	95.5	89.9	95.1	93.9	89.8	97.5	95.7	88.4	95.5	93.8
03100	CONCRETE FORMS & ACCESSORIES	96.9	65.7	70.1	100.2	56.1	62.3	89.7	65.0	68.5	96.7	39.6	47.7	90.3	86.3	86.8	85.0	44.5	50.2
03200	CONCRETE REINFORCEMENT	99.3	65.7	80.2	95.7	66.2	79.0	97.9	65.6	79.6	94.0	41.7	64.3	94.0	116.1	106.5	92.9	96.7	95.1
03300	CAST-IN-PLACE CONCRETE	95.8	63.7	82.4	97.4	58.8	81.3	99.2	63.1	84.2	85.5	48.0	69.9	85.5	74.2	80.8	85.5	51.5	71.3
03	CONCRETE	95.0	65.7	80.2	90.1	59.8	74.9	97.0	65.1	80.9	104.5	44.5	74.3	103.5	87.0	95.2	103.5	56.3	79.8
04	MASONRY	102.3	60.5	76.3	102.1	48.9	68.9	126.4	59.9	84.9	92.1	44.6	62.5	108.8	65.6	81.9	116.0	42.6	70.3
05	METALS	100.4	76.6	91.9	87.6	74.4	82.9	98.7	75.7	90.4	97.0	73.8	88.7	96.8	84.5	92.4	95.2	57.9	81.8
06	WOOD & PLASTICS	96.8	67.0	81.4	102.1	58.0	79.3	89.2	66.5	77.4	98.9	38.0	67.4	91.9	87.0	89.4	86.3	42.5	63.7
07	THERMAL & MOISTURE PROTECTION	101.4	66.6	85.1	95.8	57.9	78.1	101.5	65.8	84.8	100.9	42.4	73.5	100.8	64.1	83.6	100.7	50.8	77.3
08	DOORS & WINDOWS	101.9	69.2	93.9	98.2	63.0	89.0	104.1	69.7	95.7	105.2	46.4	90.9	105.2	78.9	98.8	105.1	39.4	89.1
09200	PLASTER & GYPSUM BOARD	100.2	66.8	79.5	94.8	57.6	71.8	98.2	66.2	78.4	96.8	35.1	58.6	93.8	85.7	88.8	92.0	39.7	59.6
09500	CEILINGS	100.8	66.8	78.5	102.5	57.6	73.0	102.1	66.2	78.6	91.1	35.1	54.3	89.7	85.7	87.0	89.7	39.7	56.9
09600	FLOORING	108.9	61.8	97.4	118.7	54.6	103.1	104.8	61.2	94.1	96.4	52.3	85.7	94.3	52.3	84.1	92.4	52.3	82.7
09900	PAINTS & COATINGS	96.8	68.2	80.1	95.9	50.7	69.5	96.8	64.2	78.0	93.5	29.2	55.9	93.5	40.4	62.4	93.5	29.8	56.3
09	FINISHES	102.8	65.2	83.6	102.7	55.0	78.3	101.2	64.3	82.4	95.1	40.0	66.9	93.7	75.1	84.2	92.8	43.2	67.5
10 - 14	TOTAL DIV. 10000 - 14000	100.0	71.6	94.0	100.0	69.1	93.5	100.0	71.4	94.0	100.0	66.0	92.8	100.0	89.0	97.7	100.0	67.3	93.1
15	MECHANICAL	100.1	63.2	83.4	100.0	59.0	81.4	96.9	62.4	81.2	99.9	31.7	69.0	99.9	81.8	91.7	96.7	63.7	81.8
16	ELECTRICAL	97.8	70.5	79.2	96.4	71.6	79.5	93.9	64.8	74.1	104.4	81.0	88.4	101.0	89.7	93.3	97.6	37.4	56.6
01 - 16	WEIGHTED AVERAGE	100.5	69.2	85.3	96.8	63.9	80.9	100.8	67.6	84.7	99.8	55.8	78.5	100.0	82.9	91.7	99.0	55.0	77.7

City Cost Indexes

DIVISION		MAINE																	
		HOULTON 047			KITTERY 039			LEWISTON 042			MACHIAS 046			PORTLAND 040 - 041			ROCKLAND 048		
		MAT.	INST.	TOTAL	MAT.	INST.	TOTAL	MAT.	INST.	TOTAL	MAT.	INST.	TOTAL	MAT.	INST.	TOTAL	MAT.	INST.	TOTAL
02	SITE CONSTRUCTION	90.0	95.6	94.3	84.6	94.8	92.4	88.0	97.5	95.3	89.4	95.5	94.1	87.4	97.5	95.1	85.9	94.8	92.7
03100	CONCRETE FORMS & ACCESSORIES	94.6	78.3	80.6	85.6	51.5	56.3	96.8	86.0	87.6	91.1	61.6	65.8	96.1	86.0	87.5	92.3	61.9	66.2
03200	CONCRETE REINFORCEMENT	94.0	99.5	97.1	92.3	53.8	70.5	115.5	116.0	115.8	94.0	82.4	87.4	115.5	116.0	115.8	94.0	82.4	87.5
03300	CAST-IN-PLACE CONCRETE	85.5	55.9	73.2	85.5	56.8	73.5	95.2	74.1	86.4	85.5	57.4	73.8	88.6	74.1	82.5	87.2	57.7	74.9
03	CONCRETE	104.6	73.0	88.7	98.4	54.9	76.5	106.9	86.8	96.8	104.1	64.0	83.9	103.6	86.8	95.2	100.7	64.3	82.4
04	MASONRY	92.1	51.0	66.5	105.3	48.1	69.7	93.2	65.6	76.0	92.1	51.2	66.6	91.0	65.6	75.2	85.8	51.3	64.3
05	METALS	95.4	57.3	81.7	95.1	74.9	87.9	100.1	84.2	94.4	95.4	69.4	86.1	100.1	84.2	94.4	95.3	70.0	86.2
06	WOOD & PLASTICS	96.6	86.0	91.1	85.6	51.1	68.4	98.9	87.0	92.8	92.9	61.9	76.9	98.9	87.0	92.8	94.2	61.9	77.5
07	THERMAL & MOISTURE PROTECTION	100.9	53.9	78.9	100.3	50.5	77.0	100.7	64.1	83.5	100.8	49.6	76.8	100.5	64.1	83.5	100.5	54.5	79.0
08	DOORS & WINDOWS	105.2	62.6	94.8	105.1	55.6	93.1	108.3	78.9	101.1	105.2	55.8	93.2	108.3	78.9	101.1	105.2	55.8	93.1
09200	PLASTER & GYPSUM BOARD	95.9	84.6	88.9	92.0	48.6	65.1	98.5	85.7	90.5	94.5	59.7	73.0	98.5	85.7	90.5	94.7	59.7	73.1
09500	CEILINGS	91.1	84.6	86.8	89.7	48.6	62.7	99.4	85.7	90.4	91.1	59.7	70.5	99.4	85.7	90.4	89.7	59.7	70.0
09600	FLOORING	95.6	52.3	85.1	92.7	52.3	82.8	96.4	52.3	85.7	94.6	25.7	77.8	96.4	52.3	85.7	94.9	52.3	84.5
09900	PAINTS & COATINGS	93.5	40.4	62.4	93.5	41.0	62.8	93.5	40.4	62.4	93.5	30.6	56.7	93.5	40.4	62.4	93.5	40.4	62.4
09	FINISHES	94.7	71.1	82.7	92.6	50.0	70.8	96.6	75.1	85.6	94.2	52.1	72.7	96.6	75.1	85.6	93.8	57.8	75.4
10 - 14	TOTAL DIV. 10000 - 14000	100.0	75.0	94.7	100.0	69.3	93.5	100.0	88.8	97.6	100.0	73.4	94.4	100.0	88.8	97.6	100.0	73.4	94.4
15	MECHANICAL	96.7	50.1	75.6	96.7	55.3	77.9	99.9	81.8	91.7	96.7	52.8	76.8	99.9	81.8	91.7	96.7	52.8	76.8
16	ELECTRICAL	104.4	34.3	56.6	97.6	57.5	70.2	104.4	55.3	70.9	104.4	85.4	91.4	104.6	55.3	71.0	104.3	85.4	91.4
01 - 16	WEIGHTED AVERAGE	98.8	59.9	79.9	97.7	60.2	79.5	101.0	77.1	89.4	98.6	65.9	82.7	100.5	77.1	89.2	97.7	66.7	82.7

DIVISION		MAINE			MARYLAND														
		WATERVILLE 049			ANNAPOLIS 214			BALTIMORE 210 - 212			COLLEGE PARK 207 - 208			CUMBERLAND 215			EASTON 216		
		MAT.	INST.	TOTAL	MAT.	INST.	TOTAL	MAT.	INST.	TOTAL	MAT.	INST.	TOTAL	MAT.	INST.	TOTAL	MAT.	INST.	TOTAL
02	SITE CONSTRUCTION	90.1	95.1	94.0	94.5	88.0	89.5	95.3	92.5	93.2	101.1	90.2	92.8	87.6	87.9	87.8	93.9	86.1	87.9
03100	CONCRETE FORMS & ACCESSORIES	84.3	39.6	45.9	101.0	67.3	72.1	101.1	75.2	78.9	78.9	72.5	73.5	89.3	81.5	82.6	86.5	42.9	49.1
03200	CONCRETE REINFORCEMENT	94.0	41.7	64.3	97.8	87.5	92.0	97.8	87.6	92.0	101.0	72.8	85.0	86.5	78.6	82.0	85.7	25.7	51.7
03300	CAST-IN-PLACE CONCRETE	85.5	48.0	69.9	103.8	79.0	93.4	103.8	79.9	93.8	117.5	75.6	100.0	92.7	83.9	89.0	102.9	50.6	81.1
03	CONCRETE	105.2	44.5	74.7	102.4	76.6	89.4	102.4	80.5	91.4	107.5	75.2	91.2	91.3	82.9	87.1	99.1	44.4	71.6
04	MASONRY	103.0	44.6	66.6	89.8	73.6	79.7	89.8	73.7	79.7	103.8	72.1	84.0	87.1	76.1	80.2	101.9	46.8	67.6
05	METALS	95.4	73.8	87.6	94.2	99.3	96.0	95.1	100.0	96.9	92.5	97.3	94.2	94.0	95.3	94.5	94.2	69.0	85.2
06	WOOD & PLASTICS	85.4	38.0	60.9	101.9	65.7	83.2	101.9	76.3	88.7	76.8	73.4	75.1	89.7	82.4	85.9	86.6	43.4	64.3
07	THERMAL & MOISTURE PROTECTION	100.8	42.4	73.5	94.7	76.2	86.0	94.7	77.6	86.7	93.7	78.5	86.6	94.2	83.2	89.0	94.3	50.4	73.7
08	DOORS & WINDOWS	105.2	46.4	90.9	92.5	78.9	89.2	92.5	81.8	89.9	97.3	73.9	91.6	93.7	87.3	92.1	91.8	39.7	79.1
09200	PLASTER & GYPSUM BOARD	91.8	35.1	56.7	101.8	65.4	79.2	101.8	76.2	85.9	93.5	72.4	80.5	98.6	82.6	88.7	97.6	42.4	63.4
09500	CEILINGS	91.1	35.1	54.3	90.9	65.4	74.2	90.9	76.2	81.2	95.6	72.4	80.4	90.9	82.6	85.5	90.9	42.4	59.1
09600	FLOORING	92.2	52.3	82.5	88.3	82.4	86.9	88.3	82.4	86.9	85.8	80.4	84.5	84.1	79.1	82.9	83.2	20.0	67.8
09900	PAINTS & COATINGS	93.5	29.2	55.9	97.9	79.9	87.4	97.9	79.9	87.4	98.9	82.9	89.6	97.9	61.3	76.5	97.9	32.3	59.6
09	FINISHES	93.1	40.0	66.0	89.8	70.6	80.0	89.9	76.8	83.2	90.2	74.9	82.4	87.7	78.8	83.1	87.5	38.6	62.5
10 - 14	TOTAL DIV. 10000 - 14000	100.0	66.0	92.8	100.0	79.4	95.6	100.0	81.3	96.1	100.0	83.1	96.4	100.0	83.7	96.6	100.0	64.0	92.4
15	MECHANICAL	96.7	31.7	67.3	100.0	80.7	91.3	100.0	80.7	91.3	96.8	81.5	89.9	96.8	76.3	87.5	96.8	35.4	69.0
16	ELECTRICAL	104.3	81.0	88.4	102.0	87.6	92.2	103.3	94.1	97.0	98.0	92.7	94.4	99.7	86.1	90.4	98.8	64.6	75.5
01 - 16	WEIGHTED AVERAGE	99.1	55.8	78.1	96.7	81.5	89.3	97.0	84.6	91.0	97.3	82.7	90.3	93.9	82.8	88.5	95.5	52.4	74.6

DIVISION		MARYLAND														MASSACHUSETTS		
		ELKTON 219			HAGERSTOWN 217			SALISBURY 218			SILVER SPRING 209			WALDORF 206			BOSTON 020 - 022, 024		
		MAT.	INST.	TOTAL	MAT.	INST.	TOTAL	MAT.	INST.	TOTAL	MAT.	INST.	TOTAL	MAT.	INST.	TOTAL	MAT.	INST.	TOTAL
02	SITE CONSTRUCTION	82.4	87.2	86.1	85.7	88.2	87.6	93.9	86.2	88.0	89.5	82.6	84.2	95.8	82.9	85.9	95.5	104.2	102.2
03100	CONCRETE FORMS & ACCESSORIES	94.7	79.6	81.7	88.1	76.2	77.9	105.4	40.2	49.4	88.5	71.4	73.8	97.5	66.7	71.0	102.8	135.0	130.5
03200	CONCRETE REINFORCEMENT	85.7	75.7	80.0	86.5	78.5	82.0	85.7	75.5	79.9	99.9	74.8	85.7	100.6	74.7	85.9	114.2	169.9	145.8
03300	CAST-IN-PLACE CONCRETE	83.3	59.2	73.3	88.2	65.8	78.9	102.9	49.6	80.7	120.3	77.5	102.5	134.8	72.4	108.8	110.3	142.8	123.8
03	CONCRETE	84.0	72.5	78.2	87.5	74.3	80.8	100.5	51.4	75.8	105.8	75.6	90.6	117.0	71.5	94.1	117.8	142.8	130.4
04	MASONRY	88.1	51.5	65.3	94.0	76.1	82.8	100.8	43.8	65.3	101.9	72.6	83.6	86.4	68.0	74.9	111.8	147.4	134.0
05	METALS	94.2	82.2	89.9	94.1	94.7	94.3	94.2	71.9	86.2	98.2	95.0	97.1	98.2	90.4	95.4	102.9	126.8	111.5
06	WOOD & PLASTICS	95.2	88.7	91.8	88.4	76.1	82.1	106.6	38.4	71.4	84.4	72.3	78.1	93.6	68.5	80.6	103.3	134.4	119.4
07	THERMAL & MOISTURE PROTECTION	93.9	61.3	78.6	93.9	77.0	86.0	94.7	68.0	82.2	94.7	82.8	88.7	94.6	78.6	87.1	101.2	141.3	120.0
08	DOORS & WINDOWS	91.8	72.7	87.1	91.7	77.9	88.4	92.0	33.2	77.7	90.5	75.1	86.8	91.3	66.9	85.3	102.2	138.3	111.0
09200	PLASTER & GYPSUM BOARD	99.9	89.1	93.2	98.1	76.2	84.5	103.2	37.2	62.3	98.1	72.4	82.2	100.8	68.5	80.8	98.5	134.6	120.8
09500	CEILINGS	90.9	89.1	89.7	90.9	76.2	81.2	90.9	37.2	55.7	102.5	72.4	82.8	102.5	68.5	80.2	99.4	134.6	122.5
09600	FLOORING	85.9	76.3	83.6	83.8	79.1	82.6	89.8	76.3	86.5	91.6	80.4	88.8	95.6	76.3	90.9	96.2	152.6	110.0
09900	PAINTS & COATINGS	97.9	90.5	93.6	97.9	49.5	69.6	97.9	36.5	62.0	104.2	82.9	91.7	104.2	85.2	93.1	94.2	143.5	123.0
09	FINISHES	88.2	81.7	84.9	87.4	73.8	80.4	90.5	45.5	67.5	90.4	74.0	82.0	92.3	70.1	81.0	96.3	139.3	118.3
10 - 14	TOTAL DIV. 10000 - 14000	100.0	71.8	94.0	100.0	82.9	96.4	100.0	62.6	92.1	100.0	75.7	94.9	100.0	68.5	93.3	100.0	127.0	105.7
15	MECHANICAL	96.8	76.2	87.5	100.0	84.2	92.9	96.8	53.5	77.2	96.8	84.1	91.0	96.8	81.5	89.9	99.9	125.0	111.3
16	ELECTRICAL	102.0	81.7	88.1	99.3	86.1	90.3	96.6	77.6	83.6	95.4	92.7	93.6	91.3	92.7	92.3	99.5	128.8	119.4
01 - 16	WEIGHTED AVERAGE	93.0	76.1	84.8	94.2	82.0	88.3	96.0	59.9	78.5	96.7	82.2	89.7	97.5	79.2	88.7	103.0	131.1	116.6

City Cost Indexes

		MASSACHUSETTS																	
		BROCKTON			BUZZARDS BAY			FALL RIVER			FITCHBURG			FRAMINGHAM			GREENFIELD		
DIVISION		023			025			027			014			017			013		
		MAT.	INST.	TOTAL	MAT.	INST.	TOTAL	MAT.	INST.	TOTAL	MAT.	INST.	TOTAL	MAT.	INST.	TOTAL	MAT.	INST.	TOTAL
02	SITE CONSTRUCTION	93.2	100.4	98.7	84.3	100.2	96.5	92.1	100.5	98.6	85.5	100.3	96.9	82.8	99.5	95.6	89.5	99.0	96.8
03100	CONCRETE FORMS & ACCESSORIES	102.4	119.4	117.0	99.2	115.1	112.9	102.4	115.6	113.8	92.1	113.8	110.7	100.6	113.8	112.0	89.9	103.2	101.3
03200	CONCRETE REINFORCEMENT	115.5	169.5	146.1	92.6	150.5	125.4	115.5	154.3	137.5	92.6	162.9	132.5	92.5	169.2	136.0	96.5	133.7	117.6
03300	CAST-IN-PLACE CONCRETE	105.0	135.2	117.6	87.2	138.2	108.5	101.5	139.0	117.2	87.2	131.6	105.7	91.5	122.1	104.2	94.0	111.2	101.2
03	CONCRETE	114.8	133.2	124.0	97.1	128.8	113.0	113.1	130.0	121.6	92.8	128.0	110.5	98.4	126.1	112.3	99.1	110.9	105.0
04	MASONRY	106.9	139.2	127.0	98.1	142.8	126.0	106.7	142.7	129.1	94.4	138.3	121.8	100.5	135.4	122.2	98.8	110.7	106.2
05	METALS	99.9	123.4	108.4	94.9	116.5	102.7	99.9	118.6	106.6	94.8	117.3	102.9	94.8	121.5	104.4	97.2	104.8	99.9
06	WOOD & PLASTICS	102.4	115.9	109.3	99.0	110.6	105.0	102.4	111.0	106.8	94.0	109.2	101.9	101.5	108.9	105.3	91.8	101.1	96.6
07	THERMAL & MOISTURE PROTECTION	101.0	134.0	116.4	100.2	130.0	114.2	100.9	129.1	114.1	100.3	124.2	111.5	100.4	131.9	115.2	100.3	110.5	105.1
08	DOORS & WINDOWS	102.2	128.3	108.5	97.7	119.1	102.9	102.2	120.2	106.6	107.2	123.1	111.1	98.2	124.5	104.6	107.0	109.1	107.5
09200	PLASTER & GYPSUM BOARD	98.5	115.4	109.0	95.2	110.0	104.4	98.5	110.0	105.6	93.8	108.6	103.0	96.1	108.6	103.8	95.3	100.1	98.3
09500	CEILINGS	99.4	115.4	109.9	89.7	110.0	103.0	99.4	110.0	106.3	89.7	108.6	102.1	89.7	108.6	102.1	99.4	100.1	99.9
09600	FLOORING	96.6	152.6	110.3	95.4	152.6	109.4	96.4	152.6	110.1	94.3	152.6	108.5	95.7	152.6	109.6	93.6	112.3	98.2
09900	PAINTS & COATINGS	93.6	131.4	115.7	93.6	131.4	115.7	93.6	131.4	115.7	93.5	131.4	115.7	94.2	131.4	116.0	93.5	98.2	96.2
09	FINISHES	96.4	126.4	111.7	93.3	123.3	108.6	96.4	123.6	110.3	92.8	122.5	108.0	93.4	122.2	108.1	94.8	103.9	99.5
10 - 14	TOTAL DIV. 10000 - 14000	100.0	123.4	104.9	100.0	122.7	104.8	100.0	123.7	105.0	100.0	108.7	101.8	100.0	121.6	104.6	100.0	111.2	102.4
15	MECHANICAL	99.9	105.6	102.5	96.7	106.0	100.9	99.9	106.1	102.7	97.1	103.1	99.8	97.1	110.4	103.1	97.1	99.5	98.2
16	ELECTRICAL	99.2	99.3	99.3	94.3	99.8	98.0	98.6	99.8	99.4	105.2	99.6	101.4	99.3	125.4	117.1	105.2	85.2	91.6
01 - 16	WEIGHTED AVERAGE	101.8	117.2	109.3	96.2	115.6	105.6	101.5	116.1	108.6	97.4	114.1	105.5	97.0	120.1	108.2	99.0	101.6	100.3

		MASSACHUSETTS																	
		HYANNIS			LAWRENCE			LOWELL			NEW BEDFORD			PITTSFIELD			SPRINGFIELD		
DIVISION		026			019			018			027			012			010 - 011		
		MAT.	INST.	TOTAL	MAT.	INST.	TOTAL	MAT.	INST.	TOTAL	MAT.	INST.	TOTAL	MAT.	INST.	TOTAL	MAT.	INST.	TOTAL
02	SITE CONSTRUCTION	89.7	100.2	97.7	93.9	100.4	98.9	93.0	100.3	98.6	91.8	100.5	98.5	94.1	99.1	98.0	93.3	99.3	97.9
03100	CONCRETE FORMS & ACCESSORIES	92.0	115.1	111.9	102.3	119.4	117.0	98.6	119.5	116.6	102.4	115.6	113.8	98.5	98.8	98.6	98.7	106.4	105.3
03200	CONCRETE REINFORCEMENT	92.6	150.5	125.4	114.5	150.9	135.1	115.5	142.9	131.0	115.5	154.3	137.5	95.7	128.6	114.4	115.5	140.7	129.8
03300	CAST-IN-PLACE CONCRETE	95.8	138.2	113.5	105.3	139.6	119.6	96.3	139.6	114.3	98.6	139.0	115.4	105.0	104.5	104.8	98.6	111.5	104.0
03	CONCRETE	104.3	128.8	116.6	114.8	131.2	123.0	105.0	129.7	117.4	111.7	130.0	120.9	106.7	105.5	106.1	106.1	113.6	109.9
04	MASONRY	105.9	142.8	128.9	106.5	139.2	126.9	93.2	138.3	121.3	106.4	142.7	129.1	93.9	102.3	99.2	93.4	112.3	105.2
05	METALS	96.5	116.5	103.7	97.5	116.9	104.5	97.5	112.7	103.0	99.9	118.6	106.6	97.3	99.8	98.2	99.9	105.1	101.7
06	WOOD & PLASTICS	91.6	110.6	101.4	102.4	115.9	109.3	101.1	115.9	108.7	102.4	111.0	106.8	101.1	96.7	98.8	101.1	104.9	103.1
07	THERMAL & MOISTURE PROTECTION	100.4	130.0	114.3	100.9	134.6	116.7	100.7	133.3	116.0	100.9	129.1	114.1	100.8	106.8	103.6	100.7	111.4	105.7
08	DOORS & WINDOWS	98.4	119.1	103.4	102.2	123.8	107.4	108.3	123.0	111.9	102.2	120.2	106.6	108.3	104.0	107.2	108.3	111.2	109.0
09200	PLASTER & GYPSUM BOARD	92.9	110.0	103.5	98.5	115.4	109.0	98.5	115.4	109.0	98.5	110.0	105.6	98.5	95.6	96.7	98.5	104.1	102.0
09500	CEILINGS	89.7	110.0	103.0	99.4	115.4	109.9	99.4	115.4	109.9	99.4	110.0	106.3	99.4	95.6	96.9	99.4	104.1	102.5
09600	FLOORING	93.2	152.6	107.7	96.4	152.6	110.1	96.4	152.6	110.1	96.4	152.6	110.1	96.6	102.4	98.1	96.3	114.2	100.6
09900	PAINTS & COATINGS	93.6	130.9	115.4	93.6	131.4	115.7	93.5	131.4	115.7	93.6	131.4	115.7	93.5	98.2	96.2	94.8	98.2	96.8
09	FINISHES	92.5	123.3	108.2	96.4	126.4	111.7	96.3	126.4	111.7	96.3	123.6	110.2	96.4	98.8	97.6	96.4	106.7	101.7
10 - 14	TOTAL DIV. 10000 - 14000	100.0	122.7	104.8	100.0	123.4	105.0	100.0	123.4	105.0	100.0	123.7	105.0	100.0	107.4	101.6	100.0	109.7	102.0
15	MECHANICAL	99.9	106.0	102.7	99.9	105.1	102.3	99.9	119.6	108.8	99.9	106.1	102.7	99.9	92.0	96.3	99.9	98.8	99.4
16	ELECTRICAL	94.3	99.8	98.0	103.5	109.1	107.3	104.4	109.0	107.6	99.5	99.8	99.7	104.4	81.5	88.8	104.4	93.2	96.8
01 - 16	WEIGHTED AVERAGE	98.5	115.6	106.8	101.8	117.8	109.5	100.6	119.8	109.9	101.4	116.1	108.5	100.8	96.6	98.8	101.1	104.0	102.5

		MASSACHUSETTS			MICHIGAN															
		WORCESTER			ANN ARBOR			BATTLE CREEK			BAY CITY			DEARBORN			DETROIT		
DIVISION		015 - 016			481			490			487			481			482		
		MAT.	INST.	TOTAL	MAT.	INST.	TOTAL	MAT.	INST.	TOTAL	MAT.	INST.	TOTAL	MAT.	INST.	TOTAL	MAT.	INST.	TOTAL
02	SITE CONSTRUCTION	93.0	100.3	98.6	79.7	93.9	90.6	83.2	88.6	87.4	68.8	93.2	87.5	79.4	94.1	90.7	97.7	95.9	96.3
03100	CONCRETE FORMS & ACCESSORIES	99.2	113.8	111.7	96.1	116.0	113.2	95.0	91.5	92.0	96.2	97.3	97.1	96.0	118.6	115.4	98.1	118.7	115.8
03200	CONCRETE REINFORCEMENT	115.5	153.8	137.2	95.4	116.1	107.1	98.2	89.3	93.2	95.4	115.0	106.5	95.4	116.1	107.1	94.8	116.1	106.8
03300	CAST-IN-PLACE CONCRETE	96.3	136.6	111.8	95.4	123.6	107.1	98.0	110.1	103.1	91.2	101.7	95.6	93.2	124.5	106.2	102.3	124.5	111.6
03	CONCRETE	105.0	127.2	116.1	90.1	119.1	104.7	96.7	96.7	96.7	88.1	103.1	95.6	89.0	120.6	104.9	93.5	119.0	106.4
04	MASONRY	93.2	138.3	121.3	103.2	117.6	112.2	99.1	78.4	86.2	102.9	91.9	96.0	103.0	120.5	113.9	101.9	120.5	113.5
05	METALS	99.9	117.2	106.1	98.0	125.2	107.7	96.3	86.6	92.9	98.6	121.9	106.9	98.0	125.4	107.8	98.6	101.4	99.6
06	WOOD & PLASTICS	101.7	109.2	105.6	92.5	116.2	104.8	94.7	91.1	92.8	92.5	98.2	95.4	92.5	117.9	105.6	94.8	117.9	106.8
07	THERMAL & MOISTURE PROTECTION	100.7	124.5	111.8	94.6	117.7	105.4	92.3	88.3	90.4	92.9	95.0	93.9	93.8	123.8	107.9	92.1	123.8	106.9
08	DOORS & WINDOWS	108.3	123.1	111.9	97.2	113.0	101.1	93.7	83.9	91.3	97.2	101.2	98.2	97.2	114.0	101.3	96.9	114.1	101.1
09200	PLASTER & GYPSUM BOARD	98.5	108.6	104.7	113.1	114.9	114.2	103.3	84.7	91.8	113.1	96.3	102.7	113.1	116.7	115.3	113.1	116.7	115.3
09500	CEILINGS	99.4	108.6	105.4	102.6	114.9	110.7	95.1	84.7	88.3	102.6	96.3	98.5	102.6	116.7	111.8	102.6	116.7	111.8
09600	FLOORING	96.4	146.6	108.7	92.9	125.3	100.8	96.2	79.2	92.1	92.7	72.2	87.7	92.5	119.4	99.0	92.6	119.4	99.2
09900	PAINTS & COATINGS	93.5	121.6	109.9	95.3	116.8	107.9	97.4	90.0	93.1	95.3	88.2	91.2	95.3	113.6	106.0	97.3	113.6	106.9
09	FINISHES	96.3	120.2	108.5	100.3	117.7	109.2	96.4	88.7	92.5	99.9	91.9	95.8	100.1	118.3	109.4	101.0	118.3	109.8
10 - 14	TOTAL DIV. 10000 - 14000	100.0	112.9	102.7	100.0	112.3	102.6	100.0	114.9	103.1	100.0	105.9	101.2	100.0	113.8	102.9	100.0	113.8	102.9
15	MECHANICAL	99.9	103.1	101.4	100.0	104.2	101.9	100.1	86.5	94.0	100.0	90.8	95.9	100.0	111.9	105.4	100.0	115.0	106.8
16	ELECTRICAL	104.4	99.6	101.1	94.7	104.2	101.2	93.5	91.4	92.1	94.2	89.6	91.1	94.7	112.3	106.7	96.0	112.2	107.0
01 - 16	WEIGHTED AVERAGE	100.9	113.8	107.2	97.2	111.0	103.9	96.6	89.1	93.0	96.6	96.5	96.5	97.0	114.7	105.6	98.2	113.1	105.4

617

City Cost Indexes

DIVISION		MICHIGAN																	
		FLINT 484 - 485			GAYLORD 497			GRAND RAPIDS 493,495			IRON MOUNTAIN 498 - 499			JACKSON 492			KALAMAZOO 491		
		MAT.	INST.	TOTAL	MAT.	INST.	TOTAL	MAT.	INST.	TOTAL	MAT.	INST.	TOTAL	MAT.	INST.	TOTAL	MAT.	INST.	TOTAL
02	SITE CONSTRUCTION	67.4	93.3	87.3	79.1	86.6	84.9	82.8	88.3	87.0	86.5	95.7	93.5	100.7	87.2	90.4	83.5	88.6	87.4
03100	CONCRETE FORMS & ACCESSORIES	97.3	98.8	98.6	92.3	76.8	79.0	96.0	80.8	83.0	87.9	91.3	90.9	89.5	93.4	92.9	95.0	91.4	91.9
03200	CONCRETE REINFORCEMENT	95.4	115.2	106.6	91.5	106.0	99.7	98.2	88.4	92.6	91.1	101.5	97.0	88.8	115.3	103.9	98.2	89.7	93.4
03300	CAST-IN-PLACE CONCRETE	96.0	104.7	99.6	97.7	91.4	95.1	98.3	90.7	95.1	115.8	89.7	104.9	97.6	96.9	97.3	100.0	110.1	104.2
03	CONCRETE	90.5	104.9	97.7	93.1	88.7	90.8	96.9	85.2	91.0	103.2	92.7	98.0	87.9	99.7	93.9	99.9	96.7	98.3
04	MASONRY	103.2	98.0	99.9	111.8	77.0	90.1	96.8	60.7	74.3	96.6	95.5	95.9	90.1	94.8	93.1	99.2	85.9	90.9
05	METALS	98.0	122.5	106.8	97.7	111.9	102.8	96.1	84.3	91.9	97.1	95.0	96.3	97.9	120.9	106.1	96.3	87.2	93.1
06	WOOD & PLASTICS	92.5	98.2	95.4	88.4	74.5	81.2	96.2	81.7	88.7	87.8	90.5	89.2	87.0	93.7	90.5	94.7	91.1	92.8
07	THERMAL & MOISTURE PROTECTION	92.9	99.0	95.8	91.3	74.4	83.4	92.6	68.7	81.4	93.9	88.7	91.5	90.5	100.7	95.4	92.7	98.3	95.3
08	DOORS & WINDOWS	97.2	101.2	98.2	95.4	75.4	90.6	93.6	74.6	89.0	100.2	81.0	95.5	94.4	96.1	94.8	93.7	80.9	90.5
09200	PLASTER & GYPSUM BOARD	113.1	96.3	102.7	106.3	71.6	84.8	103.3	75.0	85.8	70.2	90.4	82.9	105.1	91.4	96.6	103.3	84.7	91.8
09500	CEILINGS	102.6	96.3	98.5	92.3	71.6	78.7	95.1	75.0	81.9	92.0	90.7	91.1	93.7	91.4	92.2	95.1	84.7	88.3
09600	FLOORING	92.7	90.8	92.2	89.6	76.2	86.3	96.2	53.6	85.8	108.7	98.8	106.3	87.8	99.8	90.7	96.2	62.1	87.9
09900	PAINTS & COATINGS	95.3	93.3	94.1	92.8	70.2	79.6	97.4	53.6	71.8	116.3	58.5	82.5	92.8	105.6	100.3	97.4	86.8	91.2
09	FINISHES	99.6	96.5	98.0	96.9	75.3	85.9	96.4	72.5	84.2	98.2	89.2	93.6	97.4	95.6	96.5	96.4	84.8	90.5
10 - 14	TOTAL DIV. 10000 - 14000	100.0	107.1	101.5	100.0	94.2	98.8	100.0	111.4	102.4	100.0	100.2	100.0	100.0	105.1	101.1	100.0	114.9	103.1
15	MECHANICAL	100.0	100.0	100.0	97.0	82.0	90.2	100.0	65.7	84.5	96.9	94.7	95.9	97.0	90.3	93.9	100.0	94.0	97.3
16	ELECTRICAL	94.7	98.0	97.0	89.4	75.7	80.1	93.6	75.6	81.3	99.7	86.4	90.6	95.9	104.1	101.5	93.2	88.5	90.0
01 - 16	WEIGHTED AVERAGE	96.8	101.3	98.9	96.2	83.5	90.0	96.5	75.9	86.5	98.2	92.0	95.2	95.4	98.3	96.8	97.0	90.3	93.7

DIVISION		MICHIGAN																MINNESOTA		
		LANSING 488 - 489			MUSKEGAN 494			ROYAL OAK 480,483			SAGINAW 486			TRAVERSE CITY 496			BEMIDJI 566		
		MAT.	INST.	TOTAL	MAT.	INST.	TOTAL	MAT.	INST.	TOTAL	MAT.	INST.	TOTAL	MAT.	INST.	TOTAL	MAT.	INST.	TOTAL
02	SITE CONSTRUCTION	87.5	93.3	92.0	83.3	88.7	87.5	85.4	93.1	91.3	69.7	93.2	87.7	72.7	95.5	90.2	79.1	97.9	93.5
03100	CONCRETE FORMS & ACCESSORIES	99.8	96.7	97.2	96.4	89.8	90.7	91.7	117.4	113.7	96.2	97.1	96.9	87.6	73.7	75.7	81.7	92.4	90.9
03200	CONCRETE REINFORCEMENT	95.4	115.3	106.7	98.1	89.7	93.3	86.5	115.7	103.0	95.4	114.9	106.5	92.5	75.0	82.6	102.6	96.8	99.3
03300	CAST-IN-PLACE CONCRETE	95.4	102.7	98.4	98.0	109.4	102.8	83.5	111.8	95.3	94.2	101.6	97.3	90.3	80.0	86.0	98.6	93.7	96.6
03	CONCRETE	90.4	103.3	96.9	96.8	95.8	96.3	78.0	113.8	96.0	89.5	103.0	96.3	84.6	76.7	80.6	91.3	95.0	93.2
04	MASONRY	96.3	99.2	98.1	99.1	87.7	92.0	96.4	113.2	106.9	104.6	91.9	96.7	94.7	77.3	83.9	102.8	90.4	95.0
05	METALS	97.3	122.4	106.3	95.5	87.2	92.5	97.7	97.5	97.6	98.1	121.7	106.5	97.0	83.7	92.3	93.3	120.1	102.9
06	WOOD & PLASTICS	96.1	96.1	96.1	96.2	87.7	91.8	88.2	117.9	103.5	92.5	98.2	95.4	87.4	72.0	79.4	73.9	93.7	84.2
07	THERMAL & MOISTURE PROTECTION	94.2	98.0	96.0	92.3	84.2	88.5	91.4	120.5	105.0	93.4	95.0	94.2	92.8	72.0	83.1	97.5	90.2	94.0
08	DOORS & WINDOWS	97.2	100.1	97.9	94.5	87.3	92.7	97.4	114.0	101.5	97.2	101.2	98.2	100.2	66.9	92.1	103.8	115.7	106.7
09200	PLASTER & GYPSUM BOARD	116.1	94.2	102.6	104.4	81.2	90.0	111.0	116.7	114.5	113.1	96.3	102.7	70.2	71.6	71.0	110.9	94.1	100.5
09500	CEILINGS	102.6	94.2	97.1	100.6	81.2	87.9	101.3	116.7	111.4	102.6	96.3	98.5	92.0	71.6	78.6	141.0	94.1	110.2
09600	FLOORING	92.9	79.4	89.6	96.2	77.4	91.6	90.0	104.9	93.6	92.9	72.2	87.9	108.7	76.2	100.7	106.5	127.2	111.5
09900	PAINTS & COATINGS	95.3	88.2	91.2	97.4	77.7	85.9	97.3	113.5	106.8	95.3	88.2	91.2	116.3	70.2	89.4	104.7	88.9	95.5
09	FINISHES	101.2	92.0	96.5	97.6	85.5	91.4	99.0	115.0	107.1	100.1	91.2	95.5	97.4	73.5	85.2	111.6	98.4	104.9
10 - 14	TOTAL DIV. 10000 - 14000	100.0	106.5	101.4	100.0	115.3	103.2	100.0	112.4	102.6	100.0	105.9	101.2	100.0	76.3	95.0	100.0	91.7	98.2
15	MECHANICAL	100.0	97.0	98.7	100.0	95.7	98.1	97.0	106.7	101.4	100.0	90.6	95.8	96.9	72.6	85.9	97.3	91.7	94.8
16	ELECTRICAL	93.5	100.8	98.5	93.5	77.6	82.7	98.2	111.4	107.2	97.0	89.6	92.0	91.9	74.8	80.2	100.5	96.8	98.0
01 - 16	WEIGHTED AVERAGE	97.0	100.4	98.6	96.7	89.0	93.0	94.8	108.8	101.6	97.0	96.4	96.7	94.7	77.2	86.2	98.3	97.8	98.0

DIVISION		MINNESOTA																	
		BRAINERD 564			DETROIT LAKES 565			DULUTH 556 - 558			MANKATO 560			MINNEAPOLIS 553 - 555			ROCHESTER 559		
		MAT.	INST.	TOTAL	MAT.	INST.	TOTAL	MAT.	INST.	TOTAL	MAT.	INST.	TOTAL	MAT.	INST.	TOTAL	MAT.	INST.	TOTAL
02	SITE CONSTRUCTION	79.6	104.4	98.6	77.5	98.2	93.3	80.2	103.3	97.9	76.8	104.6	98.1	80.0	109.6	102.7	79.4	102.9	97.4
03100	CONCRETE FORMS & ACCESSORIES	83.1	90.1	89.1	77.9	79.4	79.2	97.3	121.1	117.8	93.6	97.0	96.5	98.0	138.4	132.6	97.8	104.2	103.3
03200	CONCRETE REINFORCEMENT	101.5	96.8	98.9	102.6	96.7	99.3	91.4	100.6	96.6	99.3	103.7	101.8	91.6	115.8	105.3	91.4	114.8	104.6
03300	CAST-IN-PLACE CONCRETE	107.2	101.1	104.6	95.7	98.1	96.7	103.8	110.8	106.7	98.8	94.2	96.8	102.3	126.2	112.3	100.3	103.2	101.5
03	CONCRETE	95.2	96.5	95.9	88.9	90.8	89.8	93.4	114.1	103.8	90.8	98.6	94.7	94.1	130.0	112.2	91.7	106.9	99.3
04	MASONRY	126.0	99.2	109.3	125.2	98.9	108.8	108.2	111.8	110.5	112.0	112.7	112.4	108.3	135.8	125.5	108.0	105.4	106.4
05	METALS	94.7	120.0	103.8	93.2	119.4	102.6	94.3	121.0	103.9	94.5	126.1	105.8	96.3	131.7	109.0	94.2	129.1	106.7
06	WOOD & PLASTICS	88.1	85.9	86.9	70.0	72.8	71.4	102.6	123.0	113.1	99.8	93.9	96.7	103.1	138.5	121.4	103.1	101.9	102.5
07	THERMAL & MOISTURE PROTECTION	96.1	105.6	100.6	97.3	94.7	96.1	97.0	120.9	108.2	96.7	104.3	100.3	98.1	135.5	115.5	96.8	108.2	102.1
08	DOORS & WINDOWS	90.1	111.4	95.3	103.8	96.9	102.1	100.4	127.6	107.1	95.3	119.5	101.2	101.6	145.2	112.2	100.4	126.3	106.7
09200	PLASTER & GYPSUM BOARD	84.9	86.2	85.7	109.5	72.5	86.6	96.5	124.0	113.8	88.8	94.5	92.3	96.5	140.4	123.7	96.5	102.8	100.4
09500	CEILINGS	68.6	86.2	80.2	141.0	72.5	96.0	89.4	124.5	112.4	68.6	94.5	85.6	89.4	140.4	122.9	89.4	102.8	98.2
09600	FLOORING	105.1	127.2	110.5	104.9	127.2	110.4	114.2	127.2	117.4	107.3	127.2	112.2	111.4	121.4	113.8	114.0	95.3	109.5
09900	PAINTS & COATINGS	100.1	88.9	93.6	104.7	80.3	90.4	104.0	110.2	107.7	109.2	78.8	91.4	111.2	116.7	114.4	106.7	101.8	103.9
09	FINISHES	93.7	95.9	94.8	110.9	87.3	98.8	100.7	121.8	111.5	95.4	100.1	97.8	100.2	133.9	117.4	100.8	102.4	101.6
10 - 14	TOTAL DIV. 10000 - 14000	100.0	89.2	97.7	100.0	92.2	98.4	100.0	108.7	101.8	100.0	102.9	100.6	100.0	111.5	102.4	100.0	100.0	100.0
15	MECHANICAL	96.5	98.4	97.3	97.3	95.9	96.7	99.8	105.5	102.4	96.5	96.6	96.5	99.8	115.8	107.0	99.8	94.7	97.5
16	ELECTRICAL	96.7	102.3	100.5	100.5	59.4	72.5	97.9	102.3	100.9	107.0	94.1	98.2	99.0	105.5	103.5	97.9	94.1	95.3
01 - 16	WEIGHTED AVERAGE	96.3	101.5	98.8	99.0	90.4	94.8	98.1	111.4	104.5	96.6	103.1	99.8	98.6	122.8	110.3	97.8	104.0	100.8

City Cost Indexes

DIVISION		MINNESOTA															MISSISSIPPI		
		SAINT PAUL			ST. CLOUD			THIEF RIVER FALLS			WILLMAR			WINDOM			BILOXI		
		550 - 551			563			567			562			561			395		
		MAT.	INST.	TOTAL	MAT.	INST.	TOTAL	MAT.	INST.	TOTAL	MAT.	INST.	TOTAL	MAT.	INST.	TOTAL	MAT.	INST.	TOTAL
02	SITE CONSTRUCTION	82.1	103.9	98.8	75.1	105.6	98.5	77.0	97.9	93.0	74.9	102.4	96.0	69.7	101.1	93.8	109.7	87.4	92.6
03100	CONCRETE FORMS & ACCESSORIES	88.1	130.6	124.6	79.3	99.3	96.5	80.7	91.5	90.0	79.0	74.4	75.1	84.9	69.4	71.6	99.9	54.5	60.9
03200	CONCRETE REINFORCEMENT	88.2	115.7	103.8	100.9	114.8	108.8	104.3	96.6	100.0	100.0	102.6	101.5	100.0	102.1	101.2	95.8	67.8	79.9
03300	CAST-IN-PLACE CONCRETE	103.2	122.5	111.3	94.6	112.5	102.0	94.7	71.0	84.8	96.1	90.0	93.5	83.1	68.7	77.1	107.9	55.3	85.9
03	CONCRETE	95.5	125.3	110.5	86.8	108.0	97.4	88.6	86.7	87.7	86.7	86.7	86.7	78.1	77.0	77.5	97.1	58.9	77.9
04	MASONRY	119.3	135.8	129.6	111.0	111.1	111.1	102.5	90.4	94.9	114.8	88.2	98.2	124.8	66.7	88.6	88.1	49.5	64.1
05	METALS	93.0	131.4	106.7	95.1	130.0	107.6	93.3	119.1	102.6	94.5	115.7	102.1	94.4	115.0	101.8	98.9	84.9	93.9
06	WOOD & PLASTICS	92.6	128.1	110.9	84.0	91.2	87.7	72.7	93.7	83.6	83.6	73.9	78.6	90.2	73.9	81.8	99.8	55.3	76.8
07	THERMAL & MOISTURE PROTECTION	96.7	135.6	114.9	96.3	122.4	108.5	97.3	79.9	89.1	96.1	71.0	84.3	96.1	61.5	79.9	96.9	56.3	77.9
08	DOORS & WINDOWS	97.5	139.5	107.8	95.3	113.1	99.6	103.8	115.7	106.7	92.7	89.5	91.9	97.4	89.5	95.5	97.3	56.0	87.3
09200	PLASTER & GYPSUM BOARD	92.7	129.8	115.7	83.5	91.8	88.7	110.4	94.1	100.3	83.5	74.0	77.6	85.8	74.0	78.5	109.2	54.4	75.3
09500	CEILINGS	86.6	129.8	115.0	68.6	91.8	83.8	141.0	94.1	110.2	68.6	74.0	72.1	68.6	74.0	72.1	96.3	54.4	68.8
09600	FLOORING	107.1	121.4	110.6	101.3	127.2	107.6	106.1	127.2	111.2	103.2	57.7	92.1	105.9	51.1	92.5	114.5	51.0	99.0
09900	PAINTS & COATINGS	111.2	115.9	113.9	109.2	88.9	97.3	104.7	80.5	90.4	104.7	84.6	93.0	104.7	82.2	91.5	106.3	45.7	70.9
09	FINISHES	98.0	127.7	113.1	92.6	101.9	97.4	111.3	97.4	104.2	92.9	71.7	82.1	93.7	67.1	80.2	105.3	52.7	78.5
10 - 14	TOTAL DIV. 10000 - 14000	100.0	109.7	102.1	100.0	101.3	100.3	100.0	91.5	98.2	100.0	81.3	96.0	100.0	76.2	95.0	100.0	71.2	93.9
15	MECHANICAL	99.8	118.4	108.2	99.7	112.0	105.2	97.3	89.8	93.9	96.5	83.8	90.7	96.5	75.0	86.7	99.9	62.8	83.1
16	ELECTRICAL	96.9	100.2	99.2	96.7	94.1	94.9	94.0	59.4	70.4	96.7	87.3	90.3	107.0	87.3	93.5	95.8	58.9	70.7
01 - 16	WEIGHTED AVERAGE	98.0	120.0	108.6	95.6	108.0	101.6	97.3	89.6	93.6	94.7	88.2	91.5	95.5	82.0	89.0	98.9	63.3	81.7

DIVISION		MISSISSIPPI																	
		CLARKSDALE			COLUMBUS			GREENVILLE			GREENWOOD			JACKSON			LAUREL		
		386			397			387			389			390 - 392			394		
		MAT.	INST.	TOTAL	MAT.	INST.	TOTAL	MAT.	INST.	TOTAL	MAT.	INST.	TOTAL	MAT.	INST.	TOTAL	MAT.	INST.	TOTAL
02	SITE CONSTRUCTION	107.3	86.5	91.3	109.1	86.9	92.1	114.2	87.2	93.5	111.1	86.2	92.0	106.1	87.2	91.6	115.7	85.4	92.5
03100	CONCRETE FORMS & ACCESSORIES	82.1	24.2	32.4	79.5	39.4	45.1	77.6	45.2	49.8	94.3	26.6	36.2	94.2	51.7	57.7	79.8	30.9	37.9
03200	CONCRETE REINFORCEMENT	103.1	22.1	57.2	102.4	40.0	67.0	103.8	56.3	76.9	103.1	60.1	78.7	95.8	61.1	76.1	103.1	39.3	66.9
03300	CAST-IN-PLACE CONCRETE	105.2	32.6	74.9	110.0	45.4	83.1	108.4	48.4	83.4	113.2	32.4	79.5	105.8	51.4	83.1	107.6	36.4	77.9
03	CONCRETE	94.7	29.4	61.8	98.0	43.8	70.8	99.4	50.4	74.8	101.1	37.3	69.1	95.6	55.2	75.3	99.8	36.7	68.1
04	MASONRY	88.0	26.4	49.6	108.5	31.2	60.3	128.8	46.9	77.8	88.6	24.7	48.8	91.1	46.9	63.6	106.4	32.0	59.9
05	METALS	96.5	66.0	85.5	96.3	73.1	88.0	97.3	80.3	91.2	96.5	78.2	89.9	98.9	82.1	92.8	96.4	71.2	87.4
06	WOOD & PLASTICS	79.9	22.8	50.4	77.1	41.2	58.5	75.2	44.8	59.5	93.1	25.7	58.3	93.2	53.3	72.6	78.1	32.4	54.5
07	THERMAL & MOISTURE PROTECTION	96.4	34.7	67.5	96.6	34.6	67.6	96.8	51.1	75.4	96.9	29.0	65.1	96.6	52.7	76.0	96.7	33.0	66.8
08	DOORS & WINDOWS	96.8	27.0	79.8	96.7	36.3	82.0	96.8	46.7	84.6	96.8	37.4	82.3	97.7	52.4	86.7	93.5	31.4	78.4
09200	PLASTER & GYPSUM BOARD	103.5	20.9	52.3	102.6	39.8	63.7	102.1	43.6	65.9	108.1	23.9	55.9	109.2	52.3	74.0	101.7	30.8	57.8
09500	CEILINGS	90.8	20.9	44.9	90.8	39.8	57.3	90.8	43.6	59.8	90.8	23.9	46.8	96.3	52.3	67.4	90.8	30.8	51.4
09600	FLOORING	107.6	29.7	88.6	106.2	25.6	86.6	105.2	43.6	90.2	114.5	26.7	93.1	114.5	50.5	98.9	104.2	24.1	84.7
09900	PAINTS & COATINGS	106.3	19.2	55.4	106.3	28.3	60.7	106.3	45.0	70.5	106.3	26.2	59.5	106.3	45.0	70.5	106.3	28.3	60.7
09	FINISHES	101.1	24.5	62.0	100.7	36.3	67.8	100.6	44.7	72.1	104.3	26.5	64.6	105.3	50.9	77.5	100.3	30.2	64.5
10 - 14	TOTAL DIV. 10000 - 14000	100.0	56.4	90.8	100.0	60.1	91.6	100.0	68.9	93.4	100.0	56.8	90.9	100.0	70.1	93.7	100.0	35.9	86.4
15	MECHANICAL	98.5	24.9	65.2	98.2	33.9	69.0	99.9	45.5	75.3	98.5	26.2	65.7	99.9	45.1	75.1	98.2	28.9	66.8
16	ELECTRICAL	94.8	20.5	44.2	91.8	35.6	53.5	94.8	49.0	63.6	94.8	19.0	43.1	96.4	49.0	64.1	94.2	26.0	47.7
01 - 16	WEIGHTED AVERAGE	97.1	35.8	67.4	98.3	45.2	72.5	100.4	54.9	78.3	98.5	38.3	69.3	98.8	56.7	78.4	98.3	39.8	70.0

DIVISION		MISSISSIPPI									MISSOURI								
		MCCOMB			MERIDIAN			TUPELO			BOWLING GREEN			CAPE GIRARDEAU			CHILLICOTHE		
		396			393			388			633			637			646		
		MAT.	INST.	TOTAL	MAT.	INST.	TOTAL	MAT.	INST.	TOTAL	MAT.	INST.	TOTAL	MAT.	INST.	TOTAL	MAT.	INST.	TOTAL
02	SITE CONSTRUCTION	100.6	86.3	89.7	104.9	87.2	91.3	105.3	86.5	90.9	96.1	93.5	94.1	96.9	93.3	94.1	98.4	89.6	91.7
03100	CONCRETE FORMS & ACCESSORIES	79.5	29.6	36.6	75.9	43.9	48.4	78.3	31.4	38.0	94.6	86.8	87.9	85.0	84.9	84.9	90.9	68.2	71.4
03200	CONCRETE REINFORCEMENT	103.8	18.7	55.5	102.4	60.6	78.7	101.0	60.9	78.3	98.8	93.7	95.9	100.0	93.4	96.3	98.6	80.8	88.5
03300	CAST-IN-PLACE CONCRETE	95.5	34.4	70.0	102.2	51.4	81.0	105.2	47.2	81.0	88.4	93.8	90.7	87.5	92.0	89.4	94.8	78.7	88.1
03	CONCRETE	87.8	31.7	59.6	92.2	51.6	71.8	94.1	44.7	69.3	85.8	92.2	89.0	84.6	90.7	87.7	96.4	75.5	85.9
04	MASONRY	110.4	26.4	58.1	87.7	38.2	56.9	116.7	40.6	69.3	119.3	90.0	101.1	113.7	81.2	93.5	103.4	73.4	84.7
05	METALS	96.5	64.4	85.0	97.2	81.2	91.5	96.4	80.3	90.6	94.7	116.6	102.6	95.5	116.4	103.0	93.2	94.4	93.7
06	WOOD & PLASTICS	77.1	30.1	52.8	73.3	43.0	57.6	75.9	30.1	52.2	95.2	87.5	91.2	85.0	86.5	85.8	90.5	67.1	78.4
07	THERMAL & MOISTURE PROTECTION	96.1	32.1	66.1	96.2	49.5	74.3	96.4	43.4	71.6	91.9	90.5	91.3	91.7	84.6	88.4	97.8	75.7	87.5
08	DOORS & WINDOWS	96.8	30.2	80.6	96.8	45.8	84.4	96.8	35.4	81.8	97.9	88.1	95.5	97.9	87.5	95.4	93.0	67.1	86.7
09200	PLASTER & GYPSUM BOARD	102.6	28.4	56.7	101.2	41.7	64.4	102.1	28.4	56.5	101.6	86.7	92.4	98.4	85.7	90.5	107.8	65.7	81.8
09500	CEILINGS	90.8	28.4	49.9	90.8	41.7	58.6	90.8	28.4	49.9	92.5	86.7	88.7	92.5	85.7	88.0	94.3	65.7	75.6
09600	FLOORING	106.2	29.7	87.6	104.2	35.4	87.4	105.6	32.7	87.8	103.7	92.6	101.0	99.7	92.6	98.0	94.1	66.5	87.4
09900	PAINTS & COATINGS	106.3	26.7	59.8	106.3	44.4	70.1	106.3	45.0	70.5	95.6	100.5	98.5	95.6	93.3	94.3	91.6	70.4	79.2
09	FINISHES	100.2	29.7	64.2	99.6	41.7	70.0	100.3	32.7	65.8	94.5	89.1	91.7	92.7	86.9	89.7	96.6	68.2	82.1
10 - 14	TOTAL DIV. 10000 - 14000	100.0	57.4	91.0	100.0	68.8	93.4	100.0	58.2	91.1	100.0	84.0	96.6	100.0	82.7	96.3	100.0	77.1	95.2
15	MECHANICAL	98.2	23.1	64.2	99.9	45.1	75.1	98.7	32.5	68.7	96.9	96.9	96.9	100.1	94.5	97.6	97.1	57.8	79.3
16	ELECTRICAL	89.4	23.6	44.5	94.2	52.6	65.9	94.3	32.4	52.1	95.6	80.7	85.4	95.6	106.2	102.8	91.0	54.0	65.8
01 - 16	WEIGHTED AVERAGE	96.7	36.9	67.7	96.8	54.4	76.2	98.4	45.6	72.8	96.3	92.5	94.4	96.4	94.7	95.6	96.0	70.2	83.5

City Cost Indexes

		\multicolumn{18}{c}{MISSOURI}												
		\multicolumn{3}{c}{COLUMBIA}	\multicolumn{3}{c}{FLAT RIVER}	\multicolumn{3}{c}{HANNIBAL}	\multicolumn{3}{c}{HARRISONVILLE}	\multicolumn{3}{c}{JEFFERSON CITY}	\multicolumn{3}{c}{JOPLIN}
	DIVISION	\multicolumn{3}{c}{652}	\multicolumn{3}{c}{636}	\multicolumn{3}{c}{634}	\multicolumn{3}{c}{647}	\multicolumn{3}{c}{650 - 651}	\multicolumn{3}{c}{648}
		MAT.	INST.	TOTAL	MAT.	INST.	TOTAL	MAT.	INST.	TOTAL	MAT.	INST.	TOTAL	MAT.	INST.	TOTAL	MAT.	INST.	TOTAL
02	SITE CONSTRUCTION	98.1	94.9	95.7	99.1	93.5	94.8	93.1	93.4	93.3	90.0	91.4	91.1	100.3	95.0	96.2	98.1	96.4	96.8
03100	CONCRETE FORMS & ACCESSORIES	85.6	64.6	67.6	103.4	86.1	88.6	92.3	85.5	86.5	86.9	82.3	83.0	104.1	71.3	76.0	105.1	60.9	67.2
03200	CONCRETE REINFORCEMENT	97.7	95.0	96.1	100.0	99.0	99.4	98.3	93.6	95.6	98.2	86.4	91.5	96.5	105.1	101.3	101.7	76.3	87.3
03300	CAST-IN-PLACE CONCRETE	85.8	74.3	81.0	91.3	93.8	92.3	83.6	93.1	87.6	97.0	85.7	92.3	89.8	77.9	84.8	102.6	75.4	91.3
03	CONCRETE	80.3	76.1	78.1	88.7	92.9	90.8	82.2	91.4	86.8	93.2	85.3	89.2	84.2	82.1	83.1	96.7	70.5	83.5
04	MASONRY	131.1	67.9	91.7	116.4	84.4	96.4	110.9	87.3	96.2	98.3	80.3	87.1	91.9	67.9	76.9	94.6	63.7	75.4
05	METALS	94.1	115.8	101.9	94.6	118.7	103.2	94.7	116.5	102.5	93.2	100.2	95.7	93.3	120.3	103.0	97.0	95.1	96.3
06	WOOD & PLASTICS	88.6	60.0	73.8	104.9	86.5	95.4	92.8	86.5	89.5	85.7	81.9	83.7	108.6	68.3	87.8	103.7	58.1	80.1
07	THERMAL & MOISTURE PROTECTION	91.7	73.4	83.2	92.2	87.8	90.1	91.7	85.7	88.9	96.7	84.7	91.1	102.0	74.7	83.9	98.4	68.3	84.3
08	DOORS & WINDOWS	93.8	81.4	90.7	97.9	89.0	95.7	97.9	87.5	95.4	93.4	82.4	90.7	90.4	88.6	89.9	94.2	71.9	88.8
09200	PLASTER & GYPSUM BOARD	103.9	58.4	75.7	104.8	85.7	92.9	100.7	85.7	91.4	100.4	81.0	88.4	109.9	66.9	83.3	112.0	56.4	77.6
09500	CEILINGS	90.9	58.4	69.6	92.5	85.7	88.0	92.5	85.7	88.0	94.3	81.0	85.6	90.9	66.9	75.1	97.1	56.4	70.4
09600	FLOORING	108.0	66.2	97.8	107.5	92.6	103.9	102.7	92.6	100.3	90.1	86.1	89.1	115.7	66.2	103.6	118.0	62.3	104.4
09900	PAINTS & COATINGS	95.3	66.0	78.2	95.6	93.3	94.3	95.6	100.5	98.5	95.9	87.6	91.1	95.3	66.0	78.2	90.9	55.1	70.0
09	FINISHES	94.5	63.0	78.4	96.3	87.6	91.9	93.9	88.2	91.0	94.0	83.5	88.7	98.0	67.8	82.6	103.6	59.5	81.0
10 - 14	TOTAL DIV. 10000 - 14000	100.0	94.0	98.7	100.0	84.0	96.6	100.0	83.4	96.5	100.0	87.7	97.4	100.0	95.1	99.0	100.0	82.0	96.2
15	MECHANICAL	100.2	72.9	87.8	96.9	96.3	96.6	96.9	95.6	96.3	96.9	84.5	91.3	100.2	89.5	95.3	100.3	60.4	82.2
16	ELECTRICAL	93.6	80.7	84.8	102.6	106.2	105.1	93.7	80.6	84.8	102.2	89.2	93.4	96.7	80.7	85.8	88.7	70.8	76.5
01 - 16	WEIGHTED AVERAGE	96.2	79.8	88.2	97.4	96.1	96.8	95.1	91.6	93.4	95.6	87.0	91.4	94.9	85.2	90.2	97.7	71.6	85.1

		\multicolumn{18}{c}{MISSOURI}												
		\multicolumn{3}{c}{KANSAS CITY}	\multicolumn{3}{c}{KIRKSVILLE}	\multicolumn{3}{c}{POPLAR BLUFF}	\multicolumn{3}{c}{ROLLA}	\multicolumn{3}{c}{SEDALIA}	\multicolumn{3}{c}{SIKESTON}
	DIVISION	\multicolumn{3}{c}{640 - 641}	\multicolumn{3}{c}{635}	\multicolumn{3}{c}{639}	\multicolumn{3}{c}{654 - 655}	\multicolumn{3}{c}{653}	\multicolumn{3}{c}{638}
		MAT.	INST.	TOTAL	MAT.	INST.	TOTAL	MAT.	INST.	TOTAL	MAT.	INST.	TOTAL	MAT.	INST.	TOTAL	MAT.	INST.	TOTAL
02	SITE CONSTRUCTION	91.1	94.6	93.8	89.3	89.9	89.7	75.0	93.7	89.4	98.1	94.7	95.5	87.8	91.0	90.2	78.5	93.8	90.2
03100	CONCRETE FORMS & ACCESSORIES	104.8	92.5	94.3	82.6	72.6	74.1	82.5	72.7	74.1	95.8	52.5	58.6	93.2	64.6	68.6	83.9	73.0	74.6
03200	CONCRETE REINFORCEMENT	96.6	96.2	96.4	98.8	76.7	86.2	101.8	77.0	87.7	98.2	70.0	82.2	96.6	76.2	85.1	101.1	77.0	87.4
03300	CAST-IN-PLACE CONCRETE	95.4	96.8	96.0	91.3	80.4	86.7	70.1	76.1	72.6	87.8	58.3	75.5	91.7	76.8	85.5	74.9	78.4	76.4
03	CONCRETE	92.9	95.5	94.2	96.8	77.4	87.0	75.4	76.0	75.7	82.4	60.3	71.3	93.2	72.5	82.8	79.0	77.0	78.0
04	MASONRY	101.3	97.4	98.8	122.4	66.0	87.3	113.6	73.1	88.4	107.7	58.9	77.3	113.9	60.5	80.7	114.0	74.0	89.0
05	METALS	100.4	106.9	102.7	93.0	96.0	94.1	95.5	97.1	96.1	93.3	100.7	95.9	91.4	94.7	92.6	95.4	97.1	96.0
06	WOOD & PLASTICS	104.0	90.6	97.1	79.5	72.7	76.0	78.3	72.7	75.4	99.5	49.9	73.9	92.9	60.9	76.3	79.6	72.7	76.0
07	THERMAL & MOISTURE PROTECTION	97.2	96.4	96.8	95.8	77.7	87.3	95.3	78.6	87.5	92.0	57.7	76.0	95.4	76.2	86.4	95.5	79.0	87.8
08	DOORS & WINDOWS	98.4	94.6	97.5	101.9	76.1	95.6	102.9	69.1	94.7	93.8	53.9	84.0	97.4	67.4	90.1	102.9	69.1	94.7
09200	PLASTER & GYPSUM BOARD	107.1	90.0	96.5	96.0	71.4	80.7	96.5	71.4	80.9	107.2	48.0	70.5	99.0	59.3	74.4	97.0	71.4	81.1
09500	CEILINGS	102.6	90.0	94.3	89.7	71.4	77.7	92.5	71.4	78.6	90.9	48.0	62.7	89.5	59.3	69.7	92.5	71.4	78.6
09600	FLOORING	96.2	96.4	96.2	79.8	68.2	77.0	94.2	62.3	86.5	112.9	58.1	99.0	87.0	86.1	86.8	94.9	62.3	86.9
09900	PAINTS & COATINGS	95.9	96.9	96.5	91.2	55.1	70.1	90.6	65.5	75.9	95.3	43.2	64.9	95.3	87.6	90.8	90.6	65.5	75.9
09	FINISHES	98.5	93.1	95.7	90.0	69.2	79.4	92.1	69.2	80.4	96.4	51.0	73.2	91.4	70.1	80.5	92.6	69.4	80.7
10 - 14	TOTAL DIV. 10000 - 14000	100.0	94.9	98.9	100.0	77.0	95.1	100.0	77.2	95.2	100.0	74.7	94.6	100.0	83.2	96.4	100.0	77.5	95.2
15	MECHANICAL	100.1	96.9	98.6	97.0	84.1	91.1	97.0	93.1	95.2	97.0	90.2	93.9	96.9	81.8	90.1	97.0	93.5	95.4
16	ELECTRICAL	102.8	103.4	103.2	94.3	80.6	85.0	94.5	106.2	102.5	91.0	36.4	53.8	92.4	92.4	92.4	93.1	106.2	102.0
01 - 16	WEIGHTED AVERAGE	98.9	97.9	98.4	97.2	80.0	88.8	94.4	86.7	90.7	94.5	67.8	81.6	95.6	80.1	88.1	94.9	87.0	91.1

		\multicolumn{9}{c}{MISSOURI}	\multicolumn{9}{c}{MONTANA}
		\multicolumn{3}{c}{SPRINGFIELD}	\multicolumn{3}{c}{ST. JOSEPH}	\multicolumn{3}{c}{ST. LOUIS}	\multicolumn{3}{c}{BILLINGS}	\multicolumn{3}{c}{BUTTE}	\multicolumn{3}{c}{GREAT FALLS}
	DIVISION	\multicolumn{3}{c}{656 - 658}	\multicolumn{3}{c}{644 - 645}	\multicolumn{3}{c}{630 - 631}	\multicolumn{3}{c}{590 - 591}	\multicolumn{3}{c}{597}	\multicolumn{3}{c}{594}
		MAT.	INST.	TOTAL	MAT.	INST.	TOTAL	MAT.	INST.	TOTAL	MAT.	INST.	TOTAL	MAT.	INST.	TOTAL	MAT.	INST.	TOTAL
02	SITE CONSTRUCTION	90.3	94.5	93.5	92.5	91.4	91.6	96.3	97.0	96.8	80.6	98.8	94.6	86.5	97.5	94.9	89.8	98.2	96.3
03100	CONCRETE FORMS & ACCESSORIES	103.7	59.3	65.5	104.7	75.5	79.6	102.9	107.5	106.9	94.8	87.9	88.8	80.7	90.5	89.1	101.3	88.2	90.1
03200	CONCRETE REINFORCEMENT	93.9	95.0	94.5	95.4	86.2	90.2	91.5	106.8	100.2	97.7	101.8	100.0	106.0	101.9	103.7	97.7	102.0	100.1
03300	CAST-IN-PLACE CONCRETE	97.0	66.4	84.3	95.3	91.4	93.7	87.5	110.7	97.2	114.9	83.9	102.0	118.3	79.1	102.0	125.0	72.8	103.2
03	CONCRETE	92.7	70.0	81.3	92.7	84.2	88.4	84.8	109.7	97.3	100.9	89.4	95.1	100.7	88.9	94.8	106.3	85.7	96.0
04	MASONRY	87.4	68.9	75.9	100.4	75.0	84.6	97.3	110.4	105.4	122.7	94.3	105.0	121.0	96.8	105.9	124.9	99.9	109.4
05	METALS	97.5	100.8	98.7	99.0	99.7	99.2	98.0	124.6	107.5	97.0	98.6	97.6	96.4	98.0	97.0	97.4	98.5	97.8
06	WOOD & PLASTICS	102.5	56.5	78.7	104.5	73.7	88.6	104.0	105.6	104.8	96.7	88.4	92.4	83.1	93.7	88.6	105.6	88.1	96.6
07	THERMAL & MOISTURE PROTECTION	96.0	68.6	83.2	97.9	80.4	89.7	91.7	104.8	97.9	97.2	88.1	92.9	96.8	87.0	92.0	97.4	87.3	92.7
08	DOORS & WINDOWS	99.5	68.2	91.9	99.2	83.0	95.2	94.8	111.2	98.8	103.3	86.9	99.3	99.9	89.7	97.4	103.6	86.8	99.5
09200	PLASTER & GYPSUM BOARD	109.2	54.7	75.4	112.8	72.5	87.9	105.2	105.4	105.3	97.2	88.5	91.8	89.7	94.0	92.4	97.2	88.2	91.6
09500	CEILINGS	96.4	54.7	69.0	101.3	72.5	82.4	96.6	105.4	102.4	111.4	88.5	96.4	108.7	94.0	99.1	111.4	88.2	96.2
09600	FLOORING	110.4	62.3	98.7	98.8	76.4	93.3	107.2	101.0	105.7	111.4	64.9	100.0	103.2	51.6	90.6	111.4	70.1	101.3
09900	PAINTS & COATINGS	90.2	73.1	80.2	91.6	82.3	86.1	95.6	110.0	104.0	104.7	81.0	90.9	104.7	62.5	80.1	104.7	70.4	84.7
09	FINISHES	99.3	60.3	79.4	99.7	75.6	87.4	97.0	106.1	101.7	105.8	82.7	94.0	101.4	80.5	90.7	105.9	82.8	94.1
10 - 14	TOTAL DIV. 10000 - 14000	100.0	84.7	96.8	100.0	89.2	97.7	100.0	107.0	101.5	100.0	88.0	97.5	100.0	88.0	97.5	100.0	88.8	97.6
15	MECHANICAL	100.3	68.5	85.9	100.3	77.4	89.9	100.1	109.6	104.4	100.3	89.8	95.5	100.3	81.3	91.7	100.3	91.0	96.1
16	ELECTRICAL	100.0	71.5	80.5	102.2	76.0	84.3	99.3	117.0	111.4	92.8	86.1	88.2	104.2	64.4	77.1	92.8	83.1	86.2
01 - 16	WEIGHTED AVERAGE	97.6	74.0	86.2	98.9	81.6	90.5	96.4	110.4	103.2	100.8	90.2	95.6	100.5	84.8	92.9	102.0	89.9	96.1

City Cost Indexes

MONTANA

DIVISION		HAVRE 595			HELENA 596			KALISPELL 599			MILES CITY 593			MISSOULA 598			WOLF POINT 592		
		MAT.	INST.	TOTAL	MAT.	INST.	TOTAL	MAT.	INST.	TOTAL	MAT.	INST.	TOTAL	MAT.	INST.	TOTAL	MAT.	INST.	TOTAL
02	SITE CONSTRUCTION	93.6	98.3	97.2	91.5	98.2	96.7	78.1	97.4	92.9	83.8	98.3	94.9	71.0	97.5	91.3	99.4	98.3	98.5
03100	CONCRETE FORMS & ACCESSORIES	71.9	90.8	88.1	101.4	88.7	90.5	85.2	85.2	85.2	94.0	90.9	91.3	85.2	85.3	85.3	84.9	88.7	88.1
03200	CONCRETE REINFORCEMENT	106.9	101.9	104.0	101.1	101.8	101.5	108.9	109.2	109.0	106.3	101.8	103.8	107.8	109.2	108.6	107.9	101.8	104.5
03300	CAST-IN-PLACE CONCRETE	127.4	82.3	108.6	127.3	85.6	110.0	102.7	83.6	94.7	112.5	82.0	99.8	87.2	83.6	85.7	125.9	75.9	105.1
03	CONCRETE	108.3	90.1	99.1	107.9	90.3	99.1	90.8	89.1	90.0	98.0	90.1	94.0	80.0	89.4	84.7	111.8	87.0	99.3
04	MASONRY	122.0	100.4	108.5	121.9	99.5	108.0	120.2	85.3	98.4	127.9	98.4	109.5	148.1	85.3	109.0	128.9	100.4	111.1
05	METALS	96.2	98.0	96.8	96.6	98.0	97.1	96.1	92.7	94.9	95.5	98.1	96.4	96.5	98.2	97.1	95.6	97.9	96.4
06	WOOD & PLASTICS	73.4	90.8	82.4	105.8	88.1	96.6	88.2	88.1	88.1	96.0	90.8	93.3	88.2	88.1	88.1	86.2	88.1	87.2
07	THERMAL & MOISTURE PROTECTION	96.8	87.9	92.7	97.4	89.1	93.5	96.1	92.6	94.4	96.5	93.3	95.0	95.6	92.6	94.2	97.4	86.0	92.1
08	DOORS & WINDOWS	100.0	88.2	97.1	103.0	87.0	99.1	100.0	84.2	96.1	99.4	88.2	96.7	99.9	87.9	97.0	99.4	86.7	96.3
09200	PLASTER & GYPSUM BOARD	87.0	91.1	89.5	96.6	88.2	91.4	91.6	88.2	89.5	96.4	91.1	93.1	91.6	88.2	89.5	92.7	88.2	89.9
09500	CEILINGS	108.7	91.1	97.1	108.7	88.2	95.2	108.7	88.2	95.2	107.3	91.1	96.6	108.7	88.2	95.2	107.3	88.2	94.8
09600	FLOORING	99.6	56.8	89.2	111.4	67.8	100.8	105.0	80.7	99.1	111.1	63.7	99.6	105.0	80.7	99.1	106.9	56.8	94.7
09900	PAINTS & COATINGS	104.7	69.5	84.1	104.7	78.1	89.2	104.7	70.4	84.7	104.7	65.7	81.9	104.7	70.4	84.7	104.7	65.7	81.9
09	FINISHES	100.3	82.1	91.0	105.4	83.6	94.3	101.9	82.8	92.1	104.6	83.0	93.6	101.5	82.8	92.0	103.6	80.0	91.6
10 - 14	TOTAL DIV. 10000 - 14000	100.0	79.9	95.7	100.0	79.3	95.6	100.0	75.1	94.7	100.0	79.8	95.7	100.0	75.1	94.7	100.0	79.5	95.7
15	MECHANICAL	97.1	92.1	94.8	100.3	80.4	91.3	97.1	83.6	90.9	97.1	91.9	94.7	100.3	83.7	92.8	97.1	92.1	94.8
16	ELECTRICAL	92.8	83.1	86.2	92.8	83.1	86.2	99.3	86.0	90.3	92.8	86.0	88.2	100.9	86.0	90.8	92.8	86.0	88.2
01 - 16	WEIGHTED AVERAGE	99.9	90.5	95.3	101.8	88.3	95.3	98.0	87.1	92.7	99.2	91.0	95.3	98.8	87.9	93.5	101.2	90.2	95.9

NEBRASKA

DIVISION		ALLIANCE 693			COLUMBUS 686			GRAND ISLAND 688			HASTINGS 689			LINCOLN 683 - 685			MCCOOK 690		
		MAT.	INST.	TOTAL	MAT.	INST.	TOTAL	MAT.	INST.	TOTAL	MAT.	INST.	TOTAL	MAT.	INST.	TOTAL	MAT.	INST.	TOTAL
02	SITE CONSTRUCTION	88.1	97.3	95.1	88.5	90.9	90.3	92.6	92.0	92.1	91.9	91.6	91.7	83.7	92.0	90.0	92.7	90.9	91.3
03100	CONCRETE FORMS & ACCESSORIES	89.1	32.0	40.1	100.5	29.8	39.8	99.8	58.7	64.5	104.5	55.9	62.8	105.3	51.4	59.0	96.6	37.9	46.2
03200	CONCRETE REINFORCEMENT	110.3	54.8	78.8	102.8	61.0	79.1	102.3	74.5	86.5	102.3	61.0	78.9	93.9	74.3	82.8	104.1	60.9	79.6
03300	CAST-IN-PLACE CONCRETE	103.3	43.8	78.4	105.8	55.3	84.7	112.1	65.7	92.7	112.1	61.0	90.8	101.2	68.4	87.5	112.1	43.7	83.5
03	CONCRETE	110.2	41.6	75.7	97.9	46.7	72.1	102.1	65.5	83.7	102.5	60.1	81.2	94.8	63.2	78.9	102.1	46.4	74.1
04	MASONRY	107.9	36.3	63.3	114.1	37.5	66.4	105.2	56.2	74.6	113.9	49.2	73.6	95.4	53.4	69.2	104.3	36.2	61.9
05	METALS	102.8	64.1	88.9	92.7	80.6	88.3	94.0	87.2	91.6	92.7	81.0	88.5	97.0	86.8	93.4	92.9	85.0	90.1
06	WOOD & PLASTICS	86.9	30.7	57.9	99.6	27.4	62.3	98.7	55.0	76.1	103.5	55.0	78.5	104.2	45.5	73.9	95.6	38.2	65.9
07	THERMAL & MOISTURE PROTECTION	102.2	41.9	74.0	96.1	38.5	69.1	96.3	64.0	81.2	96.3	55.8	77.3	96.7	56.8	78.0	96.2	42.5	71.1
08	DOORS & WINDOWS	93.3	34.8	79.0	93.0	41.1	80.3	93.0	57.0	84.2	93.0	56.0	84.0	98.6	54.0	87.7	93.0	43.4	80.9
09200	PLASTER & GYPSUM BOARD	100.5	28.2	55.8	106.4	24.7	55.8	106.0	53.2	73.3	107.4	53.2	73.8	109.2	43.4	68.4	105.5	35.8	62.4
09500	CEILINGS	111.1	28.2	56.7	89.5	24.7	47.0	89.5	53.2	65.7	89.5	53.2	65.7	96.4	43.4	61.6	89.5	35.8	54.3
09600	FLOORING	99.8	35.5	84.1	99.8	45.1	86.5	99.6	43.3	85.9	101.2	45.1	87.6	101.5	45.1	87.7	98.5	35.5	83.1
09900	PAINTS & COATINGS	134.2	27.2	71.7	87.6	62.3	72.8	87.6	51.6	66.5	87.6	62.3	72.8	87.6	52.0	66.8	87.6	29.6	53.7
09	FINISHES	98.5	31.4	64.3	94.3	34.9	64.0	94.3	53.9	73.7	95.1	54.5	74.3	96.5	48.8	72.1	93.9	35.9	64.3
10 - 14	TOTAL DIV. 10000 - 14000	100.0	62.0	92.0	100.0	59.3	91.4	100.0	83.1	96.4	100.0	68.6	93.4	100.0	81.9	96.2	100.0	60.7	91.7
15	MECHANICAL	97.0	33.3	68.1	97.1	30.3	66.8	100.3	84.2	93.0	97.1	47.9	74.8	100.3	84.2	93.0	97.1	51.8	76.5
16	ELECTRICAL	94.2	23.0	45.7	93.0	54.7	66.9	90.8	54.7	66.2	89.8	54.7	65.9	100.3	54.7	69.2	97.0	23.1	46.6
01 - 16	WEIGHTED AVERAGE	99.7	42.9	72.1	96.3	49.7	73.7	97.2	69.5	83.8	96.9	60.0	79.0	97.7	67.8	83.2	96.7	49.3	73.7

NEBRASKA / NEVADA

DIVISION		NORFOLK 687			NORTH PLATTE 691			OMAHA 680 - 681			VALENTINE 692			CARSON CITY 897			ELKO 898		
		MAT.	INST.	TOTAL	MAT.	INST.	TOTAL	MAT.	INST.	TOTAL	MAT.	INST.	TOTAL	MAT.	INST.	TOTAL	MAT.	INST.	TOTAL
02	SITE CONSTRUCTION	75.1	90.2	86.7	93.4	91.6	92.0	75.8	90.6	87.1	79.0	93.9	90.4	65.3	102.6	93.9	60.4	102.0	92.3
03100	CONCRETE FORMS & ACCESSORIES	84.0	56.3	60.2	99.6	56.3	62.4	100.1	71.2	75.3	85.2	28.5	36.6	99.5	96.3	96.8	115.4	87.9	91.8
03200	CONCRETE REINFORCEMENT	102.9	62.4	79.9	103.5	61.0	79.4	98.3	75.7	85.5	104.1	49.2	73.1	108.6	117.3	113.5	107.3	79.6	91.6
03300	CAST-IN-PLACE CONCRETE	106.4	67.5	90.2	112.1	61.2	90.8	105.0	77.4	93.5	98.6	51.5	78.9	117.3	95.4	108.2	105.3	89.1	98.6
03	CONCRETE	96.2	62.2	79.1	102.3	60.6	81.3	96.1	74.5	85.2	98.5	42.0	70.1	113.6	99.7	106.6	104.9	86.9	95.9
04	MASONRY	119.8	74.4	91.5	90.5	49.2	64.8	101.0	78.6	87.0	102.7	37.6	62.2	134.3	85.4	103.9	133.4	80.4	100.4
05	METALS	96.9	74.1	88.7	94.2	85.9	91.2	99.6	78.3	92.0	110.0	64.7	93.8	104.1	101.6	103.2	103.4	88.9	98.2
06	WOOD & PLASTICS	82.0	54.2	67.6	98.5	55.0	76.0	100.2	70.9	85.1	82.5	26.6	53.6	95.2	95.8	95.5	112.4	88.9	100.3
07	THERMAL & MOISTURE PROTECTION	94.0	65.7	80.7	96.2	55.9	77.3	91.9	74.6	83.8	94.7	37.6	68.0	104.1	90.8	97.9	104.1	86.4	95.8
08	DOORS & WINDOWS	95.4	56.9	86.0	92.2	59.1	84.2	102.5	67.7	94.0	95.5	33.3	80.4	94.5	104.9	97.0	97.1	89.9	95.4
09200	PLASTER & GYPSUM BOARD	107.7	53.2	74.0	106.2	53.2	73.4	115.0	70.4	87.4	112.8	24.7	58.2	89.0	95.4	93.0	95.0	88.3	90.9
09500	CEILINGS	124.2	53.2	77.6	90.9	53.2	66.2	135.2	70.4	92.7	133.4	24.7	62.0	97.7	95.4	96.2	97.7	88.3	91.5
09600	FLOORING	124.8	54.2	107.6	99.5	45.1	86.3	130.3	54.2	111.8	127.7	54.1	109.8	106.7	85.8	101.6	112.9	85.8	106.3
09900	PAINTS & COATINGS	139.8	62.3	94.5	87.6	62.3	72.8	139.8	62.3	94.5	139.8	41.8	82.5	111.0	91.6	99.7	111.0	94.5	101.3
09	FINISHES	116.5	55.8	85.5	94.6	54.5	74.1	121.4	66.6	93.4	120.3	33.9	76.2	98.7	93.2	95.9	101.3	88.3	94.6
10 - 14	TOTAL DIV. 10000 - 14000	100.0	79.3	95.6	100.0	68.6	93.4	100.0	83.3	96.5	100.0	50.1	89.5	100.0	124.7	105.2	100.0	94.2	98.8
15	MECHANICAL	96.7	64.3	82.0	100.3	61.0	82.5	99.9	83.2	92.4	96.5	29.8	66.3	100.0	97.9	99.0	98.1	95.3	96.8
16	ELECTRICAL	89.1	68.8	75.2	94.1	54.7	67.2	88.2	86.1	86.7	87.9	54.6	65.2	93.7	91.7	92.3	92.8	86.9	88.8
01 - 16	WEIGHTED AVERAGE	98.8	68.2	83.9	96.7	63.2	80.5	100.3	79.4	90.2	100.6	47.2	74.7	102.0	96.9	99.5	100.9	90.0	95.6

City Cost Indexes

DIVISION		NEVADA ELY 893 MAT.	INST.	TOTAL	LAS VEGAS 889-891 MAT.	INST.	TOTAL	RENO 894-895 MAT.	INST.	TOTAL	NEW HAMPSHIRE CHARLESTON 036 MAT.	INST.	TOTAL	CLAREMONT 037 MAT.	INST.	TOTAL	CONCORD 032-033 MAT.	INST.	TOTAL
02	SITE CONSTRUCTION	66.1	102.0	93.6	65.5	103.9	95.0	65.7	102.6	94.0	87.4	95.2	93.4	81.4	95.2	92.0	94.1	96.4	95.8
03100	CONCRETE FORMS & ACCESSORIES	104.0	88.1	90.4	95.6	107.9	106.2	100.0	96.4	96.9	82.8	47.1	52.1	90.7	47.1	53.3	84.6	74.0	75.5
03200	CONCRETE REINFORCEMENT	106.1	79.6	91.1	100.3	124.9	114.2	100.3	124.4	113.9	92.3	73.6	81.7	92.3	73.6	81.7	92.3	101.2	97.3
03300	CAST-IN-PLACE CONCRETE	113.1	91.9	104.3	111.7	111.4	111.5	119.5	95.4	109.5	102.3	62.4	85.6	94.0	62.4	80.8	99.4	89.7	95.3
03	CONCRETE	114.0	88.0	100.9	109.4	111.8	110.6	113.6	101.1	107.3	108.6	58.1	83.2	99.7	58.1	78.8	105.1	84.5	94.7
04	MASONRY	138.4	80.4	102.3	128.3	101.5	111.6	134.4	85.4	103.9	88.0	52.8	66.1	88.1	52.8	66.2	93.1	97.8	96.0
05	METALS	103.5	89.1	98.4	104.8	106.8	105.5	104.7	104.3	104.5	94.8	73.4	87.1	94.8	73.4	87.1	94.9	88.2	92.5
06	WOOD & PLASTICS	99.2	88.9	93.8	90.0	105.2	97.8	95.6	95.8	95.7	84.1	46.5	64.7	92.5	46.5	68.7	85.9	68.6	77.0
07	THERMAL & MOISTURE PROTECTION	104.7	88.9	97.3	103.7	102.1	102.9	104.1	90.0	97.9	99.8	55.3	79.0	99.8	55.3	78.9	100.4	98.2	99.4
08	DOORS & WINDOWS	97.1	89.9	95.3	94.8	111.8	99.0	94.8	105.5	97.4	105.0	50.1	91.1	106.4	50.1	92.7	106.3	76.3	99.0
09200	PLASTER & GYPSUM BOARD	91.8	88.3	89.6	89.2	105.1	99.1	90.1	95.4	93.4	90.6	43.8	61.6	93.4	43.8	62.7	91.1	66.6	75.9
09500	CEILINGS	97.7	88.3	91.5	103.2	105.1	104.5	103.2	95.4	98.1	89.7	43.8	59.6	89.7	43.8	59.6	89.7	66.6	74.6
09600	FLOORING	110.3	95.9	106.8	106.7	102.1	105.5	106.7	85.8	101.6	91.2	49.4	81.0	93.8	49.4	83.0	91.8	104.8	95.0
09900	PAINTS & COATINGS	111.0	94.5	101.3	111.0	117.0	114.5	111.0	91.6	99.7	93.5	40.4	62.4	93.5	40.4	62.4	93.5	84.8	88.4
09	FINISHES	100.4	90.4	95.3	99.7	107.3	103.6	99.8	93.2	96.5	91.2	46.0	68.1	92.1	46.0	68.5	92.3	79.8	85.9
10 - 14	TOTAL DIV. 10000 - 14000	100.0	94.2	98.8	100.0	114.2	103.0	100.0	124.7	105.2	100.0	60.2	91.6	100.0	60.2	91.6	100.0	96.2	99.2
15	MECHANICAL	98.1	95.3	96.8	100.0	112.7	105.7	100.0	98.0	99.1	96.7	49.8	75.4	96.7	49.8	75.4	99.9	82.9	92.2
16	ELECTRICAL	92.8	86.9	88.8	95.8	108.2	104.2	93.7	91.7	92.3	100.3	45.5	62.9	100.3	45.5	62.9	104.4	80.5	88.1
01 - 16	WEIGHTED AVERAGE	102.3	90.5	96.6	101.5	108.3	104.8	102.2	97.3	99.9	98.0	57.0	78.2	97.1	57.0	77.7	99.4	85.9	92.8

DIVISION		NEW HAMPSHIRE KEENE 034 MAT.	INST.	TOTAL	LITTLETON 035 MAT.	INST.	TOTAL	MANCHESTER 031 MAT.	INST.	TOTAL	NASHUA 030 MAT.	INST.	TOTAL	PORTSMOUTH 038 MAT.	INST.	TOTAL	NEW JERSEY ATLANTIC CITY 082,084 MAT.	INST.	TOTAL
02	SITE CONSTRUCTION	94.8	95.2	95.1	81.4	94.8	91.7	93.1	96.4	95.6	95.3	96.4	96.1	89.2	95.9	94.3	98.1	99.9	99.5
03100	CONCRETE FORMS & ACCESSORIES	88.8	48.4	54.2	101.6	54.3	61.0	98.3	73.9	77.4	98.7	73.9	77.4	84.6	70.9	72.8	109.9	126.2	123.9
03200	CONCRETE REINFORCEMENT	92.3	87.7	89.7	93.1	51.0	69.2	115.5	101.2	107.4	115.5	101.2	107.4	92.3	101.1	97.3	88.0	118.2	105.1
03300	CAST-IN-PLACE CONCRETE	102.8	62.8	86.1	92.3	55.5	76.9	99.8	89.7	95.6	97.3	89.7	94.1	92.3	85.7	89.5	87.2	125.9	103.3
03	CONCRETE	108.4	61.5	84.8	99.2	54.6	76.8	109.3	84.4	96.8	108.1	84.4	96.2	98.9	81.7	90.2	101.6	123.1	112.5
04	MASONRY	93.4	52.8	68.1	100.5	51.0	69.7	94.3	97.8	96.5	94.7	97.8	96.6	89.1	91.1	90.4	97.0	123.1	113.2
05	METALS	94.8	79.7	89.4	94.9	64.2	83.9	99.9	88.1	95.7	99.9	88.1	95.7	96.5	86.8	93.0	94.6	101.1	96.9
06	WOOD & PLASTICS	90.7	46.5	67.8	102.5	53.4	77.1	101.2	68.6	84.3	101.1	68.6	84.3	85.9	68.6	77.0	113.4	126.6	120.2
07	THERMAL & MOISTURE PROTECTION	100.4	57.5	80.3	99.9	60.8	81.6	100.6	98.2	99.4	100.9	98.2	99.6	100.4	96.5	98.6	100.6	122.4	110.8
08	DOORS & WINDOWS	103.5	61.4	93.2	107.5	49.8	93.4	108.3	76.3	100.5	108.3	76.3	100.5	109.3	69.2	99.6	106.0	120.7	109.6
09200	PLASTER & GYPSUM BOARD	92.9	43.8	62.5	103.5	50.9	70.9	98.5	66.6	78.8	98.5	66.6	78.8	91.1	66.6	75.9	100.7	126.5	116.6
09500	CEILINGS	89.7	43.8	59.6	89.7	50.9	64.2	99.4	66.6	77.9	99.4	66.6	77.9	89.7	66.6	74.6	89.7	126.5	113.8
09600	FLOORING	93.2	80.9	90.2	102.4	49.4	89.5	96.6	104.8	98.6	96.4	104.8	98.5	91.8	104.8	95.0	100.1	127.0	106.7
09900	PAINTS & COATINGS	93.5	40.4	62.4	93.5	78.2	84.5	93.5	84.8	88.4	93.5	84.8	88.4	93.5	48.9	67.4	93.5	136.5	118.6
09	FINISHES	93.2	52.4	72.4	96.3	55.6	75.6	96.5	79.8	88.0	96.7	79.8	88.1	92.0	74.1	82.9	96.9	128.0	112.8
10 - 14	TOTAL DIV. 10000 - 14000	100.0	85.8	97.0	100.0	63.5	92.3	100.0	96.2	99.2	100.0	96.2	99.2	100.0	93.5	98.6	100.0	119.5	104.1
15	MECHANICAL	96.7	53.3	77.0	96.7	62.9	81.4	99.9	82.9	92.2	99.9	82.9	92.2	99.9	79.2	90.6	99.7	118.6	108.2
16	ELECTRICAL	100.3	45.5	62.9	102.1	72.1	81.7	104.6	75.8	85.0	104.4	75.8	84.9	101.1	80.5	87.0	95.8	123.1	114.4
01 - 16	WEIGHTED AVERAGE	98.6	60.6	80.2	98.6	64.0	81.8	101.5	85.1	93.6	101.5	85.1	93.6	98.6	82.9	91.0	99.3	118.3	108.5

DIVISION		NEW JERSEY CAMDEN 081 MAT.	INST.	TOTAL	DOVER 078 MAT.	INST.	TOTAL	ELIZABETH 072 MAT.	INST.	TOTAL	HACKENSACK 076 MAT.	INST.	TOTAL	JERSEY CITY 073 MAT.	INST.	TOTAL	LONG BRANCH 077 MAT.	INST.	TOTAL
02	SITE CONSTRUCTION	98.6	100.4	100.0	110.4	101.1	103.2	114.0	100.6	103.7	110.8	100.7	103.0	98.6	100.7	100.3	103.8	100.9	101.6
03100	CONCRETE FORMS & ACCESSORIES	98.7	126.1	122.2	94.7	118.9	115.5	110.1	115.0	114.3	94.7	125.6	121.2	98.7	121.4	118.2	99.5	126.0	122.2
03200	CONCRETE REINFORCEMENT	115.5	112.8	114.0	88.9	119.7	106.4	88.9	119.6	106.3	88.9	119.6	106.3	115.5	123.4	120.0	88.9	123.1	108.3
03300	CAST-IN-PLACE CONCRETE	84.6	122.5	100.4	108.5	130.6	117.7	93.2	124.2	106.1	106.0	130.4	116.2	84.6	130.7	103.8	94.0	124.6	106.8
03	CONCRETE	101.9	120.9	111.5	109.8	122.3	116.1	105.4	118.3	111.9	107.7	125.1	116.5	101.9	123.8	112.9	107.4	123.7	115.6
04	MASONRY	86.9	122.9	109.3	90.9	127.3	113.6	108.1	119.2	115.0	95.4	127.3	115.3	86.9	127.3	112.0	99.7	121.4	113.2
05	METALS	99.7	99.0	99.5	94.6	109.6	99.9	96.2	108.7	100.7	94.6	109.0	99.8	99.8	106.5	102.2	94.7	105.8	98.6
06	WOOD & PLASTICS	101.1	126.6	114.3	100.0	116.9	108.6	117.4	116.7	117.0	100.0	126.6	113.7	101.1	120.0	110.9	102.3	126.5	114.8
07	THERMAL & MOISTURE PROTECTION	100.4	118.6	108.9	100.0	122.3	110.9	101.2	118.9	109.5	100.6	123.1	111.1	100.4	114.9	107.2	100.5	121.9	110.5
08	DOORS & WINDOWS	108.3	118.6	110.8	111.7	121.4	114.0	109.5	121.4	112.4	108.2	126.1	113.1	108.3	123.2	111.9	104.3	125.5	109.5
09200	PLASTER & GYPSUM BOARD	98.5	126.5	115.8	95.6	116.3	108.4	101.6	116.3	110.7	95.6	126.5	114.7	98.5	119.7	111.6	97.0	126.5	115.2
09500	CEILINGS	99.4	126.5	117.1	89.7	116.3	107.2	89.7	116.3	107.2	89.7	126.5	113.8	99.4	119.7	112.7	89.7	126.5	113.8
09600	FLOORING	96.4	126.7	103.8	95.5	122.0	102.0	100.6	68.3	92.7	95.5	68.3	88.9	96.4	137.0	106.3	96.7	126.8	104.0
09900	PAINTS & COATINGS	93.5	134.1	117.2	93.4	139.1	120.1	93.4	139.1	120.1	93.4	139.1	120.1	93.5	139.1	120.1	93.5	135.5	118.0
09	FINISHES	97.0	127.7	112.7	94.8	121.1	108.2	97.7	107.9	102.9	94.7	115.7	105.4	97.0	126.1	111.9	95.6	127.4	111.8
10 - 14	TOTAL DIV. 10000 - 14000	100.0	119.5	104.1	100.0	100.9	100.2	100.0	94.9	98.9	100.0	99.5	99.9	100.0	101.4	100.3	100.0	101.9	100.4
15	MECHANICAL	99.9	119.1	108.6	99.7	126.0	111.6	99.9	110.7	104.8	99.7	125.9	111.6	99.9	125.0	111.3	99.7	122.7	110.1
16	ELECTRICAL	104.4	123.1	117.1	98.0	123.8	115.6	99.1	120.4	113.6	98.0	131.9	121.1	106.0	131.9	123.6	97.4	118.2	111.6
01 - 16	WEIGHTED AVERAGE	100.4	117.7	108.8	100.8	119.7	109.9	101.9	113.0	107.3	100.4	121.0	110.4	100.5	121.2	110.6	100.0	118.4	108.9

City Cost Indexes

NEW JERSEY

DIVISION		NEWARK 070-071			NEW BRUNSWICK 088-089			PATERSON 074-075			POINT PLEASANT 087			SUMMIT 079			TRENTON 085-086		
		MAT.	INST.	TOTAL	MAT.	INST.	TOTAL	MAT.	INST.	TOTAL	MAT.	INST.	TOTAL	MAT.	INST.	TOTAL	MAT.	INST.	TOTAL
02	SITE CONSTRUCTION	117.9	100.7	104.7	111.9	100.9	103.5	112.3	100.7	103.4	113.3	100.4	103.4	111.5	100.0	102.7	99.8	101.0	100.7
03100	CONCRETE FORMS & ACCESSORIES	97.0	126.7	122.5	102.8	125.8	122.6	97.5	126.4	122.3	95.9	126.0	121.7	98.5	114.9	112.5	97.8	126.6	122.5
03200	CONCRETE REINFORCEMENT	115.5	123.5	120.0	88.9	123.1	108.3	115.5	123.5	120.0	88.9	119.3	106.1	88.9	119.6	106.3	115.5	118.1	117.0
03300	CAST-IN-PLACE CONCRETE	95.3	131.1	110.2	107.7	124.6	114.8	107.8	130.7	117.4	107.7	125.3	115.1	90.1	125.8	105.0	95.2	125.5	107.9
03	CONCRETE	107.0	126.5	116.8	120.8	123.6	122.2	113.1	126.2	119.7	120.2	123.3	121.8	101.6	118.8	110.3	107.0	123.3	115.2
04	MASONRY	96.2	128.1	116.1	93.8	121.4	111.0	91.3	127.3	113.7	87.3	122.9	109.5	93.6	119.2	109.5	87.9	122.9	109.7
05	METALS	99.8	109.7	103.4	94.6	105.6	98.6	99.8	109.7	103.4	94.6	105.5	98.5	94.6	108.5	99.6	100.6	103.8	101.7
06	WOOD & PLASTICS	103.0	126.6	115.2	106.0	126.5	116.6	103.0	126.6	115.2	98.3	126.0	112.6	104.2	116.7	110.7	101.2	126.5	114.1
07	THERMAL & MOISTURE PROTECTION	100.6	123.6	111.4	100.7	121.9	110.6	101.1	123.1	111.4	100.7	122.4	110.9	101.5	118.9	109.6	99.2	122.1	109.9
08	DOORS & WINDOWS	114.4	126.7	117.4	100.1	126.7	106.6	114.4	126.7	117.4	102.3	124.3	107.7	117.2	121.4	118.2	108.3	123.9	112.1
09200	PLASTER & GYPSUM BOARD	98.5	126.5	115.8	98.4	126.5	115.8	98.5	126.5	115.8	95.6	126.0	114.4	97.0	116.3	109.0	98.5	126.2	115.7
09500	CEILINGS	99.4	126.5	117.1	89.7	126.5	113.8	99.4	126.5	117.1	89.7	126.0	113.5	89.7	116.3	107.2	99.4	126.2	117.0
09600	FLOORING	96.6	137.0	106.5	97.8	88.6	95.6	96.4	137.0	106.3	95.5	126.8	103.1	96.8	68.3	89.8	96.6	126.8	104.0
09900	PAINTS & COATINGS	93.4	139.1	120.1	93.5	139.1	120.1	93.4	139.1	120.1	93.5	135.5	118.0	93.4	139.1	120.1	93.5	139.1	120.1
09	FINISHES	97.6	130.2	114.2	96.8	121.2	109.3	97.2	130.0	113.9	95.7	127.5	111.9	95.6	107.9	101.9	97.1	128.1	112.9
10 - 14	TOTAL DIV. 10000 - 14000	100.0	102.5	100.5	100.0	102.0	100.4	100.0	102.2	100.5	100.0	102.5	100.5	100.0	94.8	98.9	100.0	119.3	104.1
15	MECHANICAL	99.9	126.0	111.7	99.7	120.0	108.9	99.9	125.9	111.7	99.7	123.5	110.4	99.7	110.7	104.7	99.9	123.6	110.6
16	ELECTRICAL	105.8	126.9	120.2	97.1	127.4	117.7	106.0	123.8	118.2	95.8	118.2	111.1	99.1	113.0	108.6	104.6	128.7	121.0
01 - 16	WEIGHTED AVERAGE	103.0	122.3	112.4	101.2	118.7	109.7	103.3	121.6	112.2	100.7	118.6	109.4	100.8	111.7	106.1	101.2	120.7	110.7

DIVISION		NEW JERSEY VINELAND 080,083			NEW MEXICO ALBUQUERQUE 870-872			CARRIZOZO 883			CLOVIS 881			FARMINGTON 874			GALLUP 873		
		MAT.	INST.	TOTAL	MAT.	INST.	TOTAL	MAT.	INST.	TOTAL	MAT.	INST.	TOTAL	MAT.	INST.	TOTAL	MAT.	INST.	TOTAL
02	SITE CONSTRUCTION	102.8	100.3	100.9	86.1	110.4	104.8	109.1	110.2	109.9	94.8	110.2	106.6	91.5	110.2	105.8	101.8	110.2	108.3
03100	CONCRETE FORMS & ACCESSORIES	92.6	125.5	120.8	97.9	72.3	75.9	97.9	70.1	74.0	97.9	72.0	75.7	97.9	70.2	74.1	97.9	70.6	74.5
03200	CONCRETE REINFORCEMENT	88.0	118.2	105.1	100.9	71.8	84.4	108.3	71.5	87.4	109.4	62.1	82.6	110.2	71.6	88.3	105.6	71.6	86.3
03300	CAST-IN-PLACE CONCRETE	94.0	125.1	107.0	108.9	78.7	96.8	101.8	77.7	91.7	101.7	78.3	92.0	110.5	77.7	96.8	103.9	78.3	93.2
03	CONCRETE	106.8	122.5	114.7	108.8	75.5	92.0	124.5	74.1	99.2	111.9	73.5	92.6	112.5	74.2	93.2	119.4	74.5	96.9
04	MASONRY	83.6	121.7	107.4	110.6	73.3	87.4	115.1	67.7	85.6	115.1	73.2	89.0	125.1	69.5	90.5	115.2	70.5	87.4
05	METALS	94.5	100.9	96.8	105.9	92.0	100.9	104.3	90.8	99.5	104.0	88.4	98.4	105.0	91.1	100.0	104.4	91.1	99.6
06	WOOD & PLASTICS	94.7	126.6	111.2	96.1	73.5	84.5	96.1	71.6	83.5	96.1	73.5	84.5	96.1	71.6	83.5	96.1	71.6	83.5
07	THERMAL & MOISTURE PROTECTION	100.2	121.8	110.3	103.0	79.5	92.0	105.6	77.7	92.5	104.0	79.9	92.7	103.9	78.1	91.8	105.3	78.6	92.7
08	DOORS & WINDOWS	101.8	120.0	106.2	94.8	74.7	89.9	93.8	71.7	88.4	94.0	70.9	88.4	97.8	71.7	91.4	97.8	71.7	91.4
09200	PLASTER & GYPSUM BOARD	94.3	126.5	114.2	90.1	72.1	79.0	88.5	70.1	77.1	88.5	72.1	78.3	88.5	70.1	77.1	88.5	70.1	77.1
09500	CEILINGS	89.7	126.5	113.8	103.2	72.1	82.8	94.9	70.1	78.6	94.9	72.1	79.9	94.9	70.1	78.6	94.9	70.1	78.6
09600	FLOORING	94.5	127.0	102.4	106.7	75.6	99.1	106.7	68.5	97.4	106.7	68.5	97.4	106.7	75.6	99.1	106.7	75.6	99.1
09900	PAINTS & COATINGS	93.5	80.3	85.8	111.0	67.5	85.6	111.0	71.7	88.0	111.0	71.7	88.0	111.0	71.7	88.0	111.0	71.7	88.0
09	FINISHES	94.5	121.3	108.2	100.3	72.3	86.0	100.3	70.0	84.8	99.2	71.4	85.0	98.9	71.4	84.8	99.8	71.7	85.4
10 - 14	TOTAL DIV. 10000 - 14000	100.0	119.0	104.0	100.0	82.3	96.3	100.0	76.4	95.0	100.0	77.1	95.2	100.0	76.4	95.0	100.0	76.8	95.1
15	MECHANICAL	99.7	117.9	107.9	99.9	77.8	89.9	97.9	74.0	87.1	97.9	76.1	88.0	99.9	73.6	87.9	97.8	74.1	87.1
16	ELECTRICAL	95.8	123.1	114.4	88.2	79.1	82.0	88.2	79.1	82.0	84.5	79.1	80.8	85.5	79.1	81.1	84.3	79.1	80.7
01 - 16	WEIGHTED AVERAGE	98.3	117.2	107.5	100.7	81.1	91.2	102.8	78.9	91.3	100.4	79.9	90.4	102.0	79.3	91.0	102.1	79.6	91.2

NEW MEXICO

DIVISION		LAS CRUCES 880			LAS VEGAS 877			ROSWELL 882			SANTA FE 875			SOCORRO 878			TRUTH/CONSEQUENCES 879		
		MAT.	INST.	TOTAL	MAT.	INST.	TOTAL	MAT.	INST.	TOTAL	MAT.	INST.	TOTAL	MAT.	INST.	TOTAL	MAT.	INST.	TOTAL
02	SITE CONSTRUCTION	99.0	85.8	88.9	90.6	110.2	105.6	96.7	110.1	107.0	85.4	110.4	104.6	87.7	110.0	104.8	111.6	85.5	91.6
03100	CONCRETE FORMS & ACCESSORIES	95.4	70.0	73.6	97.9	70.2	74.1	97.9	67.1	71.4	97.9	72.3	75.9	97.9	67.2	71.6	95.4	65.4	69.6
03200	CONCRETE REINFORCEMENT	104.3	65.2	82.1	107.4	71.6	87.1	109.4	61.8	82.4	108.3	71.8	87.6	109.4	71.6	88.0	103.1	61.5	79.5
03300	CAST-IN-PLACE CONCRETE	96.1	67.8	84.3	107.4	77.7	95.0	101.7	76.2	91.1	103.8	78.7	93.3	105.3	76.3	93.2	115.3	65.6	94.6
03	CONCRETE	88.7	69.1	78.8	109.6	74.2	91.8	113.0	70.5	91.6	106.8	75.5	91.0	108.1	72.4	90.1	101.6	65.5	83.5
04	MASONRY	110.8	65.1	82.3	115.5	69.5	86.9	126.2	67.0	89.4	115.2	73.3	89.1	115.4	69.0	86.5	112.8	64.9	83.0
05	METALS	99.2	79.1	92.0	104.1	91.0	99.4	105.0	86.3	98.3	105.0	92.0	100.3	104.4	90.9	99.5	98.3	77.0	90.6
06	WOOD & PLASTICS	89.4	71.9	80.4	96.1	71.6	83.5	96.1	68.0	81.6	96.1	73.5	84.5	96.1	68.0	81.6	89.4	66.4	77.5
07	THERMAL & MOISTURE PROTECTION	88.9	71.0	80.5	103.4	78.1	91.6	104.1	77.0	91.4	103.3	79.5	92.1	103.3	77.5	91.2	89.6	70.1	80.5
08	DOORS & WINDOWS	87.0	72.0	83.3	94.0	71.7	88.6	93.8	66.7	87.2	94.0	74.7	89.3	93.8	69.7	87.9	86.9	65.8	81.8
09200	PLASTER & GYPSUM BOARD	90.4	72.1	79.1	88.5	70.1	77.1	88.5	66.4	74.8	88.5	72.1	78.3	88.5	66.4	74.8	90.4	66.4	75.5
09500	CEILINGS	98.4	72.1	81.1	94.9	70.1	78.6	94.9	66.4	76.2	94.9	72.1	79.9	94.9	66.4	76.2	98.4	66.4	77.4
09600	FLOORING	139.9	51.2	118.3	106.7	75.6	99.1	106.7	68.5	97.4	106.7	75.6	99.1	106.7	75.6	99.1	139.9	68.5	122.5
09900	PAINTS & COATINGS	104.1	47.9	71.3	111.0	65.7	84.5	111.0	71.7	88.0	111.0	64.3	83.7	111.0	65.7	84.5	104.1	71.7	85.2
09	FINISHES	116.8	63.9	89.8	98.8	70.7	84.4	99.3	67.7	83.2	98.6	72.0	85.0	98.7	68.5	83.3	117.3	66.5	91.4
10 - 14	TOTAL DIV. 10000 - 14000	100.0	77.6	95.3	100.0	76.4	95.0	100.0	75.7	94.9	100.0	82.3	96.3	100.0	75.7	94.9	100.0	71.0	93.9
15	MECHANICAL	100.1	77.2	89.7	97.8	74.0	87.0	99.9	73.6	88.0	99.9	77.8	89.9	97.8	73.2	86.6	97.8	72.7	86.4
16	ELECTRICAL	86.7	66.4	72.8	88.2	79.1	82.0	86.6	79.1	81.5	88.2	79.1	82.0	85.0	80.1	81.6	91.0	80.0	83.5
01 - 16	WEIGHTED AVERAGE	98.0	72.2	85.5	100.3	79.3	90.1	101.9	77.4	90.0	100.3	81.1	91.0	99.8	78.5	89.5	99.7	72.7	86.6

City Cost Indexes

		NEW MEXICO			NEW YORK													
		TUCUMCARI			ALBANY			BINGHAMTON			BRONX			BROOKLYN			BUFFALO		
DIVISION		884			120 - 122			137 - 139			104			112			140 - 142		
		MAT.	INST.	TOTAL	MAT.	INST.	TOTAL	MAT.	INST.	TOTAL	MAT.	INST.	TOTAL	MAT.	INST.	TOTAL	MAT.	INST.	TOTAL
02	SITE CONSTRUCTION	94.4	110.1	106.5	73.9	108.3	100.3	97.9	91.0	92.6	132.2	127.0	128.2	125.8	130.4	129.3	102.1	93.0	95.1
03100	CONCRETE FORMS & ACCESSORIES	97.9	67.1	71.5	99.6	90.4	91.7	105.7	77.7	81.6	104.5	161.7	153.6	113.8	161.6	154.8	103.0	115.1	113.4
03200	CONCRETE REINFORCEMENT	107.4	61.9	81.6	100.6	90.4	94.8	99.6	89.8	94.1	95.7	183.4	145.4	100.8	183.4	147.6	104.1	106.0	105.2
03300	CAST-IN-PLACE CONCRETE	101.7	76.2	91.1	81.6	103.6	90.8	103.3	93.6	99.2	109.3	158.4	129.8	105.2	157.3	126.9	106.6	122.6	113.3
03	CONCRETE	111.3	70.5	90.8	99.5	96.3	97.9	101.0	88.0	94.4	112.6	162.8	137.9	114.4	162.4	138.5	103.4	115.1	109.3
04	MASONRY	130.0	68.7	91.8	90.7	98.5	95.6	106.6	85.3	93.3	99.6	160.6	137.6	117.3	160.5	144.2	104.4	119.5	113.8
05	METALS	104.0	86.4	97.7	99.9	110.5	103.7	95.0	122.8	105.0	96.7	137.3	111.3	106.6	137.6	117.7	98.3	96.7	97.8
06	WOOD & PLASTICS	96.1	68.0	81.6	101.7	88.3	94.8	111.8	74.0	92.3	107.7	162.7	136.2	116.3	162.6	140.2	104.1	115.1	109.7
07	THERMAL & MOISTURE PROTECTION	104.0	77.4	91.5	92.2	96.0	94.0	101.9	87.4	95.1	107.5	153.4	129.0	107.0	152.9	128.5	103.9	110.0	106.8
08	DOORS & WINDOWS	93.7	66.7	87.2	98.7	85.3	95.4	93.6	76.6	89.4	95.8	164.3	112.5	92.9	164.2	110.2	93.4	108.5	97.1
09200	PLASTER & GYPSUM BOARD	88.5	66.4	74.8	100.1	87.6	92.3	106.2	72.5	85.3	98.5	164.5	139.4	101.7	164.5	140.6	96.8	115.1	108.1
09500	CEILINGS	94.9	66.4	76.2	99.4	87.6	91.6	99.4	72.5	81.8	85.7	164.5	137.4	82.8	164.5	136.4	92.0	115.1	107.1
09600	FLOORING	106.7	68.5	97.4	83.6	88.6	84.8	93.6	87.3	92.0	94.2	145.9	106.8	100.1	145.9	111.3	91.7	121.2	98.9
09900	PAINTS & COATINGS	111.0	71.7	88.0	80.6	82.5	81.7	82.5	75.5	78.4	98.1	148.6	127.6	104.8	148.6	130.4	80.5	114.6	100.4
09	FINISHES	99.2	67.7	83.1	95.4	88.6	91.9	95.8	78.2	86.8	103.5	157.8	131.2	109.1	157.6	133.9	93.3	116.9	105.3
10 - 14	TOTAL DIV. 10000 - 14000	100.0	75.7	94.9	100.0	95.7	99.1	100.0	91.5	98.2	100.0	136.2	107.7	100.0	135.7	107.5	100.0	106.8	101.4
15	MECHANICAL	97.9	73.6	86.9	100.4	93.3	97.2	100.6	89.9	95.8	100.4	161.2	127.9	100.0	164.6	129.3	100.0	99.1	99.6
16	ELECTRICAL	88.2	79.1	82.0	102.8	97.2	99.0	103.8	81.4	88.5	98.7	163.2	142.6	103.3	163.2	144.1	101.9	103.7	103.1
01 - 16	WEIGHTED AVERAGE	101.4	77.5	89.8	98.1	97.1	97.6	99.0	88.7	94.1	102.1	155.0	127.7	105.2	155.9	129.8	99.3	106.3	102.7

		NEW YORK																
		ELMIRA			FAR ROCKAWAY			FLUSHING			GLENS FALLS			HICKSVILLE			JAMAICA		
DIVISION		148 - 149			116			113			128			115,117,118			114		
		MAT.	INST.	TOTAL	MAT.	INST.	TOTAL	MAT.	INST.	TOTAL	MAT.	INST.	TOTAL	MAT.	INST.	TOTAL	MAT.	INST.	TOTAL
02	SITE CONSTRUCTION	97.7	89.4	91.3	130.8	130.4	130.5	130.8	130.4	130.5	64.3	107.8	97.7	118.0	132.1	128.8	124.0	130.4	128.9
03100	CONCRETE FORMS & ACCESSORIES	81.7	76.5	77.3	94.8	161.6	152.1	100.1	161.6	152.9	82.4	84.6	84.3	90.0	160.5	150.6	100.1	161.6	152.9
03200	CONCRETE REINFORCEMENT	105.3	92.5	98.1	100.8	183.4	147.6	102.6	183.4	148.4	100.3	90.3	94.7	100.8	183.4	147.6	100.8	183.4	147.6
03300	CAST-IN-PLACE CONCRETE	99.7	92.3	96.6	113.8	157.3	132.0	113.8	157.3	132.0	78.5	101.5	88.1	96.6	157.1	121.8	105.2	157.3	126.9
03	CONCRETE	96.7	87.7	92.1	120.7	162.4	141.7	121.3	162.4	142.0	90.4	93.0	91.7	105.0	162.1	133.7	113.4	162.4	138.0
04	MASONRY	100.9	83.4	90.0	120.2	160.5	145.3	114.1	160.5	143.0	91.7	95.4	94.0	109.1	158.1	139.6	118.4	160.5	144.7
05	METALS	94.0	124.8	105.1	106.6	137.6	117.7	106.6	137.6	117.7	96.5	109.5	101.1	109.3	141.4	120.8	106.6	137.6	117.7
06	WOOD & PLASTICS	82.2	73.8	77.9	95.4	162.6	130.1	101.3	162.6	133.0	83.1	82.5	82.8	90.3	162.6	127.6	101.3	162.6	133.0
07	THERMAL & MOISTURE PROTECTION	104.5	85.1	95.4	106.7	152.9	128.3	106.8	152.9	128.4	90.9	94.1	92.4	106.1	151.4	127.4	106.6	152.9	128.3
08	DOORS & WINDOWS	96.2	77.5	91.7	91.2	164.2	109.0	91.2	164.2	109.0	92.2	82.2	89.7	91.2	164.2	109.0	91.2	164.2	109.0
09200	PLASTER & GYPSUM BOARD	93.8	72.5	80.6	95.3	164.5	138.1	97.1	164.5	138.8	90.7	81.6	85.1	93.9	164.5	137.6	97.1	164.5	138.8
09500	CEILINGS	89.2	72.5	78.2	82.8	164.5	136.4	82.8	164.5	136.4	91.1	81.6	84.9	82.8	164.5	136.4	82.8	164.5	136.4
09600	FLOORING	82.0	88.0	83.5	94.6	145.9	107.1	96.2	145.9	108.3	75.6	88.6	78.8	93.3	108.4	96.9	96.2	145.9	108.3
09900	PAINTS & COATINGS	78.5	83.1	81.2	104.8	148.6	130.4	104.8	148.6	130.4	80.6	82.9	81.9	104.8	148.6	130.4	104.8	148.6	130.4
09	FINISHES	88.9	78.4	83.5	106.7	157.6	132.7	107.5	157.6	133.1	89.1	84.6	86.8	105.3	149.0	127.6	107.1	157.6	132.9
10 - 14	TOTAL DIV. 10000 - 14000	100.0	92.2	98.4	100.0	135.7	107.5	100.0	135.7	107.5	100.0	80.5	95.9	100.0	134.5	107.3	100.0	135.7	107.5
15	MECHANICAL	97.1	88.4	93.1	96.6	164.6	127.4	96.6	164.6	127.4	97.1	87.6	92.8	99.8	150.6	122.9	96.6	164.6	127.4
16	ELECTRICAL	98.1	87.6	91.0	116.8	163.2	148.4	116.8	163.2	148.4	95.5	93.3	94.0	103.3	159.8	141.8	101.5	163.2	143.5
01 - 16	WEIGHTED AVERAGE	96.2	89.3	92.9	105.8	155.9	130.1	105.7	155.9	130.0	93.3	93.4	93.3	102.9	151.8	126.6	103.6	155.9	129.0

		NEW YORK																
		JAMESTOWN			KINGSTON			LONG ISLAND CITY			MONTICELLO			MOUNT VERNON			NEW ROCHELLE		
DIVISION		147			124			111			127			105			108		
		MAT.	INST.	TOTAL	MAT.	INST.	TOTAL	MAT.	INST.	TOTAL	MAT.	INST.	TOTAL	MAT.	INST.	TOTAL	MAT.	INST.	TOTAL
02	SITE CONSTRUCTION	99.1	88.8	91.2	126.1	129.4	128.6	127.5	130.4	129.7	120.9	129.4	127.4	143.1	126.9	130.7	141.4	126.9	130.3
03100	CONCRETE FORMS & ACCESSORIES	81.8	86.1	85.5	84.2	113.6	109.5	106.1	161.6	153.8	94.2	113.6	110.8	91.2	129.4	124.0	117.1	129.3	127.5
03200	CONCRETE REINFORCEMENT	105.5	96.3	100.3	100.7	133.2	119.1	100.8	183.4	147.6	99.9	133.2	118.8	94.8	147.8	124.8	94.9	147.7	124.9
03300	CAST-IN-PLACE CONCRETE	103.3	97.4	100.8	106.0	115.4	109.9	108.6	157.3	128.9	99.2	115.4	106.0	122.2	131.3	126.0	122.2	131.3	126.0
03	CONCRETE	99.6	91.6	95.6	114.1	117.6	115.9	117.0	162.4	139.8	108.7	117.6	113.2	124.5	133.3	128.9	124.4	132.9	128.7
04	MASONRY	108.5	93.5	99.2	107.8	130.5	122.0	114.3	160.5	143.1	100.0	130.5	119.0	105.9	131.5	121.9	105.9	131.5	121.9
05	METALS	94.0	86.8	91.4	106.6	116.5	110.1	106.6	137.6	117.7	106.5	116.4	110.1	96.4	136.1	110.7	96.7	128.2	108.0
06	WOOD & PLASTICS	81.2	80.6	80.9	84.6	109.5	97.5	108.1	162.6	136.2	95.5	109.5	102.8	93.8	126.8	110.9	122.3	126.8	124.6
07	THERMAL & MOISTURE PROTECTION	103.7	93.0	98.7	108.7	131.0	119.1	106.8	152.9	128.4	108.3	130.7	118.7	106.3	136.5	121.9	109.4	136.5	122.1
08	DOORS & WINDOWS	96.1	86.9	93.8	96.6	123.5	103.2	91.2	164.2	109.0	91.3	123.5	99.1	95.8	144.8	107.8	95.9	139.4	106.5
09200	PLASTER & GYPSUM BOARD	89.3	79.5	83.2	91.8	109.8	102.9	99.4	164.5	139.7	95.0	109.8	104.2	93.6	127.5	114.6	102.8	127.5	118.1
09500	CEILINGS	86.4	79.5	81.9	81.4	109.8	100.0	82.8	164.5	136.4	81.4	109.8	100.0	84.3	127.5	112.6	84.3	127.5	112.6
09600	FLOORING	84.8	99.2	88.3	87.4	113.8	93.8	97.9	145.9	109.6	90.2	113.8	96.0	86.5	145.9	101.0	93.2	145.9	106.0
09900	PAINTS & COATINGS	80.5	85.3	83.3	102.1	105.8	104.3	104.8	148.6	130.4	102.1	105.8	104.3	95.4	146.5	125.3	95.4	146.5	125.3
09	FINISHES	88.8	87.7	88.2	103.2	111.7	107.5	108.1	157.6	133.4	104.2	111.7	108.0	100.7	133.6	117.5	104.0	133.6	119.1
10 - 14	TOTAL DIV. 10000 - 14000	100.0	100.2	100.0	100.0	114.7	103.1	100.0	135.7	107.5	100.0	114.7	103.1	100.0	126.5	105.6	100.0	126.5	105.6
15	MECHANICAL	96.8	93.0	95.1	97.1	112.6	104.1	100.0	164.6	129.3	97.1	110.6	103.2	97.3	134.5	114.1	97.3	134.3	114.1
16	ELECTRICAL	97.3	88.8	91.5	99.8	96.9	97.8	102.4	163.2	143.8	99.8	91.7	94.3	96.3	136.2	123.5	96.3	136.2	123.5
01 - 16	WEIGHTED AVERAGE	96.8	90.3	93.7	103.3	115.2	109.1	105.0	155.9	129.6	101.6	114.0	107.6	102.9	133.8	117.9	103.6	132.8	117.7

City Cost Indexes

		NEW YORK																	
		NEW YORK			NIAGARA FALLS			PLATTSBURGH			POUGHKEEPSIE			QUEENS			RIVERHEAD		
	DIVISION	100 - 102			143			129			125 - 126			110			119		
		MAT.	INST.	TOTAL	MAT.	INST.	TOTAL	MAT.	INST.	TOTAL	MAT.	INST.	TOTAL	MAT.	INST.	TOTAL	MAT.	INST.	TOTAL
02	SITE CONSTRUCTION	142.9	128.3	131.7	101.6	90.8	93.3	89.6	103.7	100.4	122.0	129.4	127.7	121.6	130.4	128.3	118.2	130.7	127.8
03100	CONCRETE FORMS & ACCESSORIES	110.0	169.4	161.0	81.7	114.0	109.5	88.4	81.7	82.6	84.2	140.6	132.7	90.3	161.6	151.5	94.7	160.1	150.8
03200	CONCRETE REINFORCEMENT	101.3	183.6	147.9	104.1	105.8	105.1	104.6	90.1	96.4	100.7	133.4	119.2	102.6	183.4	148.4	102.8	183.3	148.4
03300	CAST-IN-PLACE CONCRETE	123.1	160.8	138.8	107.0	120.8	112.8	96.0	99.8	97.6	102.6	115.8	108.1	100.0	157.3	123.9	94.8	156.4	120.5
03	CONCRETE	125.0	167.2	146.2	102.2	113.7	108.0	105.0	89.2	97.1	111.0	129.6	120.4	108.4	162.4	135.5	104.0	161.5	132.9
04	MASONRY	112.4	161.9	143.2	118.0	123.0	121.1	89.4	91.2	90.5	101.9	130.5	119.7	108.9	160.5	141.0	116.4	157.9	142.2
05	METALS	109.6	142.1	121.2	94.1	92.3	93.4	103.0	83.5	96.0	106.6	117.4	110.5	106.6	137.6	117.7	109.4	139.2	120.1
06	WOOD & PLASTICS	113.2	171.7	143.4	81.1	111.0	96.6	90.9	78.3	84.4	84.6	144.9	115.8	90.5	162.6	127.2	95.3	162.6	130.1
07	THERMAL & MOISTURE PROTECTION	108.0	155.6	130.3	103.8	106.9	105.3	99.3	91.3	95.6	108.6	134.6	120.8	106.2	152.9	128.1	106.2	151.4	127.4
08	DOORS & WINDOWS	99.5	169.3	116.5	96.1	103.3	97.8	102.1	79.9	96.7	96.6	142.6	107.8	91.2	164.2	109.0	91.2	164.2	109.0
09200	PLASTER & GYPSUM BOARD	103.9	173.8	147.2	89.3	110.8	102.6	108.9	76.6	88.9	91.8	146.3	125.6	93.9	164.5	137.6	95.3	164.5	138.1
09500	CEILINGS	103.7	173.8	149.7	86.4	110.8	102.4	100.4	76.6	84.8	81.4	146.3	124.0	82.8	164.5	136.4	82.8	164.5	136.4
09600	FLOORING	95.4	163.2	112.0	84.7	116.7	92.5	97.5	84.2	94.3	87.4	134.9	99.0	93.3	145.9	106.1	94.6	108.4	98.0
09900	PAINTS & COATINGS	98.1	155.8	131.8	80.5	111.0	98.3	98.2	78.2	86.5	102.1	105.6	104.3	104.8	148.6	130.4	104.8	148.6	130.4
09	FINISHES	108.6	166.9	138.4	88.9	114.8	102.1	97.6	80.8	89.0	103.0	137.0	120.4	105.4	157.6	132.1	105.8	149.0	127.9
10 - 14	TOTAL DIV. 10000 - 14000	100.0	147.3	110.0	100.0	109.0	101.9	100.0	73.3	94.3	100.0	119.4	104.1	100.0	135.7	107.5	100.0	128.9	106.1
15	MECHANICAL	100.4	161.4	128.0	96.8	102.2	99.3	97.0	87.7	92.8	97.1	113.7	104.6	100.0	164.6	129.3	99.8	150.6	122.8
16	ELECTRICAL	112.0	177.7	156.8	95.0	104.5	101.5	95.0	88.2	90.4	99.8	96.3	97.4	103.3	163.2	144.1	104.4	159.8	142.2
01 - 16	WEIGHTED AVERAGE	108.6	160.5	133.8	97.6	106.0	101.7	99.0	88.0	93.7	102.5	121.4	111.6	103.0	155.9	128.6	103.4	151.2	126.6

		NEW YORK																	
		ROCHESTER			SCHENECTADY			STATEN ISLAND			SUFFERN			SYRACUSE			UTICA		
	DIVISION	144 - 146			123			103			109			130 - 132			133 - 135		
		MAT.	INST.	TOTAL	MAT.	INST.	TOTAL	MAT.	INST.	TOTAL	MAT.	INST.	TOTAL	MAT.	INST.	TOTAL	MAT.	INST.	TOTAL
02	SITE CONSTRUCTION	77.6	107.3	100.4	73.9	108.3	100.3	148.2	127.0	132.0	137.4	126.0	128.6	96.5	108.2	105.5	72.0	107.3	99.0
03100	CONCRETE FORMS & ACCESSORIES	102.7	97.9	98.6	105.2	90.4	92.5	90.6	161.5	151.5	103.0	111.2	110.0	104.5	88.0	90.3	105.7	76.9	81.0
03200	CONCRETE REINFORCEMENT	104.9	93.2	98.3	99.3	90.4	94.3	95.7	183.4	145.4	94.9	133.3	116.7	100.6	90.2	94.7	100.6	82.1	90.1
03300	CAST-IN-PLACE CONCRETE	101.4	105.8	103.2	94.8	103.6	98.5	122.2	158.3	137.3	118.5	122.2	120.0	94.8	96.9	95.7	87.6	91.1	89.1
03	CONCRETE	109.6	100.8	105.2	106.2	96.3	101.2	126.6	162.7	144.7	120.2	118.9	119.5	104.4	92.8	98.6	102.9	84.5	93.6
04	MASONRY	100.7	101.3	101.1	91.6	98.6	96.0	115.0	160.6	143.4	106.7	128.7	120.4	97.0	94.9	95.7	90.2	88.5	89.2
05	METALS	101.5	111.2	105.0	99.9	110.5	103.7	96.7	137.1	111.2	96.7	116.6	103.9	99.7	109.7	103.3	98.0	106.4	101.0
06	WOOD & PLASTICS	100.4	98.3	99.3	108.5	88.3	98.1	92.3	162.7	128.7	106.8	106.4	106.6	108.5	86.4	97.1	108.5	74.3	90.8
07	THERMAL & MOISTURE PROTECTION	100.6	100.1	100.3	92.4	96.0	94.1	107.9	153.4	129.2	109.0	131.4	119.5	100.3	95.1	97.8	92.4	91.0	91.7
08	DOORS & WINDOWS	98.9	94.6	97.9	98.7	85.3	95.4	95.8	164.3	112.5	95.9	121.8	102.2	96.6	82.4	93.1	98.7	75.8	93.1
09200	PLASTER & GYPSUM BOARD	98.3	98.1	98.1	100.1	87.6	92.3	93.5	164.5	137.4	97.8	106.4	103.1	100.1	85.6	91.1	100.1	73.1	83.4
09500	CEILINGS	94.7	98.1	96.9	99.4	87.6	91.6	85.7	164.5	137.4	84.3	106.4	98.8	99.4	85.6	90.4	99.4	73.1	82.1
09600	FLOORING	81.0	103.2	86.4	83.6	88.6	84.8	90.5	145.9	104.0	89.5	75.4	86.1	85.4	85.0	85.3	83.6	80.7	82.9
09900	PAINTS & COATINGS	77.2	105.6	93.8	80.6	82.5	81.7	98.1	148.6	127.6	95.4	109.6	103.7	85.8	88.5	87.4	80.6	88.5	85.2
09	FINISHES	94.2	100.0	97.1	95.1	88.7	91.8	102.7	157.8	130.8	101.9	102.3	102.1	96.6	87.0	91.7	95.1	78.1	86.4
10 - 14	TOTAL DIV. 10000 - 14000	100.0	98.9	99.8	100.0	95.8	99.1	100.0	136.2	107.7	100.0	122.5	104.8	100.0	90.9	98.1	100.0	91.1	98.1
15	MECHANICAL	100.0	93.4	97.0	100.4	93.3	97.2	100.4	147.8	121.9	97.3	123.7	109.3	100.4	92.4	96.8	100.4	88.6	95.0
16	ELECTRICAL	107.0	95.2	99.0	103.3	96.1	98.4	98.7	163.2	142.6	109.5	122.7	118.5	103.3	88.9	93.5	103.2	85.5	91.1
01 - 16	WEIGHTED AVERAGE	100.6	99.7	100.1	99.0	96.9	98.0	104.9	152.4	127.9	103.6	120.3	111.7	99.9	94.2	97.1	98.2	89.2	93.8

		NEW YORK									NORTH CAROLINA								
		WATERTOWN			WHITE PLAINS			YONKERS			ASHEVILLE			CHARLOTTE			DURHAM		
	DIVISION	136			106			107			287 - 288			281 - 282			277		
		MAT.	INST.	TOTAL	MAT.	INST.	TOTAL	MAT.	INST.	TOTAL	MAT.	INST.	TOTAL	MAT.	INST.	TOTAL	MAT.	INST.	TOTAL
02	SITE CONSTRUCTION	80.6	109.4	102.7	131.6	126.9	128.0	142.3	126.4	130.1	101.7	73.3	79.9	102.0	73.3	80.0	101.7	83.4	87.6
03100	CONCRETE FORMS & ACCESSORIES	84.7	86.5	86.2	110.8	129.2	126.6	109.8	138.6	134.5	92.9	46.8	53.3	101.6	46.8	54.6	95.1	46.9	53.7
03200	CONCRETE REINFORCEMENT	101.2	70.4	83.7	94.9	147.7	124.9	98.7	107.2	103.5	93.8	44.7	66.0	94.2	44.7	66.2	94.2	50.9	69.7
03300	CAST-IN-PLACE CONCRETE	102.0	99.3	100.9	108.4	131.3	117.9	121.4	131.4	125.5	99.5	54.0	80.6	102.1	54.0	82.0	99.8	54.1	80.7
03	CONCRETE	116.1	89.6	102.8	112.3	133.2	122.8	124.0	130.5	127.3	99.2	51.1	75.0	100.4	51.2	75.6	98.8	52.4	75.5
04	MASONRY	91.3	101.7	97.8	106.0	131.5	121.9	112.1	131.8	124.4	78.6	40.6	54.9	80.6	40.6	55.7	79.1	40.6	55.1
05	METALS	98.0	109.3	102.1	99.5	135.9	112.6	106.4	135.8	117.0	93.7	82.4	89.6	94.9	82.4	90.4	95.0	85.1	91.4
06	WOOD & PLASTICS	85.5	82.0	83.7	115.4	126.8	121.3	113.9	139.5	127.1	94.8	47.5	70.4	105.7	47.5	75.7	97.6	47.5	71.7
07	THERMAL & MOISTURE PROTECTION	92.6	97.2	94.7	108.9	135.7	121.4	109.3	137.9	122.7	94.9	47.6	72.7	95.0	48.7	73.3	95.6	48.2	73.4
08	DOORS & WINDOWS	98.7	82.5	94.8	95.9	144.8	107.8	99.5	151.8	112.3	92.7	46.3	81.4	96.6	46.3	84.4	96.6	48.6	84.9
09200	PLASTER & GYPSUM BOARD	92.8	81.1	85.6	100.5	127.5	117.2	103.6	140.5	126.5	99.5	45.7	66.2	105.8	45.7	68.6	105.8	45.7	68.6
09500	CEILINGS	99.4	81.1	87.4	84.3	127.5	112.6	102.3	140.5	127.4	88.0	45.7	60.2	92.1	45.7	61.7	92.1	45.7	61.7
09600	FLOORING	76.3	82.1	77.7	91.5	145.9	104.8	91.2	145.9	104.5	89.5	49.6	79.7	92.7	49.6	82.2	92.9	49.6	82.4
09900	PAINTS & COATINGS	80.6	88.1	85.0	95.4	146.5	125.3	95.4	146.5	125.3	103.6	47.2	70.7	103.6	47.2	70.7	103.6	47.2	70.7
09	FINISHES	92.3	84.9	88.5	102.5	133.6	118.4	106.8	141.1	124.3	91.0	47.1	68.6	93.7	47.1	69.9	93.8	47.1	70.0
10 - 14	TOTAL DIV. 10000 - 14000	100.0	96.9	99.3	100.0	126.5	105.6	100.0	138.1	108.1	100.0	69.7	93.6	100.0	69.7	93.6	100.0	69.7	93.6
15	MECHANICAL	100.4	89.4	95.4	100.6	124.9	111.6	100.6	126.0	112.1	100.1	52.2	78.4	100.1	52.2	78.4	100.1	52.3	78.5
16	ELECTRICAL	103.2	88.3	93.0	96.3	136.2	123.5	109.5	138.4	129.2	100.8	49.6	66.0	100.3	44.7	62.4	98.7	47.6	63.9
01 - 16	WEIGHTED AVERAGE	99.6	93.7	96.7	102.7	131.9	116.9	107.7	133.8	120.3	95.9	54.8	76.0	97.1	54.0	76.2	96.7	56.0	77.0

City Cost Indexes

	DIVISION	NORTH CAROLINA																	
		ELIZABETH CITY			FAYETTEVILLE			GASTONIA			GREENSBORO			HICKORY			KINSTON		
		279			283			280			270,272 - 274			286			285		
		MAT.	INST.	TOTAL	MAT.	INST.	TOTAL	MAT.	INST.	TOTAL	MAT.	INST.	TOTAL	MAT.	INST.	TOTAL	MAT.	INST.	TOTAL
02	SITE CONSTRUCTION	106.4	84.2	89.4	99.8	83.4	87.2	101.4	73.3	79.8	101.6	83.4	87.6	101.2	82.2	86.6	100.0	83.0	86.9
03100	CONCRETE FORMS & ACCESSORIES	76.9	30.2	36.8	90.1	46.9	53.0	101.8	46.7	54.5	95.1	46.9	53.7	86.9	29.2	37.4	82.2	29.3	36.8
03200	CONCRETE REINFORCEMENT	93.3	53.9	71.0	93.3	50.9	69.3	94.2	44.7	66.2	94.2	50.9	69.7	93.8	24.3	54.4	93.3	24.4	54.2
03300	CAST-IN-PLACE CONCRETE	99.9	38.3	74.2	96.3	54.1	78.7	97.3	54.0	79.2	98.9	54.1	80.2	99.5	41.2	75.2	96.0	42.4	73.7
03	CONCRETE	98.8	40.1	69.3	96.3	52.4	74.2	98.1	51.1	74.5	98.4	52.4	75.3	98.8	35.3	66.8	95.5	35.9	65.5
04	MASONRY	92.0	32.7	55.1	83.1	40.6	56.6	80.2	40.6	55.5	78.9	40.6	55.0	68.4	32.5	46.0	74.6	32.9	48.6
05	METALS	94.5	84.5	90.9	93.9	85.1	90.8	95.8	82.3	90.9	95.8	85.1	91.9	93.1	75.0	86.6	93.4	76.7	87.4
06	WOOD & PLASTICS	77.2	30.6	53.1	91.4	47.5	68.7	105.7	47.5	75.7	97.6	47.5	71.7	87.5	29.7	57.7	82.7	29.7	55.3
07	THERMAL & MOISTURE PROTECTION	94.8	31.2	65.1	95.2	48.2	73.2	95.3	48.7	73.5	95.6	48.2	73.4	95.7	34.4	66.7	95.0	35.0	66.9
08	DOORS & WINDOWS	93.6	34.3	79.2	92.8	48.6	82.1	96.6	46.3	84.4	96.6	48.6	84.9	92.8	29.5	77.4	92.8	30.0	77.5
09200	PLASTER & GYPSUM BOARD	96.2	27.3	53.5	99.7	45.7	66.3	105.8	45.7	68.6	105.8	45.7	68.6	99.5	27.3	54.8	97.4	27.3	54.0
09500	CEILINGS	92.1	27.3	49.6	89.4	45.7	60.7	92.1	45.7	61.7	92.1	45.7	61.7	88.0	27.3	48.2	89.4	27.3	48.6
09600	FLOORING	83.7	19.5	68.1	89.6	49.6	79.9	92.9	34.1	78.6	92.9	49.6	82.4	89.4	17.7	71.9	86.3	26.6	71.8
09900	PAINTS & COATINGS	103.6	16.1	52.5	103.6	47.2	70.7	103.6	47.2	70.7	103.6	47.2	70.7	103.6	18.9	54.1	103.6	20.8	55.2
09	FINISHES	89.6	25.9	57.1	91.3	47.1	68.8	93.7	44.4	68.6	93.8	47.1	70.0	91.1	25.3	57.5	89.9	27.6	58.1
10 - 14	TOTAL DIV. 10000 - 14000	100.0	68.4	93.3	100.0	69.7	93.6	100.0	69.7	93.6	100.0	69.7	93.6	100.0	64.7	92.5	100.0	64.7	92.5
15	MECHANICAL	96.9	33.8	68.3	100.1	52.3	78.5	100.1	52.2	78.4	100.1	52.3	78.5	96.9	23.2	63.5	96.9	24.1	63.9
16	ELECTRICAL	101.6	36.7	57.4	93.8	47.6	62.3	99.8	44.7	62.2	99.8	47.6	64.2	96.6	36.7	55.8	96.2	36.7	55.7
01 - 16	WEIGHTED AVERAGE	95.9	44.5	71.0	95.2	56.0	76.2	96.9	53.7	76.0	96.8	56.0	77.0	94.1	40.4	68.1	93.9	41.3	68.4

	DIVISION	NORTH CAROLINA															NORTH DAKOTA		
		MURPHY			RALEIGH			ROCKY MOUNT			WILMINGTON			WINSTON-SALEM			BISMARCK		
		289			275 - 276			278			284			271			585		
		MAT.	INST.	TOTAL	MAT.	INST.	TOTAL	MAT.	INST.	TOTAL	MAT.	INST.	TOTAL	MAT.	INST.	TOTAL	MAT.	INST.	TOTAL
02	SITE CONSTRUCTION	103.4	72.1	79.4	102.8	83.4	87.9	104.8	82.3	87.5	102.7	73.3	80.1	102.0	83.4	87.7	83.0	96.6	93.4
03100	CONCRETE FORMS & ACCESSORIES	102.6	29.2	39.6	98.3	46.9	54.1	85.5	34.9	42.0	94.7	46.9	53.7	96.6	46.8	53.9	93.6	58.9	63.8
03200	CONCRETE REINFORCEMENT	93.3	24.1	54.1	94.2	50.9	69.7	93.3	24.3	54.2	94.5	50.9	69.8	94.2	44.7	66.2	106.0	71.0	86.1
03300	CAST-IN-PLACE CONCRETE	102.9	36.6	75.2	105.6	54.1	84.1	97.7	35.5	71.7	99.1	54.1	80.3	102.1	54.0	82.1	103.7	63.2	86.8
03	CONCRETE	102.5	33.6	67.9	101.9	52.4	77.0	99.9	35.8	67.6	99.0	52.4	75.6	100.0	51.2	75.5	96.4	63.5	79.8
04	MASONRY	70.5	29.5	45.0	86.3	40.6	57.8	74.8	34.9	49.9	68.4	40.6	51.1	79.2	40.6	55.1	108.3	59.5	77.9
05	METALS	92.9	74.6	86.4	95.0	85.0	91.4	93.5	74.8	86.8	94.0	85.1	90.8	94.9	82.4	90.5	94.2	75.3	87.4
06	WOOD & PLASTICS	106.7	29.7	66.9	101.5	47.5	73.6	86.4	36.5	60.6	96.8	47.5	71.3	97.6	47.5	71.7	86.1	56.6	70.9
07	THERMAL & MOISTURE PROTECTION	95.2	36.7	67.8	95.4	47.6	73.0	95.5	33.4	66.4	94.9	48.2	73.1	95.6	48.2	73.4	97.9	60.0	80.2
08	DOORS & WINDOWS	92.7	34.1	78.4	93.5	48.6	82.6	92.9	32.1	78.1	92.9	48.6	82.1	96.6	48.6	84.5	104.1	53.9	91.9
09200	PLASTER & GYPSUM BOARD	105.0	27.3	56.9	108.8	45.7	69.7	98.4	34.2	58.6	100.2	45.7	66.5	105.8	45.7	68.6	115.3	55.8	78.4
09500	CEILINGS	88.0	27.3	48.2	92.1	45.7	61.7	89.4	34.2	53.1	89.4	45.7	60.7	92.1	45.7	61.7	142.3	55.8	85.5
09600	FLOORING	93.3	28.3	77.5	92.9	49.6	82.4	87.7	18.7	70.9	90.1	49.6	80.2	92.9	49.6	82.4	111.9	57.3	98.6
09900	PAINTS & COATINGS	103.6	21.5	55.6	103.6	47.2	70.7	103.6	35.9	64.1	103.6	47.2	70.7	103.6	47.2	70.7	104.7	48.5	71.9
09	FINISHES	93.1	27.5	59.6	94.3	47.1	70.2	90.9	32.4	61.0	91.5	47.1	68.9	93.8	47.1	70.0	114.5	57.1	85.2
10 - 14	TOTAL DIV. 10000 - 14000	100.0	64.8	92.5	100.0	69.7	93.6	100.0	66.5	92.9	100.0	69.7	93.6	100.0	69.7	93.6	100.0	62.9	92.1
15	MECHANICAL	96.9	23.3	63.6	100.1	52.3	78.5	96.9	23.3	63.5	100.1	52.3	78.5	100.1	52.3	78.4	100.5	66.4	85.1
16	ELECTRICAL	102.4	36.7	57.6	99.4	47.6	64.1	105.1	36.7	58.5	100.6	47.6	64.5	99.8	47.6	64.2	94.8	74.1	80.7
01 - 16	WEIGHTED AVERAGE	95.5	39.4	68.3	97.3	56.0	77.3	95.3	41.8	69.4	95.4	55.0	75.8	96.9	55.5	76.9	100.2	68.5	84.8

	DIVISION	NORTH DAKOTA																	
		DEVILS LAKE			DICKINSON			FARGO			GRAND FORKS			JAMESTOWN			MINOT		
		583			586			580 - 581			582			584			587		
		MAT.	INST.	TOTAL	MAT.	INST.	TOTAL	MAT.	INST.	TOTAL	MAT.	INST.	TOTAL	MAT.	INST.	TOTAL	MAT.	INST.	TOTAL
02	SITE CONSTRUCTION	87.9	96.4	94.4	94.7	96.4	96.0	82.9	96.5	93.4	90.7	96.4	95.0	87.0	96.4	94.2	88.4	96.5	94.6
03100	CONCRETE FORMS & ACCESSORIES	102.9	53.7	60.7	84.8	53.7	58.1	94.2	59.3	64.3	90.0	53.7	58.9	86.8	53.8	58.5	84.4	58.7	62.4
03200	CONCRETE REINFORCEMENT	106.0	64.2	82.3	107.1	64.2	82.8	97.7	71.0	82.5	104.3	64.2	81.6	106.6	67.4	84.4	108.0	71.0	87.0
03300	CAST-IN-PLACE CONCRETE	114.6	63.9	93.5	103.8	63.9	87.1	106.2	65.9	89.4	103.7	63.9	87.1	113.2	63.9	92.7	103.7	61.4	86.1
03	CONCRETE	105.1	60.2	82.5	103.7	60.2	81.8	96.6	64.5	80.5	101.0	60.2	80.5	103.1	60.8	81.8	99.5	62.7	81.0
04	MASONRY	114.0	55.1	77.3	115.2	55.1	77.8	108.4	58.9	77.6	108.3	55.1	75.2	125.0	55.2	81.5	108.2	59.6	77.9
05	METALS	94.4	72.6	86.5	94.2	72.6	86.5	94.1	75.1	87.3	94.2	72.6	86.5	94.3	73.7	86.9	94.6	75.1	87.6
06	WOOD & PLASTICS	96.1	52.0	73.3	76.9	52.0	64.1	86.1	57.5	71.3	82.4	52.0	66.7	79.2	52.0	65.1	76.6	56.6	66.3
07	THERMAL & MOISTURE PROTECTION	98.5	58.0	79.5	99.0	58.0	79.8	98.4	60.0	80.4	98.6	58.0	79.6	98.1	58.0	79.3	98.3	59.9	80.3
08	DOORS & WINDOWS	104.2	49.8	90.9	104.1	49.8	90.9	104.1	54.4	92.0	104.1	49.8	90.9	104.1	50.6	91.1	104.3	53.9	92.0
09200	PLASTER & GYPSUM BOARD	119.0	51.0	76.9	112.1	51.0	74.3	115.3	56.6	79.0	113.9	51.0	75.0	113.0	51.0	74.6	112.1	55.8	77.2
09500	CEILINGS	142.3	51.0	82.4	142.3	51.0	82.4	142.3	56.6	86.1	142.3	51.0	82.4	142.3	51.0	82.4	142.3	55.8	85.5
09600	FLOORING	116.1	56.4	101.5	107.8	56.4	95.3	111.7	56.4	98.2	110.2	56.4	97.1	108.8	56.4	96.1	107.6	57.5	95.4
09900	PAINTS & COATINGS	104.7	37.1	65.2	104.7	37.1	65.2	104.7	44.6	69.6	104.7	37.1	65.2	104.7	37.1	65.2	104.7	40.6	67.2
09	FINISHES	116.7	52.0	83.7	113.6	52.0	82.1	114.4	57.0	85.1	114.3	52.0	82.5	113.4	52.0	82.0	113.0	56.3	84.0
10 - 14	TOTAL DIV. 10000 - 14000	100.0	60.6	91.7	100.0	60.6	91.7	100.0	62.9	92.1	100.0	60.6	91.7	100.0	60.6	91.7	100.0	62.9	92.1
15	MECHANICAL	97.3	57.7	79.4	97.3	57.7	79.4	100.5	59.9	82.1	100.5	57.7	81.1	97.3	57.8	79.4	100.5	59.8	82.1
16	ELECTRICAL	94.0	60.0	70.8	107.6	60.0	75.1	94.7	53.3	66.5	99.4	60.0	72.5	94.0	60.0	70.8	104.4	74.1	83.8
01 - 16	WEIGHTED AVERAGE	101.3	62.4	82.5	101.9	62.4	82.7	100.2	63.8	82.6	101.3	62.4	82.4	101.1	62.6	82.5	101.3	67.0	84.7

City Cost Indexes

		NORTH DAKOTA			OHIO													
	DIVISION	WILLISTON			AKRON			ATHENS			CANTON			CHILLICOTHE			CINCINNATI		
		588			442 - 443			457			446 - 447			456			451 - 452		
		MAT.	INST.	TOTAL	MAT.	INST.	TOTAL	MAT.	INST.	TOTAL	MAT.	INST.	TOTAL	MAT.	INST.	TOTAL	MAT.	INST.	TOTAL
02	SITE CONSTRUCTION	89.3	96.4	94.7	102.4	109.7	108.0	93.6	95.6	95.1	102.5	108.6	107.2	83.0	111.5	104.8	79.8	110.7	103.5
03100	CONCRETE FORMS & ACCESSORIES	92.1	53.7	59.2	101.6	99.6	99.9	91.0	72.8	75.3	101.6	88.3	90.2	93.9	97.2	96.8	96.6	86.8	88.2
03200	CONCRETE REINFORCEMENT	109.0	64.2	83.6	98.4	97.2	97.7	92.1	71.4	80.4	98.4	82.2	89.2	89.7	84.0	86.5	95.1	84.3	89.0
03300	CAST-IN-PLACE CONCRETE	103.7	63.9	87.1	101.1	108.2	104.1	107.1	85.2	98.0	102.1	105.3	103.4	97.2	97.8	97.5	89.6	90.5	89.9
03	CONCRETE	101.1	60.2	80.5	98.5	101.0	99.7	102.2	76.5	89.3	99.0	92.3	95.6	96.7	94.7	95.7	91.3	87.7	89.5
04	MASONRY	103.4	55.1	73.3	94.5	100.2	98.0	77.6	76.1	76.7	95.3	90.2	92.1	86.6	96.5	92.7	86.1	90.1	88.6
05	METALS	94.5	72.6	86.7	91.3	82.3	88.1	103.7	69.1	91.3	91.3	75.7	85.7	93.3	86.8	90.9	95.2	87.1	92.3
06	WOOD & PLASTICS	84.6	52.0	67.8	97.8	98.8	98.3	84.2	71.6	77.7	97.8	87.5	92.5	94.9	96.2	95.6	97.7	84.8	91.0
07	THERMAL & MOISTURE PROTECTION	98.6	58.0	79.5	107.9	100.0	104.2	107.3	79.7	94.4	108.4	95.5	102.4	95.0	96.3	95.7	92.5	93.0	92.7
08	DOORS & WINDOWS	104.2	49.8	91.0	107.5	98.8	105.4	97.8	68.7	90.7	101.8	81.5	96.9	92.3	87.3	91.1	98.2	82.1	94.3
09200	PLASTER & GYPSUM BOARD	114.8	51.0	75.3	102.4	98.0	99.7	103.1	70.2	82.7	102.4	86.3	92.4	101.9	96.4	98.5	103.0	84.6	91.6
09500	CEILINGS	142.3	51.0	82.4	99.5	98.0	98.5	99.7	70.2	80.3	99.5	86.3	90.8	90.8	96.4	94.5	92.1	84.6	87.2
09600	FLOORING	111.2	56.4	97.9	111.5	105.5	110.0	108.9	85.1	103.1	111.7	89.5	106.3	89.0	89.7	89.1	89.8	101.9	92.7
09900	PAINTS & COATINGS	104.7	37.1	65.2	104.5	116.4	111.4	107.0	64.0	81.8	104.5	93.1	97.8	103.0	97.8	99.9	103.0	93.5	97.5
09	FINISHES	114.7	52.0	82.7	104.7	102.5	103.6	95.8	73.8	84.6	104.8	88.4	96.5	92.1	96.1	94.2	92.6	90.3	91.4
10 - 14	TOTAL DIV. 10000 - 14000	100.0	60.6	91.7	100.0	99.7	99.9	100.0	77.6	95.3	100.0	92.3	98.4	100.0	94.2	98.8	100.0	88.6	97.6
15	MECHANICAL	97.3	57.7	79.4	100.1	100.9	100.5	96.6	73.9	86.3	100.1	88.3	94.8	97.4	98.1	97.7	99.8	90.9	95.8
16	ELECTRICAL	100.1	60.0	72.7	98.6	94.9	96.1	100.2	70.5	80.0	97.1	95.6	96.0	97.7	91.3	93.4	95.4	81.1	85.6
01 - 16	WEIGHTED AVERAGE	100.5	62.4	82.0	99.8	99.0	99.4	98.1	75.6	87.2	99.1	91.1	95.2	94.7	95.8	95.2	95.3	89.9	92.7

		OHIO																
	DIVISION	CLEVELAND			COLUMBUS			DAYTON			HAMILTON			LIMA			LORAIN		
		441			430 - 432			453 - 454			450			458			440		
		MAT.	INST.	TOTAL	MAT.	INST.	TOTAL	MAT.	INST.	TOTAL	MAT.	INST.	TOTAL	MAT.	INST.	TOTAL	MAT.	INST.	TOTAL
02	SITE CONSTRUCTION	102.4	109.3	107.7	82.5	100.6	96.4	78.7	110.3	102.9	78.9	110.5	103.2	88.1	96.1	94.3	101.8	109.0	107.3
03100	CONCRETE FORMS & ACCESSORIES	101.7	106.6	105.9	95.4	88.4	89.4	96.5	85.9	87.4	96.5	85.9	87.4	90.9	88.0	88.4	101.6	89.1	90.9
03200	CONCRETE REINFORCEMENT	98.8	97.6	98.1	104.7	85.8	94.0	95.1	82.4	87.9	95.1	84.3	89.0	92.1	82.3	86.5	98.4	97.3	97.8
03300	CAST-IN-PLACE CONCRETE	99.2	116.5	106.4	92.8	96.8	94.4	81.6	89.9	85.0	89.3	89.2	89.3	98.4	97.6	98.1	96.3	98.0	97.0
03	CONCRETE	97.6	107.2	102.4	94.7	90.3	92.5	87.4	86.1	86.8	91.2	86.9	89.0	95.3	89.9	92.6	96.1	92.9	94.5
04	MASONRY	99.2	110.7	106.4	89.9	93.9	92.4	85.4	89.0	87.6	86.0	87.9	87.2	106.1	85.8	93.4	91.2	99.5	96.4
05	METALS	92.7	86.0	90.3	95.4	80.6	90.1	94.6	76.9	88.2	94.6	87.0	91.9	103.8	80.3	95.3	91.8	83.3	88.8
06	WOOD & PLASTICS	97.1	103.9	100.6	95.6	87.2	91.2	99.7	85.5	92.4	97.7	84.8	91.0	84.1	87.4	85.8	97.8	85.9	91.7
07	THERMAL & MOISTURE PROTECTION	106.6	113.5	109.8	103.4	96.0	99.9	97.4	91.9	94.8	95.2	92.7	94.0	109.3	95.7	102.9	108.3	102.6	105.6
08	DOORS & WINDOWS	96.1	102.2	97.6	100.3	83.4	96.2	98.5	81.8	94.4	98.1	82.1	94.2	97.8	76.6	92.7	101.8	92.4	99.5
09200	PLASTER & GYPSUM BOARD	101.7	103.3	102.6	105.0	86.5	93.5	103.0	85.3	92.1	103.0	84.6	91.6	103.1	86.5	92.8	102.4	84.6	91.4
09500	CEILINGS	98.2	103.3	101.5	95.1	86.5	89.4	92.1	85.3	87.7	92.1	84.6	87.2	98.2	86.5	90.5	99.5	84.6	89.8
09600	FLOORING	111.3	115.7	112.3	96.1	95.8	96.0	92.2	90.8	91.9	89.8	93.3	90.6	108.3	90.7	104.0	111.7	110.8	111.5
09900	PAINTS & COATINGS	104.5	122.1	114.8	99.2	97.8	98.4	103.0	91.5	96.3	103.0	89.7	95.2	107.0	87.8	95.8	104.5	122.1	114.8
09	FINISHES	104.4	109.8	107.1	95.6	90.6	93.0	93.4	87.3	90.3	92.6	87.8	90.1	95.0	88.4	91.6	104.8	96.1	100.4
10 - 14	TOTAL DIV. 10000 - 14000	100.0	105.0	101.1	100.0	91.9	98.3	100.0	87.3	97.3	100.0	87.7	97.4	100.0	95.9	99.1	100.0	99.1	99.8
15	MECHANICAL	100.1	109.7	104.4	100.1	94.8	97.7	100.8	87.9	94.9	100.6	89.4	95.7	96.6	89.8	93.5	100.1	93.1	96.9
16	ELECTRICAL	97.7	107.0	104.1	102.1	87.7	92.3	94.1	87.6	89.6	94.1	83.1	86.6	100.7	86.3	90.9	97.1	95.2	95.8
01 - 16	WEIGHTED AVERAGE	98.6	106.4	102.4	97.4	91.1	94.4	95.0	88.7	91.9	95.3	89.3	92.4	98.6	88.2	93.6	98.6	95.4	97.0

		OHIO																
	DIVISION	MANSFIELD			MARION			SPRINGFIELD			STEUBENVILLE			TOLEDO			YOUNGSTOWN		
		448 - 449			433			455			439			434 - 436			444 - 445		
		MAT.	INST.	TOTAL	MAT.	INST.	TOTAL	MAT.	INST.	TOTAL	MAT.	INST.	TOTAL	MAT.	INST.	TOTAL	MAT.	INST.	TOTAL
02	SITE CONSTRUCTION	97.9	108.5	106.1	78.4	101.1	95.8	79.2	110.1	102.9	115.2	110.0	111.2	81.7	101.3	96.8	102.3	109.8	108.0
03100	CONCRETE FORMS & ACCESSORIES	88.1	87.5	87.6	91.2	83.7	84.7	96.5	86.6	88.0	91.0	92.7	92.4	95.4	98.8	98.3	101.6	90.3	91.9
03200	CONCRETE REINFORCEMENT	89.5	82.3	85.4	96.5	83.8	89.3	95.1	82.4	87.9	94.1	90.6	92.2	104.7	88.8	95.6	98.4	90.8	94.1
03300	CAST-IN-PLACE CONCRETE	93.7	96.1	94.7	84.7	91.5	87.6	85.5	89.8	87.3	92.0	100.8	95.7	92.7	104.9	97.8	100.2	107.0	103.0
03	CONCRETE	90.6	88.7	89.7	86.9	85.9	86.4	89.4	86.4	87.9	90.9	94.1	92.5	94.7	98.5	96.6	98.0	95.4	96.7
04	MASONRY	93.8	93.3	93.5	93.3	86.6	89.1	85.7	88.8	87.7	80.5	95.6	89.9	99.5	97.3	98.2	94.6	96.4	95.7
05	METALS	92.4	75.3	86.3	96.5	77.5	89.7	94.5	76.8	88.2	91.3	78.6	86.8	95.2	86.1	91.9	91.3	80.1	87.3
06	WOOD & PLASTICS	83.5	85.9	84.7	91.1	82.0	86.4	99.7	86.6	92.9	87.2	90.7	89.0	95.6	100.4	98.1	97.8	88.3	92.9
07	THERMAL & MOISTURE PROTECTION	107.5	95.0	101.7	102.7	90.4	96.9	97.4	91.9	94.9	115.7	95.7	106.3	105.3	102.7	104.1	108.6	97.7	103.5
08	DOORS & WINDOWS	101.0	76.7	95.1	96.8	73.4	91.1	98.5	78.7	93.6	97.5	94.2	96.7	100.2	94.0	98.7	101.8	88.7	98.6
09200	PLASTER & GYPSUM BOARD	98.4	84.6	89.9	104.1	81.1	89.9	103.0	86.5	92.8	105.5	89.4	95.5	105.0	100.1	102.0	102.4	87.1	92.9
09500	CEILINGS	102.3	84.6	90.7	97.8	81.1	86.9	92.1	86.5	88.4	90.3	89.4	89.7	93.5	100.1	97.9	99.5	87.1	91.4
09600	FLOORING	106.0	101.0	104.8	94.4	93.7	94.2	92.2	90.8	91.9	123.1	100.4	117.6	95.3	76.1	90.6	111.7	96.4	108.0
09900	PAINTS & COATINGS	104.5	72.9	86.0	99.2	72.9	83.8	103.0	91.5	96.3	116.5	92.6	102.5	99.2	103.2	101.6	104.5	102.5	103.3
09	FINISHES	102.6	88.4	95.3	95.2	84.1	89.5	93.4	88.0	90.6	112.2	93.7	102.8	95.0	95.0	95.0	104.9	92.2	98.4
10 - 14	TOTAL DIV. 10000 - 14000	100.0	93.4	98.6	100.0	86.5	97.1	100.0	87.4	97.3	100.0	97.6	99.5	100.0	97.5	99.5	100.0	97.6	99.5
15	MECHANICAL	96.9	92.3	94.8	96.9	94.4	95.8	100.8	87.8	94.9	97.3	99.4	98.3	100.1	100.0	100.1	100.1	94.4	97.5
16	ELECTRICAL	93.1	89.1	90.4	93.5	89.1	90.5	94.1	87.7	89.7	85.1	105.8	99.2	102.8	99.3	100.4	97.1	89.2	91.7
01 - 16	WEIGHTED AVERAGE	96.6	90.4	93.6	94.8	88.3	91.6	95.3	88.6	92.1	96.6	97.4	97.0	97.9	97.5	97.7	99.0	93.7	96.4

City Cost Indexes

		OHIO			OKLAHOMA												
		ZANESVILLE			ARDMORE			CLINTON			DURANT			ENID			GUYMON		
	DIVISION	437 - 438			734			736			747			737			739		
		MAT.	INST.	TOTAL	MAT.	INST.	TOTAL	MAT.	INST.	TOTAL	MAT.	INST.	TOTAL	MAT.	INST.	TOTAL	MAT.	INST.	TOTAL
02	SITE CONSTRUCTION	81.1	101.0	96.3	107.5	89.3	93.6	108.9	87.7	92.6	101.4	87.8	91.0	110.2	88.1	93.3	114.3	87.1	93.4
03100	CONCRETE FORMS & ACCESSORIES	87.2	85.6	85.8	91.3	56.8	61.7	89.8	48.6	54.4	80.3	56.7	60.0	94.3	47.7	54.3	99.5	29.7	39.6
03200	CONCRETE REINFORCEMENT	95.9	84.4	89.4	95.5	77.3	85.2	96.2	77.3	85.5	96.4	69.6	81.2	95.4	83.6	88.7	96.2	41.5	65.1
03300	CAST-IN-PLACE CONCRETE	89.3	90.8	89.9	100.8	55.3	81.8	97.4	55.3	79.9	94.3	54.8	77.8	97.5	59.0	81.4	97.5	36.0	71.8
03	CONCRETE	90.4	86.6	88.5	91.9	60.0	75.9	91.2	56.4	73.7	86.5	58.3	72.3	91.8	58.4	75.0	94.3	34.6	64.2
04	MASONRY	88.6	85.7	86.8	93.9	56.8	70.8	119.3	56.8	80.4	92.4	57.7	70.8	100.6	57.7	73.9	97.6	29.1	54.9
05	METALS	96.5	77.8	89.8	92.3	65.9	82.8	92.4	65.9	82.9	92.4	61.0	81.1	93.5	66.9	84.0	92.8	44.3	75.4
06	WOOD & PLASTICS	87.1	85.2	86.1	92.0	57.6	74.2	90.5	46.4	67.7	80.4	57.6	68.6	95.5	45.1	69.5	101.1	29.4	64.0
07	THERMAL & MOISTURE PROTECTION	102.8	93.0	98.2	96.7	65.2	81.9	96.8	64.0	81.5	96.2	64.3	81.3	97.0	64.1	81.6	97.4	35.7	68.5
08	DOORS & WINDOWS	96.8	81.5	93.1	96.7	64.6	88.9	96.7	58.6	87.4	96.7	61.3	88.1	96.7	58.1	87.3	96.9	32.1	81.1
09200	PLASTER & GYPSUM BOARD	102.7	84.5	91.4	91.3	57.4	70.3	90.4	45.9	62.8	87.7	57.4	68.9	91.8	44.5	62.5	93.4	28.3	53.1
09500	CEILINGS	97.8	84.5	89.1	94.2	57.4	70.0	94.2	45.9	62.5	94.2	57.4	70.0	94.2	44.5	61.6	95.6	28.3	51.4
09600	FLOORING	92.5	89.7	91.8	115.2	67.2	103.5	113.4	64.0	101.4	108.9	69.3	99.2	115.8	64.0	103.2	118.4	37.9	98.8
09900	PAINTS & COATINGS	99.2	62.9	78.0	95.9	43.8	65.5	95.9	64.5	77.5	95.9	64.5	77.5	95.9	64.5	77.5	95.9	24.3	54.0
09	FINISHES	94.5	83.7	89.0	100.0	56.6	77.9	99.5	51.9	75.2	97.2	59.4	77.9	100.5	51.0	75.2	102.2	30.6	65.6
10 - 14	TOTAL DIV. 10000 - 14000	100.0	90.8	98.0	100.0	71.1	93.9	100.0	69.7	93.6	100.0	68.2	93.3	100.0	69.5	93.5	100.0	63.8	92.3
15	MECHANICAL	96.9	85.6	91.8	96.8	68.9	84.1	96.8	68.9	84.1	96.8	68.7	84.1	100.0	68.9	85.9	96.8	34.5	68.6
16	ELECTRICAL	92.5	81.4	84.9	93.2	73.7	79.9	94.9	73.7	80.4	97.6	70.5	79.1	94.9	73.7	80.4	97.6	25.4	48.4
01 - 16	WEIGHTED AVERAGE	94.9	85.8	90.5	96.0	67.3	82.1	97.4	65.7	82.0	94.9	66.1	81.0	97.5	66.0	82.2	97.4	38.7	69.0

		OKLAHOMA																
		LAWTON			MCALESTER			MIAMI			MUSKOGEE			OKLAHOMA CITY			PONCA CITY		
	DIVISION	735			745			743			744			730 - 731			746		
		MAT.	INST.	TOTAL	MAT.	INST.	TOTAL	MAT.	INST.	TOTAL	MAT.	INST.	TOTAL	MAT.	INST.	TOTAL	MAT.	INST.	TOTAL
02	SITE CONSTRUCTION	105.6	89.7	93.4	93.4	87.6	88.9	96.0	85.1	87.6	96.0	84.3	87.1	105.8	90.3	93.9	102.4	88.1	91.4
03100	CONCRETE FORMS & ACCESSORIES	98.7	65.6	70.3	77.6	59.7	62.2	94.6	67.6	71.4	100.5	43.3	51.4	100.2	56.1	62.4	88.7	57.5	62.0
03200	CONCRETE REINFORCEMENT	95.7	83.6	88.8	96.2	60.6	76.0	94.7	77.4	84.9	95.4	46.4	67.6	95.7	83.6	88.8	95.4	77.3	85.2
03300	CAST-IN-PLACE CONCRETE	94.3	59.0	79.5	82.6	58.2	72.4	86.6	57.1	74.3	87.7	44.9	69.8	94.3	57.6	79.0	96.8	48.2	76.5
03	CONCRETE	88.4	66.3	77.3	77.7	59.0	68.3	82.3	66.5	74.4	83.8	45.6	64.6	88.6	61.6	75.0	88.8	57.9	73.3
04	MASONRY	97.4	57.7	72.7	109.9	56.6	76.7	95.8	58.6	72.7	108.4	52.5	73.6	99.3	59.5	74.5	85.6	57.9	68.4
05	METALS	95.6	66.9	85.3	92.3	55.0	79.0	92.3	83.1	89.0	93.5	64.5	83.1	95.8	66.9	85.4	92.3	66.5	83.1
06	WOOD & PLASTICS	99.8	69.4	84.1	77.3	62.6	69.7	95.5	71.7	83.2	102.0	44.6	72.2	102.1	55.6	78.0	89.6	58.5	73.5
07	THERMAL & MOISTURE PROTECTION	96.7	66.6	82.6	95.7	63.5	80.6	96.4	67.4	82.8	96.5	52.6	75.9	95.9	65.9	81.9	96.5	64.0	81.3
08	DOORS & WINDOWS	98.2	71.3	91.6	96.7	61.3	88.1	96.7	72.1	90.7	96.7	43.9	83.8	98.2	63.8	89.8	96.7	65.6	89.1
09200	PLASTER & GYPSUM BOARD	94.8	69.6	79.2	86.7	62.6	71.8	92.2	71.8	79.6	94.1	43.8	63.0	94.8	55.3	70.3	90.4	58.3	70.5
09500	CEILINGS	102.5	69.6	80.9	94.2	62.6	73.4	94.2	71.8	79.5	94.2	43.8	61.1	102.5	55.3	71.5	94.2	58.3	70.6
09600	FLOORING	118.9	64.0	105.5	107.5	64.0	96.9	116.9	69.3	105.3	120.0	51.6	103.3	118.9	64.0	105.5	112.9	64.0	101.0
09900	PAINTS & COATINGS	95.9	43.8	65.5	95.9	60.1	75.0	95.9	56.0	72.6	95.9	43.7	65.4	95.9	64.5	77.5	95.9	64.5	77.5
09	FINISHES	103.1	62.8	82.5	96.1	60.7	78.0	99.9	66.8	83.0	101.2	45.2	72.6	103.2	57.5	79.9	99.0	59.1	78.6
10 - 14	TOTAL DIV. 10000 - 14000	100.0	72.7	94.2	100.0	71.8	94.0	100.0	73.2	94.3	100.0	40.5	87.4	100.0	71.6	94.0	100.0	71.6	94.0
15	MECHANICAL	100.0	68.9	85.9	96.8	48.5	74.9	96.8	68.1	83.8	100.0	36.7	71.4	100.0	69.9	86.4	96.8	65.2	82.5
16	ELECTRICAL	97.6	64.1	74.8	95.0	63.8	73.7	97.4	63.8	74.5	94.3	43.6	59.7	96.4	73.7	80.9	94.3	67.8	76.3
01 - 16	WEIGHTED AVERAGE	97.8	68.0	83.3	94.2	60.8	78.0	94.9	69.5	82.6	96.6	49.7	73.9	97.8	68.2	83.5	94.9	65.7	80.8

		OKLAHOMA												OREGON					
		POTEAU			SHAWNEE			TULSA			WOODWARD			BEND			EUGENE		
	DIVISION	749			748			740 - 741			738			977			974		
		MAT.	INST.	TOTAL	MAT.	INST.	TOTAL	MAT.	INST.	TOTAL	MAT.	INST.	TOTAL	MAT.	INST.	TOTAL	MAT.	INST.	TOTAL
02	SITE CONSTRUCTION	76.1	83.7	81.9	105.6	88.0	92.1	102.7	85.2	89.2	109.5	87.7	92.8	108.6	106.2	106.7	96.2	106.3	104.0
03100	CONCRETE FORMS & ACCESSORIES	85.6	57.4	61.4	80.2	55.5	59.0	100.0	57.1	63.1	89.9	48.5	54.4	106.4	109.8	109.3	101.8	109.8	108.7
03200	CONCRETE REINFORCEMENT	96.6	77.4	85.7	95.4	66.8	79.2	95.7	83.6	88.8	95.4	77.4	85.2	99.9	98.2	98.9	109.9	102.2	105.6
03300	CAST-IN-PLACE CONCRETE	86.7	56.6	74.1	99.9	54.3	80.9	95.6	57.1	79.5	97.5	55.3	79.9	100.9	108.8	104.2	97.7	109.0	102.4
03	CONCRETE	83.9	61.9	72.8	90.4	57.1	73.6	89.2	62.9	76.0	91.5	56.4	73.8	124.3	107.0	115.6	112.8	107.8	110.3
04	MASONRY	96.8	57.8	72.5	111.2	55.9	76.7	96.5	58.7	73.0	90.2	57.7	69.9	114.5	107.6	110.2	112.4	107.7	109.5
05	METALS	92.3	82.9	89.0	92.3	59.7	80.6	96.2	82.8	91.4	92.5	65.9	83.0	94.3	101.8	97.0	95.3	102.4	97.8
06	WOOD & PLASTICS	85.8	58.6	71.7	80.1	57.3	68.3	100.6	57.3	78.2	90.7	46.3	67.8	102.1	108.8	105.6	97.3	108.8	103.3
07	THERMAL & MOISTURE PROTECTION	96.4	65.4	81.9	96.4	64.7	81.6	96.5	64.3	81.4	96.9	64.2	81.6	111.7	102.5	107.4	110.6	100.3	105.8
08	DOORS & WINDOWS	96.7	66.2	89.3	96.7	57.0	87.0	98.2	64.1	89.9	96.8	58.5	87.4	101.1	111.6	103.7	101.5	111.6	104.0
09200	PLASTER & GYPSUM BOARD	89.5	58.3	70.2	87.7	57.1	68.7	94.8	56.9	71.3	90.7	45.7	62.9	95.2	109.1	103.8	94.1	109.1	103.4
09500	CEILINGS	94.2	58.3	70.6	94.2	57.1	69.8	102.5	56.9	72.5	95.6	45.7	62.9	115.4	109.1	111.2	116.8	109.1	111.7
09600	FLOORING	112.1	69.3	101.6	108.7	51.0	94.7	118.7	66.7	106.0	113.5	67.2	102.2	121.3	104.9	117.3	119.5	104.9	115.9
09900	PAINTS & COATINGS	95.9	62.5	76.4	95.9	54.3	71.6	95.9	64.5	77.5	95.9	64.5	77.5	122.9	75.9	95.4	122.9	85.7	101.2
09	FINISHES	97.3	59.7	78.1	97.4	54.2	75.3	102.6	59.1	80.4	99.9	52.3	75.6	117.4	105.3	111.2	116.0	106.4	111.1
10 - 14	TOTAL DIV. 10000 - 14000	100.0	68.7	93.4	100.0	70.4	93.7	100.0	74.5	94.6	100.0	69.6	93.6	100.0	110.9	102.3	100.0	110.9	102.3
15	MECHANICAL	96.8	64.3	82.1	96.8	63.3	81.6	100.0	64.8	84.1	96.8	68.9	84.1	96.9	111.0	103.3	100.2	111.0	105.1
16	ELECTRICAL	94.5	63.8	73.6	97.8	73.7	81.4	97.6	56.9	69.8	97.4	73.7	81.2	106.0	102.7	103.8	103.9	102.7	103.1
01 - 16	WEIGHTED AVERAGE	94.0	66.5	80.7	96.6	64.4	81.0	97.8	65.7	82.3	96.1	65.8	81.4	105.3	106.5	105.9	103.9	106.8	105.3

City Cost Indexes

		OREGON															
	DIVISION	KLAMATH FALLS			MEDFORD			PENDLETON			PORTLAND			SALEM			VALE		
		976			975			978			970 - 972			973			979		
		MAT.	INST.	TOTAL	MAT.	INST.	TOTAL	MAT.	INST.	TOTAL	MAT.	INST.	TOTAL	MAT.	INST.	TOTAL	MAT.	INST.	TOTAL
02	SITE CONSTRUCTION	113.4	106.2	107.9	105.1	106.2	106.0	99.6	98.6	98.9	97.3	106.3	104.2	96.4	106.3	104.0	86.3	98.5	95.6
03100	CONCRETE FORMS & ACCESSORIES	97.3	109.6	107.8	96.1	109.6	107.7	99.7	109.9	108.4	103.4	110.0	109.1	103.3	110.0	109.1	108.1	109.2	109.0
03200	CONCRETE REINFORCEMENT	99.9	98.2	98.9	101.5	98.2	99.6	98.6	96.0	97.2	105.1	102.4	103.6	111.0	102.4	106.2	96.3	95.9	96.1
03300	CAST-IN-PLACE CONCRETE	100.9	108.6	104.1	100.9	108.6	104.1	101.7	110.4	105.3	104.9	109.0	106.6	101.3	109.0	104.5	80.0	110.1	92.6
03	CONCRETE	128.2	106.8	117.4	120.3	106.8	113.5	100.8	107.3	104.1	115.7	107.9	111.8	114.8	107.9	111.3	83.7	106.8	95.3
04	MASONRY	131.1	107.6	116.4	109.5	107.6	108.3	119.3	110.9	114.1	114.0	110.9	112.1	114.6	110.9	112.3	117.6	110.9	113.5
05	METALS	94.3	101.4	96.9	94.8	101.4	97.2	107.6	102.5	105.7	94.0	102.8	97.1	95.9	102.8	98.3	107.3	101.8	105.3
06	WOOD & PLASTICS	92.4	108.8	100.9	91.1	108.8	100.3	95.8	109.0	102.6	99.2	108.8	104.2	99.2	108.8	104.2	105.0	109.0	107.1
07	THERMAL & MOISTURE PROTECTION	112.0	99.2	106.0	111.5	99.2	105.7	99.1	98.5	98.8	110.5	105.2	108.0	110.4	105.2	108.0	98.3	98.2	98.3
08	DOORS & WINDOWS	101.2	111.6	103.7	104.3	111.6	106.1	97.8	109.7	100.7	98.9	111.6	102.0	100.6	111.6	103.3	97.7	106.7	99.9
09200	PLASTER & GYPSUM BOARD	92.0	109.1	102.6	91.6	109.1	102.4	76.6	109.1	96.7	93.2	109.1	103.0	95.1	109.1	103.7	79.8	109.1	97.9
09500	CEILINGS	115.4	109.1	111.2	115.4	109.1	111.2	72.9	109.1	96.6	116.8	109.1	111.7	126.4	109.1	115.0	72.9	109.1	96.6
09600	FLOORING	117.8	104.9	114.7	117.1	104.9	114.1	82.6	104.9	88.0	119.5	101.3	115.1	119.5	104.9	115.9	85.0	104.9	89.8
09900	PAINTS & COATINGS	122.9	72.5	93.5	122.9	72.5	93.5	100.9	92.2	95.8	122.9	85.7	101.2	122.9	85.7	101.2	100.9	75.9	86.3
09	FINISHES	116.1	104.9	110.4	115.2	104.9	109.9	78.4	107.2	93.1	115.9	105.6	110.7	117.8	106.4	112.0	78.9	105.4	92.4
10 - 14	TOTAL DIV. 10000 - 14000	100.0	110.9	102.3	100.0	110.9	102.3	100.0	111.3	102.4	100.0	111.0	102.3	100.0	110.9	102.3	100.0	111.3	102.4
15	MECHANICAL	96.9	111.0	103.3	100.2	111.0	105.1	95.9	112.8	103.5	100.2	111.0	105.1	100.2	111.0	105.1	95.9	93.5	94.8
16	ELECTRICAL	103.9	95.1	97.9	109.5	95.1	99.7	91.9	102.1	98.8	103.7	109.1	107.4	103.5	102.7	103.0	91.9	80.7	84.2
01 - 16	WEIGHTED AVERAGE	106.4	105.0	105.7	105.5	105.0	105.3	98.2	106.5	102.2	103.9	108.2	106.0	104.4	107.3	105.8	95.7	98.7	97.2

		PENNSYLVANIA															
	DIVISION	ALLENTOWN			ALTOONA			BEDFORD			BRADFORD			BUTLER			CHAMBERSBURG		
		181			166			155			167			160			172		
		MAT.	INST.	TOTAL	MAT.	INST.	TOTAL	MAT.	INST.	TOTAL	MAT.	INST.	TOTAL	MAT.	INST.	TOTAL	MAT.	INST.	TOTAL
02	SITE CONSTRUCTION	95.5	108.1	105.1	100.1	108.0	106.2	98.7	106.8	104.9	95.2	108.1	105.1	90.9	108.5	104.4	89.8	106.5	102.6
03100	CONCRETE FORMS & ACCESSORIES	104.4	107.1	106.7	82.7	92.1	90.8	84.2	92.9	91.6	86.2	89.0	88.6	84.8	107.1	104.0	86.3	92.1	91.3
03200	CONCRETE REINFORCEMENT	100.6	108.3	105.0	97.6	97.9	97.7	103.7	81.6	91.1	99.6	92.3	95.5	98.2	97.0	97.6	98.0	87.6	92.1
03300	CAST-IN-PLACE CONCRETE	86.8	99.2	92.0	96.8	90.1	94.0	102.6	91.2	97.9	92.4	88.4	90.7	85.4	100.3	91.6	93.8	93.2	93.6
03	CONCRETE	99.2	105.7	102.5	93.3	94.0	93.7	97.5	91.7	94.6	100.8	91.1	95.9	84.7	104.3	94.6	107.3	93.2	100.2
04	MASONRY	95.2	95.4	95.3	97.1	84.7	89.4	95.1	90.4	92.2	92.8	83.5	87.0	99.2	99.5	99.4	95.7	92.0	93.4
05	METALS	97.9	124.9	107.6	91.6	119.4	101.6	98.9	113.7	104.2	95.8	118.8	104.0	91.4	126.9	104.1	95.8	115.8	102.9
06	WOOD & PLASTICS	108.5	109.3	108.9	81.7	93.0	87.8	84.0	93.1	88.7	88.6	86.8	87.6	83.9	108.8	96.7	93.0	93.0	93.0
07	THERMAL & MOISTURE PROTECTION	100.3	112.7	106.1	99.1	95.3	97.3	96.0	97.6	96.8	100.1	94.6	97.5	98.7	103.1	100.8	99.0	88.9	94.3
08	DOORS & WINDOWS	96.6	108.1	99.4	91.0	102.2	93.8	92.7	99.3	94.3	97.1	96.0	96.8	91.0	114.6	96.8	93.4	91.6	92.9
09200	PLASTER & GYPSUM BOARD	100.1	109.2	105.8	85.2	92.5	89.7	87.7	92.5	90.7	92.1	86.0	88.3	85.6	108.7	99.9	95.7	92.5	93.7
09500	CEILINGS	99.4	109.2	105.9	93.8	92.5	92.9	95.4	92.5	93.5	91.1	86.0	87.8	93.8	108.7	103.6	91.1	92.5	92.0
09600	FLOORING	85.4	89.6	86.4	79.8	71.0	77.7	84.8	72.1	81.7	79.1	79.7	79.3	80.7	96.3	84.5	83.1	71.9	80.4
09900	PAINTS & COATINGS	85.8	102.8	95.7	81.1	107.6	96.6	89.4	107.6	100.0	85.8	91.0	88.8	81.1	107.5	96.5	85.8	89.0	87.7
09	FINISHES	96.6	103.7	100.2	91.8	89.3	90.5	94.9	89.9	92.4	91.9	86.6	89.2	91.7	105.6	98.8	93.0	87.9	90.4
10 - 14	TOTAL DIV. 10000 - 14000	100.0	102.2	100.5	100.0	102.1	100.4	100.0	103.2	100.7	100.0	103.1	100.7	100.0	106.7	101.4	100.0	94.7	98.9
15	MECHANICAL	100.4	102.1	101.2	99.8	89.4	95.1	96.6	90.5	93.8	97.1	97.1	97.1	96.6	98.8	97.6	97.1	94.7	96.1
16	ELECTRICAL	103.2	90.1	94.3	88.4	104.6	99.4	90.4	104.6	100.1	92.5	104.6	100.7	89.5	84.2	85.9	90.3	82.1	84.7
01 - 16	WEIGHTED AVERAGE	98.8	103.4	101.1	94.7	97.7	96.2	96.1	97.6	96.8	96.4	98.0	97.2	92.8	102.6	97.6	96.7	94.1	95.5

		PENNSYLVANIA															
	DIVISION	DOYLESTOWN			DUBOIS			ERIE			GREENSBURG			HARRISBURG			HAZLETON		
		189			158			164 - 165			156			170 - 171			182		
		MAT.	INST.	TOTAL	MAT.	INST.	TOTAL	MAT.	INST.	TOTAL	MAT.	INST.	TOTAL	MAT.	INST.	TOTAL	MAT.	INST.	TOTAL
02	SITE CONSTRUCTION	111.6	90.0	95.0	104.0	107.0	106.3	96.4	108.4	105.6	94.6	107.5	104.5	83.8	106.6	101.3	88.5	107.5	103.1
03100	CONCRETE FORMS & ACCESSORIES	81.5	124.0	118.0	83.3	93.2	91.8	102.7	93.8	95.0	93.0	107.3	105.3	95.4	88.8	89.7	78.6	91.5	89.7
03200	CONCRETE REINFORCEMENT	97.2	97.3	97.3	103.0	92.7	97.1	99.6	97.3	98.3	103.0	97.2	99.7	100.6	91.4	95.4	97.7	105.4	102.1
03300	CAST-IN-PLACE CONCRETE	82.0	105.6	91.9	99.0	97.4	98.3	95.2	96.3	95.6	95.3	100.5	97.5	85.2	91.5	87.8	82.0	91.9	86.1
03	CONCRETE	94.3	112.9	103.6	98.4	96.1	97.3	92.5	96.8	94.6	92.6	104.5	98.6	97.6	91.8	94.7	90.7	95.8	93.3
04	MASONRY	98.0	117.4	110.1	95.3	86.8	90.0	88.5	100.2	95.8	106.8	103.3	104.6	92.5	90.3	91.2	106.1	96.4	100.0
05	METALS	94.9	120.9	104.2	98.9	118.9	106.1	91.8	120.0	101.9	98.8	126.4	108.7	99.6	116.9	105.8	97.6	122.7	106.6
06	WOOD & PLASTICS	82.6	125.9	105.0	82.9	93.1	88.2	84.7	91.3	97.5	93.5	108.9	101.5	99.9	88.2	93.9	80.5	91.1	86.0
07	THERMAL & MOISTURE PROTECTION	98.3	123.9	110.3	96.3	98.3	97.2	99.2	100.4	99.8	95.9	104.8	100.1	103.4	106.2	104.7	99.5	106.3	102.7
08	DOORS & WINDOWS	98.2	132.0	106.5	92.7	102.9	95.2	91.2	92.6	91.5	92.7	114.7	98.0	96.6	97.6	96.8	97.3	103.6	98.8
09200	PLASTER & GYPSUM BOARD	85.0	126.4	110.6	87.3	92.5	90.5	94.1	90.7	92.0	90.2	108.7	101.7	100.1	87.5	92.3	91.6	90.4	90.9
09500	CEILINGS	95.2	126.4	115.7	95.4	92.5	93.5	99.4	90.7	93.7	94.0	108.7	103.7	99.4	87.5	91.6	98.0	90.4	93.0
09600	FLOORING	70.2	125.4	83.7	84.4	74.7	82.0	86.3	93.0	87.9	88.8	96.3	90.6	85.6	85.6	85.6	76.5	85.4	78.7
09900	PAINTS & COATINGS	85.1	109.8	99.5	89.4	107.5	100.0	89.1	91.0	90.2	89.4	107.6	100.0	85.8	89.0	87.7	85.8	98.4	93.2
09	FINISHES	85.6	123.1	104.7	95.1	90.4	92.7	96.6	92.8	94.7	96.1	105.6	101.0	95.5	87.7	91.5	91.8	90.4	91.1
10 - 14	TOTAL DIV. 10000 - 14000	100.0	108.6	101.8	100.0	103.2	100.7	100.0	104.9	101.0	100.0	107.1	101.5	100.0	93.4	98.6	100.0	94.4	98.8
15	MECHANICAL	96.6	120.7	107.5	96.6	90.6	93.9	99.8	98.9	99.4	96.6	95.2	96.0	100.4	93.9	97.4	97.1	91.9	94.8
16	ELECTRICAL	92.4	113.0	106.4	91.4	104.6	100.4	90.9	89.2	89.7	91.4	104.6	100.4	102.3	82.2	88.6	93.9	86.1	88.6
01 - 16	WEIGHTED AVERAGE	95.5	115.6	105.2	96.5	98.5	97.5	95.0	99.1	97.0	96.3	105.6	100.8	98.2	94.4	96.4	96.0	97.0	96.5

City Cost Indexes

		PENNSYLVANIA																	
DIVISION		INDIANA			JOHNSTOWN			KITTANNING			LANCASTER			LEHIGH VALLEY			MONTROSE		
		157			159			162			175 - 176			180			188		
		MAT.	INST.	TOTAL	MAT.	INST.	TOTAL	MAT.	INST.	TOTAL	MAT.	INST.	TOTAL	MAT.	INST.	TOTAL	MAT.	INST.	TOTAL
02	SITE CONSTRUCTION	92.7	107.6	104.1	99.3	107.2	105.4	93.7	108.7	105.2	81.1	106.7	100.7	93.0	108.0	104.5	91.7	107.8	104.0
03100	CONCRETE FORMS & ACCESSORIES	85.1	95.6	94.1	83.3	93.7	92.2	84.8	107.5	104.3	89.0	89.4	89.4	94.9	107.1	105.4	79.7	93.0	91.1
03200	CONCRETE REINFORCEMENT	102.2	97.2	99.4	103.7	105.1	104.5	98.2	97.3	97.7	97.7	91.2	94.0	97.7	101.9	100.1	102.4	105.3	104.0
03300	CAST-IN-PLACE CONCRETE	93.4	100.5	96.4	103.6	91.9	98.7	88.8	100.5	93.7	79.8	92.9	85.3	88.8	98.9	93.0	87.1	93.8	89.9
03	CONCRETE	90.4	99.3	94.9	98.1	96.8	97.4	87.4	104.6	96.0	94.1	92.6	93.4	97.9	104.6	101.3	95.9	97.1	96.5
04	MASONRY	92.2	103.3	99.1	93.0	87.3	89.5	102.3	103.3	102.9	101.3	91.5	95.2	95.3	96.8	96.2	95.2	97.0	96.3
05	METALS	98.9	126.6	108.8	98.9	124.1	107.9	91.4	127.4	104.3	95.8	116.5	103.2	97.5	125.3	107.5	95.9	122.6	105.5
06	WOOD & PLASTICS	85.0	93.1	89.2	82.9	93.1	88.2	83.9	108.8	96.7	95.9	88.2	91.9	98.0	109.3	103.9	81.8	91.1	86.6
07	THERMAL & MOISTURE PROTECTION	95.7	102.7	98.9	96.0	98.4	97.1	98.8	105.1	101.8	98.5	107.5	102.7	100.0	111.8	105.5	99.5	107.1	103.0
08	DOORS & WINDOWS	92.7	106.2	96.0	92.7	105.5	95.8	91.0	114.6	96.8	93.4	91.5	92.9	97.3	114.2	101.4	93.6	103.6	96.0
09200	PLASTER & GYPSUM BOARD	88.2	92.5	90.8	87.0	92.5	90.4	85.6	108.7	99.9	96.6	87.5	91.0	96.2	109.2	104.3	90.7	90.4	90.5
09500	CEILINGS	95.4	92.5	93.5	94.0	92.5	93.0	93.8	108.7	103.6	91.1	87.5	88.7	98.0	109.2	105.4	91.1	90.4	90.7
09600	FLOORING	85.3	109.8	91.3	84.4	90.5	85.9	80.7	109.6	87.8	84.2	85.6	84.5	82.0	125.4	92.6	77.0	90.1	80.2
09900	PAINTS & COATINGS	89.4	107.6	100.0	89.4	107.6	100.0	81.1	107.6	96.6	85.8	90.9	88.8	85.8	102.8	95.7	85.8	98.4	93.2
09	FINISHES	94.8	99.0	96.9	94.5	94.0	94.2	91.8	108.1	100.1	93.0	87.8	90.4	94.5	109.8	102.3	90.8	92.5	91.7
10 - 14	TOTAL DIV. 10000 - 14000	100.0	105.0	101.1	100.0	103.4	100.7	100.0	106.6	101.4	100.0	94.5	98.8	100.0	99.6	99.9	100.0	99.6	99.9
15	MECHANICAL	96.6	90.8	94.0	96.6	91.1	94.1	96.6	101.5	98.8	97.1	95.4	96.4	97.1	103.4	100.0	97.1	93.8	95.6
16	ELECTRICAL	91.4	104.6	100.4	91.4	104.6	100.4	88.4	104.6	99.5	92.7	72.7	79.1	93.9	98.9	97.3	92.5	90.2	90.9
01 - 16	WEIGHTED AVERAGE	95.0	102.7	98.7	96.1	99.8	97.9	93.4	107.3	100.1	95.4	93.1	94.3	96.9	106.0	101.3	95.3	98.7	96.9

		PENNSYLVANIA																	
DIVISION		NEW CASTLE			NORRISTOWN			OIL CITY			PHILADELPHIA			PITTSBURGH			POTTSVILLE		
		161			194			163			190 - 191			150 - 152			179		
		MAT.	INST.	TOTAL	MAT.	INST.	TOTAL	MAT.	INST.	TOTAL	MAT.	INST.	TOTAL	MAT.	INST.	TOTAL	MAT.	INST.	TOTAL
02	SITE CONSTRUCTION	91.4	108.7	104.7	101.2	96.7	97.8	89.7	108.1	103.8	107.7	95.9	98.7	97.7	110.4	107.4	84.3	106.6	101.4
03100	CONCRETE FORMS & ACCESSORIES	84.8	103.4	100.8	83.0	125.4	119.4	84.8	88.0	87.5	105.8	130.8	127.2	103.5	104.4	104.2	77.9	88.8	87.3
03200	CONCRETE REINFORCEMENT	97.0	99.4	98.4	96.0	108.4	103.0	98.2	98.9	98.6	99.1	122.1	112.1	104.2	109.3	107.1	96.8	93.5	95.0
03300	CAST-IN-PLACE CONCRETE	86.3	100.3	92.1	79.5	120.8	96.7	83.7	94.3	88.1	97.5	124.4	108.7	99.0	102.1	100.3	85.0	96.5	89.8
03	CONCRETE	85.2	102.9	94.1	94.2	120.6	107.5	83.4	93.9	88.7	107.4	126.5	117.0	96.3	105.8	101.0	98.0	94.1	96.0
04	MASONRY	96.5	99.6	98.4	105.8	118.4	113.7	98.1	92.6	94.7	92.4	127.7	114.4	89.7	103.8	98.5	95.0	90.8	92.4
05	METALS	91.4	122.5	102.6	96.3	120.2	104.9	91.4	119.5	101.5	99.6	121.6	107.5	99.9	127.3	109.7	96.0	120.2	104.7
06	WOOD & PLASTICS	83.9	104.2	94.4	86.0	125.7	106.5	83.9	84.8	84.4	110.2	130.7	120.8	104.8	104.3	104.5	83.8	88.2	86.1
07	THERMAL & MOISTURE PROTECTION	98.7	102.4	100.4	98.4	126.0	111.3	98.6	96.8	97.8	99.5	132.0	114.8	96.3	105.2	100.5	98.5	105.1	101.6
08	DOORS & WINDOWS	91.0	105.1	94.5	92.3	129.4	101.4	91.0	89.7	90.7	98.4	133.9	107.0	92.9	112.9	97.7	93.4	99.9	95.0
09200	PLASTER & GYPSUM BOARD	85.6	104.0	97.0	88.5	126.4	111.9	85.6	84.0	84.6	94.9	131.5	117.6	93.7	104.0	100.1	92.5	87.5	89.4
09500	CEILINGS	93.8	104.0	100.5	99.9	126.4	117.3	93.8	84.0	87.4	99.9	131.5	120.6	95.4	104.0	101.0	91.1	87.5	88.7
09600	FLOORING	80.7	89.5	82.9	74.5	125.4	86.9	80.7	74.7	79.2	82.2	136.0	95.3	93.4	106.8	96.7	79.9	95.2	83.6
09900	PAINTS & COATINGS	81.1	107.5	96.5	88.3	133.8	114.9	81.1	107.5	96.5	88.3	139.4	118.1	89.4	112.8	103.1	85.8	74.8	79.4
09	FINISHES	91.7	101.2	96.6	90.4	126.3	108.7	91.6	86.6	89.1	94.0	132.8	113.9	98.5	105.4	102.0	91.2	88.0	89.5
10 - 14	TOTAL DIV. 10000 - 14000	100.0	106.2	101.3	100.0	124.3	105.1	100.0	103.4	100.7	100.0	126.2	105.5	100.0	106.5	101.4	100.0	96.0	99.1
15	MECHANICAL	96.6	98.4	97.4	96.8	121.8	108.1	96.6	97.9	97.2	100.0	123.8	110.8	99.8	103.6	101.6	97.1	90.8	94.3
16	ELECTRICAL	89.5	92.4	91.5	90.6	113.0	105.9	92.0	91.3	91.6	97.9	131.9	121.1	95.1	104.6	101.6	89.6	84.4	86.0
01 - 16	WEIGHTED AVERAGE	92.8	102.3	97.4	95.6	118.3	106.6	92.8	97.0	94.8	99.7	124.8	111.9	97.4	107.6	102.4	95.1	95.0	95.0

		PENNSYLVANIA																	
DIVISION		READING			SCRANTON			STATE COLLEGE			STROUDSBURG			SUNBURY			UNIONTOWN		
		195 - 196			184 - 185			168			183			178			154		
		MAT.	INST.	TOTAL	MAT.	INST.	TOTAL	MAT.	INST.	TOTAL	MAT.	INST.	TOTAL	MAT.	INST.	TOTAL	MAT.	INST.	TOTAL
02	SITE CONSTRUCTION	105.1	113.6	111.7	96.0	108.2	105.3	86.5	106.9	102.1	90.5	107.6	103.6	95.0	107.7	104.7	93.3	107.6	104.2
03100	CONCRETE FORMS & ACCESSORIES	105.2	89.5	91.7	104.5	93.4	95.0	85.1	92.3	91.3	87.3	87.1	87.1	91.9	84.3	85.4	75.5	107.3	102.8
03200	CONCRETE REINFORCEMENT	97.4	98.8	98.2	100.6	105.6	103.5	98.8	98.1	98.4	100.9	101.3	101.1	99.6	89.1	93.7	103.0	97.2	99.7
03300	CAST-IN-PLACE CONCRETE	71.2	95.0	81.1	90.7	94.6	92.3	87.5	89.1	88.2	85.4	91.4	87.9	93.0	91.4	92.3	93.4	100.5	96.4
03	CONCRETE	94.3	94.8	94.6	101.2	97.7	99.4	101.2	93.8	97.5	94.9	93.1	94.0	101.7	89.4	95.5	89.8	104.5	97.2
04	MASONRY	94.5	91.9	92.9	95.5	103.7	100.6	100.1	89.5	93.5	92.4	90.3	91.1	92.9	90.6	91.4	106.7	103.3	104.6
05	METALS	96.6	120.6	105.2	99.7	124.1	108.5	95.6	119.9	104.3	97.6	122.6	106.6	95.7	116.5	103.2	98.6	126.7	108.7
06	WOOD & PLASTICS	110.5	86.8	98.1	108.5	91.1	99.5	91.7	93.0	92.4	89.8	84.7	87.2	94.7	82.6	88.5	74.6	108.9	92.3
07	THERMAL & MOISTURE PROTECTION	99.5	108.8	103.9	100.1	108.9	104.2	98.7	105.4	101.8	99.8	86.3	93.4	100.0	103.5	101.6	95.5	104.8	99.9
08	DOORS & WINDOWS	98.4	93.1	97.1	96.6	103.6	98.3	93.3	102.2	95.5	97.3	90.0	95.6	93.5	94.5	93.8	92.6	114.7	98.0
09200	PLASTER & GYPSUM BOARD	102.4	86.2	92.4	100.1	90.4	94.1	95.0	92.5	93.4	93.9	83.9	87.7	93.6	81.7	86.2	84.7	108.7	99.6
09500	CEILINGS	102.6	86.2	91.9	99.4	90.4	93.5	89.7	92.5	91.5	98.0	83.9	88.7	89.7	81.7	84.4	94.0	108.7	103.7
09600	FLOORING	80.4	88.3	82.3	85.4	90.1	86.5	82.6	77.4	81.4	79.6	79.9	79.6	81.0	66.1	77.4	81.1	99.7	85.6
09900	PAINTS & COATINGS	86.3	98.4	93.4	85.8	98.4	93.2	85.8	107.6	98.5	85.8	102.8	95.7	85.8	98.4	93.2	89.4	107.8	100.2
09	FINISHES	95.0	89.4	92.1	96.5	92.7	94.6	92.2	90.7	91.5	93.3	87.0	90.1	92.5	81.9	87.1	92.7	106.1	99.5
10 - 14	TOTAL DIV. 10000 - 14000	100.0	98.3	99.6	100.0	97.6	99.5	100.0	94.0	98.7	100.0	98.5	99.7	100.0	95.2	99.0	100.0	107.1	101.5
15	MECHANICAL	100.3	101.1	100.7	100.4	94.2	97.6	97.1	93.8	95.6	97.1	93.5	95.5	97.1	92.2	94.9	96.6	93.2	95.1
16	ELECTRICAL	100.5	84.4	89.5	103.3	90.2	94.4	91.2	104.6	100.3	93.9	81.6	85.5	90.1	85.1	86.7	86.5	104.6	98.8
01 - 16	WEIGHTED AVERAGE	98.1	97.9	98.0	99.4	99.7	99.5	96.1	99.2	97.6	96.1	94.1	95.1	96.0	93.4	94.8	94.9	105.3	100.0

City Cost Indexes

		PENNSYLVANIA																	
	DIVISION	WASHINGTON			WELLSBORO			WESTCHESTER			WILKES-BARRE			WILLIAMSPORT			YORK		
		153			169			193			186 - 187			177			173 - 174		
		MAT.	INST.	TOTAL	MAT.	INST.	TOTAL	MAT.	INST.	TOTAL	MAT.	INST.	TOTAL	MAT.	INST.	TOTAL	MAT.	INST.	TOTAL
02	SITE CONSTRUCTION	93.3	107.6	104.3	99.6	107.6	105.7	107.7	94.5	97.6	88.3	108.0	103.4	85.5	106.9	101.9	84.6	106.7	101.6
03100	CONCRETE FORMS & ACCESSORIES	85.3	107.5	104.4	85.4	83.6	83.9	92.5	120.7	116.7	90.7	92.6	92.4	87.3	82.7	83.3	82.5	88.4	87.5
03200	CONCRETE REINFORCEMENT	103.0	97.1	99.7	98.8	105.2	102.4	95.1	107.5	102.1	99.6	105.4	102.9	98.8	89.0	93.3	99.6	91.3	94.9
03300	CAST-IN-PLACE CONCRETE	93.4	100.6	96.4	91.7	89.8	90.9	88.0	118.6	100.8	82.0	93.8	86.9	78.7	81.6	79.9	85.4	93.1	88.6
03	CONCRETE	90.5	104.5	97.5	103.6	91.6	97.6	102.9	117.0	110.0	91.9	97.0	94.5	88.5	85.3	86.9	99.2	92.2	95.7
04	MASONRY	93.4	100.5	97.9	99.8	87.7	92.3	100.7	118.0	111.5	107.0	99.4	102.3	85.7	87.0	86.5	93.8	91.5	92.3
05	METALS	98.6	125.2	108.1	95.7	123.2	105.6	96.3	106.7	100.1	95.8	123.6	105.8	95.8	116.1	103.1	97.0	116.8	104.1
06	WOOD & PLASTICS	85.1	108.9	97.4	87.8	82.6	85.1	96.2	121.4	109.3	93.5	91.1	92.2	89.8	82.6	86.1	88.8	86.3	87.5
07	THERMAL & MOISTURE PROTECTION	95.7	104.1	99.1	91.6	104.1	96.6	98.9	123.4	110.4	99.5	107.4	103.2	99.3	102.2	100.7	98.6	107.3	102.7
08	DOORS & WINDOWS	92.6	113.6	97.7	97.1	99.0	97.5	92.3	108.6	96.3	93.6	97.5	94.5	93.5	83.7	91.1	93.4	96.5	94.2
09200	PLASTER & GYPSUM BOARD	87.9	108.7	100.8	91.8	81.7	85.5	91.2	121.9	110.2	93.4	90.4	91.6	92.5	81.7	85.8	94.3	85.5	88.9
09500	CEILINGS	94.0	108.7	103.7	89.7	81.7	84.4	99.9	121.9	114.3	91.1	90.4	90.7	91.1	81.7	84.9	91.1	85.5	87.5
09600	FLOORING	85.4	105.0	90.2	78.9	79.1	79.0	77.8	125.4	89.4	80.7	79.5	80.4	79.6	77.4	79.0	81.6	85.6	82.6
09900	PAINTS & COATINGS	89.4	107.8	100.2	85.8	98.4	93.2	88.3	133.8	114.9	85.8	98.4	93.2	85.8	98.4	93.2	85.8	81.1	83.0
09	FINISHES	94.5	107.5	101.2	91.8	83.4	87.5	92.2	123.3	108.1	92.2	90.5	91.3	91.5	82.9	87.1	92.0	86.5	89.2
10 - 14	TOTAL DIV. 10000 - 14000	100.0	107.1	101.5	100.0	94.2	98.8	100.0	123.5	105.0	100.0	97.1	99.4	100.0	91.4	98.2	100.0	94.3	98.8
15	MECHANICAL	96.6	101.5	98.8	97.1	92.8	95.2	96.8	121.2	107.9	97.1	93.5	95.5	97.1	92.2	94.9	100.4	95.4	98.1
16	ELECTRICAL	90.4	104.6	100.1	92.5	74.6	80.3	90.4	102.1	98.4	93.9	86.1	88.6	90.8	69.3	76.1	92.7	78.6	83.1
01 - 16	WEIGHTED AVERAGE	94.9	106.6	100.6	97.2	92.7	95.0	96.8	113.1	104.7	95.7	97.7	96.7	93.7	89.3	91.6	96.4	94.1	95.3

		RHODE ISLAND						SOUTH CAROLINA											
	DIVISION	NEWPORT			PROVIDENCE			AIKEN			BEAUFORT			CHARLESTON			COLUMBIA		
		028			029			298			299			294			290 - 292		
		MAT.	INST.	TOTAL	MAT.	INST.	TOTAL	MAT.	INST.	TOTAL	MAT.	INST.	TOTAL	MAT.	INST.	TOTAL	MAT.	INST.	TOTAL
02	SITE CONSTRUCTION	88.3	99.3	96.7	87.6	99.5	96.8	115.0	81.5	89.3	110.1	81.7	88.3	93.2	82.4	84.9	93.0	82.4	84.9
03100	CONCRETE FORMS & ACCESSORIES	102.4	111.2	109.9	101.4	111.2	109.8	98.6	35.0	44.0	96.4	35.6	44.2	95.1	45.6	52.6	100.6	48.2	55.6
03200	CONCRETE REINFORCEMENT	115.5	128.4	122.8	115.5	128.4	122.8	95.3	14.0	49.2	94.4	36.3	61.5	94.2	47.3	67.6	94.2	46.9	67.4
03300	CAST-IN-PLACE CONCRETE	85.4	115.2	97.9	90.3	115.4	100.8	75.0	36.7	59.0	75.0	47.2	63.4	84.8	56.5	73.0	83.1	52.3	70.2
03	CONCRETE	105.2	115.3	110.3	107.5	115.4	111.5	102.1	34.0	67.9	99.1	42.1	70.4	91.5	51.7	71.5	91.0	51.4	71.1
04	MASONRY	100.1	116.1	110.1	102.5	116.1	111.0	71.9	23.1	41.5	84.7	31.3	51.4	85.9	43.9	59.7	84.9	38.4	55.9
05	METALS	99.9	112.3	104.4	99.9	112.4	104.4	93.1	64.0	82.7	93.2	73.0	85.9	95.0	79.5	89.4	94.9	78.8	89.2
06	WOOD & PLASTICS	102.3	109.6	106.1	102.3	109.6	106.1	101.0	36.0	67.4	99.1	36.0	66.5	97.6	45.3	70.6	104.5	49.3	76.0
07	THERMAL & MOISTURE PROTECTION	100.9	108.7	104.5	99.9	108.7	103.9	96.5	31.2	66.0	96.2	40.0	69.9	95.1	51.6	74.7	95.1	49.4	73.7
08	DOORS & WINDOWS	102.2	116.7	105.7	102.2	116.7	105.7	92.7	30.9	77.7	92.8	36.0	79.0	96.6	43.5	83.7	96.6	45.7	84.2
09200	PLASTER & GYPSUM BOARD	98.5	109.0	105.0	98.5	109.0	105.0	101.3	33.8	59.5	105.4	33.8	61.1	105.8	43.4	67.2	105.8	47.5	69.7
09500	CEILINGS	99.4	109.0	105.7	99.4	109.0	105.7	88.0	33.8	52.4	88.0	33.8	52.4	92.1	43.4	60.2	92.1	47.5	62.8
09600	FLOORING	96.4	124.9	103.4	96.4	124.9	103.4	91.9	21.0	74.6	93.4	38.3	80.0	92.9	48.9	82.2	92.7	48.9	82.0
09900	PAINTS & COATINGS	93.6	119.2	108.6	93.6	119.2	108.6	103.6	32.4	62.0	103.6	32.4	62.0	103.6	47.6	70.9	103.6	47.6	70.9
09	FINISHES	96.4	114.7	105.8	96.3	114.7	105.7	94.0	32.7	62.7	94.8	35.7	64.6	94.0	46.1	69.5	93.9	48.4	70.7
10 - 14	TOTAL DIV. 10000 - 14000	100.0	113.8	102.9	100.0	113.8	102.9	100.0	65.5	92.7	100.0	49.5	89.3	100.0	68.6	93.3	100.0	69.1	93.5
15	MECHANICAL	99.9	102.7	101.2	99.9	102.7	101.2	96.9	38.6	70.5	96.9	32.2	67.6	100.1	54.1	79.3	100.1	45.6	75.4
16	ELECTRICAL	99.5	97.4	98.1	99.0	97.4	97.9	96.7	19.8	44.3	102.6	32.9	55.1	99.7	41.2	59.8	100.9	43.4	61.7
01 - 16	WEIGHTED AVERAGE	100.2	107.8	103.9	100.5	107.8	104.0	95.6	39.4	68.4	96.3	43.4	70.7	96.0	54.6	75.9	96.0	53.1	75.2

		SOUTH CAROLINA												SOUTH DAKOTA					
	DIVISION	FLORENCE			GREENVILLE			ROCK HILL			SPARTANBURG			ABERDEEN			MITCHELL		
		295			296			297			293			574			573		
		MAT.	INST.	TOTAL	MAT.	INST.	TOTAL	MAT.	INST.	TOTAL	MAT.	INST.	TOTAL	MAT.	INST.	TOTAL	MAT.	INST.	TOTAL
02	SITE CONSTRUCTION	103.5	82.4	87.3	98.7	82.0	85.9	96.8	81.7	85.2	98.5	82.0	85.9	79.4	95.8	92.0	77.2	95.8	91.4
03100	CONCRETE FORMS & ACCESSORIES	78.5	48.2	52.5	95.0	48.3	54.9	92.9	34.1	42.5	99.3	48.3	55.5	93.9	46.7	53.4	92.8	46.7	53.2
03200	CONCRETE REINFORCEMENT	93.9	47.2	67.4	93.8	47.3	67.4	94.7	22.5	53.7	93.8	47.3	67.4	104.6	64.2	81.7	103.9	64.2	81.4
03300	CAST-IN-PLACE CONCRETE	75.0	56.4	67.2	75.0	56.4	67.2	75.0	41.5	61.0	75.0	56.4	67.2	98.9	58.2	81.9	96.0	58.2	80.3
03	CONCRETE	93.0	52.9	72.8	91.7	52.9	72.2	90.1	36.9	63.3	92.1	52.9	72.4	94.6	55.2	74.8	92.4	55.2	73.7
04	MASONRY	71.9	43.9	54.5	70.0	43.9	53.8	92.5	28.8	52.8	71.9	43.9	54.5	108.8	61.4	79.3	96.3	61.4	74.6
05	METALS	93.9	79.3	88.6	93.7	79.3	88.6	93.2	67.6	84.0	93.8	79.3	88.6	105.7	74.1	94.4	104.9	74.1	93.8
06	WOOD & PLASTICS	78.8	49.3	63.6	97.0	49.3	72.3	94.8	34.7	63.8	101.9	49.3	74.7	96.6	44.4	69.6	95.4	44.4	69.0
07	THERMAL & MOISTURE PROTECTION	95.3	52.0	75.0	95.4	52.0	75.1	95.2	34.8	66.9	95.5	52.0	75.1	96.2	55.6	77.2	95.9	55.6	77.0
08	DOORS & WINDOWS	92.8	45.7	81.3	92.7	45.7	81.3	92.8	32.1	78.0	92.8	45.7	81.3	99.1	48.1	86.6	97.3	48.1	85.3
09200	PLASTER & GYPSUM BOARD	95.6	47.5	65.8	100.4	47.5	67.7	99.9	32.5	58.2	101.8	47.5	68.2	98.6	43.2	64.3	97.9	43.2	64.0
09500	CEILINGS	89.4	47.5	61.9	88.0	47.5	61.4	88.0	32.5	51.6	88.0	47.5	61.4	109.5	43.2	66.0	108.2	43.2	65.5
09600	FLOORING	84.2	48.9	75.6	90.5	49.9	80.6	89.8	37.8	77.1	92.2	49.9	81.8	112.2	72.3	102.5	111.7	72.3	102.1
09900	PAINTS & COATINGS	103.6	47.6	70.9	103.6	47.6	70.9	103.6	32.4	62.0	103.6	47.6	70.9	104.7	42.5	68.3	104.7	42.5	68.3
09	FINISHES	90.0	48.4	68.8	92.2	48.6	69.9	91.8	34.9	62.7	93.0	48.6	70.3	105.8	50.7	77.7	105.1	50.7	77.4
10 - 14	TOTAL DIV. 10000 - 14000	100.0	69.0	93.4	100.0	69.1	93.5	100.0	65.4	92.7	100.0	69.1	93.5	100.0	62.1	92.0	100.0	62.1	92.0
15	MECHANICAL	100.1	45.6	75.4	100.1	45.8	75.5	96.9	23.0	63.4	100.1	45.8	75.5	100.1	43.6	74.5	96.9	43.6	72.8
16	ELECTRICAL	96.6	24.8	47.7	100.1	39.5	58.8	100.1	24.4	48.5	100.1	39.5	58.8	100.2	50.1	66.1	97.7	50.1	65.3
01 - 16	WEIGHTED AVERAGE	94.3	50.8	73.2	94.5	53.3	74.5	94.6	38.8	67.6	94.8	53.3	74.7	100.5	57.7	79.8	98.2	57.7	78.6

City Cost Indexes

		SOUTH DAKOTA															TENNESSEE		
		MOBRIDGE			PIERRE			RAPID CITY			SIOUX FALLS			WATERTOWN			CHATTANOOGA		
	DIVISION	576			575			577			570 - 571			572			373 - 374		
		MAT.	INST.	TOTAL	MAT.	INST.	TOTAL	MAT.	INST.	TOTAL	MAT.	INST.	TOTAL	MAT.	INST.	TOTAL	MAT.	INST.	TOTAL
02	SITE CONSTRUCTION	77.1	95.8	91.4	77.8	95.7	91.6	78.1	95.5	91.4	79.1	97.7	93.4	77.0	95.8	91.4	106.0	98.0	99.9
03100	CONCRETE FORMS & ACCESSORIES	81.3	46.7	51.6	92.1	49.2	55.2	105.3	44.8	53.3	92.1	47.1	53.5	76.8	46.7	51.0	96.4	53.9	59.9
03200	CONCRETE REINFORCEMENT	106.6	64.2	82.6	104.0	72.9	86.4	97.7	73.1	83.8	97.7	73.7	84.1	101.1	64.2	80.2	93.8	50.4	69.2
03300	CAST-IN-PLACE CONCRETE	96.0	58.2	80.3	96.0	53.8	78.4	95.3	50.6	76.7	99.3	58.4	82.3	96.0	58.2	80.3	103.6	55.7	83.6
03	CONCRETE	91.9	55.2	73.4	92.3	56.4	74.2	92.1	53.3	72.6	93.1	57.2	75.0	90.8	55.2	72.9	93.4	55.8	74.5
04	MASONRY	106.7	61.4	78.5	106.7	51.8	72.5	106.9	48.8	70.7	103.9	64.3	79.2	130.4	61.4	87.5	95.7	53.4	69.4
05	METALS	105.1	74.1	94.0	105.6	76.8	95.3	107.7	77.1	96.7	108.2	78.3	97.5	104.9	74.1	93.8	97.4	83.9	92.6
06	WOOD & PLASTICS	82.9	44.4	63.0	94.5	48.0	70.5	105.6	44.9	74.2	94.5	44.4	68.6	78.0	44.4	60.6	96.1	55.0	74.9
07	THERMAL & MOISTURE PROTECTION	96.0	55.6	77.0	96.3	53.3	76.2	96.8	52.1	75.9	96.3	58.4	78.5	95.6	55.6	76.9	102.9	55.8	80.9
08	DOORS & WINDOWS	100.6	48.1	87.8	102.6	52.5	90.4	103.6	50.8	90.7	103.6	50.5	90.7	97.3	48.1	85.3	100.1	54.3	89.0
09200	PLASTER & GYPSUM BOARD	93.6	43.2	62.4	97.4	46.9	66.1	98.5	43.7	64.6	98.5	43.2	64.3	91.5	43.2	61.6	92.4	54.2	68.7
09500	CEILINGS	109.5	43.2	66.0	108.2	46.9	67.9	113.7	43.7	67.7	113.7	43.2	67.4	108.2	43.2	65.5	93.4	54.2	67.6
09600	FLOORING	106.3	72.3	98.0	111.4	51.5	96.8	111.4	60.5	99.0	111.4	77.3	103.1	104.4	72.3	96.5	98.1	60.8	89.0
09900	PAINTS & COATINGS	104.7	42.5	68.3	104.7	42.5	68.3	104.7	42.5	68.3	104.7	42.5	68.3	104.7	42.5	68.3	105.8	51.6	74.1
09	FINISHES	103.0	50.7	76.3	105.0	48.6	76.2	106.2	47.1	76.0	106.2	51.8	78.4	101.8	50.7	75.7	97.5	55.0	75.8
10 - 14	TOTAL DIV. 10000 - 14000	100.0	62.1	92.0	100.0	62.6	92.1	100.0	59.8	91.5	100.0	62.2	92.0	100.0	62.1	92.0	100.0	67.7	93.2
15	MECHANICAL	96.9	43.6	72.8	100.1	43.7	74.5	100.1	40.5	73.1	100.1	43.8	74.6	96.9	43.6	72.8	100.2	52.4	78.6
16	ELECTRICAL	100.2	50.1	66.1	94.0	54.6	67.2	94.7	54.6	67.4	93.7	68.9	76.8	96.4	50.1	64.9	102.8	66.6	78.1
01 - 16	WEIGHTED AVERAGE	98.9	57.7	79.0	99.9	57.9	79.5	100.7	56.2	79.1	100.5	62.3	82.0	99.2	57.7	79.1	98.8	63.6	81.7

		TENNESSEE																	
		COLUMBIA			COOKEVILLE			JACKSON			JOHNSON CITY			KNOXVILLE			MCKENZIE		
	DIVISION	384			385			383			376			377 - 379			382		
		MAT.	INST.	TOTAL	MAT.	INST.	TOTAL	MAT.	INST.	TOTAL	MAT.	INST.	TOTAL	MAT.	INST.	TOTAL	MAT.	INST.	TOTAL
02	SITE CONSTRUCTION	98.7	84.8	88.0	105.7	85.3	90.0	106.9	96.0	98.5	116.0	86.7	93.5	93.3	86.7	88.2	104.9	85.5	90.0
03100	CONCRETE FORMS & ACCESSORIES	78.1	52.1	55.8	78.3	29.9	36.7	86.7	39.0	45.7	79.8	56.1	59.5	95.3	56.3	61.8	88.8	29.5	37.9
03200	CONCRETE REINFORCEMENT	99.3	52.3	72.6	99.3	23.8	56.5	99.6	49.8	71.4	97.8	50.6	71.0	93.8	50.6	69.3	100.3	23.5	56.7
03300	CAST-IN-PLACE CONCRETE	93.9	56.8	78.4	106.4	37.6	77.7	103.8	39.4	76.9	83.3	62.2	74.5	97.3	57.0	80.5	104.0	38.1	76.5
03	CONCRETE	92.9	55.7	74.2	102.8	34.0	68.2	94.7	43.5	68.9	98.1	59.1	78.5	90.5	57.3	73.8	101.8	34.0	67.7
04	MASONRY	106.6	49.4	71.0	101.2	26.8	54.9	107.0	32.1	60.3	103.8	53.2	72.3	72.9	53.2	60.6	104.3	43.9	66.6
05	METALS	94.0	83.2	90.1	94.1	69.1	85.1	95.9	80.4	90.4	95.1	83.5	91.0	97.7	83.7	92.7	94.1	70.5	85.6
06	WOOD & PLASTICS	68.8	57.8	63.2	69.1	30.8	49.3	83.7	41.3	61.8	71.6	58.4	64.8	87.9	58.4	72.6	80.5	29.9	54.4
07	THERMAL & MOISTURE PROTECTION	93.7	51.8	74.1	94.2	34.4	66.2	102.1	41.1	73.0	97.0	56.7	78.1	94.7	56.1	76.6	94.3	37.0	67.5
08	DOORS & WINDOWS	92.2	54.1	82.9	92.2	28.7	76.7	100.7	45.3	87.2	96.4	57.9	87.0	93.1	57.9	84.5	92.2	28.5	76.7
09200	PLASTER & GYPSUM BOARD	100.2	57.0	73.5	100.2	29.2	56.2	98.8	40.0	62.4	102.8	57.6	74.8	109.2	57.6	77.3	103.9	28.2	57.1
09500	CEILINGS	88.0	57.0	67.7	88.0	29.2	49.4	96.9	40.0	59.6	83.4	57.6	66.5	90.3	57.6	68.8	88.0	28.2	48.8
09600	FLOORING	88.2	31.6	74.4	88.3	29.4	74.0	89.6	29.4	75.0	91.9	62.5	84.7	97.2	62.5	88.7	91.9	60.5	84.2
09900	PAINTS & COATINGS	94.1	30.4	56.8	94.1	28.3	55.6	95.4	33.7	59.4	103.6	66.1	81.7	103.6	66.1	81.7	94.1	32.3	58.0
09	FINISHES	96.7	47.1	71.4	97.1	29.7	62.7	96.3	37.3	66.2	98.4	58.6	78.1	92.9	58.6	75.4	98.7	35.8	66.6
10 - 14	TOTAL DIV. 10000 - 14000	100.0	63.2	92.2	100.0	29.3	85.0	100.0	61.1	91.8	100.0	69.2	93.5	100.0	69.4	93.5	100.0	27.8	84.7
15	MECHANICAL	98.2	47.9	75.4	98.2	28.1	66.4	100.1	46.2	75.7	99.8	55.6	79.8	99.8	59.6	81.6	98.2	25.6	65.3
16	ELECTRICAL	90.4	34.9	52.6	93.2	26.7	47.8	98.4	38.1	57.3	90.4	40.7	56.5	99.7	65.4	76.3	92.8	22.1	44.6
01 - 16	WEIGHTED AVERAGE	95.6	54.6	75.7	97.0	38.4	68.6	98.8	50.3	75.3	98.0	59.8	79.5	94.9	64.5	80.2	97.3	39.7	69.4

		TENNESSEE						TEXAS											
		MEMPHIS			NASHVILLE			ABILENE			AMARILLO			AUSTIN			BEAUMONT		
	DIVISION	375,380 - 381			370 - 372			795 - 796			790 - 791			786 - 787			776 - 777		
		MAT.	INST.	TOTAL	MAT.	INST.	TOTAL	MAT.	INST.	TOTAL	MAT.	INST.	TOTAL	MAT.	INST.	TOTAL	MAT.	INST.	TOTAL
02	SITE CONSTRUCTION	100.6	92.5	94.4	96.1	100.7	99.7	103.1	84.4	88.8	103.2	85.5	89.6	90.3	86.5	87.4	98.5	82.8	86.4
03100	CONCRETE FORMS & ACCESSORIES	96.3	61.3	66.2	96.0	68.6	72.4	96.6	54.3	60.3	100.3	60.8	66.4	99.0	65.5	70.2	106.0	71.6	76.5
03200	CONCRETE REINFORCEMENT	93.6	61.0	75.1	93.4	61.7	75.4	95.7	63.3	77.3	95.7	56.9	73.7	93.6	64.9	77.3	93.8	59.3	74.2
03300	CAST-IN-PLACE CONCRETE	97.4	67.7	85.0	91.0	66.7	80.9	99.1	50.9	79.0	102.8	60.2	85.0	89.2	62.5	78.1	92.8	68.5	82.7
03	CONCRETE	90.8	65.4	78.0	87.7	68.3	77.9	90.7	55.8	73.1	92.7	60.6	76.6	81.2	64.9	73.0	88.7	68.6	78.6
04	MASONRY	82.5	64.3	71.2	84.6	62.6	70.9	99.8	54.4	71.5	103.5	52.6	71.8	100.0	59.3	74.7	103.0	71.3	83.2
05	METALS	97.2	93.0	95.7	99.8	91.2	96.7	96.8	73.7	88.5	96.8	70.9	87.5	96.9	74.3	88.8	97.1	73.2	88.5
06	WOOD & PLASTICS	94.6	62.4	77.9	91.2	70.9	80.7	96.4	55.6	75.3	99.7	63.3	80.9	97.1	68.4	82.3	107.2	74.2	90.1
07	THERMAL & MOISTURE PROTECTION	99.9	66.0	84.0	97.2	64.8	82.0	96.6	60.1	79.5	98.7	55.6	78.5	94.9	64.3	80.6	96.6	71.0	84.6
08	DOORS & WINDOWS	99.8	66.1	91.6	93.8	69.3	87.8	93.6	58.4	85.0	93.6	58.0	84.9	96.2	67.6	89.2	98.1	67.5	90.7
09200	PLASTER & GYPSUM BOARD	101.2	61.5	76.6	100.1	70.5	81.8	94.8	55.1	70.2	94.8	63.2	75.2	95.9	68.3	78.8	96.9	74.4	83.0
09500	CEILINGS	90.4	61.5	71.4	91.8	70.5	77.8	102.5	55.1	71.4	102.5	63.2	76.7	90.3	68.3	75.8	106.9	74.4	85.6
09600	FLOORING	93.4	55.5	84.2	99.5	68.1	91.8	118.9	60.7	104.7	118.7	54.3	103.0	98.7	62.4	89.8	118.3	78.1	108.5
09900	PAINTS & COATINGS	98.0	67.0	79.9	107.3	62.2	81.0	94.7	70.7	80.7	94.7	50.5	68.9	99.3	56.7	74.4	90.5	68.1	77.4
09	FINISHES	94.9	60.4	77.3	101.7	68.1	84.6	102.6	57.1	79.4	102.6	58.6	80.2	94.1	64.1	78.8	99.0	72.8	85.6
10 - 14	TOTAL DIV. 10000 - 14000	100.0	73.2	94.3	100.0	73.3	94.4	100.0	70.9	93.8	100.0	64.3	92.4	100.0	67.7	93.2	100.0	75.7	94.9
15	MECHANICAL	100.0	66.2	84.7	100.0	65.1	84.2	100.0	46.7	75.9	100.0	56.3	80.2	99.9	61.9	82.7	99.9	64.9	84.0
16	ELECTRICAL	98.6	77.5	84.2	100.8	59.5	72.6	97.6	49.4	64.8	98.5	59.9	72.2	97.4	67.5	77.0	94.5	75.0	81.2
01 - 16	WEIGHTED AVERAGE	96.8	72.3	84.9	96.9	71.0	84.4	97.7	58.1	78.5	98.3	62.0	80.7	95.5	67.3	81.9	97.5	71.8	85.1

City Cost Indexes

		\multicolumn{18}{c}{TEXAS}
	DIVISION	\multicolumn{3}{c}{BROWNWOOD}	\multicolumn{3}{c}{BRYAN}	\multicolumn{3}{c}{CHILDRESS}	\multicolumn{3}{c}{CORPUS CHRISTI}	\multicolumn{3}{c}{DALLAS}	\multicolumn{3}{c}{DEL RIO}
		\multicolumn{3}{c}{768}	\multicolumn{3}{c}{778}	\multicolumn{3}{c}{792}	\multicolumn{3}{c}{783 - 784}	\multicolumn{3}{c}{752 - 753}	\multicolumn{3}{c}{788}
		MAT.	INST.	TOTAL	MAT.	INST.	TOTAL	MAT.	INST.	TOTAL	MAT.	INST.	TOTAL	MAT.	INST.	TOTAL	MAT.	INST.	TOTAL
02	SITE CONSTRUCTION	110.7	83.4	89.8	88.3	83.1	84.3	116.3	84.1	91.6	125.9	81.6	92.0	129.2	83.1	93.9	111.5	85.2	91.3
03100	CONCRETE FORMS & ACCESSORIES	93.9	39.1	46.9	79.6	53.0	56.8	94.2	59.7	64.5	102.5	52.1	59.2	96.3	68.6	72.5	90.7	34.4	42.4
03200	CONCRETE REINFORCEMENT	97.6	34.7	62.0	98.0	65.1	79.4	96.4	55.2	73.1	91.8	54.6	70.7	95.1	65.3	78.2	93.6	24.1	54.2
03300	CAST-IN-PLACE CONCRETE	106.8	46.7	81.7	74.4	71.6	73.2	101.6	58.5	83.6	105.8	57.5	85.6	103.8	67.2	88.5	113.8	40.5	83.3
03	CONCRETE	97.7	42.1	69.8	73.9	62.7	68.3	98.3	59.1	78.6	91.2	56.6	73.8	91.4	69.1	80.2	105.8	35.9	70.7
04	MASONRY	125.0	45.7	75.6	139.7	64.9	93.1	103.5	51.5	71.1	90.7	53.6	67.6	94.2	67.8	77.8	105.9	39.8	64.7
05	METALS	94.9	58.1	81.7	97.4	75.9	89.7	94.7	68.8	85.4	96.4	87.6	93.2	97.7	91.1	95.3	95.6	52.5	80.1
06	WOOD & PLASTICS	93.7	38.0	64.9	73.9	49.2	61.1	94.6	63.3	78.4	107.2	52.9	79.1	101.3	70.4	85.3	86.0	35.4	59.9
07	THERMAL & MOISTURE PROTECTION	96.8	48.4	74.1	91.0	66.9	79.7	97.4	54.6	77.3	97.5	57.9	79.0	93.5	70.0	82.5	95.1	39.0	68.8
08	DOORS & WINDOWS	89.7	38.7	77.3	100.5	58.2	90.2	90.7	58.5	82.9	103.4	52.9	91.1	107.3	66.7	97.5	97.1	29.9	80.7
09200	PLASTER & GYPSUM BOARD	93.5	37.0	58.5	88.7	48.6	63.9	93.1	63.2	74.5	97.0	52.0	69.2	98.2	70.4	81.0	93.1	34.2	56.6
09500	CEILINGS	98.3	37.0	58.1	102.8	48.6	67.2	98.3	63.2	75.2	95.8	52.0	67.1	105.1	70.4	82.3	87.5	34.2	52.5
09600	FLOORING	117.5	51.8	101.5	88.9	67.8	83.7	116.3	54.3	101.2	110.7	59.0	98.1	111.2	72.0	101.6	95.7	24.8	78.4
09900	PAINTS & COATINGS	94.7	40.6	63.1	85.9	75.7	80.0	94.7	53.7	70.7	114.1	50.6	77.0	100.5	66.1	80.4	99.3	35.7	62.1
09	FINISHES	101.7	41.3	70.8	87.0	56.8	71.6	101.8	58.5	79.6	100.6	53.0	76.3	104.0	69.1	86.2	93.4	33.5	62.8
10 - 14	TOTAL DIV. 10000 - 14000	100.0	57.9	91.1	100.0	65.9	92.8	100.0	63.3	92.2	100.0	71.3	93.9	100.0	75.5	94.8	100.0	54.6	90.4
15	MECHANICAL	96.8	39.8	71.0	96.7	74.6	86.7	96.8	54.6	77.7	99.9	47.1	76.0	99.9	67.7	85.3	96.6	26.0	64.6
16	ELECTRICAL	91.0	37.6	54.6	90.5	72.8	78.4	97.6	51.3	66.1	94.3	62.0	72.3	99.1	71.3	80.2	97.4	26.7	49.3
01 - 16	WEIGHTED AVERAGE	98.1	47.1	73.4	94.9	69.3	82.5	97.7	59.5	79.2	96.6	60.5	80.1	100.1	72.6	86.8	98.4	39.3	69.7

		\multicolumn{18}{c}{TEXAS}
	DIVISION	\multicolumn{3}{c}{DENTON}	\multicolumn{3}{c}{EASTLAND}	\multicolumn{3}{c}{EL PASO}	\multicolumn{3}{c}{FORT WORTH}	\multicolumn{3}{c}{GALVESTON}	\multicolumn{3}{c}{GIDDINGS}
		\multicolumn{3}{c}{762}	\multicolumn{3}{c}{764}	\multicolumn{3}{c}{798 - 799,885}	\multicolumn{3}{c}{760 - 761}	\multicolumn{3}{c}{775}	\multicolumn{3}{c}{789}
		MAT.	INST.	TOTAL	MAT.	INST.	TOTAL	MAT.	INST.	TOTAL	MAT.	INST.	TOTAL	MAT.	INST.	TOTAL	MAT.	INST.	TOTAL
02	SITE CONSTRUCTION	111.1	77.1	85.0	114.2	83.5	90.7	103.6	84.4	88.8	103.0	84.8	89.0	129.9	80.6	92.1	97.3	85.7	88.4
03100	CONCRETE FORMS & ACCESSORIES	105.6	51.5	59.2	97.2	46.2	53.4	97.3	41.5	49.4	99.4	68.1	72.6	90.0	77.1	79.0	89.1	46.6	52.4
03200	CONCRETE REINFORCEMENT	97.3	53.5	72.5	98.3	57.2	75.0	95.7	56.9	73.7	95.7	64.2	78.2	97.7	66.1	79.8	96.5	58.2	74.8
03300	CAST-IN-PLACE CONCRETE	82.6	62.2	74.1	112.9	47.6	85.7	101.5	45.7	78.2	97.4	61.8	82.6	98.8	72.8	87.9	96.5	48.3	76.4
03	CONCRETE	77.5	57.8	67.6	102.4	49.5	75.8	91.9	47.0	69.3	90.0	65.9	77.9	90.4	75.0	82.6	87.0	50.0	68.4
04	MASONRY	132.4	52.3	82.5	98.9	45.9	65.8	98.6	52.3	69.7	95.1	60.1	73.3	98.2	65.0	77.5	112.8	45.8	71.0
05	METALS	94.5	86.9	91.8	94.8	62.0	83.0	96.6	68.9	86.7	96.6	74.8	88.8	99.1	92.8	96.8	95.1	62.5	83.4
06	WOOD & PLASTICS	109.1	52.9	80.1	101.9	47.8	73.9	97.8	40.0	67.9	104.4	70.3	86.8	88.1	81.6	84.8	85.9	48.0	66.3
07	THERMAL & MOISTURE PROTECTION	95.9	57.7	78.0	97.3	48.9	74.6	96.2	58.3	78.5	97.1	64.1	81.6	89.4	70.1	80.4	95.5	49.1	73.8
08	DOORS & WINDOWS	106.5	55.7	94.2	78.3	43.7	69.9	93.6	44.6	81.6	87.8	66.7	82.6	106.1	76.0	98.8	96.3	43.8	83.5
09200	PLASTER & GYPSUM BOARD	98.0	52.2	69.6	93.5	47.2	64.8	94.8	39.0	60.3	94.8	70.4	79.7	93.4	81.9	86.3	91.9	47.2	64.2
09500	CEILINGS	109.4	52.2	71.8	98.3	47.2	64.7	102.5	39.0	60.8	102.5	70.4	81.4	105.6	81.9	90.0	86.1	47.2	60.6
09600	FLOORING	109.1	60.4	97.2	153.0	37.3	124.8	118.9	60.8	104.7	154.3	60.4	131.4	104.6	67.8	95.6	94.8	37.3	80.8
09900	PAINTS & COATINGS	105.9	41.1	68.0	95.9	42.4	64.6	94.7	48.5	67.7	95.9	72.0	81.9	99.4	67.8	80.9	99.3	42.4	66.0
09	FINISHES	98.5	51.8	74.7	113.6	44.6	78.4	102.6	45.0	73.2	114.3	67.4	90.4	97.3	74.9	85.8	92.0	44.5	67.8
10 - 14	TOTAL DIV. 10000 - 14000	100.0	60.5	91.6	100.0	58.7	91.3	100.0	62.7	92.1	100.0	75.5	94.8	100.0	76.9	95.1	100.0	56.9	90.9
15	MECHANICAL	96.8	52.6	76.8	96.8	40.2	71.2	100.0	45.8	75.5	100.0	66.0	84.6	96.6	71.0	85.0	96.7	40.2	71.1
16	ELECTRICAL	93.4	61.0	71.3	91.0	46.9	60.9	96.4	62.1	73.0	96.4	65.6	75.4	93.3	71.1	78.2	92.3	46.9	61.3
01 - 16	WEIGHTED AVERAGE	97.9	60.6	79.8	97.3	50.8	74.8	97.7	55.8	77.4	97.9	68.5	83.6	98.2	74.8	86.9	95.4	51.1	74.0

		\multicolumn{18}{c}{TEXAS}
	DIVISION	\multicolumn{3}{c}{GREENVILLE}	\multicolumn{3}{c}{HOUSTON}	\multicolumn{3}{c}{HUNTSVILLE}	\multicolumn{3}{c}{LAREDO}	\multicolumn{3}{c}{LONGVIEW}	\multicolumn{3}{c}{LUBBOCK}
		\multicolumn{3}{c}{754}	\multicolumn{3}{c}{770 - 772}	\multicolumn{3}{c}{773}	\multicolumn{3}{c}{780}	\multicolumn{3}{c}{756}	\multicolumn{3}{c}{793 - 794}
		MAT.	INST.	TOTAL	MAT.	INST.	TOTAL	MAT.	INST.	TOTAL	MAT.	INST.	TOTAL	MAT.	INST.	TOTAL	MAT.	INST.	TOTAL
02	SITE CONSTRUCTION	120.1	78.4	88.1	126.9	80.7	91.5	108.9	81.5	87.9	90.4	86.3	87.2	112.0	86.3	92.3	133.4	82.7	94.5
03100	CONCRETE FORMS & ACCESSORIES	85.2	38.6	45.2	92.8	79.6	81.4	89.5	48.3	54.2	90.4	52.4	57.8	79.1	39.9	45.5	98.6	56.4	62.4
03200	CONCRETE REINFORCEMENT	97.8	34.8	62.1	96.4	66.3	79.3	99.2	39.6	65.4	93.6	54.9	71.7	97.0	31.8	60.0	94.9	63.0	76.8
03300	CAST-IN-PLACE CONCRETE	94.4	46.2	74.3	95.7	76.1	87.5	102.3	56.8	83.3	81.5	61.9	73.3	109.9	42.7	81.9	99.3	60.3	83.0
03	CONCRETE	84.5	42.8	63.6	88.1	77.2	82.6	95.7	50.6	73.0	80.9	57.1	68.9	98.2	40.6	69.2	89.4	61.0	75.1
04	MASONRY	136.9	44.3	79.2	97.9	74.0	83.0	139.7	46.0	81.4	99.4	58.4	74.0	134.3	40.9	76.1	99.4	48.6	67.7
05	METALS	95.4	73.4	87.5	101.5	93.2	98.5	97.3	60.9	84.2	97.7	69.5	87.6	88.5	57.1	77.2	100.8	89.7	96.9
06	WOOD & PLASTICS	89.3	38.0	62.8	91.1	81.6	86.2	84.9	51.1	67.4	85.7	52.7	68.7	81.1	41.2	60.5	100.3	58.3	78.6
07	THERMAL & MOISTURE PROTECTION	93.1	49.6	72.7	89.3	74.7	82.4	92.3	50.5	72.7	93.8	61.6	78.7	93.4	42.6	69.6	89.8	59.4	75.6
08	DOORS & WINDOWS	108.1	38.7	91.2	106.4	76.0	99.0	100.5	43.6	86.6	96.9	52.0	85.9	97.1	37.4	82.5	105.9	57.6	94.1
09200	PLASTER & GYPSUM BOARD	94.4	37.0	58.9	95.1	81.9	86.9	92.0	50.6	66.3	93.6	52.0	67.9	91.3	40.5	59.9	95.3	57.8	72.1
09500	CEILINGS	102.3	37.0	59.5	109.7	81.9	91.4	102.8	50.6	68.5	90.3	52.0	65.2	98.2	40.5	60.3	105.2	57.8	74.1
09600	FLOORING	106.0	60.4	94.9	105.8	67.8	96.6	92.6	37.3	79.1	95.6	59.0	86.7	112.7	53.2	98.2	109.8	51.3	95.5
09900	PAINTS & COATINGS	100.5	35.0	62.2	99.4	75.7	85.6	85.9	43.5	61.1	99.3	56.1	74.0	91.3	46.6	65.2	106.5	45.3	70.7
09	FINISHES	100.8	42.0	70.8	105.4	77.3	91.0	89.6	46.3	67.5	92.9	53.5	72.8	106.2	43.2	74.0	104.1	53.9	78.5
10 - 14	TOTAL DIV. 10000 - 14000	100.0	57.2	90.9	100.0	79.3	95.6	100.0	56.5	90.8	100.0	69.9	93.6	100.0	31.4	85.5	100.0	71.7	94.0
15	MECHANICAL	96.7	39.0	70.6	99.8	77.9	89.9	96.7	37.9	70.1	99.8	55.3	79.7	96.6	38.5	70.3	99.8	56.1	80.0
16	ELECTRICAL	94.1	37.6	55.6	96.2	72.8	80.3	90.5	46.0	60.1	97.4	68.4	77.6	96.5	57.9	70.2	96.1	60.2	71.6
01 - 16	WEIGHTED AVERAGE	99.6	47.9	74.5	100.0	78.1	89.4	98.6	50.3	75.2	95.3	62.6	79.5	99.2	49.2	74.9	100.3	62.7	82.1

City Cost Indexes

TEXAS

DIVISION		LUFKIN 759			MCALLEN 785			MCKINNEY 750			MIDLAND 797			ODESSA 797			PALESTINE 758		
		MAT.	INST.	TOTAL	MAT.	INST.	TOTAL	MAT.	INST.	TOTAL	MAT.	INST.	TOTAL	MAT.	INST.	TOTAL	MAT.	INST.	TOTAL
02	SITE CONSTRUCTION	105.1	86.4	90.8	130.9	81.6	93.1	114.9	79.3	87.6	138.0	82.7	95.6	103.1	85.0	89.2	114.5	85.9	92.5
03100	CONCRETE FORMS & ACCESSORIES	83.4	56.5	60.3	100.9	49.0	56.4	83.7	59.3	62.7	104.0	54.2	61.2	96.6	54.0	60.0	71.7	46.1	49.8
03200	CONCRETE REINFORCEMENT	98.6	44.1	67.7	93.2	54.6	71.3	97.8	57.7	75.1	100.6	63.0	79.3	97.9	62.9	78.0	96.3	58.3	74.7
03300	CAST-IN-PLACE CONCRETE	98.3	50.1	78.2	115.0	55.3	90.1	88.6	63.9	78.3	105.5	56.6	85.1	99.1	55.3	80.8	89.8	46.7	71.8
03	CONCRETE	90.9	53.0	71.8	97.0	54.4	75.6	80.2	62.4	71.2	94.8	58.7	76.6	90.9	57.1	73.9	90.3	49.3	69.7
04	MASONRY	107.0	54.0	74.0	104.2	53.6	72.7	145.8	57.0	90.5	116.0	48.7	74.1	99.8	48.6	67.9	100.9	48.0	67.9
05	METALS	95.6	68.7	85.9	95.4	84.9	91.6	95.3	86.2	92.1	99.1	89.3	95.6	97.0	73.0	88.4	95.4	61.4	83.2
06	WOOD & PLASTICS	89.5	59.4	73.9	104.9	49.2	76.1	87.8	60.5	73.7	106.3	55.5	80.0	96.4	55.3	75.2	76.6	47.6	61.7
07	THERMAL & MOISTURE PROTECTION	93.1	55.2	75.4	97.6	55.0	77.7	92.9	63.4	79.1	90.2	56.5	74.4	96.6	55.5	77.3	93.6	52.0	74.1
08	DOORS & WINDOWS	78.2	54.6	72.4	101.9	50.0	89.3	108.1	59.8	96.4	104.9	55.5	92.9	93.6	55.4	84.3	78.1	44.5	69.9
09200	PLASTER & GYPSUM BOARD	92.5	59.3	72.0	96.0	48.2	66.4	94.0	60.3	73.1	96.4	54.9	70.7	94.8	54.9	70.1	88.4	47.2	62.9
09500	CEILINGS	92.6	59.3	70.8	93.0	48.2	63.6	102.3	60.3	74.7	101.1	54.9	70.8	102.5	54.9	71.2	92.6	47.2	62.8
09600	FLOORING	152.1	49.7	127.2	109.9	59.6	97.7	105.5	60.4	94.5	111.7	50.6	96.8	118.9	50.6	102.2	141.5	37.3	116.1
09900	PAINTS & COATINGS	91.3	52.5	68.6	114.1	40.1	70.8	100.5	65.4	80.0	106.5	44.8	70.4	94.7	45.3	65.8	91.3	42.4	62.7
09	FINISHES	117.9	55.0	85.8	99.9	49.7	74.3	100.3	60.2	79.8	104.4	52.2	77.8	102.6	52.1	76.9	114.5	44.1	78.5
10 - 14	TOTAL DIV. 10000 - 14000	100.0	65.8	92.8	100.0	69.8	93.6	100.0	62.8	92.1	100.0	66.1	92.8	100.0	65.6	92.7	100.0	58.2	91.2
15	MECHANICAL	96.6	45.0	73.2	96.7	43.5	72.6	96.7	59.7	80.0	96.6	50.3	75.6	100.0	50.2	77.5	96.6	40.2	71.1
16	ELECTRICAL	98.8	64.3	75.3	93.0	47.5	62.0	94.1	65.3	74.4	96.1	56.1	68.8	97.6	56.0	69.3	92.1	53.1	65.5
01 - 16	WEIGHTED AVERAGE	97.0	59.0	78.6	99.0	56.2	78.3	99.3	65.4	82.8	101.0	60.1	81.1	97.7	58.5	78.7	95.8	52.3	74.7

TEXAS

DIVISION		SAN ANGELO 769			SAN ANTONIO 781 - 782			TEMPLE 765			TEXARKANA 755			TYLER 757			VICTORIA 779		
		MAT.	INST.	TOTAL	MAT.	INST.	TOTAL	MAT.	INST.	TOTAL	MAT.	INST.	TOTAL	MAT.	INST.	TOTAL	MAT.	INST.	TOTAL
02	SITE CONSTRUCTION	106.3	84.5	89.6	90.0	90.3	90.3	92.8	84.2	86.2	98.2	86.6	89.3	111.3	86.6	92.4	136.1	77.3	91.0
03100	CONCRETE FORMS & ACCESSORIES	94.3	50.9	57.0	90.4	71.0	73.7	101.0	49.0	56.4	93.1	47.6	54.0	85.8	45.9	51.5	90.7	56.9	61.7
03200	CONCRETE REINFORCEMENT	96.9	62.8	77.6	93.6	61.0	75.1	98.0	58.7	75.7	96.1	64.3	78.1	97.0	64.6	78.6	92.8	44.1	65.2
03300	CAST-IN-PLACE CONCRETE	100.8	53.3	80.9	79.9	66.3	74.2	82.6	61.3	73.7	90.5	54.2	75.4	107.9	52.7	84.9	110.9	51.9	86.3
03	CONCRETE	93.1	54.9	73.9	80.1	67.9	74.0	80.2	56.1	68.1	85.0	54.1	69.5	97.9	52.9	75.3	97.4	55.0	76.1
04	MASONRY	122.0	46.1	74.7	99.6	64.5	77.7	131.9	57.7	85.7	149.4	46.3	85.2	140.0	50.8	84.5	114.7	54.1	77.0
05	METALS	95.1	70.7	86.4	98.3	73.1	89.3	94.8	70.2	86.0	88.3	72.7	82.7	95.2	73.3	87.4	97.3	86.0	93.3
06	WOOD & PLASTICS	94.0	52.9	72.7	85.7	73.9	79.6	105.7	47.8	75.8	96.5	47.0	70.9	91.9	44.3	67.3	91.8	59.7	75.2
07	THERMAL & MOISTURE PROTECTION	96.6	52.6	76.0	93.8	68.2	81.8	96.1	58.2	78.3	92.9	54.6	75.0	93.5	57.2	76.5	93.7	56.3	76.2
08	DOORS & WINDOWS	89.7	54.0	81.0	96.9	68.2	89.9	76.2	48.6	69.5	97.0	52.4	86.2	78.1	50.9	71.5	105.9	56.9	94.0
09200	PLASTER & GYPSUM BOARD	93.5	52.3	68.0	93.6	73.8	81.3	94.9	47.2	65.3	95.6	46.5	65.2	93.0	43.7	62.5	93.1	59.3	72.2
09500	CEILINGS	98.3	52.3	68.1	90.3	73.8	79.5	98.3	47.2	64.7	96.8	46.5	63.8	92.6	43.7	60.5	111.1	59.3	77.1
09600	FLOORING	117.6	45.2	100.0	95.6	59.8	86.9	156.1	62.1	133.2	122.5	48.1	104.3	154.2	55.2	130.1	104.0	49.7	90.7
09900	PAINTS & COATINGS	94.7	46.8	66.7	99.3	56.1	74.0	95.9	46.7	67.2	91.3	54.7	69.9	91.3	54.7	69.9	99.6	52.5	72.0
09	FINISHES	101.5	49.2	74.8	92.9	67.5	79.9	113.7	50.4	81.4	109.2	47.6	77.7	119.0	47.5	82.5	96.2	55.3	75.3
10 - 14	TOTAL DIV. 10000 - 14000	100.0	63.5	92.3	100.0	75.4	94.8	100.0	58.8	91.3	100.0	69.1	93.5	100.0	68.7	93.4	100.0	62.6	92.1
15	MECHANICAL	96.8	45.7	73.6	99.8	74.8	88.5	96.8	47.9	74.6	96.6	52.3	76.5	96.6	64.0	81.9	96.7	45.0	73.3
16	ELECTRICAL	97.6	44.7	61.5	97.4	68.4	77.6	93.1	56.8	68.4	98.6	61.3	73.2	96.5	58.2	70.4	100.5	64.3	75.9
01 - 16	WEIGHTED AVERAGE	97.7	54.4	76.7	95.3	71.9	84.0	95.8	57.9	77.4	98.6	58.6	79.2	99.7	60.6	80.8	100.4	60.0	80.8

TEXAS / UTAH

DIVISION		WACO 766 - 767			WAXAHACKIE 751			WHARTON 774			WICHITA FALLS 763			LOGAN 843			OGDEN 842,844		
		MAT.	INST.	TOTAL	MAT.	INST.	TOTAL	MAT.	INST.	TOTAL	MAT.	INST.	TOTAL	MAT.	INST.	TOTAL	MAT.	INST.	TOTAL
02	SITE CONSTRUCTION	101.9	84.7	88.7	118.1	79.2	88.2	144.7	79.0	94.3	102.7	84.4	88.7	99.1	100.0	99.8	85.6	100.0	96.7
03100	CONCRETE FORMS & ACCESSORIES	98.8	53.5	59.9	83.7	60.0	63.4	83.0	51.4	55.9	98.8	56.6	62.6	102.6	69.1	73.8	102.6	69.1	73.8
03200	CONCRETE REINFORCEMENT	95.7	64.7	78.1	97.8	59.3	76.0	97.2	61.9	77.2	95.7	56.6	73.5	100.7	73.5	85.3	100.7	73.5	85.3
03300	CAST-IN-PLACE CONCRETE	89.4	63.9	78.8	93.5	51.6	76.0	114.0	62.2	92.4	95.6	58.5	80.1	92.8	77.3	86.3	94.2	77.3	87.2
03	CONCRETE	86.1	60.2	73.1	83.5	58.7	71.1	100.9	59.1	79.9	89.1	58.3	73.6	112.6	73.1	92.8	101.4	73.1	87.2
04	MASONRY	97.1	61.6	75.0	137.4	58.5	88.3	99.3	47.2	66.9	97.5	57.6	72.7	120.9	70.4	89.5	113.9	70.4	86.8
05	METALS	96.7	74.3	88.7	95.4	86.3	92.1	99.1	87.4	94.9	96.7	74.3	88.7	105.5	78.0	95.7	106.0	78.0	95.9
06	WOOD & PLASTICS	103.2	51.4	76.4	87.8	63.3	75.1	80.7	54.0	66.9	103.2	58.2	79.9	90.2	69.3	79.4	90.2	69.3	79.4
07	THERMAL & MOISTURE PROTECTION	97.3	60.6	80.1	93.1	59.8	77.5	89.7	60.9	76.2	97.3	61.1	80.4	103.7	76.7	91.1	102.3	76.7	90.3
08	DOORS & WINDOWS	87.8	52.1	79.1	108.1	61.4	96.8	106.1	55.2	93.7	87.8	59.7	80.9	89.7	67.5	84.3	89.7	67.5	84.3
09200	PLASTER & GYPSUM BOARD	94.8	50.8	67.6	94.0	63.2	74.9	91.1	53.4	67.8	94.8	57.8	71.9	90.1	68.1	76.5	90.1	68.1	76.5
09500	CEILINGS	102.5	50.8	68.6	102.3	63.2	76.6	105.6	53.4	71.3	102.5	57.8	73.2	103.2	68.1	80.2	103.2	68.1	80.2
09600	FLOORING	154.5	47.0	128.3	105.5	49.7	91.9	101.9	37.3	86.1	155.0	65.9	133.3	106.7	74.3	98.8	106.7	74.3	98.8
09900	PAINTS & COATINGS	95.9	46.7	67.2	100.5	52.5	72.4	99.4	43.5	66.7	97.9	74.1	84.0	111.0	59.7	81.0	111.0	59.7	81.0
09	FINISHES	114.4	50.7	81.9	100.5	57.7	78.7	96.8	48.2	72.0	114.7	60.1	86.8	101.5	68.7	84.7	100.4	68.7	84.2
10 - 14	TOTAL DIV. 10000 - 14000	100.0	73.0	94.3	100.0	63.4	92.3	100.0	60.2	91.6	100.0	62.7	92.1	100.0	79.6	95.7	100.0	79.6	95.7
15	MECHANICAL	100.0	59.0	81.4	96.7	56.3	78.4	96.6	37.1	69.7	100.0	56.2	80.2	99.9	74.5	88.4	99.9	74.5	88.4
16	ELECTRICAL	97.6	66.7	76.6	94.0	74.5	80.7	99.9	46.0	63.2	102.4	66.7	78.1	93.7	74.8	80.8	93.7	74.8	80.8
01 - 16	WEIGHTED AVERAGE	97.6	63.7	81.2	99.4	65.6	83.0	100.3	54.7	78.2	98.4	63.8	81.6	102.0	76.0	89.4	99.7	76.0	88.2

COST INDEXES

634

City Cost Indexes

		UTAH								VERMONT								
	DIVISION	PRICE			PROVO			SALT LAKE CITY		BELLOWS FALLS			BENNINGTON			BRATTLEBORO		
		845			846 - 847			840 - 841		051			052			053		
		MAT.	INST.	TOTAL	MAT.	INST.	TOTAL	MAT.	INST.	TOTAL	MAT.	INST.	TOTAL	MAT.	INST.	TOTAL	MAT.	INST.	TOTAL
02	SITE CONSTRUCTION	96.8	97.1	97.0	95.1	98.1	97.4	85.1	100.0	96.5	84.1	93.7	91.5	83.5	93.4	91.1	85.0	93.7	91.7
03100	CONCRETE FORMS & ACCESSORIES	105.7	40.1	49.4	104.4	69.1	74.1	101.2	69.2	73.7	98.5	33.7	42.8	95.5	30.4	39.6	98.9	33.7	42.9
03200	CONCRETE REINFORCEMENT	108.1	70.7	86.9	108.9	73.5	88.9	98.7	73.6	84.4	92.3	66.4	77.6	92.3	21.8	52.3	91.4	66.4	77.2
03300	CAST-IN-PLACE CONCRETE	92.9	54.9	77.1	92.9	77.3	86.4	101.9	77.4	91.6	99.2	49.2	78.3	99.2	40.2	74.6	102.3	49.2	80.2
03	CONCRETE	113.8	52.2	82.8	112.1	73.2	92.5	104.8	73.2	88.9	104.8	46.1	75.3	104.6	33.3	68.8	107.4	46.1	76.6
04	MASONRY	126.0	52.1	80.0	126.0	70.4	91.4	132.7	70.4	93.9	93.1	38.4	59.1	101.0	38.0	61.8	100.5	38.4	61.9
05	METALS	102.2	75.4	92.6	102.9	78.1	94.0	104.7	78.1	95.2	95.0	65.9	84.6	95.0	49.7	78.7	95.0	65.9	84.6
06	WOOD & PLASTICS	93.5	37.8	64.7	92.2	69.3	80.4	90.4	69.3	79.5	101.1	31.7	65.2	97.8	29.1	62.3	101.5	31.7	65.4
07	THERMAL & MOISTURE PROTECTION	106.0	62.8	86.8	106.0	76.7	92.9	105.1	76.7	91.8	99.9	52.9	77.9	99.9	51.3	77.1	100.0	52.9	77.9
08	DOORS & WINDOWS	94.1	48.2	82.9	94.1	67.5	87.6	89.7	67.5	84.3	105.1	39.4	89.1	105.1	24.7	85.5	105.1	39.4	89.1
09200	PLASTER & GYPSUM BOARD	89.7	35.6	56.2	89.2	68.1	76.2	90.1	68.1	76.5	96.6	28.6	54.5	95.2	25.8	52.3	96.6	28.6	54.5
09500	CEILINGS	96.3	35.6	56.4	96.3	68.1	77.8	103.2	68.1	80.2	89.7	28.6	49.6	89.7	25.8	47.8	89.7	28.6	49.6
09600	FLOORING	107.9	55.2	95.0	107.4	74.3	99.4	106.2	74.3	98.4	96.4	46.6	84.3	95.4	46.6	83.5	96.6	46.6	84.4
09900	PAINTS & COATINGS	111.0	59.7	81.0	111.0	66.1	84.7	111.0	66.1	84.7	93.5	19.8	50.4	93.5	18.5	49.6	93.5	19.8	50.4
09	FINISHES	101.0	43.4	71.6	100.7	69.4	84.7	100.7	69.4	84.7	93.6	33.4	62.9	93.0	31.6	61.7	93.7	33.4	62.9
10 - 14	TOTAL DIV. 10000 - 14000	100.0	70.7	93.8	100.0	79.6	95.7	100.0	79.7	95.7	100.0	56.9	90.9	100.0	56.3	90.8	100.0	56.9	90.9
15	MECHANICAL	98.0	51.0	76.7	99.9	74.5	88.4	100.0	74.5	88.5	96.7	40.1	71.1	96.7	19.8	61.9	96.7	40.1	71.1
16	ELECTRICAL	101.5	56.3	70.7	94.2	74.8	81.0	94.6	77.3	82.8	104.4	22.3	48.4	104.4	25.7	50.8	104.4	22.3	48.4
01 - 16	WEIGHTED AVERAGE	102.5	58.6	81.2	102.2	75.9	89.5	101.2	76.5	89.2	98.5	45.3	72.7	98.8	37.9	69.3	99.3	45.3	73.1

		VERMONT																
	DIVISION	BURLINGTON			GUILDHALL			MONTPELIER			RUTLAND			ST. JOHNSBURY			WHITE RIVER JCT.		
		054			059			056			057			058			050		
		MAT.	INST.	TOTAL	MAT.	INST.	TOTAL	MAT.	INST.	TOTAL	MAT.	INST.	TOTAL	MAT.	INST.	TOTAL	MAT.	INST.	TOTAL
02	SITE CONSTRUCTION	86.5	95.7	93.6	83.2	94.1	91.6	85.7	95.5	93.2	86.4	95.7	93.6	83.2	94.1	91.6	87.2	93.7	92.2
03100	CONCRETE FORMS & ACCESSORIES	87.3	59.0	63.0	95.0	36.3	44.6	92.1	59.0	63.7	99.1	59.0	64.7	92.1	36.4	44.3	92.1	30.4	39.2
03200	CONCRETE REINFORCEMENT	115.5	67.8	88.5	93.1	43.2	64.8	91.9	67.8	78.2	115.5	67.8	88.5	91.4	66.7	77.4	92.3	32.1	58.1
03300	CAST-IN-PLACE CONCRETE	101.9	63.9	86.0	96.1	56.8	79.8	102.3	63.9	86.3	97.1	63.9	83.3	96.1	56.8	79.8	102.3	54.5	82.4
03	CONCRETE	109.5	62.7	86.0	102.1	45.6	73.7	107.0	62.6	84.7	108.0	62.7	85.2	101.7	49.6	75.5	109.3	40.2	74.6
04	MASONRY	94.2	72.2	80.5	100.8	45.2	66.2	82.5	72.2	76.1	82.5	72.2	76.1	127.4	45.2	76.2	112.7	38.0	66.1
05	METALS	100.6	70.7	89.9	95.0	58.7	82.0	95.0	70.7	86.3	99.9	70.7	89.4	95.0	58.8	82.0	95.0	53.5	80.1
06	WOOD & PLASTICS	87.2	58.1	72.2	95.4	34.7	64.0	90.5	58.1	73.7	101.7	58.1	79.2	90.5	34.7	61.7	94.0	28.8	60.3
07	THERMAL & MOISTURE PROTECTION	100.3	64.8	83.7	99.7	45.8	74.4	99.7	64.8	83.6	100.1	64.8	83.6	99.6	45.9	74.4	100.0	39.5	71.7
08	DOORS & WINDOWS	108.3	53.7	95.0	105.1	35.6	88.2	105.1	53.7	92.6	108.3	53.7	95.0	105.1	35.6	88.2	105.1	27.8	86.3
09200	PLASTER & GYPSUM BOARD	98.5	55.8	72.1	100.2	31.7	57.8	104.4	55.8	74.3	98.5	55.8	72.1	104.4	31.7	59.4	93.8	25.5	51.5
09500	CEILINGS	99.4	55.8	70.8	89.7	31.7	51.6	89.7	55.8	67.5	99.4	55.8	70.8	89.7	31.7	51.6	89.7	25.5	47.6
09600	FLOORING	96.4	84.2	93.4	99.6	46.6	86.7	103.2	84.2	98.5	96.4	84.2	93.4	103.2	46.6	89.4	94.3	46.6	82.7
09900	PAINTS & COATINGS	93.5	48.5	67.2	93.5	32.4	57.8	93.5	48.5	67.2	93.5	48.5	67.2	93.5	32.4	57.8	93.5	25.8	53.9
09	FINISHES	95.8	62.6	78.9	95.1	36.9	65.4	97.0	62.6	79.4	95.7	62.6	78.8	96.8	36.9	66.2	92.7	32.3	61.9
10 - 14	TOTAL DIV. 10000 - 14000	100.0	65.0	92.6	100.0	57.8	91.1	100.0	65.0	92.6	100.0	65.0	92.6	100.0	57.8	91.1	100.0	56.3	90.8
15	MECHANICAL	99.9	61.9	82.7	96.7	45.2	73.4	96.7	61.9	80.9	99.9	61.9	82.7	96.7	45.2	73.4	96.7	31.4	67.1
16	ELECTRICAL	105.7	48.9	67.0	104.4	32.3	55.3	104.4	48.9	66.6	104.4	48.9	66.6	104.4	32.3	55.3	104.4	35.1	57.2
01 - 16	WEIGHTED AVERAGE	101.3	64.9	83.7	98.7	48.0	74.1	98.5	64.8	82.2	100.5	64.9	83.2	100.2	48.6	75.2	100.0	42.8	72.3

		VIRGINIA																
	DIVISION	ALEXANDRIA			ARLINGTON			BRISTOL			CHARLOTTESVILLE			CULPEPER			FAIRFAX		
		223			222			242			229			227			220 - 221		
		MAT.	INST.	TOTAL	MAT.	INST.	TOTAL	MAT.	INST.	TOTAL	MAT.	INST.	TOTAL	MAT.	INST.	TOTAL	MAT.	INST.	TOTAL
02	SITE CONSTRUCTION	113.5	84.2	91.1	123.6	83.9	93.1	106.5	82.0	87.7	111.0	84.8	90.9	112.2	83.4	90.1	122.9	83.9	93.0
03100	CONCRETE FORMS & ACCESSORIES	91.5	73.2	75.8	88.3	72.0	74.3	82.2	39.2	45.3	79.9	47.7	52.3	76.7	66.4	67.9	80.5	71.8	73.0
03200	CONCRETE REINFORCEMENT	83.8	89.6	87.1	94.4	79.5	86.0	94.5	76.3	84.2	93.9	76.2	83.8	94.4	72.2	81.8	94.4	79.5	85.9
03300	CAST-IN-PLACE CONCRETE	103.0	80.9	93.8	100.2	76.4	90.3	99.8	48.2	78.3	103.8	61.2	86.0	102.6	49.3	80.4	100.2	76.0	90.1
03	CONCRETE	98.8	80.1	89.4	106.4	76.1	91.2	102.4	51.3	76.7	103.0	59.6	81.2	99.6	63.0	81.2	105.8	76.0	90.8
04	MASONRY	85.0	72.0	76.9	95.7	69.5	79.4	84.3	37.9	55.4	106.5	51.8	72.5	96.2	42.2	62.6	95.7	68.9	79.0
05	METALS	95.0	98.4	96.2	93.9	93.3	93.7	93.0	86.9	90.8	93.2	91.6	92.6	93.3	89.0	91.7	93.3	96.6	94.5
06	WOOD & PLASTICS	96.4	73.3	84.5	91.6	73.3	82.2	82.5	39.1	60.1	80.0	47.1	63.0	78.5	73.3	75.8	82.5	73.3	77.8
07	THERMAL & MOISTURE PROTECTION	95.3	80.1	88.2	96.0	76.9	87.1	95.5	37.4	71.9	95.3	53.3	75.7	95.2	54.3	76.0	95.9	76.7	86.9
08	DOORS & WINDOWS	96.6	75.4	91.5	94.6	72.3	89.2	97.9	46.2	85.3	95.9	50.6	84.9	96.3	70.9	90.1	94.6	74.5	89.8
09200	PLASTER & GYPSUM BOARD	105.8	72.3	85.1	101.1	72.3	83.3	98.1	37.0	60.3	96.7	44.4	64.3	97.4	72.3	81.9	98.4	72.3	82.2
09500	CEILINGS	92.1	72.3	79.1	89.4	72.3	78.2	88.0	37.0	54.5	88.0	44.4	59.4	89.4	72.3	78.2	89.4	72.3	78.2
09600	FLOORING	92.9	81.3	90.1	91.4	82.2	89.2	87.4	41.1	76.1	86.0	48.5	76.9	86.0	81.3	84.9	87.8	70.9	83.7
09900	PAINTS & COATINGS	114.7	80.5	94.7	114.7	96.5	104.1	103.6	40.2	66.5	103.6	66.1	81.7	114.7	48.0	75.7	114.7	83.9	96.7
09	FINISHES	94.5	75.0	84.6	93.7	76.6	84.9	90.5	39.0	64.2	89.8	48.9	68.9	90.6	68.3	79.2	92.1	73.0	82.3
10 - 14	TOTAL DIV. 10000 - 14000	100.0	86.7	97.2	100.0	85.6	96.9	100.0	69.5	93.6	100.0	71.5	94.0	100.0	70.4	93.7	100.0	85.3	96.9
15	MECHANICAL	100.1	85.3	93.4	100.1	83.5	92.6	96.9	54.0	77.7	96.9	68.2	83.9	96.9	47.7	74.6	96.9	83.3	90.7
16	ELECTRICAL	99.9	93.6	95.6	96.3	91.2	92.8	99.8	37.6	57.4	99.7	72.2	80.9	104.0	60.5	74.4	101.7	91.2	94.6
01 - 16	WEIGHTED AVERAGE	97.5	84.0	90.9	98.5	81.9	90.5	96.3	53.2	75.4	97.4	66.4	82.4	96.9	63.1	80.5	97.8	81.7	90.0

City Cost Indexes

VIRGINIA

DIVISION		FARMVILLE 239 MAT.	INST.	TOTAL	FREDERICKSBURG 224-225 MAT.	INST.	TOTAL	GRUNDY 246 MAT.	INST.	TOTAL	HARRISONBURG 228 MAT.	INST.	TOTAL	LYNCHBURG 245 MAT.	INST.	TOTAL	NEWPORT NEWS 236 MAT.	INST.	TOTAL
02	SITE CONSTRUCTION	107.3	84.4	89.7	111.7	83.4	90.0	104.4	81.5	86.8	119.9	83.5	92.0	105.1	82.1	87.5	105.5	85.3	90.0
03100	CONCRETE FORMS & ACCESSORIES	96.2	32.7	41.7	80.5	54.1	57.8	86.2	29.3	37.4	75.3	33.6	39.5	82.2	43.1	48.6	95.1	54.8	60.5
03200	CONCRETE REINFORCEMENT	94.5	58.4	74.0	95.2	78.4	85.7	93.2	58.6	73.6	94.4	72.0	81.7	93.9	76.3	83.9	94.2	75.1	83.4
03300	CAST-IN-PLACE CONCRETE	103.8	53.7	82.9	101.7	55.4	82.4	99.8	42.7	76.0	100.2	69.5	87.4	99.8	54.3	80.8	100.8	60.7	84.1
03	CONCRETE	102.4	46.6	74.3	99.3	61.0	80.0	101.1	41.5	71.2	103.2	55.6	79.3	100.9	55.2	77.9	99.3	62.4	80.7
04	MASONRY	91.7	39.8	59.4	94.7	43.5	62.8	86.5	35.4	54.7	95.4	40.4	61.1	100.1	41.2	63.4	87.4	53.1	66.1
05	METALS	93.2	71.7	85.5	93.3	93.3	93.3	93.0	74.0	86.2	93.2	88.9	91.7	93.2	87.2	91.0	95.0	90.8	93.5
06	WOOD & PLASTICS	99.1	31.4	64.1	82.5	55.5	68.5	86.7	28.8	56.8	76.9	27.6	51.5	82.5	42.6	61.9	97.6	55.6	75.9
07	THERMAL & MOISTURE PROTECTION	95.2	40.4	69.5	95.2	53.5	75.7	95.4	38.1	68.6	95.6	61.4	79.6	95.2	47.0	72.6	95.1	53.2	75.5
08	DOORS & WINDOWS	96.3	31.4	80.5	96.0	64.0	88.2	97.9	30.6	81.5	96.3	46.2	84.1	96.3	48.1	84.5	96.6	56.1	86.7
09200	PLASTER & GYPSUM BOARD	105.0	28.1	57.4	98.4	53.9	70.8	99.0	26.4	54.1	96.7	25.2	52.4	98.1	40.6	62.5	105.8	53.0	73.1
09500	CEILINGS	88.0	28.1	48.7	89.4	53.9	66.1	88.0	26.4	47.6	88.0	25.2	46.8	88.0	40.6	56.9	92.1	53.0	66.5
09600	FLOORING	92.9	66.3	86.4	87.8	81.3	86.2	89.0	30.3	74.7	85.5	81.3	84.5	87.4	45.0	77.1	92.9	48.0	82.0
09900	PAINTS & COATINGS	103.6	42.5	67.9	114.7	48.0	75.7	103.6	40.2	66.5	114.7	52.3	78.2	103.6	40.2	66.5	103.6	51.4	73.1
09	FINISHES	93.1	39.7	65.8	91.3	57.9	74.2	91.0	30.5	60.1	90.6	42.6	66.1	90.3	42.4	65.9	93.8	52.8	72.9
10-14	TOTAL DIV. 10000 - 14000	100.0	58.2	91.2	100.0	72.4	94.2	100.0	55.4	90.6	100.0	69.2	93.5	100.0	71.0	93.9	100.0	72.8	94.2
15	MECHANICAL	96.9	33.2	68.0	96.9	78.1	88.4	96.9	33.4	68.1	96.9	48.8	75.1	96.9	67.5	83.6	100.1	63.1	83.3
16	ELECTRICAL	95.0	55.5	68.1	96.6	99.6	98.7	99.8	25.9	49.4	100.1	46.4	63.5	101.6	36.8	57.4	99.8	60.4	73.0
01-16	WEIGHTED AVERAGE	96.6	49.4	73.7	96.3	74.0	85.5	96.3	42.0	70.0	97.2	55.4	76.9	96.8	56.9	77.5	97.4	64.7	81.5

VIRGINIA

DIVISION		NORFOLK 233-235 MAT.	INST.	TOTAL	PETERSBURG 238 MAT.	INST.	TOTAL	PORTSMOUTH 237 MAT.	INST.	TOTAL	PULASKI 243 MAT.	INST.	TOTAL	RICHMOND 230-232 MAT.	INST.	TOTAL	ROANOKE 240-241 MAT.	INST.	TOTAL
02	SITE CONSTRUCTION	104.7	86.5	90.7	109.3	86.0	91.5	103.9	85.0	89.4	103.8	81.5	86.7	106.1	86.0	90.7	102.3	82.1	86.8
03100	CONCRETE FORMS & ACCESSORIES	100.2	54.9	61.3	86.6	62.7	66.1	81.2	54.7	58.4	86.2	29.3	37.4	96.0	62.7	67.4	94.9	42.9	50.2
03200	CONCRETE REINFORCEMENT	94.2	75.1	83.4	93.9	76.4	84.0	93.9	72.2	81.9	93.2	58.6	73.6	94.2	76.4	84.1	94.2	76.3	84.1
03300	CAST-IN-PLACE CONCRETE	103.8	60.8	85.9	103.8	62.7	86.7	99.9	62.8	84.4	99.8	42.7	76.0	107.3	62.7	88.7	104.6	54.2	83.6
03	CONCRETE	101.2	62.4	81.7	103.1	66.7	84.8	97.8	62.5	80.1	101.1	41.5	71.2	102.5	66.7	84.5	101.1	55.0	77.9
04	MASONRY	95.1	53.1	68.9	99.4	56.9	72.9	92.1	53.0	67.8	81.7	35.4	52.8	83.9	56.9	67.0	86.9	41.0	58.3
05	METALS	94.2	91.0	93.1	93.3	92.3	92.9	94.2	89.4	92.5	93.0	74.0	86.2	95.0	92.3	94.0	94.8	87.0	92.0
06	WOOD & PLASTICS	104.1	55.6	79.0	87.4	65.4	76.0	81.8	55.6	68.2	86.7	28.8	56.8	99.0	65.4	81.6	97.6	42.6	69.2
07	THERMAL & MOISTURE PROTECTION	94.9	53.2	75.4	95.1	57.4	77.4	95.0	53.2	75.4	95.4	38.1	68.6	94.7	57.4	77.2	95.0	48.5	73.3
08	DOORS & WINDOWS	96.6	58.9	87.4	96.0	60.4	87.3	96.7	57.5	87.1	97.9	30.6	81.5	96.6	60.4	87.8	96.6	48.1	84.8
09200	PLASTER & GYPSUM BOARD	105.8	53.0	73.1	99.3	63.2	76.9	97.1	53.0	69.8	99.0	26.4	54.1	105.8	63.2	79.4	105.8	40.6	65.4
09500	CEILINGS	92.1	53.0	66.5	89.4	63.2	72.2	92.1	53.0	66.5	88.0	26.4	47.6	92.1	63.2	73.1	92.1	40.6	58.3
09600	FLOORING	92.7	48.0	81.8	88.7	71.3	84.4	85.4	48.0	76.2	89.0	30.3	74.7	92.7	71.3	87.5	92.9	43.0	80.8
09900	PAINTS & COATINGS	103.6	53.7	74.4	103.6	66.1	81.7	103.6	45.1	69.4	103.6	40.2	66.5	103.6	66.1	81.7	103.6	40.2	66.5
09	FINISHES	93.8	53.1	73.0	91.3	64.9	77.8	90.1	52.1	70.7	91.0	30.5	60.1	93.7	64.9	79.0	93.7	42.1	67.3
10-14	TOTAL DIV. 10000 - 14000	100.0	72.8	94.2	100.0	74.9	94.7	100.0	72.8	94.2	100.0	55.4	90.6	100.0	74.9	94.7	100.0	71.0	93.9
15	MECHANICAL	100.1	64.4	83.9	96.9	69.6	84.5	100.1	59.9	81.9	96.9	33.4	68.1	100.1	69.6	86.3	100.1	49.1	77.0
16	ELECTRICAL	99.8	65.2	76.2	100.1	72.2	81.1	96.9	65.2	75.3	99.8	25.9	49.4	100.9	72.2	81.3	99.8	43.0	61.1
01-16	WEIGHTED AVERAGE	97.9	66.0	82.5	97.2	71.0	84.5	96.5	64.7	81.1	96.0	42.0	69.8	97.7	71.0	84.7	97.5	54.4	76.6

VIRGINIA / WASHINGTON

DIVISION		STAUNTON 244 MAT.	INST.	TOTAL	WINCHESTER 226 MAT.	INST.	TOTAL	CLARKSTON 994 MAT.	INST.	TOTAL	EVERETT 982 MAT.	INST.	TOTAL	OLYMPIA 985 MAT.	INST.	TOTAL	RICHLAND 993 MAT.	INST.	TOTAL
02	SITE CONSTRUCTION	107.8	84.6	90.0	118.7	82.9	91.2	93.9	90.5	91.3	90.8	118.6	112.1	94.0	118.6	112.9	94.7	91.2	92.0
03100	CONCRETE FORMS & ACCESSORIES	85.9	42.4	48.5	78.4	45.1	49.8	111.5	87.1	90.5	108.0	103.8	104.4	90.9	104.2	102.3	111.7	87.2	90.7
03200	CONCRETE REINFORCEMENT	93.9	58.5	73.8	93.8	72.1	81.5	106.7	94.6	99.8	108.5	96.4	101.6	111.4	96.1	102.8	101.7	94.4	97.6
03300	CAST-IN-PLACE CONCRETE	103.8	54.6	83.3	100.2	68.2	86.9	107.9	91.3	101.0	92.9	108.9	99.6	97.5	111.7	103.4	108.2	91.8	101.4
03	CONCRETE	102.1	51.8	76.8	102.7	60.1	81.3	109.8	89.7	99.7	104.6	103.4	104.0	107.9	104.6	106.2	109.5	89.9	99.6
04	MASONRY	98.0	42.6	63.5	92.4	40.3	59.9	114.4	93.4	101.2	134.1	106.0	116.6	126.9	102.4	111.7	114.1	91.1	99.8
05	METALS	93.2	85.4	90.4	93.3	88.1	91.4	94.2	87.0	91.6	101.2	89.7	97.1	100.8	89.3	96.7	93.3	86.7	90.9
06	WOOD & PLASTICS	86.7	42.6	63.9	80.4	43.5	61.3	101.5	85.7	93.4	110.6	102.9	106.6	91.6	102.9	97.4	101.8	85.7	93.5
07	THERMAL & MOISTURE PROTECTION	95.0	47.9	73.0	95.8	62.9	80.4	169.2	86.8	130.6	104.2	102.5	103.4	103.8	101.4	102.6	170.8	87.3	131.7
08	DOORS & WINDOWS	96.3	46.1	84.1	98.0	54.8	87.5	116.1	87.9	109.3	103.4	98.3	102.1	103.2	98.3	102.1	116.2	82.7	108.0
09200	PLASTER & GYPSUM BOARD	99.0	39.7	62.3	97.9	41.5	63.0	127.7	85.2	101.4	98.9	102.6	101.2	95.7	102.6	100.0	127.7	85.2	101.4
09500	CEILINGS	88.0	39.7	56.3	89.4	41.5	57.9	119.5	85.2	97.0	112.3	102.6	106.0	110.9	102.6	105.5	119.5	85.2	97.0
09600	FLOORING	88.5	43.4	77.5	87.0	81.3	85.6	114.7	74.9	105.0	126.4	97.8	119.5	118.8	101.4	114.6	115.0	55.0	100.4
09900	PAINTS & COATINGS	103.6	33.2	62.4	114.7	33.9	67.4	120.4	90.1	102.7	117.4	90.1	101.4	117.4	93.0	103.1	120.4	80.0	96.8
09	FINISHES	90.8	41.7	65.8	91.4	49.5	70.0	132.5	84.6	108.0	119.0	101.0	109.8	115.9	102.4	109.0	132.6	79.6	105.5
10-14	TOTAL DIV. 10000 - 14000	100.0	72.8	94.2	100.0	66.7	93.0	100.0	100.3	100.1	100.0	105.9	101.3	100.0	106.5	101.4	100.0	95.9	99.1
15	MECHANICAL	96.9	42.8	72.4	96.9	74.9	86.9	97.3	92.3	95.1	100.0	108.0	103.6	100.1	102.7	101.3	100.6	104.2	102.3
16	ELECTRICAL	97.8	47.0	63.2	97.3	42.0	59.6	101.3	99.2	99.9	105.5	97.4	100.0	105.2	100.2	101.8	104.0	99.2	100.8
01-16	WEIGHTED AVERAGE	96.8	53.6	75.8	97.0	61.4	79.8	107.9	91.4	99.9	105.4	103.3	104.4	104.9	102.7	103.8	108.7	92.7	100.9

City Cost Indexes

	DIVISION	WASHINGTON																	
		SEATTLE			SPOKANE			TACOMA			VANCOUVER			WENATCHEE			YAKIMA		
		980 - 981,987			990 - 992			983 - 984			986			988			989		
		MAT.	INST.	TOTAL	MAT.	INST.	TOTAL	MAT.	INST.	TOTAL	MAT.	INST.	TOTAL	MAT.	INST.	TOTAL	MAT.	INST.	TOTAL
02	SITE CONSTRUCTION	94.9	116.3	111.3	95.5	91.2	92.2	93.6	118.7	112.8	106.5	103.4	104.1	105.8	117.7	114.9	96.9	117.8	112.9
03100	CONCRETE FORMS & ACCESSORIES	91.0	104.6	102.7	120.1	87.3	92.0	91.0	104.4	102.5	95.5	100.1	99.5	97.4	82.9	85.0	93.5	100.5	99.5
03200	CONCRETE REINFORCEMENT	102.4	96.5	99.1	102.4	94.6	97.9	102.4	96.4	99.0	105.4	96.3	100.2	106.7	95.8	100.5	105.0	95.7	99.7
03300	CAST-IN-PLACE CONCRETE	97.5	111.8	103.5	112.1	91.8	103.6	95.5	111.7	102.3	106.7	105.1	106.0	97.6	89.5	94.2	102.1	93.0	98.3
03	CONCRETE	106.8	105.0	105.9	111.9	90.0	100.9	105.8	104.7	105.2	118.1	100.8	109.4	117.6	87.4	102.4	111.4	96.3	103.8
04	MASONRY	126.8	107.4	114.7	115.9	94.2	102.4	126.7	107.4	114.7	128.8	106.2	114.7	129.4	91.9	106.1	118.3	91.0	101.3
05	METALS	102.7	91.9	98.8	93.0	87.0	90.9	102.3	89.8	97.8	100.7	91.9	97.6	100.6	86.9	95.7	101.2	86.4	95.9
06	WOOD & PLASTICS	92.8	102.9	98.0	112.4	85.7	98.6	91.6	102.9	97.4	90.2	98.8	94.6	98.4	80.8	89.3	94.4	102.9	98.8
07	THERMAL & MOISTURE PROTECTION	104.0	103.8	103.9	168.5	87.9	130.8	103.8	103.3	103.6	107.3	97.7	102.8	104.8	84.8	95.5	104.0	87.9	96.5
08	DOORS & WINDOWS	103.5	98.3	102.2	115.9	81.7	107.6	103.9	98.3	102.5	100.5	96.8	99.6	103.5	78.7	97.5	103.5	88.7	99.9
09200	PLASTER & GYPSUM BOARD	93.8	102.6	99.3	130.9	85.2	102.6	96.5	102.6	100.3	97.9	98.8	98.5	98.0	79.8	86.7	96.6	102.6	100.3
09500	CEILINGS	116.4	102.6	107.4	119.5	85.2	97.0	115.1	102.6	106.9	123.5	98.8	107.3	110.9	79.8	90.5	110.9	102.6	105.5
09600	FLOORING	118.2	104.9	114.9	118.3	82.7	109.6	118.8	104.9	115.4	124.5	101.2	118.8	122.2	82.7	112.5	120.2	82.7	111.0
09900	PAINTS & COATINGS	117.4	93.0	103.1	120.4	84.8	99.6	117.4	93.0	103.1	126.1	78.9	98.5	117.4	80.0	95.5	117.4	84.8	98.3
09	FINISHES	116.5	103.1	109.7	134.2	85.8	109.5	116.8	103.1	109.8	117.1	97.9	107.3	118.2	81.8	99.6	116.7	95.9	106.0
10 - 14	TOTAL DIV. 10000 - 14000	100.0	106.5	101.4	100.0	95.9	99.1	100.0	106.5	101.4	100.0	102.3	100.5	100.0	95.1	99.0	100.0	103.7	100.8
15	MECHANICAL	100.0	113.3	106.0	100.6	93.7	97.5	100.1	102.7	101.3	100.3	102.3	101.2	96.9	88.5	93.1	100.1	104.6	102.2
16	ELECTRICAL	105.2	105.1	105.2	94.5	90.0	91.5	105.2	100.2	101.8	118.0	106.1	109.9	106.6	86.4	92.8	110.0	99.2	102.7
01 - 16	WEIGHTED AVERAGE	105.2	106.2	105.7	108.6	90.2	99.6	105.0	103.4	104.2	107.5	101.4	104.5	106.3	89.9	98.3	105.5	98.8	102.2

	DIVISION	WEST VIRGINIA																	
		BECKLEY			BLUEFIELD			BUCKHANNON			CHARLESTON			CLARKSBURG			GASSAWAY		
		258 - 259			247 - 248			262			250 - 253			263 - 264			266		
		MAT.	INST.	TOTAL	MAT.	INST.	TOTAL	MAT.	INST.	TOTAL	MAT.	INST.	TOTAL	MAT.	INST.	TOTAL	MAT.	INST.	TOTAL
02	SITE CONSTRUCTION	98.5	85.2	88.3	98.3	85.2	88.2	104.2	85.0	89.5	100.9	85.9	89.4	104.8	85.0	89.7	101.8	85.6	89.4
03100	CONCRETE FORMS & ACCESSORIES	80.8	96.3	94.1	82.7	95.5	93.7	81.7	92.2	90.7	103.6	97.4	98.2	78.4	92.4	90.4	81.0	92.4	90.7
03200	CONCRETE REINFORCEMENT	92.8	65.8	77.5	92.8	65.7	77.5	93.4	83.6	87.8	94.2	83.7	88.3	93.4	83.5	87.8	93.4	80.7	86.2
03300	CAST-IN-PLACE CONCRETE	97.6	106.6	101.3	97.6	96.6	97.2	97.3	94.0	95.9	98.7	105.3	101.4	106.6	94.1	101.4	101.9	103.9	102.7
03	CONCRETE	96.5	95.2	95.8	96.7	91.1	93.9	100.0	92.1	96.1	98.9	98.2	98.5	103.8	92.3	98.0	100.1	95.1	97.6
04	MASONRY	87.5	88.5	88.1	83.3	88.5	86.5	93.6	92.3	92.8	85.7	90.6	88.7	94.8	92.3	93.3	96.0	88.0	91.0
05	METALS	93.2	99.1	95.3	93.2	93.8	93.4	93.3	102.4	96.6	95.0	100.4	96.9	93.3	103.5	97.0	93.3	101.9	96.4
06	WOOD & PLASTICS	82.0	98.6	90.6	84.2	98.6	91.6	83.2	92.4	87.9	107.5	98.0	102.6	79.5	92.4	86.1	82.3	92.4	87.5
07	THERMAL & MOISTURE PROTECTION	95.1	91.3	93.3	95.2	86.5	91.1	95.5	89.3	92.6	95.1	92.1	93.7	95.4	89.3	92.5	95.1	90.0	92.7
08	DOORS & WINDOWS	97.3	88.5	95.1	98.1	85.4	95.0	98.2	86.3	95.3	97.8	88.2	95.5	98.2	92.8	96.8	96.1	86.1	93.7
09200	PLASTER & GYPSUM BOARD	95.8	98.4	97.4	96.3	98.4	97.6	96.5	92.0	93.7	105.8	97.7	100.8	95.6	92.0	93.3	96.1	92.0	93.5
09500	CEILINGS	88.0	98.4	94.8	88.0	98.4	94.8	89.4	92.0	91.1	92.1	97.7	95.8	89.4	92.0	91.1	89.4	92.0	91.1
09600	FLOORING	84.7	90.7	86.2	85.4	90.7	86.6	85.0	101.8	89.1	92.7	93.9	93.0	83.7	101.8	88.1	84.8	91.8	86.5
09900	PAINTS & COATINGS	103.6	58.0	77.0	103.6	58.0	77.0	103.6	93.6	97.7	103.6	100.1	101.6	103.6	93.6	97.7	103.6	100.1	101.6
09	FINISHES	88.9	91.8	90.4	89.1	91.8	90.5	89.7	94.6	92.2	93.8	97.6	95.8	89.1	94.6	91.9	89.3	93.2	91.3
10 - 14	TOTAL DIV. 10000 - 14000	100.0	89.2	97.7	100.0	89.0	97.7	100.0	97.1	99.4	100.0	98.9	99.8	100.0	97.1	99.4	100.0	97.1	99.4
15	MECHANICAL	96.9	82.2	90.2	96.9	75.0	87.0	96.9	96.1	96.5	100.1	88.3	94.7	96.9	96.3	96.6	96.9	97.2	97.0
16	ELECTRICAL	93.5	84.1	87.1	98.0	56.4	69.6	100.3	85.1	90.0	99.8	81.1	87.0	100.3	85.1	90.0	100.3	83.2	88.7
01 - 16	WEIGHTED AVERAGE	94.8	88.7	91.8	95.1	81.4	88.4	96.5	92.1	94.3	97.4	91.1	94.3	96.9	92.5	94.8	96.2	91.8	94.1

	DIVISION	WEST VIRGINIA																	
		HUNTINGTON			LEWISBURG			MARTINSBURG			MORGANTOWN			PARKERSBURG			PETERSBURG		
		255 - 257			249			254			265			261			268		
		MAT.	INST.	TOTAL	MAT.	INST.	TOTAL	MAT.	INST.	TOTAL	MAT.	INST.	TOTAL	MAT.	INST.	TOTAL	MAT.	INST.	TOTAL
02	SITE CONSTRUCTION	102.6	85.8	89.7	113.4	85.2	91.8	102.0	83.2	87.6	99.2	85.4	88.6	106.8	85.9	90.8	98.5	85.4	88.4
03100	CONCRETE FORMS & ACCESSORIES	95.8	99.6	99.1	78.9	96.0	93.6	80.8	41.0	46.6	78.9	92.5	90.6	84.1	96.0	94.3	83.1	82.4	82.5
03200	CONCRETE REINFORCEMENT	94.2	90.5	92.1	93.4	65.8	77.7	92.8	64.8	77.0	93.4	91.6	92.4	92.8	88.9	90.6	92.8	69.6	79.7
03300	CAST-IN-PLACE CONCRETE	106.5	106.2	106.3	97.7	106.5	101.3	102.2	54.7	82.4	97.3	94.1	96.0	99.5	96.8	98.3	97.3	91.9	95.1
03	CONCRETE	102.1	100.8	101.4	106.5	94.8	100.6	100.2	52.6	76.3	96.3	93.7	95.0	101.9	95.6	98.7	96.5	84.6	90.5
04	MASONRY	87.1	101.0	95.7	86.0	88.5	87.5	90.5	53.6	67.5	112.7	92.3	100.0	74.6	94.5	87.0	90.1	92.3	91.5
05	METALS	95.0	103.6	98.1	93.3	94.3	93.6	93.5	93.0	93.3	93.4	103.6	97.0	93.8	102.5	96.9	93.4	98.8	95.3
06	WOOD & PLASTICS	97.6	99.8	98.7	79.9	98.6	89.6	82.0	37.8	59.2	79.9	92.4	86.3	84.5	95.6	90.2	84.5	79.7	82.0
07	THERMAL & MOISTURE PROTECTION	95.3	95.4	95.4	96.2	91.3	93.9	95.4	63.4	80.4	95.2	89.3	92.4	95.4	91.1	93.4	95.3	81.5	88.8
08	DOORS & WINDOWS	96.6	90.6	95.2	98.1	85.4	95.0	99.4	51.2	87.7	99.5	92.8	97.9	97.2	85.5	94.4	99.5	78.5	94.4
09200	PLASTER & GYPSUM BOARD	105.8	99.6	101.9	95.6	98.4	97.3	96.1	35.6	58.7	95.6	92.0	93.3	98.8	95.3	96.6	96.5	78.9	85.6
09500	CEILINGS	92.1	99.6	97.0	89.4	98.4	95.3	89.4	35.6	54.1	89.4	92.0	91.1	89.4	95.3	93.2	89.4	78.9	82.5
09600	FLOORING	92.7	110.3	97.0	83.9	90.7	85.5	84.7	83.6	84.4	83.9	101.8	88.2	88.1	97.7	90.5	85.5	101.8	89.5
09900	PAINTS & COATINGS	103.6	98.4	100.6	103.6	58.0	77.0	103.6	68.5	83.1	103.6	93.6	97.7	103.6	102.4	102.9	103.6	67.1	82.3
09	FINISHES	93.7	102.2	98.0	89.8	91.8	90.8	89.3	50.8	69.6	88.8	94.6	91.7	91.0	97.4	94.3	89.5	84.2	86.8
10 - 14	TOTAL DIV. 10000 - 14000	100.0	99.7	99.9	100.0	89.0	97.7	100.0	72.0	94.1	100.0	95.0	98.9	100.0	99.0	99.8	100.0	95.4	99.0
15	MECHANICAL	100.1	91.1	96.0	96.9	82.0	90.2	96.9	50.9	76.1	96.9	96.3	96.6	100.1	95.4	98.0	96.9	97.8	97.3
16	ELECTRICAL	99.8	91.4	94.1	93.5	56.4	68.2	103.4	38.0	58.8	100.7	85.1	90.1	100.6	98.0	98.8	106.1	80.4	88.6
01 - 16	WEIGHTED AVERAGE	97.6	95.7	96.7	96.6	83.4	90.2	96.6	57.0	77.4	96.9	92.7	94.9	96.6	95.3	95.9	96.3	88.4	92.4

City Cost Indexes

		WEST VIRGINIA						WISCONSIN												
	DIVISION	ROMNEY			WHEELING			BELOIT			EAU CLAIRE			GREEN BAY			KENOSHA		
		267			260			535			547			541 - 543			531		
		MAT.	INST.	TOTAL	MAT.	INST.	TOTAL	MAT.	INST.	TOTAL	MAT.	INST.	TOTAL	MAT.	INST.	TOTAL	MAT.	INST.	TOTAL
02	SITE CONSTRUCTION	101.3	85.3	89.0	107.5	85.4	90.5	82.6	104.5	99.4	80.2	102.4	97.2	82.2	98.2	94.5	86.7	102.1	98.5
03100	CONCRETE FORMS & ACCESSORIES	77.8	81.8	81.2	86.3	93.3	92.3	97.6	94.3	94.7	96.4	93.8	94.1	116.6	93.6	96.8	111.7	102.5	103.8
03200	CONCRETE REINFORCEMENT	93.4	69.4	79.8	92.2	91.6	91.9	103.9	108.0	106.2	101.2	97.2	98.9	97.7	94.8	96.1	110.9	100.2	104.9
03300	CAST-IN-PLACE CONCRETE	101.9	89.0	96.5	99.5	100.3	99.8	104.2	95.7	100.7	96.7	93.0	95.2	100.0	96.8	98.7	113.4	99.6	107.7
03	CONCRETE	99.9	83.2	91.5	101.9	96.2	99.0	98.1	97.4	97.8	93.5	94.2	93.9	95.2	95.0	95.1	104.5	101.0	102.8
04	MASONRY	86.3	81.5	83.3	96.7	93.9	95.0	105.4	101.5	103.0	94.2	98.9	97.1	115.1	97.9	104.4	101.0	107.1	104.8
05	METALS	93.4	96.2	94.4	94.0	103.8	97.5	95.3	101.5	97.5	92.7	95.5	93.7	95.2	94.7	95.0	96.4	100.2	97.8
06	WOOD & PLASTICS	78.8	79.7	79.3	86.7	92.4	89.6	109.2	92.9	100.7	106.5	92.6	99.3	124.5	92.6	108.0	120.5	101.1	110.5
07	THERMAL & MOISTURE PROTECTION	95.3	77.9	87.2	95.5	90.7	93.3	96.3	102.0	99.0	95.7	89.8	93.0	98.4	89.8	94.4	96.7	98.8	97.7
08	DOORS & WINDOWS	99.4	73.0	93.0	98.1	92.8	96.8	106.3	99.2	104.6	104.8	90.6	101.3	103.4	88.7	99.9	101.6	101.8	101.6
09200	PLASTER & GYPSUM BOARD	95.1	78.9	85.1	99.3	92.0	94.7	105.2	93.2	97.8	109.6	92.9	99.3	93.0	92.9	92.9	94.7	101.7	99.0
09500	CEILINGS	89.4	78.9	82.5	89.4	92.0	91.1	94.7	93.2	93.7	107.3	92.9	97.8	102.1	92.9	96.1	85.6	101.7	96.1
09600	FLOORING	83.6	82.1	83.2	89.0	101.8	92.1	91.3	91.3	91.3	93.4	73.0	88.4	112.4	99.5	109.2	108.1	92.9	104.4
09900	PAINTS & COATINGS	103.6	67.1	82.3	103.6	94.6	98.3	92.1	92.4	92.3	94.0	85.0	88.8	104.7	76.8	88.4	103.4	96.7	99.5
09	FINISHES	88.8	80.1	84.4	91.4	95.1	93.3	97.8	93.6	95.6	100.2	88.8	94.4	104.0	93.1	98.4	100.7	100.3	100.5
10 - 14	TOTAL DIV. 10000 - 14000	100.0	95.4	99.0	100.0	99.5	99.9	100.0	85.2	96.9	100.0	97.1	99.4	100.0	96.1	99.2	100.0	100.2	100.0
15	MECHANICAL	96.9	97.4	97.1	100.1	97.2	98.8	100.0	94.8	97.7	100.1	90.9	95.9	100.5	92.6	97.0	100.4	98.3	99.4
16	ELECTRICAL	104.8	80.4	88.1	95.8	94.8	95.1	91.4	97.0	95.2	102.7	88.4	93.0	94.0	89.6	91.0	93.3	103.7	100.4
01 - 16	WEIGHTED AVERAGE	96.3	86.0	91.3	97.6	95.1	96.4	98.7	97.7	98.2	98.0	93.1	95.6	99.7	93.6	96.7	99.6	101.4	100.5

		WISCONSIN																	
	DIVISION	LA CROSSE			LANCASTER			MADISON			MILWAUKEE			NEW RICHMOND			OSHKOSH		
		546			538			537			530,532			540			549		
		MAT.	INST.	TOTAL	MAT.	INST.	TOTAL	MAT.	INST.	TOTAL	MAT.	INST.	TOTAL	MAT.	INST.	TOTAL	MAT.	INST.	TOTAL
02	SITE CONSTRUCTION	74.6	102.4	95.9	82.1	103.7	98.7	82.0	105.1	99.7	83.2	93.8	91.4	77.3	103.3	97.3	75.6	98.8	93.4
03100	CONCRETE FORMS & ACCESSORIES	78.8	93.9	91.7	96.5	96.3	96.3	98.3	102.3	101.7	101.3	107.1	106.3	88.6	92.4	91.9	88.3	92.5	91.9
03200	CONCRETE REINFORCEMENT	99.7	89.4	93.9	112.5	74.9	91.2	103.9	89.7	95.9	103.9	100.5	102.0	98.1	97.1	97.5	98.4	92.1	94.8
03300	CAST-IN-PLACE CONCRETE	87.0	92.7	89.4	103.6	86.3	96.4	102.2	96.5	99.8	102.2	103.0	102.6	100.7	96.2	98.8	92.8	95.3	93.9
03	CONCRETE	84.7	92.7	88.7	98.7	88.5	93.5	97.2	97.8	97.5	97.6	103.6	100.6	91.6	94.6	93.1	86.8	93.5	90.2
04	MASONRY	93.4	99.2	97.0	105.4	82.9	91.4	105.2	102.7	103.7	105.2	111.7	109.3	118.8	96.7	105.0	107.0	95.6	99.9
05	METALS	92.6	92.3	92.5	93.2	79.1	88.1	96.8	93.9	95.7	98.4	90.5	95.6	95.1	93.1	94.4	93.2	93.3	93.2
06	WOOD & PLASTICS	87.0	92.6	89.8	108.1	103.3	105.6	109.3	103.3	106.2	112.9	106.0	109.3	94.3	92.1	93.2	93.0	92.3	92.7
07	THERMAL & MOISTURE PROTECTION	95.0	89.5	92.4	96.0	75.2	86.2	95.8	98.6	97.1	95.1	103.3	99.0	96.4	100.5	98.3	97.1	88.8	93.2
08	DOORS & WINDOWS	104.7	85.0	99.9	101.4	85.0	97.4	106.3	99.2	104.6	106.2	104.5	105.8	92.2	87.6	91.1	98.3	89.1	96.1
09200	PLASTER & GYPSUM BOARD	103.3	92.9	96.9	103.9	103.9	103.9	105.2	103.9	104.4	105.2	106.5	106.0	86.8	92.6	90.4	84.3	92.6	89.5
09500	CEILINGS	103.1	92.9	96.4	87.8	103.9	98.4	94.7	103.9	100.7	94.7	106.5	102.5	68.6	92.6	84.4	98.0	92.6	94.5
09600	FLOORING	86.3	98.4	89.3	90.8	78.1	87.7	91.3	92.6	91.6	94.1	109.8	97.9	105.2	66.9	95.9	103.2	83.3	98.4
09900	PAINTS & COATINGS	94.0	85.0	88.8	92.1	92.4	92.3	92.1	92.4	92.3	94.7	105.6	101.1	109.2	85.0	95.1	104.7	71.8	85.5
09	FINISHES	95.9	94.1	95.0	96.1	94.5	95.3	97.7	99.0	98.4	98.9	107.5	103.3	94.5	86.7	90.5	98.7	88.6	93.5
10 - 14	TOTAL DIV. 10000 - 14000	100.0	97.2	99.4	100.0	71.5	94.0	100.0	96.5	99.3	100.0	102.0	100.4	100.0	96.8	99.3	100.0	89.7	97.8
15	MECHANICAL	100.1	90.7	95.9	96.8	66.2	82.9	100.0	95.7	98.1	100.0	99.0	99.5	96.5	89.7	93.4	97.3	87.0	92.6
16	ELECTRICAL	103.1	88.4	93.1	91.4	88.4	89.4	92.1	96.3	95.0	93.0	106.5	102.2	100.4	88.4	92.2	101.0	89.0	92.8
01 - 16	WEIGHTED AVERAGE	96.0	93.0	94.6	97.0	84.3	90.8	98.8	93.8	98.6	99.4	102.3	100.8	96.2	92.5	94.4	96.0	91.2	93.7

		WISCONSIN															WYOMING		
	DIVISION	PORTAGE			RACINE			RHINELANDER			SUPERIOR			WAUSAU			CASPER		
		539			534			545			548			544			826		
		MAT.	INST.	TOTAL	MAT.	INST.	TOTAL	MAT.	INST.	TOTAL	MAT.	INST.	TOTAL	MAT.	INST.	TOTAL	MAT.	INST.	TOTAL
02	SITE CONSTRUCTION	73.8	104.4	97.2	82.3	106.0	100.4	86.1	97.5	94.9	74.5	104.3	97.3	71.9	98.3	92.2	86.5	99.7	96.6
03100	CONCRETE FORMS & ACCESSORIES	86.8	101.8	99.7	98.0	102.5	101.9	85.1	90.1	89.4	86.7	95.8	94.5	87.9	93.0	92.2	100.4	52.3	59.1
03200	CONCRETE REINFORCEMENT	113.2	89.4	99.7	103.9	100.2	101.8	99.7	89.8	94.1	99.3	90.6	94.3	99.7	89.2	93.8	107.4	55.4	77.9
03300	CAST-IN-PLACE CONCRETE	88.9	98.5	92.9	102.2	99.2	101.0	105.2	84.3	96.5	94.9	98.9	96.6	86.5	88.5	87.3	104.5	67.1	88.9
03	CONCRETE	86.8	98.2	92.5	97.1	100.9	99.0	98.3	87.9	93.1	86.4	95.9	91.2	82.2	90.8	86.5	107.2	58.8	82.8
04	MASONRY	104.3	100.3	101.8	105.2	107.1	106.4	125.3	82.4	98.6	118.1	101.0	107.4	106.5	94.6	99.1	108.0	48.3	70.8
05	METALS	93.3	93.1	93.6	96.8	100.2	98.0	93.3	86.1	90.7	96.0	93.2	95.0	93.2	91.7	92.7	105.3	67.6	91.8
06	WOOD & PLASTICS	96.9	103.9	100.2	109.5	101.1	105.2	89.4	92.3	90.9	92.1	94.7	93.5	92.2	92.6	92.4	96.5	51.0	73.0
07	THERMAL & MOISTURE PROTECTION	95.2	97.9	96.5	96.3	98.7	97.5	98.0	80.6	89.8	96.0	99.0	97.6	96.8	82.4	90.0	103.8	64.3	85.3
08	DOORS & WINDOWS	101.5	99.2	101.0	106.3	101.8	105.2	98.4	86.2	95.4	92.1	92.4	92.2	98.6	88.5	96.1	93.7	52.2	83.6
09200	PLASTER & GYPSUM BOARD	100.5	103.9	102.6	105.2	101.7	103.0	83.4	92.6	89.1	86.6	95.4	92.0	83.9	92.9	89.5	88.8	49.2	64.3
09500	CEILINGS	89.2	103.9	98.9	94.7	101.7	99.3	98.0	92.6	94.5	69.9	95.4	86.6	98.0	92.9	94.6	96.3	49.2	65.4
09600	FLOORING	87.1	104.5	91.3	91.3	92.9	91.7	102.2	83.5	97.7	106.6	103.6	105.8	103.0	93.8	100.8	106.7	52.0	93.3
09900	PAINTS & COATINGS	92.1	92.4	92.3	92.0	94.0	93.2	104.7	83.8	92.5	100.1	103.0	101.8	104.7	88.7	95.3	111.0	69.4	86.7
09	FINISHES	94.3	102.0	98.2	97.7	100.0	98.9	98.9	88.9	93.8	94.3	98.2	96.3	98.3	93.2	95.7	99.1	53.3	75.7
10 - 14	TOTAL DIV. 10000 - 14000	100.0	95.9	99.1	100.0	100.2	100.0	100.0	82.1	96.2	100.0	97.6	99.5	100.0	96.0	99.2	100.0	79.6	95.7
15	MECHANICAL	96.8	94.3	95.7	100.0	98.3	99.3	97.3	87.1	92.7	96.5	94.5	95.6	97.3	88.4	93.3	99.9	66.3	84.7
16	ELECTRICAL	96.6	94.3	95.0	91.1	102.0	98.5	100.0	87.4	91.4	108.1	99.1	101.9	102.8	84.1	90.1	93.7	63.4	73.1
01 - 16	WEIGHTED AVERAGE	95.4	97.7	96.5	98.8	101.5	100.1	98.7	87.7	93.4	96.1	97.5	96.8	95.5	90.5	93.1	100.5	64.5	83.1

City Cost Indexes

		WYOMING																	
		CHEYENNE			NEWCASTLE			RAWLINS			RIVERTON			ROCK SPRINGS			SHERIDAN		
	DIVISION	820			827			823			825			829 - 831			828		
		MAT.	INST.	TOTAL	MAT.	INST.	TOTAL	MAT.	INST.	TOTAL	MAT.	INST.	TOTAL	MAT.	INST.	TOTAL	MAT.	INST.	TOTAL
02	SITE CONSTRUCTION	86.2	99.7	96.6	86.9	99.2	96.4	104.6	99.2	100.5	97.3	99.4	98.9	89.3	99.3	96.9	94.2	99.7	98.4
03100	CONCRETE FORMS & ACCESSORIES	100.4	52.4	59.2	93.5	37.0	45.0	100.1	37.0	45.9	91.7	43.2	50.0	106.2	37.0	46.8	92.2	52.2	57.9
03200	CONCRETE REINFORCEMENT	100.7	55.5	75.1	108.9	50.8	76.0	108.5	50.8	75.8	109.4	50.9	76.3	109.4	50.8	76.2	109.4	55.5	78.9
03300	CAST-IN-PLACE CONCRETE	104.5	67.1	88.9	105.0	50.2	82.1	105.1	50.2	82.2	105.0	51.6	82.8	105.0	50.2	82.1	105.0	67.0	89.2
03	CONCRETE	106.3	58.9	82.4	107.1	45.4	76.1	122.0	45.4	83.5	116.0	48.6	82.1	108.1	45.5	76.6	113.6	58.7	86.0
04	MASONRY	107.1	51.0	72.1	108.0	39.4	65.3	108.0	39.4	65.3	108.1	42.6	67.3	173.1	39.4	89.8	108.1	51.2	72.6
05	METALS	106.6	67.9	92.7	104.7	64.8	90.4	104.7	64.8	90.4	104.8	65.1	90.6	105.3	66.0	91.2	104.8	67.7	91.5
06	WOOD & PLASTICS	95.5	51.0	72.5	89.2	36.1	61.8	96.4	36.1	65.2	87.5	42.9	64.4	103.3	36.1	68.5	87.9	51.0	68.8
07	THERMAL & MOISTURE PROTECTION	103.8	65.0	85.6	104.2	48.7	78.2	106.4	48.7	79.4	105.5	52.4	80.6	104.5	48.7	78.4	105.2	65.0	86.4
08	DOORS & WINDOWS	94.4	52.2	84.1	100.4	43.1	86.4	100.0	43.1	86.2	100.2	46.8	87.2	100.1	43.9	86.9	100.6	53.7	89.2
09200	PLASTER & GYPSUM BOARD	90.1	49.2	64.8	86.5	33.8	53.9	88.8	33.8	54.8	86.0	40.9	58.1	91.5	33.8	55.8	86.0	49.2	63.2
09500	CEILINGS	103.2	49.2	67.8	96.3	33.8	55.3	96.3	33.8	55.3	96.3	40.9	59.9	96.3	33.8	55.3	96.3	49.2	65.4
09600	FLOORING	106.7	73.3	98.5	103.9	51.0	91.0	106.6	51.0	93.0	103.2	51.0	90.5	109.0	51.0	94.9	103.4	57.3	92.2
09900	PAINTS & COATINGS	110.8	69.4	86.6	111.0	42.1	70.7	111.0	42.1	70.7	111.0	38.6	68.7	111.0	42.1	70.7	111.0	69.4	86.7
09	FINISHES	100.5	57.7	78.6	97.8	39.2	67.9	100.4	39.2	69.2	98.4	43.4	70.3	100.3	39.2	69.1	98.2	54.4	75.9
10 - 14	TOTAL DIV. 10000 - 14000	100.0	79.6	95.7	100.0	63.0	92.2	100.0	63.0	92.2	100.0	64.9	92.6	100.0	63.0	92.2	100.0	79.6	95.7
15	MECHANICAL	99.9	55.3	79.7	98.0	39.0	71.3	98.0	39.0	71.3	98.0	49.8	76.2	99.9	39.0	72.3	98.0	53.3	77.7
16	ELECTRICAL	93.6	66.1	74.9	95.2	57.0	69.2	95.2	57.0	69.2	95.2	57.0	69.2	92.6	57.0	68.3	95.2	57.0	69.2
01 - 16	WEIGHTED AVERAGE	100.7	63.7	82.8	100.7	52.3	77.2	103.4	52.3	78.6	102.1	56.0	79.8	105.2	52.4	79.6	101.8	61.4	82.2

		WYOMING						CANADA											
		WHEATLAND			WORLAND			YELLOWSTONE NAT'L PA			BARRIE, ONTARIO			BATHURST, NEW BRUNSWICK			BRANDON, MANITOBA		
	DIVISION	822			824			821											
		MAT.	INST.	TOTAL	MAT.	INST.	TOTAL	MAT.	INST.	TOTAL	MAT.	INST.	TOTAL	MAT.	INST.	TOTAL	MAT.	INST.	TOTAL
02	SITE CONSTRUCTION	93.0	99.4	97.9	88.1	99.3	96.7	88.2	99.6	96.9	113.5	97.6	101.3	93.2	95.8	95.2	110.6	96.5	99.8
03100	CONCRETE FORMS & ACCESSORIES	96.2	46.3	53.4	96.3	37.0	45.4	96.3	42.9	50.4	120.3	102.4	104.9	91.7	70.5	73.5	119.1	81.4	86.8
03200	CONCRETE REINFORCEMENT	108.9	50.9	76.0	109.4	50.8	76.2	111.4	50.9	77.1	179.5	99.2	134.0	147.5	70.7	103.9	160.8	66.2	107.2
03300	CAST-IN-PLACE CONCRETE	114.5	52.0	88.4	105.0	50.2	82.1	105.0	66.0	88.7	174.7	94.1	141.1	144.0	65.8	111.4	162.8	80.9	128.6
03	CONCRETE	115.1	50.1	82.4	107.1	45.4	76.1	107.6	53.4	80.4	154.4	98.4	126.2	136.9	69.4	102.9	139.0	78.9	108.8
04	MASONRY	108.8	43.7	68.2	108.1	39.4	65.3	108.1	46.7	69.8	154.9	115.7	130.5	144.5	72.8	99.8	147.9	78.0	104.4
05	METALS	104.6	65.2	90.5	104.8	66.0	90.9	105.4	65.3	91.0	106.3	90.1	100.5	100.6	76.8	92.1	100.8	83.0	94.4
06	WOOD & PLASTICS	92.2	46.7	68.7	92.3	36.1	63.2	92.3	39.9	65.2	120.1	101.5	110.5	90.3	70.8	80.2	116.4	82.7	99.0
07	THERMAL & MOISTURE PROTECTION	104.6	54.0	80.9	104.3	48.7	78.3	103.7	56.2	81.5	105.0	101.7	103.4	103.8	70.2	88.1	104.4	81.6	93.7
08	DOORS & WINDOWS	99.0	48.9	86.8	100.6	43.9	86.8	93.9	45.2	82.0	91.9	94.4	92.5	86.1	62.8	80.4	92.1	73.7	87.6
09200	PLASTER & GYPSUM BOARD	87.4	44.8	61.0	87.4	33.8	54.2	87.6	37.8	56.8	169.1	101.6	127.3	190.9	69.9	116.0	133.6	82.2	101.8
09500	CEILINGS	96.3	44.8	62.5	96.3	33.8	55.3	97.7	37.8	58.4	102.5	101.6	101.9	103.8	69.9	81.6	102.5	82.2	89.2
09600	FLOORING	105.2	51.0	92.0	105.2	51.0	92.0	105.2	63.0	94.9	132.0	111.9	127.1	114.3	52.6	99.3	132.2	78.3	119.0
09900	PAINTS & COATINGS	111.0	42.1	70.7	111.0	42.1	70.7	111.0	42.1	70.7	110.2	103.9	106.5	109.5	58.8	79.8	109.6	68.1	85.3
09	FINISHES	98.7	46.7	71.9	98.4	39.2	68.2	98.7	45.3	71.4	124.3	104.9	114.4	120.2	66.0	92.5	118.0	79.8	98.5
10 - 14	TOTAL DIV. 10000 - 14000	100.0	74.5	94.6	100.0	63.0	92.2	100.0	71.2	93.9	100.0	87.7	97.4	100.0	74.8	94.7	100.0	77.4	95.2
15	MECHANICAL	98.0	46.1	74.5	98.0	39.0	71.3	98.0	49.2	75.9	102.7	100.4	101.7	102.6	69.5	87.6	102.6	84.5	94.4
16	ELECTRICAL	95.2	57.0	69.2	95.2	57.0	69.2	93.6	57.0	68.7	127.7	102.6	110.6	123.4	77.4	92.1	128.6	88.6	101.3
01 - 16	WEIGHTED AVERAGE	101.8	56.3	79.8	100.9	52.4	77.4	100.1	57.3	79.4	115.6	100.8	108.4	109.8	73.9	92.3	111.7	83.6	98.1

		CANADA																	
		BRANTFORD, ONTARIO			CALGARY, ALBERTA			CAP-DE-LA-MADELEINE, PQ			CHARLESBOURG, QUEBEC			CHARLOTTETOWN, PEI			CHICOUTIMI, QUEBEC		
	DIVISION	MAT.	INST.	TOTAL	MAT.	INST.	TOTAL	MAT.	INST.	TOTAL	MAT.	INST.	TOTAL	MAT.	INST.	TOTAL	MAT.	INST.	TOTAL
02	SITE CONSTRUCTION	112.6	98.1	101.5	107.1	97.7	99.9	98.9	98.0	96.4	90.9	98.0	96.4	104.4	95.2	97.3	89.8	98.0	96.1
03100	CONCRETE FORMS & ACCESSORIES	120.1	111.2	112.5	119.0	82.6	87.8	128.8	98.9	103.1	128.8	98.9	103.1	84.8	63.9	66.8	128.7	98.7	103.0
03200	CONCRETE REINFORCEMENT	174.2	97.6	130.8	160.8	65.6	106.9	147.5	99.5	120.3	147.5	99.5	120.3	147.5	57.9	96.7	111.4	99.4	104.6
03300	CAST-IN-PLACE CONCRETE	170.4	116.5	147.9	162.8	84.6	130.2	137.5	105.1	124.0	137.5	105.1	124.0	166.1	64.7	123.8	131.4	105.0	120.4
03	CONCRETE	151.6	110.1	130.7	146.0	80.6	113.1	132.6	101.0	116.7	132.6	101.0	116.7	153.9	63.7	108.6	125.0	100.8	112.8
04	MASONRY	149.9	121.0	131.9	147.9	83.9	108.0	144.0	100.6	117.0	144.0	100.6	117.0	146.1	70.3	98.9	142.8	100.6	116.5
05	METALS	100.7	98.2	99.8	100.8	84.0	94.8	101.0	95.3	98.9	101.0	95.3	98.9	100.7	72.3	90.5	99.1	95.1	97.7
06	WOOD & PLASTICS	118.6	110.2	114.3	116.6	81.9	98.7	132.3	99.0	115.1	132.3	99.0	115.1	81.1	63.4	72.0	132.3	99.0	115.1
07	THERMAL & MOISTURE PROTECTION	104.6	108.2	106.3	104.4	82.7	94.2	104.6	101.8	103.3	104.6	101.8	103.3	105.1	67.3	87.4	104.6	101.8	103.3
08	DOORS & WINDOWS	91.3	106.0	94.8	92.1	77.8	88.6	92.1	93.1	92.3	92.1	93.1	92.3	94.4	55.9	85.0	91.7	85.7	90.2
09200	PLASTER & GYPSUM BOARD	141.9	110.6	122.5	146.1	81.4	106.0	212.7	99.0	142.3	212.7	99.0	142.3	186.1	62.3	109.4	212.4	99.0	142.2
09500	CEILINGS	102.5	110.6	107.8	102.5	81.4	88.6	102.5	99.0	100.2	102.5	99.0	100.2	123.5	62.3	76.1	101.1	99.0	99.7
09600	FLOORING	132.2	111.9	127.2	132.2	85.5	120.8	132.2	112.8	127.4	132.2	112.8	127.4	110.4	70.0	100.5	132.2	112.8	127.4
09900	PAINTS & COATINGS	109.5	114.0	112.1	109.4	80.7	92.7	109.5	106.1	107.5	109.5	106.1	107.5	109.5	48.8	74.0	109.5	106.1	107.5
09	FINISHES	118.9	112.3	115.5	119.0	83.3	101.2	128.7	102.7	115.4	128.7	102.7	115.4	118.5	63.7	90.5	128.4	102.7	115.3
10 - 14	TOTAL DIV. 10000 - 14000	100.0	90.6	98.0	100.0	92.3	98.4	100.0	97.9	99.6	100.0	97.9	99.6	100.0	73.9	94.5	100.0	97.9	99.6
15	MECHANICAL	102.6	104.0	103.3	102.6	75.7	90.5	102.6	92.0	97.8	102.6	92.0	97.8	102.6	63.8	85.1	102.6	92.0	97.8
16	ELECTRICAL	126.8	107.0	113.3	113.6	83.7	93.2	123.4	92.6	102.4	123.4	92.6	102.4	123.4	65.8	84.2	123.4	92.6	102.4
01 - 16	WEIGHTED AVERAGE	113.3	106.7	110.1	111.6	83.1	97.8	111.2	96.8	104.2	111.2	96.8	104.2	112.9	68.7	91.5	109.8	96.5	103.3

City Cost Indexes

CANADA

DIVISION		CORNER BROOK, NFLD			CORNWALL, ONTARIO			DALHOUSIE, NB			DARTMOUTH, NOVA SCOTIA			EDMONTON, ALBERTA			FORT MCMURRAY, ALBERTA		
		MAT.	INST.	TOTAL	MAT.	INST.	TOTAL	MAT.	INST.	TOTAL	MAT.	INST.	TOTAL	MAT.	INST.	TOTAL	MAT.	INST.	TOTAL
02	SITE CONSTRUCTION	112.4	95.6	99.6	110.4	97.6	100.6	93.2	95.8	95.2	96.4	97.0	96.9	108.6	97.7	100.2	106.1	97.7	99.6
03100	CONCRETE FORMS & ACCESSORIES	95.0	67.4	71.3	116.6	102.6	104.6	91.7	70.5	73.5	82.8	78.7	79.3	116.2	82.6	87.3	116.2	82.6	87.3
03200	CONCRETE REINFORCEMENT	147.5	60.3	98.0	174.2	97.2	130.6	147.5	70.7	103.9	147.5	60.0	97.9	160.8	65.6	106.9	160.8	65.6	106.9
03300	CAST-IN-PLACE CONCRETE	179.6	74.1	135.6	153.3	106.2	133.7	144.0	65.8	111.4	178.4	77.8	136.4	163.6	84.6	130.6	144.3	84.6	119.4
03	CONCRETE	164.5	69.2	116.6	143.0	102.7	122.7	136.9	69.4	102.9	150.7	75.4	112.8	146.2	80.6	113.2	136.8	80.6	108.5
04	MASONRY	147.2	71.7	100.2	148.6	111.1	125.3	144.5	72.8	99.8	147.1	81.1	106.0	147.9	83.9	108.0	146.5	83.9	107.5
05	METALS	102.1	77.4	93.2	100.7	96.5	99.2	100.6	76.8	92.1	100.0	80.7	93.1	100.8	84.0	94.8	100.8	84.0	94.8
06	WOOD & PLASTICS	91.9	66.4	78.7	116.5	102.4	109.2	90.3	70.8	80.2	80.8	77.9	79.3	112.1	81.9	96.5	112.1	81.9	96.5
07	THERMAL & MOISTURE PROTECTION	105.9	69.1	88.6	104.3	101.0	102.7	103.8	70.2	88.1	103.1	79.4	92.0	104.3	82.7	94.2	104.3	82.7	94.2
08	DOORS & WINDOWS	98.4	63.8	90.0	92.1	98.6	93.7	86.1	62.8	80.4	82.6	73.1	80.3	92.1	77.8	88.6	92.1	77.8	88.6
09200	PLASTER & GYPSUM BOARD	191.7	65.4	113.5	262.5	102.6	163.5	190.9	69.9	116.0	185.6	77.3	118.6	143.4	81.4	105.0	143.4	81.4	105.0
09500	CEILINGS	105.2	65.4	79.1	102.5	102.6	102.5	103.8	69.9	81.6	102.5	77.3	86.0	102.5	81.4	88.6	102.5	81.4	88.6
09600	FLOORING	115.1	62.9	102.3	132.2	110.3	126.8	114.3	52.6	99.3	110.3	75.0	101.7	132.2	85.5	120.8	132.2	85.5	120.8
09900	PAINTS & COATINGS	109.5	69.3	86.0	109.5	106.2	107.6	109.5	58.8	79.8	109.5	73.0	88.1	109.6	80.7	92.7	109.6	80.7	92.7
09	FINISHES	121.6	66.7	93.6	136.6	105.0	120.2	120.2	66.0	92.5	117.7	77.5	97.2	119.6	83.3	101.1	119.6	83.3	101.1
10 - 14	TOTAL DIV. 10000 - 14000	100.0	74.7	94.7	100.0	87.1	97.3	100.0	74.8	94.7	100.0	76.6	95.0	100.0	87.8	97.4	100.0	87.8	97.4
15	MECHANICAL	102.6	70.9	88.3	102.6	101.2	102.0	102.6	69.5	87.6	102.6	84.5	94.4	102.6	75.7	90.4	102.6	75.7	90.4
16	ELECTRICAL	123.4	72.8	88.9	123.4	103.3	109.7	123.4	77.4	92.1	123.4	81.2	94.7	113.6	83.7	93.2	113.6	83.7	93.2
01 - 16	WEIGHTED AVERAGE	115.6	73.3	95.1	113.9	101.9	108.1	109.6	73.9	92.3	110.7	81.6	96.6	111.6	83.0	97.7	110.3	83.0	97.1

CANADA

DIVISION		FREDERICTON, NB			GATINEAU, QUEBEC			HALIFAX, NOVA SCOTIA			HAMILTON, ONTARIO			KAMLOOPS, BC			KINGSTON, ONTARIO		
		MAT.	INST.	TOTAL	MAT.	INST.	TOTAL	MAT.	INST.	TOTAL	MAT.	INST.	TOTAL	MAT.	INST.	TOTAL	MAT.	INST.	TOTAL
02	SITE CONSTRUCTION	92.8	95.8	95.1	87.2	98.0	95.5	96.4	97.0	96.9	110.9	97.9	100.9	112.7	98.9	102.1	110.4	97.6	100.6
03100	CONCRETE FORMS & ACCESSORIES	118.5	70.8	77.5	128.8	98.8	103.0	82.8	78.8	79.4	118.4	107.3	108.9	101.4	102.7	102.5	116.6	102.7	104.7
03200	CONCRETE REINFORCEMENT	147.5	70.8	104.0	156.3	99.4	124.1	147.5	60.2	98.0	174.2	99.3	131.7	116.0	99.7	106.7	174.2	97.2	130.6
03300	CAST-IN-PLACE CONCRETE	137.1	65.9	107.4	108.1	105.0	106.8	178.4	80.0	137.4	157.2	109.2	137.2	114.1	106.6	110.9	153.3	106.2	133.6
03	CONCRETE	135.6	69.5	102.4	119.4	100.9	110.1	150.7	76.3	113.3	145.0	106.1	125.4	136.5	103.2	119.7	143.0	102.7	122.7
04	MASONRY	143.9	74.5	100.7	142.5	100.6	116.4	147.1	83.9	107.7	149.7	115.1	128.1	143.4	107.8	121.2	148.6	111.2	125.3
05	METALS	100.2	77.3	92.0	101.0	95.1	98.9	100.0	81.0	93.2	100.9	97.3	99.6	99.9	95.7	98.4	100.7	96.4	99.2
06	WOOD & PLASTICS	121.1	70.8	95.1	132.3	99.0	115.1	80.8	77.9	79.3	117.4	106.7	111.8	94.1	101.3	97.9	116.5	102.6	109.3
07	THERMAL & MOISTURE PROTECTION	104.3	71.4	88.9	104.6	101.8	103.3	103.1	80.2	92.4	104.4	104.2	104.3	105.0	95.9	100.7	104.3	100.5	102.5
08	DOORS & WINDOWS	86.0	61.6	80.0	92.1	87.7	91.0	82.6	73.1	80.3	92.1	102.1	94.5	87.9	98.6	90.5	92.1	96.3	93.1
09200	PLASTER & GYPSUM BOARD	206.7	69.9	122.0	143.2	99.0	115.9	185.6	77.3	118.6	206.3	106.9	144.8	148.8	101.5	119.5	262.5	102.8	163.6
09500	CEILINGS	102.5	69.9	81.1	102.5	99.0	100.2	102.5	77.3	86.0	102.5	106.9	105.4	102.5	101.5	101.8	102.5	102.8	102.7
09600	FLOORING	127.4	52.6	109.2	132.2	112.8	127.4	110.3	75.0	101.7	132.2	111.9	127.2	126.6	63.9	111.3	132.2	110.3	126.8
09900	PAINTS & COATINGS	109.5	74.9	89.3	109.5	106.1	107.5	109.5	73.0	88.1	109.5	114.0	112.1	109.5	96.1	101.6	109.5	108.0	108.6
09	FINISHES	126.6	67.8	96.5	118.5	102.7	110.4	117.7	77.5	97.2	128.3	109.3	118.6	119.7	95.3	107.2	136.6	105.3	120.6
10 - 14	TOTAL DIV. 10000 - 14000	100.0	74.9	94.7	100.0	97.9	99.6	100.0	76.6	95.1	100.0	114.2	103.0	100.0	109.2	102.0	100.0	87.1	97.3
15	MECHANICAL	102.6	78.5	91.7	102.6	92.0	97.8	102.6	84.5	94.4	102.6	105.6	104.0	102.6	95.3	99.3	102.6	101.3	102.0
16	ELECTRICAL	122.9	96.5	104.9	123.4	92.6	102.4	123.4	87.4	98.9	126.8	106.2	112.7	126.4	104.5	111.5	123.4	102.0	108.9
01 - 16	WEIGHTED AVERAGE	110.4	79.2	95.3	108.3	96.6	102.6	110.7	83.1	97.3	113.6	105.6	109.7	110.4	100.0	105.4	113.9	101.7	108.0

CANADA

DIVISION		KITCHENER, ONTARIO			LAVAL, QUEBEC			LETHBRIDGE, ALBERTA			LLOYDMINSTER, ALBERTA			LONDON, ONTARIO			MEDICINE HAT, ALBERTA		
		MAT.	INST.	TOTAL	MAT.	INST.	TOTAL	MAT.	INST.	TOTAL	MAT.	INST.	TOTAL	MAT.	INST.	TOTAL	MAT.	INST.	TOTAL
02	SITE CONSTRUCTION	98.2	97.5	97.6	91.3	98.0	96.5	104.8	97.7	99.3	106.1	97.7	99.6	110.9	97.6	100.7	104.8	97.7	99.3
03100	CONCRETE FORMS & ACCESSORIES	111.2	98.2	100.1	128.8	98.8	103.0	119.0	82.6	87.8	116.2	82.6	87.3	118.3	101.8	104.1	119.0	82.6	87.8
03200	CONCRETE REINFORCEMENT	110.1	99.1	103.9	156.3	99.4	124.1	160.8	65.6	106.9	160.8	65.6	106.9	122.4	97.3	108.2	160.8	65.6	106.9
03300	CAST-IN-PLACE CONCRETE	146.7	88.1	122.3	140.1	105.0	125.5	144.3	84.6	119.4	144.3	84.6	119.4	157.2	106.4	136.0	144.3	84.6	119.4
03	CONCRETE	124.4	94.6	109.4	135.1	100.9	117.9	137.0	80.6	108.6	136.8	80.6	108.5	138.3	102.3	120.2	137.0	80.6	108.6
04	MASONRY	147.1	110.0	124.0	144.9	100.6	117.3	146.5	83.9	107.5	146.5	83.9	107.5	148.9	111.6	125.7	146.5	83.9	107.5
05	METALS	100.2	93.6	97.8	101.0	95.1	98.9	100.8	84.0	94.8	100.8	84.0	94.8	99.1	96.8	98.3	100.8	84.0	94.8
06	WOOD & PLASTICS	110.0	97.0	103.3	132.3	99.0	115.1	116.6	81.9	98.7	112.1	81.9	96.5	117.4	100.5	108.6	116.6	81.9	98.7
07	THERMAL & MOISTURE PROTECTION	102.9	98.8	101.0	104.6	101.8	103.3	104.4	82.7	94.2	104.3	82.7	94.2	104.4	101.9	103.2	104.4	82.7	94.2
08	DOORS & WINDOWS	83.5	92.9	85.8	92.1	87.7	91.0	92.1	77.8	88.6	92.1	77.8	88.6	92.9	97.6	94.0	92.1	77.8	88.6
09200	PLASTER & GYPSUM BOARD	146.9	97.0	116.0	143.2	99.0	115.9	146.1	81.4	106.0	143.4	81.4	105.0	206.3	100.5	140.8	146.1	81.4	106.0
09500	CEILINGS	102.5	97.0	98.9	102.5	99.0	100.2	102.5	81.4	88.6	102.5	81.4	88.6	102.5	100.5	101.2	102.5	81.4	88.6
09600	FLOORING	127.9	111.9	124.0	132.2	112.8	127.4	132.2	85.5	120.8	132.2	85.5	120.8	132.2	111.9	127.2	132.2	85.5	120.8
09900	PAINTS & COATINGS	109.5	103.9	106.2	109.5	106.1	107.5	109.4	80.7	92.7	109.6	80.7	92.7	109.5	114.0	112.1	109.4	80.7	92.7
09	FINISHES	117.6	101.6	109.4	118.5	102.7	110.4	119.8	83.3	101.2	119.6	83.3	101.1	128.3	105.2	116.5	119.8	83.3	101.2
10 - 14	TOTAL DIV. 10000 - 14000	100.0	111.4	102.4	100.0	97.9	99.6	100.0	92.3	98.4	100.0	87.8	97.4	100.0	112.5	102.6	100.0	92.3	98.4
15	MECHANICAL	102.6	101.0	101.9	102.6	92.0	97.8	102.6	75.7	90.5	102.6	75.7	90.4	102.6	101.4	102.1	102.6	75.7	90.5
16	ELECTRICAL	126.8	102.1	110.0	123.4	92.6	102.4	113.6	83.7	93.2	113.6	83.7	93.2	126.8	102.3	110.1	113.6	83.7	93.2
01 - 16	WEIGHTED AVERAGE	108.2	100.1	104.3	110.4	96.6	103.7	110.4	83.1	97.1	110.3	83.0	97.1	112.6	102.4	107.6	110.4	83.1	97.1

COST INDEXES

City Cost Indexes

		CANADA																	
DIVISION		MONCTON, NEW BRUNSWICK			MONTREAL, QUEBEC			MOOSE JAW, SASKATCHEWAN			NEWCASTLE, NB			NEW GLASGOW, NOVA SCOTIA			NORTH BAY, ONTARIO		
		MAT.	INST.	TOTAL	MAT.	INST.	TOTAL	MAT.	INST.	TOTAL	MAT.	INST.	TOTAL	MAT.	INST.	TOTAL	MAT.	INST.	TOTAL
02	SITE CONSTRUCTION	92.6	95.8	95.0	91.3	98.0	96.5	108.0	96.0	98.8	93.2	95.8	95.2	96.4	97.0	96.9	112.6	97.3	100.9
03100	CONCRETE FORMS & ACCESSORIES	91.7	70.5	73.5	128.8	98.8	103.0	96.6	67.7	71.8	91.7	70.5	73.5	82.8	78.7	79.3	120.1	99.8	102.7
03200	CONCRETE REINFORCEMENT	147.5	70.7	103.9	158.9	99.4	125.2	113.6	79.0	93.9	147.5	70.7	103.9	147.5	60.0	97.9	174.2	96.7	130.3
03300	CAST-IN-PLACE CONCRETE	138.7	65.8	108.3	140.1	105.0	125.5	152.8	76.8	121.1	144.0	65.8	111.4	178.4	77.8	136.4	170.4	92.3	137.8
03	CONCRETE	134.2	69.4	101.6	135.4	100.9	118.1	126.7	73.5	99.9	136.9	69.4	102.9	150.7	75.4	112.8	151.6	96.5	123.9
04	MASONRY	144.1	72.8	99.7	144.9	100.6	117.3	146.3	74.0	101.2	144.5	72.8	99.8	147.1	81.1	106.0	149.9	106.4	122.8
05	METALS	100.6	76.8	92.1	101.0	95.1	98.9	97.9	81.2	91.9	100.6	76.8	92.1	100.0	80.7	93.1	100.7	96.3	99.1
06	WOOD & PLASTICS	90.3	70.8	80.2	132.3	99.0	115.1	92.4	66.1	78.8	90.3	70.8	80.2	80.8	77.9	79.3	118.6	100.8	109.4
07	THERMAL & MOISTURE PROTECTION	103.8	70.2	88.1	104.6	101.8	103.3	103.6	73.3	89.4	103.8	70.2	88.1	103.1	79.4	92.0	104.6	98.3	101.6
08	DOORS & WINDOWS	86.1	62.8	80.4	92.1	87.7	91.0	87.0	65.1	81.6	86.1	62.8	80.4	82.6	73.1	80.3	91.3	95.8	92.4
09200	PLASTER & GYPSUM BOARD	190.9	69.9	116.0	143.2	99.0	115.9	120.6	65.0	86.2	190.9	69.9	116.0	185.6	77.3	118.6	141.9	100.9	116.5
09500	CEILINGS	103.8	69.9	81.6	102.5	99.0	100.2	99.7	65.0	77.0	103.8	69.9	81.6	102.5	77.3	86.0	102.5	100.9	101.5
09600	FLOORING	114.3	52.6	99.3	132.2	112.8	127.4	120.0	71.2	108.1	114.3	52.6	99.3	110.3	75.0	101.7	132.2	110.3	126.8
09900	PAINTS & COATINGS	109.5	58.8	79.8	109.5	106.1	107.5	109.5	76.0	89.9	109.5	58.8	79.8	109.5	73.0	88.1	109.5	105.4	107.1
09	FINISHES	120.2	66.0	92.5	118.5	102.7	110.4	111.6	68.9	89.8	120.2	66.0	92.5	117.7	77.5	97.2	118.9	103.0	110.8
10 - 14	TOTAL DIV. 10000 - 14000	100.0	74.8	94.7	100.0	97.9	99.6	100.0	74.2	94.6	100.0	74.8	94.7	100.0	76.6	95.0	100.0	85.4	96.9
15	MECHANICAL	102.6	69.5	87.6	102.6	92.0	97.8	102.6	76.2	90.7	102.6	69.5	87.6	102.6	84.5	94.4	102.6	98.9	100.9
16	ELECTRICAL	123.4	77.4	92.1	123.4	92.6	102.4	128.6	79.2	94.9	123.4	77.4	92.1	123.4	81.2	94.7	126.8	103.2	110.7
01 - 16	WEIGHTED AVERAGE	109.3	73.9	92.1	110.5	96.6	103.7	108.1	77.0	93.0	109.6	73.9	92.3	110.7	81.2	96.6	113.3	99.7	106.7

		CANADA																	
DIVISION		OSHAWA, ONTARIO			OTTAWA, ONTARIO			OWEN SOUND, ONTARIO			PETERBOROUGH, ONTARIO			PORTAGE LA PRAIRIE, MB			PRINCE ALBERT, SS		
		MAT.	INST.	TOTAL	MAT.	INST.	TOTAL	MAT.	INST.	TOTAL	MAT.	INST.	TOTAL	MAT.	INST.	TOTAL	MAT.	INST.	TOTAL
02	SITE CONSTRUCTION	112.5	97.7	101.1	111.2	97.7	100.8	113.5	97.4	101.2	112.6	97.6	101.1	109.1	96.5	99.5	104.0	96.1	97.9
03100	CONCRETE FORMS & ACCESSORIES	120.1	102.9	105.4	116.6	102.9	104.8	120.3	97.9	101.1	120.1	100.9	103.6	119.1	81.4	86.8	96.4	67.6	71.6
03200	CONCRETE REINFORCEMENT	174.2	100.0	132.1	174.2	97.3	130.6	179.5	99.2	134.0	174.2	97.3	130.6	160.8	66.2	107.2	118.5	79.0	96.1
03300	CAST-IN-PLACE CONCRETE	169.7	94.4	138.3	159.6	106.3	137.4	174.7	88.0	138.5	170.4	94.2	138.6	151.2	80.9	121.9	145.1	76.7	116.6
03	CONCRETE	151.2	99.1	125.0	146.0	102.8	124.3	154.4	94.5	124.3	151.6	97.8	124.5	133.4	78.9	106.0	123.5	73.4	98.3
04	MASONRY	149.9	114.2	127.7	149.9	111.1	125.7	154.9	112.9	128.8	149.9	114.2	127.7	147.0	78.0	104.0	145.8	74.0	101.1
05	METALS	100.7	97.1	99.4	100.9	96.8	99.4	106.3	93.5	101.7	100.7	96.7	99.3	100.8	83.0	94.4	98.0	81.1	91.9
06	WOOD & PLASTICS	118.6	101.5	109.8	116.5	102.4	109.2	120.1	97.0	108.2	118.6	99.3	108.6	116.4	82.7	99.0	92.4	66.1	78.8
07	THERMAL & MOISTURE PROTECTION	104.4	101.3	102.9	104.3	101.0	102.7	105.0	99.2	102.3	104.6	101.9	103.3	104.4	81.6	93.7	103.5	72.1	88.8
08	DOORS & WINDOWS	91.3	98.9	93.1	92.1	98.6	93.7	91.9	92.9	92.1	91.3	97.8	92.8	92.1	73.7	87.6	85.5	65.1	80.5
09200	PLASTER & GYPSUM BOARD	151.0	101.6	120.4	262.5	102.6	163.5	169.1	97.0	124.5	141.9	99.3	115.6	133.6	82.2	101.8	120.6	65.0	86.2
09500	CEILINGS	102.5	101.6	101.9	102.5	102.6	102.5	102.5	97.0	98.9	102.5	99.3	100.4	102.5	82.2	89.2	99.7	65.0	77.0
09600	FLOORING	132.2	114.8	127.9	132.2	110.3	126.8	132.2	111.9	127.3	132.2	110.3	126.8	132.2	78.3	119.0	120.0	71.2	108.1
09900	PAINTS & COATINGS	109.5	124.4	118.2	109.5	106.2	107.6	110.2	103.9	106.5	109.5	108.0	108.6	109.6	68.1	85.3	109.5	64.8	83.4
09	FINISHES	120.2	107.9	113.9	136.6	105.0	120.5	124.3	101.6	112.7	118.9	103.8	111.2	118.0	79.8	98.5	111.6	67.7	89.1
10 - 14	TOTAL DIV. 10000 - 14000	100.0	113.1	102.8	100.0	110.1	102.1	100.0	86.0	97.0	100.0	87.4	97.3	100.0	77.4	95.2	100.0	74.2	94.6
15	MECHANICAL	102.6	103.1	102.9	102.6	101.7	102.2	102.7	99.0	101.0	102.6	102.9	102.8	102.6	84.5	94.4	102.6	76.2	90.7
16	ELECTRICAL	126.8	103.0	110.6	123.4	103.3	109.7	128.7	101.9	110.4	126.8	103.0	110.6	128.6	88.6	101.3	128.6	79.2	94.9
01 - 16	WEIGHTED AVERAGE	113.4	103.1	108.4	114.4	102.6	108.7	115.7	99.3	107.7	113.3	101.7	107.7	111.0	83.6	97.7	107.4	76.8	92.6

		CANADA																	
DIVISION		PRINCE GEORGE, BC			QUEBEC, QUEBEC			RED DEER, ALBERTA			REGINA, SASKATCHEWAN			SAINT JOHN, N B			SARNIA, ONTARIO		
		MAT.	INST.	TOTAL	MAT.	INST.	TOTAL	MAT.	INST.	TOTAL	MAT.	INST.	TOTAL	MAT.	INST.	TOTAL	MAT.	INST.	TOTAL
02	SITE CONSTRUCTION	119.2	98.9	103.6	92.7	98.0	96.8	104.2	97.7	99.2	109.3	96.0	99.1	93.4	95.8	95.2	110.9	97.7	100.8
03100	CONCRETE FORMS & ACCESSORIES	101.4	102.7	102.5	128.8	98.9	103.1	119.0	82.6	87.8	96.6	67.7	71.8	118.5	70.8	77.5	118.2	109.3	110.6
03200	CONCRETE REINFORCEMENT	116.0	99.7	106.7	147.5	99.5	120.3	160.8	65.6	106.9	113.6	79.0	94.0	147.5	70.8	104.0	122.4	102.6	111.2
03300	CAST-IN-PLACE CONCRETE	164.8	106.6	140.5	151.5	105.1	132.2	139.7	84.6	116.7	162.8	76.8	126.9	141.9	65.9	110.2	157.2	108.0	136.7
03	CONCRETE	161.3	103.2	132.1	139.5	101.0	120.1	134.8	80.6	107.5	131.6	73.5	102.4	138.0	69.5	103.5	138.3	107.2	122.6
04	MASONRY	147.2	107.8	122.7	145.0	100.6	117.4	146.1	83.9	107.3	147.0	74.5	101.8	144.3	74.5	100.9	148.9	117.4	129.3
05	METALS	99.9	95.7	98.4	101.0	95.3	98.9	100.8	84.0	94.8	97.9	81.3	92.0	100.2	77.3	92.0	98.7	98.5	98.6
06	WOOD & PLASTICS	94.1	101.3	97.9	132.3	99.0	115.1	116.0	81.9	98.7	92.4	66.1	78.8	121.1	70.8	95.1	117.4	109.0	113.1
07	THERMAL & MOISTURE PROTECTION	105.0	95.9	100.7	104.6	101.8	103.3	104.4	82.7	94.2	103.6	73.4	89.5	104.3	71.4	88.9	104.4	107.4	105.8
08	DOORS & WINDOWS	87.9	98.6	90.5	92.1	93.1	92.3	92.1	77.8	88.6	87.0	65.1	81.6	86.0	61.6	80.0	92.9	103.1	95.4
09200	PLASTER & GYPSUM BOARD	148.8	101.5	119.5	212.7	99.0	142.3	146.1	81.4	106.0	120.6	65.0	86.2	206.7	69.9	122.0	206.3	109.3	146.3
09500	CEILINGS	102.5	101.5	101.8	102.5	99.0	100.2	102.5	81.4	88.6	99.7	65.0	77.0	102.5	69.9	81.1	102.5	109.3	107.0
09600	FLOORING	126.6	63.9	111.3	132.2	112.8	127.4	132.2	85.5	120.8	120.0	71.2	108.1	127.4	52.6	109.2	132.2	116.6	128.4
09900	PAINTS & COATINGS	109.5	96.1	101.6	109.5	106.1	107.5	109.4	80.7	92.7	109.5	76.0	89.9	109.5	74.9	89.3	109.5	118.6	114.8
09	FINISHES	119.7	95.3	107.2	128.7	102.7	115.4	119.8	83.3	101.2	111.6	68.9	89.8	126.6	67.8	96.5	128.3	112.4	120.2
10 - 14	TOTAL DIV. 10000 - 14000	100.0	109.2	102.0	100.0	97.9	99.6	100.0	92.3	98.4	100.0	74.2	94.6	100.0	74.9	94.7	100.0	89.4	97.8
15	MECHANICAL	102.6	95.3	99.3	102.6	92.0	97.8	102.6	75.7	90.5	102.6	76.2	90.7	102.6	78.5	91.7	102.6	109.0	105.5
16	ELECTRICAL	126.4	104.5	111.5	123.4	92.6	102.4	113.6	83.7	93.2	128.6	79.2	94.9	122.9	96.5	104.9	126.8	111.3	116.3
01 - 16	WEIGHTED AVERAGE	113.9	100.0	107.1	112.1	96.8	104.7	110.1	83.1	97.0	108.8	77.0	93.4	110.7	79.2	95.4	112.5	107.5	110.1

City Cost Indexes

		\multicolumn{18}{c}{CANADA}																
		\multicolumn{3}{c}{SASKATOON, SASKATCHEWAN}	\multicolumn{3}{c}{SHERBROOKE, QUEBEC}	\multicolumn{3}{c}{ST CATHARINES, ONTARIO}	\multicolumn{3}{c}{ST JOHNS, NEWFOUNDLAND}	\multicolumn{3}{c}{SUDBURY, ONTARIO}	\multicolumn{3}{c}{SUMMERSIDE, PEI}
DIVISION		MAT.	INST.	TOTAL	MAT.	INST.	TOTAL	MAT.	INST.	TOTAL	MAT.	INST.	TOTAL	MAT.	INST.	TOTAL	MAT.	INST.	TOTAL
02	SITE CONSTRUCTION	104.5	96.1	98.0	91.3	98.0	96.5	98.6	97.4	97.7	112.4	95.6	99.6	98.5	97.3	97.6	104.6	95.2	97.4
03100	CONCRETE FORMS & ACCESSORIES	96.4	67.6	71.6	128.8	98.8	103.0	108.4	100.4	101.5	95.0	67.4	71.3	103.4	100.0	100.5	84.8	63.9	66.8
03200	CONCRETE REINFORCEMENT	118.5	79.0	96.1	156.3	99.4	124.1	111.0	99.2	104.3	147.5	60.3	98.0	112.0	96.7	103.3	147.5	57.9	96.7
03300	CAST-IN-PLACE CONCRETE	149.1	76.7	118.9	140.1	105.0	125.5	140.1	93.0	120.5	179.6	74.1	135.6	140.1	92.4	120.2	168.0	64.7	124.9
03	CONCRETE	125.5	73.4	99.3	135.1	100.9	117.9	121.1	97.3	109.1	164.5	69.2	116.6	120.9	96.6	108.7	154.8	63.7	109.0
04	MASONRY	146.1	74.5	101.5	144.9	100.6	117.3	146.6	106.8	121.8	147.2	71.7	100.2	146.6	106.4	121.6	146.3	70.3	98.9
05	METALS	98.0	81.2	92.0	101.0	95.1	98.9	97.4	93.8	96.1	102.1	77.4	93.2	97.6	96.5	97.2	100.7	72.3	90.5
06	WOOD & PLASTICS	92.4	66.1	78.8	132.3	99.0	115.1	106.9	102.2	104.5	91.9	66.4	78.7	101.3	100.8	101.1	81.1	63.4	72.0
07	THERMAL & MOISTURE PROTECTION	103.5	72.3	88.9	104.6	101.8	103.3	102.9	100.2	101.7	105.9	69.1	88.6	102.9	98.3	100.7	105.1	67.3	87.4
08	DOORS & WINDOWS	85.5	65.1	80.5	92.1	87.7	91.0	83.1	97.7	86.6	98.4	63.8	90.0	84.0	95.8	86.9	94.4	55.9	85.0
09200	PLASTER & GYPSUM BOARD	120.6	65.0	86.2	143.2	99.0	115.9	144.5	102.4	118.4	191.7	65.4	113.5	141.7	100.9	116.5	186.1	62.3	109.4
09500	CEILINGS	99.7	65.0	77.0	102.5	99.0	100.2	99.7	102.4	101.5	105.2	65.4	79.1	99.7	100.9	100.5	102.5	62.3	76.1
09600	FLOORING	120.0	71.2	108.1	132.2	112.8	127.4	126.3	111.9	122.8	115.1	62.9	102.3	123.5	110.3	120.3	110.4	70.0	100.5
09900	PAINTS & COATINGS	109.5	64.8	83.4	109.5	106.1	107.5	109.5	114.0	112.1	109.5	69.3	86.0	109.5	105.4	107.1	109.5	48.8	74.0
09	FINISHES	111.6	67.6	89.1	118.5	102.7	110.4	116.2	104.7	110.3	121.6	66.7	93.6	114.9	103.0	108.8	118.5	63.7	90.5
10 - 14	TOTAL DIV. 10000 - 14000	100.0	74.2	94.6	100.0	97.9	99.6	100.0	84.8	96.8	100.0	74.7	94.7	100.0	110.8	102.3	100.0	73.9	94.5
15	MECHANICAL	102.6	76.2	90.7	102.6	92.0	97.8	102.6	98.9	101.0	102.6	70.9	88.3	102.6	95.7	99.5	102.6	63.8	85.1
16	ELECTRICAL	128.6	79.2	94.9	123.4	92.6	102.4	126.8	103.4	110.8	123.4	72.8	88.9	126.8	103.2	110.7	123.4	65.8	84.2
01 - 16	WEIGHTED AVERAGE	107.7	76.9	92.7	110.4	96.6	103.7	107.1	100.0	103.6	115.6	73.3	95.1	107.0	99.7	103.5	113.0	68.7	91.5

		\multicolumn{18}{c}{CANADA}
		\multicolumn{3}{c}{SYDNEY, NOVA SCOTIA}	\multicolumn{3}{c}{THUNDER BAY, ONTARIO}	\multicolumn{3}{c}{TORONTO, ONTARIO}	\multicolumn{3}{c}{TROIS RIVIERES, QUEBEC}	\multicolumn{3}{c}{VANCOUVER, B C}	\multicolumn{3}{c}{VICTORIA, B C}
DIVISION		MAT.	INST.	TOTAL	MAT.	INST.	TOTAL	MAT.	INST.	TOTAL	MAT.	INST.	TOTAL	MAT.	INST.	TOTAL	MAT.	INST.	TOTAL
02	SITE CONSTRUCTION	91.1	97.0	95.6	104.6	97.5	99.2	111.8	98.3	101.5	90.9	98.0	96.4	104.7	99.2	100.5	119.2	98.9	103.6
03100	CONCRETE FORMS & ACCESSORIES	82.8	78.7	79.3	118.9	101.7	104.1	120.1	115.4	116.1	128.8	98.9	103.1	110.6	103.5	104.5	101.4	102.7	102.5
03200	CONCRETE REINFORCEMENT	147.5	60.0	97.9	99.2	98.4	98.7	170.5	104.0	132.8	147.5	99.5	120.3	160.8	99.8	126.2	116.0	99.7	106.7
03300	CAST-IN-PLACE CONCRETE	137.4	87.8	112.5	155.4	106.4	134.9	164.5	117.5	144.9	137.5	105.1	124.0	159.6	106.9	137.6	164.8	106.6	140.5
03	CONCRETE	130.6	75.4	102.9	131.3	102.5	116.8	148.2	113.5	130.7	132.6	101.0	116.7	150.0	103.7	126.7	161.3	103.2	132.1
04	MASONRY	144.0	81.1	104.8	147.6	107.8	122.8	150.3	125.1	134.6	144.0	100.6	117.0	147.6	107.8	122.8	147.2	107.8	122.7
05	METALS	100.0	80.7	93.1	96.7	95.7	96.3	100.9	100.8	100.8	101.0	95.3	98.9	101.1	96.5	99.4	99.9	95.7	98.4
06	WOOD & PLASTICS	80.8	77.9	79.3	118.6	102.7	110.4	118.6	114.3	116.4	132.3	99.0	115.1	103.8	101.3	102.5	94.1	101.3	97.9
07	THERMAL & MOISTURE PROTECTION	103.1	79.4	92.0	103.4	99.4	101.5	104.7	112.7	108.4	104.6	101.8	103.3	104.2	103.4	103.8	105.0	95.9	100.7
08	DOORS & WINDOWS	82.6	73.1	80.3	82.1	97.7	85.9	91.3	109.2	95.6	92.1	93.1	92.3	92.1	98.9	93.7	87.9	98.6	90.5
09200	PLASTER & GYPSUM BOARD	185.6	77.3	118.6	150.2	102.8	120.9	141.9	114.8	125.1	212.7	99.0	142.3	154.3	101.5	121.6	148.8	101.5	119.5
09500	CEILINGS	102.5	77.3	86.0	98.3	102.8	101.3	102.5	114.8	110.6	102.5	99.0	100.2	102.5	101.5	101.8	102.5	101.5	101.8
09600	FLOORING	110.3	75.0	101.7	132.2	64.4	115.6	132.2	118.6	128.8	132.2	112.8	127.4	132.2	110.1	126.8	126.6	63.9	111.3
09900	PAINTS & COATINGS	109.5	73.0	88.1	109.5	108.4	108.9	112.5	124.4	119.4	109.5	106.1	107.5	109.4	120.0	115.6	109.5	96.1	101.6
09	FINISHES	117.7	77.5	97.2	119.0	96.8	107.7	119.1	117.5	118.3	128.7	102.7	115.4	121.1	106.0	113.4	119.7	95.3	107.2
10 - 14	TOTAL DIV. 10000 - 14000	100.0	76.6	95.0	100.0	86.2	97.1	100.0	117.1	103.6	100.0	97.9	99.6	100.0	109.2	102.0	100.0	109.2	102.0
15	MECHANICAL	102.6	84.5	94.4	102.6	99.9	101.4	102.6	111.9	106.8	102.6	92.0	97.8	102.6	103.6	103.1	102.6	95.3	99.3
16	ELECTRICAL	123.4	81.2	94.7	126.8	101.4	109.5	126.8	112.6	117.1	123.4	92.6	102.4	130.7	104.5	112.8	126.4	104.5	111.5
01 - 16	WEIGHTED AVERAGE	107.9	81.6	95.2	108.8	99.8	104.5	113.0	111.8	112.4	111.2	96.8	104.2	113.3	103.3	108.4	113.9	100.0	107.1

		\multicolumn{15}{c}{CANADA}
		\multicolumn{3}{c}{WHITEHORSE, YUKON}	\multicolumn{3}{c}{WINDSOR, ONTARIO}	\multicolumn{3}{c}{WINNIPEG, MANITOBA}	\multicolumn{3}{c}{YARMOUTH, NOVA SCOTIA}	\multicolumn{3}{c}{YELLOWKNIFE, NWT}
DIVISION		MAT.	INST.	TOTAL	MAT.	INST.	TOTAL	MAT.	INST.	TOTAL	MAT.	INST.	TOTAL	MAT.	INST.	TOTAL
02	SITE CONSTRUCTION	104.9	96.0	98.1	94.5	97.7	97.0	108.5	96.5	99.3	96.4	97.0	96.9	104.9	96.1	98.1
03100	CONCRETE FORMS & ACCESSORIES	96.4	65.7	70.0	118.9	102.4	104.8	119.1	81.4	86.8	82.8	78.7	79.3	96.4	65.9	70.2
03200	CONCRETE REINFORCEMENT	118.5	75.5	94.1	108.7	97.5	102.3	160.8	66.2	107.2	147.5	60.0	97.9	118.5	75.2	93.9
03300	CAST-IN-PLACE CONCRETE	152.8	74.5	120.1	143.4	107.9	128.6	146.7	80.9	119.3	178.4	77.8	136.4	152.8	74.4	120.1
03	CONCRETE	127.3	71.2	99.2	123.2	103.2	113.1	131.2	78.9	104.9	150.7	75.4	112.8	127.3	71.2	99.1
04	MASONRY	146.4	71.2	99.5	146.8	115.8	127.5	146.6	78.0	103.9	147.1	81.1	106.0	146.4	71.1	99.5
05	METALS	98.0	79.6	91.4	96.8	97.1	96.9	100.8	83.0	94.4	100.0	80.7	93.1	98.0	79.4	91.3
06	WOOD & PLASTICS	92.4	64.2	77.8	118.6	100.9	109.5	116.4	82.7	99.0	80.8	77.9	79.3	92.4	64.5	78.0
07	THERMAL & MOISTURE PROTECTION	103.5	70.1	87.9	103.0	103.9	103.4	104.4	81.6	93.7	103.1	79.4	92.0	103.5	70.0	87.8
08	DOORS & WINDOWS	85.5	62.7	80.0	81.8	97.5	85.6	92.1	73.7	87.6	82.6	73.1	80.3	85.5	62.8	80.0
09200	PLASTER & GYPSUM BOARD	120.6	63.2	85.1	150.2	101.0	119.7	133.6	82.2	101.8	185.6	77.3	118.6	120.6	63.4	85.2
09500	CEILINGS	99.7	63.2	75.7	98.3	101.0	100.1	102.5	82.2	89.2	102.5	77.3	86.0	99.7	63.4	75.9
09600	FLOORING	120.0	68.1	107.4	132.2	112.7	127.4	132.2	78.3	119.0	110.3	75.0	101.7	120.0	67.7	107.3
09900	PAINTS & COATINGS	109.5	63.5	82.6	109.5	107.8	108.5	109.6	68.1	85.3	109.5	73.0	88.1	109.5	63.2	82.4
09	FINISHES	111.6	65.5	88.1	118.7	105.3	111.8	118.0	79.8	98.5	117.7	77.5	97.2	111.6	65.6	88.1
10 - 14	TOTAL DIV. 10000 - 14000	100.0	72.9	94.3	100.0	87.7	97.4	100.0	77.4	95.2	100.0	76.6	95.0	100.0	73.9	94.5
15	MECHANICAL	102.6	74.9	90.1	102.6	103.2	102.9	102.6	84.5	94.4	102.6	84.5	94.4	102.6	74.7	90.0
16	ELECTRICAL	128.6	78.1	94.2	126.8	103.7	111.0	128.6	88.6	101.3	123.4	81.2	94.7	128.6	77.8	94.0
01 - 16	WEIGHTED AVERAGE	107.9	75.2	92.1	107.4	103.0	105.3	110.7	83.6	97.6	110.7	81.6	96.6	107.9	75.1	92.0

Location Factors

Costs shown in *Means cost data publications* are based on National Averages for materials and installation. To adjust these costs to a specific location, simply multiply the base cost by the factor and divide by 100 for that city. The data is arranged alphabetically by state and postal zip code numbers. For a city not listed, use the factor for a nearby city with similar economic characteristics.

STATE/ZIP	CITY	MAT.	INST.	TOTAL
ALABAMA				
350-352	Birmingham	96.7	77.7	87.5
354	Tuscaloosa	96.4	62.2	79.8
355	Jasper	97.6	52.8	75.9
356	Decatur	96.4	67.5	82.4
357-358	Huntsville	96.3	70.1	83.6
359	Gadsden	97.2	66.3	82.2
360-361	Montgomery	97.2	60.6	79.5
362	Anniston	95.7	53.5	75.2
363	Dothan	96.9	59.0	78.6
364	Evergreen	96.2	60.6	78.9
365-366	Mobile	97.2	69.4	83.7
367	Selma	96.5	58.9	78.3
368	Phenix City	97.3	60.4	79.4
369	Butler	96.6	58.9	78.3
ALASKA				
995-996	Anchorage	134.4	115.4	125.2
997	Fairbanks	130.2	118.5	124.5
998	Juneau	132.9	115.6	124.5
999	Ketchikan	144.0	115.4	130.1
ARIZONA				
850,853	Phoenix	100.5	77.8	89.5
852	Mesa/Tempe	100.4	68.2	84.8
855	Globe	101.3	72.1	87.1
856-857	Tucson	99.2	75.4	87.7
859	Show Low	101.5	73.2	87.8
860	Flagstaff	103.0	75.9	89.8
863	Prescott	100.3	72.3	86.7
864	Kingman	98.9	73.8	86.7
865	Chambers	98.9	73.2	86.4
ARKANSAS				
716	Pine Bluff	95.7	63.8	80.3
717	Camden	94.2	46.8	71.2
718	Texarkana	94.9	52.6	74.4
719	Hot Springs	93.3	46.1	70.5
720-722	Little Rock	96.2	64.4	80.8
723	West Memphis	95.5	63.8	80.2
724	Jonesboro	95.5	63.8	80.1
725	Batesville	94.2	58.2	76.7
726	Harrison	95.7	58.2	77.5
727	Fayetteville	92.7	40.1	67.2
728	Russellville	94.4	56.1	75.8
729	Fort Smith	96.3	62.0	79.7
CALIFORNIA				
900-902	Los Angeles	104.5	114.9	109.6
903-905	Inglewood	100.8	114.0	107.2
906-908	Long Beach	102.6	114.0	108.1
910-912	Pasadena	100.7	114.0	107.1
913-916	Van Nuys	104.6	113.9	109.1
917-918	Alhambra	103.3	114.2	108.6
919-921	San Diego	105.0	108.2	106.6
922	Palm Springs	102.5	110.8	106.5
923-924	San Bernardino	100.0	112.0	105.8
925	Riverside	104.5	111.3	107.8
926-927	Santa Ana	102.3	111.5	106.8
928	Anaheim	104.8	113.5	109.0
930	Oxnard	105.3	113.5	109.3
931	Santa Barbara	104.5	112.6	108.4
932-933	Bakersfield	104.3	106.9	105.5
934	San Luis Obispo	106.3	112.0	109.1
935	Mojave	102.9	108.7	105.7
936-938	Fresno	105.3	111.7	108.4
939	Salinas	107.3	117.9	112.4
940-941	San Francisco	111.8	136.6	123.8
942,956-958	Sacramento	107.4	113.5	110.4
943	Palo Alto	105.5	129.2	117.0
944	San Mateo	108.6	128.8	118.4
945	Vallejo	106.1	125.2	115.3
946	Oakland	110.5	127.4	118.7
947	Berkeley	110.1	127.3	118.5
948	Richmond	109.9	125.2	117.3
949	San Rafael	111.9	125.1	118.3
950	Santa Cruz	110.9	118.0	114.4

STATE/ZIP	CITY	MAT.	INST.	TOTAL
CALIFORNIA (CONT'D)				
951	San Jose	110.1	129.1	119.3
952	Stockton	105.8	114.1	109.8
953	Modesto	106.0	114.9	110.3
954	Santa Rosa	107.3	127.9	117.3
955	Eureka	108.6	109.8	109.2
959	Marysville	107.4	112.3	109.8
960	Redding	108.4	112.0	110.2
961	Susanville	108.6	112.0	110.2
COLORADO				
800-802	Denver	102.1	85.9	94.2
803	Boulder	100.2	69.5	85.3
804	Golden	102.7	80.5	91.9
805	Fort Collins	103.4	79.5	91.8
806	Greeley	100.4	69.5	85.4
807	Fort Morgan	100.9	80.5	91.0
808-809	Colorado Springs	100.8	81.6	91.5
810	Pueblo	102.8	79.6	91.6
811	Alamosa	105.1	70.0	88.1
812	Salida	105.1	70.1	88.1
813	Durango	105.7	65.9	86.4
814	Montrose	104.0	63.3	84.3
815	Grand Junction	107.1	63.6	86.0
816	Glenwood Springs	105.1	75.4	90.7
CONNECTICUT				
060	New Britain	102.2	105.2	103.7
061	Hartford	102.5	105.8	104.1
062	Willimantic	103.0	104.8	103.8
063	New London	99.5	105.2	102.3
064	Meriden	102.2	105.5	103.8
065	New Haven	102.4	105.4	103.9
066	Bridgeport	103.6	104.2	103.9
067	Waterbury	103.0	105.8	104.3
068	Norwalk	102.9	102.8	102.9
069	Stamford	103.1	111.3	107.0
D.C.				
200-205	Washington	99.4	91.9	95.7
DELAWARE				
197	Newark	99.4	100.1	99.7
198	Wilmington	98.7	100.2	99.4
199	Dover	99.3	100.1	99.7
FLORIDA				
320,322	Jacksonville	98.7	67.5	83.6
321	Daytona Beach	98.8	74.6	87.1
323	Tallahassee	99.3	57.4	79.0
324	Panama City	99.8	44.6	73.1
325	Pensacola	99.4	67.0	83.7
326,344	Gainesville	100.3	63.5	82.4
327-328,347	Orlando	100.7	69.4	85.5
329	Melbourne	100.8	73.7	87.7
330-332,340	Miami	98.2	73.1	86.0
333	Fort Lauderdale	98.0	74.4	86.6
334,349	West Palm Beach	97.0	67.8	82.8
335-336,346	Tampa	100.0	66.2	83.6
337	St. Petersburg	101.8	65.9	84.4
338	Lakeland	98.8	65.9	82.8
339,341	Fort Myers	98.4	62.8	81.2
342	Sarasota	100.2	63.2	82.3
GEORGIA				
300-303,399	Atlanta	96.8	80.9	89.1
304	Statesboro	97.1	36.5	67.7
305	Gainesville	96.0	51.3	74.4
306	Athens	95.1	66.9	81.4
307	Dalton	97.0	37.5	68.1
308-309	Augusta	95.7	62.2	79.5
310-312	Macon	95.9	66.9	81.8
313-314	Savannah	98.0	65.6	82.3
315	Waycross	98.0	50.9	75.2
316	Valdosta	97.5	55.0	76.9
317	Albany	97.5	61.1	79.9
318-319	Columbus	97.7	59.8	79.3

643

Location Factors

STATE/ZIP	CITY	MAT.	INST.	TOTAL
HAWAII				
967	Hilo	115.6	128.8	122.0
968	Honolulu	116.2	128.8	122.3
STATES & POSS.				
969	Guam	115.4	46.1	81.8
IDAHO				
832	Pocatello	102.1	85.2	93.9
833	Twin Falls	104.3	53.5	79.7
834	Idaho Falls	101.4	64.7	83.6
835	Lewiston	113.8	90.5	102.5
836-837	Boise	102.1	86.1	94.3
838	Coeur d'Alene	115.3	64.7	90.8
ILLINOIS				
600-603	North Suburban	97.3	117.8	107.3
604	Joliet	97.3	113.4	105.1
605	South Suburban	97.3	117.2	106.9
606	Chicago	97.6	124.7	110.7
609	Kankakee	93.7	105.4	99.4
610-611	Rockford	96.8	110.1	103.3
612	Rock Island	94.6	99.5	97.0
613	La Salle	96.1	100.7	98.3
614	Galesburg	95.7	100.0	97.8
615-616	Peoria	98.2	103.0	100.5
617	Bloomington	95.0	103.8	99.3
618-619	Champaign	98.7	100.6	99.6
620-622	East St. Louis	93.4	105.9	99.5
623	Quincy	95.2	95.2	95.2
624	Effingham	94.5	99.9	97.1
625	Decatur	95.9	98.6	97.2
626-627	Springfield	95.6	100.9	98.2
628	Centralia	92.7	103.5	98.0
629	Carbondale	92.4	99.0	95.6
INDIANA				
460	Anderson	96.7	88.1	92.5
461-462	Indianapolis	99.6	89.1	94.5
463-464	Gary	97.5	101.8	99.6
465-466	South Bend	95.9	84.5	90.4
467-468	Fort Wayne	96.9	86.0	91.6
469	Kokomo	95.2	85.4	90.4
470	Lawrenceburg	93.3	82.9	88.2
471	New Albany	95.0	80.1	87.7
472	Columbus	97.1	86.0	91.7
473	Muncie	97.3	86.6	92.1
474	Bloomington	99.1	84.5	92.0
475	Washington	95.4	90.3	92.9
476-477	Evansville	96.9	91.2	94.2
478	Terre Haute	97.3	89.6	93.6
479	Lafayette	96.9	83.6	90.4
IOWA				
500-503,509	Des Moines	97.8	84.9	91.5
504	Mason City	96.2	63.6	80.4
505	Fort Dodge	96.5	58.0	77.8
506-507	Waterloo	97.6	66.6	82.6
508	Creston	97.0	72.1	85.0
510-511	Sioux City	98.4	67.4	83.4
512	Sibley	97.1	57.6	77.9
513	Spencer	99.3	57.5	79.1
514	Carroll	96.0	64.6	80.8
515	Council Bluffs	99.1	76.0	87.9
516	Shenandoah	96.0	55.8	76.5
520	Dubuque	97.7	74.8	86.6
521	Decorah	97.1	62.5	80.3
522-524	Cedar Rapids	99.0	84.1	91.8
525	Ottumwa	97.2	73.4	85.7
526	Burlington	96.0	75.7	86.2
527-528	Davenport	98.0	89.6	93.9
KANSAS				
660-662	Kansas City	97.8	91.1	94.5
664-666	Topeka	97.3	72.0	85.1
667	Fort Scott	96.9	70.2	83.9
668	Emporia	96.5	64.0	80.8
669	Belleville	98.6	64.9	82.2
670-672	Wichita	96.5	72.9	85.1
673	Independence	98.3	59.7	79.6
674	Salina	98.0	63.8	81.4
675	Hutchinson	93.6	58.6	76.6
676	Hays	97.8	64.9	81.8
677	Colby	98.6	64.9	82.3

STATE/ZIP	CITY	MAT.	INST.	TOTAL
KANSAS (CONT'D)				
678	Dodge City	99.3	64.9	82.6
679	Liberal	97.5	53.9	76.4
KENTUCKY				
400-402	Louisville	96.0	85.7	91.0
403-405	Lexington	96.7	72.8	85.2
406	Frankfort	96.0	78.3	87.4
407-409	Corbin	94.5	51.7	73.8
410	Covington	95.3	89.9	92.7
411-412	Ashland	93.4	99.3	96.3
413-414	Campton	95.4	51.7	74.2
415-416	Pikeville	96.2	68.7	82.9
417-418	Hazard	94.7	51.7	73.9
420	Paducah	93.4	87.4	90.5
421-422	Bowling Green	95.4	82.4	89.1
423	Owensboro	95.3	82.7	89.2
424	Henderson	92.9	91.2	92.1
425-426	Somerset	92.4	51.7	72.7
427	Elizabethtown	92.1	83.7	88.0
LOUISIANA				
700-701	New Orleans	100.5	69.2	85.3
703	Thibodaux	100.8	67.6	84.7
704	Hammond	97.7	66.1	82.4
705	Lafayette	99.9	63.0	82.0
706	Lake Charles	100.0	65.7	83.4
707-708	Baton Rouge	99.7	63.6	82.2
710-711	Shreveport	96.8	63.9	80.9
712	Monroe	96.7	60.0	78.9
713-714	Alexandria	96.8	58.0	78.0
MAINE				
039	Kittery	97.7	60.2	79.5
040-041	Portland	100.5	77.1	89.2
042	Lewiston	101.0	77.1	89.4
043	Augusta	99.8	55.8	78.5
044	Bangor	100.0	82.9	91.7
045	Bath	99.0	55.0	77.7
046	Machias	98.6	65.9	82.7
047	Houlton	98.8	59.9	79.9
048	Rockland	97.7	66.7	82.7
049	Waterville	99.1	55.8	78.1
MARYLAND				
206	Waldorf	97.5	79.2	88.7
207-208	College Park	97.3	82.7	90.3
209	Silver Spring	96.7	82.2	89.7
210-212	Baltimore	97.0	84.6	91.0
214	Annapolis	96.7	81.5	89.3
215	Cumberland	93.9	82.8	88.5
216	Easton	95.5	52.4	74.6
217	Hagerstown	94.2	82.0	88.3
218	Salisbury	96.0	59.9	78.5
219	Elkton	93.0	76.1	84.8
MASSACHUSETTS				
010-011	Springfield	101.1	104.0	102.5
012	Pittsfield	100.8	96.6	98.8
013	Greenfield	99.0	101.6	100.3
014	Fitchburg	97.4	114.1	105.5
015-016	Worcester	100.9	113.8	107.2
017	Framingham	97.0	120.1	108.2
018	Lowell	100.6	119.8	109.9
019	Lawrence	101.8	117.8	109.5
020-022, 024	Boston	103.0	131.1	116.6
023	Brockton	101.8	117.2	109.3
025	Buzzards Bay	96.2	115.6	105.6
026	Hyannis	98.5	115.6	106.8
027	New Bedford	101.4	116.1	108.5
MICHIGAN				
480,483	Royal Oak	94.8	108.8	101.6
481	Ann Arbor	97.2	111.0	103.9
482	Detroit	98.2	113.1	105.4
484-485	Flint	96.8	101.3	98.9
486	Saginaw	97.0	96.4	96.7
487	Bay City	96.6	96.5	96.5
488-489	Lansing	97.0	100.4	98.6
490	Battle Creek	96.6	89.1	93.0
491	Kalamazoo	97.0	90.3	93.7
492	Jackson	95.4	98.3	96.8
493,495	Grand Rapids	96.5	75.9	86.5
494	Muskegan	96.7	89.0	93.0

Location Factors

STATE/ZIP	CITY	MAT.	INST.	TOTAL
MICHIGAN (CONT'D)				
496	Traverse City	94.7	77.2	86.2
497	Gaylord	96.2	83.5	90.0
498-499	Iron mountain	98.2	92.0	95.2
MINNESOTA				
550-551	Saint Paul	98.0	120.0	108.6
553-555	Minneapolis	98.6	122.8	110.3
556-558	Duluth	98.1	111.4	104.5
559	Rochester	97.8	104.0	100.8
560	Mankato	96.6	103.1	99.8
561	Windom	95.5	82.0	89.0
562	Willmar	94.7	88.2	91.5
563	St. Cloud	95.6	108.0	101.6
564	Brainerd	96.3	101.5	98.8
565	Detroit Lakes	99.0	90.4	94.8
566	Bemidji	98.3	97.8	98.0
567	Thief River Falls	97.3	89.6	93.6
MISSISSIPPI				
386	Clarksdale	97.1	35.8	67.4
387	Greenville	100.4	54.9	78.3
388	Tupelo	98.4	45.6	72.8
389	Greenwood	98.5	38.3	69.3
390-392	Jackson	98.8	56.7	78.4
393	Meridian	96.8	54.4	76.2
394	Laurel	98.3	39.8	70.0
395	Biloxi	98.9	63.3	81.7
396	Mccomb	96.7	36.9	67.7
397	Columbus	98.3	45.2	72.5
MISSOURI				
630-631	St. Louis	96.4	110.4	103.2
633	Bowling Green	96.3	92.5	94.4
634	Hannibal	95.1	91.6	93.4
635	Kirksville	97.2	80.0	88.8
636	Flat River	97.4	96.1	96.8
637	Cape Girardeau	96.4	94.7	95.6
638	Sikeston	94.9	87.0	91.1
639	Poplar Bluff	94.4	86.7	90.7
640-641	Kansas City	98.9	97.9	98.4
644-645	St. Joseph	98.9	81.6	90.5
646	Chillicothe	96.0	70.2	83.5
647	Harrisonville	95.6	87.0	91.4
648	Joplin	97.7	71.6	85.1
650-651	Jefferson City	94.9	85.2	90.2
652	Columbia	96.2	79.8	88.2
653	Sedalia	95.6	80.1	88.1
654-655	Rolla	94.5	67.8	81.6
656-658	Springfield	97.6	74.0	86.2
MONTANA				
590-591	Billings	100.8	90.2	95.6
592	Wolf Point	101.2	90.2	95.9
593	Miles City	99.2	91.0	95.3
594	Great Falls	102.0	89.9	96.1
595	Havre	99.9	90.5	95.3
596	Helena	101.8	88.3	95.3
597	Butte	100.5	84.8	92.9
598	Missoula	98.8	87.8	93.5
599	Kalispell	98.0	87.1	92.7
NEBRASKA				
680-681	Omaha	100.3	79.4	90.2
683-685	Lincoln	97.7	67.8	83.2
686	Columbus	96.3	49.7	73.7
687	Norfolk	98.8	68.2	83.9
688	Grand Island	97.2	69.5	83.8
689	Hastings	96.9	60.0	79.0
690	Mccook	96.7	49.3	73.7
691	North Platte	96.7	63.2	80.5
692	Valentine	100.6	47.2	74.7
693	Alliance	99.7	42.9	72.1
NEVADA				
889-891	Las Vegas	101.5	108.3	104.8
893	Ely	102.3	90.5	96.6
894-895	Reno	102.2	97.3	99.9
897	Carson City	102.0	96.9	99.5
898	Elko	100.9	90.0	95.6
NEW HAMPSHIRE				
030	Nashua	101.5	85.1	93.6
031	Manchester	101.5	85.1	93.6

STATE/ZIP	CITY	MAT.	INST.	TOTAL
NEW HAMPSHIRE (CONT'D)				
032-033	Concord	99.4	85.9	92.8
034	Keene	98.6	60.6	80.2
035	Littleton	98.6	64.0	81.8
036	Charleston	98.0	57.0	78.2
037	Claremont	97.1	57.0	77.7
038	Portsmouth	98.6	82.9	91.0
NEW JERSEY				
070-071	Newark	103.0	122.3	112.4
072	Elizabeth	101.9	113.0	107.3
073	Jersey City	100.5	121.2	110.6
074-075	Paterson	103.3	121.6	112.2
076	Hackensack	100.4	121.0	110.4
077	Long Branch	100.0	118.4	108.9
078	Dover	100.8	119.7	109.9
079	Summit	100.8	111.7	106.1
080,083	Vineland	98.3	117.2	107.5
081	Camden	100.4	117.7	108.8
082,084	Atlantic City	99.3	118.3	108.5
085-086	Trenton	101.2	120.7	110.7
087	Point Pleasant	100.7	118.6	109.4
088-089	New Brunswick	101.2	118.7	109.7
NEW MEXICO				
870-872	Albuquerque	100.7	81.1	91.2
873	Gallup	102.1	79.6	91.2
874	Farmington	102.0	79.3	91.0
875	Santa Fe	100.3	81.1	91.0
877	Las Vegas	100.3	79.3	90.1
878	Socorro	99.8	78.5	89.5
879	Truth/Consequences	99.7	72.7	86.6
880	Las Cruces	98.0	72.2	85.5
881	Clovis	100.4	79.8	90.4
882	Roswell	101.9	77.4	90.0
883	Carrizozo	102.8	78.9	91.3
884	Tucumcari	101.4	77.5	89.8
NEW YORK				
100-102	New York	108.6	160.5	133.8
103	Staten Island	104.9	152.4	127.9
104	Bronx	102.1	155.0	127.7
105	Mount Vernon	102.9	133.8	117.9
106	White Plains	102.7	131.9	116.9
107	Yonkers	107.7	133.8	120.3
108	New Rochelle	103.6	132.8	117.7
109	Suffern	103.6	120.3	111.7
110	Queens	103.0	155.9	128.6
111	Long Island City	105.0	155.9	129.6
112	Brooklyn	105.2	155.9	129.8
113	Flushing	105.7	155.9	130.0
114	Jamaica	103.6	155.9	129.0
115,117,118	Hicksville	102.9	151.8	126.6
116	Far Rockaway	105.8	155.9	130.1
119	Riverhead	103.4	151.2	126.6
120-122	Albany	98.1	97.1	97.6
123	Schenectady	99.0	96.9	98.0
124	Kingston	103.3	115.2	109.1
125-126	Poughkeepsie	102.5	121.4	111.6
127	Monticello	101.6	114.0	107.6
128	Glens Falls	93.3	93.4	93.3
129	Plattsburgh	99.0	88.0	93.7
130-132	Syracuse	99.9	94.2	97.1
133-135	Utica	98.2	89.2	93.8
136	Watertown	99.6	93.7	96.7
137-139	Binghamton	99.0	88.7	94.1
140-142	Buffalo	99.3	106.3	102.7
143	Niagara Falls	97.6	106.0	101.7
144-146	Rochester	100.6	99.7	100.1
147	Jamestown	96.8	90.3	93.7
148-149	Elmira	96.2	89.3	92.9
NORTH CAROLINA				
270,272-274	Greensboro	96.8	56.0	77.0
271	Winston-Salem	96.9	55.5	76.9
275-276	Raleigh	97.3	56.0	77.3
277	Durham	96.7	56.0	77.0
278	Rocky Mount	95.3	41.8	69.4
279	Elizabeth City	95.9	44.5	71.0
280	Gastonia	96.9	53.7	76.0
281-282	Charlotte	97.1	54.0	76.2
283	Fayetteville	95.2	56.0	76.2
284	Wilmington	95.4	55.0	75.8
285	Kinston	93.9	41.3	68.4

645

Location Factors

STATE/ZIP	CITY	MAT.	INST.	TOTAL
NORTH CAROLINA (CONT'D)				
286	Hickory	94.1	40.4	68.1
287-288	Asheville	95.9	54.8	76.0
289	Murphy	95.5	39.4	68.3
NORTH DAKOTA				
580-581	Fargo	100.2	63.8	82.6
582	Grand Forks	101.3	62.4	82.4
583	Devils Lake	101.3	62.4	82.5
584	Jamestown	101.1	62.6	82.5
585	Bismarck	100.2	68.5	84.8
586	Dickinson	101.9	62.4	82.7
587	Minot	101.3	67.0	84.7
588	Williston	100.5	62.4	82.0
OHIO				
430-432	Columbus	97.4	91.1	94.4
433	Marion	94.8	88.3	91.6
434-436	Toledo	97.9	97.5	97.7
437-438	Zanesville	94.9	85.8	90.5
439	Steubenville	96.6	97.4	97.0
440	Lorain	98.6	95.4	97.0
441	Cleveland	98.6	106.4	102.4
442-443	Akron	99.8	99.0	99.4
444-445	Youngstown	99.0	93.7	96.4
446-447	Canton	99.1	91.1	95.2
448-449	Mansfield	96.6	90.4	93.6
450	Hamilton	95.3	89.3	92.4
451-452	Cincinnati	95.3	89.9	92.7
453-454	Dayton	95.0	88.7	91.9
455	Springfield	95.3	88.6	92.1
456	Chillicothe	94.7	95.8	95.2
457	Athens	98.1	75.6	87.2
458	Lima	98.6	88.2	93.6
OKLAHOMA				
730-731	Oklahoma City	97.8	68.2	83.5
734	Ardmore	96.0	67.3	82.1
735	Lawton	97.8	68.0	83.3
736	Clinton	97.4	65.7	82.0
737	Enid	97.5	66.0	82.2
738	Woodward	96.1	65.8	81.4
739	Guymon	97.4	38.7	69.0
740-741	Tulsa	97.8	65.7	82.3
743	Miami	94.9	69.5	82.6
744	Muskogee	96.6	49.7	73.9
745	Mcalester	94.2	60.8	78.0
746	Ponca City	94.9	65.7	80.8
747	Durant	94.9	66.1	81.0
748	Shawnee	96.6	64.4	81.0
749	Poteau	94.0	66.5	80.7
OREGON				
970-972	Portland	103.9	108.2	106.0
973	Salem	104.4	107.3	105.8
974	Eugene	103.9	106.8	105.3
975	Medford	105.5	105.0	105.3
976	Klamath Falls	106.4	105.0	105.7
977	Bend	105.3	106.5	105.9
978	Pendleton	98.2	106.5	102.2
979	Vale	95.7	98.7	97.2
PENNSYLVANIA				
150-152	Pittsburgh	97.4	107.6	102.4
153	Washington	94.9	106.6	100.6
154	Uniontown	94.9	105.3	100.0
155	Bedford	96.1	97.6	96.8
156	Greensburg	96.3	105.6	100.8
157	Indiana	95.0	102.7	98.7
158	Dubois	96.5	98.5	97.5
159	Johnstown	96.1	99.8	97.9
160	Butler	92.8	102.6	97.6
161	New Castle	92.8	102.3	97.4
162	Kittanning	93.4	107.3	100.1
163	Oil City	92.8	97.0	94.8
164-165	Erie	95.0	99.1	97.0
166	Altoona	94.7	97.7	96.2
167	Bradford	96.4	98.0	97.2
168	State College	96.1	99.2	97.6
169	Wellsboro	97.2	92.7	95.0
170-171	Harrisburg	98.2	94.4	96.4
172	Chambersburg	96.7	94.1	95.5
173-174	York	96.4	94.1	95.3
175-176	Lancaster	95.4	93.1	94.3

STATE/ZIP	CITY	MAT.	INST.	TOTAL
PENNSYLVANIA (CONT'D)				
177	Williamsport	93.7	89.3	91.6
178	Sunbury	96.0	93.4	94.8
179	Pottsville	95.1	95.0	95.0
180	Lehigh Valley	96.9	106.0	101.3
181	Allentown	98.8	103.4	101.1
182	Hazleton	96.0	97.0	96.5
183	Stroudsburg	96.1	94.1	95.1
184-185	Scranton	99.4	99.7	99.5
186-187	Wilkes-Barre	95.7	97.7	96.7
188	Montrose	95.3	98.7	96.9
189	Doylestown	95.5	115.6	105.2
190-191	Philadelphia	99.7	124.8	111.9
193	Westchester	96.8	113.1	104.7
194	Norristown	95.6	118.3	106.6
195-196	Reading	98.1	97.9	98.0
RHODE ISLAND				
028	Newport	100.2	107.8	103.9
029	Providence	100.5	107.8	104.0
SOUTH CAROLINA				
290-292	Columbia	96.0	53.1	75.2
293	Spartanburg	94.8	53.3	74.7
294	Charleston	96.0	54.6	75.9
295	Florence	94.3	50.8	73.2
296	Greenville	94.5	53.3	74.5
297	Rock Hill	94.6	38.8	67.6
298	Aiken	95.6	39.4	68.4
299	Beaufort	96.3	43.4	70.7
SOUTH DAKOTA				
570-571	Sioux Falls	100.5	62.3	82.0
572	Watertown	99.2	57.7	79.1
573	Mitchell	98.2	57.7	78.6
574	Aberdeen	100.5	57.7	79.8
575	Pierre	99.9	57.9	79.5
576	Mobridge	98.9	57.7	79.0
577	Rapid City	100.7	56.2	79.1
TENNESSEE				
370-372	Nashville	96.9	71.0	84.4
373-374	Chattanooga	98.8	63.6	81.7
375,380-381	Memphis	96.8	72.3	84.9
376	Johnson City	98.0	59.8	79.5
377-379	Knoxville	94.9	64.5	80.2
382	Mckenzie	97.3	39.7	69.4
383	Jackson	98.8	50.3	75.3
384	Columbia	95.6	54.6	75.7
385	Cookeville	97.0	38.4	68.6
TEXAS				
750	Mckinney	99.3	65.4	82.8
751	Waxahackie	99.4	65.6	83.0
752-753	Dallas	100.1	72.6	86.8
754	Greenville	99.6	47.9	74.5
755	Texarkana	98.6	58.6	79.2
756	Longview	99.2	49.2	74.9
757	Tyler	99.7	60.6	80.8
758	Palestine	95.8	52.3	74.7
759	Lufkin	97.0	59.0	78.6
760-761	Fort Worth	97.9	68.5	83.6
762	Denton	97.9	60.6	79.8
763	Wichita Falls	98.4	63.8	81.6
764	Eastland	97.3	50.8	74.8
765	Temple	95.8	57.9	77.4
766-767	Waco	97.6	63.7	81.2
768	Brownwood	98.1	47.1	73.4
769	San Angelo	97.7	54.4	76.7
770-772	Houston	100.0	78.1	89.4
773	Huntsville	98.6	50.3	75.2
774	Wharton	100.3	54.7	78.2
775	Galveston	98.2	74.8	86.9
776-777	Beaumont	97.5	71.8	85.1
778	Bryan	94.9	69.3	82.5
779	Victoria	100.4	60.0	80.8
780	Laredo	95.3	62.6	79.5
781-782	San Antonio	95.3	71.9	84.0
783-784	Corpus Christi	98.6	60.5	80.1
785	Mc Allen	99.0	56.2	78.3
786-787	Austin	95.5	67.3	81.9
788	Del Rio	98.4	39.3	69.7
789	Giddings	95.4	51.1	74.0
790-791	Amarillo	98.3	62.0	80.7

Location Factors

STATE/ZIP	CITY	MAT.	INST.	TOTAL
TEXAS (CONT'D)				
792	Childress	97.7	59.5	79.2
793-794	Lubbock	100.3	62.7	82.1
795-796	Abilene	97.7	58.1	78.5
797	Midland	101.0	60.1	81.1
798-799,885	El Paso	97.7	55.8	77.4
UTAH				
840-841	Salt Lake City	101.2	76.5	89.2
842,844	Ogden	99.7	76.0	88.2
843	Logan	102.0	76.0	89.4
845	Price	102.5	58.6	81.2
846-847	Provo	102.2	75.9	89.5
VERMONT				
050	White River Jct.	100.0	42.8	72.3
051	Bellows Falls	98.5	45.3	72.7
052	Bennington	98.8	37.9	69.3
053	Brattleboro	99.3	45.3	73.1
054	Burlington	101.3	64.9	83.7
056	Montpelier	98.5	64.8	82.2
057	Rutland	100.5	64.9	83.2
058	St. Johnsbury	100.2	48.6	75.2
059	Guildhall	98.7	48.0	74.1
VIRGINIA				
220-221	Fairfax	97.8	81.7	90.0
222	Arlington	98.5	81.9	90.5
223	Alexandria	97.5	84.0	90.9
224-225	Fredericksburg	96.3	74.0	85.5
226	Winchester	97.0	61.4	79.8
227	Culpeper	96.9	63.1	80.5
228	Harrisonburg	97.2	55.4	76.9
229	Charlottesville	97.4	66.4	82.4
230-232	Richmond	97.7	71.0	84.7
233-235	Norfolk	97.9	66.0	82.5
236	Newport News	97.4	64.7	81.5
237	Portsmouth	96.5	64.7	81.1
238	Petersburg	97.2	71.0	84.5
239	Farmville	96.6	49.4	73.7
240-241	Roanoke	97.5	54.4	76.6
242	Bristol	96.3	53.2	75.4
243	Pulaski	96.0	42.0	69.8
244	Staunton	96.8	53.6	75.8
245	Lynchburg	96.8	56.9	77.5
246	Grundy	96.3	42.0	70.0
WASHINGTON				
980-981,987	Seattle	105.2	106.2	105.7
982	Everett	105.4	103.3	104.4
983-984	Tacoma	105.0	103.4	104.2
985	Olympia	104.9	102.7	103.8
986	Vancouver	107.5	101.4	104.5
988	Wenatchee	106.3	89.9	98.3
989	Yakima	105.5	98.8	102.2
990-992	Spokane	108.6	90.2	99.6
993	Richland	108.7	92.7	100.9
994	Clarkston	107.9	91.4	99.9
WEST VIRGINIA				
247-248	Bluefield	95.1	81.4	88.4
249	Lewisburg	96.6	83.4	90.2
250-253	Charleston	97.4	91.1	94.3
254	Martinsburg	96.6	57.0	77.4
255-257	Huntington	97.6	95.7	96.7
258-259	Beckley	94.8	88.7	91.8
260	Wheeling	97.6	95.1	96.4
261	Parkersburg	96.6	95.3	95.9
262	Buckhannon	96.5	92.1	94.3
263-264	Clarksburg	96.9	92.5	94.8
265	Morgantown	96.9	92.7	94.9
266	Gassaway	96.2	91.8	94.1
267	Romney	96.3	86.0	91.3
268	Petersburg	96.3	88.4	92.4
WISCONSIN				
530,532	Milwaukee	99.4	102.3	100.8
531	Kenosha	99.6	101.4	100.5
534	Racine	98.8	101.5	100.1
535	Beloit	98.7	97.7	98.2
537	Madison	98.8	98.3	98.6
538	Lancaster	97.0	84.3	90.8
539	Portage	95.4	97.7	96.5
540	New Richmond	96.2	92.5	94.4

STATE/ZIP	CITY	MAT.	INST.	TOTAL
WISCONSIN (CONT'D)				
541-543	Green Bay	99.7	93.6	96.7
544	Wausau	95.5	90.5	93.1
545	Rhinelander	98.7	87.7	93.4
546	La Crosse	96.0	93.0	94.6
547	Eau Claire	98.0	93.1	95.6
548	Superior	96.1	97.5	96.8
549	Oshkosh	96.0	91.2	93.7
WYOMING				
820	Cheyenne	100.7	63.7	82.8
821	Yellowstone Nat'l Park	100.1	57.3	79.4
822	Wheatland	101.8	56.3	79.8
823	Rawlins	103.4	52.3	78.6
824	Worland	100.9	52.4	77.4
825	Riverton	102.1	56.0	79.8
826	Casper	100.5	64.5	83.1
827	Newcastle	100.7	52.3	77.2
828	Sheridan	101.8	61.4	82.2
829-831	Rock Springs	105.2	52.4	79.6
CANADIAN FACTORS (reflect Canadian currency)				
ALBERTA				
	Calgary	111.6	83.1	97.8
	Edmonton	111.6	83.0	97.7
	Fort McMurray	110.3	83.0	97.1
	Lethbridge	110.4	83.1	97.1
	Lloydminster	110.3	83.0	97.1
	Medicine Hat	110.4	83.1	97.1
	Red Deer	110.1	83.1	97.0
BRITISH COLUMBIA				
	Kamloops	110.4	100.0	105.4
	Prince George	113.9	100.0	107.1
	Vancouver	113.3	103.3	108.4
	Victoria	113.9	100.0	107.1
MANITOBA				
	Brandon	111.7	83.6	98.1
	Portage la Prairie	111.0	83.6	97.7
	Winnipeg	110.7	83.6	97.6
NEW BRUNSWICK				
	Bathurst	109.6	73.9	92.3
	Dalhousie	109.6	73.9	92.3
	Fredericton	110.4	79.2	95.3
	Moncton	109.3	73.9	92.1
	Newcastle	109.6	73.9	92.3
	Saint John	110.7	79.2	95.4
NEWFOUNDLAND				
	Corner Brook	115.6	73.3	95.1
	St. John's	115.6	73.3	95.1
NORTHWEST TERRITORIES				
	Yellowknife	107.9	75.1	92.0
NOVA SCOTIA				
	Dartmouth	110.7	81.6	96.6
	Halifax	110.7	83.1	97.3
	New Glasgow	110.7	81.6	96.6
	Sydney	107.9	81.6	95.2
	Yarmouth	110.7	81.6	96.6
ONTARIO				
	Barrie	115.6	100.8	108.4
	Brantford	113.3	106.7	110.1
	Cornwall	113.9	101.9	108.1
	Hamilton	113.6	105.6	109.7
	Kingston	113.9	101.7	108.0
	Kitchener	108.2	100.1	104.3
	London	112.6	102.4	107.6
	North Bay	113.3	99.7	106.7
	Oshawa	113.4	103.1	108.4
	Ottawa	114.4	102.6	108.7
	Owen Sound	115.7	99.3	107.7
	Peterborough	113.3	101.7	107.7
	Sarnia	112.5	107.5	110.1
	St. Catharines	107.1	100.0	103.6
	Sudbury	107.0	99.7	103.5
	Thunder Bay	108.8	99.8	104.5
	Toronto	113.0	111.8	112.4
	Windsor	107.4	103.0	105.3

Location Factors

STATE/ZIP	CITY	MAT.	INST.	TOTAL
PRINCE EDWARD ISLAND				
	Charlottetown	112.9	68.7	91.5
	Summerside	113.0	68.7	91.5
QUEBEC				
	Cap-de-la-Madeleine	111.2	96.8	104.2
	Charlesbourg	111.2	96.8	104.2
	Chicoutimi	109.8	96.5	103.3
	Gatineau	108.3	96.6	102.6
	Laval	110.4	96.6	103.7
	Montreal	110.5	96.6	103.7
	Quebec	112.1	96.8	104.7
	Sherbrooke	110.4	96.6	103.7
	Trois Rivieres	111.2	96.8	104.2
SASKATCHEWAN				
	Moose Jaw	108.1	77.0	93.0
	Prince Albert	107.4	76.8	92.6
	Regina	108.8	77.0	93.4
	Saskatoon	107.7	76.9	92.7
YUKON				
	Whitehorse	107.9	75.2	92.1

Abbreviations

A	Area Square Feet; Ampere	Cab.	Cabinet	d.f.u.	Drainage Fixture Units
ABS	Acrylonitrile Butadiene Stryrene; Asbestos Bonded Steel	Cair.	Air Tool Laborer	D.H.	Double Hung
		Calc	Calculated	DHW	Domestic Hot Water
A.C.	Alternating Current; Air-Conditioning; Asbestos Cement; Plywood Grade A & C	Cap.	Capacity	Diag.	Diagonal
		Carp.	Carpenter	Diam.	Diameter
		C.B.	Circuit Breaker	Distrib.	Distribution
		C.C.A.	Chromate Copper Arsenate	Dk.	Deck
A.C.I.	American Concrete Institute	C.C.F.	Hundred Cubic Feet	D.L.	Dead Load; Diesel
AD	Plywood, Grade A & D	cd	Candela	DLH	Deep Long Span Bar Joist
Addit.	Additional	cd/sf	Candela per Square Foot	Do.	Ditto
Adj.	Adjustable	CD	Grade of Plywood Face & Back	Dp.	Depth
af	Audio-frequency	CDX	Plywood, Grade C & D, exterior glue	D.P.S.T.	Double Pole, Single Throw
A.G.A.	American Gas Association			Dr.	Driver
Agg.	Aggregate	Cefi.	Cement Finisher	Drink.	Drinking
A.H.	Ampere Hours	Cem.	Cement	D.S.	Double Strength
A hr.	Ampere-hour	CF	Hundred Feet	D.S.A.	Double Strength A Grade
A.H.U.	Air Handling Unit	C.F.	Cubic Feet	D.S.B.	Double Strength B Grade
A.I.A.	American Institute of Architects	CFM	Cubic Feet per Minute	Dty.	Duty
AIC	Ampere Interrupting Capacity	c.g.	Center of Gravity	DWV	Drain Waste Vent
Allow.	Allowance	CHW	Chilled Water; Commercial Hot Water	DX	Deluxe White, Direct Expansion
alt.	Altitude			dyn	Dyne
Alum.	Aluminum	C.I.	Cast Iron	e	Eccentricity
a.m.	Ante Meridiem	C.I.P.	Cast in Place	E	Equipment Only; East
Amp.	Ampere	Circ.	Circuit	Ea.	Each
Anod.	Anodized	C.L.	Carload Lot	E.B.	Encased Burial
Approx.	Approximate	Clab.	Common Laborer	Econ.	Economy
Apt.	Apartment	C.L.F.	Hundred Linear Feet	EDP	Electronic Data Processing
Asb.	Asbestos	CLF	Current Limiting Fuse	EIFS	Exterior Insulation Finish System
A.S.B.C.	American Standard Building Code	CLP	Cross Linked Polyethylene	E.D.R.	Equiv. Direct Radiation
Asbe.	Asbestos Worker	cm	Centimeter	Eq.	Equation
A.S.H.R.A.E.	American Society of Heating, Refrig. & AC Engineers	CMP	Corr. Metal Pipe	Elec.	Electrician; Electrical
		C.M.U.	Concrete Masonry Unit	Elev.	Elevator; Elevating
A.S.M.E.	American Society of Mechanical Engineers	CN	Change Notice	EMT	Electrical Metallic Conduit; Thin Wall Conduit
		Col.	Column		
A.S.T.M.	American Society for Testing and Materials	CO_2	Carbon Dioxide	Eng.	Engine, Engineered
		Comb.	Combination	EPDM	Ethylene Propylene Diene Monomer
Attchmt.	Attachment	Compr.	Compressor		
Avg.	Average	Conc.	Concrete	EPS	Expanded Polystyrene
A.W.G.	American Wire Gauge	Cont.	Continuous; Continued	Eqhv.	Equip. Oper., Heavy
AWWA	American Water Works Assoc.	Corr.	Corrugated	Eqlt.	Equip. Oper., Light
Bbl.	Barrel	Cos	Cosine	Eqmd.	Equip. Oper., Medium
B. & B.	Grade B and Better; Balled & Burlapped	Cot	Cotangent	Eqmm.	Equip. Oper., Master Mechanic
		Cov.	Cover	Eqol.	Equip. Oper., Oilers
B. & S.	Bell and Spigot	C/P	Cedar on Paneling	Equip.	Equipment
B. & W.	Black and White	CPA	Control Point Adjustment	ERW	Electric Resistance Welded
b.c.c.	Body-centered Cubic	Cplg.	Coupling	E.S.	Energy Saver
B.C.Y.	Bank Cubic Yards	C.P.M.	Critical Path Method	Est.	Estimated
BE	Bevel End	CPVC	Chlorinated Polyvinyl Chloride	esu	Electrostatic Units
B.F.	Board Feet	C.Pr.	Hundred Pair	E.W.	Each Way
Bg. cem.	Bag of Cement	CRC	Cold Rolled Channel	EWT	Entering Water Temperature
BHP	Boiler Horsepower; Brake Horsepower	Creos.	Creosote	Excav.	Excavation
		Crpt.	Carpet & Linoleum Layer	Exp.	Expansion, Exposure
B.I.	Black Iron	CRT	Cathode-ray Tube	Ext.	Exterior
Bit.; Bitum.	Bituminous	CS	Carbon Steel, Constant Shear Bar Joist	Extru.	Extrusion
Bk.	Backed			f.	Fiber stress
Bkrs.	Breakers	Csc	Cosecant	F	Fahrenheit; Female; Fill
Bldg.	Building	C.S.F.	Hundred Square Feet	Fab.	Fabricated
Blk.	Block	CSI	Construction Specifications Institute	FBGS	Fiberglass
Bm.	Beam			F.C.	Footcandles
Boil.	Boilermaker	C.T.	Current Transformer	f.c.c.	Face-centered Cubic
B.P.M.	Blows per Minute	CTS	Copper Tube Size	f'c.	Compressive Stress in Concrete; Extreme Compressive Stress
BR	Bedroom	Cu	Copper, Cubic		
Brg.	Bearing	Cu. Ft.	Cubic Foot	F.E.	Front End
Brhe.	Bricklayer Helper	cw	Continuous Wave	FEP	Fluorinated Ethylene Propylene (Teflon)
Bric.	Bricklayer	C.W.	Cool White; Cold Water		
Brk.	Brick	Cwt.	100 Pounds	F.G.	Flat Grain
Brng.	Bearing	C.W.X.	Cool White Deluxe	F.H.A.	Federal Housing Administration
Brs.	Brass	C.Y.	Cubic Yard (27 cubic feet)	Fig.	Figure
Brz.	Bronze	C.Y./Hr.	Cubic Yard per Hour	Fin.	Finished
Bsn.	Basin	Cyl.	Cylinder	Fixt.	Fixture
Btr.	Better	d	Penny (nail size)	Fl. Oz.	Fluid Ounces
BTU	British Thermal Unit	D	Deep; Depth; Discharge	Flr.	Floor
BTUH	BTU per Hour	Dis.;Disch.	Discharge	F.M.	Frequency Modulation; Factory Mutual
B.U.R.	Built-up Roofing	Db.	Decibel		
BX	Interlocked Armored Cable	Dbl.	Double	Fmg.	Framing
c	Conductivity, Copper Sweat	DC	Direct Current	Fndtn.	Foundation
C	Hundred; Centigrade	DDC	Direct Digital Control	Fori.	Foreman, Inside
C/C	Center to Center, Cedar on Cedar	Demob.	Demobilization	Foro.	Foreman, Outside

649

Abbreviations

Fount.	Fountain	J.I.C.	Joint Industrial Council	M.C.P.	Motor Circuit Protector
FPM	Feet per Minute	K	Thousand; Thousand Pounds; Heavy Wall Copper Tubing, Kelvin	MD	Medium Duty
FPT	Female Pipe Thread			M.D.O.	Medium Density Overlaid
Fr.	Frame	K.A.H.	Thousand Amp. Hours	Med.	Medium
F.R.	Fire Rating	KCMIL	Thousand Circular Mils	MF	Thousand Feet
FRK	Foil Reinforced Kraft	KD	Knock Down	M.F.B.M.	Thousand Feet Board Measure
FRP	Fiberglass Reinforced Plastic	K.D.A.T.	Kiln Dried After Treatment	Mfg.	Manufacturing
FS	Forged Steel	kg	Kilogram	Mfrs.	Manufacturers
FSC	Cast Body; Cast Switch Box	kG	Kilogauss	mg	Milligram
Ft.	Foot; Feet	kgf	Kilogram Force	MGD	Million Gallons per Day
Ftng.	Fitting	kHz	Kilohertz	MGPH	Thousand Gallons per Hour
Ftg.	Footing	Kip.	1000 Pounds	MH, M.H.	Manhole; Metal Halide; Man-Hour
Ft. Lb.	Foot Pound	KJ	Kiljoule	MHz	Megahertz
Furn.	Furniture	K.L.	Effective Length Factor	Mi.	Mile
FVNR	Full Voltage Non-Reversing	K.L.F.	Kips per Linear Foot	MI	Malleable Iron; Mineral Insulated
FXM	Female by Male	Km	Kilometer	mm	Millimeter
Fy.	Minimum Yield Stress of Steel	K.S.F.	Kips per Square Foot	Mill.	Millwright
g	Gram	K.S.I.	Kips per Square Inch	Min., min.	Minimum, minute
G	Gauss	kV	Kilovolt	Misc.	Miscellaneous
Ga.	Gauge	kVA	Kilovolt Ampere	ml	Milliliter, Mainline
Gal.	Gallon	K.V.A.R.	Kilovar (Reactance)	M.L.F.	Thousand Linear Feet
Gal./Min.	Gallon per Minute	KW	Kilowatt	Mo.	Month
Galv.	Galvanized	KWh	Kilowatt-hour	Mobil.	Mobilization
Gen.	General	L	Labor Only; Length; Long; Medium Wall Copper Tubing	Mog.	Mogul Base
G.F.I.	Ground Fault Interrupter			MPH	Miles per Hour
Glaz.	Glazier	Lab.	Labor	MPT	Male Pipe Thread
GPD	Gallons per Day	lat	Latitude	MRT	Mile Round Trip
GPH	Gallons per Hour	Lath.	Lather	ms	Millisecond
GPM	Gallons per Minute	Lav.	Lavatory	M.S.F.	Thousand Square Feet
GR	Grade	lb.; #	Pound	Mstz.	Mosaic & Terrazzo Worker
Gran.	Granular	L.B.	Load Bearing; L Conduit Body	M.S.Y.	Thousand Square Yards
Grnd.	Ground	L. & E.	Labor & Equipment	Mtd.	Mounted
H	High; High Strength Bar Joist; Henry	lb./hr.	Pounds per Hour	Mthe.	Mosaic & Terrazzo Helper
		lb./L.F.	Pounds per Linear Foot	Mtng.	Mounting
H.C.	High Capacity	lbf/sq.in.	Pound-force per Square Inch	Mult.	Multi; Multiply
H.D.	Heavy Duty; High Density	L.C.L.	Less than Carload Lot	M.V.A.	Million Volt Amperes
H.D.O.	High Density Overlaid	Ld.	Load	M.V.A.R.	Million Volt Amperes Reactance
Hdr.	Header	LE	Lead Equivalent	MV	Megavolt
Hdwe.	Hardware	LED	Light Emitting Diode	MW	Megawatt
Help.	Helpers Average	L.F.	Linear Foot	MXM	Male by Male
HEPA	High Efficiency Particulate Air Filter	Lg.	Long; Length; Large	MYD	Thousand Yards
		L & H	Light and Heat	N	Natural; North
Hg	Mercury	LH	Long Span Bar Joist	nA	Nanoampere
HIC	High Interrupting Capacity	L.H.	Labor Hours	NA	Not Available; Not Applicable
HM	Hollow Metal	L.L.	Live Load	N.B.C.	National Building Code
H.O.	High Output	L.L.D.	Lamp Lumen Depreciation	NC	Normally Closed
Horiz.	Horizontal	L-O-L	Lateralolet	N.E.M.A.	National Electrical Manufacturers Assoc.
H.P.	Horsepower; High Pressure	lm	Lumen		
H.P.F.	High Power Factor	lm/sf	Lumen per Square Foot	NEHB	Bolted Circuit Breaker to 600V.
Hr.	Hour	lm/W	Lumen per Watt	N.L.B.	Non-Load-Bearing
Hrs./Day	Hours per Day	L.O.A.	Length Over All	NM	Non-Metallic Cable
HSC	High Short Circuit	log	Logarithm	nm	Nanometer
Ht.	Height	L.P.	Liquefied Petroleum; Low Pressure	No.	Number
Htg.	Heating	L.P.F.	Low Power Factor	NO	Normally Open
Htrs.	Heaters	LR	Long Radius	N.O.C.	Not Otherwise Classified
HVAC	Heating, Ventilation & Air-Conditioning	L.S.	Lump Sum	Nose.	Nosing
		Lt.	Light	N.P.T.	National Pipe Thread
Hvy.	Heavy	Lt. Ga.	Light Gauge	NQOD	Combination Plug-on/Bolt on Circuit Breaker to 240V.
HW	Hot Water	L.T.L.	Less than Truckload Lot		
Hyd.;Hydr.	Hydraulic	Lt. Wt.	Lightweight	N.R.C.	Noise Reduction Coefficient
Hz.	Hertz (cycles)	L.V.	Low Voltage	N.R.S.	Non Rising Stem
I.	Moment of Inertia	M	Thousand; Material; Male; Light Wall Copper Tubing	ns	Nanosecond
I.C.	Interrupting Capacity			nW	Nanowatt
ID	Inside Diameter	M²CA	Meters Squared Contact Area	OB	Opposing Blade
I.D.	Inside Dimension; Identification	m/hr; M.H.	Man-hour	OC	On Center
I.F.	Inside Frosted	mA	Milliampere	OD	Outside Diameter
I.M.C.	Intermediate Metal Conduit	Mach.	Machine	O.D.	Outside Dimension
In.	Inch	Mag. Str.	Magnetic Starter	ODS	Overhead Distribution System
Incan.	Incandescent	Maint.	Maintenance	O.G.	Ogee
Incl.	Included; Including	Marb.	Marble Setter	O.H.	Overhead
Int.	Interior	Mat; Mat'l.	Material	O & P	Overhead and Profit
Inst.	Installation	Max.	Maximum	Oper.	Operator
Insul.	Insulation/Insulated	MBF	Thousand Board Feet	Opng.	Opening
I.P.	Iron Pipe	MBH	Thousand BTU's per hr.	Orna.	Ornamental
I.P.S.	Iron Pipe Size	MC	Metal Clad Cable	OSB	Oriented Strand Board
I.P.T.	Iron Pipe Threaded	M.C.F.	Thousand Cubic Feet	O. S. & Y.	Outside Screw and Yoke
I.W.	Indirect Waste	M.C.F.M.	Thousand Cubic Feet per Minute	Ovhd.	Overhead
J	Joule	M.C.M.	Thousand Circular Mils	OWG	Oil, Water or Gas

Abbreviations

Oz.	Ounce	S.	Suction; Single Entrance; South	THHN	Nylon Jacketed Wire
P.	Pole; Applied Load; Projection	SCFM	Standard Cubic Feet per Minute	THW.	Insulated Strand Wire
p.	Page	Scaf.	Scaffold	THWN;	Nylon Jacketed Wire
Pape.	Paperhanger	Sch.; Sched.	Schedule	T.L.	Truckload
P.A.P.R.	Powered Air Purifying Respirator	S.C.R.	Modular Brick	T.M.	Track Mounted
PAR	Weatherproof Reflector	S.D.	Sound Deadening	Tot.	Total
Pc., Pcs.	Piece, Pieces	S.D.R.	Standard Dimension Ratio	T-O-L	Threadolet
P.C.	Portland Cement; Power Connector	S.E.	Surfaced Edge	T.S.	Trigger Start
P.C.F.	Pounds per Cubic Foot	Sel.	Select	Tr.	Trade
P.C.M.	Phase Contract Microscopy	S.E.R.;	Service Entrance Cable	Transf.	Transformer
P.E.	Professional Engineer;	S.E.U.	Service Entrance Cable	Trhv.	Truck Driver, Heavy
	Porcelain Enamel;	S.F.	Square Foot	Trlr	Trailer
	Polyethylene; Plain End	S.F.C.A.	Square Foot Contact Area	Trlt.	Truck Driver, Light
Perf.	Perforated	S.F.G.	Square Foot of Ground	TV	Television
Ph.	Phase	S.F. Hor.	Square Foot Horizontal	T.W.	Thermoplastic Water Resistant Wire
P.I.	Pressure Injected	S.F.R.	Square Feet of Radiation		
Pile.	Pile Driver	S.F. Shlf.	Square Foot of Shelf	UCI	Uniform Construction Index
Pkg.	Package	S4S	Surface 4 Sides	UF	Underground Feeder
Pl.	Plate	Shee.	Sheet Metal Worker	UGND	Underground Feeder
Plah.	Plasterer Helper	Sin.	Sine	U.H.F.	Ultra High Frequency
Plas.	Plasterer	Skwk.	Skilled Worker	U.L.	Underwriters Laboratory
Pluh.	Plumbers Helper	SL	Saran Lined	Unfin.	Unfinished
Plum.	Plumber	S.L.	Slimline	URD	Underground Residential Distribution
Ply.	Plywood	Sldr.	Solder		
p.m.	Post Meridiem	SLH	Super Long Span Bar Joist	US	United States
Pntd.	Painted	S.N.	Solid Neutral	USP	United States Primed
Pord.	Painter, Ordinary	S-O-L	Socketolet	UTP	Unshielded Twisted Pair
pp	Pages	sp	Standpipe	V	Volt
PP; PPL	Polypropylene	S.P.	Static Pressure; Single Pole; Self-Propelled	V.A.	Volt Amperes
P.P.M.	Parts per Million			V.C.T.	Vinyl Composition Tile
Pr.	Pair	Spri.	Sprinkler Installer	VAV	Variable Air Volume
P.E.S.B.	Pre-engineered Steel Building	spwg	Static Pressure Water Gauge	VC	Veneer Core
Prefab.	Prefabricated	Sq.	Square; 100 Square Feet	Vent.	Ventilation
Prefin.	Prefinished	S.P.D.T.	Single Pole, Double Throw	Vert.	Vertical
Prop.	Propelled	SPF	Spruce Pine Fir	V.F.	Vinyl Faced
PSF; psf	Pounds per Square Foot	S.P.S.T.	Single Pole, Single Throw	V.G.	Vertical Grain
PSI; psi	Pounds per Square Inch	SPT	Standard Pipe Thread	V.H.F.	Very High Frequency
PSIG	Pounds per Square Inch Gauge	Sq. Hd.	Square Head	VHO	Very High Output
PSP	Plastic Sewer Pipe	Sq. In.	Square Inch	Vib.	Vibrating
Pspr.	Painter, Spray	S.S.	Single Strength; Stainless Steel	V.L.F.	Vertical Linear Foot
Psst.	Painter, Structural Steel	S.S.B.	Single Strength B Grade	Vol.	Volume
P.T.	Potential Transformer	sst	Stainless Steel	VRP	Vinyl Reinforced Polyester
P. & T.	Pressure & Temperature	Sswk.	Structural Steel Worker	W	Wire; Watt; Wide; West
Ptd.	Painted	Sswl.	Structural Steel Welder	w/	With
Ptns.	Partitions	St.; Stl.	Steel	W.C.	Water Column; Water Closet
Pu	Ultimate Load	S.T.C.	Sound Transmission Coefficient	W.F.	Wide Flange
PVC	Polyvinyl Chloride	Std.	Standard	W.G.	Water Gauge
Pvmt.	Pavement	STK	Select Tight Knot	Wldg.	Welding
Pwr.	Power	STP	Standard Temperature & Pressure	W. Mile	Wire Mile
Q	Quantity Heat Flow	Stpi.	Steamfitter, Pipefitter	W-O-L	Weldolet
Quan.; Qty.	Quantity	Str.	Strength; Starter; Straight	W.R.	Water Resistant
Q.C.	Quick Coupling	Strd.	Stranded	Wrck.	Wrecker
r	Radius of Gyration	Struct.	Structural	W.S.P.	Water, Steam, Petroleum
R	Resistance	Sty.	Story	WT., Wt.	Weight
R.C.P.	Reinforced Concrete Pipe	Subj.	Subject	WWF	Welded Wire Fabric
Rect.	Rectangle	Subs.	Subcontractors	XFER	Transfer
Reg.	Regular	Surf.	Surface	XFMR	Transformer
Reinf.	Reinforced	Sw.	Switch	XHD	Extra Heavy Duty
Req'd.	Required	Swbd.	Switchboard	XHHW; XLPE	Cross-Linked Polyethylene Wire Insulation
Res.	Resistant	S.Y.	Square Yard		
Resi.	Residential	Syn.	Synthetic	XLP	Cross-linked Polyethylene
Rgh.	Rough	S.Y.P.	Southern Yellow Pine	Y	Wye
RGS	Rigid Galvanized Steel	Sys.	System	yd	Yard
R.H.W.	Rubber, Heat & Water Resistant; Residential Hot Water	t.	Thickness	yr	Year
		T	Temperature; Ton	Δ	Delta
rms	Root Mean Square	Tan	Tangent	%	Percent
Rnd.	Round	T.C.	Terra Cotta	~	Approximately
Rodm.	Rodman	T & C	Threaded and Coupled	$\emptyset$	Phase
Rofc.	Roofer, Composition	T.D.	Temperature Difference	@	At
Rofp.	Roofer, Precast	T.E.M.	Transmission Electron Microscopy	#	Pound; Number
Rohe.	Roofer Helpers (Composition)	TFE	Tetrafluoroethylene (Teflon)	<	Less Than
Rots.	Roofer, Tile & Slate	T. & G.	Tongue & Groove; Tar & Gravel	>	Greater Than
R.O.W.	Right of Way				
RPM	Revolutions per Minute	Th.; Thk.	Thick		
R.R.	Direct Burial Feeder Conduit	Thn.	Thin		
R.S.	Rapid Start	Thrded	Threaded		
Rsr	Riser	Tilf.	Tile Layer, Floor		
RT	Round Trip	Tilh.	Tile Layer, Helper		

Index

A

Abandon catch basin	34
Abatement asbestos	369
ABC extinguisher	310
Abrasive aggregate	90
floor	112
floor tile	274
nosing	166, 167
silicon carbide	112
stair tread concrete	102
terrazzo	276
tile	274
tread	275
ABS DWV pipe	397
Absorption testing	10
water chillers	423
Accelerator set	90
sprinkler system	377
Access control	373
door & panels	238
door basement	116
door duct	430
door fire rated	238
door floor	238
door metal	238
door roof	222
door stainless steel	238
floor	304
road and parking area	17
security	373
Accessories bath	316
bathroom	273, 316
door and window	258
drywall	273
duct	430
plaster	268
reinforcing steel	103-105
roof	222
Accessory drainage	220
fireplace	305
formwork	98-102
furniture	345
masonry	120, 121
reinforcing	103
Accordion door	235
Acid proof floor	278, 282
resistant pipe	394
Acoustic ceiling board	276
Acoustical batt	284
block	134
booth	358
ceiling	276, 277, 353
door	242, 314
enclosure	356, 358
folding partition	314
glass	259
metal deck	156
panel	285, 313
partition	313, 314
phone booth	315
room	354
sealant	225, 271
spray	223
treatment	284
underlayment	285
wall tile	278
wallboard	271
window wall	262
Acrylic ceiling	277
floor	281
latex	90
rubber roofing	215
sign	308
wall coating	293
wallcovering	284
wood block	279
Act. tools & fast. powder	143
Adhesive cement	281
EPDM	200
neoprene	200
PVC	200
roof	213
wallpaper	284
Adjustable astragal	258
Adjustment factor	6
Admixture	120
cement	90, 113
concrete	90
water reducing	90
Adobe brick	129
masonry	129
Aeration sewage	62
Aerator	76
Aerial lift	22
photography	10
seeding	85
survey	7
Aggregate abrasive	90
concrete	90
exposed	112
lightweight	90
marble	91
masonry	120
panels exposed	211
plaster	90
roof	91
stone	84
testing	10
Air balancing	436
cleaner electronic	436
compressor	22
compressor control system	374
compressor dental	341
conditioner central station	425
conditioner cooling & heating	426
conditioner direct expansion	428
conditioner fan coil	428
conditioner gas heat	426
conditioner packaged terminal	426
conditioner receptacle	453
conditioner removal	390
conditioner rooftop	426
conditioner self-contained	426
conditioner thru-wall	426
conditioner ventilating	430
conditioner wiring	455
conditioning	567
conditioning computer	425
conditioning fan	431
conditioning ventilating	374, 423, 424, 427, 431-433, 436
control	402
cooled belt drive condenser	424
curtain	241
curtain door	241
entraining agent	90
extractor	430
filter	436
filter roll type	436
filtration	369
handling fan	431
handling troffer	464
handling unit	425
hose	22
lock	352
make-up unit	425
register	435
return grille	434
sampling	369
spade	22
supply pneumatic control	374
supply register	435
supported building	12, 352
supported enclosure	363
supported storage tank cover	352
supported structures	352, 557
tool	22
tube system	386
vent automatic	402
vent roof	223
wall	314
Air-compressor mobilization	57
Airfoil fan centrifugal	432
Airless sprayer	370
Airplane hangar	361
hangar door	241
Alarm burglar	375
exit control	375
residential	454
sprinkler	375
standpipe	375
valve sprinkler	378
Alley bowling	336
Alloy steel chain	168
Altar	321
Alteration fee	6
Aluminium fence	77
Aluminum astragal	257
bench	81
blind	345
cable tray	446
ceiling tile	278
column	146, 193
column base	168
conduit	446
coping	128
cross	322
curtain wall	262
diffuser perforated	434
directory board	307
dome	359
door	242, 244, 245
door frame	244
downspout	219
drip edge	220
ductwork	430
entrance	244
expansion joint	170, 222
extrusion	152
fence	76
flagpole	306
flashing	211, 217, 408
foil	202, 206, 285
framing	152
grandstand	360
grating	165
gravel stop	220
grille	434
gutter	221
handrail	304
ladder	163
louver	302, 434
mansard	209
nail	173
operating louver	435
pipe	67, 68
register	435
reglet	221
rivet	143
roof	209
salvage	391
sash	247
screw	211
service entrance cable	444
shade	346
sheet metal	219
shelter	363
shingle	206
shore	101
siding	211
siding paint	290
sign	308
stair	164
stair tread	102, 167
steeple	322
storefront	245
storm door	237
structural	152
tile	208, 275
transom	244
trench cover	167
tube frame	262
weatherstrip	257
weld rod	144
window	247
window demolition	41
wire	445
Ammeter	458
Analysis petrographic	10
sieve	10, 11
Anchor bolt	98, 121, 138, 142
brick	121
buck	121
bumper	328
chemical	140
dovetail	99
expansion	140
framing	174
hollow wall	140
joist	174
lead screw	140
machine	142
masonry	121
nailing	140
partition	121
rafter	174
rigid	121
screw	141
sill	174
steel	121
stone	121
super high-tensile	55
wall	139
wedge	141
Anechoic chamber	354
Angle corner	303
fiberglass	195
framing	148
valve	400
Antenna system	468
Anti-siphon water	405
Apartment call system	468
S.F. & C.F.	470
Appliance	334
plumbing	417, 418
residential	325, 334, 455
Approach ramp	304
Apron wood	186
Arch laminated	185
radial	185
Architectural equipment	334
fee	6, 487
panel	182
precast	115
woodwork	189
Area clean-up	370
wall	258
Arena sport S.F. & C.F.	476
Ark	321
Armored cable	442
Arrow	71
Artificial turf	74
Asbestos abatement	369

Index

demolition ... 369
disposal ... 371
felt ... 214
removal ... 368
removal process ... 559
As-built drawings ... 27
Ash conveyor ... 329
receiver ... 316, 345
urn ... 345
Ashlar stone ... 131
veneer ... 132, 532
Asphalt base sheet ... 213
block ... 73
block floor ... 73
coating ... 200
curb ... 71
cutting ... 40
distributor ... 23
expansion joint ... 100
felt ... 213, 214
flashing ... 218
flood coat ... 213
mix design ... 10
paper ... 206
pavement ... 74
paver ... 23
primer ... 216, 281
sheathing ... 184
shingle ... 206
sidewalk ... 72
testing ... 10
Asphaltic binder ... 70
concrete ... 30, 70
emulsion ... 84
pavement ... 70
paving ... 492
wearing course ... 70
Aspirator dental ... 341
Astragal adjustable ... 258
aluminum ... 257
magnetic ... 257, 258
molding ... 187
one piece ... 257
overlapping ... 257
rubber ... 257, 258
split ... 257
steel ... 257
Astronomy observation dome ... 359
Athletic equipment ... 82, 337
paving ... 74
pole ... 82
post ... 82
Atomizer water ... 370
Attic stair ... 335
Audio masking ... 356
Audiometric room ... 354
Auditorium chair ... 348
S.F. & C.F. ... 470
Auger ... 19
boring ... 33
hole ... 32
Augering ... 56
Auto park drain ... 406
Automatic fire suppression ... 378
flush ... 411
opener ... 252
operator ... 252
teller ... 321
washing machine ... 335
Automotive equipment ... 326, 327
lift ... 326
sales S.F. & C.F. ... 470
Autopsy equipment ... 340
Awning ... 311
window ... 247, 248
Axial flow fan ... 431

B

Backer rod ... 100, 225
Backerboard ... 200, 269, 270
Backfill ... 47, 51
compaction ... 496
planting pit ... 86
trench ... 52
Backflow preventer ... 405
Backhoe ... 48
excavation ... 51
Backsplash ... 344
countertop ... 172
Backstop baseball ... 82
basketball ... 82, 337
electric ... 337
handball court ... 82
squash court ... 82
tennis court ... 82
Backup block ... 125
Backwater valve ... 407
Baffle roof ... 223
Bag disposable ... 370
glove ... 368
Baked enamel door ... 228
enamel frame ... 230
Balanced door ... 246
Balancing air ... 436
water ... 437
Balcony fire escape ... 164
Bale hay ... 54
Baler ... 329
Ball valve ... 398, 400, 401
wrecking ... 25
Ballast fixture ... 465
railroad ... 169
Baluster ... 192
Balustrade painting ... 289
Band joist framing ... 161
molding ... 186
riser ... 323
Banding grating ... 166
iron ... 100
Bank counter ... 321
equipment ... 320, 321
S.F. & C.F. ... 470
window ... 320
Bankrun gravel ... 30, 90
Banquet booth ... 350
Baptistry ... 321
Bar bell ... 337
front ... 329
grab ... 316
grouted ... 108
joist ... 154
joist painting ... 155
panic ... 256
parallel ... 340
restaurant ... 348
tie ... 104
touch ... 256
towel ... 316
tub ... 316
ungrouted ... 108
Zee ... 273
Barbed wire fence ... 76
Barber equipment ... 324
Bark mulch ... 84
mulch redwood ... 84
Barrel ... 22
bolts ... 253
Barricade ... 17, 22, 80
Barrier delineator ... 80
impact ... 80
median ... 80
moisture ... 201
parking ... 80
separation ... 369
waterstop ... 100
X-ray ... 357
Barriers and enclosures ... 17
Base cabinet ... 189, 190
carpet ... 283
ceramic tile ... 273
column ... 146, 151, 168, 174
course ... 69
cove ... 280
form ... 102
gravel ... 70
light ... 98
masonry ... 129
molding ... 186
plate column ... 151
quarry tile ... 274
resilient ... 280
road ... 70
sheet ... 213, 214
sign ... 98
sink ... 189
stone ... 70, 131
terrazzo ... 275, 276
vanity ... 191
wood ... 186, 194
Baseball backstop ... 82
scoreboard ... 338
Baseboard demolition ... 39
heat ... 428
heat electric ... 429
Basement stair ... 116
Basketball backstop ... 82, 337
scoreboard ... 338
Batch trial ... 11
Bath accessories ... 316
paraffin ... 340
steam ... 356
whirlpool ... 341, 414
Bathroom ... 414
accessories ... 273, 316
exhaust fan ... 432
faucet ... 411
fixture ... 414-416
heater & fan ... 433
soaking ... 414
Bathtub ... 414
bar ... 316
enclosure ... 302
removal ... 390
Batt acoustical ... 284
insulation ... 203
thermal ... 284
Battery light ... 466
Bead blast demo ... 37
board insulation ... 355
casing ... 273
corner ... 268
parting ... 187
Beam & girder framing ... 176
and girder formwork ... 92
bolster ... 103
bond ... 120, 125
bondcrete ... 269
bottom ... 92
box ... 182
castellated ... 151
ceiling ... 193
concrete ... 109-111
drywall ... 270, 271
fiberglass wide flange ... 195
fireproofing ... 223
formwork ... 92
grade ... 95, 111
hanger ... 174
laminated ... 185
mantel ... 193
plaster ... 268, 269
precast ... 114
reinforcing ... 106
removal ... 37
side ... 92
soldier ... 46
spandrel ... 92
steel ... 151
tee ... 115
test ... 11
wide flange ... 149, 151
wood ... 176, 177
Bearing pad ... 143
Bed molding ... 186
Bedding brick ... 73
pipe ... 47
Beech tread ... 193
Belgian block ... 72
Bell ... 321
& spigot pipe ... 393
bar ... 337
caisson ... 59
church ... 322
system ... 467
Bellow expansion joint pipe ... 402
Belt material handling ... 385
Bench aluminum ... 81
fiberglass ... 81
folding ... 337
greenhouse ... 360
park ... 81
planter ... 87
players ... 81
wood ... 81
work ... 339
Bentonite ... 46, 201
Berm pavement ... 71
road ... 71
Bevel glass ... 259
siding ... 212
Bicycle rack ... 82
trainer ... 337
Bi-fold door ... 229, 234
Bin parts ... 315
Binder asphaltic ... 70
Bi-passing closet door ... 234
Birch door ... 231, 234, 235
molding ... 187
paneling ... 188
stair ... 192
wood frame ... 237
Bit drill ... 33
Bituminous block ... 73
coating ... 87, 200
expansion joint ... 100
paver ... 23
paving ... 492
Blackboard ... 314
Blank leaf door ... 236
Blanket curing ... 114
insulation ... 203, 391
sound attenuation ... 284
Blast demo bead ... 37
floor shot ... 37
Blaster shot ... 24
Blasting ... 47
cap ... 48
mat ... 48
Bleacher ... 360
outdoor ... 82
stadium ... 82
telescoping ... 337
Blind exterior ... 194
venetian ... 345

INDEX

Index

window 345
Block acoustical 134
asphalt 73
asphalt floor 73
backup 125
belgian 72
bituminous 73
Block, brick and mortar 530
Block cap 129
chimney 125
column 134
concrete 125-128, 134, 135
concrete bond beam 125
concrete exterior 127
corner 128
decorative concrete 126, 127
filler 293
floor 279
glass 129, 532
glazed 128, 129
grooved 126
ground face 126
hexagonal face 127
high strength 127
insulation 121
interlocking 128
lightweight 134
lintel 127
manhole 68
partition 134, 135
patio 131
planter 73
profile 126, 127
reflective 129
removal 34
scored 127
scored split face 126
sill 128
slotted 134
slump 126
solar screen 128
split rib 126
wall 531
wall removal 34
Blocking carpentry 176
wood 176, 179
Blockout slab 96
Blood pressure unit 340
Bloodbank refrigeration 340
Blower pneumatic tube 386
Blown in cellulose 201
in fiberglass 201
in insulation 201
Blueboard 270
partition 272
Bluegrass sod 85
Bluestone 130
sidewalk 72
sill 130
step 72
Board & batten fence 78
& batten siding 212
bulletin 299
ceiling 276
control 299, 323
directory 307
dock 328
drain 201
gypsum 270
insulation 202
paneling 189
ridge 180
sheathing 183
valance 190
verge 187
Boat dock 54
Boiler 419
control system 374
demolition 390
electric 419
electric steam 419
feed pump 409
gas fired 419
gas/oil combination 419
general 390
hot water 419, 420
insulation 391
mobilization 57
oil fired 420
steam 419, 420
Bollards pipe 81
Bolster beam 103
slab 103
Bolt 139
anchor 121, 138, 142
expansion 140
flush 253
steel 139, 174
toggle 140
wedge 141
Bolt-on circuit-breaker 459
Bolts barrel 253
Bond beam 120, 125
performance 8, 486
Bondcrete 269
Bonding agent 90
surface 120
Bookcase 189, 190
Bookshelf 322
Boom lift 22
Booster fan 431
Boot pile 58
Booth acoustical 358
banquet 350
fixed 350
mounted 350
painting 327
portable 358
telephone 315
ticket 358
Border light 323
Bored pile 58
Boring 32
auger 33
cased 32
horizontal 55
Borosilicate pipe 393
Borrow 30, 51
Bottle storage 330
Bottom beam 92
Boulder excavation 48
Bow window 248
Bowling alley 336
Bowstring truss 185
Box 449
beam 182
buffalo 61
distribution 64
electrical 450
explosionproof 450
locker 309
mail 311
out 96
out opening formwork 94
pull 450
safety deposit 321
stair 192
steel mesh 54
storage 12
termination 468
vent 122
Boxed headers & beams 158, 161
Boxes & wiring device 450
Brace cross 148
Bracing 176
let-in 176
metal joists 160
stud walls 157
Braided bronze hose 403
Branch circuit fusible switch 461
Brass hinge 254
pipe 393
salvage 391
screw 174
valve 398
Brazed connection 441
Break glass station 376
Breaker circuit 459
vacuum 405
Breeching draft inducer 423
insulation 391
Brick 124, 529
adobe 129
anchor 121
bedding 73
Brick, block and mortar 530
Brick cart 23
catch basin 68
chimney simulated 304
chimneys 527
cleaning 136
concrete 128
demolition 37, 38
economy 123
engineer 124
face 123
floor 73
forklift 23
molding 186
paving 73
removal 34
saw 136
shelf 96
sidewalk 73
sill 130
simulated 133
stair 136
step 72
testing 11
veneer 122, 123, 530
veneer demolition 39
vent 434
vent box 122
wall 135
wall panel 136
wash 136
Bricklaying 527
Bridge 81
concrete 81
cranes 386
pedestrian 81
sidewalk 15
Bridging 155, 176
metal joists 160
steel 176
stud walls 157
wood 176
Broiler 330
Bronze body strainer 403
cross 322
expansion joint 170
letter 308
plaque 308
push-pull plate 256
stair tread 167
valve 398
Broom cabinet 190
finish concrete 112
Brown coat 276
Brownstone 131
Brush clearing 42
cutter 20
mowing 42
Brush-off blast metal 152
Bubbler 413
drinking 413
Buck anchor 121
rough 179
Bucket concrete 19
crane 20
Buffalo box 61
Buggy concrete 19, 112
Builder's risk insurance 480
Building air supported 352
directory 307, 468
disposal 40
greenhouse 360
hangar 361
hardware 253
insulation 201
model 6
moving 88
paper 206
permit 8
portable 12
prefabricated 360
slipform 98
sprinkler 377
temporary 12
tension 358
Built-in range 334
shower 415
Built-up roof 213
roofing 548
Bulbtee subpurlin 145
Bulk bank measure excavating 48
storage dome 359
Bulkhead door 238
formwork 94-96
Bull winch 99
Bulldozer 21, 47
Bullet resistant window 321
Bulletin board 299
board cork 299
Bulletproof glass 260
Bullnose block 127
Bumper car 80, 169, 303
dock 327
door 253, 254, 259
parking 80
rail 80
railroad 169
wall 254, 303
Burglar alarm 375
alarm indicating panel 375
Burlap curing 114
rubbing 113
Bus-duct cable tap box 462
copper 462
fitting 462
plug in 462, 463
switch 463
tap box 462
Bush hammer 113
hammer concrete 113
Butt fusion machine 22
Butterfly valve 60, 400
Buttress formwork 96
Butyl caulking 225
expansion joint 222
flashing 218
waterproofing 200

INDEX

Index

C

Cabinet base 189, 190
   broom 190
   casework 190
   corner base 189
   corner wall 190
   current transformer 451
   demolition 39
   door 190, 191
   electrical 450
   electrical hinged 450
   fire equipment 310
   hardboard 189
   hardware 191
   hinge 191
   hose rack 310
   hotel 317
   key 344
   kitchen 189, 344
   laboratory 339
   medicine 317
   metal 344
   oven 190
   portable 330
   school 344
   shower 301
   stain 285
   storage 339
   strip 468
   transformer 451
   valve 310
   varnish 285
   wall 190
   wardrobe 318
Cable armored 442
   control 443
   copper 443
   copper shielded 444
   crimp termination 443
   electric 442, 445
   guide rail 80
   heating 429
   jack 26
   mineral insulated 444
   PVC jacket 443
   sheathed nonmetallic 444
   sheathed romex 444
   shielded 444
   tap box bus-duct 462
   termination 443
   tray 446
   tray aluminum 446
   tray ladder type 446
   wire rope 152
Cafe door 233
Caisson 58, 499
   bell 59
   concrete 58
   displacement 60
   foundation 58
Calcium chloride 54, 90
Call system apartment 468
   system emergency 467
   system nurse 467
Camera TV closed circuit 468
Canopy 311
   entrance 311, 362
   framing 146
   wood 180
Cant roof 180
Cantilever retaining wall 79
Canvas awning 311
Cap blasting 48
   block 129
   column 146

pile 95, 111, 112
post 174
Capacitor indoor 458
Capital column 93
Car bumper 80, 169, 303
   motorized 385
Carbon black 90
   dioxide extinguisher .. 310, 378
Carborundum 113
Card catalog file 323
Carillon 321
Carousel compactor 328
   telephone 315
Carpentry finish 186, 189
   rough 176
Carpet 282
   base 283
   computer room 304
   felt pad 283
   floor 282
   maintenance 283
   nylon 282
   padding 283
   removal 37
   stair 283
   urethane pad 283
   wool 283
Carport 311
Carrel 323
Carrier ceiling 267
   channel 266
   fixture 409
   fixture support 409
Cart brick 23
   concrete 19, 112
   mounted extinguisher 310
Carving stone 131
Case display 344
   exhibit 344
   refrigerated 324
   work 190
Cased boring 32
Casement window 246-249
Casework cabinet 190
   custom 189
   demolition 39
   ground 181
   manufactured 344
   metal 344
   painting 285
   varnish 285
Cash register 324
Casing bead 273
   wood 186
Cast in place concrete .. 109, 110
   in place pile 55, 56
   in place terrazzo 276
   iron bench 81
   iron casting 168
   iron damper 305
   iron drain 406
   iron manhole cover 69
   iron miscellaneous 168
   iron pipe 393
   iron pull box 450
   iron radiator 428
   iron stair 164
   iron stair tread 102, 164, 167
   iron trap 407
   iron tread 164
   iron weld rod 144
   trim lock 255
Castellated beam 151
Casting 168
   construction 168
Castings fiberglass 195

Catch basin 68
   basin brick 68
   basin frame and cover 69
   basin masonry 68
   basin precast 68
   basin removal 34
   door 191
Catwalk 15
Caulking 225, 226
   lead 393
   oakum 393
   polyurethane 226
   sealant 225
Cavity wall grout 120
   wall insulation 202
   wall reinforcing 121
Cedar closet 188
   fence 78
   paneling 189
   post 194
   roof deck 184
   roof plank 183
   shingle 207
   siding 212
   stair 192
Ceiling acoustical 276, 277
   beam 193
   board 276
   board acoustic 276
   board fiberglass 276
   bondcrete 269
   carrier 267
   demolition 35
   diffuser 434
   drill 140
   drywall 270
   eggcrate 277
   expansion joint 170
   fan 432
   framing 177
   furring 181, 266, 267
   hatch 238
   heater 334
   insulation 201
   integrated 353
   lath 267, 268
   luminous 277, 466
   molding 186
   painting 292, 293
   plaster 268, 269
   polystyrene 355
   radiant 353
   register 435
   stair 335
   stressed skin 149, 182
   support 146
   suspended 267, 270, 277, 353
   tile 277
   woven wire 312
Ceilings 276
ceilings 276
Cell prison 329
Cellar door 238
   wine 330
Cellular concrete 110
   deck concrete 115
   fill 117
   metal deck 156
Cellulose blown in 201
   insulation 202
Cement adhesive 281
   admixture 90, 113
   color 90, 120
   content 513
   flashing 217
   grout 44, 117, 120

gunite 113
gypsum 120
keene 268
liner 87
masonry 91, 120
masonry unit ... 125-128, 134, 135
mortar 274, 526
parging 200
plaster 354
Portland 91
testing 11
vermiculite 224
Cementitious deck 115-117
   roof deck 116
   waterproofing 201
Center bulb waterstop 103
Central station air handling unit . 425
   vacuum 320
Centrifugal airfoil fan 432
   fan 431
   liquid chiller 423
   pump 24
Ceramic coating 293
   mulch 84
   tile 273, 274
   tile countertop 172
   tile demolition 37
   tile floor 273
Certification welding 12
Chain alloy steel 168
   core rope 348
   hoist 26, 386
   hoist door 242
   link fence 17, 76-78
   link fence paint 288
   link fence removal 34
   link gate 77
   saw 24
   steel 168
   trencher 21
Chair 348
   barber 324
   dental 341
   hydraulic 341
   library 323
   life guard 364
   locker bench 309
   molding 187
   movie 322, 326, 348
   portable 323
   rail demolition 39
   seating 348
Chalkboard 298
   electric 298
   freestanding 298
   liquid chalk 299
   metal 314
   sliding 298
   swing leaf 298
   wall hung 299
Chamber anechoic 354
   decontamination 369
   echo 356
Chamfer strip 99
   strip galvanized 99
   strip PVC 99
   strip wood 99
Channel carrier 266
   frame 228
   framing 148
   furring 266, 273
   siding 212
   slab concrete 117
   slotted 148
   steel 268
Channelizing traffic 80

Index

Charges disposal 371
Charging desk 323
Chase formwork 102
Check floor 254
   swing valve 398, 400
   valve 378, 401
Checkered plate 166, 167
   plate nosing 166
Checkout counter 324
   scanner 324
   supermarket 324
Chemical anchor 140
   dry extinguisher 310
   water closet 62
Chest 349
Chilled water coil 428
Chiller water 423
Chimney 134
   accessories 304
   block 125
   brick 134
   concrete 134
   demolition 38
   flue 133
   foundation 109
   metal 134, 304, 422
   screen 305
   simulated brick 304
   vent 422
China cabinet 189
Chipper brush 20
Chipping hammer 22
   stump 42
Chips wood 84
Chlorination system 364
Church bell 322
   equipment 321
   pew 322
   S.F. & C.F. 471
Chute linen 385
   mail 311
   package 385
   refuse 385
   rubbish 40
Chutes 385
Cinema equipment 325
C.I.P. concrete 109
Circuit-breaker 459
   bolt-on 459
   feeder section 461
Circular saw 24
   slipforms 502
Circulating pump 408
City hall S.F. & C.F. 477
Cladding 211
Clamp water pipe ground 441
Clamshell 48, 53
   bucket 20
Clapboard painting 290
Classroom heating & cooling ... 428
   seating 348
Clay fire 133
   masonry 122
   pipe 64, 67
   roofing tile 208
   tennis court 74
   tile 208
   tile coping 128
Clean room 354
   steam 136
   tank 31
Cleaner steam 24
Cleaning brick 136
   face brick 532
   masonry 136
   steam 136
   up 27
Cleanout door 305
   floor 405
   pipe 405
   PVC 406
   tee 406
Clean-up area 370
Clear & grub 42
Clearing brush 42
   site 42
Clevis hooks 168
Climbing crane 25
   jack 26
Clinic S.F. & C.F. 474
Clip insulation 202
   plywood 174
   tie 104
Clock 374
   & bell system 375
   system 375
   timer 454
Closed circuit T.V. 321
   deck grandstand 360
Closer concealed 253
   door 253
   electronic 253
   floor 253
   holder 253
Closet cedar 188
   door 229, 234, 235
   pole 187
   rod 189
   water 416
Closure trash 82
Cloth hardware 78
Clothes dryer commercial 325
   dryer residential 334
Club country S.F. & C.F. 471
CMU 125
Coal tar pitch 214, 216
Coat brown 276
   glaze 292
   hook 316
   rack 190, 318
   scratch 276
Coated reinforcing 106
Coating bituminous 87, 200
   ceramic 293
   epoxy 87, 113, 293
   flood 214
   glazed 293
   polyethylene 87
   roof 216
   rubber 201
   silicone 201
   spray 200
   structural steel 533
   trowel 200
   wall 293
   water repellent 201
   waterproofing 200
Cocks gas 398
Coffee urn 331
Cofferdam excavation 49
Cohesion testing 10
Coil cooling 428
   flanged 428
Coiling door & grilles 239
Coin operated gate 309
Cold applied roofing 214
   roofing 214
   storage door 240
Coldformed framing 157
   joists 160
   metal joists 162
Collected stone 132
Collection box lobby 312
Collector solar energy system ... 372, 373
College S.F. & C.F. 471
Colonial door 231, 234
   wood frame 236, 237
Color concrete 109
   floor 113
   wheel 323
Column 146, 193
   aluminum 146
   base 146, 151, 168, 174
   base plate 151
   block 134
   bondcrete 269
   brick 134
   cap 146
   capital 93
   concrete 109, 111, 114, 116
   demolition 38
   drywall 270, 271
   fireproof 147, 223
   form 99
   formwork 92
   lally 146
   laminated wood 185
   lath 267
   pipe 146
   plaster 268, 269
   precast 114
   reinforcing 106
   removal 37, 38
   steel 146
   stone 131
   wood 177, 178, 193
Columns formwork 92
Combination device 452
   storm door 230, 234
Comfort station 358
Command dog 18
Commercial door 228, 236, 238
   folding partition 314
   greenhouse 360
   gutting 38
   refrigeration 354
   water heater 418
   window glass 552
Commissioning 27
Common brick 122
   nail 173
Communicating lockset 255
Communication system 467
Community center S.F. & C.F. .. 471
Compact fill 51
Compaction 47
   backfill 496
   soil 47
   test Proctor 11
   vibroflotation 495
Compactor 328
   earth 20
   residential 334
   waste 328
Compartments shower 301
   toilet 300
Compensation workers' 8
Component sound system 467
   sprinkler system 377
Composite door 228
   insulation 205
   joist 156
   metal deck 156
Composition flooring 279, 281
   flooring removal 37
Compressive strength 11
Compressor air 22
tank mounted 374
Computer air conditioning 425
   floor 304
   room carpet 304
Concealed closer 253
Concrete 116, 510-515, 517-519
   admixture 90
   aggregate 90
   asphaltic 30, 70
   beam 109-111
   block 125-128, 134, 135, 531
   block back-up 125
   block bond beam 125
   block decorative 125-127
   block demolition 36
   block exterior 127
   block foundation 127
   block grout 120
   block insulation 202
   block paver 73
   block planter 73
   brick 128
   bridge 81
   broom finish 112
   bucket 19
   buggy 112
   bush hammer 113
   caisson 58
   cart 19, 112
   cast in place 109
   catch basin 68
   cellular 110, 117
   cellular deck 115
   channel slab 117
   chimney 134
   C.I.P. 109
   color 109
   column 109, 111, 116
   conversion 514
   conveyer 19
   coping 128
   core 32
   cost 515
   cribbing 79
   curb 71
   curing 71, 90, 114
   curtain wall 116
   cutout 36
   cutting 41
   cylinder 11
   darby finish 112
   demo 35
   demolition 36, 37
   drill 32, 139
   elevated slab 111
   field mix 109, 516
   finish 71, 112, 113, 520
   flagpole 306
   float 19
   float finish 112
   floor 90, 520
   footing 110, 111
   forms 501, 503, 504
   formwork 92
   foundation 110
   furring 181
   granolithic finish 112
   gravel 90
   grout 120
   hand hole 65
   hand trowel finish 112
   hardener 90, 113
   hydrodemolition 34
   in place 517-519
   insulation 114, 117
   integral finish 112

656

Index

joist	110, 114
lift slab	520
lightweight	109, 110, 117, 525
lintel	116
manhole	65, 68
materials	513
mix design	11
mixer	19
mixes	513, 514
monolithic finish	112
patching	117
paver	23
paving	71
perlite	90
pier	109
pile	55
pipe	62, 63, 66, 500
pipe removal	35
placing	111, 516
plank	117
planter	87
post	80
poured	90
precast	114, 116
prestressed	522
prestressed precast	521
protection	91
pump	19, 111
ready mix	109
reinforcement	103
removal	35
retaining wall	79
roof	115
roof deck	115
sand	90
saw	19, 40
sealer	90
septic tank	64
shingle	208
sidewalk	72
siding	115
sill	130
silo	363
slab	110, 112
slabs X-ray	12
spreader	24
stair	111
stair tread	102
stamping	113
stengths	514
structural	109
tank	365
testing	10
tie	169
tile	208
tilt up	524
topping	112
trowel	19
truck	19
utility vault	30
vibrator	19
wall	110, 112
wheeling	112
Condensate pump	408
steam meter	409
Condenser	424
air cooled belt drive	424
Conductive floor	276, 280, 282
Conductor	442
& grounding	442-445
wire	445
Conductors	570
Conduit & fitting flexible	456
aluminum	446
electrical	446
fitting	447
flexible metallic	447
greenfield	447
high installation	447
in slab	447
in slab PVC	447
in trench electrical	447
in trench steel	447
intermediate	446
intermediate steel	446
raceway	446
rigid in slab	447
rigid steel	446
Cone traffic	17
Confessional	322
Connection brazed	441
motor	456
standpipe	376
Connector flexible	402
joist	174
screw-on	443
stud	143
timber	174
water copper tubing	403
wire	443
Consolidation test	11
Construction aids	13
casting	168
cost index	7
management fee	6
photography	10
sump hole	43
temporary	13
time	6, 8, 480
Contaminated soil	31
Contingencies	7
Continuous hinge	257
Contractor equipment	19, 320
overhead	14, 27
pump	24
scale	317
Control & motor starter	459
air	402
board	299, 323
cable	443
expansion	170
hand-off-automatic	459
joint	121
joint PVC	121
joint rubber	121
radiator supply	403
station	459
station stop-start	459
system air compressor	374
system boiler	374
system cooling	374
system electronic	374
system split system	374
system ventilator	374
tower	358
traffic	80
valve heating	403
Controller time system	375
Convection oven	332
Convector cover	172
heating unit	428
Conveyor	19, 385
ash	329
door	241
material handling	385
system	384
vertical	385
Cooking equipment	330, 334
range	334
Cooler	355
beverage	331
door	241
water	413
Cooling & heating A/C	426
coil	428
control system	374
tower	424
tower louver	434
towers fiberglass	424
towers stainless	424
Coping	128, 131
aluminum	128
clay tile	128
concrete	128
removal	39
terra cotta	125, 128
Copper bus-duct	462
cable	442, 443
downspout	219
drum trap	408
DWV tubing	394
flashing	218, 408
gutter	221
pipe	393
reglet	221
rivet	143
roof	217
salvage	391
shielded cable	444
wall covering	283
wire	445
Corbel formwork	96
Core drill	11, 19, 32
testing	11
Coreboard	271
Cork expansion joint	100
floor	280
insulation	355
tile	280
wall tile	283
Corner base cabinet	189
bead	268
block	128
guard	303
protection	303
wall cabinet	190
Cornice drain	406
molding	186
painting	289
stone	131
Correspondence lift	384
Corrosion resistant fan	432
resistant pipe	394
Corrugated metal pipe	67
pipe	67
roof tile	208
siding	209, 210, 212
Cost control	10
mark-up	9
Costs of trees	500
Cot prison	329
Counter bank	321
checkout	324
door	239
flashing	221
hospital	344
top	172, 339
top demolition	39
window	311
Countertop backsplash	172
sink	415
Country club S.F. & C.F.	471
Course base	69
wearing	70
Court air supported	352
handball	356
paddle tennis	83
racquetball	356
squash	355
tennis	74
Courthouse S.F. & C.F.	471
Cove base	280
base ceramic tile	273
base terrazzo	275
molding	186
molding scotia	186
Cover convector	172
ground	84
manhole	69
pool	364
stadium	352
stair tread	13
tank	352
trench	167
walkway	311
Covers expansion	170
CPVC pipe	397
valve	401
Crane	386
bucket	20
climbing	25
crawler	25
hydraulic	25
material handling	25, 386
mobilization	57
rail	386
scale	317
tower	25
Crawler crane	25
Crematory	329
Crew	7
survey	6
Cribbing concrete	79
Crimp termination cable	443
Critical path schedule	9
Cross brace	148
wall	322
Crossing pedestrian	83
Crown molding	186, 187
Crushed gravel	90
stone sidewalk	72
Cubic foot building cost	470
Cubicle	302
curtain	302
detention	329
shower	415
toilet	300
Culvert end	63
formwork	93
reinforced	62
Cupola	306
Cupric oxychloride floor	282
Curb	71
and gutter	71
asphalt	71
concrete	71
edging	72, 147
formwork	94, 95
granite	72, 130
inlet	72
precast	72
removal	35
roof	180
seal	73
terrazzo	275, 276
Curing blanket	114
concrete	71, 90, 114
paper	206
Current transformer	458
transformer cabinet	451
Curtain air	241
cubicle	302
damper fire	431
divider	338

gymnasium divider 338
rod 316
sound 356
track 323
wall 262
wall concrete 116
wall glass 262
Curved stair 192
Custom casework 189
Cut stone trim 131
Cutoff pile 56
Cutout 35
counter 172
slab 35
Cutter brush 20
Cutting asphalt 40
block 172
concrete 41
saw 40
steel 41, 145
torch 24
tree 42
Cylinder concrete 11
lockset 255
recore 255, 256

D

Dairy case 324
Damper 431
fireplace 305
foundation vent 122
multi-blade 431
Dampproofing 201
Darby finish concrete 112
Darkroom 354
door 336
equipment 335
revolving 354
sink 336
Dasher hockey 365
Data safe 320
Day gate 239
Deadbolt 254
Deadlocking latch 254
Deciduous shrub 86
tree 86
Deciduous tree 86
Deck cementitious 115-117
concrete cellular 115
drain 406
edge form 157
metal 156
roof 156, 182, 183
steel 156
wood 183, 184
Decontamination chamber 369
enclosure 369
equipment 370
Decorative block 126
Decorator device 451
switch 451
Deep freeze 334
seal trap 408
therapy room 357
well pump 62
Dehumidifier 334
Delicatessen case 324
Delineator barrier 80
Delivery charge 46
Deluge valve assembly sprinkler . 378
Demo concrete 35
Demolition 30, 33, 34, 37, 39, 41, 88, 390
asbestos 369
baseboard 39
boiler 390
brick 37, 38
cabinet 39
casework 39
ceiling 35
ceramic tile 37
chimney 38
column 38
concrete 36, 37
concrete block 36
door 36
drywall 35
ductwork 390
electric 440, 441
enclosure 370
explosive 33
fireplace 39
flooring 37
framing 38
glass 41
granite 39
gutter 39
hammer 20
house 33
HVAC 390
implosive 33
joist 38
lath 35
masonry 36, 38, 41
metal deck 39
metal stud 41
millwork 39
paneling 39
partition 41
pavement 34, 35
plaster 35, 41
plumbing 390
plywood 35, 39, 41
post 38
rafter 38
railing 39
roofing 39, 40, 369
selective 35
siding 39
site 33, 34
steel window 41
stucco 41
terra cotta 41
terrazzo 37
tile 35
trim 39
wall 41
walls and partitions 41
window 41
wood 35
Demountable partition 312
partitions 312
Dental equipment 341
metal interceptor 407
Department store S.F. & C.F. 471
Depository night 321
Derail 169
Derrick 25
guyed 25
stiffleg 25
Desk 348, 349
Detection system 375
tape 30
Detector infra-red 375
motion 375
smoke 375
temperature rise 375
ultrasonic motion 375
Detention cubicle 329
Detention doors & frames 239
Detention equipment 329
Developing tank 336
Device & box wiring 450
combination 452
decorator 451
exit 256
GFI 453
panic 256
receptacle 453
residential 451
wiring 456
Dewater 26, 43, 44
pumping 43
Dewatering 43
Diamond lath 267
Diaphragm pump 24
Diesel fire pump 377
generator 457
hammer 20
Diffuser ceiling 434
opposed blade damper 434
perforated aluminum 434
rectangular 434
steel 434
T-bar mount 434
Dimensionaire ceiling 353
Dimensions & wts. of sheet steel 542
Dimmer switch 451, 456
Direct expansion A/C 428
Directional sign 83
Directory 307
board 307
building 307, 468
shelf 315
Disappearing stairs 335
Discharge hose 23
Dishwasher 331, 334
Diskette safe 320
Dispenser hot water 411
napkin 316
product 326
soap 316
toilet tissue 316
towel 316
Dispersion nozzle 378
Displacement caisson 60
Display case 344
Disposable bag 370
Disposal 34, 40
asbestos 371
building 40
charges 371
field 64
garbage 334
waste 328
Distiller medical 340
water 341
Distribution box 64
section electric 461
Distributor asphalt 23
Ditching 43, 45
Divider curtain 338
strip terrazzo 276
Diving board 364
Dock board 328
board magnesium 328
boat 54
bumper 327
leveler 328
loading 327
shelter 328
truck 327
Doctor register 467
Dog command 18
Dome drain 407
fiberglass 94
geodesic 359, 360
metal 94
roof 263
steel 153
structures 558
wood 182
Domes 359
Door & grilles coiling 239
& panels access 238
accordion 235
acoustical 242, 314
air curtain 241
air lock 353
aluminum 244, 245
automatic entrance 245
balanced 246
bell residential 454
bell system 467
bi-fold 229, 234
bi-passing closet 234
birch 235
blind 194
bulkhead 238
bullet resistant 321
bumper 253, 254, 259
cabinet 190, 191
cafe 233
catch 191
cleanout 305
closer 253
closet 229, 235
cold storage 240
combination storm 234
commercial 228, 236
composite 228
conveyor 241
cooler 241
counter 239
darkroom 336
demolition 36
double 229
duct access 430
dutch 234
dutch oven 305
entrance 234, 245
fire 229, 235, 236, 240, 375
fire rated access 238
flexible 242
floor 238, 239
flush 231, 232, 234
flush wood 231
folding 314
folding accordion 235
frame 228, 230, 236, 362
frame grout 120
frame interior 237
frame lead lined 357
glass 237, 242, 245
handle 191
hangar 241, 361
hardware 252
industrial 240
kennel 242
kick plate 254
knob 255
labeled 229, 235, 236
metal 228, 242, 362
metal access 238
mirror 261
molding 187
moulded 233
opener 240, 243, 252
operator 252
overhead 242, 243
overhead commercial 242
panel 233

INDEX

Index

paneled	229
partition	313
passage	231, 235
plastic	230
prefinished	232
pre-hung	230
prison	329
protection	259
pull	256
refrigerator	240
release	468
removal	36
residential	229, 234, 237
residential garage	243
revolving	246, 336, 353
revolving entrance	246
rolling	240
roof	222
sauna	355
seal	328
shower	301
sill	187, 236, 258
sliding	230, 237
special	238, 242
stain	287
stainless steel (access)	238
steel	228, 229
stop	254, 262
storm	237
swing	242
switch burglar alarm	375
telescoping	243
threshold	237
traffic	244
varnish	287
vault	239
vertical lift	244
weatherstrip	258
weatherstrip garage	258
wire partition	312
wood	230-232, 236
Doors & frames detention	239
Doors & windows interior paint	286
Dormer gable	181
Dormitory S.F. & C.F.	472
Double acting door	242
brick	124
hung window	246, 248, 249
tee beam	115
wall pipe	64, 65
weight hinge	257
Dovetail anchor	99
Dowel support	95
Dowels reinforcing	106
Downspout	219, 220, 311
aluminum	219
copper	219
elbow	220
lead coated copper	219
steel	220
strainer	219
Dozer	21, 47
excavating bulk	49
excavation	49
Draft inducer breeching	423
Dragline	48, 51
bucket	20
Drain	406, 407
board	201
cast iron	406
deck	406
dome	407
floor	406
main	407
pipe	43
scupper	407
sediment bucket	406
shower	407
trench	407
Drainage accessory	220
mat	345
pipe	62, 63, 67, 392, 394
site	68
trap	407, 408
Drains roof	407
Drapery hardware	346
Drawer	191
kitchen	189
track	191
type cabinet	344
wood	191
Drawings as-built	27
Dredge	53
hydraulic	53
Dressing unit	349
Drill auger	33
bit	33
ceiling	140
concrete	32, 139
console dental	
core	11, 19, 32
drywall	140
earth	32, 58
hammer	23
hole	55
plaster	140
rig	32
rig mobilization	57
rock	32, 47
shop	339
steel	22
track	22
wall	139
wood	174
Drilled pier	59
Drilling horizontal	55
quarry	48
rock bolt	48
Drinking bubbler	413
fountain	413
fountain deck rough-in	413
fountain floor rough-in	413
fountain handicap	413
fountain support	409
fountain wall rough-in	413
Drip edge	220
edge aluminum	220
Drive pin	143
Driven pile	55
Driver sheeting	22
Drive-up window	321
Driveway	72
removal	35
Drop pan ceiling	277
panel formwork	94
Drum trap	408
trap copper	408
Dry chemical extinguisher	310
pipe sprinkler head	377
transformer	457
wall leaded	357
Drycleaner	325
Dryer clothes	334
commercial clothes	325
darkroom	336
hand	316
receptacle	453
residential	325
vent	335
Drywall	270
accessories	273
ceiling	270
column	271
cutout	36
demolition	35, 41
drill	140
frame	230
gypsum	270
nail	173
painting	292
partition	272
prefinished	271
removal	41
screw	273
Duck tarpaulin	17
Duckboard	345
Duct access door	430
accessories	430
connection fabric flexible	431
electric	448, 462
electrical underfloor	448
fitting trench	448
fitting underfloor	448
flexible insulated	430
flexible noninsulated	430
furnace	421
grille	304
heater electric	427
humidifier	427
HVAC	430
insulation	391
liner	391
mechanical	430
silencer	431
steel trench	448
trench	448
underfloor	448
underground	66
utility	66
Ductile iron fitting	61
iron pipe cement lined	60
Ductwork	430, 568
aluminum	430
demolition	390
fabric coated flexible	430
fabricated	430
galvanized	430
laboratory	339
rectangular	430
rigid	430
Dumbbell	337
waterstop	102
Dumbwaiter	380
electric	380
manual	380
Dump charge	40
truck	21, 53
Dumpster	40
partition	40
Duplex pump	409
receptacle	456
Dust collector shop	339
Dustproofing	90, 113
Dutch door	234
oven door	305
DWV pipe ABS	397
PVC pipe	397
tubing copper	394

E

Earth compactor	20
drill	32, 58
rolling	47
vibrator	47, 52
Earthwork	46
equipment	19, 47
haul	51
Eave overhang	362
Ecclesiastical equipment	321
Echo chamber	356
Economy brick	123
Edge drip	220
form	94, 96, 157
Edging curb	72, 147
Education religious S.F. & C.F.	475
Educational TV studio	468
Efflorescence testing	11
Eggcrate ceiling	277
EIFS	205
Ejector pump	417
Elastomeric roof	215
waterproofing	200
Elbow downspout	220
pipe	396
Elderly housing S.F. & C.F.	473
Electric appliance	334
backstop	337
ballast	465
baseboard heater	429
bed	349
boiler	419
cable	442, 445
capacitor	458
chalkboard	298
curing	114
demolition	440, 441
duct	448, 462
duct heater	427
dumbwaiter	380
feeder	462
fire pump	377
fixture	463, 465
furnace	421
generator	23
generator set	456
heater	334
heating	429
hinge	256
hoist	386
lamp	466
log	305
metallic tubing	446
meter	458
panelboard	460
pool heater	420
service	459
stair	335
switch	456, 459
unit heater	429
utility	66
water heater	417
wire	445
Electrical box	450
cabinet	450
conduit	446
conduit greenfield	447
fee	6
laboratory	340
pole	65
site work	65
tubing	446
Electronic air cleaner	436
closer	253
control system	374
Elementary school S.F. & C.F.	475
Elevated conduit	447
floor	109
slab	109, 502
slab concrete	111
slab formwork	93
water tank	365
Elevating scraper	50

Index

Elevator 380, 561-563
   construction 26
   fee 6
   holeless 380
   options 382
   shaft wall 273
Embankment fill 51
Embossed print door 233
Emergency call system 467
   lighting 466
Employer liability 8
EMT 446
Emulsion adhesive 281
   asphaltic 84
   pavement 73
   penetration 70
   sprayer 23
Encapsulation 371
   pipe 371
Encasement pile 58
Enclosure acoustical 356
   bathtub 302
   decontamination 369
   demolition 370
   shower 302
   swimming pool 363
   telephone 315
End culvert 63
Energy circulator air solar 373
   system controller solar 373
   system heat exchanger solar ... 373
Engineer brick 124
Engineering fee 6, 488
Entrance aluminum 244
   and storefront 244
   canopy 311, 362
   door 231, 234, 245
   frame 236
   lock 254
   screen 300
   sliding 245
   strip 244
EPDM adhesive 200
   roof 215
Epoxy 44
   coating 87, 113, 293
   dustproofing 90
   fiberglass wound pipe 394
   floor 276, 282
   grating 196
   grout 44, 117, 274
   lined silo 363
   resin 90
   terrazzo 276, 282
   wall coating 293
Equipment 19, 324
   athletic 82
   bank 320, 321
   barber 324
   cinema 325
   contractor 320
   darkroom 335
   dental 341
   detention 329
   earthwork 19, 47
   ecclesiastical 321
   fire 376
   formwork 95
   foundation 95
   general 9
   gymnasium 337
   health club 336
   hospital 340
   installation 338
   insurance 8
   laboratory 339

laundry 325
library 322
loading dock 327
maintenance 320
medical 340
mobilization 46, 57, 60
parking 327
playfield 82
playground 82
rental 19, 25, 490
security and vault 320
service station 326
shop 338
stage 323
swimming pool 364
waste handling 328
Erosion control 54
Escalator 384
Escalators 563
Escape fire 164
Estimate electrical heating 429
Estimating 480
Evergreen shrub 85
   tree 85
Excavating bulk bank measure 48
   bulk dozer 49
   equipment 498
Excavation 47, 48, 53, 497
   backhoe 51
   bulk scraper 50
   cofferdam 49
   dozer 49
   hand 47, 50, 52
   hauling 53
   machine 47
   planting pit 86
   rock 47
   scraper 50
   septic tank 64
   structural 50, 51
   trench 51-53
Excavator rental 19
Exchanger heat 425
Exercise equipment 337
   ladder 337
   rope 338
   weight 337
Exhaust hood 334, 339
   vent 435
Exhauster roof 433
Exhibit case 344
Exit and emergency lighting 466
   control alarm 375
   device 256
   light 466
Expanded aluminum grating 166
   steel grating 166
Expansion anchor 140
   bolt 140
   control 170
   covers 170
   joint 100, 170, 222, 268, 402
   joint asphalt 100
   joint bituminous 100
   joint cork 100
   joint cover 170
   joint neoprene 100
   joint polyethylene 100
   joint polyurethane 100
   joint premolded 100
   joint PVC 100
   joint roof 222
   joint rubber 100
   shield 140
   tank 402
   tank steel 402

Expense office 9
Explosionproof box 450
   fixture 464
   lighting 464
Explosive demolition 33
Explosives 48
Exposed aggregate 72, 112
   aggregate coating 293
   aggregate panels 211
Exterior blind 194
   concrete block 127
   door frame 236
   floodlamp 465
   insulation 205
   insulation finish system 205
   lighting 465
   lighting fixture 465
   molding 186
   plaster 269
   pre-hung door 230
   residential door 234
   sprinkler 74
   tile 274
   wood frame 236
Extinguisher ABC 310
   carbon dioxide 310
   chemical dry 310
   fire 310
   installation 310
   pressurized fire 310
   standard 310
Extra work 9
Extractor 20
   air 430
   industrial 325
Extrusion aluminum 152
Eye wash fountain 414
   wash portable 414

F

Fabric flashing 218
   stabilization 70
   stile 257, 258
   structure 352
   waterproofing 200
   welded wire 107
   wire 78
Fabricated ductwork 430
Fabrication metal 163
Fabrications plastic 195
Face brick 123, 528
   brick cleaning 532
   wash fountain 414
Faceted glass 259
Facing panel 132
   panel concrete 115
   stone 131
   tile structural 124
Factory S.F. & C.F. 472
Fan 431
   air conditioning 431
   air handling 431
   axial flow 431
   bathroom exhaust 432
   booster 431
   ceiling 432
   centrifugal 431
   coil air conditioning 428
   corrosion resistant 432
   induced draft 423
   in-line 431
   kitchen exhaust 433
   low sound 431
   paddle 455

   propeller exhaust 432
   residential 455
   roof 433
   utility set 432, 433
   vane-axial 432
   ventilation 455
   ventilator 431
   wall exhaust 433
   wiring 455
Farm type siding 211
Fascia board demolition 38
   metal 211
   wood 180, 186
Fast food equipment 331
Fastener metal 170
   steel 142
   timber 173
   wood 174
Fastenings metal 138
Faucet & fitting 411
   bathroom 411
Fee architectural 6
   engineering 6
Feeder electric 462
   section 461
   section branch circuit 461
   section circuit-breaker 461
   section frame 461
Felt 214
   asphalt 213, 214
   carpet pad 283
   tarred 214
   waterproofing 200
Fence aluminum 77
   and gate 76
   auger, truck mounted 19
   board & batten 78
   cedar 78
   chain link 17, 76, 78
   helical topping 78
   mesh 76
   metal 77
   picket paint 288
   plywood 17
   removal 34
   residential 77
   security 76
   snow 77
   steel 76
   temporary 17
   tennis court 78
   treated pine 78
   wire 17, 76
Fences 288
Fencing wire 78
Fiber cement shingle 207
   cement siding 211
   optic cable 444
   optics 443
   reinforcing 108
   tube formwork 92
   wood 116
Fiberboard insulation 204
Fiberglass tank 367
Fiberglass 195
   angle 195
   area wall cap 258
   bench 81
   blown in 201
   castings 195
   ceiling board 276
   cooling towers 424
   cross 322
   cupola 306
   dome 94
   door 242

Index

flagpole 306
flat sheet 195
formboard 117
formwork 92
grating 195
handrail 195
insulation 201-203, 362, 391
panel 13, 210, 285, 363
planter 87
reinforced ceiling 355
round bar 195
round tube 195
shade 346
shower stall 301
square bar 195
square tube 195
stair tread 167, 195
steeple 322
threaded rod 195
trash receptacle 81
wall lamination 293
waterproofing 201
wide flange beam 195
wool 202
Field disposal 64
  mix concrete 109, 516
  office 12
  office expense 12
  seeding 85
Fieldstone 132
Fill 47, 51
  cellular 117
  embankment 51
  floor 110
  gravel 30, 51
Filler block 293
  strip 257
Fillet weld 145
Film equipment 325, 336
  polyethylene 84
Filter air 436
  grille 434
  mechanical media 436
  stone 54
  swimming pool 364
Filtration air 369
  equipment 364
Fine grade 47, 85
Finish carpentry 186, 189
  concrete 71, 112, 113
  float 113
  floor 280, 293
  hardware 253
  keene cement 268
  lime 120
  nail 173
  refrigeration 354
  wall 113
Finisher floor 19
Finishes wall 283
Fir column 193
  floor 279
  molding 187
  roof deck 184
  roof plank 183
Fire alarm 375
  brick 133
  call pullbox 376
  clay 133
  damper curtain type 431
  door 229, 235, 236, 240, 375
  door frame 230
  equipment 376
  equipment cabinet 310
  escape 164
  escape disappearing 335
extinguisher 310
extinguisher portable 310
extinguishing system 310, 378
horn 375
hose 376
hose adapter 376
hose nozzle 376
hose storage cabinet 310
hose valve 377
hydrant 30, 376
protection 310
protection kitchen 332
protection system 376
pump 377
rated closer 253, 254
rated tile 124
resistant drywall 271
resistant wall 272
retardant lumber 172
retardant pipe 394
retardant plywood 172
signal bell 375
station S.F. & C.F. 472
suppression automatic 378
Firebrick 133
Fireplace accessory 305
  box 133
  built-in 305
  damper 305
  demolition 39
  form 305
  free standing 304
  mantel 193
  masonry 133
  prefabricated 304
Fireproof column 147
Fireproofing 223
  plaster 223
  plastic 223
  spray 223
Firestop gypsum 267
  wood 178
Firestopping 224, 549
Fitting bus-duct 462
  conduit 447
  greenfield 447
  grooved joint 396
  grooved joint pipe 396
  poke-thru 449
  tee 61
  underfloor duct 448
  vent chimney 422
  water pipe 61
  wye 61
Fixed blade louver 435
  booth 350
  roof tank 366
Fixture ballast 465
  bathroom 414-416
  carrier 409
  electric 463
  explosionproof 464
  explosionproof mercuryvapor . 465
  exterior mercury vapor ... 465
  fluorescent 463, 464
  incandescent 464
  incandescent vaportight .. 464
  interior light 463, 464
  mercury vapor 464
  plumbing ... 62, 407-409, 413, 415, 416
  removal 390
  residential 454
  sodium high pressure 465
  support handicap 409
  vandalproof 465
Flagging 73, 278
  slate 73
Flagpole 306
  aluminum 306
  fiberglass 306
  foundation 307
  wall 307
Flange clip 104
Flanged coil 428
Flashing 217, 218, 221
  aluminum 211, 217, 408
  asphalt 218
  butyl 218
  cement 217
  copper 218, 408
  counter 221
  fabric 218
  lead coated 218
  masonry 217
  membrane 214
  metal 362
  PVC 218
  stainless 219
  valley 206
  vent 408
Flat plate formwork 93
  sheet fiberglass 195
  slab formwork 94
Flatbed truck 22
Flexible conduit & fitting .. 456
  connector 402
  door 242
  duct connection fabric ... 431
  ductwork fabric coated ... 430
  insulated duct 430
  metal hose 403
  metallic conduit 447
  plastic netting 196
Float concrete 19
  finish 113
  finish concrete 112
  glass 259
Floater equipment 8
Floating dock 54
  pin 254
  roof tank 366
Flood coating 214
Floodlamp exterior 465
Floodlight 23
  pole mounted 465
Floor 278
  abrasive 112
  access 304
  acid proof 278
  asphalt block 73
  brick 73, 278
  carpet 282
  ceramic tile 274
  check 254
  cleaning 27
  cleanout 405
  closer 253
  color 113
  concrete 90
  conductive 276, 280, 282
  cork 280
  deck steel 156
  door 238, 239
  drain 406
  elevated 109
  epoxy 276, 282
  expansion joint 170
  fill 110
  finish 293
  finisher 19
  flagging 73
framing removal 36
grating 165, 166, 196
hatch 238
marble 131
mastic 282
mat 345
metal deck 156
nail 173
neoprene 282
oak 279
paint 289
pans 502
parquet 279
patching concrete 117
pedestal 304
plank 178
plywood 184
polyacrylate 282
polyester 282
polyethylene 280
polyurethane 281
portable 323
quarry tile 274
refrigeration 355
register 435
removal 33
rubber 281
sander 24
scale 317
scupper 407
slab formwork 94
slate 132
sleeper 180
stain 289
stone 72
stressed skin 182
subfloor 184
terrazzo 276, 555
tile terrazzo 275
topping 113
underlayment 184
varnish 289
vinyl 281
wood 279
wood composition 279
Flooring 278
  composition 281
  demolition 37
Flow meter 404
Flue chimney 133
  chimney metal 422
  liner 133, 134
  prefab metal 422
  screen 305
  tile 133
Fluid heat transfer 373
Fluorescent fixture 463, 464
  lamp 466
Fluoroscopy room 357
Flush automatic 411
  bolt 253
  door 231, 232, 234
  tube framing 262
  valve 411
  wood door 231
Flying truss 101
FM200 fire extinguisher 378
Foam glass insulation 203
  insulation 202
  pipe covering 392
  urethane 226
Foil aluminum 202, 206, 285
  back insulation 204
  barrier 285
Folded plate roof 182
Folder laundry 325

Index

Folding accordion door 235
  accordion partition 314
  bench 337
  door 314
  door shower 301
  gate 312
  partition 314
  table 348
  window 247
Food mixer 331
  service equipment 330
  warmer 331
Foot valve 401
Football goalpost 82
  scoreboard 338
Footing concrete 110, 111
  formwork 95
  pressure injected 60
  reinforcing 106
  removal 34
  spread 95
Forged valves 401
Forklift brick 23
Form base 102
  edge 94, 96, 157
  fireplace 305
  liner 97
  material 501
  patch 102
  polystyrene 97
  release 90
  slab 156
Formbloc 126
Formboard 117
Forms 503, 504
  concrete 503, 504
  slab 502
Formwork 92, 93, 96
  accessory 98-102
  beam 92
  beam and girder 92
  box out opening 94
  bulkhead 95, 96
  buttress 96
  chase 102
  columns 92
  concrete 92
  corbel 96
  culvert 93
  curb 94, 95
  drop panel 94
  elevated slab 93
  equipment 95
  fiber tube 92
  flat plate 93
  flat slab 94
  floor slab 94
  footing 95
  foundation 95
  gang wall 97
  girder 92
  hanger 100
  insert 101
  interior beam 92
  labor hours 505
  liner 97
  lintel 97
  oil 102
  pilaster 98
  plywood 92, 93, 96
  retaining wall 97
  sill 97
  sleeve 102
  snap tie 102
  stair 96
  steel frame 93
  structural 92
  trench 96
  upstanding beam 92
  void 94
  wall 96
Fossil fuel boiler 420
Foundation caisson 58
  chimney 109
  concrete 110
  concrete block 127
  equipment 95
  flagpole 307
  formwork 95
  mat 110, 111
  pile 56
  scale pit 317
  underpin 44
  vent 122
  wall 127
Fountain 76
  drinking 413
  eye wash 414
  face wash 414
  lighting 76
  wash 412
Frame baked enamel 230
  door 228, 230, 236, 362
  drywall 230
  entrance 236
  fire door 230
  grating 166
  labeled 230
  metal 229
  metal butt 230
  roof 148
  space 153
  steel 229
  welded 230
  window 251
  wood 190, 236
Framing aluminum 152
  anchor 174
  angle 148
  band joist 161
  beam & girder 176
  canopy 146
  ceiling 177
  channel 148
  coldformed 157
  demolition 38
  heavy 177
  joists 162
  laminated 185
  lightweight 148
  metal 145
  metal joists 162
  metal partitions 158
  metal studs 158
  opening 362
  partition 266
  removal 38
  roof rafters 179
  roof truss 185
  sleepers 180
Framing, stair stringer 179
Framing steel 151
  suspended ceiling 177
  timber 177
  tube 262
  wall 181
  window 362
  window wall 148, 262
  wood 176-181
Fraternal club S.F. & C.F. 471
Fraternity house S.F. & C.F. ... 472
Freestanding chalkboard 298

Freeze deep 334
Freezer 324, 331, 355
  door 241
Freight elevators 560
Friction collar 99
  pile 56, 58
Front end loader 48
  shovel 21
  vault 239
Frp panel 284
Fryer 331
Fuel storage tank 366
  tank 367
Full vision door 245
  vision glass 259
Fume hood 339
Funeral home S.F. & C.F. 472
Furnace duct 421
  electric 421
  gas 421
  gas fired 421
  hot air 421
  oil fired 421
  wall 421
Furnishings site 81
Furniture 347
  accessory 345
  hospital 349
  hotel 348
  library 349
  office 347
  restaurant 348
  school 348
Furring and lathing 266
  ceiling 181, 266, 267
  channel 266, 273
  metal 266
  steel 266
  wall 181, 267
Fusible link closer 253
  switch branch circuit 461
Fusion machine butt 22

G

Gabion 54, 79
Gable dormer 181
Galvanized chamfer strip 99
  ductwork 430
  roof 210
  steel reglet 221
  structural steel 152
  wire gabion 79
Galvanizing 152
Gang wall formwork 97
Gantry crane 386
Garage 359
  door 242, 243
  door weatherstrip 258
  residential 359
  S.F. & C.F. 472
Garbage disposal 334
Garden house 360
Gas cocks 398
  connector 402
  fired boiler 419
  fired furnace 421
  fired infra-red heater 429
  fired pool heater 420
  fired space heater 421
  furnace 421
  generator 457
  heat air conditioner 426
  incinerator 329
  log 305

pipe 65
station formwork 98
station piping 64
station tank 367
stop valve 398
vent 422
water heater 418
Gasket neoprene 226
Gasoline generator 457
  pump 326
  tank 366
Gate chain link 77
  day 239
  fence 77
  folding 312
  security 312
  swing 77
  valve 400, 401
  valve soldered 398
Gauging plaster 268
General contractor's overhead .. 481
  equipment 9
Generator construction 23
  diesel 457
  emergency 456
  gas 457
  gasoline 457
  set 457
  steam 341
Geodesic dome 359, 360
GFI receptacle 453
Girder formwork 92
  reinforcing 106
  removal 37
  wood 176, 178
Glass 259, 261
  acoustical 259
  and glazing 259
  bead 262
  bead molding 187
  bevel 259
  block 129, 532
  block skylight 252
  bulletin board 300
  bulletproof 260
  curtain wall 262
  demolition 41
  door 237, 242, 245
  door astragal 257
  door shower 301
  faceted 259
  float 259
  full vision 259
  heat reflective 260
  insulating 260
  laminated 260
  lead 357
  lined water heater 335
  low emissivity 259
  masonry 129
  mirror 261, 316
  mosaic 275
  obscure 260
  patterned 260
  pipe 392, 393
  railing 169
  reflective 260
  safety 260
  sandblast 260
  sheet 260
  shower stall 301
  spandrel 260
  tempered 259
  tile 261
  tinted 259
  vinyl 261

Index

window	261
window wall	262
wire	261
Glassware washer	339
Glasweld	182
Glaze coat	292
Glazed block	128, 129
brick	123
ceramic tile	273
coating	293
wall coating	293
Glazing	259
labor	552
plastic	261
polycarbonate	261
Globe valve	399, 401
Glove bag	368
box	339
Glued laminated	185
laminated construction	185
Goalpost	82
football	82
soccer	82
Golf shelter	82
tee surface	280
Gore line	71
Grab bar	316
Gradall	20, 48
Grade beam	95, 111
fine	47, 85
wall	110
Grader motorized	20
Grading	47, 53
Grandstand	82, 360
Granite	130
building	130
chips	84
conductive floor	282
curb	72, 130
demolition	39
indian	72
paver	130
paving block	73
sidewalk	73
Granolithic finish	520
finish concrete	112
Grass cloth wallpaper	284
lawn	85
seed	85
sprinkler	74
Grating	164
aluminum	165
area wall	258
area way	258
banding	166
fiberglass	195
floor	165, 166, 196
frame	166
steel	167
Gravel bankrun	30
base	70
concrete	90
crushed	90
fill	30, 51
pack well	61
pea	84
roof	90
stop	220
Gravity retaining wall	79
Grease interceptor	407
Greenfield conduit	447
fitting	447
Greenhouse	360
air supported	352
cooling	360
Grid spike	174

Griddle	331
Grille air return	434
aluminum	434
convector cover	172
decorative wood	193
duct	304
filter	434
painting	289
plastic	434
roll up	239
top coiling	239
window	251
Grinder pump system	416
shop	339
terrazzo	19
Grooved block	126
joint fitting	396
joint pipe	395
Ground	181
box hydrant	412
clamp water pipe	441
cover	84
face block	126
fault protection	458
rod	441
socket	82
water monitoring	12
Grounding	441
& conductor	442-445
wire brazed	441
Group shower	415
wash fountain	412
Grout	120
Grout cavity wall	120
cement	117, 120
column base	117
concrete	120
concrete block	120
door frame	120
epoxy	44, 117, 274
non-shrink	117
pressure	44
tile	273
topping	118
underlayment	118
wall	120
Grouted bar	108
strand	108
Guard corner	303
gutter	221
lamp	466
service	18
wall	303
window	168
Guardrail temporary	17
Guards wall & corner	303
Guidance inertial	7
Guide rail	80
rail cable	80
rail removal	34
rail timber	80
Guide/guard rail	80
Gunite	113
mesh	113
Gutter	221, 362
aluminum	221
copper	221
demolition	39
guard	221
lead coated copper	221
monolithic	71
stainless	221
steel	221
strainer	221
swimming pool	364
vinyl	221

wood	221
Gutting	38
Guyed derrick	25
tower	66
Guying tree	87
Gym floor underlayment	279
mat	338
Gymnasium divider curtain	338
equipment	337
floor	279, 281
floor expansion joint	170
S.F. & C.F.	473
squash court	355
Gypsum block demolition	36
board	270
board accessories	273
board leaded	357
board system	272
cement	120
drywall	270
fabric wallcovering	284
firestop	267
lath	267, 553
lath nail	173
partition	272, 312
plaster	268, 554
poured	117
roof	117
roof plank demolition	39
shaft wall	273
sheathing	184
weatherproof	183

H

H pile	57
Hair interceptor	407
Half round molding	187
Hammer bush	113
chipping	22
demolition	20
drill	23
hydraulic	23
pile	20
pile mobilization	57
Hand carved door	233
clearing	42
dryer	316
excavation	47, 50, 52
hole	30
hole concrete	65
rail	165
split shake	208
trowel finish concrete	112
Handball court	356
court backstop	82
Handicap drinking fountain	413
fixture support	409
lever	254
opener	252
ramp	110
water cooler	414
Handle door	191
Handling material	386
rubbish	40
unit air	425
waste	328
Hand-off-automatic control	459
Handrail	165, 304
aluminum	304
and railing	165
fiberglass	195
wood	187, 191
Hangar	361
door	241

Hanger beam	174
formwork	100
joist	174
Hanging wire	266
Hardboard cabinet	189
overhead door	243
paneling	188
siding	213
soffit	188
tempered	188
underlayment	184
Hardener concrete	90, 113
Hardware	252
cabinet	191
cloth	78
door	252
drapery	346
finish	253
ooor	252
panic	256
rough	176
window	252
Hardwood carrel	323
floor stage	323
grille	193
Hasp	254
Hat and coat strip	316
rack	318
Hatch ceiling	238
floor	238
roof	222, 223
smoke	223
Haul earthwork	51
Hauling excavation	53
truck	53
Hay	84
bale	54
Hazardous waste cleanup	30
waste disposal	30
waste handling	328
Head sprinkler	377
Header pipe	26
wood	181
Headers & beams boxed	158, 161
Health club equipment	336
Hearth	133
Heat baseboard	428
electric baseboard	429
exchanger	425
exchanger plate	425
greenhouse	360
loss	566
pump	427
pump residential	455
radiant	353, 429
recovery	329
recovery air to air	429
reflective glass	260
temporary	12, 91
therapy	340
transfer fluid	373
transfer package	418
Heated doorway	241
Heater & fan bathroom	433
contractor	23
electric	334
electric residential	429
electric wall	429
floor mounted space	421
infra-red	429
sauna	355
swimming pool	420
terminal	429
unit	421, 428
warm air	421
water	335, 417, 418

Index

Heating . 366, 373, 391, 392, 419-422, 425 , 565
& cooling classroom 428
cable 429
electric 429
estimate electrical 429
hot air 421, 428
hydronic 420, 428
insulation 391
subsoil 365
unit convector 428
Heavy construction 81
framing 177
lifting 491
Hemlock column 194
Hex nut 139
nuts steel 139
Hexagonal face block 127
High bay lighting 464
build coating 293
chair 104
early strength cement 91
installation conduit 447
intensity discharge lighting 464, 465
rib lath 267
rise glazing 259
school S.F. & C.F. 476
strength block 126, 127
strength bolts 141, 533
strength concrete 90
strength steel 151, 537
temperature vent chimney 422
Highway paver 73
paving 81
sign 83
Hinge brass 254
cabinet 191
continuous 257
electric 256
hospital 257
paumelle 256
prison 257
residential 254
school 257
security 256
special 256
steel 254
Hinges 551
Hip rafter 180
Historical Cost Index 7
Hockey dasher 365
scoreboard 338
Hoist 386
and tower 26
automotive 326
chain 386
contractor 26
electric 386
lift equipment 25
overhead 386
personnel 26
Holder closer 253
Holding tank 62
Holdown 174
Hole drill 55
drilling 139
Hollow core door 231, 235
metal 229
metal door 228
metal frames 549
metal stud partition 268
wall anchor 140
Honed block 127
Hood exhaust 339
fire protection 332
fume 339
range 334
Hook coat 316
robe 316
Hooks clevis 168
Hopper refuse 385
Horizontal auger 19
boring 55
drilling 55
shore 101
Horn fire 375
Hose adapter fire 376
air 22
braided bronze 403
discharge 23
equipment 376
fire 376
metal flexible 403
nozzle 376
rack 376
rack cabinet 310
suction 23
valve cabinet 310
water 23
Hospital cabinet 344
door hardware 253
equipment 340
furniture 349
hinge 257
kitchen equipment 333
partition 302
S.F. & C.F. 473
tip pin 254
Hot air furnace 421
air heating 421, 428
water boiler 419, 420
water dispenser 411
water heating 419, 428
water-steam exchange 418, 425
water-water exchange 425
Hotel cabinet 317
furniture 348
lockset 255
House demolition 33
garden 360
telephone 468
Housewrap 206
Housing elderly S.F. & C.F. 473
public S.F. & C.F. 473
S.F. & C.F. 473
HTRW 368
Hubbard tank 340
Humidification equipment 360
Humidifier 335
duct 427
room 427
Humidifiers 427
Humus peat 84
HVAC demolition 390
duct 430
piping specialties 402, 404
Hydrant fire 30, 376
ground box 412
removal 34
wall 412
water 412
Hydrated lime 120
Hydraulic chair 341
crane 25, 386
dredge 53
hammer 23
jack 26
jacking 491
seeding 85
Hydrocumulator 416
Hydrodemolition concrete 34
Hydronic heating 420, 428
Hypalon neoprene roofing 215

I

Ice rink 365
skating equipment 365
skating rink S.F. & C.F. 473
Icemaker 331, 335
Impact barrier 80
wrench 22
Implosive demolition 33
Improvement site 82
Incandescent fixture 464
Incinerator gas 329
municipal 329
waste 329
Inclined ladder 165
ramp 384
Increaser vent chimney 422
Incubator 340
Index construction cost 7
Indian granite 72
Indicating panel burglar alarm 375
Induced draft fan 423
Industrial address system 467
chimneys 527
door 238
equipment 338
equipment installation 338
folding partition 314
lighting 464
railing 165
safety fixture 414
window 246
Inert gas 31
Inertial guidance 7
Infra-red broiler 330
detector 375
heater 429
heater gas fired 429
Inlet curb 72
In-line fan 431
Insecticide 54
Insert formwork 101
slab lifting 105
Inspection technician 11
Installation extinguisher 310
Instrument switchboard 458
Insulated glass spandrel 261
panels 205
wall panels 116
Insulating glass 260
Insulation 201, 202, 391, 392
batt 203
blanket 203, 391
board 202
boiler 391
breeching 391
building 201
cavity wall 202
cellulose 202
composite 205
concrete 114, 117
duct 391
exterior 205
fasteners 202
fiberglass 201-203, 362, 391
foam 202
foam glass 203
insert 121
isocyanurate 203
masonry 202
mineral fiber 204
pipe 392
polystyrene 202, 203
refrigeration 355
removal 368
roof 204, 362
roof deck 204
sandwich panel 211
screw 202
shingle 206
spray 202
subsoil 365
urethane 202
vapor barrier 206
vermiculite 202
wall 121, 203
Insurance 8, 480
builder risk 8
equipment 8
public liability 8
Intake-exhaust louver 435
vent 436
Integral finish concrete 112
floor finish 520
waterproofing 90, 91
Integrated ceiling 353
Interceptor 407
grease 407
metal recovery 407
oil 407
Intercom 467
Interior beam formwork 92
door frame 237
light fixture 463, 464
pre-hung door 231
residential door 234
wood frame 237
Interlocking block 128
Intermediate conduit 446
Interval timer 451
Intrusion system 375
Invert manhole 69
Inverted bucket steam trap 403
tee beam 114
Investigation subsurface 32
Iron alloy mechanical joint pipe 394
body valve 400
wrought 169
Ironer laundry 325
Ironing center 344
Ironspot brick 278
Irrigation system 74
Island formwork 98
Isocyanurate insulation 203
I.V. track system 302

J

Jack cable 26
hydraulic 26
mud 19
post 147
roof 180
screw 45
Jackhammer 22, 48
Jacking 55
Jail equipment 329
S.F. & C.F. 473
Jalousie 247
Jet water system 62
Jetting pump 26
Jeweler safe 320
Job condition 7
Jockey pump fire 377
Joint control 121
expansion 100, 170, 222, 268, 402
reinforcing 121

INDEX

664

Index

roof 222
sealer 225
Jointer shop 339
Joist anchor 174
   bar 154
   composite 156
   concrete 110, 114
   connector 174
   demolition 38
   hanger 174
   metal 154
   open web 154, 155
   steel 154
   web stiffeners 162
   wood 178, 184
Joists coldformed 160
   framing 162
Jumbo brick 123, 124
Jute mesh 54

K

Kalamein door 230
Keene cement 268
Kennel door 242
   fence 76
Kettle 332
Kettle/pot tar 24
Key cabinet 344
   keeper 311
Keyless lock 256
Keyway 95
Kick plate 165, 166, 254
   plate door 254
Kiln dried lumber 172
   vocational 339
King brick 124
Kiosk 361
Kitchen appliance 334
   cabinet 189, 344
   equipment 330
   equipment fee 6
   exhaust fan 433
   heat recovery 429
   sink 416
   sink faucet 411
K-lath 267
Knob door 255
Kraft paper 206

L

Labeled door 229, 235, 236
   frame 230
Labor formwork 505
   index 8
Laboratory cabinet 339
   equipment 339
   research S.F. & C.F. ... 475
   sink 339
   table 339
Ladder 23, 163
   building 163, 165
   exercise 337
   monkey 82
   rolling 15
   swimming pool 364
   towel 316
   type cable tray 446
Lag screw 141
   screws 142
Lagging 46
Lally column 146
Laminated beam 185

construction glued 185
countertop 172
countertop plastic 172
epoxy & fiberglass 293
framing 185
glass 260
lead 357
roof deck 184
veneer members 185, 186
wood 185
Lamp fluorescent 466
   guard 466
   metal halide 466
   post 168, 465
Lampholder 456
Lamphouse 325
Landfill fees 40
Landing newel 193
   stair 164, 276
Landscape fee 6
   surface 280
Laser 23
Latch deadlocking 254
   set 255
Latex acrylic 90
   caulking 226
   underlayment 281
Lath demolition 35
   gypsum 267, 553
   metal 267, 269, 553
   rib 267
Lathe shop 339
Lattice molding 187
Lauan door 235
Laundry equipment 325
   faucet 411
   sink 414
   tray 414
Lava stone 132
Lavatory 415
   faucet 411
   support 409
   vanity top 415
   wall hung 415
Lawn grass 85
   seed 85
Lazy susan 190
Leaching field chamber 64
   pit 64
Lead barrier 285
   caulking 393
   coated copper downspout .. 219
   coated copper gutter 221
   coated downspout 219
   coated flashing 218
   flashing 218
   glass 357
   gypsum board 357
   lined darkroom 354
   lined door frame 357
   paint encapsulation ... 372
   paint remediation methods ... 559
   paint removal 371
   pile 20
   plastic 357
   roof 217
   salvage 391
   screw anchor 140
   sheets 357
   shielding 357
   wool 100
Lean-to type greenhouse ... 360
Lectern 322
Lecture hall seating 348
Ledger wood 180
Lens movie 325

Let-in bracing 176
Letter sign 308
   slot 311
Leveler dock 328
   truck 328
Lever handicap 254
Lexan 261
Liability employer 8
   insurance 481
Library chair 323
   equipment 322
   furniture 349
   S.F. & C.F. 474
   shelf 322
Life guard chair 364
Lift 384
   aerial 22
   automotive 326
   correspondence 384
   parcel 384
   scissor 386
   slab 110, 520
   wheelchair 380
Lifter platform 328
Light base 98
   border 323
   dental 341
   exit 466
   fixture interior .. 463-465
   loading dock 328
   nurse call 467
   pole 65, 465
   pole aluminum 465
   pole steel 465
   portable 22
   shield 80
   stand 370
   strobe 323
   support 146
   temporary 12
   tower 23
   traffic 83
   underwater 364
Lighting 463-465
   ceiling 353
   darkroom 336
   emergency 466
   exit and emergency 466
   explosionproof 464
   exterior 465
   fixture exterior 465
   fountain 76
   high intensity discharge .. 464, 465
   incandescent 464
   industrial 464
   mercury vapor 464
   outdoor 22
   outlet 454
   residential 454
   roadway 465
   stage 323
   strip 464
   surgical 341
Lightning suppressor 451
Lightweight aggregate 90
   block 134
   concrete 109, 110, 117, 525
   concrete plank 117
   floor fill 110
   framing 148
   natural stone 132
Lime finish 120
   hydrated 120
Limestone 85, 130, 131
   coping 128
Line gore 71

Linen chute 385
   collector 385
   wallcovering 284
Liner cement 87
   duct 391
   flue 133
   formwork 97
   pipe 87
Link mat 345
Lint collector 325
Lintel 116, 131, 148
   block 127
   concrete 116
   formwork 97
   precast 116
Liquid chiller centrifugal . 423
Load test pile 57
Loadbearing studs 157
Loadcenter residential 451
Loader front end 48
   tractor 21
   vacuum 370
Loading dock 327
   dock equipment 327
   dock light 328
Loam 30, 47
Lobby collection box 312
Lock electric release 256
   entrance 254
   keyless 256
   tile 345
   time 239
   tubular 254
Locker metal 309
   steel 309
   wall mount 309
   wire mesh 309
Locking receptacle 453
Lockset communicating 255
   cylinder 255
   hotel 255
   mortise 255
Log electric 305
   gas 305
Louver 302, 434
   aluminum 302, 434
   aluminum operating 435
   coating 435
   cooling tower 434
   fixed blade 435
   intake-exhaust 435
   midget 302
   mullion type 435
   redwood 193
   silencer 431
   ventilation 193
   wall 303
   wood 193
Louvered door 231, 235
Low-voltage silicon rectifier .. 456
   switching 456
   switchplate 456
   transformer 456
Lube equipment 327
Lug terminal 443
Lumber 544, 545
   core paneling 188
   kiln dried 172
   treated 180
   treatment 172
Luminaire ceiling 353
Luminous ceiling 277, 466
   panel 277

Index

M

Macadam	70
penetration	70
Machine anchor	142
excavation	47
trowel finish	112
welding	25
Machines screws	142
Magazine shelving	323
Magnesium dock board	328
oxychloride	223
Magnetic astragal	257, 258
motor starter	459
particle test	12
Mahogany door	233
Mail box	311
box call system	467
chute	311
slot	311
Main drain	407
office expense	9
Maintenance	27
carpet	283
equipment	320
Make-up air unit	425
Mall front	245
Management fee construction	6
Manhole	68
concrete	65, 68
cover	69
electric service	65
invert	69
precast	68
raise	69
removal	34
step	69
Mansard aluminum	209
Mantel beam	193
fireplace	193
Manual dumbwaiter	380
Manufactured casework	344
Map rail	299
Maple countertop	172
floor	279
wall	356
Marble	131
aggregate	91
chips	84
coping	128
countertop	172
floor	131
screen	300
shower stall	301
sill	130
soffit	131
stair	131
synthetic	278
tile	278
Marina small boat	54
Mark-up cost	9
Masonry accessory	120, 121
adobe	129
aggregate	120
anchor	121
base	129
brick	134, 278
catch basin	68
cement	91, 120
clay	122
cleaning	136
color	120
demolition	36, 38, 41
fireplace	133
flashing	217
furring	181
glass	129
insulation	202
manhole	68
nail	173
needle	136
panel	136
point	136
reinforcing	121, 527
removal	34, 38
restoration	136
saw	24
sill	130
step	72
testing	11
toothing	36
unit	122
ventilator	122
wall	80, 125
wall tie	121
waterproofing	120
Massage table	337
Master clock system	375
Mastic floor	282
Mat blasting	48
floor	345
foundation	95, 110, 111
gym	338
wall	338
Material handling	385, 386
handling belt	385
handling system	385
hoist	26
index	8
Materials concrete	513
Meat case	324
Mechanical	564
duct	430
equipment demolition	390
fee	6
media filter	436
Median barrier	80
precast	80
Medical clinic S.F. & C.F.	474
distiller	340
equipment	340
office S.F. & C.F.	474
X-ray	341
Medicine cabinet	317
Melting snow	429
Membrane curing	114
flashing	214
roofing	213
waterproofing	200
Mercury vapor exterior fixture	465
vapor fixture	464
vapor fixture vaportight	464
vapor lamp	466
Mesh fence	76
gunite	113
partition	312
stucco	267
wire	78
Metal bookshelf	322
brush-off blast	152
butt frame	230
cabinet	344
casework	344
chalkboard	298, 314
chimney	134, 304, 422
deck	156
deck composite	156
deck demolition	39
dome	94
door	228, 242, 245, 362
door frame	244, 362
door residential	229
ductwork	430
fabrication	163
facing panel	209
fascia	211
fastener	170
fastenings	138
fence	77
flashing	362
flexible hose	403
flue chimney	422
frame	229
framing	145
furring	266
gutter	362
halide lamp	466
hollow	229
joist	154
joists bracing	160
joists bridging	160
joists coldformed	162
joists framing	162
lath	267, 269, 553
locker	309
miscellaneous	258
molding	268
ornamental	169, 258
overhead door	243
pan	94
pan ceiling	277
panel	211
partitions framing	158
pipe	60, 67
recovery interceptor	407
roof	209, 210
sash	246
screen	246
sheet	217
shelf	314
shingle	207
siding	209, 211
sign	308
soffit	212
stair	164
stair tread	102
steam cleaning	152
stud	266, 269
stud demolition	41
stud partition	158, 266
stud walls	158
studs	158
studs framing	158
support assemblies	266
threshold	258
tile	275
toilet partition	300
trash receptacle	345
truss	151
water blasting	152
window	246, 247, 363
Metallic foil	202
hardener	90, 113
waterproofing	90
Meter electric	458
steam condensate	409
venturi flow	404
water supply	404
water supply domestic	404
Metric conversion	514
rebar specs	506
Microphone	467
Microtunneling	55
Microwave oven	334
Middle school S.F. & C.F.	476
Mill construction	177
Millwork	186, 190
demolition	39
Mineral fiber ceiling	277
fiber insulation	204
insulated cable	444
roof	216
siding	182
wool blown in	201
Mirror	261, 316
ceiling board	277
door	261
glass	261
plexiglass	261
wall	261
Miscellaneous metal	258
Mix design asphalt	10
planting pit	86
Mixer concrete	19
food	331
mortar	19, 23
plaster	23
transit	19
Mixes concrete	513, 514
Mixing valve	411
Mobile shelving	323
X-ray	341
Mobilization	46, 59
air-compressor	57
equipment	45, 57, 60
Model building	6
Modification to cost	7
Modified bitumen roofing	548
Modular office system	312, 313
playground	82
Modulus of elasticity	11
Moil point	22
Moisture barrier	201
content test	11
Molding	186
base	186
brick	186
ceiling	186
chair	187
cornice	186
exterior	186
hardboard	188
metal	268
pine	186
trim	187
wood	192
wood transition	280
Monel rivet	143
Money safe	320
Monitor support	146
Monkey ladder	82
Monolithic finish concrete	112
gutter	71
terrazzo	276
Monorail	385
Monument survey	6
Mop holder strip	316
roof	214
sink	416
Mortar	120, 526
Mortar, brick and block	530
Mortar cement	274
masonry cement	120
mixer	19, 23
point	120
Portland cement	120
sand	120
testing	11
thinset	274
Mortise lockset	255
Mortuary refrigeration	340
Mosaic glass	275
Moss peat	84
Motel S.F. & C.F.	474

Index

Motion detector 375
Motor connection 456
   generator 23
   starter 459
   starter & control 459
   starter magnetic 459
   starter w/circuit protector .. 459
   starter w/fused switch 459
   support 146
Motorized car 385
   roof 363
Motors 571
Moulded door 233
Mounted booth 350
Mounting board plywood 183
Movable louver blind 346
   office partition 312
Movie equipment 325
   lens 325
   projector 326
   screen 325
   S.F. & C.F. 476
Moving building 88
   ramp 384
   ramps and walks 563
   shrub 84
   structure 88
   tree 84
   walk 384
Mowing brush 42
Mud jack 19
   sill 180
Mulch 84
   bark 84
   ceramic 84
   stone 84
Mullion type louver 435
   vertical 262
Multi-blade damper 431
Multi-channel rack enclosure . 444
Multizone air cond. rooftop .. 426
Municipal incinerator 329
Muntin window 251
Mushroom ventilator stationary . 436
Music room 354
Mylar tarpaulin 17

N

Nail 173
   lead head 357
   stake 102
Nailable concrete plank 117
Nailer 23
   pneumatic 22
   wood 178
Napkin dispenser 316
Needle masonry 136
Neoprene adhesive 200
   expansion joint 100, 222
   flashing 218
   floor 282
   gasket 226
   roof 215
   vibration pad 143
   waterproofing 200
Net safety 13
   tennis court 74
Netting flexible plastic 196
Newel 192
Newspaper rack 323
Night depository 321
No hub pipe 393
Non-destructive testing 11
Non-metallic hardener 90

Non-removable pin 254
Non-shrink grout 117
Norwegian brick 123
Nosing checkered plate 166
   mat 345
   rubber 280
   safety 280
   stair 102, 164, 167, 276, 280
Nozzle dispersion 378
   fire hose 376
   fog 376
Nurse call light 467
   call system 467
   speaker station 467
   station cabinet 344
Nursing home bed 349
   home S.F. & C.F. 474
Nut hex 139
Nylon carpet 282

O

Oak door frame 236
   floor 279
   molding 187
   paneling 188
   stair tread 193
   threshold 237
Oakum caulking 393
Obscure glass 260
Observation dome 359
   well 61
Off highway truck 22
Office expense 9
   field 12
   floor 304
   furniture 347
   medical S.F. & C.F. 474
   partition 312
   partition movable 312
   safe 320
   S.F. & C.F. 474, 477
   system modular 313
   trailer 12
Oil filled transformer 458
   fired boiler 420
   fired furnace 421
   fired space heater 421
   formwork 102
   heater temporary 23
   interceptor 407
   storage tank 366
   tank 402
   water heater 418
Olive knuckle hinge 256
Omitted work 9
One piece astragal 257
Onyx 84
Ooor hardware 252
Open web joist 154
Opener automatic 252
   door 240, 243, 252
   handicap 252
Opening framing 362
Operable partition 314
Operating cost 19
   cost equipment 19
Operator automatic 252
Options elevator 382
Ornamental metal 169, 258
   railing 169
OSHA testing 370
Outdoor bleacher 82
   lighting 22
Outlet box steel 449

lighting 454
Outrigger wall pole 307
Oven 330, 334
   cabinet 190
   convection 332
   microwave 334
Overbed table 349
Overhaul 40, 53
Overhead & profit 9, 485
   commercial door 242
   contractor 14, 27, 481
   door 242, 243
   hoist 386
   support 239
Overlapping astragal 257
Overlay face door 232, 235, 236
Overpass 81
Overtime 7, 9, 487
Oxygen lance cutting 41

P

P trap 408
   trap running 407
Package chute 385
   receiver 321
Packaging waste 371
Pad bearing 143
   vibration 143
Padding carpet 283
Paddle blade air circulator .. 432
   fan 455
   tennis court 83
Paint 556
   aluminum siding 290
   chain link fence 288
   doors & windows interior .. 286
   encapsulation lead 372
   fence picket 288
   floor 289
   floor concrete 288, 289
   floor wood 289
   removal 371
   siding 290
   trim, exterior 291
   walls masonry, exterior ... 292
Painting 556
   balustrade 289
   bar joist 155
   booth 327
   casework 285
   ceiling 292, 293
   clapboard 290
   cornice 289
   decking 289
   drywall 292
   grille 289
   parking stall 71
   pavement 71
   pipe 289
   plaster 292
   reflective 71
   siding 290
   sprayer 23
   stair tread 167
   steel 152
   steel siding 290
   stucco 290
   swimming pool 364
   temporary road 71
   tennis court 74
   thermoplastic 71
   trim 289
   truss 290
   wall 292, 293

   window 287, 288
Paints & coatings 285
Palladian windows 250
Pallet rack 315
Pan metal 94
   shower 218
   slab 110
   stair 102
Panel acoustical 285, 313
   architectural 182
   brick wall 136
   ceramic tile 274
   divider 313
   door 233, 234
   facing 132
   fiberglass 13, 210
   fire 375
   FRP 284
   luminous 277
   masonry 136
   metal 211
   metal facing 209
   portable 313
   precast 116
   sandwich 210
   sound 356
   spandrel 261
   structural 182
   wall 110, 115, 133
   woven wire 312
Panelboard 460
   electric 460
   w/circuit-breaker 459, 460
Paneled door 229
   pine door 235
Paneling 188
   board 189
   cutout 36
   demolition 39
   hardboard 188
   plywood 188
   wood 188
Panelized shingle 208
Panels insulated 205
   prefabricated 205
Panic bar 256
   device 256
   device door hardware 253
Paper building 206
   curing 114
   sheathing 206
   tubing 105
Paperhanging 283
Paperholder 316
Paraffin bath 340
Parallel bar 338, 340
Parcel lift 384
Pargeting 547
Parging cement 200
Park bench 81
Parking barrier 80
   barrier precast 81
   bumper 80
   control equipment 327
   equipment 327
   garage 359
   garage S.F. & C.F. 473
   lots paving 70
   stall painting 71
Parquet floor 279
Particle board siding 213
   board underlayment 184
   core door 232
Parting bead 187
Partition 312, 313
   acoustical 313, 314

Index

anchor	121
block	134, 135
blueboard	272
bulletproof	321
demolition	41
demountable	312
door	313
drywall	272
dust	40
folding	314
folding leaf	314
framing	266
gypsum	272, 312
hospital	302
mesh	312
movable office	312
office	312
operable	314
plaster	269
portable	314
refrigeration	355
sheetrock	272
shower	131, 301
steel	313, 314
stud	268
support	146
thin plaster	272
tile	125
toilet	131, 300
wall	272
wire	312
wood frame	179
woven wire	312
Partitions demountable	312
wire mesh	312
Parts bin	315
Passage door	231, 235
Passenger elevators	563
Patch core hole	11
form	102
roof	216
Patching asphalt	30
concrete	117
concrete floor	117
concrete wall	117
Patient nurse call	467
Patio	72
block	131, 223
door	237
Patterned glass	260
Paumelle hinge	256
Pavement	73
asphalt	74
asphaltic	70
berm	71
demolition	34, 35
emulsion	73
markings	71
painting	71
replacement	70
sealer	74
slate	132
Paver bituminous	23
concrete	23
concrete block	73
floor	278
highway	73
Paving asphaltic	492
athletic	74
block granite	73
brick	73
concrete	71
highway	81
parking lots	70
Pea gravel	84
stone	91

Peastone	84
Peat humus	84
moss	84
Pedestal floor	304
type seating	348
Pedestrian bridge	81
crossing	83
traffic control	348
Pegboard	188
Penetration macadam	70
test	10
Penthouse roof louver	435
Perforated aluminum pipe	67
ceiling	277
pipe	67
PVC pipe	67
Performance bond	8, 486
Perlite concrete	90, 117
insulation	203
plaster	268, 554
sprayed	293
Permit building	8
Personal respirator	370
Personnel hoist	26
protection	13
Petrographic analysis	10
Pew church	322
Phone booth	315
Photography	10
aerial	10
construction	10
time lapse	10
Physician's scale	340
PIB roof	215
Pickup truck	25
Picture window	246, 247, 250
Pier brick	134
concrete	109
Pilaster formwork	98
toilet partition	301
wood column	194
Pile	56
boot	58
bored	58
cap	95, 111, 112
cast in place	55
concrete	55
cutoff	56
driven	55
driver	20
driving	490
driving mobilization	57
encasement	58
foundation	56
friction	58
H	46, 57
hammer	20
high strength	45
lightweight	45
load test	56
mobilization hammer	57
pipe	56, 57
point	57
pile point	57
Pile point heavy duty	58
precast	56
prestressed	56
sod	86
splice	57, 58
steel	57
steel sheet	44
step tapered	57
testing	56
treated	58
wood	58, 499
wood sheet	45

Piling sheet	44, 495
Pillow tank	367
Pin floating	254
non-removable	254
Pine door	234
door frame	236
fireplace mantel	193
floor	279
molding	186
roof deck	184
shelving	189
siding	213
stair	192
stair tread	192
Pipe	68
& fittings	393-395, 397, 399-401, 405, 407, 408, 410-412
acid resistant	394
aluminum	67, 68
and fittings	564
bedding	47
bellow expansion joint	402
bollards	81
brass	393
bumper	303
cast iron	393
clay	64, 67
cleanout	405
column	146
concrete	62, 63, 66, 500
copper	393
corrosion resistant	394
corrugated	67
corrugated metal	67
covering	391
covering fiberglass	391, 392
CPVC	397
double wall	64, 65
drain	43
drainage	62, 63, 67, 394
DWV ABS	397
DWV PVC	397
elevated installation	390
encapsulation	371
epoxy fiberglass wound	394
fire retardant	394
fitting grooved joint	396
gas	65
glass	392, 393
grooved joint	395
header	26
insulation	392
insulation removal	368
iron alloy mechanical joint	394
liner	87
metal	67
no hub	393
painting	289
perforated	67
perforated aluminum	67
pile	56, 57
plastic	397
plastic, FRP	397
polyethylene	65
polypropylene	394
proxylene	394
PVC	63
quick coupling	26
rail	165
reinforced concrete	62
relay	43
removal	35, 390
sewage	62, 63, 67
shock absorber	405
single hub	393
soil	393

stainless	165
stainless steel	396, 397
steel	67, 147, 394, 395
subdrainage	66, 67
support framing	149
vitrified clay	63, 67
weld joint	395
wrapping	87
Piping gas station	64
specialties HVAC	402, 404
subdrainage	66
Pit cover	167
excavation	50
leaching	64
scale	317
sump	43
test	33
Pitch coal tar	214
emulsion tar	73
pocket	222
pockets	222
Pivoted window	246
Placing concrete	111, 516
reinforcment	507
Plain tube framing	262
Planer shop	339
Plank concrete	117
floor	178
grating	166
roof	183
sheathing	183
Plant bed preparation	86
power S.F. & C.F.	475
screening	21
Planter	87, 350
bench	87
concrete	87
fiberglass	87
Planting	84
Plants and planters	350
Plaque	308
Plaster	266
accessories	268
beam	268
ceiling	269
cement	354
column	268
cutout	36
demolition	35, 41
drill	140
gauging	268
ground	182
gypsum	268
mixer	23
painting	292
partition	269
partition thin	272
perlite	268
thincoat	269
vermiculite	268
wall	269
Plasterboard	270
Plastic convector cover	172
door	230
fabrications	195
faced hardboard	188
fireproofing	223
Plastic, FRP pipe	397
Plastic glazing	261
grating	196
grille	434
laminate door	232, 236
laminated countertop	172
lead	357
pipe	397
roof ventilator	432

Index

sign 308	closet 187	Posts sign 83	fence 76
skylight 252	cross arm 66	Post-tensioned concrete 522	hinge 257
valve 400	electrical 65	Potable water softener 417	S.F. & C.F. 473
Plastics structural 195	light 465	water treatment 417	toilet 329
Plate 149	portable decorative 348	Potters wheel 339	Process air handling fan 431
checkered 166, 167	steel light 465	Poured concrete 90	Proctor compaction test 11
glass greenhouse 360	utility 65, 465	gypsum 117	Produce case 324
heat exchanger 425	Police station S.F. & C.F. 475	insulation 202	Product dispenser 326
push-pull 256	Polyacrylate floor 282	Powder act. tools & fast. 143	piping 65
shear 174	Polycarbonate glazing 261	Power plant S.F. & C.F. 475	Profile block 126, 127
wall switch 456	Polyester floor 282	system & capacitor 457	Progress schedule 9
wood 181	Polyethylene coating 87	temporary 12	Project overhead 9
Platform checkered plate 167	expansion joint 100	wiring 459	sign 18
grating 166	film 84	Preblast survey 48	size modifier 572
lifter 328	floor 280	Precast architectural 115	Projected window 246, 247
telescoping 323	pipe 65	beam 114	Projection screen ... 298, 325, 326
tennis 83	pool cover 364	bridge 81	Projector movie 326
trailer 24	septic tank 64	catch basin 68	Propeller exhaust fan 432
Plating zinc 173	tarpaulin 17	column 114	Property line survey 6
Players bench 81	waterproofing 200	concrete 114-116	Protection corner 303
Playfield equipment 82	Polypropylene pipe 394	coping 128	door 259
Playground equipment 82	shower 415	curb 72	fire 310
modular 82	valve 401	lintel 116	slope 54
slide 82	Polystyrene blind 194	manhole 68	stile 257, 258
surface 281	ceiling 277, 355	median 80	termite 54
whirler 82	ceiling panel 277	members 521	winter 13
Plenum silencer 431	insulation ... 202, 203, 355	panel 116	worker 370
Plexiglass 261	Polysulfide caulking 226	parking barrier 81	Proxylene pipe 394
mirror 261	Polyurethane caulking 226	pile 56	P&T relief valve 399
Plinth 95	expansion joint 100	receptor 301	Public address system 467
Plug in circuit-breaker 463	floor 281, 282	roof 117	garage 558
in switch 463	roof 215	septic tank 64	housing S.F. & C.F. 473
wall 122	varnish 293	sill 130	Pull box 450
Plugmold raceway 449	Polyvinyl chloride roof .. 215, 216	stair 116	box explosionproof 450
Plumbing 405	soffit 212	tee 115	door 191
appliance 417, 418	tarpaulin 17	terrazzo 275	Pulpit church 322
demolition 390	Pool accessory 364	tilt-up 116	Pump 62, 409, 416
fixture . 62, 407-409, 413, 415, 416,	cover 364	wall 523	circulating 408
565	cover polyethylene 364	Pre-engineered steel building ... 361	concrete 19, 111
fixtures removal 390	heater electric 420	steel buildings 557	condensate return 408
laboratory 340	heater gas fired 420	structure 358	contractor 43
Plywood 546	swimming 364	Prefabricated building 360	diaphragm 24
clip 174	swimming S.F. & C.F. ... 476	comfort station 358	duplex 409
demolition 35, 39, 41	therapeutic 365	fireplace 304	fire 377
fence 17	Porcelain tile 274	panels 205	gasoline 326
floor 184	Porch molding 187	Prefinished door 232	heat 427
formwork 92, 93, 96	Portable air compressor 22	drywall 271	jetting 26
joist 184	booth 358	floor 279	operator 44
mounting board 183	building 12	shelving 189	pressure booster 416
paneling 188	cabinet 330	Preformed roof panel 209	rental 24
sheathing, roof and walls ... 183	eye wash 414	roofing & siding 209	sewage ejector 417
shelving 189	fire extinguisher 310	Pre-hung door 230	shallow well 62
sidewalk 17	floor 279	Premolded expansion joint 100	staging 13, 489
siding 213	heater 23	Preparation plant bed 86	submersible 24, 417
sign 308	light 22	surface 293, 294	sump 335, 417
soffit 188	panel 313	Pressure booster pump 416	trash 24
stressed skin 182	partition 314	grout 44	water 24, 26, 61
subfloor 184	post 348	grouting cement 44	wellpoint 26
treatment 172	scale 317	injected footing 60, 499	Pumping 43
underlayment 184	stage 323	regulator oil 403	dewater 43
Pneumatic control system 374	Portland cement 91	regulator steam 403	station 330
nailer 22	Post 347	relief valve 399	Purlin roof 178
tube 321	athletic 82	switch switchboard 461	Push button lock 255
tube system 386	cap 174	valve relief 399	Push-pull plate 256
Pocket door frame 237	cedar 194	wash 294	Putlog 15
pitch 222	concrete 80	Pressurized fire extinguisher ... 310	Putting surface 281
Pockets pitch 222	demolition 38	Prestressed concrete 115, 522	Puttying 293
Point heavy duty pile 58	fence 77	pile 56	PVC adhesive 200
masonry 136	jack 147	precast concrete 521	blind 346
moil 22	lamp 168, 465	Prestressing steel 108	chamfer strip 99
mortar 120	office S.F. & C.F. 475	Pretreatment termite 54	cleanout 406
pile 57, 58	portable 348	Prime coat 70	conduit in slab 447
Poisoning soil 54	recreational 82	Primer asphalt 216, 281	control joint 121
Poke-thru fitting 449	tennis court 74	steel 152	DWV pipe 397
Pole aluminum light 465	tensioned concrete 115	Prison door 329	expansion joint 100
athletic 82	Postal specialty 311	equipment 329	flashing 218

Index

gravel stop 221
pipe 61, 63, 397
sheet 200, 281
siding 212, 213
underground duct 66
valve 400, 401
waterstop 102
Pyrex pipe 393

Q

Quarried stone 132
Quarry drilling 48
 tile 274
Quarter round molding 187
Quartz 84
Quoins 131

R

Raceway 446, 448
 conduit 446
 plugmold 449
 surface 449
 wiremold 449
 wireway 449
Rack bicycle 82
 coat 190, 318
 hat 318
 hose 376
 pallet 315
Racquetball court 356
Radial arch 185
 wall 97
Radiant ceiling 353
 heat 353, 429
Radiator cast iron 428
 supply control 403
 thermostat control sys. ... 374
Radio frequency shielding 358
 tower 66
Radiography test 12
Rafter anchor 174
 demolition 38
 hip 180
 tie 180
 valley 180
 wood 180
Rail bumper 80
 crane 386
 crash 304
 dock shelter 328
 guide 80
 guide/guard 80
 hand 165
 map 299
 pipe 165
 trolley 303
 wall 165
Railing aluminum 165
 church 322
 demolition 39
 industrial 165
 ornamental 169
 wood 187, 191-193
Railroad bumper 169
 siding 169, 543
 track 169
 track removal 35
 trackwork 169
 work 169
Raise manhole 69
 manhole frame 69
Raised floor 304

Ramp approach 304
 handicap 110
 moving 384
 temporary 17
Ranch plank floor 279
Range cooking 334
 hood 334
 receptacle 453, 456
 restaurant 332
 shooting 338
Ratio water cement 11
Razor wire 78
Reading table 323
Ready mix 515
 mix concrete 109, 515, 516
 mix cost 515
Receiver ash 345
 trash 345
Receptacle air conditioner 453
 device 453
 dryer 453
 duplex 456
 GFI 453
 locking 453
 range 453, 456
 telephone 454
 television 454
 trash 81
 waste 317
 weatherproof 453
Receptor shower 301, 415
Recessed mat 345
Reciprocating water chiller .. 423
Recirculating chemical toilet .. 62
Recorder videotape 468
Recore cylinder 255, 256
Recreational post 82
Rectangular diffuser 434
 ductwork 430
 structural tubing 147
Rectifier low-voltage silicon .. 456
Redwood bark mulch 84
 cupola 306
 louver 193
 paneling 189
 siding 212, 213
 tank 367
 trim 187
 wine cellar 330
Refinish floor 280
Reflective block 129
 glass 260
 insulation 202
 painting 71
 sign 83
Refrigerant removal 371
Refrigerated case 324
 wine cellar 330
Refrigeration 354
 bloodbank 340
 commercial 324, 332
 equipment 365
 floor 355
 insulation 355
 mortuary 340
 panel fiberglass 355
 partition 355
 residential 335
 walk-in 355
Refrigerator door 240
Refuse chute 385
 hopper 385
Register air supply 435
 cash 324
 doctor 467
 in-out 467

return 435
 steel 435
 wall 435
Reglet 221
 aluminum 221
 galvanized steel 221
Regulator oil pressure 403
 steam pressure 403
Reinforced brick 529
 concrete pipe 62, 63
 culvert 62
 PVC roof 215
Reinforcement 507
 concrete 103
 welded wire 509
Reinforcing 106, 507, 508
 accessory 103
 beam 106
 coated 106
 column 106
 cost 507
 dowels 106
 extras 507
 fiber 108
 footing 106
 girder 106
 joint 121
 masonry 121
 metric 506
 prices 508
 slab 106
 splice 107
 steel 103, 506, 507
 steel accessory 104, 105
 steel cost 508
 testing 11
 wall 106
Relay pipe 43
 railroad track 169
Release door 468
 form 90
Relief vent ventilator 436
 water vacuum 405
Religious education S.F. & C.F. .. 475
Removal air conditioner 390
 asbestos 368
 block wall 34
 catch basin 34
 chain link fence 34
 concrete 35
 concrete pipe 35
 curb 35
 driveway 35
 fence 34
 floor 33
 guide rail 34
 hydrant 34
 insulation 368
 masonry 34
 paint 371
 pipe 35, 390
 pipe insulation 368
 plumbing fixtures 390
 railroad track 35
 refrigerant 371
 shingle 40
 sidewalk 35
 sod 86
 steel pipe 35
 stone 34
 stump 43
 tank 31
 tree 42, 43
 vat 369
 window 41
Rendering 6

Rental equipment 19, 25
 equipment rate 14, 19
 generator 23
Repellent water 293
Replacement pavement 70
 sash 251
Resturant roof 217
Reshoring 101
Residential alarm 454
 appliance 325, 334, 455
 closet door 229
 device 451
 door 229, 234, 237
 door bell 454
 dryer 325
 elevator 380
 fan 455
 fence 77
 fixture 454
 folding partition 314
 garage 359
 garage door 243
 greenhouse 360
 gutting 38
 heat pump 455
 heater electric 429
 hinge 254
 lighting 454
 loadcenter 451
 lock 255
 overhead door 243
 refrigeration 335
 roof jack 433
 service 451
 smoke detector 454
 stair 192
 storm door 237
 switch 451
 transition 433
 wall cap 433
 washer 325
 water heater 417, 455
 wiring 451, 455
Resilient floor 280
 flooring 555
 pavement 74
Resin epoxy 90
Respirator 370
 personal 370
Resquared shingle 207
Restaurant furniture 348
 range 332
 S.F. & C.F. 475
Restoration masonry 136
 window 371
Resurfacing railroad 169
Retail store S.F. & C.F. ... 475
Retaining wall 79, 80, 111
 wall formwork 97
 wall segmental 79
Retarder vapor 206
Return register 435
Revolving darkroom 354
 dome 359
 door 246, 336, 353
 entrance door 246
Rewind table 326
Rib lath 267
Ribbed siding 211
 waterstop 102
Ridge board 180
 cap 206, 209
 flashing 362
 roll 210
 vent 222, 303
Rig drill 32

Index

Rigid anchor 121	gravel 90	tub 414	wood 251
conduit in-trench 447	gypsum 117	Round bar fiberglass 195	Sauna 355
in slab conduit 447	hatch 222	diffuser 434	door 355
insulation 202	insulation 204, 362	rail fence 78	Saw 24
Ring split 174	jack 180	table 348	brick 136
toothed 174	jack residential 433	tube fiberglass 195	chain 24
Rink ice 365	joint 222	Rowing machine 337	circular 24
Rip rap 54	lead 217	Rubber astragal 257, 258	concrete 19, 40
Riser rubber 280	metal 209	base 280	cutting 40
stair 275	mineral 216	coating 201	masonry 24
terrazzo 275	modified bitumen 216	control joint 121	shop 339
wood stair 193	mop 214	dock bumper 327	table 339
Rivet 143	nail 173	expansion joint 100	Scaffold specialties 15, 16
tool 143	panel preformed 209	floor 281	steel tubular 14
Road base 70	patch 216	floor tile 281	Scaffolding 489
berm 71	PIB 215	mat 345	tubular 14
sign 83	plywood 182	nosing 280	Scale 317
temporary 17	polyvinyl chloride ... 215, 216	pavement 74	contractor 317
Roadway lighting 465	precast 117	pipe insulation 392	crane 317
Robe hook 316	purlin 178	riser 280	floor 317
Rock bolt drilling 48	rafters framing 179	sheet 281	physician's 340
bolts 55	reinforced PVC 215	stair 280	pit 317
drill 32, 47	restaurant 217	threshold 258	portable 317
excavation 47	roll 213	tile 280	truck 317
Rod backer 100, 225	rolling 263	tired roller 21	warehouse 317
closet 189	sheathing 183	waterproofing 200	Scanner checkout 324
curtain 316	sheet metal 217	waterstop 103	Schedule 9
ground 441	shingle 206	Rubbing wall 113	board 299
tie 45, 148	single-ply 215	Rubbish chute 40	critical path 9
weld 144	skylight 252	handling 40	progress 9
Roll ridge 210	slate 207, 548	Rumble strip 80	School cabinet 344
roof 213	specialties, prefab 219	Running track 74, 279	crosswalk 84
roofing 216	stainless steel 217	trap 407	door hardware 253
type air filter 436	steel 210	Rupture testing 11	elementary S.F. & C.F. 475
up grille 239	stressed skin 182	Rustication strip 97	equipment 337
Roller compaction 47	tile 208		furniture 348
earth 20	tile corrugated 208		hinge 257
sheepsfoot 47	tile spanish 208	**S**	S.F. & C.F. 476
Rolling door 240, 245	truss 151, 178, 185		T.V. 468
earth 47	vent 222, 363	Safe data 320	Scissor gate 312
ladder 15	ventilator 223	diskette 320	lift 386
roof 263	ventilator plastic 432	office 320	Scoreboard baseball 338
service door 240	walkway 223	Safety deposit box 321	Scored block 127
tower 15	wood 182	equipment laboratory 340	split face block 126
Roman brick 124	zinc 217	fixture industrial 414	SCR brick 124
Romex copper 444	Roofing & siding preformed ... 209	glass 260	Scraper 21
Roof accessories 222	cold 214	glass nosing 168	elevating 50
adhesive 213	demolition 39, 40, 369	net 13	excavation 50
aggregate 91	membrane 213	nosing 280	excavation bulk 50
aluminum 209	modified bitumen 216	shower 414	mobilization 46
baffle 223	roll 216	switch 459	self propelled 50
beam 185	sheet metal 217	Salamander 24	Scratch coat 276
built-up 213	single ply 215	Sales tax 8, 486	Screed 96
cant 180, 214	Rooftop air conditioner 426	Salt treatment lumber 172	base 105
clay tile 208	multizone air cond. 426	Sampling air 369	holder 105
coating 216	Room acoustical 354	Sand 120	Screen chimney 305
concrete 115	audiometric 354	concrete 90	entrance 300
copper 217	clean 354	fill 30	fence 78
CSPE 215	humidifier 427	screened 120	metal 246
curb 180	Rope decorative 348	Sandblast glass 260	molding 187
deck 156, 182, 183	exercise 338	Sandblasting equipment 24	projection 298, 325, 326
deck concrete 115	wire 152	Sander floor 24	security 168, 246
deck insulation 204	Rosewood door 233	Sanding 293	sight 313
deck laminated 184	Rosin paper 206	floor 280	squirrel and bird 305
deck wood 184	Rotary hammer drill 23	Sandstone 131	urinal 300
dome 263	Rough buck 179	flagging 73	window 246, 251
drains 407	carpentry 176	Sandwich panel 210	wood 251
elastomeric 215	hardware 176	panel aluminum 211	Screened sand 120
EPDM 215	stone wall 132	panel skylight 252	Screening plant 21
exhauster 433	Rough-in drinking fountain deck 413	wall panel 212	Screw aluminum 211
expansion joint 170, 222	drinking fountain floor 413	Sanitary base cove 273	anchor 105, 141
fiberglass 210	drinking fountain wall 413	Sash 363	anchor bolt 105
fill 525	sink countertop 416	aluminum 247	brass 174
folded plate 182	sink raised deck 416	metal 246	drywall 273
frame 148	sink service floor 416	replacement 251	eye bolt 105
framing removal 36	sink service wall 416	steel 246	insulation 202

Index

jack	45
lag	141
sheet metal	173
steel	174
wood	174
Screw-on connector	443
Screws lag	142
machines	142
Scrub station	340
Scupper drain	407
floor	407
Seal curb	73
door	328
pavement	73
security	257
Sealant	200, 206
acoustical	271
caulking	225
tape	226
Sealcoat	73
Sealer concrete	90
joint	225
pavement	74
Seamless floor	282
Seating	348
church	322
movie	326
Secondary treatment plant	62
Section steel	147
Sectional door	242, 243
overhead door	243
Security access	373
and vault equipment	320
fence	76
gate	312
hinge	256
sash	246
screen	168, 246
seal	257
vault door	239
Sediment bucket drain	406
strainer Y valve	401
Seeding	85, 500
See-saw	82
Segmental retaining wall	79
Selective demolition	35
Self propelled crane	25
Self-closing relief valve	399
Self-contained air conditioner	426
Self-propelled scraper	50
Self-supporting tower	66
Sentry dog	18
Separation barrier	369
Septic system	64
tank	64
tank concrete	64
tank precast	64
Service door	240
electric	459
entrance cable aluminum	444
residential	451
sink	416
sink faucet	411
station equipment	326
Set accelerator	90
Sewage aeration	62
ejector pump	417
holding tank	62
municipal waste water	62
pipe	62, 63, 67
pumping station	330
treatment plant	62
Shade	346
Shaft wall	272
Shake wood	208
Shear connectors weld	143
plate	174
test	11
wall	183
Sheathed nonmetallic cable	444
romex cable	444
Sheathing	183
asphalt	184
gypsum	184
paper	206
roof	183
Sheathing, roof & walls ply.	183
Sheepsfoot roller	47
Sheet base	214
floor	282
glass	260
metal	217
metal aluminum	219
metal roofing	217
metal screw	173, 211
piling	44, 495
Sheeting	44
driver	22
open	46
tie back	46
wale	45
wood	43, 45, 494
Sheetrock	270
partition	272
removal	41
Sheets lead	357
Shelf bathroom	316
bin	315
brick	96
directory	315
library	322
metal	314
Shellac door	287
Shelter aluminum	363
dock	328
golf	82
rail dock	328
Shelving	189
mobile	323
pine	189
plywood	189
prefinished	189
refrigeration	355
steel	315
storage	314
wood	189
Shield expansion	140
light	80
Shielded cable	444
copper cable	444
Shielding lead	357
radio frequency	358
Shift work	7
Shingle	206
aluminum	206
asphalt	206
concrete	208
metal	207
panelized	208
removal	40
roof	206
strip	206
wood	207
Shipping door	241
Shock absorber	405
absorber pipe	405
absorbing door	242
Shooting range	338
Shop drill	339
equipment	338
Shoring	44, 45, 101
frame system	101
horizontal	101
installation	101
temporary	45
vertical	101
Shot blast floor	37
blaster	24
Shovel	48
front	21
Shower arm	415
built-in	415
by-pass valve	411
compartments	301
cubicle	415
door	301
drain	407
enclosure	302
glass door	301
group	415
pan	218
partition	131, 301
polypropylene	415
receptor	301, 415
safety	414
stall	415
surround	302
Shredder	329
compactor	328
Shrinkage test	11
Shrub broadleaf evergreen	85
deciduous	86
evergreen	85
moving	84
Shutter	346
Siamese	376
Side beam	92
Sidelight	237
Sidewalk	72, 73, 278
asphalt	72
brick	73
bridge	15
concrete	72
removal	35
temporary	17
Siding aluminum	211
concrete	115
demolition	39
fiber cement	211
fiberglass	210
hardboard	213
metal	209, 211
mineral	182
nail	173
paint	290
painting	290
plywood	213
railroad	169, 543
redwood	213
removal	40
ribbed	211
steel	212
vinyl	212
wood	212
wood product	213
Sieve analysis	10, 11
Sign	18, 83, 308
aluminum	308
base	98
directional	83
letter	308
posts	83
project	18
reflective	83
road	83
street	308
traffic	83
Signal bell fire	375
traffic	83
Silencer duct	431
louver	431
plenum	431
Silicon carbide abrasive	112
carbide aggregate	90
Silicone caulking	226
coating	201
water repellent	201
Sill	131
anchor	174
block	128
door	187, 236, 258
formwork	97
masonry	130
mud	180
precast	130
quarry tile	275
stone	130, 132
window	251
wood	180
Sillcock	411
Silo	363
slipform	98
Silt fence	54
Simulated brick	133
stone	133
Single hub pipe	393
hung window	247
ply roofing	215
tee beam	115
track	543
zone rooftop unit	426
Single-ply roof	215
Sink	415
barber	324
base	189
countertop	415
countertop rough-in	416
darkroom	336
kitchen	416
laboratory	339
laundry	414
lavatory	415
raised deck rough-in	416
removal	390
service	416
service floor rough-in	416
service wall rough-in	416
support	410
Site clearing	42
demolition	33, 34
drainage	68
furnishings	81
improvement	74, 81, 82
plan	6
preparation	32
work electrical	65
Skating mat	345
rink S.F. & C.F.	473
Skin stressed	149
Skirtboard	193
Skylight	252, 363
removal	40
roof	252
Skyroof	263
Slab blockout	96
bolster	103
concrete	110, 112
cutout	35
elevated	109
form	156
lift	110, 520
lifting insert	105
on grade	95, 110
on grade removal	35

672

Index

pan 110
precast 115
reinforcing 106
support 15
textured 110
waffle 110
Slate 132
  flagging 73
  pavement 132
  removal 40
  roof 207
  shingle 207
  sidewalk 73
  sill 130
  stair 132
  tile 278
Sleeper clip 105
  floor 180
  wood 180
Sleepers framing 180
Sleeve formwork 102
Slide gate 77
  playground 82
  swimming pool 364
Sliding chalkboard 298
  door 230, 237
  door shower 301
  entrance 245
  glass door 237
  mirror 317
  panel door 245
  window 247, 250
Slipform 98
  building 98
  silo 98
Slipforms 502
Slop sink 414
Slope protection 54
Slot letter 311
Slotted block 134
  channel 148
  pipe 63
Slump block 126
Slurry trench 45
Small tools 17
Smoke detector 375
  hatch 223
  vent 223
  vent chimney 422
Smokestack 134
Snap tie formwork 102
  tie hanger 100
Snow fence 77
  melting 429
Soaking bathroom 414
Soap dispenser 316
  holder 316
  tank 316
Soccer goalpost 82
Social club S.F. & C.F. 471
Socket ground 82
Sod 85
  tennis court 74
Sodium low pressure fixture .. 465
Soffit drywall 270, 271
  marble 131
  metal 212
  plaster 269
  plywood 188
  steel 362
  stucco 269
  vent 303
  vinyl 212
  wood 180, 187
Softener water 335
Soil compaction 47

compactor 20
decontamination 31
pipe 393
poisoning 54
sample 33
stabilization 54
tamping 47
test 10, 11
treatment 54
Solar energy 372, 373
  energy circulator air 373
  energy system collector 373
  energy system controller 373
  energy system heat exchanger 373
  film glass 260
  heating 560
  screen block 128
Soldier beam 46
Solid core door 232
  wood door 233
Sorority house S.F. & C.F. 472
Sound attenuation 284
  curtain 356
  movie 326
  panel 356
  proof enclosure 354
  system component 467
  system speaker 467
Space frame 153, 182
  heater floor mounted 421
  heater rental 23
Spacer 105
Spade air 22
Spandrel beam 92
  glass 260
Spanish roof tile 208
Speaker movie 326
  sound system 467
  station nurse 467
Special construction 356
  door 238, 242
  hinge 256
  purpose room 353
  systems 467, 468
Specialties 316
  & accessories, roof 219
  piping HVAC 402, 404
  scaffold 15, 16
Specific gravity 11
  gravity testing 10
Speed bump 80
Spike grid 174
  railroad 169
Spinner ventilator 436
Spiral stair 164, 192
Spire church 322
Splice pile 57, 58
  reinforcing 107
Splicer movie 326
Split astragal 257
  rail fence 78
  rib block 126
  ring 174
  system control system 374
Splitter damper assembly 431
Sport arena S.F. & C.F. 476
Spotlight 323
Spotter 53
Spray acoustical 223
  coating 200
  fireproofing 223
  insulation 202
  substrate 370
Sprayer airless 370
  emulsion 23
Spread footing 95, 110

Spreader concrete 24
Spring bolt astragal 257
  bronze weatherstrip 257
  hinge 257
Sprinkler alarm 375
  grass 74
  head 377
  system 74
  system accelerator 377
  system component 377
Square bar fiberglass 195
  foot building cost 470
  structural tubing 147
  tube fiberglass 195
Squash court 355
  court backstop 82
Stabilization fabric 70
  soil 54
Stacked bond block 127
Stadium 360
  bleacher 82
  cover 352
Stage equipment 323
  lighting 323
  portable 323
Staging aids 16
  pump 13
  swing 16
Stain cabinet 285
  door 287
  floor 289
  lumber 185
  siding 213
  truss 290
Stainless cooling towers 424
  duct 430
  flashing 219
  gutter 221
  pipe 165
  reglet 221
  screen 300
  sign 308
  steel bolt 139, 141
  steel corner guard 303
  steel cot 329
  steel cross 322
  steel downspout 220
  steel grate 166
  steel gravel stop 221
  steel hinge 254
  steel pipe 396, 397
  steel rivet 143
  steel roof 217
  steel shelf 316
  steel storefront 245
  steel weld rod 144
Stair 193
  basement 116, 192
  brick 136
  carpet 283
  ceiling 335
  climber 380
  concrete 111
  electric 335
  finish 113
  fire escape 164
  formwork 96
  grating 166
  landing 164, 276
  marble 131
  metal 164
  nosing 102, 164, 167, 276, 280
  pan 102
  pan form 164
  precast 116
  railroad tie 72

removal 38
residential 192
riser 275
riser vinyl 280
rubber 280
slate 132
spiral 192
steel 164
stone 83
stringer 276
stringer framing, ... 179
temporary protection 13
terrazzo 276
tread 102, 130, 167, 280
tread aluminum 167
tread fiberglass 195
tread terrazzo 275
tread tile 274
tread wood 193
wood 192
Stairlift wheelchair 380
Stairs disappearing 335
  fire escape 164
Stairway door hardware 253
Stairwork and handrails 191
Stake nail 102
Stall shower 415
  toilet 300
  type urinal 412
  urinal 412
Stamp time 375
Stamping concrete 113
  texture 113
Standard extinguisher 310
Standpipe alarm 375
  connection 376
  steel 366
Starter board & switch 460-462
  motor 459
  wall 95
Starting newel 192
Station control 459
  hospital 340
  transfer 329
  weather 13
Stationary ventilator 436
  ventilator mushroom 436
Steam bath 356
  bath residential 356
  boiler 419, 420
  boiler electric 419
  clean 136
  cleaner 24
  cleaning 136
  cleaning metal 152
  condensate meter 409
  humidifier 427
  jacketed kettle 332
  pressure valve 403
  regulator pressure 403
  trap 403
Steamer 333
Steel anchor 121
  astragal 257
  beam hanger 100
  bin wall 79
  blind 345
  blocking 176
  bolt 139, 174
  bridge 81
  bridging 176
  building components 540
  building pre-engineered 361
  chain 168
  channel 268
  conduit in slab 447

673

Index

conduit in trench	447
conduit intermediate	446
conduit rigid	446
corner guard	303
cutting	41, 145
deck	156
diffuser	434
dome	153, 359
door	228, 229, 239, 242
downspout	220
drill	22
estimating	535
expansion tank	402
fastener	142
fence	76, 77
form	93
frame	229
frame formwork	93, 98
furring	266
grating	166, 167
gravel stop	221
gutter	221
hex nuts	139
hinge	254
joist	154
ladder	163
lath	267
lintel	149
locker	309
lockers	557
mesh box	54
metal deck	156
painting	152
partition	313, 314
pile	57
pipe	67, 147, 394, 395
pipe downspout	220
pipe removal	35
prestressing	108
primer	152
protective coating	152
register	435
reinforcement	506
reinforcing	103, 506, 507
roof	210
salvage	391
sash	246, 550
scaffold tubular	14
screw	174
section	147
sections	538
sheet pile	44
shelving	315
shingle	207
shore	101
siding	212
silo	363
stair	164
stair tread	167
standpipe	366
structural	145, 146, 149, 151
stud	266
tank	366, 402
testing	10
tower	66
tube	147
underground duct	66
valve	401
weld rod	144
window	246, 363
window demolition	41
wire rope	152
Steeple	322
aluminum	322
tip pin	254
Step	83
bluestone	72
brick	72
manhole	69
masonry	72
stone	131
tapered pile	57
Sterilizer barber	324
dental	341
medical	340
Stiffleg derrick	25
Stile fabric	257, 258
protection	257, 258
Stone	130
aggregate	84, 90
anchor	121
ashlar	131
base	70, 131
cast	133
collected	132
column	131
fill	30
filter	54
floor	72, 131
ground cover	84
mulch	84
paver	73, 130
pea	91
quarried	132
removal	34
sill	130, 132
simulated	133
stair	83
step	131
stool	132
threshold	130
tread	130
trim cut	131
wall	80, 132
Stool cap	187
doctor	341
stone	132
window	130, 131
Stop door	187
gravel	220
wheel	169
Stop-start control station	459
Storage bottle	330
box	12
building S.F. & C.F.	477
cabinet	339
dome	359
dome bulk	359
shelving	314
tank	365, 367
tank cover	352
trailer	24
Store department S.F. & C.F.	471
retail S.F. & C.F.	475
Storefront	246
aluminum	245
Storm door	237
window	251, 363
Stove	305, 332
wood burning	305
Strainer bronze body	403
downspout	219
gutter	221
roof	219
wire	220
Y type	403
Y type iron body	404
Strand grouted	108
ungrouted	108
Stranded wire	445
Strap tie	174
Straw	84
Street sign	308
Streetlight	465
Strength bolts high	141
compressive	11
Stressed skin	149, 182
skin roof	182
Stressing tendons	108
Stringer stair	276
stair terrazzo	276
Strip cabinet	468
chamfer	99
entrance	244
filler	257
floor	279
footing	110
lighting	464
rumble	80
shingle	206
soil	47
Stripping topsoil	43
Strobe light	323
Structural aluminum	152
concrete	109, 517-519
excavation	50, 51
face tile	124
facing tile	124
fee	6
formwork	92
framing	536
panel	182
plastics	195
steel	145-147, 149, 151, 534
steel extras	539
tile	124
tubing	147
welding	145, 151
Structure fabric	352
moving	88
pre-engineered	358
tension	358
Stub switchboard	462
Stucco	269, 554
demolition	41
mesh	267
painting	290
Stud connector	143
demolition	38
metal	266, 269
partition	179, 268
partitions metal	158
steel	266
wall	179, 181, 266, 272
walls bracing	157
walls bridging	157
walls metal	158
wood	181, 269
Studs	553
loadbearing	157
metal	158
weld	144
Stump chipping	42
removal	43
Subcontractor O & P	9
Subdrainage pipe	66, 67
piping	66
system	66
Subfloor	184
plywood	184
wood	184
Subgrade chair	105
stake	106
Submersible pump	24, 61, 417
Subpurlin	145
bulbtee	145
truss tee	145
Subpurlins	541
Subsoil heating	365
Subsurface exploration	33
investigation	32
Suction hose	23
Sump hole construction	43
pit	43
pump	335, 417
Super high-tensile anchor	55
Supermarket checkout	324
scanner	324
S.F. & C.F.	476
Support carrier fixture	409
ceiling	146
dowel	95
drinking fountain	409
framing pipe	149
lavatory	409
light	146
monitor	146
motor	146
partition	146
sink	410
slab	15
urinal	410
water closet	410
water cooler	411
Suppressor lightning	451
Surface bonding	120
landscape	280
playground	281
preparation	293, 294
raceway	449
treatment	74
Surfacing	73
Surfactant	370
Surgery equipment	341
table	341
Surgical lighting	341
Surround shower	302
tub	302
Surveillance system	321
system TV	468
Survey aerial	7
crew	6
monument	6
preblast	48
property line	6
stake	16
topographic	6
Suspended ceiling	267, 270, 277, 353
ceiling framing	177
Suspension system ceiling	266
Swell testing	10
Swimming pool	364
pool blanket	364
pool enclosure	363
pool equipment	364
pool heater	420
pool hydraulic lift	364
pool ladder	364
pool painting	364
pool ramp	365
pool S.F. & C.F.	476
pools	558
Swing	82
check valve	398
clear hinge	257
door	242
gate	77
staging	16
Swing-up overhead door	243
Switch box	449
bus-duct	463
decorator	451
dimmer	451, 456
electric	456, 459

674

Index

general duty	459
residential	451
safety	459
timber	169
toggle	456
Switchboard electric	459
instrument	458
pressure switch	461
stub	462
w/bus bar	461
w/CT compartment	461
Switching low-voltage	456
Switchplate low-voltage	456
Synthetic marble	278
turf	281
System antenna	468
clock	375
control	374
fire extinguishing	310, 378
grinder pump	416
irrigation	74
septic	64
sprinkler	74, 377
subdrainage	66
tube	386
T.V.	468
T.V. surveillance	468
UHF	468
VHF	468

T

Table folding	348
laboratory	339
massage	337
overbed	349
physical therapy	341
reading	323
rewind	326
round	348
saw	339
surgery	341
top	172
Tamper	22, 52
Tamping soil	47
Tandem roller	20
Tank clean	31
cover	352
darkroom	336
developing	336
disposal	31
expansion	402
fiberglass	367
fixed roof	366
gasoline	366
holding	62
hubbard	340
oil & gas	402
pillow	367
removal	31
septic	64
soap	316
steel	366, 402
storage	365, 367
testing	12
water	365, 366
water storage solar	373
Tap box bus-duct	462
Tape detection	30
sealant	226
temporary	71
underground	30
Tar kettle/pot	24
paper	84
pitch emulsion	73

roof	216
Target range	338
Tarpaulin	13, 17
duck	17
mylar	17
polyethylene	17
polyvinyl	17
Tarred felt	214
Tax	8
sales	8
social security	8
unemployment	8
Taxes	486
T-bar mount diffuser	434
Teak floor	279
molding	187
paneling	188
Technician inspection	11
Tee beam	114, 115
cleanout	406
fitting	61
pipe	396
precast	115
Telephone booth	315
enclosure	315
exchange S.F. & C.F.	476
house	468
manhole	65
pole	65
receptacle	454
Telescoping bleacher	337
door	243
platform	323
Television equipment	468
receptacle	454
Teller automatic	321
window	321
Temperature relief valve	399
rise detector	375
Tempered glass	259
glass greenhouse	360
hardboard	188
Tempering valve	400
Temporary building	12
construction	12, 13, 18
facility	17
fence	17
guardrail	17
heat	12, 91
light	12
oil heater	23
road painting	71
shoring	45
tape	71
toilet	24
utility	12
Tendons stressing	108
Tennis court	74
court air supported	352
court backstop	82
court clay	74
court fence	78
court net	74
court post	74
court sod	74
court surface	281
Tensile test	11
Tension structure	358
Terminal A/C packaged	426
heater	429
lug	443
Termination box	468
cable	443
Termite pretreatment	54
protection	54
Terne coated flashing	219

Terra cotta	125
cotta coping	128
cotta demolition	36, 41
Terrazzo	275, 276
abrasive	276
base	275, 276
cove base	275
curb	276
demolition	37
epoxy	276, 282
floor	276, 555
monolithic	276
precast	275
receptor	301
stair	276
stringer stair	276
venetian	276
wainscot	276
Test beam	11
load pile	57
moisture content	11
pile load	56
pit	33
soil	11
ultrasonic	12
well	61
Testing	10
OSHA	370
pile	56
sulfate soundness	10
tank	12
Texture stamping	113
Textured slab	110
Theater S.F. & C.F.	476
Therapeutic pool	365
Thermal batt	284
Thermometer	341
Thermoplastic painting	71
Thermostat control sys. radiator	374
integral	429
wire	456
Thin plaster partition	272
shell construction	151, 182
Thincoat plaster	269
Thinning tree	42
Thinset ceramic tile	273
mortar	274
Threaded rod fiberglass	195
Threshold	258
door	237
stone	130, 131
wood	187
Thru-wall air conditioner	426
Ticket booth	358
Tie back hanger	100
back sheeting	46
bar	104
clip	104
concrete	169
formwork snap	102
hanger	100
rafter	180
rod	45, 148
strap	174
wall	121
wire	106
wood	169
Tier locker	309
Tile	273, 281
abrasive	274
aluminum	208
ceiling	277, 353
ceramic	273, 274
clay	208
concrete	208
cork	280

cork wall	283
demolition	35
exterior	274
fire rated	124
flue	133
glass	261, 275
grout	273
marble	278
metal	275
partition	125
porcelain	274
quarry	274
roof	208
rubber	280
slate	278
stainless steel	275
stair tread	274
structural	124
terra cotta	125
vinyl	281
wall	273
window sill	275
Tilt up concrete	524
up construction	116
Tilt-up precast	116
Timber connector	174
fastener	173
framing	177
guide rail	80
laminated	185
roof deck	183
switch	169
Time lapse photography	10
lock	239
stamp	375
system controller	375
Timer clock	454
interval	451
Tin clad door	231
clad fire door	550
Tinted glass	259
Toaster	333
Toe plate	166
Toggle bolt	140
switch	456
Toilet	62
accessory	316
bowl	416
compartments	300
door hardware	253
partition	131, 300
partition removal	41
prison	329
stall	300
temporary	24
tissue dispenser	316
trailer	24
Tool air	22
rivet	143
van	25
Tools small	17
Toothed ring	174
Toothing masonry	36
Top coiling grille	239
demolition counter	39
dressing	47
table	172
Topographic survey	6
Topping concrete	112
epoxy	282
floor	113
grout	118
Topsoil	30, 47
stripping	43
Torch cutting	24, 41, 145
Touch bar	256

675

Index

Tour station watchman 375
Towel bar 316
   dispenser 316
Tower control 358
   cooling 424
   crane 25, 490
   hoist 26
   light 23
   radio 66
   rolling 15
Town hall S.F. & C.F. 477
Track curtain 323
   drawer 191
   drill 22
   railroad 169
   running 74, 279
   surface 281
   traverse 346
   work 169
Trackwork railroad 169
Tractor 21, 47, 52
   loader 21
   truck 25
Traffic channelizing 80
   cone 17
   control 80
   control pedestrian 348
Traffic door 244
Traffic line 71
   sign 83
   signal 83
Trailer office 12
   platform 24
   storage 24
   toilet 24
   truck 22, 24
Trainer bicycle 337
Training 27
Transceiver 444
Transfer station 329
Transformer 457, 458
   & bus-duct 457, 463
   cabinet 451
   current 458
   dry type 457
   fused potential 458
   low-voltage 456
   oil filled 458
Transit 24
   mixer 19
Transition molding wood 280
   residential 433
Transom aluminum 244
   lite frame 230
Transplanting 84
Trap cast iron 407
   deep seal 408
   drainage 407, 408
   grease 407
   inverted bucket steam 403
   P 408
   rock surface 113
   steam 403
Trash closure 82
   pump 24
   receiver 345
   receptacle 81
   receptacle fiberglass 81
Traverse 346
   track 346
Travertine 131
Tray cable 446
   laundry 414
Tread abrasive 275
   cast iron 164
   rubber 280

stair 102, 130, 167
   stone 131, 132
   vinyl 280
   wood 193
Treadmill 337, 338
Treated lumber 180
   pile 58
   pine fence 78
Treatment acoustical 284
   lumber 172
   plant secondary 62
   plywood 172
   potable water 417
   wood 172
Tree 86
   cutting 42
   deciduous 86
Tree evergreen 85
   guying 87
   moving 84
   removal 42, 43
   thinning 42
Trench backfill 52
   box 24
   cover 166, 167
   drain 407
   duct 448
   duct fitting 448
   duct steel 448
   excavation 51-53
   formwork 96
   slurry 45
   utility 52
Trencher 21
   chain 21
Trenching 43, 47
Trial batch 11
Trim demolition 39
   exterior 186
Trim, exterior paint 291
Trim molding 187
   painting 289
   redwood 187
   tile 273
   wood 186
Triple brick 124
   weight hinge 257
Troffer air handling 464
Trolley rail 303
Trowel coating 200
   concrete 19
Truck concrete 19
   dock 327
   dump 21
   flatbed 22
   hauling 53
   leveler 328
   loading 51
   mounted crane 25
   off highway 22
   pickup 25
   rental 22
   scale 317
   tractor 25
   trailer 22, 24
   vacuum 25
Trucking 53
Truss bowstring 185
   flying 101
   metal 151
   painting 290
   plate 175
   roof 151, 178, 185
   stain 290
   tee subpurlin 145
   varnish 290

Tub bar 316
   rough-in 414
   soaking 414
   surround 302
Tube framing 262
   pneumatic 321
   steel 147
   system 386
   system air 386
   system pneumatic 386
Tubing copper 393
   electric metallic 446
   electrical 446
   structural 147
Tubular fence 76
   lock 254
   scaffolding 14
   steel door 242
   steel joist 185
Tumbler holder 316
Tunnel support system 55
Tunneling 55
Turf artificial 74
   synthetic 281
Turned column 193
Turnout 169
Turnstile 308
TV closed circuit camera 468
   system 468
Two piece astragal 257

U

UHF system 468
Ultrasonic cleaner 341
   motion detector 375
   test 11, 12
Underdrain 69
Undereave vent 303
Underfloor duct 448
   duct electrical 448
   duct fitting 448
Underground duct 66
   storage tank removal 493
   tank 367
   tape 30
Underlayment 184
   acoustical 285
   grout 118
   gym floor 279
   hardboard 184
   latex 281
Underpin foundation 44
Underwater light 364
Undisturbed soil 11
Unemployment tax 8
Ungrouted bar 108
   strand 108
Unit heater 421, 428
   heater electric 429
   masonry 122
Upstanding beam formwork 92
Urethane foam 226, 285
   insulation 202, 355
   wall coating 293
Urinal 412
   removal 390
   screen 300
   stall 412
   stall type 412
   support 410
   wall hung 412
Urn ash 345
Utensil washer medical 340
Utility brick 123, 124

duct 66
electric 66
pole 65, 465
pump 409
set fan 432, 433
sitework 65
temporary 12
trench 52
vault 30

V

Vacuum breaker 405
   central 320
   cleaning 320
   loader 370
   relief water 405
   truck 25
Valance board 190
Valley flashing 206
   gutter 362
   rafter 180
Valve 398-401
   angle 400
   assembly sprinkler deluge . 378
   backwater 407
   ball 398, 400, 401
   brass 398
   bronze 398
   butterfly 400
   cabinet 310
   check 378, 401
   check swing 400
   CPVC 401
   fire hose 377
   flush 411
   foot 401
   gas stop 398
   gate 400, 401
   globe 399, 401
   hot water radiator 403
   iron body 400
   mixing 411
   plastic 400
   polypropylene 401
   PVC 400
   relief pressure 399
   shower by-pass 411
   soldered gate 398
   sprinkler alarm 378
   steam pressure 403
   steel 401
   swing check 398
   tempering 400
   water pressure 399
   Y sediment strainer 401
Valves forged 401
Vandalproof fixture 465
Vane-axial fan 432
Vanity base 191
   top lavatory 415
Vapor barrier 206, 355
   barrier sheathing 184
   retarder 206
Vaportight fixture incandescent . 464
   mercury vapor fixture 464
Varnish cabinet 285
   casework 285
   door 287
   floor 289, 293
   polyurethane 293
   truss 290
Vat removal 369
Vault 356
   door 239

Index

front 356
utility 30
Vaulting side horse 338
Vct removal 37
Veneer ashlar 132, 532
brick 122, 123, 530
core paneling 188
granite 130
members laminated 185, 186
removal 39
Venetian blind 345
terrazzo 275, 276
Vent automatic air 402
box 122
brick 434
chimney 422
chimney all fuel 422
chimney fitting 422
chimney high temperature 422
chimney increaser 422
dryer 335
exhaust 435
flashing 408
foundation 122
intake-exhaust 436
metal chimney 422
metal deck 156
ridge 222, 303
ridge strip 303
roof 222, 363
silencer 431
smoke 223
soffit 303
Ventilating air conditioning 374, 423, 424, 427, 430-433, 436
Ventilation 567
fan 455
louver 193
Ventilator control system 374
fan 431
masonry 122
mushroom stationary 436
relief vent 436
roof 223
spinner 436
Venturi flow meter 404
Verge board 187
Vermiculite cement 224
insulation 202
plaster 268
Vertical blinds 346
conveyor 385
lift door 244
shore 101
VHF system 468
Vibration pad 143
Vibrator concrete 19
earth 21, 47, 52
plate 47
Vibratory equipment 20
roller 20
Vibroflotation 45
compaction 495
Videotape recorder 468
Vinyl blind 194
coated fence 77
composition floor 281
downspout 220
faced wallboard 270
floor 281
glass 261
gutter 221
lead barrier 285
mat 345
plastic waterproofing 200
roof 215
siding 212
soffit 212
stair riser 280
stair tread 280
tile 281
tread 280
wall coating 293
wallpaper 284
Vitrified clay pipe 63, 67
Vocational kiln 339
school S.F. & C.F. 476
shop equipment 339
Void formwork 94
Voltmeter 458
Volume control damper 431

W

Waffle slab 110
Wainscot ceramic tile 273
molding 187
quarry tile 275
terrazzo 276
Wale sheeting 45
Wales 45
Walk 72
moving 384
Walk-in refrigeration 355
Walkway cover 311
roof 223
Wall & corner guards 303
aluminum 211
anchor 139, 140
area 258
block 531
brick 135
bumper 254, 303
cabinet 190, 339, 344
canopy 311
cap residential 433
ceramic tile 274
coating 293
concrete 110, 112
covering 555
cross 322
curtain 262
cutout 36
demolition 41
drill 139
drywall 270
exhaust fan 433
expansion joint 170
finish 113
Wall finishes 283
Wall flagpole 307
forms 501
formwork 96
foundation 127
framing 181
framing removal 36
furnace 421
furring 181, 267
grout 120
guard 303
heater 334
heater electric 429
hung lavatory 415
hung urinal 412
hydrant 376, 412
insulation 121, 202, 203, 362
lath 267
louver 303
masonry 80, 125
mat 338
mirror 261
painting 292, 293
panel 110, 115, 133
panel brick 136
panel ceramic tile 274
panel woven wire 312
paneling 188
panels insulated 116
partition 272
patching concrete 117
plaster 268, 269
plug 122
precast concrete 523
radial 97
rail 165
register 435
reinforcing 106, 507
removal 34
retaining 79, 80, 111
rubbing 113
shaft 272
shear 183
sheathing 183
siding 213
starter 95
steel bin 79
stone 132
stucco 269
stud 179, 181, 266, 269, 272
switch plate 456
tie 121
tie masonry 121
tile 273
tile cork 283
urn ash receiver 316, 317
window 262
Wallboard acoustical 271
Wallcovering 283
acrylic 284
Wallguard 303
Wallpaper 283, 284
Walls and partitions demolition 41
Walnut door frame 236
floor 279
Wardrobe 318
cabinet 318
wood 190
Warehouse 352
scale 317
S.F. & C.F. 477
Warm air heater 421
Wash bowl 415
brick 136
fountain 412
fountain group 412
Washer 175
commercial 325
darkroom 336
residential 325, 335
Washing machine automatic 335
Waste compactor 328
disposal 328
handling 328
handling equipment 328
incinerator 329
packaging 371
receptacle 317
Wastewater treatment system 330
Watchdog 18
Watchman service 18
tour station 375
Water anti-siphon 405
atomizer 370
balancing 437
blasting metal 152
cement ratio 11
chiller 423
chillers absorption 423
closet 416
closet chemical 62
closet removal 390
closet support 410, 411
coil chilled 428
cooler 413, 565
cooler handicap 414
cooler support 411
copper tubing connector 403
dispenser hot 411
distiller 341
distribution system 60
effect testing 10
fountain removal 390
heater 335, 418, 455
heater commercial 418
heater electric 417
heater gas 418
heater oil 418
heater removal 391
heater residential 417
heating hot 419, 428
hose 23
hydrant 412
pipe ground clamp 441
pressure relief valve 399
pressure valve 399
pump 24, 26, 61, 335, 408
pump fire 377
pumping 43
repellent 293
repellent coating 201
repellent silicone 201
softener 335
softener potable 417
storage solar tank 373
supply domestic meter 404
supply meter 404
tank 365, 366
tank elevated 365
tank sprayer 24
tempering valve 399
trailer 24
treatment 330
vacuum relief 405
well 61
Waterproofing 90, 200, 201
butyl 200
cementitious 201
coating 200
elastomeric 200
integral 91
masonry 120
membrane 200
neoprene 200
rubber 200
Waterstop 102
barrier 100
center bulb 103
dumbbell 102
fitting 103
PVC 102
ribbed 102
rubber 103
Watt meter 458
Wearing course 70
Weather station 13
Weatherproof receptacle 453
Weatherstrip 240, 258
aluminum 257
and seals 257
door 258
Weathervanes 306
Web stiffeners joist 162
Wedge anchor 141

INDEX

677

Index

bolt 141
Weight exercise 337
   lifting multi station 338
Weld fillet 145
   joint pipe 395
   rod 144
   rod aluminum 144
   shear connectors 143
   studs 144
Welded frame 230
   shear connector 143
   structural steel 533
   wire fabric 107, 509
Welder arc 339
Welding 145
   certification 12
   machine 25
   structural 145, 151
Well 43, 61
   gravel pack 61
   pump deep 62
   pump shallow 62
   water 61
Wellpoint 44, 493
   equipment 26
   header pipe 26
   pump 26
Wheel color 323
   guard 303
   potters 339
   stop 169
Wheelbarrow 25
Wheelchair lift 380
   stairlift 380
Whirler playground 82
Whirlpool bath 341, 414
Wide flange beam 151
   flange beam fiberglass 195
   throw hinge 257
Winch bull 99
Window 246
window 246
Window 247
   aluminum 247
   awning 248
   bank 320
   blind 194, 345
   bullet resistant 321
   casement 248, 249
   casing 186
   counter 311
   demolition 41
   double hung 249
   drive-up 321
   frame 251
   framing 362
   glass 261
   grille 251
   guard 168
   hardware 252
   metal 247, 363
   muntin 251
   painting 287, 288
   picture 250
   removal 41
   restoration 371
   screen 246, 251
   sill 130
   sill marble 131
   sill tile 275
   sliding 250
   steel 246, 363
   stool 130, 131
   storm 251
   teller 321
   trim set 187

wall 262
wall framing 148, 262
wood 248-251
Wine cellar 330
Winter protection 13, 91
Wire 168, 569
   aluminum 445
   connector 443
   copper 445
   electric 445
   fabric welded 107
   fence 17, 76
   fencing 78
   glass 261
   ground 441
   hanging 266
   mesh 78, 269
   mesh locker 309
   mesh partitions 312
   partition 312
   razor 78
   reinforcement 509
   rope 152
   rope steel 152
   strainer 220
   thermostat 456
   THW 445
   THWN-THHN 445
   window guard 168
Wiremold raceway 449
Wireway raceway 449
Wiring air conditioner 455
   device 450, 456
   device & box 450
   fan 455
   methods 442
   power 459
   residential 451, 455
Wood base 186, 194
   beam 176, 177
   bench 81
   blind 194, 346
   block floor 278
   block floor demolition 37
   blocking 176, 179
   bridge 81
   bridging 176
   cabinet 344
   canopy 180
   casing 186
   chamfer strip 99
   chips 84
   column 177, 178, 193
   cupola 306
   deck 183, 184
   demolition 35
   dome 182, 359
   door 230-232, 234, 236
   doors 550
   drawer 191
   fascia 180, 186
   fastener 174
   fiber 116, 117
   fiber insulation 202
   fiber sheathing 184
   fiber soffit 187
   fiber subfloor 184
   fiber underlayment 184
   firestop 178
   flagpole 307
   floor 278, 279
   floor demolition 37
   folding partition 314
   frame 190, 236
   framing 176, 178, 181
   furring 181

girder 176
gutter 221
handrail 187
header 181
joist 178, 184
laminated 185
ledger 180
louver 193
molding 192
nailer 178
overhead door 243
panel door 233
paneling 188
partition 179
pile 58, 499
planter 87
plate 181
pole 66
product siding 213
rafter 180
railing 191, 193
roof 182
roof deck 183
roof deck demolition 39
roof truss 185
roof trusses 547
sash 251
screen 251
screw 174
shade 346
shake 208
sheathing 183
sheathing cutting 41
sheet piling 494
sheeting 43, 45, 494
shelving 189
shingle 207
sidewalk 72
siding 212
siding demolition 40
sill 180
sleeper 180
soffit 180, 187
stair 192, 547
storm door 230
stud 181, 269
subfloor 184
tank 367
threshold 187
tie 169
tread 193
treatment 172
trim 186
veneer wallpaper 283
wardrobe 190
window 248-251, 551
window demolition 41
Woodwork architectural 189
Wool carpet 283
   fiberglass 202
   lead 100
Work extra 9
Worker protection 370
Workers' compensation 8, 482-484
Woven wire partition 312
Wrapping pipe 87
Wrecking ball 25
Wrench impact 22
Wrestling mat 338
Wrought iron 169
Wye fitting 61

X

X-ray barrier 357

concrete slabs 12
dental 341
medical 341
mobile 341
protection 357

Y

Y sediment strainer valve 401
   type iron body strainer 404
Yellow pine floor 280
YMCA club S.F. & C.F. 471

Z

Z bar suspension 266
Zee bar 273
Zinc divider strip 276
   plating 173
   roof 217
   terrazzo strip 276
   weatherstrip 258

CMD Group...

R.S. Means Company, Inc., a CMD Group company, the leading provider of construction cost data in North America, supplies comprehensive construction cost guides, related technical publications and educational services.

CMD Group, a leading worldwide provider of proprietary construction information, is comprised of three synergistic product groups crafted to be the complete resource for reliable, timely and actionable construction market data. In North America, CMD Group encompasses:
- Architects' First Source
- Construction Market Data (CMD)
- Associated Construction Publications
- Manufacturer's Survey Associates (MSA)
- R.S. Means
- CMD Canada
- BIMSA/Mexico
- Worldwide, CMD Group includes Byggfakta Scandinavia (Denmark, Estonia, Finland, Norway and Sweden) and Cordell Building Information Services (Australia).

First Source for Products, available in print, on the Means CostWorks CD, and on the Internet, is a comprehensive product information source. In alliance with The Construction Specifications Institute (CSI) and Thomas Register, Architects' First Source also produces CSI's SPEC-DATA® and MANU-SPEC® as well as CADBlocks.℠ Together, these products offer commercial building product information for the building team at each stage of the construction process.

CMD Exchange, a single electronic community where all members of the construction industry can communicate, collaborate, and conduct business, better, faster, easier.

Construction Market Data provides complete, accurate and timely project information through all stages of construction. Construction Market Data supplies industry data through productive leads, project reports, contact lists, market penetration analysis and sales evaluation reports. Any of these products can pinpoint a county, look at a state, or cover the country. Data is delivered via paper, e-mail or the Internet.

Construction Market Data Canada serves the Canadian construction market with reliable and comprehensive information services that cover all facets of construction. Core services include: Buildcore Product Source, a preliminary product selection tool available in print and on the Internet; CMD Building Reports, a national construction project lead service; CanaData, statistical and forecasting information; Daily Commercial News, a construction newspaper reporting on news and projects in Ontario; and Journal of Commerce, reporting news in British Columbia and Alberta.

Manufacturers' Survey Associates is a quantity survey and specification service whose experienced estimating staff examines project documents throughout the bidding process and distributes edited information to its clients. Material estimates, edited plans and specifications and distribution of addenda form the heart of the Manufacturers' Survey Associates product line, which is available in print and on CD-ROM.

Associated Construction Publications, founded in 1930, are a group of 14 regional magazines focused on highway and heavy construction and cost data. With one of the world's largest editorial staffs dedicated to construction coverage, the magazine group focuses on local and regional news of projects in all phases of construction, precise project volumes, plus material and labor costs. The magazines have a cumulative ABC and BPA audited circulation of 115,000, an estimated pass-along readership of one-half million, and more than 25,000 pages of annual advertising.

Clark Reports is the premier provider of industrial construction project data, noted for providing earlier, more complete project information in the planning and pre-planning stages for all segments of industrial and institutional construction.

Byggfakta Scandinavia AB, founded in 1936, is the parent company for the leaders of customized construction market data for Denmark, Estonia, Finland, Norway and Sweden. Each company fully covers the local construction market and provides information across several platforms including subscription, ad-hoc basis, electronically and on paper.

Cordell Building Information Services, with its complete range of project and cost and estimating services, is Australia's specialist in the construction information industry. Cordell provides in-depth and historical information on all aspects of construction projects and estimation, including several customized reports, construction and sales leads, and detailed cost information among others.

For more information, please visit our website at www.cmdg.com

CMD Group Corporate Offices
30 Technology Parkway South #100
Norcross, GA 30092-2912
(770) 417-4000
(770) 417-4002 (fax)
www.cmdg.com

Means Project Cost Report

By filling out and returning the Project Description, you can receive a discount of $20.00 off any one of the Means products advertised in the following pages. The cost information required includes all items marked (✔) except those where no costs occurred. The sum of all major items should equal the Total Project Cost.

$20.00 Discount per product for each report you submit.

DISCOUNT PRODUCTS AVAILABLE—FOR U.S. CUSTOMERS ONLY—STRICTLY CONFIDENTIAL

Project Description (No remodeling projects, please.)

- ✔ Type Building _____
- ✔ Location _____
- Capacity _____
- ✔ Frame _____
- ✔ Exterior _____
- ✔ Basement: full ☐ partial ☐ none ☐ crawl ☐
- ✔ Height in Stories _____
- ✔ Total Floor Area _____
- Ground Floor Area _____
- ✔ Volume in C.F. _____
- % Air Conditioned _____ Tons _____
- Comments _____

- Owner _____
- Architect _____
- General Contractor _____
- ✔ Bid Date _____
- Typical Bay Size _____
- ✔ Labor Force: _____ % Union _____ % Non-Union
- ✔ Project Description (Circle one number in each line)
  1. Economy 2. Average 3. Custom 4. Luxury
  1. Square   2. Rectangular 3. Irregular 4. Very Irregular

		Item	$	Unit
	✔	**Total Project Cost**	$	
A	✔	**General Conditions**	$	
B	✔	**Site Work**	$	
BS		Site Clearing & Improvement		
BE		Excavation	(	C.Y.)
BF		Caissons & Piling	(	L.F.)
BU		Site Utilities		
BP		Roads & Walks Exterior Paving	(	S.Y.)
C	✔	**Concrete**	$	
C		Cast in Place	(	C.Y.)
CP		Precast	(	S.F.)
D	✔	**Masonry**	$	
DB		Brick	(	M)
DC		Block	(	M)
DT		Tile	(	S.F.)
DS		Stone	(	S.F.)
E	✔	**Metals**	$	
ES		Structural Steel	(	Tons)
EM		Misc. & Ornamental Metals		
F	✔	**Wood & Plastics**	$	
FR		Rough Carpentry	(	MBF)
FF		Finish Carpentry		
FM		Architectural Millwork		
G	✔	**Thermal & Moisture Protection**	$	
GW		Waterproofing-Dampproofing	(	S.F.)
GN		Insulation	(	S.F.)
GR		Roofing & Flashing	(	S.F.)
GM		Metal Siding/Curtain Wall	(	S.F.)
H	✔	**Doors and Windows**	$	
HD		Doors	(	Ea.)
HW		Windows	(	S.F.)
HH		Finish Hardware		
HG		Glass & Glazing	(	S.F.)
HS		Storefronts	(	S.F.)

		Item	$	Unit
J	✔	**Finishes**	$	
JL		Lath & Plaster	(	S.Y.)
JD		Drywall		S.F.)
JM		Tile & Marble		S.F.)
JT		Terrazzo		S.F.)
JA		Acoustical Treatment	(	S.F.)
JC		Carpet	(	S.Y.)
JF		Hard Surface Flooring		S.F.)
JP		Painting & Wall Covering		S.F.)
K	✔	**Specialties**	$	
KB		Bathroom Partitions & Access.	(	S.F.)
KF		Other Partitions	(	S.F.)
KL		Lockers	(	Ea.)
L	✔	**Equipment**	$	
LK		Kitchen		
LS		School		
LO		Other		
M	✔	**Furnishings**	$	
MW		Window Treatment		
MS		Seating	(	Ea.)
N	✔	**Special Construction**	$	
NA		Acoustical	(	S.F.)
NB		Prefab. Bldgs.	(	S.F.)
NO		Other		
P	✔	**Conveying Systems**	$	
PE		Elevators	(	Ea.)
PS		Escalators	(	Ea.)
PM		Material Handling		
Q	✔	**Mechanical**	$	
QP		Plumbing (No. of fixtures		)
QS		Fire Protection (Sprinklers)		
QF		Fire Protection (Hose Standpipes)		
QB		Heating, Ventilating & A.C.		
QH		Heating & Ventilating (BTU Output		)
QA		Air Conditioning	(	Tons)
R	✔	**Electrical**	$	
RL		Lighting	(	S.F.)
RP		Power Service		
RD		Power Distribution		
RA		Alarms		
RG		Special Systems		
S	✔	**Mech./Elec. Combined**	$	

Product Name _____

Product Number _____

Your Name _____

Title _____

Company _____

☐ Company
☐ Home   Street Address _____

City, State, Zip _____

☐ Please send _____ forms.

Please specify the Means product you wish to receive. Complete the address information as requested and return this form with your check (product cost less $20.00) to address below.

R.S. Means Company, Inc.,
Square Foot Costs Department
100 Construction Plaza, P.O. Box 800
Kingston, MA 02364-9988

R.S. Means Company, Inc... a tradition of excellence in Construction Cost Information and Services since 1942.

For more information visit Means Web Site at www.rsmeans.com

Table of Contents
Annual Cost Guides, Page 2
Reference Books, Page 6
Seminars, Page 11
Consulting Services, Page 13
Electronic Data, Page 14
New Titles, Page 15
Order Form, Page 16

Book Selection Guide

The following table provides definitive information on the content of each cost data publication. The number of lines of data provided in each unit price or assemblies division, as well as the number of reference tables and crews is listed for each book. The presence of other elements such as an historical cost index, city cost indexes, square foot models or cross-referenced index is also indicated. You can use the table to help select the Means' book that has the quantity and type of information you most need in your work.

Unit Cost Divisions	Building Construction Costs	Mechanical	Electrical	Repair & Remodel.	Square Foot	Site Work Landsc.	Assemblies	Interior	Concrete Masonry	Open Shop	Heavy Construc.	Residential	Light Commercial	Facil. Construc.	Plumbing	Western Construction Costs
1	1108	587	588	780		1035		488	1000	1098	1075	430	625	1524	670	1095
2	3126	1460	506	2292		9120		1150	1636	3339	5584	1152	1217	5043	1819	3100
3	1516	90	68	764		1320		193	1935	1489	1492	263	208	1356	43	1493
4	875	28	0	675		770		642	1193	857	677	320	405	1143	0	853
5	1743	176	173	829		780		837	681	1710	1054	634	661	1733	95	1726
6	1396	95	82	1297		482		1317	334	1364	590	1515	1454	1393	59	1749
7	1365	178	91	1331		492		545	454	1366	353	803	1037	1359	188	1367
8	1898	60	0	1900		335		1818	723	1878	9	1131	1180	2030	0	1898
9	1610	48	0	1466		184		1706	322	1556	113	1250	1350	1805	48	1603
10	978	58	32	551		204		823	197	982	0	253	454	979	251	978
11	1157	434	212	621		152		915	35	1051	96	117	239	1220	369	1050
12	352	0	0	48		226		1536	29	343	0	71	68	1537	0	343
13	1384	1217	493	582		443		1071	106	1355	302	230	1331	965	1080	1335
14	364	43	0	261		36		331	0	365	39	9	18	363	17	363
15	2518	14708	764	2249		1802		1471	77	2518	2088	1065	1549	12986	10913	2547
16	1517	801	10627	1195		1098		1289	60	1529	911	703	1286	10207	708	1453
17	459	369	459	0		0		0	0	460	0	0	0	459	369	459
Totals	23366	20352	14095	16841		18479		16132	8782	23260	14383	9946	13082	46102	16629	23412

Assembly Divisions	Building Construction Costs	Mechanical	Electrical	Repair & Remodel.	Square Foot	Site Work Landsc.	Assemblies	Interior	Concrete Masonry	Open Shop	Heavy Construc.	Residential	Light Commercial	Facil. Construc.	Plumbing	Western Construction Costs
1		0	0	202	131	738	775	0	689		752	844	131	89	0	
2		0	0	40	33	35	48	0	49		0	589	32	0	0	
3		0	0	443	1114	0	3064	228	1245		0	1546	806	174	0	
4		0	0	714	1353	0	3172	255	1273		0	2046	1171	26	0	
5		0	0	288	227	0	459	0	0		0	1082	223	25	0	
6		0	0	1006	845	0	1295	1723	152		0	1082	738	330	0	
7		0	0	42	89	0	179	159	0		0	723	33	71	0	
8		2257	149	998	1691	0	2720	926	0		0	1694	1191	1114	2084	
9		0	1346	355	357	0	1283	306	0		0	241	358	319	0	
10		0	0	0	0	0	0	0	0		0	0	0	0	0	
11		0	0	365	465	0	724	153	0		0	0	466	201	0	
12		501	160	539	84	2392	699	0	698		541	0	84	119	850	
Totals		2758	1655	4992	6389	3165	14418	3750	4106		1293	9847	5233	2468	2934	

Reference Section	Building Construction Costs	Mechanical	Electrical	Repair & Remodel.	Square Foot	Site Work Landsc.	Assemblies	Interior	Concrete Masonry	Open Shop	Heavy Construc.	Residential	Light Commercial	Facil. Construc.	Plumbing	Western Construction Costs
Tables	150	46	84	66	4	84	223	59	83	146	55	49	70	86	49	148
Models					102							32	43			
Crews	408	408	408	389		408		408	408	391	408	391	391	389	408	408
City Cost Indexes	yes	yes	yes	yes	yes	yes	yes	yes	yes	yes	yes	yes	yes	yes	yes	yes
Historical Cost Indexes	yes	yes	yes	yes	yes	yes	yes	yes	yes	yes	yes	no	yes	yes	yes	yes
Index	yes	yes	yes	yes	no	yes	yes	yes	yes	yes	yes	yes	yes	yes	yes	yes

Annual Cost Guides

For more information visit Means Web Site at www.rsmeans.com

Means Building Construction Cost Data 2000

Available in Both Softbound and Looseleaf Editions

The "Bible" of the industry comes in the standard softcover edition or the looseleaf edition.

Many customers enjoy the convenience and flexibility of the looseleaf binder, which increases the usefulness of *Means Building Construction Cost Data 2000* by making it easy to add and remove pages. You can insert your own cost information pages, so everything is in one place. Copying pages for faxing is easier also. Whichever edition you prefer, softbound or the convenient looseleaf edition, you'll get the *DesignBuildIntelligence* newsletter at no extra cost.

$85.95 per copy, Softbound
Catalog No. 60010

$109.95 per copy, Looseleaf
Catalog No. 61010

Means Building Construction Cost Data 2000

Offers you unchallenged unit price reliability in an easy-to-use arrangement. Whether used for complete, finished estimates or for periodic checks, it supplies more cost facts better and faster than any comparable source. Over 23,000 unit prices for 2000. The City Cost Indexes cover over 930 areas, for indexing to any project location in North America. Order and get the *DesignBuildIntelligence* newsletter sent to you FREE. You'll have year-long access to the Means Estimating **Hotline** FREE with your subscription. Expert assistance when using Means data is just a phone call away.

$85.95 per copy
Over 650 pages, illustrated, available Oct. 1999
Catalog No. 60010

Means Building Construction Cost Data 2000
Metric Version

The Federal Government has stated that all federal construction projects must now use metric documentation. The *Metric Version* of *Means Building Construction Cost Data 2000* is presented in metric measurements covering all construction areas. Don't miss out on these billion dollar opportunities. Make the switch to metric today.

$89.95 per copy
Over 650 pages, illustrated, available Nov. 1999
Catalog No. 63010

For more information
visit Means Web Site
at www.rsmeans.com

Annual Cost Guides

Means Mechanical Cost Data 2000

• HVAC • Controls

Total unit and systems price guidance for mechanical construction . . . materials, parts, fittings, and complete labor cost information. Includes prices for piping, heating, air conditioning, ventilation, and all related construction.

Plus new 2000 unit costs for:
• Over 2500 installed HVAC/controls assemblies
• "On Site" Location Factors for over 930 cities and towns in the U.S. and Canada
• Crews, labor and equipment

$89.95 per copy
Over 600 pages, illustrated, available Oct. 1999
Catalog No. 60020

Means Plumbing Cost Data 2000

Comprehensive unit prices and assemblies for plumbing, irrigation systems, commercial and residential fire protection, point-of-use water heaters, and the latest approved materials. This publication and its companion, *Means Mechanical Cost Data*, provide full-range cost estimating coverage for all the mechanical trades.

$89.95 per copy
Over 500 pages, illustrated, available Oct. 1999
Catalog No. 60210

Means Electrical Cost Data 2000

Pricing information for every part of electrical cost planning: More than 15,000 unit and systems costs with design tables; clear specifications and drawings; engineering guides and illustrated estimating procedures; complete labor-hour and materials costs for better scheduling and procurement; the latest electrical products and construction methods.

• A Variety of Special Electrical Systems including Cathodic Protection
• Costs for maintenance, demolition, HVAC/mechanical, specialties, equipment, and more

$89.95 per copy
Over 450 pages, illustrated, available Oct. 1999
Catalog No. 60030

Means Electrical Change Order Cost Data 2000

You are provided with electrical unit prices exclusively for pricing change orders—based on the recent, direct experience of contractors and suppliers. Analyze and check your own change order estimates against the experience others have had doing the same work. It also covers productivity analysis and change order cost justifications. With useful information for calculating the effects of change orders and dealing with their administration.

$89.95 per copy
Over 450 pages, available Oct. 1999
Catalog No. 60230

Means Facilities Maintenance & Repair Cost Data 2000

Published in a looseleaf format, *Means Facilities Maintenance & Repair Cost Data* gives you a complete system to manage and plan your facility repair and maintenance costs and budget efficiently. Guidelines for auditing a facility and developing an annual maintenance plan. Budgeting is included, along with reference tables on cost and management and information on frequency and productivity of maintenance operations.

The only nationally recognized source of maintenance and repair costs. Developed in cooperation with the Army Corps of Engineers.

$199.95 per copy
Over 600 pages, illustrated, available Dec. 1999
Catalog No. 60300

Means Square Foot Costs 2000

It's Accurate and Easy To Use!

• **Updated 2000 price information**, based on nationwide figures from suppliers, estimators, labor experts and contractors.

• "How-to-Use" Sections, with **clear examples** of commercial, residential, industrial, and institutional structures.

• Realistic graphics, offering true-to-life illustrations of building projects.

• Extensive information on using square foot cost data, including **sample estimates** and **alternate pricing methods.**

$99.95 per copy
Over 450 pages, illustrated, available Nov. 1999
Catalog No. 60050

Annual Cost Guides

For more information visit Means Web Site at www.rsmeans.com

Means Repair & Remodeling Cost Data 2000
Commercial/Residential

You can use this valuable tool to estimate commercial and residential renovation and remodeling.

Includes: New costs for hundreds of unique methods, materials and conditions that only come up in repair and remodeling. PLUS:
- Unit costs for over 16,000 construction components
- Installed costs for over 90 assemblies
- Costs for 300+ construction crews
- Over 930 "On Site" localization factors for the U.S. and Canada.

$79.95 per copy
Over 600 pages, illustrated, available Oct. 1999
Catalog No. 60040

Means Facilities Construction Cost Data 2000

For the maintenance and construction of commercial, industrial, municipal, and institutional properties. Costs are shown for new and remodeling construction and are broken down into materials, labor, equipment, overhead, and profit. Special emphasis is given to sections on mechanical, electrical, furnishings, site work, building maintenance, finish work, and demolition. More than 45,000 unit costs plus assemblies and reference sections are included.

$209.95 per copy
Over 1150 pages, illustrated, available Nov. 1999
Catalog No. 60200

Means Residential Cost Data 2000

Contains square foot costs for 30 basic home models with the look of today—plus hundreds of custom additions and modifications you can quote right off the page. With costs for the 100 residential systems you're most likely to use in the year ahead. Complete with blank estimating forms, sample estimates and step-by-step instructions.

$74.95 per copy
Over 550 pages, illustrated, available Dec. 1999
Catalog No. 60170

Means Light Commercial Cost Data 2000

Specifically addresses the light commercial market, which is an increasingly specialized niche in the industry. Aids you, the owner/designer/contractor, in preparing all types of estimates, from budgets to detailed bids. Includes new advances in methods and materials. Assemblies section allows you to evaluate alternatives in the early stages of design/planning.

Over 13,000 unit costs for 2000 ensure you have the prices you need... when you need them.

$76.95 per copy
Over 600 pages, illustrated, available Dec. 1999
Catalog No. 60180

Means Assemblies Cost Data 2000

Means Assemblies Cost Data 2000 takes the guesswork out of preliminary or conceptual estimates. Now you don't have to try to calculate the assembled cost by working up individual components costs. We've done all the work for you.

Presents detailed illustrations, descriptions, specifications and costs for every conceivable building assembly—240 types in all—arranged in the easy-to-use UniFormat system. Each illustrated "assembled" cost includes a complete grouping of materials and associated installation costs including the installing contractor's overhead and profit.

$139.95 per copy
Over 550 pages, illustrated, available Oct. 1999
Catalog No. 60060

Means Site Work & Landscape Cost Data 2000

Means Site Work & Landscape Cost Data 2000 is organized to assist you in all your estimating needs. Hundreds of fact-filled pages help you make accurate cost estimates efficiently.

Updated for 2000!
- Demolition features—including ceilings, doors, electrical, flooring, HVAC, millwork, plumbing, roofing, walls and windows
- State-of-the-art segmental retaining walls
- Flywheel trenching costs and details
- Updated Wells section
- Thousands of landscape materials, flowers, shrubs and trees

$89.95 per copy
Over 600 pages, illustrated, available Nov. 1999
Catalog No. 60280

For more information
visit Means Web Site
at www.rsmeans.com

Annual Cost Guides

Means Open Shop Building Construction Cost Data 2000

The latest costs for accurate budgeting and estimating of new commercial and residential construction... renovation work... change orders... cost engineering. *Means Open Shop BCCD* will assist you to...
· Develop benchmark prices for change orders
· Plug gaps in preliminary estimates, budgets
· Estimate complex projects
· Substantiate invoices on contracts
· Price ADA-related renovations

$89.95 per copy
Over 650 pages, illustrated, available Oct. 1999
Catalog No. 60150

Means Heavy Construction Cost Data 2000

A comprehensive guide to heavy construction costs. Includes costs for highly specialized projects such as tunnels, dams, highways, airports, and waterways. Information on different labor rates, equipment, and material costs is included. Has unit price costs, systems costs, and numerous reference tables for costs and design. Valuable not only to contractors and civil engineers, but also to government agencies and city/town engineers.

$89.95 per copy
Over 450 pages, illustrated, available Nov. 1999
Catalog No. 60160

Means Building Construction Cost Data 2000
Western Edition

This regional edition provides more precise cost information for western North America. Labor rates are based on union rates from 13 western states and western Canada. Included are western practices and materials not found in our national edition: tilt-up concrete walls, glu-lam structural systems, specialized timber construction, seismic restraints, landscape and irrigation systems.

$89.95 per copy
Over 600 pages, illustrated, available Dec. 1999
Catalog No. 60220

Means Heavy Construction Cost Data 2000
Metric Version

Make sure you have the Means industry standard metric costs for the federal, state, municipal and private marketplace. With thousands of up-to-date metric unit prices in tables by CSI standard divisions. Supplies you with assemblies costs using the metric standard for reliable cost projections in the design stage of your project. Helps you determine sizes, material amounts, and has tips for handling metric estimates.

$89.95 per copy
Over 450 pages, illustrated, available Nov. 1999
Catalog No. 63160

Means Construction Cost Indexes 2000

Who knows what 2000 holds? What materials and labor costs will change unexpectedly? By how much?
· Breakdowns for 305 major cities.
· National averages for 30 key cities.
· Expanded five major city indexes.
· Historical construction cost indexes.

$198.00 per year/$49.50 individual quarters
Catalog No. 60140

Means Interior Cost Data 2000

Provides you with prices and guidance needed to make accurate interior work estimates. Contains costs on materials, equipment, hardware, custom installations, furnishings, labor costs . . . every cost factor for new and remodel commercial and industrial interior construction, including updated information on office furnishings, plus more than 50 reference tables. For contractors, facility managers, owners.

$89.95 per copy
Over 550 pages, illustrated, available Oct. 1999
Catalog No. 60090

Means Concrete & Masonry Cost Data 2000

Provides you with cost facts for virtually all concrete/masonry estimating needs, from complicated formwork to various sizes and face finishes of brick and block, all in great detail. The comprehensive unit cost section contains more than 8,500 selected entries. Also contains an assemblies cost section, and a detailed reference section which supplements the cost data.

$79.95 per copy
Over 450 pages, illustrated, available Nov. 1999
Catalog No. 60110

Means Labor Rates for the Construction Industry 2000

Complete information for estimating labor costs, making comparisons and negotiating wage rates by trade for over 300 cities (United States and Canada). With 46 construction trades listed by local union number in each city, and historical wage rates included for comparison. No similar book is available through the trade.

Each city chart lists the county and is alphabetically arranged with handy visual flip tabs for quick reference.

$189.95 per copy
Over 300 pages, available Dec. 1999
Catalog No. 60120

Reference Books

For more information visit Means Web Site at www.rsmeans.com

Project Scheduling & Management for Construction
New Revised Edition
By David R. Pierce, Jr.

A comprehensive, yet easy-to-follow guide to construction project scheduling and control—from vital project management principles through the latest scheduling, tracking, and controlling techniques. The author is a leading authority on scheduling with years of field and teaching experience at leading academic institutions. Spend a few hours with this book and come away with a solid understanding of this essential management topic.

$64.95 per copy
Over 250 pages, illustrated, Hardcover
Catalog No. 67247A

Means Landscape Estimating Methods
New 3rd Edition
By Sylvia H. Fee

This revised edition offers expert guidance for preparing accurate estimates for new landscape construction and grounds maintenance. Includes a complete project estimate featuring the latest equipment and methods, and **two totally new chapters—Life Cycle Costing and Landscape Maintenance Estimating.**

$62.95 per copy
Over 300 pages, illustrated, Hardcover
Catalog No. 67295A

Total Productive Facilities Management
By Richard W. Sievert, Jr.

A New Operational Standard for Facilities Management.

The TPFM program incorporates the best of today's cost and project management, quality, and value engineering principles. The book includes:
- Realistic ways to collect and use benchmarking data.
- Value Engineering to find economical answers to the organization's needs.
- Selecting the best scheduling, control, and contracting methods for construction projects.

$79.95 per copy
Over 270 pages, illustrated, Hardcover
Catalog No. 67321

Cyberplaces: The Internet Guide for Architects, Engineers & Contractors
By Paul Doherty

Internet applications for business and project management. Includes Book, CD-ROM and Web Site.

The CD-ROM offers built-in links to web sites, tours of captured sites, FREE browser software and a document workshop, plus a test for Continuing Education Credits. **The Web Site** keeps the book current with updates on new technologies, interactive workshops, links to new tools and sites, and reports from top firms.

$59.95 per copy
Over 700 pages, illustrated, Softcover
Catalog No. 67317

Value Engineering: Practical Applications
...For Design, Construction, Maintenance & Operations
By Alphonse Dell'Isola, P.E., leading authority on VE in construction

A tool for immediate application—for engineers, architects, facility managers, owners and contractors. Includes: Making the Case for VE—The Management Briefing, Integrating VE into Planning and Budgeting, Conducting Life Cycle Costing, Integrating VE into the Design Process, Using VE Methodology in Design Review and Consultant Selection, Case Studies, VE Workbook, and a Life Cycle Costing program on disk.

$79.95 per copy
Over 450 pages, illustrated, Hardcover
Catalog No. 67319

Means Environmental Remediation Estimating Methods
By Richard R. Rast

The first-ever guide to estimating any size environmental remediation project... anywhere in the country.

Field-tested guidelines for estimating 50 standard remediation technologies. This resource will help you: prepare preliminary budgets, develop detailed estimates, compare costs, select solutions, estimate liability, review quotes, and negotiate settlements.

A valuable support tool for *Means Environmental Remediation Unit Price* and *Assemblies* books.

$99.95 per copy
Over 600 pages, illustrated, Hardcover
Catalog No. 64777

For more information visit Means Web Site at www.rsmeans.com

Reference Books

Cost Planning & Estimating for Facilities Maintenance

In this unique book, a team of facilities management authorities shares their expertise at:
- Evaluating and budgeting maintenance operations
- Maintaining & repairing key building components
- Applying *Means Facilities Maintenance & Repair Cost Data* to your estimating

With the special maintenance requirements of the 10 major building types.

$82.95 per copy
Over 475 pages, Hardcover
Catalog No. 67314

Facilities Planning & Relocation

By David D. Owen

An A-Z working guide—complete with the checklists, schematic diagrams and questionnaires to ensure the success of every office relocation. Complete with a step-by-step manual, 100-page technical reference section, over 50 reproducible forms in hard copy and on computer diskettes.
Winner of the International Facility Managers Assoc. "Distinguished Author of the Year" award.

$109.95 per copy, Textbook-384 pages,
Forms Binder-146 pages, illustrated, Hardcover
Catalog No. 67301

Means Facilities Maintenance Standards

Unique features of this one-of-a-kind working guide for facilities maintenance

A working encyclopedia that points the way to solutions to every kind of maintenance and repair dilemma. With a labor-hours section to provide productivity figures for over 180 maintenance tasks. Included are ready-to-use forms, checklists, worksheets and comparisons, as well as analysis of materials systems and remedies for deterioration and wear.

$159.95 per copy, 600 pages, 205 tables, checklists and diagrams, Hardcover
Catalog No. 67246

HVAC: Design Criteria, Options, Selection
Expanded Second Edition

By William H. Rowe III, AIA, PE

Includes Indoor Air Quality, CFC Removal, Energy Efficient Systems and Special Systems by Building Type. Helps you solve a wide range of HVAC system design and selection problems effectively and economically. Gives you clear explanations of the latest ASHRAE standards.

$84.95 per copy
Over 600 pages, illustrated, Hardcover
Catalog No. 67306

The Facilities Manager's Reference

By Harvey H. Kaiser, PhD

The tasks and tools the facility manager needs to accomplish the organization's objectives, and develop individual and staff skills. Includes Facilities and Property Management, Administrative Control, Planning and Operations, Support Services, and a complete building audit with forms and instructions, widely used by facilities managers nationwide.

$86.95 per copy, over 250 pages,
with prototype forms and graphics, Hardcover
Catalog No. 67264

Facilities Maintenance Management

By Gregory H. Magee, PE

Now you can get successful management methods and techniques for all aspects of facilities maintenance. This comprehensive reference explains and demonstrates successful management techniques for all aspects of maintenance, repair and improvements for buildings, machinery, equipment and grounds. Plus, guidance for outsourcing and managing internal staffs.

$86.95 per copy
Over 280 pages with illustrations, Hardcover
Catalog No. 67249

Understanding Building Automation Systems

- Direct Digital Control
- Security/Access Control
- Energy Management
- Life Safety
- Lighting

By Reinhold A. Carlson, PE & Robert Di Giandomenico

The authors, leading authorities on the design and installation of these systems, describe the major building systems in both an overview with estimating and selection criteria, and in system configuration-level detail.

Now $39.98 per copy, limited quantity
Over 200 pages, illustrated, Hardcover
Catalog No. 67284

The ADA in Practice

By Deborah S. Kearney, PhD

Helps you meet and budget for the requirements of the Americans with Disabilities Act. Shows how to do the job right by understanding what the law requires, allows, and enforces. Includes a "buyers guide" for 70 ADA-compliant products.

*Winner of IFMA's "Distinguished Author of the Year" award.

See page 9 for another ADA resource.

$72.95 per copy
Over 600 pages, illustrated, Softcover
Catalog No. 67147A

Reference Books

For more information visit Means Web Site at www.rsmeans.com

Basics for Builders: Plan Reading & Material Takeoff
By Wayne J. DelPico

For Residential and Light Commercial Construction

A valuable tool for understanding plans and specs, and accurately calculating material quantities. Step-by-step instructions and takeoff procedures based on a full set of working drawings.

$35.95 per copy
Over 420 pages, Softcover
Catalog No. 67307

Means Illustrated Construction Dictionary Unabridged Edition

Written in contractor's language, the information is adaptable for report writing, specifications or just intelligent discussion. Contains over 13,000 construction terms, words, phrases, acronyms and abbreviations, slang, regional terminology, and hundreds of illustrations.

$99.95 per copy
Over 700 pages, Hardcover
Catalog No. 67292

Superintending for Contractors:
How to Bring Jobs in On-time, On-Budget
By Paul J. Cook

This book examines the complex role of the superintendent/field project manager, and provides guidelines for the efficient organization of this job. Includes administration of contracts, change orders, purchase orders, and more.

$35.95 per copy
Over 220 pages, illustrated, Softcover
Catalog No. 67233

Means Forms for Contractors
For general and specialty contractors

Means editors have created and collected the most needed forms as requested by contractors of various-sized firms and specialties. This book covers all project phases. Includes sample correspondence and personnel administration.
Full-size forms on durable paper for photocopying or reprinting.

$79.95 per copy
Over 400 pages, three-ring binder, more than 80 forms
Catalog No. 67288

Risk Management for Building Professionals
By Thomas E. Papageorge, RA

Outlines a master plan for phasing a risk management system into each company function. Guides the manager in writing a detailed Risk Management Plan for each project. Contains sample procedures manual.

Now $29.98 per copy, limited quantity
Over 200 pages, illustrated, Hardcover
Catalog No. 67254

Means Heavy Construction Handbook

Informed guidance for planning, estimating and performing today's heavy construction projects. Provides expert advice on every aspect of heavy construction work including hazardous waste remediation and estimating. To assist planning, estimating, performing, or overseeing work.

$74.95 per copy
Over 430 pages, illustrated, Hardcover
Catalog No. 67148

Estimating for Contractors:
How to Make Estimates that Win Jobs
By Paul J. Cook

Estimating for Contractors is a reference that will be used over and over, whether to check a specific estimating procedure, or to take a complete course in estimating.

$35.95 per copy
Over 225 pages, illustrated, Softcover
Catalog No. 67160

HVAC Systems Evaluation
By Harold R. Colen, PE

You get direct comparisons of how each type of system works, with the relative costs of installation, operation, maintenance and applications by type of building. With requirements for hooking up electrical power to HVAC components. Contains experienced advice for repairing operational problems in existing HVAC systems, ductwork, fans, cooling coils, and much more!

$84.95 per copy
Over 500 pages, illustrated, Hardcover
Catalog No. 67281

Quantity Takeoff for Contractors:
How to Get Accurate Material Counts
By Paul J. Cook

Contractors who are new to material takeoffs or want to be sure they are using the best techniques will find helpful information in this book, organized by CSI MasterFormat division.

Now $17.98 per copy, limited quantity
Over 250 pages, illustrated, Softcover
Catalog No. 67262

Roofing: Design Criteria, Options, Selection
By R.D. Herbert, III

This book is required reading for those who specify, install or have to maintain roofing systems. It covers all types of roofing technology and systems. You'll get the facts needed to intelligently evaluate and select both traditional and new roofing systems.

Now $31.48 per copy
Over 225 pages with illustrations, Hardcover
Catalog No. 67253

For more information visit Means Web Site at www.rsmeans.com

Reference Books

Means Estimating Handbook

This comprehensive reference covers a full spectrum of technical data for estimating, with information on sizing, productivity, equipment requirements, codes, design standards and engineering factors.
Means Estimating Handbook will help you: evaluate architectural plans and specifications, prepare accurate quantity takeoffs, prepare estimates from conceptual to detailed, and evaluate change orders.

$99.95 per copy
Over 900 pages, Hardcover
Catalog No. 67276

Means Repair and Remodeling Estimating Third Edition
By Edward B. Wetherill & R.S. Means

Focuses on the unique problems of estimating renovations of existing structures. It helps you determine the true costs of remodeling through careful evaluation of architectural details and a site visit.
New section on disaster restoration costs.

$69.95 per copy
Over 450 pages, illustrated, Hardcover
Catalog No. 67265A

Successful Estimating Methods:
From Concept to Bid
By John D. Bledsoe, PhD, PE

A highly practical, all-in-one guide to the tips and practices of today's successful estimator. Presents techniques for all types of estimates, *and* advanced topics such as life cycle cost analysis, value engineering, and automated estimating.
Estimate spreadsheets available at Means Web site.

$64.95 per copy
Over 300 pages, illustrated, Hardcover
Catalog No. 67287

Means Productivity Standards for Construction
Expanded Edition *(Formerly Man-Hour Standards)*

Here is the working encyclopedia of labor productivity information for construction professionals, with labor requirements for thousands of construction functions in CSI MasterFormat.
Completely updated, with over 3,000 new work items.

$159.95 per copy
Over 800 pages, Hardcover
Catalog No. 67236A

Means Electrical Estimating Methods Second Edition

Expanded version includes sample estimates and cost information in keeping with the latest version of the CSI MasterFormat. Contains new coverage of Fiber Optic and Uninterruptible Power Supply electrical systems, broken down by components and explained in detail. A practical companion to *Means Electrical Cost Data*.

$64.95 per copy
Over 325 pages, Hardcover
Catalog No. 67230A

Means Mechanical Estimating Methods Second Edition

This guide assists you in making a review of plans, specs and bid packages with suggestions for takeoff procedures, listings, substitutions and pre-bid scheduling. Includes suggestions for budgeting labor and equipment usage. Compares materials and construction methods to allow you to select the best option.

$64.95 per copy
Over 350 pages, illustrated, Hardcover
Catalog No. 67294

Means Scheduling Manual
Third Edition
By F. William Horsley

Fast, convenient expertise for keeping your scheduling skills right in step with today's cost-conscious times. Covers bar charts, PERT, precedence and CPM scheduling methods. Now updated to include computer applications.

$64.95 per copy
Over 200 pages, spiral-bound, Softcover
Catalog No. 67291

Maintenance Management Audit

This annual audit program is essential for every organization in need of a proper assessment of its maintenance operation. The forms in this easy-to-use workbook allow managers to identify and correct problems and enhance productivity. Includes electronic forms on disk.

Now $32.48 per copy, limited quantity
125 pages, spiral bound, illustrated, Hardcover
Catalog No. 67299

Means Square Foot Estimating Methods Second Edition
By Billy J. Cox and F. William Horsley

Proven techniques for conceptual and design-stage cost planning. Steps you through the square foot cost process, demonstrating faster, better ways to relate the design to the budget. Now updated to the latest version of UniFormat.

$69.95 per copy
Over 300 pages, illustrated, Hardcover
Catalog No. 67145A

Means ADA Compliance Pricing Guide

Gives you detailed cost estimates for budgeting modification projects, including estimates for each of 260 alternates. You get the 75 most commonly needed modifications for ADA compliance, each an assembly estimate with detailed cost breakdown.

$72.95 per copy
Over 350 pages, illustrated, Softcover
Catalog No. 67310

Reference Books

For more information visit Means Web Site at www.rsmeans.com

Means Unit Price Estimating Methods
2nd Edition
$59.95 per copy
Catalog No. 67303

Legal Reference for Design & Construction
By Charles R. Heuer, Esq., AIA
Now $54.98 per copy, limited quantity
Catalog No. 67266

Means Plumbing Estimating Methods
New 2nd Edition
By Joseph J. Galeno & Sheldon T. Greene
$59.95 per copy
Catalog No. 67283

Structural Steel Estimating
By S. Paul Bunea, PhD
Now $39.98 per copy, limited quantity
Catalog No. 67241

Business Management for Contractors
How to Make Profits in Today's Market
By Paul J. Cook
Now $17.98 per copy, limited quantity
Catalog No. 67250

Understanding Legal Aspects of Design/Build
By Timothy R. Twomey, Esq., AIA
$79.95 per copy
Catalog No. 67259

Construction Paperwork
An Efficient Management System
By J. Edward Grimes
Now $26.48 per copy, limited quantity
Catalog No. 67268

Contractor's Business Handbook
By Michael S. Milliner
Now $21.48 per copy, limited quantity
Catalog No. 67255

Basics for Builders: How to Survive and Prosper in Construction
By Thomas N. Frisby
$34.95 per copy
Catalog No. 67273

Successful Interior Projects Through Effective Contract Documents
By Joel Downey & Patricia K. Gilbert
Now $34.98 per copy, limited quantity
Catalog No. 67313

The Building Professional's Guide to Contract Documents
By Waller S. Poage, AIA, CSI, CCS
$64.95 per copy
Catalog No. 67261

Illustrated Construction Dictionary, Condensed
$59.95 per copy
Catalog No. 67282

Managing Construction Purchasing
By John G. McConville, CCC, CPE
Now $31.48 per copy, limited quantity
Catalog No. 67302

Hazardous Material & Hazardous Waste
By Francis J. Hopcroft, PE, David L. Vitale, M. Ed., & Donald L. Anglehart, Esq.
Now $44.98 per copy, limited quantity
Catalog No. 67258

Fundamentals of the Construction Process
By Kweku K. Bentil, AIC
Now $34.98 per copy, limited quantity
Catalog No. 67260

Construction Delays
By Theodore J. Trauner, Jr., PE, PP
$59.95 per copy
Catalog No. 67278

Basics for Builders: Framing & Rough Carpentry
By Scot Simpson
$24.95 per copy
Catalog No. 67298

Interior Home Improvement Costs
6th Edition
$19.95 per copy
Catalog No. 67308B

Exterior Home Improvement Costs
6th Edition
$19.95 per copy
Catalog No. 67309B

Concrete Repair and Maintenance Illustrated
By Peter H. Emmons
$69.95 per copy
Catalog No. 67146

How to Estimate with Metric Units
Now $24.98 per copy, limited quantity
Catalog No. 67304

For more information visit Means Web Site at www.rsmeans.com

Seminars

Developing Facility Assessment Programs

This two-day program concentrates on the management process required for planning, conducting, and documenting the physical condition and functional adequacy of buildings and other facilities. Regular facilities condition inspections are one of the facility department's most important duties. However, gathering reliable data hinges on the design of the overall program used to identify and gauge deferred maintenance requirements. Knowing where to look . . . and reporting results effectively are the keys. This seminar is designed to give the facility executive essential steps for conducting facilities inspection programs.

Inspection Program Requirements • Where is the deficiency? • What is the nature of the problem? • How can it be remedied? • How much will it cost in labor, equipment and materials? • When should it be accomplished? • Who is best suited to do the work?

Note: Because of its management focus, this course will not address trade practices and procedures.

Repair and Remodeling Estimating

Repair and remodeling work is becoming increasingly competitive as more professionals enter the market. Recycling existing buildings can pose difficult estimating problems. Labor costs, energy use concerns, building codes, and the limitations of working with an existing structure place enormous importance on the development of accurate estimates. Using the exclusive techniques associated with Means' widely acclaimed **Repair & Remodeling Cost Data**, this seminar sorts out and discusses solutions to the problems of building alteration estimating. Attendees will receive two intensive days of eye-opening methods for handling virtually every kind of repair and remodeling situation . . . from demolition and removal to final restoration.

Mechanical and Electrical Estimating

This seminar is tailored to fit the needs of those seeking to develop or improve their skills and to have a better understanding of how mechanical and electrical estimates are prepared during the conceptual, planning, budgeting and bidding stages. Learn how to avoid costly omissions and overlaps between these two interrelated specialties by preparing complete and thorough cost estimates for both trades. Featured are order of magnitude, assemblies, and unit price estimating. In combination with the use of **Means Mechanical Cost Data**, **Means Plumbing Cost Data** and **Means Electrical Cost Data**, this seminar will ensure more accurate and complete Mechanical/Electrical estimates for both unit price and preliminary estimating procedures.

Unit Price Estimating

This seminar shows how today's advanced estimating techniques and cost information sources can be used to develop more reliable unit price estimates for projects of any size. It demonstrates how to organize data, use plans efficiently, and avoid embarrassing errors by using better methods of checking.

You'll get down-to-earth help and easy-to-apply guidance for:
- making maximum use of construction cost information sources
- organizing estimating procedures in order to save time and reduce mistakes
- sorting out and identifying unusual job requirements to improve estimating accuracy.

Square Foot Cost Estimating

Learn how to make better preliminary estimates with a limited amount of budget and design information. You will benefit from examples of a wide range of systems estimates with specifications limited to building use requirements, budget, building codes, and type of building. And yet, with minimal information, you will obtain a remarkable degree of accuracy.

Workshop sessions will provide you with model square foot estimating problems and other skill-building exercises. The exclusive Means building assemblies square foot cost approach shows how to make very reliable estimates using "bare bones" budget and design information.

Scheduling and Project Management

This seminar helps you successfully establish project priorities, develop realistic schedules, and apply today's advanced management techniques to your construction projects. Hands-on exercises familiarize participants with network approaches such as the Critical Path Method. Special emphasis is placed on cost control, including use of computer-based systems. Through this seminar you'll perfect your scheduling and management skills, ensuring completion of your projects *on time* and *within budget*. Includes hands-on application of **Means Scheduling Manual** and **Means Building Construction Cost Data.**

Facilities Maintenance and Repair Estimating

With our Facilities Maintenance and Repair Estimating seminar, you'll learn how to plan, budget, and estimate the cost of ongoing and preventive maintenance and repair for all your buildings and grounds. Based on R.S. Means' groundbreaking cost estimating book, this two-day seminar will show you how to decide to either contract out or retain maintenance and repair work in-house. In addition, you'll learn how to prepare budgets and schedules that help cut down on unplanned and costly emergency repair projects. Facilities Maintenance and Repair Estimating crystallizes what facilities professionals have learned over the years, but never had time to organize or document. This program covers a variety of maintenance and repair projects, from underground storage tank removal, roof repair and maintenance, exterior wall renovations, and energy source conversions, to service upgrades and estimating energy-saving alternatives.

Managing Facilities Construction and Maintenance

In you're involved in new facility construction, renovation or maintenance projects and are concerned about getting quality work done on time and on or below budget, in Means' seminar **Managing Facilities Construction and Maintenance** you'll learn management techniques needed to effectively plan, organize, control and get the most out of your limited facilities resources.

Learn how to develop budgets, reduce expenditures and check productivity. With the knowledge gained in this course, you'll be better prepared to successfully sell accurate project budgets, timing and manpower needs to senior management . . . plus understand how to evaluate the impact of today's facility decisions on tomorrow's budget.

Call 1-800-448-8182 for more information

Seminars

For more information visit Means Web Site at www.rsmeans.com

2000 Means Seminar Schedule

Location	Dates
Las Vegas, NV	March 13 - 16
Dallas, TX	April 10-13
Washington, DC	April 17-20
Denver, CO	May 22-25
San Francisco, CA	June 12-15
Cape Cod, MA	September 11-14
Washington, DC	September 25-28
San Diego, CA	October TBD
Atlantic City, NJ	November TBD
Orlando, FL	November 13-16

Registration Information

Register Early... Save up to $150! Register 30 days before the start date of a seminar and save $150 off your total fee. *Note: This discount can be applied only once per order.*

How to Register Register by phone today! Means toll-free number for making reservations is: **1-800-448-8182**.

Individual Seminar Registration Fee $875 To register by mail, complete the registration form and return with your full fee to: Seminar Division, R.S. Means Company, Inc., 63 Smiths Lane, Kingston, MA 02364.

Federal Government Pricing All Federal Government employees save 25% off regular seminar price. Other promotional discounts cannot be combined with Federal Government discount.

Team Discount Program Two to four seminar registrations: $760 per person—Five or more seminar registrations: $710 per person—Ten or more seminar registrations: Call for pricing.

Consecutive Seminar Offer One individual signing up for two separate courses at the same location during the designated time period pays only $1,400. You get the second course for only $525 (**a 40% discount**). Payment must be received at least ten days prior to seminar dates to confirm attendance.

Refunds Cancellations will be accepted up to ten days prior to the seminar start. There are no refunds for cancellations postmarked later than ten working days prior to the first day of the seminar. A $150 processing fee will be charged for all cancellations. Written notice or telegram is required for all cancellations. Substitutions can be made at any time before the session starts. **No-shows are subject to the full seminar fee.**

AACE Approved Courses The R.S. Means Construction Estimating and Management Seminars described and offered to you here have each been approved for 14 hours (1.4 recertification credits) of credit by the AACE International Certification Board toward meeting the continuing education requirements for re-certification as a Certified Cost Engineer/Certified Cost Consultant.

AIA Continuing Education R.S. Means is registered with the AIA Continuing Education System (AIA/CES) and is committed to developing quality learning activities in accordance with the CES criteria. R.S. Means seminars meet the AIA/CES criteria for Quality Level 2. AIA members will receive (28) learning units (LUs) for each two day R.S. Means Course.

Daily Course Schedule The first day of each seminar session begins at 8:30 A.M. and ends at 4:30 P.M. The second day is 8:00 A.M.–4:00 P.M. Participants are urged to bring a hand-held calculator since many actual problems will be worked out in each session.

Continental Breakfast Your registration includes the cost of a continental breakfast, a morning coffee break, and an afternoon break. These informal segments will allow you to discuss topics of mutual interest with other members of the seminar. (You are free to make your own lunch and dinner arrangements.)

Hotel/Transportation Arrangements R.S. Means has arranged to hold a block of rooms at each hotel hosting a seminar. To take advantage of special group rates when making your reservation be sure to mention that you are attending the Means Seminar. You are of course free to stay at the lodging place of your choice. (**Hotel reservations and transportation arrangements should be made directly by seminar attendees.**)

Important Class sizes are limited, so please register as soon as possible.

Registration Form

Please register the following people for the Means Construction Seminars as shown here. Full payment or deposit is enclosed, and we understand that we must make our own hotel reservations if overnight stays are necessary.

☐ Full payment of $ _____ enclosed.

☐ Bill me

Name of Registrant(s)
(To appear on certificate of completion)

P.O. #: _____
GOVERNMENT AGENCIES MUST SUPPLY PURCHASE ORDER NUMBER

Call 1-800-448-8182 to register or FAX 1-800-632-6732

Firm Name _____

Address _____

City/State/Zip _____

Telephone No. _____ Fax No. _____

E-Mail Address _____

Charge our registration(s) to: ☐ MasterCard ☐ VISA ☐ American Express ☐ Discover

Account No. _____ Exp. Date _____

Cardholder's Signature _____

Seminar Name	City	Dates

Please mail check to: R.S. MEANS COMPANY, INC., 63 Smiths Lane, P.O. Box 800, Kingston, MA 02364 USA

Consulting Services Group

Solutions For Your Construction and Facilities Cost Management Problems

We are leaders in the cost engineering field, supporting the unique construction and facilities management costing challenges of clients from the Federal Government, Fortune 1000 Corporations, Building Product Manufacturers, and some of the World's Largest Design and Construction Firms.

Developers of the Dept. of Defense Tri-Services Estimating Database

Recipient of SEARS 1997 Chairman's Award for Innovation

Research...

- **Custom Database Development**—Means expertise in construction cost engineering and database management can be put to work creating customized cost databases and applications.
- **Data Licensing & Integration**—To enhance applications dealing with construction, any segment of Means vast database can be licensed for use, and harnessed in a format compatible with a previously developed proprietary system.
- **Cost Modeling**—Pre-built custom cost models provide organizations that expend countless hours estimating repetitive work with a systematic time-saving estimating solution.
- **Database Auditing & Maintenance**—For clients with in-house data, Means can help organize it, and by linking it with Means database, fill in any gaps that exist and provide necessary updates to maintain current and relevant proprietary cost data.

Estimating...

- **Estimating Service**—Means expertise is available to perform construction cost estimates, as well as to develop baseline schedules and establish management plans for projects of all sizes and types. Conceptual, budget and detailed estimates are available.
- **Benchmarking**—Means can run baseline estimates on existing project estimates. Gauging estimating accuracy and identifying inefficiencies improves the success ratio, precision and productivity of estimates.
- **Project Feasibility Studies**—The Consulting Services Group can assist in the review and clarification of the most sound and practical construction approach in terms of time, cost and use.
- **Litigation Support**—Means is available to provide opinions of value and to supply expert interpretations or testimony in the resolution of construction cost claims, litigation and mediation.

Training...

- **Core Curriculum**—Means educational programs, delivered on-site, are designed to sharpen professional skills and to maximize effective use of cost estimating and management tools. On-site training cuts down on travel expenses and time away from the office.
- **Custom Curriculum**—Means can custom-tailor courses to meet the specific needs and requirements of clients. The goal is to simultaneously boost skills and broaden cost estimating and management knowledge while focusing on applications that bring immediate benefits to unique operations, challenges, or markets.
- **Staff Assessments and Development Programs**—In addition to custom curricula, Means can work with a client's Human Resources Department or with individual operating units to create programs consistent with long-term employee development objectives.

MEMBER:

www.eas.asu.edu/joc/

For more information and a copy of our capabilities brochure, please call 1-800-448-8182 and ask for the Consulting Services Group, or reach us at www.rsmeans.com

MeansData™

CONSTRUCTION COSTS FOR SOFTWARE APPLICATIONS
Your construction estimating software is only as good as your cost data.

Software Integration

A proven construction cost database is a mandatory part of any estimating package. We have linked MeansData™ directly into the industry's leading software applications. The following list of software providers can offer you MeansData™ as an added feature for their estimating systems. Visit them on-line at *www.rsmeans.com/demo/* for more information and free demos. Or call their numbers listed below.

ACT
Applied Computer Technologies
Facility Management Software
919-851-7172

AEPCO, Inc.
301-670-4642

ArenaSoft ESTIMATING
888-370-8806

ASSETWORKS, Inc.
Facility Management Software
800-659-9001

BSD
Building Systems Design, Inc.
888-BSD-SOFT

CDCI
Construction Data Controls, Inc.
800-285-3929

CMS
Computerized Micro Solutions
800-255-7407

CONAC GROUP
604-273-3463

CONSTRUCTIVE COMPUTING, Inc.
800-456-2113

ESTIMATING SYSTEMS, Inc.
800-967-8572

G2 Estimator
A Div. of Valli Info. Syst., Inc.
800-627-3283

GEAC COMMERCIAL SYSTEMS, Inc.
800-554-9865

GRANTLUN CORPORATION
602-897-7750

HCI SYSTEMS, Inc.
800-750-4424

IQ BENECO
801-565-1122

MC^2
Management Computer Controls
800-225-5622

PRISM COMPUTER CORPORATION
Facility Management Software
800-774-7622

PYXIS TECHNOLOGIES
888-841-0004

QUEST SOLUTIONS, Inc.
800-452-2342

RICHARDSON ENGINEERING SERVICES, Inc.
602-497-2062

SANDERS SOFTWARE, Inc.
800-280-9760

STN, Inc.
Workline Maintenance Systems
800-321-1969

TIMBERLINE SOFTWARE CORP.
800-628-6583

TMA SYSTEMS, Inc.
Facility Management Software
800-862-1130

US COST, Inc.
800-955-1385

VERTIGRAPH, Inc.
800-989-4243

WENDLWARE
714-895-7222

WINESTIMATOR, Inc.
800-950-2374

DemoSource™ One-stop shopping for the latest cost estimating software for just $19.95. This evaluation tool includes product literature and demo diskettes for ten or more estimating systems, all of which link to MeansData™. Call 1-800-334-3509 to order.

**FOR MORE INFORMATION ON ELECTRONIC PRODUCTS CALL
1-800-448-8182 OR FAX 1-800-632-6732.**

MeansData™ is a registered trademark of R.S. Means Co., Inc., *A CMD Group* Company.

For more information
visit Means Web Site
at www.rsmeans.com

New Titles

Facilities Operations & Engineering Reference
An all-in-one technical reference for planning & managing facility projects & solving day-to-day operations problems

Ten major sections address:

- **Management Skills**
- **Economics** (budgeting/cost control, financial analysis, VE, etc.)
- **Civil Engineering & Construction Practices**
- **Maintenance** (detailed staffing guidance and job descriptions, CMMS, planning/scheduling, training, work orders, preventive/predictive maintenance)
- **Energy Efficiencies** (optimizing energy use, including heating, cooling, lighting, and water)
- **HVAC**
- **Mechanical Engineering**
- **Instrumentation & Controls**
- **Electrical Engineering**
- **Environmental, Health & Safety**

Produced jointly by R.S. Means and the Association for Facilities Engineering.

$99.95 per copy
Over 700 pages, Illustrated, Hardcover
Catalog No. 67318

Planning & Managing Interior Projects New Second Edition
by Carol E. Farren, CFM

Revised edition addresses the major changes in technology and business that have affected all aspects of interiors projects. Guides you through every step in managing interior design and construction for commercial endeavors, from initial client meetings to post-project administration. Corporate downsizing, company mergers, and advances in information technology have revolutionized the workplace. This book addresses all of these changes, showing how to plan effectively for current and future organizational needs. Includes new sections on:

- Evaluating space requirements and selecting the right site
- Alternative work models (telecommuting, hoteling, etc.)
- Records and document management
- Telecommunications and data issues
- Working with consultants
- Environmental considerations

$69.95 per copy
Over 400 pages, Illustrated, Hardcover
Catalog No. 67245A

Residential & Light Commercial Construction Standards
A Unique Collection of Industry Standards That Define Quality in Construction

For Contractors & Subcontractors, Owners, Developers, Architects & Engineers, Attorneys & Insurance Personnel

Compiled from the nation's major building codes, and from scores of publications and reports from professional institutes and other authorities, this one-of-a-kind resource enables you to:

- Set a standard for subcontractors and employees
- Protect yourself against defect claims
- Substantiate your own claim for inferior workmanship
- Resolve disputes
- Overview installation methods
- Answer client questions with an authoritative reference

$59.95 per copy
Over 500 pages, Illustrated, Softcover
Catalog No. 67322

2000 Order Form

ORDER TOLL FREE 1-800-334-3509 OR FAX 1-800-632-6732.

Qty.	Book No.	COST ESTIMATING BOOKS	Unit Price	Total
	60060	Assemblies Cost Data 2000	$139.95	
	60010	Building Construction Cost Data 2000	85.95	
	61010	Building Const. Cost Data-Looseleaf Ed. 2000	109.95	
	63010	Building Const. Cost Data-Metric Version 2000	89.95	
	60220	Building Const. Cost Data-Western Ed. 2000	89.95	
	60110	Concrete & Masonry Cost Data 2000	79.95	
	50140	Construction Cost Indexes 2000	198.00	
	60140A	Construction Cost Index-January 2000	49.50	
	60140B	Construction Cost Index-April 2000	49.50	
	60140C	Construction Cost Index-July 2000	49.50	
	60140D	Construction Cost Index-October 2000	49.50	
	60310	Contr. Pricing Guide: Framing/Carpentry 2000	36.95	
	60330	Contr. Pricing Guide: Resid. Detailed 2000	36.95	
	60320	Contr. Pricing Guide: Resid. Sq. Ft. 2000	39.95	
	64020	ECHOS Assemblies Cost Book 2000	149.95	
	64010	ECHOS Unit Cost Book 2000	99.95	
	54000	ECHOS (Combo set of both books)	214.95	
	60230	Electrical Change Order Cost Data 2000	89.95	
	60030	Electrical Cost Data 2000	89.95	
	60200	Facilities Construction Cost Data 2000	209.95	
	60300	Facilities Maintenance & Repair Cost Data 2000	199.95	
	60160	Heavy Construction Cost Data 2000	89.95	
	63160	Heavy Const. Cost Data-Metric Version 2000	89.95	
	60090	Interior Cost Data 2000	89.95	
	60120	Labor Rates for the Const. Industry 2000	189.95	
	60180	Light Commercial Cost Data 2000	76.95	
	60020	Mechanical Cost Data 2000	89.95	
	60150	Open Shop Building Const. Cost Data 2000	89.95	
	60210	Plumbing Cost Data 2000	89.95	
	60040	Repair and Remodeling Cost Data 2000	79.95	
	60170	Residential Cost Data 2000	74.95	
	60280	Site Work & Landscape Cost Data 2000	89.95	
	60050	Square Foot Costs 2000	99.95	
		REFERENCE BOOKS		
	67147A	ADA in Practice	72.95	
	67310	ADA Pricing Guide	72.95	
	67298	Basics for Builders: Framing & Rough Carpentry	24.95	
	67273	Basics for Builders: How to Survive and Prosper	34.95	
	67307	Basics for Builders: Plan Reading & Takeoff	35.95	
	67261	Building Prof. Guide to Contract Documents	64.95	
	67312	Building Spec Homes Profitably	29.95	
	67250	Business Management for Contractors	17.98	
	67146	Concrete Repair & Maintenance Illustrated	69.95	
	67278	Construction Delays	59.95	
	67268	Construction Paperwork	26.48	
	67255	Contractor's Business Handbook	21.48	
	67314	Cost Planning & Est. for Facil. Maint.	82.95	
	67317	Cyberplaces: The Internet Guide for A/E/C	59.95	
	67230A	Electrical Estimating Methods-2nd Ed.	64.95	
	64777	Environmental Remediation Est. Methods	99.95	
	67160	Estimating for Contractors	35.95	
	67276	Estimating Handbook	99.95	
	67249	Facilities Maintenance Management	86.95	

Qty.	Book No.	REFERENCE BOOKS (Con't)	Unit Price	Total
	67246	Facilities Maintenance Standards	$159.95	
	67264	Facilities Manager's Reference	86.95	
	67318	Facilities Operations & Engineering Reference	99.95	
	67301	Facilities Planning & Relocation	109.95	
	67231	Forms for Building Const. Profess.	94.95	
	67288	Forms for Contractors	79.95	
	67260	Fundamentals of the Construction Process	34.98	
	67258	Hazardous Material & Hazardous Waste	44.98	
	67148	Heavy Construction Handbook	74.95	
	67308B	Home Improvement Costs-Interior Projects	19.95	
	67309B	Home Improvement Costs-Exterior Projects	19.95	
	67304	How to Estimate with Metric Units	24.98	
	67306	HVAC: Design Criteria, Options, Select.-2nd Ed.	84.95	
	67281	HVAC Systems Evaluation	84.95	
	67282	Illustrated Construction Dictionary, Condensed	59.95	
	67292	Illustrated Construction Dictionary, Unabridged	99.95	
	67295A	Landscape Estimating-3rd Ed.	62.95	
	67266	Legal Reference for Design & Construction	54.98	
	67299	Maintenance Management Audit	32.48	
	67302	Managing Construction Purchasing	31.48	
	67294	Mechanical Estimating-2nd Ed.	64.95	
	67245A	Planning and Managing Interior Projects, 2nd Ed.	69.95	
	67283A	Plumbing Estimating Methods, 2nd Ed.	59.95	
	67236A	Productivity Standards for Constr.-3rd Ed.	159.95	
	67247A	Project Scheduling & Management for Constr.	64.95	
	67262	Quantity Takeoff for Contractors	17.98	
	67265A	Repair & Remodeling Estimating-3rd Ed.	69.95	
	67322	Residential & Light Commercial Const. Stds.	59.95	
	67254	Risk Management for Building Professionals	29.98	
	67253	Roofing: Design Criteria, Options, Selection	31.48	
	67291	Scheduling Manual-3rd Ed.	64.95	
	67145A	Square Foot Estimating Methods-2nd Ed.	69.95	
	67241	Structural Steel Estimating	39.98	
	67287	Successful Estimating Methods	64.95	
	67313	Successful Interior Projects	34.98	
	67233	Superintending for Contractors	35.95	
	67321	Total Productive Facilities Management	79.95	
	67284	Understanding Building Automation Systems	39.98	
	67259	Understanding Legal Aspects of Design/Build	79.95	
	67303	Unit Price Estimating Methods-2nd Ed.	59.95	
	67319	Value Engineering: Practical Applications	79.95	

MA residents add 5% state sales tax

Shipping & Handling**

Total (U.S. Funds)*

Prices are subject to change and are for U.S. delivery only. *Canadian customers may call for current prices. **Shipping & handling charges: Add 7% of total order for check and credit card payments. Add 9% of total order for invoiced orders.

Send Order To: ADDV-1001

Name (Please Print) _____

Company _____

☐ Company
☐ Home   Address _____

_____

City/State/Zip _____

Phone # _____ P.O. # _____

Mail To: **R.S. Means Company, Inc.,** P.O. Box 800, Kingston, MA 02364-0800   (Must accompany all orders being billed)